Regenerative Medicine Applications in Organ Transplantation

Regenerative Medicine Applications in Organ Transplantation

Giuseppe Orlando, MD, PhD, Marie Curie Fellow

Department of Surgery
Section of Transplantation
Wake Forest Baptist Hospital
Wake Forest Institute for Regenerative Medicine
Wake Forest School of Medicine
Winston Salem, NC

Jan Lerut, MD, PhD, FACS

Starzl Unit of Abdominal Transplantation
University Hospitals Saint-Luc Université catholique Louvain
Brussels, Belgium

Shay Soker, PhD

Wake Forest Institute for Regenerative Medicine
Wake Forest School of Medicine
Winston Salem, NC

Robert J. Stratta, MD, FACS

Department of Surgery, Section of Transplantation
Wake Forest Baptist Hospital
Wake Forest School of Medicine
Winston Salem, NC

AMSTERDAM • BOSTON • HEIDELBERG • LONDON
NEW YORK • OXFORD • PARIS • SAN DIEGO
SAN FRANCISCO • SINGAPORE • SYDNEY • TOKYO

Academic Press is an Imprint of Elsevier

Acquiring Editor: Christine Minihane
Development Editor: Catherine Mullane
Project Managers: Karen East and Kirsty Halterman
Designer: Matthew Limbert

Academic Press is an imprint of Elsevier
32 Jamestown Road, London NW1 7BY, UK
225 Wyman Street, Waltham, MA 02451, USA
525 B Street, Suite 1800, San Diego, CA 92101-4495, USA

First edition 2014

Notice

No responsibility is assumed by the publisher for any injury and/or damage to persons or property as
a matter of products liability, negligence or otherwise, or from any use or operation of any methods,
products, instructions or ideas contained in the material herein.

Because of rapid advances in the medical sciences, in particular, independent verification of diagnoses and
drug dosages should be made.

British Library Cataloguing-in-Publication Data
A catalogue record for this book is available from the British Library

Library of Congress Cataloging-in-Publication Data
A catalog record for this book is available from the Library of Congress

ISBN: 978-0-12-398523-1

For information on all Academic Press publications
visit our website at elsevierdirect.com

Typeset by MPS Limited, Chennai, India
www.adi-mps.com

Printed and bound in USA

14 15 16 17 18   10 9 8 7 6 5 4 3 2 1

Working together
to grow libraries in
developing countries

www.elsevier.com • www.bookaid.org

In this book, we lay the groundwork for a new frontier in transplantation. History reveals that one can never know for certain when and whether such endeavors will bear fruit. Nonetheless, through our unwavering commitment to patients, and through the love of family and friends, nature and people, we relentlessly continue to push the boundaries of medicine with the aim of bettering people's lives. From bench to bedside, our pursuits are endowed with meaning:

"End and goal. Not every end is the goal. The end of a melody is not its goal; and yet: as long as the melody has not reached its end, it also hasn't reached its goal. A parable."

(Nietzsche, The Wanderer and His Shadow)

Contents

Part II
Kidney

24. Liver Regeneration: The Bioengineering Approach 333

Yeonhee Kim, Sinan Ozer, and Basak E. Uygun

25. Liver Regeneration: The Developmental Biology Approach 353

David A. Rudnick

26. Liver Regeneration: The Stem Cell Approach 375

Syeda H. Afroze, Kendal Jensen, Kinan Rahal, Fanyin Meng, Gianfranco Alpini, and Shannon S. Glaser

Part VII
Lung

Contents

Giuseppe Orlando, MD, PhD (*Editor-in-Chief*), Marie Curie Fellow, is an abdominal organ transplant surgeon scientist at the Wake Forest School of Medicine in Winston, Salem, USA. Dr. Orlando specialized in regenerative medicine at the Wake Forest Institute for Regenerative Medicine and in transplant immunology at the Transplant Research Immunology Group, University of Oxford, UK. His main fields of investigation are: renal, pancreas, liver, and small bowel bioengineering and regeneration; the regenerative medicine approach to clinical tolerance; and minimal immunosuppression and clinical operational tolerance after abdominal organ transplantation. He has authored more than 130 research papers, review articles, and book chapters, and is a member of numerous transplant societies, and is on the board, and acts as a regular reviewer for a number of relevant journals.

Dr. Orlando conceived and designed the book, and was responsible for the primary undertaking, completion, and supervision of the whole editorial process.

Jan P. Lerut, MD, PhD (*co-editor*), is ordinary Professor of Surgery, Director of the Abdominal Transplant Unit of the University Hospital Saint-Luc, and Director of the UCL Transplant Center in Brussels, Belgium. He has served as chairman of the Department of Abdominal and Transplantation Surgery of the University Hospital Saint-Luc in Brussels. He has published over 260 peer-reviewed articles, 24 book chapters, and 24 scientific films. He has made more than 600 communications on national and international congresses, most of them devoted to liver transplantation. His research interests focus on the development of technical refinements in liver transplantation and on the use of minimal immunosuppression and tolerance induction in liver transplantation.

Shay Soker (*co-editor*) is a Professor of Regenerative Medicine at the Wake Forest School of Medicine. Among his contributions are the integration of molecular and cellular biology principles in regenerative medicine applications. Dr. Soker's research interests are in vascular biology, identification of stem and progenitor cells that are needed for tissue damage repair and regeneration, and the biochemical nature of natural materials that can be used for tissue engineering. Dr. Soker has published more that 100 research manuscripts in scientific journals. Some of Dr. Soker's projects are now being discussed with industry collaborators in order to create new regenerative medicine products.

Robert J. Stratta, MD (*co-editor*), is Professor of Surgery and Director of Transplantation at Wake Forest Baptist Health, which is currently one of the largest kidney and pancreas transplant centers in the Southeastern United States. His areas of research interest and expertise include kidney and pancreas allocation and transplantation, immunosuppressive strategies, infection prophylaxis, organ donation and preservation, expanded criteria donors, quality of life, and economic aspects of transplantation. He is a member of 25 medical societies, several editorial and physician review boards, and numerous local, regional, and national committees including UNOS activities. He has lectured worldwide on kidney and pancreas transplantation, is board certified in General Surgery, and has been selected as one of the Best Doctors in America and one of America's Top Surgeons and Top Doctors. He is co-author of four books, over 30 book chapters, 650 abstracts, 375 peer-reviewed articles, 100 internet and 390 oral presentations dealing predominantly with transplantation.

Syeda H. Afroze   Division of Gastroenterology, Department of Medicine, Scott & White Healthcare and Texas A&M Health Science Center, Temple, TX

Gianfranco Alpini   Division of Gastroenterology, Department of Medicine, Scott & White Healthcare and Texas A&M Health Science Center, Temple, TX; Research, Central Texas Veterans Health Care System

Eric C. Anderson   Division of Hematology and Medical Oncology, Oregon Health & Science University, Portland, OR

Anthony Atala   Wake Forest School of Medicine, Winston-Salem, NC

Stephen F. Badylak   Department of Surgery, University of Pittsburgh, Pittsburgh, PA; McGowan Institute for Regenerative Medicine, Pittsburgh, PA

Silvia Baiguera   BIOAIRLab, Florence, Italy

Leonardo Baiocchi   Tor Vergata University of Rome, U.O.C. Medicina Interna, Policlinico Tor Vergata, Viale Oxford, Rome, Italy

Pedro M. Baptista   Wake Forest Institute for Regenerative Medicine, Wake Forest School of Medicine, Winston-Salem, NC

Lorena Bejarano-Pineda   Division of Transplantation, Department of Surgery, University of Illinois at Chicago, Chicago, IL

Enrico Benedetti   Division of Transplantation, Department of Surgery, University of Illinois at Chicago, Chicago, IL

Rosanna Beraldi   General Internal Medicine, Mayo Clinic, Rochester, MN; Todd and Karen Wanek Family Program for Hypoplastic Left Heart Syndrome, Mayo Clinic, Rochester, MN

Alexandra Berdichevski   Faculty of Biomedical Engineering, Technion—Israel Institute of Technology, Haifa, Israel

Mohammod Bhuyan   Department of Mechanical Engineering and Biomedical Engineering Program, University of Texas, El Paso, TX

Alex G. Bishop   Transplantation Laboratory, Royal Prince Alfred Hospital, Sydney, Australia

Khalil N. Bitar   Wake Forest Institute for Regenerative Medicine, Wake Forest School of Medicine, Winston-Salem, NC

Kory J. Blose   Department of Bioengineering, University of Pittsburgh, Pittsburgh, PA

Lauren Brasile   BREONICS, Inc., Albany, NY

Faouzi Braza   INSERM, UMR 1064, Nantes, France; Université de Nantes, Faculté de Médecine, Nantes, France

Christopher K. Breuer   Tissue Engineering Program and Surgical Research, Nationwide Children's Hospital, Columbus, OH

Sophie Brouard   INSERM, UMR 1064, Nantes, France; Université de Nantes, Faculté de Médecine, Nantes, France

Bryan N. Brown   McGowan Institute for Regenerative Medicine, Pittsburgh, PA

George W. Burke   The DeWitt Daughtry Family Department of Surgery, The Lillian Jean Kaplan Renal Transplant Center, Division of Transplantation and The Miami Transplant Institute, University of Miami, Miami, FL

Peter E. Butler   Department of Plastic and Reconstructive Surgery, Royal Free Hampstead NHS Trust Hospital, London, United Kingdom; UCL Centre for Nanotechnology and Regenerative Medicine, Division of Surgery and Interventional Science, London, UK

Colin R. Butler   Lungs for Living Research Centre, Division of Medicine, University College London, UK

Damelys Calderon   Institut National de la Santé et de la Recherché Médicale, Paris, France

Ranieri Cancedda   Dipartimento di Medicina Sperimentale, Università di Genova & AOU San Martino-Istituto Nazionale per la Ricerca sul Cancro, Genova, Italy

Susana Cantero Peral   General Internal Medicine, Mayo Clinic, Rochester, MN; Todd and Karen Wanek Family Program for Hypoplastic Left Heart Syndrome, Mayo Clinic, Rochester, MN

Marco Carbone   Liver Unit, Addenbrooke's Hospital, Cambridge; Organ Donation and Transplantation, National Health Service Blood and Transplant, Bristol, UK

Diego Castanares-Zapatero   Critical Care Unit, St Luc University Hospital, Université Catholique de Louvain, Brussels, Belgium

Edward Castro Santa   Starzl Abdominal Transplant Unit, Department of Abdominal and Transplantation Surgery, University Hospitals St. Luc, Université Catholique Louvain (UCL), Brussels, Belgium

Lucienne Chatenoud   Institut National de la Santé et de la Recherché Médicale, Paris, France

Reema Chawla   UCL Centre for Nanotechnology and Regenerative Medicine, Division of Surgery and Interventional Science, London, United Kingdom

Linda Chen   The DeWitt Daughtry Family Department of Surgery, The Lillian Jean Kaplan Renal Transplant Center, Division of Transplantation and The Miami Transplant Institute, University of Miami, Miami, FL

Valeria Chiono   Department of Mechanical and Aerospace Engineering, Politecnico di Torino, Torino, Italy

Gaetano Ciancio   The DeWitt Daughtry Family Department of Surgery, The Lillian Jean Kaplan Renal Transplant Center, Division of Transplantation and The Miami Transplant Institute, University of Miami, Miami, FL

Gianluca Ciardelli   Department of Mechanical and Aerospace Engineering, Politecnico di Torino, Torino, Italy

Olga Ciccarelli   Starzl Unit Abdominal Transplantation, St Luc University Hospital, Université Catholique de Louvain, Brussels, Belgium

Christine Collienne   Critical Care Unit, St Luc University Hospital, Université Catholique de Louvain, Brussels, Belgium

Francesca Corradini   Center for Regenerative Medicine "Stefano Ferrari", University of Modena and Reggio Emilia, Modena, Italy

Joaquin Cortiella   Mario Negri Institute for Pharmacological Research, Bergamo, Italy

Paolo Cravedi   IRCCS-Istituto di Ricerche Farmacologiche Mario Negri, Bergamo, Italy

Marcelo Cypel   Division of Thoracic Surgery, Toronto Lung Transplant Program, Toronto General Hospital, University Health Network, University of Toronto, Toronto, ON, Canada

Stefano Da Sacco   Saban Research Institute, Children's Hospital Los Angeles, Los Angeles, CA

Hiroshi Date   Department of Thoracic Surgery, Kyoto University Graduate School of Medicine, Kyoto, Japan

Paige S. Davies   Department of Dermatology, Oregon Health & Science University, Portland, OR

Paolo De Coppi   Centre for Stem Cells and Regenerative Medicine, University College London, London; Surgery Unit, Institute of Child Health and Great Ormond Street Hospital, University College London, London, UK

Michele De Luca   Center for Regenerative Medicine "Stefano Ferrari", University of Modena and Reggio Emilia, Modena, Italy

Ethan W. Dean   Yale University School of Medicine, New Haven, CT

Nicolas Degauque   INSERM, UMR 1064, Nantes, France; Université de Nantes, Faculté de Médecine, Nantes, France

Juan Domínguez-Bendala   Diabetes Research Institute, University of Miami Miller School of Medicine, Miami, FL

Jean Michel Dubernard   Department of Transplantation, Hopital Edouard Herriot, Lyon, France

Rania M. ElBackly   Dipartimento di Medicina Sperimentale, Università di Genova & AOU San Martino-Istituto Nazionale per la Ricerca sul Cancro, Genova, Italy

Karen English   Cellular Immunology Group, Institute of Immunology, National University of Ireland Maynooth, Co. Kildare, Ireland

Fred Fändrich   Department of Applied Cellular Medicine, University Schleswig-Holstein, Kiel, Germany

Alan C. Farney   Transplant Service, Department of General Surgery, Wake Forest Baptist Health Medical Center Blvd, Winston-Salem, NC; Wake Forest Institute for Regenerative Medicine, Wake Forest School of Medicine, Winston-Salem, NC

Denver M. Faulk   Department of Bioengineering, University of Pittsburgh, Pittsburgh, PA

José E. García-Arrarás Biology Department, University of Puerto Rico, Rio Piedras, Puerto Rico

Pierre Gianello Laboratory of Experimental Surgery and Transplantation, Institut de Recherche Expérimentale et Clinique, Université Catholique de Louvain, Brussels, Belgium

Adam Giangreco Lungs for Living Research Centre, Division of Medicine, University College London, London, UK

Shannon S. Glaser Division of Gastroenterology, Department of Medicine, Scott & White Healthcare and Texas A&M Health Science Center, Temple, TX; Research, Central Texas Veterans Health Care System

Ryan Gobble Brigham and Women's Hospital, Division of Plastic Surgery, Boston, MA

Aaron S. Goldstein Department of Chemical Engineering, Virginia Polytechnic Institute and State University, Blacksburg, VA; School of Biomedical Engineering and Sciences, Virginia Polytechnic Institute and State University, Blacksburg, VA

Christa N. Grant Division of Pediatric Surgery, Saban Research Institute, Children's Hospital, Los Angeles, CA

Tracy C. Grikscheit Division of Pediatric Surgery, Saban Research Institute, Children's Hospital, Los Angeles, CA

Angelika C. Gruessner College of Public Health, University of Arizona, Tucson, AZ

Rainer W.G. Gruessner Department of Surgery, University of Arizona, Tucson, AZ

Pascale V. Guillot Institute of Reproduction and Developmental Biology, Imperial College London, London

Nicholas Hamilton Lungs for Living Research Centre, Division of Medicine, University College London, London, UK

Philippe Hantson Critical Care Unit, St Luc University Hospital, Université Catholique de Louvain, Brussels, Belgium

Sij Hemal Wake Forest School of Medicine, Winston Salem, NC

Peiman Hematti Department of Medicine, University of Wisconsin-Madison, School of Medicine and Public Health, Madison, WI

Robert E. Hynds Lungs for Living Research Centre, Division of Medicine, University College London, London, UK

Luca Inverardi Diabetes Research Institute, University of Miami Miller School of Medicine, Miami, FL

Luc M. Jacquet Cardiovascular Intensive Care, Cliniques Universitaires Saint-Luc, Université Catholique de Louvain, Brussels, Belgium

Sam M. Janes Lungs for Living Research Centre, Division of Medicine, University College London, London, UK

Gavin Jell UCL Centre for Nanotechnology and Regenerative Medicine, Division of Surgical and Interventional Sciences, University College London, London, UK

Kendal Jensen Division of Gastroenterology, Department of Medicine, Scott & White Healthcare and Texas A&M Health Science Center, Temple, TX

Ina Jochmans Clinical Department of Abdominal Transplant Surgery, University Hospitals Leuven, Laboratory of Abdominal Transplant Surgery, Department of Microbiology and Immunology, KU Leuven, Leuven, Belgium

Nobuo Kanai Institute of Advanced Biomedical Engineering and Science (TWIns), Tokyo Women's Medical University, Shinjuku-ku, Tokyo, Japan

Ravi Katari Wake Forest School of Medicine, Winston-Salem, NC

Shaf Keshavjee Division of Thoracic Surgery, Toronto Lung Transplant Program, Toronto General Hospital, University Health Network, University of Toronto, Toronto, ON, Canada

Jaehyup Kim Department of Medicine, University of Wisconsin-Madison, School of Medicine and Public Health, Madison, WI

Yeonhee Kim Center for Engineering in Medicine, Massachusetts General Hospital, Harvard Medical School and Shriners Hospitals for Children, Boston, MA

Nancy M.P. King Department of Social Sciences & Health Policy and Wake Forest Institute for Regenerative Medicine, Wake Forest School of Medicine; Center for Bioethics, Health, & Society and Graduate Program in Bioethics, Wake Forest University

Eiji Kobayashi Division of Development of Advanced Therapy, Center for Development of Advanced Medical Technology, Jichi Medical University, Japan

Takaaki Koshiba Department of Disaster and Comprehensive Medicine, Fukushima Medical University, Fukushima, Japan

Jeffrey T. Krawiec Department of Bioengineering, University of Pittsburgh, Pittsburgh, PA

Hirotsugu Kurobe   Tissue Engineering Program and Surgical Research, Nationwide Children's Hospital, Columbus, OH

Quirino Lai   Starzl Abdominal Transplant Unit, Department of Abdominal and Transplantation Surgery, University Hospitals St. Luc, Université Catholique Louvain (UCL), Brussels, Belgium

Giacomo Lanzoni   Diabetes Research Institute, University of Miami Miller School of Medicine, Miami, FL

Pierre-François Laterre   Critical Care Unit, St Luc University Hospital, Université Catholique de Louvain, Brussels, Belgium

Jan P. Lerut   Starzl Abdominal Transplant Unit, Department of Abdominal and Transplantation Surgery, University Hospitals St. Luc, Université Catholique Louvain (UCL), Brussels, Belgium

Toni Lerut   Faculty of Medicine, KU Leuven, Leuven, Belgium

Ou Li   Transplantation Research Immunology Group, Nuffield Department of Surgical Sciences, John Radcliffe Hospital, University of Oxford, Oxford, UK

Paolo Macchiarini   Advanced Center for Translational Regenerative Medicine (ACTREM), Karolinska Institutet, Stockholm, Sweden

Panagiotis Maghsoudlou   Surgery Unit, Institute of Child Health and Great Ormond Street Hospital, University College London, London

Nizam Mamode   Consultant Transplant Surgeon and Reader in Transplant Surgery, Guy's and St Thomas' NHS Foundation Trust, London

Tommaso M. Manzia   Tor Vergata University of Rome, U.O.C. Chirurgia dei Trapianti, Policlinico Tor Vergata, Viale Oxford, Rome, Italy

Almudena Martinez-Fernandez   Departments of Medicine, Mayo Clinic, Rochester, MN; Marriott Heart Disease Research Program, Mayo Clinic, Rochester, MN; Medical Genetics, Mayo Clinic, Rochester, MN

Gleb Martovetsky   Department of Pediatrics, University of California at San Diego, La Jolla, CA

Maddalena Mastrogiacomo   Dipartimento di Medicina Sperimentale, Università di Genova & AOU San Martino-Istituto Nazionale per la Ricerca sul Cancro, Genova, Italy

Shigeo Masuda   Gene Expression Laboratory, The Salk Institute for Biological Studies, La Jolla, CA

John P. McQuilling   Wake Forest Institute for Regenerative Medicine, Wake Forest School of Medicine, Winston-Salem, NC

Mandeep R. Mehra   The Center for Advanced Heart Disease, Brigham and Women's Hospital and Harvard Medical School, Boston, MA

Philippe Menasché   Université Paris Descartes, Sorbonne Paris Cité, Faculté de Médecine, Paris, France

Fanyin Meng   Division of Gastroenterology, Department of Medicine, Scott & White Healthcare and Texas A&M Health Science Center, Temple, TX; Research, Central Texas Veterans Health Care System

Iris Mironi-Harpaz   Faculty of Biomedical Engineering, Technion—Israel Institute of Technology, Haifa, Israel

Majid Mirzazadeh   Wake Forest School of Medicine, Winston-Salem, NC

Naiem S. Moiemen   Department of Burns and Plastic Surgery, Queen Elizabeth Hospital, University Hospitals Birmingham NHS Foundation Trust, Edgbaston, Birmingham, UK

Emma Moran   Wake Forest Institute for Regenerative Medicine, Wake Forest School of Medicine, Winston-Salem, NC

Paolo Muiesan   The Liver Unit, University Hospitals Birmingham NHS Trust, Birmingham, UK

Tiziana Nardo   Department of Mechanical and Aerospace Engineering, Politecnico di Torino, Torino, Italy

Timothy J Nelson   General Internal Medicine, Mayo Clinic, Rochester, MN; Molecular Pharmacology and Experimental Therapeutics, Mayo Clinic, Rochester, MN; Todd and Karen Wanek Family Program for Hypoplastic Left Heart Syndrome, Mayo Clinic, Rochester, MN

Paolo A. Netti   Centre for Advanced Biomaterials for Health Care, IIT@CRIB, Istituto Italiano di Tecnologia, Largo Barsanti e Matteucci, Napoli, Italy

James M. Neuberger   Organ Donation and Transplantation, National Health Service Blood and Transplant, Bristol; Liver Unit, Queen Elizabeth Hospital, Birmingham, UK

Sanjay K. Nigam   Department of Pediatrics, University of California at San Diego, La Jolla, CA

Hidenori Ohe   Department of Surgery, Graduate School of Medicine, Kyoto University, Kyoto, Japan

Teruo Okano   Institute of Advanced Biomedical Engineering and Science (TWIns), Tokyo Women's Medical University, Shinjuku-ku, Tokyo, Japan

Emmanuel C. Opara   Institute for Regenerative Medicine, Wake Forest School of Medicine, Medical Center Blvd., Winston-Salem, NC; Wake Forest Institute for Regenerative Medicine, Wake Forest School of Medicine, Winston-Salem, NC

Dennis P. Orgill   Brigham and Women's Hospital, Division of Plastic Surgery, Boston, MA; Harvard Medical School, Boston, MA

Jean-Bernard Otte   Université Catholique de Louvain, Cliniques Saint-Luc, Brussels, Belgium

Sinan Ozer   Center for Engineering in Medicine, Massachusetts General Hospital, Harvard Medical School and Shriners Hospitals for Children, Boston, MA

Rajesh A. Pareta   Wake Forest Institute for Regenerative Medicine, Wake Forest School of Medicine, Winston-Salem, NC

Timil Patel   Wake Forest School of Medicine, Winston-Salem, NC

Graziella Pellegrini   Center for Regenerative Medicine "Stefano Ferrari", University of Modena and Reggio Emilia, Modena, Italy

Andrea Peloso   IRCCS Policlinico San Matteo, Department of General Surgery, University of Pavia, Pavia, Italy

Laura Perin   Saban Research Institute, Children's Hospital Los Angeles, Los Angeles, CA

Palmina Petruzzo   Department of Transplantation, Hopital Edouard Herriot, Lyon, France

Rafael S. Pinheiro   Starzl Abdominal Transplant Unit, Department of Abdominal and Transplantation Surgery, University Hospitals St. Luc, Université Catholique Louvain (UCL), Brussels, Belgium

Jacques Pirenne   Clinical Department of Abdominal Transplant Surgery, University Hospitals Leuven, Laboratory of Abdominal Transplant Surgery, Department of Microbiology and Immunology, KU Leuven, Leuven, Belgium

Lauren K. Poindexter   Virginia Tech Carillion School of Medicine and Research Institute, Roanoke, VA

Michel Pucéat   Université Paris Descartes, Sorbonne Paris Cité, Faculté de Médecine, Paris, France

Alberto Pugliese   Diabetes Research Institute, University of Miami, Miller School of Medicine, Miami, FL

Shreya Raghavan   Wake Forest Institute for Regenerative Medicine, Wake Forest School of Medicine, Winston-Salem, NC

Kinan Rahal   Division of Gastroenterology, Department of Medicine, Scott & White Healthcare and Texas A&M Health Science Center, Temple, TX

Paolo Rama   San Raffaele Scientific Institute, Ophthalmology Unit, Milan, Italy

Andrea Remuzzi   IRCCS-Istituto di Ricerche Farmacologiche Mario Negri, Bergamo, Italy

Giuseppe Remuzzi   Farmacologiche Mario Negri and Unit of Nephrology, Dialysis and Transplantation, Azienda Ospedaliera Papa Giovanni XXIII, Bergamo, Italy

Daniel Reyna-Soriano   Department of Mechanical Engineering and Biomedical Engineering Program, University of Texas, El Paso, TX

Juan M. Rico Juri   Starzl Abdominal Transplant Unit, Department of Abdominal and Transplantation Surgery, University Hospitals St. Luc, Université Catholique Louvain (UCL), Brussels, Belgium

Camillo Ricordi   Diabetes Research Institute, University of Miami Miller School of Medicine, Miami, FL

Keith J. Roberts   The Liver Unit, University Hospitals Birmingham NHS Trust, Birmingham, UK

Kevin A. Rocco   Department of Biomedical Engineering, Yale University, New Haven, CT

Jorge I. Rodriguez-Devora   Department of Mechanical Engineering and Biomedical Engineering Program, University of Texas, El Paso, TX

Jeffrey Rogers   Department of General Surgery, Section of Transplantation, Wake Forest Baptist Medical Center, Winston-Salem, NC

David A. Rudnick   Departments of Pediatrics and Developmental, Regenerative, and Stem Cell Biology, Washington University School of Medicine, St. Louis, MO

Piero Ruggenenti   IRCCS-Istituto di Ricerche Farmacologiche Mario Negri, Unit of Nephrology, Dialysis and Transplantation, Azienda Ospedaliera Papa Giovanni XXIII, Bergamo, Italy

Junichiro Sageshima   The DeWitt Daughtry Family Department of Surgery, The Lillian Jean Kaplan Renal Transplant Center, Division of Transplantation and The Miami Transplant Institute, University of Miami, Miami, FL

Marcus Salvatori   Newcastle School of Medicine, University of Newcastle-upon-Tyne, Newcastle upon Tyne, UK

Satyavrata Samavedi   Department of Chemical Engineering, Virginia Polytechnic Institute and State University, Blacksburg, VA

Alessandro Sannino   Department of Engineering for Innovation, University of SalentoVia per Monteroni—Complesso Ecotekne, Lecce, Italy

Irene Scalera   The Liver Unit, University Hospitals Birmingham NHS Trust, Birmingham, UK

Tina Sedaghati   UCL Centre for Nanotechnology and Regenerative Medicine, Division of Surgical and Interventional Sciences, University College London, London, UK

Sargis Sedrakyan   Saban Research Institute, Children's Hospital Los Angeles, Los Angeles, CA

Alexander M. Seifalian   Department of Plastic and Reconstructive Surgery, Royal Free Hampstead NHS Trust Hospital, London, UK; UCL Centre for Nanotechnology and Regenerative Medicine, Division of Surgery and Interventional Science, London, UK

Amelia Seifalian   UCL Centre for Nanotechnology and Regenerative Medicine, Division of Surgery and Interventional Science, London, UK

Dror Seliktar   Faculty of Biomedical Engineering, Technion—Israel Institute of Technology, Haifa, Israel

Gennaro Selvaggi   University of Miami Miller School of Medicine, Miami Transplant Institute, Miami, FL

Keren Shapira-Schweitzer   Faculty of Biomedical Engineering, Technion—Israel Institute of Technology, Haifa, Israel

Toshiharu Shinoka   Department of Cardiothoracic Surgery, The Heart Center, Nationwide Children's Hospital, Columbus, OH

Thomas Shupe   Wake Forest Institute for Regenerative Medicine, Wake Forest School of Medicine, Winston-Salem, NC

Nicholas R. Smith   Department of Dermatology, Oregon Health & Science University, Portland, OR

Jean-Paul Soulillou   INSERM, UMR 1064, Nantes, France; Université de Nantes, Faculté de Médecine, Nantes, France

Valentina Spinelli   Surgery Unit, Institute of Child Health, University College London and Great Ormond Street Hospital, London, UK

David E.R. Sutherland   Division of Transplantation, Department of Surgery, University of Minnesota, Minneapolis, MN

Shuhei Tara   Tissue Engineering Program and Surgical Research, Nationwide Children's Hospital, Columbus, OH

Takumi Teratani   Division of Development of Advanced Therapy, Center for Development of Advanced Medical Technology, Jichi Medical University, Japan

Andre Terzic   Departments of Medicine, Mayo Clinic, Rochester, MN; Marriott Heart Disease Research Program, Mayo Clinic, Rochester, MN; Medical Genetics, Mayo Clinic, Rochester, MN

Sunu S. Thomas   The Center for Advanced Heart Disease, Brigham and Women's Hospital and Harvard Medical School, Boston, MA

Giuseppe Tisone   Head of Transplantation Surgery, Tor Vergata University of Rome, U.O.C. Chirurgia dei Trapianti, Policlinico Tor Vergata, Viale Oxford, Rome, Italy

Giorgia Totonelli   Surgery Unit, Institute of Child Health and Great Ormond Street Hospital, University College London, London, UK

Andreas Tzakis   Cleveland Clinic Foundation, Weston, FL

Ivo G. Tzvetanov   Division of Transplantation, Department of Surgery, University of Illinois at Chicago, Chicago, IL

Brooks V. Udelsman   Yale University School of Medicine, New Haven, CT

Basak E. Uygun   Center for Engineering in Medicine, Massachusetts General Hospital, Harvard Medical School and Shriners Hospitals for Children, Boston, MA

Joseph P. Vacanti   Laboratory of Tissue Engineering and Organ Fabrication, Massachusetts General Hospital, Boston, MA

Olivier Van Caenegem   Cardiovascular Intensive Care, Heart Transplant Unit, Cliniques Universitaires Saint-Luc, Université Catholique de Louvain, Brussels Belgium

Mark Van Dyke   School of Biomedical Engineering and Sciences, Virginia Polytechnic Institute and State University, Blacksburg, VA

Dirk E. Van Raemdonck   Division of Experimental Thoracic Surgery, Department of Clinical and Experimental Medicine, KU Leuven, Belgium

David A. Vorp   Department of Bioengineering, University of Pittsburgh, Pittsburgh, PA; McGowan Institute for Regenerative Medicine, University of Pittsburgh, Pittsburgh, PA

Dipen Vyas   Wake Forest Institute for Regenerative Medicine, Wake Forest School of Medicine, Winston-Salem, NC

Kun Wang   Department of Orthopedic Surgery, Third Affiliated Hospital, Sun Yat-sen University, Guangzhou, Guangdong Province, People's Republic of China

Justin S. Weinbaum   Department of Bioengineering, University of Pittsburgh, Pittsburgh, PA; McGowan Institute for Regenerative Medicine, University of Pittsburgh, Pittsburgh, PA

Jason A. Wertheim   Comprehensive Transplant Center, Department of Surgery, Northwestern University Feinberg School of Medicine, Institute for BioNanotechnology in Medicine and Chemistry of Life Processes Institute, Northwestern University, Chicago, IL

David F. Williams   Wake Forest Institute of Regenerative Medicine, Winston-Salem, NC

Xavier Wittebole   Critical Care Unit, St Luc University Hospital, Université Catholique de Louvain, Brussels, Belgium

Melissa H. Wong   Department of Dermatology, Oregon Health & Science University, Portland, OR; Oregon Stem Cell Center, Oregon Health & Science University, Portland, OR

Kathryn J. Wood   Transplantation Research Immunology Group, Nuffield Department of Surgical Sciences, John Radcliffe Hospital, University of Oxford, Oxford, UK

Tao Xu   Department of Mechanical Engineering, Tsinghua University, Beijing, People's Republic of China

Masayuki Yamato   Institute of Advanced Biomedical Engineering and Science (TWIns), Tokyo Women's Medical University, Shinjuku-ku, Tokyo, Japan

Sylvaine You   Institut National de la Santé et de la Recherché Médicale, Paris, France

Yuyu Yuan   Medprin Regenerative Technologies Co. Ltd, Guangzhou, Guangdong Province, People's Republic of China

Joao Paulo Zambon   Wake Forest School of Medicine, Winston-Salem, NC

Michela Zattoni   Center for Regenerative Medicine "Stefano Ferrari", University of Modena and Reggio Emilia, Modena, Italy

Lei Zhu   Department of Plastic Surgery, Third Affiliated Hospital, Sun Yat-sen University, Guangzhou, Guangdong Province, People's Republic of China

Solid Organ Transplantation in the Regenerative Medicine Era

This new book is an up-to-date consideration of organ transplantation and regenerative medicine which together make a huge, wide-ranging subject extending from basic molecular biology to ethics.

It is now more than 50 years since the first successful clinical kidney transplant was performed by Murray and his colleagues in Boston. This demonstrated the surgical feasibility of organ grafting and the ability of a kidney to withstand the trauma of being removed from the donor and suffer ischemia before being revascularized in a recipient. This was an extremely important advance in surgery and precipitated an intense, world-wide interest in immunology in general, and immunology of transplant rejection in particular. With the exception of identical twins, control of rejection proved to be difficult. The early results of chemical immunosuppression with azathioprine and steroids produced some excellent, long-term results but there were also many failures. The introduction of cyclosporine improved the 1-year functional survival of kidney grafts in the clinic from around 50% to more than 80%. This was a watershed in organ transplantation as previously the few centers performing transplants were regarded with skepticism by the medical profession, but following

cyclosporine immunosuppression, organ grafting became respectable and instead of around 10 centers world-wide carrying out transplants, there were soon more than 1000 and the shortages of organs for transplantation started to become a serious problem. This has produced an increasing worry ever since, putting pressure on traditional medical ethics in a manner that never occurred before. Hippocrates's advocacy of "first do no harm" became obsolete, since healthy people were subjected to major and sometimes dangerous operations to procure organs.

In this book, the current status of transplantation of all the major organs is discussed, and the next phase of development, namely, regeneration of tissues using stem cell techniques and gene therapy, is now at an early stage but shows encouraging signs of becoming clinically useful after a long period of little progress.

This book should be a valuable source of reference to this whole, wide-ranging, and rapidly advancing subject, and I wish the editors and contributors every success with what must have been a very hard task to bring all the chapters together in one volume.

Sir Roy Calne, FRS
Emeritus Professor of Surgery, University of Cambridge, UK

Introduction: Regenerative Medicine and Solid Organ Transplantation from a Historical Perspective

Joseph P. Vacanti[a], Jean-Bernard Otte[b], and Jason A. Wertheim[c]

[a]Laboratory of Tissue Engineering and Organ Fabrication, Massachusetts General Hospital, Boston, MA, [b]Université Catholique de Louvain, Cliniques Saint-Luc, Brussels, Belgium, [c]Comprehensive Transplant Center, Department of Surgery, Northwestern University Feinberg School of Medicine, Institute for BioNanotechnology in Medicine and Chemistry of Life Processes Institute, Northwestern University, Chicago, IL

Chapter Outline

1.1 INTRODUCTION

As a therapeutic modality, solid organ transplantation has been available to patients in need for the past half-century, however, it is only within the past 30 years that the development of immunosuppression has allowed for long-term graft survival with minimal risk of rejection. The history of organ transplantation is a remarkable journey of scientific experimentation, progress and failures, miraculous discoveries, and personal sacrifice on the part of those early donors and recipients of transplanted organs. The first modern-day studies in transplantation began at the turn of the twentieth century in which skin grafting between animals initiated the first series of experiments that eventually led to a scientific understanding of graft rejection. This, in turn, opened the door to the modern era of human transplantation with the first successful kidney transplant between two identical siblings in 1954 by Joseph Murray at the Peter Bent Brigham Hospital in Boston. Since that time, advances in immunosuppression, development of complex operative techniques, and patient/donor selection have made organ transplantation a successful clinical modality that touches tens of thousands of patients worldwide every year.

Regenerative medicine, on the other hand, is a relatively newer concept in medical care that builds upon the principles of stem cell technology and regrowth of organs and tissues to provide new treatments to old ailments. The ability of the body to regenerate, or regrow, itself is first evident in nature. Salamanders have the ability to regrow their tails [1]. Sea cucumbers, if divided into pieces, become new, individual entities. Some spiders regrow legs, starfish can regenerate new arms. Perhaps an early interpretation of regenerative medicine may have been embodied in the concept of the "fountain of youth." Herodotus, an ancient Greek historian born in the fifth century BC first described a special liquid that gave the people of Ethiopia longevity. In the sixteenth century, the conquistador Juan Ponce de León explored the newly discovered Americas and was charged with finding the fountain of youth. However, these historical interpretations of magical water restoring youth are not entirely in line with the modern-day concept of "regenerative medicine." Perhaps the most accurate, modern-day definition is that of Leland Kaiser who wrote in 1992 that regenerative medicine "attempts to change the course of chronic disease" to restore normal function to tissues, cells, and organs [2]. This concept was later promoted by William Haseltine of Human Genome Sciences (Rockville, MD).

The earliest references to organ transplantation and regeneration are embodied in the Chimera and the Hydra in ancient Greek mythology. The Chimera was first introduced in Homer's the *Odyssey* as a collection of different body parts taken from a goat, lion, and dragon. The Hydra was a multiheaded monster that was thought impossible to defeat because once severed, a new head regrew in its place. Hercules eventually defeated the Hydra by burning its neck before a new head grew back as part of his 12 Labors.

Ancient Chinese texts documented the first human-to-human transplant and recount the exchange of two hearts between soldiers by surgeon Tsin Yue-Jen. This is reportedly the first known description of body to body transplantation [3]. In the third century AD, Christian mythology described the first cadaveric transplant when Saints Damien and Cosmas transplanted a cadaver leg onto a soldier (Figure 1.1). The first report in a medical journal describing a human-to-human skin transplant first appeared in 1881. The recipient was burned while leaning against a metal door during a lightning strike.

Transplantation and regenerative medicine perhaps both have their most significant, early contributions to skin grafting, and soft tissue reconstruction. Sir Harold Giles was the first surgeon to take skin from a healthy portion of the body and transfer it onto a burned segment. This was the case of British soldier Walter Yeo who sustained deep facial burns around his eyelids and eyes while maintaining guns aboard the British ship HMS Worspite in 1916. As medicine progressed throughout the early twentieth century, vascular bypass was established as a means to circumvent blocked arteries in the heart, abdomen, and lower extremity. It is commonly known that the body's own vascular conduits in the form of small arteries and veins perform much better when taken from their

original location and used as a bypass conduit compared to synthetically constructed, inert bypass grafts. The development of new vascular channels is another important component of regenerative medicine and shares common themes with transplantation. In 1960 at Albert Einstein College of Medicine—Bronx Municipal Hospital Center, Robert Goetz and his team performed the first coronary artery bypass surgery in 1960. Three years later, Thomas Starzl performed the first human liver transplantation in 1963. And, in the 50 years that have postdated that achievement, organ transplantation has expanded throughout the United States and the world due to revolutionary advancements in immunosuppression, surgical technique, and patient/donor selection. Now, over 117,000 patients are waiting for an organ for transplantation in the United States alone and this increasing trend has never been reversed. The promise of organ transplantation in the regenerative medicine era is the hope that one day organs and tissues can be developed "on demand" to supply all patients in need of organ and tissue replacement.

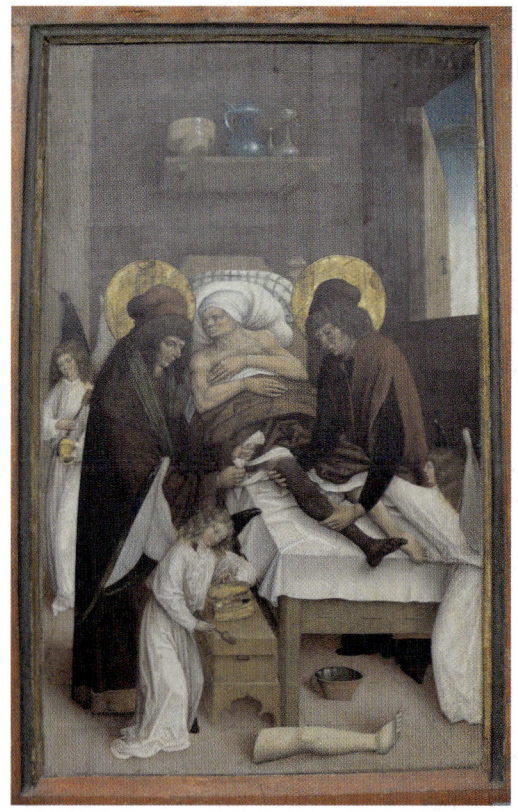

FIGURE 1.1 In the third century AD, Christian mythology described the first cadaveric transplant when Saints Damien and Cosmas transplanted a cadaver leg onto a soldier. *Legendary transplantation of a leg by Saints Cosmas and Damian, assisted by Angels (Württembergisches Landesmuseum Stuttgart), available at: http://en.wikipedia.org/wiki/File: Beinwunder_Cosmas_und_Damian.jpg.*

1.2 MAJOR ADVANCES IN ORGAN TRANSPLANTATION

1.2.1 Vascular Reconstruction

Every solid organ transplant requires precise vascular reconstruction. The technique of triangular vascular anastomosis was devised by Alexis Carrel at the turn of the twentieth century at the University of Lyon and further refined with Charles Claude Guthrie at the University of Chicago. This technique was used by Carrel to develop renal transplantation. The Noble Prize in Physiology was awarded in 1912 to Carrel for this pioneering work.

1.2.2 Organ Preservation

Carrel and Charles Lindbergh pioneered *ex vivo* pulsatile organ perfusion at the Rockefeller Institute in the 1930s. In 1964, Lillehei et al. published an early paper about *in vitro* preservation of the kidney and the heart by a combination of hypothermia and hyperbaric oxygenation [4].

Dr Francis O. Belzer developed *ex vivo* renal preservation by perfusion of recovered kidneys with cryoprecipitated plasma allowing preservation of dog kidneys for up to 72 h and successful renal human transplantation after 17 h of storage [5]. Later, he developed a miniature portable perfusion machine, the precursor of the device that is now used worldwide to pump kidneys with cold perfusate.

The first solution allowing cold organ preservation after initial perfusion was formulated by Collins et al. [6]. Although the duration of preservation of the kidney and liver was somewhat limited, Collins' solution was used for nearly 20 years.

The University of Wisconsin (UW) solution developed by Belzer et al. permitted the extension of the duration of cold storage. It has revolutionized cold storage of the liver and the pancreas [7].

1.2.3 Artificial Support

Patients with chronic renal disease need extracorporeal dialysis while waiting for a donated kidney. The artificial kidney was devised by the Dutch physician Willem Johan Kolff in 1943, under the Nazis' occupation of the Netherlands. The first patient with acute renal failure was saved by Willem Kolff in 1945. He continued his research on renal dialysis at the Mount Sinai Hospital which was granted one of the first five machines built by Kolff [8].

After his move to the University of Utah, Kolff devised the first artificial heart that was named after a coworker, Robert Jarvis. This research paved the way for artificial cardiac support to become a bridge to definitive heart transplantation for patients awaiting a transplant, either acutely with extracorporeal circulation with ECMO or chronically with an implantable left ventricular assist device.

Renal replacement in the form of hemodialysis requires access to an artery and a vein. The remarkable achievement of Belding Scribner at the University of Washington was to develop an original arterial-venous shunt made of Teflon that was used worldwide for hemodialysis access in patients with end-stage renal disease for decades prior to the introduction of the arterial-venous fistula [9].

1.2.4 Immunosuppression

The first attempts at immunosuppression included total body irradiation and steroids but were eventually unsuccessful. Pharmacological immunosuppression became the standard of care with the development of 6-mercaptopurine (Purinethol) [10] followed by azathioprine (Imuran) in the early 1960s [11]. George Hitchings was later awarded the Nobel Prize for medicine and physiology in 1988 for his work on purine chemistry. After the first successful series of renal transplantations performed in Denver by Thomas E. Starzl, the combination of azathioprine and steroids came into widespread use for the next 20 years.

The major breakthrough came with the discovery of the immunosuppressive power of cyclosporine A by Borel et al. [12] and its first clinical study, as single immunosuppressant, performed in Cambridge, UK, by Calne et al. [13]. Since none of the English patients had normal kidney function due to the nephrotoxicity of the drug, Thomas E. Starzl, in Denver, decided to decrease the dosage and use it in combination with steroids in a successful series of 22 kidney recipients [14]. As a method to facilitate lymphocyte depletion, 11 patients were prepared with thoracic duct drainage without significant advantage [15,16]. Later on, a similar protocol was used in liver transplant patients [17,18].

In the late nineteenth century, Metchnikov discovered the role of lymphocytes in the pathogenesis of autoimmune disorders [19]. The first trial of antilymphoid globulin (ALG), in combination with azathioprine and steroids, was conducted by Starzl et al. in Denver in 1966 [20]. "Rejection was practically eliminated... if ALG therapy was started at the time of transplantation. If ALG treatment was delayed, it could be used effectively to reverse established rejection. The amounts of Imuran and prednisone (especially the latter) were reduced. This was the beginning of the triple antirejection therapy..." wrote Starzl in his book *The Puzzle People: Memoirs of a Transplant Surgeon*[15]. Refinements to Starzl's protocol followed during the ensuing years and included the development of hybridoma technology by Georges Kohler and Cesar Milstein of Cambridge University (Nobel laureates in 1984) that allowed for the production of OKT3 [21].

FK506, later renamed tacrolimus, was discovered by scientists of Fujisawa Pharmaceutical Corporation in Japan and was noted to rescue liver transplants from uncontrollable rejection in the late 1980s [22]. Despite controversy about its toxicity observed in animals by Calne's group in Cambridge, extensive use in several organ transplant recipients in Pittsburgh revealed its immunosuppressive power [23]. This new member of the calcineurin inhibitor family rapidly supplanted cyclosporin in clinical transplantation.

1.2.5 Operational Tolerance

Induction of tolerance in clinical transplantation is still the Holy Grail. The occasional occurrence of perfect immune tolerance of a transplanted organ is well known, as observed occasionally in recipients who unwisely discontinued their immunosuppression and in some children taken off immunosuppression because of posttransplant lymphoproliferative disease. Unfortunately, no reliable biomarker has been identified yet to detect tolerant patients despite intense research in this field [24]. A practical approach was developed by Calne et al. with the definition of *prope* tolerance [25] leading to protocols for the *minimization* of immunosuppression [26–28]. Immunosuppression-free transplantation in the modern era of regenerative medicine is reviewed by several leading investigators in the field of tolerance in Chapter 16.

1.2.6 Organ Recovery and Definition of Brain Death

For decades, organs were procured from nonheart beating donors and quality was affected by resulting ischemic injury. The paradigm of organ donation was drastically changed with the definition of brain death allowing safe procurement of organs while the heart continued to beat.

The "coma dépassé" was defined by Pierre Mollaret's team [29] confronted with patients in intensive care units that required respiratory assistance though they had lost cerebral function. "Brain death" was defined as a new criterion of death by the Ad Hoc Committee of Harvard Medical School in 1968 [30]. The very first kidney procurement from a brain dead donor was performed in 1963 at the Université Catholique de Louvain, Belgium, by Morelle and Alexandre [31]. The growing organ shortage later triggered the need to revisit the strategy of organ procurement from nonheart beating donors with the definition of the Maastricht categories [32]. An in-depth discussion about the current organ donation process is presented in Chapter 4.

1.2.7 Living Donation

Living donor kidney transplantation began more than five decades ago by the Boston team with the first renal transplantation between identical twins performed in 1954 at the Peter Bent Brigham Hospital by Joseph Murray (Nobel laureate in 1990) [33] followed by the first successful mother-to-child transplant performed in Denver in 1962 with a combination treatment composed of irradiation, Imuran and prednisone [15]. None of the world's 24 kidney transplant recipients who had survived more than a quarter century as of 1989 had received an unrelated donor kidney [34].

During the last three decades, live organ donation has been extended to the liver, the lung, the pancreas, and the intestine. Medical guidelines and ethical aspects were addressed for the kidney by the Amsterdam forum [35] and for the other organs by the Vancouver forum [36].

1.2.8 HLA Typing and Crossmatch

The credit for the discovery of the first human leukocyte antigen (HLA) goes to Jean Dausset (Nobel laureate in 1980), who analyzed sera from patients who had received multiple blood transfusions [37]. Gerhard Opelz, using the data of the 20-year Collaborative Transplant Study [38], showed that HLA mismatches significantly influence the long-term outcome of kidney transplants, supporting his recommendation to continue kidney exchange programs for the purpose of achieving better HLA matches. Other studies have shown that the influence of HLA matching in renal transplantation has diminished in recent years, which is explained by improved immunosuppression medications and protocols. For other solid organs, HLA typing and crossmatch are either not relevant or cannot be implemented due to the short time interval required between recovery and implantation. Quoting Starzl: "It was a relief to know that the selection of donors with random tissue matching would not result in an intolerable penalty" [15].

1.3 REGENERATIVE MEDICINE

1.3.1 A New Term in Transplantation: Regenerative Medicine

As a defined term, regenerative medicine has been used in the medical literature since the turn of this most recent century. Regenerative medicine describes a process by which cell, tissue, or organ damage can be reversed through the use of stem cell technology. An example of this would be the infusion of stem cells into the heart after a myocardial infarction to minimize myocardial damage, limit fibrosis, and enhance healing. However, inherent in this definition is the notion that cells may be rejuvenated, or repaired, through growth factors or cell reprogramming, potentially through the delivery of specific genes or telomere modification.

Individual or colonies of cells do not have form or structure and therefore the field of tissue engineering has developed a close relationship with regenerative medicine (and sometimes the terms are used interchangeably) by the former's ability to develop solid scaffolds and structures to give shape to developing cells. The early concept of three-dimensional cell growth was first reported in 1951 by Wilton Earl and colleagues who noted that cells proliferated on cellophane or glass surfaces to create sheets of cells. He hypothesized that these materials could be fabricated into beads or be designed to contain tiny pores where cells could grow and still receive nutrients from the surrounding media [39]. Growth of cells on natural gels made from collagen was first reported in 1956 [40] and was later used by Eugene Bell at the Massachusetts Institute of Technology to describe the mechanics of fibroblasts as they move and contract to form a model for wound healing [41]. The seminal paper published by one of these authors (JPV) along with Robert Langer in 1993 first coined the term "tissue engineering" in the literature [42]. As outlined in this paper, tissue engineering incorporated: (i) the isolation and manipulation of individual cells or cell substitutes for therapeutic infusion, (ii) the identification of tissue inducing substances, such as growth factors and their appropriate delivery, and (iii) cells placed on or within matrices which permit the delivery of nutrients but protect cells from immunological destruction [42].

Whether the terms used to describe the development of novel tissues to treat disease or restore function are called "regenerative medicine" or "tissue engineering," the common themes of developmental biology and engineering and their relationship to these disciplines are as true today as they were in 1993 and before. Perhaps the most difficult question to answer in developmental biology is the ability to describe the complex cascade and specialized presentation of growth factors that allow for cellular development and renewal [43]. The engineering challenge is to deliver growth factors to a growing tissue in the correct spatiotemporal sequence as seen in nature. Important engineering principles that are critical to the development of laboratory grown tissues are mass transfer and diffusion, which govern how nutrients are delivered to living cells within organs that become more critical as tissues become larger over time.

Stem cells have the ability of self-renewal and differentiation. Over the last 10 years, we have seen significant advances in stem cell technology. Embryonic stem cells are derived from the inner cell mass of an embryo. These cells are pluripotent and can differentiate into any cell type or lineage. Induced pluripotent stem cells are created by reprogramming the genetic information in mature, somatic cells to develop an immature, pluripotent cell [44−46]. These cells may then be differentiated into almost any cell type given the correct growth factors, culture conditions, and genetic milieu. Though Shinya Yamanaka shared in the Nobel Prize for medicine or physiology in 2012 for his discovery of this technology, induced pluripotent stem cells are still new and challenges remain to characterize their genetic stability and potential to recapitulate the function of normal, adult cells of the body. Finally, the discovery of adult stem cells that are immature, though committed toward a certain differentiation pathway, has been found in the brain, muscle, liver, and blood. Taken together, new discoveries in stem cell technology, as briefly outlined above, may hold the cure for genetic diseases and may lead to the identification of the initial cell population from which tissues are grown.

1.3.2 Characteristics of Regenerative Tissues

If regenerative medicine has a slightly increased focus on the application of stem cells and growth factors to human health, then tissue engineering can provide cells form with the goal of improving function. Engineered tissues must be made from natural or synthetic components without toxic chemicals or deleterious interactions with surrounding tissues. In general, the micro- and macrostructural architecture of the regenerated tissue must display some similarity to the target organ. This provides optimal trafficking of blood cells, nutrients, and small molecules both between cells and with the body through the bloodstream. Regenerated tissues can be tough structures, such as bone and cartilage or softer, more compliant organs, such as muscles, tendons, intestines, and livers.

Though mechanical strength and rigidity may vary between structures, in order to effectively recapitulate tissues and organs in a laboratory their structure and function must mimic that of the native tissue. The chemical function, the ability to effectively secrete nutrients and bioactive molecules, of an engineered tissue will be largely dependent upon the cellular make-up of the tissue and its ability to modulate the release kinetics of these molecules. This is particularly seen in engineered livers and encapsulated β cells whose regulatory processes are closely controlled.

1.3.3 Polymer Development

A biomaterial may be any material, natural or man-made, i.e., biocompatible, without eliciting a deleterious immunologic or toxic response once introduced into the specific site of application. Under that definition, glass, ceramics, metals, and polymers have been used for biological applications and their specific site of use is determined by their unique physical and chemical properties. Polymers are macromolecules that repeat a basic core molecular structure. The benefit of these molecules is that their core structure can be modified to incorporate functional groups and the length of their chains can be controlled to further tailor the properties to the medical application. Early biomaterials utilized already available formulations that were initially intended for other applications like textiles. Later, during the last 30 years or more material scientists began engineering new materials and modifying existing ones. One incredibly important breakthrough was the development of biodegradable polymers. These large macromolecules are degraded by the body's natural environment. An example of biodegradable polymers is polyanhydrides, which are hydrolyzed (broken apart) by water molecules within the tissue environment. The degradation products are either absorbed by tissues or excreted from the patient.

Biodegradable polymers revolutionized the material science field by allowing materials to be implanted within the body without the need for an additional operation to retrieve them. Moreover, polymers could be engineered with a quick degradation time, allowing them to be used for short-term applications. A second critical property of polymers is the ability to create precise structures with a high porosity, like a sponge, with multiple pores to trap fluid and cells.

Poly-methyl-methacrylate was one of the first biocompatible polymers used for a variety of applications. Bone tissue engineering and prosthesis stabilization were the focus of many studies developing polymers for *in vivo* use [47]. Important characteristics were strength, stiffness, and energy absorption. When polymers were combined with cells, a high porosity and interconnectivity are important to establish uniform cell seeding and incorporation of the material into the surrounding tissue. Blocks of poly-L-lactic acid in combination with polyvinyl alcohol were created by David Mooney and colleagues at a high porosity as high as 90–95%. This allowed for a high level of uniform seeding of hepatocytes in early studies on liver tissue engineering [48]. A further, in-depth discussion of biomaterials is presented in Chapter 6.

1.3.4 Early Tissue Development

The first tissues that were attempted to be re-created in the laboratory were bone, cartilage, and skin. Charles Vacanti was the first to combine chondrocytes with biodegradable, synthetic polymers to construct specific shapes as a future application toward reconstructive surgery [49]. Later, others combined chondrocytes with fibrous meshes of polyglycolic acid to create 10 mm in diameter small cartilage scaffolds. These were grown within a rotating wall bioreactor to simulate microgravity conditions by Freed et al. in the mid-1990s [50]. When the cell–scaffold composite was grown in a simulated microgravity environment, more glycosaminoglycans and collagen were produced. Glycosaminoglycans, for cartilage tissue engineering, are particularly important because it is the major molecule that withstands compressive forces during loadbearing. Tissue-engineered cartilage has even been grown in space within the Mir Space Station. Chondrocytes cultured within a polyglycolic acid scaffold showed inferior mechanical properties compared to Earth grown cartilage, likely due to the need for physical forces to assist in the remodeling process [51].

Bone tissue engineering began in the early 1990s out of a long-standing effort to develop new mechanical prostheses, a better understanding of biomechanical processes, and improved biocompatible materials. One of the earliest reports of the use of bone grafts came in 1892 where M'Gregor described that bone fragments lived after "transplantation" and facilitated vascularization and healing [52]. In the early 1900s, there were several attempts at hip arthroplasty (replacement) using materials, such as pig bladders, gold foil, glass, and acrylic [53]. Moore and Bohlman were the first surgeons to perform a total hip replacement with a metallic prosthesis in 1940 [54]. The patient had previously sustained a fracture of the neck of the right femur and had multiple operations including nailing of the fracture which was eventually complicated by a giant cell tumor. The tumor was resected as well as a long segment of bone. A metal prosthesis taking the shape of natural bone was constructed preoperatively and was later inserted at the operation. The prosthesis was made from a cobalt–chromium alloy and appeared intact after one-and-a-half years when the patient passed away from cardiac failure. Moore improved on his prosthesis by placing a gentle curve in the neck to well approximate the bone with fenestrations on the prosthesis that allowed ingrowth of bone grafts to enhance healing, nourishment, and blood supply [54].

Further refinements in material science enhanced the structure and compatibility of bioactive materials and were later used to coat prostheses, allowing inert metals to become more biologically active. Ideal materials may be biologically inert, biodegradable, and induce bone and vascular ingrowth with enough strength to withstand compressive forces. Bone substitutes may be used to heal large sections of bone removed due to tumor ingrowth or may be used for dental applications and periodontal surgery. Hydroxyapatite, a bioactive glass, is an inorganic material that has reliable mechanical properties and can integrate well into the body, forming a minimally inflammatory bone–material interface. Bioactive glasses have been used to carry molecules like growth factors to enhance bone formation [55]. Bioactive polymers are also used to deliver cells and growth factors to joint surfaces. Charles Vacanti and colleagues report the use of polyglycolic acid constructs to deliver chondrocytes to the distal femoral joint of rabbits whose joint surfaces have been denuded of articular cartilage. After 7 weeks, new cartilage was seen in 11 of 12 experimental animals [56–58]. A new class of materials termed porous inorganic metals, such as porous magnesium and porous iron, have also been investigated as a biocompatible material for bone tissue engineering [59].

Bioengineered skin is another early tissue that was recapitulated in the laboratory in the mid-1990s, but since then has received US Food and Drug Administration approval for the treatment of full-thickness diabetic foot ulcers. Neonatal dermal fibroblasts were obtained from human skin and expanded in culture. Cells were later grown onto thin meshes made from poly-lactide-co-glycolide, a biodegradable suture material. The new tissues could be frozen and produced vascular endothelial growth factor and other bioactive agents after thawing that enhance angiogenesis and wound healing capability [60]. This human fibroblast-derived dermal substitute is now marketed under the name Dermagraft®. Composite tissue allotransplantation is a new field encompassing extremity and face transplantation. This modality combines skin, bone, cartilage, blood vessel, and muscle, and is reviewed in Chapter 15.

One common theme that runs throughout these early tissue-engineered structures is their geometrical simplicity, as their two-dimensional planar structure, small size, and high porosity is enough to allow delivery of nutrients, proteins, and host cells to the transplanted tissue. Two other tissues that follow this paradigm and have recently been engineered include bladders [61] and tracheas [62]. Bladders were constructed from biodegradable polymers comprised of polyglycolic acid and collagen and seeded with smooth muscle and urothelial cells. Scaffolds were grown in culture and implanted into three patients. Reconstructive surgery using these engineered tissues yielded a 1.58-fold increase in bladder capacity postoperatively. The tissue-engineered trachea is processed by removing cells from a human donor trachea. Initially, exogenous cells were added prior to transplantation but recently a shorter culture time allowed for earlier transplantation with improved homing of recipient cells to the tissue-engineered graft after implantation [63]. Airway development is discussed by Paolo Macchiarini in Chapter 8 and engineered urinary conduits are reviewed by Anthony Atala in Chapter 7.

1.3.5 Cell Source

Perhaps one of the most elusive challenges in tissue engineering is to select the ideal cell source for tissue and organ regeneration. The determination of cell choice will likely depend upon the type of tissue that is to be engineered. Some tissues or organs with a large number of specialized and differentiated cells may be optimally developed with precursor cells that differentiate into individual, specialized cells depending upon their location within an organ—one example of this would be the kidney. Alternatively, mature cells may be used to form organs and tissues with slightly less cellular variation, but which require a large number of cells, such as muscle or liver. Different categories of candidate cells include embryonic stem cells, adult stem/progenitor cells, differentiated cells of either fetal/neonatal/adult origin, or induced pluripotent stem cells.

Embryonic stem cells are attractive because they can be expanded while still maintaining pluripotency, and then may be differentiated to produce cells from any embryonic germ cell layer. Research into human embryonic stem cells has accelerated over the early part of the century to encompass laboratories from 41 countries in 2006 compared to only 27 in 2002 [64]. With this escalated interest, new small molecules to aid in the differentiation process have been characterized to allow for increased efficiency and generation of larger numbers of mature cells, which will be needed as tissue-engineered organs and animal models become larger [64]. Vunjak-Novakovic et al. recently described engineering bone scaffolds containing cells differentiated from human embryonic stem cells [65]. Myocardium has also been engineered from human embryonic stem cells [66].

Adult stem and progenitor cells are perhaps a more optimal cell origin from an ethical standpoint and may be easily obtained from autologous or cadaveric sources. One drawback of adult stem/progenitor cells is their slow proliferative rate that may increase the number of progenitor cells needed or the time that it takes to repopulate an engineered tissue or organ.

Terminally differentiated adult cells may be reprogrammed by exogenous delivery of genes encoded in viruses, plasmids, or other specially designed vehicles. Genes that are typically used to create human-induced pluripotent stem cells are OCT3/4, Sox2, Klf4, cMyc, Nanog, and Lin28 [44–46]. Conceptually, induced pluripotent stem cells can be used to investigate drug interactions with donor tissues without the need for invasive biopsies. Induced pluripotent stem cells can be created from skin or mononuclear cells and then differentiated into specific tissues of the body. Likewise, cells may be expanded in culture after being obtained from an intended organ recipient and used to engineer donor organs in the laboratory with the hope of eliminating or minimizing immunosuppression, because the donor organ cells could be made from the same genetic background as the recipient. Even though the 2002 Nobel Prize for medicine or physiology was awarded to Shinya Yamanaka for his discovery of induced pluripotent stem cell technology, challenges remain to characterize genetic stability and potential to recapitulate the function of normal, adult cells of the body. The use of stem cells to modulate the body's immune response to transplanted tissues is discussed by Kathryn Wood in Chapter 16.

1.3.6 Soft Tissue Regeneration as Applied to Battlefield Medicine

Blast wounds to the extremities are the leading sites of major battlefield injuries incurred by combat soldiers and are typically localized to the lower extremities. However, with the increased use of improvised explosive devices (IEDs), upper extremity injuries are now more common in soldiers participating in modern-day combat theaters like Operation Iraqi Freedom. Of 935 warriors wounded in this theatre from 2004 to 2005, 71% sustained extremity injuries with the majority being localized to the upper extremity [67]. The Armed Forces Institute for Regenerative Medicine (AFIRM) began in 2008 and was designed as a 5-year $300 million initiative to create multi-institutional (academic and industrial) consortiums. The traditional paradigm of research funding was to approve individual projects for governmental support through NIH style grants. AFIRM, on the other hand, promoted collaboration and sharing of ideas, resources and data among collaborative members to fast track laboratory discoveries into clinical realities to improve the lives of servicemen and women. The five critical focus areas for AFIRM are: (i) limb and digit salvage, (ii) craniofacial reconstruction, (iii) scarless wound healing, (iv) burned repair, and (v) compartment syndrome. Within the first 3 years of AFIRM, over 30 patients were enrolled and treated in clinical trials and further clinical investigations were made through collaborations with industry [68]. Some projects included within the AFIRM initiative are: craniofacial reconstruction, composite tissue allografting including hand transplantation, burn repair with skin substitutes, limb and digit regenerative medicine, and bioscaffolds to promote wound healing.

1.3.7 Planar Acellular Dermal Matrices

The intrinsic function of cells, including their ability to proliferate, differentiate, and perform specialized functions, can often be influenced by the specific microenvironment that they inhabit. These microenvironments, commonly called "niches," are thought to provide cells with a context in which to properly function. These cellular cues are a combination of growth factors (both soluble and immobilized) and extracellular matrix molecules (such as fibronectin, laminin, and collagen) that present specialized ligands (like the RGD peptide) to cell surface

adhesion molecules in tissue-specific three-dimensional orientations. Stroma support cells also function to secrete growth factors and present functional ligands. In the late 1970s, Swarm et al. described a mouse poorly differentiated chondrosarcoma in which cancer cells were found imbedded within a dense extracellular matrix similar to basement membrane [69]. This cancer line was later termed the Englebreth−Holm−Swarm tumor and was later characterized to contain type IV collagen, laminin, and proteoglycans [70]. By the late 1980s, this matrix was termed Matrigel, and was able to influence the growth and function of fibroblasts and endothelial cells [71,72]. This extracellular matrix formulation was later found to maintain stem and progenitor cells in a relatively undifferentiated state and promote self-renewal. The protein and growth factor milieu of Matrigel has been further characterized [73] and the effect on various cells has been evaluated using highly sophisticated methods to determine transcription factor enhancement or repression extracellular cues presented by this matrix [74].

One of the drawback of Englebreth−Holm−Swarm tumor-derived matrix is that it lacks the specific three-dimensional orientation of an organ and that it is produced by a tumor-derived cell source. Accordingly, alternative acellular matrix components were investigated in the late 1990s and since that time porcine derived, glutaraldehyde-fixed heart valves became clinically available in addition to human- and animal-derived acellular dermal substitutes. In 1995, Wainwright et al. described the use of AlloDerm, a processed human acellular dermal matrix [75]. That report describes the de-epithelialization of the donor tissue and removal of fibroblasts from the dermis. This was later used to cover a portion of tissue overlaid by a section of split-thickness skin grafts after burned skin had earlier been excised in two patients. Wainwright notes that the acellular dermal matrix supported fibroblast infiltration, neovascularization, and epithelialization. AlloDerm (Life Cell, Branchburg, NJ) is currently used clinically for abdominal wall reconstruction.

In an effort to find alternative sources of acellular tissues, the small intestinal submucosa (SIS) and urinary bladder matrix (UBM) have been prepared in a similar manner to remove adjacent cellular compartments and wash the submucosa in gentle solutions that both sterilize and decellularize [76]. One of the first investigations of acellular SIS was as a vascular graft in dogs. Stephen Badylak and colleagues implanted autogenous SIS in the infra-renal aorta of dogs in a report published in 1989. After 52 weeks, two of 12 dogs remained alive with patent grafts. Nine of the 12 dogs were sacrificed within the first 52 weeks and the luminal surface of the graft contained a dense connective tissue matrix. In contrast to Matrigel, SIS and UBS can be utilized in a three-dimensional matrix environment to support cell growth and form larger tissues [77]. SIS was evaluated in abdominal wall reconstruction of dogs and rats [78] and humans with infected hernias [79]. Porcine SIS is now marketed for hernia repair under the name Surgisis (Cook Medical, Bloomington, IN). Acellular UBM is isolated by removing the serosa and smooth muscle layers and the transitional epithelium is removed by either mechanical agitation or treatment with salt solutions. The resulting submucosa tissue is then treated with dilute paracetic acid and washed with saline and water solutions. UBM has been investigated as a treatment for esophageal perforation in a dog model [80]. Porcine UBM is marketed under the name MatriStem (ACell, Columbia, MD) for wound care and pelvic floor, hernia and esophageal repair, as well as other surgical applications.

The ability of acellular two-dimensional, planar tissue scaffolds to promote cellular infiltration with minimal scarring is particularly attractive in infected wounds and tissue defects, where the implantation of synthetic, foreign body material is less desirable. This is, in part, likely due to turnover of the extracellular matrix that is produced by new, infiltrating cells and facilitated by the excretion of matrix metalloproteinases. Over the last several years, new strategies to remove cells from tissues and organs have been refined. Physical methods to remove cells include freezing and mechanical agitation. Chemicals that have been utilized in the laboratory to remove cells include ionic detergents (e.g., sodium dodecyl sulfate), nonionic detergents (e.g., Triton-X), zwitterionic detergents (e.g., CHAPS), and enzymes (e.g., trypsin, endonucleases) [81].

1.3.8 Vascularization of Tissue-Engineered Organs

The initial building block of any engineered tissue, or any functioning, living organ for that matter, is the individual cell. The first single-celled organism was present on the Earth over 3 billion years ago. It took another 2 billion years for multicellular organisms to appear on our planet. During this time, evolution allowed organisms to develop and adapt a process by which nutrients and waste material were delivered to and removed from individual cellular components of the evolving multicellular organism. Likewise, Judah Folkman's theory of tumor growth emphasizes the dependence of the growing cell mass upon angiogenesis to deliver nutrients. The development of larger organs for tissue engineering follows a similar paradigm in which large clusters of cells outgrow their nutritional supply as the organ becomes larger unless new blood vessels are developed to feed the growing biomass. This fundamental constraint is illustrated by a growing volume of cells that increases as a function of the radius to the third power, whereas the surface area used for

nutritional exchange increases as a function of the radius to the second power.

The solution to a growing organ is the incorporation of a vascular network to feed cells and return waste material and venous blood to the recipient's circulatory system. One of our early attempts at liver tissue engineering utilized a branching network of vascular channels using microfluidic etchings in silicon and incorporated a vascular tree into a tissue-engineered organ [82]. We later improved upon this system by using a computer model of microfluidic flow to design a micro-electro-mechanical system (MEMS) in which HepG2 hepatocytes were grown in between the capillary channels that connected to the circulatory system of a rat to support the growing hepatocytes over a short time period (Figure 1.2) [84]. Vascularized systems have also been used to facilitate gas exchange in tissue-engineered lungs [85].

1.3.9 Perfusion Decellularization, Using Nature's Vascular Network

The success of acellular two-dimensional, planar tissues to allow for cellular growth and differentiation with minimal inflammation was applied to the problem of developing a vascular tree to deliver nutrients to large organs. In 2008, Ott and Taylor described the technique of "perfusion deceleration" in which hearts from 12-week-old Fisher rats were perfused with an ionic detergent (sodium dodecyl sulfate) followed by a nonionic solution (Triton-X 100) [86]. As a surrogate for cell removal, DNA content of the decellularized heart was <4% of the native heart, indicating efficient decellularization. Scaffolds contained fibronectin, laminin, and collagens I and III and could be

FIGURE 1.2 A brass mold with vascular branching conduits that direct flow from the outer, portal triad region toward the inner, central vein for *in vitro* liver tissue engineering. A similar illustration was depicted in Vacanti et al. [83].

recellularized with rat neonatal cardiac cells using an intraparenchymal injection technique. The vasculature was reconstituted with rat aortic endothelial cells. The myocardium of the recellularized heart contracted upon electrical stimulation. This was the initial report of using an organ's vasculature as a conduit to deliver solutions to render an organ acellular. Since then, similar techniques have been used to remove cells from rat lungs (Figure 1.3) [87], livers [88–90], and porcine kidneys [91,92]. As noted with two-dimensional tissues, perfusion decellularization retains critical extracellular growth factors linked to the matrix. Soto-Gutierrez demonstrated that rat livers undergoing perfusion decellularization with trypsin and Triton-X 100 contained 50% hepatocyte growth factor and approximately 40% of basic fibroblast grown factor compared to the normal organ and Badylak showed that basic fibroblast grown factor and vascular endothelial growth factor were also retained in adipose tissue using Triton-X 100 or trypsin [93]. It should be noted that the latter study relies on fluid agitation and not perfusion through a vascular tree, however, the finding that growth factors persist after decellularization is particularly relevant to the concept that acellular microenvironment is biologically active, although devoid of living cells.

1.3.10 Specific Organ Transplantation History and Link to Engineering

1.3.10.1 Liver

An early model of liver transplantation in dogs was published by Staudacher et al. in Italy in 1952 [94]. The liver was transplanted at the orthotopic position, yet vascular connections were secured with stents instead of sutures, the donor hepatic artery was not used and instead the portal vein was arterialized by the recipient hepatic artery [95]. Later, the technique of total, orthotopic replacement of the liver was further developed simultaneously in dogs by Starzl et al. in Chicago [96] and Moore et al. in Boston [97]. Prerequisites for success were cool preservation of the donor organ and decompression of the splanchnic bed with external venovenous bypass.

Auxiliary liver concept was first described in dogs by Welch [98]. This model played an important role in clarifying the need of portal blood to prevent graft atrophy [99,100]. The King's College team successfully applied the concept to *in situ*, auxiliary partial liver transplantation for the treatment of acute liver failure and metabolic diseases in children [101,102].

In human orthotopic liver transplantation, technical refinements included over the years a flexible and safe procedure for multiple cadaveric organ procurement to avoid damage to the kidneys [103,104], the use of an external portal–caval–jugular bypass during the anhepatic phase which allowed for stressless vascular reconstruction and training of new generations of surgeons [105], preservation of the retrohepatic vena cava eliminating the need of a venous bypass [106,107], and appropriate techniques of biliary reconstruction [108].

Liver replacement in children poses specific questions related to the young age and low weight of most recipients and to the scarcity of size-matched organ donors. This was the trigger for developing variant techniques,

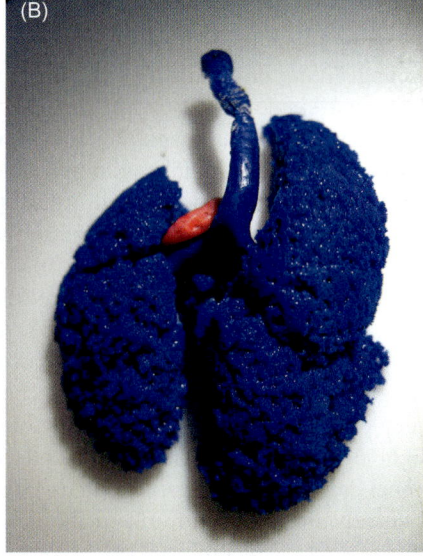

FIGURE 1.3   (A) A decellularized rodent lung and (B) a corrosion cast showing the preservation of tiny vascular channels after decellularization. *Reprinted with permission from Vacanti et al. [83]*

including the reduced size and split liver transplants. The former approach helped in launching liver transplantation in small children on a large scale [109—111], however, it has been dropped from routine practice because it does not extend the donor pool but simply diverts donor organs from the adult to the pediatric population. The first four cases of *ex situ* split transplantation were performed in 1988, three in Europe [112—114], one in the United States [115], with three long-term survivors (>15 years). The split liver transplant procedure remains somehow confined to highly experienced centers because of its technical complexity, the need for reinforced local resources and the difficulties of interinstitutional collaboration. Liver tissue engineering is reviewed in Chapter 10.

1.3.10.2 Kidney

The kidney has been the launching ramp of solid organ transplantation. From the technical point of view, no significant changes have been made since the original description by René Kuss, a French urologist [116].

As recalled earlier, the first success with kidney transplantation was between identical twins [117]. Besides this privileged setting, multiple attempts without immunosuppression in France [118], the United States [119], and the United Kingdom [120] failed. Progress came with the introduction of immunosuppression, Starzl being the first to perform a successful series of cases [14]. The kidney may be one of the most challenging organs to re-create in the laboratory and the development of bioartificial kidneys is discussed in Chapter 9.

1.3.10.3 Heart

The experimental model of heart transplantation was explored by Shumway and Lower in San Francisco [121]. The first heart transplantation in an adult patient was performed on December 3, 1967, by Barnard in Cape Town [122], followed 3 days later by Kantrowitz et al. in New York in a baby [123] and 1 month later by Shumway's team [124]. Similarly to the liver, consistent successes awaited the advent of cyclosporine in the early 1980s [125,126]. Cardiac regenerative medicine is discussed in an in-depth manner in Chapter 11.

1.3.10.4 Lung

The first lung transplantation in man was performed on November 18, 1968, in Ghent, Belgium, by Derom et al. [127] with a 10-month survival. The first long-term success in lung transplantation occurred at Stanford in 1981, with the transplantation of the heart and both lungs [128]. Bilateral lobar transplantation from a live donor has been developed in a few centers. Pulmonary engineering is presented in Chapter 14.

1.3.10.5 Pancreas

The first pancreas transplantation was performed in 1966 by the team of Kelly et al. [129], 3 years after the first kidney transplantation. Implantation of the pancreatic duct into the urinary bladder [130] competed with enteric drainage, the latter one taking the lead. Intraductal injection of neoprene, to obstruct the exocrine secretion, was introduced by Dubernard et al. in the 1970s [131] but did not stand the test of time. Most often, the pancreas is combined with the kidney, except in Minneapolis. Isolated pancreas transplantation is still facing reluctance from endocrinologists, because immunosuppression competes poorly with insulin injection. The ability to re-create pancreatic tissue and encapsulate islets of Langerhans is presented in Chapter 13.

1.3.10.6 Intestine

The first formally published case of intestinal transplantation was performed by Lillehei et al. in Minneapolis in 1967 [132]. Poor results were obtained using conventional immunosuppression. Of the 15 isolated bowel transplants performed between 1985 and 1990 using cyclosporine as the basic immunosuppressant, one adult transplanted in Kiel in 1988 [133] and one infant transplanted in Paris in 1989 [134] became the first long-term survivors with sufficient graft function. The first recipient of a combined liver—intestinal graft, performed in London, Ontario, in 1988, lived for several years [135]. Intestinal transplantation and regeneration is discussed in Chapter 12.

1.3.11 Concluding Remarks

Transplantation is an old discipline with references in ancient historical texts. Nonetheless, through scientific achievements, the modern era of solid organ transplantation has witnessed the development of the discipline from an experimental application to an accepted and optimal treatment for organ failure. The inability of cadaveric organs for transplantation to keep pace with the growing wait list of patients needing a replacement organ to sustain life and improve quality has prompted the transplant community to look for alternative sources of organs for transplantation. To be sure, other methods to achieve this, through organs from living donors and split liver transplantation, are important measures to increase the donor pool, yet these are unlikely to ultimately overcome the gap between the supply of organs for transplantation and those waiting for them.

Comparatively, regenerative medicine and tissue engineering are newer disciplines. Transplantation has become the accepted therapy of choice for patients because scientific progress and experimental investigations have allowed it to move forward. Such is true with regenerative medicine. The chapters contained within this text illustrate the current state of art in regenerative medicine and organ/tissue engineering with current and future applications to organ transplantation and tissue reconstructive surgery. For some applications, regenerative medicine and transplantation have already met and for others the scientific method of experimentation is still ongoing to one day intersect as a future therapy for patients in need.

REFERENCES

[1] Carlson BM. Principles of regenerative biology. Amsterdam/Burlington, MA: Elsevier/Academic Press; 2007.

[2] Kaiser LR. The future of multihospital systems. Top Health Care Financ 1992;18:32–45.

[3] Tilney NL. Transplant: from myth to reality. New Haven: Yale University Press; 2003.

[4] Lillehei RC, Manax WG, Bloch JH, Eyal Z, Hidalgo F, Longerbeam JK. *In vitro* preservation of whole organs by hypothermia and hyperbaric oxygenation. Cryobiology 1964;1:181–93.

[5] Belzer FO, Ashby BS, Gulyassy PF, Powell M. Successful seventeen-hour preservation and transplantation of human-cadaver kidney. N Engl J Med 1968;278:608–10.

[6] Collins GM, Bravo-Shugarman M, Terasaki PI. Kidney preservation for transportation. Initial perfusion and 30 hours' ice storage. Lancet 1969;2:1219–22.

[7] Jamieson NV, Sundberg R, Lindell S, Claesson K, Moen J, Vreugdenhil PK, et al. Preservation of the canine liver for 24–48 hours using simple cold storage with UW solution. Transplantation 1988;46:517–22.

[8] Kolff WJ. The artificial kidney. J Mt Sinai Hosp NY 1947;14:71–9.

[9] Quinton W, Dillard D, Scribner BH. Cannulation of blood vessels for prolonged hemodialysis. Trans Am Soc Artif Intern Organs 1960;6:104–13.

[10] Calne RY. The rejection of renal homografts. Inhibition in dogs by 6-mercaptopurine. Lancet 1960;1:417–8.

[11] Calne RY, Alexandre GP, Murray JE. A study of the effects of drugs in prolonging survival of homologous renal transplants in dogs. Ann NY Acad Sci 1962;99:743–61.

[12] Borel JF, Feurer C, Gubler HU, Stahelin H. Biological effects of cyclosporin A: a new antilymphocytic agent. Agents Actions 1976;6:468–75.

[13] Calne RY, Rolles K, White DJ, Thiru S, Evans DB, McMaster P, et al. Cyclosporin A initially as the only immunosuppressant in 34 recipients of cadaveric organs: 32 kidneys, 2 pancreases, and 2 livers. Lancet 1979;2:1033–6.

[14] Starzl TE, Weil III R, Iwatsuki S, Klintmalm G, Schroter GP, Koep LJ, et al. The use of cyclosporin A and prednisone in cadaver kidney transplantation. Surg Gynecol Obstet 1980;151:17–26.

[15] Starzl TE. The puzzle people: memoirs of a transplant surgeon. Pittsburgh: University of Pittsburgh Press; 1992.

[16] Starzl TE, Weil R, Koep LJ, Iwaki Y, Terasaki PI, Schroter GP. Thoracic duct drainage before and after cadaveric kidney transplantation. Surg Gynecol Obstet 1979;149:815–21.

[17] Starzl TE, Iwatsuki S, Klintmalm G, Schroter GP, Weil III R, Koep LJ, et al. Liver transplantation, 1980, with particular reference to cyclosporin A. Transplant Proc 1981;13:281–5.

[18] Lerut J, Stieber AC, Makowka L, Esquivel CO, Iwatsuki S, Gordon RD, et al. Long-term results of orthotopic liver transplantation during the cyclosporin era. 393 orthotopic liver transplantations accomplished in 313 consecutive patients at the Pittsburgh Transplantation Center. Helv Chir Acta 1989;56:405–20.

[19] Metchnikov I. Etudes sur la résorption des cellules. Ann Inst Pasteur 1899;13:737–69.

[20] Starzl TE, Marchioro TL, Porter KA, Iwasaki Y, Cerilli GJ. The use of heterologous antilymphoid agents in canine renal and liver homotransplantation and in human renal homotransplantation. Surg Gynecol Obstet 1967;124:301–8.

[21] Kohler G, Milstein C. Continuous cultures of fused cells secreting antibody of predefined specificity. Nature 1975;256:495–7.

[22] Starzl TE. Fk-506—a potential breakthrough in immunosuppression. Transplant Proc 1987;19:103.

[23] Starzl TE, Todo S, Fung J, Demetris AJ, Venkataramman R, Jain A. FK 506 for liver, kidney, and pancreas transplantation. Lancet 1989;2:1000–4.

[24] Londono MC, Danger R, Giral M, Soulillou JP, Sanchez-Fueyo A, Brouard S. A need for biomarkers of operational tolerance in liver and kidney transplantation. Am J Transplant 2012;12:1370–7.

[25] Calne R, Friend P, Moffatt S, Bradley A, Hale G, Firth J, et al. Prope tolerance, perioperative campath 1H, and low-dose cyclosporin monotherapy in renal allograft recipients. Lancet 1998;351:1701–2.

[26] Reding R, Gras J, Bourdeaux C, Wieers G, Truong QD, Latinne D, et al. Stepwise minimization of the immunosuppressive therapy in pediatric liver transplantation. A conceptual approach towards operational tolerance. Acta Gastroenterol Belg 2005;68:320–2.

[27] Londono MC, Lopez MC, Sanchez-Fueyo A. Minimization of immunosuppression in adult liver transplantation: new strategies and tools. Curr Opin Organ Transplant 2010 Sep 30 [Epub ahead of print].

[28] Sarwal MM. Out with the old, in with the new: immunosuppression minimization in children. Curr Opin Organ Transplant 2008;13:513–21.

[29] Mollaret P. Le coma dépassé. Rev Neurol 1959;101:3–15.

[30] A definition of irreversible coma. Report of the ad hoc committee of the Harvard Medical School to examine the definition of brain death. J Am Med Assoc 1968;205:337–40.

[31] Morelle J, Alexandre G, Michielsen P, Vanypersele C. The problem of renal graft. Bull Acad R Med Belg 1964;4:157–75.

[32] Kootstra G, Daemen JH, Oomen AP. Categories of non-heart-beating donors. Transplant Proc 1995;27:2893–4.

[33] Merrill JP, Murray JE, Harrison JH, Guild WR. Successful homotransplantation of the human kidney between identical twins. J Am Med Assoc 1956;160:277–82.

[34] Starzl TE, Schroter GP, Hartmann NJ, Barfield N, Taylor P, Mangan TL. Long-term (25-year) survival after renal homotransplantation—the world experience. Transplant Proc 1990;22:2361–5.

[35] Delmonico F, Council of the Transplantation. A report of the Amsterdam Forum on the care of the live kidney donor: data and medical guidelines. Transplantation 2005;79:53–66.

[36] Barr ML, Belghiti J, Villamil FG, Pomfret EA, Sutherland DS, Gruessner RW, et al. A report of the Vancouver Forum on the care of the live organ donor: lung, liver, pancreas, and intestine data and medical guidelines. Transplantation 2006;81: 1373–85.

[37] Dausset J. Iso-leuko-antibodies. Acta Haematol 1958;20:156–66.

[38] Opelz G, Dohler B. Effect of human leukocyte antigen compatibility on kidney graft survival: comparative analysis of two decades. Transplantation 2007;84:137–43.

[39] Earle WR, Schilling EL, Shannon Jr JE. Growth of animal tissue cells on three-dimensional substrates. J Natl Cancer Inst 1951;12:179–93.

[40] Ehrmann RL, Gey GO. The growth of cells on a transparent gel of reconstituted rat-tail collagen. J Natl Cancer Inst 1956;16:1375–403.

[41] Bell E, Ivarsson B, Merrill C. Production of a tissue-like structure by contraction of collagen lattices by human fibroblasts of different proliferative potential *in vitro*. Proc Natl Acad Sci USA 1979;76:1274–8.

[42] Langer R, Vacanti JP. Tissue engineering. Science 1993;260:920–6.

[43] Petit-Zeman S. Regenerative medicine. Nat Biotechnol 2001;19:201–6.

[44] Takahashi K, Tanabe K, Ohnuki M, Narita M, Ichisaka T, Tomoda K, et al. Induction of pluripotent stem cells from adult human fibroblasts by defined factors. Cell 2007;131:861–72.

[45] Takahashi K, Yamanaka S. Induction of pluripotent stem cells from mouse embryonic and adult fibroblast cultures by defined factors. Cell 2006;126:663–76.

[46] Yu J, Vodyanik MA, Smuga-Otto K, Antosiewicz-Bourget J, Frane JL, Tian S, et al. Induced pluripotent stem cell lines derived from human somatic cells. Science 2007;318:1917–20.

[47] Tencer AF, Mooney V, Brown KL, Silva PA. Compressive properties of polymer coated synthetic hydroxyapatite for bone grafting. J Biomed Mater Res 1985;19:957–69.

[48] Mooney DJ, Park S, Kaufmann PM, Sano K, McNamara K, Vacanti JP, et al. Biodegradable sponges for hepatocyte transplantation. J Biomed Mater Res 1995;29:959–65.

[49] Kim WS, Vacanti JP, Cima L, Mooney D, Upton J, Puelacher WC, et al. Cartilage engineered in predetermined shapes employing cell transplantation on synthetic biodegradable polymers. Plast Reconstr Surg 1994;94:233–7; discussion 238–40.

[50] Freed LE, Vunjak-Novakovic G. Microgravity tissue engineering. *In vitro* cellular & developmental biology. Animal 1997;33:381–5.

[51] Freed LE, Langer R, Martin I, Pellis NR, Vunjak-Novakovic G. Tissue engineering of cartilage in space. Proc Natl Acad Sci USA 1997;94:13885–90.

[52] M'Gregor AN. The repair of bone, with special reference to transplantation and other artificial aids. J Anat Physiol 1892;26:220–30.

[53] Gomez PF, Morcuende JA. Early attempts at hip arthroplasty—1700s to 1950s. Iowa Orthop J 2005;25:25–9.

[54] Moore AT, Bohlman HR. Metal hip joint. A case report. J Bone Joint Surg 1943;25:688–92.

[55] Hum J, Boccaccini AR. Bioactive glasses as carriers for bioactive molecules and therapeutic drugs: a review. J Mater Sci Mater Med 2012;23:2317–33.

[56] Vacanti CA, Kim W, Schloo B, Upton J, Vacanti JP. Joint resurfacing with cartilage grown *in situ* from cell–polymer structures. Am J Sports Med 1994;22:485–8.

[57] Vacanti CA, Upton J. Tissue-engineered morphogenesis of cartilage and bone by means of cell transplantation using synthetic biodegradable polymer matrices. Clin Plast Surg 1994;21:445–62.

[58] Vacanti CA, Vacanti JP. Bone and cartilage reconstruction with tissue engineering approaches. Otolaryngol Clin North Am 1994;27:263–76.

[59] Yusop AH, Bakir AA, Shaharom NA, Abdul Kadir MR, Hermawan H. Porous biodegradable metals for hard tissue scaffolds: a review. Int J Biomater 2012;2012:641430.

[60] Mansbridge J, Liu K, Patch R, Symons K, Pinney E. Three-dimensional fibroblast culture implant for the treatment of diabetic foot ulcers: metabolic activity and therapeutic range. Tissue Eng 1998;4:403–14.

[61] Atala A, Bauer SB, Soker S, Yoo JJ, Retik AB. Tissue-engineered autologous bladders for patients needing cystoplasty. Lancet 2006;367:1241–6.

[62] Macchiarini P, Jungebluth P, Go T, Asnaghi MA, Rees LE, Cogan TA, et al. Clinical transplantation of a tissue-engineered airway. Lancet 2008;372:2023–30.

[63] Jungebluth P, Bader A, Baiguera S, Moller S, Jaus M, Lim ML, et al. The concept of *in vivo* airway tissue engineering. Biomaterials 2012;33:4319–26.

[64] Ben-David U, Kopper O, Benvenisty N. Expanding the boundaries of embryonic stem cells. Cell Stem Cell 2012;10:666–77.

[65] Marolt D, Campos IM, Bhumiratana S, Koren A, Petridis P, Zhang GP, et al. Engineering bone tissue from human embryonic stem cells. Proc Natl Acad Sci USA 2012;109:8705–9.

[66] Soong PL, Tiburcy M, Zimmermann WH. Cardiac differentiation of human embryonic stem cells and their assembly into engineered heart muscle. Curr Protoc Cell Biol 2012;23:23–8.

[67] Dougherty AL, Mohrle CR, Galarneau MR, Woodruff SI, Dye JL, Quinn KH. Battlefield extremity injuries in operation Iraqi freedom. Injury 2009;40:772–7.

[68] Dean W. The armed forces institute of regenerative medicine: a collaborative approach to department of defense-relevant research. Regen Med 2011;6:71–4.

[69] Orkin RW, Gehron P, McGoodwin EB, Martin GR, Valentine T, Swarm R. A murine tumor producing a matrix of basement membrane. J Exp Med 1977;145:204–20.

[70] Kleinman HK, McGarvey ML, Liotta LA, Robey PG, Tryggvason K, Martin GR. Isolation and characterization of type IV procollagen, laminin, and heparan sulfate proteoglycan from the EHS sarcoma. Biochemistry 1982;21:6188–93.

[71] McGuire PG, Orkin RW. Isolation of rat aortic endothelial cells by primary explant techniques and their phenotypic modulation by defined substrata. Lab Invest 1987;57:94–105.

[72] Emonard H, Grimaud JA, Nusgens B, Lapiere CM, Foidart JM. Reconstituted basement-membrane matrix modulates fibroblast activities *in vitro*. J Cell Physiol 1987;133:95–102.

[73] Hughes CS, Postovit LM, Lajoie GA. Matrigel: a complex protein mixture required for optimal growth of cell culture. Proteomics 2010;10:1886–90.

[74] Weiss MS, Penalver Bernabe B, Bellis AD, Broadbelt LJ, Jeruss JS, Shea LD. Dynamic, large-scale profiling of transcription factor activity from live cells in 3D culture. PloS One 2010;5:14026.

[75] Wainwright JM, Czajka CA, Patel UB, Freytes DO, Tobita K, Gilbert TW, et al. Preparation of cardiac extracellular matrix from an intact porcine heart. Tissue Eng Part C Methods 2010; 16:525–32.

[76] Gilbert TW, Freund JM, Badylak SF. Quantification of DNA in biologic scaffold materials. J Surg Res 2009;152:135–9.

[77] Badylak SF, Record R, Lindberg K, Hodde J, Park K. Small intestinal submucosa: a substrate for *in vitro* cell growth. J Biomater Sci Polym Ed 1998;9:863–78.

[78] Badylak SF, Kokini K, Tullius B, Simmons-Byrd A, Morff R. Morphologic study of small intestinal submucosa as a body wall repair device. J Surg Res 2002;103:190–202.

[79] Franklin Jr ME, Gonzalez Jr JJ, Michaelson RP, Glass JL, Chock DA. Preliminary experience with new bioactive prosthetic material for repair of hernias in infected fields. Hernia: J Hernias Abdom Wall Surg 2002;6:171–4.

[80] Badylak S, Meurling S, Chen M, Spievack A, Simmons-Byrd A. Resorbable bioscaffold for esophageal repair in a dog model. J Pediatr Surg 2000;35:1097–103.

[81] Gilbert TW, Sellaro TL, Badylak SF. Decellularization of tissues and organs. Biomaterials 2006;27:3675–83.

[82] Kaihara S, Borenstein J, Koka R, Lalan S, Ochoa ER, Ravens M, et al. Silicon micromachining to tissue engineer branched vascular channels for liver fabrication. Tissue Eng 2000;6:105–17.

[83] Vacanti JP. Tissue engineering and the road to whole organs. Br J Surg 2012;99:451–3.

[84] Hsu WM, Carraro A, Kulig KM, Miller ML, Kaazempur-Mofrad M, Weinberg E, et al. Liver-assist device with a microfluidics-based vascular bed in an animal model. Ann Surg 2010;252:351–7.

[85] Hoganson DM, Pryor II HI, Vacanti JP. Tissue engineering and organ structure: a vascularized approach to liver and lung. Pediatr Res 2008;63:520–6.

[86] Ott HC, Matthiesen TS, Goh SK, Black LD, Kren SM, Netoff TI, et al. Perfusion-decellularized matrix: using nature's platform to engineer a bioartificial heart. Nat Med 2008;14:213–21.

[87] Ott HC, Clippinger B, Conrad C, Schuetz C, Pomerantseva I, Ikonomou L, et al. Regeneration and orthotopic transplantation of a bioartificial lung. Nat Med 2010;16:927–33.

[88] Soto-Gutierrez A, Zhang L, Medberry C, Fukumitsu K, Faulk D, Jiang H, et al. A whole-organ regenerative medicine approach for liver replacement. Tissue Eng Part C Methods 2011;17:677–86.

[89] Uygun BE, Soto-Gutierrez A, Yagi H, Izamis ML, Guzzardi MA, Shulman C, et al. Organ reengineering through development of a transplantable recellularized liver graft using decellularized liver matrix. Nat Med 2010;16:814–20.

[90] Baptista PM, Siddiqui MM, Lozier G, Rodriguez SR, Atala A, Soker S. The use of whole organ decellularization for the generation of a vascularized liver organoid. Hepatology. 2010.

[91] Orlando G, Farney AC, Iskandar SS, Mirmalek-Sani SH, Sullivan DC, Moran E, et al. Production and implantation of renal extracellular matrix scaffolds from porcine kidneys as a platform for renal bioengineering investigations. Ann Surg 2012;256:363–70.

[92] Sullivan DC, Mirmalek-Sani SH, Deegan DB, Baptista PM, Aboushwareb T, Atala A, et al. Decellularization methods of porcine kidneys for whole organ engineering using a high-throughput system. Biomaterials 2012;33:7756–64.

[93] Brown BN, Freund JM, Han L, Rubin JP, Reing JE, Jeffries EM, et al. Comparison of three methods for the derivation of a biologic scaffold composed of adipose tissue extracellular matrix. Tissue Eng Part C Methods 2011;17:411–21.

[94] Staudacher V. Transplantation of an organ with vascular anastomoses. Riforma Med 1952;66:1060.

[95] Busuttil RW, De Carlis LG, Mihaylov PV, Gridelli B, Fassati LR, Starzl TE. The first report of orthotopic liver transplantation in the Western world. Am J Transplant 2012;12:1385–7.

[96] Starzl TE, Kaupp HA, Brock DR, Lazarus RE, Johnson RV. Reconstructive problems in canine liver homotransplantation with special reference to the postoperative role of hepatic venous flow. Surg Gynecol Obstet 1960;111:733–43.

[97] Moore FD, Wheele HB, Demissianos HV, Smith LL, Balankura O, Abel K, et al. Experimental whole-organ transplantation of the liver and of the spleen. Ann Surg 1960;152:374–87.

[98] Welch CS. A note on transplantation of the whole liver in dogs. Transplant Bull 1955;2:54–5.

[99] Marchioro TL, Porter KA, Dickinson TC, Faris TD, Starzi TE. Physiologic requirements for auxiliary liver homotransplantation. Surg Gynecol Obstet 1965;121:17–31.

[100] Starzl TE, Francavilla A, Halgrimson CG, Francavilla FR, Porter KA, Brown TH, et al. The origin, hormonal nature, and action of hepatotrophic substances in portal venous blood. Surg Gynecol Obstet 1973;137:179–99.

[101] Faraj W, Dar F, Bartlett A, Melendez HV, Marangoni G, Mukherji D, et al. Auxiliary liver transplantation for acute liver failure in children. Ann Surg 2010;251:351–6.

[102] Rela M, Muiesan P, Andreani P, Gibbs P, Mieli-Vergani G, Mowat AP, et al. Auxiliary liver transplantation for metabolic diseases. Transplant Proc 1997;29:444–5.

[103] Starzl TE, Hakala TR, Shaw BW, Hardesty RL, Rosenthal TJ, Griffith BP, et al. A flexible procedure for multiple cadaveric organ procurement. Surg Gynecol Obstet 1984;158:223–30.

[104] Starzl TE, Miller C, Broznick B, Makowka L. An improved technique for multiple organ harvesting. Surg Gynecol Obstet 1987;165:343–8.

[105] Griffith BP, Shaw BW, Hardesty RL, Iwatsuki S, Bahnson HT, Starzl TE. Veno-venous bypass without systemic anticoagulation for transplantation of the human liver. Surg Gynecol Obstet 1985;160:270–2.

[106] Tzakis A, Todo S, Starzl TE. Orthotopic liver transplantation with preservation of the inferior vena cava. Ann Surg 1989;210:649–52.

[107] Starzl TE, Iwatsuki S, Esquivel CO, Todo S, Kam I, Lynch S, et al. Refinements in the surgical technique of liver transplantation. Semin Liver Dis 1985;5:349–56.

[108] Lerut J, Gordon RD, Iwatsuki S, Esquivel CO, Todo S, Tzakis A, et al. Biliary tract complications in human orthotopic liver transplantation. Transplantation 1987;43:47−51.

[109] Bismuth H, Houssin D. Reduced-sized orthotopic liver graft in hepatic transplantation in children. Surgery 1984; 95:367−70.

[110] Broelsch CE, Emond JC, Thistlethwaite JR, Rouch DA, Whitington PF, Lichtor JL. Liver transplantation with reduced-size donor organs. Transplantation 1988;45:519−24.

[111] Otte JB, de Ville de Goyet J, Sokal E, Alberti D, Moulin D, de Hemptinne B, et al. Size reduction of the donor liver is a safe way to alleviate the shortage of size-matched organs in pediatric liver transplantation. Ann Surg 1990;211:146−57.

[112] Bismuth H, Morino M, Castaing D, Gillon MC, Descorps Declere A, Saliba F, et al. Emergency orthotopic liver transplantation in two patients using one donor liver. Br J Surg 1989;76:722−4.

[113] Pichlmayr R, Ringe B, Gubernatis G, Hauss J, Bunzendahl H. Transplantation of a donor liver to 2 recipients (splitting transplantation)—a new method in the further development of segmental liver transplantation. Langenbecks Archiv Chirurgie 1988;373:127−30.

[114] Otte JB, de Ville de Goyet J, Alberti D, Balladur P, de Hemptinne B. The concept and technique of the split liver in clinical transplantation. Surgery 1990;107:605−12.

[115] Emond JC, Whitington PF, Thistlethwaite JR, Cherqui D, Alonso EA, Woodle IS, et al. Transplantation of two patients with one liver. Analysis of a preliminary experience with "split-liver" grafting. Ann Surg 1990;212:14−22.

[116] Kuss R, Teinturier J, Milliez P. Some attempts at kidney transplantation in man. Mem Acad Chir (Paris) 1951;77:755−64.

[117] Merrill JP, Murray JE, Harrison JH, Guild WR. Successful homotransplantation of the human kidney between identical twins. J Am Med Assoc 1984;251:2566−71.

[118] Michon L, Hamburger J, Oeconomos N, Delinotte P, Richet G, Vaysse J, et al. An attempted kidney transplantation in man: medical and biological aspects. Presse Med 1953;61:1419−23.

[119] Hume DM, Merrill JP, Miller BF. Homologous transplantation of human kidneys. J Clin Invest 1952;31:640−1.

[120] Woodruff MF, Robson JS, McWhirter R, Nolan B, Wilson TI, Lambie AT, et al. Transplantation of a kidney from a brother to sister. Br J Urol 1962;34:3−14.

[121] Lower RR, Shumway NE. Studies on orthotopic homotransplantation of the canine heart. Surg Forum 1960;11:18−9.

[122] Barnard CN. The operation. A human cardiac transplant: an interim report of a successful operation performed at Groote Schuur Hospital, Cape Town. S Afr Med J 1967;41:1271−4.

[123] Kantrowitz A, Haller JD, Joos H, Cerruti MM, Carstensen HE. Transplantation of the heart in an infant and an adult. Am J Cardiol 1968;22:782−90.

[124] Stinson EB, Dong E, Schroeder JS, Harrison DC, Shumway NE. Initial clinical experience with heart transplantation. Am J Cardiol 1968;22:791−803.

[125] Dein JR, Oyer PE, Stinson EB, Starnes VA, Shumway NE. Cardiac retransplantation in the cyclosporine era. Ann Thorac Surg 1989;48:350−5.

[126] Cabrol C, Gandjbakhch I, Pavie A, Cabrol A, Mattei MF, Lienhart A, et al. Heart transplantation. experience at La Pitie hospital. Apropos of 82 cases. Arch Mal Coeur Vaiss 1984;77:1427−33.

[127] Derom F, Barbier F, Ringoir S, Versieck J, Rolly G, Berzsenyi G, et al. Ten-month survival after lung homotransplantation in man. J Thorac Cardiovasc Surg 1971;61:835−46.

[128] Shumway NE. Thoracic transplantation. World J Surg 2000;24:811−4.

[129] Kelly WD, Lillehei RC, Merkel FK, Idezuki Y, Goetz FC. Allotransplantation of the pancreas and duodenum along with the kidney in diabetic nephropathy. Surgery 1967; 61:827−37.

[130] Cook K, Sollinger HW, Warner T, Kamps D, Belzer FO. Pancreaticocystostomy: an alternative method for exocrine drainage of segmental pancreatic allografts. Transplantation 1983;35:634−6.

[131] Dubernard JM, Traeger J, Neyra P, Touraine JL, Tranchant D, Blanc-Brunat N. A new method of preparation of segmental pancreatic grafts for transplantation: trials in dogs and in man. Surgery 1978;84:633−9.

[132] Lillehei RC, Idezuki Y, Kelly WD, Najarian JS, Merkel FK, Goetz FC. Transplantation of the intestine and pancreas. Transplant Proc 1969;1:230−8.

[133] Deltz E, Schroeder P, Gebhardt H, Gundlach M, Engemann R, Timmermann W. First successful clinical small intestine transplantation. Tactics and surgical technic. Chirurg 1989;60:235−9.

[134] Goulet O, Revillon Y, Brousse N, Jan D, Canion D, Rambaud C, et al. Successful small bowel transplantation in an infant. Transplantation 1992;53:940−3.

[135] Grant D, Abu-Elmagd K, Reyes J, Tzakis A, Langnas A, Fishbein T, et al. 2003 report of the intestine transplant registry: a new era has dawned. Ann Surg 2005;241:607−13.

Chapter 2

Solid Organ Transplantation: Has the Promise Been Kept and the Needs Met?

Marco Carbone[a,b] and James M. Neuberger[b,c]

[a]Liver Unit, Addenbrooke's Hospital, Cambridge, UK, [b]Organ Donation and Transplantation, NHS Blood and Transplant, Bristol, [c]Liver Unit, Queen Elizabeth Hospital, Birmingham, UK

Chapter Outline

2.1 INTRODUCTION

Organ transplantation (OTx) represents the only curative treatment for end-stage liver, heart, and lung diseases but it may also represent a lifesaving option following failure of other organs. Although end-stage kidney disease patients can be effectively treated with renal replacement therapy (like hemodialysis), kidney transplantation (KT) is generally accepted as the best treatment for both quality of life and cost effectiveness in the great majority of those with end-stage kidney disease. The failure of the endocrine pancreas, with consequent diabetes mellitus, is generally treated by insulin therapy; however, some patients with hypoglycemia unawareness and brittle diabetes may benefit from pancreas transplantation. In the same way, the standard treatment for patients with intestinal failure is parenteral nutrition. However, for those patients who develop life-threatening complications such as recurrent infections, thrombosis/blockage of major blood vessels, and intestinal failure-associated liver disease, bowel transplantation is a lifesaving option.

The road to successful organ grafting in humans has been long and fraught with problems. Over the past half century, there have been major advances in the field of transplantation including improved surgical techniques, anesthesia, microbiology, interventional radiology, immunosuppression (IS), and per/post-operative care. All these elements have substantially improved patient and graft survival (particularly for kidney and liver transplantation) although outcomes for both intestinal and pancreas transplantation are still poor (Table 2.1). However, this success is only partial: long-term survival of patients is reduced compared with age- and sex-matched controls; and the shortage of organs compared with need, which translates in the mortality of those who do not receive the organ. Kidney transplant (KT) has improved both short- and long-term survival, although there are remarkable disparities between countries [1]. Risk-adjusted survival after liver transplantation (LT) has improved primarily because of a reduction in early/postoperative deaths. Adult liver recipients who survive the first postoperative year have an estimated 7.7-year reduction in life expectancy compared with an age- and sex-matched population. The extent of reduced life expectancy in LT depends on several factors, including recipient's age, gender, and indication [2]. Patient and graft survival after OTx is limited by many factors, including the recurrence of the original disease: this includes not only malignancy but also viral infections (like hepatitis C) after LT and recurrence of autoimmune diseases leading to graft failure (such as recurrent glomerular disease in KT or autoimmune hepatitis in LT). Furthermore, the consequences of long-term IS may limit survival because of the increased risk of fatal infections, malignancies, cerebrovascular and cardiovascular diseases, and also drug-related side-effects, such as calcineurin inhibitor (CNI)-related diabetes mellitus and chronic renal failure.

Another major problem in OTx is organ shortage. The gap between available organs and potential recipients is increasing (Figure 2.1) and it must be stressed that the numbers on the waiting list underestimate the need since many patients who might benefit are not listed. This shortage gives rise to serious ethical and practical dilemmas in balancing the competing needs of equity of access, to this limited yet lifesaving resource. Organs can be allocated according to need, benefit, or utility. The selection of which approach is adopted will determine who receives and who

is denied access to this form of treatment. The increased risk of death on the waiting list has altered the balance of risk when higher risk organs are being considered, so more high risk organs will be used.

The major challenges in OTx remain in expanding the pool of organs suitable for donation and improving the outcomes of those organs that are offered. Developing effective methods to induce transplant tolerance, as a means to eliminate the requirement for long-term IS, will help to reduce the premature loss of life following successful OTx.

2.2 TOLERANCE IN ORGAN TRANSPLANTATION

The currently available armamentarium of immunosuppressive (IS) drugs includes agents which target different steps in the cascade of events leading to allograft rejection, such as the inhibition of antigen-presenting cells, the development and maturation of B cells and T cells (corticosteroids, sirolimus/everolimus), the blockade of production of interleukin (IL)-2 and other cytokines which play a critical role in the regulation of growth, the differentiation and survival of activated T cells (calcineurin inhibitors [CNI] cyclosporin and tacrolimus), the blockade of response of leukocytes to growth factors (sirolimus/everolimus, IL-2 receptor alpha-chain-specific antibodies), and the inhibition of DNA synthesis (azathioprine, mycophenolate). There are also newer IS drugs, such as belatacept, a fusion protein that inhibits T-cell activation by binding to CD80 and CD86, recently approved by the US Food and Drug Administration (FDA) for KT. Bortezomib, a proteasome inhibitor approved for the treatment of multiple myeloma, is under investigation in KT for desensitization and for the

TABLE 2.1 Patient Survival after Adult Transplant from DBD (2004–2006)

	% Patient Survival (95% Confidence Interval)			
	One Year		**Five Years**	
Kidney	97	96–97	90	88–91
Liver	89	88–91	77	75–80
Heart	82	77–85	72	67–77
Lungs	77	72–81	54	48–59
Pancreas	98	84–100	94	79–99
Intestine	75	52–88	–	–

DBD, donor after brain death.
Data from transplant Activity Report 2011–2012—organ donation and transplantation—NHS Blood and Transplant, United Kingdom, 2011.

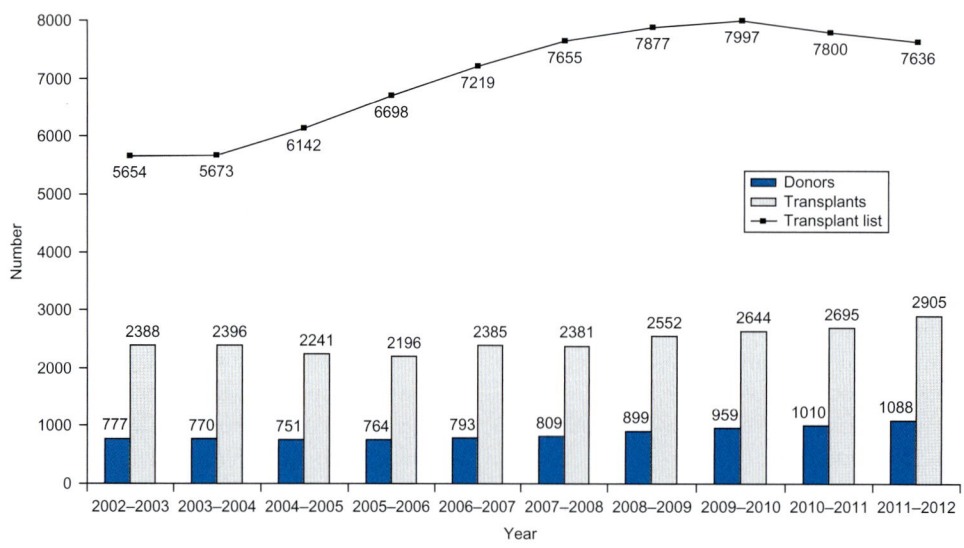

FIGURE 2.1   Number of deceased donors and transplants in the United Kingdom, April 1, 2002–March 31, 2012, and patients on the active transplant lists at March 31, 2012.

treatment of early acute antibody-mediated rejection. There is also tofacitinib, a Janus kinase (JAK)-inhibitor which showed potential for CNI-free IS.

Although these agents have contributed to a remarkable improvement of survival after OTx, they are associated with certain problems. First, the adverse consequence of the relatively nonspecific IS property of these agents is the increased risk of opportunistic infections, mostly in the early period, and malignancies in the long term. IS drugs have a high risk of toxicity, such as nephrotoxicity associated with CNI, hypertension, and cardiovascular disease associated with corticosteroids. Moreover, IS is not completely successful in preventing chronic rejection which may be antibody- or lymphocyte-mediated and represents a major cause of graft loss. Therefore, it appears essential to develop strategies which allow the minimization of IS drugs and/or induce and maintain transplant tolerance, a state in which the allograft is specifically accepted without the need for long-term IS.

Different tissues and organs have different 'tolerability,' with the liver being the most tolerogenic, followed by heart, kidney, pancreatic islet cells, and skin. Over the past two decades, studies primarily conducted in rodents extended to large animals and primates (and in different tissue and organs) have shown that certain immunologic interventions induce transplantation tolerance [3]. Unfortunately, translation of these strategies into the clinic has been remarkably difficult [3−5]. However, human transplant recipients occasionally develop spontaneous operational tolerance (SOT) to their graft, a phenomenon defined by the maintenance of stable graft function in the absence of harmful immune responses in recipients receiving no IS therapy [6−11].

The exact mechanisms (such as immunological tolerance, ignorance, immunodeficiency, or graft accommodation) responsible for the successful discontinuation of IS drugs in human organ transplant recipients are currently largely unknown. There are currently no approved tools to detect operationally tolerant recipients prior to drug withdrawal. Effective clinical tolerance has been reported more frequently in liver transplant recipients than after transplantation of other solid organs, suggesting that the liver is an immune-privileged organ; this is illustrated by the relatively low incidence of graft loss due to acute or chronic rejection and by its comparative resistance to antibody-mediated injury. Accordingly, LT is an ideal setting for considering IS withdrawal in selected recipients. However, although IS drug doses are routinely minimized, complete withdrawal is seldom attempted in LT. Lifelong IS maintenance still represents the standard therapeutic approach for the majority of transplant patients in the long term.

Tolerance is infrequently achieved outside of LT in humans and when it is recognized it is often discovered serendipitously because of noncompliance or physician-driven IS withdrawal for severe adverse effects or malignancy. Operational tolerance in organ transplant patients continues to represent an elusive clinical goal, and has stimulated a broad variety of approaches.

2.2.1 Tolerogenic Approaches

2.2.1.1 T-Cell Depletion

Many strategies have focused on T-cell depletion to eliminate alloreactive T cells and 'reset' the immune system. The underlying hypothesis is that the removal of circulating T lymphocytes through the administration of a polyclonal antibody and subsequent low doses of IS drugs in the early post-transplant period will increase the intrinsic tolerogenic properties of the graft. Several T-cell-depleting agents are currently available for use in humans: CD3-specific monoclonal antibody and anti-lymphocyte globulin (ALG) clear T cells from the peripheral blood, however there is no evidence that these agents deplete T cells from lymph nodes and spleen. The CD52-specific monoclonal antibody, alemtuzumab (Campath) causes peripheral T-cell depletion for up to 1 year following a short course of therapy, and therefore it is generally assumed that it also depletes cells from lymph nodes and spleen [12]. Alemtuzumab has been assessed in over 40 clinical trials for LT and KT and compared to other induction regimens, is associated with less cellular rejection [13,14]. Alemtuzumab has been associated with rapid homeostatic proliferation of memory T cells after depletion, increased B-cell activating factor, and higher rates of alloantibody production and humoral rejection [15−18]. Current approaches are to combine alemtuzumab with costimulation blockade, regulatory T-cell infusion, and donor stem cell transfusion.

2.2.1.2 Costimulation Blockade

Alloreactive T-cell activation requires antigen-specific engagement of the T-cell receptor with major histocompatibility complex molecules, followed by antigen nonspecific ligation of a variety of receptor−ligand combinations. Blockade of costimulation effectively prevents T-cell activation and so reduces the risk of allograft rejection. Costimulatory signals of the CD28-B7 (receptor for CD80 [B7.1] and CD86 [B7.2]) of the immunoglobulin superfamily and CD40−CD154 (receptor for CD40L) of the tumor necrosis factor (TNF)-receptor super family are the most studied and potentially most important activating costimulation pathways, although other costimulatory pathways have been identified as targets for therapeutic interventions (such as inducible costimulator (ICOS) and CD134) [12].

2.2.1.3 B-Cell Therapies

B cells play a major role in chronic rejection [19], as donor-specific alloantibodies (DSAs) have been causally linked to chronic rejection and long-term graft failure. Kidney recipients with pretransplant class I and II DSA have a 10-year graft survival of 30% compared to 72% of those without pretransplant class I and II DSA [20]. The mechanisms through which B cells mediate tolerance are unclear. There are several studies highlighting that the selective use or pairing of B-cell depleting agents can generate tolerance-promoting B-cell phenotypes and reduce factors leading to chronic rejection [21–23]. Other strategies consist of selective targeting of B-cell activation and signaling pathways, such as the blockade of the B-cell-activating factor (BAFF), a member of the TNF family involved in B-cell survival, proliferation, and maturation, which might help to overcome the problems of desensitization, in those with human leukocyte antigen (HLA) donor-specific antibodies.

2.2.1.4 Chimerism

Chimerism occurs when cells of different donor origin coexist in the same organism. Chimerism itself can be defined into two broad categories: 'mixed' or 'micro-chimerism' and 'full' or 'macro-chimerism.' Mixed chimerism is defined as the presence of both donor and recipient cell lineages coexisting in the recipient bone marrow. Full chimerism implies complete elimination of recipient hematopoietic lineages and population of the recipient bone marrow by 100% donor cells. Promotion of mixed chimerism may be a viable strategy for inducing tolerance in solid organ recipients. However, important hurdles, such as conditioning of donors and recipients to produce an environment where both donor and host hematopoietic cells can coexist, need to be overcome. Also, donor cells that could attack the host and cause graft-versus-host disease (GVHD) also need to be eliminated while preserving the recipient's ability to produce immune populations that can defend against infections.

Several groups demonstrated in animal and nonhuman primate studies that the partial irradiation of the recipient bone marrow with peripheral deletion of recipient T cells allowed for the development of both donor and recipient hematopoietic cells and the induction of tolerance to donor tissue without the need for full myoablation [24–27]. Limited, though encouraging, results come from the clinical application of these concepts [5,28].

2.2.1.5 Regulatory T Cells

There is strong evidence for the importance of regulatory T cells (Treg) in transplantation tolerance and many current strategies focus on Treg development [29–32]. *Ex vivo* expansion of Tregs, using a mixture of antibodies to CD3 plus CD28 in combination with IL-2, has been successfully used to generate a large number of Tregs capable of preventing autoimmunity in a mouse model of diabetes [33]. *In vivo* expansion of antigen-specific Tregs has also been described in a mouse model [31]. The initial clinical trials utilizing Treg immunotherapy for hematopoietic stem cell transplantation have shown promising results [34–36], thus making it an attractive prospect in tolerogenic therapy.

2.2.1.6 Maintenance IS

Maintenance IS agents have different immunoregulatory properties. CNIs inhibit the calcineurin-driven pathways of IL-2 and Interferon-gamma (IFN-γ) transcription, which inhibits T-cell activation; however, IL-2 is critical for the expression of FOXP3, a master regulator in the development and function of Tregs, and the survival and proliferation of CD4$^+$ CD25^{high} FOXP3$^+$ Tregs, which are important in the mechanism of tolerance induction [37–39]. Conversely, the mammalian target of rapamycin (mTOR) inhibitors (sirolimus, everolimus) and possibly the antimetabolites (mycophenolic acid) not only suppress alloreactive T cells but also enhance Treg generation *in vitro*, and do not inhibit initial IL-2 transcription. In the clinical setting, limited data have shown that liver and kidney transplant recipients treated with mTOR agents have higher percentages of phenotypic Tregs versus those recipients on CNI therapy [40,41]. Thus the use of CNIs might be one of the possible reasons for the low success rates in prior weaning studies. It might be that conversion from CNI to putative 'tolerogenic' IS therapies like mTOR inhibitors may increase the proportion of tolerant patients. However, this approach remains at a very early stage and needs to be rigorously tested in prospective fashion against the standard agents.

2.2.2 Spontaneous Operational Tolerance

There are several reports on the achievement of SOT, which allows successful IS discontinuation, in the setting of lymphoproliferative disorder (LPD) [42,43], planned withdrawal, or patient noncompliance. The liver has intrinsic tolerogenic properties so it is not surprising that the intentional discontinuation of IS has been attempted primarily in the liver transplant recipients.

These attempts have led to successful weaning in almost 20% of highly selected recipients [6,9,11,44–46], although the true prevalence of SOT in unselected recipients is still unknown. Although this strategy is associated with a high rate of acute rejection, these episodes tend to be mild and easily resolved. The low rate of successful weaning may be due to known CNI mechanisms inhibiting immunoregulation and the lack of available, well-defined immune monitoring to predict and detect tolerance. Most of the studies of withdrawal of IS have focused on nonviral and nonimmune diseases, although Tisone et al. [11] have examined the impact of IS withdrawal in 34 Hepatitis C virus (HCV)-positive liver allograft recipients. Complete and permanent IS withdrawal was achieved in eight patients (24%), who showed at mean follow-up of 45.5 months stabilization/improvement of histological fibrosis compared to those who were intolerant of weaning. However, a follow-up study almost 3 years later did not show any significant histological differences between HCV-positive transplant recipients in whom IS was withdrawn and those in whom it was maintained [47]. Studies designed to look at long-term histological data and to determine the consequences of a chronic absence of IS therapy are needed. Very close monitoring of liver function tests (LFTs), protocol biopsies to detect subclinical rejection with normal LFTs, and slow withdrawal of IS appear to be key features for successful withdrawal.

2.2.3 Biomarkers of Operationally Tolerant Transplant Recipients

Currently, IS withdrawal is not a feasible option in routine clinical practice. Without predictive tools or clinical guidance, the risks are considered to outweigh the benefits. If a signature of tolerance could be identified in operationally tolerant individuals, then it would allow identification of those recipients amenable to drug minimization or withdrawal; so many studies today are focused on the characterization of these signatures. The use of transcriptional profiling techniques using peripheral blood has been applied in operationally tolerant and nontolerant liver recipients. A number of gene expression classifiers that could accurately identify tolerant recipients have been reported. The natural killer cell signaling pathway seems to be a major molecular pathway associated with tolerance. Studies employing flow cytometry showed an increase in various peripheral blood mononuclear cell subsets (CD4$^+$ CD25$^+$ FOXP3$^+$ T cells, Vδ^+ $\gamma\delta$ T cells, and plasmacytoid dendritic cells) in tolerant liver recipients [48−50].

The main limitations of these studies are their design (cross-sectional, case−control studies), the small sample size (due to the paucity of operationally tolerant recipients), the difficulty in accounting for the confounding effects of pharmacological IS in the control group of recipients (i.e., stable transplant recipients on maintenance IS or recipients undergoing graft rejection), and the use of only peripheral blood samples. The key for future research is setting up of prospective clinical trials, including large populations, where biomarkers of tolerance will be sought in the allograft tissue.

2.3 ORGAN SHORTAGE

The major limitation to the widespread use of transplantation is the scarcity of organs. In March 2011, the United

Network for Organ Sharing (UNOS) reported a solid organ transplant waiting list of 110,600: 16,133 patients were listed for liver, 1389 for pancreas, and 262 for intestine transplants. The number of transplants in 2010, in contrast, was 6291, 351, and 151 for liver, pancreas, and intestine respectively. These figures clearly illustrate that many patients will never benefit from transplantation, and die while on the waiting list [51]. The problem is even greater than these figures suggest, as not all those who might benefit from transplantation are listed. In the United Kingdom, patients with chronic liver disease are listed only when estimated likelihood of death without transplant is greater than likelihood of death following transplantation. This estimation of survival is calculated based on validated models using objective laboratory data incorporated in scores such as developed by the model for end-stage liver disease (MELD) or United Kingdom End-Stage Liver Disease (UKELD). The UKELD score includes, besides the MELD parameters, serum creatinine, bilirubin, and INR, also serum sodium and is more accurate than MELD to predict mortality on the waiting list. A UKELD score of over 49 predicts a >9% 1-year mortality without LT; this is the minimum criteria for joining the UK waiting list [52]. Candidates for LT must also have a >50% probability of surviving for 5-year posttransplant with a quality of life acceptable to the patient [53]. In the United States, survival benefit starts when MELD score exceeds 17, unless the patient has other comorbid factors like liver cancer. Although not used as a criterion for selection or allocation, the development of the concept of transplant survival benefit, i.e., the extra years of life attributable to transplant, might facilitate more effective use of scarce organs and exclude access to those whose lives will be extended minimally or not at all. However, it has proved very difficult to develop a robust model on which to base such a benefit-based approach to liver allocation.

In the case of diabetes mellitus, the number of patients on waiting lists also underestimates the need for transplantation, as transplantation is offered primarily to those in need of a combined pancreas−kidney transplantation. Transplantation for diabetes alone is largely restricted to some patients with hypoglycemia unawareness and brittle diabetes. The adverse consequences of IS need to be balanced against the potential benefits of improved glycemic control. Hence it is not clear which patients with diabetes mellitus might benefit from transplantation.

2.3.1 Strategies to Increase the Donor Pool

Deceased donor rates vary considerably between countries because of cultural, legal and logistic issues, and range from 2 per million population in Greece to 35 per million population in Spain [54]. The success of public health initiatives, leading to better awareness of vascular diseases such as hypertension and reduction in fatal road accidents have contributed to a fall in the number of 'traditional ideal' (young post-traumatic) organ donors. Therefore the potential donor pool is now smaller, with donors being older and heavier. Since the number and quality of organs suitable for transplantation decreases with donor age and weight, clinicians have had to use more 'high risk' donors [55]. The increasing donor risk profile partially reduces the benefits made by better surgery and peritransplant care.

2.3.1.1 Increasing the Donation Rates

Several approaches have been adopted to increase donor rates. These include the development of national advertising campaigns, the establishment of organ donor registers, the promotion of both opt-in and opt-out approaches, incentives to donation, prioritization in organ allocation to those who are donors, and even restricting allocation to those who have agreed to donation. These approaches have met with varying degrees of success in different countries, and it is clear that there is no one simple solution to this problem.

2.3.1.2 Greater Use of Deceased Potential Donors

To increase the number of transplantable organs, newer techniques and policies have been adopted. Use of extended criteria donors (ECDs), donation following circulatory death (DCD) [also referred to as non-heart-beating donors (NHBD)], use of split liver transplantation (SLT), and organs from living donors [LDLT (living donor liver transplantation)] can contribute to enlargement of the donor pool. However, use of such organs is not without additional risk and raises medical and ethical concerns as the welfare of both the living donor and recipient may be compromised. All the allografts from deceased donors carry an increased risk of primary nonfunction (PNF), early or delayed graft dysfunction (DGF) and possibly a greater risk transmission of infection and cancers [56]. It has been clearly shown that ECD organs, defined as organs originating from donors dying from cerebrovascular accident, donation after circulatory death, longer cold ischemia time (CIT), older age and moderate—severe steatosis compromise outcome.

In order to further expand the donor pool, organs from DCD donors are increasingly used in LT, KT, and pancreas transplantation, especially in the UK and the Netherlands. These grafts are usually restricted to those originating from controlled donors—those donors in Maastricht category III (awaiting cardiac arrest) [57]. However, legal constraints do not allow use of NHBD in all countries [58]. Although DCD LT can have good outcomes, their use is associated with a significantly higher risk of graft failure [59], severe biliary complications, and higher costs [60—62]. However,

increasing understanding of the pathophysiology of the events surrounding DCD and better selection and timing may improve outcome in the future. Warm ischemic injury in KT increases the incidence of DGF suggesting that kidneys from DCD donors are inferior to those from donation after brain death (DBD) donors. However, Summers et al. showed in a cohort of 9134 kidney transplant recipients in the UK that for controlled DCD donors, graft survival and function (estimated glomerular filtration rate [eGFR]) did not differ from kidneys from DBD donors at 5 years after transplantation [63]. Several reports described successful islet isolation and transplantation from DCD donors [64—68]. These donors could provide an important resource for islet transplantation if used under strict criteria and in multiple transplantation, particularly in countries where heart-beating donors are not readily available. Bowel from DCD donors is not used. Both the CIT and prolonged rewarming time during allograft implantation have been associated, in LT, with DGF, and biliary complications, particularly for allografts from donors >60 years old, and severe recurrent disease and reduced graft and patient survival, in recipients with hepatitis C infection. The mechanisms that determine this association are probably multifactorial. In these cases, optimal function is achieved when cold ischemia is under 8 h. Prolonged CIT is an important risk factor in KT for the development of DGF and acute rejections which are both significant determinants of short- and long-term graft survival. CIT has an important influence on the outcomes of ECD transplantation. Summers et al. [63] showed in a UK cohort that a CIT of <12 h was strongly associated with superior kidney graft survival, but few kidneys from DCD donors had such a short CIT. Therefore, increased efforts are needed to restrict the CIT of these kidneys by reduction of the time taken for cross-match testing before transplantation, shortening of organ transport times, and ensuring adequate infrastructure in transplant centers.

As discussed above, donor demographics have substantially changed in recent years, so the use of elderly donors has significantly increased. For livers procured from older donors, there is a significant increase in the risk of graft failure, particularly for donors >60 years old [69]. The deleterious effect of increased donor age on transplant outcome is in part dependent on indication and is seen especially in those grafted with HCV infection [70] and is associated with increased rates of HCV disease progression, with a reduction in graft and patient survival. Increasing age of the donor is also a well-established independent risk factor for poor graft outcome in recipients of kidneys from DBD and DCD donors. In the latter, eGFR reduces as age of the donor increases, presumably because these kidneys have less functional reserve, have more vasculopathy, and are less able to withstand the ischemia and reperfusion injury that accompanies transplantation than kidneys from younger donors.

Severe liver allograft steatosis, defined as >60% fatty infiltration, is associated with a greater risk of PNF and lower patient and graft survival [71]. Liver grafts with more than 30% steatosis have been reported to be safe in low-risk recipients but associated with more risk in recipients with MELD >30 [72]. Clinical estimation of the degree of fatty infiltration correlates poorly with histologic assessment and microvesicular steatosis is more associated with nonfunction than macrovesicular steatosis. The widespread clinical use of histological assessment by liver biopsy of the donor graft is limited by a high interobserver variability, difficulties to distinguish between micro- and macrosteatosis and the poor reliability of frozen tissue samples.

Models investigating the interaction between donor and recipient risk profiles have been developed to predict the likelihood of graft and patient survival after LT. Feng et al. recently identified, in a large donor cohort study, eight factors (higher age, lower height, donation after circulatory death donors, split liver grafts, black race, cause of death from cerebrovascular accident, regional sharing, and longer CIT) which were independently associated with graft failure. As a corollary, a donor risk index (DRI) predicting the effect of these variables on graft survival was developed [69]. It is clear that it will become increasingly important to match donor risk scores with recipient etiology and disease severity, in order to assure an acceptable outcome for the recipient.

2.3.1.3 Splitting Livers

SLT has been developed as a strategy to increase the number of liver grafts by creating two grafts from one donated liver. The bipartition of a liver is especially important in the small group of pediatric patients for whom size-matched whole liver allografts are scarce. Indeed the use of split grafts has been associated with a reduction in the risk of death on the pediatric waiting list so that although some centers have reported an increased risk of graft failure, the split procedure for adult—pediatric pair is now accepted as a valuable technical variant in pediatric LT [69,72]. Donor selection for splitting, technical and logistic expertise to decrease total ischemia time are all important factors for a successful outcome of the procedure. This technique is much less successful in the adult—adult split constellation.

2.3.1.4 Living Donation

With respect to transplantation of organs of the gastrointestinal system, living donation is essentially confined to LT. Better understanding of the anatomy and increasing surgical skills has allowed LDLT to become a routine procedure in some centers. LDLT has been widely adopted in Asia, accounting for over 95% of liver transplants, because of the very low rates of deceased organ donation. In 2008,

a Chinese series of 234 right-liver living donor liver transplants showed 1-, 3-, and 5-year overall survival rates of 93.2%, 85.7%, and 82.4%, respectively, comparable with deceased donor liver transplant (DDLT) outcomes [73]. Good outcomes have been shown even when using grafts with a graft-to-recipient weight ratio (GRWR) < 0.8%, with a rate of small-for-size syndrome similar to those receiving graft with a GRWR >0.8%, provided the recipient is receiving the graft from a young donor [74]. In the western world, LDLT is practiced much less frequently not only because of the greater availability of deceased donors but also because of major concerns with donor mortality, especially when transplanting the right lobe which is associated with an estimated risk of donor death of 0.08% and a morbidity around 20%. The reported morbidity and mortality data underestimate the real risk. There have been anecdotal reports of donors requiring a transplant for hepatic failure. The outcome of LDLT in Western countries is good with 1-, 5- and 10-year graft survival of 81%, 70% and 68%, respectively [75−78]. The survival rates after LDLT are better than full size DDLT in children but somewhat lower in adults [77].

Living kidney transplantation represents today the predominant form of KT in Western countries [79]. In the United States, the annual number of living kidney donations (LKDs) has surpassed the number of deceased donations since 2001, although the absolute number of transplants from deceased donors still outnumbers those from living donors [80,81]. This source of kidney grafts may help to overcome in part the organ donor shortage of cadaveric donors. The advantages of live versus deceased donor transplantation are readily apparent as live donation affords earlier transplantation and the best long-term survival. LKD has also been fostered by the technical advance of laparoscopic nephrectomy and immunologic maneuvers that can overcome biologic obstacles, such as HLA disparity and ABO or cross-match incompatibility. Congressional legislation has provided an important model to remove financial disincentives to being a live donor. Federal employees are now given paid leave and coverage for travel expenses. Candidates for renal transplantation seem aware of these developments, and have become less hesitant to ask family members, spouses, or friends to become live kidney donors. Living donation has become safe with minimal immediate and long-term risk for the donor. However, the situation in the future may not stay the same, as western society is becoming increasingly obese and developing associated health problems.

Living donor bowel transplantation (LDBT) has been reported as an additional resource for patients with intestinal failure with total parenteral nutrition (TPN)-related life-threatening complications [82]. However, the very limited data from the Intestinal Transplant Registry (ITR) do not demonstrate a clear advantage of living donor intestine donation over deceased donor intestine transplant [83–85]. The early outcomes of combined intestinal and liver transplantation using living donors are promising and the elimination of the high mortality on the cadaver waiting list (30%) for this category of patients represents a substantial advantage [86].

2.3.1.5 Improved Use of Organs

Current technology for improvement in the maintenance of organ function is allowing organs to be better maintained. The development of improved perfusion fluids has been of some value, but greater advances are likely to come from machine perfusion. While most techniques focused originally on hypothermic perfusion, normothermic perfusion is now increasingly used. This may be applied after retrieval either in the donor or the recipient hospital. Although such technologies are still being developed and require expensive consumables and skilled personnel, it is likely that such approaches will not only improve the quality of those organs that are retrieved but may also render some organs that were previously considered ungraftable, suitable for transplantation.

2.3.1.6 Xenotransplantation

Research in xenotransplantation has grown in the last decades [87,88]. The use of knockout pigs with multiple gene modifications has reduced the frequency of hyperacute rejection, which was a major problem in earlier models [89,90]. Many physiological restraints, as evidenced by a systemic inflammatory response involving the innate immune system, by platelet, leukocyte and complement activation, and by coagulation dysfunction associated with coagulation–anticoagulation incompatibilities of primates and pigs, remain. The transplantation of porcine organs has been carried out in nonhuman primates with better outcomes with hearts or kidneys, compared with liver. The main problem with the latter is a coagulation dysfunction with thrombocytopenia leading to spontaneous bleeding [91]. However, pig livers may provide sufficient function to maintain short-term support and might so be used in patients with acute liver failure, either until the native liver recovers or as a bridge to liver allograft. Only one clinical pig-to-human LT has been reported so far by a surgical team in Los Angeles headed by Makowka, in a patient with fulminant hepatitis. The patient underwent preoperative plasmaphoresis to remove circulating xenoantibodies and the porcine liver graft was placed in a heterotopic position. The pig liver was rejected in few hours and the patient died before a human liver became available [92].

Significant obstacles such as the immunologic hurdle of cross species transplantation and transmission of infections, particularly endogenous retroviruses, need to be overcome before pig organ xenotransplantation can become a clinical reality. Furthermore, the physiological impact of xenotransplantation remains unclear as shown by the immunogenicity and uncertain physiological functioning of pig proteins in the maintenance of homeostasis.

Progress is being made in this difficult field of transplantation as shown by the report of encouraging outcome of pig hepatocyte xenotransplantation with the benefit of the lack of an acute humoral xenograft rejection, the immediate restoration of the liver function, and the resistance to specific human viruses [93]. Equally encouraging are results of xenotransplantation in the field of pancreatic islet. Encapsulated pig islet cells prevent antibody or T-cell contact with islets while allowing through insulin to reach the systemic circulation [94]. However, xenotransplantation remains illegal at this time.

2.4 CONCLUSIONS

Many lessons have been learned in organ transplantation during the past half-century and the overall outcome of transplant patients has dramatically improved compared to the early days of this clinical practice. However, the future of organ transplantation still presents challenges. Strategies aimed at completely withdrawing IS therapy are promising but are difficult to apply in routine clinical practice. Many strategies are also in place to address the problem of organ shortage with contradictory results.

Regenerative medicine (RM) technology and tissue engineering (TE) applied to organ transplantation have the potential to meet two major needs: namely, the identification of a potentially inexhaustible source of organs and an IS-free state. The attraction of a failing organ being replaced by a bioengineered organ generated from a decellularized scaffold and seeded with autologous stem cells is obvious but not without limitations. Many critical aspects of RM and TE, such as long-term safety, tolerability, and efficacy in the clinical setting and the development of an European Medicines Evaluation Agency (EMEA)/FDA approved product need to be addressed, before they become the protagonists of a new scientific era.

ABBREVIATIONS

ALG	anti-lymphocyte globulin
ATG	antithymocyte globulin
BAFF	B-cell-activating factor
CNI	calcineurin inhibitor
DBD	donation after brain death
DDLT	deceased donor liver transplant
DGF	delayed graft function
DRI	donor risk index
EMEA	European medicines evaluation agency

FDA	food and drug administration
GVHD	graft-versus-host disease
HCV	hepatitis C virus
HLA	human leukocyte antigen
ICOS	inducible costimulator
IFN	Interferon
IL	interleukin
IS	immunosuppression
JAK	Janus kinase
KT	kidney transplantation
LDLT	living donor liver transplantation
LFT	liver function test
MELD	model for end-stage liver disease
OTx	organ transplantation
mTOR	mammalian target of rapamycin
SOT	spontaneous operational tolerance
PNF	primary nonfunction
TPN	total parenteral nutrition
UKELD	UK End-Stage Liver Disease
UNOS	United Network for Organ Sharing

REFERENCES

[1] Gondos A, Döhler B, Brenner H, Opelz G. Kidney graft survival in Europe and the United States: strikingly different long-term outcomes. Transplantation 2012;95:267−74.

[2] Barber K, Blackwell J, Collett D, Neuberger J; UK Transplant Liver Advisory Group. Life expectancy of adult liver allograft recipients in the UK. Gut 2007;56(2):279−82.

[3] Sanchez-Fueyo A, Strom TB. Immunologic basis of graft rejection and tolerance following transplantation of liver or other solid organs. Gastroenterology 2011;140:51−64.

[4] Turka LA, Lechler RI. Towards the identification of biomarkers of transplantation tolerance. Nat Rev Immunol 2009;9:521−6.

[5] Kawai T, Cosimi AB, Spitzer TR, Tolkoff-Rubin N, Suthanthiran M, Saidman SL, et al. HLA-mismatched renal transplantation without maintenance immunosuppression. N Engl J Med 2008;358:353−61.

[6] Devlin J, Doherty D, Thomson L, Wong T, Donaldson P, Portmann B, et al. Defining the outcome of immunosuppression withdrawal after liver transplantation. Hepatology 1998;27:926−33.

[7] Lerut J, Sanchez-Fueyo A. An appraisal of tolerance in liver transplantation. Am J Transplant 2006;6:1774−80.

[8] Brouard S, Mansfield E, Braud C, Li L, Giral M, Hsieh SC, et al. Identification of a peripheral blood transcriptional biomarker panel associated with operational renal allograft tolerance. Proc Natl Acad Sci USA 2007;104:15448−53.

[9] Mazariegos GV, Reyes J, Marino I, Flynn B, Fung JJ, Starzl TE. Weaning of immunosuppression in liver transplant recipients. Transplantation 1997;63:243−9.

[10] Pons JA, Yélamos J, Ramírez P, Oliver-Bonet M, Sánchez A, Rodríguez-Gago M, et al. Endothelial cell chimerism does not influence allograft tolerance in liver transplant patients after withdrawal of immunosuppression. Transplantation 2003;75:1045−7.

[11] Tisone G, Orlando G, Cardillo A, Palmieri G, Manzia TM, Baiocchi L, et al. Complete weaning off immunosuppression in HCV liver transplant recipients is feasible and favourably impacts on the progression of disease recurrence. J Hepatol 2006;44:702−9.

[12] Page EK, Dar WA, Knechtle SJ. Tolerogenic therapies in transplantation. Front Immunol. 2012;3:198.

[13] Knechtle SJ, Fernandez LA, Pirsch JD, Becker BN, Chin LT, Becker YT, et al. Campath-1H in renal transplantation. The University of Wisconsin Experience 2004. Surgery 2004 Oct;136(4):754−60.

[14] Hanaway MJ, Woodle ES, Mulgaonkar S, Peddi VR, Kaufman DB, First MR, et al. Alemtuzumab induction in renal transplantation. N Engl J Med 2011;364:1909−19.

[15] Knechtle SJ, Pirsch JD, Fechner Jr. J, Becker BN, Friedl A, Colvin RB, et al. Campath-1H induction plus rapamycin monotherapy for renal transplantation: results of a pilot study. Am J Transplant 2003;3:722−30.

[16] Pearl JP, Parris J, Hale DA, Hoffmann SC, Bernstein WB, Mccoy KL, et al. Immunocompetent T cells with a memory-like phenotype are the dominant cell type following antibody mediated T-cell depletion. Am J Transplant 2005;5:465−74.

[17] Trzonkowski P, Zilvetti M, Chapman S, Wieckiewicz J, Sutherland A, Friend P, et al. Homeostatic repopulation by CD28−CD8+ T cells in alemtuzumab-depleted kidney transplant recipients treated with reduced immunosuppression. Am J Transplant 2008;8:338−47.

[18] Thompson SA, Jones JL, Cox AL, Compston DA, Coles AJ. B-cell reconstitution and BAFF after alemtuzumab (Campath-1H) treatment of multiple sclerosis. J Clin Immunol 2010;30:99−105.

[19] Kwun J, Knechtle SJ. Overcoming chronic rejection—can it B? Transplantation 2009;88:955−61.

[20] Otten HG, Verhaar MC, Borst HPE, Hene RJ, VanZuilen AD. Pretransplant donor-specific HLA class-I and -II antibodies are associated with an increased risk of kidney graft failure. Am J Transplant 2012;12:1618−23.

[21] Liu C, Noorchashm H, Sutter JA, Naji M, Prak EL, Boyer J, et al. B lymphocyte-directed immunotherapy promotes longterm islet allograft survival in non human primates. Nat Med 2007;13:1295−8.

[22] Kelishadi SS, Azimzadeh AM, Zhang T, Stoddard T, Welty E, Avon C, et al. Preemptive CD20+ B-cell depletion attenuates cardiac allograft vasculopathy in cyclosporine-treated monkeys. J Clin Invest 2010;120:1275−84.

[23] Kopchaliiska D, Zachary AA, Montgomery RA, Leffell MS. Reconstitution of peripheral allospecific CD19+ B-cell subsets after B-lymphocyte depletion therapy in renal transplant patients. Transplantation 2009;87:1394−401.

[24] Ildstad ST, Sachs DH. Reconstitution with syngeneic plus allogeneic or xenogeneic bone marrow leads to specific acceptance of allografts or xenografts. Nature 1984;307:168−70.

[25] Sharabi Y, Sachs DH. Mixed chimerism and permanent specific transplantation tolerance induced by a non lethal preparative regimen. J Exp Med 1989;169:493−502.

[26] Kaufman CL, Ildstad ST. Induction of donor-specific tolerance by transplantation of bone marrow. Ther Immunol 1994;1:101−11.

[27] Colson YL, Wren SM, Schuchert MJ, Patrene KD, Johnson PC, Boggs SS, et al. A non lethal conditioning approach to achieve durable multilineage mixed chimerism and tolerance across major minor and hematopoietic histocompatibility barriers. J Immunol 1995;155:4179−88.

[28] Spitzer TR, Sykes M, Tolkoff-Rubin N, Kawai T, Mcafee SL, Dey BR, et al. Long-term follow-up of recipients of combined human leukocyte antigen-matched bone marrow and kidney transplantation for multiple myeloma with end-stage renal disease. Transplantation 2011;91:672−6.

[29] Levings MK, Sangregorio R, Roncarolo MG. Human CD25(+) CD4(+) T-regulatory cells suppress naive and memory T-cell proliferation and can been expanded *in vitro* without loss of function. J Exp Med 2001;193:1295–302.

[30] Godfrey WR, Ge YG, Spoden DJ, Levine BL, June CH, Blazar BR, et al. *In vitro*-expanded human CD4(+) CD25(+) T-regulatory cells can markedly inhibit allogeneic dendritic cell stimulated MLR cultures. Blood 2004;104:453–61.

[31] Nishimura E, Sakihama T, Setoguchi R, Tanaka K, Sakaguchi S. Induction of antigen-specific immunologic tolerance by *in vivo* and *in vitro* antigen specific expansion of naturally arising Foxp3$^+$ CD25$^+$ CD4$^+$ regulatory T cells. Int Immunol 2004;16:1189–201.

[32] Yamazaki S, Iyoda T, Tarbell K, Olson K, Velinzon K, Inaba K, et al. Direct expansion of functional CD25$^+$ CD4$^+$ regulatory T cells by antigen processing dendritic cells. J Exp Med 2003;198:235–47.

[33] Tang Q, Henriksen KJ, Bi M, Finger EB, Szot G, Ye J, et al. *In vitro*-expanded antigen-specific regulatory T cells suppress autoimmune diabetes. J Exp Med. 2004;199:1455–65.

[34] Edinger M, Hoffmann P. Regulatory T cells in stem cell transplantation: strategies and first clinical experiences. Curr Opin Immunol 2011;23:679–84.

[35] Brunstein CG, Miller JS, Cao Q, Mckenna DH, Hippen KL, Curtsinger J, et al. Infusion of *ex vivo* expanded T regulatory cells in adults transplanted with umbilical cord blood: safety profile and detection kinetics. Blood 2011;117:1061–70.

[36] Di Ianni M, Falzetti F, Carotti A, Terenzi A, Castellino F, Bonifacio E, et al. Tregs prevent GVHD and promote immune reconstitution in HLA-haplo identical transplantation. Blood 2011;117:3921–8.

[37] Woltman AM, de Fijter JW, Kamerling SW, Paul LC, Daha MR, van Kooten C. The effect of calcineurin inhibitors and corticosteroids on the differentiation of human dendritic cells. Eur J Immunol 2000;30:1807–12.

[38] Szabo G, Gavala C, Mandrekar P. Tacrolimus and cyclosporine A inhibit allostimulatory capacity and cytokine production of human myeloid dendritic cells. J Investig Med 2001;49:442–9.

[39] Gao W, Lu Y, El Essawy B, Oukka M, Kuchroo VK, Strom TB. Contrasting effects of cyclosporine and rapamycin in *de novo* generation of alloantigen-specific regulatory T cells. Am J Transplant 2007;7:1722–32.

[40] Levitsky J, Mathew JM, Flaa CW, Rosen A, Tambur AR, Miller J. Immunoregulatory effects of conversion from tacrolimus to sirolimus in liver transplant recipients [Abstract]. Hepatology 2009;50:568A.

[41] Hendrikx TK, Velthuis JH, Klepper M, van Gurp E, Geel A, Schoordijk W, et al. Monotherapy rapamycin allows an increase of CD4 CD25 FoxP3 T cells in renal recipients. Transpl Int 2009;22:884–91.

[42] Hurwitz M, Desai DM, Cox KL, Berquist WE, Esquivel CO, Millan MT. Complete immunosuppressive withdrawal as a uniform approach to post-transplant lymphoproliferative disease in pediatric liver transplantation. Pediatr Transplant 2004;8:267–72.

[43] Birkeland SA, Hamilton-Dutoit S, Bendtzen K. Long term follow-up of kidney transplant patients with posttransplant lymphoproliferative disorder: duration of posttransplant lymphoproliferative disorder-

induced operational graft tolerance, interleukin-18 course, and results of retransplantation. Transplantation 2003;76:153–8.

[44] Takatsuki M, Uemoto S, Inomata Y, Sakamoto S, Hayashi M, Ueda M, et al. Analysis of alloreactivity and intragraft cytokine profiles in living donor liver transplant recipients with graft acceptance. Transpl Immunol 2001;8:279–86.

[45] Eason JD, Cohen AJ, Nair S, Alcantera T, Loss GE. Tolerance: is it worth the risk? Transplantation 2005;79:1157–9.

[46] Assy N, Adams PC, Myers P, Simon V, Minuk GY, Wall W, et al. Randomized controlled trial of total immunosuppression withdrawal in liver transplant recipients: role of ursodeoxycholic acid. Transplantation 2007;83:1571–6.

[47] Orlando G, Manzia T, Baiocchi L, Sanchez-Fueyo A, Angelico M, Tisone G. The Tor Vergata weaning off immunosuppression protocol in stable HCV liver transplant patients: the updated follow up at 78 months. Transpl Immunol 2008;20:43–7.

[48] Martínez-Llordella M, Puig-Pey I, Orlando G, Ramoni M, Tisone G, Rimola A, et al. Multiparameter immune profiling of operational tolerance in liver transplantation. Am J Transplant 2007;7:309–19.

[49] Li Y, Koshiba T, Yoshizawa A, Yonekawa Y, Masuda K, Ito A, et al. Analyses of peripheral blood mononuclear cells in operational tolerance after pediatric living donor liver transplantation. Am J Transplant 2004;4:2118–25.

[50] Mazariegos GV, Zahorchak AF, Reyes J, Chapman H, Zeevi A, Thomson AW. Dendritic cell subset ratio in tolerant, weaning and non-tolerant liver recipients is not affected by extent of immunosuppression. Am J Transplant 2005;5:314–22.

[51] Available from: http://www.unos.org/. Based on OPTN data as of June 21, 2013.

[52] Neuberger J, Gimson A, Davies M, Akyol M, O'Grady J, Burroughs A, Liver Advisory Group, et al. UK blood and transplant. Gut 2008;57(2):252–7.

[53] Neuberger J, James O. Guidelines for selection of patients for liver transplantation in the era of donor-organ shortage. Lancet 1999;354:1636–9.

[54] Council of Europe. Available from: <http://ec.europa.eu/health/ph_threats/human_substance/oc_organs/docs/fact_figures.pdf. > [accessed 24.06.11].

[55] Summers DM, Counter C, Johnson RJ, Murphy PG, Neuberger JM, Bradley JA. Is the increase in DCD organ donors in the United Kingdom contributing to a decline in DBD donors? Transplantation 2010;90:1506–10.

[56] Alkofer B, Samstein B, Guarrera JV, Kin C, Jan D, Bellemare S, et al. Extended-donor criteria liver allografts. Semin Liver Dis 2006;26:221–33.

[57] Kootstra G, Daemen JH, Oomen AP. Categories of non-heart-beating donors. Transplant Proc 1995;27:2893–4.

[58] De Vera ME, Lopez-Solis R, Dvorchik I, Campos S, Morris W, Demetris AJ, et al. Liver transplantation using donation after cardiac death donors: long-term follow-up from a single center. Am J Transplant 2009;9:773–81.

[59] Domínguez-Gil B, Haase-Kromwijk B, Van Leiden H, Neuberger J, Coene L, Morel P, , et al.European Committee (Partial Agreement) on Organ Transplantation. Council of Europe (CD-P-TO) Current

situation of donation after circulatory death in European countries. Transpl Int 2011;24:676—86.

[60] Chan EY, Olson LC, Kisthard JA, Perkins JD, Bakthavatsalam R, Halldorson JB, et al. Ischemic cholangiopathy following liver transplantation from donation after cardiac death donors. Liver Transpl 2008;14:604—10.

[61] Grewal HP, Willingham DL, Nguyen J, Hewitt WR, Taner BC, Cornell D, et al. Liver transplantation using controlled donation after cardiac death donors: an analysis of a large single-center experience. Liver Transpl 2009;15:1072—82.

[62] Yamamoto S, Wilczek HE, Duraj FF, Groth CG, Ericzon BG. Liver transplantation with grafts from controlled donors after cardiac death: a 20-year follow-up at a single center. Am J Transplant 2010;10:602—11.

[63] Summers DM, Johnson RJ, Allen J, Fuggle SV, Collett D, Watson CJ, et al. Analysis of factors that affect outcome after transplantation of kidneys donated after cardiac death in the UK: a cohort study. Lancet 2010;376:1303—11.

[64] Markmann JF, Deng S, Desai NM, Huang X, Velidedeoglu E, Frank A, et al. The use of non-heart-beating donors for isolated pancreatic islet transplantation. Transplantation 2003;75:1423.

[65] Zhao M, Muiesan P, Amiel SA, Srinivasan P, Asare-Anane H, Fairbanks L, et al. Human islets derived from donors after cardiac death are fully biofunctional. Am J Transplant 2007;7:2318.

[66] Matsumoto S, Tanaka K. Pancreatic islet cell transplantation using non-heart-beating donors (NHBDs). J Hepatobiliary Pancreat Surg 2005;12:227.

[67] Matsumoto S, Okitsu T, Iwanaga Y, Noguchi H, Nagata H, Yonekawa Y, et al. Successful islet transplantation from non-heart-beating donor pancreas using modified Ricordi islet isolation method. Transplantation 2006;82:460.

[68] Saito T, Gotoh M, Satomi S, Uemoto S, Kenmochi T, Itoh T, et al. Working members of the Japanese Pancreas and Islet Transplantation Association. Islet transplantation using donors after cardiac death: report of the Japan Islet Transplantation Registry. Transplantation. 2010;15;90(7):740—7.

[69] Feng S, Goodrich NP, Bragg-Gresham JL, Dykstra DM, Punch JD, DebRoy MA, et al. Characteristics associated with liver graft failure: the concept of a donor risk index. Am J Transplant 2006;6:783—90.

[70] Verran D, Kusyk T, Painter D, Fisher J, Koorey D, Strasser S, et al. Clinical experience gained from the use of 120 steatotic donor livers for orthotopic liver transplantation. Liver Transpl 2003;9:500—5.

[71] Briceno J, Padillo J, Rufian S, Solorzano G, Pera C. Assignment of steatotic livers by the Mayo model for end-stage liver disease. Transpl Int 2005;18:577.

[72] Broering DC, Topp S, Schaefer U, Fischer L, Gundlach M, Sterneck M, et al. Split transplantation and risk to the adult recipient: analysis using matched pairs. J Am Coll Surg 2002;195:648—57.

[73] Chan SC, Fan ST, Lo CM, Liu CL, Wei WI, Chik BH, et al. A decade of right liver adult-to-adult living donor liver transplantation: the recipient mid-term outcomes. Ann Surg 2008;248:411—9.

[74] Moon JI, Kwon CH, Joh JW, Jung GO, Choi GS, Park JB, et al. Safety of small-for-size grafts in adult-to-adult living donor liver transplantation using the right lobe. Liver Transpl 2010;16:864—9.

[75] Hashikura Y, Ichida T, Umeshita K, Kawasaki S, Mizokami M, Mochida S, et al. Japanese liver transplantation society. Donor complications associated with living donor liver transplantation in Japan. Transplantation. 2009 Jul 15;88(1):110—4.

[76] Renz JF, Emond JC, Yersiz H, Ascher NL, Bussutil RW. Split liver transplantation in the United States: outcomes of a national survey. Ann Surg 2004;239:172—81.

[77] Adam R, Karam V, Delvart V, Kilic M, Paul A, Fisher L. et al. ELITA. Living donor liver transplantation: a European liver transplant registry (ELTR) report on 2634 cases. The 2009 international congress of ILTS, New York, USA, July 8—11, 2009.

[78] Gillespie BW, Merion RM, Ortiz-Rios E, Tong L, Shaked A, Brown RS, et al. Database comparison of the adult-to-adult living donor liver transplantation cohort study (A2ALL) and the SRTR U.S. Transplant Registry. Am J Transplant 2010;10:1621—33.

[79] Price D. Living kidney donation in Europe: legal and ethical perspectives—the EUROTOLD project. Transpl Int 1994;7:665—7.

[80] OPTN/SRTR Annual Report. Available from: <http://optn.transplant.hrsa.gov/>. Based on OPTN data as of June 21, 2013.

[81] Delmonic FL, Sheehy E, Marks WH, Baliga P, McGowan JJ, Magee JC. Organ donation and utilization in the United States. Am J Transplant 2004;5:862—73.

[82] Testa G, Holterman M, Abcarian H, Iqbal R, Benedetti E. Simultaneous or sequential combined living donor-intestine transplantation in children. Transplantation 2008;85:713—7.

[83] The Intestinal Transplant Registry. Available from: http://www.optn.org>.

[84] Benedetti E, Holterman M, Asolati M, Di Domenico S, Oberholzer J, Sankary H, et al. Living related segmental bowel transplantation: from experimental to standardized procedure. Ann Surg 2006;244:694—9.

[85] Mazariegos GV, Superina R, Rudolph J, Cohran V, Burns RC, Bond GJ, et al. Current status of pediatric intestinal failure, rehabilitation, and transplantation: summary of a colloquium. The sixth international pediatric intestinal failure and rehabilitation symposium, September 9—11, 2010, Chicago. Transplantation; 2010.

[86] Gangemi A, Tzvetanov IG, Beatty E, Oberholzer J, Testa G, Sankary HN, et al. Lessons learned in pediatric small bowel and liver transplantation from living-related donors. Transplantation 2009;87:1027—30.

[87] Ekser B, Ezzelarab M, Hara H, van der Windt DJ, Wijkstrom M, Bottino R, et al. Clinical xenotransplantation: the next medical revolution? Lancet 2011;379:672—83.

[88] Hammer C. Xenotransplantation—will it bring the solution to organ shortage. Ann Transplant. 2004;9(1):7—10.

[89] Kuwaki K, Tseng YL, Dor FJ, Shimizu A, Houser SL, Sanderson TM, et al. Heart transplantation in baboons using α1,1-galactosyltransferase gene-knockout pigs as donors: initial experience. Nat Med 2005;11:29—31.

[90] Yamada K, Yazawa K, Shimizu A, Iwanaga T, Hisashi Y, Nuhn M, et al. Marked prolongation of porcine renal xenograft survival in baboons through the use of α1,1-galactosyltransferase gene-knockout donors and the cotransplantation of vascularized thymic tissue. Nat Med 2005;11:32—4.

[91] Ekser B, Echeverri GJ, Hassett AC, Yazer MH, Long C, Meyer M, et al. Hepatic function after genetically engineered pig liver transplantation in baboons. Transplantation 2010;90:483—93.

[92] Makowka L, Cramer DV, Hoffman A, Breda M, Sher L, Eiras-Hreha G, et al. The use of a pig liver xenograft for temporary support of a patient with fulminant hepatic failure. Transplantation 1995;59:1654.

[93] Bonavita AG, Quaresma K, Cotta-de-Almeida V, Pinto MA, Saraiva RM, Alves LA. Hepatocyte xenotransplantation for treating liver disease. Xenotransplantation 2010;17:181—7.

[94] Dufrane D, Goebbels RM, Gianello P. Alginate macroencapsulation of pig islets allows correction of streptozotocin-induced diabetes in primates up to 6 months without immunosuppression. Transplantation 2010;90:1054—62.

Part I

Principles of Regenerative Medicine and Cell, Tissue, and Organ Bioengineering

Chapter 3

Strategies for the Specification of Tissue Engineering Biomaterials

David F. Williams

Wake Forest Institute of Regenerative Medicine, Winston-Salem, NC

Chapter Outline

3.1 INTRODUCTION: THE ENGINEERING APPROACH TO TISSUE REGENERATION

In order to discuss tissue engineering biomaterials, we have to consider the engineering context in which regenerative medicine has evolved and the unique types of characteristics that we require in these materials. The reason why there is the word "engineering" in the term "tissue engineering" is not really obvious. Most definitions of engineering invoke the use of scientific knowledge to solve practical problems and/or the systematic analysis of data to yield useful end products. Neither of these concepts is readily translated into the paradigms of tissue engineering, which does have practical end products but is far more related to cell, molecular and developmental biology than to the physical sciences that normally underpin classical engineering. Another meaning of engineering, however, which is best appreciated when we consider that the origin of the term is the Latin *ingenium*, is that it is creativeness that is really at the heart of the subject. This is not a matter of semantics but of immense importance in both the philosophy and practical development of tissue engineering, and the broader area of regenerative medicine.

Tissues and organs suffer from a wide variety of diseases and injuries, as a result of which they lose some degree of function. Primarily these conditions are associated with acute injury or chronic degenerative changes. Without any medical intervention, the response of the body is quite limited and mainly restricted to repair processes. Repair may lead to the restoration of continuity in the affected part by the synthesis of collagenous scar tissue, which is not reminiscent of the indigenous damaged tissue. This may be an effective front line response to injury but does not lead to the restoration of normal structure and function and may, if uncontrolled, lead to detrimental effects in the patient.

The logical conclusion to discussions which emphasize that repair is not an effective outcome, and that replacement has serious limitations with respect to logistics and lack of biological functionality, is the consideration that tissue regeneration is the only possible alternative, which is aimed at restoring normal structure and function through the production of new tissue that does replicate exactly that which has been lost. If we wish to persuade the human adult to regenerate whole organs or tissues that do not spontaneously regenerate, then we have to give them some cues or signals, and superimpose on them a mechanism that is not the natural response to those conditions. Induced regeneration is the essence of tissue engineering, which is, of course, very different to either repair or replacement of tissues. Tissue engineering is, therefore, a matter of the creation of new tissue and to engineer here is, quite simply, to create.

Clearly it is not a trivial process persuading cells to produce new tissue under circumstances in which they do not normally do so. Moreover, it is of the utmost importance that, during this process, exactly the right type of tissue is generated, that the signals given to the cells can be switched off when the process is complete, and that the resulting tissue is fully functional. The process of tissue engineering starts with the sourcing of the relevant cells and ends with the full incorporation of the functional regenerated tissue into the host. The pathway between these two points can take many forms, but is essentially represented by the central tissue engineering paradigm. The types of cells include those derived from autologous, allogeneic, or xenogeneic sources, and they may be fully differentiated cells or stem/progenitor cells. The degree of cell manipulation will depend on the origin of the cells and the complexity of the tissue, and may be dependent on gene transfer in order to optimize processes of, e.g., cell expansion, or to control phenotype under these abnormal circumstances. Normally the cells will require some supporting structure, often referred to a scaffold or a matrix, within or on which they will express the new tissue. They will be persuaded to do so by molecular signals provided by relevant cytokines, growth factors, or other molecules, and by mechanical signals transmitted via the support and the fluid medium. The environment in which this takes place may be described as a bioreactor. The tissue that forms will, if generated *ex vivo*, have to be placed within the host, where it has to be fully and functionally incorporated into indigenous tissues. This should take into account the responses that should be avoided, such as excessive inflammation, immune responses and carcinogenicity or teratogenicity, and also the responses that may be required, such as vascularization and innervation, and indeed the further development and maturation of the tissue itself. It should be borne in mind that this paradigm does not have to be rigidly followed, and many tissue engineering processes are evolving with, for example, much of the regeneration actually occurring *in vivo* rather than *ex vivo*.

Having set out the framework of the generic tissue engineering approach, we have to identify the scientific and infrastructure factors that control the development of tissue engineering. It has to be recognized here that tissue engineering processes are complex and, as yet, have not been effectively translated into widely used, clinically acceptable procedures, or indeed commercial successes. There are several reasons for this but probably the most important is the difficulty of integrating all of these components into a coherent system that is able to accommodate the requirements and specifications for each phase of this paradigm into an efficient and cost-effective process within a quality-validated, clinically oriented environment, and which takes into account the impositions of regulatory, ethical, and reimbursement schemes. A systems engineering approach to regenerative medicine appears to be an essential element of future developments with respect to this integration. It has become clear that it is the dynamic interactions of molecules and cells that give rise to biological function, and that knowledge about individual biological components, from genes to proteins, subcellular components, cells, tissues, organs, and whole organisms does not in itself lead to an understanding of cell and organ function. It is rather the understanding of the inter- and intracellular processes that will do this, leading, for example, to a far greater appreciation of disease causation and drug design. So it is within tissue engineering and regenerative medicine. The paradigm discussed in this chapter is not hierarchical but temporal, based on the practical transition from cell derivation to tissue construct integration.

3.2 PRINCIPLES OF BIOMATERIALS SELECTION IN TISSUE ENGINEERING

We now come to the main theme of this chapter, that is, the biomaterials that are used in the processes of regenerative medicine, and especially, tissue engineering. It is tempting here to provide a list of those materials that have been used over the course of the last several decades for these processes, with a rationale for their selection and a discussion of their known performance. However, that would not be too helpful as a list of previously used materials does not provide a good perspective on the specifications for those biomaterials that are likely to be successful in the future. The selection of so-called scaffolds and matrices for tissue engineering processes was initially pragmatic. It was natural that consideration was given to the regulatory approval that would be required for their use in human patients. This was already a formidable task when just considering the issues involving the manipulation of cells outside of the body, let alone the use of radically new materials to assist in that manipulation. Early stage products were almost wholly based on biomaterials that had prior regulatory approval in other nontissue engineering applications.

With hindsight, this was unfortunate. Since regulatory approval with respect to biomaterial-based devices is, to a large extent, based on the issues of biological safety it was assumed that data on safety that had been clearly demonstrated in one nontissue engineering application could simply be transferred to specific tissue engineering applications. However, biological safety is just a component of biocompatibility and, in fact, the performance of biomaterials with respect to biocompatibility varies from one application to another. The predicate biomaterials for tissue engineering scaffolds were assumed to be those materials that had been safely used in implantable medical devices or drug delivery systems. These materials, however, had been designed to perform some nonbiologic function in those devices without having any undesirable effect on the host, which usually meant having no biological activity at all.

In tissue engineering processes, certain cells, recruited for this purpose, have to be directed to express new tissue, and they require a combination of molecular and mechanical cues in order to do so. The biomaterial is the support or template within which these processes take place. We cannot expect, therefore, that these processes will be optimized, or even take place at all, if we use materials that cannot take part in any biological activity. Instead, the biomaterials that we use must be designed to facilitate cell performance and fully participate in the cell signaling and tissue expression processes.

3.2.1 The Concept of Tissue Engineering Templates

One problem here is that some terminology has already been established which attempts to describe these materials and structures but where the words do not adequately express what is happening. The two most common are scaffold and matrix. It would not make any sense to ignore these since they are in common usage. It is very important to recognize that the performance of the structures that we use in tissue engineering processes depends on both the material chemistry and its physical form, and indeed the latter may be the more important. The description of a "scaffold material" therefore falls far short of the characterization and specification that are required. The use of the composite noun "injectable scaffold" shows just how difficult the terminology discussion has become.

I therefore prefer the use of the overarching term "template" to describe the biomaterial-based constructs that are used in tissue engineering. This embraces the external shape or form, the internal architecture, the material of construction, any surface-bound molecules or structures, and any added biomolecules.

3.2.2 The Objectives of Tissue Engineering Templates

Consider what we are really trying to do with tissue engineering biomaterials. We start with a single cell that, in its preexisting state, does not have the ability to express any extracellular matrix (ECM) molecules. The internal characteristics of that cell and/or its environmental characteristics have to be changed in order for the cell to recapitulate or gain that ability. Obviously practical processes are unlikely to work with just a single cell and so it is necessary to turn attention to a population of identical cells and try to change their collective characteristics so that together they express multiple ECM molecules or other molecules that are of interest to us in a regenerative therapy. It may well be possible to do this with the cells in suspension in culture, and, if the molecules they are releasing are the sole players in the therapy, all we have to do is deliver this suspension to the requisite site. This is the basis of cell therapy and could relate, for example, to the delivery of dopamine-producing cells to the brain for treatment of Parkinson's disease or stem cells that facilitate axon regeneration in the treatment of spinal cord injury. Templates and biomaterials may not have any role here.

However, in most situations it will be necessary to have spatial, and probably temporal, control over the behavior of these cells, and this is the control that the template should exercise. The following are the components of this control:

1. There will need to be control over the volume and, in many cases, shape of the region in which the cells are operating.
2. There may need to be control of the mechanical signaling to the cells, which is unlikely to happen in most situations where they are simply injected into tissue.
3. Cell-to-cell contact and signaling may be required in order to optimize their performance.
4. Under *ex vivo* conditions, the supply of nutrients, including oxygen, needs to be regulated.
5. Equally, molecular signaling through growth factors or alteration of cell phenotype through gene transfer may be required, and this may need spatio-temporal control.
6. In cases of complex tissue regeneration, more than one cell type is required for the expression of different types of ECM and this will definitely need spatio-temporal control; this becomes even more complex when we consider whole organ printing.
7. In some situations, allogeneic or even xenogeneic cells may be used and they need protection from the host immune system through the use of some membrane template.
8. The behavior of cells, whether stem cells or fully differentiated adult cells, and whether individually or in colonies, may have to be optimized through control over their interactions with substrates, usually best achieved through carefully engineered substrates, often at the nanoscale.

These, then, are the generic objectives of tissue engineering templates, which allows the consideration of the form that such templates can take in practice. The first step is to produce a classification of tissue engineering templates based on the current portfolio of experimental and commercial structures. This classification refers to the characteristics of the templates themselves and not just the materials of their construction.

3.2.3 Classification of Tissue Engineering Templates

Templates may take several different forms. They may be simple homogeneously porous structures, into which cells are seeded under *ex vivo* culture conditions, and from which implantable constructs arise. They may be gels that attempt to mimic the environment of cells or they may be derived from naturally occurring ECM tissues. They may be used solely *ex vivo* to direct cell behavior or may provide *in vivo* protection to cells. The following system represents a broad classification of these templates.

Class I: Homogeneous, isotropic, unmodified porous structures; for example, a tricalcium phosphate block of uniform pore distribution, which could be implanted in a critical size defect in bone in order to facilitate osteoconduction, possibly assisted by prior cell seeding.

Class II: Homogeneous, isotropic porous structures with chemically modified surfaces, such as a similar construct to that in Class I but with bone morphogenetic protein (BMP)-2 attached to the pore surface.

Class III: Homogeneous, anisotropic, unmodified porous structures, for example, an electrospun homopolymer with oriented fibers prepared as a tubular template for a blood vessel regeneration.

Class IV: Homogeneous, anisotropic porous structures with chemically modified surfaces, such as an RGD-modified electrospun homopolymer with oriented fibers.

Class V: Homogeneous, isotropic porous structures with biomolecule delivery, for example, a porous polymer made by salt leaching that incorporates vascular endothelial cell growth factor (VEGF).

Class VI: Homogeneous, anisotropic porous structures with biomolecule delivery, as in a polymer with aligned porosity that incorporates a neurotrophic factor for delivery to a nerve.

Class VII: Heterogeneous, multiphase, unmodified porous structures, such as blends of silk and elastin fibers.

Class VIII: Heterogeneous, multiphase, porous structures with chemically modified surfaces, including an RGD-functionalized chitosan—alginate porous solid.

Class IX: Heterogeneous, multiphase, porous structures with biomolecule delivery, for example, a porous PCL-PHB structure containing growth factors.

Class X: Derivatized ECM like decellularized small intestine submucosa (known as SIS).

Class XI: Injectable, *in situ* polymerizing or setting substance, for example, a photopolymerizable methacrylate modified Dextran hydrogel.

Class XII: Noninjectable hydrogel like a thermoresponsive polymer for cell sheet engineering.

Class XIII: Membranes for cell encapsulation, typically consisting of alginate microcapsules.

Class XIV: Micro-engineered, patterned surfaces for cell culture, for example, microfabricated platform of polyethylene glycol (PEG) microwell arrays.

3.2.4 Specifications for Template Materials

Templates should facilitate, or even optimize, the delivery of molecular and mechanical signals to the target cells. They should accommodate those cells within an environment that replicates, as far as possible, the environment in which they normally reside, as with the stem cell niche, in order to maximize the chance of those cells expressing new tissue. The template should be responsive and adaptable to time sensitive changes in the environment. It should allow for optimal flow of nutrients and gases. It should not have any deleterious chemical effects on the cells. It should usually give shape and volume to the new tissue construct. It should have an architecture that facilitates innervation and vascularization. It will normally be expected to degrade during and after tissue regeneration without stimulating significant inflammation or other undesirable effects within the tissue.

This is a formidable list of requirements for the template. These are translated into the following specification for tissue engineering materials, which can be separated into the mandatory features and those which are application specific.

3.2.4.1 Mandatory Specifications

1. The material should be capable of recapitulating the architecture of the niche of the target cells.
2. It should be capable of adapting to the constantly changing microenvironment.
3. The material should have elastic properties that favor mechanical signaling to the target cells in order to optimize differentiation, proliferation, and gene expression.
4. It should have optimal surface or interfacial energy characteristics to facilitate cell adhesion and function.
5. The material should be capable of orchestrating molecular signaling to the target cells, either by

directing endogenous molecules or delivering exogenous molecules.

6. The material should be of a physical form that provides appropriate shape and size to the regenerated tissue.

7. The material should be capable of forming an architecture that optimizes cell, nutrient, gas, and biomolecule transport, and facilitates blood vessel and nerve development.

8. It should be intrinsically noncytotoxic and nonimmunogenic, and minimally proinflammatory.

3.2.4.2 Optional/Application-Specific Requirements

1. The material should be degradable if that is desired with appropriate degradation kinetics and appropriate degradation profiles.

2. The material should be injectable if that is desired with the appropriate rheological characteristics and transformation mechanisms and kinetics.

3. Where necessary, the material should be compatible with the processing techniques that simultaneously pattern both the material and living cells.

4. Where multiple cell types are involved, the material properties should be tunable in order to accommodate variable cellular requirements with spatio-temporal control as appropriate.

5. When used in a significantly stressed *in vivo* environment, the material must have sufficient strength and toughness.

6. In those situations where the biomaterial encapsulates cells, optimal diffusion characteristics concerning key molecules are required.

3.3 SPECIFIC TYPES OF TEMPLATE MATERIAL

If the essential features of template materials are extracted from the long lists of template classes, requirements, and specifications given above, a simple hierarchy of structures can be identified. There are, in fact, three groups of template biomaterials, the porous solids, the hydrogels, and the decellularized tissues. There is some overlap between these groups since some polymers can be presented as a porous solid or a gel, and there are significant subdivisions within them, for example, preformed and injectable gels, and soft and hard porous solids. Nevertheless, these three groups have sufficient distinctive features that allow these groupings to be robust and scientifically meaningful.

3.3.1 Porous Solids

Many of the first attempts to use biomaterials in tissue engineering involved therapies for skin regeneration and cartilage repair. The former were targeted toward burns and ulcers and the material was required to be in the form of a simple porous flat sheet. Cartilage lesions were a little different and required more of a cylindrical shape. The tissues being addressed here were soft and it was reasonably logical to turn to porous polymers as a basis for cell seeding and regeneration of tissue at the affected site; it was thought desirable for the polymer to degrade as the new tissue was formed and it was assumed that since several biodegradable polymers had already been used successfully in medical devices, with regulatory approval, these would be ideal for the so-called tissue engineering scaffolds.

Thus tissue engineering products emerged with various porous polyester formulations, including polylactic and polyglycolic acids and their copolymers, and polycaprolactone. It is fair to say that these scaffold materials have not been very successful, although it is also true that many more factors other than the material have been at fault.

Considering the mandatory specifications, porous synthetic biodegradable polymers do not resemble the cell niche, either physically, chemically, or mechanically. There have been some commendable modifications to the porous architecture, including the development of nanoarchitecture, but it is essentially impossible to replicate that niche with manufactured synthetic fibers or open foam structures. It is hard to replicate tissue elasticity with such porous structures and very difficult to achieve the optimal hydrophilicity. These mean that the delivery of molecular and mechanical signaling is not optimal with these materials and structures.

Having said that, some success in experimental systems with such polymers has been obtained. Some of the deficiencies can be addressed by surface modification of the polymers, for example, by the attachment of RGD sequences to fiber surfaces or plasma treatments, and a wide variety of copolymerization techniques have been used in order to tune degradation profiles. Within hard tissues, limited success has been achieved with porous calcium phosphate materials.

In many ways, a far better option than using synthetic degradable polymers in soft tissue engineering involves the use of natural biopolymers. There has been a significant move towards proteins (collagen, elastin, silk, etc.) and polysaccharides (alginates, chitosan, hyaluronan, etc.) all of which may be prepared in solid porous form. Several of these are clearly involved in signaling processes in their normal biological state and they may be associated directly with the cell niche. The use of copolymers and blends allows for fine-tuning of many of the properties, including elasticity and biodegradation. They may also be formulated as gels.

3.3.2 Hydrogels

An indication of the potential for gels to facilitate the culturing of cells, including stem cells, can be seen with the ubiquitous use of the material Matrigel™ in laboratories. This is a commercially available gel that allows cells to replicate many of their normal complex behavior patterns under culture conditions, yielding phenomena that are not replicated with synthetic polymer substrates. The material is a gel derived from a tissue-cultured mouse sarcoma line and contains laminin, type IV collagen, entactin, heparan sulfate proteoglycans, and growth factors.

Hydrogels are 3D networks formed from hydrophilic polymers which are usually soft and elastic with strong thermodynamic compatibility with water. The cross-linked structure is characterized by junctions formed from strong chemical bonds, physical entanglements, micro-crystallite formation, or weak interactions. The network morphology can be amorphous, semicrystalline, supramo-lecular, or colloidal aggregates. The polymers can be combined in the form of blends, copolymers, and interpenetrating networks and can be neutral, cationic, anionic, or ampholytic, determined by the pendant groups.

Hydrogels have been used in regenerative medicine and have also been used as drug or growth factor depots, as cell encapsulants, and as adhesives. They are potentially attractive template material since they may be tailored to mimic properties of the ECM like mechanical properties. Cells may be adherent to or suspended within the gel network. It is recognized that the chemicals used in the preparation of hydrogels may display some toxicity and care has to be taken if the degree of conversion is not 100%; initiators, organic solvents, stabilizers, emulsifiers, unreacted monomers, cross-linking agents, and other substances have to be considered in this light and such chemicals may need to be removed.

Among the more significant synthetic hydrogels are poly(2-hydroxethyl methacrylate) (PHEMA), PEG, and poly(vinyl alcohol) (PVA), but only PEG has significant potential in tissue engineering. PEG homopolymer is a polyether that can be polymerized from ethylene oxide, the resulting chains possessing terminal hydroxyl groups. These may be derivatized to make PEG macromers for use in a wide variety of reaction schemes. Photocurable PEG-based gels are used to encapsulate cells. They may also incorporate bioactive peptides attached to PEG chains thereby influencing cellular behavior.

The biopolymer, or naturally derived, hydrogels include those based on hyaluronic acid, alginates, collagen, fibrin, and peptides. They tend to be considered as superior to synthetic gels with respect to biocompatibility since they may offer better molecular and morphological cues to cells. Hyaluronic acid, or hyaluranon (HA), is a high molecular weight glycosaminoglycan (GAG) with repeating disaccharide units composed of D-glucuronic acid and N-acetyl-D-glucosamine which is found in several soft connective tissues, including skin, umbilical cord, synovial fluid, and vitreous humor. This polysaccharide offers multiple sites for cross-linking and property modification using its carboxyl and hydroxyl groups. It may also be modified with peptides to enhance cell attachment, spreading, and proliferation. These peptide functionalized gels may also be used as *in situ* gelling injectable constructs for *in vivo* tissue engineering.

Alginate is a linear block copolymer of D-mannuronic acid (M) and L-guluronic acid (G) residues that has been widely used in medical applications, including wound healing and cell encapsulation. The viscosity of alginates and the stiffness after gelling depend on the concentration of the polymer and its molecular weight distribution. Cross-linking between polymer chains can be arranged through multivalent cations and with carboxylic acid groups in the sugars. Alginates have generally good biocompatibility and can be prepared as an injectable ionic solution. They do have poorly controlled degradation and variable cell adhesion characteristics.

Collagen may be prepared in various forms of gel for tissue engineering applications, including those with *in situ* formation capability. Many of these applications involve unmodified collagen, chemical cross-linkers can be used to inhibit degradation and resorption when necessary.

An increasingly important class of hydrogels for regenerative medicine comprises those made from self-assembled peptides. These are polypeptides that assemble under specific conditions to form nanoscale structures. For example, a group of self-assembled peptides made from amphiphilic molecules is derived from polypeptides linked to a polycarbon chain. The polypeptide region is typically hydrophilic while the hydrocarbon chain is hydrophobic. They can self-assemble into rod structures due to the assembly of the hydrophobic regions as well as the charge shielding of the hydrophilic end groups by ionic molecules in the solution. These molecules can be decorated with functional groups to facilitate cellular adhesion and signaling.

The significance of these engineered peptide hydrogels is that they epitomize this direction towards materials that can replicate cell niches, referred to above as the most important specification for tissue engineering templates. Considering the stem cell niche in particular, these contain ECM components, such as laminin and hyaluronan, which present cell adhesion ligands, and also soluble factors such as cytokines and growth factors, with a constantly replenished supply of differentiation cues. Peptide materials can be designed at a molecular level to give combined structural and biological activity characteristics that address these niche characteristics. These engineered, self-assembled peptides contain relatively short chains of amino acids. Through the careful choice of amino acid monomer sequences, the peptides can fold into secondary structures like β sheets, which themselves self-assemble into hierarchical structures, such as fibers and micelles. These fibrous hydrogels replicate the required cell niches far more than other materials.

3.3.3 ECM-Derived Materials

The ECM is, in principle, an obvious starting point for the development of a tissue engineering template since it comprises the structure in which cells reside in the various tissues and organs of the body and stands a good chance of providing the right type of environmental cues that drive cells into the regeneration process. The ECM in fact is in a state of dynamic equilibrium with the cells of that particular tissue, where it provides the best structural environment for its cells, which in turn provide the biochemical and molecular environment to support the matrix.

The ECM has a complex structure and composition, which varies to some extent from tissue to tissue and from species to species. There are some common major features, however, that control the normal, natural homeostasis mechanisms in the tissue and should play some role if that ECM was to be used as a template in tissue engineering. The most structurally important and indeed most abundant protein in the ECM is collagen. Type I collagen is present in the largest quantities but many of the other 20 or so types will be present, their presence being important since they modulate both mechanical properties and ligand-mediated interactions with cells. For example, fibronectin is the second most abundant protein and is especially prevalent in submucosal structures, interstitial tissues, and basement membranes. It has multiple ligands to enable adhesion to many cell types and since it is a dominant protein in developing embryos, it should play an important part in regenerative processes. Similarly, laminin is an adhesion protein with an important role in tissue development. In addition to these and other proteins, various types of GAGs are present in ECM, their nature and abundance varying quite considerably from tissue to tissue. The main GAGs present in ECM are hyaluronic acid, heparan sulfate, heparin, and some forms of chondroitin sulfate.

Many of these substances found in the ECM have been used separately as biomaterials in various forms of tissue engineering template. One very fundamental question here is whether they are better used individually or collectively. One advantage of the individual use is the greater uniformity with which they can be manufactured, and the greater control that can be exerted on their biological properties, including immunogenicity. However, the totality of the ECM has a much better chance of recapitulating the ideal environment for tissue regeneration since the ECM also contains growth factors and cytokines, the activity of which should powerfully assist the cells that are effecting the regeneration. These growth factors include VEGF, platelet-derived growth factor (PDGF), and BMPs.

3.4 CONCLUDING COMMENTS

This discussion has attempted to explain and rationalize the author's current position on the design and specification for tissue engineering templates. It is absolutely clear that the conventional approach of using simple synthetic materials that have prior regulatory approval for medical device-oriented devices as tissue engineering "scaffolds" cannot be expected to meet the specifications for the complex templates that will provide the optimal environment in which target cells, including stem cells, express new tissue. It is of no surprise here that the outcome of the process of exploring these specifications results in structures that tend to recapitulate the overall architecture of solid organs that constitute the basis for viable organ transplantation, emphasizing the conceptual similarity between transplantation and regenerative approaches to major diseases.

AUTHOR'S NOTE

This essay is based on several papers by the author published in the last few years concerned with the evolving nature of biomaterials and tissue engineering [1−5]. Parts of this essay summarize detailed analyses of biomaterials performance and specifications that will be published in monograph form.

REFERENCES

[1] Williams DF. Tissue engineering: the multidisciplinary epitome of hope and despair. In: McNamara L, editor. Multidisciplinary approaches to theory in medicine. Amsterdam: Elsevier; 2006.

[2] Williams DF. To engineer is to create. Trends Biotechnol. 2006;24:4−8.

[3] Williams DF. On the mechanisms of biocompatibility. Biomaterials 2008;29:2941−53.

[4] Williams DF. On the nature of biomaterials. Biomaterials 2009;30:5897−909.

[5] Williams DF. Essential biomaterials science. Cambridge University Press, in press.

Chapter 4

Principles of Stem Cell Biology

Valentina Spinelli[a], Pascale V. Guillot[b], and Paolo De Coppi[a]

[a]Surgery Unit, Institute of Child Health, University College London and Great Ormond Street Hospital, London, [b]Institute of Reproduction and Developmental Biology, Imperial College London, London

4.1 THE ORIGIN OF THE CELL REPROGRAMMING

Nuclear reprogramming is a term that describes a switch in nuclear gene expression of one kind of cell to another cell type.

In the early embryo of vertebrates, totipotent cells have the potential to differentiate and give rise to all the embryonic and the extraembryonic tissues, such as the placenta, amniotic fluid, or cord blood (CB). This process of cell specification is controlled by the interplay of endogenous and exogenous factors. At the blastocyst stage of the early embryo, the cells of the inner cell mass (from which embryonic stem (ES) cell lines are derived) are pluripotent: they are able to form only the embryonic tissues and cells derived from the three germ layers—the endoderm, ectoderm, and mesoderm. Eventually, cells that are committed to each of these germ layers specialize to give rise to the tissues of the adult body, such as the brain, intestine, or cardiac muscle [1]. Ultimately, multipotent stem cells are able to give rise to several cell types but those types are limited in number and are generally referred to by their tissue or germ layer origin (mesenchymal stem cell (MSC), adipose-derived stem cell, endothelial stem cell, etc.) [2].

Several classic studies [3,4] demonstrated the "plastic" fate of "committed" cells of the embryo, indeed the fate of these cells can change when they are explanted and exposed to a different microenvironment [1]. These observations have led to the development of various approaches to nuclear reprogramming in the last 60 years: nuclear transfer, cell fusion, and transcription factor transduction (induced pluripotent stem cells (iPSCs) generation). The latter has more recently led to many ways to generate integration-free iPSCs, such as plasmids [5], Sendai virus [6], adenovirus [7], synthesized RNAs [8], proteins [9], and histone deacetylase inhibitors (HDACis) like valproic acid (VPA) [10]. However, the nonintegrative methods still have pitfalls, being mainly associated with poor efficiency of pluripotent cells generation [11].

Generally, all of these approaches have shown that the fate of a specialized cell type (a cell that has been carefully determined to be differentiated) can be reversed, returning to an embryonic state [1].

The timeline shows an overview of the milestones contributing to the conception of vertebrate developmental biology and the initiation of nuclear reprogramming (Figure 4.1); all the nuclear reprogramming methods used are extensively described in the following section.

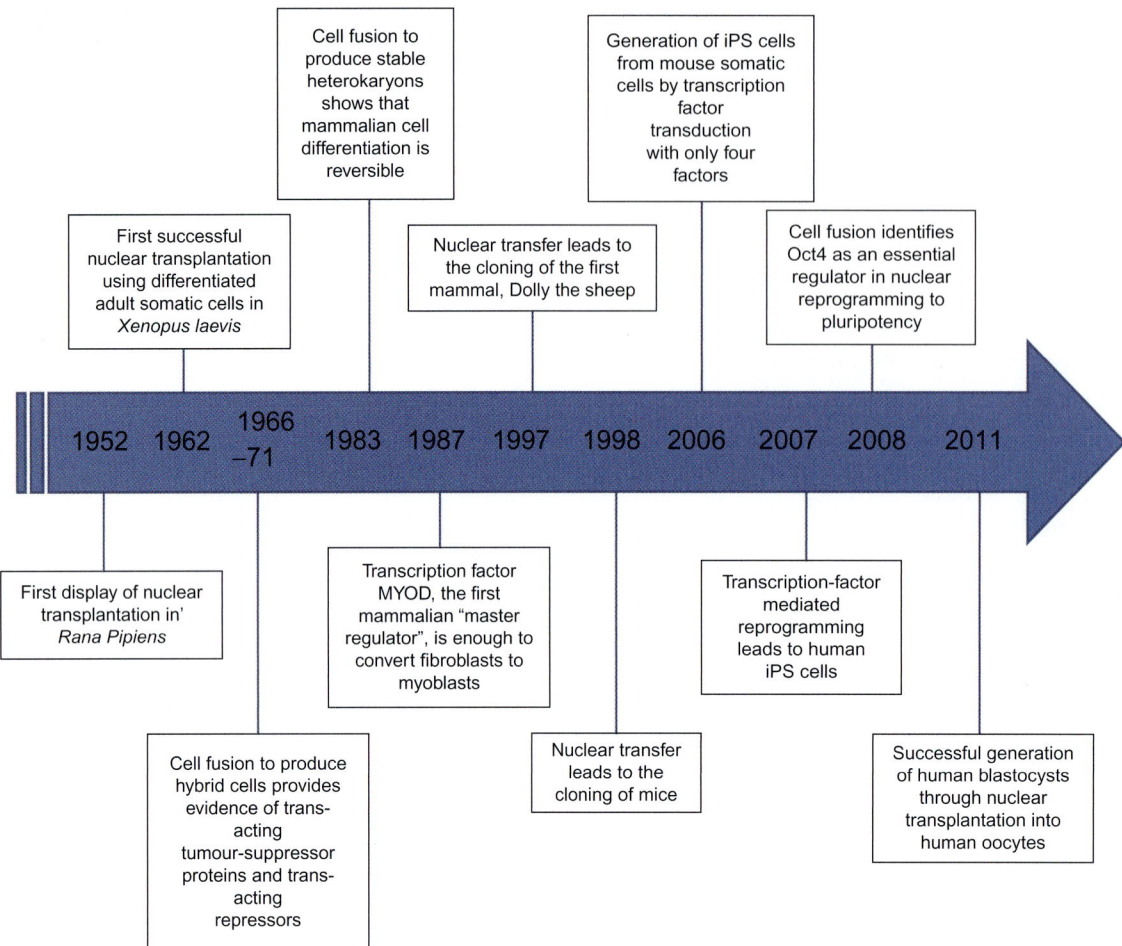

FIGURE 4.1   Timeline shows the discoveries in nuclear reprogramming since 1952.

4.2 SOMATIC CELL NUCLEAR TRANSFER

The transplantation of a nucleus from a differentiated somatic cell, such as an epithelial cell, into an enucleated oocyte, leads to the generation of an entire individual, which is a genetically identical clone of the original somatic cell. The process of transplanting a donor nucleus into an oocyte has now been termed somatic cell nuclear transfer (SCNT) (Figure 4.2) and it is the first mechanism of cell reprogramming performed from around the 1950s.

4.2.1 SCNT in Amphibians

The earliest evidence for the experimental reversal of cell differentiation came from the transplantation of a viable cell nucleus into an enucleated frog egg. In 1952, Briggs and King successfully transplanted a nucleus from a *Rana pipiens* embryo into an enucleated oocyte to produce adult organisms [12]. Despite their first innovative discovery, in a later study, Briggs and King showed that the transfer of nuclei from slightly older (gastrula) embryos resulted only

in abnormal development and they concluded that cell differentiation was likely to involve irreversible nuclear changes. Similar experiments were carried out in *Xenopus laevis* [13]; nuclei from terminally differentiated intestinal epithelial cells were successfully transplanted into irradiated oocytes and developed into fertile male and female frogs [14]. In this case, entirely normal and fertile male and female frogs were obtained. These results, which led Sir John B. Gordon to share the Nobel Price for Medicine in 2012 with Prof. Shina Yamanaha, showed that the process of cell differentiation can be fully reversed and does not require irreversible nuclear changes. In fact, it involves changes in nuclear gene expression but not in gene content [15].

The SCNT, as previously described, was first used in 1958 to create pluripotent cells from adult somatic cells [13]. This finding was followed by several studies in the following years which confirmed the initial finding in different species. Indeed, a series of successful transplantations of nuclei from different spectrum of cell origins, such as kidney, lung, and skin [16,17], were demonstrated.

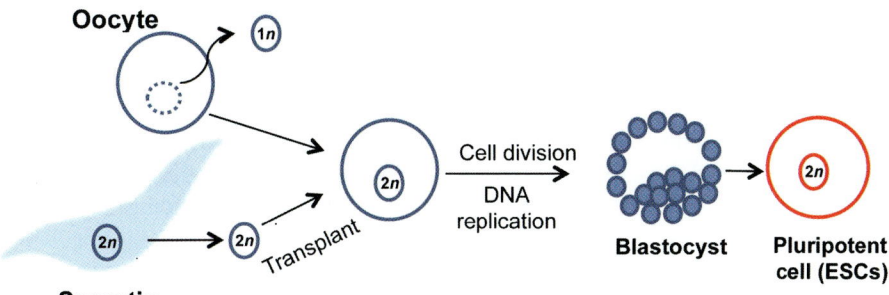

FIGURE 4.2 Somatic cell nuclear transfer. In this method, the nucleus of a somatic cell (which is diploid, 2n) is transplanted into an enucleated oocyte. The somatic nucleus is reprogrammed thus the cells that derived from it are pluripotent. A blastocyst is generated from the oocyte and ES cells are isolated from the blastocyst.

Moreover, nondividing erythrocytes from *R. pipiens* were demonstrated to produce normal larvae upon serial transplantations [18]. Nuclear transfer from myotomes could also effectively produce swimming tadpoles at a frequency of 2% [19]. Transcription of muscle differentiation markers ceased upon entry into the oocyte and reactivated once gastrulation occurred. This study demonstrated that the chemical environment in the oocyte is able to perturb the transcriptional machinery of the donor cell and reverse its differentiation status [15].

4.2.2 SCNT in Mammals

Gurdon's reports dominated the field of amphibians SCNT for more than three decades until cloning of mammal species became possible [20]. The first successfully cloned mammal was Dolly the sheep, made by fusing a mammary cell with an enucleated oocyte. In this study, Wilmut et al. performed the nuclear transfer using a fusogenic electrical pulse and an unfertilized oocytes as recipients together with donor cells that had been induced to exit the cell cycle by serum deprivation in culture. They postulated that the serum deprivation would change the chromatin structure leading to a nuclear reprogramming [1].

The following year, in 1998, Wakayama et al. using an enucleation pipette designed to deliver piezoelectric pulses, performed the cloning of mice [21]. This instrument facilitated the removal of the nucleus from the mouse oocyte in a more delicate manner compared to the instrument used for the sheep and replaced with the nucleus of a somatic cell. This particular process of SCNT was soon adopted in many laboratories and became extremely useful for generating transgenic mice [1].

However, it was only in 2002 that Jaenisch et al. finally emphasized that the "reprogramming" using SCNT resulted from the presence of stem cells or progenitor cells [22]. The authors obtained evidence that the fate of nuclei from specialized olfactory neurons or from B cells in which the immunoglobulin locus had been rearranged could be reversed to produce a mouse clone. SCNT is possible not only with oocytes but also with fertilized eggs (or zygotes), showing that the reprogramming factors are still present at this stage of development, and not only before the fusion of the two gamete cells [1]. This discovery opened the way to the concept of totipotent stem cells which, as previously described, are present in the zygote and are able to generate the entire organism comprising the extraembryonic tissues. In the following years, a wide range of species have been successfully cloned using SCNT in addition to sheep and mice, specifically dogs, goats, mules to wild animals such as African wildcats and wolves [23].

Furthermore, some experiments showed the possibility to cryopreserve the cells to employ in the SCNT [24,25] showing that nuclei obtained from frozen tissues could be transplanted into enucleated oocytes also after a decade of tissue freezing. These studies were carried out with the purpose of a therapeutic use of cryopreserved cells.

Beside the interesting and fascinating discoveries obtained with the previously mentioned studies, some aspects of SCNT remain to be clarified. A low efficiency of nuclear cloning (1−2%) typical of mice and the frequency of abnormalities in cloned animals generated by nuclear transfer suggests that nuclear reprogramming is incomplete and a better understanding of the mechanisms of gene regulation, particularly at the epigenetic level, is required. For example, cloned mice with apparently normal gross anatomy could hide numerous abnormalities, which raise concerns about the fidelity of SCNT for generating cloned organisms or cells without phenotypic defects. Some common abnormalities include aberrant gene expression in embryos, telomere elongation, obesity in adults, impaired immune systems and, often, increased cancer susceptibility and premature death [1]. In order to avoid these problems, different technical modifications have been adopted for SCNT in mice, including chemical activation of oocytes to make them more responsive, changes in the time of enucleation and cytokinesis inhibition [23]. Despite all the attempts, these alterations have led to only modest increase in the frequency of cloned animals, which still remains around 1−3% [23].

The gold standard for the completeness of cell reprogramming has been described as the formation of a fertile adult animal containing every type of functional cells. Nevertheless, it would not be therapeutically useful to supply a patient with a specific organ disease with replacement cells of every kind. In the case of SCNT, it is important to determine the efficiency of obtaining a particular differentiated cell type by using the transplanted nucleus of an entirely unrelated cell type. It has been shown that the success of nuclear reprogramming decreases as donor cells become more differentiated [26]. In this prospective, cells derived from a nuclear transplant blastocyst can be used to derive ES cells, whose differentiation potential can later be evaluated by transplanting these cells into normal host embryos. Although, the ethical concerns about obtaining human unfertilized eggs have lead to the possible use of animal eggs, such as those of cows, mice, or rabbits, to generate ES cells from transplanted human somatic nuclei [26]. Nevertheless, eggs produced by transfers between very different species, such as human and mouse, cow, or pig, generally die before the 32-cell stage [21]. In conclusion, there is no confirmed evidence that proliferating ES cells can be obtained from such distant combinations, including human nuclei in monkey cytoplasm.

4.3 CELL FUSION

Pluripotent cells can also be obtained by fusion of two or more cell types to form a single entity cell, an approach called cell fusion (Figure 4.3).

The fusion between two different genomes gives rise to the existence of *trans*-acting repressors and tumor-suppressor proteins able to regulate the gene expression after the cellular fusion. Fusion experiments performed nearly four decades ago showed that when a differentiated cell such as a mouse fibroblast was fused with a hamster melanocyte or a rat hepatocyte, mouse melanin and tyrosine aminotransferase, respectively, ceased to be

synthesized [27,28]. These studies provided evidence on the gene expression, which is regulated not only by acting DNA elements but also by acting DNA repressors. Subsequently other evidences regarding the differentiated state of mammalian somatic cells as a reversible and flexible state were generated by some studies conducted in the late 1970s [29–31]. Indeed, the genome is not irreversible and fixed but it requires continuous regulation and a good balance between the gene regulators. Moreover, the cell fusion experiments showed that the pluripotent state can dominate the differentiated state under certain conditions, leading to activation or repression of some genes that were previously silent or activate. On the other hand, the cell fusion approach is now considered as a potent way of elucidating some epigenetic mechanisms like DNA demethylation that are required for nuclear reprogramming [1].

Using the cell fusion approach, two different entities can be generated: hybrids, characterized by the ability to proliferate after nuclei fusion and heterokaryons, in which the nuclei remain separated and do not proliferate. Some experiments of cell fusion performed later reported that in cell hybrids there is cancerogenesis regulation activated after the nuclei fusion; it was shown that *trans*-acting tumor-suppressor proteins exist in the hybrids, because in some cases in which noncancerous cells and tumor cells were fused, the "normal" state dominated the transformed state, preventing tumor formation. Nevertheless, the suppression of malignancy did not lead to the oncogene loss, because after several cell proliferations the transformed phenotype re-emerged [32]. Moreover, from these preliminary observations it is not clear if the observed gene activation was caused by the loss of a gene encoding a repressor or the action of an activator.

In 1983, the production of heterokaryons in mammalian cells gave the first definitive evidence that some genes previously silent could be activated [29]. The nuclear reprogramming can be assessed when the cell types come from different species because their gene

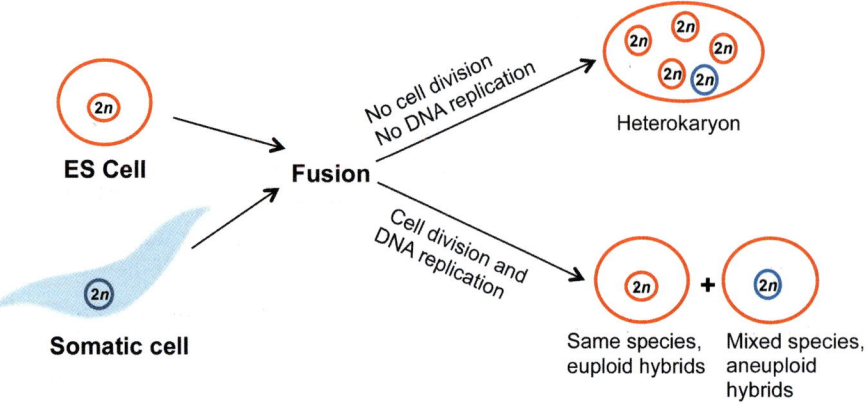

FIGURE 3. Cell fusion. In this approach, one somatic cell is combined with a pluripotent stem cell: the resultant fused cells can be heterokaryons or hybrids.

products can be easily distinguished. This demonstration of nuclear reprogramming generated incredulity, because differentiated state of mammalian cells was thought to be fixed and irreversible, but it confirmed what was previously described in the hybrids [27,28].

However, it is important to alight on a difference between hybrids and heterokaryons: by the using of the second ones the problems of chromosome loss and rearrangement that were typical of hybrids were overcome, because in the absence of proliferation the nuclei of the two cell types remained distinct and intact [1].

Early experiments on heterokaryons performed before 1983 had shown nuclear swelling and DNA and RNA synthesis, but not the activation of silent genes [33], presumably due to the cell type chosen for the cell fusion reprogramming. These studies involved chicken erythrocytes, which have a nucleus but are among the most specialized and difficult cells to reprogram. As explained above, the activation of previously silent genes was first detected by Blau et al. [29] using heterokaryons of muscle cells and amniotic cells. For the first time, this study alighted on a particular type of fetal cells, the human amniotic cells. The authors used mouse muscle cells as the fusion partner in order to increase gene dosage and avoid cell division and also because they are naturally multinucleate and postmitotic; human amniotic cells were chosen as the other fusion partner for their embryonic origin. This choice indicated that human amniotic cells might be more plastic than other cell types. The results obtained from the heterokaryons showed directed differentiation and several human muscle proteins expression, indicating that muscle genes had been activated in nonmuscle cells [29,30]. With other experiments subsequently carried out, Blau et al. also demonstrated that in heterokaryons formed from mouse muscle syncytia and diverse cell types, including human fibroblasts (which arise from the mesoderm), hepatocytes (from the endoderm) and keratinocytes (from the ectoderm), silent muscle genes could be activated in cells representative of all three embryonic lineages [29,30]. Some important evidences emerged from these and other studies: (i) the relative ratio of the nuclei, or the gene dosage, contributed by the two cell types dictated the direction of reprogramming [29,34]; (ii) DNA replication was not required and DNA methylation status was crucial to the outcome in heterokaryon studies [35,36]; and (iii) the frequency and kinetics of reprogramming also differed among cell types [30]. The proof that silent genes could be activated in muscle-cell-containing heterokaryons was rapidly confirmed by others in other cell types [37]. For example, erythroid-specific and hepatocyte-specific genes were activated in fibroblast-derived nuclei present in nondividing mixed-species heterokaryons [38,39]. Altogether, these experiments established that the differentiated state was continuously regulated by the balance of regulators present at any given time, providing strong evidence for nuclear plasticity.

The interest in cell fusion as a reprogramming method to study the pluripotency and the regulatory mechanisms that are involved was reestablished during the late 1990s; in particular, to avoid the problems of aneuploidy, all the studies on cell fusion were performed using the same species [1]. In 1997, Tada et al. showed for the first time the nuclear reprogramming of somatic cells in proliferative hybrids [40]. They fused pluripotent stem cells derived from primordial germ cells (PGCs) with thymocytes isolated from adult mice. Subsequently, they analyzed the demethylation of the DNA (an epigenetic modification of the genome) by using DNA methylation-sensitive restriction enzymes, and whether certain imprinted and nonimprinted genes from the somatic genome had been activated. Moreover, they showed that the fused tetraploid cells that resulted from the experiment were pluripotent and they could contribute to the three germ layers (mesoderm, ectoderm, and endoderm) in chimeric embryos. This was an innovative finding that opened the way to another work carried out by the same authors, in which they could confirm that somatic cells can acquire a pluripotent state after fusion with ES cells [41]. They fused ES cells from male mice with thymocytes from female mice containing a GFP reporter transgene driven by the promoter of mouse *Oct4* (a pluripotency transcription factor also known as *Pou5f1*), and they found that genes on the inactive X chromosome and the *Oct4−GFP* reporter construct of thymocytes were reactivated. The difference related to the use of ES cells as a fusion partner, in contrast to germ cells (described above), was related to the fact that the imprinted genes in fused tetraploid cells were not demethylated.

After these set of experiments, the same group generated hybrid cells between subspecies with *Mus musculus domesticus* ES cells and *M. musculus molossinus* thymocytes [42]. The presence of frequent DNA sequence polymorphisms between the genomes of these two subspecies allowed the researchers to monitor the origin of the RNA and DNA in hybrid clones. Using this elegant approach, they showed that the reprogrammed somatic genome in hybrids with ES cells becomes hyperacetylated at histones H3 and H4, whereas the lysine residue at position 4 of H3 becomes globally hyper-dimethylated and hyper-trimethylated, an indication that the epigenome was converted to a pluripotent state. Usually the trimethylation of the histone H3 on the lysine 4 is an epigenetic modification able to mark transcriptionally active genes.

One year later, Cowan et al. showed that nuclear reprogramming of human somatic cells could be realized by fusing them with human ES cells in tetraploid hybrids in a 1:1 ratio [43]. In 2006, Smith et al. subsequently demonstrated that, in mice, overexpression of *Nanog*, a pluripotency transcription factor, substantially enhanced fusion-based nuclear reprogramming [44], which clearly showed that regulators of pluripotency can reverse regulators of cell differentiation.

Therefore, heterokaryons derived from cell fusion of different species permit more comprehensive studies of nuclear reprogramming than same-species cell fusions, as all gene expression changes can be monitored on the basis of species-specific differences. In addition, chromosome loss, rearrangement and aneuploidy, which modify the analysis of the results obtained with mixed-species hybrids, in mixed-species heterokaryons, do not occur. However, the use of heterokaryons for studying the induction of pluripotency in somatic cells seems to be counterintuitive, because ES cells divide rapidly and extensively. Therefore, it seemed probable that the induction of growth arrest that follows cell fusion in heterokaryons, which are nondividing, would lead to a loss of pluripotency and to differentiation.

Since the scientific discoveries cannot be definitive, in contrast to that which was shown some years before, results from two laboratories using mixed-species heterokaryons have demonstrated that pluripotency genes, such as *Oct4* and *Nanog*, are activated and their promoters demethylated when mouse ES cells are fused with human B cells or with human fibroblasts [45,46].

As previously mentioned, the molecular events that underlie nuclear reprogramming can be ideally analyzed using heterokaryons. Indeed, the secret of gene expression analyses is the possibility of purifying a small proportion of the population ($\sim$1%) immediately after fusion, by flow cytometry. Besides the knowledge that heterokaryons are characterized by low proliferative capacity, in the experiments carried out so far, hybrids have proliferated extensively and been drug selected before being analyzed, so it has not been possible to assess the earliest changes in gene expression.

The efficiency of inducing pluripotency still remains the main challenge for all of the nuclear reprogramming methods and it depends on the cell types used; for example, in the data shown above, the efficiency to form the heterokaryons was 16% for B cells and 70% for fibroblasts [45,46] and differences in the timing of gene activation and DNA demethylation were observed.

In contrast to that which was observed in the hybrids, the rapid rate of reprogramming in heterokaryons is making them useful for elucidating the molecular mechanisms that are required for initiating reprogramming to a pluripotent state, by using *loss-of-function* (gene silencing) and *gain-of-function* (gene activation) approaches. For instance, when mouse ES cells that have lost *Oct4* expression are fused with human B cells to form heterokaryons, the B cells are not reprogrammed, showing that Oct4 is required for reprogramming toward a pluripotent state

[45]. In another study, using heterokaryons, Bhutani et al. recently elucidated a new role for the DNA cytidine deaminase activation-induced deaminase (AID), an enzyme known to deaminate cytosine residues. They revealed an active mechanism that is essential for DNA demethylation and for the induction of nuclear reprogramming of fibroblasts toward pluripotency [46].

The notion that has been suggested from these new data is represented by the idea that the DNA repair mechanism may operate in mammals, in which a methylated base or nucleotide is exchanged for an unmethylated one [1]. These studies demonstrate the potential of cell fusion to exemplify mechanistic knowledge of genome modification, such as DNA demethylation, to the successful reprogramming of somatic cells by SCNT or transcription factor transduction.

4.3.1 The Transcription Factors Induce Cell Lineage Switch

Cell fusion experiments have clearly shown that there are specific components within each cell lineage, defending it from external influences and protecting its designed function. In particular, these intrinsic cell-specific properties can be used to redirect differentiation signals. In 1989, using the evidence that DNA demethylating agent, 5-azacytidine (AZA) enables efficient reprogramming into muscle lineage [36], Weintraub et al. identified a gene as a main regulator of myogenesis: MyoD. They showed that its ectopic expression forced recipient nonmuscle cells to convert into myocytes [47].

As previously described in many studies, cells from three germ layers were successfully reprogrammed; nevertheless, an efficient genome conversion was performed starting with the material originating from the mesoderm, whereas cells from endoderm and ectoderm lineages retained their original gene signatures. This has suggested the necessity of additional factors to induce a complete reform of the preexisting gene expression profile.

The general idea is that the forced expression of transcription factors that naturally occurs during developmental programs can incite lineage switches. As described above, expression of MyoD can redirect nonmuscle cells into the muscle lineage [47]. To reinforce this hypothesis, in 1994 it was demonstrated that the overexpression of adipocyte-specific nuclear hormone receptor, peroxisome proliferator-activated receptor (PPAR)γ2, triggers adipogenesis in fibroblasts [48]. Although single genetic factors have been described to control cell fate in adipocytes and myocytes, one genetic factor is not sufficient to induce the transdifferentiation in all cell types. For example, dermal fibroblasts require a combination of Gata4, Mef2c, and Tbx5 to be converted into cardiomyocytes [49]. Similarly, ectopic expression of separate sets of transcription factors has successfully triggered transdifferentiation of dermal fibroblasts into neuron-like cells [50], blood progenitors [51], insulin secreting β cells [52], and brown adipose cells [53]. Further exploration in the lineage switching has placed emphasis on the importance of spatial, temporal, and quantitative control of transcription factor expression in the generation of an array of cell types [15].

In conclusion, cell lineage switching could be considered as an acquisition of the pluripotency closely related with SCNT and cell fusion which occurs without the need for cell division. Moreover, markers of progenitor cells are often not detected, suggesting that reprogramming may take place in a direct fashion without the presence of an undifferentiated intermediate. However, the possibility that a transient intermediate state could be present is yet to be excluded [15].

4.4 TRANSCRIPTION FACTOR TRANSDUCTION: iPS CELL GENERATION

As extensively reported above, the overexpression of a single transcription factor in somatic cells was unexpectedly found to activate cohorts of genes that were typical of other somatic cell types, leading to remarkable changes in cell fate. Furthermore, it was discovered that pluripotency can be reacquired by numerous differentiated somatic cell types through the overexpression of four transcription factor-encoding genes. Besides ES cells, these cells, known as iPS cells, represent the strongest example of the plasticity of cells in response to a disruption in the stoichiometry of their transcriptional regulators. Human iPS cells, avoiding the ethical and immunological issues linked to the use of ES cells, represent an interesting candidate to employ in regenerative medicine and cell therapy. Moreover, because iPSCs could be generated from a patient's own cells (autogenous derivation), *in vitro* models of human diseases could be generated allowing drug screening in an unprecedented manner.

This section provides an overview on the recent development in the field of the iPS cells from their discovery to the novel methods of transcription factor transduction both in mouse and in human.

4.4.1 iPSCs from Mouse and Human Somatic Cells

The conversion of somatic cells to a pluripotent state is a highly complex process that might involve the cooperation of more than 100 factors, such as proteins, transcriptional factors, or other regulators of the genome. After the fusion experiments performed by Tada et al. [41] where they clearly showed that ES cells and embryonic germ cells contain factors that can induce reprogramming and

pluripotency in somatic cells, many efforts have been made to identify the "master regulators" of the ES cell state but, unfortunately, without success. Only in 2006, pluripotent stem cells could be generated from mouse fibroblasts by ectopic expression of the OKSM factors, i.e., Oct3/4, Sox2, C-Myc, and Klf4 [54]. These cells, designated as iPS cells, have similar, but not identical, morphology, growth properties, and genetic profile of mES cells [1] (Figure 4.4).

At the beginning, the results evoked excitement as well as scepticism but they alighted on an intensive work in this field that is still in continuous expansion. They used retroviral vectors to introduce into mouse embryonic and adult fibroblasts a mini-complementary-DNA library of 24 genes expressed by ES cells, and these genes were then tested for their ability to induce pluripotency. The pluripotent state was evaluated by examining for activation of a reporter gene construct containing the promoter of Fbx15, a gene previously identified as being specific to ES cells [55]. Clones in which the Fbx15 promoter was activated produced a reporter protein that rendered them resistant to the drug neomycin, and these drug-resistant clones had similar morphology, growth properties, and gene expression characteristics to ES cells. The pluripotency of iPS cells was assessed *in vivo* by their capacity to form teratomas (tumors that include cells of all three germ layers) following subcutaneous transplantation into immunocompromised mice and to contribute to embryonic development following injection into blastocysts [1]. Nevertheless, because of the simplicity of the method, and because iPS cells initially failed to produce adult chimeric mice, many scientists formulated doubts about the validity of iPSCs as stem cells with ES-cell-like properties [1].

The doubts were dispelled 1 year later following the derivation of iPS cells using retroviral transduction of the OKSM factors in human dermal fibroblasts [56]. Impressively, transgenes encoding OKSM factors need to be present only when iPS cells are being generated. When these iPS cells are established, the retroviral transgenes

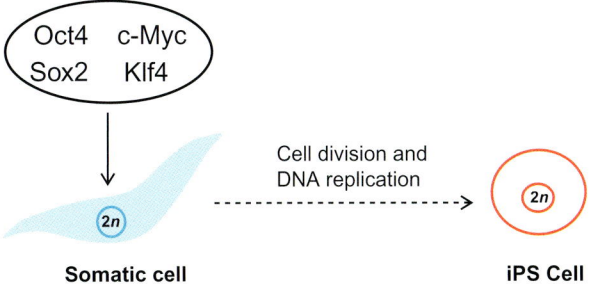

FIGURE 4.4 Transcription factor transduction. This method can be used to form iPS cells from almost any cell type of the body (adult and fetal) through the introduction of four genes (Oct4, Sox2, Klf4, and c-Myc) by using retroviruses; iPS cells show properties similar to ES cells.

become silenced, and the endogenous genes encoding the four factors are activated. Therefore, the self-renewal of iPS cells and maintenance of endogenous expression of the genes encoding Oct4, Sox2, Klf4, and c-Myc indicate that cells are reprogrammed to pluripotency. At the same time, James Thomson's group also reported the generation of human iPSCs from somatic cells using a different combination of factors. They identified another set of four genes (OSNL factors), i.e., Oct4, Sox2, Nanog, and Lin28, capable of reprogramming somatic cells to full pluripotency at a clonal level [1].

4.4.1.1 The Hidden Fate of iPSCs Derived from Somatic Cells

The similarity of the phenotype of iPS derived from human somatic cells compared to hES cells is a challenge. It is well accepted that the iPS cells are not identical to ES cells; with increasing evidence showing that iPS cells are distinct from hES cells, albeit both being pluripotent, as defined by their capacity to differentiate into lineages of the three germ layers (Table 4.1) [1]. For example, differences have been described at the level of gene expression, DNA methylation, and stability of the pluripotent phenotype over time, as well as the epigenetic memory. These may be attributed to somatic mutations [57], copy number variations (CNVs) [58], and immunogenicity [59], which could be altered in iPS cells. Moreover, the factor combinations, gene delivery methods, and culture conditions might also contribute to the differences obtained between the different iPS cell populations generated. Finally, some variations may be attributed to stochastic events during reprogramming, which cannot be controlled [60]. Thus, increasing efforts now focus on finding the "best candidate parental population" to generate iPS cells for *in vitro* studies and future clinical applications. The sequencing of the majority of the protein-coding exons of 22 human iPS lines and the nine parental fibroblast revealed that some of the reprogramming-associated mutations were likely to preexist in the starting fibroblast cultures, while the others occurred during reprogramming and subsequent culturing [57]. The comparison of CNVs of different passages of human iPS cells with their fibroblast cell population and with ES cells showed that the reprogramming process is associated with high mutation rates characterized by increased levels of CNVs and genetic mosaics in particular during early passage of human iPS lines [58]. Immunologically, it has been observed that mouse embryonic fibroblasts (MEFs), reprogrammed into iPSCs by either retroviral approach (ViPS) cells or a novel episomal approach (EiPS cells) generated an immunoresponse when transplanted in B6 mice. In contrast to B6-derived ES cells, teratomas formed by B6 (ViPS) cells transplantation were mostly immune-rejected by B6 recipients, and the majority of teratomas formed by B6-derived iPS cells were immunogenic with T-cell infiltration and apparent tissue damage observed in a small fraction of teratomas [59].

In this context, fetal stem cells (FSCs) have emerged as an intermediate phenotype between embryonic and adult stem cells. FSCs are neither fully pluripotent nor limited to

TABLE 4.1 Number of ESC and Somatic-Derived iPSC Clones Compared in Published Studies. The Table Summarizes the Conclusion Reported by Different Studies About the Relationship Between ESC and iPSC with the Author's Name, the Year of the Article Publication and the Number of the Clones Analysed.

Relationship Between ESCs and iPSCs	Author	Year	Numbers of Clone	
			ESC	iPSC
It is difficult to discriminate between ESCs and iPSCs	A.M. Newman	2010	23	68
	M.G. Guenther	2010	36	54
	C. Bock	2011	20	12
There are significant differences between ESCs and iPSCs	M. Chin	2009	3	5
	C.M. Marchetto	2009	2	2
	J. Deng	2009	3	4
	Z. Ghosh	2010	6	4
	A. Doi	2011	3	9
	Y. Ohi	2011	3	9
	K. Kim	2011	6	12
	R. Lister	2011	2	5

multipotency, when compared with their adult counterparts, FSCs appear to be more primitive, with higher growth kinetics, smaller cell size, active telomerase, and greater plasticity, whilst lacking tumorogenicity. These features may represent an advantage for regenerative medicine because they might be easier to reprogram [1].

Moreover, one of the major limitations related to iPS cell generation has been the use of retroviruses or lentiviruses, which could cause mutagenesis leading to a risk of teratogenesis and other adverse effects like those seen in some attempts at gene therapy [61]. Therefore, it has been reported that to ameliorate the efficiency of iPS generated from somatic cells, some groups have tried to modulate key components of the cell cycle like repression of the Ink4a/Arf locus or downregulation of the p53−p21 pathway; nevertheless, p53 suppression can lead to increased levels of DNA damage and genomic instability [58]. FSCs represent an alternative source for cell reprogramming and regenerative medicine since they are easily achievable, they show high proliferation rate, negligible immunogenity and demonstrate no evidence for teratoma formation and no ethical concerns [1]. The next section is dedicated to the generation of iPS cells from FSCs and their future applications in regenerative medicine. Since FSCs are a promising source for nuclear reprogramming, an accurate analysis of their origin, their characteristics and their pluripotency potential is required.

4.4.2 iPS Cells Derived from Fetal and Extraembryonic Tissues

In recent years, FSCs have emerged as a potential reservoir for cellular reprogramming and an alternative cell type in regenerative medicine. Indeed, FSCs can be isolated both from fetal tissues, such as blood, liver, bone marrow (BM), pancreas, spleen, and kidney [62], and from the extraembryonic tissues, such as CB, placenta, umbilical cord, and amniotic fluid. FSCs from BM or liver can be obtained from termination of pregnancy or during pregnancy with a direct biopsy of the fetus. Less invasively, first-trimester placenta and mid-trimester amniotic fluid samples can be obtained during amniocentesis and chorionic villus sampling performed for the prenatal screening.

FSCs are heterogeneous populations which include stromal stem cells/MSCs, hematopoietic stem cells (HSCs), and pro-pluripotent cells isolated from the extraembryonic tissues. In particular, their phenotypic feature, their properties and cell markers expression depend on their tissue of origin and gestational age. The pluripotent fate and the therapeutic potential of FSCs are related to their origin. Table 4.2 shows some applications of FSCs isolated from fetal and extraembryonic tissues.

Since the FSCs are represented by an extensive cells type, the following paragraphs are focused on analyzing the characteristics of some specific kind of FSCs, such as human fetal HSCs, placental stem cells, and amniotic fluid stem cells (AFSCs) and their use for nuclear reprogramming.

4.4.2.1 iPS from Human Fetal HSCs

HSCs are multipotent stem cells responsible for the maintenance of hematopoiesis throughout fetal and adult life by the derivation of all the hematopoietic lineages [63]. HSCs express some typical markers, such as CD34 and CD45 antigens, but not CD38 and human leukocyte antigen (HLA)/DRE [64]. Human fetal HSCs localize in the BM "niche" which are able to perform self-renewal and

TABLE 4.2

Origin and Cell Type	Treated Recipient	Regenerated Tissue	Cell Delivery Method	Disease Treated	Reference
Human extra-embryonic cord blood	Human	Bone marrow	Systemic administration	Malignant and non-malignant blood disorders	Broxmeyer et al, 2010
Human amniotic fluid	Mouse	Lung	Systemic administration	Lung injury	Carraro et al, 2008
Human amniotic fluid	Rat	Heart	Intramyocardial injection	Myocardial infarction	Yeh et al, 2010
Human amniotic fluid	Rat	Smooth muscle	Intramuscolar injection	Wound healing bladder injury	De Coppi et al, 2007
Human placenta	Mouse	Brain	Intracranial injection	Parkinson's disease	Kong et al, 2008
Human fetal tissue blood and kidney	Mouse	Bone	Intrauterine transplantation	Osteogenesis Imperfecta	Guillot et al, 2008

differentiation throughout adult life. Interestingly, during the development of the embryo and the fetus, before reaching the BM, newly formed HSCs migrate to the fetal liver, which is the main hematopoietic organ before birth [65]. The migration of HSCs is accompanied by a modification in the percentage of the cells in each region of hematopoiesis: first-trimester fetal blood contains more CD34$^+$ cells than term gestation blood, in which CD34$^+$ cells constitute 4% of cells in blood, 16.5% in BM, 6% in liver, 5% in spleen, and 1.1% in the thymus [66]. Moreover, during the third trimester the frequency of CD34$^+$ cells in the blood gradually decreases probably because the marrow is the primary site of hematopoiesis [67]. Compared to HSCs from adult BM, fetal HSCs have a greater proliferative capacity, lower immunological reactivity, and lower risk of graft-versus-host disease (GVHD) [68]; therefore, thanks to their repopulating capacity after intra-bone injection of severe combined immunodeficiency mice [69,70], HSCs can be used as an alternative cell source in respect to adult BM HSCs.

Some advantages derived from the use of blood HSCs to generate iPSCs compare to dermal fibroblasts, such as the more convenient and less invasive procedure to obtain peripheral blood (PB) than skin biopsy [1]. A recent study reported the generation of iPSCs from human immature mononuclear cells (MNCs) CD34$^+$ or CD133$^+$ isolated from umbilical cord blood (UCB), adult PB and BM using a set of EBNA1/OriP plasmids using inclusion of the EBNA1 gene and the OriP DNA sequence from the Epstein−Barr virus (EBV) enables a plasmid, after one-time DNA transfection, to replicate extra-chromosomally in many types of primate cells as a circular episome [71]. In this particular experiment of nuclear reprogramming, they adopted two sets of plasmids to transduce the cells. In the first set, EBNA1/OriP plasmid (called pEB-C5) and the reprogramming factors Oct4, Sox2, Klf4, c-Myc, and LIin28 are expressed as a single polycistronic unit; in the second one, EBNA1/OriP plasmid, SV40 large T antigen (Tg), Nanog, or a small hairpin RNA targeting p53 (p53shRNA) is individually expressed. They observed a highly efficient reprogramming of blood MNCs. Within 14 days of one-time transfection by one plasmid, up to 1000 iPSC-like colonies per two million transfected CB MNCs were generated. Although the efficiency of deriving iPSCs from adult PB MNCs was approximately 50-fold lower, it could be enhanced by inclusion of a second EBNA1/OriP plasmid for transient expression of additional genes like SV40 T antigen. The time of obtaining iPSC colonies from adult PB MNCs was reduced to half (∼14 days) as compared to adult fibroblastic cells (28−30 days). More than nine human iPSC lines derived from PB or CB cells are extensively characterized, including those from PB MNCs of an adult patient with sickle cell disease [71]. This method represents an innovative and useful instrument to generate human iPSCs from blood MNCs that will accelerate their use in both research and future clinical applications.

Around 1% of fetal HSCs isolated from UCB at birth are characterized for the positive expression of the CD34 surface marker and negative expression of CD38. Following cytokine mobilization, the frequency of CD34$^+$ cells in CB is higher than that of adult BM or PB and compared with BM cells, CD34$^+$/CD38$^-$ UCB cells proliferate more rapidly and generate larger numbers of progeny cells [72]. A possible explanation for the greater proliferative capacity of UCB could be the longer telomere lengths compared to the adult counterpart [72]. It was also demonstrated that UCB HSCs express neuronal proteins and can differentiate into neuronal-like cells or glial cells [73]. Altogether, these properties designate that UCB has an alternative source of HSCs for transplantation. Despite this, obtaining an adequate cell number from a single UCB unit is difficult, also because the homing and engraftment capacity of HSCs seems to be dependent on cytokines release, molecular and cellular factors [74].

Other transduction systems were employed to generate iPS from human CD34$^+$ UCB cells; for example, a brief report described the iPS production from fresh CB and CB cryopreserved for 58 years [69]. A lentiviral vector encoding Oct4, KLF-4, Sox2, and c-Myc was used to transduce CD34$^+$ cells; iPS cell colonies stained positive for Oct4, Nanog, Tra-1-60, SSEA-4, alkaline phosphatase (AP) and after quantitative RT-PCR, they expressed endogenous Oct4, Sox2, and Nanog. The expression of ectodermal, mesodermal, and endodermal proteins was confirmed *in vitro* by embryoid bodies (EBs) and *in vivo* by teratoma formation. The authors also showed that the generation of iPS from frozen CB produced cells expressing Tra-1-60, SSEA-4, Nanog, and Oct4 and able to form teratomas with endoderm, mesoderm, and ectoderm markers expression; moreover, the efficiency of iPS cell from thawed CB ranged from 0.027% to 0.05% per CD34$^+$ cell, similar to cultured CD34$^+$ cells from freshly isolated or shorter term frozen CB [69]. Therefore, since it was reported that the absence of the p53 gene results in spontaneous reversion to a pluripotent state of germ cell stem cells, Takenaka et al. investigated if the repression of p53 could mediate the induction of pluripotency in CD34$^+$ CB cells [70,74]. A shTP53 RNA construct expressing a short-hairpin RNA (shRNA) sequence able to reduce the amount of endogenous p53 transcripts was introduced in addition to Sox2, Oct3/4, Klf4, and c-Myc factors. With this system, they obtained a large number of bona fide iPS cell clones from 2×10^4 virus-infected cells.

4.4.2.2 iPS from Placental Stem Cells

The placenta is the organ involved in maintaining fetal tolerance and allows nutrient uptake and gas exchange with the mother, but it is now clear that progenitors and stem cells are also present [75]. Amnion, chorion (fetal sides), and deciduas (the maternal side) are the three parts that constitute the placenta, each of these are characterized by the presence of different stem cell populations. Amniotic epithelial cells (AECs) and amniotic mesenchymal cells (AMSCs) are isolated from amniotic membrane; an evidence suggests that AECs and AMSCs express pluripotency markers and have the ability, *in vitro*, to form xenogeneic chimera with mouse ES cells [76]. AMSCs and stem cells derived from chorion (CSCs) can be isolated throughout gestation from first trimester to delivery. They show a differentiation potential toward mesodermal lineages and express some markers, such as SSEA-4, TRA-1-61, and TRA-1-80, typically present in ES cells. Nevertheless, some differences exist between AMSCs and CSCs regarding their differentiation potential; indeed, AMSCs seem to be more directed to the adipogenic lineage whereas CMSCs are directed more to chrondo-, osteo-, myo-, and neurogenic [77]. Finally, another recent study [78] comparing the phenotype of first-trimester and term fetal placental chorionic stem cells (e-CSC and l-CSC, respectively) has shown that e-CSC are smaller cells with faster growth kinetics and higher levels of pluripotency marker expression. The authors also found that e-CSC expressed Oct4A variant 1 and had potential to differentiate into lineages of the three germ layers *in vitro* and high capacity of tissue repair *in vivo*. In addition e-CSC and l-CSC express markers associated with PGCs and thus may share a developmental origin with these cells.

Human amnion-derived cells (hADCs) are described as a heterogeneous group of multipotent progenitor cells that can be easily derived from placental tissue after delivery [75]. Recently, the nuclear reprogramming of hADCs has been demonstrated: hADCs are able to give rise to iPS using lentivirus encoding Oct4, Sox2, and Nanog as transduction system. hADC-iPS colonies showed positive expression of AP, Oct4, Sox2, Nanog, SSEA-3, SSEA-4, Tra-1-60, and Tra-1-81; moreover, hADc-iPS could form *in vitro* EBs expressing markers of the three embryonic germ layers. Therefore, teratoma-like masses containing mesoderm, ectoderm, and endoderm proteins were observed 6−8 weeks after the injection of hADc-iPS into immunodeficient mice [79]. In conclusion, hADCs could be an ideal source to efficiently reprogram into individual-specific iPS cells.

4.4.2.3 iPS from AFSCs

Human amniotic fluid (hAF) is known to contain multiple cell types and recently, it was demonstrated that lines of broadly multipotent cells (hAFS cells) are present in AF that can give rise to adipogenic, osteogenic, myogenic,

endothelial, neurogenic, and hepatic lineages, inclusive of all embryonic germ layers [80]. This study carefully analyzed hAFSCs showing that they are characterized by an easy growth in culture and maintenance of a stable phenotype and genotype. Approximately 1% of AF cells express the surface antigen c-Kit (CD117) and a number of mesenchymal and/or neural stem cells markers, including CD44 (hyaluronan receptor), CD73, CD105 (endoglin); 90% of hAFSC express the pluripotency markers Oct4, Nanog, and SSEA-4 but they did not express other ES cells markers as SSEA-3 and Tra-1-81 [80]. As mentioned above, hAFSC had multipotent properties and exhibited the intrinsic capacity to differentiate into cell types indicative of the three germ layers. Therefore, since these cells did not form teratomas upon transplantation into mice, they could be considered for therapeutic applications [81].

Table 4.2 shows some applications of FSCs isolated from hAF in tissue engineering and cell therapies. In particular, regarding tissue engineering several studies indicated that AFSCs have the capacity to attach and proliferate on biodegradable scaffolds; an example was offered by the use of ovine mesenchymal AF-derived cells for the generation of cartilage grafts [82,83] and tendon grafts to employ in diaphragmatic hernia repair [82,84].

Another interesting example of the use of AFSC regards congenital malformations of the heart; hAFSCs were selected for the expression of CD133 surface marker in order to obtain the two cell types found in heart valves, namely myofibroblast like cells (CD133$^-$) and endothelial cells (CD133$^+$) and they were used to regenerate the functionality of the heart valves. After seeding of these cells on the valve scaffolds, heart valve opening and closing capability has been shown [85].

The phenotypic fate and the genes expression profile that characterized hAFSC draws these cells as "precursor" stem cells and, since the "precursor" state could be rapidly and efficiently reprogrammed, hAFSCs seem to be a good source for cell reprogramming using the transduction system. Therefore, samples of amniotic fluid can be easily obtained (after written and informed consent from pregnant women) from the routinely amniocentesis performed for prenatal screening during mid-trimester of pregnancy (around 18 weeks) or during the intervention of amnioreduction. Amniocytes isolated from the specimens could be easier to reprogram the iPS cell state than the other fetal or somatic cell types because their transcriptional and epigenetic states that make them more similar to early ES cells [86]. Thanks to their embryonic origin, amniocytes may have accumulated less genetic damage or somatic mutation than older cell types or somatic cells, previously described. Moreover, amniocytes are autologous to the fetus and semiallogeneic to each parent, thereby expanding the potential utility of AFiPS cells to other family members [86].

Several studies have recently reported the capacity of hAFSCs to generate iPSCs using different protocols; but, since one of the main aims of the researchers is to identify the right protocol to guarantee maximum efficiency and viability of iPS colonies, several small molecular drugs, such as histone methyltransferase inhibitors [87], L-channel calcium agonist [87,88], Wnt inhibitors [88], and vitamin C [89], have been used during cells transduction.

Some examples of nuclear transduction reprogramming are listed as follows. CD34$^+$ subpopulation cells isolated from hAFCs could generate iPS cell lines after infection with lentiviral constructs encoding only Oct4. Immunofluorescence staining revealed high levels of Oct4, Sox2, Nanog, Rex1, and AP. The expression of these stem cells markers was ∼5−120-fold higher in human iPS cells than in hAFCs after qRT-PCR analysis. Moreover, the *in vivo* study demonstrated that the injection of iPS cells into the hind leg of mice gave rise to teratomas contained cellular type representatives of all the three germ layers [90]. In an other recent article, iPS cells were derived by transduction of hAFSCs with a retroviral cocktail consisting of Oct4, Sox2, Klf4, and c-Myc (OKSM). AFiPSCs were characterized by analysis of AP activity, expression of several markers of the undifferentiated state, including Nanog, Oct4, Sox2, SSEA-4, Tra-1-60, Tra-1-81; therefore, after passages in culture AFiPSCs exhibited a normal karyotype their genetic relationship to primary AFC cells was confirmed by DNA fingerprinting analysis. AFiPSCs were able to form derivatives of the three embryonic germ layers and also of the extraembryonic trophoblast lineage activating of BMP signaling cascades and blocking of TGFbeta/activin/nodal signaling [91].

Not long ago, a fascinating analysis was performed between AFiPSCs, iPS derived from fibroblasts cells line (FiPSCs) and hESC in order to evaluate which genes were lost, acquired, or retained after reprogramming both in fetal than in somatic cells. As a result, the lost genes included HOXB7, HOXA9, HOXA10, PAX8, DSCR1, Myc in AFiPSCs and EMX2, FOXF2, FOXF1, Myc, Klf4 in FiPSCs. Concerning the acquired genes, AFiPSCs and FiPSCs showed some common self-renewal genes (Oct4, Sox2, Nanog), and other iPSC type-dependent genes (SIX6, EGR2 for AFiPSCs or PKNOX2, HOXD4, HOXD10 in FiPSCs). Instead, retained gene expression included genes such as PKNOX2 for AFiPSCs, HMBOX1, MGA for FiPSCs, or RAXL1 for both [92]. Viral integrations are probably the cause of the partially inconsistent gene expression patterns observed in AFiPSCs and FiPSCs respect to ES cell.

4.4.3 Conclusions on iPS from Fetal Tissues

The studies reported above have demonstrated the possibility of iPSCs generation from fetal and extraembryonic tissues and their advantages in respect to the reprogramming

of adult somatic cells. Notably, the iPS cells derived from hFSCs have the advantage of being easily obtainable and rapidly expandable to an adequate number required for the clinical application; they do not display tumorigenic potential and immunogenicity and they are not linked to ethical concerns [81]. For example, hAFSCs are easily reprogrammed by primary infection within 5−6 days, compared with about 10 days to induce iPS cell colonies from keratinocytes and 2 weeks or more from MEFs. Compared to other somatic cell types, such as adult human fibroblasts, MEFs, blood cells, adipose stem cells, and keratinocytes, hAFSCs present also the advantage of being a safer source for nuclear reprogramming. For instance, skin keratinocytes, used by several groups for obtaining disease- and patient-specific iPSC lines, could have some hidden risks: (i) they have a considerably higher probability of harboring silent genetic aberrations and (ii) the long procedure to establish keratinocyte or fibroblast cultures from patient skin biopsy specimens could allow the accumulation and enrichment of cellular subpopulations harboring mutations that may either hinder subsequent reprogramming or encourage clonal dominance [93]. Because of their early embryonic origin, amniocytes have probably accumulated less genetic damage or somatic mutation compared to the older FSCs. Amniotic fluid, among other sources of fetal and extra ES cells, can be taken during gestation with minimal risks both for the fetuses and the mother. Moreover, from the therapeutic point of view, isolation of hAFSCs from fetal with chromosomal anomalies, such as Trisomy 13, Trisomy 18, or Trisomy 21, the consequent generation of iPSCs and their *in vitro* differentiation, could be an interesting model to study the outcome of these pathologies and to develop new therapeutic strategies to improve the quality of life of affected newborns [81].

4.4.4 Novel Findings in iPS Cell Generation

Many ways to ameliorate the iPSC efficiency and to generate integration-free iPSCs have been recently tested; the OKSM transcriptional factors, initially identified by Takahashi in 2006 can now be substituted with different factors: plasmids [5], Sendai virus [6], adenovirus [7], synthesized RNAs [8], cell permeant protein (CPPs) [94], or molecules able to regulate the gene transcription [10].

However, the nonintegrative methods still have pitfalls, being mainly associated with poor efficiency of iPS generation thus the original finding—that a set of factors is required—remains firm, and particular key factors like Oct4 cannot be missed.

Nevertheless, many efforts have been directed in order to ameliorate the potency of the retroviral vector as transduction system. For example, exposure to a hypoxic environment or ascorbic acid has been shown to increase the frequency of iPS cell generation [94]. As already cited in the section concerning iPS from somatic cells, the disruption of the signaling pathways mediated by the tumor-suppressor protein p53 or the cell cycle regulator INK4A removes cell cycle control checkpoints and permits a more rapid generation of iPS cells [58].

It is known that integration of foreign DNA into the host genome could silence or induce dysregulation of indispensable genes [54] and moreover the viral infection could activate innate immunity by interaction with toll-like receptors (TLRs) (TLRs recognize pathogen-associated molecular patterns (PAMPs) binding to viral protein, lipopolysaccharides, DNA, or RNA). Interestingly, a recent study has shown a clear function of innate immunity system in nuclear reprogramming and the crucial role of the TLR3 (one isoform of TLRs) in augmenting its efficiency [95].

First, the authors compared retroviral vectors transduction system encoding Sox2 and Oct4 genes (pMX-Oct4 and pMX-Sox2) versus the cell permeant protein system encoding the same transcriptional factors (CPP-Oct4 and CPP-Sox2) in human fibroblast cells. They showed that CPP-Oct4 or CPP-Sox2 constructs alone are unable to induce nuclear reprogramming, but when a vector containing GFP protein (pMX-GFP) is combined with CPP-Sox2 or CPP-Oct4, the expression of the pluripotency genes is accelerated, mimicking the effect observed with the only retroviral vector. Second, in order to establish whether TLRs pathway might be involved in nuclear reprogramming using pMX-Oct4 or pMX-Sox2 they knocked down TLR3 or TRIF (the adaptor of TLR3 in the signaling TLR3-mediated) showing that TLR3 is required for full induction of pluripotency genes expression and for an efficient generation of human iPSC colonies with the retrovirus vector approach. In addition, they demonstrated that polyinosinic—polycytidylic acid (poly I:C) (an agonist of TLR3 receptor) could accelerate the gene expression after CPP transduction in both fibroblast cells and iPSC colonies. In the end, the authors hypothesized that activation of the TLR3 pathway might enhance early transcriptional activation; thus, they analyzed two epigenetic modification to mark transcriptionally active genes and transcriptionally silenced genes, such as the trimethylation of histone H3 on lysine 4 (H3K4) and the trimethylation of histone H3 on lysine 9 (H3K9), respectively. They could conclude that an epigenetic mechanism exists to explain the effect of TLR3 activation in nuclear reprogramming, in particular H3K4 trimethylation leads to transcriptional genes activating TLR3 dependent instead H3K9 trimethylation leads to transcriptional genes silencing TLR3 dependent [95].

The novelty of this study is it was not only to demonstrate that an interaction between innate immunity and nuclear reprogramming exists but also that the reprogramming could be efficiently performed using a viral-free transduction system (Poly I:C-CPPs).

Importantly, the use of a transduction viral-free could acquire an interesting role in the regenerative medicine and cells/tissues transplantation, avoiding the problems related with the genome modification that the viruses can induce.

In this way, another study recently showed for the first time that functional iPSCs from hAFSCs expressing Oct4, Sox2, Klf4, C-Myc, and hESC-specific surface antigens can be generated without ectopic reprogramming factors by culture on Matrigel in hESC medium supplemented with the HDACi like VPA [10]. In this fascinating work, the authors demonstrated that mid-trimester hAFSCs, when expanded under MSC conditions, expressed the MSC markers CD105, CD90, CD73, CD44, and CD29 along with a subset of cells expressing Oct4, c-Myc, and SSEA-4. Rather than, AFSCs cultured in ES conditions and combined with VPA added for 5 days showed an upregulation of Oct4, Sox2, c-Myc, and Klf4, with cells expressing Nanog, SSEA-3, SSEA-4, Tra-1-60, and Tra-1-81. Moreover, EBs and teratoma formation competency was confirmed in reprogrammed VPA cells gain, showing that a chemical approach can also be used in this cell type to obtain functional iPSCs.

4.5 NOVELTY IN NUCLEAR REPROGRAMMING

It is clear that the OKSM reprogramming approach introduced to generate iPS cells led to a shift from the initial aim to identify and use a single "master" regulatory transcription factor to the use of multiple factors in the reprogramming of mammalian cell fate [1]. By using the system discovered by Takahashi et al. for generating iPS cells, two research groups have deciphered the individual contributions of a full array of transcription factors; they identified three factors able to convert one cell type to another. In particular, the three transcription factors Ngn3 (also known as Neurog3), Pdx1, and Mafa could reprogram differentiated pancreatic exocrine cells in adult mice into cells that closely resemble insulin-producing endocrine cells (β cells) *in vivo* [50]. On the other hand, Ascl1, Brn2 (also called Pou3f2), and Myt1l could rapidly and efficiently convert mouse embryonic and postnatal fibroblasts into functional neurons *in vitro* and fibroblasts into excitatory neurons *in vitro* [53].

The knowledge of the role of the large numbers of transcription factors and the identification of sets that can be used to convert one type of somatic cell to another type or to a progenitor or stem cell became the first aim of many groups in the application for tissue engineering. A better understanding of the nuclear reprogramming mechanism, its impact on the epigenetic regulation and how the epigenetic memory is established could reduce the possibility of generating pluripotent but tumorigenic cells and therefore increase the usefulness of iPS cells, supplying safer sources for future cell therapy [1].

4.6 COMPARISON BETWEEN THE THREE APPROACHES OF NUCLEAR REPROGRAMMING

Around 60 years of research have characterized the history of nuclear reprogramming, during which elaborated tools have been developed and have become available. However, the three previously explained approaches are distinct to each other, each finding is synergistic and informative to the others [1] and a comparison of the three approaches shows common features.

Remarkably, there are two characteristics of cell differentiation at the base of nuclear reprogramming. One is that every cell seems to express those genes whose products determine its state of differentiation. For example, a muscle to maintain its differentiated state will provide by auto activation a high enough content of MyoD (the key protein in the regulation of muscle differentiation). The second and last characteristic common to all the three approaches is that the rearrangement of gene expression becomes increasingly difficult as cells become more differentiated, probably due to a shutdown of inappropriate lineages. These sentences can be better understood by evaluating the hypothesis of "fleeting access" formulated by Gurdon et al. in 2008. They suggested the presence of an increasing firm association among the combination of DNA binding or chromosomal proteins with the regulatory regions of inactive genes [26]. The concept of "fleeting access" is based on the thought that most proteins dissociate from DNA at frequent intervals of seconds or a few minutes and in few cases for longer, during this process a multicomponent complex as an entirity may have a very long dwell time on inactive genes. It will be a very rare event for a sufficient number of individual proteins in a complex to dissociate from a chromosome at the same time for a gene region to be accessible to reprogramming factors. On the point of view of this theory, the probability of reprogramming taking place in somatic nuclear transfer, cell fusion, and transcription factor transduction (iPSCs) seems to depend on the access frequency of gene regulatory regions and the concentration and duration of other transcription factors. Indeed, a more successful reprogramming would be performed in large cells, such as eggs or myotubes, because of their high and heterogeneous content of factors.

In more detailed analysis, these approaches reveal other common peculiarities that include the demethylation of pluripotency gene promoters and the activation of *Oct4*, *Nanog*, and some factors expressed in ES cells. Moreover, the telomeres length and the increased activity

of the telomerase demonstrate the activation of cell "rejuvenation" after reprogramming [96]. And to conclude, some studies have shown that in female cells, genes on the inactive X chromosome are reactivated, although this reversal of relatively stable methylated states is often incomplete [97,98].

On the other hand, many differences exist among the three approaches in technical feasibility, cells availability, time required for reprogramming, efficiency of pluripotency gene induction at the single cell level, and cell yield. Somatic nuclear transfer, compared with the others, permits a more rapid reprogramming and it is ideally suited to explaining the crucial principles of early embryonic development and reproductive biology [1]. Cell fusion is technically easier than SCNT and iPSCs generation and it is also suited to elucidating the molecular mechanisms that control the onset of nuclear reprogramming. When cell fusion is used to form mixed-species heterokaryons, pluripotency genes are activated quickly and efficiently. Despite this, the method does not yield clinically useful cells [1].

In contrast, transcription factor transduction system is currently unparalleled because of the simplicity and abundance of iPS cells generation, their potential for studying the mechanisms underlying human disease, and their usefulness for drug discovery and cell therapy [1]. Regarding the molecular process that underlines this method, a passive and stochastic DNA demethylation mechanism has been suggested: the cells expansion is associated with a progressive DNA demethylation, thus the methylation is not maintained after DNA replication [99]. This is in contrast with what has been reported for nuclear transfer and heterokaryons, in which an active DNA demethylation mechanism is present: pluripotency genes are activated and their promoters demethylated in the absence of DNA replication or cell division [100].

4.7 CONCLUSIONS

From its discovery until the present day, nuclear reprogramming has represented a fascinating and attracting field able to generate great interest in researchers which are still searching for the right combination between cells type, transcription factors, and efficient reprogramming methods.

Particularly, the aim of all the reprogramming approaches is to provide replacement cells for patients and also to create long-lasting cell lines from patients with genetic diseases (Trisomy 13, Trisomy 18, or Trisomy 21), in order to test drugs or other treatments. Nevertheless, for future therapeutic application of reprogrammed cells sufficient numbers of replacement cells, a clear knowledge of their function into the host tissues and their capacity to produce the correct amount of their product will be necessary [26].

Regarding the cells number required for the transplantation, generally, the starting success rate of iPS cells is 10^{-4}; for example, since a human adult liver contains about 10^{14} cells, to create this number of hepatocyte cells differentiated from iPSCs, an enormous number of cell divisions in culture are needed. On the other hand, it is true that many parts of the human body necessitate a smaller number of cells to ameliorate the organs or tissues function. This is confirmed by the human eye retina, in which only 10^5 cells are required for therapeutic benefit [26]. As mentioned above, it is fundamental to understand if transplanted cells can be useful even if they are not "properly" integrated into the host recipient. In this sense, it is convenient to emphasize that organs like the pancreas contain different cell types: exocrine (acinar) cells, ductal cells, and at least four kinds of hormone-secreting cells in the endocrine islet. Replacement endocrine cells can provide useful therapeutic benefit even if not incorporated into the normal complex pancreas cell configuration [50]. To conclude, it is not yet well defined whether the reprogrammed cells will be able to produce the desired product, such as proteins, enzymes, and other factors in a regulated manner, after their implantation. Future studies will enhance the current findings of nuclear reprogramming and cell differentiation, which are essential to better analyze the mechanisms of human diseases *in vitro* and to discover novel therapeutic agents to improve cell transplantation.

REFERENCES

[1] Yamanaka S, Blau Helen M. Nuclear reprogramming to a pluripotent state by three approaches. Nature 2010;465:704−12.

[2] Barrilleaux B, et al. Review: *ex vivo* engineering of living tissues with adult stem cells. Tissue Eng 2006;12:3007−19.

[3] Hadorn E. Constancy, variation and type of determination and differentiation in cells from male genitalia rudiments of *Drosophila melanogaster* in permanent culture *in vivo*. Dev Biol 1966;13:424−509.

[4] Gehring W. Clonal analysis of determination dynamics in cultures of imaginal disks in *Drosophila melanogaster*. Dev Biol 1967;16:438−56.

[5] Aoi T, et al. Generation of pluripotent stem cells from adult mouse liver and stomach cells. Science 2008;321:699−702.

[6] Fusaki N, et al. Efficient induction of transgene-free human pluripotent stem cells using a vector based on Sendai virus, an RNA virus that does not integrate into the host genome. Proc Jpn Acad Ser B Phys Biol Sci 2009;85:348−62.

[7] Stadtfeld M, et al. Induced pluripotent stem cells generated without viral integration. Science (New York, NY) 2008;322:945−9.

[8] Warren L, et al. Highly efficient reprogramming to pluripotency and directed differentiation of human cells with synthetic modified mRNA. Cell Stem Cell 2010;7:618−30.

[9] Kim D, et al. Generation of human induced pluripotent stem cells by direct delivery of reprogramming proteins. Cell Stem Cell 2009;4:472−6.

[10] Moschidou D, et al. Valproic acid confers functional pluripotency to human amniotic fluid stem cells in a transgene-free approach. Mol Ther J Am Soc Gene Ther 2012;20:1953−67.

[11] Pozzobon M, Ghionzoli M, De Coppi P. ES, iPS, MSC, and AFS cells. Stem cells exploitation for pediatric surgery: current research and perspective. Pediatr Surg Int 2010;26:3−10.

[12] Briggs R, King TJ. Transplantation of living nuclei from blastula cells into enucleated frogs' eggs. Proc Natl Acad Sci USA 1952;38:455−63.

[13] Gurdon JB, Elsdale TR, Fishberg M. Sexually mature individuals of *Xenopus laevis* from the transplantation of single somatic nuclei. Nature 1958;182:64−5.

[14] Gurdon JB. The developmental capacity of nuclei taken from intestinal epithelium cells of feeding tadpoles. J Embryol Exp Morphol 1962;10:622−40.

[15] Ooi J, Liu P. Delineating nuclear reprogramming. Protein Cell 2012;3:329−45.

[16] Laskey RA, Gurdon JB. Genetic content of adult somatic cells tested by nuclear transplantation from cultured cells. Nature 1970;228:1332−4.

[17] Gurdon JB, Laskey RA, Reeves OR. The developmental capacity of nuclei transplanted from keratinized skin cells of adult frogs. J Embryol Exp Morphol 1975;34:93−112.

[18] DiBerardino MA, Hoffner NJ. Gene reactivation in erythrocytes: nuclear transplantation in oocytes and eggs of Rana. Scienc (New York, NY) 1983;219:862−4.

[19] Gurdon JB, et al. Transcription of muscle-specific actin genes in early Xenopus development: nuclear transplantation and cell dissociation. Cell 1984;38:691−700.

[20] Wilmut I, et al. Viable offspring derived from fetal and adult mammalian cells. Nature 1997;386:810−3.

[21] Wakayama T, et al. Full-term development of mice from enucleated oocytes injected with cumulus cell nuclei. Nature 1998;394:369−74.

[22] Hochedlinger K, Jaenisch R. Monoclonal mice generated by nuclear transfer from mature B and T donor cells. Nature 2002;415:1035−8.

[23] Thuan N, Van Kishigami S, Wakayama T. How to improve the success rate of mouse cloning technology. J Reprod Dev 2010;56:20−30.

[24] Wakayama S, et al. Production of healthy cloned mice from bodies frozen at $-20°C$ for 16 years. Proc Natl Acad Sci USA 2008;105:17318−22.

[25] Yang X. Nuclear reprogramming of cloned embryos and its implications for therapeutic cloning. Nat Genet 2007;39:295−302.

[26] Gurdon JB, Melton DA. Nuclear reprogramming in cells. Science (New York, NY) 2008;322:1811−5.

[27] Davidson RL, Ephrussi B, Yamamoto K. Regulation of pigment synthesis in mammalian cells, as studied by somatic hybridization. Proc Natl Acad Sci USA 1966;56:1437−40.

[28] NR R. Cell hybrids. Academic; 1977.

[29] Blau HM, Chiu CP, Webster C. Cytoplasmic activation of human nuclear genes in stable heterocaryons. Cell 1983;32:1171−80.

[30] Blau HM, et al. Plasticity of the differentiated state. Science 1985;230:758−66.

[31] Blau HM, Baltimore D. Differentiation requires continuous regulation. J Cell Biol 1991;112:781–3.

[32] Harris H, et al. Suppression of malignancy by cell fusion. Nature 1969;223:363–8.

[33] Harris H, et al. Artificial heterokaryons of animal cells from different species. J Cell Sci 1966;1:1–30.

[34] Pavlath GK, Blau HM. Expression of muscle genes in heterokaryons depends on gene dosage. J Cell Biol 1986;102:124–30.

[35] Chiu CP, Blau HM. Reprogramming cell differentiation in the absence of DNA synthesis. Cell 1984;37:879–87.

[36] Chiu CP, Blau HM. 5-Azacytidine permits gene activation in a previously noninducible cell type. Cell 1985;40:417–24.

[37] Wright WE. Induction of muscle genes in neural cells. J Cell Biol 1984;98:427–35.

[38] Baron MH, Maniatis T. Rapid reprogramming of globin gene expression in transient heterokaryons. Cell 1986;46:591–602.

[39] Spear BT, Tilghman SM. Role of alpha-fetoprotein regulatory elements in transcriptional activation in transient heterokaryons. Mol Cell Biol 1990;10:5047–54.

[40] Tada M, et al. Embryonic germ cells induce epigenetic reprogramming of somatic nucleus in hybrid cells. EMBO J 1997;16:6510–20.

[41] Tada M, et al. Nuclear reprogramming of somatic cells by *in vitro* hybridization with ES cells. Curr Biol 2001;11:1553–8.

[42] Kimura H, et al. Histone code modifications on pluripotential nuclei of reprogrammed somatic cells. Mol Cell Biol 2004;24:5710–20.

[43] Cowan CA, et al. Nuclear reprogramming of somatic cells after fusion with human embryonic stem cells. Science (New York, NY) 2005;309:1369–73.

[44] Silva J, et al. Nanog promotes transfer of pluripotency after cell fusion. Nature 2006;441:997–1001.

[45] Pereira CF, et al. Heterokaryon-based reprogramming of human B lymphocytes for pluripotency requires Oct4 but not Sox2. PLoS Genet 2008;4:e1000170.

[46] Bhutani N, et al. Reprogramming towards pluripotency requires AID-dependent DNA demethylation. Nature 2010;463:1042–7.

[47] Weintraub H, et al. Activation of muscle-specific genes in pigment, nerve, fat, liver, and fibroblast cell lines by forced expression of MyoD. Proc Natl Acad Sci USA 1989;86:5434–8.

[48] Tontonoz P, Hu E, Spiegelman BM. Stimulation of adipogenesis in fibroblasts by PPAR gamma 2, a lipid-activated transcription factor. Cell 1994;79:1147–56.

[49] Ieda M, et al. Direct reprogramming of fibroblasts into functional cardiomyocytes by defined factors. Cell 2010;142:375–86.

[50] Zhou Q. *In vivo* reprogramming of adult pancreatic exocrine cells to beta cells. Nature 2008;455:627–32.

[51] Kajimura S, et al. Initiation of myoblast to brown fat switch by a PRDM16-C/EBP-beta transcriptional complex. Nature 2009;460:1154–8.

[52] Szabo E, et al. Direct conversion of human fibroblasts to multilineage blood progenitors. Nature 2010;468:521–6.

[53] Vierbuchen T, et al. Direct conversion of fibroblasts to functional neurons by defined factors. Nature 2010;463:1035–41.

[54] Takahashi K, Yamanaka S. Induction of pluripotent stem cells from mouse embryonic and adult fibroblast cultures by defined factors. Cell 2006;126:663–76.

[55] Tokuzawa Y, et al. Fbx15 is a novel target of Oct3/4 but is dispensable for embryonic stem cell self-renewal and mouse development. Mol Cell Biol 2003;23:2699–708.

[56] Takahashi K, et al. Induction of pluripotent stem cells from fibroblast cultures. Nat Protoc 2007;2:3081–9.

[57] Gore A, et al. Somatic coding mutations in human induced pluripotent stem cells. Nature 2011;471:63–7.

[58] Hussein SM. Copy number variation and selection during reprogramming to pluripotency. Nature 2011;471:58–62.

[59] Zhao T, et al. Immunogenicity of induced pluripotent stem cells. Nature 2011;474:212–5.

[60] Yamanaka S. Induced pluripotent stem cells: past, present, and future. Cell Stem Cell 2012;10:678–84.

[61] Hacein-Bey-Abina S, et al. LMO2-associated clonal T cell proliferation in two patients after gene therapy for SCID-X1. Science (New York, NY) 2003;302:415–9.

[62] Marcus AJ, Woodbury D. Fetal stem cells from extra-embryonic tissues: do not discard. J Cell Mol Med 2008;12:730–42.

[63] Weissman IL, Shizuru JA. The origins of the identification and isolation of hematopoietic stem cells, and their capability to induce donor-specific transplantation tolerance and treat autoimmune diseases. Blood 2008;112:3543–53.

[64] Huss R. Perspectives on the morphology and biology of CD34-negative stem cells. J Hematother Stem Cell Res 2000;9:783–93.

[65] Bigas A, D'Altri T, Espinosa L. The Notch pathway in hematopoietic stem cells. Curr Topics Microbiol Immunol 2012;360:1–18.

[66] Lim FTH, Kanhai HHH, Falkenburg JHF. Characterization of the human CD34$^+$ hematopoietic progenitor cell compartment during the second trimester of pregnancy. Haematologica 2005;90:173–9.

[67] Wagers AJ, Allsopp RC, Weissman IL. Changes in integrin expression are associated with altered homing properties of Lin(-/lo)Thy1.1(lo)Sca-1(+)c-kit(+) hematopoietic stem cells following mobilization by cyclophosphamide/granulocyte colony-stimulating factor. Exp Hematol 2002;30:176–85.

[68] Broxmeyer HE, et al. Hematopoietic stem/progenitor cells, generation of induced pluripotent stem cells, and isolation of endothelial progenitors from 21- to 23.5-year cryopreserved cord blood. Blood 2011;117:4773–7.

[69] Mazurier F, et al. Rapid myeloerythroid repopulation after intrafemoral transplantation of NOD-SCID mice reveals a new class of human stem cells. Nat Med 2003;9:959–63.

[70] Wang J, et al. SCID-repopulating cell activity of human cord blood-derived CD34$^-$ cells assured by intra-bone marrow injection. Blood 2003;101:2924–31.

[71] Chou BK, et al. Efficient human iPS cell derivation by a non-integrating plasmid from blood cells with unique epigenetic and gene expression signatures. Cell Res 2011;21:518–29.

[72] Delaney C, Ratajczak MZ, Laughlin MJ. Strategies to enhance umbilical cord blood stem cell engraftment in adult patients. Expert Rev Hematol 2010;3:273–83.

[73] McGuckin CP, et al. Umbilical cord blood stem cells can expand hematopoietic and neuroglial progenitors *in vitro*. Exp Cell Res 2004;295:350–9.

[74] Takenaka C, et al. Effective generation of iPS cells from CD34$^+$ cord blood cells by inhibition of p53. Exp Hematol 2010;38:154–62.

[75] Pipino C, Shangaris P, Resca E, Zia S, Deprest J, Sebire NJ, et al. Placenta as a reservoir of stem cells: an underutilized resource? Br Med Bull 2013;105:43−68.

[76] Tamagawa T, et al. Differentiation of mesenchymal cells derived from human amniotic membranes into hepatocyte-like cells *in vitro*. Human Cell 2007;20:77−84.

[77] Bieback K, Brinkmann I. Mesenchymal stromal cells from human perinatal tissues: from biology to cell therapy. World J Stem Cells 2010;2:81−92.

[78] Jones GN, et al. Ontological differences in first compared to third trimester human fetal placental chorionic stem cells. PLoS One 2012;7:e43395.

[79] Zhao H. Rapid and efficient reprogramming of human amnion-derived cells into pluripotency by three factors OCT4/SOX2/NANOG. Differentiation 2010;80:123−9.

[80] De Coppi P, et al. Isolation of amniotic stem cell lines with potential for therapy. Nat Biotechnol 2007;25:100−6.

[81] Cananzi M, De Coppi P. CD117(+) amniotic fluid stem cells: state of the art and future perspectives. Organogenesis 2012;8:77−88.

[82] Kunisaki SM, et al. Diaphragmatic repair through fetal tissue engineering: a comparison between mesenchymal amniocyte- and myoblast-based constructs. J Pediatric Surg 2006;41:34−9.

[83] Kunisaki SM, et al. A comparative analysis of cartilage engineered from different perinatal mesenchymal progenitor cells. Tissue Eng 2007;13:2633−44.

[84] Fuchs JR, et al. Diaphragmatic reconstruction with autologous tendon engineered from mesenchymal amniocytes. J Pediatric Surg 2004;39:834−8.

[85] Schmidt D, et al. Prenatally fabricated autologous human living heart valves based on amniotic fluid derived progenitor cells as single cell source. Circulation 2007;116:64−70.

[86] Anchan RM, et al. Amniocytes can serve a dual function as a source of iPS cells and feeder layers. Human Mol Genet 2011;20:962−74.

[87] Shi Y, et al. Induction of pluripotent stem cells from mouse embryonic fibroblasts by Oct4 and Klf4 with small-molecule compounds. Cell Stem Cell 2008;3:568−74.

[88] Park J, et al. Reprogramming of mouse fibroblasts to an intermediate state of differentiation by chemical induction. Cell Reprog 2011;13:121−31.

[89] Wang Q, et al. Lithium, an anti-psychotic drug, greatly enhances the generation of induced pluripotent stem cells. Cell Res 2011;21:1424−35.

[90] Esteban MA, et al. Porcine induced pluripotent stem cells may bridge the gap between mouse and human iPS. IUBMB Life 2010;62:277−82.

[91] Liu T, et al. High efficiency of reprogramming CD34$^+$ cells derived from human amniotic fluid into induced pluripotent stem cells with Oct4. Stem Cell Dev 2012;21:2322−32.

[92] Wolfrum K, et al. The LARGE principle of cellular reprogramming: lost, acquired and retained gene expression in foreskin and amniotic fluid-derived human iPS cells. PLoS One 2010;5:13703.

[93] Mukherjee S, Thrasher AJ. iPSCs: Unstable origins? Mol Ther J Am Soc Gene Ther 2011;19:1188−90.

[94] Yoshida Y, et al. Hypoxia enhances the generation of induced pluripotent stem cells. Cell Stem Cell 2009;5:237−41.

[95] Lee J, et al. Activation of innate immunity is required for efficient nuclear reprogramming. Cell 2012;151:547−58.

[96] Marion RM, et al. Telomeres acquire embryonic stem cell characteristics in induced pluripotent stem cells. Cell Stem Cell 2009;4:141−54.

[97] Eggan K, et al. X-Chromosome inactivation in cloned mouse embryos. Science (New York, NY) 2000;290:1578−81.

[98] Nolen LD, et al. X chromosome reactivation and regulation in cloned embryos. Dev Biol 2005;279:525−40.

[99] Hanna J, et al. Direct cell reprogramming is a stochastic process amenable to acceleration. Nature 2009;462:595−601.

[100] Simonsson S, Gurdon J. DNA demethylation is necessary for the epigenetic reprogramming of somatic cell nuclei. Nat Cell Biol 2004;6:984−90.

Chapter 5

Principles of Cell Sheet Technology

Nobuo Kanai, Masayuki Yamato, and Teruo Okano
Institute of Advanced Biomedical Engineering and Science (TWIns), Tokyo Women's Medical University, Shinjuku-ku, Tokyo, Japan

Chapter Outline

5.1 INTRODUCTION

5.1.1 Tissue Engineering for Regenerative Medicine

Regeneration of tissues and organs offers an innovative approach to the treatment of injury and disease with substituted artificial organs and organ transplantations. The term "tissue engineering" was first utilized in the late 1980s based on the pioneering work of Langer and Vacanti [1,2], the use of cultured cells as a therapeutic modality was initiated by Howard Green and associates, who transplanted cultured sheets of autologous epidermis to patients resulting from severe burns [3], giant congenital nevi [4], and skin ulcers [5]. The key technology is the use of biodegradable polymer scaffolds, preformed in the target tissue shape, for seeding cells, as demonstrated in the well-publicized reconstruction of cartilage tissues for the growth of human ears on mouse back. More recently, as the field has advanced and incorporated newer technologies, such as drug and gene delivery and nanotechnology, traditional tissue engineering has become integrated into the overarching discipline of regenerative medicine. Since the mid-1990s, significant progress from basic science has resulted in clinical applications of tissue engineering for the replacement of a variety of tissues and organs. For example, tissue-engineered bone using autologous osteoblast cells from the periosteum seeded into porous coral scaffolds regenerated the thumb of a patient who had lost the appendage in a traumatic accident [6]. Similarly, tissues created using autologous cells isolated from peripheral blood vessels and polycaprolactone-polyglycolic acid copolymers, have also been used to successfully replace occluded pulmonary arteries [7]. Vascular smooth muscle and endothelial cells that expressed telomerase were used to reconstruct tissue-engineered arteries [8], introducing an alternative method for autologous treatments using cells from elderly patients. In addition, tissue-engineered bladders created using urothelial and smooth muscle cells seeded into collagen-based scaffolds have also been used to treat patients who would normally require bladder augmentation [9]. These therapeutic applications and achievements of tissue engineering over the past two decades, have therefore spurred significant interest into the field of regenerative medicine.

5.1.2 Cell-Sheet Engineering and Temperature-Responsive Culture Surface

An innovative approach to tissue engineering using a thermo-responsive culture surface has been developed from the 1990s [10]. Poly(*N*-isopropylacrylamide) (PIPAAm) exhibits a lower critical solution temperature (LCST) at around 32°C in aqueous media [11]. While they hydrate and form an expanded structure in an aqueous media below the LCST, they dehydrate and form a compact structure above the LCST. Such a conformational change in response to temperature has been extensively used to modulate the physicochemical properties of polymeric thin surfaces. At 37°C, PIPAAm-grafted surface is slightly hydrophobic, allowing cells to proliferate under normal conditions and to become a confluent cell sheet, which is regularly found on usual tissue culture polystyrene. A decrease in temperature lower than 32°C, however, results in the hydration of the polymer surface, giving the spontaneous detachment of the cells as a monolithic tissue-like cell sheet for less than 1 h without any enzymes such as trypsin. Since PIPAAm is covalently immobilized onto the culture surfaces, PIPAAm remains bound to the surfaces even after cell-sheet detachment, realizing the noninvasive harvest of cultured cells as an intact cell sheet having deposited extracellular matrix (ECM). This technology allows us to transplant cell sheets to host tissues without using biodegradable scaffolds. PIPAAm thickness on the order of nanometers is necessary for expressing such interesting properties as temperature-controlled cell attachment or detachment [12−14]. For example, a PIPAAm layer of ~20 nm is optimal for cell adhesion and detachment properties in response to temperature change for a PIPAAm-modified tissue culture polystyrene system. Since the PIPAAm-grafted surfaces facilitate spontaneous cell detachment, the use of conventional proteolytic enzymes such as dispase, trypsin, and collagenase can be avoided. With noninvasive cell harvest, cell-to-cell junction and ECM proteins can therefore be maintained [15] (Figure 5.1).

Numerous cell types including epidermal keratinocytes [16], vascular endothelial cells [17], renal epithelial cells [18,19], periodontal ligaments (PDLs) [20,21], and cardiomyocytes [22,23] have shown the maintenance of differentiated functions after low-temperature cell-sheet harvest, due to the preservation of cell surface proteins, such as growth factor receptors, ion channels, and cell-to-cell junction proteins. Additionally, due to the presence of deposited ECM that is produced during *in vitro* incubation, cell sheets can be easily transplanted and attached to sites such as culture dishes and even host tissues. To fabricate thick tissues, cell sheets can be stacked in layers because cell sheets connect each other in a short time. A study showed that bilayer cardiomyocyte sheets were completely coupled 46 ± 3 min (mean thickness determined by scanning electron) after initial layering [24], suggesting that multilayered cell sheets

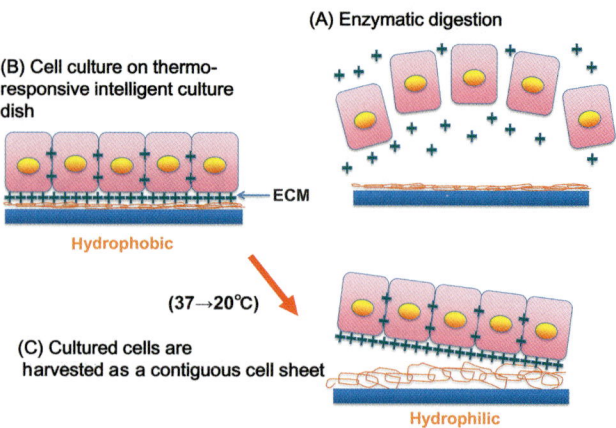

FIGURE 5.1 Cell-sheet detachment from temperature-responsive culture surfaces. (A) Enzymatic digestion. (B) Cell culture on a thermo-responsive intelligent culture dish. (C) Cultured cells are harvested as a contiguous cell sheet.

can communicate and be synchronized as functional tissues. Based on this study, multilayered transplantation was performed [25]. When more than three cardiomyocyte sheets were layered and transplanted into the subcutaneous space, the appearance of fibrosis and disordered vasculature indicated the presence of fibrotic areas within the transplanted laminar structures. Although the rapid establishment of microvascular networks occurred within the engineered tissues, this formation of new vessels was not able to rescue tissues when the thickness was above 80 μm. Multiple-step transplantation at one- or two-day intervals established the rapid neovascularization of the engineered myocardial tissues with more than 1 mm thickness [25], and these results directed us to fabricate prevascularized cell sheets. Recent studies showed that the combination use of different types of cells, for example an endothelial cell sheet sandwiched with other types of cell sheets, induced prevascularization *in vitro*, which may allow the graft to survive and function [26,27]. Three-dimensional manipulation of fibroblast cell sheets and micropatterned endothelial cells with the gelatin-coated stacking manipulator-induced microvascular-like network within 5-day culture *in vitro* [28]. Nonpatterned endothelial cell sheets and other types of cell sheets with a fibrin gel manipulator also induced prevascular networks both *in vitro* [29] and *in vivo* [30].

Therefore, cell-sheet technology enables us to avoid scaffolds, fixation, or sutures that are needed by conventional tissue engineering approaches using isolated cell injections and scaffold-based technologies, which is the limitation of applicability. In the following sections, the specific regenerative applications of cell-sheet technology are described.

5.2 FUNCTIONAL TISSUE REGENERATION USING CELL SHEETS

5.2.1 Corneal Regeneration

Limbal stem-cell deficiency by ocular trauma or diseases causes corneal opacification and visual loss. To recruit limbal stem cells, a novel cell-sheet manipulation technology using temperature-responsive culture surfaces was developed [31]. The results showed that multilayered corneal epithelial cell sheets were successfully fabricated and their characteristics were similar to those of native tissues. Transplantation of these cell sheets induced the corneal surface reconstruction in rabbits. For patients who suffer from unilateral limbal stem deficiency, corneal epithelial cell sheets can be cultured from autologous limbal stem cells. In case the objective is the repair of bilateral corneal stem-cell deficiency, autologous oral mucosal epithelial cells are utilized to create the oral mucosal epithelial cell sheets. The cell sheets contain both cell-to-cell junctions and ECM proteins, and can be transplantable without the need for any carrier substrate or sutures. Therefore, oral mucosal epithelial sheets were examined as an alternative cell source to expand the opportunity of autologous transplantation [32]. Autologous transplantation to rabbit corneal surfaces successfully reconstructed the corneal surface with restoration of transparency. Four weeks after transplantation, epithelial stratification was similar to that in the normal corneal epithelium, although the keratin expression profile retained characteristics of the oral mucosal epithelium.

The first clinical trial of cell-sheet engineering was the corneal reconstruction using autologous mucosal epithelial cells, and the results were published in 2004 [33]. Oral mucosal tissue was harvested from four patients with bilateral total corneal stem-cell deficiencies. Then cells were cultured for 2 weeks on mitomycin C-treated 3T3 feeder layer and transplanted directly to the denuded corneal surfaces without sutures. Results showed that complete reepithelialization of the corneal surfaces occurred and vision of all patients was restored. Recently, autologous oral mucosal epithelial cell sheets cultured with the UpCell-Insert technology (CellSeed, Tokyo, Japan) without feeder layer were transplanted in 25 patients for the treatment of corneal limbal epithelial deficiency in France. The safety of the products was proved during the follow-up (360 days), and the results suggested its efficacy for reconstructing the ocular surface [34] (Figure 5.2).

5.2.2 Cardiac Regeneration

To enhance the function of cardiac tissue, neonatal rat cardiomyocyte sheets were fabricated and their characteristics were examined. When four sheets were layered, spontaneous beating of engineered constructs was observed. When they were transplanted into subcutaneously, heart tissue-like structures and neovascularization within contractile tissues were observed [35]. Long-term survival of pulsatile cardiac grafts was confirmed over more than 1 year [36]. The next study was performed to create thick tissue [25]. However, the thickness limit for layered cell sheets in subcutaneous tissue was ∼80 μm (three layers). To overcome this limitation, repeated transplantation of triple-layer grafts was performed and multistep transplantation created ∼1 mm thick myocardium with a well-organized microvascular network. Other types of cell sheets were also examined to improve cardiac functions. Adipose-derived mesenchymal stem cells [37] and skeletal myoblasts [38–40] were transplanted as cell sheets and results showed the efficacy for cardiac repairs.

The first clinical trial in the field of cardiac regeneration was started in 2007. Autologous myoblast cells from the patient's femoris muscle were fabricated as cell sheets, and these cell sheets have been transplanted to end-stage dilated cardiomyopathy patients who need left ventricular assist systems in Osaka University, Japan. The myoblastic cell sheets were transplanted into the affected part of the heart in patients. The first patient was successfully treated and discharged from hospital without requiring a ventricular assist device [41] (Figure 5.3).

5.2.3 Esophageal Regeneration

Endoscopic treatments for early esophageal cancer and Barrett's esophagus with high-grade dysplasia have gained widespread acceptance as minimally invasive therapies [42–44]. In particular, endoscopic submucosal dissection (ESD) makes it possible to resect superficial cancer *en bloc* regardless of size and allows for an accurate histological assessment for diagnosis. However, severe postoperative esophageal stricture is inevitably observed when ESD is performed for widespread superficial neoplasms that remove a large area of mucosa, i.e., >75% of the esophageal lumen [45]. For solving the demand, our laboratory has developed a method combining ESD with the endoscopic transplantation of autologous oral mucosal epithelial cell sheets [43,46]. Results showed the effectiveness of a novel combined endoscopic approach for the potential treatment of esophageal cancers that can effectively enhance wound healing and possibly prevent postoperative esophageal stenosis. In addition, we reported using fabricated autologous skin epidermal cell sheet, as another epithelial cell source, prevent severe stricture following full-circumferential ESD in a porcine model [47]. Furthermore, several studies revealed

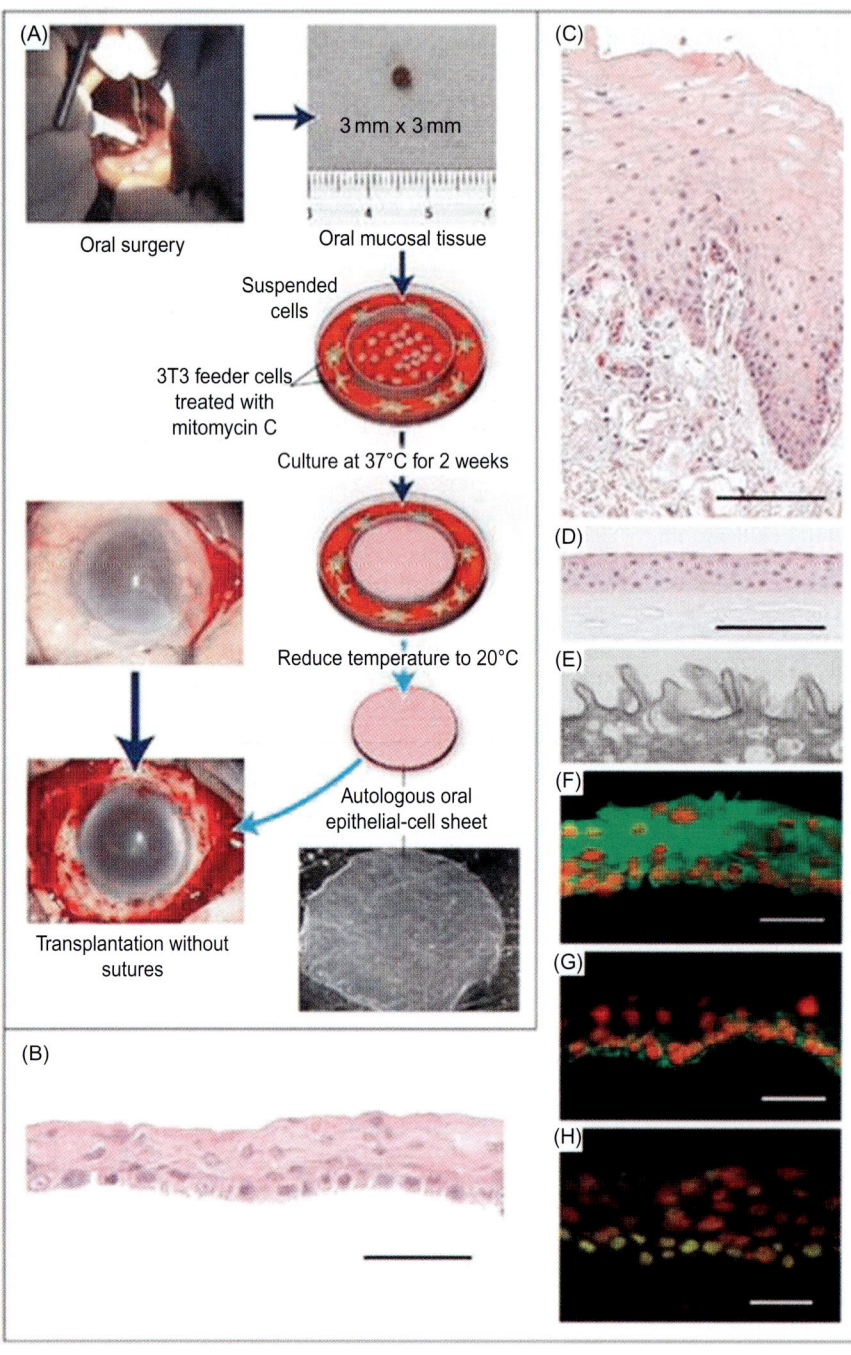

FIGURE 5.2 Ocular surface reconstruction using tissue-engineered sheets of oral mucosal epithelium. From a small biopsy of the patient's own oral mucosa, autologous epithelial cells can be isolated for *ex vivo* expansion on temperature-responsive culture surfaces in the presence of 3T3 feeder layers. After 2 weeks in culture, the autologous oral mucosal epithelial cell sheets can be harvested by temperature-reduction to 20°C for 30 min. After removal of the damaged tissue from the corneal surface, the harvested oral mucosal epithelial cell sheet can be transplanted directly to the surface of the denuded corneal stroma, without sutures. *Reprinted with permission from [33], Copyright © 2004 Massachusetts Medical Society. All rights reserved.*

epithelial cell sheets can be fabricated with temperature-responsive culture inserts without feeder layers [48,49], which can exclude xenogeneic factors for animal-free cell transplantation.

Based on these animal studies, human autologous oral mucosal epithelial cell sheets are fabricated with the UpCell-Insert technology [50]. The first clinical trial was esophageal regeneration using autologous oral mucosal cell sheets after a large-size removal ESD for superficial

carcinoma corneal, and the results were published in 2012 [51]. Results showed that early reepithelialization of the ulcer site and the results suggested its efficacy for preventing stricture in case of circumferential ESD (Figure 5.4).

5.2.4 Periodontal Regeneration

Periodontitis is a bacterial infection-induced inflammation that causes the destruction of attachment apparatus of

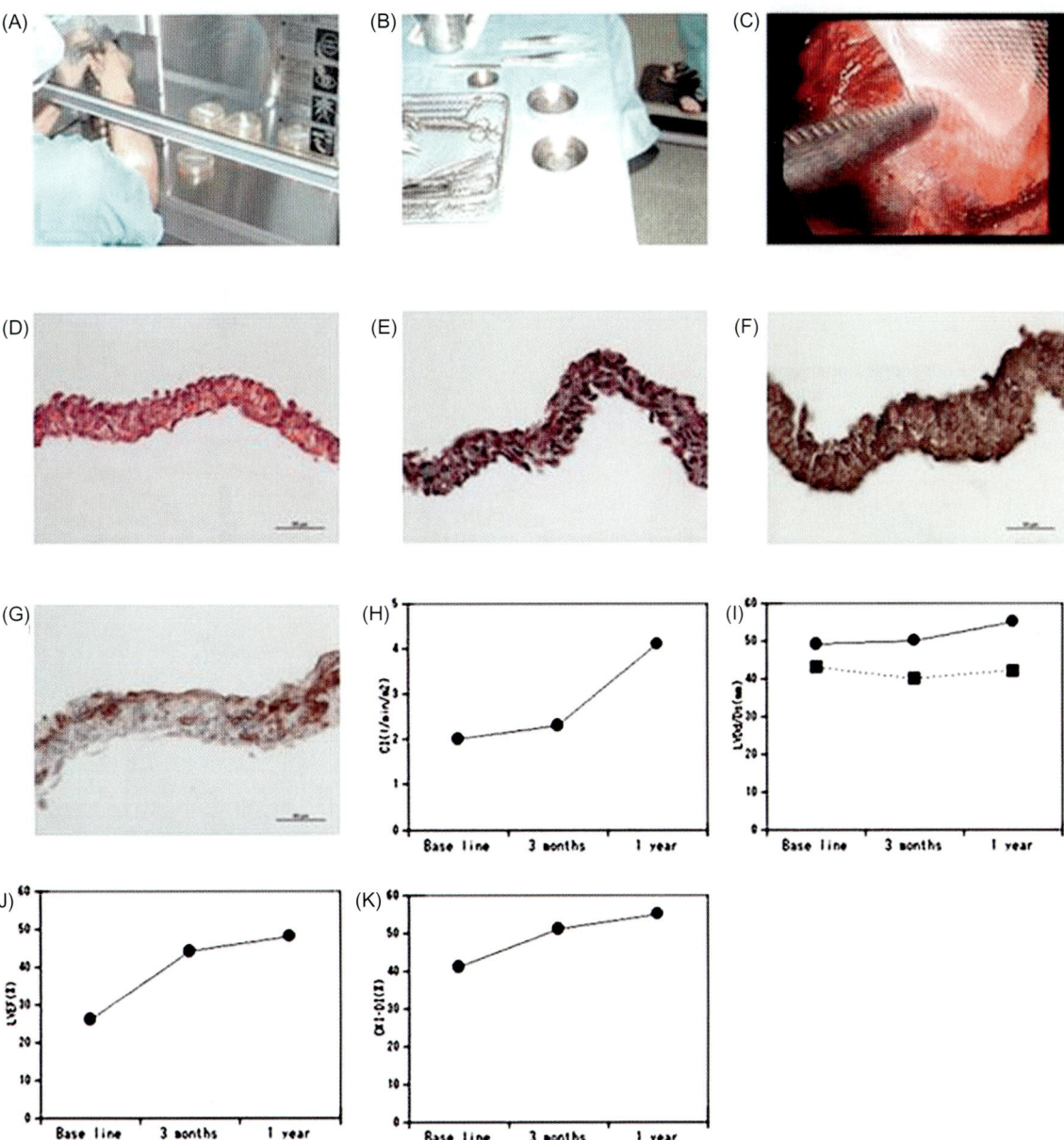

FIGURE 5.3 The histology of the myoblast sheet and the changes in hemodynamic variables from baseline (before myoblast-sheet implantation), and at 3 months and 1 year follow-up examinations. Hemodynamic variables at baseline and at the 3-month follow-up were obtained at the examination for left ventricle assisting system (LVAS) removal. (A and B) Preparation of myoblast sheets in the operating room. (C) Implantation of a myoblast sheet via a left lateral thoracotomy. (D) H&E staining of a myoblast sheet; the myoblast sheet included many cells, and its thickness was about 50 μm. (E) Masson-Trichrome staining; the myoblast sheet consisted of cells and ECM. (F) Desmin staining; nearly all of the cells in the myoblast sheets were desmin positive. (G) Alpha-myosin heavy chain staining; some cells in the myoblast sheets were alpha-myosin heavy chain positive. (H) Cardiac index. (I) left ventricular end-diastolic diameter (LVDd) (*solid line*) and left ventricular end-systolic diameter (LVDs) (*dotted line*). (J) Left ventricle (LV) ejection fraction. (K) Color kinesis diastolic indices (CK-DI). *Reprinted with permission from [41], Copyright © 2012 Springer.*

dental roots. Several materials, such as bone graft materials, barrier membranes, and protein products had been developed and utilized to treat periodontal defects clinically, however, it is difficult to regenerate complete periodontal tissue. First, the cell sheets of human PDL cells were successfully created with the temperature-responsive dishes, and the characteristics of human PDL cell sheets were investigated [21]. And then, triple-layered PDL cell sheets supported with woven poly(glycolic acid) are transplanted to dental root surfaces having three-wall periodontal defects in an autologous manner, and bone defects are filled with porous β-tricalcium phosphate [52]

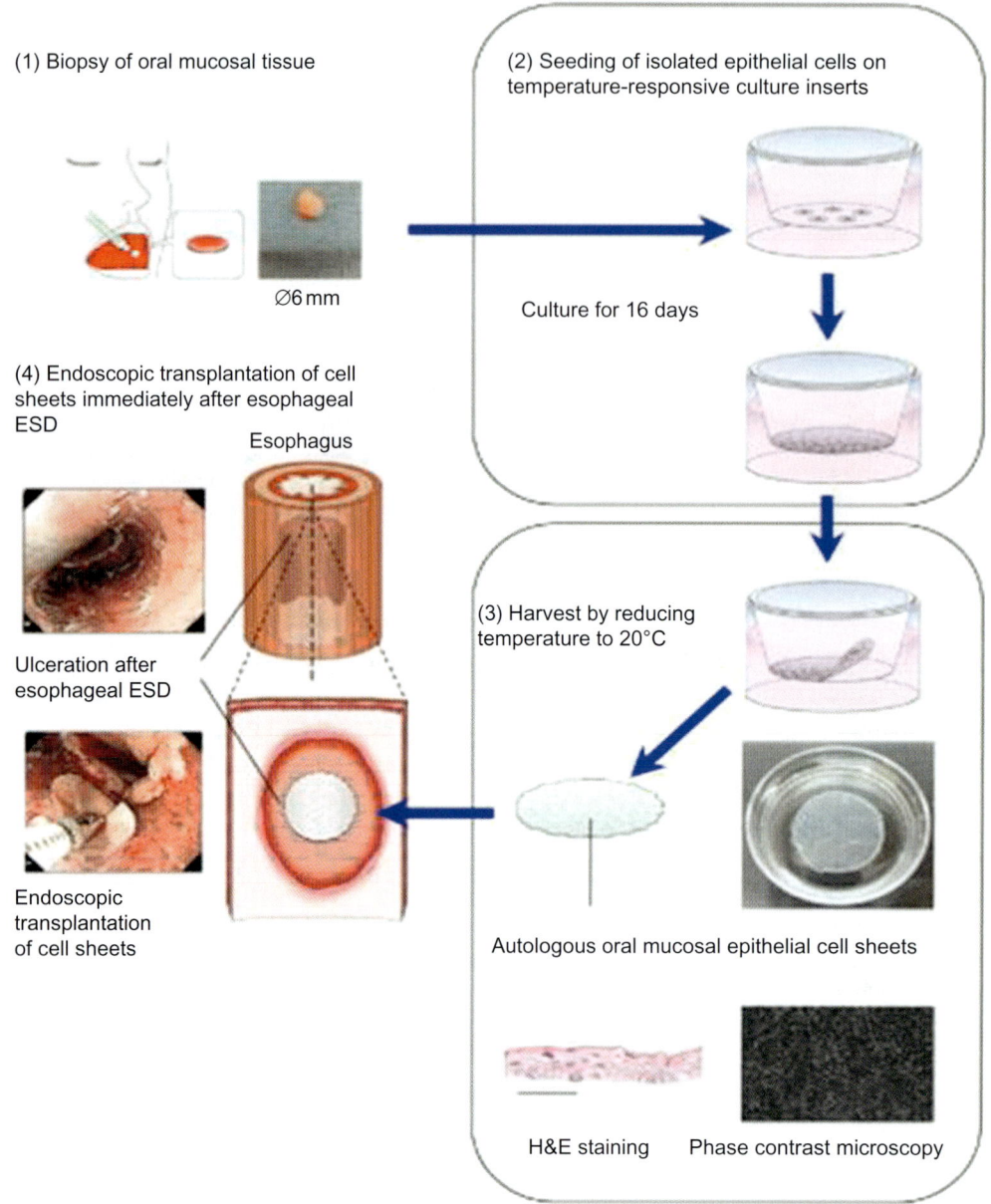

FIGURE 5.4   Treatment of the artificial ulceration after esophageal ESD by transplantation of autologous oral mucosal epithelial cell sheets fabricated on temperature-responsive culture inserts. (1) Biopsies were taken from the patient's own oral buccal mucosal tissue. Oral epithelial cells were isolated from the tissue by proteolytic enzyme treatment. (2) The epithelial cells were seeded onto temperature-responsive culture inserts without a 3T3 feeder layer, and cultured with autologous serum for 16 days at 37°C. (3) Oral mucosal epithelial cell sheets (23 mm in diameter) were harvested by reducing the culture temperature to 20°C. Bar = 50 μm. (4) Autologous oral mucosal epithelial cell sheets on a support membrane were transplanted with endoscopic forceps onto the bed of the esophageal ulceration immediately after ESD. *Reprinted with permission from [51], Copyright ©️ 2012 Elsevier. All rights reserved.*

(Figure 5.5). Cell-sheet transplantation regenerates both new bone and cementum connecting with well-oriented collagen fibers, while only limited bone regeneration is observed in the control group without cell-sheet transplantation. In addition, cell-sheet transplantation was performed using three kinds of mesenchymal tissue (PDL, alveolar periosteum, and bone marrow)-derived cells for comparing the differences between cell sources in a canine severe defect model (one-wall intrabony defect) [53]. Eight weeks after the transplantation, periodontal regeneration was significantly observed with both newly formed cementum and well-oriented PDL fibers more in the PDL cell group than in the other groups. Furthermore, nerve filaments were also observed in the regenerated PDL tissue

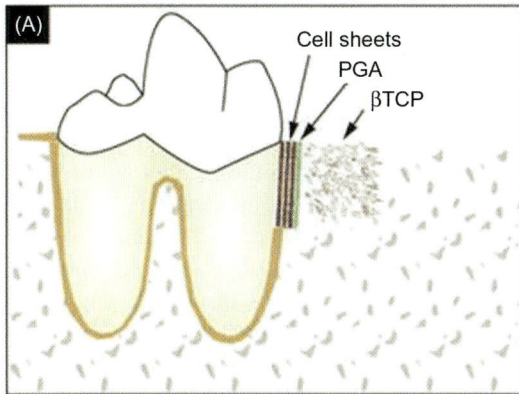

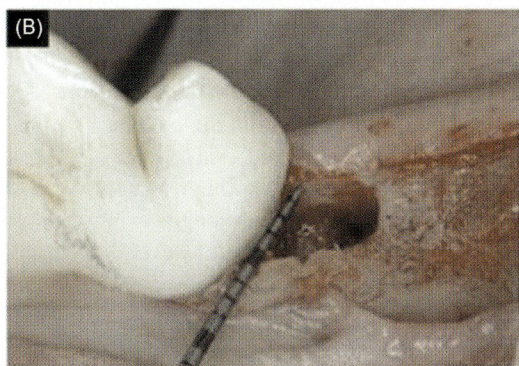

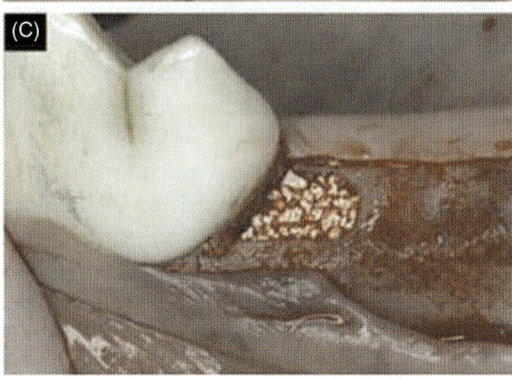

FIGURE 5.5 **Three-wall infrabony defect in dog.** (A) Schematic illustration of artificial periodontal defects and three-layered cell-sheet transplantation. After three-wall bone defect was created, root surface was curetted to remove all PDLs and cementum. Then, three-layered PDL cell sheets supported by woven polyglycolic acid (PGA) were transplanted to the root surface. *Note*: Cell sheets were directly attached to the naked surface. (B) After the flap was raised, 3-wall infrabony defects ($5 \times 5 \times 4 \, mm^3$, depth, mesio-distal width, and bucco-lingual width, respectively) were surgically created on the mesial side of bilateral mandibular first molars of dogs. (C) Three-layered PDL cell sheets were trimmed and applied to the mechanically and chemically planed root surface. After transplantation of cell sheets, porous βTCP (Osferion®) was filled in the infrabony defect. *Reprinted with permission from [52], Copyright © 2012 Elsevier. All rights reserved.*

only in the PDL cell group. The amount of alveolar bone regeneration was highest in the PDL cell group, although the amount was unable to show a statistical significance among the groups. Recently, we reported that PDL cell

sheets combined with β-TCP/collagen scaffold serve as a promising tool for periodontal regeneration.

In order to evaluate the efficacy of PDL cell-sheet transplantation, the clinical trial of periodontal regeneration was started in 2011 at Tokyo Women's Medical University, Japan.

5.2.5 Cartilage Regeneration

Chondrocyte sheets applicable to cartilage regeneration were prepared with cell-sheet technique using temperature-responsive culture dishes. The layered chondrocyte sheets were able to maintain the phenotype of cartilage, and could be attached to the sites of cartilage damage. The cell sheets acted as a barrier to prevent a loss of proteoglycan from these sites and to protect them from catabolic factors in the joint [54]. Chondrocyte sheets with a consistent cartilaginous phenotype and adhesive properties were confirmed and it may lead to a new strategy for cartilage regeneration [55,56] (Figure 5.6).

The clinical trial of cartilage regeneration was started in 2011 at the Tokai University, Japan. Autologous chondrocytes and synoviocytes are cocultured with UpCell-Insert technology. After a certain period for cultivation cocultured cell sheets were three-layered and transplanted into the defective cartilage of patients.

5.2.6 Others

Hepatic tissue sheets transplanted into the subcutaneous space resulted in efficient engraftment to the surrounding cells, with the formation of two-dimensional hepatic tissues that stably persisted for longer than 200 days [57]. The engineered hepatic cell sheet also showed several characteristics of liver-specific functionality and bilayered sheets enhanced the effects more.

In thoracic surgery, the development of postoperative air leaks is the most common cause of prolonged hospitalization. To seal the lung leakage, autologous fibroblast sheets were put on the defects and showed to be effective in permanently sealing air leaks in a dynamic fashion [58]. Using almost same procedures, pleural defects were also closed by fibroblast sheets [59].

Postoperative adhesions often cause severe complications such as bowel obstruction and abdominopelvic pain. Mesothelial cell sheets were examined to ascertain whether they can prevent the postoperative adhesions in a canine model [60]. Mesothelial cells were harvested from tunica vaginalis [46] and cell sheets were fabricated on a fibrin gel. The results showed that mesothelial cell sheet is effective for preventing postoperative adhesion formation.

The cells from rat thyroid were spread on the temperature-responsive culture dishes, and cell sheets

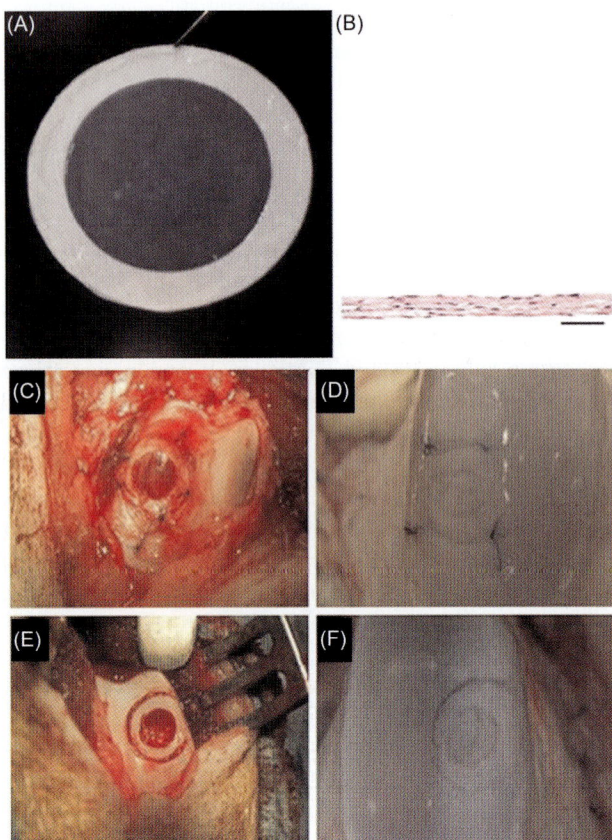

FIGURE 5.6 (A) Layered chondrocyte sheet. Chondrocytes cultured on a temperature-responsive surface can be released from the dish (4.2 cm^2) by reducing the temperature without the need for proteolytic enzymes. Confluent cultured chondrocytes were harvested as a single contiguous cell sheet retaining cell—cell junctions, and ECM was deposited on the basal side. (B) Chondrocyte sheets can be layered and thereafter adhere to other cell sheets. Culture of five-layered chondrocyte sheets can be continued for up to 1 week (scale bar = 150 μm).
Gross appearance (C—F) in the group receiving layered chondrocyte sheets (C and D) and the control group (E and F). Gross appearance is shown at transplantation (C and E) and 3 months later (D and F). The defect in the transplantation group (D) was filled with cartilaginous tissue, but filling in the defect of the control group (F) was insufficient, and the subchondral bone was exposed partially. The defect of 6 mm in diameter and 5 mm deep was made, and outside circle of the defect is 8 mm in diameter, and is used for suturing chondrocyte sheet. *Reprinted with permission from [56], Copyright © 2012 Elsevier. All rights reserved.*

were created [61]. Rats were exposed to total thyroidectomy as hypothyroidism models and received thyroid cell-sheet transplantation 1 week after total thyroidectomy. Transplantation of the thyroid cell sheets was able to restore the thyroid function 1 week after the cell-sheet transplantation and the improvement was maintained.

To establish a novel approach for diabetes mellitus, pancreatic islet cell sheets were fabricated and transplanted [62]. Laminin-5 was coated on the temperature-responsive dishes to enhance the initial cell attachment and specific molecules, such as insulin and glucagon, were positive in the recipient site.

5.3 CONCLUSIONS

In this review, we described the technology of "cell-sheet engineering" and its current applications for both animal models and human clinical trials. The limitation of tissue engineering using biodegradable scaffolds and injection of cell suspension has been disclosed, and alternative strategies have been proposed recently. The aspect of cell-sheet engineering is the application of temperature-responsive polymer to the surface of cell culture dishes to overcome the problems. Temperature-responsive dishes, which are now commercially available as UpCell™ Surface (Thermo Scientific), enable harvesting of cells without enzymes, and it permits the cell sheets to be readily manipulated, transferred, layered, or fabricated, because they adhere rapidly to other surfaces, cell sheets, and recipient sites. Cell-sheet engineering has already been tested in clinical settings for several incurable diseases. To create cell sheets, cells should be confluent before transplantation, or the cell sheets tend to be fragile. In addition, this technology enables us to harvest sticky cells, which are considered to be declined in function by enzymatic digestion, such as microglia [63], osteoclasts [64], and macrophages [65]. It is also possible to analyze native cell surface proteins of adherent cells, which are conventionally analyzed after enzymatic treatments. Thus, our temperature-responsive polymer chemistry can also contribute to the basic science fields. We believe that methods that can effectively apply cell-sheet engineering will provide new possibilities in the field of regenerative medicine.

ACKNOWLEDGMENT

This study was supported by Creation of Innovation Centers for Advanced Interdisciplinary Research Areas Program in the Project for Developing Innovation Systems "Cell Sheet Tissue Engineering Center (CSTEC)" from the Ministry of Education, Culture, Sports, Science and Technology (MEXT), Japan.

REFERENCES

[1] Vacanti JP. Beyond transplantation. Third Annual Samuel Jason Mixter lecture. Arch Surg 1988;123:545—9.

[2] Langer R, Vacanti JP. Tissue engineering. Science 1993;260:920—6.

[3] Gallico 3rd GG, O'Connor NE, Compton CC, Kehinde O, Green H. Permanent coverage of large burn wounds with autologous cultured human epithelium. N Engl J Med 1984;311:448—51.

[4] Gallico 3rd GG, O'Connor NE, Compton CC, Remensnyder JP, Kehinde O, Green H. Cultured epithelial autografts for giant congenital nevi. Plast Reconstr Surg 1989;84:1—9.

[5] Phillips TJ, Kehinde O, Green H, Gilchrest BA. Treatment of skin ulcers with cultured epidermal allografts. J Am Acad Dermatol 1989;21:191—9.

[6] Vacanti CA, Bonassar LJ, Vacanti MP, Shufflebarger J. Replacement of an avulsed phalanx with tissue-engineered bone. N Engl J Med 2001;344:1511—4.

[7] Shin'oka T, Imai Y, Ikada Y. Transplantation of a tissue-engineered pulmonary artery. N Engl J Med 2001;344:532—3.

[8] Poh M, Boyer M, Solan A, Dahl SL, Pedrotty D, Banik SS, et al. Blood vessels engineered from human cells. Lancet 2005;365:2122—4.

[9] Atala A, Bauer SB, Soker S, Yoo JJ, Retik AB. Tissue-engineered autologous bladders for patients needing cystoplasty. Lancet 2006;367:1241—6.

[10] Okano T, Yamada N, Sakai H, Sakurai Y. A novel recovery system for cultured cells using plasma-treated polystyrene dishes grafted with poly(N-isopropylacrylamide). J Biomed Mater Res 1993;27:1243—51.

[11] Heskins M, Guillet JE. Solution properties of poly(N-isopropylacrylamide). J Macromol Sci Part A Pure Appl Chem 1968;2:1441—55.

[12] Akiyama Y, Kikuchi A, Yamato M, Okano T. Ultrathin poly(N-isopropylacrylamide) grafted layer on polystyrene surfaces for cell adhesion/detachment control. Langmuir 2004;20:5506—11.

[13] Fukumori K, Akiyama Y, Yamato M, Kobayashi J, Sakai K, Okano T. Temperature-responsive glass coverslips with an ultra-thin poly(N-isopropylacrylamide) layer. Acta Biomater 2009;5:470—6.

[14] Fukumori K, Akiyama Y, Kumashiro Y, Kobayashi J, Yamato M, Sakai K, et al. Characterization of ultra-thin temperature-responsive polymer layer and its polymer thickness dependency on cell attachment/detachment properties. Macromol Biosci 2010;10:1117—29.

[15] Kushida A, Yamato M, Konno C, Kikuchi A, Sakurai Y, Okano T. Decrease in culture temperature releases monolayer endothelial cell sheets together with deposited fibronectin matrix from temperature-responsive culture surfaces. J Biomed Mater Res 1999;45:355—62.

[16] Yamato M, Utsumi M, Kushida A, Konno C, Kikuchi A, Okano T. Thermo-responsive culture dishes allow the intact harvest of multilayered keratinocyte sheets without dispase by reducing temperature. Tissue Eng 2001;7:473—80.

[17] Yamato M, Okuhara M, Karikusa F, Kikuchi A, Sakurai Y, Okano T. Signal transduction and cytoskeletal reorganization are required for cell detachment from cell culture surfaces grafted with a temperature-responsive polymer. J Biomed Mater Res 1999;44:44—52.

[18] Kushida A, Yamato M, Isoi Y, Kikuchi A, Okano T. A noninvasive transfer system for polarized renal tubule epithelial cell sheets using temperature-responsive culture dishes. Eur Cell Mater 2005;10:23—30 [discussion 23—30].

[19] Kushida A, Yamato M, Kikuchi A, Okano T. Two-dimensional manipulation of differentiated Madin-Darby canine kidney (MDCK) cell sheets: the noninvasive harvest from temperature-responsive culture dishes and transfer to other surfaces. J Biomed Mater Res 2001;54:37—46.

[20] Akizuki T, Oda S, Komaki M, Tsuchioka H, Kawakatsu N, Kikuchi A, et al. Application of periodontal ligament cell sheet for periodontal regeneration: a pilot study in beagle dogs. J Periodontal Res 2005;40:245—51.

[21] Hasegawa M, Yamato M, Kikuchi A, Okano T, Ishikawa I. Human periodontal ligament cell sheets can regenerate periodontal ligament tissue in an athymic rat model. Tissue Eng 2005;11:469—78.

[22] Shimizu T, Yamato M, Akutsu T, Shibata T, Isoi Y, Kikuchi A, et al. Electrically communicating three-dimensional cardiac tissue mimic fabricated by layered cultured cardiomyocyte sheets. J Biomed Mater Res 2002;60:110—7.

[23] Shimizu T, Yamato M, Kikuchi A, Okano T. Two-dimensional manipulation of cardiac myocyte sheets utilizing temperature-responsive culture dishes augments the pulsatile amplitude. Tissue Eng 2001;7:141—51.

[24] Haraguchi Y, Shimizu T, Yamato M, Kikuchi A, Okano T. Electrical coupling of cardiomyocyte sheets occurs rapidly via functional gap junction formation. Biomaterials 2006;27:4765—74.

[25] Shimizu T, Sekine H, Yang J, Isoi Y, Yamato M, Kikuchi A, et al. Polysurgery of cell sheet grafts overcomes diffusion limits to produce thick, vascularized myocardial tissues. FASEB J 2006;20:708—10.

[26] Haraguchi Y, Shimizu T, Sasagawa T, Sekine H, Sakaguchi K, Kikuchi T, et al. Fabrication of functional three-dimensional tissues by stacking cell sheets in vitro. Nat Protoc 2012;7:850—8.

[27] Pirraco RP, Obokata H, Iwata T, Marques AP, Tsuneda S, Yamato M, et al. Development of osteogenic cell sheets for bone tissue engineering applications. Tissue Eng Part A 2011;17:1507—15.

[28] Tsuda Y, Shimizu T, Yamato M, Kikuchi A, Sasagawa T, Sekiya S, et al. Cellular control of tissue architectures using a three-dimensional tissue fabrication technique. Biomaterials 2007;28:4939—46.

[29] Asakawa N, Shimizu T, Tsuda Y, Sekiya S, Sasagawa T, Yamato M, et al. Pre-vascularization of in vitro three-dimensional tissues created by cell sheet engineering. Biomaterials 2010;31:3903—9.

[30] Sasagawa T, Shimizu T, Sekiya S, Haraguchi Y, Yamato M, Sawa Y, et al. Design of prevascularized three-dimensional cell-dense tissues using a cell sheet stacking manipulation technology. Biomaterials 2010;31:1646—54.

[31] Nishida K, Yamato M, Hayashida Y, Watanabe K, Maeda N, Watanabe H, et al. Functional bioengineered corneal epithelial sheet grafts from corneal stem cells expanded ex vivo on a temperature-responsive cell culture surface. Transplantation 2004;77:379—85.

[32] Hayashida Y, Nishida K, Yamato M, Watanabe K, Maeda N, Watanabe H, et al. Ocular surface reconstruction using autologous rabbit oral mucosal epithelial sheets fabricated ex vivo on a temperature-responsive culture surface. Invest Ophthalmol Vis Sci 2005;46:1632—9.

[33] Nishida K, Yamato M, Hayashida Y, Watanabe K, Yamamoto K, Adachi E, et al. Corneal reconstruction with tissue-engineered cell sheets composed of autologous oral mucosal epithelium. N Engl J Med 2004;351:1187—96.

[34] Burillon C, Huot L, Justin V, Nataf S, Chapuis F, Decullier E, et al. Cultured autologous oral mucosal epithelial cell sheet (CAOMECS) transplantation for the treatment of corneal limbal epithelial stem cell deficiency. Invest Ophthalmol Vis Sci 2012;53:1325—31.

[35] Shimizu T, Yamato M, Isoi Y, Akutsu T, Setomaru T, Abe K, et al. Fabrication of pulsatile cardiac tissue grafts using a novel 3-dimensional cell sheet manipulation technique and temperature-responsive cell culture surfaces. Circ Res 2002;90:e40.

[36] Shimizu T, Sekine H, Isoi Y, Yamato M, Kikuchi A, Okano T. Long-term survival and growth of pulsatile myocardial tissue grafts engineered by the layering of cardiomyocyte sheets. Tissue Eng 2006;12:499−507.

[37] Miyahara Y, Nagaya N, Kataoka M, Yanagawa B, Tanaka K, Hao H, et al. Monolayered mesenchymal stem cells repair scarred myocardium after myocardial infarction. Nat Med 2006;12:459−65.

[38] Kondoh H, Sawa Y, Miyagawa S, Sakakida-Kitagawa S, Memon IA, Kawaguchi N, et al. Longer preservation of cardiac performance by sheet-shaped myoblast implantation in dilated cardiomyopathic hamsters. Cardiovasc Res 2006;69:466−75.

[39] Hata H, Matsumiya G, Miyagawa S, Kondoh H, Kawaguchi N, Matsuura N, et al. Grafted skeletal myoblast sheets attenuate myocardial remodeling in pacing-induced canine heart failure model. J Thorac Cardiovasc Surg 2006;132:918−24.

[40] Hoashi T, Matsumiya G, Miyagawa S, Ichikawa H, Ueno T, Ono M, et al. Skeletal myoblast sheet transplantation improves the diastolic function of a pressure-overloaded right heart. J Thorac Cardiovasc Surg 2009;138:460−7.

[41] Sawa Y, Miyagawa S, Sakaguchi T, Fujita T, Matsuyama A, Saito A, et al. Tissue engineered myoblast sheets improved cardiac function sufficiently to discontinue LVAS in a patient with DCM: report of a case. Surg Today 2012;42:181−4.

[42] Gotoda T, Kondo H, Ono H, Saito Y, Yamaguchi H, Saito D, et al. A new endoscopic mucosal resection procedure using an insulation-tipped electrosurgical knife for rectal flat lesions: report of two cases. Gastrointest Endosc 1999;50:560−3.

[43] Ono H, Kondo H, Gotoda T, Shirao K, Yamaguchi H, Saito D, et al. Endoscopic mucosal resection for treatment of early gastric cancer. Gut 2001;48:225−9.

[44] Seewald S, Akaraviputh T, Seitz U, Brand B, Groth S, Mendoza G. Circumferential EMR and complete removal of Barrett's epithelium: a new approach to management of Barrett's esophagus containing high-grade intraepithelial neoplasia and intramucosal carcinoma. Gastrointest Endosc 2003;57:854−9.

[45] Ono S, Fujishiro M, Niimi K, Goto O, Kodashima S, Yamamichi N, et al. Predictors of postoperative stricture after esophageal endoscopic submucosal dissection for superficial squamous cell neoplasms. Endoscopy 2009;41:661−5.

[46] Asano T, Takazawa R, Yamato M, Kageyama Y, Kihara K, Okano T. Novel and simple method for isolating autologous mesothelial cells from the tunica vaginalis. BJU Int 2005;96:1409−13.

[47] Kanai N, Yamato M, Ohki T, Yamamoto M, Okano T. Fabricated autologous epidermal cell sheets for the prevention of esophageal stricture after circumferential ESD in a porcine model. Gastrointest Endosc 2012;76:873−81.

[48] Murakami D, Yamato M, Nishida K, Ohki T, Takagi R, Yang J, et al. Fabrication of transplantable human oral mucosal epithelial cell sheets using temperature-responsive culture inserts without feeder layer cells. J Artif Organs 2006;9:185−91.

[49] Murakami D, Yamato M, Nishida K, Ohki T, Takagi R, Yang J, et al. The effect of micropores in the surface of temperature-responsive culture inserts on the fabrication of transplantable canine oral mucosal epithelial cell sheets. Biomaterials 2006;27:5518−23.

[50] Takagi R, Yamato M, Murakami D, Kondo M, Yang J, Ohki T, et al. Preparation of keratinocyte culture medium for the clinical applications of regenerative medicine. J Tissue Eng Regen Med 2011;5:63−73.

[51] Ohki T, Yamato M, Ota M, Takagi R, Murakami D, Kondo M, et al. Prevention of esophageal stricture after endoscopic submucosal dissection using tissue-engineered cell sheets. Gastroenterology 2012;143:582−8.

[52] Iwata T, Yamato M, Tsuchioka H, Takagi R, Mukobata S, Washio K, et al. Periodontal regeneration with multi-layered periodontal ligament-derived cell sheets in a canine model. Biomaterials 2009;30:2716−23.

[53] Tsumanuma Y, Iwata T, Washio K, Yoshida T, Yamada A, Takagi R, et al. Comparison of different tissue-derived stem cell sheets for periodontal regeneration in a canine 1—wall defect model. Biomaterials 2011;32:5819−25.

[54] Kaneshiro N, Sato M, Ishihara M, Mitani G, Sakai H, Mochida J. Bioengineered chondrocyte sheets may be potentially useful for the treatment of partial thickness defects of articular cartilage. Biochem Biophys Res Commun 2006;349:723−31.

[55] Mitani G, Sato M, Lee JI, Kaneshiro N, Ishihara M, Ota N, et al. The properties of bioengineered chondrocyte sheets for cartilage regeneration. BMC Biotechnol 2009;9:17.

[56] Ebihara G, Sato M, Yamato M, Mitani G, Kutsuna T, Nagai T, et al. Cartilage repair in transplanted scaffold-free chondrocyte sheets using a minipig model. Biomaterials 2012;33:3846−51.

[57] Ohashi K, Yokoyama T, Yamato M, Kuge H, Kanehiro H, Tsutsumi M, et al. Engineering functional two- and three-dimensional liver systems in vivo using hepatic tissue sheets. Nat Med 2007;13:880−5.

[58] Kanzaki M, Yamato M, Yang J, Sekine H, Kohno C, Takagi R, et al. Dynamic sealing of lung air leaks by the transplantatiozn of tissue engineered cell sheets. Biomaterials 2007;28:4294−302.

[59] Kanzaki M, Yamato M, Yang J, Sekine H, Takagi R, Isaka T, et al. Functional closure of visceral pleural defects by autologous tissue engineered cell sheets. Eur J Cardiothorac Surg 2008;34:864−9.

[60] Asano T, Takazawa R, Yamato M, Takagi R, Iimura Y, Masuda H, et al. Transplantation of an autologous mesothelial cell sheet prepared from tunica vaginalis prevents post-operative adhesions in a canine model. Tissue Eng 2006;12:2629−37.

[61] Arauchi A, Shimizu T, Yamato M, Obara T, Okano T. Tissue-engineered thyroid cell sheet rescued hypothyroidism in rat models after receiving total thyroidectomy comparing with non-transplantation models. Tissue Eng Part A 2009;15:3943−9.

[62] Shimizu H, Ohashi K, Utoh R, Ise K, Gotoh M, Yamato M, et al. Bioengineering of a functional sheet of islet cells for the treatment of diabetes mellitus. Biomaterials 2009;30:5943−9.

[63] Nakajima K, Honda S, Nakamura Y, Lopez-Redondo F, Kohsaka S, Yamato M, et al. Intact microglia are cultured and non-invasively harvested without pathological activation using a novel cultured cell recovery method. Biomaterials 2001;22:1213−23.

[64] Ishii KA, Fumoto T, Iwai K, Takeshita S, Ito M, Shimohata N, et al. Coordination of PGC-1beta and iron uptake in mitochondrial biogenesis and osteoclast activation. Nat Med 2009;15:259−66.

[65] Collier TO, Anderson JM, Kikuchi A, Okano T. Adhesion behavior of monocytes, macrophages, and foreign body giant cells on poly(N-isopropylacrylamide) temperature-responsive surfaces. J Biomed Mater Res 2002;59:136−43.

Chapter 6

Principles of Bioprinting Technology

Tao Xu[a,b], Jorge I. Rodriguez-Devora[b], Daniel Reyna-Soriano[b], Mohammod Bhuyan[b], Lei Zhu[c], Kun Wang[d], and Yuyu Yuan[e]

[a]Department of Mechanical Engineering, Tsinghua University, Beijing, People's Republic of China, [b]Department of Mechanical Engineering and Biomedical Engineering Program, University of Texas, El Paso, TX, [c]Department of Plastic Surgery, Third Affiliated Hospital, Sun Yat-sen University, Guangzhou, Guangdong Province, People's Republic of China, [d]Department of Orthopedic Surgery, Third Affiliated Hospital, Sun Yat-sen University, Guangzhou, Guangdong Province, People's Republic of China, [e]Medprin Regenerative Technologies Co. Ltd, Guangzhou, Guangdong Province, People's Republic of China

6.1 INTRODUCTION

Since the first successful organ transplantation with a kidney in 1954 [1], scientists have maintained the dream of being able to fabricate organs per request. Creation of living tissues and organs from artificial manipulation of cells, materials, growth factors (GFs), and other organ elements has been waiting for appropriate fabrication technologies. These technologies should be capable of rebuilding compositional and structural complexities of human tissues and organs. The recent development of bioprinting technologies (known by their high resolution and high-speed construction) has revived interest in applying those emerging methods for organ construction. The term "organ printing" has become standard since the 2000s [2–4], referring to the line of investigations related to the development of these technologies on the construction of three-dimensional (3D) structures based on the deposition of different cell lines and biochemical promoters.

Although individual tissue systems have been successfully engineered for various applications using basic tissue engineering approaches, fabrication of complex tissues that consist of multiple cells and tissue components has not been established. This is due to various challenges encountered in the tissue building process. One of the challenges has been the inability to recreate the well-defined cellular configurations and functions of a native tissue. Living tissues contain multiple cell types and various extracellular materials arranged in specific patterns that are difficult to replicate *in vitro*. Thus, one important goal of tissue engineering and regenerative medicine is to develop a tissue fabrication method that allows for specific control over the placement of various cells and matrices in three dimensions in order to mimic the complexity of native tissue architecture. Emerging "organ printing" or "bioprinting" methodologies are being investigated to create tissue-engineered constructs that initially have more defined spatial organization. The underlying hypothesis is that these biomimetic constructs can achieve ideal therapeutic outcomes.

The following sections provide a brief and concise introduction to basic elements and current trends as they pertain to cell and organ printing, including: bio-blueprint, bio-ink, bio-paper, bio-printers, and different printing methods. Organ printing stands at the forefront of tissue engineering based on recent studies that have demonstrated the feasibility and benefits of implementing bioprinting of biomolecules and cells for tissue engineering and biological applications.

6.2 BIO-BLUEPRINT

As a preprocessing step in organ printing technology, bio-blueprint depicts the complete pictorial feature of organ structure along with the required biological data and serves the following basic functions: (i) describes anatomy, geometry, and internal architecture of an organ of interest, including tissue heterogeneity, individual tissue geometry, and boundary distinction within the organ of interest and (ii) defines a vascular network of an organ of interest.

The bio-blueprint is generated using medical imaging data in order to replicate organ/tissue complexity. As an example, Figure 6.1 shows a blueprint of a heart, depicting only five cross section and side views, however, the real life bio-blueprint consists of innumerable numbers of cross sections (equal to the number of cell layers). The blueprint will specify by color, the type of cells in each layer to print using the bioprinting method. In general, an image-based Bio-CAD modeling process involves the following four major steps: (i) noninvasive image acquisition, (ii) imaging process and 3D reconstruction to form voxel-based volumetric image representation, (iii) construction of CAD-based model, and (iv) viewing operations and display.

6.2.1 Data Acquisition

Modern imaging and reverse engineering techniques have been extensively used in a variety of biomedical engineering applications including clinical medicine, customized medical implant design, and tissue engineering. The primary imaging modalities capable of producing 3D views, such as computed tomography (CT), magnetic resonance imaging (MRI), optical microscopy, and single photon emission computed tomography (SPECT), are used for anatomic data acquisition to construct a Bio-CAD model. The bio-model provides the needed biological data for organ anatomy, tissue heterogeneity, and vascular networking, which can be used to introduce and facilitate the design or manufacturing intent.

6.2.1.1 CT and Micro-CT

CT provides nondestructive 3D visualization and characterization, creating images that map the variation of X-ray attenuation within objects, which relates closely to density. X-ray CT measures the spatially varying X-ray attenuation [5] to show internal structures. CT or micro-CT scans require exposure of a sample to small quantities of ionizing radiation and the absorption of which is detected and imaged. This results in a series of two-dimensional (2D) images displaying a density map of the sample. Stacking these images creates a 3D representation of the scanned area. In particular, the latest development of micro-CT technology has been successfully used to quantify the microstructure-function relationship of tissues and the designed tissue structures, including the characterization of the microarchitecture of tissue scaffolds to help with the design and fabrication of tailored tissue microstructures, and determine tissue morphologies and internal physical activities.

6.2.1.2 Magnetic Resonance Imaging

MRI provides images for soft tissues as well as for hard tissues, and as such is vastly superior in differentiating soft tissue types and recognizing border regions of tissues of similar density. MRI shows excellent contrast between varieties of soft tissues, however, the variety of surfaces presents a challenge to 3D surface construction and required techniques for selective extraction and display.

6.2.1.3 Optical Microscopy

Optical microscopy has limited applications to 3D bio-tissue modeling due to intensive data manipulation. For example, to examine a sample with high resolution using optical microscopy, it must be physically sectioned onto very thin slides. The division into these slides is a labor-intensive process, and the resulting images of the target organ would be thousands of 2D images that must be digitally stacked into 3D columns. Due to excessive computation and a memory-intensive process, it is a significant challenge to train computers to identify individual cells by their visual characteristics, even with the aid of complex staining. However, up to date differentiating tissue down to the level of the individual cell may still be only possible by using optical microscopy.

6.2.1.4 Single Photon Emission Computed Tomography

SPECT measures the emission of gamma rays. The sources of these rays are radioisotopes distributed within the body. SPECT can show the presence of

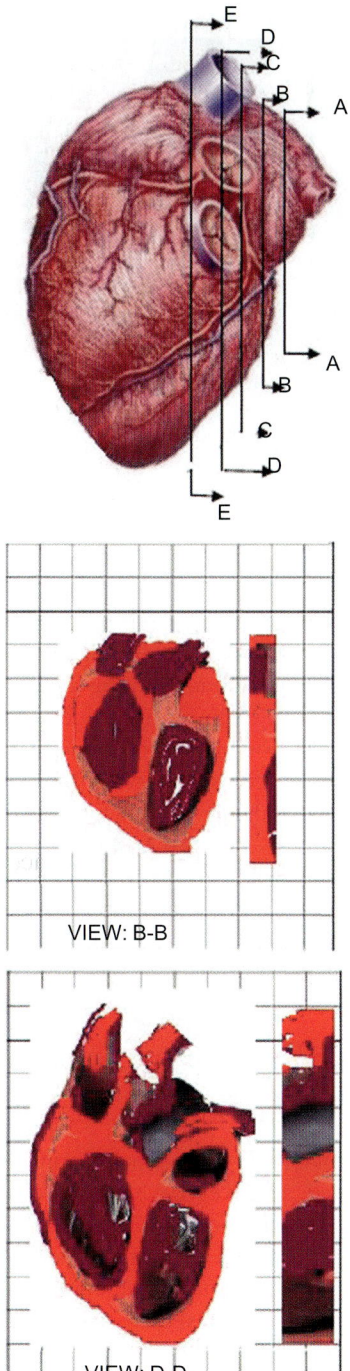

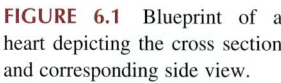

FIGURE 6.1 Blueprint of a heart depicting the cross section and corresponding side view.

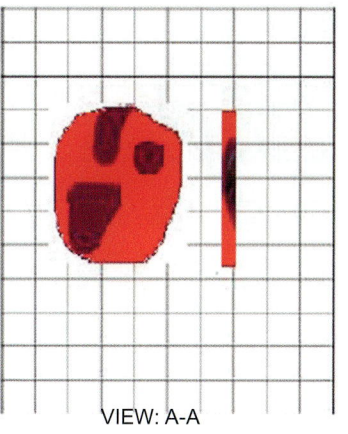

VIEW: A-A

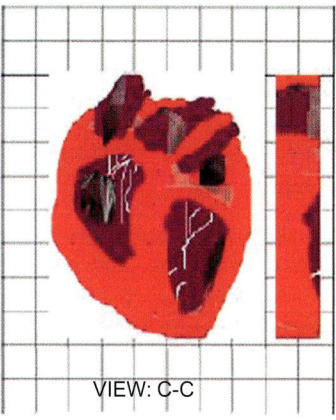

VIEW: B-B   VIEW: C-C

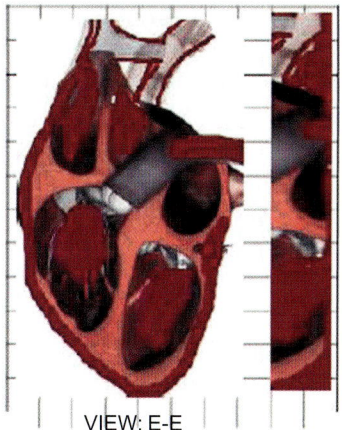

VIEW: D-D   VIEW: E-E

blood in structures with a much lower dose than that required by CT.

6.2.2 Bio-Blueprint Modeling

The framework of development of a biomimetic model is outlined through the following major steps: (i) development of a computer modeling representation of a 3D organ, (ii) development of a 3D vascularization network, and (iii) development of a CAD-based organ bio-blueprint model.

The development of a bio-blueprint model starts from 3D reconstruction of organ anatomy (without vascular system) from the given modalities (medical imaging data) and a CAD-based modeling representation that can be used to explicitly describe organ geometry, topology, and

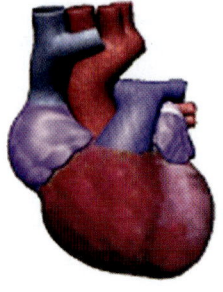 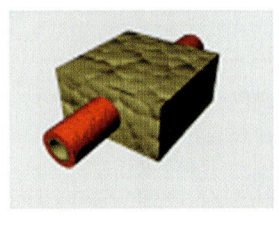  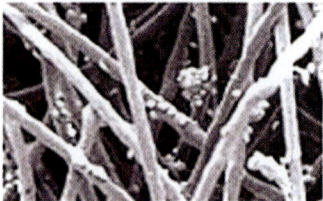

FIGURE 6.2   Hierarchical scales of organ structure [8].

individual tissues. The model development involves using state-of-the-art 3D reconstruction, reverse engineering, CAD platforms, and in-house heterogeneous modeling algorithms [6,7]. A bio-blueprint model is generated from noninvasive medical imaging for capturing and replicating organ anatomy including detailed internal and external morphology, geometry, vascularization, and tissue identification. Therefore, all the following three hierarchical scales, as shown in Figure 6.2, need to be considered in the construction of the bio-blueprint model:

1. the scale of the organ (to consider the organ's macrostructure tissue types, vasculature, ducting, and anatomical compatibility);
2. the scale of the tissue or suborgan (to consider the heterogeneity of tissue with appropriate type of cells and their interaction);
3. the scale at the cellular level [to consider the selection of extracellular matrix (ECM) or scaffold materials and to divide bio-blueprint model into small blocks to enable local definition of ECM/scaffold materials].

The bio-blueprint model is represented in a CAD format, which adopts "boundary representation" by which an organ or tissue anatomy and topology can be explicitly described by the enclosed and adjacent boundaries through mathematically defined NURBS functions [8]. Reducing memory usage of the bio-blueprint or of interpolating structures not visible in medical imaging, repetitive or patterned structures can be done by reducing the combinations of feature primitives. This method also smoothens out irregularities in continuous features caused by noise in the imaging modality. For instance, a feature primitive-based reconstruction method for vascular networks is used to generate a 3D biological vascular system for organ growth. In this primitive feature modeling approach, the basic vascular primitive characteristic parameters are determined from patient-specific CT/MRI images, and further use of Boolean operation algebra forms a high-level vessel assembly. The vascular feature primitives are represented as NURBS bases, and the

parameters in the NURBS equations can be determined through measuring the spatial positions of the vascular CT/MRI images at different projections [8].

In summary, bio-blueprint is a crucial step for the organ printing process because it provides a description and representation of details of organ anatomy, morphology, tissue heterogeneity, and vascular systems at different organ organizational scales. Reverse engineering plays an important role for the organ printing process, transforming a real organ into engineering models and concepts. Current methods to obtain the organ modeling data are the noninvasive data acquisition, which are CT scan, MRI scan, and optical methods. Bio-blueprint is an early stage and more research is required to improve the current limitations. The future goal is to realize an intelligent artificial organ using reverse engineering to create an optimal bio-blueprint model not only to define the dimensions and parameters, but also to focus in mimetic the functionality.

6.3 BIO-INK

As one of the required elements to pursue organ printing, the bio-inks and associated additives have been developed. It has being called "bio-ink" for the fact that inkjet technologies have been early found with the most intimate characteristics to create tissue and organ constructs. While inkjet technologies were initially used for printing applications (texts and pictures), they have found their way towards more interesting applications, such as the construction of electronic components and biomedical applications. For printing applications, the ink has been developed for years to optimize its characteristics to improve printing quality (precision and accuracy) and speed by creating a controlled flow of ink and preventing clogging and drying of ink at the print head [9–14]. These results and founding are also important to the desired organ printing devices. Therefore, intensive research has been focused on developing the bio-inks to advance toward organ printing.

Strategies based on the types of materials utilized have been pursued. Starting from all kinds of living cells to small proteins and plasmids, they have been utilized by different techniques to create *in vitro* 3D structures. In general, the requirements for a viable suspension for a bio-ink include highly biocompatibility, low sedimentation, high control of viscosity, ability to be arranged three dimensionally, and so forth, justifying the focus on improving one or more of these characteristics by different research groups worldwide. In the following paragraphs, a description divided by suspension source is provided to illustrate the strategies surveyed by researchers.

6.3.1 Cell Suspensions

The human body is formed by many different tissues, such as muscles, bones, nerves, vessels, and skin. Each tissue is composed of different and numerous cell types to achieve its function. If the intention of organ printing consists of the fabrication of living structures to overcome lack of organ donation, it is expected that most of these types of cells are tested to evaluate the engineering capabilities of assembling viable 3D structures. To date, many of these cell types have been used for tissue engineering applications [15]. Within these experiments, we can identify two strategies. First, by using differentiated cells such as epithelial, muscle, bone, nerve, and so forth. However, therapeutic rejection of such cells is common as implanted cells have to perfectly match the ones contained within the patient's body. On the other hand, cell proliferation of patient's biopsied cells can be approached as an alternative solution (autologous treatment). As a second strategy, the pluripotency of stem cells sets them as a promising approach for creating human-made organs. Even though ethical concerns are still present regarding the acquisition of stem cells, research using them has initiated and continued to grow.

6.3.2 GFs/Cytokines

Along with the utilization of differentiated and stem cells, other factors have been found important to artificially create living tissues. The natural organization of living tissues is governed by environmental edifying signals such as biochemical cues (i.e., GFs, cytokines). For instance, VEGF (vascular endothelial growth factor), a GF well known by its angiogenesis property, represents a potential factor to improved vascularization of printed organs, emphasizing the need for the investigation of these cues to further advance towards organ printing reality.

By employing different GFs, Campbell's group has achieved recent progress in experiments eliciting musculoskeletal tissues using inkjet printing technologies to accurately dispense the GF-loaded bio-inks [16]. Spatial patterns of GFs were built over oriented sub-micron fibers (mimicking musculoskeletal ECM), which were evaluated *in vitro* to demonstrate its control over cell fate. Further studies used bone morphogenetic protein-2 to guide stem cells towards osteoblast differentiation by the same inkjet technology [17]. This protein was patterned over 200-μm (average)-thick scaffolds which are promising for controlling cell differentiation in 3D structures. These results represent the first steps towards fabrication of more complicated structures.

6.3.3 Other Additives

Other materials have been used to modify bio-ink characteristics to improve its printability capacities and assembly features. Printability capacities are highly related to viscosity and surface tension, while assembly features require the contrary characteristics by means of quick solidification (i.e., via gelation) and structural properties to enable 3D organ-like structures. Very few studies have focused on the intrinsic characteristics of the bio-ink as enhanced agents, which represent a potential promising field for broad investigations. One of the first agents investigated for improving sedimentation and aggregation rates is surfactants. Surfactants are characterized by decreasing shear stress via diminishing surface tension. For instance, pluronic has been known by decreasing cell aggregation [18]. In a recent study, pluronic has been proven biocompatible, and more importantly decreases cell aggregation and sedimentation when dispensed by a piezoelectric inkjet printing system [19]. Another alternative for improving bio-ink printability characteristics has been polyethylene glycol (PEG). Studies have shown that 10% concentration on cell suspensions effectively improves bio-ink circulation time and biocompatibility [20,21].

ECM is crucial for cell homeostasis *in vivo*. The cell containing bio-ink should be composed of an ECM fraction accordingly. Consistent with the layer-by-layer 3D building strategy, the solidification of the bio-ink onto the substrate is necessary and should be controlled for at least three reasons: (i) to stabilize the printed 2D pattern, (ii) to mechanically support the subsequent bio-ink layer, and (iii) to mimic cell type specific ECM with the ability to regulate cell fate [22]. The solidification process of the bio-ink should not be harmful to the cells. In laser-based printing methods [23], cell suspensions have been supplemented with 1% (w/v) alginate hydrogel as a preliminary approach to mimic ECM. Cell adhesion proteins have been printed successfully since Klebe's pioneering work [24]. Fibrin is a versatile biopolymer, which presents potential for tissue engineering applications [25]. Cui and Boland have recently printed a cellularized pattern reminiscent of microvasculature using an inkjet bio-printer to print cells in combination with fibrin hydrogel [26].

In summary, a key characteristic in the elicitation of organ printing is the selection of a mixture of cells, additives, cytokinesis, and GFs. These components will provide unique properties, rendering on different effects of the final product.

6.4 BIO-PAPER

Efficacious organ printing requires appropriate substrates or scaffolds as bio-paper that are complemented with concerned organ constituents, to contribute biological cues to extract explicit cellular responses and direct new tissue formation [14,27]. Bio-paper is usually used as a scaffolding material in organ printing, contributing a biological and structural support for cells to attach, proliferate, and differentiate. Bio-papers are printed for the accommodation of cells, the supervision of cellular augmentation, and the revival of 3D tissues with proper support and functions. The most important task of a bio-paper is to guarantee a solid support to body structures to allow the stress transfer over time to implanted sites, and promote tissue regeneration on the paper structure. Some general properties for biomaterials, including biocompatibility, and some special properties we should concern for the materials used for bioprinting application, including viscosity, mechanical, and physical requirements, are the main criterions when choosing materials for organ printing applications. Bio-paper is chosen from biomaterials and the commonly available material used for bioprinting can be classified into two major categories, including natural polymers and synthetic polymers.

6.4.1 Natural Polymers Used as Bio-Paper

One of the common materials used as bio-paper is natural polymers due to their excellent properties with ease of handling and shaping under certain conditions. Natural biomaterials are enormously diverse and complex, and are obtained from proteins and carbohydrates. These proteins and carbohydrates have evolved to execute very explicit biological, mechanical, biochemical, and structural roles. Natural biodegradable polymers, such as alginate, gelatin, collagen, silk fibroin, and chitosan, show potential advantages over synthetic polymers because of their complimentary properties, including splendid biocompatibility, biodegradability, and biorestorability.

6.4.2 Synthetic Polymers

Synthetic biopolymers are the combination of two or more materials, whether by design or by natural processes. Compared to natural polymers, synthetic bio-papers are more promising for organ printing applications, as their block structures, gelation dynamics, crosslinkings, pore structures, and mechanical properties can be manipulated and reproducible in the laboratory. Polylactic-co-glycolic acid (PLGA), polycaprolacton (PCL), PEG are common synthetic polymers used for organ printing applications.

6.5 ORGAN BIO-PRINTERS AND BIOPRINTING METHODS

Bioprinting is an emerging field represented by various biologically applied deposition and assembling systems, which range from direct writing, microstamping, photolithography, laser writing, electro-printing, microfluidics, stereolithography, extrusion, to inkjet deposition. However, much of the bioprinting work has focused on 2D patterning for basic biological studies. Development and adaptation of fabrication methods to 3D cellular or tissue constructs is required to advance the road towards organogenesis. To date, those technologies inspired by rapid prototyping based on layer-by-layer fabrications for 3D structures are more attractive to researchers. Based on that, extrusion, inkjet (thermal and piezoelectric), and laser printing devices will be the main focus for organ printing in this chapter, as these techniques have made significant progress in creating 3D living biological structures.

The ideal organ bio-printer has specific system requirements, which include high resolution, high throughput, ability to dispense various materials simultaneously, biocompatibility, cell viability, process repeatability, and ability to control dispensability of bio-inks by modification of properties, such as viscosity, surface tension, and density. Resolution is a critical factor as high detail is required for the reconstruction of the organ structures. For instance, layers of vessels are under the range of a few micrometers in thickness. The throughput or process speed is also important as the main material used for organ fabrication is living cells, and viability can be compromised if the process takes a lengthy time during fabrication. However, an alternative solution can be the use of a special bio-reactor that provides an environment required to maintain cells alive for a long period.

Organs are naturally composed of diverse types of cells and structures, which encourage researchers to develop the organ bio-printer to assemble different cells and materials. The organ bio-printer has to be safe in the sense of avoiding any potential cross-contamination by isolating material reservoirs. Furthermore, the organ bio-printer should dispense materials where they are designed to be allocated; as a result biocompatibility and cell viability will be ensured within the system. As discussed in previous sections, manipulation of bio-ink properties is essential for the future organ bio-printer. This system is not allowed to accommodate clogging and/or changes in dispensing rates; therefore, bio-ink conditions have to be monitored and controlled to avoid these potential drawbacks. Another concern is the possibility that the printing process could adversely affect cellular phenol- and genotypes, which calls for the need for a specific detection method to assure cell integrity through the bio-printing process.

The great opportunity and potential financial interest which the organ printing technology can bring has motivated the biomedical industry to develop and start commercializing 3D bioprinting systems. Even though actual printing of organs is far from being accomplished, the launching of the prototyping systems supports technical investigations to move this emerging technology forward. Currently, the commercially available 3D bio-printers are produced by EnvisionTEC, Organovo, Tengion, Sciperio, and Neatco. As shown in Figure 6.3, examples of these systems are given which have implications on drug discovery, biological assays, experimental biological depositions, stem cell experiments, and other biomedical applications.

There are different organ printing strategies with distinct focuses. Based on the assembling process of cells and materials, these strategies have been categorized as follows [15]: (i) structural: organ structure is foreseen to be achieved by dispensing a mixture of cells and structural entities by the same bioprinting system (either by different heads or a single head); (ii) conformal: in this approach, prefabricated scaffolds (i.e., hydrogel layers) are set as substrate to conformed sites of deposition of cell suspension; and (iii) aggregation: as its name implies,

cells are expected to be aggregated (organized, analog to "bricks") in such manner that "bricks" are utilized to fabricate macro-scaled biological structures. In this approach, genetic cell modification is needed to maintain "brick" integrity. Meanwhile, organ printing strategies based on the basic work mechanism can also be elucidated, where the most promising technologies include extrusion, inkjet, and laser deposition-based systems. Since they meet many key characteristics required by the ideal organ-printer, it is worth reviewing them in this chapter.

6.5.1 Extrusion Printing

The pressure or extrusion-based method had been historically used for quite a long time, mainly for material transportation. The work mechanism is better explained by the syringe mechanism, where the plunger exerts pressure over liquid enclosed in the barrel to dispense a controlled volume through the needle. Since tissue engineering was introduced by Langer and Vacanti in the last century [28], this technology has also started to be investigated as a promising technique for creating living tissues [29−31]. Researchers have worked on creating and adapting more complex systems that can accommodate tissue and organ construction [32−34]. The main advantage consists in its ability to deposit highly viscous materials, however, the major drawback comes from its low resolution and high droplet volume. One example is approximately 600-μm-tall hepatic tissue built by Tsinghua University's group in China [35]. In this study, hepatocyte cells were entrapped in an artificial matrix (mixture of gelatin and chitosan

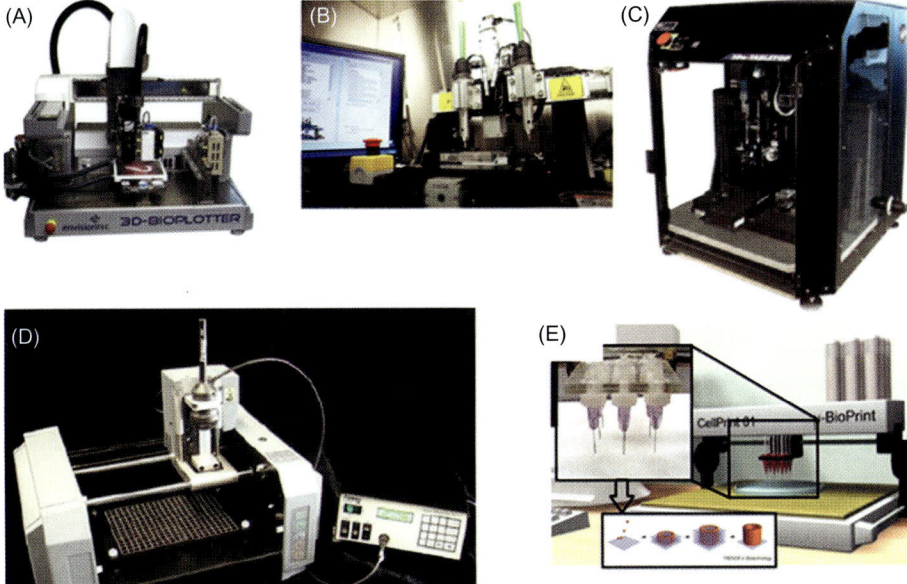

FIGURE 6.3 Examples of available bio-printers: (A) EnvisionTEC 3D bio-plotter, (B) NovoGen MMX Bioprinter™, (C) Sciperio-nScript bio-printer, (D) Neatco bio-printer, and (E) Boland's group at Clemson University developed Bio-plotter [3].

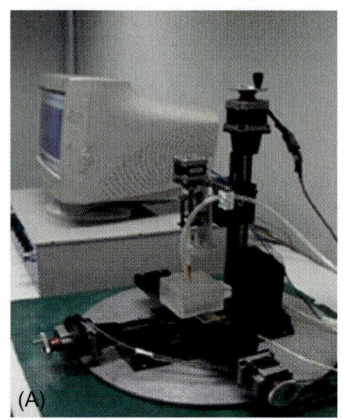
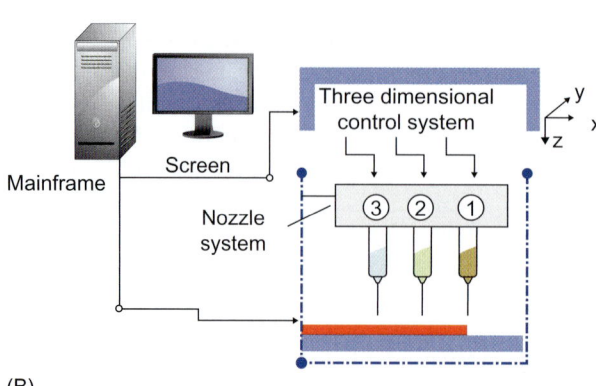

FIGURE 6.4 Tsinghua University group's cell assembling machine [35]: (A) actual machine and (B) schematic of cell assembling device.

hydrogels) and were proven to remain viable and functional after 3D biofabrication process. Their developed cell assembling machine is described in Figure 6.4. Moreover, studies using the same cell assembling system were applied for traumatic brain injury [36]. This study fabricated porous scaffolds made of gelatin and a mixture of gelatin and hyaluronan hydrogels. These scaffolds were implanted in a rat's brain and evaluated for their biocompatibility and biodegradation properties. Results indicate that the printed scaffolds improved tissue regeneration by infiltration observed in the study.

Direct cell writing (DCW), another example of the extrusion systems, has recently combined with the microfluidic devices technique for the fabrication of a novel pharmacokinetic model [4]. The DCW process is integrated with a microfluidic device to fabricate 3D tissue/organ constructs/chambers, as opposed to producing 2D cell monolayers. Biological studies reveal the unstable cellular phenotype and reduced tissue-specific gene expression with conventional monolayer *in vitro* culture techniques [29−31,37]. A 3D tissue model will, in contrast, foster improved retention of cell-specific function. The study resulted in tissue structures that are more closely resembled to their *in vivo* physiological state where cell−cell communication either from direct contact or paracrine signaling is important for proper cellular behavior, differentiation, and proliferation, along with the concomitant ECM produced by the neighboring cells. Further, optimization of process parameters (nozzle pressure, motion arm velocity, nozzle tip size, etc.) and material parameters (biopolymer viscosity, crosslinking agent concentrations, etc.) was done to achieve high-fidelity 3D structures and seamless integration onto microfluidic tissue micro-organ chambers.

Recently, the extrusion technique has been investigated to pursue the construction of skin-like tissue construction by Yoo's group. A 10-layered scaffold has been constructed with embedded human skin fibroblasts (FBs) and keratinocytes (KCs) at intermediate layers, which

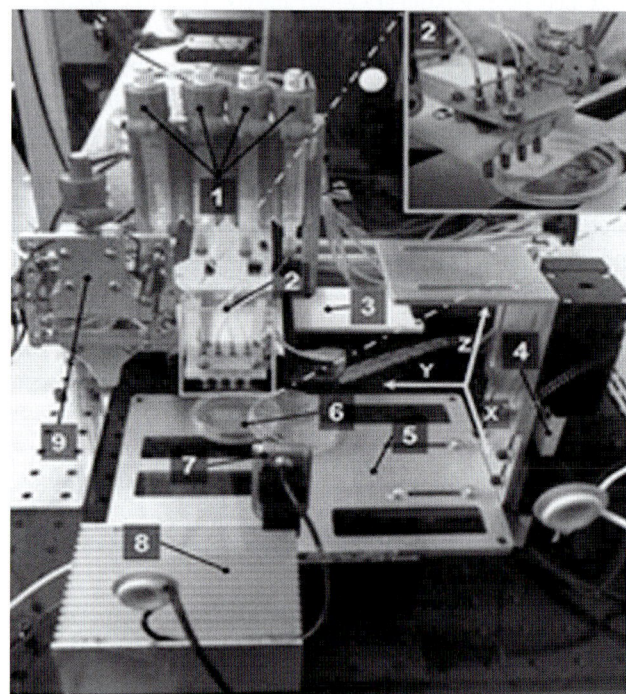

FIGURE 6.5 Picture of the modular tissue printing platform: (1) four syringes as "cartridges" to load cell suspensions and hydrogel precursors; (2) an array of four-channel dispensers; (3) horizontal stage; (4) vertical stage; (5) target platform; (6) target substrate; (7) camera; (8) stage heater, vertical stage heater; and (9) independent heated/cooled dispenser [39].

imitate the skin's natural structure [38]. As illustrated in Figure 6.5, the system utilized consists of four microvalves connected to reservoirs (5−10 mL disposable syringes) with camera-based monitoring systems to assure printing quality. Moreover, one of the dispensing units and the vertical stage were temperature controlled (operating temperature between 5 and 40°C). In this system, more than 30 human and animal cell lines [39] have been dispensed at high survival rates (above 95%). These results suggest that this technique can be generally applied to most of the cell types and to demonstrate the ability to print and culture multilayered cell−hydrogel composites on planar and nonplanar surfaces.

6.5.2 Nozzle-Based Printing

Two main types of nozzle-based printers at the forefront of this technology are inkjet and piezoelectric printers. The inkjet printer is the most common [40], primarily because the parts are readily available and simple modifications and changes to the ink cartridge and printer itself make it ready to dispense biomaterials [40]. The piezoelectric printer is the next most common, as it too can be an off the shelf item.

In both technologies, ink is deposited from the cartridge reservoir through small nozzles arranged linearly to finally reach its target. Triggering of each nozzle could be possible with the development of personal computers, which happened along this event. The thermal branch uses a heater element in each nozzle to cause heat, causing a bubble within the ink overpassing surface tension. These characteristics offer the main advantage of this type of technology, which consists in the small volume of each droplet dispensed (in the picoliter range, 2−70 pL) [41]. On the other side, the piezoelectric branch uses a vibrating element (crystal exhibiting piezoelectric effect) in the body of each nozzle resulting in the same effects as the thermal branch. However, this method does not expose ink to high temperatures, making it a promising technique when moving on to cell suspensions as bio-inks. Some other interesting characteristics for inkjet printing include additive operation enabling control over concentration of given biochemical entities, ability to process different materials at the same time by usage of different compartments for each bio-ink, flexible patterning through a computer-controlled system enabling rapid buildup, large-area printing, cost-effectiveness, and its noncontact nature that reduces possibility of cross-contamination during printing.

Studies using the piezoelectric branch of inkjet printing have been presented by Derby's group [42−44]. A piezoelectric-actuated inkjet printing was used to characterize the effect of amplitude and time pulse on cell stresses. Results indicate that shear stresses elicited during the printing process did not significantly affect cell survival rate. As the driving voltage was increased for droplet ejection, droplet velocity and the forces generated during drop generation and impact increased.

For thermal inkjet printing, a modified HP Desktop printer (HP 550C) and a modified HP 51626a ink cartridge were used to print the CHO cells and motoneurons directly into specific patterns [14]. Moreover, by using the modified commercial HP Deskjet 550 thermal inkjet printer,

the microscopic and macroscopic contractile cardiomyocyte constructs were built *in vitro*[45]. Results suggest that thermal inkjet bio-printers are suitable constructions of functional cardiac pseudo-tissues [45].

An alternative system has recently been developed to overcome limitations in inkjet printing by Demirci's group [46]. This system utilizes mechanical valves that enable higher viscous materials to be printed. They have demonstrated that this new printing system is able to create hydrogel structures characterized by (i) printing of multilayered 3D cell-laden hydrogel structures (16.2 μm thick per layer) with controlled spatial resolution (proximal axis: 18.0 ± 7.0 μm and distal axis: 0.5 ± 4.9 μm), (ii) high-throughput droplet generation (1 s per layer, 160 droplets = s), (iii) cell seeding uniformity (26 ± 2 cells/mm^2 at 1 million cells/mL, 122 ± 20 cells/mm^2 at 5 million cells/mL, and 216 ± 38 cells/mm^2 at 10 million cells/mL), and (iv) long-term viability in culture ($>90\%$, 14 days).

6.5.3 Laser Printing

Laser printing has gained a great interest since it was first introduced into the biomedical engineering field because of its major benefits, such as high resolution and throughput [23,47−49]. The principle of laser bioprinting is shown in Figure 6.6. Laser-based biological deposition is based on ejection of materials via laser powered beam. Different materials layers are disposed of in such manner that the first layer absorbs a great part of the beam energy but just letting enough of the energy to push adjacent layers (biological materials of interest) towards the desired substrate in a precise location. The laser beam is extremely small which provides this method with a great precision character. Moreover, this laser can be computer-controlled to create complex patterns. The nozzle free nature of the method enables the usage of high viscosity materials, an advantage that makes it worth being investigated, since inkjet and extrusion systems have limited viscosity. Creation of stable gels will certainly increase the chances of constructing 3D organ-like structures in a layer-by-layer basis.

Laser-guided direct writing (LGDW) is an example of laser printing, which is able to deposit cells with micrometer accuracy on arbitrary matrices, including soft gels such as collagen or Matrigel™ [49−52]. The unique ability of LGDW to micropatterned cells on the basement membrane Matrigel™ allows researchers to use the intrinsic ability of endothelial cells to self-assemble into vascular structures [53,54] for the assembly of liver tissue. To avoid optical damage to cells, LGDW uses the near-infrared part of the electromagnetic spectrum (700−1000 nm), which is past the absorption spectra of most proteins and before the infrared absorption of water.

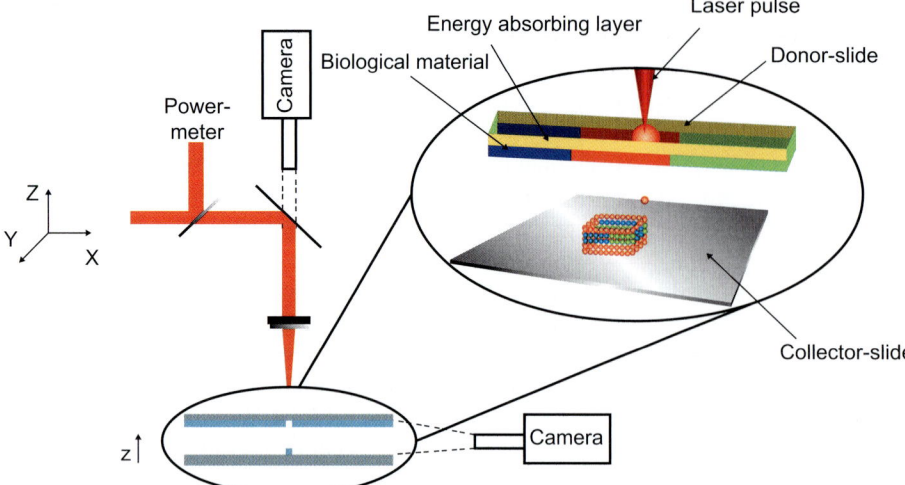

FIGURE 6.6   A schematic diagram of laser printing.

Photons in this range of the spectrum lack the energy to create free radicals (as in ultraviolet) and are not absorbed by DNA, suggesting that the laser has little chance of creating mutations or causing cellular death although precise wavelength optimization remains an open issue [55,56]. LGDW has been used to pattern living embryonic chick neurons [51,52], multipotent adult progenitor cells [50], endothelial cells [49], bacteria and other particles [52,57,58] with no apparent deleterious effect on cellular viability and function. The LGDW variant, optical trapping, has been commercially available for the past decade and has been used extensively in the field.

Laser-induced forward transfer (LIFT) is another example of laser printing, which has been investigated to transfer small amounts of various materials (including living cells) in 2D or 3D patterns. LIFT has previously been proven for printing single to tens of cells simultaneously without any observable damage to pheno- and genotype [59]. This laser jet printing has successfully been used to generate cellular patterns either by printing cell-adhesive molecules for subsequent cell seeding or by directly depositing cells onto substrates with predesigned patterns. Koch et al. have used this approach to transfer living cells in a 2D structure [60]. Further, Barron et al. [61] and Othon et al. [62] have transferred single cells also in a 3D structure by using layer-by-layer technique. The cells survived the transfer procedure with a rate of $98 \pm 1\%$ (skin cells) and $90 \pm 10\%$ (human musculoskeletal cells), respectively. These studies demonstrate LIFT as a suitable technique for unharmed computer-controlled positioning of different cell types and a promising tool for future applications in the *ex vivo* generation of tissue replacements [60].

As stated previously, nozzle-based cell printing has the limitation of printer head clogging issues due to cell concentration of the bio-ink superior to 10^7 cells/mL. On the contrary, recent studies have demonstrated that the laser-based printing, in particular laser-assisted bio-printer (LAB) system [23] can print a high cell fraction per droplet volume (i.e., cells with little surrounding bio-ink volume). The LAB has proven to be able to print cells one by one from a high cell concentration bio-ink (10^8 cells/mL) and viscosity (above 100 mPa.s) to fabricate a tissue-engineered product with comparable organization and cell density with living tissues, in which cells are in physical contact with each other [23]. Further development may focus on the implementation of a cell recognition scanning technology onto the ribbon prior to printing, so that the laser beam could exactly aim one single cell per pulse. Moreover, previous studies have shown that LAB is allowed to print undamaged mammalian cells in terms of cell viability and DNA damage and function [62–66]. However, current investigations are in progress to confirm the functional integrity of printed cell suspensions [23]. More information regarding the LAB system can be found in the previous report [67].

6.6 SUMMARY

Organ printing is the natural evolution of tissue engineering, manufacturing, biology, and medicine. Recent study in this field shows promising progress towards printing living tissues. Even though organ printing is not achievable today, estimations give a 15- to 20-year window before initiation of biomedical companies producing organs or organ parts ready for implantation. For wider *in vivo* applications of bioprinting, it is necessary to develop new materials and methods to fabricate 3D layer-by-layer printed bio-paper that allow manipulation and implantation. The next step also includes the improvement of bio-ink formulation for the printed cells with minimal damage during printing as well as maintenance of their specific genotype and phenotype.

Computer-aided layer-by-layer assembly of biological tissues and organs is currently feasible, fast-evolving and predicted to be a major technology in tissue engineering. Organ printing uses the principle of cellular self-assembly into tissues, similar to the way in which the embryonic-like tissues sort and fuse into functional forms dictated by the rules of the developmental biology. At present, several groups are actively working on the improvement of organ inkjet printing technology. Inkjet bioprinting technology can benefit enormously due to being affordable. Moreover, from the long-term point of view, the existence of profitable and well-established inkjet printer technology with global companies, such as HP, Lexmark, Canon, and Xerox, represents an important advantage especially when the technology becomes mature enough for commercialization. Besides the application of organ transplantation, bioprinting of 3D perfused and vascularized human tissues (or structural–functional units of human organs) can become a popular screening assay for drug discovery and other biomedical research. It is safe to predict that in the 21st century specific cell and organ printers will be as widely used as biomedical research tools as the electron microscope was in the 20th century.

ACKNOWLEDGMENTS

The authors would like to thank NIH-National Heart, Lung, and Blood Institute (NHLBI) (grant No. 1SC2HL107235-01), National Science Foundation (grant No. CBET0936238), and Department of Education (grant No. P116V090013) for funding this project. Dr Lei Zhu and Dr Kun Wang want to thank the Science and Technology Program of Guangdong Province of China (grant No. 2009B06070045) for financial support. This work is also supported by Guangdong Innovative Research Team Program (No. 2011S055).

REFERENCES

[1] Murray JE. The 50th anniversary of the first successful human organ transplant. Rev Invest Clin 2005;57:118–9.

[2] Mironov V. Toward human organ printing: Charleston bioprinting symposium. ASAIO J 2006;52:27–30.

[3] Mironov V, Boland T, Trusk T, Forgacs G, Markwald RR. Organ printing: computer-aided jet-based 3D tissue engineering. Trends Biotechnol 2003;21:157–61.

[4] Chang R, Nam J, Sun W. Direct cell writing of 3D microorgan for in vitro pharmacokinetic model. Tissue Eng C Methods 2008;14:157–66.

[5] Bates RHT, Garden KL, Peters TM. Overview of computerized tomography with emphasis on future-developments. Proc IEEE 1983;71:356–72.

[6] Sun W. Multi-volume CAD modeling for heterogeneous object design and fabrication. J Comput Sci Technol 2000;15:27–36.

[7] Sun W, Hu X. Reasoning Boolean operation based modeling for heterogeneous objects. Comput Aided Des 2002;34:481–8.

[8] Sun W, Darling A, Starly B, Nam J. Computer-aided tissue engineering: overview, scope and challenges. Biotechnol Appl Biochem 2004;39:29–47.

[9] Boland T, Cui X, Chaubey A, Burg TC, Groff RE, Burg KJL. Precision printing of cells and biomaterials onto 3D matrices. Proceedings of the ASME international conference on manufacturing science and engineering; 2007. p. 77–81.

[10] Boland T, Tao X, Damon BJ, Manley B, Kesari P, Jalota S, et al. Drop-on-demand printing of cells and materials for designer tissue constructs. Mater Sci Eng C Biomim Supramol Syst 2007;27:372–6.

[11] Cui XF, Boland T. Simultaneous deposition of human microvascular endothelial cells and biomaterials for human microvasculature fabrication using inkjet printing. Nip24/digital fabrication 2008: 24th international conference on digital printing technologies, technical program and proceedings; 2008. p. 480–83.

[12] Kesari P, Xu T, Boland T. Layer-by-layer printing of cells and its application to tissue engineering. Nanoscale Mater Sci Biol Med 2005;845:111–7.

[13] Xu T, Gregory CA, Molnar P, Cui X, Jalota S, Bhaduri SB, et al. Viability and electrophysiology of neural cell structures generated by the inkjet printing method. Biomaterials 2006;27:3580–8.

[14] Xu T, Jin J, Gregory C, Hickman JJ, Boland T. Inkjet printing of viable mammalian cells. Biomaterials 2005;26:93–9.

[15] Narayan R, Boland T, Lee YS. Printed biomaterials: novel processing and modeling techniques for medicine and surgery. Biological and medical physics, biomedical engineering. London; New York: Springer; 2010.

[16] Ker EDP, Nain AS, Weiss LE, Wang J, Suhan J, Amon CH, et al. Bioprinting of growth factors onto aligned sub-micron fibrous scaffolds for simultaneous control of cell differentiation and alignment. Biomaterials 2011;32:8097–107.

[17] Cooper GM, Miller ED, DeCesare GE, Usas A, Lensie EL, Bykowski MR, et al. Inkjet-based biopatterning of bone morphogenetic protein-2 to spatially control calvarial bone formation. Tissue Eng Part 2010;16:1749–59.

[18] Ma NN, Chalmers JJ, Aunins JG, Zhou WC, Xie LZ. Quantitative studies of cell-bubble interactions and cell damage at different pluronic F-68 and cell concentrations. Biotechnol Progr 2004;20:1183–91.

[19] Parsa S, Gupta M, Loizeau F, Cheung KC. Effects of surfactant and gentle agitation on inkjet dispensing of living cells. Biofabrication 2010;2:025003.

[20] Holtsberg FW, Ensor CM, Steiner MR, Bomalaski JS, Clark MA. Poly(ethylene glycol) (PEG) conjugated arginine deiminase: effects of PEG formulations on its pharmacological properties. J Control Release 2002;80:259–71.

[21] Bomalaski JS, Holtsberg FW, Ensor CM, Clark MA. Uricase formulated with polyethylene glycol (uricase-PEG 20): biochemical rationale and preclinical studies. J Rheumatol 2002;29:1942–9.

[22] Engler AJ, Sen S, Sweeney HL, Discher DE. Matrix elasticity directs stem cell lineage specification. Cell 2006;126:677–89.

[23] Guillotin B, Souquet A, Catros S, Duocastella M, Pippenger B, Bellance S, et al. Laser assisted bioprinting of engineered tissue with high cell density and microscale organization. Biomaterials 2010;31:7250–6.

[24] Klebe RJ. Cytoscribing: a method for micropositioning cells and the construction of two- and three-dimensional synthetic tissues. Exp Cell Res 1988;179:362–73.

[25] Ahmed TA, Dare EV, Hincke M. Fibrin: a versatile scaffold for tissue engineering applications. Tissue Eng B Rev 2008;14:199–215.

[26] Cui XF, Boland T. Human microvasculature fabrication using thermal inkjet printing technology. Biomaterials 2009;30:6221–7.

[27] Fedorovich NE, Alblas J, de Wijn JR, Hennink WE, Verbout AJ, Dhert WJA. Hydrogels as extracellular matrices for skeletal tissue engineering: state-of-the-art and novel application in organ printing. Tissue Eng 2007;13:1905–25.

[28] Langer R, Vacanti JP. Tissue engineering. Science 1993;260:920–6.

[29] Roskelley CD, Desprez PY, Bissell MJ. Extracellular matrix-dependent tissue-specific gene expression in mammary epithelial cells requires both physical and biochemical signal transduction. Proc Natl Acad Sci USA 1994;91:12378–82.

[30] Mooney DJ, Sano K, Kaufmann PM, Majahod K, Schloo B, Vacanti JP, et al. Long-term engraftment of hepatocytes transplanted on biodegradable polymer sponges. J Biomed Mater Res 1997;37:413–20.

[31] Knight B, Laukaitis C, Akhtar N, Hotchin NA, Edlund M, Horwitz AR. Visualizing muscle cell migration in situ. Curr Biol 2000;10:576–85.

[32] Chang R, Nam J, Sun W. Effects of dispensing pressure and nozzle diameter on cell survival from solid freeform fabrication-based direct cell writing. Tissue Eng A 2008;14:41–8.

[33] Lee W, Pinckney J, Lee V, Lee JH, Fischer K, Polio S, et al. Three-dimensional bioprinting of rat embryonic neural cells. Neuroreport 2009;20:798–803.

[34] Hamid Q, Snyder J, Wang C, Timmer M, Hammer J, Guceri S, et al. Fabrication of three-dimensional scaffolds using precision extrusion deposition with an assisted cooling device. Biofabrication 2011;3:034109.

[35] Yan Y, Wang X, Pan Y, Liu H, Cheng J, Xiong Z, et al. Fabrication of viable tissue-engineered constructs with 3D cell-assembly technique. Biomaterials 2005;26:5864–71.

[36] Zhang T, Yan YN, Wang XH, Xiong Z, Lin F, Wu RD, et al. Three-dimensional gelatin and gelatin/hyaluronan hydrogel structures for traumatic brain injury. J Bioact Compat Polym 2007;22:19–29.

[37] Patz TM, Doraiswamy A, Narayan RJ, He W, Zhong Y, Bellamkonda R, et al. Three-dimensional direct writing of B35 neuronal cells. J Biomed Mater Res B Appl Biomater 2006;78:124–30.

[38] Lee W, Debasitis JC, Lee VK, Lee JH, Fischer K, Edminster K, et al. Multi-layered culture of human skin fibroblasts and keratinocytes through three-dimensional freeform fabrication. Biomaterials 2009;30:1587–95.

[39] Ringeisen BR, Spargo BJ, Wu PK. Cell and organ printing. 1st ed. New York: Springer; 2010.

[40] Hon KKB, Li L, Hutchings IM. Direct writing technology—advances and developments. CIRP Ann Manuf Technol 2008;57:601.

[41] Lim T, Han S, Chung J, Chung JT, Ko S, Grigoropoulos CP. Experimental study on spreading and evaporation of inkjet printed pico-liter droplet on a heated substrate. Int J Heat Mass Transfer 2009;52:431–41.

[42] Saunders R, Derby B, Gough J, Reis N. Ink-jet printing of human cells. Archit Appl Biomater Biomol Mater 2004;1:95–7.

[43] Saunders R, Gough J, Derby B. Ink jet printing of mammalian primary cells for tissue engineering applications. Nanoscale Mater Sci Biol Med 2005;845:57–62.

[44] Saunders RE, Gough JE, Derby B. Delivery of human fibroblast cells by piezoelectric drop-on-demand inkjet printing. Biomaterials 2008;29:193–203.

[45] Xu T, Baicu C, Aho M, Zile M, Boland T. Fabrication and characterization of bio-engineered cardiac pseudo tissues. Biofabrication 2009;1:035001.

[46] Moon S, Hasan SK, Song YS, Xu F, Keles HO, Manzur F, et al. Layer by layer three-dimensional tissue epitaxy by cell-laden hydrogel droplets. Tissue Eng C Methods 2010;16:157–66.

[47] Duocastella M, Fernandez-Pradas JM, Morenza JL, Zafra D, Serra P. Novel laser printing technique for miniaturized biosensors preparation. Sens Actuators B Chem 2010;145:596–600.

[48] Kattamis N, Brown M, Arnold CB. Incident beam shape effects on thick-film laser induced forward transfer. 2010 conference on lasers and electro-optics (CLEO) and quantum electronics and laser science conference (QELS); 2010.

[49] Nahmias Y, Schwartz RE, Verfaillie CM, Odde DJ. Laser-guided direct writing for three-dimensional tissue engineering. Biotechnol Bioeng 2005;92:129–36.

[50] Nahmias YK, Gao BZ, Odde DJ. Dimensionless parameters for the design of optical traps and laser guidance systems. Appl Opt 2004;43:3999–4006.

[51] Odde DJ, Renn MJ. Laser-guided direct writing for applications in biotechnology. Trends Biotechnol 1999;17:385–9.

[52] Odde DJ, Renn MJ. Laser-guided direct writing of living cells. Biotechnol Bioeng 2000;67:312–8.

[53] Kubota Y, Kleinman HK, Martin GR, Lawley TJ. Role of laminin and basement membrane in the morphological differentiation of human endothelial cells into capillary-like structures. J Cell Biol 1988;107:1589–98.

[54] Vernon RB, Angello JC, Iruela-Arispe ML, Lane TF, Sage EH. Reorganization of basement membrane matrices by cellular traction promotes the formation of cellular networks in vitro. Lab Invest 1992;66:536–47.

[55] Neuman KC, Chadd EH, Liou GF, Bergman K, Block SM. Characterization of photodamage to *Escherichia coli* in optical traps. Biophys J 1999;77:2856–63.

[56] Liang H, Vu KT, Krishnan P, Trang TC, Shin D, Kimel S, et al. Wavelength dependence of cell cloning efficiency after optical trapping. Biophys J 1996;70:1529–33.

[57] Renn MJ, Montgomery D, Vdovin O, Anderson DZ, Wieman CE, Cornell EA. Laser-guided atoms in hollow-core optical fibers. Phys Rev Lett 1995;75:3253–6.

[58] Renn MJ, Pastel R. Particle manipulation and surface patterning by laser guidance. J Vac Sci Technol B 1998;16:3859–63.

[59] Gruene M, Deiwick A, Koch L, Schlie S, Unger C, Hofmann N, et al. Laser printing of stem cells for biofabrication of scaffold-free autologous grafts. Tissue Eng C Methods 2010.

[60] Koch L, Kuhn S, Sorg H, Gruene M, Schlie S, Gaebel R, et al. Laser printing of skin cells and human stem cells. Tissue Eng C Methods 2010;16:847−54.

[61] Barron JA, Wu P, Ladouceur HD, Ringeisen BR. Biological laser printing: a novel technique for creating heterogeneous 3-dimensional cell patterns. Biomed Microdevices 2004;6:139−47.

[62] Othon CM, Wu X, Anders JJ, Ringeisen BR. Single-cell printing to form three-dimensional lines of olfactory ensheathing cells. Biomed Mater 2008;3:034101.

[63] Chen CY, Barron JA, Ringeisen BR. Cell patterning without chemical surface modification: cell−cell interactions between printed bovine aortic endothelial cells (BAEC) on a homogeneous cell-adherent hydrogel. Appl Surf Sci 2006;252:8641−5.

[64] Hopp B, Smausz T, Kresz N, Barna N, Bor Z, Kolozsvari L, et al. Survival and proliferative ability of various living cell types after laser-induced forward transfer. Tissue Eng 2005;11:1817−23.

[65] Barron JA, Ringeisen BR, Kim HS, Spargo BJ, Chrisey DB. Application of laser printing to mammalian cells. Thin Solid Films 2004;453:383−7.

[66] Ringeisen BR, Kim H, Barron JA, Krizman DB, Chrisey DB, Jackman S, et al. Laser printing of pluripotent embryonal carcinoma cells. Tissue Eng 2004;10:483−91.

[67] Guillemot F, Souquet A, Catros S, Guillotin B, Lopez J, Faucon M, et al. High-throughput laser printing of cells and biomaterials for tissue engineering. Acta Biomater 2010;6:2494−500.

Chapter 7

Synthetic Biomaterials for Regenerative Medicine Applications

Satyavrata Samavedi[a], Lauren K. Poindexter[b], Mark Van Dyke[c], and Aaron S. Goldstein[a,c]

[a]Department of Chemical Engineering, Virginia Polytechnic Institute and State University, Blacksburg, VA, [b]Virginia Tech Carillion School of Medicine and Research Institute, Roanoke, VA, [c]School of Biomedical Engineering and Sciences, Virginia Polytechnic Institute and State University, Blacksburg, VA

Chapter Outline

7.1 INTRODUCTION

Synthetic materials have been used for the reconstruction and regeneration of a variety of tissues and organs for the last 100 years. Beginning with inert implantable materials that primarily provided replacement of mechanical function, the field of regenerative medicine (RM) has become a field unto itself with essential ties to engineering, materials, and biology. RM has seen the use of a wide spectrum of materials ranging from metals and ceramics to polymers whose properties can be appropriately designed for specific applications. Materials in RM applications may be used to augment existing tissue (e.g., engraftment of stem cells), promote the growth of neotissue (e.g., engineered blood vessels), or engineer a tissue equivalent for full organ replacement. Thus, a regenerative strategy mandates the use of degradable materials that serve as temporary implants and subsequently degrade once tissue is regenerated or engrafted *in vivo*. In contrast, relatively long-lasting implants may be used in cases where the implants are expected to replace the function of damaged tissue. While RM in its strictest sense encompasses the former strategy—leveraging the human body's intrinsic ability to regenerate damaged tissue—nondegradable materials for tissue replacement will also be discussed in this chapter.

Primary considerations for the use of synthetic biomaterials are biocompatibility, degradation characteristics, mechanical, chemical, and increasingly, biological properties. While the first generation of biomaterials was designed to be inert and primarily meant for replacing tissue mechanical function, the second and third generations have incorporated degradability and bioactivity (e.g., cell-responsive cues), respectively. The third generation of materials in particular has utilized polymeric vehicles for the delivery of therapeutics. From the perspective of long-term implants, materials such as metals and ceramics have shown a lot of promise for hard tissue replacement due to their mechanical properties. In contrast, polymers offer much more flexibility in terms of chemical properties (e.g., degradation rates, surface chemistry, glass transition temperature (T_g)) that make them attractive for soft tissue repair and RM. Thus, material properties drive their application and these properties can be suitably designed to elicit a specific host response.

This chapter discusses important classes of synthetic biomaterials used in RM applications. Section 7.2 describes biologically relevant properties of synthetic biomaterials, such as biocompatibility, degradation, and mechanical properties, that influence their use in specific applications. This section also provides a brief overview of design criteria for implants and scaffolds based on synthetic biomaterials. Section 7.3 briefly discusses classes of nondegradable materials that have traditionally been used for replacement of tissue and organs. While Section 7.4 forms the bulk of this chapter and addresses a whole array of degradable polymers that have been used for in RM applications, Section 7.5 describes hydrogels. Both Sections 7.4 and 7.5 include a description of key material properties, their merits and limitations, and a summary of applications in which materials have been employed. (Here, discussions are limited to clinical trials and animal models.) Finally, Section 7.6 provides insights on current and future directions for the use of synthetic biomaterials in RM applications.

7.2 BIOLOGICALLY RELEVANT PROPERTIES OF POLYMERIC BIOMATERIALS

Materials used in RM applications possess certain essential properties, such as biocompatibility and appropriate mechanical/chemical properties [1]. In cases where biomaterials are used as temporary implants or as delivery systems, the materials are degradable and the products of degradation are nontoxic. In addition to these generic properties, materials used in specialized applications possess other specific properties. For example, in the case of ocular applications, implants possess suitable optical properties [2]. In several other applications such as wound healing or controlling cell fate *in situ*, biologically inert biomaterials can be functionalized via chemical treatments, coatings, and surface grafting [1]. Herein, some significant properties of these materials relevant to RM applications are discussed.

7.2.1 Biocompatibility

The definition of the term "biocompatibility" has undergone several changes over the years [3]. In its most general sense, biocompatibility may be defined as the ability of a material, device, or system to perform without a clinically significant host response in a specific application [4]. While this definition extends the concept of biocompatibility to mean more than mere biological safety to the patient, it is still broad and general. For example, this generality does not account for interactions of the biomaterial with the body or vice versa in applications where the biomaterial is intended to elicit a specific biological response as in the case of wound healing. To overcome such ambiguities, the use of the phrase "biological performance" to describe the concept of biocompatibility has been suggested [5]. Biological performance is in turn divided into two subcategories, namely host response and material response. While host response refers to the local and systemic response of the biological system to the implanted material, material response refers to the effect of the material on the biological system. In light of these ideas, biocompatibility may be redefined as the ability of a material, device, or system to perform appropriately in an intended application and elicit a host response suitable to that application. This umbrella definition of biocompatibility includes physical/chemical properties of the biomaterial, its bioactivity, degradability, and products of degradation, all of which are critical in determining the appropriateness of material performance as well as the host response.

7.2.2 Mechanical Properties

Typically, the design of biomaterials is inspired by the nature of the tissue to be regenerated. The biomaterial implant is expected to match the mechanical properties of target tissue and confer a certain degree of functionality until complete regeneration occurs [6]. For example, biomaterials used in the regeneration of ligaments are viscoelastic and thus promote joint stability, allowing for joint flexibility. In RM applications such as the regeneration of load-bearing stiff tissues, mechanical properties such as stiffness and compressive moduli of the biomaterials play a major role in influencing the eventual success of the implant [7]. In other applications like engineering blood vessels, the biomaterial is robust enough to flex under cyclic hydrostatic pressure as well as hydrodynamic shear stress. In the regeneration of cartilage tissue, hydrogels are typically used because they mimic the mechanical properties of cartilage and allow for effective load transfer [8]. In the reconstruction of complex organs such as the bladder, the mechanical properties of biomaterials must be carefully chosen, keeping in mind the various mechanical deformations and forces the organ undergoes under normal physiological conditions [9]. In addition, the mechanical properties of the biomaterial will change as the material degrades and new tissue replaces it; moreover, mechanical demands on the biomaterial may change on a short-term scale (i.e., hours to days) as physiologic loading occurs. All of these considerations must be kept in mind when selecting materials for a particular application.

7.2.3 Degradability

In cases where biomaterials are used as replacements (such as in total hip replacement), degradability can be detrimental to the functionality of the biomaterial. In contrast, degradability is a desired property in most RM applications where tissue regeneration is the ultimate goal [6]. Degradation rate is typically chosen to match with the regeneration rate of neotissue [10] (Figure 7.1). In this context, degradable polymers have great value in RM because their degradation rates can be tuned to specific applications. The degradation rate of the biomaterial depends on the backbone chemistry of the polymer as well as the side groups attached to the backbone. Properties of biomaterials, such as hydrophobicity, molecular weight, and crystallinity, strongly affect degradation (Figure 7.2). The degradability of polymeric biomaterials can be exploited to load and deliver therapeutics, such as drugs, vaccines, and proteins, to host tissue [6]. In other words, the polymeric biomaterial can aid the controlled release of therapeutics as it degrades. While degradable polymers are very useful in RM applications, a vital requirement is that the products of degradation be nontoxic and nonimmunogenic [11]. Products should be readily cleared by the excretory system and not accumulate in the liver, kidney, and spleen.

7.2.4 Bioactivity

Biochemical properties of materials play a crucial role in guiding tissue regeneration *in vivo*. Here, two distinct strategies have been employed. In a cell-free approach, biomaterial scaffolds are used to enhance the intrinsic ability of the tissue for regeneration via release of a biochemical agent that directs homing and recruitment of progenitor cells. In a tissue engineering approach, biomaterial scaffolds are seeded with cells, scaffolds are subsequently conditioned in a bioreactor before implantation into a patient [12]. In both the approaches, the biomaterials used for the fabrication of scaffolds can be appropriately "bioactivated" [13]. For example, biomaterials used for wound healing applications can be designed to incorporate chemoattractants to promote cell migration and subsequent healing of injured tissue. Certain other applications require nonfouling surfaces, for which biomaterials can be coated with poly(ethylene glycol) (PEG) [14]. In applications like bone tissue implants, inert titanium (Ti) plates are often coated with a layer of calcium phosphate (CaP) mineral in order to promote osseointegration with host tissue [15]. Thus, biomaterials can be suitably

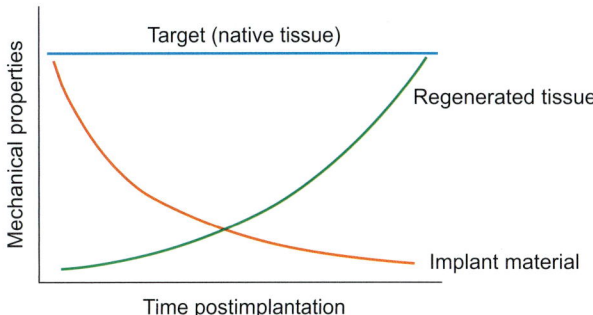

FIGURE 7.1 Immediately following implantation, the mechanical properties of the implant material match the properties of the target tissue. As the material degrades with time, the mechanical properties of the regenerated tissue slowly increase due to tissue maturation within the implant. This process leaves behind biologically functional neotissue.

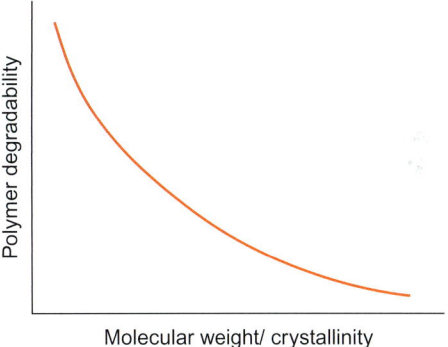

FIGURE 7.2 The degradability of polymeric materials decreases with an increase in molecular weight and crystallinity. Polymers possessing high molecular weights and crystallinity degrade slowly *in vivo*.

designed to promote specific cell responses in various applications.

7.3 CLASSIC MATERIALS

7.3.1 Metals

Metals have been traditionally used in surgical procedures since the seventeenth century. Due to their superior mechanical properties as compared to other biomaterials, metals have been extensively used in load-bearing applications, such as the reconstruction of knees and hips. Conventional metals used for these applications include stainless steel (SS), cobalt–chromium (Co–Cr)-based alloys and Ti (and its alloys) [16]. The properties of these metals are briefly discussed in the following paragraphs.

7.3.1.1 Stainless Steel

SS materials are cost-effective and exhibit mechanical strength sufficient for orthopedic applications [17]. They possess moduli close to 200 GPa and ultimate tensile strengths in excess of 500 MPa depending on the processing method. They are also resistant to corrosion due to the presence of a small amount of molybdenum and a high percentage of chromium. Other elements, when present in trace amounts, confer various types of functionalities. For example, low phosphorous content is usually tied to improved ductility, while carbon—usually restricted to very low amounts—improves corrosion resistance [17]. The presence of nickel in SS materials can be a cause for concern because nickel is a known allergen and its release from SS implants can cause severe allergies in patients with a known history of nickel allergy [17,18]. Therefore, nickel-free SS materials with the inclusion of nitrogen have been developed [19]. These materials exhibit the excellent mechanical and corrosion resistance properties of SS in addition to being nickel-free. SS materials suffer from poor wear resistance, leading to severe wear and debris formation especially in prostheses where excessive contact between two metallic implants occurs [20]. As a result, SS is usually used in devices such as fracture plates and hip nails, and much less in applications like joint prostheses.

7.3.1.2 Co−Cr-Based Alloys

Co−Cr-based alloy implants help overcome the limitations encountered with SS implants. In addition to exhibiting corrosion resistance like SS materials, these alloys also resist wear better than SS implants [16]. Co−Cr-based alloys also possess a high elastic modulus similar to that of SS and have higher fatigue strength (in excess of 600 MPa) in air compared to SS (<400 MPa) [21]. They also possess higher resistance to corrosion even in alkaline environments. As with SS materials, the presence of nickel is a concern due to its potential as an allergen. Therefore, Ni-free and low-C Co−Cr-based alloys have been more recently investigated [18]. Due to their high ultimate tensile strength and fatigue strength, these alloys may cause stress shielding, a phenomenon by which the absence of mechanical loading on bone results in a local resorption of the tissue [22]. Moreover, ions from these alloys can leach out of the implant and accumulate in tissue over a period of time [23]. Nevertheless, Co−Cr-based alloys have been used in joint prosthesis, hip prosthesis, and in artificial disc replacement systems.

7.3.1.3 Titanium

Ti-based implants have been explored for orthopedic applications primarily due to their moderately high elastic modulus of around 110 GPa and low density compared to the metals previously described [16]. As compared to SS-based and Co−Cr-based materials, Ti-based alloys exhibit lower cyclic fatigue [21]. Ti-based implants are also

favored because of their superior corrosion-resistant properties compared to SS-based and Co−Cr-based materials and a relatively lower amount of ions released from the implant [24]. Ti derives its corrosion resistance from the ability to form an inert oxide layer on its surface, which allows for passivation. Ti-based implants can readily ossointegrate, allowing them to function better in the long term than traditional SS implants. Ti alloys like Ti−6Al−4V, which incorporate varying amounts of other metals, such as aluminum (Al) and vanadium (V), have been manufactured to help improve the properties of the base metal. However, V and Al ions released from the alloys have been found to be cytotoxic and detrimental to the long-term health of patients [25]. Moreover, Ti suffers from poor wear resistance and is not easily amenable to processing. To overcome these limitations, alloys such as Ti−6Al−7Nb and Ti−5Al−2.5Fe, which include niobium (Nb) and iron (Fe) instead of V, have been synthesized [18]. Ti alloys have also been fabricated with Young's moduli closer to that of bone [25]. Ti−Ni-based alloys (nitinol) have been studied for application in orthopedics due to their shape-memory properties (superelasticity) [26]. Despite possessing attractive properties, the use of nitinol has been restricted due to the allergy and toxicity associated with Ni. Therefore, Ni-free alternatives such as Ti−Nb−Al and Ti−Mo−Al have also been explored. Several studies have investigated the role of surface properties, such as roughness, charge, and ceramic/biochemical coating, toward improving their osseointegration with host tissue [16].

7.3.2 Ceramics

Ceramics are biocompatible materials typically used in orthopedics and dentistry, particularly as components in joint prostheses and bone fillers. Due to their popular use, they have evolved considerably over the years: first-generation materials were designed to replace bone, second-generation materials were intended to repair bone, and third-generation materials aim to regenerate bone [27]. Common ceramic materials include CaPs as well as silica-based (e.g., Bioglass 45S5®) and alumina-based compounds.

7.3.2.1 Calcium Phosphate

CaP materials are biocompatible, osteoconductive, and can be engineered to possess a wide variety of physical and chemical properties [28,29]. Specific formulations of synthetic CaP include hydroxyapatite (HAP), β-tricalcium phosphate (TCP), and biphasic calcium phosphate (varying ratios of HAP and TCP). Some forms of CaP readily osseointegrate (i.e., achieve intimate biological fixation) with natural bone due to their chemical and physical similarities with carbonated HAP crystals. However, CaP biomaterials are brittle, and therefore perform best *in vivo* in a limited array of forms or in conjunction with stronger materials such as a nanoscale coating on Ti rods, screws, and plates, among others [27,30,31]. Commercially available FDA-cleared CaP products range from moldable putties to presized wedges. Moreover, injectable CaP cements are widely employed in surgery because of their high compressive strength and ease of defect filling, despite inconclusive documentation of their slow resorption rates *in vivo*[29,32].

7.3.2.2 Bioglass

Melt-derived bioactive silicate glass (Bioglass 45S5®) is an inorganic, modified silica network comprised of 45% SiO_2, 24.5% Na_2O, 24.5% CaO, and 6% P_2O_5[33]. Due to its ability to bond chemically with surrounding native tissue, this material promotes excellent repair of bony defects [27,33] and does not stimulate an immune-derived fibrous capsule around the implant. However, histological evaluation of long-term indwelling Bioglass® prostheses reveals partial degradation, fragmentation, and connective tissue invasion of the implant and implant site [34]. An alternative sol–gel process for glass formation results in a bioactive glass that, unlike melt-derived Bioglass 45S5®, is able to degrade *in vivo*[35,36]. In particular, a $70SiO_2/30CaO$ (mol%) composition with or without additional compounds like P_2O_5 is now a popular material for third-generation tissue regeneration applications [33]. Recently, the sol–gel process has been applied to Bioglass 45S5®[37,38]. To date, sol–gel silica glasses have been manufactured in a large spectrum of sizes and shapes (scaffolds, foams, powders, nanoparticles, etc.) for a vast array of orthopedic and dental applications. While the bioglass foam structure approximates that of cancellous bone, few studies demonstrate adequate strength (relative to cancellous bone) of bioglass scaffolds pre- and postimplantation for tissue engineering [39].

7.3.2.3 Alumina

Alumina (Al_2O_3) has been widely used in the manufacture of artificial prostheses for total hip arthroplasty (THA) since the 1970s. Although this inert oxide ceramic component (first generation) performed marginally well initially, improvements in manufacturing, design, and surgical technique have significantly improved its success rate as a hip prosthesis [40]. Fine-grain alumina is a desirable material for orthopedic applications because a nearly perfect spherical alumina surface, well-lubricated with joint fluid, exhibits minimal friction and dramatically reduced metallic wear, as compared to other joint prosthesis materials [41]. One example of an FDA-approved alumina-on-alumina model (Al–Al) is the Stryker® Trident® Ceramic System. This prosthesis is composed of a Ti acetabular component and an alumina insert that articulates with an alumina ball. Two separate follow-up studies on Al–Al THAs show mid-term (5–9 years) prosthetic survival rates above 90% for patients [42,43]. Despite its wide use in orthopedic applications, alumina-based biomaterials can suffer from squeaking, component wear from edge-loading, chipping and fracture [40].

7.3.3 Nondegradable Polymers

7.3.3.1 Polymethylmethacrylate

Polymethylmethacrylate (PMMA) is a strong but lightweight polymer possessing a compressive strength between 85 and 110 MPa and a tensile strength between 30 and 50 MPa [44]. PMMA possesses a relatively high coefficient of thermal expansion, and during polymerization *in situ*, temperatures can reach values as high as 40 and 56°C [45,46]. Consequently, the curing process *in vivo* can result in a shrinkage of around 6–7% [45] and cause tissue necrosis. The mechanical properties of PMMA can be tuned by varying the mixing ratios of monomer and initiator during polymerization. PMMA can absorb water over several weeks at body temperature [45] and properties such as tensile strength and fatigue strength, decrease upon water absorption. Despite this seemingly poor stability, PMMA degrades very little in aqueous environments. In addition to strength and stability in aqueous environments, PMMA possesses excellent optical properties [47].

Due to its biocompatible nature and tunable mechanical and optical properties, PMMA has been widely used in bone cements, as intraocular lenses and as screws in bone fixation [48]. Although PMMA's use as bone cement has certain disadvantages, such as causing thermally induced necrosis, chemically induced necrosis due to the release of unreacted monomer, shrinkage after setting and inflammation due to particles released from the cement, bone cement continues to enjoy popularity in orthopedic surgery and is used under six commercial formulations in the United States [48]. Concurrently, bone cements have also been used extensively in percutaneous vertebroplasty [49]. Since PMMA does not play a role in inducing new bone formation, recent work has investigated the addition of small quantities of bioactive glass, ceramics, and osteogenic growth factors to PMMA resulting in the formation of "bioactive" bone cements [49]. Apart from use as bone cement, PMMA has also found widespread application as hard contact lenses due to its excellent optical properties [2]. However, PMMA's use as long-term lenses is limited due to its low oxygen permeability. This limitation can be overcome by copolymerizing methylmethacrylate with methacrylate-functionalized siloxanes. PMMA has also been used in intraocular implants and as intrastromal corneal rings due to its superior mechanical and optical properties and biocompatibility [50,51].

7.3.3.2 Poly(ethylene terephthalate)

Poly(ethylene terephthalate) (PET) is a thermoplastic semicrystalline polymer, first commercialized under the trade name Dacron [52]. Its T_g is generally accepted to be 70°C, while its melting point (T_m) is ~ 250°C. Although it is lightweight, PET possesses a yield strength of around 40 MPa [53] and a tensile strength of around 170 MPa. In addition, it is extremely resistant to wear and has a high flexural modulus. PET is quite impermeable to most gases and liquids, and this property can be improved by either increasing its crystallinity or by copolymerizing it with other monomers [54]. Once molded and set, PET does not absorb water easily due to its crystalline nature, and is also resistant to dissolution by common solvents.

PET is biocompatible and is FDA approved for the repair of large blood vessels and other soft tissues. It has been widely used in procedures, such as axillofemoral bypass [55], rhinoplasty [56], and ligament reconstruction [57], particularly in the form of woven/knitted structures. Although Dacron has been used successfully in reconstruction surgery for a number of decades, it has several limitations. For example, it is known to elongate, creep, and undergo permanent deformation under sustained loads; this behavior can be detrimental to reconstructed blood vessels or ligaments where the shape of the device is critical for its proper function. Arterial procedures based on Dacron have been found to result in graft dilation [58] and multiple aneurysms, while carotid patching procedures have been reported in a small number of instances to result in pseudoaneurysms and infection [59].

7.3.3.3 Poly(tetrafluoroethylene) (PTFE)

Poly(tetrafluoroethylene) (PTFE), more commonly known by its trade name Teflon, is yet another thermoplastic crystalline polymer that is highly resistant to thermal treatment [52]. It possesses a T_g value between -110 and -97°C, while its T_m value is 330°C. Teflon is highly hydrophobic and is also resistant to attack by most commonly used chemicals and solvents. Its inert properties stem from the strong carbon–fluoride bonds and the absence of virtually any branching. Teflon is generally thought to possess a Young's modulus of around 500 MPa, while its tensile strength is between 14 and 23 MPa. Although strong and highly resistant to wear, Teflon can also undergo creep, which can pose a problem when used in medical applications.

Similar to PET, Teflon has been used in a number of reconstructive procedures due to its biocompatibility, inertness, and mechanical properties. Specifically, woven/knitted Teflon-based materials have been used for small artery replacement [60] as synthetic ligament prostheses (under the name Gore-tex) [61] and as prosthetic patches for congenital diaphragmatic hernia repair [62]. Teflon, in its injectable form, has also been used for vocal-fold augmentation [63] and in the treatment of vesicoureteric reflux in young adults and children [64]. However, the performance of Teflon has been shown to deteriorate over time when used in the reconstruction of ligaments [61]. In addition, in certain cases, it has been reported to migrate away from site of injection [65] and has been reported to result in granulomas and subglottic overfilling when used in vocal-fold augmentation [63].

7.3.3.4 Silicones

Silicones, or polysiloxanes, are a class of polymers that exhibit a wide variety of properties that range from being liquids to gels and elastomers. The specific properties of silicones depend upon the side chains and the extent of cross-linking. Liquid silicones are biocompatible and nondegradable [66]. They possess very low surface tension values and have the ability to repel water. In fact, they are considered to possess the highest surface tension values for biocompatible elastomers. Silicones have low thermal conductivity and are almost nonreactive chemically. They are very stable and resistant to degradation by heat or ultraviolet radiation. Due to a large free volume, silicones demonstrate high permeability to gases, such as oxygen and nitrogen. They are considered nondegradable and are nontoxic. Silicones are also biocompatible and do not support the growth of microbes. Given their broad properties, silicones have been used for a wide variety of applications, including intraocular lenses, bandages, scar treatment sheets, and breast implants (liquid silicone).

7.3.3.5 Polyurethanes

Polyurethanes (PUs) are a class of elastomers that are typically produced by the reaction between a diol (soft segment) diisocyanate and a chain extender (hard segments) [67]. The wide variety of choices available for each of these monomers allows for the fabrication of PUs possessing a range of mechanical and chemical properties. PUs are usually very flexible and elastic, displaying minimal hysteresis over repeated cycles of cyclic strain. For example, an ether-based PU has been shown to possess Young's modulus between 4.7 and 7.4 MPa (depending on the humidity and temperature), and to exhibit rate-dependent viscoelastic behavior [68]. Also, PUs possess better tear and abrasion resistance than silicones.

Although classic PUs synthesized from aromatic diisocyanates are considered nondegradable, materials synthesized from degradable monomers, such as polylactide and polycaprolactone (PCL), can undergo hydrolysis. Moreover, the use of nonaromatic diisocyanates, such as lysine ethyl ester diisocyanate and 1,4 diisocyanatobutane, has resulted in the fabrication of degradable PUs that can be used for regeneration of tissues [69,70]. Finally, the use of nonaromatic diisocyanates mitigates toxicity issues typically encountered with the degradation of aromatic diisocyanate-based PUs. Due to their elastomeric nature and blood compatibility, PUs have been used for the reconstruction of blood vessels (vascular grafts), as breast implants, as wound healing materials, and in facial reconstruction [71,72].

7.4 DEGRADABLE POLYMERS

Degradable polymers comprise several different families possessing a wide variety of properties that eventually determine their specific application. The following section discusses clinically important properties of some of these polymers and examines their applications in RM. Table 7.1 provides a list of these polymers, along with their applications.

7.4.1 Polyesters

Polyesters are a class of polymers that have been used widely in RM applications due to their biocompatibility and biodegradability. These polymers derive their biodegradability from an ability of the ester bonds to undergo hydrolysis in aqueous environments. Among the various types of polyesters, poly(lactic acid) (PLA), poly(glycolic acid) (PGA), and PCL have been widely used. The properties and potential applications of these polymers are discussed in detail below.

7.4.1.1 Poly(lactic) Acid: Properties

PLA exists in several isoforms given the inherent chirality of the pendant methyl groups. The L-isoform (PLLA) and the racemic DL-isoform (PDLLA) in particular have been routinely used in RM applications. PLLA is a semicrystalline polymer with a T_g value of 60−65°C, while PDLLA is an amorphous polymer possessing a T_g value of 55−60°C. Enantiomerically pure PLA possesses a T_m value of around 180°C [100]. In general, properties such as crystallinity and T_m depend on the molecular weight of the polymer as well as its thermal history [101,102]. However, these properties can be varied by copolymerization with other monomers, such as glycolide and caprolactone. PLA and its isoforms are hydrophobic and typically soluble in chlorinated and fluorinated organic solvents. Moreover, the racemic mixture is soluble in solvents, such as acetone, tetrahydrofuran, xylene, and dimethylsulfoxide. However, PLA and its isoforms are not soluble in water and alcohols. PLA possesses a high tensile modulus (~ 3 GPa) and strength ($\sim 50-70$ MPa), and as a result, it is commonly used as a fixative in anchoring bone implants [100,103]. Mechanical properties, such as tensile modulus and elongation at break, depend upon molecular weight and crystallinity. For example, increasing the molecular weight of PLLA has been shown to result in superior mechanical properties; however, this effect appears to plateau at very high molecular weights [104]. In aqueous environments, pure PLA has a slow degradation rate, i.e., over months to years due to its high crystallinity, which hinders water penetration. However, crystallinity can be reduced by the addition of small amounts of D-lactide resulting in amorphous PDLLA that degrades faster [105]. Degradation rate also depends on molecular weight, with low molecular weight polymers degrading more quickly. The mechanism of degradation occurs essentially via the hydrolysis of ester bonds, resulting in progressive loss of molecular weight. However, the acidic degradation products can cause damage to surrounding tissue when used in RM applications. Moreover, when used as scaffolds in traditional tissue engineering applications, acidic by-products can be detrimental to seeded cells.

7.4.1.2 Poly(glycolic) Acid: Properties

In contrast, PGA is a crystalline polymer with a T_g value between 35 and 40°C, and a T_m value between 224 and 227°C. PGA can be up to 55% crystalline [103]. Using a variety of synthesis routes, a range of molecular weights can be obtained for PGA [106]. Due to its high crystallinity, PGA is insoluble in common organic solvents, such as chloroform, tetrahydrofuran, and acetone, but soluble in highly fluorinated solvents like hexafluoroisopropanol [73]. PGA is more hydrophilic than PLA and thus more easily degraded in aqueous environments. However, PGA has a higher tensile modulus (~ 7 GPa) than PLA, although its mechanical properties can decrease considerably over time due to bulk degradation [73]. As with PLA, high molecular weights and crystallinity of PGA can result in higher mechanical properties, while also reducing degradation rates. This can be useful depending on the nature of certain applications. PGA is very unstable in aqueous environments and can degrade over a period of a few weeks. Due to its hydrolytic instability, it is typically used in temporary sutures. The products of degradation are biocompatible and can enter the acetic acid cycle within the body, and be metabolized to carbon

TABLE 7.1 Degradable Polymers Used in RM Applications

Polymer Class	Applications	References
Polyesters	• Bone fixation • Absorbable sutures • Long-term contraception device	[73−75]
Polyorthoesters	Drug delivery for • Opthalmic applications • Periodontitis • Pain management (analgesics)	[76−80]
Polyanhydrides	Drug delivery: • To treat glioblastoma • For osteomyelitis (antibiotics) • Anesthetics • For prevention of restenosis (heparin) • In ophthalmic surgery (5-FU)	[81−89]
Polyphosphazenes	Drug delivery for • Periodontal diseases • Bone regeneration • Nerve regeneration • Inflammation • Diabetes	[90−95]
Polyphosphoesters	Drug delivery for • Tumor treatment (paclitaxel) • Nerve regeneration	[96−99]

dioxide and water. However, major concerns with PGA include the sudden drop in mechanical properties upon absorption of water, subsequent degradation, and the accumulation of acidic degradation products. Moreover, foreign body responses have been noted when PGA has been used in animal studies [107,108].

7.4.1.3 Poly(lactic-co-glycolic acid): Properties

Poly(lactic-co-glycolic acid) (PLGA) is a random copolymer of lactic and glycolic acid groups that combines the attractive properties of PLA and PGA [109]. Physical and chemical properties (e.g., crystallinity, Young's modulus, degradation rate) can be tuned by varying the ratio of monomer units. However, copolymer properties are not linearly related to monomer properties [110,111]. Rather, copolymers generally degrade more rapidly than corresponding homopolymers [112,113], are completely amorphous unlike either homopolymer, and possess T_g values of around 40−60°C. Finally, PLGA is also soluble in a wide range of organic solvents, including acetone,

tetrahydrofuran, choloroform, and other chlorinated/fluorinated organic solvents [114].

7.4.1.4 Polycaprolactone: Properties

PCL is another semicrystalline polyester that has been widely used for a variety of RM applications. It has a T_g value of −60°C and a T_m value of 55−60°C [115,116]. As with other polyesters, the mechanical and physical properties of the polymer depend upon its molecular weight and degree of crystallinity. Like PLGA, PCL is soluble in a wide variety of chlorinated/fluorinated organic solvents, as well as partially soluble in acetone and dimethylformamide; however, it is insoluble in alcohols and water [117]. The polymer possesses a tensile strength of around 16−500 MPa and a Young's modulus of around 400 MPa [104,115,116]. Due to its semicrystalline nature, PCL possesses relatively low degradability in aqueous environments and can degrade over a period of 2−4 years depending on its molecular weight and degree of crystallinity [118]. It is easily amenable to polymerization with a variety of other polymers resulting in copolymers with a host of tunable characteristics. Copolymer blends may be useful in drug delivery applications where degradation and subsequent release of drug is critical. PCL has also been copolymerized with polyethylene oxide (PEO) to yield polymers with both hydrophobic and hydrophilic microdomains [119].

7.4.1.5 Poly(propylene fumarate): Properties

Poly(propylene fumarate) (PPF) is an unsaturated linear polyester that can be chemically modified and/or cross-linked to polymers with a wide variety of degradation characteristics. As with other polyesters, degradation occurs by the hydrolysis of ester bonds, and the products of degradation (fumaric acid and propylene glycol) are nontoxic and easily processed by the human body [120]. Subcutaneous implantation studies in rats indicate that PPF induces a mild initial inflammatory response, but no any long-term negative effects [121]. Degradation and mechanical properties heavily depend on the molecular weight, cross-linking duration, and cross-linking chemistry [122].

7.4.1.6 Applications of Polyesters

The different types of degradable polyesters have been employed in a wide variety of RM applications. Since the products of degradation from PLA, PGA, and PLGA are nontoxic and readily processed by the human body [73], these polymers are approved by the FDA for application in bone fixation and sutures. For example, PLA has been used extensively as a bone fixative, while PGA is commonly used for the preparation of absorbable sutures (e.g., DEXON [74]) and fixatives (e.g., Biox) possessing degradation periods <3 months. Like its monomers, PLGA has also been used extensively as bone fixation screws and absorbable sutures. In contrast, PCL is often used in applications such as long-term drug delivery and in slow-degrading sutures due to its much slower rate of degradation as compared to PLGA. PCL has also been used in a long-term contraception device Capronor for the release of levonorgestrel over extended periods of time [75].

7.4.2 Polyorthoesters

Polyorthoesters (POEs) are a class of surface-eroding polymers that can be processed into materials that exhibit near zero-order release of drugs and therapeutics [123]. Their surface-eroding properties stem from the presence of a hydrolytically degradable backbone within a hydrophobic polymer. These polymers can be formed into dense hydrophobic matrices that subsequently degrade into smaller water-soluble polymers [124]. Four families of POEs, namely POE I, POE II, POE III, and POE IV have been synthesized and used for various applications.

7.4.2.1 POE I, II, and III: Properties

Although POE I, II, and III have been developed with a wide range of properties, they suffer from inherent limitations [74,124]. For example, POE I was initially used in several applications, such as the treatment of burns [125], for drug delivery [126], and in bone regeneration [127]. Later, its usage was discontinued due to a rather low T_g value and the acidic degradation products that also promoted autocatalytic degradation. In the case of POE II, the polymer's hydrophobicity hinders its degradation under normal physiological conditions. Although POE III, possessing a flexible backbone, was developed as an improvement over the first two families of POEs, its use has been limited due to long synthesis periods and a lack of reproducibility [74].

7.4.2.2 POE IV: Properties

POE IV has shown promise for commercialization primarily as a result of the presence of latent acidic chain segments (based on lactic and glycolic acids) within the backbone of the polymer, which allows for controlled degradation rates from a few weeks to a few months [128]. The polymer degrades with a linear decrease in weight over time, primarily driven by surface erosion and to a smaller extent by bulk degradation [129]. Moreover, the products of degradation are neither toxic nor acidic. The thermal and mechanical properties of POE IV can be controlled independent of erosion rates by appropriate choice of diols during synthesis [128]. These polymers

show excellent stability at room temperature and are also relatively stable upon irradiation [130].

7.4.2.3 Applications of POE II, III, and IV

POEs have been shown to be useful for the controlled delivery of drugs and other therapeutics. For example, POE III has been routinely used in ophthalmic applications [76]. In particular, it has been shown to be biocompatible in rabbit ophthalmic models and has been used (in a semi-solid form) in intravitreal injections for the delivery of the anti-inflammatory/antiscarring agent 5-fluorouracil (FU) [77]. The biocompatibility of viscous semisolid POE IV has been demonstrated in rabbits during subconjunctival, intracameral, intravitreal, and suprachoiroidal injections [78]. Importantly, the lifetime of POE IV was found to be much higher *in vivo* than POE III. Inflammatory reactions were found to be none to minimal. POEs have also been used for the delivery of bupivacaine for the management of postsurgical pain [79] and for the delivery of tetracycline into periodontal pockets in humans [80]. In the latter case, patients suffering from periodontitis tolerated the POE formulations with no observable signs of irritation or discomfort. Furthermore, an improved clinical procedure ensured the longer retention of the formulations in the periodontal pockets, which allowed for the release of therapeutic doses of tetracycline over a period of 11 days.

7.4.3 Polyanhydrides

7.4.3.1 Properties

Polyanhydrides (PAHs) have been investigated extensively as biomaterials for the delivery of drugs and other therapeutics [131]. These polymers are biocompatible and degrade to release their respective diacids, which are nontoxic and nonmutagenic [132]. Due to a hydrophobic backbone, these polymers degrade via surface erosion, thus allowing for close to zero-order release kinetics over a period of weeks [131]. While aliphatic anhydrides are prone to degradation (and require storage under moisture-free conditions), aromatic anhydrides are very stable in hydrolytic environments [74]. Aliphatic PAHs usually possess T_m values <100°C, while the aromatic PAHs possess T_m values over 100°C. The properties of PAHs can be tuned by adjusting polymer composition. For example, the inclusion of PEG improves their hydrophilicity [133], while copolymerization of sebacic acid with fatty acid dimers results in hydrophobic polymers with lower rates of degradation [131]. PAHs are generally crystalline, but their crystallinity can be tuned based on the choice of monomers [131]. PAHs suffer from poor mechanical properties. For example, poly 1,6 carboxy-phenoxy hexane has a Young's modulus of about 1.3 MPa [134]. However, copolymers like polyanhydrides-co-imides have been shown to possess improved mechanical properties [135]. In particular, compressive strengths in the range of 50−60 MPa have been reported for polyanhydrides-co-imides [136]. Mechanical properties and degradation rates can also be varied by the use of cross-linkable PAHs. Polymers based on dimethacrylated anhydride have been synthesized [137], and the mechanical properties of these polymers can potentially be tuned based on monomer−cross-linker ratios and the choice of monomers and cross-linkers [73]. Specifically, such cross-linked anhydrides have been shown to possess compressive strengths of 30−40 MPa and tensile strengths of 15−27 MPa [138].

7.4.3.2 Applications

Like POEs, PAHs have traditionally been used for the localized delivery of drugs and therapeutics in RM applications [81]. Polycarboxy phenoxy propane sebacic acid (PCPP-SA), a type of PAH, has been shown to be biocompatible [82] and has also been approved by the FDA for the delivery of a therapeutic for treating brain cancer (glioblastoma) under the commercial name Gliadel. This system has been used for the delivery of the drug carmustine (BCNU) to residual tumor cells after surgical removal of the tumor [83]. Studies of delivery in a monkey brain revealed a local reaction at the site of implantation, including subacute inflammation and local edema; however, no negative effects were observed after 72 days [84]. In another study with a rat malignant glioma model, paclitaxel delivery from a PAH-based delivery system was found to significantly increase the survival rates of rats compared to controls [85]. PAH-based systems have also been used for the delivery of carboplatin [86] and methotrexate [83], and have been shown to increase median survival times in rat glioma models. In addition, poly(fatty acid dimer-sebacic acid) has been used for the delivery of cisplatin to treat human squamous cell carcinoma xenografts in an animal model [87]. Apart from the delivery of drugs to treat cancer, a PAH-based system, under the commercial name Septacin, has been used for the delivery of antibiotics to treat osteomyelitis, for the delivery of local anesthetics [81], for the delivery of heparin to prevent restenosis in an animal model [88] and for the delivery of 5-FU in prolonging the success of glaucoma filtering surgery in a rabbit model [89].

7.4.4 Polyphosphazenes

7.4.4.1 Properties

Polyphosphazenes (PNEs) are high molecular weight linear polymers that possess an alternating nitrogen and phosphorous backbone [139]. Most PNEs are soluble in commonly used organic solvents, which allows for easy processing [74]. The properties of PNEs can vary very widely depending on the nature of their organic side chains. For example, the use of glycine side groups results in PNEs with a T_g value of $-40°C$, while phenylalanine results in a T_g value of $41.6°C$ [140], consistent with the hypothesis that a bulkier side chain results in a higher T_g value. The mechanical properties and crystallinity of these polymers can be varied based on the nature of the side groups as well. While methoxy and ethoxy side groups result in thermoplastics, fluoroalkoxy and chloroalkoxy side chains result in the formation of elastomers [139]. Elastomeric PNEs possessing T_g values as low as $-60°C$ have been manufactured and used in a wide variety of applications. By appropriate choice of side groups, PNE-based hydrogels with pH- and temperature-dependent properties have been synthesized [141,142]. Side groups, such as amines and imidazoles, can confer hydrolytic instability, making these polymers degrade more quickly. In contrast, the substitution of hydrophobic side groups has the opposite effect on degradation rates. In fact, the degradation of PNEs can be tuned from weeks to months depending on the molecular weight, nature of the side groups, and degree of substitution [143,144]. Products of degradation include phosphate, ammonia, and side groups, which are nontoxic [145].

7.4.4.2 Applications

PNEs have been used for the controlled delivery of drugs and therapeutics for a wide variety of applications [90]. Initial studies revealed that these polymers could be used to release progesterone in Sprague-Dawley rats at a sustained rate [91]. PNE-based polymeric vehicles have also been used for the delivery of naproxen and the sustained therapeutic efficacy in a rat model of acute inflammation and chronic arthritis has been demonstrated [92]. PNE-based microspheres prepared by a double-emulsion method were found to release insulin over a sustained period of time and kept glucose levels low in a diabetic mouse model for over 1000 h [93]. PNE membranes and microspheres have also been investigated as delivery vehicles in the treatment of periodontal diseases and in bone regeneration [94]. The release of trimethoprim and succinlysulfathiazole was tested in a rat model and the results indicated that the PNE membranes were able to release these drugs locally to surrounding gingival tissue over a period at least 110 h. The release of naproxen, an anti-inflammatory drug, from PNE-based membranes and microspheres was also tested in the same model and the blood concentration levels of the drug were found to be therapeutically satisfactory. In the same study, significant bone healing was noted after 1 month in a rabbit tibial bone defect model, when the defects were implanted with PNE-based membranes. Moreover, no signs of inflammation were detected. Polyorganophosphazenes have also been used for the regeneration of neural tissue [95]. Specifically, PNE-based nerve guides were found to promote the formation of tissue cable bridging a rat sciatic nerve defect, 45 days postimplantation.

7.4.5 Polyphosphoesters

7.4.5.1 Properties

Polyphosphoesters (PPEs) are nucleic acid/techoic acid analogs that possess phosphorous atoms in the backbone attached to an oxygen atom and a side group. Based on the nature of the side group, these polymers are further classified as polyphosphites, polyphosphonates, and polyphosphates [74,96]. PPEs possess a hydrolytically degradable phosphoester backbone, resulting in the formation of phosphate, alcohols, and diols upon degradation. Like PNEs, the degradation of PPEs can be tuned by changing the molecular weight, crystallinity, backbone chemistry, as well as the chemistry of the side groups. The presence of amino pendant groups has been shown to result in PPEs that can degrade within a period of days to weeks [146]. PPEs can also undergo enzymatic degradation via the action of intracellular enzymes like phosphodiesterase I [147]. The side groups and the organic moieties within the backbone can be varied to also tune the mechanical properties of PPEs, including the production of crystalline or amorphous solids, liquids, and elastomers [96]. Copolymers of PPEs have been investigated widely for the production of polymers possessing an interesting and useful blend of properties. For example, blending PPE with PDLA to result in poly(lactide-co-ethyl phosphate) has been shown to alter T_g, degradation rates and degradation mechanism of the copolymer based on the relative amounts of the monomers [148]. The synthesis of photo-polymerizable hydrogels based on phosphoester−PEG copolymers has been investigated for alkaline phosphatase-mediated degradation [149]. More recently, the production of thermosensitive PPEs has been reported. In particular, block copolymers based on PEG and PPE or PCL and PPE have been shown to confer thermosensitive properties like tunable lower critical solution temperatures (LCSTs) [150,151]. The LCST values depend on the molecular weights of the blocks, their hydrophobicity and ratios of the blocks.

7.4.5.2 Applications

PPEs have been used for a variety of applications, including controlled delivery of drugs as well as in tissue engineering due to their biocompatible nature and degradation properties [96]. Copolymers of PLA and PPE have been used as a drug delivery system under the name PACLIMER [97]. Specifically, the release of paclitaxel from the polymeric system was found to reduce tumor volume as well as significantly increase in tumor doubling time as compared to intratumoral administration of paclitaxel in nude mice possessing human NSCLC cell lines. The release of paclitaxel from PACLIMER microspheres, when injected intraperitoneally into dogs, was found to reach a much lower peak plasma concentration as compared to bolus injections of paclitaxel [96]. The microspheres also helped maintain therapeutic levels of paclitaxel in the plasma over extended periods of time. In a similar manner, PACLIMER microspheres were found to be more active against OVCAR-3, a relevant animal model for ovarian cancer, compared to Taxol injections. Finally, a phase I clinical trial involving PACLIMER microspheres to treat ovarian cancer, demonstrated that the PACLIMER system was able to maintain therapeutic levels of paclitaxel (below toxic threshold levels) in the plasma over a 2-month period [98]. PPE-based systems also have been used for the regeneration of neural tissues. In a study by Wang et al., two PPE-based nerve conduits were found to result in the formation of a thin fibrous capsule upon implantation in the sciatic nerve of rats [99]. Although tube breakage was observed 5 days postimplantation, regeneration of a gap defect in the rat sciatic nerve model was observed. Moreover, the conduits demonstrated negligible swelling and no crystallization upon implantation.

7.5 HYDROGELS

Hydrogels differ from other degradable polymeric biomaterials in their hydrophilicity and capacity to absorb many times their weight in water. Consequently, most hydrogel materials are chemically cross-linked to prevent their dissolution in aqueous environments. Although several hydrogel materials have been tested for RM applications, the following polymers have been extensively used clinically. Table 7.2 provides a list of these hydrogels, along with their applications.

7.5.1 Polyacrylamide

7.5.1.1 Properties

Polyacrylamide (PAAM) forms soft, cationic nondegradable hydrogels consisting of entangled polymer chains tethered to one another via cross-linkers [175]. Its hydrophilic nature results in a high capacity for swelling and tendency for cell (bacterial and mammalian) adhesion and migration, granuloma formation, and fibrosis *in vivo* [152,153,176]. The extent of swelling is inversely related

TABLE 7.2 Hydrogels Used in RM Applications

Hydrogel	Applications	References
Polyacrylamide	• Tissue bulking procedures • Subcutaneous injections for aesthetic purposes • Muscle regeneration	[152–156]
Poly(2-hydroxyethyl methacrylate)	• Opthalmic applications: corneal prosthesis • Controlled release of proteins • Cardiac tissue regeneration • Axonal regeneration in spinal cord injury • Replacement of intervertebral discs	[157–162]
Polyvinyl alcohol	• Surgical sponges • Wound dressings • Hyaline cartilage reconstructions • Meniscal implant • Injectable intravascular emboli	[163–168]
Polyethyelene glycol	• Antimicrobial/low-fouling coatings • Wound dressings • Osmotic laxatives • Reducing the formation of fibrin bands between adjacent tissues	[169–174]

to the extent of cross-linking. However, lightly cross-linked PAAM gels—composed of up to 99% water by weight—still maintain their physical, chemical, and mechanical stabilities [154]. Alterations in the physical properties of PAAM have been achieved primarily by varying the degree of cross-linking and by copolymerization with other monomers: these modifications change hydrogel swelling capacity and stiffness, respectively [176]. Specifically, copolymerization—with synthetic or biologic monomers—allows the modulation of material properties, including stiffness, charge, and the densities of amide and ester side groups for surface modification [176,177]. Although PAAM is a versatile hydrogel, it requires purification prior to clinical use, as acrylamide monomers *in vivo* degrade to glycidamide, a genotoxic compound [178,179].

7.5.1.2 Applications

Popular surgical applications of PAAM hydrogels include tissue bulking procedures and subcutaneous injections for aesthetic purposes [152,155,156]. As a bulking agent,

Bulkamid® has been injected into urethral submucosa for the creation of artificial urethral cushions to treat stress urinary incontinence (or mixed urinary incontinence) [155]. Furthermore, Aquamid® is an alternative to silicone oil for soft tissue augmentation (e.g., face, breasts) with few reports of severe adverse reactions [153,156]. Additionally, the cationic nature of PAAM allows it to deform under electric fields, allowing its potential use in muscle regeneration [154]. However, PAAM has not been approved for use in the United States, despite its use in plastic surgery in Europe and Asia for over 10 years. At present Aquamid® and Bulkamid® are undergoing FDA clinical trial and premarket approval, respectively.

7.5.2 Poly(2-hydroxyethylmethacrylate)

7.5.2.1 Properties

Poly(2-hydroxyethylmethacrylate) (pHEMA) is an inert, water-stable, nondegradable hydrogel [180] with high transparency. The physical properties of pHEMA (e.g., swelling, stiffness, rheology) can be tuned by varying cross-linking density, incorporating different chemistries through copolymerization, and introducing mesoscopic pores. Specifically, a reduction in cross-linking density results in a softer, more malleable hydrogel [157] that may be better suited for soft tissue regeneration. Moreover, copolymerization with acetic acid, methylmethacrylate, or dextran can adjust the permanence, hydrophilicity, and cellular adhesion *in vivo*[158,181,182]. Finally, the introduction of mesoscopic porogens can facilitate vascular ingrowth, improve cellular attachment, and overcome limited permeability [159,183]. Although pHEMA is considered nondegradable (which makes it ideally suited for long-term applications *in vivo*), degradable pHEMA copolymers have been fabricated by the integration of enzymatically susceptible monomers (e.g., dextran) or cross-linking agents [158]. These degradable materials show promise for controlled release of pharmaceuticals and proteins [158,160,184].

7.5.2.2 Applications

Due to its excellent optical properties, pHEMA has primarily been used in ophthalmic applications [157] under the generic names etafilcon A and vifilcon A. In addition, it has been examined for controlled release of proteins and drugs [158,161], engineering of cardiac tissue [159], axonal regeneration in spinal cord injury [160], and replacement of intervertebral discs [162]. However, two limitations of pHEMA are its propensity for calcification and the toxicity of the 2-hydroxyethylmethacrylate monomers. Phase I testing of pHEMA for corneal prostheses (keratoprosthesis) revealed calcium salt deposition within

2.5 years after implantation [180,181]. At the same time, residual HEMA monomer can compromise the mechanical properties of the hydrogel, and leach into surrounding tissue with toxic effects [185,186].

7.5.3 Polyvinyl Alcohol

7.5.3.1 Properties

Polyvinyl alcohol (PVA) is a biodegradable, biocompatible, and protein-resistant polymer that gels spontaneously in aqueous media. Improved stability has been achieved by freeze-thawing [163,187] and covalent cross-linking (e.g., glutaraldehyde [188] and gamma irradiation [163,180,189]). Both physically and chemically cross-linked PVA can be processed into a wide range of forms, including solutions, pellets, sheets, sponges, foams, coatings, meshes, and microspheres [187,190]. The degradation rate of PVA hydrogels can be controlled by varying polymer molecular weight, polymer concentration, and the combination of physical and chemical cross-links [190]. By copolymerizing and blending with natural and synthetic polymers, PVA has been used to create a vast array of hydrogels with diverse characteristics [191−195]. These blends allow for a greater diversity of behaviors and applications than PVA alone.

7.5.3.2 Applications

PVA's resistance to cell adhesion and protein absorption prevents bacterial infection, granuloma formation, and postoperative adhesions. Therefore, PVA-based hydrogels have been extensively used in a wide variety of RM applications. Such FDA-approved applications, include contact lenses [164], eye wetting drops, surgical sponges, wound dressings [165], catheter coatings, barrier films, and injectable intravascular emboli. Cartiva™ (not currently FDA approved) is a PVA-based implant to replace degraded hyaline cartilage in osteoarthritis. Additionally, research is under way to develop a PVA meniscal implant [166−168].

7.5.4 Polyethylene Glycol

7.5.4.1 Properties

Cross-linked PEG materials are widely considered biocompatible and biodegradable, suitable to be used in a vast array of medical applications [180]. PEG hydrogels biodegrade *in situ* via hydrolysis and low molecular weight PEG oligomers excreted in the urine. Due to the scarcity of side chains and overall hydrophilicity, PEG hydrogels are considered to be "stealth materials." Specifically, they are surrounded by an impenetrable water layer *in vivo* that permits diffusion, but prevents protein adsorption and cell adhesion [196]. This property reduces both the formation of biofilms on PEG-coated devices and the metabolism/degradation of PEG-encapsulated drug particles *in vivo*[169,197]. When taken orally, PEG is toxic only at high doses [198], and very limited immunological reaction to indwelling PEG devices has been reported [199].

7.5.4.2 Applications

Currently, PEG is employed in several applications, including osmotic laxatives (e.g., MiraLAX®, GoLYTELY®) [170], as antimicrobial coatings/low-fouling coatings on medical devices [169], and as surgical sealants, adhesion barriers, and wound dressings. PEG-based liquid sealants (e.g., ProGEL™, CoSeal™, and DuraSeal™) have shown positive results as adjuncts to tissue closure for pulmonary, vascular, and neurological surgeries. Moreover, PEG's use as an adhesion barrier in cardiothoracic (e.g., REPEL-CV®), abdominal, and gynecologic surgeries has demonstrated reduced formation of fibrin bands between adjacent tissues [171−173]. Lastly, PEG hydrogels are well suited as wound dressings as they provide a moist environment and a barrier to bacterial infiltration. Amerigel® Hydrogel Saturated Gauze, which contains a PEG400/3350 mix and proprietary oak-extract compound (Oakin®; an antimicrobial agent) embedded in a blended polyester−rayon gauze, is an FDA-approved product for wound dressings. The PEG-Oakin® mix is highly absorbent, pliable, and inexpensive compared to equally efficacious silver-impregnated dressings [174].

7.6 SUMMARY AND FUTURE DIRECTIONS

Past generations of biomaterials typically arose from industrial materials and were predominantly used to perform mechanical functions as medical implants. Such devices have been enormously relevant to reconstructive surgery, but not necessarily transplant surgery because the level of sophistication of these devices was not physiologic in nature. More recently, as the technology of biomaterials has evolved, new "smart" materials have appeared, taking on a higher level of sophistication by indirectly or directly performing organ-like function. Innovations that now allow greater cell interaction, and even control over cell behavior, have taken inspiration from nature and in the case of RM, developmental biology. New biomaterials are in advanced stages of preclinical development that can support, facilitate, and even direct cell adhesion, migration, assembly, and growth. Technologies that mimic organoid function, such as pancreas, liver, and kidney, are being developed through tissue engineering approaches. Cell delivery using novel biomaterial carriers to augment organ function and rescue diseased organs is beginning to enter clinical trials and small trials of tissue-engineered organs are becoming less sporadic events. These technologies, envisioned decades ago and only now starting to come to fruition are on the horizon of transplant surgery and have great potential to provide the next generation of treatment modalities.

REFERENCES

[1] Roach P, Eglin D, Rohde K, Perry CC. Modern biomaterials: a review-bulk properties and implications of surface modifications. J Mater Sci Mater Med 2007;18:1263−77.

[2] Lloyd AW, Faragher RGA, Denyer SP. Ocular biomaterials and implants. Biomaterials 2001;22:769−85.

[3] Williams DF. On the mechanisms of biocompatibility. Biomaterials 2008;29:2941−53.

[4] Consensus Conference on Biocompatibility. Nephrol Dial Transplant 1994;9(Suppl. 2):1−186 [Consensus Development Conference].

[5] Black J. Biological performance of materials: fundamentals of biocompatibility. 4th ed. Boca Raton; London: CRC Taylor & Francis; 2006.

[6] Kohane DS, Langer R. Polymeric biomaterials in tissue engineering. Pediatr Res 2008;63:487−91.

[7] Hutmacher DW. Scaffolds in tissue engineering bone and cartilage. Biomaterials 2000;21:2529−43.

[8] Spiller KL, Maher SA, Lowman AM. Hydrogels for the repair of articular cartilage defects. Tissue Eng Part B Rev 2011;17:281−99.

[9] Atala A. Tissue engineering of human bladder. Br Med Bull 1997;97:81−104.

[10] Dhandayuthapani B, Yoshida Y, Maekawa T, Kumar DS. Polymeric scaffolds in tissue engineering application: a review. Int J Polym Sci 2011; Article ID 290602.

[11] Griffith LG. Emerging design principles in biomaterials and scaffolds for tissue engineering. Ann N Y Acad Sci 2002;961:83−95.

[12] Guilak F. Homing in on a biological joint replacement. Stem Cell Res Ther 2010;1:40.

[13] Shin H, Jo S, Mikos AG. Biomimetic materials for tissue engineering. Biomaterials 2003;24:4353−64.

[14] Kingshott P, Griesser HJ. Surfaces that resist bioadhesion. Curr Opin Solid State Mater Sci 1999;4:403−12.

[15] Ripamonti U, Roden LC, Renton LF. Osteoinductive hydroxyapatite-coated titanium implants. Biomaterials 2012;33:3813−23.

[16] Navarro M, Michiardi A, Castano O, Planell JA. Biomaterials in orthopaedics. J R Soc Interface 2008;5:1137−58.

[17] Disegi JA, Eschbach L. Stainless steel in bone surgery. Injury Int J Care Injured 2000;31:2−6.

[18] Niinomi M. Metallic biomaterials. J Artif Organs 2008;11:105−10.

[19] Uggowitzer PJ, Bahre WF, Wohlfromm H, Speidel MO. Nickel-free high nitrogen austenitic stainless steels produced by metal injection moulding. High Nitrogen Steels 1999;:663−71.

[20] Case CP, Langkamer VG, James C, Palmer MR, Kemp AJ, Heap PF, et al. Widespread dissemination of metal debris from implants. J Bone Joint Surg Br 1994;76:701−12.

[21] Teoh SH. Fatigue of biomaterials: a review. Int J Fatigue 2000;22:825−37.

[22] Bauer TW, Schils J. The pathology of total joint arthroplasty. II. Mechanisms of implant failure. Skeletal Radiol 1999;28:483−97.

[23] Gotman I. Characteristics of metals used in implants. J Endod 1997;11:383−9.

[24] Okazaki Y, Gotoh E. Comparison of metal release from various metallic biomaterials in vitro. Biomaterials 2005;26:11−21.

[25] Geetha M, Singh AK, Asokamani R, Gogia AK. Ti based biomaterials, the ultimate choice for orthopaedic implants—a review. Prog Mater Sci 2009;54:397−425.

[26] Bansiddhi A, Sargeant TD, Stupp SI, Dunand DC. Porous NiTi for bone implants: a review. Acta Biomater 2008;4:773−82.

[27] Vallet-Regí M, Ruiz-Hernández E. Bioceramics: from bone regeneration to cancer nanomedicine. Adv Mater (Deerfield Beach, FL) 2011;23:5177−218.

[28] LeGeros RZ. Calcium phosphate-based osteoinductive materials. Chem Rev 2008;108:4742−53.

[29] Vallet-Regí M, González-Calbet JM. Calcium phosphates as substitution of bone tissues. Prog Solid State Chem 2004;32:1−31.

[30] Jimbo R, Coelho PG, Vandeweghe S, Schwartz-Filho HO, Hayashi M, Ono D, et al. Histological and three-dimensional evaluation of osseointegration to nanostructured calcium phosphate-coated implants. Acta Biomater 2011;7:4229−34.

[31] Narayanan R, Seshadri SK, Kwon TY, Kim KH. Calcium phosphate-based coatings on titanium and its alloys. J Biomed Mater Res Part B Appl Biomater 2008;85:279−99.

[32] De Long Jr. WG, Einhorn TA, Koval K, McKee M, Smith W, Sanders R, et al. Bone grafts and bone graft substitutes in orthopaedic trauma surgery. A critical analysis. J Bone Joint Surg 2007;89:649−58.

[33] Hench LL. The story of bioglass. J Mater Sci Mater Med 2006;17:967−78.

[34] Bahmad JF, Merchant SN. Histopathology of ossicular grafts and implants in chronic otitis media. Ann Otol Rhinol Laryngol 2007;116:181−91.

[35] Li R, Clark AE, Hench LL. An investigation of bioactive glass powders by sol-gel processing. J Appl Biomater 1991;2:231−9.

[36] Nandi SK, Kundu B, Datta S, De DK, Basu D. The repair of segmental bone defects with porous bioglass: an experimental study in goat. Res Veter Sci 2009;86:162−73.

[37] Cacciotti I, Lombardi M, Bianco A, Ravaglioli A, Montanaro L. Sol−gel derived 45S5 bioglass: synthesis, microstructural evolution and thermal behaviour. J Mater Sci Mater Med 2012;23:1849−66.

[38] Chen QZ, Thouas GA. Fabrication and characterization of sol−gel derived 45S5 Bioglass®-ceramic scaffolds. Acta Biomater 2011;7:3616−26.

[39] Chen QZ, Thompson ID, Boccaccini AR. 45S5 Bioglass-derived glass−ceramic scaffolds for bone tissue engineering. Biomaterials 2006;27:2414−25.

[40] Jeffers JRT, Walter WL. Ceramic-on-ceramic bearings in hip arthroplasty: state of the art and the future. J Bone Joint Surg Br Vol 2012;94:735−45.

[41] Hannouche D, Zaoui A, Zadegan F, Sedel L, Nizard R. Thirty years of experience with alumina-on-alumina bearings in total hip arthroplasty. Int Orthop 2011;35:207−13.

[42] Garcia-Rey E, Cruz-Pardos A, Garcia-Cimbrelo E. Alumina-on-alumina total hip arthroplasty in young patients: diagnosis is more important than age. Clin Orthop Relat Res 2009;467:2281−9.

[43] Bizot P, Larrouy M, Witvoet J, Sedel L, Nizard R. Press-fit metal-backed alumina sockets: a minimum 5-year followup study. Clin Orthop Relat Res 2000;:134–42.

[44] Jaeblon T. Polymethylmethacrylate: properties and contemporary uses in orthopaedics. J Am Acad Orthop Surg 2010;18:297–305.

[45] Kuehn KD, Ege W, Gopp U. Acrylic bone cements: composition and properties. Orthop Clin North Am 2005;36:17–28.

[46] Webb JCJ, Spencer RF. The role of polymethylmethacrylate bone cement in modern orthopaedic surgery. J Bone Joint Surg Br 2007;89:851–7.

[47] Chehade M, Elder MJ. Intraocular lens materials and styles: a review. Aust NZ J Ophthalmol 1997;25:255–63.

[48] Frazer RQ, Byron RT, Osborne PB, West KP. PMMA: an essential material in medicine and dentistry. J Long Term Eff Med Implants [Comp Study Rev] 2005;15:629–39.

[49] Provenzano MJ, Murphy KPJ, Riley LH. Bone cements: review of their physiochemical and biochemical properties in percutaneous vertebroplasty. Am J Neuroradiol 2004;25:1286–90.

[50] Hoffmann F, Kruse H, Schuler A. Mechanical methods in refractive corneal surgery. Curr Opin Ophthalmol 1993;4:84–90.

[51] Nose W, Neves RA, Schanzlin DJ, Belfort Jr. R. Intrastromal corneal ring—one-year results of first implants in humans: a preliminary nonfunctional eye study. Refract Corneal Surg 1993;9:452–8.

[52] Puskas JE, Chen Y. Biomedical application of commercial polymers and novel polyisobutylene-based thermoplastic elastomers for soft tissue replacement. Biomacromolecules 2004;5:1141–54.

[53] Arefazar A, Biddlestone F, Hay JN, Haward RN. The effect of physical aging on the properties of poly(ethylene-terephthalate). Polymer 1983;24:1245–51.

[54] Polyakova A, Liu RYF, Schiraldi DA, Hiltner A, Baer E. Oxygen-barrier properties of copolymers based on ethylene terephthalate. J Polym Sci Polym Phys 2001;39:1889–99.

[55] el-Massry S, Saad E, Sauvage LR, Zammit M, Davis CC, Smith JC, et al. Axillofemoral bypass with externally supported, knitted Dacron grafts: a follow-up through twelve years. J Vasc Surg [Comparative Study] 1993;17:107–14.

[56] Fanous N, Samaha M, Yoskovitch A. Dacron implants in rhinoplasty: a review of 136 cases of tip and dorsum implants. Arch Facial Plast Surg 2002;4:149–56.

[57] Barrett GR, Line LL, Shelton WR, Manning JO, Phelps R. The Dacron ligament prosthesis in anterior cruciate ligament reconstruction—a 4-year review. Am J Sport Med 1993;21:367–73.

[58] Nunn DB, Freeman MH, Hudgins PC. Postoperative alterations in size of Dacron aortic grafts: an ultrasonic evaluation. Ann Surg 1979;189:741–5.

[59] Borazjani BH, Wilson SE, Fujitani RM, Gordon I, Mueller M, Williams RA. Postoperative complications of carotid patching: pseudoaneurysm and infection. Ann Vasc Surg 2003;17:156–61.

[60] Massell TB, Heringman EC, Greenstone SM. Woven Dacron and woven teflon prostheses. Use for small artery replacement. Arch Surg 1962;84:73–9.

[61] Roolker W, Patt TW, van Dijk CN, Vegter M, Marti RK. The gore-tex prosthetic ligament as a salvage procedure in deficient knees. Knee Surg Sport Tr A 2000;8:20–5.

[62] Grethel EJ, Cortes RA, Wagner AJ, Clifton MS, Lee H, Farmer DL, et al. Prosthetic patches for congenital diaphragmatic hernia repair: surgisis vs Gore-Tex. J Pediatr Surg 2006;41:29–32.

[63] Nakayama M, Ford CN, Bless DM. Teflon vocal fold augmentation: failures and management in 28 cases. Otolaryngol Head Neck Surg 1993;103:493–8.

[64] O'Donnell B, Puri P. Treatment of vesicoureteric reflux by endoscopic injection of Teflon. Br Med J 1984;289:7–9.

[65] Bhatti HA, Khattak H, Boston VE. Efficacy and causes of failure of endoscopic subureteric injection of Teflon in the treatment of primary vesicoureteric reflux. Br J Urol 1993;71:221–5.

[66] Narins RS, Beer K. Liquid injectable silicone: a review of its history, immunology, technical considerations, complications, and potential. Plast Rec Surg 2006;118:77–84.

[67] Lamba NMK, Woodhouse KA, Cooper SL, Lelah MDPim. Polyurethanes in biomedical applications. Boca Raton; London: CRC; 1998.

[68] Kanyanta V, Ivankovic A. Mechanical characterisation of polyurethane elastomer for biomedical applications. J Mech Behav Biomed Mater 2010;3:51–62.

[69] Guelcher SA, Gallagher KM, Didier JE, Klinedinst DB, Doctor JS, Goldstein AS, et al. Synthesis of biocompatible segmented polyurethanes from aliphatic diisocyanates and diurea diol chain extenders. Acta Biomater 2005;1:471–84.

[70] Spaans CJ, De Groot JH, Belgraver VW, Pennings AJ. A new biomedical polyurethane with a high modulus based on 1,4-butanediisocyanate and epsilon-caprolactone. J Mater Sci Mater Med 1998;9:675–8.

[71] Burke A, Hasirci N. Polyurethanes in biomedical applications. Adv Exp Med Biol 2004;553:83–101.

[72] Zdrahala RJ, Zdrahala IJ. Biomedical applications of polyurethanes: a review of past promises, present realities, and a vibrant future. J Biomater Appl 1999;14:67–90.

[73] Gunatillake PA, Adhikari R. Biodegradable synthetic polymers for tissue engineering. Eur Cells Mater 2003;5:1–16.

[74] Nair LS, Laurencin CT. Polymers as biomaterials for tissue engineering and controlled drug delivery. Tissue Eng I Scaffold Syst Tissue Eng 2006;102:47–90.

[75] Ueda H, Tabata Y. Polyhydroxyalkanonate derivatives in current clinical applications and trials. Adv Drug Deliver Rev 2003;55:501–18.

[76] Heller J. Ocular delivery using poly(ortho esters). Adv Drug Deliver Rev 2005;57:2053–62.

[77] Bernatchez SF, Merkli A, Minh TL, Tabatabay C, Anderson JM, Gurny R. Biocompatibility of a new semisolid bioerodible poly (ortho ester) intended for the ocular delivery of 5-fluorouracil. J Biomed Mater Res 1994;28:1037–46.

[78] Einmahl S, Ponsart S, Bejjani RA, D'Hermies F, Savoldelli M, Heller J, et al. Ocular biocompatibility of a poly(ortho ester) characterized by autocatalyzed degradation. J Biomed Mater Res A 2003;67:44–53.

[79] Heller J, Barr J, Ng SY, Shen HR, Schwach-Abdellaoui K, Gurny R, et al. Development and applications of injectable poly(ortho esters) for pain control and periodontal treatment. Biomaterials 2002;23:4397–404.

[80] Schwach-Abdellaoui K, Loup PJ, Vivien-Castioni N, Mombelli A, Baehni P, Barr J, et al. Bioerodible injectable poly(ortho ester) for tetracycline controlled delivery to periodontal pockets: preliminary trial in humans. AAPS Pharmsci 2002;4(4):14–20.

[81] Jain JP, Modi S, Domb AJ, Kumar N. Role of polyanhydrides as localized drug carriers. J Control Release 2005;103:541–63.

[82] Laurencin C, Domb A, Morris C, Brown V, Chasin M, Mcconnell R, et al. Poly(anhydride) administration in high-doses *in vivo* studies of biocompatibility and toxicology. J Biomed Mater Res 1990;24:1463−81.

[83] Wang PP, Frazier J, Brem H. Local drug delivery to the brain. Adv Drug Deliver Rev 2002;54:987−1013.

[84] Brem H, Tamargo RJ, Olivi A, Pinn M, Weingart JD, Wharam M, et al. Biodegradable polymers for controlled delivery of chemotherapy with and without radiation-therapy in the monkey brain. J Neurosurg 1994;80:283−90.

[85] Walter KA, Cahan MA, Gur A, Tyler B, Hilton J, Colvin OM, et al. Interstitial taxol delivered from a biodegradable polymer implant against experimental malignant glioma. Cancer Res 1994;54:2207−12.

[86] Olivi A, Ewend MG, Utsuki T, Tyler B, Domb AJ, Brat DJ, et al. Interstitial delivery of carboplatin via biodegradable polymers is effective against experimental glioma in the rat. Cancer Chemother Pharm 1996;39:90−6.

[87] Shikani AH, Eisele DW, Domb AJ. Polymer delivery of chemotherapy for squamous-cell carcinoma of the head and neck. Arch Otolaryngol 1994;120:1242−7.

[88] Orloff LA, Glenn MG, Domb AJ, Esclamado RA. Prevention of venous thrombosis in microvascular surgery by transmural release of heparin from a polyanhydride polymer. Surgery 1995;117:554−9.

[89] Lee DA, Leong KW, Panek WC, Eng CT, Glasgow BJ. The use of bioerodible polymers and 5-fluorouracil in glaucoma filtration surgery. Invest Ophth Vis Sci 1988;29:1692−7.

[90] Lakshmi S, Katti DS, Laurencin CT. Biodegradable polyphosphazenes for drug delivery applications. Adv Drug Deliver Rev 2003;55:467−82.

[91] Laurencin CT, Koh HJ, Neenan TX, Allcock HR, Langer R. Controlled release using a new bioerodible polyphosphazene matrix system. J Biomed Mater Res 1987;21:1231−46.

[92] Conforti A, Bertani S, Lussignoli S, Grigolini L, Terzi M, Lora S, et al. Anti-inflammatory activity of polyphosphazene-based naproxen slow-release systems. J Pharm Pharmacol 1996;48:468−73.

[93] Caliceti P, Veronese FM, Lora S. Polyphosphazene microspheres for insulin delivery. Int J Pharm 2000;211:57−65.

[94] Veronese FM, Marsilio F, Lora S, Caliceti P, Passi P, Orsolini P. Polyphosphazene membranes and microspheres in periodontal diseases and implant surgery. Biomaterials 1999;20:91−8.

[95] Langone F, Lora S, Veronese FM, Caliceti P, Parnigotto PP, Valenti F, et al. Peripheral-nerve repair using a poly(organo) phosphazene tubular prosthesis. Biomaterials 1995;16:347−53.

[96] Zhao Z, Wang J, Mao HQ, Leong KW. Polyphosphoesters in drug and gene delivery. Adv Drug Deliver Rev 2003;55:483−99.

[97] Harper E, Dang WB, Lapidus RG, Garver RI. Enhanced efficacy of a novel controlled release paclitaxel formulation (PACLIMER Delivery System) for local-regional therapy of lung cancer tumor nodules in mice. Clin Cancer Res 1999;5:4242−8.

[98] Armstrong DK, Fleming GF, Markman M, Bailey HH. A phase I trial of intraperitoneal sustained-release paclitaxel microspheres (Paclimer(R)) in recurrent ovarian cancer: a gynecologic oncology group study. Gynecol Oncol 2006;103:391−6.

[99] Wang S, Wan ACA, Xu XY, Gao SJ, Mao HQ, Leong KW, et al. A new nerve guide conduit material composed of a biodegradable poly(phosphoester). Biomaterials 2001;22:1157−69.

[100] Sodergard A, Stolt M. Properties of lactic acid based polymers and their correlation with composition. Prog Polym Sci 2002;27:1123−63.

[101] Jamshidi K, Hyon SH, Ikada Y. Thermal characterization of polylactides. Polymer 1988;29:2229−34.

[102] Migliaresi C, Cohn D, Delollis A, Fambri L. Dynamic mechanical and calorimetric analysis of compression-molded plla of different molecular-weights—effect of thermal treatments. J Appl Polym Sci 1991;43:83−95.

[103] Middleton JC, Tipton AJ. Synthetic biodegradable polymers as orthopedic devices. Biomaterials 2000;21:2335−46.

[104] Engelberg I, Kohn J. Physicomechanical properties of degradable polymers used in medical applications—a comparative study. Biomaterials 1991;12:292−304.

[105] Li SM, Garreau H, Vert M. Structure property relationships in the case of the degradation of massive aliphatic poly-(alpha-hydroxy acids) in aqueous media. 1. Poly(DL-lactic acid). J Mater Sci Mater Med 1990;1:123−30.

[106] Singh V, Tiwari M. Structure-processing-property relationship of poly(glycolic acid) for drug delivery systems. 1. Synthesis and catalysis. Int J Polym Sci 2010 Article ID 652719. Available from: http://dx.doi.org/10.1155/2010/652719.

[107] Bostman O, Partio E, Hirvensalo E, Rokkanen P. Foreign-body reactions to polyglycolide screws—observations in 24/216 malleolar fracture cases. Acta Orthop Scand 1992;63:173−6.

[108] Bostman O, Paivarinta U, Partio E, Vasenius J, Manninen M, Rokkanen P. Degradation and tissue replacement of an absorbable polyglycolide screw in the fixation of rabbit femoral osteotomies. J Bone Joint Surg Am Vol 1992;74:1021−31.

[109] Griffith LG. Polymeric biomaterials. Acta Mater 2000;48:263−77.

[110] Miller RA, Brady JM, Cutright DE. Degradation rates of oral resorbable implants (polylactates and polyglycolates)—rate modification with changes in Pla−Pga copolymer ratios. J Biomed Mater Res 1977;11:711−9.

[111] Gilding DK, Reed AM. Biodegradable polymers for use in surgery—polyglycolic−poly(acetic acid) homopolymers and copolymers. Polymer 1979;20:1459−64.

[112] Park TG. Degradation of poly(lactic-co-glycolic acid) microspheres—effect of copolymer composition. Biomaterials 1995;16:1123−30.

[113] Lu L, Peter SJ, Lyman MD, Lai HL, Leite SM, Tamada JA, et al. *In vitro* and *in vivo* degradation of porous poly(DL-lactic-co-glycolic acid) foams. Biomaterials 2000;21:1837−45.

[114] Wu XS, Wang N. Synthesis, characterization, biodegradation, and drug delivery application of biodegradable lactic/glycolic acid polymers. Part II: biodegradation. J Biomat Sci Polym E 2001;12:21−34.

[115] Van de Velde K, Kiekens P. Biopolymers: overview of several properties and consequences on their applications. Polym Test 2002;21:433−42.

[116] Ikada Y, Tsuji H. Biodegradable polyesters for medical and ecological applications. Macromol Rapid Commun 2000;21:117−32.

[117] Sinha VR, Bansal K, Kaushik R, Kumria R, Trehan A. Poly-epsilon-caprolactone microspheres and nanospheres: an overview. Int J Pharm 2004;278:1−23.

[118] Woodruff MA, Hutmacher DW. The return of a forgotten polymer polycaprolactone in the 21st century. Prog Polym Sci 2010;35:1217−56.

[119] Li SM, Chen XH, Gross RA, McCarthy SP. Hydrolytic degradation of PCL/PEO copolymers in alkaline media. J Mater Sci Mater Med 2000;11:227–33.

[120] Fisher JP, Holland TA, Dean D, Mikos AG. Photoinitiated cross-linking of the biodegradable polyester poly(propylene fumarate). Part II. *In vitro* degradation. Biomacromolecules 2003;14:1335–42.

[121] Peter SJ, Miller ST, Zhu GM, Yasko AW, Mikos AG. *In vivo* degradation of a poly(propylene fumarate) beta-tricalcium phosphate injectable composite scaffold. J Biomed Mater Res 1998;41:1–7.

[122] Peter SJ, Kim P, Yasko AW, Yaszemski MJ, Mikos AG. Crosslinking characteristics of an injectable poly(propylene fumarate)/beta-tricalcium phosphate paste and mechanical properties of the crosslinked composite for use as a biodegradable bone cement. J Biomed Mater Res 1999;44:314–21.

[123] Heller J. Poly (ortho esters). Adv Polym Sci 1993;107:41–92.

[124] Heller J, Barr J, Ng SY, Abdellauoi KS, Gurny R. Poly(ortho esters): synthesis, characterization, properties and uses. Adv Drug Deliver Rev 2002;54:1015–39.

[125] Vistnes LM, Schmitt EE, Ksander GA, Rose EH, Balkenhol WJ, Coleman CL. Evaluation of a prototype therapeutic system for prolonged, continuous topical delivery of homosulfanilamide in the management of pseudomonas burn wound sepsis. Surgery 1976;79:690–6.

[126] Benagiano G, Schmitt E, Wise D, Goodman M. Sustained-release hormonal preparations for the delivery of fertility-regulating agents. J Polym Sci Part C Polym Symp 1979;66 (1):129–48.

[127] Solheim E, Pinholt EM, Andersen R, Bang G, Sudmann E. The effect of a composite of polyorthoester and demineralized bone on the healing of large segmental defects of the radius in rats. J Bone Joint Surg Am Vol 1992;74:1456–63.

[128] Heller J, Barr J. Poly(ortho esters)—from concept to reality. Biomacromolecules 2004;5:1625–32.

[129] Schwach-Abdellaoui K, Heller J, Gurny R. Hydrolysis and erosion studies of autocatalyzed poly(ortho esters) containing lactoyl–lactyl acid dimers. Macromolecules 1999;32:301–7.

[130] Ng SY, Shen HR, Lopez E, Zherebin Y, Barr J, Schacht E, et al. Development of a poly(ortho ester) prototype with a latent acid in the polymer backbone for 5-fluorouracil delivery. J Control Release 2000;65:367–74.

[131] Kumar N, Langer RS, Domb AJ. Polyanhydrides: an overview. Adv Drug Deliver Rev 2002;54:889–910.

[132] Leong KW, Damore P, Marletta M, Langer R. Bioerodible polyanhydrides as drug-carrier matrices. 2. Biocompatibility and chemical reactivity. J Biomed Mater Res 1986;20:51–64.

[133] Jiang HL, Zhu KJ. Preparation, characterization and degradation characteristics of polyanhydrides containing poly(ethylene glycol). Polym Int 1999;48:47–52.

[134] Uhrich KE, Thomas TT, Laurencin CT, Langer R. *In vitro* degradation characteristics of poly(anhydride-imides) containing trimellitylimidoglycine. J Appl Polym Sci 1997;63:1401–11.

[135] Muggli DS, Burkoth AK, Anseth KS. Crosslinked polyanhydrides for use in orthopedic applications: degradation behavior and mechanics. J Biomed Mater Res 1999;46:271–8.

[136] Uhrich KE, Gupta A, Thomas TT, Laurencin CT, Langer R. Synthesis and characterization of degradable poly(anhydride-co-imides). Macromolecules 1995;28:2184–93.

[137] Muggli DS, Burkoth AK, Keyser SA, Lee HR, Anseth KS. Reaction behavior of biodegradable, photo-cross-linkable polyanhydrides. Macromolecules 1998;31:4120–5.

[138] Anseth KS, Svaldi DC, Laurencin CT, Langer R. Photopolymerization of novel degradable networks for orthopedic applications. Photopolymerization 1997;673:189–202.

[139] Allcock HR. Polyphosphazene elastomers, gels, and other soft materials. Soft Matter 2012;8:7521–32.

[140] Deng M, Kumbar SG, Wan YQ, Toti US, Allcock HR, Laurencin CT. Polyphosphazene polymers for tissue engineering: an analysis of material synthesis, characterization and applications. Soft Matter 2010;6:3119–32.

[141] Allcock HR, Ambrosio AM. Synthesis and characterization of pH-sensitive poly(organophosphazene) hydrogels. Biomaterials 1996;17:2295–302.

[142] Allcock HR, Pucher SR, Turner ML, Fitzpatrick RJ. Poly(organophosphazenes) with poly(alkyl ether) side groups—a study of their water solubility and the swelling characteristics of their hydrogels. Macromolecules 1992;25:5573–7.

[143] Singh A, Krogman NR, Sethuraman S, Nair LS, Sturgeon JL, Brown PW, et al. Effect of side group chemistry on the properties of biodegradable L-alanine cosubstituted polyphosphazenes. Biomacromolecules 2006;7:914–8.

[144] Andrianov AK, Marin A. Degradation of polyaminophosphazenes: effects of hydrolytic environment and polymer processing. Biomacromolecules 2006;7:1581–6.

[145] Kumbar SG, Bhattacharyya S, Nukavarapu SP, Khan YM, Nair LS, Laurencin CT. *In vitro* and *in vivo* characterization of biodegradable poly(organophosphazenes) for biomedical applications. J Inorg Organomet Polym Mater 2006;16:365–85.

[146] Wang J, Mao HQ, Leong KW. A novel biodegradable gene carrier based on polyphosphoester. J Am Chem Soc 2001;123:9480–1.

[147] Wang YC, Yuan YY, Du JZ, Yang XZ, Wang J. Recent progress in polyphosphoesters: from controlled synthesis to biomedical applications. Macromol Biosci 2009;9:1154–6114.

[148] Chaubal MV, Su G, Spicer E, Dang WB, Branham KE, English JP, et al. *In vitro* and *in vivo* degradation studies of a novel linear copolymer of lactide and ethylphosphate. J Biomat Sci Polym E 2003;14:45–61.

[149] Wang DA, Williams CG, Yang F, Cher N, Lee H, Elisseeff JH. Bioresponsive phosphoester hydrogels for bone tissue engineering. Tissue Eng 2005;11:201–13.

[150] Wang YC, Li Y, Yang XZ, Yuan YY, Yan LF, Wang J. Tunable thermosensitivity of biodegradable polymer micelles of poly (epsilon-caprolactone) and polyphosphoester block copolymers. Macromolecules 2009;42:3026–32.

[151] Wang YC, Tang LY, Li Y, Wang J. Thermoresponsive block copolymers of poly(ethylene glycol) and polyphosphoester: thermo-induced self-assembly, biocompatibility, and hydrolytic degradation. Biomacromolecules 2009;10:66–73.

[152] Lee S, Son Y, Kim C, Lee J, Kim S, Koh Y. Voice outcomes of polyacrylamide hydrogel injection laryngoplasty. Laryngoscope 2007;117:1871–5.

[153] Christensen LH, Breiting VB, Aasted A, Jorgensen A, Kebuladze I. Long-term effects of polyacrylamide hydrogel on human breast tissue. Plast Reconstr Surg 2003;111:1883–90.

[154] Bassil M, Davenas J, El Tahchi M. Electrochemical properties and actuation mechanisms of polyacrylamide hydrogel for

artificial muscle application. Sens Actuators B Chem 2008;134:496–501.

[155] Toozs-Hobson P, Al-Singary W, Fynes M, Tegerstedt G, Lose G. Two-year follow-up of an open-label multicenter study of polyacrylamide hydrogel (Bulkamid®) for female stress and stress-predominant mixed incontinence. Int Urogynecol J 2012;23:1373–8.

[156] Pallua N, Wolter TP. A 5-year assessment of safety and aesthetic results after facial soft-tissue augmentation with polyacrylamide hydrogel (Aquamid): a prospective multicenter study of 251 patients. Plast Reconstr Surg 2010;125:1797–804.

[157] Refojo MF, Yasuda H. Hydrogels from 2-hydroxycthylmcthacrylatc and propylcne glycol monoacrylatc. J Appl Polym Sci 1965;9:2425–35.

[158] Meyvis T, De Smedt S, Stubbe B, Hennink W, Demeester J. On the release of proteins from degrading dextran methacrylate hydrogels and the correlation with the rheologic properties of the hydrogels. Pharm Res 2001;18:1593–9.

[159] Madden LR, Ratner BD, Mortisen DJ, Sussman EM, Dupras SK, Fugate JA, et al. Proangiogenic scaffolds as functional templates for cardiac tissue engineering. Proc Natl Acad Sci USA 2010;107:15211–6.

[160] Hejcl A, Lesny P, Pradny M, Sedy J, Zamecnik J, Jendelova P, et al. Macroporous hydrogels based on 2-hydroxyethyl methacrylate. Part 6: 3D hydrogels with positive and negative surface charges and polyelectrolyte complexes in spinal cord injury repair. J Mater Sci Mater Med 2009;20:1571–7.

[161] Gupta H, Aqil M. Contact lenses in ocular therapeutics. Drug Discov Today 2012;17:522–7.

[162] Gloria A, De Santis R, Ambrosio L, Causa F, Tanner KE. A multi-component fiber-reinforced PHEMA-based hydrogel/HAPEX device for customized intervertebral disc prosthesis. J Biomater Appl 2011;25:795–810.

[163] Peppas NA, Hilt JZ, Khademhosseini A, Langer R. Hydrogels in biology and medicine: from molecular principles to bionanotechnology. Adv Mater 2006;18:1345–60.

[164] Hyon SH, Cha WI, Ikada Y, Kita M, Ogura Y, Honda Y. Poly (vinyl alcohol) hydrogels as soft contact-lens material. J Biomat Sci Polym E 1994;5:397–406.

[165] Masters KS, Leibovich SJ, Belem P, West JL, Poole-Warren LA. Effects of nitric oxide releasing poly(vinyl alcohol) hydrogel dressings on dermal wound healing in diabetic mice. Wound Repair Regen 2002;10:286–94.

[166] Kelly BT, Robertson W, Potter HG, Deng XH, Turner AS, Lyman S, et al. Hydrogel meniscal replacement in the sheep knee. Am J Sports Med 2007;35:43.

[167] Kobayashi M, Toguchida J, Oka M. Preliminary study of polyvinyl alcohol-hydrogel (PVA-H) artificial meniscus. Biomaterials 2003;24:639–47.

[168] Kobayashi M, Chang YS, Oka M. A two year *in vivo* study of polyvinyl alcohol-hydrogel (PVA-H) artificial meniscus. Biomaterials 2005;26:3243–8.

[169] Saldarriaga Fernandez IC, Mei HCvd, Metzger S, Grainger DW, Engelsman AF, Nejadnik MR, et al. *In vitro* and *in vivo* comparisons of staphylococcal biofilm formation on a cross-linked poly (ethylene glycol)-based polymer coating. Acta Biomater 2010;6:1119–24.

[170] Guenaga KF, Matos D, Wille-Jorgensen P. Mechanical bowel preparation for elective colorectal surgery. Cochrane Database Syst Rev 2011;7(9):CD001544. Available from: http://dx.doi.org/10.1002/14651858.CD001544.pub4.

[171] Haensig M, Mohr FW, Rastan AJ. Bioresorbable adhesion barrier for reducing the severity of postoperative cardiac adhesions: focus on REPEL-CV((R)). Med Devices (Auckl) 2011;4:17–25.

[172] ten Broek RPG, Kok-Krant N, Verhoeve HR, van Goor H, Bakkum EA. Efficacy of polyethylene glycol adhesion barrier after gynecological laparoscopic surgery: results of a randomized controlled pilot study. Gynecol Surg 2012;9:29–35.

[173] Suzuki S, Ikada Y. Barriers to prevent tissue adhesion. Totowa, NJ: Humana Press; 201291–130

[174] Moore J, Perkins A. Evaluating antimicrobial efficacy and cost of 3 dressings containing silver versus a novel antimicrobial hydrogel impregnated gauze dressing containing Oakin, an oak extract. Adv Skin Wound Care 2010;23:544–51.

[175] Lev B, Alexander Yu,G, Eriko Sato M, Yasuo S, Toyoichi T. Dependency of swelling on the length of subchain in poly(N,N-dimethylacrylamide)-based gels. J Chem Phys 1997;106:2906–10.

[176] Baker BA, Murff RL, Milam VT. Tailoring the mechanical properties of polyacrylamide-based hydrogels. Polymer 2010;51:2207–14.

[177] Kulicke WM, Kniewske R, Klein J. Preparation, characterization, solution properties and rheological behaviour of polyacrylamide. Prog Polym Sci 1982;8:373–468.

[178] Rice JM. The carcinogenicity of acrylamide. MutRes Genetic Toxicol Environ Mutagen 2005;580:3–20.

[179] Xi TF, Fan CX, Feng XM, Wan ZY, Wang CR, Chou LL. Cytotoxicity and altered c-myc gene expression by medical polyacrylamide hydrogel. J Biomed Mater Res Part A 2006;78:283–90.

[180] Slaughter BV, Khurshid SS, Fisher OZ, Khademhosseini A, Peppas NA. Hydrogels in regenerative medicine. Adv Mater 2009;21:3307–29.

[181] Lai JY, Wang TP, Li YT, Tu IH. Synthesis, characterization and ocular biocompatibility of potential keratoprosthetic hydrogels based on photopolymerized poly(2-hydroxyethyl methacrylate)-co-poly(acrylic acid). J Mater Chem 2012;22:1812–23.

[182] Lahooti S, Sefton MV. Microencapsulation of normal and transfected L929 fibroblasts in a HEMA–MMA copolymer. Tissue Eng 2000;6:139–49.

[183] Bryant SJ, Cuy JL, Hauch KD, Ratner BD. Photo-patterning of porous hydrogels for tissue engineering. Biomaterials 2007;28:2978–86.

[184] Denizli A, Tuncel A, Olcay M, Sarnatskaya V, Sergeev V, Nikolaev VG, et al. Biologically modified PHEMA beads for hemoperfusion: preliminary studies. Clin Mater 1992;11:129–37.

[185] Saini R, Bajpai J, Bajpai AK. Synthesis of poly(2-hydroxyethyl methacrylate) (PHEMA) based nanoparticles for biomedical and pharmaceutical applications. Methods Mol Biol 2012;906:321.

[186] Pawlowska E, Poplawski T, Ksiazek D, Szczepanska J, Blasiak J. Genotoxicity and cytotoxicity of 2-hydroxyethyl methacrylate. MutRes Genetic Toxicol Environ Mutagen 2010;696:122–9.

[187] Baker MI, Walsh SP, Schwartz Z, Boyan BD. A review of polyvinyl alcohol and its uses in cartilage and orthopedic applications. J Biomed Mater Res B Appl Biomater 2012;100:1451–7.

[188] Peppas NA, Wright SL. Drug diffusion and binding in ionizable interpenetrating networks from poly(vinyl alcohol) and poly (acrylic acid). Eur J Pharm Biopharm 1998;46:15–29.

[189] Hassan CM, Peppas NA. Structure and applications of poly(vinyl alcohol) hydrogels produced by conventional crosslinking or by freezing/thawing methods. Berlin: Springer-Verlag; 2000;37–65

[190] Alves MH, Jensen BEB, Smith AAA, Zelikin AN. Poly(vinyl alcohol) physical hydrogels: new vista on a long serving biomaterial. Macromol Biosci 2011;11:1293–313.

[191] Koyano T, Minoura N, Nagura M, Kobayashi K. Attachment and growth of cultured fibroblast cells on PVA/chitosan-blended hydrogels. J Biomed Mater Res 1998;39:486–90.

[192] Wang M, Li Y, Wu J, Xu F, Zuo Y, Jansen JA. *In vitro* and *in vivo* study to the biocompatibility and biodegradation of hydroxyapatite/poly(vinyl alcohol)/gelatin composite. J Biomed Mater Res Part A 2008;85:418–26.

[193] Lang RA, Grüntzig PM, Weisgerber C, Weis C, Odermatt EK, Kirschner MH. Polyvinyl alcohol gel prevents abdominal adhesion formation in a rabbit model. Fertil Steril 2007;88:1180–6.

[194] Kokabi M, Sirousazar M, Hassan ZM. PVA–clay nanocomposite hydrogels for wound dressing. Eur Polym J 2007;43:773–81.

[195] Lin W, Yu D, Yang M. Blood compatibility of novel poly (gamma-glutamic acid)/polyvinyl alcohol hydrogels. Colloids Surf B Biointerfaces 2006;47:43–9.

[196] Lee JH, Lee HB, Andrade JD. Blood compatibility of polyethylene oxide surfaces. Prog Polym Sci 1995;20:1043–79.

[197] Gref R, Domb A, Quellec P, Blunk T, Müller RH, Verbavatz JM, et al. The controlled intravenous delivery of drugs using PEG-coated sterically stabilized nanospheres. Adv Drug Delivery Rev 1995;16:215–33.

[198] Webster R, Elliott V, Park BK, Walker D, Hankin M, Taupin P. PEG and PEG conjugates toxicity: towards an understanding of the toxicity of PEG and its relevance to PEGylated biologicals. Basel: Birkhauser Basel; 2009;127–146.

[199] Thoma DS, Subramani K, Weber FE, Luder HU, Hämmerle CHF, Jung RE. Biodegradation, soft and hard tissue integration of various polyethylene glycol hydrogels: a histomorphometric study in rabbits. Clin Oral Implants Res 2011;22:1247–54.

Chapter 8

Natural Biomaterials for Regenerative Medicine Applications

Denver M. Faulk[a] and Stephen F. Badylak[a,b,c]

[a]*Department of Bioengineering, University of Pittsburgh, Pittsburgh, PA,* [b]*Department of Surgery, University of Pittsburgh, Pittsburgh, PA,*
[c]*McGowan Institute for Regenerative Medicine, University of Pittsburgh, PA*

Chapter Outline

8.1 INTRODUCTION

8.1.1 Naturally Occurring Biomaterials

Naturally occurring biomaterials are commonly used as surgical mesh devices for a number of clinical applications and are increasingly recognized in regenerative medicine for their inductive properties in tissue and organ engineering. These materials are often composed of extracellular matrix (ECM), a material which consists of the secreted structural and functional components of resident cells. The specific composition and ultrastructure of ECM will vary depending on the source tissue. The composition of ECM scaffolds includes a complex mixture of molecules arranged in unique three-dimensional (3D) patterns that are ideally suited to the tissue from which the ECM is harvested. The ECM provides signals which cue cell migration, proliferation, and differentiation [1−7]. Maintaining the native ultrastructure and composition of the ECM during the tissue/organ decellularization process is essential for optimal outcomes with biologic scaffolds [8−12]. ECM scaffold materials are derived from a variety of tissues and organs including blood vessels [13−15], ligaments [16], heart valves [17−23], skin [24], nerves [25,26], skeletal muscle, tendons [27,28], small intestinal submucosa (SIS) [29−31], heart [32,33], urinary bladder [34−36], and liver [37]. Although a number of studies have been conducted with many different ECM materials, the most comprehensive studies regarding structure−function relationships have been reported for urinary bladder matrix (UBM) and SIS. ECM scaffold materials are biodegradable unless processed with crosslinking agents, and their degradation products have been shown to be important bioactive contributors to the constructive remodeling process [38,39]. Furthermore, incomplete removal of cellular material from the source tissue results in an undesirable host remodeling response [40]. A more detailed discussion of the mechanisms by which naturally occurring biologic materials support constructive remodeling can be found later in this chapter.

8.1.2 Synthetic Biomaterials

New generations of synthetic biomaterials which attempt to mimic the native ECM are being developed at a rapid pace for use as 3D scaffolds. Poly(glycolic acid) (PGA), poly(lactic acid) (PLA), and the copolymer PLGA have been extensively used as synthetic 3D scaffold biomaterials. Biomimetic synthetic polymers have been created to elicit specific cellular functions and to direct cell—cell interactions. The use of synthetic materials for regenerative medicine applications can be effective, but there are inherent limitations to the use of synthetic materials. Of note, synthetic or chemically crosslinked biologic materials invariably elicit a foreign body response.

Generally, the presence of an absorbable component as part or the entirety of a synthetic device or a natural device results in a less severe foreign body response than a nondegradable device. The effective ability of the host to effectively degrade and remodel an implanted biomaterial facilitates the integration of the implanted device with surrounding tissue and minimizes the formation of dense capsule formation.

8.1.3 Utilizing Nature's Engineered Scaffold Material

Individual ECM components or combinations of specific ECM components have been used as substrates for *in vitro* cell culture systems for several decades. Although cell culture substrates composed of proteins, such as collagen type I, laminin, and fibronectin, have facilitated cell attachment, proliferation, and differentiation, these systems are far from physiologically relevant. The lack of an intact ECM structure negatively affects cell—cell interaction, cell—ECM interaction, physical—chemical influences, and the effects of mechanical stimuli upon the cultured cells. As a result, the cells in culture eventually lose their functional cell phenotype and initiate a dedifferentiation process. Studies have shown that intact ECM provides a more favorable substrate than the use of individual components of ECM for the growth and differentiation of various cell types. Squamous epithelial, fibroblastic (Swiss 3T3), glandular epithelial (adenocarcinoma), and smooth muscle-like (urinary bladder) cells cultured on SIS-ECM maintained superior expression of tissue-specific phenotype than those same cells cultured on plastic, Vitrogen, or Matrigel. In a study that used human islet cells, SIS enhanced the insulin producing function of the cells *in vitro* better than islets cultured on standard islet substrates [41—43]. Lin et al. compared hepatocytes cultured on L-ECM to the well-characterized hepatocyte culture models, double-gel (sandwich) cultures, and adsorbed collagen monolayers. Hepatocytes survived up to 45 days on L-ECM, and several liver-specific functions, such as albumin synthesis, urea production, and P-450 IA1 activity, were significantly greater than the growth and metabolism of cells cultured on collagen [44].

In summary, there is an increasing body of evidence that tissue-specific signals are present within the ECM from each tissue and organ. This property may be important in selected regenerative medicine strategies.

8.2 ECM COMPOSITION AND TISSUE-SPECIFIC ULTRASTRUCTURE

The ECM provides a physical substratum for the attachment and spatial organization of cells [45]. Well-known biophysical properties, such as ECM composition and 3D surface topography, play an important role in cell phenotype and behavior [46—50]. The 3D surface topography of ECM derived from different tissues includes variation in pore diameter, distance between the pores [51], surface texture [49,52], hydrophobicity, hydrophilicity, and the presence or absence of a basement membrane. In addition to surface topography, the spatial presentation of the surface proteins contacting the cells can alter cell attachment, migration, proliferation, and differentiation (Figure 8.1).

Studies have suggested that tissue- and organ-specific ECM can promote site-appropriate differentiation of progenitor cells [53,54] and maintain site appropriate phenotype in *in vitro* culture systems [55]. Since the ECM of each tissue and organ is produced by the resident cells and logically represents the ideal scaffold or substrate for these cells, it is intuitive that a substrate composed of a tissue-specific ECM would be favorable for those tissue cells. It has been shown that biologic ECM scaffolds support tissue-specific cell phenotype. Hepatic sinusoidal endothelial cells maintained their differentiated phenotype longer when cultured on ECM derived from the liver compared to sinusoidal endothelial cells cultured on ECM derived from small intestine or urinary bladder [55]. These results suggest that ECM biologic scaffolds provide a set of unique, tissue-specific signals that are dependent upon the tissue from which an ECM is derived. Lin et al. compared rat hepatocytes cultured on porcine liver ECM biologic scaffolds to well-characterized hepatocyte culture models (type I collagen sandwich configuration or a single layer of type I collagen) [44]. Hepatocytes survived up to 45 days on a sheet form of porcine liver ECM and several liver-specific functions, such as albumin synthesis, urea production, and P-450 IA1 activity, were markedly enhanced compared with the growth and metabolism of cells cultured on a single layer of type I collagen.

Zeisberg et al. isolated liver-derived basement membrane matrix from human or bovine liver, and used the substrate for culture of human hepatocytes [56]. Human hepatocytes adhered more efficiently to liver-derived basement membrane matrix and expressed lower levels of vimentin and cytokeratin-18, which are markers of

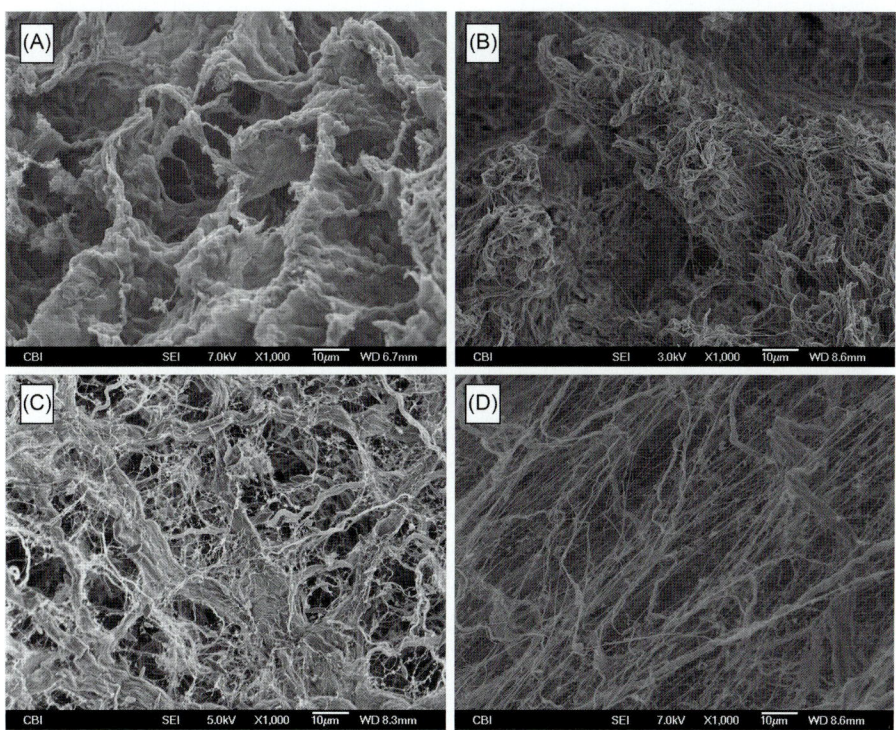

FIGURE 8.1 SEM images (1000×) of naturally occurring ECM scaffold derived from (A) adrenal, (B) dermal, (C) liver, and (D) pancreas tissues. These images demonstrate a distinct, tissue-specific architecture.

hepatocyte dedifferentiation, compared with hepatocytes cultured on Matrigel or type I collagen. However, the maintenance of liver-specific functions *in vitro* was not reported [56].

Biologic scaffolds composed of ECM provide a more physiologically relevant culture substrate compared to reconstituted ECM proteins and have unique biochemical profiles that are dependent upon the tissue from which an ECM scaffold is derived. Minimally processed ECM scaffolds and gels like PLECM provide a combination of ECM proteins derived from the liver that are in physiologically relevant amounts (e.g., Types I, IV, VI, III, XI, XIX, heparin sulfate proteoglycan, ECM-bound growth factors, laminin, biglycan, tenascin, fibronectin) [52,57,58].

8.3 PRODUCTION OF NATURALLY OCCURRING BIOMATERIALS FROM DECELLULARIZED TISSUES

Decellularization of anatomically distinct tissues to produce an ECM bioscaffold has been reported for hollow organs, such as small intestinal, urinary bladder [59,60], and dermis [61], as well as solid organs, such as liver [62] and heart [33]. The use of these bioscaffolds to support tissue reconstruction varies from bridging repairs to structural replacement (i.e., heart valves [63,64]). There have been several comprehensive reviews of commonly

used decellularization methods, which will be briefly discussed below [60,65,66].

While the specific methods used to decellularize the tissue can vary greatly, the physical form of the resulting ECM can vary greatly. Hollow organs, such as the urinary bladder or small intestine, are typically incised and split open, thus forming a sheet-like structure. The sheet form can be air dried or lyophilized and further processed into a comminuted powdered form. The sheet form is typically very thin; therefore, multiple sheets can be laminated together by vacuum pressing to increase strength of the resulting multilayer construct. Lamination can also be used to create more complex 3D shapes like tubes by molding the sheets around a mandrel. Lamination, however, is limited by the nature of the starting sheets and by casting requirements of particular shapes.

The sheet form of ECM has many advantages, not least of which is that it is usually rapidly produced, often requiring only a couple of processing steps and easily scaled for mass production. The sheet or laminate sheet form lends itself well to patch, bridging, or soft tissue reinforcement applications for which robust strength is usually required. The comminuted form in contrast is well suited for filling of volumetric injuries and can be applied by minimally invasive methods like injection.

The maintenance of 3D organ morphology has recently emerged as a method for organ engineering [67]. Organs, such as the heart, liver, kidney, and lung, are

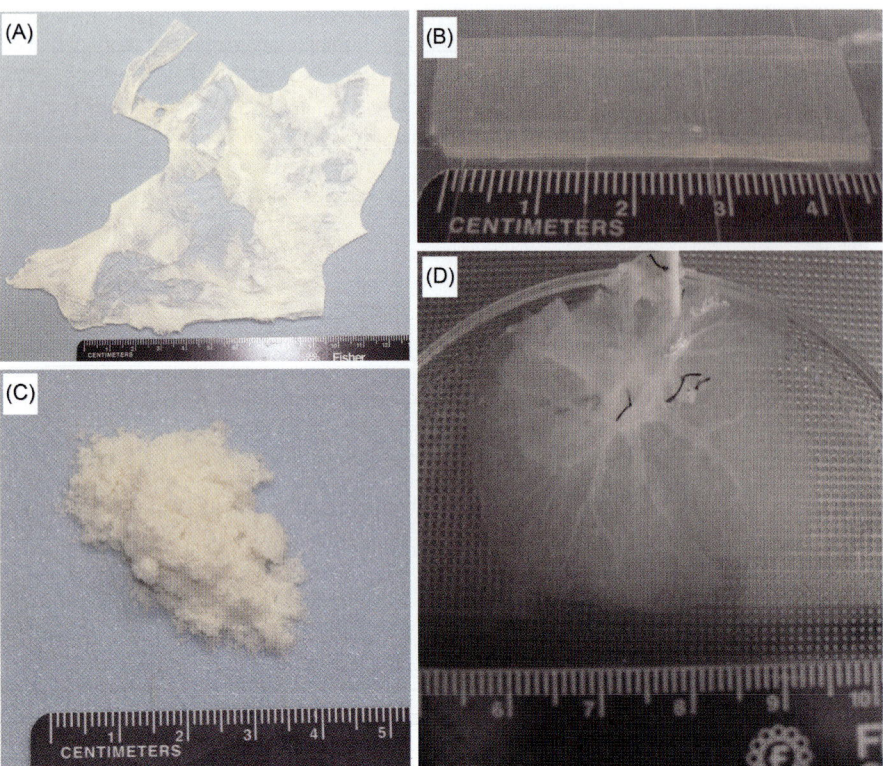

FIGURE 8.2 Naturally occurring scaffolds processed in various formats: (A) sheet, (B) hydrogel, (C) powder, and (D) 3D whole organ.

decellularized by using the native vasculature to deliver decellularizing agents. This method leaves the organ as semitransparent in appearance while retaining the shape of the native organ with intact vasculature. The benefit of this method is the ability to subsequently connect the vasculature of the decellularized organ to the recipient circulatory system. Re-endothelialization of the 3D ECM construct is necessary to prevent thrombosis (Figure 8.2).

8.3.1 Techniques Used to Decellularize Tissue

The most effective agents for decellularization of each tissue and organ will depend upon many factors, including the tissue cellularity (e.g., liver versus tendon), density (e.g., dermis versus adipose tissue), lipid content (e.g., brain versus urinary bladder), and thickness (e.g., dermis versus pericardium). It should be understood that all cell removal agents and methods will disrupt ECM composition and ultrastructure. Minimization of these undesirable effects rather than complete avoidance is the objective of optimal decellularization.

The optimal recipe of decellularization agents is dependent upon tissue characteristics, such as thickness and density, the agents being used, and the intended clinical application of the decellularized tissue. Prior to applying decellularization agents, undesirable excess tissue may be removed to simplify the cell removal process. Cell removal may focus on the retention of key ECM components like the basement membrane. Mechanical or physical methods may be used to facilitate decellularization. For thin tissue structures, such as urinary bladder, intestine, pericardium, and amnion, the most commonly used decellularization techniques include freezing and thawing, mechanical removal of undesirable layers such as muscle or submucosa, and relatively brief exposure to easily removed detergents or acids followed by rinsing. Thicker tissue structures like dermis may require more extensive biochemical exposure and longer rinse times. Fatty, amorphous organs and tissues such as adipose tissue, brain, and pancreas often require the addition of lipid solvents like alcohols. The complexity and length of the decellularization protocol is usually proportional to the degree of geometric and biologic conservation desired for the postprocessed tissue (e.g., macrostructure, ultrastructure, matrix and basement membrane proteins, growth factors, etc.), especially for composite tissues and whole organs.

Regardless of the decellularization method chosen, minimizing the disruption of matrix molecules is of paramount importance. The method used to decellularize the tissue can be broadly divided into physical, enzymatic, or

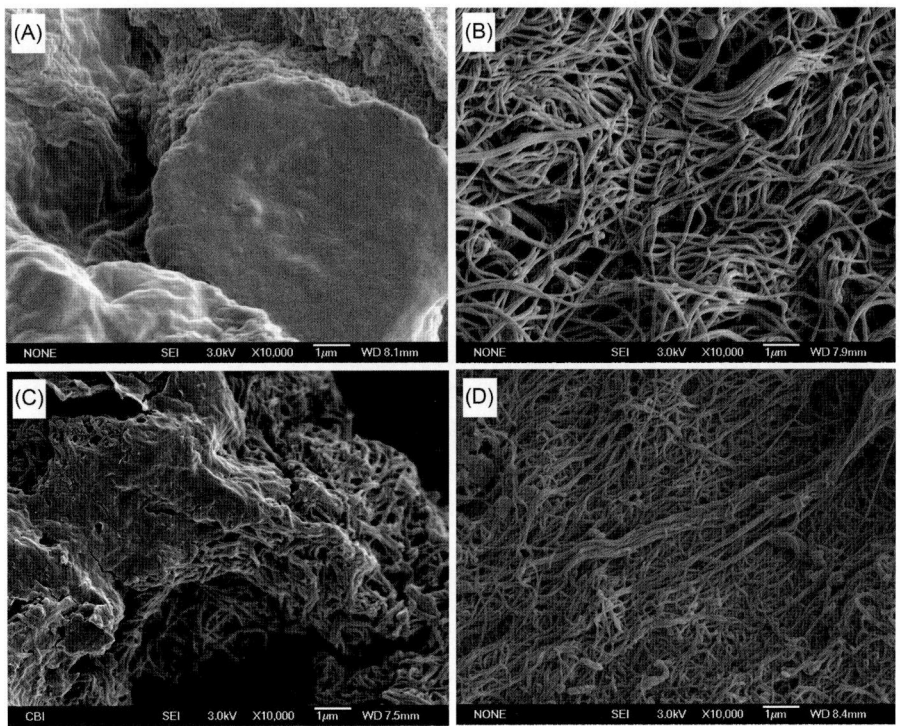

FIGURE 8.3 SEM images (10,000 ×) of porcine urinary bladder tissue decellularized with (A) 8 mM CHAPS, (B) 4% deoxycholate, (C) 1% SDS, and (D) 3% Triton X-100 have unique ECM fiber networks. Note the marked differences in architecture as a function of decellularizing agents.

chemical techniques with many protocols mixing various methods and agents to optimize decellularization while maintaining native structure and composition. Physical methods not only typically involve the separation of the epithelial and muscular layer of an organ from the adjacent submucosal layers but can also include freezing or osmotic shock with hyper- and/or hypotonic solutions.

While there are many potential enzymatic methods of disrupting cells, trypsin is one of the most commonly used and effective enzymes. Trypsin cleaves peptide chains mainly at the carboxyl side of the amino acids lysine or arginine, except when either is followed by proline. This results in cellular detachment from the matrix and cell lysis. One of the disadvantages of using enzymes is the destruction of matrix proteins, especially with prolonged exposure which can also disrupt the native matrix architecture. Following enzymatic treatment, the tissue must be thoroughly rinsed to remove or inactivate any remnant enzyme and to remove cellular material. Enzymes are often used in conjunction with other methods in a stepwise process to facilitate dislodgement of the cells while follow-up methods target the remaining, often damaged, cells.

Chemical methods of decellularization are varied in nature and mechanism of action. Detergents, changes of pH, and solvents or chelators can all be effectively used to decellularize tissue. Detergents comprise some of the most commonly used chemicals for decellularization, including sodium dodecyl sulfate (SDS), deoxycholate, and Triton X-100. SDS and deoxycholate are ionic

detergents that not only solubilize cellular membranes but also disrupt protein−protein interactions that may damage the ECM ulstrastructure and protein integrity. Triton X-100 is a nonionic detergent that disrupts lipid interactions while leaving protein interactions intact. Zwitterionic detergents like CHAPS mix properties of both ionic and nonionic detergents but are less commonly used.

Placing tissues in acidic or alkaline solutions can not only effectively disrupt cells but can also irreversibly damage ECM proteins. Peracetic acid (PAA) is a relatively weak acid that has been widely used as a decellularizing agent and a disinfectant, especially with ECMs, such as SIS and UBM, which have a thin cross-sectional area (Figure 8.3).

Following cell lysis, the membranes and cytoplasmic debris must be removed from the matrix, which is usually accomplished by aggressive rinsing in sterile water. This washing procedure also serves to remove any remnant detergents or enzymes from the decellularization process. The use of iso-, hyper- or hypotonic washing solutions can facilitate additional cell lysis as a physical method of decellularization as described above.

8.3.2 Criteria for Decellularization

Following decellularization, the tissue usually assumes a pale or translucent quality. However, macroscopic appearance alone is insufficient to determine the extent of decellularization. While there is no universal consensus for criteria for adequate decellularization, standard metrics are beginning to emerge. It has been proposed that three relatively stringent criteria must be met to establish sufficient decellularization: (i) tissue must have <50 ng of double stranded DNA (dsDNA) per mg of dry weight, (ii) DNA fragments <200 bp in length, and (iii) no visible nuclear material in histologic analysis with DAPI or H&E [66,68,69].

Failure to completely decellularize a tissue leads to negative outcomes upon *in vivo* implantation, including a pro-inflammatory response with recruitment of M1 macrophages and subsequent fibrosis [40]. Such a reaction is likely caused in part by damage-associated molecular pattern (DAMP) molecules and can lead to seroma formation, sterile abscess formation, and chronic inflammation [70].

8.3.3 Chemical Crosslinking and Sterilization

Following decellularization, further processing of the ECM can include dehydration, lyophilization, and/or comminution. Once comminuted, the powder can be enzymatically solubilized (e.g., pepsin) to produce a liquid form that can be subsequently induced to gel [32,61,71–73]. This soluble form allows for minimally invasive applications, such as needle or catheter-based injections [32]. The viscosity of the soluble ECM can be adjusted so that it can be induced to conform to the contours of native tissue or irregularly shaped tissue defects. Although the solubilization of an ECM scaffold obviously destroys the 3D architecture, entrapped growth factors and newly created bioactive cryptic peptides may be released.

Terminal sterilization of an ECM bioscaffold is required prior to *in vivo* implantation. Gamma irradiation, e-beam, glutaraldehyde, ethylene oxide, and PAA have all been used as methods of sterilization and have been extensively evaluated for their effect on bioscaffold mechanical and biological integrities. In addition to sterilization, glutaraldehyde effectively crosslinks ECM proteins. Other chemical agents, such as carbodiimide and genipin, are also crosslinking agents prior to sterilization. Chemical crosslinking of ECM proteins (e.g., collagen) stabilizes and strengthens the ECM structure and severely inhibits *in vivo* degradation. Chemically crosslinked and nondegradable ECM elicits a foreign body response very similar to nondegradable synthetic polymer scaffolds (e.g., polypropylene).

8.3.4 Source Material of ECM

Surgical mesh materials composed of ECM are harvested from a variety of allogeneic or xenogeneic tissue sources, including dermis, urinary bladder, small intestine, mesothelium, pericardium, and heart valves, and from several different species. Although there have been no reported differences in the host remodeling response as a function of

species source, there are clear differences with respect to age [74], anatomic location of the source tissue, and other factors [69]. The potential for disease transmission to humans is much less for a xenogeneic ECM compared to allogeneic (human) ECMs. Although a certain amount of biologic variability is unavoidable with naturally occurring biomaterials, such variability can be minimized by controlling for factors, such as age, weight, breed, and diet of the source animals.

8.4 MECHANISMS BY WHICH BIOMATERIALS SUPPORT CONSTRUCTIVE REMODELING

ECM scaffolds can change the default wound healing response from the well-described pro-inflammatory and scarring events toward a more constructive remodeling response [75]. Site appropriate formation of functional tissue has been shown for a variety of anatomic locations including dermis [76], esophagus [77], skeletal muscle [78–80], heart [32,81–85], and temporomandibular joint meniscus, among others. There currently exists strong evidence for three mechanisms by which biomaterials support constructive remodeling: (i) contribution from bioactive cryptic peptides released during the process of *in vivo* ECM degradation [86,87], (ii) recruitment of endogenous stem and progenitor cells to the site of ECM remodeling [86,87], and (iii) modulation of the host immune response toward an M2-Th2 phenotype [38,70,88,89].

8.4.1 Cryptic Peptides and Bioactive Molecules

The host response to an ECM that is not chemically crosslinked is distinctly different from the response to synthetic scaffold materials and from the response to either purified components of ECM, such as collagen I or chemically crosslinked forms of ECM. The purified components elicit a specific response to the particular molecule, for example, angiogenesis in response to Vascular Endothelial Growth Factor (VEGF) or bone formation in response to Bone morphogenetic protein (BMP). The specific response may be desirable for particular medical/surgical needs but typically lacks the complex constructive wound healing response observed when the intact ECM is used as a reparative scaffold. It has been shown that matrix metalloproteinases (MMPs) can cause release of matricryptic peptides from ECM, such as endostatin [90], restin [91], and arrestin [92]. Macrophages can release matricryptic angiostatic factors from ECM [93]. Degradation and remodeling of the ECM by heparanases can release bioactive matricryptic peptides [94,95].

ECM scaffolds in preclinical studies have been shown to result in constructive tissue remodeling, promote angiogenesis, and to resist deliberate bacterial infection [75,96]. *In vitro* studies have shown that the degradation of ECM generates low molecular weight peptides with biological properties, such as chemotaxis, angiogenesis, and antimicrobial activity [97,98]. In contrast, intact ECM does not possess such activity [99], suggesting that these biological activities are associated with and dependent upon the products of ECM degradation, rather than molecules present in intact ECM. Low molecular weight peptides isolated from acid-hydrolyzed small intestinal submucosa (SIS-ECM) have been shown to possess chemotactic activity for primary murine adult liver, heart, and kidney endothelial cells and to promote vascularization *in vivo* in Matrigel plug assays [100]. A recent study has shown that *in vivo* degradation of urinary bladder-derived ECM can produce bioactive matricryptic peptides that cause chemotaxis of multipotential progenitor cells [101]. Previous studies have also shown that subcutaneous implantation of porcine SIS-ECM induces migration of bone marrow-derived cells to the site of constructive remodeling [102], and that this phenomenon is associated with complete degradation of the ECM scaffold.

The concept of functional matricryptic peptides is not new. During ECM degradation, many large insoluble molecules present within the matrix are reduced to fragments which possess biological activities that are not possessed by the parent molecules. In addition to proteolysis-generated bioactive fragments, functional sites of the parent molecules that are hidden and inactive within the ECM can also become active due to conformational changes [103]. Some of the ECM molecules that have been shown to possess this property are collagen [91,104,105], fibronectin [106], laminin [107−109], and hyaluronan [110], all of which are present in ECM scaffold materials [75].

8.4.2 Recruitment of Progenitor Cells to the Site of Injury

The second mechanism of site-specific ECM scaffold remodeling involves the recruitment of multipotent or lineage-directed progenitor cells by scaffold degradation products. Both the age and species of the tissue from which the ECM is harvested have an effect upon this chemoattractant potential. Transitional ECM instructs cell behavior in both the blastema and mammalian skeletal muscle regeneration, especially tenascin-C, hyaluronic acid, and fibronectin [111].

Studies have shown degradation products of human fetal skin-derived ECM possess stronger chemoattractant activity for skin-specific lineage-directed stem and progenitor cells than do degradation products of human adult skin-derived ECM. In addition, degradation products of porcine adult skin-derived ECM showed stronger chemoattractant activity than degradation products of human adult skin-derived ECM. These results suggest that ECM degradation products from younger tissue sources may have more potent chemoattractant activity for local tissue stem or progenitor cells than older tissue sources, and that the species of the tissue source also has an effect on chemoattractant activity.

The early gestation human fetus (<24 weeks gestation), unlike the later-gestation fetus or adult, is able to heal incisional skin wounds without scarring [112]. This scarless wound healing ability of fetal skin may be due to the fetal cells and/or the fetal ECM. Fetal platelets, inflammatory cells, and fibroblasts all demonstrate differences from their adult counterparts that may contribute to the scarless wound healing phenomenon. Fetal ECM contains a higher proportion of type III collagen and a greater concentration of hyaluronic acid than ECM of adult skin. Unwounded fetal skin at gestational ages associated with scarless wound healing has also been shown to express low levels of Transforming Growth Factor (TGF)β1, high levels of TGFβ3, and increased expression of MMPs, a family of proteases associated with ECM degradation and remodeling [113]. Although studies have not addressed the relative contribution of cells versus ECM to scarless fetal wound healing, it appears clear that the ECM, which represents the secreted product of local cells, contains signaling molecules that can affect stem and progenitor cell activity.

The results of a previous study show that degradation products of ECM possess chemoattractant activity for local tissue progenitor and stem cells, and that this chemoattractant activity may decline as a function of the age of the tissue from which the ECM is harvested [114], and that ECM may vary between different species. These findings add a new perspective to the role of ECM in wound healing and the differences between fetal and adult wound healing. Though fetal ECM is not a likely candidate for tissue engineering/regenerative medicine applications because of the scarce availability of the raw material, these findings could be applied to the development of methods to induce migration of lineage-directed progenitor cells to tissue sites in need of repair, thereby facilitating a regenerative tissue response rather than default scar tissue formation.

8.4.3 Modulation of the Host Immune Response—Push Towards an M2 Constructive Remodeling Macrophage Phenotype

The exact mechanisms by which certain biologic mesh materials are capable of modulating the host macrophage population towards a more constructive remodeling phenotype are not fully understood. However, it has been shown that the presence of large amounts of cellular remnant material as a result of ineffective tissue decellularization and chemical crosslinking has detrimental effects upon the host remodeling response [38,76,115]. This response is not surprising in the case of xenogeneic cellular components like the α-Gal epitope, which may be recognized by the host immune system and elicit an immune response following implantation. Other molecules, including those associated with cell death, are also known to have potent immunomodulatory effects. These cell death-associated molecules, collectively termed DAMPs, are recognized by pattern recognition receptors on cells of the innate immune system. Therefore, large amounts of these molecules within biologic materials that derive from mammalian tissues as a result of inefficient removal during processing or due to cellular death upon implantation may have detrimental effects upon the ability of ECM scaffold materials to promote constructive tissue remodeling.

Macrophages are a heterogeneous subset of the mononuclear cell population involved in the host response to implanted materials. Macrophages are activated in response to tissue damage or infection, causing an increase in the production of cytokines, chemokines, and other inflammatory molecules to which they are exposed. Recently, macrophage phenotype has been characterized based on distinct functional properties, surface markers, and the cytokine profile of the microenvironment. Polarized macrophages are referred to as either M1 or M2 cells, mimicking the Th1/Th2 nomenclature.

Macrophages are a plastic cell population capable of sequentially changing their polarization in response to local stimuli during the process of wound healing. The macrophages participating in the host response to an implanted material are exposed to multiple stimuli including cytokines and effector molecules secreted by cells including other macrophages that are participating in the host response, microbial agents, epitopes associated with the implanted biomaterial, and the degradation products of the biomaterial, among others. Therefore, it is logical to assume that the host macrophage response after implantation of a biomaterial is modulated via "crosstalk" between macrophages and the other cells involved in the host response as well as factors within the local microenvironment.

Considerable variability has been seen in the host immune response to xenogeneic biologic scaffolds depending upon the source of the ECM, method of decellularization, and the presence of any modifications like chemical crosslinking. Crosslinking of ECM with carbodiimide has been correlated with delayed degradation, a

chronic mononuclear cell accumulation around the device and a foreign body reaction.

Pro-inflammatory responses toward biomaterials are typically associated with encapsulation and a foreign body reaction. However, the bioactive and bioinductive molecules within the ECM that induce polarization are unclear, although it is likely that cellular remnants like DAMPs retained within the scaffold may play a role in the investigation of the immunomodulatory effects of common ECM scaffolds. Results showed that tissue source, decellularization method, and chemical crosslinking modifications affect the presence of the well-characterized DAMP—HMGB1 (High-mobility group box-1). In addition, these factors were correlated with differences in cell proliferation, death, secretion of the chemokines CCL2 and CCL4, and upregulation of the pro-inflammatory signaling receptor TLR4 (toll-like receptor 4). Inhibition of HMGB1 with glycyrrhizin increased the pro-inflammatory response, increasing cell death and upregulating chemokine and TLR4 mRNA expression. This suggests the importance of HMGB1 and other DAMPs as bioinductive molecules within the ECM scaffold. Identification and evaluation of other ECM bioactive molecules will be an area of future interest for new biomaterial development.

REFERENCES

[1] Vorotnikova E, McIntosh D, Dewilde A, Zhang J, Reing JE, Zhang L, et al. Extracellular matrix-derived products modulate endothelial and progenitor cell migration and proliferation in vitro and stimulate regenerative healing in vivo. Matrix Biol 2010;29:690–700.

[2] Barkan D, Green JE, Chambers AF. Extracellular matrix: a gatekeeper in the transition from dormancy to metastatic growth. Eur J Cancer 2010;46:1181–8.

[3] Nelson CM, Bissell MJ. Of extracellular matrix, scaffolds, and signaling: tissue architecture regulates development, homeostasis, and cancer. Annu Rev Cell Dev Biol 2006;22:287–309.

[4] Taylor KR, Gallo RL. Glycosaminoglycans and their proteoglycans: host-associated molecular patterns for initiation and modulation of inflammation. FASEB J 2006;20:9–22.

[5] Nagase H, Visse R, Murphy G. Structure and function of matrix metalloproteinases and TIMPs. Cardiovasc Res 2006;69:562–73.

[6] Werner S, Grose R. Regulation of wound healing by growth factors and cytokines. Physiol Rev 2003;83:835–70.

[7] Bornstein P, Sage EH. Matricellular proteins: extracellular modulators of cell function. Curr Opin Cell Biol 2002;14:608–16.

[8] Ott HC, Matthiesen TS, Goh SK, Black LD, Kren SM, Netoff TI, et al. Perfusion-decellularized matrix: using nature's platform to engineer a bioartificial heart. Nat Med 2008;14:213–21.

[9] Uygun BE, Soto-Gutierrez A, Yagi H, Izamis ML, Guzzardi MA, Shulman C, et al. Organ reengineering through development of a transplantable recellularized liver graft using decellularized liver matrix. Nat Med 2010;16:814–20.

[10] Petersen TH, Calle EA, Zhao L, Lee EJ, Gui L, Raredon MB, et al. Tissue-engineered lungs for *in vivo* implantation. Science 2010;329:538−41.

[11] Nakayama KH, Batchelder CA, Lee CI, Tarantal AF. Decellularized rhesus monkey kidney as a three-dimensional scaffold for renal tissue engineering. Tissue Eng Part A 2010;16:2207−16.

[12] Allen RA, Seltz LM, Jiang H, Kasick RT, Sellaro TL, Badylak SF, et al. Adrenal extracellular matrix scaffolds support adrenocortical cell proliferation and function *in vitro*. Tissue Eng Part A 2010;16:3363−74.

[13] Conklin BS, Richter ER, Kreutziger KL, Zhong DS, Chen C. Development and evaluation of a novel decellularized vascular xenograft. Med Eng Phys 2002;24:173−83.

[14] Schmidt CE, Baier JM. Acellular vascular tissues: natural biomaterials for tissue repair and tissue engineering. Biomaterials 2000;21:2215−31.

[15] Uchimura E, Sawa Y, Taketani S, Yamanaka Y, Hara M, Matsuda H, et al. Novel method of preparing acellular cardiovascular grafts by decellularization with poly(ethylene glycol). J Biomed Mater Res A 2003;67:834−7.

[16] Woods T, Gratzer PF. Effectiveness of three extraction techniques in the development of a decellularized bone-anterior cruciate ligament-bone graft. Biomaterials 2005;26:7339−49.

[17] Bader A, Schilling T, Teebken OE, Brandes G, Herden T, Steinhoff G, et al. Tissue engineering of heart valves—human endothelial cell seeding of detergent acellularized porcine valves. Eur J Cardiothorac Surg 1998;14:279−84.

[18] Booth C, Korossis SA, Wilcox HE, Watterson KG, Kearney JN, Fisher J, et al. Tissue engineering of cardiac valve prostheses I: development and histological characterization of an acellular porcine scaffold. J Heart Valve Dis 2002;11:457−62.

[19] Grauss RW, Hazekamp MG, Oppenhuizen F, van Munsteren CJ, Gittenberger-de Groot AC, DeRuiter MC. Histological evaluation of decellularised porcine aortic valves: matrix changes due to different decellularisation methods. Eur J Cardiothorac Surg 2005;27:566−71.

[20] Kasimir MT, Rieder E, Seebacher G, Silberhumer G, Wolner E, Weigel G, et al. Comparison of different decellularization procedures of porcine heart valves. Int J Artif Organs 2003;26:421−7.

[21] Korossis SA, Booth C, Wilcox HE, Watterson KG, Kearney JN, Fisher J, et al. Tissue engineering of cardiac valve prostheses II: biomechanical characterization of decellularized porcine aortic heart valves. J Heart Valve Dis 2002;11:463−71.

[22] Rieder E, Kasimir MT, Silberhumer G, Seebacher G, Wolner E, Simon P, et al. Decellularization protocols of porcine heart valves differ importantly in efficiency of cell removal and susceptibility of the matrix to recellularization with human vascular cells. J Thorac Cardiovasc Surg 2004;127:399−405.

[23] Schenke-Layland K, Vasilevski O, Opitz F, Konig K, Riemann I, Halbhuber KJ, et al. Impact of decellularization of xenogeneic tissue on extracellular matrix integrity for tissue engineering of heart valves. J Struct Biol 2003;143:201−8.

[24] Chen RN, Ho HO, Tsai YT, Sheu MT. Process development of an acellular dermal matrix (ADM) for biomedical applications. Biomaterials 2004;25:2679−86.

[25] Hudson TW, Liu SY, Schmidt CE. Engineering an improved acellular nerve graft via optimized chemical processing. Tissue Eng 2004;10:1346−58.

[26] Kim BS, Yoo JJ, Atala A. Peripheral nerve regeneration using acellular nerve grafts. J Biomed Mater Res A 2004;68:201−9.

[27] Borschel GH, Dennis RG, Kuzon Jr. WM. Contractile skeletal muscle tissue-engineered on an acellular scaffold. Plast Reconstr Surg 2004;113:595−602 discussion 603−604

[28] Tanaka T, Sun YL, Zhao C, Zobitz ME, An KN, Amadio PC. Effect of curing time and concentration for a chemical treatment that improves surface gliding for extrasynovial tendon grafts *in vitro*. J Biomed Mater Res A 2006;79:451−5.

[29] Badylak SF, Lantz GC, Coffey A, Geddes LA. Small intestinal submucosa as a large diameter vascular graft in the dog. J Surg Res 1989;47:74−80.

[30] Badylak SF, Tullius R, Kokini K, Shelbourne KD, Klootwyk T, Voytik SL, et al. The use of xenogeneic small intestinal submucosa as a biomaterial for Achilles tendon repair in a dog model. J Biomed Mater Res 1995;29:977−85.

[31] Kropp BP, Eppley BL, Prevel CD, Rippy MK, Harruff RC, Badylak SF, et al. Experimental assessment of small intestinal submucosa as a bladder wall substitute. Urology 1995;46:396−400.

[32] Singelyn JM, Sundaramurthy P, Johnson TD, Schup-Magoffin PJ, Hu DP, Faulk DM, et al. Catheter-deliverable hydrogel derived from decellularized ventricular extracellular matrix increases endogenous cardiomyocytes and preserves cardiac function post-myocardial infarction. J Am Coll Cardiol 2012;59:751−63.

[33] Wainwright JM, Czajka CA, Patel UB, Freytes DO, Tobita K, Gilbert TW, et al. Preparation of cardiac extracellular matrix from an intact porcine heart. Tissue Eng Part C Methods 2010;16:525−32.

[34] Chen F, Yoo JJ, Atala A. Acellular collagen matrix as a possible "off the shelf" biomaterial for urethral repair. Urology 1999;54:407−10.

[35] Freytes DO, Badylak SF, Webster TJ, Geddes LA, Rundell AE. Biaxial strength of multilaminated extracellular matrix scaffolds. Biomaterials 2004;25:2353−61.

[36] Gilbert TW, Stolz DB, Biancaniello F, Simmons-Byrd A, Badylak SF. Production and characterization of ECM powder: implications for tissue engineering applications. Biomaterials 2005;26:1431−5.

[37] Sellaro TL, Ranade A, Faulk DM, McCabe GP, Dorko K, Badylak SF, et al. Maintenance of human hepatocyte function *in vitro* by liver-derived extracellular matrix gels. Tissue Eng Part A 2010;16:1075−82.

[38] Brown BN, Valentin JE, Stewart-Akers AM, McCabe GP, Badylak SF. Macrophage phenotype and remodeling outcomes in response to biologic scaffolds with and without a cellular component. Biomaterials 2009;30:1482−91.

[39] Reing JE, Zhang L, Myers-Irvin J, Cordero KE, Freytes DO, Heber-Katz E, et al. Degradation products of extracellular matrix affect cell migration and proliferation. Tissue Eng Part A 2009;15:605−14.

[40] Keane TJ, Londono R, Turner NJ, Badylak SF. Consequences of ineffective decellularization of biologic scaffolds on the host response. Biomaterials 2012;33:1771−81.

[41] Tian XH, Xue WJ, Pang XL, Teng Y, Tian PX, Feng XS. Effect of small intestinal submucosa on islet recovery and function *in vitro* culture. Hepatobiliary Pancreat Dis Int 2005;4:524−9.

[42] Tian XH, Xue WJ, Ding XM, Pang XL, Teng Y, Tian PX, et al. Small intestinal submucosa improves islet survival and function during *in vitro* culture. World J Gastroenterol 2005;11:7378—83.

[43] Xiaohui T, Wujun X, Xiaoming D, Xinlu P, Yan T, Puxun T, et al. Small intestinal submucosa improves islet survival and function *in vitro* culture. Transplant Proc 2006;38:1552—8.

[44] Lin P, Chan WC, Badylak SF, Bhatia SN. Assessing porcine liver-derived biomatrix for hepatic tissue engineering. Tissue Eng 2004;10:1046—53.

[45] Hubbell JA. Materials as morphogenetic guides in tissue engineering. Curr Opin Biotechnol 2003;14:551—8.

[46] Dow JA, Clark P, Connolly P, Curtis AS, Wilkinson CD. Novel methods for the guidance and monitoring of single cells and simple networks in culture. J Cell Sci Suppl 1987;8:55—79.

[47] Clark P, Connolly P, Curtis AS, Dow JA, Wilkinson CD. Topographical control of cell behaviour. II. Multiple grooved substrata. Development 1990;108:635—44.

[48] Clark P, Connolly P, Curtis AS, Dow JA, Wilkinson CD. Topographical control of cell behaviour. I. Simple step cues. Development 1987;99:439—48.

[49] den Braber ET, de Ruijter JE, Smits HT, Ginsel LA, von Recum AF, Jansen JA. Quantitative analysis of cell proliferation and orientation on substrata with uniform parallel surface micro-grooves. Biomaterials 1996;17:1093—9.

[50] Chehroudi B, Soorany E, Black N, Weston L, Brunette DM. Computer-assisted three-dimensional reconstruction of epithelial cells attached to percutaneous implants. J Biomed Mater Res 1995;29:371—9.

[51] Zeltinger J, Sherwood JK, Graham DA, Mueller R, Griffith LG. Effect of pore size and void fraction on cellular adhesion, proliferation, and matrix deposition. Tissue Eng 2001;7:557—72.

[52] Brown B, Lindberg K, Reing J, Stolz DB, Badylak SF. The basement membrane component of biologic scaffolds derived from extracellular matrix. Tissue Eng 2006;12:519—26.

[53] Cepko CL. The roles of intrinsic and extrinsic cues and bHLH genes in the determination of retinal cell fates. Curr Opin Neurobiol 1999;9:37—46.

[54] Jadhav AP, Mason HA, Cepko CL. Notch 1 inhibits photoreceptor production in the developing mammalian retina. Development 2006;133:913—23.

[55] Sellaro TL, Ravindra AK, Stolz DB, Badylak SF. Maintenance of hepatic sinusoidal endothelial cell phenotype *in vitro* using organ-specific extracellular matrix scaffolds. Tissue Eng 2007;13:2301—10.

[56] Zeisberg M, Kramer K, Sindhi N, Sarkar P, Upton M, Kalluri R. De-differentiation of primary human hepatocytes depends on the composition of specialized liver basement membrane. Mol Cell Biochem 2006;283:181—9.

[57] Voytik-Harbin SL, Brightman AO, Kraine MR, Waisner B, Badylak SF. Identification of extractable growth factors from small intestinal submucosa. J Cell Biochem 1997;67:478—91.

[58] Hodde JP, Badylak SF, Brightman AO, Voytik-Harbin SL. Glycosaminoglycan content of small intestinal submucosa: a bioscaffold for tissue replacement. Tissue Eng 1996;2:209—17.

[59] Sandusky Jr. GE, Badylak SF, Morff RJ, Johnson WD, Lantz G. Histologic findings after *in vivo* placement of small intestine submucosal vascular grafts and saphenous vein grafts in the carotid artery in dogs. Am J Pathol 1992;140:317—24.

[60] Badylak SF, Freytes DO, Gilbert TW. Extracellular matrix as a biological scaffold material: structure and function. Acta Biomater 2009;5:1—13.

[61] Wolf MT, Daly KA, Brennan-Pierce EP, Johnson SA, Carruthers CA, D'Amore A, et al. A hydrogel derived from decellularized dermal extracellular matrix. Biomaterials 2012;33:7028—38.

[62] Soto-Gutierrez A, Zhang L, Medberry C, Fukumitsu K, Faulk D, Jiang H, et al. A whole-organ regenerative medicine approach for liver replacement. Tissue Eng Part C Methods 2011;17:677—86.

[63] Bloch O, Erdbrugger W, Volker W, Schenk A, Posner S, Konertz W, et al. Extracellular matrix in deoxycholic acid decellularized aortic heart valves. Med Sci Monit 2012;18:487—92.

[64] Kneib C, von Glehn CQ, Costa FD, Costa MT, Susin MF. Evaluation of humoral immune response to donor HLA after implantation of cellularized versus decellularized human heart valve allografts. Tissue Antigens 2012;80:165—74.

[65] Crapo PM, Gilbert TW, Badylak SF. An overview of tissue and whole organ decellularization processes. Biomaterials 2011;32:3233—43.

[66] Gilbert TW, Sellaro TL, Badylak SF. Decellularization of tissues and organs. Biomaterials 2006;27:3675—83.

[67] Badylak SF, Weiss DJ, Caplan A, Macchiarini P. Engineered whole organs and complex tissues. Lancet 2012;379:943—52.

[68] Gilbert TW, Freund JM, Badylak SF. Quantification of DNA in biologic scaffold materials. J Surg Res 2009;152:135—9.

[69] Badylak SF, Gilbert TW. Immune response to biologic scaffold materials. Semin Immunol 2008;20:109—16.

[70] Daly KA, Liu S, Agrawal V, Brown BN, Johnson SA, Medberry CJ, et al. Damage associated molecular patterns within xenogeneic biologic scaffolds and their effects on host remodeling. Biomaterials 2012;33:91—101.

[71] Singelyn JM, Christman KL. Modulation of material properties of a decellularized myocardial matrix scaffold. Macromol Biosci 2011;11:731—8.

[72] Freytes DO, Martin J, Velankar SS, Lee AS, Badylak SF. Preparation and rheological characterization of a gel form of the porcine urinary bladder matrix. Biomaterials 2008;29:1630—7.

[73] Medberry CJ, Crapo PM, Siu BF, Carruthers CA, Wolf MT, Nagarkar SP, et al. Hydrogels derived from central nervous system extracellular matrix. Biomaterials 2013;34:1033—40.

[74] Sicari BM, Johnson SA, Siu BF, Crapo PM, Daly KA, Jiang H, et al. The effect of source animal age upon the *in vivo* remodeling characteristics of an extracellular matrix scaffold. Biomaterials 2012;33:5524—33.

[75] Badylak SF. The extracellular matrix as a scaffold for tissue reconstruction. Semin Cell Dev Biol 2002;13:377—83.

[76] Daly KA, Liu S, Agrawal V, Brown BN, Huber A, Johnson SA, et al. The host response to endotoxin-contaminated dermal matrix. Tissue Eng Part A 2012;18:1293—303.

[77] Londono R, Jobe BA, Hoppo T, Badylak SF. Esophagus and regenerative medicine. World J Gastroenterol 2012;18:6894—9.

[78] Sicari BM, Agrawal V, Siu BF, Medberry CJ, Dearth CL, Turner NJ, et al. A murine model of volumetric muscle loss and a regenerative medicine approach for tissue replacement. Tissue Eng Part A 2012;18:1941–8.

[79] Turner NJ, Badylak JS, Weber DJ, Badylak SF. Biologic scaffold remodeling in a dog model of complex musculoskeletal injury. J Surg Res 2012;176:490–502.

[80] Turner NJ, Badylak SF. Regeneration of skeletal muscle. Cell Tissue Res 347:759–74.

[81] Wainwright JM, Hashizume R, Fujimoto KL, Remlinger NT, Pesyna C, Wagner WR, et al. Right ventricular outflow tract repair with a cardiac biologic scaffold. Cells Tissues Organs 195:159–170.

[82] Badylak SF, Kochupura PV, Cohen IS, Doronin SV, Saltman AE, Gilbert TW, et al. The use of extracellular matrix as an inductive scaffold for the partial replacement of functional myocardium. Cell Transplant 2006;15(Suppl. 1):29–40.

[83] Kochupura PV, Azeloglu EU, Kelly DJ, Doronin SV, Badylak SF, Krukenkamp IB, et al. Tissue-engineered myocardial patch derived from extracellular matrix provides regional mechanical function. Circulation 2005;112:144–9.

[84] Kastner T, Berkner P, DeSouza T, Wight D, Waran S. Rett syndrome and metabolic disorder. J Am Acad Child Adolesc Psychiatry 1992;31:567–8.

[85] Rane AA, Christman KL. Biomaterials for the treatment of myocardial infarction: a 5-year update. J Am Coll Cardiol 2011;58:2615–29.

[86] Agrawal V, Tottey S, Johnson SA, Freund JM, Siu BF, Badylak SF. Recruitment of progenitor cells by an extracellular matrix cryptic peptide in a mouse model of digit amputation. Tissue Eng Part A 2011;17:2435–43.

[87] Agrawal V, Kelly J, Tottey S, Daly KA, Johnson SA, Siu BF, et al. An isolated cryptic peptide influences osteogenesis and bone remodeling in an adult mammalian model of digit amputation. Tissue Eng Part A 2011;17:3033–44.

[88] Brown BN, Londono R, Tottey S, Zhang L, Kukla KA, Wolf MT, et al. Macrophage phenotype as a predictor of constructive remodeling following the implantation of biologically derived surgical mesh materials. Acta Biomater 2012;8:978–87.

[89] Brown BN, Ratner BD, Goodman SB, Amar S, Badylak SF. Macrophage polarization: an opportunity for improved outcomes in biomaterials and regenerative medicine. Biomaterials 2012;33:3792–802.

[90] O'Reilly MS, Boehm T, Shing Y, Fukai N, Vasios G, Lane WS, et al. Endostatin: an endogenous inhibitor of angiogenesis and tumor growth. Cell 1997;88:277–85.

[91] Ramchandran R, Dhanabal M, Volk R, Waterman MJ, Segal M, Lu H, et al. Antiangiogenic activity of restin, NC10 domain of human collagen XV: comparison to endostatin. Biochem Biophys Res Commun 1999;255:735–9.

[92] Colorado PC, Torre A, Kamphaus G, Maeshima Y, Hopfer H, Takahashi K, et al. Anti-angiogenic cues from vascular basement membrane collagen. Cancer Res 2000;60:2520–6.

[93] Houghton AM, Grisolano JL, Baumann ML, Kobayashi DK, Hautamaki RD, Nehring LC, et al. Macrophage elastase (matrix metalloproteinase-12) suppresses growth of lung metastases. Cancer Res 2006;66:6149–55.

[94] Vlodavsky I, Goldshmidt O, Zcharia E, Atzmon R, Rangini-Guatta Z, Elkin M, et al. Mammalian heparanase: involvement in cancer metastasis, angiogenesis and normal development. Semin Cancer Biol 2002;12:121–9.

[95] Roy M, Marchetti D. Cell surface heparan sulfate released by heparanase promotes melanoma cell migration and angiogenesis. J Cell Biochem 2009;106:200–9.

[96] Badylak SF, Coffey AC, Lantz GC, Tacker WA, Geddes LA. Comparison of the resistance to infection of intestinal submucosa arterial autografts versus polytetrafluoroethylene arterial prostheses in a dog model. J Vasc Surg 1994;19:465–72.

[97] Sarikaya A, Record R, Wu CC, Tullius B, Badylak S, Ladisch M. Antimicrobial activity associated with extracellular matrices. Tissue Eng 2002;8:63–71.

[98] Brennan EP, Reing J, Chew D, Myers-Irvin JM, Young EJ, Badylak SF. Antibacterial activity within degradation products of biological scaffolds composed of extracellular matrix. Tissue Eng 2006;12:2949–55.

[99] Holtom PD, Shinar Z, Benna J, Patzakis MJ. Porcine small intestine submucosa does not show antimicrobial properties. Clin Orthop Relat Res 2004;:18–21.

[100] Li F, Li W, Johnson S, Ingram D, Yoder M, Badylak S. Low-molecular-weight peptides derived from extracellular matrix as chemoattractants for primary endothelial cells. Endothelium 2004;11:199–206.

[101] Beattie AJ, Gilbert TW, Guyot JP, Yates AJ, Badylak SF. Chemoattraction of progenitor cells by remodeling extracellular matrix scaffolds. Tissue Eng Part A 2009;15:1119–25.

[102] Badylak SF, Park K, Peppas N, McCabe G, Yoder M. Marrow-derived cells populate scaffolds composed of xenogeneic extracellular matrix. Exp Hematol 2001;29:1310–8.

[103] Schenk S, Quaranta V. Tales from the crypt[ic] sites of the extracellular matrix. Trends Cell Biol 2003;13:366–75.

[104] Ortega N, Werb Z. New functional roles for non-collagenous domains of basement membrane collagens. J Cell Sci 2002;115:4201–14.

[105] Mott JD, Werb Z. Regulation of matrix biology by matrix metalloproteinases. Curr Opin Cell Biol 2004;16:558–64.

[106] Ambesi A, Klein RM, Pumiglia KM, McKeown-Longo PJ. Anastellin, a fragment of the first type III repeat of fibronectin, inhibits extracellular signal-regulated kinase and causes G(1) arrest in human microvessel endothelial cells. Cancer Res 2005;65:148–56.

[107] Ponce ML, Hibino S, Lebioda AM, Mochizuki M, Nomizu M, Kleinman HK. Identification of a potent peptide antagonist to an active laminin-1 sequence that blocks angiogenesis and tumor growth. Cancer Res 2003;63:5060–4.

[108] Ponce ML, Kleinman HK. Identification of redundant angiogenic sites in laminin alpha1 and gamma1 chains. Exp Cell Res 2003;285:189–95.

[109] Schenk S, Hintermann E, Bilban M, Koshikawa N, Hojilla C, Khokha R, et al. Binding to EGF receptor of a laminin-5 EGF-like fragment liberated during MMP-dependent mammary gland involution. J Cell Biol 2003;161:197–209.

[110] Chen WY, Abatangelo G. Functions of hyaluronan in wound repair. Wound Repair Regen 1999;7:79—89.

[111] Calve S, Odelberg SJ, Simon HG. A transitional extracellular matrix instructs cell behavior during muscle regeneration. Dev Biol 344:259—271.

[112] Adzick NS, Longaker MT. Scarless fetal healing. Therapeutic implications. Ann Surg 1992;215:3—7.

[113] Bullard KM, Longaker MT, Lorenz HP. Fetal wound healing: current biology. World J Surg 2003;27:54—61.

[114] Brennan EP, Tang XH, Stewart-Akers AM, Gudas LJ, Badylak SF. Chemoattractant activity of degradation products of fetal and adult skin extracellular matrix for keratinocyte progenitor cells. J Tissue Eng Regen Med 2008;2:491—8.

[115] Brown BN, Freund JM, Han L, Rubin JP, Reing, JE, Jeffries EM, et al. Comparison of three methods for the derivation of a biologic scaffold composed of adipose tissue extracellular matrix. Tissue Eng Part C Methods 17:411—421.

Chapter 9

Bioartificial Biomaterials for Regenerative Medicine Applications

Valeria Chiono, Tiziana Nardo, and Gianluca Ciardelli

Department of Mechanical and Aerospace Engineering, Politecnico di Torino, Torino, Italy

Chapter Outline

9.1 RELEVANCE OF BIOARTIFICIAL MATERIALS IN REGENERATIVE MEDICINE

Tissue engineering (TE) strategies are generally aimed at tissue regeneration, mediated by purposely designed three-dimensional (3D) matrices (scaffolds) regulating cell function. TE methods for tissue regeneration require the design of (i) appropriate scaffolding biomaterials, (ii) suitable 3D scaffolds, and (iii) the presence of molecules on the scaffold/biomaterial able to impart haptotactic signals to cells.

In any organ and tissue, the extracellular matrix (ECM) components provide the means by which adjacent cells communicate with each other and the external environment, therefore the ECM organization and composition impact organ development and function [1]. The vital interface between cells and ECM components is mainly provided by transmembrane cell surface receptors, integrins, which ensure communication and mediate bidirectional signaling across the cell membrane. Binding of integrins to ECM results in the activation of a cascade of events named "outside-in" signaling. In doing so, integrins influence a wide range of activities including cell morphology, proliferation, and survival [2].

A promising approach to regulate cell behavior is to engineer "biomimetic" scaffolds modulating cell response, in a similar manner as the ECM does *in vivo*.

The most important requirement for a biodegradable polymer to be used in medical applications is its cell and tissue compatibility in a specific environment, together with the noncytotoxicity of its degradation products [3]. Aliphatic polyesters, such as polylactide (PLA), polyglycolide (PGA), poly(lactide-co-glycolide) (PLGA), and poly(ε-caprolactone) (PCL), are among the few synthetic polymers that meet most of these requirements and have therefore been used in the engineering of tissues, such as cartilage [4−7], bone [8−10], tendon [11], skin [12], liver [13], and heart valves [14]. Synthetic polymers are advantageous for the fabrication of TE scaffolds, due to their usually performing mechanical properties, reproducible chemistry, and times of degradation in physiological conditions that match tissue regeneration processes. However, cell affinity towards synthetic polymers is generally poor as a consequence of their low hydrophilicity and lack of surface cell recognition sites [15,16]. In Table 9.1, the main water-unsoluble synthetic polymers used in TE applications are collected, together with their key properties and applications.

Polymers listed in Table 9.1 degrade principally by simple hydrolysis of the ester bonds in the polymer backbone. Partial chain scission degrades the polymer to $10-40$ micron particles, capable of being phagocytosed

TABLE 9.1 Main Biocompatible Water-Unsoluble Polymers for TE Applications and Their Properties

Synthetic Polymer	Molecular Formula	Main Characteristics	Applications
Poly(α-hydroxyacids)			
PLA		• Biocompatible • Bioresorbable (degradation time > 24 months) • Optically active polymer due to the presence of chiral centers • Poly(L-lactic acid) (PLLA): semicrystalline polymer with glass transition temperature (T_g) between 50 and 80°C and melting temperature (T_m) between 173 and 178°C • Racemic PLA: amorphous polymer with T_g between 55 and 60°C	Suture, bone TE [17], drug delivery [18], other TE applications [19], such as bioresorbable stents [20]
PGA		• Biocompatible • Bioresorbable (degradation time: 6–12 months) • Highly crystalline (T_g: 35–40°C; T_m: 225–230°C) • Rigid (elastic modulus, E: 7 GPa) • Not soluble in most organic solvents due to high crystallinity • Soluble in highly fluorinated solvents	Sutures, bone TE [21]
PLGA		• Biocompatible • Bioresorbable (degradation time: 1–16 months) • Amorphous with T_g of 45–60°C • Mechanical and degradation properties can be finely tuned through variations in molecular weight and copolymer ratio • No linear relationship between the physical properties of homopolymers and monomer composition	Drug release [22–24]
PCL		• Biocompatible • Bioresorbable (degradation time > 24 months) • Semicrystalline with T_g of around −65°C and T_m of 58–63°C	Drug delivery [25], various TE applications [26,27]
PLCL		• Biocompatible • Bioresorbable • Amorphous with T_g of around 16°C • Mechanical and degradation properties can be altered by changing the monomer content	Various TE applications, including nerve [28], cartilage, and bone regeneration [29,30]
Polyhydroxyalkanoates			
Poly (hydroxybutyrate) (PHB)		• Biocompatible • Bioresorbable (by surface erosion) • Produced by bacteria (such as *Ralstonia eutrophus* or *Bacillus megaterium*) fermentation • High cost	Various TE applications, such as nerve [31,32] and bone [33] regeneration

(Continued)

TABLE 9.1 (Continued)

Synthetic Polymer	Molecular Formula	Main Characteristics	Applications
		• Optically active polymer due to the presence of chiral centers • Semicrystalline (high crystallinity) with T_g of 4–10°C and T_m of around 175°C • Brittleness and rigidity • High thermal degradation rate at higher temperature than melting point	
Poly (hydroxybutyrate-co-hydroxyvalerate) (PHBHV)		• Biocompatible • Bioresorbable (by surface erosion) • Produced by bacteria fermentation • Remarkable mechanical, physical, and thermoplastic properties depending on monomer ratio • Tenacity • High cost	Various TE applications including nerve [34], skin [35], and vascular [36] regeneration
Poly(anhydride)s			
Poly(anhydride)s		• Degradation into nontoxic diacid monomers • Degradation by surface erosion	Various TE applications depending on their chemistry [37,38]

TABLE 9.2 Main Biocompatible Water-Soluble Polymers for TE Applications and Their Properties

Synthetic Polymer	Molecular Formula	Main Characteristics	Applications
Poly (ethylene glycol) (PEG)		• Not biodegradable • Antiadhesive and antifouling properties • Semicrystalline	Antiadhesive and antifouling coating [39], constituent of TE hydrogels [40], and copolymers [41], porogen for scaffold production [42,43]
Poly(vinyl pyrrolidone) (PVP)		• Amorphous polymer with T_g of 130–175°C • Soluble in polar solvents • Able to absorb up to 40% of its weight in atmospheric water	Drug delivery systems [44], hydrogels [45], various TE applications [46,47]
Poly(vinyl alcohol) (PVA)		• Atactic • Semicrystalline with T_m of 230°C and 180–190°C for the fully hydrolyzed and partially hydrolyzed grades, respectively • It decomposes rapidly above 200°C • Excellent film forming, emulsifying, and adhesive properties	Drug delivery systems [48], emulsifying agent, hydrogels [49]

and metabolized to carbon dioxide and water. The degradation time is a function of the chemical structure of the polymer and its molecular weight.

Table 9.2 collects the main water-soluble synthetic polymers used for TE applications, together with their main properties and applications.

Typical natural polymers for TE include polysaccharides and proteins. Polysaccharides [chitosan (CS), chondroitin sulfate, heparin sulfate, heparin, hyaluronic acid, etc.] are attractive materials for TE applications since their carbohydrate moieties are components of membrane proteoglycans and extracellular matrix molecules (glycosaminoglycans, proteoglycans, glycoproteins). On the other hand, proteins, such as collagen (or its derivative, gelatin), fibronectin (FN), laminin, fibrin, and silk fibroin, are natural components of the ECM or display functional motifs (peptides) which are ligands for specific cell surface receptors.

Tables 9.3 and 9.4 report the basic features of proteins and polysaccharides and their potential applications in the biomedical field. Collected data are mainly derived from Refs. [98—100].

The cell communicating activity of proteins is mainly developed through short portions of their macromolecular chain, therefore this property may be obtained by biomimetic peptides, with which the material surface can be functionalized and that will interact with cell surface

TABLE 9.3 Proteins for TE Applications and Their Properties

Protein	Biological Role	Molecular Structure	Main Properties	Applications
Collagen	Structural protein in tissues, such as connective tissue, tendon, skin, bone, and cartilage	• Primary structure: Gly—X—Y triplet (X and Y are usually proline and 4-hydroxyproline, respectively) • Secondary structure: alpha helix • Tertiary structure: triple helix	• Biodegradability • Low antigenecity • Biocompatibility • High cost	Scaffolds for TE applications, drug, and cell delivery [50,51]
Gelatin	Protein obtained by thermal or chemical degradation of collagen	• Type A gelatin, from porcine skin: mainly random coil and residual triple helices • Type B gelatin, from bovine skin: random coil	• Biodegradability • Biocompatibility • Low antigenecity • Water soluble at higher temperature than 37°C • It forms a gel at lower temperature than 37°C • Low cost	Scaffolds for TE applications [52,53], drug and cell delivery
Elastin	ECM structural protein in mammals Main component of skin, blood vessels, such as the aorta, and lung tissues	• Primary structure: mainly composed of Gly, Pro, Ala. • Formed by covalently crosslinked molecules of its precursor tropoelastin, a 67 kDa soluble, nonglycosylated highly hydrophobic protein • Crosslinking through desmosine and isodesmosine amino acids • Around 60—70 amino acids between two crosslinking points	• Water insoluble • Elastomeric properties in its native form • Due to elastin insolubility, artificial proteins incorporating elastin-like peptides are of interest for the development of new protein-based materials	TE applications including vascular graft [54], hydrogels [55], bone repair [56], and drug delivery [57]
Fibronectin	ECM glycoprotein with functions, such as structural support and signaling for cell survival, migration, contractility, and differentiation	• Two polypeptide chains linked by disulfide bonds • These subunits are formed by three types of repeating modules, named types I, II and III	• Binding motifs for cell integrins, collagen, fibrin, and heparin ECM components	Scaffold functionalization [58—61]
Laminin	ECM glycoprotein, main component of the basal lamina, connecting cells to ECM by integrin binding	• Three polypeptide chains (α, β, γ chains), organized to form a crossed structure stabilized by disulfide bonds	• Binding motifs for cell integrins, type IV collagen, heparin, entactin, and proteoglycans ECM components	Cell adhesion/scaffold functionalization [62—64]
Fibrinogen	• Protein synthesized by the liver, which freely circulates in the bloodstream and has a major role in hemostasis	• 340 kDa glycoprotein consisting of a pair of three polypeptide chains: 2Aα, 2Bβ, and 2γ	• Biocompatibility • Integrin biding motifs (e.g., RGD) • High affinity binding with various growth factors, such	Cell adhesion/scaffold functionalization [65] Fibrin is used as medical sealant [66],

(Continued)

TABLE 9.3 (Continued)

Protein	Biological Role	Molecular Structure	Main Properties	Applications
	• During coagulation, fibrinogen is cleaved by thrombin and converted into fibrin, which spontaneously aggregates into fibrin protofibrils, assembling into larger fibers. The loosely assembled clot is then stabilized by covalent crosslinks by transglutaminase catalysis (factor XIIIa)		as fibroblast growth factor (FGF), vascular endothelial growth factor (VEGF), and several other cytokines. The presence of these growth factors may promote angiogenesis and enhance cell chemotaxis and mitogenesis at the implant site	in TE [67] and for drug delivery [68]
Silk fibroin	It forms the core of natural silk fibers. Silk produced by the silkworm species *Bombyx mori* is the most studied	• Primary structure: mainly composed of Ala—Gly sequences • It consists of heavy and light chains (350 and 25 kDa) linked together by a disulfide bond • Secondary structure: β-sheets alternated to random coils, α-helices, and β-turns	• High crystallinity • High elasticity, strength, toughness • Biocompatibility • Self-assembly	Wound dressing [69], sutures [70], scaffolds for TE [71—74], and drug delivery [75]

TABLE 9.4 Polysaccharides for TE Applications and Their Properties

Polysaccharides	Biological Role	Molecular Structure	Main Characteristics	Applications
Chitin	Component of cell walls of fungi, the exoskeletons of arthropods such as crustaceans (e.g., crabs, lobsters, and shrimps) and insects, the radulas of mollusks, and the beaks and internal shells of cephalopods, including squid and octopuses It has a structural function	Homopolymer based on 2-acetamido-2-deoxy-β-D-glucopyranose units	• Biocompatibility • Biodegradability • Wound healing ability	Scaffolds for TE applications [76], wound dressing [77], drug delivery [77], and cancer diagnosis [77]
Chitosan	Derived from chitin by deacetylation	Mainly composed of 2-amino-2-deoxy-β-D-glucopyranose units, with some residual 2-acetamido-2-deoxy-β-D-glucopyranose units	• Biocompatibility • High charge density • Nontoxicity • Mucoadhesion • Antibacterial activity	Scaffolds for TE applications [78,79], wound dressing [77], drug delivery [77], and cancer diagnosis [77]
Hyaluronic acid	GAG component of connective tissue, synovial fluid of vertebrates, and the vitreous humor of the eye Functions: • Increased viscosity of the synovial fluid • Responsible for the resilience of articular cartilage • Involved in skin tissue repair • Contributes to cell proliferation and migration	It is composed of alternating disaccharide units (250—25,000) of D-glucuronic acid and *N*-acetyl-D-glucosamine with β(1 → 4) interglycosidic linkages	• Lubricating properties • Water sorption • Cell attachment, migration, and proliferation • Biocompatibility	Scaffolds for TE of various tissues [80—85] and drug delivery [86]

(Continued)

TABLE 9.4 (Continued)

Polysaccharides	Biological Role	Molecular Structure	Main Characteristics	Applications
Chonodroitin sulfate	Major component of aggrecan, the most abundant GAG found in the proteoglycans of articular cartilage	GAG composed of alternating disaccharide units of N-acetylgalactosamine and D-glucuronic acid	• Anti-inflammatory activity • Improved wound healing • Polyanion	Scaffolds for TE of various tissues, such as skin, cartilage, and bone [79,85,87,88]
Alginate	High molecular mass polysaccharide extracted from various species of kelp. It gives strength and flexibility to the algal tissue, and regulates the water content in the seaweed. It is also produced extracellularly by *Pseudomonas aeruginosa* and *Azotobacter vinilandii*	Linear anionic block copolymers of heteropolysaccharides of β-D-mannuronic acid and α-L-guluronic acid residues, which can be arranged in different proportions and sequences along the polymer chain	• Slow degradation • Poor mechanical properties • It forms hydrogels by ionotropic gelation in the presence of multivalent ions like Ca^{2+} • Able of absorbing 200–300 times its own weight in water	Cell [89] and drug delivery [90], scaffolds for TE applications [91–93]
Agar	Cell wall polysaccharides extracted from the *Gelidiaceae* and *Gracilariaceae* families of seaweeds	Mainly composed of alternating (1–4)-D-galactose and (1–3)-3,6 anhydro-L-galactose repeating units. The disaccharide may be substituted by sulfate esters and methoxyl and may also carry pyruvic acid residues	• Melt on heating and reset on cooling • Gelling ability depending on chemical features of molecules and their molecular weight • Nonadhesive for cells	Drug delivery [94,95], gel-forming agent, thickener, water-holding agent
Starch	Major form of stored carbohydrate in plants, such as corn, wheat, potatoes, and rice	It is a mixture of two polymers of D-glucose: linear amylase and highly branched amylopectin. Amylose contains 200–20,000 glucose units with helix conformation. Amylopectin has 30 glucose unit side chains attached to the main glucose chain every 20–30 glucose units. It contains up to 2 million glucose units	• Easy availability • Biodegradability • Low cost	Scaffolds for TE, in particular of bone and cartilage tissue [96,97]

receptors and drive a specific information to adherent cells. Table 9.5 collects the main biomimetic peptides used for the aforementioned purpose.

The RGD (Arg−Gly−Asp) sequence was identified by Pierschbacher and Rouslahti in 1984 [103] as the minimal motif in FN promoting cell adhesion. Subsequently, it has been found in many other cell adhesion proteins, such as vitronectin, fibrinogen, collagen, and laminin. Around half of the 24 existing integrins show RGD-binding affinity. For this reason, RGD is a nonselective cell adhesion motif. DGEA (Asp-Gly-Glu-Ala), LDV (Leu-Asp-Val), and REDV (Arg-Glu-Asp-Val) are other integrin-binding peptide sequences: while DGEA promotes platelet adhesion and neurite outgrowth, LDV and REDV are selective towards endothelial cell attachment and avoid platelet adhesion, representing the ideal candidate for material functionalization in vascular applications [102]. Laminin-derived peptides, such as IKVAV (Ile-Lys-Val-Ala-Val), YIGSR (Tyr-Ile-Gly-Ser-Arg), RNIAEIIKDI (Arg-Asn-Ile-Ala-Glu-Ile-Ile- Lys-Asp-Ile), bind to nonintegrin receptors like LR67 and are known to favor neurite outgrowth [102].

Heparin-binding motifs like FHRRIKA (Phe-His-Arg-Arg-Ile-Lys-Ala) bind to proteoglycan receptors and contribute to the formation of focal adhesion. Finally, enzyme-cleavable peptide sequences (poly(A) (poly

TABLE 9.5 Immobilized Ligands Used in TE (from [101,102])

Biomimetic Peptide	ECM Molecule Source	Applications
RGD	FN, laminin, vitronectin, collagen, and thrombospondin	• Bone and cartilage regeneration • Neurite outgrowth • Myoblast and endothelial cell adhesion, proliferation, and differentiation
DGEA	Collagen	• Platelet adhesion • Neural cell adhesion
LDV, REDV	FN	• Promote the adhesion of endothelial cells and fibroblasts but inhibit platelet adhesion
IKVAV, YIGSR, RNIAEIIKDI	Laminin	• Neurite outgrowth
FHRRIKA	Heparin-binding motif	• Cell adhesion mediated by proteoglycan receptors • Improve osteoblastic mineralization
Poly(A)	Elastin	• Proteolytic degradation mediated by elastase
APGL	Collagen	• Proteolytic degradation mediated by collagenase
VRN	Fibrin	• Proteolytic degradation mediated by plasmin

materials will be referred to as "bioartificial materials" and they include any combination between natural and synthetic biocompatible polymers, such as bioartificial blends, copolymers, hydrogels, and materials obtained by the coating of synthetic polymers with natural polymers.

Bioartificial materials can be designed at different scale levels. In blends of natural and synthetic polymers (bioartificial blends), micrometric domains of the natural polymer are generally introduced into a synthetic polymer matrix. On the other hand, in the case of bioartificial copolymers and hydrogels, natural polymer monomers or oligomers are introduced into the polymer chains and material modification is performed at the molecular level. Finally, coating of synthetic substrates with natural polymers usually involves a surface modification of the material at the nano- or microscale, by covalent grafting, affinity bonding, or noncovalent functionalization through electrostatic interactions or weak van der Waals forces (physical adsorption).

9.2 DIFFERENT LEVELS OF BIOMIMICKRY IN BIOARTIFICIAL MATERIALS

Currently, regenerative medicine research is addressed to the functionalization of scaffolds with polysaccharides or biomimetic proteins/peptides.

Polysaccharides promote cell response through their binding to nonintegrin receptors. On the contrary, proteins/peptides bind to integrin or nonintegrin laminin receptors, leading to focal adhesion and a cascade of intracellular signaling directing cell behavior.

Functionalization through proteins or peptides is associated with different concerns [103]. First of all, proteins are generally isolated from other organisms, therefore their use in implants can potentially transmit immune reactions and infections. In addition, proteins can be easily degraded by proteolytic enzymes in vivo, and their degradation rate can be accelerated by inflammatory reaction and infections. Finally, proteins should arrange in a proper conformation to efficiently expose the epitope responsible for cell recognition. Biomaterial surface features (such as charge, wettability, and topography) influence protein conformation: for instance, proteins interact with hydrophobic surfaces through their hydrophobic amino acids, and can change their native conformation up to denaturation.

On the other hand, peptides are generally more stable against sterilization conditions, thermal treatment, pH variations, and storage as compared to proteins [103]. Due to their low molecular weight, peptides have a relatively stable conformation and can be easily identified and characterized by physicochemical analysis techniques [103]. Proteins contain different peptide sequences and are thus multifunctional; on the contrary, peptides drive specific signaling to cells. In addition, cyclic peptides are

(Ala)), APGL (Ala-Pro-Gly-Leu), and VRN (Val-Arg-Asn)) have been identified in ECM components: their introduction into biomaterials allows the obtainment of biodegradable substrates [102].

Materials derived from natural sources are advantageous for biomedical applications, due to their inherent biorecognition properties; drawbacks include poor mechanical properties and lack of water stability, as well as complexities associated with purification and immunogenicity concerns.

Thus, the current approach of regenerative medicine is to design custom-based tissue-specific biomimetic materials by the incorporation of the key characteristics of naturally derived materials into purposely selected or synthesized synthetic polymers [104,105]. In this chapter, such particularly resistant to enzymatic degradation [103]. Finally, chemistry of bioartificial materials containing peptides can be more easily controlled as compared to that of those containing proteins, due to stable conformation of shorter peptide sequences [103].

TE research efforts are addressed to find the best strategy for biomimetic functionalization of biomaterials. Recent investigations on ECM biology have shown that cell—ECM interactions are complex, highly coordinated, and dynamic; for instance, ECM contains cryptic bioactive sites for cell adhesion which become accessible as a function of time. In this context, functionalization with proteins is promising as they provide complex and dynamic multiple cell-binding sites [106]. On the other hand, functionalization strategies employing peptides can be improved by the use of peptide combinations, obtaining a multifunctional and modular material, which preserves a simple chemistry (leading to simpler validation and certification procedures) although avoiding immune response and disease transmission [106].

Protein/peptide functionalization has some limitations. First, the use of adhesive ECM proteins, such as collagen, laminin, and FN, or peptide combination is unsuitable for close mimicking the signal complexity of natural ECM. Second, the combination of proteins and proteoglycans in each ECM is tissue specific. Third, the presence of pathologies may alter ECM composition. Therefore, a third approach has been developed in the last years for biomimetic scaffold design, based on the assumption that the chemical composition of tissue-specific natural ECM represents the *ideal material* for a biomimetic scaffold aimed at the regeneration of that tissue. Decellularized ECM has thus been proposed as acellular and biodegradable scaffold preserving the biomechanical and physical architecture of the original tissue [107,108]. Several decellularized products, including dermis (Alloderm®; LifeCell), small intestine (Surgi SIS®; Cook Biotech, Inc.; Restore®, DePuy Orthopaedics, Inc.), heart valves (Synergraft; Cryolife, Inc.), or urinary bladder (ACell, Inc.) have received regulatory approval for clinical use [109].

As each ECM has a different composition, the optimal decellularized scaffold for a tissue should be based on its own ECM. These biological scaffolds are obtained through a lysis mechanism in an iper- or ipotonic environment leading to cell membrane disruption [107,108]. Then, a detergent-enzymatic method is applied for cell debris removal [107,108]. In this procedure, enzymes are used to degrade cell DNA fragments, but they can also damage ECM components. For this reason, some authors have proposed the combined application of enzymatic inhibitors like trypsin in combination with enzymes. As the performed treatments may modify the ECM biological and mechanical characteristics, reported decellularization protocols are variable, differing for used detergents and extraction times.

Main disadvantages in the use of decellularized ECM as biological scaffold derive from its nonautologous origin, which may lead to an adverse host response and disease transmission. Moreover, detergents can alter the ECM structure, decreasing the ability of the scaffold to provide mechanical support during regeneration.

Decellularized ECM can also be solubilized by mild treatments, obtaining a gel-based ECM [107,110,111]. This stock material can be used in combination with synthetic polymers to obtain bioartificial solid structures or applied as an injectable scaffold. However, drawbacks of this alternative strategy are due to the nonautologous origin of decellularized ECM.

A further improvement of this method can be achieved by using an ECM produced by *in vitro* culture of autologous cells, isolated from the patient and further expanded *in vitro*. This innovative approach, taking advantage of the abundant synthesis of human ECM proteins *in vitro*, opens the possibility for the fabrication of custom-made, patient-specific scaffolds for regenerative medicine approaches.

9.3 BIOARTIFICIAL MATERIALS BY BLENDING

Polymer blends are commonly prepared by mixing polymers in a melt state or coprecipitation from a common solvent [99]. However, natural polymers are generally water soluble, while most of the synthetic polymers are soluble in organic solvents, therefore solution blending is either not feasible or requires potentially toxic solvents. Bioartificial blends prepared by solution mixing have been widely used in the form of electrospun matrices, as reviewed by Gunn and Zhang [112].

On the other hand, natural polymer chains strongly interact with each other either by hydrogen bonding or electrostatic interactions, therefore they chemically degrade before melting, when heated. For this reason, bioartificial blends prepared by melt mixing are generally based on the compounding of a synthetic polymer melt with natural polymer unmelt particles.

As an alternative, solutions of natural and synthetic polymers, respectively, in water-based and organic solvents are mixed at high speed, obtaining an emulsion, which is then stabilized by freezing and further lyophilization (emulsion mixing).

Processing conditions for blending, such as the application of high temperatures and shear stresses (melt mixing) or the use of organic solvents (solution mixing and emulsion mixing), may denature proteins or degrade the natural polymers.

Table 9.6 collates the advantages and disadvantages of the methods for bioartificial blends preparation.

TABLE 9.6 Advantages and Disadvantages of Methods for Bioartificial Blend Preparation

Blending Methods	Advantages	Disadvantages
Solution mixing	• Effective method for blending in the case of water-soluble synthetic polymers	• Use of potentially toxic solvents in the case of water unsoluble synthetic polymers (e.g., hexafluoro-propanol) • Organic solvents may cause protein denaturation
Emulsion mixing	• Use of low temperatures • Single process for blending and scaffold formation	• Phase separation between blend components • Use of potentially toxic solvent for synthetic polymer dissolution
Melt mixing	• Scalable method for blend preparation avoiding the use of solvents	• High temperature may induce protein denaturation or natural polymer degradation • Phase separation between blend components • During processing, natural polymers do not melt and simply disperse in the synthetic polymer matrix as solid particles

9.3.1 Bioartificial Blends Based on Proteins and Water-Unsoluble Synthetic Polymers

Bioartificial blends have been prepared by combining proteins (Table 9.3) and synthetic polymers (Table 9.1), using various blending methods (Table 9.6). Some examples of bioartificial blends based on proteins are described below.

Collagen has been widely used in bioartificial blends, aimed at the development of drug delivery devices and scaffolds for the regeneration of various tissues, such as skin, bone, cartilage, tendons, blood vessels, cornea, and nerve [99,113].

Bioartificial blends based on collagen have been commonly prepared in the form of nanofibers (NFs) by electrospinning, using a fluorinated common solvent like 1,1,1,3,3,3-hexafluoro-2-propanol [112]. In this context, PCL/collagen blend NFs with different fiber orientation have been prepared by electrospinning to engineer muscle tissue [114]. Human skeletal muscle cells (hSkMCs) have been seeded onto the electrospun PCL/collagen NF meshes and analyzed for cell adhesion, proliferation, and organization. Aligned NFs have been found to induce muscle cell alignment and myotube formation as compared to randomly oriented NFs.

Aligned NF blends of PLGA and collagen with different PLGA/collagen ratios have been prepared from solutions in 1,1,1,3,3,3-hexafluoro-2-propanol for bone TE, identifying the 80/20 wt/wt composition as the most suitable for this application [115]. Similarly, electrospun tertiary blends have been prepared from PCL, collagen, and elastin for vascular TE, obtaining a three-layered conduit [116].

Bioartificial blends based on gelatin have been prepared using a fluorinated common solvent for applications, such as bone [117,118] and nerve [119] regeneration.

As an alternative to the solution mixing method, Chiono et al. have prepared tubular conduits based on a binary blend between PCL and gelatin (10 wt% content) obtained by melt mixing, for the regeneration of peripheral nerves [120]. Two blending methods have been applied. In the first method, gelatin microparticles crosslinked by Transglutaminase have been prepared and then premixed with PCL pellets; finally, the mixed material has been extruded. The second method was based on the extrusion of premixed (uncrosslinked) gelatin powder and PCL pellets. In both cases, extrusion allowed the fabrication of tubular conduits. Prepared PCL/gelatin blends were not miscible and phase separation between the components was observed. However, the use of natural polymer microparticles allowed the obtainment of bioartificial blends with enhanced dispersion degree of the natural polymer into the synthetic polymer matrix. *In vitro* cell tests using S5Y5 neuroblastoma cells showed a similar cell response for PCL and PCL/gelatin blends, suggesting that gelatin is not a specific substrate for promoting nerve cell adhesion. When poly(L-lysine) was grafted by transglutaminase on the exposed gelatin domains of the inner guide surface, S5Y5 neuroblastoma cell attachment was significantly enhanced, due to the specificity of this polypeptide in promoting nerve regeneration.

9.3.2 Bioartificial Blends Based on Polysaccharides and Water-Unsoluble Synthetic Polymers

Among the polysaccharides listed in Table 9.4, CS, a naturally derived polysaccharide with structural similarity to various glycosaminoglycans (GAGs), has been widely used for the preparation of bioartificial blends. It is composed of glucosamine and *N*-acetyl glucosamine units linked by $\beta(1-4)$ glycosidic bonds and derives from the deacetylation of chitin. CS has been found to support chondrogenic activity [121–123] and peripheral nerve regeneration [124]. Moreover, CS possesses antibacterial, hemostatic, fungistatic, antitumoral, and anticholesteremic properties [124]. Main disadvantages of CS are due to its brittleness in wet state (40–50% of strain at break) and poor water stability in the noncrosslinked form. Blending CS with a synthetic hydrophobic polymer could be a simple technique to balance its highly hydrophilic character and poor mechanical properties.

PCL (Table 9.1) is a Food and Drug Administration (FDA) approved polymer for biomedical applications with ductile mechanical properties (elastic modulus: 0.4 GPa; elongation at break > 1000%). Several attempts have been performed to prepare blends between PCL and CS by dissolution in common solvents, such as 1,1,1,3,3,3-hexa-fluoro-2-propanol (HFlP) [125,126] or acetic acid [127,128].

An alternative method to produce PCL/CS blend films by solvent casting was based on blending between low concentration solutions of CS in 0.5 M acetic acid and PCL in glacial acetic acid [127]. The authors have concluded that blends were miscible for all compositions and that the lack of improvement in the mechanical properties was due to the low concentration of starting solutions.

Cruz et al. have adopted a similar method [128] based on the addition of 2% (w/v) CS solution in 0.1 M acetic acid into 1.8% (w/v) PCL solution in glacial acetic acid, to obtain blends containing 10, 20, 30, and 40 wt% CS. The authors have found that in blends with CS amount ≥ 20 wt%, CS was the continuous phase. In each blend, phase separation between PCL and CS was observed.

The biological response of PCL/CS blends prepared according to this method has been analyzed by the authors by culturing primary human chondrocytes on the substrates [129]. Cell proliferation was decreased in PCL/CS blends as compared to PCL: this behavior was attributed to the conformation of proteins adsorbed from the culture medium onto the substrate. For instance, laminin was found to adopt a globular or aligned conformation on PCL and CS substrates, respectively. Protein conformation affects cell accessibility to protein integrin-binding domains.

PCL/CS blend fibers have been prepared by wet spinning by Malheiro et al. [130]: PCL and CS were dissolved in a common solvent mixture (acetone/formic acid 70:30 vol%) and extruded into a methanol containing coagulation bath. Blend fibers containing 0, 25, 50, 75 and 100 wt% CS were prepared. Phase separation between CS and PCL was observed, although PCL distribution into CS matrix was at the micrometric scale (< 10 μm).

Neves et al. have recently prepared blends between PCL and CS at various compositions (100, 75, 50 wt% CS) by dissolution in 100% formic acid. Micrometric fibers were prepared by wet spinning using methanol as a nonsolvent.

Fibers were folded into a cylindrical mold and a thermal treatment was applied at 60°C for 3 h to obtain porous scaffolds for cartilage regeneration. Phase separation between CS and PCL was detected by thermal (Differential Scanning Calorimetry, DSC) and infrared (Fourier Transform Infrared Spectroscopy Attenuated Total Reflectance, FTIR-ATR and chemical mapping) analyses. However, PCL domains were homogeneously distributed in the CS matrix. The blend containing 75 wt% CS showed the best compromise of biological and physicochemical properties, as it showed reduced swelling degree and improved mechanical properties as compared to pure CS and promoted GAG production by chondrocytes [131].

In the above-reported studies, HFlP has been frequently used as common solvent for blend preparation, however, it is toxic, carcinogenic, expensive, and difficult to remove [132]. Alternatively, diluted acetic acid as well as acetone/formic acid 70:30 vol% and formic acid solutions have led to phase separation [128,131].

Emulsion mixing methods have also been proposed to prepare bioartificial blends based on CS. For instance, Moshfeghian et al. [133] have prepared blends between CS and PLGA (LA:GA = 50:50 mol:mol) by emulsifying a CS acetic acid solution (0.2 M) with a PLGA solution in chloroform, methylene chloride, dimethyl sulfoxide, or benzene. As a stabilizer of the emulsion, 1,2-dimyristoyl-sn-glycero-3-phosphocholine (DMPC) was used.

Beside CS, starch represents one of the mainly used polysaccharides for the preparation of bioartificial blends due to its biodegradability and low cost [134]. Starch (Table 9.4) is a polysaccharide produced by higher plants as energy storage and is composed of two polymers of D-glucose: amylose and amylopectin. Amylose is a lightly branched polymer, whereas amylopectin is highly branched with an overall tree-like structure.

Ciardelli et al. have prepared bioartificial blends based on a medium molecular weight PCL (MW: 45,000 Da) and 9.1 wt% starch (S), dextran (D) and gellan (G), by a solution-precipitation technique, based on material dissolution in dimethyl sulfoxide and further precipitation in methanol [105]. This technique can be applied for the preparation of blends containing low or medium molecular weight PCL. Blend microparticles with lower size than 125 μm were prepared by cryogenical milling and used to prepare two-dimensional scaffolds in the form of square-meshed grids, by selective laser sintering (SLS) through a CO_2 laser. The distance between the sintered stripes in the square-meshed grids was 2 mm, each stripe was 700 μm wide and 300 μm deep. Sintering has been performed on slurries of particles in demineralized water, after selecting the optimal sintering parameters (power: 2 W; beam speed: 5 mm/s). Morphological analysis of fractured sections of sintered scaffolds has shown that sintering was not optimal on the nonexposed side, probably due to the irregular morphology of microparticles. FTIR chemical imaging analysis has demonstrated that the polysaccharide phase was homogeneously distributed within the PCL matrix, except for PCL/D blend. *In vitro* cell tests using NIH-3T3 fibroblasts have shown that cell adhesion was promoted by blending PCL with G or S.

The starch granule is partially crystalline and the melting temperature of dry starch is around 220−240°C, whereas the onset degradation temperature is around 220°C. For these reasons, starch has to be plasticized before melt processing [135].

Chiono et al. have prepared binary blends between a high molecular weight PCL (Mn: 80,000 Da) and a natural polymer (S or CS) with different contents of the natural polymer (5−30 wt%) [104]. For the preparation of PCL/S blends, starch has been plasticized and then the polymers have been blended in a Brabender Plasticorder static mixer at a higher temperature than the PCL melting point. PCL/S blends have been manually reduced into pellets and, then, fed into a single-screw extruder to fabricate tubular conduits for nerve repair. On the other hand, a double precipitation method has been applied to obtain PCL/CS blends. A dispersion of CS microparticles in a water solution has been prepared by CS dissolution in 1% acetic acid solution (0.125%, w/v) and further addition of 10% (v/v) Na_2SO_4 solution. PCL solution in chloroform (5%, w/v) has been dropped into the CS dispersion (after the addition of an emulsifier). The system has been kept under vigorous stirring. After solvent evaporation, mixed microparticles have been collected by centrifugation and then used for the melt extrusion of tubular guides. Physicochemical analysis has evidenced phase separation between blend components. However, a homogeneous composition and morphology has been found at low polysaccharide content (≤10 wt%). *In vitro* cell tests on extruded guides have been performed using NIH-3T3 mouse fibroblasts. Cell attachment has followed this trend: PCL/S blends > PCL > PCL/CS blends. For each blend type, cell adhesion has been found to decrease as a function of the natural polymer content, probably due to the decreasing degree of structural morphological homogeneity of samples. *In vitro* cell tests using S5Y5 neuroblastoma cells have been carried out on blend conduits containing low amount of natural polymer (5−10 wt%), showing the absence of cytotoxicity toward this cell type.

Blends between starch and PLLA or PCL have been frequently used for cartilage and bone regeneration [96,97,136−138]. Part of research on starch-based blends has been carried out using commercially available blends [96,97,136−140].

3D scaffolds made of corn starch and PCL (70 wt% PCL, supplied from Novamont), incorporating Si-OH groups, have been fabricated by molding wet-spun fibers prepared using a calcium silicate solution as a coagulation bath [136]. A control blend scaffold has also been prepared from fibers obtained using methanol as a coagulation bath. *In vitro* studies have been performed by seeding and culturing wet-spun fiber mesh scaffolds, with human adipose stem cells (hASCs), either in a flow perfusion bioreactor or under static conditions. During 14 days of culture in dynamic conditions, scaffolds incorporating Si-OH groups sustained hASC proliferation and differentiation into the osteogenic lineage.

Ghosh et al. have used a PLLA/S blend with 50 wt% S, supplied for Novamont, for the fabrication of bilayered scaffolds for TE of osteochondral defects, by compression molding and particulate leaching (using NaCl particles) [137]. PLLA/S blend has been used as cartilage side layer, due to its adequate hydration capability. PLLA reinforced with hydroxyapatite has been used as a bone layer, due to its stiffness and mechanical strength. Bone layer has been found to induce the deposition of calcium phosphate crystals after incubation in simulated body fluid; on the contrary, the cartilage layer has not exhibited any ability for calcification.

9.3.3 Bioartificial Blends Based on Solubilized ECM

At a higher biomimickry level, bioartificial blends based on solubilized ECM produced *in vitro* by cell culturing are currently under investigation and their potentiality is a promising field to be explored in the near future.

Schenke-Layland et al. have cultured human foreskin fibroblasts for 4 weeks on silicon-based nanopatterned surface fabricated by interference lithography and deep reactive ion etching, with the aim to produce ECM sheets [141]. ECM were decellularized using 0.25% sodium deoxycholate and 0.25% Triton X-100 treatment for 6 h. Then, ECM protein extracts were solubilized in hexafluoro-2-propanol, alone or in combination with PCL for electrospinning nanofibrous substrates. Fibrous scaffolds were found to support adipose-derived stem cells attachment and migration.

9.4 BIOARTIFICIAL HYDROGELS

Hydrogels are 3D polymeric networks, able to absorb a considerable water amount, displaying high swelling degree. Hydrogels can be prepared via physical or chemical crosslinking of hydrophilic homopolymers, copolymers, or macromers (preformed macromolecular chains). They are classified into chemical or physical hydrogels, depending on the nature of the network crosslinking bonds: covalent bonding (chemical hydrogels) or inter- and intramolecular interactions by hydrogen bonding or van der Waals forces (physical hydrogels) [142]. Hydrogel water content, soft nature, and porous structure mimic biological tissues and make them suitable to accommodate cells and to encapsulate and release water-soluble compounds in a controlled way. Several hydrogel types with vastly different chemical and physical properties have been developed over the last several decades from a wide variety of chemical building blocks and using an array of synthetic techniques [143]. Among the various synthetic polymers for preparing hydrogels for biomedical applications, PEG, PVA, PVP (Table 9.2), and polyacrylates (PAs) have been widely used, owing to their biocompatibility characteristics [144,145]. Improvements in the biological response of hydrogels can be achieved by the introduction of natural polymers or biomimetic peptides. Bioartificial hydrogels can be prepared from blends of natural and synthetic water-soluble polymers or from bioartificial copolymers.

9.4.1 Bioartificial Hydrogels by Blending

Hydrogels derived from blends of natural and synthetic polymers have been an increasing interest in recent years for their potentiality in the biomedical field, such as drug delivery systems, hemodialysis, or scaffold membranes [146−149]. Bioartificial hydrogels combine the mechanical/rheological and aging properties, typical of synthetic polymers with the biocompatibility and adequate cell response promoted by the natural polymer.

Typical synthetic polymers in bioartificial hydrogels are vinylic polymers [137−139]. The first prepared bioartificial hydrogels contained protein macromolecules, such as fibrin or collagen [150], then polysaccharides have also been used [140,142,151,152].

A few examples of bioartificial hydrogels obtained by blending are described below.

Cascone et al. have prepared blends of PVA with different biological macromolecules (gelatin, dextran, and hyaluronic acid) to obtain bioartificial hydrogels, with the aim to improve the biocompatibility properties of PVA [149]. The effect exerted by each biological component on hydrogel pore size and distribution has been investigated. It has been found that when a natural macromolecule is added to PVA, the hydrogel internal structure changes: hydrogels containing 20% of the biological component displayed the most regular structure and the lowest total porosity; samples with the highest content of natural polymer (40%) showed the less regular structure and the highest total porosity.

Blends between PVA and CS have been recently prepared with various compositions [153]. Computational tools based on molecular mechanics and dynamics techniques have been combined with experimental tests with the aim of predicting mechanical properties and diffusion of small molecules like urea within the hydrogel as a function of composition. This work has demonstrated that computer simulations can be a valid tool to reduce the

material characterization efforts allowing the evaluation of the dependency of the properties on the blend composition.

9.4.2 Bioartificial Hydrogels Based on Bioartificial Copolymers

An alternative approach to obtain bioartificial hydrogels is based on its covalent functionalization with peptides or proteins/polysaccharides.

The nonionic hydrophilic PEG gel systems have been increasingly used as matrices for controlling drug delivery and as cell delivery vehicles for promoting tissue regeneration [154]. The versatility of the PEG macromer chemistry [155], together with its excellent biocompatibility, have caused the development of numerous intelligently designed hydrogel systems for regenerative medicine applications. To direct cell response, PEG-based hydrogels can be modified by the covalent incorporation of bioactive peptides or natural polymers, obtaining bioartificial hydrogels.

Hubbell and coworkers have been the first to produce cell-adhesive, proteolytically sensitive hybrid biomaterial for tissue regeneration. They have used a PEG backbone modified with RGD oligopeptide, and crosslinked with short oligopeptides containing plasmin or collagenase degradation substrate [156,157]. Similarly, West and co-workers have also studied a proteolytically sensitive PEG−peptide biomaterial, functionalized with adhesion ligands and growth factors for enhancing smooth muscle cell migration, proliferation, and new matrix production [158−160]. PEG−peptide hydrogels need the use of peptide combinations to achieve both cell adhesion and hydrogel biodegradation. As an alternative, PEG−protein approach has been proposed as a simpler method to obtain a cell-adhesive biodegradable hydrogel. Some authors have end functionalized PEG−OH obtaining PEG−diacrylate molecules [161,162]. Fibrinogen has been selected as a bioactive molecule for hydrogel functionalization, as it contains several cell signaling domains including a protease degradation substrate and cell adhesion motifs. A Michael-type addition reaction has then been used to form ester bonds between the free thiols in the denatured fibrinogen cysteines and acrylate end groups on the PEG−DA (PEGylation) (Figure 9.1). After PEGylation, unreacted acrylates on the difunctional PEGs have been used to crosslink the fibrinogen backbone into a hydrogel network using photopolymerization (Figure 9.1).

Similarly, PEG-based bioartificial hydrogels have been prepared from branched PEG−OH molecules, by proper functionalization of end groups with reactive groups, able to further react with crosslinker molecules containing bioactive peptides [163]. For instance, PEG−OH molecules have been reacted with an excess of divinylsulfone, obtaining PEG−vinylsulfone branched

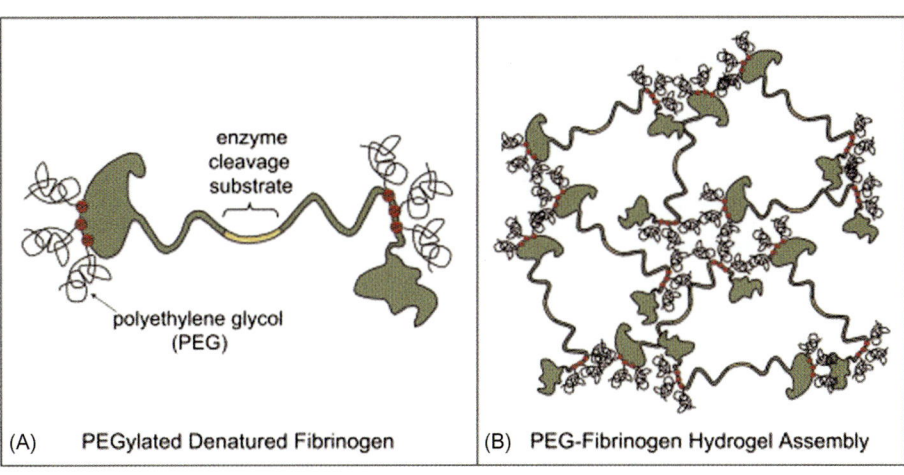

(A)　PEGylated Denatured Fibrinogen

(B)　PEG-Fibrinogen Hydrogel Assembly

FIGURE 9.1 Schematic representation of the PEG–fibrinogen hydrogel assembly. PEGylated fibrinogen fragments (A) contain a natural protease cleavage site (yellow) and multiple unpaired thiols (red) for covalent conjugation of functionalized PEG by Michael-type addition reaction. PEG–fibrinogen hydrogel assembly (B) is accomplished by photoinitiation of unreacted PEG-diacrylate, resulting in a hydrogel network of PEGylated fibrinogen. *Derived with permission from Ref. [161].*

molecules. Subsequently, a chemical hydrogel has been obtained by reacting the PEG–vinylsulfone branched molecules with RGDSP (Arg-Gly-Asp-Ser-Pro)- and PQGIW-containing oligopeptides through Michael-type addition reaction [163]. The proposed method has allowed a modulation of hydrogel elasticity, cell adhesion, and proteolytic susceptibility nearly independently. P19 embryonal carcinoma cells have been used as a model cell line to study their differentiation *in vitro* towards a cardiac lineage. Hydrogel elasticity has been found to direct the first stages of cell differentiation, while terminal cell differentiation has required the presence of ligands for $\alpha_v\beta_3$ and $\alpha_5\beta_1$ integrins and protease-mediated enzymatic cleavage.

Hydrogels that are sensitive to environmental stimuli, such as temperature, pH, and electric field, are called "smart hydrogels." Among thermoresponsive hydrogels, poly(ethylene oxide) (PEO)–poly(propylene oxide) (PPO)–poly(ethylene oxide) (PEO) copolymers, often called Pluronics or Poloxamers, are commercially available for various biomedical applications, including the use as injectable matrices for drug delivery [164].

Pluronics have been covalently modified using both natural polymers or bioactive peptides [65,165,166].

Cha et al. have modified Pluronic F68, by reacting its terminal hydroxyl group with oligo-lactides (oligo-LA) or oligo-caprolactones (oligo-CL), with the aim to increase copolymer hydrophobicity [165]. The resulting copolymer has then been modified with 4-methacryloxyethyl trimellitic anhydride (4-META) and a peptide ligand (RGD). Figure 9.2 shows a schematic representation of the modification steps of Pluronic F68. Water solutions of Pluronic F68 showed a sol–gel–sol transition with increasing temperature: due to the polymer hydrophilicity, sol–gel transition occurred at higher temperature than 36°C and higher solution concentration than 30% (w/v). Water solutions of the synthesized copolymers showed a different rheological behavior with gel–sol–gel transition with increasing temperature.

Park et al. have modified Pluronic F127 by end-grafting of CS (Figure 9.3) [166]. Monocarboxylated Pluronic F127 has first been synthesized using succinic anhydride as a reagent. Then, carbodiimide chemistry has been exploited for the grafting of CS amino groups with the carboxylic groups of monocarboxylated Pluronic F127. Solutions of the copolymer have been found to undergo sol-to-gel transition at higher temperature than 15°C, depending on their concentration. The main driving force for the formation of the hydrogel has been attributed to hydrophobic interactions of PPO groups and dehydrated CS at high temperature. The introduction of CS has increased the mechanical characteristics of the hydrogels: G′ storage modulus of Pluronic F127 (20%, w/v) was lower than 10^4 Pa, whereas G′ values of copolymer solutions were 20 and 40 kPa, for 16 and 20% (w/v) hydrogels, respectively. Chondrocytes have been inserted into copolymer hydrogels: after 30 days, the number of viable cells was around 6×10^5 and 5×10^5 for copolymer and alginate hydrogels, respectively. Moreover, cells deposited type II collagen and aggrecan as a function of time: the aggrecan expression was higher than for alginate control hydrogels.

Shachaf et al. have modified Pluronic F127 with fibrinogen to obtain a material for 3D cell encapsulation and TE [65]. Figure 9.4 shows a schematic representation of the synthesis procedure. Pluronic F127 was first functionalized with acrylate groups and then reacted with denatured fibrinogen through a Michael-type addition reaction. These copolymers were used to obtain 3D hydrogels by free radical polymerization using light activation (photopolymerization). These materials showed a reversible sol–gel transition driven by hydrophobic interactions and an irreversible light-activated chemical crosslinking.

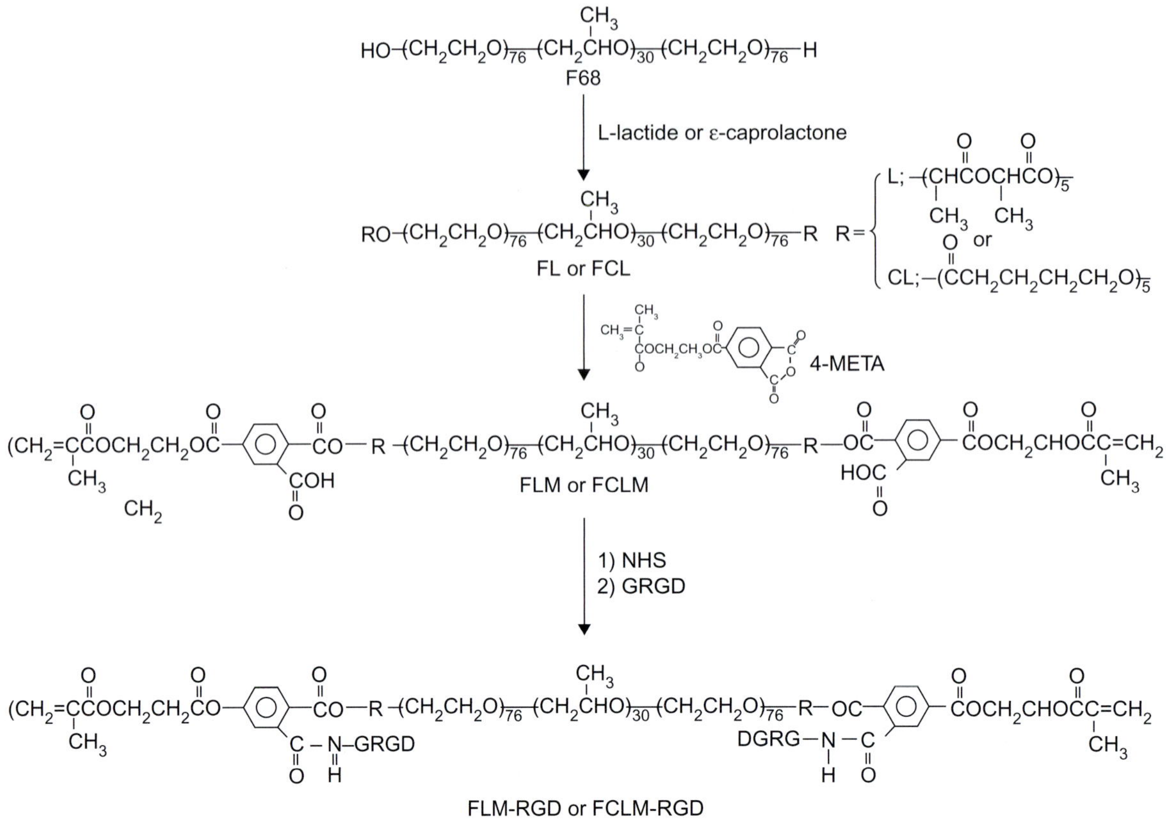

FIGURE 9.2 Schematic illustration of the synthesis steps for the development of Pluronic F68-based copolymers, modified with oligo(lactic acid) FL or oligo(caprolactone) (FCL) and functionalized with RGD peptide. 4-META: 4-methacryloxyethyl trimellitic anhydride; NHS: N-hydroxysuccinimide; FLM: FL functionalised with 4-META; FCLM: FCL functionalised with 4-META. *Derived with permission from Ref. [165].*

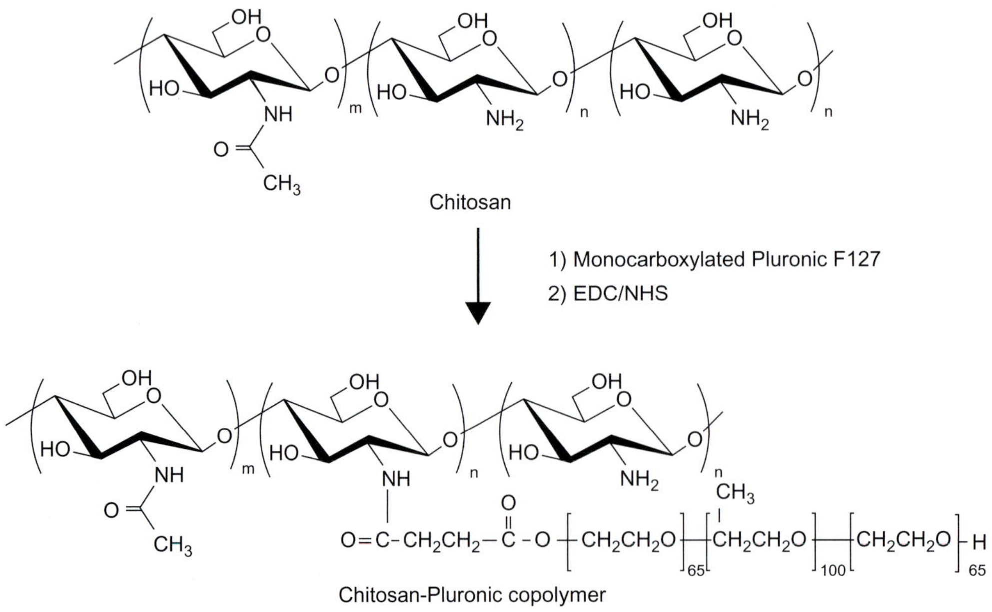

FIGURE 9.3 Schematic illustration of the modification of monocarboxylated Pluronic F127 with CS molecules. EDC/NHS: 1-Ethyl-3-(3-dimethyla-minopropyl)carbodiimide/N-hydroxysuccinimide. *Derived with permission from Ref. [166].*

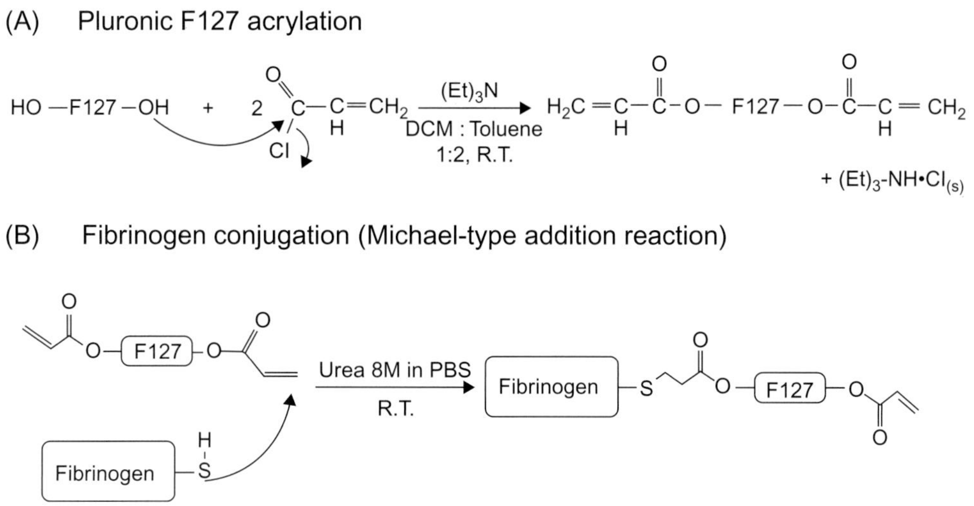

(A)   Pluronic F127 acrylation

(B)   Fibrinogen conjugation (Michael-type addition reaction)

FIGURE 9.4 Schematic representation of the synthesis steps for the preparation of Pluronic F127 functionalized with fibrinogen: (A) acrylation of Pluronic F127 and (B) Michael-type addition reaction for fibrinogen conjugation. DCM: dichloromethane. *Derived with permission from Ref. [65].*

As an alternative to the modification of commercially available thermosensitive polymers, bioactive reverse thermal polymers may be synthesized with properties tailored to the specific application. Polyurethanes are a class of block copolymers with a versatile chemistry, which composition can be modulated to obtain thermosensitive and biomimetic properties [167].

Park et al. have synthesized an amine-functionalized ABA block copolymer, poly(ethylene glycol)−poly(serinol hexamethylene urethane) [168]. This reverse thermal gel consisted of a hydrophobic block (B): poly(serinol hexamethylene urethane) and a hydrophilic block (A): poly(ethylene glycol). The copolymer was biocompatible and degradable mainly by hydrolysis of ester bonds. Its solutions in phosphate buffered saline underwent gelification at around 32°C forming a relatively soft gel. The presence of reactive amino groups allowed the functionalization with the hexapeptide, Ile−Lys−Val−Ala−Val−Ser (IKVAVS).

9.5 BIOARTIFICIAL MATERIALS THROUGH COATING STRATEGIES

9.5.1 Coating of Synthetic Polymers with Natural Polymers/Peptides

Many surface modification techniques have been developed to improve biomaterial biocompatibility. Surface modification techniques provide accessible functional groups for the immobilization of drugs, enzymes, antibodies, or other biologically active species for a variety of biomedical applications [169]. Through surface functionalization, bulk properties of the biomaterial are not altered. Biomacromolecules, such as proteins, polysaccharides, proteoglycans and their derivatives, are known to act as

biological cues for adherent cells. The initial cell adhesion and spread are obtained through ECM proteins [170], such as FN, vitronectin, fibrinogen, and collagen, which have the ability to prompt cell adhesion. Therefore, if biomaterial surfaces are functionalized with these bioactive macromolecules, the biocompatibility can be significantly improved.

Several approaches can be adopted to modify the surface properties of biocompatible polymers with biomacromolecules, such as covalent attachment, physical adsorption, and affinity bonding (Table 9.7).

The surface modification techniques, as briefly summarized in Table 9.7, have been described in many reviews and book chapters [169,183−186]. In particular, an exhaustive list of crosslinkers for the immobilization of bioactive molecules to a functionalized substrate has been reported by Goddard and Hotchkiss [187].

Surface coating by the three above-mentioned approaches is often proceeded by a surface prefunctionalization step via surface chemical etching (in acidic or basic solution [188]), chemical functionalization (e.g., by aminolysis [174,189,190]), or plasma treatment [191,192].

Chemical etching involves the partial degradation of the surface material by the corrosive action of a liquid or gaseous acid or an alkali. Etching treatment of aliphatic polyesters results in the hydrolysis of exposed macromolecules, leading to the formation of carboxylic and hydroxylic moieties [187].

Aminolysis is another widely applied technique for surface functionalization of polyesters with amino groups, which allow the grafting of biomolecules. During the aminolysis reaction, a diamine reagent reacts with exposed ester groups of polyesters, reducing the molecular weight of surface macromolecules and forming amide bonds. As a result, amino and hydroxyl groups form on the material surface [193,194].

TABLE 9.7 Surface Modification Approaches

Modifications	Mechanisms	Methods
Chemical grafting	Covalent grafting of bioactive peptides, proteins, and polysaccharides to improve cell adhesion and biocompatibility	Carbodiimide [171–174], glutaraldehyde [175,176], hydroxyethylmethacrylate [177,178]
Physical adsorption	Noncovalent immobilization of bioactive macromolecules and growth factors (by hydrogen bonding, van der Waals interactions, or electrostatic forces) to promote cell interaction and response, respectively. LbL self-assembly technique belongs to physical adsorption techniques	Physical adsorption [179–181]
Affinity bonding	Streptavidin (avidin)/biotin non-covalent interaction	Surface functionalization with a biotin containing molecule, able to interact with streptavidin (avidin), then binding to a biotin-functionalized bioactive molecule [182]

Plasma treatment can introduce hydroxyl, carboxyl, and amino groups on polymer surfaces using different carrier reactive gases, such as argon, oxygen, nitrogen, hydrogen, and ammonia. When these gases are exposed to plasma, they dissociate and react with the surface, creating different chemical functional groups on the surface. A plasma, which can be regarded as the fourth state of matter, is composed of highly excited atomic, molecular, ionic, and radical species. It is typically obtained when gases are excited into energetic states by radio frequency, microwave, or electrons from a hot filament discharge [183]. Plasma surface modification involves the interaction of the plasma-generated excited species with a polymeric interface. The plasma process results in a physical and/or chemical modification of the first few molecular layers of the polymer surface. The effectiveness of the treatment is determined by the plasma source gases, the configuration of the plasma system, and the plasma operating parameters.

Below some examples of successful surface modification strategies are reported, based on a premodification step, followed by surface functionalization.

In order to prepare PLLA surfaces, which favor cell attachment and spreading, Chen and Su have grafted cationized gelatin (CG) onto oxygen plasma-treated PLLA NFs, using water-soluble carbodiimide as a coupling agent [191]. PLLA NF membranes have first been treated with oxygen plasma to introduce −COOH groups on the surface. Then, plasma-treated membranes have been exposed to water-soluble carbodiimide to activate the carboxylic groups, followed by covalent grafting of CG molecules onto the fiber surface. The physicochemical characterization of PLLA and CG-PLLA NF membranes has demonstrated a successful grafting of CG on the surface of electrospun PLLA NF by oxygen plasma pretreatment. Both *in vitro* and *in vivo* experiments have shown that rabbit articular chondrocytes seeded in CG-PLLA NF membranes could grow into tissue-like constructs with high viability and abundant deposition of GAGs and type II collagen.

Zhang et al. [180] have investigated the bioactivity of FN immobilized on polyethylene terephthalate (PET) using a physical adsorption method (Figure 9.5). Clean PET surface aminated by ethylenediamine solution was immersed in freshly prepared glutaraldehyde solution, and then reacted at room temperature. After the surface was blocked by ethanolamine, FN was added into the system for physical adsorption on the surface. The study demonstrated that the adsorption method preserved the compact conformation of FN, reaching saturation when a monolayer of FN was formed. Biological characterization by adhesion of baby hamster kidney 21 (BHK21) cells and enzyme-linked immunosorbent assay (ELISA) for active RGD domains revealed that an active RGD domain was maintained, thereby promoting cell adhesion.

Nanocoatings can be also fabricated using the layer-by-layer (LbL) self-assembly technique [195]. The buildup of LbL multilayers is mainly driven by the electrostatic attraction between oppositely charged polyelectrolytes [196]. Generally, LbL self-assembly proceeds as schematically reported in Figure 9.6. A charged substrate is immersed in a solution of an oppositely charged polyelectrolyte to electrostatically attract the first monolayer. A washing step follows to remove unbound material and preclude contamination of the subsequent oppositely charged polyelectrolyte solution, in which the coated substrate is then submerged to deposit the second layer. A washing step is then performed again. These four steps are repeated several times to obtain a multilayered coating. LbL advantages include: (i) the simplicity of the technique, (ii) the availability of an abundance of natural and synthetic polyelectrolytes, (iii) the flexible application to objects with irregular shapes and sizes, (iv) the use

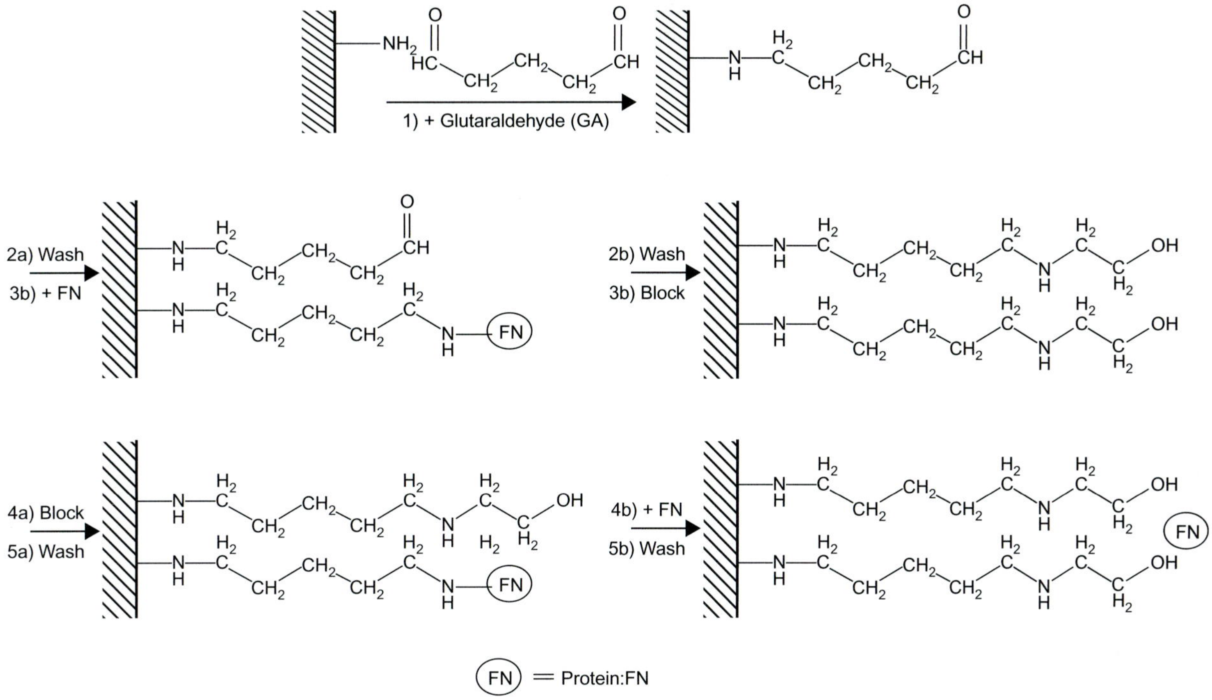

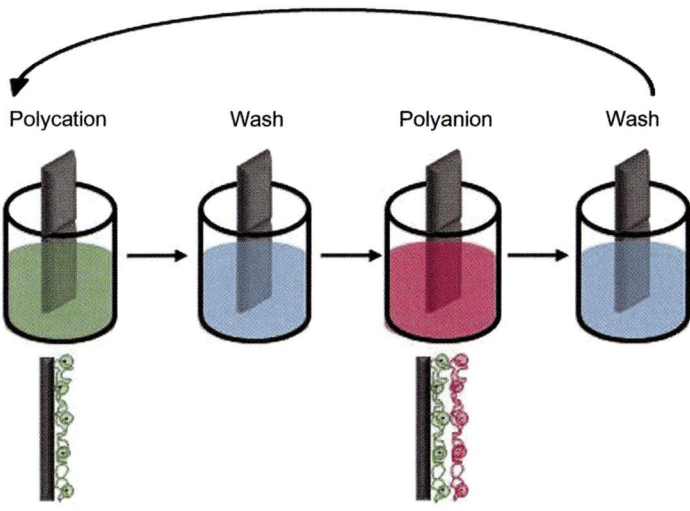

FIGURE 9.5 Schematic representation of the steps for FN adsorption on aminated PET surface. *Derived with permission from Ref. [180].*

FIGURE 9.6 A schematic illustration of the alternate adsorption of the polyelectrolytes to produce a multilayered structure.

of mild processing conditions, and (v) the control of the coating multilayer thickness.

To illustrate the applicability of LbL assembly on surface modification of porous scaffolds, Gong et al. [181] selected chondroitin sulfate A (CHS) and collagen type I (Col) as the building blocks to build biomimetic coatings on PLLA scaffolds. The high negative polarity resulting from SO_4^{2-} and COO^- groups allows CHS to be used as a polyanion, on which positively charged Col can be adsorbed.

PLLA porous scaffolds surface was activated through the aminolysis technique. Then, the samples were subsequently incubated in CHS solution for 20 min to adsorb a layer of CHS, and then rinsed with water containing NaCl three times. In the next step, positively charged Col was absorbed by incubating the CHS-enriched PLLA membrane in Col solution for 20 min, followed by rinsing with acetic solution (containing NaCl) and then water. The process was repeated until the desired number of layers had been deposited. The LbL-modified PLLA scaffolds were finally dried under reduced pressure.

UV–vis spectroscopy and ninhydrin analysis verified the deposition of CHS/Col multilayers on the aminolyzed PLLA scaffolds. Moreover, confocal laser scanning microscopy (CLSM) observation and hydroxyproline quantification revealed the successful LBL assembly of CHS/Col multilayers in the interior of PLLA porous scaffolds. *In vitro* chondrocyte culture showed that the presence of CHS and Col greatly improved the cytocompatibility of the PLLA scaffolds in terms of cell attachment, proliferation, viability, and GAG secretion. The study has proposed a practical and simple technique, i.e., aminolysis and subsequent LbL assembly, through which a cytocompatible polymeric material can be easily fabricated.

Surface functionalization techniques have also been employed to graft bioactive peptides on biomaterial surfaces [103]. As an example, Santiago et al. have covalently attached IKVAV, RGD, and YIGSR peptide sequences on the surface of a PCL substrate, by an aminolysis treatment followed by a grafting reaction mediated by carbodiimide chemistry [174]. Peptide functionalization needs the contemporary presence of antifouling molecules (e.g., PEG oligomers or PEG-modified peptides) to avoid surface fouling by plasma proteins, affecting surface bioactivity, and to contrast bacterial infection [197,198].

9.5.2 Bioartificial Materials Through Innovative Biomimetic Coatings

A completely innovative approach for surface coating of biomaterials consists of *in vitro* cell culturing on the surface of porous scaffolds, leading to ECM deposition, followed by scaffold decellularization. This approach has been recently described [199]. A synthetic poly(ester urethane) (DegraPol®; Ab Medica, Italy) has been processed by thermally induced phase separation to obtain porous cylindrical scaffolds. Human mesenchymal stem cells (hMSCs) from bone marrow aspirates, obtained from the iliac crest of healthy donors, have been seeded into the porous scaffolds in a bioreactor using a differentiative or proliferative culture medium. With the aim to produce hybrid constructs, a two phase perfusion culture regime has been followed consisting of a proliferative phase to increase cell number followed by a differentiative phase to stimulate ECM production and its mineralization for a total culture period of 28 days.

The constructs have then been decellularized by a two-step procedure based on (i) various freeze and thaw cycles with intermediate rinse in hypotonic solution to devitalize cells and (ii) perfusion of an isotonic solution to remove cellular debris. This protocol effectively removed cell remnants (94.2 ± 6.0% DNA reduction) but preserved collagen and calcium phosphate deposited by cells. Subsequently, *in vitro* culture experiments using hMSCs showed that scaffolds functionalized with cell-deposited ECM demonstrated the ability to promote cell differentiation, at a similar extent as for hydroxyapatite control scaffolds.

9.6 CONCLUSIONS AND FUTURE PERSPECTIVE

Current regenerative medicine strategies are addressed at the design of biomimetic materials by the combination of biocompatible synthetic polymer, providing structural and mechanical properties, and natural polymers or peptides, able to direct cell response. Many strategies have been investigated for the realization of bioartificial biomimetic materials although four distinctive routes can be delineated:

- The incorporation of minimal cell-binding motifs (bioactive peptides) into biomaterials to specifically direct cell behavior. Peptides can be introduced into the biomaterial bulk (obtaining bioartificial copolymer hydrogels) or grafted on biomaterial surface by a series of available approaches.

- The functionalization of biomaterials with proteins providing multifunctional materials with bioactive properties. Proteins can be introduced into the biomaterials bulk by blending (bioartificial blends or hydrogels) or grafting methods (bioartificial hydrogels), or they can be coated on the biomaterial surface through various approaches. Due to their multifunctionality, proteins are not able to drive a specific and selective cell response, but can improve scaffold biocompatibility and biomimickry.

- The functionalization of biomaterials with biomimetic polysaccharides to improve material biocompatibility and enhancing cell–material interactions. Cells interact with polysaccharides by nonintegrin receptors; moreover, proteins in the biological fluids are easily absorbed by exposed polysaccharides, then promoting cell–biomaterial communication mediated by integrin receptors [200]. Similarly to proteins, polysaccharides can be introduced into the biomaterials bulk by blending (bioartificial blends or hydrogels) or grafting methods (bioartificial hydrogels), or they can be coated on the biomaterial surface through various approaches. Biomimetic polysaccharides are not able to promote a specific cell response, although they increase biomaterial biocompatibility.

- The development of innovative strategies based on the incorporation of decellularized ECM into biomaterials. *In vitro* culturing of patient-derived cells on preformed scaffolds allows the deposition of ECM on the scaffold surface; a subsequent decellularization step leaves an ECM-coated scaffold. As an alternative, ECM prepared by *in vitro* patient-derived cells culture can be decellularized and solubilized. Solubilized ECM can then be used as a coating layer or as a stock material for the preparation of bioartificial blends or hydrogels. Intact or solubilized ECMs preserve the complexity of natural ECM. Although this approach is not able to address a specific and selective cell response, it allows the obtainment of a close microenvironment to natural niches, potentially directing adult stem cell behavior.

The use of decellularized ECM has been introduced only recently and demonstrates that current regenerative medicine strategies are addressed towards the development of biomimetic patient-specific scaffolds for tissue regeneration.

On the other hand, functionalization through biomimetic peptides continues to be of great interest for the development of smart devices driving specific signaling to selected cell phenotypes (e.g., vascular scaffolds).

ACKNOWLEDGMENTS

"Starigen" FIRB2010 Project and "Biodress" MANUNET-ERANET Project are acknowledged.

REFERENCES

[1] Hynes RO. The extracellular matrix: not just pretty fibrils. Science 2009;326:1216−9.

[2] Humphries MJ, Travis MA, Clark K, Mould AP. Mechanisms of integration of cells and extracellular matrices by integrins. Biochem Soc Trans 2004;32:822−5.

[3] Pachence JM, Kohn J. Biodegradable polymers. In: Lanza RP, Langer R, Chick WL, editors. Principles of tissue engineering. 2nd ed. San Diego, CA: Academic Press; 2000. p. 263−77.

[4] Agrawal CM, Ray RB. Biodegradable polymeric scaffolds for musculoskeletal tissue engineering. J Biomed Mater Res 2001;55:141−50.

[5] Pitt C. Poly-ε-caprolactone and its copolymers. In: Chasin M, Langer R, editors. Biodegradable polymers as drug delivery systems. New York: Marcel-Dekker; 1990. p. 71−120.

[6] Ma PX, Langer R. Morphology and mechanical function of long-term *in vitro* engineered cartilage. J Biomed Mater Res 1999;44:217−21.

[7] Hutmacher DW. Scaffolds in tissue engineering bone and cartilage. Biomaterials 2000;21:2529−34.

[8] Marra KG, Szem JW, Kumta PN, Di Milla PA, Weiss LE. *In vitro* analysis of biodegradable polymer blend/hydroxyapatite composites for bone tissue engineering. J Biomed Mater Res 1999;47:324−35.

[9] Winet H, Bao JY. Fibroblast growth factor-2 alters the effect of eroding polylactide−polyglycolide on osteogenesis in the bone chamber. J Biomed Mater Res 1998;40:567−76.

[10] Whang K, Tsai DC, Nam EK, Aitken M, Sprague SM, Patel PK, et al. Ectopic bone formation via rhBMP-2 delivery from porous bioabsorbable polymer scaffolds. J Biomed Mater Res 1998;42: 491−9.

[11] Cao YL, Vacanti JP, Ma PX, Paige KJ, Chowanski I, Schloo B, et al. Generation of neo-tendon using synthetic polymers seeded with enocytes. Transplant Proc 1993;25:1019−21.

[12] Zacchi V, Soranzo C, Cortivo R, Radice M, Brun P, Abatangelo G. *In vitro* engineering of human skin-like tissue. J Biomed Mater Res 1997;36:17−28.

[13] Kaufmann P, Heimarath S, Kim B, Mooney DJ. Highly porous polymer matrices as a three-dimensional culture systems for hepatocytes. Cell Transplant 1997;6:463−8.

[14] Zund G, Breuer CK, Shinoka T, Ma PX, Langer R, Mayer JE, et al. The *in vitro* construction of a tissue engineered bioprosthetic heart valve. Eur J Cardiothorac Surg 1997;11:493−7.

[15] Cai Q, Wan Y, Bei J, Wang S. Synthesis and characterization of biodegradable polylactide-grafted dextran and its application as compatilizer. Biomaterials 2003;24:3555−62.

[16] Cai Q, Yang J, Bei J, Wang S. A novel porous cells scaffold made of polylactide−dextran blend by combining phase-separation and particle-leaching techniques. Biomaterials 2002;23:4483−92.

[17] Nampoothiri KM, Nair NR, John RP. An overview of the recent developments in polylactide (PLA) research. Bioresour Technol 2010;101:8493−501.

[18] Zhao YM, Wang ZY, Wang J, Mai HZ, Yan B, Yang F. Direct synthesis of poly(D,L-lactic acid) by melt polycondensation and its applications in drug delivery. J Appl Polym Sci 2004;91:2143−50.

[19] Mehta R, Kumar V, Bhunia H, Upadhyay SN. Synthesis of poly(lactic acid): a review. J Macromol Sci Polym Rev 2005;45:325−49.

[20] Garg S, Serruys PW. Coronary stents: looking forward. J Am Coll Cardiol 2010;56:43−78.

[21] Dai TT, Jiang ZH, Li S, Zhoua GD, Kretlow JD, Cao WG, et al. Reconstruction of lymph vessel by lymphatic endothelial cells combined with polyglycolic acid scaffolds: a pilot study. J Biotechnol 2010;150:182−9.

[22] Cao Y, Mitchell G, Messina A, Priced L, Thompson E, Penington A, et al. The influence of architecture on degradation and tissue ingrowth into three-dimensional poly(lactic-co-glycolic acid) scaffolds *in vitro* and *in vivo*. Biomaterials 2006;27:2854−64.

[23] Fredenberg S, Wahlgren M, Reslow M, Axelsson A. The mechanisms of drug release in poly(lactic-co-glycolic acid)-based drug delivery systems—a review. Int J Pharm 2011;415:34−52.

[24] Danhier F, Ansoren E, Silva JM, Coco R, Le Breton A, Préat V. PLGA-based nanoparticles: an overview of biomedical applications. J Control Release 2012;161:505−22.

[25] Chang H-I, Wang Y, Perrie Y, Coombes AGA. Microporous polycaprolactone matrices for drug delivery and tissue engineering: the release behaviour of bioactives having extremes of aqueous solubility. J Drug Delivery Sci Technol 2010;20:207−12.

[26] Chiono V, Vozzi G, Vozzi F, Salvadori C, Dini F, Carlucci F, et al. Melt-extruded guides for peripheral nerve regeneration. Part I: poly(ε-caprolactone). Biomed Microdevices 2009;11:1037−50.

[27] Elomaa L, Teixeira S, Hakala R, Korhonen H, Grijpma DW, Seppälä JV. Preparation of poly(ε-caprolactone)-based tissue engineering scaffolds by stereolithography. Acta Biomater 2011;7:3850−6.

[28] Meek MF, Jansen K, Steendam R, van Oeveren W, van Wachem PB, van Luyn MJA. *In vitro* degradation and biocompatibility of poly(DL-lactide-co-caprolactone) nerve guides. J Biomed Mater Res A 2004;68A:43−51.

[29] Li C, Wang L, Yang Z, Kim G, Chen H, Ge Z. A viscoelastic chitosan-modified three-dimensional porous poly(L-lactide-co-ε-caprolactone) scaffold for cartilage tissue engineering. J Biomater Sci 2012;23:405—24.

[30] Gentile P, Chiono V, Tonda-Turo C, Ferreira AM, Ciardelli G. Polymeric membranes for guided bone regeneration. J Biotechnol 2011;6:1187—97.

[31] Novikova LN, Pettersson J, Brohlin M, Wiberg M, Novikov LN. Biodegradable poly-beta-hydroxybutyrate scaffold seeded with Schwann cells to promote spinal cord repair. Biomaterials 2008;29:1198—206.

[32] Khorasani MT, Mirmohammadi SA, Irani S. Polyhydroxybutyrate (PHB) scaffolds as a model for nerve tissue engineering application: fabrication and in vitro assay. Int J Polym Mater 2011;60:562—75.

[33] Sombatmankhong K, Sanchavanakit N, Pavasant P, Supaphol P. Bone scaffolds from electrospun fiber mats of poly(3-hydroxybutyrate), poly(3-hydroxybutyrate-co-3-hydroxyvalerate) and their blend. Polymer 2007;48:1419—27.

[34] Chiono V, Ciardelli G, Vozzi G, Sotgiu MG, Vinci B, Domenici C, et al. Poly(3-hydroxybutyrate-co-3-hydroxyvalerate)/poly(ε-caprolactone) blends for tissue engineering applications in the form of hollow fibers. J Biomed Mater Res A 2008;85:938—53.

[35] Veleirinho B, Coelho DS, Dias PF, Maraschin M, Ribeiro-do-Valle RM, Lopes-da-Silva JA. Nanofibrous poly(3-hydroxybutyrate-co-3-hydroxyvalerate)/chitosan scaffolds for skin regeneration. Int J Biol Macromol 2012;51:343—50.

[36] Del Gaudio C, Fioravanzo L, Folin M, Marchi F, Ercolani E, Bianco A. Electrospun tubular scaffolds: on the effectiveness of blending poly(ε-caprolactone) with poly(3-hydroxybutyrate-co-3-hydroxyvalerate). J Biomed Mater Res B 2012;100:1883—98.

[37] Shi Q, Zhong S, Chen Y, Whitaker W. Crosslinking copolymers based polyanhydride and 1G olyamidoamine-methacrylamide as bone tissue engineering: synthesis, characterization, and in vitro degradation. Polym Degrad Stab 2010;95:1961—8.

[38] Griffin J, Carbone A, Delgado-Rivera R, Meiners S, Uhrich KE. Design and evaluation of novel polyanhydride blends as nerve guidance conduits. Acta Biomater 2010;6:1917—24.

[39] Dalsin JL, Messersmith PB. Bioinspired antifouling polymers. Mater Today 2005;8:38—46.

[40] Park JH, Bae YH. Hydrogels based on poly(ethylene oxide) and poly(tetramethylene oxide) or poly(dimethyl siloxane): synthesis, characterization, in vitro protein adsorption and platelet adhesion. Biomaterials 2002;23:1797—808.

[41] Pourcelle V, Freichels H, Stoffelbach F, Auzély-Velty R, Jéroôme C, Marchand-Brynaert J. Light induced functionalization of PCL—PEG block copolymers for the covalent immobilization of biomolecules. Biomacromolecules 2009;10:966—74.

[42] Sun XF, Jing Z, Wang G. Preparation and swelling behaviors of porous hemicellulose-g-polyacrylamide hydrogels. J Appl Polym Sci 2012;128:1861—70.

[43] Tonda-Turo C, Audisio C, Gnavi S, Chiono V, Gentile P, Raimondo S, et al. Porous poly(ε-caprolactone) nerve guide filled with porous gelatin matrix for nerve tissue engineering. Adv Eng Mater 2011;13:151—64.

[44] Comolli N, Donaldson O, Grantier N, Zhukareva V, Tom VJ. Polyvinyl alcohol—polyvinyl pyrrolidone thin films provide local short-term release of anti-inflammatory agents post spinal cord injury. J Biomed Mater Res B 2012;100B:1867—73.

[45] Saha N, Saarai A, Roy N, Kitano T, Saha P. Polymeric biomaterial based hydrogels for biomedical applications. J Biomater Nanobiotechnol 2011;2:85—90.

[46] Ahn MY, Hwang IT, Jung CH, Nho YC, Choi JH, Huh KM. Cell patterning on a poly(N-vinyl pyrrolidone)-patterned polystyrene substrate by using ion implantation. J Ind Eng Chem 2010;16:87—90.

[47] Hwang S, Jeong S. Electrospun nano composites of poly(vinyl pyrrolidone)/nano-silver for antibacterial materials. J Nanosci Nanotechnol 2011;11:610—3.

[48] Basak P, Adhikari B. Poly(vinyl alcohol) hydrogels for pH dependent colon targeted drug delivery. J Mater Sci Mater Med 2009;20:137—46.

[49] Zhang H, Xia H, Zhao Y. Poly(vinyl alcohol) hydrogel can autonomously self-heal. ACS Macro Lett 2012;1:1233—6.

[50] Ferreira AM, Gentile P, Sartori S, Pagliano C, Cabrele C, Chiono V, et al. Biomimetic soluble collagen purified from bones. J Biotechnol 2012;7:1386—94.

[51] Ferreira AM, Gentile P, Chiono V, Ciardelli G. Collagen for bone tissue regeneration. Acta Biomater 2012;8:3191—200.

[52] Chiono V, Pulieri E, Vozzi G, Ciardelli G, Ahluwalia A, Giusti P. Genipin-crosslinked chitosan/gelatin blends for biomedical applications. J Mater Sci Mater Med 2008;19:889—98.

[53] Pulieri E, Chiono V, Ciardelli G, Vozzi G, Ahluwalia A, Domenici C, et al. Chitosan/gelatin blends for biomedical applications. J Biomed Mater Res A 2008;86:311—22.

[54] Wise SG, Byrom MJ, Waterhouse A, Bannon PG, Ng MK, Weiss AS. A multilayered synthetic human elastin/polycaprolactone hybrid vascular graft with tailored mechanical properties. Acta Biomater 2010;7:295—303.

[55] Dinerman AA, Cappello J, El-Sayed M, Hoag SW, Ghandehari H. Influence of solute charge and hydrophobicity on partitioning and diffusion in a genetically engineered silk-elastin-like protein polymer hydrogel. Macromol Biosci 2010;10:1235—47.

[56] Rocha LB, Adam RL, Leite NJ, Metze K, Rossi MA. Biomimeralization of polyanionic collagen-elastin matrices during cavarian bone repair. J Biomed Mater Res A 2006;79:237—45.

[57] Bessa PC, Machado R, Nürnberger S, Dopler D, Banerjee A, Cunha AM, et al. Thermoresponsive self-assembled elastin-based nanoparticles for delivery of BMPs. J Control Release 2010;142:312—8.

[58] Rexeisen EL, Fan W, Pangburn TO, Taribagil RR, Bates FS, Lodge TP, et al. Self-assembly of fibronectin mimetic peptide-amphiphile nanofibers. Langmuir 2009;26:1953—9.

[59] Amaral IF, Unger RE, Fuchs S, Mendonça AM, Sousa SR, Barbosa MA, et al. Fibronectin-mediated endothelisation of chitosan porous matrices. Biomaterials 2009;30:5465—75.

[60] Custódio CA, Alves CM, Reis RL, Mano JF. Immobilisation of fibronectin in chitosan substrates improves cell adhesion and proliferation. J Tissue Eng Regener Med 2010;4:316—23.

[61] Wittmer CR, Phelps JA, Saltzman WM, Tassel PRV. Fibronectin terminated multilayer films: protein adsorption and cell attachment studies. Biomaterials 2007;28:851—60.

[62] Kikkava Y, Takahashi N, Matsuda Y, Miwa T, Akizuki T, Kataoka A, et al. The influence of synthetic peptides derived from the laminin α1 chain on hepatocyte adhesion and gene expression. Biomaterials 2009;30:6888—95.

[63] Hozumi K, Akizuki T, Yamada Y, Hara T, Urushibada S, Katagiri F, et al. Cell adhesive peptide screening of mouse laminin α1 chain G domain. Arch Biochem Biophys 2010;503:213—22.

[64] Sarfati G, Dvir T, Elkabets M, Apte RN, Cohen S. Targeting of polymeric nanoparticles to lung metastases by surface attachment of YIGSR peptide from laminin. Biomaterials 2011;32:152−61.

[65] Shachaf Y, Gonen-Wadmany M, Seliktar D. The biocompatibility of Pluronic® F127 fibrinogen-based hydrogels. Biomaterials 2010;31:2836−47.

[66] Spotniz WD. Fibrin sealant: past, present and future: a brief review. World J Surg 2010;34:632−4.

[67] Ahmann KA, Weinbaum JS, Johnson SL, Tranquillo RT. Fibrin degradation enhances vascular smooth muscle cell proliferation and matrix deposition in fibrin-based tissue constructs fabricated *in vitro*. Tissue Eng Part A 2010;16:3261−70.

[68] Le YB, Polio S, Lee W, Dai G, Menon L, Carroll RS, et al. Bioprinting of collagen and VEGF-releasing fibrin gel scaffolds for neural stem cell culture. Exp Neurol 2010;223:645−52.

[69] Baoyong L, Jian Z, Denglong C, Min L. Evaluation of a new type of wound dressing made from recombinant spider silk protein using rat models. Burns 2010;36:891−6.

[70] Omenetto FG, Kaplan DL. New opportunities for an ancient material. Science 2010;329:528−31.

[71] Altman GH, Diaz F, Jakuba C, Calabro T, Horan LR, Chen J, et al. Silk-based biomaterials. Biomaterials 2003;24:401−16.

[72] Fan H, Liu H, Toh SL, Goh JCH. Enhanced differentiation of mesenchymal stem cells co-cultured with ligament fibroblasts on gelatin/silk fibroin hybrid scaffold. Biomaterials 2008;29:1017−27.

[73] Sell SA, McClure MJ, Garg K, Wolfe PS, Bowlin GL. Electrospinning of collagen/biopolymers for regenerative medicine and cardiovascular tissue engineering. Adv Drug Delivery Rev 2009;61:1007−19.

[74] Kluge JA, Rabotyagova O, Leisk GG, Kaplan DL. Spider silks and their applications. Trends Biotechnol 2008;26:244−51.

[75] Lammel AS, Hu X, Park SH, Kaplan DL, Scheibel TR. Controlling silk fibroin particle features for drug delivery. Biomaterials 2010;31:4583−91.

[76] Kumar PTS, Srinivasan S, Lakshmanan VK, Tamura H, Nair SV, Jayakumar R. Chitin hydrogel/nano hydroxyapatite composite scaffolds for tissue engineering applications. Carbohydr Polym 2011;85:584−91.

[77] Jayakumar R, Menona D, Manzoor K, Naira SV, Tamura H. Review biomedical applications of chitin and chitosan based nanomaterials—a short review. Carbohydr Polym 2010;82:227−32.

[78] Kim IY, Seo SJ, Moon HS, Yoo MK, Park IY, Kim BC, et al. Chitosan and its derivatives for tissue engineering applications. Biotechnol Adv 2008;26:1−21.

[79] Muzzarelli RAA, Greco F, Busilacchi A, Sollazzo V, Gigante A. Chitosan, hyaluronan and chondroitin sulfate in tissue engineering for cartilage regeneration: a review. Carbohydr Polym 2012;89:723−39.

[80] Wang X, He J, Wang Y, Cui FZ. Hyaluronic acid-based scaffold for central neural tissue engineering. Interface Focus 2012;2:278−91.

[81] Unterman SA, Gibson M, Lee JH, Crist J, Chansakul T, Yang EC, et al. Hyaluronic acid-binding scaffold for articular cartilage repair. Tissue Eng Part A 2012;18:2497−506.

[82] Iwasaki N, Kasahara Y, Yamane S, Igarashi T, Minami A, Nisimura SI. Chitosan-based hyaluronic acid hybrid polymer fibers as a scaffold biomaterial for cartilage tissue engineering. Polymer 2011;3:100−13.

[83] Kim IL, Mauck RL, Burdick JA. Hydrogel design for cartilage tissue engineering: a case study with hyaluronic acid. Biomaterials 2011;32:8771−82.

[84] Liu LS, Thompson AY, Heidaran MA, Poser JW, Spiro RC. An osteoconductive collagen hyaluronate matrix for bone regeneration. Biomaterials 1999;20:1097−108.

[85] Park SN, Lee HJ, Lee KH, Suh H. Biological characterization of EDC-crosslinked collagen−hyaluronic acid matrix in dermal tissue restoration. Biomaterials 2003;24:1631−41.

[86] Luo Y, Kirker KR, Prestwich DG. Cross-linked hyaluronic acid hydrogel films: new biomaterials for drug delivery. J Control Release 2000;69:169−84.

[87] Liang WH, Kienitz BL, Penick KJ, Welter JF, Zawodzinski TA, Baskaran H. Concentrated collagen-chondroitin sulfate scaffolds for tissue engineering applications. J Biomed Mater Res A 2010;15:1050−60. doi:10.1002/jbm.a.32774.

[88] Wollenweber M, Domaschke H, Hanke T, Boxberger S, Schmack G, Gliesche K, et al. Mimicked bioartificial matrix containing chondroitin sulphate on a textile scaffold of poly(3-hydroxybutyrate) alters the differentiation of adult human mesenchymal stem cells. Tissue Eng 2006;12:345−59.

[89] Popa EG, Gomes ME, Reis RL. Cell delivery systems using alginate−carrageenan hydrogel beads and fibers for regenerative medicine applications. Biomacromolecules 2011;12:3952−61.

[90] Tønnesen HH, Karlsen J. Alginate in drug delivery systems. Drug Dev Ind Pharm 2002;28:621−30.

[91] Wong M. Alginates in tissue engineering. In: Hollander AP, Hatton PV, editors. Biopolymer methods in tissue engineering, vol. 238. New York: Humana Press; 2004. p. 77−86.

[92] Augst AD, Kong HJ, Mooney DJ. Alginate hydrogels as biomaterials. Macromol Biosci 2006;6:623−33.

[93] Eiselt P, Yeh J, Latvala RK, Shea LD, Mooney DJ. Porous carriers for biomedical applications based on alginate hydrogels. Biomaterials 2000;21:1921−7.

[94] Lead JR, Starchev K, Wilkinson KJ. Diffusion coefficient of humic substances in agarose gel and in water. Environ Sci Technol 2003;37:482−7.

[95] Sumathi S, Ray AR. Release behavior of drugs from tamarind seed polysaccharide tablets. J Pharm Sci 2002;5:12−8.

[96] Duarte AR, Mano JF, Reis RL. Preparation of starch-based scaffolds for tissue engineering by supercritical immersion precipitation. J Supercrit Fluids 2009;49:279−85.

[97] Duarte AR, Mano JF, Reis RL. Enzymatic degradation of 3D scaffolds of starch−poly-(ε-caprolactone) prepared by supercritical fluid technology. Polym Degrad Stab 2010;95:2110−7.

[98] Sell SA, Wolfe PS, Garg K, McCool JM, Rodriguez IA, Bowlin GL. The use of natural polymers in tissue engineering: a focus on electrospun extracellular matrix analogues. Polymer 2010;2:522−53.

[99] Sionkowska A. Current research on the blends of natural and synthetic polymers as new biomaterials: review. Prog Polym Sci 2011;36:1254−76.

[100] Nandagiri VK, Chiono V, Gentile P, Montevecchi FM, Ciardelli G. Biomaterials of natural origin in regenerative medicine. In: Dumitriu S, Popa V, editors. Polymeric biomaterials: structure and function, vol. 1. CRC Press/Taylor and Francis Group, Boca Raton, FL; 2013. p. 38.

[101] Tessmar JK, Göpferich AM. Matrices and scaffolds for protein delivery in tissue engineering. Adv Drug Delivery Rev 2007;59:274–91.

[102] Shin H, Jo S, Mikos AG. Biomimetic materials for tissue engineering. Biomaterials 2003;24:4353–64.

[103] Hersel U, Dahmen C, Kessler H. RGD modified polymers: biomateriale for stimulated cell adhesion and beyond. Biomaterials 2003;24:4385–415.

[104] Chiono V, Vozzi G, D'Acunto M, Brinzi S, Domenici C, Vozzi F, et al. Characterisation of blends between poly(ε-caprolactone) and polysaccharides for tissue engineering applications. Mater Sci Eng C 2009;29:2174–87.

[105] Ciardelli G, Chiono V, Vozzi G, Pracella M, Ahluwalia A, Barbani N, et al. Blends of poly-(ε-caprolactone) and polysaccharides in tissue engineering applications. Biomacromolecules 2005;6:1961–76.

[106] Williams DF. The role of short synthetic adhesion peptides in regenerative medicine: the debate. Biomaterials 2011;32:4195–7.

[107] Barnes CA, Brison J, Michel R, Brown BN, Castner DG, Badylak SF, et al. The surface molecular functionality of decellularized extracellular matrices. Biomaterials 2011;32:137–43.

[108] Ott HC, Matthiesen TS, Goh S-K, Black LD, Kren SM, Netoff TI, et al. Perfusion-decellularized matrix: using nature's platform to engineer a bioartificial heart. Nat Med 2008;14:213–21.

[109] Badylak DF, Freytes DO, Gilbert TW. Extracellular matrix as a biological scaffold material: structure and function. Acta Biomater 2009;5:1–13.

[110] Okada M, Payne TR, Oshima H, Momoi N, Tobita K, Huard J. Differential efficacy of gels derived from small intestinal submucosa as an injectable biomaterial for myocardial infarct repair. Biomaterials 2010;31:7678–83.

[111] Singelyn JM, DeQuach JA, Seif-Naraghi SB, Littlefield RB, Schup-Magoffin PJ, Christman KL. Naturally derived myocardial matrix as an injectable scaffold for cardiac tissue engineering. Biomaterials 2009;30:5409–16.

[112] Gunn J, Zhang M. Polyblend nanofibers for biomedical applications: perspectives and challenges. Trends Biotechnol 2010;28:189–97.

[113] Dai NT, Williamson MR, Khammo N, Adams EF, Coombes AGA. Composite cell support membranes based on collagen and polycaprolactone for tissue engineering of skin. Biomaterials 2004;25:4263–71.

[114] Choi JS, Lee SJ, Christ GJ, Atala A, Yoo JJ. The influence of electrospun aligned poly(3-caprolactone)/collagen nanofiber meshes on the formation of self-aligned skeletal muscle myotubes. Biomaterials 2008;29:2899–906.

[115] Moncy JV, Vinoy T, Dean DR, Nyairo E. Fabrication and characterization of aligned nanofibrous PLGA/collagen blends as bone tissue scaffolds. Polymer 2009;50:3778–85.

[116] McClure MJ, Sell SA, Simpson DG, Walpoth BH, Bowlin GL. A three-layered electrospun matrix to mimic arterial architecture using polycaprolactone, elastin and collagen: a preliminary study. Acta Biomater 2010;6:2422–33.

[117] Ji W, Yang F, Ma F, Bouma MJ, Boerman OC, Chen Z, et al. Incorporation of stromal cell-derived factor-1a in PCL/gelatin electrospun membranes for guided bone regeneration. Biomaterials 2013;34: 753–745

[118] Alvarez Perez MA, Guarino V, Cirillo V, Ambrosio L. *In vitro* mineralization and bone osteogenesis in poly(ε-caprolactone)/gelatin nanofibers. J Biomed Mater Res A 2012;100A:3008–19.

[119] Gupta D, Venugopal J, Prabhakaran MP, Dev VRG, Low S, Choon AT, et al. Aligned and random nanofibrous substrate for the *in vitro* culture of Schwann cells for neural tissue engineering. Acta Biomater 2009;5:2560–9.

[120] Chiono V, Ciardelli G, Vozzi G, Cortez J, Barbani N, Gentile P, et al. Enzymatically modified melt-extruded guides for peripheral nerve repair. Eng Life Sci 2008;8:226–37.

[121] Di Martino A, Sittinger M, Risbud MV. Chitosan: a versatile biopolymer for orthopaedic tissue engineering. Biomaterials 2005;26:5983–90.

[122] Kuo YC, Lin CY. Effect of genipin-crosslinked chitin–chitosan scaffolds with hydroxyapatite modifications on the cultivation of bovine knee chondrocytes. Biotechnol Bioeng 2006;95:132–44.

[123] Kim SE, Park JH, Cho YW, Chung H, Jeong SY, Lee EB, et al. Porous chitosan scaffold containing microspheres loaded with transforming growth factor-b1: implications for cartilage tissue engineering. J Control Release 2003;91:365–74.

[124] Ciardelli G, Chiono V. Materials for peripheral nerve regeneration. Macromol Biosci 2006;6:13–26.

[125] Honma T, Senda T, Inoue Y. Thermal properties and crystallization behaviour of blends of poly(epsilon-caprolactone) with chitin and chitosan. Polym Int 2003;52:1839–46.

[126] Senda T, He Y, Inoue Y. Biodegradable blends of poly(epsilon-caprolactone) with alpha-chitin and chitosan: specific interactions, thermal properties and crystallization behavior. Polym Int 2002;51:33–9.

[127] Sarasam A, Madihally SV. Characterization of chitosan–polycaprolactone blends for tissue engineering applications. Biomaterials 2005;26:5500–8.

[128] Cruz DMG, Ribelles JLG, Sanchez MS. Blending polysaccharides with biodegradable polymers. I. Properties of chitosan/polycaprolactone blends. J Biomed Mater Res B 2008;85:303–13.

[129] Cruz DMG, Coutinho DF, Costa Martinez E, Mano JF, Gòmez Ribelles JL, Salmeròn Sànchez M. Blending polysaccharides with biodegradable polymers. II. Structure and biological response of chitosan/polycaprolactone blends. J Biomed Mater Res B 2008;87B:544–54.

[130] Malheiro VN, Caridade SG, Alves NM, Mano JF. New poly(ε-caprolactone)/chitosan blend fibers for tissue engineering applications. Acta Biomater 2010;6:418–28.

[131] Neves SC, Moreira Teixeira LS, Moroni L, Reis RL, Van Blitterswijk CA, Alves NM, et al. Chitosan/poly(3-caprolactone) blend scaffolds for cartilage repair. Biomaterials 2011;32:1068–79.

[132] Nielsen GD, Abraham MH, Hansen LF, Hammer M, Cooksey CJ, Andonian-Haftvan J, et al. Sensory irritation mechanisms investigated from model compounds: trifluoroethanol, hexafluoroisopropanol and methyl hexafluoroisopropyl ether. Arch Toxicol 1996;70:319–28.

[133] Moshfeghian A, Tillman J, Madihally SV. Characterization of emulsified chitosan–PLGA matrices formed using controlled rate freezing and lyophilization technique. J Biomed Mater Res A 2006;79A:418–30.

[134] Mano JF, Koniarova D, Reis RL. Thermal properties of thermoplastic starch/synthetic polymer blends with potential biomedical applicability. J Mater Sci Mater Med 2003;14:127–35.

[135] Avèrous L. Biodegradable multiphase systems based on plasticized starch: a review. J Macromol Sci C 2005;44:231–74.

[136] Rodrigues AI, Gomes ME, Leonor IB, Reis RL. Bioactive starch-based scaffolds and human adipose stem cells are a good combination for bone tissue engineering. Acta Biomater 2012;8:3765−76.

[137] Ghosh S, Viana JC, Reis RL, Mano JF. Bi-layered constructs based on poly(L-lactic acid) and starch for tissue engineering of osteochondral defects. Mater Sci Eng C 2008;28:80−6.

[138] Pavlov MP, Mano JF, Neves NM, Reis RL. Fibers and 3D mesh scaffolds from biodegradable starch-based blends: production and characterization. Macromol Biosci 2004;4:776−84.

[139] Sobral JM, Caridade SG, Sousa RA, Mano JF, Reis RL. Three-dimensional plotted scaffolds with controlled pore size gradients: effect of scaffold geometry on mechanical performance and seeding efficiency. Acta Biomater 2011;7:1009−18.

[140] Neves NM, Kouyumdzhiev A, Reis RL. The morphology, mechanical properties and ageing behavior of porous injection molded starch-based blends for tissue engineering scaffolding. Mater Sci Eng C 2005;25:195−200.

[141] Schenke-Layland K, Rofail F, Heydarkhan S, Gluck JM, Ingle NP, Angelis E, et al. The use of three-dimensional nanostructures to instruct cells to produce extracellular matrix for regenerative medicine strategies. Biomaterials 2009;30:4665−75.

[142] Hoffman A. Hydrogels for biomedical applications. Adv Drug Delivery Rev 2002;43:3−12.

[143] Slaughter BV, Khurshid SS, Fisher OZ, Khademhosseini A, Peppas NA. Hydrogels in regenerative medicine. Adv Mater 2009;21:3307−29.

[144] Jin R, Dijkstra PJ. Hydrogels for tissue engineering applications. In: Ottenbrite RM, Park K, Okano T, editors. Biomedical applications of hydrogels handbook. Milan: Springer-Verlag; 2010. p. 203−21.

[145] Zavan B, Cortivo R, Abatangelo G. Hydrogels and tissue engineering. In: Barbucci R, editor. Hydrogels: biological properties and applications. Milan: Springer-Verlag; 2009. p. 1−8.

[146] Mangala E, Suresh Kumar T, Baska S, Panduranga RK. Development of chitosan/poly(vinyl alcohol) blend membranes as burn dressings. Trends Biomater Artif Organ 2003;17:34−40.

[147] Alberti M, Snakenborg D, Lopacinska JM, Dufa M, Kutter JP. Characterization of a patch-clamp microchannel array towards neuronal networks analysis. Microfluid Nanofluid 2010;9:963−70.

[148] Cascone MG, Barbani N, Cristallini C, Giusti P, Ciardelli G, Lazzeri L. Bioartificial polymeric materials based on polysaccharides. J Biomater Sci Polym Ed 2001;3:267−81.

[149] Cascone MG, Lazzeri L, Sparvoli E, Scatena M, Serino LP, Danti S. Morphological evaluation of bioartificial hydrogels as potential tissue engineering scaffolds. J Mater Sci Mater Med 2004;15:1309−13.

[150] Giusti P, Lazzeri L, De Petris S, Palla M, Cascone MG. Collagen-based new bioartificial polymeric materials. Biomaterials 1994;15:1229−33.

[151] Shokrgozar M, Mottaghitalab FM, Vahid FM. Fabrication of porous chitosan/poly(vinyl alcohol) reinforced single-walled carbon nanotube nanocomposites for neural tissue engineering. J Biomed Nanotechnol 2011;7:276−84.

[152] Bispo VM, Mansur AP, Barbosa-Stancioli EF, Mansur HS. Biocompatibility of nanostructured chitosan/poly(vinyl alcohol) blends chemically crosslinked with genipin for biomedical applications. J Biomed Nanotechnol 2010;6:166−75.

[153] Ionita M, Iovu H. Mechanical properties, urea diffusion, and cell cultural response of poly(vinyl alcohol)−chitosan bioartificial membranes via molecular modelling and experimental investigation. Composites Part B 2012;43:2464−70.

[154] Peppas NA, Hilt JZ, Khademhosseini A, Langer R. Hydrogels in biology and medicine: from molecular principles to bionanotechnology. Adv Mater 2006;18:1345−60.

[155] Nuttelman CR, Rice MA, Rydholm AE, Salinas CN, Shah DN, Anseth KS. Macromolecular monomers for the synthesis of hydrogel niches and their application in cell encapsulation and tissue engineering. Prog Polym Sci 2008;33:167−79.

[156] Lutolf MP, Lauer-Fields JL, Schmoekel HG, Metters AT, Weber FE, Fields GB, et al. Synthetic matrix metalloproteinase-sensitive hydrogels for the conduction of tissue regeneration: engineering cell invasion characteristics. Proc Natl Acad Sci USA 2003;100:5413−8.

[157] Pratt AB, Weber FE, Schmoekel HG, Muller R, Hubbell JA. Synthetic extracellular matrices for *in situ* tissue engineering. Biotechnol Bioeng 2004;86:27−36.

[158] Mann BK, Gobin AS, Tsai AT, Schmedlen RH, West JL. Smooth muscle cell growth in photopolymerized hydrogels with cell adhesive and proteolytically degradable domains: synthetic ECM analogs for tissue engineering. Biomaterials 2001;22:3045−51.

[159] Gobin AS, West JL. Cell migration through defined, synthetic ECM analogs. FASEB J 2002;16:751−3.

[160] Mann BK, Schmedlen RH, West JL. Tethered-TGF-beta increases extracellular matrix production of vascular smooth muscle cells. Biomaterials 2001;22:439−44.

[161] Almany L, Seliktar D. Biosynthetic hydrogel scaffolds made from fibrinogen and polyethylene glycol for 3D cell cultures. Biomaterials 2005;26:2467−77.

[162] Dikovsky D, Bianco-Peleda H, Seliktar D. The effect of structural alterations of PEG−fibrinogen hydrogel scaffolds on 3-D cellular morphology and cellular migration. Biomaterials 2006;27:1496−506.

[163] Kraehenbuehl TP, Zammaretti P, Van der Vlies AJ, Schoenmakers RG, Lutolf MP, Jaconi ME, et al. Three-dimensional extracellular matrix-directed cardioprogenitor differentiation: systematic modulation of a synthetic cell-responsive PEG hydrogel. Biomaterials 2008;29:2757−66.

[164] Alexandridis P, Hatton TA. Poly(ethylene oxide)−poly(propylene oxide)−poly(ethylene oxide) block copolymer surfactants in aqueous solutions and at interfaces: thermodynamics, structure, dynamics, and modeling. Colloids Surf A 1995;96:1−46.

[165] Cha MH, Choi J, Choi BG, Park K, Kim H, Jeong B, et al. Synthesis and characterization of novel thermo-responsive F68 block copolymers with cell-adhesive RGD peptide. J Colloid Interface Sci 2011;360:78−85.

[166] Park KM, Lee SY, Joung YK, Na JS, Lee MC, Park KD. Thermosensitive chitosan−pluronic hydrogel as an injectable cell delivery carrier for cartilage regeneration. Acta Biomater 2009;5:1956−65.

[167] Zdrahala RJ, Zdrahala IJ. Biomedical applications of polyurethanes: a review of past promises, present realities, and a vibrant future. J Biomater Appl 1999;16:67−90.

[168] Park D, Wu W, Wang Y. A functionalizable reverse thermal gel based on a polyurethane/PEG block copolymer. Biomaterials 2011;32:777−86.

[169] Jiao YP, Cui FZ. Surface modification of polyester biomaterials for tissue engineering. Biomed Mater 2007;2:24—37.

[170] Di Toro R, Betti V, Spampinato S. Biocompatibility and integrin-mediated adhesion of human osteoblasts to poly(DL-lactide-co-glycolide) copolymers. Eur J Pharm Sci 2004;21:161—9.

[171] Grøndahl L, Chandler-Temple A, Trau M. Polymeric grafting of acrylic acid onto poly(3-hydroxybutyrate-co-3-hydroxyvalerate): surface functionalization for tissue engineering applications. Biomacromolecules 2005;6:2197—203.

[172] Kang G, Yu H, Liu Z, Cao Y. Surface modification of a commercial thin film composite polyamide reverse osmosis membrane by carbodiimide-induced grafting with poly(ethylene glycol) derivatives. Desalination 2011;275:252—9.

[173] Russo L, Zanini S, Giannoni P, Landi E, Villa A, Sandri M, et al. The influence of plasma technology coupled to chemical grafting on the cell growth compliance of 3D hydroxyapatite scaffolds. J Mater Sci Mater Med 2012;23:2727—38.

[174] Santiago LY, Nowal RW, Rubin JP, Marra KG. Peptide surface modification of poly(caprolactone) with laminin-derived sequences for adipose-derived stem cell applications. Biomaterials 2006;27:2962—9.

[175] Zhu YB, Gao CY, Liu YX, Shen JC. Surface modification of polycaprolactone membrane via aminolysis and biomacromolecule immobilization for promoting cytocompatibility of human endothelial cells. Biomacromolecules 2002;3:1312—9.

[176] Zhu YB, Gao CY, Liu YX, Shen JC. Endothelial cell functions *in vitro* cultured on poly(L-lactic acid) membranes modified with different methods. J Biomed Mater Res A 2004;69A:436—43.

[177] Lao HK, Renard E, Linossier I, Langlois V, Vallée-Rehel K. Modification of poly(3-hydroxybutyrate-co-3-hydroxyvalerate) film by chemical graft copolymerization. Biomacromolecules 2007;8:416—23.

[178] dos Santos KS, Coelho JF, Ferreira P, Pinto I, Lorenzetti SG, Ferreira EI, et al. Synthesis and characterization of membranes obtained by graft copolymerization of 2-hydroxyethyl methacrylate and acrylic acid onto chitosan. Int J Pharm 2006;310:37—45.

[179] Guo C, Gemeinhart RA. Understanding the adsorption mechanism of chitosan onto poly(lactide-co-glycolide) particles. Eur J Pharm Biopharm 2008;70:597—604.

[180] Zhang Y, Chai C, Jiang XS, Teoh SH, Leong KW. Fibronectin immobilized by covalent conjugation or physical adsorption shows different bioactivity on aminated-PET. Mater Sci Eng C 2007;27:213—9.

[181] Gong Y, Zhu Y, Liu Y, Ma Z, Gao C, Shen J. Layer-by-layer assembly of chondroitin sulfate and collagen on aminolyzed poly (L-lactic acid) porous scaffolds to enhance their chondrogenesis. Acta Biomater 2007;3:677—85.

[182] Schmidt TGM, Koepke J, Frank R, Skerra A. Molecular interaction between the strep-tag affinity peptide and its cognate target, streptavidin. J Mol Biol 1995;255:753—66.

[183] Chu PK, Chen JY, Wang LP, Huang N. Plasma-surface modification of biomaterials. Mater Sci Eng 2002;R36:143—206.

[184] de Villiers MM, Otto DP, Strydom SJ, Lvov YM. Introduction to nanocoatings produced by layer-by-layer (LbL) self-assembly. Adv Drug Delivery Rev 2011;63:701—15.

[185] Ma Z, Mao Z, Gao C. Surface modification and property analysis of biomedical polymers used for tissue engineering. Colloids Surf B 2007;60:137—57.

[186] Chiono V, Descrovi E, Sartori S, Gentile P, Ballarini M, Giorgis F, et al. Biomimetic tailoring of the surface properties of polymers at the nanoscale: medical applications. In: Bhushan B, editor. Scanning probe microscopy in nanoscience and nanotechnology, Part 3. Berlin/Heidelberg: Springer-Verlag; 2011. p. 645—89. Chapter 22.

[187] Goddard JM, Hotchkiss JH. Polymer surface modification for the attachment of bioactive compounds. Prog Polym Sci 2007;32:698—725.

[188] Klokkevold PR, Nishimura RD, Adachi M, Caputo A. Osseointegration enhanced by chemical etching of the titanium surface. A torque removal study in the rabbit. Clin Oral Implants Res 1997;8:442—7.

[189] Croll TI, O'Connor AJ, Stevens GW, Cooper-White JJ. Controllable surface modification of poly(lactic-co-glycolic acid) (PLGA) by hydrolysis or aminolysis. I. Physical, chemical, and theoretical aspects. Biomacromolecules 2004;5:463—73.

[190] Zhu Y, Gao C, Liu X, Shen J. Surface modification of polycaprolactone membrane via aminolysis and biomacromolecule immobilization for promoting cytocompatibility of endothelial cells. Biomacromolecules 2002;3:1312—9.

[191] Chen JP, Su CH. Surface modification of electrospun PLLA nanofibers by plasma treatment and cationized gelatin immobilization for cartilage tissue engineering. Acta Biomater 2011;7:234—43.

[192] Sartori S, Rechichi A, Vozzi G, D'Acunto M, Heine E, Giusti P, et al. Surface modification of a synthetic polyurethane by plasma glow discharge: preparation and characterization of bioactive monolayers. React Funct Polym 2008;68:809—21.

[193] Wu JD, Tan HP, Li LH, Gao CY. Covalently immobilized gelatin gradients within three-dimensional porous scaffolds. Chin Sci Bull 2009;54:3174—80.

[194] Lin Y, Wang L, Zhang P, Wang X, Chen X, Jing X, et al. Surface modification of poly(L-lactic acid) to improve its cytocompatibility via assembly of polyelectrolytes and gelatin. Acta Biomater 2006;2:155—64.

[195] Kotov NA, Podsiadlo P, Tang Z, Wang Y. Biomedical applications of layer-by-layer assembly: from biomimetics to tissue engineering. Adv Mater 2006;18:3203.

[196] Ariga K, Hill JP, Ji Q. Layer-by-layer assembly as a versatile bottom-up nanofabrication technique for exploratory research and realistic application. Phys Chem Chem Phys 2007;21:2319—40.

[197] Subbiahdoss G, Pidhatika B, Coullerez G, Charnley M, Kuijer R, van der Mei HC, et al. Bacterial biofilm formation versus mammalian cell growth on titanium-based mono- and bi-functional coatings. Eur Cell Mater 2010;19:205—13.

[198] Roberts MJ, Bentley MD, Harris JM. Chemistry for peptide and protein PEGylation. Adv Drug Delivery Rev 2002;54:459—76.

[199] Sadr N, Pippenger BE, Scherberich A, Wendt D, Mantero S, Martin I, et al. Enhancing the biological performance of synthetic polymeric materials by decoration with engineered, decellularized extracellular matrix. Biomaterials 2012;33: 5085—93.

[200] Alves CM, Reis RL, Hunt JA. Preliminary study on human protein adsorption and leukocyte adhesion to starch-based biomaterials. J Mater Sci Mater Med 2003;14:157—65.

Chapter 10

Bioactivated Materials for Cell and Tissue Guidance

Paolo A. Netti

Centre for Advanced Biomaterials for Health Care, IIT@CRIB, Istituto Italiano di Tecnologia, Napoli, Italy

Chapter Outline

10.1 INTRODUCTION

Tissue engineering (TE) and regenerative medicine aim to restore and repair damaged tissue using three basic ingredients, namely scaffolds, bioactive signals (molecular and biophysical), and cells [1,2]. Recent clinical evidence indicates that the success of any TE approach strongly relies on the complex and dynamic interplay among these three components and that functional tissuegenesis is achieved only upon their sapient integration [3,4].

The general strategy is to seed cells within a scaffold that fits the size and the shape of the damage and offers the appropriate molecular microenvironment to promote tissue regeneration [5,6]. Scaffolds should not only provide correct shape, sufficient porosity for nutrients and cell trafficking and adequate mechanical and structural support, but should also actively guide and control cell attachment, migration, proliferation, and differentiation. The integration of these important features has been attempted by conjugating biologically active molecules, such as growth factors or integrin-binding peptides, within the scaffold to provide the correct molecular milieu to guide and direct cell functions [1,7]. However, albeit potent tissue morphogenic and tissue modulator molecules have been discovered and isolated in the past decade, their use in TE strategies has often disappointedly led to the formation of dysfunctional or ectopic tissues [8–13].

Bone morphogenetic protein (BMP)-2, for instance, is an osteogenic growth factor frequently used in clinics for bone generation. Interspersing BMP-2 in collagen scaffolds has been proved to stimulate bone formation and help in healing bone defects [14] but, at the same time, it has been shown to induce dysfunctional tissues [8–11]. Analogously, vascular endothelial growth factor (VEGF) has been commonly used or proposed to promote scaffold vascularization. However, animal studies and preclinical tests have demonstrated that the use of exogenous VEGF often leads to abnormal vascularization and dysfunctional tissue growth [15–21]. Endogenously secreted growth factors (GFs) elicit their action at a low dose and few nanograms per gram of extracellular matrix (ECM) are generally sufficient to guide tissue repairing mechanisms. Furthermore, since the GFs are continuously synthesized by cells seated at the periphery of the damaged tissue, their action of GFs is prolonged in time and confined to specific regions. Therefore, the clinical failure of the use of GFs interspersed in scaffolds could be attributable to the poor control of the spatial–temporal presentation of these factors. To overcome these limitations, a more complex and sophisticated scaffold able to tune and tightly control the sequestration and delivery of relevant bioactive molecules should be designed. Dysfunctional tissue growth probably results from an incorrect presentation of these signals that, in order to elicit the right morhogenic program, must be presented in the right form, at the right time, and at the right site.

Apart from GFs, also matricellular cues, or integrin-associated signals, play an important role in controlling and guiding cell functions and fate [22,23]. During the last decade it has been shown that cells sense biochemical, topographical, and mechanical cues on their substrata and translate them into commands that regulate activity and fate [24]. These findings have led to the proposition that the interplay between the focal adhesion (FA) dynamics and the cytoskeleton assembly is at the base cell—material interaction and transduction mechanism. Therefore, any materials feature that might interfere with the dynamics of FA can potentially influence cell behavior and fate. The formation of FAs and complexes is mediated by integrin recognition of the contact surface and occurs on a length scale that spans from few tens of nanometres (focal complexes) to micron (mature FA) [25]. Hence, nanometric patterns of biochemical, topographical, and mechanical features embossed in a three-dimensional (3D) scaffold could provide specific instructive environment to control and guide cell behavior. The effect of pattern of these integrin-associated signals has been already proved to be an effective tool to control cell adhesion, migration, and differentiation [26—29], and their integration in cell-instructive TE scaffolds is required to provide the correct microenvironment for tissue growth.

The concept of tissue and cell guidance is rapidly evolving as more information on the biological control of the extracellular microenvironment on cellular function and tissue morphogenesis becomes available. These findings have burst a novel concept in bioactive material design based on nanometric control of structural and functional features to recapitulate the spatio-temporal molecular regulatory program and the 3D architecture of the native ECM. Micro- and nanostructured scaffolds able to sequester and deliver in a tightly spatial and temporal controlled manner biomolecular moieties have been proposed as highly effective in tissue repairing, in guiding functional angiogenesis, and in controlling stem cell differentiation [30]. Although these materials are a first attempt to mimic the complex and dynamic microenvironment presented *in vivo*, an increased symbiosis among material engineering, micro- and nanotechnology, drug delivery, and cell and molecular biology is needed to fabricate biomaterials that encode the whole array of biosignals to guide and control developmental process in tissue- and organ-specific differentiation and morphogenesis.

Recent advances in micro- and nanofabrication technologies make it possible to engineer scaffolds with a well-defined stereoregulated architecture and defined structural and topographical features at the nanometric level providing a control of cellular spatial organization, mimicking the microarctitectural organization of cells in native tissues [31—37]. Furthermore, the biological activity of these matrices can be enriched by encoding in them the capability to expose an array of biological signals at a given site, with an adequate dose and for a desired time frame. Through the tight control over time and space of tiny quantities of multiple biomacromolecular factors and of their gradients along with the structural features, it would be virtually possible to program these materials to induce and guide any desired morphogenetic pathway and to lead the process of neotissue formation, inducing on demand different pathways to cell response. In this chapter, the evolution in scaffold design derived from the passage from shape to cell guidance concept will be overviewed along with the most important technological and scientific achievements. The most advanced attempt to engineer cell and tissue guiding programmable scaffolds will be presented and their impact on the modern TE and regenerative medicine scheme outlined.

10.2 EVOLUTION OF SCAFFOLD DESIGN CONCEPT

The principle of scaffolds design and manufacturing has been continuously revised during the last 25 years passing from almost inert temporary material permissive to cells and tissue growth to proactive cell-instructive material able to control and guide tissue morphogenesis (Figure 10.1). At the beginning of the 90s, scaffolds were conceived as a temporary structure in which seeded cells could proliferate and deposit *de novo* synthesized the tissue. They were essentially organic or inorganic sponges with a high degree of pore interconnectivity designed to guide tissue geometrical features (i.e., size and shape) and chemical properties (i.e., degradation rate). Indeed, the tissue regeneration scheme relied on the match between scaffolds degradation and neotissue deposition rate and, on the fit between scaffolds and tissue geometric features. According to the modern concept, the ideal scaffold should guide the process of tissue repairing by presenting a repertoire of cues—chemical, biochemical, and biophysical—able to control and promote specific events at the cellular and tissue level. Spatio-temporal presentation of biological cues must be integrated with microstructural and mechanical properties to provide a proper cell-instructive environment not only within the scaffolds but also at the interface with the native tissues. Over the last three decades, the concept of cell guidance in tissue regeneration has been extensively discussed and progressively revised as new knowledge of the complex features of cell—material interaction has been disclosed and

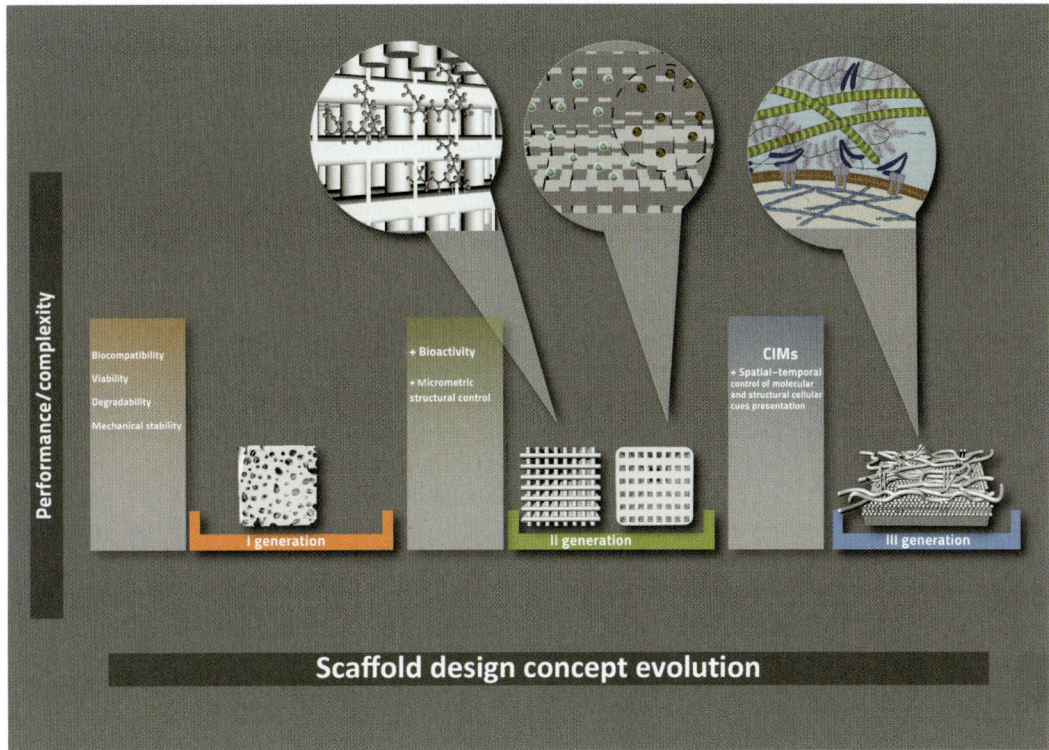

FIGURE 10.1 Evolution of the concept of scaffold design. During the last 25 years, scaffold design has witnessed a radical transformation passing from simple biocompatible and biodegradable sponges to complex proactive materials encoding sophisticated chemical and structural strategies to control and guide cell and tissue processes. The first generation of scaffolds was simple highly porous materials engineered to temporarily replace the function of damaged tissues while providing a permissive environment to cell and tissue growth. The concept of guidance had only a geometrical significance that is to form the scaffolds with same shape and dimension of the damaged tissue. The extension of biomaterial capability to influence cell behavior and tissuegenesis processes by enriching them with bioactive molecules has introduced new concepts and has opened the way to the second generation of scaffolds with stretched function and biological activity. In this stage, there has been the flourishment of the bioactivation concept that aimed to entrap bioactive signals, both molecular and physical, within the scaffolds 3D structure to provide a proactive environment for controlling and guiding complex cellular and tissue processes. The concept of scaffolds guidance is extended and revised and encompasses the control over cell attachment, migration, differentiation, vascularization, and specific tissuegenesis. Furthermore, the fast advancements of the microfabrication and materials printing techniques also lead to an enhanced control over microstructural features, such as porosity size, shape, and interconnectivity. The need for a tight spatial–temporal control of presentation of an exact amount of multiple signals—both molecular and physical—with a nanometric resolution call for a new paradigm shift in scaffold design. The next generation of scaffolds is envisaged as programmable materials able to display in a highly controlled manner a complex array of signals at the right site with the right dose, possibly with a cell-demanding logic, recapitulating in a simplified fashion, the complex cascade of signals occurring *in vivo* during a tissue regeneration process. These materials should also be capable to recruit specific cells from the surrounding tissue and expand them within their physical domain, avoiding the *ex vivo* cell manipulation.

elucidated [38]. This new concept of molecular and cell biology, along with advances in material science and nanotechnology, have been gradually and constantly integrated in the design strategies leading to a costant increase in complexity and sophistication of scaffold architecture and function. The general trend is to recreate the ECM function in simplified form in a synthetic material by embossing a complex array of cellular cues with a spatial and temporal controlled presentation. Following the stages depicted in Figure 10.1, the most significant milestones of this evolution will be discussed and future trends in scaffold design will be highlighted.

Natural, synthetic, semisynthetic, and hybrid materials have been proposed and used as scaffolds for TE [39].

Purified ECM components or decellularized ECMs derived from animals have been widely used in TE. Indeed, even if they are subject to purification and sterilization treatments, these materials retain important characteristics of the physical and chemical structure of native ECM. Owing to their similarity to the ECM, natural polymers may also avoid the stimulation of chronic inflammation or immunological reactions and toxicity, often detected with synthetic polymers [40]. Although decellularized ECM has been successfully used as a scaffold for soft tissue applications [41], single purified ECM components, such as collagen, hyaluronic acid, and fibrin, can be combined appropriately to create more controlled and standardized materials that are potentially less immunogenic and have a similar structure to native ECM [42]. Animal- or vegetal-derived proteins have been shown to have potential to be used as scaffolds for TE applications. Silk proteins, for example, contain a high content of β-sheet sequences that make this polymer particularly suitable for high-strength and slow degradation purposes [43]. Alginate and chitosan, which are glycans extracted from brown algae and the exoskeleton of shellfish, respectively, have gained popularity because of their biocompatibility, ease of processing, and ability to encapsulate cells and bioactive molecules [40,44]. Natural proteins, such as gelatin and zein, have also been investigated as biomaterial scaffolds for applications spanning from soft to hard tissue regeneration [45,46].

Albeit materials derived from biological tissues present several advantages, notably biocompatibility and biological recognition, synthetic materials have attracted an ever-growing attention due to the high degree of control of their properties and to the possibility to tailor their performance on the specific application. Furthermore, on top of being chemically programmable and reproducible, synthetic materials do not suffer from immunogenic and purification issues related to materials of animal or human origin. On the other hand, the lack of a molecular environment recognizable by cells impairs and limits the use of synthetic materials as tissue regeneration scaffolds. To overcome the limitation of both natural and synthetic materials, hybrid strategies have been proposed and implemented. For instance, the need to improve natural materials' physical and chemical performance prompted their modification to generate their semisynthetic counterparts [47] with the expectation to impart better and tunable physiochemical properties with undiminished biological activity. However, poor processability and unsatisfactory mechanical properties have restricted their use in TE. A more successful strategy has sought to extend the biological performance of synthetic materials by encoding biomolecular cues in their structure to obtain hybrid proactive materials capable to promote and control cell interaction and functions. The development of synthetic material able to present a complex array of bioactive signals with a defined time and space program is at the frontier of biomaterials science for the realization of artificial ECM (α-ECM) replica and it will be discussed in details in the following section.

The first generation of scaffold was conceived as biocompatible and biodegradable foams with a high degree of pore interconnection permissive to cell growth and able to stimulate transplanted cells to regenerate biological tissues with defined size and shape [48]. The main properties required from a scaffold included (i) biocompatibility: intended as the capability to perform with an appropriate host response in a specific application; (ii) biodegradability without producing toxic degradation by-products; (iii)

processability to manufacture scaffolds of desired internal structure and external shape; (iv) sterilizability by using process technologies appropriate for biological uses, and finally (v) provide mechanical properties tailored for the required application. The scaffold's porosity network was indented to provide cell and tissue invasion and nutrient trafficking, while its degradation and mechanical properties were tailored to provide a temporary support function to damaged tissue and progressively remove as the new tissue was deposited and able to restore the original function. Sponges of biopolymer (collagen and collagen—proteoglycans) were among the first scaffolds used in TE for the regeneration of skin-equivalent tissue and other connective tissues [49]. Porous scaffolds made of a wide range of natural and synthetic polymers, bioactive glasses and their composites, were copiously proposed and used for bone, cartilage, blood vessels, nerve, and derma. A variety of biodegradable synthetic polymers, including poly(α-esther)s, such as polyglycolide, polylactides, polycaprolactone and their copolymers, polyanhydrides and poly(propylene fumarate) have been extensively investigated for their use as scaffolds [42,44]. The application of synthetic biodegradable polymers in the TE field provides several advantages. First of all, these materials can be synthesized in a variety of chemical structures, enabling the possibility to easily tailor their microstructural features and degradation behavior.

Also inorganic materials like calcium phosphates (e.g., hydroxyapatite and α- and β-tricalciumphosphate) have been widely investigated and used as scaffolds for bone TE [50]. These materials show several advantages, such as bioactivity, osteoconductivity, and ability to bond directly to bone. Ceramic implants for osteogenesis are based mainly on hydroxyapatite, since this is the inorganic component of bone [51]. Bioactive glasses containing SiO_2, Na_2O, CaO, and P_2O_5 have been shown as equally suitable for TE applications [50,52]. Indeed, they have been proved to be osteoinductive and, when exposed to biological fluids, showed the ability to form a carbonated hydroxyapatite layer, which serves as a bonding interface between implant and surrounding bone. Silica-rich scaffolds have revealed excellent bone forming ability and resorption rates directly dependent on the silica content [51].

The advent of the TE approach has called for scaffolds with defined porosity, mechanical properties, and degradation rate; it has led to a tremendous advancement and enrichment in technologies to fabricate porous scaffolds with controlled porosity during the decade across the year 2000. First technologies used for manufacturing porous scaffolds were adapted from conventional processes, including textile technologies, porogen leaching, phase separation method of polymeric solutions, freeze-drying, gas foaming, and sintering. Textile technologies were among the first technologies to be adopted; they consisted of woven or knitted synthetic fibers made of polyglycolic acid, polylactic acid or their copolymers and have been used for nerve, skin, ligament, and cartilage regeneration [53]. Fiber bonding technologies have also been proposed and used to increase the structural stability of textile scaffolds [54]. Reverse templating was a widely used method to manufacture highly porous (up to 93%) biodegradable polylactic acid foams [55]. The process is based on the use of a microparticle template (commonly sodium chloride, sodium tartrate, or sodium citrate particle) that is dispersed in a polymeric solution or melt and a successive removal of the template after the setting of the polymer. This technique has been applied to producing scaffolds with classes of biomaterials, such as synthetic and natural polymers [56,57] and ceramics [58]. More recently, an enhancement of control of pore shape and mechanical properties has been achieved by means of the use of continuous porogens [59]. In gas foaming, phase separation and freeze-drying techniques, porosity is induced by the attainment of thermodynamic instabilities in multiphase polymeric systems. In the gas foaming process, the depressurization of a gas-loaded polymer phase induced nucleation and growth of pores inside the material [60,61]. Processing variables, such as temperature, pressure, and flow, can control the extension and the degree of interconnectivity of the porous network. In the phase separation and freeze-drying methods, the porosity is created by bringing a polymer solution into a thermodynamically instable state and letting the polymer separate abruptly from the solvent, generally by decreasing the temperature or adding a nonsolvent, leading to the formation of a multiphase system characterized by polymer-rich and polymer-lean phases. The subsequent removing of the solvent from the system induces the crystallization of the polymer-rich phase and the formation of the porous network in the polymer-lean phase. Depending on the polymer-solvent choice and the phase separation conditions, highly porous scaffolds, up to 99%, with random or oriented pores can be produced [56].

All the above methods, and also some of their combinations, have been used to better control the microstructural architecture of scaffolds in terms of porosity degree, interconnection, and morphology [62]. However, a highly interconnected porosity often leads to mechanical properties depreciation and therefore composite strategies have been pursued to enhance the load-bearing features of porous scaffolds especially of those designed for hard TE. Several examples of organic/inorganic composite scaffolds made, for instance, by interspersing micro- and nanohydroxyapatite particles inside a polymeric matrix to improve mechanical properties, degradation kinetic and bioactivity are reported in the literature [63,64]. Furthermore, the interaction between ionic dissolution products of ceramics and cell metabolic activity has also

been proved to be an extra beneficial outcome of these composite scaffolds to promote cell differentiation and tissue neovascularization [50,65]. Also natural soft macromolecular gels have been enriched with ceramic nanoparticles both to enhance their mechanical performances and to improve their osseointegration characteristics [51]. Finally, long and short fibre reinforced composites have also been proposed and tailored to enhance the mechanical properties of scaffolds and to mimic the anisotropy occurring in natural tissues [66].

The improved control over scaffolds microstructural features has, however, not always been followed by a positive *in vitro* and *in vivo* tissue regeneration outcome. Indeed, while a highly interconnected porous network embossed in a biocompatible material was clearly proved to facilitate cell and tissue invasion along with a free nutrient trafficking, tissue growth within the biological inhert porosity resulted far from optimal and often dysfunctional. As a consequence of these disappointing results, during the early years of this century it was commonly recognized that a cell seated in a cavity within a 3D matrix would not receive all the microenvironmental cues essential to direct and guide the tissuegenesis process. This notion has motivated the scientific community towards the design of the next generation of bioactivated multifunctional scaffolds that, on top of microstructural, chemical, and mechanical features, were enriched with biomolecular cues to extend the function of the scaffolds also to provide control of the cellular microenviroment. In second-generation scaffolds, the presentation of bioactive molecules, such as growth factors, cytokines, potent morphogens and integrin-associated matricellular cues, was included in their design and manufacturing strategies [1,7].

The regenerative potency of TE scaffolds has been dramatically improved by integrating mechanisms of sequestration and delivery of GFs [67–69]. Signaling molecules have been integrated within scaffolds by simply interspersing them in the matrix or by using smart approaches to embed mechanisms of sequestration and delivery. Simple interdispersion of GFs within scaffolds, although far from ideal, has been widely used in the literature. Bioactive factors have been dispersed in naturally derived (collagen, fibrin, chitosan) [47,70] or synthetic hydrogels (poly(ethylene glycol), PEG) [47,71–73], or included in hydrolytically degradable hydrogels in which chemical or physical cross-linkings offer the possibility to control the diffusion of solubilized hydrophilic macromolecules [74–76]. However, direct interspersion of GFs within hydrogel does not guarantee the preservation of their biological activity and the control over the dose and time of their presentation [77]. A more sophisticated approach is represented by the attempt of integrating responsive logic based on enzymatically triggered release of GFs [23]. Proteolytically sensitive PEG-based networks have been proved to be effective in controlling the presentation of bioactive agents based on a cell-demanded mechanism recapitulating the dual-reciprocity process occurring in a native ECM [78].

Protein-loaded synthetic solid biodegradable materials, suitable for hard tissue repair, also present serious limitations regarding protein leaching and stability [67]. Direct encapsulation of proteins in solid scaffolds is often impaired by the drastic environmental condition required by their process and should be preferentially produced under mild process, such as gas foaming and electrospinning, possibly combined with particulate leaching [77]. Angiogenic factor enriched PLGA scaffolds have been proposed to engineer bone tissues [79] and vascular beds [80] with alternate success. Composite scaffolds made by microparticle-loaded GFs included in a 3D matrix represent a more controlled and viable approach [81−84].

In this case, GFs are loaded in a stable form in polymeric depots which are then included into various biomaterials to enable a sustained and controlled point source release while preserving bioactivity even for a long time [67,83,85−88]. Some relevant results following this approach have been achieved by Mooney and co-workers [89], who developed PLGA scaffolds for the sequential release of multiple GFs by mixing free VEGF with empty and platelet-derived growth factor (PDGF)-loaded polymer particles and subsequently assembling them into a porous scaffold [84] or anisotropic system based on a porous bi-layered PLGA scaffold able to expose only VEGF in one spatial region, and deliver VEGF and PDGF in an adjacent region [90]. In a similar attempt, PLA microparticles plasticized with PEG were sintered into scaffolds formed by protein-free and protein-loaded layers, thus allowing a release of different bioactive molecules restricted to specific regions within the scaffold [91]. These scaffolds might be particularly effective in applications where GF gradient or a region-dependent tissue growth is required.

An alterative route to control the time presentation of GFs within a 3D scaffold is based on the strategy of sequestration, by immobilization or complexation, of signals. Polymer scaffolds can be modified to interact with signaling molecules, thereby hindering their diffusion out of the polymer platform and prolonging release. Signal immobilization can occur through reversible association with the scaffold (i.e., binding/debinding kinetics), irreversible binding to the polymer, or signals can be immobilized with release dependent upon degradation of a linking tether or the matrix itself [92,93]. One of the most common approaches to improve release kinetics of the immobilized molecule relies on the use of heparin-immobilized scaffolds [94−100]. Also heparinized cross-linked collagens for *in vivo* endothelial cell seeding have been studied with respect to GFs binding and release

[95,100]. Collagen matrices [95,100] and synthetic PLGA-based scaffold [98,99] have been enriched with heparin and demonstrated to be able to attract host blood vessels and osteoblasts through a sustained delivery of angoigenic factors and bone morphogenic proteins.

On top of soluble GFs, scaffolds should also provide suitable signals to promote cell attachment, spreading, and migration. Indeed, in the natural context the extracellular space is populated by integrin-binding signals, including fibronectin, laminin, and vitronectin that control the maintenance and development of cell function within the tissue. Through these signals the extracellular space transmits information across the cell membrane via cytoskeleton; consequently, they are key regulators of cell adhesion and migration [24,101]. The identification of small integrin-binding oligopeptide sequences along the macromolecular component of the ECM opened up the possibility to recreate cell adhesion pathways artificially by conjugating small peptides in the scaffold structure with controlled density and spatial distribution [102,103]. Gly−Arg−Gly−Asp−Ser (GRGDS)-based peptides are the most widespread signals used to date as an adhesive motif, even if other non-RGD peptides, as well as peptidomimetics, have also been deeply studied and used in prototype multifunctional scaffolds. Ligand density in 2D as well as in 3D, spatial distribution, co-regulation, gradient and nature of material substrate, all affect cell behavior and have also been included in scaffold design guidelines [24,104]. PEG-enriched adhesive peptide like RGD was among the first examples of hybrid materials synthetized as a simplified replica of the ECM [31]. Many other materials have been since decorated with adhesive peptides to enhance cell adhesion, spreading, and migration [105] and have become today the standard for TE scaffolds.

Apart from biochemical signals, it has also been understood that topographical and biophysical signals play an important role in providing tissue and cell guidance [24]. The scaffold's pore size and shape, for instance, could elicit a specific cell-signaling pathway and therefore play a more relevant biological role than expected. Indeed, pore size alone, as a signal, can trigger and possibly regulate cell phenotype, control the morphogenic process occurring within the scaffold, and ultimately enhance or repress specific cellular functions. As a result, a scaffold must possess a definite distribution of porosity for any specific application. For instance, average pore size should be about 35 mm for stimulating vascularization, between 20 and 125 mm for enhancing fibroblast ingrowth and skin regeneration, between 100 and 350 mm for bone regeneration [106−108]. Also pore shape and orientation might play an important role in guiding and boosting the regeneration of a tissue. Several tissues, including nerve, muscle, tendon, ligament, blood vessel, and bone have anisotropic properties and their regeneration could need an oriented and anisotropic porous network [109]. Given the importance of microstructural features in controlling and guiding the tissuegenesis process, novel methods to fabricate scaffolds with predefined reliable and reproducible microstructural architecture at micrometric level were introduced and optimized. Solid free-form fabrication (SFF) technology is a collective term for a group of technologies that can be used to manufacture objects in a layer-by-layer fashion from a 3D computer-generated image of the object. This layer-by-layer mode enables the manufacturing of objects with a complex internal architecture, not possible with traditional manufacturing methods. The versatility of SFF in internal architectural and external shape control is the main advantage of this technique. Commercially available SFF systems may be categorized into three major groups based on the way materials are deposited [110]: (i) laser-based machines that either photopolymerize liquid monomer or sinter powdered materials; (ii) printing of a chemical binder onto powdered material or directly printing wax; and (iii) nozzle-based systems, such as bioplotter, which are able to print biological cells as well as a range of biomaterials. These technologies have been successfully implemented to manufacture highly defined microstructured scaffolds made out of polymers, hydrogels, ceramic, and even metal biomaterials [110–112]. Due to their detailed and defined microstructural features, SFF scaffolds often allow for a better mechanical and biological performance than those displayed by scaffolds obtained using traditional approaches [110]. The increasing demand for nano- and micrometric-controlled features of scaffolds has led to the development of integrated approaches, which combine high-resolution 2D structure manufacturing with layer-by-layer assembly. These approaches differ from those of SFF because the "multilayer scaffolds" are built in a semi-automated two-step process. This approach has been used for constructing, among others, complex 3D microfluidic scaffolds for tissue vascularization [113] as well as biomimetic and bioactive scaffolds with precise microarchitecture and surface, micro- and nanotextures for controlled cell ingrowth and differentiation [114].

Despite the impressive enhancement in tissue guidance and regeneration offered by the bioactivated scaffolds depicted above, preclinical and clinical studies indicate that several limitations have yet to be resolved. TE has yet to reproduce complex tissue architectures for tissues in which the structure is either relatively homogeneous, such as cartilage, or develops naturally, such as skin. More intricate tissue architectures not only demand appropriate morphological characteristics from the scaffold but, probably, also require changing molecular signals during the entire process of neotissue formation. The ability to program the physical location and the lifetime of biomolecular signals in a coordinated way is therefore crucial for a successful TE scaffold. Bioactivated scaffolds have been designed to deliver potent biological signals to control and guide the morphogenic and tissuegenic processes. However, to elicit the desired cell response, a given signal must be presented at the right time, with the right dose, and for the right time frame; these control strategies in signal presentation are generally missing in bioactivated scaffolds. They include the tight control over time and space of tiny quantities of multiple biomacromolecular factors and of their gradients within the interstitial space of the scaffold, as well as at the scaffold–tissue interface. Moreover, there is a paucity of studies regarding the effective dose in the local microenvironment, the magnitude of the spatial and temporal gradients, and the development of technological strategies to integrate and position drug delivery devices with a submicrometric spatial resolution within the scaffolds. Therefore, there is today a common thinking that bioactivated scaffolds do not provide a correct presentation of the signals since they miss the spatial and temporal control of presentation. This presents a significant challenge, since several features of the processes that control cell guidance in 3D materials still remain to be defined despite the enormous progress made over the last decades. Tissue regeneration, whenever it occurs—for example, in response to physiological conditions or to trauma—is a result of a complex cascade of events. These events are coordinated in spatial and temporal modalities, and each of them is governed by biophysical and biochemical signals. In turn, these signals are triggered by the extracellular microenvironment. Therefore, the final outcome of cell life for most connective tissues depends on the dynamic and reciprocal interaction between the cell and the ECM. It is this interaction that ultimately determines cell fate and ECM degradation or remodeling [101]. These conclusions point to the need to increase the functionalities of these already complex scaffolds towards the engineering of programmable materials with encoded strategies to control the space and time presentation of bioactive moieties and to control and guide specific events at the cellular scale. The potential to pattern material properties with nanometric precision paves the way to the next scaffold generation—referred to as cell-instructive materials (CIMs)—with extended functionality and bioactivity. CIMs are envisioned as nanofeatured materials expressly programmed to impart even complex commands or instructions to cells with the aim of directing, guiding, and controlling their fate. The realization of these attractive materials relies upon a deep understanding of the mechanisms that regulate cell–material interactions and in particular upon the disclosing of the complex molecular machinery of recognition and decoding that occurs at the interface between cell membrane and materials. This next scaffold generation represents the synthetic replica of the

ECM in which the presentation of bioactive signals can be programmed, possibly with a cell-demanding logic, in terms of site, time, and dose.

10.3 BIOACTIVATED PROGRAMMABLE CELL-INSTRUCTIVE SCAFFOLDS

ECM, the natural medium in which cells grow, differentiate, and migrate, is the gold standard material for tissue regeneration [115]. The interaction between cells and their ECM is specific and biunivocal. Cells synthesize, assemble, and degrade ECM components responding to specific signals and, at the same time, ECM sequesters and presents signals that control and guide specific cell functions. The continuous crosstalk between cells and ECM is essential for tissue and organ development and repair, providing both a structural (i.e., mechano and topological cues) and molecular (presentation of molecular cues) guidance at cell level. ECM is a highly organized dynamic biomolecular environment in which many proliferation—adhesion—differentiation motifs, governing cell behaviors, are continuously generated, sequestered, and released, inducing matrix synthesis and degradation. These molecular motifs are locally released according to cellular stimuli, generally occurring upon degradation of the adhesion sites binding them to the ECM [116]. Integrin receptors, for instance, are recruited in microdomains of cell membrane forming adhesion plaques, and in these areas integrins communicate with structural and signaling molecules influencing transport, degradation, and secretion of ECM molecules, endocytosis, and cellular fate [116—118]. Moreover, solid-state, structural ECM molecules, such as heparin, act as reservoirs for secreted signaling molecules for their on demand release [119—121]. GFs are locally secreted by ECM, in which they are stored in insoluble/latent forms through specific binding with glycosaminoglycans (GAGs) (e.g., heparins), and can elicit their biological activity once presented in their active form. During tissue morphogenesis, the presence of GFs guides cellular behaviors, thus governing neotissue formation and organization. The sequestration of GFs within ECM in inert form is necessary for rapid signal transduction, allowing extracellular signal processing to take place in time frames similar to those inside cells. Moreover, concentration gradients of GFs play a major role in ECM maintenance and equilibrium because they are able to direct cell adhesion, migration, and differentiation deriving from given progenitor cells and organize patterns of cells into complex structures, such as vascular networks and nervous system [122—124]. Thus, spatial patterns in tissues are dictated by both the architectural features of ECM and concentration profiles/gradients of diffusible bioactive factors [125].

Recent scaffold development has been driven by biomimickry-inspired design to recapitulate in a simplified form the essential microenviroment architecture and dual reciprocity existing in the ECM. To this end, several micro- and nanofabrication strategies have been applied in an attempt to mimic the spatial distribution of the fibrillar structure of the ECM, which provides essential guidance for cell organization, survival, and function [32,36]. These technologies include molecular and nanoparticulate self-assembly, micro and nanoprinting, electrospinning, molecular and nanotemplating [5,6,32,35—37,126]. Patterning at the nanometric scale of bioactive molecules controlling the fibrillar hierarchy from nano to microlevel, and controlling the time presentation of signals, might all be possible features that can be included within prefabricated scaffolds. Topographical and stereomorphological cellular cues can be provided by controlling fiber dimension and arrangement [127]. Chrono- and spatial-programmed presentation of bioactive moieties can be encoded by placing morphogenic factor-loaded degradable microparticles in predefined regions of the scaffold [6,30], and finally the exposition of matricellular cues can be controlled, eventually in a dynamic fashion, by properly decorating scaffold structure with integrin adhesive motifs [22] (Figure 10.2). Furthermore, the scaffold should not only provide a controlled administration of relevant bioactive molecules and their gradients but also present them in a suitable conformation state, mimicking ECM—GFs binding (Figure 10.2). Proper engineering of all these features would make it possible to present a complex array of biochemical, topological, and mechanical signals to every cell seated within the scaffold.

The tight control over the microenviromental cues at cell level denotes the key shift from the concept of shape to cell guidance that is accompanying the modern scaffold design strategies. However, the control over space, time, and molecular arrangement of the complex cascade of signals necessary to guide the process of tissue or organ repair is, albeit theoretically possible, practically and economically nonpursuable [5]. The recapitulation of the complex molecular events occurring within the extracellular space during the process of tissue repair and regeneration should be captured in the most essential features and reproduced by the use of simplified strategies and molecules. For instance, ECM possesses a vast repertoire of integrin-binding motifs within its fibrillar components, each of them equipped with a specific function and activity [22,24]. Most of these motifs have been molecularly identified and their corresponding peptides synthetically reproduced [24]. The availability of an entire library of integrin-binding peptide allows to mimic the matricellular integrin-mediated crosstalk by inserting small molecular units within the scaffold instead of the whole fibrillar protein, such as collagen, fibronectin, or laminin,

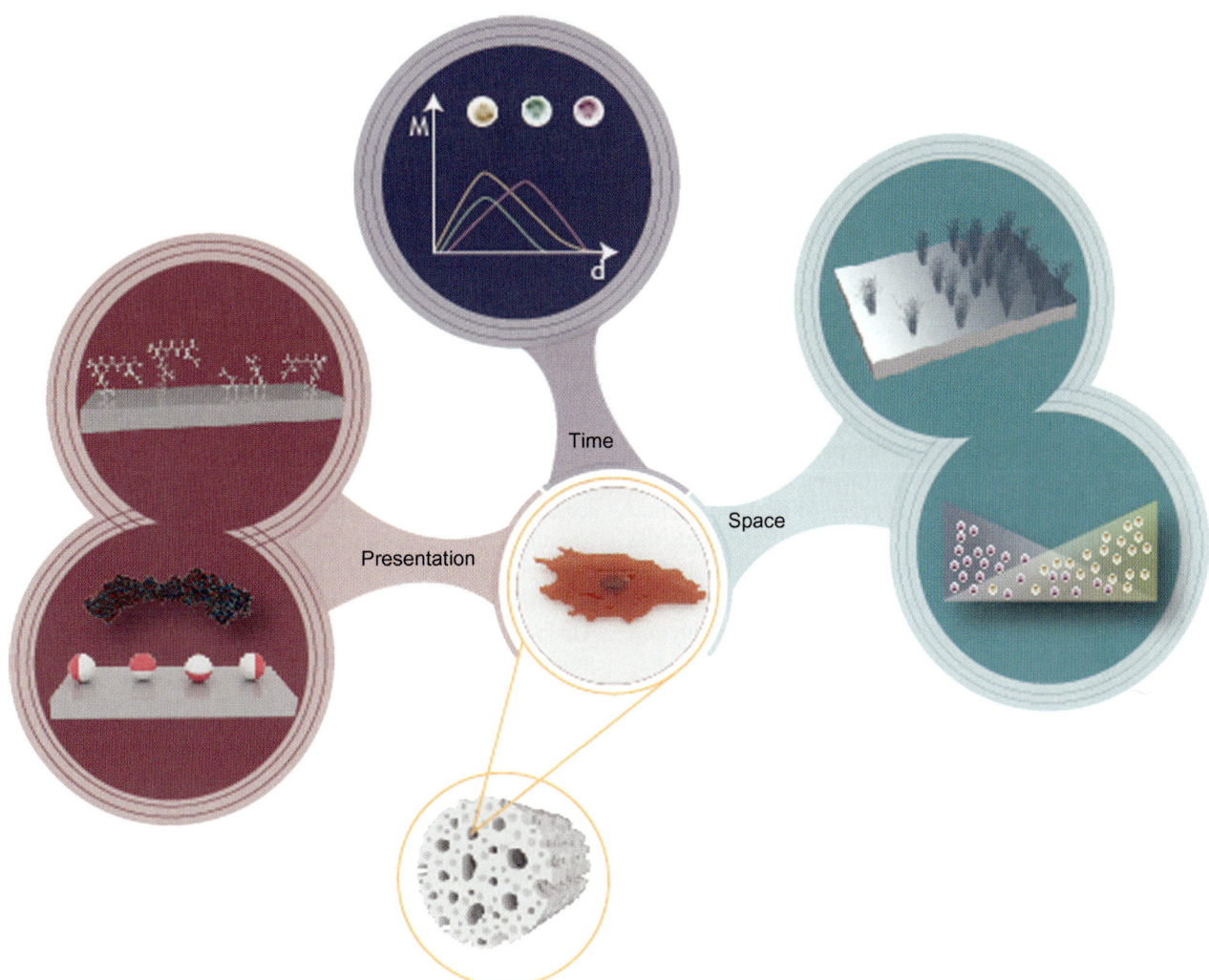

FIGURE 10.2 Control of presentation of bioactive signals within multifunctional bioactive scaffolds. The scaffold should place cells in the correct microenviromental context by displaying an array of bioactive signals—molecular and physical—depending on the site and on the time of evolution of the tissuegenesis process. Clockwise, from bottom: matricellular, GFs, or topographical signals must be presented with the correct conformation to elicit the desired cell response. Signals presented with incorrect stereoregular arrangement elicit no or improper cell response. GFs, for instance, often present an ECM-binding domain within their structure and elicit their function when bound to the extracellular space [5]. If a given GF is presented in an unbound form or bound to the scaffold materials through a different binding domain, then its signaling pathways could be different and could possibly elicit a nonphysiological response [137]. Casting within the scaffolds degradable GF-loaded microparticles, time evolution of multiple GFs presentation could be programmed [83]. The rate of GFs delivery from the single polymer degradable depot can be modulated by controlling their rate of degradation to program the GFs supply over time for the control of the specific morphogenetic process. Signals should also be presented with a controlled spatial distribution. Gradient of bioactive signals can be embossed within the 3D scaffolds structure using different strategies. GFs gradient could be built by micropositioning degradable microparticles loaded GFs according to a predefined spatial distribution in such a way that the continuous release of the GFs from the microparticle can sustain GFs gradient over time. An alternative way to present a gradient of GFs is to bind them to the scaffold surface in a specific orientation. This can be attained by properly conjugating the surface of the scaffolds with molecules, like heparin, that can bind to GFs [5].

providing a great simplification in scaffold bioactivation. The inclusion of all possible matricellular cues mimicking peptides in a scaffold, however, would represent a more faithful replica of the natural molecular niche for cells, on one hand, but would make the scaffold design impracticably, and perhaps unnecessarily, complex on the other. Furthermore, GFs are continuously produced within the extracellular space and dynamically presented over cell surface during any tissue repairing process. The use of very complex and generally labile proteins within the scaffold architecture requires sophisticated technologies and elaborate strategies to preserve their activity for a medium-long time period [83]. The use of peptides capable of eliciting comparable morphogenic activity instead of the original protein allows a dramatic reduction of the level of sophistication of the architecture and a massive simplification in scaffold complexity. Even if these peptides elicit the appropriate cell response generally at a higher dose compared to their natural or recombinant counterparts, they provide a viable alternative in terms of cost and handling. QK peptide has been proved to be effective in eliciting angiogenic response and has already been exploited as an alternative to VEGF in promoting scaffold [128]. Analogously, BMP mimetic peptide already been proved a potent osteogenic active molecule [129]. These small molecules, as their natural counterparts, often impart a potentiated biological response if bound to a solid substrate. Indeed, it has been shown that molecularly decorated materials enhance tissue formation through the modulation of the interaction between protein signaling and biomaterials; it appears to be fundamental to provide a better integration of the scaffold with the neoforming tissue [130]. In natural ECM, GAGs provide binding domains for GFs and this mechanism of action could be encoded within artifical ECM by introducing specific binding domain for the mimicking peptides. Alginate and poly(acrylamide) gel, for instance, have been sulphated to enhance the binding affinity to some GFs, including VEGF, PDGF, and hepatocyte growth factor, potentiating the angiogenic activity and extending the flexibility of the scaffold for growth factor presentation and preservation [131–133]. Furthermore, the modulation of binding affinity within the scaffold structure provides a viable strategy to control stable gradients of GFs (Figure 10.2), or their mimicking peptides, which are proved to be essential in controlling and guiding morphogenetic processes [134].

In natural ECM, there is a continuos production of GFs that are sequestred within even remote molecular recess and then used upon cell request. Continuous supply of GFs, or their mimicking peptides, at a specific location within a synthetic scaffold can be endowed with the use of micro- or nanoparticles loaded with bioactive moieties and programmed to deliver a certain dose according to a definite profile (Figure 10.2). Integration of GF loaded microparticles programmed to sequentially release multiple GFs has already been discussed in the literature [30,84,89]. Following this approach, it is possible to control the spatial distribution and the gradients of bioactive agents at different locations within the scaffold [30,83,90,91]. A more advanced approach to create microsphere-integrated scaffolds able to regulate both temporally and spatially GFs release kinetics may take advantage of micromanipulation-based techniques. Possible developments and advancement include the control over the presentation of relevant signals not only within the physical domain of the scaffolds but also within the host surrounding tissues. Engineered template embedding microspheres releasing GFs at known release rates in a predetermined and optimized spatial distribution within the scaffold may benefit from advances in micro- and nanotechnologies. A microdepot acting as a single point source may be micropositioned by 3D printing and soft lithography to obtain highly regulated structures able to trigger the extent, and possibly the architecture/structure of tissue formation [33,135]. The combination of micropositioning systems and mathematical modeling describing the complex and multiple mechanisms governing the release kinetics from single microspheres within the scaffold can be of help in creating scaffolds with a highly controlled architecture by computer-aided scaffold design (CASD) programs [33,136].

These novel bioactivated multifunctional computer-aided designed scaffolds offer the potentiality to be programmed for recruiting stem cells from the surrounding tissue, expanding them within their structure, guiding their fate, and promoting functional tissue regeneration. This approach requires the control of delivery and presentation of chemoattractants for recruiting target cell (including endothelial cell), presentation of mitogenic factor for cell expansion and finally control of the presentation of relevant morphogenic factors for controlling and guiding tissue formation. The possibility to program materials, through a tight control of biological cues presentation, to recruit specific host cells and guide *in situ* their fate hold the promise to circumvent severe limitations in the current tissue engineering scheme. Indeed, using cell-free scaffolds able to direct the tissue regeneration process via endogenous stem cell homing avoids the expensive and complex stage of autologous cell *ex vivo* manipulation.

10.4 CONCLUSION AND REMARKS

Over the past 25 years, the concept of biomaterials has been revolutionized, passing from a space filler material to the modern version of programmable bioactive material able to guide and control complex cellular processes. By a fortunate integration of the recent discovery of cellular and molecular biology and the advancement in material science and nanotechnology, future scaffolds will be a simplified, yet effective, replica of the natural ECM with the potentiality to make TE a real clinical success. It is expected that the next generation of biomaterials will integrate more biology able to display with a defined spatial and temporal orchestration a complex array of bioactive signals—chemically and physically—to guide the tissue regeneration process at single cell level. These biomaterials hold the promise that in the near future, the classical TE scheme could be simplified by avoiding the *ex vivo* manipulation of the autologous cell by driving the repair process promoting the recruitment of resident stem cells from surrounding tissues.

ABBREVIATIONS

BMP	bone morphogenetic protein
CASD	computer-aided scaffold design
CIMs	cell-instructive materials
ECM	extracellular matrix
GFs	growth factors
HGF	hepatocyte growth factor
GAGs	glycosaminoglycans
PDGF	platelet-derived growth factor
SFF	solid free-form fabrication
TE	tissue engineering
VEGF	vascular endothelial growth factor

REFERENCES

[1] Mikos AG, et al. Engineering complex tissues. Tissue Eng 2006;12:3307–39.

[2] Lavik E, Langer R. Tissue engineering: current state and perspectives. Appl Microbiol Biotechnol 2004;65:1–8.

[3] Goldberg M, Langer R, Jia X. Nanostructured materials for applications in drug delivery and tissue engineering. J Biomater Sci Polym Ed 2007;18:241–68.

[4] Niklason LE, Langer R. Prospects for organ and tissue replacement. J Am Med Assoc 2001;285:573–6.

[5] Place ES, Evans ND, Stevens MM. Complexity in biomaterials for tissue engineering. Nat Mater 2009;8:457–70.

[6] Manav Mehta KS, Georg B, Duda N, Mooney DJ. Biomaterial delivery of morphogens to mimic the natural healing cascade in bone. Adv Drug Delivery Rev 2012;64:1257–76.

[7] Matsumoto T, Mooney DJ. Cell instructive polymers. In: Lee K, Kaplan D, editors. Tissue engineering. I. Scaffold systems for tissue engineering. 2006. p. 113–37.

[8] Carragee EJ, Hurwitz EL, Weiner BK. A critical review of recombinant human bone morphogenetic protein-2 trials in spinal surgery: emerging safety concerns and lessons learned. Spine J Off J N Am Spine Soc 2011;11:471–91.

[9] Buchowski M, Nussenbaum B. In reference to acute airway obstruction in cervical spinal procedures with bone morphogenetic proteins. Laryngoscope 2011;121(11):2501.

[10] Yaremchuk MST, Somers ML, Peterson E. Acute airway obstruction in cervical spinal procedures with bone morphogenetic proteins. Laryngoscope 2010;120:1954–7.

[11] Epstein NE. Pros, cons, and costs of infuse in spinal surgery. Surg Neurol Int 2011;2:10.

[12] Service RF. Tissue engineers build new bone. Science 2000;289:1498–500.

[13] Shields LB, Glassman SD, Campbell M, Vitaz T, Harpring J, Shields CB. Adverse effects associated with high-dose recombinant human bone morphogenetic protein-2 use in anterior cervical spine fusion. Spine 2006;31:542–7.

[14] Li G, Luppen C, Li XJ, Wood M, Seeherman HJ, Wozney JM, et al. Bone consolidation is enhanced by rhBMP-2 in a rabbit model of distraction osteogenesis. J Orthop Res 2002;20:779–88.

[15] Street J, deGuzman L, Bunting S, Peale Jr. FV, Ferrara N, Steinmetz H, et al. Vascular endothelial growth factor stimulates bone repair by promoting angiogenesis and bone turnover. Proc Natl Acad Sci USA 2002;99:9656–61.

[16] Kimoto T, Kubo T, Maeda M, Sano A, Akagawa Y. Continuous administration of basic fibroblast growth factor (FGF-2) accelerates bone induction on rat calvaria—an application of a new drug delivery system. J Dent Res 1998;77:1965–9.

[17] Ennett AB, Mooney DJ. Temporally regulated delivery of VEGF *in vitro* and *in vivo*. J Biomed Mater Res A 2006;79:176–84.

[18] Santo VE, Carida M, Cancedda R, Gomes ME, Mano JF, Reis RL. Carrageenan-based hydrogels for the controlled delivery of PDGF-BB in bone tissue engineering applications. Biomacromolecules 2009;10:1392–401.

[19] Chen FM, Chen R, Wang XJ, Sun HH, Wu ZF. In vitro cellular responses to scaffolds containing two microencapsulated growth factors. Biomaterials 2009;30:5215–24.

[20] Patel ZS, Young S, Tabata Y, Jansen JA, Wong ME, Mikos AG. Dual delivery of an angiogenic and an osteogenic growth factor for bone regeneration in a critical size defect model. Bone 2008;43:931–40.

[21] Hasirci V, Yilgor P, Hasirci N. Sequential BMP-2/BMP-7 delivery from polyester nanocapsules. J Biomed Mater Res A 2010;93 (2):528–36.

[22] Causa F, Netti PA, Ambrosio L. A multi-functional scaffold for tissue regeneration: the need to engineer a tissue analogue. Biomaterials 2007;28(34):5093–9.

[23] Lutolf MP, Hubbell JA. Synthetic biomaterials as instructive extracellular microenvironment for morphogenesis in tissue engineering. Nat Biotechnol 2005;23:47–55.

[24] Ventre M, Causa F, Netti PA. Determinants of cell-material crosstalk at the interface: towards engineering of cell instructive materials. J R Soc Interface 2012;9(74):2017–32.

[25] Pelham RJ, Wang YL. Cell locomotion and focal adhesions are regulated by substrate flexibility (vol. 94, p. 13661, 1997). Proc Natl Acad Sci USA 1998;95(20):12070.

[26] Kantawong F, et al. Whole proteome analysis of osteoprogenitor differentiation induced by disordered nanotopography and mediated by ERK signalling. Biomaterials 2009;30(27):4723–31.

[27] Biggs MJP, et al. Interactions with nanoscale topography: adhesion quantification and signal transduction in cells of osteogenic and multipotent lineage. J Biomed Mater Res Part A 2009;91A (1):195–208.

[28] Biggs MJP, et al. The use of nanoscale topography to modulate the dynamics of adhesion formation in primary osteoblasts and ERK/MAPK signalling in STRO-1 + enriched skeletal stem cells. Biomaterials 2009;30(28):5094–103.

[29] Chen CS, et al. Cell shape provides global control of focal adhesion assembly. Biochem Biophys Res Commun 2003;307(2):355–61.

[30] Luciani A, et al. PCL microspheres based functional scaffolds by bottom-up approach with predefined microstructural properties and release profiles. Biomaterials 2008;29(36):4800–7.

[31] Lutolf MP, Hubbell JA. Synthetic biomaterials as instructive extracellular microenvironments for morphogenesis in tissue engineering. Nat Biotechnol 2005;23(1):47–55.

[32] Sachlos E, Czernuszka JT. Making tissue engineering scaffolds work. Review: The application of solid free-form fabrication technology to the production of tissue engineering scaffolds. Eur Cells Mater 2003;5:29–39 [discussion 39–40]

[33] Sun W, et al. Computer-aided tissue engineering: overview, scope and challenges. Biotechnol Appl Biochem 2004;39:29–47.

[34] Boland T, et al. Application of inkjet printing to tissue engineering. Biotechnol J 2006;1(9):910–7.

[35] Teo W-E, He W, Ramakrishna S. Electrospun scaffold tailored for tissue-specific extracellular matrix. Biotechnol J 2006;1(9):918–29.

[36] Guarino V, Causa F, Ambrosio L. Bioactive scaffolds for bone and ligament tissue. Expert Rev Med Devices 2007;4 (3):405–18.

[37] Hutmacher DW. Scaffold design and fabrication technologies for engineering tissues—state of the art and future perspectives. J Biomater Sci Polym Ed 2001;12(1):107–24.

[38] Hacker MC, Mikos AG. Trends in tissue engineering research. Tissue Eng 2006;12:2049–57.

[39] Nair LS, Laurencin CT. Polymers as biomaterials for tissue engineering and controlled drug delivery. In: Lee K, Kaplan, D, editors. Tissue engineering. I. Scaffold systems for tissue engineering, editors; 2006. p. 47–90.

[40] Mano JF, et al. Natural origin biodegradable systems in tissue engineering and regenerative medicine: present status and some moving trends. J R Soc Interface 2007;4:999–1030.

[41] Voytik-Harbin SL, et al. Small intestinal submucosa: a tissue-derived extracellular matrix that promotes tissue-specific growth and differentiation of cells *in vitro*. Tissue Eng 1998;4:157–74.

[42] Chan G, Mooney DJ. New materials for tissue engineering: towards greater control over the biological response. Trends Biotechnol 2008;26:382–92.

[43] Rockwood DN, et al. Materials fabrication from Bombyx mori silk fibroin. Nat Protoc 2011;6(10):1612–31.

[44] Nair LS, Laurencin CT. Biodegradable polymers as biomaterials. Prog Polym Sci 2007;32(8–9):762–98.

[45] Chang WH, et al. A genipin-crosslinked gelatin membrane as wound-dressing material: in vitro and in vivo studies. J Biomater Sci Polym Ed 2003;14(5):481–95.

[46] Wang H-J, et al. In vivo biocompatibility and mechanical properties of porous zein scaffolds. Biomaterials 2007;28(27):3952–64.

[47] Langer R, Tirrell DA. Designing materials for biology and medicine. Nature 2004;428(6982):487–92.

[48] Langer R, Vacanti JP. Tissue engineering. Science 1993;260 (5110):920–6.

[49] Bell E, et al. Living tissue formed in vitro and accepted as skin-equivalent tissue of full thickness. Science (New York, NY) 1981;211(4486):1052–4.

[50] Hoppe A, Gueldal NS, Boccaccini AR. A review of the biological response to ionic dissolution products from bioactive glasses and glass ceramics. Biomaterials 2011;32(11):2757–74.

[51] Karageorgiou V, Kaplan D. Porosity of 3D biomaterial scaffolds and osteogenesis. Biomaterials 2005;26(27):5474–91.

[52] Hench LL. Bioceramics. J Am Ceram Soc 1998;81(7):1705–28.

[53] Chen GP, Ushida T, Tateishi T. Scaffold design for tissue engineering. Macromol Biosci 2002;2(2):67–77.

[54] Mikos AG, et al. Preparation of poly(glycolic acid) bonded fiber structures for cell attachment and transplantation. J Biomed Mater Res 1993;27(2):183–9.

[55] Mikos AG, et al. *Preparation and characterization of poly(L-lactic acid) foams.* Polymer 1994;35(5):1068–77.

[56] Guarino V, et al. Design and manufacture of microporous polymeric materials with hierarchal complex structure for biomedical application. Mater Sci Technol 2008;24(9):1111–7.

[57] Gomes ME, et al. Alternative tissue engineering scaffolds based on starch: processing methodologies, morphology, degradation and mechanical properties. Mater Sci Eng CBiomimetic Supramol Syst 2002;20(1–2):19–26.

[58] Chevalier E, et al. Fabrication of porous substrates: a review of processes using pore forming agents in the biomaterial field. J Pharm Sci 2008;97(3):1135–54.

[59] Salerno A, et al. Design of porous polymeric scaffolds by gas foaming of heterogeneous blends. J Mater Sci Mater Med 2009;20 (10):2043–51.

[60] Salerno A, et al. Engineering of foamed structures for biomedical application. J Cell Plast 2009;45(2):103–17.

[61] Mathieu LM, et al. Architecture and properties of anisotropic polymer composite scaffolds for bone tissue engineering. Biomaterials 2006;27(6):905–16.

[62] Salerno A, et al. Engineered mu-bimodal poly(epsilon-caprolactone) porous scaffold for enhanced hMSC colonization and proliferation. Acta Biomater 2009;5(4):1082–93.

[63] Murugan R, Ramakrishna S. Development of nanocomposites for bone grafting. Compos Sci Technol 2005;65 (15-16):2385–406.

[64] Salerno A, et al. *Novel 3D porous multi-phase composite scaffolds based on PCL, thermoplastic zein and ha prepared via supercritical CO₂foaming for bone regeneration.* Compos Sci Technol 2010;70(13):1838–46.

[65] Gerhardt L-C, et al. The pro-angiogenic properties of multifunctional bioactive glass composite scaffolds. Biomaterials 2011;32 (17):4096–108.

[66] Jandt KD. Evolutions, revolutions and trends in biomaterials science—A perspective. Adv Eng Mater 2007;9(12):1035–50.

[67] Tessmar JK, Goepferich AM. Matrices and scaffolds for protein delivery in tissue engineering. Adv Drug Delivery Rev 2007;59 (4–5):274–91.

[68] Chen RR, Mooney DJ. Polymeric growth factor delivery strategies for tissue engineering. Pharm Res 2003;20(8):1103–12.

[69] Holland TA, Mikos AG. Review: Biodegradable polymeric scaffolds. Improvements in bone tissue engineering through controlled drug delivery. In: Lee K, Kaplan D, editors. Tissue engineering. I. Scaffold systems for tissue engineering. 2006. p. 161–85.

[70] Malafaya PB, Silva GA, Reis RL. Natural-origin polymers as carriers and scaffolds for biomolecules and cell delivery in tissue engineering applications. Adv Drug Delivery Rev 2007;59 (4–5):207–33.

[71] Lee KY, Mooney DJ. Hydrogels for tissue engineering. Chem Rev 2001;101(7):1869–79.

[72] Anseth KS, Bowman CN, BrannonPeppas L. Mechanical properties of hydrogels and their experimental determination. Biomaterials 1996;17(17):1647–57.

[73] Sokolsky-Papkov M, et al. Polymer carriers for drug delivery in tissue engineering. Adv Drug Delivery Rev 2007;59 (4–5):187–206.

[74] Peppas NA, et al. Poly(ethylene glycol)-containing hydrogels in drug delivery. J Control Release 1999;62(1–2):81–7.

[75] Hoffman AS. Hydrogels for biomedical applications. Adv Drug Delivery Rev 2012;64:18—23.

[76] Van Tomme SR, Hennink WE. Biodegradable dextran hydrogels for protein delivery applications. Expert Rev Med Devices 2007;4(2):147—64.

[77] Biondi M, et al. Controlled drug delivery in tissue engineering. Adv Drug Delivery Rev 2008;60(2):229—42.

[78] Rizzi SC, et al. Recombinant protein-co-PEG networks as cell-adhesive and proteolytically degradable hydrogel matrixes. Part II. Biofunctional characteristics. Biomacromolecules 2006;7 (11):3019—29.

[79] Murphy WL, et al. Sustained release of vascular endothelial growth factor from mineralized poly(lactide-co-glycolide) scaffolds for tissue engineering. Biomaterials 2000;21(24):2521—7.

[80] Peters MC, Polverini PJ, Mooney DJ. Engineering vascular networks in porous polymer matrices. J Biomed Mater Res 2002;60 (4):668—78.

[81] Ungaro F, et al. Microsphere-integrated collagen scaffolds for tissue engineering: effect of microsphere formulation and scaffold properties on protein release kinetics. J Control Release 2006;113(2):128—36.

[82] Mollica F, et al. Mathematical modelling of the evolution of protein distribution within single PLGA microspheres: prediction of local concentration profiles and release kinetics. J Mater Sci Mater Med 2008;19(4):1587—93.

[83] Borselli C, et al. Induction of directional sprouting angiogenesis by matrix gradients. J Biomed Mater Res Part A 2007;80A (2):297—305.

[84] Richardson TP, et al. Polymeric system for dual growth factor delivery. Nat Biotechnol 2001;19(11):1029—34.

[85] Kretlow JD, Klouda L, Mikos AG. Injectable matrices and scaffolds for drug delivery in tissue engineering. Adv Drug Delivery Rev 2007;59(4—5):263—73.

[86] Zhang G, Suggs LJ. Matrices and scaffolds for drug delivery in vascular tissue engineering (vol. 59, p. 360, 2007). Adv Drug Delivery Rev 2009;61(14):1386.

[87] Lee S-H, Shin H. Matrices and scaffolds for delivery of bioactive molecules in bone and cartilage tissue engineering. Adv Drug Delivery Rev 2007;59(4—5):339—59.

[88] Ivana d'Angelo, Oliviero O, Ungaro F, Quaglia, F, Netti PA. *Engineering strategies to control VEGF stability and levels in a collagen matrix for angiogenesis: the role of heparin sodium salt and the PLGA-based microsphere approach.* Acta Biomater 2013;9 (7):7389—98.

[89] Saltzman WM, Olbricht WL. Building drug delivery into tissue engineering. Nat Rev Drug Discov 2002;1(3):177—86.

[90] Chen RR, et al. Spatio-temporal VEGF and PDGF delivery patterns blood vessel formation and maturation. Pharm Res 2007;24(2):258—64.

[91] Suciati T, et al. Zonal release of proteins within tissue engineering scaffolds. J Mater Sci Mater Med 2006;17 (11):1049—56.

[92] Salvay DM, Shea LD. Inductive tissue engineering with protein and DNA-releasing scaffolds. Mol Biosyst 2006;2 (1):36—48.

[93] Chung HJ, Park TG. Surface engineered and drug releasing pre-fabricated scaffolds for tissue engineering. Adv Drug Delivery Rev 2007;59(4—5):249—62.

[94] Sakiyama-Elbert SE, Hubbell JA. Development of fibrin derivatives for controlled release of heparin-binding growth factors. J Controll Release 2000;65(3):389—402.

[95] Wissink MJB, et al. Binding and release of basic fibroblast growth factor from heparinized collagen matrices. Biomaterials 2001;22(16):2291—9.

[96] Lee AC, et al. Controlled release of nerve growth factor enhances sciatic nerve regeneration. Exp Neurol 2003;184 (1):295—303.

[97] Taylor SJ, McDonald JW, Sakiyama-Elbert SE. Controlled release of neurotrophin-3 from fibrin gels for spinal cord injury. J Control Release 2004;98(2):281—94.

[98] Yoon JJ, et al. Heparin-immobilized biodegradable scaffolds for local and sustained release of angiogenic growth factor. J Biomed Mater Res Part A 2006;79A(4):934—42.

[99] Jeon O, et al. *Enhancement of ectopic bone formation by bone morphogenetic protein-2 released from a heparin-conjugated poly(L-lactic-co-glycolic acid) scaffold.* Biomaterials 2007;28 (17):2763—71.

[100] Nillesen STM, et al. Increased angiogenesis and blood vessel maturation in acellular collagen-heparin scaffolds containing both FGF2 and VEGF. Biomaterials 2007;28(6):1123—31.

[101] Kleinman HK, Philp D, Hoffman MP. Role of the extracellular matrix in morphogenesis. Curr Opin Biotechnol 2003;14 (5):526—32.

[102] Harris BP, et al. Photopatterned polymer brushes promoting cell adhesion gradients. Langmuir 2006;22(10):4467—71.

[103] Guarnieri D, et al. Covalent immobilized RGD gradient on PEG hydrogel scaffold influences cell migration parameters. Acta Biomater 2010;6(7):2532—9.

[104] Harbers GM, Healy KE. The effect of ligand type and density on osteoblast adhesion, proliferation, and matrix mineralization. J Biomed Mater Res Part A 2005;75A(4):855—69.

[105] Hersel U, Dahmen C, Kessler H. RGD modified polymers: biomaterials for stimulated cell adhesion and beyond. Biomaterials 2003;24(24):4385—415.

[106] Ranucci CS, Moghe PV. Polymer substrate topography actively regulates the multicellular organization and liver-specific functions of cultured hepatocytes. Tissue Eng 1999; 5(5):407—20.

[107] O'Brien FJ, et al. The effect of pore size on cell adhesion in collagen-GAG scaffolds. Biomaterials 2005;26(4):433—41.

[108] Salerno A, Guarnieri D, Iannone M, Zeppetelli S, Netti PA. Effect of micro- and macroporosity of bone tissue three-dimensional-poly(ε-caprolactone) scaffold on human mesenchymal stem cells invasion, proliferation, and differentiation in vitro. Tissue Eng 2010;16:2661—3.

[109] Ma PX, Zhang RY. Microtubular architecture of biodegradable polymer scaffolds. J Biomed Mater Res 2001; 56(4):469—77.

[110] Hollister SJ. Porous scaffold design for tissue engineering (vol. 4, p. 518, 2005). Nat Mater 2006;5(7):590.

[111] Li JP, et al. Porous Ti6Al4V scaffolds directly fabricated by 3D fibre deposition technique: effect of nozzle diameter. J Mater Sci Mater Med 2005;16:1159—63.

[112] Simon JL, et al. In vivo bone response to 3D periodic hydroxyapatite scaffolds assembled by direct ink writing. J Biomed Mater Res Part A 2007;83A(3):747—58.

[113] Vacanti JP, Shin YM, Ogilvie J, Svy A, Maemura T, Ishii O, et al. Fabrication of vascularized tissue using microfabricated two-dimensional molds. US Patent; 2010.

[114] Mata A, et al. A three-dimensional scaffold with precise microarchitecture and surface micro-textures. Biomaterials 2009;30 (27):4610–7.

[115] Bosman FT, Stamenkovic I. Functional structure and composition of the extracellular matrix. J Pathol 2003;200(4):423–8.

[116] Katz E, Streuli CH. The extracellular matrix as an adhesion checkpoint for mammary epithelial function. Int J Biochem Cell Biol 2007;39(4):715–26.

[117] Fittkau MH, et al. The selective modulation of endothelial cell mobility on RGD peptide containing surfaces by YIGSR peptides. Biomaterials 2005;26(2):167–74.

[118] Stupack DG, Cheresh DA. ECM remodeling regulates angiogenesis: endothelial integrins look for new ligands. Sci STKE 2002;2002(119):pe7.

[119] Rapraeger AC. Syndecan-regulated receptor signaling. J Cell Biol 2000;149(5):995–7.

[120] Wijelath ES, et al. Novel vascular endothelial growth factor binding domains of fibronectin enhance vascular endothelial growth factor biological activity. Circulation Res 2002;91(1):25–31.

[121] Taipale J, KeskiOja J. Growth factors in the extracellular matrix. FASEB J 1997;11(1):51–9.

[122] Gurdon JB, et al. Activin signaling and response to a morphogen gradient. Nature 1994;371(6497):487–92.

[123] Tanabe Y, Jessell TM. Diversity and pattern in the developing spinal cord (vol. 274, p. 1115, 1996). Science 1997;276 (5309):21.

[124] Burgess BT, Myles JL, Dickinson RB. Quantitative analysis of adhesion-mediated cell migration in three-dimensional gels of RGD-grafted collagen. Ann Biomed Eng 2000;28(1):110–8.

[125] Kong HJ, Mooney DJ. Microenvironmental regulation of biomacromolecular therapies. Nat Rev Drug Discov 2007;6(6):455–63.

[126] Beniash E, et al. Self-assembling peptide amphiphile nanofiber matrices for cell entrapment. Acta Biomater 2005;1(4):387–97.

[127] Brown TD, Dalton PD, Hutmacher DW. Direct writing by way of melt electrospinning. Adv Mater 2011;23(47):5651.

[128] Finetti F, et al. Functional and pharmacological characterization of a VEGF mimetic peptide on reparative angiogenesis. Biochem Pharmacol 2012;84(3):303–11.

[129] Zouani OF, et al. Differentiation of pre-osteoblast cells on poly (ethylene terephthalate) grafted with RGD and/or BMPs mimetic peptides. Biomaterials 2010;31(32):8245–53.

[130] Wang D-A, et al. Multifunctional chondroitin sulphate for cartilage tissue-biomaterial integration. Nat Mater 2007;6(5):385–92.

[131] Merkel TC, et al. Ultrapermeable, reverse-selective nanocomposite membranes. Science 2002;296(5567):519–22.

[132] Rouet V, et al. A synthetic glycosaminoglycan mimetic binds vascular endothelial growth factor and modulates angiogenesis. J Biol Chem 2005;280(38):32792–800.

[133] Chaterji S, Gemeinhart RA. Enhanced osteoblast-like cell adhesion and proliferation using sulfonate-bearing polymeric scaffolds. J Biomed Mater Res Part A 2007;83A(4):990–8.

[134] Griffith LG, Swartz MA. Capturing complex 3D tissue physiology in vitro. Nat Rev Mol Cell Biol 2006;7(3):211–24.

[135] Whitesides GM, et al. Soft lithography in biology and biochemistry. Ann Rev Biomed Eng 2001;3:335–73.

[136] Hutmacher DW, Sittinger M, Risbud MV. Scaffold-based tissue engineering: rationale for computer-aided design and solid free-form fabrication systems. Trends Biotechnol 2004;22:354–62.

[137] Causa F, et al. Surface investigation on biomimetic materials to control cell adhesion: the case of RGD conjugation on PCL. Langmuir 2010;26:9875–84.

Chapter 11

Biocompatibility and Immune Response to Biomaterials

Bryan N. Brown[a,b] and Stephen F. Badylak[a,b,c]

[a]McGowan Institute for Regenerative Medicine, Pittsburgh, PA, [b]Department of Bioengineering, Pittsburgh, PA, [c]Department of Surgery, University of Pittsburgh, Pittsburgh, PA

Chapter Outline

11.1 INTRODUCTION

The mechanisms of the host innate and humoral response to whole organ transplantation are reasonably well understood. Xenogeneic and allogeneic cellular antigens are recognized by the host, elicit immune activation, and cause the production of proinflammatory mediators with downstream cytotoxicity and transplant tissue/organ rejection. There are a number of important and recognized considerations which determine success or failure of the transplant procedure. Among these considerations are the quality and functionality of the harvested organ, time to transplantation, and patient–donor human leukocyte antigen (HLA) matching.

Tissue engineering and regenerative medicine strategies for organ transplantation offer the potential of custom-designed organs composed of the patients' own cells, presumably avoiding the complications of allogeneic antigens. However, although each strategy may differ in the specific combination of cells, biomaterials, and bioactive factors selected for the application, it is not possible to entirely circumvent the host response, and, in fact, the host response should not be circumvented. It is well understood that the manner in which the host responds to the selected intervention will dictate long-term success or failure. Depending on the source of the cells used, the type of host response to an engineered organ will vary. When autologous cells are delivered on a synthetic or biologic scaffold material, it is more likely that the host innate immune response will play a dominant role rather than the acquired response.

This chapter will briefly review the key considerations in the host response to transplanted organs and to tissue engineering and regenerative medicine strategies for organ transplant including transplant rejection, the host response to tissue injury, the foreign body

Regenerative Medicine Applications in Organ Transplantation.

reaction, and newly defined paradigms surrounding the host macrophage response. While these considerations are applicable to all tissue engineering and regenerative medicine strategies, the present chapter is intended as a companion to the previous chapter. Therefore, a specific emphasis is placed on the role of the host immune response to extracellular matrix (ECM)-based scaffold materials in determining the ability to support constructive and functional remodeling and the implications of this response in the success or failure of strategies employing whole organ decellularization and recellularization.

11.2 TRANSPLANT REJECTION

Rejection is a well-studied and well-understood consequence of the transplantation of tissue or organs from a nonself donor. Largely, rejection is driven via the recognition of nonself alloantigen [primarily MHC (major histocompatibility class) I] on cells of the donor organ and activation of the acquired immune response (CD4$^+$ and CD8$^+$ T cells). In the absence of immune suppression, this response occurs within 10–13 days posttransplant (acute rejection). Subsequent tissue or organ transplants from the same donor will be subject to a accelerated rejection response within 6–8 days posttransplant (accelerated rejection) due to a memory-type immune response. With the understanding that MHC (also referred to as HLA in humans) is a potent driver of the acquired immune response to transplanted organs, much effort is now devoted to HLA matching between donors prior to transplant. However, an exact HLA match is only possible between related individuals, and the majority of transplants (with the exception of identical twins) between related donors will also be rejected, although more slowly, due to mismatch at other genetic loci. Therefore, it can be stated that the success of transplantation is largely due to advances in immunosuppression rather than advances in HLA matching. Further, the scarcity of donor organs is a major limiting factor in transplant medicine. As the transplant rejection response has been described in detail elsewhere (readers are referred to any of a number of basic immunobiology texts including [1,2]), it is described only briefly herein as a basis for understanding and distinguishing additional responses which may occur following the implantation of tissue-engineered organs.

Briefly, alloreactive T cells are activated by antigen presenting cells (APCs) bearing both alloantigen and costimulatory molecules. This T-cell accumulation can occur by one of two commonly recognized pathways including direct and indirect allorecognitions. Direct allorecognition involves the migration of donor-derived APCs to the lymph nodes and subsequent activation of a T-cell response, while the indirect pathway involves uptake of alloantigen by host APCs and subsequent presentation to T cells. The activation of a rejection response via the direct pathway can be reduced by depletion of donor APCs within the organ prior to transplant or in cases of transplantation of organs without lymphatic drainage. The relative contributions of the direct and indirect pathways of recognition and T-cell activation in transplant rejection are unclear, but the direct pathway is largely responsible for the acute rejection response.

Antibodies also play an important role in the rejection response. Pre-existing alloantibodies which recognize nonself antigens can initiate a complement-dependent cascade within minutes of transplantation resulting in hyperacute rejection of the organ. Hyperacute rejection results from the interaction of antibodies and the subsequent complement cascade with the vascular endothelium leading to clotting and loss of blood supply to the transplanted organ. Hyperacute rejection can typically be avoided by screening of blood types and for the presence of reactive alloantibodies prior to transplant; however, various methods to desensitize patients are available. For these same reasons, xenografts are unsuitable for human transplant applications. For example, the α-Gal epitope on the cells of other mammals is rapidly recognized by preexisting antibodies in humans, leading to hyperacute rejection.

The availability of immunosuppression largely allows for the success of transplantation in the short term. In the long term, transplanted organs fail not only for a number of reasons including alloreactivity but also due to the recruitment of innate immune cells (monocytes and macrophages) by activated T cells leading to chronic inflammation, scarring and subsequent graft failure. It should be noted that, regardless of transplant acceptance or rejection, activation of T cells is commonly observed. However, the phenotype of the activated cells (CD4$^+$ T-helper cells) has been observed to be different in cases of rejection versus acceptance. Briefly, the role of T lymphocytes, especially the Th1 and Th2 lymphocyte phenotypes, in cell-mediated immune responses to allografts and xenografts has been widely studied [3,4]. Th1 lymphocytes produce cytokines, such as interleukin (IL)-2, interferon (IFN)-γ, and tumor necrosis factor (TNF)-β leading to macrophage activation, stimulation of complement fixing antibody isotypes and differentiation of CD8$^+$ cells to a cytotoxic phenotype [5,6]. Activation of this pathway is associated with both allogeneic and xenogeneic transplant rejections [3,4,7]. Th2 lymphocytes produce IL-4, IL-5, IL-6, and IL-10, cytokines that do not activate macrophages and that lead to the production of noncomplement fixing antibodies. Activation of the Th2 pathway is commonly associated with transplant acceptance [8–10]. Note that

T-cell activation occurs in both rejection and acceptance conditions, however, the phenotype of the activated T cell is the critical difference. The Th1/Th2 paradigm is discussed in the context of tissue engineering and regenerative medicine in more detail below.

Tissue engineering and regenerative medicine approaches which utilize autologous cell sources provide promise for the development of whole organ or tissue transplants without the complications of the rejection-type response as described above. However, as will be discussed below, these engineered organs will be subject to other potentially detrimental responses which are dependent upon the noncellular components used to construct the organ. Nonetheless, it is important to appreciate the transplant rejection response as aspects of the acquired immune system may be activated depending on the type of cells used and their source (i.e., nonautologous), and the type of bioscaffold used. Further, and as will be discussed below, the activation of components of the acquired immune response can be observed, even in the case of acellular, scaffold-only based approaches to tissue reconstruction.

11.3 THE HOST RESPONSE TO TISSUE INJURY

The default mammalian response to tissue injury represents a protective mechanism designed to prevent further insult to the host by hemorrhage, potential pathogens, loss of function, or by a prolonged inflammatory response. This default response, commonly referred to as wound healing, has been studied in-depth in many tissues and organ systems. It is well understood that, due to the need for creation of a surgical defect, this response is an unavoidable phenomenon following both organ transplant and the placement of materials or tissue-engineered constructs within the body. The mechanisms of the wound healing response have been described at length elsewhere [11−13], so are only reviewed briefly herein as a basis from which to initiate discussion of the response to biomaterials implantation.

The host response to tissue injury is generally considered to occur in four overlapping stages, eventually resulting in the deposition of scar tissue consisting of dense fibrous connective tissue.

11.3.1 Hemostasis

Following tissue injury and the associated damage to the vasculature, platelets are activated by exposure to subendothelial structures and released tissue factor resulting in the activation of a cascade of clotting factors that causes the formation of a provisional fibrin clot with entrapped erythrocytes. The provisional matrix provides a substrate for further cell migration into the site of injury and a medium for cell signaling. In addition to their role in hemostasis and provisional matrix formation, platelets also release cytokines including platelet-derived growth factor (PDGF), transforming growth factor beta (TGF)-β, chemokine C-X-C ligand 4 (C-X-C L4), IL-1β, and the CD47 ligand thrombospondin [14−18]. These factors, among others, contribute to the initial repair process via recruitment of multiple cell types, including neutrophils, macrophages, fibroblasts, and other tissue-specific cells to the injury site [16].

11.3.2 Inflammation

Neutrophils are the first inflammatory cell type to arrive at the wound site. Neutrophils phagocytose and destroy foreign material, bacteria, or cell debris that may be present and provide additional signaling molecules that recruit macrophages to the injury site [14]. Mast cells participate in the early stages of wound healing by releasing granules rich in enzymes, histamine, and other factors that modulate the inflammatory response [12,19]. By 48−72 h postinjury, however, macrophages typically dominate the cell population at the site of injury [20]. These cells are of a predominantly proinflammatory phenotype and secrete cytokines and chemokines that promote the further recruitment of leukocytes to the site of injury [14,16]. Macrophages also remove apoptotic neutrophils, the phagocytosis of which may lead to a change toward a more reparative macrophage phenotype (discussed in further detail below) and the resolution of the inflammatory phase of wound healing [21−23]. The T-lymphocyte population also plays an important late regulatory role in the resolution of the inflammatory process through local secretion of cytokines and chemokines, many of which are known to affect macrophage polarization [24].

11.3.3 Proliferative Phase

The proliferative phase of wound healing involves cellular proliferation, angiogenesis, new ECM deposition, and the formation of granulation tissue—processes that are largely mediated via the effects of the local microenvironment, including pH and oxygen tension, and the cytokine milieu secreted by macrophages, T lymphocytes, and other cells within the wound site [16,25,26]. These cytokines include epidermal growth factor (EGF), basic fibroblast growth factor (b-FGF), transforming growth factor (TGF-α, TGF-β), vascular endothelial growth factor (VEGF), and others depending on the nature of the injured tissue [12]. Importantly, macrophage participation and phenotypic polarization during this proliferative stage or the injury response may have significant downstream remodeling effects.

11.3.4 Remodeling Phase

Following the deposition of significant amounts of new ECM which is rich in collagen types I and III during the proliferative phase, the remodeling phase of wound healing begins. This phase is characterized by matrix metalloproteinase (MMP)- and tissue inhibitor of metalloproteinase (TIMP)-mediated degradations and remodeling of the newly deposited matrix components with eventual scar tissue formation/maturation [12,27]. In some cases, dysregulation of the default healing process can result in prolonged inflammation and remodeling leading to fibrosis or hypertrophic scar formation [12,28–30].

11.4 THE FOREIGN BODY REACTION

The foreign body reaction is a well-described tissue response in the context of biomaterials [31,32], particularly those materials which are composed of nondegradable synthetic and metallic components and intended for long-term implantation. Although the foreign body response is expected and even accepted for many such biomaterials, it is typically associated with negative implications for material longevity and local tissue structure/function relationships. While many of the materials used in the fabrication of tissue-engineered whole organs are transient (degradable) or composed of biologic materials which are not necessarily subject to the foreign body response as described below, it is none-the-less important to fully understand the foreign body response. In the following sections, additional responses which do not fall under the category of a foreign body response will be discussed along with their implications for tissue engineering and regenerative medicine strategies.

The host response following the implantation of a nondegradable synthetic or metallic biomaterial involves a series of overlapping processes that include: blood—material interaction with the deposition of a protein film on the biomaterial, provisional matrix formation, acute inflammation, chronic inflammation, granulation tissue formation, foreign body reaction, and fibrosis and capsule development [32,33]. While many of these processes are similar to those described above for the default host response to tissue injury, there are a number of key differences.

11.4.1 Blood—Material Interaction and Provisional Matrix Formation

The surgical implantation of a biomaterial is invariably associated with tissue damage and disruption of the vasculature at the surgical site. Release of blood into the wound site results in degranulation of platelets, formation of a provisional matrix as described above for wound healing, and signaling that recruits inflammatory cells (i.e., neutrophils and macrophages) to the surgical site. Blood contact also results in adsorption of proteins to the surface of the biomaterial within seconds of implantation [34]. The proteins that adsorb to a biomaterial may include components of the coagulation system (fibrinogen and tissue factors), complement cascade (C5), and other plasma-derived proteins (albumin and IgG) [32,35]. These proteins provide a substrate with which the inflammatory cells arriving at the site of injury interact at the surface of the biomaterial. The specific proteins that attach and the behavior of the attached cells are dependent on the physical and functional nature of the biomaterial surface and on an adsorption/desorption process that is governed by the affinity of the proteins for the biomaterial surface (known as the Vroman Effect) [34,35]. As described briefly below, interactions of cells with the proteins adsorbed to the surface of the biomaterial may lead to a variety of cellular responses including adherence, activation, or triggering of phagocytic pathways, among others, depending on the cell type and the proteins involved [36,37].

11.4.2 Acute Inflammation

Acute inflammation, consisting of the emigration of neutrophils from the vasculature into the implant site, follows the formation of the provisional matrix and the release of chemoattractant factors by platelets and other cells within the inflammatory site, much like the process described above for default wound healing. However, upon arrival within the wound site, neutrophils interact with the proteins adsorbed onto the biomaterial surface through integrin receptors specific for the adsorbed proteins [32]. For example, the adsorption of fibronectin and IgG plays significant roles in the Mac-1-mediated attachment of neutrophils and macrophages to biomaterial surfaces during the acute phase of inflammation [38]. Complement and serum immunoglobulin adsorption to a pathogen (termed opsonization) leads to phagocytosis by neutrophils and/or macrophages, or destruction of the pathogen via the complement pathway. In comparison, an opsonized biomaterial elicits either phagocytosis from neutrophils (and later macrophages) or will be subjected to frustrated phagocytosis, depending on the nature of the biomaterial and its size [32,39]. The process of frustrated phagocytosis involves the extracellular release of microbicidal contents at the surface of a foreign body. This release may cause the erosion of implanted materials, and may eventually lead to the failure of the material to perform as intended.

11.4.3 Chronic Inflammation

The chronic inflammation phase associated with the implantation of a biomaterial is typically characterized by the presence of activated macrophages. This process of macrophage accumulation may occur for a period of days to months depending on the nature of the implanted material and the adsorbed proteins. A meshwork of new ECM is deposited around the biomaterial and an accompanying angiogenic process is prominent. The continued presence of macrophages at the site of biomaterial implantation is often the precursor to the formation of granulation tissue, the foreign body giant cell response, and the eventual encapsulation of the biomaterial within a dense layer of collagenous connective tissue as described below.

11.4.4 Granulation Tissue Formation, Foreign Body Reaction, and Tissue Encapsulation

Chronic inflammation can progress to a granulation tissue phase, in which the deposition of new ECM and the robust angiogenesis into the implantation site are conspicuous. The persistence of granulation tissue combined with the presence of a nondegradable biomaterial is eventually associated with the formation of foreign body giant cells. The classic histologic description of a foreign body reaction consists of macrophages and foreign body giant cells, formed through fusion of macrophages, which are typically located at or near the interface of the host tissue with the biomaterial [32,40]. There are a number of factors including the chemical composition and surface topography that determine protein adsorption to the biomaterial, and the subsequent degree to which a material elicits a foreign body giant cell response [41−43]. As previously stated, macrophages generally interact with protein-adsorbed surfaces through cell surface integrin receptors, the ligation of which induces intracellular signaling cascades that regulate macrophage behavior. Depending on the type of signaling elicited and the immunologic microenvironment, macrophages may fuse with adjacent macrophages forming foreign body giant cells. The exact mechanisms of foreign body giant cell formation are highly complex and have yet to be fully described. An in-depth discussion of the process of foreign body giant cell formation is beyond the scope of this chapter, however, the topic of foreign body giant cell formation as it relates to biomaterials has been reviewed elsewhere [32]. In the final stage of the host response following the implantation of a biomaterial, an increasingly dense layer of collagenous connective tissue is deposited around the surface of the material, thus isolating or "encapsulating" it from the surrounding healthy tissue.

This view of the "classical" host response to biomaterials was developed in the context of the response to primarily synthetic materials intended for long-term implantable applications. With some exceptions, the materials used in tissue engineering and regenerative medicine approaches to whole organ transplantation are composed of naturally derived or degradable synthetic components intended as a transient scaffold for the delivery of cells or as an inductive template for constructive remodeling. The interaction of host innate immune cells with these materials has long been viewed as a detrimental occurrence and a precursor to the response described above (i.e., foreign body reaction and tissue encapsulation with loss of function). However, it is increasingly being recognized that host innate immune cells are not singular inflammatory mediators, and can play important and determinant roles in constructive remodeling following tissue injury. A description of those roles is provided in the following sections.

11.5 MACROPHAGE POLARIZATION

Initially described in the contexts of the host response to pathogen and cancer immunology, macrophages have been shown to have heterogeneous phenotypes ranging from M1 (classically activated, proinflammatory) to M2 (anti-inflammatory, homeostatic, wound healing) [44−49], mimicking the Th1/Th2 pathways as described above. M1 macrophages are activated by well-known proinflammatory signals, such as IFN-γ and LPS; produce characteristic pro-inflammatory cytokines, such as IL-1β, IL-6, IL-12, IL-23, and TNF-α; have low production of anti-inflammatory cytokines like IL-10; produce high levels of reactive oxygen species (ROS); are efficient APCs; and cause the formation of inducer and effector cells in the Th1 pathway. In contrast, M2 cells are activated by molecular cues, such as IL-4, IL-13, IL-10, and immune complexes; produce high levels of IL-10 and have increased the expression of scavenger, mannose, and galactose receptors; produce ornithine and polyamines in place of ROS; and are involved in polarized Th2-type reactions. It should be noted that the term M2 macrophage has evolved to encompass virtually all macrophages that do not fit the M1, classically activated, description [48−50]. However, segregation into two distinct phenotypes is a marked simplification of the *in vivo* reality. Some have described M2 macrophages as falling into one of three distinct subsets termed M2a, M2b, and M2c. These subsets, based upon their observed functions, have been described as alternative, Type II, and deactivated respectively [45]. Others, however, have described macrophages as falling into three categories which include classically activated, wound-healing, and regulatory macrophages [49]. It is likely that, with an increasing

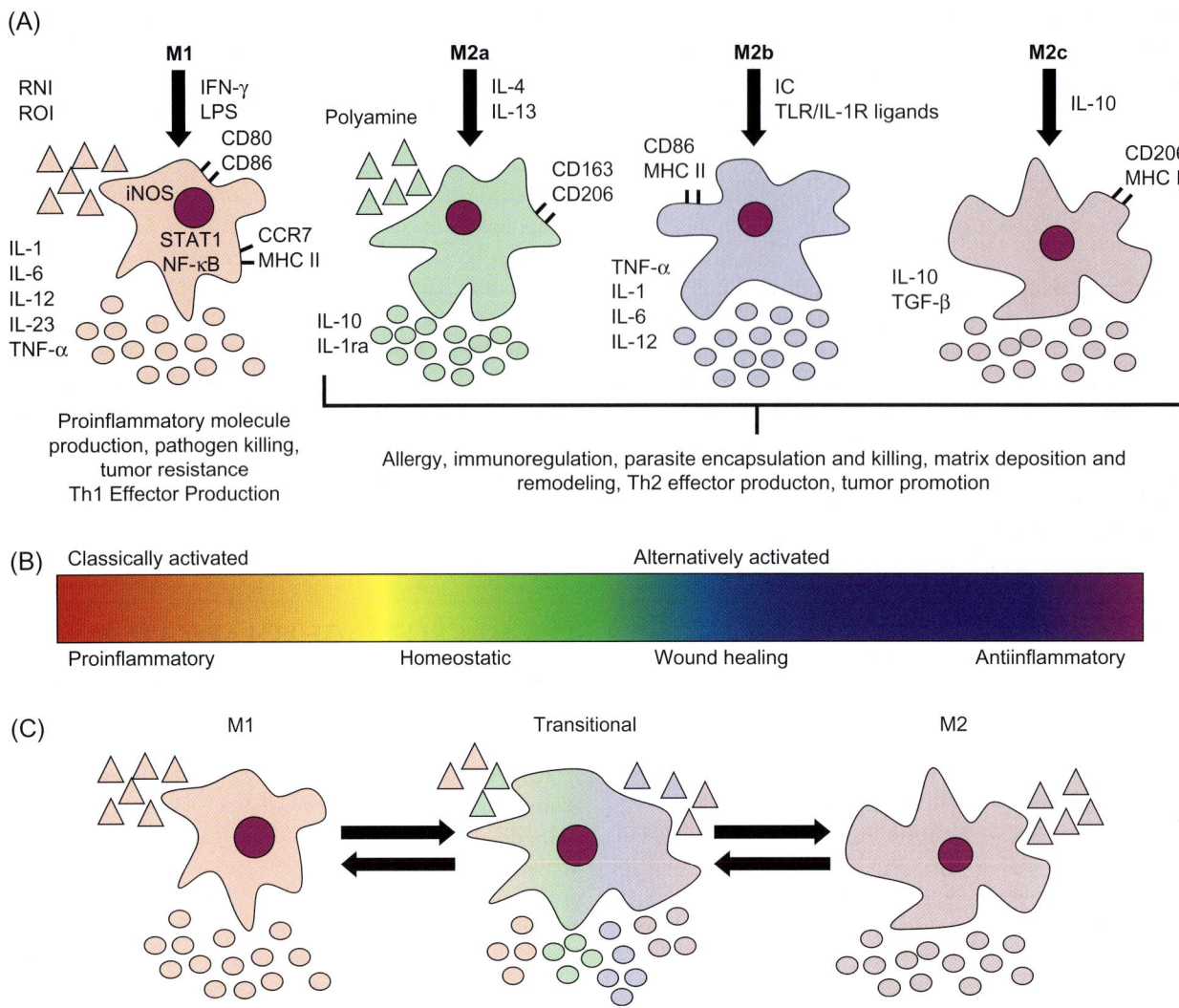

FIGURE 11.1   Macrophage polarization paradigm. (A) Common inducers, surface molecule expression, cytokine and effector molecule production, nuclear translocation, and functional characteristics of recognized macrophage subsets as described by Mantovani et al. [45]. (B) However, this represents a simplified view of macrophage polarization and Mosser et al. have described macrophage polarization as occurring on a spectrum between M1 and M2 extremes [49]. (C) Macrophage phenotype is plastic and can change with paracrine and autocrine signals. Therefore, it is logical that macrophages may adopt a transitional phenotype with characteristics and functions of both M1 and M2 subsets. Triangles represent secreted reactive species and circles represent secreted cytokines. Arg, arginase; C/EPBβ, CCAAT/enhancer-binding protein beta; CCR,C−C chemokine receptor; CD, cluster of differentiation; IC, immune complex; IFN-γ, interferon gamma; IL, interleukin; iNOS, inducible nitric oxide synthase; LPS, lipopolysaccharide; NF-κB, nuclear factor kappa B; MIIC, major histocompatibility class; RNI, reactive nitrogen intermediates; ROI, reactive oxygen intermediates; STAT, signal transducer and activator of transcription; TGF-β, transforming growth factor beta; TLR, toll-like receptor; TNF-α, tumor necrosis alpha. *Reprinted from [51] with permission from Elsevier*

understanding of macrophage phenotype, the number of generally accepted phenotypes will expand. The properties of M1, M2a, M2b, and M2c macrophages, their activating signals, and functional activities are presented in more detail in Figure 11.1A. Macrophage phenotype is plastic and determined by highly complex microenvironments, and therefore likely more accurately considered as a spectrum between the M1 and M2 extremes, where any given cell may express certain components of multiple M1 or M2 phenotypes (Figure 11.1B).

Macrophages, unlike T cells, appear to possess remarkable plasticity once activated (Figure 11.1C). M1 to M2 and M2 to M1 phenotype switching have been observed in multiple studies [52−55]. This plasticity likely represents a protective mechanism by which the host not only can mount an appropriate host response to pathogen but also effectively resolve such a response without excessive local or systemic damage. Dysfunction of macrophage phenotype and dysfunction of macrophage plasticity in particular have been proposed as an underlying mechanism to a number of diseases, including cancer, atherosclerosis, insulin resistance, inflammatory bowel disease, and fibrosis [28−30,46,56−62]. Macrophage phenotype has also been found to be an important modulator of the tissue remodeling process which occurs following injury in skin, skeletal muscle, cardiac tissue, and the central nervous system among others [63−72]. In general, an initial M1-type response is mounted to destroy potential pathogens within the wound site and to debride the wound site of dead cells and damaged tissue. Transition to an M2 phenotype is associated with tissue remodeling resulting in either scar tissue or constructive, functional remodeling as an outcome depending on the timing of the phenotype switch. Prolonged M1 polarization or overly exuberant transition to an M2 phenotype may lead to excessive scarring or a delay in wound healing, respectively. A description of the unique role of macrophage polarization in each of these diseases and tissue remodeling in each of these organs has been reviewed elsewhere [28,56], but it is important to note that an effective and timely switch in macrophage polarization is almost always a key component of a positive outcome.

As a result of the increased understanding of the necessary and determinant role of macrophages in tissue remodeling following injury, a number of studies have begun to apply similar paradigms to the outcomes observed in tissue engineering and regenerative medicine approaches to tissue reconstruction. Briefly, it has been observed that strategies which promote a transition from an initially M1-type response to a more "friendly" M2-type response are better able to promote constructive tissue remodeling and recovery of function than those which promote only an M1 response or lead to a foreign body reaction. An example of this is provided below in the context of ECM scaffold materials for tissue engineering and regenerative medicine.

11.6 THE HOST RESPONSE TO ECM-BASED SCAFFOLDS

As was described in the preceding chapter, ECM scaffold materials have been used in a wide variety of tissue engineering and regenerative medicine-based approaches to tissue and organ restoration. The success of these materials in preclinical and clinical applications appears to be attributable, in large part, to their ability to modulate the default mechanisms of wound healing as described above towards a more functional and constructive remodeling outcome. It is increasingly understood that the modulation of the host immune response, in particular, is essential to the formation of new, functional host tissues following ECM scaffold placement. Therefore, the host response to

ECM-based scaffold materials in tissue engineering and regenerative medicine applications is described herein as an example which demonstrates the importance of the immune response in determining outcomes in tissue engineering and regenerative medicine.

11.6.1 Potential Immune Activating Molecules Within ECM Scaffolds

The preparation of ECM scaffolds for tissue engineering and regenerative medicine applications involves the decellularization of the tissue or organ from which the ECM is to be harvested [73]. The removal of the cellular component produces a different type of "tissue graft" than is typically presented with autogeneic, allogeneic, or xenogeneic whole organ grafts. An ECM scaffold ideally consists primarily of the ECM constituent molecules, many of which have been found to be conserved across species [74], thus mitigating many adverse components of the host immune response [75]. However, many ECM scaffolds have been shown to contain a number of components that are thought to induce adverse host immune and/or rejection-type responses when present in large quantities. These components include the α-Gal epitope and DNA. The α-Gal epitope is known to cause hyperacute rejection of organ transplants [76−79]. However, studies of α-Gal positive ECM scaffold implantation have not shown adverse responses that can be attributed to the α-Gal epitope [80,81]. A recent study investigated the effects of the presence of the α-Gal epitope upon the remodeling of ECM scaffolds in a nonhuman primate model [80]. The study compared the host response to ECM derived from allogeneic, xenogeneic porcine, and xenogeneic α-Gal −/− porcine sources. The results of the study showed that although those animals implanted with an ECM scaffold containing the α-Gal epitope exhibited an increase in serum anti-Gal antibodies, there were no adverse effects of the α-Gal epitope upon the remodeling response. Several studies have shown the presence of DNA fragments remaining within ECM scaffolds following the decellularization and sterilization processes [82−84]. A relevant study examined the presence of DNA within a number of commercially available ECM scaffolds [85]. The results of the study showed that, although all of the products tested contained small amounts of DNA, the remnants generally consisted of fragments of <300 bp. Despite the presence of small amounts of both the α-Gal epitope and DNA within ECM scaffolds, adverse clinical effects have not been observed. The absence of an adverse host response is likely due to the minute amounts of these components present and the rapid degradation of the ECM scaffold. It has been shown that the presence of large amounts of cellular material within an implanted ECM scaffold leads to prolonged inflammation and scar tissue formation as opposed to the modulation of the host response towards a constructive remodeling outcome [86]. Therefore, it is probable that there is a threshold amount of these components required to induce adverse effects upon the remodeling response.

11.6.2 Innate Immune Response to ECM Scaffolds

In general, innate immune cells (neutrophils and macrophages) are the first cells to encounter and respond to implanted biomaterials. The immediate cellular response observed following the implantation of an ECM scaffold consists almost exclusively of neutrophils, as one might expect, but there is also a significant mononuclear cell component. In the absence of large amounts of cellular debris within the scaffold, chemical crosslinking, or contaminants like endotoxin, the neutrophil infiltrate diminishes almost entirely within 72 h and is replaced by a mononuclear cell population. This type of response, characterized by a large infiltration of innate immune cells, has been conventionally interpreted as either acute or chronic inflammation with associated negative implications. However, the presence of these cells, especially mononuclear macrophages, has been shown to be essential to the formation of the type of constructive remodeling response that has been observed following the implantation of ECM scaffolds [86–89].

A histologically similar population of neutrophils and macrophages is observed following the implantation of ECM scaffolds which either have or have not been processed using chemical cross-linking agents, such as glutaraldehyde or carbodiimide; however, the tissue remodeling outcome observed following the implantation of chemically cross-linked ECM scaffolds is distinctly different than is observed with the use of noncross-linked scaffolds [90]. The host tissue response typically observed following the implantation of an acellular ECM scaffold that has not been chemically cross-linked is characterized by a dense infiltration of neutrophils at early time points changing to primarily mononuclear cells thereafter. This infiltrate of innate immune cells is accompanied by rapid degradation of the ECM scaffold and replacement with organized, site-specific, functional host tissue. If the scaffold has been processed using chemical cross-linking agents, such as glutaraldehyde or carbodiimide, the host response is characterized by a similar presence of a large number of neutrophils and macrophages, but results in a persistent pro-inflammatory response consisting of dense fibrous tissue encapsulation and the prolonged presence of a multinucleate giant cell population [90]. Although histomorphologically similar populations of neutrophils

and macrophages are present in the host response to either scaffold type, studies have linked the differences observed in remodeling outcomes, in part, to differences in the phenotype of the host innate immune cells which participate in the host response to implanted ECM scaffolds [87,89].

Recent studies of the M1/M2 profile of the macrophages responding to implanted ECM scaffolds have shown that acellular, noncross-linked ECM scaffolds elicit a predominantly M2-type macrophage response and result in constructive tissue remodeling [87,89]. Chemically cross-linked ECM scaffolds, however, elicit a predominantly M1-type macrophage response and result in a more typical foreign body type of response that includes the deposition of dense collagenous connective tissue and a lack of constructive remodeling. Autograft controls also exhibited a predominance of the M1 phenotype and resulted in scarring. Examples of the histologic response and macrophage phenotype following the implantation of noncross-linked, cross-linked, and autologous test articles can be seen in Figure 11.2. The exact mechanisms by which acellular noncross-linked ECM scaffolds are capable of modulating the default host macrophage response are as yet unknown. However, it is increasingly clear that the M1/M2 polarization profile of the macrophages that participate in the host response to ECM scaffolds is related to the downstream outcome associated with their implantation. Further, the characterization and control of the M1/M2 phenotype may provide a tool by which a constructive and functional tissue remodeling outcome can be predicted and/or promoted.

11.6.3 T-Cell-Mediated Immune Response to ECM Scaffolds

In addition to eliciting a robust, but friendly, host innate immune response, acellular noncross-linked ECM scaffolds have consistently been shown to evoke a Th2-type T-cell response [75,91]. The Th2 response is generally associated with transplant acceptance. One study utilized a mouse model of subcutaneous implantation to examine the T-cell response to xenogeneic muscle tissue, syngeneic muscle tissue, and an acellular ECM scaffold [75]. Results showed that the xenogeneic tissue implant was associated with a response consistent with rejection. That is, the xenogeneic muscle implant showed signs of necrosis, granuloma formation, and encapsulation. The syngeneic tissue and the ECM scaffold elicited an acute inflammatory response that resolved with time and resulted in organized tissue morphology at the remodeling site. Tissue cytokine analysis revealed that the ECM group elicited expression of IL-4 and suppressed the expression of IFN-γ compared to the xenogeneic tissue implants. The ECM group elicited the production of an

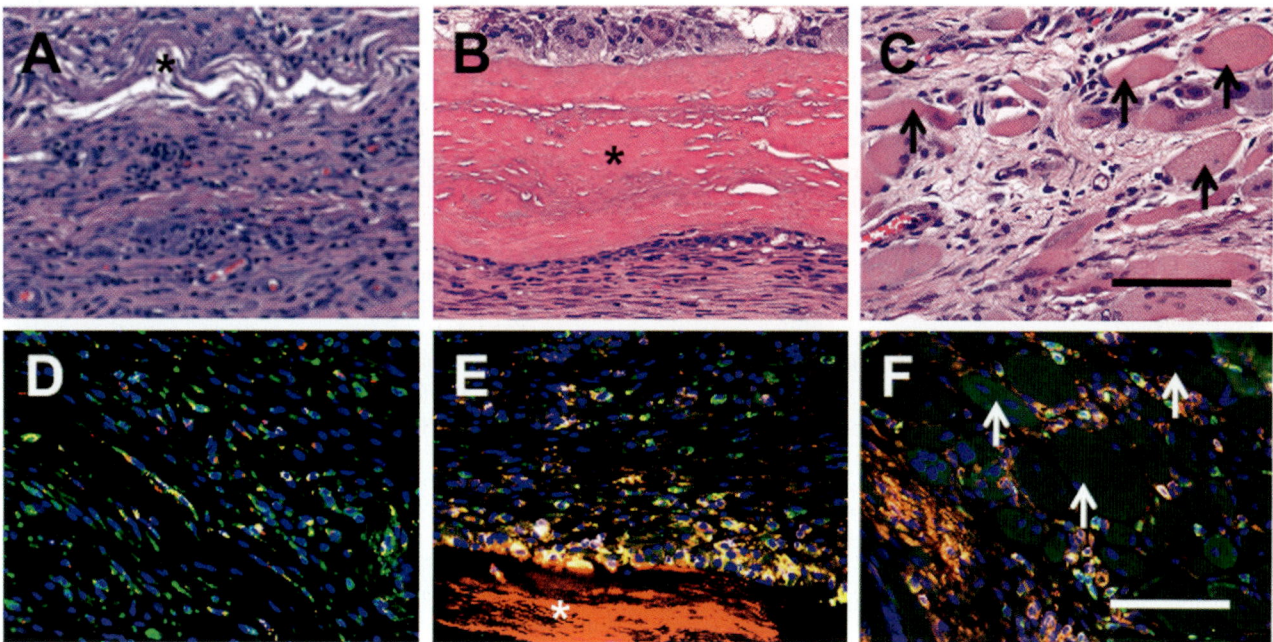

FIGURE 11.2 Host response to noncross-linked ECM (A, D), carbodiimide crosslinked ECM (B, E), and autograft muscle tissue (C, E) implanted within a rat abdominal wall defect at 14 days postimplantation. Macrophages can be observed within the noncross-linked implant (A). A similar population is present around the crosslinked implant (B) and between necrotic bundles of skeletal muscle within the autologous tissue test article (C). The host macrophage response to the noncross-linked ECM test article is predominantly of the M2 (CD206$^+$, green) phenotype, and a lesser number of M1 cells are observed (D). The host macrophage response to the crosslinked ECM test article is predominantly of the M1 (CCR7$^+$, orange) phenotype at the interface with the implant. The host macrophage response to the autologous tissue test article is a mix of both M1 and M2 macrophages. * represents test article remnants and arrow heads represent necrotic skeletal muscle within the autologous tissue test article. Magnification = 40×, scale bars = 100 µm.

ECM-specific antibody response; however, it was restricted to the IgG1 isotype. Reimplantation of the mice with another ECM scaffold led to a secondary anti-ECM antibody response that was also restricted to the IgG1 isotype and there was no evidence of the formation of a Th1-type response. Further investigation confirmed that the observed responses were in fact T-cell dependent. Finally, it has been shown that, while both T and B cells respond to ECM scaffolds, they are not required for acceptance or constructive remodeling of an ECM implant [75]. This tolerable absence of T- or B-cell presence further indicates the importance of the host innate immune response in driving/determining the downstream remodeling outcome following the implantation of an ECM scaffold.

11.6.4 Application to Whole Organ Decellularization and Whole Organ Engineering

The exact mechanisms by which ECM-based scaffold materials are able to modulate the host innate and adaptive immune responses are not yet understood. However,

it is likely that interplay between innate and adaptive arms of the immune response is crucial. It is clear, though, that ECM scaffolds must be prepared properly in order for this immunomodulatory response to occur. For example, cells and associated molecules (DNA, α-Gal epitope) must be removed to avoid the activation of a detrimental immune response [86]. Further, the manner in which the material is processed following decellularization (i.e., chemical-crosslinking) will also play an important role [87,90]. The observation that ECM scaffold materials which have been chemically crosslinked possess an altered surface chemistry and are prevented from degradation, thereby preventing the release of bioactive matricryptic peptides, and also elicit a primarily M1-type immune response with scar tissue formation suggests that both the surface "ligand landscape" and molecules released during scaffold degradation may have potential immunomodulatory effects [87,92,93]. Therefore, and as was discussed in the previous chapter, it will be essential to maintain the composition and structure of the native ECM to the degree possible while removing potentially immunogeneic cellular components. The manner in which this is achieved in whole organ decellularization approaches differs in a number of respects from nonwhole organ approaches to tissue decellularization [94,95]. However, the underlying mechanisms by which whole organ ECM-based approaches may modulate the immune response are likely the same. Additionally, the efficiency and techniques of reseeding of decellularized whole organs will undoubtedly be a significant contributor to their future success. Therefore, careful consideration of both the decellularization process and the re-seeding process is warranted. Those approaches which maintain cellular viability to the highest degree will prevent potentially inflammatory responses to cellular necrosis and will likely result in the most functional outcomes.

11.7 CONCLUSION

While the mechanisms by which the host responds to a tissue-engineered whole organ will likely differ from the mechanisms associated with success or failure of allogeneic whole organ transplantation, the success or failure of tissue engineering and regenerative medicine-based approaches will in turn depend upon the manner in which the host responds to the intervention. The aspects of the host innate or acquired immune system which are activated will depend largely on the constituent cells, biomaterials, and bioactive factors included in the strategy of choice. Approaches which include a logical and informed understanding of the host response will likely result in better and more functional outcomes. Specifically, it is increasingly clear that the modulation and not suppression of the host immune response is essential for constructive remodeling. In particular, and as has been described above in the context of ECM-based approaches, those strategies which elicit more "friendly" Th2 and M2 outcomes will meet with greater success and those which promote a predominantly Th1- and M1-type response will result in poor outcomes.

REFERENCES

[1] Murphy K, et al. 7th ed. Janeway's immunobiology, XXI. New York: Garland Science; 2008. p. 887.

[2] Abbas AK, Lichtman AH, Pillai S. 6th ed. Cellular and molecular immunology, VIII. Philadelphia: Saunders/Elsevier; 2010. p. 566

[3] Strom TB, Roy-Chaudhury P, Manfro R, Zheng XX, Nickerson PW, Wood K, et al. The Th1/Th2 paradigm and the allograft response. Curr Opin Immunol 1996;8:688–93.

[4] Zhai Y, Ghobrial RM, Busuttil RW, Kupiec-Weglinski JW. Th1 and Th2 cytokines in organ transplantation: paradigm lost? Crit Rev Immunol 1999;19:155–72.

[5] Abbas AK, Murphy KM, Sher A. Functional diversity of helper T lymphocytes. Nature 1996;383:787–93.

[6] Matsumiya G, Shirakura R, Miyagawa S, Izutani H, Nakata S, Matsuda H. Assessment of T-cell subsets involved in antibody production and cell-mediated cytotoxicity in rat-to-mouse cardiac xenotransplantation. Transplant Proc 1994;26:1214–6.

[7] Chen N, Gao Q, Field EH. Prevention of Th1 response is critical for tolerance. Transplantation 1996;61:1076–83.

[8] Bach FH, Ferran C, Hechenleitner P, Mark W, Koyamada N, Miyatake T, et al. Accommodation of vascularized xenografts: expression of "protective genes" by donor endothelial cells in a host Th2 cytokine environment. Nat Med 1997;3:196–204.

[9] Chen N, Field EH. Enhanced type 2 and diminished type 1 cytokines in neonatal tolerance. Transplantation 1995;59:933–41.

[10] Piccotti JR, Chan SY, VanBuskirk AM, Eichwald EJ, Bishop DK. Are Th2 helper T lymphocytes beneficial, deleterious, or irrelevant in promoting allograft survival? Transplantation 1997;63:619–24.

[11] Robbins SL, Kumar V, Cotran RS. 8th ed. Robbins and Cotran pathologic basis of disease, XIV. Philadelphia, PA: Saunders/Elsevier; 20108th ed. Robbins and Cotran pathologic basis of disease, XIV. Philadelphia, PA: Saunders/Elsevier; 2010. p. 1450

[12] Diegelmann RF, Evans MC. Wound healing: an overview of acute, fibrotic and delayed healing. Front Biosci 2004;9:283–9.

[13] Guo S, Dipietro LA. Factors affecting wound healing. J Dent Res 2010;89:219–29.

[14] Barrientos S, Stojadinovic O, Golinko MS, Brem H, Tomic-Canic M. Growth factors and cytokines in wound healing. Wound Repair Regen 2008;16:585–601.

[15] Broughton II G, Janis JE, Attinger CE. The basic science of wound healing. Plast Reconstr Surg 2006;117:12–34.

[16] Werner S, Grose R. Regulation of wound healing by growth factors and cytokines. Physiol Rev 2003;83:835–70.

[17] Isenberg JS, Pappan LK, Romeo MJ, Abu-Asab M, Tsokos M, Wink DA, et al. Blockade of thrombospondin-1-CD47 interactions prevents necrosis of full thickness skin grafts. Ann Surg 2008;247:180–90.

[18] Sweetwyne MT, Murphy-Ullrich JE. Thrombospondin1 in tissue repair and fibrosis: TGF-beta-dependent and independent mechanisms. Matrix Biol 2012;31:178–86.

[19] Artuc M, Hermes B, Steckelings UM, Grützkau A, Henz BM. Mast cells and their mediators in cutaneous wound healing—active participants or innocent bystanders? Exp Dermatol 1999;8:1–16.

[20] Adamson R. Role of macrophages in normal wound healing: an overview. J Wound Care 2009;18:349–51.

[21] Erwig LP, Henson PM. Immunological consequences of apoptotic cell phagocytosis. Am J Pathol 2007;171:2–8.

[22] Fadok VA, McDonald PP, Bratton DL, Henson PM. Regulation of macrophage cytokine production by phagocytosis of apoptotic and post-apoptotic cells. Biochem Soc Trans 1998;26:653–6.

[23] Sylvia CJ. The role of neutrophil apoptosis in influencing tissue repair. J Wound Care 2003;12:13–6.

[24] Keen D. A review of research examining the regulatory role of lymphocytes in normal wound healing. J Wound Care 2008;17:218–20.

[25] Knighton DR, Hunt TK, Scheuenstuhl H, Halliday BJ, Werb Z, Banda MJ. Oxygen tension regulates the expression of angiogenesis factor by macrophages. Science 1983;221:1283–5.

[26] LaVan FB, Hunt TK. Oxygen and wound healing. Clin Plast Surg 1990;17:463–72.

[27] Gill SE, Parks WC. Metalloproteinases and their inhibitors: regulators of wound healing. Int J Biochem Cell Biol 2008;40:1334–47.

[28] Murray PJ, Wynn TA. Protective and pathogenic functions of macrophage subsets. Nat Rev Immunol 2011;11:723–37.

[29] Wynn TA, Barron L. Macrophages: master regulators of inflammation and fibrosis. Semin Liver Dis 2010;30:245–57.

[30] Wynn TA. Cellular and molecular mechanisms of fibrosis. J Pathol 2008;214:199–210.

[31] Anderson JM, Jones JA. Phenotypic dichotomies in the foreign body reaction. Biomaterials 2007;28:5114–20.

[32] Anderson JM, Rodriguez A, Chang DT. Foreign body reaction to biomaterials. Semin Immunol 2008;20:86–100.

[33] Ratner BD. 2nd ed. Biomaterials science: an introduction to materials in medicine, XII. Amsterdam, Boston: Elsevier Academic Press; 2004. p. 851

[34] Horbett TA. The role of adsorbed proteins in tissue response to biomaterials. In: Ratner BD, et al., editors. Biomaterials science: an introduction to biomaterials in medicine. San Diego, CA: Elsevier Academic Press; 2004. p. 237–46.

[35] Wilson CJ, Clegg RE, Leavesley DI, Pearcy MJ. Mediation of biomaterial–cell interactions by adsorbed proteins: a review. Tissue Eng 2005;11:1–18.

[36] Jenney CR, Anderson JM. Adsorbed serum proteins responsible for surface dependent human macrophage behavior. J Biomed Mater Res 2000;49:435–47.

[37] Brodbeck WG, Colton E, Anderson JM. Effects of adsorbed heat labile serum proteins and fibrinogen on adhesion and apoptosis of monocytes/macrophages on biomaterials. J Mater Sci Mater Med 2003;14:671–5.

[38] Hu WJ, Eaton JW, Ugarova TP, Tang L. Molecular basis of biomaterial-mediated foreign body reactions. Blood 2001;98:1231–8.

[39] Henson PM. Mechanisms of exocytosis in phagocytic inflammatory cells (Parke-Davis Award Lecture). Am J Pathol 1980;101:494–511.

[40] McNally AK, Anderson JM. Macrophage fusion and multinucleated giant cells of inflammation. Adv Exp Med Biol 2011;713:97–111.

[41] Dadsetan M, Jones JA, Hiltner A, Anderson JM. Surface chemistry mediates adhesive structure, cytoskeletal organization, and fusion of macrophages. J Biomed Mater Res A 2004;71:439–48.

[42] MacEwan MR, Brodbeck WG, Matsuda T, Anderson JM. Monocyte/lymphocyte interactions and the foreign body response: in vitro effects of biomaterial surface chemistry. J Biomed Mater Res A 2005;74:285–93.

[43] Chen S, Jones JA, Xu Y, Low HY, Anderson JM, Leong KW. Characterization of topographical effects on macrophage behavior in a foreign body response model. Biomaterials 2010;31:3479–91.

[44] Gordon S, Taylor PR. Monocyte and macrophage heterogeneity. Nat Rev Immunol 2005;5:953–64.

[45] Mantovani A, Sica A, Sozzani S, Allavena P, Vecchi A, Locati M. The chemokine system in diverse forms of macrophage activation and polarization. Trends Immunol 2004;25:677–86.

[46] Mantovani A, Sozzani S, Locati M, Allavena P, Sica A. Macrophage polarization: tumor-associated macrophages as a paradigm for polarized M2 mononuclear phagocytes. Trends Immunol 2002;23:549–55.

[47] Mills CD, Kincaid K, Alt JM, Heilman MJ, Hill AM. M-1/M-2 macrophages and the Th1/Th2 paradigm. J Immunol 2000;164:6166–73.

[48] Mosser DM. The many faces of macrophage activation. J Leukoc Biol 2003;73:209–12.

[49] Mosser DM, Edwards JP. Exploring the full spectrum of macrophage activation. Nat Rev Immunol 2008;8:958–69.

[50] Gordon S. Alternative activation of macrophages. Nat Rev Immunol 2003;3:23–35.

[51] Brown BN, Badylak SF. Expanded applications, shifting paradigms and an improved understanding of host-biomaterial interactions. Acta Biomater 2013;9:4948–55.

[52] Stout RD, Watkins SK, Suttles J. Functional plasticity of macrophages: in situ reprogramming of tumor-associated macrophages. J Leukoc Biol 2009;86:1105–9.

[53] Porcheray F, Viaud S, Rimaniol AC, Léone C, Samah B, Dereuddre-Bosquet N, et al. Macrophage activation switching: an asset for the resolution of inflammation. Clin Exp Immunol 2005;142:481–9.

[54] Stout RD, Jiang C, Matta B, Tietzel I, Watkins SK, Suttles J. Macrophages sequentially change their functional phenotype in response to changes in microenvironmental influences. J Immunol 2005;175:342–9.

[55] Lawrence T, Natoli G. Transcriptional regulation of macrophage polarization: enabling diversity with identity. Nat Rev Immunol 2011;11:750–61.

[56] Brown BN, Ratner BD, Goodman SB, Amar S, Badylak SF. Macrophage polarization: an opportunity for improved outcomes in biomaterials and regenerative medicine. Biomaterials 2012;33:3792–802.

[57] Johnson JL, Newby AC. Macrophage heterogeneity in atherosclerotic plaques. Curr Opin Lipidol 2009;20:370–8.

[58] Mantovani A, Garlanda C, Locati M. Macrophage diversity and polarization in atherosclerosis: a question of balance. Arterioscler Thromb Vasc Biol 2009;29:1419–23.

[59] Mantovani A, Sica A. Macrophages, innate immunity and cancer: balance, tolerance, and diversity. Curr Opin Immunol 2010;22:231–7.

[60] Moore KJ, Tabas I. Macrophages in the pathogenesis of atherosclerosis. Cell 2011;145:341–55.

[61] Odegaard JI, Chawla A. Alternative macrophage activation and metabolism. Annu Rev Pathol 2011;6:275–97.

[62] Olefsky JM, Glass CK. Macrophages, inflammation, and insulin resistance. Annu Rev Physiol 2010;72:219–46.

[63] Rodero MP, Khosrotehrani K. Skin wound healing modulation by macrophages. Int J Clin Exp Pathol 2010;3:643–53.

[64] Deonarine K, Panelli MC, Stashower ME, Jin P, Smith K, Slade HB, et al. Gene expression profiling of cutaneous wound healing. J Transl Med 2007;5:11.

[65] Tidball JG, Villalta SA. Regulatory interactions between muscle and the immune system during muscle regeneration. Am J Physiol Regul Integr Comp Physiol 2010;298:1173–87.

[66] Arnold L, Henry A, Poron F, Baba-Amer Y, van Rooijen N, Plonquet A, et al. Inflammatory monocytes recruited after skeletal

muscle injury switch into antiinflammatory macrophages to support myogenesis. J Exp Med 2007;204:1057–69.

[67] Villalta SA, Rinaldi C, Deng B, Liu G, Fedor B, Tidball JG. Interleukin-10 reduces the pathology of mdx muscular dystrophy by deactivating M1 macrophages and modulating macrophage phenotype. Hum Mol Genet 2011;20:790–805.

[68] Villalta SA, Nguyen HX, Deng B, Gotoh T, Tidball JG. Shifts in macrophage phenotypes and macrophage competition for arginine metabolism affect the severity of muscle pathology in muscular dystrophy. Hum Mol Genet 2009;18:482–96.

[69] Kigerl KA, Gensel JC, Ankeny DP, Alexander JK, Donnelly DJ, Popovich PG. Identification of two distinct macrophage subsets with divergent effects causing either neurotoxicity or regeneration in the injured mouse spinal cord. J Neurosci 2009;29:13435–44.

[70] Rolls A, Shechter R, Schwartz M. The bright side of the glial scar in CNS repair. Nat Rev Neurosci 2009;10:235–41.

[71] Schwartz M. "Tissue-repairing" blood-derived macrophages are essential for healing of the injured spinal cord: from skin-activated macrophages to infiltrating blood-derived cells? Brain Behav Immun 2010;24:1054–7.

[72] Anzai A, Anzai T, Nagai S, Maekawa Y, Naito K, Kaneko H, et al. Regulatory role of dendritic cells in postinfarction healing and left ventricular remodeling. Circulation 2012;125:1234–45.

[73] Gilbert TW, Sellaro TL, Badylak SF. Decellularization of tissues and organs. Biomaterials 2006;27:3675–83.

[74] van der Rest M, Garrone R. Collagen family of proteins. FASEB J 1991;5:2814–23.

[75] Allman AJ. Xenogeneic extracellular matrix grafts elicit a TH2-restricted immune response. Transplantation 2011;71:1631–40.

[76] Collins BH. Mechanisms of injury in porcine livers perfused with blood of patients with fulminant hepatic failure. Transplantation 1994;58:1162–71.

[77] Cooper DK, Good AH, Koren E, Oriol R, Malcolm AJ, Ippolito RM, et al. Identification of alpha-galactosyl and other carbohydrate epitopes that are bound by human anti-pig antibodies: relevance to discordant xenografting in man. Transpl Immunol 1993;1:198–205.

[78] Galili U, Macher BA, Buehler J, Shohet SB. Human natural anti-alpha-galactosyl IgG. II. The specific recognition of alpha (1−3)-linked galactose residues. J Exp Med 1985;162:573–82.

[79] Oriol R, Ye Y, Koren E, Cooper DK. Carbohydrate antigens of pig tissues reacting with human natural antibodies as potential targets for hyperacute vascular rejection in pig-to-man organ xenotransplantation. Transplantation 1993;56:1433–42.

[80] Daly K, Stewart-Akers AM, Hara H, Ezzelarab M, Long C, Cordero K, et al. Effect of the alphaGal Epitope on the response to small intestinal submucosa extracellular matrix in a nonhuman primate model. Tissue Eng Part A 2009;15:3877–88.

[81] Raeder RH, Badylak SF, Sheehan C, Kallakury B, Metzger DW. Natural anti-galactose alpha1,3 galactose antibodies delay, but do not prevent the acceptance of extracellular matrix xenografts. Transpl Immunol 2002;10:15–24.

[82] Derwin KA, Baker AR, Spragg RK, Leigh DR, Iannotti JP. Commercial extracellular matrix scaffolds for rotator cuff tendon repair. Biomechanical, biochemical, and cellular properties. J Bone Joint Surg Am 2006;88:2665–72.

[83] Roberts R, Gallagher J, Spooncer E, Allen TD, Bloomfield F, Dexter TM. Heparan sulphate bound growth factors: a mechanism for stromal cell mediated haemopoiesis. Nature 1988;332:376–8.

[84] Zheng MH, Chen J, Kirilak Y, Willers C, Xu J, Wood D. Porcine small intestine submucosa (SIS) is not an acellular collagenous matrix and contains porcine DNA: possible implications in human implantation. J Biomed Mater Res B Appl Biomater 2005;73:61–7.

[85] Gilbert TW, Freund JM, Badylak SF. Quantification of DNA in biologic scaffold materials. J Surg Res 2009;152:135–9.

[86] Brown BN, Valentin JE, Stewart-Akers AM, McCabe GP, Badylak SF. Macrophage phenotype and remodeling outcomes in response to biologic scaffolds with and without a cellular component. Biomaterials 2009;30:1482–91.

[87] Brown BN, Londono R, Tottey S, Zhang L, Kukla KA, Wolf MT, et al. Macrophage phenotype as a predictor of constructive remodeling following the implantation of biologically derived surgical mesh materials. Acta Biomater 2012;8:978–87.

[88] Valentin JE, Stewart-Akers AM, Gilbert TW, Badylak SF. Macrophage participation in the degradation and remodeling of extracellular matrix scaffolds. Tissue Eng Part A 2009;15:1687–94.

[89] Badylak SF, Valentin JE, Ravindra AK, McCabe GP, Stewart-Akers AM. Macrophage phenotype as a determinant of biologic scaffold remodeling. Tissue Eng Part A 2008; 14:1835–42.

[90] Valentin JE, Badylak JS, McCabe GP, Badylak SF. Extracellular matrix bioscaffolds for orthopaedic applications. A comparative histologic study. J Bone Joint Surg Am 2006; 88:2673–86.

[91] Allman AJ, McPherson TB, Merrill LC, Badylak SF, Metzger DW. The Th2-restricted immune response to xenogeneic small intestinal submucosa does not influence systemic protective immunity to viral and bacterial pathogens. Tissue Eng 2002;8:53–62.

[92] Brown BN, Barnes CA, Kasick RT, Michel R, Gilbert TW, Beer-Stolz D, et al. Surface characterization of extracellular matrix scaffolds. Biomaterials 2010;31:428–37.

[93] Reing JE, Zhang L, Myers-Irvin J, Cordero KE, Freytes DO, Heber-Katz E, et al. Degradation products of extracellular matrix affect cell migration and proliferation. Tissue Eng Part A 2009;15:605–14.

[94] Badylak SF, Taylor D, Uygun K. Whole-organ tissue engineering: decellularization and recellularization of three-dimensional matrix scaffolds. Annu Rev Biomed Eng 2011;13:27–53.

[95] Crapo PM, Gilbert TW, Badylak SF. An overview of tissue and whole organ decellularization processes. Biomaterials 2011;32:3233–43.

Chapter 12

Harnessing Regenerative and Immunomodulatory Properties of Mesenchymal Stem Cells in Transplantation Medicine

Jaehyup Kim[a] **and Peiman Hematti**[b]

[a]*Department of Medicine, University of Wisconsin-Madison, School of Medicine and Public Health, Madison, WI,* [b]*University of Wisconsin Carbone Cancer Center, Madison, WI*

12.1 INTRODUCTION

The cells currently known as mesenchymal stem (or stromal) cells (MSCs) were first described by Friedenstein et al. as an adherent, fibroblast-like population of cells cultured from the bone marrow (BM) of rodent animals more than four decades ago [1]. Through a series of elegant experiments, he showed that when implanted (e.g., under the kidney capsule), these cells generated rudimentary bone tissue capable of supporting hematopoiesis. Friedenstein's method was subsequently used to derive similar cells from the human BM [2] and further research showed that these cells were capable of supporting hematopoiesis in long-term culture assays [3]. Later, Caplan proposed that these cells were stem cells for the mesenchymal tissues, and therefore coined the term "mesenchymal stem cells" [4]. He also

proposed that these MSCs could have therapeutic potential in the regeneration of various tissues. While isolation and *ex vivo* expansion of MSCs from small amounts of tissue are relatively easy, MSCs are postulated to comprise only a small population (<0.01%) of adult BM cells [5]. The scarcity of MSCs *in vivo* give rise to one of the major challenges in this field which is the lack of any reliable or widely accepted marker for direct isolation of MSCs from the BM aspirates. Several markers, such as Stro-1, CD271, and CD146, have been proposed for direct isolation of MSCs, but there is still no consensus regarding which marker is the most representative of cells present in their *in situ* BM environment. The low number of MSCs in BM also means that for any *in vivo* use, be it experimental or clinical, the cells have to be expanded *ex vivo* to generate a sufficient number of cells. MSCs from the BM are most commonly isolated using density gradient separation and plating of BM mononuclear cells in culture plates and allowing the attached cells to grow and expand. Upon reaching near confluency the adherent cells are passaged, which leads to expansion of a homogenous, fibroblast-looking population of cells [6]. In 2006 to provide harmonization in the field, the International Society for Cellular Therapy (ISCT) proposed criteria to characterize MSCs: (a) expression of a certain set of markers (i.e., CD105, CD73, CD90) and lack of expression of markers associated with hematopoietic lineages (i.e., CD45, CD34, CD14, CD11b, CD79a, CD19, and HLA-DR) and (b) the ability to differentiate into osteoblasts, adipocytes, and chondroblasts *in vitro*[7]. These criteria are now widely accepted and effectively define a seemingly homogenous population of cells, but still it is widely recognized that the population of cells defined in this manner are functionally heterogeneous and are comprised of various subpopulations with different differentiation capabilities at a clonal level [8]. This chapter discusses the use of *ex vivo* culture-expanded MSCs in different clinical settings with special focus on transplantation medicine and solid organ transplantation. It also addresses the rationale behind the use of MSCs, their immunomodulatory and tissue regenerative properties, practical issues to consider in clinical applications of MSCs, and evolving concepts in this dynamic field.

12.2 USE OF MSCS IN HEMATOPOIETIC STEM CELL TRANSPLANTATION

12.2.1 Background

Hematopoietic stem cell (HSC) transplantation was the first and remains the only form of stem cell therapy accepted as standard of care for certain conditions. The sources of HSCs include the BM aspirate, HSCs pharmacologically mobilized into peripheral blood followed by their isolation using leukopheresis, and HSCs from umbilical cord blood units [9]. Autologous or allogeneic HSCs are used to replace diseased HSCs of the recipients, usually with the goal of treating hematologic malignancies or immunodeficiency disorders. However, despite decades of clinical experience, HSC transplantation continues to be a high-risk procedure with a high rate of morbidity and mortality related to the toxic effects of pretransplant preparative regimen, HSC graft failure and/or rejection, and immunologically mediated phenomenon of graft-versus-host disease (GVHD). Indeed, the latter is the main reason preventing the use of HSC transplantation as a way to induce tolerance after solid organ transplantation.

Since MSCs are generally considered to be an essential constituent of the BM stromal microenvironment with a critical role in supporting hematopoiesis [10], HSC transplant physicians were the first to test the use of MSCs in a clinical setting. The Lazarus team was the first group to conduct a phase I clinical trial to show the safety of intravenous infusion of autologous BM-derived MSCs that were culture-expanded *ex vivo* over several weeks from small volume of BM aspirates [11]. This seminal clinical study, which showed that intravenous administration of autologous *ex vivo* generated MSCs is safe and does not cause any adverse reactions, paved the way for further use of MSCs in the context of HSC transplantation. Furthermore, recently there has been an exponential increase in use of MSCs in a wide variety of other disorders and pathologies.

12.2.2 MSCs for Supporting Hematopoiesis

Expediting recovery of hematopoiesis after HSC transplantation is important for decreasing morbidity and mortality associated with the procedure and has been the subject of intensive investigation for decades. MSCs have emerged as an attractive cell therapy for this purpose, as MSCs are presumed to play a major role in supporting hematopoiesis in the BM [12]. Thus, it was natural that promoting engraftment of HSCs was one of the earliest indications for which MSCs were investigated. Koc et al. reported the result of a phase I–II clinical trial on breast cancer patients where *ex vivo* culture-expanded autologous MSCs were infused at the time of autologous HSC transplantation [13]. In another study by Lazarus' group, mononuclear cells isolated from BM of human leukocyte antigen (HLA)-identical sibling donors were used to drive allogeneic MSCs, which were then culture-expanded for infusion into the respective allogeneic HSC transplant recipients [14]. The result from these studies provided further reassurance to the scientific community that the use of culture-expanded autologous or allogeneic MSCs as a form of cellular therapy is feasible and safe, although their efficacy in promoting engraftment of HSCs could not be concluded in a definitive matter from these studies. Subsequent studies have continued to investigate the use of autologous or allogeneic *ex vivo* culture-expanded, BM-derived MSCs in the context of HSC transplantation for reducing the risk of graft failure, preventing repeat rejections, or rescuing graft failure [9]. Most of these studies have provided encouraging results but the clinical effectiveness of MSCs in these clinical settings has not been conclusively proven through a randomized clinical trial [15,16]. Furthermore, due to the very low risk of graft failure in recent studies, the significant decrease in morbidity during the posttransplant cytopenic phase, and the use of less toxic conditioning regimens, one could argue that the use of MSCs to accelerate posttransplant recovery might not be a cost-effective approach anymore.

Nevertheless, there are still many reports appearing in the literature on the ability of MSCs to support hematopoiesis in different settings [17,18].

12.2.3 MSCs for Treatment of GVHD

In addition to their role as elements of hematopoietic supportive stroma, interaction of MSCs with various immune cells and their ability to modulate the immune responses *in vitro* and *in vivo* have been the subject of extensive investigation over the last decade [19–21]. MSCs have been shown to suppress proliferation of activated T lymphocytes while increasing the number of regulatory T lymphocytes. MSCs also decrease activation and proliferation of B lymphocytes while suppressing cytotoxicity of natural killer (NK) cells and maturation of dendritic cells. They also modulate activity of neutrophils and changes in the immunophenotype of macrophages. Among these, suppression of T-cell proliferation and activation received much attention due to the importance of T lymphocytes in various immune function [21]. It is important to note that these immunosuppressive effects of MSCs are independent of the HLA compatibility status of the donor and recipient cells [22], supporting the notion that these cells could be used clinically without worrying about histocompatibility issues which enable banking large number of cells from a single donor for use in a multitude of patients later. Thus, the effect of MSCs in the setting of acute GVHD, which is mainly a T-cell-mediated process, was among the first target applications for MSCs to receive serious attention [23]. Le Blanc et al. were the first to investigate the potential of MSC infusion for the treatment of refractory GVHD in a 9-year-old boy who had received a HLA-matched, unrelated donor HSC transplant for leukemia [24]. Infusion of *ex vivo* expanded MSCs generated from the patient's mother, not the original HSC donor, resulted in the resolution of GVHD symptoms that were refractory to several lines of treatments and high dose corticosteroids. Subsequent reports on larger studies from the same group verified the safety of MSC administration to patients with steroid-refractory GVHD from HLA-identical siblings, haplo-identical family donors, and unrelated mismatched donors [25]. These studies showed that infusions of MSCs resulted in a significantly improved survival rate in responders, and responses were again shown to be independent of the HLA compatibility between the donors and recipients. The latter point is very important, as the generation of patient-specific MSCs is very time-consuming, costly, and in many instances impractical due to the urgent nature of the need for their use. The use of MSCs from a donor without the need for HLA typing allows the use of MSCs as an off-the-shelf, universal cell product, making them a very popular type of stem cell therapeutic.

After these exciting initial reports, various groups started using BM-derived MSCs for the treatment of GVHD in diverse patient populations. It must be noted that in these trials different doses of MSCs and frequencies of administration were tested along with various methodologies for their production [9]. What has emerged from these studies is the fact that there is now almost unanimous agreement regarding the safety of the use of MSCs in these very sick patients. However, there is no consensus on the clinical efficacy of those cells, as the results are based on mostly small and nonrandomized clinical trials, so occasionally the results can even be contradictory. The largest clinical trials performed with MSCs to date have been two Phase III, double-blind, placebo-controlled, randomized trials evaluating a proprietary formulation of MSCs (trade name Prochymal) derived from the marrow of a single third-party donor as a first-line treatment for acute GVHD or for the treatment of refractory acute GVHD. These multicenter, international clinical trials sought to assess the safety and efficacy of Prochymal using a double-blinded, placebo-controlled (in a 2:1 actual treatment to placebo ratio) design. Highly anticipated results from these studies failed to provide data supporting the clinical efficacy of Prochymal in GVHD with regard to survival as the primary end point. This result was received with great surprise by most transplant physicians and investigators in the field. It must be noted that several potentially confounding factors, which include the differences in treatment regimens administered to patients in conjunction with Prochymal, might be responsible for the observed lack of effect. However, the use of the same Prochymal formulation in pediatric patients with severe refractory acute GVHD resulted in more promising results [26] and approval of Prochymal for pediatric patients with GVHD in Canada and New Zealand as the first form of "off-the-shelf universal stem cell therapy product." Prochymal has not yet been approved in the United States.

12.3 USE OF MSCS IN NON-HSC TRANSPLANT SETTINGS

Over the last decade, MSCs have generated a lot of excitement in other disciplines of medicine as well, and their use has now expanded beyond those related to HSC transplantation. Indeed, it can be argued that very few cell types have attracted so much attention in the field of cellular therapy as have MSCs. Originally much of this enthusiasm stemmed from the assumption that MSCs are capable of differentiating not only into the mesenchymal tissues, such as bone and cartilage, but also into other types of cells, such as cardiomyocytes, hepatocytes, and pancreatic islets [27]. Several groups reported the unorthodox plasticity of MSCs both *in vitro* and in preclinical models, and these results seemed to provide support for the ability of MSCs to transdifferentiate. However, subsequent research failed to replicate those findings, and it is now accepted that those early observations of transdifferentiation of MSCs into various tissues were due to imperfect experimental tools, the use of animal models that failed to adequately represent human biology, or other unusual mechanisms like cell fusion [28]. By the time scientific community reached a consensus regarding the lack of transdifferentiation potential of MSCs, hundreds of patients had already been recruited for numerous clinical trials with many of them reporting promising preliminary results. One reason for the initial successful results was the small scale and the nonrandomized format of those trials, but these positive results might be explained by other modes of action for MSCs that initially were not appreciated. Indeed, the discovery of these hitherto unrecognized mechanisms of actions precipitated a major paradigm shift in the field. Currently, tropism of MSCs for migration into sites of tissue damage or inflammation, their capability to support and stimulate proliferation and/or survival of resident tissue progenitor cells through secretion of a variety of cytokines and chemokines, and their contribution to angiogenesis in tissues are proposed to explain some of the observed effects of MSCs [29,30]. These newly discovered modes of action, in addition to their previously recognized immunomodulatory and anti-inflammatory properties, form the basis for the current continuation of investigating MSCs for use in a growing number of human pathologies. At this point, the list of phase I–II trials involving MSCs includes a variety of nonhematological, indications including myocardial infarction, chronic obstructive pulmonary disease, amyotrophic lateral sclerosis, stroke, multiple sclerosis, Crohn's disease, diabetes mellitus, systemic sclerosis, systemic lupus erythematosus, and refractory wounds among others [31,32]. In many of these studies, allogeneic third-party MSCs were used without any immunosuppression, as MSCs are assumed to work across major histocompatibility complex barriers in humans and escape destruction by cytotoxic T cells or NK cells [33]. Again, this assumption has been the cornerstone for the use of "off-the-shelf *ex vivo* culture-expanded BM-derived MSCs" from "third-party" donors, and is supported by the large body of clinical experience that has confirmed the safety of such MSCs. This use of "off-the-shelf," ready-to-use MSCs as a therapeutic entity allows large-scale production and commercialization of cellular therapeutics by biopharmaceutical entities, which is virtually impractical with patient-specific MSCs due to cost and logistical concerns [34]. This also provides the capability to deliver MSCs in acute conditions, such as myocardial infraction or stroke.

12.4 ROLE OF MSCS IN SOLID ORGAN TRANSPLANTATION

12.4.1 The Immunomodulatory Properties of MSCs

Solid organ transplantation is one of the major triumphs of modern medicine and a lifesaving therapy for many patients with end-stage organ damage. The development of immunosuppressive drugs for prevention of rejection of transplanted tissues and organs could be considered as a turning point for this field that converted it from an experimental procedure to a standard of care. However, long-term administration of immunosuppressive medications has potential unfavorable consequences including increased susceptibility to opportunistic infections and malignancies, and damaging effects on different organ functions like kidney toxicity. Thus, the development of nontoxic immunosuppressive-sparing strategies could have a major impact on improving the long-term outcome for transplant recipients. A growing body of basic, preclinical, and clinical research studies suggests that MSCs may fulfill this need due to a combination of immunomodulatory and tissue protective effects.

MSCs interact, directly or indirectly, with a wide range of cell types that play a role in various inflammatory and immune-mediated diseases. While MSCs have been reported to interact with almost all different cells of the immune system, their effect on T cells had been considered as the main pathway by which MSCs exert their immunomodulatory effects due to the prominent role of T cells in adaptive immune responses in general. This also seems to be the case with solid organ transplantation, as expansion of T cells and activation of their effector functions play a paramount role in transplant rejection. MSCs have been shown to suppress the proliferation of effector T cells in response to various cellular and nonspecific mitogen stimulation [35] and hinder antigen recognition of naïve and memory T lymphocytes [36] via various factors, including indoleamine 2,3-dioxygenase (IDO), prostaglandin E2 (PGE2), transforming growth factor beta (TGF-beta), and hepatocyte growth factor (HGF). There are also reports about the role of contact inhibition in MSC-mediated suppression of lymphocyte proliferation [37], and Augello et al. proposed that direct cell-to-cell contact and activation via programmed death ligand 1 (PD-L1) mediates the inhibition of lymphocyte proliferation observed in cocultures with MSCs [38]. However, it is unlikely that a single pathway will be able to explain this whole phenomenon, and thus a combination of a multitude of factors, both cell-to-cell contact-mediated and paracrine-related, is likely to be responsible. No doubt that further research in this area will yield a wealth of information that could be potentially targeted for therapeutic applications.

Because of the important role of regulatory T cells in transplantation biology, modulation of their activity by MSCs has also received major attention. MSCs have been shown to increase the level of regulatory $CD4^+$ CD25 high $FoxP3^+$ T cells and the potential mediators of this effect are suggested to be CCR8, HLA-G, PGE2, and TGF-beta [39]. English et al. also suggested that in cocultures of MSCs with purified T cells, cell-to-cell contact was required, and coculturing MSCs with unpurified mononuclear cells increased the number of regulator T cells which suggest that multicellular interactions play a role in upregulation of regulatory T cells [39]. In a study using a mouse kidney transplantation model, MSCs were shown to increase regulatory T cells, which in turn prolonged the survival of the graft via induction of IDO and associated increase of $D4^+CD25^+FoxP3^+$ regulatory T cells [40].

While the interaction of MSCs with T cells has attracted the most attention, MSCs have been shown to interact with other immune cells as well. MSCs have been shown to inhibit B-cell differentiation and proliferation which is similar to what have been observed with T lymphocytes. Asari et al. reported that MSCs suppressed the terminal differentiation of B cells as shown by decreased percentage and increased apoptosis of CD138 (+) cells along with increased production of IgM [41], and a similar finding was reported by Rasmusson et al. [42]. However, the latter also reported that immunoglobulin G production could either increase or decrease based on the context (such as LPS stimulation, cytomegalovirus, or varicella zoster virus stimulation), which suggests that MSCs may not be fully suppressive in certain situations. Decreased proliferation of B cells was found to be mediated through cell cycle arrest [43], and this could serve as another mechanism of immunomodulation by MSCs in addition to secretion of cytokines or growth factors. The effect of MSC on B lymphocytes is important in organ transplantation since antibody-mediated rejection is increasingly recognized as a major player in transplant rejection [44], and while it was first observed in the kidney transplantation, it is now accepted that this phenomenon could occur in almost all types of transplantation [45]. The potential role of B-cell contribution to organ rejection and pathology as well as the modulation of B-cell function by MSCs could support the use of MSCs as a strategy for immune modulation after solid organ transplantation. However, MSCs have also been shown to increase the viability of B cells, so the effect of MSCs may not be entirely inhibitory on B cells [43]. For example, Traggiari et al. reported that MSCs induced proliferation and maturation of B cells in contrast to the reports mentioned above [46], showing that the effect might be dependent on the context of interaction and caution is necessary in extrapolating *in vitro* experiments to *in vivo*

since multitudes of cells will be interacting an in actual clinical setting. All these findings point to the uncertainty regarding the ultimate role and potential mechanisms of action of MSCs in solid organ transplantation.

The effects of MSCs on immune cells other than lymphocytes have also been studied, but to a much lesser degree, so further work is required in this area. MSCs have been shown to inhibit proliferation and terminal differentiation of NK cells [47,48], but the inhibition of proliferation was more pronounced for resting NK cells than activated NK cells. Again, factors known to be involved in interaction of MSCs with other immune cells, such as TGF-beta, PGE2, and IDO, are thought to mediate the effect of MSCs on NK cells too [49,50] as blocking of those factors restored the function of NK cells [47]. Dendritic cells play a key role in transplantation as antigen-presenting cells and are important component of immune response in general. So far researchers have shown that MSCs affect dendritic cells in a variety of ways, including induction of regulatory phenotypes and suppression of differentiation [51,52]. MSCs can also suppress the maturation of monocytes into dendritic cells [53], which provide another venue for suppression of dendritic cell function by blocking supply of new dendritic cells. Jiang et al. reported that dendritic cells cocultured with MSCs were less efficient in stimulating allogeneic T cells [54] and studies by Zhang et al. suggest that MSCs drive dendritic cells toward Jagged-2 dependent regulatory phenotype, possibly via cell-to-cell contact [55].

Lastly, the interaction of MSCs and macrophages deserves special attention because more recent studies suggest that the beneficial effects of MSCs could be mediated via their effects on macrophages. In a mouse model of sepsis, Nemeth et al. showed MSC increase survival of mice with sepsis induced by colon perforation, but depletion of macrophages eliminated this survival benefit [20]. This study provided evidence for the increased level of IL-10 mediating improved survival. We have also shown that coculturing macrophages with MSCs increased the levels of IL-6 and IL-10 and decreased the levels of IL-12 and TNF-alpha, as measured by intracellular staining of $CD14^+$ monocyte-derived macrophages [19]. Other researchers reported similar findings where MSCs from BM or gingiva induced a regulatory phenotype in macrophages in mouse models [56,57], lending support to the idea that MSCs polarize the immunophenotype of macrophages toward a phenotype similar to alternatively activated or regulatory macrophages. As macrophages are present in almost all the tissues in the body and are considered to be major mediators of inflammation and its subsequent resolution, the interactions between MSCs and macrophages could have significant implications in the therapeutic potential of MSCs in transplantation medicine.

It has to be recognized that differences in biological properties of MSCs derived from different species, differences in immunological assays used, and different animal transplantation models could contribute to the occasional contradictory results among reported studies. The ultimate potential of MSCs for any indication and their effect on immune systems could only be realized through systematic investigation of changes in the immune repertoire of patients after MSC transplantation which is especially important since many reports show that the effect of MSCs can be dependent on the context and unlike *in vitro* system where two cell populations can be isolated and studied, interactions *in vivo* probably entail involvement of several types of cells at the same time.

12.4.2 The Tissue Regenerative Properties of MSCs

MSCs are thought to secrete various growth factors and cytokines and provide supportive signals that could help to regenerate damaged tissues and restore tissue homeostasis. Indeed, several groups reported beneficial effects of MSCs in the setting of damaged tissue repair. For example, Peng et al. used a pig myocardial infarction model to study the effect of autologous MSCs injected into the coronary artery and observed improvement of cardiac function using various imaging modalities [58]. MSCs were found to have beneficial effects in chronic obstructive pulmonary disease as well using a chronic thromboembolic pulmonary hypertension rat model where intratracheal injection of allogeneic MSCs resulted in improvement of clinical parameters associated with lung function and hemodynamics in addition to changes in proteome which suggested tissue reconstruction [59]. In another study, Hoffman et al. reported the derivation of lung MSCs from mice and compared them to BM MSCs. They found several differences between these two types of MSCs and reported that while there are differences in gene expression between BM MSCs and lung MSCs, both were equally active in reducing tissue injury induced by elastase [60]. The question regarding whether MSCs derived from the same organ targeted for therapy is superior to commonly used MSCs like BM MSC remains to be determined in future research and will be discussed in a later section of this chapter. Beneficial effect of MSCs in the setting of tissue damage has also been observed in the several kidney injury models in nontransplant settings. For example, Morigi et al. reported that injection of human cord blood MSCs into a cisplatin-induced acute kidney injury mouse model prolonged survival by protecting renal function and preventing tubular damage [61]. Similar findings have been reported by Altun et al., where a rat ischemia/reperfusion kidney injury model was used

[62]. They found that MSCs and darbopoietin were beneficial and the combination of both provided additional benefit in preventing tissue damage. These results point to the beneficial role of MSCs on organ function and increased interest in the use of MSCs not only to induce favorable modulation of immune responses but also to contribute toward tissue regeneration and restoration of function via nonimmunologically mediated effects.

12.4.3 Preclinical Data Supporting Use of MSCs in Solid Organ Transplantation

Available *in vivo* data support the notion that infusion of MSCs could prolong survival of transplanted organs in various animal models. For example, Bartholomew et al. reported that MHC mismatched MSCs prolonged survival of autologous, donor, and third-party skin grafts in a baboon model potentially via suppression of lymphocyte proliferation [63]. In the case of heart transplantation, Ge et al. reported that MSCs injected 24 h after heart transplantation from C57BL/6 to balb/c mice almost doubled the survival of transplanted organs and when rapamycin was added to MSCs, there was an even more prolonged survival of over 100 days [64]. One important observation from this group is that the recipients developed tolerance toward the donor skin but not to skin from third-party origin, suggesting that MSCs induced tolerance specifically toward the donor organs. A similar observation was reported by Popp et al. in a rat model, where donor MSCs conferred only a small increase in survival of transplanted hearts but combination of donor MSCs with mycophenolate increased survival in a significant way to achieve long-term engraftment for more than 100 days [65]. They observed gradual loss of allogeneic MSCs over time and suggested that dendritic cells are a potential mediator of tolerance induction.

As mentioned earlier, the effects of MSCs may not solely rely on their immunomodulatory properties. For example, in addition to immunomodulation and improved graft function, MSCs might help control nonimmune-mediated organ damage that is inevitable in transplantation. Rackham et al. used a pancreatic islet transplantation model where kidney derived MSCs co-injected with islets under the kidney capsule of streptozotocin induced diabetic C57BL/6 mice improved vascular engraftment and helped retain islet structure [66]. Moreover, 92% of co-transplanted mice achieved normoglycemia compared to 42% in the islet only group. Since these islets were of syngenic origin, the effects observed strongly suggest that MSCs might directly improve functioning of the transplanted organs, such as increasing angiogenesis or preventing apoptosis. Kidney transplantation is also prone to various delayed tissue damages due to the heavy metabolic demand of the kidney. Franquesa et al. reported that injection of MSCs prevented interstitial fibrosis and tubular atrophy in a rat kidney transplantation model, but the effect of MSCs on the regeneration of damaged tissue was not observed [67]. In other words, the primary role of MSCs might be prevention of nonimmune damage to tissues, such as hypoxia and reperfusion injury, and this may determine when MSCs should be used in the setting of organ transplantation.

12.4.4 Clinical Experience with MSCs in Solid Organ Transplantation

Similar to the aforementioned putative role of MSCs in the setting of HSC transplantation, potential beneficial effects of MSCs in the setting of solid organ transplantation are supported by immune modulation, anti-inflammatory properties, and paracrine effects via secretion of various growth and angiogenic factors. However, despite the numerous trials on use of MSCs in solid organ transplantation listed on www.clinicaltrials.gov, there is a paucity of studies published in peer reviewed journals.

Perico et al. were the first to report the use of autologous MSC injection in the context of solid organ transplantation [68]. In this study, autologous MSCs were generated from sternal BM aspirates according to good manufacturing practice procedures 4 months prior to patients receiving living, related donor kidney transplants. Three patients who received a living-related kidney prior to these two patients were used as the control group. Induction regimen consisted of basiliximab pretransplant and on day 4 posttransplant, and low-dose rabbit antithymocyte globulin daily from day 0 to day 6 posttransplant. Interestingly, in both MSC-treated patients serum creatinine levels increased 7–14 days after MSC administration. However, at the time of publication (360 and 180 days posttransplant) both patients had stable graft function. Importantly, the level of regulatory T cells (CD4$^+$, CD25 high, FoxP3$^+$, and CD127$^-$) increased progressively, while that of memory T cells (CD45RO$^+$, CD45RA$^-$, and CD8$^+$) showed a decrease compared to the control group. This study is important because it showed that infusion of MSCs in kidney transplant recipients is feasible.

More recently, Tan et al. reported results from a large, prospective, open-label, randomized clinical trial on end-stage renal disease patients that received a kidney from a living, related donor. In this study, the use of autologous MSCs after kidney transplantation reduced the incidence of acute rejection and preserved kidney function after 1 year [69]. In this clinical trial, autologous MSCs generated from BM were used at the time of kidney reperfusion and 2 weeks later. They used the combination of MSCs

with either regular or low-dose calcineurin inhibitor and compared the results to a control group given a standard dose of calcineurin inhibitor and anti-IL-2 receptor antibody. The results showed that BM-derived autologous MSC injection resulted in about a 7% rejection rate regardless of the dose of calcineurin inhibitor, but the control group experienced a rejection rate of about 21%. In addition to decrease in the rate of rejection, MSC injection expedited the recovery of estimated glomerular filtration rate and decreased rate of opportunistic injection, although the effect on kidney function does not seem to last in the long run. While it is not clear whether allogeneic MSCs will have a similar effect on the function and survival of the kidney engraft, these results are very encouraging. However, further clinical trials and verification of these results by other groups are warranted before any definite conclusions can be derived.

12.5 PRACTICAL ISSUES

12.5.1 How to Generate MSCs for Clinical Applications

In clinical medicine, a confounding factor in interpreting the results of clinical trials of pharmaceuticals is the heterogeneity of patient physiology, which affects the absorption, metabolism, and pharmacokinetics of the drug, and the heterogeneity of the targeted diseases, which affects the potential responsiveness of the disease to the administered drug. However, in clinical trials of cellular therapeutics, particularly MSCs, another layer of complexity exists and that is the enormous heterogeneity in the final cellular product. Indeed, the lack of standard culture methodology for generating MSCs poses a major challenge to the field [70]. It is well known that culturing MSCs using different types of growth media (containing fetal bovine serum, serum free media, autologous serum, fresh frozen plasma, or human platelet lysates) could affect their phenotype, rate of growth, and their functional properties [71]. Even the specific batch of fetal bovine serum used could have a major impact on the properties of the end product [72,73]. All these subtle differences could potentially result in clinically significant changes in the ultimate biological characteristics of the cellular product. However, delineating the potential correlation between the effects of different culture methodologies on the final product and clinical outcomes will be extremely challenging. Other factors that add to the complexity of interpreting the results of clinical trials are the fact that MSCs have been used at different doses and frequencies, of different passages, either fresh or frozen, given at different stages of disease, and given alone or with different combinations of other medications and immunosuppressive treatments.

A major challenge in harmonizing the culture conditions is the fact that although these variations in the culture methodologies could affect the immunomodulatory and regenerative properties of MSCs, there is no clinically applicable potency assay that is widely accepted for MSCs. It should be also recognized that one potency assay could not be suitable for assessing the effect of MSCs for so many different indications. For example, the same MSCs that are used for the treatment of GVHD have also been used for the treatment of myocardial infarction or chronic obstructive pulmonary disease [32]. However, the latter disorders are not T-cell-mediated disorders, so potency assays depending on the T-cell suppression capability of MSCs are probably not suitable for functional analyses of MSCs for these disorders.

12.5.2 Donor Source for MSCs

MSCs could be derived from tissues of autologous or allogeneic donors. However, MSCs generated from tissues collected from the patients might not be the appropriate source, as these MSCs may also be afflicted by the patient's primary disease processes [74]. However, it can also be argued that autologous MSCs could avoid issues, such as tissue incompatibility and rejection. However, even in the case of normal healthy allogeneic donors, we do not know if there are donor-specific characteristics that make certain allogeneic donors potentially more suitable for the production of MSCs, such as donors of younger age or special physical attributes. Additionally, in the case of allogeneic MSCs, we still do not know the appropriate tissue sources of MSCs that will be most effective for specific indications (e.g., BM versus adipose tissues). Furthermore, the generation of MSCs, which usually takes several weeks, might not be practical for certain diseases due to the urgency of the specific clinical scenarios. However, it is reassuring that use of MSCs from allogeneic donors has been shown in a myriad of studies to be safe and potentially efficacious.

12.5.3 Tissue Source for MSCs

Although MSCs were originally isolated from BM, cells with similar phenotype, differentiation potential, and biological characteristics have now been derived from almost all adult tissues, including adipose tissue [75], pancreas and heart [76], neonatal tissues like placenta [77], fetal tissues such as lung, liver, and blood [78], and even embryonic stem cells [79]. Furthermore, it has been repeatedly shown that MSCs derived from non-BM tissues have immunomodulatory properties very similar to BM-derived MSCs [80,81]. Indeed, based on experimental data showing that adipose tissue-derived MSCs possess immunological characteristics similar to BM-derived

MSCs, they have been used for treatment of GVHD after HSC transplantation [82]. Adipose tissue MSCs are expected to find their highest clinical applications in the fields of plastic and reconstructive surgery [83,84], and indeed use of fat graft, lipo-aspirates, or stromal vascular fraction (SVF) derived from lipo-aspirates could be considered as a form of MSC therapy because such tissues contain a large number of stromal cells. Nevertheless, use of lipo-aspirates or SVF is not equal to the use of MSCs because MSCs by definition are products of *ex vivo* culture expansion. Other types of MSCs that have reached the clinic include placenta-derived MSCs for the treatment of GVHD [85] and umbilical cord-derived MSCs for the treatment of severe and refractory systemic lupus erythematosus [86]. It should be no surprise to see other novel sources of MSCs to be tested for specific clinical settings based on their tissue of origin as there are reports regarding differences among MSCs derived from distinct tissues. For example, MSCs derived from BM and adipose tissues have been reported to show certain differences, while similar in terms of their immunomodulatory capacity [87]. While further research efforts are required, it is possible that the tissue of origin determines the properties of the MSCs, making them useful for reconstructing a microenvironment specific to that tissue. For example, we proposed that MSCs derived from pancreatic islets could be of more value for indications, such as the protection of transplanted islets after cadaveric transplantation [88], or MSCs generated from the lung could be a better candidate for use in lung disorders or after lung transplantation.

12.5.4 Timing of the Use of MSCs

While the use of MSCs for promotion of survival of transplanted organs seems promising, several crucial details need to be worked out. For example, Casiraghi et al. reported that the timing of MSC infusion could affect the survival of a transplanted kidney [89] in a mouse kidney transplant model. In this study, when MSCs were infused before kidney transplantation from balb/c mice to C57 recipient mice, survival of kidney grafts was significantly prolonged via regulatory T-cell-dependent mechanisms. However, the infusion of MSCs 2 days after transplantation was associated with neutrophil and complement activation, leading to premature rejection. Nevertheless, more studies are needed to determine whether delayed use of MSCs confers any detrimental or beneficial effects.

Another issue with the timing of MSC infusion is to determine whether repeated injection of MSCs would provide additional benefit. Until now, no systematic comparison has been performed to study the effect of number of injections or the timing of injection on the clinical outcome. In the context of solid organ transplantation, it is not clear if MSCs should be used prophylactically or only if there is a rejection episode.

12.5.5 Combination of MSCs with Immunosuppressive Medications

All transplant patients receive immunosuppressive medications, so it is very important to fully understand the effect of these drugs on MSC potency and function. A number of studies have investigated the effect of immunosuppressive drugs on immunomodulatory properties of MSCs *in vitro* or the effect of their administration in combination with MSCs on *in vivo* models of organ transplantation. Hoogduijn et al. showed that in the presence of MSCs, tacrolimus and rapamycin were less effective in suppressing the proliferation of peripheral blood mononuclear cells (PBMCs). Buron et al. approached the effect of immunosuppressants on MSCs and reported that calcineurin inhibitors and rapamycin suppressed the ability of MSCs to inhibit proliferation of allogeneic lymphocytes. Furthermore, Eggenhofer et al. showed that the administration of MSCs alone did not have much effect on the survival of transplanted hearts in a murine heart transplantation model; however, when an immunosuppressant was used in combination with MSCs, the survival of the graft increased significantly [90]. However, in a kidney transplantation model from Dark Agouti rat into Lewis strain, the use of BM MSCs from the Lewis rat alone was able to suppress inflammation [91]. It is unclear whether these observed discrepancies are due to organ-specific effects or the specific drug used. Indeed, there are reports about MSCs antagonizing the effects of immunosuppressants in certain cases [92,93].

Extrapolating from these animal models to human clinical trials will be very challenging; nevertheless, it is obvious that further research is required in order to define the optimal combination of MSCs and immunosuppressive agents. It is highly unlikely that in any clinical trial MSCs could be used alone for treatment of rejection episodes. Rejection of transplanted organs, especially vital organs such as hearts and lungs, carries such dire consequences that it would be unethical to use MSCs alone and thus potentially deprive patients of lifesaving antirejection treatments. This in itself makes it difficult to discern the role of MSCs versus coadministered treatments in many cases. Another major challenge in this field is the fact that current immunosuppressive induction regimens and antirejection treatments are so effective, especially in the case of kidney transplantation, that showing any potential benefits would require the recruitment of hundreds and even thousands of patients in randomized trials [88]. On the other hand, in the case of pancreatic islet transplantation due to the high rate of graft failure, it could be argued that the potential beneficial effects of MSC could be discerned much more easily [81].

12.6 EVOLVING CONCEPTS

MSCs were originally promoted in the field of regenerative medicine mainly for their potential to differentiate into many different types of cells and to replace lost or damaged cells. However, these expectations were never realized in human studies. Indeed, despite seemingly encouraging positive results in numerous clinical trials, the durability of the infused MSCs is now a matter of great debate [94], and the new wave of studies report that the extent of MSC engraftment is usually very minimal and could not explain the observed clinical benefits [95]. Indeed, the lack of engraftability of BM MSCs has been well known to HSC transplant physicians for a long time, as most reports have shown that in BM transplant recipients, the recipient MSCs remain and are not replaced by the donor MSCs carried in BM grafts [96]. However, it could be argued that in these BM transplant scenarios these cells were transplanted in low numbers and therefore were different from *ex vivo* culture-expanded MSCs. In any circumstance, there have been significant changes in our understanding of the potential mechanisms for MSC to exert their beneficial effects (Figure 12.1). This new paradigm proposes that MSCs exert their beneficial effects through mechanisms, such as immune modulation and paracrine effects, prior to their demise [97]. Thus, instead of "replacing" damaged cells, MSCs contribute to the "regeneration" of damaged cells and tissues indirectly, mainly via indirect paracrine effects. This low level of survival after administration of MSCs could also explain why repeated infusions of MSCs might be needed to achieve a clinical effect. It is ironic that one of the original attractive properties of MSCs was assumed to be their presumed lack of immunogenicity [33], and thus feasibility of transplanting across HLA barriers without concern for rejection. However, it could be argued that the lack of durability and persistence of MSCs, especially third-party MSCs, could be a desirable property of MSCs, as it could preclude any chance of tumorigenicity in the future.

A major question that has preoccupied the minds of clinicians and basic researchers alike is why these fibroblast-looking cells [98], derived mostly from BM, could have a therapeutic effect in such a wide range of conditions with seemingly unrelated pathophysiology, such as GVHD, myocardial infarction, heart failure, chronic obstructive pulmonary disease, Crohn's disease, diabetes, osteoarthritis, systemic lupus erythematosus, kidney transplantation, stroke, nonhealing wounds, systemic sclerosis, cirrhosis, and others. We propose that the common pathophysiological theme for all these disorders

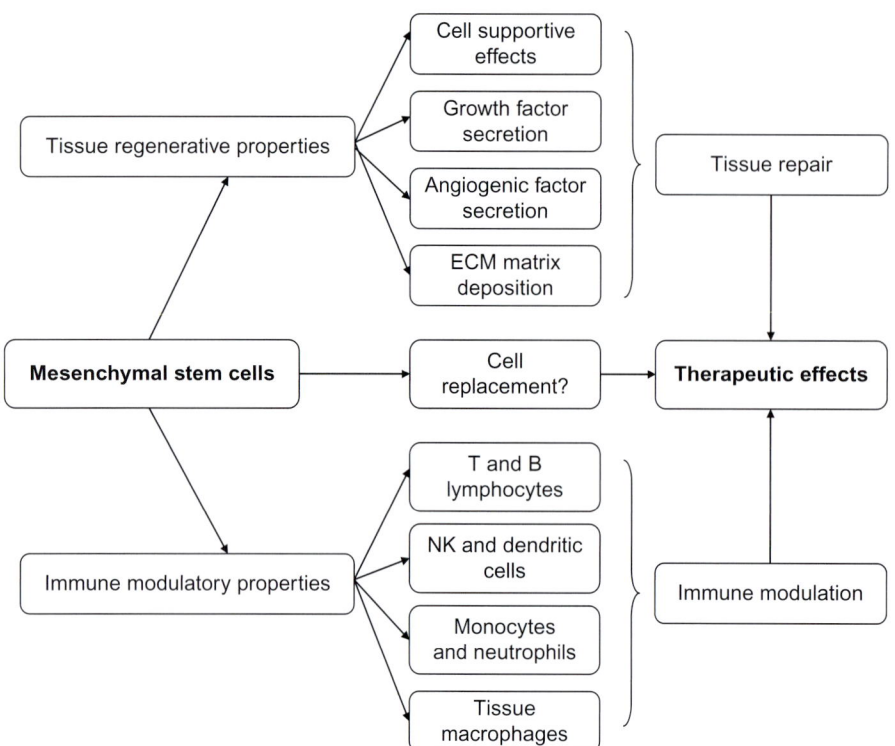

FIGURE 12.1 Potential mechanisms of action of mesenchymal stem cells in immune modulation and tissue regeneration relevant to the solid organ transplantation.

is the contribution of "inflammation," and the major way that MSCs exert their effects is through their interactions with macrophages. This paradigm could provide a unifying rationale for the continued investigation of these cells in such a wide range of applications including solid organ transplantation.

candidate for improving the durability of transplanted organs in certain clinical scenarios. Clinical trials utilizing MSC in solid organ transplantation are currently in their infancy and only systematic and well-designed clinical trials could provide convincing conclusions regarding the potential role of MSCs in the field of solid organ transplantation.

12.7 CONCLUSIONS

The original clinical trials of use of MSCs in the setting of HSC transplantation not only provided the initial safety data but also generated much excitement, encouraging other disciplines of medicine to take advantage of the potential therapeutic effects of MSCs. A major factor in the expansion of MSC trials is the lack of any documented toxicity or long-term side effects and the unmet need for novel therapies for many of these diseases. However, due to the inherent shortcomings of small and nonrandomized clinical trials in which MSCs have been investigated, variability in the manufacturing methodology, cell doses and frequency of administration, the use of multiple concomitant medications, and the heterogeneity of the clinical conditions treated with MSCs, a conclusive clinical benefit has been difficult to discern, and important questions remain to be addressed. Nevertheless, considering their immunomodulatory activity and supportive effects on tissue repair, MSCs seem to be an ideal

REFERENCES

[1] Friedenstein AJ, Petrakova KV, Kurolesova AI, Frolova GP. Heterotopic of bone marrow. Analysis of precursor cells for osteogenic and hematopoietic tissues. Transplantation 1968;6:230−47.

[2] Castro-Malaspina H, Gay RE, Resnick G, Kapoor N, Meyers P, Chiarieri D, et al. Characterization of human bone marrow fibroblast colony-forming cells (CFU-F) and their progeny. Blood 1980;56:289−301.

[3] Dexter TM. Stromal cell associated haemopoiesis. J Cell Physiol 1982;1:87−94.

[4] Caplan AI. Mesenchymal stem cells. J Orthop Res 1991;9:641−50.

[5] Caplan AI. Adult mesenchymal stem cells for tissue engineering versus regenerative medicine. J Cell Physiol 2007;213:341−7.

[6] Pittenger MF, Mackay AM, Beck SC, Jaiswal RK, Douglas R, Mosca JD, et al. Multilineage potential of adult human mesenchymal stem cells. Science 1999;284:143−7.

[7] Dominici M, Le Blanc K, Mueller I, Slaper-Cortenbach I, Marini F, Krause D, et al. Minimal criteria for defining multipotent mesenchymal stromal cells. The international society for cellular therapy position statement. Cytotherapy 2006;8:315−7.

[8] Ho AD, Wagner W, Franke W. Heterogeneity of mesenchymal stromal cell preparations. Cytotherapy 2008;10:320–30.

[9] Battiwalla M, Hematti P. Mesenchymal stem cells in hematopoietic stem cell transplantation. Cytotherapy 2009;11:503–15.

[10] Raaijmakers MH, Scadden DT. Evolving concepts on the microenvironmental niche for hematopoietic stem cells. Curr Opin Hematol 2008;15:301–6.

[11] Lazarus HM, Haynesworth SE, Gerson SL, Rosenthal NS, Caplan AI. Ex vivo expansion and subsequent infusion of human bone marrow-derived stromal progenitor cells (mesenchymal progenitor cells): implications for therapeutic use. Bone Marrow Transpl 1995;16:557–64.

[12] Almeida-Porada G, Flake AW, Glimp HA, Zanjani ED. Cotransplantation of stroma results in enhancement of engraftment and early expression of donor hematopoietic stem cells in utero. Exp Hematol 1999;27:1569–75.

[13] Koc ON, Gerson SL, Cooper BW, Dyhouse SM, Haynesworth SE, Caplan AI, et al. Rapid hematopoietic recovery after coinfusion of autologous-blood stem cells and culture-expanded marrow mesenchymal stem cells in advanced breast cancer patients receiving high-dose chemotherapy. J Clin Oncol 2000;18:307–16.

[14] Lazarus HM, Koc ON, Devine SM, Curtin P, Maziarz RT, Holland HK, et al. Cotransplantation of HLA-identical sibling culture-expanded mesenchymal stem cells and hematopoietic stem cells in hematologic malignancy patients. Biol Blood Marrow Transpl 2005;11:389–98.

[15] Le Blanc K, Samuelsson H, Gustafsson B, Remberger M, Sundberg B, Arvidson J, et al. Transplantation of mesenchymal stem cells to enhance engraftment of hematopoietic stem cells. Leukemia 2007;21:1733–8.

[16] Ning H, Yang F, Jiang M, Hu L, Feng K, Zhang J, et al. The correlation between cotransplantation of mesenchymal stem cells and higher recurrence rate in hematologic malignancy patients: outcome of a pilot clinical study. Leukemia 2008;22:593–9.

[17] Li X, Wang D, Liang J, Zhang H, Sun L. Mesenchymal SCT ameliorates refractory cytopenia in patients with systemic lupus erythematosus. Bone Marrow Transpl 2013;48:544–50.

[18] Sanchez-Guijo FM, Lopez-Villar O, Lopez-Anglada L, Villaron EM, Muntion S, Diez-Campelo M, et al. Allogeneic mesenchymal stem cell therapy for refractory cytopenias after hematopoietic stem cell transplantation. Transfusion 2012;52:1086–91.

[19] Kim J, Hematti P. Mesenchymal stem cell-educated macrophages: a novel type of alternatively activated macrophages. Exp Hematol 2009;37:1445–53.

[20] Nemeth K, Leelahavanichkul A, Yuen PS, Mayer B, Parmelee A, Doi K, et al. Bone marrow stromal cells attenuate sepsis via prostaglandin E(2)-dependent reprogramming of host macrophages to increase their interleukin-10 production. Nat Med 2009;15:42–9.

[21] Keating A. How do mesenchymal stromal cells suppress T cells? Cell Stem Cell 2008;2:106–8.

[22] Le Blanc K, Tammik L, Sundberg B, Haynesworth SE, Ringden O. Mesenchymal stem cells inhibit and stimulate mixed lymphocyte cultures and mitogenic responses independently of the major histocompatibility complex. Scand J Immunol 2003;57:11–20.

[23] Koc ON, Lazarus HM. Mesenchymal stem cells: heading into the clinic. Bone Marrow Transpl 2001;27:235–9.

[24] Le Blanc K, Rasmusson I, Sundberg B, Gotherstrom C, Hassan M, Uzunel M, et al. Treatment of severe acute graft-versus-host disease with third party haploidentical mesenchymal stem cells. Lancet 2004;363:1439–41.

[25] Le Blanc K, Frassoni F, Ball L, Locatelli F, Roelofs H, Lewis I, et al. Mesenchymal stem cells for treatment of steroid-resistant, severe, acute graft-versus-host disease: a phase II study. Lancet 2008;371:1579–86.

[26] Prasad VK, Lucas KG, Kleiner GI, Talano JA, Jacobsohn D, Broadwater G, et al. Efficacy and safety of ex vivo cultured adult human mesenchymal stem cells (Prochymal) in pediatric patients with severe refractory acute graft-versus-host disease in a compassionate use study. Biol Blood Marrow Transpl 2011;17:534–41.

[27] Prockop DJ. Repair of tissues by adult stem/progenitor cells (MSCs): controversies, myths, and changing paradigms. Mol Ther 2009;17:939–46.

[28] Phinney DG, Prockop DJ. Concise review: mesenchymal stem/multipotent stromal cells: the state of transdifferentiation and modes of tissue repair—current views. Stem Cells 2007;25:2896–902.

[29] Devine SM, Cobbs C, Jennings M, Bartholomew A, Hoffman R. Mesenchymal stem cells distribute to a wide range of tissues following systemic infusion into nonhuman primates. Blood 2003;101:2999–3001.

[30] Prockop DJ, Olson SD. Clinical trials with adult stem/progenitor cells for tissue repair: let's not overlook some essential precautions. Blood 2007;109:3147–51.

[31] Deda H, Inci MC, Kurekci AE, Sav A, Kayihan K, Ozgun E, et al. Treatment of amyotrophic lateral sclerosis patients by autologous bone marrow-derived hematopoietic stem cell transplantation: a 1-year follow-up. Cytotherapy 2009;11:18–25.

[32] Taupin P. OTI-010 osiris therapeutics/JCR pharmaceuticals. Curr Opin Investig Drugs 2006;7:473–81.

[33] Tse WT, Pendleton JD, Beyer WM, Egalka MC, Guinan EC. Suppression of allogeneic T-cell proliferation by human marrow stromal cells: implications in transplantation. Transplantation 2003;75:389–97.

[34] McKernan R, McNeish J, Smith D. Pharma's developing interest in stem cells. Cell Stem Cell 2010;6:517–20.

[35] Di Nicola M, Carlo-Stella C, Magni M, Milanesi M, Longoni PD, Matteucci P, et al. Human bone marrow stromal cells suppress T-lymphocyte proliferation induced by cellular or nonspecific mitogenic stimuli. Blood 2002;99:3838–43.

[36] Krampera M, Glennie S, Dyson J, Scott D, Laylor R, Simpson E, et al. Bone marrow mesenchymal stem cells inhibit the response of naive and memory antigen-specific T cells to their cognate peptide. Blood 2003;101:3722–9.

[37] Beyth S, Borovsky Z, Mevorach D, Liebergall M, Gazit Z, Aslan H, et al. Human mesenchymal stem cells alter antigen-presenting cell maturation and induce T-cell unresponsiveness. Blood 2005;105:2214–9.

[38] Augello A, Tasso R, Negrini SM, Amateis A, Indiveri F, Cancedda R, et al. Bone marrow mesenchymal progenitor cells inhibit lymphocyte proliferation by activation of the programmed death 1 pathway. Eur J Immunol 2005;35:1482–90.

[39] English K, Ryan JM, Tobin L, Murphy MJ, Barry FP, Mahon BP. Cell contact, prostaglandin E(2) and transforming growth factor beta 1 play non-redundant roles in human mesenchymal stem cell induction of $CD4^+CD25(High)$ forkhead box $P3^+$ regulatory T cells. Clin Exp Immunol 2009;156:149–60.

[40] Ge W, Jiang J, Arp J, Liu W, Garcia B, Wang H. Regulatory T-cell generation and kidney allograft tolerance induced by mesenchymal stem cells associated with indoleamine 2,3-dioxygenase expression. Transplantation 2010;90:1312−20.

[41] Asari S, Itakura S, Ferreri K, Liu CP, Kuroda Y, Kandeel F, et al. Mesenchymal stem cells suppress B-cell terminal differentiation. Exp Hematol 2009;37:604−15.

[42] Rasmusson I, Le Blanc K, Sundberg B, Ringden O. Mesenchymal stem cells stimulate antibody secretion in human B cells. Scand J Immunol 2007;65:336−43.

[43] Tabera S, Perez-Simon JA, Diez-Campelo M, Sanchez-Abarca LI, Blanco B, Lopez A, et al. The effect of mesenchymal stem cells on the viability, proliferation and differentiation of B-lymphocytes. Haematologica 2008;93:1301−9.

[44] Sis B, Mengel M, Haas M, Colvin RB, Halloran PF, Racusen LC, et al. Banff '09 meeting report: antibody mediated graft deterioration and implementation of Banff working groups. Am J Transpl 2010;10:464−71.

[45] Crew RJ, Ratner LE. ABO-incompatible kidney transplantation: current practice and the decade ahead. Curr Opin Org Transplant 2010;15:526−30.

[46] Traggiari E, Volpi S, Schena F, Gattorno M, Ferlito F, Moretta L, et al. Bone marrow-derived mesenchymal stem cells induce both polyclonal expansion and differentiation of B cells isolated from healthy donors and systemic lupus erythematosus patients. Stem Cells 2008;26:562−9.

[47] Spaggiari GM, Capobianco A, Abdelrazik H, Becchetti F, Mingari MC, Moretta L. Mesenchymal stem cells inhibit natural killer-cell proliferation, cytotoxicity, and cytokine production: role of indoleamine 2,3-dioxygenase and prostaglandin E2. Blood 2008;111:1327−33.

[48] Spaggiari GM, Capobianco A, Becchetti S, Mingari MC, Moretta L. Mesenchymal stem cell-natural killer cell interactions: evidence that activated NK cells are capable of killing MSCs, whereas MSCs can inhibit IL-2-induced NK-cell proliferation. Blood 2006;107:1484−90.

[49] Aggarwal S, Pittenger MF. Human mesenchymal stem cells modulate allogeneic immune cell responses. Blood 2005;105:1815−22.

[50] Sotiropoulou PA, Perez SA, Gritzapis AD, Baxevanis CN, Papamichail M. Interactions between human mesenchymal stem cells and natural killer cells. Stem Cells 2006;24:74−85.

[51] Nauta AJ, Kruisselbrink AB, Lurvink E, Willemze R, Fibbe WE. Mesenchymal stem cells inhibit generation and function of both CD34^{+}-derived and monocyte-derived dendritic cells. J Immunol 2006;177:2080−7.

[52] Ramasamy R, Fazekasova H, Lam EW, Soeiro I, Lombardi G, Dazzi F. Mesenchymal stem cells inhibit dendritic cell differentiation and function by preventing entry into the cell cycle. Transplantation 2007;83:71−6.

[53] Maccario R, Podesta M, Moretta A, Cometa A, Comoli P, Montagna D, et al. Interaction of human mesenchymal stem cells with cells involved in alloantigen-specific immune response favors the differentiation of CD4^{+} T-cell subsets expressing a regulatory/suppressive phenotype. Haematologica 2005;90:516−25.

[54] Jiang XX, Zhang Y, Liu B, Zhang SX, Wu Y, Yu XD, et al. Human mesenchymal stem cells inhibit differentiation and function of monocyte-derived dendritic cells. Blood 2005;105:4120−6.

[55] Zhang B, Liu R, Shi D, Liu X, Chen Y, Dou X, et al. Mesenchymal stem cells induce mature dendritic cells into a novel Jagged-2-dependent regulatory dendritic cell population. Blood 2009;113:46−57.

[56] Zhang QZ, Su WR, Shi SH, Wilder-Smith P, Xiang AP, Wong A, et al. Human gingiva-derived mesenchymal stem cells elicit polarization of m2 macrophages and enhance cutaneous wound healing. Stem Cells 2010;28:1856−68.

[57] Maggini J, Mirkin G, Bognanni I, Holmberg J, Piazzon IM, Nepomnaschy I, et al. Mouse bone marrow-derived mesenchymal stromal cells turn activated macrophages into a regulatory-like profile. PLoS ONE 2010;5:e9252.

[58] Peng C, Yang K, Xiang P, Zhang C, Zou L, Wu X, et al. Effect of transplantation with autologous bone marrow stem cells on acute myocardial infarction. Int J Cardiol 2013;162:158−65.

[59] Jungebluth P, Luedde M, Ferrer E, Luedde T, Vucur M, Peinado VI, et al. Mesenchymal stem cells restore lung function by recruiting resident and non-resident proteins. Cell Transplant 2011;20:1561−74.

[60] Hoffman AM, Paxson JA, Mazan MR, Davis AM, Tyagi S, Murthy S, et al. Lung-derived mesenchymal stromal cell post-transplantation survival, persistence, paracrine expression, and repair of elastase-injured lung. Stem Cells Dev 2011;20:1779−92.

[61] Morigi M, Rota C, Montemurro T, Montelatici E, Lo Cicero V, Imberti B, et al. Life-sparing effect of human cord blood-mesenchymal stem cells in experimental acute kidney injury. Stem Cells 2010;28:513−22.

[62] Altun B, Yilmaz R, Aki T, Akoglu H, Zeybek D, Piskinpasa S, et al. Use of mesenchymal stem cells and darbepoetin improve ischemia-induced acute kidney injury outcomes. Am J Nephrol 2012;35:531−9.

[63] Bartholomew A, Sturgeon C, Siatskas M, Ferrer K, McIntosh K, et al. Mesenchymal stem cells suppress lymphocyte proliferation *in vitro* and prolong skin graft survival *in vivo*. Exp Hematol 2002;30:42−8.

[64] Ge W, Jiang J, Baroja ML, Arp J, Zassoko R, Liu W, et al. Infusion of mesenchymal stem cells and rapamycin synergize to attenuate alloimmune responses and promote cardiac allograft tolerance. J Am Soc Transpl Am Soc Transpl Surg 2009;9:1760−72.

[65] Popp FC, Eggenhofer E, Renner P, Slowik P, Lang SA, Kaspar H, et al. Mesenchymal stem cells can induce long-term acceptance of solid organ allografts in synergy with low-dose mycophenolate. Transpl Immunol 2008;20:55−60.

[66] Rackham CL, Chagastelles PC, Nardi NB, Hauge-Evans AC, Jones PM, King AJ. Co-transplantation of mesenchymal stem cells maintains islet organisation and morphology in mice. Diabetologia 2011;54:1127−35.

[67] Franquesa M, Herrero E, Torras J, Ripoll E, Flaquer M, Goma M, et al. Mesenchymal stem cell therapy prevents interstitial fibrosis and tubular atrophy in a rat kidney allograft model. Stem Cells Dev 2012;21:3125−35.

[68] Perico N, Casiraghi F, Introna M, Gotti E, Todeschini M, Cavinato RA, et al. Autologous mesenchymal stromal cells and kidney transplantation: a pilot study of safety and clinical feasibility. Clin J Am Soc Nephrol 2011;6:412−22.

[69] Tan J, Wu W, Xu X, Liao L, Zheng F, Messinger S, et al. Induction therapy with autologous mesenchymal stem cells in

living-related kidney transplants: a randomized controlled trial. J Am Med Assoc 2012;307:1169—77.

[70] Samuelsson H, Ringden O, Lonnies H, Le Blanc K. Optimizing *in vitro* conditions for immunomodulation and expansion of mesenchymal stromal cells. Cytotherapy 2009;11:129—36.

[71] Bieback K, Hecker A, Kocaomer A, Lannert H, Schallmoser K, Strunk D, et al. Human alternatives to fetal bovine serum for the expansion of mesenchymal stromal cells from bone marrow. Stem Cells 2009;27:2331—41.

[72] Shahdadfar A, Fronsdal K, Haug T, Reinholt FP, Brinchmann JE. *In vitro* expansion of human mesenchymal stem cells: choice of serum is a determinant of cell proliferation, differentiation, gene expression, and transcriptome stability. Stem Cells 2005;23:1357—66.

[73] Duggal S, Brinchmann JE. Importance of serum source for the *in vitro* replicative senescence of human bone marrow derived mesenchymal stem cells. J Cell Physiol 2011;226:2908—15.

[74] Kastrinaki MC, Sidiropoulos P, Roche S, Ringe J, Lehmann S, Kritikos H, et al. Functional, molecular and proteomic characterisation of bone marrow mesenchymal stem cells in rheumatoid arthritis. Ann Rheum Dis 2008;67:741—9.

[75] Zuk PA, Zhu M, Mizuno H, Huang J, Futrell JW, Katz AJ, et al. Multilineage cells from human adipose tissue: implications for cell-based therapies. Tissue Eng 2001;7:211—28.

[76] Lushaj EB, Anstadt E, Haworth R, Roenneburg D, Kim J, Hematti P, et al. Mesenchymal stromal cells are present in the heart and promote growth of adult stem cells *in vitro*. Cytotherapy 2011;13:400—6.

[77] In't Anker PS, Scherjon SA, Kleijburg-van der Keur C, de Groot-Swings GM, Claas FH, Fibbe WE, et al. Isolation of mesenchymal stem cells of fetal or maternal origin from human placenta. Stem Cells 2004;22:1338—45.

[78] In't Anker PS, Noort WA, Scherjon SA, Kleijburg-van der Keur C, Kruisselbrink AB, van Bezooijen RL, et al. Mesenchymal stem cells in human second-trimester bone marrow, liver, lung, and spleen exhibit a similar immunophenotype but a heterogeneous multilineage differentiation potential. Haematologica 2003;88:845—52.

[79] Hematti P. Human embryonic stem cell-derived mesenchymal progenitors: an overview. Methods Mol Biol 2011;690:163—74.

[80] Trivedi P, Hematti P. Derivation and immunological characterization of mesenchymal stromal cells from human embryonic stem cells. Exp Hematol 2008;36:350—9.

[81] Kim J, Breunig MJ, Escalante LE, Bhatia N, Denu RA, Dollar BA, et al. Biologic and immunomodulatory properties of mesenchymal stromal cells derived from human pancreatic islets. Cytotherapy 2012;14:925—35.

[82] Fang B, Song Y, Liao L, Zhang Y, Zhao RC. Favorable response to human adipose tissue-derived mesenchymal stem cells in steroid-refractory acute graft-versus-host disease. Transpl Proc 2007;39:3358—62.

[83] Hanson SE, Gutowski KA, Hematti P. Clinical applications of mesenchymal stem cells in soft tissue augmentation. Aesthet Surg 2010;30:838—42.

[84] Hanson SE, Bentz ML, Hematti P. Mesenchymal stem cell therapy for nonhealing cutaneous wounds. Plast Rec Surg 2010;125:510—6.

[85] Brooke G, Rossetti T, Pelekanos R, Ilic N, Murray P, Hancock S, et al. Manufacturing of human placenta-derived mesenchymal stem cells for clinical trials. Br J Haematol 2009;144:571—9.

[86] Gu Z, Akiyama K, Ma X, Zhang H, Feng X, Yao G, et al. Transplantation of umbilical cord mesenchymal stem cells alleviates lupus nephritis in MRL/lpr mice. Lupus 2010;19:1502—14.

[87] Ikegame Y, Yamashita K, Hayashi S, Mizuno H, Tawada M, You F, et al. Comparison of mesenchymal stem cells from adipose tissue and bone marrow for ischemic stroke therapy. Cytotherapy 2011;13:675—85.

[88] Hematti P. Role of mesenchymal stromal cells in solid organ transplantation. Transpl Rev 2008;22:262—73.

[89] Casiraghi F, Azzollini N, Todeschini M, Cavinato RA, Cassis P, Solini S, et al. Localization of mesenchymal stromal cells dictates their immune or proinflammatory effects in kidney transplantation. Am J Transpl 2012;12:2373—83.

[90] Eggenhofer E, Renner P, Soeder Y, Popp FC, Hoogduijn MJ, Geissler EK, et al. Features of synergism between mesenchymal stem cells and immunosuppressive drugs in a murine heart transplantation model. Transpl Immunol 2011;25:141—7.

[91] Hara Y, Stolk M, Ringe J, Dehne T, Ladhoff J, Kotsch K, et al. *In vivo* effect of bone marrow-derived mesenchymal stem cells in a rat kidney transplantation model with prolonged cold ischemia. Transplant Int 2011;24:1112—23.

[92] Hoogduijn MJ, Crop MJ, Korevaar SS, Peeters AM, Eijken M, Maat LP, et al. Susceptibility of human mesenchymal stem cells to tacrolimus, mycophenolic acid, and rapamycin. Transplantation 2008;86:1283—91.

[93] Buron F, Perrin H, Malcus C, Hequet O, Thaunat O, Kholopp-Sarda MN, et al. Human mesenchymal stem cells and immunosuppressive drug interactions in allogeneic responses: an *in vitro* study using human cells. Transplant Proc 2009;41:3347—52.

[94] Uccelli A, Moretta L, Pistoia V. Mesenchymal stem cells in health and disease. Nat Rev 2008;8:726—36.

[95] von Bahr L, Batsis I, Moll G, Hagg M, Szakos A, Sundberg B, et al. Analysis of tissues following mesenchymal stromal cell therapy in humans indicate limited long-term engraftment and no ectopic tissue formation. Stem Cells 2012;30:1575—8.

[96] Koc ON, Peters C, Aubourg P, Raghavan S, Dyhouse S, DeGasperi R, et al. Bone marrow-derived mesenchymal stem cells remain host-derived despite successful hematopoietic engraftment after allogeneic transplantation in patients with lysosomal and peroxisomal storage diseases. Exp Hematol 1999;27:1675—81.

[97] Prockop DJ, Kota DJ, Bazhanov N, Reger RL. Evolving paradigms for repair of tissues by adult stem/progenitor cells (MSCs). J Cell Mol Med 2010;14:2190—9.

[98] Hematti P. Mesenchymal stromal cells and fibroblasts: a case of mistaken identity? Cytotherapy 2012;14:516—21.

Chapter 13

Bioreactors for Tissue Engineering Purposes

Kory J. Blose[a], Jeffrey T. Krawiec[a], Justin S. Weinbaum[e], and David A. Vorp[a,b,c,d,e]

[a]Department of Bioengineering, University of Pittsburgh, Pittsburgh, PA, [b]Department of Cardiothoracic Surgery, University of Pittsburgh, Pittsburgh, PA, [c]Department of Surgery, University of Pittsburgh, Pittsburgh, PA, [d]Center for Vascular Remodeling and Regeneration, University of Pittsburgh, Pittsburgh, PA, [e]McGowan Institute for Regenerative Medicine, University of Pittsburgh, Pittsburgh, PA

Tissue engineering focuses on developing ideal tissue replacements that adequately mimic, or are functionally equivalent to, native tissues. Generally these are intended to be autologous tissues which require being matured in culture for an extended period of time before being placed back within the patient. It is this culture period where bioreactors can be useful. When trying to recreate functional tissues one must consider the complex array of stimuli that are seen *in vivo* which have been used to achieve the level of diversity seen in the body. Additionally, as tissues being developed for functional replacement are typically three dimensional, the standard static two-dimensional cell culture procedure is obsolete and unable to provide the necessary stimuli to fully mature them prior to implant. Bioreactors can deliver both a productive three-dimensional culture environment and instructive stimuli.

When considering tissue engineered approaches, metabolic demand, mechanical stimuli, and chemical stimuli need to be accounted for. Tissue engineering bioreactors are *in vitro* cell culture systems that have been designed to alter the following basic physiological phenomena: cell survival along with tissue structure, organization, and function. To illustrate the principle that complex stimuli can mature tissues differently, consider the case of blood vessels. Arteries are constantly subjected to specific pressures and flows depending on their proximity to the heart. When arteries are close to the heart, where pressures and flows are highest, they are thick with additional musculature, highly elastic, and require additional sources of nutrition in the form of small capillaries in the wall known as the vasa vasorum. Blood vessels distal to the heart, where flow and pressure are low, lack a vasa vasorum, have less musculature and show lower elasticity. The following sections will describe design principles related to the two main types of bioreactors used for tissue engineering, using the paradigm of vascular engineering for illustration. First we will examine mixing bioreactors. Following that, bioreactors designed to guide tissue structure and function will be discussed.

13.1 MIXING BIOREACTORS

Typical flask culture can be both labor intensive and inappropriate for the three-dimensional constructs commonly seen in tissue engineering. In this type of culture, the media sits stagnant while the cells rapidly deplete it of nutrients and accumulate toxic waste products.

Additionally, concerns can arise about toxic degradation products accumulating around the cells when synthetic degradable scaffolds are used [1]. Replenishing nutrients and removing waste products require manually replacing the culture media. Mixing bioreactors (also known as dynamic culture bioreactors) alleviate both the concerns of supplying nutrients and removing waste by consistently circulating the media, whether through stirring, rocking, or perfusion.

From a mass transport perspective, mixing bioreactors also allow for convection instead of relying on diffusion. In static cell culture, diffusion is the main source of transport where solutes will move from high to low concentrations, and their velocity of movement is inversely related to their size [2]. Convection on the other hand utilizes fluid flow for transport which provides a much stronger mixing phenomenon. This distinction is key, especially in cell culture which utilizes a very small concentration of various chemical constituents like growth factors. With their large size and low concentrations, growth factors are not likely to diffuse at a high rate or distribute evenly throughout the media. By providing convection, uniform distribution of growth factors is more easily achievable. Beyond these considerations from a basic nutrition perspective, oxygen diffusion is limited in static media [3]; therefore, mixing can improve cellular metabolism.

Additionally, when creating tissue engineered constructs scaffolding materials are routinely used which have a substantial thickness and may provide resistance to mass transport, which is dependent on thickness and pore size. By providing convection, cells on the interior of the construct more easily receive nutrients. Without convection, cells would migrate outward towards the media, creating nonuniformity, and would not proliferate as well within the scaffold [4].

Despite the advantages of mixing, the degree of shear stress caused by this fluid motion must also be considered and has been characterized in some studies [5,6].

The following section will provide examples of some common mixing bioreactors, such as spinner flasks, rocker platforms, and perfusion systems.

13.1.1 Spinner Flasks

Spinner flasks are culture flasks that have a magnetic stir bar attached to an impeller to circulate media providing a dynamic flow (Figure 13.1A). These have been used for expanding various stem cell populations as well as for engineering cartilage [7–9], bone [10,11], and cardiovascular [4,12] tissues. Generally, spinner flask culture increases cell proliferation and tissue generation over static culture. During culture, constructs can either be fixed to remain stationary within the spinner flask [7] or allowed to float freely with the fluid flow [4] with the latter exposing cells to less shear. The shear cells experience is dependent on the size of the impeller diameter and rotation speed of the impeller. While an optimal shear may vary depending on cell population, a high level of shear is very likely to destroy the cells [13]; therefore, lower rotation speeds are recommended and utilized.

13.1.2 Rocker Platforms

Rocker platforms, which can also be known as wave bioreactors, provide a gentle side to side motion producing waves within the media to cause mixing (Figure 13.1B). These bioreactors share many of the same principles that are apparent in the spinner flask: nutrients are mixed via convection and the level of shear must be monitored. While rocker platforms are commonly used in regular laboratory chemical assays, preparation of biomaterials [14–16], and cell culturing [17], their use is limited [18] in tissue engineering due to the higher sophistication of other approaches.

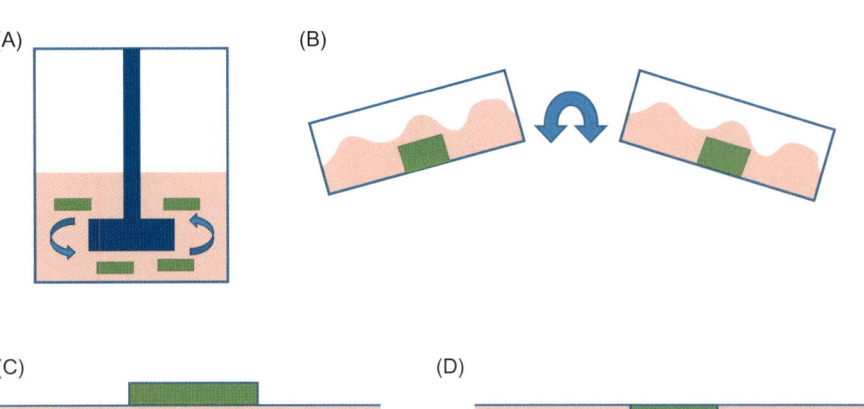

FIGURE 13.1 Mechanisms of action of different mixing bioreactors (tissue construct represented in green): (A) spinner flask, (B) rocker platform, (C) perfusion across the surface of a construct, and (D) perfusion through a construct.

13.1.3 Perfusion Systems

Perfusion systems can be designed in two ways. The first design drives media past the constructs; this media movement can be achieved via axial or parallel plate flow and is commonly used in tissue engineering of tubular tissues like blood vessels (Figure 13.1C). In the second design, tissue constructs are placed in line with the fluid flow forcing the fluid through the construct (Figure 13.1D). Perfusion systems use a pressure gradient which provides better mixing over diffusion. Constructs matured in a perfusion system show better cell distribution and tissue-specific protein expression than constructs stimulated in a spinner flask bioreactor [19,20]. Since a flow loop can be used for these systems they can be connected to a large media reservoir allowing constant media exchange. As with all the other mixing systems, perfusion systems impart shear on the cells which must be considered; at higher perfusion rates a decrease in cell growth is seen and excessive fluid flow can remove paracrine growth-promoting factors that cells secrete [13].

13.1.4 Cell Seeding Bioreactors

While the main function of many mixing bioreactors is to culture cells in such a way to promote cell growth, a secondary function of bioreactors is to distribute cells optimally throughout a scaffold prior to long-term culture. In order to accomplish this optimal distribution, many approaches have been developed. One simple approach is to apply cell suspensions [21] directly onto the scaffold, allowing the cells to settle and potentially infiltrate on their own; however, this method can be suboptimal resulting in a nonuniform distribution of cells through the thickness of the scaffold. To combat nonuniform cell distribution, other approaches have been developed using various physical phenomena, such as vacuums [22], electromagnetism [23], and acoustic waves [24], to force cells

throughout the thickness of scaffolds. Additional considerations can be made for more complicated geometries like tubes. Flow-mediated recruitment, where a cell suspension is infused and cells attach to the inner wall as they float by [25], can be used for cell seeding. Additionally, rotation can be used to help achieve a more uniform circumferential distribution and overcome the effects of settling from gravity [26]. One cell seeding bioreactor (Figure 13.2) for tubular constructs [27] combines vacuum, rotation, and flow to distribute cells uniformly in a three-dimensional porous tubular scaffold creating a distribution closely resembling a native artery. Other advanced techniques such as cell spraying [28], where a cell suspension is turned into an aerosol and projected at a surface, or microintegration [29], where polymers are synthesized in a cell suspension, can also be used. More in depth reviews of seeding techniques can be found elsewhere [30].

13.2 GUIDING TISSUE STRUCTURE AND FUNCTION

While one main function of bioreactors is to provide sufficient nutrition for cell survival, they also can be used to guide tissue structure, organization, and ultimately function. To guide cells in this way, two broad categories of stimulation can be used: chemical and mechanical. Chemical stimulation represents a large variety of bioactive factors including growth factors, hormones, and chemical additives, which are added to the cell culture media to induce effects, such as proliferation or differentiation. Typically, these factors can be added to media in any tissue culture system and therefore do not fall under the category of bioreactor design; for this reason, chemical stimulation will not be reviewed here. Mechanical stimulation, however, is a key component of bioreactor design which can have a variety of beneficial effects on engineered tissue development.

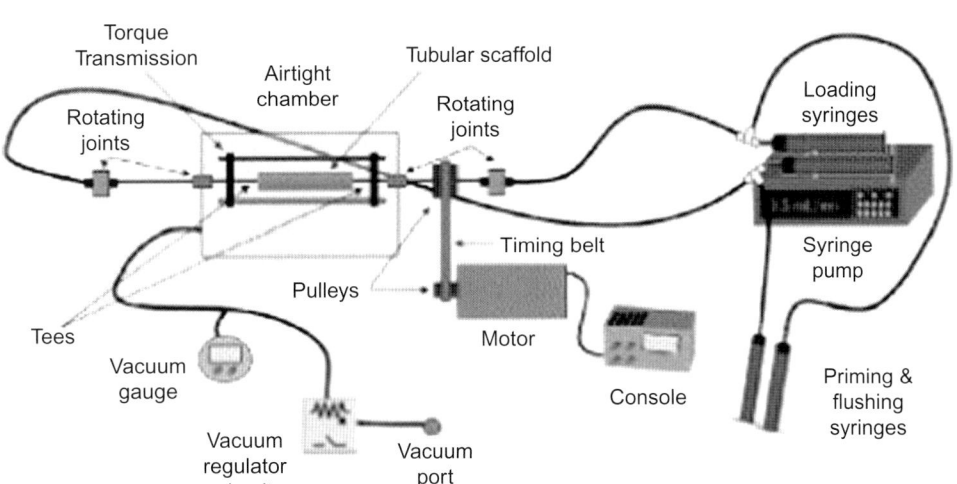

FIGURE 13.2 Schematic illustration of rotational vacuum seeding device. *From Ref. [27] with permission.*

Simply defined, mechanical stimulation encompasses any type of mechanical loading imposed on a tissue. When designing bioreactors, engineers typically look to the body as a guide for finding appropriate mechanical loads. For example, muscles, tendons, ligaments, and vascular tissues all undergo stretch but to different levels. Endothelial cells (ECs) along the inner wall of blood vessels experience shear and pressure loads while bone tissues will experience compression, bending, and torsion under normal physiological conditions. This variety in physiological loading highlights the need to accurately select the desired type of mechanical loading when designing bioreactors for tissues of interest.

Mechanical stimulation can guide the organization and structure of tissue formation at a cellular level by mechanotransduction, a process in which cells convert mechanical stimuli into a cellular signaling response [31]. While some baseline level of mechanical stimulation is required for the maintenance of many tissues, imparting loads above or below this baseline level can lead to adaptation within the tissue. Adaptation was first described for bone tissue by the nineteenth century German surgeon Julius Wolff. Wolff's law states that healthy bone will adapt or remodel in response to imposed loads [32]. Wolff's law has been confirmed experimentally in studies with elevated [33] and decreased loading [34]. While Wolff's law was specifically described for bone tissue, it has a corollary for soft tissues as well known as Davis' law. The law is attributed to the nineteenth century American orthopedic surgeon Henry Gassett Davis. Davis, in his book *Conservative Surgery*[35], outlined the idea of ligaments or any soft tissue remodeling while under various loads, growing in response to applied loads and shortening in response to laxity. These laws of adaptive remodeling can be used to a tissue engineer's advantage when designing bioreactors.

Altering cellular signaling and downstream gene transcription is not the only advantage of using mechanical stimuli in bioreactors. The direction of loading also plays a part in organizing and assembling aligned protein complexes in the extracellular matrix (ECM). Studies have shown that ECM fibers organize along the axis of loading in tissue engineered vascular constructs [36,37]. The organization of the ECM fibers plays an important role in the functionality of the tissue. When designing a bioreactor for tissue engineering purposes, this fiber organization and structure needs to be considered. For example, arterial walls and tendons are anisotropic tissues that contain fiber alignment to support their physiological function. Conversely, liver tissue may be considered mechanically isotropic, meaning that fiber alignment will be less important in creating an artificial liver.

Mechanical stimulation can also be used to guide stem cell differentiation. Many tissue engineering approaches

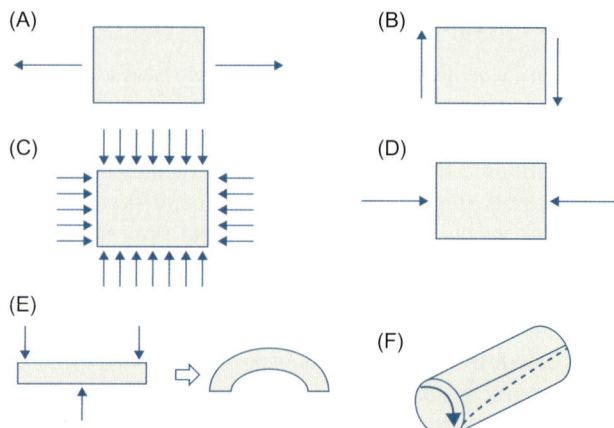

FIGURE 13.3   Loading conditions used in bioreactors for tissue engineering: (A) uniaxial stretch, (B) shear stress, (C) pressure, (D) compression, (E) bending, and (F) torsion.

utilize undifferentiated stem cells as a starting point. Stem cells are advantageous for engineering purposes because they are self-renewing and multipotent. The presence of certain mechanical loading conditions increases stem cell proliferation [38–41], thus enhancing the inherent self-renewal ability of stem cells. Stem cell fate can be guided by mechanical loading through a process known as mechanodifferentiation [40]. Mechanical stimulation is therefore a key design consideration for bioreactors that are being used for stem cell-based tissue engineered tissues.

Though many complex loading patterns can exist within the body, most loading patterns used in bioreactors can be described simply through some combination of stretch (Figure 13.3A), shear (Figure 13.3B), pressure (Figure 13.3C), compression (Figure 13.3D), bending (Figure 13.3E), and torsion (Figure 13.3F).

13.2.1 Stretch

Stretch is defined as the displaced length divided by the initial length. This displacement can be in any direction and sometimes even in multiple directions at the same time. In vascular tissue engineering, cyclic stretch has been shown to increase structural (ECM) protein content, increase contractile protein expression, and differentiate stem cells to a more contractile phenotype [42]. Stretch bioreactors have wide applications throughout tissue engineering: tendons [43], ligaments [44,45], heart valves [46], cardiac muscle [47,48], and smooth muscle [49–52] tissues all have benefited from bioreactor culture. The application of stretch in these contexts leads to tissue engineered constructs with ECM and cell alignment similar to that of the native tissues.

13.2.2 Shear

Shear stress is defined as the component of stress that acts parallel to a material cross section. The most common source of shear stress occurs when forces are applied directly parallel to a surface like the fluid shear stress that occurs in vascular tissue from flowing blood interacting with the vessel wall. This shear stress, as well as changes in it, can be detected by the ECs lining the blood vessel wall. ECs send paracrine signals to smooth muscle cells (SMCs) in the vascular tunica media layer maintaining a homeostatic level of shear stress. When shear on the ECs is increased, they release vasodilators causing the SMCs to relax and increase the vessel's diameter thereby returning shear stress to homeostatic levels. When the shear stress on ECs is lowered, they release vasoconstrictors causing the SMCs to contract and decrease the vessel diameter. Bioreactors that impart shear stress on cells have been used to differentiate mesenchymal stem cells to a more endothelial-like cell that expresses some endothelial markers [40,53], and enhance the endothelialization of scaffold surfaces [54] (Figure 13.4).

While most of the discussion above focuses on shear stress for blood vessels it also plays an important role in how heterogeneous tissues like bone are formed. The flow of fluid and marrow through the pores in trabecular bone

or scaffolds can affect developing bone tissue [11,55]. While shear can be explicitly designed into bioreactors, it may often go unnoticed in perfusion and mixing bioreactors which impart some level of shear stress on the tissue; therefore, the bioreactor designer should consider the effects of this additional mechanical stimulus.

13.2.3 Pressure

Pressure is defined as a load that is evenly distributed over an area. Our bodies are constantly exposed to this stimulus. Skin, for example, is constantly subjected to atmospheric pressure while arteries are constantly under pressure loads from blood. Stem cells in bioreactors change their shape in response to pressure, assuming a cobblestone morphology similar to ECs [40,41]. Bioreactors have also been designed to expose tissue engineered tubes to pressure at the luminal surface [49−52]. Additionally, these pressures can cause stretching in the circumferential direction of the tube and could be considered a type of stretch bioreactor (Figure 13.5).

Consider the case of engineered tissues for heart valve repair. In the heart valve, the pressure differences across the valve provide the mechanical stimulus to open and close the valve resulting in a complex strain pattern.

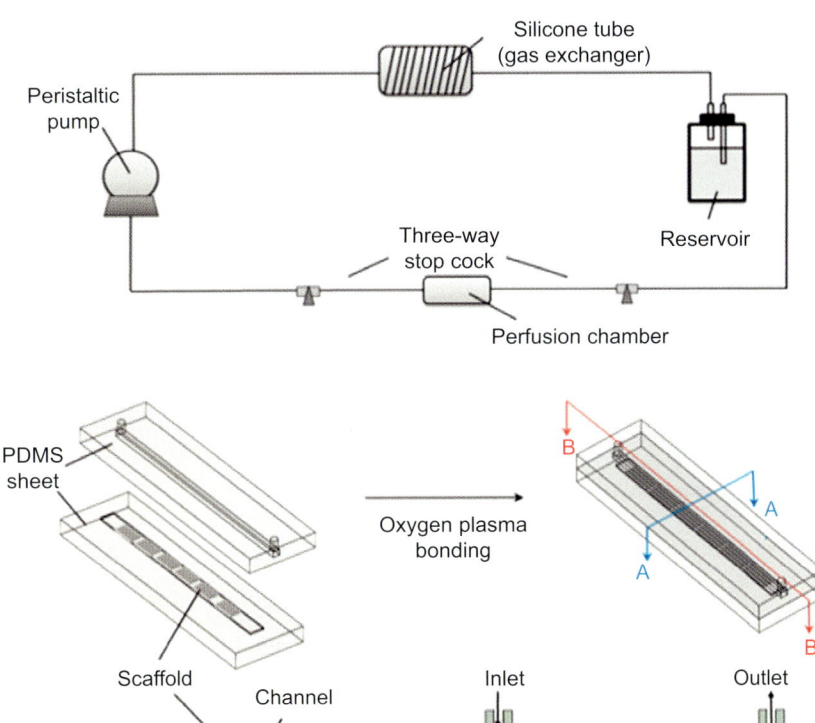

FIGURE 13.4 Schematic illustration bioreactor providing shear stress via parallel flow. *From Ref. [54] with permission.*

Bioreactors have been created to deliver this pressure gradient to tissue engineered heart valves [56–58], and have produced tissues with mechanical properties closer to that of native tissues when compared to statically cultured counterparts.

13.2.4 Compression

Compression is defined as a directional force leading to a stretch value <1 in the loading direction. Many tissues in the body experience some amount of compression—forces pushing inward. These forces are directional. For example, blood vessels experience compression in the radial direction due to blood pressure acting on the lumen; however, the most obvious tissues under high compression are weight bearing tissues surrounding joints like cartilage in the knee joint. Engineered cartilage has been created using bioreactors that subject the constructs

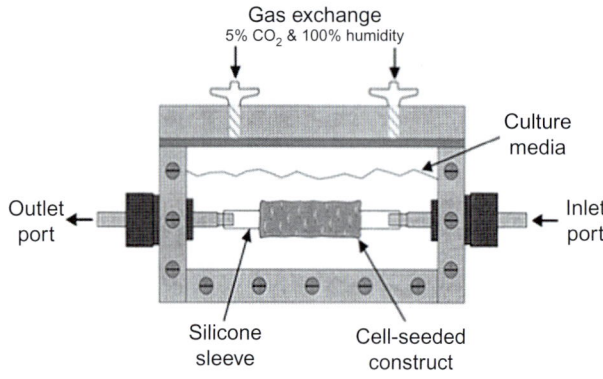

FIGURE 13.5 Schematic illustration of a cyclic pressure bioreactor which also provides circumferential stretch in the construct. *From Ref. [36] with permission.*

to compression [59–61]; dynamic compression results in engineered cartilage with a more organized structure and higher glycosaminoglycan content. Since cartilage is a heterogeneous tissue, it is important to understand the microscale loads caused by macroscale loading. For example, when cartilage tissue is compressed at the macroscopic level, fluid within this tissue will flow and impart shear on the interstitial components at the microscopic level (Figure 13.6).

13.2.5 Bending/Torsion

Sometimes tissues are subjected to even more complex loading conditions than previously discussed. The coronary artery is a prime example. This artery is wrapped around the heart causing bending due to its geometrical constraints. Additional bending and torsion result from displacements created while the heart expands and contracts. These loading conditions have motivated the creation of bioreactors that impart cyclic bending or flexure [62] and cyclic twisting or torsion and extension [63] on arterial segments. Musculoskeletal tissues represent another example of tissues under bending and torsion moments, and these loading conditions should be considered designing bioreactors for the tissue engineering of bone, cartilage, tendon, and ligament tissues (Figure 13.7).

13.3 CONCLUSION

The design principles behind bioreactors used for tissue engineering can be categorized as elements which promote tissue health and elements which guide tissue function. A number of different mixing bioreactor designs can be utilized to ensure sufficient nutrition delivery and cell

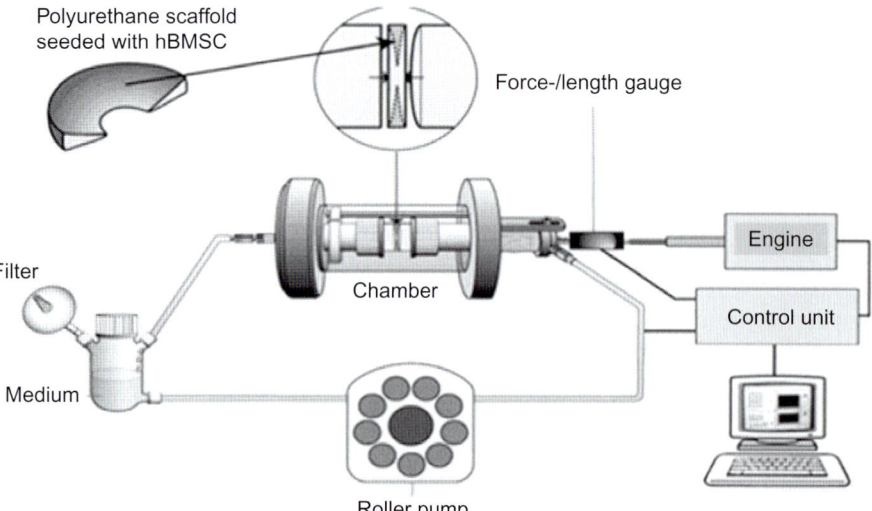

FIGURE 13.6 Schematic illustration of a cyclic pressure bioreactor. *From Ref. [61] with permission.*

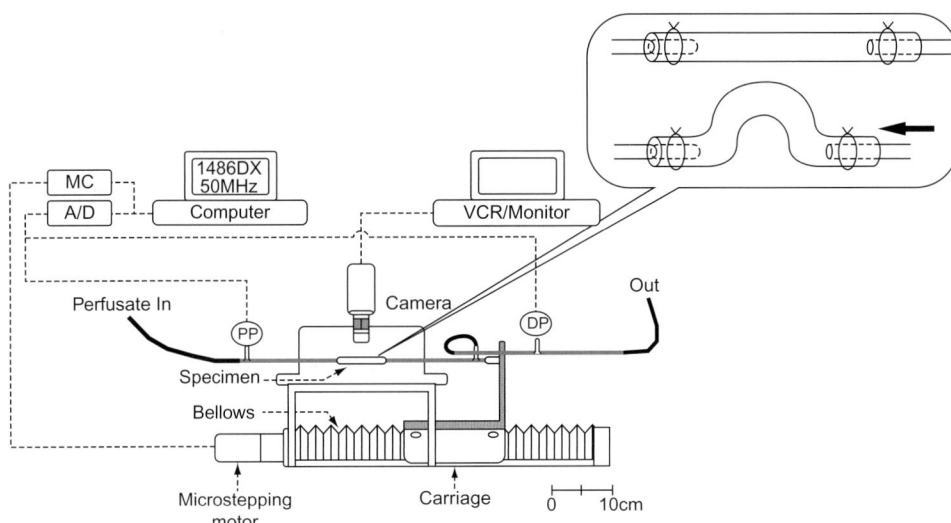

FIGURE 13.7 Schematic illustration of a cyclic bending. *From Ref. [62] with permission.*

survival. Tissue structure, organization, and function can be formed through a variety of bioreactors that utilize appropriate mechanical loading. Proper implementation of these bioreactor design principles can lead to better tissue engineered products like artificial organs for transplantation.

REFERENCES

[1] Gunatillake PA, Adhikari R. Biodegradable synthetic polymers for tissue engineering. Eur Cell Mater 2003;5:1−16.

[2] Bird RB, Stewart WE, Lightfoot EN. Transport phenomena. Hoboken, New Jersey: Wiley; 2006.

[3] Griffith LG, Swartz MA. Capturing complex 3D tissue physiology *in vitro*. Nat Rev Mol Cell Biol 2006;7:211−24.

[4] Nieponice A, Soletti L, Guan J, Deasy BM, Huard J, Wagner WR, et al. Development of a tissue-engineered vascular graft combining a biodegradable scaffold, muscle-derived stem cells and a rotational vacuum seeding technique. Biomaterials 2008;29:825−33.

[5] Sucosky P, Osorio DF, Brown JB, Neitzel GP. Fluid mechanics of a spinner-flask bioreactor. Biotechnol Bioeng 2004;85:34−46.

[6] Cherry RS, Papoutsakis ET. Physical mechanisms of cell damage in microcarrier cell culture bioreactors. Biotechnol Bioeng 1988;32:1001−14.

[7] Almarza AJ, Athanasiou KA. Seeding techniques and scaffolding choice for tissue engineering of the temporomandibular joint disk. Tissue Eng 2004;10:1787−95.

[8] Chang C-H, Liu H-C, Lin C-C, Chou CH, Lin F-H. Gelatin−chondroitin−hyaluronan tri-copolymer scaffold for cartilage tissue engineering. Biomaterials 2003;24:4853−8.

[9] Vunjak-Novakovic G, Freed LE, Biron RJ, Langer R. Effects of mixing on the composition and morphology of tissue-engineered cartilage. AIChE J 1996;42:850−60.

[10] Sikavitsas VI, Bancroft GN, Mikos AG. Formation of three-dimensional cell/polymer constructs for bone tissue engineering in a spinner flask and a rotating wall vessel bioreactor. J Biomed Mater Res 2002;62:136−48.

[11] Meinel L, Karageorgiou V, Fajardo R, Snyder B, Shinde-Patil V, Zichner L, et al. Bone tissue engineering using human mesenchymal stem cells: effects of scaffold material and medium flow. Ann Biomed Eng 2004;32:112−22.

[12] Bursac N, Papadaki M, Cohen R, Schoen F, Eisenberg S, Carrier R, et al. Cardiac muscle tissue engineering: toward an *in vitro* model for electrophysiological studies. Am J Physiol Heart Circ Physiol 1999;277:433−44.

[13] King JA, Miller WM. Bioreactor development for stem cell expansion and controlled differentiation. Curr Opin Chem Biol 2007;11:394−8.

[14] DeKosky BJ, Dormer NH, Ingavle GC, Roatch CH, Lomakin J, Detamore MS, et al. Hierarchically designed agarose and poly (ethylene glycol) interpenetrating network hydrogels for cartilage tissue engineering. Tissue Eng Part C Methods 2010;16:1533−42.

[15] Ungaro F, Biondi M, d'Angelo I, Indolfi L, Quaglia F, Netti PA, et al. Microsphere-integrated collagen scaffolds for tissue engineering: effect of microsphere formulation and scaffold properties on protein release kinetics. J Control Release 2006;113:128−36.

[16] Massia SP, Holecko MM, Ehteshami GR. *In vitro* assessment of bioactive coatings for neural implant applications. J Biomed Mater Res Part A 2004;68A:177−86.

[17] Just L, Kursten A, Borth-Bruhns T, Lindenmaier W, Rohde M, Dittmar K, et al. Formation of three-dimensional fetal myocardial tissue cultures from rat for long-term cultivation. Dev Dyn 2006;235:2200−9.

[18] Chen P, Marsilio E, Goldstein RH, Yannas IV, Spector M. Formation of lung alveolar-like structures in collagen-glycosaminoglycan scaffolds *in vitro*. Tissue Eng 2005;11:1436−48.

[19] Martin I, Wendt D, Heberer M. The role of bioreactors in tissue engineering. Trends Biotechnol 2004;22:80−6.

[20] Carrier RL, Rupnick M, Langer R, Schoen FJ, Freed LE, Vunjak-Novakovic G. Perfusion improves tissue architecture of engineered cardiac muscle. Tissue Eng 2002;8:175–88.

[21] Shirota T, He H, Yasui H, Matsuda T. Human endothelial progenitor cell-seeded hybrid graft: proliferative and antithrombogenic potentials *in vitro* and fabrication processing. Tissue Eng 2003;9:127–36.

[22] Van Wachem P, Stronck J, Koers-Zuideveld R, Dijk F, Wildevuur CR. Vacuum cell seeding: a new method for the fast application of an evenly distributed cell layer on porous vascular grafts. Biomaterials 1990;11:602–6.

[23] Ito A, Ino K, Hayashida M, Kobayashi T, Matsunuma H, Kagami H, et al. Novel methodology for fabrication of tissue-engineered tubular constructs using magnetite nanoparticles and magnetic force. Tissue Eng 2005;11:1553–61.

[24] Li H, Friend JR, Yeo LY. A scaffold cell seeding method driven by surface acoustic waves. Biomaterials 2007;28:4098–104.

[25] Alvarez-Barreto JF, Linehan SM, Shambaugh RL, Sikavitsas VI. Flow perfusion improves seeding of tissue engineering scaffolds with different architectures. Ann Biomed Eng 2007; 35:429–42.

[26] Mazzucotelli JP, Roudière JL, Bernex F, Bertrand P, Léandri J, Loisance D. A new device for endothelial cell seeding of a small-caliber vascular prosthesis. Artif Organs 1993;17:787–90.

[27] Soletti L, Nieponice A, Guan J, Stankus JJ, Wagner WR, Vorp DA. A seeding device for tissue engineered tubular structures. Biomaterials 2006;27:4863–70.

[28] Stankus JJ, Soletti L, Fujimoto K, Hong Y, Vorp DA, Wagner WR. Fabrication of cell microintegrated blood vessel constructs through electrohydrodynamic atomization. Biomaterials 2007; 28:2738–46.

[29] Grassl ED, Oegema TR, Tranquillo RT. A fibrin-based arterial media equivalent. J Biomed Mater Res Part A 2003;66:550–61.

[30] Villalona GA, Udelsman B, Duncan DR, McGillicuddy E, Sawh-Martinez RF, Hibino N, et al. Cell-seeding techniques in vascular tissue engineering. Tissue Eng Part B: Rev 2010;16:341–50.

[31] Duncan R, Turner C. Mechanotransduction and the functional response of bone to mechanical strain. Calcif Tissue Int 1995; 57:344–58.

[32] Wolff J. Das Gesetz der transformation der knochen/A. Berlin, Germany: Hirshwald; 1892.

[33] Goodship A, Lanyon L, McFie H. Functional adaptation of bone to increased stress. J Bone Joint Surg A 1979;61:539–46.

[34] Vico L, Chappard D, Palle S, Bakulin A, Novikov V, Alexandre C. Trabecular bone remodeling after seven days of weightlessness exposure (BIOCOSMOS 1667). Am J Physiol Regul Integr Comp Physiol 1988;255:243–7.

[35] Davis HG. Conservative surgery. New York: Appleton; 1867.

[36] Seliktar D, Black RA, Vito RP, Nerem RM. Dynamic mechanical conditioning of collagen-gel blood vessel constructs induces remodeling *in vitro*. Ann Biomed Eng 2000;28:351–62.

[37] Stickler P, De Visscher G, Mesure L, Famaey N, Martin D, Campbell JH, et al. Cyclically stretching developing tissue *in vivo* enhances mechanical strength and organization of vascular grafts. Acta Biomater 2010;6:2448–56.

[38] Kurpinski K, Chu J, Hashi C, Li S. Anisotropic mechanosensing by mesenchymal stem cells. Proc Natl Acad Sci USA 2006;103:16095–100.

[39] Lee W-C, Maul T, Vorp D, Rubin JP, Marra K. Effects of uniaxial cyclic strain on adipose-derived stem cell morphology, proliferation, and differentiation. Biomech Model Mechanobiol 2007;6:265–73.

[40] Maul T, Chew D, Nieponice A, Vorp D. Mechanical stimuli differentially control stem cell behavior: morphology, proliferation, and differentiation. Biomech Model Mechanobiol 2011;10:939–53.

[41] Maul TM, Hamilton DW, Nieponice A, Soletti L, Vorp DA. A new experimental system for the extended application of cyclic hydrostatic pressure to cell culture. J Biomech Eng 2007; 129:110–6.

[42] Nieponice A, Maul TM, Cumer JM, Soletti L, Vorp DA. Mechanical stimulation induces morphological and phenotypic changes in bone marrow-derived progenitor cells within a three-dimensional fibrin matrix. J Biomed Mater Res Part A 2007;81:523–30.

[43] Saber S, Zhang AY, Ki SH, Lindsey DP, Smith RL, Riboh J, et al. Flexor tendon tissue engineering: bioreactor cyclic strain increases construct strength. Tissue Eng Part A 2010;16:2085–90.

[44] Lee CH, Shin HJ, Cho IH, Kang Y-M, Kim I, Park KD, et al. Nanofiber alignment and direction of mechanical strain affect the ECM production of human ACL fibroblast. Biomaterials 2005;26:1261–70.

[45] Langelier E, Rancourt D, Bouchard S, Lord C, Stevens PP, Germain L, et al. Cyclic traction machine for long-term culture of fibroblast-populated collagen gels. Ann Biomed Eng 1999;27:67–72.

[46] Mol A, Bouten C, Zund G, Gunter C, Visjager J, Turina M, et al. The relevance of large strains in functional tissue engineering of heart valves. Thorac Cardiovasc Surg 2003;51:78–83.

[47] Gonen-Wadmany M, Gepstein L, Seliktar D. Controlling the cellular organization of tissue-engineered cardiac constructs. Ann N Y Acad Sci 2006;1015:299–311.

[48] Fink C, Ergun S, Kralisch D, Remmers U, Weil J, Eschenhagen T. Chronic stretch of engineered heart tissue induces hypertrophy and functional improvement. FASEB J 2000;14:669–79.

[49] Syedain ZH, Meier LA, Bjork JW, Lee A, Tranquillo RT. Implantable arterial grafts from human fibroblasts and fibrin using a multi-graft pulsed flow-stretch bioreactor with noninvasive strength monitoring. Biomaterials 2011;32:714–22.

[50] Williams C, Wick TM. Perfusion bioreactor for small diameter tissue-engineered arteries. Tissue Eng 2004;10:930–41.

[51] Hoerstrup SP, Sodian R, Sperling JS, Vacanti JP, Mayer Jr JE. New pulsatile bioreactor for *in vitro* formation of tissue engineered heart valves. Tissue Eng 2000;6:75–9.

[52] Jeong SI, Kwon JH, Lim JI, Cho S-W, Jung Y, Sung WJ, et al. Mechano-active tissue engineering of vascular smooth muscle using pulsatile perfusion bioreactors and elastic PLCL scaffolds. Biomaterials 2005;26:1405–11.

[53] Wang H, Riha GM, Yan S, Li M, Chai H, Yang H, et al. Shear stress induces endothelial differentiation from a murine embryonic mesenchymal progenitor cell line. Arterioscler Thromb Vasc Biol 2005;25:1817–23.

[54] Kang TY, Hong JM, Kim BJ, Cha HJ, Cho DW. Enhanced endothelialization for developing artificial vascular networks with a natural vessel mimicking the luminal surface in scaffolds. Acta Biomater 2013;9:4716–25.

[55] Yu X, Botchwey EA, Levine EM, Pollack SR, Laurencin CT. Bioreactor-based bone tissue engineering: the influence of dynamic flow on osteoblast phenotypic expression and matrix mineralization. Proc Natl Acad Sci USA 2004;101:11203−8.

[56] Engelmayr GC, Sales VL, Mayer JE, Sacks MS. Cyclic flexure and laminar flow synergistically accelerate mesenchymal stem cell-mediated engineered tissue formation: implications for engineered heart valve tissues. Biomaterials 2006;27:6083−95.

[57] Engelmayr GC, Hildebrand DK, Sutherland FW, Mayer JE, Sacks MS. A novel bioreactor for the dynamic flexural stimulation of tissue engineered heart valve biomaterials. Biomaterials 2003;24:2523−32.

[58] Mol A, Driessen NJ, Rutten MC, Hoerstrup SP, Bouten CV, Baaijens FP. Tissue engineering of human heart valve leaflets: a novel bioreactor for a strain-based conditioning approach. Ann Biomed Eng 2005;33:1778−88.

[59] Hung CT, Mauck RL, Wang CC-B, Lima EG, Ateshian GA. A paradigm for functional tissue engineering of articular cartilage via applied physiologic deformational loading. Ann Biomed Eng 2004;32:35−49.

[60] Seidel J, Pei M, Gray M, Langer R, Freed L, Vunjak-Novakovic G. Long-term culture of tissue engineered cartilage in a perfused chamber with mechanical stimulation. Biorheology 2004; 41:445−58.

[61] Liu C, Abedian R, Meister R, Haasper C, Hurschler C, Krettek C, et al. Influence of perfusion and compression on the proliferation and differentiation of bone mesenchymal stromal cells seeded on polyurethane scaffolds. Biomaterials 2012;33: 1052−64.

[62] Vorp DA, Peters DG, Webster MW. Gene expression is altered in perfused arterial segments exposed to cyclic flexure *ex vivo*. Ann Biomed Eng 1999;27:366−71.

[63] Vorp DA, Severyn DA, Steed DL, Webster MW. A device for the application of cyclic twist and extension on perfused vascular segments. Am J Physiol Heart Circ Physiol 1996; 270:787−95.

Part II

Kidney

Chapter 14

Current Status of Renal Transplantation

Jeffrey Rogers

Department of General Surgery, Section of Transplantation, Wake Forest Baptist Medical Center, Winston-Salem, NC

Chapter Outline

Kidney transplantation continues to be the preferred modality of renal replacement therapy for appropriately selected candidates with advanced chronic kidney disease or end-stage renal disease (ESRD), providing superior quality and length of life in comparison to untransplanted candidates awaiting transplantation. Advances in immunosuppression and in our understanding of mechanisms of acute and chronic rejection, as well as other causes of graft failure, have resulted in decreasing rates of acute rejection and improved short-term graft survival. Unfortunately, the transplant waiting list continues to grow in the face of a relatively stagnant deceased donor organ pool, resulting in longer patient waiting times and an increased likelihood of waitlist mortality. A variety of strategies have evolved in response to the ongoing organ shortage, including increasing use of "marginal" or "suboptimal" organs, efforts to improve the national donation conversion rate, and methods to increase living donation. Despite these efforts, the demand for kidney transplantation continues to outpace the available supply of donor organs. A substantial revision of the current kidney allocation policy is now being proposed in an attempt to optimize kidney utilization and transplant outcomes. The aim of this chapter is to review the current status of kidney transplantation in the United States, with specific attention to outcomes, efforts to overcome the barriers and limitations outlined above, and outlook for the future.

14.1 THE WAITING LIST

The number of candidates on the kidney transplant waiting list continues to grow annually with over 95,000 candidates registered [1]. A major change in Organ Procurement and Transplantation Network (OPTN) policy in 2003 allowed inactive patients to accrue waiting time on the list. Because of the increasing time between wait listing and transplant, a consequence of this policy change has been an increase in the number of inactive registrants, most commonly due to incomplete candidate work up, insurance issues, or severity of illness precluding transplantation [2]. The number of candidates who were inactive within 7 days of wait listing increased from 718 in 2003 to 9628 in 2011. The number of active candidates on the waiting list has risen steadily from 7404 in 2003 to 32,501 in 2011 [2]. The growing number of candidates on the waiting list in the face of relatively stable organ donation rates has resulted in a steady decrease in transplant rates for adult candidates awaiting kidney transplantation since 1998. The deceased donor transplant rate decreased from 20.6 transplants per 100 waitlist years in 1998 to 11.4 transplants per 100 waitlist years in 2011 [2]. Consequently, more than 20,000 candidates were removed from the waiting list over the last 3 years because they died or became too sick to undergo transplant [2]. The wait listed population continues to age, with patients aged 50 years or older representing a steadily rising proportion of wait listed candidates. Despite this fact, there has been a trend towards decreasing pretransplant mortality in wait listed candidates. The transplant rate within 5 years of wait listing for patients with panel reactive antibody (PRA) 80–100% is 30.5%, which is comparable to that of patients with PRA <1%, as a result of allocation priority points provided to highly sensitized patients [2].

14.2 DONATION

Rates of deceased donation remained flat in 2010, with 21.8 donors per million population and 2.4 donations per 100,000 deaths in 2009–2010 [3]. Even more concerning, an increasing number of kidneys recovered for transplant are discarded. The discard rate has increased steadily from 12.7% in 2002 to 17.9% in 2011. The most common donor-specific reasons for kidney discard after procurement were biopsy findings (37.3%), poor organ function (9.2%), and anatomic abnormalities (7.1%) [2]. Furthermore, in 2011, kidneys were not recovered from 9% of donors in whom at least one other organ was recovered for transplant, primarily due to poor organ function and donor medical history. The increasing rates of kidney discard and nonrecovery undoubtedly reflect a deceased donor pool which is clearly aging and has more associated risks for graft failure. Improvements in organ donation, utilization, and allocation are needed to maximize the benefits of transplant from an aging and relatively stagnant donor pool.

14.3 OUTCOMES

Patient survival after kidney transplantation remains excellent, with 1-year unadjusted survival rates of 95% for deceased donor recipients and 98% for living donor recipients [4]. Five-year patient survival is higher for recipients of living donor kidneys (90%), compared to recipients of standard criteria deceased donor (SCD) kidneys (83%) and expanded criteria deceased donor (ECD) kidneys (69%) [4]. Over the past decade, there has been consistent improvement in 90-day, 6-month, and 1-, 3-, and 5-year graft survival, with a suggestion of improvement in 10-year outcomes for both living and deceased donor recipients (Table 14.1, only deceased donor data shown). Over the past 10 years, graft half-life has improved by

TABLE 14.1 Death-Censored Deceased Donor Kidney Graft Survival.

Year	30 Days	6 Months	1 Year	3 Years	5 Years	10 Years
1998	94.8%	94%	92.5%	86.3%	79.2%	66.4%
1999	95.2%	94.5%	92.7%	86.9%	80.5%	68.2%
2000	95%	94.4%	92.6%	86.2%	80.2%	68.6%
2001	95.5%	94.9%	93.2%	87.7%	81.9%	70.4%
2002	95.5%	94.7%	93.1%	87%	81.2%	
2003	95.6%	95.1%	93.5%	87.7%	82.6%	
2004	95.9%	95.1%	93.6%	87.4%	82.4%	
2005	96.3%	95.6%	94%	88.7%	83.9%	
2006	96.6%	95.9%	94.3%	89.2%	84%	
2007	96.9%	96.1%	94.8%	89.9%		
2008	96.8%	96.2%	94.9%	90.6%		
2009	97.3%	96.5%	95.3%			
2010	96.9%	96.3%	95.2%			
2011	99.1%	97.1%				

Source: Data from Matas et al. [2]

TABLE 14.2 Deceased Donor Kidney Graft Half-Life for Grafts Functioning Beyond the First Year.

Year	Half-Life (Years)
1999	9.8
2000	9.7
2001	9.9
2002	10.2
2003	10.2
2004	10.5
2005	10.7
2006	10.6
2007	10.7
2008	10.7
2009	10.7

Source: Data from Matas et al. [2]

about 1 year, for grafts functioning beyond the first year in both living and deceased donor recipients (Table 14.2, only deceased donor data shown). This improvement can be attributed to a decrease in the rate of graft failure and return to dialysis. The rate of death with functioning graft has not changed significantly. As of June 30, 2011, 164,200 adults in the United States were alive with a functioning kidney transplant, about twice as many as in the previous decade [2]. The incidence of acute rejection in the first year posttransplant has declined by approximately 50% over the past 10 years, reported as 11% in deceased donors and 10% in living donor recipients in 2009 [3]. Despite these improvements, there is a persistent, chronic, long-term deterioration in graft survival, with 5-year graft survival rates of 80% for living donor kidneys and 68% for deceased donor kidneys [4].

14.4 CAUSES OF GRAFT LOSS

According to registry data, "chronic rejection" is the most common etiology of late graft failure; however, this terminology is a misnomer since it implies that late allograft scarring is related to T-cell-mediated alloimmune injury [4]. The term "chronic allograft nephropathy" (CAN) was subsequently introduced to dispel this misconception. In 2007, CAN was removed from the Banff classification for kidney allograft pathology because the use of this terminology was thought to undermine the identification of histopathologic findings associated with specific causes of chronic allograft dysfunction. Recent efforts have focused on identifying specific etiologies which lead to the lesions of interstitial fibrosis and tubular atrophy (IF/TA) and chronic glomerular injury [4]. These lesions are nonspecific responses to injury; consequently, a wide variety of potential etiologies have been postulated, including antibody-mediated endothelial activation, calcineurin inhibitor (CNI) toxicity, recurrence of primary renal disease, chronic inflammation, innate immune mechanisms, hypertension, and diabetes [4]. It is likely that a combination of immunologic and nonimmunologic causes of tubular injury results in a final common pathway leading to graft fibrosis, although the exact mechanisms responsible for this process are still being investigated. One of the consequences of heightened immunosuppression in the current era has been the emergence of polyomavirus (BK virus) as a significant cause of graft failure, with as many as 10% of new transplants affected in the last decade [5,6]. It is believed that the net state of immunosuppression, rather than a specific immunosuppressive agent, is responsible for viral reactivation. The mainstay of treatment is reduction of immunosuppression, although this predisposes to acute rejection and chronic graft dysfunction. Over 50% of patients with biopsy proven polyomavirus nephropathy develop graft failure. Treatment modalities include cidofovir, intravenous immunoglobulin (IVIG), leflunomide, and quinolones, all of which have been used in uncontrolled trials in conjunction with immunosuppression reduction. There is evidence that an approach of prospective viral load monitoring with immunosuppression reduction upon detection of viremia prevents the development of polyomavirus nephropathy without an increased incidence of rejection [7]. Successful retransplantation after graft loss secondary to polyomavirus is possible.

14.5 MECHANISMS OF REJECTION

Acute allograft rejection is an immune-mediated process involving T cells, antibody, complement, and other cell types, and primarily affects tubular and endothelial cells [8]. Acute T-cell-mediated rejection typically manifests as a mononuclear interstitial infiltrate, largely consisting of T cells and macrophages, with varying degrees of tubulitis and/or arteritis [9]. Although the precise mechanism of T-cell-mediated graft injury has not been defined, it is thought to be related to cell-mediated cytotoxicity and cytokine release. $CD8^+$ class I reactive T cells act on target cells via perforin and granzyme B or Fas/FasL cytotoxicity pathways [4]. These theories are supported by recent gene expression analysis of human kidney allograft biopsies and urine specimens obtained during acute rejection episodes, which revealed selective expression of mRNA transcripts for cytokines and molecules associated with cytotoxic lymphocytes [10,11]. Acute antibody-mediated rejection (AMR) can occur concurrently with T-cell-mediated rejection or independently, and manifests as neutrophilic and mononuclear infiltration of the peritubular and glomerular capillaries [12]. The availability of C4d staining has significantly facilitated identification of AMR and correlates strongly with histologic features. Approximately 90% of patients with positive C4d staining on biopsy in the setting of acute allograft dysfunction have human leukocyte antigen (HLA) class I or II antibodies compared to <10% in C4d negative rejection. C4d positivity in the absence of identified circulating HLA antibody may be due to antibody absorption by the graft or the presence of non-HLA antibodies. Conversely, histologic changes of AMR may exist in the absence of C4d deposition but in the presence of circulating antibody [4]. The histologic findings associated with chronic AMR have become better defined in recent years. Biopsy findings include transplant glomerulopathy, peritubular capillaritis, transplant arteriopathy, and IF/TA [13]. These histologic findings, in combination with C4d deposition in the peritubular capillaries and circulating donor-specific antibodies, are diagnostic of chronic AMR and are associated with poor graft survival. It is unclear whether modification or augmentation of immunosuppression is beneficial in this setting. Antibodies to MICA as well as other non-HLA antibodies are also believed to contribute to late graft failure [14].

14.6 EXPANDING THE DECEASED DONOR ORGAN POOL

The burgeoning crisis in organ supply challenges the transplant community to maximize and optimize the use of organs from all consented deceased donors. The persistent scarcity of available deceased donor kidneys in the face of a growing waitlist mandates an ongoing reappraisal of the limits of acceptability when considering whether or not to transplant a recovered kidney. The annual mortality of a patient on the waiting list is 6% and as high as 10% in diabetic candidates. When the risk of waitlist mortality is considered in the context of a potential transplant recipient's age and other comorbidities, it is clear that the amount of time patients can "afford" to wait for a kidney transplant is highly variable. More aggressive utilization of kidneys which may be collectively termed "marginal" takes into account that, in appropriately selected recipients, more timely transplantation and improved patient survival in exchange for decreased long-term graft survival may represent a worthwhile trade-off. The next several sections focus on various strategies which are being employed in ongoing efforts to maximize utilization of deceased donor kidneys.

14.7 EXPANDED CRITERIA DONORS

Over the past decade, there has been an increasing, yet disproportionate shift toward increasing numbers of older donors and recipients in kidney transplantation, with cerebrovascular events now the leading cause of brain death in deceased donation [15]. The aging recipient and donor population have inevitably resulted in increased transplantation of kidneys from older donors. The value of transplanting such kidneys was previously questioned because of concerns over decreased graft survival and poorer predicted outcomes [16−18]. In October 2007, The United Network of Organ Sharing (UNOS) defined ECD as donor age >60 years or 50−59 years with at least two of the following criteria: history of hypertension, terminal serum creatinine >1.5 mg/dL, and cerebrovascular cause of death. This definition was derived from a Scientific Registry of Transplant Recipients (SRTR) retrospective analysis of primary deceased donor adult kidney transplants. Based on this analysis, ECD kidneys had a 70% higher risk of graft failure compared to a reference group of kidneys from donors aged 10 to 39, without hypertension, cerebrovascular cause of death, or terminal serum creatinine >1.5 mg/dL [16,18]. Optimal use of ECD kidneys continues to be debated; however, there is general agreement that these kidneys should be used to improve access to transplantation for patients whose life expectancy is less than their predicted waiting time for a kidney, particularly older patients and diabetics [19].

Although there are reports of inferior graft survival and poorer intermediate-term results, Stratta et al. have contended that ECD kidneys are defined by suboptimal nephron mass and that appropriate donor and recipient selection may optimize outcomes with ECD kidney transplantation [15]. Rather than transplanting ECD kidneys indiscriminately, this group preferentially transplants ECD kidneys into "low risk" and "low functional need" recipients, specifically older patients with low body mass index (BMI) and low immunologic risk. By profiling recipients in this way, acceptable intermediate-term outcomes with ECD kidney transplantation were achieved, with graft survival comparable to a group of concurrently transplanted SCD kidneys. ECD kidney graft survival has improved with in this era, possibly due to introduction of mycophenolate and subsequent decreases in CNI doses and target levels. Unfortunately, kidney quality and graft survival is highly variable within the ECD category, with some ECD kidneys outperforming SCD kidneys. Consequently, a continuous kidney donor risk index (KDRI) was proposed in 2009, combining donor and transplant variables to quantify graft failure risk [20]. There was a considerable overlap in KDRI distribution according to ECD and SCD classification. The graded impact of KDRI on graft outcome is thought to provide a more useful decision-making tool compared to simply categorizing donors as ECD or SCD. A modified version of KDRI is utilized in the recent new kidney allocation proposal currently under review. This is discussed in further detail later in this chapter.

14.8 DONATION AFTER CARDIAC DEATH

Although donation rates have not changed significantly, the percentage of transplants performed from donation after cardiac death (DCD) donors has increased steadily from 1.4% in 1998 to 15.8% in 2011 [2]. Compared with kidneys recovered from brain dead donors, DCD kidneys are subjected to variable periods of warm ischemia after withdrawal of life support, followed by declaration of death by cardiocirculatory arrest with subsequent organ procurement [21−23]. Warm ischemia is known to be associated with acute tubular necrosis, irreversible cell damage, and reduced graft survival after kidney transplantation. Despite the warm ischemia associated with DCD donation, numerous studies show comparable short- and intermediate-term graft survival rates between brain dead non-ECD kidneys and DCD non-ECD kidneys, although delayed graft function (DGF) is more common with DCD kidneys, with an incidence of 25−90% [24]. Studies have shown that older donor age, and more specifically ECD status, negatively impacts graft survival. In a study by Farney et al., DCD−ECD kidney graft survival at 3 years was 48% compared to 79% for non-ECD–DCD [24]. It has been suggested that utilizing DCD–ECD kidneys as dual transplants could potentially improve outcomes. Interestingly, several studies have shown that DGF does not have the same detrimental effect on graft survival after transplantation of DCD kidneys as it does after transplantation of donation after brain death kidneys [25–27]. Two recent studies of machine preservation reported opposite conclusions regarding the efficacy of machine preservation in reducing the incidence of DGF after DCD kidney transplantation, with an incidence of DGF > 50% in both studies [28,29]. Early experience with extracorporeal support to reduce ischemic injury and DGF is promising, although application of this technology is limited to date.

14.9 DUAL KIDNEY TRANSPLANTS

Concerns over the limited life span of ECD or other "marginal" kidneys have resulted in an increased prevalence of simultaneous transplantation of both kidneys from a donor into a single recipient. Dual kidney transplantation (DKT) is performed to optimize outcomes by providing adequate nephron mass using kidneys thought to be unsuitable for single use which might otherwise be discarded. DKT has been shown to have comparable outcomes to ideal single kidney transplantation, including similar rates of DGF and graft survival. Criteria for allocating kidneys for DKT vary, although primary considerations include donor age, donor creatinine clearance, machine preservation characteristics, cold ischemia time, and preimplantation donor biopsy characteristics [30]. A recent multicenter prospective analysis showed excellent short- and long-term results after allocation of kidneys from donors older than 60 years of age based on a preimplantation histologic scoring system [31]. While DKT graft survival was similar to solitary kidney transplant graft survival in this study, it is possible that some DKT could have had comparable results had they been transplanted as single kidneys, since the histologic scoring system allocates kidneys with only mild chronic changes to DKT. Furthermore, while donor arteriosclerosis was included in the allocation criteria, the severity and extent of these lesions were not well defined. A recent report from Kayler et al. examined outcomes of donor kidneys with moderate (> 25%) arteriosclerosis on preimplantation biopsy when transplant as single transplants or DKT [30]. In this study, single kidney transplantation from donors with moderate arteriosclerosis resulted in poor 1-year graft survival (71%) when compared to a 95% 1-year graft survival as DKT, despite the use of older and more ECD kidneys in the DKT group. Appropriate recipient selection is critical to achieving good outcomes with DKT. While it is possible to perform DKT in elderly recipients, it is important to select patients with adequate cardiac reserve and without extensive iliac artery atherosclerosis. Selecting patients with BMI < 30 mg/kg is preferable from a technical standpoint and may also help improve functional outcomes by maximizing nephron mass in recipients with lower muscle mass and decreased metabolic demand. Over the past decade, DKT has evolved from a bilateral iliac fossa approach (either intraperitoneal or extraperitoneal) to an extraperitoneal unilateral approach. This shortens operating time by limiting iliac artery and vein dissection to one side and preserves the contralateral iliac vessels, should retransplantation be required in the future.

14.10 CENTERS FOR DISEASE CONTROL AND PREVENTION HIGH-RISK KIDNEYS

The use of kidneys from donors designated by the Centers for Disease Control and Prevention (CDC) as being at increased risk for transmission of viral infections is being increasingly accepted as another approach to expanding the donor pool. The risk of transplanting organs from such donors is not the potential for decreased graft survival associated with "marginal" donors; rather, it is the possibility of transmitting potentially life-threatening viral infections, such as human immunodeficiency virus, hepatitis C, and hepatitis B. Unfortunately, transplantation of kidneys from CDC high-risk donors continues to be controversial because data on risks and outcomes have been limited. The reported incidence of viral transmission with CDC high-risk donors is exceptionally low [32,33], and may actually be lower than the risk of viral infection associated with long-term hemodialysis [34]. Despite these facts, CDC high-risk donors remain underutilized because of physician and patient perception that the risk of viral transmission is greater than it actually is. Introduction of nucleic acid testing has significantly shortened the window between the time of infection and detection, and broader availability of this testing should further minimize the risks of viral transmission with transplantation. A recent, single center, retrospective study of 50 recipients of kidneys from CDC high-risk donors revealed no instances of seroconversion with nearly 1 year of follow-up [35]. Donors in this study were younger and less like to be ECD. Willingness to accept a kidney from a CDC high-risk donor was associated with significantly shorter time on the waitlist prior to transplantation. Education of transplant physicians, surgeons, and prospective recipients regarding the actual risks and benefits associated with using kidneys from CDC high-risk donors is necessary so that informed decisions can be made and so that potentially lifesaving organs are appropriately utilized.

14.11 PEDIATRIC KIDNEYS

Kidneys from pediatric donors are still regarded controversially with respect to long-term graft survival and function, particularly kidneys from donors younger than 12 months of age. Furthermore, there is also general reluctance to separate pediatric kidneys for transplantation into two recipients. Several recent studies have demonstrated excellent results with pediatric *en bloc* kidney transplantation. Thomusch et al. performed a single center, retrospective analysis of 78 pediatric *en bloc* kidney transplants between 1989 and 2008 [36]. Mean donor age was 15 months in the pediatric *en bloc* group and 38 years in a matched control group of SCD kidney recipients. Mean follow-up was 9 years. One-, 5-, and 10-year graft survival were 83%, 76%, and 74% for the pediatric *en bloc* group and 90%, 79%, and 58% for the matched control group, respectively. One-, 5-, and 10-year serum creatinine were 1.0, 0.8, and 1.1 mg/dL for the pediatric *en bloc* group and 1.5, 1.7, and 1.6 mg/dL for the matched control group, respectively. Sharma et al. showed excellent long-term outcomes after pediatric *en bloc* kidney transplantation from donors weighing ≤15 kg, with 5-year graft survival comparable to that of living donor kidney transplantation (88%) [37]. Functional outcomes at 1 year were also similar. A recent analysis of UNOS data demonstrated that long-term outcomes of pediatric *en bloc* kidney transplants (donor <5 years of age) were superior to matched recipients of solitary pediatric kidney transplants (donor <5 years of age) and SCD adult kidney recipients [38]. The ongoing disparity between static supply and increasing demand for organs has prompted some surgeons to explore the limits of acceptability of single kidney transplantation from small pediatric donors into two recipients. Laurence et al. constructed a decision analysis model to predict outcome in life years for patients with ESRD on the waiting list, depending on whether they received *en bloc* or solitary pediatric transplants [39]. At all recipient ages, the projected life years of both recipients of a solitary transplant exceeded the projected life years of an *en bloc* kidney transplant. Only recipients of solitary kidney transplants from donors weighing <10 kg had an estimated net loss of life years. In a review of OPTN data, Sureshkumar et al. also found that the graft failure risk of solitary pediatric kidneys was consistently lower when the donor weight exceeded 10 kg [40]. However, in this study, pediatric *en bloc* kidneys outperformed solitary pediatric kidneys for all donor weight groups. In a review of SRTR data from 1995 to 2007, Kayler et al. found that the graft failure risk of solitary pediatric kidney transplants from donors weighing more than 35 kg was similar to ideal SCD kidney transplants (donor age 18–39 without other risk factors), whereas solitary kidneys from pediatric donors weighing 10–35 kg performed more like nonideal SCD kidneys [41]. In conclusion, these studies indicate that more liberal transplantation of kidneys from pediatric donors may expand the donor pool. Solitary, rather than *en bloc* transplantation of kidneys from donors weighing more than 10 kg offers more cumulative graft years and maximizes organ utilization without compromising outcomes.

14.12 KIDNEYS FROM DONORS WITH ACUTE RENAL FAILURE

Until recently, kidneys from donors with acute renal failure (ARF) were generally refused for transplantation because of the anticipated poor outcome. In 2006, Anil Kumar et al. reported a series of 55 kidneys transplanted from donors with ARF [42]. Outcome was compared with 55 concurrent and matched recipients of SCD kidneys and 55 concurrent and matched recipients of ECD kidneys. ARF kidneys were accepted from donors aged < 50 years with a negative history of kidney disease and a pretransplant kidney biopsy negative for chronic changes. Three-year patient and graft survival was 90% and 90% in the ARF group, 100% and 89% in the SCD group, and 83% and 66% in the ECD group. Biopsy proven acute rejection rates were comparable among groups; however, there was significantly more CAN in the ECD group. Mean serum creatinine levels were 1.9, 1.9, and 2.2 mg/dL in the ARF, SCD, and ECD groups, respectively (SCD and ARF vs. ECD, $P = 0.04$). It was concluded that transplantation of kidneys from selected donors with ARF provides comparable graft survival and function to non-ARF donors, despite a higher DGF rate, and that transplantation of kidneys from ARF donors may help expand the donor pool. In 2009, Kayler et al. reviewed SRTR data from 1995 to 2007 to study the outcome and utilization of kidneys from deceased donors with ARF [43]. For ECD recipients, the relative risk for graft failure significantly increased with increasing serum creatinine. Among potential SCD donors, elevated serum creatinine was a strong independent risk factor for kidney discard. However, when kidney transplantation was performed, elevated donor terminal creatinine was not a risk factor for graft loss. This study clearly underscores the fact that potentially transplantable kidneys are being discarded on the basis of elevated serum creatinine, and that a more aggressive approach to transplanting kidneys from donors with terminal ARF may increase the number of deceased donor kidneys available for transplantation. Others have also reported good results with transplantation of kidneys from donors with ARF, with 5-year graft survival rates similar to non-ARF kidneys despite a significantly higher incidence of DGF with ARF kidneys [44,45].

14.13 LIVING DONATION

Living donation is covered in detail in a separate chapter of this book; however, a brief overview of recent developments is provided herein. A number of significant innovations and advances over the past decade have increased access to living donation for kidney transplant candidates, particularly ABO incompatible donor-recipient pairs and highly sensitized recipients. HLA sensitization is a major problem, limiting access to transplantation for 30% of patients awaiting kidney transplantation [46]. Current desensitization protocols include high dose IVIG, plasmapheresis with low dose IVIG, and IVIG combined with rituximab [47]. A recent study by Montgomery et al. analyzed the outcomes of 211 highly sensitized patients who underwent live donor renal transplantation after undergoing desensitization with plasmapheresis and IVIG, compared with two carefully matched control groups of patients on the waiting list for kidney transplantation who continued to undergo dialysis (dialysis-only group) or who underwent either dialysis or HLA-compatible transplantation (dialysis or transplantation group) [48]. Live donor transplantation after desensitization provided a significant survival advantage for sensitized patients compared to waiting for a compatible organ. This survival advantage is more than doubled at 8 years. Technological advances in antibody characterization have permitted a more comprehensive assessment of anti-HLA antibody activity and have provided new insights into the clinical effects of HLA antibody class and specificity [49]. Although desensitization protocols have helped to overcome incompatibility barriers in live donor transplantation, with satisfactory early- to intermediate-term graft survival, AMR remains a significant issue, occurring in 20–50% of HLA-incompatible transplants and having a significant impact on graft survival [49]. Emerging therapies to prevent and treat AMR include proteosome inhibitors aimed at plasma cells and modifiers of complement-mediated injury [47]. ABO incompatible transplantation can now be performed successfully without high intensity immunomodulation. Current protocols involve only a brief escalation in immunosuppression using plasmapheresis and IVIG without long-term B-cell suppression from splenectomy or anti-CD20 treatment with rituximab [50]. AMR rates are low (<15%) and long-term graft survival is excellent, as high as 88% at 5 years in a recent report. A newer development has been the growth of paired kidney donation as an alternative for kidney transplant candidates with willing and medically suitable live donors who cannot donate to their intended recipient due to incompatibility of blood type, HLA crossmatch, or both. In the most basic form of paired donation, the incompatibility problems with two donor-recipient pairs can be solved by exchanging donors [51].

Using advanced software, several organizations have successfully completed paired kidney donations involving three or more pairs. Another permutation of paired kidney donor protocols is the "nondirected" or nonsimultaneous, extended, altruistic donor chain, which is initiated by an altruistic donor and completed when the last paired donor in the chain donates to an unpaired recipient on the deceased donor waiting list [51]. The proliferation of paired kidney donation programs has increased transportation of live donor kidneys and has reduced geographic barriers to paired kidney donation and live donor transplantation [52]. Recent implementation of a national paired kidney donation program should continue to promote growth of this practice. Despite the introduction of desensitization protocols, paired kidney donation, and nondirected donations, national living donation rates have remained essentially flat over the last decade [2]. Although national donation rates have not changed significantly, donation rates have increased in some areas of the country and decreased by 5–10% in other areas. Comparison of living donation rates with deceased donation rates by state reveals interesting differences. Some states have high rates for both; others low rates for both; still others, high rates for one and low rates for the other [2]. Reasons for these geographic variations require further investigation.

14.14 IMMUNOSUPPRESSION: CURRENT STATUS AND EMERGING THERAPIES

The combination of CNI, antimetabolite, and steroids remains the most prevalent maintenance immunosuppression regimen at the time of hospital discharge among US transplant centers; however, changes have occurred over the past decade. Whereas in 1997, 77% of patients were discharged on cyclosporine, in 2011 86% were discharged on tacrolimus [2]. This shift was based on a multitude of mostly uncontrolled, and often conflicting, trials demonstrating lower acute rejection rates, improved graft function, and better blood pressure control with tacrolimus, at the expense of a higher incidence of new onset diabetes, neurologic and gastrointestinal side effects. Mycophenolate mofetil remains the antimetabolite of choice based on several large multicenter trials demonstrating significant reduction of acute rejection rates compared to azathioprine in patients receiving the older formulation of cyclosporine [53,54]. More recent reports have questioned whether this benefit persists in patients receiving the microemulsion formulation of cyclosporine. Increased use of mycophenolate sodium has occurred, despite controlled trials failing to show a reduction in gastrointestinal side effects with this agent. Mammalian target of rapamycin (MTOR) inhibitor use peaked in 2001 and has steadily declined since that time, largely due to higher acute rejection rates, questionable improvement in graft function, poor tolerability, synergistic nephrotoxicity with CNI, and association with proteinuria [4]. Over the past decade, steroid use at the time of hospital discharge declined from 97% to 68%, whereas the use of induction therapy increased from 35% to 79%, with usage most prevalent in patients not placed on maintenance steroids [2]. As of 2010, 58% of recipients received T-cell depleting induction with either thymoglobulin (48%) or alemtuzumab (10%). Twenty-one percent of patients received IL-2 receptor antagonist induction (IL-2RA) [2], with possible use declining since daclizumab is no longer available. Randomized prospective trials with thymoglobulin have shown lower acute rejection rates but higher adverse even rates compared to IL-2RA in patients at high risk for DGF and acute rejection [55]. Alemtuzumab has been used as a preconditioning agent with tacrolimus monotherapy and spaced weaning [56] as well as an induction agent in combination with CNI, mycophenolate, and/or steroid maintenance immunosuppression. A recent single center randomized trial of alemtuzumab versus thymoglobulin induction in renal and pancreas transplantation showed similar safety profiles but a lower incidence of biopsy proven rejection in the alemtuzumab group in short-term follow-up [57]. Contrary to some earlier reports, late acute rejection episodes were uncommon with alemtuzumab in this study. In an effort to reduce immunosuppression-related toxicity and adverse effects, numerous approaches toward CNI and steroid minimization or elimination have been studied. Cyclosporine withdrawal in patients with stable graft function in both the azathioprine and mycophenolate eras has uniformly resulted in increased acute rejection rates with worse graft survivals in most studies [58], whereas CNI reduction in patients with deteriorating graft function had a beneficial effect on graft function. CNI avoidance in *de novo* renal transplantation by substituting low dose rapamycin in the SYMPHONY study resulted in higher rejection rates and worse graft survival compared to the CNI arm [59]. Results from the CONVERT study indicated that conversion from CNI to rapamycin could be safely performed after 6 months if baseline glomerular filtration rate was > 40 mL/min and urinary protein was normal [60]. However, the benefits of conversion on graft survival were small, and this approach has not yet been shown to improve graft survival. Conversion was associated with an increase in adverse events, although a significantly lower incidence of malignancy was observed. Recent FDA approval of belatacept, which works by costimulation blockade, marked the availability of the first significantly novel therapeutic agent for use in transplantation in decades. It is approved for use in combination with basiliximab induction, mycophenolate mofetil, and steroids. In low to moderate immunologic risk patients, short-term patient and graft survival appear comparable to that seen with cyclosporine, with improved renal function despite higher early rejection rates [61]. Preliminary data from phase 2 steroid avoidance and conversion trials suggest that better renal function, acceptable rejection rates, and comparable patient and graft survival may be achieved with belatacept compared with CNI. Other benefits of belatacept include favorable cardiovascular and metabolic profiles as well as the lack of requirement for therapeutic drug monitoring. Downsides include a higher incidence of posttransplant lymphoproliferative disorder (which may be decreased by avoiding transplantation of kidneys from EBV seropositive donors into EBV seronegative recipients) and the need for scheduled outpatient intravenous infusions. At this time, it is unclear if the significantly increased drug costs associated with belatacept will be offset by long-term improvements in patient and graft survival. Additional clinical experience with belatacept will be needed to determine its place in the immunosuppression armamentarium. Several uncontrolled studies in the current immunosuppression era have reported acceptable acute rejection rates and excellent short-term graft survival after steroid minimization. However, two recent randomized controlled trials have demonstrated a significant increase in biopsy proven rejection with steroid avoidance [62,63]. One of these studies showed no difference in graft function at 5 years, although a significant increase in biopsy proven CAN was observed in subsequent analysis. Although improvements in metabolic profile were seen, longer follow-up is needed to determine if these changes significantly affect patient survival.

14.15 PROGRESS TOWARD TOLERANCE

Operational tolerance after transplantation involves prevention of acute and chronic rejection, normal graft function, and indefinite graft survival without the need for maintenance immunosuppression in an immunocompetent recipient [64]. Tolerance is widely regarded as the ultimate goal of transplantation, and while it has remained largely elusive, recent progress towards achieving tolerogenic treatment regimens has been achieved. In 2008, Kawai et al. reported on a series of five patients with ESRD who received combined bone marrow and kidney transplants from HLA single-haplotype mismatched living related donors, with the use of a nonmyeloablative preparative regimen [65]. Transient chimerism was observed in all recipients. Irreversible AMR and graft loss occurred in one patient. In the other four recipients, it was possible to discontinue all immunosuppressive therapy 9−14 months after transplantation and renal function has remained stable for 2.0−5.3 years since transplantation. *In vitro* testing of T cells from these four recipients showed donor-specific unresponsiveness. In 2012, Scandling et al. reported on a series of 16 patients conditioned with total lymphoid irradiation and antithymocyte globulin who received kidney transplants and an injection of $CD34^+$ hematopoietic progenitor cells and T cells from HLA-matched donors in a tolerance induction protocol [66]. Fifteen patients developed multilineage chimerism without graft versus host disease and eight with chimerism for at least 6 months were withdrawn from immunosuppression for $1-3$ years without subsequent episodes of acute rejection. Four chimeric patients had completed or were in the process of immunosuppression withdrawal at the time of publication, and four patients did not undergo drug withdrawal due to rejection or recurrence of primary disease. Blood cells from all patients showed early high ratios of $CD4^+CD5^+$ regulatory T cells and NKT cells versus conventional naïve $CD4^+$ T cells. Patients who underwent successful immunosuppression withdrawal all demonstrated donor-specific unresponsiveness.

14.16 NEW KIDNEY ALLOCATION POLICY PROPOSAL IN THE UNITED STATES

Many features of the existing kidney allocation system have functioned well for nearly 30 years without substantial modifications. However, there are several areas where the system is not designed well and require substantial revision. The OPTN/UNOS Kidney Committee has recently proposed a series of improvements to enhance recipient survival, make better use of available kidneys, and increase transplant opportunities for highly sensitized patients [67]. The current policy proposal is not a policy about to go into effect, but is the culmination of more than a decade of work by the Kidney Committee and is the product of a consensus driven process over that period of time. The proposal was issued for public comment in late 2012. Substantive concerns will be addressed by the Kidney Committee, which will present its recommendations to the OPTN/UNOS Board of Directors for approval. The information below is intended to provide the reader with an overview of the proposed policy at the time of public comment and may be subject to change prior to ultimate implementation. The major areas of concern are as follows: (1) some potentially transplantable kidneys are discarded because of how donors are classified and how organs are offered. Although an increase in donation rates is certainly desirable, more transplants can still be performed by making better use of existing organs. (2) There is no maximum life year benefit in the current system in which there are significant mismatches between the projected longevity of kidneys and the likely survival of recipients. Consequently, some patients with long expected survival may require retransplantation, thereby lowering the chances for others to receive a first transplant. (3) Access to transplant varies among candidates. In many cases, highly sensitized patients wait much longer than average for a transplant. The key components to the proposal are (1) kidney donor profile index (KDPI) and estimated posttransplant survival (EPTS)—The KDPI is a formula that has been established to estimate the expected graft survival of a donor kidney. EPTS is a second formula which will be used to estimate the likely benefit a specific patient would derive from transplantation. The proposal would give consideration for the 20% of kidneys with the best KDPI (longest estimated graft survival) to the 20% of candidates estimated by EPTS to have the longest time to benefit from a kidney transplant. For the remaining 80% of candidates, organ allocation would be similar to the existing system unless the patient falls into one of the "hard-to-match" categories. KDPI provides more detailed information than current designations of SCD and ECD, and it is anticipated that a more graded rather than binary assessment of expected graft survival will improve kidney utilization and increase transplant rates. (2) Blood type subgroup matching—Blood type B is relatively uncommon and is present in only about 16% of kidney transplant candidates, many of whom are ethnic minorities. Although patients with blood type B could receive kidneys from blood type O donors, this practice would reduce access to deceased donor kidneys for blood type O recipients who can only receive blood type O kidneys. It has been shown that donors with the A2 subtype of blood type A are compatible with blood type B recipients and provide good outcomes after kidney transplantation. It is proposed that kidneys from donors with blood type A2 be offered to blood type B candidates on a national basis in order to improve access and reduce waiting time for blood type B candidates. (3) Immune sensitivity matching—For many years, the kidney allocation system has offered some additional priority in the form of extra allocation points for candidates with a PRA of 80% or higher, since these patients tend to wait longer for transplant because of increased difficulty obtaining an organ from an HLA-compatible donor. However, no extra points are awarded as PRA increases, nor are points given to moderately sensitized patients. Based on the results of a time-to-kidney offer analysis according to PRA, the proposed policy will give priority points to sensitized patients using a sliding scale, starting at a PRA of 20%. The most highly sensitized patients (PRA 98%, 99%, and 100%) will receive significantly more priority than they had previously. It is expected that this will provide more equitable access to transplant for a group of patients with the longest waiting times as a result of compatibility obstacles. (4) Dialysis waiting time—In the current system, the duration of time on the waiting list is the predominant factor in prioritization of organ offers. Whereas some patients have regular access to medical care and are promptly placed on the waiting list when they begin dialysis or have a glomerular filtration rate of <20 mL/min, other patients reach ESRD and may be on dialysis for many years prior to being placed on the waiting list. Since waiting time is a priority factor, patients who are listed late for a transplant will likely be on dialysis longer than patients listed at an earlier stage of renal failure. Many patients who are not listed promptly have socioeconomic disadvantages. Therefore, it is proposed that waiting time priority begins at the point at which a patient begins dialysis or meets a defined stage of declining kidney function.

14.17 CONCLUSION

Ongoing efforts continue to expand the deceased donor kidney pool by maximizing organ donation and utilization as well as improving organ allocation. Novel approaches to overcoming ABO and HLA-incompatible barriers have also improved access to living kidney donation. Ideally, these efforts should increase transplant rates, optimize life years gained from transplantation, and minimize the need for retransplantation. In reality, the demand for kidney transplantation rapidly continues to outpace organ availability despite these recent accomplishments. Advances in immunosuppression and in our understanding of immunologic and nonimmunologic mechanisms of allograft injury have resulted in improved graft survival. Nevertheless, the immunosuppression armamentarium has changed only modestly over the past decade. Continued reliance upon nephrotoxic immunosuppressive regimens contributes to an inexorable decline of graft function over time. Tolerance induction, cell transplantation, bioartificial organ technology, and other advances in regenerative medicine are undoubtedly the frontiers of transplantation science, and may represent the best prospects for clinical solutions which may ultimately mitigate the widening disparity between organ supply and demand.

REFERENCES

[1] UNOS/OPTN Data. <http://optn.transplant.hrsa.gov/latestData/rptData.asp>.
[2] Matas AJ, Smith JM, Skeans MA, Lamb KE, Gustafson SK, Samana CJ, et al. OPTN/SRTR 2011 annual data report: kidney. Am J Transplant 2013;13:11–46.
[3] 2012 United States Renal Data System Annual Report. 2012 Atlas of CKD & ESRD. <http://www.usrds.org/atlas.aspx>. p. 283–94.
[4] Womer KL, Kaplan B. Recent developments in kidney transplantation—a critical assessment. Am J Transplant 2009;9:1265–71.
[5] Bohl DL, Brennan DC. BK virus nephropathy and kidney transplantation. Clin J Am Soc Nephrol 2007;2:36–46.
[6] Randhawa P, Brennan DC. BK virus infection in transplant recipients: an overview and update. Am J Transplant 2006;6:2000–5.
[7] Brennan DC, Agha I, Bohl DL, Schnitzler MA, Hardinger KL, Lockwood M, et al. Incidence of BK with tacrolimus versus cyclosporine and impact of preemptive immunosuppression reduction. Am J Transplant 2005;5:582–94.
[8] Solez K, Colvin RB, Racusen LC, Haas M, Sis B, Mengel M, et al. Banff 07 classification of renal allograft pathology: updates and future directions. Am J Transplant 2008;8:753–60.
[9] Cornell LD, Smith RN, Colvin RB. Kidney transplantation: mechanisms of rejection and acceptance. Annu Rev Pathol 2008;3:189–220.
[10] Desvaux D, Schwarzinger M, Pastural M, Baron C, Abtahi M, Berrehar F, et al. Molecular diagnosis of renal-allograft rejection: correlation with histopathologic evaluation and antirejection-therapy resistance. Transplantation 2004;78:647–53.
[11] Sarwal M, Chua MS, Kambham N, Hsieh SC, Satterwhite T, Masek M, et al. Molecular heterogeneity in acute renal allograft rejection identified by DNA microarray profiling. N Engl J Med 2003;349:125–38.
[12] Colvin RB. Antibody-mediated renal allograft rejection: diagnosis and pathogenesis. J Am Soc Nephrol 2007;18:1046–56.
[13] Solez K, Colvin RB, Racusen LC, Sis B, Halloran PF, Birk PE, et al. Banff '05 meeting report: differential diagnosis of chronic allograft injury and elimination of chronic allograft nephropathy (CAN). Am J Transplant 2007;7:518–26.
[14] Zou Y, Stastny P, Susal C, Dohler B, Opelz G. Antibodies against MICA antigens and kidney-transplant rejection. N Engl J Med 2007;357:1293–300.
[15] Stratta RJ, Rohr MS, Sundberg AK, Farney AC, Hartmann EL, Moore PS, et al. Intermediate-term outcomes with expanded criteria deceased donors in kidney transplantation: a spectrum or specter of quality? Ann Surg 2006;243:594–601 [discussion 601–3].
[16] Metzger RA, Delmonico FL, Feng S, Port FK, Wynn JJ, Merion RM. Expanded criteria donors for kidney transplantation. Am J Transplant 2003;3:114–25.
[17] Ojo AO, Hanson JA, Meier-Kriesche H, Okechukwu CN, Wolfe RA, Leichtman AB, et al. Survival in recipients of marginal cadaveric donor kidneys compared with other recipients and wait-listed transplant candidates. J Am Soc Nephrol 2001;12:589–97.
[18] Port FK, Bragg-Gresham JL, Metzger RA, Dykstra DM, Gillespie BW, Young EW, et al. Donor characteristics associated with reduced graft survival: an approach to expanding the pool of kidney donors. Transplantation 2002;74:1281–6.
[19] Pascual J, Zamora J, Pirsch JD. A systematic review of kidney transplantation from expanded criteria donors. Am J Kidney Dis 2008;52:553–86.
[20] Rao PS, Schaubel DE, Guidinger MK, Andreoni KA, Wolfe RA, Merion RM, et al. A comprehensive risk quantification score for deceased donor kidneys: the kidney donor risk index. Transplantation 2009;88:231–6.
[21] Abt PL, Fisher CA, Singhal AK. Donation after cardiac death in the US: history and use. J Am Coll Surg 2006;203:208–25.
[22] Howard RJ, Schold JD, Cornell DL. A 10-year analysis of organ donation after cardiac death in the United States. Transplantation 2005;80:564–8.
[23] Perico N, Cattaneo D, Sayegh MH, Remuzzi G. Delayed graft function in kidney transplantation. Lancet 2004;364:1814–27.

[24] Farney AC, Hines MH, al-Geizawi S, Rogers J, Stratta RJ. Lessons learned from a single center's experience with 134 donation after cardiac death donor kidney transplants. J Am Coll Surg 2011;212:440–51 [discussion 451–3].

[25] Ojo AO, Wolfe RA, Held PJ, Port FK, Schmouder RL. Delayed graft function: risk factors and implications for renal allograft survival. Transplantation 1997;63:968–74.

[26] Singh RP, Farney AC, Rogers J, Zuckerman J, Reeves-Daniel A, Hartmann E, et al. Kidney transplantation from donation after cardiac death donors: lack of impact of delayed graft function on post-transplant outcomes. Clin Transplant 2011;25:255–64.

[27] Yarlagadda SG, Coca SG, Formica Jr. RN, Poggio ED, Parikh CR. Association between delayed graft function and allograft and patient survival: a systematic review and meta-analysis. Nephrol Dial Transplant 2009;24:1039–47.

[28] Jochmans I, Moers C, Smits JM, Leuvenink HG, Treckmann J, Paul A, et al. Machine perfusion versus cold storage for the preservation of kidneys donated after cardiac death: a multicenter, randomized, controlled trial. Ann Surg 2010;252:756–64.

[29] Watson CJ, Wells AC, Roberts RJ, Akoh JA, Friend PJ, Akyol M, et al. Cold machine perfusion versus static cold storage of kidneys donated after cardiac death: a UK multicenter randomized controlled trial. Am J Transplant 2010;10:1991–9.

[30] Kayler LK, Mohanka R, Basu A, Shapiro R, Randhawa PS. Single versus dual renal transplantation from donors with significant arteriosclerosis on pre-implant biopsy. Clin Transplant 2009;23:525–31.

[31] Remuzzi G, Cravedi P, Perna A, Dimitrov BD, Turturro M, Locatelli G, et al. Long-term outcome of renal transplantation from older donors. N Engl J Med 2006;354:343–52.

[32] Kucirka LM, Ros RL, Subramanian AK, Montgomery RA, Segev DL. Provider response to a rare but highly publicized transmission of HIV through solid organ transplantation. Arch Surg 2011;146:41–5.

[33] Kucirka LM, Singer AL, Segev DL. High infectious risk donors: what are the risks and when are they too high? Curr Opin Organ Transplant 2011;16:256–61.

[34] Freeman RB, Cohen JT. Transplantation risks and the real world: what does "high risk" really mean? Am J Transplant 2009;9:23–30.

[35] Lonze BE, Dagher NN, Liu M, Kucirka LM, Simpkins CE, Locke JE, et al. Outcomes of renal transplants from Centers for Disease Control and Prevention high-risk donors with prospective recipient viral testing: a single-center experience. Arch Surg 2011;146:1261–6.

[36] Thomusch O, Tittelbach-Helmrich D, Meyer S, Drognitz O, Pisarski P. Twenty-year graft survival and graft function analysis by a matched pair study between pediatric en bloc kidney and deceased adult donors grafts. Transplantation 2009;88:920–5.

[37] Sharma A, Fisher RA, Cotterell AH, King AL, Maluf DG, Posner MP. En bloc kidney transplantation from pediatric donors: comparable outcomes with living donor kidney transplantation. Transplantation 2011;92:564–9.

[38] Bhayana S, Kuo YF, Madan P, Mandaym S, Thomas PG, Lappin JA, et al. Pediatric en bloc kidney transplantation to adult recipients: more than suboptimal? Transplantation 2010;90:248–54.

[39] Laurence JM, Sandroussi C, Lam VW, Pleass HC, Eslick GD, Allen RD. Utilization of small pediatric donor kidneys: a decision analysis. Transplantation 2011;91:1110–3.

[40] Sureshkumar KK, Patel AA, Arora S, Marcus RJ. When is it reasonable to split pediatric en bloc kidneys for transplantation into two adults? Transplant Proc 2010;42:3521–3.

[41] Kayler LK, Magliocca J, Kim RD, Howard R, Schold JD. Single kidney transplantation from young pediatric donors in the United States. Am J Transplant 2009;9:2745–51.

[42] Anil Kumar MS, Khan SM, Jaglan S, Heifets M, Moritz MJ, Saeed MI, et al. Successful transplantation of kidneys from deceased donors with acute renal failure: three-year results. Transplantation 2006;82:1640–5.

[43] Kayler LK, Garzon P, Magliocca J, Fujita S, Kim RD, Hemming AW, et al. Outcomes and utilization of kidneys from deceased donors with acute kidney injury. Am J Transplant 2009;9:367–73.

[44] Greenstein SM, Moore N, McDonough P, Schechner R, Tellis V. Excellent outcome using "impaired" standard criteria donors with elevated serum creatinine. Clin Transplant 2008;22:630–3.

[45] Lin NC, Yang AH, King KL, Wu TH, Yang WC, Loong CC. Results of kidney transplantation from high-terminal creatinine donors and the role of time-zero biopsy. Transplant Proc 2010;42:3382–6.

[46] Montgomery RA, Warren DS, Segev DL, Zachary AA. HLA incompatible renal transplantation. Curr Opin Organ Transplant 2012;17:386–92.

[47] Reinsmoen NL, Lai CH, Vo A, Jordan SC. Evolving paradigms for desensitization in managing broadly HLA sensitized transplant candidates. Discov Med 2012;13:267–73.

[48] Montgomery RA, Lonze BE, King KE, Kraus ES, Kucirka LM, Locke JE, et al. Desensitization in HLA-incompatible kidney recipients and survival. N Engl J Med 2011;365:318–26.

[49] Gloor J, Stegall MD. Sensitized renal transplant recipients: current protocols and future directions. Nat Rev Nephrol 2010;6:297–306.

[50] Montgomery RA, Locke JE, King KE, Segev DL, Warren DS, Kraus ES, et al. ABO incompatible renal transplantation: a paradigm ready for broad implementation. Transplantation 2009;87:1246–55.

[51] Rees MA, Kopke JE, Pelletier RP, Segev DL, Rutter ME, Fabrega AJ, et al. A nonsimultaneous, extended, altruistic-donor chain. N Engl J Med 2009;360:1096–101.

[52] Segev DL, Veale JL, Berger JC, Hiller JM, Hanto RL, Leeser DB, et al. Transporting live donor kidneys for kidney paired donation: initial national results. Am J Transplant 2011;11:356–60.

[53] Cravedi P, Perna A, Ruggenenti P, Remuzzi G. Mycophenolate mofetil versus azathioprine in organ transplantation. Am J Transplant 2009;9:2856–7.

[54] Remuzzi G, Lesti M, Gotti E, Ganeva M, Dimitrov BD, Ene-Iordache B, et al. Mycophenolate mofetil versus azathioprine for prevention of acute rejection in renal transplantation (MYSS): a randomised trial. Lancet 2004;364:503–12.

[55] Brennan DC, Daller JA, Lake KD, Cibrik D, Del Castillo D. Rabbit antithymocyte globulin versus basiliximab in renal transplantation. N Engl J Med 2006;355:1967–77.

[56] Shapiro R, Ellis D, Tan HP, Moritz ML, Basu A, Vats AN, et al. Alemtuzumab pre-conditioning with tacrolimus monotherapy in pediatric renal transplantation. Am J Transplant 2007;7:2736–8.

[57] Farney AC, Doares W, Rogers J, Singh R, Hartmann E, Hart L, et al. A randomized trial of alemtuzumab versus antithymocyte globulin induction in renal and pancreas transplantation. Transplantation 2009;88:810–9.

[58] Srinivas TR, Meier-Kriesche HU. Minimizing immunosuppression, an alternative approach to reducing side effects: objectives and interim result. Clin J Am Soc Nephrol 2008;3:101–16.

[59] Ekberg H, Tedesco-Silva H, Demirbas A, Vitko S, Nashan B, Gurkan A, et al. Reduced exposure to calcineurin inhibitors in renal transplantation. N Engl J Med 2007;357:2562–75.

[60] Schena FP, Pascoe MD, Alberu J, del Carmen Rial M, Oberbauer R, Brennan DC, et al. Conversion from calcineurin inhibitors to sirolimus maintenance therapy in renal allograft recipients: 24-month efficacy and safety results from the CONVERT trial. Transplantation 2009;87:233–42.

[61] Su VC, Harrison J, Rogers C, Ensom MH. Belatacept: a new biologic and its role in kidney transplantation. Ann Pharmacother 2012;46:57–67.

[62] Vincenti F, Schena FP, Paraskevas S, Hauser IA, Walker RG, Grinyo J. A randomized, multicenter study of steroid avoidance, early steroid withdrawal or standard steroid therapy in kidney transplant recipients. Am J Transplant 2008;8:307–16.

[63] Woodle ES, First MR, Pirsch J, Shihab F, Gaber AO, Van Veldhuisen P. A prospective, randomized, double-blind, placebo-controlled multicenter trial comparing early (7 day) corticosteroid cessation versus long-term, low-dose corticosteroid therapy. Ann Surg 2008;248:564–77.

[64] Salama AD, Womer KL, Sayegh MH. Clinical transplantation tolerance: many rivers to cross. J Immunol 2007;178:5419–23.

[65] Kawai T, Cosimi AB, Spitzer TR, Tolkoff-Rubin N, Suthanthiran M, Saidman SL, et al. HLA-mismatched renal transplantation without maintenance immunosuppression. N Engl J Med 2008;358:353–61.

[66] Scandling JD, Busque S, Dejbakhsh-Jones S, Benike C, Sarwal M, Millan MT, et al. Tolerance and withdrawal of immunosuppressive drugs in patients given kidney and hematopoietic cell transplants. Am J Transplant 2012;12:1133–45.

[67] Proposal to substantially revise the National Kidney Allocation System (Kidney Transplantation Committee). <http://optn.transplant.hrsa.gov/PublicComment/pubcommentPropSub_311.pdf>; 2012. p. 1–59.

Chapter 15

Living Donor Renal Transplantation: Progress, Pitfalls, and Promise

Nizam Mamode

Consultant Transplant Surgeon and Reader in Transplant Surgery, Guy's and St Thomas' NHS Foundation Trust, London

Chapter Outline

15.1 INTRODUCTION

Although the first successful solid organ transplant was a living donor renal transplant performed by Joseph Murray in 1954 [1], it is only in the twenty-first century that living donation (LDT) has been widely practiced. The imperative for the rapid development of LDT was the decline in available deceased donor organs in the1980s, coupled with a rise in the transplant waiting list, but the tremendous success of LDT, with 1-year graft survival rates of over 95%, has meant that in many centers LDT is the first choice for renal replacement therapy. At the same time, the introduction of minimally invasive donor surgery in the late 1990s has encouraged both potential donors and transplant centers to undertake donor nephrectomy, with minimal mortality and morbidity [2].

The advantages of LDT include the ability to plan surgery electively, thus optimizing the condition of the recipient, and the ability to undertake transplantation early in the development of renal failure, often "pre-emptively," that is, before any renal replacement has been instituted. This means that the detrimental effects of chronic dialysis such as cardiovascular impairment can be minimized. These considerations also mean that LDT may be more appropriate for recipients who are at high perioperative risk from transplantation, for example, due to ischemic heart disease or severe obesity, in whom controlled elective surgery carries a lower risk. Furthermore, some would hold the view that when a transplant fails in a high-risk recipient after deceased donation, a valuable and limited resource (the kidney) has been lost which could have been given to another recipient, whereas this is not the case in LDT. Therefore, LDT in the high-risk situation only requires willingness from donor, recipient, and the transplant team to proceed, but not a consideration of rationing of limited organs.

Finally, LDT represents a highly significant emotional event for most donors and recipients. The opportunity for living donors to directly help improve the health of someone close to them can be an intense and transformative experience, and recipients inevitably feel a tremendous gratitude towards their donor. The profound nature of this experience will remain with most, even after the organ has failed.

15.2 THE DONOR OPERATION

It is less than two decades since a minimally invasive approach to living donor nephrectomy was first described,

TABLE 15.1 Techniques for Donor Nephrectomy

Technique	Advantages	Disadvantages
Hand-assisted laparoscopic	Safe, fast	Less cosmetic incision
Totally laparoscopic	Better cosmesis	Less control of acute bleeding, longer surgery, difficult to learn
Retroperinoteoscopic	Fast, less bowel injury/obstruction	Plane not always achieved
Single-port laparoscopic (LESS)	Better cosmesis, quicker recovery?	Longer warm time
Natural orifice surgery (NOTES)	Better cosmesis	Technique not yet evaluated fully
Mini-incision surgery	Easy for nonlaparoscopic surgeons to learn	More pain than laparoscopic nephrectomy
Open surgery	Was standard of care	More pain and slower recovery than laparoscopic nephrectomy

but in the United States and the United Kingdom this has now become the technique of choice. A variety of different surgical procedures are employed and these, along with their potential advantages and disadvantages, are summarized in Table 15.1. There are seven randomized controlled trials and two meta-analyses comparing laparoscopic donor nephrectomy (LDN) with open surgery, generally concluding that there is less acute and chronic pain, a shorter hospital stay and a quicker return to work after LDN [3–11], although the mini-incision approach may be useful in open surgery; only one small randomized trial compares totally laparoscopic nephrectomy with the hand-assisted approach (HALDN) [12], making it difficult to draw firm conclusions, but many centers believe that the safety and speed afforded by HALDN make this approach preferable. If HALDN is to be employed the option of a retroperitoneal approach may be advantageous, with quicker surgery and less bowel injury. This is currently being tested in the ongoing hand-assisted retroperitoneoscopic nephrectomy (HARP) study.

Robotically assisted donor nephrectomy has been performed [13], using the da Vinci robotic system—in this setting, the surgeon sits at a console and controls laparoscopic instruments remotely, whilst looking through a binocular eyepiece. The potential advantages of the robot include scaling of movement, so that surgery may be more precise, three-dimensional vision, which may be beneficial in the laparoscopic setting, and increased flexibility of the laparoscopic instruments. The last of these results from the fact that the robotic arm can move in six different directions ("degrees of freedom"), rather than four in conventional laparoscopy, and thus matches natural movements of the human arm more closely. However, no controlled studies have been performed of robotically assisted donor nephrectomy against

conventional laparoscopy, and there is no clear evidence of clinical benefit using this technology in this setting. Furthermore, disadvantages of robotic surgery include cost (over $1 million per robot) and the lack of haptic feedback; the operator has no sense of resistance from the tissues, and therefore could potentially cause significant trauma. Despite these caveats, it is likely that new robotic systems will be available in the near future, which will be cheaper, include haptic feedback and be less cumbersome, and robotic donor nephrectomy may well become the standard of care.

15.3 PERIOPERATIVE RISKS OF LIVING DONOR NEPHRECTOMY

There is good evidence regarding the perioperative risk of living donor nephrectomy; over 20 years ago, perioperative mortality rates were found to be 0.03% [14], and subsequent large studies, with many thousands of participants, have found similar results [15]. This risk is similar to that of death after appendicectomy or in childbirth. The introduction of laparoscopic nephrectomy might have been expected to increase the risk of complications related to hemorrhage, but a study of 2509 consecutive donors using UK registry data showed that major morbidity was 4.5% after laparoscopic donation and 5.1% after open surgery, whilst the rate of any morbidity was 10.3% and 15.7%, respectively ($P = 0.001$) [2].

15.4 POTENTIAL PROBLEMS WITH LIVING DONORS

Candidates for living donor nephrectomy are extensively screened for the presence of underlying conditions which

TABLE 15.2 Contraindications to Donation

Absolute Contraindications	Relative Contraindications
Inadequate renal function on measured GFR	Severe obesity (BMI > 35 kg/m^2)
Renal or urinary tract disease	Bilateral nephrolithiasis
Diabetes mellitus	Steatosis
Significant cardiovascular disease	Previous malignancy
Uncontrolled hypertension	Family history of renal disease

might preclude donation. A thorough clinical history, with particular attention to previous urinary tract infections, hematuria or other evidence of nephrolithiasis, is essential. Diabetes or significant cardiovascular disease will mean that the donor cannot proceed in most centers. A family history, to rule out familial causes of end-stage renal disease, must be carefully elicited. Common contraindications to donation are shown in Table 15.2.

Investigations will include a renal ultrasound, a cardiac stress test for older donors (exercise tolerance test or myocardial perfusion scan), urine assessment for the presence of microscopic hematuria or proteinuria, and isotope quantification of GFR.

Donors with hypertension may be considered if there is no evidence of end organ damage, and blood pressure is stable and controlled on one or two antihypertensive medications. For those with borderline hypertension, a 24-h ambulatory blood pressure record is obtained.

Most centers will perform annual follow-up of donors indefinitely, with screening of blood, urine, and blood pressure. Perhaps because of this, long-term survival rates after donor nephrectomy are excellent, with good evidence that nephrectomy does not shorten life or lead to an increased risk of renal failure [16]. However, it is important to consider that donors from developing countries, who travel to donate to a relative, may not have easy access to regular follow-up, or indeed to intervention if a problem is detected. This is particularly pertinent for those with previous nephrolithiasis (see below), and transplant teams should carefully consider whether a recurrent stone in the remaining kidney could be adequately treated in this setting.

There are three common issues which create difficulties when deciding whether to accept potential donors for nephrectomy; obesity, nephrolithiasis, and malignancy.

The epidemic of obesity in Western societies has meant that in the last decade increasing numbers of potential donors are severely obese, with a body mass index

(BMI) over 30 kg/m^2. In the United States, a study of over 80,000 donors over 15 years found around one-fifth of donors to be clinically obese [15]. An increased rate of perioperative complications, particularly wound infections, and a longer operating time, have been found in obese donors [17,18]. However, a meta-analysis which included 484 obese donors, with a mean BMI of 34 kg/m^2 showed that although these differences were statistically significant, they were not clinically significant [19]. In summary, therefore, it seems that although the perioperative risks for obese donors may be a little higher, if donors are counseled appropriately these risks should not in themselves preclude donation. Concerns have, however, been raised about the long-term risks of uni-nephrectomy in the obese. After 10 years, obese patients have an increased risk of proteinuria and renal insufficiency after uni-nephrectomy [20], although registry data from the United States in over 5000 living donors suggests that the degree of renal insufficiency at 6 months was not related to obesity [21]. There are, however, few long-term data available on the very obese (BMI > 35 kg/m^2). The conclusion from these and other studies is that there may be an increased long-term risk in obese donors, and LDT in this group should be approached with caution.

A history of nephrolithiasis in potential living donors has previously been thought to preclude donation, due to the risk of developing a stone in the remaining kidney, and a subsequent risk of a renal failure as a result. There is a 50% risk of developing a further stone over 5 years in those who present with a symptomatic calcium oxalate stone. However, in selected cases, this may not be an absolute contraindication. In our practice, such donors undergo biochemical screening of blood and urine for raised calcium, oxalate, and citrate. If this metabolic screen is normal, and there is a unilateral stone, donors can be counseled about the potential long-term risks and nephrectomy performed, with retrieval of the kidney which contains the stone. Attempts can be made to remove the stone on the bench, and in this scenario the risk to the recipient is low.

Finally, a history of recent malignancy is, in general, a contraindication to donation. However, in deceased donation, organs are increasingly used from donors with documented malignancy, where the risk of spread to the kidney and transmission to the recipient is low [22], and consideration can be given to living donors with previous malignancy on a case by case basis. In this respect, the Israel Penn registry can be helpful in attempting to quantify risk.

15.5 OUTCOMES AFTER LDT

There is good evidence that outcomes after LDT are significantly better than those after deceased donor transplantation. The decline in the numbers of standard criteria deceased donors which most countries have experienced in recent years, due to improved selection criteria for intensive care treatment after head injury or stroke, and the decline in head injuries after road traffic accidents has meant that organs retrieved from deceased donors are commonly of poor quality. In the United Kingdom, 40% of organ donors in 2011/12 were donors after cardiac death [23], while in the United States, 22% of deceased donors in 2011 were categorized as "extended criteria donors" [24], which may lead to poorer outcomes. Thus it is likely current outcomes after LDT may be even better than the advantage shown in comparison to deceased donation in historic data.

The 2010/11 report from NHS Blood and Transplant in the United Kingdom shows that 1-year graft and patient survival after LDT are excellent, being 97% (95% CI 96−97%) and 99% (98−99%), respectively [25]. Five-year survival rates are 92% (90−93%) and 96% (95−97%). However, after deceased donation, 5-year graft and patient survival rates are 84% (82−85%) and 89% (88−90%).

There is therefore a clear benefit in receiving a living donor kidney transplant, and median graft survival time is around 25 years after LDT, compared with 15 years for deceased donor recipients. This benefit occurs irrespective of HLA matching—even completely mismatched living donor kidneys will do better than well-matched deceased donor organs [26]. Indeed, in an analysis of UK registry data, the degree of HLA mismatching had no effect on graft survival after LDT in the 2000−2007 cohort of living donor recipients [27].

For children, outcomes are similar, but good long-term graft survival is even more important, since second or third transplants will become necessary and will potentially be immunologically more difficult due to sensitization (the development of anti-HLA antibodies) as a result of the first transplant.

The implication of these data is that a living donor should be sought as the first option for a patient with end-stage renal disease, and assuming there are no donor issues, or contraindications due to antibody incompatibility (see below), should be considered as a better option than deceased donation.

15.6 PROBLEMS WITH LDT

One of the commonest issues in LDT is antibody incompatibility, when the recipient has antibodies which would react against the endothelial cells of the donor kidney. Approximately 30% of recipients presenting with a living donor will have antibody incompatibility [28], and in most cases these will preclude transplantation without specific intervention, due to hyperacute or accelerated

early rejection. There are two key types of antibody which cause incompatibility: anti-HLA and isohemagglutinin (or blood group) antibodies. HLA antibodies arise for three reasons: previous transplantation, pregnancy, and blood transfusion, and patients with these antibodies are termed "sensitized." It is often stated that the degree of sensitization is related to the cause of sensitization, with previous transplantation leading to the greatest problems and blood transfusion the least, but it is increasingly apparent that transfusions can lead to severe and intractable sensitization, especially when leukocyte-depleted blood is not used. The combination of transplantation and transfusion leads to sensitization in over 70% of patients, although interestingly antibodies tend to be generated against the antigens present in the transplant rather than the blood [29].

The presence of anti-HLA antibodies has several implications: the transplant may not proceed due to a positive flow cytometric cross-match (FXCM), the risk of rejection even if transplantation does take place may be high, and waiting time on the deceased donor list will be much longer, or even indefinite, if living donor transplantation does not proceed.

Hemagglutinins arise due to cross-reaction between antibodies generated against gut bacteria and blood group antigens, which are present on a number of tissues apart from red cells. Indeed the expression of these antigens varies according to tissue type, with the kidney having much greater expression than the liver or heart. For this reason, blood group incompatible transplantation (ABOi) is most difficult in renal transplantation. The fact that the antibodies arise from exposure to intestinal flora has two important implications: first, antibody levels may vary over time and second children tend to have low antibody levels—indeed, we have shown that 45% of those on the pediatric waiting list have low titers of blood group antibodies [30]. The distribution of blood groups in Western populations is shown in Table 15.3, and the clear implication is that the commonest incompatibility will be with a blood group A donor and group O recipient. Group A and B antigens are formed from the action of glycosyltransferase on the substance H (glycoprotein) core chain. There are four types of core chain, and differences in the utilization of these results in the division of A into A_1 and A_2 subtypes; group A_2 is less antigenic than A_1, so that ABOi using kidneys from these donors is easier.

15.7 HLA-INCOMPATIBLE TRANSPLANTATION

Living donor transplants can be performed when there is HLA incompatibility (HLAi), but this requires additional therapy. First, antibody must be removed or suppressed

TABLE 15.3 Blood Group Antigens and Antibodies

Phenotype	Genotype	Antigens	Antibodies	Frequency in the United Kingdom
O	OO	None	Anti-A and anti-B	44%
A	AA or AO	A	Anti-B	45%
B	BB or BO	B	Anti-A	8%
AB	AB	A and B	None	3%

so that a negative cross-match is achieved. Many centers will require a negative FXCM, but some will transplant if the complement-dependent cytotoxic (CDC) cross-match is negative, even if the FXCM is positive. Antibody removal is normally undertaken using plasmapheresis or variants of this technique like double-filtration plasmapheresis (DFPP). Blood is removed from the recipient, and the plasma is separated and either completely replaced or filtered, with the filtrate replaced, using albumin or fresh frozen plasma. Typically several volumes of blood are processed in each session and multiple treatments may be required in order to sufficiently reduce the antibody. In some cases, antibody levels remain high and transplantation cannot proceed.

Immunoabsorption columns are an alternative means of removing HLA antibody. Typically sheep-derived polyclonal antihuman antibody is bound to a sepharose matrix, and passage of plasma through the column results in antibody depletion. This has the advantage of having a minimal effect on clotting factors, unlike plasmapheresis, which is an important consideration in the perioperative period. However, immunoabsorption is usually more expensive than plasmapheresis and is therefore not used by all centers.

As well as antibody removal, the administration of IvIgG is often used in these recipients; this is a collection of pooled antibody from healthy volunteers, which acts through several mechanisms. Downregulation of antibody production occurs, especially when it is given after plasmapheresis, and an anti-idiotypic effect results in binding and neutralization of antidonor antibody. IvIgG may be given in low dose (0.5 g/kg) after each plasmapheresis or in less frequent high doses (2 g/kg).

In HLAi, induction therapy is often more powerful, in an attempt to reduce the risk of rejection. Antithymocyte globulin (ATG—a polyclonal depleting antibody) or alemtuzumab (a monoclonal depleting antibody) is commonly used in place of the IL-2 inhibitor basiliximab. Rituximab, a chimeric B-cell depleting antibody is given by some centers 1 month prior to transplantation—this does not result in depletion of antibody levels in itself, but may have an effect on T-cell activation through the antigen-presenting

role and cytokine release of B cells, and thus may help to reduce the risk of subsequent antibody-mediated rejection.

When HLAi transplantation does occur, outcomes are certainly worse when compared with standard LDT, with 1-year graft survival rates of around 85% [31]. Rejection rates are around 40% and severe early antibody-mediated rejection may result in graft loss. However, US registry data has shown that HLAi results in better patients survival when compared with remaining on the waiting list for a deceased donor [32].

In summary, HLAi will entail more treatment and higher risks, and for this reason a compatible donor should always be sought. If none is available, alternatives to HLAi (see below) may be considered, but if these have failed direct HLAi is generally preferable to remaining on the waiting list for a deceased donor.

15.8 BLOOD GROUP INCOMPATIBLE TRANSPLANTATION

ABOi is generally more feasible than HLAi and in some centers has become routine. Similar principles apply as in HLAi; antibody is removed, using plasmapheresis or immunoabsorption prior to transplantation. However, specific anti-A or anti-B columns are available, giving a highly specific and effective reduction in antibody. Typically, antibody titers of 1 in 8 or less are required for transplantation to proceed.

Rituximab is used widely in ABOi, and has replaced splenectomy which was previously a routine part of regimens, particularly in Japan where large numbers of ABOi have been performed for many years.

Outcomes after ABOi are widely accepted to be excellent. One-year graft survival rates are equivalent to compatible living donor transplantation [33,34] and long-term outcomes are similarly very good [35]. However, recent analysis of US registry data has shown that there is a small excess early risk of graft loss in ABOi, which is probably related to antibody-mediated rejection [36].

For recipients with low baseline hemagglutinin titers, direct ABOi may be undertaken in the absence of a compatible donor. Those with higher titers may consider the alternatives detailed below, but only 15% of group O recipients with an A donor will achieve a transplant through a pooled scheme, and thus direct ABOi is likely to be necessary after a period within such a scheme.

15.9 ALTERNATIVES TO ANTIBODY-INCOMPATIBLE TRANSPLANTATION

In recent years, several countries, including the United States, the United Kingdom, and the Netherlands, have introduced pooled or paired exchange schemes. In these schemes, antibody-incompatible living donor pairs are registered in the scheme, and a computerized algorithm is run to seek compatible matches with other pairs. Transplantation typically takes place on the same day, with the organ from one donor being given to the recipient from the other pair and vice versa. There may be two-way or three-way swaps, as shown in Figure 15.1.

The advantage of a paired exchange scheme is that compatible transplants are achieved; not only can the transplant take place but also both short- and long-term risks are reduced. The disadvantages include the need to maintain anonymity, in order to protect one pair from unwanted attention from another, the logistical issues in coordinating simultaneous surgery, and most importantly the fact that certain groups will have a low chance of success in the scheme. Group O recipients tend to be over-represented in paired schemes, and as noted above A into O incompatible pairs have a low chance of success. Similarly, the most highly sensitized recipients will find it difficult to obtain a match; 61% of those in the UK paired scheme have a calculated reaction frequency (the percentage of the last 10,000 donors against whom they are sensitized) of 95−100%, and the transplantation rate in this group is 5%. Attempts have been made to overcome these difficulties, with the inclusion of voluntary compatible pairs within the schemes in the United States and the United Kingdom, and the use of nondirected donors.

Nondirected (or altruistic) donors are individuals who decide to donate a kidney in the same way as someone might donate blood. This currently occurs in only three countries: the United Kingdom, the United States, and the Netherlands. Rigorous psychological screening, as well as physical assessment, takes place before considering donation. A review of the first 117 nondirected donors in the United Kingdom unsurprisingly found not only excellent physical outcomes but also a very low level of regret and a quicker recovery when compared with conventional donors [37]. Organs from nondirected donors are typically used in two ways. They may be treated as a deceased donor organ and offered through the normal matching algorithm, or they may be used to "prime" a chain. In this scenario (shown in Figure 15.1), the organ from the nondirected donor is given to the recipient of an incompatible pair, whose donor then donates to a second recipient and so on. These transplants are often performed sequentially (hence are known as nonsimultaneous extended altruistic donation—NEAD) and may result in very long chains of 30 or more transplants [38].

Finally, paired schemes and desensitization may be combined, so that recipients are entered into the scheme with the attention of carrying out plasmapheresis prior to transplantation—this may allow the list of unacceptable antigens in the donor to be reduced, thus increasing the chances of transplantation in the most highly sensitized, but does run the risk that the transplant might not be able to proceed.

15.10 FUTURE DIRECTIONS IN LIVING DONOR TRANSPLANTATION

15.10.1 Minimally Invasive Transplantation

Although the donor operation has seen a number of minimally invasive approaches, recipient surgery has remained largely unchanged over the last 60 years. Recently, however, attempts have been made to improve this. A mini-incision approach has been used by a Norwegian group, with good results in a small number of patients [39]. However, several groups have begun to use laparoscopic techniques for implantation. In a recent report of 72 cases, operating time increased from 176 to 224 min and warm ischemic time from 30 to 60 min when those undergoing open surgery were compared with the laparoscopic group [40]. For many these disadvantages would outweigh the benefits of laparoscopic surgery, and these reflect the difficulty of suturing vascular anastomoses laparoscopically. This can be partially overcome by use of the da Vinci robot, notwithstanding the issues noted above. The Chicago group have described

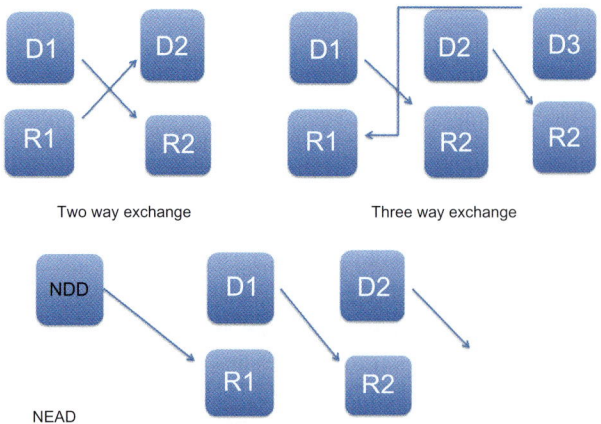

FIGURE 15.1 Two- and three-way exchanges and NEAD. NDD = nondirected donor.

good outcomes when performing robot laparoscopic implantation in 39 obese recipients, who would normally be expected to have a high complication rate [41]. At present, however, laparoscopic transplantation remains experimental.

15.10.2 Antibody-Incompatible Transplantation

Two new drugs have recently become available which may help HLAi. Bortezomib is a protease inhibitor which, by acting on plasma cells may reduce HLA antibody and thus help in both desensitization regimens and the treatment of rejection. So far reports are conflicting, and the most appropriate use of this drug is yet to be defined [42,43].

A more exciting development is the terminal complement inhibitor eculizumab [44]. This monoclonal antibody does not attempt to suppress or remove HLA antibody; rather, it inhibits formation of the membrane attack complex, and thus prevents the antibody acting on the endothelial cell. An initial report in 26 patients at the Mayo Clinic showed that using eculizumab prophylactically in HLAi living donor transplants resulted in a 7% rate of antibody-mediated rejection, against an expected rate of 40% [45]. A multicenter, randomized trial is currently underway and will determine whether the outcomes can be significantly improved in this difficult group of patients.

15.10.3 Chronic Rejection

Although 1-year graft survival rates are excellent after LDT, as noted above, the rate of graft loss over subsequent years has remained fairly constant over the last 20 years (Figure 15.2). The main cause of late graft loss, after death with a functioning graft, is chronic rejection, and at present the only treatment for those with such features on biopsy is to increase levels of immunosuppression. Two randomized trials are attempting to find an alternative solution, using B-cell depletion. The REMIND study involves using rituximab 1 month prior to transplantation in compatible living donor recipients, in the hope that as B cells reconstitute after a year, donor-specific clones will not appear and chronic rejection will be decreased. The RituxiCan study uses rituximab in those with chronic rejection who are not responding to optimization of immunosuppression. Both studies are ongoing.

15.10.4 Tolerance

This will be discussed in more detail elsewhere, but a very exciting development has been the use of nonmyeloablative

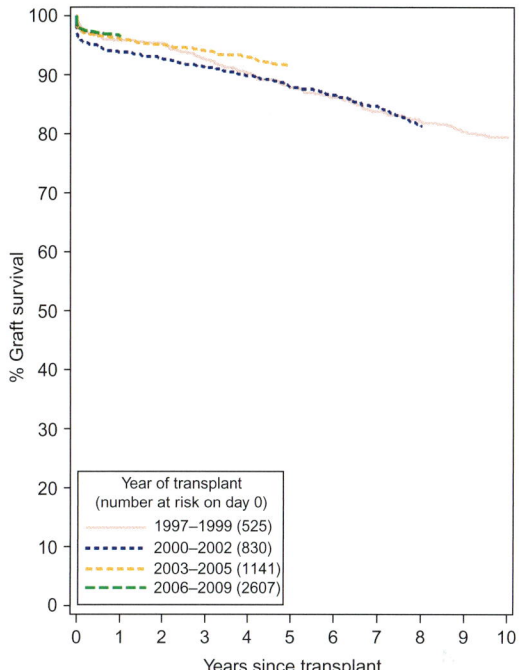

FIGURE 15.2 Long-term graft survival after first adult living donor transplant in the United Kingdom. *NHSBT Activity Report 2011 [25]*

hematopoietic stem cell transplantation in combination with renal transplantation to induce tolerance in conventional living donor recipients. Stanford has recently reported outcomes in 16 fully HLA-matched recipients, using total lymphoid irradiation and infusion of donor-derived CD34 and CD3 cells, with successful tolerance in eight patients and a further four undergoing weaning [46]. This approach may represent a major advance in living donor transplantation and further studies are awaited.

15.11 CONCLUSION

Living donor renal transplantation has been one of the great success stories of the late twentieth century. However, many problems remain, including antibody incompatibility and late graft loss through chronic rejection. A combination of new drugs, new logistical approaches, and stem cell therapies offer tremendous hope in the twenty-first century, and it is likely that our current techniques will look very primitive when the centenary of the first transplant is reached. However, we should always consider that the most exciting technical or immunological advances will not concern those individuals donating or receiving a kidney as much as the fact that they have undergone one of the most meaningful experiences of their lives. To be given the opportunity to participate in and facilitate that experience remains a tremendous privilege.

REFERENCES

[1] Merrill JP, Murray JE, Harrison JH, Guild WR. Successful homo-transplantation of the human kidney between identical twins. J Am Med Assoc 1956;160:277–82.

[2] Hadjianastassiou VG, Johnson RJ, Rudge CJ, Mamode N. 2509 living donor nephrectomies, morbidity and mortality, including the UK introduction of laparoscopic donor surgery. Am J Transplant Off J Am Soc Transplant Am Soc Transplant Surg 2007;7:2532–7.

[3] Kok NF, Lind MY, Hansson BM, Pilzecker D, Mertens zur Borg IR, Knipscheer BC, et al. Comparison of laparoscopic and mini incision open donor nephrectomy: single blind, randomised controlled clinical trial. BMJ 2006;333:221.

[4] Wolf Jr. JS, Merion RM, Leichtman AB, Campbell Jr. DA, Magee JC, Punch JD, et al. Randomized controlled trial of hand-assisted laparoscopic versus open surgical live donor nephrectomy. Transplantation 2001;72284–90.

[5] Nicholson ML, Kaushik M, Lewis GR, Brook NR, Bagul A, Kay MD, et al. Randomized clinical trial of laparoscopic versus open donor nephrectomy. Br J Surg 2010;97:21–8.

[6] Oyen O, Andersen M, Mathisen L, Kvarstein G, Edwin B, Line PD, et al. Laparoscopic versus open living-donor nephrectomy: experiences from a prospective, randomized, single-center study focusing on donor safety. Transplantation 2005;79:1236–40.

[7] Simforoosh N, Basiri A, Tabibi A, Shakhssalim N, Hosseini Moghaddam SM. Comparison of laparoscopic and open donor nephrectomy: a randomized controlled trial. BJU Int 2005;95:855.

[8] Basiri A, Simforoosh N, Heidari M, Moghaddam SM, Otookesh H. Laparoscopic v open donor nephrectomy for pediatric kidney recipients: preliminary report of a randomized controlled trial. J Endourol Endourol Soc 2007;21:1033–6.

[9] Antcliffe D, Nanidis TG, Darzi AW, Tekkis PP, Papalois VE. A meta-analysis of mini-open versus standard open and laparoscopic living donor nephrectomy. Transplant Int Off J Eur Soc Organ Transplant 2009;22:463–74.

[10] Nanidis TG, Antcliffe D, Kokkinos C, Borysiewicz CA, Darzi AW, Tekkis PP, et al. Laparoscopic versus open live donor nephrectomy in renal transplantation: a meta-analysis. Ann Surg 2007;247:58–70.

[11] Hofker HS, Nijboer WN, Niesing J, Krikke C, Seelen MA, van Son WJ, et al. A randomized clinical trial of living donor nephrectomy: a plea for a differentiated appraisal of mini-open muscle splitting incision and hand-assisted laparoscopic donor nephrectomy. Transplant Int Off J Eur Soc Organ Transplant 2012;25:976–86.

[12] Bargman V, Sundaram CP, Bernie J, Goggins W. Randomized trial of laparoscopic donor nephrectomy with and without hand assistance. J Endourol 2006;20:717–22.

[13] Renoult E, Hubert J, Ladriere M, Billaut N, Mourey E, Feuillu B, et al. Robot-assisted laparoscopic and open live-donor nephrectomy: a comparison of donor morbidity and early renal allograft outcomes. Nephrol Dial Transplant 2006;21:472–7.

[14] Najarian JS, Chavers BM, McHugh LE, Matas AJ. 20 years or more of follow-up of living kidney donors. Lancet 1992;340:807–10.

[15] Segev DL, Muzaale AD, Caffo BS, Mehta SH, Singer AL, Taranto SE, et al. Perioperative mortality and long-term survival following live kidney donation. J Am Med Assoc 2010;303:959–66.

[16] Ibrahim HN, Foley R, Tan L, Rogers T, Bailey RF, Guo H, et al. Long-term consequences of kidney donation. N Engl J Med 2009;360:459–69.

[17] Heimbach JK, Taler SJ, Prieto M, Cosio FG, Textor SC, Kudva YC, et al. Obesity in living kidney donors: clinical characteristics and outcomes in the era of laparoscopic donor nephrectomy. Am J Transplant Off J Am Soc Transplant Am Soc Transplant Surg 2005;5:1057–64.

[18] Patel S, Cassuto J, Orloff M, Tsoulfas G, Zand M, Kashyap R, et al. Minimizing morbidity of organ donation: analysis of factors for perioperative complications after living-donor nephrectomy in the United States. Transplantation 2008;85:561–5.

[19] Young A, Storsley L, Garg AX, Treleaven D, Nguan CY, Cuerden MS, et al. Health outcomes for living kidney donors with isolated medical abnormalities: a systematic review. Am J Transplant Off J Am Soc Transplant Am Soc Transplant Surg 2008;8:1878–90.

[20] Praga M, Hernandez E, Herrero JC, Morales E, Revilla Y, Diaz-Gonzalez R, et al. Influence of obesity on the appearance of proteinuria and renal insufficiency after unilateral nephrectomy. Kidney Int 2000;58:2111–8.

[21] Reese PP, Feldman HI, Asch DA, Thomasson A, Shults J, Bloom RD. Short-term outcomes for obese live kidney donors and their recipients. Transplantation 2009;88:662–71.

[22] Desai R, Collett D, Watson CJ, Johnson P, Evans T, Neuberger J. Cancer transmission from organ donors-unavoidable but low risk. Transplantation 2012;94:1200–7.

[23] http://www.organdonation.nhs.uk/statistics/latest_statistics/.

[24] http://www.unos.org/donation/index.php?topic = data_resources.

[25] http://www.organdonation.nhs.uk/statistics/transplant_activity_report/archive_activity_reports/pdf/ukt/activity_report_2010_11.pdf.

[26] Terasaki PI, Cecka JM, Gjertson DW, Takemoto S. High survival rates of kidney transplants from spousal and living unrelated donors. N Engl J Med 1995;333:333–6.

[27] United Kingdom guidelines for living donor kidney transplantation. 2011 http://www.bts.org.uk/Documents/Guidelines/Active/UK%20Guidelines%20for%20Living%20Donor%20Kidney%20July%202011.pdf.

[28] Segev DL, Gentry SE, Warren DS, Reeb B, Montgomery RA. Kidney paired donation and optimizing the use of live donor organs. J Am Med Assoc 2005;293:1883–90.

[29] Scornik JC, Meier-Kriesche HU. Blood transfusions in organ transplant patients: mechanisms of sensitization and implications for prevention. Am J Transplant Off J Am Soc Transplant Am Soc Transplant Surg 2011;11:1785–91.

[30] Barnett AN, Hudson A, Hadjianastassiou VG, Marks SD, Reid CJ, Maggs TP, et al. Distribution of ABO blood group antibody titers in pediatric patients awaiting renal transplantation: implications for organ allocation policy. Transplantation 2012;94:362–8.

[31] Stegall MD, Gloor J, Winters JL, Moore SB, Degoey S. A comparison of plasmapheresis versus high-dose IVIG desensitization in renal allograft recipients with high levels of donor specific alloantibody. Am J Transpl Off J Am Soc Transplant Am Soc Transplant Surg 2006;6:346–51.

[32] Montgomery RA, Lonze BE, King KE, Kraus ES, Kucirka LM, Locke JE. Desensitization in HLA-incompatible kidney recipients and survival. N Engl J Med 2011;365:318–26.

[33] Ishida H, Miyamoto N, Shirakawa H, Shimizu T, Tokumoto T, Ishikawa N, et al. Evaluation of immunosuppressive regimens in ABO-incompatible living kidney transplantation—single center analysis. Am J Transplant Off J Am Soc Transplant Am Soc Transplant Surg 2007;7:825−31.

[34] Mamode N, Dorling A, Kenchayikoppad S, Barnett N. Successful outcomes after minimising antibody modulation in blood group incompatible transplantation. Glasgow: European Society of Transplantation; 2011.

[35] Wilpert J, Fischer KG, Pisarski P, Wiech T, Daskalakis M, Ziegler A. Long-term outcome of ABO-incompatible living donor kidney transplantation based on antigen-specific desensitization. An observational comparative analysis. Nephrol Dialysis Transplant Off Publ Europ Dialysis Transplant Assoc Eur Renal Assoc 2010;25:3778−86.

[36] Montgomery JR, Berger JC, Warren DS, James NT, Montgomery RA, Segev DL. Outcomes of ABO-incompatible kidney transplantation in the United States. Transplantation 2012;27:603−9.

[37] Maple H, Burnapp L, Santhouse A, Gibbs P, Weinman J, Mamode N. Assessment of psychosocial characteristics and outcomes in 117 non-directed altruistic donors in the UK. Bournemouth, UK: British Transplantation Society; 2013.

[38] Montgomery RA. Renal transplantation across HLA and ABO antibody barriers: integrating paired donation into desensitization protocols. Am J Transplant Off J Am Soc Transplant Am Soc Transplant Surg 2010;10:449−57.

[39] Oyen O, Scholz T, Hartmann A, Pfeffer P. Minimally invasive kidney transplantation: the first experience. Transplant Proc 2006;38:2798−802.

[40] Modi P, Pal B, Modi J, Singla S, Patel C, Patel R. Retroperitoneoscopic living-donor nephrectomy and laparoscopic kidney transplantation: experience of initial 72 cases. Transplantation 2013;95:100−5.

[41] Oberholzer J, Giulianotti P, Danielson KK, Spaggiari M, Bejarano-Pineda L, Bianco F, et al. Minimally invasive robotic kidney transplantation for obese patients previously denied access to transplantation. Am J Transplant Off J Am Soc Transplant Am Soc Transplant Surg 2013;13(3):721−8 [Epub 27 February 2013].

[42] Schmidt N, Alloway RR, Walsh RC, Sadaka B, Shields AR, Girnita AL, et al. Prospective evaluation of the toxicity profile of proteasome inhibitor-based therapy in renal transplant candidates and recipients. Transplantation 2012;94:352−61.

[43] Sberro-Soussan R, Zuberl J, Suberbielle-Boissel C, Legendre C. Bortezomib alone fails to decrease donor specific anti-HLA antibodies: even after one year post-treatment. Clin Transplant 2010;409−14.

[44] Barnett AN, Asgari E, Chowdhury P, Sacks SH, Dorling A, Mamode N. The use of eculizumab in renal transplantation. Clin Transplant 2013;27(3):E216−29.

[45] Stegall MD, Diwan T, Raghavaiah S, Cornell LD, Burns J, Dean PG, et al. Terminal complement inhibition decreases antibody-mediated rejection in sensitized renal transplant recipients. Am J Transplant Off J Am Soc Transplant Am Soc Transplant Surg 2011;11:2405−13.

[46] Scandling JD, Busque S, Dejbakhsh-Jones S, Benike C, Sarwal M, Millan MT, et al. Tolerance and withdrawal of immunosuppressive drugs in patients given kidney and hematopoietic cell transplants. Am J Transplant Off J Am Soc Transplant Am Soc Transplant Surg 2012;12:1133−45.

Chapter 16

Machine Perfusion of Kidneys Donated After Circulatory Death: The Carrel and Lindbergh Legacy

Ina Jochmans[a,b] and Jacques Pirenne[a,b]

[a]Clinical Department of Abdominal Transplant Surgery, University Hospitals Leuven, Leuven, Belgium, [b]Laboratory of Abdominal Transplant Surgery, Department of Microbiology and Immunology, KU Leuven, Belgium

Chapter Outline

16.1 INTRODUCTION

After decades of experimental work optimizing surgical techniques—the most important being the development of surgical methods for vascular anastomosis by Alexis Carrel (1873–1944) [1]—the era of modern kidney transplantation started in 1954. On December 23, Joseph Murray, John Merril, and their team successfully transplanted a kidney between identical twins [2]. Both Carrel and Murray later won the Nobel prize for their pioneering work. However, in the 1950s and early 1960s, the majority of attempts at allotransplantation were defeated by rejection. It took until 1963, when Roy Calne brought Azathioprine with him to the Peter Bent Brigham Hospital, for kidney transplantation of deceased donors to become a clinical reality [3]. After "mastering" both surgical and immunological challenges, the next issue was preservation of the kidney grafts. This again is an area that was already explored by Carrel earlier on in the twentieth century; Carrel indeed also studied tissue and organ preservation by both cold storage and machine perfusion [1].

16.2 CONCEPT OF MACHINE PERFUSION

Alexis Carrel was one of the first to observe that hypothermia (1–2°C), contrary to normothermic conditions, slows down tissue destruction in an anoxic environment. At the time, this observation led to the cold preservation of deceased bodies, used in experimental settings as donors of grafts and tissues. As such, blood vessels preserved for several days in Locke's solution could be transplanted. Soon thereafter, other tissues such as skin, cornea, and omentum were successfully transplanted clinically after having been cold stored. Despite the successful short-term preservation of a variety of tissues, Carrel was soon confronted with the limitations of cold storage preservation: limited in time and imperfect. Therefore, organ perfusion, a technique that would allow

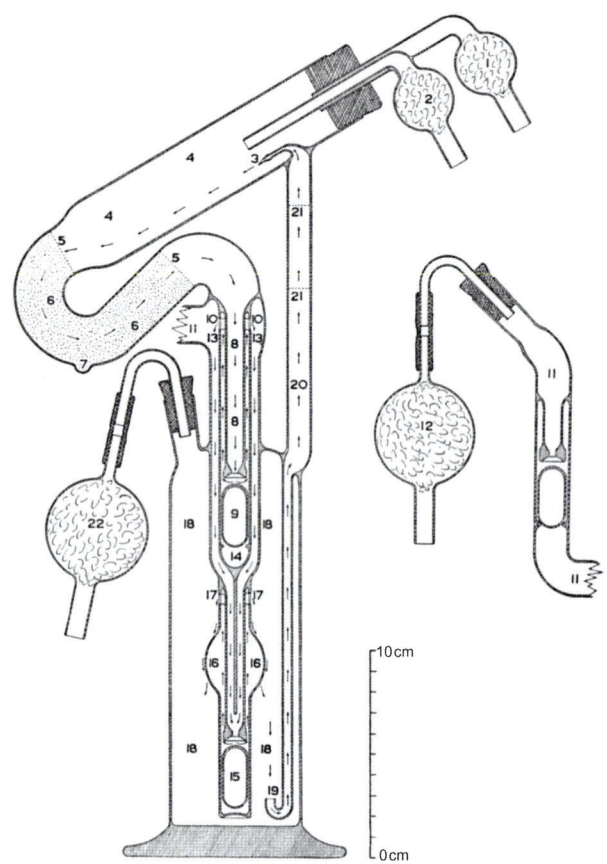

FIGURE 16.1 Cross section of the pulsating perfusion pump designed by Lindbergh. *Reprinted with permission from Ref. [5]. © 1935 Rockefeller University Press. Originally published in Journal of Experimental Medicine;62(3):409–31. Epub 1935/08/31.*

extracorporeal survival of organs for longer periods of time became a new focus of his research. When Carrel met the American engineer and pilot Charles Lindbergh, organ perfusion became more than an elusive idea envisioned by Julien-Jean-Cesar Le Gallois in 1813. This latter had already said *"if one could replace for the heart some kind of injection of artificial blood, either natural or artificially made ... one could succeed easily in maintaining alive indefinitely any part of the body..."* [4].

In 1935, Charles Lindbergh described a device designed to maintain a sterile, pulsating circulation of fluid through living organs [5]. Three glass chambers—an organ chamber at the top, a pressure-equalizing chamber in the middle, and a fluid reservoir at the bottom (Figure 16.1)—allowed a sterile pulsating circulation through the organ studied. A gas mixture (oxygen, carbon dioxide, and nitrogen) was transmitted under a pulsating pressure to the perfusion fluid. The system pressures could be maintained within adjustable minimum and maximum limits. The power for the pulsating gas pressure

was obtained from compressed air, operated by a small motor. As such, the nutritious fluid could be driven from the lowest fluid reservoir up to a glass tube connected to the organ's artery. The design also allowed the aseptical removal of gas and perfusion fluid, to adjust temperature range, and to directly observe the perfused organ. In fact, it is quite extraordinary that this first perfusion machine already contains all the essential elements we find in machine perfusion devices today.

16.3 MACHINE PERFUSION INTRODUCED IN THE CLINICS

Although a substantial amount of work on experimental isolated organ preservation and transplantation continued in the Soviet Union in the 1950s by Lapchinsky et al. it was not until Folkert O Belzer, from Wisconsin, USA, got interested in machine preservation that a truly clinically applicable machine was designed in the 1960s. In a series of experiments, one more successful than the other, Belzer discovered that cryoprecipitated plasma was superior to either whole blood or plasma to perfuse canine kidneys [6]. This preclinical work resulted in the first successful human transplantation of a hypothermically machine-perfused kidney in 1968 [9]. Technological improvements led to the construction of smaller, portable machines. This relatively miniaturized preservation machine was used clinically for the first time to transport a kidney over the Atlantic Ocean, from New York to Leiden, the Netherlands, on Christmas Eve of 1971 where, after 37 h of hypothermic preservation, it was successfully transplanted into a 42-year-old truck driver with polycystic kidney disease [10].

16.4 MACHINE PERFUSION DISAPPEARS TO THE BACKGROUND

Although hypothermic machine perfusion was used by transplant centers in both the United States and Europe in the 1970s, this technology disappeared to the background for the next few decades. Indeed, around the same time, the first reports on successful long-term static cold storage appeared. Geoffrey Collins published that canine kidney grafts could be transplanted successfully after 30 h of preservation in a solution containing a high potassium and glucose concentration [11]. Obviously, compared to machine perfusion, organ storage in a preservation solution on melting ice was by far an easier and cheaper method of organ preservation and allowed easy transportation between transplant centers. The research interest in static cold storage increased and this led to the development of newer preservation solutions such as the

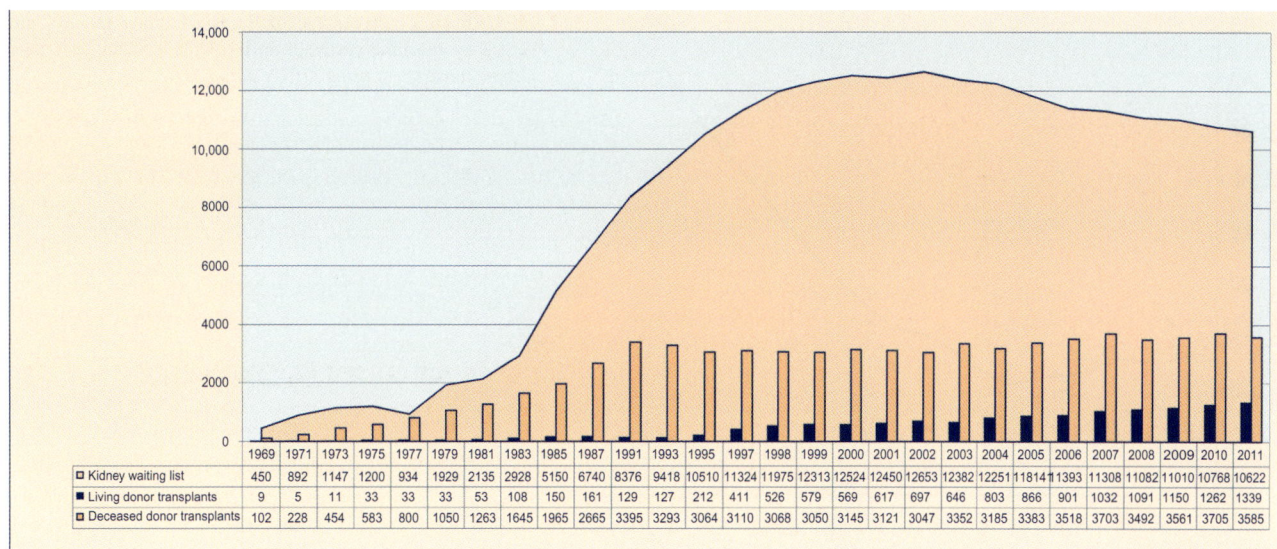

	1969	1971	1973	1975	1977	1979	1981	1983	1985	1987	1991	1993	1995	1997	1998	1999	2000	2001	2002	2003	2004	2005	2006	2007	2008	2009	2010	2011
Kidney waiting list	450	892	1147	1200	934	1929	2135	2928	5150	6740	8376	9418	10510	11324	11975	12313	12524	12450	12653	12382	12251	11814	11393	11308	11082	11010	10768	10622
Living donor transplants	9	5	11	33	33	33	53	108	150	161	129	127	212	411	526	579	569	617	697	646	803	866	901	1032	1091	1150	1262	1339
Deceased donor transplants	102	228	454	583	800	1050	1263	1645	1965	2665	3395	3293	3064	3110	3068	3050	3145	3121	3047	3352	3185	3383	3518	3703	3492	3561	3705	3585

FIGURE 16.2 Dynamics of the Eurotransplant kidney transplant waiting list between 1969 and 2011. *Reprinted with permission from Eurotransplant [27].*

EuroCollins' solution that emerged in 1976 as a modified Collins' solution [12]. At the same time, the deceased donor population was also changing: from kidneys retrieved from donors that died after a circulatory arrest (donation after circulatory death or DCD) to brain-dead donors (donation after brain death or DBD). Indeed, soon after the description of "irreversible coma," a legal definition of brain death was published [13,14]. Kidneys were no longer exposed to detrimental warm ischemia, but came from young donors that had died from head trauma. These pristine kidneys tolerated cold ischemia in the—at that time—available preservation solutions relatively well. It seemed that the complex use of relatively cumbersome hypothermic machine perfusion units was no longer necessary.

Nevertheless, despite the increased use of static cold storage, Belzer continued his work on hypothermic machine perfusion. In the 1980s, he developed a purely synthetic machine perfusion solution that could preserve canine kidneys for up to 5 days [15,16]. This Belzer's machine perfusion solution is still used today. Furthermore, it is this solution that, with a few modifications, gave rise to the University of Wisconsin cold storage preservation solution, still the gold standard for abdominal organ preservation [17].

We have described the history of organ perfusion in transplantation in detail elsewhere [18].

16.5 MACHINE PRESERVATION REVIVAL

The deteriorating quality of available deceased donor kidneys in the twenty-first century announces a new dawn on machine perfusion. Indeed, continued organ shortage, as illustrated for Eurotransplant (the international organ-exchange organization of Austria, Belgium, Croatia, Germany, Luxemburg, the Netherlands, and Slovenia) in Figure 16.2, has led to another change in the profile of the deceased donor pool.

As mentioned earlier, DCD donation had almost come to a stop after the introduction of brain death criteria and the superior results achieved with DBD kidneys [19]. Indeed, compared with DBD kidneys, DCDs have considerably higher incidences of delayed graft function (DGF) and primary nonfunction (PNF) (13−35% and 1−10% vs. 28−88% and 1−18%, respectively) and inferior graft outcome [20,21]. However, the success of clinical transplantation made it clear that the DBD pool was vastly insufficient to sustain the increasing demand for kidney grafts. It was estimated that the DCD pool was substantially larger than the DBD pool and that DCD donation could potentially double or even quadruple the number of deceased donor kidney transplantations [22]. For this reason DCD programs were set up again. In addition, a few landmark papers at the turn of the century showed that excellent long-term graft survival, equivalent to DBD kidneys, could be achieved DCD kidneys, and this despite a higher rate of DGF [23,24]. These early reports were later confirmed in larger series [20,25,26]. The excellent results of DCD kidney transplantation combined with the growing organ shortage have led to a steady increase of DCD kidney transplant activity in countries with the required legal framework. DCD kidney transplantation now reaches up to 30−40% of deceased donor kidney transplantations in the United Kingdom (UK) and the Netherlands [25,27].

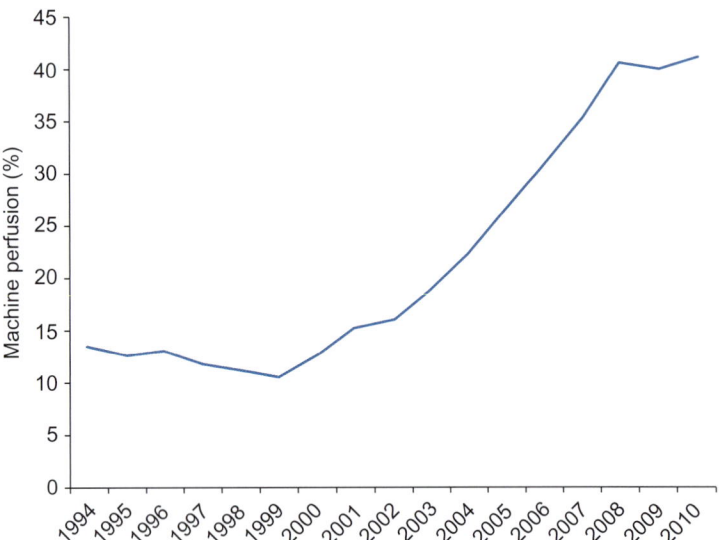

FIGURE 16.3 Percentage of deceased donor kidneys preserved by machine perfusion in the United States each year. *Graph constructed using OPTN data [33].*

The continued organ shortage has urged transplant physicians to expand the acceptance criteria of DBD donation as well. The use of so-called "expanded criteria donor" (ECD) kidneys—as opposed to "standard criteria donors" (SCD)—is no longer an exception. Compared to SCDs, ECDs have an inherent higher risk for the development of DGF (15−25% vs. up to 40%) and PNF after transplantation [28,29]. This is because ECDs are donors older than 60 years of age or 50−59 years of age with at least 2 of the following characteristics: a history of hypertension, death caused by a cerebrovascular accident and a final preprocurement serum creatinine >1.5 mg/dL, resulting in an overall risk of graft failure of 1.7 compared to SCDs [30]. Nevertheless, transplantation of ECD kidneys has a significant survival benefit compared with continued dialysis treatment [31]. About one-third of deceased kidney transplant activity in the United States is now performed with kidneys from ECDs and DCDs [32].

Because both DCD and ECD kidneys are more prone to preservation-induced damage, posttransplant DGF and PNF, it was thought that machine perfusion preservation, that is closer to mimicking the physiological status, might improve the results after transplantation of this type of kidneys. This belief, together with technological evolutions over the years, has led to an enormous increase of the use of hypothermic machine perfusion preservation. In fact most centers that restarted active DCD programs use machine perfusion. Furthermore, machine perfusion creates a window of opportunity to assess the graft before transplantation and to assess its viability or quality, possibly allowing the selection of grafts with better outcome after transplantation. Data from the Organ Procurement and Transplantation Network (OPTN) from the United States show that in 2010 almost 45% of deceased donor kidneys were machine perfused (Figure 16.3) [33].

16.6 WHAT IS THE EVIDENCE?

16.6.1 Does Machine Perfusion Provide Better Preservation?

Ever since static cold storage solutions became available a "cold war" exists between advocates of hypothermic machine perfusion and static cold storage to determine which preservation method is the best in terms of outcome. Today, in times of financial crises, the cost-−benefit of either storage method is as important to consider. So, where should our alliance go considering the currently available evidence?

16.6.1.1 Data from the Twentieth Century

Between 1971 and 2001, 20 papers (reporting on 16 studies) were published that prospectively compared the outcome of transplantation of kidneys preserved by either hypothermic machine perfusion or static cold storage [34,35]. In a metaanalysis of these studies, Wight et al. report that the use of machine perfusion is associated with a relative risk of DGF of 0.80 (95% confidence interval 0.67−0.96) compared to cold storage [34]. No effect of machine perfusion on 1-year graft survival was detected in this metaanalysis. However, all studies, even when aggregated, were severely underpowered with respect to the likely impact on graft survival [34]. They conclude their report by indicating that in the long term machine perfusion would be expected to be cost-effective compared to cold storage for DBD and DCD kidney transplants. However, a definite study establishing the effect of machine perfusion on DGF and long-term graft survival, together with an economic evaluation would be of great value. In an attempt to provide these additional

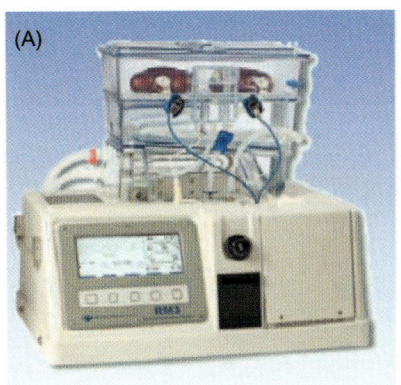
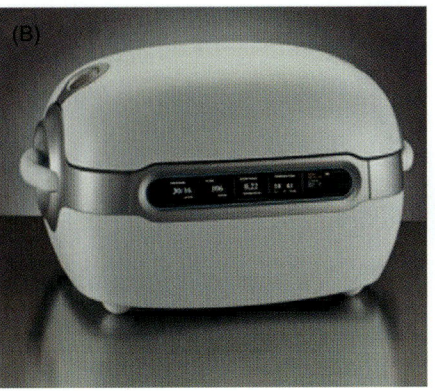
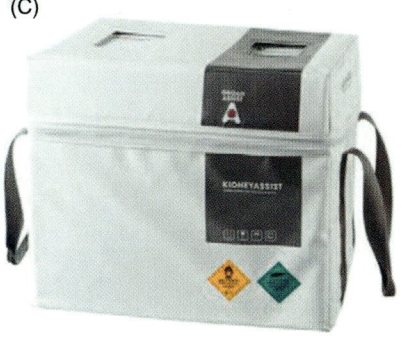

FIGURE 16.4 Currently available hypothermic machine perfusion devices for clinical kidney preservation. Panel A: RM3 (Waters Medical Systems, Birmingham, AL), www.wtrs.com; panel B: LifePort Kidney Transporter (Organ Recovery Systems, Itasca, IL), www.organ-recovery.com; panel C: Kidney Assist (Organ Assist, Groningen, the Netherlands), www.organ-assist.nl.

data, a few randomized controlled trials were set up in Europe. We review these trials below.

16.6.1.2 Machine Perfusion Preservation of Deceased Donor Kidneys

In 2005, the Machine Preservation Trial (MP Trial)—an international randomized controlled trial within the Eurotransplant region—started recruitment in the Netherlands, Belgium, and the federal state of North Rhine-Westphalia in Germany [36]. From each consecutive, eligible donor, one kidney was randomized to machine perfusion and the contralateral organ to cold storage. The MP Trial included kidneys from all deceased donor types (SCD, ECD, DCD), aged 16 years or older and maintained a strict paired design in which kidney pairs were only included if both organs were transplanted into two different recipients. Combined transplantations (e.g., kidney–pancreas or kidney–liver) were excluded. DCD kidneys had to be Maastricht category III (awaiting circulatory death after withdrawal of treatment) or category IV (circulatory death in a brain-dead donor) [37]. Surgical teams were allowed to switch preservation methods only if the kidney assigned to machine perfusion had an aortic patch that was too small or if it had too many renal arteries for a reliable connection to the machine perfusion device; this switch in preservation methods changed the initial randomization. As such, data were analyzed "per protocol."

LifePort Kidney Transporter machines (Organ Recovery Systems, Itasca, IL) (Figure 16.4) were used to preserve kidneys randomized to machine perfusion. The LifePort delivered a pulsatile flow of University of Wisconsin machine preservation solution (Kidney Preservation Solution-1) [38] at 1–8°C from organ procurement until transplantation. The systolic perfusion pressure was fixed at 30 mm Hg and was not changed during the preservation period. Intravascular renal resistance and flow readings, recorded by the LifePort, were never revealed to the transplantation team in order to prevent bias in clinical decisions about transplanting or discarding an organ. Kidneys that were randomized to cold storage followed the standard cold storage protocols of Eurotransplant. After an initial vascular washout, kidneys were submerged in the preservation solution and stored on melting ice. As such, cold-stored kidneys were preserved by either University of Wisconsin solution or Histidine–Tryptophan–Ketoglutarate solution. The trial's primary end point was DGF, defined as the requirement for dialysis during the first week after transplantation. Secondary end points included the duration of DGF, PNF, and 1-year graft and patient survival. Additionally, the incidence of "functional DGF" was assessed. Functional DGF is a more refined surrogate marker for early kidney graft function than DGF. It is defined as the absence of a decrease in the serum creatinine level of at least 10% per day for at least 3 consecutive days in the first week after transplantation, not including patients in whom acute rejection, toxicity of the calcineurin inhibitor, or both developed within the first week [39,40].

From November 1, 2005, through October 31, 2006, there were 654 potential deceased kidney donors 16 years of age or older in the three trial regions. 336 kidney pairs (672 recipients) were included in the analysis. In 25 donors (4.6%), preservation methods were switched because of aberrant vascular anatomy of the kidney assigned to machine perfusion. Vascular anomalies were not observed to have a significant effect on DGF. DGF occurred in 70 recipients in the machine perfusion group (20.8%) as compared with 89 patients in the cold storage group (26.5%). Logistic regression showed that machine perfusion significantly reduced the risk of DGF (adjusted odds ratio 0.57; 95% confidence interval 0.36–0.88;

TABLE 16.1 Multivariate Analysis of the Risk of DGF and Graft Failure[a]

Variable	Odds Ratio (95% CI)	Hazard Ratio (95% CI)	P Value
Delayed graft function			
Machine perfusion versus cold storage	0.57 (0.36−0.88)		0.01
Panel-reactive antibody level (%)	1.01 (0.99−1.02)		0.29
Recipient age (years)	1.01 (0.99−1.03)		0.28
Donor age (years)	1.03 (1.00−1.06)		0.04
ECD donor versus SCD donor[b]	1.04 (0.46−2.34)		0.92
Cold ischemic time (h)	1.08 (1.03−1.14)		0.003
HLA mismatches (no.)	1.13 (0.94−1.37)		0.18
Duration of pretransplantation dialysis (years)	1.16 (1.03−1.31)		0.01
Second or later transplantation versus first transplantation	3.01 (1.75−5.18)		<0.001
DCD donor versus DBD donor	17.2 (8.16−36.2)		<0.001
Graft failure within 1 year after transplantation[c]			
Machine perfusion versus cold storage		0.52 (0.29−0.93)	0.03
DCD donor versus DBD donor		0.90 (0.28−2.92)	0.87
Recipient age (years)		0.97 (0.95−1.00)	0.02
Duration of pretransplantation dialysis (years)		1.00 (0.87−1.15)	0.97
Panel-reactive antibody level (%)		1.01 (0.99−1.03)	0.31
Cold ischemic time (h)		1.04 (0.97−1.11)	0.25
Donor age (years)		1.05 (1.01−1.10)	0.02
ECD donor versus SCD donor[b]		1.18 (0.42−3.27)	0.76
HLA mismatches (no.)		1.23 (0.98−1.55)	0.08
Second or later transplantation versus first transplantation		1.72 (0.88−3.35)	0.11

[a]A logistic regression model was used to determine the odds ratio for DGF and a Cox proportional hazards model was used to determine the hazard ratio for graft failure. Odds ratios and hazard ratios are associated with a 1-unit increase in each covariate. CI denotes confidence interval, DBD donation after brain death, DCD donation after cardiocirculatory death, ECD expanded criteria donation, and SCD standard criteria donation.
[b]Since donor age was a separate covariate in these models and donor age was also part of the ECD definition, the effect of ECD versus SCD on DGF and the risk of graft failure may appear to be less pronounced than commonly reported.
[c]Data on graft survival were censored at the time of death in patients who died with a functioning allograft.
Source: Reprinted with permission from Ref. [36] © 2009 Massachusetts Medical Society.

$P = 0.01$) (Table 16.1). If DGF developed, its duration was 3 days shorter after machine perfusion as compared with cold storage (10 days vs. 13 days, $P = 0.04$). Functional DGF occurred in 22.9% of recipients in the machine perfusion group and in 30.1% of recipients in the cold storage group ($P = 0.03$). PNF occurred in 2.1% of machine-perfused kidneys compared to 4.8% of cold-stored kidneys; this difference did not reach statistical significance ($P = 0.08$).

Graft survival was also significantly improved by machine perfusion compared to cold storage (94% vs. 90%, $P = 0.04$). (Figure 16.5). Cox regression analysis (Table 16.1) showed that machine perfusion significantly reduced the risk of graft failure in the first year after

transplantation, with a hazard ratio of 0.52 (95% confidence interval 0.29−0.93; $P = 0.03$). In 2012, the 3-year follow-up data of the MP Trial were published [41]. Overall, 3-year graft survival was better for machine-perfused kidneys (91% vs. 87%, adjusted hazard ratio 0.60; 95% confidence interval 0.37−0.97; $P = 0.04$) (Figure 16.5).

16.6.1.3 Machine Perfusion Preservation of ECD Kidneys

As mentioned earlier, ECD kidneys are increasingly used because of organ shortage. Unfortunately, ECD kidneys have a higher incidence of DGF and a more complicated

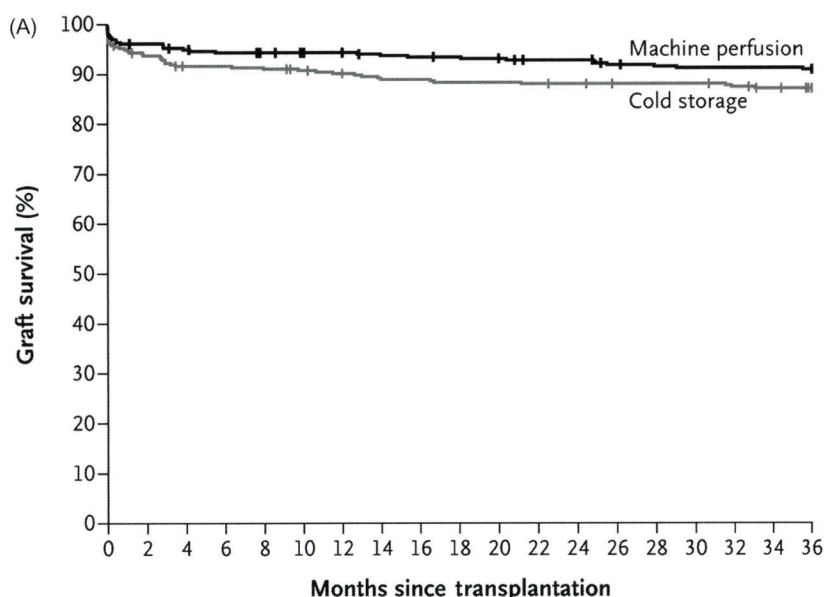

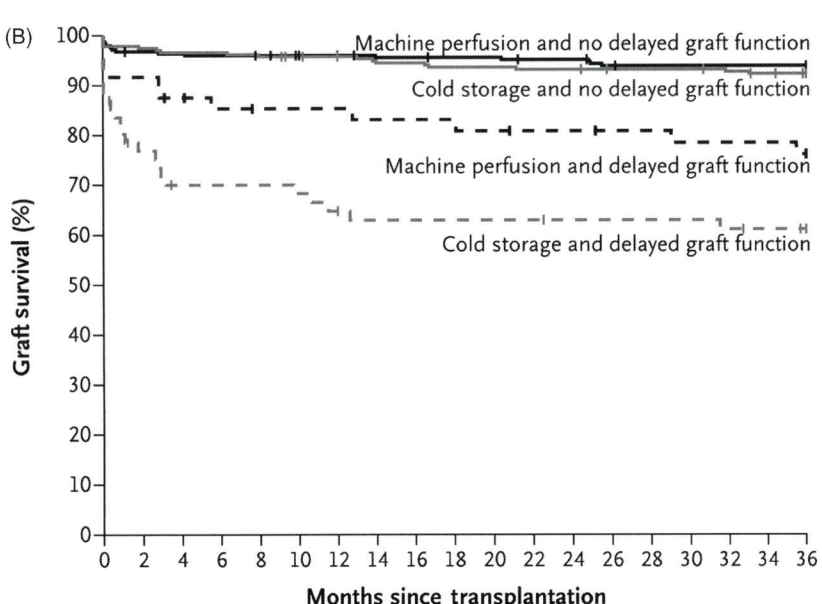

FIGURE 16.5 Three-year graft survival of deceased donor kidneys included in the MP Trial. Panel A shows graft survival in 672 kidney recipients, with a hazard ratio for graft failure in the machine perfusion group of 0.60 (95% confidence interval 0.37−0.97; *P* = 0.04). Panel B shows the *post hoc* analysis of a subgroup of 588 recipients of kidneys donated after brain death, with data split according to whether DGF developed in the recipient. DGF was defined as the need for dialysis in the first week after transplantation. *Reprinted with permission from Ref. [41] © 2012 Massachusetts Medical Society.*

postoperative course, resulting in an inferior long-term graft survival [30,42−45]. The beneficial effect of machine perfusion on the occurrence of DGF and the overall graft survival might be more pronounced in ECD kidneys [30,46]. Of the 336 kidney pairs included in the MP Trial, 91 were ECDs [47]. 22.0% of machine-perfused ECD kidneys developed DGF compared to 29.7% of cold-stored ECD kidneys (*P* = 0.27). Multivariate logistic regression showed that machine perfusion significantly reduced the risk of DGF compared with cold storage (adjusted odds ratio 0.46; 95% confidence interval 0.21−0.99; *P* = 0.047). The incidence of PNF of cold-stored ECD kidneys was significantly higher

compared to machine-perfused kidneys (12% vs. 3%, *P* = 0.04). The incidence of functional DGF was 29.7% after cold storage and 20.8% after machine perfusion (*P* = 0.31).

One-year death censored graft survival was significantly higher in machine-perfused kidneys compared with cold-stored kidneys (92.3% vs. 80.2%, *P* = 0.02) (Figure 16.6A). The presence or absence of DGF seems to have an impressive effect on graft survival, at least in cold-stored ECD kidneys. Although there was a difference of nearly 10% for 1-year graft survival if DGF occurred compared with kidneys with immediate function in the machine preservation group, this difference was

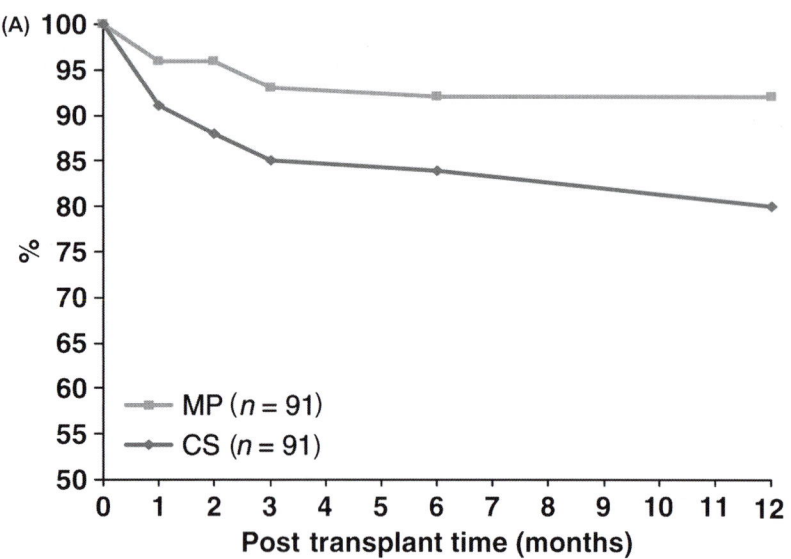

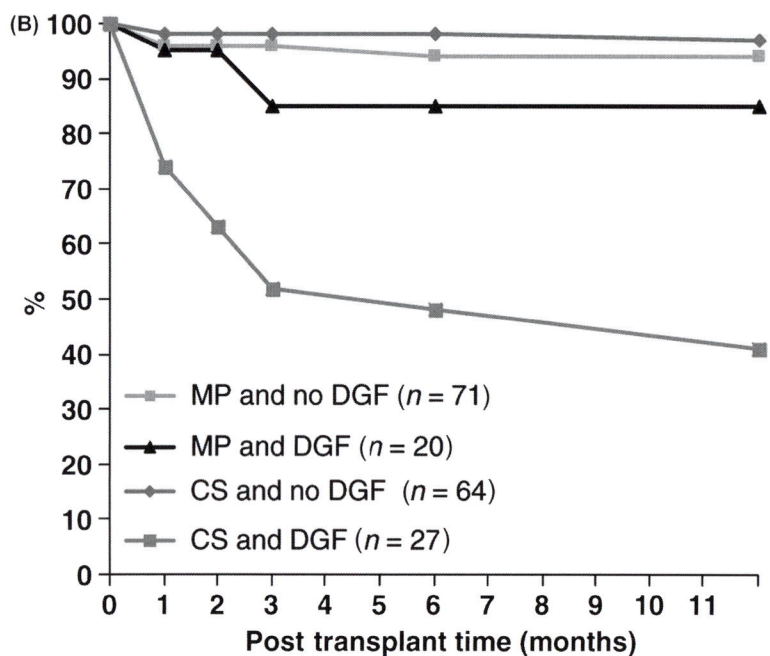

FIGURE 16.6 Panel A: Posttransplant graft survival rates of 182 ECD kidney recipients in the MP Trial. Logrank test of equality machine perfusion (MP) versus cold storage (CS), $P = 0.02$. Panel B: Within CS group DGF versus no DGF, $P < 0.0001$. Within MP group DGF versus no DGF, $P = 0.164$. Within no DGF group MP versus CS, $P = 0.48$. Within DGF group MP versus CS, $P = 0.003$. MP denotes machine perfusion, CS cold storage. *Reprinted with permission from Ref. [47]*

not statistically significant (94% vs. 85%, $P = 0.16$). However, in cold-stored kidneys that developed DGF, graft survival was significantly worse compared to when the graft functioned immediately (41% vs. 97%, $P < 0.0001$). When only recipients of grafts that developed DGF were analyzed, there was a significant difference in 1-year graft survival between machine-perfused kidneys and cold-stored kidneys (85% vs. 41%, $P = 0.003$) (Figure 16.6B). Cox regression analysis showed that machine perfusion significantly reduced the risk of graft failure in the first year with an adjusted hazard ratio of 0.35 (95% confidence interval 0.15–0.86;

$P = 0.02$). The 3-year graft survival advantage after machine perfusion was maintained for ECD kidneys (86% vs. 76%, adjusted hazard ratio 0.38; 95% confidence interval 0.18–0.80; $P = 0.01$) [41].

16.6.1.4 Machine Perfusion Preservation of Kidneys Donated After Circulatory Death

Because DCD kidneys are exposed to warm ischemia during the period of donor circulatory arrest, they are particularly vulnerable to the development of DGF. DGF occurs in 28–88% of cases, depending on the report compared to 13−35% in DBD donation [20,21,48]. Because of this high incidence of DGF, it was thought that especially DCD kidneys would benefit from machine perfusion preservation and would allow a safer and wider use of this potentially large donor source.

Indeed, previous studies have suggested that machine perfusion of DCD kidneys results in better early function and improved graft survival compared to static cold storage. However, other studies do not support this conclusion [49−53]. A comprehensive metaanalysis failed to show a statistically significant risk reduction of DGF in machine-perfused versus static cold-stored DCD kidneys [34,35].

The previously mentioned MP Trial included DCD donors but 87.5% of inclusions were DBD donors. An extension of this trial specifically addressed the issue of machine perfusion in DCD kidneys. The trial only included Maastricht category III (circulatory arrest after withdrawal of treatment) DCD donors aged 16 years or older and a 5 min "no-touch" period was always respected [54]. Data from 82 recipients in each study group were analyzed. DCD kidneys had a median warm ischemic time of 16 (6−38) min and a cold ischemic time of 15.0 (4.3−28.9) h in the machine perfusion group and 15.9 (8.6−46.6) h in the cold storage group.

Machine perfusion preservation significantly reduced the risk for the development of DGF (adjusted odds ratio 0.43; 95% confidence interval 0.20−0.89; $P = 0.025$). Of machine-perfused DCD kidneys, 53.7% developed DGF compared to 69.5% of the cold-stored DCD kidney ($P = 0.007$). Functional DGF was even more reduced by machine perfusion (19.5% vs. 51.2%; $P < 0.0001$). PNF occurred in only two cases in each study group.

Unlike for DBD kidneys, machine perfusion did not result in an increased 1-year graft survival (93.9% for machine-perfused kidneys versus 95.1% for cold-stored kidneys). At analysis of the 3-year data, this trend was confirmed. Machine perfusion did not offer a benefit to graft survival, despite its effect on reducing the incidence of DGF [41]. Although this might seem contradictory, this finding is in line with an increasing number of reports showing similar medium-term graft survival for DCD and DBD kidneys despite higher rates of DGF in DCD kidneys [20,48,55]. DGF does not influence graft survival after DCD kidney transplantation in the same way it does after DBD kidney transplantation.

At the same time the MP trial started including patients, a similar trial focusing on the effect of machine perfusion in DCD kidneys was set up in the UK [56]. In this trial, all adult Maastricht Category III DCD donors at five participating centers in the UK were eligible for inclusion. Here as well, a 5-min "no-touch" period was always respected. DCD kidneys were not shared through the national sharing scheme and were nearly always transplanted at the center that performed the procurement. *In situ* perfusion of kidneys was done with University of Wisconsin preservation solution. Cold-stored kidneys were preserved in this solution on melting ice. Kidneys allocated to machine perfusion were perfused on the LifePort Kidney Transporter machines and perfused with Kidney Preservation Solution-1 at a pulsatile mode with a pressure of 30 mm Hg. In this trial, similar to the MP trial, data on flow and resistance were not used to determine whether a kidney should or should not be transplanted. Where organ procurement occurred away from the base transplant hospital, the kidney randomized to machine perfusion could first undergo a period of CS during transport to the base hospital. In this case, machine perfusion was started when the kidney arrived at the transplant center. The trial followed a sequential design, allowing it to be stopped when analysis after a predetermined number of patients did not show a difference between the study arms. The trial was indeed stopped when data for 45 DCD kidney pairs were available. Kidneys were exposed to a mean of 15 (4.0−35.0) min of warm ischemia followed by 13.9 (6.7−24.2) h of cold storage in the machine preservation group and 14.3 (7.0−30.1) h in the cold storage group. Of the machine-perfused kidneys, 58% developed DGF compared to 56% of cold-stored DCD kidneys ($P = 0.99$). In multivariate analysis, this leads to an odds ratio of 1.14 (95% confidence interval 0.38—3.49). One kidney in the machine preservation group suffered from PNF.

One-year graft survival was similar in both groups (93.3% in the machine preservation group and 97.8% in the cold storage group, $P = 0.3$).

The controversial results between the two trials, currently the largest trials performed and the best available evidence, remains unexplained. One explanation might be that machine perfusion was started immediately after procurement in the Eurotransplant MP Trial whereas it was delayed in a subset of kidneys in the UK trial. It may be that pumping kidneys throughout the entire preservation period is necessary to benefit from machine perfusion, which obviously has important logistical consequences [57]. In support of this hypothesis, Nicholson et al. showed, in a porcine kidney transplant model, that the beneficial effect of machine perfusion disappears when kidneys are not pumped immediately after procurement [58]. Another noticeable difference is the higher DGF rate in the cold storage arm of the MP Trial. Increased ischemic injury, and perhaps other factors inherent to large multicentric international studies, may account for this difference. Indeed, DCD kidneys were exposed to longer *mean* warm (*plus* 2.6 min) and cold (*plus* 3 h) ischemia times in the Eurotransplant compared with the UK trial.

16.6.2 Does Machine Perfusion Predict Graft Viability?

In theory, accurately predicting DGF, PNF, and graft survival before transplantation would allow the selection of kidneys with a high likelihood of favorable outcome and would help in tailoring peri- and postoperative care. However, kidney graft quality cannot be assessed when kidney grafts are cold stored on melting ice. On the contrary, hypothermic machine perfusion may not only provide superior preservation but also offer the opportunity to study perfusion characteristics like renal vascular resistance (RR) in real time. Furthermore, injury markers in the perfusate can also be studied.

16.6.2.1 The Value of RR

Retrospective evidence suggests that RR and flow rate during machine perfusion correlate with kidney graft function [34]. However, in most of these studies, a selection bias was introduced because kidneys were systematically discarded based on arbitrarily defined perfusion parameter thresholds. Today, "poor" perfusion dynamics are still frequently used to discard kidneys for transplantation even though their true prognostic value on graft outcome had never been studied until recently. Indeed, more than 15% of machine-perfused kidneys are discarded annually in the United States, partly based on elevated renal resistance [59]. In the aforementioned MP Trial, the preservation method was not revealed at the time of organ offer and in case of machine perfusion clinicians had no knowledge of the RR value. The decision to accept a given kidney was based solely on traditional donor data, making it possible to elucidate the real prognostic value of RR on PNF, DGF, and 1-year graft survival [60]. To date this is the only reported prospective study available to determine the true value of RR.

Analysis of RR of all kidneys included in the MP Trial showed that RR at the end of machine perfusion was an independent risk factor for the development of DGF, independent of donor type (SCD vs. ECD vs. DCD). Univariable analysis showed that RR was a risk factor for the development of DGF at 30 min, 2 and 4 h and at the end of machine perfusion. In multivariable analysis, only RR at the end of MP proved to be an independent risk factor of DGF (adjusted odds ratio 38.1; 95% confidence interval 1.56−93.4; $P = 0.03$). However, the c-statistic of the receiver−operator characteristic curve for RR at the end of machine perfusion was only 0.58. This means that, despite the association of RR and DGF, RR has a limited value in the prediction of DGF for a specific donor−recipient pair.

Interestingly and contrary to common belief, renal resistance values of PNF cases were not the highest renal resistance values measured. Although the number of PNF cases was low ($n = 7$)—and a definite relationship between renal resistance and PNF needs cautious interpretation—this observation together with the poor predictive power of renal resistance for DGF indicates that kidneys should not be discarded based on RR criteria alone. Some previous reports, analyzing data in retrospect, also recommend caution in using RR in the assessment of kidney quality. Indeed, Sonnenday et al. stressed the importance of considering not only the perfusion parameters but also donor factors when assessing graft quality. These authors could successfully transplant 11 out of 14 kidneys with favorable donor characteristics that had been turned down by other centers due to "poor" perfusion parameters [61]. Mozes et al. analyzed 336 consecutive machine-perfused ECD kidneys and showed that the outcome of kidneys with "poor" perfusion parameters (0.40 mm Hg/mL/min < RR < 0.60 mm Hg/mL/min) was similar to the kidneys with "good" perfusion parameters [62]. Guarrera et al. reported acceptable short- and long-term results in a small series of deceased donor kidneys with "poor" perfusion parameters (flow < 80 mL/min/100 g and RR > 0.40 mm Hg/mL/min/100 g) but no other high donor risk factors [63].

In the analysis of the MP Trial, RR was also a risk factor for 1-year graft failure in both unadjusted and adjusted Cox regression analyses (adjusted hazard ratio 12.3, 95% confidence interval 1.11−136.9; $P = 0.004$).

The analysis of prospectively collected RR values of kidneys stored by machine perfusion showed that RR is an independent risk factor for both DGF and 1-year graft failure. These findings suggest that RR is an important additional objective tool to be used in kidney graft quality assessment. Nevertheless, because of its low predictive accuracy, RR cannot be used as a stand-alone viability parameter to accept or discard a given kidney.

16.6.2.2 The Value of Perfusate Biomarkers

Retrospective data also suggest that biomarker concentrations measured in the perfusate of machine-perfused kidneys correlate with graft outcome. The groups of Newcastle and Maastricht have previously determined the perfusate concentration of total glutathione-S-transferase (GST), heart-type fatty acid binding protein (H-FABP), and alanine-aminopeptidase (Ala-AP) and found higher concentrations in perfusate of discarded kidneys [64−66]. Furthermore, lactate dehydrogenase (LDH) and GST concentrations correlated with warm ischemia time, and GST, H-FABP, and Ala-AP levels were higher in uncontrolled DCD kidneys (Maastricht category II) compared to controlled DCD kidneys (Maastricht category III). However, these data are confronted with the same methodological design and selection bias as the perfusion dynamics data, and none of the previous studies investigated whether perfusate biomarkers were independently associated with graft outcome. Despite absence of prospective analysis of the real predictive value of perfusate biomarkers, some centers are still using them in their current decision algorithms.

During the MP Trial, perfusate was prospectively collected to determine whether GST, H-FABP, Ala-AP, LDH, aspartate aminotransferase (AST), and N-acetyl-β-D-glucosaminidase (NAG) concentrations were independent predictors of PNF, DGF, and 1-year graft survival [67]. Only GST, NAG, and H-FABP were independent risk factors of DGF, but not of PNF or graft survival. The predictive power of these three biomarkers was also moderate at best with c-statistics of 0.67, 0.64, and 0.64, respectively. So, similar to RR, perfusate biomarkers alone should not lead to kidney discard.

There are multiple donor, preservation, and recipient factors influencing graft outcome. It is therefore not surprising that RR and perfusate biomarkers are not sufficient on their own to be used as sole predictors of outcome. However, given the fact that they do independently correlate with DGF (RR, biomarkers) and with graft survival (RR), they are valuable in assisting clinicians in decision making. In the future, it will be interesting to study whether inclusion of RR and perfusate biomarkers into existing multifactorial scoring systems can increase their predictive value.

16.6.3 Is Machine Perfusion Cost Effective?

In times of economic difficulties, another important issue is not only whether a specific treatment is "better" than another but also whether it is cost effective. The meta-analysis by Wight et al., assessing the cost effectiveness of machine perfusion, concluded that the published economic evidence is of poor quality and not based on randomized studies [34]. In 2009, Bond et al. performed an updated metaanalysis on machine perfusion versus cold storage, including preliminary data from both the MP Trial and the UK DCD Trail. As always, with only few trials available, the authors conclude that the cost effectiveness of machine perfusion is difficult to determine and requires more research [68].

Data of the MP Trial were used to perform an economic evaluation of machine perfusion versus cold storage [69]. The economic evaluation combined the 1-year results based on the empirical data from the study with a Markov model with a 10-year time horizon. Direct medical costs of hospital stay, dialysis treatment, and complications were included. Data regarding long-term survival, quality of life, and long-term costs were derived from the literature. Short-term evaluation showed that machine perfusion results in lower average costs in the first year after

transplantation compared to cold storage. The costs of dialysis and the costs of readmission were mainly responsible for the difference in costs. This is applicable for the overall dataset (including DBDs and DCDs). A subset analysis showed the same benefit for ECDs, however, because the number of graft failures in the DCD population ($n = 5$) was too small, firm conclusions concerning cost effectiveness of machine perfusion for DCDs could not be drawn.

The long-term analyses showed that machine perfusion results in lower costs and better outcomes in the mixed overall population of deceased donor kidneys irrespective of donor type (DBD or DCD) and for both SCDs and ECDs. The Markov model revealed cost savings of \$86.750 per life year gained in favor of machine preservation. The corresponding incremental cost–utility ratio was − \$496.223 per quality-adjusted life year (QALY) gained. The authors conclude that life years and QALYs can be gained while reducing costs at the same time, when kidneys are preserved by machine perfusion instead of cold storage. Furthermore, a sensitivity analyses showed that only substantial increases in costs of disposables could affect the short-term cost effectiveness of machine perfusion over cold storage.

16.7 MACHINE PERFUSION IN THE TWENTY-FIRST CENTURY—ACTIVE ORGAN REPAIR DURING MACHINE PERFUSION

Although machine perfusion has been around for over 50 years, the technology can still be greatly improved and many potential applications remain to be explored.

16.7.1 An Improved Preservation Solution

Currently, Belzer's machine perfusion solution (amongst others available as Kidney Preservation Solution-1) is used to perfuse kidneys. This solution is a gluconate-based perfusate that contains hydroxylethyl starch and, contrary to the static preservation solution, has a low potassium concentration to avoid vasoconstriction. The constitution of the perfusion solution is likely to play an important role in the outcome of the graft after transplantation. Keeping in mind the mechanisms of ischemia–reperfusion injury, it is possible that improved solutions or the addition of specific reagents targeting these mechanisms to the preservation solution could be developed. New solutions that resemble cell culture media have been developed (AQIX RS-I and Lifor Organ Preservation Medium) [70,71]. They contain amino acids, metabolic substrates, vitamins, salts, and organic buffers that make them ideal potential new solutions for machine perfusion. Furthermore, additives in the perfusion solution, such as radical scavengers (e.g., curcumin) [72,73], antioxidants (e.g., trimetazidine) [74,75], and many others, might decrease the detrimental effects of warm and cold ischemic injury by counteracting several key events in the cascade of ischemia—reperfusion injury.

16.7.2 The Addition of Oxygen

As can be made up from the construction of the Lindbergh apparatus that included a gas mixture of carbon dioxide, nitrogen, and oxygen, the latter was considered to be a vital part of kidney preservation from the start of preservation technology [5]. Although the rationale of using oxygen to sustain metabolic cell processes makes perfect sense, it is quite paradoxical that the majority of clinically applied preservation methods nowadays do not use oxygen. However, with the availability of high-tech perfusion platforms, the deteriorating quality of donor kidneys and the need to "recondition" or "repair" these higher risk kidneys prior to transplantation, oxygenation is finding its way into transplant research. Increasing evidence suggests that adding oxygen to the preservation solution is beneficial, especially when it comes to restoring cellular levels of adenosine triphosphate after the kidney has been exposed to ischemia [76]. On the other hand, the presence of oxygen could potentially increase the production of radical oxygen species and thereby causing increased injury [77]. Any preservation solution will need to keep the delicate balance between active oxygenation and antioxidant properties of the preservation solution into account.

Oxygen can be delivered in a number of ways: by simple retrograde persufflation of oxygen directly through the renal vein (without necessarily using machine perfusion); during machine perfusion where oxygen can be dissolved into the perfusate; or by adding artificial oxygen carriers (e.g., acellular solution based on perfluorocarbons by Brasile et al. [78—80], Hemo2Life [81]). The use of oxygen during kidney preservation is reviewed by Hosgood et al. [76].

16.7.3 Normothermic Machine Perfusion

Nevertheless, one might wonder whether oxygenation and resuscitation in a cold environment is the most optimal way to "repair" or "recondition" kidney grafts. Indeed, during hypothermic conditions the cellular metabolism is—intentionally—lowered to approximately 10%, also reducing the need for oxygen. At the same time, hypothermia prevents active repair mechanisms from taking place and limits the effect of (metabolically active) drugs intervening on ischemia—reperfusion injury. In this light, warmer, i.e., subnormothermic or normothermic, machine perfusion creates a near-to-normal physiological environment in which kidneys could be "reconditioned" prior to their transplantation. This has been shown in a preclinical model of kidney transplantation in which kidneys exposed to 2 h of warm ischemia were successfully transplanted after normothermic preservation with an acellular preservation solution while cold storage or hypothermic machine preservation resulted in graft failure [79]. Although (sub)normothermic machine perfusion and preservation has already been applied successfully for the preservation of human heart and lungs, it is only recently that normothermic machine perfusion of human kidneys has been proven possible [82]. The team of Nicholson in Leicester developed an *ex vivo* kidney perfusion circuit that has been used in numerous preclinical studies [83—85]. The circuit consists of a centrifugal blood pump, a heat exchanger, PVC tubing, and a membrane oxygenator. The first case of normothermic *ex vivo* perfusion was performed in 2011: an ECD kidney was first cold stored for 11 h after which it was perfused at 33.9°C during 35 min immediately before transplantation [82]. The *ex vivo* circuit was primed with Ringers' solution, one unit of cross-matched packed red cells from the blood bank, mannitol, dexamethasone, sodium bicarbonate, heparin, prostacyclin, glucose 5%, parental nutrition, insulin, and multivitamins. Until now the team of Leicester has reported 17 cases in which the kidneys were successfully transplanted after being perfused on this *ex vivo* normothermic perfusion circuit following a period of cold storage, clearly showing the feasibility of the technique [86]. A randomized controlled trial needs to determine whether normothermic preservation is superior to hypothermic preservation and which type of kidneys might benefit most.

In addition to its potential superiority in preserving and reconditioning kidneys, normothermia may also open the door to true viability assessment. Indeed, true graft quality assessment in the current setting of static cold storage is probably elusive. Cold preservation, even with machine perfusion, remains cold, precluding real-time assessment of organ function at a normal metabolic rate. On the contrary, during warmer, oxygenated preservation, metabolism is sustained and true viability assessment may become feasible. In addition, viability assessment during cold storage is always biased by the fact that evaluation takes place prior to ischemia—reperfusion injury in the recipient. In (sub)normothermic machine perfusion, the anticipated repercussions of ischemia—reperfusion injury may either be minimal or—in any case—be observed (and potentially treated) before transplantation. Standard parameters of kidney function like diuresis and calculating creatinine clearance, as has been done in experimental settings of *ex vivo* kidney reperfusion [85] would become a possibility. It is possible that RR and perfusate injury biomarkers in this setting will correlate better with graft outcome than during cold perfusion. Warmer machine perfusion is also likely to result in greater differences in protein and gene expression profiles. These might lead to the discovery of new markers of injury and outcome and to the identification of new targets of interest for possible intervention.

Another field of interest is *in situ* normothermic recirculation making use of the setup of extracorporeal membrane oxygenation. After circulatory arrest and the diagnosis of death the DCD donor is connected to a reperfusion circuit reinstituting oxygenation normothermic perfusion. The thoracic cavity is excluded from the normothermic recirculation so the heart will not resuscitate. This method of recirculation theoretically shortens the exposure to warm ischemia and allows more time for logistics involved in the procurement procedure. This is especially interesting in uncontrolled Maastricht Category II DCDs (failed resuscitation). Preclinical studies have shown the benefit of normothermic recirculation in kidney transplantation [87]. There are also emerging reports on the clinical use with good results for both liver [88] and kidney [89]. Randomized controlled trials evaluating normothermic recirculation remain to be performed.

16.7.4 The Potential of *Ex Vivo* Reconditioning with Mesenchymal Stem Cells

Mesenchymal stem cells (MSCs) are multipotent, self-renewing cells isolated from whole bone marrow. They have the potential to differentiate into different cell types but their effector function is mostly based on paracrine effect and cross-talk with other cells within diseased tissues [90]. Indeed, they have immunomodulatory properties—interaction with innate immunity and suppression of T-cell responses—and they secrete soluble factors [91]. Both properties make them particularly interesting to use as a cellular therapy in solid organ transplantation [92,93].

In kidney, the administration of MSC enhances the recovery from ischemia—reperfusion induced acute kidney injury in rats [94]. While the operating mechanisms are not yet identified with certainty, they seem to involve anti-inflammatory properties and the facilitation of tissue repair through endocrine and paracrine interactions and cellular cross-talk between MSC, dendritic cells, macrophages and other cell types. MSCs are now considered as a new therapeutic tool in patients with acute kidney injury because of their renoprotective properties and their capacity to promote tissue repair after injury. Ongoing trials are testing their effect [95]. The role of autologous MSCs as an induction therapy to promote graft acceptance has also been studied in a randomized controlled trial after living-related kidney transplantation. This study demonstrated a decreased incidence in acute rejection, a lower risk of opportunistic infections, and a better estimated renal function at 1 year after transplantation compared with the IL-2 receptor antibody control group [96].

Machine perfusion offers a unique platform to selectively administer MSCs directly and selectively to the donor organ, overcoming issues such as homing, trafficking, and safety. While donor-derived MSCs might not be useful in deceased organ transplantation (because of the extended time required for their isolation after retrieval from the donor bone marrow), allogeneic MSCs are more attractive because of their wide availability at the time of organ transplant. From an experimental point of view, direct administration of MSCs to an organ during machine perfusion in an isolated circuit *ex vivo* would provide a unique platform to study the potential therapeutic effects of these cells and their operating mechanisms. Perfusing an organ directly with MSC may allow us to better study the trafficking of these cells, mechanisms of homing to particular organs, and the factors influencing this process; their effect on the immunogenicity of the perfused organ (MHC expression, effect on endothelial cells and passenger antigen-presenting cells); their effect on intragraft innate immunity and local inflammation (cytokine, adhesion molecule, chemokine expression, cellular infiltration, etc.), and their mode of action. Administration of MSC during machine perfusion carries many potentials: the possibility to promote repair of damage sustained prior to preservation due to donor-related injuries and the possibility to downregulate local intragraft inflammation and reduce graft immunogenicity in order to make grafts less vulnerable to ischemia—reperfusion injury and to rejection in the recipient [90].

16.8 CONCLUSIONS

The concept of machine perfusion of organs is as old as transplantation itself. Over the past century, it has been transformed from sketches on a drawing table to a clinical reality. The ever deteriorating donor profile has encouraged the use of machine perfusion once again. Simpler and transportable perfusion systems have been developed and strong evidence showing the superiority of hypothermic machine perfusion over static cold storage has emerged. As a consequence, machine perfusion is increasingly used. The use of machine perfusion as a viability predictor is still under investigation. However, the more we learn about hypothermic machine perfusion, the more we are confronted with its shortcomings. Therefore, new ways of using the machine as a preservation and assessment method (e.g., warmer temperatures, oxygenation, newer preservation solutions, pharmacological additions to the perfusate, stem cell research, etc.) are being explored. The pioneering work by Carrell and Lindbergh continues in the twenty-first century and is likely to change the way we preserve kidneys for transplantation.

REFERENCES

[1] Carrel A. Landmark article of Nov 14,1908, results of the transplantation of blood vessels, organs and limbs. JAMA 1983;250:944–53.

[2] Harrison JH, Merrill JP, Murray JE. Renal homotransplantation in identical twins. Surg Forum 1956;6:432–6.

[3] Calne R, Loughridge LW, MacGillivray JB, Zilva JF, Levi AJ. Renal transplantation in man: a report of five cases, using cadaveric donors. Br Med J 1963;2:645–51.

[4] Glass Heart. Time 1935;41–2.

[5] Lindbergh CA. An apparatus for the culture of whole organs. J Exp Med 1935;62:409–31.

[6] Belzer FO, Park HY, Vetto RM. Factors influencing renal blood flow during isolated perfusion. Surg Forum 1964;15:222–4.

[7] Belzer FO, Ashby BS, Dunphy JE. 24-hour and 72-hour preservation of canine kidneys. Lancet 1967;2:536–8.

[8] Belzer FO, Ashby BS, Huang JS, Dunphy JE. Etiology of rising perfusion pressure in isolated organ perfusion. Ann Surg 1968;168:382–91.

[9] Belzer FO, Ashby BS, Gulyassy PF, Powell M. Successful seventeen-hour preservation and transplantation of human-cadaver kidney. N Engl J Med 1968;278:608–10.

[10] Belzer FO. Organ preservation: a personal perspective. In: Terasaki PI, ed. History of transplantation: thirty-five recollection. Los Angeles, CA: UCLA Tissue Typing Laboratory; 1991. p. 597–613.

[11] Collins GM, Bravo-Shugarman M, Terasaki PI. Kidney preservation for transportation. Initial perfusion and 30 hours' ice storage. Lancet 1969;2:1219–22.

[12] Watkins GM, Prentiss NA, Couch NP. Successful 24-hour kidney preservation with simplified hyperosmolar hyperkalemic perfusate. Transplant Proc 1971;3:612–5.

[13] Beecher HA. Definition of irreversible coma. Report of the Ad Hoc Committee of the Harvard Medical School to examine the definition of brain death. JAMA 1968;205:337–40.

[14] Selby R, Selby MT. Status of the legal definition of death. Neurosurgery 1979;5:535–40.

[15] Hoffmann RM, Southard JH, Lutz M, Mackety A, Belzer FO. Synthetic perfusate for kidney preservation. Its use in 72-hour preservation of dog kidneys. Arch Surg 1983;118:919–21.

[16] McAnulty JF, Ploeg RJ, Southard JH, Belzer FO. Successful five-day perfusion preservation of the canine kidney. Transplantation 1989;47:37–41.

[17] Ploeg RJ, Goossens D, McAnulty JF, Southard JH, Belzer FO. Successful 72-hour cold storage of dog kidneys with UW solution. Transplantation 1988;46:191–6.

[18] Monbaliu D, Liu Q, Vekemans K, Pirenne J. History of organ perfusion in organ transplantation. In: Talbot D, D'Alessandro A, editors. Organ donation and transplantation after cardiac death. Oxford, UK: Oxford University press; 2009. p. 31–60.

[19] Kootstra G. History of non-heart-beating donation. In: Talbot D, D'Alessandro A, eds. Organ donation and transplantation after cardiac death. Oxford, UK: Oxford University Press; 2009. p. 1–6.

[20] Kokkinos C, Antcliffe D, Nanidis T, Darzi AW, Tekkis P, Papalois V. Outcome of kidney transplantation from nonheart-beating versus heart-beating cadaveric donors. Transplantation 2007;83:1193–9.

[21] Moers C, Leuvenink HG, Ploeg RJ. Non-heart beating organ donation: overview and future perspectives. Transpl Int 2007;20:567–75.

[22] Daemen JW, Oomen AP, Kelders WP, Kootstra G. The potential pool of non-heart-beating kidney donors. Clin Transplant 1997;11:149–54.

[23] Wijnen RM, Booster MH, Stubenitsky BM, de Boer J, Heineman E, Kootstra G. Outcome of transplantation of non-heart-beating donor kidneys. Lancet 1995;345:1067–70.

[24] Weber M, Dindo D, Demartines N, Ambuhl PM, Clavien PA. Kidney transplantation from donors without a heartbeat. N Engl J Med 2002;347:248–55.

[25] Summers DM, Johnson RJ, Allen J, Fuggle SV, Collett D, Watson CJ, et al. Analysis of factors that affect outcome after transplantation of kidneys donated after cardiac death in the UK: a cohort study. Lancet 2010;376:1303–11.

[26] Barlow AD, Metcalfe MS, Johari Y, Elwell R, Veitch PS, Nicholson ML. Case-matched comparison of long-term results of non-heart beating and heart-beating donor renal transplants. Br J Surg 2009;96:685–91.

[27] Eurotransplant. Eurotransplant annual report. Leiden, the Netherlands: Eurotransplant International Foundation; 2011. Available from: <www.eurotransplant.org>; [accessed November 2012].

[28] Saidi RF, Elias N, Kawai T, Hertl M, Farrell ML, Goes N, et al. Outcome of kidney transplantation using expanded criteria donors and donation after cardiac death kidneys: realities and costs. Am J Transplant 2007;7:2769–74.

[29] Fraser SM, Rajasundaram R, Aldouri A, Farid S, Morris-Stiff G, Baker R, et al. Acceptable outcome after kidney transplantation using "expanded criteria donor" grafts. Transplantation 2010;89:88–96.

[30] Metzger RA, Delmonico FL, Feng S, Port FK, Wynn JJ, Merion RM. Expanded criteria donors for kidney transplantation. Am J Transplant 2003;3:114–25.

[31] Ojo AO, Hanson JA, Meier-Kriesche H, Okechukwu CN, Wolfe RA, Leichtman AB, et al. Survival in recipients of marginal cadaveric donor kidneys compared with other recipients and wait-listed transplant candidates. J Am Soc Nephrol 2001;12:589–97.

[32] OPTN. OPTN/SRTR annual report [09/2011]. Available from: <http://optn.transplant.hrsa.gov/ar2009/>; 2009.

[33] OPTN. 2009 Annual report of the U.S. Organ Procurement and Transplantation Network and the scientific registry of transplant recipients: transplant data 1999–2008. Rockville, MD: U.S. Department of Health and Human Services, Health Resources and Services Administration, Healthcare Systems Bureau, Division of Transplantation; 2009. Available from: <http://optn.transplant.hrsa.gov> [accessed 4.1.11].

[34] Wight J, Chilcott J, Holmes M, Brewer N. The clinical and cost-effectiveness of pulsatile machine perfusion versus cold storage of kidneys for transplantation retrieved from heart-beating and non-heart-beating donors. Health Technol Assess 2003;7:1–94.

[35] Wight JP, Chilcott JB, Holmes MW, Brewer N. Pulsatile machine perfusion vs. cold storage of kidneys for transplantation: a rapid and systematic review. Clin Transplant 2003;17:293−307.

[36] Moers C, Smits JM, Maathuis MH, Treckmann J, Van Gelder F, Napieralski BP, et al. Machine perfusion or cold storage in deceased-donor kidney transplantation. N Engl J Med 2009;360:7−19.

[37] Kootstra G, Daemen JH, Oomen AP. Categories of non-heart-beating donors. Transplant Proc 1995;27:2893−4.

[38] Belzer FO, Glass NR, Sollinger HW, Hoffmann RM, Southard JH. A new perfusate for kidney preservation. Transplantation 1982;33:322−3.

[39] Boom H, Mallat MJ, de Fijter JW, Zwinderman AH, Paul LC. Delayed graft function influences renal function, but not survival. Kidney Int 2000;58:859−66.

[40] Boom H, Paul L, de Fijter J. Delayed graft function in renal transplantation. Transplant Rev 2004;18:139−52.

[41] Moers C, Pirenne J, Paul A, Ploeg RJ. Machine perfusion or cold storage in deceased-donor kidney transplantation. N Engl J Med 2012;366:770−1.

[42] Pascual J, Zamora J, Pirsch JD. A systematic review of kidney transplantation from expanded criteria donors. Am J Kidney Dis 2008;52:553−86.

[43] Daly PJ, Power RE, Healy DA, Hickey DP, Fitzpatrick JM, Watson RW. Delayed graft function: a dilemma in renal transplantation. BJU Int 2005;96:498−501.

[44] McLaren AJ, Jassem W, Gray DW, Fuggle SV, Welsh KI, Morris PJ. Delayed graft function: risk factors and the relative effects of early function and acute rejection on long-term survival in cadaveric renal transplantation. Clin Transplant 1999;13:266−72.

[45] Troppmann C, Gillingham KJ, Benedetti E, Almond PS, Gruessner RW, Najarian JS, et al. Delayed graft function, acute rejection, and outcome after cadaver renal transplantation. The multivariate analysis. Transplantation 1995;59:962−8.

[46] Polyak MM, Arrington BO, Stubenbord WT, Boykin J, Brown T, Jean-Jacques MA, et al. The influence of pulsatile preservation on renal transplantation in the 1990s. Transplantation 2000;69:249−58.

[47] Treckmann J, Moers C, Smits JM, Gallinat A, Maathuis MH, van Kasterop-Kutz M, et al. Machine perfusion versus cold storage for preservation of kidneys from expanded criteria donors after brain death. Transpl Int 2011;24:548−54.

[48] Kootstra G, van Heurn E. Non-heartbeating donation of kidneys for transplantation. Nat Clin Pract Nephrol 2007;3:154−63.

[49] Plata-Munoz JJ, Muthusamy A, Quiroga I, Contractor HH, Sinha S, Vaidya A, et al. Impact of pulsatile perfusion on postoperative outcome of kidneys from controlled donors after cardiac death. Transpl Int 2008;21:899−907.

[50] Moustafellos P, Hadjianastassiou V, Roy D, Muktadir A, Contractor H, Vaidya A, et al. The influence of pulsatile preservation in kidney transplantation from non-heart-beating donors. Transplant Proc 2007;39:1323−5.

[51] St Peter SD, Imber CJ, Friend PJ. Liver and kidney preservation by perfusion. Lancet 2002;359:604−13.

[52] van der Vliet JA, Kievit JK, Hene RJ, Hilbrands LB, Kootstra G. Preservation of non-heart-beating donor kidneys: a clinical prospective randomised case-control study of machine perfusion versus cold storage. Transplant Proc 2001;33:847.

[53] Opelz G, Terasaki PI. Advantage of cold storage over machine perfusion for preservation of cadaver kidneys. Transplantation 1982;33:64−8.

[54] Jochmans I, Moers C, Smits JM, Leuvenink HGD, Treckmann J, Paul A, et al. Machine perfusion versus cold storage for the preservation of kidneys donated after cardiac death: a multicenter randomized controlled trial. Ann Surg 2010;252:756−64.

[55] Brook NR, Waller JR, Nicholson ML. Nonheart-beating kidney donation: current practice and future developments. Kidney Int 2003;63:1516−29.

[56] Watson CJE, Wells AC, Roberts RJ, Akoh JA, Friend PJ, Akyol M, et al. Cold machine perfusion versus static cold storage of kidneys donated after cardiac death: a UK Multicenter Randomized Controlled Trial. Am J Transplant 2010;10:1991−9.

[57] Jochmans I, Moers C, Ploeg RJ, Pirenne J. To perfuse of to not perfuse kidneys donated after cardiac death. Am J Transplant 2011;11:409−10.

[58] Hosgood SA, Mohamed IH, Bagul A, Nicholson ML. Hypothermic machine perfusion after static cold storage does not improve the preservation condition in an experimental porcine kidney model. Br J Surg 2011;98:943−50.

[59] Sung RS, Christensen LL, Leichtman AB, Greenstein SM, Distant DA, Wynn JJ, et al. Determinants of discard of expanded criteria donor kidneys: impact of biopsy and machine perfusion. Am J Transplant 2008;8:783−92.

[60] Jochmans I, Moers C, Smits JM, Leuvenink HG, Treckmann J, Paul A, et al. The prognostic value of renal resistance during hypothermic machine perfusion of deceased donor kidneys. Am J Transplant 2011;11:2214−20.

[61] Sonnenday CJ, Cooper M, Kraus E, Gage F, Handley C, Montgomery RA. The hazards of basing acceptance of cadaveric renal allografts on pulsatile perfusion parameters alone. Transplantation 2003;75:2029−33.

[62] Mozes MF, Skolek RB, Korf BC. Use of perfusion parameters in predicting outcomes of machine-preserved kidneys. Transplant Proc 2005;37:350−1.

[63] Guarrera JV, Goldstein MJ, Samstein B, Henry S, Reverte C, Arrington B, et al. 'When good kidneys pump badly': outcomes of deceased donor renal allografts with poor pulsatile perfusion characteristics. Transpl Int 2010;23:444−6.

[64] Gok MA, Pelsers M, Glatz JF, Shenton BK, Peaston R, Cornell C, et al. Use of two biomarkers of renal ischemia to assess machine-perfused non-heart-beating donor kidneys. Clin Chem 2003;49:172−5.

[65] Gok MA, Pelzers M, Glatz JF, Shenton BK, Buckley PE, Peaston R, et al. Do tissue damage biomarkers used to assess machine-perfused NHBD kidneys predict long-term renal function post-transplant? Clin Chim Acta 2003;338:33−43.

[66] Daemen JW, Oomen AP, Janssen MA, van de Schoot L, van Kreel BK, Heineman E, Kootstra G, et al. Glutathione-S-transferase as predictor of functional outcome in transplantation of machine-preserved non-heart-beating donor kidneys. Transplantation 1997;63:89−93.

[67] Moers C, Varnav OC, van Heurn E, Jochmans I, Kirste GR, Rahmel A, et al. The value of machine perfusion perfusate biomarkers for predicting kidney transplant outcome. Transplantation 2010;90:966−73.

[68] Bond M, Pitt M, Akoh J, Moxham T, Hoyle M, Anderson R. The effectiveness and cost-effectiveness of methods of storing donated kidneys from deceased donors: a systematic review and economic model. Health Technol Assess 2009;13:1–156.

[69] Groen H, Moers C, Smits JM, Treckmann J, Monbaliu D, Rahmel A, et al. Cost-effectiveness of hypothermic machine preservation versus static cold storage in renal transplantation. Am J Transplant 2012;12:1824–30.

[70] Kay MD, Hosgood SA, Harper SJ, Bagul A, Waller HL, Nicholson ML. Normothermic versus hypothermic *ex vivo* flush using a novel phosphate-free preservation solution (AQIX) in porcine kidneys. J Surg Res 2011;171:275–82.

[71] Gage F, Leeser DB, Porterfield NK, Graybill JC, Gillern S, Hawksworth JS, et al. Room temperature pulsatile perfusion of renal allografts with Lifor compared with hypothermic machine pump solution. Transplant Proc 2009;41:3571–4.

[72] Bayrak O, Uz E, Bayrak R, Turgut F, Atmaca AF, Sahin S, et al. Curcumin protects against ischemia/reperfusion injury in rat kidneys. World J Urol 2008;26:285–91.

[73] Thuillier R, Trieu MTN, Giraud S, Lathelize H, Marchand E, Parkkinen J, et al. A therapy to make borderline grafts suitable: curcumin supplementation optimises function and recovery and outcome. Transpl Int 2011;24:51.

[74] Hauet T, Goujon JM, Vandewalle A, Baumert H, Lacoste L, Tillement JP, et al. Trimetazidine reduces renal dysfunction by limiting the cold ischemia/reperfusion injury in autotransplanted pig kidneys. J Am Soc Nephrol 2000;11:138–48.

[75] Jayle C, Favreau F, Zhang K, Doucet C, Goujon JM, Hebrard W, et al. Comparison of protective effects of trimetazidine against experimental warm ischemia of different durations: early and long-term effects in a pig kidney model. Am J Physiol Renal Physiol 2007;292:1082–93.

[76] Hosgood SA, Nicholson HF, Nicholson ML. Oxygenated kidney preservation techniques. Transplantation 2012;93:455–9.

[77] Fuller BJ, Lee CY. Hypothermic perfusion preservation: the future of organ preservation revisited? Cryobiology 2007;54:129–45.

[78] Brasile L, Stubenitsky BM, Booster MH, Arenada D, Haisch C, Kootstra G. Hypothermia—a limiting factor in using warm ischemically damaged kidneys. Am J Transplant 2001;1:316–20.

[79] Brasile L, Stubenitsky BM, Booster MH, Lindell S, Araneda D, Buck C, et al. Overcoming severe renal ischemia: the role of *ex vivo* warm perfusion. Transplantation 2002;73:897–901.

[80] Brasile L, Stubenitsky BM, Haisch CE, Kon M, Kootstra G. Repair of damaged organs *in vitro*. Am J Transplant 2005;5:300–6.

[81] Thuillier R, Dutheil D, Trieu MT, Mallet V, Allain G, Rousselot M, et al. Supplementation with a new therapeutic oxygen carrier reduces chronic fibrosis and organ dysfunction in kidney static preservation. Am J Transplant 2011;11:1845–60.

[82] Hosgood SA, Nicholson ML. First in man renal transplantation after *ex vivo* normothermic perfusion. Transplantation 2011;92:735–8.

[83] Bagul A, Hosgood SA, Kaushik M, Kay MD, Waller HL, Nicholson ML. Experimental renal preservation by normothermic resuscitation perfusion with autologous blood. Br J Surg 2008;95:111–8.

[84] Hosgood SA, Bagul A, Yang B, Nicholson ML. The relative effects of warm and cold ischemic injury in an experimental model of nonheartbeating donor kidneys. Transplantation 2008;85:88–92.

[85] Hosgood SA, Barlow AD, Yates PJ, Snoeijs MG, van Heurn EL, Nicholson ML. A pilot study assessing the feasibility of a short period of normothermic preservation in an experimental model of non heart beating donor kidneys. J Surg Res 2011;171:283–90.

[86] Hosgood SA, Nicholson HF. The first clinical series of normothermic perfusion in marginal donor kidney transplantation. Transpl Int 2012;25:21.

[87] Rojas-Pena A, Reoma JL, Krause E, Boothman EL, Padiyar NP, Cook KE, et al. Extracorporeal support: improves donor renal graft function after cardiac death. Am J Transplant 2010;10:1365–74.

[88] Fondevila C, Hessheimer AJ, Ruiz A, Calatayud D, Ferrer J, Charco R, et al. Liver transplant using donors after unexpected cardiac death: novel preservation protocol and acceptance criteria. Am J Transplant 2007;7:1849–55.

[89] Billault C, Godfroy F, Thibaut F, Bart S, Arzouk N, Van Glabeke E, et al. Organ procurement from donors deceased from cardiac death: a single-center efficiency assessment. Transplant Proc 2011;43:3396–7.

[90] Van Raemdonck D, Neyrinck AP, Rega FR, Devos T, Pirenne J. Machine perfusion in organ transplantation: a tool for *ex vivo* graft conditioning with mesenchymal stem cells? Curr Opin Organ Transplant 2013;18:24–33.

[91] English K, Mahon BP. Allogeneic mesenchymal stem cells: agents of immune modulation. J Cell Biochem 2011;112:1963–8.

[92] Hoogduijn MJ, Popp FC, Grohnert A, Crop MJ, van Rhijn M, Rowshani AT, et al. Advancement of mesenchymal stem cell therapy in solid organ transplantation (MISOT). Transplantation 2010;90:124–6.

[93] Roemeling-van Rhijn M, Weimar W, Hoogduijn MJ. Mesenchymal stem cells: application for solid-organ transplantation. Curr Opin Organ Transplant 2012;17:55–62.

[94] Lange C, Togel F, Ittrich H, Clayton F, Nolte-Ernsting C, Zander AR, et al. Administered mesenchymal stem cells enhance recovery from ischemia/reperfusion-induced acute renal failure in rats. Kidney Int 2005;68:1613–7.

[95] Togel FE, Westenfelder C. Mesenchymal stem cells: a new therapeutic tool for AKI. Nat Rev Nephrol 2010;6:179–83.

[96] Tan J, Wu W, Xu X, Liao L, Zheng F, Messinger S, et al. Induction therapy with autologous mesenchymal stem cells in living-related kidney transplants: a randomized controlled trial. JAMA 2012;307:1169–77.

Chapter 17

Renal Regeneration: The Bioengineering Approach

Marcus Salvatori[b], Andrea Peloso[a,c], Timil Patel[a], Sij Hemal[a], Joao Paulo Zambon[a], Ravi Katari[a], and Giuseppe Orlando[a]

[a]*Wake Forest School of Medicine, Winston-Salem,* [b]*Newcastle School of Medicine, University of Newcastle-upon-Tyne,* [c]*School of Medicine, University of Pavia, Pavia, Italy*

17.1 INTRODUCTION

The prevalence of chronic kidney disease (CKD) continues to outpace the development of effective treatment strategies. Best practice guidelines for CKD suggest supportive measures that minimize comorbidities but fail to arrest progression. Although tight blood pressure control may help fewer than 40% of patients achieve and maintain therapeutic targets [1,2]. Patients with advanced disease will eventually require lifelong renal replacement therapy in the form of peritoneal or hemodialysis. Although lifesaving, dialysis approximates the filtration functions of the kidney at considerable cost and inconvenience, and fails to restore the homoeostatic, resorptive, metabolic, endocrine, and immunomodulatory functions. As a result of these shortcomings, patients with CKD remain at an increased risk of cardiovascular and all-cause mortality [3].

Allogeneic transplantation remains the only restorative treatment available for advanced CKD with progression to end-stage renal disease (ESRD). Renal transplantation, pioneered by Dr. Joseph Murray [4], is associated with reduced healthcare costs, longer life expectancy, and improved quality of life compared to dialysis [5,6]. The majority of patients are unable to realize these advantages, however, due to the critical shortage of organs that has emerged over recent decades (Figure 17.1). Furthermore, the side effects of lifelong immunosuppressive therapy dramatically impact the overall outcome and significantly limit clinical application.

Emerging technologies in the field of regenerative medicine (RM) seek to address the limitations of current treatment strategies. Technologies that are currently being implemented to bioengineer or regenerate kidneys for transplant purposes can be broadly classified into the following categories: (i) cell-based strategies, (ii) developmental biology strategies, (iii) bioartificial kidney strategies, and (iv) whole-organ bioengineering strategies. Collaborative efforts across these fields aim to produce a bioengineered kidney capable of restoring renal function in patients with end-stage disease. In this chapter, we review recent advancements in RM as applied to kidney bioengineering.

17.2 OVERVIEW OF ORGAN BIOENGINEERING

Organ bioengineering technologies are based on recellularization protocols designed to seed cells onto scaffolds

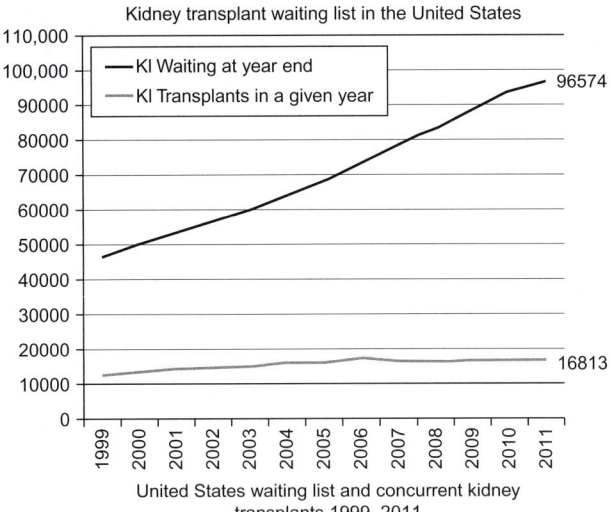

Kidney transplant waiting list in the United States

United States waiting list and concurrent kidney transplants 1999–2011

FIGURE 17.1 Official, open access organ procurement and transplantation network data as of August 24, 2012, reporting on the number of patients in the waiting list for a deceased donor renal transplant and the effective number of transplant performed. This work was supported in part by Health Resources and Services Administration contract 234-2005-370011C. The content is the responsibility of the authors alone and does not necessarily reflect the views or policies of the Department of Health and Human Services, nor does mention of trade names, commercial products, or organizations imply endorsement by the US Government.

capable of driving cellular proliferation and differentiation. Seeding populations currently include pluripotent embryonic stem cells, pluripotent adult stem cells, multipotent adult progenitor cells, and induced pluripotent stem cells, generated via retroviral infection [7], as well as adult, already committed cells. The scaffolds can be derived from native, animal extracellular matrix (ECM) or synthetic biomaterial carriers designed to mimic the structure of natural ECM. The use of native ECM has been made possible through the refinement of decellularization protocols capable of removing DNA, cellular material, and cell surface antigens while leaving the attachment sites, structural integrity, and vascular channels intact [8]. This approach, termed "cell-scaffold technology," is based heavily upon principles of developmental biology that suggest that the ECM is a biologically active structure that influences cell differentiation, growth, survival, and function.

The basic sequence of events required for successful engraftment is: scaffold attachment → growth and division → expansion → maturation. These processes are conducted in a bioreactor, a device produced to recreate the biologic environment required for cell growth, including temperature, flow, oxygenation, and nutrient supply. These conditions allow the construct to effectively mature and dynamically prepare for the *in vivo* environment that follows implantation. Theoretically, this sequence is

intended to recreate the events occurring *in utero* during organ ontogenesis, though many of the specifics remain unknown. Therefore, an in-depth knowledge of the mechanisms of organogenesis is essential for successful bioengineering, as well as details regarding the regeneration and endogenous repair mechanisms of specific organs.

17.3 KIDNEY ECM

Since Grobstein's early characterization of renal organogenesis, researchers have recognized that the ECM plays a crucial role in kidney development and repair [9–11]. The ECM is a three-dimensional framework of structural and functional proteins in a state of dynamic reciprocity with intracellular cytoskeletal and nuclear elements [12]. Over the last decade, research has shown that the ECM plays a fundamental role in the welfare of cells, tissues, and organs. The role of the ECM extends far beyond mechanical support and architecture to influence molecular composition, cell adhesion, signaling, and binding of growth factors. In addition, the mechanical stiffness and deformability of the ECM contribute significantly to the determination, differentiation, proliferation, survival, polarity, migration, and behavior of cells [13]. The proliferation of societies specializing in ECM biology, such as the American Society for Matrix Biology (www.asmb.net), the British Society for Matrix Biology (www.bsmb.ac.uk), the Federation of European Connective Tissue Societies, the International Society of Matrix Biology (www.ismb.org/), and the Matrix Biology Society of Australia and New Zealand (www.mbsanz.org/), is a testament to the rising relevance of ECM technology in the context of biological sciences that have been traditionally focused on cellular elements. The success of organ bioengineering and RM depends on a deep understanding of cell–ECM interactions during ontogenesis and adult life.

ECM molecules and their receptors influence organogenesis and repair by (i) providing a 3D scaffold for the spatial organization of cells, (ii) secreting and storing growth factors and cytokines, and (iii) regulating signal transduction [14]. The components of the ECM include type IV collagen, laminins, proteoglycans, nephronectin, tenascin, fibronectin, and vitronectin [15]. The composition and 3D spatial arrangement of the ECM components are crucial for the phenotype of the collecting system, as shown by cell culture models of renal tubulogenesis [16]. Furthermore, disruptions of these ECM components have been shown to underlie major kidney pathologies, such as CKD [17], Goodpasture's syndrome [18], Alport syndrome [19], nephrotic syndrome [20], and diabetic glomerulosclerosis [21].

17.3.1 Laminins

The laminins are a family of trimeric glycoproteins expressed by the ECM that interact with specific receptors on the cell surface termed "integrins." Through these laminin—integrin interactions, the ECM has been shown to influence several key mechanisms of renal development, including branching morphogenesis. Ureteric branching morphogenesis is a fundamental developmental process required for the amplification of the epithelial surface area that serves as the interface for gaseous and molecular exchange [22]. In the kidney, branching morphogenesis begins with the formation of the ureteric bud (UB) from the Wolffian duct. Reciprocal interactions between the UB and the metenaphric mesenchyme result in the development of renal vesicles, epithelial collections that serve as the precursor to functional nephrons [23]. Mutant knockout mice lacking the laminin-α5 chain exhibit deficient UB branching and aberrant glomerulogenesis [23]. The importance of the laminin-α5 chain for renal organogenesis is further evidenced by the finding that 20% of knockout embryos shows complete, bilateral renal agenesis [23]. The mechanism of developmental failure remains unknown, although the authors theorized that the laminin-α5 chain is an essential component of the signaling cascade required for UB outgrowth, mesenchymal differentiation, or branching morphogenesis. Similarly, laminin-γ1 knockouts show a complete absence of basement membrane, resulting in a lethal arrest of renal development during embryogenesis [24]. Studies investigating selective inhibition of laminin-γ1 expression in the UB showed a disordered basement membrane and disorganized ampullary epithelia [25].

17.3.2 Integrins

The integrins are a family of cell surface receptors responsible for cell—matrix adhesion and the transduction of external signals to the cytoskeleton [26]. Several integrins, including α3β1, α6β4, and α8β1, have been shown to be crucial for metanephric proliferation, branching, and epithelialization during organogenesis [27,28]. Renal papillae cells from α3 knockout mice show a reduced number of collecting ducts in culture, suggesting a deficiency in branching morphogenesis [29]. By contrast, α6 knockout mice show normal renal development, although α6 integrin blocking antibodies reduced branching morphogenesis in both whole kidney and isolated UB culture system [28]. These findings correlate with the aforementioned laminin studies, as laminins-5 is a ligand for both integrins α3 and α6. α8 deficient mice show similarly disordered renal development with defective branching and epithelial—mesenchymal interaction during kidney morphogenesis [27]. α8β1 has been shown to function as a receptor for several ECM elements, including nephronectin, tenascin, fibronectin, and vitronectin [15]. 50% of α8 knockout mice show bilateral renal agenesis caused by failure of UB development or invasion into the metanephric mesenchyme [27].

17.3.3 Proteoglycans

Heparan sulfate proteoglycans (HSPGs) are key constituents of both cell membranes and the ECM. HSPGs are thought to play a role in several molecular signaling pathways, including fibroblast growth factors (FGFs), Wingless/Wnt, transforming growth factor-β (TGF-β), and Hedgehog families [30]. Cell surface HSPGs contribute to the presentation of these secreted signaling molecules to their target receptors. Alternatively, the HSPGs in the ECM have been shown to bind and concentrate growth factors [31]. The signaling and storage functions of the HSPGs have been shown to play a vital role in the branching morphogenesis of the embryonic kidney [32]. Mice lacking the enzyme required for the sulfation of heparan show deficiencies in branching morphogenesis, resulting in complete renal agenesis despite normal UB outgrowth [30].

17.4 ECM IN RENAL BIOENGINEERING: BIOARTIFICIAL SCAFFOLDS

The constituent elements of the ECM are thus essential for renal development and function. The necessity of an extracellular scaffold is widely recognized in the bioengineering literature, as anchorage-dependent cells deprived of attachment sites quickly undergo apoptosis induced by loss of cellular adhesion, termed anoikis. Researchers have sought to fulfill the anchorage requirement artificially via biomaterial carriers that mimic the structure of native ECM. These carriers serve the functions of delivery platform, transitory structural support, and mechanical immune barrier, thus enabling renal cell transplantation into heterotopic sites. The properties of these carriers can also be precisely manipulated, allowing for the study of individual ECM components. Several bioartificial carriers have been reported in the literature, including polyglycolic acid (PGA), poly-lactic-co-glycolic acid (PLGA), polyvinyl alcohol (PVA), poly(ethylene glycol) (PEG), poly(N-isopropylacrylamide) (NIPA), and biopolymer films. These carriers vary in their structure, strength, stability, rigidity, biocompatibility, growth factor binding capacity, and amenability to manipulation.

Successful attachment, proliferation, and function of renal cells on bioartificial carriers would enable the development of advanced renal replacement therapies that incorporate conventional hemofiltration with functional renal bioreactors termed renal tubule assisted devices

(RADs) [33]. The addition of the bioreactor provides the homoeostatic, resorptive, metabolic, endocrine, and immunomodulatory functions of the kidney that are lacking from current renal replacement therapies. Furthermore, these units are designed to be portable or implantable providing continuous renal support with minimal lifestyle disruption.

In their study on the histocompatibility of nuclear transfer generated cells, Lanza et al. report the first successful *in vivo* reconstitution and structural remodeling of functional renal tissues from cloned embryonic renal cells seeded on bioartificial scaffolds [34]. Renal cells were harvested from an embryonic cloned metanephros and expanded *in vitro*. In addition to renal cell specific markers, these cells expressed erythropoietin and 1,25-dihydroxyvitamin D, confirming endocrine functionality. Following expansion and characterization, the primitive renal cells were seeded onto collagen-coated PGA tubular delivery vehicles designed to biodegrade in 8–12 weeks. PGA polymers are biodegradable, biocompatible, and only mildly antigenic. These devices were fitted with collecting systems that enabled the detection, quantification, and analysis of produced filtrate. The constructs were then implanted into the donor animal from which the cells were originally cloned and maintained *in vivo* for 12 weeks. Upon explanation, the devices were found intact and showed straw-yellow fluid within the reservoirs consistent with unidirectional secretion and concentration of urea nitrogen and creatinine. The electrolyte levels and specific gravity of the fluid suggested that the renal cells developed filtration, reabsorption, and secretory capabilities. Additionally, the explants showed vascularization, self-assembled glomeruli, and tubule-like structures lined with cuboid epithelial cells. These findings suggest that bioartificial carriers are capable of fulfilling some of the roles of the native ECM, supporting the attachment, survival, differentiation, and function of embryonic renal cells.

Kim et al. [35] reported similar success after seeding PGA scaffolds with renal segments isolated from postnatal rats. The renal segments were composed of various cell types, including nephron epithelial cells, endothelial cells, vascular smooth muscle cells, and stromal cells. The scaffolds were composed of nonwoven meshes of PGA fibers designed to provide a three-dimensional platform for reconstitution. The porosity of the scaffolds provided an efficient substrate for cellular adhesion, as scaffold adherence was observed within 1 day postseeding. The seeded constructs were implanted in the subcutaneous tissue of recipient mice and maintained for 2 or 4 weeks *in vivo*. After 2 weeks, histological analysis revealed tubular structures and glomerular structures with glomerular tufts and hollow centers. Immunohistochemistry confirmed that the newly generated constructs expressed cell markers specific to the glomerular endothelium. Although this study demonstrates that a bioartificial scaffold can structurally support developing renal structures, the use of postnatal renal segments as the seeding population precludes the ability to determine whether bioartificial scaffolds provide the signals necessary to drive cellular differentiation. Furthermore, the authors did investigate whether these constructs were capable of performing the physiological functions of the native kidney.

Recently, Joraku et al. [36] reported the successful generation of 3D renal structures *in vitro* using primary renal cells in a rigid collagen matrix. The authors homogenized and digested the donor renal tissues from 4-week-old mice in order to produce a heterogeneous seeding population. The heterogeneous seeding population was expected to provide additional signals required for cellular differentiation, such as epithelial–mesenchymal cell interactions. The cells were then plated and allowed to attach and expand until reaching 90–95% confluence. The heterogeneous renal cell population was then harvested and suspended in a 3D matrix composed of collagen gel. The matrix gel was composed of rat tail collagen type 1 diluted in Medium 199 at a 9:1 ratio to produce a firm yet flexible matrix permissible to the diffusion of nutrients. The cell matrix was then plated and incubated in a 37°C incubator for 20 min for solidification. Growth medium was added to each well and the incubation was continued for 7 days. Within 7 days, the authors reported the formation of glomerular- and tubule-like structures within the solidified matrix. Immunohistochemical analysis confirmed that the tubule-like structures expressed Tam-Horsfall protein, a cell marker expressed by cells of the descending loop of Henle and distal convoluted tubule. Similar to the Kim et al. study, these methods did not allow the authors to determine whether bioartificial scaffolds are sufficient to support renal cell differentiation or function.

17.5 ECM IN RENAL BIOENGINEERING: NATIVE ECM

Although biomaterial carriers allow for well-characterized, controlled investigative conditions, they may not reflect the complete complement of native ECM components required for *in vivo* renal cell attachment, survival, and function. A suitable scaffold for functional tissue engineering should (i) be easily explanted and decellularized, (ii) retain native structural relationships and vascular channels, (iii) be amenable to repopulation via immersion or perfusion, and (iv) provide the cues necessary for cellular adhesion and proliferation. Cutting-edge technologies in RM have recently allowed researchers to exploit and appreciate the advantages of preserving the innate ECM for organ bioengineering investigations [37–41]. Indeed, innate

ECM represents a biochemically, geometrically, and spatially ideal platform for such investigations, because it is biocompatible [42], it has both basic components (proteins and polysaccharides) and matrix-bound growth factors and cytokines preserved and at physiological levels [43], it retains an intact and patent vasculature which sustains physiologic blood pressure when implanted *in vivo*[42], and is able to drive differentiation of progenitor cells into an organ-specific phenotype [44,45]. In simplified terms, the *in natura* innate ECM represents the requisite environment for cell welfare because it contains all indispensable information for growth and function [13]. RM is now exploring the possibility of using intact ECM from animal or human whole organs for bioengineering purposes.

17.5.1 Decellularization—Recellularization Treatment

ECM scaffolds from whole animal or human-cadaveric organs can be generated through detergent-based decellularization [38—40,46] (Figure 17.2). Current decellularization techniques are capable of removing DNA, cellular material, and cell surface antigens from the ECM scaffold while preserving attachment sites, structural integrity, and vascular channels [8]. Decellularization protocols involve the repeated irrigation of cadaveric tissues with detergents or acids through the innate vasculature, although organs with higher fat content, like the pancreas, often require the addition of lipid solvents like alcohol [47]. Typical protocols employ a combination of ionic and nonionic detergents, enzymatic nucleases, and antimicrobials, although the optimal composition and concentration of the detergent solution varies with the size, age, and density of the organ.

Effective decellularization protocols are designed to achieve a series of key outcomes, including disruption of the cell membrane, cell lysis, removal of cytoplasmic contents, induction of endogenous nucleases, disruption of nuclear membranes, and degradation of nuclear material. It remains unclear whether detergent decellularization damages essential components of the ECM, although irrigation through the existing vasculature is thought to limit potentially disruptive ECM exposure. Complete decellularization is essential as residual cellular material may contain antigenic epitopes that trigger inflammatory responses [48] and compromise subsequent

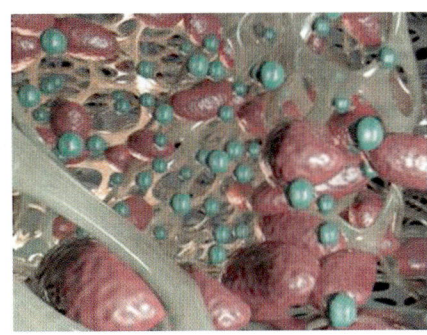 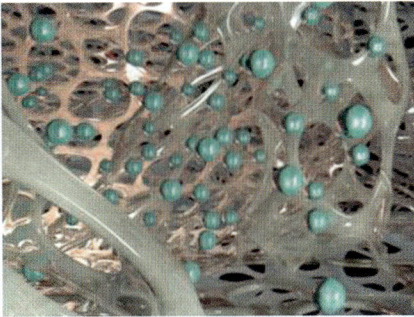 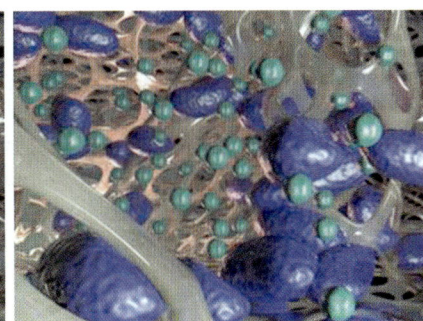

Decellularization:

During this phase, the cellular compartment is cleared with detergent-based solutions. Clearance of ≥95% of the cellular/nuclear Content is the target outcome

Recellularization:

Cells are delivered through intravascular or transmural injection. The main objectives of this phase are the reconstitution of the endothelium to allow implantation and of the organ-specific cellular compartment, namely parenchynal cells

 = Growth factors

= Native animal cells destined to be removed

 = Patient's autologous cells

FIGURE 17.2 Principles of whole-organ bioengineering. Animal or human organs are processed with detergent-based solutions to remove cells. In the case of the kidney, the higher density of the renal texture requires treatment with very strong detergents like SDS, whereas less dense organs, such as the heart, liver, intestine, pancreas, and lung, may be successfully decellularized with milder detergents, such as Triton or sodium deoxycholate. Once cells are removed, the so-obtained extracellular matrix scaffold represents a formidable platform for organ bioengineering and regeneration investigations. Next step is the reconstitution of the different cellular compartments, namely the parenchymal compartment to allow function and the endothelium to allow implantation. Unfortunately, while the decellularization process is quite do-able for all organs of all sizes and species, the repopulation of the scaffolds remains far from the realm of the possible and represents the greatest challenge for the years to come.

recellularization [49]. Following decellularization, gamma irradiation [50], ethylene oxide [51], or paracetic acid [52] have been shown to effectively sterilize the ECM without denaturing the ECM proteins or growth factors, although the risk of viral contamination remains [37].

The decellularized, sterilized ECM serves as the scaffold on which stem cells or renal progenitors are seeded with the intent to reconstitute the cellular compartment (recellularization). The successful recellularization of ECM scaffolds has been reported in several organ systems, including liver [53], respiratory tract [54], nerve [47], tendon [55], valve [56], bladder [57], and mammary gland [58]. These results demonstrate the potential of RM to dramatically impact organ transplantation, with the possibility of upscaling to more complex, modular organs. Organs bioengineered from autologous cells may enable surgeons to successfully address the two major obstacles currently facing organ transplantation: (i) the need for a new, ideally inexhaustible source of organs and (ii) the achievement of an immunosuppression-free state posttransplantation.

17.5.2 Bioreactors

A critical advancement in the field of organ bioengineering has been the development of suitable bioreactors. Bioreactors provide a dynamic, controlled, and reproducible culture environment designed to support structures that are capable of taking part in a specific biologic process, from which the products can be harvested or extracted [59]. These devices allow for the continuous support, monitoring, and manipulation of the biological, biochemical, and biophysical processes involved in organ bioengineering. Bioreactors facilitate the even distribution of seed populations on scaffolds while providing effective nutrient supply, waste removal, and hydrodynamic shear stress [59,60]. Flow-dependent shear stress is a critical mediator of cellular development, as the mechanical stimulus activates signaling pathways that influence the cellular activity, differentiation, and function of the newly engrafted cells [61]. Theoretically, bioreactors provide an environment where the reseeded scaffolds can mature under "*in vivo*-like" conditions in preparation for implantation.

The bioengineering literature boasts a myriad of bioreactor designs, often tailored to suit the needs of the specific organ under investigation. Despite this variety, several key features are common to all bioreactor designs. An ideal bioreactor device must allow (i) precision control of environmental factors, (ii) perfusion of innate channels with the seed population, (iii) automated operation allowing for sterile conditions, (iv) real-time monitoring, and (v) modifiable levels of shear stress. Perfusion bioreactors are essential for uniform cellular dispersion, as static culture conditions result in poor

cellular migration, dispersion, and adherence. As a result, static recellularization protocols are associated with a disproportionate deposition of cells around the organ exterior with a paucity of cells within the inner parenchyma [62]. Poor seeding of the inner layers compromises organ integrity as it (i) fails to recapitulate the 3D structure of the native organ and (ii) results in poor deposition of cells around the vascular channels. Close proximity to the arterial network is a necessity for the successful engraftment, proliferation, and long-term viability of the seed population, due to the high metabolic requirements of differentiating cells. Perfusion bioreactors overcome this obstacle by distributing cells uniformly throughout the organ while providing shear stress via fluid flow [61]. However, large, complex, and modular organs pose additional challenges, as perfusion pressure must be finely regulated to penetrate the full thickness of the parenchyma without damaging the delicate vascular network.

At our institute, we employ a 37°C single-compartment perfusion bioreactor assembly that allows continuous, controlled flow through innate vascular channels in a controlled environment. The bioreactor serves as the environment for both seeding and scaffold maturation, during which attachment, proliferation, and differentiation are expected to occur. The organ is suspended by the arterial cannulas and submerged in culture medium to prevent mechanical compression of the innate vasculature. The organ and reservoir are maintained on a magnetic mixer to minimize the deposition and accumulation of nutrients and cellular debris. The growth media is circulated continuously in a closed circuit, driven by a peristaltic pump. A pulse dampener placed in series between the pump and the organ minimizes pressure fluctuations and ensures steady versus pulsatile flow. Pluripotent cells are suspended in growth media and introduced directly into the arterial cannula via discrete injections and allowed to recirculate. We have found that multiple, small-volume seedings increase cell attachment and viability. The absence of filtration not only affords the cells multiple opportunities for attachment but also increases the likelihood of microvessel occlusion. In addition, it is highly probable that the mechanical stress associated with recirculation compromises cell viability. The culture medium is therefore changed every 48 h to remove accumulated cellular debris and replenish the nutrient supply. Perfusion pressures during seeding are typically $4-6\times$ those used during the maturation phase, although pressures must be titrated according to the robustness of the cell population. The maturation phase is continued for approximately 1 week prior to collection and analysis. Microscopic, histological, and immunohistochemical analyses are used to determine the number, distribution, and differentiation of the engrafted cells. At present, our bioreactor systems do not allow for the control of oxygenation or fluid pH, although these measures may be introduced as required.

17.5.3 Vascularization

To date, several investigators have attempted to implant and maintain bioengineered constructs in live recipients. However, the vast majority of these constructs have been implanted without reconnection to the systemic vascular system, thereby exposing the new organ to ischemia and risk of graft failure. Transplanted organs require an efficient vascular supply, as cells can only survive within 1−3 mm of nutrient and oxygen supply [63]. Therefore, RM cannot neglect the necessity of vascular reconnection for organs larger than 0.3 cm. Supporting evidence comes from Macchiarini's group, who observed proximal ventral collapse of the proximal 1 cm of a tracheal graft within 8 months of implantation (unpublished results). The authors speculated that this complication was due to the pulsatile compression from the aortic arch and migration of the stem-cell-derived chondrocytes into the endoluminal surface of the graft. However, the possibility remains that graft failure was caused by ischemia, as the new organ did not have a vascular pedicle and was not reconnected to systemic circulation. The bioengineering of complex modular organs, such as kidneys, livers, and hearts, will clearly necessitate the reconnection of the vascular pedicle to the systemic circulation. For this reason, natural, decellularized ECM scaffolds represent the ideal platform for future transplantation investigation, due to the preservation of the intrinsic vascular tree.

17.5.4 Decellularization−Recellularization Technology Efforts in the Kidney

Recent advances have allowed researchers to apply the principles of decellularization−recellularization technology (DRT) to complex organs, including the kidney [42,44,50,64,65]. Ross et al. [44] were the first to report the successful recellularization of whole-organ, rat kidney scaffolds with xenotransplanted murine embryonic stem cells (ESC) perfused through the innate vascular and ureteric networks. The Ross protocol involved the harvesting of kidneys from systemically anticoagulated rats, followed by cannulation of the renal artery and ureter. The kidneys were arterially perfused *in situ* using a saline/nitroprusside solution to remove blood remaining in renal circulation. These preharvesting measures were employed to minimize the likelihood of vasoconstriction and clotting that could impair the patency of the channels required for even perfusion. Blood clearance was confirmed visually, through the observation of uniform blanching, which also served to confirm the patency of

the innate vasculature. The harvested kidneys were suspended and gravity perfused for 5 days through an arterial cannula at a constant physiological fluid pressure of 100 mmHg. The perfusion fluid was continuously filtered and recirculated to remove cellular debris, further minimizing the likelihood of vascular and tubular blockage. The decellularization solutions contained a combination of ionic and nonionic detergents, including Trition-X-100, 5 mM calcium chloride, 5 mM magnesium sulfate, 1 M sodium chloride, sodium azide, 0.0025% deoxyribonuclease 1, sodium acetate, sodium deoxycholate, and sodium dodecyl sulfate (SDS). These components were chosen to degrade the native cellular elements through solubilization, as well as enzymatic and osmotic disruption. Decellularization protocols produced a transparent whole-organ scaffold that was easily compressible with sponge-like reexpansion. Compared with sodium deoxycholate, SDS-based detergent protocols showed a higher degree of cell clearance. Cell removal, ECM preservation, and scaffold morphology were confirmed through scanning electron microscopy (SEM), histological analysis, and immunohistochemistry. The decellularized matrix was shown to retain the essential morphological features and protein complement of the native ECM, including laminin and collagen IV. ESC cells were chosen as a seeding population due to their high doubling capacity, pluripotency, and potential to differentiate and integrate into primordial kidney cultures. Prior to implantation, ES cells were cultured and maintained in an undifferentiated state on gelatin-coated dishes, bathed in a serum solution containing antibiotics and recombinant mouse leukemia inhibitory factor. Following harvest, the undifferentiated ES cells were suspended in a growth media devoid of prodifferentiation agents. Approximately 2×10^6 of these ES cells were manually injected through the arterial or ureteral cannula. Following confirmation of even cellular dispersion via fluorescence microscopy, the whole-organ scaffolds with engrafted ES cells were cultured for 12 h to promote cellular adherence. ES cell doubling time was approximately 24 h. The authors then attempted three incubation protocols to investigate the long-term viability and differentiation of the engrafted cells: (i) *static-state whole-organ scaffold immersion in growth media.* Static, whole-organ suspension proved inadequate for long-term cell viability, as cell death was commonly observed at 4 days, prior to ES cell differentiation. (ii) *Transverse sectioning and culture of the preseeded scaffold.* Transverse sectioning of the preseeded scaffold allowed for observation of cellular response, transient differentiation patterns, gross cell localization and migration, but precluded the ability to identify the key features and obstacles to whole-organ repopulation. (iii) *Dynamic bioreactor perfusion.* This protocol employed an automated whole-organ perfusion system designed to recreate the dynamic, *in situ* physiological environment of the native kidney. This system utilized a peristaltic pump to deliver cells and media at physiological pressures up to 120/80 mmHg with a periodicity of 270–300 beats per minute. The entire perfusion system and organ scaffold was maintained in a temperature-controlled incubator supplied with sterile gases, with CO_2 titrated to keep media pH at 7.4. Histochemical and microscopic evaluation of the engrafted ES cells was performed after a range of time intervals (3–10 days) to assess the distribution, migration, grouping tendencies, and cell–ECM interactions. The engrafted cells lost their primitive ES cytology by day 4, and began to adopt the morphological appearance of glomerular and epithelial cells, depending on the site of engraftment. Longer incubation times were associated with central necrosis within tubular structures, suggestive of lumen formation. Developmental protein markers were also localized to assess the pattern and degree of differentiation. The decellularized kidney scaffold successfully supported the growth and migration of the xenotransplanted ESCs within glomerular, vascular, and tubular structures while inducing differentiation down renal cell lines. Within 10 days of seeding, the ESCs showed gross morphological changes consistent with epithelial maturation, as well as immunohistochemical markers of renal and epithelial differentiation, including Pax2 [66], Ksp-cadherin [67], and pan-cytokeratin [68].

Although the importance of their seminal findings cannot be overstated, Ross et al. identified several lines of investigation that must be satisfied in order to advance the field. The authors theorize that pretreatment of the ESC seed population with prodifferentiation agents, such as retinoic acid, activin-A, and bone morphogenetic protein 7 (BMP7) [69], may promote engraftment and proliferation by providing a more kidney-specific lineage. Furthermore, although seeding via innate vascular channels provides even cell distribution in the renal cortex, the renal collecting system is omitted. Retrograde seeding through the ureter was attempted to address this problem, but this route was associated with uneven cell dispersion, possibly due to the papillary architecture of the rat kidney. Furthermore, murine scaffolds do not adequately reflect the size and structure of the adult human kidney. The authors suggest that the kidney architecture of higher order mammals may thus be better suited as a platform for kidney bioengineering. Furthermore, larger pig and primate organs have the physiological and structural capacity to support the critical mass of nephrons required to meet human renal requirements.

The application of DRT protocols to larger, complex organs can be challenging as perfusion decellularization relies in part on diffusion. Larger organs with greater parenchymal mass require higher perfusion pressures, stronger detergents, and prolonged detergent exposure that may damage the native architecture and ECM proteins [42]. Nakayama et al. [65] first performed decellularization studies on sectioned nonhuman primate kidneys. The age range of the kidneys included fetal (T2/T3), infant (birth–1 year), juvenile (1–3 years), or adult (4–12 years). Sections were bath perfused in detergent solutions of 1% SDS or 1% Triton X-100 at 4 or 37°C. Decellularization solutions were changed 8 h after initial submersion, and every 48 h thereafter for 7–10 days until the tissues were transparent. Microscopy, histology, and immunohistochemistry confirmed effective bath decellularization with preservation of native ECM architecture and protein complement. Although both concentrations of SDS effectively removed the cellular compartment, the group treated with 1% SDS at 4°C showed the greatest preservation of the native ECM architecture. At 4°C, the glomerular basement membrane and tubules remained intact with a tightly organized collagen matrix and preserved tissue volume. Interestingly, their results showed that decellularization rate correlated inversely with the animal age due in part to the greater presence of polysaccharide chains in fetal ECM. The authors also theorize that the ECM of the immature kidneys is susceptible to scaffold collapse following removal of the cellular compartment, which prevents perfusion and decellularization of the innermost cells. Higher incubation temperatures during decellularization (37°C) were associated with gross morphological changes and a loss of scaffold volume. Similarly, scaffolds treated with Triton X-100 showed greater disruption of the basement membrane and disorganization of the collagen matrix, with associated increases in Bowman's space, incomplete decellularization, and the accumulation of cellular debris. Overall, higher incubation temperatures during decellularization caused greater ECM disruption and loss of volume. Subsequent immunohistochemistry confirmed the expression of key ECM proteins, including HSPG, fibronectin, collagen I and IV, and laminins. Biomechanical testing was also conducted to assess the preservation of scaffold integrity. The acellular scaffolds showed a decrease in stiffness and resistance to deformation when compared to native tissues, although the implications of these changes for later recellularization and transplantation remain unclear. The authors further tested recellularization via direct extension by culturing acellular scaffolds layered with age-matched fetal kidney explants from unrelated donors for a maximum of 5 days. The explants fused with the scaffold within 3 days of culture. Although previous studies reported poor cell migration through SDS-treated tissues, histological analysis showed local migration of renal cells from the explant to the explant-scaffold border to a maximum depth of 300 μm. Immunohistochemistry for the mesenchymal marker vimentin, epithelial marker cytokeratin, renal marker Pax2, and transcription factor WT1 was used to identify the phenotype of the migrating cells. Two discrete cell populations were identified among the migrating cells: (i) dispersed, vimentin/Pax2 double-positive cells of probable mesenchymal origin and (ii) clustered, vimentin/Pax2/cytokeratin triple positive cells of probable UB origin. These findings served as effective proof-of-principle, demonstrating that primate kidneys are amenable to DRT protocols.

More recently, intact, whole-organ ECM scaffolds have been produced from porcine kidneys [42,50]. Sullivan et al. [50] successfully scaled up their murine decellularization methods for use in porcine kidneys using a high-throughput system. The authors tested different detergents in order to identify a decellularization protocol capable of reliably and rapidly producing multiple acellular scaffolds from large, complex organs. The authors prestated criteria for successful decellularization included: (i) complete or near complete removal of native cellular materials (DNA content at or below 50 ng DNA/mg tissue), with any residual DNA fragments below 200 dp in length, (ii) preservation of the native vascular architecture of the decellularized organ, and (iii) preservation of native extracellular components and ultrastructure of the decellularized organ. Multiple 200 L tanks were batch-filled with either (i) PBS, (ii) 0.2 g/L potassium chloride, (iii) 1.15 g/L sodium phosphate dibasic anhydrous, (iv) 0.2 g/L potassium phosphate monobasic anhydrous in deionized water, or (v) detergent. These tanks were arranged to allow for the simultaneous decellularization of up to 24 organs. The importance of this modular and scalable design cannot be overstated, as the widespread clinical implementation of bioengineered organs will require translation of research-scale product designs to large-scale operations capable of mass production. Flow from these reservoirs to the kidneys was driven by peristaltic pumps and system valves that allowed individualized perfusion from each of the reservoirs as required. Kidneys were harvested from adult pigs and heparinized via renal artery cannula. The kidneys were then arterially perfused with either (i) 0.25% SDS, (ii) 0.5% SDS, or (iii) 1% Triton X-100/0.1% ammonium hydroxide for a total of 36 h, followed by DNase and 10 mM magnesium chloride overnight. Although the SDS-treated kidneys showed minimal changes within the first 12 h of detergent perfusion, they appeared largely translucent by visual inspection at 36 h. Translucency is widely recognized as a coarse indicator of cellular clearance in the RM literature. The scaffold perfused with 0.5% SDS solution appeared more translucent than those perfused with 0.25% SDS, although decellularization occurred within the same time frame overall. These results were achieved with the same SDS concentration used on smaller, simpler organs, suggesting that organ size and complexity does not critically influence outcome as long as the innate vasculature can accommodate and withstand detergent perfusion. By contrast, the kidneys treated with Triton X-100/

0.1% ammonium hydroxide showed rapid visual changes, with translucency apparent within 6 h of perfusion. Despite these early results, the Triton-treated kidneys showed little subsequent change and did not achieve the same final degree of translucency as the SDS group. However, quantification of residual DNA and histological analysis showed no significant differences between the three decellularization groups. Importantly, only the 0.5% SDS achieved the prestated criteria of residual DNA content amounting to 50 ng/mg or less. Despite these high rates of clearance, the native kidney architecture was maintained in all three groups, with preservation of glomerular and tubular structures observed under microscopy. A key outcome for these decellularization protocols was the preservation of the innate vasculature, as the arterial perfusion of detergents can compromise vessel integrity, leading to extravasation and microvessel occlusion. Contrast CT imaging confirmed the preservation of the arterial-venous renal system of the 0.5% SDS group, with demonstrable continuity of the vascular circuit. Histological analysis confirmed the preservation of essential ECM elements, including glycosaminoglycans and collagens, in the decellularized scaffolds. As suggested by visual inspection, the Triton-treated kidneys showed a greater amount of residual cellular cytoplasmic components compared to the SDS groups. Following perfusion decellularization, the scaffolds were rinsed and sterilized using 10.0 kGy gamma irradiation, which results in continued sterility for 6 months or more. Small scaffold biopsies were then statically seeded with primary human renal cells to assess the degree of soluble and contact toxicity. Because even trace amounts of residual detergent could compromise the membrane integrity of the seeded population, a 3-day PBS rinsing protocol was employed to ensure maximal clearance. Although biopsies from the SDS-treated scaffolds supported the growth of the primary human renal cells over 4 days, the Triton-treated biopsies showed high rates of necrosis and poor cell viability. These findings were confirmed by DNA quantitation, which showed that the Triton-treated biopsies were devoid of cells by day 4. These findings suggest that the Triton X-100 detergent is retained in the renal matrix to a degree sufficient to compromise cell integrity. The authors conclude that the effective, reproducible decellularization of relevantly sized kidneys was possible, establishing the innate ECM as the ideal platform for bioengineering investigation.

Orlando et al. [42] advanced the field further by transplanting and maintaining decellularized porcine kidneys for extended observation in live hosts. The kidneys were retrieved and rinsed with a simple 0.9% isotonic solution, without the preceding anticoagulation or vasodilatation described above by Ross et al. The kidneys were decellularized using a series of steps, including

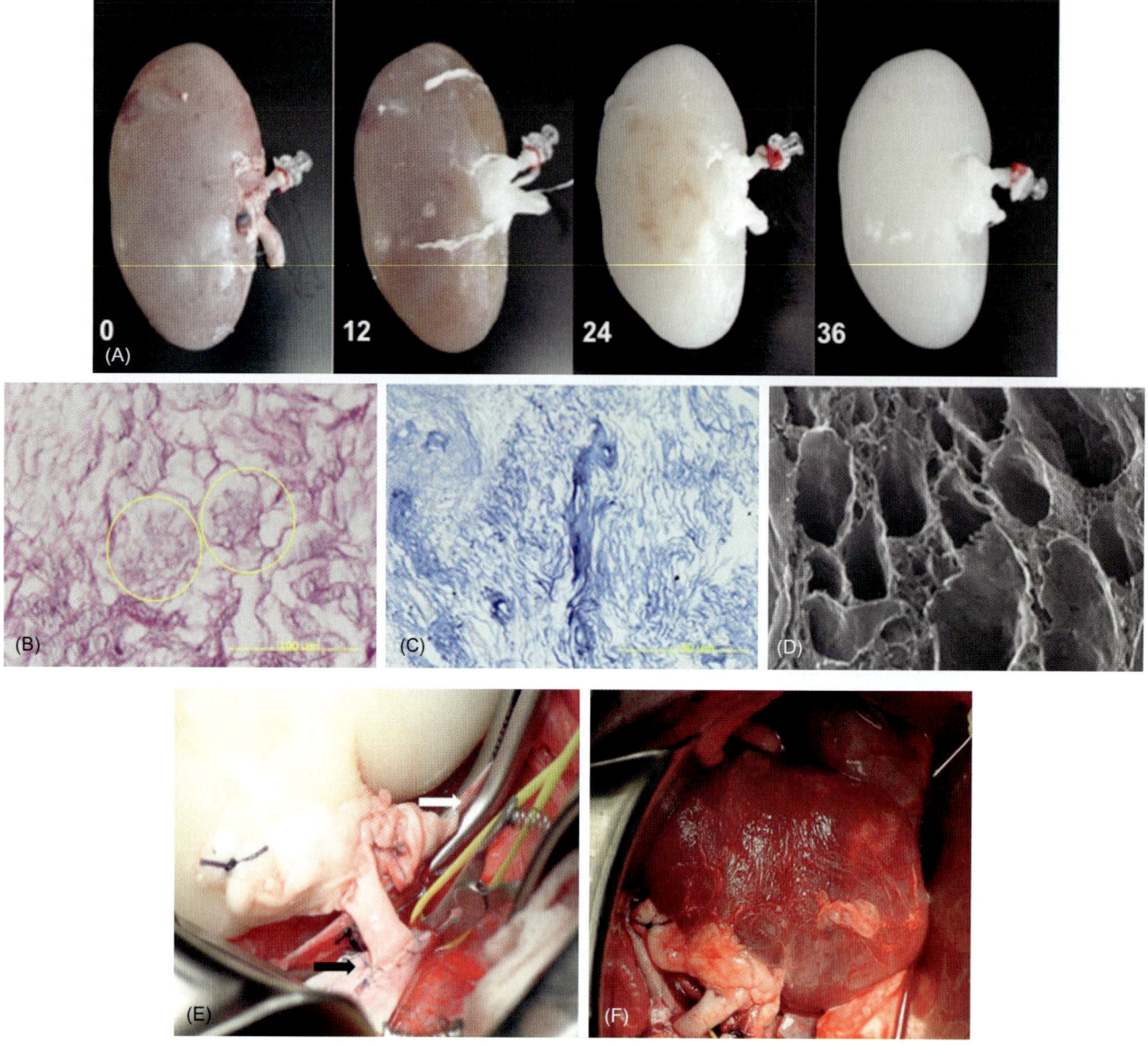

FIGURE 17.3 Porcine kidneys were decellularized with SDS-based solution. Cell clearance was quite spectacular. After decellularization, the renal ECM scaffold acquires a whitish gross appearance (A). H&E shows total clearance of nuclear and cellular material (B), while collagen (stained pink) remains intact. Masson trichrome (C) and electron microscopy (D) confirms cell clearance and persistence of collagen fibers. Scaffolds were implanted in pigs (E and F). The renal vein of the scaffold was implanted on the vena cava right above the iliac bifurcation while the renal artery was implanted on the aorta before the bifurcation of the iliac axis. *From Orlando et al. [42] with permission*

distilled water at 12 mL/min for 12 h to osmotically disrupt cellular integrity, SDS at 12 mL/min for 48 h to denature cellular debris, and phosphate buffered saline at 6 mL/min for 120 h to rinse the channels of toxic chemicals (Figure 17.3). Similar to Sullivan et al. [50], the authors report that both 1% and 5% Triton X-100 nonionic detergents resulted in poor clearance of the cellular compartment, despite demonstrated efficacy in liver, pancreas, and intestine bioengineering [70]. 4′,6-diamidino-2-phenylindole staining for integrin cell surface receptors $\alpha 3$ and $\beta 1$ showed complete cell

clearance, while methenamine silver staining of the basement membrane showed an intact and continuous black outline along the contours of the glomerular structure. SEM further confirmed the absence of cellular material within the scaffold with preservation of the ultrastructural architecture of the glomeruli, tubules, and vessels. The dense, fibrous matrix showed reticular collagen fibers with pores sized $10-50\,\mu$m. The integrity of the vascular tree was confirmed using X-ray fluoroscopy, performed by the injection of contrast media through the renal artery. Importantly, the decellularization protocol preserved the patency and hierarchical branching structures of the vascular network, which is imperative for subsequent transplant, perfusion, and recellularization. The contrast media was observed to flow from larger vessels to progressively smaller capillaries, eventually draining from the renal vein. Extravasation of the contrast media within the scaffold parenchyma was not observed. The patency and strength of the innate vascular network was further confirmed via pressure and flow testing, which showed a physiological increase in renal arterial pressures with increased flow. Four sterilized, acellular, whole-organ scaffolds were then implanted and surgically reconnected in age-and-weight-matched recipients to determine whether the intact porcine scaffolds could maintain physiological blood pressures for an extended period *in vivo*. Satisfactory blood flow was directly observed for 60 min following implantation before surgical closure. Scaffold implantation was well tolerated and no adverse reactions were observed during surgery or follow up. Blood pressure was effectively maintained, and no blood extravasation was observed over the course of the study, proving the integrity of the intrinsic vasculature. The recipients were euthanized and scaffolds were retrieved after 2 weeks. At retrieval, one of four scaffolds was encased within a foreign-body reaction-like fibrous capsule, suggesting that the matrix had retained some antigenic properties. The other three scaffolds were mobile and only slightly adherent to surrounding tissues. However, the renal artery and vein were obstructed by massive thrombi despite strong anticoagulation prophylaxis. The observed coagulation was an expected by-product of the interaction of whole blood with the de-endothelialized vascular network. Further dissection showed a preservation of the gross anatomy with clear demarcation of both the cortex and medulla. However histological staining showed massive nonspecific inflammatory infiltrate in the pericapsular zone, though the animals did not show any observable signs of fever or infection during the recovery period.

Despite these advances, kidney bioengineering and regeneration still lags behind other organs in the field. Although effective protocols have been established for whole-organ kidney decellularization in pigs and rats, similar studies involving nonhuman primates have not been attempted. Nonhuman primate studies will become increasingly important as the field advances, as porcine kidneys show marked differences in vascular architecture [71]. Second, the attempted recellularization of higher order mammalian kidneys has not been reported. It thus remains unclear whether larger kidneys can support the retrograde ureteric perfusion needed to recellularize the renal collecting system. These studies are currently underway at our institute. Additionally, the most effective

pluripotent seeding population remains to be determined. Likewise, it is unclear whether the decellularized kidney ECM can support the proliferation and differentiation of stem cells into the approximately 26 requisite cell types that comprise the mature human kidney [72]. Even after reliable recellularization protocols have been established, the homoeostatic, resorptive, metabolic, endocrine, and immunomodulatory functions must be reestablished by the bioengineered kidney. Finally, it is unclear whether perfusion recellularization is sufficient to prevent the thrombogenic effects of the collagen scaffold observed upon reimplantation.

17.6 ORGAN BIOENGINEERING AND REGENERATION TECHNOLOGY: MOVING FORWARD

At present, bioengineered organs have been implanted in at least 160 human recipients worldwide. As the field advances, it is becoming imperative that detailed case information is recorded and shared between centers performing such procedures. Information sharing among the RM community promotes collaboration and provides the clinical data required for analysis and audit. Therefore, we believe that the time has come for an International Registry of Implanted Bioengineered Organs. Previous experiences with solid organ and composite tissue transplantation have shown the importance of such registries. Indeed, registries allow for the ongoing assessment of new technologies as they are implemented in the clinical setting. This allows detailed analysis of procedural obstacles and outcomes, providing a critical instrument for the evaluation of risk-to-benefit ratio [73]. In addition, registries represent an ideal interface with the public, providing accurate, up-to-date information to the media and to the scientific community [74].

17.7 CONCLUSION

We have reviewed key aspects of RM technologies as they relate to the kidney. Advances in stem cell technology, renal replacement therapy, and organ bioengineering and regeneration are intersecting with promise to resolve the dire shortage of transplantable organs. Despite steady progress, renal bioengineering lags behind other organs in the field due to the complex architecture and physiology of the native kidney, and a poor understanding of the interactions between ECM, renal cells, and growth factors. While current research endeavors are promising, the transition to safe and effective clinical implementation faces significant obstacles. Collaborative efforts and investigation are required to drive the field towards the production of a bioengineered kidney capable of restoring renal function in patients with end-stage disease.

REFERENCES

[1] Peralta CA, Hicks LS, Chertow GM, Ayanian JZ, Vittinghoff E, Lin F, et al. Control of hypertension in adults with chronic kidney disease in the United States. Hypertension 2005;45:1119−24.

[2] Sarafidis PA, Li S, Chen SC, Collins AJ, Brown WW, Klag MJ, et al. Hypertension awareness, treatment, and control in chronic kidney disease. Am J Med 2008;121:332−40.

[3] Tonelli M, Wiebe N, Culleton B, House A, Rabbat C, Fok M, et al. Chronic kidney disease and mortality risk: a systematic review. J Am Soc Nephrol 2006;17:2034−47.

[4] Guild WR, Harrison JH, Merrill JP, Murray J. Successful homo-transplantation of the kidney in an identical twin. Trans Am Clin Climatol Assoc 1955;67:167−73.

[5] Jassal SV, Krahn MD, Naglie G, Zaltzman JS, Roscoe JM, Cole EH, et al. Kidney transplantation in the elderly: a decision analysis. J Am Soc Nephrol 2003;14:187−96.

[6] Laupacis A, Keown P, Pus N, Krueger H, Ferguson B, Wong C, et al. A study of the quality of life and cost-utility of renal transplantation. Kidney Int 1996;50:235−42.

[7] Yokote S, Yamanaka S, Yokoo T. *De novo* kidney regeneration with stem cells. J Biomed Biotechnol 2012;2012:453−519.

[8] Gilbert TW, Sellaro TL, Badylak SF. Decellularization of tissues and organs. Biomaterials 2006;27:3675−83.

[9] Cuppage FE, Tate A. Repair of the nephron following injury with mercuric chloride. Am J Pathol 1967;51:405−29.

[10] Oliver J. Correlations of structure and function and mechanisms of recovery in acute tubular necrosis. Am J Med 1953;15:535−57.

[11] Haagsma BH, Pound AW. Mercuric chloride-induced tubulonecrosis in the rat kidney: the recovery phase. Br J Exp Pathol 1980;61:229−41.

[12] Bissell MJ, Hall HG, Parry G. How does the extracellular matrix direct gene expression? J Theor Biol 1982;99:31−68.

[13] Hynes RO. The extracellular matrix: not just pretty fibrils. Science 2009;326:1216−9.

[14] Lelongt B, Ronco P. Role of extracellular matrix in kidney development and repair. Pediatr Nephrol 2003;18:731−42.

[15] Schnapp LM, Hatch N, Ramos DM, Klimanskaya IV, Sheppard D, Pytela R. The human integrin alpha 8 beta 1 functions as a receptor for tenascin, fibronectin, and vitronectin. J Biol Chem 1995;270:23196−202.

[16] Santos OF, Nigam SK. HGF-induced tubulogenesis and branching of epithelial cells is modulated by extracellular matrix and TGF-beta. Dev Biol 1993;160:293−302.

[17] Liu Y. Epithelial to mesenchymal transition in renal fibrogenesis: pathologic significance, molecular mechanism, and therapeutic intervention. J Am Soc Nephrol 2004;15:1−12.

[18] Okada H, Inoue T, Kanno Y, Kobayashi T, Ban S, Kalluri R, et al. Renal fibroblast-like cells in Goodpasture syndrome rats. Kidney Int 2001;60:597−606.

[19] Hahm K, Lukashev ME, Luo Y, Yang WJ, Dolinski BM, Weinreb PH, et al. Alphav beta6 integrin regulates renal fibrosis and inflammation in Alport mouse. Am J Pathol 2007;170:110−25.

[20] Jones CL, Buch S, Post M, McCulloch L, Liu E, Eddy AA. Renal extracellular matrix accumulation in acute puromycin aminonucleoside nephrosis in rats. Am J Pathol 1992;141:1381−96.

[21] Qian Y, Feldman E, Pennathur S, Kretzler M, Brosius 3rd FC. From fibrosis to sclerosis: mechanisms of glomerulosclerosis in diabetic nephropathy. Diabetes 2008;57:1439−45.

[22] Kim HY, Nelson CM. Extracellular matrix and cytoskeletal dynamics during branching morphogenesis. Organogenesis 2012;8:56−64.

[23] Miner JH, Li C. Defective glomerulogenesis in the absence of laminin alpha5 demonstrates a developmental role for the kidney glomerular basement membrane. Dev Biol 2000;217:278−89.

[24] Smyth N, Vatansever HS, Murray P, Meyer M, Frie C, Paulsson M, et al. Absence of basement membranes after targeting the LAMC1 gene results in embryonic lethality due to failure of endoderm differentiation. J Cell Biol 1999;144:151−60.

[25] Yang DH, McKee KK, Chen ZL, Mernaugh G, Strickland S, Zent R, et al. Renal collecting system growth and function depend upon embryonic gamma1 laminin expression. Development 2011;138:4535−44.

[26] Hynes RO. Integrins: versatility, modulation, and signaling in cell adhesion. Cell 1992;69:11−25.

[27] Muller U, Wang D, Denda S, Meneses JJ, Pedersen RA, Reichardt LF. Integrin alpha8 beta1 is critically important for epithelial-mesenchymal interactions during kidney morphogenesis. Cell 1997;88:603−13.

[28] Zent R, Bush KT, Pohl ML, Quaranta V, Koshikawa N, Wang Z, et al. Involvement of laminin binding integrins and laminin-5 in branching morphogenesis of the ureteric bud during kidney development. Dev Biol 2001;238:289−302.

[29] Kreidberg JA, Donovan MJ, Goldstein SL, Rennke H, Shepherd K, Jones RC, et al. Alpha 3 beta 1 integrin has a crucial role in kidney and lung organogenesis. Development 1996;122:3537−47.

[30] Bullock SL, Fletcher JM, Beddington RS, Wilson VA. Renal agenesis in mice homozygous for a gene trap mutation in the gene encoding heparan sulfate 2-sulfotransferase. Genes Dev 1998;12:1894−906.

[31] Forsten-Williams K, Chu CL, Fannon M, Buczek-Thomas JA, Nugent MA. Control of growth factor networks by heparan sulfate proteoglycans. Ann Biomed Eng 2008;36:2134−48.

[32] Reichsman F, Smith L, Cumberledge S. Glycosaminoglycans can modulate extracellular localization of the wingless protein and promote signal transduction. J Cell Biol 1996;135:819−27.

[33] Tasnim F, Deng R, Hu M, Liour S, Li Y, Ni M, et al. Achievements and challenges in bioartificial kidney development. Fibrogenesis Tissue Repair 2010;3:14.

[34] Lanza RP, Chung HY, Yoo JJ, Wettstein PJ, Blackwell C, Borson N, et al. Generation of histocompatible tissues using nuclear transplantation. Nat Biotechnol 2002;20:689−96.

[35] Kim SS, Park HJ, Han J, Choi CY, Kim BS. Renal tissue reconstitution by the implantation of renal segments on biodegradable polymer scaffolds. Biotechnol Lett 2003;25:1505−8.

[36] Joraku A, Stern KA, Atala A, Yoo JJ. *In vitro* generation of three-dimensional renal structures. Methods 2009;47:129−33.

[37] Song JJ, Ott HC. Organ engineering based on decellularized matrix scaffolds. Trends Mol Med 2011;17:424−32.

[38] Orlando G, Baptista P, Birchall M, De Coppi P, Farney A, Guimaraes-Souza NK, et al. Regenerative medicine as applied to solid organ transplantation: current status and future challenges. Transpl Int 2011;24:223−32.

[39] Orlando G, Wood KJ, Stratta RJ, Yoo JJ, Atala A, Soker S. Regenerative medicine and organ transplantation: past, present, and future. Transplantation 2011;91:1310−7.

[40] Orlando G, Wood KJ, De Coppi P, Baptista PM, Binder KW, Bitar KN, et al. Regenerative medicine as applied to general surgery. Ann Surg 2012;255:867−80.

[41] Badylak SF, Weiss DJ, Caplan A, Macchiarini P. Engineered whole organs and complex tissues. Lancet 2012;379:943−52.

[42] Orlando G, Farney AC, Iskandar SS, Mirmalek-Sani SH, Sullivan DC, Moran E, et al. Production and implantation of renal extracellular matrix scaffolds from porcine kidneys as a platform for renal bioengineering investigations. Ann Surg 2012;256:363−70.

[43] Wang Y, Cui CB, Yamauchi M, Miguez P, Roach M, Malavarca R, et al. Lineage restriction of human hepatic stem cells to mature fates is made efficient by tissue-specific biomatrix scaffolds. Hepatology 2011;53:293−305.

[44] Ross EA, Williams MJ, Hamazaki T, Terada N, Clapp WL, Adin C, et al. Embryonic stem cells proliferate and differentiate when seeded into kidney scaffolds. J Am Soc Nephrol 2009;20:2338−47.

[45] Ng SL, Narayanan K, Gao S, Wan AC. Lineage restricted progenitors for the repopulation of decellularized heart. Biomaterials 2011;32:7571−80.

[46] Badylak SF, Taylor D, Uygun K. Whole-organ tissue engineering: decellularization and recellularization of three-dimensional matrix scaffolds. Annu Rev Biomed Eng 2011;13:27−53.

[47] Crapo PM, Medberry CJ, Reing JE, Tottey S, van der Merwe Y, Jones KE, et al. Biologic scaffolds composed of central nervous system extracellular matrix. Biomaterials 2012;33:3539−47.

[48] Badylak SF, Gilbert TW. Immune response to biologic scaffold materials. Semin Immunol 2008;20:109−16.

[49] Brown BN, Valentin JE, Stewart-Akers AM, McCabe GP, Badylak SF. Macrophage phenotype and remodeling outcomes in response to biologic scaffolds with and without a cellular component. Biomaterials 2009;30:1482−91.

[50] Sullivan DC, Mirmalek-Sani SH, Deegan DB, Baptista PM, Aboushwareb T, Atala A, et al. Decellularization methods of porcine kidneys for whole organ engineering using a high-throughput system. Biomaterials 2012;33:7756−64.

[51] Reing JE, Brown BN, Daly KA, Freund JM, Gilbert TW, Hsiong SX, et al. The effects of processing methods upon mechanical and biologic properties of porcine dermal extracellular matrix scaffolds. Biomaterials 2010;31:8626−33.

[52] Brown B, Lindberg K, Reing J, Stolz DB, Badylak SF. The basement membrane component of biologic scaffolds derived from extracellular matrix. Tissue Eng 2006;12:519−26.

[53] Baptista PM, Siddiqui MM, Lozier G, Rodriguez SR, Atala A, Soker S. The use of whole organ decellularization for the generation of a vascularized liver organoid. Hepatology 2011;53:604−17.

[54] Song JJ, Kim SS, Liu Z, Madsen JC, Mathisen DJ, Vacanti JP, et al. Enhanced *in vivo* function of bioartificial lungs in rats. Ann Thorac Surg 2011;92:998−1005.

[55] Martinello T, Bronzini I, Volpin A, Vindigni V, Maccatrozzo L, Caporale G, et al. Successful recellularization of human tendon scaffolds using adipose-derived mesenchymal stem cells and collagen gel. J Tissue Eng Regen Med 2012; Jun 19 [Epub ahead of print].

[56] Honge JL, Funder J, Hansen E, Dohmen PM, Konertz W, Hasenkam JM. Recellularization of aortic valves in pigs. Eur J Cardiothorac Surg 2011;39:829−34.

[57] Loai Y, Yeger H, Coz C, Antoon R, Islam SS, Moore K, et al. Bladder tissue engineering: tissue regeneration and neovascularization of HA-VEGF-incorporated bladder acellular constructs in mouse and porcine animal models. J Biomed Mater Res A 2010;94:1205−15.

[58] Wicha MS, Lowrie G, Kohn E, Bagavandoss P, Mahn T. Extracellular matrix promotes mammary epithelial growth and differentiation *in vitro*. Proc Natl Acad Sci USA 1982;79:3213−7.

[59] Baiguera S, Birchall MA, Macchiarini P. Tissue-engineered tracheal transplantation. Transplantation 2010;89:485−91.

[60] Rauh J, Milan F, Gunther KP, Stiehler M. Bioreactor systems for bone tissue engineering. Tissue Eng Part B Rev 2011;17:263−80.

[61] Chen HC, Hu YC. Bioreactors for tissue engineering. Biotechnol Lett 2006;28:1415−23.

[62] Martin I, Obradovic B, Freed LE, Vunjak-Novakovic G. Method for quantitative analysis of glycosaminoglycan distribution in cultured natural and engineered cartilage. Ann Biomed Eng 1999;27:656−62.

[63] Folkman J, Hochberg M. Self-regulation of growth in three dimensions. J Exp Med 1973;138:745−53.

[64] Liu CX, Liu SR, Xu AB, Kang YZ, Zheng SB, Li HL. Preparation of whole-kidney acellular matrix in rats by perfusion. Nan Fang Yi Ke Da Xue Xue Bao 2009;29:979−82.

[65] Nakayama KH, Batchelder CA, Lee CI, Tarantal AF. Decellularized rhesus monkey kidney as a three-dimensional scaffold for renal tissue engineering. Tissue Eng Part A 2010;16:2207−16.

[66] Narlis M, Grote D, Gaitan Y, Boualia SK, Bouchard M. Pax2 and pax8 regulate branching morphogenesis and nephron differentiation in the developing kidney. J Am Soc Nephrol 2007;18:1121−9.

[67] Shao X, Johnson JE, Richardson JA, Hiesberger T, Igarashi P. A minimal Ksp-cadherin promoter linked to a green fluorescent protein reporter gene exhibits tissue-specific expression in the developing kidney and genitourinary tract. J Am Soc Nephrol 2002;13:1824−36.

[68] Gupta S, Verfaillie C, Chmielewski D, Kren S, Eidman K, Connaire J, et al. Isolation and characterization of kidney-derived stem cells. J Am Soc Nephrol 2006;17:3028−40.

[69] Kim D, Dressler GR. Nephrogenic factors promote differentiation of mouse embryonic stem cells into renal epithelia. J Am Soc Nephrol 2005;16:3527−34.

[70] Baptista PM, Orlando G, Mirmalek-Sani SH, Siddiqui M, Atala A, Soker S. Whole organ decellularization—a tool for bioscaffold fabrication and organ bioengineering. Conf Proc IEEE Eng Med Biol Soc 2009;2009:6526−9.

[71] Bagetti Filho HJ, Pereira-Sampaio MA, Favorito LA, Sampaio FJ. Pig kidney: anatomical relationships between the renal venous arrangement and the kidney collecting system. J Urol 2008;179:1627−30.

[72] Al-Awqati Q, Oliver JA. Stem cells in the kidney. Kidney Int 2002;61:387−95.

[73] Petruzzo P, Lanzetta M, Dubernard JM, Landin L, Cavadas P, Margreiter R, et al. The international registry on hand and composite tissue transplantation. Transplantation 2010;90:1590−4.

[74] Petruzzo P, Lanzetta M, Dubernard JM, Margreiter R, Schuind F, Breidenbach W, et al. The international registry on hand and composite tissue transplantation. Transplantation 2008;86:487−92.

Chapter 18

Renal Regeneration: The Stem Cell Biology Approach

Stefano Da Sacco, Laura Perin, and Sargis Sedrakyan
Saban Research Institute, Children's Hospital Los Angeles, Los Angeles, CA

18.1 INTRODUCTION

End-stage renal disease (ESRD) is a condition of kidney failure that affects more than 400,000 patients in the United States and is characterized by loss of primary renal function incompatible with sustaining normal physiology [1]. The rising occurrence of ESRD in the last years and the growing number of patients affected by this disease have created a strong demand in terms of therapies and new clinical approaches. Origin of ESRD can be identified in both acute and chronic injuries. Acute kidney failure (AKF) is characterized by a rapid decrease in renal functionality often accompanied by increase over several days of physiological parameters, including proteinuria and creatinine. AKF can be caused by a wide variety of pathologies, including inadequate renal perfusion, hemorrhage and loss of intravascular fluid, low cardiac output, low systemic vascular resistance, acute tubular injury, glomerulonephritis, as well as urinary obstruction. On the other hand, chronic kidney disease (CKD) is characterized by a longstanding and progressive deterioration of the renal function that slowly develops into ESRD. CKD can be originated by a wide variety of pathological situations, including diabetic nephropathy (DN), the most common cause of CKD in the United States, hypertensive nephroangiosclerosis, various glomerulopathies, such as IgA nephropathy and hereditary nephropathies like policystic kidney disease.

Despite the large amount of patients affected by ESRD and the rising costs for the health system, a definitive cure is not available. The current therapeutic strategies employed to delay ESRD are focused on the control of blood pressure and decrease of proteinuria by administration of angiotensin converting enzyme (ACE) inhibitors to antagonize the deleterious effects of an over-activated renal angiotensin system [2]. However, treatment with ACE inhibitors is not sufficient to grant long-term protection of the kidney and while the use of antihypertensive medication slows the disease progression, dialysis is eventually required. Nonetheless, morbidity and mortality of these patients remain high, due to filtration proprieties of dialysis, the suboptimal quality of life for a large number of these patients, and the very high treatment costs. Renal transplantation is still considered the option of choice for the treatment of most ESRD patients. However, only a small percentage of these patients are recipients of organ transplantation due to the limited number of organ donors [3]. Moreover, frequent complications with immunosuppressive drugs underscore the need for possibly better alternative therapies for ESRD patients.

Absence of efficient and cost-effective therapies has created a strong sense of urgency for both scientists and clinicians, driving the research of novel approaches to improve the outcome of renal diseases [4].

The regenerative potential of the kidney itself is limited as all the nephrons are formed during embryonic development. The kidney is a complex organ comprised of more than 30 terminally differentiated cell types with distinct functions [5]. This structural complexity of the kidney makes the regeneration of renal tissue one of the most challenging tasks to be overcome. Trying to identify an appropriate source of stem cells that can efficiently recapitulate the morphological and functional features of these renal cells in a diseased organ is a very difficult task. Nevertheless, a substantial amount of scientific evidence currently suggests that stem cells play a beneficial role and have the potential to rescue acute and chronic renal phenotypes.

In the last decade, the study of stem cells for the treatment of kidney disease has increased exponentially and a wide variety of approaches are being attempted to establish whether cell administration may be suitable as a therapeutic treatment of patients with ESRD. In particular, to effectively treat ESRD a cure to promote endogenous renal cell regeneration, replace the damaged cells, or eventually prevent fibrosis, a consequence of end-stage disease, must be considered. A wide variety of cell types, with different origins, degrees of differentiation and potentiality have been employed in studies to determine cell therapy feasibility. In the following paragraph, differences and analogies for these different cell types, along with a discussion on the different animal models, from acute to chronic injuries are discussed.

18.2 ACUTE KIDNEY FAILURE

AKF is characterized by a sudden decline in renal function over a period of hours or days that leads to a rapid fall in glomerular filtration rate (GFR), retention of nitrogenous waste products and, sometimes, retention of sodium, water and development of metabolic acidosis and hyperkalemia [6].

A wide range of factors predispose to acute kidney injury (AKI) and include, but are not limited to, cardiovascular pathologies such as hemodynamic instability, hypovolemia, atherosclerosis, congestive cardiac failure and other conditions such as peritonitis, ileus obstruction, diabetes mellitus, hypoxia, ischemia and reperfusion (I/R), preeclampsia/eclampsia, sepsis, major burns, pancreatitis, and jaundice. Also, several clinical and surgical treatments have been linked to an increased risk of AKI, including drug treatments, major vascular surgery, diuretic therapy, preoperative starvation, and biliary

surgery [7]. Advanced age is another factor that predisposes to AKI [8].

AKI is usually classified as prerenal, intrinsic, or postrenal to identify its etiology. Prerenal AKI occurs when there is reduction of renal perfusion due to intravascular volume depletion. Common causes of prerenal AKI are ischemia, shock, or hypovolemia caused by dehydration, fever, vomiting, and diarrhea [8].

Intrinsic renal failure is caused by damage to the tubular, glomerular, interstitial, or vascular compartment. The most common cause of intrinsic AKI is acute tubular necrosis, in which tubular cells massively start to die and consist of three phases identified as initiation, maintenance (7–14 days), and recovery, characterized by increased diuresis and slow return of kidney function [8]. Postrenal AKI is caused by an increase in pressure within the collecting system caused by obstruction [7]. However, no matter what the cause of AKI is, they all share the same rapid decrease in GFR [7] and the same high risk.

Mortality rate of patients with AKI ranges between 25 and 70%, and, in patients requiring renal replacement therapy, may be as high as 80%, despite the use of various pharmacologic agents [7]. No specific pharmacologic therapy is effective in patients with established AKI, and the care of such patients is limited to supportive treatment, including renal replacement therapy. The current treatment of choice for AKI consists of intermittent dialysis. However, risk of hypotension is common and about 10% of the patients cannot be treated because of hemodynamic instability [6,9–12]. For this reason, the necessity to find alternative approaches for the treatment of AKI is urgent.

18.3 STEM CELLS AND AKF

In recent years, cell-based therapy, and in particular stem cells, have become a focus for a large number of scientists and clinicians around the globe. Different sources of stem cells and progenitor cells are being employed in regenerative medicine, ranging from embryonic stem cells (ESCs) to hematopoietic stem cells (HSCs) to adipose tissue cells. However, for the purpose of kidney regeneration, bone marrow (BM) stem cells and amniotic fluid stem cells (AFSCs) [13] are the most used exogenous cell types. Originally, stem cells were thought to engraft and differentiate within the tissue participating in the recovery of the damaged organ; however, all recent studies have reported that the frequency of integration ranges between 3 and 22% [14–18] and these numbers are too low to justify the positive outcomes obtained in animal models. Further evidence has suggested that stem cells act as modulators of the microenvironment in the damaged organ [13] via autocrine or paracrine action.

18.3.1 Ischemia Reperfusion

Ischemia reperfusion injury remains one of the prevalent causes of AKI in both native and transplanted kidneys [19]. Renal injury caused by ischemia reperfusion is usually associated with acute tubular necrosis and death of tubular epithelial cells [20−22]. Prevalence of necrosis or apoptosis determines the severity of the ischemic insult. The pathological events involve cytoskeleton disruption, alteration in adenine nucleotide metabolism and intracellular calcium, loss of cell−cell/cell−ECM attachments, generation of reactive oxygen molecules and inflammatory responses following activation of endothelium leading to recruitment and infiltration of inflammatory cells and apoptosis [23−28]. The damage is mainly caused by the production of reactive oxygen species (ROS) during the ischemia that increase even more the injury to endothelial cells during the reperfusion phase [29]. Recovery from ischemia reperfusion damage is generally obtained through cell dedifferentiation and proliferation [19].

Stem cell therapy for the treatment of ischemia reperfusion has been widely studied in recent years. Experimentally, ischemia reperfusion is usually induced by clamping the renal artery for a time ranging from 30 to 60 min, depending on the animal model, followed by reperfusion. Renal failure occurs rapidly, generally within 24 h [7].

The pro-regenerative and anti-inflammatory mechanism of action of stem cells has been widely confirmed. In particular, injection of bone marrow mesenchymal stem cells (BMSCs) has shown a marked reduction in inflammatory cytokines within the milieu of damaged kidneys along with a decreased occurrence of apoptosis and necrosis in mouse [30] and rat [31−36]. A significant effect on expression and secretion of growth factors, including VEGF, Hepatocyte Growth Factor (HGF), and Insulin Growth Factor 1 (IGF-1), was confirmed in rats by Patel et al. [36]. Using a BMSC population knockout for VEGF, Westenfelder et al. [37] confirmed an overall amelioration of kidney functionality. However, the cells were less effective compared to cells in which VEGF was normally expressed [37] proving that this growth factor plays a major role in the mediation of the beneficial effect of BMSCs. Modulation of the milieu and homing in the renal cortex were also reported in a study by Shi et al. [38] using BMSCs expressing Glial Derived Growth Factor (GDNF). Generally, upon injection, the major favorable effects included an increase in renal function with secretion of VEGF, promoting regeneration and a decrease in damaging processing, like for example oxidative stress. Similar results were obtained by combined therapy of BMSCs with growth factors and other therapeutic molecules [39−43].

In contrast, research by Jiang et al. found that BM cells were able to replace damaged renal cells after ischemia reperfusion without any immunomodulatory effect [44].

However, independently of the mechanisms of action and the inflammatory factors that were modulated by BMSCs, all the studies reported amelioration of the renal function, a decrease in cell death, and a higher cell proliferation, that suggest an improved clinical outcome.

While BMSCs generally exert their action exclusively through a paracrine effect, stem cells from different sources show a higher predisposition for homing within the renal tissue and, at the same time, modulate the microenvironment of the damaged organ [45−49]. In some studies, stem cells from alternative sources have been found to act through a mechanism of action that has been linked solely with homing and differentiation of the injected stem cells rather than modulation of the organ milieu. This was demonstrated, for example, by human umbilical endothelial cells [50], skeletal muscle-derived stem cells capable of differentiating into endothelial cells [51], and HSC transdifferentiating into tubular cells [52].

Taken together, these results suggest that the main action of pluripotent stem cells in ischemia reperfusion injury is the stimulation of residing progenitor cells rather than active engraftment and proliferation that appears to be restricted to few stem cell lines.

18.3.2 Glycerol-Induced Tubular Necrosis

Rhabdomyolysis-induced AKI is developed following skeletal muscle trauma caused by a variety of events, including infection, metabolic disorders, toxins, ischemia, or physical traumas [53]. The destruction of the muscle fibers releases into the circulation myoglobin and other intracellular proteins that are able to cause AKI in about 10−50% of the patients, while other complications include disseminated intravascular coagulation, hyperkalemia, or other metabolic imbalances and acute cardiomyopathy. Injury of the kidney causes acute inflammation, apoptosis, and oxidative stress.

Injection of glycerol within the skeletal muscle is an effective way of obtaining rhabdomyolysis, and the subsequent renal injury in experimental animal models [53]. Several groups have looked at the effect of stem cells for the treatment of glycerol-induced AKI. In particular, a protective effect of stem cells was first shown in a model of glycerol-induced acute tubular necrosis by Herrera in 2004 and 2007 [54,55]. After infusion, BMSCs were able to integrate and differentiate into tubular cells and, at the same time, stimulated a significant increase in tubular epithelial proliferation helping kidney recovery.

In 2009, Bruno et al. confirmed that the administration of BMSCs improves recovery from glycerol-induced AKI and suggested that their paracrine action on tubular cell proliferation was mediated by release of microvesicles [56].

Using AFSCs in a mouse model of glycerol-induced AKI, Perin et al. demonstrated that the cells were able to provide a protective effect with a marked amelioration of acute tubular necrosis, decreased number of damaged tubules, and proliferation of tubular epithelial cells [57]. The suggested mechanism of action for AFSC has been identified in a paracrine effect with a prominent role on the modulation of inflammatory cytokines. At the same time, no evidence was found that AFSC can participate directly to tissue regeneration [57]. These results obtained with AFSC were also confirmed by Camussi et al. [58]. Along with the study of extrarenal stem cells for the treatment of AKI, the use of endogenous progenitor cells has been attempted in order to understand their therapeutic potential and gain new knowledge about their role in the damaged organs.

In 2005, Bussolati et al. described a population of renal progenitor cells, identified by the expression of the surface marker CD133, capable of differentiating into tubule cells *in vitro* and able to home and integrate within the tubules in a mouse model of glycerol-induced AKI [59]. A year later, Sagrinati et al. reported the isolation of a $CD24^+CD133^+$ population [60]. After infusion in glycerol-induced AKI mice, the cells repopulated the tubules and provided amelioration of both morphology and renal function.

In summary, the role of exogenous stem cells in the protection of renal following glycerol-induced AKI appears to reside in their ability to secrete a broad range of renoprotective factors. On the other hand, the few studies on endogenous renal cells seem to highlight the higher predisposition of these cells to engraft and actively participate in the repairing processes by means of proliferation and differentiation.

18.3.3 Cisplatin-Induced AKI

Cisplatin (*cis*-diaminedichloroplatinum(II)), is an antitumoral drug, widely used for the treatment of a broad range of cancers including ovarian, head and neck carcinomas, and germ cell tumors. Nephrotoxicity of cisplatin is well reported and is the major limiting factor for its use in clinical therapy, since significant kidney damage occurs at high doses [7]. Lower doses for several days are generally less aggressive, however, significant changes in serum creatinine, creatinine clearance and higher concentration of urine N-acetyl-*b*-D-glucosaminidase (an indicator of tubular damage) levels have been reported [61]. Mechanisms of cisplatin toxicity have been identified in direct tubular toxicity in the form of apoptosis and necrosis that is mediated through inflammation, ROS, calcium overload, phospholipase activation, depletion of reduced glutathione, inhibition of mitochondrial respiratory chain function, induction of apoptosis, opening of mitochondrial permeability transition pore (MPTP) and ATP depletion [61–65].

A wide variety of pluripotent cells has been tested for their regenerative potential in cisplatin-induced AKI. BMSC has been shown to accelerate recovery of AKI and prolong survival in mice [17,30] following nephrotoxic cisplatin damage of tubular cells. In an *in vitro* experiment, Imberti et al. [66] showed that BMSCs provided a protective effect, possibly mediated by IGF-1, by promoting cell proliferation in proximal tubular epithelial cells exposed to cisplatin. To test whether VEGF plays a role in mediating the healing of renal tubular epithelial cells, Yuan et al. cocultured renal cells with VEGF-upregulated BMSCs confirming that, compared to tubular cells cocultured with nonmodified mesenchymal stem cells (MSCs), viability and proliferation were increased while apoptosis was slowed down [67]. However, VEGF is not the only secreted molecule that can explain BMSC effects. In 2008, Bi et al. suggested that erythropoietin (EPO) secreting BMSCs might be involved in the protection of tubular cells diminishing the decline of renal function associated with cisplatin administration [68]. Microvesicles released by BMSCs have been shown to have a beneficial effect on renal cells, enhancing the pro-survival capabilities of tubular epithelial cells. In particular, it has been shown *in vitro* that tubular epithelial cells following exposure to microvesicles from BMSCs upregulated expression of several antiapoptotic genes, including bcl-xl, bcl2, and BIRC8, while downregulating proapoptotic effectors [69]. An innovative mechanism of action for BMSCs, studied on a mouse model of cisplatin-induced AKI, has been recently suggested by Benigni et al. [70]. They identified the release of microparticles and exosomes enriched with mRNAs for IGF-1 and suggested that the transfer of mRNAs to tubular cells might explain, at least in part, the effect of BMSCs on tubular cell proliferation. Since paracrine secretion of growth factors and cytokines appears to be the main mechanism of action of pluripotent cells in an acutely injured kidney, the administration of conditioned medium, instead of live cells, has been considered extremely appealing. In 2007, Bi et al. observed that injection of adipocyte-derived stromal cells conditioned medium protects the kidney from toxic injury by enhancing secretion of factors that limit apoptosis and enhance proliferation of the endogenous tubular cells, suggesting that transplantation of the cells themselves is not necessary [71]. Unfortunately, in 2011 Gheisari et al. testing conditioned media from both MSCs and human umbilical cord blood cells could not confirm those extremely positive results and concluded that the use of conditioned media was not sufficient to promote renal repair [72]. However, it might be worth noting that the use of different mouse strains and the adoption of different cell lines in the two experiments might account for the differences in the outcome of the two studies.

With a different approach, and following the discovery that human adipose tissue-derived MSCs exert a beneficial effect on kidney cells and recovery of renal function following cisplatin administration in rats [73], Yasuda et al. proposed the use of nonexpanded autologous adipose tissue-derived MSCs for protection from nephrotoxic injury and were able to show that the cells exerted a significantly positive effect by ameliorating renal function with decreased tubular damage by producing VEGF and HGF after 14 days from administration in the renal subcapsula [74].

Along with MSCs, other pluripotent stem cells have been investigated for their ability to provide protection during acute settings. In particular, a positive effect on recovery after injury has been shown by injection of AFSC with a beneficial effect on lifespan, renal function, and tubular damage. Confirming the previous data on AFSC in the glycerol-induced acute models [57,58], AFSC did not show any differentiation into tubular cells, suggesting again that a paracrine effect is the main mechanism of action [75]. Along with unstimulated cells, the authors also tested a population of AFSC preconditioned with GDNF, demonstrating that a higher ability to ameliorate the damage was reached by increasing production of growth factors and enhancing survival [75].

In a recent study, Luo et al. described the use of BMSCs derived from differentiation of ESCs for the treatment of cisplatin-induced AKI [76]. Specifically, the transplanted cells were able to engraft within the renal tissue and promote tubular cell proliferation while counteracting apoptosis caused by the nephrotoxic agent. In addition, upregulation of anti-inflammatory cytokines was confirmed. The novelty of this study lies in the fact that the BMSCs used for this study were differentiated from ESCs, proving that cells differentiated *in vitro* before the administration can promote renal repair.

In 2005, Hishikawa et al. described the use of an endogenous kidney side cell population that, following infusion, contributed to the amelioration of the renal function and improved renal regeneration by secreting renoprotective factors and LIF in a cisplatin-induced model of AKI. The same cell population fails to deliver any benefit in chronic settings [77].

Again, the use of exogenous or endogenous stem cells for the treatment or the prevention of cisplatin-induced AKI proves that they are a feasible alternative that might prove beneficial effects, in particular in patients undergoing chemotherapy with nephrotoxic drugs. In summary, the adoption of pluripotent stem cells from a broad variety of sources seems to be a feasible and useful approach for the treatment of AKI. Independently from the cause of the acute injury, the main mechanism of action is mainly being identified in the ability of pluripotent cells to modulate the milieu of the acute damaged organ by reducing inflammation and cell death while enhancing functional recovery and endogenous regeneration. However, despite the enormous advances in the field and the profoundly encouraging results, understanding the specific mechanisms of action of the cells and their fate after the injury has been resolved, are essential before translating these experimental therapies into clinical approaches for the treatment of AKI.

18.4 CHRONIC RENAL DISEASES

CKD is a global health problem. It is associated with gradual decline in renal function, which develops into ESRD and culminates in renal failure. Many different etiological factors are involved in the initiation and progression of CKD. The two most prevailing medical reasons for the development of progressive kidney disease are diabetes and hypertension. Current knowledge classifies and characterizes progressive renal disease in addition to the clinical signs and symptoms by expansion of the glomerulo-tubulo-interstitium and accumulation of extracellular matrix within these tissue compartments. However, depending on the type of etiology, CKD can initiate in different compartments or structures of the kidney, such as in the glomeruli, in the tubuli, or in the renal vessels. Several key mechanisms variably drive CKD progression at cellular level, which include mesangial and fibroblast activation, epithelial to mesenchymal transition (EMT), monocyte/macrophage, and T-cell infiltration. Regardless of the site of the initial insult, all forms of progressive kidney diseases converge into a final common pathway characterized by glomerulosclerosis and tubule-interstitial fibrosis. Accumulation of fibrotic scar tissue gradually replaces healthy functional nephrons. In the beginning, nephrons that are little damaged or nonaffected adapt and functionally compensate the dysfunctional nephrons until over 60% of the nephrons become dysfunctional after which kidneys become incapable of cleaning the blood from waste products [78,79].

18.5 STEM CELLS AND CKD

The emergence of regenerative medicine and the potential use of stem cells for renal regeneration have generated vast interest and many novel opportunities to develop new therapies to treat kidney disease. At present many efforts are being made to apply this approach for kidney repair and regeneration using several types of stem or progenitor cells, such as MSCs, ESCs, renal progenitor cells, AFSCs, and induced pluripotent stem cells. The feasibility of stem cell therapy to treat renal disease has been extensively studied in models of AKI [5,80−83], but only a few investigations have addressed the use of stem cells in models of CKD. Therefore, our current knowledge of potential beneficial outcomes of stem cell treatments of CKD and the underlying mechanisms of interaction between stem cells and chronically damaged renal tissue is still a very new area of investigation.

18.5.1 Animal Models of CKD

A range of animal models of genetic or experimentally induced chronic kidney damage have been applied to investigate molecular pathogenesis as well as to understand mechanisms of stem cell actions in CKD, such as models of Alport syndrome, 5/6 nephrectomy, cystinosis, DN, and adriamycin-induced CKD. Most reports have so far been concentrated on transplantation of whole BM or fractions thereof like BMSCs. Similar to cell therapy in AKI models, in most of these studies only little integration of exogenously administered cells has been observed, suggesting that integration and differentiation of donor cells might be of minor importance and that instead cell therapy may involve mechanisms that interfere with the inflammatory and fibrotic pathways activated during chronic progression [84].

18.5.2 Alport Syndrome

Alport syndrome is a hereditary disorder caused by mutations in the a3a4a5(IV) collagen network genes resulting in structural defects in the glomerular basement membrane (GBM) early during development leading subsequently to the breakdown of the filtration barrier, development of renal fibrosis and kidney failure. Two distinct mouse models of Alport syndrome have been described thus far: COL4A3 knockout mice on C57BL/6 background and on 129Sv background, respectively, and COL4A5 knockout mice on C57BL/6 background that differ with respect to disease progression due to differences in compensatory mechanisms [85–88]. A handful of studies mostly using BM transplantation and in rare cases AFSCs have demonstrated the beneficial role of stem cells, nevertheless, the mechanisms by which stem cells may be contributing to such observations are still highly debatable. In particular, two independent studies [89,90] have demonstrated that transplantation of whole BM derived from wild-type (WT) mice into irradiated COL4A3 −/− mice on C57BL/6 background led to a partial restoration of the defective a3a4a5(IV) collagen network accompanied by significant improvements in renal morphology and function. In one study transplanted cells (Lac-Z tagged) [89] were shown to express podocyte and mesangial markers, indicating transdifferentiation of BM-derived progenitor cells into kidney cells without ruling out the possibility of cell fusion. Such observations led investigators of both studies to the conclusion that transplantation of whole BM improves Alport disease by providing WT cells that are able to repopulate the damaged glomeruli, via differentiation or cell fusion, and to at

least in part restore glomerular a3(IV) expression. However, because of the small number of integrated cells, Prodromidi et al. also speculated that paracrine actions could contribute to the beneficial effects of whole BM [90]. These results were later challenged by Katayama et al. who demonstrated that irradiation alone can prolong the lifespan and improve physiologic parameters of COL4A3 −/− mice on 129Sv background without the need for BM transplantation [91]. Such contrasting outcomes might be attributed to the differences in the mouse background (C57BL/6 vs. 129Sv) and the timing and dose of irradiation. In addition, studies using nonradiated COL4A3 −/− mice on C57BL/6 background that received multiple injection of WT BM at a late stage of disease not only demonstrated significant improvement in renal morphology and function but also presented with *de novo* expression of a3(IV) collagen in the GBM [92]. Unfortunately, this study was conducted with a very small number of mice per experimental group ($n = 3$), and the authors never assessed the animal lifespan or whether transplanted cells integrated into damaged kidneys. Positive results were also obtained using blood transfusion and undifferentiated mouse embryonic stem cells (mESCs), suggesting that plasticity could be an important property of the so far unidentified therapeutic cell type in BM and blood.

Whether BMSCs have any contribution in treating Alport disease is still questionable. Neither single injection of BMSCs in COL4A3 −/− mice on the C57BL/6 background [90] nor multiple injections of BMSCs in COL4A3 −/− mice on the 129Sv background [93] showed any improvement in physiological and functional parameters. In the former study, these observations led the authors to the conclusion that HSCs rather than BMSCs are the effective component in BM. Multiple injections, however, prevented the loss of peritubular capillaries and reduced interstitial renal fibrosis but not glomerulosclerosis with no effect on animal lifespan. In addition, VEGF upregulation in kidneys was detected in BMSC-injected Alport mice, suggesting protection of peritubular capillaries via paracrine effects.

Similar to BM cells, AFSCs have shown to provide therapeutic benefits in COL4A5 −/− mice on C57BL/6 background. Single injection of AFSC at an early stage before the onset of proteinuria was shown to promote antifibrotic cytokine expression profile and macrophage infiltration and activation [94]. In addition, treated mice demonstrated less interstitial and glomerular fibrosis with improved renal structure and function and increase in animal lifespan. Unlike in studies with BMSCs, no *de novo* production of Col4a5(IV) collagen was detected with AFSC injection, leading the authors to conclude that AFSC might work mainly through endocrine/paracrine mechanisms to delay disease progression.

Even if results reported between groups are different, the evidence suggests that the immunomodulatory role of donor cells acting via autocrine and paracrine pathways is likely the predominant mechanism. Additional studies designed to answer this particular question will be necessary to determine the final mechanism(s).

18.5.3 Nephritic Cystinosis

Nephritic cystinosis (NS) is the only other genetic model of CKD currently used to study stem cell potential for possible therapies. NS is a rare lysosomal storage disease (LSD), characterized by accumulation of cystine in lysosomes in multiple organs including kidney tubular cells. Cystinosis ($Ctns^{-/-}$) mice represent a good model for CKD, with renal dysfunction onset at 6 months of age leading to chronic renal insufficiency at about a year thereafter [95,96]. Stem cell therapy holds promise to ameliorate some of the symptoms associated with this condition. The best approach for treatment in this model would be the effective engraftment and integration of donor cells carrying the functional protein that would ultimately reverse cystine accumulation in the tissue, allowing functional recovery. In a proof of principle study, Syres et al. [97] demonstrated that a single injection of both BMCs and BMSCs from a WT donor at 2 months of age improved levels of serum creatinine at 4 months after treatment. Overall, BMC showed better outcomes with abundant engraftment in the renal interstitium, with some colocalization in the tubules and glomeruli [97]. In a similar but long-term study, the same research team reported significant improvement in renal function well over 6 months posttransplantation of BMC. No difference was found between mice transplanted before and after 6 months of age, suggesting that WT BM cells can protect the kidney from further damage even at later stages of kidney damage. Strong correlation was found between preservation of kidney function and high level of donor-derived WT BMC engraftment. Most of the donor BM-derived cells localized in the interstitium within the kidney, acquiring lymphoid, dendritic, fibroblastic/myofibroblastic phenotype. Hence, it was concluded that with over 50% engraftment of donor cells in the kidney the progression of CKD in the mouse model of cystinosis can be effectively treated [98]. Mice with <50% engraftment showed similar or worse kidney function like the untreated controls. Although, this study indicates donor cell engraftment as a mechanism for renal protection in this particular model, nevertheless, additional supporting evidence will be necessary to confirm such claims.

18.5.4 5/6 Nephrectomy

Surgical 5/6 nephrectomy (renal mass reduction) is one of the most common ways inducing experimental uremia.

One important feature of renal injury in this particular model is hypertension, which alone perpetuates endothelial activation, inflammation, and proliferation, followed by vascular obliteration and glomerulosclerosis, a process that involves growth factor and cytokine signaling and influx of interstitial and glomerular inflammatory cells [99]. To date, several different stem/progenitor cells have been used to evaluate whether cell therapy could halt progressive fibrosis in the remnant kidney model, such as BMSCs, endothelial progenitor cells (EPCs), and fetal kidney precursor cells.

A single injection of BMSC was shown capable of reducing interstitial fibrosis and glomerulosclerosis, accompanied by reduction in the expression of the proinflammatory cytokines TNFα and IL-6, markers of EMT and matrix remodeling with an increase in the anti-inflammatory cytokines IL-4 and IL-10 [100]. However, because donor cells only temporarily integrated into kidney tissue, these observations had only a transient effect and led to the speculation that the protective effect of BMSCs was mediated via paracrine mechanisms. In order to increase the efficacy of BMSC therapy, new studies were conducted based on a speculation that the effectiveness of BMSC therapy of CKD might depend on administration frequency [101]. Lee et al. later successfully demonstrated that weekly administration of BMSCs in this model ameliorated functional parameters and led to a more significant improvement in morphology. Other progenitor cell lines, such as BM-derived EPCs [102] and lineage negative (Lin −) BM cells [103] produced similar immunomodulatory paracrine effects in mice undergoing 5/6 nephrectomy.

Alternatively to their paracrine role, BMSCs are hypothesized to also trigger reparative/regenerative response in the kidney. In a study by Villanuera et al., Sprague-Dawley rats receiving single BMSC infusions post 5/6 nephrectomy presented reduced expression of ED-1 and a-SMA associated with tissue damage. Similar to the previous reports, donor cell engraftment was minimal at 35 days after BMSC delivery. Meanwhile, significant increase in developmental markers, such as Pax-2, basic Fibroblast Growth Factor (bFGF), Bone Morphogenic Protein 7 (BMP-7), associated with tubular development as well as VEGF and Tie-2, important transcriptional factors in angiogenesis were reported. BMSC associated elevation of Pax-2, bFGF, and BMP-7 could in turn activate EMT, supporting a reparation and regeneration hypothesis underlying the effects of BMSCs or otherwise activate cellular repair mechanisms [104].

Interesting outcomes were also reported with fetal precursor cells isolated from metanephroi of E17.5 rat fetuses containing populations expressing mesenchymal, hematopoietic, and embryonic surface markers. Direct injection of these precursor cells under the renal capsule post 5/6 reduction injury resulted in significant attenuation of glomerulosclerosis and improvement in survival rates with donor cells reconstituting tubules and glomeruli, respectively [105]. However, the feasibility of this approach in humans is limited due to limited availability and ethical considerations regarding human fetal cells.

18.5.5 Adriamycin Nephropathy

Adriamycin nephropathy is a rodent model of CKD, which mirrors human primary focal segmental glomerulosclerosis (FSGS) and is characterized by podocyte injury followed by glomerulosclerosis, tubulointerstitial inflammation and fibrosis [106,107]. It has been demonstrated that BMSCs have beneficial effects on adriamycin-damaged podocytes both *in vitro* and *in vivo*. *In vitro*, BMSCs rescued immortalized podocytes from adriamycin-induced apoptosis, whereas *in vivo* they were shown to improve glomerular sclerosis, but did not have an effect on functional parameters, such as proteinuria and serum creatinine [108]. Additionally, multiple injections of BMSCs decreased podocyte loss and apoptosis with partial preservation of nephrin and CD2AP, which are significantly lost in adriamycin-damaged podocytes. Furthermore, attenuation of glomerulosclerosis was associated with decreased formation of glomerular podocyte−PEC bridges and normalized distribution of Neural Cell Adhesion Molecule (NCAM)-progenitor cells along the Bowman's capsule [109].

Kidney side populations (SPs) represent stem and progenitor cells within the kidney which are characterized by low fluorescence after staining with Hoechst 33342, have been shown to improve albumin creatinine ratio within 1 week of adriamycin treatment when injected systemically and directly into the renal cortex in an adriamycin nephropathy mouse model [110]. *In vitro* they successfully incorporate into the developing kidney in a metanephric organ culture system [110].

Adriamycin-induced renal injury is hypothesized to also impair EPC function [111]. Adaptive transfer of intact EPCs is shown to improve GFR and proteinuria and significantly reduce mortality in adriamycin-treated mice. These functional improvements were associated with a decrease in plasma concentration of proinflammatory cytokines, IL-1a and -b, and granulocyte-colony stimulating factor (G-CSF), decrease in VEGF, and correlated with improvement in renal vascular density and reduction of apoptosis. Such observations seem to suggest that reversal of kidney disease may not require that stem cells differentiate to replace damaged local tissue cells, but rather they may contribute to the regeneration of endogenous tissue by improving local microvasculature with enhanced perfusion and oxygenation [112].

18.5.6 Diabetic Nephropathy

DN is a complication of diabetes mellitus caused by angiopathy of capillaries supplying the glomeruli [113]. It is considered as the most common cause of progressive kidney disease worldwide [114−117], characterized by glomerular hyperfiltration, followed by thickening of the GBM, glomerular hypertrophy, and mesangial expansion, ultimately leading to proteinuria, renal fibrosis, and ESRD. The main promoting mechanisms leading to nephropathy include hyperglycemia-induced activation of ROS, which in turn result in upregulation of profibrotic cytokines like TGFβ [118,119]. Progression occurs in both type 1 and type 2 diabetes. Finding effective therapies for treatment or prevention of diabetes and DN is crucial to stop the rapidly increasing need for donor organs and dialysis treatment. Stem cell-based therapies provide new possibilities to treat DN. Unlike the currently available pharmacotherapies which target only a single aspect of the disease, such as ACE inhibitors, stem cells usually act through multiple mechanisms and hence are more suitable for this complex disease. Human and murine BMSCs have been used for the treatment of DN in NOD/SCID and C57BL/6 mice which succumb to DN after application of multiple low doses of streptozotocin (STZ) [113,120]. Data obtained from similar studies using the above models clearly indicate that donor cells efficiently engraft in the damaged kidneys, undergo renal differentiation and through paracrine and immune modulation provide structural and functional improvement to the kidney in diabetic animals [113,120−124]. However, data also suggest that donor cells are not able to proliferate after engraftment, as 1 month after MSC treatment, only a few human MSCs were detected in kidneys [120]. Interestingly, systemic administration of BMSCs at the onset of proteinuria did not reverse hyperglycemia but protected kidneys from progression to macroalbuminuria and prevented development of tubular dilatations, such as sclerosis, mesangial expansion, and podocyte loss [113]. In contrast, when syngeneic BMC transplantation was administered at an early stage after onset of diabetes, blood glucose levels were significantly reduced [120] in addition to reduction of renal oxidative stress, monocyte chemoattractant protein (MCP-1), and improvement in renal morphology and hypertrophy. Therefore, it is reasonable to conclude that the cell therapy is most effective when it is administered at an early stage after the onset of diabetes. Taken together, these studies may provide proof of the therapeutic potential of BMC and MBSC transplantation in the prevention and intervention of DN.

18.6 CONCLUSION

Based on what is currently known from studies on the animal models discussed in this chapter, it can be concluded that downstream effects of stem cell treatments can vary greatly depending on the damage site and the tissue microenvironment characteristic of each disease. Collectively, many studies indicate that BMCs, BMSCs, EPCs, etc. act as paracrine agents secreting anti-inflammatory and antifibrotic factors protecting the kidney from progressive damage. Taken together it is reasonable to conclude that stem cell injections provide a transient benefit to the kidney, delaying disease progression but in most cases complete regression does not occur yet. Despite this, we strongly believe that stem cells have a great therapeutic potential in regenerative medicine for kidney disease.

REFERENCES

[1] US Renal Data System. Available from: <http://www.usrds.org> [accessed July 2010].

[2] Kidney Disease Outcomes Quality I. K/DOQI clinical practice guidelines on hypertension and antihypertensive agents in chronic kidney disease. Am J Kidney Dis 2004;43:1−290.

[3] Rastogi A, Nissenson AR. Technological advances in renal replacement therapy: five years and beyond. Clin J Am Soc Nephrol 2009;4:132−6.

[4] Yeagy BA, Cherqui S. Kidney repair and stem cells: a complex and controversial process. Pediatr Nephrol 2011;26:1427−34.

[5] Yokoo T, Kawamura T, Kobayashi E. Stem cells for kidney repair: useful tool for acute renal failure? Kidney Int 2008;74:847−9.

[6] Tolwani A. Continuous renal-replacement therapy for acute kidney injury. N Engl J Med 2012;367:2505−14.

[7] Singh AP, Junemann A, Muthuraman A, Jaggi AS, Singh N, Grover K, et al. Animal models of acute renal failure. Pharmacol Rep 2012;64:31−44.

[8] Needham E. Management of acute renal failure. Am Fam Physician 2005;72:1739−46.

[9] Emili S, Black NA, Paul RV, Rexing CJ, Ullian ME. A protocol-based treatment for intradialytic hypotension in hospitalized hemodialysis patients. Am J Kidney Dis 1999;33:1107−14.

[10] Conger J. Dialysis and related therapies. Semin Nephrol 1998;18:533−40.

[11] Briglia A, Paganini EP. Acute renal failure in the intensive care unit. Therapy overview, patient risk stratification, complications of renal replacement, and special circumstances. Clin Chest Med 1999;20:347−66.

[12] Paganini EP, Sandy D, Moreno L, Kozlowski L, Sakai K. The effect of sodium and ultrafiltration modelling on plasma volume changes and haemodynamic stability in intensive care patients receiving haemodialysis for acute renal failure: a prospective, stratified, randomized, cross-over study. Nephrol Dial Transplant 1996;11:32−7.

[13] Perin L, Da Sacco S, De Filippo RE. Regenerative medicine of the kidney. Adv Drug Deliv Rev 2011;63:379−87.

[14] Kale S, Karihaloo A, Clark PR, Kashgarian M, Krause DS, Cantley LG. Bone marrow stem cells contribute to repair of the ischemically injured renal tubule. J Clin Invest 2003;112:42−9.

[15] Togel F, Zhang P, Hu Z, Westenfelder C. VEGF is a mediator of the renoprotective effects of multipotent marrow stromal cells in acute kidney injury. J Cell Mol Med 2009;13:2109−14.

[16] Lin F, Moran A, Igarashi P. Intrarenal cells, not bone marrow-derived cells, are the major source for regeneration in postischemic kidney. J Clin Invest 2005;115:1756−64.

[17] Morigi M, Introna M, Imberti B, Corna D, Abbate M, Rota C, et al. Human bone marrow mesenchymal stem cells accelerate recovery of acute renal injury and prolong survival in mice. Stem Cells 2008;26:2075−82.

[18] Fang TC, Alison MR, Cook HT, Jeffery R, Wright NA, Poulsom R. Proliferation of bone marrow-derived cells contributes to regeneration after folic acid-induced acute tubular injury. J Am Soc Nephrol 2005;16:1723−32.

[19] Bagul A, Frost JH, Drage M. Stem cells and their role in renal ischaemia reperfusion injury. Am J Nephrol 2013;37:16−29.

[20] Dai RP, Dheen ST, Tay SS. Induction of cytokine expression in rat post-ischemic sinoatrial node (SAN). Cell Tissue Res 2002;310:59−66.

[21] Suzuki S, Maruyama S, Sato W, Morita Y, Sato F, Miki Y, et al. Geranylgeranylacetone ameliorates ischemic acute renal failure via induction of Hsp70. Kidney Int 2005;67:2210−20.

[22] Wei Q, Wang MH, Dong Z. Differential gender differences in ischemic and nephrotoxic acute renal failure. Am J Nephrol 2005;25:491−9.

[23] Molitoris BA, Marrs J. The role of cell adhesion molecules in ischemic acute renal failure. Am J Med 1999;106:583−92.

[24] Sáenz-Morales D, Escribese MM, Stamatakis K, García-Martos M, Alegre L, Conde E, et al. Requirements for proximal tubule epithelial cell detachment in response to ischemia: role of oxidative stress. Exp Cell Res 2006;312:3711−27.

[25] Boros P, Bromberg JS. New cellular and molecular immune pathways in ischemia/reperfusion injury. Am J Transplant 2006;6:652−8.

[26] Thurman JM. Triggers of inflammation after renal ischemia/reperfusion. Clin Immunol 2007;123:7−13.

[27] Devarajan P. Update on mechanisms of ischemic acute kidney injury. J Am Soc Nephrol 2006;17:1503−20.

[28] Bonventre JV. Dedifferentiation and proliferation of surviving epithelial cells in acute renal failure. J Am Soc Nephrol 2003;14:55−61.

[29] Munshi R, Hsu C, Himmelfarb J. Advances in understanding ischemic acute kidney injury. BMC Med 2011;9:11.

[30] Morigi M, Imberti B, Zoja C, Corna D, Tomasoni S, Abbate M, et al. Mesenchymal stem cells are renotropic, helping to repair the kidney and improve function in acute renal failure. J Am Soc Nephrol 2004;15:1794−804.

[31] Togel F, Hu Z, Weiss K, Isaac J, Lange C, Westenfelder C. Administered mesenchymal stem cells protect against ischemic acute renal failure through differentiation-independent mechanisms. Am J Physiol Renal Physiol 2005;289:31−42.

[32] Togel F, Weiss K, Yang Y, Hu Z, Zhang P, Westenfelder C. Vasculotropic, paracrine actions of infused mesenchymal stem cells are important to the recovery from acute kidney injury. Am J Physiol Renal Physiol 2007;292:1626−35.

[33] Semedo P, Wang PM, Andreucci TH, Cenedeze MA, Teixeira VP, Reis MA, et al. Mesenchymal stem cells ameliorate tissue damages triggered by renal ischemia and reperfusion injury. Transplant Proc 2007;39:421−3.

[34] Semedo P, Palasio CG, Oliveira CD, Feitoza CQ, Goncalves GM, Cenedeze MA, et al. Early modulation of inflammation by mesenchymal stem cell after acute kidney injury. Int Immunopharmacol 2009;9:677−82.

[35] Hara Y, Stolk M, Ringe J, Dehne T, Ladhoff J, Kotsch K, et al. *In vivo* effect of bone marrow-derived mesenchymal stem cells in

a rat kidney transplantation model with prolonged cold ischemia. Transpl Int 2011;24:1112—23.

[36] Patel J, Pancholi N, Gudehithlu KP, Sethupathi P, Hart PD, Dunea G, et al. Stem cells from foreign body granulation tissue accelerate recovery from acute kidney injury. Nephrol Dial Transplant 2012;27:1780—6.

[37] Westenfelder C, Togel FE. Protective actions of administered mesenchymal stem cells in acute kidney injury: relevance to clinical trials. Kidney Int Suppl 2011;1:103—6.

[38] Shi H, Patschan D, Dietz GP, Bahr M, Plotkin M, Goligorsky MS. Glial cell line-derived neurotrophic growth factor increases motility and survival of cultured mesenchymal stem cells and ameliorates acute kidney injury. Am J Physiol Renal Physiol 2008;294:229—35.

[39] Tian H, Lu Y, Shah SP, Wang Q, Hong S. 14S,21R-dihydroxy-docosahexaenoic acid treatment enhances mesenchymal stem cell amelioration of renal ischemia/reperfusion injury. Stem Cells Dev 2012;21:1187—99.

[40] Altun B, Yilmaz R, Aki T, Akoglu H, Zeybek D, Piskinpasa S, et al. Use of mesenchymal stem cells and darbepoetin improve ischemia-induced acute kidney injury outcomes. Am J Nephrol 2012;35:531—9.

[41] Nafar M, Parvin M, Sadeghi P, Ghoraishian M, Soleimani M, Tabibi A, et al. Effects of stem cells and granulocyte colony stimulating factor in reperfusion injury. Iran J Kidney Dis 2010;4:207—13.

[42] Huls M, Russel FG, Masereeuw R. Insights into the role of bone marrow-derived stem cells in renal repair. Kidney Blood Press Res 2008;31:104—10.

[43] Hagiwara M, Shen B, Chao L, Chao J. Kallikrein-modified mesenchymal stem cell implantation provides enhanced protection against acute ischemic kidney injury by inhibiting apoptosis and inflammation. Hum Gene Ther 2008;19:807—19.

[44] Jiang H, Qu L, Li Y, Gu L, Shi Y, Zhang J, et al. Bone marrow mesenchymal stem cells reduce intestinal ischemia/reperfusion injuries in rats. J Surg Res 2011;168:127—34.

[45] Wang Y, Hu F, Wang ZJ, Wang GX, Zhang ZH, Xie P, et al. Administration of bone marrow-derived stem cells suppresses cellular necrosis and apoptosis induced by reperfusion of ischaemic kidneys in rats. Chin Med J (Engl) 2008;121:268—71.

[46] Wang PH, Schwindt TT, Barnabe GF, Motta FL, Semedo P, Beraldo FC, et al. Administration of neural precursor cells ameliorates renal ischemia—reperfusion injury. Nephron Exp Nephrol 2009;112:20—8.

[47] Patschan D, Patschan S, Wessels JT, Becker JU, David S, Henze E, et al. Epac-1 activator 8-O-cAMP augments renoprotective effects of syngeneic [corrected] murine EPCs in acute ischemic kidney injury. Am J Physiol Renal Physiol 2010;298:78—85.

[48] Furuichi K, Shintani H, Sakai Y, Ochiya T, Matsushima K, Kaneko S, et al. Effects of adipose-derived mesenchymal cells on ischemia—reperfusion injury in kidney. Clin Exp Nephrol 2012;16:679—89.

[49] Chen Y, Sun C, Lin Y, Chang L, Chen Y, Tsai T, et al. Adipose-derived mesenchymal stem cell protects kidneys against ischemia—reperfusion injury through suppressing oxidative stress and inflammatory reaction. J Transl Med 2011;9:51.

[50] Brodsky SV, Yamamoto T, Tada T, Kim B, Chen J, Kajiya F, et al. Endothelial dysfunction in ischemic acute renal failure:

rescue by transplanted endothelial cells. Am J Physiol Renal Physiol 2002;282:1140—9.

[51] Arriero M, Brodsky SV, Gealekman O, Lucas PA, Goligorsky MS. Adult skeletal muscle stem cells differentiate into endothelial lineage and ameliorate renal dysfunction after acute ischemia. Am J Physiol Renal Physiol 2004;287:621—7.

[52] Lin F, Cordes K, Li L, Hood L, Couser WG, Shankland SJ, et al. Hematopoietic stem cells contribute to the regeneration of renal tubules after renal ischemia—reperfusion injury in mice. J Am Soc Nephrol 2003;14:1188—99.

[53] Korrapati MC, Shaner BE, Schnellmann RG. Recovery from glycerol-induced acute kidney injury is accelerated by suramin. J Pharmacol Exp Ther 2012;341:126—36.

[54] Herrera MB, Bussolati B, Bruno S, Fonsato V, Romanazzi GM, Camussi G. Mesenchymal stem cells contribute to the renal repair of acute tubular epithelial injury. Int J Mol Med 2004;14:1035—41.

[55] Herrera MB, Bussolati B, Bruno S, Morando L, Mauriello-Romanazzi G, Sanavio F, et al. Exogenous mesenchymal stem cells localize to the kidney by means of CD44 following acute tubular injury. Kidney Int 2007;72:430—41.

[56] Bruno S, Grange C, Deregibus MC, Calogero RA, Saviozzi S, Collino F, et al. Mesenchymal stem cell-derived microvesicles protect against acute tubular injury. J Am Soc Nephrol 2009;20:1053—67.

[57] Perin L, Sedrakyan S, Giuliani S, Da Sacco S, Carraro G, Shiri L, et al. Protective effect of human amniotic fluid stem cells in an immunodeficient mouse model of acute tubular necrosis. PLoS One 2010;5:9357.

[58] Hauser PV, De Fazio R, Bruno S, Sdei S, Grange C, Bussolati B, et al. Stem cells derived from human amniotic fluid contribute to acute kidney injury recovery. Am J Pathol 2010;177:2011—21.

[59] Bussolati B, Bruno S, Grange C, Buttiglieri S, Deregibus MC, Cantino D, et al. Isolation of renal progenitor cells from adult human kidney. Am J Pathol 2005;166:545—55.

[60] Sagrinati C, Netti GS, Mazzinghi B, Lazzeri E, Liotta F, Frosali F, et al. Isolation and characterization of multipotent progenitor cells from the Bowman's capsule of adult human kidneys. J Am Soc Nephrol 2006;17:2443—56.

[61] Willox JC, McAllister EJ, Sangster G, Kaye SB. Effects of magnesium supplementation in testicular cancer patients receiving cis-platin: a randomised trial. Br J Cancer 1986;54:19—23.

[62] Arany I, Safirstein RL. Cisplatin nephrotoxicity. Semin Nephrol 2003;23:460—4.

[63] Muthuraman A, Sood S, Singla SK, Rana A, Singh A, Singh A, et al. Ameliorative effect of flunarizine in cisplatin-induced acute renal failure via mitochondrial permeability transition pore inactivation in rats. Naunyn Schmiedebergs Arch Pharmacol 2011;383:57—64.

[64] Kawai Y, Nakao T, Kunimura N, Kohda Y, Gemba M. Relationship of intracellular calcium and oxygen radicals to Cisplatin-related renal cell injury. J Pharmacol Sci 2006;100:65—72.

[65] Buzzi FC, Fracasso M, Filho VC, Escarcena R, del Olmo E, San Feliciano A. New antinociceptive agents related to dihydrosphin-gosine. Pharmacol Rep 2010;62:849—57.

[66] Imberti B, Morigi M, Tomasoni S, Rota C, Corna D, Longaretti L, et al. Insulin-like growth factor-1 sustains stem cell mediated renal repair. J Am Soc Nephrol 2007;18:2921—8.

[67] Yuan L, Wu MJ, Sun HY, Xiong J, Zhang Y, Liu CY, et al. VEGF-modified human embryonic mesenchymal stem cell implantation enhances protection against cisplatin-induced acute kidney injury. Am J Physiol Renal Physiol 2011;300:207–18.

[68] Bi B, Guo J, Marlier A, Lin SR, Cantley LG. Erythropoietin expands a stromal cell population that can mediate renoprotection. Am J Physiol Renal Physiol 2008;295:1017–22.

[69] Bruno S, Grange C, Collino F, Deregibus MC, Cantaluppi V, Biancone L, et al. Microvesicles derived from mesenchymal stem cells enhance survival in a lethal model of acute kidney injury. PLoS One 2012;7:33115.

[70] Tomasoni S, Longaretti L, Rota C, Morigi M, Conti S, Gotti E, et al. Transfer of growth factor receptor mRNA via exosomes unravels the regenerative effect of mesenchymal stem cells. Stem Cells Dev 2013;22:772–80.

[71] Bi B, Schmitt R, Israilova M, Nishio H, Cantley LG. Stromal cells protect against acute tubular injury via an endocrine effect. J Am Soc Nephrol 2007;18:2486–96.

[72] Gheisari Y, Ahmadbeigi N, Naderi M, Nassiri SM, Nadri S, Soleimani M. Stem cell-conditioned medium does not protect against kidney failure. Cell Biol Int 2011;35:209–13.

[73] Kim JH, Park DJ, Yun JC, Jung MH, Yeo HD, Kim HJ, et al. Human adipose tissue-derived mesenchymal stem cells protect kidneys from cisplatin nephrotoxicity in rats. Am J Physiol Renal Physiol 2012;302:1141–50.

[74] Yasuda K, Ozaki T, Saka Y, Yamamoto T, Gotoh M, Ito Y, et al. Autologous cell therapy for cisplatin-induced acute kidney injury by using non-expanded adipose tissue-derived cells. Cytotherapy 2012;14:1089–100.

[75] Rota C, Imberti B, Pozzobon M, Piccoli M, De Coppi P, Atala A, et al. Human amniotic fluid stem cell preconditioning improves their regenerative potential. Stem Cells Dev 2012;21:1911–23.

[76] Luo J, Zhao X, Tan Z, Su Z, Meng F, Zhang M. Mesenchymal-like progenitors derived from human embryonic stem cells promote recovery from acute kidney injury via paracrine actions. Cytotherapy 2013;15(6):649–62.

[77] Hishikawa K, Marumo T, Miura S, Nakanishi A, Matsuzaki Y, Shibata K, et al. Musculin/MyoR is expressed in kidney side population cells and can regulate their function. J Cell Biol 2005;169:921–8.

[78] Goligorsky MS. Regenerative nephrology. 1st ed. Amsterdam, Boston: Academic; 2011.

[79] Metcalfe W. How does early chronic kidney disease progress? a background paper prepared for the UK consensus conference on early chronic kidney disease. Nephrol Dial Transplant 2007;22:26–30.

[80] Li L, Black R, Ma Z, Yang Q, Wang A, Lin F. Use of mouse hematopoietic stem and progenitor cells to treat acute kidney injury. Am J Physiol Renal Physiol 2012;302:9–19.

[81] Lee PT, Lin HH, Jiang ST, Lu PJ, Chou KJ, Fang HC, et al. Mouse kidney progenitor cells accelerate renal regeneration and prolong survival after ischemic injury. Stem Cells 2010;28:573–84.

[82] Morigi M, Benigni A, Remuzzi G, Imberti B. The regenerative potential of stem cells in acute renal failure. Cell Transplant 2006;15:111–7.

[83] Saito A, Aung T, Sekiguchi K, Sato Y, Vu DM, Inagaki M, et al. Present status and perspectives of bioartificial kidneys. J Artif Organs 2006;9:130–5.

[84] Sedrakyan S, Angelow S, De Filippo RE, Perin L. Stem cells as a therapeutic approach to chronic kidney diseases. Curr Urol Rep 2012;13:47–54.

[85] Cosgrove D, Meehan DT, Grunkemeyer JA, Kornak JM, Sayers R, Hunter WJ, et al. Collagen COL4A3 knockout: a mouse model for autosomal Alport syndrome. Genes Dev 1996;10:2981–92.

[86] Miner JH, Sanes JR. Molecular and functional defects in kidneys of mice lacking collagen alpha 3(IV): implications for Alport syndrome. J Cell Biol 1996;135:1403–13.

[87] Rheault MN, Kren SM, Thielen BK, Mesa HA, Crosson JT, Thomas W, et al. Mouse model of X-linked Alport syndrome. J Am Soc Nephrol 2004;15:1466–74.

[88] Kang JS, Wang XP, Miner JH, Morello R, Sado Y, Abrahamson DR, et al. Loss of alpha3/alpha4(IV) collagen from the glomerular basement membrane induces a strain-dependent isoform switch to alpha5alpha6(IV) collagen associated with longer renal survival in Col4a3 −/− Alport mice. J Am Soc Nephrol 2006;17:1962–9.

[89] Sugimoto H, Mundel TM, Sund M, Xie L, Cosgrove D, Kalluri R. Bone-marrow-derived stem cells repair basement membrane collagen defects and reverse genetic kidney disease. Proc Natl Acad Sci USA 2006;103:7321–6.

[90] Prodromidi EI, Poulsom R, Jeffery R, Roufosse CA, Pollard PJ, Pusey CD, et al. Bone marrow-derived cells contribute to podocyte regeneration and amelioration of renal disease in a mouse model of Alport syndrome. Stem Cells 2006;24:2448–55.

[91] Katayama K, Kawano M, Naito I, Ishikawa H, Sado Y, Asakawa N, et al. Irradiation prolongs survival of Alport mice. J Am Soc Nephrol 2008;19:1692–700.

[92] LeBleu V, Sugimoto H, Mundel TM, Gerami-Naini B, Finan E, Miller CA, et al. Stem cell therapies benefit Alport syndrome. J Am Soc Nephrol 2009;20:2359–70.

[93] Ninichuk V, Gross O, Segerer S, Hoffmann R, Radomska E, Buchstaller A, et al. Multipotent mesenchymal stem cells reduce interstitial fibrosis but do not delay progression of chronic kidney disease in collagen4A3-deficient mice. Kidney Int 2006;70:121–9.

[94] Sedrakyan S, Da Sacco S, Milanesi A, Shiri L, Petrosyan A, Varimezova R, et al. Injection of amniotic fluid stem cells delays progression of renal fibrosis. J Am Soc Nephrol 2012;23:661–73.

[95] Cherqui S, Sevin C, Hamard G, Kalatzis V, Sich M, Pequignot MO, et al. Intralysosomal cystine accumulation in mice lacking cystinosin, the protein defective in cystinosis. Mol Cell Biol 2002;22:7622–32.

[96] Nevo N, Chol M, Bailleux A, Kalatzis V, Morisset L, Devuyst O, et al. Renal phenotype of the cystinosis mouse model is dependent upon genetic background. Nephrol Dial Transplant 2010;25:1059–66.

[97] Syres K, Harrison F, Tadlock M, Jester JV, Simpson J, Roy S, et al. Successful treatment of the murine model of cystinosis using bone marrow cell transplantation. Blood 2009;114:2542–52.

[98] Yeagy BA, Harrison F, Gubler MC, Koziol JA, Salomon DR, Cherqui S. Kidney preservation by bone marrow cell transplantation in hereditary nephropathy. Kidney Int 2011;79:1198–206.

[99] Hobo A, Yuzawa Y, Kosugi T, Kato N, Asai N, Sato W, et al. The growth factor midkine regulates the renin−angiotensin system in mice. J Clin Invest 2009;119:1616−25.

[100] Semedo P, Correa-Costa M, Antonio Cenedeze M, Costa Malheiros MA, Antonia dos Reis D, Shimizu M, et al. Mesenchymal stem cells attenuate renal fibrosis through immune modulation and remodeling properties in a rat remnant kidney model. Stem Cells 2009;27:3063−73.

[101] Lee SR, Lee SH, Moon JY, Park JY, Lee D, Lim SJ, et al. Repeated administration of bone marrow-derived mesenchymal stem cells improved the protective effects on a remnant kidney model. Ren Fail 2010;32:840−8.

[102] Sangidorj O, Yang SH, Jang HR, Lee JP, Cha RH, Kim SM, et al. Bone marrow-derived endothelial progenitor cells confer renal protection in a murine chronic renal failure model. Am J Physiol Renal Physiol 2010;299:325−35.

[103] Alexandre CS, Volpini RA, Shimizu MH, Sanches TR, Semedo P, di Jura VL, et al. Lineage-negative bone marrow cells protect against chronic renal failure. Stem Cells 2009;27:682−92.

[104] Villanueva S, Ewertz E, Carrion F, Tapia A, Vergara C, Cespedes C, et al. Mesenchymal stem cell injection ameliorates chronic renal failure in a rat model. Clin Sci (Lond) 2011;121:489−99.

[105] Kim SS, Park HJ, Han J, Gwak SJ, Park MH, Song KW, et al. Improvement of kidney failure with fetal kidney precursor cell transplantation. Transplantation 2007;83:1249−58.

[106] Pippin JW, Brinkkoetter PT, Cormack-Aboud FC, Durvasula RV, Hauser PV, Kowalewska J, et al. Inducible rodent models of acquired podocyte diseases. Am J Physiol Renal Physiol 2009;296:213−29.

[107] Lee VW, Harris DC. Adriamycin nephropathy: a model of focal segmental glomerulosclerosis. Nephrology (Carlton) 2011;16:30−8.

[108] Magnasco A, Corselli M, Bertelli R, Ibatici A, Peresi M, Gaggero G, et al. Mesenchymal stem cells protective effect in adriamycin model of nephropathy. Cell Transplant 2008;17:1157−67.

[109] Zoja C, Garcia PB, Rota C, Conti S, Gagliardini E, Corna D, et al. Mesenchymal stem cell therapy promotes renal repair by limiting glomerular podocyte and progenitor cell dysfunction in adriamycin-induced nephropathy. Am J Physiol Renal Physiol 2012;303:1370−81.

[110] Challen GA, Bertoncello I, Deane JA, Ricardo SD, Little MH. Kidney side population reveals multilineage potential and renal functional capacity but also cellular heterogeneity. J Am Soc Nephrol 2006;17:1896−912.

[111] Yasuda K, Park HC, Ratliff B, Addabbo F, Hatzopoulos AK, Chander P, et al. Adriamycin nephropathy: a failure of endothelial progenitor cell-induced repair. Am J Pathol 2010;176:1685−95.

[112] Fine LG. First heal thyself: rescue of dysfunctional endothelial progenitor cells restores function to the injured kidney. Am J Pathol 2010;176:1586−7.

[113] Ezquer FE, Ezquer ME, Parrau DB, Carpio D, Yanez AJ, Conget PA. Systemic administration of multipotent mesenchymal stromal cells reverts hyperglycemia and prevents nephropathy in type 1 diabetic mice. Biol Blood Marrow Transplant 2008;14:631−40.

[114] Ritz E, Rychlik I, Locatelli F, Halimi S. End-stage renal failure in type 2 diabetes: a medical catastrophe of worldwide dimensions. Am J Kidney Dis 1999;34:795−808.

[115] Viswanathan V. Type 2 diabetes and diabetic nephropathy in India—magnitude of the problem. Nephrol Dial Transplant 1999;14:2805−7.

[116] Parving HH. Diabetic nephropathy: prevention and treatment. Kidney Int 2001;60:2041−55.

[117] Remuzzi G, Schieppati A, Ruggenenti P. Clinical practice. Nephropathy in patients with type 2 diabetes. N Engl J Med 2002;346:1145−51.

[118] Dronavalli S, Duka I, Bakris GL. The pathogenesis of diabetic nephropathy. Nat Clin Pract Endocrinol Metab 2008;4:444−52.

[119] Ezquer ME, Ezquer FE, Arango-Rodriguez ML, Conget PA. MSC transplantation: a promising therapeutic strategy to manage the onset and progression of diabetic nephropathy. Biol Res 2012;45:289−96.

[120] Lee RH, Seo MJ, Reger RL, Spees JL, Pulin AA, Olson SD, et al. Multipotent stromal cells from human marrow home to and promote repair of pancreatic islets and renal glomeruli in diabetic NOD/SCID mice. Proc Natl Acad Sci USA 2006;103:17438−43.

[121] Tolar J, Nauta AJ, Osborn MJ, Panoskaltsis Mortari A, McElmurry RT, Bell S, et al. Sarcoma derived from cultured mesenchymal stem cells. Stem Cells 2007;25:371−9.

[122] Ouyang J, Hu G, Wen Y, Zhang X. Preventive effects of syngeneic bone marrow transplantation on diabetic nephropathy in mice. Transpl Immunol 2010;22:184−90.

[123] Le Blanc K, Pittenger M. Mesenchymal stem cells: progress toward promise. Cytotherapy 2005;7:36−45.

[124] Ezquer F, Ezquer M, Simon V, Pardo F, Yanez A, Carpio D, et al. Endovenous administration of bone-marrow-derived multipotent mesenchymal s0tromal cells prevents renal failure in diabetic mice. Biol Blood Marrow Transplant 2009;15:1354−65.

Chapter 19

Renal Regeneration: The Developmental Approach

Gleb Martovetsky[a,b] and Sanjay K. Nigam[a,b,c,d]

[a]Department of Pediatrics, University of California at San Diego, La Jolla, CA, [b]Department of Biomedical Sciences, University of California at San Diego, La Jolla, CA, [c]Department of Medicine, University of California at San Diego, La Jolla, CA, [d]Department of Cellular and Molecular Medicine, University of California at San Diego, La Jolla, CA

Chapter Outline

19.1 NORMAL KIDNEY DEVELOPMENT

Because the tissue engineering strategies that will be discussed are based on normal kidney development, it is worthwhile to first discuss how the kidney develops in the embryo. Over half a century of research has greatly advanced the understanding of the cellular and molecular basis of kidney development. However, there is much work to be done to understand how two progenitor subpopulations of intermediate mesodermal cells [the ureteric bud (UB) cells and metanephric mesenchymal (MM) cells], over the span of fetal metanephrogenesis, form a highly complex organ consisting of over 25 spatially and largely functionally distinct cell types. Most of the experimentation has been performed with rodents, in which, other than a few notable differences, such as timing and scale, kidney development appears quite similar to that of humans.

The functional unit of the kidney is the nephron, composed of the glomerulus, proximal tubule, loop of Henle, and distal tubule. A mature human kidney is generally thought to have approximately 0.5−1 million nephrons. The distal tubule of each nephron connects to a network of branched collecting ducts, which feed into

the ureter on each side. Kidney organogenesis begins when the Wolffian duct (WD), sometimes referred to as the nephric duct, is induced by the MM to "bud"; the resulting outgrowth is referred to as the UB. This process takes place at around 4 weeks of gestation in humans, embryonic day 11 in mice, and day 12 in rats. In turn, the UB interacts with the surrounding MM to establish a group of cells that will later form parts of the nephron. After the initial budding event, the UB goes through numerous rounds of dichotomous branching, essentially becoming an iterative tip-stalk generator, while at the same time inducing mesenchymal-to-epithelial transition (MET) and subsequent nephron formation in the mesenchyme. Through this mutual induction process about which much is now understood, these two multicellular structures—the UB and the MM—will form the collecting duct system and the MM-derived parts of the nephron, respectively (Figure 19.1). Gene knockout models and *in vitro* tissue culture systems have implicated many growth factors, extracellular matrix molecules, transcription factors, and signaling pathways in various aspects of nephrogenesis. The details are beyond the scope of this chapter but have been detailed elsewhere [2]. Those molecules, particularly growth factors, which are relevant

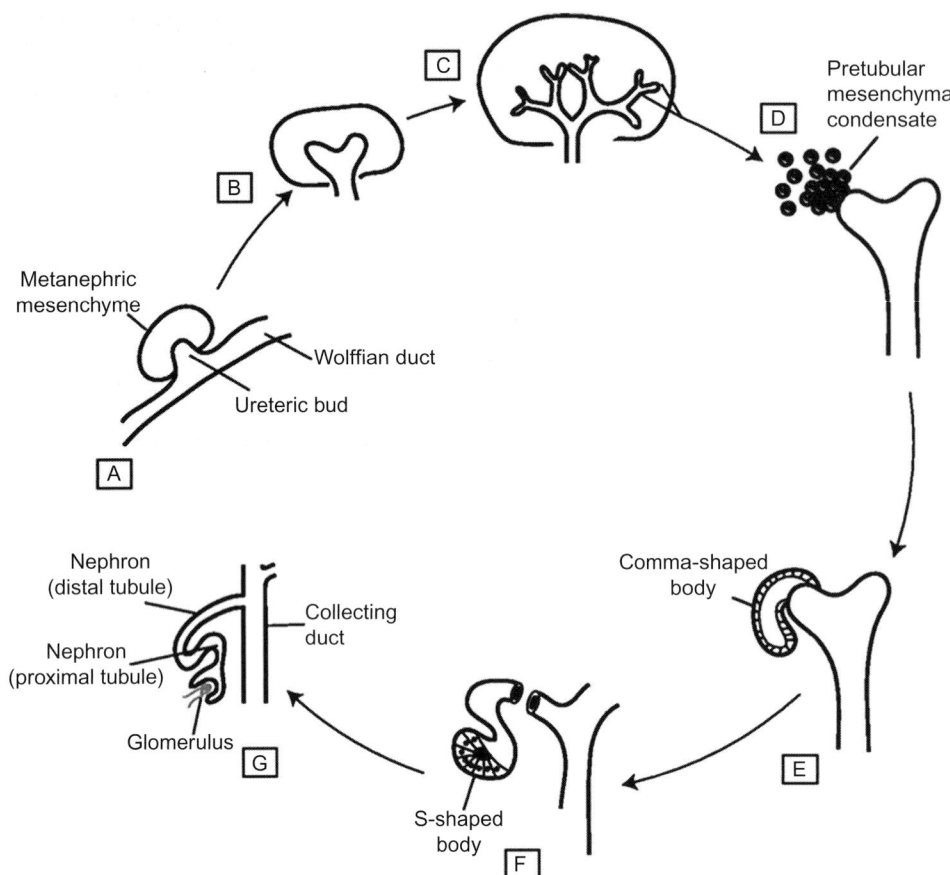

FIGURE 19.1 Summary of kidney development. (A) Kidney development initiates when the intermediate mesoderm-derived MM induces the outgrowth of the UB from the WD. (B and C) The UB then undergoes multiple rounds of iterative branching while invading the expanding mesenchyme. (D) Cells near branch tips are induced to condense into cellular aggregates, which then proceed to undergo nephrogenesis. (E and F) The pretubular condensate goes through precisely choreographed morphogenesis on the way to becoming a nephron, including intermediate stages called comma-shaped and s-shaped bodies. (G) The final result is an MM-derived epithelial nephron attached to the UB-derived collecting duct. The origin of the endothelial cells in the glomerulus is still debated. *Modified from Shah et al. [1]*

to tissue engineering strategies, are described in the following sections.

There is also the separate and still debated matter of vascular and nervous system development within the kidney. Whether vascular cells differentiate within the mesenchyme, migrate from outside, or both, is not yet clear. At approximately the 36th week of gestation in humans (though it continues for 1−2 weeks after birth in rats), the kidney is thought to have developed its full "endowment" of nephrons. Nevertheless, there is a substantial period of postnatal growth and maturation, and significant changes occur through sexual maturity.

One can examine the processes involved in kidney development morphologically, through analysis of markers of various cell types [3], or by analyzing gene expression patterns using microarrays and related gene expression profiling technologies. For example, to examine the transcriptional dynamics of kidney development, a study has profiled the genome-wide transcription in rat kidneys at many developmental time-points [4]. The resulting expression data at each time-point was analyzed with hierarchical clustering, and used to generate self-organized maps (Figure 19.2). These analytical approaches revealed that, based on gene expression patterns, kidney development might occur in distinct stages, opposed to a smooth continuum. This possibly suggests transcriptional enrichment of genes involved in biological processes prioritized at that stage in development.

19.2 MODELING DEVELOPMENTAL PROGRAMS USING ORGAN CULTURE

A great deal of current research is aimed at developing strategies to preserve kidney function. Nevertheless, engineering an implantable organ that can completely replace kidney function remains a major goal. Some of these strategies, including the ones emphasized in this chapter, are developmentally based. They can be traced back to early attempts to recapitulate *in vivo* morphogenetic events of renal development in *ex vivo* and *in vitro* settings. In 1953, two pioneering studies by Grobstein and colleagues [5,6] demonstrated that early kidney rudiments, comprised of the UB and surrounding

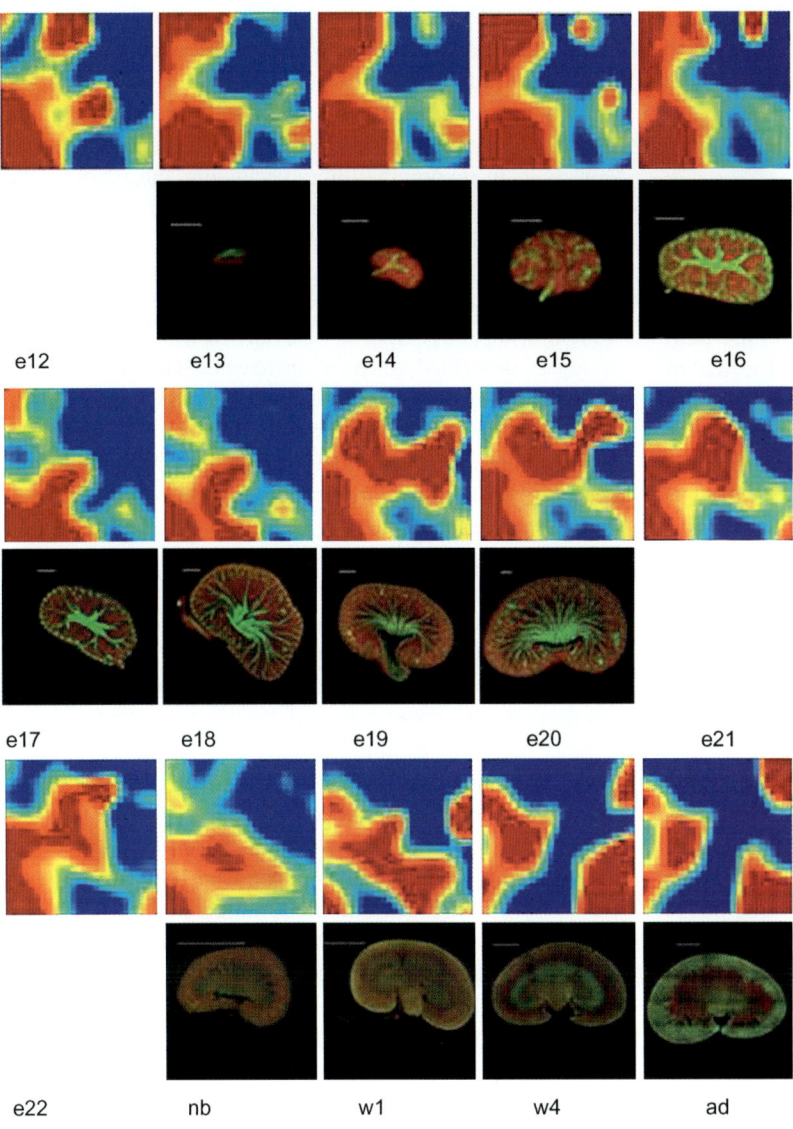

FIGURE 19.2 Self-organized maps of gene expression reveal transcriptional dynamics throughout kidney development. Metagene portraits and corresponding kidney slices of developing rat kidneys are portrayed for the indicated time-points: 12–22 days postcoitum, newborn, 1 week old, 4 weeks old, and adult (8 weeks old). Inspection of transitions in gene expression revealed a stage-wise progression corresponding to development, rather than a continuum. Pictures of rat kidney slices at e12, e21, and e22 were not obtained. Slices were stained with dolichos biflorus lectin (green), which marks the collecting ducts in embryonic kidneys, and peanut agglutinin (red). *Modified from Tsigelny et al. [4]*

MM, can be dissected from mice and cultured on top of a permeable membrane. These cultures were able to grow substantially in size, and recapitulated a significant portion of development, including the formation of an extensive collecting system via branching morphogenesis and nephron-like structures (after an MET). Unlike in submerged tissue culture, cohesive forces cause a thin layer of media to form around the tissue, allowing for more efficient gas exchange. While this technique is still limited to small embryonic tissues, tissue culture on these transwell filters has emerged as a major technique to this date for exploring kidney tissue engineering strategies (described below). Grobstein et al. also went on to show that after enzymatically separating the UB and the MM, these two tissues can continue much of their natural course of development when recombined (i.e., placed side by side in a culture system). Over the years, variations of these culture models have been devised in order to investigate the potential of embryonic kidney tissues to induce and undergo various developmental processes.

After it became clear that embryonic kidney cultures could recapitulate many aspects of *in vivo* development with minimal supplementation, efforts were made to differentiate between the intrinsic properties of the tissues and those that require inductive stimuli. The results of these early experiments are summarized elsewhere [7]. If simply cultured in isolation, the UB and MM fail to grow. However, by coculturing these rudiments with other embryonic tissues, it became apparent that many different tissues are capable of inducing the UB to branch and the MM to undergo various degrees of nephrogenesis.

Spinal cord was found to be a particularly potent inducer of the MM. Thus, the inductive signals are not unique to the MM and the UB, and their specific responses appear to be at least in part preprogrammed within the renal progenitor tissues.

19.3 MODELING RENAL MORPHOGENESIS WITH MODIFIED ORGAN CULTURES

The aforementioned studies raised the question of whether direct cell contact between the UB and the MM was indeed necessary for the morphogenesis of either of those tissues to occur. It was argued that direct cell contact might be necessary. In an attempt to reproduce an environment with the necessary soluble inductive factors using cell-based systems, cells were derived from the UB and MM, immortalized, and cultured to produce conditioned media. It was shown that concentrated conditioned media from UB cells was indeed able to induce morphogenesis in isolated mesenchyme cultures [8]. This indicated that even in the absence of cell contact with the inducing tissue, soluble factors were sufficient to induce nephrogenesis. However, conditioned media contains thousands of protein species, generating the question: "What is the minimal set of specific factors sufficient for induction?" To address this issue, individual proteins were purified from this conditioned media and tested for inductive capacity. One of the purified factors was identified as leukemia inhibitory factor (LIF), and turned out to be sufficient to induce a significant degree of nephrogenesis [9]. Later, other factors were discovered that were also sufficient to induce early nephron formation or synergistic when used in combination [10]. These findings suggest that complex developmental processes can be driven by a relatively small number of soluble inputs. One can imagine that if the correct cells are delivered to a target area, nephrons can be induced to form locally with a combination of soluble factors. Alternatively, these strategies can be applied to refine the development of engineered renal tissues.

Unlike the MM, the UB does not respond consistently when exposed to conditioned media from MM cells on a two-dimensional (2D) filter. In this case, a three-dimensional (3D) structural environment containing extracellular matrix proteins plays a key role in supporting branching morphogenesis of the UB. (The extracellular environment is also likely to be vital for supporting nephrogenesis in the mesenchyme cultures but is provided by the surrounding stromal cells.) To overcome the UB's need for structural support, various 3D gels composed of extracellular matrix proteins were used in an attempt to recreate the native environment. When UBs were embedded in a complex gel scaffold in the presence of MM-conditioned media and glial cell derived neurotrophic factor (GDNF), the UB was able to develop and branch [11]. Using this method, it was also discovered that if the UB is induced to branch in culture, and a tip is then cut off and subcultured, it will propagate and maintain the capacity to induce the MM similar to the early UB [12]. Moreover, in recombination experiments, the nascent collecting ducts and distal tubules were able to form direct connections. This could be done for several generations. Thus, once a branching UB-derived structure can be generated, this tissue is "replenishable" without losing developmental capacity. This is relevant to tissue engineering strategies described below. Considering that mass transfer is a major hurdle in engineering large tissues, one solution might be to subdivide an engineered kidney culture into numerous small modules, which might then be recombined to form a large organ once an endothelial network is established.

Morphologically, cultured whole embryonic kidneys or renal progenitor tissues are capable of undergoing morphogenesis similar to endogenous development and express markers that are representative of specific renal structures. However, it was important to establish the functionality of these cultured tissues. One indicator of kidney function is the capacity to transport specific solutes. The proximal tubule of the kidney plays an essential role in clearing many widely administered pharmaceuticals, exogenous toxins, and harmful endogenous metabolites. These functions depend on specific transporters, which are absent in progenitor renal tissues. One method to test functional transport is by using a fluorescent substrate known to be specifically taken up by proximal tubule cells: 6-carboxyfluorescein (6CF). This is largely mediated by organic anion transporters (e.g., Oat1, Oat3)—major drug, toxin, and metabolite transporters in the kidney—and is inhibitable by the drug probenecid. Using this technique, it was demonstrated that nephron-like structures were capable of probenecid-sensitive 6CF uptake in both embryonic kidney cultures and recombined renal tissue [13]. Expression analysis of these culture systems confirmed the expression of transporters known to be involved in organic anion transport, which increased with culture time.

19.4 TISSUE ENGINEERING

After it was demonstrated that nephrogenesis could be broken down into modular entities that recapitulate key developmental processes, such as WD budding, UB branching, MET, and nephron formation, methods initially used to understand kidney development started being explored for their potential to engineer renal tissue. Considering the UB and MM can mutually induce each other to undergo proper morphogenesis when recombined, it was hypothesized that it might be possible to create functional engineered kidneys by generating and propagating these two progenitor tissues, and somehow integrating vasculature. In the culture methods involving embryonic kidney tissues discussed above, there is no blood supply to form the elaborate network of capillaries in the glomerulus and around the remaining nephron. However, earlier studies had explored this issue, and discovered that implanted developing kidneys have an intrinsic capacity to recruit blood supply from nearby endothelial networks, akin to nephrons during normal kidney development. For example, when the embryonic metanephros is implanted into a rabbit eye, under the renal capsule, or on a chicken chorioallantoic membrane, vasculature invades the developing tissue [14]. Therefore, it might be possible to support the vascularization of engineered renal tissue by exposing it to an active endothelial network throughout the culture process.

If a "building block approach" is used for generating an engineered kidney, the source of the UB and mesenchyme must be considered. The MM is an early derivative of germ layer tissue; while to form the UB, mesodermal cells must first undergo epithelialization and tubulogenesis to form the WD, of which a small section undergoes reorganization upon induction by the MM or its soluble factors to form the UB. Thus, it was of interest to investigate whether the precursor to the UB—the WD—can be used as the starting point for generating the UB and the nascent collecting duct system, which is derived from the UB. Ever since the first transgenic mouse model was generated, many knockout models have been created which exhibit kidney agenesis or varying degrees of developmental/functional defects. One of these models revealed that the growth factor GDNF and its receptor c-Ret are required for the UB to bud from the WD [15]. It was also shown, if WDs are isolated from rodent embryos and grown on Transwells, addition of purified GDNF plus a fibroblast growth factor (FGF) is sufficient to induce ectopic budding [17]. After this finding, the required building blocks necessary to recapitulate the foundation of kidney development were potentially reduced to a WD-like tubule and a source of mesenchyme.

Understanding the effects on cell and tissue behavior in response to the modulation of signaling pathways is important for engineering renal tissue, because this allows for the replacement of developmental cues that are lost *ex vivo*. There seems to be no *a priori* reason why the engineering kidneys cannot deviate from the strictly choreographed cascade of biochemical events that take place during development *in vivo*. Due to the high interconnectivity of cellular signaling pathways, it might be possible to tissue engineer a functional kidney by sometimes modulating alternate or redundant biological pathways. In such a scenario, manipulation of signaling pathways would be critical to enhance morphogenesis and/or differentiation. To alter intracellular signaling, the agent must either somehow enter into the cells, or alter activity via activating cell surface receptors. Because of this, the manipulation of various modified organ cultures is highly amenable to small molecules, which can diffuse freely into cells, and proteins that act on extracellular receptors (e.g., extracellular matrix proteins, growth factors, cytokines, membrane-bound ligands). One example was the use of an activin A-blocking antibody and the growth factor GDNF to induce ectopic budding of the isolated WD [17]. While it was known that GDNF is required *in vivo*, it was not sufficient *in vitro*. However, by attenuating the inhibitory functions of activin signaling, budding was permitted. This possibly recreated endogenous signaling cues which are likely provided *in vivo* by surrounding cells but absent in culture. Another example was the use of the factors LIF, transforming growth factor beta (TGFβ), and FGF2 to induce nephrogenesis in isolated mesenchyme [9]. These factors appear to recreate a minimal set of inductive stimuli that are normally provided by the UB. However, more than one combination of factors, as well as a variety of small molecules not present endogenously, can induce nephrogenesis, indicating that nonendogenous stimuli can still be used to achieve the desirable effect. Future advances in targeted drug delivery would allow for signals to be delivered only to selected cells, expanding the potential to control development even further.

Once the foundation for a feasible renal tissue engineering strategy is established, it might be possible to manipulate the extracellular environment to instruct or enhance development. One example involved the use of a soluble matrix protein—hyaluronic acid (HA) [18]—to modulate nephrogenesis *ex vivo*. It has been shown that culturing embryonic kidneys in the presence of HA of various concentration and molecular weight had different morphogenetic consequences. Namely, one of the conditions was found to promote tubulogenesis and differentiation, as measured by gene expression and morphometry, thereby enhancing the baseline capacity of cultured embryonic kidneys to follow initial stages of organogenesis. Another study aimed to improve embryonic kidney development in culture by providing structural support via various 3D hydrogels made up of recombinant extracellular matrix proteins [19]. It was determined that, compared to Transwells, a scaffold made up purely of collagen IV greatly enhanced the spatial organization of cultured metanephroi (Figure 19.3). Interestingly, collagen IV is one of the predominant components of renal tubular basal lamina. It might be possible that more complex scaffolds will be able to recreate specific extracellular microenvironments, which could then go beyond structural support and play a part in instructing cellular behavior.

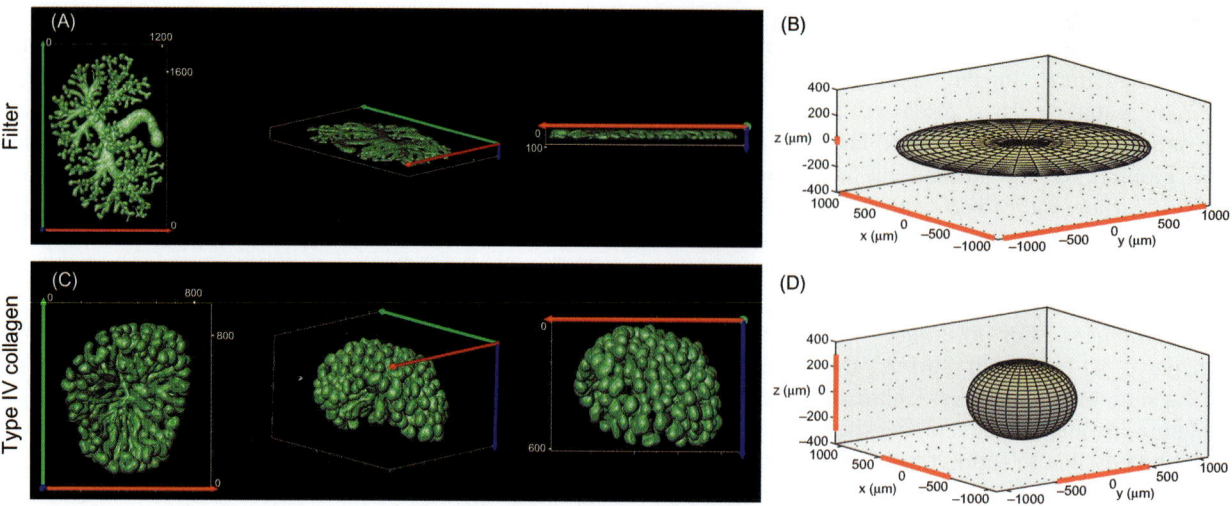

FIGURE 19.3 3D scaffold made of collagen IV enhances the architecture of cultured embryonic mouse kidneys. Kidney rudiments were removed from e12 mouse embryos from a HoxB7−GFP reporter line, which marks the collecting duct system in the kidney, and cultured for 7 days. (A and B) 3D projections of a kidney grown on a Transwell revealed relatively flat morphology. (C and D) When cultured on a Transwell in a gel made of collagen IV, the kidneys were able to develop a 3D structure that more accurately mimicks endogenous morphology. *Modified from Rosines et al. [19].*

It appears that, if engineered renal tissue develops within reach of an endothelial network, vascularization and subsequent functionality may follow with minimal input. Collection and removal of the filtrate would need to be considered, but the major limitation of this approach is that the tissue must remain viable prior to establishing a blood supply. Presumably, the starting tissue must be small enough to exchange nutrients and gases exclusively by diffusion and integrate vasculature as it develops (e.g., in the patient, surrogate animal, or some sort of bioreactor). Considering that the UB and MM meet these criteria, and are sufficient to undergo multiple stages of kidney organogenesis, this strategy appears to be theoretically feasible. Unfortunately, the UB and MM cease to exist as entities once kidney development has begun. So, while kidney development has been gradually broken down into fundamental building blocks, most of the tissue engineering strategies discussed so far have required progenitor renal tissues as the starting material, which are not easily replenishable and do not match the host's genetic identity. There are, however, strategies to generate cells with a patient's genetic background, such as somatic-cell nuclear transfer and programming of induced pluripotent stem cells (iPSCs). Thus, the fundamental obstacle to generating histocompatible renal tissues could be overcome if the UB and MM could be constructed from single cells.

19.5 *EX VIVO/IN VIVO* APPROACHES

In 2007, a study combined all of these previously developed concepts to create, in a modular fashion, a rudimentary, implantable engineered kidney tissue [20]. In this

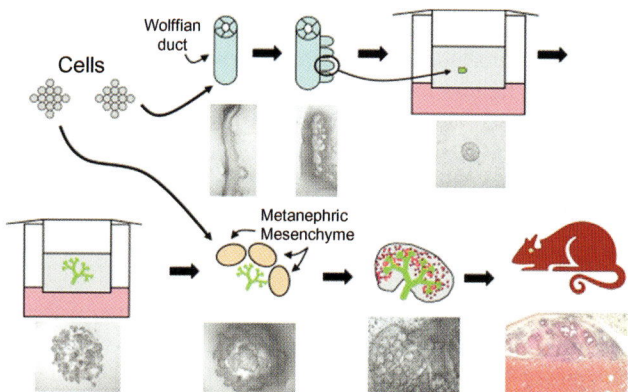

FIGURE 19.4 Strategy to generate implantable engineered renal tissue. WDs were first dissected from rat embryos, and induced to bud *in vitro*. These buds were then isolated, and induced to grow and branch in culture. Resulting branched structures were then recombined with freshly isolated MM and cocultured for 4−6 days. Final tissue constructs were then implanted under the renal capsule of adult rats, where they recruited vasculature and formed glomeruli. While cells were not used as the starting material for this study, the potential for cells to replace the requirement for the WD and MM is indicated. *Modified from Rosines et al. [20].*

study, kidney-like tissue was engineered using a WD and uninduced MM as the starting materials (Figure 19.4). First, a WD was made to bud in culture. This bud was then propagated and induced to branch. These branched structures were then recombined with freshly isolated MM, inducing nephrogenesis and further branching morphogenesis. To investigate the resemblance of the engineered tissue to the composition of native kidneys, genome-wide expression of the tissue constructs was

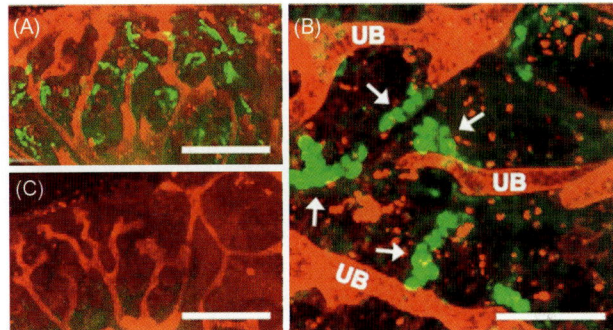

FIGURE 19.5 Recombined renal tissue constructs are capable of functional organic anion transport. Recombined tissue was stained with dolichos biflorus (red), which marks the UB-derived collect duct system, and exposed to organic anion 6CF (green) in the absence or presence of organic anion transport inhibitor—probenecid. (A and B) Accumulation of 6CF can be seen in MM-derived tubules. (C) In the presence of probenecid, 6CF uptake was inhibited, confirming that accumulation is dependent on specific transport. (Scale bars: A and C, 500 μm; B, 200 μm) *Modified from Rosines et al. [20]*

examined and compared to existing time-series data of kidney development. With this approach, it was determined that recombined tissues roughly resembled the transcriptional profile of E18 rat kidneys, at which time proximal tubules have begun to form. To examine functional transport properties of the engineered tissue, the probenecid-sensitive capability to take up the fluorescent organic anion 6CF was examined. Indeed, the engineered renal tissues were able to specifically uptake organic anions, confirming that transcriptional changes reflect functional characteristics (Figure 19.5). Furthermore, this showed that the engineered renal tissue had established functional properties without further manipulation. The resulting tissue was implanted into rats under the renal capsule (thin, fibrous membrane surrounding the kidney) and examined after 2 weeks. The tubular structures in this tissue resembled nephrons and appeared to recruit vasculature. This study thus demonstrated that, following a developmental strategy and using modifications of the *in vitro/ex vivo* culture systems for the isolated WD, isolated UB and UB−MM recombinations, a complex, implantable renal tissue can be generated from relatively simple starting materials using a potentially scalable modular strategy.

While the cultured nephrogenic structures appeared to recruit vasculature when implanted *in vivo*, their functional capacity to concentrate urine has not yet been determined. However, a previous study suggests that they indeed might have had some filtration and transport capabilities, or at least that achieving this is quite feasible [21]. In this study, bovine fibroblasts were fused with enucleated donor oocytes, followed by activation and gestation, resulting in embryos that have the genetic makeup of the animal from which the fibroblast was derived. MM cells were then derived from these embryos, and seeded on a scaffold consisting of biodegradable polymer tubes, which were connected to nondegradable catheters leading to a sealed reservoir. The final construct was then implanted into original fibroblast donors. Inspection after 12 weeks of culture in the subcutaneous space revealed that the polymer tubes had been replaced by vascularized nephron-like structures, as judged by several markers and histology. The reservoir contained yellow "urine-like" fluid. Analysis of this fluid provided some evidence that these structures were able to secrete urea nitrogen and creatinine, although the extent of glomerular filtration and collecting duct concentrating ability was unclear.

19.6 CELL-BASED APPROACHES

Constructing organs from single cells has long been a goal in the tissue engineering field. This may be more difficult to achieve with a kidney than with less complex tissues; however, advances have been recently made that demonstrate that this might be possible. Initial studies of cultured embryonic kidney-derived and mature renal cells explored their capacity to provide and respond to developmental cues. We have already mentioned that UB and MM-derived cell lines secrete factors that induce the morphogenesis of the MM and UB, respectively. Interestingly, UB cells were also found to respond to soluble factors. For example, specific factors or conditioned media were able to induce tubulogenesis and branching morphogenesis in culture [22]. Considering that the WD is ultimately a tubular cellular structure, it might be possible to form this structure from cultured cells, as long as appropriate cells are induced to undergo tubulogenesis. This structure can then be used to generate one or more UBs. Alternatively, it may be possible to induce cells primed to form renal collecting ducts to undergo branching morphogenesis, thereby circumventing the requirement for an UB.

In 2010, a study addressed the potential of cells to recreate the functions of the UB and mesenchyme [19]. When the UB or MM are enzymatically disassociated and cultured, the resulting cells do not retain their phenotype for more than a few passages. Ideally, the cell sources for renal tissue engineering should be stable and indefinitely replenishable. To simulate this effect, cell lines of UB and MM origin from genetically modified mice were employed; these mice expressed the SV40 large T-antigen oncogene, allowing the derivation of immortalized cells. Using a hanging drop technique, aggregates of UB-derived cells were able to induce isolated MM while undergoing branching morphogenesis. Interestingly, mature collecting duct cells derived from the same mouse line behaved in a similar fashion, suggesting that differentiated cells from mature organs should be considered as a cell source for future approaches. When immortalized MM cells were aggregated by the same hanging drop method, they were able to induce isolated UBs to undergo branching morphogenesis. This recapitulates the finding that conditioned media from these immortalized mesenchymal cells is able to induce branching of the UB. However, these mesenchymal aggregates did not undergo nephrogenesis. More recently, it has been demonstrated that 3D cultured structures from cultured renal-derived cells can be engrafted onto the chick chorioallantoic membrane to enhance differentiation [23]. Overall, these types of studies demonstrate the feasibility of using cell populations to generate organized renal tissue.

While immortalized cell lines were able to partially recapitulate kidney development, the cell lines of mesenchymal origin were not able to behave as the intact MM. The approaches used in previously described studies to immortalize the UB and MM cells can cause significant changes to cellular identity, potentially explaining why MM-derived cells cannot recapitulate the functions of a native MM. To test this notion, one group partially simulated the initial steps of kidney organogenesis using primary cells derived from disassociated UBs and MMs. In this case, primary cells from embryonic tissues served as surrogates for endogenous cellular identities, which presumably can be generated with the advance of cell reprogramming. When a mixture of UB and MM cells were aggregated by centrifugation, and cultured on a Transwell filter, they were able to form structures that resemble those found in whole embryonic kidney cultures [24]. A key component to this strategy was the inclusion of a Rho kinase inhibitor, whose role was initially established *in vitro*[25], which appeared to support the survivability of the primary cells until they formed cohesive structures and stabilized. However, the complex architecture and nephron segmentation, including glomeruli, were lacking. Nevertheless, the implantation of such engineered tissue, together with treatment with the angiogenic factor vascular endothelial growth factor (VEGF), lead to vascularization and the formation of distinguishable glomeruli, as well as the development of several functional properties of mature nephrons [26]. This is another example of how cells can be used to "engineer" 3D renal-like tissue and supports the view that engineered kidney-like tissue is capable of further development and maturation *in vivo*.

Whereas a heterogeneous mixture of UB and MM cells failed to pattern correctly *in vitro*, the UB and MM can instruct each other to undergo spatiotemporally organized morphogenesis when recombined in culture. With this in consideration, in a follow-up article to that discussed above [24], the strategy was to recreate these structures separately from dissociated embryonic kidney cells and refine culture conditions [27]. In earlier attempts, when a mix of UB and MM cells was aggregated and cultured, distinct UB-like structures were observable. This strategy was repeated, but this time the UB-like cyst was separated out from the surrounding cells, and served as the starting point for tissue culture. Pure UB cells can also be aggregated and used to induce the MM [19], but the presence of surrounding mesenchymal cells possibly aids in the patterning of the UB cells, resulting in more organized branching morphogenesis [28]. Primary MM cells were then aggregated and placed on top of the isolated UB-like cyst, and cultured using a novel low-volume culture technique [27]. The resulting tissues looked similar to cultures of whole embryonic kidneys. In summary, these recent findings from multiple studies demonstrate the feasibility of renal tissue engineering starting from cells.

19.7 CONCLUSION

We have focused on some recent developmental approaches to tissue engineering of the kidney. A number of other types of approaches have also been discussed in the literature, including 3D bioprinting, bioartificial devices, stem cell-based, and other methods. All of these methods hold a great deal of promise, and it seems likely that future work will combine developmental approaches with these other methods in the hope of creating tissue that can be used for transplantation in humans or, at the very least, for functional and toxicity studies in 3D renal-like tissue.

REFERENCES

[1] Shah MM, Sampogna RV, Sakurai H, Bush KT, Nigam SK. Branching morphogenesis and kidney disease. Development 2004;131(7):1449—62.

[2] Nigam SK, Wu W, Bush KT. Organogenesis forum lecture: *in vitro* kidney development, tissue engineering and systems biology. Organogenesis 2008;4(3):137—43.

[3] Brunskill EW, Aronow BJ, Georgas K, Rumballe B, Valerius MT, Aronow J, et al. Atlas of gene expression in the developing kidney at microanatomic resolution. Dev Cell 2008;15(5):781—91.

[4] Tsigelny IF, Kouznetsova VL, Sweeney DE, Wu W, Bush KT, Nigam SK. Analysis of metagene portraits reveals distinct transitions during kidney organogenesis. Sci Signal 2008;1(49):RA16.

[5] Grobstein C. Morphogenetic interaction between embryonic mouse tissues separated by a membrane filter. Nature 1953;172 (4384):869—70.

[6] Grobstein C. Inductive epitheliomesenchymal interaction in cultured organ rudiments of the mouse. Science 1953;118 (3054):52—5.

[7] Saxén L. Organogenesis of the kidney. Developmental and cell biology series, viii. Cambridge Cambridgeshire; New York: Cambridge University Press; 1987. p. 173

[8]   Barasch J, Pressler L, Connor J, Malik A. A ureteric bud cell line induces nephrogenesis in two steps by two distinct signals. Am J Physiol 1996;271(1 Pt 2):F50−61.

[9]   Barasch J, Yang J, Ware CB, Taga T, Yoshida K, Erdjument-Bromage H, et al. Mesenchymal to epithelial conversion in rat metanephros is induced by LIF. Cell 1999;99(4):377−86.

[10]   Plisov SY, Yoshino K, Dove LF, Higinbotham KG, Rubin JS, Perantoni AO. TGF beta 2, LIF and FGF2 cooperate to induce nephrogenesis. Development 2001;128(7):1045−57.

[11]   Qiao JH, Sakurai, Nigam SK. Branching morphogenesis independent of mesenchymal−epithelial contact in the developing kidney. Proc Natl Acad Sci USA 1999;96(13):7330−5.

[12]   Steer DL, Bush KT, Meyer TN, Schwesinger C, Nigam SK. A strategy for in vitro propagation of rat nephrons. Kidney Int 2002;62(6):1958−65.

[13]   Sweet DH, Eraly SA, Vaughn DA, Bush KT, Nigam SK. Organic anion and cation transporter expression and function during embryonic kidney development and in organ culture models. Kidney Int 2006;69(5):837−45.

[14]   Hammerman MR. Renal organogenesis from transplanted metanephric primordia. J Am Soc Nephrol 2004;15:1126−32.

[15]   Costantini F, Shakya R. GDNF/Ret signaling and the development of the kidney. Bioessays 2006;28(2):117−27.

[16]   Maeshima A, Sakurai H, Choi Y, Kitamura S, Vaughn DA, Tee JB, et al. Glial cell-derived neurotrophic factor independent ureteric bud outgrowth from the Wolffian duct. J Am Soc Nephrol 2007;18(12):3147−55.

[17]   Maeshima A, Vaughn DA, Choi Y, Nigam SK. Activin A is an endogenous inhibitor of ureteric bud outgrowth from the Wolffian duct. Dev Biol 2006;295(2):473−85.

[18]   Rosines E, Schmidt HJ, Nigam SK. The effect of hyaluronic acid size and concentration on branching morphogenesis and tubule differentiation in developing kidney culture systems: potential applications to engineering of renal tissues. Biomaterials 2007;28 (32):4806−17.

[19]   Rosines E, Johkura K, Zhang X, Schmidt HJ, Decambre M, Bush KT, et al. Constructing kidney-like tissues from cells based on programs for organ development: toward a method of in vitro tissue engineering of the kidney. Tissue Eng Part A 2010;16(8):2441−55.

[20]   Rosines E, Sampogna RV, Johkura K, Vaughn DA, Choi Y, Sakurai H, et al. Staged in vitro reconstitution and implantation of engineered rat kidney tissue. Proc Natl Acad Sci USA 2007;104 (52):20938−43.

[21]   Lanza RP, Chung HY, Yoo JJ, Wettstein PJ, Blackwell C, Borson N, et al. Generation of histocompatible tissues using nuclear transplantation. Nat Biotechnol 2002;20(7):689−96.

[22]   Sakurai H, Barros EJ, Tsukamoto T, Barasch J, Nigam SK. An in vitro tubulogenesis system using cell lines derived from the embryonic kidney shows dependence on multiple soluble growth factors. Proc Natl Acad Sci USA 1997;94(12):6279−84.

[23]   Buzhor E, Harari-Steinberg O, Omer D, Metsuyanim S, Jacob-Hirsch J, Noiman T, et al. Kidney spheroids recapitulate tubular organoids leading to enhanced tubulogenic potency of human kidney-derived cells. Tissue Eng Part A 2011;17(17−18):2305−19.

[24]   Unbekandt M, Davies JA. Dissociation of embryonic kidneys followed by reaggregation allows the formation of renal tissues. Kidney Int 2010;77(5):407−16.

[25]   Meyer TN, Schwesinger C, Sampogna RV, Vaughn DA, Stuart RO, Steer DL, et al. Rho kinase acts at separate steps in ureteric bud and metanephric mesenchyme morphogenesis during kidney development. Differentiation 2006;74(9−10):638−47.

[26]   Xinaris C, Benedetti V, Rizzo P, Abbate M, Corna D, Azzollini N, et al. In vivo maturation of functional renal organoids formed from embryonic cell suspensions. J Am Soc Nephrol 2012;23(11):1857−68.

[27]   Chang CH, Davies JA. An improved method of renal tissue engineering, by combining renal dissociation and reaggregation with a low-volume culture technique, results in development of engineered kidneys complete with loops of henle. Nephron Exp Nephrol 2012;121(3−4):e79−85.

[28]   Shah MM, Tee JB, Meyer T, Meyer-Schwesinger C, Choi Y, Sweeney DE, et al. The instructive role of metanephric mesenchyme in ureteric bud patterning, sculpting, and maturation and its potential ability to buffer ureteric bud branching defects. Am J Physiol Renal Physiol 2009;297(5):F1330−1341.

Part III

Liver

Chapter 20

Current Status of Liver Transplantation

Quirino Lai, Rafael S. Pinheiro, Juan M. Rico Juri, Edward Castro Santa, and Jan P. Lerut

Starzl Abdominal Transplant Unit, Department of Abdominal and Transplantation Surgery, University Hospitals St. Luc, Université Catholique Louvain (UCL), Brussels, Belgium

Chapter Outline

20.1 HISTORICAL NOTE

At the beginning, liver transplantation (LT) was considered an impossible medical and surgical task. Transforming this idea into a reality represented one of the most extraordinary adventures the medical sciences faced during the twentieth century. When LT was introduced in clinical practice, many clinical and fundamental research projects were needed in order to respond to five, different but interconnected, questions related to: (a) the study of the metabolic interactions between the different intraabdominal organs, (b) the refinement of the liver and multivisceral transplantation as well as procurement and preservation techniques in order to make the procedure feasible, (c) the appropriate use of prophylactic and therapeutic immunosuppression (IS) to "control" the immune response to the liver allograft, (d) the problems interfering with early and long-term outcomes, and finally (e) the impact of this procedure on social, ethical, legal, and public issues [1].

The first human LT was carried out by Thomas E. Starzl on March 26, 1963, Denver, CO [2]. The recipient died due to intraoperative bleeding reflecting all shortcomings of procurement, organ preservation as well as peritransplant care. Similarly, all nine cases performed in the first year of clinical activity had poor results; indeed no patient survived more than 23 days [1,3]. Due to these appalling results, a 3.5-year long voluntary moratorium was observed, the LT procedure labeled at that time as "the impossible operation." In 1967, the same team performed the worldwide 10th attempt at human LT. This time the operation was successful. The recipient, a 1.5-year infant presenting a biliary atresia complicated with the development of a huge hepatocellular cancer, died of tumor recurrence 400 days later [4]. The stage for this incredible medico-surgical adventure was set by Starzl's team.

20.2 THE STATUS IN 1983

Following this first success, it took 20 years more to develop a safely applicable surgical procedure. In the beginning of the 80s, the introduction of cyclosporine A was a wonderful "aid" allowing a sudden increase in success rates; the 1-year survival rate indeed raised from <50% during the period 1976–1979, to more than 75% in the period 1980–1984. These encouraging results led to a steep increase in the number of transplant procedures (from 59 to 394, respectively), essentially performed in Starzl's team after his move to the University of Pittsburgh (PA).

In June 1983, the National Institute of Health Consensus Development Conference reviewed the worldwide LT experience comprising 540 patients transplanted during the period 1963–83 in four centers only. One was America, Denver–Pittsburgh, the three other programs were located in Europe, namely Groningen, Hannover, and Cambridge. The Consensus Conference concluded that LT was "a promising alternative to current therapy in the management of late phase of several forms of serious liver diseases" and that it had the potential to become a "clinical service" as opposed to an experimental procedure [5]. However, this potential was recognized only on the condition to restrict the procedure to very well-selected patients according to the forwarded 10 indications and 10 absolute and 5 relative contraindications (Table 20.1).

20.3 THE STATUS 30 YEARS LATER...

Only 6 years after this conference, Starzl (who else?) opened his *New England Journal of Medicine* paper with the following statement: "the conceptual appeal of liver transplantation is so great that the procedure may come to mind as a last resort for virtually every patient with lethal hepatic disease" [6,7]. This visionary sentence perfectly synthesized what happened in the last three decades in this fascinating field of medicine and surgery. Indeed progresses were spectacular and number of procedures grew exponentially, reaching now, exactly 50 years later, the 250,000 mark!

During these three decades of development, all but one, namely, active sepsis outside the hepatobiliary system, of the contraindications to LT forwarded by the Consensus Conference have been eliminated [8,9]. Today, LT is "the" answer to more than 50 different liver pathologies (Table 20.2).

20.4 INDICATIONS FOR LT IN THE TWENTY-FIRST CENTURY

20.4.1 Acute Liver Failure

20.4.1.1 Toxic Acute Liver Failure

Many drugs induce severe irreversible liver injury. Besides non steroidal inflammatory drugs (NSAID) and other antibiotics, acetaminophen (or paracetamol) hepatotoxicity is without any doubt the most common cause of acute liver failure (ALF), especially in the United Kingdom and the United States. The great majority of acetaminophen-induced ALFs occur in the setting of deliberate ingestion with suicidal intent, often in the context of psychiatric illness and substance or alcohol abuse [10]. Improved intensive care and medical management, mainly gathered by experience liver (transplant) units, markedly improved the outcome of these patients. Acetaminophen-induced ALF has a distinct clinical course characterized by rapid clinical deterioration, which contrasts with its relatively good prognosis (Figure 20.1A). The most widely used criteria to justify indication for LT in such patients are the King's College criteria originally described by O'Grady in 1989 and afterwards refined by the same group in 1993 [11]. These criteria rely essentially on the combination of severe encephalopathy and acidosis. LT largely revolutionized the management of ALF allowing to refine medical treatment allowing nowadays to rescue patients

TABLE 20.1 Indications and Contraindications for LT Identified at the 1983 National Institute of Health Consensus Conference

Indications	Contraindications	
	Absolute	Relative
1. Young patient <50 years	1. Age >55 years	1. Age >50 years
2. No viral infection	2. HBsAg—HBeAg positive state	2. HBsAg positive state
3. No alcohol and drug abuse	3. Active alcoholism	3. Intrahepatic or biliary sepsis
4. Ability to accept procedure/understand its nature	4. Inability to accept procedure or understand its nature or costs	4. Advanced alcoholic liver disease in abstinent alcoholic
5. Ability to accept costs	5. Sepsis outside hepatobiliary system	5. Prior abdominal surgery[a]
6. Normal vessel state	6. Portal vein thrombosis	6. Portal hypertension surgery
7. No CP or renal disease	7. Advanced CP or renal disease	
8. No prior abdominal surgery	8. Severe hypoxemia (R to L shunts)	
9. No infection	9. Metastatic HB malignancy	
10. No (advanced) malignancy	10. Primary malignant disease outside hepatobiliary system	

Abbreviations: CP, cardiopulmonary; HB, hepatitis B; R, right; L, left.
[a]*Especially in the right upper quadrant.*

TABLE 20.2 Todays Accepted Indications for LT

1. Acute liver failure
Hepatitis A virus
Hepatitis B virus
Hepatitis C virus
Hepatitis D virus
Hepatitis E virus
Acetaminophen
Other drugs
Postoperative failure
Posttraumatic failure
Wilson disease
Budd–Chiari syndrome
Autoimmune hepatitis
Cryptogenic
Fatty infiltration—acute fatty liver of pregnancy
Reye syndrome
2. Cirrhosis from chronic liver disease
Chronic hepatitis B virus infection
Chronic hepatitis C virus infection
Chronic hepatitis B and D virus infection
Chronic hepatitis E virus infection
Cirrhosis virus-related (other viruses)
Cirrhosis drug-related
Alcoholic liver disease
Autoimmune hepatitis
Cryptogenic liver disease
Nonalcoholic fatty liver disease
3. Cholestatic liver diseases
Primary biliary cirrhosis
Primary sclerosing cholangitis
Caroli disease
Secondary biliary cirrhosis
Alagille syndrome
Byler disease
Choledocal cyst
Congenital biliary fibrosis
Extrahepatic biliary atresia
4. Metabolic liver diseases
Wilson disease
Hereditary hemochromatosis
Alpha-1 antitrypsin deficiency
Glycogen storage disease I and IV
Familial homozygous hypercholesterolemia
Tyrosinemia
Familiar amyloidotic polyneuropathy
Hyperoxaluria Type 1
Protoporphyria
Other types of porphyria
Crigler–Najjar syndrome
Cystic fibrosis
Galactosemia
Hemophilia A and B
5. Benign liver diseases
Hepatic adenoma
Adenomatosis
Giant hemangioma
Focal nodular hyperplasia
Nodular regenerative hyperplasia
6. Malignant liver diseases
Hepatocellular carcinoma on cirrhosis
Hepatocellular carcinoma on noncirrhotic liver
Intrahepatic cholangiocarcinoma
Biliary tract carcinoma (Klatskin)
Epithelioid hemangioendothelioma
Hepatoblastoma
NET metastasis
Colorectal tumor metastasis[a]
7. Vascular liver diseases
Budd–Chiari syndrome
Hereditary hemorrhagic telangiectasia
Veno-occlusive disease
8. Miscellaneous
Polycystic liver disease
Alveolar echinococcosis
Cystic echinococcosis
Hepatic trauma
Schistosomiasis
Severe graft-versus-host disease
Sarcoidosis

[a]Only in pilot study.

without the need for LT. Patients transplanted because of acetaminophen-related ALF nowadays experience a 5-year patient survival of 75%, results that are almost similar to those obtained in patients electively transplanted for chronic liver diseases [10].

20.4.1.2 Other Causes of ALF

Various other conditions may lead to ALF (Table 20.2). A great heterogeneity throughout the reported studies dealing with nonacetaminophen-induced ALF has been observed. ALF caused by an identified hepatotrophic virus accounts for 15–50% of cases in Europe, with an additional 20% of cases related to hepatitis of unknown etiology; in these cases viral etiology is frequently presumed. It should be mentioned that hepatitis C viral infection, rarely responsible for ALF in Western countries, accounts for a higher proportion of cases in Japan and that hepatitis E infection, commonly observed in the Indian subcontinent, may also lead to ALF particularly in pregnant women. Uncommon nonviral-related causes of ALF are represented by the acute presentation of Wilson disease (WD) or poisoning after ingestion of the mushroom amanita phalloides.

Analyzing the various etiologies of ALF, hepatitis A or B viral infections have a relatively good prognosis,

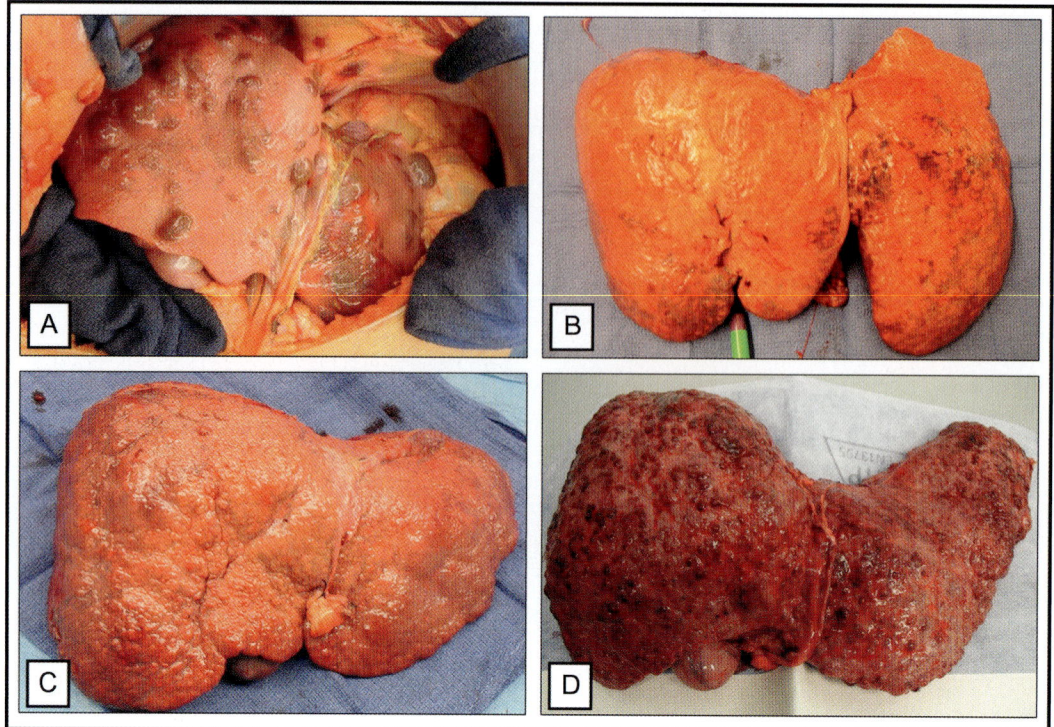

FIGURE 20.1 Hepatectomy specimens: (A) ALF; (B) HBV-related cirrhosis; (C) HCV-related cirrhosis complicated by hepatocellular cancer; and (D) alcohol-related cirrhosis.

whilst seronegative hepatitis infection has a more insidious clinical course and higher mortality rate (<30%) without transplantation. LT allows obtaining cumulative 5-year patient survival rates in these patients superior to 70% [10].

As the King's College Criteria have a low negative predictive value for nonacetaminophen-induced ALF, the Clichy Criteria were developed, with the intent to better identify those patients presenting with viral infection in need for LT. These criteria essentially based on the combination of severe encephalopathy and coagulation disturbances (expressed by the FV level of <20 or 30% related to the age of the patient) were able to predict mortality without LT with a positive predictive value of 82% and a negative predictive value of 98%. However, as several studies reported contrasting data, no definitive conclusions have been obtained on the best selection criteria to be adopted in such patients [12].

The recently developed Drug-Induced Liver Injury Network (DILIN) in the United States aims at collecting prospectively all cases of drug-induced liver injuries and at investigating more appropriately the role of different culprit drugs in the development of ALF [13]. Finally, the gained knowledge in the field of LT was of great help not only to improve the care of ALF but also to develop artificial liver support devices. The potential impact of such devices on

the need for LT and outcome after transplantation and resection surgery is promising. Today these costly devices are mostly used as a "bridge" to LT rather than to a transplant-free survival.

20.4.2 Chronic Parenchymal Liver Diseases

20.4.2.1 Viral Diseases

Hepatitis B Viral Infection

Worldwide, hepatitis B viral (HBV) infection causes not only 600,000 deaths each year but it is the leading cause, mainly in Eastern countries, of cirrhosis and development of liver cancer. In Europe, about 14 million people are chronically infected (Figure 20.1B). An aggressive vaccination policy of the population is the most effective strategy to prevent the disease [14]. The introduction during the last decade of potent antiviral drugs dramatically improved the evolution of the disease without LT as well as the long-term results of LT itself with 5-year patient survival rates superior to 80%. Lifelong prophylactic therapy with specific IV or IM anti-HB immunoglobulins (HBIG), eventually reinforced with nucleos(t)ide analogs, such as lamivudine, adefovir, and tenofovir, in case of active replication at moment of LT, has nowadays reduced the rate of allograft reinfection to <5%. As HBV recurrence after 3-year post-LT is extremely rare, the costly immunoprophylaxis has recently been successfully switched in selected cases toward the oral nucleoside analogs only [15].

Hepatitis D Viral Infection

Hepatitis D viral (HDV) infection concerns a distinct subgroup of simultaneously HBV-infected individuals usually presenting with severe chronic liver disease. HDV is a defective RNA agent needing the presence of HBV for its life cycle. Two major specific patterns of infection can occur, namely HDV−HBV coinfection or HDV superinfection of a chronic HBV carrier status. Prevention and therapeutic strategies for HDV are similar to those applied in HBV recipients [16]. As most of these recipients present with a nonreplicating HBV status, results of LT are outstanding with survival rates exceeding 90%.

Hepatitis C Viral Infection

Worldwide, Hepatitis C viral (HCV) infection causes 350,000 deaths each year. In Europe, approximately 9 million people are chronically infected. In many Mediterranean countries, HCV-related cirrhosis with or without hepatocellular carcinoma (HCC) is now the primary indication for LT [14]. Unfortunately, this virus is not eliminated by LT so graft reinfection is the rule. The natural history of recurrent allograft HCV reinfection hepatitis is speeded up after LT resulting in cirrhosis and allograft failure in 20−30% and 10% of cases 5−10 years after LT (Figure 20.1C). This unfortunate evolution is influenced by several host (e.g., IS, HLA, genotype, viral load, etc.) as well as donor (marginal, steatotic, and old grafts) factors. Adequate immunoprophylaxis does not exist due to the great variability of viral genotypes and antiviral therapy is still insufficient in most liver recipients. Combination of pegylated interferon and ribavirin, still the actual therapy of choice, allows obtaining a sustained viral response in only 20−30% of recipients. The efficacy of this therapy is greatly hampered due to the appearance of major side effects of the antivirals, such as pronounced anemia, development of *de novo* autoimmune hepatitis, and rejection. Delayed retransplantation can be a good last resort option in well-selected patients experiencing allograft failure [17]. Newer antiviral drugs which will be implemented soon in clinical transplant praxis as well as the search for an optimal immunosuppressive scheme will change the outlook of these liver recipients in a very near future.

Human Immunodeficiency Viral infection

Human immunodeficiency virus (HIV) has been considered for a long time as an absolute contraindication for LT. The success of highly active antiretroviral treatments (HAART) allowed to bring these patients, which are frequently present a HCV coinfection or hepatocellular cancer, to LT. Improved knowledge about handling of the combination of antiviral and immunosuppressive drugs has also contributed to improve the results of LT. To date, differences between HIV-positive and -negative patients in relation to indications and survival after LT are fading away on the condition that HIV viral load at the time of transplant is undetectable and $CD4^+$ count is >200/mL. Despite this outcome of HIV−HCV coinfected patients still remains compromised due to the HCV allograft reinfection. More effective antiviral therapies at an early stage after LT are here also underway [18].

20.4.2.2 Alcohol-Related Liver Disease

Alcoholic liver disease represents the first and second most common indication for LT throughout Europe and United States (Figures 20.1D and 20.2A). This indication still remains a controversial subject in terms of public attitude toward the responsibility of the patient for his self-inflicted hepatic disease. The main problem resides in the fact that relapse after transplant cannot be excluded despite adequate pre-LT psychological evaluation and post-LT follow-up. The 6-month abstinence "rule" considered as a "safety belt" in most transplant centers reveals to be an unreliable selection criterion to justify LT. Recidivism after LT remains an issue of concern, the exact incidence of which remains unknown [19]. Probably the best selection parameter is to be assured that the patient with an "alcohol dependent" cirrhosis remains integrated in his familial, professional, as well as social environment. Evidently the general practitioner, usually knowing best the "real" context of his patient, should be consulted in the decision when considering LT as a therapeutic option.

The "social debate" about LT and alcoholic liver disease has recently been fueled by the Lille experience which proposed to perform LT even in case of severe alcoholic hepatitis not responding to medical therapy [20]. Results of medical treatment based on glucocorticoid administration are indeed very poor; most patients die within 2 months and 6-month survival rate only reaches 30% in patients having a Lille score of ≥ 0.45 after 7 days of treatment The Lille group therefore designed a prospective study proposing early LT in alcoholic hepatitis patients selected on the basis of a "beneficial" Lille score. LT allowed to obtain a significantly higher 6-month survival rate than medical treatment (77 vs. 23%, $P < 0.001$); moreover only 11% (of 26) patients resumed alcoholism post-LT [20].

Despite the fact that LT represents the best therapy for both alcoholic cirrhosis and severe alcoholic hepatitis not responding to medical therapies, alcohol abstinence represents a mandatory objective to be obtained,

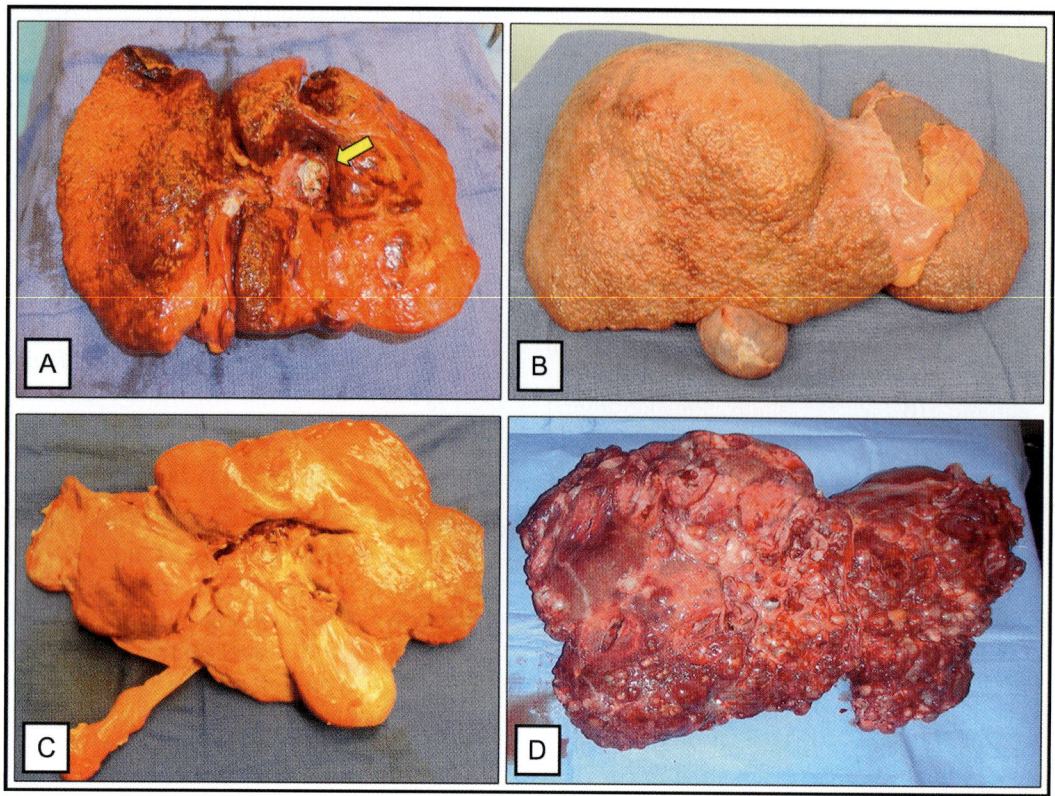

FIGURE 20.2   Hepatectomy specimens: (A) alcohol-related cirrhosis complicated by hepatocellular cancer (yellow arrow: TIPSS placed at the level of the right hepatic vein); (B) NASH-related cirrhosis; (C) PSC; and (D) polycystic liver disease.

interconnecting patient compliance and public reassurance. One should also not forget that in the absence of alcohol use, this is one of the rare diseases that do not recur after LT. This patient group therefore represents a "fertile" research domain in the field of immunosuppressive minimization or clinical operational tolerance (COT) program protocols.

20.4.2.3 Nonalcoholic Steatohepatitis or Cirrhosis

Nonalcoholic fatty liver disease (NAFLD) and nonalcoholic steatohepatitis (NASH) are becoming the most frequent liver diseases of our modern society. By the year 2025, it is estimated that over 25 million Americans will have NASH-related liver disease. NASH is already the third most common indication for LT in the United States...the way to the highest step on the podium is coming closer (Figure 20.2B). The recent study of the American Scientific Registry of Transplant Recipients (SRTR) showed that the percentage of patients undergoing LT for NASH increased from 1.2% in 2001 to already 9.7% in 2009 [21]. NASH moreover affects all phases of LT, comprising (a) impaired quality of liver allografts, as steatosis is responsible for an increased risk of allograft

nonfunction or dysfunction; (b) prolonged waiting time due to the frequent need to treat several comorbidities linked to the metabolic syndrome and insulin resistance, and finally (c) compromised long-term posttransplant course due to comorbidity and allograft recurrent disease [22]. A study of 88 recipients showed that not only recurrent disease is common (39%) within the first 5-year post-LT but also follow-up is rendered complicated due to all different expressions of the metabolic syndrome. Recurrence as such was not associated with higher total mortality, cardiovascular mortality, and morbidity is of major concern [23].

20.4.2.4 Autoimmune Hepatitis or Cirrhosis

In 1950, Waldenström first described a chronic form of hepatitis in young women, called "lupoid hepatitis," termed as such because of its association with other autoimmune disease manifestations. In 1965, this disease was termed "autoimmune hepatitis" and in the 70s and 80s, several autoantibodies directed against proteins of the endoplasmatic reticulum expressed in liver and kidney and against soluble liver antigens were identified in these patients. The immunoscrological and genetic heterogeneity of this pathology has been well determined explaining the great variability of the disease expression in relation to race, geographical distribution, and genetic predisposition [24]. A recent ELTR–ELITA study about this disease showed a 5-year patient survival rate of 73% after LT. The increased risk of dying of infectious complications in the early postoperative period, especially in recipients aged over 50 years, significantly affected patient survival [25].

20.4.3 Cholestatic Liver Diseases

20.4.3.1 Primary Biliary Cirrhosis

LT represents the gold standard in the treatment of patients with end-stage liver disease or in patients presenting intractable symptoms related to primary biliary cirrhosis (PBC). Indications for LT are impaired quality of life or anticipated death in <1 year. Despite a great number of prognostic models, serum bilirubin level of 4 mg/dL provides the simplest guide for the timing of LT [26]. PBC has a more favorable outcome after LT, compared to viral hepatitis and alcohol-associated liver disease. As these recipients seem more prone to develop chronic rejection, they are poor candidates to withdraw IS. Following LT, antimitochondrial antibodies usually persist and histological features of recurrent allograft PBC may be seen in up to 50% of recipients after 10 years [27]. Recurrence, which may occur in the presence of normal liver tests, is often discovered on protocol biopsies. In the medium term, disease recurrence rarely causes clinical problems. During follow-up, osseous complications (bone fractures) and flaring up of several other disease manifestations, such as polyarthritis, sklerodermia, and thyroiditis, may compromise the quality of life and thus the success of LT.

20.4.3.2 Primary Sclerosing Cholangitis (PSC)

Primary sclerosing cholangitis (PSC) is another well-known indication for LT (Figure 20.2C). Patient selection and timing for LT are not always easy. Up to 70% of patients present other major disease manifestations like inflammatory bowel disease (mostly ulcerative colitis or less frequently Crohn's disease). Up to 20% of patients develop cholangiocellular cancer (CCC). Good predictors of patient survival are difficult to identify although serum bilirubin, splenomegaly, and duration of disease have been proposed as predictors of outcome [28]. Pre-LT screening with MRI, Pet-scan, and brush cytology of the biliary tract is indicated to detect early biliary cancer.

Early patient and graft survival rates following LT are excellent with 1-year patient survival exceeding 90%. The late outcome of these patients is, however, frequently compromised as expressed by 5- and 10-year survival rates of around 50% only. Recurrent PSC affects 20–40% of all recipients and many patients will need further surgical treatment of their GI disease [29].

20.4.3.3 Cystic Biliary Tract Disease or Caroli Disease

Caroli disease (CD) is a rare inherited disorder which may cause severe, life-threatening, cholangitis, or even lead to biliary cancer. The disease is characterized by the development of gross segmental, mostly bilobar, grape fruit-like dilatations of the intrahepatic bile ducts. CD can be associated to congenital hepatic fibrosis which is a different entity, or, less frequently, to cystic renal disease; this constellation is termed as Caroli's syndrome. LT should be proposed in patients presenting severe recurring cholangitis or suspicion of CCC (7–14%). In patients with concomitant end-stage renal disease, combined liver–kidney transplantation is the best therapeutic option. The ELTR–ELITA study performed on 110 patients showed a 5-year patient survival of 86% [30].

20.4.4 Metabolic Liver Diseases

LT is an accepted treatment for various hepatic-based inborn errors of metabolism. LT can be considered in this context as a means of gene therapy. In some cases, the liver is directly involved by the pathology leading to a cirrhogenic evolution. Conversely, in metabolic diseases in which the main symptoms are extrahepatic the liver appears completely normal [31].

20.4.4.1 Cirrhogenic Metabolic Liver Diseases

Wilson Disease

WD is related to an abnormal copper metabolism. This pathology has two different clinical patterns of presentation; an ALF or a chronic hepatic disease ultimately leading to cirrhosis. LT represents the treatment of choice for both acute and end-stage chronic diseases unresponsive to or not timely treated with copper chelating agents. Indication for LT is also justified even in patients with progressive and severe neurological involvement. LT corrects copper metabolism and complications. Excellent 1- and 5-year survivals can be obtained [32].

Hereditary hemochromatosis

Hereditary hemochromatosis (HH), a disorder of iron metabolism, is a rather uncommon indication for LT. These patients present a higher risk to develop cardiac complications, due to iron deposition in the myocardium as well as a propensity to develop (aggressive) hepatocellular cancer. Transplanted patients have a higher morbidity and mortality related to a higher incidence of cardiac, diabetic, and infectious complications as well as recurrent cancer [33].

Alpha-1 Antitrypsin Deficiency

This disease is characterized by a pathologic reduction of the serum concentration of alpha-1 antitrypsin, the most important antiprotease in man. Typically, alpha-1 antitrypsin deficiency (A1ATD) is related to lung emphysema, especially in the lower lobes. Homozygote patients often develop neonatal cholestatic hepatitis; in contrast, only few adult patients develop chronic liver disease or cirrhosis. Besides experimental therapeutic approaches, liver disease can only be treated by LT. Excellent 5-year patient survivals of 75% have been reported [34].

Cystic Fibrosis

Secondary cirrhosis, affecting 4—10% of cystic fibrosis (CF) individuals, represents the third leading cause of death in these patients. LT is an accepted therapy for severe liver disease. Results after LT are excellent. Controversial results have been obtained in relation to the beneficial impact of LT on pulmonary infections and function. Some reports showed a positive impact of LT on the lung function explained by the improved nutritional status after successful LT; conversely a recent larger series showed a progressive decline of the pulmonary function in comparison to non-transplanted patients [35]. In exceptional cases, combined liver—lung transplantation may be necessary.

20.4.4.2 Noncirrhogenic Metabolic Liver Diseases

LT is curative and lifesaving for a wide range of inherited noncirrhogenic liver-based metabolic diseases, such as type I hyperoxalosis, glycogen storage diseases, and familial amyloidotic polyneuropathy (FAP) [36].

Familial amyloidotic polyneuropathy

FAP is a fatal disease, belonging to a group of systemic disorders caused by a transthyretin variant. Up to now more than 80 genetic variants have been identified; the most frequent one is the Val30Met variant. LT eliminates the source of the variant molecule and therefore represents the only known curative treatment of this disease as of today. A fascinating consequence of this treatment is the possibility to implant the removed "normal" FAP liver into another non-FAP recipient (Figure 20.3A and B). The domino or sequential LT procedure was first described by Furtado in Coimbra in 1993. The FAP World Transplant and Domino Liver Transplantation Registries, founded respectively in 1995 and 1999 at Karolinska Institute in Stockholm, document very well

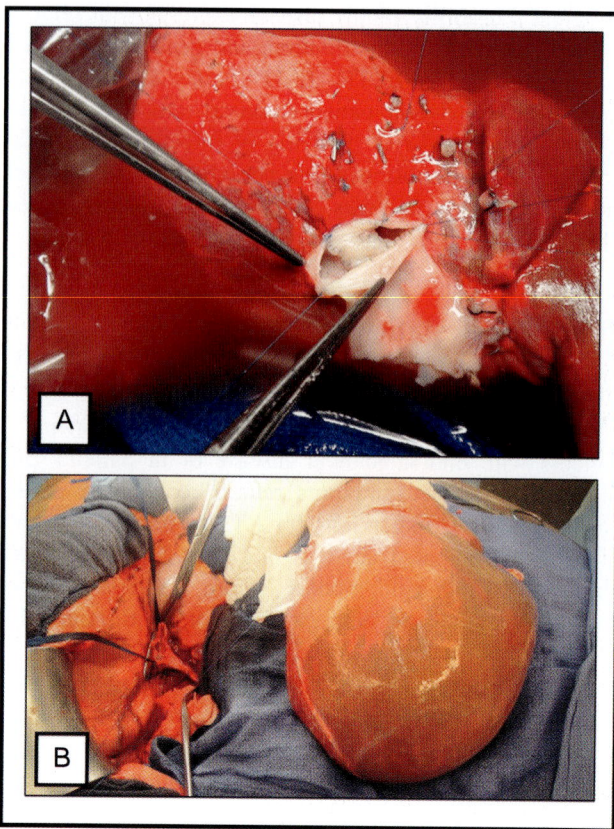

FIGURE 20.3 (A) FAP liver. Hepatic veins have been reconstructed with an iliac venous patch for performing domino transplantation. (B) FAP liver during the phase of implantation of domino transplantation.

the interest of the sequential LT procedure. Up to December 2010, 977 domino LT were performed in 58 hospitals in 21 countries. Patient survival after LT is similar to the results obtained in other chronic liver disorders [37]. Although "transfer" of polyneuropathy by the domino graft has been documented in very rare cases, FAP patients represent a valuable source to enlarge the liver allograft pool. Maple syrup disease represents the only other metabolic disease which can be considered for safe sequential LT. Indeed experiences with LT using allografts originating from oxalotic patients were very disappointing as the recipients rapidly developed renal insufficiency due to oxalic acid overload.

20.4.5 Benign Liver Diseases

Several benign liver diseases can be cured by LT. The main problem in the evaluation of the real impact of LT in the treatment of these pathologies is many times the small number of reported cases [38]. Prospective, systematic, collection of long-term results in such patients would be of great interest for the medical community [39].

20.4.5.1 Polyadenomatosis and Adenoma

Malignant transformation of liver cell adenoma is present in about 5% of patients. LT has been exceptionally reported in the treatment of badly located or ruptured giant (>5 cm) adenomas, elective surgical resection remains however the treatment of choice, even if complicated by bleeding or malignant transformation [38].

Liver cell adenomatosis is a different clinical entity occurring in 10–24% of all patients with liver cell adenomas. Adenomatosis, characterized by the presence of multiple adenomas (from 3 or 10), is related to modified vascularization caused by a modified hepatic parenchyma. The exact pathogenetic mechanism of this disease still remains unknown (Figure 20.4A). LT has been reported in 6% of published series of liver adenomatosis and about 5% of transplants performed for benign liver tumors in the United States and Europe [38]. LT is indicated in case of progressive, symptomatic growth of the remaining adenomas after previous hepatectomy or when malignancy is suspected [40]. Degeneration is exceptional except in the context of a "metabolic background" such as glycogenosis.

20.4.5.2 Giant Hemangioma

Cavernous hepatic hemangioma is the most frequent benign liver tumor with prevalence in autopsy and imaging studies of up to 7%. Almost all hepatic hemangiomas can be managed conservatively; when surgical treatment is indicated, tumor enucleation is the technique of choice. LT has been reported in exceptional cases of complicated huge hemangiomas or hemangiomatosis. According to a recent review, <15 cases have been reported in the English literature and most of them were associated with the presence of Kasabach–Merritt syndrome characterized by coagulopathy, thrombocythemia, hypofibrinogenemia, and fibrinolysis [41]. In the United States, only three patients received LT for hepatic hemangiomas in the period 2000–2009 [38]. Infantile hepatic hemangioma is a clinically different entity which must be distinguished from hemangioma in adults. 10–20% of patients are refractory to medical therapy; when they develop life-threatening complications, LT is the only suitable therapeutic option.

20.4.6 Malignant Liver Diseases

20.4.6.1 Hepatocellular Carcinoma

HCC in Cirrhosis

HCC represents the sixth most common malignant tumor worldwide and the third leading cause of cancer-related death. LT is the optimal strategy as it simultaneously removes the tumor and the underlying liver disease. During the 80s, the first wave of enthusiasm led to an attitude to see LT as a "last chance" in desperate, nonresectable tumors. As could be expected, results were very poor due to rapid tumor recurrence. In 1996, the introduction of Milan Criteria (MC) (one lesion smaller than 5 cm or up to three lesions smaller than 3 cm, without extrahepatic manifestations or vascular invasion) and the progressive improvement of locoregional therapies (LRT) represented real breakthroughs in transplant oncology (Figure 20.5A and B) [42,43]. Nowadays, wide implementation of MC and pre-LT LRT allows obtaining excellent disease-free survival 5-year rates of 85%. Worldwide experiences have confirmed the solidity of the

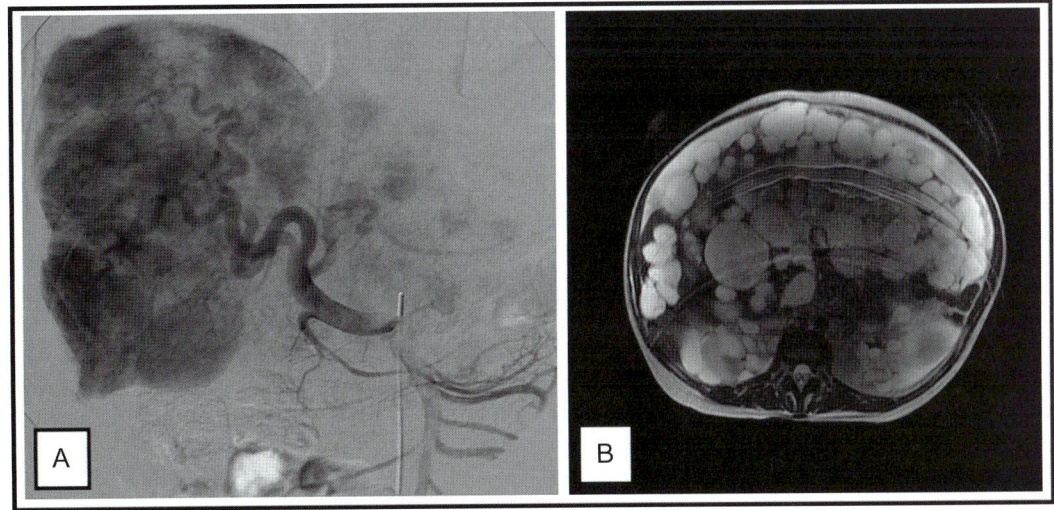

FIGURE 20.4   (A) Arteriography showing a polyadenomatosis in a patient with a Rendu–Osler–Weber disease and (B) MRI showing a polycystic kidney and liver disease.

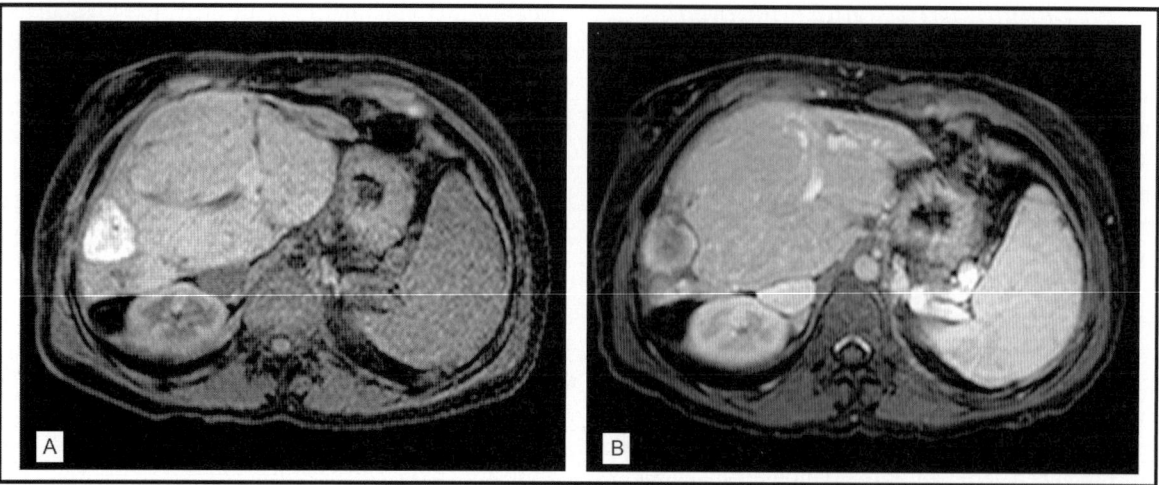

FIGURE 20.5 Patient with Alagille syndrome complicated with a huge hepatocellular cancer: (A) MRI showing a large HCC involving segment 6 (November 2011) and (B) MRI showing the same lesion after downstaging following LRT (May 2012): the tumor is almost necrotic.

MC. Several groups now extended these (too) restrictive MC as they unjustifiably denied access to LT for many patients. A prudent increase in the selection criteria has been proposed by different Western and Eastern centers, the most important ones being San Francisco, Kyoto, Seoul, and again Milan with the up-to-seven Metroticket concept. The introduction of immunosuppressants harboring antitumor properties like mTOR inhibitors will undeniably further foster the field of transplant oncology [44]. The respective roles of liver resection and transplantation are still under debate. Both surgical approaches should, however, be seen as complementary rather than competitive therapies [45]. Partial hepatectomy should be adopted in case of resectable tumors in livers with preserved function, whilst LT should be used in cases of unresectable HCC or in the presence of HCC in an advanced liver disease. Due to the high recurrence rate after hepatectomy, reported in up to 60% of patients after 3 years, LT can also play a role as salvage therapy [46,47]. Although tumor recurrence can reduce the applicability of salvage LT, an evident benefit derived from the "resection first strategy" is the opportunity to obtain a precise pathological examination of the resected specimen, allowing to identify risk factors for recurrence, such as microvascular invasion and presence of satellite nodules [48]. The recent Zurich Consensus Conference reviewed the current practice of LT in HCC patients in order to streamline internationally accepted guidelines with the aim of rational use of the scarce allograft resource in HCC patients [49]. Despite extensive and expert review at this meeting, many areas, however, still remain controversial, such as impact and type of IS in relation to HCC and LT or the role of neoadjuvant and adjuvant therapies. There is now more and more evidence that appropriately applied LRT may increase the chance of cure especially in case of initially advanced (this means outside MC) tumors.

HCC in Noncirrhotic Liver

Only a small number of HCC (5–15%) develop in normal, noncirrhotic, or nonfibrotic livers (NC-HCC). Typically, NC-HCC occur in young and healthy female patients in their 30s. The diagnosis is frequently made at an advanced stage in the absence of a clear etiological factor. Only surgical approaches can be considered curative for this type of cancer. It is commonly thought that partial liver resection represents the therapeutic gold standard in these patients. Review of the literature, however, shows unexpectedly high recurrence rates (up to 70%) after 5 years. LT seems therefore to become an important player in two different therapeutic scenarios: first in the initially nonresectable tumor and second in case of tumor recurrence after previous R0 resection (salvage LT) [45]. A recent ELTR–ELITA analysis, dealing with 62 patients (only) with "nonresectable" NC-HCC presented excellent 60% 5-year long-term disease-free survival rates. Forty-three patients who underwent salvage LT for intrahepatic tumor recurrence after partial liver resection for NC-HCC had 5-year overall and disease-free survival rates of 58% and 48% [50]. Macrovascular invasion, lymph node invasion, and an interval of <12 months between the previous partial resection and recurrence were the significant risk factors for poor outcome, while tumor diameter and MC were not. Larger studies are needed to better explore the value of LT in these patients.

20.4.6.2 Cholangiocellular Carcinoma

CCC is a primary neoplasm arising from malignant transformation of the biliary epithelium. Chronic biliary tree inflammation such as observed in PSC is a risk factor for its development. CCC is a very aggressive tumor which has been for a long time considered an absolute contraindication to LT [51]. The Nebraska and Mayo Clinic transplant teams pioneered a strategy of neoadjuvant radiochemotherapy followed by LT for patients with unresectable hilar CCC. The Mayo Clinic protocol is based on a strict, although cumbersome multidisciplinary therapeutic project beginning with external beam radiation with 5-FU, followed by brachytherapy via an endoscopic approach. After negative surgical laparoscopic exploration, capecitabine is started until (living donor) LT. This protocol is only indicated in highly selected patients presenting with a tumor <3 cm in absence of extrahepatic metastasis including negative nodes. Despite the complexity of the therapeutic algorithm, the high dropout and the very restrictive patient selection, this protocol has generated excellent outcomes with 5-year recurrence-free survival reaching 68%. In the most recent update concerning 199 patients, elevated CA 19-9 (>120 UI/mL), portal vein encasement, and residual tumor on explant were highly significant predictors of CCC recurrence [52].

20.4.6.3 Vascular Tumors

The place of LT in the treatment of vascular tumors has for a long time been underestimated, due to scarce and often contradictory reported data. The recent ELTR−ELITA study showed that hepatic epithelioid hemangioendothelioma (EHE) can currently be considered a very good indication for LT with excellent patient and disease-free 5-year survival rates of 83% and 82%. Good results can even be obtained in case of (limited) extrahepatic disease, estimated to be present in around 20% of patients [53]. Conversely infantile EHE is a more difficult indication for LT especially due to the often difficult differential diagnosis with hemangiosarcoma. Hemangiosarcoma is an absolute contraindication for LT as universal tumor recurrence is observed within 6 months and no single recipient survival has been reported after 2 years [53].

20.4.6.4 Secondary Liver Metastases

Neuroendocrine Metastases

LT for nonresectable metastatic neuroendocrine metastases (NET) is another controversial area in the field of LT. Refining selection criteria is key in this field of transplant oncology. The French multicenter study comprising 85 patients reported a 5-year survival of 47%. In these series primary location of the tumor was unknown in 16% of cases. Independent factors of poor prognosis were concomitant upper abdominal exenteration and primary tumor in duodenum or pancreas with accompanying tumor hepatomegaly. Recipients presenting with these unfavorable prognostic factors had significantly worse results (5-year survival rates of 12% vs. 68%) [54]. The Milan group showed that LT can benefit well-selected patients with nonresectable NET metastases. Low proliferation indexes (Ki < 5%), delay between R0 resection of the primary tumor and LT of more than 6 months, tumor location outside the portal venous draining system, adapted IS and neo- and adjuvant surgical are all key factors in the multidisciplinary treatment of these patients [55]. Adherence to these criteria allowed obtaining 85% disease-free survival after LT. Without any doubt, LT will take a greater place in the treatment of these patients in the near future as LT is indeed the only therapy that allows obtaining a long-term disease-free survival state.

Colorectal Cancer Liver Metastases

Colorectal cancer (CRC) liver metastases were rapidly identified as absolute contraindications for LT due to the very poor 5-year patient survival of 18%. The recent retrospective ELTR−ELITA reevaluation of the originally transplanted patients underlined that near half of these "seminal patients" died due to causes unrelated to tumor recurrence. Nowadays, better imaging techniques and improved surgical and medical multimodality treatment will lead to a change in relation to the impact of LT in these patients. The preliminary data from the Norwegian SECA (secondary cancer) study looking at the value of LT in patients, not responding anymore to the most updated chemotherapeutic treatments, go along with this reasoning. Although two-thirds of their patients presented disease recurrence, a survival rate of 94% after a median follow-up of 25 months was obtained. The obtained results allowed already to have a look at selection criteria and to adapt (neo)-adjuvant therapies and immunosuppressive handling in the context of LT...promising results are in the air [56].

20.4.7 Vascular Liver Diseases

20.4.7.1 Budd−Chiari Syndrome

Budd−Chiari syndrome (BCS) is a rare pathology, described 100 years ago, caused by the occlusion of at least two of the three major hepatic veins. In half of the cases the etiology of BCS is unknown; in the remaining ones hematologic diseases, including myeloproliferative disorders and thrombophilic status (due to, e.g., protein C or S deficiency, factor V Leiden mutation and antiphospholipid syndrome), are present. LT is indicated for the treatment of both severe acute or chronic forms of BCS with liver failure. Small series of LT have been reported for the treatment of BCS [57,58]. Despite higher rates of vascular complications having been reported after LT, comprising portal thrombosis and recurrence of BCS, survival rates are similar compared to other indication groups. The large ELTR—ELITA study analyzing 248 LT recipients reported an overall 5-year patient survival of 71%; four patients died due to BCS recurrence. Anticoagulation is therefore an obligatory compound of the LT treatment. The pre-LT predictors of mortality on multivariate analysis were impaired renal function and a history of previous shunt procedures [59].

20.4.7.2 Hereditary Hemorrhagic Telangiectasia

Hereditary hemorrhagic telangiectasia (HHT) or Rendu—Osler—Weber disease is a rare hereditary disease characterized by the presence of nose bleeding and arteriovenous malformations. Hepatic involvement can lead to life-threatening conditions. The largest reported transplant series in the treatment of hepatic-based HHT (40 pats) has been reported in the ELTR—ELITA study [60]. Indications for LT related to the three phenotypic expressions of the disease: cardiac failure, biliary necrosis causing hepatic failure, and severe portal hypertension. Pulmonary arterial hypertension, present in a majority of cases, is usually reversed after successful LT. Besides the excellent 5-year survival rates of 82.5%, it should be noted that LT markedly improved the performance status of these patients due to the elimination of hepatobiliary sepsis and reversal of cardiopulmonary changes. LT should be proposed earlier in the course of symptomatic, usually life-threatening, conditions. Palliative interventions, especially on the hepatic artery, should be avoided in view of their high (infectious) complication rate due to the development of liver necrosis.

20.4.8 Miscellaneous

20.4.8.1 Adult Polycystic Liver Disease

Autosomal dominant polycystic disease is a genetically heterogeneous disease. Two mutations in two distinct genes (PKD1 and PKD2) predispose to combined polycystic liver and kidney disease (Figures 20.2D and 20.4B); isolated PCLD without renal involvement is rare. The most common complications are intracystic hemorrhage, infection, and posttraumatic rupture, hepatic failure with ascites and encephalopathy is exceptional. LT offers the best chance of definitive treatment in patients presenting with complicated cyst disease and especially in those presenting an invalidating hepatomegaly. The severity of the disease often correlates with the severity of renal cystic disease and degree of renal dysfunction. In cases of coexisting renal failure (creatinin clearance of <20 mL/min/1.73 m^2 or in case of already ongoing hemodialysis) combined liver—kidney transplantation represents the optimal treatment [61]. Three major advantages of combined liver—kidney transplantation are important to underline. Indeed the liver allograft offers major immunological advantages to the recipient of a double graft: the half-life time of the concomitantly transplanted kidney is doubled, the renal transplantation can be done safely despite the presence of a positive crossmatch or a highly immunized status and finally there is usually no need for a reinforced (renal) immunosuppressive therapy.

Despite the huge hepatomegaly and the increased risks for bleeding during the total hepatectomy, meticulous surgery and the application of the cava sparing LT technique have been shown to be of importance in order to reduce postoperative morbidity and mortality [62]. Extensive cyst fenestrations, better known as the Lin procedure, should be avoided at any price in these patients, as this surgery may seriously compromise later transplant surgery. A recent study from ELTR—ELITA database analyzed 58 adult polycystic liver disease (APCLD) patients treated with LT [63] showed that severe abdominal pain due to the "liver mass" effect (75%) and portal hypertension (35%) were the most common indications for LT. In 17% of patients, LT was rendered more difficult due the prior liver surgery (cyst fenestration). The 5-year patient survival rate was 92.3% and quality of life as well as psychological well-being of most patients improved dramatically.

20.4.8.2 Alveolar and Cystic Echinococcosis

The human cystic and alveolar echinococcoses are zoonotic diseases caused by larval stages of the tapeworms *Echinococcus granulosus* and *Echinococcus multilocularis*. In man, liver involvement represents the most common form of the diseases. This pathology is lethal in the absence of appropriate therapeutic management. Complete surgical resection of the parasite at an early stage of infection provides favorable prospects for cure, but, due to a long clinical latency, many cases are diagnosed at an advanced stage which reduces the possibility to perform partial liver resection to 35% of patients. LT has been considered for the treatment of incurable liver echinococcosis. A series of 47 European patients, transplanted during the period 1985 and 2002, had 5-year patient and disease-free survival rates of 71% and 58%, respectively [64]. Five late deaths were related to echinococcus recurrence, located in the brain in three cases. LT for incurable alveolar echinococcus, although very complex due to involvement of many neighboring organs and major vessels, can be discussed in highly symptomatic cases. However, before LT, a careful evaluation of possible distant metastases should be done.

20.4.8.3 Hepatic Trauma

Severe liver trauma is a very rare indication for LT usually because of uncontrollable bleeding followed by liver necrosis and/or failure. LT may be a lifesaving procedure in selected cases of surgically uncontrollable liver injury despite the fact that LT for this indication has been occasionally described as a "waste of organs." In one monocentric retrospective series only 6 (0.04%) of 1.529 LT were performed because of uncontrollable liver trauma caused by traffic accident [65]. Five deceased donor LT and one living donor LT were performed. Two patients survived after a median follow-up of 33 months [10−55,66]. Another US study reported about four LT done because of penetrating or blunt liver trauma; three patients had a good outcome [67].

20.4.9 Immunosuppression

20.4.9.1 Minimal IS

Because of the markedly improved short-term results of LT and persistently high number of long-term complications, the attention of transplant physicians has nowadays been shifted toward minimizing protocols of IS. The development of more selective IS drugs contributed largely to the high success rates in terms of both patient and graft survival. However, the many adverse effects of IS seriously affect long-term outcome of LT recipients. Complications related to IS are responsible for the majority of deaths in patients surviving more than 1 year. The search for an optimal IS regimen is thus of paramount importance. Progressive reduction of IS has now been adopted in several centers with the intent to minimize the side effects caused by their lifelong administration [68]. The minimization IS strategy is now more and more applied in many centers as exemplified by the "10 UCL commandments" used in Brussels (Table 20.3).

20.4.9.2 Steroid Withdrawal

The role of steroid-based IS has been largely investigated. Steroids, the cornerstone IS in the first decades of transplantation, are well known to be responsible for a substantial morbidity and mortality post-LT. Hence, withdrawal of steroids can be done safely in most patients resulting in a great benefit in terms of metabolic and cardiovascular complications. During recent years, long-term "steroid-free" regimens have been widely implemented. Most recipients with stable graft function can be easily

TABLE 20.3 The 10 "UCL Commandments" of Prophylactic and Therapeutic Immunotherapy in LT

1. Every IS, even minimal, is still too great.
2. Toxicity (e.g., neuro/nephrotoxicity) indicates too heavy IS.
3. Immunosuppressive therapy should not be changed continuously as this renders the patient insecure and noncompliant.
4. If the patient is on multiple immunosuppressive drugs, only one drug should be changed at any time when the treatment has to be modified.
5. The immunosuppressive medication should have maximum priority in the posttransplant care of the patient. The more drugs the patient takes, the less efficiently the immunosuppressive treatment is followed.
6. The treatment of different medical complications, such as arterial hypertension, hyperuricemia, and hypercholesterolemia, should not consist of adding more medications, but first of reducing or even withdrawing the responsible immunosuppressive drug.
7. In case of transplantation for HCV-related disease, IS should never be changed significantly as this measure may cause a major viral reactivation.
8. One should always explain to the patient why a modification of therapy is necessary, in order to obtain maximum compliance.
9. Detailed follow-up of the patient, including biochemistry and a problem-list follow-up sheet, is necessary in order to have a good immunosuppressive follow-up by all the different medical caregivers.
10. Every major modification of IS must be transmitted and discussed with the clinical transplant coordinator—the preferred contact person for the recipient and his family—in order to assure an adequate follow-up of such modifications.

maintained using a single drug IS already from 6 or 12 months after LT. A recent paper from our group shows that excellent results can be obtained by applying a steroid-free minimal IS, without compromising graft and patient survival [69].

20.4.9.3 Steroid Avoidance

As a consequence of successful steroid withdrawal, steroid avoidance was a next hurdle to take in clinical practice of LT. In a recent meta-analysis comparing steroid-free vs. steroid-based IS lower rates of hypertension, hypercholesterolemia, cytomegalovirus infection, and HCV recurrence were observed after steroid avoidance [70]. Due to the heterogeneity of the trials performed to date, clinical guidelines in this field have not yet been clearly defined.

20.4.9.4 Infratherapeutic Monotherapy

The liver is an "immunologically privileged" organ, capable in several animal models to be accepted as an allograft with very few IS or even without any intervention on the immune system of the recipient. In some humans, liver allograft acceptance has been observed after IS withdrawal. Identification of those recipients in which such strategies would work out well is still in its developmental phase [71]. Experiences with monotherapy "*ab initio*" have proven to be effective. A randomized study confirmed the validity of a regimen based on early withdrawal of steroids and monotherapy with tacrolimus (TAC) versus cyclosporin (CYA): in this study of comparison between TAC and CYA, both the IS regimens showed good patient survivals (actuarial 5-year survivals of 68% and 70%, respectively). TAC monotherapy "*ab initio*" was associated with lower rejection rates and renal complications, compared to CYA [72]. In our randomized controlled prospective study, TAC placebo was compared with TAC low-dose short-term (64 days) steroid IS. One-year patient survival was similar in the two groups (87% and 95%, respectively); the same was true for 1-year acute rejection rates (23% and 20%, respectively) [73]. These data clearly show that TAC monotherapy can be achieved safely without compromising patient survival after LT. Further large-scale minimization studies are needed to better investigate this approach.

20.4.9.5 Induction Therapy

In contrast to other solid organ transplants, the role of antibody induction IS in LT remains controversial. When taking into consideration the previous paragraphs, it seems curious that induction IS for LT recipients increased from 13.3% in 1999 to 26.7% in 2008 in the United States. The rationale for adopting induction therapy is the attempt to avoid use of CYA, TAC, or steroids in the early post-LT period, preventing in this way the development or aggravation of renal dysfunction, hypertension, and diabetes. Different drugs are used as induction agents; anti-IL-2 receptors (e.g., Basiliximab, Novartis, CH) and the polyclonal agents (r-ATG, Fresenius Biotech-G) represent the most commonly adopted in today's clinical practice [74].

20.4.9.6 Clinical Operational Tolerance

COT is a condition characterized by normal graft function in the absence of any maintenance IS. COT has been reported more frequently in LT recipients than in other transplants. Cautious and carefully supervised weaning of IS drugs has been tried in controlled trials monitored by protocol biopsies [75]. The worldwide experience demonstrates that COT can be achieved safely in one quarter of *very well-selected* individuals, irrespective of the immunological background of donor and recipient, patient age, indication for LT, study endpoint, length of the weaning period,

and pre-/postweaning follow-up, presence or absence of chimerism [76]. The investigation of COT in LT has been hampered by confusion regarding the definitions of rejection and tolerance as well as by the absence of prospective studies correlating results of immune monitoring assays and clinical outcome [77]. Only very recently a pilot study has been performed in pediatric patients. A higher percentage (60%) of recipients remained off IS therapy for at least 1 year with normal graft function and stable allograft histology, possibly demonstrating that children can more easily achieve a tolerant status than adults [78]. The lack of tests capable to safely identify tolerant patients among recipients receiving maintenance IS still represents the main problem. Although many candidates such as higher number of circulating regulatory T-cell subsets (CD4$^+$ CD25$^+$ T cells and Vdelta1$^+$ T cells) have been identified, up to now all of these "markers" of COT remain largely insufficient [79]. Intensive collaboration and interaction between clinicians, immunologists, and basic scientists is needed to go forward in this difficult field of transplantation. The systematic application of these principles should allow further improvements in quality of life and long-term survival after LT [80].

CONCLUSION

It is without any doubt that LT has revolutionized modern medical practice. Progress in this field has been spectacular over the last half century. Indications have become clear and early, long-term results are excellent and, most importantly, quality of life of successfully transplanted recipients is excellent. The further improvement of several medical therapies such as those directed toward viral and metabolic liver diseases, actually the main indications for LT, will lead to a more prominent role of LT in the field of oncology. The development of living liver donation in the western world will be another important factor in this future development. Larger long-term experiences will constitute an important research tool to unravel, slowly but surely, the mystery of clinical, operational tolerance.

ACKNOWLEDGMENTS

Q. Lai—Recipient of ESOT Travel Grant 2012 and ILTS Grant 2012, R.S. Pinheiro—recipient of grant Université Catholique Louvain—University of Sao Paulo, J.M. Rico Juri and E. Castro Santa—recipient of "Henri Bismuth Institute" Master in Hepatobiliary and Pancreatic Surgery.

REFERENCES

[1] Starzl TE, Fung JJ. Themes of liver transplantation. Hepatology 2010;51:1869–84.

[2] Starzl TE, Marchioro TL, Vonkaulla KN, Hermann G, Brittain RS, Waddell WR. Homotransplantation of the liver in humans. Surg Gynecol Obstet 1963;117:659–76.

[3] Starzl TE. Experience in liver transplantation, **9**. Philadelphia: *WB Saunders Company*; 1969.

[4] Starzl TE, Groth CG, Brettschneider L, Penn I, Fulginiti VA, Moon JB, et al. Orthotopic homotransplantation of the human liver. Ann Surg 1968;168:392–415.

[5] National Institutes of Health Consensus Development Conference Statement: liver transplantation—June 20–23, 1983. Hepatology (Suppl.):107S–110S.

[6] Starzl TE, Demetris AJ, Van Thiel D. Liver transplantation. N Engl J Med 1989;321:1014–22.

[7] Starzl TE, Demetris AJ, Van Thiel D. Liver transplantation. N Engl J Med 1989;321:1092–9.

[8] Starzl TE, Groth C, Makowka L. Clio Chirurgica: liver transplantation, 363. Austin, TX: Silvergirl Books; 1988.

[9] Starzl TE, Demetris AJ. Liver transplantation: a 31-year perspective (current problems in surgery classic), 224. Chicago, Year Book Medical Publishers; 1990.

[10] Karvellas CJ, Safinia N, Auzinger G, Heaton N, Muiesan P, O'Grady J, et al. Medical and psychiatric outcomes for patients transplanted for acetaminophen-induced acute liver failure: a case–control study. Liver Int 2010;30:826–33.

[11] O'Grady J. Modern management of acute liver failure. Clin Liver Dis 2007;11:291–303.

[12] Pauwels A, Mostefa-Kara N, Florent C, Lévy VG. Emergency liver transplantation for acute liver failure. Evaluation of London and Clichy criteria. J Hepatol 1993;17:124–7.

[13] Fontana RJ, Watkins PB, Bonkovsky HL, Chalasani N, Davern T, Serrano J, DILIN Study Group, et al. Drug-Induced Liver Injury Network (DILIN) prospective study: rationale, design and conduct. Drug Saf 2009;32:55–68.

[14] Hatzakis A, Wait S, Bruix J, Buti M, Carballo M, Cavaleri M, et al. The state of hepatitis B and C in Europe: report from the hepatitis B and C summit conference. J Viral Hepat 2011;18:1–16.

[15] Samuel D. The option of liver transplantation for hepatitis B: where are we? Dig Liver Dis 2009;41:185–9.

[16] Moradpour D, Negro F. Hepatitis D: forgotten but not gone. Rev Med Suisse 2010;6:1656–9.

[17] Rubín A, Aguilera V, Berenguer M. Liver transplantation and hepatitis C. Clin Res Hepatol Gastroenterol 2011;35:805–12.

[18] Duclos-Vallée JC. Liver transplantation in HIV infected patients: indications and results. Acta Gastroenterol Belg 2010;73:380–2.

[19] Mathurin P, Ehrhard F. Management of alcohol dependence in transplant candidates: we are far away from the objective line. Liver Transpl 2011;17:492–3.

[20] Mathurin P, Moreno C, Samuel D, Dumortier J, Salleron J, Durand F, et al. Early liver transplantation for severe alcoholic hepatitis. N Engl J Med 2011;365:1790–800.

[21] Charlton MR, Burns JM, Pedersen RA, Watt KD, Heimbach JK, Dierkhising RA. Frequency and outcomes of liver transplantation for nonalcoholic steatohepatitis in the United States. Gastroenterology 2011;141:1249–53.

[22] Burke A, Lucey MR. Non-alcoholic fatty liver disease, non-alcoholic steatohepatitis and orthotopic liver transplantation. Am J Transplant 2004;4:686–93.

[23] Dureja P, Mellinger J, Agni R, Chang F, Avey G, Lucey M, et al. NAFLD recurrence in liver transplant recipients. Transplantation 2011;91:684–9.

[24] Manns MP, Vogel A. Autoimmune hepatitis, from mechanisms to therapy. Hepatology 2006;43:132–44.

[25] Schramm C, Bubenheim M, Adam R, Karam V, Buckels J, O'Grady JG, European Liver Intestine Transplant Association, et al. Primary liver transplantation for autoimmune hepatitis: a comparative analysis of the European Liver Transplant Registry. Liver Transpl 2010;16:461–9.

[26] Carbone M, Neuberger J. Liver transplantation in PBC and PSC: indications and disease recurrence. Clin Res Hepatol Gastroenterol 2011;35:446–54.

[27] Schreuder TC, Hübscher SG, Neuberger J. Autoimmune liver diseases and recurrence after orthotopic liver transplantation: what have we learned so far? Transpl Int 2009;22:144–52.

[28] Brandsaeter B, Broomé U, Isoniemi H, Friman S, Hansen B, Schrumpf E, et al. Liver transplantation for primary sclerosing cholangitis in the Nordic countries: outcome after acceptance to the waiting list. Liver Transpl 2003;9:961–9.

[29] Bjøro K, Brandsaeter B, Foss A, Schrumpf E. Liver transplantation in primary sclerosing cholangitis. Semin Liver Dis 2006;26:69–79.

[30] De Kerckhove L, De Meyer M, Verbaandert C, Mourad M, Sokal E, Goffette P, et al. The place of liver transplantation in Caroli's disease and syndrome. Transpl Int 2006;19:381–8.

[31] Meyburg J, Hoffmann GF. Liver transplantation for inborn errors of metabolism. Transplantation 2005;80:135–7.

[32] Catana AM, Medici V. Liver transplantation for Wilson disease. World J Hepatol 2012;4:5–10.

[33] Dar FS, Faraj W, Zaman MB, Bartlett A, Bomford A, O'Sullivan A, et al. Outcome of liver transplantation in hereditary hemochromatosis. Transpl Int 2009;22:717–24.

[34] Hughes Jr MG, Khan KM, Gruessner AC, Sharp H, Hill M, Jie T, et al. Long-term outcome in 42 pediatric liver transplant patients with alpha 1-antitrypsin deficiency: a single-center experience. Clin Transplant 2011;25:731–6.

[35] Miller MR, Sokol RJ, Narkewicz MR, Sontag MK. Pulmonary function in individuals with cystic fibrosis from the U.S. cystic fibrosis foundation registry who had undergone liver transplant. Liver Transpl 2012;:10.1002/lt.23389.

[36] Moini M, Mistry P, Schilsky ML. Liver transplantation for inherited metabolic disorders of the liver. Curr Opin *Organ Transplant* 2010;15:269–76.

[37] Ericzon BG, Larsson M, Herlenius G, Wilczek HE, Familial Amyloidotic Polyneuropathy World Transplant Registry. Report from the Familial Amyloidotic Polyneuropathy World Transplant Registry (FAPWTR) and the Domino Liver Transplant Registry (DLTR). Amyloid 2003;10:67–76.

[38] Ercolani G, Grazi GL, Pinna AD. Liver transplantation for benign hepatic tumours: a systematic review. Dig Surg 2010;27:68–75.

[39] Clavien PA. Is there a place for liver transplantation for 'non HCC' tumours? J Hepatol 2007;47:454–5.

[40] Veteläinen R, Erdogan D, de Graaf W, ten Kate F, Jansen PL, Gouma DJ, et al. Liver adenomatosis: re-evaluation of aetiology and management. Liver Int 2008;28:499–508.

[41] Lerut J, Weber M, Orlando G, Dutkowski P. Vascular and rare liver tumours: a good indication for liver transplantation? J Hepatol 2007;47:454–75.

[42] Mazzaferro V, Bhoori S, Sposito C, Bongini M, Langer M, Miceli R, et al. Milan criteria in liver transplantation for hepatocellular carcinoma: an evidence-based analysis of 15 years of experience. Liver Transpl 2011;17:44–57.

[43] Ciccarelli O, Lai Q, Goffette P, Finet P, De Reyck C, Roggen F, et al. Liver transplantation for hepatocellular cancer: UCL experience in 137 adult cirrhotic patients. Alpha-foetoprotein level and locoregional treatment as refined selection criteria. Transpl Int 2012;25:867–75.

[44] Lerut J. The modernized treatment of hepatocellular cancer: time to think twice!. Updates Surg 2011;63:229–31.

[45] Hwang S, Moon DB, Lee SG. Liver transplantation and conventional surgery for advanced hepatocellular carcinoma. Transpl Int 2010;23:723–7.

[46] Ng KK, Lo CM, Liu CL, Poon RT, Chan SC, Fan ST. Survival analysis of patients with transplantable recurrent hepatocellular carcinoma: implications for salvage liver transplant. Arch Surg 2008;143:68–74.

[47] Lai Q, Avolio AW, Lerut J, Singh G, Chan SC, Berloco PB, et al. Recurrence of hepatocellular cancer after liver transplantation: the role of primary resection and of salvage transplantation in east and west. J Hepatol 2012;57:974–9.

[48] Fuks D, Dokmak S, Paradis V, Diouf M, Durand F, Belghiti J. Benefit of initial resection of hepatocellular carcinoma followed by transplantation in case of recurrence: an intention-to-treat analysis. Hepatology 2012;55:132–40.

[49] Clavien PA, Lesurtel M, Bossuyt PM, Gores GJ, Langer B, Perrier A, OLT for HCC Consensus Group. Recommendations for liver transplantation for hepatocellular carcinoma: an international consensus conference report. Lancet Oncol 2012;13:11–22.

[50] Lerut J, Mergental H, Kahn D, Albuquerque L, Marrero J, Vauthey JN, et al. Place of liver transplantation in the treatment of hepatocellular carcinoma in the normal liver. Liver Transpl 2011;17:90–7.

[51] Rosen CB, Heimbach JK, Gores GJ. Liver transplantation for cholangiocarcinoma. Transpl Int 2010;23:692–7.

[52] Murad SD, Ray Kim W, Therneau T, Gores GJ, Rosen CB, Martenson JA, et al. Predictors of pre-transplant dropout and post-transplant recurrence in patients with perihilar cholangiocarcinoma. Hepatology 2012; Available from: http://dx.doi.org/10.1002/hep.25629.

[53] Bonaccorsi-Riani E, Lerut JP. Liver transplantation and vascular tumours. Transpl Int 2010;23:686–91.

[54] Le Treut YP, Grégoire E, Belghiti J, Boillot O, Soubrane O, Mantion G, et al. Predictors of long-term survival after liver transplantation for metastatic endocrine tumours: an 85-case French multicentric report. Am J Transplant 2008;8:1205–13.

[55] Bonaccorsi-Riani E, Apestegui C, Jouret-Mourin A, Sempoux C, Goffette P, Ciccarelli O, et al. Liver transplantation and neuroendocrine tumours: lessons from a single centre experience and from the literature review. Transpl Int 2010;23:668–78.

[56] Foss A, Adam R, Dueland S. Liver transplantation for colorectal liver metastases: revisiting the concept. Transpl Int 2010;23:679–85.

[57] Chinnakotla S, Klintmalm GB, Kim P, Tomiyama K, Klintmalm E, Davis GL, et al. Long-term follow-up of liver transplantation for Budd–Chiari syndrome with antithrombotic therapy based on the etiology. Transplantation 2011;92:341–5.

[58] Ulrich F, Pratschke J, Neumann U, Pascher A, Puhl G, Fellmer P, et al. Eighteen years of liver transplantation experience in patients with advanced Budd–Chiari syndrome. Liver Transpl 2008;14:144–50.

[59] Mentha G, Giostra E, Majno PE, Bechstein WO, Neuhaus P, O'Grady J, et al. Liver transplantation for Budd–Chiari syndrome: a European study on 248 patients from 51 centres. J Hepatol. 2006;44:520–8.

[60] Lerut J, Orlando G, Adam R, Sabbà C, Pfitzmann R, Klempnauer J, European Liver Transplant Association, et al. Liver transplantation for hereditary hemorrhagic telangiectasia: report of the European liver transplant registry. Ann Surg 2006;244:854–62.

[61] Rossi M, Spoletini G, Bussotti A, Lai Q, Travaglia D, Ferretti S, et al. Combined liver–kidney transplantation in polycystic disease: case reports. Transpl Proc 2008;40:2075–6.

[62] Lerut J, Ciccarelli O, Rutgers M, Orlando G, Mathijs J, Danse E, et al. Liver transplantation with preservation of the inferior vena cava in case of symptomatic adult polycystic disease. Transpl Int 2005;18:513–8.

[63] van Keimpema L, Nevens F, Adam R, Porte RJ, Fikatas P, Becker T, European Liver and Intestine Transplant Association, et al. Excellent survival after liver transplantation for isolated polycystic liver disease: an European Liver Transplant Registry study. Transpl Int 2011;24:1239–45.

[64] Bresson-Hadni S, Koch S, Miguet JP, Gillet M, Mantion GA, Heyd B, European Group of Clinicians, et al. Indications and results of liver transplantation for Echinococcus alveolar infection: an overview. Langenbecks Arch Surg 2003;388:231–8.

[65] Heuer M, Kaiser GM, Lendemans S, Vernadakis S, Treckmann JW, Paul A. Transplantation after blunt trauma to the liver: a valuable option or just a 'waste of organs'? Eur J Med Res 2010;15:169–73.

[66] McPhail MJ, Wendon JA, Bernal W. Meta-analysis of performance of Kings's College Hospital Criteria in prediction of outcome in non-paracetamol-induced acute liver failure. J Hepatol 2010;53:492–9.

[67] Delis SG, Bakoyiannis A, Selvaggi G, Weppler D, Levi D, Tzakis AG. Liver transplantation for severe hepatic trauma: experience from a single center. World J Gastroenterol 2009;15:1641–4.

[68] Londoño MC, López MC, Sánchez-Fueyo A. Minimization of immunosuppression in adult liver transplantation: new strategies and tools. Curr Opin Organ Transplant 2010;15:685–90.

[69] Segev DL, Sozio SM, Shin EJ, Nazarian SM, Nathan H, Thuluvath PJ, et al. Steroid avoidance in liver transplantation: meta-analysis and meta-regression of randomized trials. Liver Transpl 2008;14:512–25.

[70] Lerut J, Bonaccorsi-Riani E, Finet P, Gianello P. Minimization of steroids in liver transplantation. Transpl Int 2009;22:2–19.

[71] Raimondo ML, Burroughs AK. Single-agent immunosuppression after liver transplantation: what is possible? Drugs 2002;62:1587–97.

[72] Cholongitas E, Shusang V, Germani G, Tsochatzis E, Raimondo ML, Marelli L, et al. Long-term follow-up of immunosuppressive monotherapy in liver transplantation: tacrolimus and microemulsified cyclosporin. Clin Transplant 2011;25:614–24.

[73] Lerut J, Mathys J, Verbaandert C, Talpe S, Ciccarelli O, Lemaire J, et al. Tacrolimus monotherapy in liver transplantation: one-year

results of a prospective, randomized, double-blind, placebo-controlled study. Ann Surg 2008;248:956—67.

[74] Wiesner RH, Fung JJ. Present state of immunosuppressive therapy in liver transplant recipients. Liver Transpl 2011;17:1—9.

[75] Tisone G, Orlando G, Angelico M. Operational tolerance in clinical liver transplantation: emerging developments. Transpl Immunol 2007;17:108—13.

[76] Orlando G, Soker S, Wood K. Operational tolerance after liver transplantation. J Hepatol 2009;50:1247—57.

[77] Lerut J, Sanchez-Fueyo A. An appraisal of tolerance in liver transplantation. Am J Transplant 2006;6:1774—80.

[78] Feng S, Ekong UD, Lobritto SJ, Demetris AJ, Roberts JP, Rosenthal P, et al. Complete immunosuppression withdrawal and subsequent allograft function among pediatric recipients of parental living donor liver transplants. JAMA 2012;307:283—93.

[79] Martínez-Llordella M, Puig-Pey I, Orlando G, Ramoni M, Tisone G, Rimola A, et al. Multiparameter immune profiling of operational tolerance in liver transplantation. Am J Transplant 2007;7:309—19.

[80] Starzl TE, Murase N, Abu-Elmagd K, Gray EA, Shapiro R, Eghtesad B, et al. Tolerogenic immunosuppression for organ transplantation. Lancet 2003;361:1502—10.

Chapter 21

Living-Related Liver Transplantation: Progress, Pitfalls, and Promise

Quirino Lai and Jan P. Lerut

Starzl Abdominal Transplant Unit, Department of Abdominal and Transplantation Surgery, University Hospitals St. Luc, Université Catholique Louvain (UCL), Brussels, Belgium

Chapter Outline

21.1 HISTORICAL NOTE

On March 1, 1963, Starzl first described in detail a case of deceased-donor liver transplantation (DDLT), establishing the technical feasibility of liver transplantation in humans [1]. Only several years later, could long recipient survival be achieved [2]. Nowadays liver transplantation (LT) has become, after half a century of clinical application, the standard treatment modality for many acute and chronic end-stage liver diseases. Due to its success, the need for LT rapidly outstripped the supply of liver grafts. The even more pronounced shortage of pediatric donors leads to the development in clinical practice of reduced-size [3] and split-graft LT [4]. The experience obtained from these technical variants of whole liver grafting and the marked progresses obtained during the last three decades in conventional liver surgery paved the way to living-donor liver transplantation (LDLT), an idea already proposed by Smith as early as 1969 [5]. The first attempt of an adult-to-child LDLT was done by S. Raia in Sao Paulo Brazil in December 1988 [6]; the first success was achieved somewhat later by R. Strong in Sydney, Australia [7] in July 1989. After a well thought preparation including an in-depth ethical evaluation, the Chicago group, led by Ch. Broelsch, launched the first real adult-to-child LDLT program. Afterwards several small series of adult-to-child LDLT were reported from the United States [8] as well as from Europe [9]. In Asia, where the problem of deceased-donor liver graft shortage is particularly important due to religious and cultural matters [10], LDLT literally exploded [11,12]. In November 1993, the first adult-to-adult LDLT (ALDLT) was performed by M. Makuuchi in Tokyo, Japan using a left liver in a small primary biliary cirrhotic woman [13]. Because the left liver usually represents an insufficiently sized graft, adult LDLT developed at a much slower pace than pediatric LDLT. The most important pediatric LDLT program was developed in Kyoto under the guidance of K. Tanaka. In 1993, this team had to switch during an LDLT procedure for a 9-year-old recipient to the unprogrammed use of a right hemiliver [14]. This was the start for many teams to set up an LDLT program using right

TABLE 21.1 Landmark Publications of Surgical Evolution in Liver Transplantation in Chronological Order

Author	Discovery and Application	Year	Location
Starzl et al. [1]	First attempts of DDLT in human	1963	United States
Starzl et al. [2]	First long survival DDLT recipients	1967–1968	United States
Smith [5]	Conceptualization of LDLT	1969	United States
Bismuth and Houssin [3]	First adult-to-child reduced size DDLT	1981	France
Pichlmayr et al. [4]	Split-graft DDLT for two recipients	1988	Germany
Raia et al. [6]	First attempt of LDLT	December 1988	Brazil
Strong et al. [7]	First successful LDLT from adult to child	July 1989	Australia
Yamaoka et al. [14]	First report of right liver graft from adult to child	1992	Japan
Hashikura et al. [13]	First successful left liver ALDLT	November 1993	Japan
Lo et al. [15]	First successful ALDLT using right liver	May 1996	Hong Kong
Cherqui et al. [17]	First laparoscopic left lobectomy from adult to child	2001	France
Lee et al. [18]	First adult LDLT using dual left grafts from two donors	2001	Korea
Koffron et al. [19]	First hand-assisted laparoscopic adult-to-adult right lobectomy	2006	United States
Eguchi et al. [20]	First LDLT by hybrid hand-assisted laparoscopy	July 2010	Japan
Giulianotti et al. [21]	First robotic adult-to-adult right lobectomy	2011	United States

Abbreviations: DDLT, deceased-donor liver transplantation; LDLT, living-donor liver transplantation.

livers. The first right liver LDLT program was started up by Fan at the Queen Mary University Hospital in Hong Kong, on May 10, 1996 [15]. This surgical innovation indeed opened the way for many adult recipients toward a timely scheduled, adequately sized and functioning liver graft [16].

Starting from these historical backgrounds (Table 21.1), LDLT nowadays represents a justified answer to the deceased-donor liver graft shortage. At the turn of the twenty-first century, Tanaka highlighted the still many unresolved problems as well as perspectives for the future development of LDLT [22]. It is clear that living donation, especially of the right liver, has high technical and ethical challenges. This surely explains why mainly in Western countries, the medical community and society still consider living liver donation a risky surgical procedure to do with caution under very well specified conditions [23–25].

21.2 ACTUAL STATUS: EASTERN VERSUS WESTERN EXPERIENCES

In 2003, an extensive report about the first decade of LDLT performed in the five most important centers in Asia was published. The experience with 1500 cases reflected the necessity of LDLT in Asia as a consequence of the severe shortage of postmortem grafts. The pioneering Tokyo and Kyoto experiences of the early 90s led to the development in Hong Kong, Kaoshiung (Taiwan), and Seoul (South Korea) of a technically very demanding and new field of hepatobiliary surgery [26]. The rapid evolution of LDLT can be easily illustrated by some numbers. By the end of 2004, 2667 patients underwent LDLT in Japan and Tanaka reported about the first 1000 LDLT done in Kyoto [27]. Between 2001 and 2003, 985 LDLT were performed in South Korea; in June 2005, Lee reported the first 1000 consecutive LDLT from the Asan Medical Center in Seoul [28]. The China Liver Transplant Registry reported about 643 LDLT performed in the period 1995–March 2008; interestingly 588 of them (91%) were realized during the last 3 years. Recently numerable programs have been started up throughout Asia, confirming that in this part of the world, LDLT is the routine approach to organ shortage [29–31].

The Western experience is much more humble; this as a consequence of the higher availability of postmortem donors and the ethical problems related to living donation. After the initial wave of enthusiasm in the 90s, 49 centers performed at least one LDLT in the United States. This enthusiasm unfortunately has been dampened by the first cases of donor mortality reported in 2002. Since then, the number of patients who have undergone LDLT has dramatically declined. By July 2005, 2734 LDLT cases, 1761 of them in adults, were reported. A recent study from the Adult-to-Adult Living-Donor Liver Transplantation (A2ALL) Cohort Study reported 2366 cases of ALDLT performed during the period January 1998−December 2007 [32]. The European experience parallels the US one. During a first decade of optimism (October 1991−December 2001), the European Liver Transplant Registry (ELTR) reported 806 LDLT performed in 46 centers in 15 different European countries [33]. Until December 2009, the number of LDLT has increased to 3622 (2.59%) on a total of 93,634 transplants, performed in 74 centers.

Also in Europe, the experience with ALDLT is limited; again the occurrence of some donor deaths played an important role in this evolution. In contrast pediatric experiences fare well [34], only a small number of centers display a high volume experience in adult donation [35].

21.3 CONTROVERSIAL ISSUES

21.3.1 High Urgency LDLT

Use of LDLT in case of acute or acute-on chronic liver failure (high urgency condition) represents a real challenge from both clinical and ethical points of view. Despite the fact that early experiences showed inferior surgical outcomes of LDLT in the high urgency situation [36−38], the paucity of postmortem donors triggered the development of this type of program, especially in Asia [39]. Lo showed that application of LDLT in this context significantly increased the 1-year survival rate in patients waiting for emergency LT in comparison with patients waiting for a postmortem graft (85% versus 11%) [39]. With more experience, the outcome of urgent LDLT improved substantially in adult as well as pediatric recipients [40−42]. An important question to raise is when a patient becomes too sick to be transplanted [43]. An European study evaluating the concept of "high-risk" patients, which showed that the sickest patients are those who benefit most from LDLT when compared to "low-risk" recipients, stimulated the medical community to consider LDLT in urgent patients [44]. However, a good predictive tool to better discriminate between high-risk and low-risk patients does not exist yet. Although the introduction in clinical practice of the model for end-stage liver disease score (MELD) surely improved selection of potential liver recipients, an effective threshold value of "liver sickness state not to be transplanted" has not been validated yet [45,46]. Besides these "physical" criteria, one should be aware that the psychosocial evaluation of the potential donor in case of urgent donation should be done very carefully avoiding thereby any coercion. The Live Organ Donor Consensus Group has largely underlined this point of view [47]. The donor selection for LDLT is indeed a complex, multidisciplinary process

resulting (too?) many times in donor exclusion due to anatomic abnormalities, blood type incompatibility, or even psychosocial conditions. In a recent paper from Israel, an urgent selection model based on an expeditious protocol was evaluated in donation because of fulminant liver disease. Despite a high number of potential candidates, only four donor procedures could be finalized; in all other cases the recipient died waiting for the transplantation [48]. Centers performing LDLT for acute liver failure should therefore have an infrastructure allowing rapid completion of the donor evaluation including blood testing, ECG, chest X-ray, pulmonary function test, echocardiography, imaging of the liver, psychological assessment, and evaluation by the ethical board within 24−48 h.

21.3.2 Hepatocellular Cancer and LDLT

LT represents the gold standard therapy for hepatocellular cancer (HCC). Milan Criteria [49] and University of California San Francisco Criteria [50] are the worldwide most commonly adopted selection criteria for these cancer patients. Tumor size and number are considered as surrogate markers of biological behavior and tumor aggressiveness. LDLT has the great advantage of eliminating the factor (waiting) time for LT, consequently reducing the potential risk of tumor growth from moment of registration on the waiting list. The Hong Kong group showed that LDLT improved intention-to-treat survival compared to postmortem donor LT (4-year survival: 66% versus 31%, $P = 0.029$) [51] and the Singapore group showed that waiting times of more than 8 months favor pre-LT tumor growth [52]. However, LDLT for HCC also has a drawback as a higher recurrence rate has been reported in several Eastern as well as Western series [53,54]. Many explanations have been put forward to explain these observations, such as the higher regeneration rate and ischemia/reperfusion injury of small grafts providing a favorable environment for HCC cell implantation and growth [55] and the preservation of the inferior vena cava responsible for more manipulation, tumor compression, and dissemination of cancer cells [56]. Recently, however, a different explanation has been put forward. Patients receiving a postmortem LT have gone through a severe selection process in which only candidates presenting with a slowly growing, and thus less aggressive, HCC are able to wait for DDLT. On the opposite side, patients undergoing fast-tracking LDLT for HCC have higher recurrence rates [57]. In a series of 56 recipients with HCC who underwent LDLT in Japan, patient and recurrence-free survival rates were significantly worse if the Milan criteria were not met [58] and in the Hong Kong experience, worse results were observed even in the subgroup of Milan-in patients [53]. The large multicenter A2ALL Cohort Study Group showed that the higher tumor recurrence observed after LDLT can be explained by differences in tumor characteristics, pre-transplant HCC management, and length of waiting time [54]. A recent paper comparing Western and Eastern experiences confirmed all these observations; patients treated with "salvage [for HCC recurrence] LT" after previous liver resection had a much worse outcome; previous liver resection was the main reason for the significantly higher recurrence rate of HCC after LT reported in the Eastern series. Tumors beyond conventional selection criteria and shorter selection periods go along with a higher incidence of HCC recurrence, this independently of the use of a living or postmortem donor LT [59].

21.3.3 Hepatitis C Virus and LDLT

Hepatitis C viral (HCV) cirrhosis represents, despite its universal and early recurrence, an acceptable indication for LDLT. Moreover reports from Western centers raised concerns that HCV recurrence may occur earlier and more severely after LDLT leading to an even more frequent graft loss than after postmortem LT [60–62]. These initial observations were explained by the more pronounced regeneration of partial liver grafts facilitating thereby the entry of HCV into hepatocytes or promoting the HCV replication due to specific cellular changes occurring during this vigorous proliferative response [63]. The increased genetic similarity and the higher degree of HLA matching in intrafamilial LDLT may be another explanation for the more aggressive reinfection of the allograft [64]. On the contrary, LDLT has the advantage to be able to use grafts with shorter ischemia times and from younger donors, factors known to positively influence the outcome of LT for HCV disease. The heterogeneity of reported results and a recent large review about this matter finally led to the conclusion that results of LDLT and DDLT for HCV cirrhosis are quite similar [65–68]. The appropriate timing for LT and the use in clinical practice of more effective medical antiviral treatments against HCV allograft reinfection require further investigation.

21.4 TECHNICAL CHALLENGES AND CONTROVERSIES

21.4.1 Pediatric LDLT: A Standardized Procedure

Pediatric LDLT represents a standard surgical procedure mostly requiring the procurement of the left liver lobe containing Couinauds' segments II and III for implantation. The weight of the left liver lobe is always above 150 mg, making implantation of such graft in children having a weight up to 15 kg a safe procedure. LT results

are optimal due to the lower number of anatomical abnormalities and a lower number of reconstructions is required at graft implantation. Several large series have been reported. The Kyoto group, directed by K. Tanaka and S. Uemoto, reported about 600 pediatric LDLT done during the period June 1990–December 2003, including 568 primary LT and 32 re-LT. One-, 5-, and 10-year patient survival rates were 84.6%, 82.4%, and 77.2%; graft survival rates were 84.1%, 80.9%, and 74.5%. Fifty-five (9.7%) patients have been completely weaned off immunosuppression (IS) [69]. The Chang Gung Memorial Hospital in Kaoshiung (Taiwan), directed by Ch.-L. Cheng, showed an extraordinary 5-year patient survival of 98% in children with biliary atresia [70]. Similar results are obtained in Western series. The Saint-Luc Hospital in Brussels based on 100 pediatric LDLT performed during the period July 1993–April 2002 reported 5-year patient and graft survival rates of 92% and 89%, respectively [34].

Despite the excellent reported results, pediatric LDLT may sometimes be a real technical challenge because of the small caliber of the anatomical structures. This problem is relevant in the case of biliary reconstruction, in which duct-to-duct reconstruction or the Roux-en-Y hepaticojejunostomy can be performed. The first approach, currently the standard procedure in ALDLT, has rarely been reported in pediatric series. The Kyoto center reported about 60 recipients transplanted between November 2005 and June 2008; 14 were treated with duct-to-duct reconstruction and 46 with Roux-en-Y hepaticojejunostomy. The first group presented a higher rate of biliary strictures (28.6% versus 10.9%) confirming that Roux-en-Y hepaticojejunostomy remains the preferable option of bile duct reconstruction in pediatric LDLT [71].

Routine use of microsurgical technique for both biliary and arterial reconstructions has represented a technical innovation leading to decreased early anastomotic complications. The Taiwanese experience in 85 consecutive LDLT performed using microscopic reconstruction done by a single microsurgeon decreased the risk of biliary complications by 2.5 times [72]. A large experience from Hong Kong about 28 pediatric and 124 adult LDLT with hepatic arterial anastomosis using microvascular technique reported an overall complication rate of only 2% [73].

21.4.2 Adult Liver Donation and the Dilemma of the Middle Hepatic Vein

Graft size is one of the most important factors for the success of LDLT. The importance of a good venous drainage of the liver graft as such and of the anterior sector of the right hemiliver graft in particular are, however, even in the case of a sufficient liver graft mass, crucial for the postoperative liver function in ALDLT [74].

The first reported case of an ALDLT was performed in November 1993 by M. Makuuchi; the left lobe from a 25-year-old man was transplanted into his 53-year-old mother suffering from primary biliary cirrhosis [13]. Despite the excellent results obtained in this first experience, an immediate evolution from left to right hemiliver use was observed to overcome the high incidence of small-for-size liver grafts. In 1994, Tanaka first reported about the use of a right liver in order to transplant a great 9-year-old child [14] and in 1996, Lo described the first series of eight ALDLT realized at the University of Hong Kong [15]. These first surgical experiences rapidly unravelled that the use of right hemiliver implied some anatomical controversies. Indeed the venous congestion of Couinauds' segments V and VIII of the right graft was frequently observed if the middle hepatic vein (MHV) was not included in the liver graft. This technique implies that the MHV tributaries from these segments are ligated at procurement [75]. The consequences of such compromised venous outflow become most evident after portal vein reperfusion only. The segments V and VIII many times become swollen and turn bluish, in some cases even graft rupture has been reported due to severe congestion of the anterior sector after reperfusion [76]. The poor venous allograft outflow can also increase the risk of hepatic artery thrombosis, impair graft regeneration by elevating sinusoidal pressure, and disrupt sinusoidal endothelium [77]. All these hemodynamic consequences may lead to a functional "small-for-size" graft syndrome despite the adequate transferred liver mass. In order to minimize these problems and to ensure satisfactory postoperative outcome, different surgical strategies have been advocated in various centers, ranging from routine inclusion of the MHV in the graft to selective reconstruction of the venous drainage on the basis of specific criteria. In Hong Kong, the MHV is routinely included in the right liver graft [78]. Irrespective of the venous drainage pattern of segment IV of the remnant left liver, the segment 4B hepatic vein is preserved, when draining into the MHV. The outflow capacity of the graft is optimized by a back-table venoplasty joining MHV and right hepatic vein into one single cuff [79]. Several centers adopt the original technique proposed by Tanaka, in which MHV was not included in the graft [80], others include or exclude the MHV in the allograft according to peculiar donor and/or recipient characteristics. The Tokyo group excludes the MHV in the allograft based on intraoperative ultrasound findings [81]. After clamping of the MHV at the end of the parenchymal transection, the flow pattern in the right anterior sector portal vein is studied using intraoperative ultrasonography; in case of reversed flow in the portal vein reconstruction of segment V and VIII

branches is considered to be necessary [81,82]. Clamping of the right hepatic artery and MHV at donor operation is another test that has been proposed; if the right anterior sector becomes dusky, hepatic vein reconstruction is needed [83].

The Kyoto group includes the MHV in the graft when the MHV is dominant, when graft-to-recipient weight ratio (GRWR) is <1%, and in all cases when the remnant left liver in the recipient is larger than 35% [84]. The Kaoshiung group proposes an algorithm based on donor–recipient body weight ratio, right hemiliver-to-recipient standard liver volume estimation, and donor hepatic venous anatomy in order to determine if it is necessary to include the MHV in the right graft. The MHV is not included in the graft if the donor is bigger than the recipient and if: (a) the estimated graft volume by computed tomographic volumetry is >50% of the standard liver volume after correction for steatosis; (b) the RHV is large; and (c) the segment V and VIII hepatic veins are <5 mm in size [85].

Recently, several studies focused on the MHV reconstruction using various venous or arterial interposition grafts [86]. Prominent (>5 mm) segment V and VIII hepatic vein branches are anastomosed using a homologous or cryopreserved vein graft to the cuff of middle and/or left hepatic veins of the recipient. Reconstruction of the MHV tributaries using recipient's autogenous interposition vein grafts (including external iliac vein or saphenous vein) or even artificial grafts obtained excellent results [86,87]. Despite all these various criteria adopted by different investigators for selective inclusion of the MHV in the liver graft or selective reconstruction of the segment V and VIII venous tributaries, there is yet no real consensus on how to handle this problem. Some authors even proposed a real "back to the past," simplifying the procedure as much as possible. The improved computerized analysis of the liver vasculature and volume determination of the liver segments will undoubtedly play a more and more important role in the choice of procedure.

In 2001, Lee first described an ALDLT using a dual left graft from two different donors [18]. In 2007, 226 such dual grafts were reported by the same group [88]. The idea of adopting a double graft was based on the necessity to minimize the risk of "small-for-size" syndrome in the recipient and to reduce the risk in the donor by performing mainly a left hepatectomy.

Recently, experimental and clinical research in relation to portal inflow modulation has led to renewed interest in the use of a single left liver graft in ALDLT. In 2004, Kawasaki from Shinshu University, Japan, reported a series of 97 adult recipients transplanted using left liver graft with or without the left portion of the caudate lobe only; 5-year patient survival rate reached 84% [89]. In 2012, the same authors reported about their last 42 such consecutive ALDLT cases. In the absence of splenectomy and portocaval shunt (used to modulate portal inflow), an excellent 5-year patient survival rate of 91% was obtained validating thereby this surgical approach in LDLT [90]. Soejima from the Kyushu University in Fukuoka confirmed the value of this approach in a series of 200 consecutive ALDLT using a left lobe [91,92]. The 5-year patient survival rate was 77.9% with an incidence of small-for-size syndrome (SFSS) of 19.5%. Donor postoperative liver tests and length of hospital stay were significantly better in comparison to those of right liver donors [92]. Very recently, the use of the left liver for LDLT has been evaluated retrospectively in the United States. This procedure minimizes donor morbidity and mortality [93]. During the period 1998–2010, 154 (5.4%) of 2844 ALDLT were done using a left liver. Again reported results are discordant; the number of left liver LDLT increased during recent years despite the increased risk for allograft failure and an inferior patient survival.

21.4.3 Laparoscopy in LDLT: A New Challenge

Laparoscopic living donation was initially developed in the field of kidney transplantation aiming at reducing donor postoperative pain, morbidity rate, and length of hospital stay [94]. The introduction of this surgical approach in the field of LDLT was appealing as open living donation mostly requires a large abdominal incision. This aspect, which is especially important in Eastern countries, combined with more postoperative pain, longer hospital stay, and recovery, represents a barrier to donation, especially in young women [17]. Clearly, donor risk increases with the type of hepatectomy, the right liver donation being more complex than the left one. Experimental models demonstrated the feasibility of laparoscopic living donation using the available technology [95]. Cherqui reported about the first series of laparoscopic LDLT in 2001 for pediatric donation. The technique provided similar or even better short-term graft function and long-term survival rates compared to postmortem LT [17]. In 2006, Soubrane reported the safety of laparoscopic left lateral sectionectomy (LLS or left lobectomy including segments II and III), in 16 consecutive live donors compared to conventional open procurement [96]. Several series have been reported worldwide since then [97,98]. Kim proposed as an alternative the adoption of a small upper midline incision using a standard open technique for adult donation. Twenty-three consecutive donors underwent a right

hepatectomy (RH) using an epigastric midline incision; operative time and periods of analgesic use were shorter and the complaints about wound pain were much less [99]. Recently, the same author compared 11 laparoscopic to 11 open LLS; the laparoscopic group displayed a significantly shorter hospital stay, whilst the duration of surgery, blood loss, warm ischemia time, and out-of-pocket medical costs was comparable between both groups [100]. A similar study from Washington DC compared laparoscopic or laparoscopic-assisted left hepatectomy (LH) and RH for liver donation with 15 open procedures. No substantial differences were found in terms of early graft function, allograft biliary, and vascular complications and survival rates (1-year graft and patient survival: 100% versus 93% in laparoscopic and open group) [101]. Koffron described in 2006 the first hand-assisted laparoscopic RH for adult-to-adult live donation [19]. Kurosaki reported in the same period about 13 consecutive video-assisted adult-to-adult laparoscopic hepatectomies (3 RH and 10 LH with or without segment I) [102]. The surgical manipulation was obtained via ports or via a 12 cm incision whilst the exposure combined direct and laparoscopic visions. Median operation time was 363 ± 33 min and median blood loss was 302 ± 191 mL. No complications were reported, the restoration of liver function was smooth and the use of analgesics was inferior compared to a historical control group. In 2008, the Samsung Seoul group lead by Kuh reported about the first series of hand-assisted laparoscopic-modified RH preserving the middle hepatic vein; two cases of laparoscopic RH and seven cases of laparoscopic-assisted RH with a hand-port device were reported [103,104]. Hilar dissection and parenchymal transection were done under pneumoperitoneum or through a minilaparotomy. The graft was extracted through the site of the hand-port device or the minilaparotomy. Operative time was 765 min in the laparoscopic RH patients and ranged from 310 to 575 min in the laparoscopy-assisted surgery. At Northwestern University in Chicago, Baker et al. retrospectively compared 33 open to 33 laparoscopic living-donor RH. Laparoscopy had an equivalent safety, resource utilization and effectiveness, with several adjunctive physical and psychological benefits [105]. Donor operative times were shorter for the laparoscopic group (265 min versus 316 min); blood loss and length of stay and total hospitalization costs were comparable. In another Seoul study, single-port laparoscopic-assisted donor RH ($n = 40$) was compared to laparoscopic-assisted donor RH ($n = 20$) and open donor RH ($n = 90$). Postoperative complications and reoperation rates were not significantly different, the single-port group showing the lowest level of postoperative pain [106]. In 2011, the first right lobe donor hepatectomy using robotic surgical technology was performed, further underlining the safety of the minimally invasive technology [21]. Finally, the first nine cases of LDLT through a short midline incision combined with hand-assisted laparoscopic surgery have been reported in Japan [20]. All patients were cirrhotic with a median MELD score of 14. Total hepatectomy was carried out through a hand-assisted laparoscopic approach with an 8-cm upper midline incision. Retrieval of the diseased liver was obtained through the upper midline incision which was extended to 12−15 cm. Afterwards the partial liver grafts were implanted through the upper midline incision. Median surgical time was 741 min and the median blood loss was 3940 mL.

Till now no mortalities have been encountered in laparoscopic living donation. The preliminary report of laparoscopic LT represents an extraordinary innovation, opening new perspectives in this fascinating surgical field. Further evolutions related to the use of minimal access surgery in LT are expected in upcoming years, but only centers with both expertise in hepatic minimal access surgery and LDLT should embark on such programs [107] (Figures 21.1−21.5).

21.5 RECIPIENT MORBIDITY AND MORTALITY

21.5.1 Small-for-Size Syndrome: A Physiological Understanding

Post-LDLT recipient survival is dependent on adequate graft size in relation to recipient body size. The Kyoto group demonstrated that the use of small-for-size grafts was responsible for a lower graft survival (58% versus 93%). The cutoff value of 0.8% of GRWR should be respected in order to avoid a graft dysfunction and/or loss [108]. The Tokyo group favors the expression of graft weight in function of the recipient standard liver volume; here the cutoff value should be 40. Several formulas, in the Western and in the Eastern hemispheres, aiming at measuring precisely graft volume in relation to body weight or standard liver volume of the recipient have been put forward [109,110] (Table 21.2). Typically 35−40% of the estimated standard liver weight is the minimum requirement for a successful LT [115]. Despite these well-defined cutoff values, several successes using very small

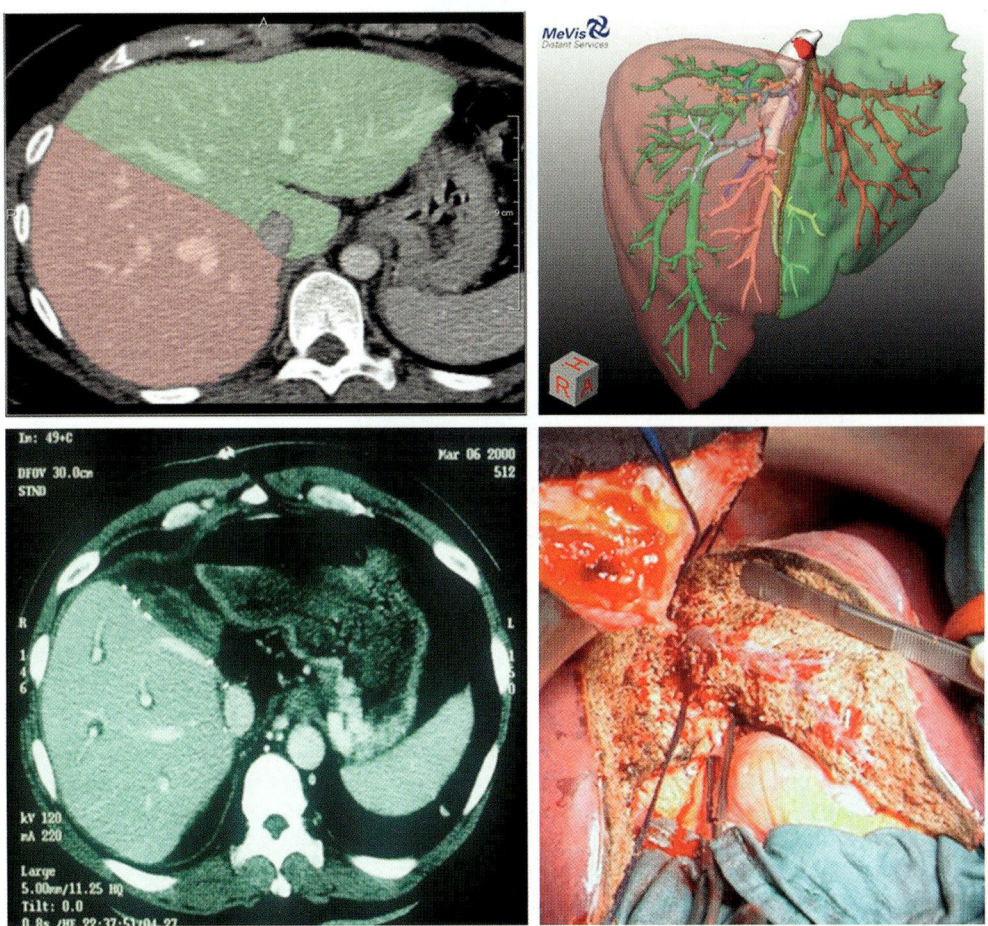

FIGURE 21.1   Mevis analysis allowing to plan "perfectly" the living donation procedure.

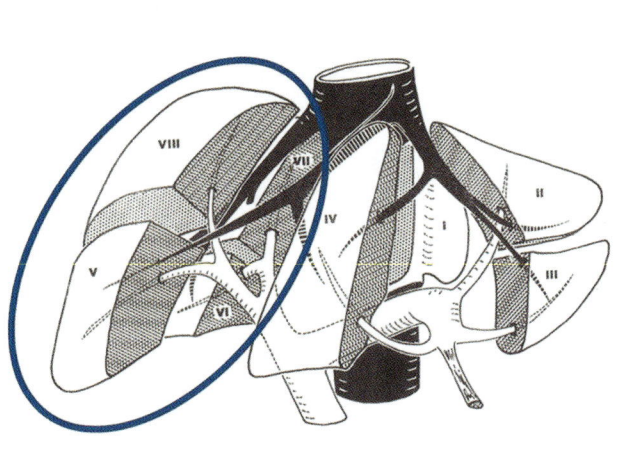

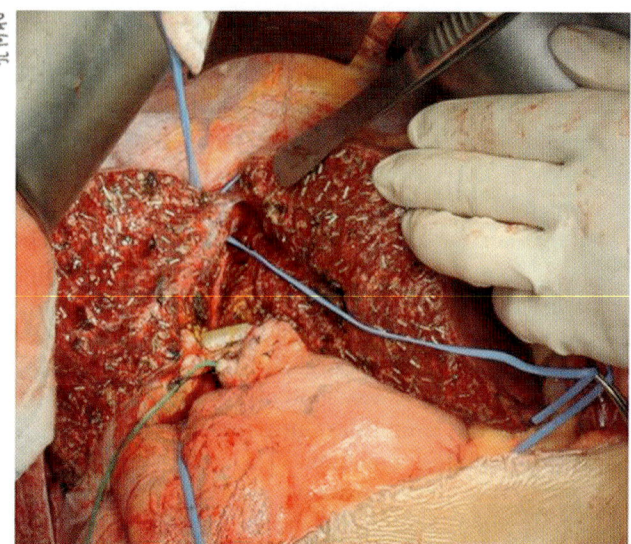

FIGURE 21.2   Operative view of a right liver donation procedure.

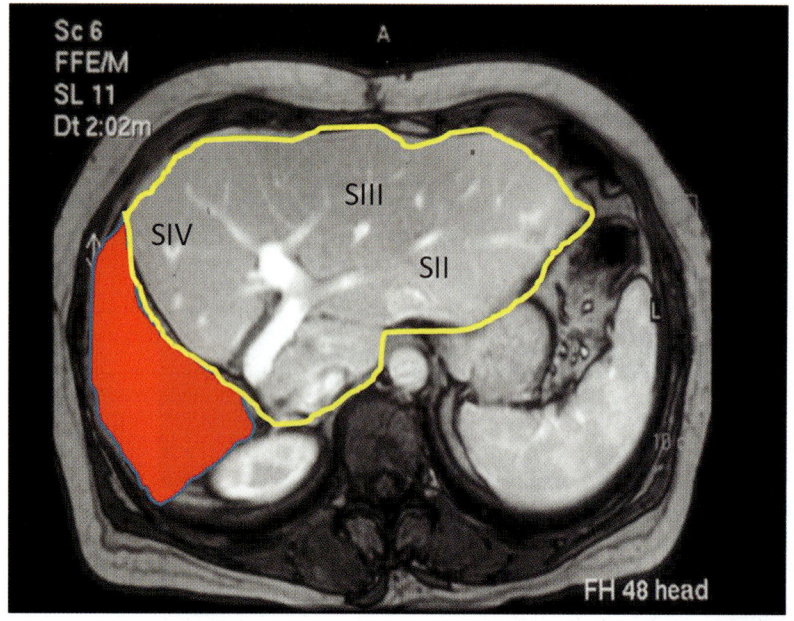

FIGURE 21.3   Full regeneration of left liver after right liver donation.

grafts for ALDLT (going as low as 25% of the ratio of GR/SLV and even 20% but then with portal flow manipulation) have been reported [116,117]. Recent Japanese experiences even report similar results in the absence of portal flow modulation [91].

The poorer results when not respecting these mentioned cutoff values of GRWR are explained by the (further) loss of functional liver mass due to portal overperfusion (and thus portal hypertension) of the graft leading to enhanced parenchymal cell injury, reduced metabolic and synthetic capacity. The ensuing SFSS includes hepatocyte ballooning, steatosis, centrilobular necrosis, and parenchymal cholestasis [118]. These histologic findings are translated in the clinical evolution by pronounced jaundice, encephalopathy, coagulation disturbances, and production of massive quantities of ascites during a prolonged time period. SFSS cannot always be foreseen as it is also dependent on different recipient as well as graft factors. Several approaches have been adopted in order to overcome the problem [119]. Troisi reported that, in addition to technical and graft-related factors, the metabolic requirements of the recipient due to his disease status, the degree of portal hypertension, and the hyperdynamic status all play a

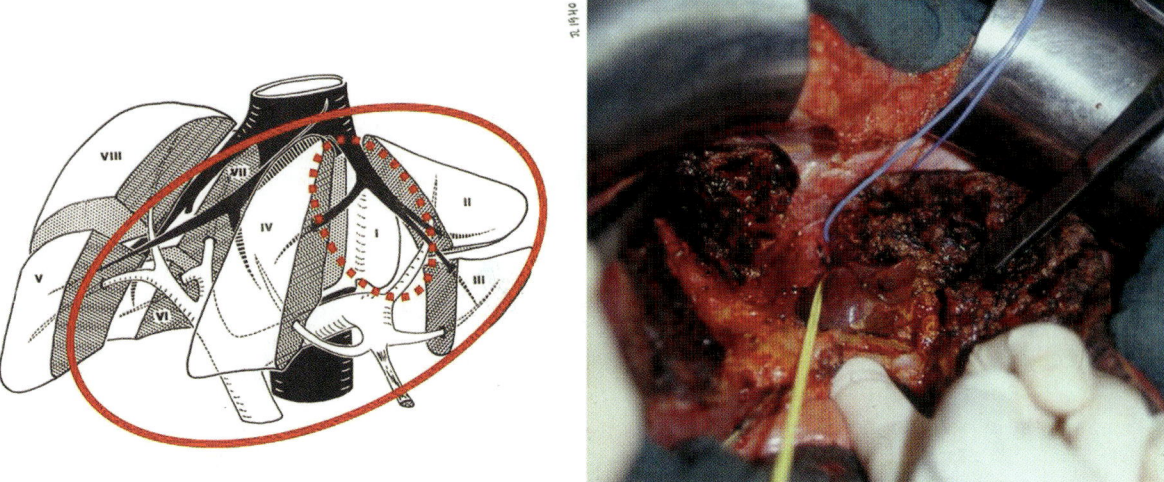

FIGURE 21.4   Operative view of a left liver, including segment I, donation LDLT.

Before

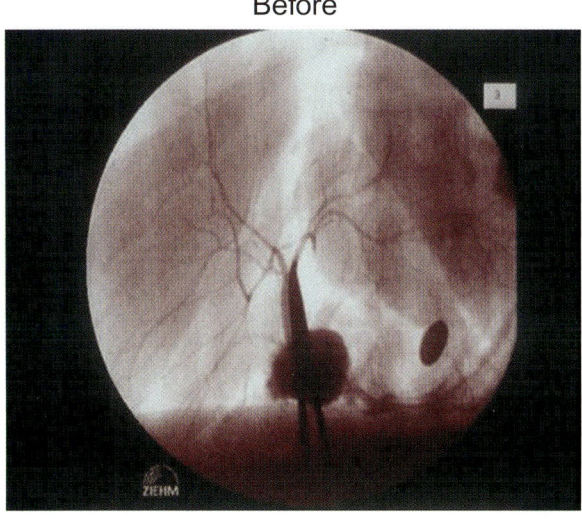

After

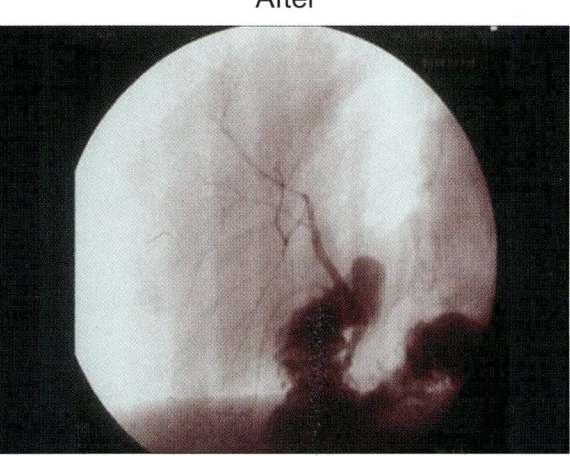

FIGURE 21.5   Intraoperative cholangiography after retrieval of the right liver graft.

role in the amount of stress a small-for-size graft can tolerate. The sicker the recipient, the higher the needed liver mass should be! As portal hyperperfusion is the main reason for the development of SFSS, several techniques aiming at the modulation of the portal flow have been described, including mesocaval [117,120] hemiportocaval shunting [121], inflow modulation by splenic artery ligation or splenectomy [122]. Portal flow can also be influenced by pharmacologic means. An experimental model from Hong Kong investigated the role of low-dose nitric oxide donor FK 409 in attenuating small-for-size graft injury [123]. New combined strategies of pharmacological flow and pressure gradient modulation and pharmacological protection aiming at a reduction of the ischemia/reperfusion injury will also play an important role in the near future, especially as left livers will probably be used increasingly in order to make the donor operation safer [124].

21.5.2 Biliary Complications and Their Management

Biliary complications remain a major problem in both adult and pediatric LDLT. Several, mainly Asian, experiences focalized on the incidence of biliary complications and their management. In the Taiwanese pediatric report with 157 LDLT, 10 (6.3%) biliary complications were diagnosed. These were divided into three groups: bile leakage and biliary strictures with or without vascular complication. The bile leakages recovered after interventional radiology; the biliary strictures were treated using interventional radiology or surgery [125]. In 265 adult LDLT performed in Hong Kong, 55 (20.8%) patients presented a biliary anastomotic stricture. Cold ischemia time

TABLE 21.2 Reported Formulae for ESLV and ESLW

Author	Report Date	Formula	Material Used (Race, Number)
Urata et al. [111]	1995	$ESLV = 706.2 \times BSA + 2.4$	CT Volumetry (Japanese, 96)
Heinemann et al. [112]	1999	$ESLV = 1072.8 \times BSA - 345.7$	Autopsy (Caucasian, 1332)
Vauthey et al. [109]	2002	$ESLV = 18.51 \times BW + 191.8$	CT Volumetry (Western, 292)
Lee et al. [113]	2002	$ESLV = 691 \times BSA + 95$	LDLT (Korea, 311)
Chouker et al. [114]	2004	$ESLW = 452 + 16.34 \times BW + 11.85 \times age - 166 \times gender$ (F = 1, M = 0)	Autopsy (Caucasian, 728)
Chan et al. [110]	2006	$ESLW = 218.32 + BW \times 12.29 + gender \times 50.74$ (M = 1, F = 0)	LDLT (Chinese, 159)

Abbreviations: ESLV, estimated standard liver volume; ESLW, estimated liver weight; BSA, body surface area; BW, body weight; CT, computed tomography; LDLT, Living donor liver transplantation.

TABLE 21.3 Clavien Classification of Surgical Complications Adapted for Living Liver Donors

Grade 1: Nonlife-threatening complications
Require interventions only at the bedside, postoperative bleeding of <4 units of packed red blood cells, never associated with prolongation of ICU or hospital stay longer than twice the median of the population in study.

Grade 2: No residual disability
2a: Require only use of medication or four or more units of packed red blood cells.
2b: Require therapeutic interventions, readmission to the hospital or ICU, or prolongation of regular ICU stay for more than 5 days.
2c: Any potential donor who has an aborted surgery. Donor surgery does not result in transplantation.

Grade 3: Residual disability
3a: There is low risk of death that results in permanent but not progressive disability.
3b: There is lasting disability that is either difficult to control or has a significant risk of death or liver failure.

Grade 4: Liver failure or death
4a: Lead to liver transplantation.
4b: Lead to donor death.

and acute rejection were significant factors for their development; graft survival rates of patients with or without biliary stricture were comparable [126]. The Korean experience in 339 LDLT observed 121 (35.7%) cases with biliary complications, 95 (78.5%) of them appeared within 1 year after LT. An intensive program of interventional procedures using stent insertion or stricture dilatation was successful in 80% of patients. Poor outcomes after endoscopic treatments were reported in case of nonanastomotic strictures and of stenosed hepatic artery [127].

21.6 DONOR MORBIDITY AND MORTALITY

21.6.1 Donor Morbidity

The incidence of complications associated with LDLT varies widely since a uniform definition of complications in this setting was lacking. Recently, the use of the Clavien system to record living-donor complications has been proposed (Table 21.3) [128]. Although results of American, European, and Asian series differ, it can be stated that the right liver donation has a high morbidity (ranging from 20% to 60%) and a higher incidence of severe complications than left liver or lobe donation. Brown et al. reported about 449 ALDLT performed in 42 centers. Donor complications were more frequent in centers performing fewer numbers of LDLT; they included biliary complications requiring interventional procedures (6%) or reoperation (4.5%); the mortality was 0.2%. Biliary and vascular complications occurred in 22% and 9.8%, respectively, of recipients [129].

Abecassis reported on 760 living donors accepted for surgery for ALDLT. Twenty procedures were aborted, 740 were completed. Forty percent of donors had mostly Clavien grades 1 (232) and 2 (269) complications (557 complications among 296 donors). Grade 3 (residual disability, $n = 5$) and grade 4 (leading to death, $n = 3$)

TABLE 21.4 Reported Cases of Donor Death After Living Donor Hepatectomy During Period 1988–2012

No. Death	Cause of Death	Lobe	Timing Death	Location
Intraoperative				
01	Bleeding/cardiac failure/cardiac arrest	R	0	North America
Early postoperative (<60 days)				
02	Anaphylaxis	LLS	1 day	North America
03	Pulmonary embolism	LLS	2 days	Europe
04	Cardiac arrest/vegetative state	R	2 days	Asia
05	Cardiac arrhythmia	R	2 days	South America
06	Gastric necrosis (*C. perfringens*)	R	3 days	North America
07	Cardiac arrest	R	4 days	North America
08	Massive bleeding	R	4 days	Europe
09	Subarachnoid hemorrhage	R	7 days	South America
10	Myocardial infarction	R	10 days	Asia
11	Unknown	Unknown	10 days	Asia
12	Sepsis/MOF	R	11 days	Europe
13	Sepsis/MOF	R	21 days	Europe
14	Bile leak/sepsis/MOF	R	21 days	North America
15	Fall at home	R	28 days	Asia
16	Cardiac failure/LT	R	32 days	Europe
17	Berardinelli-Seip/LT/cardiac failure	R	32 days	Europe
18	Subarachnoid hemorrhage	R	42 days	Asia
19	MOF	R	49 days	Europe
20	Complication of multiple myeloma	R	56 days	Europe
21	Bile peritonitis/sepsis/MOF	R	60 days	Middle East
22	Suicide	L	60 days	North America
23	Pulmonary embolism	L	Unknown	North America
Late postoperative (>60 days)				
24	Duodenal-IVC fistula/air embolism	R	2.3 months	Asia
25	NASH/liver failure/liver transplant	R	9 months	Asia
26	Lung cancer	R	22 months	Asia
27	Suicide	R	22 months	North America
28	Suicide	R	23 months	North America
29	Lung cancer	R	3.4 years	Asia
30	Suicide	R	4 years	South America
31	Suicide	LLS	5 years	South America
32	Asthma	R	5 years	Asia
33	Myocardial infarction	L	6 years	Asia
34	Acute Budd–Chiari	Unknown	Unknown	Europe

Abbreviations: R, right lobe; L, left lobe; LLS, left lateral sectionectomy; MOF, multiorgan failure; LT, liver transplantation; IVC, inferior vena cava; NASH, nonalcoholic steatohepatitis.

complications were rare [130]. A monocentric French experience with 91 consecutive adult living donations displayed 53 (47%) complications occurring in 43 donors [112]. Nineteen (37%) complications were Clavien Grade III or more category. The biliary fistula (14%) was the most common complication; there was no donor mortality. A large multicenter survey from Asia based on 1508 LDLT (766 adults and 742 children) reported a donor complication rate of 15.8%; 1.1% of donors needed surgery[131]. Donor blood loss was <1 L in 94.1% of the cases. Complication rate was higher in right lobe (28%) than in left lobe (9.3%) and left liver (7.5%) donors. In particular, right lobe donors had more serious complications, such as cholestasis (7.3%), bile leakage (6.1%), biliary stricture (1.1%), portal vein thrombosis (0.5%), intraabdominal bleeding (0.5%), and pulmonary embolism (0.5%). A large survey from Japan on 3565 LDLT by the end of December 2006 in 38 Japanese centers reported preoperative problems in two donors, intraoperative problems in 27, and postoperative complications in 270 patients [132]. In total, 299 (8.4%) donors had complications related to the procedure. Postoperative complications included biliary problems in 3%, reoperation in 1.3%, severe sequelae in two (0.06%), and death (related to donor surgery) in one donor (0.03%). The incidence of postoperative complications in left and right liver donors was 8.7% and 9.4%, respectively.

21.6.2 Donor Mortality

LDLT represents a lifesaving alternative for end-stage liver disease patients who have no chance to receive a timely deceased-donor organ. However, donor safety is and must remain central in the LDLT project. When LDLT application was extended from children to adults and from left to right liver grafts, the dilemma between recipient benefit and donor risk came more and more into the spotlight [133,134]. Different, widely publicized, donor deaths have been reported during the last years. Among them the most appealing were the male donor in New York, who succumbed to gas gangrene caused by *Clostridium perfringens* 3 days after donor RH [135], the hypertensive Japanese lady who died from liver failure after right liver donation with a residual left liver of 28% of total liver volume and presenting a nonalcoholic steatohepatitis [136]; the fatal pulmonary embolism in a left liver donor in Germany [9] and the donor mother, who had a history of substance abuse, who died from drug overdose 2 months after donation to her 3-year-old son [137]. Very recently, a large web-based worldwide survey of liver transplant programs has been performed aiming to improve the knowledge about the actual incidence of fatal events [138]. Indeed reported percentages about donor adverse events and death widely vary and potentially life-threatening events, such as aborted donor hepatectomy or intraoperative hemorrhage ("near miss" events), are rarely reported in cases of successful management. According to this survey, 23 (0.2%) donor deaths out of 11,553 donor hepatectomy procedures have been reported, but an additional 11 donor deaths not captured in the survey have been reported in the literature! All the reported cases with fatal outcome are displayed in Table 21.4.

A total of 136 (1.16%) aborted donor hepatectomies has been reported. Such mishap was mostly due to the unexpected vascular or biliary anomalies or quality of the liver parenchyma. In 126 (1.1%) cases, a "near miss" event occurred; bleeding requiring surgical reintervention was most frequent, followed by thrombotic events, biliary reconstruction procedures, life-threatening sepsis, and iatrogenic injury of the bowel or vessels. The overall reported donor morbidity rate was 23.9%. Five donors needed a transplantation themselves; four required a liver and one a kidney.

21.7 CONCLUSION

Great progress in LDLT has been made during the last decades, thanks to a better understanding of the liver physiology and surgical techniques. Many efforts have been made in order to especially optimize venous in- and outflow of the graft. Morbidity and mortality rates have been progressively lowered in LDLT recipients. LDLT in pediatric recipients has proven its outstanding value. In adults the situation still remains more complex due to the higher perioperative risks both in donors and recipients. It can be foreseen that there will be in the near future a revived interest in implantation of left livers in adult recipients as this approach has now been validated in a few monocentric series. This evolution could have a positive impact on the development of LDLT, especially in Western countries where this procedure is still considered as a too risky and many times questionable intervention. Balancing donor safety and recipient benefit should be the master switch in the further development of LDLT. Moreover the decision to go for a LDLT procedure should always be based on a sound medical, surgical, and ethical judgment.

ACKNOWLEDGMENT

Q. Lai—Recipient of ESOT Travel Grant 2012 and ILTS Grant 2012.

REFERENCES

[1] Starzl TE, Marchioro TL, Vonkaulla KN, Hermann G, Rittain RS, Addel WR. Homotransplantation of the liver in humans. Surg Gynecol Obstet 1963;117:659—76.

[2] Starzl TE, Groth CG, Brettschneider L, Penn I, Fulginiti VA, Moon JB, et al. Orthotopic homotransplantation of the human liver. Ann Surg 1968;168:392−415.

[3] Bismuth H, Houssin D. Reduced-sized orthotopic liver graft in hepatic transplantation in children. Surgery 1984;95:367−70.

[4] Pichlmayr R, Ringe B, Gubernatis G, Hauss J, Bunzendahl H. Transplantation of a donor liver to 2 recipients (splitting transplantation)—a new method in the further development of segmental liver transplantation. Langenbecks Arch Chir 1988;373:127−30.

[5] Smith B. Segmental liver transplantation from a living donor. J Pediatr Surg 1969;4:126−32.

[6] Raia S, Nery JR, Mies S. Liver transplantation from live donors. Lancet 1989;2:497.

[7] Strong RW, Lynch SV, Ong TH, Matsunami H, Koido Y, Balderson GA. Successful liver transplantation from a living donor to her son. N Engl J Med 1990;322:1505−7.

[8] Broelsch CE, Whitington PF, Emond JC, Heffron TG, Thistlethwaite JR, Stevens L, et al. Liver transplantation in children from living related donors. Surgical techniques and results. Ann Surg 1991;214:428−37 discussion 437−9

[9] Malago M, Rogiers X, Burdelski M, Broelsch CE. Living related liver transplantation: 36 cases at the University of Hamburg. Transplant Proc 1994;26:3620−1.

[10] Nudeshima J. Obstacles to brain death and organ transplantation in Japan. Lancet 1991;338:1063−4.

[11] Makuuchi M, Kawarazaki H, Iwanaka T, Kamada N, Takayama T, Kumon M. Living related liver transplantation. Surg Today 1992;22:297−300.

[12] Ozawa K, Uemoto S, Tanaka K, Kumada K, Yamaoka Y, Kobayashi N, et al. An appraisal of pediatric liver transplantation from living relatives. Initial clinical experiences in 20 pediatric liver transplantations from living relatives as donors. Ann Surg 1992;216:547−53.

[13] Hashikura Y, Makuuchi M, Kawasaki S, Matsunami H, Ikegami T, Nakazawa Y, et al. Successful living-related partial liver transplantation to an adult patient. Lancet 1999;343:1233−4.

[14] Yamaoka Y, Washida M, Honda K, Tanaka K, Mori K, Shimahara Y, et al. Liver transplantation using a right lobe graft from a living related donor. Transplantation 1994;57:1127−30.

[15] Lo CM, Fan ST, Liu CL, Wei WI, Lo RJ, Lai CL, et al. Adult-to-adult living donor liver transplantation using extended right lobe grafts. Ann Surg 1997;226:261−9.

[16] Lo CM, Fan ST, Liu CL, Lo RJ, Lau GK, Wei WI, et al. Extending the limit on the size of adult recipient in living donor liver transplantation using extended right lobe graft. Transplantation 1997;63:1524−8.

[17] Cherqui D, Soubrane O, Husson E, Barshasz E, Vignaux O, Ghimouz M, et al. Laparoscopic living donor hepatectomy for liver transplantation in children. Lancet 2002;359:392−6.

[18] Lee S, Hwang S, Park K, Lee Y, Choi D, Ahn C, et al. An adult-to-adult living donor liver transplant using dual left lobe grafts. Surgery 2001;129:647−50.

[19] Koffron AJ, Kung R, Baker T, Fryer J, Clark L, Abecassis M. Laparoscopic-assisted right lobe donor hepatectomy. Am J Transplant 2006;6:2522−5.

[20] Eguchi S, Takatsuki M, Soyama A, Hidaka M, Tomonaga T, Muraoka I, et al. Elective living donor liver transplantation by hybrid hand-assisted laparoscopic surgery and short upper midline laparotomy. Surgery 2011;150:1002−5.

[21] Giulianotti PC, Tzvetanov I, Jeon H, Bianco F, Spaggiari M, Oberholzer J, et al. Robot-assisted right lobe donor hepatectomy. Transpl Int 2012;25:5−9.

[22] Kiuchi T, Tanaka K. Living donor adult liver transplantation: status quo in Kyoto and perspectives in the new millennium. Acta Chir Belg 2000;100:279−83.

[23] Singer PA, Siegler M, Whitington PF, Lantos JD, Emond JC, Thistlethwaite JR, et al. Ethics of liver transplantation with living donors. N Engl J Med 1989;321:620−2.

[24] Cronin 2nd DC, Millis JM, Siegler M. Transplantation of liver grafts from living donors into adults—too much, too soon. N Engl J Med 2001;344:1633−7.

[25] Surman OS. The ethics of partial-liver donation. N Engl J Med 2002;346:1038.

[26] Chen CL, Fan ST, Lee SG, Makuuchi M, Tanaka K. Living-donor liver transplantation: 12 years of experience in Asia. Transplantation 2003;75:6−11.

[27] Takada Y, Suzukamo Y, Oike F, Egawa H, Morita S, Fukuhara S, et al. Long-term quality of life of donors after living donor liver transplantation. Liver Transpl 2012;18:1343−52.

[28] Hwang S, Lee SG, Lee YJ, Sung KB, Park KM, Kim KH, et al. Lessons learned from 1000 living donor liver transplantations in a single center: how to make living donations safe. Liver Transpl 2006;12:920−7.

[29] Soin AS, Kakodkar R. Living donor liver transplantation in India. Trop Gastroenterol 2007;28:96−8.

[30] Jy L, Ln Y. Transfusion rate for 500 consecutive living donor liver transplantations: experience of one liver transplantation center. Transplantation 2012;94:66−7.

[31] Lee VT, Yip CC, Ganpathi IS, Chang S, Mak KS, Prabhakaran K, et al. Expanding the donor pool for liver transplantation in the setting of an "opt-out" scheme: 3 years after new legislation. Ann Acad Med Singapore 2009;38:315−7.

[32] Olthoff KM, Abecassis MM, Emond JC, Kam I, Merion RM, Gillespie BW, et al. Adult-to-adult living donor liver transplantation cohort study group. outcomes of adult living donor liver transplantation: comparison of the adult-to-adult living donor liver transplantation cohort study and the national experience. Liver Transpl 2011;17:789−97.

[33] Adam R, McMaster P, O'Grady JG, Castaing D, Klempnauer JL, Jamieson N, et al. Evolution of liver transplantation in Europe: report of the European liver transplant registry. Liver Transpl. 2003;9:1231−43.

[34] Bourdeaux C, Darwish A, Jamart J, Tri TT, Janssen M, Lerut J, et al. Living-related versus deceased donor pediatric liver transplantation: a multivariate analysis of technical and immunological complications in 235 recipients. Am J Transplant 2007;7:440−7.

[35] Azoulay D, Bhangui P, Andreani P, Salloum C, Karam V, Hoti E, et al. Short- and long-term donor morbidity in right lobe living donor liver transplantation: 91 consecutive cases in a European center. Am J Transplant 2011;11:101−10.

[36] Zieniewicz K, Skwarek A, Nyckowski P, Pawlak J, Micha-lowicz B, Patkowski W, et al. Comparison of the results of liver transplantation for elective versus urgent indications. Transplant Proc 2003;35:2262−4.

[37] Uemoto S, Inomata Y, Sakurai T, Egawa H, Fujita S, Kiuchi T, et al. Living donor liver transplantation for fulminant hepatic failure. Transplantation 2000;70:152−7.

[38] Abt PL, Mange KC, Olthoff KM, Markmann JF, Reddy KR, Shaked A. Allograft survival following adult-to-adult living donor liver transplantation. Am J Transplant 2004;4:1302−7.

[39] Lo CM, Fan ST, Liu CL, Wei WI, Chan JK, Lai CL, et al. Applicability of living donor liver transplantation to high-urgency patients. Transplantation 1999;67:73−7.

[40] Liu CL, Fan ST, Lo CM, Yong BH, Fung AS, Wong J. Right-lobe live donor liver transplantation improves survival of patients with acute liver failure. Br J Surg 2002;89:317−22.

[41] Liu CL, Fan ST, Lo CM, Wong J. Living-donor liver transplantation for high-urgency situations. Transplantation 2003;75:33−6.

[42] Mack CL, Ferrario M, Abecassis M, Whitington PF, Superina RA, Alonso EM. Living donor liver transplantation for children with liver failure and concurrent multiple organ system failure. Liver Transpl 2001;7:890−5.

[43] Merion RM. When is a patient too well and when is a patient too sick for a liver transplant? Liver Transpl 2004;10:69−73.

[44] Durand F, Belghiti J, Troisi R, Boillot O, Gadano A, Francoz C, et al. Living donor liver transplantation in high-risk vs. low-risk patients: optimization using statistical models. Liver Transpl 2006;12:231−9.

[45] Saab S, Wang V, Ibrahim AB, Durazo F, Han S, Farmer DG, et al. MELD score predicts 1-year patient survival post-orthotopic liver transplantation. Liver Transpl 2003;9:473−6.

[46] Olthoff KM, Merion RM, Ghobrial RM, Abecassis MM, Fair JH, Fisher RA, et al. Outcomes of 385 adult-to-adult living donor liver transplant recipients: a report from the A2ALL Consortium. Ann Surg 2005;242:314−23.

[47] Abecassis M, Adams M, Adams P, Arnold RM, Atkins CR, Barr ML, et al. Consensus statement on the live organ donor. JAMA 2000;284:2919−26.

[48] Ben-Haim M, Carmiel M, Lubezky N, Keidar R, Katz P, Blachar A, et al. Donor recruitment and selection for adult-to-adult living donor liver transplantation in urgent and elective circumstances. Isr Med Assoc J 2005;7:169−73.

[49] Mazzaferro V, Regalia E, Doci R, Andreola S, Pulvirenti A, Bozzetti F, et al. Liver transplantation for the treatment of small hepatocellular carcinomas in patients with cirrhosis. N Engl J Med 1996;334:693−9.

[50] Yao FY, Ferrell L, Bass NM, Watson JJ, Bacchetti P, Venook A, et al. Liver transplantation for hepatocellular carcinoma: expansion of the tumor size limits does not adversely impact survival. Hepatology 2001;33:1394−403.

[51] Lo CM, Fan ST, Liu CL, Chan SC, Wong J. The role and limitation of living donor liver transplantation for hepatocellular carcinoma. Liver Transpl 2004;10:440−7.

[52] Mak KS, Tan KC. Liver transplantation for hepatocellular carcinoma: an Asian perspective. Asian J Surg 2002;25:271−6.

[53] Ng KK, Lo CM, Chan SC, Chok KS, Cheung TT, Fan ST. Liver transplantation for hepatocellular carcinoma: the Hong Kong experience. J Hepatobiliary Pancreat Sci 2010;17:548−54.

[54] Kulik LM, Fisher RA, Rodrigo DR, Brown Jr RS, Freise CE, Shaked A, et al. Outcomes of living and deceased donor liver transplant recipients with hepatocellular carcinoma: results of the A2ALL cohort. Am J Transplant 2012;12:2997−3007.

[55] Lo CM, Fan ST, Liu CL, Chan SC, Ng IO, Wong J. Living donor versus deceased donor liver transplantation for early irresectable hepatocellular carcinoma. Br J Surg 2007;94:78−86.

[56] Lee S, Ahn C, Ha T, Moon D, Choi K, Song G, et al. Liver transplantation for hepatocellular carcinoma: Korean experience. J Hepatobiliary Pancreat Sci 2010;17:539−47.

[57] Kulik L, Abecassis M. Living donor liver transplantation for hepatocellular carcinoma. Gastroenterology 2004;127:277−82.

[58] Kaihara S, Kiuchi T, Ueda M, Oike F, Fujimoto Y, Ogawa K, et al. Living-donor liver transplantation for hepatocellular carcinoma. Transplantation 2003;75:37−40.

[59] Lai Q, Avolio AW, Lerut J, Singh G, Chan SC, Berloco PB, et al. Recurrence of hepatocellular cancer after liver transplantation: the role of primary resection and salvage transplantation in East and West. J Hepatol 2012;57:974−9.

[60] Thuluvath PJ, Yoo HY. Graft and patient survival after adult live donor liver transplantation compared to a matched cohort who received a deceased donor liver transplantation. Liver Transpl 2004;10:1263−8.

[61] Schiano TD, Gutierrez JA, Walewski JL, Fiel MI, Cheng B, Bodenheimer H, et al. Accelerated hepatitis C virus kinetics but similar survival rates in recipients of liver grafts from living versus deceased donors. Hepatology 2005;42:1420−8.

[62] Garcia-Retortillo M, Forns X, Llovet JM, Navasa M, Feliu A, Massaguer A, et al. Hepatitis C recurrence is more severe after living donor compared to cadaveric liver transplantation. Hepatology 2004;40:699−707.

[63] Zimmerman MA, Trotter J. Living donor liver transplantation in patients with hepatitis C. Liver Transpl 2003;9:53−7.

[64] Russo MW, Shrestha R. Is severe recurrent hepatitis C more common after adult living donor liver transplantation? Hepatology 2004;40:524−6.

[65] Schmeding M, Neumann UP, Puhl G, Bahra M, Neuhaus R, Neuhaus P. Hepatitis C recurrence and fibrosis progression are not increased after living donor liver transplantation: a single center study of 289 patients. Liver Transpl 2007;13:687−92.

[66] Gallegos-Orozco JF, Yosephy A, Noble B, Aqel BA, Byrne TJ, Carey EJ, et al. Natural history of post-liver transplantation hepatitis C: a review of factors that may influence its course. Liver Transpl 2009;15:1872−81.

[67] Jain A, Singhal A, Kashyap R, Safadjou S, Ryan CK, Orloff MS. Comparative analysis of hepatitis C recurrence and fibrosis progression between deceased-donor and living-donor liver transplantation: 8-year longitudinal follow-up. Transplantation 2011;92:453−60.

[68] Takada Y, Uemoto S. Living donor liver transplantation for hepatitis C. Surg Today 2012;43:709−14.

[69] Ueda M, Oike F, Ogura Y, Uryuhara K, Fujimoto Y, Kasahara M, et al. Long-term outcomes of 600 living donor liver transplants for pediatric patients at a single center. Liver Transpl 2006;12:1326−36.

[70] Chen CL, Concejero AM, Cheng YF. More than a quarter of a century of liver transplantation in Kaohsiung Chang Gung memorial hospital. Clin Transpl 2011;213−21.

[71] Tanaka H, Fukuda A, Shigeta T, Kuroda T, Kimura T, Sakamoto S, et al. Biliary reconstruction in pediatric live donor liver transplantation: duct-to-duct or Roux-en-Y hepaticojejunostomy. J Pediatr Surg 2010;45:1668−75.

[72] Lin TS, Concejero AM, Chen CL, Chiang YC, Wang CC, Wang SH, et al. Routine microsurgical biliary reconstruction decreases early anastomotic complications in living donor liver transplantation. Liver Transpl 2009;15:1766−75.

[73] Wei WI, Lam LK, Ng RW, Liu CL, Lo CM, Fan ST, et al. Microvascular reconstruction of the hepatic artery in live donor liver transplantation: experience across a decade. Arch Surg 2004;139:304−7.

[74] Liu CL, Lo CM, Fan ST. What is the best technique for right hemiliver living donor liver transplantation? With or without the middle hepatic vein? Duct-to-duct biliary anastomosis or Roux-en-Y hepaticojejunostomy? Forum Liver Transplant J Hepatol 2005;43:13−37.

[75] Marcos A, Orloff M, Mieles L, Olzinski AT, Renz JF, Sitzmann JV. Functional venous anatomy for right-lobe grafting and techniques to optimize outflow. Liver Transpl 2001;7:845−52.

[76] Marcos A, Fisher RA, Ham JM, Olzinski AT, Shiffman ML, Sanyal AJ, et al. Emergency portacaval shunt for control of hemorrhage from a parenchymal fracture after adult-to-adult living donor liver transplantation. Transplantation 2000;69:2218−21.

[77] Man K, Fan ST, Lo CM, Liu CL, Fung PC, Liang TB, et al. Graft injury in relation to graft size in right lobe live donor liver transplantation: a study of hepatic sinusoidal injury in correlation with portal hemodynamics and intragraft gene expression. Ann Surg 2003;237:256−64.

[78] Fan ST, Lo CM, Liu CL, Wang WX, Wong J. Safety and necessity of including the middle hepatic vein in the right lobe graft in adult-to-adult live donor liver transplantation. Ann Surg 2003;238:137−48.

[79] Lo CM, Fan ST, Liu CL, Wong J. Hepatic venoplasty in living-donor liver transplantation using right lobe graft with middle hepatic vein. Transplantation 2003;75:358−60.

[80] Campsen J, Hendrickson RJ, Zimmerman MA, Wachs M, Bak T, Russ P, et al. Adult right lobe live donor liver transplantation without reconstruction of the middle hepatic vein: a single-center study of 109 cases. Transplantation 2008;85:775−7.

[81] Cescon M, Sugawara Y, Sano K, Ohkubo T, Kaneko J, Makuuchi M. Right liver graft without middle hepatic vein reconstruction from a living donor. Transplantation 2002;73:1164−6.

[82] Sano K, Makuuchi M, Miki K, Maema A, Sugawara Y, Imamura H, et al. Evaluation of hepatic venous congestion: proposed indication criteria for hepatic vein reconstruction. Ann Surg 2002;236:241−7.

[83] Sugawara Y, Makuuchi M, Sano K, Imamura H, Kaneko J, Ohkubo T, et al. Vein reconstruction in modified right liver graft for living donor liver transplantation. Ann Surg 2003;237:180−5.

[84] Kasahara M, Takada Y, Fujimoto Y, Ogura Y, Ogawa K, Uryuhara K, et al. Impact of right lobe with middle hepatic vein graft in living-donor liver transplantation. Am J Transplant 2005;5:1339−46.

[85] de Villa VH, Chen CL, Chen YS, Wang CC, Lin CC, Cheng YF, et al. Right lobe living donor liver transplantation-addressing the middle hepatic vein controversy. Ann Surg 2003;238:275−82.

[86] Lee S, Park K, Hwang S, Lee Y, Choi D, Kim K, et al. Congestion of right liver graft in living donor liver transplantation. Transplantation 2001;71:812−4.

[87] Hwang S, Jung DH, Ha TY, Ahn CS, Moon DB, Kim KH, et al. Usability of ringed polytetrafluoroethylene grafts for middle hepatic vein reconstruction during living donor liver transplantation. Liver Transpl 2012;18:955−65.

[88] Lee SG, Hwang S, Park KM, Kim KH, Ahn CS, Lee YJ, et al. Seventeen adult-to-adult living donor liver transplantations using dual grafts. Transplant Proc 2001;33:3461−3.

[89] Hashikura Y, Kawasaki S. Living donor liver transplantation: issues regarding left liver grafts. HPB (Oxford) 2004;6:99−105.

[90] Ishizaki Y, Kawasaki S, Sugo H, Yoshimoto J, Fujiwara N, Imamura H. Left lobe adult-to-adult living donor liver transplantation: should portal inflow modulation be added?. Liver Transpl 2012;18:305−14.

[91] Taketomi A, Kayashima H, Soejima Y, Yoshizumi T, Uchiyama H, Ikegami T, et al. Donor risk in adult-to-adult living donor liver transplantation: impact of left lobe graft. Transplantation 2009;87:445−50.

[92] Soejima Y, Shirabe K, Taketomi A, Yoshizumi T, Uchiyama H, Ikegami T, et al. Left lobe living donor liver transplantation in adults. Am J Transplant 2012;12:1877−85.

[93] Saidi RF, Jabbour N, Li Y, Shah SA, Bozorgzadeh A. Is left lobe adult-to-adult living donor liver transplantation ready for widespread use? The US experience (1998−2010). HPB (Oxford) 2012;14:455−60.

[94] Flowers JL, Jacobs S, Cho E, Morton A, Rosenberger WF, Evans D, et al. Comparison of open and laparoscopic live donor nephrectomy. Ann Surg 1997;226:483−9.

[95] Lin E, Gonzalez R, Venkatesh KR, Mattar SG, Bowers SP, Fugate KM, et al. Can current technology be integrated to facilitate laparoscopic living donor hepatectomy? Surg Endosc 2003;17:750−3.

[96] Soubrane O, Cherqui D, Scatton O, Stenard F, Bernard D, Branchereau S, et al. Laparoscopic left lateral sectionectomy in living donors: safety and reproducibility of the technique in a single center. Ann Surg 2006;244:815−20.

[97] Troisi R, Debruyne R, Rogiers X. Laparoscopic living donor hepatectomy for pediatric liver transplantation. Acta Chir Belg 2009;109:559−62.

[98] Coelho JC, Freitas AC, Mathias JE. Laparoscopic resection of the left lateral segment of the liver in living donor liver transplantation. Rev Col Bras Cir 2009;36:537−8.

[99] Kim SH, Cho SY, Lee KW, Park SJ, Han SS. Upper midline incision for living donor right hepatectomy. Liver Transpl 2009;15:193−8.

[100] Kim KH, Jung DH, Park KM, Lee YJ, Kim DY, Kim KM, et al. Comparison of open and laparoscopic live donor left lateral sectionectomy. Br J Surg 2011;98:1302−8.

[101] Thenappan A, Jha RC, Fishbein T, Matsumoto C, Melancon JK, Girlanda R, et al. Liver allograft outcomes after laparoscopic-assisted and minimal access live donor hepatectomy for transplantation. Am J Surg 2011;201:450−5.

[102] Kurosaki I, Yamamoto S, Kitami C, Yokoyama N, Nakatsuka H, Kobayashi T, et al. Video-assisted living donor hemihepatectomy through a 12-cm incision for adult-to-adult liver transplantation. Surgery 2006;139:695−703.

[103] Suh KS, Yi NJ, Kim J, Shin WY, Lee HW, Han HS, et al. Laparoscopic hepatectomy for a modified right graft in adult-to-adult living donor liver transplantation. Transplant Proc 2008;40:3529−31.

[104] Suh KS, Yi NJ, Kim T, Kim J, Shin WY, Lee HW, et al. Laparoscopy-assisted donor right hepatectomy using a hand port system preserving the middle hepatic vein branches. World J Surg 2009;33:526−33.

[105] Baker TB, Jay CL, Ladner DP, Preczewski LB, Clark L, Holl J, et al. Laparoscopy-assisted and open living donor right hepatectomy: a comparative study of outcomes. Surgery 2009;146:817−23.

[106] Choi HJ, You YK, Na GH, Hong TH, Shetty GS, Kim DG. Single-port laparoscopy-assisted donor right hepatectomy in living donor liver transplantation: sensible approach or unnecessary hindrance? Transplant Proc 2012;44:347−52.

[107] Lai Q, Pinheiro RS, Levi Sandri GB, Spoletini G, Melandro F, Guglielmo N, et al. Laparoscopy in liver transplantation: the future has arrived. HPB Surg 2012;148387.

[108] Kiuchi T, Kasahara M, Uryuhara K, Inomata Y, Uemoto S, Asonuma K, et al. Impact of graft size mismatching on graft prognosis in liver transplantation from living donors. Transplantation 1999;67:321−7.

[109] Vauthey JN, Abdalla EK, Doherty DA, Gertsch P, Fenstermacher MJ, Loyer EM, et al. Body surface area and body weight predict total liver volume in Western adults. Liver Transpl 2002;8:233−40.

[110] Chan SC, Liu CL, Lo CM, Lam BK, Lee EW, Wong Y, et al. Estimating liver weight of adults by body weight and gender. World J Gastroenterol 2006;12:2217−22.

[111] Urata K, Kawasaki S, Matsunami H, Hashikura Y, Ikegami T, Ishizone S, et al. Calculation of child and adult standard liver volume for liver transplantation. Hepatology 1995;21:1317−21.

[112] Heinemann A, Wischhusen F, Püschel K, Rogiers X. Standard liver volume in the Caucasian population. Liver Transpl Surg 1999;5:366−8.

[113] Lee SG, Park KM, Hwang S, Lee YJ, Kim KH, Ahn CS, et al. Adult-to-adult living donor liver transplantation at the Asan medical center, Korea. Asian J Surg 2002;25:277−84.

[114] Chouker A, Martignoni A, Dugas M, Eisenmenger W, Schauer R, Kaufmann I, et al. Estimation of liver size for liver transplantation: the impact of age and gender. Liver Transpl 2004;10:678−85.

[115] Fan ST, Lo CM, Liu CL, Yong BH, Wong J. Determinants of hospital mortality of adult recipients of right lobe live donor liver transplantation. Ann Surg 2003;238:864−9.

[116] Lo CM, Fan ST, Chan JK, Wei W, Lo RJ, Lai CL. Minimum graft volume for successful adult-to-adult living donor liver transplantation for fulminant hepatic failure. Transplantation 1996;62:696−8.

[117] Masetti M, Siniscalchi A, De Pietri L, Braglia V, Benedetto F, Di Cautero N, et al. Living donor liver transplantation with left liver graft. Am J Transplant 2004;4:1713−6.

[118] Kiuchi T, Onishi Y, Nakamura T. Small-for-size graft: not defined solely by being small for size. Liver Transpl 2010;16:815−7.

[119] Kiuchi T, Tanaka K, Ito T, Oike F, Ogura Y, Fujimoto Y, et al. Small-for-size graft in living donor liver transplantation: how far should we go? Liver Transpl 2003;9:29−35.

[120] Boillot O, Delafosse B, Mechet I, Boucaud C, Pouyet M. Small-for-size partial liver graft in an adult recipient; a new transplant technique. Lancet 2002;359:406−7.

[121] Troisi R, Ricciardi S, Smeets P, Petrovic M, Van Maele G, Colle I, et al. Effects of hemi-portocaval shunts for inflow modulation on the outcome of small-for-size grafts in living donor liver transplantation. Am J Transplant 2005;5:1397−404.

[122] Lo CM, Liu CL, Fan ST. Portal hyperperfusion injury as the cause of primary nonfunction in a small-for-size liver graft-successful treatment with splenic artery ligation. Liver Transpl 2003;9:626−8.

[123] Man K, Lee TK, Liang TB, Lo CM, Fung PC, Tsui SH, et al. FK 409 ameliorates small-for-size liver graft injury by attenuation of portal hypertension and down-regulation of Egr-1 pathway. Ann Surg 2004;240:159−68.

[124] Troisi RI, Sainz-Barriga M. Successful transplantation of small-for-size grafts: a reappraisal. Liver Transpl 2012;18:270−3.

[125] Lu CH, Tsang LL, Huang TL, Chen TY, Ou HY, Yu CY, et al. Biliary complications and management in pediatric living donor liver transplantation for underlying biliary atresia. Transplant Proc 2012;44:476−7.

[126] Chok KS, Chan SC, Cheung TT, Sharr WW, Chan AC, Lo CM, et al. Bile duct anastomotic stricture after adult-to-adult right lobe living donor liver transplantation. Liver Transpl 2011;17:47−52.

[127] Chang JH, Lee IS, Choi JY, Yoon SK, Kim DG, You YK, et al. Biliary stricture after adult right-lobe living-donor liver transplantation with duct-to-duct anastomosis: long-term outcome and its related factors after endoscopic treatment. Gut Liver 2010;4:226−33.

[128] Barr ML, Belghiti J, Villamil FG, Pomfret EA, Sutherland DS, Gruessner RW, et al. A report of the Vancouver Forum on the care of the live organ donor: lung, liver, pancreas, and intestine data and medical guidelines. Transplantation 2006;81:1373−85.

[129] Brown Jr. RS, Russo MW, Lai M, Shiffman ML, Richardson MC, Everhart JE, et al. A survey of liver transplantation from living adult donors in the United States. N Engl J Med 2003;348:818−25.

[130] Abecassis MM, Fisher RA, Olthoff KM, Freise CE, Rodrigo DR, Samstein B, et al. Complications of living donor hepatic lobectomy—a comprehensive report. Am J Transplant 2012;12:1208−17.

[131] Lo CM. Complications and long-term outcome of living liver donors: a survey of 1,508 cases in five Asian centers. Transplantation 2003;75:12−5.

[132] Hashikura Y, Ichida T, Umeshita K, Kawasaki S, Mizokami M, Mochida S, et al. Donor complications associated with living donor liver transplantation in Japan. Transplantation 2009;88:110−4.

[133] Beavers KL, Sandler RS, Shrestha R. Donor morbidity associated with right lobectomy for living donor liver transplantation to adult recipients: a systematic review. Liver Transpl 2002;8:110−7.

[134] Ringe B, Strong RW. The dilemma of living liver donor death: to report or not to report? Transplantation 2008;85:790−3.

[135] Miller C, Florman S, Kim-Schluger L, Lento P, De La Garza J, Wu J, et al. Fulminant and fatal gas gangrene of the stomach in a healthy live liver donor. Liver Transpl 2004;10:1315−9.

[136] Akabayashi A, Slingsby BT, Fujita M. The first donor death after living-related liver transplantation in Japan. Transplantation 2004;77:634.

[137] Ringe B, Petrucci RJ, Soriano HE, Reynolds JC, Meyers WC. Death of a living liver donor from illicit drugs. Liver Transpl 2007;13:1193−4.

[138] Cheah YL, Simpson MA, Pomposelli JJ, Pomfret EA. Incidence of death and potentially life-threatening "near miss" events in living donor hepatic lobectomy: a world-wide survey. Liver Transpl 2012;19:499−506.

Chapter 22

Donation After Cardiac Death in Liver Transplantation

Keith J. Roberts, Irene Scalera, and Paolo Muiesan

Departments of Liver Transplantation and Hepatobiliary and Pancreatic Surgery, University Hospitals Birmingham NHS Trust, Birmingham, UK

22.1 INTRODUCTION

Donation after circulatory death (DCD) arguably represents the greatest potential way to provide organs for transplantation. The vast majority of people die of causes that do not lead to brain death (98% [1]) and thus DCD could meet the needs of organ transplantation based upon the actual numbers of potential donors alone. Furthermore, DCD liver transplantation has played a unique role in the genesis of solid organ transplantation and paved the way for early pioneers of the speciality.

There are many challenges, however, to successful DCD programs. In contemporary practice, despite large numbers of potential donors, DCD is not the panacea to the shortfall of organs for transplantation due to complications relating primarily to ischemia—reperfusion injury which are responsible for primary and/or delayed graft dysfunction or failure. The liver is particularly sensitive to ischemia—reperfusion injury and displays organ-specific injury in terms of ischemic biliary damage. Ethical issues, societal idiosyncrasies, and religious beliefs have a huge impact upon organ donation. These vary between nations, and within due to multiculturism, and can be so extreme as to prevent or limit DCD. Some of the more important issues that affect DCD follow. Controlled DCD is not possible in some countries where withdrawal of life-sustaining treatment is prohibited; as all religious scriptures were created before organ donation no religion states organ donation should be prohibited though some individuals perceive this to be so [2]; there is a perception by some people that efforts at resuscitation, provided by healthcare professionals, would be negatively influenced if uncontrolled DCD was a possible outcome for the patient in cardiac arrest [3]; certain noninvasive and invasive procedures are required during uncontrolled DCD before consent for donation can be sought. Finally, there are difficult logistic issues surrounding liver procurement, particularly for uncontrolled DCD. Thus the process of DCD is presented with strong challenges from society, individuals, and the technical and logistical process of liver procurement itself.

22.2 CLASSIFICATION OF DCD

DCD can be classified using the modified Maastricht criteria [4] (Table 22.1). In Maastricht category I, death is declared outside of hospital and the potential donor is brought to the hospital without resuscitation attempts. In type II, cardiocirculatory arrest occurs unexpectedly and resuscitation attempts are unsuccessful. In category III, cardiac arrest is induced by withdrawal of life support therapy (WLST) from a patient with brain damage, though not meeting criteria for brain death certification. In type IV, brain death is either declared before unpredicted cardiac arrest or, to respect family wishes, donation proceeds just after cardiac arrest. Thus category IV can be either uncontrolled or controlled. Categories I and II are considered uncontrolled and category III controlled.

Therefore uncontrolled DCD occurs following the unanticipated cardiac arrest of a patient; due to logistical reasons and the associated degree of ischemic injury only deaths occurring at a center with established organ retrieval teams and pathways are suitable for donation of liver grafts (category II). It is possible to overcome some of these logistical challenges by directing healthcare resources outside of the hospital; in order to maximize rates of uncontrolled DCD and optimize donor management mobile medical teams are tasked to patients in out-of-hospital cardiac arrest. Controlled DCD occurs in the presence of organ retrieval teams and limits the ischemic injury associated with death. The process of dying in type III DCD, however, may be associated with a prolonged premortem period of hypotension and/or hypoxia which is ultimately responsible for ischemic injury that may prevent organ donation or be accountable for graft dysfunction or nonfunction of the transplanted organ. In this respect, it is crucial that we recognize that long before cardiocirculatory arrest happens, there is inadequate arterial and portal blood flow through the liver [5].

22.3 DCD LIVER TRANSPLANTATION: THE PAST, A HISTORICAL PERSPECTIVE

The first organ transplants were from DCD subjects. In 1906, a goat kidney and a pig kidney were transplanted into two patients with acute renal failure by Jaboulay [6]; the outcome is obvious to us now but this process utilized anatomical principles that underpin transplantation today and more importantly demonstrate the ingenuity, imagination, and accomplishment that has brought organ transplantation to where it is at present and will continue to take it forward in the future. In 1936, six patients in Kiev received kidney transplants for mercury poisoning. All patients died, presumably due to ischemic injury of the graft as organ recovery occurred hours after death. Appreciation of the importance of ischemic injury associated with DCD led to live donation of kidneys. This was followed by evolution of the surgical technique with implantation of the donor kidney to the iliac vessels and bladder in 1951; this technique remains the standard today.

Due to procedural complexity and recipient morbidity associated with liver failure, attempts at liver transplantation followed renal transplantation by over half a century. The first canine and human transplants were associated with no survivors beyond 2 months. Canine models utilized living or deceased donors whilst human transplants utilized DCD subjects exclusively. Starzl acknowledged the role of ischemic injury associated with DCD and described biochemical evidence of hepatocellular damage, postoperative coagulopathy, and biliary necrosis in transplanted grafts at autopsy [7]. The importance of limiting ischemic injury by hypothermic perfusion and the time from donor death to revascularization was highlighted in these early experiments. Calne, also utilizing DCD donors, recommended limiting the time from death to cold perfusion to <15 min [8]. He also addressed some ethical challenges associated with pairing a DCD organ donor with a suitable recipient by taking the transplant team to the site of the donor hospital as opposed to moving the donor for the purposes of transplantation.

Early liver transplant programs experienced many problems relating to peri- and postoperative care, immunosuppression and poor graft and patient survival. The recognition of brain death [9–11], clinically and legally, led to interest in recovering organs following donation after brain death (DBD). DBD limited ischemic injury and

TABLE 22.1 Modified Maastricht classification of DCD [4a]. Controlled forms of DCD limit ischaemic injury and are associated with the highest quality DCD organs for transplantation. Type 1 DCD yields tissue, not organs, for donation due to the ischaemic injury.

Category	Description	Type of DCD	Location
I	Dead on arrival (at hospital)	Uncontrolled	ED
II	Unsuccessful resuscitation	Uncontrolled	ED
III	Anticipated cardiac arrest (after withdrawal of treatment)	Controlled	ICU and ED
IV	Cardiac arrest in a brain dead donor	Controlled	Theatre
V	Unexpected cardiac arrest in an ICU patient	Uncontrolled	ICU

thus became, and remains, the standard by which cadaveric liver grafts are retrieved and led to the cessation of almost all DCD programs. With the advent of cyclosporine in the 1980s improved graft survival was observed and led to widespread acceptance of liver transplantation for patients with end-stage liver disease. Consequently, the number of potential recipients increased at a rate exceeding the number of brain dead donors (a problem exacerbated by progress in road safety and treatment of conditions that may lead to brain death). Thus the demand for donor liver organs was not, and will never be met, solely by DBD. The reemergence of DCD as a viable organ source has occurred as a direct result of increased demand for donor organs. In the United Kingdom, there has been a 10-fold increase in DCD donors between 2000 and 2010 [12] (Figure 22.1). Early enthusiasm with DCD was directed toward renal transplantation where long-term renal function has been demonstrated comparable to organs obtained following DBD [13−16]. Similar outcomes were not observed in early liver DCD programs where a high incidence of graft dysfunction and failure was observed in addition to biliary complications [17−20]. The Pittsburgh group published early experience

of both uncontrolled and controlled DCD liver transplantation in a small number of subjects [18]. Only one of six uncontrolled DCD liver grafts survived longer than a month. Postreperfusion biopsies demonstrated large areas of centrilobular and/or periportal hemorrhagic necrosis. All controlled DCD grafts functioned though two were lost due to early hepatic artery thrombosis. Equivalent 1-year patient survival between recipients of DBD or DCD liver grafts was observed by D'Alessandro et al. but at the expense of higher rates of retransplantation due to primary nonfunction (PNF) in the DCD group. One-year DCD graft survival was 54% versus 81% for DBD grafts [19]. In a review of United Network for Organ Sharing (UNOS) data independent predictors of early DCD graft failure were prolonged cold ischemic time (CIT) and recipient life support, highlighting the importance of recipient variables in addition to donor variables in the use of DCD liver grafts [20]. By reducing the CIT and with careful donor and recipient selection, however, graft and patient survival has subsequently been demonstrated equivalent between recipients of DBD and DCD livers [21,22]. In 2000, Reich et al. published 100% patient and graft survival amongst eight recipients of controlled DCD livers

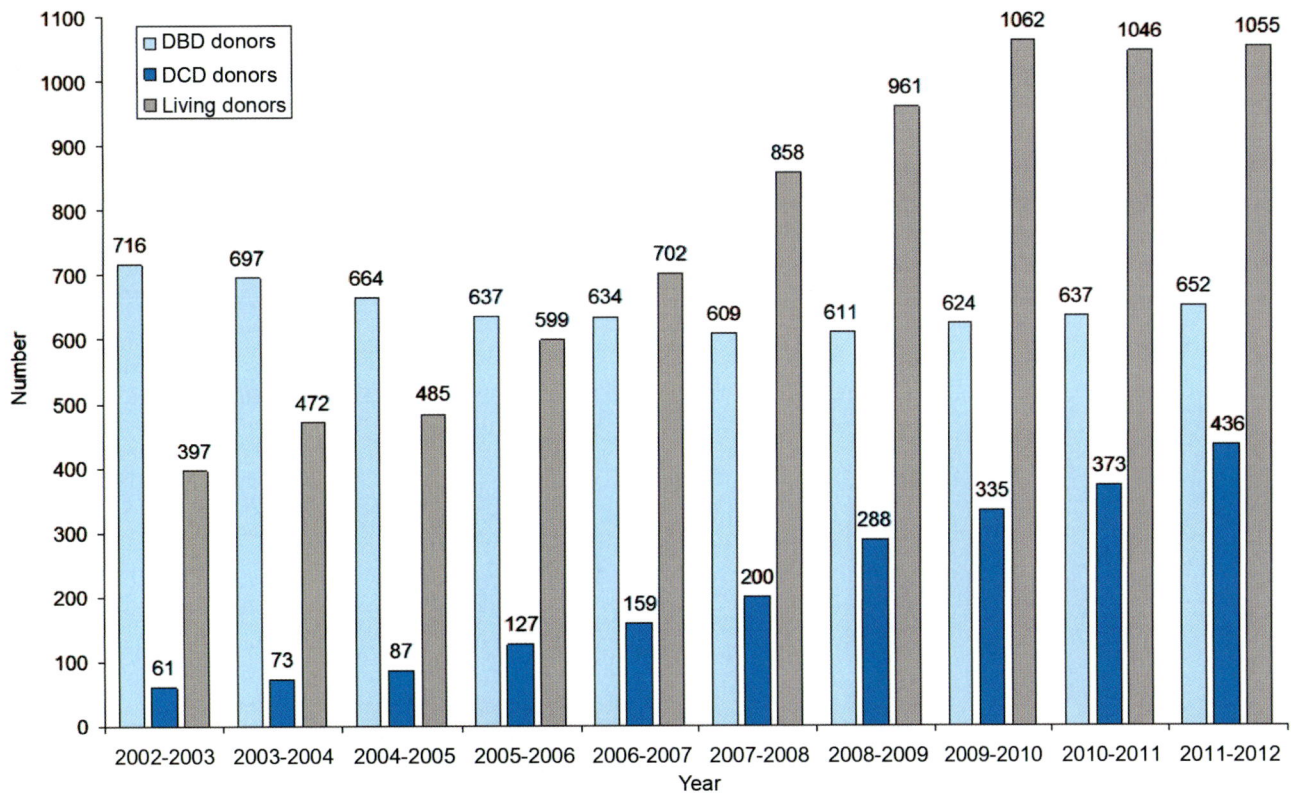

FIGURE 22.1 Steps of controlled donation after circulatory death.

[22]. Although the authors statement that their DCD program significantly expanded the donor pool was a little overstated (5% or eight donors over 3 years) there is no doubt that their excellent graft and patient survival reflects an emphasis upon careful donor selection and in particular a focus upon short premortem warm ischemic intervals. It is interesting to note that over half of the donors were over 50 years old. Subsequently larger numbers of successful DCD liver transplantation programs have demonstrated graft and patient survival equivalent to DBD recipients [21]. Short warm and cold ischemic times in addition to careful donor and recipient selection appear responsible for this equivalency.

A problem almost unique to liver DCD transplantation is the recognition of associated ischemic-type biliary strictures (ITBS) (Figure 22.2). This was first described in 1995 [18] and subsequently recognized to be associated with an increased duration of DWIT [23−25].

22.4 DCD LIVER TRANSPLANTATION: THE PRESENT

The reemergence of DCD as a viable source of liver grafts for transplantation presents challenges, some of which were not present at the beginning of organ transplantation. In the present climate, great emphasis is placed upon clinical outcomes; 5- and 10-year graft and patient survival of recipients transplanted with DBD organs is well established. In health care, interventions yielding results below what is considered a standard of care are not usually tolerated. However, patients continue to die on organ donor waiting lists and thus the use of DCD organs appears justified given the paucity of DBD organs and the inherent risk of donor morbidity and mortality that is associated with living liver donation. Therefore the present technical challenges that face DCD liver transplantation relate to preserving organ function

and achieving outcomes similar to those from the use of DBD grafts. Careful donor and recipient selection is key as the main complications that arise from the use of organs from DCD subjects (delayed graft function, PNF, and other organ-specific complications like ITBS) relate directly to ischemic injury during the donor warm ischemic time (DWIT) and cold storage. The focus of current research aims to limit this ischemic injury.

Ethical and legal considerations are also intrinsically associated with DCD. One contemporary argument in favor of DCD is that it provides deceased patients and their families with the chance of organ donation if brain death is not the cause of death. This is essential to permit the ethical and legal justification for the process of DCD [26,27]. It also benefits those individuals who do not recognize brain death and in whom death can only be accepted following the cessation of the heart beat. The process associated with DCD, however, raises ethical and legal concerns [28−30]; healthcare professionals may feel a conflict of interest and be placed in particularly challenging situations when faced with patients in whom ongoing treatment is futile and who may be suitable DCD donors [31]. Further uncertainty exists on what are acceptable interventions pre- and postmortem prior to DCD, particularly in the case of uncontrolled DCD of the liver, and what conditions must be met in order to confirm death in DCD donors.

22.4.1 Controlled DCD

The current challenges faced by controlled DCD liver transplantation surround ethical issues, achieving equivalent graft and patient survival to recipients of DBD grafts *whilst* expanding the DCD donor pool and decreasing the rate of ITBS. Furthermore there is concern within the transplant community that an active controlled DCD program may impact negatively upon rates of DBD.

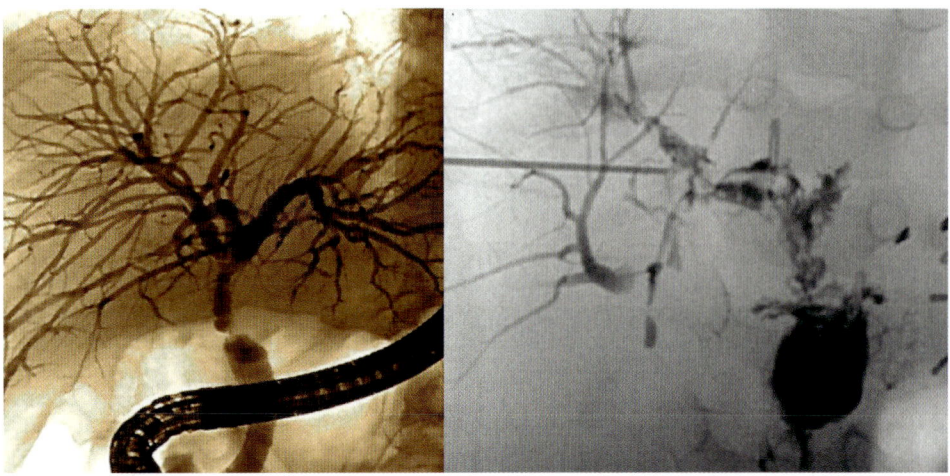

FIGURE 22.2 Progressive increase of DCD donors in the UK 2002−2012. http://www.nhsbt.nhs. uk/. *Source: Transplant activity in the UK, 2011−2012, NHS Blood and Transplant.*

22.4.2 Controlled DCD: An Ethical Perspective

It is essential from an ethical perspective that the decision to undertake WLST is made without consideration to the possibility of DCD. A team separate from that which is responsible for the ongoing treatment of the patient will assess the suitability of donation. The decision to withdraw treatment precedes and must be completely independent from that to consider organ donation. This process is transparent to the public and no conflict of interest can be perceived. When donation is considered possible the family or next of kin are approached and consent for DCD is sought. Organ retrieval teams are dispatched to the donor hospital in a timely manner where WLST subsequently occurs. This process of events with key points is summarized in Figure 22.3. The vast majority of potential DCD donors lack capacity to consent to donation during their final illness. In the United Kingdom providing medical care to patients deemed to be near to and unrecoverable from an end of life state has been limited to what is considered to be in the patient's medical best interests. Thus any procedure or intervention deemed not required for that patient's ongoing medical care could be considered unethical or even unlawful. A more holistic approach to the patient recognizes that an overall benefit arising from interventions may be appropriate when taking into consideration wider emotional, cultural, family, and religious issues [27]. This broader approach to assessing the best interest of an individual has been upheld by the courts and is fundamental in UK law [32–34]. This is an essential requirement for controlled DCD. A person's wish in regards to organ donation can be assessed by discussion with family and by review of national organ donor registries. When organ donation is considered to have been a dying patient's wish then intervention may be appropriate. Appropriate interventions include discussing donation with family, reviewing a patient's past medical history for the purposes of donation, taking blood and serum, maintenance of life-sustaining treatment, delaying WLST until the surgical team is on site and changing a patients location for purposes of WLST and subsequent donation. Controversial interventions include those that can be considered as anything that places the person at risk of harm or distress and include systemic heparinization where this might accelerate death, placement of femoral catheters, and cardiopulmonary resuscitation (CPR) prior to the commencement of organ recovery processes [35].

Within the transplant community, there is growing concern that by promoting controlled DCD some clinicians supporting donation may avoid prolonged neurological testing that is required for DBD. In the United Kingdom, nearly twice as many organs are recovered from the average DBD donor compared to a DCD donor and given the association of delayed graft function with DCD organs this must be avoided. Controlled DCD aims to expand the donor pool and support the already existing DBD programs that cannot meet the needs of potential recipients. The actual situation is unclear with some authors not supporting this view [36] suggesting improvements in road safety and the management of primary brain

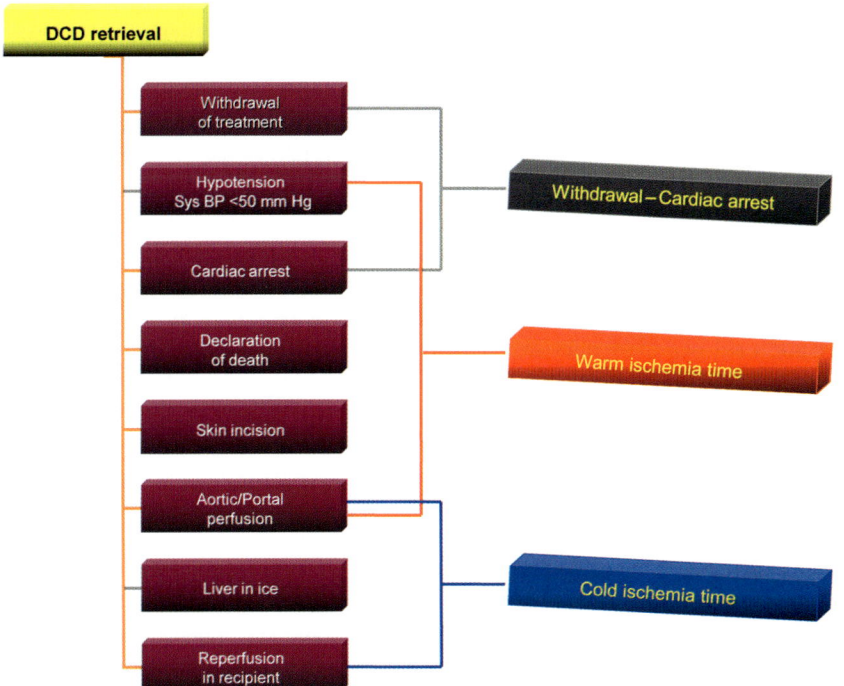

FIGURE 22.3 ERCP and PTC images of ischemic cholangiopathy after DCD liver transplantation

injury are responsible for the fall in DBD rates. A review of organ donation within 23 European countries, however, identified that 18 had demonstrated an increase in DBD rates between 2000 and 2009 [37]. Three of the five countries with decreased rates of DBD were those with expanding and outstanding controlled DCD programs (Belgium, the Netherlands and the United Kingdom).

The location of WLST is controversial. Ideally, to keep the DWIT to a minimum for the purposes of organ donation, this should occur in the theater complex where the organ recovery procedure will occur [22]. In situations where prolonged and protracted travel from the donor's previous site of care to theatre occur this may be essential. At most institutions, this argument cannot be made and thus choosing the best site of WLST is a compromise between providing the final treatment to the patient and their family and respecting the needs of the donor and subsequent recipient. The issues in conflict with treating the needs of the donor include the dying patient's right to dignity and privacy; providing continuity of care by intensive care or emergency department staff; dignity, privacy, and access of family and loved ones. This plan must also take into consideration the outcome if donation does not occur.

22.4.3 Controlled DCD: Donor Selection

The increased incidence of graft dysfunction and nonfunction associated with DCD organs has resulted in great attention being placed upon donor selection. A very strict selection policy of the very best controlled DCD grafts has achieved excellent results so that some of these liver grafts have been reduced to allow transplantation in children [38].

This process is dynamic and is influenced by events following WLST; the key component that is considered is the *functional* donor warm ischemic time (FDWIT). FDWIT was originally defined in the United Kingdom as the duration of time from when a patients systolic blood pressure falls below 50 mm Hg or, it was later added, when the arterial oxygen saturation falls below 70%, whichever comes first, and it ends at the commencement of cold perfusion [35]. A similar definition has been proposed in the United States and makes a distinction between the total DWIT (time from WLST and cold perfusion) and the true DWIT (from when systolic blood pressure falls below 50 mm Hg and cold perfusion).

The tolerance to warm ischemic injury is organ dependent with the kidney being most resistant and hearts least so. Current practice in the United Kingdom is based upon a maximum FDWIT as follows: liver and pancreas, 30 min; lungs, 60 min, and kidneys, 120 min. These times are not absolute but typically represent the maximum duration that would be tolerated. Other factors that affect this time (to decrease it) include advanced donor age, obese body habitus, comorbidity, and other premortem

events. Typically in suboptimal controlled DCD when considering liver donation, the upper limit of FDWIT is reduced to 20 min [35]. Characteristics describing "optimal" and "suboptimal" criteria for donors of liver grafts and other organs are shown in Table 22.2.

Identifying which patients will die following WLST, and ideally how rapidly this occurs, is attractive as this would optimize resource allocation, limit family distress, and identify those donors most likely to provide organs with minimal ischemic injury. A UNOS committee presented five criteria to predict death within 60 min from WLST [39]. These criteria are summarized in Table 22.3. The UNOS criteria have subsequently been validated in an external cohort [40]. There are obvious limitations of these criteria; the 60-min end point does not consider the duration of FDWIT within this time frame and is not organ specific, that is to say potential liver and pancreas donors will be identified in addition to those that do die but outside of the 30-min window specific to those organs; similarly for kidney donors a 60-min target will miss potential donors who die beyond 60 min but within standard or extended criteria for that organ. Furthermore several of the criteria within the UNOS score are uncommon in clinical practice, such as ventricular assist devices and extracorporeal membrane support and thus of limited value. Despite this when three or more criteria are present the likelihood of death within 60 min from WLST exceeds 80% [40]. The original University of Wisconsin DCD evaluation tool utilizes five separate components (Table 22.4) to predict death at 60 min after WLST with nearly 90% accuracy [41] (82% when repeated in an external cohort [40]); these are probably of more value than the UNOS criteria as all criteria are applicable to all potential donors. Although these criteria again identify death within 60 min they can also be used to identify death within 120 min. Other authors have focused upon neurological criteria to predict death in DCD patients [42,43] and again validated in external cohorts [44]. It must be noted that these criteria have not been validated outside of the United States and do not identify which donors will have an FDWIT of 30 min. More recent work has focused upon predicting death in Maastricht-type 3 donors and specifically liver graft usage from a United Kingdom group [45]. Independent predictors of death in these potential donors were age under 40, use of inotopes, and the absence of a cough/gag reflex. Independent predictors that the liver graft would not be subsequently implanted were donor age in excess of 50, BMI in excess of 30, FDWIT in excess of 25 min, an intensive care stay exceeding 7 days, and an ALT exceeding 4× the normal limit. Formulas to predict death and of graft nonusage were created from this data and subsequently successfully tested upon a validation cohort. Use of these formulas could be used to identify suitable donors for DCD liver

TABLE 22.2 Current Criteria to Identify Optimal and Suboptimal DCD Organs

Organ	Criteria	Optimal DCD donor	Suboptimal DCD donor
Liver	Age	< 50 years	> 50 years
	Weight	< 100 Kg	> 100 Kg
	ICU stay	< 5 days	> 5 days
	FWIT	< 20 minutes	> 20 but < 30 minutes
	CIT	< 8 hours	> 8 but < 12 hours
	Steatosis	< 10%	> 15%
Kidney	Age	< 50 years	> 50 years
	Hypertension	None	Present/treated
	CVA	None	Present/previous
	Terminal creatinine	< 133 µmol/l	> 133 µmol/l
	FWIT	< 60 minutes	> 60 minutes
Pancreas	Age	< 45 years	> 45 but < 60 years
	BMI	< 30 (whole organ use)	> 30 (whole organ use)
	FWIT	< 20 minutes	> 20 but < 30 minutes
	Steatosis	Subjective assessment	Subjective assessment
Lung*	Age	< 55 years	
	Blood gas oxygenation	Reasonable with FiO$_2$ 40%	
	Thoracic surgery	No significant previous	
	Chest x-ray	Clear in previous 24 hours	
	Serology/Tissue type	Available at time of retrieval	

*ICU, intensive care unit; FWIT, functional warm ischaemic time (see text); CIT, cold ischaemic time; CVA, cerebrovascular accident; BMI, body mass index; * these are current referral criteria for consideration of lung DCD.*
Modified from [35].

TABLE 22.3 UNOS Criteria for Identifying Potential DCD Patients

Respiratory Failure	Cardiac Support	Respiratory Support	Cardiovascular Support 1	Cardiovascular Support 2
Apnea	LVAD	PEEP ≥ 10 and SaO2 ≤ 92%	Norepinephrine, epinephrine or phenylephrine ≥ 0.2 lg/kg/min	IABP 1:1 OR dobutamine or dopamine ≥ 10 µg/kg/min and CI ≤ 2.2 L/min/m^2
RR < 8	RVAD	FiO2 ≥ 0.5 and SaO2 ≤ 92%	Dopamine ≥ 15 lg/kg/min	IABP 1:1 and CI ≤ 1.5 L/min/m2
RR > 30 during trial off mechanical ventilation	V-A ECMO	V-V ECMO		
	Pacemaker with unassisted rhythm < 30			

RR = respiratory rate; LVAD = left ventricular assist device; RVAD = right ventricular assist device; V-A ECMO = venoarterial extracorporeal membrane oxygenation; PEEP = positive end-expiratory pressure; SaO2 = arterial oxygen saturation; FiO2 = fraction of inspired oxygen; V-V ECMO = venovenous extracorporeal membrane oxygenation; IABP = intra-aortic ballon pump; CI = cardiac index.
Modified from [39].

TABLE 22.4 The University of Wisconsin Donation after Cardiac Death Evaluation Tool [41] for identifying potential DCD patients.

Criteria	Assigned Points	Pt. Score
Spontaneous respirations after 10 min.		
Rate > 12	1	
Rate < 12	3	
TV > 200 cc	1	
TV < 200 cc	3	
NIF < 20	3	
NIF > 20	1	
No spontaneous respirations	9	
Vasopressors/inotropes		
None	1	
Single drug	2	
Multiple drugs	3	
Patient age		
0–30	1	
31–50	2	
51 +	3	
Intubation		
Endotracheal tube	3	
Tracheostomy	1	
Oxygenation after 10 minutes		
SO_2 > 90%	1	
SO_2 80–89%	2	
SO_2 < 79%	3	
	Final Score	
	Time from extubation to expiration	

TV = tidal volume; NIF = negative inspiratory force; SO_2 = oxygen saturation

transplantation and avoid unnecessary family distress, donor team activation and costs. Given that the number of DCD offers in the United Kingdom is increasing by nearly 100% each year evidence-based donor selection may be increasingly utilized [45].

22.4.4 Controlled DCD: Surgical Procedure

Casavilla et al. originally described the super-rapid organ recovery technique for controlled DCD [18]. A midline laparotomy is followed by rapid isolation, cannulation, and perfusion of the distal abdominal aorta followed by venting of the abdominal vena cava, sternotomy, clamping of the descending thoracic aorta, and venting of the supradiaphragmatic inferior vena cava. In order to facilitate rapid cooling of the liver, the portal system is cannulated and perfused followed by packing of the abdominal cavity with ice cold fluid and ice. Depending upon planned pancreatic organ recovery and local practice the inferior mesenteric vein, the superior mesenteric vein at the base of the colonic mesentery, or the portal vein above the duodenum are all suitable sites for portal cannulation. Supraduodenal portal perfusion is preferred in cases of pancreas recovery where it is usual to vent the portal vein as it appears from under the duodenum. Heparin is added to both aortic and portal perfusion fluid given that it is unlawful to administer heparin prior to certification of death in the United Kingdom. Aortic perfusion using gravity alone is insufficient to produce rapid cooling so aortic fluids are pressurized to 200 mm Hg. Careful surgical technique is required to recover the liver during DCD where the hepatic artery is at particular risk to injury due to the frequent occurrence of replaced or accessory hepatic arteries and absence of pulse.

An alternative technique of donor organ recovery consists of *en bloc* removal of the abdominal viscera. Following *in situ* hypothermic perfusion these steps are performed: the distal stomach is divided, the lesser curve vessels are preserved, and the greater omentum and short gastric vessels divided and the stomach is reflected into the thorax; the colon is reflected downwards off the pancreas and the small bowel mesentery is divided; the ureters are identified within the pelvis, divided and reflected cranially; division of the vena cava and aorta at its bifurcation is performed as well as the supracoeliac aorta, suprahepatic vena cava, and wide division of the diaphragm; the organ complex is mobilized off the spine in a cranial direction to include the kidneys, pancreas, spleen, liver and associated soft tissues. On the back table the various organs are dissected. The main advantage is that this procedure is performed viewing the organs posterior surfaces including the aorta and the various arterial ostia. This technique is well suited to DCD organ procurement given that the whole dissection will occur in the cold pulseless phase whether it is *in situ* or *ex situ*. Supporters of this procedure have published case series with no cases of iatrogenic hepatic arterial injury [46,47]. Others report lower hepatic transaminases following *en bloc* dissection compared to standard *in situ* dissection during DBD organ retrieval [48]. The dissection time is less in DCD organ recovery, even in the presence of aberrant hepatic arterial anatomy, with *en bloc* resection [49,50], where possible surgeons can review premortem computerized tomographic imaging to identify normal and aberrant hepatic vascular anatomy [51].

22.4.5 Choice of Preservation Solution

Rapid cooling of the liver with synchronous removal of blood from the hepatic microvasculature is essential to limit warm ischemic injury and should be the focus of the organ recovery team/procedure. For example, using supraduodenal portal perfusion, the correct placement of the cannula is essential. The length of portal vein from the point of cannula insertion to its bifurcation is approximately 5 cm. It is easy to imagine the situation where, in the excitement of the controlled DCD process, a surgeon places the portal perfusion cannula tip too distal within the portal system so that the catheter tip lies within the right or left portal system without appreciating the importance of correct placement. The preservation fluid, however, also plays a role. University of Wisconsin (UW) solution is typically preferred for perfusion

as data from the UNOS database from over 17,000 liver transplants associates this solution with improved graft function compared to histidine–tryptophan–ketoglutarate (HTK) solution. This effect was particularly pronounced in cases of controlled DCD where HTK was associated with early graft loss with an odds ratio of 1.63 compared to UW solution [52]. In an early randomized study of these two solutions, there was no difference in clinical outcomes following liver DBD transplantation though only 60 patients were recruited [53]. There is no randomized evidence from which to directly influence practice in DCD liver transplantation. UW is not the ideal solution, however, as it is viscous, will therefore cool more slowly than less viscous solutions and contains more potassium than other solutions.

22.4.6 Controlled DCD: Recipient Selection and Procedure

It is essential to limit the CIT. This requires clear communication between donor and recipient transplant teams to permit rapid and seamless progression from organ recovery, transport, assessment, preparation, and implantation. An early assessment of the donor graft as to its suitability for transplantation is essential. This may not be without risk. A reassuring opinion on the DCD liver graft by the donor surgeon may drive the recipient team to start the transplant early with the aim of reducing CIT; in case such assessment is inaccurate it may lead to the implantation of a marginal DCD with potentially dramatic consequences. That is why many surgeons are willing to compromise slightly on CIT in order to personally inspect the graft at its arrival before sending the recipient to the operating room. In cases where there is likely to be a protracted recipient hepatectomy the expected duration of CIT must be known. If the hepatectomy is more difficult than anticipated the implanting surgeon may find him/herself committed to hepatectomy having performed irreversible operative procedures but with a CIT in excess of 8 h. Furthermore following the implantation of DCD grafts recipients typically display more physiological stress than those of a DBD graft. This can manifest itself as cardiovascular instability followed by coagulopathy during the remaining operative procedure and as postoperative organ dysfunction which may affect the liver as well as other organs. The requirement for renal support is higher following DCD liver implantation compared to DBD liver grafts [54]. Consequently patients with severe comorbidity, extensive previous upper abdominal surgery, those with previous spontaneous bacterial peritonitis or those requiring a redo liver transplant are typically considered unsuitable recipients of a DCD liver.

22.4.7 Uncontrolled DCD

Uncontrolled DCD presents unique challenges. As the timing of death is uncontrolled skilled resources must be directed rapidly to potential donors to limit DWIT and preserve organ function. This process is resource heavy and is directly responsible for the majority of ethical issues inherent with uncontrolled DCD. It is usual for a potential donor's family not to be present when procedures to preserve organ function are required. These must therefore be performed with presumed consent. There are also issues with obtaining consent from the coroner for organ donation.

Early experience of uncontrolled DCD demonstrated very poor outcomes. All of Starzl's early patients died within 30 days of transplantation [7]; donors were a combination of controlled and uncontrolled DCD and the organ recovery process differed from today's process making comparison difficult. Recognition of warm ischemic injury during organ recovery prevented further efforts at uncontrolled DCD at that time. In 1995, Casavilla et al. reported on the outcome of 14 uncontrolled DCD subjects all of whom suffered in-hospital cardiac arrest [18]. Six organs were transplanted but only one graft survived past 2 months. Two grafts were lost to PNF. All donors were young with a mean age of 24 years; CPR was commenced at the time of circulatory arrest and continued into theater until the commencement of organ retrieval. The mean duration of CPR was 37 min. Therefore despite careful donor selection and relatively short periods of DWIT in the presence of CPR, outcomes were poor. Subsequent series of uncontrolled DCD predominately originate from

Spain where the legal framework supports uncontrolled DCD programs (Table 22.5). Contemporary uncontrolled DCD consists of advanced cardiorespiratory support, femoral cannula placement, extracorporeal membrane oxygenation (ECMO) support and finally organ recovery (described below). Even with these processes and careful donor selection, PNF still affects 10−18% of grafts [55−58]. Graft and patient survival at 1 year is 49−90% and 62−90%, respectively [55,56,58,59]. In the largest published series, there is evidence of improving outcomes; 6-month graft survival was 53% in the first half of the series compared to 88% in the second cohort ($P = 0.024$) [59]. An evolution of technical processes and a better understanding of recipient selection appear responsible for this improvement. Published outcomes of liver uncontrolled DCD series are given in Table 22.5.

There are theoretical advantages to the use of uncontrolled DCD compared to controlled DCD donors. If resuscitation is implemented early following cardiac arrest in uncontrolled DCD donors the duration of warm ischemic injury may be less than that of a controlled DCD donor who suffers a prolonged period of perimortem FDWIT. Uncontrolled DCD donors, if well selected, may be healthier at the time of donation compared to controlled DCD donors who have experienced ICU care up to the point of donation. Finally uncontrolled donors will never become controlled DCD or DBD donors and their inclusion within donor programs truly expands the donor pool.

From certain ethical perspectives, uncontrolled DCD presents fewer challenges than controlled DCD. The process of dying in uncontrolled DCD is a natural

TABLE 22.5 Published Series of Uncontrolled Liver Transplantation with Outcomes

Author	LTx Centre	Year	N uDCD	PNF (%)	HAT (%)	BC (%)	Re-LTx (%)	Graft Survival(%)	Patient Survival (%)	Follow up (months)
Casavailla et al.	Pittsburgh	1995	6	33	17	−	50	17	67	12
Busuttil and Tanaka	UCLA	2003	16	6.25	−	−	−	75	88	12
Otero et al.	La Coruna	2003	20	25	0	5	25	55	80	24
Quintela et al.	La Coruna	2005	10	10	0	0	10	90	90	57
Fondevila et al.	Barcelona	2007	10	10	10	10	25	50	70	23
Suarez et al.	La Coruna	2008	27	18	3.6	25	−	49	62	60
Jiminez − Galanes et al.	Madrid	2009	20	10	0	5	15	80	85.5	12
Fondevila et al. *	Barcelona	2012	34	−	−	12	−	82	70	24

BC−biliary complications; HAT−hepatic artery thrombosis; LTx−liver transplant; PNF−primary non function; * the 34 patients contain 10 patients from the 2007 publication.

spontaneous event and has not occurred following the WLST [60,61]. In some countries where the WLST is prohibited uncontrolled DCD may be the only viable method of DCD. Certain aspects of uncontrolled DCD do, however, raise issues of ethical concern. Postmortem procedures, such as systemic heparinization and cannulation, are required typically without specific consent. In Spanish law, following declaration of death steps to ensure organ viability for the purpose of donation should be made [62]. No consideration of consent is required at this stage although it is for the process of organ retrieval and transplantation. In other countries, the law is less clear. In America, donor cards are considered as legal documents permitting postmortem organ recovery (Uniform Anatomical Gift Act, UAGA; 49).

22.4.8 Uncontrolled DCD: Donor Selection

Due to the established warm ischemic injury and time required to mobilize the organ recovery teams to the donor usually kidney only donation is possible of donors presenting to hospitals with established organ recovery pathways. Uncontrolled DCD donors of livers that are used for transplantation are typically younger, have a lower body mass index and have a more favorable biochemical profile when on ECMO [59].

Due to the risks of PNF and graft dysfunction careful selection of donors is key and resource intensive in uncontrolled liver DCD. The Barcelona experience over 8 years yielded 400 uncontrolled DCD protocol activations. 110 patients were excluded during the process of cardio-respiratory support, then 145 further during ECMO to leave 145 organ donors from which 34 liver transplants were performed [59].

22.4.9 Uncontrolled DCD: Donor Procedure

The sequence of events that lead to liver donation in uncontrolled DCD is unique and includes (i) advanced cardiorespiratory support, (ii) normothermic ECMO (NECMO), and (iii) organ recovery.

1. It is essential that the cardiac arrest is witnessed so that the DWIT is known. Bystander basic life support is provided until advanced life support by medical/paramedical professionals. The duration of resuscitation that is required before death is declared varies according to local protocol; in centers with established programs this is 20 min [59]. Once this has occurred and the donor fulfills basic criteria, such as age limits and mode of death, a no touch period of 5 min is required to confirm death in the absence of cardiac and respiratory function before restarting chest compressions for the purpose of donation. This may be provided by external chest compression devices. Donor coordinators begin the process of organ donation; blood samples are obtained, heparin is given and the organ recovery team mobilized. Perfusion catheters are placed in the femoral artery and vein.

2. A large bore Fogarty catheter is introduced via the contralateral femoral artery and inserted so that the balloon lies within the aorta above the origin of the coeliac axis or in the thoracic aorta. The balloon contains radioopaque contrast so the position is confirmed on X-ray, balloon inflated and the cerebral circulation is isolated from warm perfusion. NECMO is commenced to maintain pump flow in excess of 1.7 L/min. Certain physiological criteria must be met during NECMO including satisfactory pH and temperature. During this period an assessment of the biochemical liver function tests can be performed and reperfusion with warm blood permits an assessment of the liver in near physiological conditions.

3. A full abdominal exploration is performed in the warm phase during NECMO perfusion prior to cold portal and aortic perfusion and final organ recovery. This process is similar to that during DBD organ recovery.

22.4.10 DCD Liver Transplantation: The Future

The present use of DCD liver grafts represents the tip of the iceberg. Presently many controlled DCD grafts are declined due to various adverse factors, including the duration of warm ischemia, macroscopic appearance, and donor comorbidity. The actual fate of these grafts, had they been implanted, is never known. Thus one strategy to increase the usage of these organs would be to use NECMO as an intermediate step between the donation process and implantation. This would, in theory, prevent the ischemic/reperfusion injury and hypothermic injury associated with traditional cold storage and, in addition, allow analysis of ischemic injury by pathological and biochemical assessments before implantation. In a porcine model of extreme warm ischemic injury NECMO was associated with a much improved survival over simple cold storage. Animals experienced 90 min of asystole followed by either NECMO for 1 h or not and then followed by 4 h of cold storage. PNF affected 6/6 animals in the no NECMO group and 1/6 in the NECMO group [63]. A period of NECMO permits an assessment of organ function where NECMO is associated with improved bile production, hemodynamic and biochemical parameters over static cold perfusion [64]. Whether assessment during this period could discriminate between viable and nonviable human grafts remains to be established.

Ex vivo organ perfusion has demonstrated promising results in animal models of both controlled and uncontrolled DCD. Following a 90-min period of warm ischemia pigs were randomized to receive NECMO followed by either of cold storage or normothermic machine perfusion (NMP). NMP was associated with significantly reduced histological features of ischemia−reperfusion injury and reduced expression of pro-inflammatory genes and cytokines compared to cold storage. Interestingly, at the time of euthanasia, the biliary epithelium was normal in NMP group whilst in the cold storage group cholangiocytes were swollen with peripherally displaced nuclei [63]. One of six animals suffered PNF in the cold storage group whilst none did in the NMP group. NMP is potentially more beneficial than hypothermic machine perfusion (HMP) as it provides substrates at body temperature with which hepatocytes can ameliorate previous ischemic injury [65,66]. Furthermore both Kupffer cells and hepatic endothelial cells are sensitive to hypothermic injury [67−71]. NMP has not been trialed in humans to date, however, HMP has. Only the group led by Guarrera used HMP for grafts that were *a priori* marginal. According to the authors, HMP preservation significantly reduces pro-inflammatory cytokine expression when compared with cold storage. In a case-controlled study, HMP was associated with a lower rate of primary dysfunction compared to recipients of standard cold storage grafts (5 vs. 25%, $P = 0.08$) and significantly lower peak levels of AST, ALT, bilirubin, and creatinine [72].

The introduction of uncontrolled DCD could expand the donor pool in countries/regions without an existing program without impacting upon existing controlled DCD or DBD programs. There are, however, many barriers to uncontrolled DCD programs. The provision of trained donor coordinators, local pathways, organ recovery surgeons, and ECMO facilities are required at each site of uncontrolled DCD. This clearly presents logistical and financial challenges. Having met these a program would need to provide sufficient numbers of donors to create a genuine impact. Even at centers with dedicated facilities the contribution of uncontrolled DCD grafts to liver transplant programs is under 5% [59]. Developing links with local prehospital teams and services is essential. Developing uncontrolled DCD organ recovery at centers served by well-developed helicopter emergency medical services (HEMS) would be a logical process. HEMS teams cover large geographical areas, are tasked to a high proportion of patients in cardiac arrest and are trained to a high skill level. In the United Kingdom, it is estimated that by using HEMS teams alone there could be an additional 300−528 potential donors per year [73]. This estimate was based upon patients with a witnessed cardiac arrest, short WIT, and under 55 years old.

A novel technique which aims to decrease the incidence of ITBS following DCD liver transplantation is the combined arterial and portal reperfusion of the graft. It is possible that staged portal perfusion followed by arterial perfusion prolongs the duration of warm ischemia of the biliary tissue. In theory, this strategy would limit a second warm hypoxic injury specific to the hepatic arterial circulation of the biliary tree. In a prospective nonrandomized study of 40 DBD recipients, a lower rate of intrahepatic biliary strictures was observed in the recipients who received combined reperfusion compared to those with sequential reperfusion (0% vs. 21%). The use of DBD grafts, the unusually high incidence of intrahepatic structures in the control group and nonrandomized design of this study limit its applicability. There is clear potential, however, to review this strategy in future studies of recipients of DCD grafts.

22.5 CONCLUSIONS

Liver transplantation using DCD grafts increases the donor pool. However, due to concerns over graft dysfunction, nonfunction and ischemic biliary injury careful donor and recipient selection is essential. Outcomes equivalent to DBD liver transplantation are achievable in such a setting but many potential donor organs are not transplanted as a consequence. Uncontrolled DCD may expand the donor pool further but it is likely that this will be at the expense of reduced graft survival.

Whether controlled DCD programs negatively impact upon rates of DBD is unclear, recent evidence suggests this may be the case. It is essential that the transplant community works alongside allied medical professionals and lay people to maximize all forms of donation.

Strategies to decrease or limit ischemic injury, such as normothermic ECMO and machine perfusion (hypothermic or normothermic), have shown great promise in experimental models. High quality randomized trials are required to investigate their effects in humans. Addressing ethical and legal concerns, and logistical issues are prerequisites for any group wishing to implement an uncontrolled DCD liver transplant program.

REFERENCES

[1] Centers for Disease Control and Prevention. Vital statistics report. 2002.
[2] Randhawa G. Death and organ donation: meeting the needs of multiethnic and multifaith populations. Br J Anaesth 2012;108:88−91.
[3] Volk ML, Warren GJ, Anspach RR, Couper MP, Merion RM, Ubel PA. Attitudes of the American public toward organ donation after uncontrolled (sudden) cardiac death. Am J Transplant 2010;10: 675−80.
[4] Kootstra G, Kievit J, Nederstigt A. Organ donors: heartbeating and non-heartbeating. World J Surg 2002;26:181−4.
[4a] Kootstra G, Daemen JH, Oomen AP. Categories of non-heart beating organ donors. Transplant Proc 1995;27:2893−4.

[5] Hernandez-Alejandro R, Caumartin Y, Chent C, Levstik MA, Quan D, Muirhead N, et al. Kidney and liver transplants from donors after cardiac death: initial experience at the London health sciences centre. Can J Surg 2010;53:93−102.

[6] Morris PJ. Transplantation—a medical miracle of the 20th century. N Engl J Med 2004;351:2678−80.

[7] Starzl TE, Marchioro TL, Huntley RT, Rifkind D, Rowlands Jr DT, Dickinson TC, et al. Experimental and clinical homotransplantation of the liver. Ann N Y Acad Sci 1964;120:739−65.

[8] Calne RY, Williams R. Liver transplantation in man. I. Observations on technique and organization in five cases. Br Med J 1968;4: 535−40.

[9] A Definition of Irreversible Coma. Report of the Ad Hoc Committee of the Harvard Medical School to examine the definition of brain death. JAMA 1968;205:337−40.

[10] Diagnosis of Brain Death. Statement issued by the honorary secretary of the conference of medical royal colleges and their faculties in the United Kingdom. Br Med J 1976;2:1187−8.

[11] Diagnosis of Brain Death. Lancet 1976;2:1069−70.

[12] Manara AR, Murphy PG, O'Callaghan G. Donation after circulatory death. Br J Anaesth 2012;108:108−21.

[13] Weber M, Dindo D, Demartines N, Ambuhl PM, Clavien PA. Kidney transplantation from donors without a heartbeat. N Engl J Med 2002;347:248−55.

[14] Akoh JA, Denton MD, Bradshaw SB, Rana TA, Walker MB. Early results of a controlled non-heart-beating kidney donor programme. Nephrol Dial Transplant 2009;24:1992−6.

[15] Summers DM, Johnson RJ, Allen J, Fuggle SV, Collett D, Watson CJ, et al. Analysis of factors that affect outcome after transplantation of kidneys donated after cardiac death in the UK: a cohort study. Lancet 2010;376:1303−11.

[16] Thomas I, Caborn S, Manara AR. Experiences in the development of non-heart beating organ donation scheme in a regional neurosciences intensive care unit. Br J Anaesth 2008;100:820−6.

[17] Foley DP, Fernandez LA, Leverson G, Chin LT, Krieger N, Cooper JT, et al. Donation after cardiac death: the University of Wisconsin experience with liver transplantation. Ann Surg 2005;242:724−31.

[18] Casavilla A, Ramirez C, Shapiro R, Nghiem D, Miracle K, Bronsther O, et al. Experience with liver and kidney allografts from non-heart-beating donors. Transplantation 1995;59:197−203.

[19] D'Alessandro AM, Hoffmann RM, Knechtle SJ, Odorico JS, Becker YT, Musat A, et al. Liver transplantation from controlled non-heart-beating donors. Surgery 2000;128:579−88.

[20] Abt PL, Desai NM, Crawford MD, Forman LM, Markmann JW, Olthoff KM, et al. Survival following liver transplantation from non-heart-beating donors. Ann Surg 2004;239:87−92.

[21] Muiesan P, Girlanda R, Jassem W, Melendez HV, O'Grady J, Bowles M, et al. Single-center experience with liver transplantation from controlled non-heartbeating donors: a viable source of grafts. Ann Surg 2005;242:732−8.

[22] Reich DJ, Munoz SJ, Rothstein KD, Nathan HM, Edwards JM, Hasz RD, et al. Controlled non-heart-beating donor liver transplantation: a successful single center experience, with topic update. Transplantation 2000;70:1159−66.

[23] Abt P, Crawford M, Desai N, Markmann J, Olthoff K, Shaked A. Liver transplantation from controlled non-heart-beating donors: an increased incidence of biliary complications. Transplantation 2003;75:1659−63.

[24] Kaczmarek B, Manas MD, Jaques BC, Talbot D. Ischemic cholangiopathy after liver transplantation from controlled non-heart-beating donors-a single-center experience. Transplant Proc 2007;39: 2793−5.

[25] Monbaliu D, Van GF, Troisi R, De HB, Lerut J, Reding R, et al. Liver transplantation using non-heart-beating donors: belgian experience. Transplant Proc 2007;39:81−1484.

[26] Coggon J, Brazier M, Murphy P, Price D, Quigley M. Best interests and potential organ donors. BMJ 2008;336:1346−7.

[27] Richards B, Rogers WA. Organ donation after cardiac death: legal and ethical justifications for antemortem interventions. Med J Aust 2007;187:168−70.

[28] Bell MD. Non-heart beating organ donation: old procurement strategy—new ethical problems. J Med Ethics 2003;29:176−81.

[29] Bell MD. Non-heart beating organ donation: in urgent need of intensive care. Br J Anaesth 2008;100:738−41.

[30] Gardiner D, Riley B. Non-heart-beating organ donation-solution or a step too far? Anaesthesia 2007;62:431−3.

[31] Mandell MS, Zamudio S, Seem D, McGaw LJ, Wood G, Liehr P, et al. National evaluation of healthcare provider attitudes toward organ donation after cardiac death. Crit Care Med 2006;34:2952−8.

[32] Ahsan v University Hospitals Leicester NHS Trust. PIQR 1993: P19.

[33] Airedale NHS Trust v Bland. AC 2007;:789.

[34] Mental Capacity Act 2005.

[35] Department of Health. Organ Donation after Circulatory Death. Report of a consensus meeting. Intensive Care Society, NHS Blood and Transplant, and British Transplantation Society. Available from: <http://www.ics.ac.uk/intensive_care_professional/standards_and_guidelines/dcd>; 2010.

[36] Summers DM, Counter C, Johnson RJ, Murphy PG, Neuberger JM, Bradley JA. Is the increase in DCD organ donors in the United Kingdom contributing to a decline in DBD donors? Transplantation 2010;90:1506−10.

[37] Dominguez-Gil B, Haase-Kromwijk B, Van LH, Neuberger J, Coene L, Morel P, et al. European Committee (Partial Agreement) on Organ Transplantation. Council of Europe (CD-P-TO). Current situation of donation after circulatory death in European countries. Transplant Int 2011;24:676−86.

[38] Muiesan P, Jassem W, Girlanda R, Steinberg R, Vilca-Melendez H, Mieli-Vergani G, et al. Segmental liver transplantation from non-heart beating donors—an early experience with implications for the future. Am J Transplant 2006;6:1012−6.

[39] Bernat JL, D'Alessandro AM, Port FK, Bleck TP, Heard SO, Medina J, et al. Report of a national conference on donation after cardiac death. Am J Transplant 2006;6:281−91.

[40] Devita MA, Brooks MM, Zawistowski C, Rudich S, Daly B, Chaitin E. Donors after cardiac death: validation of identification criteria (DVIC) study for predictors of rapid death. Am J Transplant 2008;8:432−41.

[41] Lewis J, Peltier J, Nelson H, Snyder W, Schneider K, Steinberger D, et al. Development of the University of Wisconsin donation after cardiac death evaluation tool. Prog Transplant 2003;13: 265−73.

[42] Rabinstein AA, Yee AH, Mandrekar J, Fugate JE, de Groot YJ, Kompanje EJ, et al. Prediction of potential for organ donation after cardiac death in patients in neurocritical state: a prospective observational study. Lancet Neurol 2012;11:414−9.

[43] Yee AH, Rabinstein AA, Thapa P, Mandrekar J, Wijdicks EF. Factors influencing time to death after withdrawal of life support in neurocritical patients. Neurology 2010;74:1380−5.

[44] de Groot YJ, Lingsma HF, Bakker J, Gommers DA, Steyerberg E, Kompanje EJ. External validation of a prognostic model predicting time of death after withdrawal of life support in neurocritical patients. Crit Care Med 2012;40:233−8.

[45] Davila D, Ciria R, Jassem W, Briceno J, Littlejohn W, Vilca-Melendez H, et al. Prediction models of donor arrest and graft utilization in liver transplantation from Maastricht—3 donors after circulatory death. Am J Transplant 2012;12:3414−24.

[46] Boggi U, Vistoli F, Del CM, Signori S, Pietrabissa A, Costa A, et al. A simplified technique for the *en bloc* procurement of abdominal organs that is suitable for pancreas and small-bowel transplantation. Surgery 2004;135:629−41.

[47] de Ville de GJ, Reding R, Hausleithner V, Lerut J, Otte JB. Standardized quick *en bloc* technique for procurement of cadaveric liver grafts for pediatric liver transplantation. Transplant Int 1995;8:280−5.

[48] Imagawa DK, Olthoff KM, Yersiz H, Shackleton CR, Colquhoun SD, Shaked A, et al. Rapid *en bloc* technique for pancreas-liver procurement. Improved early liver function. Transplantation 1996;61:1605−9.

[49] Jeon H, Ortiz JA, Manzarbeitia CY, Alvarez SC, Sutherland DE, Reich DJ. Combined liver and pancreas procurement from a controlled non-heart-beating donor with aberrant hepatic arterial anatomy. Transplantation 2002;74:1636−9.

[50] Sindhi R, Fox IJ, Heffron T, Shaw Jr. BW, Langnas AN. Procurement and preparation of human isolated small intestinal grafts for transplantation. Transplantation 1995;60:771−3.

[51] Roberts KJ, Malde DJ, Adams B, Hodson J, Sheridan M, Hidalgo E. Assessment of hepatic arterial anatomy prior to organ recovery. Open J Organ Transplant Surg. 2012;2(4):28−31.

[52] Stewart ZA, Cameron AM, Singer AL, Montgomery RA, Segev DL. Histidine−tryptophan−ketoglutarate (HTK) is associated with reduced graft survival in deceased donor livers, especially those donated after cardiac death. Am J Transplant 2009;9:286−93.

[53] Erhard J, Lange R, Scherer R, Kox WJ, Bretschneider HJ, Gebhard MM, et al. Comparison of histidine−tryptophan−ketoglutarate (HTK) solution versus University of Wisconsin (UW) solution for organ preservation in human liver transplantation. A prospective, randomized study. Transplant Int 1994;7:177−81.

[54] Leithead JA, Tariciotti L, Gunson B, Holt A, Isaac J, Mirza DF, et al. Donation after cardiac death liver transplant recipients have an increased frequency of acute kidney injury. Am J Transplant 2012;12:965−75.

[55] Fondevila C, Hessheimer AJ, Ruiz A, Calatayud D, Ferrer J, Charco R, et al. Liver transplant using donors after unexpected cardiac death: novel preservation protocol and acceptance criteria. Am J Transplant 2007;7:1849−55.

[56] Jimenez-Galanes S, Meneu-Diaz MJ, Elola-Olaso AM, Perez-Saborido B, Yiliam FS, Calvo AG, et al. Liver transplantation using uncontrolled non-heart-beating donors under normothermic extracorporeal membrane oxygenation. Liver Transplant 2009;15:1110−8.

[57] Otero A, Gomez-Gutierrez M, Suarez F, Arnal F, Fernandez-Garcia A, Aguirrezabalaga J, et al. Liver transplantation from Maastricht category 2 non-heart-beating donors. Transplantation 2003;76:1068−73.

[58] Quintela J, Gala B, Baamonde I, Fernandez C, Aguirrezabalaga J, Otero A, et al. Long-term results for liver transplantation from non-heart-beating donors maintained with chest and abdominal compression-decompression. Transplant Proc 2005;37:3857−8.

[59] Fondevila C, Hessheimer AJ, Flores E, Ruiz A, Mestres N, Calatayud D, et al. Applicability and results of Maastricht type 2 donation after cardiac death liver transplantation. Am J Transplant 2012;12:162−70.

[60] Huddle TS, Schwartz MA, Bailey FA, Bos MA. Death, organ transplantation and medical practice. Philos Ethics Humanit Med 2008;3:5.

[61] Kaufman BJ, Wall SP, Gilbert AJ, Dubler NN, Goldfrank LR. Success of organ donation after out-of-hospital cardiac death and the barriers to its acceptance. Crit Care 2009;13:189.

[62] Royal Decree 2070/1999, December 30, 1999, Article 10, Section I: Protocol of diagnosis and certification of death for the extraction of organs from deceased donors. Available from: <www.noticias.juridicas.com/base_datos/Admin/rd2070-1999.html#anexo1>; 2012.

[63] Fondevila C, Hessheimer AJ, Maathuis MH, Munoz J, Taura P, Calatayud D, et al. Superior preservation of DCD livers with continuous normothermic perfusion. Ann Surg 2011;254:1000−7.

[64] Gong J, Lao XJ, Wang XM, Long G, Jiang T, Chen S. Preservation of non-heart-beating donor livers in extracorporeal liver perfusion and histidine−trytophan−ketoglutarate solution. World J Gastroenterol 2008;14:2338−42.

[65] Butler AJ, Rees MA, Wight DG, Casey ND, Alexander G, White DJ, et al. Successful extracorporeal porcine liver perfusion for 72 h. Transplantation 2002;73:1212−8.

[66] Imber CJ, Peter St SD, Lopez dCI, Pigott D, James T, Taylor R, et al. Advantages of normothermic perfusion over cold storage in liver preservation. Transplantation 2002;73:701−9.

[67] Belzer FO, Ashby BS, Huang JS, Dunphy JE. Etiology of rising perfusion pressure in isolated organ perfusion. Ann Surg 1968;168:382−91.

[68] Hansen TN, Dawson PE, Brockbank KG. Effects of hypothermia upon endothelial cells: mechanisms and clinical importance. Cryobiology 1994;31:101−6.

[69] Jain S, Xu H, Duncan H, Jones Jr. JW, Zhang JX, Clemens MG, et al. *Ex vivo* study of flow dynamics and endothelial cell structure during extended hypothermic machine perfusion preservation of livers. Cryobiology 2004;48:322−32.

[70] Maathuis MH, Manekeller S, van der Plaats A, Leuvenink HG, 't Hart NA, Lier A,B, et al. Improved kidney graft function after preservation using a novel hypothermic machine perfusion device. Ann Surg 2007;246:982−8.

[71] Xu H, Lee CY, Clemens MG, Zhang JX. Inhibition of TXA synthesis with OKY-046 improves liver preservation by prolonged hypothermic machine perfusion in rats. J Gastroenterol Hepatol 2008;23:212−20.

[72] Guarrera JV, Henry SD, Samstein B, Odeh-Ramadan R, Kinkhabwala M, Goldstein MJ, et al. Hypothermic machine preservation in human liver transplantation: the first clinical series. Am J Transplant 2010;10:372−81.

[73] Roberts KJ, Bramhall S, Mayer D, Muiesan P. Uncontrolled organ donation following prehospital cardiac arrest: a potential solution to the shortage of organ donors in the United Kingdom? Transpl Int 2011;24:477−81.

Chapter 23

Artificial Liver Support

Xavier Wittebole[a], Diego Castanares-Zapatero[a], Christine Collienne[a], Olga Ciccarelli[b], Philippe Hantson[a], and Pierre-François Laterre[a]

[a]Critical Care Unit, St Luc University Hospital, Université Catholique de Louvain, Brussels, Belgium, [b]Starzl Unit Abdominal Transplantation, St Luc University Hospital, Université Catholique de Louvain, Brussels, Belgium

Chapter Outline

23.1 INTRODUCTION

Liver failure is a major health problem across the world. It results from a loss in liver cell mass leading to various clinical and biological symptoms. Three major presentations are described in the literature [1]. The term fulminant liver failure was initially used by Charles Trey to describe a potentially reversible condition resulting from an insult to the liver and leading to encephalopathy within 8 weeks of symptoms [2]. This definition was later reviewed by O'Grady et al. [3], who introduced the notion of hyperacute, acute, and subacute liver failure according to the interval between the development of jaundice and encephalopathy, two major characteristics already described by Trey. In parallel, Bernuau et al. divided the syndrome of acute liver failure (ALF) into fulminant and subfulminant liver failure [4]. ALF, developed in an otherwise healthy liver, is a rare condition and is mainly due to toxic substances like paracetamol (acetaminophen) in western countries while virus (A, B, and E) account for the principal etiology in other parts of the world [5]. Acute-on-chronic liver failure (AoCLF) is a more recently defined syndrome [6]. It occurs when a precipitating event, such as variceal bleeding, infection, acute alcohol abuse, adverse drug reaction, or surgery, precipitates rapid liver decompensation in a patient with stable chronic liver disease. Finally, liver failure may occur as a result of end-stage chronic liver failure.

While sophisticated critical care protocols have been developed and have contributed to the improved outcome of those patients, liver transplantation (OLT) is often viewed as the last and only effective treatment for ALF and AoCLF not responding to medical treatment. Despite never being demonstrated for ALF in a randomized trial nor a carefully constructed case comparison study, this approach has to be viewed as an integral management plan of those patients [7]. Indeed, and despite increased age for the donors, outcome for OLT in ALF has improved over the last 20 years [8]. Improved surgical techniques as well as better use of immunosuppressive agents are certainly part of this success. However, as for some other medical resources, there is no balance between the number of patients in need of OLT and the available resources [9,10]. The use of marginal donors and other so-called extended criteria for donation and living donation were all developed to counteract this existing gap. However those measures are not sufficient. Furthermore,

Regenerative Medicine Applications in Organ Transplantation.

new indications for liver transplantation, such as acute alcoholic hepatitis [11], chronic steatohepatitis-related cirrhosis [12], or liver cancers [13], are either emerging or increasing. Hence, the shortage of organs is even more pronounced. Therefore, there is an urgent need for a therapy to treat all these patients with various kinds of liver failure.

On the other hand, the liver displays the extraordinary capacity of regeneration. This is another valuable reason to provide temporary hepatic support to patients, while waiting for this spontaneous recovery. This could avoid unnecessary liver transplantation, decreasing treatment costs and making organs available for those patients whose liver does not recover, while limiting the need for immunosuppressive therapy.

Having all those reasons in mind, and because medical treatment of liver failure is mainly supportive, many teams over the world developed over the last 60 years various systems and techniques aiming at providing liver support during the time patients were suffering ALF or AoCLF. Those systems may be classified as artificial or nonbiological (without any living material) and bioartificial techniques. The first techniques do include various plasmapheresis and dialysis techniques, and the concept of albumin dialysis. The latter techniques are based on the use of biological resources, such as blood, liver cells, or even whole liver. The potential for those liver support techniques and devices to have an impact on the need for OLT and outcomes thereafter certainly remains an exciting and major challenge in medicine [14].

23.2 LIVER SUPPORT TECHNIQUES

23.2.1 Exchange Transfusion and Total Body Washout

Based upon the idea that hepatic coma is metabolic in origin and because of the extraordinary capacity of the liver to regenerate after an insult, various authors hypothesized that exchange transfusion (ET) in patients with acute yellow atrophy might remove sufficient toxic substances to allow the liver to recover. The effect of ET was assessed as soon as 1916 for the treatment of carbon monoxide poisoning. This technique was later described as a therapeutic tool for hemolytic disease of the newborn [15]. In the particular setting of liver failure, ET was initially reported in the late 50s [16] and was further promoted by Trey et al. [17]. In a series of seven cases of ALF complicated by coma, they observed recovery from coma in all patients while five survived the episode of acute hepatic failure. Complications described in this case series include septicemia and fatal bronchospasm. Other side effects of this treatment did include anaphylaxis and hypotension. In the Boston Hepatic Surveillance Study, the survival rate was 20% out of 168 patients while it was

only 8% out of 116 patients treated conventionally [18]. The dramatic improvement in outcome was not confirmed in other series despite some improvement in neurological parameters like EEG findings [19]. Total body washout (TBW) represents an extreme in the use of ET for ALF. In this technique proposed by Klebanoff et al. [20,21], patients undergo complete exsanguination in conjunction with profound hypothermia. Case reports and case series display potential beneficial effects in patients mainly with Reye's syndrome [22–25]. During the procedure, severe cardiac arrhythmias including ventricular fibrillation may occur. In a dog model of liver insufficiency, the TBW technique was found to be superior to liver perfusion to clear bilirubin [26]. Beyond the nondemonstrated effect on mortality in randomized trials, the main issue with those two techniques lies within the amount of blood required to allow adequate ET and TBW. This may rise to 58 L of whole blood for a single patient [19] and therefore places a heavy and unacceptable burden on the hospital blood bank.

23.2.2 Cross Circulation

To avoid the above-mentioned need for a large amount of blood products, parabiotic cross circulation was developed. It was demonstrated to reverse the clinical, biochemical, and hematological findings characteristic of the hepatic coma in a dog model of total hepatectomy-related ALF [27]. Cross circulation in man was performed through arteriovenous forearm canula interconnected artery to vein [28]. Cross circulation directly through the portal vein after umbilical vein catheterization has also been reported [29]. Partners for this technique were either cancer patients with normal liver function [29,30], normal healthy volunteers [30], or brain dead patients with persistent cardiac activity. Despite a decreased use of blood products as opposed to ET, the potential risk for transmission of viral hepatitis as well as other hematogenous pathogens to healthy subjects limited this technique to toxic-related ALF or alcoholic hepatitis-related liver failure. Therefore, cross circulation with baboon partners was developed in South Africa [31,32] and further used by others [33]. Animal's blood volume was washed out with Ringer's Lactate solution under hypothermia conditions before it was replaced with human blood. The animal was then used for cross circulation in patients with severe hepatic coma. Some improvement in neurological status could be achieved and some authors reported complete liver function recovery in alcoholic hepatitis for instance [34]. Cross circulation with baboons is, as it is for ET, a procedure requiring a large amount of blood; indeed, to avoid any hemolysis in the patient due to immunological incompatibilities, the baboon's blood is replaced before the procedure by human blood. The method was therefore found unpractical and was abandoned.

Another technique also referred to as cross perfusion involves dialysis of the patient's blood against the blood of another volunteer or an animal (xeno-cross dialysis) [35, reported in 36]. The use of animal liver tissue preparations in this particular context is also reported [37]. While beneficial to the patients described in the literature, those techniques were barely reported and abandoned.

23.2.3 Plasmapheresis

Developed by Lepore et al. in New York because they could not find compatible blood for their patient in fulminant hepatic failure-related coma, the authors proposed to perform plasmapheresis [38,39]. The concept was based upon the idea that red blood cells probably had little significance in the development of encephalopathy. Since plasma was found to be more easily obtainable than fresh whole blood and more conveniently stored for a long period of time, plasmapheresis progressively replaced whole blood ET. Furthermore, and as opposed to ET and cross perfusion, another advantage of this technique includes the fact that no third party is involved in the procedure. This treatment was also applied with some success in chronic liver failure and cirrhosis [40].

While still in use in some centers, plasmapheresis is usually limited to ALF related to Wilsonian crisis [41,42], acute humoral rejection [43,44] and early liver dysfunction after living donor liver transplantation [45] or large hepatectomy [46,47]. Many bioartificial devices also use the principle of plasmapheresis in their systems (see the chapter on those devices).

The procedure further evolved to the so-called "high volume plasma exchange" in which exchanges of up to 8 L of plasma may be performed over a 7-h period [48]. Neurological, hemodynamic, and metabolic parameter improvements are reported [48−51]. Likewise, total plasma exchange (calculated to exchange one plasma volume) was reported in a series of 39 patients suffering ALF [52]. In this retrospective analysis, liver function tests improved significantly and renal function tended to improve.

To increase its potential efficacy in the removal of cytokines and to reduce potential side effects, plasmapheresis has also been associated in series or in parallel with continuous hemofiltration [53] and hemodiafiltration [54]. Only case series were reported.

23.2.4 Extracorporeal Liver Perfusion

Extracorporeal whole liver perfusion (ECLP) is another concept aimed at providing support to patients with hepatic failure. Actually, the story of liver perfusion starts in 1855 when Claude Bernard used such preparation to

demonstrate the role of the liver in regulating blood sugar levels [55]. The improvement of blood pump oxygenators for cardiac surgery in the 50s allowed searchers to develop extracorporeal organ perfusion of many kinds. Thus, about one century after Claude Bernard's landmark study, Otto et al. demonstrated a significant decrease in ammonia levels in hepatectomized dogs treated with an allogenic ECLP system as compared to control animals [56]. The first human experiment was conducted at the University of Bombay, India, in 1964 [57]. Five patients were treated with a cadaveric human liver ECLP system. Four patients died while one 27-year-old man completely recovered from ALF. At the same time, Eiseman et al. started this procedure in seven alcohol-related cirrhotic patients whose clinical situation was characterized by ALF secondary to gastrointestinal hemorrhage due to variceal or ulcer bleeding [58]. A 17-year-old boy suffering from acute viral hepatitis was also treated with the same procedure. Eleven procedures using pig liver in a perfusion chamber were performed on those eight patients. As described in the cross circulation technique using liver of baboons, livers used for ECLP had to be asanguineous and washed from their own blood before they were perfused with patient's blood. Beyond the fact the procedure was globally well tolerated, the authors observed a sometimes dramatic improvement in neurological condition during or after, in at least eight procedures; some patients displayed clear signs of awakenings and were even able to talk coherently despite the fact they were in deep coma prior to the procedure. One patient developed hypotension and some of them presented with various degree of bleeding since heparin was used as an anticoagulant. The observed beneficial effects were thought to be related to the clearance of some toxins like ammonia; however, the authors also hypothesized that the new liver might also add some needed metabolites.

By 1975, approximately 100 patients were treated with a whole liver perfusion technique [59] and by early 2000, 270 patients in acute, subacute, or chronic liver failure had undergone 527 liver perfusions in 49 centers around the world, with livers from various different species (human, nonhuman primates, pigs, cow, and calves) [60].

Since animal livers (mainly from pigs) were used for the technique, immunologic reactions were thought to be an issue; however, severe reactions were only reported in five patients, including three presenting with anaphylactic reactions [60]. Immunohistologic examination of the used livers demonstrated some damage not consistent with acute rejection [61] while other authors described signs of humoral rejection, such as deposition of immunoglobulins, complement, and fibrinogen [62]. Another major issue with the use of pig livers for ECLP was the potential transmission of the porcine endogenous retrovirus (PERV).

Two patients treated with transgenic porcine liver perfusion before liver transplantation remained negative for serial polymerase chain reaction analysis of their peripheral blood polymorphonuclear cells for PERV [63].

Age below 40 years, coma stage lower than III and IV, total perfusion time over 10 h, hepatitis B as the cause of ALF as well as the use of baboon and human livers were identified as independent positive prognostic markers for improved survival [60].

Actually, in terms of outcome, ECLP achieved its better results when used as a bridge to liver transplant. Started in the early 90s at Johns Hopkins Medical Institution, Baltimore, the procedure was only reported on a few times [64]. Interestingly, some authors used human livers found not suitable for liver transplantation as the liver for ECLP before patients were transplanted with adequate liver [65]. Despite the poor quality of the liver used, however, they observed some neurological and biological improvements. However, the combined ECLP—liver transplantation procedure did not appear to have better efficacy than other bioartificial support methods in use until 2000 [60].

Unavailability of whole organs, their limited viability, and the development of other techniques, especially those directed towards utilizing cell cultures, led to a decreased interest in whole liver perfusion techniques.

Recently, Bikhchandani et al. developed an ECLP system with a semipermeable membrane (Evaclio EC4A, Kuraray Medical Inc., Tokyo, Japan), which separates the two blood compartments of the system [66]. On one side, we find patient's blood, on the other porcine blood which will perfuse the extracorporeal liver. The separation of those two compartments allows a decreased risk of PERV transmission. In an *in vitro* experiment, the system was found to be effective on galactose elimination, ammonia clearance, and *para*-aminobenzoic acid metabolism. The use of this system in humans has not been reported so far and it is certainly too early to propose this system in a clinical setting.

23.2.5 Hemodialysis and Related Techniques

Because of its success in treating patients with renal failure, unmodified hemodialysis was proposed during the late 50s and 60s as a way to treat patients with ALF. Indeed, ammonia, a molecule known to be involved in hepatic coma for a long time [67], was demonstrated to be efficiently removed by hemodialysis in patient with hepatic coma, leading to improvement of patient's clinical condition when dialysis was effective [68]. Peritoneal dialysis has also been proposed in those patients with hepatic coma [69], introducing the debate on which dialysis technique to use in liver failure patients.

However, despite some clearance of ammonia, serum ammonia levels were not substantially lowered with hemodialysis [70]. The lack of hemodialysis in improving hepatic encephalopathy led to the development of other types of filtration systems, including the so-called sorbent hemoperfusion techniques in which blood circulates over a sorbent material for the purpose of removal of toxins from the patient's blood. Three classes of sorbents were used (and still are in some devices).

Charcoal hemoperfusion was used for a long time as an add-on therapy in renal failure [71] as well as in some cases of acute intoxication with some substances, such as barbiturates or gluthethimide. Indeed, it efficiently removes molecules in the 1000—1500 kDa range but neither efficiently removes ammonia nor protein-bound molecules. To avoid complement activation and decreased platelet and neutrophil count, common side effects of this technique, charcoal is usually coated on albumin or encapsulated in hydrophilic gels like cellulose [72]. While found beneficial in a dog model of ischemic ALF in term of prolonged survival [73], in two human randomized trials, neither the duration of charcoal hemoperfusion (10 h versus 5 h) nor the technique itself as compared to standard usual care could demonstrate any superiority in treating patients with various causes of fulminant hepatic failure [74].

Synthetic resin and anion exchange resin hemoperfusion, on the other hand, are more effective at removing lipid-soluble and protein-bound molecules. Resins may be used in hemoperfusion devices (with whole blood passing through the resin) as well as with plasma perfusion devices (solely the plasma separated from the whole blood passes through the resin). Initially described by Willson et al. [75], the technique was tested at King's College Hospital, London [76]. Perfusion of blood from ALF dogs through the Amberlite-XAD2 resin was associated with a fall in serum concentration of bilirubin and the complete clearance from the blood of the ^{14}C-labeled sodium glycolate [76]. However, mortality in those dogs was not affected. Those effects on bilirubin levels were further confirmed with other types of resins [77—79]. Some beneficial effects were described in humans with fulminant hepatic failure but those results came from non-randomized trials [80]. Some cytokines (IL-1, TNFα, but not IL-6) and endotoxin were also affected by *in vitro* and *in vivo* hemoperfusion in man [81]. Interestingly, the porous anion exchange resin Medisorba BL-300 (Kuraray Medical Inc., Tokyo, Japan) is in use in Japan where it is recommended for the treatment of hyperbilirubinemia caused by many diseases [82].

Those three techniques (hemodialysis, perfusion on charcoal, and perfusion on resins) may be used altogether within the same device (some of the bioartificial devices, the molecular adsorbent recirculating system, . . .) in order to improve efficacy in treating patients with liver failure. Working as the Biologic-HD system used to treat uremia, the BioLogic DT system (also known as HemoCleanse, HemoCleanse, Inc., Lafayette, IN) is one of those devices. It includes a cellulosic plate dialyzer and 1.5−2 L sorbent suspension, on the dialysate side. This suspension contains powdered charcoal of high surface area, sodium-loaded IRP-69 (a cation exchanger), polymeric flow inducing agents and salts. Since the plate dialyzer serves as a pump, withdrawing and returning blood, a single blood access is sufficient. The general principle of this technique is called fluidized bed adsorbent system and displays several theoretical advantages [83]. In preclinical [84] and clinical studies [85,86], this system was found to allow effective mass transfer for selected dialyzable toxins (such as aromatic amino acids, glutamine, mercaptans, bile acids and others) [87] and many drugs, even those with high protein binding. In a series of 15 patients presenting with acute hepatic failure and coma treated for 8 to 12 h a day with the system, neurological and physiological status both improved [86].

The system further evolved to the BioLogic-DTPF (detoxifier/plasma filter) system in which two plasma filters are added downstream from the plate dialyzer, in order to allow plasma to pass out of the blood, contact powdered charcoal in suspension, and then return to the blood [87]. Clinical data with this system are rather limited and do confirm neurological improvement [88]; however, a positive effect on survival could never be demonstrated.

23.2.6 Albumin Dialysis with the Molecular Adsorbent Recirculating System

Albumin, a long-time and widely used molecule as volume expander, has been shown to display, beyond its oncotic properties, many other biological activities [89]. A high capacity for transport of various molecules, including water nonsoluble molecules, free radical scavenging and catalysis of S-nitrosothiols leading to vascular tone control, and modulatory effect on neutrophil function are the major characteristics involved in the concept of albumin dialysis with the Molecular Adsorbent Recirculating System (MARS™, Gambro).

In summary, the patient's blood is dialyzed against an albumin containing solution across a high-flux permeable membrane with albumin-related binding sites that have a 50- to 60-kDa limit, thus avoiding endogenous albumin, hormones, and carrier proteins to pass through the membrane. Protein-bound and albumin toxins that accumulate during liver failure are released from the protein binding by physicochemical interactions with the membrane and cleared from blood by diffusion. Purified albumin

molecules pick up the toxins on the dialysate side by specific binding, thereby creating a continuous concentration gradient for these molecules. The albumin dialysate is recycled and cleaned through a charcoal column and an anion exchange resin. Water-soluble substances, such as urea and plasma creatinine, are removed from the albumin dialysate by conventional hemodialysis or hemodiafiltration, using the mechanisms of diffusion and/or convection. Hence, the system has three different compartments. The blood circuit is an extracorporeal circuit like any other circuit used for people with renal failure for instance. The main difference lies within the high-flux permeable membrane used in the MARS™. The second circuit is an albumin circuit filled in with about 600 mL of an albumin 20% solution. Finally, there is the dialysate circuit for usual dialysis or hemofiltration. Therefore, the use of MARS™ requires a standard dialysis or hemofiltration device on top of the system used for the albumin solution, as well as the system used for the circulation of blood in the first compartment.

The use of this device is supposed to block the "auto-intoxication in liver failure following a first hit to the liver" hypothesis [90]. MARS™ is certainly the system for which the most extensive literature exists. Developed in the early 90s by Stange et al. in Rostock, Germany [91,92], the MARS™ technique was applied for the first time in humans in 1996. The initial report demonstrated, in patients with a life-threatening exacerbation of chronic liver failure not responding to "state-of-the-art" standard therapy, an effective clearance of bile acids, bilirubin, creatinine, and ammonia [93]. An unexpected high survival rate of 69% in those 13 patients was observed with improvement in various clinical endpoints, such as the Child-Turcotte-Pugh score and the grade of hepatic encephalopathy.

In the first randomized clinical trial performed, the same team evaluated the efficacy of MARS™ in treating patients with type I hepatorenal syndrome as compared to continuous hemodiafiltration on top of standard medical treatment [94]. MARS™ treatment was carried out for 6−8 h a day, for a maximum of 10 days. A significant prolongation of survival was observed in the MARS™ group: mortality rate was 100% at day 7 in the control group while it was 62.5% in the MARS™ group at the same time, and 75% at day 30. MARS™ treatment was well tolerated throughout all single sessions. Because of slow enrollment, the study was prematurely stopped after enrolling 13 out of the 28 patients initially planned. None of the patients included in the trial were treated with vasopressin analogs or placement of a transjugular intrahepatic portosystemic shunt, a procedure contraindicated in the studied population because of its high severity (bilirubin level above 15 mg/dL and severe hepatic encephalopathy).

Over the next years, various clinical and biological responses were observed and confirmed by using MARS™, mainly in patients with AoCLF. Unfortunately, numerous data come from case reports, case series, and observational studies and firm conclusions on the potential benefit of MARS™ on mortality for instance are difficult to define.

23.2.6.1 Effects on Mortality

Indeed, the effect of MARS™ on survival was evaluated only in a small number of randomized studies. A prolonged survival was observed in the above-mentioned study [94]. However, even in the MARS™-treated group, mortality at 30 days remained extremely high. In another German randomized trial evaluating the effect of albumin dialysis in 23 patients with cirrhosis, superimposed ALF and bilirubin level above 20 mg/dL, the 30 days mortality was decreased as compared to the group receiving standard medical therapy (SMT) alone (1/12 vs. 5/11, $P < 0.05$) [95]. A trend towards improved survival was observed in a study on patients with hypoxic liver failure after cardiogenic shock [96]. The improved survival of 50% in the MARS™ group (while it was only 32% in the control group) did not reach statistical significance. In a nonrandomized trial, Di Campli et al. compared the observed mortality at 3 months in 20 patients with ALF and AoCLF as opposed to the expected mortally assessed by the MELD score [97]. A reduction of mortality was found for those patients with a MELD score between 20 and 29 as well as 30 and 39. However, those positive results on mortality were not confirmed by all authors. Indeed, in their randomized study, Sen et al. (18 patients randomized) as well as Hassanein et al. (70 patients randomized) were not able to demonstrate any survival benefit with MARS™ treatment in alcohol-related liver failure and hepatic encephalopathy in decompensated cirrhosis of various etiologies, respectively [98,99].

Therefore, meta-analyses were performed. Likewise, a positive effect on survival could not be confirmed. Even the most recent meta-analysis, that included nine randomized controlled trials and one nonrandomized study, was not able to detect any beneficial effect on mortality [100].

Finally, the most recent and largest study with MARS™ in AoCLF patients did not find a difference in survival as compared to SMT. The results of this large multicenter, randomized, so-called RELIEF trial to be published in 2013 [101] were presented at the 2010 Meeting of the European Association for the Study of the Liver in Vienna, Austria. The study enrolled 189 patients with decompensated cirrhosis with hepatic encephalopathy, hepatorenal syndrome and/or progressive hyperbilirubinemia. While similar severity of patients in both groups is reported, there was a trend toward an increased number of patients with a MELD score above 20 in the MARS™ group as well as an increased number of patients with spontaneous bacterial peritonitis. A more pronounced decrease in creatinine and bilirubin values was observed in the MARS™-treated group and a trend towards improved grade of encephalopathy was also described. However, mortality at day 30 was not affected. Nevertheless, the authors conclude that MARS™ has an acceptable safety profile.

23.2.6.2 Clinical Effects

Improvement in general hemodynamics is usually reported. Indeed, an increase in mean arterial pressure (MAP) and systemic vascular resistance index as well as a stable or decreased cardiac output are described in fulminant liver failure [102] and AoCLF [103,104]. A clearance of vasoactive mediators is hypothesized. However in the study showing a decreased level of nitric oxide (a potent vasodilator) with MARS™, the MAP was not affected [105].

Local hemodynamics are also affected by MARS™ treatment. A decrease in portal pressure leading to a decreased hepatic venous gradient is reported as soon as 6 h after the onset of treatment and persistent up to 18 h after it was stopped [106]. However, the exact mechanism of this effect remains elusive.

Cerebral perfusion also increases with MARS™ [107,108] and increased cerebral blood flow and cerebral oxygen consumption as assessed by a decrease in saturation in the venous jugular bulb ($SvjO_2$) are also described. An effect on brain edema and brain water content has only been reported in an animal model of liver failure [109].

Renal blood flow also increases [90] and this could be related to the various hormonal changes (such as rennin activity, aldosterone, and others) observed during treatment [104].

Finally, hepatic encephalopathy and Glasgow Coma Scale are consistently reported to be improved with MARS™. This was confirmed in AoCLF [98,99] as well as in ALF [110]. The mechanisms proposed to explain this effect on patient's neurological condition are numerous and do include removal of ammonia, aromatic amino acids and endozepines, as well as increased albumin-binding capacity and modulation of cerebral hemodynamics by changes in endogenous vasoactive substances [111]. Of importance, an effect on intracranial pressure is also reported in animal models of ALF [109,112] and this was correlated to intracranial ammonia levels as measured by microdialysis probe [112]. This decrease in intracranial pressure was confirmed in patients suffering ALF by some authors [113] while others did not report any changes [114].

23.2.6.3 Biological Effects

Total bile acid pool, total and conjugated bilirubin, creatinine and urea values, and ammonia are consistently decreased with treatment. Despite MARS™ not being a cell-based system allowing possible synthesis of various liver proteins, liver capacity to synthetize coagulation factors and other proteins is described. This is, however, not an uniform finding.

23.2.6.4 Effects on Renal Function

The first randomized trial on MARS™ therapy demonstrated its efficacy in treating type 1 hepatorenal syndrome (HRS) [94]. A large amount of literature does confirm decreased levels of creatinine and urea values. However, in a most recent study in patients with HRS who did not improve their kidney function with vasopressin analogs, the authors could not confirm the previous studies [105]. Indeed, while they also observed decreased creatinine values, kidney function, assessed by inulin and para-aminohippurate clearances, was not altered. Hence, the effects of MARS™ on biological markers of renal failure are most probably related to the filtration process of the procedure.

23.2.6.5 Effects in ALF

In patients with fulminant liver failure, MARS™ has been proposed as a bridge to liver transplantation. In a retrospective analysis, native liver recovery was more frequent in those patients treated with MARS™ and improved survival was described in the transplanted population pretreated with MARS™ [115]. This improved liver recovery was confirmed in a French case series of 18 patients with fulminant hepatic failure treated with MARS™ for at least 15 h, and compared to a French liver transplant database [110].

Finally, the results of a French study that evaluated the effect of MARS™ in patients with fulminant and subfulminant liver failure were presented at the AASLD meeting in San Francisco in 2008 [116]. After listing for emergency liver transplantation, patients were randomized to MARS™ treatment or medical therapy. A trend towards improved survival at 6 months was observed in the 102 patients included in the intention to treat analysis and in the 88 patients included in the per protocol analysis. This was mainly related to a statistically significant decrease in mortality in the paracetamol-related fulminant liver failure group. While the overall results of the study could be found negative, we have to remember that about 75% of the patients underwent liver transplantation within 24 h after inscription on the urgent waiting list, hence limiting the interpretation of the possible effects of MARS™ treatment. Furthermore, the transplant-free 6-month survival was significantly increased in patients who underwent at least three sessions of MARS™.

23.2.6.6 Other Potential Interest

Patients suffering from medical-resistant pruritus secondary to liver cholestatic diseases, such as primary sclerosing cholangitis or primary biliary cirrhosis, have been proposed to undergo MARS™ treatment. Documented by Visual Analogic Scale, a positive effect on this symptom could be observed in various studies [117,118] and this treatment has even been proposed as an outpatient procedure in this particular scenario [119]. This positive effect may last for several days, weeks, or even months after two or three sessions of MARS™. For those patients responding to therapy, this meant an improvement in quality of life. Unfortunately, some patients remain resistant.

Interestingly, the concept of albumin dialysis with MARS™ was also applied to patients admitted in critical care after acute intoxication with various substances sharing the characteristic of high protein binding. Those substances include phenytoin, lamotrigine, theophyllin, calcium channel blockers, and heavy metals such as copper and chromium [120]. While the concept of removing highly bound substances from the blood looks interesting on a toxico-cinetic point of view, further randomized trials are required before we can draw any conclusion on a possible interest of MARS™ in acute intoxication without secondary liver failure.

MARS™ has also been used in patients receiving extracorporeal membrane oxygenation and developing multiple organ failure, including hyperbilirubinemia above 300 μmom/L [121]. Improvement in mortality was observed when the five patients were compared with historical controls.

23.2.7 Plasma Separation and Adsorption and Dialysis with Prometheus™

The initial report on the use of fractionated plasma separation and adsorption (FPSA) with the Prometheus™ system (Fresenius Medical Care, Bad Homburg, Germany) describes a 27-year-old patient with fulminant hepatic failure secondary to massive combined cocaine/MDMA ingestion [122]. Despite massive cerebral edema seen on CT, and because of his history of drug abuse, the patient was not found eligible for liver transplant. He was started on Prometheus™ and treated for the next 6 days. His neurological evolution was excellent and the patient was discharged completely recovered. Since this initial report, increasing interest led to more than a 1000 treatments performed worldwide in patients with ALF and AoCLF. Developed at the Danube University Krems, Austria, the system combines a typical dialysis procedure with an adsorber treatment. The Prometheus™ membrane (Albuflow®) has a molecular weight cutoff of 250 kDa, keeping on one side blood cells and large molecules, but allowing albumin and albumin-bound toxins to cross the membrane to the "plasma circuit" which is cleaned by passing on two adsorber resins. Total blood is reformed before it passes through a hemodialysis filter then returning to the patient.

As opposed to MARS™, in the particular setting of ALF, only uncontrolled data exists with Prometheus™. In the first series of 13 patients with ALF of various etiologies, FPSA induced significant changes in total bilirubin, ammonia, and aminotransferase levels [123]. The procedure also equalized acid−base balance and water−electrolytes homeostasis. Furthermore, the system afforded body temperature adjustment. Those results were later expanded to a higher number of patients [124]. Other authors confirmed in a series of 11 patients the same decrease of biological markers [125]. In this study, the authors also report on some decrease in TNFα, C-reactive protein, procalcitonin, and α-fetoprotein while levels of hepatocyte growth factor increased with treatment. In a series of eight patients with mushroom poisoning related ALF, one to four sessions were performed on top of SMT [126]. Improvement of various biochemical parameters was also observed after the first treatment. Out of eight, one patient, admitted in hospital with severe encephalopathy, died while the other seven survived and were discharged. In a series of 27 patients, of whom 23 had ALF, treatment induced the same kind of biological effects already described [127]. Procedures were well tolerated and the system was found safe. In AoCLF, a recent European randomized controlled trial was reported [128]. Over a period of about 30 months, 145 cirrhotic patients with a precipitating event, such as infection or variceal bleeding, were randomized and received a total of 545 treatments (average: 8.1 treatment/patient). Approximately 80% of the patients were encephalopathic and roughly 55% had renal failure. The study was prematurely ended because a survival advantage could not be demonstrated. A trend towards decreased mortality at day 90 was described (47% versus 37% with Prometheus™; $P = 0.35$). However, the subgroup analyses of patients with a MELD score above 30 showed beneficial effect of treatment with Prometheus™. In univariate analyses, type 1 hepatorenal syndrome and alcoholic cirrhosis were also found to be positively influenced by the system. This, however, disappeared in multivariate analyses. As compared to SMT, the major biological change was a decrease in bilirubin. All other laboratory parameters were unaffected.

In general, the treatment was well tolerated in the various studies reported. A transient decrease in the MAP was described for some patients, this side effect being usually not reported with the use of MARS™.

As for MARS™, a potential effect of FPSA in acute intoxication of protein-bound substance may be hypothesized. Finally, the positive effect on pruritus described with MARS™ has also been reported with Prometheus™ in seven patients suffering various cholestatic diseases, the benefit lasting for up to 4 weeks after treatment [129].

The question whether clinicians should use Prometheus™ versus MARS™ remains a matter of debate. None of these systems could ever demonstrate any effect on survival. However, those systems would probably be reevaluated in some particular setting with a much higher number of patients. Indeed, in the study by Kribben et al. [128], a 9% difference in mortality at day 90, in favor of the Prometheus™ system, was described. To confirm this statistically significant 9% benefit on mortality, a study with 1000 patients should be required. Both systems were compared in various studies. In summary, both systems achieved clearance of water-soluble and protein-bound toxins, Prometheus™ being more efficient [130]. On the other hand, the MARS™ device, but not the Prometheus™ system, was able to significantly attenuate the hyperdynamic circulation observed in patients with AoCLF, presumably by a better removal of vasoactive agents, such as plasma rennin activity, aldosterone, norepinephrin, and vasopressin [104]. Hence, the debate remains open.

23.2.8 Single Pass Albumin Dialysis

Initially described as a model to study the general principle of albumin dialysis [131], single pass albumin dialysis (SPAD) is another system designed to remove protein-bound toxins using the carrier properties of albumin. As opposed to MARS™, the albumin dialysate is not regenerated but well discarded after a single passage in the high-flux albumin impermeable dialyzer. Efficacy of the two techniques was compared *in vitro* and favored MARS™ treatment for the clearance of bilirubin, bile acids, and sulfobromophtalein [132]. While in this study, improved efficiency was documented for SPAD at a higher cost (with increased albumin concentration), other authors have suggested similar detoxification capacities for those two systems in the removal of ammonia, bilirubin, bile acid, and creatinine [133]. In a case control study of 13 patients with paracetamol-induced ALF, the six patients treated with SPAD did not achieve better survival or significant improvement in clinical and biological parameters [134]. The 21 sessions performed were well tolerated. Nevertheless, because of its simplicity, SPAD has been used in patients with HRS [135] or fulminant Wilsonian liver failure [136].

Interestingly, Rosa Diez et al. proposed to combine SPAD with extended daily dialysis (EDD) in a patient suffering acute rejection 1 year after liver transplant

[137]. This allowed better removal of bilirubin than EDD alone. Likewise, Chawla et al. combined SPAD with continuous venovenous hemodiafiltration to treat a patient with sepsis-related severe liver failure with high bilirubin levels [138]. At the present time, human randomized controlled trials with SPAD are lacking.

23.2.9 Another System Using the Concept of Albumin Dialysis: The Hepa Wash®

Other devices were tested in preclinical and sometimes clinical conditions. Among them the Hepa Wash is an ongoing clinical study in critical care patients with hepatic dysfunction (NCT01079104) and AoCLF patients (NCT01079091). It is a liver and renal support system based on the use of recycled albumin dialysate, which has shown a high detoxification capacity in *in vitro* and preclinical studies. Precise description of the system and results of those studies are, however, not published at the present time.

23.2.10 Selective Plasma Filtration Therapy and Plasma Diafiltration Techniques

At the Cedars Sinai Medical Center, Los Angeles, CA, Rozga et al. developed a specific hollow fiber filter for selective plasma filtration therapy (SEPET). This membrane is included in a standard hemodialysis system and has a cutoff pore size of 100 kDa, allowing albumin to cross the filter while larger molecules cannot, the idea being that removal of pro-inflammatory cytokines and TGFβ1 while keeping hepatocyte growth factor would be of interest [139]. The filtered plasma is discarded while fresh frozen plasma (FFP) is transfused as replacement solution. In animals with ischemic liver failure, this technique was found safe and allowed prolonged survival while it effectively and actively removed ammonia, aromatic amino acids, creatinine, cytokines, and C3a. A phase I trial was started in humans with grade II–III hepatic encephalopathy. Results are not published so far.

Two other techniques are named selective plasma filtration and are different from SEPET. The first technique was described by Ho and includes a plasma filter, which separates plasma from blood, and a selective plasma filter which separates the plasma components according to their molecular weight. Total blood is reconstituted by the original cell bloods, the "cleaned" plasma, and fresh plasma which replaces the discarded plasma from the selective plasma filter [140]. In an animal model of D-galactosamine-induced fulminant liver failure, this system improved liver function, the regeneration index of hepatocytes and survival.

The other technique with the Evacure® filter (Kuraray Medical Inc., Tokyo, Japan) was initially described by Mori et al. as an attempt to bridge to transplant in a patient suffering drug-induced ALF [141]. This filter is a membrane plasma separator and has been used as a secondary membrane in plasma exchange procedures. The procedure requires dialysate, which compensates for the removal of middle weight molecular substances, and restitution fluid (albumin solution or FFP).

This technique was also used in a patient with HRS [142], fulminant hepatitis [143,144], ALF [144], and postoperative liver failure [145]. None of these reports were randomized nor controlled. Therefore and despite improved biological parameters and, as stated by the authors, an improved efficacy as compared to MARS™, definite conclusions are difficult to drawn.

23.2.11 Cell-Based Liver Assist Devices

As discussed earlier, extracorporeal liver perfusion using human, baboon, and pig livers has been shown to effectively support patients with ALF, sometimes for several days. However, this technique is rather impractical for wider use, mainly because of the limited availability of human livers and the lack of quality control and consistency of animal livers. Nevertheless, the idea of using hepatocytes rather than complete livers emerged. Using a system with living cells was based on the idea that those hepatocytes would add their own functions to the system, leading to better detoxification, improved metabolic functions, and synthesis of proteins and other molecules. Various hepatocyte lines have been used and do include rabbit, pig, and human hepatocytes, either cryopreserved, freshly isolated, or from culture.

The first application of a bioartificial liver device was reported in Japan in 1987, in a patient with inoperable hepatic cholangiocarcinoma-related ALF [146]. It was based on the principle of dialysis against a suspension of 10×10^9 functioning rabbit hepatocytes. A cellulose membrane (permeable to low and middle weight molecules) separates patient's blood from the hepatocytes solution. The patient survived after two treatments of 5 and 4.5 h, respectively. Rabbit hepatocytes have been used in another system (the artificial liver support system or ALSS developed in Tsukuba, Japan), in which they were entrapped in a calcium alginate hydrogel [147]. Results of *in vitro* experiments demonstrated urea synthesis and ammonium metabolism. To our knowledge, there are no human reports with this system.

In 1989, a clinical trial on the use of a bioartificial liver device with porcine hepatocytes was reported from Riga, Latvia [148]. Their device was made of a capsule that contained hepatocytes, activated charcoal, and inorganic quartz glass, and was placed in a forearm arteriovenous fistula. The capsule was changed every hour during the 6-h treatment period. In this large randomized trial, 59 patients treated with this system were compared to a control group of 67 patients. The cause of liver failure included B-virus hepatitis, nonA-nonB hepatitis, acute toxic hepatitis, active liver cirrhosis, sepsis, leptospirosis, and long-term subhepatic jaundice. In the group of patients treated with those capsules, overall mortality was 39% while it was 59% in the SMT group. While those results look encouraging, a subanalysis of patients in coma showed a mortality of 100% despite treatment (after improvement of neurological function for 25% of them) and mortality was 18% in those patients treated with the system at the time they were in precoma. The mortality results for the medical treatment group were 90% and 38%, respectively.

There were no further reports on the use of those three initial devices.

Many other systems were developed all over the world and some of them are listed in Table 23.1.

The *extracorporeal liver assist device* (ELAD) was developed by Sussman et al. in Houston, TX [149]. It is a bioartificial liver device in which human hepatocytes are used. The C3A cell line is a clonal derivate of the hepatoblastoma HepG2 cell line with reduced tumorigenic potential and increased capacity of synthesis of albumin and alpha-foetoprotein. Blood is pumped into a series of two cartridges (giving a theoretical mass of approximately 400 g hepatocytes) with a standard dialysis pump unit. A

negative pressure is applied through the cartridges membrane in order to allow the so-formed ultrafiltrate to pass over the cells before returning to the patient. In the phase I trial, the device was considered to be well tolerated but the positive effects reported were limited to improvement of some biological values [150]. One patient underwent treatment for 168 h. Another phase I randomized trial was later conducted in 25 patients with ALF and published as an abstract [151]. There was a trend toward improved survival at day 30 in those patients listed for liver transplant and receiving ELAD treatment as compared to those who received SMT alone (survival 83% versus 43%; $P = 0.12$) and patients who underwent ELAD treatment were more frequently bridged to OLT (92% versus 43%; $P < 0.05$). In another trial of a slightly modified ELAD system developed in Chicago, five patients underwent treatment until liver transplantation [152]. The device remained metabolically active throughout the study period as documented by oxygen use. A phase II study (NCT00030225) was started but stopped and positive effects on mortality with the ELAD were never published.

The *HepatAssist BAL device* was developed at the Cedars Sinaï Medical Center in Los Angeles, CA by Demetriou et al. The "bioartificial liver" (BAL) is comprised of approximately seven billion porcine hepatocytes housed within a hollow fiber bioreactor. The system also includes a perfusion pump, a charcoal column, a combined oxygenator/blood warmer, and custom tubing that connects the various components to a commercially

TABLE 23.1 Some of the Existing Bio-Artificial Liver- Assist Devices According to the Type of Cells used in the Bioreactor

Rabbit Hepatocytes			
Dialysis against hepatocytes in suspension		Japan	[146]
Artificial Liver Support System (ALSS)	ALSS	Japan	[147]
Porcine Hepatocytes			
Bioartificial Liver		Latvia	[148]
Hepat Assist BAL device		California	[153−156]
Academic Medical Center Bioartificial Liver	AMC-BAL	Netherlands	[159−160]
Bioartificial Liver Support System	BLSS	Pennsylvania	[165−167]
Radial Flow Bioreactor		Italy	[168−169]
Modular Extracorporeal Liver Support	MELS	Germany	[175]
Hybrid Bioartificial Liver	HBAL	China	[176−177]
Hybrid Artificial Liver Support System	HALSS	China	[178]
Bioartificial Hepatic Support	BHS	Italy	[179]
Human Hepatocytes			
Extra Corporeal Liver Assist Device (C3A Cell line)	ELAD	Texas	[149]
Academic Medical Center Bioartificial Liver (cBAL111 cell line)	AMC-BAL	Netherlands	[163]
Academic Medical Center Bioartificial Liver (HepaRG cell line)	AMC-BAL	Netherlands	[164]
Radial Flow Bioreactor (FLC-4 cell line)	RFB	Japan	[170]
Modular Extracorporeal Liver Support (Cells harvested from human livers)	MELS	Germany	[171−174]

available plasmapheresis machine. During the process, plamsmapheresis is performed and the separated plasma is pumped into the BAL, moving the plasma into the lumens of the fibers that are surrounded by a suspension of porcine hepatocytes. The pore size of the membrane is 0.15 μm which prevents the passing of hepatocytes and cell debris into the patient's blood. Preliminary uncontrolled clinical studies evaluated the effects of this system in patients with ALF [153−155]. Improvement of neurological function, reduced intracranial pressure, and increased cerebral perfusion pressure were described. In the phase I trial, the system was evaluated in three groups of patients: ALF patients who were candidates for liver transplant, posttransplantation patients with primary nonfunction-related liver failure, and AoCLF [154]. Most of the patients in groups 1 and 2 were successfully bridged to transplantation and the procedure was found to be safe. Therefore, a phase II−III randomized trial was started and randomized 171 patients with fulminant or subfulminant liver failure (the largest study on a bioartificial liver device so far) [156]. Overall, a trend toward improved survival at day 30 was observed (71% versus 62%; $P = 0.26$) and the subanalysis of patients admitted because of acetaminophen-induced ALF was in favor of the HepatAssist system. As for extracorporeal pig liver perfusion, one of the feared drawbacks is the potential transmission of PERV. This could, however, not be established [157,158]. Further development and studies were supposed to occur with 15−20 billion cells (rather than seven billion) with a more efficient platform, but were unfortunately never achieved.

The *AMC-Bioartificial Liver* (AMC-BAL), developed in Amsterdam, the Netherlands, consists of a hollow fiber bioreactor (filled in with about 10×10^9 pig hepatocytes attached in a three-dimensional configuration to the hydrophilic polyester matrix) and a plasmapheresis system. Plasma obtained after plasma filtration is perfused through the bioreactor before it is reunited with the patient's blood cells. The initial assessment was performed in an anhepatic pig model and led to a prolonged survival time and evidence of a renewed liver function [159]. The phase I clinical trial was performed in Naples and Roma, Italy [160]. Seven patients diagnosed with hyperacute and ALF according to the Crepaldi criteria underwent treatment. This was associated with improved neurological status, hemodynamic stabilization, and improved urine output in those patients with superimposed renal failure. One patient recovered after two AMC-BAL treatments while the six others underwent liver transplantation. As for the HepatAssist BAL system, transmission of PERV was never demonstrated and the follow-up reached 8.7 years after exposure to the device [161,162]. To avoid this potential drawback, other liver cell lines were recently assessed for use in the AMC-BAL. Those cell lines

include the human fetal liver cell line (cBAL111), which displays liver-specific functionality when cultured inside the AMC-BAL [163] and the human hepatoma cell line HepaRG, a liver progenitor cell line [164]. Those latter cells were tested within the HepaRG-AMC-BAL in a model of complete liver ischemia related ALF in rats. Prolonged survival time and delayed hepatic encephalopathy were characteristics of those animals in the treated group. No human data are available at the present time.

The *Bioartificial Liver Support System* (BLSS; Excorp Medical Inc., St Paul, MN), developed in Pittsburgh, PA, uses approximately 100 g porcine hepatocytes, mixed with collagen, and housed in an hollow fiber bioreactor. The system also includes a blood pump, a heat exchanger, and an oxygenator. Preclinical evaluation was performed in a dog model of D-galactosamine-induced ALF and prolonged survival was observed when the treatment was started within 16−18 h of the liver insult [165]. Initially tested in an African-American woman with ALF, the procedure induced slight biological changes indicative of liver support [166]. The further safety phase I/II study confirmed in four patients that the treatment was well tolerated and induced moderate biochemical response with a decrease in ammonia (−33%) and total bilirubin (−6%) levels [167]. Neither complete results of this phase I study nor further reports were ever published.

The *Radial Flow Bioreactor* is a system developed in Ferrara, Italy, in which the fluid perfuses the module from the center to the periphery, after having diffused through a space occupied by a three-dimensional structure filled with porcine hepatocytes. *In vitro* experiments confirmed the system was metabolically active [168]. The phase I trial confirmed in seven patients suffering ALF that treatment was well tolerated and allowed an improvement in the level of encephalopathy, and a decrease in serum ammonia, bilirubin, and prothrombin time [169]. Again, short-term follow-up for the detection of PERV transmission was negative.

Another radial flow bioreactor was developed in Tokyo, Japan [170]. There are two differences with the previous devices. The cells are a highly functional human hepatocarcinoma cell line (FLC-4) immediately cultured in the radial flow bioreactor. The second difference is the plasma flow which, in this module, goes from the periphery to the center. In a pig model of α-amanitin and lipopolysaccharide-induced ALF, the use of this bioreactor allowed improvement in electroencephalogram findings and prolonged survival [170].

In Berlin, Germany, Sauer et al. developed the *Modular Extracorporeal Liver Support* (MELS) which includes the CellModule, a specific bioreactor charged with primary human hepatocytes harvested from various livers found not suitable for organ donation. This bioreactor is associated with the DetoxModule which enables albumin dialysis, and if necessary with a DialysisModule for continuous hemofiltration if hepatorenal syndrome is present [171]. *In vitro* comparison with the AMC-BAL system concluded to similar efficacy [172]. Eight patients underwent treatment for ALF of various etiologies without any adverse events [173]. Another patient with primary graft nonfunction was successfully treated with this system [174]. The bioreactor of this system was also tested with porcine hepatocytes in eight patients as bridge to liver transplant [175]. Continuous liver support over a period of 8–46 h was safely administered and well tolerated by all patients. The eight patients were still alive 3 years after successful liver transplantation.

From Nanjing, China, Ding et al. developed the *Hybrid BioArtificial Liver* (HBAL), which includes a plasma separator allowing plasma to pass through a charcoal column or a bilirubin adsorption column, before it passes through a bioreactor inoculated with porcine hepatocytes. The reacted plasma is then reconstituted with blood cells and returned to the patient [176]. Twelve patients with acute hepatitis B were treated with the system. Improvement in ammonia levels, prothrombin time, and bilirubin were observed mainly in the group of patients treated with the so-called "simultaneous treatment" (BAL with concomitant plasma exchange or bilirubin adsorption column). The same team developed a HBAL based on multilayer flat plate bioreactor with cocultured pig hepatocytes and bone marrow mesenchymal stem cells, in order to mimic the *in vivo* microenvironment [177]. This HBAL is combined with an anionic resin absorption column. In a dog model of D-galactosamine-induced ALF, improvement in biological parameters as well as prolonged survival was demonstrated. Human data with this device are lacking so far.

In Suzhou, China, Gan et al. tested another *Hybrid Artificial Liver Support System* (HALSS) in 10 patients suffering severe liver failure [178]. This system differs from HBAL in that it includes a hemo-adsorba column, to remove bilirubin and toxins and because the bioreactor is filled with spheroid porcine hepatocytes. The authors randomized 30 patients with ALF in three groups of 10: the HALSS group, the plasmapheresis group, and the conventional treatment group. An improved neurological advantage as well as an improved survival was described in the 10 patients from the HALSS group as compared to the others.

The use of the *Bioartificial Hepatic Support* (BHS), developed in Udine, Italy, was reported in one patient suffering AoCLF [179]. After plasmapheresis, the plasma of the patients passes a charcoal column before it enters a bioreactor filled with 15 billion porcine hepatocytes and collagen-coated dextran microcarriers. Improvement in biological parameters as well as grade of encephalopathy

was described. There are no further reports on this system.

Finally, the concept of fluidized adsorbent bed system, as described in the Biologic-DT system has also been applied to bioartificial liver devices. The *fluidized bed bioreactor*, as described by Doré et al, is based on microencapsulated hepatocytes [180]. They provide immunoprotection and a high surface area that facilitates maximal material transport. Those systems were shown to have beneficial effects in terms of preventing the increase in ammonia and intracranial pressure in a pig model of ALF [181,182]. However, with the use of this concept, it seems that PERV transmission could be an issue [183]. *In vitro* human study was carried out with plasma from liver failure patients [184]. Alginate-encapsulate HepG2 cells maintained viability, metabolic, synthetic, and detoxificatory activities, indicating that this could be used as the biological component of a bioreactor to treat patients with ALF. Note that the use of microencapsulated porcine hepatocytes has also been described in a mice model of fulminant hepatic failure treated with transplantation of those capsules leading to improved survival [185]. Those hepatocytes maintained metabolic functions after cryopreservation and encapsulation.

23.3 FUTURE PERSPECTIVES AND CONCLUSIONS

For more than 60 years, researchers all over the world tried to develop a promising liver assist device to support those patients suffering liver failure, a really dramatic condition that carries out, in the absence of OLT, an unacceptably high mortality rate. Unfortunately, we have to admit results are quite unsatisfactory. A systematic review published in 2004 evaluated the effect of artificial and bioartificial liver devices on outcome [186]. Overall, those systems did not affect mortality or bridging to liver transplantation. However, a significant improvement in liver encephalopathy was described. In subgroup analysis, liver support systems decreased mortality in AoCLF (RR: 0.67; 95% IC: 0.51–0.9) but not in ALF (RR: 0.95; 95% IC: 0.71–1.29). Most trials included in this systematic review had unclear methodological evidence and the strength of evidence was found insufficient. To the opposite, a meta-analysis, covering randomized trials published between 1995 and 2010, demonstrated survival advantages in patients suffering ALF (RR: 0.7; 95% IC: 0.49–1) but not AoCLF (RR: 0.87; 95% IC: 0.64–1.18) [187]. From the 74 potentially relevant abstracts identified, eight randomized controlled trials with sufficient information on outcome were retained. From those eight abstracts, only three addressed the issue of ALF (for a total number of 198 patients) and five evaluated the outcome in AoCLF (total number of patients: 157). With this small number of studies and patients, firm conclusions on a potential benefit on mortality are rather difficult to obtain, because of a risk of underpowered analysis. Nevertheless, this meta-analysis confirmed an acceptable safety profile.

Despite encouraging preclinical results, those conflicting results demonstrate how difficult it is to develop an universal liver assist device. Considering the complexity of liver functions and the heterogeneity of patients to be treated, this is not surprising. Therefore, defining a more specific trial design according to the system to evaluate as well as the population to be studied should certainly be of interest. Second, as discussed earlier, to avoid underpowered analysis, a higher number of patients should certainly be enrolled in randomized trials. This, however, substantially raises the costs of those trials, leading the pharmaceutical industry to some reluctance, especially if the system is already on the market. Finally, the primary objective of a trial should be discussed. Mortality is certainly of importance but other endpoints could be of interest. Hemodynamics or encephalopathy improvements before going to OLT are certainly endpoints that could be addressed with a more limited number of patients. However some could argue that if those endpoints do not translate into outcome benefit, they would be of limited value.

Some important issues encountered in patients with liver failure should be kept in mind when developing a new system. For instance, none of the artificial devices addresses the issue of specific cytokine removal or endotoxin clearance, two pathophysiologic mechanisms involved in liver failure and its consequences [1].

While the artificial devices, such as MARS™ or Prometheus™, are not expected to be modified in the near future, recent advances in bioreactors, cell production, and cell transplantation could be of interest for the future.

The development of an ideal bioreactor faces many challenges. The ideal cell line for use in liver bioreactors has still to be defined. This cell line should combine the following characteristics: full functionality of adult human hepatocytes, unlimited lifespan, potentials for *in vitro* proliferative capacity, no risk for metastatic tumor formation, nor zoonosis transmission or immunogenicity [188]. More than 200 articles focused on the ideal cell sources of hepatocytes in bioreactors [189]. Until now, rabbit, pig, primary human hepatocytes, and liver tumor-derived cells have been used in bioreactors. Because of the fear of zoonosis (such as PERV transmission), potential immunogenicity, the fact pig hepatocytes do not display total metabolic similarity with human hepatocytes (for instance, pig hepatocytes do not synthetize compatible coagulation factors [190]), and the poor quality of primary human hepatocytes, researchers did focus on other

nonprimary cell sources, such as other human tumor-derived cell lines (GS-HepG2, HepG2-GS-3A4, FLC-4), immortalized fetal human hepatocytes (for instance cBAL111), immortalized adult human hepatocytes, human stem cell-derived hepatocytes from liver progenitor cells, hematopoietic stem cells, and embryonic origin stem cells. None of these cells have been tested in the clinical setting so far. A second key point in choosing the optimal cell line is the potential coculture of different cell lines together with hepatocytes. Indeed, liver cells do communicate with each other to maintain the various liver functions. Therefore some authors advocate the use of cocultivation in order to promote the activities and functions of hepatocytes by forming spherical aggregates, hence increasing direct contact between the liver cells and other cells [191]. This approach was recently documented in *in vitro* and *in vivo* experiments [192]. *In vitro*, both fibroblasts and mesenchymal stem cells (MSCs) supported long-term hepatocyte functions, MSC coculture allowing better hepatocyte detoxification capacities. Furthermore, the presence of MSCs improved hepatocytes survival when exposed to galactosamine. Finally, a liver assist device with both hepatocytes and MSC was superior to other devices in terms of survival of rats exposed to galactosamine. A sinusoid-like structure was demonstrated when hepatocytes were cocultured with MSCs [193], reaching better metabolic control than isolated hepatocyte cultures.

The bioreactor is another key point of the bioartificial liver. The ideal bioreactor should provide an *in vivo* like environment for optimal viability and functionality of hepatocytes [188]. Hepatocytes oxygenation and bile secretion are two major issues for development of the device. Until now, four types of bioreactors exist: hollow fiber bioreactor, flat plate bioreactor, perfused bed or scaffold bioreactor, and the encapsulated suspension bioreactor. Each of them displays advantages and disadvantages [191]. One of the most recent developments is the possibility to generate a matrix from decellularized livers before it is perfused with hepatocytes and/or other cell types. This "reconstructed liver" could be used in a bioreactor or be transplanted as it was demonstrated in animals [194].

Other methods to support the failing liver are under development. Among them, hepatocytes transplantation certainly deserves attention. It has been used in inborn metabolic errors as well as chronic liver fibrosis and in ALF of various etiologies [195]. The total number of treated patients is relatively low and results in ALF are limited to some improvements in encephalopathy and biological variables. Two other techniques may help in liver support [196]. The first technique is described here above. It consists of creating a scaffold and seeding cells on the scaffold. The second technique is manipulating the cell pathways to induce liver regeneration of dead or resected tissue.

Those techniques are beyond the scope of this review describing the available liver assist devices. Despite 60 years of research, a lot of work mainly in the development of bioreactors has still to be done. Meanwhile, the use of albumine dialysis systems, such as MARS™ or Prometheus™, allows improvement in encephalopathy and hemodynamics. Those techniques should be used in specialized units where liver transplantation is available.

REFERENCES

[1] Tritto G, Davies NA, Jalan R. Liver replacement therapies. Semin Respir Crit Care Med 2012;33(1):70−9.

[2] Trey C, Davidson CS. The management of fulminant hepatic failure. Prog Liver Dis 1970;3:282−98.

[3] O'Grady JG, Schalm SW, Williams R. Acute liver failure: redefining the syndromes. Lancet 1993;342(8866):273−5.

[4] Bernuau J, Rueff B, Benhamou JP. Fulminant and subfulminant liver failure: definitions and causes. Semin Liver Dis 1986;6 (2):97−106.

[5] Bernal W, Auzinger G, Dhawan A, Wendon J. Acute liver failure. Lancet 2010;376(9736):190−201.

[6] Jalan R, Williams R. Acute-on-chronic liver failure: pathophysiological basis of therapeutic options. Blood Purif 2002;20 (3):252−61.

[7] O'Grady J. Liver transplantation for acute liver failure. Best Pract Res Clin Gastroenterol 2012;26(1):27−33.

[8] Germani G, Theocharidou E, Adam R, Karam V, Wendon J, O'Grady J, et al. Liver transplantation for acute liver failure in Europe: outcomes over 20 years from the ELTR registry. J Hepatol 2012;57(2):288−96.

[9] Abouna GM. Organ shortage crisis: problems and possible solution. Transplant Proc 2008;40(1):34−8.

[10] Arulraj R, Neuberger J. Liver transplantation: filling the gap between supply and demand. Clin Med 2011;11(2):94−8.

[11] Mathurin P, Moreno C, Samuel D, Dumortier J, Salleron J, Durand F, et al. Early liver transplantation for severe alcoholic hepatitis. N Engl J Med 2011;365(19):1790−800.

[12] Agopian VG, Kaldas FM, Hong JC, Whittaker M, Holt C, Rana A, et al. Liver transplantation for nonalcoholic steatohepatitis: the new epidemic. Ann Surg 2012;256(4):624−33.

[13] Khorsandi SE, Heaton N. Contemporary strategies in the management of hepatocellular carcinoma. HPB Surg 2012;2012:154056.

[14] O'Grady J. Modern management of acute liver failure. Clin Liver Dis 2007;11(2):291−303.

[15] Wallerstein H. Treatment of severe erythoblastosis by simultaneous removal and replacement of blood of newborn infant. Science 1946;105(2680):583.

[16] Lee C, Tink A. Exchange transfusion in hepatic coma: report of a case. Med J Aust 1958;45(2):40−2.

[17] Trey C, Burns DG, Saunders SJ. N Engl J Med 1966;274 (9):473−81.

[18] Trey C. The fulminant hepatic failure surveillance study. Brief review of the effects of presumed etiology and age of survival. Can Med Assoc J 1972;106:525−8.

[19] Jones EA, Clain D, Clink HM, MacGillivray M, Sherlock S. Hepatic coma due to acute hepatic necrosis treated by exchange blood transfusion. Lancet 1967;290(7508):169−72.

[20] Klebanoff G, Armstrong RG, Cline RE, Powell JR, Bedingfield JR. Resuscitation of a patient in stage IV hepatic coma using total body washout. J Surg Res 1972;13(4):159−65.

[21] Cline RE, Klebanoff G, Armstrong RG, Stanford W. Extracorporal circulation in hypothermia as used for total body washout in stage IV hepatic coma. Ann Thorac Surg 1973;16(1):44−51.

[22] Talmage EA, Thomas JM, Weeks JH. Total blood washout for Reye's syndrome. Anesth Analg 1973;52(4):563−9.

[23] Lansky LL, Fixley M, Romig DA, Keitges PW, Boggan M, Reis RL. Letter: hypothermic total body-washout with survival in Reye's syndrome. Lancet 1974;2(7887):1019.

[24] Cooper Jr GN, Karlson KE, Clowes GH, Martin H, Randall HT. Total blood washout and exchange. A valuable tool in acute hepatic coma and Reye's syndrome. Am J Surg 1977;133(4):522−30.

[25] Lansky LL, Kalavsky SM, Brackett CE, Wallas CH, Reis RL. Hypothermic total body washout and intracranial pressure monitoring in stage IV Reye syndrome. J Pediatr 1977;90(4):639−40.

[26] Wakabayashi A, Kubo T, Gilman P, Charney K, Connolly JE. Total body washout and *ex vivo* liver perfusion in acute hepatic failure: a comparative study. Arch Surg 1974;109(1):52−6.

[27] Joyeuse R, Ivanisevic B, Longmire Jr WP, Maloney Jr. JV. The treatment of experimental coma by parabiotic cross circulation. Surg Forum 1962;13:334−6.

[28] Burnell JM, Thomas ED, Ansell JS, Cross HE, Dillard DH, Epstein RB, et al. Observations on cross circulation in man. Am J Med 1965;38:832−41.

[29] Swift JE, Ghent WR, Beck IT. Direct transhepatic cross-circulation in hepatic coma in man. Can Med Assoc J 1967;97(24):1435−45.

[30] Burnell JM, Dawborn JK, Epstein RB, Gutman RA, Leinbach GE, Thomas ED, et al. Acute hepatic coma treated by cross-circulation or exchange transfusion. N Engl J Med 1967;276 (17):935−43.

[31] Bosman SC, Terblanche J, Saunders SJ, Harrison GG, Barnard CN. Cross-circulation between man and baboon. Lancet 1968;2 (7568):583−5.

[32] Saunders SJ, Terblanche J, Bosman SC, Harrison GG, Walls R, Hickman R, et al. Acute hepatic coma treated by cross circulation with a baboon and by repeated exchange transfusions. Lancet 1968;2(7568):585−8.

[33] Gayle Jr WE, Williams GM, Hume DM. Immunologic consequences of human being and baboon cross circulation. Surg Forum 1969;20:354−5.

[34] Hollander D, Klebanoff G, Osteen RT. Human−baboon cross-circulation for hepatic coma. JAMA 1971;218(1):67−71.

[35] Kimoto S. The artificial liver: experiments and clinical application. Trans Am Soc Artif Intern Organs 1959;5:102−12.

[36] van de Kerkhove MP, Hoekstra R, Chamuleau RA, van Gulik TM. Clinical application of bioartificial liver support systems. Ann Surg 2004;240(2):216−30.

[37] Nose Y, Mikami J, Kasai Y, Sasaki E, Agishi T, Danjo Y. An experimental artificial liver utilizing extracorporeal metabolism with sliced or granulated canine liver. Trans Am Soc Artif Intern Organs 1963;9:358−62.

[38] Lepore MJ, Martel AJ. Plasmapheresis in hepatic coma. Lancet 1967;290(7519):771−2.

[39] Lepore MJ, Martel AJ. Plasmapheresis and plasma exchange in hepatic coma. Methods and results in five patients with acute hepatic necrosis. Ann Intern Med 1970;72(2):165–74.

[40] Sabin S, Merritt JA. Treatment of hepatic coma in cirrhosis by plasmapheresis and plasma infusion (plasma exchange). Ann Intern Med 1968;68(1):1–7.

[41] Jhang JS, Schilsky ML, Lefkowitch JH, Schwartz J. Therapeutic plasmapheresis as a bridge to liver transplantation in fulminant Wilson disease. J Clin Apher 2007;22(1):10–4.

[42] Asfaha S, Almansori M, Qarni U, Gutfreund KS. Plasmapheresis for hemolytic crisis and impending acute liver failure in Wilson disease. J Clin Apher 2007;22(5):295–8.

[43] Kamar N, Lavayssière L, Muscari F, Selves J, Guilbeau-Frugier C, Cardeau I, et al. Early plasmapheresis and rituximab for acute humoral rejection after ABO-compatible liver transplantation. World J Gastroenterol 2009;15(27):3426–30.

[44] Chan KM, Lee CS, Wu TJ, Lee CF, Chen TC, Lee WC. Clinical perspective of acute humoral rejection after blood type-compatible liver transplantation. Transplantation 2011;91(5): e29–30.

[45] Park CS, Hwang S, Park HW, Park YH, Lee HJ, Namgoong JM, et al. Role of plasmapheresis as liver support for early graft dysfunction following adult living donor liver transplantation. Transplant Proc 2012;44(3):749–51.

[46] Hwang S, Ha TY, Ahn CS, Kim KH, Lee SG. Reappraisal of plasmapheresis as a supportive measure in a patient with hepatic failure after major hepatectomy. Case Rep Gastroenterol 2007;1 (1):162–7.

[47] Sotiropoulos GC, Lang H, Herget-Rosenthal S, Molmenti EP, Baba HA, Karaliotas C, et al. Salvage plasmapheresis for small for size syndrome following hepatic resection for colorectal liver metastasis. Int J Colorectal Dis 2008;23(5):553.

[48] Kondrup J, Almdal T, Vilstrup H, Tygstrup N. High volume plasma exchange in acute hepatic failure. Int J Artif Organs 1992;15(11):669–76.

[49] Larsen FS, Ejlersen E, Hansen BA, Mogensen T, Tygstrup N, Secher NH. Systemic vascular resistance during high volume plasmapheresis in patients with fulminant hepatic failure: relationship with oxygen consumption. Eur J Gastroenterol Hepatol 1995;7 (9):887–92.

[50] Clemmesen JO, Larsen FS, Ejlersen E, Schiodt FV, Ott P, Hansen BA. Haemodynamic changes after high-volume plasmapheresis in patients with chronic and acute liver failure. Eur J Gastroenterol Hepatol 1997;9(1):55–60.

[51] Clemmesen JO, Kondrup J, Nielsen LB, Larsen FS, Ott P. Effects of high-volume plasmapheresis on ammonia, urea and amino acids in patients with acute liver failure. Am J Gastroenterol 2001;96(4):1217–23.

[52] Akdogan M, Camci C, Gurakar A, Gilcher R, Alamian S, Wright H. The effect of total plasma exchange on fulminant hepatic failure. J Clin Apher 2006;21(2):96–9.

[53] Matsubara S. Combination of plasma exchange and continuous hemofiltration as temporary metabolic support for patients with acute liver failure. Artif Organs 1994;18(5):363–6.

[54] Sadahiro T, Hirasawa H, Oda S, Shiga H, Nakanishi K, Kitamura N, et al. Usefulness of plasma exchange plus continuous hemodiafiltration to reduce adverse effects associated with plasma exchange in patients with acute liver failure. Crit Care Med 2001;29(7):1386–92.

[55] Bernard C. Sur le mécanisme de la formation du sucre dans le foie. CR Acad Sci (Paris) 1855;41:461.

[56] Otto J, Pender JC, Cleary JH, Sensenig DM, Welch CS. The use of a donor liver in experimental animals with elevated blood ammonia. Surgery 1958;43(2):301–9.

[57] Sen PK, Bhalerao RA, Parulkar GP, Samsi AB, Shah BK, Kinare SG. Use of isolated perfused cadaveric liver in the management of acute hepatic failure. Surgery 1966;59(5):774.

[58] Eiseman B. Treatment of hepatic coma by extracorporeal liver perfusion. Ann R Coll Surg Engl 1966;38(6):329–48.

[59] Slapak M. Fulminant liver failure: clinical and experimental study. Ann R Coll Surg Engl 1975;57(5):234–47.

[60] Pascher A, Sauer IM, Hammer C, Gerlach JC, Neuhaus P. Extracorporeal liver perfusion as hepatic assist in acute liver failure: a review of world experience. Xenotransplantation 2002;9:309–24.

[61] Collins BH, Chari RS, Magee JC, Harland RC, Lindman BJ, Logan JS, et al. Immunopathology of porcine livers perfused with blood of humans with fulminant hepatic failure. Transplant Proc 1995;27(1):280–1.

[62] Abouna GM. Simultaneous liver hemoperfusion and hemodialysis for treatment of hepatic coma and hepatorenal failure. Surgery 1973;73(4):541–9.

[63] Levy MF, Crippin J, Sutton S, Netto G, McCormack J, Curiel T, et al. Liver allotransplantation after extracorporeal hepatic support with transgenic (hCD55/hCD59) porcine livers: clinical results and lack of pig-to-human transmission of the porcine endogenous retrovirus. Transplantation 2000;69 (2):272–80.

[64] Fair J, Maley W, Stephenson R, Kittur D, Wiener C, Klein A. Extracorporeal porcine liver perfusion as a successful bridge to orthotopic liver transplantation (OLT). Gastroenterology 1993;104:809A [abstract].

[65] Horslen SP, Hammel JM, Fristoe LW, Kangas JA, Collier DS, Sudan DL, et al. Extracorporeal liver perfusion using human and pig livers for acute liver failure. Transplantation 2000;70 (10):1472–8.

[66] Bikhchandani J, Metcalfe M, Illouz S, Puls F, Dennison A. Extracorporeal liver perfusion system for artificial liver support across a membrane. J Surg Res 2011;171(1):e139–47.

[67] Schwartz R, Phillips GB, Gabuzda Jr GJ, Davidson CS. Blood ammonia and electrolytes in hepatic coma. J Lab Clin Med 1953;42:499–508.

[68] Kiley JE, Pender JC, Welch HF, Welch CS. Ammonia intoxication treated by hemodialysis. N Engl J Med 1956;259 (24):1156–61.

[69] Jones RC, Strader Jr LD, Berry WC. Peritoneal dialysis in hepatic coma. US Armed Forces Med J 1959;10:977–82.

[70] Keynes WM. Haemodialysis in the treatment of liver failure. Lancet 1968;2(7580):1236–8.

[71] Yatzidis H. A convenient haemoperfusion microapparatus over charcoal for the treatment of endogenous and exogenous intoxication. I. Its use as an effective artificial kidney. Proc Eur Dial Transplant Ass 1964;1:83–5.

[72] Barshes NR, Gay AN, Williams B, Patel AJ, Awad SS. Support for the acutely failing liver: a comprehensive review of historic and contemporary strategies. J Am Coll Surg 2005;201(3):458–76.

[73] Toledo-Perevra LH. Role of activated carbon hemoperfusion in the recovery of livers exposed to ischemic damage. Arc Surg 1985;120(4):462−5.

[74] O'Grady JG, Gimson AE, O'Brien CJ, Pucknell A, Hughes RD, Williams R. Controlled trials of charcoal hemoperfusion and prognostic factors in fulminant hepatic failure. Gastroenterology 1988;94(5Pt1):1186−92.

[75] Willson RA, Webster KH, Hofmann AF, Summerskill WH. Toward an artificial liver: *in vitro* removal of unbound and protein-bound plasma compounds related to hepatic failure. Gastroenterology 1972;62(6):1191−9.

[76] Weston MJ, Gazzard BG, Buxton BH, Winch J, Machado AL, Flax H, et al. Effects of hemoperfusion through charcoal or XAD-2 resin on an animal model of fulminant liver failure. Gut 1974;15(6):482−6.

[77] Lopukhin UM, Molodenkev MN, Levkin UA, Gorchakov VD, Evseev NG, Ryabov AV. Removal of plasma bilirubin by hemoperfusion in dogs. Am J Gastroenterol 1977;68(4):345−53.

[78] Ton HY, Hughes RD, Silk DB, Williams R. Albumin coated Amberlite XAD-7 resin for hemoperfusion in acute liver failure. Part I. Adsorption studies. Artif Organs 1979;3(1):20−2.

[79] Chamuleau RA, Schoemaker LP, Schmit EM. *In vitro* adsorption of possible aetiological factors of hepatic encephalopathy. Int J Artif Organs 1979;2(6):284−8.

[80] Bihari D, Hughes RD, Gimson AE, Langley PG, Ede RJ, Eder G, et al. Effects of serial resin hemoperfusion in fulminant hepatic failure. Int J Artif Organs 1983;6(6):299−302.

[81] Wang YJ, Wang ZW, Luo BW, Liu HL, Wen HW. Assessment of resin perfusion in hepatic failure *in vitro* and *in vivo*. World J Gastroenterol 2004;10(6):837−40.

[82] Nakaji S, Hayashi N. Bilirubin adsorption column Medisorba BL 300. Ther Apher Dial 2003;7(1):98−103.

[83] Falkenhagen D, Brandl M, Hartmann J, Kellner KH, Posnicek T, Weber V. Fluidized bed adsorbent systems for extracorporeal liver support. Ther Apher Dial 2006;10 (2):154−9.

[84] Barile RG, Wang NH, Blake DE, Belcastro PF, Gupta S, Regnier FE, et al. A reciprocating, single needle hemodialyzer with bidirectional flow of sorbent suspension. Artif Organs 1982;6(3):267−79.

[85] Shihab-Eldeen AA, Peck GE, Ash SR, Kaufman G. Evaluation of the sorbent reciprocating dialyzer in the treatment of overdose of paracetamol and phenobarbitone. J Pharm Pharmacol 1988;40 (6):381−7.

[86] Ash SR, Blake DE, Carr DJ, Carter C, Howard T, Makowka L. Neurologic improvement of patients with hepatic failure and coma during sorbent suspension dialysis. ASAIO Trans 1991;37 (3):M332−4.

[87] Ash SR, Blake DE, Carr DJ, Harker KD. Push−pull sorbent based pheresis for treatment of acute hepatic failure: the BioLogic-detoxifier-plasma filter system. ASAIO J 1998;44 (3):129−39.

[88] Ellis AJ, Hughes RD, Nicholl D, Langley PG, Wendon JA, O'Grady JG, et al. Temporary extracorporeal liver support for severe alcoholic hepatitis using the BioLogig-DT. Int J Artif Organs 1999;22(1):27−34.

[89] Quinlan GJ, Martin GS, Evans TW. Albumin: biochemical properties and therapeutic potential. Hepatology 2005;41(6):1211−9.

[90] Mitzner SR. Extracorporeal liver support-albumin dialysis with the molecular adsorbent recirculating system (MARS). Ann Hepatol 2011;10(S1):S21−8.

[91] Stange J, Mitzner S, Ramlow W, Gliesche T, Hickstein H, Schmidt R. A new procedure for the removal of protein bound drugs and toxins. ASAIO J 1993;39(3):M621−5.

[92] Stange J, Ramlow W, Mitzner S, Schmidt R, Klinkmann H. Dialysis against a recycle albumin solution enables the removal of albumin bound toxins. Artif Organs 1993;17:809−13.

[93] Stange J, Mitzner S, Risler T, Erley CM, Lauchart W, Goehl H, et al. Molecular adsorbent recycling system (MARS): clinical results of a new membrane-based blood purification system for bioartificial liver support. Artif Organs 1999;23(4):319−30.

[94] Mitzner SR, Stange J, Klammt S, Risler T, Erley CM, Bader BD, et al. Improvement of hepatorenal syndrome with extracorporeal albumin dialysis MARS: results of a prospective randomized controlled clinical trial. Liver Transpl 2000;6(3):277−86.

[95] Heemann U, Treichel U, Loock J, Philipp T, Gerken G, Malago M, et al. Albumin dialysis in cirrhosis with superimposed acute liver injury: a prospective, controlled study. Hepatology 2002;36 (4 Pt 1):949−58.

[96] El Banayosy A, Kizner L, Schueler V, Berqmeiers S, Cobaugh D, Koerfer R. First use of the molecular adsorbent recirculating system technique on patients with hypoxic liver failure after cardiogenic shock. ASAIO J 2004;50(4):332−7.

[97] Di Campli C, Santoro MC, Gaspari R, Merra G, Zileri Dal Verme L, Zocco MA, et al. Catholic university experience with molecular adsorbent recycling system in patients with severe liver failure. Transplant Proc 2005;37(6):2547−50.

[98] Sen S, Davies NA, Mookerjee RP, Cheshire LM, Hodges SJ, Williams R, et al. Pathophysiological effects of albumin dialysis in acute on chronic liver failure: a randomized controlled study. Liver Transpl 2004;10(9):1109−19.

[99] Hassanein TI, Tofteng F, Brown Jr RS, McGuire B, Lynch P, Mehta R, et al. Randomized controlled study of extracorporeal albumin dialysis for hepatic encephalopathy in advanced cirrhosis. Hepatology 2007;46(6):1853−62.

[100] Vaid A, Chweich H, Balk EM, Jaber BL. Molecular adsorbent recirculating system as artificial support therapy for liver failure: a meta-analysis. ASAIO J 2012;58(1):51−9.

[101] Banares R, Nevens F, Larsen FS, Jalan R, Albillos A, Dollinger M, et al. Extracorporeal albumin dialysis with the molecular adsorbent recirculating system in acute-on-chronic liver failure. Hepatology 2012;: doi: 10.2002/hep.26185. [Epub ahead of print].

[102] Schmidt LE, Wang LP, Hansen BA, Larsen FS. Systemic hemodynamic effects of treatment with the molecular adsorbent recirculating system in patients with hyperacute liver failure: a prospective controlled trial. Liver Transpl 2003;9(3):290−7.

[103] Mitzner SR, Stange J, Klammt S, Peszynski P, Schmidt R, Nöldge-Schomburg G. Extracorporeal detoxification using the molecular adsorbent recirculating system for critically ill patients with liver failure. J Am Soc Nephrol 2001;12(S7):S75−82.

[104] Laleman W, Wilmer A, Evenepoel P, Elst IV, Zeegers M, Zaman Z, et al. Effects of the molecular adsorbent recirculating system and the Prometheus devices on systemic haemodynamics and vasoactive agents in patients with acute-on-chronic alcoholic liver failure. Crit Care 2008;10(4):R108.

[105] Wong F, Raina N, Richardson R. Molecular adsorbent recirculating system is ineffective in the management of type 1 hepatorenal syndrome in patient with cirrhosis with ascites who have failed vasoconstrictor treatment. Gut 2010;59(3):381−6.

[106] Sen S, Mookerjee RP, Cheshire LM, Davies NA, Williams R, Jalan R. Albumine dialysis reduces portal pressure acutely in patients with severe alcoholic hepatitis. J Hepatol 2005;43(1):142−8.

[107] Sorkine P, Ben Abraham R, Szold O, Biderman P, Kidron A, Merchav H, et al. Role of the molecular adsorbent recirculating system (MARS) in the treatment of patients with acute exacerbation of chronic liver failure. Crit Care Med 2001;29(7):1332−6.

[108] Mitzner S, Loock J, Peszynski P, Klammt S, Majcher-Peszynska J, Gramowski A, et al. Improvement in central nervous system functions during treatment of liver failure with albumin dialysis MARS: a review of clinical, biochemical and electrophysiological data. Metab Brain Dis 2002;17(4):463−75.

[109] Sen S, Rose C, Ytrebo LM, Davies NA, Nedredal GI, Drevland SS, et al. Effect of albumin dialysis on intracranial pressure increase in pigs with acute liver failure: a randomized controlled study. Crit Care Med 2006;34(1):158−64.

[110] Camus C, Lavoué S, Gacouin A, Compagnon P, Boudjéma K, Jacquelinet C, et al. Liver transplantation avoided in patients with fulminant hepatic failure who received albumin dialysis with the molecular adsorbent recirculating system while on the waiting list: impact of the duration of therapy. Ther Apher Dial 2009;13(6):549−54.

[111] Wauters J, Wilmer A. Albumin dialysis: current practice and future options. Liver Int 2011;31(S3):9−12.

[112] Zwirner K, Thiel C, Thiel K, Morgalla MH, Königsrainer A, Schenk M. Extracellular brain ammonia levels in association with arterial ammonia, intracranial pressure and the use of albumin dialysis devices in pigs with acute liver failure. Metab Brain Dis 2010;25(4):407−12.

[113] Pugliese F, Ruberto F, Perrella SM, Cappannoli A, Bruno K, Martelli S, et al. Modifications of intracranial pressure after molecular adsorbent recirculating system treatment in patients with acute liver failure: case reports. Transplant Proc 2007;39(6):2042−4.

[114] Lai WK, Haydon G, Mutimer D, Murphy N. The effect of molecular adsorbent recirculating system on pathophysiological parameters in patients with acute liver failure. Intensive Care Med 2005;31(11):1544−9.

[115] Kantola T, Koivusalo AM, Höckerstedt K, Isionemi H. The effect of molecular adsorbent recirculating system treatment on survival, native liver recovery and need for liver transplantation in acute liver failure patients. Transplant Int 2008;21(9):857−66.

[116] Saliba F, Camus C, Durand F, Mathurin P, Delafosse B, Barange K, et al. Randomized multicenter controlled trial evaluating the efficacy and safety of albumin dialysis with MARS® in patients with fulminant and subfulminant hepatic failure. Hepatology 2008;48(Suppl.):377A.

[117] Parés A, Herrera M, Avilés J, Sanz M, Mas A. Treatment of resistant pruritus from cholestasis with albumin dialysis: combined analysis of patients from 3 centers. J Hepatol 2010;53(2):307−12.

[118] Chazouillères O. MARS: the ultimate warrior against pruritus of cholestasis? J Hepatol 2010;53(2):228−9.

[119] Leckie P, Tritto G, Proven A, Cheshire L, Mookerjee RP, Jones DE, et al. Out-patient albumin dialysis for itch in chronic cholestatic liver disease patients is feasible, safe and efficacious alternative to liver transplantation. Hepatology 2009;50:376A [abstract 155]

[120] Wittebole X, Hantson P. Use of the molecular adsorbent recirculating system (MARS™) for the management of acute poisoning with or without liver failure. Clin Toxicol 2011;49(9):782−93.

[121] Peek GJ, Killer HM, Sosnowski MA, Firmin RK. Modular extracorporeal life support for multiple organ failure patients. Liver 2002;22(S2):69−71.

[122] Kramer L, Bauer E, Schenk P, Steininger R, Vigl M, Mallek R. Successful treatment of refractory cerebral edema in ecstasy/cocaine-fulminant hepatic failure using a new high-efficacy liver detoxification device (FPSA-Prometheus). Wien Klin Wochenschr 2003;115(15−16):599−603.

[123] Skwarek A, Grodzicki M, Nyckowski P, Kotulski M, Zieniewicz K, Michalowicz B, et al. The use Prometheus FPSA system in the treatment of acute liver failure: preliminary results. Transplant Proc 2006;38(1):209−11.

[124] Grodzicki M, Kotulski M, Leonowicz D, Zieniewicz K, Krawczyk M. Results of treatment of acute liver failure patients with use of the Prometheus FPSA system. Transplant Proc 2009;41(8):3079−81.

[125] Rocen M, Kieslichova E, Merta D, Uchytilova E, Pavlova Y, Cap J, et al. The effect of Prometheus device on laboratory markers of inflammation and tissue regeneration in acute liver failure management. Transplant Proc 2010;42(9):3606−11.

[126] Vardar R, Gunsar F, Ersoz G, Akarca US, Karasu Z. Efficacy of fractionated plasma separation and adsorption system (Prometheus) for treatment of liver failure due to mushroom poisoning. Hepatogastroenterology 2010;57(99−100):573−7.

[127] Sentürk E, Esen F, Özcan PE, Rifai K, Pinarbasi P, Cakar N, et al. The treatment of acute liver failure with fractionated plasma separation and adsorption system: experience in 85 applications. J Clin Apher 2010;25(4):195−201.

[128] Kribben A, Gerken G, Haag S, Herget-Rosenthal S, Treichel U, Betz C, et al. Effects of fractionated plasma separation and adsorption on survival in patients with acute on chronic liver failure. Gastroenterology 2012;142(4):782−9.

[129] Rifai K, Hafer C, Rosenau J, Athmann C, Haller H, Peter Manns M, et al. Treatment of severe refractory pruritus with fractionated plasma separation and adsorption (Prometheus). Scand J Gastroenterol 2006;41(10):1212−7.

[130] Evenepoel P, Laleman W, Wilmer A, Claes K, Kuypers D, Bammens B, et al. Prometheus versus molecular adsorbents recirculating system: comparison of efficiency in two different liver detoxification devices. Artif Organs 2006;30(4):276−84.

[131] Awad SS, Rich PB, Kolla S, Younger JG, Reickert CA, Downing VP, et al. Characteristics of an albumin dialysate hemodiafiltration system for the clearance of unconjugated bilirubin. ASAIO J 1997;43(5):M745−9.

[132] Peszynski P, Klammt S, Peters E, Mitzner S, Stange J, Schmidt R. Albumin dialysis: single pass vs. recirculation. Liver 2002;22(S2):40−2.

[133] Sauer IM, Goetz M, Steffen I, Walter G, Kehr DC, Schwartlander R, et al. In vitro comparison of the molecular adsorbent recirculating system (MARS) and single pass albumin dialysis (SPAD). Hepatology 2004;39(5):1408−14.

[134] Karvellas CJ, Bagshaw SM, McDermid RC, Stollery DE, Bain VG, Gibney RT. A case—control study of single-pass albumin dialysis for acetaminophen-induced acute liver failure. Blood Purif 2009;28(3):151—8.

[135] Rahman E, Al Suwaida AK, Askar A. Single pass albumin dialysis in hepatorenal syndrome. Saudi J Kidney Dis Transpl 2008;19(3):479—84.

[136] Collins KL, Roberts EA, Adeli K, Bohn D, Harvey EA. Single pass albumin dialysis (SPAD) in fulminant Wilsonian liver failure: a case report. Pediatr Nephrol 2008;23(6):1013—6.

[137] Rosa Diez G, Greloni G, Gadano A, Giannasi S, Crucelegui M, Trillini M, et al. Combined extended haemodialysis with single pass albumin dialysis (SPAED). Nephrol Dial Transplant 2007;22(9):2731—2.

[138] Chawla LS, Georgescu F, Abell B, Seneff MG, Kimmel PL. Modification of continuous venovenous hemodialfiltration with single-pass albumin dialysate allows for removal of serum bilirubin. Am J Kidney Dis 2005;45(3):e51—6.

[139] Rozga J, Umehara Y, Trofimenko A, Sadahiro T, Demetriou AA. A novel plasma filtration therapy for hepatic failure: preclinical studies. Ther Apher Dial 2006;10(2):138—44.

[140] Ho DW, Fan ST, To J, Woo YH, Zhang Z, Lau C, et al. Selective plasma filtration for treatment of fulminant hepatic failure induced by D-galactosamine in a pig model. Gut 2002;50 (6):869—76.

[141] Mori T, Eguchi Y, Shimizu T, Endo Y, Yoshioka T, Hanasawa K, et al. A case of acute hepatic insufficiency treated with novel plasmapheresis plasma diafiltration for bridge use until liver transplantation. Ther Apher 2002;6(6):463—6.

[142] Nakae H, Igarashi T, Tajimi K, Kusano T, Shibata S, Kume M, et al. A case report of hepatorenal syndrome treated with plasma diafiltration (selective plasma filtration with dialysis). Ther Apher Dial 2007;11(5):391—5.

[143] Nakae H, Igarashi T, Tajimi K, Noguchi A, Takahashi I, Tsuchida S, et al. A case report of pediatric fulminant hepatitis treated with plasma diafiltration. Ther Apher Dial 2008;12 (4):329—32.

[144] Nakae H, Eguchi Y, Saotome T, Yoshioka T, Yoshimura N, Kishi Y, et al. Multicenter study of plasma diafiltration in patients with acute liver failure. Ther Apher Dial 2010;14 (5):444—50.

[145] Nakae H, Eguchi Y, Yoshioka T, Yoshimura N, Isono M. Plasma diafiltration therapy in patients with post operative liver failure. Ther Apher Dial 2011;15(4):406—10.

[146] Matsumura KN, Guevara GR, Huston H, Hamilton WL, Rikimaru M, Yamasaki G, et al. Hybrid bioartificial liver in hepatic failure: preliminary clinical report. Surgery 1987;101 (1):99—103.

[147] Yanagi K, Ookawa K, Mizuno S, Ohshima N. Performance of a new hybrid artificial liver support system using hepatocytes entrapped within a hydrogel. ASAIO Trans 1989;35(3):570—2.

[148] Margulis MS, Erukhimov EA, Andreiman LA, Viksna LM. Temporary organ substitution by hemoperfusion through suspension of active donor hepatocytes in a total complex of intensive therapy in patients with acute hepatic insufficiency. Resuscitation 1989;18(1):85—94.

[149] Sussman NL, Chong MG, Koussayer T, He DE, Shang TA, Whisennand HH, et al. Reversal of fulminant hepatic failure using an extracorporeal liver assist device. Hepatology 1992;16 (1):60—5.

[150] Ellis AJ, Hughes RD, Wendon JA, Dunne J, Langley PG, Kelly JH, et al. Pilot-controlled trial of the extracorporeal liver assist device in acute liver failure. Hepatology 1996;24(6):1446—51.

[151] Millis JM, Kramer DJ, O'Grady J, Heffron TG, Caldwell S, Hart M, et al. Results of phase I trial of the extracorporeal liver assist device for patients with fulminant hepatic failure. Am J Transplant 2001;1(S1):391 A1012

[152] Millis JM, Cronin DC, Johnson R, Conjeevaram H, Conlin C, Trevino S, et al. Initial experience with the modified extracorporeal liver-assist device for patients with fulminant hepatic failure: system modifications and clinical impact. Transplantation 2002;74(12):1735—46.

[153] Mullon C, Pitkin Z. The Hepatassist bioartificial liver support system: clinical study and pig hepatocyte process. Expert Opin Investig Drugs 1999;8(3):229—35.

[154] Watanabe FD, Mullon CJ, Hewitt WR, Arkadopoulos N, Kahaku E, Egushi S, et al. Clinical experience with a bioartificial liver in the treatment of severe liver failure. A phase I clinical trial. Ann Surg 1997;225(5):484—91.

[155] Detry O, Arkadopoulos N, Ting P, Kahaku E, Watanabe FD, Rozga J, et al. Clinical use of a bioartificial liver in the treatment of acetaminophen-induced fulminant hepatic failure. Am Surg 1999;65(10):934—8.

[156] Demetriou AA, Brown Jr RS, Busuttil RW, Fair J, McGuire BM, Rosenthal P, et al. Prospective, randomized, multicenter, controlled trial of a bioartificial liver in treating acute liver failure. Ann Surg 2004;239:5.

[157] Pitkin Z, Mullon C. Evidence of absence of porcine endogenous retrovirus (PERV) infection in patients treated with a bioartificial liver support system. Artif Organs 1999;23(9):829—33.

[158] Paradis K, Langford G, Long Z, Heneine W, Sandstrom P, Switzer WM, et al. Search for cross-species transmission of porcine endogenous retrovirus in patients treated with living pig tissues. Science 1999;285(5431):1236—41.

[159] Sosef MN, Abrahamse LSL, van de Kerkhove MP, Hartman R, Chamuleau RA, van Gulik TM. Assessment of the AMC-bioartificial liver in the anhepatic pig. Transplantation 2002;73 (2):204—9.

[160] van de Kerkhove MP, Di Florio E, Scuderi V, Mancini A, Belli A, Bracco A, et al. Phase I clinical trial with the AMC-Bioartificial liver. Int J Artif Organs 2002;25(10):950—9.

[161] Di Nicuolo G, van de Kerkhove MP, Hoekstra R, Beld MG, Amoroso P, Battisti S, et al. No evidence of in vitro and in vivo porcine endogenous retrovirus infection after plasmapheresis through the AMC-bioartificial liver. Xenotransplantation 2005;12(4):286—92.

[162] Di Nicuolo G, D'Alessandro A, Andria B, Scuderi V, Scognamiglio M, Tammaro A, et al. Long-term absence of porcine endogenous retrovirus infection in chronically immunosuppressed patients after treatment with the porcine cell-based Academic Medical Center bioartificial liver. Xenotransplantation 2010;17(6):431—9.

[163] Poyck PP, van Wijck AC, van der Hoeven TV, de Waart DR, Chalumeau RA, van Gulik TM, et al. Evaluation of a new immortalized human fetal liver cell line (cBAL111) for application in bioartificial liver. J Hepatol 2008;48(2):266—75.

[164] Nibourg GA, Chamuleau RA, van der Hoeven TV, Maas MA, Ruiter AF, Lamers WH, et al. Liver progenitor cell line HepaRG differentiated in a bioartificial liver effectively supplies liver support to rats with acute liver failure. PLoS One 2012;7(6): e38778.

[165] Patzner II JF, Mazariegos GV, Lopez R, Bioartificial Liver Program Investigators. Preclinical evaluation of the Excorp Medical Inc. bioartificial liver support system. J Am Coll Surg 2002;195(3):299—310.

[166] Mazariegos GV, Patzer II JF, Lopez RC, Giraldo M, Devera ME, Grogan TA, et al. First clinical use of a novel bioartificial liver support system (BLSS). Am J Transplant 2002;2(3):260—6.

[167] Mazariegos GV, Kramer DJ, Lopez RC, Shakil AO, Rosenbloom AJ, DeVera ME, et al. Safety observations in phase I clinical evaluation of the Excorp Medical bioartificial liver support system after the first four patients. ASAIO J 2001;47(5):471—5.

[168] Morsiani E, Puviani AC, Brogli M, Pazzi P, Tosatti S, Valieri L, et al. In vitro experimentation of a new model of radial flow bioreactor containing isolate hepatocytes. Ann Ital Chir 2000;71 (3):337—45.

[169] Morsiani E, Pazzi P, Puviani AC, Brogli M, Valieri L, Gorini P, et al. Early experiences with a porcine hepatocyte-based bioartificial liver in acute hepatic failure patients. Int J Artif Organs 2002;25(3):192—202.

[170] Kanai H, Marushima H, Kimura N, Iwaki T, Saito M, Maehashi H, et al. Extracorporeal bioartificial liver using the radial-flow bioreactor in treatment of fatal experimental hepatic encephalopathy. Artif Organs 2007;31(2):148—51.

[171] Sauer IM, Gerlach JC. Modular extracorporeal liver support. Artif Organs 2002;26(8):703—6.

[172] Poyck PP, Pless G, Hoekstra R, Roth S, Van Wijck AC, Schwartländer R, et al. In vitro comparison of two bioartificial liver support system: MELS Cellmodule and AMC-BAL. Int J Artif Organs 2007;30(3):183—91.

[173] Sauer IM, Zeilinger K, Obermayer N, Pless G, Gruenwald A, Pascher A, et al. Primary human liver cells as source for modular extracorporeal liver support—a preliminary report. Int J Artif Organs 2002;25(10):1001—5.

[174] Sauer IM, Zeilinger K, Pless G, Kardassis D, Theruvath T, Pascher A, et al. Extracorporeal liver support based on primary human liver cells and albumin dialysis—treatment of a patient with primary graft non function. J Hepatol 2003;39(4):649—53.

[175] Sauer IM, Kardassis D, Zeilinger K, Pascher A, Gruenwald A, Pless G, et al. Clinical extracorporeal hybrid liver support— phase I study with porcine liver cells. Xenotransplantation 2003;10(5):460—9.

[176] Ding YT, Qiu YD, Chen Z, Xu QX, Zhang HY, Tang Q, et al. The development of a new bioartificial liver and its application in 12 acute liver failure patients. World J Gastroenterol 2003;9 (4):829—32.

[177] Shi XL, Zhang Y, Chu XH, Han B, Gu JY, Xiao JQ, et al. Evaluation of a novel hybrid bioartificial liver based on a multi-layer flat-plate bioreactor. World J Gastroenterol 2012;18 (28):3752—60.

[178] Gan JH, Zhou XQ, Qin AL, Luo EP, Zhao WF, Yu H, et al. Hybrid artificial liver support system for treatment of severe liver failure. World J Gastroenterol 2005;11 (6):890—4.

[179] Donini A, Baccarani U, Risaliti A, Degrassi A, Bresadola F. Temporary neurological improvement in a patient with acute on chronic liver failure treated with a bioartificial liver device. Am J Gastroenterol 2000;95(4):1102—4.

[180] Doré E, Legallais C. A new concept of bioartificial liver based on a fluidized bed bioreactor. Ther Apher 1999;3(3):264—7.

[181] Hwang YJ, Kim YI, Lee JG, Lee JW, Kim JW, Chung JM. Development of bioartificial liver system using a fluidized-bed bioreactor. Transplant Proc 2000;32(7):2349—51.

[182] Hwang YJ, Chang SK, Kim JY, Kim SG, Yun YG, Kim YI. Bioartificial liver system using a fluidized-bed bioreactor. Liver Transpl 2008;14(S1): S188-A430

[183] Yang Q, Liu F, Pan XP, Lv G, Zhang A, Yu CB, et al. Fluidized-bed bioartificial liver assist devices (BLADs) based on microencapsulated primary porcine hepatocytes have risk of porcine endogenous retrovirus transmission. Hepatol Int 2010;4(4):757—61.

[184] Coward SM, Legallais C, David B, Thomas M, Foo Y, Mavri-Damelin D, et al. Alginate-encapsulated HepG2 cells in a fluidized bed bioreactor maintain function in human liver failure plasma. Artif Organs 2009;33(12):1117—26.

[185] Mei J, Sgroi A, Mai G, Baertschiger R, Gonelle-Gispert C, Serre-Beinier V, et al. Improved survival of fulminant liver failure by transplantation of microencapsulated cryopreserved porcine hepatocytes in mice. Cell Transplant 2009;18(1):101—10.

[186] Liu JP, Gluud LL, Als-Nielsen B, Gluud C. Artificial and bioartificial support systems for liver failure. Cochrane Database Syst Rev 2004;1:CD003628.

[187] Strutchfield BM, Simpson K, Wigmore SJ. Systematic review and meta-analysis of survival following extracorporeal liver support. Br J Surg 2011;98(5):623—31.

[188] Zhao LF, Pan XP, Li LJ. Key challenges to the development of extracorporeal bioartificial liver support systems. Hepatobiliary Pancreat Dis Int 2012;11(3):243—9.

[189] Pan XP, Li LJ. Advances in cell sources of hepatocytes for bioartificial liver. Hepatobiliary Pancreat Dis Int 2012;11 (6):594—605.

[190] Cowan PJ, d'Apice AJ. The coagulation barrier in xenotransplantation: incompatibilities and strategies to overcome them. Curr Opin Organ Transplant 2008;13(2):178—83.

[191] Ding YT, Shi XL. Bioartificial liver devices: perspectives on the state of the art. Front Med 2011;5(1):15—9.

[192] Yagi H, Parekkadan B, Suganuma K, Soto-Gutierrez A, Tompkins RG, Tilles AW, et al. Long-term superior performance of a stem cell/hepatocyte device for the treatment of acute liver failure. Tissue Eng Part A 2009;15(11):3377—88.

[193] Soto-Gutierrez A, Navarro-Alvarez N, Yagi H, Nahmias Y, Yarmush ML, Kobayashi N. Engineering of an hepatic organoid to develop liver assist device. Cell Transplant 2010;19 (6):815—22.

[194] Uygun BE, Soto-Gutierrez A, Yagi H, Izamis ML, Guzzardi MA, Schulman C, et al. Organ reengineering through development of transplantable recellularized liver graft using decellularized matrix. Nat Med 2010;16(7):814—20.

[195] Hughes RD, Mitry RR, Dhawan A. Current status of hepatocyte transplantation. Transplantation 2012;93(4):342—7.

[196] Booth C, Soker T, Baptista P, Ross CL, Soker S, Farooq U, et al. Liver bioengineering: current status and future perspectives. World J Gastroenterol 2012;18(47):6926—34.

Liver Regeneration: The Bioengineering Approach

Yeonhee Kim, Sinan Ozer, and Basak E. Uygun

Center for Engineering in Medicine, Massachusetts General Hospital, Harvard Medical School and Shriners Hospitals for Children, Boston, MA

Chapter Outline

24.1 INTRODUCTION

The liver is the largest internal organ, constituting 2% of the weight of an adult human [1,2]. It is located on the right side of the abdominal cavity, just below the diaphragm and above the stomach. It is composed of two lobes, left lobe and right lobe, and these lobules are separated by the falciform ligament [3]. Two large vessels, hepatic artery and portal vein, supply the liver with blood at a rate of approximately 1.5 L/min [4]. The portal vein provides 75% of hepatic blood flow, and the hepatic artery contributes to the other 25% [5]. The portal blood circulates through the stomach, spleen, and intestine before entering the liver designating liver as the primary processing unit for ingested substances before being distributed throughout the body [6,7].

The liver performs many metabolic functions that are essential for the maintenance of health of the body. There are more than 500 functions attributed to the liver [8]. The major role of liver is the regulation of metabolism at molecular, cellular, and organ levels. It serves as an endocrine gland by secreting albumin and urea into the blood, as an exocrine gland by secreting bile [5]. One of its essential roles is in the metabolism of carbohydrates, hormones, and lipids [9]. It is the organ that is responsible for detoxification of drugs and xenobiotics [10]. It also plays a role in the body's response to trauma and injury by secreting acute phase proteins and by assuming a hypermetabolic state [11]. Liver is also capable of regenerating after an injury provided that at least 20% of its original size is preserved [12].

The major cell type responsible for liver functions is the hepatocyte. It represents 80% of liver cell mass and 60% of total cell population [13]. The rest is called the nonparenchymal cell fraction and is composed of bile duct epithelial cells, liver sinusoidal endothelial cells, hepatic stellate cells, Kupffer cells, and pit cells. The cells of the liver are arranged in basic functional units called acinus [14]. Plates of hepatocytes radiate from the central vein toward the portal triad with blood flow between the hepatocytes in structures called the liver sinusoids [2]. Liver sinusoids are lined by liver sinusoidal endothelial cells that are separated from hepatocytes by an extracellular matrix (ECM)-rich space called the space of Disse. Hepatic stellate cells reside in the space of Disse. The blood in the sinusoids provides 2000 nmol/mL of oxygen to the surrounding hepatocytes. The nutrient and oxygen gradients formed along the sinusoids are responsible for the zonation of hepatocytes [15–17].

Liver failure is responsible for 25,000 deaths per year in the United States [18]. Given that liver is capable of regeneration, an acute liver failure may resolve by itself within a couple of weeks. In cases where the liver cannot recover, the treatment requires orthotopic liver transplantation. Similarly, the only solution to chronic liver failure which may be caused by sustained exposure of toxic substances or cancer is orthotopic liver transplantation. However, donor organs suitable for transplantation are limited [19,20] and alternative treatments for liver transplantation are being thoroughly investigated. Here we will discuss engineering approaches for replacing failing liver functions as organ replacement therapy or as a bridge to transplantation. These include development of bioartificial liver (BAL) devices, cell-based therapies, and tissue engineered liver. We will conclude this chapter by discussing the novel whole liver engineering approach.

24.2 DEVELOPMENT OF BAL DEVICES

One approach to compensate for the functions of a failing liver is to use an extracorporeal liver support device. These devices are intended for use to bridge liver failure patients until a suitable donor liver becomes available or provide hepatic support until the liver recovers from an acute liver failure [21]. Depending on whether or not cells are included in the design, the device can provide artificial or bioartificial support. An artificial liver support device aims to replace the detoxification functions of the liver by using membrane separation techniques that involve adsorption by charcoal, anion or cation exchange resins. The separation selectively removes toxins and regenerates the dialysate. A BAL device aims to replace not only the detoxification functions but also synthetic functions of the liver by incorporating hepatocytes into the design of the device. Although it sounds promising, over three decades of work on the development of a BAL device has not yet resulted in a device that demonstrates clinical efficacy [22]. An ideal BAL device should be able to perform human-specific liver synthetic and detoxification functions in a sustainable fashion on the order of a couple weeks while the native liver regenerates [23].

The efficacy of a BAL device is determined by the use of an appropriate cell source and the design of a bioreactor system that maintains the function of those cells. The ideal cell type for use in a BAL device is primary adult human hepatocytes but their use is limited by the scarcity of high quality human cells [19]. Several alternatives exist including porcine hepatocytes and hepatic cancer cell lines but they are far from ideal due to poor functionality and the high inherent risk of zoonosis, immunogenicity, metastasis formation, and virus transfer [24]. The search for an ideal cell type persists through studies on derivation of human hepatocyte-like cells from pluripotent stem cells; however, practical and scalable methods to produce hepatocyte-like cells with mature phenotype have not been established yet [25].

The engineering perspective on the BAL device design focuses on the design of the bioreactor. The bioreactor is the part of the device where the cells are cultured and its design directly affects the viability and the function of the cells, hence the efficiency of the device [26]. The general operation of a BAL device is such that hepatocytes are seeded in the bioreactor and the patient's blood or plasma is circulated through the device. The treated blood or plasma is then returned to the patient. The bioreactor design should maximize mass transfer to the hepatocytes hence maximizing the delivery of oxygen, nutrients, and toxins from the patient's blood to the hepatocytes. This task requires a large surface area for mass transfer as well as uniform hepatocyte and flow distribution [27]. Oxygen delivery is especially essential for cell viability because hepatocytes are highly metabolically active cells with very high demand for oxygen [28]. For a clinically successful device, the bioreactor should mimic the native liver's microstructure and microenvironment to maintain the differentiated state of the hepatocytes. Physiologically, hepatocytes are not exposed to flow conditions as they are protected by an endothelial cell layer in the liver sinusoids. Therefore, the bioreactor should be designed in such a way that the shear stress on the hepatocytes is minimized since they are susceptible to shear-induced injury. Finally, the bioreactor should be able to maintain high density hepatocyte cultures, approximately 10^{10} cells, to support life during an acute liver failure [29].

There are four main types of bioreactor design proposed for BAL devices (Table 24.1). Hollow fiber bioreactors utilize capillary size hollow fiber cartridges housed in a cylindrical casing. The cells are loaded inside the cartridge and the patient's plasma or blood is circulated in the lumen of the casing [30–32]. The hollow fibers are made of a semipermeable membrane which allows for the selective transport of molecules from the patient's plasma and the device. Due to the large diameter of fibers used in these designs as well as the resistance to mass transport through the fibers, hollow fiber bioreactor design is associated with issues related to substrate limitations [33–37]. Oxygen delivery is limited by diffusion through the semipermeable membrane, therefore a hollow fiber bioreactor requires an extra means to enhance oxygen delivery to the cells for survival and optimal function. Mathematical modeling studies of oxygen distribution aided in determining the optimized parameters used in the design of hollow fiber bioreactors that resulted in improved cell distribution and function [33–35]. A comprehensive mathematical evaluation of design parameters for oxygen profiles led to the conclusion that it is not possible to maintain physiological oxygen levels throughout the bioreactor (70–25 mm Hg) with the current design parameters [38]. It is necessary to provide supraphysiological levels of oxygen at the bioreactor inlet. Early

TABLE 24.1 Characteristics of Bioreactor Designs in BAL Devices

Type of Bioreactor	Advantages	Disadvantages
Hollow fiber	Cells are protected from shear stress. Cells are immunoprotected.	Limited oxygenation of cells. Nonuniform cell distribution.
Packed bed	3D environment. Easy scale-up.	Cells exposed to shear stress. Nonuniform perfusion.
Fluidized bed	3D environment. Potential for immunoprotection. Large surface to volume ratio for better mass transfer.	Risk of microcapsule degradation.
Flat plate	Uniform cell distribution and microenvironment. Scale-up possible.	Cells exposed to shear stress.

experimental attempts to enhance oxygen delivery to the cells involved incorporating hollow fibers as conduits for oxygenation that were shown to help hepatocytes express differentiated function over several weeks [39−41]. Later studies focused on increasing the oxygen delivery to the cells through the use of oxygen carriers in the perfusate, such as red blood cells [42] or perfluorocarbons [28,43].

The second type of bioreactor used in BAL devices is the packed bed bioreactor. In this type of design, hepatocytes are seeded in porous three-dimensional (3D) polymer scaffolds and constructs are cultured in packed bed bioreactors [44]. Different materials were used in the design of the packed beds, such as polyvinyl formal resins [45−47] and plant-based glycosylated chitosan [48]. Although hepatocyte densities up to 10^7 cells/cm^3 could be achieved in one study [49], complications of flow dynamics such as channeling and clogging due to poorly defined pore structures in the scaffolds may affect device function [44]. Direct exposure of cells to perfused plasma may cause xenoimmunogenic reactions [50].

In order to protect cells from toxic exposures to perfused blood, microencapsulated hepatocytes have been used in fluidized bed-type bioreactors for BAL support recently [51,52]. The advantage of fluidized bed-type bioreactors is the large mass transfer area provided by spherical capsules that contain the cells. The encapsulation also protects the cells from liquid shear forces reducing the risk of shear stress injury. However, there is the risk of degradation of microcapsules over time, resulting in decline in the performance of the bioreactor.

The fourth type of bioreactor design in BAL devices, called the flat plate bioreactor, is based on flat surface geometry on which the hepatocytes are attached. In such a design, it is easier to control the internal flow distribution and adequately perfuse and oxygenate the cells. In a flat plate bioreactor, hepatocytes are cultured in the monolayer form rather than in the aggregate form, hence oxygen transfer to the cells is not dependent on diffusion

but convection which is more efficient. However, one disadvantage of such a design is to achieve sufficient concentration of cells without increasing the dead volume in the bioreactor beyond physiologically possible. In a flat plate bioreactor, the chamber volume is directly proportional to the channel height, and decreasing the channel height to decrease the volume increases the shear stress exerted by the flow on the hepatocytes. In an attempt to adequately oxygenate pig hepatocytes and protect the cells from shear stress in a flat plate bioreactor, De Bartolo et al. cultured liver cells between two gas permeable membranes and showed that the cells maintained differentiated functions for 18 days under flow conditions [53,54]. A comparison of a radial flow flat plate design to a hollow fiber bioreactor revealed that cells achieved polarity faster and expressed higher hepatic functions in the radial flow flat plate bioreactor due to the maximal contact of hepatocytes with the fresh perfusate in the radial flow configuration. The radial flow design also permits scale-up without sacrificing the survival and the function of the cells [55]. Maringka et al. achieved high hepatocyte density cultures by stacking 50 flat membrane modules that contain 2×10^8 cells [56]. Each module was equipped with their own oxygenation source, therefore the cultures were kept viable and functional for 12 days. The same group later showed that the same device has the potential to biotransform diazepam for 20 days even in the presence of human plasma [57]. The duration for efficient BAL device operation was later extended to 35 days when hepatocytes were cultured on a self-assembling peptide nanofiber scaffold [58]. The clinical efficacy of the stacked flat membrane design BAL device remains to be shown in an animal model of acute liver failure.

Similarly, using microfabrication techniques, a microchannel flat plate bioreactor with an internal gas permeable membrane through which oxygen is supplied is shown in Figure 24.1[59]. The oxygen distribution was found to be sufficient to maintain high hepatocyte survival

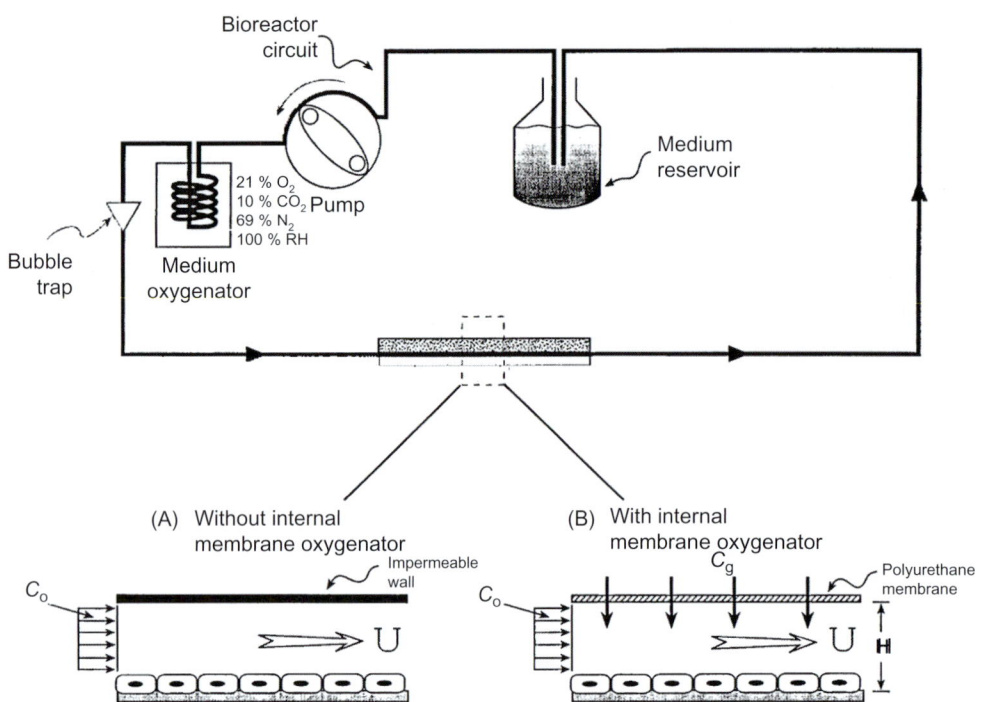

FIGURE 24.1 Schematic diagram of the BAL device for improved oxygen delivery. Flow circuit perfusion system and the bioreactors (A) without and (B) with internal membrane oxygenator. *Reproduced with the permission of John Wiley and Sons [59].*

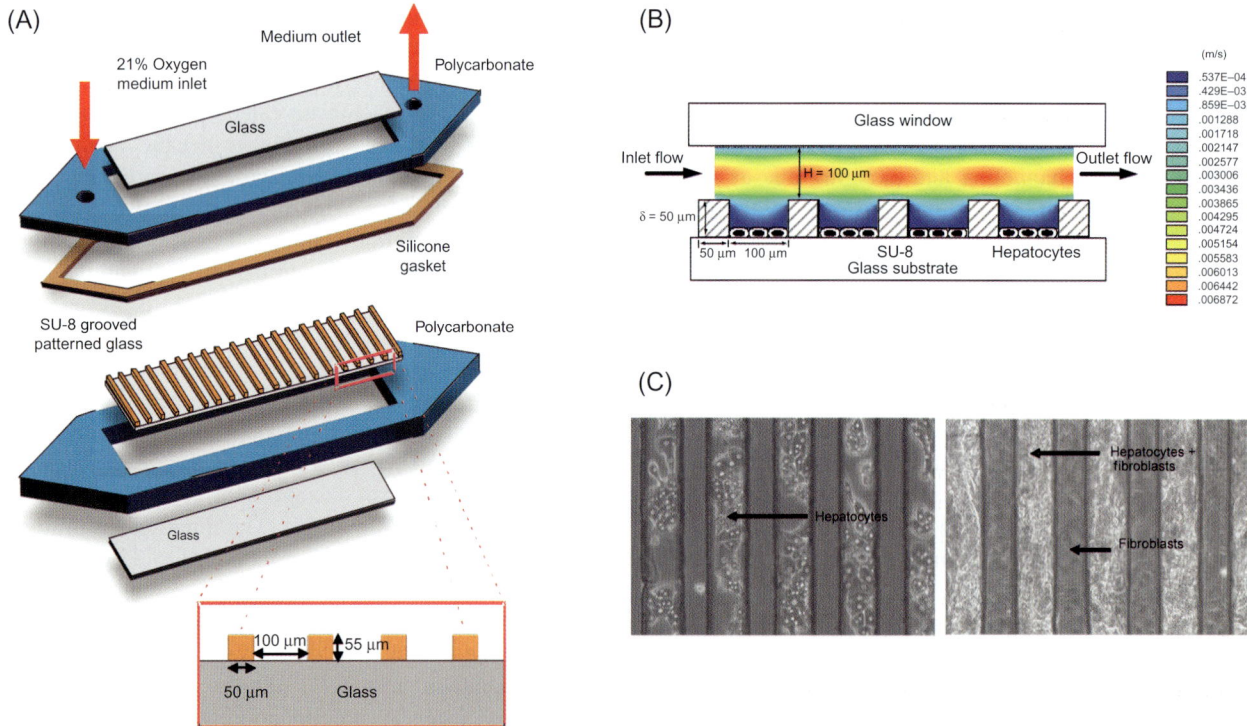

FIGURE 24.2 BAL device designed for minimal shear stress. (A) Schematic representation of bioreactor with microgrooved glass substrate. (B) Contour plot of velocity vector field in channel of the microgrooved substrate bioreactor using Finite Element Analysis (FEA) modeling for volumetric flow rate set at 0.7 mL/min. (C) Phase contrast images of hepatocytes and hepatocytes/fibroblasts cultured on microgrooved substrate 1 day after seeding. . *Reproduced with the permission of John Wiley and Sons [64].*

and albumin and urea functions for 4 days of perfusion [60−62]. In addition, the BAL device with the internal membrane oxygenator was shown to rescue animals undergoing D-galactosamine-induced fulminant hepatic failure [63]. Further improvement to this design was made by incorporating microgrooves on the substrate plate. Hepatocytes were seeded in the microgrooves to protect them from detrimental shear stress (Figure 24.2). With the microgroove design, the cell viability was much higher resulting in a more stable production of albumin and urea over a 4-day culture period when compared to the flat substrate bioreactors [64]. In order to increase the cellular capacity of bioreactors without introducing oxygen deprivation, later studies were performed using stacked microgrooved plates in radial flow configuration [65].

24.3 ENGINEERING PERSPECTIVE ON HEPATOCYTE TRANSPLANTATION

Hepatocyte transplantation is a promising alternative to orthotopic liver transplantation to treat liver-based metabolic disorders, acute, and chronic liver failures [66−68]. Hepatocyte transplantation may act as a bridge to orthotopic liver transplantation or alternatively trigger regeneration of the acutely injured native liver [20]. It involves the transfer of healthy hepatocytes into diseased liver, by injecting isolated hepatocytes into the spleen or splenic artery or infusing directly into the portal vein with the expectation that transplanted cells engraft into the liver and perform the functions of the failing liver (Figure 24.3). Although clinical success has been demonstrated for a limited number of metabolic liver diseases [70,71], there are several challenges associated with the approach. The first is the source of human hepatocytes. The use of primary adult human hepatocyte is limited by shortage of healthy donor liver tissue for hepatocyte isolation and the reduced proliferative capacity of primary hepatocytes [19]. Therefore, there have been many efforts to identify alternative cell sources that are sustainable and readily available for transplantation. Additional efforts aim to improve the quality of primary human hepatocytes that are isolated from resected liver tissue. The second challenge associated with hepatocyte transplantation is the low engraftment and survival of transplanted cells [19]. There are several factors that may contribute to this problem. The transplanted cells may become the target of the host immune system and be eliminated from the system. Additionally, the lack of homotypic interactions as well as structural support for the anchorage-dependent hepatocytes in a system where cell transplantation is simply done by injecting a suspension of cells may result in their poor survival. In the following sections, we will discuss the potential engineering solutions for hepatocyte transplantation to improve the clinical outcome.

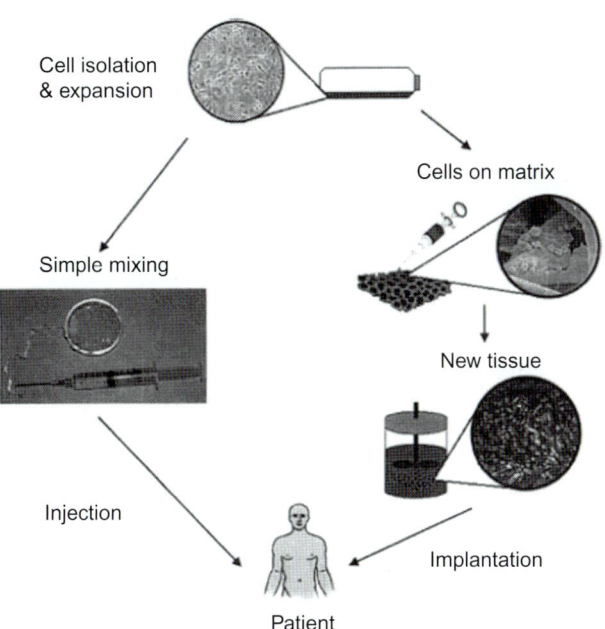

FIGURE 24.3 Current strategy for cell transplantation. Cells are expanded *ex vivo* and injected into the damaged tissue or systemic circulation as suspension (left). Tissue engineering approach involves seeding of cells in a 3D biomaterial scaffold and culturing the construct in the presence of mechanical and chemical stimuli *ex vivo*. The newly formed tissue can then be implanted into the patient to replace the failing organ functions (right). *Adapted with permission from [69]. Copyright 2001 American Chemical Society.*

24.3.1 Cell Sourcing

The selection of the ideal cell source for cell transplantation studies depends on several factors. The cells should perform the liver-specific functions, be readily available and have reduced risk of zoonosis, immunogenicity, and tumor formation. Various cell types that have been suggested for hepatocyte transplantation include primary human hepatocytes, porcine hepatocytes, stem cell-derived hepatocyte-like cells, fetal progenitor cells, and oval cells. The comparison of different cell types in terms of clinical efficacy and their associated risks has been well presented [19]. Here we will focus on the primary hepatocytes as the ideal cell source for transplantation and discuss engineering approaches to increase their quality and availability. We will also include studies that utilize engineering tools, such as microfabrication and tissue engineering, to improve the differentiation efficiency of pluripotent stem cells into hepatocyte-like cells.

24.3.1.1 Adult Primary Hepatocytes

Hepatocytes, the principal cell type of liver, are a natural cell source that can restore deteriorated functions and regenerate damaged tissue. Human hepatocytes are isolated from liver tissues which are donated but rejected or deemed unsuitable for organ transplantation, including livers from nonheart-beating donors or those with long cold ischemic times. In addition, a split liver segment or section IV that can be used to isolation, but not for transplantation, can be a valuable source for hepatocyte transplantation. A collagenase perfusion technique is used to digest donor liver tissue, and the cells are purified by density gradient method [72]. The number and viability of good quality hepatocytes usable to transplantation are evaluated using trypan blue exclusion method [19,73]. The first clinical trial of hepatocyte autotransplantation was performed on 10 patients in 1992 [74]. The recipients' native cirrhotic left lateral liver segment was used as the cell source for hepatocyte transplantation and the results showed improved encephalopathy in two of the patients. Since then, hepatocyte transplantation has shown therapeutic potential for treating inborn errors of metabolism [70,75–77], acute [78], and chronic liver diseases [79,80]. For example, cryopreserved hepatocytes were injected into spleen of a patient with end-stage liver disease as a "bridging" technique to sustain the patient until orthotopic liver transplantation could be conducted [79]. Also, hepatocyte transplantation was performed in three patients with alcoholic liver cirrhosis in 1996 and all of them were alive in 2000 [80]. In addition, transplantation of cryopreserved hepatocytes, of which the advantages are ease of use in emergency and effective quarantining of cells against bacterial or fungal contamination was performed as the clinical trials [81–83].

Although primary hepatocytes are the most suitable cell type for transplantation, this cell source relies largely on the limited availability of donor livers for the cell isolation. In addition, it has been shown that the energy sources of donor livers were depleted during cold storage which led to a decreased number of viable hepatocytes isolated from these livers [84]. In an effort to increase the viable pool of donor organs, machine perfusion of marginally ischemic livers has been demonstrated as a promising approach [85,86]. In this technique, marginal livers taken from nonheart-beating donors and usually considered unsuitable for transplantation undergo continuous hypothermic (4°C) [87–89] or subnormothermic (20°C) [90–92] perfusion with oxygen and other nutrients at physiological rates for a period of a couple of hours. The controlled perfusion regime promotes a washout of blood and other toxic substances, allows for repletion of ATP reserves of cells leading to increased metabolic activity of the organ [85]. Interestingly, subnormothermic (i.e., 20°C) perfusion has also been shown to improve the yield and the quality primary hepatocytes isolated from marginal rat livers [93]. The hepatocyte yields from perfusion recovered ischemic livers were 25-fold higher than those obtained from livers that were not perfused and not significantly different from fresh livers.

The hepatocytes remained viable and functional *in vitro* during the 2-week culture period. Although the efficacy of the technique remains to be shown for obtaining viable hepatocytes from marginal human livers, these results hold the promise to extend the availability of high quality human hepatocytes for cell transplantation.

Another challenge to overcome in using primary hepatocytes in cell transplantation is the dramatic loss of viability, hepatic phenotype, and functional properties *in vitro* following cryopreservation [94]. Successful cryopreservation will allow establishment of cell banks with full characterization of hepatocytes. It will also improve the availability of human hepatocytes in a timely manner, so that they can be used in emergency situations of acute liver failure and repeated or planned administrations to treat patients with chronic liver failure and metabolic disorders [95]. The primary reason for cryoinjury to the cells is the disruption of plasma membrane during freezing due to intracellular ice crystal formation. Several attempts have been made to improve the viability and function of hepatocytes upon thawing by incubating the cells with antioxidants and cryoprotectants prior to freezing [95,96]. One interesting study reported that loading of a nonmetabolizable glucose, 3-O-methyl glucose, into the cells prior to cryopreservation improved the viability and function of the cells by at least 50% upon thawing. These cells were functional up to 2 weeks in culture and showed cytosolic enzyme activity at the levels similar to nonfrozen controls [97]. Another strategy to eliminate the formation of ice crystals is cryopreservation by vitrification [95]. Vitrification is transition from liquid phase to a glassy state by rapid cooling usually in the presence of cryoprotectants. Two recent studies report preservation of primary hepatocytes by vitrification both in suspension and in attached cells [98,99]. The fraction of apoptotic cells post thaw in vitrification samples were similar to nonfrozen controls while conventional frozen samples had a higher apoptosis rate [98]. The scale-up of the vitrification protocol has also been presented making this an interesting alternative to conventional freezing preservation [98]. As a separate but parallel approach to cryopreservation, simple cold storage (2–8°C) has been suggested for human hepatocytes [100]. Modified with iron chelators and other supplements, organ cold storage solutions (e.g., Custodial-N) were used to store monolayers of human hepatocytes at 4°C for up to 2 weeks. The cold-induced injury assessed by lactate dehydrogenase release was reduced to 20%. Although these results need to be reproduced, the fact that alternatives to conventional cryopreservation methods are under investigation raises the promise for better technologies for storage of primary hepatocytes that can be used in cell transplantation therapies.

24.3.1.2 Pluripotent Stem Cell-Derived Hepatocytes

Embryonic stem (ES) cells and induced pluripotent stem (iPS) cells have the great capacity to proliferate and differentiate into almost any cell type making them an attractive cell source for cell transplantation therapy [101]. There are numerous studies that report the generation of hepatocyte-like cells from pluripotent stem cells [25,102] and demonstrate their use in cell transplantation studies [103−105]. Protocols, capable of generating ES/iPS cell-derived hepatocytes, are usually based on signaling events in the early embryo, known to initiate specification of definitive endoderm [106−109]. Treatment with growth factors or cytokines or chemicals in determined sequence has led to direct differentiation of ES/iPS cells toward endoderm and further toward hepatic progenitor state exhibiting molecular features found in primary hepatocytes [110−115]. After intravenous transplantation of murine ES cell-derived hepatic cells to injured liver of three recipient mice, significant suppression of liver fibrosis, and enhancement of liver functions and survival rate were observed [116,117]. Transplanted mature hepatic cells (ES cell-derived hepatocytes) were able to integrate into the injured liver of immunodeficient mice with [110] or without a 50% hepatectomy [118] and to express hepatic phenotypic functions similar to those of primary human hepatocytes. ES cell studies remain at a preclinical stage owing to the risk of tumor formation [119] from undifferentiated ES cells and low levels of engraftment.

Recently, iPS cells were shown to differentiate into human hepatocyte-like cells that exhibited major liver functions and integrated into the hepatic parenchyma *in vivo* [120,121]. More recently, human hepatic cells derived from iPS cells regardless of their parental epigenetic memory could successfully repopulate the liver tissue of mice with liver cirrhosis and secrete human-specific liver proteins comparable to those in primary human hepatocytes [122]. The advent of iPS cell technology in mid-2006 has made possible for the first time the generation of patient-specific pluripotent stem cells for cell-based therapy. These cells should circumvent the need for immune suppression and at the same time alleviate the ethical considerations of using human ES cells [123−125]. However, investigation of iPS cell-based applications for liver regeneration is still in infancy with various risks, such as teratoma formation and retention of epigenetic memory prior to reprogramming.

One engineering strategy to drive efficient stem cell differentiation into mature phenotype while facilitating cell transplantation is to employ 3D biomaterials to culture and differentiate pluripotent stem cells mimicking the *in vivo* microenvironment [126,127]. The cell−cell and cell−ECM interactions within the 3D microenvironment provide additional cues to direct efficient differentiation of pluripotent stem cells [128]. To this effect, ES cells were cultured in collagen scaffolds [129], in Algimatrix 3D culture platform [130] and on synthetic nonbiodegradable polyamide nanofibers [131] for hepatic differentiation. 3D culture conditions resulted in higher expression levels of hepatocyte markers when compared to 2D culture conditions. In order to increase the yield of hepatocyte-like cells, one study reported culture and differentiation of ES cells under perfusion bioreactor conditions [132]. Although these studies represent preliminary results, applying tissue engineering techniques to generate hepatocyte-like cells from ES cells holds great promise to address the cell source problem for hepatocyte transplantation therapies.

24.3.2 Engineering Cells to Improve Engraftment

Although several clinical successes of hepatocyte transplantation via direct delivery to patients with liver failure have been reported, insufficient engraftment and low survival of transplanted cells is still a major challenge. The sustained function and therapeutic efficacy through hepatocyte transplantation can be enhanced by appropriate means of cell delivery which will provide a protected environment for the transplanted cells from the recipient immune system. In addition, a 3D structural support that mimics the native microenvironment can improve the survival of transplanted cells by recapitulating the mechanical and chemical cues present *in vivo*.

24.3.2.1 Cell Encapsulation

Immune rejection of transplanted cells is a major issue in cell transplantation that results in low survival of transplanted cells. Cell encapsulation has been suggested as a means to protect the cells from the hostile microenvironment in the recipient tissue and provide an opportunity for the cells to survive [133,134]. Encapsulation is a technique where a number of cells are entrapped in a spherical semipermeable polymeric membrane. Different techniques have been used for microencapsulation of cells within 0.3−1.5 mm size range. These include microencapsulation by polyelectrolyte complexation, microencapsulation by agarose, encapsulation based on interfacial phase inversion, and encapsulation by *in situ* polymerization [135]. This technique has been used for hepatocyte transplantation to improve cell survival and subsequent engraftment. For example, intrasplenic transplantation of rat hepatocytes encapsulated within polyvinylidene difluoride hollow fibers showed survival for 28 days [136]. Transplantation of encapsulated hepatocytes within alginate—poly-L-lysine microcapsules in rat fulminant hepatic failure model showed decrease in mortality and improvement in liver functions [137]. Alginate-based encapsulation techniques have been extensively studied as alginate—poly-L-lysine—alginate microcapsules [138,139] or as alginate—chitosan microcapsules [138,140,141] or as calcium—alginate—poly-L-lysine matrix (CAP) [142]. Rat hepatocytes encapsulated in CAP were transplanted in animal models of Crigler—Najjar syndrome, resulting in the reduced bilirubin levels and no rejection of the cells [142]. Later studies with animal models of liver failure also showed immunoprotection of hepatocytes upon transplantation when the cells were encapsulated [141,143,144]. Multilayered microcapsules, capable of selective permeation, have addressed mechanical stability, selective permeability, and a favorable microenvironment for improving cellular functions. Taken together, encapsulation helped circumvent the immune barrier and sustained the transplanted cells. In addition, cotransplantation of hepatocytes with bone marrow mesenchymal stem cells [145—147] or with Sertoli cells [148,149] or with endothelial cells [150] could be performed via the coencapsulation approach, highlighting improved functions and long-term survival of the cells. Despite numerous preclinical studies demonstrating the efficacy of the microencapsulated hepatocyte transplantation, the lack of clinical grade polymers and the difficulty in producing uniform microcapsules hindered the advancement of the field to the clinical setting [133].

24.3.2.2 Polymer Scaffolds for Cell Transplantation

An early study investigated the transplantation of hepatocytes seeded into bioerodible polymeric materials and showed that cell survival and function was improved as compared to hepatocytes transplanted without the biomaterial [151]. It has also been suggested that the impact of transplanted cells may be significantly increased if the cells are transplanted within a carrier material that has the biological cues necessary for cell survival and provide the mechanical support for the proper cell function [152]. This approach forms the basis of hepatic tissue engineering and has been an attractive field of study for over two decades. The types of materials used to harbor cells as well as the work on preclinical transplantation of tissue engineered hepatic constructs will be discussed in detail later in the chapter.

24.3.2.3 Sites of Cell Transplantation

While the focus in cell transplantation for liver tissue repair and regeneration is on developing delivery mechanisms to improve efficiency of cell survival and engraftment, one other important aspect is the site of hepatocyte transplantation. The traditional route to deliver hepatic cells to failing liver is the hepatic portal vein [153,154] or hepatic artery [20,155,156]. The hepatic portal vein drains blood from the gastrointestinal tract and spleen into the liver. It is known from animal studies that infused hepatocytes disperse with the portal blood flow, translocate to the hepatic sinusoids and eventually integrate into the parenchyma [157]. During intercellular contacts with host cells, transplanted hepatocytes start to form gradually increasing cell clusters and finally repopulate the recipient liver. In some cases, pulmonary infiltrates have been reported, suggesting that some cells lodge in the pulmonary capillary bed [158]. The number of hepatocytes that may be transplanted through the portal vein at a time is also limited by portal pressure. Therefore, direct injection of cells into the liver circulation may be used in basic research but not in the clinical setting.

The other major route of injecting cells into liver is via spleen, usually the splenic pulp. The spleen is a natural location for transplanting cells for treating liver diseases [153,159—161] since it has a rich blood supply which is accessible to hepatic portal circulation, leading to the translocation of the transplanted cells to the hepatic sinusoids. In early studies, 40% of the spleen was replaced by transplanted hepatocytes and donor hepatocytes repopulated up to 97% of the host liver in mice with genetically induced liver disease [79,162,163]. Transplanted hepatocytes had high capability to replicate hepatocyte foci with an average of 12 cell doublings [162]. Direct intrasplenic injection was suggested to be a more feasible method to transplant hepatocytes compared to the splenic artery infusion, since the latter gave rise to vascular occlusion with hepatocytes, gastric erosion, and large areas of splenic necrosis [20,163,164]. There have been reported cases of intrasplenic transplantation leading to local embolus and large numbers of the transplanted cells retained in the spleen pulp [159,165].

The intraperitoneal route has been used to accommodate a large number of transplanted cells to sustain function in acute liver failure [166]. Prior to transplantation, cells were attached to collagen-based microcarriers [167] or encapsulated in biodegradable polymers [151] or in prevascularized poly(vinyl alcohol) sponges [168] to prevent the transplanted cells from the attack of the host immune system and to provide a scaffold to sustain survival and function. The immune response and the delivery distance between injection and target site are the main challenges with this mode of delivery. However, in a recent study, porcine hepatocytes encapsulated in hydrogel could be xenotransplanted into rats survive and function for 30 days without any immune suppression [169].

Kidney capsule could be used as a site for hepatocyte transplantation [170], but it resulted in low survival of transplanted cells and it had insufficient space for transplantation of a meaningful cell mass. When hepatocytes mixed in the ECM obtained from murine Engelbreth—Holm—Swarm tumor cell lines gel were transplanted under the bilateral kidney capsule spaces, they resulted in a liver tissue formation that possessed active regeneration potential [171,172]. In a more recent study, 3D liver organoids comprised of hepatocytes and nonparenchymal cells (sinusoidal endothelial cells and hepatic stellate cells) were transplanted under kidney capsule and showed enhanced hepatic functions and survival for 8 weeks after transplantation [173].

Taken together, the choice of sites is one of critical factors to determine the efficiency of hepatocyte transplantation. Sites that can prolong transplanted cells' survival and enhance the efficiency of engraftment into recipient's liver are of significant interest. Modulation of the host liver environment to boost the regenerative capacity of endogenous hepatocytes might be triggered by transplantation sites in and near the liver. Environmental cues, such as ECM, nutrients, growth factors [174], can also stimulate essential interactions of the transplanted cells with nonparenchymal cells of the host liver to support the engraftment and prolonged survival of the transplanted cells [175].

24.4 HEPATIC TISSUE ENGINEERING

The ultimate goal of tissue engineering is to create structures that mimic the native organ architecture with the aims of replacing or restoring lost organ functions. Tissues are composed of repeating 3D units in *in vivo* (e.g., liver lobules, islets, nephrons). On the macroscopic scale, the 3D architecture of tissues underlies the coordination of multicellular processes, emergent mechanical properties of the tissue, and the integration with other organ systems through microcirculation. Microscopically, the three dimensionality allows the formation of the cellular "microenvironment" that presents the biochemical, physical, and cellular stimuli that orchestrate cellular processes, such as proper functions, proliferation, differentiation, migration, and apoptosis [176]. Thus tissue engineering should address the creation of 3D tissue constructs both macroscopically and microscopically. Traditional tissue engineering approach involves fabrication of 3D polymer scaffolds in which the parenchymal cells are seeded and cultured in the presence of biochemical and mechanical stimuli. However, this has proven to be insufficient since over two decades of work in tissue engineering only resulted in fewer than 10 products that are used clinically [177]. The success for liver has been even less, mainly due to inefficiency of supplying nutrients and oxygen for hepatocytes that are seeded inside the tissue construct. In addition, the early work on fabricating hepatic tissue constructs focused on creating a large

hepatic mass and neglected the fact that proper hepatocyte function requires the presence of nonparenchymal cells, such as endothelial and stellate cells as well as the native ECM cues [178]. Liver tissue engineering should assume an organotypic approach, aiming to create the hepatic microenvironment microscopically, while providing the mechanical structure and the architecture to allow for the microcirculation of oxygen and other nutrients for the survival of the tissue. In this section, we will summarize the work in hepatic tissue engineering and describe the current work with an emphasis on the organotypic approach.

24.4.1 Cellular Constructs

Adult primary hepatocytes can successfully be isolated using the two-step perfusion technique described by Seglen [179]. The isolation protocol yields high purity hepatocytes ($>95\%$) with a very high viability ($>90\%$) [180]. Although the isolation protocol was established almost four decades ago, it has been a great challenge to maintain differentiated hepatocyte function and morphology *in vitro* [181]. A successful way to rescue differentiated hepatocyte function *in vitro* is to overlay a monolayer of primary hepatocytes with an ECM gel (e.g., collagen) creating a sandwich configuration. This technique rescues differentiated hepatocyte function at 7 days of culture and stabilizes synthetic functions (e.g., albumin and urea secretion) for over 6 weeks [182]. While this model provides a robust system for the study of hepatocyte functions *in vitro*, it is not a scalable system and not suitable for use as an organ replacement therapy. Another approach of sustaining hepatocyte functions *in vitro* is 3D spheroid cultures. Hepatocytes cultured on a basement membrane protein blend (Matrigel) form into spherical aggregates called spheroids within 48 h and display differentiated hepatocyte functions for a week [183]. Hepatocytes form spheroids on other protein surfaces [184,185], nonadhesive surfaces, such as poly(2-hydroxyethyl methacrylate) (poly-HEMA)-coated surfaces [186,187] and Primaria culture plates [188]. A number of studies report that hepatocyte spheroids remained functional for up to 6 weeks displaying sustained albumin production and cytochrome P450 activity [186,187]. The homotypic interactions result in the deposition of basement membrane proteins, such as laminin and fibronectin as well as in the formation of the bile ductular structures within the spheroids [186]. One disadvantage of spheroid hepatocyte cultures is the limited transport of oxygen and other nutrients. The difficulty of size control in the formation of spheroids with conventional techniques may result in spheroids too large so that the cells in the center of the structure become necrotic [181]. The advances in microfabrication techniques allow for precise control over the size of the spheroids [189−191]. Nevertheless, the primary hepatocytes in spheroid configuration do not display any proliferative capacity, hence it is unlikely that this technique will be used as scalable method for organ replacement therapy.

The presence of heterotypic interactions as well as homotypic interactions is essential for maintaining liver-specific functions in vitro. In vivo, two-thirds of the liver is composed of parenchymal cells, and the rest of liver is composed of nonparenchymal cells [192]. It has been shown that cocultivation of parenchymal and mesenchymal cell types, such as endothelial or fibroblasts, promotes retention of hepatocyte viability and liver-specific functions that would normally be lost and leads to improved viability, function, and tissue stability [193]. When cultured with murine 3T3-J2 fibroblasts, it was found that there was a dose-dependent increase in liver-specific function as the number of fibroblasts were increased in the culture [194]. Similarly, 3D cocultures of hepatocytes with stellate cells have been shown to maintain liver-specific functions longer than monocultures [195−197]. In an effort to build the liver microarchitecture in vitro, micropatterning techniques have been used to coculture hepatocytes with nonparenchymal cell types. These techniques include laser-guided direct writing (LGDW) [198] and micropatterning using stencils [199,200]. LGDW is a patterning technique which allows deposition of cells with precision on matrices, including soft gels such as collagen and Matrigel [201−204]. Micropatterning of endothelial cells done by LGDW creates an environment for the endothelial cells to self-assemble into vascular structures. The purpose of the organization of these vascular structures is to overcome the difficulties faced by hepatic tissue engineering and to form heterotypic cell−cell interactions found in the adult liver, which is important for maintaining liver function, in the engineered liver tissue [205,206]. The method itself is also suited for studying properties of tissue architecture and effects of heterotypic cell−cell interactions on morphogenesis, differentiation, and angiogenesis [198]. The single mode laser is weakly focused to a spot about 5 μm in radius. The cells that are in the laser beam path experience a radiation force and they are pushed in the opposite direction by the force of the scattering photons from the cell−media interface, which allows them to be deposited on the surface [203].

One other approach for creating cellular constructs for liver tissue engineering is cell sheet engineering. Hepatocyte sheet engineering employs stacking layered cellular constructs to make stratified architecture that mimics in vivo structural features [207]. It has been reported that the shape of the hepatocytes is closely related with the cellular functions, dedifferentiation, cell proliferative growth, and cell survival [208]. Therefore,

the purpose of cell sheets is the regulation of cell shape while creating a transplantable liver tissue that can be implanted with minimally invasive procedures (e.g., local anesthesia and small skin incision) [209]. Cell sheets are generated by growing cells in a cell culture dish at a certain confluence, which allows for development of cell−cell interactions and deposition of ECM over time. An example of cell sheet engineering is hepatocytes cultured on a temperature-responsive polymer, poly(N-isopropylacrylamide) [210]. Poly(N-isopropylacrylamide) is hydrophobic in water, but hydrophilic at temperatures below 32°C. Using this feature of the polymer, cells can be seeded at 37°C, a temperature at which the polymer is hydrophobic, and allowed to adhere, spread, and proliferate. Afterwards, the temperature is lowered to 20°C, forcing the polymer to become hydrophilc and removing the cultured cells as an intact cell sheet [211]. Hepatocyte sheets displayed higher liver-specific functions than single hepatocytes and the engineered sheets were transplanted successfully in the subcutaneous space for over 250 days [210]. Cell sheet engineering is an attractive approach to fabricate small hepatic constructs which can be transplanted into sites with limited space [209].

24.4.2 Cell-Scaffold Hybrid Hepatic Constructs

One of the first examples of cell-scaffold hybrid constructs for hepatic tissue engineering was shown using prevascularized polyvinyl alcohol discs. The discs were repopulated with rat hepatocytes and transplanted into animals for up to 1 year. Promisingly, even though initial cell engraftment was low, the discs contained viable cells after 1 year of transplantation [212,213]. Later studies focused on demonstrating optimal hepatocyte function and viability in vitro using tissue engineering techniques. The first biomaterials that were tested for liver tissue engineering include those that are already approved for use in biomedical applications, such as polylactic acid (PLA), polyglycolic acid (PGA) and copolymers of lactic and glycolic acid (PLGA) [214]. These materials are biodegradable, biocompatible, and can be fabricated into highly porous structures that may accommodate highly oxygen demanding hepatocytes [151,215]. Improved hepatocyte function due to spheroid formation has been shown in PLA-based discs [216−218] and PLA-based scaffolds [195]. A natural biopolymer, chitosan, has been the focus of several liver tissue engineering studies. Chitosan is a linear polysaccharide with randomly distributed β-(1−4)-linked D-glucosamine and N-acetyl-D-glucosamine units. It is a desirable material for biomedical applications because it is biodegradable, biocompatible [219], and easy to chemically modify for manipulating its biological and mechanical properties [220]. Chitosan-based biomaterials that have been tested for hepatocyte culture and function so far include chitosan/alginate and chitosan/collagen blends [221], galactosylated chitosan/alginate and galactosylated chitosan/alginate/heparin blends [222,223], chitosan/galactosylated hyaluronic acid (HA) composites [224] and nanofibrous galactosylated chitosan [225]. Another natural polymer that was investigated for hepatic tissue engineering is HA. HA is a nonsulfated, linear polysaccharide with the repeating disaccharide, β-1,4-D-glucuronic acid–β-1,3-N-acetyl-D-glucosamine. It is an essential component of the ECM and its structural and biological properties mediate cellular signaling, wound repair, morphogenesis, and matrix organization [226–228]. Several studies have demonstrated the benefits of using HA-based scaffolds in liver tissue engineering [229,230]. Fetal liver cell-loaded HA sponge scaffolds could be transplanted into a rat model of Wilson's disease, a genetic disorder resulting in accumulation of copper. As a result, this approach prevented jaundice and showed a significant reduction in blood copper concentration, which consequently led to restoring liver functions [230]. Another study has proven the beneficial effects of the use of HA-based scaffolds for long-term culture and efficient formation of hepatic aggregates *in vivo* after transplantation [229]. Recently, peptide-based biomaterials have drawn interest in tissue engineering applications. A well-defined synthetic hydrogel forming peptide, PuraMatrix™, was investigated for its capacity to support hepatocyte functions in a 3D environment for rat [185] and porcine hepatocyte cultures [231]. Even though albumin and urea production as well as cytochrome P450 activities in PuraMatrix™ cultures were similar to those observed in collagen sandwich and Matrigel cultures, the advantage of using PuraMatrix™ was the possibility of achieving high functional capacity through the high density cell cultures [185]. It also serves as a delivery vehicle for cell transplantation for which it has been shown to be highly effective in a number of animal models of liver failure [232].

One limitation of using polymer scaffolds for liver tissue engineering is the lack of inherent vascular network that would allow circulation of nutrients and removal of waste from the cells that are seeded inside the scaffold. Current technology in the fabrication of 3D scaffolds limits the volume of hepatic constructs one can achieve since nutrient delivery can occur by diffusion only. A promising approach is to reconstruct the artificial liver tissue by building the tissue from ground up using microfabrication techniques [178,233–235]. Even though it has not been realized so far, this promising approach involves a silicone or pyrex platform with microchannels that allow continuous perfusion seeded with endothelial cells to form a well-sealed vascular network. Once the vascular network is established, the system is expected to be repopulated with hepatocytes and biliary epithelial cells to form the liver tissue [236].

24.5 WHOLE LIVER TISSUE ENGINEERING

The first-generation tissue engineering materials have found limited use since they had limited ability to modulate the repair and regeneration of the host tissue. Consequently, material and host tissue interactions, such as controlling specific cell-binding interactions and responding to environmental cues, became important parameters in the design of new materials [237]. Material microarchitecture is another factor that plays an important role in tissue morphogenesis. Considering these design constraints, whole organ decellularization is an attractive technique to prepare tissue engineering scaffolds, as the technique retains relevant aspects of complex structure and chemical composition of the organ's ECM [238] as well as the native 3D structure of the organ [239]. The primary advantage of using whole organ scaffolds is that the resulting construct possesses the native vascular bed which facilitates connection. The feasibility of the technique was first described for the heart [240], which was quickly followed by reports for the liver [241] and lung [242–244], and preliminary results for the kidney and pancreas have appeared in the literature [245]. Figure 24.4 below depicts the basics of whole liver tissue engineering.

Our group was to first report a transplantable liver graft using perfusion decellularized liver matrix [241]. The decellularization process preserved the structural and functional characteristics of the native microvascular network as shown by dye perfusion and corrosion cast molding (Figure 24.5) [246]. The preserved vasculature allowed subsequent recellularization with adult rat hepatocytes with over 90% efficiency (Figure 24.6). The *in vitro* perfusion culture of the recellularized liver graft showed that the graft supported liver-specific function at approximately 30% functional capacity of normal liver, including albumin secretion, urea synthesis, and cytochrome P450 expression *in vitro*. Furthermore, the recellularized liver grafts were transplanted in rats, supporting hepatocyte survival and function with minimal ischemic damage for a duration of 8 h. This report was soon followed by successful creation of humanized rat [247] and mouse [248] liver grafts using human fetal hepatoblasts, scale-up to porcine grafts [249], and long-term transplantation in rats [250]. Still, the process of creating reengineered liver grafts remains a multiplex of several complicated procedures, with many hurdles remaining, such as immunological issues related to organ source, difficulty of recapturing the liver microarchitecture inside

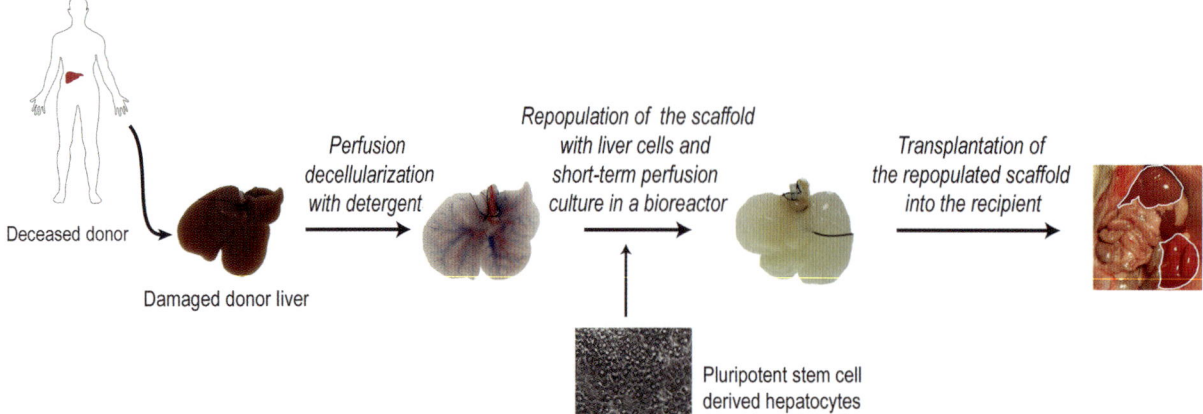

FIGURE 24.4 Principles of whole liver engineering. The liver unsuitable for transplantation is removed from a deceased donor. The organ can be decellularized with a detergent and the process retains the ECM, which is a scaffold that can be repopulated with hepatocytes. The repopulated graft can then be transplanted into a patient. *Permission obtained from Nature Publishing Group Ltd [241].*

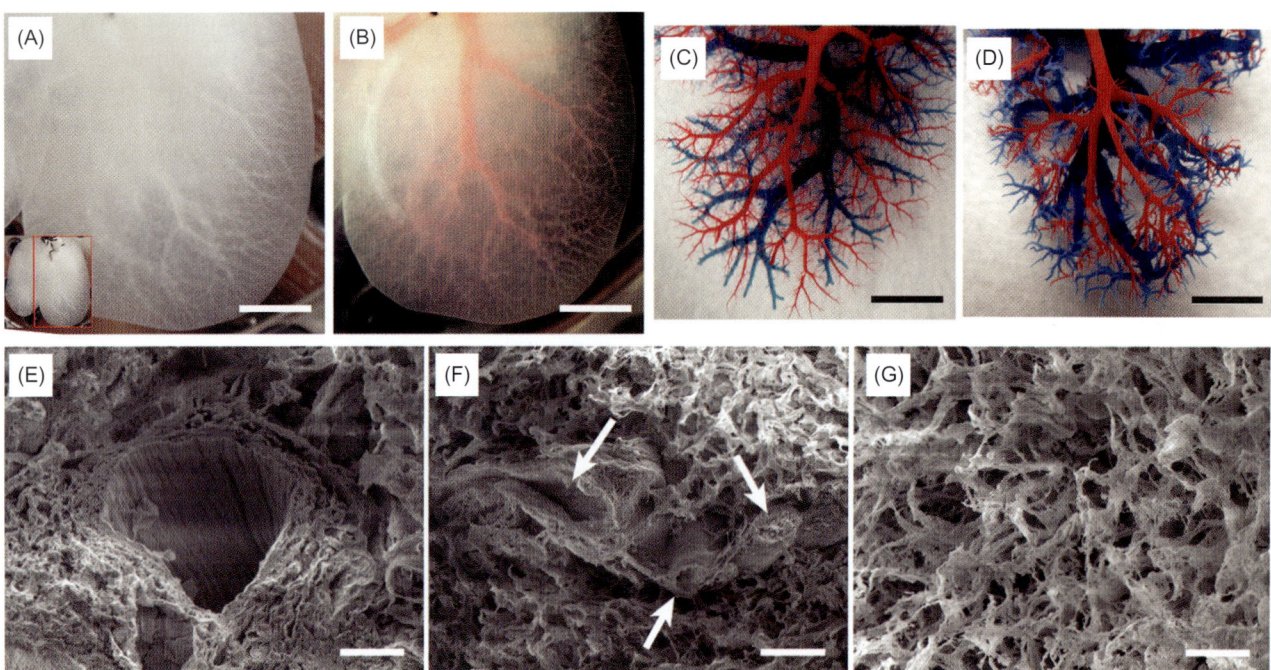

FIGURE 24.5 Decellularized liver matrix scaffold retains intact lobular structure and vascular bed upon perfusion decellularization. (A) Representative photograph of decellularized left lateral and median lobes of rat liver, with the vascular tree visible. (B) The vascular tree, after perfusion with Allura Red AC dye. Corrosion cast model of left lobe of (C) a normal liver and (D) the DLM, portal (red) and venous (blue) vasculature. Scale bars (A and B) 10 mm, (C and D) 5 mm. *Permission obtained from Nature Publishing Group Ltd [241].*

the scaffold, and the risk of blood clot formation with incomplete reendothelialization [251,252].

24.6 CONCLUSION

Even though a healthy liver has the capacity to regenerate, damage to the organ may result in mortality if the hepatic function is not supported. Since the current treatment methods for liver failure are limited, tissue engineering has offered tools to develop alternative treatment options, such as providing temporary support in terms of BAL device, techniques to sustain long-term differentiated function of adult primary hepatocytes, and development of biomaterials as delivery vehicles of hepatocytes for cell transplantation. Furthermore, the tissue engineering tools made development of transplantable liver grafts possible from a wide range of synthetic and biological materials. With the recent advances in whole organ engineering, large-scale

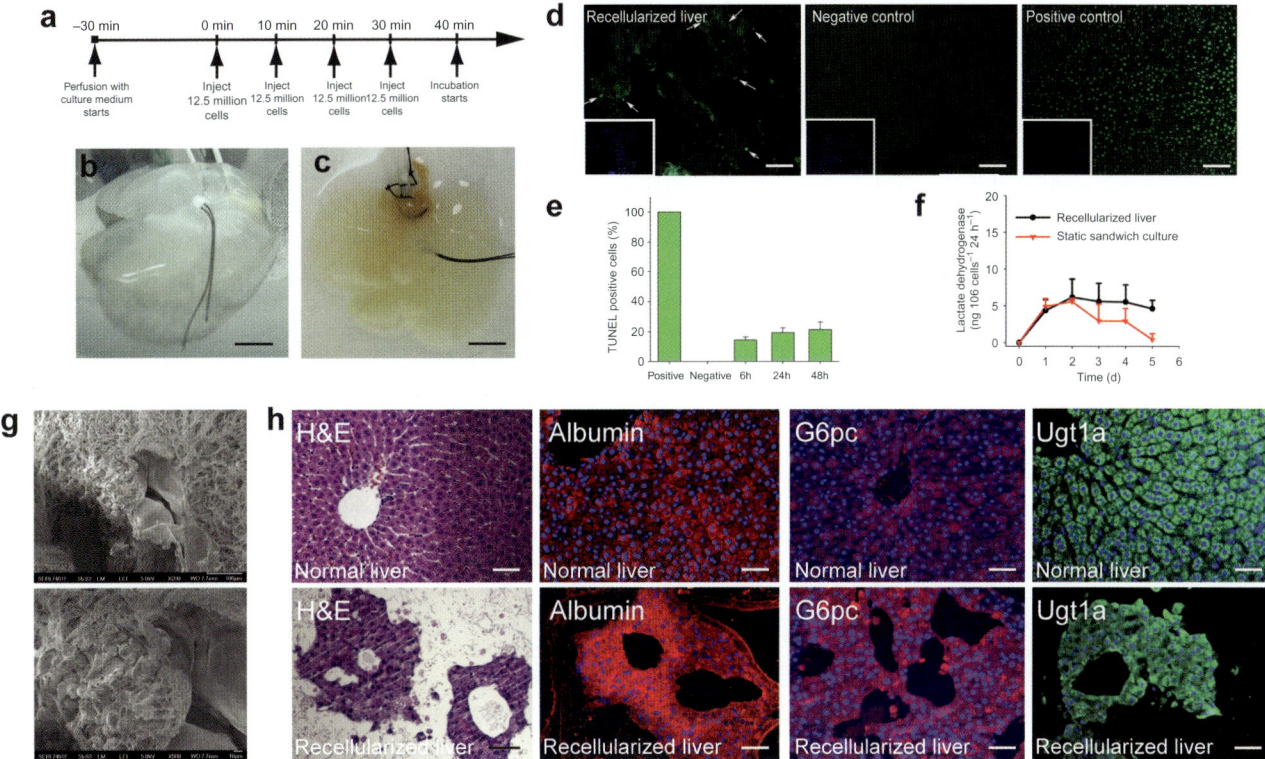

FIGURE 24.6 Repopulation of decellularized liver matrix with adult primary hepatocytes. (A) Decellularized whole liver matrix, (B) same liver after recellularization with about 50 million hepatocytes. (C) Hematoxylin and eosin staining of hepatocyte repopulated decellularized liver matrix. Scale bars (A and B) 20 mm, (C) 100 μm. *Permission obtained from Nature Publishing Group Ltd [241].*

transplantable liver grafts could be developed. Whole liver grafts allow for sustained hepatocyte function within the scaffold, and the presence of the vascular network facilitates their reconnection to the blood torrent *in vivo*. The future work in the development of a transplantable liver graft requires the addition of the nonparenchymal component of the liver into the decellularized liver matrix scaffold, such as endothelial cells and biliary epithelial cells. The clinical translation requires overcoming hurdles, such as immunological issues and blood clot formation, due to incomplete reendothelialization.

REFERENCES

[1] Zakim D, Boyer TD. 3rd ed. Hepatology: a textbook of liver disease, 2 vols. Philadelphia: Saunders; 1996.

[2] Grisham JW. Organizational principles of the liver. In: Arias IM, editor. The liver, biology and pathobiology. 4th ed. Philadelphia: Lippincott Williams & Wilkins; 2001.

[3] Kuntz E, Kuntz HS. Hepatology: textbook and atlas. 3rd ed. Berlin: Springer; 2008.

[4] Cotran RS, Kumar V, Collins T. Robbins pathologic basis of disease. 6th ed. Philadelphia: Saunders; 1998.

[5] Kibble JD, Halsey CR. *The Big Picture: Medical Physiology.* 1st ed. New York: McGraw-Hill; 2009.

[6] Bismuth H. Surgical anatomy and anatomical surgery of the liver. In: Blumgart LH, editor. Surgery of the liver and the biliary tract. Edinburgh: Churchill Livingstone; 1988. p. 1–9.

[7] Dawson JL, Tan KC. Anatomy of the liver. In: Millward-Sadler GH, Wright R, Arthur MJP, editors. Wright's liver and biliary disease. Pathophysiology, diagnosis and management. 3rd ed. London: WB Saunders; 1992. p. 1.

[8] Kuntz E, Kuntz HD. Hepatology: principles and practice: history, morphology, biochemistry, diagnostics, clinic, therapy. 2nd ed. Heidelberg: Springer; 2006.

[9] Taub R. Liver regeneration: from myth to mechanism. Nat Rev Mol Cell Biol 2004;5:836–47.

[10] Dutton GJ. Glucuronication of drugs and other compounds. Boca Raton: CRC Press, Inc.; 1980.

[11] Ramadori G, Christ B. Cytokines and the hepatic acute-phase response. Semin Liver Dis 1999;19:141–55.

[12] DeFrances MC, Michalopoulos GK. Liver regeneration and partial hepatectomy. In: Häussinger D, editor. Liver Regen. Dusseldorf: de Gruyter; 2011.

[13] Kmiec Z. Cooperation of liver cells in health and disease. Adv Anat Embryol Cell Biol 2001;161(III–XIII):1–151.

[14] LeCluyse EL, Witek RP, Andersen ME, Powers MJ. Organotypic liver culture models: meeting current challenges in toxicity testing. Crit Rev Toxicol 2012;42:501–48.

[15] Gebhardt R. Metabolic zonation of the liver: regulation and implications for liver function. Pharmacol Ther 1992;53:275–354.

[16] Katz NR. Metabolic heterogeneity of hepatocytes across the liver acinus. J Nutr 1992;122:843–9.

[17] Jungermann K, Kietzmann T. Zonation of parenchymal and nonparenchymal metabolism in liver. Annl Rev Nutr 1996;16:179–203.

[18] Popovic JR, Kozak LJ. National hospital discharge survey: annual summary, 1998. Vital Health Stat 2000;13:1–194.

[19] Dhawan A, Puppi J, Hughes RD, Mitry RR. Human hepatocyte transplantation: current experience and future challenges. Nat Rev Gastroenterol Hepatol 2010;7:288–98.

[20] Zhang W, Tucker-Kellogg L, Narmada BC, Venkatraman L, Chang S, Lu Y, et al. Cell-delivery therapeutics for liver regeneration. Adv Drug Del Rev 2010;62:814–26.

[21] Carpentier B, Gautier A, Legallais C. Artificial and bioartificial liver devices: present and future. Gut 2009;58:1690–702.

[22] Zhao L-F, Pan XP, Li L-J. Key challenges to the development of extracorporeal bioartificial liver support systems. Hepatobiliary Pancreat Dis Int 2012;11:243–9.

[23] Roll C, Ballauff A, Lange R, Erhard J. Heterotopic auxiliary liver transplantation in a 3-year-old boy with acute liver failure and aplastic anemia. Transplantation 1997;64:658–60.

[24] Pless G, Maurel P. Bioartificial liver support systems. Method Mol Biol 2010;640:511–23.

[25] Zhang Z, Liu J, Liu Y, Li Z, Gao W-Q, He Z. Generation, characterization and potential therapeutic applications of mature and functional hepatocytes from stem cells. J Cell Physiol 2013;228:298–305.

[26] Yu CB, Pan XP, Li LJ. Progress in bioreactors of bioartificial livers. Hepatobiliary Pancreat Dis Int 2009;8:134–40.

[27] Tilles AW, Berthiaume F, Yarmush ML, Tompkins RG, Toner M. Bioengineering of liver assist devices. J Hepatobiliary Pancreat Surg 2002;9:686–96.

[28] Nahmias Y, Kramvis Y, Barbe L, Casali M, Berthiaume F, Yarmush ML. A novel formulation of oxygen-carrying matrix enhances liver-specific function of cultured hepatocytes. FASEB J 2006;20:2531–3.

[29] van de Kerkhove M-P, Hoekstra R, van Gulik TM, Chamuleau RAFM. Large animal models of fulminant hepatic failure in artificial and bioartificial liver support research. Biomaterials 2004;25:1613–25.

[30] Ellis AJ, Hughes RD, Wendon JA, Dunne J, Langley PG, Kelly JH, et al. Pilot-controlled trial of the extracorporeal liver assist device in acute liver failure. Hepatology 1996;24:1446–51.

[31] Watanabe FD, Shackleton CR, Cohen SM, Goldman DE, Arnaout WS, Hewitt W, et al. Treatment of acetaminophen-induced fulminant hepatic failure with a bioartificial liver. Transplant Proc 1997;29:487–8.

[32] Kamohara Y, Rozga J, Demetriou AA. Artificial liver: review and Cedars-Sinai experience. J Hepatobiliary Pancreat Surg 1998;5:273–85.

[33] Kleinstreuer C, Agarwal SS. Analysis and simulation of hollow-fiber bioreactor dynamics. Biotechnol Bioeng 1986;28:1233–40.

[34] Piret JM, Cooney CL. Model of oxygen transport limitations in hollow fiber bioreactors. Biotechnol Bioeng 1991;37:80–92.

[35] Hay PD, Veitch AR, Smith MD, Cousins RB, Gaylor JD. Oxygen transfer in a diffusion-limited hollow fiber bioartificial liver. Artif Org 2000;24:278–88.

[36] Abu-Absi SF, Seth G, Narayanan RA, Groehler K, Lai P, Anderson ML, et al. Characterization of a hollow fiber bioartificial liver device. Artif Org 2005;29:419–22.

[37] Consolo F, Fiore GB, Truscello S, Caronna M, Morbiducci U, Montevecchi FM, et al. A computational model for the optimization of transport phenomena in a rotating hollow-fiber bioreactor for artificial liver. Tissue Eng Part C Methods 2009;15:41–55.

[38] Sullivan JP, Gordon JE, Palmer AF. Simulation of oxygen carrier mediated oxygen transport to C3A hepatoma cells housed within a hollow fiber bioreactor. Biotechnol Bioeng 2006;93:306–17.

[39] Gerlach JC, Encke J, Hole O, Muller C, Ryan CJ, Neuhaus P. Bioreactor for a larger scale hepatocyte in vitro perfusion. Transplantation 1994;58:984–8.

[40] Flendrig LM, la Soe JW, Jorning GG, Steenbeek A, Karlsen OT, Bovee WM, et al. In vitro evaluation of a novel bioreactor based on an integral oxygenator and a spirally wound nonwoven polyester matrix for hepatocyte culture as small aggregates. J Hepatol 1997;26:1379–92.

[41] Jasmund I, Langsch A, Simmoteit R, Bader A. Cultivation of primary porcine hepatocytes in an OXY-HFB for use as a bioartificial liver device. Biotechnol Prog 2002;18:839–46.

[42] Gordon J, Palmer AF. Impact of increased oxygen delivery via bovine red blood cell supplementation of culturing media on select metabolic and synthetic functions of C3A hepatocytes maintained within a hollow fiber bioreactor. Artif Cell Blood Substit Immobil Biotechnol 2005;33:297–306.

[43] Chen G, Palmer AF. Mixtures of hemoglobin-based oxygen carriers and perfluorocarbons exhibit a synergistic effect in oxygenating hepatic hollow fiber bioreactors. Biotechnol Bioeng 2010;105:534–42.

[44] Tzanakakis ES, Hess DJ, Sielaff TD, Hu WS. Extracorporeal tissue engineered liver-assist devices. Ann Rev Biomed Eng 2000;2:607–32.

[45] Miyoshi H, Yanagi K, Ohshima N, Fukuda H. Long-term continuous culture of hepatocytes in a packed-bed reactor utilizing porous resin. Biotechnol Bioeng 1994;43:635–44.

[46] Ohshima N, Yanagi K, Miyoshi H. Packed-bed type reactor to attain high density culture of hepatocytes for use as a bioartificial liver. Artif Org 1997;21:1169–76.

[47] Ehashi T, Ohshima N, Miyoshi H. Three-dimensional culture of porcine fetal liver cells for a bioartificial liver. J Biomed Mater Res Part A 2006;77A:90–6.

[48] Chen JP, Lin TC. High-density culture of hepatocytes in a packed-bed bioreactor using a fibrous scaffold from plant. Biochem Eng J 2006;30:192–8.

[49] Yanagi K, Miyoshi H, Ohshima N. A high-density culture of hepatocytes using porous substrate for use as a bioartificial liver. J Artif Org 1999;2:124–8.

[50] Matsushita T, Ijima H, Funatsu K, Hamazaki K, Koide N. Development of a hybrid type artificial liver using PUF/spheroid culture system of adult rat hepatocytes-estimation of hepatic functions in blood plasma. Jpn J Artif Org 1994;23:469–72.

[51] Yang Q, Liu F, Pan XP, Lv G, Zhang A, Yu CB, et al. Fluidized-bed bioartificial liver assist devices (BLADs) based on microencapsulated primary porcine hepatocytes have risk of porcine endogenous retroviruses transmission. Hepatol Int 2010;4:757–61.

[52] Lv G, Zhao L, Zhang A, Du W, Chen Y, Yu C, et al. Bioartificial liver system based on choanoid fluidized bed bioreactor improve the survival time of fulminant hepatic failure pigs. Biotechnol Bioeng 2011;108:2229–36.

[53] De Bartolo L, Jarosch-Von Schweder G, Haverich A, Bader A. A novel full-scale flat membrane bioreactor utilizing porcine hepatocytes: cell viability and tissue-specific functions. Biotechnol Prog 2000;16:102—8.

[54] De Bartolo L, Bader A. Review of a flat membrane bioreactor as a bioartificial liver. Ann Transplant 2001;6:40—6.

[55] Morsiani E, Galavotti D, Puviani AC, Valieri L, Brogli M, Tosatti S, et al. Radial flow bioreactor outperforms hollow-fiber modules as a perfusing culture system for primary porcine hepatocytes. Transplant Proc 2000;32:2715—8.

[56] Maringka M, Giri S, Bader A. Preclinical characterization of primary porcine hepatocytes in a clinically relevant flat membrane bioreactor. Biomaterials 2010;31:156—72.

[57] Maringka M, Giri S, Nieber K, Acikgöz A, Bader A. Biotransformation of diazepam in a clinically relevant flat membrane bioreactor model using primary porcine hepatocytes. Fundam Clin Pharmacol 2011;25:343—53.

[58] Giri S, Acikgöz A, Pathak P, Gutschker S, Kürsten A, Nieber K, et al. Three dimensional cultures of rat liver cells using a natural self-assembling nanoscaffold in a clinically relevant bioreactor for bioartificial liver construction. J Cell Physiol 2012;227:313—27.

[59] Tilles AW, Baskaran H, Roy P, Yarmush ML, Toner M. Effects of oxygenation and flow on the viability and function of rat hepatocytes cocultured in a microchannel flat-plate bioreactor. Biotechnol Bioeng 2001;73:379—89.

[60] Roy P, Baskaran H, Tilles AW, Yarmush ML, Toner M. Analysis of oxygen transport to hepatocytes in a flat-plate microchannel bioreactor. Ann Biomed Eng 2001;29:947—55.

[61] Roy P, Washizu J, Tilles AW, Yarmush ML, Toner M. Effect of flow on the detoxification function of rat hepatocytes in a bioartificial liver reactor. Cell Transplant 2001;10:609—14.

[62] Shito M, Kim NH, Baskaran H, Tilles AW, Tompkins RG, Yarmush ML, et al. In vitro and in vivo evaluation of albumin synthesis rate of porcine hepatocytes in a flat-plate bioreactor. Artif Org 2001;25:571—8.

[63] Shito M, Tilles AW, Tompkins RG, Yarmush ML, Toner M. Efficacy of an extracorporeal flat-plate bioartificial liver in treating fulminant hepatic failure. J Surg Res 2003;111:53—62.

[64] Park J, Berthiaume F, Toner M, Yarmush ML, Tilles AW. Microfabricated grooved substrates as platforms for bioartificial liver reactors. Biotechnol Bioeng 2005;90:632—44.

[65] Park J, Li Y, Berthiaume F, Toner M, Yarmush ML, Tilles AW. Radial flow hepatocyte bioreactor using stacked microfabricated grooved substrates. Biotechnol Bioeng 2008;99:455—67.

[66] Allen KJ, Soriano HE. Liver cell transplantation: the road to clinical application. J Lab Clin Med 2001;138:298—312.

[67] Dhawan A, Mitry RR, Hughes RD. Hepatocyte transplantation for liver-based metabolic disorders. J Inherited Metabolic Dis 2006;29:431—5.

[68] Fisher RA, Strom SC. Human hepatocyte transplantation: worldwide results. Transplantation 2006;82:441—9.

[69] Lee KY, Mooney DJ. Hydrogels for tissue engineering. Chem Rev 2001;101:1869—79.

[70] Fox IJ, Chowdhury JR, Kaufman SS, Goertzen TC, Chowdhury NR, Warkentin PI, et al. Treatment of the Crigler—Najjar syndrome type I with hepatocyte transplantation. N Engl J Med 1998;338:1422—6.

[71] Horslen SP, Fox IJ. Hepatocyte transplantation. Transplantation 2004;77:1481—6.

[72] Shulman M, Nahmias Y. Long-term culture and coculture of primary rat and human hepatocytes. Method Mol Biol 2012;103:287—302.

[73] Mitry RR, Hughes RD, Aw MM, Terry C, Mieli-Vergani G, Girlanda R, et al. Human hepatocyte isolation and relationship of cell viability to early graft function. Cell Transplant 2003;12:69—74.

[74] Mito M, Kusano M, Kawaura Y. Hepatocyte transplantation in man. Transplant Proc 1992;24:3052—3.

[75] Sokal EM, Smets F, Bourgois A, Van Maldergem L, Buts JP, Reding R, et al. Hepatocyte transplantation in a 4-year-old girl with peroxisomal biogenesis disease: technique, safety, and metabolic follow-up. Transplantation 2003;76:735—8.

[76] Muraca M, Gerunda G, Neri D, Vilei MT, Granato A, Feltracco P, et al. Hepatocyte transplantation as a treatment for glycogen storage disease type 1a. Lancet 2002;359:317—8.

[77] Horslen SP, McCowan TC, Goertzen TC, Warkentin PI, Cai HB, Strom SC, et al. Isolated hepatocyte transplantation in an infant with a severe urea cycle disorder. Pediatrics 2003;111:1262—7.

[78] Habibullah CM, Syed IH, Qamar A, Taher-Uz Z. Human fetal hepatocyte transplantation in patients with fulminant hepatic failure. Transplantation 1994;58:951—2.

[79] Strom SC, Fisher RA, Thompson MT, Sanyal AJ, Cole PE, Ham JM, et al. Hepatocyte transplantation as a bridge to orthotopic liver transplantation in terminal liver failure. Transplantation 1997;63:559—69.

[80] Kobayashi N, Noguchi H, Watanabe T, Matsumura T, Totsugawa T, Fujiwara T, et al. Establishment of a tightly regulated human cell line for the development of hepatocyte transplantation. Human Cell 2000;13:7—13.

[81] Stephenne X, Najimi M, Smets F, Reding R, de Ville de Goyet J, Sokal EM. Cryopreserved liver cell transplantation controls ornithine transcarbamylase deficient patient while awaiting liver transplantation. Am J Transplant 2005;5:2058—61.

[82] Puppi J, Tan N, Mitry RR, Hughes RD, Lehec S, Mieli-Vergani G, et al. Hepatocyte transplantation followed by auxiliary liver transplantation—a novel treatment for ornithine transcarbamylase deficiency. Am J Transplant 2008;8:452—7.

[83] Meyburg J, Das AM, Hoerster F, Lindner M, Kriegbaum H, Engelmann G, et al. One liver for four children: first clinical series of liver cell transplantation for severe neonatal urea cycle defects. Transplantation 2009;87:636—41.

[84] Berendsen TA, Izamis M-L, Xu H, Liu Q, Hertl M, Berthiaume F, et al. Hepatocyte viability and adenosine triphosphate content decrease linearly over time during conventional cold storage of rat liver grafts. Transplant Proc 2011;43:1484—8.

[85] Balfoussia D, Yerrakalva D, Hamaoui K, Papalois V. Advances in machine perfusion graft viability assessment in kidney, liver, pancreas, lung, and heart transplant. Exp Clin Transplant 2012;10:87—100.

[86] Guarrera JV. Assist devices: machine preservation of extended criteria donors. Liver Transplant 2012;18:31—3.

[87] Guarrera JV, Henry SD, Chen SWC, Brown T, Nachber E, Arrington B, et al. Hypothermic machine preservation attenuates ischemia/reperfusion markers after liver transplantation: preliminary results. J Surg Res 2011;167:365—73.

[88] Monbaliu D, Liu Q, Libbrecht L, De Vos R, Vekemans K, Debbaut C, et al. Preserving the morphology and evaluating the quality of liver grafts by hypothermic machine perfusion: a proof-of-concept study using discarded human livers. Liver Transplant 2012;18:1495−507.

[89] Shigeta T, Matsuno N, Obara H, Mizunuma H, Kanazawa H, Tanaka H, et al. Functional recovery of donation after cardiac death liver graft by continuous machine perfusion preservation in pigs. Transplant Proc 2012;44:946−7.

[90] Xu H, Tim B, Kim K, Soto-Gutierrez A, Berthiaume F, Yarmush ML, et al. Excorporeal normothermic machine perfusion resuscitates pig DCD livers with extended warm ischemia. J Surg Res 2012;173:e83−8.

[91] Gringeri E, Bonsignore P, Bassi D, D'Amico FE, Mescoli C, Polacco M, et al. Subnormothermic machine perfusion for non−heart-beating donor liver grafts preservation in a swine model: a new strategy to increase the donor pool? Transplant Proc 2012;44:2026−8.

[92] Tolboom H, Izamis M-L, Sharma N, Milwid JM, Uygun BE, Berthiaume F, et al. Subnormothermic machine perfusion at both 20°C and 30°C recovers ischemic rat livers for successful transplantation. J Surg Res 2012;175:149−56.

[93] Izamis M-L, Calhoun C, Uygun BE, Guzzardi MA, Luitje M, Price G, et al. Simple Machine Perfusion Significantly Enhances Hepatocyte Yields of Ischemic and Fresh Rat Livers. Cell Med 2013;4:24−109.

[94] Hughes RD, Mitry RR, Dhawan A. Current status of hepatocyte transplantation. Transplantation 2012;93:342−7.

[95] Stéphenne X, Najimi M, Sokal EM. Hepatocyte cryopreservation: is it time to change the strategy? World J Gastroenterol 2010;16:1.

[96] Terry C, Dhawan A, Mitry RR, Lehec SC, Hughes RD. Optimization of the cryopreservation and thawing protocol for human hepatocytes for use in cell transplantation. Liver Transplant 2010;16:229−37.

[97] Sugimachi K, Roach KL, Rhoads DB, Tompkins RG, Toner M. Nonmetabolizable glucose compounds impart cryotolerance to primary rat hepatocytes. Tissue Eng Part A 2006;12:579−88.

[98] Magalhaes R, Nugraha B, Pervaiz S, Yu H, Kuleshova LL. Influence of cell culture configuration on the post-cryopreservation viability of primary rat hepatocytes. Biomaterials 2012;33:829−36.

[99] Wang X, Magalhaes R, Wu Y, Wen F, Gouk SS, Watson PF, et al. Development of a modified vitrification strategy suitable for subsequent scale-up for hepatocyte preservation. Cryobiology 2012;65:289−300.

[100] Pless G, Sauer IM, Rauen U. Improvement of the cold storage of isolated human hepatocytes. Cell Transplant 2012;21:23−37.

[101] Ho PJ, Yen ML, Yet SF, Yen BL. Current applications of human pluripotent stem cells: possibilities and challenges. Cell Transplant 2012;21:801−14.

[102] Yi F, Liu GH, Izpisua Belmonte JC. Human induced pluripotent stem cells derived hepatocytes: rising promise for disease modeling, drug development and cell therapy. Protein Cell 2012;3:246−50.

[103] Yin Y, Lim YK, Salto-Tellez M, Ng SC, Lin CS, Lim SK. AFP (+), ESC-derived cells engraft and differentiate into hepatocytes *in vivo*. Stem Cells 2002;20:338−46.

[104] Heo J, Factor VM, Uren T, Takahama Y, Lee JS, Major M, et al. Hepatic precursors derived from murine embryonic stem cells contribute to regeneration of injured liver. Hepatology 2006;44:1478−86.

[105] Ishikawa T, Banas A, Teratani T, Iwaguro H, Ochiya T. Regenerative cells for transplantation in hepatic failure. Cell Transplant 2012;21:387−99.

[106] Zaret KS. Hepatocyte differentiation: from the endoderm and beyond. Curr Opin Genet Dev 2001;11:568−74.

[107] Kubo A, Shinozaki K, Shannon JM, Kouskoff V, Kennedy M, Woo S, et al. Development of definitive endoderm from embryonic stem cells in culture. Development 2004;131:1651−62.

[108] D'Amour KA, Agulnick AD, Eliazer S, Kelly OG, Kroon E, Baetge EE. Efficient differentiation of human embryonic stem cells to definitive endoderm. Nat Biotechnol 2005;23:1534−41.

[109] Hay DC, Fletcher J, Payne C, Terrace JD, Gallagher RC, Snoeys J, et al. Highly efficient differentiation of hESCs to functional hepatic endoderm requires ActivinA and Wnt3a signaling. Proc Natl Acad Sci USA 2008;105:12301−6.

[110] Basma H, Soto-Gutierrez A, Yannam GR, Liu L, Ito R, Yamamoto T, et al. Differentiation and transplantation of human embryonic stem cell-derived hepatocytes. Gastroenterology 2009;136:990−9.

[111] Song Z, Cai J, Liu Y, Zhao D, Yong J, Duo S, et al. Efficient generation of hepatocyte-like cells from human induced pluripotent stem cells. Cell Res 2009;19:1233−42.

[112] Duan Y, Ma X, Zou W, Wang C, Behbahan IS, Ahuja TP, et al. Differentiation and characterization of metabolically functioning hepatocytes from human embryonic stem cells. Stem Cell 2010;28:674−86.

[113] Behbahan IS, Duan Y, Lam A, Khoobyari S, Ma X, Ahuja TP, et al. New approaches in the differentiation of human embryonic stem cells and induced pluripotent stem cells toward hepatocytes. Stem Cell Rev Rep 2011;7:748−59.

[114] Takayama K, Inamura M, Kawabata K, Sugawara M, Kikuchi K, Higuchi M, et al. Generation of metabolically functioning hepatocytes from human pluripotent stem cells by FOXA2 and HNF1α transduction. J Hepatol 2012;57:628−36.

[115] Wang A, Sander M. Generating cells of the gastrointestinal system: current approaches and applications for the differentiation of human pluripotent stem cells. J Mol Med 2012;90:763−71.

[116] Teratani T, Yamamoto H, Aoyagi K, Sasaki H, Asari A, Quinn G, et al. Direct hepatic fate specification from mouse embryonic stem cells. Hepatology 2005;41:836−46.

[117] Kumashiro Y, Asahina K, Ozeki R, Shimizu-Saito K, Tanaka Y, Kida Y, et al. Enrichment of hepatocytes differentiated from mouse embryonic stem cells as a transplantable source. Transplantation 2005;79:550−7.

[118] Cai J, Zhao Y, Liu Y, Ye F, Song Z, Qin H, et al. Directed differentiation of human embryonic stem cells into functional hepatic cells. Hepatology 2007;45:1229−39.

[119] Cao F, van der Bogt KE, Sadrzadeh A, Xie X, Sheikh AY, Wang H, et al. Spatial and temporal kinetics of teratoma formation from murine embryonic stem cell transplantation. Stem Cell Dev 2007;16:883−91.

[120] Espejel S, Roll GR, McLaughlin KJ, Lee AY, Zhang JY, Laird DJ, et al. Induced pluripotent stem cell-derived hepatocytes

have the functional and proliferative capabilities needed for liver regeneration in mice. J Clin Invest 2010;120:3120−6.

[121] Si-Tayeb K, Noto FK, Nagaoka M, Li J, Battle MA, Duris C, et al. Highly efficient generation of human hepatocyte-like cells from induced pluripotent stem cells. Hepatology 2010;51:297−305.

[122] Liu H, Kim Y, Sharkis S, Marchionni L, Jang YY. In vivo liver regeneration potential of human induced pluripotent stem cells from diverse origins. Sci Trans Med 2011;3:82ra39.

[123] Nishikawa S, Goldstein RA, Nierras CR. The promise of human induced pluripotent stem cells for research and therapy. Nat Rev Mol Cell Biol 2008;9:725−9.

[124] Cheng X, Gadue P. Liver regeneration from induced pluripotent stem cells. Mol Ther 2010;18:2044−5.

[125] Robinton DA, Daley GQ. The promise of induced pluripotent stem cells in research and therapy. Nature 2012;481:295−305.

[126] Lund AW, Yener B, Stegemann JP, Plopper GE. The natural and engineered 3D microenvironment as a regulatory cue during stem cell fate determination. Tissue Eng Part B Rev 2009;15:371−80.

[127] Kraehenbuehl TP, Langer R, Ferreira LS. Three-dimensional biomaterials for the study of human pluripotent stem cells. Nat Method 2011;8:731−6.

[128] Sharma R, Greenhough S, Medine CN, Hay DC. Three-dimensional culture of human embryonic stem cell derived hepatic endoderm and its role in bioartificial liver construction. J Biomed Biotechnol 2010;2010:1−13.

[129] Baharvand H, Hashemi SM, Kazemi Ashtiani S, Farrokhi A. Differentiation of human embryonic stem cells into hepatocytes in 2D and 3D culture systems in vitro. Int J Dev Biol 2006;50:645−52.

[130] Ramasamy TS, Yu JSL, Selden C, Hodgson H, Cui W. Application of three-dimensional culture conditions to human embryonic stem cell-derived definitive endoderm cells enhances hepatocyte differentiation and functionality. Tissue Eng Part A 2013;19:360−7.

[131] Farzaneh Z, Pournasr B, Ebrahimi M, Aghdami N, Baharvand H. Enhanced functions of human embryonic stem cell-derived hepatocyte-like cells on three-dimensional nanofibrillar surfaces. Stem Cell Rev Rep 2010;6:601−10.

[132] Miki T, Ring A, Gerlach J. Hepatic differentiation of human embryonic stem cells is promoted by three-dimensional dynamic perfusion culture conditions. Tissue Eng Part C Method 2011;17:557−68.

[133] Orive G, Hernandez RM, Gascon AR, Calafiore R, Chang TM, De Vos P, et al. Cell encapsulation: promise and progress. Nat Med 2003;9:104−7.

[134] Hernandez RM, Orive G, Murua A, Pedraz JL. Microcapsules and microcarriers for in situ cell delivery. Adv Drug Delivery Rev 2010;62:711−30.

[135] Uludag H, De Vos P, Tresco PA. Technology of mammalian cell encapsulation. Adv Drug Del Rev 2000;42:29−64.

[136] Aoki T, Umehara Y, Ferraresso C, Sugiyama N, Middleton Y, Avital I, et al. Intrasplenic transplantation of encapsulated cells: a novel approach to cell therapy. Cell Transplant 2002;11:553−61.

[137] Aoki T, Jin Z, Nishino N, Kato H, Shimizu Y, Niiya T, et al. Intrasplenic transplantation of encapsulated hepatocytes decreases mortality and improves liver functions in fulminant hepatic failure from 90% partial hepatectomy in rats. Transplantation 2005;79:783−90.

[138] Gaserod O, Sannes A, Skjak-Braek G. Microcapsules of alginate-chitosan. II. A study of capsule stability and permeability. Biomaterials 1999;20:773−83.

[139] Rokstad AM, Holtan S, Strand B, Steinkjer B, Ryan L, Kulseng B, et al. Microencapsulation of cells producing therapeutic proteins: optimizing cell growth and secretion. Cell Transplant 2002;11:313−24.

[140] Haque T, Chen H, Ouyang W, Martoni C, Lawuyi B, Urbanska A, et al. Investigation of a new microcapsule membrane combining alginate, chitosan, polyethylene glycol and poly-L-lysine for cell transplantation applications. Int J Artif Org 2005;28:631−7.

[141] Haque T, Chen H, Ouyang W, Martoni C, Lawuyi B, Urbanska AM, et al. In vitro study of alginate−chitosan microcapsules: an alternative to liver cell transplants for the treatment of liver failure. Biotechnol Lett 2005;27:317−22.

[142] Bruni S, Chang TM. Hepatocytes immobilised by microencapsulation in artificial cells: effects on hyperbilirubinemia in Gunn rats. Biomater Artif Cell Artif Org 1989;17:403−11.

[143] Balladur P, Crema E, Honiger J, Calmus Y, Baudrimont M, Delelo R, et al. Transplantation of allogeneic hepatocytes without immunosuppression: long-term survival. Surgery 1995;117:189−94.

[144] Hamazaki K, Doi Y, Koide N. Microencapsulated multicellular spheroid of rat hepatocytes transplanted intraperitoneally after 90% hepatectomy. Hepatogastroenterology 2002;49:1514−6.

[145] Liu ZC, Chang TM. Increased viability of transplanted hepatocytes when hepatocytes are co-encapsulated with bone marrow stem cells using a novel method. Artif Cell Blood Substit Immobil Biotechnol 2002;30:99−112.

[146] Liu ZC, Chang TM. Coencapsulation of hepatocytes and bone marrow stem cells: in vitro conversion of ammonia and in vivo lowering of bilirubin in hyperbilirubemia Gunn rats. Int J Artif Org 2003;26:491−7.

[147] Shi XL, Zhang Y, Gu JY, Ding YT. Coencapsulation of hepatocytes with bone marrow mesenchymal stem cells improves hepatocyte-specific functions. Transplantation 2009;88:1178−85.

[148] Rahman TM, Diakanov I, Selden C, Hodgson H. Co-transplantation of encapsulated HepG2 and rat Sertoli cells improves outcome in a thioacetamide induced rat model of acute hepatic failure. Transplant Int 2005;18:1001−9.

[149] Zheng MH, Lin HL, Qiu LX, Cui YL, Sun QF, Chen YP. Mixed microencapsulation of rat primary hepatocytes and Sertoli cells improves the metabolic function in a D-galactosamine and lipopolysaccharide-induced rat model of acute liver failure. Cytotherapy 2009;11:326−9.

[150] Qiu L, Wang J, Wen X, Wang H, Wang Y, Lin Q, et al. Transplantation of co-microencapsulated hepatocytes and HUVECs for treatment of fulminant hepatic failure. Int J Artif Org 2012;35:458−65.

[151] Vacanti JP, Morse MA, Saltzman WM, Domb AJ, Perez-Atayde A, Langer R. Selective cell transplantation using bioabsorbable artificial polymers as matrices. J Pediatric Surg 1988;23:3−9.

[152] Mooney DJ, Vandenburgh H. Cell delivery mechanisms for tissue repair. Cell Stem Cell 2008;2:205−13.

[153] Overturf K, Al-Dhalimy M, Tanguay R, Brantly M, Ou CN, Finegold M, et al. Hepatocytes corrected by gene therapy are selected *in vivo* in a murine model of hereditary tyrosinaemia type I. Nat Genet 1996;12:266−73.

[154] Gunsalus JR, Brady DA, Coulter SM, Gray BM, Edge AS. Reduction of serum cholesterol in watanabe rabbits by xenogeneic hepatocellular transplantation. Nat Med 1997;3:48−53.

[155] Gordon MY, Levicar N, Pai M, Bachellier P, Dimarakis I, Al-Allaf F, et al. Characterization and clinical application of human CD34+ stem/progenitor cell populations mobilized into the blood by granulocyte colony-stimulating factor. Stem Cell 2006;24:1822−30.

[156] Fitzpatrick E, Mitry RR, Dhawan A. Human hepatocyte transplantation: state of the art. J Int Med 2009;266:339−57.

[157] Koenig S, Stoesser C, Krause P, Becker H, Markus PM. Liver repopulation after hepatocellular transplantation: integration and interaction of transplanted hepatocytes in the host. Cell Transplant 2005;14:31−40.

[158] Bilir BM, Guinette D, Karrer F, Kumpe DA, Krysl J, Stephens J, et al. Hepatocyte transplantation in acute liver failure. Liver Transplant 2000;6:32−40.

[159] Mito M, Ebata H, Kusano M, Onishi T, Saito T, Sakamoto S. Morphology and function of isolated hepatocytes transplanted into rat spleen. Transplantation 1979;28:499−505.

[160] Kay MA, Baley P, Rothenberg S, Leland F, Fleming L, Ponder KP, et al. Expression of human alpha 1-antitrypsin in dogs after autologous transplantation of retroviral transduced hepatocytes. Proc Natl Acad Sci USA 1992;89:89−93.

[161] Kobayashi N, Ito M, Nakamura J, Cai J, Gao C, Hammel JM, et al. Hepatocyte transplantation in rats with decompensated cirrhosis. Hepatology 2000;31:851−7.

[162] Weglarz TC, Degen JL, Sandgren EP. Hepatocyte transplantation into diseased mouse liver. Kinetics of parenchymal repopulation and identification of the proliferative capacity of tetraploid and octaploid hepatocytes. Am J Pathol 2000;157:1963−74.

[163] Nagata H, Ito M, Shirota C, Edge A, McCowan TC, Fox IJ. Route of hepatocyte delivery affects hepatocyte engraftment in the spleen. Transplantation 2003;76:732−4.

[164] Nussler A, Konig S, Ott M, Sokal E, Christ B, Thasler W, et al. Present status and perspectives of cell-based therapies for liver diseases. J Hepatol 2006;45:144−59.

[165] Vroemen JP, Buurman WA, Heirwegh KP, van der Linden CJ, Kootstra G. Hepatocyte transplantation for enzyme deficiency disease in congenic rats. Transplantation 1986;42:130−5.

[166] Henne-Bruns D, Kruger U, Sumpelmann D, Lierse W, Kremer B. Intraperitoneal hepatocyte transplantation: morphological results. Virchows Archiv A Pathol Anat Histopathol 1991;419:45−50.

[167] Demetriou AA, Levenson SM, Novikoff PM, Novikoff AB, Chowdhury NR, Whiting J, et al. Survival, organization, and function of microcarrier-attached hepatocytes transplanted in rats. Proc Natl Acad Sci USA 1986;83:7475−9.

[168] Uyama S, Kaufmann PM, Takeda T, Vacanti JP. Delivery of whole liver-equivalent hepatocyte mass using polymer devices and hepatotrophic stimulation. Transplantation 1993;55:932−5.

[169] Baldini E, Cursio R, De Sousa G, Margara A, Honiger J, Saint-Paul MC, et al. Peritoneal implantation of cryopreserved encapsulated porcine hepatocytes in rats without immunosuppression: viability and function. Transplant Proc 2008;40:2049−52.

[170] Ohashi K, Marion PL, Nakai H, Meuse L, Cullen JM, Bordier BB, et al. Sustained survival of human hepatocytes in mice: a model for *in vivo* infection with human hepatitis B and hepatitis delta viruses. Nat Med 2000;6:327−31.

[171] Ohashi K, Kay MA, Yokoyama T, Kuge H, Kanehiro H, Hisanaga M, et al. Stability and repeat regeneration potential of the engineered liver tissues under the kidney capsule in mice. Cell Transplant 2005;14:621−7.

[172] Ohashi K, Koyama F, Tatsumi K, Shima M, Park F, Nakajima Y, et al. Functional life-long maintenance of engineered liver tissue in mice following transplantation under the kidney capsule. J Tissue Eng Regen Med 2010;4:141−8.

[173] Saito R, Ishii Y, Ito R, Nagatsuma K, Tanaka K, Saito M, et al. Transplantation of liver organoids in the omentum and kidney. Artif Org 2011;35:80−3.

[174] Gupta S, Gorla GR, Irani AN. Hepatocyte transplantation: emerging insights into mechanisms of liver repopulation and their relevance to potential therapies. J Hepatol 1999;30:162−70.

[175] Gupta S, Chowdhary JR. Hepatocyte transplantation: back to the future. Hepatology 1992;15:156−62.

[176] Griffith LG, Swartz MA. Capturing complex 3D tissue physiology *in vitro*. Nat Rev Mol Cell Biol 2006;7:211−24.

[177] Lysaght MJ, Hazlehurst AL. Tissue engineering: the end of the beginning. Tissue Eng 2004;10:309−20.

[178] Tsang VL, Bhatia SN. Fabrication of three-dimensional tissues. Adv Biochem Eng Biotechnol 2007;103:189−205.

[179] Seglen PO. Preparation of isolated rat liver cells. Method Cell Biol 1976;13:29−83.

[180] Strain AJ. Isolated hepatocytes: use in experimental and clinical hepatology. Gut 1994;35:433−6.

[181] Nahmias Y, Berthiaume F, Yarmush ML. Integration of technologies for hepatic tissue engineering. Adv Biochem Eng Biotechnol 2007;103:309−29.

[182] Berthiaume F, Moghe PV, Toner M, Yarmush ML. Effect of extracellular matrix topology on cell structure, function, and physiological responsiveness: hepatocytes cultured in a sandwich configuration. FASEB J 1996;10:1471−84.

[183] Schuetz EG, Li D, Omiecinski CJ, Muller-Eberhard U, Kleinman HK, Elswick B, et al. Regulation of gene expression in adult rat hepatocytes cultured on a basement membrane matrix. J Cell Physiol 1988;134:309−23.

[184] Koide N, Shinji T, Tanabe T, Asano K, Kawaguchi M, Sakaguchi K, et al. Continued high albumin production by multicellular spheroids of adult rat hepatocytes formed in the presence of liver-derived proteoglycans. Biochem Biophys Res Commun 1989;161:385−91.

[185] Wang S, Nagrath D, Chen PC, Berthiaume F, Yarmush ML. Three-dimensional primary hepatocyte culture in synthetic self-assembling peptide hydrogel. Tissue Eng Part A 2008;14:227−36.

[186] Landry J, Bernier D, Ouellet C, Goyette R, Marceau N. Spheroidal aggregate culture of rat liver cells: histotypic reorganization, biomatrix deposition, and maintenance of functional activities. J Cell Biol 1985;101:914−23.

[187] Tong JZ, De Lagausie P, Furlan V, Cresteil T, Bernard O, Alvarez F. Long-term culture of adult rat hepatocyte spheroids. Exp Cell Res 1992;200:326−32.

[188] Wu FJ, Friend JR, Remmel RP, Cerra FB, Hu WS. Enhanced cytochrome P450 IA1 activity of self-assembled rat hepatocyte spheroids. Cell Transplant 1999;8:233−46.

[189] Napolitano AP, Dean DM, Man AJ, Youssef J, Ho DN, Rago AP, et al. Scaffold-free three-dimensional cell culture utilizing micro-molded nonadhesive hydrogels. Biotechniques 2007;43:494−6.

[190] Mori R, Sakai Y, Nakazawa K. Micropatterned organoid culture of rat hepatocytes and HepG2 cells. J Biosci Bioeng 2008;106:237−42.

[191] Ong S-M, Zhang C, Toh Y-C, Kim SH, Foo HL, Tan CH, et al. A gel-free 3D microfluidic cell culture system. Biomaterials 2008;29:3237−44.

[192] Naughton BA. The importance of stromal cells. In: Bronzino JD, editor. Biomedical engineering handbook. Boca Raton: CRC Press; 1995. p. 1710−22.

[193] Bhatia SN, Balis UJ, Yarmush ML, Toner M. Effect of cell−cell interactions in preservation of cellular phenotype: cocultivation of hepatocytes and nonparenchymal cells. FASEB J 1999;13:1883−900.

[194] Bhatia SN, Balis UJ, Yarmush ML, Toner M. Microfabrication of hepatocyte/fibroblast co-cultures: role of homotypic cell inter-actions. Biotechnol Prog 1998;14:378−87.

[195] Riccalton-Banks L, Liew C, Bhandari R, Fry JR, Shakesheff K. Long-term culture of functional liver tissue: three-dimensional coculture of primary hepatocytes and stellate cells. Tissue Eng Part A 2003;9:401−10.

[196] Hansen LK, Hu W-S. Three-dimensional co-culture of hepato-cytes and stellate cells. Cytotechnology 2004;45:125−40.

[197] Thomas RJ, Bhandari R, Barrett DA, Bennett AJ, Fry JR, Powe D, et al. The effect of three-dimensional co-culture of hepato-cytes and hepatic stellate cells on key hepatocyte functions *in vitro*. Cells Tissues Org 2005;181:67−79.

[198] Nahmias Y, Odde DJ. Micropatterning of living cells by laser-guided direct writing: application to fabrication of hepatic−endothelial sinusoid-like structures. Nat Protoc 2006;1:2288−96.

[199] Khetani SR, Bhatia SN. Microscale culture of human liver cells for drug development. Nat Biotechnol 2008;26:120−6.

[200] Cho CH, Park J, Tilles AW, Berthiaume F, Toner M, Yarmush ML. Layered patterning of hepatocytes in co-culture systems using microfabricated stencils. Biotechniques 2010;48:47−52.

[201] Odde DJ, Renn MJ. Laser-guided direct writing for applications in biotechnology. Trends Biotechnol 1999;17:385−9.

[202] Odde DJ, Renn MJ. Laser-guided direct writing of living cells. Biotechnol Bioeng 2000;67:312−8.

[203] Nahmias YK, Gao BZ, Odde DJ. Dimensionless parameters for the design of optical traps and laser guidance systems. Appl Opt 2004;43:3999−4006.

[204] Nahmias Y, Schwartz RE, Verfaillie CM, Odde DJ. Laser-guided direct writing for three-dimensional tissue engineering. Biotechnol Bioeng 2005;92:129−36.

[205] Bhatia SN, Balis UJ, Yarmush ML, Toner M. Probing hetero-typic cell interactions: hepatocyte function in microfabricated co-cultures. J Biomater Sci Polym Ed 1998;9:1137−60.

[206] Griffith LG, Naughton G. Tissue engineering—current chal-lenges and expanding opportunities. Science 2002;295:1009−14.

[207] Elloumi-Hannachi I, Yamato M, Okano T. Cell sheet engineering: a unique nanotechnology for scaffold-free tissue reconstruction with clinical applications in regenerative medi-cine. J Int Med 2010;267:54−70.

[208] Singhvi R, Kumar A, Lopez GP, Stephanopoulos GN, Wang DI, Whitesides GM, et al. Engineering cell shape and function. Science 1994;264:696−8.

[209] Takagi S, Ohno M, Ohashi K, Utoh R, Tatsumi K, Okano T. Cell shape regulation based on hepatocyte sheet engineering technologies. Cell Transplant 2012;21:411−20.

[210] Ohashi K, Yokoyama T, Yamato M, Kuge H, Kanehiro H, Tsutsumi M, et al. Engineering functional two- and three-dimensional liver systems *in vivo* using hepatic tissue sheets. Nat Med 2007;13:880−5.

[211] Haraguchi Y, Shimizu T, Sasagawa T, Sekine H, Sakaguchi K, Kikuchi T, et al. Fabrication of functional three-dimensional tis-sues by stacking cell sheets *in vitro*. Nat Protoc 2012;7:850−8.

[212] Kaufmann PM, Kneser U, Fiegel HC, Kluth D, Herbst H, Rogiers X. Long-term hepatocyte transplantation using three-dimensional matrices. Transplant Proc 1999;31:1928−9.

[213] Fiegel HC, Kaufmann PM, Bruns H, Kluth D, Horch RE, Vacanti JP, et al. Hepatic tissue engineering: from transplanta-tion to customized cell-based liver directed therapies from the laboratory. J Cell Mol Med 2008;12:56−66.

[214] Davis MW, Vacanti JP. Toward development of an implantable tissue engineered liver. Biomaterials 1996;17:365−72.

[215] Mikos AG, Sarakinos G, Lyman MD, Ingber DE, Vacanti JP, Langer R. Prevascularization of porous biodegradable polymers. Biotechnol Bioeng 1993;42:716−23.

[216] Torok E, Pollok JM, Ma PX, Kaufmann PM, Dandri M, Petersen J, et al. Optimization of hepatocyte spheroid formation for hepatic tissue engineering on three-dimensional biodegradable polymer within a flow bioreactor prior to implantation. Cells Tissues Org 2001;169:34−41.

[217] Torok E, Pollok JM, Ma PX, Vogel C, Dandri M, Petersen J, et al. Hepatic tissue engineering on 3-dimensional biodegradable polymers within a pulsatile flow bioreactor. Digestive Surg 2001;18:196−203.

[218] Torok E, Vogel C, Lutgehetmann M, Ma PX, Dandri M, Petersen J, et al. Morphological and functional analysis of rat hepatocyte spheroids generated on poly(L-lactic acid) polymer in a pulsatile flow bioreactor. Tissue Eng 2006;12:1881−90.

[219] VandeVord PJ, Matthew HW, DeSilva SP, Mayton L, Wu B, Wooley PH. Evaluation of the biocompatibility of a chitosan scaffold in mice. J Biomed Mater Res 2002;59:585−90.

[220] Kim IY, Seo SJ, Moon HS, Yoo MK, Park IY, Kim BC, et al. Chitosan and its derivatives for tissue engineering applications. Biotechnol Adv 2008;26:1−21.

[221] Elcin YM, Dixit V, Gitnick G. Hepatocyte attachment on bio-degradable modified chitosan membranes: *in vitro* evaluation for the development of liver organoids. Artif Org 1998;22:837−46.

[222] Seo SJ, Choi YJ, Akaike T, Higuchi A, Cho CS. Alginate/galac-tosylated chitosan/heparin scaffold as a new synthetic extracellu-lar matrix for hepatocytes. Tissue Eng 2006;12:33−44.

[223] Seo SJ, Kim IY, Choi YJ, Akaike T, Cho CS. Enhanced liver functions of hepatocytes cocultured with NIH 3T3 in the alginate/galactosylated chitosan scaffold. Biomaterials 2006;27:1487—95.

[224] Fan J, Shang Y, Yuan Y, Yang J. Preparation and characterization of chitosan/galactosylated hyaluronic acid scaffolds for primary hepatocytes culture. J Mater Sci Mater Med 2010;21:319—27.

[225] Feng ZQ, Chu X, Huang NP, Wang T, Wang Y, Shi X, et al. The effect of nanofibrous galactosylated chitosan scaffolds on the formation of rat primary hepatocyte aggregates and the maintenance of liver function. Biomaterials 2009;30:2753—63.

[226] Toole BP. Hyaluronan: from extracellular glue to pericellular cue. Nat Rev Cancer 2004;4:528—39.

[227] Allison DD, Grande-Allen KJ. Review. Hyaluronan: a powerful tissue engineering tool. Tissue Eng 2006;12:2131—40.

[228] Prestwich GD. Hyaluronic acid-based clinical biomaterials derived for cell and molecule delivery in regenerative medicine. J Control Release 2011;155:193—9.

[229] Zavan B, Brun P, Vindigni V, Amadori A, Habeler W, Pontisso P, et al. Extracellular matrix-enriched polymeric scaffolds as a substrate for hepatocyte cultures: in vitro and in vivo studies. Biomaterials 2005;26:7038—45.

[230] Katsuda T, Teratani T, Ochiya T, Sakai Y. Transplantation of a fetal liver cell-loaded hyaluronic acid sponge onto the mesentery recovers a Wilson's disease model rat. J Biochem 2010;148:281—8.

[231] Navarro-Alvarez N, Soto-Gutierrez A, Rivas-Carrillo JD, Chen Y, Yamamoto T, Yuasa T, et al. Self-assembling peptide nanofiber as a novel culture system for isolated porcine hepatocytes. Cell Transplant 2006;15:921—7.

[232] Navarro-Alvarez N, Soto-Gutierrez A, Chen Y, Caballero-Corbalan J, Hassan W, Kobayashi S, et al. Intramuscular transplantation of engineered hepatic tissue constructs corrects acute and chronic liver failure in mice. J Hepatol 2010;52:211—9.

[233] Leclerc E, Furukawa KS, Miyata F, Sakai Y, Ushida T, Fujii T. Fabrication of microstructures in photosensitive biodegradable polymers for tissue engineering applications. Biomaterials 2004;25:4683—90.

[234] Borenstein JT, Weinberg EJ, Orrick BK, Sundback C, Kaazempur-Mofrad MR, Vacanti JP. Microfabrication of three-dimensional engineered scaffolds. Tissue Eng 2007;13:1837—44.

[235] Baudoin R, Corlu A, Griscom L, Legallais C, Leclerc E. Trends in the development of microfluidic cell biochips for in vitro hepatotoxicity. Toxicol In Vitro 2007;21:535—44.

[236] Ochoa ER, Vacanti JP. An overview of the pathology and approaches to tissue engineering. Ann New York Acad Sci 2002;979:10—26 [discussion 35—38]

[237] Lutolf MP, Hubbell JA. Synthetic biomaterials as instructive extracellular microenvironments for morphogenesis in tissue engineering. Nat Biotechnol 2005;23:47—55.

[238] Gilbert TW, Sellaro TL, Badylak SF. Decellularization of tissues and organs. Biomaterials 2006;27:3675—83.

[239] Badylak SF, Taylor D, Uygun K. Whole-organ tissue engineering: decellularization and recellularization of three-dimensional matrix scaffolds. Ann Rev Biomed Eng 2011;13:27—53.

[240] Ott HC, Matthiesen TS, Goh SK, Black LD, Kren SM, Netoff TI, et al. Perfusion-decellularized matrix: using nature's platform to engineer a bioartificial heart. Nat Med 2008;14:213—21.

[241] Uygun BE, Soto-Gutierrez A, Yagi H, Izamis ML, Guzzardi MA, Shulman C, et al. Organ reengineering through development of a transplantable recellularized liver graft using decellularized liver matrix. Nat Med 2010;16:814—20.

[242] Ott HC, Clippinger B, Conrad C, Schuetz C, Pomerantseva I, Ikonomou L, et al. Regeneration and orthotopic transplantation of a bioartificial lung. Nat Med 2010;16:927—33.

[243] Petersen TH, Calle EA, Zhao L, Lee EJ, Gui L, Raredon MB, et al. Tissue-engineered lungs for in vivo implantation. Science 2010;329:538—41.

[244] Price AP, England KA, Matson AM, Blazar BR, Panoskaltsis-Mortari A. Development of a decellularized lung bioreactor system for bioengineering the lung: the matrix reloaded. Tissue Eng Part A 2010;16:2581—91.

[245] Song JJ, Ott HC. Organ engineering based on decellularized matrix scaffolds. Trends Mol Med 2011;17:424—32.

[246] Uygun BE, Price G, Saedi N, Izamis ML, Berendsen T, Yarmush M, et al. Decellularization and recellularization of whole livers. J Visualized Exp JoVE 2011;13:27—53.

[247] Baptista PM, Siddiqui MM, Lozier G, Rodriguez SR, Atala A, Soker S. The use of whole organ decellularization for the generation of a vascularized liver organoid. Hepatology 2011;53:604—17.

[248] Zhou P, Lessa N, Estrada DC, Severson EB, Lingala S, Zern MA, et al. Decellularized liver matrix as a carrier for the transplantation of human fetal and primary hepatocytes in mice. Liver Transplant 2011;17:418—27.

[249] Barakat O, Abbasi S, Rodriguez G, Rios J, Wood RP, Ozaki C, et al. Use of decellularized porcine liver for engineering humanized liver organ. J Surg Res 2012;173:11—25.

[250] Bao J, Shi Y, Sun H, Yin X, Yang R, Li L, et al. Construction of a portal implantable functional tissue-engineered liver using perfusion-decellularized matrix and hepatocytes in rats. Cell Transplant 2011;20:753—66.

[251] Uygun BE, Yarmush ML, Uygun K. Application of whole-organ tissue engineering in hepatology. Nat Rev Gastroenterol Hepatol 2012;9:738—44.

[252] Soto-Gutierrez A, Wertheim JA, Ott HC, Gilbert TW. Perspectives on whole-organ assembly: moving toward transplantation on demand. J Clin Invest 2012;122:3817—23.

Liver Regeneration: The Developmental Biology Approach

David A. Rudnick

Departments of Pediatrics and Developmental, Regenerative, and Stem Cell Biology, Washington University School of Medicine, St. Louis, MO

Chapter Outline

25.1 INTRODUCTION

Liver diseases have substantial impact on human morbidity and mortality. While some disease-specific therapies exist, host survival and recovery from all liver injuries depends on the liver's remarkable capacity to regenerate from such insults. Liver regeneration is also essential for recovery after hepatic tumor resection and living-donor liver transplantation (LDLT). Based on these considerations, hepatic regeneration has been subjected to decades of rigorous experimental investigation (reviewed in [1−4]), with hope that mechanistic insights provided by such research will lead to novel pro-regenerative therapies

with which to improve the management of human liver diseases. Such analyses show that hepatic regenerative capacity is conserved in all vertebrates where it has been examined, presumably because of the essential detoxification, metabolic, and synthetic functions subserved by this organ, and which includes animals from fish to human. While other body structures also regenerate in lower vertebrates, e.g., the zebra fish's amputated fin [5], the liver is unique amongst mammalian visceral organs in the ability to recover from injury by regeneration instead of scar formation. Thus, elucidating the mechanisms of hepatic regeneration might also inform efforts to promote the regeneration of other human organs [6].

Although various experimental models have been used to study liver regeneration [7], the most commonly employed paradigm has been rodent partial hepatectomy, in which a rat or mouse undergoes surgical removal of a portion of liver. Afterwards, a tissue-specific regenerative response ensues. This response begins with stereotypical changes in hepatic and extrahepatic metabolism and physiology [8], and the activation of specific extra- and intracellular signaling events, followed by alterations in gene and protein expression. Together, these events induce normally quiescent hepatocytes and other cells in the liver to reenter the cell cycle. Such cell proliferation leads to restoration of the preresection liver-to-body-mass ratio and normalization of hepatic function. Afterwards, hepatic lobular architecture, temporarily distorted by the regenerative response, is remodeled, and the liver returns to its preregenerative state of proliferative inactivity [1−4]. Nonsurgical models, including those of toxin- and genetically induced hepatocellular injury, and xenobiotic-stimulated hepatocellular proliferation have also been employed to investigate the regulation of hepatocellular proliferation and liver regeneration [9,10]. Those studies provide further evidence for the precision with which liver mass is regulated in health and restored by regeneration following hepatic injury. Taken together, such analyses infer the existence of master regulator of liver-to-body-mass ratio, i.e., a so-called "*hepatostat*" [1]. Nevertheless, the nature and identities of the most proximal events that initiate hepatic regeneration and those distal signals that terminate this response are still incompletely defined, and the essence of this hepatostat remains essentially unknown.

25.2 HISTORICAL PERSPECTIVES

The regenerative potential of the liver might have been recognized as early as the eighth century BCE, based on the reference in Hesiod's *Theogony* to the regeneration of Prometheus' liver [11]. However, similar mythological references to the regeneration of other organs suggest the Ancients may have lacked a specific understanding of the liver's unique regenerative capability amongst mammalian organs [12,13]. Accounts of experimental, hepatic resection-induced liver regeneration first appeared in the scientific literature in the late nineteenth century, but a detailed, quantitative description of the rodent partial hepatectomy model was not published until 1931 [14]. In that manuscript, Higgins and Anderson reported on the recovery of hepatic mass following the removal of the left and median liver lobes in rat. Those lobes constitute ∼65% of rat liver mass. Thus, this approach, which has also been used to study liver regeneration in mouse and other species, is referred to as the "two-thirds partial hepatectomy" model. Non-surgical models have subsequently been used to study liver regeneration as well; nevertheless, since its

initial description, partial hepatectomy has been the most commonly employed paradigm for investigating the biology of liver regeneration. Investigations using these models have defined many of the physiological, cellular, and molecular signals that control liver regeneration, while pharmacological and genetic manipulation of those signals has permitted assessment of their necessity and sufficiency for induction of normal liver regeneration [1−4].

25.3 EXPERIMENTAL MODELS OF LIVER REGENERATION

25.3.1 Partial Hepatectomy

The best characterized and most extensively utilized experimental strategy for investigating liver regeneration has been the rodent partial hepatectomy paradigm [14]. In the typical ("two-thirds") version of this model, the anesthetized rodent is subjected to mid-ventral laparotomy with exposure of normal hepatic lobar structures, followed by ligation and resection of the left and median hepatic lobes (leaving the right and caudate lobes intact), closure of the surgical wounds, and postoperative recovery. Variations, including "one-third" partial hepatectomy (in which only the median lobes are removed [15]) and subtotal hepatectomy (in which the left, median, and a portion of the right lobe are resected [16]), have also been employed. The metrics with which regeneration is quantified, in these as well as the other, nonsurgical models, include assessments of hepatocellular proliferation, cell volume, and liver-to-body-mass ratio at serial times after the surgery (Figures 25.1 and 25.2).

The principle advantage of partial hepatectomy over other experimental models of regeneration is that the hepatic remnant is not intentionally injured by the surgically induced regenerative stimulus. However, unintentional injury to such structures can occur. Damage to the extrahepatic bile duct or perihepatic vascular structures can cause biliary infarcts or centrilobular necrosis, which are technical complications whose frequency decreases with experience. Choice of anesthesia might also influence the likelihood of technical complications [17]. These and other considerations related to technical aspects of the partial hepatectomy model are well summarized in a recent review of this system [18].

25.3.2 Toxin and Genetically Based Models of Liver Injury

Models of liver injury induced by various toxins (e.g., carbon tetrachloride (CCl_4), thioacetamide, acetaminophen, and D-galactosamine [7,9]) or genetic manipulations (e.g., liver-specific transgenic expression of the human α1-antitrypsin PiZ allele [10,19,20], fumarylacetoacetate hydrolase (FAH) deficiency [21]) have also been employed to study liver regeneration. Unlike partial hepatectomy, in those models the entire liver is subjected to injury-induced inflammation and necrosis, which affects the ensuing regenerative response. Nevertheless, because hepatocellular inflammation and necrosis occur in and probably influence recovery from many human liver diseases, these models provide valuable additional information about the mechanisms of regeneration.

25.3.3 Xenobiotic-Induced Hepatocellular Proliferation

Experimental chemical (xenobiotic)-induced liver growth has also been employed to investigate the regulation of liver

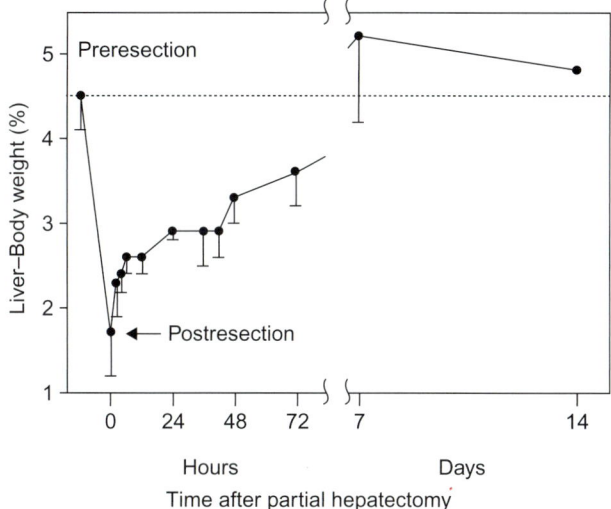

FIGURE 25.1   Recovery of % liver-to-body weight after partial hepatectomy in mice. The liver-to-body-mass ratio in unoperated, 2- to 3-month-old C57BL6 mice is ~4.5%. The recovery of liver-to-body mass is illustrated as a function of time after two-third partial hepatectomy.

mass and hepatocellular proliferation. Administration of certain chemicals [e.g., phenobarbitol, hypolipidemic fibrates, 1,4-bis(2-(3,5-dichloropyridoxyloxy)) benzene (TCPOBOP), and others] to rodents induces hepatomegaly without liver injury [22]. The increase in liver mass results from a combination of chemically induced hepatocellular hypertrophy and hyperplasia. Withdrawal of the inciting agent is followed by a return to "normal" liver mass, providing additional evidence for the "hepatostat." In many cases, specific nuclear receptor transcription factors that mediate the liver growth promoting activities of individual xenobiotics have been identified. Moreover, partial hepatectomy-induced liver regeneration is affected in some models in which those factors are genetically or pharmacologically altered. These observations suggest that the mechanisms that regulate xenobiotic-induced liver growth and surgically induced liver regeneration overlap to some degree. However, other studies show that the disruption of certain molecular signals implicated in the regulation of partial hepatectomy-induced regeneration does not suppress xenobiotic-dependent changes in liver mass [23−25], suggesting that at least some xenobiotics act downstream of or in parallel to those signals. These points are further considered in Section 25.5.3.

25.3.4 Primary Hepatocyte Cell Culture

This *in vitro* approach permits the assessment of the direct effects of specific signals on proliferation of hepatocytes. Although such evaluation is limited by the relatively poor proliferative activity and phenotypic instability of primary hepatocytes in culture, important observations have been made using this paradigm [26]. Indeed, this approach has been used to distinguish those signals associated with the initiation of liver regeneration *in vivo* that also promote primary hepatocyte proliferation

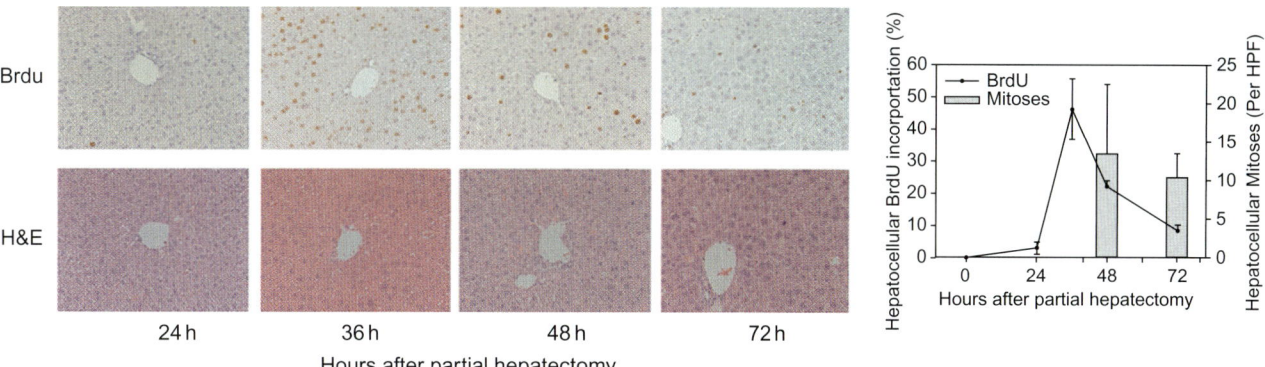

FIGURE 25.2   Hepatocellular BrdU incorporation and mitotic frequency after partial hepatectomy. Immunohistochemical analysis of hepatocellular BrdU incorporation (brown-stained nuclei) and liver histology (hematoxylin and eosin, H&E, staining, left panels) and a representative summary of quantification of hepatocellular BrdU incorporation and mitoses (right panel; HPF, high-powered field) at serial times after partial hepatectomy in wild-type C57BL6 mice are shown. For this analysis, mice were subjected to partial hepatectomy, recovered, and then injected with BrdU 1 h before animal sacrifice and tissue harvest for these analyses.

in serum free medium [e.g., hepatocyte growth factor (HGF) and epidermal growth factor receptor (EGF-R) ligands] from those (many other signals) that do not independently promote such proliferation *in vitro*.

25.4 GENERAL PRINCIPLES OF LIVER REGENERATION

Experimental analyses using the models described above have defined several fundamental properties of the hepatic regenerative response to liver injury.

25.4.1 Liver Regeneration is Subject to Humoral Regulation

The regulation of liver regeneration by circulating factors was first established by parabiosis experiments in which cross-circulation was established between pairs of rats. When one rat was subjected to partial hepatectomy (but not sham surgery) increased hepatic DNA synthesis was detectable in the other, unoperated animal [27,28]. Humoral regulation of regeneration is also suggested by the observation that after partial hepatectomy periportal hepatocytes (closest to the afferent arterial and portal blood supplies of the liver) proliferate prior to centrilobular hepatocytes (which are furthest from those vascular supplies [29]). These sentinel observations led to subsequent efforts to discover the circulating growth factors, cytokines, metabolites, and other biologically active molecules that regulate liver regeneration. The relative increase in intrahepatic portal blood flow that occurs after partial hepatectomy, which results from postsurgical delivery of such flow through a fraction of the original tissue mass, has also been suggested to be a pro-regenerative stimulus contributing to the initiation of regeneration [1,2].

25.4.2 Liver Mass is Regulated in Proportion to Body Mass

This conclusion is based on clinical and experimental evidence showing that the liver-to-body-mass ratio is preserved in health and restored by regeneration after liver injury. For example, autopsy- and imaging-based analyses of people without liver disease demonstrate a linear relationship between liver size and body mass [30,31]. Restoration of the preresection liver-to-body-mass ratio following rodent partial hepatectomy was reported in the original description of that model [14], and has subsequently been replicated by many investigators (Figure 25.1). This phenomenon is also observed in humans recovering from partial hepatic resection (e.g., for tumor removal, see Figure 25.3 and [32]), or LDLT

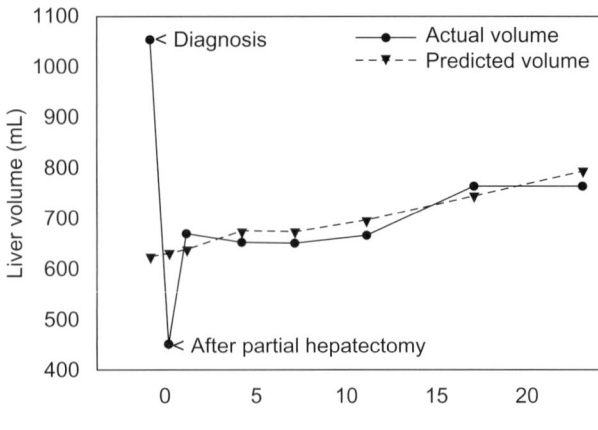

FIGURE 25.3   Recovery of liver volume after liver tumor resection in a patient. The recovery of liver volume, assessed by quantitative analysis of serial abdominal CT scans, is shown along with predicted liver volume (based on [31]) before and at serial times after liver tumor resection. The initial liver volume exceeded the predicted volume by an amount equal to the tumor volume.

[33,34]). Furthermore, both animal model and clinical observations show that after transplantation of small- or large-for-host-size livers, the size of the transplanted organ adjusts until the original liver-to-body-mass set point is restored [35,36]. Similarly and as already noted, xenobiotic-induced hepatomegaly reverses upon discontinuation of the inciting chemical [22]. Finally, one-third partial hepatectomy induces a less robust hepatocellular proliferative response than two-thirds hepatectomy [15], suggesting that, at least over a certain range, hepatic regeneration occurs in proportion to the degree of liver injury. Nevertheless, despite such evidence, the specific extrahepatic signals that regulate liver-to-body mass remain enigmatic. One clue about those signals was recently provided by the observation that mice with genetically induced hypertrophy of skeletal muscle (i.e., myostatin-null mice) exhibit diminished liver-to-body-mass ratio compared to their wild-type littermates [37]. Those data demonstrate that muscle mass is not the extrahepatic compartment against which liver mass is positively regulated, and they suggest a previously unrecognized degree of extrahepatic tissue specificity to the regulation of liver-to-body mass.

25.4.3 Regenerative Recovery of Liver Mass Does Not Require a Stem Cell

Partial hepatectomy induces a sequence of events that triggers the initiation of proliferation of many cell types within the affected liver. Hepatocytes, which constitute the bulk of the cellular mass and are responsible for many physiological functions of the liver, proliferate first [38]. In rats, peak hepatocellular proliferation [as assessed by

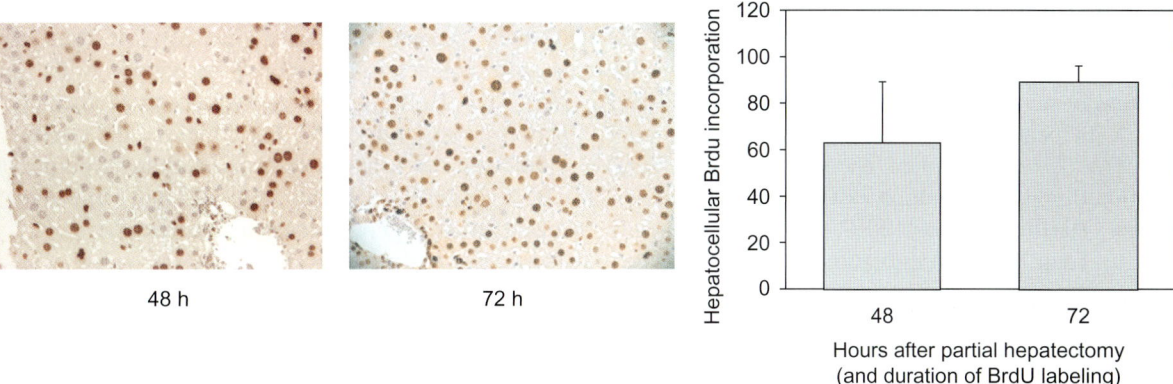

FIGURE 25.4 Cumulative hepatocellular BrdU incorporation after partial hepatectomy in mice. Mice were implanted with a sustained release BrdU-containing osmotic pump at the time of partial hepatectomy. In this approach, all hepatocytes that replicate their DNA throughout the time-course of the experiment (48 or 72 h) exhibit hepatocellular BrdU incorporation. The data show that almost all hepatocytes replicate their DNA at least once by 72 h after partial hepatectomy.

nuclear bromodeoxyuridine (BrdU) incorporation] occurs 24 h after surgery. In contrast, peak hepatocellular BrdU labeling in mice occurs 36 h after liver resection, with subsequent rounds of proliferation occurring in approximately 24-h intervals [39]. This difference in the kinetics of rat versus mouse liver regeneration is a cell-autonomous property of hepatocytes, as shown by a study in which hepatocellular proliferation after partial hepatectomy was examined in immunodeficient, urokinase plasminogen activator (uPA)-transgenic mice whose livers were repopulated with either mouse or rat hepatocytes [40]. Despite this specificity with which the time from liver resection to peak hepatocellular DNA synthesis (S phase) is regulated, experimental analyses in which the time-of-day of partial hepatectomy is varied show that subsequent progression from S phase to mitosis depends on circadian influences [41,42]. This observation, that hepatocellular DNA synthesis and mitosis are differentially regulated during liver regeneration, is also illustrated by the effects of partial hepatectomy on hepatocellular ploidy. The abundance of polyploid hepatocytes, which are present in quiescent liver, increases after partial hepatectomy [43,44]. Moreover, polyploid (and aneuploid) hepatocytes can revert to normal ploidy, prompting speculation that generation and reversal of such abnormal ploidy might serve as an adaptive mechanism to repeatedly generate hepatocellular genetic diversity with which to combat chemically induced and other forms of liver injury [45]. Towards the end of the hepatocellular proliferative response to partial hepatectomy, which occurs ~5 days after surgery [39], a small hepatocellular apoptotic response has been observed [46]. This response might correct "over-shoot" of liver mass recovery after partial hepatectomy (Figure 25.1). Of note, the author has also observed such hepatocellular apoptosis anecdotally in a patient with increased liver:body mass

after large-graft-for-host-size LDLT (D.A.R., unpublished observations). The broader relevance of these observations for liver-to-body-mass regulation requires further investigation.

Virtually all of the previously quiescent hepatocytes in the remnant liver re-enter the cell cycle and replicate their DNA at least once after "two-thirds" partial hepatectomy (Figure 25.4), indicating that the typical regenerative response to this stimulus does not depend upon a stem cell population. Recent lineage-tracing analyses using Cre-dependent, cell type-specific marker expression strategies provide further evidence that mature hepatocytes in the remnant are the predominant source of new hepatocytes after partial hepatectomy and toxin-induced liver injury [47]. Nevertheless, a bipotential hepatic progenitor cell population can be detected in response to a regenerative stimulus when the proliferation of mature hepatocytes is simultaneously prevented. For example, rats treated with 2-acetylaminofluorene (AAF), which alkylates and prevents the replication of hepatocellular DNA, then subjected to partial hepatectomy (or exposed to other specific liver injuries) cannot regenerate their livers through the proliferation of mature hepatocytes in the liver remnant. Instead, those animals exhibit delayed expansion of small, periportal, oval-shaped cells which exhibit immunohistochemical evidence of both biliary and hepatocytic properties. Eventually these "oval" cells become mature hepatocytes or cholangiocytes, restoring normal liver mass and histology [48–53]. Cells with similar histological and immunohistochemical characteristics to rodent oval cells have been identified in human livers affected by many different forms of liver injury, thus, in at least some cases, these so-called progenitor or ductular hepatocytes might contribute to restoration of hepatocellular mass after liver injury in humans [54,55].

Following hepatocellular proliferation, other cell types in the regenerating liver also proliferate. Regrowth of the intrahepatic biliary tree after hepatectomy occurs via the proliferation of mature cholangiocytes present in the liver remnant, with peak proliferation occurring 3 days after surgery [56]. Liver sinusoidal endothelial cell (LSEC) proliferation during regeneration exhibits similar kinetics, with bone marrow-derived LSEC precursors recently reported to contribute to restoration of sinusoidal networks in regenerating liver tissue [57]. Kupffer and stellate cells have also been reported to proliferate during and exert complex effects on the regenerative response to partial hepatectomy [58].

25.4.4 Hepatocyte Replicative Potential

The remarkable capacity for (uninjured) hepatocytes to replicate has been demonstrated both by subjecting experimental animals to serial partial hepatectomy [59] and by analyses of experimental repopulation of diseased livers after hepatocyte transplantation in susceptible animal models. In the latter studies, wild-type hepatocytes transplanted into the livers of mice with genetically based hepatocellular injury replace and restore hepatocellular mass in those animals over time. For example, such repopulation has been demonstrated in mouse models of FAH deficiency (a model of hereditary tyrosinemia type I [60,61]), transgenic hepatic uPA overexpression [62], and transgenic PiZ expression-induced α1-antitrypsin deficiency liver disease [63]. In each case, a relatively small number of wild-type hepatocytes ($\sim 0.1-10\%$ of estimated normal hepatocellular mass) were transplanted (by intrasplenic injection), then engrafted in the diseased liver, proliferated, and replaced the majority of hepatocellular mass of those livers. These data illustrate a competitive proliferative advantage of transplanted normal hepatocytes over endogenous diseased cells, which implies that hepatocellular proliferative competence is reduced under certain pathological conditions. Such transplanted wild-type hepatocytes can be serially retransplanted and, thereby, repopulate many more livers of affected mice [61]. These observations also provide experimental support for efforts to develop therapeutic hepatocellular transplantation strategies for patients with metabolic liver diseases.

25.4.5 Hypertrophy Versus Hyperplasia

Some studies have examined the relative importance of hepatocellular hypertrophy versus hyperplasia during liver regeneration. In one study, recovery of liver mass after partial hepatectomy in rodents in whom both hepatocellular and oval cell proliferation were coincidentally suppressed occurred via hepatocyte hypertrophy [64].

Similarly, hepatocellular hypertrophy was reported to mediate pregnancy-induced recovery of the impaired hepatic regenerative response associated with old age [65]. Such hypertrophy was also observed after one-third partial hepatectomy [66]. Nevertheless, the relative magnitude and mechanisms of hepatocellular hyperplasia versus hypertrophy during normal and impaired liver regeneration require further characterization.

25.5 MOLECULAR REGULATION OF LIVER REGENERATION

Experimental analyses using partial hepatectomy and other models have identified many signals that are regulated during and important for normal liver regeneration. These signals include humoral and other extracellular regulators, which activate intracellular signaling pathways and alter transcriptional programs, inducing normally quiescent cellular populations of the liver to re-enter the cell cycle and proliferate until liver mass is restored. Afterwards, other signals terminate the regenerative response.

25.5.1 Humoral, Extracellular, and Matrix-Derived Signals

Recognition of the humoral regulation of liver regeneration led to the search for circulating factors that promote hepatic regeneration *in vivo*. Amongst the identified factors, HGF and EGF-R ligands also stimulate primary hepatocytes to proliferate in culture in the absence of other factors [38]. Other humoral, cellular, and matrix-based signals "prime" hepatocytes to respond to these primary mitogens or cooperate in other ways to promote regeneration *in vitro* and *in vivo*, but do not augment hepatocellular proliferation by themselves. Tumor necrosis factor α (TNFα) is an example of such a factor [38].

25.5.1.1 Hepatocyte Growth Factor

Circulating levels of HGF increase after partial hepatectomy and HGF is mitogenic in primary hepatocyte culture [67]. This growth factor, which is also expressed and has important functions outside of the liver [68,69], has been studied extensively for its role during liver regeneration. Those analyses show that inactive HGF protein, present in the extracellular matrix (ECM) of quiescent liver (and other organs), is proteolytically activated following partial hepatectomy by uPA and perhaps other proteases [70,71]. Activated HGF binds to c-Met to induce autophosphorylation-dependent activation of this integral membrane receptor tyrosine kinase (RTK [72]). Such activation is detectable shortly after partial hepatectomy and induces downstream intracellular signaling events that promote liver mass restoration [73]. Induction of hepatic

HGF expression in endothelial [74] and perhaps stellate [75] cells also occurs in response to humoral factors induced during early regeneration including vascular endothelial growth factor (VEGF [76,77]) and perhaps interleukin 6 (IL6 [78]). HGF signaling is functionally important during liver regeneration based on studies showing that the exogenous administration of this growth factor induces rodent hepatocellular proliferation *in vivo* [79,80] and more recent studies demonstrating that gene-knockout (KO) and small inhibitory RNA (siRNA)-based disruption of hepatic c-Met expression inhibit resection- and toxin-induced regeneration [81−83]. Genetic disruption of HGF/c-Met signaling in regenerating liver impairs downstream intracellular signaling events including the activation of Akt and p42/44 mitogen-activated protein kinases (MAPKs) and liver regeneration-associated changes in hepatic gene expression, and delays hepatocellular G_1-S and G_2-M cell cycle progression [81−84].

25.5.1.2 EGF-R Ligands

EGF family members also promote liver regeneration, as suggested by their mitogenic activity on hepatocytes *in vivo* [85] and *in vitro*[1,2,38]. These ligands include EGF itself, which is produced by salivary and duodenal Brunner's glands [86,87], as well as transforming growth factor α (TGFα, expressed in liver, [88]), heparin-binding EGF (Hb-EGF, produced by endothelial and Kupffer cells, [89]), and amphiregulin (expressed in gastrointestinal epithelium and other tissues, [90]). Hepatocellular proliferation *in vivo* is increased in the absence of partial hepatectomy by administration of EGF [85] and in mice transgenic for TGFα [91]. TGFα KO mice regenerate their livers normally [92], however, genetic disruption of Hb-EGF, amphiregulin, or EGF-R, which (like c-Met) is a ligand-activated RTK, results in impaired regeneration after partial hepatectomy [93−96]. Disruption of EGF-R expression results in impaired activation of nuclear factor kappa B (NFκB) and delayed induction of cell cycle progression machinery after partial hepatectomy [95,96].

25.5.1.3 TNFα and IL6

These cytokines are regulated during and important for normal liver regeneration, however, neither factor appears to induce hepatocyte proliferation directly. Circulating levels of TNFα and IL6 rise after partial hepatectomy [97,98] with Kupffer cells [99] and lymphocytes are likely contributors. Disruption of TNFα signaling by administration of neutralizing anti-TNFα antibody [100] or genetic disruption of the TNFα receptor subtype TNFR1 (but not TNFR2 [97,101,102]) results in impaired regeneration after partial hepatectomy or CCl_4 exposure. TNFR1-null mice also exhibit impaired hepatic activation of the transcription factors NFκB and STAT3 (signal transducer and activator of transcription 3) and diminished elevation of circulating IL6 in these models. Similarly, liver regeneration is also suppressed and NFκB and STAT3 activation deranged after resection- or hepatotoxin-induced liver injury in IL6-KO mice [98,103]. However, these mice exhibit increased mortality after partial hepatectomy, with surviving mice regenerating their livers relatively normally [104]. Such data highlights the challenges presented by animal models of coincident impairment of liver regeneration and increased mortality, in which reduced animal survival might be secondary to the incompetent regenerative response or, alternatively, reduced regeneration might be caused by events associated with increased mortality.

25.5.1.4 Other Humoral, Extracellular, and Liver Matrix-Based Regulators

A multitude and expanding list of other humoral and extracellular factors have also been implicated as regulated during and important for normal liver regeneration. Space limitations preclude comprehensive review here of all such signals. The interested reader is referred to several excellent recent reviews on this subject [1−4], as well as primary literature concerning some of these factors, which include insulin and glucagon [105−107], growth hormone [108,109], fibroblast growth factors [110,111], VEGF [112], serotonin [113,114], catecholamines [115,116], prostaglandins [117,118], complement cascade components [17,119−121], leptin [122−125], and many other factors. Wnt-dependent stabilization of β-catenin and Jagged-1 stimulation of Notch signaling are also regulated during and appear to modulate liver regeneration. These pathways are reviewed together with other transcriptional regulators of liver regeneration in Section 25.5.3. Metabolites, derived from the metabolic alterations that occur in response to hepatic insufficiency and whose circulating levels are altered in models of liver regeneration, represent yet another category of extracellular signals that appear to regulate liver regeneration [8]. This possibility is considered in greater detail in Section 25.5.4.

As mentioned previously, HGF protein is activated and released from ECM during early regeneration. Tranforming growth factor β1 (TGFβ1), which is implicated in the termination of regeneration (see Section 25.5.6), is also present in the matrix of quiescent liver and released into the circulation (but inactivated) after partial hepatectomy. These data suggest that ECM itself participates in the regulation of regeneration [126]. Consistent with this idea, the hepatic expression of enzymes that regulate matrix degradation, including matrix metalloproteinases (MMPs) 2 and 9 and tissue inhibitor of MMP (TIMP) 1, is induced during liver regeneration [127], and regeneration is impaired in MMP9-null mice [128].

25.5.2 Secondary Messengers

The molecules and signaling pathways discussed above activate specific integral membrane receptors expressed on hepatocytes and other liver cells to induce intracellular secondary messenger signaling pathways. Many of these pathways are also known to be regulated during and, in some cases, have been shown to be essential for normal liver regeneration [129]. For example, RTKs, such as c-Met and EGF-R, activate phospholipase Cγ (PLCγ) to induce inositol phosphate pathway-dependent changes in calcium signaling and protein kinase B (i.e., Akt) and C (PKC) activities. RTKs also activate MAPKs. Similarly, G-protein coupled receptors, including those that bind glucagon, catecholamines, and prostaglandins, stimulate ligand-dependent activation of adenylyl cyclase (AC), PLCβ, and cytosolic Rho GTPases. PLCβ and γ [130], PKC [131], MAPKs [132,133], cyclic AMP (cAMP [134,135]), calcium [136], small molecular weight GTP-binding proteins [73], and other downstream signals have also been shown to be regulated in various models of liver regeneration. In many cases, the functional importance of such regulation has been examined using genetically or pharmacologically modified mouse models. For example, mice treated with cyclooxygenase 2 (COX2)-specific inhibitors exhibit impaired activation of the cAMP-dependent binding protein (CREB) transcriptional regulator and impaired liver regeneration [117,118,137]. Regeneration is also impaired in mice in which the cAMP-response element modifier (CREM) transcription factor [138] or phosphoinositide 3 (PI3) kinase [139] is deleted, and disruption of cytosolic [140] or mitochondrial [141] calcium regulation also impairs regeneration. Glycogen synthase kinase (GSK) 3β [142−144], which phosphorylates many targets in addition to glycogen synthase [145], and Akt [139,146−151], which, like GSK3β, plays a key role in metabolic regulation and other processes, have also been implicated in the regulation of liver regeneration

Other intracellular signaling pathways relevant to liver biology are also induced during and important for normal liver regeneration. One example is the Wnt/β-catenin pathway [152]. Under basal conditions, cytosolic β-catenin is proteolytically degraded by a process dependent upon its association with other proteins, including axin, adenomatous polyposis coli (APC), and GSK3β. Following partial hepatectomy, hepatocellular β-catenin is stabilized and undergoes nuclear translocation [153]. The functional importance of this signal, which results in β-catenin-dependent changes in gene expression, was demonstrated by the impairment in regeneration seen in rodents in which hepatic β-catenin expression is disrupted [154,155] and by evidence for cross talk between this pathway and HGF- and EGF-R dependent signaling [156−158]. Another such example is

that of Jagged/Notch signaling. Notch-1 is an integral membrane protein induced to undergo proteolytic-cleavage of its intracellular domain upon binding to Jagged-family ligands. The cleaved fragment, known as the notch intracellular domain (NICD), undergoes nuclear translocation and affects gene expression [159]. Genetic mutations in Jagged or Notch cause Alagille syndrome, which is characterized by paucity of intrahepatic bile ducts and extrahepatic abnormalities. Partial hepatectomy induces hepatocellular NICD nuclear translocation, and disruption of Notch-1 or Jagged-1 expression impairs liver regeneration [160].

25.5.3 Transcriptional Regulation of Liver Regeneration

Signals induced in response to partial hepatectomy and other regenerative stimuli affect the expression and activities of many transcriptional regulators, which, in turn, modulate patterns of hepatic gene expression to promote cell proliferation. For example, TNFα activates NFκB through well-characterized pathways involving disruption of NFκB binding to the inhibitor of kappa B (IκB). Such activation prevents TNFα-induced apoptosis. Consistent with that consideration, adenoviral vector-based hepatic expression of a dominant-negative mutant IκB resulted in partial hepatectomy-induced hepatocellular apoptosis [161]. However, transgenic mice engineered for hepatocyte-specific expression of dominant-negative IκB regenerated normally, suggesting that non-parenchymal hepatic NFκB expression and activation might be necessary for prevention of TNFα-induced hepatocellular apoptosis in regenerating liver [162]. Mice deleted for the p50 subunit of NFκB also regenerate their livers normally, implying some degree of redundancy in the transcriptional effects of specific NFκB subunits during regeneration [163]. Similarly, IL6 induces phosphorylation-dependent activation of STAT3, which was also implicated as important for normal liver regeneration by analyses of KO mice [164]. However, as with IL6 nulls, impaired regeneration in STAT3 KOs is associated with increased mortality [165]. CREM and CREB are also activated during early liver regeneration [135]. These and other events induce an immediate-early gene expression program [166,167], which, in turn, directs changes in the expression of additional transcription factors (including increased levels of c-Fos, c-Jun, c-Myc, C/EBPβ, Egr-1 and decreased expression of C/EBPα [168−172]) and cell cycle regulators that promote hepatocellular cell cycle progression. Inactivation of hepatocyte nuclear factor (HNF) 4α, which is expressed in and contributes to the differentiated phenotype of mature, quiescent hepatocytes, might also promote liver regeneration based on recent data showing increased hepatocellular proliferation in HNF4α null mice [173].

The development of comprehensive gene expression profiling strategies has added to current understanding of the transcriptional regulation of liver regeneration. The earliest such studies used subtraction hybridization approaches to identify growth-, cell cycle-, and post-growth phase-regulated patterns of gene expression in the regenerating liver [166,167,174]. Subsequent analyses using microarray profiling defined temporal patterns of change in mRNA expression of transcription factors, cell cycle and apoptosis regulators, matrix and structural components, inflammation-associated, signal transduction, angiogenic, metabolic, and other genes involved in the regulation of liver regeneration [124,175−180]. Array-based approaches have recently been used to further characterize cell-specific patterns of mRNA expression [99,181,182] and the regulation of microRNA (miRNA) expression during regeneration [183,184]. miRNAs regulate gene expression by affecting mRNA stability and translation. Their role in liver regeneration has been suggested by demonstration of impaired regeneration in mice with global disruption of miRNA processing and in mice with targeted disruption of miRNA21 [185,186].

Regulation of liver mass and regeneration by xenobiotic-activated nuclear receptor transcriptional regulators, including peroxisome proliferator-activated receptor (PPAR) α, PPARγ, constitutive androstane receptor (CAR), pregnane X receptor (PXR), farnesoid X receptor (FXR), and liver X receptor (LXR), has also been studied. The specific importance of PPARα expression during normal liver regeneration remains uncertain based on conflicting reports about the magnitude of derangement of the regenerative response in PPARα-null mice [187−190], which are resistant to the hepatomegaly-inducing activity of PPARα ligands. Nevertheless, recent reports implicating endogenous lipid metabolites as ligand activators of PPARα [191,192] raise the hypothesis that naturally occurring xenobiotic receptor ligands might influence the hepatic regenerative response. Additional support for this provocative idea comes from other data showing that global disruption of expression of FXR, a bile acid-activated transcriptional regulator, results in marked impairment of liver regeneration [193]. FXR is expressed in liver and intestine, each of which participate in the entero-hepatic circulation of bile acids, and recent data shows impairment of liver regeneration in both liver- and intestine-specific FXR-null mice [194,195]. Unoperated, bile acid-fed, wild-type animals exhibit FXR-dependent increases in hepatocellular mitoses and liver mass, suggesting that proportionately increased portal delivery of bile acids to the liver remnant after partial hepatectomy might initiate liver regeneration. However, wild-type mice subjected to partial hepatectomy show no increase in serum bile acids and reduced hepatic bile acid content. Thus, bile acids were probably not the circulating factor that promoted hepatocellular proliferation in the parabiotic experiments discussed earlier. Mice also develop hepatomegaly when treated with CAR- and PXR-activating ligands, including phenobarbitol and TCPOBOP; however, CAR- and PXR-null mice show only modestly impaired liver regeneration after partial hepatectomy [193,196]. In contrast, PPARγ null mice exhibit mildly accelerated regeneration [197], and pharmacological activation of PPARγ [198] or of LXR [199] suppresses regeneration. Together, these observations suggest a novel "metabolic model" of liver regeneration in which alterations in systemic metabolism that occur in response to hepatic insufficiency (induced by partial hepatectomy or other liver injuries) lead to accumulation of metabolites that activate pro-regenerative xenobiotic receptors or reduction of metabolites that stimulate anti-regenerative receptors in the regenerating liver.

25.5.4 Metabolism and Liver Regeneration

The liver is centrally involved in the regulation of intermediary metabolism [200]. In addition to the discussion above, various other observations have suggested that the metabolic response to hepatic insufficiency might contribute essential signals to the regulation of liver regeneration. These data are considered here.

25.5.4.1 Glycemia and Regeneration

Consistent with the liver's important role in gluconeogenesis and glycogen metabolism, rodents subjected to partial hepatectomy or exposed to hepatotoxins develop hypoglycemia [8,201]. Moreover, provision of supplemental dextrose to these animals suppresses the hepatocellular proliferative responses to these regenerative stimuli [8,201−209]. LXR, which as mentioned above have been implicated as a negative regulator of liver regeneration, has also been suggested to function as a glucose sensor [210]. Conversely, dietary caloric restriction accelerates the initiation of hepatocellular proliferation after various hepatic regenerative stimuli [211,212]. Not surprisingly, circulating insulin levels decline in response to partial hepatectomy-induced hypoglycemia (and are augmented by dextrose supplementation [201]). In addition, toxin-induced liver regeneration is augmented and accelerated in mice with streptozotocin-induced insulin-deficient diabetes [213,214]. However, systemic diversion of portal circulation (e.g., portacaval shunting, which diverts pancreas-derived insulin away from the liver) causes atrophy of the liver lobe from which such flow is diverted, and insulin supplementation reverses such atrophy [215−218]. Thus, the functional role of glycemia-induced changes in hepatic insulin signaling during liver regeneration requires further clarification. Nevertheless, these observations suggest that such alterations in response to hepatic insufficiency contribute to the regulation of regeneration.

25.5.4.2 Systemic Catabolism During Regeneration

After onset of hypoglycemia and prior to initiation of regenerative hepatocellular proliferation, mice subjected to partial hepatectomy exhibit a stereotypical decline in extrahepatic lean and fat tissue stores and a subsequent rise in circulating and hepatic levels of metabolites, including free fatty acids and specific amino acids [15,188,219]. Those alterations in metabolism, like the regenerative response itself, occur in proportion to the degree of induced hepatic insufficiency. For example, two-thirds partial hepatectomy results in greater reduction of systemic adipose mass (and more robust hepatocellular proliferation) than does one-third hepatectomy. Systemic fat depletion also occurs in toxin-induced models of liver regeneration [15,220], suggesting that catabolism of peripheral tissues is a common response to regeneration-inducing liver injuries.

25.5.4.3 Hepatic Lipid Metabolism and Regeneration

The early regenerating liver transiently accumulates lipid [221–224]. Such steatosis, which follows onset of partial hepatectomy-induced hypoglycemia and resolves as regeneration proceeds, can be suppressed by various genetic and pharmacological strategies. For example, fatty liver dystrophy (fld) mice, which exhibit a paucity of systemic adipose, exhibit reduced hepatic fat accumulation after partial hepatectomy compared to control animals [15]. Those data suggest that the lipid that accumulates in regenerating liver derives from systemic fat stores, a conclusion supported by the observation that genetic disruption of de novo hepatic lipogenesis, as occurs in liver-specific fatty acid synthase null (FAS-KOL) mice, does not suppress fat accumulation in regenerating liver [188]. Of note, in several models in which hepatic fat accumulation is suppressed either pharmacologically (e.g., by clofibrate [225], leptin [124], or propranolol [226]) or genetically (e.g., as in fld mice or by liver-specific disruption of glucocorticoid receptor expression [124]), liver regeneration is impaired. However, other studies seem to contradict these results. For example, liver regeneration in caveolin 1-null mice, in which partial hepatectomy-induced hepatic steatosis is reduced, has been variably reported as impaired in one study and normal in another [227,228]. Furthermore, hepatic fat accumulation is diminished but regeneration proceeds

normally following partial hepatectomy in liver fatty acid binding protein (L-Fabp) KO mice [188] and in mice with intestine-specific deletion of the microsomal triglyceride transfer protein (MTP-IKO). Liver resection-induced fat accumulation was not completely abrogated in either of those mouse models [188], leading those investigators to speculate about the existence of a "threshold of adaptive lipogenesis" essential for regeneration and not affected in those mice. Together with data showing that fat accumulates concomitantly with cell proliferation in primary hepatocyte culture [229], these observations raise the possibility that alterations in systemic adipose metabolism and hepatocellular accumulation of lipid in response to hepatic insufficiency might promote hepatocyte proliferation at least under certain circumstances. Nevertheless, the molecular mechanisms responsible for this link remain uncertain. One possibility is suggested by older studies reporting increased dependency of regenerating liver on β-oxidation of fatty acids for energy production [205,230]. Indeed, it has been speculated that the inhibitory effect of dextrose supplementation on liver regeneration [201] might be secondary to the suppressive effects of such supplementation on the release of free fatty acids from systemic adipose stores [205,231], and both dietary and parenteral administration of various lipid-based formulations accelerate resection- [205] and toxin- [232] induced hepatocellular proliferation. However, these findings are difficult to reconcile with other data showing that diminished hepatic adenine nucleotide content [233] and increased AMP kinase (AMPK) activity [234,235] promote liver regeneration.

25.5.4.4 Hepatic Methionine Metabolism During Regeneration

The considerations discussed above raise the intriguing possibility, noted previously, that accumulation of specific metabolites in response to hepatic insufficiency might promote regeneration by affecting signaling. Indirect support for this idea also comes from analyses of methionine metabolism during liver regeneration. The liver plays a central role in methionine metabolism, and such metabolism is altered in both regenerating liver and chronic liver disease [236]. Accumulation of a transaminated product of methionine catabolism, α-NH$_2$-butyrate (Aab [237,238]) was recently reported in regenerating liver [219]. Moreover, liver regeneration is impaired in methionine-adenosyl transferase 1a (Mat1a) KO mice, in which methionine metabolism is disrupted [239]. Together, these findings indicate that metabolic flux from methionine to Aab is a marker of normal liver regeneration, and they suggest such metabolism might also mediate this response.

25.5.5 Cell Cycle Progression

The events discussed above lead quiescent liver cells to re-enter the cell cycle, proliferate, and thereby restore hepatocellular mass. Such cell cycle progression is regulated by expression and activation of specific cyclin−cyclin-dependent kinase (CDK) complexes (reviewed in [240]). For example, progression of quiescent (G_0) hepatocytes into G_1 is characterized by increased expression of cyclin D1, which complexes with CDK4 or CDK6 to induce cyclin E1 expression. Cyclin D-CDK4/6 and cyclin E/CDK2 complexes phosphorylate retinoblastoma (Rb) protein, releasing Rb-bound E2F transcription factors to direct E2F-dependent changes in gene expression that promote entry into the DNA synthesis (S) phase of the cell cycle. Cyclins A and B are subsequently and sequentially induced with cyclin A-CDK2 driving S-G_2 progression and cyclin B-CDK1 regulating entry into and progression through mitosis. Hepatic expression of cyclin-dependent kinase inhibitors (CDKIs), including $p21^{Cip1}$ and $p27^{Kip1}$, is also regulated during the liver regeneration, with these factors influencing hepatocellular cell cycle progression by transiently suppressing the activities of specific cyclin−CDK complexes. These positive and negative molecular mediators of cell cycle progression are not unique to hepatocellular proliferation. Nevertheless, several intriguing observations about their regulation and function during liver regeneration have been made. For example, $p21^{Cip1}$ is not expressed in quiescent liver, but, rather, induced following partial hepatectomy and other liver injuries [241,242]. It has been speculated that low levels of $p21^{Cip1}$ expression might facilitate cyclin D-CDK4/6 complex formation and, thereby, promote hepatocellular G_1 progression. However, increased hepatic $p21^{Cip1}$ expression is observed in many models of impaired regeneration, including that of dextrose supplementation in which genetic disruption of $p21^{Cip1}$ expression rescued the impairment [201]. In contrast to $p21^{Cip1}$, $p27^{Kip1}$ protein is expressed in quiescent liver and its level declines as regeneration proceeds [198]. As with $p21^{Cip1}$, increased $p27^{Kip1}$ is also reported in models of suppressed regeneration. Moreover, genetic disruption of these Cip/Kip CDKIs accelerates liver regeneration in wild-type mice [242−245], and deletion of $p18^{INK4c}$, a member of the INK4 CDKI subfamily, cooperatively enhances such acceleration [244]. Degradation of $p21^{Cip1}$ and $p27^{Kip1}$ protein is regulated by the E3 ubiquitin ligase Skp2 [246], and interestingly, Skp2 KO mice recover liver mass by hepatocellular hypertrophy without induction of hepatocellular proliferation after partial hepatectomy [247].

25.5.6 Termination of Regeneration

Once initiated, regeneration of the liver continues until the preresection liver-to-body-mass ratio is restored (Figure 25.1 and [14]). Although these data indicate the precision with which termination of regeneration is also regulated, the molecular mechanisms responsible for such control have generally been less studied than those involved in initiation of regeneration. Nevertheless, several signals have been implicated as important in such regulation. For example, TGBβ family members and their receptors appear to participate in the termination of liver regeneration. This conclusion is suggested by several observations: pharmacological administration of TGFβ delays initiation of regeneration *in vivo* [248] and suppresses hepatocyte proliferation in culture [249]. Furthermore, mice with liver-specific disruption of TGFβ receptor type II (TGFβRII) expression and pharmacological suppression of signaling by activin (a TGFβ-superfamily member) exhibit prolonged hepatocellular proliferation after partial hepatectomy [250]. Finally, genetic disruption of hepatic TGFβRII- and activin receptor type II-dependent signaling increases hepatocellular proliferation in rats whose livers were otherwise uninjured [251]. As already noted, TGFβ is present in the ECM of quiescent liver and its circulating levels rise during early regeneration. Those data suggest that liver ECM not only influences initiation but also termination of liver regeneration. Recent studies implicate the integrin-linked kinase (ILK), which mediates cellular transduction of ECM signals, in modulation of hepatocellular proliferation *in vivo*, with ILK-null mice shown to exhibit increased proliferation at baseline and prolonged proliferation after partial hepatectomy [252−254]. Expression of HGF and activation of the Hippo kinase signaling cascade, which is also implicated in liver and other organ size control, were dysregulated during regeneration in the ILK KO. Finally, glypican 3, which is a proteoglycan component of ECM, has also been suggested to participate in the termination of liver regeneration. Hepatic glypican 3 expression is induced in late regenerating liver, and loss of glypican 3 function, which occurs in a human overgrowth (Simpson−Golabi−Behmel) syndrome, is associated with enhanced growth of hepatocytes in culture [255]. Moreover, transgenic overexpression of glypican 3 impairs regeneration after partial hepatectomy [256] and decreases xenobiotic-induced hepatocellular proliferation and hepatomegaly [257]. Taken together, these data implicate ECM-derived signaling by TGFβ family members, glypican 3, and ILK in the termination of liver regeneration.

25.6 CLINICAL IMPLICATIONS

The regenerative capacity of the liver has obvious clinical relevance to human health. For example, liver regeneration is essential for recovery after intrahepatic tumor resection and LDLT. Indeed, such regeneration is required for survival and recovery from any serious liver injury. Thus, reports of impaired regeneration in experimental models of fatty liver disease, aging, fulminant liver failure and other conditions have potential implications for evaluation and management of patients with liver disease. Some of these issues are considered here.

25.6.1 Fatty Liver

As already discussed, transient steatosis in response to hepatic insufficiency occurs during and might be important for normal liver regeneration. However, chronic hepatic steatosis has been associated with impaired resection- and toxin-induced regeneration in many experimental animal models, including leptin-deficient (ob/ob) [122,125,258,259] and -resistant (db/db) [123,260], diabetic KK-A(y) [261], and "Western" [262] and high-fructose [263] diet-fed mice, and leptin-resistant obese Zucker rats [264,265]. In contrast, liver regeneration is not impaired in rats with milder hepatic steatosis, as occurs with orotic acid- [265] and choline- [266] deficient diet exposure, leading some investigators to speculate that the degree of steatosis is important in determining its effect on liver regeneration. Consistent with that interpretation, liver regeneration is variably affected in animals administered a methionine–choline deficient (MCD) diet, depending on the magnitude of induced steatosis [265,267–270]. Chronic steatosis is also associated with adverse outcomes after major hepatic resection in humans. For example, in one study the risk of postoperative complications in patients with any steatosis undergoing hepatectomy (for neoplasm) was double that of their nonsteatotic counterparts, and those with excessive (>30%) steatosis had a ~threefold increased risk of death [271]. Despite these findings, which are consistent with the animal model studies discussed above, the mechanisms responsible for the distinct influences of chronic versus acute hepatic fat accumulation on liver regeneration are enigmatic and require further investigation.

25.6.2 Aging

Old age is also associated with reduced hepatic regenerative capacity in surgically induced experimental liver regeneration. First noted ~60 years ago [272,273], the molecular mechanisms responsible for this observation remain incompletely defined. Analyses using partial hepatectomy have shown that the typical suppression of C/EBPα and induction of C/EBPβ and FoxM1 expression that occur in regenerating livers of young mice [168,169,274] are deranged in older animals, and that such dysregulation is accompanied by augmented hepatic expression of p21^{Cip1}[275–278]. Interestingly,

similar molecular alterations are induced in association with the impaired regenerative response in dextrose-supplemented animals [201]. More recent studies implicate age-dependent effects on histone deacetylase (HDAC) 1-dependent epigenetic regulation in the antiregenerative effect of aging [279]. Transgenic expression of FoxM1 or administration of growth hormone, whose levels also decline with advancing age, partially reverses the effects of aging on these molecular perturbations and on cell proliferation in mice subjected to partial hepatectomy [108,278]. Surprisingly, age does not appear to suppress the hepatocellular proliferative responses to certain hepatotoxins [9] and xenobiotics (e.g., TCPOBOP [280]). The relevance of these data to human requires further investigation. In particular, whether such diminished regeneration of the aged liver contributes to the reduced survival of transplanted liver grafts from older donors seen in some [281] but not other [282] studies should be examined.

25.6.3 Fulminant Liver Failure

Experimental analyses in which 85–90% of the native liver is resected often show impaired liver regeneration and increased mortality [283]. Those observations suggest the existence of a threshold amount of remnant liver tissue mass below which regenerative recovery is unable to rescue the animal. Increased liver parenchymal injury in association with such extensive resection is often reported, however, a recent study demonstrated that such injury is not a requisite aspect of the impairment in regeneration in that model [16]. Glucose supplementation and other interventions [283,284] improve survival after experimental subtotal hepatectomy, which has been considered to be an experimental paradigm for acute liver failure (ALF) and "small-for-size" syndrome (SFSS). The recent identification of serum Aab levels as a biomarker of normal liver regeneration in the mouse partial hepatectomy model and predictor of spontaneous survival in pediatric ALF patients [219] suggests that further experimental analyses of subtotal liver resection and regeneration might lead to the discovery of more robust markers with which to predict outcomes in ALF and SFSS.

25.7 SUMMARY AND AREAS OF FUTURE INVESTIGATION

Our current understanding of the regulation of liver regeneration has advanced tremendously since the initial description of the rodent partial hepatectomy model. Nevertheless, important questions remain about this biologically fascinating and clinically important process.

25.7.1 What is the Essence of the Hepatostat?

Despite extensive knowledge of many specific aspects of the physiological, cellular, and molecular regulation of liver regeneration, the identities of the most proximal processes that initiate this response and those distal signals that terminate regeneration are still not adequately known. For example, the molecular mechanisms linking the metabolic response to hepatic insufficiency with the initiation of regeneration require further characterization.

25.7.2 Can Reliable Biomarkers of Liver Regeneration be Discovered?

Such markers would have great utility with respect to clinical decision-making algorithms in patients with ALF, SFSS, and other serious liver diseases. For example, the decision as to whether or not to proceed with liver transplantation in ALF patients would likely benefit from the development of robust, noninvasive strategies with which to assess regenerative recovery of the injured liver in real time. The recent discovery, noted above, of a serum metabolic biomarker of experimental liver regeneration whose level correlates with spontaneous recovery in pediatric ALF, suggests the enormous potential of combining experimental models of regeneration with modern genomic, proteomic, and metabolomic technologies to inform such efforts.

25.7.3 Can Liver Regeneration Be Augmented for Therapeutic Benefit in Patients with Chronic Hepatic Steatosis, Old Age, ALF, Cirrhosis, and Even Liver Cancer?

The answer to this question will ultimately depend on continued efforts to address the questions posed above. The progress made over the past century offers real hope that such research will not only offer novel insight into the mechanisms of liver regeneration, but might also be harnessed to develop effective pro-regenerative therapies that improve outcomes in patients with serious liver diseases.

25.8 ACKNOWLEDGMENTS

Dr Rudnick is grateful to all current and former members of the laboratory, some of whose work is reviewed in this chapter. Those studies were supported by grants to Dr Rudnick from NIH (DK-02900, DK-068219), the American Digestive Health Foundation, the Washington University Child Health Research Center of Excellence, the Children's Digestive Health and Nutrition Foundation, the Washington University Digestive Disease Research Core Center (P30-DK52574), March of Dimes, and the Washington University and St Louis Children's Hospital Children's Discovery Institute.

ABBREVIATIONS

Aab	α-NH_2-butyrate
AAF	acetyl aminofluorene
ALF	acute liver failure
BrdU	bromodeoxyuridine
CAR	constitutive androstane receptor
CCl$_4$	carbon tetrachloride
CREB	cAMP responsive element binding protein
CREM	cAMP response element modifier
ECM	extracellular matrix
CDK	cyclin-dependent kinase
CDKI	cyclin-dependent kinase inhibitor
EGF-R	epidermal growth factor receptor
FAH	fumarylacetoacetate hydrolase
FXR	farnesoid X receptor
GSK	glycogen synthase kinase
Hb-EGF	heparin-binding EGF
HGF	hepatocyte growth factor
IL6	interleukin 6
KO	knockout mouse
ILK	integrin-linked kinase
LDLT	living-donor liver transplantation
LSEC	liver sinusoidal endothelial cell
LXR	liver X receptor
MAPK	mitogen-activated protein kinase
MiRNA	microRNA
NFκB	nuclear factor κB
PKC	protein kinase C
PLC	phospholipase C
PPAR	peroxisome proliferator-activated receptor
PXR	pregnane X receptor
Rb	retinoblastoma
RTK	receptor tyrosine kinase
SAMe	*S*-adenosylmethionine
SiRNA	small inhibitory RNA
TGFα	transforming growth factor α
TNFα	tumor necrosis factor α
uPa	urokinase plasminogen activator
VEGF	vascular endothelial growth factor

REFERENCES

[1] Michalopoulos GK. Liver regeneration after partial hepatectomy: critical analysis of mechanistic dilemmas. Am J Pathol 2010;176 (1):2−13.

[2] Michalopoulos GK. Liver regeneration. J Cell Physiol 2007;213 (2):286−300.

[3] Riehle KJ, Dan YY, Campbell JS, Fausto N. New concepts in liver regeneration. J Gastroenterol Hepatol 2011;26(Suppl. 1):203−12.

[4] Fausto N, Campbell JS, Riehle KJ. Liver regeneration. Hepatology 2006;43(2 Suppl. 1):S45−53.

[5] Johnson SL, Bennett P. Growth control in the ontogenetic and regenerating zebrafish fin. Methods Cell Biol 1999;59:301−11.

[6] Davenport RJ. What controls organ regeneration? Science 2005;309(5731):84.

[7] Palmes D, Spiegel HU. Animal models of liver regeneration. Biomaterials 2004;25(9):1601−11.

[8] Rudnick DA, Davidson NO. Functional relationships between lipid metabolism and liver regeneration. Int J Hepatol 2012:549241. Available from: http://dx.doi.org/10.1155/2012/549241.

[9] Mehendale HM. Tissue repair: an important determinant of final outcome of toxicant-induced injury. Toxicol Pathol 2005;33(1):41–51.

[10] Rudnick DA, Perlmutter DH. Alpha-1-antitrypsin deficiency: a new paradigm for hepatocellular carcinoma in genetic liver disease. Hepatology 2005;42(3):514–21.

[11] Hesiod. Theogeny. Oxford University Press; 1966.

[12] Tiniakos DG, Kandilis A, Geller SA. Tityus: a forgotten myth of liver regeneration. J Hepatol 2010;53(2):357–61. Available from: http://dx.doi.org/10.1016/j.jhep.2010.02.032.

[13] Power C, Rasko JE. Whither prometheus' liver? Greek myth and the science of regeneration. Ann Intern Med 2008;149(6):421–6.

[14] Higgins GM, Anderson RM. Experimental pathology of the liver. 1. Restoration of the liver of the white rat following partial surgical removal. Arch Pathol 1931;12:186–202.

[15] Gazit V, Weymann A, Hartman E, Finck BN, Hruz PW, Tzekov A, et al. Liver regeneration is impaired in lipodystrophic fatty liver dystrophy mice. Hepatology 2010;52(6):2109–17.

[16] Lehmann K, Tschuor C, Rickenbacher A, Jang JH, Oberkofler CE, Tschopp O, et al. Liver failure after extended hepatectomy in mice is mediated by a p21-dependent barrier to liver regeneration. Gastroenterology 2012;143(6):1609–19.

[17] Clark A, Weymann A, Hartman E, Turmelle Y, Carroll M, Thurman JM, et al. Evidence for non-traditional activation of complement factor C3 during murine liver regeneration. Mol Immunol 2008;45(11):3125–32.

[18] Mitchell C, Willenbring H. A reproducible and well-tolerated method for 2/3 partial hepatectomy in mice. Nat Protoc 2008;3 (7):1167–70.

[19] Rudnick DA, Liao Y, An JK, Muglia LJ, Perlmutter DH, Teckman JH. Analyses of hepatocellular proliferation in a mouse model of alpha-1-antitrypsin deficiency. Hepatology 2004;39 (4):1048–55.

[20] Rudnick DA, Shikapwashya O, Blomenkamp K, Teckman JH. Indomethacin increases liver damage in a murine model of liver injury from alpha-1-antitrypsin deficiency. Hepatology 2006;44 (4):976–82.

[21] Willenbring H, Sharma AD, Vogel A, Lee AY, Rothfuss A, Wang Z, et al. Loss of p21 permits carcinogenesis from chronically damaged liver and kidney epithelial cells despite unchecked apoptosis. Cancer Cell 2008;14(1):59–67.

[22] Hall AP, Elcombe CR, Foster JR, Harada T, Kaufmann W, Knippel A, et al. Liver hypertrophy: a review of adaptive (adverse and non-adverse) changes—conclusions from the 3rd international ESTP expert workshop. Toxicol Pathol 2012;40(7):971–94.

[23] Columbano A, Ledda-Columbano GM, Pibiri M, Piga R, Shinozuka H, De LV, et al. Increased expression of c-Fos, c-Jun and LRF-1 is not required for in vivo priming of hepatocytes by the mitogen TCPOBOP. Oncogene 1997;14(7):857–63.

[24] Menegazzi M, Carcereri-De Prati A, Suzuki H, Shinozuka H, Pibiri M, Piga R, et al. Liver cell proliferation induced by nafenopin and cyproterone acetate is not associated with increases in activation of transcription factors NF-kappaB and AP-1 or with expression of tumor necrosis factor alpha. Hepatology 1997;25(3):585–92.

[25] Ledda-Columbano GM, Curto M, Piga R, Zedda AI, Menegazzi M, Sartori C, et al. In vivo hepatocyte proliferation is inducible through a TNF and IL-6-independent pathway. Oncogene 1998;17 (8):1039–44.

[26] Fausto N, Webber EM. Control of liver growth. Crit Rev Eukaryot Gene Expr 1993;3(2):117–35.

[27] Moolten FL, Bucher NLR. Regeneration of rat liver: transfer of humoral agent by cross circulation. Science 1967;158:272–4.

[28] Fisher B, Szuch P, Levine M, Fisher ER. A portal blood factor as the humoral agent in liver regeneration. Science 1971;171:575–7.

[29] Rabes HM, Wirsching R, Tuczek HV, Iseler G. Analysis of cell cycle compartments of hepatocytes after partial hepatecomy. Cell Tissue Kinet 1976;9(6):517–32.

[30] Chaib E, Morales MM, Bordalo MB, Antonio LG, Feijo LF, Ishida RY, et al. Predicting the donor liver lobe weight from body weight for split-liver transplantation. Braz J Med Biol Res 1995;28(7):759–60.

[31] Urata K, Kawasaki S, Matsunami H, Hashikura Y, Ikegami T, Ishizone S, et al. Calculation of child and adult standard liver volume for liver transplantation. Hepatology 1995;21(5):1317–21.

[32] Yamanaka N, Okamoto E, Kawamura E, Kato T, Oriyama T, Fujimoto J, et al. Dynamics of normal and injured human liver regeneration after hepatectomy as assessed on the basis of computed tomography and liver function. Hepatology 1993;18(1):79–85.

[33] Kawasaki S, Makuuchi M, Ishizone S, Matsunami H, Terada M, Kawarazaki H. Liver regeneration in recipients and donors after transplantation. Lancet 1992;339(8793):580–1.

[34] Haga J, Shimazu M, Wakabayashi G, Tanabe M, Kawachi S, Fuchimoto Y, et al. Liver regeneration in donors and adult recipients after living donor liver transplantation. Liver Transpl 2008;14(12):1718–24.

[35] Van Thiel DH, Gavaler JS, Kam I, Francavilla A, Polimeno L, Schade RR, et al. Rapid growth of an intact human liver transplanted into a recipient larger than the donor. Gastroenterology 1987;93(6):1414–9.

[36] Kam I, Lynch S, Svanas G, Todo S, Polimeno L, Francavilla A, et al. Evidence that host size determines liver size: studies in dogs receiving orthotopic liver transplants. Hepatology 1987;7(2):362–6.

[37] Huang J, Glauber M, Qiu Z, Gazit V, Dietzen DJ, Rudnick DA. The influence of skeletal muscle on the regulation of liver:body mass and liver regeneration. Am J Pathol 2012;180(2):575–82.

[38] Michalopoulos GK, DeFrances MC. Liver regeneration. Science 1997;276(5309):60–6.

[39] Zou Y, Bao Q, Kumar S, Hu M, Wang GY, Dai G. Four waves of hepatocyte proliferation linked with three waves of hepatic fat accumulation during partial hepatectomy-induced liver regeneration. PLoS One 2012;7(2):e30675.

[40] Weglarz TC, Sandgren EP. Timing of hepatocyte entry into DNA synthesis after partial hepatectomy is cell autonomous. Proc Natl Acad Sci USA 2000;97(23):12595–600.

[41] Matsuo T, Yamaguchi S, Mitsui S, Emi A, Shimoda F, Okamura H. Control mechanism of the circadian clock for timing of cell division in vivo. Science 2003;302(5643):255–9.

[42] Schibler U. Circadian rhythms. Liver regeneration clocks on. Science 2003;302(5643):234–5.

[43] Melchiorri C, Chieco P, Zedda AI, Coni P, Ledda-Columbano GM, Columbano A. Ploidy and nuclearity of rat hepatocytes after

compensatory regeneration or mitogen-induced liver growth. Carcinogenesis 1993;14(9):1825–30.

[44] Gupta S. Hepatic polyploidy and liver growth control. Semin Cancer Biol 2000;10(3):161–71.

[45] Duncan AW, Taylor MH, Hickey RD, Hanlon Newell AE, Lenzi ML, Olson SB, et al. The ploidy conveyor of mature hepatocytes as a source of genetic variation. Nature 2010;467(7316):707–10.

[46] Sakamoto T, Liu Z, Murase N, Ezure T, Yokomuro S, Poli V, et al. Mitosis and apoptosis in the liver of interleukin-6-deficient mice after partial hepatectomy. Hepatology 1999;29(2):403–11.

[47] Malato Y, Naqvi S, Schurmann N, Ng R, Wang B, Zape J, et al. Fate tracing of mature hepatocytes in mouse liver homeostasis and regeneration. J Clin Invest 2011;121(12):4850–60.

[48] Oh SH, Hatch HM, Petersen BE. Hepatic oval "stem" cell in liver regeneration. Semin Cell Dev Biol 2002;13(6):405–9.

[49] Fausto N, Campbell JS. The role of hepatocytes and oval cells in liver regeneration and repopulation. Mech Dev 2003;120 (1):117–30.

[50] Erker L, Grompe M. Signaling networks in hepatic oval cell activation. Stem Cell Res 2007;1(2):90–102.

[51] Duncan AW, Dorrell C, Grompe M. Stem cells and liver regeneration. Gastroenterology 2009;137(2):466–81.

[52] Matthews VB, Yeoh GC. Liver stem cells. IUBMB Life 2005;57 (8):549–53.

[53] Oertel M, Shafritz DA. Stem cells, cell transplantation and liver repopulation. Biochim Biophys Acta 2008;1782(2):61–74.

[54] Santoni-Rugiu E, Jelnes P, Thorgeirsson SS, Bisgaard HC. Progenitor cells in liver regeneration: molecular responses controlling their activation and expansion. APMIS 2005;113 (11–12):876–902.

[55] Bird TG, Lorenzini S, Forbes SJ. Activation of stem cells in hepatic diseases. Cell Tissue Res 2008;331(1):283–300.

[56] Lesage G, Glaser SS, Gubba S, Robertson WE, Phinizy JL, Lasater J, et al. Regrowth of the rat biliary tree after 70% partial hepatectomy is coupled to increased secretin-induced ductal secretion. Gastroenterology 1996;111(6):1633–44.

[57] Wang L, Wang X, Xie G, Wang L, Hill CK, DeLeve LD. Liver sinusoidal endothelial cell progenitor cells promote liver regeneration in rats. J Clin Invest 2012;122(4):1567–73.

[58] Malik R, Selden C, Hodgson H. The role of non-parenchymal cells in liver growth. Semin Cell Dev Biol 2002;13(6):425–31.

[59] Stocker E, Wullstein HK, Brau G. Capacity of regeneration in liver epithelia of juvenile, repeated partially hepatectomized rats. Autoradiographic studies after continous infusion of ^{3}H-thymidine (author's transl). Virchows Arch B Cell Pathol 1973;14 (2):93–103.

[60] Overturf K, Al Dhalimy M, Tanguay R, Brantly M, Ou CN, Finegold M, et al. Hepatocytes corrected by gene therapy are selected in vivo in a murine model of hereditary tyrosinaemia type I. Nat Genet 1996;12(3):266–73.

[61] Overturf K, Al Dhalimy M, Ou CN, Finegold M, Grompe M. Serial transplantation reveals the stem-cell-like regenerative potential of adult mouse hepatocytes. Am J Pathol 1997;151 (5):1273–80.

[62] Rhim JA, Sandgren EP, Palmiter RD, Brinster RL. Complete reconstitution of mouse liver with xenogeneic hepatocytes. Proc Natl Acad Sci USA 1995;92(11):4942–6.

[63] Ding J, Yannam GR, Roy-Chowdhury N, Hidvegi T, Basma H, Rennard SI, et al. Spontaneous hepatic repopulation in transgenic mice expressing mutant human alpha1-antitrypsin by wild-type donor hepatocytes. J Clin Invest 2011;121(5):1930–4.

[64] Nagy P, Teramoto T, Factor VM, Sanchez A, Schnur J, Paku S, et al. Reconstitution of liver mass via cellular hypertrophy in the rat. Hepatology 2001;33(2):339–45.

[65] Gielchinsky Y, Laufer N, Weitman E, Abramovitch R, Granot Z, Bergman Y, et al. Pregnancy restores the regenerative capacity of the aged liver via activation of an mTORC1-controlled hyperplasia/hypertrophy switch. Genes Dev 2010;24(6):543–8.

[66] Miyaoka Y, Ebato K, Kato H, Arakawa S, Shimizu S, Miyajima A. Hypertrophy and unconventional cell division of hepatocytes underlie liver regeneration. Curr Biol 2012;22(13):1166–75.

[67] Boros P, Miller CM. Hepatocyte growth factor: a multifunctional cytokine [Review]. Lancet 1995;345(8945):293–5.

[68] Nakamura T, Mizuno S. The discovery of hepatocyte growth factor (HGF) and its significance for cell biology, life sciences and clinical medicine. Proc Jpn Acad Ser B Phys Biol Sci 2010;86 (6):588–610.

[69] Nakamura T, Sakai K, Nakamura T, Matsumoto K. Hepatocyte growth factor twenty years on: much more than a growth factor. J Gastroenterol Hepatol 2011;26(Suppl. 1):188–202.

[70] Mars WM, Zarnegar R, Michalopoulos GK. Activation of hepatocyte growth factor by the plasminogen activators uPA and tPA. Am J Pathol 1993;143(3):949–58.

[71] Pediaditakis P, Lopez-Talavera JC, Petersen B, Monga SPS, Michalopoulos GK. The processing and utilization of hepatocyte growth factor/scatter factor following partial hepatectomy in the rat. Hepatology 2001;34(4 Part 1):688–93.

[72] Naldini L, Vigna E, Narsimhan RP, Gaudino G, Zarnegar R, Michalopoulos GK, et al. Hepatocyte growth factor (HGF) stimulates the tyrosine kinase activity of the receptor encoded by the proto-oncogene c-MET. Oncogene 1991;6(4):501–4.

[73] Stolz DB, Mars WM, Petersen BE, Kim TH, Michalopoulos GK. Growth factor signal transduction immediately after two-thirds partial hepatectomy in the rat. Cancer Res 1999;59 (16):3954–60.

[74] Maher JJ. Cell-specific expression of hepatocyte growth factor in liver. Upregulation in sinusoidal endothelial cells after carbon tetrachloride. J Clin Invest 1993;91(5):2244–52.

[75] Asai K, Tamakawa S, Yamamoto M, Yoshie M, Tokusashi Y, Yaginuma Y, et al. Activated hepatic stellate cells overexpress p75NTR after partial hepatectomy and undergo apoptosis on nerve growth factor stimulation. Liver Int 2006;26(5):595–603.

[76] LeCouter J, Moritz DR, Li B, Phillips GL, Liang XH, Gerber HP, et al. Angiogenesis-independent endothelial protection of liver: role of VEGFR-1. Science 2003;299(5608):890–3.

[77] Ding BS, Nolan DJ, Butler JM, James D, Babazadeh AO, Rosenwaks Z, et al. Inductive angiocrine signals from sinusoidal endothelium are required for liver regeneration. Nature 2010;468 (7321):310–5.

[78] Sun R, Jaruga B, Kulkarni S, Sun H, Gao B. IL-6 modulates hepatocyte proliferation via induction of HGF/p21cip1: regulation by SOCS3. Biochem Biophys Res Commun 2005;338(4):1943–9.

[79] Liu ML, Mars WM, Zarnegar R, Michalopoulos GK. Collagenase pretreatment and the mitogenic effects of hepatocyte growth

factor and transforming growth factor-alpha in adult rat liver. Hepatology 1994;19(6):1521−7.

[80] Patijn GA, Lieber A, Schowalter DB, Schwall R, Kay MA. Hepatocyte growth factor induces hepatocyte proliferation *in vivo* and allows for efficient retroviral-mediated gene transfer in mice. Hepatology 1998;28(3):707−16.

[81] Borowiak M, Garratt AN, Wustefeld T, Strehle M, Trautwein C, Birchmeier C. Met provides essential signals for liver regeneration. Proc Natl Acad Sci USA 2004;101(29):10608−13.

[82] Huh CG, Factor VM, Sanchez A, Uchida K, Conner EA, Thorgeirsson SS. Hepatocyte growth factor/c-met signaling pathway is required for efficient liver regeneration and repair. Proc Natl Acad Sci USA 2004;101(13):4477−82.

[83] Paranjpe S, Bowen WC, Bell AW, Nejak-Bowen K, Luo JH, Michalopoulos GK. Cell cycle effects resulting from inhibition of hepatocyte growth factor and its receptor c-Met in regenerating rat livers by RNA interference. Hepatology 2007;45(6):1471−7.

[84] Factor VM, Seo D, Ishikawa T, Kaposi-Novak P, Marquardt JU, Andersen JB, et al. Loss of c-Met disrupts gene expression program required for G2/M progression during liver regeneration in mice. PLoS One 2010;5:9.

[85] Bucher NL, Patel U, Cohen S. Hormonal factors concerned with liver regeneration. Ciba Found Symp 1977;55:95−107.

[86] Poulsen SS, Nexo E, Olsen PS, Hess J, Kirkegaard P. Immunohistochemical localization of epidermal growth factor in rat and man. Histochemistry 1986;85(5):389−94.

[87] Skov OP, Boesby S, Kirkegaard P, Therkelsen K, Almdal T, Poulsen SS, et al. Influence of epidermal growth factor on liver regeneration after partial hepatectomy in rats. Hepatology 1988;8(5):992−6.

[88] Mead JE, Fausto N. Transforming growth factor alpha may be a physiological regulator of liver regeneration by means of an autocrine mechanism. Proc Natl Acad Sci USA 1989;86(5):1558−62.

[89] Kan M, Huang JS, Mansson PE, Yasumitsu H, Carr B, McKeehan WL. Heparin-binding growth factor type 1 (acidic fibroblast growth factor): a potential biphasic autocrine and paracrine regulator of hepatocyte regeneration. Proc Natl Acad Sci USA 1989;86(19):7432−6.

[90] Busser B, Sancey L, Brambilla E, Coll JL, Hurbin A. The multiple roles of amphiregulin in human cancer. Biochim Biophys Acta 2011;1816(2):119−31.

[91] Webber EM, Wu JC, Wang L, Merlino G, Fausto N. Overexpression of transforming growth factor-alpha causes liver enlargement and increased hepatocyte proliferation in transgenic mice. Am J Pathol 1994;145(2):398−408.

[92] Russell WE, Kaufmann WK, Sitaric S, Luetteke NC, Lee DC. Liver regeneration and hepatocarcinogenesis in transforming growth factor-alpha-targeted mice. Mol Carcinog 1996;15 (3):183−9.

[93] Mitchell C, Nivison M, Jackson LF, Fox R, Lee DC, Campbell JS, et al. Heparin-binding epidermal growth factor-like growth factor links hepatocyte priming with cell cycle progression during liver regeneration. J Biol Chem 2005;280 (4):2562−8.

[94] Berasain C, Garcia-Trevijano ER, Castillo J, Erroba E, Lee DC, Prieto J, et al. Amphiregulin: an early trigger of liver regeneration in mice. Gastroenterology 2005;128(2):424−32.

[95] Natarajan A, Wagner B, Sibilia M. The EGF receptor is required for efficient liver regeneration. Proc Natl Acad Sci USA 2007;104(43):17081−6.

[96] Paranjpe S, Bowen WC, Tseng GC, Luo JH, Orr A, Michalopoulos GK. RNA interference against hepatic epidermal growth factor receptor has suppressive effects on liver regeneration in rats. Am J Pathol 2010;176(6):2669−81.

[97] Yamada Y, Kirillova I, Peschon JJ, Fausto N. Initiation of liver growth by tumor necrosis factor: deficient liver regeneration in mice lacking type I tumor necrosis factor receptor. Proc Natl Acad Sci USA 1997;94(4):1441−6.

[98] Cressman DE, Greenbaum LE, DeAngelis RA, Ciliberto G, Furth EE, Poli V, et al. Liver failure and defective hepatocyte regeneration in interleukin-6- deficient mice. Science 1996;274 (5291):1379−83.

[99] Xu CS, Jiang Y, Zhang LX, Chang CF, Wang GP, Shi RJ, et al. The role of Kupffer cells in rat liver regeneration revealed by cell-specific microarray analysis. J Cell Biochem 2012;113 (1):229−37.

[100] Akerman P, Cote P, Yang SQ, McClain C, Nelson S, Bagby GJ, et al. Antibodies to tumor necrosis factor-alpha inhibit liver regeneration after partial hepatectomy. Am J Physiol 1992;263(4 Pt 1):G579−85.

[101] Yamada Y, Fausto N. Deficient liver regeneration after carbon tetrachloride injury in mice lacking type 1 but not type 2 tumor necrosis factor receptor. Am J Pathol 1998;152(6):1577−89.

[102] Yamada Y, Webber EM, Kirillova I, Peschon JJ, Fausto N. Analysis of liver regeneration in mice lacking type 1 or type 2 tumor necrosis factor receptor: requirement for type 1 but not type 2 receptor [comment]. Hepatology 1998;28 (4):959−70.

[103] Kovalovich K, DeAngelis RA, Li W, Furth EE, Ciliberto G, Taub R. Increased toxin-induced liver injury and fibrosis in interleukin-6- deficient mice. Hepatology 2000;31(1):149−59.

[104] Blindenbacher A, Wang X, Langer I, Savino R, Terracciano L, Heim MH. Interleukin 6 is important for survival after partial hepatectomy in mice.[see comment]. Hepatology 2003;38 (3):674−82.

[105] Bucher ML, Swaffield MN. Regulation of hepatic regeneration in rats by synergistic action of insulin and glucagon. Proc Natl Acad Sci USA 1975;72(3):1157−60.

[106] Lai HS, Chung YC, Chen WJ, Chen KM. Rat liver regeneration after partial hepatectomy: effects of insulin, glucagon and epidermal growth factor. J Formos Med Assoc 1992;91(7):685−90.

[107] Hwang TL, Chen MF, Chen TJ. Augmentation of liver regeneration with glucagon after partial hepatectomy in rats. J Formos Med Assoc 1993;92(8):725−8.

[108] Krupczak-Hollis K, Wang X, Dennewitz MB, Costa RH. Growth hormone stimulates proliferation of old-aged regenerating liver through forkhead box m1b. Hepatology 2003;38(6):1552−62.

[109] Pennisi PA, Kopchick JJ, Thorgeirsson S, LeRoith D, Yakar S. Role of growth hormone (GH) in liver regeneration. Endocrinology 2004;145(10):4748−55.

[110] Kan NG, Junghans D, Izpisua Belmonte JC. Compensatory growth mechanisms regulated by BMP and FGF signaling mediate liver regeneration in zebrafish after partial hepatectomy. FASEB J 2009;23(10):3516−25.

[111] Bohm F, Speicher T, Hellerbrand C, Dickson C, Partanen JM, Ornitz DM, et al. FGF receptors 1 and 2 control chemically induced injury and compound detoxification in regenerating livers of mice. Gastroenterology 2010;139(4):1385−96.

[112] Wang L, Wang X, Wang L, Chiu JD, van de Ven G, Gaarde WA, et al. Hepatic vascular endothelial growth factor regulates recruitment of rat liver sinusoidal endothelial cell progenitor cells. Gastroenterology 2012;143(6):1555−63.

[113] Lesurtel M, Graf R, Aleil B, Walther DJ, Tian Y, Jochum W, et al. Platelet-derived serotonin mediates liver regeneration. Science 2006;312(5770):104−7.

[114] Papadimas GK, Tzirogiannis KN, Mykoniatis MG, Grypioti AD, Manta GA, Panoutsopoulos GI. The emerging role of serotonin in liver regeneration. Swiss Med Wkly 2012;142:w13548.

[115] Cruise JL, Houck KA, Michalopoulos GK. Induction of DNA synthesis in cultured rat hepatocytes through stimulation of alpha 1 adrenoreceptor by norepinephrine. Science 1985;227(4688):749−51.

[116] Cruise JL. Alpha 1-adrenergic receptors in liver regeneration. Dig Dis Sci 1991;36(4):485−8.

[117] Rudnick DA, Perlmutter DH, Muglia LJ. Prostaglandins are required for CREB activation and cellular proliferation during liver regeneration. Proc Natl Acad Sci USA 2001;98(15):8885−90.

[118] Rudnick DA, Muglia LJ. Eicosanoids and liver regeneration. In: Curtis-Prior P, editor. The Eicosanoids. West Sussex, England: John Wiley and Sons; 2004. p. 415−22.

[119] Mastellos D, Papadimitriou JC, Franchini S, Tsonis PA, Lambris JD. A novel role of complement: mice deficient in the fifth component of complement (C5) exhibit impaired liver regeneration. J Immunol 2001;166(4):2479−86.

[120] Strey CW, Markiewski M, Mastellos D, Tudoran R, Spruce LA, Greenbaum LE, et al. The proinflammatory mediators C3a and C5a are essential for liver regeneration. J Exp Med 2003;198(6):913−23.

[121] Markiewski MM, Mastellos D, Tudoran R, DeAngelis RA, Strey CW, Franchini S, et al. C3a and C3b activation products of the third component of complement (C3) are critical for normal liver recovery after toxic injury. J Immunol 2004;173(2):747−54.

[122] Leclercq IA, Field J, Farrell GC. Leptin-specific mechanisms for impaired liver regeneration in ob/ob mice after toxic injury. Gastroenterology 2003;124(5):1451−64.

[123] Yamauchi H, Uetsuka K, Okada T, Nakayama H, Doi K. Impaired liver regeneration after partial hepatectomy in db/db mice. Exp Toxicol Pathol 2003;54(4):281−6.

[124] Shteyer E, Liao Y, Muglia LJ, Hruz PW, Rudnick DA. Disruption of hepatic adipogenesis is associated with impaired liver regeneration in mice. Hepatology 2004;40(6):1322−32.

[125] Leclercq IA, Vansteenberghe M, Lebrun VB, VanHul NK, Abarca-Quinones J, Sempoux CL, et al. Defective hepatic regeneration after partial hepatectomy in leptin-deficient mice is not rescued by exogenous leptin. Lab Invest 2006;86(11):1161−71.

[126] Martinez-Hernandez A, Amenta PS. The extracellular matrix in hepatic regeneration. FASEB J 1995;9(14):1401−10.

[127] Kim TH, Mars WM, Stolz DB, Michalopoulos GK. Expression and activation of pro-MMP-2 and pro-MMP-9 during rat liver regeneration. Hepatology 2000;31(1):75−82.

[128] Olle EW, Ren X, McClintock SD, Warner RL, Deogracias MP, Johnson KJ, et al. Matrix metalloproteinase-9 is an important

factor in hepatic regeneration after partial hepatectomy in mice. Hepatology 2006;44(3):540−9.

[129] Diehl AM, Rai RM. Liver regeneration.3. Regulation of signal transduction during liver regeneraiton. FASEB J 1996;10(2):215−27.

[130] Albi E, Rossi G, Maraldi NM, Magni MV, Cataldi S, Solimando L, et al. Involvement of nuclear phosphatidylinositol-dependent phospholipases C in cell cycle progression during rat liver regeneration. J Cell Physiol 2003;197(2):181−8.

[131] Okamoto Y, Nishimura K, Nakayama M, Nakagawa M, Nakano H. Protein kinase C in the regenerating rat liver. Biochem Biophys Res Commun 1988;151(3):1144−9.

[132] Talarmin H, Rescan C, Cariou S, Glaise D, Zanninelli G, Bilodeau M, et al. The mitogen-activated protein kinase kinase/extracellular signal-regulated kinase cascade activation is a key signalling pathway involved in the regulation of G(1) phase progression in proliferating hepatocytes. Mol Cell Biol 1999;19(9):6003−11.

[133] Campbell JS, Argast GM, Yuen SY, Hayes B, Fausto N. Inactivation of p38 MAPK during liver regeneration. Int J Biochem Cell Biol 2011;43(2):180−8.

[134] Servillo G, Penna L, Foulkes NS, Magni MV, Della Fazia MA, Sassone-Corsi P. Cyclic AMP signalling pathway and cellular proliferation: induction of CREM during liver regeneration. Oncogene 1997;14(13):1601−6.

[135] Servillo G, Della Fazia MA, Sassone-Corsi P. Coupling cAMP signaling to transcription in the liver: pivotal role of CREB and CREM [Review] [131 refs]. Exp Cell Res 2002;275(2):143−54.

[136] Garcin I, Tordjmann T. Calcium signalling and liver regeneration. Int J Hepatol 2012;2012:630670.

[137] Casado M, Callejas NA, Rodrigo J, Zhao X, Dey SK, Bosca L, et al. Contribution of cyclooxygenase-2 to liver regeneration after partial hepatectomy. FASEB J. 2001;15(11):2016−8.

[138] Servillo G, Della Fazia MA, Sassone-Corsi P. Transcription factor CREM coordinates the timing of hepatocyte proliferation in the regenerating liver. Genes Dev 1998;12(23):3639−43.

[139] Jackson LN, Larson SD, Silva SR, Rychahou PG, Chen LA, Qiu S, et al. PI3K/Akt activation is critical for early hepatic regeneration after partial hepatectomy. Am J Physiol Gastrointest Liver Physiol 2008;294(6):G1401−10.

[140] Lagoudakis L, Garcin I, Julien B, Nahum K, Gomes DA, Combettes L, et al. Cytosolic calcium regulates liver regeneration in the rat. Hepatology 2010;52(2):602−11.

[141] Guerra MT, Fonseca EA, Melo FM, Andrade VA, Aguiar CJ, Andrade LM, et al. Mitochondrial calcium regulates rat liver regeneration through the modulation of apoptosis. Hepatology 2011;54(1):296−306.

[142] Sekiya S, Suzuki A. Glycogen synthase kinase 3 beta-dependent Snail degradation directs hepatocyte proliferation in normal liver regeneration. Proc Natl Acad Sci USA 2011;108(27):11175−80.

[143] Jin J, Wang GL, Shi X, Darlington GJ, Timchenko NA. The age-associated decline of glycogen synthase kinase 3beta plays a critical role in the inhibition of liver regeneration. Mol Cell Biol 2009;29(14):3867−80.

[144] Schwabe RF, Brenner DA. Role of glycogen synthase kinase-3 in TNF-alpha-induced NF-kappaB activation and apoptosis in hepatocytes. Am J Physiol Gastrointest Liver Physiol 2002;283(1):G204−11.

[145] Kockeritz L, Doble B, Patel S, Woodgett JR. Glycogen synthase kinase-3—an overview of an over-achieving protein kinase. Curr Drug Targets 2006;7(11):1377—88.

[146] Hong F, Nguyen VA, Shen X, Kunos G, Gao B. Rapid activation of protein kinase B/Akt has a key role in antiapoptotic signaling during liver regeneration. Biochem Biophys Res Commun 2000;279(3):974—9.

[147] Haga S, Ogawa W, Inoue H, Terui K, Ogino T, Igarashi R, et al. Compensatory recovery of liver mass by Akt-mediated hepato-cellular hypertrophy in liver-specific STAT3-deficient mice. J Hepatol 2005;43(5):799—807.

[148] Haga S, Ozaki M, Inoue H, Okamoto Y, Ogawa W, Takeda K, et al. The survival pathways phosphatidylinositol-3 kinase (PI3-K)/phosphoinositide-dependent protein kinase 1 (PDK1)/Akt modulate liver regeneration through hepatocyte size rather than proliferation. Hepatology 2009;49(1):204—14.

[149] Mullany LK, Nelsen CJ, Hanse EA, Goggin MM, Anttila CK, Peterson M, et al. Akt-mediated liver growth promotes induction of cyclin E through a novel translational mechanism and a p21-mediated cell cycle arrest. J Biol Chem 2007;282 (29):21244—52.

[150] Chen P, Yan H, Chen Y, He Z. The variation of Akt/TSC1—TSC1/mTOR signal pathway in hepatocytes after partial hepatectomy in rats. Exp Mol Pathol 2009;86(2):101—7.

[151] Nechemia-Arbely Y, Shriki A, Denz U, Drucker C, Scheller J, Raub J, et al. Early hepatocyte DNA synthetic response posthe-patectomy is modulated by IL-6 trans-signaling and PI3K/Akt activation. J Hepatol 2011;54(5):922—9.

[152] Thompson MD, Monga SP. WNT/beta-catenin signaling in liver health and disease. Hepatology 2007;45(5):1298—305.

[153] Monga SP, Pediaditakis P, Mule K, Stolz DB, Michalopoulos GK. Changes in WNT/beta-catenin pathway during regulated growth in rat liver regeneration. Hepatology 2001;33(5):1098—109.

[154] Sodhi D, Micsenyi A, Bowen WC, Monga DK, Talavera JC, Monga SP. Morpholino oligonucleotide-triggered beta-catenin knockdown compromises normal liver regeneration. J Hepatol 2005;43(1):132—41.

[155] Tan X, Behari J, Cieply B, Michalopoulos GK, Monga SP. Conditional deletion of beta-catenin reveals its role in liver growth and regeneration. Gastroenterology 2006;131(5):1561—72.

[156] Tan X, Apte U, Micsenyi A, Kotsagrelos E, Luo JH, Ranganathan S, et al. Epidermal growth factor receptor: a novel target of the Wnt/beta-catenin pathway in liver. Gastroenterology 2005;129(1):285—302.

[157] Apte U, Zeng G, Muller P, Tan X, Micsenyi A, Cieply B, et al. Activation of Wnt/beta-catenin pathway during hepatocyte growth factor-induced hepatomegaly in mice. Hepatology 2006;44(4):992—1002.

[158] Monga SPS, Mars WM, Pediaditakis P, Bell A, Mule K, Bowen WC, et al. Hepatocyte growth factor induces Wnt-independent nuclear translocation of beta-catenin after met-ss-catenin dissoci-ation in hepatocytes. Cancer Res 2002;62(7):2064—71.

[159] Kopan R, Ilagan MX. The canonical Notch signaling pathway: unfolding the activation mechanism. Cell 2009;137(2):216—33.

[160] Kohler C, Bell AW, Bowen WC, Monga SP, Fleig W, Michalopoulos GK. Expression of Notch-1 and its ligand Jagged-1 in rat liver during liver regeneration. Hepatology 2004;39(4):1056—65.

[161] Iimuro Y, Nishiura T, Hellerbrand C, Behrns KE, Schoonhoven R, Grisham JW, et al. NFκB prevents apoptosis and liver dysfunction during liver regeneration. J Clin Invest 1998;101:802—11.

[162] Chaisson ML, Brooling JT, Ladiges W, Tsai S, Fausto N. Hepatocyte-specific inhibition of NF-kappaB leads to apoptosis after TNF treatment, but not after partial hepatectomy. J Clin Invest 2002;110(2):193—202.

[163] DeAngelis RA, Kovalovich K, Cressman DE, Taub R. Normal liver regeneration in p50/nuclear factor kappaB1 knockout mice. Hepatology 2001;33(4):915—24.

[164] Li W, Liang X, Kellendonk C, Poli V, Taub R. STAT3 contri-butes to the mitogenic response of hepatocytes during liver regeneration. J Biol Chem 2002;277(32):28411—7.

[165] Moh A, Iwamoto Y, Chai GX, Zhang SS, Kano A, Yang DD, et al. Role of STAT3 in liver regeneration: survival, DNA synthesis, inflammatory reaction and liver mass recovery. Lab Invest 2007;87(10):1018—28.

[166] Mohn KL, Laz TM, Melby AE, Taub R. Immediate-early gene expression differs between regenerating liver, insulin-stimulated H-35 cells, and mitogen-stimulated Balb/c 3T3 cells. Liver-specific induction patterns of gene 33, phosphoenolpyruvate car-boxykinase, and the jun, fos, and egr families. J Biol Chem 1990;265(35):21914—21.

[167] Mohn KL, Laz TM, Hsu JC, Melby AE, Bravo R, Taub R. The immediate-early growth response in regenerating liver and insulin-stimulated H-35 cells: comparison with serum-stimulated 3T3 cells and identification of 41 novel immediate-early genes. Mol Cell Biol 1991;11(1):381—90.

[168] Greenbaum LE, Cressman DE, Haber BA, Taub R. Coexistence of C/EBP alpha, beta, growth-induced proteins and DNA synthe-sis in hepatocytes during liver regeneration. Implications for maintenance of the differentiated state during liver growth. J Clin Invest 1995;96(3):1351—65.

[169] Greenbaum LE, Li W, Cressman DE, Peng Y, Ciliberto G, Poli V, et al. CCAAT enhancer- binding protein beta is required for normal hepatocyte proliferation in mice after partial hepatec-tomy. J Clin Invest 1998;102(5):996—1007.

[170] Taub R, Greenbaum LE, Peng Y. Transcriptional regulatory sig-nals define cytokine-dependent and - independent pathways in liver regeneration. Semin Liver Dis 1999;19(2):117—27.

[171] Wang H, Peiris TH, Mowery A, Le Lay J, Gao Y, Greenbaum LE. CCAAT/enhancer binding protein-beta is a transcriptional regulator of peroxisome-proliferator-activated receptor-gamma coactivator-1alpha in the regenerating liver. Mol Endocrinol 2008;22(7):1596—605.

[172] Liao Y, Shikapwashya ON, Shteyer E, Dieckgraefe BK, Hruz PW, Rudnick DA. Delayed hepatocellular mitotic progression and impaired liver regeneration in early growth response-1-deficient mice. J Biol Chem 2004;279(41):43107—16.

[173] Bonzo JA, Ferry CH, Matsubara T, Kim JH, Gonzalez FJ. Suppression of hepatocyte proliferation by hepatocyte nuclear factor 4alpha in adult mice. J Biol Chem 2012;287 (10):7345—56.

[174] Haber BA, Mohn KL, Diamond RH, Taub R. Induction patterns of 70 genes during nine days after hepatectomy define the tem-poral course of liver regeneration. J Clin Invest 1993;91 (4):1319—26.

[175] Su AI, Guidotti LG, Pezacki JP, Chisari FV, Schultz PG. Gene expression during the priming phase of liver regeneration after partial hepatectomy in mice. Proc Natl Acad Sci USA 2002;99 (17):11181–6.

[176] Locker J, Tian J, Carver R, Concas D, Cossu C, Ledda-Columbano GM, et al. A common set of immediate-early response genes in liver regeneration and hyperplasia. Hepatology 2003;38(2):314–25.

[177] Arai M, Yokosuka O, Chiba T, Imazeki F, Kato M, Hashida J, et al. Gene expression profiling reveals the mechanism and pathophysiology of mouse liver regeneration. J Biol Chem 2003;278(32):29813–8.

[178] Morita T, Togo S, Kubota T, Kamimukai N, Nishizuka I, Kobayashi T, et al. Mechanism of postoperative liver failure after excessive hepatectomy investigated using a cDNA microarray. J Hepatobil Pancreat Surg 2002;9(3):352–9.

[179] Kelley-Loughnane N, Sabla GE, Ley-Ebert C, Aronow BJ, Bezerra JA. Independent and overlapping transcriptional activation during liver development and regeneration in mice. Hepatology 2002;35(3):525–34.

[180] Li J, Campbell JS, Mitchell C, McMahan RS, Yu X, Riehle KJ, et al. Relationships between deficits in tissue mass and transcriptional programs after partial hepatectomy in mice. Am J Pathol 2009;175(3):947–57.

[181] Xu CS, Chen XG, Chang CF, Wang GP, Wang WB, Zhang LX, et al. Analysis of time-course gene expression profiles of sinusoidal endothelial cells during liver regeneration in rats. Mol Cell Biochem 2011;350(1–2):215–27.

[182] Chen X, Xu C, Zhang F, Ma J. Comparative analysis of expression profiles of chemokines, chemokine receptors, and components of signaling pathways mediated by chemokines in eight cell types during rat liver regeneration. Genome 2010;53(8):608–18.

[183] Shu J, Kren BT, Xia Z, Wong PY, Li L, Hanse EA, et al. Genomewide microRNA down-regulation as a negative feedback mechanism in the early phases of liver regeneration. Hepatology 2011;54(2):609–19.

[184] Raschzok N, Werner W, Sallmon H, Billecke N, Dame C, Neuhaus P, et al. Temporal expression profiles indicate a primary function for microRNA during the peak of DNA replication after rat partial hepatectomy. Am J Physiol Regul Integr Comp Physiol 2011;300(6):R1363–72.

[185] Song G, Sharma AD, Roll GR, Ng R, Lee AY, Blelloch RH, et al. MicroRNAs control hepatocyte proliferation during liver regeneration. Hepatology 2010;51(5):1735–43.

[186] Ng R, Song G, Roll GR, Frandsen NM, Willenbring H. A microRNA-21 surge facilitates rapid cyclin D1 translation and cell cycle progression in mouse liver regeneration. J Clin Invest 2012;122(3):1097–108.

[187] Rao MS, Peters JM, Gonzalez FJ, Reddy JK. Hepatic regeneration in peroxisome proliferator-activated receptor alpha-null mice after partial hepatectomy. Hepatol Res 2002;22(1):52–7.

[188] Newberry EP, Kennedy SM, Xie Y, Luo J, Stanley SE, Semenkovich CF, et al. Altered hepatic triglyceride content after partial hepatectomy without impaired liver regeneration in multiple murine genetic models. Hepatology 2008;48(4):1097–105.

[189] Wheeler MD, Smutney OM, Check JF, Rusyn I, Schulte-Hermann R, Thurman RG. Impaired Ras membrane association and activation in PPARalpha knockout mice after partial hepatectomy. Am J Physiol Gastrointest Liver Physiol 2003;284 (2):G302–12.

[190] Anderson SP, Yoon L, Richard EB, Dunn CS, Cattley RC, Corton JC. Delayed liver regeneration in peroxisome proliferator-activated receptor-alpha-null mice. Hepatology 2002;36(3):544–54.

[191] Chakravarthy MV, Pan Z, Zhu Y, Tordjman K, Schneider JG, Coleman T, et al. "New" hepatic fat activates PPARalpha to maintain glucose, lipid, and cholesterol homeostasis. Cell Metab 2005;1(5):309–22.

[192] Chakravarthy MV, Lodhi IJ, Yin L, Malapaka RR, Xu HE, Turk J, et al. Identification of a physiologically relevant endogenous ligand for PPARalpha in liver. Cell 2009;138(3):476–88.

[193] Huang W, Ma K, Zhang J, Qatanani M, Cuvillier J, Liu J, et al. Nuclear receptor-dependent bile acid signaling is required for normal liver regeneration. Science 2006;312(5771):233–6.

[194] Zhang L, Wang YD, Chen WD, Wang X, Lou G, Liu N, et al. Promotion of liver regeneration/repair by farnesoid X receptor in both liver and intestine in mice. Hepatology 2012;56 (6):2336–43.

[195] Borude P, Edwards G, Walesky C, Li F, Ma X, Kong B, et al. Hepatocyte-specific deletion of farnesoid X receptor delays but does not inhibit liver regeneration after partial hepatectomy in mice. Hepatology 2012;56(6):2344–52.

[196] Dai G, He L, Bu P, Wan YJ. Pregnane X receptor is essential for normal progression of liver regeneration. Hepatology 2008;47 (4):1277–87.

[197] Gazit V, Huang J, Weymann A, Rudnick DA. Analysis of the role of hepatic PPARgamma expression during mouse liver regeneration. Hepatology 2012;56:1489–98.

[198] Turmelle YP, Shikapwashya O, Tu S, Hruz PW, Yan Q, Rudnick DA. Rosiglitazone inhibits mouse liver regeneration. FASEB J 2006;20:2609–11.

[199] Lo SG, Celli N, Caboni M, Murzilli S, Salvatore L, Morgano A, et al. Down-regulation of the LXR transcriptome provides the requisite cholesterol levels to proliferating hepatocytes. Hepatology 2009.

[200] Felber JP, Golay A. Regulation of nutrient metabolism and energy expenditure. Metabolism 1995;44(2 Suppl. 2):4–9.

[201] Weymann A, Hartman E, Gazit V, Wang C, Glauber M, Turmelle Y, et al. p21 is required for dextrose-mediated inhibition of mouse liver regeneration. Hepatology 2009;50:207–15.

[202] Bengmark S, Olsson R, Svanborg A. The influence of glucose supply on liver steatosis and regeneration rate after partial hepatectomy. Acta Chir Scand 1965;130:216–23.

[203] Simek J, Melka J, Pospisil M, Neradilkova M. Effect of protracted glucose infusion on the development of early biochemical changes and initiation of regeneration in rat liver after partial hepatectomy. Physiol Bohemoslov 1965;14(4):366–70.

[204] Caruana JA, Whalen Jr. DA, Anthony WP, Sunby CR, Ciechoski MP. Paradoxical effects of glucose feeding on liver regeneration and survival after partial hepatectomy. Endocr Res 1986;12 (2):147–56.

[205] Holecek M. Nutritional modulation of liver regeneration by carbohydrates, lipids, and amino acids: a review. Nutrition 1999;15 (10):784–8.

[206] Simek J, Chmelar V, Melka J, Pazderka J, Charvat Z. Influence of protracted infusion of glucose and insulin on the composition

and regeneration activity of liver after partial hepatectomy in rats. Nature 1967;213(5079):910−1.

[207] Ngala Kenda JF, de Hemptinne B, Lambotte L. Role of metabolic overload in the initiation of DNA synthesis following partial hepatectomy in the rat. Eur Surg Res 1984;16 (5):294−302.

[208] Chanda S, Mehendale HM. Nutritional impact on the final outcome of liver injury inflicted by model hepatotoxicants: effect of glucose loading. FASEB J 1995;9(2):240−5.

[209] Chanda S, Mehendale HM. Role of nutrition in the survival after hepatotoxic injury. Toxicology 1996;111(1−3):163−78.

[210] Mitro N, Mak PA, Vargas L, Godio C, Hampton E, Molteni V, et al. The nuclear receptor LXR is a glucose sensor. Nature 2007;445(7124):219−23.

[211] Cuenca AG, Cress WD, Good RA, Marikar Y, Engelman RW. Calorie restriction influences cell cycle protein expression and DNA synthesis during liver regeneration. Exp Biol Med (Maywood) 2001;226(11):1061−7.

[212] Apte UM, Limaye PB, Desaiah D, Bucci TJ, Warbritton A, Mehendale HM. Mechanisms of increased liver tissue repair and survival in diet-restricted rats treated with equitoxic doses of thioacetamide. Toxicol Sci 2003;72(2):272−82.

[213] Shankar K, Vaidya VS, Wang T, Bucci TJ, Mehendale HM. Streptozotocin-induced diabetic mice are resistant to lethal effects of thioacetamide hepatotoxicity. Toxicol Appl Pharmacol 2003;188(2):122−34.

[214] Shankar K, Vaidya VS, Apte UM, Manautou JE, Ronis MJ, Bucci TJ, et al. Type 1 diabetic mice are protected from acetaminophen hepatotoxicity. Toxicol Sci 2003;73(2):220−34.

[215] Starzl TE, Francavilla A, Halgrimson CG, Francavilla FR, Porter KA, Brown TH, et al. The origin, hormonal nature, and action of hepatotrophic substances in portal venous blood. Surg Gynecol Obstet 1973;137(2):179−99.

[216] Starzl TE, Porter KA, Putnam CW. Intraportal insulin protects from the liver injury of portacaval shunt in dogs. Lancet 1975;2 (7947):1241−2.

[217] Starzl TE, Francavilla A, Porter KA, Benichou J, Jones AF. The effect of splanchnic viscera removal upon canine liver regeneration. Surg Gynecol Obstet 1978;147(2):193−207.

[218] Starzl TE, Porter KA, Kashiwagi N. Portal hepatotrophic factors, diabetes mellitus and acute liver atrophy, hypertrophy and regeneration. Surg Gynecol Obstet 1975;141(6):843−58.

[219] Rudnick DA, Dietzen DJ, Turmelle YP, Shepherd R, Zhang S, Belle SH, et al. Serum alpha-NH-butyric acid may predict spontaneous survival in pediatric acute liver failure. Pediatr Transplant 2009;13(2):223−30.

[220] Klingensmith JS, Mehendale HM. Chlordecone-induced fat depletion in the male rat. J Toxicol Environ Health 1982;10 (1):121−9.

[221] Glende Jr. EA, Morgan WS. Alteration in liver lipid and lipid fatty acid composition after partial hepatectomy in the rat. Exp Mol Pathol 1968;8(2):190−200.

[222] Delahunty TJ, Rubinstein D. Accumulation and release of triglycerides by rat liver following partial hepatectomy. J Lipid Res 1970;11:536−43.

[223] Girard A, Roheim PS, Eder HA. Lipoprotein synthesis and fatty acid mobilization in rats after partial hepatectomy. Biochim Biophys Acta 1971;248(1):105−13.

[224] Gove CD, Hems DA. Fatty acid synthesis in the regenerating liver of the rat. Biochem J 1978;170(1):1−8.

[225] Srinivasan SR, Chow CK, Glauert HP. Effect of the peroxisome proliferator ciprofibrate on hepatic DNA synthesis and hepatic composition following partial hepatectomy in rats. Toxicology 1990;62(3):321−32.

[226] Walldorf J, Hillebrand C, Aurich H, Stock P, Hempel M, Ebensing S, et al. Propranolol impairs liver regeneration after partial hepatectomy in C57Bl/6-mice by transient attenuation of hepatic lipid accumulation and increased apoptosis. Scand J Gastroenterol 2010;45(4):468−76. Available from: http://dx.doi.org/10.3109/00365520903583848.

[227] Fernandez MA, Albor C, Ingelmo-Torres M, Nixon SJ, Ferguson C, Kurzchalia T, et al. Caveolin-1 is essential for liver regeneration. Science 2006;313(5793):1628−32.

[228] Mayoral R, Fernandez-Martinez A, Roy R, Bosca L, Martin-Sanz P. Dispensability and dynamics of caveolin-1 during liver regeneration and in isolated hepatic cells. Hepatology 2007;46 (3):813−22.

[229] Michalopoulos G, Cianciulli HD, Novotny AR, Kligerman AD, Strom SC, Jirtle RL. Liver regeneration studies with rat hepatocytes in primary culture. Cancer Res. 1982;42 (11):4673−82.

[230] Nakatani T, Ozawa K, Asano M, Ukikusa M, Kamiyama Y, Tobe T. Differences in predominant energy substrate in relation to the resected hepatic mass in the phase immediately after hepatectomy. J Lab Clin Med 1981;97(6):887−98.

[231] Holecek M, Simek J, Palicka V, Zadak Z. Effect of glucose and branched chain amino acid (BCAA) infusion on onset of liver regeneration and plasma amino acid pattern in partially hepatectomized rats. J Hepatol 1991;13(1):14−20.

[232] Chanda S, Mehendale M. Role of nutritional fatty acid and L-carnitine in the final outcome of thioacetamide hepatotoxicity. FASEB J 1994;8(13):1061−8.

[233] Crumm S, Cofan M, Juskeviciute E, Hoek JB. Adenine nucleotide changes in the remnant liver: an early signal for regeneration after partial hepatectomy. Hepatology 2008;48 (3):898−908.

[234] Vazquez-Chantada M, Ariz U, Varela-Rey M, Embade N, Martinez-Lopez N, Fernandez-Ramos D, et al. Evidence for LKB1/ AMP-activated protein kinase/ endothelial nitric oxide synthase cascade regulated by hepatocyte growth factor, S-adenosylmethionine, and nitric oxide in hepatocyte proliferation. Hepatology 2009;49 (2):608−17.

[235] Varela-Rey M, Beraza N, Lu SC, Mato JM, Martinez-Chantar ML. Role of AMP-activated protein kinase in the control of hepatocyte priming and proliferation during liver regeneration. Exp Biol Med (Maywood) 2011;236(4):402−8.

[236] Mato JM, Lu SC. Role of S-adenosyl-L-methionine in liver health and injury. Hepatology 2007;45(5):1306−12.

[237] Steele RD. Transaminative metabolism of alpha-amino-n-butyrate in rats. Metabolism 1982;31(4):318−25.

[238] Newsholme EA, Leech. AR. Amino acid metabolism. Biochemistry for the medical sciences. New York: John Wiley & Sons; 1983382−441

[239] Chen L, Zeng Y, Yang H, Lee TD, French SW, Corrales FJ, et al. Impaired liver regeneration in mice lacking methionine adenosyltransferase 1A. FASEB J 2004;18(7):914−6.

[240] Crawford DF, Piwnica-Worms H. Regulation of the eukaryotic cell cycle. In: Irwin M, Arias JL, Boyer FV, Chisari NF, Schachter D, Shafritz DA, editors. The liver: biology and pathobiology. Philadelphia: Lipincott Williams & Wilkins; 2001. p. 977–86.

[241] Albrecht JH, Meyer AH, Hu MY. Regulation of cyclin-dependent kinase inhibitor p21(WAF1/Cip1/Sdi1) gene expression in hepatic regeneration. Hepatology 1997;25(3):557–63.

[242] Albrecht JH, Poon RY, Ahonen CL, Rieland BM, Deng C, Crary GS. Involvement of p21 and p27 in the regulation of CDK activity and cell cycle progression in the regenerating liver. Oncogene 1998;16(16):2141–50.

[243] Jaime M, Pujol MJ, Serratosa J, Pantoja C, Canela N, Casanovas O, et al. The p21(Cip1) protein, a cyclin inhibitor, regulates the levels and the intracellular localization of CDC25A in mice regenerating livers. Hepatology 2002;35(5):1063–71.

[244] Luedde T, Rodriguez ME, Tacke F, Xiong Y, Brenner DA, Trautwein C. p18(INK4c) collaborates with other CDK-inhibitory proteins in the regenerating liver. Hepatology 2003;37 (4):833–41.

[245] Hayashi E, Yasui A, Oda K, Nagino M, Nimura Y, Nakanishi M, et al. Loss of p27(Kip1) accelerates DNA replication after partial hepatectomy in mice. J Surg Res 2003;111 (2):196–202.

[246] Kossatz U, Dietrich N, Zender L, Buer J, Manns MP, Malek NP. Skp2-dependent degradation of p27kip1 is essential for cell cycle progression. Genes Dev 2004;18(21):2602–7.

[247] Minamishima YA, Nakayama K, Nakayama K. Recovery of liver mass without proliferation of hepatocytes after partial hepatectomy in Skp2-deficient mice. Cancer Res 2002;62(4):995–9.

[248] Russell WE, Coffey Jr. RJ, Ouellette AJ, Moses HL. Type beta transforming growth factor reversibly inhibits the early proliferative response to partial hepatectomy in the rat. Proc Natl Acad Sci USA 1988;85(14):5126–30.

[249] Houck KA, Cruise JL, Michalopoulos G. Norepinephrine modulates the growth-inhibitory effect of transforming growth factor-beta in primary rat hepatocyte cultures. J Cell Physiol 1988;135 (3):551–5.

[250] Oe S, Lemmer ER, Conner EA, Factor VM, Leveen P, Larsson J, et al. Intact signaling by transforming growth factor beta is not required for termination of liver regeneration in mice. Hepatology 2004;40(5):1098–105.

[251] Ichikawa T, Zhang YQ, Kogure K, Hasegawa Y, Takagi H, Mori M, et al. Transforming growth factor beta and activin tonically inhibit DNA synthesis in the rat liver. Hepatology 2001;34 (5):918–25.

[252] Gkretsi V, Apte U, Mars WM, Bowen WC, Luo JH, Yang Y, et al. Liver-specific ablation of integrin-linked kinase in mice results in abnormal histology, enhanced cell proliferation, and hepatomegaly. Hepatology 2008;48(6):1932–41.

[253] Apte U, Gkretsi V, Bowen WC, Mars WM, Luo JH, Donthamsetty S, et al. Enhanced liver regeneration following changes induced by hepatocyte-specific genetic ablation of integrin-linked kinase. Hepatology 2009;50(3):844–51.

[254] Donthamsetty S, Bhave VS, Kliment CS, Bowen WC, Mars WM, Bell AW, et al. Excessive hepatomegaly of mice with hepatocyte-targeted elimination of integrin linked kinase following treatment with 1,4-bis[2-(3,5-dichaloropyridyloxy)] benzene. Hepatology 2011;53(2):587–95.

[255] Liu B, Paranjpe S, Bowen WC, Bell AW, Luo JH, Yu YP, et al. Investigation of the role of glypican 3 in liver regeneration and hepatocyte proliferation. Am J Pathol 2009;175(2):717–24.

[256] Liu B, Bell AW, Paranjpe S, Bowen WC, Khillan JS, Luo JH, et al. Suppression of liver regeneration and hepatocyte proliferation in hepatocyte-targeted glypican 3 transgenic mice. Hepatology 2010;52(3):1060–7.

[257] Lin CW, Mars WM, Paranjpe S, Donthamsetty S, Bhave VS, Kang LI, et al. Hepatocyte proliferation and hepatomegaly induced by phenobarbital and 1,4-bis [2-(3,5-dichloropyridyloxy)] benzene is suppressed in hepatocyte-targeted glypican 3 transgenic mice. Hepatology 2011;54(2):620–30.

[258] Yang SQ, Lin HZ, Mandal AK, Huang J, Diehl AM. Disrupted signaling and inhibited regeneration in obese mice with fatty livers: implications for nonalcoholic fatty liver disease pathophysiology. Hepatology 2001;34:694–706.

[259] Torbenson M, Yang SQ, Liu HZ, Huang J, Gage W, Diehl AM. STAT-3 overexpression and p21 up-regulation accompany impaired regeneration of fatty livers. Am J Pathol 2002;161 (1):155–61.

[260] Shirai M, Yamauchi H, Nakayama H, Doi K, Uetsuka K. Expression of epidermal growth factor receptor protein in the liver of db/db mice after partial hepatectomy. Exp Toxicol Pathol 2007;59(3–4):157–62.

[261] Aoyama T, Ikejima K, Kon K, Okumura K, Arai K, Watanabe S. Pioglitazone promotes survival and prevents hepatic regeneration failure after partial hepatectomy in obese and diabetic KK-A(y) mice. Hepatology 2009;49(5):1636–44.

[262] DeAngelis RA, Markiewski MM, Taub R, Lambris JD. A high-fat diet impairs liver regeneration in C57BL/6 mice through overexpression of the NF-kappaB inhibitor, IkappaBalpha. Hepatology 2005;42(5):1148–57.

[263] Tanoue S, Uto H, Kumamoto R, Arima S, Hashimoto S, Nasu Y, et al. Liver regeneration after partial hepatectomy in rat is more impaired in a steatotic liver induced by dietary fructose compared to dietary fat. Biochem Biophys Res Commun 2011;407 (1):163–8.

[264] Selzner M, Clavien PA. Failure of regeneration of the steatotic rat liver: disruption at two different levels in the regeneration pathway. Hepatology 2000;31(1):35–42.

[265] Picard C, Lambotte L, Starkel P, Sempoux C, Saliez A, Van den Berge V, et al. Steatosis is not sufficient to cause an impaired regenerative response after partial hepatectomy in rats. J Hepatol 2002;36(5):645–52.

[266] Rao MS, Papreddy K, Abecassis M, Hashimoto T. Regeneration of liver with marked fatty change following partial hepatectomy in rats. Dig Dis Sci 2001;46(9):1821–6.

[267] Zhang BH, Weltman M, Farrell GC. Does steatohepatitis impair liver regeneration? A study in a dietary model of non-alcoholic steatohepatitis in rats. Eur J Gastroenterol Hepatol 1999;14 (2):133–7.

[268] Vetelainen R, Bennink RJ, van Vliet AK, van Gulik TM. Mild steatosis impairs functional recovery after liver resection in an experimental model. Br J Surg 2007;94(8):1002–8.

[269] Vetelainen R, van Vliet AK, van Gulik TM. Severe steatosis increases hepatocellular injury and impairs liver regeneration in a rat model of partial hepatectomy. Ann Surg 2007;245 (1):44–50.

[270] Donthamsetty S, Bhave VS, Mitra MS, Latendresse JR, Mehendale HM. Nonalcoholic fatty liver sensitizes rats to carbon tetrachloride hepatotoxicity. Hepatology 2007;45 (2):391–403.

[271] de Meijer VE, Kalish BT, Puder M, Ijzermans JN. Systematic review and meta-analysis of steatosis as a risk factor in major hepatic resection. Br J Surg 2010;97(9):1331–9.

[272] Bucher NL, Glinos AD. The effect of age on regeneration of rat liver. Cancer Res 1950;10(5):324–32.

[273] Bucher NL, Swaffield MN, Ditroia JF. The influence of age upon the incorporation of thymidine-2-C14 into the DNA of regenerating rat liver. Cancer Res 1964;24:509–12.

[274] Wang X, Kiyokawa H, Dennewitz MB, Costa RH. The forkhead box m1b transcription factor is essential for hepatocyte DNA replication and mitosis during mouse liver regeneration. Proc Natl Acad Sci USA 2002;99(26):16881–6.

[275] Timchenko NA, Wilde M, Kosai KI, Heydari A, Bilyeu TA, Finegold MJ, et al. Regenerating livers of old rats contain high levels of C/EBPalpha that correlate with altered expression of cell cycle associated proteins. Nucleic Acids Res 1998;26(13):3293–9.

[276] Iakova P, Awad SS, Timchenko NA. Aging reduces proliferative capacities of liver by switching pathways of C/EBPalpha growth arrest. Cell 2003;113(4):495–506.

[277] Timchenko NA. Old livers—C/EBPalpha meets new partners. Cell Cycle 2003;2(5):445–6.

[278] Wang X, Quail E, Hung NJ, Tan Y, Ye H, Costa RH. Increased levels of forkhead box M1B transcription factor in transgenic mouse hepatocytes prevent age-related proliferation defects in regenerating liver. Proc Natl Acad Sci USA 2001;98(20):11468–73.

[279] Wang GL, Salisbury E, Shi X, Timchenko L, Medrano EE, Timchenko NA. HDAC1 cooperates with C/EBPalpha in the inhibition of liver proliferation in old mice. J Biol Chem 2008;283(38):26169–78.

[280] Ledda-Columbano GM, Pibiri M, Cossu C, Molotzu F, Locker J, Columbano A. Aging does not reduce the hepatocyte proliferative response of mice to the primary mitogen TCPOBOP. Hepatology 2004;40(4):981–8.

[281] Timchenko NA. Aging and liver regeneration.. Trends Endocrinol Metab 2009;20(4):171–6.

[282] Sirivatanauksorn Y, Taweerutchana V, Limsrichamrern S, Kositamongkol P, Mahawithitwong P, Asavakarn S, et al. Analysis of donor risk factors associated with graft outcomes in orthotopic liver transplantation. Transplant Proc 2012;44 (2):320–3.

[283] Gaub J, Iversen J. Rat liver regeneration after 90% partial hepatectomy. Hepatology 1984;4(5):902–4.

[284] Cai SR, Motoyama K, Shen KJ, Kennedy SC, Flye MW, Ponder KP. Lovastatin decreases mortality and improves liver functions in fulminant hepatic failure from 90% partial hepatectomy in rats. J Hepatol 2000;32(1):67–77.

Chapter 26

Liver Regeneration: The Stem Cell Approach

Syeda H. Afroze[a], Kendal Jensen[a], Kinan Rahal[a], Fanyin Meng[a,b,c], Gianfranco Alpini[a,b,c], and Shannon S. Glaser[a,b,c]

[a]*Division of Gastroenterology, Department of Medicine, Scott & White Healthcare and Texas A&M Health Science Center, Temple, TX,* [b]*Scott & White Digestive Disease Research Center, Temple, Texas,,* [c]*Research, Central Texas Veterans Health Care System, Temple, Texas*

26.1 LIVER STRUCTURE AND FUNCTION

To better understand the biology of hepatic stem cells, it is important to understand the liver architecture (Figure 26.1). Hepatocytes are organized into plates that are one-cell thick in mammals. These plates are separated from one another by large capillary spaces (i.e., sinusoids). Sinusoids are characterized by discontinuous, fenestrated endothelial cells with no basement membrane [1]. Sinusoids facilitate the interaction between blood and hepatocytes. These endothelial cells are separated from hepatocytes by the space of Disse, which drains lymph into lymphatics within portal tracts [2]. Hepatic plates are arranged into lobules that are considered the functional units of the liver [3]. Portal areas, that are located along the lobule perimeter, consist of a small portal vein, hepatic artery, and bile ducts. Blood enters the liver from both the hepatic artery and the portal vein and it flows through the sinusoids toward the central vein that is located in the middle of the lobule. The central veins converge to form two hepatic veins that drain the blood to the inferior vena cava. Bile canaliculus is formed by the apical faces of adjacent hepatocytes [4]. Bile is produced by hepatocytes and then is secreted into the bile canaliculus [4]. The latter drains via the canal of Hering into the bile ducts, which in turn drains into hepatic ducts that carry bile to the gut [5].

Hepatocytes are involved in the regulation of various metabolic and biochemical functions. The liver is responsible for the synthesis of serum proteins, metabolism of carbohydrates, and lipids [6]. Also, the liver plays a vital role as a detoxifying system. Remarkably, the liver has the capacity to regenerate after injury and to adjust its size to match its host. Partial hepatectomy is a procedure where lobes are removed from the liver [7,8]. After such a procedure, all populations of cells within the liver, including hepatocytes, cholangiocytes, and endothelial

Regenerative Medicine Applications in Organ Transplantation.

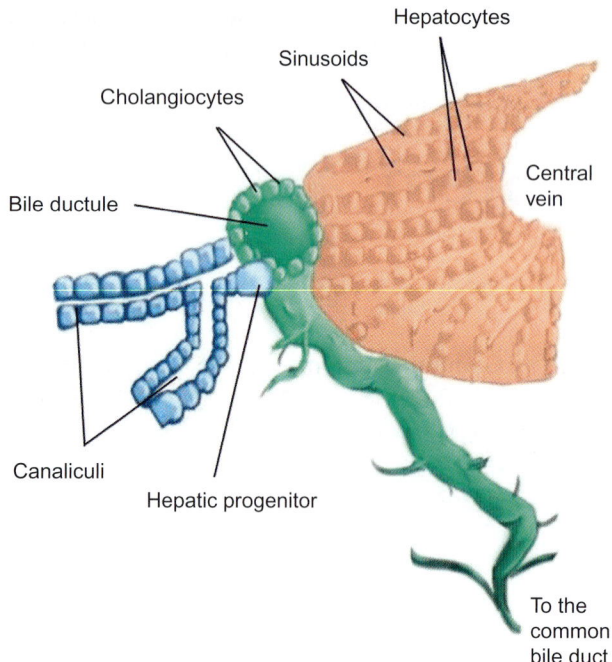

FIGURE 26.1 General structures and cells in the liver. Two major structures involved in liver physiology and damage are hepatocytic areas and bile ducts. Hepatocytes are arranged in plates that run toward the central vein (full circle not shown). The liver sinusoids facilitate the interaction between blood and hepatocytes and run into the central vein. Cholangiocytes comprise the epithelial barrier surrounding the bile duct. Canaliculi form the bile which runs into bile ductules which become bile ducts and finally join the common bile duct which empties into the duodenum. HPCs are located near the junction of the canaliculi and bile ductules.

cells proliferate to restore liver mass [8]. Cellular proliferation starts in the periportal area where proliferating hepatocytes initially form clumps [7]. These clumps later are transformed into the classical plates. Hepatocytes are highly differentiated cells but they can act as functional stem cells in the liver under pathological circumstances. After liver injury, hepatocytes regenerate to restore parenchymal liver mass. Reconstitution of the entire liver mass is completed within 5−7 days in rodents [9]. Using classic thymidine labeling studies, it has been shown that hepatocytes in the remaining liver undergo mitosis to restore the original cell number in 3−4 days [10,11]. In the case of persistent liver injury, the contribution of other cells such as hepatic progenitor cells (HPCs) and oval cells (OCs) becomes apparent [12,13]. HPCs are bipotent progenitor cells, which can fully differentiate into hepatocytes and cholangiocytes [14]. As shown by other studies, OCs can differentiate into both hepatocytes and cholangiocytes when hepatocytes are unable to sustain a proliferative response to liver injury [15,16]. Sources of exogenous stem/progenitor cells that are involved in liver injury include embryonic stem cells

(ESCs), bone marrow (BM) or fat-derived mesenchymal stem cells (MSCs), fetal stem cells, and endothelial progenitor cells [17−21]. In this chapter, we will review the liver regeneration in normal and damaged liver. We will also review the latest literature about progenitor cell isolation, identification, and their role in cell therapy.

26.2 ANIMAL MODELS OF LIVER REGENERATION

Several models (e.g., partial liver resection and acute CCl_4 administration) [8,22] have been used to evaluate the regenerative capacity of liver cells such as hepatocytes, cholangiocytes, and endothelial cells. Also, transplantation of hepatic cells into damaged livers has been used in rodents to determine the role and expansion of hepatocytes, endothelial cells, and cholangiocytes during liver regeneration to identify and characterize putative endothelial, hepatocyte, and cholangiocyte progenitors [23]. The activation of HPCs has also been achieved by retrosine treatment followed by partial hepatectomy in mice [24]. With regard to the models of liver regeneration, a study in hepatectomized rats has shown that segmental cholangitis (SC, induced by lipopolysaccharide or LPS) impairs the regeneration ability of the contralateral remnant liver [25]. This study suggests that liver resection should not be performed in patients with SC even if it occurs in the part of the liver to be resected [25].

A number of studies in models of liver repair have shown that mature hepatocytes and liver progenitor cells regulate the homeostasis of the liver after hepatic injury [26]. These experimental models include partial hepatectomy, acute injury by CCl_4, D-galactosamine (GalN), and N-nitrosomorpholine (NNM), and chemical hepatocarcinogenesis by feeding NNM at both low and high doses [26]. In the injury models of GalN and NNM, there is regeneration of the liver by proliferation of bile ductules/OCs that likely derive from the canals of Hering [26]. These proliferating bile ductules/OCs showed enhanced expression of fetal genes and alpha-fetoprotein (AFP) synthesis [26]. The study suggests that due to the same embryonic origin of bile ducts and hepatocytes, cholangiocytes and OCs have a defined role in liver regeneration as a transit and amplification compartment [26]. Human multipotent stem/progenitor cells in the biliary epithelium can differentiate into hepatocytes, cholangiocytes, and pancreatic islets [27].

A recent study has developed an important pig model of regeneration for extrahepatic bile ducts [28]. This model implies the use of the human basic fibroblast growth factor (bFGF) fused with a collagen-binding domain (CBD), a biomaterial that promotes extrahepatic bile duct regeneration at the injury site without causing structure or hepatic dysfunctions [28]. Recent evidence supports the hypothesis that: (i) progenitor cells exist in extrahepatic bile ducts and (ii) peribiliary glands can act as a local progenitor cell niche in human extrahepatic bile ducts [29].

26.3 REGENERATION OF LIVER CELL SUBPOPULATIONS

26.3.1 Hepatocytes

Hepatocytes are the major cells that replicate in response to partial hepatectomy. Regarding liver regeneration, inhibition of hemoxygenase-1 (HO-1) has been shown to improve survival after liver resection in cholestatic rats [30]. The study suggests that overexpression of HO-1 may be detrimental for liver resection during acute cholestasis [30]. Another study has shown that prospero-related homeobox 1 (Prox1) is a stable hepatocyte marker during liver development, injury, and regeneration that is absent from "OCs" [31]. Further support for the oval-cell differentiation into hepatocytes (when regeneration of parenchymal cells is hampered) in the acetylaminofluorene-treated regenerating rat liver has also been suggested by a study from Golding et al. [32]. This study supports the notion that reactive bile ductules represent an adaptive response of the liver to replenish damaged hepatocytes [32]. A recent study has demonstrated that biliary and progenitor marker epithelial cell adhesion molecule can be used to detect "regenerative clusters" of mixed cholangiocyte—hepatocyte differentiation [33]. The study has shown that the expression of epidermal growth factor receptor occurs in cholangiocytes, progenitor cells, and hepatocytes, whereas the activation of epidermal growth factor receptor is restrictive to regenerative cluster of hepatocytes.

With regard to the mechanisms of hepatocyte regeneration, a study has demonstrated the role and activation of Notch-1 and its ligand Jagged-1 in hepatocyte regeneration following 70% partial hepatectomy [34]. Since the authors have also shown that Notch and Jagged-1 are expressed not only in hepatocytes but also cholangiocytes and endothelial cells [34], the study suggests that the Notch/Jagged-1 signaling may regulate the regeneration of other hepatic cells including cholangiocytes. Also, in a murine model of liver injury, β-catenin-positive hepatocytes are able to repopulate the livers of β-catenin KO mice [35].

A recent study has shown that macrophage-derived Wnt opposes Notch signaling to determine the cell specification of HPCs during liver damage [36]. Specifically, the study has shown that during cholangiocyte regeneration the expression of Jagged-1 by myofibroblasts stimulates Notch signaling in HPCs and their cell specification to bile ducts [36]. On the contrary, during hepatocyte regeneration, there is increased expression of canonical Wnt signaling in nearby HPCs, sustaining the expression of Numb within HPCs and their specification to hepatocytes [36]. Also, in transgenic mice with conditional overexpression of the cell-cycle inhibitor p16 (INK4a), there is faster differentiation of hepatocytes and activation of OCs already in postnatal mice without negative effects on liver function [37].

26.3.2 Cholangiocytes

A number of studies have demonstrated that cholangiocytes participate in the regrowth of the biliary epithelium following partial hepatectomy [8]. Indeed, we have shown that following 70% hepatectomy both small and large cholangiocytes regenerate as early as 3 h and participate in the regrowth of the biliary epithelium by sustained biliary growth up to 28 days [8]. The enhanced regenerative capacity of cholangiocytes following partial hepatectomy was associated with increased secretin-receptor expression and secretin-stimulated biliary secretion [8], functional markers of biliary proliferation/regeneration [5,8,22,38].

Consistent with remodeling of the biliary epithelium during liver regeneration, a recent study has shown increased expression of colony-stimulating factors (GM-CSFs) and stem cell factors (SCFs) in cholangiocytes until 7 days after partial hepatectomy [39]. The study has also demonstrated significant increases of key remodeling molecules including S100 calcium-binding protein A4 (S100A4) and miR-181b, after SCF plus GM-CSF administration in small cholangiocytes [39]. Also, small cholangiocytes (lining small bile ducts) [40] secrete high levels of soluble and bound SCFs and GM-CSFs in response to transforming growth factor-beta (TGF-β). The levels of MMP-2, MMP-9, and miR-181b were upregulated in isolated cholangiocytes after 70% partial hepatectomy [39]. The data suggests that altered expression of SCF + GM-CSF after partial hepatectomy may be important in biliary remodeling (e.g., in posttransplanted conditions). Physiological and differential expression of SCF and SDF-1 has been described in several murine models of liver injury and regeneration induced by tyrosinaemia, 3,5-diethoxycarbonyl-1,4-dihydrocollidine (DDC) or liver irradiation [41]. Furthermore, granulocyte colony-stimulating factor (G-CSF) administration after partial orthotopic liver transplantation (PLTx) stimulates liver regeneration of partial graft, partly by its mobilizing hematopoietic stem cells (HSCs) into the injured liver to differentiate into hepatocytes through hepatic engraftment of OCs [42].

Regarding the mechanisms of liver regeneration, a study evaluated the effects of hyperbaric oxygenation (HBO) on regeneration of bile ducts in hepatectomized rats. The study demonstrates that HBO promoted hepatocyte and cholangiocyte proliferation by reducing *c-Met* mRNA expression [43]. Recently, fibroblast growth factor 15 (FGF15) has been identified as a novel mediator of liver regeneration since FGF15 was shown to: (i) stimulate proliferation of both hepatocytes and cholangiocytes and (ii) prevent liver injury during liver regeneration [44]. Cholangiocyte regeneration was also prevented by rapamycin by blockage of interleukin-6 (IL-6)/STAT3 signal pathway after liver transplantation [45]. With regard to the mechanisms of hepatobiliary regeneration, a recent study has shown that Foxl1 promotes liver repair following cholestatic injury in mice [46]. Specifically, the study shows that in *Foxl1*(−/−) livers there were impaired biliary and hepatocyte proliferation, and reduced bile duct mass. Wnt3a and Wnt7b expression was decreased in the livers of *Foxl1*(−/−) mice along with reduced expression of the beta-catenin target gene Cyclin D1 in *Foxl1*(−/−) cholangiocytes [46].

Regeneration of sinusoidal endothelial cells, hepatocytes, and cholangiocytes has been shown in rat models following irradiation and partial hepatectomy. Specifically, in this study, the livers of dipeptidyl peptidase IV (DPPIV)-deficient rats were preconditioned by irradiation followed by two-third resection of the right liver lobules followed by a one-third partial hepatectomy of the untreated lobule. DPPIV-positive liver cells (nonparenchymal cells containing liver sinusoidal endothelial cells, cholangiocytes, and hepatocytes) were transplanted via the spleen into the recipient livers [47]. The extent of donor cell engraftment and growth was studied over a long-term interval of 16 weeks after transplantation and demonstrated liver cell repopulation by the donor hepatocytes, sinusoidal endothelial cells, and cholangiocytes [47]. Choline-deficient diet has been shown to activate not only OCs but also regeneration of sinusoidal endothelial cells, hepatic stellate, and Kupffer cells [48].

In rats treated with 2-acetylaminofluorene and subjected to 70% partial hepatectomy (2-AAF/PH model), concomitant with reduced hepatocyte and cholangiocyte proliferation, there was activation of hepatic stem cells and/or small ductules [49] or HPCs [50]. Specifically, the study from Wang et al. has shown that the regenerating gene *(Reg)* I (a regenerative/proliferative factor for pancreatic islet cells) was expressed in proliferating bile ductules (colocalized with CK-19, OV6, and AFP) and increased during regeneration [50].

The chemokine SDF-1 and its receptor, CXCR4, are important regulators of cell function in a number of systems. A recent study has shown [51] that cholangiocytes and HPCs synthesize the chemokine, SDF-1, and express CXCR4, an axis that may be important for HPC growth/differentiation during damage-induced liver regeneration. The findings were obtained in *MxCre CXCR4* (f/null) mice that were more susceptible to severe chronic liver damage [51].

A recent study in mice has shown that epiplakin1 *(Eppk1)* is expressed in cholangiocytes in normal liver and adult progenitor cells in the injured liver [52]. During liver injury induced by a choline-deficient ethionine-supplemented diet, there was a significant increase in the number of EPPK1-positive cells [52].

A recent study also showed that spermatogenic immunoglobulin superfamily (SgIGSF) is a cell adhesion molecule that mediates the development of bile ductules [53]. In this model of liver regeneration induced by combined 2-acetylaminofluorene/partial hepatectomy model, SgIGSF was detected exclusively in "OCs" that aligned in ductal and trabecular patterns 2 weeks after partial hepatectomy [53]. Furthermore, in the regenerative model of 2-acetylaminofuorene/partial hepatectomy, there was increased expression of the human gankyrin gene product (p28GANK) in OCs and enhanced oval cell-cycle progression [54]. In support of the activation of "OCs" and their role in liver regeneration, ductular damage induced by methylene dianiline (in a rodent model of 2-acetylaminofluorene + CCl_4 or partial hepatectomy) has been shown to inhibit "oval-cell" activation and their regenerative activity [55]. Among the several players regulating biliary regeneration, a study has shown that bile ducts and portal and central veins are major producers of tumor necrosis factor-alpha in the regenerating livers of both normal and Kupffer cell-depleted rats [56]. Another study has shown that bombesin and neurotensin display antiproliferative effects on OCs and enhances the regenerative response of cholestatic rat livers [57].

26.3.3 Vascular Cells

During liver injury, BM-derived liver sinusoidal endothelial progenitor cells (BM-SPCs) repopulate the sinusoids as liver sinusoidal endothelial cells. Following liver resection, BM-SPCs secrete hepatocyte growth factor, thus playing a key role in liver regeneration. Hepatic vascular endothelial growth factor (VEGF) has been shown to favor the recruitment of rat liver sinusoidal endothelial cell progenitor cells (LSECs) 24 h after dimethylnitrosamine-induced liver injury [58]. Knockdown of hepatic VEGF prevented dimethylnitrosamine-induced proliferation of BM-SPCs, reduced their engraftment, prevented formation of fenestration after engraftment as LSECs, and exacerbated dimethylnitrosamine liver injury [58].

26.3.4 Hepatic Stellate Cells

Rat pancreatic stellate cells (PSCs), which may contribute to pancreatic fibrosis, have stem cell features [59]. The study demonstrated that transplantation of culture-activated PSCs from enhanced green fluorescent protein-expressing rats into wild-type rats after liver resection in the presence of 2-acetylaminofluorene demonstrated that PSCs participate in the regrowth of large areas of the host liver through differentiation into parenchymal cells and cholangiocytes [59].

Blockade of the renin—angiotensin system (by the use of the angiotensin-converting enzyme (ACE) inhibitor, captopril) improves the early stages of liver regeneration (by increasing the regeneration of hepatocytes and HSCs) and liver function [60].

26.4 EXTRAHEPATIC LIVER CELL PROGENITORS

Extrahepatic stem/progenitor cells have been shown to play a role in the regeneration and repair of the liver. Several extrahepatic sources have been identified as hepatic progenitors, which can differentiate into hepatic lineages both *in vitro* and *in vivo*[61]. BM, adipose tissue, umbilical cord, and peripheral blood are the most common sources of extra-HPCs [61]. In addition, the hepatic differentiation potentials of ESCs, amniotic fluid—derived stem cells, and adipose tissue—derived stem cells have also been evaluated [19,62—67]. Controversy remains regarding the capability of these progenitors of extrahepatic origin to differentiate into hepatic cell types and take up residence with liver cells of the recipient when transplanted.

Self-renewal capable stem cells are able to differentiate into specialized cell types. These cells are able to differentiate into hepatocytes-like cells [68]. There are a number of stem cells that have been identified and act as a source of liver progenitors including the BM-derived MSCs, ESCs, HSCs, fetal stem cells, and umbilical or adult stem cells [69,70]. Identification of adult liver stem/progenitor cells has not been clearly delineated, which is a cause of controversy regarding the existence and significance of these cells. Kanazawa et al. used different mouse liver injury models (an albumin-urokinase transgenic mouse and a hepatitis B transgenic mouse) to determine the ability of bone marrow cells (BMCs) to form hepatocytes in models of liver injury [71]. The findings indicated that there is little or no contribution of BMCs to the replacement of injured livers in these models [71]. On the other hand, other studies have shown that BM-derived cells are capable of transdifferentiation into hepatocytes while the BM-recipient animals were treated with CCl$_4$ to induce liver injury [70].

BM transplantation has a long-term successful history in the treatment of hematopoietic malignancies [72]. Another study has shown that BMCs can promote the repair of nonhematopoietic tissues such as skeletal muscle regeneration under certain conditions [73]. Nevertheless, several studies attempted to demonstrate that BMC-

derived cells have much more efficiency to adopt varied lineage characteristics as compared to populations enriched in HSCs. These multipotent adult progenitor cells (MAPCs) are nonhematopoietic mesenchymal cells isolated by long-term culture of adherent marrow-derived cells from humans or rodents. MAPCs are able to differentiate into muscle, cartilage, bone, neurons, and hepatocytes as well [74]. Verfaillie et al. have shown that these cells are capable of differentiating into binucleated hepatocyte-like cells that display morphological and functional characteristics such as urea synthesis, albumin secretion, and cytochrome p450 induction [75]. This finding demonstrates that MAPCs display many of the basic criteria of liver progenitors, which support the concept that MAPCs have a potential role in the hepatocyte engraftment in injured liver.

MSCs from different tissue sources have shown similar characteristics and display the ability to differentiate into cells of mesoderm lineage, like adipocytes, chondrocytes, and osteoblast [18,21,74—79]. Basically, MSCs referred to as a type of multipotent adult stem cells, which generally originate from BM. There is a beneficial site to deal with BM-derived MSCs, because of its immunosuppressive activities, which is assumed to be helpful in treating patients having certain liver disorders such as autoimmune diseases, or patients needing immune modulation [80]. Moreover, other studies have shown that transplanted BM-MSCs can not only differentiate into gastrointestinal tract multiple lineage cells but also differentiate into multiple lineage cells including brain, lung, renal, skin, and pancreatic-β cells [75,77,78]. It has also been shown that cells, produced by the cellular differentiation of primary BM-derived stem cells are positive for hepatocyte-specific markers. Moreover, these cells display hepatocyte-specific bioactivities, including urea production, albumin secretion, and glycogen storage [18,74]. It has also been assumed that transdifferentiation of MSCs may produce soluble factors, which have a direct positive impact on the regeneration and survival of hepatocellular tissue in recipient liver after MSCs transplantation. In addition to the hepatocellular protection mechanism, *in vitro* studies also showed that MSC-CM (MSC-conditioned media) has direct influence on cultured hepatocytes by anti-apoptotic and pro-mitotic properties [81]. *In vivo*, in mice with 70% partial hepatectomy, administration of concentrated MSC-CM upregulated hepatic gene expression of cytokines and growth factors relevant for cell proliferation, angiogenesis, and anti-inflammatory responses [81]. Consistent with the previous study, evidence indicates MSC-CM therapy provides trophic support to the injured liver by inhibiting hepatocellular death and stimulating regeneration in rats after D-galactosamine-induced fulminant hepatic failure [82].

Several studies have been done on MSCs collected from different sources to evaluate their capability to differentiate toward hepatic lineages *in vitro* and *in vivo*, such as BM [83,84], skin [85], umbilical cord [21,86], and adipose tissues [19,62–66]. Different types of markers have been identified to recognize different stem cells. Pittenger et al. identified the classical MSC markers CD29, CD44, CD71, CD90, CD106, CD120a, and CD124. Expression of these MSC markers is not tissue specific [76]. CD13, CD29, CD44, and CD90 are the mesenchymal cell-specific markers that have been used to detect culture selected clonal BM-derived MSCs expression, whereas CD3, CD14, CD34, and CD45 known as hematopoietic markers have been used to detect HSCs [83]. HSCs have been identified as BM stem cells, which showed highly dense population *in vivo* and *in vitro* in response to liver injury [87]. Jangs et al. demonstrated that these cells can be transformed into functional hepatocytes by nonfusion mechanism and thus contribute to the repair mechanism of injured liver [88].

26.4.1 Embryonic Stem Cells

The inner cell mass of the mammalian blastocyst gives rise to ESCs. ESCs are the best example of pluripotent cells having the capacity to differentiate into a variety of cell lineages *in vitro*. This capability helps the ESCs to act as an important source of HPCs, which can generate new hepatoblasts, hepatocytes, and cholangiocytes by cellular differentiation during liver pathology. Hepatocyte-like cells, which are originated from ESCs, are morphologically, biologically, and functionally similar to that of original liver hepatocytes [89]. It has also been shown that the ESC-derived hepatocytes improve liver functions and ameliorate liver injury in mice [90].

26.4.2 Adipose Tissue–Derived Stem Cells

Adipose tissue–derived stem cells have also been shown to differentiate into hepatic lineages [19,62–67]. Adipose tissue–derived MSCs are positive for these CD9, CD13, CD29, CD44, CD49d, CD54, CD73, CD90, CD105, CD146, CD166, osteopontin, and osteonectin cell surface markers, and are not responsive to CD45, CD34, and CD31 markers, which are known as hematopoietic and endothelial markers. Zuk et al. demonstrated that ASCs express CD13, CD29, CD44, CD71, CD90, CD105/SH2, SH3, and STRO-1. This study showed that there is no expression of the hematopoietic lineage markers CD14, CD16, CD31, CD34, CD45, CD 56, CD 61, CD 62E, CD 104, and CD106 [91].

26.4.3 Amniotic Fluid–Derived Cells

Several studies have been performed on amniotic fluid–derived progenitor cells as a source of different

organ regeneration. New findings regarding amniotic fluid might give a better understanding of the underlying mechanism of the regeneration of organ by using heterogeneous amniotic fluid–derived cells, [92] and will also help to increase potential usage of amniotic fluid–derived cells in the purpose of regenerative medicine. Since the amniotic fluid, which is contained inside the amniotic cavity, is directly dealing with the fetus, it has been considered as a relevant source for progenitor cells (undifferentiated or partially differentiated) of any organ [92]. It acts as a protective mechanism for the developing embryo later during fetal development. The volume of amniotic fluid in the amniotic cavity increases according to the gestational period. However, the cellular composition of amniotic fluid is still unclear. In addition to the role of amniotic fluid in fetal development, previous investigation has suggested that in animals, the presence of molecular composition and nutritive substances in amniotic fluid has direct influence on cell proliferation, and differentiation of various cell types like epithelial and mucosal cells [93]. Specifically, it has been stated that in the fetus, amniotic fluid is considered as a secure and dependable source for identifying genetic and congenital disease.

In the past, due to the lack of advanced techniques and proper investigation, the amniotic fluid was not available as a diagnostic tool for detecting congenital or genetic diseases such as fetal distress, hemolytic disease and fetal maturity [94], neural tube defects, and lung maturity [93]. But over time the advancement of technology helped to correlate the relation between amniotic fluid and diagnosis of different conditions such as preterm labor, infective process, and embryo disease [92], suggesting that the identification of genetic or congenital disease or to diagnose the status of health of the fetus could be easier based on the amniotic fluid corresponding to that each condition.

From the previous studies on amniotic fluid, it could be summarized that different types of cells are observed to be present in amniotic fluid during the different stages of fetal development. This suggests that in organs of the developing fetus, such as lung and gastrointestinal tract, the presence of different types of cells in the fluid also can be explained by the direct connection that prevails within amniotic fluid and compartments of the developing fetus. It has also been stated that cells shedding from the forming kidney or coming off from the fetal skin may contribute significantly to cellular composition in amniotic fluid [92]. More reports have identified the presence of mature cell lines derived from all three germ layers [95,96]. In humans, at 12th week of gestation, the presence of mesenchymal and hematopoietic progenitor cells have been observed [97], and there are some reports that cells from specific tissue types including brain, heart, and pancreas are able to express proteins as well as various genetic markers [98–101].

Furthermore, it has been published that the unfractionated mesenchymal stem cells from human samples collected between 20 and 37 weeks of gestation have been isolated successfully and have been expanded, confirming the occurrence of a multipotent mesenchymal cell type over the development of gestation [102]. In 2007, De Coppi et al. first reported the complete characterization of amniotic fluid pluripotent cell population [103]. A surface marker like c-kit and also some transcription factors distinctive of ESC such us OCT-4 and SSEA-4 are used for identifying, and characterizing stem cells of mesenchymal origin. Some hyaluronan receptors, which are also known as surface markers such as CD29, CD44, and also including CD73, CD90, and CD105 surface markers, have shown positive response to the MSCs as well as to the neural stem cells [103]. Nevertheless, a clear idea about the cellular composition of amniotic fluid is still uncertain. For the first time, Da Sacco et al. in 2010 were able to state that the cellular composition of amniotic fluid changes over the gestational period of time [101]. To evaluate the optimal capability of amniotic fluid—derived cells for hepatic lineage differentiation, a small number of works have been performed. It has been reported that under the same condition, second trimester amniotic fluid (AFSC) showed better response to hepatic lineage differentiation compared to that of BM-MSC [104]. In addition, it has also been stated that adult- and fetal-derived cells, as well as AFSC has equal potentiality of differentiation into hepatic lineage for liver regeneration [105].

26.5 ISOLATION AND IDENTIFICATION OF EXTRA-HPCs

For the last two decades, the concept of using stem cells for organ regeneration has been considered as a potential clinical approach. To date, a large number of studies have been performed based on different scientific points of view using stem cells to confirm their different lineage abilities. The use of progenitors in terms of regenerative tools has been considered in several studies from a mechanistic perspective. Due to having inconsistent results, controversy still exists [106]. There are a number of tissues that have been identified as a source of adult stem cells and MSCs. Physiologically, tissue-specific adult stem cells replace new cells during normal cell turnover or during the organ injury [107–109]. Predominantly, the presence of stem cells have been observed in epidermis, hair, HSCs, or in the gastrointestinal tract during their normal cellular turnover, indicating that stem cells play an important role in the process of cellular turnover in the organ. But the underlying mechanism of this process is not well known yet [107,110,111]. However, the isolation of these progenitors from native tissue is very important,

and is a critical step prior to transplantation of these progenitors to the recipient organ. In this section, we are going to discuss the isolation procedure of extrahepatic progenitors.

BM, adipose tissue, periodontal ligament, tendon, synovial membranes, trabecular bone, embryonic tissues, the nervous system, skin, periosteum, and muscle are considered as vital sources of isolating MSCs. MSCs are shown to have a great level of plasticity [112,113], multipotent differentiation, and also have extensive proliferative capacity. MSCs also have a common surface marker profile with other cell types. Due to the lack of distinctive surface marker for MSCs till now general approaches are used to isolate all kinds of MSCs, which has been described earlier [107]. Generally, Dulbecco's Modified Eagle Medium (DMEM) is used to dissolve collagenase. One hour and 37°C is considered, as optimal digestion criterion for isolating stem cells. Stem cell isolation procedure needs to be performed as soon as possible following euthanasia. Temperature of culture medium should not go below room temperature [107].

Compared to other MSCs, the isolation procedure of adipose-derived stem cells (ASCs) is much easier and convenient to get large amount of cells [114]. In 2001, Zuk et al. for the first time was able to isolate ASCs[91]. ASCs are able to differentiate into ectodermal, endodermal, and the mesodermal lineage [115]. Liposuction aspirates or excised fat are being performed to isolate ASCs. A total of 100–200 mL of ASCs can be taken following local anesthesia. It has also been observed that adipose tissues have higher yield (approximately 5000 stem cells/1 g of adipose tissue) capacity for the stem cell compared to the BM-derived MSCs (100–1000 MSCs/ mL of marrow) from BM. Usually, it has been observed that, approximately 2% of nucleated cells are present in processed lipoaspirate as yield of ASCs [116]. This can suggest that recently these ASCs are becoming more attractive in the field of regenerative medicine as a therapeutic tool due to its easy isolation procedure and availability of large quantity [106].

The isolation procedure of BM-MSCs differs from the other MSCs because of having extracellular matrix deposition in BM. Gentle mechanical disruption procedure is followed by repeated pipetting to create a stromal and hematopoietic cells suspension. As soon as BM-MSCs adhere to the culture plate, nonadherent hematopoietic cells need to be removed by changing the media. Since MSCs do not have unique cell surface marker, it is delicate to isolate pure stem cells from primary isolation, which contains heterogeneous BM-MSC population [110].

It is well known that mammalian skin is a highly accessible tissue source and MSCs are present in the dermal layer of skin. In 2001, Toma et al. isolated multipotent adult stem cells from juvenile and adult rodent skin. These multipotent adult stem cells show positive response to nestin and fibronectin [117]. A recent study stated that multipotent stem cells isolated from epidermal layer of human skin are able to differentiate into adipocytes, osteoblasts, chondrocytes, neurons, hepatocytes, and insulin-producing pancreatic cells [118]. These isolated skin stem cells showed positive response to CD105, CD90, CD73, CD29, CD13, and CD44 lineage markers but not showed positive response to endothelial and hematopoietic lineage markers such as CD45, CD34, CD31, CD14, and HLA DR [118]. Later on, a study stated that newborn dermis originated MSCs also showed positive response to CD59, vascular cell adhesion molecule-1 (VCAM-1), and intercellular adhesion molecule-1 (ICAM-1) [119]. Other reports have stated that SSCs expressed other surface markers like as CD49, CD166, SH2, SH4, EGFR, PDGFRA, CD271, Stro-1, CD71, CD133, and CD166 [120–122].

Amniotic fluid—derived pluripotent and multipotent stem cells have great opportunity to be used as attractive regenerative tools due to their convenient isolation procedure. Moreover, due to having a specified phenotype compared to the pluripotent stem cells, multipotent stem cells have been more considered as an attractive progenitor. Generally these cells are isolated from the collected amniotic fluid and by the procedure amniotic fluid is collected known as amniocentesis. Prior to amniocentesis, the most important thing is to perform ultrasound, to confirm overall condition of fetus. Fetal viability, gestational age, number of fetuses, placental location, volume, fetal anatomical survey, and uterine cavity abnormalities are considered as important factors to be evaluated by performing ultrasound preceding amniocentesis. Another important thing is to evaluate the best needle-insertion site. To collect the amniotic liquid a 20 cc syringe is used. The cells are seeded with specific culture media after collection and then culture should be expanded depending on the time of cell attachment and growth.

Umbilical cord is also a good source of Wharton's Jelly stem cells (WJ-MSCs). UC-MSCs can differentiate into endothelial cells, adipogenic, osteogenic, chondrogenic, neurogenic lineages [123], insulin-producing cells [124], and hepatocyte-like cells as well [125]. WJ-MSCs have a higher expression of undifferentiated human ESCs (hESCs) markers like NANOG, DNMT3B, and GABRB3 [126]. They also have positive expression for MSCs markers (CD105, CD73, and CD90) and negative for CD45, CD34, CD14, CD19, and HLA-DR [123]. Umbilical cord blood (UCB) has also been taken an excellent alternative source of HSCs for transplantation [127]. Generally, UCB has been considered as more primitive compared to other types of MSCs that can be obtained with no ethical consideration [106]. Up to 63% of 59 low-volume UCB units cells were isolated with a characteristic mesenchymal morphology and immune phenotype (MSC-like cells) under optimized isolation and culture conditions. It has been observed that these cells proliferated extensively with at least 20 population doublings within eight passages. The most important part to isolate MSC-like cells from UCB has been considered a time from collection to isolation of less than 15 h, a net volume of more than 33 mL, and an MNC count of more than 1×108 MNCs [128]. Fundamental preclinical studies evaluated a higher percentage of CD34 + CD38 − cells are present in UCB that suggested more abundance of primitive progenitors in neonatal blood [129]. Under suitable induction conditions, UCB cells can differentiate into bone, cartilage, and fat. Unexpectedly, these cells were also able to differentiate into neuroglial- and hepatocyte-like cells under suitable induction conditions. Thus it has been suggested that these cells may act more than MSCs as evidenced by their ability to differentiate into cell types of all three germ layers [130].

26.6 TRANSPLANTABLE CELLS THAT REPOPULATE THE LIVER—POTENTIAL FOR CELL THERAPY

Over the past two decades, researchers have sought to identify liver stem cells that can repopulate liver mass through transplantation [131]. Cells obtained from deceased donor livers and/or other tissues have been the targets of investigation since donor availability is limited. While it is held that these therapies allow for the replacement of hepatocytes only [131], new methods such as the use of human hepatocyte growth factor (hHGF) on OCs have implicated repopulation of the biliary system as well [132]. Since many animal models have used DNA-damaging agents including chemicals and γ-irradiation to damage the liver a selective advantage has been given to the transplanted cells over endogenous hepatocytes. Various cell types are under investigation for their ability to regenerate the liver including hepatocytes, fetal hepatoblasts, OCs, pancreatic progenitors, MSCs, and BM-derived progenitors (Figure 26.2). Each is summarized below.

26.6.1 Hepatocytes

The concept that hepatocytes can repopulate the liver stems from the understanding that differentiated hepatocytes have an extremely high potential for proliferation. Unlike the hematopoietic system, hepatocytes have been demonstrated to undergo >30 cell divisions without any loss of function [133]. Since transplantation of polyploid hepatocytes identified by fluorescence-activated cell sorting resulted in extensive proliferation, it was concluded that mature, polyploid adult hepatocytes have high regenerative potential [134]. This regenerative potential is also

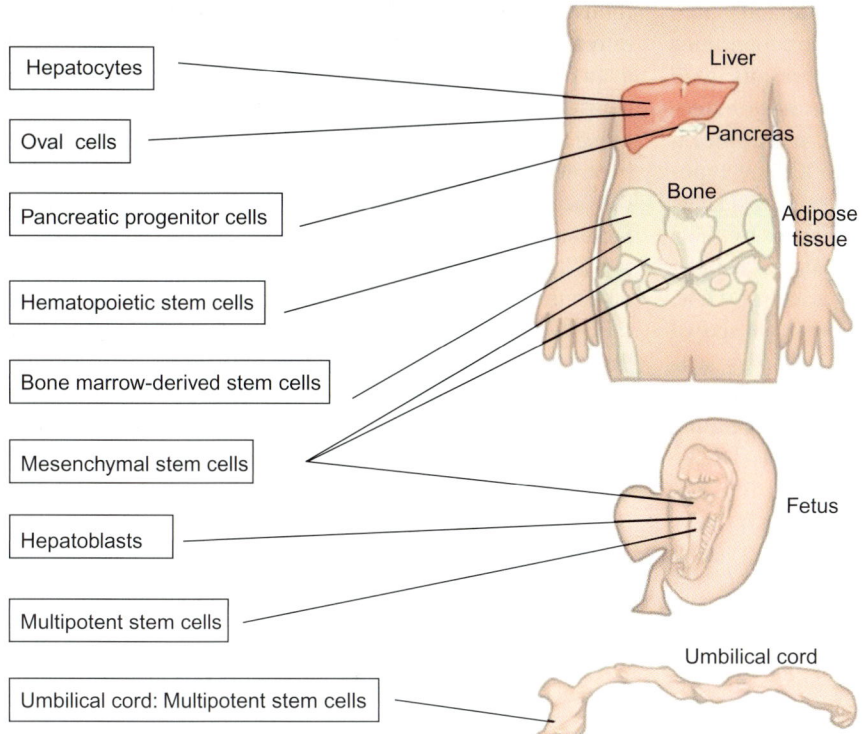

Hepatocytes

Oval cells

Pancreatic progenitor cells

Hematopoietic stem cells

Bone marrow-derived stem cells

Mesenchymal stem cells

Hepatoblasts

Multipotent stem cells

Umbilical cord: Multipotent stem cells

FIGURE 26.2 Transplantable cell sources. Transplantable cells capable of regenerating the liver can be obtained from these organs including (by organ): liver—hepatocytes and OCs; pancrease—pancreatic progenitors; bone—HSCs, BM-derived stem cells, and MSCs; fetus—MSCs hepatoblasts and multipotent stem cells; umbilical cord—multipotent stem cells.

evidenced by the clinical finding that spontaneous mutations in hepatocytes may correct the genetic defect in hereditary tyrosinemia [135]. Some studies have demonstrated that hepatocytes are bipotential and are sufficient to differentiate into cholangiocytes in addition to hepatocytes [136]. These results are supported by more recent evidence that hepatocytes undergo dedifferentiation into liver progenitor cells [137]. During this process, mature hepatocytes underwent an intermediate oval-cell-like stage expressing the oval-cell markers OV6, CK7, and GGT [137].

26.6.2 Fetal Hepatoblasts and Multipotent Stem/Progenitor Cells

During embryogenesis, the liver contains hepatoblasts that are capable of differentiating into hepatocytes and biliary epithelia. These cells have been investigated for their transplantable properties in animal models [138–142]. One study demonstrated that these transplantable cells can be separated into subpopulations based on histochemical markers that are either unipotential for hepatocytes, cholangiocytes, or bipotential for both [138]. Hepatoblasts expressing AFP and albumin, but not CK-19 were hepatocytic, hepatoblasts expressing CK-19 only were cholangiocytic, and hepatoblasts expressing AFP and CK-19 were bipotential [138]. At the same time, the cell lineages that fetal hepatoblasts extend to mesenchymal fates outside of

the liver such as cartilage and fat [13,143]. This is promising data since the liver houses a number of important mesenchymal cells such as HSCs and fibroblasts. It has been demonstrated that the regenerative potential of hepatoblasts is enhanced by using retroviruses to increase their proliferation [139]. While this may be effective, care should be taken since retroviruses contain enhancer and promoter sequences that can increase the oncogenic potential of fetal hepatoblasts [144]. Overall, the data indicate that fetal hepatoblast transplantation could be an effective method for liver regeneration. Another group has shown that fetal biliary trees contain multipotent stem/progenitor cells comparable with those the same group has identified in adults [145,146]. These multipotent stem/progenitor cells were expanded and induced *in vitro* to differentiate into liver and pancreatic mature fates [145,146]. They also engrafted and differentiated into mature cells when transplanted *in vivo* [145,146].

26.6.3 Hepatic OCs

When liver damage causes a loss in hepatocytes and cholangiocytes, hepatic oval cells (HOCs) are activated to regenerate the dying/damaged cells [147]. During this process, liver damage causes the release of chemokines, cytokines, and other chemoattractants that activate HOC differentiation and migration [148]. OCs have been tested for their transplantable abilities. It was demonstrated in rat livers that transplanted OCs can differentiate into mature hepatocytes, even without selective conditions [142]. Several others have supported these results with data demonstrating that transplantable HOCs have extensive repopulation capacities [133], and a more recent study has demonstrated that transplanted HOCs are bipotential for hepatocytes and biliary epithelium [132]. HOC migration and/or proliferation can be affected by hormones including somatostatin [149], and human hepatic growth factor [132,150]. Moreover, derivation of HOC from human cord blood has been achieved by usage of cytokines involved in OC activation [151]. It was suggested HOC-based transplantation may be developed as a therapy for treating liver damage [132]. Understanding the processes governing HOC regeneration may have important implications in the transplantation of other liver-regenerating therapies since HOCs may actually be an intermediary stage for repopulation of other transplantable cells including hepatocytes [137].

26.6.4 Pancreatic Progenitors

During embryogenesis the main cells of the liver and pancreas arise from a common endodermal precursor located in the ventral foregut. Due to this close relationship in embryogenesis, it is believed that a common hepatopancreatic precursor may exist in the adult liver and/or pancreas. This hypothesis is strengthened by the finding that during pancreatic cancer, pancreatic adenocarcinoma cells display hepatocellular markers indicative of a reversal in cellular differentiation [152]. Moreover, experiments have identified this precursor by exposing rodents to a number of stimuli/growth factors, which caused the production of hepatocytes or hepatocyte-like cells from the pancreas [153–157]. Data like these indicate a therapeutic potential for specific targeting of pancreatic cells into different liver cell fates. These cells, even nonspecifically, have been demonstrated to repopulate the damaged liver [158]. Future clinical research should determine whether pancreatic cells may be used as a therapy for liver failure in humans.

26.6.5 Mesenchymal Stem Cells

MSCs are stem cells originating from the embryonic mesoderm [18]. These cells differentiate in response to stimuli, and have been demonstrated to undergo hepatic differentiation in response to hepatocyte growth factor and oncostatin M [18]. Transplantation of MSCs isolated from human umbilical cords [159], BM [160], and adipose tissue [64] improved liver functions after acute liver damage using carbon tetrachloride. *In vitro* differentiation of MSC into hepatocyte-like cells also improved liver functions such as aspartate aminotransferase and alanine aminotransferase

following acute liver damage [63,160]. The therapeutic effects of MSC on acute liver damage are thought to be due to two reasons. First, transplanted MSCs differentiate into hepatocytes in the liver, and regenerate the number of functional hepatocytes [63,64,159,160]. Second, MSCs have been demonstrated to promote liver repair by secreting soluble factors. For instance, human MSC-CM infused into rats facilitates the recovery of acutely damaged livers [63].

While MSCs have shown potential in animal models of acute liver injury, this is not the case in chronic liver failure. Studies have demonstrated that in chronic liver failure MSCs fail to improve liver function [161], or may even accelerate the course of chronic liver disease [162,163]. For example, both mouse and human MSCs have been demonstrated to promote myofibroblast formation and worsen fibrosis in a mouse model for liver cirrhosis [162,163]. These results make a strong argument against the usage of MSCs for both acute and chronic forms of liver damage and should be taken into account when preparing for clinical trials using MSC.

26.6.6 BM-Derived Progenitors

BMCs have multiple utility for transplantation as they support stem and progenitor cells and may differentiate into a number of cell types. BMCs including MSCs and HSCs have a high potential for differentiation into liver cells. Transplantation using cross-sex BMCs into animals with chemically damaged livers resulted in regenerated hepatocytes and OCs that were donor derived [164]. Similar reports have demonstrated the appearance of donor-derived liver epithelial cells in the regenerating liver [165,166]. It is suggested that HSCs have a particular ability to differentiate into hepatocytes. In a mouse model of hereditary tyrosinemia type I, donor HSCs rescued the mouse and restored its normal liver functions [159]. Three mechanisms have been proposed to explain how BMCs form liver cells. First, BM could harbor endodermal stem cells that specialize in differentiating into hepatocytes and other epithelial cells. Second, a single stem cell progenitor could have multipotent effects allowing it to become both hepatocytes and blood cells. Third, hepatocytes could be derived by the fusion of donor BMCs with host hepatocytes [167]. This third mechanism is thought to explain the majority of BM-derived hepatocytes [131]. It is still being determined to what extent BM transplantation can be used for liver regeneration in humans.

26.7 CONCLUSION AND FUTURE DIRECTIONS

Modulating liver regeneration by manipulating endogenous mechanisms or transplantation is becoming a central goal in managing liver diseases. In this chapter, we have summarized the findings related to liver regeneration regarding the natural regenerative responses during liver injury, the cell lines that have regenerative properties, the isolation and identification of progenitor cells, and the transplantable cells of the liver. The liver houses multiple anatomical compartments that subsist in a delicate balance modulating growth and proliferation. Within these compartments lie cells capable of regenerating part or even entire livers. These cells have stem cell-like capabilities, enabling them to fulfill multiple niches when needed, and have the capacity to proliferate nearly indefinitely. Scientists have evaluated the mechanisms governing these properties, and have begun the miraculous translation into the clinic. By manipulating the conditions during liver regeneration, scientists have unlocked many key mechanisms relating to the control of liver regrowth. Such models *in vivo* include mice, rats, and pig.

The establishment of a stem cell capable of differentiating into any cell within the liver itself is a primary goal in liver-regeneration science. It is necessary that this stem cell be able to transdifferentiate into both cholangiocytes and hepatocytes having the correct anatomical and functional properties. The finding that the liver houses multipotent stem/progenitor cells is an advancement in this area. *In vitro* and *in vivo* studies using hepatocytes and cholangiocytes have demonstrated a remarkable multipotent capacity for both of these cell types. Understanding the signals that regulate their decision to adopt a multipotent fate and differentiate to fill a niche is an important future direction. OCs and HPCs are also an important area of investigation.

Extrahepatic liver cell progenitors are another area of investigation. BM, adipose tissue, umbilical cords, and peripheral blood all contain important stem cells that have regenerative capacity. While the research appears promising, there is still much that remains unknown, and important questions should be addressed such as: what signals govern the differentiation of extrahepatic progenitors to a hepatic fate? What specific disease models can be targeted by extrahepatic progenitors, and can we program progenitors to fulfill a specific niche? Additionally, research into the isolation of extra-HPCs may help in producing cells with more specific regenerative properties. While controversy remains regarding the use of human fetuses for transplantation purposes, stem cells isolated from human umbilical cords are being studied for their potential use in the near future.

While the regenerative potential for individual cells is great, it is important to note that the generation of organs involves interplay between multiple cell types. To this day, the ability to regenerate whole livers or even whole liver compartments (such as the bile duct) is still not met. Single compartments may require multiple different signaling pathways to be produced. Thus, potential studies using the partnering of multiple cells to achieve regeneration is warranted. We suggest studying regeneration using stem cells paired with natural liver cells such as hepatic progenitors/cholangiocytes/hepatocytes.

Until recently, the vast number of potential sources for liver regeneration has been underestimated. Studies involving liver regeneration have focused on identifying the markers for cells capable of a hepatic fate, and the signals that govern their regenerative properties. We suggest more research in the area of mapping the signals that govern regeneration. Finally, as further research refines the conditions involved in these systems, a more translational approach may unlock the potential for intra- and extrahepatic stem cells as a therapy for pathophysiological states of the liver. This will open room for new therapies modulating the regenerative response, and may allow for a better clinical prognosis of terminal liver diseases.

REFERENCES

[1] Wisse E, DeZanger R, Charels K, van der Smissen P, McCuskey R. The liver sieve: considerations concerning the structure and function of endothelial fenestrae, the sinusoidal wall, and the space of Disse. Hepatology 1985;5:683.

[2] Heat T, Lowden S. Pathways of interstitial fluid and lymph flow in the liver acinus of the sheep and mouse. J Anat 1998;192:351–8.

[3] Mall FP. A study of the structural unit of the liver. Am J Anat 1906;:227–308.

[4] Nathanson MH, Boyer JL. Mechanisms and regulation of bile secretion. Hepatology 1991;14:551–66.

[5] Alpini G, Lenzi R, Sarkozi L, Tavoloni N. Biliary physiology in rats with bile ductular cell hyperplasia. Evidence for a secretory function of proliferated bile ductules. J Clin Invest 1988;81:569–78.

[6] Jungermann K, Kietzmann T. Zonation of parenchymal and nonparenchymal metabolism in liver. Annu Rev Nutr 1996;16:179–203.

[7] Lee VM, Cameron RG, Archer MC. Zonal location of compensatory hepatocyte proliferation following chemically induced hepatotoxicity in rats and humans. Toxicol Pathol 1998;26:621–7.

[8] LeSage G, Glaser S, Gubba S, Robertson WE, Phinizy JL, Lasater J, et al. Regrowth of the rat biliary tree after 70% partial hepatectomy is coupled to increased secretin-induced ductal secretion. Gastroenterology 1996;111:1633–44.

[9] Straube RL, Patt HM. Regeneration following partial hepatectomy as a function of age in the mouse. Fed Proc 1961;20:286.

[10] Bucher NL, Swaffield MN. The rate of incorporation of labeled thymidine into the deoxyribonucleic acid of regenerating rat liver in relation to the amount of liver excised. Cancer Res 1964;24:1611–25.

[11] Stocker E, Pfeifer U. On the manner of proliferation of the liver parenchyma after partial hepatectomy. Autoradiography studies using 3H-thymidine. Naturwissenschaften 1965;52:663.

[12] Lowes KN, Croager EJ, Olynyk JK, Abraham LJ, Yeoh GC. Oval cell-mediated liver regeneration: role of cytokines and growth factors. J Gastroenretrol Hepatol 2003;18:4–12.

[13] Suzuki A, Zheng YW, Kaneko S, Onodera M, Fukao K, Nakauchi H, et al. Clonal identification and characterization of self-renewing pluripotent stem cells in the developing liver. J Cell Biol 2002;156:173—84.

[14] Thorgeirsson SS. Hepatic stem cells in liver regeneration. FASEB J 1996;10:1249—56.

[15] Evarts RP, Hu Z, Omori N, Omori M, Marsden ER, Thorgeirsson SS. Precursor-product relationship between oval cells and hepatocytes: comparison between tritiated thymidine and bromodeoxyuridine as tracers. Carcinogenesis 1996;17:2143—51.

[16] Lazaro CA, Rhim JA, Yamada Y, Fausto N. Generation of hepatocytes from oval cell precursors in culture. Cancer Res 1998;58:5514—22.

[17] Chien CC, Yen BL, Lee FK, Lai TH, Chen YC, Chan SH, et al. In vitro differentiation of human placenta-derived multipotent cells into hepatocyte-like cells. Stem Cells 2006;24:1759—68.

[18] Lee KD, Kuo TK, Whang-Peng J, Chung YF, Lin CT, Chou SH, et al. In vitro hepatic differentiation of human mesenchymal stem cells. Hepatology 2004;40:1275—84.

[19] Banas A, Teratani T, Yamamoto Y, Tokuhara M, Takeshita F, Quinn G, et al. Adipose tissue-derived mesenchymal stem cells as a source of human hepatocytes. Hepatology 2007;46:219—28.

[20] Cai J, Zhao Y, Liu Y, Ye F, Song Z, Qin H, et al. Directed differentiation of human embryonic stem cells into functional hepatic cells. Hepatology 2007;45:1229—39.

[21] Campard D, Lysy PA, Najimi M, Sokal EM. Native umbilical cord matrix stem cells express hepatic markers and differentiate into hepatocyte-like cells. Gastroenterology 2008;134:833—48.

[22] LeSage G, Glaser S, Marucci L, Benedetti A, Phinizy JL, Rodgers R, et al. Acute carbon tetrachloride feeding induces damage of large but not small cholangiocytes from BDL rat liver. Am J Physiol Gastrointest Liver Physiol 1999;276:G1289—301.

[23] Brilliant KE, Mills DR, Callanan HM, Hixson DC. Engraftment of syngeneic and allogeneic endothelial cells, hepatocytes and cholangiocytes into partially hepatectomized rats previously treated with mitomycin C. Transplantation 2009;88:486—95.

[24] Li WL, Su J, Yao YC, Tao XR, Yan YB, Yu HY, et al. Isolation and characterization of bipotent liver progenitor cells from adult mouse. Stem Cells 2006;24:322—32.

[25] Watanabe K, Yokoyama Y, Kokuryo T, Kawai K, Kitagawa T, Seki T, et al. Segmental cholangitis impairs hepatic regeneration capacity after partial hepatectomy in rats. HPB 2010;12:664—73.

[26] Kuhlmann WD, Peschke P. Hepatic progenitor cells, stem cells, and AFP expression in models of liver injury. Int J Exp Pathol 2006;87:343—59.

[27] Cardinale V, Wang Y, Carpino G, Cui CB, Gatto M, Rossi M, et al. Multipotent stem/progenitor cells in human biliary tree give rise to hepatocytes, cholangiocytes, and pancreatic islets. Hepatology 2011;54:2159—72.

[28] Li Q, Tao L, Chen B, Ren H, Hou X, Zhou S, et al. Extrahepatic bile duct regeneration in pigs using collagen scaffolds loaded with human collagen-binding bFGF. Biomaterials 2012;33:4298—308.

[29] Sutton ME, op den Dries S, Koster MH, Lisman T, Gouw AS, Porte RJ. Regeneration of human extrahepatic biliary epithelium: the peribiliary glands as progenitor cell compartment. Liver Int 2012;32:554—9.

[30] Scheingraber S, Bauer M, Bauer I, Bardens D, Abel K, Horn AK, et al. Inhibition of hemoxygenase-1 improves survival after liver resection in jaundiced rats. Eur Surg Res 2009;42:157—67.

[31] Dudas J, Elmaouhoub A, Mansuroglu T, Batusic D, Tron K, Saile B, et al. Prospero-related homeobox 1 (Prox1) is a stable hepatocyte marker during liver development, injury and regeneration, and is absent from "oval cells". Histochem Cell Biol 2006;126:549—62.

[32] Golding M, Sarraf CE, Lalani EN, Anilkumar TV, Edwards RJ, Nagy P, et al. Oval cell differentiation into hepatocytes in the acetylaminofluorene-treated regenerating rat liver. Hepatology 1995;22:1243—53.

[33] Hattoum A, Rubin E, Orr A, Michalopoulos GK. Expression of epidermal growth factor receptor, FAS and glypican 3 in EpCAM-positive regenerative clusters of hepatocytes, cholangiocytes, and progenitor cells in human liver failure. Human Pathol 2013;44(5):743—9.

[34] Kohler C, Bell AW, Bowen WC, Monga SP, Fleig W, Michalopoulos GK. Expression of Notch-1 and its ligand Jagged-1 in rat liver during liver regeneration. Hepatology 2004;39:1056—65.

[35] Thompson MD, Wickline ED, Bowen WB, Lu A, Singh S, Misse A, et al. Spontaneous repopulation of beta-catenin null livers with beta-catenin-positive hepatocytes after chronic murine liver injury. Hepatology 2011;54:1333—43.

[36] Boulter L, Govaere O, Bird TG, Radulescu S, Ramachandran P, Pellicoro A, et al. Macrophage-derived Wnt opposes Notch signaling to specify hepatic progenitor cell fate in chronic liver disease. Nat Med 2012;18:572—9.

[37] Ueberham E, Lindner R, Kamprad M, Hiemann R, Hilger N, Woithe B, et al. Oval cell proliferation in p16INK4a expressing mouse liver is triggered by chronic growth stimuli. J Cell Mol Med 2008;12:622—38.

[38] Glaser S, Benedetti A, Marucci L, Alvaro D, Baiocchi L, Kanno N, et al. Gastrin inhibits cholangiocyte growth in bile duct-ligated rats by interaction with cholecystokinin-B/Gastrin receptors via D-myo-inositol 1,4,5-triphosphate-, Ca(2+)-, and protein kinase C alpha-dependent mechanisms. Hepatology 2000;32:17—25.

[39] Meng F, Francis H, Glaser S, Han Y, DeMorrow S, Stokes A, et al. Role of stem cell factor and granulocyte colony-stimulating factor in remodeling during liver regeneration. Hepatology 2012;55:209—21.

[40] Alpini G, Glaser S, Robertson W, Rodgers RE, Phinizy JL, Lasater J, et al. Large but not small intrahepatic bile ducts are involved in secretin-regulated ductal bile secretion. Am J Physiol Gastrointest Liver Physiol 1997;272:G1064—74.

[41] Swenson ES, Kuwahara R, Krause DS, Theise ND. Physiological variations of stem cell factor and stromal-derived factor-1 in murine models of liver injury and regeneration. Liver Int 2008;28:308—18.

[42] Liu F, Pan X, Chen G, Jiang D, Cong X, Fei R, et al. Hematopoietic stem cells mobilized by granulocyte colony-stimulating factor partly contribute to liver graft regeneration after partial orthotopic liver transplantation. Liver Transpl 2006;12:1129—37.

[43] Idetsu A, Suehiro T, Okada K, Shimura T, Kuwano H. Hyperbaric oxygenation promotes regeneration of biliary cells and improves cholestasis in rats. WJG 2011;17:2229—35.

[44] Uriarte I, Fernandez-Barrena MG, Monte MJ, Latasa MU, Chang HC, Carotti S, et al. Identification of fibroblast growth factor 15 as a novel mediator of liver regeneration and its application in the

prevention of post-resection liver failure in mice. Gut 2013;62:899−910.

[45] Chen LP, Zhang QH, Chen G, Qian YY, Shi BY, Dong JH. Rapamycin inhibits cholangiocyte regeneration by blocking interleukin-6-induced activation of signal transducer and activator of transcription 3 after liver transplantation. Liver Transpl 2010;16:204−14.

[46] Sackett SD, Gao Y, Shin S, Esterson YB, Tsingalia A, Hurtt RS, et al. Fox11 promotes liver repair following cholestatic injury in mice. Lab Invest 2009;89:1387−96.

[47] Krause P, Rave-Frank M, Wolff HA, Becker H, Christiansen H, Koenig S. Liver sinusoidal endothelial and biliary cell repopulation following irradiation and partial hepatectomy. WJG 2010;16:3928−35.

[48] Ueberham E, Bottger J, Ueberham U, Grosche J, Gebhardt R. Response of sinusoidal mouse liver cells to choline-deficient ethionine-supplemented diet. Comp Hepatol 2010;9:8.

[49] Satoh K, Yamakawa D, Sugio H, Kida K, Sato T, Hosoi K, et al. Bile duct-bound growth of precursor cells of preneoplastic foci inducible in the initiation stage of rat chemical hepatocarcinogenesis by 2-acetylaminofluorene. Jpn J Clin Oncol 2008;38:604−10.

[50] Wang J, Koyota S, Zhou X, Ueno Y, Ma L, Kawagoe M, et al. Expression and localization of regenerating gene I in a rat liver regeneration model. Biochem Biophys Res Commun 2009;380:472−7.

[51] Tsuchiya A, Imai M, Kamimura H, Takamura M, Yamagiwa S, Sugiyama T, et al. Increased susceptibility to severe chronic liver damage in CXCR4 conditional knock-out mice. Dig Dis Sci 2012;57:2892−900.

[52] Matsuo A, Yoshida T, Yasukawa T, Miki R, Kume K, Kume S. Epiplakin1 is expressed in the cholangiocyte lineage cells in normal liver and adult progenitor cells in injured liver. Gene Expr Patterns 2011;11:255−62.

[53] Ito A, Nishikawa Y, Ohnuma K, Ohnuma I, Koma Y, Sato A, et al. SgIGSF is a novel biliary-epithelial cell adhesion molecule mediating duct/ductule development. Hepatology 2007;45:684−94.

[54] Shan YF, Zhou WP, Fu XY, Yan HX, Yang W, Liu SQ, et al. The role of p28GANK in rat oval cells activation and proliferation. Liver Int 2006;26:240−7.

[55] Petersen BE, Zajac VF, Michalopoulos GK. Bile ductular damage induced by methylene dianiline inhibits oval cell activation. Am J Pathol 1997;151:905−9.

[56] Loffreda S, Rai R, Yang SQ, Lin HZ, Diehl AM. Bile ducts and portal and central veins are major producers of tumor necrosis factor alpha in regenerating rat liver. Gastroenterology 1997;112:2089−98.

[57] Assimakopoulos SF, Tsamandas AC, Georgiou CD, Vagianos CE, Scopa CD. Bombesin and neurotensin exert antiproliferative effects on oval cells and augment the regenerative response of the cholestatic rat liver. Peptides 2010;31:2294−303.

[58] Wang L, Wang X, Chiu JD, van de Ven G, Gaarde WA, Deleve LD. Hepatic vascular endothelial growth factor regulates recruitment of rat liver sinusoidal endothelial cell progenitor cells. Gastroenterology 2012;143:1555−1563 e2.

[59] Kordes C, Sawitza I, Gotze S, Haussinger D. Stellate cells from rat pancreas are stem cells and can contribute to liver regeneration. PloS One 2012;7:e51878.

[60] Koh SL, Ager E, Malcontenti-Wilson C, Muralidharan V, Christophi C. Blockade of the renin−angiotensin system improves

the early stages of liver regeneration and liver function. J Surg Res 2013;179:66−71.

[61] Schmelzer E. Hepatic progenitors of the liver and extra-hepatic tissues, liver regeneration. In: Baptista PM, editor. Liver regeneration. Manhattan, NY: InTech; 2012.

[62] Seo MJ, Suh SY, Bae YC, Jung JS. Differentiation of human adipose stromal cells into hepatic lineage in vitro and in vivo. Biochem Biophys Res Commun 2005;328:258−64.

[63] Banas A, Teratani T, Yamamoto Y, Tokuhara M, Takeshita F, Osaki M, et al. Rapid hepatic fate specification of adipose-derived stem cells and their therapeutic potential for liver failure. J Gastroenterol Hepatol 2009;24:70−7.

[64] Banas A, Teratani T, Yamamoto Y, Tokuhara M, Takeshita F, Osaki M, et al. IFATS collection: in vivo therapeutic potential of human adipose tissue mesenchymal stem cells after transplantation into mice with liver injury. Stem Cells 2008;26:2705−12.

[65] Aurich H, Sgodda M, Kaltwasser P, Vetter M, Weise A, Liehr T, et al. Hepatocyte differentiation of mesenchymal stem cells from human adipose tissue in vitro promotes hepatic integration in vivo. Gut 2009;58:570−81.

[66] Talens-Visconti R, Bonora A, Jover R, Mirabet V, Carbonell F, Castell JV, et al. Hepatogenic differentiation of human mesenchymal stem cells from adipose tissue in comparison with bone marrow mesenchymal stem cells. WJG 2006;12:5834−45.

[67] Okura H, Komoda H, Saga A, Kakuta-Yamamoto A, Hamada Y, Fumimoto Y, et al. Properties of hepatocyte-like cell clusters from human adipose tissue-derived mesenchymal stem cells. Tissue Eng Part C Methods 2010;16:761−70.

[68] Appasani K, Appasani RK. Stem cell biology and regenerative medicine: from molecular embryology to tissue engineering. New York, NY: Springer Science + Business Media, LLC; 2011. pp. 1, online resource.

[69] Cheng T. Toward "SMART" stem cells. Gene Ther 2008;15:67−73.

[70] Zhao Q, Ren H, Zhu D, Han Z. Stem/progenitor cells in liver injury repair and regeneration. Biol Cell 2009;101:557−71.

[71] Kanazawa Y, Verma IM. Little evidence of bone marrow-derived hepatocytes in the replacement of injured liver. Proc Natl Acad Sci USA 2003;100(Suppl. 1):11850−3.

[72] Mathe G, Thomas ED, Ferrebee JW. The restoration of marrow function after lethal irradiation in man: a review. Transplant Bull 1959;6:407−9.

[73] Ferrari G, Cusella-De Angelis G, Coletta M, Paolucci E, Stornaiuolo A, Cossu G, et al. Muscle regeneration by bone marrow-derived myogenic progenitors. Science 1998;279:1528−30.

[74] Schwartz RE, Reyes M, Koodie L, Jiang Y, Blackstad M, Lund T, et al. Multipotent adult progenitor cells from bone marrow differentiate into functional hepatocyte-like cells. J Clin Invest 2002;109:1291−302.

[75] Jiang Y, Jahagirdar BN, Reinhardt RL, Schwartz RE, Keene CD, Ortiz-Gonzalez XR, et al. Pluripotency of mesenchymal stem cells derived from adult marrow. Nature 2002;418:41−9.

[76] Pittenger MF, Mackay AM, Beck SC, Jaiswal RK, Douglas R, Mosca JD, et al. Multilineage potential of adult human mesenchymal stem cells. Science 1999;284:143−7.

[77] Brazelton TR, Rossi FM, Keshet GI, Blau HM. From marrow to brain: expression of neuronal phenotypes in adult mice. Science 2000;290:1775−9.

[78] Imai E, Ito T. Can bone marrow differentiate into renal cells? Pediatr Nephrol 2002;17:790—4.

[79] Dominici M, Le Blanc K, Mueller I, Slaper-Cortenbach I, Marini F, Krause D, et al. Minimal criteria for defining multipotent mesenchymal stromal cells. The International Society for Cellular Therapy position statement. Cytotherapy 2006;8:315—7.

[80] Le Blanc K, Rasmusson I, Sundberg B, Gotherstrom C, Hassan M, Uzunel M, et al. Treatment of severe acute graft-versus-host disease with third party haploidentical mesenchymal stem cells. Lancet 2004;363:1439—41.

[81] Fouraschen SM, Pan Q, de Ruiter PE, Farid WR, Kazemier G, Kwekkeboom J, et al. Secreted factors of human liver-derived mesenchymal stem cells promote liver regeneration early after partial hepatectomy. Stem Cells Dev 2012;21:2410—9.

[82] van Poll D, Parekkadan B, Cho CH, Berthiaume F, Nahmias Y, Tilles AW, et al. Mesenchymal stem cell-derived molecules directly modulate hepatocellular death and regeneration in vitro and in vivo. Hepatology 2008;47:1634—43.

[83] Tao XR, Li WL, Su J, Jin CX, Wang XM, Li JX, et al. Clonal mesenchymal stem cells derived from human bone marrow can differentiate into hepatocyte-like cells in injured livers of SCID mice. J Cell Biochem 2009;108:693—704.

[84] Kazemnejad S, Allameh A, Soleimani M, Gharehbaghian A, Mohammadi Y, Amirizadeh N, et al. Biochemical and molecular characterization of hepatocyte-like cells derived from human bone marrow mesenchymal stem cells on a novel three-dimensional biocompatible nanofibrous scaffold. J Gastroenterol Hepatol 2009;24:278—87.

[85] De Kock J, Vanhaecke T, Biernaskie J, Rogiers V, Snykers S. Characterization and hepatic differentiation of skin-derived precursors from adult foreskin by sequential exposure to hepatogenic cytokines and growth factors reflecting liver development. Toxicol In Vitro 2009;23:1522—7.

[86] Kakinuma S, Tanaka Y, Chinzei R, Watanabe M, Shimizu-Saito K, Hara Y, et al. Human umbilical cord blood as a source of transplantable hepatic progenitor cells. Stem Cells 2003;21:217—27.

[87] Jang YY, Collector MI, Baylin SB, Diehl AM, Sharkis SJ. Hematopoietic stem cells convert into liver cells within days without fusion. Nat Cell Biol 2004;6:532—9.

[88] Harris RG, Herzog EL, Bruscia EM, Grove JE, Van Arnam JS, Krause DS. Lack of a fusion requirement for development of bone marrow-derived epithelia. Science 2004;305:90—3.

[89] Hamazaki T, Iiboshi Y, Oka M, Papst PJ, Meacham AM, Zon LI, et al. Hepatic maturation in differentiating embryonic stem cells in vitro. FEBS Lett 2001;497:15—9.

[90] Yamamoto H, Quinn G, Asari A, Yamanokuchi H, Teratani T, Terada M, et al. Differentiation of embryonic stem cells into hepatocytes: biological functions and therapeutic application. Hepatology 2003;37:983—93.

[91] Zuk PA, Zhu M, Ashjian P, De Ugarte DA, Huang JI, Mizuno H, et al. Human adipose tissue is a source of multipotent stem cells. Mol Biol Cell 2002;13:4279—95.

[92] Da Sacco S, De Filippo RE, Perin L. Amniotic fluid progenitor cells and their use in regenerative medicine. In: Wislet-Gendebien S, editor. Advances in regenerative medicine. Manhattan, NY: InTech; 2011.

[93] Underwood MA, Gilbert WM, Sherman MP. Amniotic fluid: not just fetal urine anymore. J Perinatol 2005;25:341—8.

[94] Horger III EO, Hutchinson DL. Diagnostic use of amniotic fluid. J Pediatr 1969;75:503—8.

[95] Hoehn H, Salk D. Morphological and biochemical heterogeneity of amniotic fluid cells in culture. Methods Cell Biol 1982;26:11—34.

[96] Gosden CM. Amniotic fluid cell types and culture. Br Med Bull 1983;39:348—54.

[97] Torricelli F, Brizzi L, Bernabei PA, Gheri G, Di Lollo S, Nutini L, et al. Identification of hematopoietic progenitor cells in human amniotic fluid before the 12th week of gestation. Ital J Anat Embryol 1993;98:119—26.

[98] Tsangaris G, Weitzdorfer R, Pollak D, Lubec G, Fountoulakis M. The amniotic fluid cell proteome. Electrophoresis 2005;26:1168—73.

[99] Bossolasco P, Montemurro T, Cova L, Zangrossi S, Calzarossa C, Buiatiotis S, et al. Molecular and phenotypic characterization of human amniotic fluid cells and their differentiation potential. Cell Res 2006;16:329—36.

[100] McLaughlin D, Tsirimonaki E, Vallianatos G, Sakellaridis N, Chatzistamatiou T, Stavropoulos-Gioka C, et al. Stable expression of a neuronal dopaminergic progenitor phenotype in cell lines derived from human amniotic fluid cells. J Neurosci Res 2006;83:1190—200.

[101] Da Sacco S, Sedrakyan S, Boldrin F, Giuliani S, Parnigotto P, Habibian R, et al. Human amniotic fluid as a potential new source of organ specific precursor cells for future regenerative medicine applications. J Urol 2010;183:1193—200.

[102] Kunisaki SM, Armant M, Kao GS, Stevenson K, Kim H, Fauza DO. Tissue engineering from human mesenchymal amniocytes: a prelude to clinical trials. J Pediatr Surg 2007;42:974—9, discussion; 979—80.

[103] De Coppi P, Bartsch Jr. G, Siddiqui MM, Xu T, Santos CC, Perin L, et al. Isolation of amniotic stem cell lines with potential for therapy. Nat Biotechnol 2007;25:100—6.

[104] Zheng YB, Gao ZL, Xie C, Zhu HP, Peng L, Chen JH, et al. Characterization and hepatogenic differentiation of mesenchymal stem cells from human amniotic fluid and human bone marrow: a comparative study. Cell Biol Int 2008;32:1439—48.

[105] Saulnier N, Lattanzi W, Puglisi MA, Pani G, Barba M, Piscaglia AC, et al. Mesenchymal stromal cells multipotency and plasticity: induction toward the hepatic lineage. Eur Rev Med Pharmacol Sci 2009;13(Suppl. 1):71—8.

[106] Hakan O, Morikuni T, Hiroshi M. Mesenchymal stem cells isolated from adipose and other tissues: basic biological properties and clinical applications. Stem Cells Int 2012;461718:1—9. doi:10.1155/2012/461718.

[107] da Silva Meirelles L, Chagastelles PC, Nardi NB. Mesenchymal stem cells reside in virtually all post-natal organs and tissues. J Cell Sci 2006;119:2204—13.

[108] Caplan AI. Adult mesenchymal stem cells for tissue engineering versus regenerative medicine. J Cell Physiol 2007;213:341—7.

[109] Fuchs E, Segre JA. Stem cells: a new lease on life. Cell 2000;100:143—55.

[110] Doherty MJ, Ashton BA, Walsh S, Beresford JN, Grant ME, Canfield AE. Vascular pericytes express osteogenic potential in vitro and in vivo. J Bone Miner Res 1998;13:828—38.

[111] Bianco P, Riminucci M, Gronthos S, Robey PG. Bone marrow stromal stem cells: nature, biology, and potential applications. Stem Cells 2001;19:180—92.

[112] de Bari C, Dell'accio F, Przemyslaw T. Multipotent mesenchymal stem cells from adult human synovial membrane. Arthritis Rheum 2001;44:1928−42.

[113] Young HE, Steele TA, Bray RA, Hudson J, Floyd JA, Hawkins K, et al. Human reserve pluripotent mesenchymal stem cells are present in the connective tissues of skeletal muscle and dermis derived from fetal, adult, and geriatric donors. Anat Rec 2001;264:51−62.

[114] Zuk PA, Zhu M, Mizuno H, Huang J, Futrell JW, Katz AJ, et al. Multilineage cells from human adipose tissue: implications for cell-based therapies. Tissue Eng 2001;7:211−28.

[115] Zuk PA. The adipose-derived stem cell: looking back and looking ahead. Mol Biol Cell 2010;21:1783−7.

[116] Suess Jr. RP, Martin RA, Moon ML, Dallman MJ. Rectovaginal fistula with atresia ani in three kittens. Cornell Vet 1992;82:141−53.

[117] Toma JG, Akhavan M, Fernandes KJ, Barnabe-Heider F, Sadikot A, Kaplan DR, et al. Isolation of multipotent adult stem cells from the dermis of mammalian skin. Nat Cell Biol 2001;3:778−84.

[118] Vishnubalaji R, Manikandan M, Al-Nbaheen M, Kadalmani B, Aldahmash A, Alajez NM. In vitro differentiation of human skin-derived multipotent stromal cells into putative endothelial-like cells. BMC Dev Biol 2012;12:7.

[119] Shi CM, Cheng TM. Differentiation of dermis derived multipotent cells into insulin-producing pancreatic cells *in vitro*. World J Gastroenterol 2004;10:2550−2.

[120] Shih DT, Lee DC, Chen SC, Tsai RY, Huang CT, Tsai CC, et al. Isolation and characterization of neurogenic mesenchymal stem cells in human scalp tissue. Stem Cells 2005;23:1012−20.

[121] Haniffa MA, Wang XN, Holtick U, Rae M, Isaacs JD, Dickinson AM, et al. Adult human fibroblasts are potent immunoregulatory cells and functionally equivalent to mesenchymal stem cells. J Immunol 2007;179:1595−604.

[122] Chen FG, Zhang WJ, Bi D, Liu W, Wei X, Chen FF, et al. Clonal analysis of nestin(−) vimentin(+) multipotent fibroblasts isolated from human dermis. J Cell Sci 2007;120:2875−83.

[123] Chen MY, Lie PC, Li ZL, Wei X. Endothelial differentiation of Wharton's jelly-derived mesenchymal stem cells in comparison with bone marrow-derived mesenchymal stem cells. Exp Hematol 2009;37:629−40.

[124] Wu FL, Wang NN, Liu YS, Wei X. Differentiation of Wharton's jelly primitive stromal cells into insulin-producing cells in comparison with bone marrow mesenchymal stem cells. Tissue Eng 2009;15:2865−73.

[125] Zhang YN, Lie PC, Wei X. Differentiation of mesenchymal stromal cells derived from umbilical cord Wharton's jelly into hepatocyte-like cells. Cytotherapy 2009;11:548−58.

[126] Nekanti U, Rao VB, Bahirvani AG, Jan M, Totey S, Ta M. Long-term expansion and pluripotent marker array analysis of Wharton's jelly-derived mesenchymal stem cells. Stem Cells Dev 2010;19:117−30.

[127] Grewal SS, Barker JN, Davies SM, Wagner JE. Unrelated donor hematopoietic cell transplantation: marrow or umbilical cord blood? Blood 2003;101:4233−44.

[128] Bieback K, Kern S, Kluter H, Eichler H. Critical parameters for the isolation of mesenchymal stem cells from umbilical cord blood. Stem Cells 2004;22:625−34.

[129] Broxmeyer HE, Douglas GW, Hangoc G. Human umbilical cord blood as a potential source of transplantable hematopoietic stem/progenitor cells. Proc Natl Acad Sci USA 1989;86:3828−32.

[130] Lee OK, Kuo TK, Chen W-M, Lee K-D, Hsieh S-H, Chen T-H. Isolation of multipotent mesenchymal stem cells from umbilical cord blood. Blood 2004;103:51.

[131] Duncan AW, Dorrell C, Grompe M. Stem cells and liver regeneration. Gastroenterology 2009;137:466−81.

[132] Li Z, Chen J, Li L, Ran JH, Li XH, Liu ZH, et al. Human hepatocyte growth factor (hHGF)-modified hepatic oval cells improve liver transplant survival. PLoS One 2012;7:e44805.

[133] Wang X, Foster M, Al-Dhalimy M, Lagasse E, Finegold M, Grompe M. The origin and liver repopulating capacity of murine oval cells. Proc Natl Acad Sci USA 2003;100 (Suppl. 1):11881−8.

[134] Weglarz TC, Degen JL, Sandgren EP. Hepatocyte transplantation into diseased mouse liver. Kinetics of parenchymal repopulation and identification of the proliferative capacity of tetraploid and octaploid hepatocytes. Am J Pathol 2000;157:1963−74.

[135] Kvittingen EA, Rootwelt H, Berger R, Brandtzaeg P. Self-induced correction of the genetic defect in tyrosinemia type I. J Clin Invest 1994;94:1657−61.

[136] Limaye PB, Bowen WC, Orr AV, Luo J, Tseng GC, Michalopoulos GK. Mechanisms of hepatocyte growth factor-mediated and epidermal growth factor-mediated signaling in transdifferentiation of rat hepatocytes to biliary epithelium. Hepatology 2008;47:1702−13.

[137] Chen Y, Wong PP, Sjeklocha L, Steer CJ, Sahin MB. Mature hepatocytes exhibit unexpected plasticity by direct dedifferentiation into liver progenitor cells in culture. Hepatology 2012;55:563−74.

[138] Dabeva MD, Petkov PM, Sandhu J, Oren R, Laconi E, Hurston E, et al. Proliferation and differentiation of fetal liver epithelial progenitor cells after transplantation into adult rat liver. Am J Pathol 2000;156:2017−31.

[139] Mahieu-Caputo D, Allain JE, Branger J, Coulomb A, Delgado JP, Andreoletti M, et al. Repopulation of athymic mouse liver by cryopreserved early human fetal hepatoblasts. Hum Gene Ther 2004;15:1219−28.

[140] Nierhoff D, Ogawa A, Oertel M, Chen YQ, Shafritz DA. Purification and characterization of mouse fetal liver epithelial cells with high *in vivo* repopulation capacity. Hepatology 2005;42:130−9.

[141] Oertel M, Menthena A, Chen YQ, Teisner B, Jensen CH, Shafritz DA. Purification of fetal liver stem/progenitor cells containing all the repopulation potential for normal adult rat liver. Gastroenterology 2008;134:823−32.

[142] Dabeva MD, Hwang SG, Vasa SR, Hurston E, Novikoff PM, Hixson DC, et al. Differentiation of pancreatic epithelial progenitor cells into hepatocytes following transplantation into rat liver. Proc Natl Acad Sci USA 1997;94:7356−61.

[143] Dan YY, Riehle KJ, Lazaro C, Teoh N, Haque J, Campbell JS, et al. Isolation of multipotent progenitor cells from human fetal liver capable of differentiating into liver and mesenchymal lineages. Proc Natl Acad Sci USA 2006;103:9912−7.

[144] Weber A, Touboul T, Mainot S, Branger J, Mahieu-Caputo D. Human foetal hepatocytes: isolation, characterization, and transplantation. Methods Mol Biol 2010;640:41−55.

[145] Semeraro R, Carpino G, Cardinale V, Onori P, Gentile R, Cantafora A, et al. Multipotent stem/progenitor cells in the human foetal biliary tree. J Hepatol 2012;57:987–94.

[146] Cardinale V, Wang Y, Carpino G, Mendel G, Alpini G, Gaudio E, et al. The biliary tree—a reservoir of multipotent stem cells. Nat Rev Gastroenterol Hepatol 2012;9:231–40.

[147] Oh SH, Hatch HM, Petersen BE. Hepatic oval "stem" cell in liver regeneration. Semin Cell Dev Biol 2002;13:405–9.

[148] Libbrecht L, Desmet V, Van Damme B, Roskams T. Deep intra-lobular extension of human hepatic "progenitor cells" correlates with parenchymal inflammation in chronic viral hepatitis: can "progenitor cells" migrate? J Pathol 2000;192:373–8.

[149] Jung Y, Oh SH, Witek RP, Petersen BE. Somatostatin stimulates the migration of hepatic oval cells in the injured rat liver. Liver Int 2012;32:312–20.

[150] Ishikawa T, Factor VM, Marquardt JU, Raggi C, Seo D, Kitade M, et al. Hepatocyte growth factor/c-met signaling is required for stem-cell-mediated liver regeneration in mice. Hepatology 2012;55:1215–26.

[151] Crema A, Ledda M, De Carlo F, Fioretti D, Rinaldi M, Marchese R, et al. Cord blood CD133 cells define an OV6-positive population that can be differentiated in vitro into engraftable bipotent hepatic progenitors. Stem Cells Dev 2011;20:2009–21.

[152] Hruban RH, Molina JM, Reddy MN, Boitnott JK. A neoplasm with pancreatic and hepatocellular differentiation presenting with subcutaneous fat necrosis. Am J Clin Pathol 1987;88:639–45.

[153] Scarpelli DG, Rao MS. Differentiation of regenerating pancreatic cells into hepatocyte-like cells. Proc Natl Acad Sci USA 1981;78:2577–81.

[154] Lalwani ND, Reddy MK, Qureshi SA, Reddy JK. Development of hepatocellular carcinomas and increased peroxisomal fatty acid beta-oxidation in rats fed [4-chloro-6-(2,3-xylidino)-2-pyri-midinylthio]acetic acid (Wy-14,643) in the semipurified diet. Carcinogenesis 1981;2:645–50.

[155] Rao MS, Subbarao V, Reddy JK. Induction of hepatocytes in the pancreas of copper-depleted rats following copper repletion. Cell Differ 1986;18:109–17.

[156] Krakowski ML, Kritzik MR, Jones EM, Krahl T, Lee J, Arnush M, et al. Pancreatic expression of keratinocyte growth factor leads to differentiation of islet hepatocytes and proliferation of duct cells. Am J Pathol 1999;154:683–91.

[157] Shen CN, Tosh D. Transdifferentiation of pancreatic cells to hepatocytes. Methods Mol Biol 2010;640:273–80.

[158] Wang X, Al-Dhalimy M, Lagasse E, Finegold M, Grompe M. Liver repopulation and correction of metabolic liver disease by trans-planted adult mouse pancreatic cells. Am J Pathol 2001;158:571–9.

[159] Yan Y, Xu W, Qian H, Si Y, Zhu W, Cao H, et al. Mesenchymal stem cells from human umbilical cords ameliorate mouse hepatic injury in vivo. Liver Int 2009;29:356–65.

[160] Kuo TK, Hung SP, Chuang CH, Chen CT, Shih YR, Fang SC, et al. Stem cell therapy for liver disease: parameters governing the success of using bone marrow mesenchymal stem cells. Gastroenterology 2008;134:2111–21.

[161] Avants SK, Margolin A, Kosten TR, Singer JL. Changes concur-rent with initiation of abstinence from cocaine abuse. J Subst Abuse Treat 1993;10:577–83.

[162] Russo FP, Alison MR, Bigger BW, Amofah E, Florou A, Amin F, et al. The bone marrow functionally contributes to liver fibro-sis. Gastroenterology 2006;130:1807–21.

[163] di Bonzo LV, Ferrero I, Cravanzola C, Mareschi K, Rustichell D, Novo E, et al. Human mesenchymal stem cells as a two-edged sword in hepatic regenerative medicine: engraftment and hepatocyte differentiation versus profibrogenic potential. Gut 2008;57:223–31.

[164] Petersen BE, Bowen WC, Patrene KD, Mars WM, Sullivan AK, Murase N, et al. Bone marrow as a potential source of hepatic oval cells. Science 1999;284:1168–70.

[165] Alison MR, Poulsom R, Jeffery R, Dhillon AP, Quaglia A, Jacob J, et al. Hepatocytes from non-hepatic adult stem cells. Nature 2000;406:257.

[166] Theise ND, Nimmakayalu M, Gardner R, Illei PB, Morgan G, Teperman L, et al. Liver from bone marrow in humans. Hepatology 2000;32:11–6.

[167] Forbes SJ. Myelomonocytic cells are sufficient for therapeutic cell fusion in the liver. J Hepatol 2005;42:285–6.

Chapter 27

Liver Regeneration and Bioengineering: The Role of Liver Extra-Cellular Matrix and Human Stem/Progenitor Cells

Pedro M. Baptista, Emma Moran, Dipen Vyas, Thomas Shupe, and Shay Soker

Wake Forest Institute for Regenerative Medicine, Wake Forest School of Medicine, Winston-Salem, NC

Chapter Outline

27.1 LIVER DEVELOPMENT AND THE STEM CELL NICHE

Liver development is a stepwise process that includes distinct biological events. The first step is the commitment of the foregut endoderm to the hepatic lineage. This specification process is regulated by transcription factors FoxA2 and GATA4 binding to their target sequences and facilitating the binding of C/EBPβ and nuclear factor 1 to activate transcription of albumin [1,2]. These early hepatic progenitor cells (also referred to as hepatoblasts) express albumin and alpha-fetoprotein (AFP) and have potential to differentiate into hepatocytes and bile duct epithelial cells (cholangiocytes) [5]. The progenitor cells proliferate throughout the fetal development and achieve functional maturation at various stages along with differentiating into either hepatocytes or cholangiocytes. Detailed molecular mechanisms involved in the liver development have been described recently [3,4,6−8].

For many years, human hepatoblasts were considered to be the stem cell population of the liver, but recent studies have reported isolation of a stem cell population from fetal and post natal livers that are considered to be the precursor to the hepatoblasts. These human hepatic stem cells (hHSCs) have a distinct antigenic profile from the hepatoblasts. hHSCs express neural cell adhesion molecule (NCAM), epithelial cell adhesion molecule (EpCAM), SOX9, cytokeratin (CK) 8/18/19, sonic, and Indian hedgehog. They lack AFP, albumin, intercellular adhesion molecule (ICAM-1), and early cytochrome P450 enzymes, which are expressed in hepatoblasts. The AFP$^-$/EpCAM$^+$ population of hHSCs has been obtained from fetal, neonatal, and adult livers of all ages. As mentioned earlier, hHSCs are considered to be the precursors to the hepatoblasts. The hepatoblasts are thought to be the transient amplifying progenitor cells, which are found throughout the parenchyma of fetal and neonatal livers. Maturation lineage studies demonstrated that hepatoblasts could become committed to either the hepatic or cholangiocytic lineages during the late developmental stages of the liver. These committed progenitors express either hepatic or biliary markers and do not express stem cell specific markers such as NCAM and hedgehog proteins. The committed progenitors undergo terminal differentiation to form mature hepatocytes and cholangiocytes (Figure 27.1). This process of stepwise differentiation of hepatic stem cells is accompanied by changes in gene expression, thus leading to their lineage specification into mature phenotypes [1].

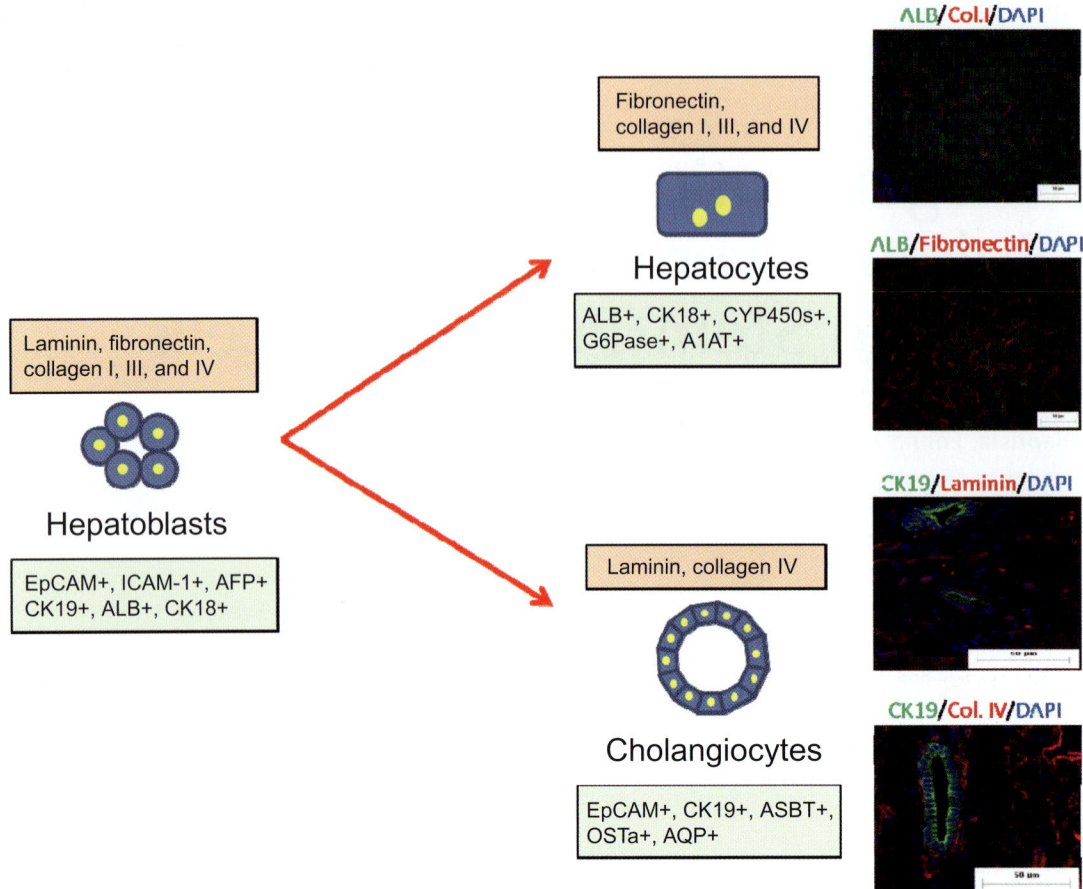

FIGURE 27.1 Schematic image showing the changes in ECM composition and phenotypic markers during hFLPC differentiation. Hepatoblasts are surrounded in the human fetal liver by an ECM composed mainly of laminin, collagen I, III, and IV, and discrete amounts of fibronectin. These bipotential progenitor cells are able to differentiate into cholangiocytes, forming the bile ducts and ductules of the adult liver and its parenchymal cells, the hepatocytes. The ECM neighboring these ductal structures is constituted mainly by collagen IV and laminin and significantly different from the parenchymal ECM, formed mainly by collagen I, III, and IV, and fibronectin. EpCAM, epithelial cell adhesion molecule; G6Pase, glucose 6 phosphatase; ASBT, apical sodium dependent bile transporter; AQP, Aquaporin; OST-α, organic solute transporter-alpha.

The stem cell microenvironment (niche) plays an important role in regulating stem cell specification and differentiation into mature cell types. Paracrine signaling between the mesenchymal and epithelial cells and interactions with the extracellular matrix (ECM) are the major components of the stem cell niche that modulates cell behavior [1,9]. Various soluble paracrine signals released from nonepithelial cells (endothelial, mesenchymal stem cells (MSCs), portal fibroblasts, hematopoeitic stem cells, stellate cells) have been shown to be critical for progenitor cell maintenance and differentiation into hepatocytes and cholangiocytes [4,6,9–11]. Bile ducts arise from the hepatoblasts lining the portal mesenchyme around the portal vein by undergoing a complex process of bile duct morphogenesis. The paracrine signaling gradient observed between the portal mesenchyme and the parenchymal space is implicated in regulating the biliary lineage specification of the hepatoblasts, as the hepatoblasts present in

the parenchymal space do not undergo biliary differentiation [6,12,13]. Two specific signaling pathways, transforming growth factor-beta (TGF-β) and Notch, have been suggested to be important for hepatoblast differentiation into cholangiocytes [14]. Jagged 1, a Notch 2 ligand expressed by myofibroblasts, activates Notch 2 in the progenitor cells and induces expression of HNF-1β, required for cholangiocyte differentiation, while suppressing HNF-1α, HNF-4α, and C/EBPα, which are required for hepatocyte differentiation [10,11]. Portal mesenchyme secretes TGF-β, thus creating a gradient from periportal to parenchymal region, with highest activity observed near the periportal mesenchyme, where the biliary ducts develop [15]. Several growth factors and soluble molecules such as hepatocyte growth factor (HGF), oncostatin M (OSM), and glucocorticoids have also been shown to be regulating hepatoblast differentiation toward hepatocytes [2,4,6]. Both stellate cells and endothelial cells have been shown to produce HGF. HGF acts via c-Met receptors and stimulates expression of transcription factors C/EBPα and HNF-4α, both required for hepatocyte differentiation [4,16]. OSM is an interleukin 6-related cytokine produced from the developing hematopoietic cell. OSM induces hepatic differentiation by promoting HNF-4α expression, and this process is suppressed by TNF-α [17]. At the latest stages of development, hepatoblasts located in the parenchyma give rise to hepatocytes, while the hepatoblasts located near the portal mesenchyme differentiate into cholangiocytes. Such lineage specification of hepatoblasts, based on their location within the liver and the presence of gradient in ECM composition, suggests an important role for the ECM in the lineage specification.

27.2 THE LIVER ECM

The liver ECM not only provides a structural framework to the liver cells, but it also has a critical role in mediating cell attachment and migration and regulating differentiation, repair, and development [18]. The intricate structure that the ECM forms in the liver can be roughly divided in two major areas: the periportal region and the pericentral region, defined by the vascular domains of the portal triad and the central vein. From one area to the other, there is a gradient of matrix molecules that can be divided in three different zones [19]. This specific organization enables and preserves the metabolic zonation observed in liver hepatocytes, with hepatocytes from the different zones having different sizes, enzymes, and major functions [20]. These matrix composition differences are mostly evident in the space of Disse (perisinusoidal space, which is the location in the liver between a hepatocyte and a sinusoid), which has also been shown to undergo changes during liver ontogenesis [21]. There is a gradient in the ECM composition in the space of Disse in the adult liver, where zone 1 (periportal region) displays fetal and neonatal ECM characteristics and zone 3 (pericentral region) displays adult characteristics [22].

The stem cell niche found in the canals of Hering has three resident stem/progenitor populations: hHSCs, angioblasts, and hepatic stellate cell precursors. The microenvironment of this niche is comprised of soluble paracrine signals and an ECM composed of hyaluronans, an integrin α6β4 binding form of laminin, collagen type III, and minimally sulfated chondroitin sulfate proteoglycans (CSPGs). This niche is devoid of collagen type I or IV or heparan sulfate proteoglycans (HSPGs). As the stem cells transition to hepatoblast stage and subsequently into successive lineage stages, changes are observed in the soluble paracrine signals and matrix composition. These changes in paracrine signals and matrix components dictate the stepwise differentiation of the stem cells to adult fates across the hepatic maturational gradient [23].

In the adult liver (Figure 27.2), the portal triad region, which includes hepatic artery, portal vein, and bile ducts, has an organized basement membrane consisting of collagen IV, laminin, entactin, and perlecan [21]. The space of Disse, which lies in between sinusoidal endothelial cells and hepatocytes, has a matrix composition unique to the liver. It lacks typical basement membrane proteins such as laminin, entactin, and perlecan but contains collagen IV. There is abundance of fibronectin, discontinuous deposits of collagen III, and continuous network of collagen I in space of Disse [8]. Liver progenitor cells demonstrate unique responses (expansion, differentiation) when grown on zone-specific ECM matrices in 2D culture system. Zone 1 matrix molecules such as laminin and collagen III and IV induce clonogenic expansion of hFLPCs, while zone 3 matrix molecule collagen I induces growth arrest and differentiation, and fibronectin inhibits cell attachment [20]. Liver epithelial cells interact with the surrounding ECM via integrin receptors found on their surface. Hepatocytes and cholangiocytes have been shown to

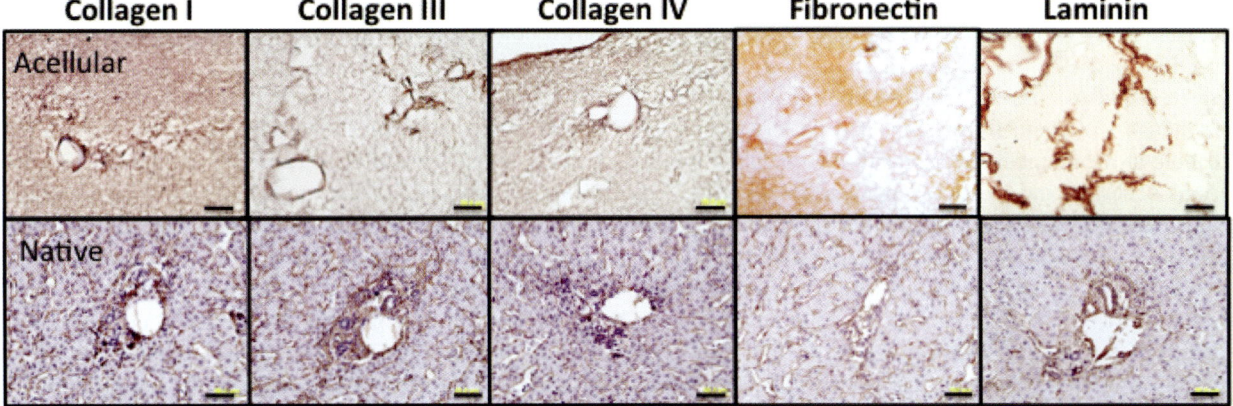

FIGURE 27.2 ECM protein localization on acellular liver scaffold and native human liver. Immunostaining for collagen I, III, IV, fibronectin, and laminin (as indicated) show similar ECM component distribution in prepared acellular liver bioscaffold and native human liver sections. Scale 50 μm.

express markedly different combinations of integrins, which correlates with the differences in ECM composition surrounding both cell types [24,25]. Cholangiocytes express variety of integrin receptors including $\alpha2\beta1$, $\alpha3\beta1$, $\alpha5\beta1$, $\alpha6\beta1$, $\alpha9\beta1$, $\alpha V\beta1$, and $\alpha6\beta4$ dimers [26], while hepatocytes express only $\alpha1\beta1$, $\alpha5\beta1$, and $\alpha9\beta1$ dimers [24]. The majority of signaling molecules implicated in ECM—integrin interactions, such as Rho guanosine triphosphatases (Rho GTPases), Raf, Ras, focal adhesion kinase (FAK), and MAPK/ERK, are ubiquitous mediators of signal transduction [27]. FAK is thought to be an important mediator of downstream signaling responsible for regulation of cell growth following integrin-dependent cell adhesion to ECM [28]. The ECM gradient in the different compartments of the liver and changes in integrin receptor subunit expression observed during the differentiation of progenitor cells, combined with the striking differences observed in the attachment efficiencies, growth rate, morphology, and differentiation of these cells on different matrix components in culture, prove that progenitor cell interaction with the ECM is vital for their maturation in the developing liver. However, the specific mechanisms and the pathways involved in this process remain to be elucidated.

27.3 THE LIVER STEM CELLS AND MATRIX MECHANOBIOLOGY

In addition to responding to biochemical signals, cell behavior is highly influenced by its mechanical environment, including the topography and stiffness of the surrounding tissue, fluid shear stress, and interstitial fluid pressure. The capability of a cell to respond to mechanical cues can lead to an array of cellular processes, including differentiation, injury response, motility, and morphological changes [29—31]. The ability of cells to sense and respond to the mechanical stiffness of its substrate, in the form of cell—matrix and cell—cell interactions is an important component of cellular mechanosensitivity. The material properties of tissue are determined by the chemical structure of the tissue as well as organization of those components. Collagen, fibronectin, and proteoglycans all contribute to mechanical strength and behavior, and modifications of any of these components lead to alterations in the mechanical properties of the tissue [32]. In most cases, the stiffness of a tissue or substrate is defined in terms of elastic or Young's modulus, a constant describing the materials ability to resist deformation, or the ratio of stress to strain. However, soft tissues, including the liver, are inherently more complex than the linear elastic modulus describes, displaying nonlinear elasticity (nonlinear stress—strain relationship) and viscoelasticity (consisting of fluid and solid components) [33]. The effects of these

more complicated material behaviors on cell mechanosensitivity are largely unknown.

Cell—matrix interactions occur at points known as focal adhesion complexes, which consist of integrins that connect the ECM to the actin—myosin cytoskeleton of the cell. Contraction of this actin—myosin cytoskeleton allows the cell to survey its mechanical environment through movement of the integrins, which pull on the ECM and then transmit that force back to the cytoskeleton. For example, if the ECM became stiffer, it would be more difficult for the cytoskeleton to contract, resulting in accumulation of more integrins, enlarged focal adhesions, and further development of the cytoskeleton [30]. Integrin movement is related to downstream signaling pathways; the main mediators of this process are Rho GTPase and the contraction of the actin—myosin cytoskeleton [32]. While less understood than cell—matrix interactions, cell—cell junctions are equally important in cell mechanosensitivity. Cell—cell interactions are thought to occur via cadherins that form a bridge between the cytoskeleton of two neighboring cells [34]. For example, the optimal stiffness for culturing and expanding any given cell type corresponds to the *in vivo* elastic modulus of its corresponding tissue. Variations in the stiffness of the tissue, a process that occurs in certain disease states, may lead to alterations in the normal behavior of a particular cell. In liver cirrhosis and fibrosis, liver tissue can increase by an order of magnitude [35], triggering an array of responses by various liver cells. When cultured on stiff substrates, hepatocytes begin to dedifferentiate and become proliferative, which is in stark contrast to their normally quiescent state [36—38]. Portal fibroblasts differentiate into myofibroblasts when supplemented with TGF-β and cultured on a stiff substrate, an important process in the early stages of biliary fibrosis [39]. Hepatic stellate cells, the major ECM producers in the liver, become increasingly transdifferentiated toward fibrogenic myofibroblasts as the stiffness of the underlying matrix is increased [40]. Disruption of HSC integrins, through disruption of $\alpha_5\beta_1$ and $\alpha_v\beta_3$ and by culturing the cells on poly-L-lysine enabling nonintegrin-dependent cell adhesion, promotes growth cycle arrest and maintains cells in a nonmyofibroblastic state [41]. Additionally, Rho GTPases, mediators of the integrin-sensing signaling pathways, are necessary for HSC transdifferentiation [42]. The transdifferentiation process of HSCs to myofibroblasts is the primary mechanism of liver fibrosis, and recent research suggests that the mechanosensitivity of HSCs plays a vital role in this process.

A groundbreaking paper by Engler et al. [29] demonstrated that stem cell differentiation could be directed toward different lineages by altering the mechanical properties of the substrate to mimic specific tissue types. MSCs cultured on soft substrates mimicking brain differentiated toward neurogenic phenotypes, MSCs cultured on matrix mimicking muscle were myogenic, and the stiffest matrix mimicking collagenous bone produced osteogenic cells. These findings are incredibly important for the field of stem cell therapy, in directing differentiation of stem cells for clinical therapies. Following this work, Lozoya et al. [43] researched a similar question pertaining to liver cells: can mimicking the physiological mechanical properties of liver tissue regulate liver stem and progenitor cells populations? As described above, the human liver contains a population of liver stem and progenitors cells that reside in a defined stem cell niche. The ECM components of the liver stem cell niche are unique from that of the rest of the liver. Due to this altered composition of ECM components, one can infer that the mechanical properties, or stiffness, of that region may also be uniquely different than the rest of the liver. In the work completed by Lozoya et al., hHSCs were cultured in 3D microenvironments constituted by hyaluronic acid (HA) hydrogels that mimicked the canals of Hering. Six different hydrogel formulations, each with their own unique set of material properties, were seeded with hHSCs and various markers were used to determine the degree of differentiation of the stem cell colonies following one week of culture. The major indicator of hHSC mechanosensitivity came from analysis of the stem cell marker CDH1, also known as E-cadherin. CDH1 is a cell surface protein that establishes cell—cell adhesions, assesses the mechanical stiffness of neighboring cells, and triggers downstream signaling pathways involved in mechanosensitivity [34]. The protein expression levels of CDH1 exhibited a dependence on stiffness, with the highest levels of expression occurring when the cells were cultured on the HA hydrogel with a shear modulus of 200 Pa. CDH1 expression on the apical side of the 200 Pa HA hydrogel seeded cells demonstrates that these exposed cells may coordinate mechanical signals to adjacent cells through CDH1. This result demonstrates a stiffness-dependent behavior of the hHSCs, suggesting that culturing the cells in their preferable mechanical environment allows them to organize themselves in the same manner observed in the stem cell niche. However, there is still a lack of convincing evidence that substrate stiffness can direct differentiation of hHSCs toward mature liver cells, which would have huge implications for liver regeneration in medicine as well as understanding the involvement of hHSCs in liver disease.

In addition to responding to the mechanical properties of the substrate, liver stem and progenitor cells and human embryonic stem cells have been shown to differentiate into mature liver cells when exposed to shear stress in perfusion bioreactor cultures [44,45]. Hepatocytes have also been shown to respond to a third type of mechanical force, parenchymal (interstitial) fluid pressure [46]. Overall, the mechanosensitivity of liver cells has important implications for regeneration in both tissue engineering and disease. In regenerative medicine, the mechanical as well as biochemical environment must be properly tuned to direct stem and progenitor cell differentiation as well as maintaining differentiated cells. The vast importance of mechanical signals in regeneration processes are just beginning to be realized and much research is yet to be done.

27.4 CELLULAR THERAPIES FOR LIVER DISEASE

Hepatocyte transplantation is certainly in the vanguard of new therapeutic strategies. The first successful hepatocyte transplantation was performed in June 1992 on a French Canadian woman with familial hypercholesterolemia. After *ex vivo* transduction with a retrovirus encoding for the human LDL receptor, the patient's hepatocytes were infused through the inferior mesenteric vein into the liver. LDL and HDL levels improved throughout the next 18 months and transgene expression was detected in a liver biopsy [47]. Following this first accomplishment, other patients were treated; however, not all patients had a clear benefit from the procedure [48]. Since then, several other metabolic diseases have been treated with hepatocyte transplantation with different degrees of success [49—53]. It has also been used as a support treatment of acute [54—56] and chronic liver diseases [55—58] by bridging severely ill patients to orthotopic liver transplantation. Low efficacy and lack of long-term therapeutic effect were common in these trials, which could be explained by the relatively small number of hepatocytes that engrafted in the recipient liver because of quality, quantity, and possibly immunosuppression protocols [59]. However, transplantation of a number of hepatocytes corresponding to 1—5% of the total liver mass has shown a positive impact in transplanted patients, even if for a limited period of time [59].

Due to the shortage of available human hepatocytes for transplantation, other cell sources have been used. Specifically, bone marrow—derived MSCs [60], hematopoietic stem cells [61,62] and fetal liver progenitor/stem cells [63] have shown to improve, to a certain extent, the condition of cirrhotic patients (Table 27.1). The latter cell type holds an enormous potential for cell/regenerative medicine therapies due to its high expansion capabilities and differentiation into hepatocytes and biliary epithelium [64].

Recent data suggests that human embryonic stem cells (hESCs) and induced pluripotent stem cells (iPSCs) hold great promise for regenerative medicine applications in every medical field. Specifically for the liver, several studies have established the required pathways to differentiate a hESC or iPSC into a hepatic fate by using defined soluble growth factor signals that mimic embryonic development [67,68]. These cells, once transplanted into rodent livers, were able to engraft and express several normal hepatic functions [69,70]. However, more extensive characterization, as well as further safety evaluation, is needed to determine whether these cells will fully function as primary adult hepatocytes.

Regardless of the origin of the cells, there is one common motif in all the clinical trials completed so far, the short-term duration of the therapeutic effect of the transplanted cells. Even when stem/progenitor cells were used, the clinical amelioration was observed for a variable and limited period of time. However, they survive well in hosts with some forms of liver disease and when native liver cell expansion is inhibited by exogenous interventions [71–73]. These data suggest that some endogenous mechanisms control the number of surviving donor hepatocytes over time. Thus, graft survival could be limited by a host-dependent survival advantage over donor hepatocytes properties. This situation would be similar to that observed in allogeneic bone marrow transplantation, where the host must undergo a preparative regimen of chemotherapy and radiotherapy to create an environment conducive to long-term engraftment. This preparation induces apoptosis of host bone marrow cells and creates a cellular vacuum for donor cell engraftment, allowing macrochimerism to take place following infusion of donor hematopoietic stem cells [74].

There are other reasons that might explain the poor engraftment observed. One major reason is the slow but steady immune destruction of the transplanted cells by the host immune system [75]. Nonoptimal immunosuppression regimens are probably the reason for this, and these regimens were improved significantly in the past decade with the introduction of new immunosuppressive drugs and protocols. The poor quality of the livers used for the isolation of hepatocyte and other cell populations is also of importance. Hepatocytes and liver progenitors are isolated usually from livers rejected for transplantation, due to the fat content and ischemia periods, and from unused segments of livers that were used for pediatric recipients. These sources are usually scarce and impact the availability of cells for transplantation, making the procedure into an experimental clinical study, rather than a concrete and valid therapeutic alternative [59,75]. Hence, quite often the harvested cells are of lower quality than cells harvested from healthier, freshly isolated livers [75]. These drawbacks do not apply to autologous bone marrow transplantation recipients with end-stage liver failure, and transient therapeutic effect is also observed [65,66].

Beyond poor viability and adverse endogenous cellular milieu that the cells encounter, one has to consider the poor engraftment and survival due to the liver microenvironment and the natural barriers that the cells encounter after their short transportation through the portal or hepatic arterial circulation to the hepatic sinusoids [76–78]. Some of the reasons that might explain these observations are related to the route that the cells take to reach the hepatic parenchyma. Crossing the endothelial cell barrier and then the basal lamina ECM, thicker in chronic liver disease, might present a significant obstacle to the millions of hepatocytes that are usually infused in a transplantation procedure [78]. This certainly limits the amount of hepatocytes and/or progenitor cells that effectively engraft, decreasing the transplantation efficiency of the whole procedure. Additionally, the ECM that the transplanted cells encounter in livers with chronic liver inflammation might not be conducive to proper engraftment, survival, and function. Actually, when ECM molecules were infused through the portal vein, as a preconditioning procedure before hepatocyte transplantation, higher engraftment efficiencies were observed [79]. This finding suggests that adequate and proper ECM molecules are necessary for efficient homing and engraftment. Hence, it is foreseeable that in the near future, new techniques of liver cell transplantation, using ECM carriers/scaffolds, will increase not only the amount of cells transplanted but also the function of the implanted cells [75]. Ectopic transplantation sites for liver cell delivery are still under investigation, but a suitable place for clinical liver cell transplantation has not been successful so far.

TABLE 27.1 List of Cell Therapy Procedures That Were Performed in Clinical Trials

Cell Type	Disease	References
Hepatocytes	Familial hypercholesterolaemia	[47,48]
Hepatocytes	Crigler-Najjar syndrome type I	[49]
Hepatocytes	Severe ornithine transcarbamylase deficiency	[50]
Hepatocytes	Crigler-Najjar syndrome type 1	[51]
Hepatocytes	Glycogen storage disease type 1a	[52]
Hepatocytes	Peroxisomal biogenesis disease	[53]
Hepatocytes	Acute liver disease	[54–56]
Hepatocytes	Chronic liver disease	[55–58]
Bone marrow mesenchymal stem cell	Chronic liver disease	[59,65,66]
Hematopoietic stem cell	Chronic liver disease	[61,62]
Fetal liver progenitor/ stem cells	Chronic liver disease	[63]

27.5 AUGMENTATION OF CELL THERAPIES BY NATURAL ECM

The ECM of the liver clearly represents a major component of the cellular microenvironment. We described above how the molecular and mechanical properties of the matrix transmit signals that regulate liver development and cell phenotype and function. Recently, several groups have published methods for the decellularization of intact liver [80–85]. Collectively, these studies have demonstrated that the composition of the resulting acellular matrix is largely intact (Figure 27.2). In fact, many of the growth factors that govern liver cell phenotype remain associated with these matrices throughout the decellularization process [81]. The ability to generate large quantities of purified liver ECM opens the possibility to incorporate this material into cell-based therapies for liver pathologies. It is difficult to imagine how natural liver matrix might be included in treatment strategies involving intravenous transplantation of hepatocytes directly into the liver. However, these matrices could be easily incorporated into extrahepatic cell transplant strategies. Decellularized liver matrix has already been shown to increase the level of functionality of hepatoblasts when used as a scaffold for cell culture [85]. This increase in cellular function capacity is maintained following transplantation into the omentum of immunodeficient mice. Studies on the treatment of acute liver injury by the transplant of alginate encapsulated hepatocytes have indicated that the efficacy of the treatment was limited by the functional capacity of the transplanted cells [86,87]. It seems likely that incorporation of natural liver ECM into these alginate capsules would support a higher level of hepatocyte function and increase the effectiveness of the therapy.

The use of natural liver ECM may also prove beneficial in the field of bioartificial livers. These extracorporeal devices provide temporary replacement of the critical functions that the liver would normally perform in a patient with a failing liver. Several iterations of bioartificial livers have been tested on animals, but very few human trials have been conducted [88–91]. The bioartificial liver designs that have advanced to clinical trials have demonstrated some ability to decrease levels of toxic substances in serum, but were ineffective at improving survival rates [88,91]. These disappointing results were attributed, in part, to insufficient functional capacity of the cellular component of the device [91–94]. Current generation bioartificial livers utilize several bioreactor designs housing hepatocytes cultured on a synthetic scaffold. Even with the use of plasma expanders, the volume of plasma that may be removed from a critically ill patient is limited. Therefore, the functional capacity for a given bioreactor volume determines the potential efficacy of the device. The use of natural liver scaffold in these systems could potentially increase the level of hepatocyte function resulting in increased biotransformative and biosynthetic capabilities.

27.6 BIOENGINEERING OF LIVERS USING LIVER ECM AS A SCAFFOLD

With the development of organ decellularization and whole liver scaffold generation, one other potential application for liver matrices is the bioengineering of human livers. As described above, our laboratory and others developed a perfusion decellularization method for the liver, reported for the first time in June 2005 [95]. We applied this technique to liver, pancreas, intestine, and kidney generating decellularized organ scaffolds for organ bioengineering [85,96]. These bioscaffolds preserve their tissue microarchitecture and an intact vascular network that can be readily used as a route for recellularization by perfusion of different cell populations with defined culture media (Figure 27.3). This organ engineering

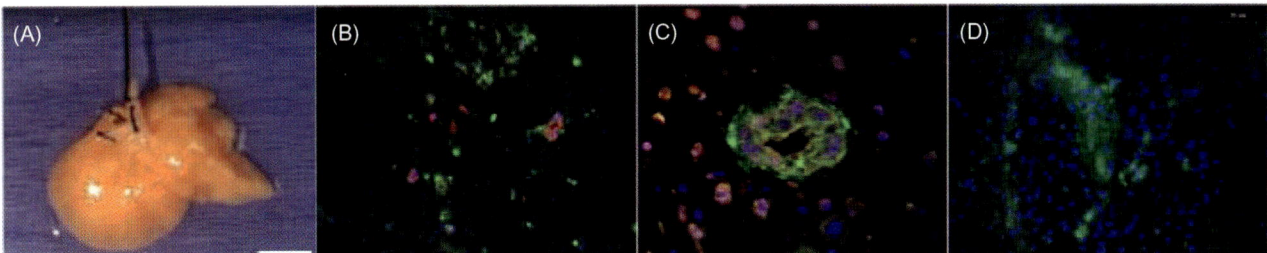

FIGURE 27.3 Bioengineered human liver. (A) Macroscopic appearance of the right lobe of a ferret liver bioscaffold seeded with human liver progenitor and endothelial cells after 7 days in a perfusion bioreactor. Scale bar = 1 cm. (B) Immunostaining for albumin (green) and cytokeratin 19 (red) shows clusters of hepatocytes synthetizing albumin and biliary tubular structures staining intensely for cytokeratin 19. (C) Bile duct immunostained with EpCAM (green) and ASBT (red) showing differentiated polarized cholangiocytes with EpCAM localized in their whole membrane with ASBT being expressed only on the luminal side of the duct. (D) Immunofluorescence staining for P450 CYP2A (green) showing human liver progenitor cells engraftment throughout the liver bioscaffold and differentiating into hepatocyte clusters.

approach has several advantages over the injection of cell suspensions into solid organs. The matrices provide sufficient volume for the transplantation of an adequate cell mass up to whole-organ equivalents [44], without oxygen and nutrient limitations, since continuous perfusion of oxygenated culture media is provided.

Using the organ scaffold technology, our laboratory recently bioengineered a humanized liver using human primary fetal liver stem/progenitor and endothelial cells seeded inside a ferret acellular liver scaffold (Figure 27.3) [85]. These bioengineered livers exhibit some of the functions of a native human liver (albumin and urea secretion, drug metabolism, etc.) and an endothelialized vascular network that prevented platelet adhesion and aggregation, critical for blood vessel patency after transplantation [85]. Hence, this technology has the potential to translate in the future into the bioengineering of human size livers, which may offer readily available organs for drug discovery and for transplantation, overcoming organ shortage.

REFERENCES

[1] Lemaigre FP. Mechanisms of liver development: concepts for understanding liver disorders and design of novel therapies. Gastroenterology 2009;137:62–79.

[2] Tanimizu N, Miyajima A. Molecular mechanism of liver development and regeneration. Int Rev Cytol 2007;259:1–48.

[3] Kinoshita T, Miyajima A. Cytokine regulation of liver development. Biochim Biophys Acta 2002;1592:303–12.

[4] Zorn AM. Liver Development, Stembook, Harvard Stem Cell Institute, Cambridge, MA 2008.

[5] McLin VA, Zorn AM. Molecular control of liver development. Clin Liver Dis 2006;10:1–25.

[6] Si-Tayeb K, Lemaigre FP, Duncan SA. Organogenesis and development of the liver. Dev Cell 2010;18:175–89.

[7] Turner R, et al. Human hepatic stem cell and maturational liver lineage biology. Hepatology 2011;53:1035–45.

[8] Martinez-Hernandez A, Amenta PS. The hepatic extracellular matrix. I. Components and distribution in normal liver. Virchows Arch A Pathol Anat Histopathol 1993;423:1–11.

[9] Wang Y, et al. Paracrine signals from mesenchymal cell populations govern the expansion and differentiation of human hepatic stem cells to adult liver fates. Hepatology 2010;52:1443–54.

[10] Boulter L, et al. Macrophage-derived Wnt opposes Notch signaling to specify hepatic progenitor cell fate in chronic liver disease. Nat Med 2012;18:572–9.

[11] Tanimizu N, Miyajima A. Notch signaling controls hepatoblast differentiation by altering the expression of liver-enriched transcription factors. J Cell Sci 2004;117:3165–74.

[12] Lemaigre FP. Development of the biliary tract. Mech Dev 2003;120:81–7.

[13] Zong Y, et al. Notch signaling controls liver development by regulating biliary differentiation. Development 2009;136:1727–39.

[14] Tanaka M, Itoh T, Tanimizu N, Miyajima A. Liver stem/progenitor cells: their characteristics and regulatory mechanisms. J Biochem 2011;149:231–9.

[15] Antoniou A, et al. Intrahepatic bile ducts develop according to a new mode of tubulogenesis regulated by the transcription factor SOX9. Gastroenterology 2009;136:2325–33.

[16] LeCouter J, et al. Angiogenesis-independent endothelial protection of liver: role of VEGFR-1. Science 2003;299:890–3.

[17] Kamiya A, et al. Fetal liver development requires a paracrine action of oncostatin M through the gp130 signal transducer. EMBO J 1999;18:2127–36.

[18] Martinez-Hernandez A, Amenta PS. The extracellular matrix in hepatic regeneration. FASEB J 1995;9:1401–10.

[19] Susick R, et al. Hepatic progenitors and strategies for liver cell therapies. Ann New York Acad Sci 2001;944:398–419.

[20] McClelland R, Wauthier E, Uronis J, Reid L. Gradients in the liver's extracellular matrix chemistry from periportal to pericentral zones: influence on human hepatic progenitors. Tissue Eng Part A 2008;14:59–70.

[21] Martinez-Hernandez A, Amenta PS. The hepatic extracellular matrix. II. Ontogenesis, regeneration and cirrhosis. Virchows Arch A Pathol Anat Histopathol 1993;423:77–84.

[22] Reid LM, Fiorino AS, Sigal SH, Brill S, Holst PA. Extracellular matrix gradients in the space of Disse: relevance to liver biology. Hepatology 1992;15:1198–203.

[23] Lozoya OA, et al. Regulation of hepatic stem/progenitor phenotype by microenvironment stiffness in hydrogel models of the human liver stem cell niche. Biomaterials 2011;32:7389–402.

[24] Volpes R, van den Oord JJ, Desmet VJ. Integrins as differential cell lineage markers of primary liver tumors. Am J Pathol 1993;142:1483–92.

[25] Couvelard A, et al. Expression of integrins during liver organogenesis in humans. Hepatology 1998;27:839–47.

[26] Volpes R, van den Oord JJ, Desmet VJ. Distribution of the VLA family of integrins in normal and pathological human liver tissue. Gastroenterology 1991;101:200–6.

[27] Miyamoto S, et al. Integrin function: molecular hierarchies of cytoskeletal and signaling molecules. J Cell Biol 1995;131:791–805.

[28] Boudreau NJ, Jones PL. Extracellular matrix and integrin signalling: the shape of things to come. Biochem J 1999;339:481–8.

[29] Engler AJ, Sen S, Sweeney HL, Discher DE. Matrix elasticity directs stem cell lineage specification. Cell 2006;126:677–89.

[30] Janmey PA, Miller RT. Mechanisms of mechanical signaling in development and disease. J Cell Sci 2011;124:9–18.

[31] Discher DE, Janmey P, Wang YL. Tissue cells feel and respond to the stiffness of their substrate. Science 2005;310:1139–43.

[32] Wells R. The role of matrix stiffness in regulating cell behavior. Hepatology 2008;47:1394–400.

[33] Suh J, DiSilvestro M. Biphasic poroviscoelastic behavior of hydrated biological soft tissue. J Appl Mech Trans ASME 1999;66:528–35.

[34] Smutny M, Yap AS. Neighborly relations: cadherins and mechanotransduction. J Cell Biol 2010;189(7):1075–7.

[35] Georges PC, et al. Increased stiffness of the rat liver precedes matrix deposition: implications for fibrosis. Am J Physiol Gastrointest Liver Physiol 2007;293(6):G1147–54.

[36] Hansen LK, Wilhelm J, Fassett JT. Regulation of hepatocyte cell cycle progression and differentiation by type I collagen structure. Current Topics Dev Biol 2006;72:205–36.

[37] Fassett J, Tobolt D, Hansen LK. Type I collagen structure regulates cell morphology and EGF signaling in primary rat

hepatocytes through cAMP-dependent protein kinase A. Mol Biol Cell 2006;17(1):345−56.

[38] Semler EJ, Ranucci CS, Moghe PV. Mechanochemical manipulation of hepatocyte aggregation can selectively induce or repress liver-specific function. Biotechnol Bioeng 2000;69(4):359−69.

[39] Li Z, et al. Transforming growth factor-beta and substrate stiffness regulate portal fibroblast activation in culture. Hepatology 2007;46(4):1246−56.

[40] Wells RG. The role of matrix stiffness in hepatic stellate cell activation and liver fibrosis. J Clin Gastroenterol 2005;39 (4 Suppl 2):S158−61.

[41] Iwamoto H, Sakai H, Nawata H. Inhibition of integrin signaling with Arg-Gly-Asp motifs in rat hepatic stellate cells. J Hepatol 1998;29(5):752−9.

[42] Yee HF. Rho directs activation-associated changes in rat hepatic stellate cell morphology via regulation of the actin cytoskeleton. Hepatology 1998;28(3):843−50.

[43] Lozoya O, et al. Regulation of hepatic stem/progenitor phenotype by microenvironment stiffness in hydrogel models of the human liver stem cell niche. Biomaterials 2011;32:7389−402.

[44] Schmelzer E, et al. Three-dimensional perfusion bioreactor culture supports differentiation of human fetal liver cells. Tissue Eng Part A 2010;16(6):2007−16.

[45] Miki T, Ring A, Gerlach J. Hepatic differentiation of human embryonic stem cells is promoted by three-dimensional dynamic perfusion culture conditions. Tissue Eng Part C-Methods 2011;17 (5):557−68.

[46] Hsu WM, et al. Liver-assist device with a microfluidics-based vascular bed in an animal model. Ann Surg 2010;252:351−7.

[47] Grossman M, et al. Successful *ex vivo* gene therapy directed to liver in a patient with familial hypercholesterolaemia. Nat Genet 1994;6:335−41.

[48] Grossman M, et al. A pilot study of *ex vivo* gene therapy for homozygous familial hypercholesterolaemia. Nat Med 1995;1:1148−54.

[49] Fox IJ, et al. Treatment of the Crigler-Najjar syndrome type I with hepatocyte transplantation. N Engl J Med 1998;338: 1422−6.

[50] Horslen SP, et al. Isolated hepatocyte transplantation in an infant with a severe urea cycle disorder. Pediatrics 2003;111:1262−7.

[51] Ambrosino G, et al. Isolated hepatocyte transplantation for Crigler-Najjar syndrome type 1. Cell Transplant 2005;14:151−7.

[52] Muraca M, et al. Hepatocyte transplantation as a treatment for glycogen storage disease type 1a. Lancet 2002;359:317−8.

[53] Sokal EM, et al. Hepatocyte transplantation in a 4-year-old girl with peroxisomal biogenesis disease: technique, safety, and metabolic follow-up. Transplantation 2003;76:735−8.

[54] Strom SC, et al. Hepatocyte transplantation as a bridge to orthotopic liver transplantation in terminal liver failure. Transplantation 1997;63:559−69.

[55] Strom SC, Chowdhury JR, Fox IJ. Hepatocyte transplantation for the treatment of human disease. Semin Liver Dis 1999;19:39−48.

[56] Strom SC, et al. Transplantation of human hepatocytes. Transplant Proc 1997;29:2103−6.

[57] Combs C, et al. Rapid development of hepatic alpha1-antitrypsin globules after liver transplantation for chronic hepatitis C. Gastroenterology 1997;112:1372−5.

[58] Mito M, Kusano M, Kawaura Y. Hepatocyte transplantation in man. Transplant Proc 1992;24:3052−3.

[59] Fisher RA, Strom SC. Human hepatocyte transplantation: worldwide results. Transplantation 2006;82:441−9.

[60] Kharaziha P, et al. Improvement of liver function in liver cirrhosis patients after autologous mesenchymal stem cell injection: a phase I-II clinical trial. Eur J Gastroenterol Hepatol 2009;21:1199−205.

[61] Salama H, et al. Autologous hematopoietic stem cell transplantation in 48 patients with end-stage chronic liver diseases. Cell Transplant 2010. p. 1475−86.

[62] Zacharoulis D, et al. Autologous infusion of expanded mobilized adult bone marrow-derived CD34 + cells into patients with alcoholic liver cirrhosis. Am J Gastroenterol 2008;103:1952−8.

[63] Khan AA, et al. Human fetal liver derived stem cell transplantation as supportive modality in the management of end stage decompensated liver cirrhosis. Cell Transplant 2010. p. 409−18.

[64] Schmelzer E, et al. Human hepatic stem cells from fetal and postnatal donors. J Exp Med 2007;204:1973−87.

[65] Peng L, et al. Autologous bone marrow mesenchymal stem cell transplantation in liver failure patients caused by hepatitis B: short-term and long-term outcomes. Hepatology 2011;54:820−8.

[66] Spahr L, et al. Autologous bone marrow mononuclear cell transplantation in patients with decompensated alcoholic liver disease: a randomized controlled trial. PLoS One 2013;8:e53719.

[67] Gouon-Evans V, et al. BMP-4 is required for hepatic specification of mouse embryonic stem cell-derived definitive endoderm. Nat Biotechnol 2006;24:1402−11.

[68] Gadue P, Huber TL, Paddison PJ, Keller GM. Wnt and TGF-beta signaling are required for the induction of an *in vitro* model of primitive streak formation using embryonic stem cells. Proc Natl Acad Sci USA 2006;103:16806−11.

[69] Basma H, et al. Differentiation and transplantation of human embryonic stem cell-derived hepatocytes. Gastroenterology 2009;136:990−9.

[70] Liu H, Kim Y, Sharkis S, Marchionni L, Jang YY. *In vivo* liver regeneration potential of human induced pluripotent stem cells from diverse origins. Sci Transl Med 2011;3:82−9.

[71] Rhim JA, Sandgren EP, Degen JL, Palmiter RD, Brinster RL. Replacement of diseased mouse liver by hepatic cell transplantation. Science 1994;263:1149−52.

[72] Laconi E, et al. Long-term, near-total liver replacement by transplantation of isolated hepatocytes in rats treated with retrorsine. Am J Pathol 1998;153:319−29.

[73] Guo D, Fu T, Nelson JA, Superina RA, Soriano HE. Liver repopulation after cell transplantation in mice treated with retrorsine and carbon tetrachloride. Transplantation 2002;73:1818−24.

[74] Latini P, et al. Radiobiological considerations of total body irradiation in bone marrow transplant conditioning: hyperfractionation of dose and early results. Rays 1987;12(81−88):110−3.

[75] Soltys KA, et al. Barriers to the successful treatment of liver disease by hepatocyte transplantation. J Hepatol 2010;53:769−74.

[76] Smets F, Najimi M, Sokal EM. Cell transplantation in the treatment of liver diseases. Pediatr Transplant 2008;12:6−13.

[77] Sukhikh GT, Shtil AA. Stem cell transplantation for treatment of liver diseases: from biological foundations to clinical experience (review). Int J Mol Med 2003;11:395−400.

[78] Gupta S, et al. Entry and integration of transplanted hepatocytes in rat liver plates occur by disruption of hepatic sinusoidal endothelium. Hepatology 1999;29:509–19.

[79] Kumaran V, Joseph B, Benten D, Gupta S. Integrin and extracellular matrix interactions regulate engraftment of transplanted hepatocytes in the rat liver. Gastroenterology 2005;129:1643–53.

[80] Uygun BE, et al. Organ reengineering through development of a transplantable recellularized liver graft using decellularized liver matrix. Nat Med 2010;16:814–20.

[81] Barakat O, et al. Use of decellularized porcine liver for engineering humanized liver organ. J Surg Res 2012;173:11–25.

[82] Lang R, et al. Three-dimensional culture of hepatocytes on porcine liver tissue-derived extracellular matrix. Biomaterials 2011;32:7042–52.

[83] Shupe T, Williams M, Brown A, Willenberg B, Petersen BE. Method for the decellularization of intact rat liver. Organogenesis 2010;6:134–6.

[84] Wang Y, et al. Lineage restriction of human hepatic stem cells to mature fates is made efficient by tissue-specific biomatrix scaffolds. Hepatology 2011;53:293–305.

[85] Baptista PM, et al. The use of whole organ decellularization for the generation of a vascularized liver organoid. Hepatology 2011;53:604–17.

[86] Sgroi A, et al. Transplantation of encapsulated hepatocytes during acute liver failure improves survival without stimulating native liver regeneration. Cell Transplant 2011;20:1791–803.

[87] Umehara Y, et al. Improved survival and ammonia metabolism by intraperitoneal transplantation of microencapsulated hepatocytes in totally hepatectomized rats. Surgery 2001;130:513–20.

[88] Carpentier B, Gautier A, Legallais C. Artificial and bioartificial liver devices: present and future. Gut 2009;58:1690–702.

[89] Demetriou AA, et al. Prospective, randomized, multicenter, controlled trial of a bioartificial liver in treating acute liver failure. Ann Surg 2004;239:660–7.

[90] van de Kerkhove MP, et al. Phase I clinical trial with the AMC-bioartificial liver. Int J Artif Org 2002;25:950–9.

[91] Watanabe FD, et al. Clinical experience with a bioartificial liver in the treatment of severe liver failure. A phase I clinical trial. Ann Surg 1997;225:484–91.

[92] Pless G. Bioartificial liver support systems. Methods Mol Biol 2010;640:511–23.

[93] Nibourg GA, et al. Liver progenitor cell line HepaRG differentiated in a bioartificial liver effectively supplies liver support to rats with acute liver failure. PLoS One 2012;7:e38778.

[94] Zhao LF, Pan XP, Li LJ. Key challenges to the development of extracorporeal bioartificial liver support systems. Hepatobiliary Pancreat Dis Int 2012;11:243–9.

[95] Baptista PM, Atala A, Soker, S. In: 3rd International society for stem cell research international meeting, San Francisco, CA; 2005.

[96] Baptista PM, et al. Whole organ decellularization—a tool for bioscaffold fabrication and organ bioengineering. Conf Proc IEEE Eng Med Biol Soc 2009;2009:6526–9.

Part IV

Heart

Chapter 28

Current Status of Heart Transplantation

Sunu S. Thomas and Mandeep R. Mehra

The Center for Advanced Heart Disease, Brigham and Women's Hospital and Harvard Medical School, Boston, MA

Chapter Outline

28.1 THE HEART FAILURE SYNDROME

There are nearly 6 million individuals in the United States with a diagnosis of heart failure. The annual incidence approximates 600,000 new cases, which is further complicated by a high hospitalization rate and yearly mortality rate in excess of 30% in patients with advanced disease. With economic costs continuing to spiral out of control, the burden and severity of this epidemic cannot be understated [1].

The cardinal feature of heart failure is the elevation of ventricular filling pressures that eventually lead to a compromised cardiac output. This may arise from poor systolic function resulting from impairment of ventricular contractility, termed *Heart Failure with Reduced Ejection Fraction* (HFrEF). Alternatively, heart failure may arise from poor ventricular compliance leading to impairment of diastolic parameters, termed *Heart Failure with Preserved Ejection Fraction* (HFpEF). Regardless of etiology, patients presenting with signs and symptoms of heart failure may do so acutely with *de novo* ventricular dysfunction, or as an exacerbation of a known chronic heart failure condition. Patients universally complain of similar symptoms including dyspnea, fatigue, presyncope, syncope, orthopnea, and paroxysmal nocturnal dyspnea reflecting the consequences of impaired circulatory flow and overall fluid retention. Physical examination typically reveals hallmark findings of volume overload including pulmonary congestion, jugular venous distension, increasing abdominal girth and peripheral edema, and signs of poor perfusion including relative hypotension, cool

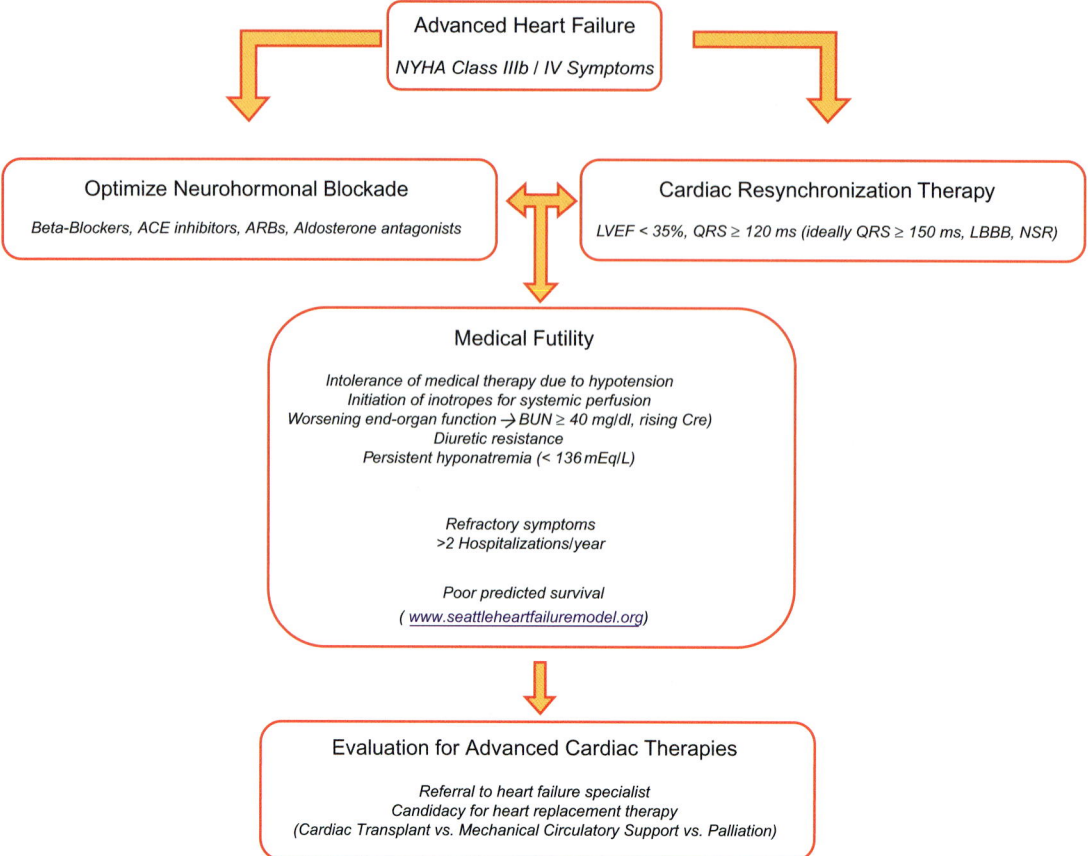

FIGURE 28.1   Algorithm for medical decision-making in patients with end-stage heart failure. *Modified from Mehra et al. [23].*

extremities, and end-organ dysfunction. These bedside findings may be further categorized into four prognostically important hemodynamic profiles on the basis of adequacy of perfusion (warm versus cold) and measure of volume overload (dry versus wet) [2]. Patients who are "cold and wet" represent the most advanced stages of disease and have a 1 year increased risk of death and need for cardiac transplantation which is also two times worse than those who are "warm and dry" [3,4].

The management of heart failure has evolved over the course of the last two decades toward an evidence-based strategy incorporating the use of beta-blockers, angiotensin converting enzyme inhibitors, or angiotensin receptor blockers and aldosterone antagonists [5–15]. In addition, survival and quality of life have significantly improved with the use of implantable defibrillators and cardiac resynchronization therapy in eligible patients. However, the clinical challenge lies in the relative ceiling effect of the benefits derived from these medical and device-based therapies [16–18]. Although mortality rates have improved by the use of disease modifying therapy, with few exceptions such as fulminant myocarditis or a peripartum cardiomyopathy, the etiology of the underlying cardiomyopathy is generally nonreversible [19,20].

It appears that regardless of inciting injury or disease process, all cardiomyopathies share a common pathophysiological pathway. While our current therapies target the mediators of neurohormonal imbalance, for reasons still poorly understood, the impaired ventricle negatively remodels through a process of dilatation and myocardial thinning [21]. Clinically, this amounts to persistence and worsening of symptoms, deteriorating ventricular function, hemodynamic and electrical instability, and frequent rehospitalizations, culminating in a medically refractory state and inevitable death [22].

Stage D heart failure, characterized by a refractory state of structural heart disease, unrelenting symptomology and grave prognosis, often requires the escalation of medical therapies [1] (Figure 28.1). This typically requires incremental use of diuretics to relieve congestion, and for the hemodynamically compromised patient, the judicious use of inotropic therapy either for temporary relief of congestion or as a longer-term bridging or palliation strategy to augment a low cardiac output state. However, symptomatic relief with such medications does not improve prognosis or reverse the downward spiral leading to disease progression, and may hasten patient mortality [24,25]. For some patients,

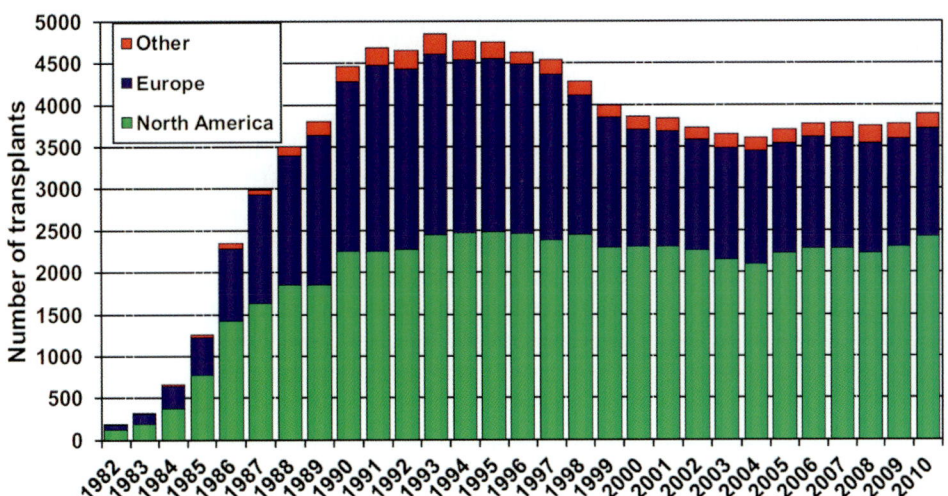

FIGURE 28.2 Cardiac transplantation trends according to time and region. *Figure reprinted from [33].*

marginal blood pressures necessitate the withdrawal of medications, such as beta-blockers and antagonists of the renin–angiotensin–aldosterone system, originally prescribed for their mortality benefit. For others too hemodynamically compromised, ventricular assist devices (VADs) have provided a necessary bridge to transplantation for mechanical circulatory support [26,27]. However, the arduous path traveled by the heart failure patient eventually reaches a crossroad at which point a decision must be made jointly by the patient and their advanced cardiac care providers as to whether to pursue a path of palliation and comfort, or to engage in the process leading to possible cardiac transplantation.

This chapter will outline our current progress and understanding of cardiac transplantation as a heart replacement therapy for selected patients with end-stage heart failure.

28.2 CARDIAC TRANSPLANTATION: HISTORICAL ACHIEVEMENT

The first successful human cardiac transplantation by Dr. Christian Baarnard in 1967 ushered in the modern age of heart replacement therapy [28]. This achievement was the result of decades of earnest investigation examining surgical technique and the biology of organ rejection. Beginning in 1905, the first case of a successful heart transplant involved the implantation of a canine heart into the neck of a canine recipient [29]. The transplant donor organ was anastomosed heterotopically to the recipient's cardiopulmonary circulation without removal of the native heart. In 1964, Dr. James Hardy performed the first successful xenotransplantation in which the heart of a chimpanzee was transplanted into a human and was functional for 90 minutes at which point it began to fail [30].

With the development of cardiopulmonary bypass, improved organ preservation with cardioplegia, and refinement of surgical technique including orthotopic transplantation, early investigators were able to improve graft survival in animal models and recognize the need for improved immunosuppression.

The success of the first human cardiac transplant was followed promptly by both a surge in optimism and in the number of surgeries to at least 100 in 17 countries, including the first US transplantation by Dr. Norman Shumway 1 month after Dr. Barnaard's achievement. However, poor posttransplant survival, attributed to poor allograft survival and opportunistic infections, resulted in contraction in both the number of transplant surgeries and programs offering them. Yet, not unlike many innovations in modern medicine requiring refinement of process, the field was revitalized with the development of endomyocardial biopsies for rejection surveillance [31], improved immunosuppression strategies primarily involving the use of cyclosporine [32], and the formation of the United Network for Organ Sharing (UNOS) to proctor the procurement and allocation of donor hearts to eligible recipients.

According to the most recent 2012 report from the Registry of the International Society for Heart and Lung Transplantation, nearly 4000 heart transplants are now performed worldwide annually, consistent with trends over the last decade (Figure 28.2) [33]. Overall, more than 100,000 heart transplants have been performed since Dr. Barnard's seminal surgery. Unfortunately, however, there continues to be great imbalance in the available donor heart pool and the number of patients requiring them, with those in need numbering close to half a million in the United States alone [1]. In fact, the number of patients active on the transplant list has grown by 19.2% since 2005 [34].

TABLE 28.1 Listing Criteria for Cardiac Transplant

Indications

Anticipated poor 1 year survival

Advanced symptoms (NYHA Class IV) refractory to medical therapy

Peak VO$_2$ < 12 mL/kg/min with beta-blockers; <14 mL/kg/min without beta-blockers

Refractory angina not amenable to revascularization

Intractable arrhythmia not amenable to pharmacotherapy, ablation, or defibrillator therapy

Contraindications

Absolute

Fixed pulmonary hypertension
    Pulmonary artery systolic pressure >60 mmHg
    Pulmonary vascular resistance >6 Wood units
    Transpulmonary gradient >15 mmHg

Concomitant illness adversely limiting life expectancy (<2 years)
    Malignancy within 5 years
    Persistent renal failure despite
      medication adjustments or
      augmentation of perfusion with inotropes or circulatory
      support

Irreversible hepatic dysfunction

Severe obstructive pulmonary disease (FEV$_1$ <1 L/min)

AIDS with recurrent opportunistic infection

Active disease with multisystem involvement, e.g., SLE, amyloidosis, sarcoidosis

Relative

Age >72 years

Morbid obesity (BMI > 35 kg/m^2)

Diabetes with end-organ involvement

Significant cerebrovascular or peripheral vascular disease

Psychosocial issues
  Active substance abuse
  Active mental illness
  Medical noncompliance
  Current smoking habit

Modified from Mehra et al. [41] and Leitz and Mancini [40].

28.3 HEART TRANSPLANT EVALUATION

The heart failure patient being evaluated for cardiac transplantation will typically have a dire prognosis. Death is anticipated within 1 year despite optimization of medical therapies that may alter symptom status, ventricular function, or electrical and hemodynamic instability. Ambulatory patients typically describe severe functional limitation with NYHA Class IIIb or IV symptoms. The severity of hemodynamic compromise is obvious in patients with cardiogenic shock requiring hemodynamic support using inotropes, intra-aortic balloon pumps, or mechanical circulatory assist devices. For such patients, a cardiac transplant may provide a solution to an otherwise irreversible process with the hope of an improved quality

and years of life. The challenge for the advanced cardiac care team is in the identification of those patients who will be both eligible and capable of benefitting from a valuable, yet limited, resource.

Based on recently published International Society of Heart and Lung Transplant (ISHLT) registry data, the typical transplant patient can be characterized as being male (76.3%), approximately 54 years of age with an underlying nonischemic cardiomyopathy (53.8%), and with medical comorbidities including diabetes (25.4%), hypertension (44.1%), and prior cardiac surgery (45.4%). The relative medical acuity of the candidate patient awaiting transplant is high, as inferred by the use of intravenous inotropes (42.6%) and nearly a majority being hospitalized at the time of transplantation (44.6%). Patients requiring left ventricular mechanical support accounted for 27.3% of transplant patients.

A challenge in prognosticating patients with heart failure is the lack of concordance between clinical symptoms, left ventricular function, and overall functional capacity. In fact, left ventricular ejection fraction is not itself a sole listing criteria for cardiac transplant. In 1991, Mancini and colleagues introduced the utility of peak oxygen consumption as an objective measure of cardiac performance and transplant eligibility [35]. In their seminal work, they found that a peak VO$_2$ less than 14 mL/kg/min was associated with a 1 year mortality rate of less than 50%. By providing a diagnostic cut-off value, they were able to conclude that cardiac transplant could be safely postponed in patients with a peak VO$_2$ greater than 14 mL/kg/min, as their 1 year survival rate was 94%. The peak VO$_2$ may be further modified by patient characteristics including age, gender, and race, and also by specific therapies such as beta-blockers [36−39]. Accordingly, cardiac transplantation may be indicated in patients with a peak VO$_2$ of less than 14 mL/kg/min, or if on concurrent beta-blocker therapy, a peak VO$_2$ less than 12 mL/kg/min.

Ultimately, a heart failure patient's candidacy for transplant is not as dependent on the burden of indication but on the identification of any contraindication that may impair graft function and the recipient's quality of life following transplant [40,41]. Table 28.1 highlights the list of indications and contraindications that may determine patient eligibility for a heart transplant. As such, each patient must proceed through a battery of clinical investigations, including a thorough evaluation of their psychological well-being and social circumstances prior to transplant listing.

28.3.1 Age and Comorbidites

Patients aged over 70 years are not typically considered for cardiac transplant, owing primarily to age-associated medical comorbidities and anticipated limited life expectancy. However, with longer waiting times for a transplant, there is a concern regarding the distinction and appropriateness of listing age versus age at transplant that will need to be addressed with a growing and aging heart failure population. Screening for medical comorbidities includes diabetes with end-organ complications, severe pulmonary disease (FEV <40%), liver cirrhosis, severe peripheral vascular disease, and osteoporosis which may complicate perioperative outcomes and limit overall life expectancy. For similar reasons, caution is exercised in listing patients with extremes of body mass index (BMI > 35 kg/m^2 and <18 kg/m^2). However, age alone is no longer an absolute contraindication for cardiac transplantation with several centers having expanded their age limits or considered using alternative lists where older or marginal donors who may otherwise not be used for traditional candidates are paired with older recipients.

28.3.2 Chronic Kidney Disease

Renal failure is common among patients with end-stage heart failure stemming from systemic disease including diabetes, hypertension, and atherosclerosis, or from complex cardiorenal interactions. Efforts to ensure that renal function is not a consequence of impaired perfusion typically employ a trial of inotropic therapy, an intra-aortic balloon pump, or mechanical circulatory support. However, patients with a persistently elevated creatinine (> 2.0 mg/dL), particularly with significant proteinuria, are deemed ineligible for a heart transplant in isolation. A key posttransplant issue is the near automatic deterioration of renal function as a consequence of particular immunosuppressive medications, particularly calcineurin inhibitors, and its associated risk with long-term survival. Therefore, collaborative efforts with renal specialists may be necessary for consideration of renal biopsies to delineate underlying kidney pathology and for potential combined listing for both heart and kidney transplant. The most recent ISHLT registry data suggest that heart–kidney transplants have increased, but still remain low with a total of only 67 such cases in 2010 [33].

28.3.3 Malignancy

All patients undergo age-related cancer screening including colonoscopy, mammography, and prostate evaluation. Patients with a history of malignancy require at least 5 years of remission prior to consideration for transplant. However, the predisposition to skin and lymphoproliferative cancers by posttransplant immunosuppressive therapies may render a patient transplant ineligible regardless

of remission status [42]. More recent guidelines suggest that at least in some situations where survival from the cancer is expected to outlast the median survival from cardiac transplantation, careful consideration to bending the 5-year remission rule may be followed in close consultation with an oncological team.

28.3.4 Infection

Patients with an active infection at the time of evaluation are typically deemed transplant ineligible. An evolving perspective in the field has involved the eligibility for patients who possess chronic viral infections, e.g., human immunodeficiency virus (HIV). The risk of opportunistic infections with immunosuppression, potential drug interactions with antiretroviral therapy, and presumed limits to HIV-related life expectancy represent a sampling of reasons explaining the relatively few cases of heart transplantation in patients with the disease [43]. However, renal transplant outcomes in HIV-positive patients are comparable to those kidney recipients not infected with the virus [44]. Personal communications from transplant centers suggest that candidacy for heart transplant may be limited by under-referral for advanced cardiac therapies stemming from prohibitive comorbidities and psychosocial circumstances.

Outcomes in cardiac transplant patients with a preexisting history of chronic hepatitis B or C viral infection may be attributed more to the progression of liver disease rather than allograft failure [45,46]. Routine transplant evaluations in such patients may involve a liver biopsy to assess for pathological cirrhosis, or a serum measurement of viral titers. In the end, however, transplant candidacy may be determined largely by center experience.

28.3.5 Pulmonary Hypertension

Prior to listing, patients undergo right heart catheterization for the assessment of pulmonary pressures. An absolute contraindication, fixed pulmonary arterial hypertension can lead to the development of acute right ventricular failure of the donor heart not accustomed to the severity of such an afterload. The degree of reversible pulmonary hypertension is evaluated using pulmonary vasodilators, including adenosine, prostacyclin, nitroprusside, or nitric oxide; or inodilators such as milrinone or dobutamine. Pulmonary hypertension is contraindicated in patients who maintain a transpulmonary gradient greater than 15, a pulmonary systolic pressure greater than 50 mmHg or the indexed pulmonary vascular resistance exceeds 6 Woods units, despite a vasodilator challenge while maintaining a systemic pressure greater than 85 mmHg [41].

TABLE 28.2 UNOS Status Definitions for Cardiac Transplant Listing According to Medical Urgency

Status	Definition

Status 1A	A. Acute inpatient mechanical circulatory support
	Left and/or right Ventricular Assist Device (VAD)[a]
	Total Artificial Heart (TAH)
	Intra-aortic balloon pump
	Extracorporeal Membrane Oxygenation (ECMO)
	B. Mechanical circulatory support with device-related complications
	Thromboembolism
	Device infection
	Mechanical failure
	Life-threatening ventricular arrhythmias
	"Other" complications with approval from a respective cardiac transplant regional review board
	C. Continuous mechanical ventilation
	D. Continuous infusion of either
	  i. a single high-dose intravenous inotrope or
	  ii. multiple intravenous inotropes
	plus continuous hemodynamic monitoring of intracardiac pressures using a pulmonary artery catheter

Status 1A Exception	Medical urgency despite failure to meet any of the aforementioned criteria Requires approval from a respective cardiac transplant regional review board

Status 1B	Transplant candidate discharged home with either
	  i. left and/or right VAD or
	  ii. continuous infusion of intravenous inotropes.

Status 2	Transplant candidate who does not meet the criteria for Status 1A or 1B
Status 7	Inactive on the transplant list despite eligibility due to prohibitive medical or social circumstance

[a]Eligible to maintain 1A status for up to 30 days postimplantation.

28.3.6 Psychosocial Functioning

Psychological health and social circumstances are major determinants of transplant eligibility, even after medical clearance. Active substance abuse is an absolute contraindication to transplant. While programs vary in their abstinence periods, patients are typically challenged to remain substance free for a period of 6 months. The data is clear in demonstrating that smoking has a deleterious impact on graft rejection and mortality posttransplant [47]. Psychological function and overall emotional stability are evaluated due to their relationship with adherence to medical regimens and self-care, which ultimately impact rejection status, morbidity, and mortality following transplant [48].

28.3.7 Sensitization

Rejection adversely affects transplant outcomes. As will be outlined, rejection may be cellular or antibody-mediated or mixed, with the consequence of graft dysfunction or more severely, overt graft failure. Antigenic exposure through blood products, pregnancy, VADs, and prior transplantation sensitizes the recipient by increasing their risk for the development of preformed antibodies against donor human leukocyte antigens (HLA). Prior to active transplant listing, programs often subject their sensitized patients to immunomodulating therapies to lessen the risk of anticipated rejection within a given donor pool. These strategies include intravenous immunoglobulin, plasmapheresis, and rituximab to reduce the burden of preformed antibodies, and in so doing, desensitize the patient.

28.4 LISTING STATUS

Once deemed eligible, patients in the United States are listed for cardiac transplant with a priority status in keeping with their medical urgency in accordance with UNOS guidelines. As highlighted in Table 28.2, Status 1A patients are considered of the highest priority due to the severity of their overall clinical acuity. These ICU-bound patients typically require hemodynamic support with a mechanical circulatory assist device, an intra-aortic balloon pump, or multiple inotropes with an *in situ* pulmonary artery catheter for hemodynamic tailoring. Status 1B may require the use of a mechanical circulatory assist device but are typically ambulatory and not admitted to hospital. Status 2 patients do not meet criteria for Status 1A or 1B listing, and Status 7 patients are typically those eligible for transplant but not active on the list due to the development of a clinical circumstance that would make either the surgery or posttransplant outcome deleterious, including active infection, stroke, or prohibitive obesity.

28.5 THE IDEAL DONOR HEART

The availability of a donor organ is typically the result of a fatal accident or self-induced trauma leading to brain death, but with preserved viability of the heart. Despite the limited number of donor organs available, appropriate selection is still necessary to ensure favorable posttransplant outcomes for the recipient. At the time of organ donation, the potential donor heart is evaluated for markers of abnormal function and the medical comorbidities of the deceased. While no single parameter ultimately disqualifies a potential donor, consensus must be reached by the transplant team taking into account the totality of such variables.

28.5.1 Donor Characteristics

Initial considerations focus on donor age, mechanism of death, high-risk social behaviors, and an ongoing assessment of the clinical requirements to ensure preservation of organ function in the deceased. Even though an upper age limit has not been defined, current guidelines recommend the use of donor hearts less than 45 years of age, namely due to the concern for age-related changes to the heart that may impact long-term graft function. However due to donor organ scarcity it is not uncommon to extend this boundary. Death of the donor typically arises from toxic overdoses or brain death resulting from trauma, anoxia, or fatal cerebrovascular accident. Although some studies suggest that traumatic brain injury may portend poorer long-term survival among transplant recipients, such a mechanism of death is nonprohibitive [49,50]. However, there is greater concern for death resulting from toxicity as the transplant outcome data are limited. Drug abuse, incarceration, and lifestyle practices that predispose to communicable disease such as HIV and hepatitis B and C, must be explored. Current consensus enables the use of donor hearts from individuals involved in such high-risk social behaviors as the long-term outcomes from smaller studies are not necessarily worse. However, due to the implicit transmissible infectious risks derived from such patients, careful screening is necessary at the time of transplant evaluation, in addition to informed consent from the recipient.

28.5.2 Evaluation of the Donor Heart

For those hearts obtained from donors older than 40 years of age, concomitant coronary artery disease should be ruled out due to its predilection for the development of coronary allograft vasculopathy and graft failure. Assessment is made by a coronary angiogram, or if an invasive evaluation is not feasible, by direct palpation of the coronary arteries during organ procurement surgery. While valvular disease is not a contraindication to transplant, surgical repair, or replacement of the affected valve can be undertaken immediately prior to transplantation. Echocardiography is an important screening tool used to evaluate the donor heart structure and function. Typically, regional wall motion abnormalities must be further evaluated for related coronary artery disease or an underlying cardiomyopathy in the donor heart. Ventricular hypertrophy has been attributed to early graft failure and mortality posttransplant, particularly if the ventricular thickness exceeds 14 mm. Even though systolic dysfunction may be considered a contraindication, its presence may reflect the transient catecholaminergic imbalance attributed to the patient's clinical condition and mechanism of death leading to global myocardial ischemia and ventricular impairment. Current recommendations prohibit the use of a potential donor heart requiring escalating inotropic therapy such as dopamine 20 μg/kg/min or equivalent doses of other catecholamines [42].

Another important consideration for an orthotopic cardiac transplantation is ensuring that the donor heart matches the size and hemodynamic needs of the recipient. Significant complications arise if a donor-recipient mismatch occurs. A large donor heart may not permit closure of the chest wall, and an undersized heart may not adequately perfuse a larger recipient. Consequently, current guidelines recommend that the weight of donor patient be no less than 30% below that of the recipient [42].

28.6 PERIOPERATIVE MANAGEMENT

28.6.1 Graft Preservation and Optimizing Ischemic Time

A major determinant of early transplant outcomes and overall right ventricular function is the duration of ischemic time. Typically, measured from the instance of aortic cross-clamping at the time of organ procurement until which time the donor heart is perfused within the recipient, there is consensus that this duration not exceed 4 h. This single variable has helped form the basis of geographical territories from which a given transplant candidate's donor pool resides.

The donor heart is immediately subjected to hypoxic-ischemic injury at the time of explantation, followed by the risk of reperfusion injury in the posttransplant period. Current strategies for graft preservation focus on curbing the perturbations in cellular, tissue, and metabolic distress associated with these surgical events. To this end, the donor organ is maintained hypothermic by direct application of ice and a cold perfusate, such as the University of Wisconsin preservation solution consisting of potassium, magnesium, and free radical scavengers, lactobionate and raffinose, to achieve homeostasis of electrolytes and limit myocardial injury [51].

Investigations are ongoing with the use of perfusion systems to improve donor organ integrity and preservation across longer ischemic times. Already approved in Europe, the Transmedics Organ Care System is currently being studied to compare its perfusion system that supplies the donor heart with warm, nutrient-enriched, oxygenated blood with that of conventional cold preservation techniques. Such strategies may potentially reduce ischemic injury and broaden the donor pool across a wider geographical region for eligible transplant candidates.

28.6.2 Early Postoperative Care

As with all high-risk surgical patients, the early posttransplant period can be marked by challenges in cardiopulmonary stability, surgical bleeding, infection, and recovery of end-organ function. Generation of an adequate cardiac output and perfusion pressure may be limited by allograft function and postoperative vasodilatory shock. In addition to preexisting comorbidities, ischemic time, and quality of graft preservation, historical evidence suggests that donor heart inotropic and chronotropic reserve may be furthered impaired by sympathetic denervation and myocardial depletion of stored catecholamines [52,53]. Invasive monitoring using pulmonary catheters may facilitate hemodynamic tailoring according to intracardiac and systemic pressures, mixed venous oxygen saturations, and calculated cardiac outputs. Epinephrine, norepinephrine, and dobutamine are classically infused to provide adrenergic support to the systemic circulation. For patients with elevated central venous pressures, milrinone, a phosphodiesterase-3 inhibitor, may be added to the inotropic regimen for enhanced myocardial contractility and reduction of right ventricular afterload due to its pulmonary vasorelaxant properties. In cases of low systemic vascular resistance, vasopressin or phenylephrine may provide necessary pressor support to maintain an adequate perfusion pressure [42].

The heart rate of the denervated heart is classically tachycardic. However, sinus node dysfunction attributed to graft ischemia, trauma, or inherent donor heart pathology may result in relative chronotropic incompetence in the early postoperative period following transplantation. Isoproterenol and theophylline have been recommended to maintain heart rates greater than 90 bpm. Atrioventricular pacing may be required in the event of medical nonresponsiveness [42].

Persistent hypotension necessitates evaluation for cardiac tamponade, graft dysfunction, and contributes to a vasodilatory state such as infection.

28.6.3 Cardiac Allograft Function

For the transplant patient, there is heightened vigilance for early graft dysfunction attributed to either (i) right ventricular failure, (ii) primary graft failure, (iii) hyperacute rejection, or (iv) acute cellular rejection.

28.6.4 Right Ventricular Failure

Right ventricular dysfunction of the donor heart may arise as a manifestation of primary graft failure. However, high pulmonary vascular pressures, excessive volume loading, and periprocedural factors including direct injury and prolonged ischemic times can compromise the right ventricular function of a normal donor heart. In addition, multiple factors can lead to an acute rise in pulmonary vascular resistance in recipients with no prior history of pulmonary arterial hypertension. Increased pulmonary vascular tone can develop from hypoxemia, unrecognized intracardiac shunts and postoperative lung atelectasis, pleural effusions, and infection. Defined hemodynamically, right ventricular failure is characterized by an elevated right atrial pressure (>20 mmHg), and a low cardiac output in the absence of high left ventricular filling pressures. Additional features include hepatic congestion marked by a transaminitis and hyperbilirubinemia, worsening renal failure, and normal to low pulmonary pressures due to poor systolic function of the right ventricle. Management is typically supportive while monitoring for ventricular recovery. Strategies range from right-sided afterload reduction with pulmonary vasodilators, including inhaled nitric oxide, prostaglandins, or phosphodiesterase inhibitors; chronotropy achieved pharmacologically with beta-agonists or with cardiac pacing, and mechanical circulatory support using right VADs.

28.6.5 Primary Graft Failure

Primary graft failure refers to a state of severe cardiac dysfunction resulting in profound circulatory insufficiency within the first 24 h following transplantation. It is defined by a low cardiac output despite adequate filling pressures, and may arise from either univentricular or biventricular failure following exclusion of conditions with similar hemodynamic profiles including hyperacute rejection and cardiac tamponade. Risk factors for the development of primary graft failure can be attributed to

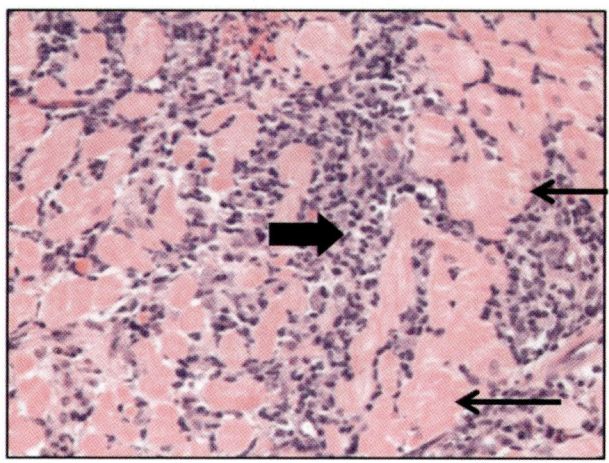

FIGURE 28.3 Acute cellular rejection. Hematoxylin-eosin staining demonstrating lymphocytic infiltration (thick arrow), and myocyte swelling and necrosis (thin arrows). *Image courtesy of Dr. Robert Padera, Brigham and Women's Hospital, Boston, MA.*

Rejection	Grade	Description
None	0	Normal myocardium
Mild	1R	≤ 1 focus of interstitial or perivascular lymphocytic infiltration with myocyte damage
Moderate	2R	≥ 2 foci of lymphocytic infiltration with myocyte damage
Severe	3R	Diffuse lymphocytic infiltration Evidence of myocyte damage, edema, hemorrhage, or vasculitis

TABLE 28.3 ISHLT Acute Cellular Rejection Grades

donor, recipient, and perioperative conditions including older age, inotropic requirements prior to surgery, renal failure, and graft ischemic time [54,55]. Primary graft failure is the main cause of 30 day mortality following transplantation accounting for 39% of all deaths within this period and nearly 20% within the first 12 months [56]. Allograft management is supportive while efforts to restore end-organ perfusion range from aggressive medical therapy using inotropes and vasopressors, to mechanical circulatory support with intra-aortic balloon pumps, extracorporeal membrane oxygenation (ECMO), or VADs to assist either left, right, or to both ventricles [57,58]. Cardiac recovery has been reported to occur within 24 h and up to 7 days following hemodynamic support [58].

28.6.6 Hyperacute Rejection

Hyperacute rejection represents the most severe of immunological reactions against the donor heart. Graft failure arises within minutes to hours following transplantation. Circulating antibodies within the recipient target the HLA class I molecules within the donor heart vascular endothelium. The consequent inflammatory response leads to near immediate graft ischemia and necrosis. While explantation of the donor heart is the only treatment option, patient mortality is inevitable.

28.6.7 Acute Graft Rejection

Acute cellular rejection is mediated by T-lymphocytes and can occur as early as within days of transplant, or as late as several years following. Diagnosis is made by tissue biopsy and can occur in up to 40% of patients within the first year of transplant. Severity is assessed using

ISHLT grades of rejection (IR, 2R, and 3R) on the basis of degree of lymphocytic infiltration and myocyte necrosis (Figure 28.3; Table 28.3). However, recent ISHLT registry data suggests that the threshold to treat rejection appears to be rising with a 10% decline over a 5-year period in the treatment of mild rejection occurring within the first year of transplant [56]. As such, treatment is typically reserved for moderate to severe rejection and may consist of pulse corticosteroids, use of cytolytic therapy, or changes in calcineurin and antiproliferative exposure.

Antibody-mediated rejection (AMR) is characterized by the adverse humoral response of preformed circulating antibodies directed against antigens within the cardiac allograft that can occur days to weeks after transplantation. Risk factors include prior blood transfusions, multiparity, pretransplant use of VADs, previous antibody exposure and retransplantation [59]. AMR is a significant risk factor for poor long-term transplant outcomes, including survival, graft failure, and the development of coronary allograft vasculopathy [60,61]. A tissue diagnosis specifically requiring immunostaining against C4D, which represents a degradation product of activated complement C4b and is a marker of antibody activity (Figure 28.4). A mechanistic approach to the treatment of AMR targets the inactivation and elimination of circulating antibodies, the suppression of T- and B-lymphocytes, plasma cell depletion, and complement inhibition (Figure 28.5) [62].

28.7 IMMUNOSUPPRESSION STRATEGIES AND REJECTION SURVEILLANCE

Immunosuppressive therapy is required to abate the cell-mediated and humoral immune responses that are directed against the donor heart. Although a thorough elaboration of transplant immunology and pharmacology extends beyond the focus of this chapter, a guiding principle governing immunosuppression aims to optimally reduce the incidence of transplant rejection without incurring the

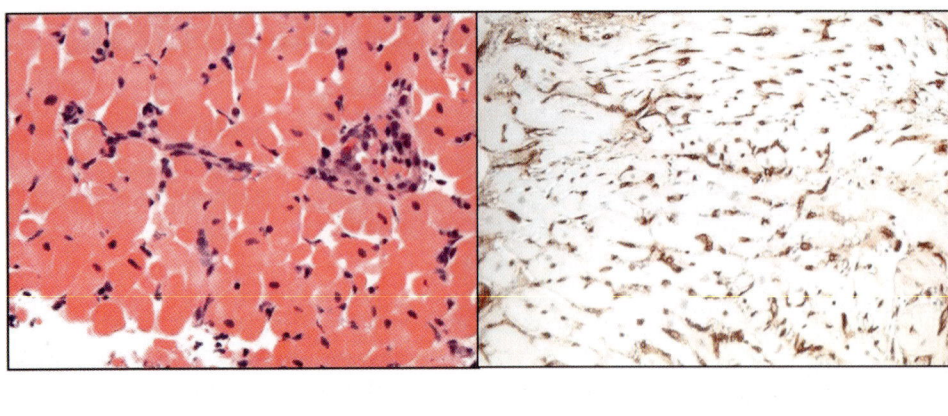

FIGURE 28.4 Antibody-mediated rejection. Hematoxylin-eosin staining (left) and C4D staining (right). *Images courtesy of Dr. Robert Padera, Brigham and Women's Hospital, Boston, MA.*

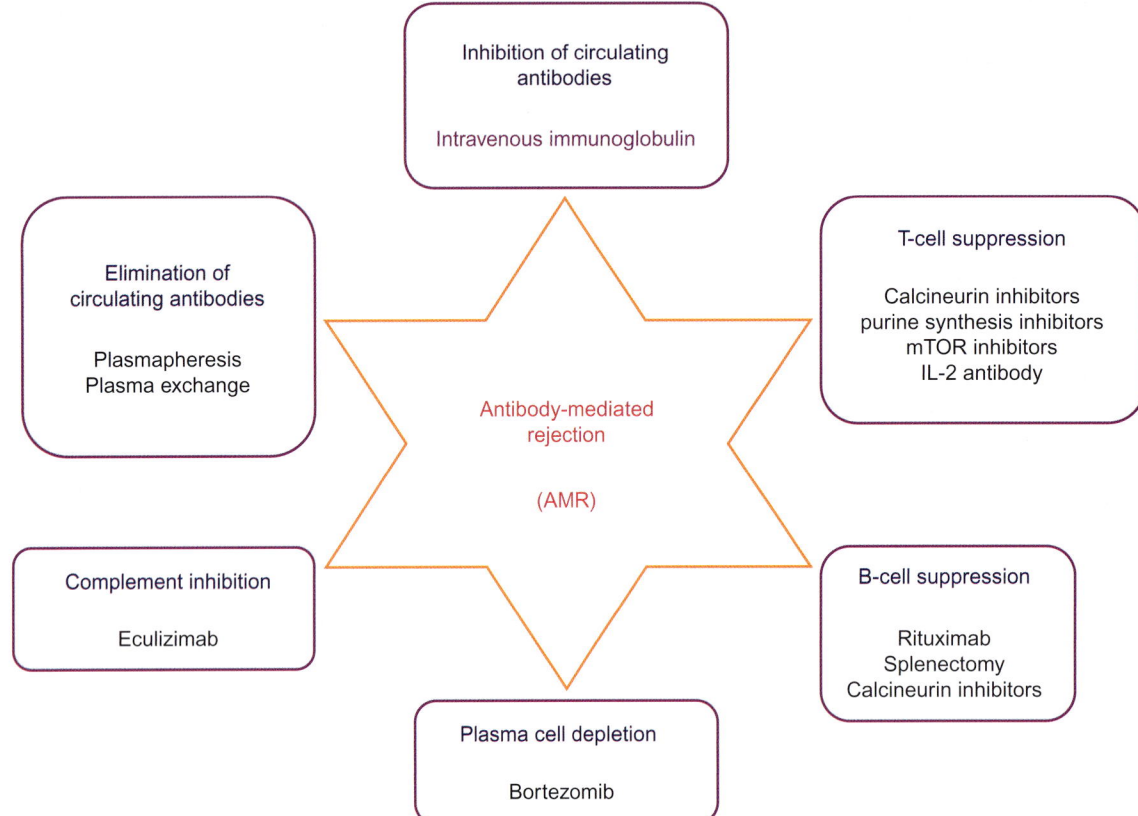

FIGURE 28.5 Targeted treatment of antibody-mediated rejection. *Modified from Nair et al. [62].*

untoward consequences of over-suppression, including opportunistic infection and malignancy risk. Current approaches to immunosuppression are guided by limited clinical trials and largely by institutional and transplant provider experience.

28.7.1 Induction Therapy

Induction therapy refers to the strategy in which an immunomodulating agent is administered to a highly sensitized recipient immediately prior to transplant to reduce the risk of acute rejection. Such patients are known to have preformed circulating antibodies, including African Americans, younger patients, and those with a history of pregnancy, prior sternotomy, or mechanical circulatory support [59]. Alternatively, induction agents have been used as an intermediary antirejection therapy in those patients in whom renal insufficiency renders initiation of a calcineurin inhibitor prohibitive. Historically, induction therapy was implemented as a panacea to promote immune tolerance; however, its contemporary use has remained controversial [63]. An observed incidence of malignancy, including posttransplant lymphoproliferative disorder (PTLD) [64], in addition to cytomegalovirus and fungal infections [65,66] have limited its general use to approximately only half of all transplant centers [56].

OKT3, a murine antithymoctye monoclonal antibody, targets the CD3 receptor of human T-cells rendering it vulnerable to opsonization and clearance by circulating macrophages. However, its use has largely been eliminated due to intolerable side effects, including a cytokine release syndrome manifesting as fever, nausea, emesis, chest pain, and dyspnea [67], and its association with the delayed emergence of PTLD and opportunistic infections in a dose-dependent manner [68]. More commonly, polyclonal antithymocyte antibodies, including the horse-derived ATGAM and rabbit-based ATG, are implemented as induction agents due to a greater potency in reducing T-lymphocytes without increasing long-term PTLD malignancy risk [69,70].

IL-2 receptor antagonists, including basiliximab and dacluzimab, also confer immune tolerance by the depletion of circulating T-lymphocytes. Basiliximab, a chimeric human/mouse antibody, has been shown to be well tolerated as compared to placebo in *de novo* transplant patients treated concomitantly with cyclosporine, mycophenolate mofetil (MMF), and steroids when administered on Day 0 and Day 4 posttransplant [71]. When compared to OKT3 in the randomized multicenter comparison of Basiliximab and Muromonab (OKT3) in heart transplantation (SIMCOR) trial, basiliximab was found to have greater tolerability with similar rates of infection, severity of rejection, and actuarial survival [72]. In comparison to polyclonal antibody therapy with antithymocyte globulin, additional studies have found basiliximab to have similar efficacy in composite endpoint and frequency of acute rejection (Grade $\geq$ 1R), and better safety endpoints including infectious death [73]. However, such results have been contradicted by other findings suggesting that while ATG may have greater infectious risk, basiliximab was not a superior prophylactic against acute rejection [74].

The induction benefit of dacluzimab was studied in a placebo-controlled randomized trial involving 434 new transplant patients who were concomitantly treated with cyclosporine, MMF, and corticosteroids [75]. Despite a reduction in acute cellular rejection, the increased rate of infection-related mortality signaled the death of dacluzimab, leading to its subsequent withdrawal from market production.

28.8 MAINTENANCE IMMUNOSUPPRESSION

Three classes of medications are typically instituted early after cardiac transplantation for maintenance immunosuppression on the basis of supportive clinical trial evidence: (i) corticosteroids, (ii) calcineurin inhibitors, (iii) antiproliferative agents. Mammalian target of rapamycin (mTOR) inhibitors, also called proliferation signal inhibitors, represent a fourth and lesser utilized class of maintenance immunosuppressive therapy.

28.8.1 Corticosteroids

Corticosteroids occupy a central role in the acute treatment of rejection, as an induction agent and for chronic maintenance immunosuppressive therapy. They provide broad immunosuppression by inhibiting the transcription of multiple mediators of the posttransplant inflammatory cascade including IL-1, IL-2, CD40 ligand, TNF-alpha, gamma interferon, GM-CSF, and related growth factors. However, concern with side effects, including psychosis, cushingoid features, hypertension, avascular necrosis, dyslipidemia, steroid-induced hyperglycemia, and adrenal insufficiency have led to the consideration of steroid-free maintenance immunosuppression regimens. Adult and pediatric studies suggest that overall rates of survival and rejection are similar to those on corticosteroids [76,77]. However, as in the case of all immunosuppression strategies, withdrawal of steroids is ultimately patient-specific depending on individual rejection history and chronicity of steroid use. Current trends demonstrate that steroid use has gradually declined further out from transplantation with more than 50% of patients being steroid-free 5 years posttransplantation [56]. Whether steroid-free immunosuppression alters late complications remains uncertain since low, relatively physiologic, doses are typically used in long-term survivors that continue maintenance steroids.

28.8.2 Calcineurin Inhibitors

Cyclosporine and tacrolimus inhibit the enzymatic action of calcineurin which plays an important role in the production of multiple cytokines including IL-2. In so doing, calcineurin inhibitors limit T-cell activation and proliferation, a fundamental tenet of an immunosuppression protocol. Cyclosporine had once been the *de facto* calcineurin inhibitor; however, its use has largely been supplanted by tacrolimus [56]. Although studies have demonstrated similar 1 year survival between cyclosporine and tacrolimus [78,79], the risk of acute rejection was found to be lower in one clinical trial [80] and further supported by a recent meta-analysis [81]. Despite similar efficacy, tacrolimus has a less adverse complication profile favoring its current use. Cyclosporine use can be complicated by hypertension, cholelithiasis, dyslipidemia, and gingival hyperplasia [78–80], in addition to significant drug interactions. Moreover, the results of the TICTAC trial have reinforced the use of tacrolimus as a pivotal agent in any immunosuppression regimen. In this trial, tacrolimus monotherapy resulted in a 6 and 12 month freedom from severe cellular rejection (ISHLT Grade ≥ 2) of 93.3% as compared to 92.9% in the tacrolimus/mycophenolate combination group [82]. In addition, the 3-arm trial comparing combination therapies of tacrolimus/MMF versus tacrolimus/sirolimus versus cyclosporine/MMF found a lower incidence of treated rejection among patients treated with tacrolimus [83]. However, tacrolimus can increase the risk of diabetes and induce neurological complications. In patients with either chronic renal insufficiency or acute kidney failure posttransplant, delayed tacrolimus initiation may be necessary and calcineurin inhibitor-based immunosuppression preempted by induction therapy.

28.8.3 Antiproliferative Agents

Antiproliferative agents, also known as antimetabolites, inhibit cell-cycle pathways to limit T- and B-cell proliferation and thereby reducing the cytotoxic response directed toward the cardiac allograft. Azathioprine, the original antiproliferative drug, mediates its effects through a purine analog metabolite that arrests DNA synthesis once incorporated into the lymphocytic nuclei. MMF and mycophenolic acid are reversible inhibitors of the enzyme inosine monophosphate dehydrogenase. This enzyme serves as an important immunosuppression target as it is both upregulated during T- and B-cell activation and forms part of an obligate pathway in purine synthesis necessary for their proliferation [84]. Contemporary practice trends highlight greater utilization of MMF and mycophenolic acid over azathioprine for maintenance immunosuppressive therapy [56]. In comparison to azathioprine, MMF promotes greater survival, less rejection, and less coronary allograft vasculopathy posttransplant [85−87]. Azathioprine and MMF both can induce significant leukopenia from profound bone marrow suppression. MMF can induce significant gastrointestinal difficulties including diarrhea, gastritis, and nausea.

28.8.4 mTOR Inhibitors

mTOR is an enzyme kinase that mediates lymphocyte growth, differentiation, and proliferation. Sirolimus and everolimus are mTOR inhibitors that in addition to their effects on T- and B-cells also inhibit the proliferation of the vascular smooth muscle wall which has broader clinical significance for the treatment of cardiac allograft vasculopathy (CAV). In comparison to azathioprine, sirolimus and everolimus have been shown to have less rejection and a lower burden of coronary allograft vasculopathy [88,89]. Sirolimus may be included as part of a maintenance immunosuppression regimen for those patients with transplant-related vasculopathy. However, general mTOR inhibitor use has been limited by their association with poor wound healing, renal failure, dyslipidemia, anemia, thrombocytopenia, and the development of both pleural and pericardial effusions [88].

28.9 REJECTION SURVEILLANCE

Acute rejection has been reported to occur in 24.5% of patients within the first year following transplant. Within 5 years, 50.9% of all transplant patients experience at least one episode of acute rejection [34]. Therefore, each institution has its own schedule and methodology for rejection monitoring. Biopsy-proven rejection, particularly those categorized as mild in severity, largely occurs in the absence of symptoms or even allograft dysfunction. Rejection may present clinically across a spectrum to also include new onset brady- or tachyarrhythmias manifesting as palpitations or more significantly as syncope, the reemergence of heart failure symptoms due to a failing graft, or even sudden cardiac death [90]. Symptom-rejection mismatch consequently results in measures of routine surveillance requiring the transplant patient to undergo a potential battery of testing modalities to monitor for cardiac allograft rejection.

28.9.1 Endomyocardial Biopsy

Historically, noninvasive markers, including cardiac enzymes or low QRS voltage on surface electrocardiogram due to myocardial edema or fibrosis, have been used to signal subclinical rejection. Endomyocardial biopsy (EMB) remains the contemporary gold standard for monitoring of graft rejection. Tissue samples are obtained from the right ventricular septum using a bioptome introduced through the right internal jugular vein, typically under fluoroscopy or by echocardiographic guidance [91]. The additional use of a right heart catheter facilitates the measurement of intracardiac pressures and cardiac output. As an invasive procedure, EMBs carry a complication risk of 0.5% in some reports [92,93]. Complications may arise from central venous access including inadvertent arterial puncture, bleeding, and pneumothorax. Alternatively, injury may result from the biopsy itself including cardiac perforation leading to tamponade, arrhythmia, and accidental sampling of tricuspid valve leaflets leading to significant valvular regurgitation [94]. Tissue biopsy yields invaluable information regarding the presence of rejection, type (cellular versus antibody), severity, and chronicity, in addition to pathological tissue changes suggestive of CAV. While institutional practice may vary, typical biopsy schedules include EMBs performed weekly for the first 4 weeks with a tapered frequency over the course of the first 12 months following transplant. Beyond the first year, EMBs are performed annually, or as motivated by patient symptoms. Some programs advocate cessation of routine biopsy surveillance after the first 2 years of transplant with similar reported outcomes to those of continued intensive surveillance.

28.9.2 Biomarker Tests

Gene expression profiling is an alternative noninvasive strategy to assess for the absence of acute cellular rejection [95–97]. With one such technique, Allomap (XDS), RNA from peripheral blood mononuclear cells is isolated and amplified using polymerase chain reaction (PCR) techniques to facilitate gene expression profiles of 11 genes known to distinguish between rejection and nonrejection in posttransplant patients. A score ranging from 0 to 40 is tabulated using an algorithm based on candidate gene expression levels. A score less than 30 has a negative predictive value for biopsy-proven cellular rejection of 99.6% [95] and is deemed to obviate the need for an EMB. Although promising, the technique has been limited by a lack of high specificity and the inability to distinguish between cellular and antibody-mediated rejection [98]. In the IMAGE trial [95], the use of this gene expression profiling biomarker was shown to decrease the number of biopsies required in surveillance while maintaining clinical outcomes. However, some have argued that the gold standard of surveillance biopsies beyond 5 months may itself be a flawed technique and appropriate comparison of the noninvasive biomarker ought to rest on its incremental value to clinical and allograft functional based monitoring. The cost-effectiveness of this approach remains uncertain as does its overall routine use in clinical practice [99].

ImmuKnow (Cylex Inc.) is an immune function assay which tests the ATP production from a transplant patient's isolated T-lymphocytes [100]. By exposing these cells to a T-cell mitogen, plant phytohemaglutinin, measured ATP can identify those patients who may be either excessively or inadequately immunosuppressed. In addition to tailoring immunosuppression, the assay may be able to predict patients at high risk for graft rejection. However, its adoption into mainstream transplant practice is pending larger-scale clinical trial data. Current observational evidence suggests that its ability to predict infection may be greater than its ability to inform on rejection.

28.10 TRANSPLANTATION OUTCOMES

One year life expectancy following heart transplant ranges from 79% to 86% depending on the underlying cardiomyopathy of the recipient, with nonischemic etiologies of the highest survivorship. In 2011, the rate of graft failure within the first 6 weeks posttransplant improved to 4.9% as compared to previous reports [34]. Risk factors for death within the first year include graft ischemic times greater than 200 min, extremes of recipient age, and a pretransplant requirement for mechanical circulatory support. When stratified according to the first 3 years posttransplant, causes for early death are predominantly associated with graft failure and overwhelming infection. Beyond 3 years, mortality is attributed to malignancy and the development of CAV [56]. Overall graft survival at 5 years approximates 74.9% regardless of medical urgency (Status 1A, 1B, or 2) or underlying heart failure etiology [34]. For those patients surviving their first year posttransplant, data from the Scientific Registry of Transplant Recipients suggest that their predicted half-life is 14.0 years. In fact, both graft and patient outcomes have continued to improve with 21,457 survivors living in 2011 as compared to 16,259 in 2001 [34], with improving trends over the last three decades (Figure 28.6).

Posttransplant morbidity is typically associated with renal failure, malignancy, acute rejection, and CAV [56]. Patients have an increasing risk of renal failure over time following transplant with an incidence of 6%, 11%, and 16% at 1, 3, and 5 years posttransplant, respectively. Risk factors include pretransplant recipient comorbidities of increasing age, history of diabetes, and the preexistence of an elevated serum creatinine. Within the first year posttransplant, mortality is typically attributed to acute rejection, infection, and graft failure. At 5 and 10 years posttransplant, the mortality burden arises from malignancy contributing to over 20% of deaths with an incidence of 6% and 15%, respectively (Figure 28.7). Skin cancer is the most common malignancy in the transplant patient. Other malignancies include lymphoproliferative disorders and cancers of the prostate, lung, bladder, kidney, breast, and colon. Risk factors include recipient age and long-term immunosuppression [42].

28.10.1 Cardiac Allograft Vasculopathy

CAV remains the most indolent long-term complication affecting the donor heart coronary vasculature following transplantation. Unlike an atherosclerotic plaque, CAV is classically characterized by the diffuse and progressive proliferation of the intimal layer of the arterial wall and can extend from epicardial vessels to deep penetrating resistance arterioles and capillaries (Figure 28.8). However, the remodeling process involves all three layers of the vascular wall and is influenced by adventitial scarring and alterations in medial tone. These hyperplastic lesions can lead to luminal narrowing sufficient to obstruct coronary blood flow. Although episodes of cellular and antibody-mediated rejection have been associated

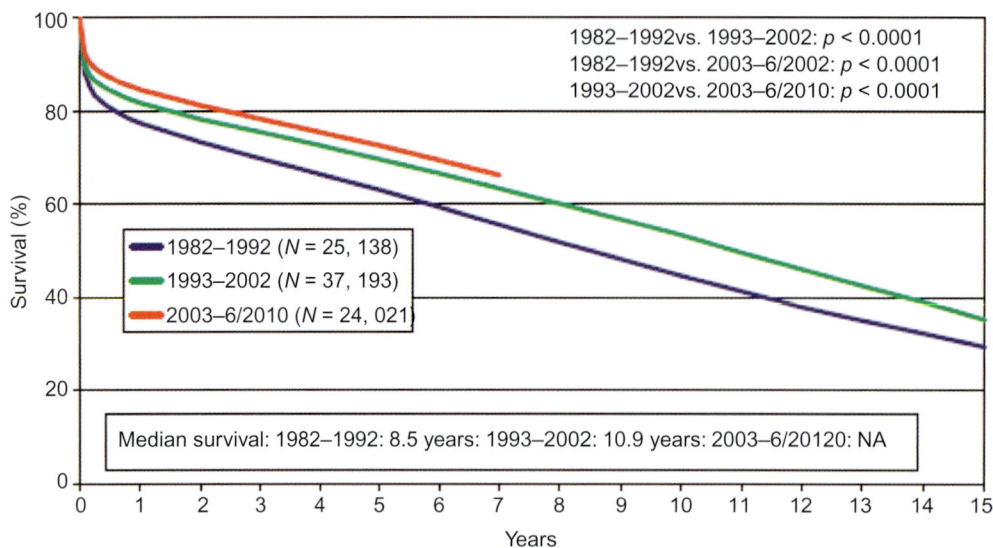

FIGURE 28.6 Cardiac transplant survival curves according to transplant era. *Figure reprinted from [33].*

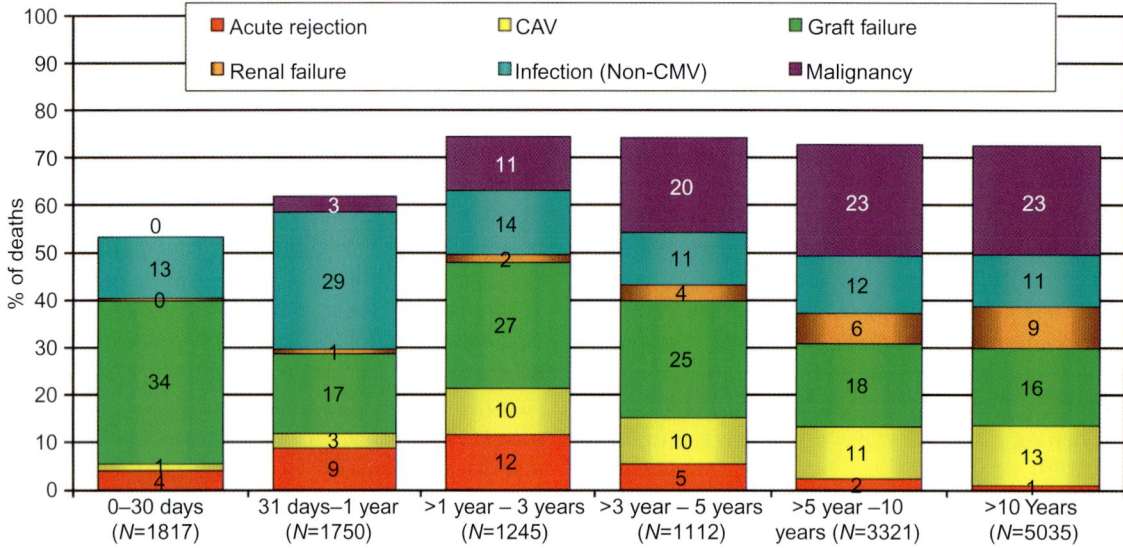

FIGURE 28.7 Causes of death posttransplant. *Figure reprinted from [33].*

with CAV, current immunosuppression protocols are limited in their ability to prevent or delay its inevitable occurrence. Numerous risk factors have been associated with the development of CAV including age, sex, preexisting coronary artery disease, and cytomegalovirus infection in both donor and recipient. Most importantly, the number and severity of acute cellular rejection predisposes to the development of CAV suggesting that its predisposition is more likely a manifestation of the immunogenicity of the cardiac allograft itself. Mechanistic studies demonstrate that CAV is the inflammatory consequence of a vascular remodeling process mediated by the complex interplay between numerous

cytokines, including interferon-γ and the cellular agents of innate immunity, such as T-cells, macrophages, and monocytes. Although 10% of patients can develop CAV as early as within the first year of transplant, it is as much a marker of graft survival with a prevalence of greater than 50% in those patients surviving at least 10 years following transplantation [56].

Coronary angiography remains the gold standard for the evaluation of CAV. Severity of CAV is classified according to the degree of luminal narrowing of the affected vessel (e.g., left main), its associated anatomic location (proximal versus distal third), and the total number of vessels involved [60]. Greater severity is ascribed

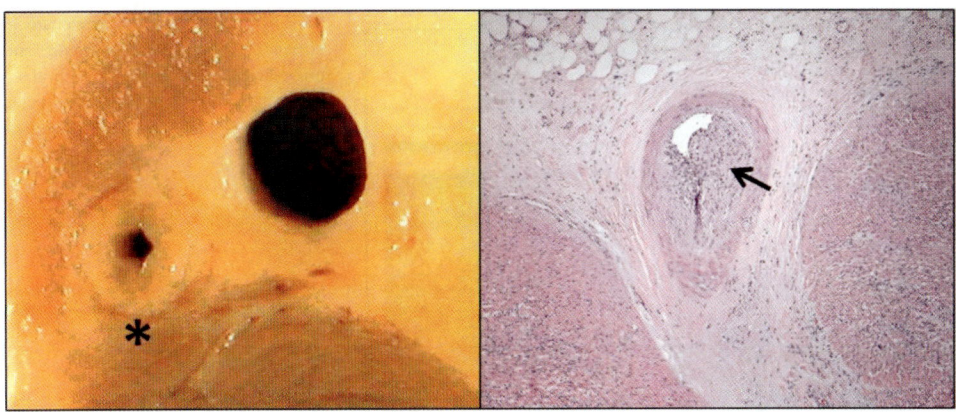

FIGURE 28.8 CAV—gross tissue specimen (left) demonstrating significant CAV-related vascular narrowing of a coronary artery (*) as compared to an unaffected adjacent larger vessel. Intimal proliferation with near luminal obliteration (arrow) by hematoxylin-eosin staining (right). *Images courtesy of Dr. Robert Padera, Brigham and Women's Hospital, Boston, MA.*

to those lesions associated with allograft dysfunction. This may be the result of systolic impairment with a measured ejection fraction of less than 45% or with hemodynamic or echocardiographic indices of restrictive physiology, if contractility is preserved. Other imaging modalities that have been employed to evaluate include intravascular ultrasound (IVUS), myocardial perfusion scans including technetium-99m sestamibi, dobutamine stress echocardiography, and multidetector computed tomography.

CAV prevention includes the use of statins predominantly for their lipid lower benefit. Small prospective clinical trials using pravastatin and simvastatin demonstrated long-term benefits in the reduction of CAV, intimal thickness, and improved patient survival [101–103]. The early initiation of diltiazem, often used to treat corticosteroid or cyclosporine-induced hypertension, was also found to be beneficial in CAV prevention [104]. However, these studies were conducted in the prestatin era of transplant management. An adjuvant approach to immunosuppression in patients with documented CAV consists of the antiproliferative drugs, everolimus and sirolimus. In addition to decreasing rejection events, these drugs have also been shown to reduce CAV [88,89] as measured by IVUS detection of intimal thickening. However, issues with poor wound healing, renal failure, marrow suppression, and drug interaction have limited their use, and specifically for everolimus, FDA approval. It is unclear whether the reduction in surrogate parameters of intimal thickening will indeed be accompanied by reduced cardiac allograft adverse events.

Revascularization is often limited due to the diffuseness of disease along the coronary vasculature. Angioplasty by percutaneous intervention may be an option for patients with single-vessel disease. However, stent patency may be at risk to restenosis by the same immune factors directed against the heart. Coronary bypass surgery is typically limited due to the lack of viable distal vessel targets due to intimal proliferation. Most importantly, revascularization does not reduce the burden of allograft rejection which is the predominant risk factor for CAV progression [105,106]. As such, retransplantation remains the definitive treatment for severe CAV in those patients deemed eligible.

28.11 MECHANICAL HORIZON: VADS AS A BRIDGE TO TRANSPLANT

VADs provide mechanical circulatory support to the failing heart. Since the initial use of the first artificial heart device in 1963 for postcardiotomy shock [107], this technology has evolved to significantly alter the therapeutic options for patients with end-stage cardiomyopathy. There has been a proliferation of their use, with over 6000 VADs implanted in the United States [108]. Technological advancements over the last five decades has led to the development of smaller VADs that can provide either pulsatile or continuous blood flow to the left, right, or both ventricles.

VADs have been implanted in patients for any of four indications. As *destination therapy* (DT), VADs provide long-term hemodynamic support for patients who are deemed transplant ineligible. *Bridge to recovery* (BTR) refers to the strategy by which VADs are implanted in patients with significant cardiac decompensation but anticipated myocardial recovery, such as postcardiotomy shock, peripartum cardiomyopathy, or fulminant myocarditis. For transplant candidates awaiting heart replacement, VADs are used as a *bridge to transplantation* (BTT). *Bridge to candidacy* (BTC) represents a growing population of patients implanted with VADs who, although initially deemed transplant ineligible, are afforded both the benefits of time and augmented perfusion to enable a reconsideration of their transplant status.

Initial US FDA approval for BTT VADs took place in 1994 with older generation devices. Since 2006, data from the Interagency Registry for Mechanically Assisted

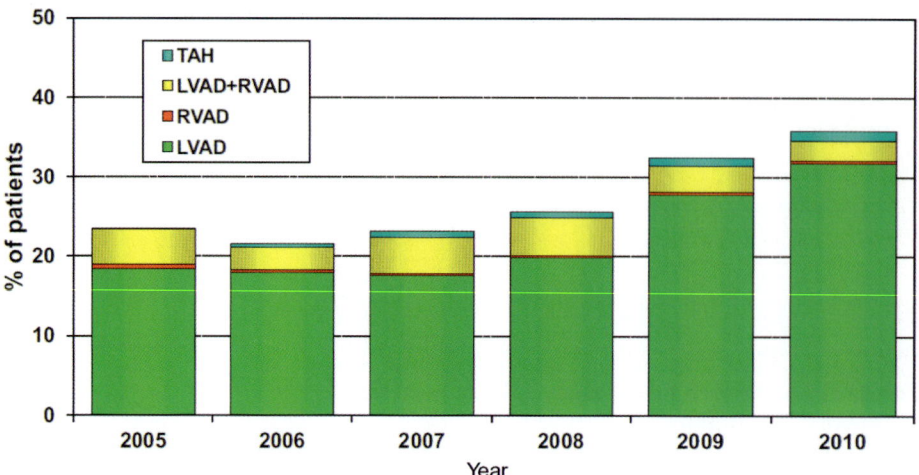

FIGURE 28.9 Use of mechanical circulatory support (biventricular and univentricular devices) as a "bridge" to cardiac transplant. *Figure reprinted from [33].*

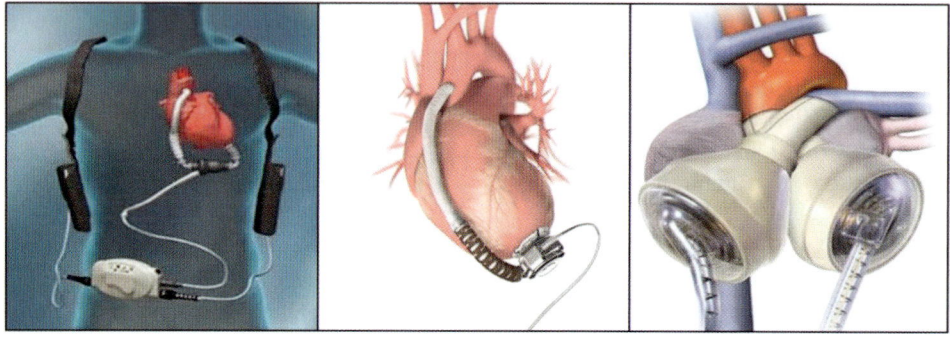

FIGURE 28.10 VADs currently FDA approved as a bridge to cardiac transplantation: Thoratec HeartMate II LVAD (Pleasanton, CA; left); HeartWare VAD (Framingham, MA; middle); Syncardia Systems Total Artificial Heart (Tuscon, AZ; right).

Circulatory Support (INTERMACS) demonstrate that at least 1600 VADs were implanted over a 5 year period of time (2006–2011) as a bridge to transplant strategy. ISHLT registry data also reveal that 36% of patients transplanted in 2010 were supported with mechanical circulatory devices reflecting a rising trend in transplant candidates with BTT VADs [109] (Figure 28.9).

The predominant contemporary VAD in use is the HeartMate II (Thoratec, Pleasanton, CA) which provides left ventricular support for patients for either DT or as a BTT. It gained FDA approval in 2008 after pivotal clinical trial findings demonstrating improved symptoms status, functional capacity, and a 68% 1-year survival rate following implantation in patients awaiting a heart transplant [110]. More recently, the Heartware Ventricular Assist Device (HVAD, Framingham, MA) was granted FDA approval as a BTT strategy following results of the ADVANCE trial that demonstrated an 86% 1-year survival rate with improved quality of life parameters in patients awaiting transplant [26]. Smaller than the HeartMate II, the HVAD can be surgically implanted within the intrapericardial space to provide left ventricular support. However, as left ventricular support devices, both require a relative preservation of right ventricular

function. Unfortunately, such devices are inadequate in patients with biventricular failure. The total artificial heart (TAH, Syncardia) provides biventricular mechanical support. It is true heart replacement therapy as the surgery requires excision of both right and left ventricles such that the device can be anastomosed to both atria. In its clinical trial, patients with the TAH had a 70% 1-year survival rate and a 79% survival to transplant [111,112] (Figure 28.10).

Despite clinical benefits and improved survivorship while waiting for transplant, VAD-related complications including device failure, infection, and cerebrovascular events due to embolic strokes or hemorrhage may compromise transplant candidacy. In addition, device maintenance requires anticoagulation which can further compound bleeding diatheses that can result from an acquired von Willebrand disease and neovascular formations that arise as consequences of nonpulsatile VAD flow [113,114]. Excessive blood loss may require transfusions that increase sensitization risks and consequent allograft rejection or failure [115]. Impact of LVAD therapy on transplant survival has been controversial [110,116]. Recent data, however, suggest patients, previously supported with continuous flow VADs, who survive beyond the initial 6 months posttransplant, have similar long-term survival outcomes as compared to conventional transplantation [56,117].

The era of mechanical circulatory support has also afforded those patients with DT devices, the opportunity for transplant reconsideration. At least 10% patients with DT VADs, originally intended a palliative therapy, were transplanted within 12 months following implantation [109]. In fact, the benefits of improved hemodynamics and organ perfusion by mechanical circulatory support have rendered the distinction between absolute and relative contraindication less dichotomous. Traditional exclusion criteria, such as renal failure and pulmonary hypertension, have been found to be "VAD-modifiable" due to mechanical unloading and improved cardiac output [118]. Improved survival with the VAD also affords valuable time for reevaluation of other modifiable contraindications including smoking, illicit drug and alcohol abuse, BMI, and patient attitude toward transplant. To this end, from 2006 to 2011, 1765 VADs were implanted in patients as a BTC allowing more individuals the opportunity for a new heart [109].

28.12 FUTURE DIRECTIONS IN CARDIAC TRANSPLANTATION

In 2010, the National Heart, Lung, and Blood Institute Working Group convened to outline research strategies that will ultimately improve the clinical decision-making required in the evolving care of cardiac transplant patients [119]. Over the next 10 years, research efforts will be encouraged to improve long-term allograft outcomes, foster tailored immunotherapy, and expand the available donor pool through improved graft preservation strategies and marginal candidate optimization. In addition, advancements in the field will require robust clinical trials that explore mechanisms and treatment of humoral and cellular-mediated rejection. This is in addition to the need for technological advances in both mechanical circulatory support and stem cell therapies that may promote true cardiac regeneration.

28.13 CONCLUSION

The management of end-stage heart disease has evolved with our improved understanding of its complex pathophysiology. Since the first heart transplant in 1969, the field of cardiac transplant medicine continues to evolve. In the end, however, the promise of heart replacement therapy is limited foremost by its availability. Until such time as more effective global donor promotion strategies are in place, our sights must rest upon a horizon in which mechanical circulatory support devices and stem cell-based therapies hold greater promise for those with end-stage heart disease.

ACKNOWLEDGMENTS

We express our sincere gratitude to Dr. Robert Padera for contributing pathological figures to supplement this manuscript.

REFERENCES

[1] Hunt SA, Abraham WT, Chin MH, Feldman AM, Francis GS, Ganiats TG, et al. Focused update incorporated into the ACC/AHA 2005 guidelines for the diagnosis and management of heart failure in adults: a report of the American College of Cardiology Foundation/American Heart Association Task Force on Practice Guidelines: developed in collaboration with the International Society for Heart and Lung Transplantation. Circulation 2009;119:391–479.

[2] Nohria A, Tsang SW, Fang JC, Lewis EF, Jarcho JA, Mudge GH, et al. Clinical assessment identifies hemodynamic profiles that predict outcomes in patients admitted with heart failure. J Am Coll Cardiol 2003;41:1797–804.

[3] Drazner MH, Hellkamp AS, Leier CV, Shah MR, Miller LW, Russell SD, et al. Value of clinician assessment of hemodynamics in advanced heart failure: the ESCAPE trial. Circ Heart Fail 2008;1:170–7.

[4] Thomas SS, Nohria A. Hemodynamic classifications of acute heart failure and their clinical application: an update. Circ J 2012;76:278–86.

[5] Pfeffer MA, Lamas GA, Vaughan DE, Parisi AF, Braunwald E. Effect of captopril on progressive ventricular dilatation after anterior myocardial infarction. N Engl J Med 1988;319:80–6.

[6] Arnold JM, Yusuf S, Young J, Mathew J, Johnstone D, Avezum A, et al. Prevention of heart failure in patients in the Heart Outcomes Prevention Evaluation (HOPE) study. Circulation 2003;107:1284–90.

[7] Flather MD, Yusuf S, Kober L, Pfeffer M, Hall A, Murray G, et al. Long-term ACE-inhibitor therapy in patients with heart failure or left-ventricular dysfunction: a systematic overview of data from individual patients. ACE-Inhibitor myocardial infarction collaborative group. Lancet 2000;355:1575–81.

[8] Granger CB, McMurray JJ, Yusuf S, Held P, Michelson EL, Olofsson B, et al. Effects of candesartan in patients with chronic heart failure and reduced left-ventricular systolic function intolerant to angiotensin-converting-enzyme inhibitors: the CHARM—Alternative trial. Lancet 2003;362:772–6.

[9] Granger CB, McMurray JJ, Yusuf S, Held P, Michelson EL, Olofsson B, et al. Effects of candesartan in patients with chronic heart failure and reduced left-ventricular systolic function taking angiotensin-converting-enzyme inhibitors: the CHARM—Added trial. Lancet 2003;362:767–71.

[10] Pfeffer MA, Swedberg K, Granger CB, Held P, McMurray JJ, Michelson EL, et al. Effects of candesartan on mortality and morbidity in patients with chronic heart failure: the CHARM—Overall programme. Lancet 2003;362:759–66.

[11] Yusuf S, Pfeffer MA, Swedberg K, Granger CB, Held P, McMurray JJ, et al. Effects of candesartan in patients with chronic

heart failure and preserved left-ventricular ejection fraction: the CHARM—Preserved Trial. Lancet 2003;362:777−81.

[12] Pitt B, Zannad F, Remme WJ, Cody R, Castaigne A, Perez A, et al. The effect of spironolactone on morbidity and mortality in patients with severe heart failure. Randomized aldactone evaluation study investigators. N Engl J Med 1999;341:709−17.

[13] Zannad F, McMurray JJ, Krum H, van Veldhuisen DJ, Swedberg K, Shi H, et al. Eplerenone in patients with systolic heart failure and mild symptoms. N Engl J Med 2011;364:11−21.

[14] Packer M, Coats AJ, Fowler MB, Katus HA, Krum H, Mohacsi P, et al. Effect of carvedilol on survival in severe chronic heart failure. N Engl J Med 2001;344:1651−8.

[15] The Cardiac Insufficiency Bisoprolol Study II (CIBIS-II): a randomised trial. Lancet 1999;353(9146):9−13.

[16] Moss AJ, Zareba W, Hall WJ, Klein H, Wilber DJ, Cannom DS, et al. Prophylactic implantation of a defibrillator in patients with myocardial infarction and reduced ejection fraction. N Engl J Med 2002;346:877−83.

[17] Solomon SD, Foster E, Bourgoun M, Shah A, Viloria E, Brown MW, et al. Effect of cardiac resynchronization therapy on reverse remodeling and relation to outcome: multicenter automatic defibrillator implantation trial: cardiac resynchronization therapy. Circulation 2010;122:985−92.

[18] Tang AS, Wells GA, Talajic M, Arnold MO, Sheldon R, Connolly S, et al. Cardiac-resynchronization therapy for mild-to-moderate heart failure. N Engl J Med 2010;363:2385−95.

[19] Felker GM, Thompson RE, Hare JM, Hruban RH, Clemetson DE, Howard DL, et al. Underlying causes and long-term survival in patients with initially unexplained cardiomyopathy. N Engl J Med 2000;342:1077−84.

[20] McCarthy 3rd RE, Boehmer JP, Hruban RH, Hutchins GM, Kasper EK, Hare JM, et al. Long-term outcome of fulminant myocarditis as compared with acute (nonfulminant) myocarditis. N Engl J Med 2000;342:690−5.

[21] Konstam MA, Kramer DG, Patel AR, Maron MS, Udelson JE. Left ventricular remodeling in heart failure: current concepts in clinical significance and assessment. JACC Cardiovasc Imaging 2011;4:98−108.

[22] Renlund DG, Kfoury AG. When the failing, end-stage heart is not end-stage. N Engl J Med 2006;355:1922−5.

[23] Mehra MR, Domanski MJ. Should left ventricular assist device should be standard of care for patients with refractory heart failure who are not transplantation candidates?: left ventricular assist devices should be considered standard of care for patients with refractory heart failure who are not transplantation candidates. Circulation 2012;126:3081−7.

[24] Cuffe MS, Califf RM, Adams Jr. KF, Benza R, Bourge R, Colucci WS, et al. Short-term intravenous milrinone for acute exacerbation of chronic heart failure: a randomized controlled trial. JAMA 2002;287:1541−7.

[25] Bayram M, De Luca L, Massie MB, Gheorghiade M. Reassessment of dobutamine, dopamine, and milrinone in the management of acute heart failure syndromes. Am J Cardiol 2005;96:47−58.

[26] Aaronson KD, Slaughter MS, Miller LW, McGee EC, Cotts WG, Acker MA, et al. Use of an intrapericardial, continuous-flow, centrifugal pump in patients awaiting heart transplantation. Circulation 2012;125:3191−200.

[27] Slaughter MS, Rogers JG, Milano CA, Russell SD, Conte JV, Feldman D, et al. Advanced heart failure treated with continuous-flow left ventricular assist device. N Engl J Med 2009;361:2241−51.

[28] Barnard CN. The operation. A human cardiac transplant: an interim report of a successful operation performed at Groote Schuur Hospital, Cape Town. S Afr Med J 1967;41:1271−4.

[29] Sade RM. Transplantation at 100 years: Alexis Carrel, pioneer surgeon. Ann Thorac Surg 2005;80:2415−8.

[30] Hardy JD, Kurrus FD, Chavez CM, Neely WA, Eraslan S, Turner MD, et al. Heart transplantation in man. Developmental studies and report of a case. JAMA 1964;188:1132−40.

[31] Caves PK, Stinson EB, Billingham ME, Rider AK, Shumway NE. Diagnosis of human cardiac allograft rejection by serial cardiac biopsy. J Thorac Cardiovasc Surg 1973;66:461−6.

[32] Modry DL, Oyer PE, Jamieson SW, Stinson EB, Baldwin JC, Reitz BA, et al. Cyclosporine in heart and heart−lung transplantation. Can J Surg 1985;28:274−80.

[33] Stehlik J, Edwards LB, Kucheryavaya AY, Benden C, Christie JD, Dipchand AI, et al. The Registry of the International Society for Heart and Lung Transplantation: 29th official adult heart transplant report—2012. J Heart Lung Transplant 2012;31:1052−64.

[34] Colvin-Adams M, Smith JM, Heubner BM, Skeans MA, Edwards LB, Waller C, et al. OPTN/SRTR 2011 annual data report: heart. Am J Transplant 2011;13:119−48.

[35] Mancini DM, Eisen H, Kussmaul W, Mull R, Edmunds Jr. LH, Wilson JR. Value of peak exercise oxygen consumption for optimal timing of cardiac transplantation in ambulatory patients with heart failure. Circulation 1991;83:778−86.

[36] Elmariah S, Goldberg LR, Allen MT, Kao A. Effects of gender on peak oxygen consumption and the timing of cardiac transplantation. J Am Coll Cardiol 2006;47:2237−42.

[37] Elmariah S, Goldberg LR, Allen MT, Kao A. The effects of race on peak oxygen consumption and survival in patients with systolic dysfunction. J Card Fail 2010;16:332−9.

[38] Peterson LR, Schechtman KB, Ewald GA, Geltman EM, de las Fuentes L, Meyer T, et al. Timing of cardiac transplantation in patients with heart failure receiving beta-adrenergic blockers. J Heart Lung Transplant 2003;22:1141−8.

[39] Goda A, Lund LH, Mancini D. The heart failure survival score outperforms the peak oxygen consumption for heart transplantation selection in the era of device therapy. J Heart Lung Transplant 2011;30:315−25.

[40] Mancini D, Lietz K. Selection of cardiac transplantation candidates in 2010. Circulation 2010;122:173−83.

[41] Mehra MR, Kobashigawa J, Starling R, Russell S, Uber PA, Parameshwar J, et al. Listing criteria for heart transplantation: international society for heart and lung transplantation guidelines for the care of cardiac transplant candidates−2006. J Heart Lung Transplant 2006;25:1024−42.

[42] Costanzo MR, Dipchand A, Starling R, Anderson A, Chan M, Desai S, et al. The International Society of Heart And Lung Transplantation guidelines for the care of heart transplant recipients. J Heart Lung Transplant 2010;29:914−56.

[43] Uriel N, Jorde UP, Cotarlan V, Colombo PC, Farr M, Restaino SW, et al. Heart transplantation in human immunodeficiency virus-positive patients. J Heart Lung Transplant 2009;28:667−9.

[44] Qiu J, Terasaki PI, Waki K, Cai J, Gjertson DW. HIV-positive renal recipients can achieve survival rates similar to those of HIV-negative patients. Transplantation 2006;81:1658–61.

[45] Hosenpud JD, Pamidi SR, Fiol BS, Cinquegrani MP, Keck BM. Outcomes in patients who are hepatitis B surface antigen-positive before transplantation: an analysis and study using the joint ISHLT/UNOS thoracic registry. J Heart Lung Transplant 2000;19:781–5.

[46] Lin MH, Chou NK, Chi NH, Chen YS, Yu HY, Huang SC, et al. The outcome of heart transplantation in hepatitis C-positive recipients. Transplant Proc 2012;44:890–3.

[47] Corbett C, Armstrong MJ, Neuberger J. Tobacco smoking and solid organ transplantation. Transplantation 2012;94:979–87.

[48] Psychosocial Outcomes Workgroup of the N, Social Sciences Council of the International Society for H, Lung T, Cupples S, Dew MA, Grady KL, et al. Report of the psychosocial outcomes workgroup of the nursing and social sciences council of the international society for heart and lung transplantation: present status of research on psychosocial outcomes in cardiothoracic transplantation: review and recommendations for the field. J Heart Lung Transplant 2006;25:716–25.

[49] Cohen O, De La Zerda DJ, Beygui R, Hekmat D, Laks H. Donor brain death mechanisms and outcomes after heart transplantation. Transplant Proc 2007;39:2964–9.

[50] Karamlou T, Shen I, Slater M, Crispell K, Chan B, Ravichandran P. Decreased recipient survival following orthotopic heart transplantation with use of hearts from donors with projectile brain injury. J Heart Lung Transplant 2005;24:29–33.

[51] Swanson DK, Pasaoglu I, Berkoff HA, Southard JA, Hegge JO. Improved heart preservation with UW preservation solution. J Heart Transplant 1988;7:456–67.

[52] Braunwald E. Innervation of the transplanted heart. N Engl J Med 1969;281(15):848–9; PubMed PMID: 4897379.

[53] Mohanty PK, Sowers JR, Thames MD, Beck FW, Kawaguchi A, Lower RR. Myocardial norepinephrine, epinephrine and dopamine concentrations after cardiac autotransplantation in dogs. J Am Coll Cardiol 1986;7(2):419–24; PubMed PMID: 3511122.

[54] Kilic A, Weiss ES, Arnaoutakis GJ, George TJ, Conte JV, Shah AS, et al. Identifying recipients at high risk for graft failure after heart retransplantation. Ann Thorac Surg 2012;93(3):712–6; PubMed PMID: 22226492.

[55] Russo MJ, Iribarne A, Hong KN, Ramlawi B, Chen JM, Takayama H, et al. Factors associated with primary graft failure after heart transplantation. Transplantation 2010;90(4):444–50; PubMed PMID: 20622755.

[56] Stehlik J, Edwards LB, Kucheryavaya AY, Aurora P, Christie JD, Kirk R, et al. The Registry of the International Society for Heart and Lung Transplantation: 27th official adult heart transplant report—2010. J Heart Lung Transplant 2010;29(10):1089–103; PubMed PMID: 20870164.

[57] Marasco SF, Esmore DS, Negri J, Rowland M, Newcomb A, Rosenfeldt FL, et al. Early institution of mechanical support improves outcomes in primary cardiac allograft failure. J Heart Lung Transplant 2005;24(12):2037–42; PubMed PMID: 16364846.

[58] Listijono DR, Watson A, Pye R, Keogh AM, Kotlyar E, Spratt P, et al. Usefulness of extracorporeal membrane oxygenation for early cardiac allograft dysfunction. J Heart Lung Transplant 2011;30(7):783–9; PubMed PMID: 21481606.

[59] Mehra MR, Uber PA, Uber WE, Scott RL, Park MH. Allosensitization in heart transplantation: implications and management strategies. Curr Opin Cardiol 2003;18(2):153–8; PubMed PMID: 12652223.

[60] Berry GJ, Angelini A, Burke MM, Bruneval P, Fishbein MC, Hammond E, et al. The ISHLT working formulation for pathologic diagnosis of antibody-mediated rejection in heart transplantation: evolution and current status (2005–2011). J Heart Lung Transplant 2011;30(6):601–11; PubMed PMID: 21555100.

[61] Prada-Delgado O, Estevez-Loureiro R, Paniagua-Martin MJ, Lopez-Sainz A, Crespo-Leiro MG. Prevalence and prognostic value of cardiac allograft vasculopathy 1 year after heart transplantation according to the ISHLT recommended nomenclature. J Heart Lung Transplant 2012;31(3):332–3; PubMed PMID: 22333404.

[62] Nair N, Ball T, Uber PA, Mehra MR. Current and future challenges in therapy for antibody-mediated rejection. J Heart Lung Transplant 2011;30(6):612–7; PubMed PMID: 21474341.

[63] Uber PA, Mehra MR. Induction therapy in heart transplantation: is there a role? J Heart Lung Transplant 2007;26(3):205–9; PubMed PMID: 17346621.

[64] Swinnen LJ, Costanzo-Nordin MR, Fisher SG, O'Sullivan EJ, Johnson MR, Heroux AL, et al. Increased incidence of lymphoproliferative disorder after immunosuppression with the monoclonal antibody OKT3 in cardiac-transplant recipients. N Engl J Med 1990;323(25):1723–8; PubMed PMID: 2100991.

[65] Costanzo-Nordin MR, Swinnen LJ, Fisher SG, O'Sullivan EJ, Pifarre R, Heroux AL, et al. Cytomegalovirus infections in heart transplant recipients: relationship to immunosuppression. J Heart Lung Transplant 1992;11(5):837–46; PubMed PMID: 1329959.

[66] Johnson MR, Mullen GM, O'Sullivan EJ, Liao Y, Heroux AL, Kao WG, et al. Risk/benefit ratio of perioperative OKT3 in cardiac transplantation. Am J Cardiol 1994;74(3):261–6; PubMed PMID: 8037132.

[67] Lindenfeld J, Miller GG, Shakar SF, Zolty R, Lowes BD, Wolfel EE, et al. Drug therapy in the heart transplant recipient: part I: cardiac rejection and immunosuppressive drugs. Circulation 2004;110(24):3734–40; PubMed PMID: 15596559.

[68] Frist WH, Merrill WH, Eastburn TE, Atkinson JB, Stewart JR, Hammon Jr. JW, et al. Unique antithymocyte serum versus OKT3 for induction immunotherapy after heart transplantation. J Heart Transplant 1990;9(5):489–94; PubMed PMID: 2121921.

[69] Kirklin JK, Bourge RC, White-Williams C, Naftel DC, Thomas FT, Thomas JM, et al. Prophylactic therapy for rejection after cardiac transplantation. A comparison of rabbit antithymocyte globulin and OKT3. J Thorac Cardiovasc Surg 1990;99(4):716–24; PubMed PMID: 2108282.

[70] Macdonald PS, Mundy J, Keogh AM, Chang VP, Spratt PM. A prospective randomized study of prophylactic OKT3 versus equine antithymocyte globulin after heart transplantation—increased morbidity with OKT3. Transplantation 1993;55 (1):110–6; PubMed PMID: 8380508.

[71] Mehra MR, Zucker MJ, Wagoner L, Michler R, Boehmer J, Kovarik J, et al. A multicenter, prospective, randomized, double-blind trial of basiliximab in heart transplantation. J Heart Lung Transplant 2005;24(9):1297–304; PubMed PMID: 16143248.

[72] Segovia J, Rodriguez-Lambert JL, Crespo-Leiro MG, Almenar L, Roig E, Gomez-Sanchez MA, et al. A randomized multicenter comparison of basiliximab and muromonab (OKT3) in heart transplantation: SIMCOR study. Transplantation 2006;81(11):1542–8; PubMed PMID: 16770243.

[73] Mattei MF, Redonnet M, Gandjbakhch I, Bandini AM, Billes A, Epailly E, et al. Lower risk of infectious deaths in cardiac transplant patients receiving basiliximab versus anti-thymocyte globulin as induction therapy. J Heart Lung Transplant 2007;26 (7):693–9; PubMed PMID: 17613399.

[74] Carrier M, Leblanc MH, Perrault LP, White M, Doyle D, Beaudoin D, et al. Basiliximab and rabbit anti-thymocyte globulin for prophylaxis of acute rejection after heart transplantation: a non-inferiority trial. J Heart Lung Transplant 2007;26(3):258–63; PubMed PMID: 17346628.

[75] Hershberger RE, Starling RC, Eisen HJ, Bergh CH, Kormos RL, Love RB, et al. Daclizumab to prevent rejection after cardiac transplantation. N Engl J Med 2005;352(26):2705–13; PubMed PMID: 15987919.

[76] Teuteberg JJ, Shullo M, Zomak R, McNamara D, McCurry K, Kormos RL. Aggressive steroid weaning after cardiac transplantation is possible without the additional risk of significant rejection. Clin Transplant 2008;22(6):730–7; PubMed PMID: 18673374.

[77] Singh TP, Faber C, Blume ED, Worley S, Almond CS, Smoot LB, et al. Safety and early outcomes using a corticosteroid-avoidance immunosuppression protocol in pediatric heart transplant recipients. J Heart Lung Transplant 2010;29(5):517–22; PubMed PMID: 20061164.

[78] Taylor DO, Barr ML, Radovancevic B, Renlund DG, Mentzer Jr. RM, Smart FW, et al. A randomized, multicenter comparison of tacrolimus and cyclosporine immunosuppressive regimens in cardiac transplantation: decreased hyperlipidemia and hypertension with tacrolimus. J Heart Lung Transplant 1999;18(4):336–45; PubMed PMID: 10226898.

[79] Reichart B, Meiser B, Vigano M, Rinaldi M, Martinelli L, Yacoub M, et al. European multicenter tacrolimus (FK506) heart pilot study: one-year results—European tacrolimus multicenter heart study group. J Heart Lung Transplant 1998;17(8):775–81; PubMed PMID: 9730426.

[80] Grimm M, Rinaldi M, Yonan NA, Arpesella G, Arizon Del Prado JM, Pulpon LA, et al. Superior prevention of acute rejection by tacrolimus vs. cyclosporine in heart transplant recipients—a large European trial. Am J Transplant 2006;6(6):1387–97; PubMed PMID: 16686762.

[81] Ye F, Ying-Bin X, Yu-Guo W, Hetzer R. Tacrolimus versus cyclosporine microemulsion for heart transplant recipients: a meta-analysis. J Heart Lung Transplant 2009;28(1):58–66; PubMed PMID: 19134532.

[82] Baran DA, Zucker MJ, Arroyo LH, Alwarshetty MM, Ramirez MR, Prendergast TW, et al. Randomized trial of tacrolimus monotherapy: tacrolimus in combination, tacrolimus alone compared (the TICTAC trial). J Heart Lung Transplant 2007;26(10):992–7; PubMed PMID: 17919618.

[83] Kobashigawa JA, Miller LW, Russell SD, Ewald GA, Zucker MJ, Goldberg LR, et al. Tacrolimus with mycophenolate mofetil (MMF) or sirolimus vs. cyclosporine with MMF in cardiac transplant patients: 1-year report. Am J Transplant 2006;6 (6):1377–86; PubMed PMID: 16686761.

[84] Staatz CE, Tett SE. Clinical pharmacokinetics and pharmacodynamics of mycophenolate in solid organ transplant recipients. Clin Pharmacokinet 2007;46(1):13–58; PubMed PMID: 17201457.

[85] Kobashigawa J, Miller L, Renlund D, Mentzer R, Alderman E, Bourge R, et al. A randomized active-controlled trial of mycophenolate mofetil in heart transplant recipients. Mycophenolate mofetil investigators. Transplantation 1998;66(4):507–15; PubMed PMID: 9734496.

[86] Kaczmarek I, Ertl B, Schmauss D, Sadoni S, Knez A, Daebritz S, et al. Preventing cardiac allograft vasculopathy: long-term beneficial effects of mycophenolate mofetil. J Heart Lung Transplant 2006;25(5):550–6; PubMed PMID: 16678034.

[87] Dandel M, Jasaityte R, Lehmkuhl H, Knosalla C, Hetzer R. Maintenance immunosuppression with mycophenolate mofetil: long-term efficacy and safety after heart transplantation. Transplant Proc 2009;41(6):2585–8; PubMed PMID: 19715979.

[88] Keogh A, Richardson M, Ruygrok P, Spratt P, Galbraith A, O'Driscoll G, et al. Sirolimus in de novo heart transplant recipients reduces acute rejection and prevents coronary artery disease at 2 years: a randomized clinical trial. Circulation 2004;110 (17):2694–700; PubMed PMID: 15262845.

[89] Eisen HJ, Tuzcu EM, Dorent R, Kobashigawa J, Mancini D, Valantine-von Kaeppler HA, et al. Everolimus for the prevention of allograft rejection and vasculopathy in cardiac-transplant recipients. N Engl J Med 2003;349(9):847–58; PubMed PMID: 12944570.

[90] Kertesz NJ, Towbin JA, Clunie S, Fenrich AL, Friedman RA, Kearney DL, et al. Long-term follow-up of arrhythmias in pediatric orthotopic heart transplant recipients: incidence and correlation with rejection. J Heart Lung Transplant 2003;22(8):889–93; PubMed PMID: 12909469.

[91] Drury JH, Labovitz AJ, Miller LW. Echocardiographic guidance for endomyocardial biopsy. Echocardiography 1997;14 (5):469–74; PubMed PMID: 11174985.

[92] Baraldi-Junkins C, Levin HR, Kasper EK, Rayburn BK, Herskowitz A, Baughman KL. Complications of endomyocardial biopsy in heart transplant patients. J Heart Lung Transplant 1993;12(Pt 1):63–7; PubMed PMID: 8443204.

[93] Saraiva F, Matos V, Goncalves L, Antunes M, Providencia LA. Complications of endomyocardial biopsy in heart transplant patients: a retrospective study of 2117 consecutive procedures. Transplant Proc 2011;43(5):1908–12; PubMed PMID: 21693299.

[94] Fiorelli AI, Coelho GH, Aiello VD, Benvenuti LA, Palazzo JF, Santos Junior VP, et al. Tricuspid valve injury after heart transplantation due to endomyocardial biopsy: an analysis of 3550 biopsies. Transplant Proc 2012;44(8):2479–82; PubMed PMID: 23026624.

[95] Pham MX, Teuteberg JJ, Kfoury AG, Starling RC, Deng MC, Cappola TP, et al. Gene-expression profiling for rejection surveillance after cardiac transplantation. N Engl J Med 2010;362 (20):1890–900; PubMed PMID: 20413602.

[96] Mehra MR, Kobashigawa JA, Deng MC, Fang KC, Klingler TM, Lal PG, et al. Clinical implications and longitudinal alteration of peripheral blood transcriptional signals indicative of future cardiac allograft rejection. J Heart Lung Transplant 2008;27(3):297–301; PubMed PMID: 18342752.

[97] Starling RC, Pham M, Valantine H, Miller L, Eisen H, Rodriguez ER, et al. Molecular testing in the management of cardiac

transplant recipients: initial clinical experience. J Heart Lung Transplant 2006;25(12):1389–95; PubMed PMID: 17178330.

[98] Kobashigawa JA. The future of heart transplantation. Am J Transplant 2012;12(11):2875–91; PubMed PMID: 22900830.

[99] Mehra MR, Parameshwar J. Gene expression profiling and cardiac allograft rejection monitoring: is IMAGE just a mirage? J Heart Lung Transplant 2010;29(6):599–602; PubMed PMID: 20497885.

[100] Kobashigawa JA, Kiyosaki KK, Patel JK, Kittleson MM, Kubak BM, Davis SN, et al. Benefit of immune monitoring in heart transplant patients using ATP production in activated lymphocytes. J Heart Lung Transplant 2010;29(5):504–8; PubMed PMID: 20133166.

[101] Kobashigawa JA, Katznelson S, Laks H, Johnson JA, Yeatman L, Wang XM, et al. Effect of pravastatin on outcomes after cardiac transplantation. N Engl J Med 1995;333(10):621–7; PubMed PMID: 7637722.

[102] Wenke K, Meiser B, Thiery J, Nagel D, von Scheidt W, Steinbeck G, et al. Simvastatin reduces graft vessel disease and mortality after heart transplantation: a four-year randomized trial. Circulation 1997;96(5):1398–402; PubMed PMID: 9315523.

[103] Wenke K, Meiser B, Thiery J, Nagel D, von Scheidt W, Krobot K, et al. Simvastatin initiated early after heart transplantation: 8-year prospective experience. Circulation 2003;107(1):93–7; PubMed PMID: 12515749.

[104] Schroeder JS, Gao SZ. Calcium blockers and atherosclerosis: lessons from the Stanford transplant coronary artery disease/diltiazem trial. Can J Cardiol 1995;11(8):710–5; PubMed PMID: 7671182.

[105] Gupta A, Mancini D, Kirtane AJ, Kaple RK, Apfelbaum MA, Kodali SK, et al. Value of drug-eluting stents in cardiac transplant recipients. Am J Cardiol 2009;103(5):659–62; PubMed PMID: 19231329.

[106] Bhama JK, Nguyen DQ, Scolieri S, Teuteberg JJ, Toyoda Y, Kormos RL, et al. Surgical revascularization for cardiac allograft vasculopathy: Is it still an option? J Thorac Cardiovasc Surg 2009;137(6):1488–92; PubMed PMID: 19464469.

[107] DeBakey ME. Left ventricular bypass pump for cardiac assistance. Clinical experience. Am J Cardiol 1971;27(1):3–11; PubMed PMID: 5538711.

[108] Kirklin JK, Naftel DC, Kormos RL, Stevenson LW, Pagani FD, Miller MA, et al. Fifth INTERMACS annual report: Risk factor analysis from more than 6000 mechanical circulatory support patients. J Heart Lung Transplant 2013;32(2):141–56; PubMed PMID: 23352390.

[109] Kirklin JK, Naftel DC, Kormos RL, Stevenson LW, Pagani FD, Miller MA, et al. The Fourth INTERMACS Annual Report:

4000 implants and counting. J Heart Lung Transplant 2012;31 (2):117–26; PubMed PMID: 22305376.

[110] Miller LW, Pagani FD, Russell SD, John R, Boyle AJ, Aaronson KD, et al. Use of a continuous-flow device in patients awaiting heart transplantation. N Engl J Med 2007;357(9):885–96; PubMed PMID: 17761592.

[111] Copeland JG, Smith RG, Arabia FA, Nolan PE, Sethi GK, Tsau PH, et al. Cardiac replacement with a total artificial heart as a bridge to transplantation. N Engl J Med 2004;351(9):859–67; PubMed PMID: 15329423.

[112] Copeland JG, Copeland H, Gustafson M, Mineburg N, Covington D, Smith RG, et al. Experience with more than 100 total artificial heart implants. J Thorac Cardiovasc Surg 2012;143(3):727–34; PubMed PMID: 22245242.

[113] Suarez J, Patel CB, Felker GM, Becker R, Hernandez AF, Rogers JG. Mechanisms of bleeding and approach to patients with axial-flow left ventricular assist devices. Circ Heart Fail 2011;4(6):779–84; PubMed PMID: 22086831.

[114] Uriel N, Pak SW, Jorde UP, Jude B, Susen S, Vincentelli A, et al. Acquired von Willebrand syndrome after continuous-flow mechanical device support contributes to a high prevalence of bleeding during long-term support and at the time of transplantation. J Am Coll Cardiol 2010;56(15):1207–13; PubMed PMID: 20598466.

[115] John R, Lietz K, Schuster M, Naka Y, Rao V, Mancini DM, et al. Immunologic sensitization in recipients of left ventricular assist devices. J Thorac Cardiovasc Surg 2003;125(3):578–91; PubMed PMID: 12658200.

[116] Patlolla V, Patten RD, Denofrio D, Konstam MA, Krishnamani R. The effect of ventricular assist devices on post-transplant mortality an analysis of the United network for organ sharing thoracic registry. J Am Coll Cardiol 2009;53(3):264–71; PubMed PMID: 19147043.

[117] John R, Pagani FD, Naka Y, Boyle A, Conte JV, Russell SD, et al. Post-cardiac transplant survival after support with a continuous-flow left ventricular assist device: impact of duration of left ventricular assist device support and other variables. J Thorac Cardiovasc Surg 2010;140(1):174–81; PubMed PMID: 20447659.

[118] Kirklin JK, Naftel DC, Kormos RL, Stevenson LW, Pagani FD, Miller MA, et al. Second INTERMACS annual report: more than 1000 primary left ventricular assist device implants. J Heart Lung Transplant 2010;29(1):1–10; PubMed PMID: 20123242.

[119] Shah MR, Starling RC, Schwartz Longacre L, Mehra MR, Working Group. Heart transplantation research in the next decade—a goal to achieving evidence-based outcomes: National Heart, Lung, and Blood Institute Working Group. J Am Coll Cardiol 2012;59:1263–9.

Chapter 29

Artificial Heart Support

Olivier Van Caenegem[a] and Luc M. Jacquet[b]

[a]*Cardiovascular Intensive Care, Heart Transplant Unit, Cliniques Universitaires Saint-Luc, Université Catholique de Louvain, Brussels Belgium,*
[b]*Cardiovascular Intensive Care, Cliniques Universitaires Saint-Luc, Université Catholique de Louvain, Brussels, Belgium*

29.1 INTRODUCTION

The introduction of cyclosporine in the 1980s [1] established heart transplantation as the gold standard therapy for terminal heart failure patients remaining symptomatic despite adequate medical therapy. Over the last decades, however, organ shortage is responsible for increasing waiting times on the transplant list and subsequent increasing mortality. In the United States, 27% of the patients listed for heart transplantation will die or be delisted as too ill at 2 years after initial UNOS (United Network for Organ Sharing) status 2 listing [2]. In the Eurotransplant area, 1-year mortality on the elective heart waiting list varies from 15% to 17% [3]. This has led to an increased use of artificial heart support as a bridge to transplantation (BTT). Thirty-six percent of the recipients transplanted worldwide in 2010 were under mechanical circulatory support (MCS) [4].

MCS has been developed to support the failing heart in various conditions including recovery of the native heart (bridge to recovery, BTR), waiting for a suitable organ for transplantation (BTT), or even permanent replacement of the failing left ventricular function (destination therapy, DT).

This chapter will describe the history of MCS, the indications for MCS, the types of MCS currently available (i.e., ventricular assist devices, VADs, and total artificial hearts, TAHs), the patient selection for MCS, the surgical procedure for MCS implantation, the complications, and the results of their use as BTT or DT.

29.2 DEFINITION

MCS is a general term making reference to any mechanical assist device supporting the circulation. It can be external and used for short periods or internal when used for prolonged periods.

The term VADs is a more specific denomination referring to mechanical pumps that support a failing ventricle. VADs are designed to support the left ventricular assist device (LVAD), right (RVAD), or both ventricles (BiVAD). VADs are connected to the heart by inflow and outflow cannulae. The inflow cannula derives blood from the heart chambers (atrium or ventricle) to fill the pump. The outflow cannula ejects blood from the pump to the aorta or the pulmonary artery.

The TAH is a device implanted in orthotopic position to replace the heart after cardiectomy.

29.3 HISTORY OF RESEARCH ON VADs AND TAH

For clarity, we will discuss separately the TAH and VADs acknowledging that progresses in both domains were strongly interrelated.

29.3.1 Total Artificial Heart

In 1958, Kolff and Akutsu from Cleveland Clinic reported their experience with a pneumatically driven plastic heart that had supported a living dog for 90 min at the American Society of Artificial Internal Organs. This was the first report of an artificial heart sustaining life of an animal. During the same session, they presented the dummy of an electromagnetically driven artificial heart [5].

In the late 1950s, several teams around the world were developing pumps for cardiac replacement. One of the most active groups was at Baylor College in Houston where DeBakey had invited Liotta, previously assistant of Kolff at Cleveland. They worked on developing a cardiac prosthesis and were able to keep animals alive for more than 24 h [6].

In 1964, the National Heart, Lung and Blood Institute (NHLBI) in the United States established an artificial heart program with special congressional approval. Contracts were awarded to support the study and development of an artificial cardiac pump. DeBakey who had played a major role in obtaining this funding remained however reluctant to proceed with human trial of the artificial heart developed at Baylor College. Probably frustrated by this decision, Liotta approached Cooley, surgeon at the nearby Texas Heart Institute and convinced him to proceed with the first clinical use. In 1969, Cooley and Liotta implanted their artificial pump in the chest of a 47-year-old man who could not be weaned from bypass after a left ventricular aneurysmectomy. The Liotta–Cooley TAH supported the patient for 64 h when a cardiac transplantation could be performed. Unfortunately, the patient died shortly thereafter from infectious complications [7].

This first implantation demonstrated that it was possible to sustain human life with an artificial pump, but many improvements were clearly still mandatory. Progress continued in the field of pump design, biomaterials, and artificial valves with several pneumatic, electric, and even nuclear-powered TAH being designed and used in animal studies. Akutsu and Cooley at Texas Heart Institute further developed new pumps and a second human implant of an Akutsu III TAH was performed in Houston in 1981 as a BTT [8]. All around the world, teams were working on projects related to the artificial heart. In Germany, Austria, Japan, and even in the former USSR, animals were supported for several months with various TAHs [9].

At the University of Utah in Salt Lake City, under the supervision of Kolff, Olsen, and Jarvik, a pneumatic TAH was developed with a uniconstructed smooth blood-contacting surface, tilting disc valves, and Dacron atrial cuffs [10]. After several years of animal studies, the designed Jarvik-7 TAH was implanted for the first time in a patient in December 1982 as a definitive therapy. The recipient, Barney Clark, survived 112 days [11]. The results of the Jarvik-7 initial trial were rather disappointing with a high rate of thromboembolic and infectious complications. Its use as a permanent TAH had to be abandoned. Nevertheless, with better medical expertise in patient management, the device continued to be used with success as a bridge to cardiac transplantation. The first of these planned interventions was performed by Copeland at the University of Arizona in 1985 [12]. The device was initially manufactured by Symbion Corporation and later by CardioWest Corporation who renamed it CardioWest C-70, with very few changes to the initial design. The CardioWest C-70 received FDA approval as BTT and, after being acquired by Syncardia, Inc., is still in use in the United States, Canada, France, and Germany. A portable driver called Freedom has now been released to power the device [13].

Meanwhile, the NHLBI continued to fund research on a totally implantable TAH. The Abiomed Company in conjunction with Texas Heart Institute and the University of Louisville was one of the NHLBI contractors and created a pump, the Abiocor, containing two ventricles and an energy converter situated between the ventricles. The energy converter contains a high-efficiency miniature centrifugal pump that rotates to pressurize a low-viscosity hydraulic fluid which is alternatively moved between the right and left ventricles resulting in alternate left and right systole. The internal controller transmits device performance data to a bedside console via radiofrequency telemetry. An internal battery is capable of powering the device for brief periods of time and is recharged by transcutaneous energy transfer [14]. The first human implantation occurred in July 2001 at Jewish Hospital, Louisville, Kentucky [15]. The initial feasibility study included 14 patients with the longest survival being 512 days. A relatively high rate of thromboembolic complications was observed requiring modifications of the pump. Research is still ongoing with several changes applied to improve the next generation Abiocor 2 (i.e., incorporation of technology acquired from Penn State University [16]).

New concepts are elaborated to design original pumps but they have not yet reached the level of clinical application. We will come back to some concepts later on.

29.3.2 Ventricular Assist Devices

Derived from the manual cardiac massage, some devices were created acting as "pneumomassage of the heart" (i.e., Bencini's cup described in 1956 [17]) but without much success. Most of the assist devices unload the natural heart by volume uptake and were, initially, derived from methods used in the heart—lung machines.

Dennis et al. developed a pump-oxygenator to replace the heart and lung and used it in clinical practice for the first time in 1951. However, realizing that the oxygenator in the circuit limited the duration of possible support, Dennis and his coworkers pursued with the idea of left heart bypass and demonstrated that left bypass could effectively decrease the workload of the normal heart in dogs [18]. They developed a cannula designed to access the left atrium by a transseptal approach. Oxygenated blood drained from the left atrium was circulated by means of a roller pump and reinfused in the femoral artery. In 1962, the first clinical application was reported in a patient suffering ventricular septal defect and cardiogenic shock after a myocardial infarction [19].

In the early 1960s, the cardiac surgery team at Baylor College in Houston, TX, concentrated its research on prolonged support for cardiogenic shock following cardiotomy while waiting for recovery of the natural heart. A tubular, pneumatically activated intrathoracic LVAD was created by Liotta and Hall and implanted in 1963 by Crawford and DeBakey. The patient however did not survive the surgical procedure [20]. In 1966, a paracorporeal Liotta—DeBakey LVAD was successfully implanted at Methodist Hospital, Houston, in a patient who could not be weaned from bypass after double valve replacement. The patient was supported for 10 days at a flow rate up to 1200 mL/min and could be weaned from the LVAD. In this early version, the pump was lined with Dacron velour to create a pseudo neoendocardium [21,22].

In Boston, Bernhard and La Farge developed a series of axiosymmetric double-valved blood pumps interposed between the left ventricular apex and the thoracic aorta. Their research work resulted in the approval of a Model X pump containing xenograft valves for use as a temporary clinical assist device [23]. At Texas Heart Institute in Houston, an abdominal LVAD interposed between the apex and the infra renal abdominal aorta was designed. This device was used for the first reported case of BTT with an LVAD in 1975 by Norman, Cooley, and Frazier [24,25].

Pierce and Donachy at Pennsylvania State University in Hershey worked on a paracorporeal, pneumatically activated pump, the Angle Port Pump [26]. They constructed a reliable pump that was acquired by Thoratec Corporation and became the Thoratec pneumatic ventricular assist device (PVAD) which is still largely used nowadays. An internal version (IVAD), activated by an external air compressor, exists since 2001.

For long-term application however, an implantable LVAD was necessary, one of the most important components being an implantable energy converter. Portner at Stanford University renewed the concept of a solenoid activator first described by Bindels in the early 1960s [27,28]. The activator was incorporated in a dual pusher-plate pump and by 1977, the first integrated left ventricular assist system (LVAS) had been designed. In 1982, the Novacor Medical was created to commercialize this LVAS. The first human implantation of an electric VAD occurred in 1984 at Stanford University Medical Center [29]. A wearable controller and rechargeable batteries for power supply were released in 1993.

In the 1970s, the possibility of using other power sources was contemplated and research was initiated on heat engine or nuclear power sources. The department of artificial organs at Cleveland Clinic (Whalen, Washizu, and Nosé) worked on a device that could be activated by a nuclear or an electrical system [30]. Outside the United States, the design of an LVAD was also the object of intense interest. In Europe, researchers in Vienna used parts of their TAH to form the ellipsoid LVAD [31]. In Japan, teams in Hiroshima, Nagaya, and Tokyo also created innovative LVAD.

Thermedics Inc., a division of Thermo Electron Corporation presented in the mid 1980s the results of their clinical experience with textured blood-contacting surfaces in a PVAD, the Thermedics Model 14 [32]. This surface results in the formation of a biologically active pseudoneointima that becomes nonthrombogenic. The technology was incorporated in the HeartMate, an implantable titanium alloy pump. The internal pneumatic HeartMate IP was implanted for the first time in 1986. A vented electric (VE) version actuated by a low-speed torque motor came into clinical use in 1991. ThermoCardio systems merged with Thoratec Corporation in 2001.

In 1999, a totally implantable LVAD including internal batteries and controller, called the LionHeart, had been created by researchers of Penn State University in collaboration with Arrow Inc. Despite encouraging results, the company ended development and sales in 2005.

Both the Novacor and HeartMate were extensively used in the 1990s as BTT for long periods of time, sometimes exceeding 1 year. This opened a completely new avenue of research on the clinical application of this therapy. A landmark study was initiated in 1998, the Randomized Evaluation of Mechanical Assistance for the Treatment of Congestive Heart Failure (REMATCH) that demonstrated the superiority of the HeartMate VE over medical treatment in patients with terminal heart failure who were not candidates for heart transplantation [33].

Meanwhile, research teams were working on new technologies to create safer devices, less cumbersome and less noisy. Continuous-flow pumps represented the most attractive alternative. In fact, back in 1960, Saxton and Andrews had already described a continuous-flow pump for use as an artificial heart [34]. Since 1964, a combined team from the University of Minnesota, Medtronic Inc. and the University of California in San Diego studied fundamental aspects of support with a centrifugal pump. They demonstrated in animal studies that pulseless flow was compatible with long-term survival [35]. Although intracorporeal implantation was the final goal, only extracorporeal pumps were developed for use during cardiac surgery or for short-term support.

In 1988, the concept of an axial pump appeared in clinical practice with the Hemopump. It had been developed by Nimbus Corporation from the idea of Wampler who applied the principle of the Archimedes screw he had seen working in Egypt for water wells. Years of research had been necessary to create this miniaturized pump, mounted on a catheter and inserted through the femoral artery or in the ascending aorta. The feasibility of sustaining life with this type of pump without excessive blood damage was demonstrated and stimulated much interest even though the Hemopump disappeared [36]. A catheter-mounted axial flow pump was later developed at Helmholtz Institute in Aachen, the Impella pump, that has since been acquired by Abiomed.

Axial flow pumps for long-term assistance were also widely investigated. The Baylor College of Medicine in Houston in collaboration with NASA developed the DeBakey-Noon pump that was implanted clinically for the first time in Berlin in 1998 [37]. Jarvik and the Texas Heart Institute designed an intraventricular axial flow pump, the Jarvik 2000. With researchers in Oxford, a skull-mounted carbon pedestal was created to transmit fine electric wires through the scalp skin which is resistant to infection [38]. The McGowan Center of the University of Pittsburgh with the Nimbus Company also initiated research on such a device. Nimbus was later acquired by Thermo Cardiosystems Inc. that merged with Thoratec. Their device, the HeartMate II (HM II), is nowadays the most widely implanted assist device for long-term support [39]. Recently published clinical trials have demonstrated the reliability of the device both as a BTT and as DT [40]. In Germany, Berlin Heart GmbH worked with the German Heart Institute on a magnetically suspended impeller which is the moving part of the Incor pump implanted for the first time in 2002 [41].

Centrifugal pumps were also developed with hydrodynamic bearings (i.e., the Ventrassist pump developed in Australia, the Eva Heart from Sinu Medical in Japan), with magnetic bearings (the HeartQuest as a collaboration between Utah Artificial Heart Institute, MedQuest

Products, and the University of Virginia; the DuraHeart developed by Terumo Heart Inc.), or hybrid pumps with passive magnets and hydrodynamic thrust bearing (HeartWare HVAD, Miami) [42].

Efforts are now concentrated in further miniaturizing the devices, especially for use in children.

29.3.3 LVAD and TAH

For years, projects have been conducted aiming at the development of a permanently implantable continuous-flow TAH.

Owing to the excellent clinical results obtained with rotary pumps as LVAD, interest has increased in using two rotary pumps either as biventricular support or as TAH. Frazier at Texas Heart Institute reported in 2006 survival of a calf for 20 days using dual continuous-flow pumps (Jarvik 2000) [43]. Biventricular continuous-flow support has now reached clinical application with several devices.

Creating a reliable, totally implantable artificial heart is much more complex than anticipated by the first researchers in this field. After years of extensive laboratory and clinical research all around the world, reliable implantable assist devices can be used in daily clinical practice for short- and long-term support. Improvements are still required and research continues in several directions (miniaturization, total implantability, clinical management, TAH), and the way is still long before we reach the Holy Grail.

29.4 INDICATIONS FOR MCS

MCS has been developed to support the circulation of patients with profound circulatory failure refractory to conventional pharmacologic therapy. Different indications are recognized according to different clinical settings. When the cause of cardiac failure is potentially reversible (e.g., postcardiotomy cardiogenic shock, acute myocarditis, drug-induced cardiac toxicity), MCS can be used until the heart recovers and can then be removed. This is called "BTR." The most widely accepted indication for many years is "BTT." In this category, patients have to be on the waiting list or considered as potential transplant candidates, and at risk of dying before a suitable donor heart is available. During the 1990s, these recipients were mostly in a critically ill condition. The benefits of MCS were not exclusively to maintain these patients alive until transplantation, but also to let them recover from possible secondary organ dysfunction like renal failure, liver failure, or muscular deconditioning. During the last decade, improvements in technical aspects and reliability of VADs led to implantation in less-severe patients. It is currently accepted that patients deteriorating during their waiting time are VAD candidates in order to avoid secondary organ failure or cardiac cachexia. Furthermore, advances in reliability opened access to "*DT*." Here, patients with refractory heart failure who are not transplant candidates because of older age or comorbidities may benefit from VAD implantation as an alternative to transplantation. Importantly, distinctions made between all those indications are very theoretical and there are many crossovers from one category to another. (For example, a patient implanted in BTR after a surgical intervention whose cardiac function does not recover can become a transplant candidate; on the contrary, a patient supported by VAD during his waiting time presenting permanent neurologic disability due to a cerebrovascular accident can be reoriented as DT.) In some cases, VAD is implanted in patients who are not transplant candidates at the time of implant (too sick), but could be listed if they improve after prolonged support, so called "*bridge to transplant candidacy*." Finally, in acute situations, short-term MCS is implanted because of emergency. Patients should then undergo careful assessment of noncardiac organ dysfunction, focusing on important sequellae of the acute heart failure. Neurologic evaluation is particularly important since low cardiac output or cardiac arrest may induce permanent cerebral disorder. If cardiac dysfunction is deemed irreversible and no life-limiting condition is observed, full LVAD support can be implemented in the hope of later providing a transplant for the patient. Short-term MCS is also used as a "*bridge to bridge*" or as a "*bridge to decision*."

29.5 TYPES OF MCS SYSTEMS

29.5.1 Short-Term Devices

Short-term mechanical circulatory systems are indicated in acute situations (e.g., cardiogenic shock due to acute myocardial infarction) or when cardiac recovery is expected (e.g., postcardiotomy heart failure). Most of these systems consist of an external pump and cannulae connected to the heart chambers or the great vessels. They are cumbersome and the patient is tethered to the device with little or no mobility. MCSs carry a risk of infection by the means of percutaneous cannulae and leads. It necessitates systemic anticoagulation since there is a hazard of thromboembolic events due to platelet and plasma protein activation in contact with device's surface. These devices have the advantage of rapid availability, ease of insertion, and low cost.

29.5.2 Nonpulsatile Pumps

29.5.2.1 Roller Pumps

A peristaltic pump is a type of positive-displacement pump used for pumping blood contained within a flexible tube fitted inside a circular pump casing. It is used in cardiopulmonary bypass circuits. Currently, roller pumps are no longer utilized for prolonged circulatory support due to coagulation disorders secondary to poor biocompatibility of blood contact surface, hemolysis induced by high shear stress, and deterioration of flexible tubing with time requiring periodic replacement.

29.5.2.2 Centrifugal Pumps

Centrifugal pumps are made of a polycarbonate housing containing rotating cones or an impeller (Figure 29.1). Rotation of the impeller or of the smooth-surfaced cones pulls the blood into the vortex of the housing. As blood flows through the pump outlet, the vortex energy is transferred to the blood in form of pressure and velocity. The pump is connected by tubings and cannulae to the patient. The cannulae are surgically inserted. Centrifugal pumps can provide left ventricular support by inserting the inflow (or drainage) cannula in the left atrium and the outflow cannula in the aorta; right ventricular support by inserting inflow cannula in the right atrium and the outflow cannula in the pulmonary artery; biventricular support by using both described circuits. Biventricular support can also be achieved by draining blood from the right atrium to the pump and reinjecting it into the aorta, but an external oxygenator is then necessary, resulting in an extracorporeal membrane oxygenation (ECMO) circuit.

The use of older generation pumps (Biomedicus©—Medtronic Biomedicus Inc., Minneapolis, MN, USA) was limited by the high incidence of thromboembolic complications despite heparinization and the risk of blood damage if the duration of support needed to be extended. Improvements were made over time by using heparin-coated circuits resulting in reduced thrombogenicity and thus reduced need for systemic anticoagulation [44]. The development of new double-lip seal bearings (Bio-Pump Plus—Medtronic, Minneapolis, MN, USA), or combined mechanical and magnetic bearings (Rotaflow©—Maquet Cardiopulmonary AG, Hirrlingen, Germany), or full magnetically levitated impellers (Centrimag©—Thoratec Europe Limited, Cambridgeshire, UK; Levitronix©—Levitronix GmbH, Zurich, Switzerland) further diminishes blood trauma and hemolysis, allowing extracorporeal support for longer periods [45,46].

29.5.2.3 Extracorporeal Membrane Oxygenation

ECMO results from the combination of an extracorporeal centrifugal pump and an oxygenator. The cannulae can be inserted percutaneously in the peripheral vessels (femoral artery and vein) or surgically in its central configuration (ascending aorta and right atrium).

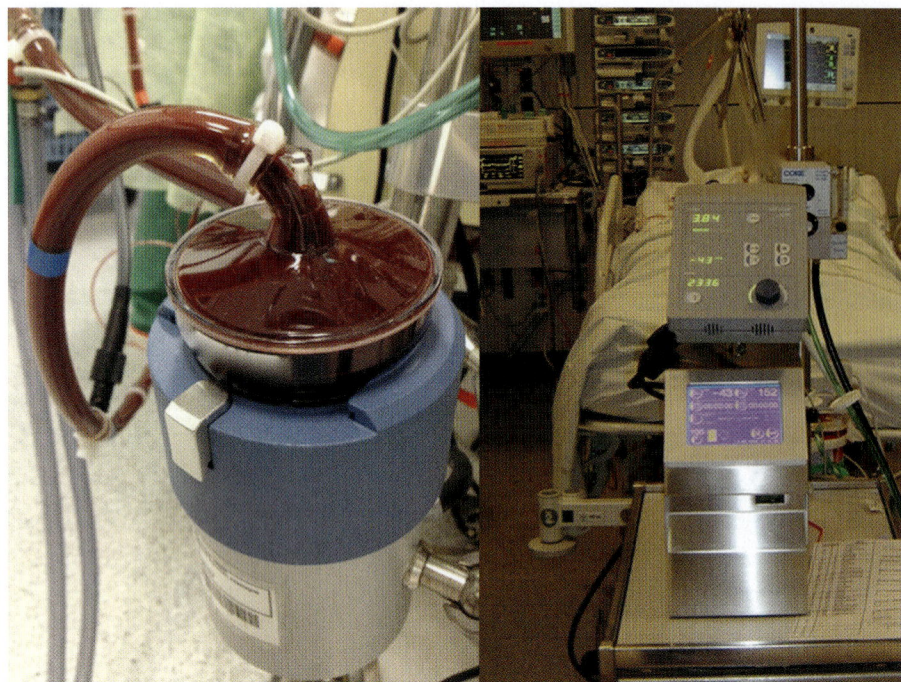

FIGURE 29.1 Paracorporeal centrifugal pump (left) and driving console (right). The rotation speed of the impeller is controlled by the driving console, which also monitors pressure into the inflow and cannula. *With the courtesy of Sorin Group.*

ECMO has different advantages. It can be inserted percutaneously. The oxygenator allows adequate oxygenation of the blood even when hypoxemia results from pulmonary edema. Both ventricles are supported. Is also has several disadvantages. The major limitation is that the left ventricle is inadequately decompressed in most cases. When left ventricular unloading is essential, decompression of the left cavities can be achieved by direct left atrial canulation through a median sternotomy or a left thoracotomy, or by introducing a transseptal catheter. Another disadvantage results from the blood/circuit surface interaction. Continuous anticoagulation is mandatory since emboli through the arterial reinjection cannula result in systemic embolic events. These events can lead to permanent disability like cerebrovascular accidents, renal failure due to infarction, and mesenteric or leg ischemia. Furthermore, the first-generation microporous membrane oxygenator caused plasma leakage and often precipitated coagulation disorders and consecutive bleeding complications. The new polymethyl pentene oxygenators have greater stability and preservation of coagulation factors and platelets, consecutively reduce consumption of blood products, reduce clot events, and have a better gas and heat exchange efficiency [47]. The use of rotational thromboelastometry (ROTEM) and platelet function analyzer (MULTIPLATE) can help to identify whether hemorrhagic complications are due to acquired coagulopathies or platelet dysfunction, the most common condition in ECMO [48]. Finally, percutaneous insertion of the outflow cannula in the femoral artery can lead to

local bleeding or leg ischemia following distal embolization or arterial occlusion due to large-sized cannulae. Leg perfusion can be achieved by insertion of a derived catheter from the ECMO circuit to the distal femoral artery. ECMO is widely used as circulatory support in the pediatric population.

The TandemHeart System© (CardiacAssist, Inc., Pittsburgh, PA, USA) is a left atrial−femoral bypass system. The circuit is made of an external centrifugal pump with an integrated motor and two cannulae. The inflow cannula is 21-French multiperforated catheter inserted through the femoral vein to the left atrium by the transseptal approach. The outflow is a 17-French cannula inserted in the femoral artery. The system provides up to 4.5−5 L/min of cardiac output. It requires anticoagulation like other paracorporeal pumps. The technical difficulty of inserting the left atrial cannula involving an experienced interventional cardiologist able to perform transseptal catheterization mainly limits its use. The pump is CE-marked and received FDA approval. The TandemHeart was compared to intra-aortic balloon pump (IABP) in cardiogenic shock with favorable hemodynamic response [49]. In the largest series of 117 patients with severe refractory cardiogenic shock supported by the TandemHeart, Kar et al. report dramatic improvements in hemodynamic parameters as well as in end-organ function, with a mortality at 30 days and 6 months of 40.2% and 45.3% respectively [50]. The TandemHeart is also being used for high-risk percutaneous coronary interventions with good results [51].

29.5.2.4 Paracorporeal Microaxial Pumps

The first percutaneous axial pump was developed during the 1970s and called the Hemopump©. The first human implant occurred in 1988 [52]. This pump is made of an impeller powered by an external electromechanical motor. The impeller is located in a 21-French catheter inserted via the femoral artery to the left ventricle through the aortic valve. The blood is drawn from the left ventricle through the transvalvular inlet cannula and pumped into the ascending aorta to provide circulatory support. The Hemopump produced flows from 2.5 to 4.5 L/min. This device is no longer manufactured.

The Impella Recover System© (Abiomed Inc., Danvers, MA, USA) is a series of three catheter-mounted microaxial rotary pumps. The Impella 2.5, which is placed percutaneously in a retrograde manner via the femoral artery across the aortic valve into the left ventricle. It can provide up to 2.5 L/min of cardiac support. The Impella 5.0 version is able to pump up to 5 L/min but needs to be surgically inserted. The Impella LD© is a catheter-based assist device inserted through the ascending aorta into the left ventricle via an open chest procedure to directly unload the left ventricle. All these pumps are FDA approved. A randomized trial compared the Impella 2.5 system to IABP in STEMI (ST segment elevation myocardial infarction) patients with cardiogenic shock and showed greater augmentation of MAP (mean arterial pressure) and CI cardiac index in the group assigned to the Impella and a more rapid decrease in serum lactates levels [53]. The 5.0 version seems to provide more efficient support than the 2.5 in profound cardiogenic shock patients [54].

A meta-analysis compared both TandemHeart and Impella to IABP in acute myocardial infarction complicated with cardiogenic shock. VAD patients were found to have higher CI and MAP, and lower PCWP (pulmonary capillary wedge pressure). However, there was no difference in 30-day mortality [55].

29.5.3 Long-Term Devices

Once circulatory support has to be extended for longer periods than a few days, intermediate or longer-term devices should be considered. Paracorporeal devices can be used for weeks or months; implantable devices for months or years.

29.5.3.1 Paracorporeal Pulsatile Devices

These devices are used for short- or medium-term support. The general principle consists of paracorporeal pneumatically driven blood pumps connected to the heart chambers or the great arteries by the means of cannulae. Sternotomy is needed for canulation regarding the size of the cannulae and the direct insertion into the heart and great arteries. Valves

between the pump chamber and the inflow and outflow cannulae ensure unidirectional anterograde blood flow. The pneumatically driven pump receives air pressure from a compressor located in an external console. Commands to adjust the pump's functioning are also on the console. Permanent anticoagulation with heparin is recommended. These VADs allow left, right, or biventricular support. The complications of such systems are thrombolic events due to thrombus into the pump or the patient's ventricles secondary to blood stasis, bleeding, infection of the exit site of the cannulae, or infection of the pump (which has to be considered as endocarditis). Technical failures are uncommon; however, cannula disinsertion by patient's mobilization has been described. The major limitation is the reduced mobility of the patient tethered to the cumbersome console.

The Abiomed BVS500© (Abiomed, Inc., Danvers, MA, USA) has been used for years worldwide to provide circulatory support up to 4 weeks. The system consisted of two cannulae, a paracorporeal pump, and a driving console. The venous cannula was surgically directly inserted in the right or the left atrium for blood drainage to the pump. The arterial cannula coming from the pump was attached to a Dacron graft inserted on the aorta or the pulmonary artery. The pump was made of two chambers. The upper filled by gravity with blood flowing from the patient's atrium. Since no vacuum was applied to assist filling, it had to be placed below the level of the patient's atrium. The lower ventricular chamber provided a pulsatile, asynchronous, flow up to 4−5 L/min. The pump was pneumatically driven by compressed room air from a console. The advantages were the facility to use a highly automated console and the different possibilities of support. The major disadvantage was the lack of mobility of the bedridden patient. This system is no longer available. Since 2005 [56], improvements came from the ABIOMED AB5000© ventricle. This new system consists of a paracorporeal pneumatically driven blood pump connected to the heart chambers by 32-French surgical cannulae and a fully automated console. The AB5000 allows flow rate up to 6 L/min. It is approved by the FDA. The small AB portable driver configured with the AB Power Pack and a small mobile cart allows independent patient ambulation. The Power Pack allows 10 h of battery.

The Thoratec PVAD© (Thoratec Corporation, Pleasanton, CA, USA) was first manufactured in 1982. More than 5000 patients have received PVAD implants over the last 30 years. This paracorporeal VAD is positioned externally, on the abdominal wall, and is designed for left, right, or biventricular support (Figure 29.2). The blood pump has a rigid polysulfone case containing a seamless elastomeric blood pumping sac composed of Thoralon®, a proprietary polyurethane multipolymer. The blood sac is compressed by air from a pneumatic driver to eject blood out of the sac. Vacuum is applied

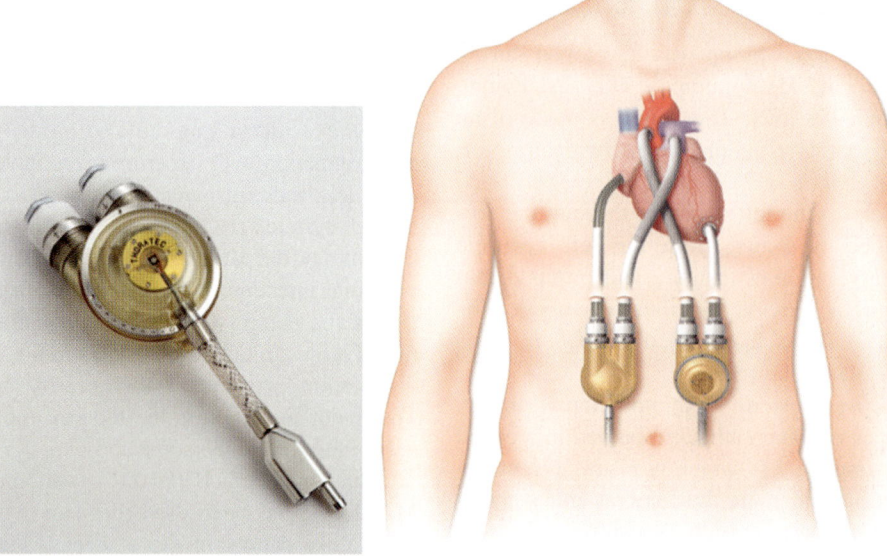

FIGURE 29.2 Thoratec PVAD paracorporeal pneumatic pulsatile pump (left) and canulation configuration for biventricular support (right). *Reprinted with the permission of Thoratec Corporation.*

during pump diastole to promote pump filling. Mechanical valves, mounted in the inflow and outflow ports of the blood pump, control the direction of blood flow. The blood pump has an effective stroke volume of 65 mL and, depending on various conditions, will pump up to 6.5 L/min at a rate of 100 beats per minute. Cannulae can be inserted in the left or right atrium or placed in the left ventricular apex to provide inflow to the VAD blood pump. Blood is returned to the patient with an arterial cannula in the aorta or the pulmonary artery depending on whether the left or right ventricle is being assisted. The cannulae are covered with external velour at the exit site to promote tissue ingrowth for the prevention of ascending infection. The blood pump is connected to the dual-drive console by flexible plastic pneumatic tubing for drive pressure and vacuum transmission and by an electrical cable for transmission of the signal from the fill switch from the pump to the driver. The TLC-II portable driver console is a compact and lightweight electrically operated pneumatic drive unit pulled on a luggage-type mobility cart (Figure 29.3). It allows patient's full ambulation and even return to home while on support [57].

The Berlin Heart Excor© (Berlin Heart GmbH, Berlin, Germany) was introduced to clinical practice in 1992 [58]. The system is very similar to the Thoratec PVAD. It is normally used as a bedside assistance with the control and drive console but can also offer mobility up to 10 h with the supplied battery pack. It is capable of

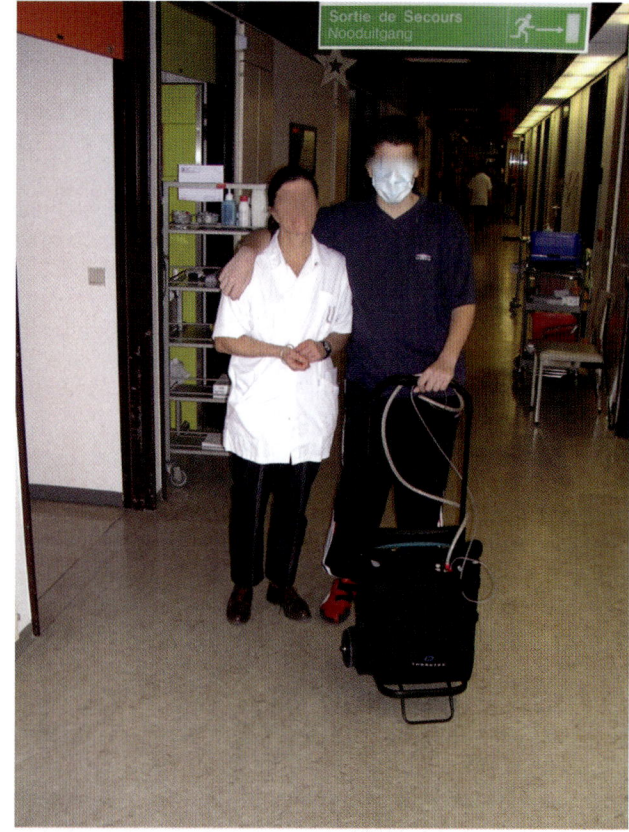

FIGURE 29.3 The TLC-II portable driver console allows complete patient ambulation. *With the permission of Thoratec Corporation.*

providing short- to long-term LV, RV, and biventricular assistance. The pumps come in six different sizes—10, 25, 30, 50, 60, and 80 mL—to suit various patient needs and have been reported with satisfactory outcomes as a pediatric support [59].

The Toyobo-NCVC VAD© (Toyobo-National Cardiovascular Center PVAD) is widely used in Japan. It was developed more than 30 years ago. Characteristics are similar to the previously described extracorporeal pneumatically driven VADs. To improve patient's mobilization, a compact wearable pneumatic drive unit has been developed [60]. The Toyobo-NCVC VAD can also support pediatric patients.

The Medos VAD-System© (MEDOS Medizintechnik AG, Stolberg, Germany) was available from 1994 until the early 2000s. The VADIII-system, like other paracorporeal VADs, consisted of cannulae, an external blood pump and a driving console. This system offered a large range of pump chambers from 10 to 80 mL, for the right or the left position. Consecutively, a specific version was developed for pediatric support, the Medos HIA-VAD-System©.

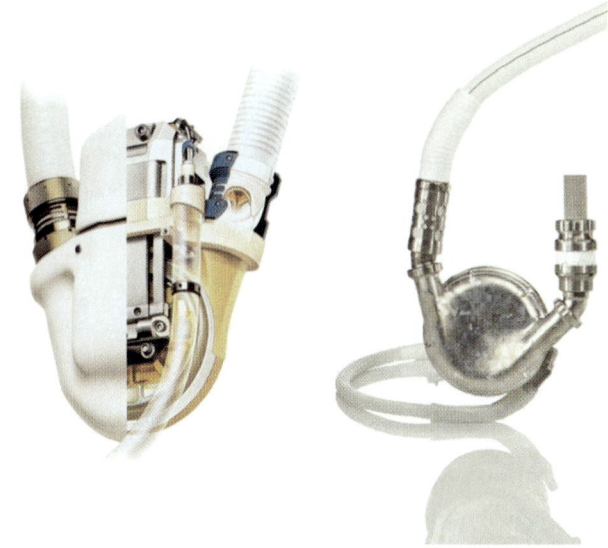

FIGURE 29.4 The Novacor LVAS (left) and HeartMate XVE (right) were the first implantable VADs allowing patients to reside outside the hospital. These systems are no longer marketed. *Reprinted with the permission of Thoratec Corporation.*

29.5.3.2 Implantable Pulsatile Devices

During the 1980s and 1990s, two implantable pulsatile volume-displacement pumps have dominated the field of long-term circulatory support: the Novacor LVAS© (Baxter Healthcare Corp, Oakland, CA, USA) and the HeartMate XVE© (Thoratec Corporation, Pleasanton, CA, USA) (Figure 29.4). These were the first electrically driven wearable LVADs that allowed patients to reside outside the hospital.

The first implant of a totally intracorporeal pump as a permanent device occurred in 1984 [61]. The Novacor LVAS was an implantable electrically driven LVAD connected to an external power source by a single percutaneous lead. The system fabricated for clinical use contained a highly efficient, electrically powered solenoid drive system coupled to a dual pusher-plate pump around a polyurethane blood sac with xenograft inflow and outflow valves. The HeartMate XVE was a positive-displacement, pusher-plate pump activated by a low-speed electrical driven torque motor derived from the HeartMate IP that was internal and pneumatically driven. Inlet and outlet conduits are fitted with porcine valves. Blood-contacting surfaces of the pumping chambers were textured to form a biological layer to interface with the blood. Both the Novacor LVAS and the HeartMate XVE were inserted surgically and placed intra-abdominally or preperitonally in a pocket under the anterior abdominal rectus muscle. The pump received blood from the left ventricle through an apical inflow cannula and pumps flow through an outflow conduit to the ascending aorta. Because of the large

volume, an extensive surgical dissection was necessary to create the pump pocket, increasing the risk for hematomas and subsequent infection. Early reoperation for bleeding, pocket drainage, or tamponade were very common. The percutaneous lead was large and stiff and also contains an air vent channel, which makes the system quite noisy and uncomfortable. This driveline connected the internal parts of the system to an external controller and power supplies. In its wearable version, two portable batteries offered 4–6 hours of power, allowing the patient to have untethered mobility. These devices provided very good circulatory support with stroke volumes of 70–80 mL and pumping rates up to 150 bpm. Antiplatelet therapy was recommended but anticoagulation was necessary only with the Novacor LVAS. The major weakness of this system was the rate of thromboembolic events resulting in transient ischemic attacks or stroke, and the rate of intracranial bleedings. In the literature, there was a large variability in rates of neurological complications following Novacor LVAS implants, ranging from 6 to 58%. This rate went down to 3–9% with the HeartMate XVE. Other serious adverse events were infections that can involve any part of the device: the driveline, the pump pocket, or the pump itself. They occurred in 18–80% of patients with pulsatile devices and were related to mortality. Finally, many moving parts and biological valves were responsible for a substantial mortality rate caused by mechanical failure of the device, especially the HeartMate XVE. The landmark

REMATCH trial showed for this device a 35% failure rate after 2 years, with a mortality of more than 10% [33]. The Novacor LVAS showed better reliable long-term performances with reported support periods longer than 4 years [61–64]. These devices are nowadays not anymore available.

Several other pulsatile LVADs have been used or tested during the past 20 years in Europe and the United States, with various clinical results.

Although the Thoratec IVAD© (Thoratec Corporation, Pleasanton, CA, USA) is described as a versatile pneumatically driven implantable VAD, only the pumps are implanted in the patient's chest. Left, right, or biventricular support can be achieved by this system. However, Thoralon® blood-contacting surfaces into the pump are designed to reduce hemolysis and thrombus, permanent anticoagulation is needed. The patient is tethered to the external TLC-II drive console by the plastic tubing and electric cables connected to the VAD. Nevertheless, the Thoratec IVAD can provide an option for extended support periods when right or biventricular support is required, or when home discharge is considered [65].

29.5.3.3 Implantable Nonpulsatile Devices

Axial Pumps

The axial pumps are composed of a rotor or impeller which spins at 8000–20,000 rpm. The impeller is the only moving part of the system. It is supported by wear-resistant bearings. The rotor is actuated by an electromagnetic power mechanism. The pump motor is located in a longitudinal housing connected by cannulae to the left ventricular apex and the ascending aorta. The blood is routed from the failing ventricle into the pump by the inflow conduit. The rotating impeller provides the power to generate blood flow through the outflow. Since blood flow is continuous rather than pulsatile, unidirectional valves are not needed. These devices provide a maximum blood flow of 5–6 L/min. They can be used in smaller patients. They do not require an abdominal pocket and can be fully inserted in the chest, reducing postoperative surgical complications. A small percutaneous driveline connects the pump to an external electric controller and power supplies. These pumps are wearable, easy to manage, and rugged, providing improved outcome compared to pulsatile devices as discussed later.

There are currently different systems available. The HeartMate II LVAD© (Thoratec Corporation, Pleasanton, CA, USA) (Figure 29.5) is the most used with more than 10,000 patients registered worldwide and hundreds of patients enrolled in different studies [66–69]. The HM II summarizes the largest clinical experience and data with continuous-flow pumps. The Jarvik 2000© (Jarvik Heart Inc., NY, USA) has a particular driveline exit site located behind the ear designed to reduce the incidence of lead infection, the pump itself being implanted in the left ventricular apex [70]. The Micromed DeBakey© is marketed in a modified fifth generation as the HeartAssist5©

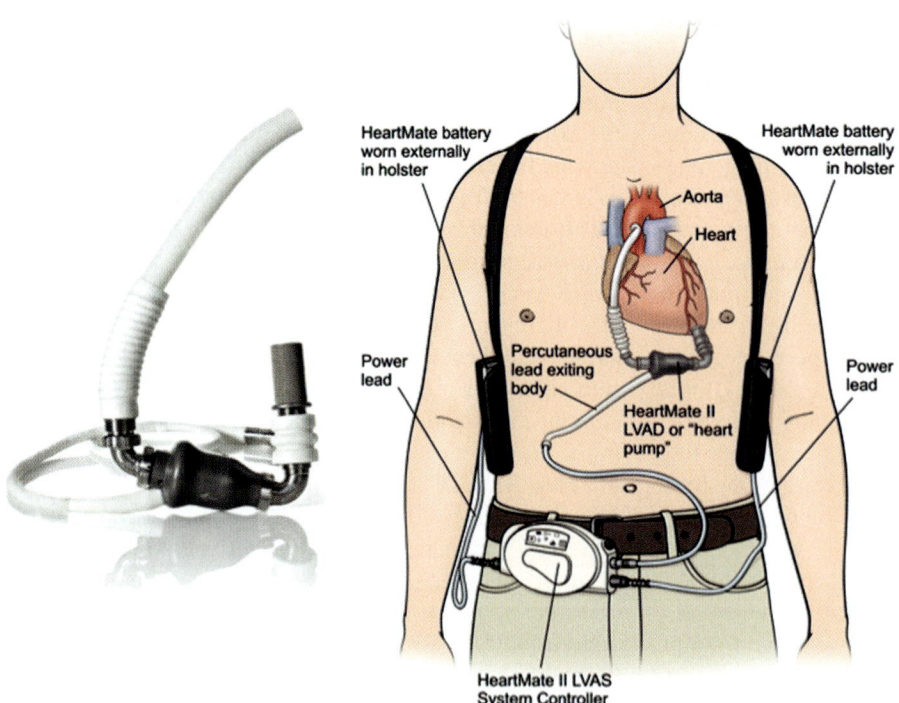

FIGURE 29.5 The HeartMate II pump (left) and the different components of the system (right). This system is the most used, with more than 10,000 patients supported by the HM II. *Reprinted with the permission of Thoratec Corporation.*

(Micromed Cardiovascular Inc., Houston, TX, USA). This system includes a flowmeter that provides direct measurements of the blood flow through the pump and remote monitoring software retrieving data from the controller from a distance. The Berlin Heart Incor© (Berlin Heart GmbH, Berlin, Germany) is distributed in Europe but not approved in the United States. It is small and placed intrapericardially. It works wear-free by means of active magnetic bearings.

Centrifugal Pumps

Centrifugal pumps have a rotating impeller which is electromagnetically driven. Rotor suspension is either hydrodynamic (VentrAssist), magnetic (DuraHeart), or hydrodynamic and magnetic (HeartWare). Their rotational speeds are slower (2000−4000 rpm) than axial pumps. These pumps tend to be easier to implant, have higher reliability, and less complications than pulsatile or axial pumps.

The DuraHeart© (Terumo Heart Inc., Ann Arbor, MI, USA) horizontal length including inflow and outflow conduits is 50−60% shorter than axial pumps. It is classically connected to the left ventricular apex and the ascending aorta, with a percutaneous driveline to an external controller and power supplies. Internal blood-contacting surfaces are heparin bounded to minimize pump clotting and thromboembolic events. Nevertheless, anticoagulation must be given. It is capable of providing flow up to 8 L/min. The system is CE-marked and commercially available in Europe but restricted to investigational use in the United States [71]. The VentrAssist© (Ventracor Limited, Sydney, Australia) was designed similarly. The company stopped their activity in 2009. The HeartWare© (Heartware International Inc., Framingham, MA, USA) is probably the smallest implantable pump currently available (Figure 29.6). It is designed to be surgically implanted above the diaphragm, intrapericardially, directly adjacent to the heart. The impeller is suspended within the pump housing through a combination of passive magnets and hydrodynamic thrust bearings. It spins at rates between 2400 and 3200 rpm, generating up to 10 L of blood flow per minute. The pump's inflow cannula is integrated with the device, ensuring proximity between the heart and the pump itself. It is CE-marked and has recently been approved by the FDA.

29.5.3.4 Total Artificial Hearts

TAHs are MCS systems that replace the two lower chambers of the heart. TAH are indicated for chronic support of patients presenting with biventricular failure, in which LVAD online is insufficient. During surgical implant, the patient's native heart is removed and replaced by a biventricular pump. Thus, in case of device failure there is no fallback mechanism to pump blood. TAH receives blood

FIGURE 29.6 The HeartWare represents the last-generation implantable centrifugal pump. The small size allows the pump to be inserted intrapericardially.

from both atria. The prosthetic ventricles are respectively connected left to the aorta and right to the pulmonary artery. Unidirectional flow is maintained by four mechanical valves. TAH are very complex devices and their implantation is challenging for surgeons. Postoperative complications like bleeding or infection are common. For the long term, anticoagulation is mandatory. The major complications are thromboembolic events, pump-related infections, and device failure. TAHs are used only in a small number of patients.

Currently, two types of TAHs are available. The CardioWest© renamed as the Syncardia TAH© (SynCardia Systems Inc., Tucson, AZ, USA) is CE-marked but still not approved by the FDA. It is a pneumatically driven pulsatile implantable biventricular pump powered by an external air compressor located in a cumbersome drive console. The patient is tethered to the console by percutaneous tubing and cables. The "Hospital Driver" is docked in a caddy allowing patients mobility in the hospital. In a current version, a wearable power supply designed to be worn in a backpack enables the patient who meets discharge criteria to leave the hospital. The Abiocor© (Abiomed Inc., Danvers, MA, USA) is a biventricular pulsatile electrically driven VAD system. It is completely contained in the chest. A battery powers the system and is charged by induction electric transfer through the skin with a special magnetic charger. Also, an implanted controller monitors and controls the pumping speed of the heart. There are two pumping chambers that function as the left and the right ventricles. The energy converter is situated between the ventricles and contains a high-efficiency miniature centrifugal pump

TABLE 29.1 Summary of the Current Available Devices and Characteristics

Duration of MCS	Device Name	Ventricle Supported	Location of Pump	Mechanism of Pump	Patient Ambulation	Expense
Short term	Biomedicus, Rotaflow, Centrimag, Levitronix	Left, right	External	Centrifugal, nonpulsatile	No	+
	ECMO	Left, right	External	Centrifugal, nonpulsatile	No	+
	TandemHeart	Left	External	Centrifugal, nonpulsatile	No	++
	Impella	Left	External	Microaxial pump, nonpulsatile	No	++
	Abiomed AB5000	Left, right	External	Pneumatic, pulsatile	No	++
Intermediate to longer term	Thoractec PVAD, Berlin Heart Excor, Toyobo-NCVC VAD	Left, right	External	Pneumatic, pulsatile	Yes	++
Long term	Thoratec IVAD	Left, right	Internal	Pneumatic, pulsatile	Yes	+++
	HeartMate II, Jarvik 2000, HeartAssist5, Berlin Heart Incor	Left	Internal	Electric, axial, nonpulsatile	Yes	+++
	HeartWare, DuraHeart	Left	Internal	Electric, centrifugal, nonpulsatile	Yes	+++
	Abiocor	Left, right	Internal	Electric, displacement, pulsatile	Yes	+++
	Syncardia	Left, right	Internal	Pneumatic, pulsatile	Yes	+++

driven by a brushless direct current motor. This centrifugal pump operates unidirectionally to pressurize a low-viscosity hydraulic fluid. A two-position switching valve is used to alternate the direction of hydraulic flow between the left and right pumping chambers. This results in alternate left and right systole.

The rate of complications, mainly thromboembolic, has limited the number of implants to less than 20.

Current available devices and their principal characteristics are summarized in Table 29.1.

29.6 PATIENT SELECTION AND TIMING OF IMPLANT

Historically, VADs were used to support the circulation of patients with profound heart failure refractory to conventional therapy. While short-term devices were implanted as a bridge to recovery or a bridge to decision, long-term support with implantable devices was used as a bridge to transplantation. Thus, the indications for chronic VAD support were transplant candidates with refractory cardiac failure pending their transplant. Those were hospitalized, under inotropic support or short-term VADs, often with secondary organ failure like renal or liver dysfunction or cachexia.

Different parameters have modified this selection process over the last years. First, the lack of organ donors increases waiting time on the transplant list. Consecutively, more patients need MCS, and for longer periods. Secondary, as the outcome of continuous-flow pumps has surpassed that of the older pulsatile pumps [68], MCS is now considered as DT, in alternative to transplantation. In consequence, any heart failure patient with signs of severity or rapid decline should be considered for referral to an advanced heart disease program to determine candidacy for transplant, MCS or both.

Increased mortality has been related to several clinical signs in heart failure patients: two or more hospitalizations for acute decompensation in a 12-month period, limitation in angiotensin-inhibitor therapy or betablocker therapy by hypotension or renal impairment, a peak oxygen consumption $<14-16$ mL/kg/min or $<50\%$ predicted value, symptoms at rest on most days (New York Heart Association, NYHA, class IV), failure to respond to resynchronization therapy, or 6-min walk test <300 m. Other high-risk clinical and laboratory features include worsening right heart failure and secondary pulmonary hypertension, diuretic refractoriness associated with worsening renal function, persistent hyponatremia, hyperuricemia, and cardiac cachexia. Intractable ventricular arrhythmia is also an indication for MCS [72]. The Seattle Heart Failure Model (SHFM) has also been applied to predict mortality after LVAD implant. Patients with anticipated mortality $>15\%$ at 1 year and $>20\%$ at

2 years based on the SHFM should be considered at high risk and evaluated for MCS [73]. The HM II risk score showed older age, higher creatinine, low albumin, and implant center inexperience to be related to post-VAD mortality [74].

Accurate evaluation of right ventricular (RV) function during the preimplant assessment is of importance since RV failure after LVAD implant increases postoperative mortality sixfold and is a major determinant for prolonged in-hospital stay [75]. Although several clinical risk factors for post-LVAD RV failure have been identified (Table 29.2) [76,77], only the more recent were developed in patients supported with continuous-flow pumps. Yet RV function plays a central role in decisions regarding uni- or biventricular support.

Chronic heart failure has serious metabolic and inflammatory effects resulting in extreme weight loss, so called "cardiac cachexia." BMI <22 has been related to early and late mortality after pulsatile VAD implant [77]. This was not confirmed after continuous-flow VAD implantation, where extreme BMI did not affect outcome [78]. Other markers of poor nutritional status including low prealbumin, low serum protein, low albumin, low cholesterol, lymphocytes count, and high C-reactive protein are associated with increased mortality and infections after cardiac surgery [72]. Low BMI should be considered during assessment of VAD candidates and adequate nutritional support, including enteral feeding, should be warranted to avoid extreme low nutritional reserve. In contrast, high BMI has not been associated with adverse outcome nor infection after LVAD implant [79].

Heart failure has a higher incidence in older patients. However, these are often considered not suitable transplant candidates. In consequence, older patient could be a target population for chronic VAD support as a DT. In the INTERMACS registry, age >70 years is consistently associated with increased early and late mortality [80]. This is probably of greater importance after pulsatile than continuous-flow VADs since age has no impact on mortality after HM II implant [81]. Frailty, which is independent of age, may be a more potent predictor of outcome than age [82].

Classical NYHA class IV symptoms do not offer adequate descriptions to allow optimal selection of patients for advanced heart failure therapies including resynchronization, transplantation, or MCS. In the early period of VADs and for many years, dependence on inotropes has been traditionally considered the threshold for circulatory support. One of the first pieces of information from the INTERMACS registry that was implemented in 2006 was to describe seven clinical profiles (Table 29.3) based on clinical and hemodynamic data. This confirmed that 80% of VAD implants occurred in the two profiles with highest levels of clinical compromise [84]. The second registry report emphasized the increased death rate of INTERMACS level 1 (cardiogenic shock) patients [85]. Worsening INTERMACS profiles have been consistently associated with higher perioperative mortality [86]. Lessons learned from those data led to a persistent decrease in the use of chronic VADs in INTERMACS level 1 patients. In the fourth report of the registry, only 14% of the implants were performed in those compromised patients, with a clear move to level 2 (41%) and level 3 (28%) patients [87]. This translates into improved survival. Whether converting a patient from level 1 to level 2 with the use of temporary MCS (e.g., Impella, TandemHeart, or ECMO) improves outcome remains untested. Nevertheless, only 18% of implants of durable VADs occur in patients not yet on intravenous inotropes. Low-implant rates in ambulatory heart failure reflect reluctance of physicians and patients. Different ongoing studies address the question of outcome after VAD implant in DT in "less sick" patients with advanced heart failure. However, waiting for end-organ failure or even for signs of debilitation before moving to LVAD support is no longer acceptable.

TABLE 29.2 Risk Scores for Right Ventricular Failure After LVAD Implantation

Right Ventricular Risk Score	Likelihood Ratio for Right Ventricular Failure
<3	0.40 (0.37−0.64)
4.0−5.0	2.8 (1.4−5.9)
>5.5	7.6 (3.4−17.1)

Right Ventricular Risk Score	Right Ventricular Failure
<5.0	11%
5.5−8.0	37%
8.5−12.0	56%
>12.5	83%

Risk score calculated by summing points awarded for the presence of a vasopressor requirement (4 points), AST ≥ 80 IU/L (2 points), bilirubin ≥ 2.0 mg/dL (2.5 points), and creatinine ≥ 2.3 mg/dL (3 points). Risk score calculated as the sum of the points assigned for the existence of each of 8 perioperative variables. DT patients 3.5 points; intra-aortic balloon counterpulsation, 4 points; pulmonary vascular resistance, quartile 1 (≤1.7 Wood units), 1 point; quartile 2 (1.8−2.7 Wood units), 2 points; quartile 3 (2.8−4.2 Wood units), 3 points; and quartile 4 (≥4.3 Wood units), 4 points; inotrope dependency, 2.5 points; obesity, 2 points; angiotensin-converting enzyme inhibitor and/or angiotensin II receptor blocker, −2.5 points; and β blocker, 2 points. Reported by Matthews et al. (upper part) [75] and Drakos et al. (lower part) [76]. Reprinted with permission.

TABLE 29.3 INTERMACS Profile Description and Time Frame for Intervention

INTERMACS Level	Level of Limitation	Time Frame for Intervention
Profile 1: Critical cardiogenic shock	Patients with life-threatening hypotension despite rapidly escalating inotropic support, critical organ hypoperfusion, often confirmed by worsening acidosis and/or lactate levels. *"Crash and burn."*	Definitive intervention needed within hours
Profile 2: Progressive decline	Patient with declining function despite intravenous inotropic support may be manifested by worsening renal function, nutritional depletion, inability to restore volume balance *"Sliding on inotropes."* Also describes declining status in patients unable to tolerate inotropic therapy.	Definitive intervention needed within a few days
Profile 3: Stable but inotrope dependent	Patient with stable blood pressure, organ function, nutrition, and symptoms on continuous intravenous inotropic support (or a temporary circulatory support device or both), but demonstrating repeated failure to wean from support due to recurrent symptomatic hypotension or renal dysfunction *"Dependent stability."*	Definitive intervention elective over a period of weeks to few months
Profile 4: Resting symptoms	Patient can be stabilized close to normal volume status but experiences daily symptoms of congestion at rest or during ADL (activities of daily life). Doses of diuretics generally fluctuate at very high levels. More intensive management and surveillance strategies should be considered, which may in some cases reveal poor compliance that would compromise outcomes with any therapy. Some patients may shuttle between 4 and 5.	Definitive intervention elective over a period of weeks to few months
Profile 5: Exertion intolerant	Comfortable at rest and with ADL but unable to engage in any other activity, living predominantly within the house. Patients are comfortable at rest without congestive symptoms, but may have underlying refractory elevated volume status, often with renal dysfunction. If underlying nutritional status and organ function are marginal, patient may be more at risk than INTERMACS 4, and require definitive intervention.	Variable urgency, depends upon maintenance of nutrition, organ function, and activity
Profile 6: Exertion limited	Patient without evidence of fluid overload is comfortable at rest and with activities of daily living and minor activities outside the home but fatigues after the first few minutes of any meaningful activity. Attribution to cardiac limitation requires careful measurement of peak oxygen consumption, in some cases with hemodynamic monitoring to confirm severity of cardiac impairment. *"Walking wounded."*	Variable, depends upon maintenance of nutrition, organ function, and activity level
Profile 7: Advanced NYHA III	A placeholder for more precise specification in future, this level includes patients who are without current or recent episodes of unstable fluid balance, living comfortably with meaningful activity limited to mild physical exertion.	Transplantation or circulatory support may not currently be indicated

Reprinted with permission from Stevenson et al. [82]

29.7 SURGICAL PROCEDURE FOR IMPLANTABLE VADs AND POSTOPERATIVE MANAGEMENT

While temporary MCS devices can be inserted percutaneously, long-term VADs implantation requires a surgical intervention. The specific technique is dependent on the device chosen and which ventricle is supported.

For pulsatile devices, the heart is exposed by median sternotomy. An abdominal incision is also required since most pumps are placed in on the left side of the abdominal wall below rectus muscles, anterior to the posterior sheath. Complete hemostasis is important to avoid a pocket hematoma and subsequent infection. Pneumatic drivelines are brought out through the anterior abdominal wall. Electric drivelines are usually brought out on the patient's right side, away from the iliac crest to avoid discomfort. For left ventricular apical VAD inflow, cardiopulmonary bypass is mandatory. The pump and components are prepared on the back table. The system is filled with heparined saline. A core of the left ventricular apex is removed and the cannula is anastomosed to the heart by the means of a sewing ring. The cannula courses between the costal margin and the diaphragm without entering the peritoneum. It is then connected to the pump placed in the abdominal pocket. After filling the device and conduit with blood, the outflow cannula is connected to the ascending aorta. After de-airing, the pump is started. Weaning of bypass machine should be done cautiously in case of LVAD. The right ventricle is not supported and at high risk of dilatation and decompensation regarding the sudden increase of venous return. Different interventions can prevent the risk of RV failure, like performing surgery on a beating heart, diminishing RV after load with vasodilators or inhaled nitric oxide in case of pulmonary hypertension and supporting the RV with inotropes. Milrinone is of particular interest since it is inotropic and it lowers pulmonary vascular resistances. Bleeding and multiple transfusions are also associated with RV failure.

The reduced size of most continuous-flow pumps allows placing them fully intrathoracically or even intra-pericardially. Median sternotomy is the classical way, but left or right thoracotomy with partial femoro-femoral bypass can be an option to avoid repeated sternotomy. The inflow cannula is inserted in the left ventricle's apex. The outflow cannula is connected to the ascending aorta. The electric driveline exits on the right side of the patient.

Transoesophageal echocardiography is routinely used to assess specific cardiac abnormalities, interatrial defects (which must be closed), or aortic regurgitation (which must be corrected to avoid recirculation). It helps also during de-airing maneuvers and to assess RV function during bypass weaning.

The most common complications following VAD implantation are reoperation for bleeding or tamponnade, thromboembolism, infection, and RV failure. Hemolysis is seen with temporary devices but is rare with chronic VADs. Renal dysfunction is common as in other cardiac interventions.

29.7.1 MCS-Associated Bleeding

As discussed earlier, meticulous hemostasis is the cornerstone of the intervention for preventing both postoperative bleeding and infection. The incidence of reoperation for bleeding or tamponnade was significantly higher with the pulsatile pumps and BiVADs (47.7% and 62% respectively) because of their large size and the extensive dissection of the chest and abdominal pocket. Continuous-flow pumps have lower incidence of reoperation (31%) [68,85].

Coagulopathy is common after VAD implantation. Most heart failure patients have liver dysfunction and subsequent coagulopathy already present before surgery. Further worsening of coagulopathy is influenced by cardiopulmonary bypass-induced platelet dysfunction, hypothermia, shock, persistent bleeding, and multiple transfusions. Accurate evaluation of the coagulation process can be obtained by new testing like ROTEM and MULTIPLATE in order to tailor the needs for platelet transfusion, plasma, fibrinogen, or coagulation factors.

29.7.2 Thromboembolism

For many years, thromboembolism has been recognized as the major problem with MCS since it can lead to permanent neurologic disability. With pulsatile devices, the incidence ranged from 10% with the HeartMate XVE to more than 50% with the Novacor LVAS [61–64]. More recent continuous-flow devices show a lower incidence, less than 10% [68,85]. Higher biocompatibility and lower thrombogenicity led to the modification of anticoagulation protocols. With pulsatile devices, anticoagulation with nonfractionated heparin was started early, as soon after the operation as possible, with respect to bleeding. Antiplatelet therapy was later associated with Coumadin (INR = 3–4). Currently, recommendations after HM II implantation minimize anticoagulation to avoid bleeding. Low-molecular heparin is started on POD1 with antiplatelet therapy. Later, Coumadin is given for an INR = 1.5–2.5.

29.7.3 Infection

Infection in patients receiving an MCS must be separated into device-related infections and nosocomial infection. Terminal heart failure patients indeed are more vulnerable to infection due to bed rest, compromised circulation, nutritional deprivation, and immunological impairment. They often have multiple catheters and invasive monitoring devices, which are potential sources for bloodstream infection.

VAD-related infections must be divided in percutaneous driveline infection, abdominal pocket infection, mediastinitis, or VAD-endocarditis. The incidence increases according to the device's size and complexity: 14–28% in continuous-flow LVADs, 38% in pulsatile flow VADs, and 32–48% in BiVADs. Among infected patients, driveline is the most common site (48%), followed by pocket (42%), endocarditis (22%), and mediastinitis (5%) [90,91]. Early mediastinitis, pocket, or pump infection are mostly the consequence of preoperative contamination, commonly by skin organisms. Bleeding and subsequent pocket hematoma greatly increases the likelihood of infection. Mediastinitis or abdominal wall pocket infection may require open debridement, evacuation of the hematoma and drainage. Rarely, infection of blood-contacting surfaces or device valves needs device replacement. In most cases, the presence of an infected large foreign body is the rationale to perform urgent cardiac transplantation, with good outcome [65,86]. In the majority of cases, driveline infection can be treated with local wound care and antibiotics.

Preventing infection is firstly related to device improvements, and secondly to patient selection and management. We discussed before the relationship between device size and infection. Another point is the improvements made in percutaneous leads. First-generation pulsatile pumps needed a large driveline containing a vent. Last-generation VADs present with a less thick and more flexible cable, resulting in fewer infections. Local care to the exit site must be performed very carefully. External fastener enhances wound healing and reduces infection rates. The use of 2% nasal mupirocin ointment (Bactroban© GSK, NC, USA) given preoperatively to carriers of nasal *Staphylococcus* decreases the rate of wound infection in cardiac surgery. Finally, there is a direct relationship between patient's status before VAD implantation and postoperative infection. The diminishing number of implants in INTERMACS level 1 patients should decrease the incidence of infection. To some extent, patient risk factors are unavoidable in patients receiving a VAD as BTT. However, patient selection for DT is pivotal to obtaining an acceptable success rate [86].

29.8 LONG-TERM OUTCOME OF IMPLANTABLE VADs

As discussed earlier, long-term results are influenced by patient selection, timing of implant, uni- or biventricular support, and type of device. As a reminder, INTERMACS level 1 patients have significantly lower survival than higher levels [83].

Historically, studies with pulsatile devices have been reported to bridge patients successfully to transplantation. The perioperative mortality was around 15−20% with an overall survival of 60−70% [63,93−95]. The high rate of thromboembolic events and infection were the main limitations of chronic pulsatile VAD support. The REMATCH trial [33] was the first randomized trial comparing best medical therapy to VAD support in terminal heart failure patients. Sixty-eight patients ineligible for heart transplantation were implanted with a HeartMate XVE and compared to sixty-one patients with optimal medical therapy. Benefit was definitively in favor of VAD support with a survival at 1 year of 52% in VAD patients and 25% in medical patients, and at 2 years of 23% versus 8%. The mean NYHA class in HeartMate patients was II versus IV in medically treated patients and quality of life was better in the VAD group. However, probability of device failure was 35% at 24 months in the trial. VAD was replaced in 10 patients during the follow-up period and total VAD-related mortality was higher

than 10%. These results raised a lot of concerns about reliability and made cardiologists reluctant to send their patients be supported with pulsatile devices.

In 2008, the FDA approved the HeartMate II as BTT. In 2010 it was approved in DT. Thoratec Inc. stopped consequently to deliver the HeartMate XVE. The HeartWare has been approved since 2012. Since January 2010, no implantable pulsatile device has been implanted in DT in the United States.

In the HM II BTT trial, survival at 1 year was 70% [66]. In the HM II DT trial, survival compared to the HM XVE was respectively 68% versus 58% at 1 year and 55% versus 24% at 2 years [69]. The overall survival after MCS reported in the INTERMACS registry is 78% at 1 year and 68% at 2 years. When stratified according to device's type, survival at 1 year is 80% after LVAD, 55% after BiVAD and 54% after TAH. At 2 years, it is 69% in LVAD patients and 53% in BiVAD patients. Continuous-flow devices have a survival of 82% at 1 year and 74% at 2-year. Pulsatile devices have a 1-year survival of 61% and 2-year survival of 43% [86,91].

The major long-term complications remain—thromboembolic events, device-related infections, and device failure. However, incidence dropped dramatically with the continuous-flow pumps. Thromboembolic events are reported to be 0.13−0.22 event/patient/year; percutaneous lead infection occurs at the rate of 0.37−0.61 event/patient/year; sepsis of 0.39−0.62 event/patient/year; and pump replacement at 0.06−0.08 event/patient/year [67,89,90].

29.9 FUTURE PERSPECTIVES

Although several improvements have been made over the last 20 years, refinements in many technical aspects of MCS are still mandatory. The ideal long-term VAD should be small, cause minimal blood components trauma and platelet activation, and have a high reliability and durability. Since the percutaneous lead is still the Achilles heel of the system, it should be totally implantable. Partial circulatory support devices are already available (CircuLite©, CircuLite Inc., Saddle Brooks, NJ, USA) or under investigation (Symphony©, Abiomed Inc., Danvers, MA, USA).

Furthermore, in the current area of evidence-based medicine, we definitely need more data about outcome. Registries like INTERMACS address this issue. A European registry is building up also. The coming Medamacs registry will help to compare outcome of VAD and medically treated heart failure patients. The REVIVE-IT trial (Evaluation of VAD InterVEntion Before Inotropic Therapy) and the Roadmap trial (Risk Assessment and Comparative Effectiveness of Left Ventricular Assist Device (LVAD) and Medical

Management) are two large randomized studies starting soon. They will compare continuous-flow pumps implantation to medical therapy in stable ambulatory heart failure patients.

Finally, MCS in younger patients is largely undeveloped. ECMO is the most used, with all its known limitations. Paracorporeal VADs like the Pedivas© (Thoratec Corporation, Pleasanton, CA, USA) or the Berlin Heart Excor (Berlin Heart GmbH, Berlin, Germany) can be used for longer support. At this moment, no really implantable VAD is available for children. The Pedimacs registry will report on outcome after MCS in children. The PUMPKIN (Pumps for Kids, Infants, and Neonates) program of the National Heart, Lung and Blood Institute in the United States has been created respond to the need for better ventricular assist options in younger patients.

REFERENCES

[1] The First International Congress on Cyclosporine: 16−19, 1983, Houston, Texas. Transplant Proc 1983;15(Suppl. 1−2):2207−3187.

[2] Dardas TF, Cowger J, Koelling TM, Aaronson KD, Pagani FD. The effectiveness of United Network of Organ Sharing status 2 transplantation in the modern era. J Heart Lung Transplant 2011;30(10):1169−74.

[3] Smits JM, Vanhaecke J, Haverich A, de Vries E, Roels L, Persijn G, Laufer G. Waiting for a thoracic transplant in Eurotransplant. Transpl Int 2006;19(1):54−66.

[4] Stehlik J, Edwards LB, Kucheryavaya AY, Benden C, Christie JD, Dipchand AI, et al. The registry of the international society for heart and lung transplantation: 29th Official Adult Heart Transplant Report—2012. J Heart Lung Transplant 2012;31(10):1052−64.

[5] Akutsu T, Kolff W. Permanent substitutes for valves and heart. Trans Am Soc Artif Intern Organs 1958;4:230−4.

[6] Hall CW, Akers WW, O'Bannon W, Liotta D, DeBakey ME. Intraventricular artificial heart. Trans Am Soc Artif Intern Organs 1965;11:263−4.

[7] Cooley D, Liotta D, Hallman G, Bloodwell RD, Leachmam RD, Milam JD. Orthotopic cardiac prosthesis for two-staged cardiac replacement. Am J Cardiol 1969;24:723−30.

[8] Cooley D, Akutsu T, Norman C, Serrato MA, Frazier OH. Total artificial heart in two-staged cardiac transplantation. Cardiovasc Dis 1981;8:305−19.

[9] Unger F. Assisted circulation. Berlin: Springer-Verlag; 1979.

[10] Lawson J, Olsen D, Kolff WJ, Liu WH, Hershgold EJ, Van Kampem K. A three months survival of a calf with an artificial heart. J Lab Clin Med 1976;87:848−58.

[11] DeVries W. Clinical use of the total artificial heart. N Eng J Med 1984;310:273−8.

[12] Copeland J, Smith R, Icenogle T, Vasu A, Rhenman B, Williams R, Cleavinger M. Orthotopic total artificial heart bridge to transplantation: preliminary results. J Heart Lung Transplant 1989;8:124−38.

[13] Discharge drivers, <http://www.syncardia.com/Medical-Professionals/discharge-drivers-us.html>2011 [accessed 22.05.11].

[14] Dowling R, Etoch S, Stevens K, Johnson AC, Gray Jr LA. Current status of the Abiocor implantable replacement heart. Ann Thorac Surg 2001;71:S147−9.

[15] Dowling R, Gray L, Etoch S. The Abiocor implantable replacement heart. Ann Thorac Surg 2003;75:S93−9.

[16] Abiocor frequently asked question, <http://www.abiomed.com/assets/2010/11/AbioCor-FAQ-FINAL.pdf> 2013 [accessed 22.05.11].

[17] Bencini A, Parola P. The pneumomassage of the heart. Surgery 1956;39:375−84.

[18] Dennis C, Hall D, Moreno J. Reduction of the oxygen utilization of the heart by left heart bypass. Circ Res 1962;10:298−305.

[19] Dennis C, Carlens E, Senning A. Clinical use of a cannula for left heart bypass without thoracotomy: experimental protection against fibrillation by left heart bypass. Ann Surg 1962;156:623−36.

[20] Liotta D, Crawford E, Cooley D, DeBakey ME, De Urquia M, Feldman L. Prolonged partial ventricular bypass by means of an intrathoracic pump implanted in the left chest. Trans Am Soc Artif Intern Organs 1962;8:90−9.

[21] DeBakey M. Left ventricular bypass pump for cardiac assistance. Am J Cardiol 1971;27:3−11.

[22] Liotta D, Hall CW, Akers W, Villanueva A, O'Neal RM RM, DeBakey ME. A pseudoendocardium for implantable blood pumps. Trans Am Soc Artif Int Organs 1966;12:129−38.

[23] Bernhard W, La Farge C, Robinson T, Yun I, Shirahige K, Kitrilakis S. An improved blood−pump interface for left ventricular bypass. Ann Surg 1968;168:750−64.

[24] Norman J. An abdominal left ventricular assist device (A-LVAD): perspectives and prospects. Cardiovasc Dis 1974;1:251−64.

[25] Norman J, Brook M, Cooley DA, Klima T, Kahan BD, Frazier OH, et al. Total support of the circulation of a patient with postcardiotomy stone-heart syndrome by a partial artificial heart (ALVAD) for 5 days followed by heart and kidney transplantation. Lancet 1978;1:1125−7.

[26] Pierce WS, Brighton JA, O'Bannon W, Doonachy JH, Phillips WM, Landis DL, et al. Complete left ventricular bypass with a paracorporeal pump: design and evaluation. Ann Surg 1974;180:418−26.

[27] Bindels J, Grigsby L. Considerations and calculations about the optimum solenoid to be used for an intrathoracic artificial heart. Trans Am Soc Artif Intern Organs 1961;7:369−72.

[28] Portner PM, Oyer PE, Miller PJ, Jassawalla JS, Ream AK, Corbin SD, Skytte KW. Evolution of the solenoid-actuated left ventricular assist system: integration with a pusher-plate pump for intra-abdominal implantation in the calf. Artif Organs 1978;2:402−12.

[29] Starnes VA, Oyer PE, Portner M, Ramasamy N, Miller PJ, Stinson EB, et al. Isolated left ventricular assist as bridge to cardiac transplantation. J Thorac Cardiovasc Surg 1988;96:62−71.

[30] Whalen RL, Molokhia FA, Jeffery D, Huffman FN, Norman JC. Current studies with simulated nuclear-powered left ventricular assist devices. Trans Am Soc Artif Intern Organs 1972;18:146−51.

[31] Wolner E, Deutsch M, Losert U, Stellwag F, Thoma H, Unger F, et al. Clinical application of the ellipsoid left heart assist device. Artif Organs 1978;2:268−72.

[32] Dasse KA, Chipman SD, Sherman CN, Levine AH, Frazier OH. Clinical experience with textured blood contacting surfaces in ventricular assist devices. Trans Am Soc Artif Intern Organs 1987;10:418−25.

[33] Rose EA, Gelijns AC, Moskowitz AJ, Heitjan DF, Stevenson LW, Dembitsky W, et al. Randomized Evaluation of Mechanical Assistance for the Treatment of Congestive Heart Failure

(REMATCH) study group. Long-term use of a left ventricular assist device for end-stage heart failure. N Engl J Med 2001;345 (20):1435—43.

[34] Saxton G, Andrew C. An ideal pump with hydrodynamic characteristics analogous to the mammalian heart. Trans Am Soc Artif Intern Organs 1960;6:288—9.

[35] Rafferty E, Kletschka H, Wynyard M, Larkin JT, Smith LV, Cheathem B. Artificial heart I. Application of nonpulsatile force-vortex principle. Minn Med 1968;51:11—6.

[36] Wampler RK, Moise JC, Frazier OH, Olsen DB. *In vivo* evaluation of a peripheral vascular access axial flow blood pump. Trans Am Soc Artif Intern Organs 1988;34:450—4.

[37] Potapov EV, Loebe M, Nasseri BA, Sinawski H, Koste A, Kuppe H. Pulsatile flow in patients with a novel nonpulsatile implantable ventricular assist device. Circulation 2000;102: III183—7.

[38] Westaby S, Katsumata Y, Evans R, Pigott D, Taggart DP, Jarvik RK. The Jarvik 2000 Oxford system: increasing the scope of mechanical circulatory support. J Thorac Cardiovasc Surg 1997;114:467—74.

[39] Griffith B, Kormos R, Borovetz H, Litwak K, Antaki JF, Poirier VL, Butler KC. HeartMate II left ventricular assist system: from concept to first clinical use. Ann Thorac Surg 2001;71:S116—20.

[40] Slaughter MS, Rogers JG, Milano CA, Russel SD, Conte JV, Feldman D, et al. Advanced heart failure treated with continuous flow assist device. N Engl J Med 2009;361:2241—51.

[41] Hetzer R, Weng Y, Potapov EV, Pasic M, Drews T, Jurmann M. First experiences with a novel magnetically suspended axial flow left ventricular assist device. Eur J Cardiothorac Surg 2004;25:964—70.

[42] LaRose J, Tanez D, Ashenuga M, Reyes C. Design concepts and principle of the HeartWare ventricular assist system. ASAIO J 2010;56:285—9.

[43] Frazier O, Tuzun E, Cohn WE, Conger JL, Fadipasaoglu KA, et al. Total heart replacement using dual intracorporeal continuous-flow pump in a chronic bovine model: a feasibility study. ASAIO J 2006;52:145—9.

[44] Weerwind PW, van der Veen FH, Lindhout T, de Jong DS, Cahalan PT. *Ex vivo* testing of heparin-coated extracorporeal circuits: bovine experiments. Int J Artif Organs 1998;21(5):291—8.

[45] Bennett M, Horton S, Thuys C, Augustin S, Rosenberg M, Brizard C. Pump-induced haemolysis: a comparison of short-term ventricular assist devices. Perfusion 2004;19(2):107—11.

[46] Sobieski MA, Giridharan GA, Ising M, Koenig SC, Slaughter MS. Blood trauma testing of CentriMag and RotaFlow centrifugal flow devices: a pilot study. Artif Organs 2012;36(8):677—82.

[47] Khoshbin E, Roberts N, Harvey C, Machin D, Killer H, Peek GJ, Sosnowski AW, Firmin RK. Poly-methyl pentene oxygenators have improved gas exchange capability and reduced transfusion requirements in adult extracorporeal membrane oxygenation. ASAIO J 2005;51(3):281—7.

[48] Oliver WC. Anticoagulation and coagulation management for ECMO. Semin Cardiothorac Vasc Anesth 2009;13(3):154—75.

[49] Burkhoff D, Cohen H, Brunckhorst C, O'Neill WW, TandemHeart Investigators Group. A randomized multicenter clinical study to evaluate the safety and efficacy of the TandemHeart percutaneous ventricular assist device versus conventional therapy with intraaortic balloon pumping for treatment of cardiogenic shock. Am Heart J 2006;152(3):469 e1—e8.

[50] Kar B, Gregoric ID, Basra SS, Idelchik GM, Loyalka P. The percutaneous ventricular assist device in severe refractory cardiogenic shock. J Am Coll Cardiol 2011;57(6):688—96.

[51] Thomas JL, Al-Ameri H, Economides C, Shareghi S, Abad DG, Mayeda G, Burstein S, Shavelle DM. Use of a percutaneous left ventricular assist device for high-risk cardiac interventions and cardiogenic shock. J Invasive Cardiol 2010;22(8):360—4.

[52] Frazier OH, Wampler RK, Duncan JM, Dear WE, Macris MP, Parnis SM, Fuqua JM. First human use of the Hemopump, a catheter-mounted ventricular assist device. Ann Thorac Surg 1990;49(2):299—304.

[53] Seyfarth M, Sibbing D, Bauer I, Fröhlich G, Bott-Flügel L, Byrne R, et al. A randomized clinical trial to evaluate the safety and efficacy of a percutaneous left ventricular assist device versus intra-aortic balloon pumping for treatment of cardiogenic shock caused by myocardial infarction. J Am Coll Cardiol 2008;52 (19):1584—8.

[54] Engström AE, Cocchieri R, Driessen AH, Sjauw KD, Vis MM, Baan J, Serruys PW, et al. The Impella 2.5 and 5.0 devices for ST-elevation myocardial infarction patients presenting with severe and profound cardiogenic shock: the Academic Medical Center intensive care unit experience. Crit Care Med 2011;39 (9):2072—9.

[55] Cheng JM, den Uil CA, Hoeks SE, van der Ent M, Jewbali LS, van Domburg RT, Serruys PW. Percutaneous left ventricular assist devices vs. intra-aortic balloon pump counterpulsation for treatment of cardiogenic shock: a meta-analysis of controlled trials. Eur Heart J 2009;30(17):2102—8.

[56] Samuels LE, Holmes EC, Garwood P, Ferdinand F. Initial experience with the Abiomed AB5000 ventricular assist device system. Ann Thorac Surg 2005;80(1):309—12.

[57] Slaughter MS, Sobieski MA, Martin M, Dia M, Silver MA. Home discharge experience with the Thoratec TLC-II portable driver. ASAIO J 2007;53(2):132—5.

[58] Rakhorst G, Hensens AG, Verkerke GJ, Blanksma PK, Bom VJ, Elstrodt J, et al. *In vivo* evaluation of the "HIA-VAD": a new German ventricular assist device. Thorac Cardiovasc Surg 1994;42:136.

[59] Hetzer R, Alexi-Meskishvili V, Weng Y, et al. Mechanical cardiac support in the young with the Berlin Heart EXCOR pulsatile ventricular assist device: 15 years' experience. Semin Thorac Cardiovasc Surg Pediatr Card Surg Annu 2006;9:99—108.

[60] Homma A, Taenaka Y, Tatsumi E, Akagawa E, Lee H, Nishinaka T, et al. Development of a compact wearable pneumatic drive unit for a ventricular assist device. Organs 2008;11(4):182—90 Epub 2008 Dec 17.

[61] Portner PM, Oyer PE, McGregor CGA. First human use of an electrically powered implantable ventricular assist system. Artif Organs 1985;9(a):36.

[62] Schmid C, Weyand M, Nabavi DG, Hammel D, Deng MC, Ringelstein EB, Scheld HH. Cerebral and systemic embolization during left ventricular support with the Novacor N100 device. Ann Thorac Surg 1998;65(6):1703—10.

[63] Di Bella I, Pagani F, Banfi C, Ardemagni E, Capo A, Klersy C, Viganò M. Results with the Novacor assist system and evaluation

of long-term assistance. Eur J Cardiothorac Surg 2000;18 (1):112−6.

[64] Strauch JT, Spielvogel D, Haldenwang PL, Correa RK, deAsla RA, Seissler PE, Baran DA, Gass AL, Lansman SL. Recent improvements in outcome with the Novacor left ventricular assist device. J Heart Lung Transplant 2003;22(6):674−80.

[65] Frazier OH, Rose EA, Oz MC, Dembitsky WP, McCarthy PM, Radovancevic B, Poirier VL, Dasse KA. Multicenter clinical evaluation of the HeartMate; vented electric left ventricular assist system in patients awaiting heart transplantation. J Heart Lung Transplant 2001;20(2):201−2.

[66] Miller LW, Pagani FD, Russell SD, John R, Boyle AJ, Aaronson KD, et al. HeartMate II clinical investigators. Use of a continuous-flow device in patients awaiting heart transplantation. N Engl J Med 2007;357(9):885−96.

[67] John R, Kamdar F, Liao K, Colvin-Adams M, Boyle A, Joyce L. Improved survival and decreasing incidence of adverse events with the HeartMate II left ventricular assist device as bridge-to-transplant therapy. Ann Thorac Surg 2008;86(4):1227−34.

[68] Lahpor J, Khaghani A, Hetzer R, Pavie A, Friedrich I, Sander K, Strüber M, et al. European results with a continuous-flow ventricular assist device for advanced heart-failure patients. Eur J Cardiothorac Surg 2010;37(2):357−61.

[69] Slaughter MS, Rogers JG, Milano CA, Russell SD, Conte JV, Feldman D, et al. HeartMate II investigators. Advanced heart failure treated with continuous-flow left ventricular assist device. N Engl J Med 2009;361(23):2241−51.

[70] Siegenthaler MP, Martin J, Pernice K, Doenst T, Sorg S, Trummer G, et al. The Jarvik 2000 is associated with less infections than the HeartMate left ventricular assist device. Eur J Cardiothorac Surg 2003;23:748−54.

[71] Morshuis M, El-Banayosy A, Arusoglu L, Koerfer R, Hetzer R, Wieselthaler G, et al. European experience of DuraHeart magnetically levitated centrifugal left ventricular assist system. Eur J Cardiothorac Surg 2009;35(6):1020−7.

[72] Stewart GC, Givertz MM. Mechanical circulatory support for advanced heart failure: patients and technology in evolution. Circulation 2012;125(10):1304−15.

[73] Levy WC, Mozaffarian D, Linker DT, Farrar DJ, Miller LW. REMATCH investigators. Can the Seattle Heart Failure Model be used to risk-stratify heart failure patients for potential left ventricular assist device therapy?. J Heart Lung Transplant 2009;28(3):231−6.

[74] Cowger J, Sundareswaran K, Rogers JG, Park SJ, Pagani FD, Bhat G, et al. The HeartMate II risk score: predicting survival in candidates for left ventricular assist device support. J Heart lung Transplant 2011;30:S31 [Abstract 70]

[75] Fitzpatrick 3rd JR, Frederick JR, Hsu VM, Kozin ED, O'Hara ML, Howell E, et al. Risk score derived from pre-operative data analysis predicts the need for biventricular mechanical circulatory support. J Heart Lung Transplant 2008;27(12):1286−92.

[76] Matthews JC, Koelling TM, Pagani FD, Aaronson KD. The right ventricular failure risk score a pre-operative tool for assessing the risk of right ventricular failure in left ventricular assist device candidates. J Am Coll Cardiol 2008;51(22):2163−72.

[77] Drakos SG, Janicki L, Horne BD, Kfoury AG, Reid BB, Clayson S, et al. Risk factors predictive of right ventricular failure after left ventricular assist device implantation. Am J Cardiol 2010;105(7):1030−5.

[78] Butler J, Howser R, Portner PM, Pierson 3rd RN. Body mass index and outcomes after left ventricular assist device placement. Ann Thorac Surg 2005;79(1):66−73.

[79] Brewer RJ, Lanfear DE, Sai-Sudhakar CB, Sundareswaran KS, Ravi Y, Farrar DJ, Slaughter MS. Extremes of body mass index do not impact mid-term survival after continuous-flow left ventricular assist device implantation. J Heart Lung Transplant 2012;31(2):167−72.

[80] Martin SI, Wellington L, Stevenson KB, Mangino JE, Sai-Sudhakar CB, Firstenberg MS, et al. Effect of body mass index and device type on infection in left ventricular assist device support beyond 30 days. Interact Cardiovasc Thorac Surg 2010;11 (1):20−3.

[81] Kirklin JK, Naftel DC, Kormos RL, Stevenson LW, Pagani FD, Miller MA, Slaughter MS. The fourth INTERMACS annual report: 4,000 implants and counting. J Heart Lung Transplant 2012;31(2):117−26.

[82] Adamson RM, Stahovich M, Chillcott S, Baradarian S, Chammas J, Jaski B, et al. Clinical strategies and outcomes in advanced heart failure patients older than 70 years of age receiving the HeartMate II left ventricular assist device: a community hospital experience. J Am Coll Cardiol 2011;57 (25):2487−95.

[83] Afilalo J, Eisenberg MJ, Morin JF, Bergman H, Monette J, Noiseux N, et al. Gait speed as an incremental predictor of mortality and major morbidity in elderly patients undergoing cardiac surgery. J Am Coll Cardiol 2010;56(20):1668−76.

[84] Stevenson LW, Pagani FD, Young JB, Jessup M, Miller L, Kormos RL, et al. INTERMACS profiles of advanced heart failure: the current picture. J Heart Lung Transplant 2009;28 (6):535−41.

[85] Kirklin JK, Naftel DC, Kormos RL, Stevenson LW, Pagani FD, Miller MA, et al. Second INTERMACS annual report: more than 1,000 primary left ventricular assist device implants. J Heart Lung Transplant 2010;29(1):1−10.

[86] Holman WL, Kormos RL, Naftel DC, Miller MA, Pagani FD, Blume E, et al. Predictors of death and transplant in patients with a mechanical circulatory support device: a multi-institutional study. J Heart Lung Transplant 2009;28(1):44−50.

[87] Kirklin JK, Naftel DC, Kormos RL, Stevenson LW, Pagani FD, Miller MA, et al. The fourth INTERMACS annual report: 4,000 implants and counting. J Heart Lung Transplant 2012;31 (2):117−26.

[88] Krishan K, Nair A, Pinney S, Adams D, Anyanwu AC. Low incidence of bleeding-related morbidity with left ventricular assist device implantation in the current era. Artif Organs 2012;36 (8):746−51.

[89] Genovese EA, Dew MA, Teuteberg JJ, Simon MA, Kay J, Siegenthaler MP, et al. Incidence and patterns of adverse event onset during the first 60 days after ventricular assist device implantation. Ann Thorac Surg 2009;88(4):1162−70.

[90] Monkowski DH, Axelrod P, Fekete T, Hollander T, Furukawa S, Samuel R. Infections associated with ventricular assist devices: epidemiology and effect on prognosis after transplantation. Transpl Infect Dis 2007;9(2):114−20.

[91] Cleveland Jr JC, Naftel DC, Reece TB, Murray M, Antaki J, Pagani FD, et al. Survival after biventricular assist device implantation: an analysis of the interagency registry for mechanically

assisted circulatory support database. J Heart Lung Transplant 2011;30(8):862—9.

[92] Poston RS, Husain S, Sorce D, Stanford E, Kusne S, Wagener M, et al. LVAD bloodstream infections: therapeutic rationale for transplantation after LVAD infection. J Heart Lung Transplant 2003;22(8):914—21.

[93] Holman WL, Pamboukian SV, McGiffin DC, Tallaj JA, Cadeiras M, Kirklin JK. Device related infections: are we making progress? J Card Surg 2010;25(4):478—83.

[94] Kalya AV, Tector AJ, Crouch JD, Downey FX, McDonald ML, Anderson AJ, et al. Comparison of Novacor and HeartMate vented electric left ventricular assist devices in a single institution. J Heart Lung Transplant 2005;24(11):1973—5.

[95] El-Banayosy A, Arusoglu L, Kizner L, Tenderich G, Minami K, Inoue K, et al. Novacor left ventricular assist system versus HeartMate vented electric left ventricular assist system as a long-term mechanical circulatory support device in bridging patients: a prospective study. J Thorac Cardiovasc Surg 2000;119 (3):581—7.

Chapter 30

Heart Regeneration: The Bioengineering Approach

Dror Seliktar, Alexandra Berdichevski, Iris Mironi-Harpaz, and Keren Shapira-Schweitzer

Faculty of Biomedical Engineering, Technion—Israel Institute of Technology, Haifa, Israel

30.1 INTRODUCTION

Nearly 2400 Americans die of cardiovascular-related diseases (CVDs) each day—an average of one death every 37 s. CVD claims approximately as many lives each year as cancer and diabetes mellitus combined [1]. Although there have been significant improvements in cardiac medicine, the optimal solution for the failing heart remains heart transplantation. Unfortunately, there is a significant shortage in donors—in 2007 there were close to 3000 heart patients on the transplant waiting list in the United States alone. There are many more who are not considered to be candidates for heart transplantation but could benefit from a cardiac restoration therapy that can rejuvenate part of the heart muscle immediately following a heart attack and myocardial infarction (MI) [2].

MI is a progressive process starting with an occlusion and a reduction of the coronary blood supply that leads to myocardial ischemia and the imminent death of cardiac cells (i.e., cardiomyocytes). A tissue repair process is initiated at the onset of MI, immediately following the loss of the cardiomyocyte population in the injured tissue. Several events occur during this repair process, including the invasion of inflammatory cells into the infract area and the increased activity of matrix metalloproteinase (MMP) at the site of inflammation. Collagen degradation is subsequently increased, and this is accompanied by additional cell loss, wall thinning, and ventricular dilation at the infarct site. After the initial inflammatory phase, the synthesis of MMP inhibitors is elevated, leading to an accumulation of collagen in the infract zone. This fibrogenic phase ends in the formation of a fibrotic scar tissue in the infarct site as well as the formation of fibrous tissue in locations that are remote to the infarct [3].

The objective of cardiac restoration therapies is to augment the fibrotic scar tissue in the infarct with functional cardiac muscle and to prevent the unavoidable deterioration of the remaining healthy cardiac muscle. Several approaches have been suggested to achieve these goals. The earliest attempts at cardiac restoration therapy were focused on the direct injection of viable cells into the infracted myocardium. Different types of cells have been employed for this purpose, including skeletal myoblasts, neonatal cardiomyocytes, fibroblasts, smooth muscle cells, embryonic stem cells, and adult stem cells (bone marrow progenitors) [4]. Even though some studies showed an improvement in cardiac function as a result of the cell injection/infusion [5,6], the percent of surviving cells in the infarcted myocardium was generally very low [7,8]. For example, Qian et al. showed that 1 h after intracoronary delivery of autologous bone marrow mononuclear cells, only ~7% of the cells appeared in the myocardium whereas more than 90% of the cells were accumulated in the liver and the spleen [8].

The low cell survival following direct cell injection motivated the use of biomaterials in cardiac restoration therapies. Two distinct categories of biomaterial-based cardiac cell therapy were introduced: a tissue engineered cardiac patch and an injectable biomaterial/cell graft. The tissue engineering method attempts to create a tissue analog *in vitro*, which is to be sutured directly onto the infarcted myocardium. The injectable biomaterial approach is designed to locally deliver cell grafts. Generally, cardiac restoration with injectable biomaterials is focused on the use of injectable hydrogels as cell carriers that, when combined with cells, should increase the cell survival and improve the overall contractility of the infarcted myocardium. In spite of the enormous potential of cardiac cell therapy in this regard, some recent studies have focused on using biomaterials alone to stabilize the cardiac wall geometry and prevent cardiac remodeling, without cell therapy.

30.2 ACELLULAR APPROACHES

The use of an acellular approach in cardiac restoration is premised on the hypothesis that a biomaterial can prevent the destructive cardiac remodeling process following MI by mechanically stabilizing the myocardial wall following the acute infarction. In this context, both injectable and noninjectable biomaterials were tested in their capacity to stabilize the wall geometry of the injured myocardium. Such studies reported that the wall stress and paradoxical systolic bulging can be reduced by grafting biomaterials that prevent the wall thinning and cardiac dilation following MI. A number of different biomaterials were tested for this purpose, including biological polymers such as collagen, alginate, Matrigel (a basement membrane extract composed of collagen, proteoglycans, and laminin), fibrin, self-assembling polypeptides, hyaluronic acid (HA)-based hydrogels, and naturally derived myocardial matrix, as well as synthetic polymers such as poly(propylene) (Marlex) and polyester [9].

The stabilization of the cardiac geometry can likewise be accomplished by restraining the ventricular wall using an external support device such as a synthetic mesh which is implanted on the heart muscle. A nondegradable or degradable polymeric mesh can be sutured either directly to the infarct region [10,11], the left ventricle (LV) [12] or around both ventricles [13,14]. A number of clinical studies performed with LV restraints demonstrated improvement in cardiac function [14−16]. Although the studies reported encouraging results in terms of improved LV function, this technique involves a complicated surgical procedure which is not preferred for patients suffering an acute MI. A similar approach using a degradable cardiac patch—a biomaterial patch sutured on the left ventricular wall—was tested in preclinical studies [17]. Fujimoto et al. reported that a biodegradable polyester urethane urea cardiac patch

implanted directly on the infarcted LV wall of rat hearts 2 weeks after coronary ligation produced encouraging results after 2 months [17]. The patch was almost entirely absorbed after 8 weeks, but the wall thickness was preserved and capillary density was increased relative to the infarction control group. Additionally, muscle-like bundles were found juxtapose to the patch implant and these stained positive for α-smooth muscle actin, indicating a contractile phenotype.

An alternative and simpler approach to stabilizing the myocardial wall is to use an injectable biomaterial in order to passively preserve cardiac geometry. In this context, some biomaterials may even be used as active implants, designed to initiate a healing response by stimulating neovascularization, angiogenesis, and cellular recruitment to the scar tissue. Christman et al. demonstrated this approach using a fibrin glue implant; an injection of fibrin into the infarcted myocardium preserved wall thickness, induced neovascularization, and reduced the overall infarct size [18,19]. Fibrin and its degradation products, including fragment E, were shown to be proangiogenic [20]. Huang et al. compared the influence of injected fibrin, collagen, and Matrigel on the infarcted myocardial remodeling in the adult rat heart [21]. They reported that in all three materials there was an acute inflammatory reaction after 1 day. However, there was no evidence for the presence of the injected biopolymers in the infarct area after 5 weeks. Furthermore, there was no significant inflammatory response at this time point when compared to the saline control animals. Interestingly, all three biopolymers used showed an enhanced angiogenic response in the myocardium, but no significant differences were found among the different treatments. The angiogenic response elicited by the three biological polymers can be readily explained by the inherent presence of cell-binding sites in the protein's sequence, including the $\alpha_v\beta_3$ integrin binding site. This particular integrin is known to promote angiogenesis *in vivo*. The authors also demonstrated that all three biological polymers are a suitable substrate for endothelial cell spreading and migration *in vitro*. In a separate study, Dai et al. evaluated the contribution of injected collagen to the infarcted myocardium. In contrast to previous reports, they found that the collagen remained in the scar tissue for up to 6 weeks. The injected collagen increased the scar tissue thickness, prevented dilation, and improved cardiac function (LV stroke volume and ejection fraction) [22].

Landa et al. reported on the effect of an injectable alginate on cardiac remodeling and function following MI [23]. This work was performed using a two-part solution of calcium cross-linker and alginate monomer with low enough initial viscosity to allow injection of the material through a small-gauge syringe needle. The solution undergoes a phase transition into a hydrogel once it is injected and mixed together in the myocardium. The efficacy work was performed in a rat ischemic heart model with both recent and established infarcts (7 days and 60 days post-ischemia, respectively). The hydrogels were fully degraded and replaced by connective tissue in 6 weeks after the injection. The alginate implant increased the scar thickness and reduced the remodeling process in both the recent and the established infracts. Mechanistically, the authors suggested that the alginate physically stabilizes the scar tissue, thus preventing dilatation of the myocardium and reducing wall stress and paradoxical systolic bulging. In addition it has been shown that alginate has an angiogenesis effect in the infarct and it can be further increased by immobilizing RGD sequences into the biopolymer [24]. Leor et al. have further demonstrated recently that intracoronary injection of the alginate hydrogel is possible and allows the prevention of cardiac remodeling process after an infarct in a swine model [25].

Self-assembled polypeptides were also introduced as an acellular biomaterial for cardiac therapy [26]. According to this study, the polypeptide hydrogel can promote the healing process of the infarcted myocardial tissue by creating a supportive microenvironment for the recruitment of endothelial cells. The synthetic material was comprised of short peptides containing 8–16 amino acids. The peptides self-assemble into nanofibers which form a 3D hydrogel network at a physiological PH. The peptide hydrogel did not cause a major inflammatory response 1 week after implantation. The gels were, however, populated with endothelial cells at a density of 33.5 ± 1.9 cells/mm^2. After 14 days postimplantation, there was a significant decrease in the endothelial cell density (17.3 ± 0.2 cells/mm^2), which remained constant at 28 days. Eventually, the endothelial cells created capillary-like structures and became clustered after 28 days. At the later time point, smooth muscle cells were also observed forming arterioles within the injected hydrogel environment. In addition to the vascular cells, the authors reported on α-sarcomeric actin positive cells which invaded the injection site. A control with Matrigel resulted in little penetration of endothelial cells and no α-sarcomeric actin positive cells within the injected site.

More recently, a HA-based hydrogel was also suggested to be injected to the infarcted cardiac muscle [27]. This hydrogel is formed by Michael-type addition reaction between acrylated HA and polyethylene glycol (PEG)-SH$_4$ thiol groups. The gelation time is approximately 10 min. This hydrogel is characterized with very high swelling ratio (1300%). Thirty days following an injection of the HA hydrogel into rat hearts with partial ligation of the left circumflex artery, there was a significant increase in the wall thickness and a reduction of the infarct area compared with the control. An increase of 150% in the capillaries and arterioles was observed in the

border zone compared to the control. The functional measurements showed a significant recovery following the injection; for example an increase in the ejection fraction from 18.2 ± 5.4 in the control group to 42.7 ± 7.5 following the injection of the hydrogel was reported. The overall function of the heart in the HA hydrogel injection group was surprisingly similar to that of the sham group; however, the authors indicated that the infarct created by the partial ligation of the left circumflex artery was not severe (the systolic function was similar in the sham and MI group) which can explain the full recovery of the injection group.

Based on the growing body of literature with animal studies, one can generally conclude that acellular therapy using biodegradable biomaterials can physically delay cardiac remodeling and in some cases even initiate the healing process by providing a temporary supportive environment for angiogenesis. Because most of the materials are absorbed after 6–8 weeks, it should be interesting to further study the long-term results from acellular treatments before knowing the true efficacy of this simple yet promising clinical approach.

30.3 CELL-BASED APPROACHES: NON-INJECTABLE MATERIALS

A common approach to cell delivery in cardiac regeneration therapy has been to implant a cardiac patch made from a biomaterial scaffold seeded with cells. One of the first reports of an implanted cardiac patch was described by Li et al. using a gelatin scaffold seeded with fetal rat cardiac cells [28]. The cells survived in the graft for up to 5 weeks in both the subcutaneous implantation model as well as on the myocardial scar tissue of adult rats. Although cells in the graft contracted spontaneously *in vitro* and in the subcutaneous implantation, there was no measurable effect on cardiac function following implantation of the patch on the myocardium. The study indicated that the size of the implanted graft was too large in comparison to the infarct size and hence may have damaged the healthy regions of the cardiac muscle that were also covered by the patch.

Since this initial study, other materials including collagen and Matrigel have been widely used for cell delivery in myocardial regeneration [29–32]. Zimmermann et al. developed engineered heart tissues (EHTs) from neonatal rat heart cell embedded in collagen I, Matrigel, and various growth supplements. The EHTs were shaped as circular rings and subjected to mechanical strain stimulation during *in vitro* culture. A multiloop EHT was created by fashioning five rings of EHTs together in a unique geometric configuration resembling a patch. The multiloop EHT was sutured onto the infarcted heart muscle of an adult rat with an MI. They demonstrated that the multi-loop EHT was integrated and electrically coupled to the native myocardium. Consequently, they showed that the cardiac systolic and diastolic function was improved. Similarly, Kutschka et al. used collagen gels conjugated with matrigel and growth factors (VEGF, FGF) to deliver fluorescently labeled cardiomyoblasts to an adult rat heart. They used *in vivo* optical bioluminescence imaging to evaluate cell survival in the graft after implantation. Consistent with previous studies, they documented low cell survival rate when the cardiomyocytes were injected alone, whereas cells seeded on the collagen matrix showed significant improvements in the survival rate after 8 days. Only 30% of the bioluminescence signal was recorded after 1 week without the collagen, and a three-fold enhancement in the imaging signal was measured with the collagen matrix. One possible explanation for the improved cell survival was based on the hypothesis that the porous collagen matrix can better guide vessel ingrowth and thus improve perfusion to the implanted cells. Interestingly, the cell survival rate was not significantly affected by the addition of the growth factors to the collagen matrix. After 4 weeks only a small fraction of the originally implanted cells survived in the scar tissue. The surviving cells did not express connexin-43 (Cx43) and hence did not integrate with the host myocardium. An improvement in cardiac function was documented, but the inhibition of cardiac remodeling and wall thinning was the predominant mechanism for the cardiac improvements, rather than a functional contribution of the transplanted cells.

Leor et al. investigated alginate sponges as a cell scaffold for myocardial repair [33]. Alginate scaffolds seeded with fetal cardiomyocytes were implanted into the infarcted rat myocardium. The cardiomyocytes maintained viability within the scaffold and formed beating aggregates after 24 h *in vitro*. The implanted patch was examined after 9 weeks, at which time the histological evaluation showed massive neovascularization at the site of implantation. This angiogenic response was thought to contribute to the high cell viability in the graft. Although the fetal cardiomyocytes differentiated into a mature cardiac phenotype *in situ*, only a small portion of the graft was occupied by cardiomyocytes whereas most of the scaffold was filled with collagen fibers and scattered fibroblasts. Moreover, the alginate was nearly completely degraded after 9 weeks and was replaced almost entirely by the collagen fibers. The authors suggested that the implantation improved cardiac function by altering the remodeling process and preventing the imminent deterioration of the cardiac wall following the MI. Because the cardiomyocyte population in the graft was very small and not completely integrated with the host muscle fibers, it was not likely that the improvement in cardiac function

was due to the contribution of the implanted cells to the contractility of the heart muscle. It was further suggested that the changes in mechanical properties induced from the scaffold implantation and the neovascularization could have been the more relevant factor in maintaining improved cardiac performance.

A cardiac tissue patch can also be created using the cell self-assembly approach pioneered by Okano and coworkers. In this approach, cell sheets are cultivated and detached from their culture substrate without the use of a scaffold material. A temperature-responsive polymer substrate made from poly(N-isopropylacrylamide) (PIPAAm) is used instead to remove the cell sheets without causing damage to their integrity. The PIPAAm is hydrophobic at 37°C and allows the cultivation of cells under normal conditions. By lowering the temperature to 32°C, the polymer quickly becomes more hydrophilic and nonadhesive to the cell sheets. This property enables the detachment of a confluent uniform monolayer of cardiomyocytes without the use of destructive enzymatic harvesting techniques. The monolayer is maintained intact, including cell−cell connections and extracellular matrix (ECM) proteins which were structured during the *in vitro* culture, so that when the 3D functional tissue is created from stacked monolayer sheets, the tissue-like integrity is preserved [34,35]. Unfortunately, only a very limited number of the 50-μ thick layers can be stacked together before the cells require additional perfusion to survive (no more than three cell sheets). To overcome this limitation, a method of sequential transplantations of multisheet constructs was proposed whereby the first implant would be vascularized *in vivo* before new sheets would be grafted on. Using this multistep transplantation procedure, the authors reported grafting a 1 mm thick cardiac tissue onto an infarcted adult rat myocardium. The cardiomyocyte sheets became integrated with the host tissue and expression of Cx43 was documented. Electrophysiological experiments showed the loss of a branch block which is seen in the control groups (no treatment or fibroblast cell sheet grafts), an indication for conductivity improvement in the scar tissue [36].

A similar approach was tested using tissue culture dishes coated with a thin layer of thrombin-polymerized fibrin [37,38]. The use of fibrin as a culture substrate with the cell self-assembly strategy has shown promising results with skeletal muscle cells [39]. The myocytes migrated and proliferated on top and inside the fibrin gel, produced ECM proteins, and degraded the fibrin matrix after 3−4 weeks *in vitro*. The fibrin degradation rate was controllable and can be matched to the ECM production rate. Huang et al. compared the use of fibrin-coated dishes by layering cardiomyocytes on top of the fibrin or embedded inside the thin fibrin layer. The cells plated on the surface of the fibrin created a uniform monolayer that displayed synchronous contractions of the sheet after 48 h; however, in the fibrin-embedded cells, only isolated groups of cells were observed beating. The authors reported that the embedded cells did not spread or form cell–cell connections.

Synthetic scaffolds have also been used in creating cardiac tissue patches for cell-based therapy. Vunjak-Novakovic and coworkers demonstrated the ability to create a cardiac tissue analog from rat neonatal cardiac cells cultured in a 3D synthetic scaffold made from a fibrous polyglycolic acid mesh [40]. The effects of cell seeding density and tissue cultivation conditions were evaluated extensively in these studies. They found that a high cell seeding density combined with a dynamic culture environment provided by a rotating bioreactor system improved the structural and functional properties of the engineered cardiac tissue.

Radisic et al. described a biomimetic approach to cardiac tissue engineering based on a porous solid scaffold. They attempted to address one of the critical problems in cardiac tissue engineering: the lack of sufficient oxygen supply. Cardiac cells and fibroblasts were cultured on a porous poly(glycerol-sebacate) (biorubber) scaffold with an array of cubically packed parallel channels which were created using a laser cutting system. In addition to the highly perfused microchannel array, the culture medium was also supplemented with a perfluorocarbon emulsion—an oxygen carrier to mimic hemoglobin oxygen delivery. They showed that the increase in oxygen concentrations also increased cell density and functional properties of the engineered tissue. Recently, Caspi et al. reported on a vascularized cardiac tissue analog made from a scaffold composed of 50% poly-L-lactic acid and 50% polylactic-glycolic acid. The vasculature was not imprinted into the scaffold, but rather it was developed by a coculture of vascular cells seeded into the material. They seeded the construct with human embryonic stem cells-derived cardiomyocytes, human embryonic stem cell-derived endothelial cells or human umbilical vein endothelial cells, and embryonic fibroblasts. They demonstrated that the preformed capillary-like network of endothelial cells was able to connect to the normal blood supply of the host upon implantation.

30.4 INJECTABLE MATERIALS

The use of injectable biomaterials for cellular cardiomyoplasty is an emerging and promising field in cardiac tissue engineering. The injectable scaffolds for cell delivery offer many advantages over the classic cardiac tissue engineering approach using a solid scaffold. A well-designed injectable biomaterial can maintain cell survival and promote cell development toward integration with the healthy cardiac tissue, but at the same time should not impose a mechanical load on the injured cardiac muscle. Identifying an appropriate biomaterial for cardiomyocyte transplantation is therefore one of the important aspects in cardiac tissue engineering. An injectable biomaterial must undergo an *in situ* liquid-to-solid transition (polymerization) with cardiomyocytes in suspension without harming the cells or the surrounding host myocardium. After polymerization, the cells should be able to readily migrate through the material and remodel the polymer so that true engraftment is possible by natural, cell-mediated pathways. To this end, a biomaterial possessing susceptibility to tissue remodeling enzymes would be advantageous [41]. At the same time, the compliance of the polymer must not obstruct cellular remodeling [26] nor distort the myocardial geometry [42]. For this reason it is important to consider the impact that material compliance has on cardiomyocyte phenotype [43]. Finally, the biomaterial needs to be a suitable growth environment for myocardial cells to survive and express a contracting cardiac phenotype so that they can functionally integrate with the host myocardium.

Injection of a liquid biomaterial, which can then be solidified *in situ*, enables the biomaterial to take the appropriate shape of the scar tissue, rather than imposing on the heart muscle with a cardiac patch having a fixed geometry. Moreover, injecting the biomaterial into the scar tissue allows for an intimate contact between the injected cells and the host tissue, as opposed to a solid graft transplanted on the surface of the myocardium. More importantly, an injectable therapy can be administered by a less invasive arthroscopic procedure or catheterization, which has a tremendous clinical advantage over open chest surgeries. Treating patients after an MI using a simple injection into the damaged tissue with guided catheterization is much less invasive than a cardiac patch implantation performed in an open heart surgery.

Christman et al. were the first to report on a fibrin glue biomaterial as an injectable scaffold to deliver myoblasts to the ischemic myocardium [19]. Fibrin is typically cross-linked into a hydrogel by mixing fibrinogen, thrombin, and factor XIII. The cross-linking reaction is initiated with the addition of thrombin to the fibrinogen solution. The reaction kinetics require a few seconds depending on the thrombin concentration, and this in turn allows the injection of the liquid solution which then solidifies *in situ*. They reported that the injection of cells with the fibrin glue did not enhance cell retention after 24 h; however, cell survival after 5 weeks was significantly higher. The fibrin glue was maintained in the myocardium for up to 10 days before fully degrading—providing for a temporary biodegradable scaffold which allows cell attachment through fibrin's Arg-Gly-Asp (RGD) integrin binding sites. The fibrin glue injection also increased neovascularization in the ischemic tissue.

The myoblasts were typically observed surrounded by arterioles in the scar tissue, which may also explain the high survival of the injected myoblasts when delivered with the fibrin glue. Although the injection of fibrin increased microvessel formation, this difference was not statistically significant when compared to injection with bovine serum albumin. Fibrin is generally known to induce angiogenesis *in vivo*[20]; however, injection of many other materials into the myocardium is also known to induce angiogenesis, and this may explain why the fibrin did not result in a significantly higher angiogenic response. In addition to increasing cell survival, fibrin also prevents cardiac remodeling after MI [18,19].

Kofidis et al. injected Matrigel combined with mouse embryonic stem cells [42,44] into an infarcted myocardium of an adult rat. The Matrigel solidified at 37°C after a few hours *in situ*. They reported an improvement in cardiac function both with and without cells. Hence, the real contribution of the stem cells embedded in the Matrigel was not clear from this investigation. Another injectable system for cell delivery is the self-assembling peptide nanofiber matrix which was presented by Davis et al. [26]. In this study, neonatal cardiomyocytes were injected together with the self-assembling peptide to healthy mice hearts. Only a very small portion of the injected cells survived in the matrix after 7 days. The authors note that the injection of the matrix with the cells increased the recruitment of native progenitor myocytes into the polypeptide matrix. The low survival rate of the injected cardiomyocytes in this system underscores the importance of properly designing an injectable cell delivery biomaterial for the special requirements of cardiomyocytes.

Cardiomyocyte transplantation using injectable polymers, such as fibrin [45], Matrigel [44], and self-assembling peptides [26], do not resolve the unanswered questions about how composition and structure of the injectable biomaterial affect the cardiomyocyte remodeling and functional integration. To this end, a new biosynthetic material based on PEGylated fibrinogen (PF) [46], which is an injectable polymer with bioactivity similar to native fibrin and the added advantage of controllable physical properties and biodegradation [47,48], has been used for both *in vitro* and *in vivo* studies. The synthetic constituent of

the PF hydrogel is PEG—a widely used polymer in biomedical and pharmacological applications because of its inert characteristics and low immunogenicity [49,50]. The injectable hydrogel is made by conjugating PEG with denatured fibrinogen in solution to form a PF liquid precursor. The liquid precursor is assembled into a 3D hydrogel scaffold *in situ*, using nontoxic photopolymerization [51] (Figure 30.1).

The fibrinogen molecule contains functional biological sequences that are essential for cell invasion and tissue remodeling. These include proteolytic cleavage sites sensitive to MMPs and plasmin, two common proteases in cardiac remodeling, as well as adhesion sites such as RGD [52]. Moreover, fibrin degradation products induce angiogenesis by stimulating the proliferation, migration, and differentiation of endothelial cells [20]. Hence, the fibrinogen-based hydrogels not only provide the cardiac cells with a 3D environment containing adhesion and proteolytic sites but will also induce an angiogenesis response which should increase the long-term *in vivo* survival of the cellular grafts. Using these favorable characteristics of fibrinogen conjugated to PEG allows one to create a hydrogel and to control the hydrogel degradation rate, reduce the immune response to the graft, and control the mechanical properties of the hydrogel by altering the composition of the matrix [48].

Shapira et al. first validated the PF as a potential cell carrier for cardiomyocyte transplantation using *in vitro* techniques. The injectable PF precursor was gelled by photopolymerization in the presence of a suspension of isolated neonatal rat cardiomyocytes to form a tissue construct for extended *in vitro* culture. They showed that the cardiomyocytes readily reorganized the matrix into a spontaneously beating tissue, where the magnitude and pattern of contraction of the constructs was quantified by image analysis of video data to reveal an inverse correlation between contraction and hydrogel stiffness. These results underscored the need to first optimize matrix stiffness when employing hydrogels for encapsulating cardiomyocytes in cardiac tissue engineering. Shapira et al. also showed that the PF supported the maturation of cardiac cells in 3D culture as evidenced by its spontaneous beating and responsiveness to various drug treatments.

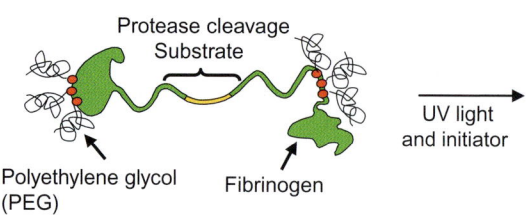

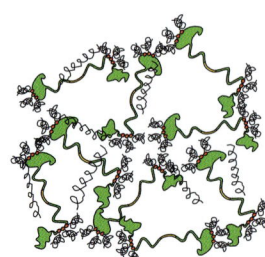

FIGURE 30.1 PEGylated fibrinogen assembles into a solid hydrogel network with UV photopolymerization. *Adapted from [46].*

Furthermore, their *in vivo* studies using the photopolymerizable and biodegradable PF hydrogels demonstrated its capabilities as an encapsulating milieu for the local delivery of cardiomyocytes to an infarcted rat heart. The hydrogel was able to act as an efficient cardiomyocyte carrier into the infarcted area as indicated by a significant attenuation of the adverse cardiac remodeling and improvements in the ventricular function of the infarcted heart.

30.5 HYDROGEL SCAFFOLDS WITH CONTROLLED MECHANICAL PROPERTIES

As underscored in previous studies, mechanical stimulation in cardiac cell therapy and tissue engineering is becoming a focal point of research and product development efforts [53]. In this context, hydrogel biomaterials are gaining favor for their use as scaffolds because they can provide both tissue-like water content and controllable mechanical properties [54]. Moreover, recent advances in hydrogel design have introduced semisynthetic materials that are capable of inductive cell signaling and biodegradation via cell-mediated proteolysis [41]. Bioactive features can be incorporated into the polymeric backbone of the hydrogel network by conventional conjugation chemistry, which provides important growth signals to resident cells via molecular interactions at the cell–biomaterial interface [55]. Inductive cell signaling may also include agonist that stimulates production of naturally secreted ECM proteins that are essential as part of any tissue repair process [56]. Concurrently, the proteolytic degradation of the material ensures a timely removal of the scaffold's basic constituents in favor of ECM laid down by resident cells [57]. Taken together, the unique combinations of bioactivity, proteolysis, and controlled mechanical properties of hydrogels provide excellent opportunities for designing inductive scaffolds for cardiac cell therapy and tissue engineering.

With mechanical properties emerging as one of the important inductive features of hydrogel scaffolds for cardiac tissue engineering, stimulating resident cells by mechanotransduction pathways is becoming a priority [58,59]. Most tissue regeneration applications that utilize hydrogel scaffolds consider the mechanical interface between resident cells and the scaffold in the context of adaptive cellular response to microenvironments [60]. Sometimes, this interface is transient, as some cell-compatible hydrogels are specifically designed to undergo biodegradation in order to accommodate cellular remodeling [61]. In this context, the transient events that occur during this remodeling process involve proteolysis, hydrolysis, and ECM production. Proteolysis and ECM synthesis are both controlled by biomolecular cell signaling, whereas material hydrolysis is mediated based on external

factors not controlled by cells [62]. Previous attempts to control scaffold remodeling have focused primarily on regulating hydrolysis using specific water-soluble synthetic polymers with known hydrolytic rates of biodegradation [63]. These strategies rely on bulk degradation of the material rather than locally mediated proteolysis associated with cell-secreted enzymes. Alternative approaches now include preparing proteolytically responsive hydrogel scaffolds containing protease-sensitive cross-linkers or other semisynthetic constituents [64]. The advantage of this strategy includes better control over local cell-mediate disassembly of the scaffold in tandem with local matrix deposition, without the bulk degradation of the material. The disadvantage of this approach is that little is known about how cell-responsive hydrogel biomaterials are degraded and restructured during cellular morphogenesis. These events will undoubtedly affect the mechanical properties of the hydrogel scaffold, which may consequently alter the mechanical signals imparted to resident cells. In the cardiac environment, where tissue deformations are prevalent to cardiac function, these alterations may have an immediate impact on the functional remodeling of the cellular grafts.

In order to take advantage of mechanotransduction using 3D hydrogel scaffolds that are proteolytically responsive, further understanding of how cells regulate matrix reorganization at their microenvironment level is required for these systems [65]. A few tissue engineering studies have recently attempted to document the process of local proteolysis in cell-responsive hydrogels using sophisticated new microscopy techniques, with some degree of success [66–68]. Nevertheless, the spatial and temporal changes in local mechanical properties of a hydrogel during cellular remodeling are still poorly understood. Moreover, the traditional instrumentation used for mapping local mechanical properties of gels, including micro-rheology and atomic force microscopy, have been difficult to apply with cell-seeded 3D hydrogel scaffolds. Given the importance of mechanical signaling in tissue engineering, more studies will be required to understand and predict the mechanical properties of the scaffold as cells reorganize the material over the course of several days in culture.

In order to investigate the relationship between mechanotransduction and changes in mechanical properties associated 3D morphogenesis in hydrogel scaffolds, Kesselman et al. documented the transient degradation and restructuring of cells seeded in hydrogel scaffolds that are undergoing active cell-mediated reorganization during 7 days in culture. A PF hydrogel matrix seeded with neonatal human dermal fibroblasts were used; rheology (*in situ* and *ex situ*) measured stiffening of the gels, and confocal laser scanning microscopy measured cell morphogenesis within the gels. The assumption that matrix modulus systematically decreases as cells locally begin to enzymatically disassemble the PF hydrogel to become spindled in the material was not supported by the bulk mechanical properties measurements. Instead, the PF hydrogels exhibited cell-mediated stiffening concurrent to their dynamic morphogenesis, as indicated by a fourfold increase in storage modulus after 1 week in culture. Fibrin hydrogels, which were used as the control biomaterial, proved similarly adaptive to cell-mediated remodeling only in the presence of exogenous serine protease inhibitor, aprotinin. Acellular and nonviable hydrogels also served as control groups to verify that transient matrix remodeling was associated entirely with cell-mediated events, including collagen deposition, cell-mediated proteolysis, and the formation of multicellular networks within the hydrogel constructs. The fact that cell network formation and collagen deposition both paralleled the transient stiffening of the PF hydrogels further reinforces the notion that cells actively balance between proteolysis and ECM synthesis when remolding proteolytically responsive hydrogel scaffolds. The fact that matrix mechanics was progressively changing in cellularized PF hydrogels underscores the need to further understand these remodeling events—including ECM turnover, cell metabolism, and proliferation—and their impact on mechanotransduction in 3D cell culture.

30.6 DONOR-TO-HOST INTEGRATION

With progress in the field of regenerative medicine relying more on approaches that deliver cells and/or bioactive factors using minimally invasive procedures, functional donor-to-host integration still remains a critical challenge. In this context, biodegradable hydrogels also provide certain advantages, such as excellent tissue compatibility, temporary protection from host inflammation, and controlled resorption based on cell-mediated degradation [54]. The integration with the host tissue is particularly influenced by the degradation and resorption characteristics of the hydrogel. Degradation of an implant, defined as the dissolution of the gel volume, and the resorption of the degradation products from the implant site can either contribute to or hamper the complex processes of tissue repair. Several recent studies have underscored the important role of implant degradation and resorption rates in enhancing the tissue regeneration [48,69−74]. Implant geometry may likewise affect the integration in the host by influencing the transport properties to and within the hydrogel. This is particularly crucial for hydrogels used in cell therapy, where strategies are constantly being developed to create vascularized networks in order to help overcome diffusion limitations [75−79]. The delivery of nutrients and removal of waste products from regions deep within a cellular hydrogel may be insufficient to sustain viability when using highly metabolic cells. These transport limitations can severely limit the use of a hydrogel patches for only the smallest of tissue grafts (i.e. >0.5 mm) [76,79−81]. Some transport limitations can be overcome by using an injectable strategy where hydrogel precursor solution is dispersed evenly and contiguously within the interstitial space of an injury prior to its in situ gelation. Injectable micro-beads of a few hundred microns also have a high surface area-to-volume ratio and may overcome some of the inherent transport and bioresorption limitations associated with patch hydrogel implants for better in vivo integration.

Optimizing the degradation properties and geometry of a hydrogel for better integration using in vitro characterization can only provide a partial outlook for the in vivo performance of the implant. One of the biggest challenges in this regard is finding a quantitative and nondestructive methodology to document the in situ integration of the implanted hydrogel. Various invasive techniques have been described for documenting implant resorption in vivo; however, these often require destructive analysis of postmortem sections, making use of distinct groups of animals and different sacrifice times for each parameter. Such invasive techniques do not provide detailed information about the in situ integration process [74,82−86]. With the optimization of a hydrogel scaffold requiring more information of the transient steps of integration, techniques to continuously monitor the degradation of a scaffold using fluorescence labeling have been described [87−89]. However, in vivo fluorescence monitoring of biomaterials often results in diffuse signal that does not allow for accurately determining the implant volume or its exact anatomical placement. Rather, fluorescence provides important information about the distribution and bioresorption of the labeled degradation products that are released from the implant.

Berdeshevski et al. reported on a noninvasive continuous monitoring of biodegradable implants used in regenerative medicine using bimodal fluorescence/MRI probes to document the subcutaneous integration of such implants. Degradable hydrogels were synthesized from a PF precursor by conjugating diethylene triamine pentaacetic acid-chelated gadolinium (GdDTPA) and Cy5.5-NHS ester dye to the polymer backbone. Open surgery was performed to implant the hydrogels in the form of 5-mm cylindrical plugs, injectable spherical micro-beads (200−500 μm), or hydrogel precursor (injected and polymerized in situ). The fluorescent Cy5.5 marker in the near-infrared spectrum provided a sensitive signal for the in vivo visualization and quantification of hydrogel resorption based on degradation products that are released from the implant. The simultaneous GdDTPA MRI monitoring provided a complete picture of the in situ hydrogel resorption through quantitative and time-resolved parameters such as residual implant size, precise anatomical location, and the fate of the degradation products. Comparisons among the three types of implant configurations, all made from the same material and the same implant volume, revealed very different resorption patterns based on these parameters. Using the quantitative aspects of bimodal imaging, they were able to document the dynamics of implant integration—underscoring a new found importance of implant geometry and implantation strategy for regenerative medicine. Moreover, they concluded that the ability to predict spatial and temporal patterns of biomaterial degradation *in vivo* is crucial to the design of engineered cell scaffolds and matrices for delivery of bioactive factors. The bimodal imaging strategy described provides important insight into the examination of *in vivo* material stability, host integration, and resorption kinetics. Although the data published in that study was based on the use of a biopolymer implant without bioactive factors, the bimodal imaging methodology is readily adapted to a variety of different resorbable biomaterial applications, including cell-seeded scaffolds for tissue engineering and controlled drug delivery systems.

30.7 FUTURE CONSIDERATIONS

Bioengineered hydrogel scaffolds are already having a dramatic impact in cardiac tissue engineering. For instance, doctors can now treat an MI injury with experimental injectable hydrogels, and stem cell researchers can make important discoveries in cardiac medicine using hydrogels in place of tissue culture plastic. As hydrogels improve through better design, their influence is likely to expand to additional clinical therapies for treating MI. For example, drug treatments may be combined with therapeutic bioactive factors and cell delivery using injectable microgels placed locally in the site of the MI. Viral gene therapy for treating MI may also be replaced with gene-delivery scaffolds made from injectable hydrogels. These approaches could thus contribute significantly to the emerging discipline of cardiovascular regenerative medicine, by providing hydrogel biomaterials as cell carriers, drug depots, and bioactive implants for when minimally invasive therapy is used to restore lost myocardial function. Thus, as the field moves to employ new biomaterial designs, it gains the ability to develop sophisticated hydrogels for these and other applications in cardiac restoration.

REFERENCES

[1] Rosamond W, Flegal K, Furie K, Go A, Greenlund K, Haase N, et al. Heart disease and stroke statistics—2008 update: a report from the American Heart Association Statistics Committee and Stroke Statistics Subcommittee. Circulation 2008;117:25—146.

[2] Akins RE. Can tissue engineering mend broken hearts? Circ Res 2002;90:120—2.

[3] Sun Y, Weber KT. Infarct scar: a dynamic tissue. Cardiovasc Res 2000;46:250—6.

[4] Murry CE, Field LJ, Menasche P. Cell-based cardiac repair: reflections at the 10-year point. Circulation 2005;112:3174—83.

[5] Britten MB, Abolmaali ND, Assmus B, Lehmann R, Honold J, Schmitt J, et al. Infarct remodeling after intracoronary progenitor cell treatment in patients with acute myocardial infarction (TOPCARE-AMI): mechanistic insights from serial contrast-enhanced magnetic resonance imaging. Circulation 2003;108:2212—8.

[6] Schachinger V, Assmus B, Britten MB, Honold J, Lehmann R, Teupe C, et al. Transplantation of progenitor cells and regeneration enhancement in acute myocardial infarction: final one-year results of the TOPCARE-AMI Trial. J Am Coll Cardiol 2004;44:1690—9.

[7] Hofmann M, Wollert KC, Meyer GP, Menke A, Arseniev L, Hertenstein B, et al. Monitoring of bone marrow cell homing into the infarcted human myocardium. Circulation 2005;111:2198—202.

[8] Qian H, Yang Y, Huang J, Gao R, Dou K, Yang G, et al. Intracoronary delivery of autologous bone marrow mononuclear cells radiolabeled by 18F-fluoro-deoxy-glucose: tissue distribution and impact on post-infarct swine hearts. J Cell Biochem 2007;102:64—74.

[9] Christman KL, Lee RJ. Biomaterials for the treatment of myocardial infarction. J Am Coll Cardiol 2006;48:907—13.

[10] Kelley ST, Malekan R, Gorman 3rd JH, Jackson BM, Gorman RC, Suzuki Y, et al. Restraining infarct expansion preserves left ventricular geometry and function after acute anteroapical infarction. Circulation 1999;99:135—42.

[11] Bowen FW, Jones SC, Narula St N, John Sutton MG, Plappert T, Edmunds Jr. LH, et al. Restraining acute infarct expansion decreases collagenase activity in borderzone myocardium. Ann Thorac Surg 2001;72:1950—6.

[12] Enomoto Y, Gorman 3rd JH, Moainie SL, Jackson BM, Parish LM, Plappert T, et al. Early ventricular restraint after myocardial infarction: extent of the wrap determines the outcome of remodeling. Ann Thorac Surg 2005;79:881—7.

[13] Chaudhry PA, Mishima T, Sharov VG, Hawkins J, Alferness C, Paone G, et al. Passive epicardial containment prevents ventricular remodeling in heart failure. Ann Thorac Surg 2000;70:1275—80.

[14] Konertz WF, Shapland JE, Hotz H, Dushe S, Braun JP, Stantke K, et al. Passive containment and reverse remodeling by a novel textile cardiac support device. Circulation 2001;104:270—5.

[15] Franco-Cereceda A, Lockowandt U, Olsson A, Bredin F, Forssell G, Owall A, et al. Early results with cardiac support device implant in patients with ischemic and non-ischemic cardiomyopathy. Scand Cardiovasc J 2004;38:159—63.

[16] Olsson A, Bredin F, Franco-Cereceda A. Echocardiographic findings using tissue velocity imaging following passive containment surgery with the Acorn CorCap cardiac support device. Eur J Cardiothorac Surg 2005;28:448—53.

[17] Fujimoto KL, Tobita K, Merryman WD, Guan J, Momoi N, Stolz DB, et al. An elastic, biodegradable cardiac patch induces contractile smooth muscle and improves cardiac remodeling and function in sub-acute myocardial infarction. J Am Coll Cardiol 2007;49:2292—300.

[18] Christman KL, Fok HH, Sievers RE, Fang Q, Lee RJ. Fibrin glue alone and skeletal myoblasts in a fibrin scaffold preserve cardiac function after myocardial infarction. Tissue Eng 2004;10:403−9.

[19] Christman KL, Vardanian AJ, Fang Q, Sievers RE, Fok HH, Lee RJ. Injectable fibrin scaffold improves cell transplant survival, reduces infarct expansion, and induces neovasculature formation in ischemic myocardium. J Am Coll Cardiol 2004;44:654−60.

[20] Bootle-Wilbraham CA, Tazzyman S, Thompson WD, Stirk CM, Lewis CE. Fibrin fragment E stimulates the proliferation, migration and differentiation of human microvascular endothelial cells in vitro. Angiogenesis 2001;4:269−75.

[21] Huang NF, Yu J, Sievers R, Li S, Lee RJ. Injectable biopolymers enhance angiogenesis after myocardial infarction. Tissue Eng 2005;11:1860−6.

[22] Dai W, Wold LE, Dow JS, Kloner RA. Thickening of the infarcted wall by collagen injection improves left ventricular function in rats: a novel approach to preserve cardiac function after myocardial infarction. J Am Coll Cardiol 2005;46:714−9.

[23] Landa N, Miller L, Feinberg MS, Holbova R, Shachar M, Freeman I, et al. Effect of injectable alginate implant on cardiac remodeling and function after recent and old infarcts in rat. Circulation 2008;117:1388−96.

[24] Yu J, Gu Y, Du KT, Mihardja S, Sievers RE, Lee RJ. The effect of injected RGD modified alginate on angiogenesis and left ventricular function in a chronic rat infarct model. Biomaterials 2009;30:751−6.

[25] Leor J, Tuvia S, Guetta V, Manczur F, Castel D, Willenz U, et al. Intracoronary injection of in situ forming alginate hydrogel reverses left ventricular remodeling after myocardial infarction in Swine. J Am Coll Cardiol 2009;54:1014−23.

[26] Davis ME, Motion JP, Narmoneva DA, Takahashi T, Hakuno D, Kamm RD, et al. Injectable self-assembling peptide nanofibers create intramyocardial microenvironments for endothelial cells. Circulation 2005;111:442−50.

[27] Yoon SJ, Fang YH, Lim CH, Kim BS, Son HS, Park Y, et al. Regeneration of ischemic heart using hyaluronic acid-based injectable hydrogel. J Biomed Mater Res B Appl Biomater 2009;91:163−71.

[28] Li RK, Jia ZQ, Weisel RD, Mickle DA, Choi A, Yau TM. Survival and function of bioengineered cardiac grafts. Circulation 1999;100:63−9.

[29] Kofidis T, de Bruin JL, Hoyt G, Ho Y, Tanaka M, Yamane T, et al. Myocardial restoration with embryonic stem cell bioartificial tissue transplantation. J Heart Lung Transplant 2005;24:737−44.

[30] Krupnick AS, Kreisel D, Engels FH, Szeto WY, Plappert T, Popma SH, et al. A novel small animal model of left ventricular tissue engineering. J Heart Lung Transplant 2002;21:233−43.

[31] Zimmermann WH, Melnychenko I, Wasmeier G, Didie M, Naito H, Nixdorff U, et al. Engineered heart tissue grafts improve systolic and diastolic function in infarcted rat hearts. Nat Med 2006;12:452−8.

[32] Thompson CA, Nasseri BA, Makower J, Houser S, McGarry M, Lamson T, et al. Percutaneous transvenous cellular cardiomyoplasty. A novel nonsurgical approach for myocardial cell transplantation. J Am Coll Cardiol 2003;41:1964−71.

[33] Leor J, Aboulafia-Etzion S, Dar A, Shapiro L, Barbash IM, Battler A, et al. Bioengineered cardiac grafts: a new approach to repair the infarcted myocardium? Circulation 2000;102:56−61.

[34] Masuda S, Shimizu T, Yamato M, Okano T. Cell sheet engineering for heart tissue repair. Adv Drug Deliv Rev 2008;60:277−85.

[35] Shimizu T, Yamato M, Kikuchi A, Okano T. Cell sheet engineering for myocardial tissue reconstruction. Biomaterials 2003;24:2309−16.

[36] Miyagawa S, Sawa Y, Sakakida S, Taketani S, Kondoh H, Memon IA, et al. Tissue cardiomyoplasty using bioengineered contractile cardiomyocyte sheets to repair damaged myocardium: their integration with recipient myocardium. Transplantation 2005;80:1586−95.

[37] Itabashi Y, Miyoshi S, Kawaguchi H, Yuasa S, Tanimoto K, Furuta A, et al. A new method for manufacturing cardiac cell sheets using fibrin-coated dishes and its electrophysiological studies by optical mapping. Artif Org 2005;29:95−103.

[38] Huang YC, Khait L, Birla RK. Contractile three-dimensional bioengineered heart muscle for myocardial regeneration. J Biomed Mater Res A 2007;80:719−31.

[39] Huang YC, Dennis RG, Larkin L, Baar K. Rapid formation of functional muscle in vitro using fibrin gels. J Appl Physiol 2005;98:706−13.

[40] Carrier RL, Papadaki M, Rupnick M, Schoen FJ, Bursac N, Langer R, et al. Cardiac tissue engineering: cell seeding, cultivation parameters, and tissue construct characterization. Biotechnol Bioeng 1999;64:580−9.

[41] Lutolf MP, Hubbell JA. Synthetic biomaterials as instructive extracellular microenvironments for morphogenesis in tissue engineering. Nat Biotechnol 2005;23:47−55.

[42] Kofidis T, Lebl DR, Martinez EC, Hoyt G, Tanaka M, Robbins RC. Novel injectable bioartificial tissue facilitates targeted, less invasive, large-scale tissue restoration on the beating heart after myocardial injury. Circulation 2005;112:173−7.

[43] McDevitt TC, Woodhouse KA, Hauschka SD, Murry CE, Stayton PS. Spatially organized layers of cardiomyocytes on biodegradable polyurethane films for myocardial repair. J Biomed Mater Res A 2003;66:586−95.

[44] Kofidis T, de Bruin JL, Hoyt G, Lebl DR, Tanaka M, Yamane T, et al. Injectable bioartificial myocardial tissue for large-scale intramural cell transfer and functional recovery of injured heart muscle. J Thorac Cardiovasc Surg 2004;128:571−8.

[45] Ye Q, Zund G, Benedikt P, Jockenhoevel S, Hoerstrup SP, Sakyama S, et al. Fibrin gel as a three dimensional matrix in cardiovascular tissue engineering. Eur J Cardiothorac Surg 2000;17:587−91.

[46] Seliktar D. Extracellular stimulation in tissue engineering. Ann N Y Acad Sci 2005;1047:386−94.

[47] Almany L, Seliktar D. Biosynthetic hydrogel scaffolds made from fibrinogen and polyethylene glycol for 3D cell cultures. Biomaterials 2005;26:2467−77.

[48] Dikovsky D, Bianco-Peled H, Seliktar D. The effect of structural alterations of PEG-fibrinogen hydrogel scaffolds on 3-D cellular morphology and cellular migration. Biomaterials 2006;27:1496−506.

[49] Barker TH, Fuller GM, Klinger MM, Feldman DS, Hagood JS. Modification of fibrinogen with poly(ethylene glycol) and its effects on fibrin clot characteristics. J Biomed Mater Res 2001;56:529−35.

[50] Bailon P, Berthold W. Polyethylene glycol-conjugated pharmaceutical proteins. Pharm Sci Technol Today 1998;1:352−6.

[51] Williams CG, Malik AN, Kim TK, Manson PN, Elisseeff JH. Variable cytocompatibility of six cell lines with photoinitiators used for polymerizing hydrogels and cell encapsulation. Biomaterials 2005;26:1211−8.

[52] Hersel U, Dahmen C, Kessler H. RGD modified polymers: biomaterials for stimulated cell adhesion and beyond. Biomaterials 2003;24:4385−415.

[53] Tibbitt MW, Anseth KS. Hydrogels as extracellular matrix mimics for 3D cell culture. Biotechnol Bioeng 2009;103:655−63.

[54] Seliktar D. Designing cell-compatible hydrogels for biomedical applications. Science 2012;336:1124−8.

[55] Hennink WE, van Nostrum CF. Novel crosslinking methods to design hydrogels. Adv Drug Deliv Rev 2002;54:13−36.

[56] Discher DE, Mooney DJ, Zandstra PW. Growth factors, matrices, and forces combine and control stem cells. Science 2009;324:1673−7.

[57] Nicodemus GD, Bryant SJ. Cell encapsulation in biodegradable hydrogels for tissue engineering applications. Tissue Eng Part B Rev 2008;14:149−65.

[58] Appelman TP, Mizrahi J, Elisseeff JH, Seliktar D. The influence of biological motifs and dynamic mechanical stimulation in hydrogel scaffold systems on the phenotype of chondrocytes. Biomaterials 2011;32:1508−16.

[59] Appelman TP, Mizrahi J, Seliktar D. A finite element model of cell-matrix interactions to study the differential effect of scaffold composition on chondrogenic response to mechanical stimulation. J Biomech Eng 2011;133(4):041010.

[60] Huebsch N, Arany PR, Mao AS, Shvartsman D, Ali OA, Bencherif SA, et al. Harnessing traction-mediated manipulation of the cell/matrix interface to control stem-cell fate. Nat Mater 2010;9:518−26.

[61] Patterson J, Martino MM, Hubbell JA. Biomimetic materials in tissue engineering. Mater Today 2010;13:14−22.

[62] Patterson J, Siew R, Herring SW, Lin AS, Guldberg R, Stayton PS. Hyaluronic acid hydrogels with controlled degradation properties for oriented bone regeneration. Biomaterials 2010;31:6772−81.

[63] Anseth KS, Metters AT, Bryant SJ, Martens PJ, Elisseeff JH, Bowman CN. In situ forming degradable networks and their application in tissue engineering and drug delivery. J Control Release 2002;78:199−209.

[64] Lutolf MP, Lauer-Fields JL, Schmoekel HG, Metters AT, Weber FE, Fields GB, et al. Synthetic matrix metalloproteinase-sensitive hydrogels for the conduction of tissue regeneration: engineering cell-invasion characteristics. Proc Natl Acad Sci USA 2003;100:5413−8.

[65] Lutolf MP, Gilbert PM, Blau HM. Designing materials to direct stem-cell fate. Nature 2009;462:433−41.

[66] Raeber GP, Lutolf MP, Hubbell JA. Molecularly engineered PEG hydrogels: A novel model system for proteolytically mediated cell migration. Biophys J 2005;89:1374−88.

[67] Raeber GP, Lutolf MP, Hubbell JA. Mechanisms of 3-D migration and matrix remodeling of fibroblasts within artificial ECMs. Acta Biomater 2007;3:615−29.

[68] Schwartz MP, Fairbanks BD, Rogers RE, Rangarajan R, Zaman MH, Anseth KS. A synthetic strategy for mimicking the extracellular matrix provides new insight about tumor cell migration. Integr Biol 2010;2:32−40.

[69] Kraehenbuehl TP, Ferreira LS, Hayward AM, Nahrendorf M, van der Vlies AJ, Vasile E, et al. Human embryonic stem cell-derived microvascular grafts for cardiac tissue preservation after myocardial infarction. Biomaterials 2011;32:1102−9.

[70] Urech L, Bittermann AG, Hubbell JA, Hall H. Mechanical properties, proteolytic degradability and biological modifications affect angiogenic process extension into native and modified fibrin matrices in vitro. Biomaterials 2005;26:1369−79.

[71] Seliktar D, Zisch AH, Lutolf MP, Wrana JL, Hubbell JA. MMP-2 sensitive, VEGF-bearing bioactive hydrogels for promotion of vascular healing. J Biomed Mater Res A 2004;68:704−16.

[72] Kloxin AM, Kasko AM, Salinas CN, Anseth KS. Photodegradable hydrogels for dynamic tuning of physical and chemical properties. Science 2009;324:59−63.

[73] Dikovsky D, Bianco-Peled H, Seliktar D. Defining the role of matrix compliance and proteolysis in three-dimensional cell spreading and remodeling. Biophys J 2008;94:2914−25.

[74] Yilgor P, Yilmaz G, Onal MB, Solmaz I, Gundogdu S, Keskil S, et al. An in vivo study on the effect of scaffold geometry and growth factor release on the healing of bone defects. J Tissue Eng Regen Med 2012 Mar 7:22396311.

[75] Bland E, Dréau D, Burg KJ. Overcoming hypoxia to improve tissue-engineering approaches to regenerative medicine. J Tissue Eng Regen Med 2012.

[76] Malda J, Klein TJ, Upton Z. The roles of hypoxia in the in vitro engineering of tissues. Tissue Eng 2007;13:2153−62.

[77] Brown DA, MacLellan WR, Dunn JC, Wu BM, Beygui RE. Hypoxic cell death is reduced by pH buffering in a model of engineered heart tissue. Artif Cells Blood Substit Immobil Biotechnol 2008;36:94−113.

[78] Griffith CK, George SC. The effect of hypoxia on in vitro prevascularization of a thick soft tissue. Tissue Eng Part A 2009;15:2423−34.

[79] Kellner K, Liebsch G, Klimant I, Wolfbeis OS, Blunk T, Schulz MB, et al. Determination of oxygen gradients in engineered tissue using a fluorescent sensor. Biotechnol Bioeng 2002;80:73−83.

[80] Sheridan MH, Shea LD, Peters MC, Mooney DJ. Bioabsorbable polymer scaffolds for tissue engineering capable of sustained growth factor delivery. J Control Release: Off J Control Release Soc 2000;64:91−102.

[81] Bland E, Dreau D, Burg KJ. Overcoming hypoxia to improve tissue-engineering approaches to regenerative medicine. J Tissue Eng Regen Med 2012.

[82] Aghion E, Levy G, Ovadia S. In vivo behavior of biodegradable Mg−Nd−Y−Zr−Ca alloy. J Mater Sci Mater Med 2012;23:805−12.

[83] Yang Z, Xu LS, Yin F, Shi YQ, Han Y, Zhang L, et al. In vitro and in vivo characterization of silk fibroin/gelatin composite scaffolds for liver tissue engineering. J Digest Dis 2012;13:168−78.

[84] Borselli C, Storrie H, Benesch-Lee F, Shvartsman D, Cezar C, Lichtman JW, et al. Functional muscle regeneration with combined delivery of angiogenesis and myogenesis factors. Proc Natl Acad Sci USA 2010;107:3287−92.

[85] Ge Y, Mei Z, Liu X. Evaluation of daidzein-loaded chitosan microspheres in vivo after intramuscular injection in rats. Yakugaku zasshi: J Pharm Soc Jpn 2011;131:1807−12.

[86] Borselli C, Cezar CA, Shvartsman D, Vandenburgh HH, Mooney DJ. The role of multifunctional delivery scaffold in the ability of cultured myoblasts to promote muscle regeneration. Biomaterials 2011;32:8905−14.

[87] Artzi N, Oliva N, Puron C, Shitreet S, Artzi S, bon Ramos A, et al. In vivo and in vitro tracking of erosion in biodegradable materials using non-invasive fluorescence imaging. Nat Mater 2011;10:704−9.

[88] Cunha-Reis C, El Haj AJ, Yang X, Yang Y. Fluorescent labeling of chitosan for use in non-invasive monitoring of degradation in tissue engineering. J Tissue Eng Regen Med 2011.

[89] Yang Y, Yiu HH, El Haj AJ. On-line fluorescent monitoring of the degradation of polymeric scaffolds for tissue engineering. Analyst 2005;130:1502−6.

Chapter 31

Heart Regeneration: The Developmental and Stem Cell Biology Approach

Almudena Martinez-Fernandez[a,b,c,d,h,i], Rosanna Beraldi[f,h,i], Susana Cantero Peral[f,h,i], Andre Terzic[a,b,c,d,e,h], and Timothy J. Nelson[d,f,g,h,i]

[a]Marriott Heart Disease Research Program, Mayo Clinic, Rochester, MN, [b]Division of Cardiovascular Diseases, Mayo Clinic, Rochester, MN, [c]Departments of Medicine, Mayo Clinic, Rochester, MN, [d]Molecular Pharmacology and Experimental Therapeutics, Mayo Clinic, Rochester, MN, [e]Medical Genetics, Mayo Clinic, Rochester, MN, [f]General Internal Medicine, Mayo Clinic, Rochester, MN, [g]Transplant Center, Mayo Clinic, Rochester, MN, [h]Center for Regenerative Medicine, Mayo Clinic, Rochester, MN, [i]Todd and Karen Wanek Family Program for Hypoplastic Left Heart Syndrome, Mayo Clinic, Rochester, MN

31.1 INTRODUCTION

In Greek mythology, the hero Prometheus was punished for stealing the fire of the gods for man. He was bound to a rock, where an eagle would tear away his liver, only to have his liver rejuvenate daily. Currently, regeneration of organs is not only the theme of mythology; it has become the foundation of new branches of medicine and surgery. Because the heart has a limited innate capability to rejuvenate, adjunctive regeneration therapies offer the promise of novel curative solutions. Therefore, in the last decade the scientific community has been focused on investigating the fascinating phenomenon of biological rejuvenation to ascertain if the heart can naturally regenerate by itself and if the innate ability can be enhanced by stem cells and regenerative medicine interventions. A series of novel discoveries and encouraging results have shown that the heart is not a postmitotic or static organ as previously described, but rather is a dynamic tissue that shows detectable signs of regeneration under physiological and pathological conditions. In fact, this paradigm shift of cardiac regeneration has revolutionized the traditional view of the heart and has fueled the hope for therapeutic applications. The new reality in myocardial biology describes the postnatal heart as a self-renewing organ that encompasses populations of stem cells or precursor cells, along with terminally differentiated cardiac cells that can reenter the cell cycle to collectively contribute to the maintenance of cardiac homeostasis. Variations of heart regeneration have been described in fish and mammalian species, demonstrating that the complex evolution of myocardial tissue has proportionally decreased the capability of the heart to self-repair. However, many of the fundamental core cardiac transcription factors are conserved across diverse species and offer a firm foundation to expand our knowledge of cardiac regeneration [1]. This chapter will summarize the current knowledge of cardiac developmental biology and innate heart regeneration in the context of ongoing clinical trials.

31.2 CARDIAC DEVELOPMENTAL BIOLOGY

The process of converting stem cells into functional tissues requires a synchronized sequence of cell fate decisions within a heterogeneous mixture of progenitor cells that behave and interact within an interdependent microenvironment. Evolutionary conserved functions are encoded within core components with the addition of regulatory mechanisms that refine the structure and function of the resulting organ system in response to environmental demands. In the following section, we will review the progression and pathways involved in the development of the heart, starting with what can be learned from zebrafish and how that has advanced our understanding of the process of cardiac formation in mammals.

31.2.1 Zebrafish Heart Development

Cardiac developmental processes in zebrafish hold great resemblance to those in mammals, with a simplified scheme corresponding to a heart with only one atrium and one ventricle. This structure and function facilitates dissection of the underlying pathways that regulate natural cardiogenesis. Zebrafish as a model system for heart regeneration has technical advantages such as rapid maturation, external postfertilization development (allowing visualization of the embryos and the developing heart at early stages), and lack of reliance on the cardiovascular system for oxygen supplies (passive diffusion is sufficient to keep embryos alive and developing for several days due to small size) [2]. These characteristics of zebrafish have made a popular model system to study cardiac development. In the late 1990s, a massive project studying embryonic phenotype after induced mutations in zebrafish set the basis for what is currently known about cardiac development in this model [3]. Collectively, these efforts have paved the way for significant scientific breakthroughs that have accelerated our understanding of cardiac developmental biology.

As little as 12 h postfertilization (hpf), two independent groups of cells form in both left and right sides of the 12-somite zebrafish embryo containing atrial and ventricular progenitors with distinct and organized locations [4]. Retinoic acid has been demonstrated to restrict the cardiac progenitor pool affecting both atrial and ventricular progenitors from an early stage [5]. The resulting specified cells migrate medially to form a common tubular structure as differentiation progresses. Once they have fused in an unique anatomical unit, the resulting hollow structure evolves and develops the defining components of the heart, such as the atrium, ventricle, valve, and conduction system.

In the process of cardiogenic specification and differentiation, several transcription factors also described to be fundamental for mammal heart development are implicated. Initially, bone morphogenetic protein (Bmp) induces transcription factor *Nkx2-5*, which is expressed at the one-to-three somite stage [6]. This fundamental regulator of early cardiogenesis was originally described in Drosophila [7,8] as the orthologue *tinman*, highlighting the hallmark of lower model organism in the process of discovery of central development regulatory elements. As seen in zebrafish, other molecules such as Nodal, Fgf8, and the transcription factor *Gata5* seem to be involved in *Nkx2-5* induction [2,9]. In particular, *Gata5* has a potent effect in the regulation of myocardium, with overexpression leading to ectopic areas of beating activity [2,10]. These experiences established the concept of transcription factor-mediated induction of cardiac specification and differentiation from pluripotent stem cells.

From the anterior—lateral regions of the zebrafish embryos, cardiac progenitors forming the two bilateral heart fields migrate and converge at the midline, giving rise to the linear heart tube. This migratory movement happens simultaneously with progression of myocardial differentiation, both processes being tightly regulated. Among the implicated factors, *Hand2* has a double function, with a cell-autonomous role in differentiation as well as a noncell autonomous regulator of fibronectin levels required for cell migration [11—13]. In fact, fibronectin arises as a key element of the migratory events, with *Hand2*, extraembryonic heparin sulfate proteoglycan Syndecan 2, and the receptor/ligand pair mil/s1p2 regulating its level and appropriate function [14,15]. Also, the G-coupled Apelin receptor and its ligand Apelin influence this developmental phase [16]. These discoveries validated the importance of microenvironments and the ability of progenitor cells to migrate into proper niche environments in order to ensure appropriate cell fate decisions.

Subsequently, cardiac structures enter into an asymmetric phase, where atrial and ventricular components, as well as the atrioventricular valve are progressively forming. Rotation of the myocardial tube positions the venous pole to the left, with the arterial pole still at the midline. This process is regulated by asymmetrical expression of Nodal and Bmp [17,18]. At the same time, the atrium and the ventricle become distinguishable, and the myocardium matures and starts beating. Retinoic acid, which originally determines the balance of cardiac versus noncardiac progenitors, extends its function into this phase restricting the number of cells that integrate into the atrium. Factors such as *Hand2*, *Fgf8*, or *Gata5* are linked to the ventricular tissue. Sarcomeric genes responsible for maturation of the contractile structures including troponin T, tropomyosin, and myosin are expressed together with transcription factors including *Mef2c*, *Gata4* and *6*, or *Tbx5*[2]. Therefore, early in the developmental process, mature cytoskeletal machinery is patterned and functional, thus establishing mechanical force and fluid dynamics as essential components for ongoing morphological maturation of the heart.

The existence of two waves of migrating cells contributing sequentially to heart development has been demonstrated for zebrafish, corresponding to the first and second heart fields described in chick, mouse, and humans. In the simplified version of the zebrafish heart, the first heart field is involved in the formation of the linear heart tube with the second heart field contributing to the arterial pole [19,20]. With cardiac chamber development underway, atrioventricular flow separation becomes necessary, and therefore, valves start developing in the interface of these structures (the atrioventricular canal). Genes regulating this process are common to those involved in amniote valve development and include *Notch, NFAT, ErbB, TgfB, and Wnt/B catenin* [21−24]. At 48 hpf the embryonic proepicardial organ can be distinguished as a mass of cells external to the heart. After 24 h, cells derived from this organ travel to the heart creating the epicardium [25,26]. This final layer of the heart plays a key role in zebrafish cardiac regeneration: upon physical injury the epicardium covers the exposed myocardium and creates new vessels that support the regenerating tissue through epithelial to mesenchyme transition of some of its cells that invade the damaged area [27]. Thus, multiple sources of distinctive cell types are interdependent during the orchestration of cardiogenesis and thereby offer multiple strategies to optimize cardiac regeneration.

Overall, development of the heart in zebrafish offers a simplified view of the mammalian process. Many of the pathways involved in the progression from specification through final differentiation are highly conserved in more complex species, highlighting the importance of the sequential events required for the conformation of this complex organ. The process of cardiogenesis is comprised of multiple cell types with cell sources working in concert with microenvironmental influences and modulated by mechanical and fluid dynamics of the maturing heart.

31.2.2 Mammalian Heart Development

During evolution, the heart has acquired additional modular anatomical structures required to adapt to advanced functional needs as imposed by new environments and physiological demands unique to each species. These challenges affecting both the anatomy and performance of the heart are responsible for the increasing intricacy of the underlying molecular pathways involved in heart development in mammalian species [1]. During this evolution, a conserved core of transcription factors is central to the development of the heart from *Drosophila* to humans. Cumulative complexity both in terms of structure and function evolves through additive layers of gene duplication and redundancy as well as cis-regulatory elements that fine-tune the expression of key genes,

restricting their anatomical location and their temporal regulation [1].

31.2.2.1 Regulation by Transcription Factors

Following the evolutionary blueprint, many of the transcription factors involved in zebrafish cardiogenesis have conserved functions during mammalian heart development, with parallel elements accounting for added complexity. For example, *Hand2* is involved in the morphogenesis of the mammalian right ventricle similarly to its function in zebrafish, but an additional *Hand1* gene is responsible for the proper development of the left ventricle in mammals. Since zebrafish only have one ventricular chamber, a duplication of the *Hand* gene happens to modularly control development of a second ventricle [28]. Similarly, several *Gata* genes are involved in mammalian cardiogenesis, often in a redundant fashion. *Gata4* plays a role in cardiomyocyte proliferation and cardiac morphogenesis [29]; *Gata5* has been described to have an impact on valve development [30] and endocardial differentiation [31]. *Gata6* regulates cardiac inflow and outflow tract development [32,33]. However, many specific functions cannot be solely attributed to only one of those factors [34]. As another example, the T-box family comprises six genes involved in mammalian heart formation, namely *Tbx1*, controlling proliferation in the second heart field [35], *Tbx18* found in epicardial and sinus venosus components [36], *Tbx5* and *20* activators of chamber myocardial programs, and *Tbx2* and *3* implicated in valvuloseptal and conduction system development [37]. Overall, these factors have undergone high degrees of diversification in order to account for the extra modules that constitute the heart of mammals, while maintaining the central core of cardiac development [1].

31.2.2.2 Regulation by the Sources of Progenitor Cells

At various stages of the developmental process, the heart receives cells from extracardiac sources that participate in creating the final structure of the organ (Figure 31.1). Cells from the cardiac neural crest migrate into the heart and contribute to septation of the outflow tract and development of the aorta [38]. Also, the proepicardial organ, a temporary organ existing only during a brief period of the embryogenesis process, sends cells to the developing heart. These cells attach and spread over the myocardium forming the epicardium. Epicardial cells invade the underlying muscular layer contributing to the noncardiomyocyte components of the heart and give rise to the entire coronary vasculature [39]. As will be discussed later in this chapter, epicardial cells are gaining

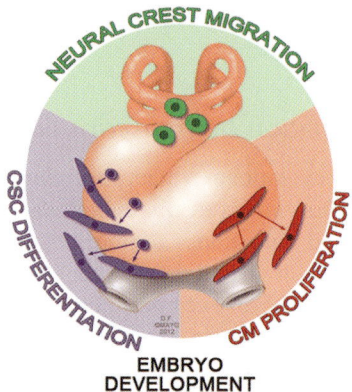

FIGURE 31.1 Cardiogenesis is the product of multiple cell sources that are both endogenous and exogenous from the cardiac tissue. Cell migration as derived from neural crest cells is the hallmark of migratory cells during cardiogenesis. Resident stem cells as noted by cardiac stem cells (CSC) contribute to early morphogenesis of the cardiac structures. Cardiomyocyte (CM) proliferation is actively contributing to hyperplasia throughout embryonic development and into perinatal life.

importance for their regenerating properties in some animal models, where they contribute actively to the formation of new tissue after cardiac damage.

31.2.2.3 Regulation by Cell Cycle and Proliferation

While cells giving rise to the vasculature and interstitial compartments migrate into the existing organ and complement its composition, cardiomyocytes develop *in situ* increasing the mass and functional capacity of the heart. C-kit myocyte progenitors have been identified at various stages of development all the way into the adult heart, where they are known as cardiac stem cells. These cells have the capacity of proliferating in an asymmetric way, maintaining a pool of undifferentiated progenitors while daughter cells specialize and differentiate (Figure 31.1). Recent studies claim that the increasing number of cardiomyocytes present during embryonic development can be explained by c-kit progenitor contribution [40]. The CSCs are also being tested in humans for their capacity to regenerate cardiac tissue even from adult diseased sources [41,42].

31.2.2.4 Regulation by miRNAs

Another layer of complexity and fine-tuning of cardiac development is created by microRNAs (miRs). These are short noncoding RNAs that can be codified in intergenic, intronic, or exonic regions of the genome and simultaneously regulate multiple target genes mainly through silencing [43]. During *in vitro* differentiation, miR-1 and miR-133 suppress endodermal and neuroectodermal tissues, with miR-1 specifically promoting cardiac

differentiation and miR-133 favoring other muscle types [43,44]. *In vivo*, precise levels of miR-1 are required to ensure proper cardiogenesis. In particular, overexpression of this regulator leads to inhibition of cardiac growth [43,45]. miR-1 has also been implicated in cardiac electrophysiology regulation as well as in cardiac hypertrophy and heart failure, scenarios in which decreased levels of expression of this miR are found. miR-133 has been shown to display several regulatory functions in the developing heart, such as regulation of the extracellular matrix [46], repression of smooth muscle formation, and inhibition of cardiomyocyte proliferation [43,47]. It is worth noticing that some of the miRs closely related to muscle (and cardiac muscle) regulation are located in intronic sequences within three muscle-specific myosins (*Myh6*, *Myh7*, and *Myh7b*). These so-called MyomiRs (miR-208a, miR-208b, and miR-499) display expression levels that mimic those of their host genes, including variations during development or in response to stress [43]. Overall, miRs are intimately involved in the development and function of the heart and contribute to the imbalance of heart function in pathological states.

31.3 INNATE CARDIAC REGENERATION

Although human heart regeneration is limited when compared to that of less complex species, a conserved developmental perspective represents a starting point to analyze the natural process of cardiac rejuvenation and to design treatment strategies for regenerative medicine. In order to supplement the cellular load of the heart, a competent contractile cytotype capable of integrating into the existing tissue without causing arrhythmias or disrupting the electrical circuitry responsible for the synchronous beating needs to be supplied. The search for progenitors fulfilling these characteristics is broad, encompassing multiple sources, delivery methods, and expected outcomes (Figure 31.2). Among many potential candidates, stem cell populations capable of differentiating toward a cardiac phenotype have been identified. Manipulation of these progenitor cells to promote their differentiation involves guiding them through a process that recapitulates normal cardiac development. In the following section, we will review the progression and pathways involved in regenerating the heart, starting with what can be learned from zebrafish and how that has advanced our understanding of the complex process of cardiac formation and regeneration in mammals.

31.3.1 Cardiac Regeneration in Zebrafish

Zebrafish have a remarkable capability to regenerate heart throughout life span. This model system has been used to decipher the mechanism underlying core features of

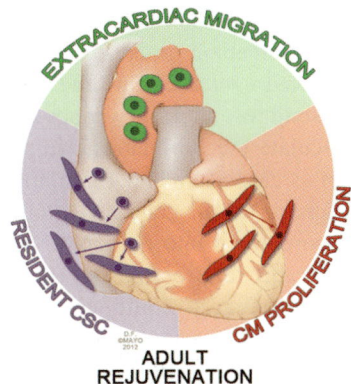

FIGURE 31.2 Cardiac regenerative strategies are the product of multiple innate rejuvenation processes designed to augment underlying cellular repair of the heart. Clinical trials have utilized cell migration as a strategy to deploy extracardiac delivery of exogenous stem cells and progenitor cells. Resident stem cells can be isolated from biopsy tissues, propagated *ex vivo*, and administered in a local and focal delivery strategy. Cardiomyocyte (CM) proliferation is an emerging strategy that could be reactivated with the delivery of small molecules or traditional pharmaceuticals. Alternatively, a combination of these strategies could offer synergistic opportunities for optimization of cardiac repair.

postnatal cardiac regeneration. Studies of this species have shown that ventricular myocardium can give rise to *de novo* cardiomyocytes through cellular proliferation and completely regenerate the structure and function of the injured tissue after surgical resection of 20% of its mass [48–50]. These reports directly oppose the traditional dogma of postmitotic cardiac tissue and have prompted substantial reexamination of the phenomenon of cardiomyocyte hyperplasia in the adult heart. Moreover, the source of proliferating cells evoked by injury has been highly investigated in the search for new mechanistic insights (Figure 31.2). The concept of *in vivo* hyperplasia is not a new finding since Lizback in 1960 and later Astorri in 1977 showed that the adult human heart can achieve an increased muscle mass under pathologic conditions by increasing the number and thickness of muscle fibers [51,52].

In deciphering the process of regeneration in lower vertebrates, ventricular resection of the native myocardium function can be reestablished by coordinated biological processes that resemble *de novo* cardiogenesis. In zebrafish, initial erythrocyte clot formation is followed by development of scar tissue that is completely replaced by new functional and contractile myocardium within 60 days. This brief period of fibrin deposition suggests a transitory event required for regeneration [49]. Due to its amenability to genetic manipulation, this model has allowed the discovery of genes and pathways involved in the regeneration process and prevention of scar formation. Specifically, cell cycle regulator *Mps1* and *Fgf* signaling

pathways have been highlighted as critically important for homeostatic repair [27,49]. How these molecular functions prevent complete regenerative response is not fully understood, yet do offer novel mechanisms to probe underlying regulatory pathways.

One of the most debated aspects of the cardiac repair process is the source of *de novo* cardiomyocytes. In zebrafish, regenerative cells have been isolated from a population of mature cardiomyocytes suggesting that reentry of the cell cycle of existing cardiomyocytes could be a predominant feature, possibly induced by cytokines or growth factors produced by the injured myocardium (Figure 31.2). To initiate this process, mature cardiomyocytes have been shown to undergo dedifferentiation with sarcomere disassembly, detachment from neighboring cells, and expression of regulators of cell-cycle progression.

Other reports have shown that the formation of new myocytes can be derived from a progenitor cell population such as the proepicardial organ or epicardial-derived cells that undergo epithelial-to-mesenchymal transition (EMT) to give rise to a new population of cardiomyocytes [27] (Figure 31.2). This mechanism resembles the cellular program activated during heart development, when undifferentiated mesodermal cells commit to the cardiac lineage and differentiate into cardiomyocytes. Indeed, Lepellina et al. have shown reactivation of the embryonic genes *Tbx18* and *Raldh2* during heart regeneration in zebrafish [27]. Taken together, these findings suggest that heart regeneration could be governed by reactivation of molecular and cellular embryonic programs originally designed for cardiac organogenesis. However, studies by Raya et al. [50] focusing on *Notch* and *Msx* genes point at the existence of specific regenerative pathways in the heart independent of developmental programs. These studies emphasize the important overlap between cardiac developmental biology and postnatal regenerative mechanisms and reveal distinctive pathways that may be required for an appropriate regenerative response.

Since zebrafish have a high capacity to recreate heart muscle, this model system remains the hallmark for advanced studies in cardiac regeneration. Moreover, low maintenance cost, rapid organogenesis, and the amenability to undergo genetic manipulations are valuable qualities that substantiate zebrafish as a gold-standard model system. However, several aspects of the zebrafish should be considered before direct comparison with a mammalian system. For example, the zebrafish heart is functionally and structurally less complex than those found in mammals. The lack of myocardial infarction (MI) models makes ventricular resection the only described surgical method to study the phenomenon of regeneration in zebrafish. Additionally, most studies in humans are derived from the study of the myocardium after an ischemic event rather than a complete removal of part of the tissue. Recently, protocols of cryoinjury that simulate a myocardial infarct in zebrafish have been proposed, and investigations are underway to determine the relevance of this model system to uncover additional mechanistic insights in to cardiac regeneration [53].

31.3.2 Cardiac Regeneration in Mammals

Limited ability of the heart to regenerate has been associated with more evolved mammalian species when compared to urodele amphibians and teleost fish. More specifically, surgical resection of the ventricular apex in perinatal mice has demonstrated the natural ability of the mammalian heart to repair itself. However, significant tissue repair is limited to mice within the first week of life, rather than the life-long capability described in zebrafish [54]. Although this regenerative potential emphasizes the evolutionary conserved natural phenomena of myocardial self-renewal, these studies also highlight the significant interspecies discrepancy in terms of efficacy of repair following heart injury. Overall, mammalian studies have shown compelling evidence that the postnatal heart harbors a limited capability of self-renewal and innate regeneration after myocardial injury (Figure 31.2). However, the low efficiency and potential enhancement of this natural phenomenon are topics of ongoing investigations.

A study by Beltrami et al. using biopsy specimens from human infarcted heart tissue provided innovative results that seemed to corroborate the regenerative capacity (although limited) of the human myocardium [55]. Proliferating myocytes were found in the infarcted tissue, indicated by the expression of Ki-67, a marker of cell proliferation. Presence of mitotic machinery components, as well as karyokinesis and cytokinesis, were also demonstrated. Former studies had previously reported similar results, but the methods used at that time to identify mitotic figures were highly contested [52]. The spatial distribution of the proliferating myocytes in Beltrami et al. was found to differ among myocardial zones, with the highest number of cells with mitotic activity bordering the area of infarction. Concomitantly, parallel studies on biopsy specimens from patients who died from nonheart related causes have demonstrated the presence of a mitotic population, although with a lower cycling activity compared to cycling cells within infarcted area of the myocardium. From these and subsequent studies, an estimation of cardiomyocyte turnover was originally established with a renewal rate of 4−5 years, which declines with age [55,56].

In 2009, Bergmann et al. addressed the same issue and advanced the field by publishing the most conclusive evidence of cardiomyocyte turnover during postnatal life in humans [57]. In this study, investigators took advantage of the radiocarbon dating method developed according to a mathematical model based on the dramatic elevation of carbon-14 released into the atmosphere after aboveground testing of nuclear weapons. This method was used to calculate the birthdate of cardiomyocytes in the human heart as cardiac cells derived after exposure to the high atmospheric levels in early 1960s would have higher levels incorporated into the DNA than cardiomyocytes born before this time period. Overall, these results corroborate the self-renewal ability of the human heart, albeit at a slower turnover rate than previously described. Indeed, Bergmann et al. estimated that approximately 50% of cardiomyocytes harbored in the heart at the time of birth are renewed by the age of 50, a percentage that slows down with age. These findings have significantly revolutionized the traditional view of the human heart as a postmitotic organ and have set the basis for a new paradigm, which establishes that the heart can undergo cellular rejuvenation (Figure 31.3).

A preponderance of reports and publications in mammalian regeneration has unleashed a wave of enthusiasm for the discovery of myocardial stem cells or cardiac progenitor cells (c-kit, Sca-1, Sca-1-SP, Isl-1) as an alternative source for *de novo* myocytes during physiological and pathological rejuvenation (Figures 31.2 and 31.3). It is not well established if these cells reside in the myocardium or if they originate in the bone marrow (BM) and migrate to the heart, where they acquire a more specialized cardiac signature to promote cardiac repair. On the other hand, BM cells (hematopoietic, mesenchymal, and

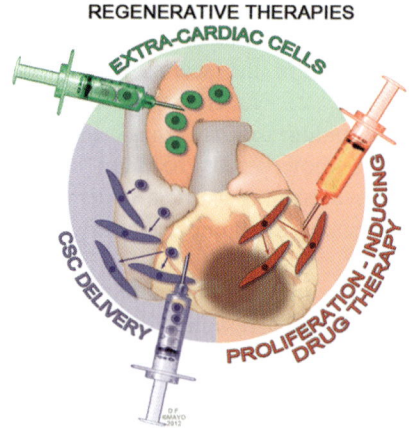

FIGURE 31.3 Cardiac rejuvenation throughout adult life is the product of multiple cell sources that are both endogenous and exogenous from the cardiac tissue. Cell migration as derived from circulating cells are the hallmark of migratory cells during cardiac rejuvenation. Resident stem cells as noted by CSCs contribute to ongoing cardiac homeostasis and cellular turn over. Cardiomyocyte (CM) proliferation is a possibility contributing to hyperplasia throughout adult life. Collectively, these processes are analogous to developmental processes and provide target pathways to engineer regenerative strategies.

endothelial cells) and CSCs could represent early and late progenitors, respectively. In this case, stem cells derived from BM may serve as a source to replenish the CSC pool, which will eventually give rise to myocytes during repair. Supporting this paradigm, cytokines or growth factors released by the injured myocardium via systemic circulation could cause the mobilization of stem cells from BM toward the damaged area, providing a pool of specialized CSCs that contributed to the continuous repair. A study in which c-kit cells were increased by recruitment from BM validated this hypothesis [58], although the existence of resident cardiac progenitor cells [54,59,60] has also been independently demonstrated. Therefore, the most useful cell to use for therapeutic interventions remains debatable. Knowledge of the cellular mechanisms that govern the physiological and pathological rejuvenation of the heart will be crucial for designing next-generation cell-based therapeutics for cardiac repair.

31.3.2.1 Resident Myocardial Progenitors

C-Kit

C-kit positive cells are the most reported source of cells for heart renewal under physiological and pathological circumstances. These resident stem cells have been isolated from rodents and human myocardium [61,62]. It has been shown in the mammalian myocardium that c-kit positive cells reside in clustered structures or niches, which are intimately connected with myocytes and fibroblasts acting as supportive cells. C-kit expressing cells are negative for CD45 and Kdr markers typically found in stem cells originating in the BM. Within these clusters, proliferative cells that stain positive for Ki-67 were found and co-expressed early cardiac stage transcription factors, *Gata4*, *Nkx2.5*, and *Mef2*. This data suggests the existence of a hierarchical cellular organization with a predetermined fate toward the cardiac lineage. In addition, sarcomeric structures were detected in a small number of cells within these clusters. Taken together, these results support the hypothesis that c-kit cells are resident CSCs giving rise to a committed cardiac lineage [62,63]. Furthermore, *in vitro* isolation and characterization of these cells from human and rodent myocardium have demonstrated self-renewal capabilities and asymmetrical cell division. Their asymmetric and symmetric cell division along with multipotency characteristics upon exposure to differentiating media and environments fulfill the criteria that define stem cell populations [61]. Moreover, when human c-kit isolated cells were injected into infarcted myocardium of immunodeficient rodents, they differentiated into myocytes, smooth muscle, and endothelial cells, resulting in increased ejection fraction, attenuated chamber dilation, and improved ventricular function as a sign of myocardial amelioration and regeneration [61,64].

Sca-1

Parallel to isolation of c-kit positive cells, a different CSC population identified by the surface stem cell antigen (Sca)-1 has been isolated from mouse myocardium [65]. Although *in vitro* analyses showed difficulty with the process of spontaneous differentiation of Sca-1 cells into myocyte lineages, treatment with 5-azacytidine, a demethylating agent, amended their ability to produce contractile cells that express Nkx2.5. Furthermore, these studies suggested that their ability to differentiate toward the myogenic lineage was dependent on the methylation/epigenetic status of components of the Bmpr1a pathway. 5-Azacytidine treatment resulted in widespread DNA demethylation and could be a limitation for their application in cell therapy. However, intravenous injection of these cells following heart injury in rodents has shown their ability to differentiate into cardiomyocytes in the ischemic microenvironment [65]. Compared to c-kit cells, Sca-1 populations have shown a propensity to fuse with host cardiomyocytes, which may limit their regenerative potential due to the loss of cellular proliferation. Current investigations have demonstrated that Sca-1 cells do not fully possess the functional properties of stem cells although they show high telomerase activity.

A side cell population positive for the Sca-1 antigen (Sca-1-SP) has also been isolated from the adult mouse myocardium. These cells express ABCG2 and MDR1 proteins, which can explain their unique ability to extrude toxic substances and Hoechst dye. They have also been shown to differentiate into structural and functional cardiomyocytes when cocultured with adult rat ventricular cardiomyocytes [66,67], but additional investigations are required to establish if this cellular subpopulation could present characteristics of stem cells and the capacity to differentiate *in vivo*.

Isl-1

The Isl-1 cell population is characterized by expression of the LIM-homeodomain transcription factor *Islet-1*. Evans et al. originally isolated these cells from mouse embryonic myocardium and showed that they were sufficient to give rise to atrial and ventricular cardiac components during cardiogenesis [68] (Figure 31.1). Subsequent studies revealed the persistence of Isl-1 cells in neonatal cardiac tissue, gradually diminishing with age. Isolation of these cells from mouse neonatal heart has shown the capability of Isl-1 cells to differentiate when cocultured with cardiomyocytes providing a potential subpopulation for regenerative applications [69] (Figure 31.2).

Cardiospheres

The field of cardiac regenerative medicine was recently expanded with the application of cardiospheres derived from human cardiac tissue as a treatment for myocardial ischemia [70]. These studies focused on the dissection of structural and functional components of the cardiospheres reporting the presence of a heterogeneous population including c-kit, Sca-1, Kdr, and CD45 positive cells. Cardiospheres have been shown to be clonogenenic, have self-renewal capability, and differentiate in beating tissue when cocultured with rat cardiomyocytes. Studies isolating cardiospheres from neonates, infants, and children with congenital heart disease have shown a linear correlation between age and number of c-kit positive cells within these structures. These results translated into a functional assay that highlighted a higher reparative potential of cardiospheres derived from neonates compared to those isolated from advanced age individuals [71].

31.3.2.2 Extracardiac Myocardial Progenitors

Transgender heart transplants where male patients received a female donor heart have been the most common example to reveal the existence and migration of endogenous stem cells from extracardiac regions (Figure 31.3). Initially, Quaini et al. reported that when female hearts were transplanted into male patients, 18% of the total cardiomyocytes contained a Y-chromosome. This percentage of *de novo* cardiomyocytes in donor hearts was contested compared to other studies in which a lower range of one male cell in 10^4 to 10^3 total cardiomyocytes was reported [72]. An explanation for the high percentage reported by Quaini could be the presence of immune cells activated following heart transplantation that produced an artificial elevation of Y-chromosome positive cells. The phenomenon of chimerism in the heart was also documented by the studies of Bayes-Genis, which showed that male fetal cells could be detected in maternal hearts affected by idiopathic dilated cardiomyopathy [73]. It is unknown why the cells were in the heart, but an attempt to cardiac regeneration by fetal cells was the evoked hypothesis. Indeed, Kara et al., encouraged by this speculation, published a study demonstrating that mouse EGFP fetal cells homed into the maternal injured heart. These cells were also noted to be capable of undergoing differentiation into cardiomyocytes, endothelial, and smooth muscle cells. However, it is still unknown if the fetal cells homed to the maternal heart in an attempt to repair [74].

To better investigate the direct involvement of BM as an extracardiac source of cells that can migrate and contribute to heart rejuvenation, cells constitutively expressing EGFP have been injected into infarcted mouse myocardium. Several weeks later, these experiments showed new EGFP-positive cardiomyocyte-like cells in the infarcted area as well as improved ventricular function in these animals. While this data generated new enthusiasm in the field, other reports contested that no significant cardiac differentiation was reported from hematopoietic stem cell transplantation [75,76]. In light of the skepticism generated from contrasting reports, BM-derived stem cells have remained under close investigation with next-generation products emerging. Mesenchymal stroma cells (MSCs) reside in the stromal compartment of the BM. The percentage of these cells is approximately 10-fold lower than their counterpart hematopoietic stem cells that are designed to give rise to circulating blood cells. MSCs do not express the canonical hematopoietic antigens CD45, CD34, and CD14, but they do carry specific adhesion molecules such as CD44 and antigens STRO-1, SH2, SH3, SH4 [77,78]. The interest for the MSCs in cardiovascular diseases was generated from studies in 1999, when it was reported that MSCs were capable of differentiating into cardiomyocytes *in vitro* [79]. Moreover, rat studies of intravenously injected LacZ MSCs showed migration into the heart in the context of MI [80]. Additionally, endothelial progenitor cells (EPCs) were identified by their expression of CD34 and CD133 [81]. This subpopulation descends from a hematopoietic stem cells family, but when cultured *in vitro*, they are capable of expressing an endothelial signature consisting of activated endothelium markers such as Flk-1 [82]. These EPCs have the capacity to migrate to the injured area mobilized by growth factors or cytokines and potentially contribute to regeneration in multiple cardiovascular diseases. Indeed several studies in animal models have shown that CD34 positive cells isolated from patients and injected intravenously were detected in the heart several days after MI. The formation of new vascular structures was the major contribution to the regenerating myocardium; however, a partial recovery of the ventricular function was shown as well [83]. New advances in the field of adult stem cells continue to provide additional cytotypes addressing target specificity for damaged tissues, with modern hypothesis implicating multiple cell type requirements for optimal regeneration.

31.3.3 Summary of Innate Cardiac Regeneration

The new paradigm recognizing the ability of the heart to rejuvenate during postnatal life also acknowledges the inherent limitations and inefficiency to repair after injury, leading to gradual deterioration of tissues in cardiovascular disease. The ability of the heart to naturally regenerate may significantly decline with age, as illustrated by the significant reduction of cardiac stem cell populations in the heart within days to weeks after birth. The oxidative environment of the heart may provide a hostile environment and lead to the reduction of resident progenitor cell populations. The reduced load of stem cells may be the underpinning of why malignancy in the heart is very rare in contrast to tissues that have high stem cell loads and harbor increased tumorigenic risk. Therefore, a conservative management of progenitor cells in the adult myocardium may have unrecognized benefits for long-term survival of the mammalian heart at the expense of decreased cardiac regenerative potential. Studies in a transgenic mouse model in which EGFP is under the control of the c-kit promoter have supported this hypothesis that progenitor cells decrease with aging. C-kit positive cardiac cells are resident in the cardiogenic mesoderm from embryonic day 6.5 and differentiate to give rise to myocyte progeny during prenatal life (Figure 31.1). In addition, these cells have been predominately detectable within the initial days of postnatal life [84]. Likewise, Isl-1$^+$ cells have been shown to be involved in heart development during embryogenesis, and their disappearance during early postnatal life could be concomitant with loss of regenerative capability in the mammalian heart. Finally, if the capacity of the heart to rejuvenate is linked to the density of cardiac stem cells, which represent a postnatal remnant of cells involved in heart development, a logical assumption could be that the process of innate rejuvenation is a "resurrection" of atavistic processes of heart development. From a molecular perspective, the well-documented reactivation of the fetal gene expression array in the failing heart could be interpreted as an attempt by the heart to regenerate by activating an embryological program [85,86]. Therefore, modulation of fetal gene expression and promoting regenerative capacity of the postnatal heart offer a wide range of therapeutic targets that could be the subject of ongoing research efforts.

31.4 CLINICAL STRATEGIES TO AUGMENT CARDIAC REGENERATION

Heart diseases remain a major cause of morbidity and mortality worldwide. Although revascularization strategies, surgical procedures, and pharmaceutical therapies are able to advert acute decompensation in many disease conditions, there is an emerging need to develop therapeutic strategies that are able to prevent or reverse the process of remodeling and thus the progression of ventricular failure. As stem cells and cell-based progenitors have the ability to self-renew and differentiate into specialized tissues, this therapeutic modality has been proposed and is being tested as an ideal class of biologics to support regeneration of damaged myocardium. There is growing evidence that such cell-based products can participate in the maturation and induction of collateral vascular growth and neovasculogenesis and may acquire phenotypic properties of neighboring cardiac myocytes [87]. However, these self-renewing processes, while sufficient to sustain normal homeostasis and turnover of cardiovascular tissues in healthy individuals, are insufficient to salvage heart muscle following massive injury or congenital heart diseases. This reality has prompted a significant effort to isolate, propagate, and reintroduce into the body a progenitor cell population necessary for a paracrine induction of endogenous stem cells or cell-autonomous differentiate into *de novo* tissue (Figure 31.2) in order to augment the underlying regenerative response in the patients with progressive disease [88]. Cardiac repair has moved rapidly from studies in experimental animals to clinical trials involving thousands of patients. The ultimate goal of cardiac repair is to regenerate the myocardium to prevent or cure progressive heart failure.

Building on encouraging preclinical animal data, stem cell therapy has been tested in a variety of clinical trials in patients with both acute and chronic ischemic myocardial injury with the goal of repairing damaged myocardium and/or inducing the growth of new blood vessels so as to improve cardiac function and symptoms (Tables 31.1 and 31.2). During this first decade of cell therapy for human heart regeneration, multiple candidate cell types have been used in preclinical animal models and in humans to repair or regenerate the injured heart either directly or indirectly (through paracrine effects), including embryonic stem cells (ESCs), induced pluripotent stem cells (iPSCs), neonatal cardiomyocytes, skeletal myoblasts (SKMs), endothelial progenitor cells (EPCs), bone marrow mononuclear cells (BMMNCs), mesenchymal stem cells (MSCs), and most recently CSCs. The ideal cell type for the treatment of heart diseases should (i) be safe, i.e., not create tumors or arrhythmias [89], (ii) improve heart function, (iii) be available "off the shelf," (iv) be delivered by minimally invasive clinical methods, and (v) be tolerated by the immune system. The vast majority of clinical trials have been conducted using autologous BMMNCs or cellular fractions thereof [90].

31.4.1 BM Cells for Cardiac Repair

Since Thomas launched the field of bone marrow transplantation (BMT) in patients with hematological disorders more than five decades ago [91], this therapeutic approach has proven the safety and feasibility of stem cell therapeutics and paved the pathway for additional regenerative applications. Subsequently, subpopulations of BM cells, such as BMMNCs, or more specifically, CD34+ or CD133+ stem cells, have been applied toward new clinical challenges outside of the hematological tissues. These applications have emerged due to the realization that mixtures of cells have the ability to contribute to nonhematological tissues. Thus, BM is a highly

TABLE 31.1 Summary of Published Major Clinical Trials Using Intracoronary Autologous BM-Derived Stem Cells in Humans in Cardiac Diseases

Study, Author, Year (Country)	No. of Patients/ Controls	Entity Cardiac Status	Cell Types and Dosage	Study Design	Timing Post-MI (Days)	Other Interventions	Cell Delivery Routes	Follow-Up (Months)
BOOST, Meyer et al., 2006; Waller et al., 2004 (Germany)	30/30	STEMI	MNC $2.46 \pm 9.4 \times 10^8$	Controlled randomized	4.8 ± 1.3	PCI and stenting	IC infusion	6, 18
REPAIR-AMI, Schachinger et al., 2006, 2009 (Germany)	101/103	STEMI	MNC 2.36×10^8	Placebo controlled	3–7	PCI and stenting	IC infusion	4, 12, 24
TOPCARE-CHD, Assmus et al., 2006 (Germany)	92	AMI at least 3 months before cell infusion	$22 \pm 11 \times 10^6$ CPC $205 \pm 110 \times 10^6$ MNC	Controlled randomized	2348–2470	ND	IC infusion	3
TOPCARE-AMI, Assmus et al., 2002; Britten et al., 2003; Schachinger et al., 2004; Leistner et al., 2011 (Germany)	29 30	AMI AMI	MNC 2.4×10^8 CPC 1.3×10^7	Non-randomized open-labeled	3–7	PCI and stenting	IC infusion	4, 12, 60 (in 55 pts)
Janssens et al., 2006 (Belgium)	33/34	STEMI	MNC 1.72×10^8	Placebo controlled	1	PCI and stenting	IC infusion	4
ASTAMI, Lunde et al., 2006; Beitnes et al., 2009 (Norway)	50/50	STEMI[a]	MNC 6.8×10^7 (median)	Randomized + placebo controlled	5~8 (median of 6)	PCI and stenting	IC infusion	6, 12, 36
IACT, Strauer et al., 2005 (Germany)	18/18	Chronic MI > 5 months	MNC 9×10^7	Controlled	5 mo to 8.5 y	PCI and stenting	IC infusion	3
Strauer et al., 2002 (Germany)	10/10	AMI	MNC 2.8×10^7	Controlled	5–9	PCI and stenting	IC infusion	3
Meluzin et al., 2006 and 2008 (Czech Republic)	44 (high and low)/ 22	STEMI	MNC High: 10^8 Low: 10^7	Randomized	5–9	PCI and stenting	IC infusion	3, 6, 12

Study	Patients	Disease	Cells/Dose	Design	Time	Procedure	Delivery	Follow-up (months)
Bartunek et al., 2005 (Belgium)	19/16	AMI	MNC CD $133 + 12.6 \times 10^6$	Cohort	11.6 ± 1.4	PCI and stenting	IC infusion	4
TCT-STAMI, Ge et al., 2006 (China)	10/10	STEMI	MNC 4×10^7	Randomized	1	PCI and stenting	IC infusion	6
Fernandez-Aviles et al., 2004 (Spain)	20/13	MI[b]	MNC $78 \pm 41 \times 10^6$	Phase I, nonrandomized	13.5 ± 5.5	PCI and stenting	IC infusion	6–21
BALANCE, Yousef et al., 2009 (Germany)	62/62	AMI	MNC 6.1×10^7	Cohort	7 ± 2	PCI and stenting	IC infusion	3, 12, 60
MYSTAR, Gyongyosi et al., 2009 (Austria)	Early 30 Late 30	STEMI	MNC IM~2×10^8, IC~1.3×10^9	Randomized	21–42 (early) 90–120 (late)	PCI	IC + Intramyocardial injection	3, 12
FINCELL, Huikuri et al., 2008 (Finland)	40/40	STEMI	MNC 4×10^8	Randomized	2–6	Thrombolysis + PCI	IC infusion	6
LateTIME, Traverse et al., 2011 (USA)	55/26	AMI	MNC 150×10^6	Randomized, double-blinded placebo controlled	14–21	PCI	IC infusion	6
TIME, Traverse et al., 2012 (USA)	3 days: 41/22 7 days: 34/15	STEMI	MNC 150×10^6	Double-blinded placebo controlled	3 vs 7	PCI	IC infusion	6

ND, not determined; G-CSF mobilization, granulocyte-colony stimulating factor 10 μg/kg; STEMI, ST-elevated myocardial infarction; AMI, acute myocardial infarction; IHD: ischemic heart disease; CHF, chronic heart failure; PCI, percutaneous coronary intervention; CABG, coronary artery bypass grafting; CPC, circulating progenitor cell; IC, intracoronary; LVEF, left ventricular ejection fraction; CHD, congestive heart disease; CABG, coronary artery bypass grafting.
[a]With cardiac functions well preserved.
[b]Extensive reperfused MI.

TABLE 31.2 Published Clinical Studies of Intramyocardial Autologous BM Stem Cells Injection for Cardiac Diseases

Study, Author, Year (Country)	No of Patients/ Controls	Entity Cardiac Status	Cell Type, Cell Number Per Injection, Volume Per Injection, No. of Injections	Site of Injection/ Other Interventions	Study Design	Follow-Up Evaluation (Months)
Perin et al., 2003, 2004 (USA)	14/7 11/9	Post-MI IHD, CHF	MNC 2×10^6 in 0.2 mL, 15 ± 2 injections	Transendocardial (catheter)/PCI and stenting or CABG	Nonrandomized, open labeled, controls	2, 4, 6, 12
FOCUS-HF, Perin et al., 2011 (USA)	20/10	Ischemic heart failure	MNC 2×10^6 in 0.2 mL, 15 ± 2 injections	Transendocardial (catheter)/PCI and stenting or CABG	Randomized, single blind	3, 6
FOCUS-CCTRN, Perin et al., 2012 (USA)	61/31	Chronic ischemic heart failure	MNC 100×10^6, 15 injections (0.2 mL each)	Endocardial injection	Randomized controlled, double blinded	6
Fuchs et al., 2003 and 2006 (USA)	27	Severe chronic myocardial ischemia	Unfractioned autologous BM $28 \pm 28 \times 10^6$ CD45 + cells/mL in 0.2 mL, 12 injections	Transendocardial (catheter based)/ PCI and stenting or CABG	Nonrandomized, open label, multicenter study	3, 12
van Ramshorst et al., 2009 (The Netherlands)	25/25	Chronic myocardial ischemia	MNC 1×10^8 total cells, 8–10 injections of 0.2–0.3 mL each	ND	Randomized, double blind, placebo controlled	3, 6
Losordo et al., 2007 (USA)	18/6	Refractory angina pectoris	CD34 + G-CSF mobilized cells, up to 5×10^5/kg, 10 injections of 0.2 mL each	Transendocardial (catheter based)/ PCI and stenting or CABG	Double blind, randomized, placebo controlled dose escalating	1, 2, 3, 6, 9, 12
Losordo et al., 2011 (USA)	111/56	Refractory angina pectoris	CD34 + G-CSF mobilized cells, 1×10^5 or 5×10^5/kg, 10 injections of 0.2 mL each	Transendocardial (catheter based)/ PCI and stenting or CABG	Double blind, randomized, placebo controlled, multicenter study	3, 6, 12
Hendrikx et al., 2006 (USA)	10/10	Post-MI	MNC 6×10^7 total cells, 0.5 mL per injection	Direct injection at the border of the infarct scar (epicardium)/CABG	Double blind, randomized, placebo controlled	4
Mocini et al., 2006 (Italy)	18/18	Post-MI (<6 months)	MNC 2.9×10^8 total cells, 26 ± 10 injections/patient, 0.2 mL per injection	Direct injection at the border of the infarct scar (epicardium)/CABG	Nonrandomized, controls	3–12
Tse et al., 2006 (China)	12	Severe coronary artery disease	MNC 10^6/injection, 12–16 injections, 0.1 mL	Transendocardial (catheter based)	Nonrandomized	3, 6, 36
PROTECT-CAD, Tse et al., 2007 (Hong Kong & Australia)	19/9	Severe coronary artery diseases	MNC Low: 1.6×10^6/0.1 mL injection High: 2×10^6/0.1 mL injection 14.6 ± 0.7 injections/ patient	Catheter-based direct endomyocardial injection	Randomized, blinded, and placebo controlled	3, 6: mean follow-up of 19 ± 9 months
Beeres et al., 2006 (Netherlands)	20	Angina pectoris	MNC $41 \pm 16 \times 10^6$/mL, 0.2 mL, 10 ± 2 injections	Direct intramyocardial injection	Cohort	1, 3, 6
Beeres et al., 2007 (Netherlands)	15	Chronic MI	MNC $9.4 \pm 1.4 \times 10^7$ (total), 0.2 mL, 10 ± 1 injections	Direct intramyocardial injection	Cohort	3, 6

(Continued)

TABLE 31.2 (Continued)

Study, Author, Year (Country)	No of Patients/ Controls	Entity Cardiac Status	Cell Type, Cell Number Per Injection, Volume Per Injection, No. of Injections	Site of Injection/ Other Interventions	Study Design	Follow-Up Evaluation (Months)
Beeres et al., 2007 (Netherlands)	20	Chronic ischemic heart disease	MNC 1×10^8/total, 0.2 mL, 10 ± 1 injections	Direct intramyocardial injection	Cohort	3, 6
Stamm et al., 2007 (Germany and USA)	20 vs 20	Chronic ischemic heart disease	Autologous BM-CD133 + median 5.8×10^6, 0.2 mL, 10 injections	Direct injection at the border of the infarct scar (epicardium)/CABG	Randomized	Up to 36
Klein et al., 2007 (Germany)	10	End-stage chronic ischemic cardiomyopathy	Autologous BM-CD133 + $1.5-9.7 \times 10^6$, $4-10$ mL total, $10-20$ injections	Direct injection (transepicardially)/ thoracotomy	Cohort	3, 6, 9
Pompilio et al., 2008 (Italy)	5	Angina pectoris	Autologous BM-CD133 + $4-12 \times 10^6$	Direct injection/ thoracotomy	Cohort	6, 12
Patel et al., 2005 (USA)	10/10	Post-MI IHD, CHF	MNC CD34 + 22×10^6 total cells, $28-30$ injections	Subepicardial (peri-infarcted) injection/ off-pump CABG	Randomized	1, 3, 6
Galiñanes et al., 2004 (UK)	14	Post-IM	Unmanipulated autologous BM, 3.1×10^6/injection 0.25 mL/injection, 26 ± 2 injections	Direct injection into the scarred myocardium/CABG	Nonrandomized	1, 10
Ang et al., 2008 (UK)	21/20	Chronic ischemic heart disease	MNC $84 \pm 56 \times 10^6$/total 20 injections of 0.5 mL	Directly into myocardial scars	Single blinded, randomized, controlled	6
Pokushalov et al., 2010 (Russia)	55/54	Chronic MI and end-stage CHF	MNC $41 \pm 16 \times 10^6$ (total), 0.2 mL, 10 injections	Border zone of MI	Randomized	1, 3, 6, 12
Zhao et al., 2008 (China)	18/18	Ischemic heart failure	MNC $6.59 \pm 5.12 \times 10^8$, 0.5 mL, 10 injections	Into the infracted and marginal areas during CABG	Randomized	Every month up to 6 months
Ahmadi et al., 2007 (Iran)	18/9	Recent MI	Autologous BM-CD133 + 1.89×10^6 (total) 0.2 mL, 10 injections	Into the peri-infarct zone, CABG	Prospective, nonrandomized, open label, controls	6, 12, 18
Yoo et al., 2008 (Canada and Korea)	5	Ischemic cardiomyopathy	MNC 1.6×10^9, $0.7-1$ mL/ injection, $10-15$ injections	Into nongraftable areas/ off-pump CABG	Cohort	2, 4
Briguori et al., 2006 (Italy)	10	Refractory angina pectoris	Unfractioned BM, at least 10^7 cells/injection, $0.5-1$ mL/injection, 11 ± 2 injections	Catheter-based direct percutaneous	Cohort	12
Li et al., 2003 (Japan)	6	Ischemic heart disease	MNC 5×10^7 to 1×10^8 cells/injection, 0.1 mL/ injection, $15-28$ injections	Into the ischemic myocardium/CABG	Cohort	Up to 12
Li et al., 2007 (Japan)	8	Ischemic heart disease	MNC $3-22 \times 10^8$ total cells, 0.1 mL/injection, $6-22$ injections	Into the ungrafted ischemic area/CABG	Cohort	1, 3, 6, 12, 24, 36, 60
Gowdak et al., 2008 (Brazil)	10	Limiting angina	MNC 1.3×10^8/total, 0.2 mL/injection, 25 injections	Into the ungrafted ischemic area/CABG	Nonrandomized, open label, phase I	1, 3, 6, 12

(Continued)

TABLE 31.2 (Continued)

Study, Author, Year (Country)	No of Patients/ Controls	Entity Cardiac Status	Cell Type, Cell Number Per Injection, Volume Per Injection, No. of Injections	Site of Injection/ Other Interventions	Study Design	Follow-Up Evaluation (Months)
Krause et al., 2009 (Germany)	20	AMI	MNC, 10×10^7/mL, 0.1 mL/injection, 20 injections	Low-voltage area (area at risk) of the MI. Acute PCI with stenting	Nonrandomized	6, 12
Silva et al., 2004 (USA)	5	Severe ischemic heart failure	MNC, cell dose not available, 0.2 mL/injection, 15 injections	Transendocardial (catheter based)	Nonrandomized, open label	2, 6
Hossne et al., 2009 (Brazil)	8	Refractory angina	MNC, 0.2 mL (2×10^6 cells)/injection, 40–90 injections/patient	Into ischemic areas of the left ventricle	Noncontrolled, open label	1, 3, 6, 12, 18
El Oakley et al., 2004 (Singapore)	7	Severe ischemic heart failure	MNC, 21×10^6/total, 0.2 mL, 5 injections	Into the anterior–lateral wall of the ventricle/ CABG	Noncontrolled, open label	9, 12

ND, not determined; G-CSF mobilization, granulocyte-colony stimulating factor 10 µg/kg; STEMI, ST-elevated myocardial infarction; AMI, acute myocardial infarction; MI, myocardial infarction; IHD, ischemic heart disease; CHF, chronic heart failure; PCI, percutaneous coronary intervention; CABG, coronary artery bypass grafting; MNC, mononuclear cells; CPC, circulating progenitor cell; IC, intracoronary; LVEF, left ventricular ejection fraction; CHD, congestive heart disease; CABG, coronary artery bypass grafting.

heterogeneous tissue, which contains different cell populations. The term "mononuclear cells" collectively refers to all cells present whose nuclei are unilobulated, rounded, and lack cytoplasmic granules. These physical features give the BMMNCs a similar shape and size making them amenable to separation by density centrifugation. Adult human BMMNCs contain hematopoietic progenitor cells at different stages of maturation in a heterogeneous mixture of lymphoid cells, monocytes, and macrophages. Additionally, nonhematopoietic lineages have been identified in this subpopulation of normal human adult BM or umbilical cord blood. Among these nonhematopoietic cells are the side population cells that have been demonstrated to function as primitive progenitor cells capable of multipotent differentiation [92], which includes mesenchymal stromal cells [93], hemangioblasts with endothelial progenitor cells [94], very small embryonic-like stem cells [95], and tissue-committed stem cells [96].

Human hematopoietic stem cells (HSCs) can traditionally be defined as CD34 + cells with the ability to regenerate the entire population of blood lineages with the emerging definition of the ability to transdifferentiate into tissue-specific lineages such as cardiomyocytes, endothelial cells (EPC), and smooth muscle cells [97]. EPCs are a subpopulation that promotes neovascularization either directly by differentiation into endothelial cells or indirectly through secretion of angiogenic cytokines. BMMNCs, isolated by density centrifugation following BM aspiration, contain a small percentage of stem cells (~2–4% HSCs/EPCs) with the majority of BMMNCs comprised of committed hematopoietic cells. In the clinical setting, autologous BMMNCs are the most frequently used cell type for treatment of heart disease. The use of these cells in cardiovascular diseases has the advantage that BM can be easily accessed, is renewable, and is an autologous source for regenerative medicine. The experience with this type of cell-based product delivered to the left ventricle dates back more than a decade [98,99]. These early clinical steps prompted a series of subsequent studies of acute and chronic heart diseases.

31.4.1.1 Preparation of Autologous BMCs for Cardiac Therapy

For cell therapy, up to 250 mL adult BM blood can be aspirated from the iliac crest under local anesthesia following standard clinical procedures. An alternative aspiration site is the sternal bone before median sternotomy at the time of cardiac surgery. In the past, the mononuclear fraction of cells was separated from the whole BM aspirate by density gradient centrifugation using osmolaric media such as ficoll [99,100]. These methods comprise open preparation procedures and need several washing steps that increase the risk of a contaminated product. Thus, the need for a good manufacturing practice process to produce a quality-controlled cell product and avoid contamination of the end product allowed for continued optimization of the product [101,102]. Recently, several new automatic systems were developed to gain nucleated or mononuclear cells from the whole BM aspirate [103]. The advantage of such systems is the possibility to separate the cells in a closed system. In these systems, the cell recovery is higher than with manual preparation [104], and with the same functional capabilities [103]. Additionally, the preparation time is definitely shorter and allows bedside preparations. The cell preparation and cell application can be done in one working process to produce a transplant-ready product. However, the functionality of the cells needs to be tested in preclinical model systems to ensure full potency is maintained with each individual separation protocol.

31.4.1.2 Cell Delivery Routes

One of the most important methodological questions refers to the optimum mechanism of cell delivery to the human heart. So far, stem cells for heart repair have been delivered clinically via three routes: systemic intravenous infusion, intracoronary infusion, and intramyocardial injection (either by direct open-chest injections as an adjunct to CABG or by transendocardial catheter-based injections).

Intravenous infusion, albeit simple in execution, can be limited by trapping of cells in the lungs during the first pass; therefore, a smaller number of cells may reach the coronary circulation and therefore the infarct target region [105]. This approach is typically not utilized as many patients have procedures that offer a more local and focal delivery.

Intracoronary delivery of cells is safe and convenient (can be performed with standard balloon catheters) and has the advantage that cells are infused into myocardial regions with preserved oxygen and nutrient supply, thus ensuring a favorable environment for cell survival. Cells are able to flow through the infarcted and peri-infarct tissue during the immediate first passage of the post-ischemic region. However, retention of cells may be suboptimal (most cells are washed away before they can migrate into the surrounding tissue), and underperfused regions of the myocardium may not be accessible.

Direct intramyocardial cell injections has also been demonstrated to be safe [106,107], and results in better retention of cells compared with intracoronary or systemic approaches. Intramyocardial stem cell injection seems to overcome the problem linked to insufficient vascularization, migration, and homing of transplanted stem cells rather than influence stem cell migration and results in a high stem cell persistence in heart muscle.

Surgical (epicardial) stem cell application is performed into well-exposed ischemic areas, allowing for multiple injections within and principally around the infarct area. First, clinical studies performed stem cell injection in combination with CABG [98]. Once the graft-coronary artery anastomosis is completed, the ischemic area is visualized, and the cells are injected into the border zone of the infarcted area [108].

Intracoronary infusion is the preferred route in acute MI, and intramyocardial cell injections are probably more suitable for patients with chronic ischemic cardiomyopathy or congenital heart disease. Less-invasive catheter-mediated methods of intramyocardial delivery (endocardial application) are being further developed (preferably employing electromechanical mapping and identification of viable myocardium) in order to avoid open-chest injections (epicardial application) (Table 31.2). This interventional approach offers intramyocardial cell delivery similar to the surgical approach while being less invasive.

31.4.1.3 Clinical Trials: Efficacy

Acute MI

In patients with acute MI, several major randomized controlled clinical trials have been reported with the use of intracoronary administration of autologous BM cells in patients who had undergone successful percutaneous coronary intervention of infarcted related artery (Table 31.1). Pilot studies of BMMNCs delivery after acute MI demonstrated consistently positive results [109,110]. Subsequently, large randomized studies have yielded varied results. REPAIR-AMI [111−113], the largest randomized, controlled clinical study reported, showed the benefits of intracoronary infusion of unfractionated BMMNCs in patients with MIs and a left ventricular ejection fraction (LVEF) no greater than 49%, while the similarly designed ASTAMI [100] showed no functional benefit. The difference between the two studies is speculated to be attributed to cellular features as measured by an *in vitro* cell migration assay. In these trials, BM cell therapy increased LVEF and reduced the 1-year combined clinical end point of death, recurrence of MI, and revascularization. In contrast to the REPAIR study, the BOOST, a randomized but not placebo-controlled study concluded that the functional benefit seen at 6 months [114] was not sustained at 18 months [115]. In other large randomized studies, functional benefits of BMMNCs therapy were more positive (FINCELL) [116].

The randomized, double-blind, placebo-controlled trial (LateTIME) has also focused on acute ischemic heart disease [117]. Patients with recent MI and a LVEF below 45% received 150 million autologous BMMNCs by intracoronary injection. Patients were recruited 2−3 weeks after their initial event. The same investigators conducted a parallel study with a recruitment window of 3−7 days after the initial acute MI (the TIME study) [118]. After 6 months' follow-up, cardiac magnetic resonance imaging data showed no effect of the treatment on either overall or regional cardiac function or infarct size. These data contrast with those from the previously published REPAIR-AMI trial, which reported a significant benefit on ejection fraction following intracoronary administration of BMMNCs at a median of 4 days after the clinical event. Taken together, these two studies suggest that BMMNC therapy may be efficacious if administered early after acute MI.

Several meta-analyses have been published, which permit general conclusions regarding the effects of BM cells in the treatment of patients with ischemic heart disease. Martin-Rendon and colleagues performed a Cochrane systematic review of 13 randomized trials with 14 different comparisons, involving 811 patients. In this review, autologous BM cell therapy was found to be safe and moderately beneficial: LVEF increased by ∼3%, end-systolic volume decreased by ∼5 mL and scar sized decreased by ∼3.5% in the cell-treated groups compared with controls. Subgroup analysis revealed that the benefit of cell therapy was greater when cells were infused within 7 days following infarction (the optimal time window seems to be 5−7 days post-MI) and when the dose administered was $>10^8$ BMMNCs [119]. A previous meta-analysis by Lipinski and colleagues included 10 studies (7 out of the 10 were randomized) of 698 patients, who were treated with percutaneous coronary angioplasty after acute MI, and then allocated to treatment with either intracoronary BM cell therapy or standard medical therapy. The patients were followed for a mean of 6 months (range: 3−18 months). In this meta-analysis, patients who received intracoronary BM cells showed statistically significant increases in LVEF (3%), decreases in end-systolic volume (7.4 mL), and reductions in infarct size (5.6%), compared to standard medical therapy. It was also associated with a nonsignificant reduction in death and rehospitalization from heart failure [120]. Similar conclusions were reached in the meta-analysis by Abdel-Latif and colleagues, which included 18 randomized and nonrandomized trials (999 patients) and showed that stem cells therapy significantly increased LVEF by 3.66% compared with control patients [121].

Chronic Myocardial Ischemia

In contrast to acute MI, there are a few small, randomized controlled clinical trials on stem cell therapy for treatment of refractory angina due to chronic myocardial ischemia or heart failure after MI. For treatment of refractory angina, clinical trials are mainly focused on the use of intramyocardial injections of BM cells (Table 31.2). Losordo et al. [106] showed an improving angina frequency and exercise capacity without any change in SPECT perfusion. PROTECT-CAD study [122] and van Ramshorst et al. [123] have shown that intramyocardial injection of BMMNCs was associated with significant improvement in exercise capacity and clinical status. Patients who received BM cell therapy also showed a modest but significant improvement in LVEF (3−5%). Of these studies, only van Ramshorst et al. demonstrated a significant improvement in SPECT perfusion after BM cell injection. The reasons for the conflicting results on SPECT remain unclear but may be related to the differences in the dosage cell, study sample size, and even the method of SPECT analysis.

The FOCUS-CCTRN trial [124] included patients with chronic ischemic left ventricular dysfunction with heart failure and/or angina that were entered into a randomized, double-blinded, controlled trial comparing 100 million autologous BMMNCs or placebo administered by catheter-mediated intramyocardial injection using an existing electromechanical mapping system to identify appropriate sites for injection. After 6 months' follow-up, there were no statistically significant differences between the groups for any of the primary endpoints and no differences in patient symptoms. In a prespecified subgroup analysis in this trial, a statistically significant treatment effect of LVEF of patients younger than 62 years was demonstrated, suggesting that age is an important consideration for inclusion in such trials.

Dondorf et al. conducted a systematic review and meta-analysis of available publications regarding the efficacy and safety of intramyocardial BM stem cell transplantation during CABG. They reported that clinical evidence suggests that intramyocardial injection of BM cells in combination with CABG is associated with improvements of functional parameters in patients with chronic ischemic heart disease and seems to be safe [125]. In 2011, another meta-analysis to provide systematic assessment of the safety and efficacy of direct BM cells transplantation in patients with ischemic heart disease was performed. Eight randomized clinical trials with 307 eligible patients were analyzed. Compared with controls, direct BM cells injection improved LVEF (8.4%). This meta-analysis suggests that intramyocardial injection is associated with moderate but significant improvements over regular therapy in cardiac functional parameters in patients with ischemic heart disease [126].

In contrast to animal models, the improvement in LVEF in most clinical trials is modest at best. However, it should be noted that several of the established therapies that have an impact on prognosis in patients with ischemic heart disease and a reduced LV function, such as β-blockers [127], thrombolytic therapy, and percutaneous coronary intervention [128,129], were initially associated with similar improvements in LVEF and have contributed to meaningful hard endpoints of morbidity and mortality.

31.4.1.4 Safety of Intracoronary and Intramyocardial Delivery

No safety concerns regarding delivery of BMMNCs through a recently reopened infarct-related coronary artery have emerged. Infusion of cells does not inflict additional ischemic damage to the myocardium. No increased incidence of arrhythmias or increased risk of arrhythmia has been reported. Published meta-analyses show no association to target-vessel restenosis or repeat revascularization in patients treated with BMMNCs. Finally, there is no evidence of increased tumorigenesis after BMMNC therapy. Overall, the safety profile of intracoronary and intramyocardial delivery of BMMNCs has consistently been confirmed across many studies [119,126].

Reviewing the accumulated data in those clinical trials executed worldwide, a number of conclusions can be drawn:

1. An excellent feasibility and safety profile has been established for intracoronary and intramyocardial delivery of BM-derived cells.
2. Clinical outcomes have been generally positive, although efficacy has been inconsistent and, overall, modest.
3. The beneficial effects of cell therapy seem to be related to paracrine effects, rather than direct new tissue formation. BM-derived stem cells produce and secrete a broad variety of cytokines, chemokines, and growth factors that are known to be involved in cardiac repair, and hypoxic stress increases the production of several of these factors. Potential effects of paracrine factors include cytoprotection of resident myocytes, upregulation of angiogenesis, modulation of inflammatory processes, improved cardiac metabolism and contractility, recruitment of endogenous stem cells, and induction of secondary humoral effect in the host tissue [130]. This mechanism of action rationalizes the persistence of benefit despite the disappearance of transplanted cell survival, so the new tissue originates from the recipient heart rather than from the transplant.

31.5 CONCLUSIONS

The developmental biology approach to cardiac regeneration is predicated on the experience that mechanisms of embryonic cardiogenesis are applicable to the development of next-generation biological products. The fetal gene expression profile that is reactivated in heart failure may be the signature of inducing the regenerative response, rather than the expression of pathological features. Therefore, promoting natural regenerative pathways, mechanisms, and cell populations is a strategy to promote functional regeneration. Furthermore, augmenting potential deficiencies in the innate regenerative capacity for individual patients may accelerate the discoveries and applications that are needed for patients with advanced heart disease or patients that are at high risk of advanced heart disease. These advances may come from improved understanding of cardiogenesis and be applied in the form of novel cell-based products, noncellular preparations of growth factors/cytokines, extracellular matrixes, or traditional small molecules that target regenerative pathways. The future of cardiac regenerative biology aimed to delay or prevent transplantation in patients with acquired or congenital heart disease continues to emerge with significant promise and opportunities to establish a disruptive technology that improves cardiovascular outcomes for transplantation medicine and surgery.

REFERENCES

[1] Olson EN. Gene regulatory networks in the evolution and development of the heart. Science 2006;313:1922–7.

[2] Stainier DYR. Zebrafish genetics and vertebrate heart formation. Nat Rev Genet 2001;2:39–48.

[3] Driever W, Solnica-Krezel L, Schier AF, Neuhauss SC, Malicki J, Stemple DL, et al. A genetic screen for mutations affecting embryogenesis in zebrafish. Development 1996;123:37–46.

[4] Keegan BR, Meyer D, Yelon D. Organization of cardiac chamber progenitors in the zebrafish blastula. Development 2004;131:3081–91.

[5] Keegan BR, Feldman JL, Begemann G, Ingham PW, Yelon D. Retinoic acid signaling restricts the cardiac progenitor pool. Science 2005;307:247–9.

[6] Kishimoto Y, Lee KH, Zon L, Hammerschmidt M, Schulte-Merker S. The molecular nature of zebrafish swirl: BMP2 function is essential during early dorsoventral patterning. Development 1997;124:4457–66.

[7] Bodmer R. The gene tinman is required for specification of the heart and visceral muscles in Drosophila. Development 1993;118:719–29.

[8] Bodmer R, Jan LY, Jan YN. A new homeobox-containing gene, msh-2, is transiently expressed early during mesoderm formation of Drosophila. Development 1990;110:661–9.

[9] Reifers F, Walsh EC, Leger S, Stainier DY, Brand M. Induction and differentiation of the zebrafish heart requires fibroblast growth factor 8 (fgf8/acerebellar). Development 2000;127:225–35.

[10] Reiter JF, Alexander J, Rodaway A, Yelon D, Patient R, Holder N, et al. Gata5 is required for the development of the heart and endoderm in zebrafish. Genes Dev 1999;13:2983–95.

[11] Garavito-Aguilar ZV, Riley HE, Yelon D. Hand2 ensures an appropriate environment for cardiac fusion by limiting Fibronectin function. Development 2010;137:3215–20.

[12] Kupperman E, An S, Osborne N, Waldron S, Stainier DYR. A sphingosine-1-phosphate receptor regulates cell migration during vertebrate heart development. Nature 2000;406:192−5.

[13] Yelon D, Ticho B, Halpern ME, Ruvinsky I, Ho RK, Silver LM, et al. The bHLH transcription factor hand2 plays parallel roles in zebrafish heart and pectoral fin development. Development 2000;127:2573−82.

[14] Arrington CB, Yost HJ. Extra-embryonic syndecan 2 regulates organ primordia migration and fibrillogenesis throughout the zebrafish embryo. Development 2009;136:3143−52.

[15] Osborne N, Brand-Arzamendi K, Ober EA, Jin SW, Verkade H, Holtzman NG, et al. The spinster homolog, two of hearts, is required for sphingosine 1-phosphate signaling in zebrafish. Curr Biol 2008;18:1882−8.

[16] Zeng X-XI, Wilm TP, Sepich DS, Solnica-Krezel L. Apelin and its receptor control heart field formation during zebrafish gastrulation. Dev Cell 2007;12:391−402.

[17] Baker K, Holtzman NG, Burdine RD. Direct and indirect roles for Nodal signaling in two axis conversions during asymmetric morphogenesis of the zebrafish heart. Proc Natl Acad Sci USA 2008;105:13924−9.

[18] de Campos-Baptista MIM, Holtzman NG, Yelon D, Schier AF. Nodal signaling promotes the speed and directional movement of cardiomyocytes in zebrafish. Dev Dyn 2008;237:3624−33.

[19] Buckingham M, Meilhac S, Zaffran S. Building the mammalian heart from two sources of myocardial cells. Nat Rev Genet 2005;6:826−37.

[20] Hami D, Grimes AC, Tsai HJ, Kirby ML. Zebrafish cardiac development requires a conserved secondary heart field. Development 2011;138:2389−98.

[21] Erter CE, Solnica-Krezel L, Wright CVE. Zebrafish nodal-related 2 encodes an early mesendodermal inducer signaling from the extraembryonic yolk syncytial layer. Dev Biol 1998;204:361−72.

[22] Feldman B, Gates MA, Egan ES, Dougan ST, Rennebeck G, Sirotkin HI, et al. Zebrafish organizer development and germ-layer formation require nodal-related signals. Nature 1998;395:181−5.

[23] Osada SI, Wright CV. Xenopus nodal-related signaling is essential for mesendodermal patterning during early embryogenesis. Development 1999;126:3229−40.

[24] Rebagliati MR, Toyama R, Haffter P, Dawid IB. Cyclops encodes a nodal-related factor involved in midline signaling. Proc Natl Acad Sci USA 1998;95:9932−7.

[25] Liao EC, Paw BH, Oates AC, Pratt SJ, Postlethwait JH, Zon LI. SCL/Tal-1 transcription factor acts downstream of cloche to specify hematopoietic and vascular progenitors in zebrafish. Genes Dev 1998;12:621−26.

[26] Zhong TP, Rosenberg M, Mohideen M-APK, Weinstein B, Fishman MC. Gridlock, an HLH gene required for assembly of the aorta in zebrafish. Science 2000;287:1820−4.

[27] Lepilina A, Coon AN, Kikuchi K, Holdway JE, Roberts RW, Burns CG, et al. A dynamic epicardial injury response supports progenitor cell activity during zebrafish heart regeneration. Cell 2006;127:607−19.

[28] McFadden DG, Barbosa AC, Richardson JA, Schneider MD, Srivastava D, Olson EN. The Hand1 and Hand2 transcription factors regulate expansion of the embryonic cardiac ventricles in a gene dosage-dependent manner. Development 2005;132: 189−201.

[29] Pu WT, Ishiwata T, Juraszek AL, Ma Q, Izumo S. GATA4 is a dosage-sensitive regulator of cardiac morphogenesis. Dev Biol 2004;275:235−44.

[30] Laforest B, Andelfinger G, Nemer M. Loss of Gata5 in mice leads to bicuspid aortic valve. J Clin Invest 2011;121:2876−87.

[31] Nemer G, Nemer M. Cooperative interaction between GATA5 and NF-ATc regulates endothelial-endocardial differentiation of cardiogenic cells. Development 2002;129:4045−55.

[32] Kodo K, Nishizawa T, Furutani M, Arai S, Yamamura E, Joo K, et al. GATA6 mutations cause human cardiac outflow tract defects by disrupting semaphorin-plexin signaling. Proc Natl Acad Sci USA 2009;106:13933−8.

[33] Tian Y, Yuan L, Goss AM, Wang T, Yang J, Lepore JJ, et al. Characterization and *in vivo* pharmacological rescue of a Wnt2-Gata6 pathway required for cardiac inflow tract development. Dev Cell 2010;18:275−87.

[34] Peterkin T, Gibson A, Patient R. Redundancy and evolution of GATA factor requirements in development of the myocardium. Dev Biol 2007;311:623−35.

[35] Rochais F, Mesbah K, Kelly RG. Signaling pathways controlling second heart field development. Circ Res 2009;104:933−42.

[36] Kraus F, Haenig B, Kispert A. Cloning and expression analysis of the mouse T-box gene Tbx18. Mech Dev 2001;100:83−6.

[37] Greulich F, Rudat C, Kispert A. Mechanisms of T-box gene function in the developing heart. Cardiovasc Res 2011;91:212−22.

[38] Kirby ML, Waldo KL. Neural crest and cardiovascular patterning. Circ Res 1995;77:211−5.

[39] Pérez-Pomares JM, de la Pompa JL. Signaling during epicardium and coronary vessel development. Circ Res 2011;109:1429−42.

[40] Ferreira-Martins J, Ogórek B, Cappetta D, Matsuda A, Signore S, D'Amario D, et al. Cardiomyogenesis in the developing heart is regulated by c-kit−positive cardiac stem cells. Circ Res 2012;110:701−15.

[41] Bolli R, Chugh AR, D'Amario D, Loughran JH, Stoddard MF, Ikram S, et al. Cardiac stem cells in patients with ischaemic cardiomyopathy (SCIPIO): initial results of a randomised phase 1 trial. Lancet 2011;378:1847−57.

[42] Makkar RR, Smith RR, Cheng K, Malliaras K, Thomson LEJ, Berman D, et al. Intracoronary cardiosphere-derived cells for heart regeneration after myocardial infarction (CADUCEUS): a prospective, randomised phase 1 trial. Lancet 2012;379:895−904.

[43] Liu N, Olson EN. MicroRNA regulatory networks in cardiovascular development. Dev Cell 2010;18:510−25.

[44] Ivey KN, Muth A, Arnold J, King FW, Yeh RF, Fish JE, et al. MicroRNA regulation of cell lineages in mouse and human embryonic stem cells. Cell Stem Cell 2008;2:219−29.

[45] Zhao Y, Samai E, Srivastava D. Serum response factor regulates a muscle-specific microRNA that targets Hand2 during cardiogenesis. Nature 2005;436:214−20.

[46] Duisters RF, Tijsen AJ, Schroen B, Leenders JJ, Lentink V, van der Made I, et al. miR-133 and miR-30 regulate connective tissue growth factor: implications for a role of microRNAs in myocardial matrix remodeling. Circ Res 2009;104:170−8.

[47] Liu H, Zhu F, Yong J, Zhang P, Hou P, Li H, et al. Generation of induced pluripotent stem cells from adult rhesus monkey fibroblasts. Cell Stem Cell 2008;3:587−90.

[48] Itou J, Oishi I, Kawakami H, Glass TJ, Richter J, Johnson A, et al. Migration of cardiomyocytes is essential for heart regeneration in zebrafish. Development 2012;139:4133–42.

[49] Poss KD, Wilson LG, Keating MT. Heart regeneration in zebrafish. Science 2002;298:2188–90.

[50] Raya Á, Koth CM, Büscher D, Kawakami Y, Itoh T, Raya RM, et al. Activation of Notch signaling pathway precedes heart regeneration in zebrafish. Proc Natl Acad Sci U S A 2003;100:11889–95.

[51] Astorri E, Bolognesi R, Colla B, Chizzola A, Visioli O. Left ventricular hypertrophy: a cytometric study on 42 human hearts. J Mol Cell Cardiol 1977;9:763–75.

[52] Linzbach AJ. Heart failure from the point of view of quantitative anatomy. Am J Cardiol 1960;5:370–82.

[53] Chablais F, Ja W, Ska A. Induction of myocardial infarction in adult zebrafish using cryoinjury. J Vis Exp 2012;62:e3666.

[54] Porrello ER, Mahmoud AI, Simpson E, Hill JA, Richardson JA, Olson EN, et al. Transient regenerative potential of the neonatal mouse heart. Science 2011;331:1078–80.

[55] Beltrami AP, Urbanek K, Kajstura J, Yan SM, Finato N, Bussani R, et al. Evidence that human cardiac myocytes divide after myocardial infarction. N Engl J Med 2001;344:1750–7.

[56] Anversa P, Kajstura J, Leri A, Bolli R. Life and death of cardiac stem cells: a paradigm shift in cardiac biology. Circulation 2006;113:1451–63.

[57] Bergmann O, Bhardwaj RD, Bernard S, Zdunek S, Barnabé-Heider F, Walsh S, et al. Evidence for cardiomyocyte renewal in humans. Science 2009;324:98–102.

[58] Fazel S, Cimini M, Chen L, Li S, Angoulvant D, Fedak P, et al. Cardioprotective c-kit + cells are from the bone marrow and regulate the myocardial balance of angiogenic cytokines. J Clin Invest 2006;116:1865–77.

[59] Quaini F, Urbanek K, Beltrami AP, Finato N, Beltrami CA, Nadal-Ginard B, et al. Chimerism of the transplanted heart. N Engl J Med 2002;346:5–15.

[60] Urbanek K, Quaini F, Tasca G, Torella D, Castaldo C, Nadal-Ginard B, et al. Intense myocyte formation from cardiac stem cells in human cardiac hypertrophy. Proc Natl Acad Sci USA 2003;100:10440–5.

[61] Bearzi C, Rota M, Hosoda T, Tillmanns J, Nascimbene A, De Angelis A, et al. Human cardiac stem cells. Proc Natl Acad Sci USA 2007;104:14068–73.

[62] Beltrami AP, Barlucchi L, Torella D, Baker M, Limana F, Chimenti S, et al. Adult cardiac stem cells are multipotent and support myocardial regeneration. Cell 2003;114:763–76.

[63] Urbanek K, Cesselli D, Rota M, Nascimbene A, De Angelis A, Hosoda T, et al. Stem cell niches in the adult mouse heart. Proc Natl Acad Sci USA 2006;103:9226–31.

[64] Dawn B, Stein AB, Urbanek K, Rota M, Whang B, Rastaldo R, et al. Cardiac stem cells delivered intravascularly traverse the vessel barrier, regenerate infarcted myocardium, and improve cardiac function. Proc Natl Acad Sci USA 2005;102:3766–71.

[65] Oh H, Bradfute SB, Gallardo TD, Nakamura T, Gaussin V, Mishina Y, et al. Cardiac progenitor cells from adult myocardium: Homing, differentiation, and fusion after infarction. Proc Natl Acad Sci USA 2003;100:12313–8.

[66] Hierlihy AM, Seale P, Lobe CG, Rudnicki MA, Megeney LA. The post-natal heart contains a myocardial stem cell population. FEBS Lett 2002;530:239–43.

[67] Martin CM, Meeson AP, Robertson SM, Hawke TJ, Richardson JA, Bates S, et al. Persistent expression of the ATP-binding cassette transporter, Abcg2, identifies cardiac SP cells in the developing and adult heart. Dev Biol 2004;265:262–75.

[68] Cai CL, Liang X, Shi Y, Chu PH, Pfaff SL, Chen J, et al. Isl1 identifies a cardiac progenitor population that proliferates prior to differentiation and contributes a majority of cells to the heart. Dev Cell 2003;5:877–89.

[69] Laugwitz KL, Moretti A, Lam J, Gruber P, Chen Y, Woodard S, et al. Postnatal isl1 + cardioblasts enter fully differentiated cardiomyocyte lineages. Nature 2005;433:647–53.

[70] Messina DN, Glasscock J, Gish W, Lovett M. An ORFeome-based analysis of human transcription factor genes and the construction of a microarray to interrogate their expression. Genome Res 2004;14:2041–7.

[71] Mishra R, Vijayan K, Colletti EJ, Harrington DA, Matthiesen TS, Simpson D, et al. Characterization and functionality of cardiac progenitor cells in congenital heart patients/clinical perspective. Circulation 2011;123:364–73.

[72] Deb A, Wang S, Skelding KA, Miller D, Simper D, Caplice NM. Bone marrow–derived cardiomyocytes are present in adult human heart. Circulation 2003;107:1247–9.

[73] Bayes-Genis A, Bellosillo B, de La Calle O, Salido M, Roura S, Ristol FS, et al. Identification of male cardiomyocytes of extracardiac origin in the hearts of women with male progeny: male fetal cell microchimerism of the heart. J Heart Lung Transplant 2005;24:2179–83.

[74] Kara RJ, Bolli P, Karakikes I, Matsunaga I, Tripodi J, Tanweer O, et al. Fetal cells traffic to injured maternal myocardium and undergo cardiac differentiation. Circ Res 2012;110:82–93.

[75] Murry CE, Soonpaa MH, Reinecke H, Nakajima H, Nakajima HO, Rubart M, et al. Haematopoietic stem cells do not transdifferentiate into cardiac myocytes in myocardial infarcts. Nature 2004;428:664–8.

[76] Nygren JM, Jovinge S, Breitbach M, Sawen P, Roll W, Hescheler J, et al. Bone marrow–derived hematopoietic cells generate cardiomyocytes at a low frequency through cell fusion, but not transdifferentiation. Nat Med 2004;10:494–501.

[77] Conget PA, Minguell JJ. Phenotypical and functional properties of human bone marrow mesenchymal progenitor cells. J Cell Physiol 1999;181:67–73.

[78] Pittenger MF, Mackay AM, Beck SC, Jaiswal RK, Douglas R, Mosca JD, et al. Multilineage potential of adult human mesenchymal stem cells. Science 1999;284:143–7.

[79] Makino S, Fukuda K, Miyoshi S, Konishi F, Kodama H, Pan J, et al. Cardiomyocytes can be generated from marrow stromal cells in vitro. J Clin Invest 1999;103:697–705.

[80] Toma C, Pittenger MF, Cahill KS, Byrne BJ, Kessler PD. Human mesenchymal stem cells differentiate to a cardiomyocyte phenotype in the adult murine heart. Circulation 2002;105:93–8.

[81] Asahara T, Murohara T, Sullivan A, Silver M, van der Zee R, Li T, et al. Isolation of putative progenitor endothelial cells for angiogenesis. Science 1997;275:964–6.

[82] Peichev M, Naiyer AJ, Pereira D, Zhu Z, Lane WJ, Williams M, et al. Expression of VEGFR-2 and AC133 by circulating human CD34 + cells identifies a population of functional endothelial precursors. Blood 2000;95:952–8.

[83] Kocher AA, Schuster MD, Szabolcs MJ, Takuma S, Burkhoff D, Wang J, et al. Neovascularization of ischemic myocardium by human bone-marrow-derived angioblasts prevents cardiomyocyte apoptosis, reduces remodeling and improves cardiac function. Nat Med 2001;7:430−6.

[84] Martin MJ, Muotri A, Gage F, Varki A. Human embryonic stem cells express an immunogenic nonhuman sialic acid. Nat Med 2005;11:228−32.

[85] Cappola TP. Molecular remodeling in human heart failure. J Am Coll Cardiol 2008;51:137−318.

[86] Kuwahara K, Nishikimi T, Nakao K. Transcriptional regulation of the fetal cardiac gene program. J Pharmacol Sci 2012;119:198−203.

[87] Bartunek J, Vanderheyden M, Hill J, Terzic A. Cells as biologics for cardiac repair in ischaemic heart failure. Heart 2010;96:792−800.

[88] Nelson TJ, Behfar A, Yamada S, Martinez-Fernandez A, Terzic A. Stem cell platforms for regenerative medicine. Clin Transl Sci 2009;2:222−7.

[89] Menasché P, Hagège AA, Vilquin JT, Desnos M, Abergel E, Pouzet B, et al. Autologous skeletal myoblast transplantation for severe postinfarction left ventricular dysfunction. J Am Coll Cardiol 2003;41:1078−83.

[90] Malliaras K, Marbán E. Cardiac cell therapy: where we've been, where we are, and where we should be headed. Br Med Bull 2011;98:161−85.

[91] Thomas ED, Lochte HL, Lu WC, Ferrebee JW. Intravenous infusion of bone marrow in patients receiving radiation and chemotherapy. N Engl J Med 1957;257:491−6.

[92] Challen GA, Little MH. A side order of stem cells: the SP phenotype. Stem Cells 2006;24:3−12.

[93] Salem HK, Thiemermann C. Mesenchymal stromal cells: current understanding and clinical status. Stem Cells 2010;28:585−96.

[94] Miyamoto Y, Suyama T, Yashita T, Akimaru H, Kurata H. Bone marrow subpopulations contain distinct types of endothelial progenitor cells and angiogenic cytokine-producing cells. J Mol Cell Cardiol 2007;43:627−35.

[95] Kucia MJ, Wysoczynski M, Wu W, Zuba-Surma EK, Ratajczak J, Ratajczak MZ. Evidence that very small embryonic-like stem cells are molized into peripheral blood. Stem Cells 2008;26:2083−92.

[96] Kucia M, Reca R, Jala VR, Dawn B, Ratajczak J, Ratajczak MZ. Bone marrow as a home of heterogenous populations of nonhematopoietic stem cells. Leukemia 2005;19:1118−27.

[97] Sussman MA, Murry CE. Bones of contention: marrow-derived cells in myocardial regeneration. J Mol Cell Cardiol 2008;44:950−3.

[98] Stamm C, Westphal B, Kleine HD, Petzsch M, Kittner C, Klinge H, et al. Autologous bone-marrow stem-cell transplantation for myocardial regeneration. Lancet 2003;361:45−6.

[99] Strauer BE, Brehm M, Zeus T, Köstering M, Hernandez A, Sorg RV, et al. Repair of infarcted myocardium by autologous intracoronary mononuclear bone marrow cell transplantation in humans. Circulation 2002;106:1913−8.

[100] Lunde K, Solheim S, Aakhus S, Arnesen H, Abdelnoor M, Egeland T, et al. Intracoronary injection of mononuclear bone marrow cells in acute myocardial infarction. N Engl J Med 2006;355:1199−209.

[101] Seeger FH, Tonn T, Krzossok N, Zeiher AM, Dimmeler S. Cell isolation procedures matter: a comparison of different isolation protocols of bone marrow mononuclear cells used for cell therapy in patients with acute myocardial infarction. Eur Heart J 2007;28:766−72.

[102] Griesel C, Heuft HG, Herrmann D, Franke A, Ladas D, Stiehler N, et al. Good manufacturing practice-compliant validation and preparation of BM cells for the therapy of acute myocardial infarction. Cytotherapy 2007;9:35−43.

[103] Hermann PC, Huber SL, Herrler T, von Hesler C, Andrassy J, Kevy SV, et al. Concentration of bone marrow total nucleated cells by a point-of-care device provides a high yield and preserves their functional activity. Cell Transplant 2007;16:1059−69.

[104] Aktas M, Radke TF, Strauer BE, Wernet P, Kogler G. Separation of adult bone marrow mononuclear cells using the automated closed separation system Sepax. Cytotherapy 2008;10:203−11.

[105] Freyman T, Polin G, Osman H, Crary J, Lu M, Cheng L, et al. A quantitative, randomized study evaluating three methods of mesenchymal stem cell delivery following myocardial infarction. Eur Heart J 2006;27:1114−22.

[106] Losordo DW, Schatz RA, White CJ, Udelson JE, Veereshwarayya V, Durgin M, et al. Intramyocardial transplantation of autologous CD34 + stem cells for intractable angina. Circulation 2007;115:3165−72.

[107] Perin EC, Dohmann HFR, Borojevic R, Silva SA, Sousa ALS, Mesquita CT, et al. Transendocardial, autologous bone marrow cell transplantation for severe, chronic ischemic heart failure. Circulation 2003;107:2294−302.

[108] Stamm C, Kleine HD, Choi YH, Dunkelmann S, Lauffs JA, Lorenzen B, et al. Intramyocardial delivery of CD133 + bone marrow cells and coronary artery bypass grafting for chronic ischemic heart disease: safety and efficacy studies. J Thorac Cardiovasc Surg 2007;133:717−25.

[109] Bartunek J, Vanderheyden M, Vandekerckhove B, Mansour S, De Bruyne B, De Bondt P, et al. Intracoronary injection of CD133-positive enriched bone marrow progenitor cells promotes cardiac recovery after recent myocardial infarction: feasibility and safety. Circulation 2005;112:178−83.

[110] Fernández-Avilés F, San Román JA, García-Frade J, Fernández ME, Peñarrubia MJ, de la Fuente L, et al. Experimental and clinical regenerative capability of human bone marrow cells after myocardial infarction. Circ Res 2004;95:742−8.

[111] Schächinger V, Assmus B, Erbs S, Elsässer A, Haberbosch W, Hambrecht R, et al. Intracoronary infusion of bone marrow-derived mononuclear cells abrogates adverse left ventricular remodelling post-acute myocardial infarction: insights from the reinfusion of enriched progenitor cells and infarct remodelling in acute myocardial infarction (REPAIR-AMI) trial. Eur J Heart Fail 2009;11:973−9.

[112] Schächinger V, Erbs S, Elsässer A, Haberbosch W, Hambrecht R, Hölschermann H, et al. Intracoronary bone marrow−derived progenitor cells in acute myocardial infarction. N Engl J Med 2006;355:1210−21.

[113] Schächinger V, Erbs S, Elsässer A, Haberbosch W, Hambrecht R, Hölschermann H, et al. Improved clinical outcome after intracoronary administration of bone-marrow-derived progenitor cells

in acute myocardial infarction: final 1-year results of the REPAIR-AMI trial. Eur Heart J 2006;27:2775—83.

[114] Wollert KC, Meyer GP, Lotz J, Ringes Lichtenberg S, Lippolt P, Breidenbach C, et al. Intracoronary autologous bone-marrow cell transfer after myocardial infarction: the BOOST randomised controlled clinical trial. Lancet 2004;364:141—8.

[115] Meyer GP, Wollert KC, Lotz J, Steffens J, Lippolt P, Fichtner S, et al. Intracoronary bone marrow cell transfer after myocardial infarction: eighteen months' follow-up data from the randomized, controlled BOOST (BOne marrOw transfer to enhance ST-elevation infarct regeneration) trial. Circulation 2006;113: 1287—94.

[116] Huikuri HV, Kervinen K, Niemelä M, Ylitalo K, Säily M, Koistinen P, et al. Effects of intracoronary injection of mononuclear bone marrow cells on left ventricular function, arrhythmia risk profile, and restenosis after thrombolytic therapy of acute myocardial infarction. Eur Heart J 2008;29:2723—32.

[117] Traverse JH, Henry TD, Ellis SG, Pepine CJ, Willerson JT, Zhao DX, et al. Effect of intracoronary delivery of autologous bone marrow mononuclear cells 2 to 3 weeks following acute myocardial infarction on left ventricular function: the latetime randomized trial. JAMA 2011;306:2110—9.

[118] Traverse JH, Henry TD, Pepine CJ, Willerson JT, Zhao DX, Ellis SG, et al. Effect of the use and timing of bone marrow mononuclear cell delivery on left ventricular function after acute myocardial infarction: the time randomized trial. JAMA 2012;308:2380—9.

[119] Martin-Rendon E, Brunskill SJ, Hyde CJ, Stanworth SJ, Mathur A, Watt SM. Autologous bone marrow stem cells to treat acute myocardial infarction: a systematic review. Eur Heart J 2008;29:1807—18.

[120] Lipinski MJ, Biondi-Zoccai GGL, Abbate A, Khianey R, Sheiban I, Bartunek J, et al. Impact of intracoronary cell therapy on left ventricular function in the setting of acute myocardial infarction: a collaborative systematic review and meta-analysis of controlled clinical trials. J Am Coll Cardiol 2007;50:1761—7.

[121] Abdel-Latif A, Bolli R, Tleyjeh IM, Montori VM, Perin EC, Hornung CA, et al. Adult bone marrow—derived cells for cardiac repair: a systematic review and meta-analysis. Arch Intern Med 2007;167:989—97.

[122] Tse HF, Thambar S, Kwong YL, Rowlings P, Bellamy G, McCrohon J, et al. Prospective randomized trial of direct endomyocardial implantation of bone marrow cells for treatment of severe coronary artery diseases (PROTECT-CAD trial). Eur Heart J 2007;28:2998—3005.

[123] van Ramshorst J, Bax JJ, Beeres SL, Dibbets-Schneider P, Roes SD, Stokkel MP, et al. Intramyocardial bone marrow cell injection for chronic myocardial ischemia: a randomized controlled trial. JAMA 301:1997—2004.

[124] Perin EC, Willerson JT, Pepine CJ, Henry TD, Ellis SG, Zhao DX, et al. Effect of transendocardial delivery of autologous bone marrow mononuclear cells on functional capacity, left ventricular function, and perfusion in chronic heart failure: the FOCUS-CCTRN trial. JAMA 2012;307:1717—26.

[125] Donndorf P, Kundt G, Kaminski A, Yerebakan C, Liebold A, Steinhoff G, et al. Intramyocardial bone marrow stem cell transplantation during coronary artery bypass surgery: a meta-analysis. J Thorac Cardiovasc Surg 2011;142:911—20.

[126] Wen Y, Meng L, Xie J, Ouyang J. Direct autologous bone marrow-derived stem cell transplantation for ischemic heart disease: a meta-analysis. Expert Opin Biol Ther 2011;11:559—67.

[127] Reffelmann T, Könemann S, Kloner RA. Promise of blood- and bone marrow-derived stem cell transplantation for functional cardiac repair: putting it in perspective with existing therapy. J Am Coll Cardiol 2009;53:305—8.

[128] Montalescot G, Barragan P, Wittenberg O, Ecollan P, Elhadad S, Villain P, et al. Platelet glycoprotein IIb/IIIa inhibition with coronary stenting for acute myocardial infarction. N Engl J Med 2001;344:1895—903.

[129] Stone GW, Grines CL, Cox DA, Garcia E, Tcheng JE, Griffin JJ, et al. Comparison of angioplasty with stenting, with or without abciximab, in acute myocardial infarction. N Engl J Med 2002;346:957—66.

[130] Gnecchi M, Zhang Z, Ni A, Dzau VJ. Paracrine mechanisms in adult stem cell signaling and therapy. Circ Res 2008;103: 1204—19.

Part V

Small Bowel

Chapter 32

Current Status of Intestinal Transplantation

Gennaro Selvaggi[a] and Andreas Tzakis[b]
[a]*University of Miami Miller School of Medicine, Miami Transplant Institute, Miami, FL,* [b]*Cleveland Clinic Foundation, Weston, FL*

Chapter Outline

32.1 INTRODUCTION

The history of intestinal transplantation goes back to the 1950s when Dr. Lillehei described the first isolated intestine transplant in a dog model [1]. Many years would pass before implementation of such a model would be possible in humans, mainly due to the significant obstacles posed by severe rejection and infections in the recipients. The first report of a successful isolated intestinal transplant was from the group led by Dr. Goulet in 1989, while the first successful long-term survivor from a multivisceral transplant was described by Dr. Margreiter in Germany in 1992 [2,3]. However, the field of clinical intestinal transplantation was greatly expanded in the late 1980s by the Pittsburgh group led by Dr. Thomas Starzl [4]. His original concept was based on the premise that the gastrointestinal tract is comprised of a "cluster" of functional units where each unit or combination of units can be transplanted as a whole or as parts of the cluster according to each patient diagnosis and requirements [5,6]. The early experience in Pittsburgh was soon followed by many other centers around the United States and the world, greatly expanding clinical intestinal transplantation to include a total number of over 35 current active centers throughout the world. In the United States

alone, over 2200 intestinal transplants have been performed since the year 1990, with over 20 active centers [7]. Moreover, the Center for Medicare and Medicaid Services approved funding for intestinal transplantation in the United States in the year 2000, further validating such procedure as scientifically sound [8]. Despite such remarkable clinical success, intestinal transplantation is still performed only in highly specialized centers and in overall low volumes as compared to any other solid organ. This is undoubtedly due to a limited number of patients with irreversible intestinal failure, but more importantly to the significant amount of posttransplant complications which make management of these patients quite complex and result in significant morbidity and mortality. The main interest of basic and clinical research, therefore, has been toward the minimization of such complications, first and foremost rejection. Currently, most studies are focused in three directions: implementation of immunosuppressive protocols which can prevent rejection without being too aggressive and cause infectious complications; search for a noninvasive marker of intestinal rejection that can replace the need for endoscopy and biopsy; lastly, an increased interest in the contribution of humoral immunity to the process of rejection, in terms of prevention, early diagnosis, and treatment.

Regenerative Medicine Applications in Organ Transplantation.

In this chapter, we will give a broad update on the current status of intestinal transplantation, with particular interest in new areas of development which in the future will interact with the upcoming contributions from the field of regenerative medicine and bio-artificial organ design.

32.2 DIAGNOSES AND INDICATIONS FOR INTESTINAL TRANSPLANT

The main reason to refer a patient for intestinal transplant is the development of irreversible intestinal failure with life-threatening complications [9,10]. Broadly speaking, patients who require intestinal transplant can be divided into three categories: patients in whom intestinal function is lost because of loss of bowel length (short gut); patients in whom the intestine presents functional abnormalities which preclude adequate absorption of nutrients (dysmotility or mucosal abnormalities); and patients with tumors or inflammatory bowel diseases (Table 32.1). The original pathological insult can be congenital (e.g., gastroschisis, intestinal atresia), developed over time (e.g., dysmotility disorders or Gardner's syndrome) or acquired (e.g., mesenteric thrombosis, trauma, tumors). Pediatric patients often present specific diagnoses which are different than those in adults. Specifically, in pediatric patients the most common causative diseases are gastroschisis, necrotizing enterocolitis, volvulus, and intestinal atresia in the short gut category; dysmotility disorders (Hirschsprung's disease) and mucosal disorders (microvillous inclusion

TABLE 32.2 Indications for Intestinal Transplantation

Accepted indications	Loss of vascular access for parenteral nutrition
	Development of parenteral nutrition-induced liver dysfunction
	Repeated episodes of line infection
	Repeated episodes of dehydration
Other indications	Extreme short bowel syndrome
	Intestinal dysmotility disorders
	Congenital mucosal absorption diseases (microvillous inclusion disease)
	Diffuse portomesenteric thrombosis
	Unresectable tumors

disease) in the functional category. Conversely, in adult patients, mesenteric ischemia, trauma, volvulus, and Crohn's disease are leading causes of shortened intestine, while tumors and dysmotility disorders are prevalent in the other categories. Lastly, re-transplantation accounts for about 8−10% of reasons for intestinal transplant, with rejection (chronic or acute) being the main responsible for graft loss in such cases.

When a patient has developed irreversible intestinal failure, there are some indications that most experts agree on, which should lead clinicians to consider listing patients for transplant (Table 32.2). Such indications include life-threatening complications such as impending loss of vascular access for parenteral nutrition, development of liver dysfunction related to long-term parenteral nutrition, repeated episodes of central line infection, and/or repeated episodes of dehydration as complications of parenteral nutrition [9,10]. Other less stringent indications are illustrated in Table 32.2, where there is no immediate danger for the patient's life, but where current medical care fails to restore appropriate function to the intestine [11]. Such indications include mucosal disorders for which no cure is available, diffuse portomesenteric thrombosis, and tumors at the base of the mesentery which cannot be removed using conventional surgical techniques.

32.3 INTESTINAL REHABILITATION

Before a patient can be considered for intestinal transplant, every attempt to promote or restore function of the native intestine needs to be exhausted. This is particularly true for those patients with short gut syndrome, where the length of the intestine has been shortened by surgical resection. In such patients, there is a possibility of avoiding intestinal transplant if the residual intestine can go through a phase of adaptation where eventually partial or full function can be reestablished.

Each patient has to be evaluated by a multidisciplinary team that includes transplant and pediatric surgeons,

TABLE 32.1 Diagnoses Leading to Intestinal Transplant

Short Gut	Functional	Other
Congenital	Mucosal abnormalities	Tumors
Intestinal atresia	Microvillous inclusion disease	Malignant
Volvulus/ malrotation	Secretory diarrhea	Benign
Gastroschisis	Pseudo obstruction	Gardner's syndrome
Acquired	Neurogenic	
Necrotizing enterocolitis	Myogenic	
Trauma	Megacystis microcolon	
Mesenteric thrombosis	Hirschsprung's disease	
Multiple resections		
Inflammatory bowel disease		
Crohn's disease		

gastroenterologists, interventional radiologists, nutrition specialists, and parenteral nutrition pharmacists [12,13].

The main areas of work in intestinal rehabilitation are enteral nutrition protocols, optimization of parenteral nutrition with minimization of the risks for hepato-toxicity, surgical bowel lengthening and reconnecting procedures, strategies in central venous lines placement and management; avoidance of episodes of sepsis and dehydration episodes [14–16].

If all the above mentioned measures fail, then referral for transplant is appropriate. At the same time, it is critically important that transplant surgeons be involved from the beginning, in order to avoid late referrals, such as when liver function has already deteriorated or venous access sites are already exhausted. Each patient should be followed longitudinally by the intestinal rehabilitation team with clinical reassessment every few months.

It also has to be remembered that mortality on the intestinal transplant list is low if a patient requires an isolated intestine transplant, but patients requiring a liver and intestinal graft have a much higher mortality than patients requiring a liver or an intestinal graft alone [17]. This holds especially true for pediatric patients. For this reason, current allocation protocols were modified to favor patients listed for liver and intestine graft by giving them more points on the liver transplant waiting list (Model of End-Stage Liver Disease). This resulted in a decrease in mortality of patients on the list which is now comparable to that of other solid organs [18].

32.4 TYPES OF INTESTINAL GRAFTS

Nomenclature in intestinal transplantation has been in constant evolution, sometimes due to the different types of organs that can be included in the graft, at other times depending on which native organ(s) would be left behind. Personal choices by the surgeons and center-specific preferences have also caused a proliferation of different techniques for similar types of procedures. As an example, the United Network of Organ Sharing does not have a category for transplantation of the stomach, and there is not a current procedural terminology for stomach transplant. Similarly, inclusion of the donor large intestine or the donor spleen, for example, is not routinely implemented by all transplant centers, with each center adopting a specific philosophy on when to include such organs as part of the intestinal or multivisceral graft. Generally speaking, most authors agree on the following definitions: *isolated intestine* (small ± large intestine) includes a graft comprised of only the small bowel (with or without the ileocecal valve and ascending colon) and whose vascular support comes from the superior mesenteric artery and vein. A *liver–intestine* graft can be composite or noncomposite; a composite liver–intestine graft includes the liver and the intestine (including the duodenum and pancreas), with a common vascular supply from the aortic patch that includes the celiac axis and the superior mesenteric artery. In this case the arterial supply to the liver and the portomesenteric venous axis of the liver are left untouched, and biliary drainage can be anatomical, if the duodenum and a part of the head of the pancreas are included. In a noncomposite liver–intestine graft, the liver and the intestine are implanted separately, each with its own vascular supply. In such cases, arterial supply to the liver comes from the celiac axis while arterial supply to the intestine is from the superior mesenteric artery. Portal vein flow to the liver comes from the native duodenum, pancreas, spleen, and stomach while portal venous drainage from the intestinal graft can be systemic (more commonly) to the inferior vena cava or indirect to the native portal system through a venous extension graft; biliary reconstruction is done with a hepato-jejunostomy. The main reason to perform a noncomposite liver–intestine graft is size discrepancy, especially when the liver graft has to be reduced in size; second, the noncomposite graft allows for the possibility of removing the intestinal graft alone if severe rejection were to occur without the need to sacrifice the liver graft at the same time. A *multivisceral* graft always includes the liver, pancreas, stomach, and small intestine; in many cases, the spleen and the large intestine are included in the graft. Arterial supply for a multivisceral graft comes from an aortic patch or conduit that includes both celiac axis and superior mesenteric artery, while venous drainage is from the hepatic veins. Lastly, a *modified multivisceral* graft is a multivisceral graft minus the liver; the transplant therefore includes the stomach, pancreas, small intestine, again with or without spleen and large intestine. Arterial supply again comes from the abdominal aorta conduit with celiac and mesenteric artery, while venous outflow is via the portal vein above the superior edge of the pancreas into the native liver portal vein.

32.5 SURGICAL PROCEDURE

Even if living-related donation for intestinal transplant has been described, the majority of intestinal transplant recipients receive organs which have been retrieved from brain dead deceased donors. Because of ischemia–reperfusion issues, there has been no significant use of donors after cardiac death. Organ retrieval techniques have been left relatively unchanged over the last few years [19,20]. Briefly, a cruciate abdominal incision is carried out and organs are exposed and mobilized. Attention is given to identify possible anatomical variants in the liver vascular supply and particular care is given to pediatric donors, especially if they weigh less than 10 pounds. The abdominal aorta is cannulated; after heparin has been given all abdominal organs are perfused with preservation solution. University of Wisconsin (UW) solution is more commonly used, but good results have also been described with histidine-trypotphane-ketoglutarate solution [21]. If the graft is a multivisceral type, all abdominal organs are removed together save for the kidneys. If an isolated intestinal graft is retrieved, division of the superior mesenteric vessels is performed at the lower border of the pancreas head. In a liver—intestine graft type, various techniques can be used, and often division of the organs is performed later at the back table. Intestinal grafts can be safely kept in cold ischemia for over 8 h and in a few cases up to 12 h. After that, the risk for ischemia—reperfusion injury and graft dysfunction rises significantly.

The recipient operation in intestinal transplant is quite often very challenging. Many patients come to transplant after multiple abdominal operations, with resulting adhesions, loss of abdominal domain in case of resections, presence of stomas, and feeding access ports (gastrostomies, colostomies, ileostomies, and so forth), as well as entero-cutaneous fistulas. Lastly, some patients present with abdominal wall tumors (such as desmoid tumors), which require extensive resection of the abdominal wall tissues. In those patients with liver failure there is significant portal hypertension, where all adhesions become vascularized and bleed easily because of coagulopathy. Access to the abdomen, therefore, is quite difficult and extensive blood loss is often unavoidable. Removal of native diseased organs is the first part of the operation, isolation of the target vessels for revascularization is the second step, vascular and gastrointestinal reconstruction is the third step, and abdominal closure is the last step. In isolated intestinal transplantation, the native intestine has often already been removed, while in liver—intestine and multivisceral transplantation not only the intestine but also the liver need to be removed. In composite liver—intestine transplant, the native stomach—pancreas—duodenum—spleen complex is preserved and drainage of the portal venous flow from such organs needs to be preserved either via a portosystemic shunt or by a portoportal connection [22]. In multivisceral transplant, devascularization of the celiac and superior mesenteric districts is necessary, and it is accomplished by clamping and subsequent over sowing of the vessels close to the aorta [23]. Technically more challenging is a modified multivisceral transplant, where the arterial supply to the liver has to be preserved with careful dissection of the different branches of the celiac axis, whereas the superior mesenteric artery is ligated. When multiple adhesions make mobilization of the native organs challenging, it is advisable to perform devascularization of the organs first and then proceed with the intestinal resection and the hepatectomy (if indicated). Contamination of the surgical field is often impossible to prevent due to the presence of ostomies, friable bowel, and

severe adhesions; it is quite important to perform thorough irrigation of the surgical field to minimize the risk of infection of the vascular anastomotic lines.

Isolation and exposure of the target vessels for revascularization are accomplished on the arterial side by exposing the infra-renal abdominal aorta. Rarely the native superior mesenteric artery can be used in isolated intestinal transplants. Venous drainage in isolated intestine grafts and in noncomposite liver—intestine grafts is through the portal system via a venous graft or directly onto the superior mesenteric vein. Alternatively, the venous drainage can be systemic onto the inferior vena cava. Such type of systemic drainage has been successfully utilized without a significant increase in morbidity or complications [24]. Modified multivisceral grafts drain portal outflow directly onto the native liver portal vein. Full multivisceral grafts and liver—intestine composite grafts provide venous outflow via the hepatic veins anastomosis which is almost always done piggyback style after native hepatectomy.

After the graft is reperfused, reconstruction of the intestinal continuity and biliary drainage are performed. Proximal continuity is reestablished as follows: in isolated intestine grafts the proximal jejunum is connected to the native jejunum or duodenum; in liver—intestine grafts there are multiple variants depending on the presence or not of the duodenum; in multivisceral and modified multivisceral grafts a gastrogastric anastomosis is performed. Pyloromyotomy or pyloroplasty is always performed when the stomach is part of the graft since total vagotomy is an anatomical necessity in gastric transplant. Distal gastrointestinal reconstruction always involves the creation of an ileostomy in all patients in order to be able to monitor the intestinal graft via endoscopy in the postoperative period. The native colon is reconnected to the distal transplant ileum or transplant colon, so to allow regular bowel movements once the ileostomy is closed.

Closure of the abdomen is another delicate and difficult step. Many patients cannot be closed primarily at the end of the transplant procedure. Reasons include loss of abdominal domain prior to the transplant, the presence of bowel edema at the end of the procedure, need for surgical resection of parts of the abdominal wall in case of tumors like desmoids, size discrepancy between donor and recipient body weight, and so forth. Since primary closure of the abdomen is not always possible, many strategies have been devised to overcome such obstacle. Temporary silicone or plastic meshes can be used and removed a few days later when edema has subsided, allowing for delayed primary closure [25]. Permanent mesh placement to bridge fascial gaps and/or to enlarge abdominal domain can be done with artificial nonabsorbable or absorbable material. Biological meshes have also been utilized with the advantage of a decrease in infection episodes [26]. Whenever possible, skin closure should be attempted by mobilizing and developing skin and subcutaneous flaps to avoid exposure of the mesh to open air. This decreases the chances for infection and fistula formation, especially when nonabsorbable mesh is used. In complex, extreme cases, transplantation of the abdominal wall as a vascularized, composite tissue graft has been described, as well as the use of the nonvascularized or vascularized fascia of the rectus muscle sheath [27,28].

32.6 IMMUNOSUPPRESSION

The intestinal graft presents a formidable challenge in terms of antigenic load and disparity of tissues that can be target for rejection, as well as the large amount of immune-competent lymphatic tissue residing within the mesentery which can trigger an immune interaction with the recipient's own immune system. Since the inception of the field of intestinal transplantation and still to this date, rejection presents the single most difficult complication to control, even though great progress has been made by means of new protocols for immunosuppression [29–32].

As new drugs have become available and experience in the field has evolved, it is now clear that the best results in terms of prevention of rejection in the postoperative period are achieved by the use of induction treatment [18,31–33]. All intestinal transplant centers currently use induction therapy, with most centers using anti-lymphocyte agents such as ATG, thymoglobuline, or alemtuzumab (Campath 1H), and a few centers using anti-IL-2 receptor antibodies such as simulect or daclizumab. Steroids are also routinely used for the first few days after transplant; however, most clinical protocols have abandoned prolonged use of steroids, and weaning is performed within a few days or a month after transplant. In addition to induction immunosuppression, maintenance immunosuppression is primarily achieved with calcineurin inhibitors, such as tacrolimus (and less often cyclosoporin A). Antimetabolite drugs such as mofetil mycophenolate or mycophenolate sodium can be added to the maintenance regimen. Target of rapamycin inhibitors such as sirolimus or everolimus are used as additional agent or in cases of significant nephrotoxicity from calcineurin inhibitors.

The most important consequence of the use of appropriate induction therapy is the possibility for a significant decrease in the doses of immunosuppression if no rejection episode is experienced in the first few months after transplant. In such cases, tacrolimus doses can be slowly weaned to achieve rejection-free maintenance of the graft with spaced dosing and very low trough drug levels [33].

When rejection develops, however, treatment needs to be instituted immediately since any delay for more than 24 h puts the graft at significant risk. Mild rejection can be treated with a steroid pulse and weaning cycle plus an increase in baseline tacrolimus immunosuppression; moderate and severe rejection episodes need to be treated with anti-lymphocyte agents.

It has become more apparent in the last few years that many episodes of rejection are associated with the development of high levels of donor-specific antibodies (DSA) [34–38]. This underlies an important, previously unrecognized contribution of the humoral response by the immune system during intestinal graft rejection. Because of that, current protocols for the treatment of rejection episodes in which elevated levels of DSA are detected include the use of medications that specifically target antibody-mediated response (humoral rejection). Such agents can be pharmacological such as monoclonal antibodies against B-lymphocytes (rituximab) or immune therapy such as intravenous immunoglobulins (IvIg) infusions and plasma exchange (plasmapheresis); newer approaches involve the use of proteasome inhibitors such as bortezomib [39].

32.7 MONITORING OF THE GRAFT

The gold standard for monitoring the intestinal graft is endoscopy with mucosal tissue biopsy. Such procedure is performed frequently (as often as twice a week) during the immediate postoperative period, and later as needed based on clinical suspicion for rejection. Histopathology analysis of the biopsies allows for identification of rejection and its severity, as well as differential diagnosis (for instance viral infections). Whenever the size of the graft allows, the use of zoom endoscopy allows visual magnification analysis with visualization of the intestinal villi structure which can provide valuable details on the mucosa appearance and has been shown to correlate well with pathology results [40].

There is no current laboratory marker that can precisely evaluate the intestinal function in a transplant graft such as there is for a kidney transplant (creatinine level) or a liver transplant (transaminases or bilirubin). Noninvasive markers for the detection of intestinal rejection are therefore an area which has been extensively studied lately though not yet widely utilized. Putative tests that have shown good promise are citrulline levels in the blood and calprotectin levels in the stool. Of these, citrulline has been studied the most. Levels of citrulline drop in the blood immediately after intestinal transplant due to the ischemia–reperfusion injury that the graft sustains during surgery. However, over time citrulline levels rise toward normal levels and stabilize to a certain baseline for each patient; a sudden drop in citrulline values in stable patients after transplant has been shown to be highly associated with episodes of acute cellular rejection (ACR), especially moderate and severe rejection [41,42]. The advantage of such a test is that it can also be performed with dry blood spot collection technique, thus facilitating collection and allowing patients who live far from the transplant center to mail their test specimens. If a stable patient exhibits a significant decrease in citrulline levels, they should be called to the transplant center and serious consideration should be given to perform endoscopy to rule out rejection.

Most recent papers have focused attention on the longitudinal monitoring of DSA levels. A sudden increase in such levels is strongly associated with development of rejection and therefore it should alert the clinician to implement endoscopy or increase baseline immunosuppression [37,38]. Additionally, patients with newly developed DSA are at a higher risk for rejection-related graft loss.

The future in graft monitoring comes from more sophisticated molecular biology techniques, such as gene expression analysis, microarray studies as well as metabolomics and proteomics techniques. These studies are still quite in the early phase but provide insights into the early events in rejection, in some cases before tissue damage has occurred [43,44].

32.8 COMPLICATIONS

32.8.1 Rejection

Intestinal graft rejection is still one of the most difficult challenges to the posttransplant management of recipients. As previously indicated, the intestine graft is made up of multiple cell types and tissues as well as containing a large amount of donor-derived immune-competent cells within the gut mucosal lymphatic system and its large mesenteric lymph node load. Interactions and antigen presentation between the donor and the recipient immune systems are therefore inevitable. The site of organ damage is primarily the intestinal mucosa, whereas the submucosa and serosa are relatively spared during acute rejection. Damage at the level of the arteries or arterioles occurs only in cases of hyperacute (antibody-mediated) rejection or during the development of chronic rejection. It is well known that the most sensitive area for the development of rejection is the distal ileum, where the largest amount of donor lymphoid tissue is present (Peyer's patches). Mucosal injury can be diffuse but at times it can present with skip areas of normal mucosa in between; it is therefore important to perform diagnostic endoscopy and evaluate a long segment of intestine in order not to miss a more proximal injury.

It is well known that the presence of a liver graft in multivisceral patients confers a significant degree of protection from rejection [44–46]. The true reasons behind such protective effect are still unknown, but recipients of liver–intestine and multivisceral grafts clearly experience significant decreased risk from developing rejection than recipients of isolated intestinal grafts. Additionally, in one study, recipients of modified multivisceral grafts also seemed to experience less rejection episodes than isolated intestine recipients, pointing to a possible protective effect of a multivisceral graft as well.

Rejection can be classified into hyperacute, acute cellular, and chronic.

Hyperacute rejection is rarely seen, usually in patients who are presensitized (e.g., re-transplant recipients), and it is characterized by immediate graft discoloration, swelling, and dysfunction on the operating table after reperfusion. The damage is at the endothelium level, caused by antibody fixation and deposition endothelium, with resulting micro- and macro-thrombosis of the graft [36]. It is possible to treat hyperacute rejection in the operating room, but most times the graft needs to be removed.

The most common form of rejection is ACR. ACR develops over the course of days and can happen any time after transplant, even though most episodes are observed during the first year [47]. ACR historically occurred in as many as 70% of patients, but improvements in induction therapy have decreased the rates to below 40%. Signs and symptoms of rejection include fever, a sudden increase in stoma output (or diarrhea), abdominal pain, and distension. Sepsis can develop rapidly, especially when the mucosal integrity has been violated and bacterial translocation occurs.

The most severe form of ACR is called exfoliative rejection, in which the process is so severe that the intestinal mucosa sloughs off [48]. In such cases, bloody stoma output and passage of gelatinous casts of intestinal mucosa occur.

In any case of suspected ACR endoscopy needs to be performed as soon as possible in order to obtain mucosal biopsies and assess the degree of rejection as well as differentiate from other forms of enteritis.

ACR is graded by pathological evaluation into indeterminate, mild, moderate, and severe, based on mucosal evaluation of the villous structure and the integrity of the crypts. The criteria are described in Table 32.3[49].

Subclinical rejection is defined when patients are diagnosed by biopsy with indeterminate or mild rejection but clinically have no symptoms, and therefore are usually not treated pharmacologically. These patients, however, are at higher risk for graft loss, usually from infectious complications [50].

Significantly associated with development of ACR is the risk for superimposed infections: during the rejection

TABLE 32.3 Grading of Rejection Based on Pathology Findings

Grade of Rejection	Inflammatory Infiltrate	Epithelial Injury	Apoptotic Bodies in Crypts	Mucosal Damage
Indeterminate	Minor	Minor	Less than 6	Intact mucosa
Mild	Mixed mononuclear infiltrate	Villous blunting and distortion	More than 6, less than 10	Edema and vascular congestion
Moderate	Heavy infiltrate	Erosion of surface, loss of villi	More than 10 apoptotic bodies, occasional confluent apoptosis, focal crypt loss	Significant edema and congestion
Severe	Granulation tissue with fibropurulent exudate	Loss of villi and architecture	Marked crypt damage and loss	Loss of mucosa with sloughing

episode, the integrity of the intestinal mucosa is violated and bacterial translocation is almost always present. Sepsis therefore can accompany symptoms or even be a heralding sign of a rejection episode. Most common involved agents are enteric bacteria, but fungal sepsis is also common. For this reason, especially in cases of moderate and severe rejection, empiric antibiotic and antifungal treatment are often used as part of the protocols of treatment of rejection.

Additionally, as a consequence of a rejection episode, the use of high doses of immunosuppression exposes patients to posttreatment infectious complications, especially viral infections such as from cytomegalovirus (CMV) and Epstein—Barr virus (EBV). The associated risk is so high that patients being treated for moderate or severe rejection should be placed on prophylaxis with antiviral medications such as ganciclovir.

Treatment of rejection consists of an increase in immunosuppression and supportive therapy. Supportive therapy includes treatment of dehydration, parenteral nutrition if the intestinal damage is severe, antibiotic therapy to treat infections. Immunosuppression treatment consists primarily of an increase of baseline immunosuppression and pulse therapy with steroids for mild episodes of rejection and use of anti-lymphocyte agents for moderate and severe rejections. In those patients in whom elevated levels of DSA are detected or in whom a humoral component of rejection is suspected, therapy with rituximab, plasma exchange, and IvIg should be considered.

If the patient does not respond to treatment, graft removal has to be considered so that immunosuppression can be halted and the patient's life saved. Graft enterectomy should be considered in patients in whom no improvement in the intestinal mucosa is detected by endoscopy after completion of therapy, as well as in patients where the intestinal graft becomes a source of persistent sepsis. Removal of the intestinal graft is not an easy surgical procedure, due to multiple adhesions and altered anatomy, but it can be facilitated in some cases by preoperative arterial embolization of the mesenteric artery of graft done in interventional radiology, in order to decrease bleeding and easily demarcate native from transplanted bowel [51].

Chronic rejection of intestinal grafts has also been described. Risk factors for developing chronic rejection are previous repeated episodes of ACR, length and duration of rejection episodes, previous infections especially viral infections [31,52,53]. Symptoms are usually slow to progress and include abdominal distension, chronic diarrhea with malabsorption and dehydration episodes. The intestine can develop areas of stricture with subocclusion symptoms. The bowel wall becomes thick, with tissue damage mostly localized in the submucosa (loss of crypts, fibrosis) and muscolaris mucosae, due to the development of arterial vasculopathy. For such reason, a mucosal biopsy can not always fully appreciate the extent of injury, and often chronic rejection is diagnosed on explanted specimens [52,53].

32.8.2 Infections

Infectious complications are overall the most common cause of morbidity and mortality in recipients of intestinal transplants. As mentioned, the intestine is a clean contaminated graft with a large physiological load of bacteria and other pathogens within its lumen. At time of transplant, ischemia—reperfusion injury causes disruption of the mucosal barrier; just as well any episode of rejection or an enteric viral infection can easily lead to translocation. The high level of immunosuppression required to prevent rejection also lends to a more favorable ground for development of infections. Gut decontamination protocols are implemented during the first few weeks after transplant by using enteral administration of nonabsorbable antibiotics, though restoration of normal intestinal flora is the goal after the initial reperfusion-injury phase has resolved.

Timing of infectious episodes shows that most episodes occur during the first months after transplant, then taper off as immunosuppression is decreased and the incidence of rejection decreases [53,54]. Incidence of viral and respiratory infections is much higher in pediatric patients, especially those younger than 1 year of age [55].

Infections can be localized or systemic, the latter often a consequence of the former (such as septicemia from a central venous catheter infection site). Localized infections are commonly observed in the respiratory system (pneumonia), urogenital system (urinary tract infections), intraabdominal (abscess, anastomotic leak), and gastrointestinal (viral, bacterial, or fungal) where the graft becomes at the same time the source and the target of the pathology. Systemic infections are most often bacterial with either enteric or respiratory flora.

Systemic fungal infections are less common, though more difficult to treat [56]. Again, patients are usually placed on fungal prophylaxis for the first few weeks, but susceptibility to fungal infections is still high. Most commonly observed pathogens are *Candida* and *Aspergillus* species, though *Cryptococcus* and other opportunistic species have also been described. Tissue invasive fungal infections are also possible, such as *Rhizopus* or *Mucor* species in soft tissues (sinuses, limbs) or *Histoplasma* and *Coccidiomycosis* in the lung or central nervous system (CNS).

Viral infections with CMV and EBV can occur either as reactivation of preexisting latent status or as *de novo* infection from the graft or from blood and blood products transfusions [57,58]. CMV prophylaxis with ganciclovir is routinely performed in all intestinal transplant recipients, with the use of hyperimmune-specific IvIg (Cytogam) reserved in cases at highest risk of transmission (positive donor to negative recipient).

Viral infections with enteric viruses are also quite common and can mimic signs and symptoms of rejection; endoscopy with biopsy and dedicated microbiology/immunology assays can help differentiate between a viral infection and a rejection episode, since they both can present with diarrhea as a heralding symptom.

Opportunistic infections with unusual pathogens are also possible, just like in all other types of solid organ transplants, such as with atypical *Mycobacteria*. Prophylaxis for *Pneumocystis* and *Nocardia* species is given to all patients with trimethoprim—sulfamethoxazole.

The most important clinical guidelines for the management of infections in intestinal transplant recipients are a low threshold of suspicion, early and repeated cultures, and adequate use of antibiotics with the help of the infectious disease specialists. Recently, the availability of short turnaround time polymerase chain reaction (PCR) testing has greatly reduced the waiting time for diagnosis of viral infections, as well as providing quick testing for a multitude of common and less common pathogens which in the past may have been underdiagnosed.

32.8.3 Graft Versus Host Disease

The presence of a large load of immune-competent, donor-derived lymphocytes in the graft increases the chances of graft versus host disease (GVHD) in recipients of intestinal transplant. Migration of donor cells in the bloodstream can be measured by the assessment of chimeric levels in the peripheral blood. Tissue damage occurs primarily in the skin, liver (native), lung, native intestine, and the recipient's bone marrow compartment. As a consequence, patients first develop a desquamating rash localized first on the palms and foot soles and then on the trunk, then diarrhea can be present if the native large intestine is affected. Late, serious signs of advanced GVHD include cholestasis (if liver is not part of the graft), lung fibrosis, but most important, however, is the involvement of the hematopoietic tissue with severe suppression of the host immune system and development of aplastic anemia. Patients who develop bone marrow suppression as a consequence of GVHD often succumb to disseminated infections.

Incidence of GVHD varies according to different series between 5—15%, with pediatric recipients being affected more frequently [59,60]. Presence of native spleen protects from GVHD while native splenectomy or inclusion of the donor spleen in the graft increases its incidence. Mortality from GVHD can be as high as 73%. Treatment is quite difficult because it involves the use of higher doses of immunosuppression. Topical therapy with tacrolimus can be used for mild forms of limited skin involvement but for more severe manifestations steroids first and then anti-lymphocyte agents need to be used. This unfortunately exposes the recipients to possible infections, especially in the setting of aplastic anemia. Acute GVHD can occasionally evolve into a more chronic form with skin acanthosis, diarrhea of malabsorption, cholestasis, and dry conjunctivitis.

32.8.4 Posttransplant Lympho-Proliferative Disease

Posttransplant lympho-proliferative disease (PTLD) encompasses a spectrum of multiple pathologies, all linked by the common thread of abnormal proliferation of lymphocytes in the transplant recipients related to the use of immunosuppression [61]. The main pathological event is usually driven by viral infection, mainly EBV infection, in the recipient lymphocytes. The most commonly affected are B-lymphocytes, though T-lymphocyte proliferation is observed in about 30% of cases, but in such cases usually EBV virus is not primarily involved. Risk factors for the development of

PTLD are previous episodes of rejection, concurrent viral infections, pediatric age, and pretransplant EBV seronegative status (as compared to preexisting EBV positive status) [62]. Timing of development of PTLD is usually within the first year after transplantation, corresponding to the period when the highest immunosuppression is onboard, as well as when the frequency of rejection episodes is at its highest.

PTLD is by definition a systemic disease affecting not only all lymphnodal stations in the body (mesenteric, axillary, neck, inguinal) but also the intestinal submucosa, organs such as liver and lungs, and in the most advanced case the CNS.

At the mild end of the spectrum of the disease there are flu-like syndromes, similar to mononucleosis, with generalized malaise, low-grade fever, localized lymphadenopathy, and occasional liver enzymes abnormalities. Moderate forms of the disease present with generalized lymphadenopathy and involve the intestine with diarrhea and weight loss. Severe forms of PTLD are comparable to aggressive lymphomas with multiorgan involvement of lungs, liver, and CNS. Solid tumors in the intestinal wall can cause obstruction, perforation, or massive bleeding, requiring surgical exploration.

Diagnosis of PTLD is based on clinical symptoms, laboratory measurements of EBV viremia, radiological interpretation of whole body computerized tomography scans, and mainly histological analysis of intestinal mucosa, lymph nodes, and other tissues biopsies. Special stains for EBV virus and lymphocyte markers (such as CD20 for B cells) are routinely utilized in the pathology laboratory.

Therapy of PTLD is based on severity of disease but includes three main actions: decrease in immunosuppression, treatment of EBV infection, and chemotherapy. A decrease in baseline immunosuppression is necessary to allow the patient's own immune system to fight the viral infection; the reduction cannot be too drastic, especially in intestinal transplant recipients, since rejection is always a possibility. Therapy for EBV is based on ganciclovir and Cytogam, given for prolonged periods of time and with longitudinal monitoring of EBV quantitative PCR in blood. Chemotherapy for B-cell forms of PTLD is based on rituximab if the tissue biopsy shows the lymphatic tissue to be positive for CD20 cells [63]; alternatively, chemotherapy for lymphoma is utilized based on current clinical protocols (e.g., cyclophosphamide, hydrodoxorubicin, oncovin, and prednisolone, CHOP). Early involvement of the hematology/oncology specialist helps in determining which best strategy to use.

Mortality from PTLD in intestinal transplant recipients used to be as high as 40% before the introduction of rituximab; it now ranges around 10−20% [18,23,31]. T-cell forms of PTLD are more aggressive and respond less successfully to chemotherapy.

32.9 RESULTS

As reported by the Intestinal Transplant Registry report at the end of 2011, just over 2600 intestinal transplants have been reported in the world, of which 1400 were in pediatric recipients [64]. In the United States, there has been a slight decrease in the number of cases performed over the last 3 years; the largest number of recipients was in 2007, with 198 patients, while in 2011 only 129 patients received an intestinal graft. Interestingly, while up to 2007 there was a majority of cases performed in pediatric recipients, the trend has inverted in the last 4 years, with the majority of patients receiving a transplant being over 18 years of age [7].

Graft and patient survivals have significantly improved over the last 10 years, especially if compared to previous decades. There are multiple reasons that account for such improvement: the development of immunosuppression protocols which use induction therapy and later minimize immunosuppression, a better monitoring of the grafts, availability of new drugs to treat severe rejection and improved survival from PTLD with the use of rituximab. In the largest centers, 1-year graft survival is now as high as 75−80% and patient survival up to 100%, with 3-year patient survival around 60−70% [31]. As previously described, graft survival is slightly better in recipients of liver-containing grafts, mainly due to a decrease in the incidence of rejection. Five-year survival, however, is still stuck at around 50%, similar to what it was even two decades ago. This late, continuous loss of grafts and patients has not yet been improved in most centers.

Mortality is mostly related to infectious complications and rejection episodes. The most common reason for graft failure is rejection, followed by infections, especially when PTLD forces significant reduction in immunosuppression with consequent development of graft failure. An interesting observation is that graft and patient loss is observed more frequently in patients who live far from the transplant center, at times in another country [65]. This is possibly due to the lack of local expertise in recognizing early signs of problems (rejection and infection, mainly) and being less aggressive in the diagnosis and treatment of such complications.

Re-transplantation is required in 10−15% of patients, with rejection (either acute or chronic) being the most common reason for re-transplantation [18,23,31]. Recipients of an isolated intestine transplant have a higher chance of requiring a second graft. In such cases, however, the type of graft may change in that many patients who first received an isolated intestinal graft end up requiring a liver−intestine or a multivisceral graft as a second transplant. Survival rates for re-transplant recipients are not as favorable as first-time grafts [66].

Long-term problems outside rejection include recurrence of disease (Crohn's disease, desmoid tumors), development of malignancies, and most notably chronic renal insufficiency. Renal failure in recipients of intestinal grafts is multifactorial in origin: the use of high dose of calcineurin inhibitors such as tacrolimus, repeated episodes of dehydration, sepsis, and use of other nephrotoxic drugs such as antibiotics. Five-year risk of chronic renal failure exceeds 20% in recipients of intestinal grafts [67].

In a large, recently published study of long-term survivors, enteral autonomy, as defined by no requirements for parenteral nutrition or intravenous hydration, was achieved in over 90% of patients who survived more than 5 years posttransplant [65,68]. Even if patients significantly improve their body mass index from before transplant, they still lag behind normal values for age and height as compared to healthy individuals. Lack of social support significantly affected long-term survival.

The field of quality of life (QOL) analysis is still in its infancy for what concerns recipients of intestinal transplants, due to low numbers of overall patients and long-term survivors as well as the different metrics that need to be used for adult or pediatric patients. In adult recipients improvements in QOL from before surgery have been described, with the exception of depression and financial obligations concerns [65]. In pediatric recipients there was a significant presence of neuropsychiatric disorders such as autism, developmental delay, attention deficit disorders, and so forth [65]. In some pediatric patients who developed food aversion from early lack of enteral intake, rehabilitation must include speech and swallow therapy to successfully bridge them to full oral intake.

32.10 CONCLUSIONS

Intestinal transplantation offers a viable option in patients with irreversible intestinal failure who cannot receive adequate intestinal rehabilitation and/or have developed life-threatening complications from chronic parenteral nutrition. The overall results are only now starting to approach the excellent rates observed in other solid organ transplants, due mainly to improvements in preventing and managing episodes of rejection and infections. There is however a significant late drop in survival rates so that 5-year survival rates are still similar to what they were years ago, when most patients were lost in the first year after transplant. This clinical observation stresses the importance of careful, attentive long-term monitoring of patients, because deadly complications leading to loss of graft and/or life can still happen beyond the first-year posttransplant, unlike most other solid organ graft recipients. By nature, intestinal transplantation even in the future is not going to be a widespread type of surgical procedure because of the limited number of patients who

actually develop irreversible intestinal failure. However, graft and patient survival keep improving over time due to the implementation of new immunosuppressive protocols which have significantly decreased incidence of rejection. Clinical research is currently focused on the relationship between preformed antibodies and risk of rejection. It is clear that humoral immunity has a much more significant role than thought in the past. The evidence that preformed antibodies at time of transplant and DSA posttransplant play a significant role in increasing the risk for rejection in intestinal transplant recipients is now clear. It is therefore suggested that patients should undergo a pretransplant virtual cross-match and if positive for donor-related HLA loci the transplant should be canceled [38]. Every attempt should be made to procure a graft with an HLA profile that excludes already known preformed antibodies. Timing of surgery in intestinal transplantation does not always allow waiting for the results of a final cross-match, but the refinement in virtual cross-match techniques does allow forecasting before transplant so to avoid undesired HLA combinations. Additionally, longitudinal monitoring of DSA titers should be performed since a rapid rise in such titers may be the first sign of an impending rejection episode and should prompt the physician to consider endoscopy and biopsy.

Lastly, there has been tremendous improvement in bowel rehabilitation techniques, in both surgical and medical management. Many patients who would have otherwise progressed to intestinal failure can now be rescued to partial or total enteral function. It is essential that any intestinal transplant center develop a multidisciplinary effort to establish protocols for intestinal rehabilitation in conjunction with gastroenterology experts, parenteral nutrition specialized pharmacists, and pediatric and adult abdominal surgeons.

REFERENCES

[1] Lillehei RC, Goott B, Miller FA. The physiological response of the small bowel of the dog to ischemia including prolonged *in vitro* preservation of the bowel with successful replacement and survival. Ann Surg 1959;150:543—60.

[2] Goulet O, Revillon Y, Brousse N, Jan D, Canion D, Rambaud C, et al. Successful small bowel transplantation in an infant. Transplantation 1992;53:940—2.

[3] Margreiter R, Konigsrainer A, Schmid A, Koller J, Kornberger R, Oberhuber G, et al. Successful multivisceral transplantation. Transplant Proc 1992;24:1226—7.

[4] Todo S, Reyes J, Furukawa H, Abu Elmagd K, Lee RG, Tzakis A, et al. Outcome analysis of 71 clinical intestinal transplantations. Ann Surg 1995;222:270—82.

[5] Starzl TE, Kaupp Jr HA, Brock DR, Butz Jr GW, Linman JW. Homotransplantation of multiple visceral organs. Am J Surg 1962;103:219—29.

[6] Todo S, Tzakis AG, Abu-Elmagd K, Reyes J, Nakamura K, Casavilla A, et al. Intestinal transplantation in composite visceral grafts or alone. Ann Surg 1992;216:223—33.

[7] Organ Procurement and Transplantation Network Website, <http://optn.transplant.hrsa.gov>; [accessed on 05.12.12].

[8] Centers for Medicare and Medicaid Services. Decision memo for intestinal and multivisceral transplantation (CAG-00036N), <http://www.cms.hhs.gov/mcd/viewdecisionmemo.asp>; October 4, 2000 [accessed on 05.12.12].

[9] Kaufmann SS, Atkinson JB, Bianchi A, Goulet OJ, Grant D, Langnas AN, et al. Indications for pediatric intestinal transplantation: a position paper of the American Society of Transplantation. Pediatr Transpl 2001;5:80—7.

[10] Abu-Elmagd KM. Intestinal transplantation for short bowel syndrome and gastrointestinal failure: current consensus, rewarding outcomes, and practical guidelines. Gastroenterology 2006;130:132—7.

[11] Mangus RS, Tector AJ, Kubal CA, Fridell JA, Vianna RM. Multivisceral transplantation: expanding indications and improving outcomes. J Gastrointest Surg 2013;17(1):179—86; discussion p.186—7. PMID: 23070622.

[12] Mazariegos GV, Superina R, Rudolph J, Cohran V, Burns RC, Bond GJ, et al. Current status of pediatric intestinal failure, rehabilitation and transplantation: summary of a colloquium. Transplantation 2011;92:1173—80.

[13] Sudan D, DiBlaise J, Torres C, Thompson J, Raynor S, Gilroy R, et al. A multidisciplinary approach to the treatment of intestinal failure. J Gastrointest Surg 2005;9:165—76.

[14] Puder M, Valim C, Meisel JA, Le HD, de Meijer VE, Robinson EM, et al. Parenteral fish oil improves outcomes in patients with parenteral nutrition-associated liver injury. Ann Surg 2009;250:395—402.

[15] Onder AM, Chandar J, Billings AA, Simon N, Diaz R, Francoeur D, et al. Comparison of early versus late use of antibiotic locks in the treatment of catheter-related bacteremia. Clin J Am Soc Nephrol 2008;3:1048—56.

[16] Kim HB, Lee PW, Garza J, Duggan C, Fauza D, Jaksic T. Serial transverse enteroplasty for short bowel syndrome: a case report. J Pediatr Surg 2003;38:881—5.

[17] Fryer J, Pellar S, Ormond D, Koffron A, Abecassis M. Mortality in candidates waiting for combined liver—intestine transplant exceeds that for other candidates waiting for liver transplants. Liver Transpl 2003;9:748—53.

[18] Mazariegos GV, Steffick DE, Horslen S, Farmer D, Fryer J, Grant D, et al. Intestine transplantation in the United States: 1999—2008. Am J Transplant 2010;10:1020—34.

[19] Abu Elmagd K, Fung J, Bueno J, Martin D, Madariaga JR, Mazariegos G, et al. Logistics and technique for procurement of intestinal, pancreatic, and hepatic grafts from the same donor. Ann Surg 2000;232:680—7.

[20] Kato T, Tzakis A, Selvaggi G, Madariaga JR. Surgical techniques used in intestinal transplantation. Curr Opin Organ Transpl 2004;9:207—13.

[21] Mangus RS, Tector AJ, Fridell JA, Kazimi M, Hollinger E, Vianna RM. Comparison of histidine-tryptophan-ketoglutarate solution and University of Wisconsin solution in intestinal and multivisceral transplantation. Transplantation 2008;86:298—302.

[22] Bueno J, Abu-Elmagd K, Mazariegos G, Madariaga J, Fung J, Reyes J. Composite liver-small bowel allografts with preservation of donor duodenum and hepatic biliary system in children. J Pediatr Surg 2000;35:291—6.

[23] Tzakis AG, Kato T, Levi DM, DeFaria W, Selvaggi G, Weppler D, et al. One hundred multivisceral transplants at a single center. Ann Surg 2005;242:480—93.

[24] Berney T, Kato T, Nishida S, Tector AJ, Mittal NK, Madariaga J, et al. Portal versus systemic drainage of small bowel allografts: comparative assessment of survival, function, rejection, and bacterial translocation. J Am Coll Surg 2002;195:804—13.

[25] Di Benedetto F, Lauro A, Masetti M, Cautero N, De Ruvo N, Quintini C, et al. Use of prosthetic mesh in difficult abdominal wall closure after small bowel transplantation. Transplant Proc 2005;37:2272—4.

[26] Asham E, Uknis ME, Rastellini C, Elias G, Cicalese L. Acellular dermal matrix provides a good option for abdominal wall closure following small bowel transplantation: a case report. Transplant Proc 2006;38:1770—1.

[27] Levi DM, Tzakis AG, Kato T, Madariaga J, Mittal NK, Nery J, et al. Transplantation of the abdominal wall. Lancet 2003;361:2173—6.

[28] Gondolesi G, Selvaggi G, Tzakis A, Rodriguez-Laiz G, Gonzalez-Campana A, Fauda M, et al. Use of the abdominal rectus fascia as a nonvascularized allograft for abdominal wall closure after liver, intestinal and multivisceral transplantation. Transplantation 2009;87:1884—8.

[29] Grant D, Abu-Elmagd K, Reyes J, Tzakis A, Langnas A, Fishbein T, et al. 2003 report of the intestine transplant registry: a new era has dawned. Ann Surg 2005;241:607—13.

[30] Abu-Elmagd K, Reyes J, Bond G, Mazariegos G, Wu T, Murase N, et al. Clinical intestinal transplantation: a decade of experience at a single center. Ann Surg 2001;234:404—17.

[31] Abu-Elmagd K, Costa G, Bond GJ, Soltys K, Sindhi R, Wu T, et al. Five hundred intestinal and multivisceral transplantations at a single center: major advances and new challenges. Ann Surg 2009;250:567—81.

[32] Tzakis AG, Kato T, Nishida S, Levi DM, Madariaga JR, Nery JR, et al. Preliminary experience with Campath 1H (C1H) in intestinal and liver transplantation. Transplantation 2003;75:1227—31.

[33] Reyes J, Mazariegos GV, Abu-Elmagd K, Macedo C, Bond GJ, Murase N, et al. Intestinal transplantation under tacrolimus monotherapy after perioperative lymphoid depletion with rabbit anti-thymocyte globulin (thymoglobulin). Am J Transplant 2005;5:1430—6.

[34] Tsai HL, Island ER, Chang JW, Gonzalez-Pinto I, Tryphonopoulos P, Nishida S, et al. Association between donor-specific antibodies and acute rejection and resolution in small bowel and multivisceral transplantation. Transplantation 2011;92:709—15.

[35] Gonzalez-Pinto IM, Tzakis AG, Tsai HL, Chang JW, Tryphonopoulos P, Nishida S, et al. Association between panel reactive antibodies and acute small bowel rejection: analysis of a series of 324 intestinal transplants. Transplant Proc 2010;42:4269—71.

[36] Ruiz P, Carreno M, Weppler D, Gomez C, Island E, Selvaggi G, et al. Immediate antibody-mediated (hyperacute) rejection in small-bowel transplantation and relationship to cross-match status

and donor-specific C4d-binding antibodies: case report. Transplant Proc 2010;42:95—9.

[37] Kato T, Mizutani K, Terasaki P, Quintini C, Selvaggi G, Thomspon J, et al. Association of emergence of HLA antibody and acute rejection in intestinal transplant recipients: a possible evidence of acute humoral sensitization. Transplant Proc 2006;38:1735—7.

[38] Abu-Elmagd KM, Wu G, Costa G, Lunz J, Martin L, Koritsky DA, et al. Preformed and *de novo* donor specific antibodies in visceral transplantation: long-term outcome with special reference to the liver. Am J Transplant 2012;12:3047—60.

[39] Island ER, Gonzalez-Pinto IM, Tsai HL, Ruiz P, Tryphonopoulos P, Gonzalez ML, et al. Successful treatment with bortezomib of a refractory humoral rejection of the intestine after multivisceral transplantation. Clin Transpl 2009;2009:465—9.

[40] Kato T, Gaynor JJ, Nishida S, Mittal N, Selvaggi G, Levi D, et al. Zoom endoscopic monitoring of small bowel allograft rejection. Surg Endosc 2006;20:773—82.

[41] Pappas PA, Tzakis AG, Gaynor JJ, Carreno MR, Ruiz P, Huijing F, et al. An analysis of the association between serum citrulline and acute rejection among 26 recipients of intestinal transplant. Am J Transplant 2004;4:1124—32.

[42] David AI, Selvaggi G, Ruiz P, Gaynor JJ, Tryphonopoulos P, Kleiner GI, et al. Blood citrulline level is an exclusionary marker for significant acute rejection after intestinal transplantation. Transplantation 2007;84:1077—81.

[43] Asaoka T, Island ER, Tryphonopoulos P, Selvaggi G, Moon J, Tekin A, et al. Characteristic immune, apoptosis and inflammatory gene profiles associated with intestinal acute cellular rejection in formalin-fixed paraffin-embedded mucosal biopsies. Transpl Int 2011;24:697—707.

[44] Andreev VP, Tryphonopoulos P, Blomberg BB, Tsironemas N, Weppler D, Neuman DR, et al. Peripheral blood gene expression analysis in intestinal transplantation: a feasibility study for detecting novel candidate biomarkers of graft rejection. Transplantation 2011;92:1385—91.

[45] Calne R, Davis H. Organ graft tolerance: the liver effect. Lancet 1994;343:67—8.

[46] Lerut J, Sanchez-Fueyo A. An appraisal of tolerance in liver transplantation. Am J Transplant 2006;6:1774—80.

[47] Selvaggi G, Gaynor JJ, Moon J, Kato T, Thomspon J, Nishida S, et al. Analysis of acute cellular rejection episodes in recipients of primary intestinal transplantation: a single center, 11 year experience. Am J Transplant 2007;7:1249—57.

[48] Kato T, Ruiz P, Tzakis A. Exfoliative bowel rejection—a dangerous loss of integrity. Pediatr Transplant 2004;8:426—7.

[49] Wu T, Abu-Elmagd K, Bond G, Nalesnik MA, Randhava P, Demetris AJ. A schema for histologic grading of small intestine allograft acute rejection. Transplantation 2003;75:1241—8.

[50] Takahashi H, Kato T, Selvaggi G, Nishida S, Gaynor JJ, Delacruz V, et al. Subclinical rejection in the initial postoperative period in small intestinal transplantation: a negative influence on graft survival. Transplantation 2007;84:689—96.

[51] Fan J, Tekin A, Nishida S, Moon J, Selvaggi G, Levi D, et al. Preoperative embolization of the graft superior mesenteric artery assists graft enterectomy in intestinal transplant recipients. Transplantation 2012;94:89—91.

[52] Parizhskaya M, Redondo C, Demetris A, Jaffe R, Reyes J, Ruppert K, et al. Chronic rejection of small bowel grafts: pediatric

and adult study of risk factors and morphologic progression. Pediatr Dev Pathol 2003;6:240—50.

[53] Tryphonopoulos P, Weppler D, Nishida S, Kato T, Levi D, Selvaggi G, et al. Mucosal fibrosis in intestinal transplant biopsies correlates positively with the development of chronic rejection. Transplant Proc 2006;38:1685—6.

[54] Guaraldi G, Cocchi S, Codeluppi M, Di Benedetto F, De Ruvo N, Masetti M, et al. Outcome, incidence, and timing of infectious complications in small bowel and multivisceral organ transplantation patients. Transplantation 2005;80:1742—8.

[55] Florescu DF, Qiu F, Langnas AN, Mercer DF, Chambers H, Hill LA, et al. Bloodstream infections during the first year after pediatric small bowel transplantation. Pediatr Infect Dis J 2012;31:700—4.

[56] Florescu DF, Islam KM, Grant W, Mercer DF, Langnas A, Botha J, et al. Incidence and outcome of fungal infections in pediatric small bowel transplant recipients. Transpl Infect Dis 2010;12:497—504.

[57] Bueno J, Green M, Kocoshis S, Furukawa H, Abu-Elmagd K, Yunis E, et al. Cytomegalovirus infection after intestinal transplantation in children. Clin Infect Dis 1997;25:1078—83.

[58] Florescu DF, Langnas AN, Grant W, Mercer DF, Botha J, Qiu F, et al. Incidence, risk factors, and outcomes associated with cytomegalovirus disease in small bowel transplant recipients. Pediatr Transplant 2012;16:294—301.

[59] Mazariegos GV, Abu-Elmagd K, Jaffe R, Bond G, Sindhi R, Martin L, et al. Graft versus host disease in intestinal transplantation. Am J Transplant 2004;4:1459—65.

[60] Wu G, Selvaggi G, Nishida S, Moon J, Island ER, Ruiz P, et al. Graft-versus-host-disease after intestinal and multivisceral transplantation. Transplantation 2011;91:219—24.

[61] Taylor AL, Marcus R, Bradley JA. Post-transplant lymphoproliferative disorders after solid organ transplantation. Crit Rev Oncol Hematol 2005;56:155—67.

[62] Quintini C, Kato T, Gaynor JJ, Ueno T, Selvaggi G, Gordon P, et al. Analysis of risk factors for the development of posttransplant lymphoproliperative disorder among 119 children who received primary intestinal transplants at a single center. Transplant Proc 2006;38:1755—8.

[63] Berney T, Delis S, Kato T, Nishida S, Mital NK, Madariaga J, et al. Successful treatment of post-transplant lymphoproliferative disease with prolonged rituximab treatment in intestinal transplant recipients. Transplantation 2002;74:1000—6.

[64] Intestinal Transplant Association. <http://www.intestinaltransplant.com>; [accessed on 05.12.12].

[65] Abu Elmagd KM, Kosmach-Park B, Costa G, Zenati M, Martin L, Koritsky D, et al. Long-term survival, nutritional autonomy, and quality of life after intestinal and multivisceral transplantation Ann Surg 2012;256: 494—508.

[66] Desai CS, Khan KM, Gruessner AC, Fishbein TM, Gruessner RW. Intestinal retransplantation: analysis of Organ Procurement and Transplantation Network database. Transplantation 2012;93:120—5.

[67] Ojo AO, Held PJ, Port FK, Wolfe RA, Leichtman AB, Young EW, et al. Chronic renal failure after transplantation of a nonrenal organ. N Engl J Med 2003;349:931—40.

[68] Sudan D. Quality of life after intestinal transplantation. Prog Transplant 2004;14:284—8.

Chapter 33

Living Related Small Bowel Transplantation: Progress, Pitfalls, and Promise

Ivo G. Tzvetanov, Lorena Bejarano-Pineda, and Enrico Benedetti

Division of Transplantation, Department of Surgery, University of Illinois at Chicago, Chicago, IL

Chapter Outline

Advances in multivisceral transplantation, and specifically intestinal transplantation, have opened new opportunities in the treatment of patients with intestine failure and life-threatening complications secondary to total parenteral nutrition (TPN). Intestinal failure can be caused by a reduced functional surface area, as a result of surgical resections or anatomic loss and mucosal dysfunction due to congenital and acquired disorders. Consequently, the insufficient absorption of nutrients entails the requirement of TPN.

The initial experience with intestinal transplant was described in the late 1960s. The first clinical transplant from a living donor (LD) was performed by Ralph Deterling in 1964. Deterling reported that the procedure was executed on an infant recipient with a segment of his mother's ileum; however, the patient died the next day due to medical complications not related to the transplant [1]. Nevertheless, the first documented intestinal transplant from an LD was reported by Fikri Alican from the University of Mississippi, Jackson, and was presented at the 11th Annual Meeting of the Society of the Alimentary Tract in Chicago, IL, in 1970. Alican et al. reported the case of an eight-year-old boy with the resection of the small bowel from the ligament of Treitz to the ileocecal valve secondary to strangulation. The transplant was performed in 1969, with approximately 3 ft of ileum transplanted from his mother. However, the recipient's procedure was complicated by thrombosis of the vena cava, and the allograft had to be subsequently removed on the ninth posttransplant day [1].

The introduction of cyclosporine distinctly changed the outcome for solid organ transplantation. Nonetheless, the use of cyclosporine did not have as much benefit for intestinal transplantation as it did for other transplanted solid organs [2]. In the cyclosporine era, only two intestinal transplants from LDs were reported by Deltz et al. [3,4], with both recipients having received a 60-cm segment of jejunoileum. First recipient was a boy 4 years of age with volvulus, who received the graft from his mother; unfortunately, the graft was removed due to an intractable rejection episode. Second recipient was a 42-year-old woman with a subtotal small bowel resection secondary to the thrombotic occlusion of mesenteric veins. Immunosuppressive therapy consisted of cyclosporine, steroids, and antithymocyte globulin. The patient was on full oral intake 2 weeks later and thereafter remained off parenteral nutrition until 1990, when chronic rejection caused the loss of the graft function. At that point in time, it was the first successful LD intestinal transplant with a long-term function of over 2 years [4].

The introduction of tacrolimus in 1990 impelled intestinal transplantation to a clinically accepted procedure. By the mid-1990s, Tesi and Jaffe reported two LD intestinal transplants; one of them was performed in a patient with pseudo-obstruction and the other in a patient with Gardner's syndrome [5,6]. However, their vascular anatomy was unfavorable, since they transplanted a segment of jejunum. Regardless of induction therapy with Muromonab-CD3 (a monoclonal antibody targeted at the CD3 receptor) and maintenance therapy with tacrolimus, both patients had several rejection episodes and were back to TPN 6 months later [2]. Morris et al. described an LD intestinal transplant in an adult patient with a desmoid tumor, whose donor was his monozygotic twin. They transplanted the distal ileum, ileocecal valve, and portion of the cecum; however, as they also removed the terminal ileum of the donor, he became vitamin B12 deficient [7].

At the University of Minnesota, during the 1990s, a group of transplant fellows at the time under the guidance of Rainer Gruessner, studied the technical aspects of LD intestinal transplantation in a pig model [8]. Consequently, they performed the first LD intestinal transplant at the University of Minnesota, from which they concluded the following: (i) the ileum was the best option due to its greater absorptive capacity of bile acids, vitamins, fat, and water [9]; (ii) the terminal ileum (30 cm), the ileocecal valve, and the cecum should remain in the donor to minimize morbidity; (iii) a vascular pedicle should be used consisting of only one artery and vein (either the ileocolic artery and vein, or the terminal branches of the superior mesentery artery (SMA) and superior mesenteric vein (SMV); (iv) the bowel continuity should be restored with a proximal bowel anastomosis and a distal ileostomy (to allow access to graft biopsy). The ileostomy should be taken down 6 months posttransplant, or 6 months after the last rejection episode [2]. After these two first successful LD intestinal transplants at the University of Minnesota, the group published the respective guidelines in 1997 [10] as a standardized technique for intestinal transplants.

TPN-associated complications, such as lack of vascular access, recurrent line infections, and liver failure, continue to become life-threatening. If it is taken into account that the pediatric population is more prone to liver disease and life-threatening complications secondary to TPN, then LD intestinal grafts could prevent the progression to end-stage liver disease and be considered life-saving when a deceased donor (DD) graft is not available. Thus, the main principles for performing an LD intestinal transplant are to improve posttransplant outcomes and to decrease the mortality of those on the waiting list by reducing their time on it.

33.1 DONOR

The existing large gap between the number of potential recipients and available cadaver donors for a liver and kidney transplant has justified the significant expansion of LD programs for those organs. This situation does not exist for adult recipients for an intestinal transplant, since the donor supply largely exceeds the current needs. However, this is not the case for potential pediatric recipients, especially those with associated liver failure. United Network for Organ Sharing (UNOS) data show that pediatric patients still have the highest mortality rate on the waiting list compared to all the other categories of solid organ transplantation (www.unos.org, 10/22/12).

To minimize the incidence of complications and increase the rate of success, it becomes necessary to choose the donor carefully. Donor selection starts with the determination of ABO blood type and the greatest degree of human leukocyte antigen (HLA) matching. It is also ideal that the donor would be Epstein−Barr virus (EBV) and cytomegalovirus (CMV) negative [2]. The selected donor must be able to tolerate an elective surgical procedure and be free of significant cardiovascular, pulmonary, and renal disease. Table 33.1 summarizes the required workup for living-related donor evaluation for LD intestinal transplant [11]. It is also important to mention that the interview with an ethicist should occur early on in the evaluation process to avoid unnecessary invasive procedures in potential donors. After consideration of the medical, psychosocial, and ethical factors, the workup should be completed by the evaluation of the vascular anatomy of the donor's small intestine and the anesthetic risk. Conventional selective angiography and abdominal CT-scan are performed to rule out unknown vascular anomalies and associated pathologies.

The technical aspects of LDIT were standardized by Gruessner and Sharp in 1997, as was previously described. This technique is based on donor safety and recipient independence from TPN [2,10]. As with other major elective bowel surgery, the pretransplant preparation of an LD takes place the afternoon before the surgery with polyethylene glycol and an electrolyte solution taken orally until the rectal effluent is clear. For intestinal decontamination, three doses of neomycin and metronidazole are given the day before the surgery [12].

The entire length of the small bowel from the ligament of Treitz to the ileocecal valve is measured. Subsequently, the cecum and the terminal ileum are identified and marked approximately 30 cm proximal from the ileocecal junction. The donor operation consists of harvesting 200 cm of distal ileum (160 cm for pediatric recipients), preserving at least 20−30 cm of terminal ileum and ileocecal valve to avoid macrocytic anemia and to shorten transit time [10,13]. The vascular pedicle of the graft is formed by the distal branches of the SMA and SMV or, alternately, by the ileocolic artery and vein; and, they are anastomosed to the infrarenal aorta and cava of the recipient, respectively.

TABLE 33.1 Donor Evaluation

Component of Evaluation	Test and Procedures
Medical evaluation	Comprehensive analysis of medical and surgical history, review of systems, physical examination, current medications, history of malignancy, and previous intestinal surgery
Immunologic evaluation	ABO compatibility, HLA[a] type, lymphocytotoxic cross-match
Nutritional evaluation	Comprehensive metabolic panel, vitamin A, D, E, K, and B12
Hepatology evaluation	Prothrombin time, partial thromboplastin time, alpha-fetoprotein, ammonia
Cardiology evaluation	Chest X-ray, electrocardiogram
Infectious disease evaluation	Serology (CMV, EBV, VZV, HIV, HCV, HBeAg, HBsAg, HBsAb[b]), complete blood count. Urine and stool cultures
Anesthesiology evaluation	Anesthesia history, surgical procedures, and drug allergies
Psychosocial evaluation	Psychiatry evaluation, social work consultation
Ethics consultation	An interview with a member of the institutional ethics committee to discuss with the potential donor about motivations and understanding of the risk involved
Abdominal anatomic evaluation	CT-scan of the abdomen or 3D-angio-CT-scan

[a]HLA, histocompatibility leukocyte antigen.
[b]CMV, cytomegalovirus; EBV, Eipstein—Barr virus; VZV, varicela zoster virus; HIV, human immunodeficiency virus; HCV, hepatitis C virus; HBeAg, hepatitis B virus early antigen; HBsAg, hepatitis B surface antigen; HBsAb, hepatitis B surface antibody.
Source: Data from Ref. [11]

If the procedure involved a combined intestinal and liver transplant, the donor operation becomes more complex. Combined intestinal and liver transplants have only been done for pediatric patients, requiring procurement of liver segments two and three, and the ileal graft. Because of the recipient's medical status, the liver should be procured and transplanted first. If the recipient remains stable after the liver is implanted, the intestinal procurement, (and consequently the transplant) can be performed; otherwise, the incision is closed and the intestinal transplant is rescheduled, preferably within the first two weeks after the liver transplant. The disadvantage of sequential liver and intestinal procurement is that the donor is exposed twice to operative stress and the intrinsic risks of complications of a major surgical procedure (e.g., thrombosis, pneumonia, surgical site infections) [2].

Ghafari et al. documented their experience with six combined intestinal/liver transplants at the University of Illinois Hospital [14]. The transplants were performed between 2004 and 2007, with a total of six children (average age 13.5 months) having received the grafts from one of their parents. Three of these recipients had a simultaneous transplantation while the other three recipients had a staged procedure, with an average interval of 6 days based on hemodynamic stability after the liver graft was implanted.

These donors were chosen according to ABO compatibility and the best HLA match. Of the six donors, five were the recipient's mother and one was the recipient's father. The mean age and weight of the donors were 25 (range 19—32) years and 76.5 kg, respectively. The mean hospital stay for donors was 8.6 days; none of the donors had any perioperative mortality or morbidity; all donors were discharged home on a regular diet. The mean follow-up period in the donors was 42 months (range 29—51 months). No donors were found to have macrocytic anemia secondary to vitamin B12 deficit; and, all donors presented increased looseness of stool after the first month of surgery. Further, none of them reported weight loss, changes in lifestyle, or psychosocial conditions after the donation. Five of the six children are still alive with adequate grafts function, whereas one recipient died due to plasmoblastic lymphoma, albeit with functioning graft [14].

It is very important that the donors have adequate follow-up care. The hospital stay in the postoperative course usually requires 3—4 days; once oral intake is tolerated, the patient can be discharged [2]. After discharge, they need to be evaluated on a monthly basis and then annually to review their eating and defecation patterns as well as any complications. The donor should also undergo vitamin B12 assays at 1, 6, and 12 months post donation to ensure adequate vitamin B12 absorption. Benedetti et al. reported a case of chronic diarrhea from among the donors, but it was resolved with medical therapy consisting of Imodium and Cholestyramine [14]. For 11 donors, out of their total cohort of LD intestinal transplants at the University of Illinois Hospital, the authors also reported a 36.4% reduction in LDL and a 22.3% decrease in total cholesterol levels when compared with their respective predonation lipid profiles, and they noted the difference was statistically significant [15]. However, a further follow-up in a greater cohort should be completed to conclude this finding.

Although the number of LD intestinal transplants is relatively small, there have been no reports of donor mortality or life-threatening complications [2]. Nevertheless, a more extensive follow-up is necessary to determine the presence of postsurgical complications, such as intestinal adhesions.

33.2 RECIPIENT

LDIT has several potential advantages, such as the elimination of waiting time, the elective nature of the procedure and better HLA matching, and short cold ischemia time. A well timed LDIT may be critically important to rescue patients who experience the loss of central venous accesses. We have also had a favorable experience in our center (University of Illinois Hospital) with young trauma victims with ultrashort bowel syndrome who achieved excellent outcomes, with complete nutritional rehabilitation and reestablishment of a normal lifestyle [16]. The indications for LD intestinal transplantation are the same as for DD intestinal transplantation. Registry data suggest that the patient and graft survival rates are similar for both LD and DD intestinal transplants. Nevertheless, using an LD can reduce the mortality rate for those on the waiting list, which is especially high for candidates less than 5 years of age (www.unos.org, 10/12).

Among the most important causes of intestine failure, volvulus and gastroschisis are leading indications for bowel transplantation in children [2]. Volvulus is caused by a malrotation of the bowel around a formed fixed point formed by congenital or adhesive bands, or from abnormal motility along the mesenteric axis. Gastroschisis is produced by the herniation of a variable length of the intestine and occasionally part of the liver, without a peritoneal sac through an abdominal wall defect in the right side of the umbilical cord [17]. Regarding the adult population, short bowel syndrome, secondary to ischemic intestinal disease, Crohn's disease, and massive intestinal resection due to trauma, are the major indications for intestinal transplantation [2,18].

TPN remains the standard therapy for chronic intestinal failure with a survival rate between 85% and 90% at 5 years for those who tolerate the therapy at home [2]. However, about 15% of patients receiving TPN for more than 1 year develop end-stage liver disease. In children, the incidences of liver disease is higher, especially in patients with less than 30−40 cm of remnant bowel [19,20]. Liver disease remains the leading indication for performing intestinal transplantation in children, but loss of central venous access to provide parenteral nutrition has become an indication for intestinal transplantation.

The indications for intestinal transplantation in pediatric patients were updated by Avitzur and Grant [21]. Intestinal transplantation is indicated for the following conditions [22]:

- loss of 50% of available central venous accesses due to thrombosis;
- recurrent septic episodes, resulting in multiorgan failure, shock, and metastatic infectious loci (more than two episodes per year);
- imminent or overt end-stage liver disease;
- ultrashort bowel syndrome;
- high risk of death attributable to the underlying disease;
- frequent hospitalization;
- severe dehydration episodes;
- lack of family support or unwillingness to accept long-term TPN.

Standard absolute contraindications include severe and irresolvable cardiac, respiratory, and neurological disease, multiorgan failure, overwhelming sepsis, and malignancy outside of the intestine. Relative contraindications are frequently center-specific and include physical debilitation, poor family support or noncompliance, advanced age, and low weight (children weighing less than 5 kg) [2].

A multidisciplinary evaluation of the patient with intestinal failure is essential to assess adequate candidacy for transplantation and to ensure best outcomes. The evaluation process must elucidate (i) the failure of TPN as compared to other surgical therapy strategies besides transplantation; (ii) the need of intestine or combined liver/intestine transplantation; (iii) the state of the remnant intestine and the patency great vessels; (iv) the absence of absolute contraindications or associated disease that can put at risk the procedure and postoperative course. This process contains, among others, medical history and assessment, laboratory tests, diagnostics procedures to define anatomy and functioning of gastrointestinal tract, nutritional state, hepatic function, vascular patency, infection history, immunological status, abdominal compliance, psychosocial issues, and child development.

The nutritional assessment includes a feeding history, assessment of caloric intake, and measurement of growth parameter in children. Caloric intake is estimated through the assessment of TPN solution and enteral feedings. Nutritional state is evaluated by laboratory parameters including electrolytes, blood urea nitrogen, creatinine, calcium, zinc, trace elements, lipid profile, and levels of vitamins A, D, E, K, and B12.

Hepatic function is evaluated to asses patients for possible TPN-induced liver disease and identify those with irreversible liver failure who require combined liver—intestine transplantation. Laboratory values related to liver function are obtained, including alanine and aspartate aminotransferase, gamma glutamyl transpeptidase, direct and indirect bilirubin, albumin, prothrombin time, partial thromboplastin time, alpha-fetoprotein, platelets, ammonia, and factors V and VII. An ultrasound of the liver and abdomen is obtained to assess liver size, vasculature, and evidence of portal hypertension. An upper endoscopy is obtained to assess the possible presence of varices in cirrhotic patients. A liver biopsy is routinely performed for patients with abnormal liver function tests.

Gastrointestinal barium studies are performed to clarify the extension and the characteristics of the remnant bowel in order to detect any abnormality that can be surgically corrected and to plan the required surgical anastomosis during the intestinal transplant. Esofagogastroduodenal endoscopy, endoscopic retrograde cholangiopancreatography, and colonoscopy may be indicated for specific clinical conditions.

Recipients may require intravenous fluids for prolonged periods of time after the transplant. Vascular evaluation requires a complete history of number of intravenous line placements, durations, and reasons for replacement. A baseline ultrasound of great vessels is obtained to evaluate the anatomy; however, for recipients with a history of previous central vein thrombosis, appropriate evaluation may require angiography, MRA, or double spiral CT. A detailed infections history is obtained to assess the etiology and frequency of previous infections.

A full psychosocial evaluation is essential to assess the patient's and family's ability to cope with the rigors of the transplant process. The evaluation should focus on family functioning, physical functioning, coping skills, and family support. Recipients and their family must acquire an understanding of posttransplant care requirements and recognize their ability to provide that care. All the aspects from the completed evaluation are summarized in Table 33.2 [24].

Although the criteria used for listing deceased and LD candidates are the same, we believe that certain patients may have a greater benefit from the LD option. Adults with an identical twin or HLA-identical sibling as a donor candidate should be transplanted without delay. In our experience, using donors with at least one haplotype match has been extremely favorable, with no acute

rejection episodes during the first year posttransplant. In children affected by ultrashort bowel syndrome with slim possibilities of successful weaning of TPN, LD intestinal transplant should be considered early in order to avoid progression to end-stage liver disease. For children who present TPN-related cirrhosis, the option of combined liver—bowel transplant from an adult donor may contribute to minimize the probability of death on the waiting list, which is extremely high in this population.

Intestinal transplant literature is scant, but according to experience to date, 180—200 cm of donor graft length for an adult recipient and about 120—150 cm for a pediatric recipient is sufficient to provide TPN independence. Undoubtedly, if LD intestinal transplantation becomes more popular, insight into immunologic and absorptive issues could be deepened.

The day before surgery, the patient is given polyethylene glycol and electrolyte solution, either orally or via a gastric tube, until the stomal effluent is clear. Decontamination of the remnant small bowel is not required, but neomycin and metronidazole could be given 12 h before surgery. In addition, the recipient also receives enemas from below, in case an ileocolostomy is performed, and the optimal ostomy site should be selected.

A midline incision is made. The ostomy from the distal portion of the remaining duodenum/jejunum is taken down, and the most distal portion is resected using the gastrointestinal anastomosis stapler. The nasograstric tube or the previously placed gastric tube or J tubes is put on suction. The liver is also inspected and a biopsy may be obtained to assess histopathologic changes before the intestinal transplant. An intraoperative decision is to be made to determine whether or not to perform a cholecystectomy in case of discovering gallstones secondary to TPN.

After the remaining small bowel is mobilized, the infrarenal aorta artery and vena cava are identified and dissected free from the takeoff of the renal vessels to their iliac bifurcations. The arterial anastomosis is done first, since it is more technically challenging due to the small diameter of donor's artery. The arteriotomy is then made between the origins of the inferior mesenteric artery and the renal arteries. Given the small size of the ileocolic artery from the donor, the end-to-side ileocolic artery-to-infrarenal aorta anastomosis is constructed in an interrupted fashion. The arteriotomy can be enlarged and optimally size-fashioned with an aortic punch. Continuing with the vein anastomosis, an appropriate site on cava is chosen for the venotomy, usually 2—3 cm proximal to the arterial anastomosis. The venous anastomosis is done with the quadrangulation technique, and the end-to-side ileocolic vein to infrarenal cava anastomosis is completed by continuous corner sutures. At the beginning of venous anastomosis, mannitol (0.5—1.0 g/kg body weight) is

TABLE 33.2 Recipient Evaluation

Component of Evaluation	Test and Procedures
Referral to a transplant center	Comprehensive analysis of medical and surgical history, review of systems, physical examination, current medications, current nutrition requirements
Immunologic status	Blood group, HLA[a] type, panel of reactive antibody
Gastroenterology evaluation	Upper and lower gastrointestinal barium study, esophagogastroduodenoscopy and colonoscopy, CT-scan abdomen and pelvis, motility studies (if indicated)
Nutritional evaluation	Height, weight, anthropometric measurements, nutritional support, comprehensive metabolic panel, zinc
Hepatology evaluation	Prothrombin time, partial thromboplastin time, alpha-fetoprotein, ammonia. Doppler ultrasound of liver and liver biopsy (if indicated)
Cardiology evaluation	Electrocardiogram, chest X-ray, echocardiogram, stress test if more than 50 years of age or with cardiac history, and risk factors (hypertension, diabetes mellitus)
Nephrology evaluation	Abdominal ultrasound with size of kidneys, triple renal scan, 24 h creatinine clearance
Venous access evaluation	Doppler ultrasound of upper and lower extremities veins
Infectious disease evaluation	History of infection episodes, immunization, serology (CMV, EBV, VZV, HIV, HCV, HBeAg, HBsAg, HBsAb[b], measles, rubella and mumps titers), complete blood count. Blood, urine, and stool cultures
Anesthesiology evaluation	Anesthesia history, surgical procedures and drug allergies
Developmental evaluation	Child life and development
Psychosocial evaluation	Psychiatry evaluation, social work consultation

[a]HLA, histocompatibility leukocyte antigen.
[b]CMV, cytomegalovirus; EBV, Eipstein—Barr virus, VZV, varicela zoster virus; HIV, human immunodeficiency virus; HCV, hepatitis C virus; HBeAg, hepatitis B virus early antigen; HBsAg, hepatitis B surface antigen; HBsAb, hepatitis B surface antibody.
Source: Data from Ref. [23]

given to the recipient to diminish the extent of bowel edema as a result of reperfusion injury. On completion of the vascular anastomoses, the vascular clamps are removed. Any bleeding sites, especially on the anastomotic site or on the cut surface of the mesentery, are identified and carefully controlled. When good arterial inflow and venous outflow are achieved, the graft immediately turns pink and shows some peristalsis.

When the proximal end of the intestinal graft is identified, the anastomosis to the remaining recipient duodenum/jejunum could be made in an end-to-end, end-to-side, or side-to-side fashion. Our preference is to perform a hand sewn, two-layer side-to-side anastomosis to the remaining recipient duodenum/jejunum. The hand-sewn technique decreases the risk of intraluminal anastomotic bleeding, as compared to stapled anastomosis. The nasogastric tube is advanced and placed across the anastomosis into the intestinal graft, and it is placed on intermittent suction.

Except with identical twins, the distal end of the donor graft should be brought out as a stoma to allow an easy access to endoscopy and biopsy. The stoma can be created in three ways: (i) an end ileostomy, without construction of a distal anastomosis at the time of the transplant (Brooke ileostomy); (ii) a loop ileostomy, with the distal end of the intestinal graft anastomosed to the recipient colon and with the diverting loop ileostomy created about 15 cm from the distal anastomosis; (iii) side-to-end fashion with the distal end of the graft brought out as a chimney spout ileostomy (Bishop loop ileostomy) [2]. A quarter-size skin hole is opened down to the fascia, two finger widths of fascia is then opened, and then the stoma is brought out to the site. The abdomen is subsequently irrigated with antibiotic- and antifungal-containing solutions. After the hemostasis is confirmed, the abdomen is closed via the standard fashion (Figure 33.1).

Recipients who are to receive a combined liver—intestine transplant should undergo the LD liver transplant fist, to correct the existing coagulopathy, by undergoing replacement of the cirrhotic native liver [2]. As described previously, the decision as to whether to perform

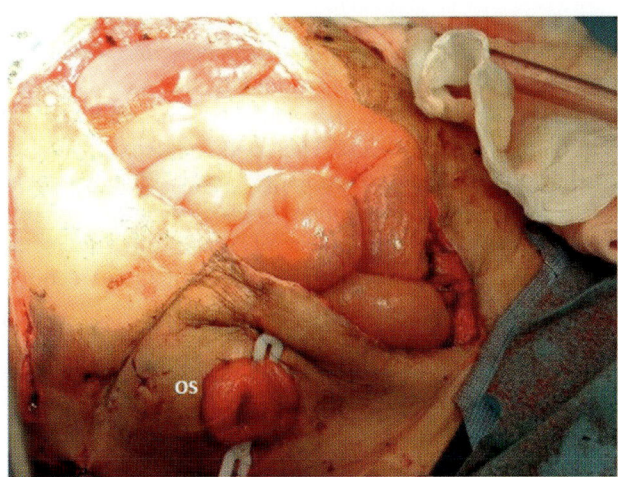

FIGURE 33.1 Recipient intestinal transplant procedure. Segmental bowel graft transplanted. The ostomy site (OS) located in the right lower quadrant.

a simultaneous or a sequential liver and intestinal transplant is made intraoperative, based on the recipient's stability after the liver is implanted. Transplantation of the LD liver segments 2 and 3 is performed. Consequently, the vascular anastomoses are completed and blood supply to the liver is restored. The bile ducts of the transplanted liver can be anastomosed to the recipient's proper hepatic duct or to the native duodenum, or whether the ducts can be stented and brought out through the skin [24]. The incision should be closed if the recipient is not hemodynamically stable to continue with the intestinal transplant. Once the intestinal transplant is performed, the cruciate incision should be closed primarily in the absence of tension. If the abdomen cannot be closed primarily, the bicostal incision should be approximated as much as possible and to close the midline incision with absorbable (polyglycolic) mesh, which subsequently could be covered with a skin graft [24].

Postoperative care for recipients of an isolated LD intestine transplant is different from the recipients of a combined LD liver and intestinal transplant. Recipients of a solitary LD intestinal transplant, after closure of the abdomen, are brought either to the postanesthesia care unit or the intensive care unit. The first 24–48 h are crucial due to surgical trauma, the degree of ischemia and reperfusion injury and onset of immunosuppression. Initially, vital signs, color of the ostomy, and laboratory parameters are monitored every 4 h. For electrolyte management, recipients are usually placed on 0.45% normal saline and dextrose infusion. Intravenous fluids are initially infused at an in/out ratio. After the first 12–24 h, the recipient goes back on TPN at a straight rate.

An important element to immediately monitor posttransplant is systemic anticoagulation: due to the small diameter of the ileocolic vessels of the LD, they are more prone to vascular thrombosis. Over the next 5–7 days, recipients are weaned off anticoagulation and started on daily baby aspirin [2]. Recipients with a medical history of thrombotic and embolic events as the principal cause of short small bowel disease remain on systemic anticoagulation for the first 6 months or indefinitely. On posttransplant day 7, a small bowel follow-through contrast study is performed to confirm intactness of the anastomosis; and, on the next day, the first graft biopsy is to be executed. Initially, endoscopy and graft biopsy are initially performed two to three times a week [25]. Within the first month of posttransplant, a marked increase in the length of intestinal villi can be noted, resulting in increase in absorptive surface [16].

After an anastomotic leak is ruled out, recipients begin a clear liquid diet; as an alternative, enteral nutrition can be started within the first 48 h posttransplant through a J tube placed beyond the proximal anastomosis. Once oral or tube feeding is tolerated, bilirubin levels and liver dysfunction start to decrease. To assess the absorptive capacity of the intestinal graft, Benedetti et al. suggest checking fat-soluble vitamin and albumin levels, quantifying fecal fat excretion and performing D-xylose absorption test at 1, 6, and 12 months posttransplant [26].

Blood, urine, stool, and sputum are collected routinely for microbiologic assessment. Antimicrobial prophylaxis usually starts with broad-spectrum antibiotics given intraoperative every 4 h, and for the first 3 days posttransplant. Antifungal medication is administered for 7 days. Antiviral medications, especially CMV prophylaxis, are first given intravenously, immediately posttransplant, and then orally when the patient tolerates oral intake. Depending on the recipient and donor CMV status, anti-CMV prophylaxis continues for up to 6 months. The recipients also receive sulfamethoxazole/trimethoprim as long-term prophylaxis against *Pneumocystis jiroveci* and *Nocardia* infections.

For recipients of a combined LD liver and intestinal transplant, postoperative care is initially dictated by the liver graft function. Once liver function has stabilized, attention can be directed to the intestinal graft function.

After the intestinal transplant has been performed, there must be a balance maintained between prevention of rejection by immunosuppression and risk of infection, so as to achieve the best outcome. The introduction of tacrolimus markedly improved the outcome of intestinal transplantation [27,28] to the point that all protocols for clinical intestinal transplant are now based on this drug. The therapeutic efficacy of tarolimus as compared to other immunosuppressants such as cyclosporine has reduced the rate of rejection and complications related to immunosuppressive therapy.

The lack of reliable markers of rejection in recipients with an intestinal transplant is a major problem, since clinical manifestations are very unspecific, and some patients may be totally asymptomatic. The most common symptom described in recipients with a rejection episode is a change in bowel habits usually with significant increase in ileostomy output or worsening diarrhea. The most effective way to diagnose rejection is by performing a videoendoscopy and biopsy of the graft. Endoscopic signs of rejection include mucosal edema, erythema, friability, and focal ulcerations. In regard to severe rejection, some of the signs are granular mucosal patterns with diffuse ulcerations and absence of peristalsis.

According to current UNOS data, thymoglobulin remains the most common immunosuppressant used as induction therapy in intestinal transplants (41.7%) (www. unos.org, 10/22/12). Induction therapy in most centers includes thymoglobulin, alemtuzumab, and baciliximab (anti-IL-2 receptor antibody). Trevizol et al. in a literature review described three different immunosuppression protocols used in transplant centers from 2006 to 2010 [29]. The protocol constituted by thymoglobulin plus rituximab as induction therapy and maintenance with tacrolimus presented the lowest infection and the higher patient survival rate, when compared with two other protocols. In our LD intestinal transplant program, we used a regimen consisting of short thymoglobulin induction accompanied by steroid bolus the day of transplantation, with a tapering off for 5 days. Tacrolimus and mycophenolate mofetil are used as a maintenance therapy for the first 2 months; if the recipient has a stable graft function, monotherapy with tacrolimus may then be considered. The tacrolimus levels for the first month, and thereafter, should be within 10−15 ng/mL and 8−12 ng/mL, respectively. Concerning maintenance therapy, tacrolimus plus steroids is the top major protocol used in intestinal transplant patients (43%); however, patients on tacrolimus, steroids and rapamycin as a maintenance therapy, had the highest graft and patient survival rates (www.unos.org, 10/22/12) [29].

Infectious complications represent a major challenge in intestinal transplantation and remain the leading cause of death in intestinal transplant recipients [29]. Intestinal grafts, unlike other transplanted abdominal organs, are directly exposed to bacterial and viral agents. Predisposing factors to bacteria translocation in intestinal transplant recipients include graft ischemia (prolonged cold ischemia time), preservation-reperfusion injury, lymphatic disruption, immunosuppression, alteration in venous drainage, and rejection episodes. In 2009, Kimura et al. studied the main sites and pathogens in recipients with bacterial infections within the first month after transplant. Their results showed that the leading sites of infection were wound and intraabdominal (ascites) infection, followed by central venous catheter associated infection;

with the principal pathogens being *Escherichia coli, Pseudomonas aeruginosa, Enterococcus*, and *Klebsiella* [30]. However, after the first month, the main sites for bacterial infection were the central venous catheter and the blood stream [30−32]. The most frequent agents isolated from the described sites were *Staphylococci, Enterococci*, and *Pseudomonas*. Also, bacterial infection was the most common cause of sepsis after intestinal transplants, representing about 73% of infectious complications [33].

The most relevant viral infections are those caused by CMV and EBV. CMV infection can cause severe enteritis leading to graft loss. The main risk factor is the combination of a CMV-positive donor with a CMV-negative recipient, which is often the case in children. The clinical manifestations of CMV enteritis are similar to those of rejection such as fever and increased stroma output. The prophylaxis and treatment of CMV infection is based on Ganciclovir and high titer anti-CMV immunoglobulin. However, there is a tendency to perform more intensive prophylaxis and monitoring in intestinal transplant recipients as compared with other solid transplants. In our center, we extend the prophylaxis with Ganciclovir for at least 1 year posttransplant in LD intestinal transplant recipients. Concerning EBV acute infection, the most dreadful related complication is posttransplant lymphoproliferative disorder. Regarding fungal infections, the most common infection in intestinal transplant recipients is *Candida albicans*, followed by *Aspergillus*. The *Aspergillus* infection usually presents within 6 months after transplantation, with a consequent high mortality rate (roughly 90%) [34]. The risk factors predisposing tan *Aspergillus* infection include a prolonged operative room time of more than 12 h, a second procedure, prolonged use of broad-spectrum antibiotics, and heavy immunosuppression therapy. Treatment options include caspifungin, voraconazole, and amphotericin.

Intestinal transplant recipients are also at high risk for the development of graft-versus-host disease (GVHD). Their predisposition is triggered because the small bowel is a major lymphoid organ laden with both mesenteric lymph nodes and gut-associated lymphoid tissue. The reported incidence of GVHD in patients with intestinal transplantation is 5−14% [35]. Many symptoms of GVHD are not specific and may be confused with other disorders such as CMV infection, drug toxicity, and allergic reactions. The diagnosis should be confirmed by tissue biopsy and demonstration of donor-type chimerism on peripheral blood cells or recipient tissue. Histopathological criteria include keratinocyte necrosis, epithelial apoptosis of the native gastrointestinal tract, epithelial cell necrosis of oral mucosa, and donor cell tissue infiltration. The management of this pathology primarily consists of control of its symptoms such as fever, diarrhea, and skin lesions, with some cases also having been successfully treated with steroid administration and optimization of tacrolimus therapy [36]. The increasing use of thymoglobulin may be an important explanation for the decreased rate of GVHD in the recipients.

Intestinal transplant recipients have the highest incidence of posttransplant lymphoproliferative disorders (PTLD) in solid organ transplantation [37]. The principal factor responsible for this trend is the aggressive immunosuppressive regimens commonly used in intestinal transplant recipients. Chronic EBV infection of seronegative naïve B cells may also result in B transformation. Because pediatric recipients are usually EBV negative, they are at increased risk of PTLD. Talisetti et al. have proposed the removal of the intestinal graft and discontinuation of immunosuppression therapy as treatment for an aggressive PTLD that is limited to the intestinal graft. Because this is not a life-supporting organ it provides an opportunity to eradicate the disease and then to retransplant the patient once the PTLD is resolved [38].

33.3 CURRENT STATUS OF INTESTINE TRANSPLANTATION

According to UNOS data, the total number of registrations on the new intestine waiting list in 2009 was 260, compared with 170 in 2000 where 54.2% of the candidates were under 18 years of age The proportion of newly listed patients who were 18 years of age or older had been increasing in the prior decade. However, the number of patients on the waiting list under 18 years of age remained higher as compared to those who were 18 years of age or older, especially for those patients under 5 years of age (71.4% and 51.8%, respectively) (www.unos.org, 10/22/12). This data further indicates that pediatric patients have a higher risk of life-threating complications secondary to TPN, when compared with adults.

During the last decade, the number of intestinal transplants increased more than twofold. In 2009, there was a total of 180 intestinal transplants, of which 94 (52%) were for recipients less than 18 years of age. This increase was due primarily to a higher number of isolated intestine transplants as well as to increased number of combined liver—intestine transplants. Through the same year, 89 (49.4%) recipients required a combined liver and intestinal transplant (www.unos.org, 10/22/12). Children are the primary candidates for intestinal transplantation, and more than 70% are affected by intestinal and liver failure. Recipients of a combined graft experience better graft survival outcomes compared to those who receive an isolated intestinal transplant [39,40].

According to UNOS, between 2000 and 2009, the rate of DD intestinal transplants had increased from 65.2 transplants per 100 patient-years on the waiting list to 80.5 per 100 patient-years. However, the rate of LD intestinal transplants remains very low, with only two transplants performed in 2009. Current data indicates that the 5-year patient survival probability was superior for LD intestinal transplant recipients as compared to intestinal/stomach and deceased intestinal transplant recipients (2011 Intestinal Transplant Registry Report, www.intestinetransplant.org; Figure 33.2). However, the experience with LD intestinal transplants remains limited, with a very small number of procedures having been performed worldwide.

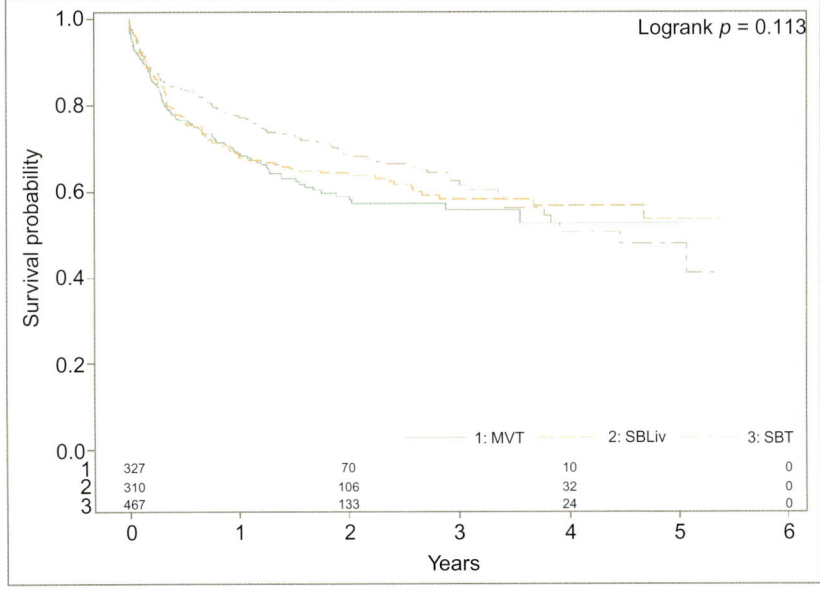

FIGURE 33.2 Survival rates of intestinal transplant recipients according to graft type from 2004 to 2009. MVT, intestine—stomach transplant; SBLiv, LD intestinal transplant; SBT, intestinal transplant. *Data from Intestinal Transplant Registry: 25 years follow-up report.*

In conclusion, LD intestinal transplantation has been perfected relative to technical details, leading results comparable with those performed with DDs. However, LD intestinal transplantation should be limited in accordance with specific indications. In particular, the best indication is probably for combined LD liver and intestinal transplantation in potential pediatric recipients with an intestinal and hepatic failure. For these potential recipients, the virtual elimination of waiting time may diminish the high mortality propability currently associated with candidates on the cadaveric waiting list. Isolated LD intestinal transplantation may further be indicated for candidates in need of an intestinal transplant with lack of central venous access as a rapid rescue strategy. Potentially, LD intestinal transplantation could be used with highly sensitized recipients, to allow the application of desensitization protocols. Finally, in the specific case of available identical twins or HLA-identical siblings, LD intestinal transplantation has a significant immunological advantage and should be offered.

33.4 REGENERATIVE MEDICINE IN ORGAN TRANSPLANTATION

The term *regenerative medicine* was defined in the late 1990s as the use of different elements coming from human resources to regenerate damaged or abnormal human tissue [41,42]. It is distinct from tissue engineering, which is defined as the manufacturing of body parts *ex vivo* by seeding cells into supporting scaffold [43]. We strongly believe, however, that the combination and collaborative efforts of these two fields can provide beneficial alternative therapies in the treatment of diseases currently presenting a pervasive gap in available resources for and knowledge on the replacement and treatment of respective dysfunctional tissue.

Concerning tissue engineered in the small intestine, some studies have shown a promising future, but there are some others that have had unsuccessful results. Nevertheless, much further research needs to be performed before applying it as a potential therapeutic approach in humans. Sala et al. in a study that was executed on two male six-week-old Yorkshire swine piglets, described the generation of a tissue-engineered small intestine and stomach from autologous tissue. A 10-cm section of jejunum was resected with a functional end-to-end anastomosis; the tissue was taken to the laboratory in which the organoid units were constituted from the resected tissue and then loaded onto biodegradable polymers. The polymers were made of polyglycolic acid tubes coated with 5% poly-L-lactic acid. Once the polymers were assembled, they were implanted into the omentum or mesentery of the autologous animal. After 7 weeks, all

the animals survived, and the implanted engineered tissue grew. The length of the villi and the depth of the crypts were similar to the native small intestine. Three types of differentiated intestinal epithelial cells were found. Enterocytes, goblet cells, and enteroendocrine cells were identified in their accustomed locations; smooth-muscle muscularis as ganglion cells between the layers of muscularis externa and submucosa were found as well. Moreover, double-cortin and CaM kinase-like-1 positive cells were identified at the base of the crypts, which is an intestinal stem cell marker [44]. This was a successful autologous animal model of an engineered small intestine from organ-specific stem cells. Totonelli et al. described a different approach, generating an acellular natural matrix preserving the intestinal architecture with the removal of donor-derived cells. They dissected the small intestine of Sprague Dawley rats from the pylorus to ileocecal valve and perfused the intestine lumen and vascular tree with phosphate-buffered saline. Subsequently, they performed a detergent enzymatic treatment (DET). Each cycle of DET was composed of deionized water, sodium deoxycholate, and 2000 kU DNase-I in 1 M NaCl. They concluded that after one cycle (roughly 31 h) a complete removal of cellular elements with the preservation of the macro and ultrastructural characteristics of the native tissue was achieved [45].

One of the investigations lacking success is described by Xu et al., who investigated the efficacy of an acellular dermal matrix for intestinal elongation. They used a Japanese white big-ear rabbit and a Wuzhishan miniature pig as donors for the elaboration of the acellular dermal matrix models and a third model using human acellular dermal matrix. They used Japanese white big-ear rabbits and Wuzhishan miniature pigs in the study. They divided them into three groups according to each type of acellular dermal matrix models. Although the animals survived after the operation, none of the groups reached a successful intestine elongation. Therefore, the authors concluded that the single use of acellular dermal matrix is not a suitable way to achieve intestinal elongation [46].

Although some investigations have been completed with successful results, we can only conclude that a better understanding of these two fields is necessary on how they can be complementarily and mutually beneficial before providing valid support and innovative strategies for patients with intestinal failure.

REFERENCES

[1] Alican F, Hardy JD, Cayirli M, Varner JE, Moynihan PC, Turner MD, et al. Intestinal transplantation: laboratory experience and report of a clinical case. Am J Surg 1971;121:150—9.

[2] Gruessner RWG, Benedetti E, editors. New York, NY: McGraw Hill; 2008.

[3] Deltz E, Mengel W, Hamelmann H. Small bowel transplantation: report of a clinical case. Prog Pediatr Surg 1990;25:90—6.

[4] Deltz E, Schroeder P, Gebhardt H, Gundlach M, Engemann R, Timmermann W. First successful clinical small intestine transplantation. Tactics and surgical technic. Chirurg 1989;60: 235—9.

[5] Tesi R, Beck R, Lambiase L, Haque S, Flint L, Jaffe B. Living-related small-bowel transplantation: donor evaluation and outcome. Transplant Proc 1997;29:686—7.

[6] Jaffe BM, Beck R, Flint L, Gutnisky G, Haque S, Lambiase L, et al. Living-related small bowel transplantation in adults: a report of two patients. Transplant Proc 1997;29:1851—2.

[7] Morris JA, Johnson DL, Rimmer JA, Kuo PC, Alfrey EJ, Bastidas JA, et al. Identical-twin small-bowel transplant for desmoid tumour. Lancet 1995;345:1577—8.

[8] Benedetti E, Pirenne J, Chul SM, Fryer J, Fasola C, Hakim NS, et al. Simultaneous en bloc transplantation of liver, small bowel and large bowel in pigs—technical aspects. Transplant Proc 1995;27:341—3.

[9] Nakhleh RE, Gruessner AC, Pirenne J, Benedetti E, Troppmann C, Gruessner RW. Colon vs small bowel rejection after total bowel transplantation in a pig model. Transpl Int 1996;9 (Suppl. 1):269—74.

[10] Gruessner RW, Sharp HL. Living-related intestinal transplantation: first report of a standardized surgical technique. Transplantation 1997;64:1605—7.

[11] Benedetti E, Holterman M, Asolati M, Di Domenico S, Oberholzer J, Sankary H, et al. Living related segmental bowel transplantation: from experimental to standardized procedure. Ann Surg 2006;244:694—9.

[12] Testa G, Panaro F, Schena S, Holterman M, Abcarian H, Benedetti E. Living related small bowel transplantation: donor surgical technique. Ann Surg 2004;240:779—84.

[13] Morris JA, Johnson DL, Rimmer JA, Kuo PC, Alfrey EJ, Bastidas JA, et al. Identical twin small bowel transplant after resection of abdominal desmoid tumor. Transplant Proc 1996;28:2731—2.

[14] Ghafari JL, Bhati C, John E, Tzvetanov IG, Testa G, Jeon H, et al. Long-term follow-up in adult living donors for combined liver/bowel transplant in pediatric recipients: a single center experience. Pediatr Transplant 2011;15:425—9.

[15] Ghafari J, Tzvetanov I, Spaggiari M, Jeon H, Oberholzer J, Benedetti E. The effect of small bowel living donation on donor lipid profile. Transpl Int 2012;25:19—20.

[16] Benedetti E, Testa G, Sankary H, Sileri P, Bogetti D, Jarzembowski T, et al. Successful treatment of trauma-induced short bowel syndrome with early living related bowel transplantation. J Trauma 2004;57:164—70.

[17] Langer JC. Abdominal wall defects. World J Surg 2003;27: 117—24.

[18] Ueno T, Fukuzawa M. Current status of intestinal transplantation. Surg Today 2010;40:1112—22.

[19] Kelly DA. Liver complications of pediatric parenteral nutrition—epidemiology. Nutrition 1998;14:153—7.

[20] Sondheimer JM, Asturias E, Cadnapaphornchai M. Infection and cholestasis in neonates with intestinal resection and long-term parenteral nutrition. J Pediatr Gastroenterol Nutr 1998;27:131—7.

[21] Avitzur Y, Grant D. Intestine transplantation in children: update 2010. Pediatr Clin North Am 2010;57:415—31.

[22] Buchman AL, Scolapio J, Fryer J. AGA technical review on short bowel syndrome and intestinal transplantation. Gastroenterology 2003;124:1111—34.

[23] Iyer KR, Iverson AK, DeVoll-Zabrocki A, Buckman S, Horslen S, Langnas A. Pediatric intestinal transplantation—review of current practice. Nutr Clin Pract 2002;17:350—60.

[24] Testa G, Holterman M, John E, Kecskes S, Abcarian H, Benedetti E. Combined living donor liver/small bowel transplantation. Transplantation 2005;79:1401—4.

[25] Sasaki T, Hasegawa T, Nakai H, Kimura T, Okada A, Musiake S, et al. Zoom endoscopic evaluation of rejection in living-related small bowel transplantation. Transplantation 2002;73:560—4.

[26] Benedetti E, Baum C, Cicalese L, Brown M, Raofi V, Massad MG, et al. Progressive functional adaptation of segmental bowel graft from living related donor. Transplantation 2001;71:569—71.

[27] Todo S, Tzakis AG, Abu-Elmagd K, Reyes J, Nakamura K, Casavilla A, et al. Intestinal transplantation in composite visceral grafts or alone. Ann Surg 1992;216:223—33.

[28] Goulet O, Michel JL, Jobert A, Damotte D, Colomb V, Cezard JP, et al. Small bowel transplantation alone or with the liver in children: changes by using FK506. Transplant Proc 1998;30: 1569—70.

[29] Trevizol AP, David AI, Dias ER, Mantovani D, Pécora R, D'Albuquerque LA. Intestinal and multivisceral transplantation immunosuppression protocols—literature review. Transplant Proc 2012;44:2445—8.

[30] Kimura T, Lauro A, Cescon M, Zanfi C, Dazzi A, Ercolani G, et al. Impact of induction therapy on bacterial infections and long-term outcome in adult intestinal and multivisceral transplantation: a comparison of two different induction protocols: daclizumab vs. alemtuzumab. Clin Transplant 2009;23:420—5.

[31] Oltean M, Herlenius G, Gäbel M, Friman V, Olausson M. Infectious complications after multivisceral transplantation in adults. Transplant Proc 2006;38:2683—5.

[32] Loinaz C, Kato T, Nishida S, Weppler D, Levi D, Dowdy L, et al. Bacterial infections after intestine and multivisceral transplantation. Transplant Proc 2003;35:1929—30.

[33] Roberts CA, Radio SJ, Markin RS, Wisecarver JL, Langnas AN. Histopathologic evaluation of primary intestinal transplant recipients at autopsy: a single-center experience. Transplant Proc 2000;32:1202—3.

[34] Green M, Michaels MG, Webber SA, Rowe D, Reyes J. The management of Epstein—Barr virus associated post-transplant lymphoproliferative disorders in pediatric solid-organ transplant recipients. Pediatr Transplant 1999;3:271—81.

[35] Pirenne J, Benedetti E, Dunn DL. Graft versus host response: clinical and biological relevance after transplantation of solid organs. Transplant Rev 1996;10:46—68.

[36] Mazariegos GV, Abu-Elmagd K, Jaffe R, Bond G, Sindhi R, Martin L, et al. Graft versus host disease in intestinal transplantation. Am J Transplant 2004;4:1459—65.

[37] Sudan DL, Kaufman SS, Shaw BW, Fox IJ, McCashland TM, Schafer DF, et al. Isolated intestinal transplantation for intestinal failure. Am J Gastroenterol 2000;95:1506—15.

[38] Talisetti A, Testa G, Holterman M, John E, Kecskes S, Benedetti E. Successful treatment of posttransplant lymphoproliferative disorder with removal of small bowel graft and subsequent second bowel transplant. J Pediatr Gastroenterol Nutr 2005;41:354—6.

[39] Gangemi A, Tzvetanov IG, Beatty E, Oberholzer J, Testa G, Sankary HN, et al. Lessons learned in pediatric small bowel and liver transplantation from living-related donors. Transplantation 2009;87: 1027−30.

[40] Testa G, Holterman M, Abcarian H, Iqbal R, Benedetti E. Simultaneous or sequential combined living donor-intestine transplantation in children. Transplantation 2008;85:713−7.

[41] Kemp P. History of regenerative medicine: looking backwards to move forwards. Regen Med 2006;1:653−69.

[42] Atala A. Engineering organs. Curr Opin Biotechnol 2009;20:575−92.

[43] Orlando G, Wood KJ, Stratta RJ, Yoo JJ, Atala A, Soker S. Regenerative medicine and organ transplantation: past, present, and future. Transplantation 2011;91:1310−7.

[44] Sala FG, Kunisaki SM, Ochoa ER, Vacanti J, Grikscheit TC. Tissue-engineered small intestine and stomach form from autologous tissue in a preclinical large animal model. J Surg Res 2009;156:205−12.

[45] Totonelli G, Maghsoudlou P, Garriboli M, Riegler J, Orlando G, Burns AJ, et al. A rat decellularized small bowel scaffold that preserves villus-crypt architecture for intestinal regeneration. Biomaterials 2012;33:3401−10.

[46] Xu HM, Wang ZJ, Han JG, Ma HC, Zhao B, Zhao BC. Application of acellular dermal matrix for intestinal elongation in animal models. World J Gastroenterol 2010;16: 2023−7.

Chapter 34

Intestinal Regeneration: The Bioengineering Approach

Khalil N. Bitar[a,b] and Shreya Raghavan[a,b]

[a]*Wake Forest Institute for Regenerative Medicine, Wake Forest School of Medicine, Winston-Salem, NC,* [b]*Virginia Tech-Wake Forest School of Biomedical Engineering and Sciences, Winston-Salem, NC*

Chapter Outline

34.1 INTRODUCTION TO THE ENTERIC NERVOUS SYSTEM

Gastrointestinal (GI) function (motility, secretion, and maintenance of fluid and electrolyte homeostasis) is controlled by three branches of the autonomic nervous system—the sympathetic, parasympathetic, and the enteric nerves. Intestinal secretomotor activity occurs in the absence of extrinsic autonomic contributions and is controlled predominantly by the enteric nervous system (ENS). Hence, the ENS is called the intrinsic innervation of the gut. The ENS assimilates information from local sensory input, muscle, and mucosa and determines the appropriate response of the gut. The ENS is the largest branch of the autonomic nervous system. The ENS is contained within the walls of the GI tract and divided into two ganglionated plexi—*Myenteric/Auerbach plexus* between the circular and longitudinal smooth muscle of the gut, *Submucosal/Meissner plexus* between the mucosal layer and circular smooth muscle of the gut. The ENS is comprised of several classes of functional neurons and enteric glia, similar to the central nervous system (CNS)[1].

The ENS demonstrates functional autonomy, capable of mediating GI secretomotor function independent of the CNS. ENS function can additionally be modulated by sympathetic and parasympathetic ganglia of the autonomic nervous system arising from the vagal, splanchnic, or the sacral ganglia. Structurally, in the mammalian gut, the neurons of the ENS (arising from either plexi) have axons >100 nm away from effector smooth muscle cells. Neurons lack a clearly defined 'postsynaptic area' and hence do not form neuromuscular junctions, as observed in other parts of the peripheral nervous system [2]. Smooth muscle cells in the intestine exist in an electrical syncytium and are innervated by hundreds of excitatory and inhibitory motor neurons to effect GI motility and the peristaltic reflex.

Major functional groups of neurons in the mammalian myenteric plexus are intrinsic sensory neurons, motor neurons, which result in smooth muscle contraction or relaxation, and interneurons. Relaxant neurotransmitters include nitric oxide, vasoactive intestinal peptide (VIP), pituitary adenylate cyclase activating peptide and purines. Contractile neurotransmitters in the gut include acetylcholine and tachykinins [3]. Depending on the species and region of the gut, the composition, size, innervation patterns and connectivity of internodal strands vary within the two plexi. The human ENS contains approximately 10^8 neurons, in par with the spinal cord [4].

The neurons and glia of the ENS are derived entirely from the neural crest. Migratory cells enter the foregut mesenchyme from the vagal axis and colonize the developing gut in a rostrocaudal fashion within 13 weeks of gestation in humans. Crest-derived cells from the sacral level also contribute to the innervation of the rectum and hindgut [5,6]. GDNF is the primary chemoattractant that mediates the migration of neural crest cells within and along the gut. The development of the ENS and neural crest cell migration are complex and dynamic processes. Multiple molecular signaling mechanisms involve neurotrophic factors (GDNF, ET3, and Notch), long-range guidance molecules (netrins, semaphorins, ephrins), and extracellular matrix components (laminins, collagens, heparan, and chondroitin sulfate proteoglycans). These cells differentiate into neurons and glia in the gut microenvironment to express a full complement of ENS neurotransmitters by 24 weeks of gestation. The ENS, however, continues to develop and differentiate up to 2 years postnatally [7,8].

34.2 DISORDERS OF THE ENS AND CURRENT THERAPEUTIC PARADIGMS

Disruption in the integrity of neural circuitry within the gut is described in several neurodegenerative conditions that result in dysfunctional motility, secretion, and inflammation. GI motor function is intimately controlled by the intramural enteric nervous system. It is a complex interplay between the smooth muscle of the muscularis externa and the two enteric neuronal plexi [6]. Depending on the region of the GI tract where nerve dysfunction occurs, a wide variety of syndromes can present themselves that include dysphagia, gastroesophageal reflux disease, altered gastric emptying, stenosis and sphincteric hypertrophy, abdominal pain, bloating, diarrhea, constipation, fecal incontinence, and megacolon.

Achalasia was originally characterized as a failure of relaxation of the lower esophageal sphincter (LES), clinically presenting itself as dysphagia and regurgitation. The main pathogeny is the absence of intrinsic inhibitory neurons in the esophagus and the LES but the preservation of cholinergic excitatory innervation, leading to tonic contraction and a lack of transient relaxation [9,10]. In achalasia of the internal anal sphincter (IAS), the rectoanal inhibitory reflex in response to rectal distention is markedly absent. Absence of nitrergic neurons

by NADPH diaphorase activity has been reported in cases of IAS achalasia [11]. Similar to achalasia, congenital hypertrophic pyloric stenosis demonstrates an inhibitory denervation of the pylorus leading to gastric obstruction. Tonic contraction maintained constantly evokes hypertrophy of the smooth muscle of the pylorus. Sphincteric achalasia has been treated with intrasphincteric injections of botulinum toxin, to temporarily inhibit the release of neuronal acetylcholine, thereby allowing transient reduction in sphincteric tone. This has been reported in both LES and IAS achalasia [12,13]. Pharmacological options include calcium channel blockers or long acting nitrates to reduce sphincteric tone. Pneumatic dilation and surgical myotomy are other options for the treatment of achalasia [14].

Neurogenic chronic intestinal pseudo-obstruction resulting in a derangement of peristalsis and intestinal failure occurs due to a noninflammatory degeneration of the ENS. Several putative pathogenic mechanisms are implicated in the loss of intrinsic neurons (altered calcium signaling, free radicals, etc.) [15]. Enteric neuropathy is also secondary to several other disorders (diabetes, Parkinson's disease, inflammation) resulting in GI dysfunction [16,17]. Neurodegeneration has been associated with the aging gut, with an age-related loss of enteric neurons resulting in altered intestinal motility and gastric emptying [18]. Pharmacological treatment is the mainstay of addressing symptoms of dysmotility in several ENS disorders. Prokinetics are documented to improve intestinal motility and control visceral sensitivity. Combination with antibiotics help reduce sepsis, nausea, and abdominal pain [19]. Surgery is not a recommended option for chronic intestinal pseudo-obstruction because only rare cases benefit from surgical resections [15].

Hirschsprung's disease or congenital megacolon is the best characterized disorder of the ENS, where enteric neurons are absent in varying lengths of distal gut. Biopsies of aganglionic segments demonstrate the absence of intrinsic inhibitory neurons leading to a lack of relaxation of the diseased segment. The aganglionic gut thus remains tonically contracted, obstructing the passage of intestinal contents leading to megacolon [20]. A pull-through surgery is essentially life-saving in Hirschsprung's disease, but the surgery can potentially disrupt anorectal innervation in the remaining ganglionic gut. Recurrent fecal soiling and fecal incontinence has been reported as a long-term outcome [21]. Enterocolitis is also a serious complication postsurgical correction for Hirschsprung's owing to the alterations in the gut microbiome and/or the loss of the intestinal mucosal barrier. It has been recently established that probiotic prophylaxis does not reduce Hirschsprung's associated enterocolitis [22].

34.3 TISSUE ENGINEERING OF THE INTESTINAL NEUROMUSCULATURE

In the following paragraphs, a current status and perspective of cell-based therapies for GI disorders is provided. The initial focus is on neural stem cell (NSC) transplantation, an emerging and exciting therapeutic paradigm to reinstate neurogenic dysfunction of the intestine. Several acellular biomaterials based approaches to intestinal regeneration and tissue engineering have been successfully applied. A future perspective on the use of NSCs in neuronal disorders of the gut is provided. Additionally, several approaches to smooth muscle replacement to remedy intestinal dysfunction are also tackled in the following paragraphs.

34.3.1 Introduction to NSC Transplantation

NSC therapy is an emerging therapeutic paradigm that ideally aims to reinstate neuronal function and thus GI motor function by repopulating the enteric plexi. The hope is to restore neuronal function, rather than pharmacologically provide palliative care. Several factors require to be taken into consideration while designing stem cell–based therapies to remedy aganglionic disorders or neurogenic disorders of the gut. Of principal importance is the source of stem cells, and if they can be reliably and efficiently isolated in adequate numbers from adults using minimally invasive procedures. A mode of delivery needs to be determined to effectively colonize and repopulate dysfunctional enteric ganglia. Furthermore, long-term viability and parameters for successful engraftment and phenotypic stability also need to be contemplated before stem cell therapy can be translated into the clinic successfully. Finally, *in vitro* research on phenotypes, trophic support, and directed differentiation could result in therapies designed to provide relief in particular GI disorders: e.g., an enriched population of nitrergic neurons could be used to relieve gastric dysfunction and gastroparesis. Such a population of neurons could also theoretically relieve sphincteric achalasia.

NSC transplantation is driven by two significant findings: (i) neuroglial progenitor cells can be isolated from adult mammalian gut, including ganglionated colon of Hirschsprung's patients [23,24] and (ii) progenitor cells can be induced to differentiate into several neuronal subtypes and glia characteristic of the ENS. Several sources of neural stem and progenitor cells have been explored— including CNS-derived stem cells, ENS-derived stem cells, and bone marrow–derived stem cells that differentiate into a neuronal lineage. The following paragraphs will expand on the various sources of neuronal stem cells, their isolation and therapeutic delivery strategies used to transplant these cells *in vivo*.

34.4 SOURCES OF SMOOTH MUSCLE AND NEURONAL STEM AND PROGENITOR CELLS

34.4.1 Isolation of Smooth Muscle Stem and Progenitor Cells

In order to avoid immunosuppressive regimens, autologous stem cell sources are ideal. Over the years, several tissue-resident organ-associated adult stem cells have been identified in the human body [25]. Bone marrow–derived stem cells have been demonstrated to assume vascular and bladder smooth muscle phenotypes and have thus been used in regenerative medicine applications [26–28]. This is an optimistic step toward producing tissue engineered smooth muscle from autologous bone marrow–derived cells. However, Hori et al. showed that bone marrow–derived mesenchymal stem cells seeded on collagen scaffolds did not regenerate a smooth muscle layer upon transplantation into the canine intestine [29]. Several populations of muscle-derived stem cells with clonal expansion and self-renewal capabilities have been identified [30,31]. These adult muscle stem cells can differentiate into myotubes as well as smooth muscle phenotypes [32]. Within the postnatal gut, interstitial cells of Cajal have been shown to transdifferentiate into smooth muscle cells upon blockade of Kit signaling, with increased expression of desmin and smooth muscle myosin [33]. Neural crest-derived nestin positive stem cells isolated from postnatal murine gut have also been demonstrated to differentiate into smooth muscle phenotypes [34].

34.4.2 Isolation of Neuronal Stem and Progenitor Cells

34.4.2.1 ENS as a Source of Neural Crest Stem and Progenitor Cells

The ENS is populated by migrating cells derived from the neural crest. Neural stem or progenitor cells have been isolated from embryonic, fetal, postnatal, and adult rodent guts. Cells have been isolated from full-thickness gut tissue, mucosal gut biopsies, or from the muscularis externa layers containing the myenteric plexus alone [34–37]. Enteric stem and progenitor cells have also been isolated from human full thickness/muscularis/mucosal gut biopsies [23,38–40]. Metzger et al. [23] have also demonstrated the reliable isolation of enteric neuronal progenitor cells from adult human gut up to 84 years of age. These cells have the ability to differentiate into a number of mature enteric neuronal subtypes. Additionally, neuroglial progenitor cells have also been isolated from ganglionated colons of Hirschsprung's patients [24]. A variety of techniques have been used to isolate neuronal stem cells from dissociated gut tissues. Embryonic mouse ENS and neonatal human

ENS-derived progenitor cells have been isolated following trypsin-EDTA dissociation of caeca [24,38]. Collagenase digestion in combination with trypsin has been used to obtain neuronal progenitor cells from rodent muscularis externa [37]. Enzymatic dissociations with a combination of collagenase/dispase combinations, as well as collagenase XI/dispase have been used to obtain human enteric neuronal progenitor cells from full thickness or mucosal biopsies [23,41]. For information on further selection of isolated ENS NSCs, refer to a review by Kulkarni et al. [43]. In a majority of these cases, isolated cells are cultured in nonadherent cultures, where these cells proliferate and aggregate to form floating spherical colonies, called neurospheres. This process of the formation of floating colonies of spherical aggregates has almost become a surrogate marker for "stemness" of NSCs.

Neural crest-derived enteric neuronal progenitor cells are positive for p75^{NTR} and have been immunoselected based on the expression of this low-affinity NGF receptor [23,24,38]. Nestin, an intermediate filament protein expressed by neuroepithelial stem cells, is also expressed in enteric neuronal progenitor cells isolated from rodent myenteric plexus [36,37,44]. Sox2 (SRY-related HMG transcription factor 2) is a marker of ENS progenitor cells and glial derivates, known to be expressed in embryonic and adult rodent gut [45]. Sox10 is another Sox transcription factor that is vital to the maintenance of progenitor status of these cells, so much so that normal Sox10 activity and endothelin signaling is essential for retrieval of sufficient numbers of enteric neuronal progenitor cells and their ability to form neurospheres. Mutations in the Sox10 gene have been implicated in the pathogenesis of congenital megacolon associated with Hirschsprung's disease [46,47]. Sox10 expression is also an indicator of gliogenic potential of neural crest-derived progenitor cells and is important for glial fate acquisition [48].

34.4.2.2 Non-ENS Sources of Neural Crest Stem and Progenitor Cells

The first report of obtaining neural crest derivatives (peripheral sensory and sympathetic neurons) from human embryonic stem cells was by Pomp et al. [50]. Lee et al. demonstrated the derivation of neural crest stem cells from human embryonic stem cells that were capable of migration and differentiation akin to neural crest identity [51]. Human embryonic stem cell–derived neuroepithelial stem cells have been transplanted into specific regions of the rodent brain, where they demonstrated capabilities of axonal growth and pathfinding [52]. Induced pluripotent stem cells offer an alternative, keeping in mind the ethical considerations of the use of human embryonic tissues. Through extensive genetic reprogramming, skin fibroblasts are routinely converted into a pluripotent

phenotype. Directed differentiation of induced pluripotent stem cells into multipotent neural crest stem cells with high efficiency, under defined media and feeder-free cultures is now well established [53]. However, induced pluripotent stem cells still retain a risk for tumorigenicity, and their therapeutic potential is severely limited by the use of retroviral transductions to derive them.

Postmigratory neural crest stem cells have been isolated from adult rodents from several distinct regions including the gut, dorsal root ganglia, sciatic nerve, and even bone marrow [54]. Recently, however, neural crest stem cells isolated from the bone marrow have been demonstrated to undergo *in vivo* tumorigenesis [55]. Neurogenesis has now been known to occur even in postnatal and adult mammalian brain [56]. Neural crest stem cells persist in restricted regions of adult brains, including the subventricular zone and the subgranular layer of the dentate gyrus and the external germinal layer of the cerebellum [57]. Similar to ENS-derived neurospheres, adult CNS-derived neural crest stem cells can be propagated in long-term cultures in the presence of epidermal and fibroblast growth factors. These cells retain their self-renewal capacity and multipotency [58]. CNS-derived neurospheres are often isolated from the subventricular zone of the rodent fetal brain. It has been demonstrated that these NSCs can be nudged into enteric-like neuronal subtypes when exposed to the gut microenvironment and gut-derived soluble factors, making them a candidate for transplantation [59]. Therefore, CNS-derived neural crest stem cells are a good alternative to ENS-derived stem cells. Additionally, it is also hypothesized that adult neural crest stem cells from various sources retain enough plasticity to assume ENS-specific phenotypes and behavior in response to the gut microenvironment. A schematic of the various sources of neuronal progenitor cells derived into enteric neurons/glia is shown in Figure 34.1.

34.5 A SUMMARY OF DELIVERY STRATEGIES FOR NSCS IN THE GI TRACT

Embryonic CNS-derived NSCs were injected bilaterally into the mid-pylorus of nNOS$^{-/-}$ mice. Grafted NSCs differentiated into neurons and glia and improved gastric emptying of liquids and improved the relaxation of the pyloric sphincter [60]. Fetal and postnatal ENS-derived NSCs were injected into the external muscle layer of the distal colon of wild-type mice. The grafted cells demonstrated several enteric neuron subtypes including nNOS and choline acetyl transferase [61]. Postnatal human enteric neurospheres were transplanted into aganglionic mouse distal hindgut where they migrated through pathways similar to neural crest cells developmentally. Furthermore, these cells differentiated into NOS and VIP neurons and glia [24]. Neuroepithelial cells were isolated from the embryonic rodent neural tube and injected into chemically denervated distal colons of rats. Grafted neuroepithelial stem cells were shown to differentiate into neurons and glia. Intraluminal pressure readings indicated an inflation stimulated contraction as well as electrical field stimulated responses [62]. Postnatal human enteric neurospheres obtained from intestinal mucosal biopsies were injected into explant cultures of human aganglionic gut or chick embryos. Grafted cells colonized aganglionic chick or human hindguts and generated ganglia-like structures with neurons and glia [39]. A recent study by Hagl et al. demonstrates that the microenvironment and smooth musculature of the aganglionic bowel of HSCR patients support NSC differentiation [63], providing promise for the development of NSC transplantation as a therapeutic tool.

34.6 SMOOTH MUSCLE INTESTINAL TISSUE ENGINEERING

Several approaches have been used to reengineer intestinal smooth musculature (either sphincteric or nonsphincteric). Conventional tissue engineering approaches initially used inert and/or prosthetic materials to replace a missing wall defect or conduit. Since then, much progress has been made toward the use of biocompatible materials that allow cellular infiltration, extensive remodeling, and adsorption. In addition to a significant repopulation of constituent cell types, ideal functional tissue engineering must also reinstate peristalsis, motility, and absorption/secretion. This requires functional smooth muscle cells and epithelial cells in addition to a functional neuronal network.

Several synthetic polymeric combinations have been demonstrated to be biocompatible, and capable of muscular ingrowth. Additionally, naturally occurring extracellular matrix-derived polymers have also been used to reconstruct segments of the GI tract. Autologous neoesophagus constructs have been engineered using human esophageal epithelial cells, aortic smooth muscle cells, and dermal fibroblasts embedded within human tendon collagen or polyglycolic acid meshes [64,65]. Saxena et al. used basement membrane matrix coated scaffolds that demonstrated muscular ingrowth of the smooth musculature as well as alignment and polarity of epithelial cells of the esophagus [66].

Several mechanical devices were used to control reflux and GERD due to sphincteric inadequacy, i.e. the lack of a sufficient high pressure from the LES. To date, the only cell-based regenerative medicine approach in repairing defects of the LES remains the injection of skeletal muscle-derived stem cells into the LES to augment sphincteric pressure in canine models. While these cells integrated within the underlying smooth musculature,

there was no demonstration of a contractile smooth muscle phenotype [67].

Collagen sponge scaffolds were used to repair gastric wall defects in which proton pump positive cells were regenerated [68]. Similar sponge scaffolds in combination with synthetic biodegradable polymers have been seeded with autologous bone marrow mesenchymal stem cells to repair gastric wall defects [69]. Gastric epithelial organoid units have been seeded on composite PLGA meshes to replace native stomach of rats [70]. The only regenerative medicine-based approach to repair defects of the pyloric sphincter is the injection of CNS-derived stem cells to reinstate nitric oxide neuronal function by Micci et al. [60].

Several studies have been published to date, reporting tissue engineering of the small and large intestines. Tissue engineering has offered an elegant solution to the bowel lengthening surgeries commonly carried out in short bowel syndrome. Autologous smooth muscle cells seeded within collagen sponge scaffolds have been used to successfully regenerate villi structures of the small intestine [71]. Tubular structure-based approaches have also had some limited success in replacing the smooth musculature of the small intestine [72–74]. Small intestinal submucosal ECM-based scaffolds have been used to replace tubular segments of bowel, where enteric neuronal plexuses were regenerated and met basic physiological demands in a canine model [75].

Intestinal organoid units generated from sigmoid colon have been used to tissue engineer tubular colonic structures [76]. Such tissue-engineered constructs have demonstrated a significant absorptive capacity when implanted. Fibrin-based hydrogels were used to generate three-dimensional concentrically aligned structures of circular smooth muscle layers of the colon [77]. Tissue engineered models of the longitudinal layer of colonic smooth muscle have also been demonstrated [78]. Individually, these structures mimic native smooth muscle alignment and maintain aspects of colonic physiology such as peristalsis and response to contractile and relaxant neurotransmitters. Recently, the use of chitosan to functionally support such tissue-engineered colonic constructs for intestinal tissue engineering was demonstrated [79].

Autologous skeletal muscle-derived stem cells were used to augment sphincteric function of the internal anal sphincter as well. However, no evidence of the development of functional smooth muscle was reported [80]. Physiologically functional tissue-engineered constructs of rodent and human internal anal sphincters have been developed, which demonstrate the spontaneous generation of myogenic basal tone [81–83]. Prewired bioengineered IAS constructs that included a functional innervation have also been bioengineered. These constructs have been implanted into athymic rodents, where they preserved neuronal networking as well as myogenic functionality [84].

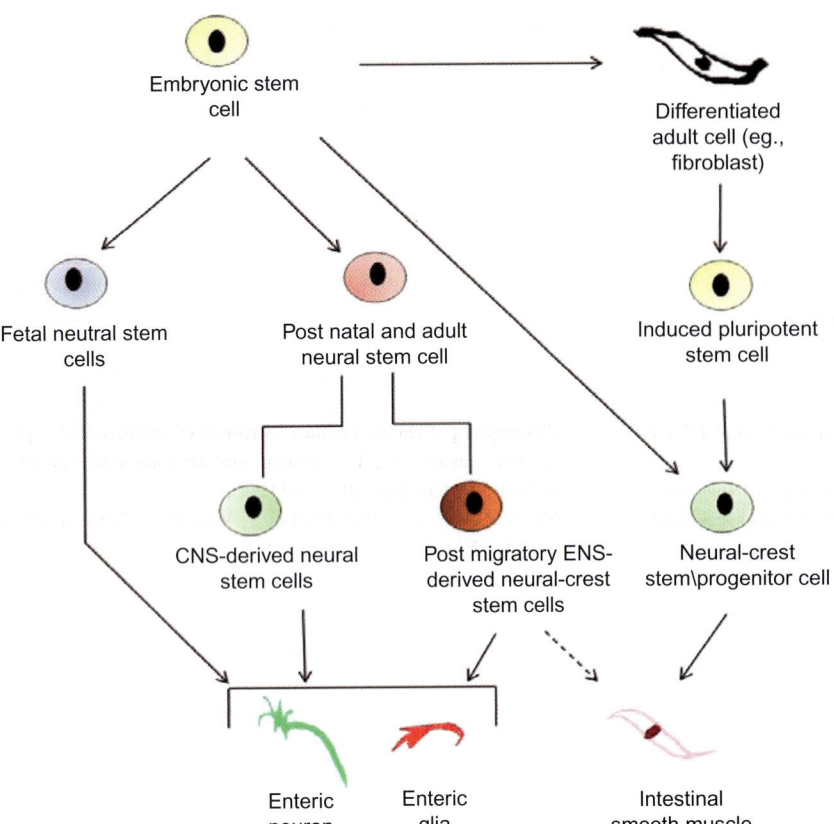

FIGURE 34.1 Schematic of the various sources of neural stem/progenitor cells. Enteric neurons and glia can be derived from various sources of stem and progenitor cells arising from different parts of the body. Adult CNS-derived as well as ENS-derived neural stem cells have been transplanted into mammalian gut, and have demonstrated differentiation into enteric neurons, enteric glia, and smooth muscle cells.

34.7 FUTURE PERSPECTIVES

Phenotypic stability, long-term survival, and posttransplant fate all remain to be optimized while moving forward with NSC transplantation for clinical use [43,44]. In order to provide trophic support and a permissive microenvironment, a more fundamental understanding of factors that affect and maintain differentiation of enteric neuronal progenitor cells is required.

As is true for all stem cells, their behavior and differentiation depends on their location, microenvironment, and the establishment or reestablishment of interactions with the extracellular matrix. Alternative to direct NSC transplantation, grafting of differentiated cells or progenitor cells committed to restricted phenotypes must be considered. This is a potential work around to avoid differentiation into undesirable phenotypes. Additionally, *in vitro* differentiation may allow tailoring of cell-based transplantation therapies to specific neurodegenerative diseases, e.g., supplying nitrergic neurons for sphincteric achalasia. Another area that needs intense focus is the identification of trophic support required for the survival of transplanted stem and/or progenitor cells. Parallel to this is the development of imaging modalities that noninvasively track or image transplanted cells. Magnetic resonance imaging

(MRI) offers a noninvasive method to image, but labeling transplanted cells with contrast agents or small paramagnetic iron oxide particles require cytotoxicity studies. Of paramount importance with cellular labeling is that the biological activity of the stem/progenitor cell is not altered, and their differentiation capability is not affected. The complex interaction between the host microenvironment and transplanted cells also needs to be considered, in order to improve survival and for cell fate determination. Biomaterial-based delivery can be used to provide trophic support to transplanted NSCs or to manipulate host inflammatory response. Extracellular matrix-based hydrogel environments can also be utilized to provide a protective microenvironment to transplanted cells. ECM molecules typical of the developing gut, similar to the developing mesenchyme, may provide a microenvironment conducive to sequestration and presentation of trophic factors, such as glial cell—derived neurotrophic factor, to influence neuronal cell survival. Lastly, as a cautionary tale from clinical studies of transplanting NSCs in several CNS-based neurological disorders, fundamental research on molecular bases of therapeutic plasticity of NSCs is important. Transplantation into animal models of GI disorders will offer a rationale for the design of future clinical studies.

ACKNOWLEDGMENT

This work was supported by *NIH RO1* DK 042876 and R01DK 071614.

REFERENCES

[1] Jessen KR, Mirsky R. Glial cells in the enteric nervous system contain glial fibrillary acidic protein. Nature 1980;286:736−7.

[2] Gabella G. Innervation of the intestinal muscular coat. J Neurocytol 1972;1:341−62.

[3] Furness JB. Types of neurons in the enteric nervous system. J Auton Nerv Syst 2000;81:87−96.

[4] Goyal RK, Hirano I. The enteric nervous system. N Engl J Med 1996;334:1106−15.

[5] Burns AJ, Douarin NM. The sacral neural crest contributes neurons and glia to the post-umbilical gut: spatiotemporal analysis of the development of the enteric nervous system. Development 1998;125:4335−47.

[6] Hansen MB. The enteric nervous system I: organisation and classification. Pharmacol Toxicol 2003;92:105−13.

[7] Young HM, Anderson RB, Anderson CR. Guidance cues involved in the development of the peripheral autonomic nervous system. Auton Neurosci 2004;112:1−14.

[8] Gariepy CE. Developmental disorders of the enteric nervous system: genetic and molecular bases. J Pediatr Gastroenterol Nutr 2004;39:5−11.

[9] Holloway RH, Dodds WJ, Helm JF, Hogan WJ, Dent J, Arndorfer RC. Integrity of cholinergic innervation to the lower esophageal sphincter in achalasia. Gastroenterology 1986;90:924−9.

[10] De Giorgio R, Di Simone MP, Stanghellini V, Barbara G, Tonini M, Salvioli B, et al. Esophageal and gastric nitric oxide synthesizing innervation in primary achalasia. Am J Gastroenterol 1999;94:2357−62.

[11] Hirakawa H, Kobayashi H, O'Briain DS, Puri P. Absence of NADPH-diaphorase activity in internal anal sphincter (IAS) achalasia. J Pediatr Gastroenterol Nutr 1995;20:54−8.

[12] Messineo A, Codrich D, Monai M, Martellossi S, Ventura A. The treatment of internal anal sphincter achalasia with botulinum toxin. Pediatr Surg Int 2001;17:521−3.

[13] Annese V, Bassotti G, Coccia G, D'Onofrio V, Gatto G, Repici A, et al. Comparison of two different formulations of botulinum toxin A for the treatment of oesophageal achalasia. The gismad achalasia study group. Aliment Pharmacol Ther 1999;13:1347−50.

[14] Pehlivanov N, Pasricha PJ. Achalasia: botox, dilatation or laparoscopic surgery in 2006. Neurogastroenterol Motil 2006;18:799−804.

[15] Antonucci A, Fronzoni L, Cogliandro L, Cogliandro RF, Caputo C, De Giorgio R, et al. Chronic intestinal pseudo-obstruction. World J Gastroenterol 2008;14:2953−61.

[16] De Giorgio R, Stanghellini V, Barbara G, Corinaldesi R, De Ponti F, Tonini M, et al. Primary enteric neuropathies underlying gastrointestinal motor dysfunction. Scand J Gastroenterol 2000;35:114−22.

[17] De Giorgio R, Guerrini S, Barbara G, Stanghellini V, De Ponti F, Corinaldesi R, et al. Inflammatory neuropathies of the enteric nervous system. Gastroenterology 2004;126:1872−83.

[18] Bitar K, Greenwood-Van Meerveld B, Saad R, Wiley JW. Aging and gastrointestinal neuromuscular function: insights from within and outside the gut. Neurogastroenterol Motil 2011;23:490−501.

[19] Rudolph CD, Hyman PE, Altschuler SM, Christensen J, Colletti RB, Cucchiara S, et al. Diagnosis and treatment of chronic intestinal pseudo-obstruction in children: report of consensus workshop. J Pediatr Gastroenterol Nutr 1997;24:102−12.

[20] Arshad A, Powell C, Tighe MP. Hirschsprung's disease. BMJ 2012;345:e5521.

[21] Catto-Smith AG, Coffey CM, Nolan TM, Hutson JM. Fecal incontinence after the surgical treatment of Hirschsprung disease. J Pediatr 1995;127:954−7.

[22] El-Sawaf M, Siddiqui S, Mahmoud M, Drongowski R, Teitelbaum DH. Probiotic prophylaxis after pullthrough for Hirschsprung disease to reduce incidence of enterocolitis: a prospective, randomized, double-blind, placebo-controlled, multicenter trial. J Pediatr Surg 2013;48:111−7.

[23] Metzger M, Bareiss PM, Danker T, Wagner S, Hennenlotter J, Guenther E, et al. Expansion and differentiation of neural progenitors derived from the human adult enteric nervous system. Gastroenterology 2009;137:2063−73.

[24] Almond S, Lindley RM, Kenny SE, Connell MG, Edgar DH. Characterisation and transplantation of enteric nervous system progenitor cells. Gut 2007;56:489−96.

[25] Weiner LP. Definitions and criteria for stem cells. Methods Mol Biol 2008;438:3−8.

[26] Tian H, Bharadwaj S, Liu Y, Ma PX, Atala A, Zhang Y. Differentiation of human bone marrow mesenchymal stem cells into bladder cells: potential for urological tissue engineering. Tissue Eng Part A 2010;16:1769−79.

[27] Kumar AH, Metharom P, Schmeckpeper J, Weiss S, Martin K, Caplice NM. Bone marrow-derived CX3CR1 progenitors contribute to neointimal smooth muscle cells via fractalkine CX3CR1 interaction. FASEB J 2010;24:81−92.

[28] Metharom P, Kumar AH, Weiss S, Caplice NM. A specific subset of mouse bone marrow cells has smooth muscle cell differentiation capacity-brief report. Arterioscler Thromb Vasc Biol 2010;30:533−5.

[29] Hori Y, Nakamura T, Kimura D, Kaino K, Kurokawa Y, Satomi S, et al. Experimental study on tissue engineering of the small intestine by mesenchymal stem cell seeding. J Surg Res 2002;102:156−60.

[30] Deasy BM, Jankowski RJ, Huard J. Muscle-derived stem cells: characterization and potential for cell-mediated therapy. Blood Cells Mol Dis 2001;27:924−33.

[31] Cao B, Huard J. Muscle-derived stem cells. Cell Cycle 2004;3:104−7.

[32] Hwang JH, Yuk SH, Lee JH, Lyoo WS, Ghil SH, Lee SS, et al. Isolation of muscle derived stem cells from rat and its smooth muscle differentiation. Mol Cells 2004;17:57−61.

[33] Torihashi S, Nishi K, Tokutomi Y, Nishi T, Ward S, Sanders KM. Blockade of kit signaling induces transdifferentiation of interstitial cells of cajal to a smooth muscle phenotype. Gastroenterology 1999;117:140−8.

[34] Suarez-Rodriguez R, Belkind-Gerson J. Cultured nestin-positive cells from postnatal mouse small bowel differentiate *ex vivo* into neurons, glia, and smooth muscle. Stem Cells 2004;22:1373−85.

[35] Schafer KH, Hagl CI, Rauch U. Differentiation of neurospheres from the enteric nervous system. Pediatr Surg Int 2003;19:340–4.

[36] Belkind-Gerson J, Carreon-Rodriguez A, Benedict LA, Steiger C, Pieretti A, Nagy N, et al. Nestin-expressing cells in the gut give rise to enteric neurons and glial cells. Neurogastroenterol Motil 2012;25:61–9.

[37] Silva AT, Wardhaugh T, Dolatshad NF, Jones S, Saffrey MJ. Neural progenitors from isolated postnatal rat myenteric ganglia: expansion as neurospheres and differentiation in vitro. Brain Res 2008;1218:47–53.

[38] Lindley RM, Hawcutt DB, Connell MG, Almond SL, Vannucchi MG, Faussone-Pellegrini MS, et al. Human and mouse enteric nervous system neurosphere transplants regulate the function of aganglionic embryonic distal colon. Gastroenterology 2008;135:205–16.

[39] Metzger M, Caldwell C, Barlow AJ, Burns AJ, Thapar N. Enteric nervous system stem cells derived from human gut mucosa for the treatment of aganglionic gut disorders. Gastroenterology 2009;136:2214–25.

[40] Rauch U, Hansgen A, Hagl C, Holland-Cunz S, Schafer KH. Isolation and cultivation of neuronal precursor cells from the developing human enteric nervous system as a tool for cell therapy in dysganglionosis. Int J Colorectal Dis 2006;21:554–9.

[41] Bondurand N, Natarajan D, Thapar N, Atkins C, Pachnis V. Neuron and glia generating progenitors of the mammalian enteric nervous system isolated from foetal and postnatal gut cultures. Development 2003;130:6387–400.

[42] Benedetti E, Holterman M, Asolati M, Di Domenico S, Oberholzer J, Sankary H, et al. Living related segmental bowel transplantation: from experimental to standardized procedure. Ann Surg 2006;244:694–9.

[43] Kulkarni S, Becker L, Pasricha PJ. Stem cell transplantation in neurodegenerative disorders of the gastrointestinal tract: future or fiction? Gut 2012;61:613–21.

[44] Schäfer Kh, Micci M-a, Pasricha PJ. Neural stem cell transplantation in the enteric nervous system: roadmaps and roadblocks. Neurogastroenterol Motil 2009;21:103–12.

[45] Heanue TA, Pachnis V. Prospective identification and isolation of enteric nervous system progenitors using Sox2. Stem Cells 2011;29:128–40.

[46] Bondurand N, Natarajan D, Barlow A, Thapar N, Pachnis V. Maintenance of mammalian enteric nervous system progenitors by SOX10 and endothelin 3 signalling. Development 2006;133:2075–86.

[47] Pingault V, Bondurand N, Kuhlbrodt K, Goerich DE, Prehu MO, Puliti A, et al. SOX10 mutations in patients with Waardenburg-Hirschsprung disease. Nat Genet 1998;18:171–3.

[48] Paratore C, Goerich DE, Suter U, Wegner M, Sommer L. Survival and glial fate acquisition of neural crest cells are regulated by an interplay between the transcription factor Sox10 and extrinsic combinatorial signaling. Development 2001;128:3949–61.

[49] Iyer KR, Iverson AK, DeVoll-Zabrocki A, Buckman S, Horslen S, Langnas A. Pediatric intestinal transplantation—review of current practice. Nutr Clin Pract 2002;17:350–60.

[50] Pomp O, Brokhman I, Ben-Dor I, Reubinoff B, Goldstein RS. Generation of peripheral sensory and sympathetic neurons and neural crest cells from human embryonic stem cells. Stem Cells 2005;23:923–30.

[51] Lee G, Kim H, Elkabetz Y, Al Shamy G, Panagiotakos G, Barberi T, et al. Isolation and directed differentiation of neural crest stem cells derived from human embryonic stem cells. Nat Biotechnol 2007;25:1468–75.

[52] Steinbeck JA, Koch P, Derouiche A, Brustle O. Human embryonic stem cell-derived neurons establish region-specific, long-range projections in the adult brain. Cell Mol Life Sci 2012;69:461–70.

[53] Menendez L, Kulik MJ, Page AT, Park SS, Lauderdale JD, Cunningham ML, et al. Directed differentiation of human pluripotent cells to neural crest stem cells. Nat Protoc 2013;8:203–12.

[54] Achilleos A, Trainor PA. Neural crest stem cells: discovery, properties and potential for therapy. Cell Res 2012;22:288–304.

[55] Wislet-Gendebien S, Poulet C, Neirinckx V, Hennuy B, Swingland JT, Laudet E, et al. In vivo tumorigenesis was observed after injection of in vitro expanded neural crest stem cells isolated from adult bone marrow. PLoS One 2012;7:e46425.

[56] Eriksson PS, Perfilieva E, Bjork-Eriksson T, Alborn AM, Nordborg C, Peterson DA, et al. Neurogenesis in the adult human hippocampus. Nat Med 1998;4:1313–7.

[57] Kriegstein A, Alvarez-Buylla A. The glial nature of embryonic and adult neural stem cells. Annu Rev Neurosci 2009;32:149–84.

[58] Gage FH, Ray J, Fisher LJ. Isolation, characterization, and use of stem cells from the CNS. Annu Rev Neurosci 1995;18:159–92.

[59] Kulkarni S, Zou B, Hanson J, Micci MA, Tiwari G, Becker L, et al. Gut-derived factors promote neurogenesis of CNS-neural stem cells and nudge their differentiation to an enteric-like neuronal phenotype. Am J Physiol Gastrointest Liver Physiol 2011;301:644–55.

[60] Micci MA, Kahrig KM, Simmons RS, Sarna SK, Espejo-Navarro MR, Pasricha PJ. Neural stem cell transplantation in the stomach rescues gastric function in neuronal nitric oxide synthase-deficient mice. Gastroenterology 2005;129:1817–24.

[61] Hotta R, Stamp LA, Foong JPP, McConnell SN, Bergner AJ, Anderson RB, et al. Transplanted progenitors generate functional enteric neurons in the postnatal colon. J Clin Invest 2013;123:1182–91.

[62] Liu W, Wu RD, Dong YL, Gao YM. Neuroepithelial stem cells differentiate into neuronal phenotypes and improve intestinal motility recovery after transplantation in the aganglionic colon of the rat. Neurogastroenterol Motil 2007;19:1001–9.

[63] Hagl CI, Rauch U, Klotz M, Heumuller S, Grundmann D, Ehnert S, et al. The microenvironment in the Hirschsprung's disease gut supports myenteric plexus growth. Int J Colorectal Dis 2012;27:817–29.

[64] Hayashi K, Ando N, Ozawa S, Kitagawa Y, Miki H, Sato M, et al. A neo-esophagus reconstructed by cultured human esophageal epithelial cells, smooth muscle cells, fibroblasts, and collagen. ASAIO J 2004;50:261–6.

[65] Sato M, Ando N, Ozawa S, Miki H, Kitajima M. An artificial esophagus consisting of cultured human esophageal epithelial cells, polyglycolic acid mesh, and collagen. ASAIO J 1994;40:389–92.

[66] Saxena AK, Kofler K, Ainodhofer H, Hollwarth ME. Esophagus tissue engineering: hybrid approach with esophageal epithelium and unidirectional smooth muscle tissue component generation in vitro. J Gastrointest Surg 2009;13:1037–43.

[67] Pasricha PJ, Ahmed I, Jankowski RJ, Micci MA. Endoscopic injection of skeletal muscle-derived cells augments gut smooth muscle sphincter function: implications for a novel therapeutic approach. Gastrointest Endosc 2009;70:1231−7.

[68] Hori Y, Nakamura T, Kimura D, Kaino K, Kurokawa Y, Satomi S, et al. Functional analysis of the tissue-engineered stomach wall. Artif Organs 2002;26:868−72.

[69] Araki M, Tao H, Sato T, Nakajima N, Sugai H, Nagayasu T, et al. Experimental study on in situ tissue engineering of stomach using new collagen sponge scaffold coated with biodegradable copolymers. Am Soc Artif Intern Organs 2006;52:5A.

[70] Maemura T, Shin M, Sato M, Mochizuki H, Vacanti JP. A tissue-engineered stomach as a replacement of the native stomach. Transplantation 2003;76:61−5.

[71] Nakase Y, Hagiwara A, Nakamura T, Kin S, Nakashima S, Yoshikawa T, et al. Tissue engineering of small intestinal tissue using collagen sponge scaffolds seeded with smooth muscle cells. Tissue Eng 2006;12:403−12.

[72] Lee M, Wu BM, Stelzner M, Reichardt HM, Dunn JC. Intestinal smooth muscle cell maintenance by basic fibroblast growth factor. Tissue Eng Part A 2008;14:1395−402.

[73] Kaihara S, Kim SS, Benvenuto M, Choi R, Kim BS, Mooney D, et al. Successful anastomosis between tissue-engineered intestine and native small bowel. Transplantation 1999;67:241−5.

[74] Kim SS, Kaihara S, Benvenuto MS, Choi RS, Kim BS, Mooney DJ, et al. Regenerative signals for intestinal epithelial organoid units transplanted on biodegradable polymer scaffolds for tissue engineering of small intestine. Transplantation 1999;67:227−33.

[75] Chen MK, Badylak SF. Small bowel tissue engineering using small intestinal submucosa as a scaffold. J Surg Res 2001;99:352−8.

[76] Grikscheit TC, Ochoa ER, Ramsanahie A, Alsberg E, Mooney D, Whang EE, et al. Tissue-engineered large intestine resembles native colon with appropriate in vitro physiology and architecture. Ann Surg 2003;238:35−41.

[77] Hecker L, Baar K, Dennis RG, Bitar KN. Development of a three-dimensional physiological model of the internal anal sphincter bioengineered in vitro from isolated smooth muscle cells. Am J Physiol Gastrointest Liver Physiol 2005;289:188−96.

[78] Raghavan S, Lam MT, Foster LL, Gilmont RR, Somara S, Takayama S, et al. Bioengineered three-dimensional physiological model of colonic longitudinal smooth muscle in vitro. Tissue Eng Part C Methods 2010;16:999−1009.

[79] Zakhem E, Raghavan S, Gilmont RR, Bitar KN. Chitosan-based scaffolds for the support of smooth muscle constructs in intestinal tissue engineering. Biomaterials 2012;33:4810−7.

[80] Kang SB, Lee HN, Lee JY, Park JS, Lee HS. Sphincter contractility after muscle-derived stem cells autograft into the cryoinjured anal sphincters of rats. Dis Colon Rectum 2008;51:1367−73.

[81] Singh J, Rattan S. Bioengineered human IAS reconstructs with functional and molecular properties similar to intact IAS. Am J Physiol Gastrointest Liver Physiol 2012;303:713−22.

[82] Somara S, Gilmont RR, Dennis RG, Bitar KN. Bioengineered internal anal sphincter derived from isolated human internal anal sphincter smooth muscle cells. Gastroenterology 2009;137:53−61.

[83] Raghavan S, Miyasaka EA, Hashish M, Somara S, Gilmont RR, Teitelbaum DH, et al. Successful implantation of physiologically functional bioengineered mouse internal anal sphincter. Am J Physiol Gastrointest Liver Physiol 2010;299:430−9.

[84] Raghavan S, Gilmont RR, Miyasaka EA, Somara S, Srinivasan S, Teitelbaum DH, et al. Successful implantation of bioengineered, intrinsically innervated, human internal anal sphincter. Gastroenterology 2011;141:310−9.

Chapter 35

Intestinal Regeneration: The Developmental Biology Approach

José E. García-Arrarás

Biology Department, University of Puerto Rico, Rio Piedras, Puerto Rico

Chapter Outline

To make an organ, it helps to know how nature does it [1]

Nature has provided two different views into organ formation. First, one can determine how the organ is made during embryological development. These events can be (and have been) studied in various animal models, however, the analysis of the important events can be confounded by the multiple ongoing events in the embryo, making it difficult to clearly determine a direct or an indirect effect. A second way of determining how "nature" builds an organ is to focus on the regenerative capacities of different species and identifying the cellular and molecular events that are important for organ regeneration. Thus, this chapter focuses on the organogenesis of the gastrointestinal tract, particularly of the small intestine, and looks at its formation during embryogenesis and during regeneration (Figure 35.1).

35.1 THE MOLECULAR BASIS OF ORGANOGENESIS

35.1.1 The Sequential Nature of Development

In order to be able to follow and understand the contents of this chapter it is necessary to provide a short introduction on how developmental biologists view the process of embryological organogenesis. What has become evident to developmental biologist is that organogenesis occurs in a sequential pathway where each differentiation step depends on both internal and exogenous factors. Thus, if we begin with the zygote, as it divides to form the cells of the embryo, the developmental pathway is initially signaled by maternal components, present within the oocyte. Later in development most of the signaling will come from the environmental milieu of the cell, including

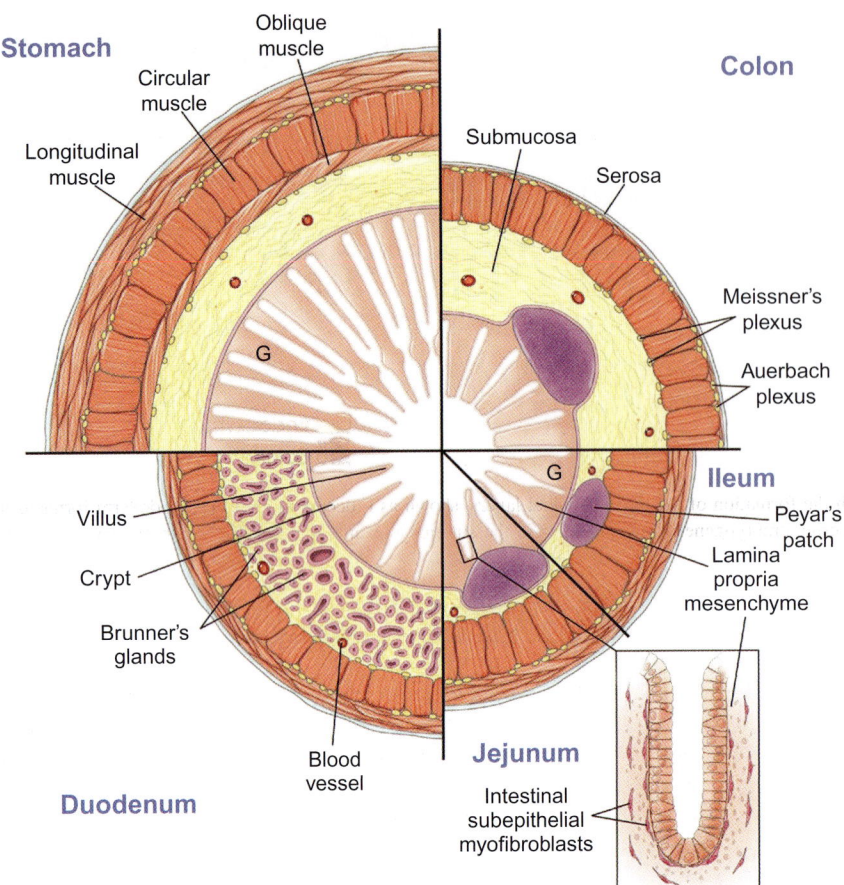

contacts with other cells, soluble factor in the extracellular space or via the components of the extracellular matrix (ECM). At each step along this path, as cells respond to a particular factor, there are changes in their profile of gene expression. These changes are mostly mediated by the activation or expression of transcription factors that bind to the cellular DNA and regulate the transcription of specific genes. The changes in gene expression can cause morphological and biochemical changes in the cell. But, more importantly, they have two obvious consequences: First, cells might stop responding to a particular factor they were able to respond to before. Second, cells will now be able to respond to a factor that they were unable to respond to previously. Thus, their gene expression pattern now changes in response to the new factors and therefore, changes occur in a stepwise manner in what classical embryologists have determined to be the stages of embryonic development.

The developmental process can be confounded by the apparent promiscuity of the differentiation-inducing factors. First, different factors might be able to induce a similar differentiation process (as determined by the gene expression changes in the cell exposed to these factors). Second, the same factor can act on different cell types to induce different differentiation outcomes or alternatively,

the same factor can act at different stages in the differentiation of a cell lineage producing different results. Take for example the effect of Wnts, which will be expanded in the following sections. In the early embryo, Wnt induces the expression of endoderm, while in latter embryonic stages it helps maintain the intestinal stem cells in a proliferating state.

An additional difficulty in the study of differentiation pathways is that there are a large number of transcription factors and many of these can be common to different cell or tissue types. For example, Choi et al. [3] did an expression survey of the transcription factors involved in intestinal development and found that an impressive high number of known transcription factors from different families (around 785) are expressed during this process. However, they found only a few differences in the transcription factor expression profile between two regions of the digestive tract (stomach and intestine).

35.2 EMBRYOLOGICAL ORGANOGENESIS

The gastrointestinal tract forms from the three main embryological layers: endoderm, mesoderm, and ectoderm. The endoderm gives rise to the luminal epithelium, the mesoderm to the muscle layers, submucosa and

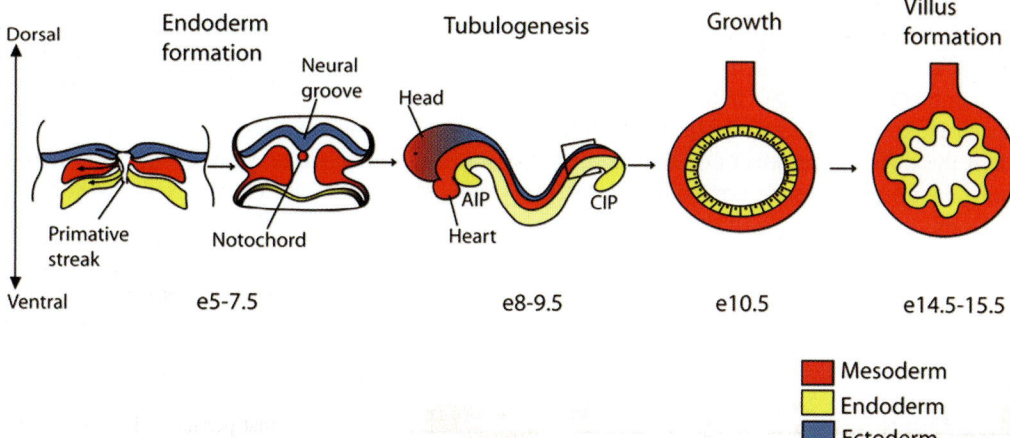

FIGURE 35.2 Intestinal development begins with the formation of the three germinal layers, shown as it occurs in mice. The endoderm forms as a single epithelial sheet underlying the mesoderm. During embryogenesis, invaginations in the anterior intestinal portal (AIP) and in the caudal intestinal portal (CIP) form a tube that is covered internally by endoderm and externally by mesoderm. This tube is innervated by cells from the neural crest that are of ectodermal origin. Thus the digestive tract (and the intestine) is made up of tissues that originate from the three germinal layers. Villus and crypt formation takes place later during gestation. *From [6].*

serosal epithelial layers, while the ectoderm gives rise to the neurons and supporting cells of the intestinal plexi (Figure 35.2). Soon after gastrulation has taken place and the three germinal layers have formed, the endoderm, which surrounds the yolk sac, establishes a close association with the overlying mesoderm (the splachnic mesoderm). The outcome of this association is the formation of a tube that extends along the anterior–posterior axis of the animal. This tube is the primordium of the digestive tract. Initially the most anterior and posterior regions of the endoderm invaginate to form the anterior intestinal portal (AIP) and the caudal intestinal portal (CIP) cavities, that will give rise to the foregut and hindgut primordia, respectively. These two cylindrical cavities extend to the midline and fuse, giving rise to the third region of the embryological digestive tract, the midgut. The formation and closure of the tube requires elongation of the endoderm and in mutants where this process does not occur, the closure of the digestive tube is hindered [4,5].

Once this has taken place the initial three divisions of the gut, the foregut, the midgut, and the hindgut, are established within the digestive tract primordium, which connects at the anterior end with the mouth and at the posterior end with the anus. Initially, the endodermal layer is homogeneous along the anterior–posterior axis, of the newly formed gut tube. Eventually the foregut will differentiate into the pharynx, esophagus, and stomach, the midgut will give rise to the small intestine while the hindgut will form the large intestine. The small and large intestines will form in the posterior region of the embryo.

The formation of the intestine involves the formation of the serosal and luminal epithelium. Between these two layers, the mesenchyme itself will form four concentric layers: the lamina propria adjacent to the luminal epithelium, the muscularis mucosa, the submucosa and the smooth muscle (circular and longitudinal) layers. As development proceeds, concomitant changes can be observed in the layers of the digestive tube, particularly in the width of the muscle and submucosal layers and in the differentiation of cells of the luminal epithelium that are associated with the formation of villi and crypts.

The formation and regionalization of the endoderm and associated mesoderm are controlled by gene regulatory networks of which several gene pathways have been characterized [7]. Studies in various vertebrate species, particularly in zebra fish, Xenopus, chicken and mouse have shown that many of the molecular pathways are highly conserved [5,8]. All of these model systems show the need of nodal/activin signaling pathway for the formation of both endoderm and mesoderm, where low doses specify mesoderm and high doses specify endoderm. The initial digestive tube shows little differences along its antero-posterior axis. However, once cells begin to differentiate into the various tissue layers the regionalization of the digestive tube will become evident. The mesodermal tissue plays an important role in the regionalization of the digestive tract. In fact, the studies of the interrelationship between the endoderm and the adjacent mesoderm provide some of the best-studied examples of epithelial–mesenchymal interactions during embryological organogenesis. During this processes, permissive and instructive signals mediate tissue morphogenesis and cell differentiation [2,8,9]. Instructive signals between the endoderm and mesoderm involve several signaling pathways; some of the best studied include Wnt, fibroblast growth factors (Fgfs), Notch, hedgehog (HH), and bone morphogenetic protein (BMP). Among these molecules, HH is probably one of the first inductive signals in intestinal development. It is expressed in the early endoderm in avians and mammals [10,11] and induces the expression of BMP4 and Hoxd genes [2,10]. In fact, it has been hypothesized that the different responses of the endoderm to the HH family members influence the formation of the different gut regions.

It is well known that Hox genes are involved in the patterning of the vertebrate digestive tract along the antero-posterior axis [10,12]. This patterning follows what is now known as the Hox code; in the developing embryo, there is a collinear expression where genes near the 3′ end of the hox gene cluster are expressed in more anterior positions while those near the 5′ end are expressed in more posterior positions [13]. Thus, a specific Hox code is expressed in each region of the developing gut. This expression is initially shown in the mesenchymal compartments and influences the development of the whole organ by the epithelial—mesenchymal interactions [14,15]. Thus, it is thought that the Hox genes provide permissive information to the mesenchyme that is the main cause of regionalization. In chickens, the expression domain of the HoxB cluster genes is associated with the formation of the midgut; duodenum, jejunum, ileum,

caeca and rectum [16]. Hoxb-6 is expressed in the jejunum and ileum, Hoxb-7 and -8 are expressed within the jejunum and Hox-9b is expressed in the posterior midgut and in the hindgut [16,17]. Similarly, in mice, regionalization depends on the Hox gene expression [17]. *In situ* studies of the expression of the different Hox clusters show that HoxA and HoxB are expressed from the duodenum, the HoxC cluster genes started from the jejunum, and HoxD cluster genes were expressed in the caudal part of the midgut, ileum, and cecum [15] (Figure 35.3).

Hox genes themselves appear to be regulated by retinoic acid (RA) in both mice and chickens [18]. RA has also been proposed to be synthesized in the endoderm [19]. Whatever its origin, it has been suggested that RA acts as a permissive signal that allows expression of Hox genes and in this way is involved in setting the anterior—posterior axis of the digestive tract [20]. The RA effects are possibly mediated by the intestinal mesenchyme and are more dramatic in the anterior part of the digestive tract but also extend to both anterior and posterior regions [21,22]. Finally, RA has also been implicated in the coiling of the intestine [23].

Once the intestinal tube has formed and the identity of each region established, the different layers undergo differentiation and growth according to the

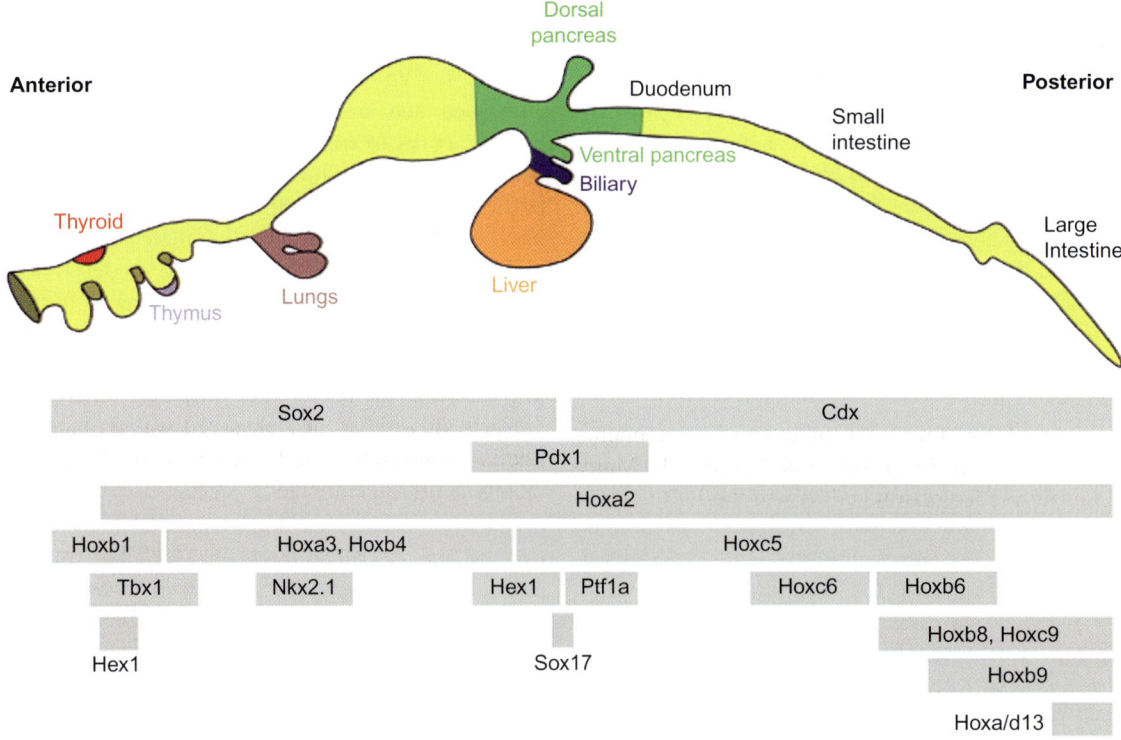

FIGURE 35.3 Expression domains of various transcription factors associated with the various regions of the digestive tract. The small intestine is characterized by the expression of Hox5c, Hox6b, Hox6c, Hox8b, Hox9, and Cdx2. Notice that different expressions of Hox genes correspond to the different regions of the small intestine. *From [8].*

parameters established during regionalization. As this takes place the radial symmetry of the intestine is established by the formation and differentiation of the various tissue layers.

35.2.1 Muscle Layers

The smooth muscle of the intestine originates from the loose mesenchymal cells of the intestinal primordium that surround the endoderm. There are spatial and temporal sequences in the formation of the muscle layers. First the circular layer is formed and this is followed by the formation of the longitudinal layer. Moreover, there is an anterior—posterior gradient where differentiation begins near the esophagus and moves caudally. The differentiation of the muscle layer can be followed by the expression of markers such as α-smooth muscle actin that is present in the early precursor myoblasts [2]. As development proceeds the myoblasts become more differentiated and begin expressing both α- and γ-smooth muscle actin.

Studies in chicken have shown that sonic hedgehog (SHH), secreted by the luminal epithelium inhibits mesenchyme from differentiating into muscle [24]. Once cells differentiate into muscle there appears to be some paracrine activity of BMPs that controls muscle development [25]. In mammals, there is early expression of BMP2 in the circular muscle layer that is apparently crucial for smooth muscle development [26]. The differentiating circular smooth muscle then produces PDGF-A that induces the differentiation of the longitudinal muscle layer [27].

Members of the HH superfamily have also been shown to be associated with the formation of the muscularis mucosa. Both indian hedgehog (IHH) and SHH have been shown to be necessary for normal formation of the muscularis mucosa and in the absence of HH activity there is mislocalization of the mesenchymal cells that would give rise to the muscularis mucosa [28].

35.2.2 Submucosa

The formation of the intestinal submucosa components has rarely been studied in an integrated approach. Many studies focus on the formation and role of the ECM, while other studies focus on the role of the mesenchymal cells, particularly the intestinal myofibroblasts (IMFs). However, these two components are closely related in view that the latter are involved in the production of ECM and can, either through the production of ECM or of soluble factors, interact with other tissue layers.

35.2.2.1 Extracellular Matrix

The ECM changes in terms of components and distribution during embryological development and these changes correlate with the transformations in intestinal morphology [29].

The role of the ECM has been well studied and there are some excellent reviews for those who would like to go into further detail [9,30]. Many ECM studies have focused on the basement membrane that lies at the base of the luminal epithelium. This basement membrane is made up by molecular components that originate both from the luminal epithelium and from the underlying mesenchymal cells. Among the molecules that have been characterized in the basement membrane are collagen, laminin, nidogen and perlecan. (This is a very simplified definition, since these proteins represent large families whose members show extensive variability. Moreover, the basement membranes themselves can be made up of different combinations of these proteins). The role of ECM molecules in the intestinal differentiation and morphological changes are best exemplified by the experimental data linking laminin isoforms to the expression of Cdx2 and the physiological maturation of the intestinal epithelium [30]. In particular, laminin α5 has been shown to be essential for the development and maintenance of the crypt-villus morphology and in its absence the intestine acquires a colon-like architecture [31]. Laminin can also play other roles during intestinal formation as it is known that it promotes neurogenesis and thus can induce the differentiation of enteric neurons [32].

The interaction between cells and the ECM is mediated via membrane-bound receptors in the cellular plasma membrane. These receptors, the most widely studied being the integrin family of receptors, recognize specific amino acid sequences in the ECM proteins and activate intracellular signaling pathways that can alter the cellular morphology, proliferation, adhesion, as well as many other cellular processes. The complexity of the integrin-ECM signaling pathway can be seen in experiments where they downregulated the β1 integrins in mice, expecting to observe changes in the adhesion of the luminal cells to the basement membrane [33]. Surprisingly, what they observed was an increase in epithelial proliferation and expansion of the mesenchymal components accompanied with a reduction in HH expression. Thus, they postulate that the integrin effect is mediated by HH effects on stromal cells that eventually cause an increase in the proliferation of the luminal epithelium.

35.2.2.2 Mesenchymal Cells

Other than the muscle cell components associated with the muscularis mucosa and the stromal lymphatic and blood vessel, the principal mesenchymal component are the fibroblasts, IMFs and the stromal stem cells [34]. The

IMFs have been subdivided into two groups, the interstitial cells of cajal (ICC) and the intestine subepithelial myofibroblasts (ISEMFs) [35,36]. The former are associated with pacemaker properties and smooth muscle control, while the later are associated with the lamina propria of the luminal layer and the regulation of luminal layer formation and homeostasis.

During embryogenesis, the cells of the original mesoderm, in their interaction with the endoderm differentiate into the muscle components. Less studied have been the interactions that give rise to the fibroblast components. It has been speculated that in the embryo IMFs originate from proto-myofibroblast, that originate from the serosal epithelium [37,38]. In contrast, the origin of adult IMFs is controversial; investigators have proposed many different origins for IMFs including mesenchymal stem cells from the bone marrow and transdifferentiation from fibroblasts [37]. Interestingly, a combination of ECM (fibronectin) and soluble factors (TGF-β) appears to be required for the differentiation of IMFs in the adult [37].

IMFs, and in particular the ISEMFs, play important roles in the formation and maintenance of the intestinal tissue [35−37]. In the human embryo, ISEMF are associated with the muscularis mucosa at the base of crypts. Their number increases during gestation to the time of birth. After birth they are mostly associated with the lower region of the crypt, where one of their main functions is the regulation of luminal epithelium proliferation and differentiation. They are also involved in crypt-villus morphogenesis. They not only produce and secrete a large range of growth factors, ECM molecules, cytokines, matrix metalloproteinases and inhibitors of metalloproteinases, but also respond to many extracellular factors secreted from the luminal epithelium or from adjacent cells within the submucosa. In fact, they are known to respond to many of the growth factors involved in the formation of the digestive tract in the embryo (IGF-I, IGF-II, IL-1B, TNF-α, PDGF, EGF, and bFGF, among others).

IMFs or their precursors, are probably responsible for most of the mesenchymal−epithelial interactions of the intestine that are described in this chapter [34,37]. These interactions can take place directly by the release of signaling molecules, or indirectly by the deposition of ECM molecules, particularly in the lamina propria, that exert an effect on the luminal epithelium. For example, disruption of HH signaling between the epithelium and the mesenchyme described elsewhere causes a disorganization in the development of myofibroblasts and malformations of the muscularis mucosa. Similarly, IMFs produce BMP antagonists that help create the stem cell niche in the luminal crypts and promote epithelial differentiation.

35.2.2.3 Lymph and Blood Vessels

The embryonic mesenchyma also forms the blood vessels during intestinal embryological organogenesis. As in many other organs, the mesoderm differentiates *in situ* into endothelial cells that form blood vessels. The origin of the lymphatic vessels is still controversial. Some evidence points to the formation of primary lymphatic endothelial cells directly from the mesenchyme, while other experiments show that they originate from embryonic venous endothelial cells [39,40]. Nonetheless, the process of vascular and lymphatic vessel formation has been studied in many organs, but there is little information specific to the intestine, thus, it will not be discussed in this review.

35.2.3 Enteric Nervous System

The enteric nervous system (ENS) is formed by neural crest cells that migrate into the gut mesenchyme of the intestinal primordium where they proliferate and differentiate forming the ENS components [41]. There are two sites of origin for the neural crest that colonize the forming gut. Anteriorly, the cranial neural crest between somites 1−7, forms the ENS of the anterior regions of the digestive tract (preumbilical). The sacral neural crest (caudal to somite 28) will form the posterior (postumbilical) ENS. Proliferation takes place in the migrating neural crest and this proliferation is essential for the colonization of the gut particularly in supplying the correct number of cells, since those in the migrating front are usually nonproliferative. Once the migrating neural crest colonizes the intestine, the cells will form ganglia between the two layers of smooth muscle (myenteric or Auerbach plexus) and in the submucosa adjacent to the luminal epithelium (submucosal or Meissner plexus). The function of the myenteric plexus is mostly associated with the regulation of contraction of the two muscle layers (the muscularis externa or tunica muscularis) in peristalsis. The function of the submucosal plexus is mainly associated with the regulation of the muscularis mucosa and the modulation of endocrine and absorptive cells within the luminal epithelium.

Among the factors that have been shown to affect ENS formation are SHH and BMPs. The ongoing hypothesis is that an endoderm−mesoderm interaction takes place where SHH induces BMP4 in the splanchnic mesoderm. SHH promotes proliferation of neural crests but inhibits migration and differentiation [42]. Mice lacking IHH or SHH either present neurons in abnormal areas of the digestive tract or lack them altogether [11,43]. In mammals, both BMP (2 and 4) and their receptors are expressed in populations of neural crest cells in the intestinal primordia [44]. Manipulations of BMP levels show that they have a large range of effects on the ENS. BMP inhibits normal migration of ENS into the mouse intestine [45,46] and later. In development induces differentiation of both neurons and glia by promoting exit of precursors from the cell cycle [42,47]. Interestingly, Hirschprung disease, a genetic disorder characterized by agangliosis (lack of ENS) in a region of the large intestine, is produced in some patients by a genetic mutation that brings about a decrease in BMP signaling [48]. Hox genes are also involved in ENS formation as described in studies showing that migrating and differentiating neural crest cells in humans express HoxB5 [49].

Other than the molecular cues that modulate the formation of most, if not all, intestinal structures, several molecules appear to play a very specific role in the formation of the ENS. Primarily among these is the RET-GDNF signaling pathway [50,51]. RET is a tyrosine kinase receptor present in the neural crest cells while GDNF (glial cell-derived neurotrophic factor) would be the ligand produced by the mesenchymal cells. GDNF expression also follows an anterior–posterior pattern. This ligand-receptor activated path is essential for the proliferation and survival of enteric neurons and glia. Moreover, it has been also shown that GDNF can serve as a chemoattractant for migrating crest cells.

A second factor that has been associated with the formation of the ENS is endothelin-3 (Edn3 or ET-3). This factor inhibits the differentiation of the neural crest cells into neurons and in its absence the digestive tract remains aganglionic [52]. It has been proposed that ET-3 promotes or maintains cell proliferation, thus inhibiting differentiation, since this requires removal from the cell cycle and the cessation of migration. Thus, ET-3 role in the developing digestive tract would be to prevent cells from moving from a proliferative state to a differentiating state before the gut has been colonized [32]. A possible molecular link with both ET-3 and GDNF is provided by the transcription factor Sox10, which is expressed in the neural crest during their migration. It has been shown that Sox10 levels are critical for maintaining the ENS progenitors in an undifferentiated state and in its absence total intestinal aganglionosis occurs [51].

In this context it is interesting to note that the submucosal ganglia forms later than the myenteric and that cell proliferation within the ganglia continues some time after it has decreased or stopped within the myenteric ganglia. It has been proposed that netrin produced by the luminal epithelium serves to attract the enteric precursors that give rise to the submucosal ganglia in a radial migration [53]. This migration of the neural crest toward the mucosa by netrin is converted to repulsion by the presence of laminin in the basement membrane. Therefore, the cells stop their migration at the submucosa–mucosa interphase, giving rise to the submucosal plexus.

Enteric neurons differ greatly in their morphological and biochemical properties as well as in the connections they make within the developing organ. Therefore, once the neural crest cells form the enteric ganglia a process of axonal growth and connections, and of neuronal differentiation takes place. Few experiments have addressed these two processes, but the available data is well summarized by Saselli et al. in their recent review [51].

35.2.4 Luminal Epithelium

In the very early stages of embryogenesis, Wnt has been associated with the formation of the endodermal tissue layer and its expression is instructive for endodermal cells to become intestinal instead of other endodermal lineages [54]. Recent studies have shown that Wnt acts in a concentration-dependent manner and that while high doses of Wnt activate endoderm gene expression patterns similar to those of the large intestine, low doses of Wnt activate endoderm gene expression patterns similar to those of the small intestine [55]. The initial luminal layer is composed of a pseudostratified epithelium that eventually is transformed into a stratified luminal epithelium. It is this stratified epithelial layer that reorganizes through the formation of secondary lumina to form a simple columnar epithelium [56,57]. This morphological reorganization takes place in a wave of morphogenetic activity that originates in the anterior part of the digestive tract and moves toward the posterior regions.

Once the endodermal epithelium is formed, the morphological and biochemical differences that characterize each region of the digestive tract must be established. For example, the luminal epithelia of the digestive tract differ in cell types and abundance, as well as on the presence of particular structures, such as crypts and villi, from the gastric or colonic epithelium. An important mediator in establishing the identity of the intestinal region is the parahox gene Cdx2. It has been proposed that an FGF gradient might serve to establish the Cdx2 expression in the endoderm [58]. Cdx2 is expressed in the endoderm from early in development and its expression is retained by the luminal epithelium of the intestine. It instructs the mucosa to acquire the characteristics of the intestinal mucosa via a cross talk between the endoderm and the underlying mesoderm [58]. In its absence, depending on the stage when it is knockout, the mucosa might acquire either gastric or esophageal properties [59]. Although Cdx2 is necessary during embryonic life for the formation of the intestinal mucosa it has also been shown that it is required in adult life for its maintenance [60,61]. When the gene is ablated in adults there are drastic changes in the intestinal mucosa that can be seen in its plasticity, cytoarchitecture and even the expression of genes not normally expressed by its cells [60,61].

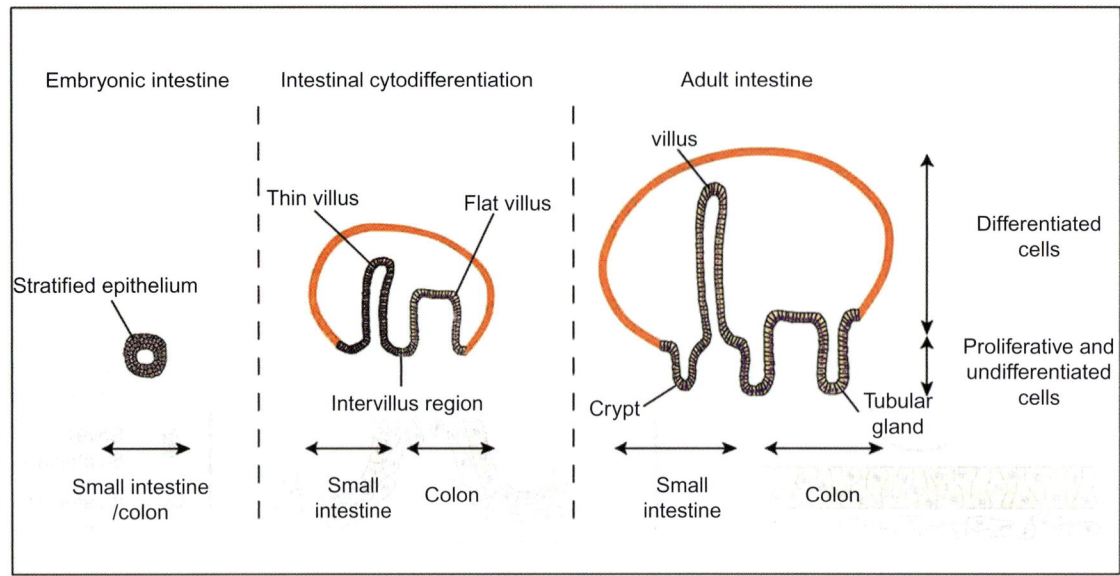

FIGURE 35.4 Differentiation of the small and large intestine. Initially, the luminal epithelium is similar all along the length of the intestinal primordium, consisting of a simple epithelial layer. As embryological development continues villi form in both small and large intestine, although with further development, crypts form in the intervillus region and villi dissapear from the large intestine. *From [15].*

The next step in the development of the intestinal mucosa is the formation of the villi and crypts. The cellular and molecular events associated with villi and crypt formation are summarized in the following section but have been presented in great detail by Spence et al. [5] (Figure 35.4).

35.2.4.1 Villi and Crypt Formation

Villi formation occurs by the invagination of the mesenchymal cells that underlie the luminal epithelia, a process that follows the anterior—posterior wave of epithelial reorganization. These mesenchymal cells condense under the basal lamina and grow in fingerlike projections toward the lumen giving rise to the villi.

Crypt formation originates within the basal part of the villi as new luminal structures form and extend into the underlying mesenchyma and eventually fuse with the intestinal lumen by an upward migration. Initially, cell proliferation occurs along most of the luminal epithelium, but once villi form it is restricted to the intervillus spaces. These are the areas where the crypts will form.

Villi and crypt formation rely on similar epithelial—mesenchymal interactions as those observed in early developmental stages of gut formation (Figure 35.5). Formation of the luminal epithelium is also affected by SHH and BMP. SHH (and IHH) are expressed at the base of the small intestinal villi. When their expression is reduced it causes the formation of abnormal villi and an increase proliferation of epithelium [11,50—62]. In both chicken and mouse embryos, BMP signaling activity has been detected within the luminal epithelium in later embryonic stages [25,63]. Inhibition of this activity in transgenics (by increasing a BMP inhibitor, noggin or mutant BMP receptors) causes abnormal villus formation and excessive crypt formation [63—65].

Wnt signaling pathways are also important in establishing and maintaining the proliferation of cells within the crypt compartments [15,66]. It has been proposed that as cell migrate out of the crypts they lose the proliferation capacity either due to dilution of the Wnt signal or the presence of other factors in the villi epithelia (such as BMP) that counteract the Wnt action [54]. Other factors are also involved in crypt and villi formation. For example, RA application induces acceleration of villus outgrowth and epithelial cell differentiation, stimulation of muscle and crypt formation [67]. These effects are also possibly meditated by the intestinal mesenchymal cells.

Initially crypts can originate from one or more stem cells. However, as the crypt matures it moves toward clonality and only one stem cell survives producing all cells within the crypt [68]. Villi on the other hand are polyclonal since they are formed by the migration of stem cell derivatives from various adjacent crypts.

An additional aspect of crypt and villi formation relates to the overall growth of the intestinal organ. As the digestive tract grows together with the growing organism, the intestine grows with it and new crypts and villi must be formed. The expansion of crypts occurs by a process named "crypt fission," where an existing crypt

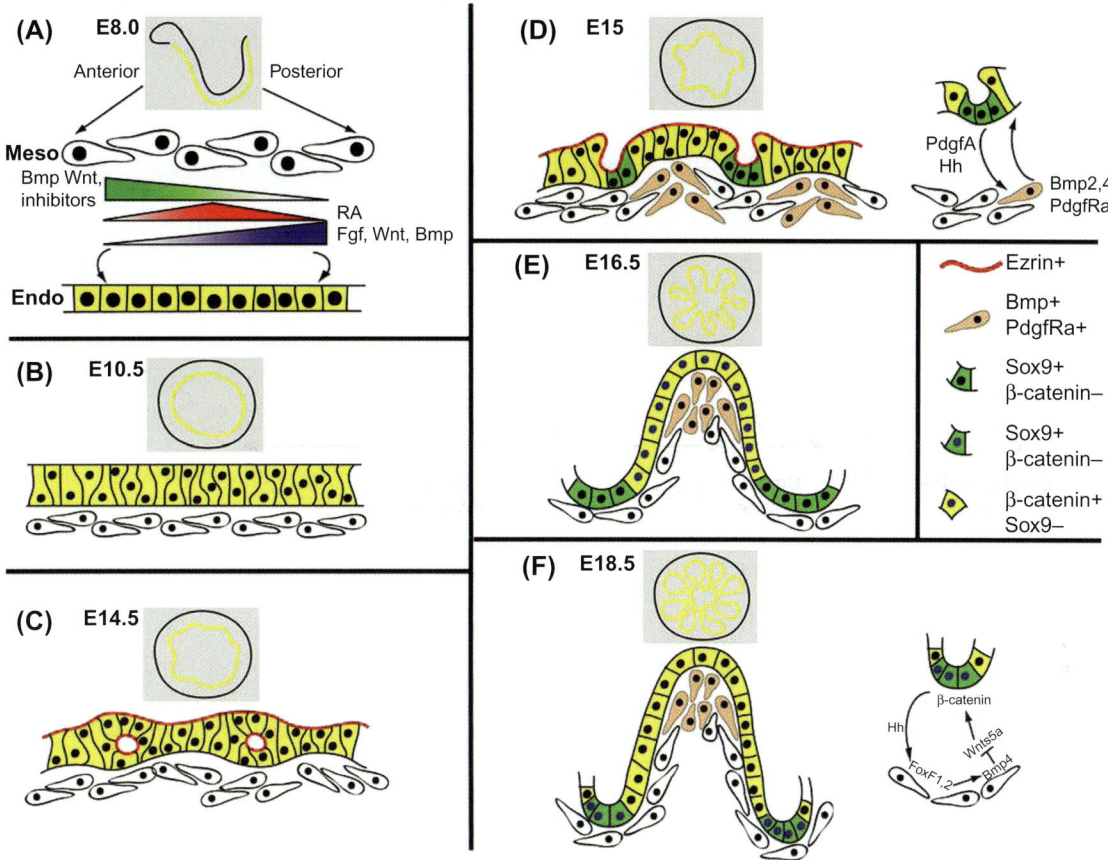

FIGURE 35.5 Mesenchymal−epithelial interactions during mammalian intestinal development. (A) Early during embryogenesis the mesoderm sends inductive signals to the endoderm. (B and C) These signals induce morphological and biochemical changes in the endoderm that eventually form a columnar epithelia. (D and E) The mesenchymal cells induce the formation of villi at later embryological stages. (F) In late gestation, the mesenchymal cells regulate the division of intestinal stem cells in the crypts. *From [69].*

divides into two new crypts. As should be expected some of the same cellular and molecular events that originally led to the formation of crypts are also involved in "crypt fission." For example, Wnt inhibition is associated with a loss of crypts while activation by Wnt agonists induces overproduction of crypts. Nonetheless, the precise mechanisms that underlie the maintenance of a "functional number" of crypts or the process of "crypt fission" are far from being determined.

35.2.4.2 Differentiation of Luminal Epithelium

The adult vertebrate small intestine is characterized by the presence of at least four different cell types that are localized in specific regions of the crypts/villi complex. There are various reviews that provide in depth information on these cells and their lineages [5,6]. The cell types have been divided according to their function and lineage into absorptive (enterocytes) or secretory (enterocrine, Goblet, Paneth) cells. The enterocytes are the most abundant cells in the small intestine, they are columnar with

apical microvilli, and play the key role associated with the intestine, that of nutrient absorption. Goblet cells are more rounded mucous-secreting cells and enteroendocrine cells are neuroendocrine cells that secrete hormones. Paneth cells secrete antimicrobial peptides, as well as several growth/maintenance factors and, in contrast to the other cell types that are mainly present in the villi, Paneth cells are found at the base of the crypts. In recent years a new, although not abundant, cell type, named tuft cells, has been found in the luminal epithelium [70−72]. These cells are chemosensitive and appear to be part of the secretory group. The different cell types begin their differentiation from the luminal endoderm at the same time as villi begin to appear. The first cells to appear are the enterocytes, enteroendocrine and Goblet cells, with Paneth cells appearing later during embryogenesis and tuft cells making a late appearance after birth. Wnt is also important in the establishment of the cells lineages. It induces the proliferation of the transit-amplifying cells but at the same time induces Paneth cell terminal differentiation [66,73].

35.2.5 Intestinal Lengthening

Once the concentric layering of tissues is established in the digestive tract, and as cellular differentiation takes place within each tissue layer, a process of intestinal lengthening must take place. This process ensures that the digestive tract grows in size as the embryo grows. However, the growth of the digestive tract in the anterior–posterior axis takes place at a faster rate than that of the embryo itself, thus resulting in the twists and curving characteristic of the digestive tract.

The events that lead to intestinal lengthening are not well understood and have received less attention than those involved in early formation of the tube itself. However, there is a growing interest in deciphering this process in view of diseases or conditions that result in Short Bowel Syndrome, where only a small region of the small intestine remains for functional absorption, and this region cannot undergo the necessary growth or lengthening to become fully functional and maintain the adult organism,

FGF9, has been shown to be involved in the elongation of the digestive tract [74]. Mice that do not express the protein have a shorter small intestine than wild type or heterozygotic mice. The effect appears to be specific to the small intestine (the colon length is normal) and to the length, since the overall caliber of the intestine and the cell type distribution also appear normal. FGF9 is expressed in the luminal epithelium and acts directly on fibroblasts [75]. Proliferation analyses using BrdU show reduced cell division in the mesenchyme of the mutant mice, once again showing the epithelial–mesenchymal cross talk common to many intestinal developmental processes [74].

Three other factors have been associated with intestinal elongation: IGF-1, epimorphin, and Wnt5a. Overexpression of IGF-1 increases the overall length of the intestine. However, this effect is not uniform as there is an increase in muscle cell thickness but not in the thickness of the mucosal layer. Epimorphin (or Stx2) is found in the mesenchymal cells, probably in the ISEMFs [20]. Its suggested role is to counterbalance the effect of FGF9, since mice that do not express the protein have a longer intestine than wild types. Wnt5a is expressed in the mesenchymal cells. Its involvement in intestinal elongation is suggested by experiments showing that the small intestine of Wnt5a null mice is reduced up to 80% length [76]. Its proposed mechanism of action is by controlling cell division during gut elongation and by regulating the intercalation of cells into the epithelium following cell division.

In addition to the involvement of these molecules in intestinal elongation other factors might be involved. For example, an inbred mice strain that shows a digestive tract with increased intestinal length has been documented, however, the molecular basis for this phenomenon remains unknown [77].

35.2.6 Intestinal Maturation

Up to now we have been mainly concerned with the formation of the intestinal organ in terms of cell types, tissue layers, and overall organ structure. Although acquiring the intestinal morphology is important, if not essential, we still need to determine when and how the functional properties of the intestine appear. Some excellent reviews provide information on the ontogeny of functionality of the luminal epithelia [78–81]. In mammals, transport mechanisms appear in the absorptive cells before birth, following intestinal organogenesis. These transporters allow absorption of carbohydrates, amino acids, and proteins from the amniotic fluid. The appearance of these transporters occurs in the later half of embryonic life and, in the fetus, can be found in both small and large intestine. Nonetheless, it is during postnatal development when large changes in absorption capacity take place to accommodate the nutrition needs of the newborn and eventually the variability in diets of the juvenile and adult.

Similarly, patterns of gut motility appear about midgestation in mammalian gut development [82,83]. These initial propagating contractions are myogenic in origin and do not depend on neurons or ICCs [84]. Contractions mediated by the nervous components appear late in fetal development, once the smooth muscle has fully matured and ICCs and neurons have established their neuronal circuitry. Similar to the ontogeny of absorptive function, it is after birth that the different and complex motility patterns of the intestine are established.

The growth and maturation of the digestive tract is also regulated by endocrine hormones [85,86]. This comes about due to the advent of the hypothalamic-pituitary axis, the beginning of function of endocrine organs, and the growing vascularization of the intestine that takes place in the later stages of gestation. Therefore, while the initial events of intestinal formation rely on regulation by autocrine and paracrine signaling between adjacent tissues, such as the mesenchymal–epithelium interactions described above, endocrine signaling plays an important role in late gestational stages and particularly postnatal. One such example is provided by the effect of two hormones, thyroid hormone and glucocorticoids. When applied exogenously, these hormones induce the precocious development of the intestinal digestive enzyme and transport functions, while removing the endocrine glands that produce them affects the normal postnatal development [87]. The glucocorticoid effects include effects on cell proliferation, regulation of enzyme expression, ion transport and vitamin metabolism [86]. Thyroid hormone also plays an integral role in postnatal intestinal development and maintenance. Its most relevant effects appear to be on the regulation of transporters in the columnar absorptive cells, but it appears to have overall effects on the different intestinal tissues. In fact, the loss of a thyroid hormone receptor (TR3α) causes particular effects in postnatal mice, including reduced intestinal diameter, lower epithelial cell mass, and a significant decrease in smooth muscle.

The mesenchyma is also essential for the response to endocrine hormones. Not only has it been shown that the mesenchymal cells have receptors for both glucocorticoids and thyroid hormones [2], but the involvement of the mesenchymal cells in the endocrine responses provides an interesting link to the maturation of the intestine. For example, it has been shown that intestinal maturation includes the deposition of new ECM components such as collagen IV, and this ECM component is made by the mesenchymal cells.

Like other developmental processes, intestinal maturation also depends on gene activation networks that regulate the expression of particular transcription factors. This is exemplified by the findings of Choi et al. [3] showing that in mice mutant for lsx, a homeobox transcription factor, a normal looking intestine develops, but at a closer look, there are problems in the expression of high density lipoprotein and cholesterol receptors.

35.2.7 Immunity–Nutrition–Microbiome

Upon birth, the intestine is challenged with new functions and changes in its environment. Factors, such as bacteria and nutrients, that are now present in the intestinal lumen are also capable of inducing differentiation and/or maturation responses in the intestinal cells, particularly those associated with the luminal layer.

The intestine is considered the largest immune organ in the vertebrate body presenting both acquired and innate immunological defenses. The complexity of the digestive tract immune system, known as gut-associated lymphoid tissue (GALT), cannot be encompassed in this chapter, but has been reviewed elsewhere [88–90]. Nonetheless it is important to emphasize that immunological defenses are put in place during embryological and fetal development, but it is with birth and the exposure to the external microbial environment that their functional capacities are tested. For example, Paneth cells which produce antimicrobial peptides are found in the embryonic intestine, but their numbers and maturation increase significantly after birth [89]. Similarly, many of the components of the immune system such as macrophages, B and T cells, Peyer's patches are present in the fetal submucosa, although in smaller quantities or in a less differentiated state than in the postnatal intestine. The maturation of the intestinal immune system in the newborn is not immediate but can take months to occur. Since, the newborn is not only exposed to nonpathogenic and pathogenic bacteria but also to a myriad of dietary antigens, the immune system must learn to respond appropriately to each component.

The type and load of nutrients themselves affect both the morphology and physiology of the intestine. Nowhere is this more evident than in snake model systems where long time intervals occur between feeding episodes, and where the ingestion of prey brings about a huge increase in the activation of enzymes and transporters, as well as in intestinal overall growth [91]. Moreover, some of the dietary effects might occur during a critical period in the newborn but can affect intestinal adaptation in later life [92].

The third element in the challenges faced by the intestine of the newborn is the establishment of the gut microbiota. The microbiota is essential to complete the development of the intestine and its effects have been shown to occur at the tissue and cellular level [93]. Studies have shown that villi capillaries fail to develop complete vascular networks in the absence of microbiota and that the upregulation of particular genes can be induced by specific bacteria. The microbiota also plays a role in providing nutrients, participate in the digestion of lipids and proteins and serve as a source of certain vitamins [94]. Finally, absence of microbiota also affects the development of the intestinal immune system [95]. This gut microbiota interaction is mediated via the activation of toll-like receptors (TLRs) and TLR-dependent pathways in the intestinal epithelial cells upon recognition of specific surface molecules in the microbiota [96].

This relationship between microbiota and intestinal development is the basis of study in certain diseases. For example, it is interesting that preterm babies are more prone to necrotizing enterocolitis, a condition where areas of the intestine die without a known cause. An unproven hypothesis is that the colonization of the gut by an abnormal microbial biota might be the cause of the failure of the intestine to develop normally and to undergo the cell death found in necrotizing enterocolitis [97].

Under normal circumstances the gut microbiota that colonizes the newborn gut originates mainly from the vaginal tract of the mother, as well as from other environmental sources. This process varies widely depending on delivery mode, feeding mode and the degree of environmental hygiene [98]. Interestingly, it has been found that the gut microbiota in babies delivered by cesarean section is widely different from those that undergo vaginal birth [99]. In addition, gut microbiota is altered with the use of antibiotics, leading to changes in the intestinal physiology. Thus, with an increment in cesarean sections and antibiotics usage in modern societies, it remains to be seen whether this different microbiota has any effect on the digestive tract morphology or physiology as the baby develops.

35.2.8 Stem Cell Differentiation *in vitro*

Developmental biologists have always considered the pathways associated with embryological development to be unidirectional, thus as cells differentiate they not only undergo morphological and molecular changes but concomitantly they lose the ability to backtrack into alternate paths that lead to different developmental lineages. The final outcome being that they become differentiated according to the series of factors that they have encountered along their differentiation route. The characterization of the factors necessary to move a cell down a specific developmental pathway, starting from the early embryo, has allowed investigators to utilize these factors in a sequential manner to direct the differentiation of pluripotent stem cells (PSCs) toward a particular cell phenotype.

The ability to isolate PSCs and culture them *in vitro* also opens the possibility of using these cells for therapeutic use. Recent experiments show the isolation of PSCs, including human PSCs, and their directed differentiation to an intestinal stem cell and then toward the several cell lineages [69,100]. It is evident that for this type of study, the accrued information from *in vivo* studies in embryos and adults is essential. The *in vitro* system not only puts into practice what is known of growth/differentiation factors involved in intestinal differentiation but

also provides a system where new experiments can be performed to determine the role of known or novel factors. Other experiments have shown the *in vitro* differentiation of human PSC into different intestinal cell phenotypes (Figure 35.6) [69]. Initially the stem cells are moved toward endoderm by applying activin A, a nodal-related TGF-β molecule. Subsequently, endoderm-derived cells were differentiated into hindgut epithelium by treating with Wnt3A and FGF4. Cdx2 expression was used as a marker to determine that the cells were indeed acquiring an intestinal identity. Interestingly, the cultures underwent morphological changes reminiscent of embryonic intestinal formation, as the cells condensed into epithelium tubes some of which budded off to form spheroid structures. When the spheroids were cultured under particular conditions they continued their development in a manner reminiscent of fetal gut development. First, the cuboidal epithelium of the spheroid differentiated morphologically into a pseudostratified epithelium surrounded by mesenchymal cells, and later into a columnar epithelium complete with villus-like structures. They also formed crypt-like structures where cell division occurred and from which the major cell types (enterocytes, Goblet, enteroendocrine, and Paneth cells) developed. It is important to highlight that in this system it is possible that the mesenchyme cells originated from a small percentage of PSCs that differentiated toward the mesoderm lineage. Therefore, these cells were able to interact in the epithelial—mesenchymal cross talk that has been shown to occur during intestinal formation in the embryo.

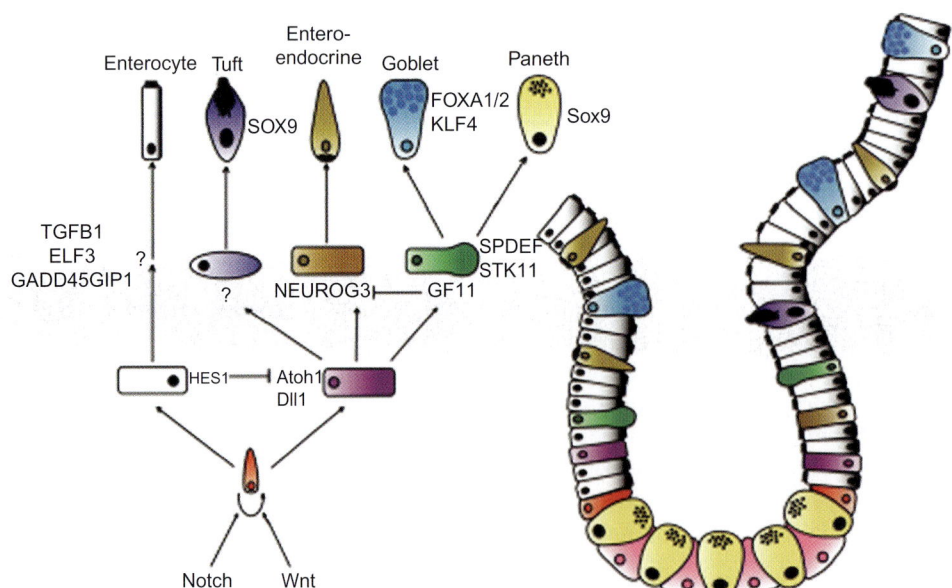

FIGURE 35.6　Intestinal organoids generated from human pluripotent stem cells (hPSCs) *in vitro*. hPSCs were grown *in vitro* initially in the presence of activin and later in the presence of FGF4 and Wnt3a. The organoids have crypt- and villus-like structures and cells within them are labeled with markers for (B) Goblet, (C) Paneth and (D) enteroendocrine cells. NUC, MUC2, CDX2, LYZ and CHGA refer to particular cell markers. *From [100].*

35.3 REGENERATIVE ORGANOGENESIS

35.3.1 Vertebrate Intestinal Regeneration in Adults

Most studies of vertebrate intestinal regeneration focus either on the regeneration of the luminal epithelium or on regeneration of deep wounds, such as those that occur following accidents or surgical manipulations that involve most intestinal tissue layers. Luminal epithelium regeneration has great biomedical relevance, since a large number of diseases or gastrointestinal problems are associated with malfunctions, abnormalities or loss of the luminal cells. The intestinal epithelium is a dynamic tissue layer, where cells are well organized in a spatial/temporal arrangement (Figure 35.7). The luminal epithelium can be said to be in a continuous regenerative state, since in order to be maintained it follows a well-defined cell proliferation, migration and differentiation process. The intestinal stem cells of the luminal epithelium have been characterized and the cellular dynamics have been well studied. In fact, this system has been used as a regeneration model exemplifying a system where a source of multipotent cells provides the basis for tissue renewal. Therefore, it bridges the embryological formation of the intestine with its long-term maintenance via a process of continuous cell replacement, proliferation and differentiation that begins in the embryo and continues in the adult up until the death of the organism.

35.3.1.1 Intestinal Epithelium Regeneration

The luminal epithelium stem cells are localized within the lower third of the crypts of Lieberkhün. These stem cells give rise to the transit-amplifying cells that proliferate and move toward the upper area of the crypt and eventually to the villi. Cell proliferation declines as cells move upward and eventually cease once they exit the crypt and enter the villi. With the termination of proliferation

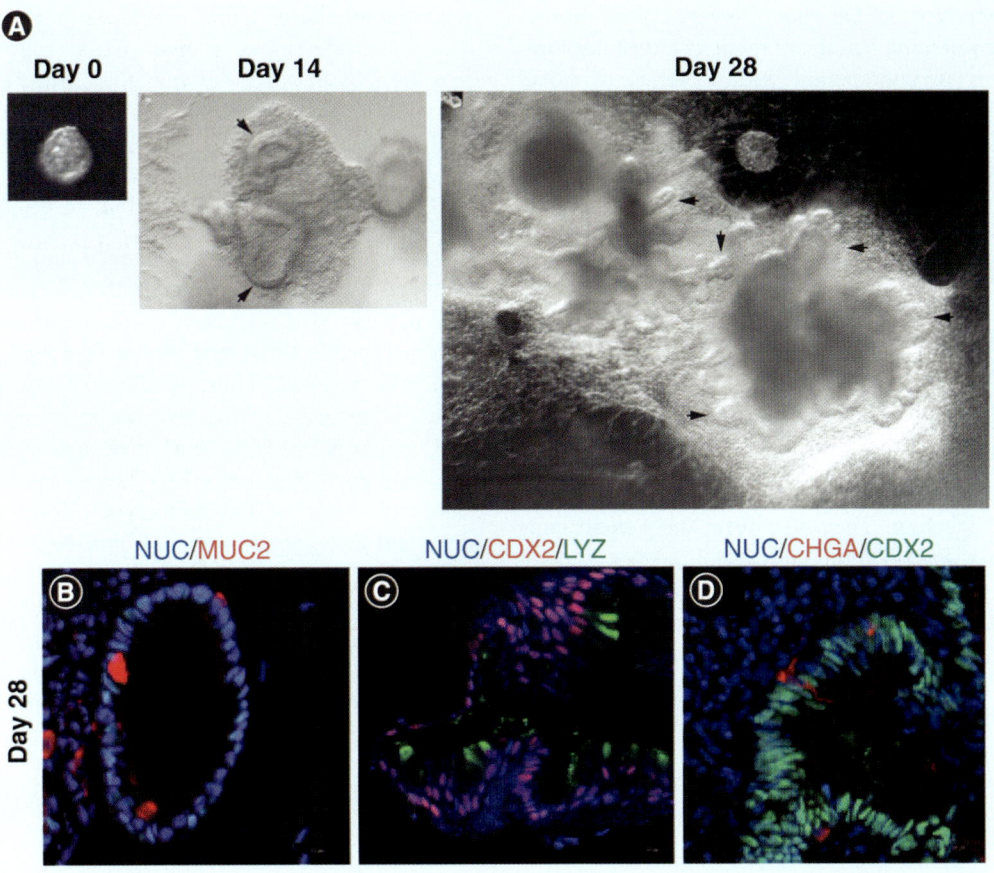

FIGURE 35.7   Proposed model for the stem cell niche and differentiation of the epithelial cell lineages. Stem cells in the crypt maintain their stem cell properties by regulation via Notch and Wnt. They are adjacent to the Paneth cells. As cell leave the niche they become transiently-dividing cells and move upward. As they leave the crypt they differentiate into the various cell lineages giving rise to all the cell types in the intestinal epithelium. The diagram includes the factors associated with the cell differentiation pathways. *From [6].*

begins the process of cell differentiation, where the transit-amplifying cells give rise to all cell phenotypes found within the intestinal villi, that is, columnar absorptive cells (enterocytes), endocrine cells and goblet cells [101]. These cells continue their movement toward the tip of the villi where eventually they are discarded off into the intestinal lumen and digested. In mammals, the time it takes from proliferation in the crypt to the tip of the villi has been documented at about 5−7 days. This makes the intestinal epithelium one of the fastest cycling tissues in the mammalian body. Stem cells also give rise to the Paneth cells, however, instead of migrating upward in the crypt, these cells move toward the basal part of the crypt where they are closely associated with the stem cells. They are mainly found within the villi but can also be found in the crypt and within the crypt base, thus it has been suggested that they may have stem cell potential, although studies have determined they are postmitotic.

The process of cell differentiation and lineages of the intestinal epithelium has been known for some years [101,102], however, many recent findings have been made regarding the regeneration of the mucosa, most of them in the laboratory of Dr. Hans Clevers. These discoveries have been possible thanks to modern molecular biology techniques in conjunction with a series of novel markers that serve to identify the various cell lineages or differentiation steps of the intestinal luminal epithelium [103]. Among these, one of the most important is the gene Lrg5 (leucine-rich repeat containing G protein-coupled receptor 5), that is expressed by intestinal stem cells [68,104]. Using this marker, investigators were able to verify the differentiation pathway that had been proposed for the formation of the intestinal mucosa. Thus, Lrg5 cells are found within the crypt and closely associated with Paneth cells. They are able to divide, and give rise to all other luminal epithelial cells. In the process, a dividing Lrg5 cell can give rise to other stem (Lrg5-labeled) cells and to proliferating cells that eventually lose the Lrg5 labeling. They also form the Paneth cells and the recently described tuft cells [68,71].

The Lrg5 marker has also allowed for the isolation of the stem cells and for extensive experimental manipulations that provide information as to their plasticity and the factors involved in modulating their developmental potential. These cells, when isolated by cell sorting and cultured, proliferate and organize themselves into structures that appeared to be "organoids" or mini-guts, with an apical/basal organization and a lumen [70] or when seeded in a particular matrix and culture media can build crypt-villus structures even in the absence of the underlying mesenchyme [105]. Moreover, after growth *in vitro*, the cells can be transplanted into the intestine of host mice, highlighting their therapeutic potential to help in the regeneration of the intestinal tissues [100].

In a completely unexpected finding, investigators culturing the luminal cells showed that all that was needed to form the stem cell niche and thus keeping stem cell proliferation was a Paneth cell. Similarly, even in the absence of mesenchymal cells, luminal epithelial cells that migrate upward from the stem cell niche have been found to regulate their expression of nuclear β-catenin, further underlying the autonomous nature of the stem cell proliferation and differentiation process. These findings have at least two important implications for intestinal functioning, first that the intestinal stem cells produce their own stem cell niche, by giving rise (and associating) with the Paneth cell. Second, that this process can take place in the absence of mesenchymal cells, thus showing an unexpected autonomy of the cellular epithelium and questioning the role of the mesenchymal cells in modulating the formation or function of the mucosa. Nonetheless, it has been noted that several factors, known to be produced by the mesenchymal cells such as laminin-rich matrix and BMP antagonists are necessary for the *in vitro* results [37]. This suggests that the mesenchymal cell role is to regulate the self-renewal and differentiation of the luminal epithelium.

35.3.1.1.1 Molecular Modulation of the Stem Cells

Among the factors involved in the modulation of stem cell proliferation and differentiation are the same factors that have been studied in other events of intestinal development, namely Wnt, HH, BMP, and TGF-β. Wnt is essential for stem cell maintenance and renewal and is produced by the epithelial cells themselves [66]. Factors that increase Wnt signaling increase the proliferation of stem cells while those that inhibit Wnt signaling decrease stem cell renewal. Thus, BMP a known Wnt inhibitor causes a decrease in the renovation of stem cells. This BMP signal is possibly being produced by the ISEMF or other mesenchymal cells [2]. Paneth cells are also the source of many of the factors that form the stem cell niche and that control stem cell proliferation and maintenance. Among the factors that have been proposed to be produced by these cell types are EGF, TGF-α, Wnt3, and the Notch ligand Dll4 [73,106]. Notch has been found to be activated in the intestinal stem cells and to be involved in the cell fate decisions toward the secretory versus the absorptive lineages.

35.3.1.1.2 Regeneration of the Luminal Epithelium following Injury

Even though the role of Lrg5 cells has been well documented in normal recycling of the intestinal mucosa, other cell types or other processes might be involved in tissue regeneration following injury. For once, when Lrg5 cells are eliminated either by target ablation [107,108] or radiation

[109,110], the intestinal epithelium can still be regenerated. In these cases, regeneration appears to depend on other cell types that represent a quiescent or slow-cycling cell population that has been described by their expression of several cell markers [111]. Although initially this quiescent cell population was thought to be in the +4 position within the intestinal crypt, more recent studies suggest that this cell population are the Paneth cells that can lose their Paneth markers in a process of dedifferentiation, enter a proliferative phase and give rise to all other cell types within the intestinal epithelium [110].

35.3.1.1.3 Role of ISEMF

The ISEMF, due to their close apposition to the basement membrane where the luminal epithelium stem cells reside, have been proposed as one of the main modulators of stem cell dynamics during normal homeostasis or during regeneration. They are viewed as the main mesenchymal components of the epithelial—mesenchymal cross talk that regulate the villi-crypt axis. Although, as explained earlier, the intestinal stem cells appear to have a large degree of autonomy, and can, when isolated, form a crypt-villi-like structure with all the derivative cell phenotypes, it is also clear that certain growth or ECM factors are needed for this to occur. In addition, ISEMF also play a role in regenerative processes following injury. As described earlier, they respond with increasing proliferation to many growth factors (IGF-I, IGF-II, IL-1B, TNF-α, PDGF, EGF, and bFGF among others). Moreover, *in vivo*, it is the ISEMF themselves that have been proposed to produce some of these molecules. Finally, as mediators of many of the stromal effects and/or actions, ISEMFs are associated with many other processes of intestinal physiology that are of importance to students of regenerative and transplant medicine. These include cancer, immune system activation, and inflammation [34,37].

35.3.1.1.4 Other Factors Associated with Wound Healing Response

Although the ISEMF certainly play a role in the normal maintenance of the luminal epithelium and in its restitution following injury, other factors have also been associated with the epithelial healing following injury. Among these are some of the factors, including nutrients, hormones, and microbiota, that are responsible for the intestinal transition from fetus to the adult [112].

35.3.2 Regeneration following Injury to Multiple Intestinal Layers

Following injury or surgical manipulations, most of the tissues of the intestine will undergo at least some type of regenerative response [113]. The best studied is the response of the luminal epithelium which shows an increased proliferative response in the adjacent tissues to a transection or resection. In addition, it has been shown that following intestinal resection in mice, there is an increase in the number of crypts [114]. This increase is probably dependent on the expansion of intestinal stem cells. Similarly, following resections or transections, there is reorganization of the ECM and regeneration of both muscularis mucosa and the smooth muscle [113]. Other studies have shown regeneration of the lymphatic vessels [115] and of the ENS [116].

Transected rat intestinal tract can form spontaneous end-to-end anastomosis in a small, though significant, number of cases [117]. When cut and left within the cavity, a mass of cells is formed from the mucosa and serosa of the cut ends that extends from one end to the other [118]. Eventually this mass gives rise to the lumen by the infiltration of luminal cells. Preliminary experiments in large intestine have used silastic tubing to extend the regenerating tissue. This procedure produces some elongation of the transected intestine with all its tissue layers, except for the presence of neural tissue.

Similar to what takes place during embryogenesis, the intestinal response to injury or to surgical/medical manipulations involves the cross talk of the luminal and mesenchymal tissue layers. This is the case following massive bowel resection where the PDFG-signaling pathway was shown to be upregulated. Moreover, animals treated with PDGF following a resection had increased intestinal cell mass and a lower loss of weight, indicating a positive effect on intestinal regeneration. The effect was most likely mediated by an overall increase in cell turn-over and a particular increase in enterocyte proliferation. More importantly, the effect appeared to be via the mesenchymal cells found in the submucosal compartment near the epithelial stem cells that give rise to the luminal epithelium.

As would be expected, many of the molecules that are associated with the formation of the intestine during embryological development are again associated with its regeneration following injury. Experiments done in the colon following injury also show the role of factors in regeneration. For example, following crypt excision in the mouse colon, new crypts form by the migration of stem cells from adjacent crypts [119]. Wnt5, probably secreted by the underlying mesenchyme plays a role in this process by potentiation of TGF-β, in a process akin to what occurs during intestinal development.

35.4 ALTERNATIVE ANIMAL MODELS

Similar to many other areas of biomedical research, scientists have used several animal species to study the process of regeneration of the digestive tract, and in particular of the intestine. The model systems used can be divided into two groups. On the one hand, we have the models that are used *in lieu* of humans. Studies in this group of animals are usually done in view of the future application to human intestinal regeneration and include mammals, such as rats, rabbits, dogs and primates, and the fruit fly Drosophila. In principle, these models share the same properties and problems (including regenerative limitations) common to most mammals. For example, regeneration of the intestine in these animals has to deal with possible bacterial contaminants and with a prolonged period during which the animal is not able to digest nutrients.

In contrast to the limited regenerative potential of injured areas or specific tissue layers, certain animal can regenerate large portions of their digestive system and/or intestine in adult stages. These animals provide model systems where the cellular and molecular basis of intestinal regeneration can be studied.

35.4.1 Drosophila

Drosophila has served as a model to study the genetic bases of intestinal stem cell regulation. Although the intestinal organization is somewhat simple, consisting of a single epithelial layer and two layers of visceral muscle, the epithelial cell layer also undergoes a process of renovation similar to what takes place in mammals [120−122]. The adult midgut epithelium turns over about once every 1−3 weeks from stem cells that reside above the basement membrane [123,124]. Upon injury, a response is induced by which cell proliferation increases (10- to 100-fold) to repair the luminal epithelia. Thus, many genetic tools available for the Drosophila system have been used to determine the signaling pathway by which cells in the midgut are induced to proliferate following lesions [120]. Not surprisingly, many of the molecular events that occur upon injury are similar if not identical to events that take place during intestinal embryogenesis, and moreover some of them are common to both mammals and flies. These events are meditated by signaling systems that are commonly activated during embryological development, such as, Notch, Wnt, BMP, EGF, and others [120−122].

35.4.2 Planaria

Planarians are known to regenerate by division and eventual differentiation of a PSC named neoblast. Intestinal regeneration in planaria is neoblast dependent [125]. Investigators have shown that the intestine originates from mesenchymal cells associated with the enteric muscle. These mesenchymal cells themselves do not divide, but depend on the neoblast precursor division for their

formation. Thus, the lineage of intestinal cells seems to be dividing stem cells giving rise to nondividing mesenchymal cells that eventually form the new enterocytes. This differentiation takes place along the entire intestine and no specific growth zones were found.

35.4.3 Tunicates

Tunicates or ascidians are deuterostomes closely related to chordates. Some of these organisms show amazing regenerative properties. For example, colonial species are known to regenerate their complete body from a group of cells named "blood cells." In terms of the digestive tract, experiments have been done where animals were cut in half and gut regeneration was determined [126]. RA inhibition either by drugs or by RNAi against RA synthesizing enzymes inhibited gut regeneration. It has been suggested that in tunicates, RA plays a role in the transdifferentiation of the atrial epithelium into the gut tissue. Nonetheless, although RA appears to be involved in initiating regeneration, its role is not exclusive to initial regeneration but might also be involved in different roles at other regeneration stages. The modulation by RA also shows the importance of RA in modulating activity of Hox genes and in establishing anterior−posterior polarity in most animals.

35.4.4 Echinoderms

Two classes of echinoderms are known to regenerate their digestive tract: starfish and holothurians (sea cucumbers). The former can regenerate their pyloric caeca and stomach, but this has been little studied following the initial description made by Anderson [127,128]. Intestinal regeneration in holothurians, on the other hand, has been widely documented both at the cellular and at the molecular levels [129,130]. In these animals the new intestine forms from the remaining tissues (Figure 35.8). An initial thickening of the distal tip of the remaining intestine is observed. This thickening mainly occurs by a combination of cellular events including the dedifferentiation of cells within the mesentery mesothelium and the proliferation of these dedifferentiated cells [131,132]. As this rudiment grows, forming a solid rod from the anterior to the posterior end of the animals, cells from the coelomic epithelium (analogous to the vertebrate serosa) differentiate forming the muscle layers [133]. The lumen itself is formed by migrating mucosal cells that originate in the two structures of the digestive tract that remain following the loss of the intestine: the esophagus and the cloaca. As these cells migrate into the connective tissue of the intestinal rudiment, they proliferate and form the lumen. Thus, the lumen forms in an anterior to posterior direction starting from the esophagus, and in a posterior to an anterior

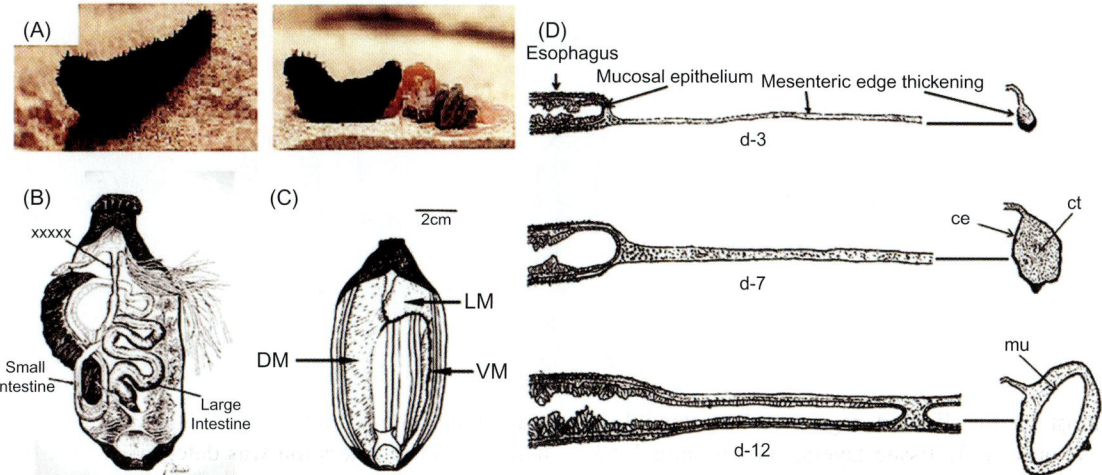

FIGURE 35.8 Holothurian model system. (A) Holothurians can undergo a process of evisceration where they eject their internal organs. (B and C) Diagram of the intestinal anatomy of *Holothuria glaberrima* showing the organs that are lost (B) and the remaining mesenteries following evisceration from where the new intestine will form. (D) Longitudinal and cross-sectional views of intestinal regeneration. Initially a solid rod forms between the esophagus and the cloaca. Epithelial cells of the esophagus migrate into the rudiment forming the lumen in the anterior region of the digestive tract and eventually fuse with migrating cells that originate from the cloaca in the posterior part of the animal, thus forming the hollow tube that will become the regenerated intestine. *DM, LM and VM refer to dorsal, lateral and ventral mesenteries. From [134].*

direction beginning from the cloaca. The two growing luminal tips join somewhat in the center of the rudiment creating a continuous lumen and thus a complete intestine. Research in holothurians has highlighted the role of cell dedifferentiation in intestinal formation, as well as the acquired potency of the new cellular precursors [135]. It is still unknown where the new nerve cells originate although some investigators have proposed that they also originate from the mesothelium, (a mesodermal derivative). Another interesting aspect of intestinal regeneration in holothurians is the possibility that in some species the mucosa itself also originates from the mesothelial cells (thus suggesting that cells from a mesodermal origin have the potential to transdifferentiate and form an endodermal derivative) [136].

More importantly, these animals are being used to determine the genes and signaling systems associated with the events that take place as the intestine regenerates [137]. It is not surprising that many of the genes and gene sequencing pathways that have been associated with holothurian intestinal regeneration are those participating in the embryological formation of the digestive tract. Therefore, genes associated with developmental processes such as Wnt, BMP, survivin, mortalin, and others have been shown to be overexpressed during the regeneration of the intestine [138–141]. Moreover, the localization of the expression of these genes mimics what is known to occur in the vertebrate embryo. For example, Wnt and BMP are expressed in the epithelium of the regenerating intestine reminiscent of the epithelial–mesenchymal interactions described previously. However, as has been

determined for other regenerative events, that is, limb regeneration in amphibians, it is not expected that intestinal regeneration is a mere recapitulation of embryological development and new genes or gene pathways might be involved. These genes can be either novel genes only associated with the regenerative process or genes that participate in other events that are activated during regeneration. Recent studies have even extended the control of intestinal regeneration to nonprotein codifying genomic sequences by suggesting that transposons might be involved in the regeneration of the intestine, as well as in other regenerative events [142].

35.4.5 Amphibians

35.4.5.1 Regeneration

Some old but highly interesting experiments performed in amphibians studied the process of intestinal regeneration after intestinal transections. They found that, in adult frogs and newts, following intestinal transections (either ileostomies or colostomies) and even in those whose intestines were ligated, the intestinal tissue reconnected and became functional again [143,144]. Most of the findings were described at the level of the tissues. The investigators showed that the initial connection was made by mucosal cells and that eventually blastemas composed of fibroblasts and mesothelial cells, formed at the cut ends of the severed intestines. Cellular dedifferentiation was observed in all successful regenerates, possibly originating from the smooth muscle and from the serosa. Cellular

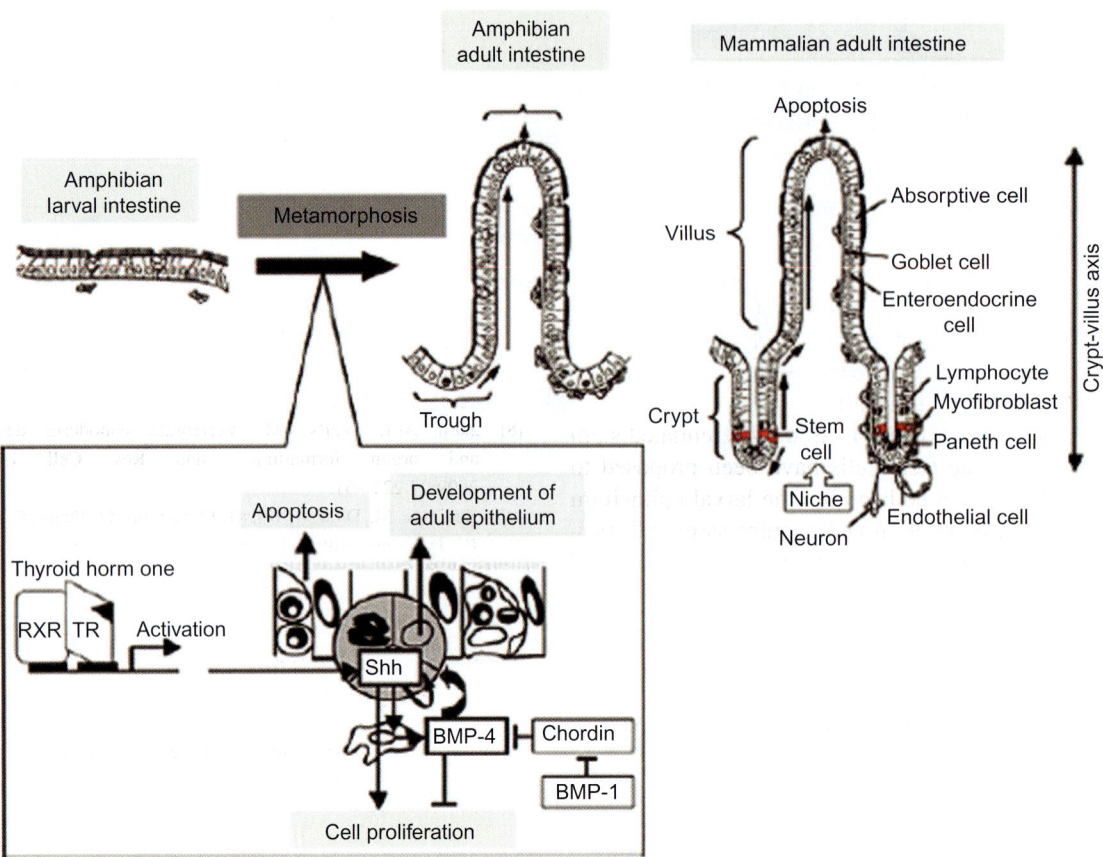

FIGURE 35.9 Changes in the intestinal epithelium of amphibians during metamorphosis. Thyroid hormone induces dramatic changes in the intestinal epithelium that are similar to those that take place during embryonic development in vertebrates. The epithelium changes form a simple layer epithelium to the formation of crypts and villi with all epithelial cell phenotypes found in vertebrates. In this system hormonal action of thyroid hormone induces the interactions between the mesenchyma and the epithelium. Gene activation takes place via the Thyroid hormone Receptor (TR) or the Retinoid-X-Receptor (RXR). *From [145].*

proliferation was also studied in newt intestine following transection showing that the initial cell proliferation takes place in the mucosal cells followed by transient proliferation of serosal and finally of muscle cells [146,147]. These investigators proposed that particular epithelia do not become totipotent but instead were only morphologically dedifferentiated during blastema formation. Nonetheless, although adult amphibians can anastomose cut ends of the intestine spontaneously, in contrast to larval amphibians, they cannot replace excised tissue but for only a few millimeters [148].

35.4.5.2 Metamorphosis

A large volume of literature is available on the changes that the intestinal tissue undergoes during amphibian metamorphosis. The changes are triggered by thyroid hormone, which controls the complete metamorphic process. This itself provides a setpoint and a molecular agent for the beginning of the changes associated with

metamorphosis. Most of this work has been done by Ishizuya-Oka et al. using *Xenopus laevis* as a model system [149,150]. In this model, the intestine of the larval stage undergoes a drastic remodeling in structure and function. The larva has a simple (one cell layer with immature connective tissue and muscle) but extended intestine that, following metamorphosis, gives rise to a shorter but much more complex intestine in the adult. The new epithelium is renewed along the trough-crest axis, the equivalent of the crypt-villus axis in mammals (Figure 35.9).

This remodeling event involves extensive apoptosis of the larval epithelial cells and proliferation and differentiation of adult stem cells. In addition, the process involves interactions between the mesenchyme and the epithelium that mimic those in the developing digestive tract. Among these interactions are changes in the ECM components that accompany the remodeling event.

Many events that take place during the formation of the amphibian adult mucosa are similar to those that take place during the formation or maintenance of the mammalian mucosa. For example, fibroblasts adjacent to the forming mucosa lamina propria secrete BMP4 suggesting that this paracrine signaling plays a key role in the differentiation of the adult mucosa. Moreover, BMP4 is apparently upregulated by SHH from the epithelium which itself is under the control of the thyroid hormone [151]. Recent studies using a dominant negative T3 receptor mutant showed that the effect of T3 is on the tissues surrounding the epithelium and that these effects are essential for the formation of the adult mucosa [150]. BMP1 and chordin have also been suggested as possible modulators of the BMP4 expression.

The new mucosa originates from undifferentiated stem cells. Interestingly, the stem cells have been proposed to arise from differentiated cells within the larval epithelium that undergo dedifferentiation and acquire stem cell markers [150,152]. More recently the intestinal stem cell marker Lgr5 was cloned from *X. laevis* and shown to be expressed transiently during metamorphosis [153]. In the intestine, the peak of Lgr5 expression correlates with the peak of adult luminal cell proliferation. Lgr5 is found in small islands of the epithelia increasing from larval to adult. The cells increased in number from the premetamorphic stage to the prometamorphic stage.

35.5 CONCLUDING REMARKS

From the information provided here, it is evident that regeneration and embryogenesis share common mechanisms in their goal to make an organ. At the same time, there are certain differences between the two processes, providing investigators with the possibility of alternate strategies in their goal to direct the regeneration of tissues and organs. Thus, it will be through the use of multiple model systems where the shared mechanisms between regeneration and embryogenesis can be discerned. This, together with the comparative analyses between regenerating and nonregenerating animals, could provide a complete understanding of regenerative processes. In conclusion, we anticipate that the information provided by experiments in intestinal regenerative and embryological organogenesis using different model systems will continue to provide valuable insights to students of regenerative medicine in their approach to present and future challenges.

REFERENCES

[1] Fountain H. A first: organs tailor-made with body's own cells. NYT 2012;15:A1.

[2] McLin VA, Henning SJ, Jamrich M. The role of the visceral mesoderm in the development of the gastrointestinal tract. Gastroenterology 2009;136:2074–91.

[3] Choi MY, Romer AI, Hu M, Lepourcelet M, Mechoor A, Yesilaltay A, et al. A dynamic expression survey identifies transcription factors relevant in mouse digestive tract development. Development 2006;133:4119–29.

[4] Garcia-Garcia MJ, Shibata M, Anderson KV. Chato, a KRAB zinc-finger protein, regulates convergent extension in the mouse embryo. Development 2008;135:3053–62.

[5] Spence JR, Lauf R, Shroyer NF. Vertebrate intestinal endoderm development. Dev Dyn 2011;240:501–20.

[6] Noah TK, Donahue B, Shroyer NF. Intestinal development and regeneration. Exp Cell Res 2011;317:2702–10.

[7] Heath JK. Transcriptional networks and signaling pathways that govern vertebrate intestinal development. Curr Top Dev Biol 2010;90:159–92.

[8] Zorn AM, Wells JM. Vertebrate endoderm development and organ formation. Annu Rev Cell Dev Biol 2009;25:221–51.

[9] Kedinger M, Duluc I, Fritsch C, Lorentz O, Plateroti M, Freund JN. Intestinal epithelial–mesenchymal cell interactions. Ann N Y Acad Sci 1998;859:1–17.

[10] Roberts DJ, Johnson RL, Burke AC, Nelson CE, Morgan BA, Tabin C. Sonic hedgehog is an endodermal signal inducing *Bmp-4* and *Hox* genes during induction and regionalization of the chick hindgut. Development 1995;121:3163–74.

[11] Ramalho-Santos M, Melton DA, McMahon AP. Hedgehog signals regulate multiple aspects of gastrointestinal development. Development 2000;127:2763–72.

[12] Zacchetti G, Duboule D, Zakany J. Hox gene function in vertebrate gut morphogenesis: the case of the caecum. Development 2007;134:3967–73.

[13] McGinnis W, Krumlauf R. Homeobox genes and axial patterning. Cell 1992;68:283–302.

[14] Kawazoe Y, Sekimoto T, Araki M, Takagi K, Araki K, Yamamura K. Region-specific gastrointestinal Hox code during murine embryological gut development. Dev Growth Differ 2002;44:77–84.

[15] De Santa Barbara P, Van den Brink GR, Roberts DJ. Development and differentiation of the intestinal epithelium. Cell Mol Life Sci 2003;60:1322–32.

[16] Sakiyama J, Yokouchi Y, Kuroiwa A. HoxA and HoxB cluster genes subdivide the digestive tract into morphological domains during chick development. Mech Dev 2001;101:233–6.

[17] Yokouchi Y, Sakiyama J, Kuroiwa A. Coordinated expression of Abd-B subfamily genes of the HoxA cluster in the developing digestive tract of the chick embryo. Dev Biol 1995;169:76–89.[16b]Sekimoto T, Yoshinobu K, Yoshida M, Kuratani S, Fujimoto S, Arake M, et al. Region-specific expression of murine *Hox* genes implies the *Hox* code-mediated patterning of the digestive tract. Genes Cells 1998;3:51–64.

[18] Huang D, Chen SW, Langston AW, Gudas LJ. A conserved retinoic acid responsive element in the murine *Hoxb-1* gene is required for expression in the developing gut. Development 1998;125:3235–46.

[19] Bayha E, Jorgensen MC, Serup P, Grapin-Botton A. Retinoic acid signaling organizes endodermal organ specification along the entire antero-posterior axis. PLoS One 2009;4:5845.

[20] Wang Y, Wang L, Iordanov H, Swietlicki EA, Zheng Q, Jiang S, et al. Epimorphin(−/−) mice have increased intestinal growth,

decreased susceptibility to dextran sodium sulfate colitis, and impaired spermatogenesis. J Clin Invest 2006;116:1535−46.

[21] Pitera JE, Smith VV, Woolf AS, Milla PJ. Embryonic gut anomalies in a mouse model of retinoic acid-induced caudal regression syndrome. Am J Pathol 2001;159:2321−9.

[22] Nadaud LD, Shelton DN, Chidester S, Yost HJ, Jones DA. The zebrafish retino dehydrogenase, rdh1l, is essentials for intestinal development and is regulated by the tumor suppressor adenomaotous polyposis coli. J Biol Chem 2005;250:30490−5.

[23] Lipscomb K, Schmitt C, Sablyak A, Yoder JA, Nascone-Yoder N. Role for retinoid signaling in left-right asymmetric digestive organ morphogenesis. Dev Dyn 2006;235:2266−75.

[24] Sukegawa A, Narita T, Kameda T, Saitoh K, Nohno T, Iba H, et al. The concentric structure of the developing gut is regulated by Sonic hedgehog derived from the endodermal epithelium. Development 2000;127:1971−80.

[25] De Santa Barbara P, Williams J, Goldstein AM, Doyle AM, Nielsen C, Winfield S, et al. Bone morphogenetic protein signaling pathway plays multiple roles during gastrointestinal tract development. Dev Dyn 2005;234:312−22.

[26] Torihashi S, Hattori T, Hasegawa H, Kurahashi M, Ogaeri T, Fujimoto T. The expression and crucial roles of BMP signaling in development of smooth muscle progenitor cells in the mouse embryonic gut. Differentiation 2009;77:277−89.

[27] Kurahashi M, Niwa Y, Cheng J, Ohsaki Y, Fujita A, Goto H, et al. Platelet-derived growth factor signals play critical roles in differentiation of longitudinal smooth muscle cells in mouse embryonic gut. Neurogastroenterol Motil 2008;20:521−31.

[28] Zacharias WJ, Madison BB, Kretovich KE, Walton KD, Richards N, Udager AM, et al. Hedgehog signaling controls homeostasis of adult intestinal smooth muscle. Dev Biol 2011;355:152 62.

[29] Simmon-Assman P, Kedinger M, Haffen K. Immunocytochemical localization of extracellular-matrix proteins in relation to rat intestinal morphogenesis. Different 1986;32:59−66.

[30] Simon-Assmann P, Spenle C, Lefebvre O, Kedinger M. The role of the basement membrane as a modulator of intestinal epithelial−mesenchymal interactions. In: Kaestner Klaus H, editor. Progress in molecular biology and translational science, vol. 96. Burlington, MA: Academic Press; 2010.ISBN:978-0-12-381280-3. p. 175−206.

[31] Mahoney ZX, Stappenbeck TS, Miner JH. Laminin alpha5 influences the architecture of the mouse small intestine mucosa. J Cell Sci 2008;121:2493−502.

[32] Gershon MD. Development determinants of the independence and complexity of the enteric nervous system. Trends Neurosci 2010;33:446−56.

[33] Jones RG, Li X, Gray PD, Kuang J, Clayton F, Samowitz WS, et al. Conditional deletion of B1 integrins in the intestinal epithelium causes a loss of Hedgehog expression, intestinal hyperplasia, and early postnatal lethality. J Cell Biol 2006;175:505−14.

[34] Pichunk IV, Mifflin RC, Saada JI, Powell DW. Intestinal mesenchymal cells. Curr Gastroenterol Rep 2010;12:310−8.

[35] Powell Mifflin RC, Valentich JD, Crowe SE, Saada JI, West AB. Myofibroblasts II Intestinal subepithelial myofibroblasts. Am J Physiol Cell Physiol 1999;277:183−201.

[36] Powell DW, Adegboyega PA, Di Mari JF, Mifflin RC. Epithelial cells and their neighbors. I. Role of intestinal myofibroblasts in development, repair, and cancer. Am J Physiol Gastrointest Liver Physiol 2005;289:2−7.

[37] Powell DW, Pinchuk IV, Saada JI, Chen X, Mifflin RC. Mesenchymal cells of the intestinal lamina propria. Annu Rev Physiol 2011;73:213−37.

[38] Artells R, Navarro A, Diaz T, Monzó M. Ultrastructural and immunohistochemical analysis of intestinal myofibroblasts during the early organogenesis of the human small intestine. Anat Rec 2011;294:462−71.

[39] Hong YK, Shin JW, Detmar M. Development of the lymphatic vascular system: a mystery unravels. Dev Dyn 2004;231:462−73.

[40] Liersch R, Detmar M. Lymphangiogenesis in development and disease. Thromb Haemost 2007;98:304−10.

[41] Le Douarin NM, Teillet MA. The migration of neural crest cells to the wall of the digestive tract in avian embryo. J Embryol Exp Morphol 1973;30:31−48.

[42] Chalazonitis A, Kessler JA. Pleitropic effect of the bone morphogenetic proteins on development of the enteric nervous system. Dev Neurobiol 2012;72:843−56.

[43] Mao J, Kim BM, Rajurkar M, Shivdasani RA, McMahon AP. Hedgehog signaling controls mesenchymal growth in the developing mammalian digestive tract. Development 2010;137:1721−9.

[44] Chalazonitis A, D'Autreaux F, Guha U, Pham TD, Faure C, Chen JJ, et al. Bone morphogenetic protein-2 and -4 limit the number of enteric neurons but promote development of a TrkC-expressing neurotrophin-3-dependent subset. J Neurosci 2004;24:4266−82.

[45] Goldstein AM, Brewer KC, Doyle AM, Nagy N, Roberts DJ. BMP signaling is necessary for neural crest cell migration and ganglion formation in the enteric nervous system. Mech Dev 2005;122:821−33.

[46] Fu M, Vohra BP, Wind D, Heuckeroth RO. BMP signaling regulates murine enteric nervous system precursor migration, neurite fasciculation, and patterning via altered Ncam1 polysialic acid addition. Dev Biol 2006;299:137−50.

[47] Chalazonitis A, Pham TD, Zhishan L, Roman D, Guha U, Gomes W, et al. Bone morphogenetic protein regulation of enteric neuronal phenotypic diversity: relationship to timing of cell cycle exit. J Comp Neurol 2008;509:474−92.

[48] Van de Putte T, Maruhashi M, Francis A, Nelles L, Kondoh H, Huylebroeck D, et al. Mice lacking ZFHX1B, the gene that codes for Smad-interacting protein- 1, reveal a role for multiple neural crest cell defects in the etiology of Hirschsprung diseasemental retardation syndrome. Am J Hum Genet 2003;72:465−70.

[49] Fu M, Hang Lui VC, Sham MH, Yin Cheung AN, Hang Tam PKH. HoxB5 expression is spatially and temporarily regulated in human embryonic gut during neural crest cell colonization and differentiation of enteric neuroblasts. Dev Dyn 2003;228:1−10.

[50] Gershon MD. Developmental determinants of the independence and complexity of the enteric nervous system. Trends Neurosci 2010;33:446−56.

[51] Sasselli V, Pachnis V, Burns AJ. The enteric nervous system. Dev Biol 2012;366:64−73.

[52] Nagy N, Goldstein AM. Endothelin-3 regulates neural crest cell proliferation and differentiation in the hindgut enteric nervous system. Dev Biol 2006;293:203−17.

[53] Young HM, Anderson RB, Anderson CR. Guidance cues involved in the development of the peripheral autonomic nervous system. Autonomic Neurosci Basic Clin 2004;112:1–14.

[54] Gregorieff A, Clevers H. Wnt signaling in the intestinal epithelium: from endoderm to cancer. Genes Dev 2005;19:877–90.

[55] Sherwood RI, Maehr R, Mazzoni EO, Melton DA. Wnt signaling specifies and patterns intestinal endoderm. Mech Dev 2011;128:387–400.

[56] Mathan M, Moxey PC, Trier JS. Morphogenesis of fetal rat duodenal villi. Am J Anat 1976;146:73–92.

[57] Matsumoto A, Hashimoto K, Yoshioka T, Otani H. Occlusion and subsequent re-canalization in early duodenal development of human embryos: integrated organogenesis and histogenesis through a possible epithelial–mesenchymal interaction. Anat Embryol 2002;205:53–65.

[58] Stringer EJ, Pritchard CA, Beck F. Cdx2 initiates histodifferentiation of the midgut endoderm. FEBS Lett 2008;582:2555–60.

[59] Grainger S, Savory JG, Lohnes D. Cdx2 regulates patterning of the intestinal epithelium. Dev Biol 2010;339:155–65.

[60] Hryniuk A, Grainger S, Savory JGA, Lohnes D. Cdx function is required for maintenance of intestinal identity in the adult. Dev Biol 2012;363:426–37.

[61] Stringer EJ, Duluc I, Saandi T, Davidson I, Bialecka M, Sato T, et al. Cdx2 determines the fate of postnatal intestinal endoderm. Dev 2012;139:465–74.

[62] Madison BB, Braunstein K, Kuizon E, Portman K, Qiao XT, Gumucio DL. Epithelial hedgehog signals pattern the intestinal crypt-villus axis. Development 2005;132:279–89.

[63] Batts LE, Polk DB, Dubois R, Kulessa H. BMP signaling is required for intestinal growth and morphogenesis. Dev Dyn 2006;235:1563–70.

[64] Haramis AP, Begthel H, van den Born M, van Es J, Jonkheer S, Offerhaus GJ, et al. De novo crypt formation and juvenile polyposis on BMP inhibition in mouse intestine. Science 2004;303:684–6.

[65] Maloum F, Allaire JM, Gagné-Sansfaçon J, Roy E, Belleviell K, Sarret P, et al. Epithelial BMP signaling is required for proper specification of epithelial cell lineages and gastric endocrine cells. Am J Physiol Gastrointes Liver Physiol 2011;300:1065–79.

[66] Clevers H. Wnt/B-catenin signaling in development and disease. Cell 2006;127:469–80.

[67] Platerotti M, Freund JN, Leberquier C, Kedinger M. Mesenchyme-mediated effects of retinoic acid during rat intestinal development. J Cell Sci 1997;110:1227–38.

[68] Simons BD, Clevers H. Stem cell self-renewal in intestinal crypt. Exp Cell Res 2011;317:2719–24.

[69] Spence JR, Mayhew CN, Rankin SA, Kuhar MF, Vallance JE, Tolle K, et al. Directed differentiation of human pluripotent stem cell into intestinal tissue *in vitro*. Nature 2011;470:105–10.

[70] Sato T, van Es JH, Snippert HJ, Stange DE, Vries RG, van den Born M, et al. Paneth cells constitute the niche for Lgr5 stem cells in intestinal crypts. Nature 2010;469:415–8.

[71] Gerbe F, van Es JH, Makrini L, Brulin B, Mellitzer G, Robine S, et al. Distinct ATOH1 and Neurog3 requirements define tuft cells as a new secretory cell type in the intestinal epithelium. J Cell Biol 2011;192:767–80.

[72] Bjerknes M, Khandanpour C, Möröy T, Fujiyama T, Hoshino M, Klisch TJ, et al. Origin of the brush cell lineage in the mouse intestinal epithelium. Dev Biol 2012;362:194–218.

[73] Farin HF, Van Es JH, Clevers H. Redundant sources of Wnt regulate intestinal stem cells and promote formation of Paneth cells. Gastroenterology 2012;143:1518–29.

[74] Geske MJ, Zhang X, Patel KK, Ornitz DM, Stappenbeck TS. Fgf9 signaling regulates small intestinal elongation and mesenchymal development. Development 2008;135:2959–68.

[75] Zhang X, Stappenbeck TS, White AC, Lavine KJ, Gordon JI, Ornitz DM. Reciprocal epithelial–mesenchymal FGF signaling is required for cecal development. Development 2006;133:173–80.

[76] Cervantes S, Yamaguchi TP, Hebrok M. Wnt5a is essential for intestinal elongation in mice. Dev Biol 2009;326:285–94.

[77] Bellier S, da Silva NR, Aubin-Houzelstein G, Elbaz C, Vanderwinden JM, Panthier JJ. Accelerated intestinal transit in inbred mice with an increased number of interstitial cells of Cajal. Am J Physiol Gastrointest Liver Physiol 2005;288:151–8.

[78] Drozdowski LA, Clandinin T, Tomson ABR. Ontogeny, growth and development of the small intestine: understanding pediatric gastroenterology. World J Gastroenterol 2010;16:787–99.

[79] Ferraris RP, Buddington RK, David ES. Ontogeny of nutrient transporters. In: 1st ed. Sanderson IR, Walker WA, editors. Development of the gastrointestinal tract, 2000. London: BC Decker Inc; 2000. p. 123–46.

[80] Pácha J. Development of intestinal transport function in mammals. Physiol Rev 2000;80:1633–67.

[81] Traber PG. Development of brushborder enzyme activity. In: 1st ed. Sanderson IR, Walker WA, editors. Development of the gastrointestinal tract, 2000. London: BC Decker Inc; 2000. p. 103–22.

[82] Bisset WM, Wingate DL, Milla PJ. Gastrointestinal motor activity in the fetus and newborn. In: 1st ed. Sanderson IR, Walker WA, editors. Development of the gastrointestinal tract, 2000. London: BC Decker Inc; 2000. p. 211–26.

[83] Burns AJ, Roberts RR, Bornstein JC, Young HM. Development of the enteric nervous system and its role in intestinal motility during fetal and early postnatal stages. Seminar Pediatr Surg 2009;18:196–205.

[84] Roberts RR, Ellis M, Gwynne RM, Bergner AJ, Lewis MD, Beckett EA, et al. The first intestinal motility patterns in fetal mice are not mediated by neurons or interstitial cells of Cajal. P Physiol 2010;588:1153–69.

[85] Ménard D, Calvert R. Fetal and postnatal development of the small and large intestine: patterns and regulation. In: 1st ed. Morisset J, Solomon TE, editors. Growth of the gastrointestinal tract: gastrointestinal hormones and growth factors, 2000. Boston, MA: CRC Press; 2000. p. 159–74.

[86] Polk DB, Barnard JA. Hormones and growth factors in intestinal development. In: 1st ed. Sanderson IR, Walker WA, editors. Development of the gastrointestinal tract, 2000. London: BC Decker Inc; 2000. p. 37–56.

[87] Thomson AB, Keelan M. The development of the small intestine. Can J Pharmacol 1986;64:13–29.

[88] Howie D, MacDonald TT. Ontogeny of T lymphocytes within the human intestine. In: 1st ed. Sanderson IR, Walker WA, editors.

Development of the gastrointestinal tract, 2000. London: BC Decker Inc; 2000. p. 165−74.

[89] Ouellette AJ, Bevins CL. Development of innate immunity in the small intestine. In: Sanderson IR, Walker WA, editors. Development of the gastrointestinal tract. 1st ed. London: BC Decker Inc; 2000. p. 147−64.

[90] Ramsay AJ, Beagley KW. Development of B lymphocytes within the mucosal immune system. In: Sanderson IR, Walker WA, editors. Development of the gastrointestinal tract. 1st ed. London: BC Decker Inc; 2000. p. 197−210.

[91] Secor SM, Diamond J. A vertebrate model of extreme physiological adaptation. Nature 1998;395:659−62.

[92] Thomson AB, Keelan M, Garg M, Clandinin MT. Evidence for critical-period programming of intestinal transport function: variations in the dietary ratio of polyunsatured to saturated fatty acids alters ontogeny of the rat intestine. Biochem Biophys Acta 1989;20:302−15.

[93] Gilbert SF. Developmental biology. 8th ed. Sunderland, MA: Sinauer Associates; 2000.

[94] Caicedo RA, Schanler RJ, Li N, Neu J. The developing intestinal ecosystem: implications for the neonate. Pediatr Res 2005;58:625−8.

[95] Rautava S, Wlaker A. Commensal bacteria and epithelial cross-talk in the developing intestine. Curr Gastroenterol Rep 2007;95:385−92.

[96] Round JL, Lee SM, Li J, Tran G, Jabri B, Chatila TA, et al. The toll-like receptor 2 pathway establishes colonization by a commensal of the human microbiota. Science 2011;332:974−7.

[97] Patel RM, Lin PW. Developmental biology of gut-probiotic interaction. Gut Microbes 2010;13:186−95.

[98] Adlerberth I, Hanson LÅ, Wold AE. Ontogeny of the intestinal flora. In: Sanderson IR, Walker WA, editors. Development of the gastrointestinal tract. 1st ed. London: BC Decker Inc; 2000. p. 279−92.

[99] Dominguez-Bello MG, Costello EK, Contreras M, Magris M, Hidalgo G, Fierer N, et al. Delivery mode shapes the acquisition and structure of the initial microbiota across multiple body habitats in newborns. Proc Natl Acad Sci USA 2010;107:11971−5.

[100] Howell JC, Wells JM. Generating intestinal tissue from stem cells: potential for research and therapy. Regen Med 2011;6:743−55.

[101] Cheng H, Leblond CP. Origin, differentiation and renewal of the four main epithelial cell types in the mouse small intestine: V. unitarian theory of the origin of the four epithelial cell types. Am J Anat 1974;141:537−61.

[102] Leblond CP, Stevens CE. The constant renewal of the intestinal epithelium in the albino rat. Anat Rec 1948;100:357−77.

[103] Shaker A, Rubin DC. Intestinal stem cells and epithelial−mesenchymal interaction in the crypt and stem cell niche. Trans Res 2010;156:180−7.

[104] Barker N, van Es JH, Kuipers J, Kujala P, van den Born M, Cozijnsen M, et al. Identification of stem cells in small intestine and colon by marker gene Lgr5. Nature 2007;449:1003−7.

[105] Sato T, Vries RG, Snippert HJ, van de Wetering M, Barker N, Stange DE, et al. Single Lgr5 stem cells build crypt-villus structures *in vitro* without a mesenchymal niche. Nature 2009;459:262−5.

[106] Lander AD, Kimble J, Clevers H, Fuchs E, Montarras D, Buckingham M, et al. What does the concept of the stem cell niche really mean today? BMC Biol 2012;10:19.

[107] van der Flier LG, van Gijn ME, Hatzis P, Kujala P, Haegebarth A, Stange DE, et al. Transcription factor achaete scute-like 2 controls intestinal stem cell fate. Cell 2009;136:903−12.

[108] Tian H, Biehs B, Warming S, Leong KG, Rangell L, Klein OD, et al. A reserve stem cell population in small intestine renders Lgr5-positive cells dispensable. Nature 2011;478:255−9.

[109] Quyn AJ, Appleton PL, Carey FA, Steele RJ, Barker N, Clevers H, et al. Spindle orientation bias in gut epithelial stem cell compartments is lost in precancerous tissue. Cell Stem Cell 2010;6:175−81.

[110] Roth S, Franken P, Sacchetti A, Kremer A, Anderson K, Sansom O, et al. Paneth cells in intestinal homeostasis and tissue injury. PLoS One 2012;7:e38965.

[111] Li L, Clevers H. Coexistence of quiescent and active adult stem cells in mammals. Science 2010;327:542−5.

[112] Iizuka M, Konno S. Wound healing of intestinal epithelial cells. World J Gastroenterol 2011;17:2161−71.

[113] Saxena SK, Thompson JS, Sharp JG. Intestinal regeneration. Austin, TX: RG Landes Co; 1943.

[114] Dekaney CM, Fong JJ, Rigby RJ, Lund PK, Henning SJ, Helmrath MA. Expansion of intestinal stem cells associated with long-term adaptation following ileocecal resection in mice. Am J Physiol 2007;293:1013−22.

[115] Shimoda H, Takahashi Y, Kato S. Regrowth of lymphatic vessels following transection of the muscle coat in the rat small intestine. Cell Tissue Res 2004;316:325−38.

[116] Wood JD. Enteric nervous system neuropathy: repair and restoration. Curr Opin Gastroenterol 2011;27:106−11.

[117] Dumont AE, Martelli AB, Iliescu H, Baron M. Spontaneous reconstitution of the mammalian intestinal tract following complete transection (40914). Proc Soc Exp Biol Med 1980;164:545−9.

[118] Dumont AE, Martelli AB, Schinella R. Regenerative lengthening of the transected rat colon 1983;143:1518−29.

[119] Miyoshi H, Ajima R, Luo CT, Yamaguchi TP, Stappenbeck TS. Wnt5a potentiates TGF-B signaling to promote colonic crypt regeneration after tissue injury. Science 2012;338:108−13.

[120] Jiang H, Edgar BA. Intestinal cells in the adult Drosophila midgut. Exp Cell Res 2011;317:2780−8.

[121] Takashima S, Mkrtchyan M, Younossi-Hartenstein A, Merriam JR, Hartestein V. The behaviour of Drosophila adult hindgut stem cells is controlled by Wnt and Hh signaling. Nature 2008;454:651−5.

[122] Wang P, Hous SX. Regulation of intestinal stem cells in mammals and Drosophila. J Cell Physiol 2009;222:33−7.

[123] Ohlstein B, Spradling A. The adult Drosophila posterior midgut is maintained by pluripotent stem cells. Nature 2006;439:470−4.

[124] Jiang H, Patel PH, Kohlmaier A, Grenley MO, McEwen DG, Edgar BA. Cytokine/Jak/Stat signaling mediates regeneration and homeostasis in the Drosophila midgut. Cell 2009;137:1343−55.

[125] Forsthoefel DJ, Park AE, Newmark PA. Stem cell-based growth, regeneration, and remodeling of the planarian intestine. Dev Biol 2011;356:445−59.

[126] Kaneko N, Katsuyama Y, Kawamura K, Fujiwara S. Regeneration of the gut requires retinoic acid in the budding ascidian Polyandrocarpa misakiensis. Dev Growth Differ 2010;52:457—68.

[127] Anderson JM. Studies on visceral regeneration in the sea-stars. I. Regeneration of pyloric caeca in *Henricia leviuscula* (Stimpson). Biol Bull 1962;122:321—42.

[128] Anderson JM. Studies on visceral regeneration in the sea-stars. II. Regeneration of pyloric caeca in Asteriidae, with notes on the source of cells in regenerating organs. Biol Bull 1965;128:1—23.

[129] García-Arrarás JE, Greenberg MJ. Visceral regeneration in holothurians. Microsc Res Tech 2001;255:438—51.

[130] Mashanov V, García-Arrarás JE. Gut regeneration in holothurians: a snapshot of recent developments. Biol Bull 2011;221:93—109.

[131] García-Arrarás JE, Estrada-Rodgers L, Santiago R, Torres II, Díaz-Miranda L, Torres-Avillán I. Cellular mechanisms of intestine regeneration in the sea cucumber, *Holothuria glaberrima* Selenka (Holothuroidea: Echinodermata). J Exp Zool 1998;281:288—304.

[132] Garcia-Arraras JE, Valentin-Tirado G, Flores JE, Rosa RJ, Rivera-Cruz A, San Miguel-Ruiz JE, et al. Cell dedifferentiation and epithelial to mesenchymal transitions during intestinal regeneration in *H. glaberrima*. BMC Dev Biol 2011;11:61.

[133] Murray G, García-Arrarás JE. Myogenesis during holothurian intestinal regeneration. Cell Tissue Res 2004;318:515—24.

[134] Quiñones JL, Rosa R, Ruiz DL, García-Arrarás JE. Extracellular matrix remodeling and metalloproteinase involvement during intestine regeneration in the sea cucumber Holthuria glaberrima. Dev Biol 2002;250:181—97.

[135] Garcia-Arraras JE, Dolmatov IY. Echinoderms: potential model systems for studies on muscle regeneration. Curr Pharm Des 2010;16:942—55.

[136] Mashanov VS, Dolmatov IY, Heinzeller T. Transdifferentiation in holothurian gut regeneration. Biol Bull 2005;209:18493.

[137] Ortiz-Pineda PA, Ramírez-Gómez F, Pérez-Ortiz J, González-Díaz S, Santiago-De Jesús F, Hernández-Pasos J, et al. Gene expression profiling of intestinal regeneration in the sea cucumber. BMC Genomics 2009;10:262.

[138] Mashanov VS, Zueva OR, Rojas-Catagena C, García-Arrarás JE. Visceral regeneration in a sea cucumber involves extensive expression of *survivin* and *mortalin* homologs in the mesothelium. BMC Dev Biol 2010;10:117.

[139] Mashanov VS, Zueva OR, García-Arrarás JE. Expression of Wnt9, TCTP, and Bmp1/Tll in sea cucumber visceral regeneration. Gene Expr Patterns 2012;12:24—35.

[140] Pasten C, Ortiz-Pineda P, García-Arrarás JE. Ubiquitin-proteasome system components are up-regulated during intestinal regeneration. Genesis 2012;50:350—65.

[141] Pasten C, Rosa R, Ortiz S, García-Arrarás JE. Characterization of proteolytic activities during intestinal regeneration of the sea cucumber *Holothuria Glaberrima*. Intl J Dev Biol 2012;56:681—91.

[142] Mashanov VS, Zueva OR, García-Arrarás JE. Retrotransposons in animal regeneration: overlooked components of the regenerative machinery? Mob Genet Elements 2012;5:1—4.

[143] Goodchild CG. Reconstitution of the intestinal tract in the adult leopard frog, *Rana pipiens* Schreber. J Exp Zool 1956;131:301—27.

[144] O'Steen WK. Regeneration of the intestine in adult urodeles. J Morphol 1958;103:435—77.

[145] Ishzuya-Oka A, Hasebe T. Sonic hedgehog and bone morphogenetic protein-4 signaling pathway involved in epithelial cell renewal along the radial axis of the intestine. Digestion 2008;77:42—7.

[146] O'Steen WK, Walker BE. Radioautographic studies of regeneration in the common newt. III. Regeneration and repair of the intestine. Anat Rec 1962;142:179—88.

[147] Grubb RB. An autoradiographic study of the origin of intestinal blastemal cells in the newt *Notophtalmus viridescens*. Dev Biol 1975;47:185—95.

[148] Dumont AE, Martelli AB, Iliescu H, Baron M. Spontaneous reconstitution of the intestinal tract following complete transection. In: Becker R, editor. Mechanisms of growth control. Springfield, IL: Charles C. Thomas; 1981. p. 394—405.

[149] Ishizuya-Oka A, Shi YB. Molecular mechanisms for thyroid hormone-induced remodeling in the amphibian digestive tract: a model for studying organ regeneration. Dev Growth Differ 2005;47:601—7.

[150] Ishizuya-Oka A, Shi YB. Thyroid hormone regulation of stem cell development during intestinal remodeling. Mol Cel Endo 2008;288:71—8.

[151] Hasebe T, Kajita M, Fu L, Shi YB, Ishizuya-Oka A. Thyroid hormone-induced Sonic hedgehog signal up-regulates its own pathway in a paracrine manner in the *Xenopus laevis* intestine during metamorphosis. Dev Dyn 2012;241:403—14.

[152] Hasebe T, Buchholz DR, Shi YB, Ishizuya-Oka A. Epithelial—connective tissue interactions induced by thyroid hormone receptor are essential for adult stem cell development in the *Xenopus laevis* intestine. Stem Cells 2011;29:154—61.

[153] Sun G, Hasebe T, Fujimoto K, Lu R, Fu L, Matsuda H, et al. Spatio-temporal expression profile of stem cell-associated gene LGR5 in the intestine during thyroid hormone-dependent metamorphosis. PLoS One 2010;5:13605.

Chapter 36

Intestinal Regeneration: The Stem Cell Approach

Christa N. Grant and Tracy C. Grikscheit

Division of Pediatric Surgery, Saban Research Institute, Children's Hospital, Los Angeles, CA

36.1 INTRODUCTION

Intestinal failure is the inability of the gastrointestinal tract to maintain nutrition and therefore health and life [1]. Short bowel syndrome (SBS) is a form of intestinal failure that results after loss of a critical length of small intestine, usually 70–75% total length [2]. Without appropriate treatment, nutritional and metabolic consequences include malnutrition, electrolyte abnormalities, weight loss, chronic diarrhea, dehydration, and failure to thrive [3]. With an incidence of 3–5 per 100,000 births per year [1], and some census based population data reporting incidence as high as 24.5 per 100,000 [4], and overall mortality ranging from 15–25% in children to 15–47% in adults [5], SBS is one of the most morbid chronic diseases, particularly in children. Additionally, the human and economic costs are high. The estimated cost for a child on home parenteral nutrition is over $500,000 in the first year of life [6]. In cases of impending liver failure, loss of vascular access sites, or sequelae of liver failure such as gastrointestinal bleeding, intestinal or multivisceral transplantation are lifesaving but morbid [7]. While bowel lengthening procedures have shown some success [8], the majority of SBS patients are maintained on parenteral nutrition during attempts at intestinal rehabilitation. Because it is difficult to maintain constant vascular access and there are essential differences in the way the body processes the nutritional content of TPN compared to enteral feeding, the 2011 Intestinal Transplant Registry Report listed the leading indication for intestinal transplantation as complications related to parenteral nutrition and central venous access, which is necessary for its delivery. Ischemia, Crohn's disease, motility disorders, volvulus, and trauma are the leading causes of SBS in adults, and congenital disorders, including gastroschisis, necrotizing enterocolitis, volvulus, and atresias predominate in children [1,9]. Advances in neonatology are accompanied by higher survival rates in premature infants, and therefore children are at risk for complications of prematurity including necrotizing enterocolitis. This will likely lead to a higher incidence of SBS. In addition, the overall survival rates of children with SBS have increased from 70% to 89%, likely secondary to increased numbers and success of

Regenerative Medicine Applications in Organ Transplantation.

multidisciplinary bowel rehabilitation programs [10]. With these advances, the prevalence of SBS will continue to outweigh donor availability. These trends describe the need for novel alternative treatments for patients with SBS.

36.2 TISSUE ENGINEERING OF THE INTESTINAL TRACT

The goal of tissue engineering is to replace the structure and function of organs or tissues that have failed or are absent. Distinct from wound repair, these engineered tissues would ideally integrate fully into the host and grow with them. Approaches can be grouped into *in vitro* and *in vivo*. While *in vivo* conditions better represent the living milieu of the intended host, it becomes difficult to monitor and control all conditions and events in the multicellular growth of the engineered tissue. In contrast, reproducing host conditions *in vitro* can prove costly and complex, and variation from homeostasis is likely. In general, *in vitro* models involve coculture of different cell types to encourage growth, or bioreactors that simulate key *in vivo* conditions such as temperature, oxygenation, flow, and tissue stress [11]. Initially quite simple, bioreactors have evolved significantly and show promise as a more malleable alternative to *in vivo* studies. One such advance is the biological vascularized scaffold, BioVasc®, which is created from decellularized porcine small bowel with the vasculature intact and allows for the pulsatile delivery of culture medium after cells of interest have been seeded into the lumen of the structure. The dynamic culture conditions have been reported to improve tissue differentiation and function by simulating the necessary biochemical and physiological regulatory signals [12].

36.3 TRANSITION TO AN *IN VIVO* MODEL

In 1988, combining both approaches, Vacanti et al. described a new method of selective cell transplantation involving the attachment and culture of individual cells to biodegradable polymer scaffolds for several days followed by implantation into several host sites in adult mice [13]. Distinct from selective cell transplantation where single cell suspensions were injected into host tissue or vasculature, this method was called chimeric morphogenesis. Despite initial poor engraftment rates, the successful growth of viable tissue masses of liver and intestine with mitotic figures and vascularization was one of the first successful applications of this *in vivo* technique.

36.4 ENGINEERED INTESTINE IS DISTINCT FROM HEALING, TISSUE TRANSFER, AND MONOLAYER CELL CULTURE

In the case of engineered intestine, the functions required of the replacement tissue include fluid and nutrient absorption, motility, and reservoir capacity. Autologous tissue is ideal, obviating the need for and eliminating the complications of immunosuppression in the host. One of the earliest investigations proving the regenerative capacity of small intestine was performed in 1973. Corroborating other results showing epithelial coverage of small intestinal defects, new jejunal mucosa was formed in dogs after enterotomy and anastomosis to the serosa of a nearby colonic segment [14]. While this technique would be classified as tissue transfer, and not true tissue engineering by current definition, the observation of epithelial cell migration from the surrounding tissue would prove to be an important one.

36.5 ORGANOID UNIT APPROACH FACILITATES ENGINEERING OF THE GASTROINTESTINAL TRACT

In the early 1990s, using structured experiments where key aspects, such as pH, CO_2, culture medium, and growth factors, were modified in turn, Evans et al. determined the ideal culture environment for rat intestinal epithelial cell primary cultures. In the same publication, a method for isolation of rat intestinal epithelium employing collagenase and dispase digestion was described [15]. The resultant isolates of epithelium consisting of villi and crypts were called organoids. Significant modifications in this technique by the Vacanti laboratory yielded greater numbers of cells, and the term organoid unit (OU) was coined. Comprised of a core of mesenchymal cells surrounded by intestinal epithelium [16], OU can be isolated from neonatal or adult mammalian tissue or from engineered intestine itself. In fact, tissue-engineered intestine has been created from OU derived from all sections of the gastrointestinal tract. OUs were generated from both neonatal and adult murine abdominal esophagus, seeded onto scaffolds, and implanted into the omentum of host Lewis rats [16]. Tissue-engineered esophagus formed in all except one implantation, and in all cases formed single lumen spheres with histological architecture similar to host esophagus. In fact, the engineered tissue acted as a conduit without complication as evaluated by fluoroscopy and normal weight gain, after anastomosis to native esophagus. Similarly, tissue-engineered small intestine (TESI) has been successfully generated, with transgenic models showing donor origin of all key cells, features of

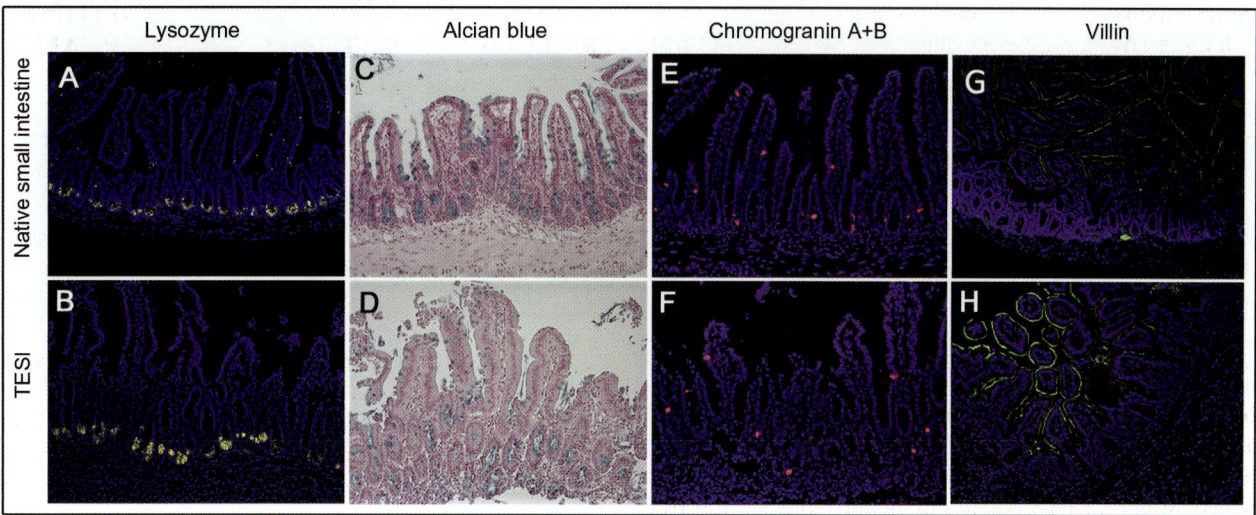

FIGURE 36.1 Morphological and histological comparison of differentiated epithelial cells in TESI compared to native small intestine. Lysozyme staining identifies Paneth cells in the crypts of native small intestine (A) and TESI (B). Alcian Blue stains goblet cells in native small intestine (C) and TESI (D). Chromogranin A staining demonstrates enteroendocrine cells in both native small intestine (E) and TESI (F). Villin stains the brush border of enterocytes in both tissues (G and H).

an intact stem cell niche, preserved epithelial–mesenchymal interactions, and terminal differentiation [17]. Histological stains have confirmed that TESI is comprised of all four differentiated intestinal cells types, enterocytes, Paneth cells, goblet cells, and enteroendocrine cells [17,18].

Defining characteristics of engineered organs include not only replication of structure but also the replacement of organ function. In a model of SBS in rats, end to end anastomosis of TESI was performed distal to the duodenum and proximal to the cecum after greater than 80% enterectomy [19]. Control animals underwent enterectomy only. As in prior investigations, histology of the TESI recapitulated that of native small intestine. While both groups lost weight initially, animals with TESI implants reached their nadir a week sooner and regained a higher percentage or preoperative body weight, 98.5%, as opposed to 76.8% in the enterectomy only group, and had higher serum levels of B12.

A necessary step toward the application of this method to human subjects was successful reproduction in a large animal model. In 2009, stomach and small intestine were both harvested from anesthetized 6-week-old Yorkshire swine. Contemporaneously, OUs were prepared, loaded onto a polymer, and implanted into the omenta of the respective donor animals [18]. After 7 weeks, all implants yielded tissue-engineered stomach or intestine. With the exception of Paneth cells, which do not stain for common Paneth cell markers in swine, all terminally differentiated epithelial cell types were present in the TESI (Figure 36.1). The tissue architecture of

both the engineered intestine and the stomach was similar to native tissue on histological sections (Figures 36.2 and 36.3). In addition to epithelial similarities, engineered intestine also had a proliferative zone in the crypts, as well as an intact lamina propria comprising of nerve and muscle elements, as well as intestinal subepithelial myofibroblasts (ISEMFs), a putative constituent of the intestinal stem cell (ISC) niche [20]. The first report of autologous OU implantation, this experiment models the ideal application of tissue-engineered intestine for human subjects. Intestine would be harvested, and OUs prepared and reimplanted in a single operation [18]. With the same technique, OUs have been isolated from the colon of Lewis rats and implanted as described above. Tissue-engineered colon (TEC) was generated after 4–16 weeks, with sizes up to 11 cm × 13 cm × 12 cm. On gross section, these constructs have circular folds as in native colon, and histology is recapitulated. TEC is generated with high fidelity even when created secondarily, i.e., OUs prepared from TEC [21]. The mechanism underlying this high proliferation is still unknown but likely linked to the elegant and regenerative ISC niche. Physiological testing using the Ussing chamber showed that TEC epithelium is composed of mature colonocytes which actively secrete chloride, participate in active ion transport, and lack SGTL1, a sodium–glucose cotransporter specific to the small intestine [21]. *In vivo* investigations of function have also proven the efficacy of TEC. After end ileostomies in two cohorts of Lewis rats, half underwent anastomosis of TEC to the ileum. Animals with ileostomies alone had 1.5 times greater weight loss,

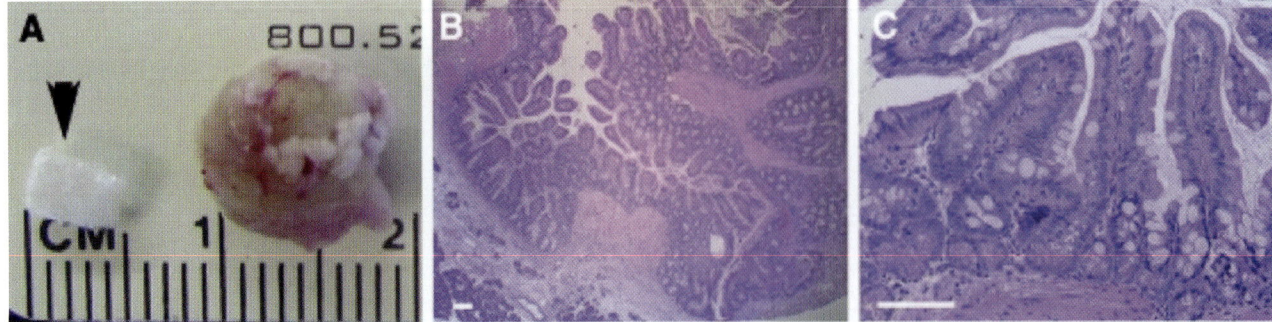

FIGURE 36.2 Morphology of TESI in the mouse model. (A) Four weeks after implantation, the tissue-engineered intestine formed a sphere about twice the size of the initial implanted polymer (arrow). (B) Hematoxylin and eosin staining at low magnification demonstrates a large amount of continuous mucosa lining the lumen of the TESI. (C) At higher magnification, the mucosa of the TESI is composed of a simple columnar epithelium forming crypt and villus structures. Goblet cells, along the crypt and villus axis, and Paneth cells, at the base of the crypts, can be identified. Scale bar: 40.0 μm. *Copyright (2011) Mary Ann Leibert, Inc. Used with permission from [17].*

longer transit times, and higher stool water content than animals with TEC anastomoses [22]. In addition, lack of TEC was associated with statistically significant hyponatremia, decreased fecal short chain fatty acid content, and increased serum urea nitrogen [22]. Continued investigations into the biological basis of TEC generation will yield insights into improved generation of other engineered tissue and understanding of the stem cells of the gastrointestinal tract.

36.6 NUTRIENT SUPPLY TO THE NEOORGAN

The blood supply is one of the well recognized barriers to the reliable generation of engineering complex tissues. In the case of skin, upper airways, and bladder, a scaffold without a defined blood supply may be employed. These tissues acquire oxygen and nutrients by imbibition and diffusion from neighboring host tissue. In contrast, complex organs comprised of multicellular and multifunctional units rapidly outweigh diffusion capacity and require the construction of a dedicated vascular supply [23]. Once implanted, all engineered tissues acquire nutrients by initial imbibition and ultimately angiogenesis. After recognizing the key importance of nonnative blood supply to increasing the volume to engineered tissue, Vacanti and colleagues at Massachusetts General Hospital partnered with the Draper laboratories at Massachusetts Institute of Technology to develop vascular molds upon which tissue could be seeded. The marriage of their respective fields of tissue engineering and materials science and bioengineering birthed new promise for the successful engineering of more complex tissues such as the liver [24]. While TESI is fed initially by imbibition

from surrounding tissues, and ultimately small-scale angiogenesis, the eventual need for a definitive blood supply remains paramount to the success of larger constructs.

36.7 CELL SIGNALING

Equally important to engineering complex tissues is cell—cell signaling. Drawing from observations in developmental biology, the use of growth factors and paracrine signals have proven essential to engineering anatomically and functionally correct constructs. Through the use of Activin A, a nodal-related transforming growth factor (TGF)-β molecule, human pluripotent stem cells were induced to differentiate into intestinal epithelium *in vitro*. This led to the formation of the embryological gut precursor, the definitive endoderm. By providing exogenous factors Wnt3a and FGF, known to be important to the differentiation of the posterior endoderm, epithelial tubes positive for midgut and hindgut markers formed from single sheets of cells [25]. Recapitulating the observation that mesenchyme—epithelial cross talk is essential for the formation of intestinal epithelium, this experiment showed that as an epithelium developed, so did a differentiated mesenchyme.

Engineering intestinal tissue by the methods described will prove time consuming and expensive, especially to generate large amounts to reverse intestinal failure. So far, *in vitro* attempts have not been successful [23] in generating the necessary blood and nutrient supply and recapitulating the important cell—cell contact and signaling on a large enough scale. Generating enough length for clinical application has proven to be one of the major hurdles in bringing human application closer. Recognizing

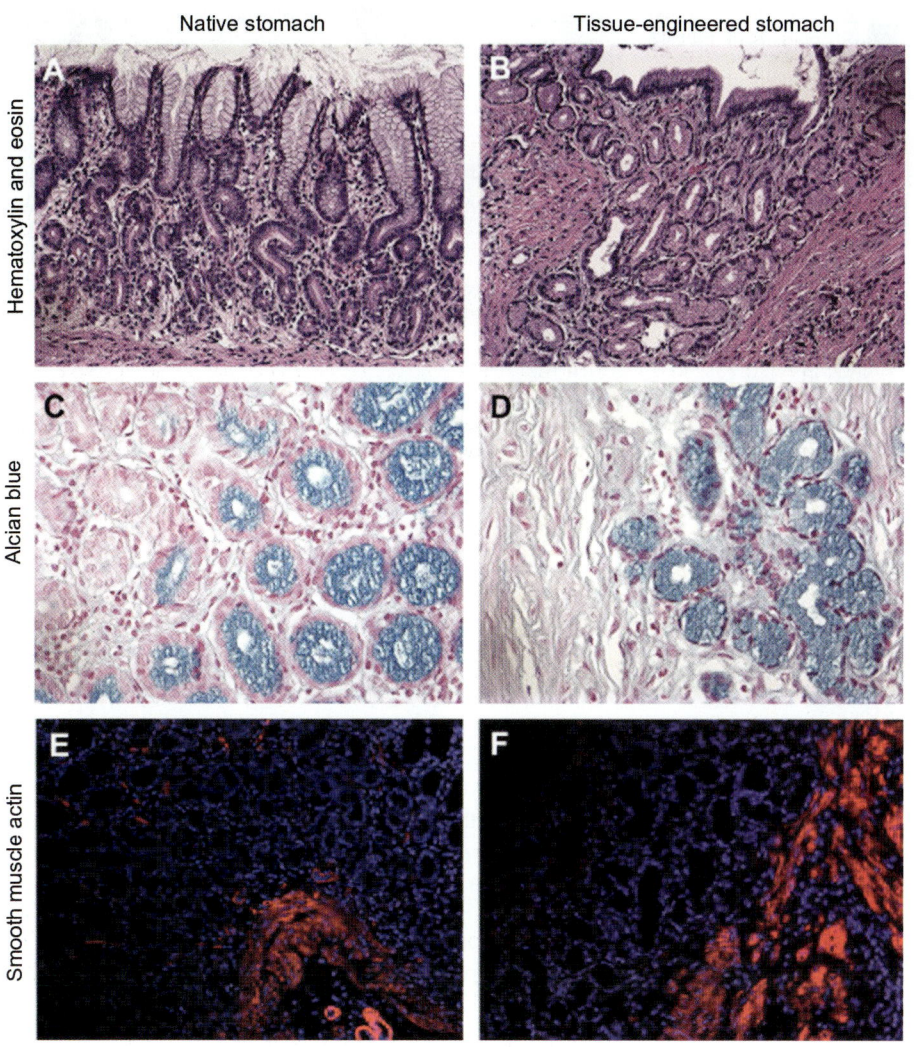

Native stomach   Tissue-engineered stomach

Hematoxylin and eosin

Alcian blue

Smooth muscle actin

FIGURE 36.3 Generation of tissue-engineered gastric mucosa. Hematoxylin and eosin staining of tissue-engineered stomach (A) and native stomach (B) (original magnification 320). (C) and (D) Alcian Blue staining of the mucous epithelial cells (original magnification 340). (E) and (F) Immunofluorescence staining of the muscularis using an anti-smooth muscle actin primary antibody (original magnification 340). *Copyright (2009) Elsevier. Used with permission from [18].*

this limitation, investigators have pursued ISCs as a plausible donor cell population. A better understanding of the location, subpopulations, and governing signals of these progenitors will direct novel applications to tissue-engineered intestine.

36.8 DEFINING THE PROGENITOR CELLS OF THE INTESTINE

The intestinal epithelium is one of the most rapidly dividing regions of cells in the human body. The mucosal lining is shed and replaced approximately every 3−4 days. This continuous self-renewal is the product of multipotent undifferentiated cells. These ISCs self-renew by giving rise to daughter stem cells and maintain epithelial homeostasis by giving rise to rapidly dividing progenitor cells. In turn, these daughter cells, also called transit amplifying cells, differentiate and give rise to all terminally

differentiated cells of the intestinal epithelium [26]. The intestine is therefore an elegant model for the study of stem and progenitor cells.

The identification of histological markers of ISCs has advanced significantly in recent years. Bjerknes and Cheng first described the proliferative zone and the stem cell zone after a series of observations in mice post-jejunal resection [27]. They observed a population at the bottom of crypts interspersed between Paneth cells, which they presumed to be the ISC. They were named crypt base columnar cells (CBCs) [28]. In a series of subsequent studies involving lineage tracing and electron microscopy, published in a series of works describing the lineage of each cell type in detail and culminating with the Unitarian Theory of origin of cell types in the small intestine. Cheng concluded that these CBCs gave rise to all four differentiated cell types of the small intestine [29]. Subsequently, most studies have agreed that there are at least two pools of stem cells, with differing

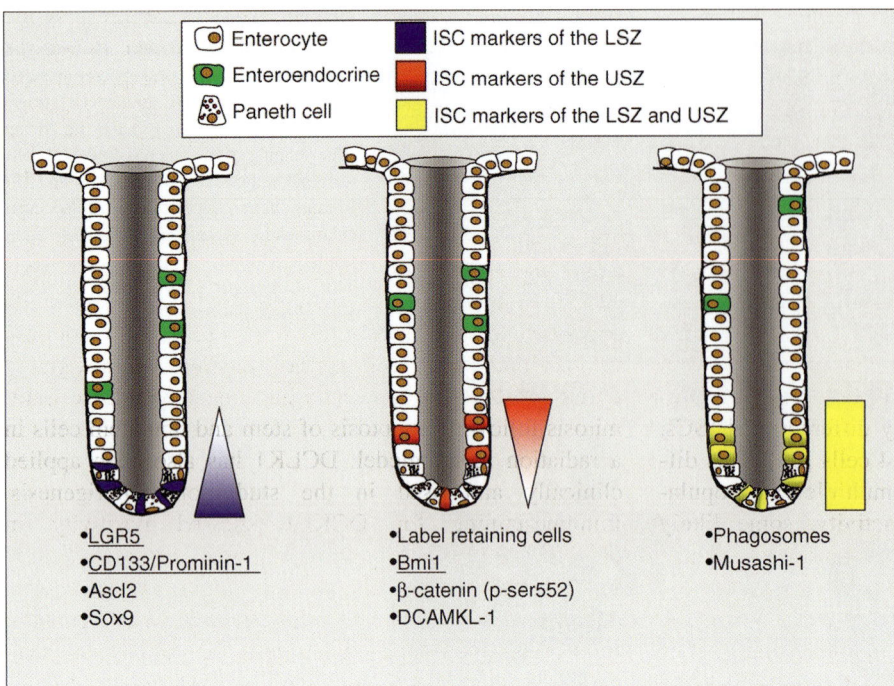

FIGURE 36.4 Markers of upper (USZ) and lower (LSZ) ISC zones. Markers represented by blue coloring are found predominantly in the LSZ but may be found in the USZ. Markers represented by red coloring are found predominantly in the USZ but may be found in the LSZ. Markers represented by yellow coloring are found throughout the USZ and LSZ. Underlined markers have been confirmed by lineage tracing. *Copyright (2009) Wolters Kluwer Health. Used with permission from [38].*

locations, namely the upper stem cell zone and the lower stem cell zone (Figure 36.4). Early studies characterized one group of ISCs by kinetic or label retention studies, most commonly BrdU. This population resides in the upper stem cell zone, on average four cell positions toward the lumen from the base of the crypt in a location designated "+4". An alternative view from studies employing lineage tracing and stem cell renewal studies suggest that the ISCs are in fact interspersed between Paneth cells at the bases of crypts [30]. Having identified several ISC markers, there now appears to be consensus that there exist both a quiescent ISC population at +4 and a faster cycling cell population at the crypt bases [31]. The proliferation, differentiation, and coexpression of molecules of interest have allowed identification of a number of markers of putative ISCs.

36.8.1 *Msi-1*

First identified in drosophila, Mushashi-1 (*Msi-1*) is a gene encoding an RNA-binding protein related to asymmetric divisions in neural progenitor cells. Okano and colleagues initially identified *Msi-1* as a marker of neural progenitor cells [32], and it was subsequently identified as one of the first markers of the ISC. In neonatal and adult irradiated mice, *Msi-1* expression was determined by immunohistochemistry. Cells in the stem cell zone of regenerating crypts stained positive across all groups [33]. In a study of mouse small intestine development, *Msi-1* shows early expression in both epithelium and mesenchyme, but by the completion of crypt development

expression was restricted to the pre-crypt regions, specifically just above the Paneth cells, as well as in the crypt base between them. This expression pattern persisted into adulthood. *Msi-1* was also found in the neck and base regions of the glandular epithelium of the stomach in postnatal mice [32].

36.8.2 Lgr5

Initially identified in colon cancer cells, leucine-rich-repeat-containing G-protein coupled receptor 5 (Lgr5), also known as GPR49, was recently found to mark the small, funnel-shaped ISCs located at the base of crypts between the Paneth cells. After noting strong downregulation following induced inhibition of the Wnt pathway, Lgr5 was identified as a *Wnt* target gene [34]. Also found in cells of the eye, brain, hair follicles, stomach, and reproductive organs, Lgr5 appeared to be isolated to the crypts of the small intestine. In addition to sparse cells at the bases of colonic crypts and stomach glands, Lgr5 expression in the small intestine was found to be limited to a small population of five to six cells at the base of each crypt, intercalated between the Paneth cells [35]. Coincident with the location and morphology (small cells with wedge-shaped nuclei and scant cytoplasm) of the CBCs identified over three decades earlier [29], these cells were thought to represent the ISC. The following observations by Clevers and colleagues support this theory [35]. Using proliferative marker Ki7 and M-phase marker phospho-histone H3, and pulse labeling with BrdU, the average cycling time of this population was

24 h. Lgr5 + cells turn over rapidly, being positive for proliferating cell nuclear antigen (PCNA) in most cases as opposed to DCLK1cells which are generally PCNA negative [26]. Employing knock-in alleles and reporter genes, lineage tracing proved all cell types were derived from these Lgr5 + CBCs [35]. To confirm that this subpopulation of ISCs was distinct from the +4 population, an injury model exploiting the knowledge of differences in radiosensitivity between the groups was employed. As previously described [36], +4 cells were exquisitely sensitive to radiation, with significant cell death after 1 Gy radiation, whereas CBCs were somewhat resistant to low doses, with apoptosis peaking after 10 Gy. Transit amplifying cells, a third group of partially differentiated ISCs, were also less radiosensitive than +4 cells [35]. This differentiation supports the theory of multiple ISC populations with varying turnover and activity, some likely participating in normal tissue homeostasis, yet others remaining quiescent until activation is necessitated by injury. While ISC–mesenchymal cross talk is thought to be mandatory for ISC maintenance and function, a provocative recent study suggests that the stem cell niche may not be entirely necessary. Single Lgr5 + stem cells isolated from small intestine crypts differentiated into budding structures representative of crypts and villi when cultured *in vitro* [35]. With the knowledge of growth factors and signals such as R-spondin, *Wnt*, and epidermal growth factor (EGF) necessary for crypt and villus development and maintenance, a supportive culture environment was created. Single Lgr5 + CBCs were isolated and plated separately, yielding organoids about 6% of the time. These organoids were composed of Paneth cell-containing crypts, as well as buds with lumens and a polarized epithelium which stained for all terminally differentiated cells [37]. Dissociation and subculture yielded secondary organoids. While remarkable, the mesenchymal niche is in fact not dispensable, as the success of these single cell cultures was likely dependent on support from the matrigel, a solubilized basement membrane preparation extracted from Engelbreth-Holm-Swarm mouse sarcoma, as well as exogenous signals known to be provided by supporting niche cells, such as EGF, R-spondin, and Noggin [30]. Successful long-term culture of single epithelial stem cells or coculture of stem cells *in vitro* could help to generate large numbers of donor cells, a major obstacle in engineering intestine. However, mesenchymal support and tissue growth will still need to be addressed.

36.8.3 DCLK1

Doublecortin and CaM kinase-like-1 (DCLK1) is a microtubule-associated kinase found in postmitotic neurons [38] and has recently been described as a putative stem cell marker of the intestine, where it has been found to be expressed on the cell surface [26]. While DCLK1-positive cells are found in the lower two-thirds of the intestinal crypt, or the stem cell zone [26] in normal intestine, these cells are likely quiescent, as they are negative for PCNA, a marker of proliferation. Though expressed in rare CBCs, DCLK1 is most frequently found in the +4 position, that is, an average of the fourth cell above the crypt base [26]. Costaining confirmed that these cells are distinct from the differentiated cell types and do not express the markers of differentiated cells except for tuft cells. These cells are also distinct, not only in location, but in cell surface markers, from Lgr5 + CBCs. DCLK1-positive cells have been shown to enter mitosis following apoptosis of stem and non-stem cells in a radiation injury model. DCLK1 has also been applied clinically and used in the study of tumorigenesis. Immunostaining for DCKL1 showed positivity in 50−78% of polyps, adenocarcinomas, and colorectal metastases. There was also an association between DCKL1 and increased cancer-specific mortality [39]. Differences in nuclear β-catenin in DCLK1 cells can also distinguish between normal and adenomatous intestinal tissues [40].

36.8.4 *mTert*

A slowly cycling subpopulation of ISCs has been found to give rise to Lgr5 + cells. These cells express mouse telomerase reverse transcriptase (*mTert*), which is thought to regulate stem cell proliferation and mobilization, a hypothesis supported by its common expression in rapidly dividing cells such as the bone marrow and testis, in addition to the intestine [41]. Consistent with known low expression levels in the intestine, labeling studies have found small numbers of *mTert*-positive cells at the bases of crypts. Using green fluorescent protein (GFP) labeling, *mTert*-positive cells were present on average between positions 5 and 8 along the crypt axis, were distinct from GPF − Lgr5 + rapidly cycling CBCs, and were generally Ki67 −, indicating their slowly cycling nature [41]. One additional distinction between *mTert* + and Lgr5 + stem cells is the comparative resistance of *mTert* + ISCs to radiation injury. This accounts for their role in regeneration after injury, with lineage tracing showing they can give rise to Lgr5 + cells. Of interest, *mTert* + cells were twice as predominant in the proximal intestine and coexpressed DCAMKL-1, Sca-1, ad BMP-R1a a small percentage of the time, indicating some overlap with other putative stem cells [41].

36.8.5 *Bmi1*

Bmi1, a gene involved in renewal of hematopoietic, neuronal, and leukemic cells, is part of the Polycomb repressing complex 1 (PRC1). The polycomb family of genes is involved in maintaining chromatin silencing. To investigate whether PRC1 is also involved in self-renewal of the ISC population, Sangiorgi et al. created transgenic mice whose *Bmi1* + cells would be labeled with yellow fluorescent protein after tamoxifen induction [42]. Present at the +4 position, *Bmi1* + cells gave rise to progeny that remained undifferentiated, and also to daughter cells that differentiated further, confirmed their identity as pluripotent ISCs. While *Bmi1* − / − knockout mice were nonviable, targeted ablation of *Bmi1* + cells using diphtheria toxin led to cell death and complete disruption of crypts, leading to the conclusion that *Bmi1* + cells are necessary for crypt maintenance [42]. In a mouse model where knock-in of diphtheria toxin at the Lgr5 locus led to complete loss of all Lgr5 + cells, crypt architecture was preserved [43]. This surprising observation suggests that Lgr5 + cells may be expendable. Further observations by these investigators proved that *Bmi1*-expressing stem cells compensate, expanding by 40% to repopulate the crypts [43]. Interestingly, Bmi1 is expressed in a descending gradient from proximal to distal in the small intestine, with very little identified in the distal ileum and colon [42]. This hints at the possibility that ISCs are more complex and may not be uniform across the organ. Regional differences may play a large role in the relative activity of stem cells, a factor that will prove important if stem cells are to be used in tissue engineering.

36.8.6 *Prom1*

Prominin 1, or CD133, has been identified as a potential cancer stem cell marker after being identified in tumors of several cell types [44]. Prominin 1 (*Prom1*) has been shown to coexpress with Lgr5 and identifies the rapidly cycling [43] population of stem cells in the lower stem cell zone [38]. In addition to closely overlapping mRNA expression, over 75% of *Prom1* + crypt cells were also Lgr5 + and were located on average at 0, 1′, and 2′; positions, the usual location of Lgr5 + CBCs [45].

36.9 THE STEM CELL NICHE

It has yet to be defined whether the self-renewing and multipotent qualities of stem cells are self-regulated or influenced by the surrounding environment. Most evidence indicates that the stem cells of the crypt are supported by physical and paracrine signals from surrounding cells in the niche, comprised of the semipermeable basement membrane and the subjacent mesenchymal, immunologic, and neural cells [30]. The canonical Wnt pathway regulates ISCs by triggering cell type−specific gene expression [46] through nuclear localization and stabilization of β-catenin. Wnt expression controls the differentiation and migration of cells along the crypts villus axis by controlling the ratio of nuclear and cytoplasmic β-catenin in cells along the crypt axis and by generating inhibitory and complementary signals. The expression of Wnt by both ISCs and cells in the surrounding niche is central to normal tissue homeostasis in the intestine. In order to prevent uncontrolled proliferation, an inhibitory mechanism must also exist. The bone morphogenic protein (Bmp) signaling pathway, which can be activated by both ISCs and mesenchymal cells, negatively regulates the proliferation of ISCs by suppressing the Wnt-β-catenin pathway [46]. By conditional inactivation of Bmpr1a, a Bmp receptor, in crypt cells, He et al. confirmed the importance of this interaction. In the absence of a functional Bmpr1a, stem and progenitor cell populations expanded rapidly, with ISC duplication leading to intestinal polyposis [47]. In order to exit senescence, ISCs are intermittently activated to enter the proliferative cycle by Bmp inhibitors such as Noggin [46]. Other Wnt inhibitors such as Dkk3 also exist in a gradient and are thought to contribute to the balance between positive and negative regulatory signals on ISCs. Signaling via the Notch pathway is thought to control these alternating gene expression patterns [46]. In addition to complex signaling pathways between ISCs and their niche, there is evidence to suggest that the balance between luminal contents and secreted peptides such as Muc2 from goblet cells and α-defensins from Paneth cells also play a role in ISC regulation [30].

Tissue resident fibroblasts are thought to play a key role in wound repair of epithelial-lined organs [48]. In the intestine, ISEMFs have been investigated closely in recent years for their role as part of the stem cell niche. With the typical appearance of myofibroblasts, ISEMFs are found throughout the lamina propria and in close conjunction with capillaries. One hypothesis is that ISEMFs and pericytes are one and the same [20]. ISEMFs are well suited for cell−cell communication and paracrine signaling and are found in close apposition to nerves with cell processes that can extend through fenestrations in the basal lamina that can come into contact with immune cells [20]. In fact, ISEMFs are thought to participate in both organogenesis and cytodifferentiation of the intestinal mucosa [49]. The Wnt family of glycoproteins serve as important signaling molecules and have been found to be expressed by ISEMFs [30].

Ootani et al. successfully cultured neonatal mouse intestinal epithelium on a collagen matrix for up to 30 days. The resultant cystic spheres had a luminal surface and a polarized epithelial monolayer. They were able to show that the spheres also contained *Bmi1*- and Lgr5-positive ISCs [50]. Spheroid growth was inhibited by the Wnt antagonist Dkk1 and stimulated by the Wnt agonist R-spondin 1, confirming the central role of Wnt signaling.

Paneth cells have been found to secrete Wnt at crypt bottoms, regulating the morphogenesis of crypts and villi in

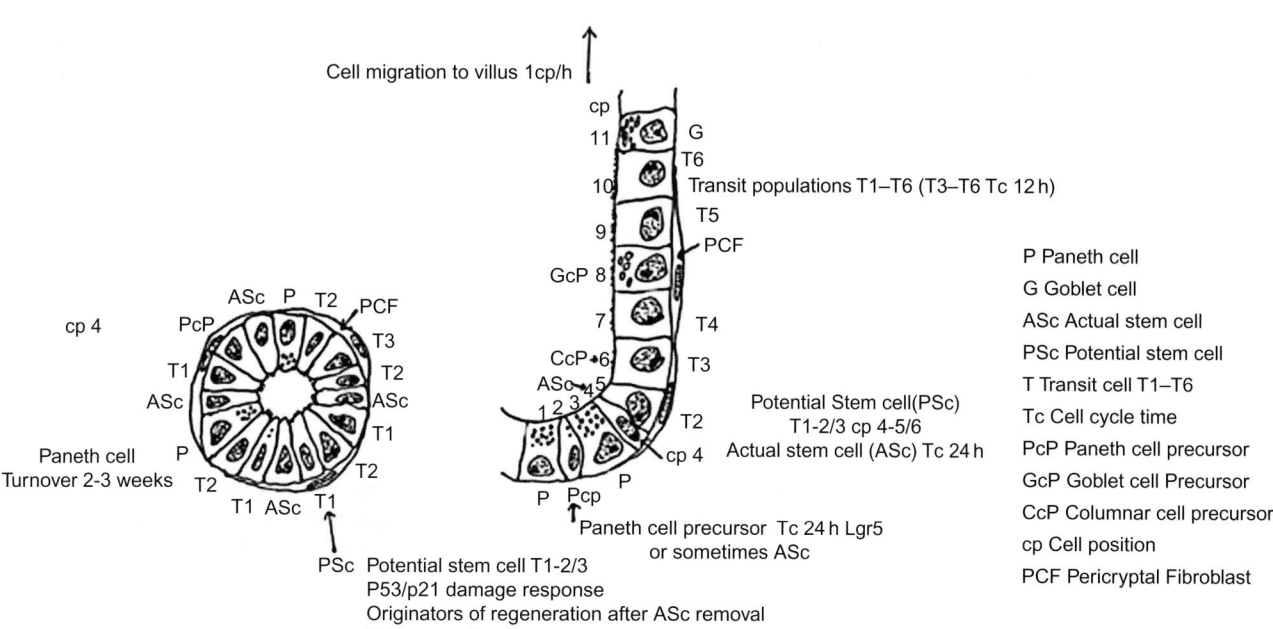

FIGURE 36.5 Schematic diagram of the lower region of small intestinal crypt in longitudinal section and a transverse section at about cell position 4. The position of the actual (ASC) and potential (PSc) stem cells is shown together with the possible relationship to the early transit generation (T1−T3). Some of the features and characteristics associated with ASCs are shown. The Paneth cell precursors (PCPs) may alternatively be regarded as ASCs, or part of the ASC population, and in the lower regions of the crypt (crypt base or cp 1−4) have been termed CBCCs. The pericryptal fibroblasts (PCF), which may undergo endoreduplication, could be important elements for the intestinal niche. *Copyright (2009) Wiley. Used with permission from [30].*

culture [51]. Also expressing EGF, TGF, and the Notch ligand Dll4 [51], it is logical that Paneth cells are often employed in coculture in support of ISCs. Isolated Paneth cells, marked by surface CD24, were cultured in matrigel alone, or with Lgr5 + sorted stem cells. Stem cells were also cultured in matrigel alone. While Paneth cells and Lgr5 + cells generated 0% and 3.3% organoids, respectively, when plated alone, 76.7% of wells plated with both Paneth cells and Lgr5 + cells yielded organoids. In addition, the small number of Lgr5 + -derived organoids disintegrated after 5 days in culture. Furthermore, in a model of Paneth cell depletion, stem cell numbers also declined, with remaining stem cells located next to few surviving Paneth cells. While these observations suggest that cell−cell contact is necessary for survival of and organoid formation from Lgr5 + cells, addition of Wnt3a to Lgr5 + cells in culture in the absence of Paneth cells yielded organoids [43]. In Figure 36.5, Potten illustrates relative positions of actual and potential stem cells and their relationship to cells in their niche, such as Paneth cells and pericryptal fibroblasts [30].

36.10 INTESTINE REGENERATION USING STEM CELLS: TESI

The era of personalized medicine is imminent [52]. In the case of TESI, tissue will ideally be harvested from a patient, and organoids prepared and implanted in a single operation, followed by anastomosis after the growth of the desired tissue substitute. While the timeline toward availability for humans is still not defined, the advances detailed above give a glimpse into the potential of this science and the impact that progress in tissue engineering the intestine may have for patients with SBS and other diseases of the entire gastrointestinal tract.

REFERENCES

[1] Squires RH, Duggan C, Teitelbaum DH, Wales PW, Balint J, Venick R, et al. Natural history of pediatric intestinal failure: initial report from the pediatric intestinal failure consortium. J Pediatr 2012;161(4):723−8.

[2] Thomson AB, Chopra A, Clandinin MT, Freeman H. Recent advances in small bowel diseases: part II. World J Gastroenterol 2012;18:3353−74.

[3] Valdovinos D, Cadena J, Montijo E, Zárate F, Cazares M, Toro E, et al. Short bowel syndrome in children: a diagnosis and management update. Rev Gastroenterol Mex 2012;77(3):130−40.

[4] Wales PW, de Silva N, Kim J, Lecce L, To T, Moore A. Neonatal short bowel syndrome: population-based estimates of incidence and mortality rates. J Pediatr Surg 2004;39:690−5.

[5] Schalamon J, Mayr JM, Höllwarth ME. Mortality and economics in short bowel syndrome. Best Pract Res Clin Gastroenterol 2003;17:931−42.

[6] Spencer AU, Kovacevich D, McKinney-Barnett M, Hair D, Canham J, Maksym C, et al. Pediatric short-bowel syndrome: the cost of comprehensive care. Am J Clin Nutr 2008;88:1552−9.

[7] Fishbein TM. Intestinal transplantation. N Engl J Med 2009;361:998−1008.

[8] Khalil BA, Ba'ath ME, Aziz A, Forsythe L, Gozzini S, Murphy F, et al. Intestinal rehabilitation and bowel reconstructive surgery: improved outcomes in children with short bowel syndrome. J Pediatr Gastroenterol Nutr 2012;54:505−9.

[9] Desai CS, Khan KM, Girlanda R, Fishbein TM. Intestinal transplantation: a review. Indian J Gastroenterol 2012;31(5):217−22.

[10] Modi BP, Langer M, Ching YA, Valim C, Waterford SD, Iglesias J, et al. Improved survival in a multidisciplinary short bowel syndrome program. J Pediatr Surg 2008;43:20−4.

[11] Grikscheit TC, Vacanti JP. The history and current status of tissue engineering: the future of pediatric surgery. J Pediatr Surg 2002;37:277−88.

[12] Schanz J, Pusch J, Hansmann J, Walles H. Vascularised human tissue models: a new approach for the refinement of biomedical research. J Biotechnol 2010;148:56−63.

[13] Vacanti JP, Morse MA, Saltzman WM, Domb AJ, Perez-Atayde A, Langer R. Selective cell transplantation using bioabsorbable artificial polymers as matrices. J Pediatr Surg 1988;23:3−9.

[14] Binnington HB, Siegel BA, Kissane JM, Ternberg JL. A technique to increase jejunal mucosa surface area. J Pediatr Surg 1973;8:765−9.

[15] Evans GS, Flint N, Somers AS, Eyden B, Potten CS. The development of a method for the preparation of rat intestinal epithelial cell primary cultures. J Cell Sci 1992;101:219−31.

[16] Grikscheit T, Ochoa ER, Srinivasan A, Gaissert H, Vacanti JP. Tissue-engineered esophagus: experimental substitution by onlay patch or interposition. J Thorac Cardiovasc Surg 2003;126:537−44.

[17] Sala FG, Matthews JA, Speer AL, Torashima Y, Barthel ER, Grikscheit TC. A multicellular approach forms a significant amount of tissue-engineered small intestine in the mouse. Tissue Eng Part A 2011;17:1841−50.

[18] Sala FG, Kunisaki SM, Ochoa ER, Vacanti J, Grikscheit TC. Tissue-engineered small intestine and stomach form from autologous tissue in a preclinical large animal model. J Surg Res 2009;156:205−12.

[19] Grikscheit TC, Siddique A, Ochoa ER, Srinivasan A, Alsberg E, Hodin RA, et al. Tissue-engineered small intestine improves recovery after massive small bowel resection. Ann Surg 2004;240:748−54.

[20] Powell DW, Mifflin RC, Valentich JD, Crowe SE, Saada JI, West AB. Myofibroblasts. II. Intestinal subepithelial myofibroblasts. Am J Physiol 1999;277:183−201.

[21] Grikscheit TC, Ochoa ER, Ramsanahie A, Alsberg E, Mooney D, Whang EE, et al. Tissue-engineered large intestine resembles native colon with appropriate *in vitro* physiology and architecture. Ann Surg 2003;238:35−41.

[22] Grikscheit TC, Ogilvie JB, Ochoa ER, Alsberg E, Mooney D, Vacanti JP. Tissue-engineered colon exhibits function *in vivo*. Surgery 2002;132:200−4.

[23] Orlando G, Baptista P, Birchall M, De Coppi P, Farney A, Guimaraes-Souza NK, et al. Regenerative medicine as applied to solid organ transplantation: current status and future challenges. Transpl Int 2011;24:223−32.

[24] Vacanti JP. Tissue engineering and the road to whole organs. Br J Surg 2012;99:451−3.

[25] Spence JR, Mayhew CN, Rankin SA, Kuhar MF, Vallance JE, Tolle K, et al. Directed differentiation of human pluripotent stem cells into intestinal tissue *in vitro*. Nature 2011;470:105−9.

[26] May R, Sureban SM, Hoang N, Riehl TE, Lightfoot SA, Ramanujam R, et al. Doublecortin and CaM kinase-like-1 and leucine-rich-repeat-containing G-protein-coupled receptor mark quiescent and cycling intestinal stem cells, respectively. Stem Cells 2009;27:2571−9.

[27] Bjerknes M, Cheng H. The stem-cell zone of the small intestinal epithelium. IV. Effects of resecting 30% of the small intestine. Am J Anat 1981;160:93−103.

[28] Cheng H, Leblond CP. Origin, differentiation and renewal of the four main epithelial cell types in the mouse small intestine. I. Columnar cell. Am J Anat 1974;141:461−79.

[29] Cheng H, Leblond CP. Origin, differentiation and renewal of the four main epithelial cell types in the mouse small intestine. V. Unitarian theory of the origin of the four epithelial cell types. Am J Anat 1974;141:537−61.

[30] Potten CS, Gandara R, Mahida YR, Loeffler M, Wright NA. The stem cells of small intestinal crypts: where are they? Cell Prolif 2009;42:731−50.

[31] Li L, Clevers H. Coexistence of quiescent and active adult stem cells in mammals. Science 2010;327:542−5.

[32] Asai R, Okano H, Yasugi S. Correlation between Musashi-1 and c-hairy-1 expression and cell proliferation activity in the developing intestine and stomach of both chicken and mouse. Dev Growth Differ 2005;47:501−10.

[33] Potten CS, Booth C, Tudor GL, Booth D, Brady G, Hurley P, et al. Identification of a putative intestinal stem cell and early lineage marker; musashi-1. Differentiation 2003;71:28−41.

[34] van de Wetering M, Sancho E, Verweij C, de Lau W, Oving I, Hurlstone A, et al. The beta-catenin/TCF-4 complex imposes a crypt progenitor phenotype on colorectal cancer cells. Cell 2002;111:241−50.

[35] Barker N, van Es JH, Kuipers J, Kujala P, van den Born M, Cozijnsen M, et al. Identification of stem cells in small intestine and colon by marker gene Lgr5. Nature 2007;449:1003−7.

[36] Potten CS, Booth C, Pritchard DM. The intestinal epithelial stem cell: the mucosal governor. Int J Exp Pathol 1997;78:219−43.

[37] Sato T, Vries RG, Snippert HJ, van de Wetering M, Barker N, Stange DE, et al. Single Lgr5 stem cells build crypt-villus structures *in vitro* without a mesenchymal niche. Nature 2009;459:262−5.

[38] Garrison AP, Helmrath MA, Dekaney CM. Intestinal stem cells. J Pediatr Gastroenterol Nutr 2009;49:2−7.

[39] Gagliardi G, Goswami M, Passera R, Bellows CF. DCLK1 immunoreactivity in colorectal neoplasia. Clin Exp Gastroenterol 2012;5:35−42.

[40] May R, Riehl TE, Hunt C, Sureban SM, Anant S, Houchen CW. Identification of a novel putative gastrointestinal stem cell and adenoma stem cell marker, doublecortin and CaM kinase-like-1, following radiation injury and in adenomatous polyposis coli/multiple intestinal neoplasia mice. Stem Cells 2008;26:630−7.

[41] Montgomery RK, Carlone DL, Richmond CA, Farilla L, Kranendonk ME, Henderson DE, et al. Mouse telomerase reverse transcriptase (mTert) expression marks slowly cycling intestinal stem cells. Proc Natl Acad Sci USA 2011;108:179−84.

[42] Sangiorgi E, Capecchi MR. Bmi1 is expressed *in vivo* in intestinal stem cells. Nat Genet 2008;40:915–20.

[43] Tian H, Biehs B, Warming S, Leong KG, Rangell L, Klein OD, et al. A reserve stem cell population in small intestine renders Lgr5-positive cells dispensable. Nature 2011;478:255–9.

[44] Miraglia S, Godfrey W, Yin AH, Atkins K, Warnke R, Holden JT, et al. A novel five-transmembrane hematopoietic stem cell antigen: isolation, characterization, and molecular cloning. Blood 1997;90:5013–21.

[45] Zhu L, Gibson P, Currle DS, Tong Y, Richardson RJ, Bayazitov IT, et al. Prominin 1 marks intestinal stem cells that are susceptible to neoplastic transformation. Nature 2009;457:603–7.

[46] Moore KA, Lemischka IR. Stem cells and their niches. Science 2006;311:1880–5.

[47] He XC, Zhang J, Tong WG, Tawfik O, Ross J, Scoville DH, et al. BMP signaling inhibits intestinal stem cell self-renewal through suppression of Wnt-beta-catenin signaling. Nat Genet 2004;36:1117–21.

[48] Stappenbeck TS, Miyoshi H. The role of stromal stem cells in tissue regeneration and wound repair. Science 2009;324:1666–9.

[49] Fritsch C, Simon-Assmann P, Kedinger M, Evans GS. Cytokines modulate fibroblast phenotype and epithelial–stroma interactions in rat intestine. Gastroenterology 1997;112:826–38.

[50] Ootani A, Li X, Sangiorgi E, Ho QT, Ueno H, Toda S, et al. Sustained *in vitro* intestinal epithelial culture within a Wnt-dependent stem cell niche. Nat Med 2009;15:701–6.

[51] Sato T, van Es JH, Snippert HJ, Stange DE, Vries RG, van den Born M, et al. Paneth cells constitute the niche for Lgr5 stem cells in intestinal crypts. Nature 2011;469:415–8.

[52] Howell JC, Wells JM. Generating intestinal tissue from stem cells: potential for research and therapy. Regen Med 2011;6:743–55.

Chapter 37

Building Blocks for Engineering the Small Intestine

Nicholas R. Smith[a], Eric C. Anderson[b], Paige S. Davies[a], and Melissa H. Wong[a,c]

[a]Department of Cell and Developmental Biology, Oregon Health & Science University, Portland, OR, [b]Division of Hematology and Medical Oncology, Oregon Health & Science University, Portland, OR, [c]Knight Cancer Institute, Oregon Health & Science University, Portland, OR

Bowel transplantation represents the current standard of care for patients suffering from severe intestinal disorders such as short bowel syndrome (SBS). Despite improvements in surgical procedures, there remains an approximate 50% 5-year survival rate for intestinal transplant patients [1]. Currently, the primary barriers to successful organ transplantation are the lack of transplantable tissue, the potential for transplant rejection, and the requirement of lifelong immunosuppression [2]. These challenges highlight the need to fully explore the developing field of intestinal epithelial stem cell biology and their ability for *ex vivo* expansion of tissue in order to reach the ultimate goal of engineering personalized transplantable intestine from patient-derived cells.

37.1 RECENT ADVANCES IN INTESTINAL TISSUE ENGINEERING

Bioengineering of functional tissues represents the next generation of medical technology. Tissue engineering and regenerative medicine harness the power of developmental and stem cell biology to reconstitute new organs that retain all aspects of normal tissue structure, physiology, and function [3]. Recently, great strides have been made in tissue engineering in many organ systems. For the bladder and liver, functional tissues were successfully created by seeding a mixture of progenitor/stem cells on an acellular scaffold, which were subsequently reconstituted *ex vivo* [4,5]. Natural scaffolds used to engineer these organs have the advantage of their immunocompatibility and capacity to provide important spatial cues to the seeded cells [6]. Similar strategies have been used to bioengineer even more complex tissues, such as the heart [2].

Clinically, autologous bioengineered intestine transplant is the paramount treatment for patients suffering from severe cases of SBS [7]. SBS is a highly morbid condition that affects both adults and children, and is characterized by the failure to adequately absorb nutrients due to lack of intestinal surface area, as a result of previous resection of diseased tissue [8]. In adults, SBS is typically caused by removal of a large portion of intestine due to Crohn's disease, ischemia, trauma, or tumors. SBS cases in children are linked to developmental defects including intestinal atresia, volvulus, and necrotizing enterocolitis, a condition most common in premature births [9]. In 1997, the estimated number of patients suffering from SBS was 3–4 per million, yet its incidence appears to be on the rise [10]. In 2004 it was estimated that 24.5 in 100,000 live births had SBS [11]. The current treatment option for SBS, total peritoneal nutrition, is insufficient as many patients fail to thrive and succumb to secondary multi-organ failure [12]. Increasing the gastrointestinal absorptive area through intestinal transplant is the only potentially curative treatment option for severe SBS cases, but how such a complex tissue can be bioengineered for successful transplant remains a scientific and clinical challenge.

The intestine is a complex organ essential for nutrient absorption and as a protective barrier from external insults. To perform these functions, the intestinal tube is composed of multiple cellular layers including the muscularis, a layer of smooth muscle cells that maintains tissue integrity and performs peristalsis, the mesenchymal stromal cell layer, and the functional layer of columnar epithelium. The mucosal layer is exposed to the external environment, therefore to protect against accumulation of DNA mutations, the epithelial layer is continuously replenished every 7–10 days [13]. Further, to increase absorptive surface area, the intestinal architecture features finger-like villus protrusions. The massive epithelial proliferation is fueled within the crypt invaginations by a population of stem cells within a niche at the base of these crypts [14]. This stem cell niche provides a microenvironment that regulates stem cell homeostasis through complex signaling and functional interactions with surrounding mesenchymal and stromal cells [15]. In order to achieve a physiologically functioning bioengineered intestine, not only must all of the normal cell types be present, but the tissue architecture and signaling mechanisms within the niche must also be successfully recapitulated.

What are the cellular and biologic components required to successfully generate viable intestine? Current strategies to bioengineer intestine are built on seminal research by the Vacanti group at Massachusetts General Hospital from the 1990s [16]. This group used porous synthetic scaffolds composed of the biodegradable extracellular matrix (ECM) molecule polyglycolic acid seeded with clusters of dissociated rat gut cells termed "organoid units" (see Table 37.1). The organoid units contained all epithelial and mesenchymal cell types normally present in the gut, presumably containing appropriate stem cell populations to regenerate both epithelial and mesenchymal compartments [17]. The seeded scaffolds were then implanted within vascular regions of host animals to regenerate the cellular components of the intestine

and to become vascularized. Here they formed small encapsulated cysts organized with villus and crypt-like regions [16]. Importantly, when these bioengineered intestinal cysts were anastomosed to the intestine of rats subjected to massive bowel resection—a model of SBS—they incorporated by creating a pouch-like structure and resulted in improved weight gain in the animals [18]. While this exciting result supports the application of engineered intestine for transplantation, it represents only the initial step in this promising process. Importantly, this approach has also shown potential application in engineered large intestine, stomach, and esophagus [19–21], as well as multiple organisms including rat, mouse, pig, and dog [17,22,23].

Despite these promising results, significant hurdles impeding progression into the clinical realm remain. Tissue integrity must be maintained in the engineered intestine in order to prove useful for transplantation. The cysts formed from seeded synthetic scaffolds, while histologically resembling mature intestine, failed to retain the tubular architecture of native intestine that will be required for extensive bowel replacement. Due to this caveat, optimization of scaffolding biomaterials has been a major research focus, including the use of natural tissue scaffolds [24]. To prepare an acellular scaffold, intestinal submucosa is decellularized resulting in the remnant ECM. This natural scaffold material retains crypt invaginations and villus protrusions, providing important spatial information for homing of seeded cells [6]. Further studies have sought to improve the efficiency of organoid unit seeding on scaffolds by testing various synthetic scaffold materials, pore sizes, mechanical properties, and inclusion of tethered growth factors [24–26].

An additional hurdle to the gut bioengineering field is the limited ability to expand intestinal epithelial cells *in vitro*. To date, only disseminated intestine formed into organoid units have been used to successfully seed scaffolds [24]. While use of these organoid units are ideal because of the minimal cellular manipulation, their use is limited by the amount of starting material required to seed sufficient lengths of scaffold. Recent and exciting advances in the ability to expand intestinal stem cells (ISCs) in culture [27–30] as well as the ability to isolate and grow patient-derived ISCs or patient-derived induced pluripotent stem cells (iPSCs) [31] represent a major and exciting advance for the field.

37.2 CURRENT UNDERSTANDING OF ISC BIOLOGY

The recent identification of novel ISC markers [32–40] and the ability to generate intestinal epithelium from iPSCs [31], coupled with a robust *in vitro* culturing system [28,30], has revitalized the ISC field. Continued

TABLE 37.1 Epithelial Cell Preparations for Use in Tissue Engineered Intestine

Cell Preparation	Cellular Composition	Expandable *in vitro?*
Organoid Unit	Epithelial and Mesenchymal cells	No
Enteroid	Epithelial cells only	Yes
Organoid	Epithelial and Mesenchymal cells	Yes

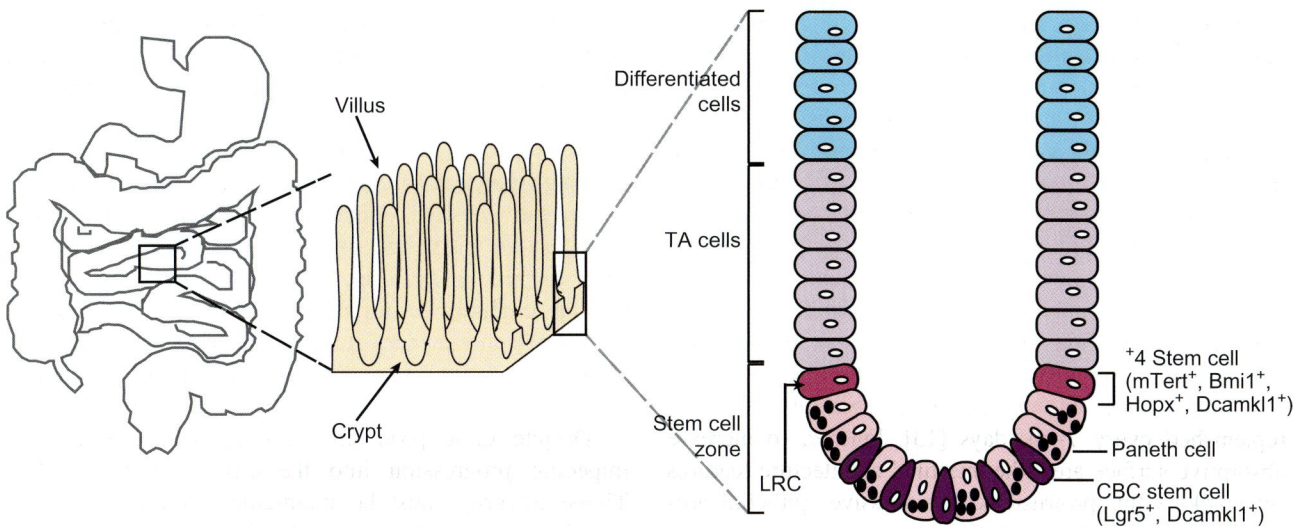

FIGURE 37.1 Architecture of the small intestine and cellular organization of the crypt. The tube of the small intestine is made up of epithelium that is compartmentalized into villus protrusions and crypt invaginations. Stem cells within the stem cell niche at the crypt base give rise to the rapidly proliferating transit-amplifying (TA) cells that differentiate as they migrate up the villus. Intestinal homeostasis is mediated by distinct stem cell populations located in the stem cell zone at the crypt base. The +4 stem cell population represents the slowly cycling, label-retaining cells (LRCs). Markers expressed by cells in this region include: mTert, Bmi1, Dcamkl1, and Hopx. The rapidly cycling crypt-based columnar stem cells (CBCs) are located in the crypt base between differentiated Paneth cells (pink). Markers of CBCs include Lgr5 and Dcamkl1.

development of these approaches now position the bioengineering field to harness these emerging technologies to achieve the ultimate goal of creating autologous patient-derived transplantable intestine. This section will highlight these exciting advances in the ISC field and their potential application to bioengineered intestine.

The adult intestinal tract displays diversity in structure and function along the cephalo-caudal axis. The small intestine is characterized by finger-like villus projections, lined with differentiated epithelial cell lineages, and adjacent crypt invaginations (crypts of Lieberkühn), that house the epithelial stem and progenitor cell populations [14]. In contrast, the adult colon is composed of epithelial-lined crypts, but lacks villi. The differentiated intestinal epithelium in both the small intestine and colon is a single layer of columnar cells composed of four primary cell lineages including absorptive enterocytes, hormone-secreting enteroendocrine cells, mucin-secreting goblet cells, and antimicrobial Paneth cells (the colon lacks Paneth cells) [14]. Under normal homeostatic conditions, the intestinal epithelial cells are regularly and rapidly turned over, with most cell types having a lifespan of approximately 7–10 days in humans—Paneth cells are longer lived, persisting for 2–3 weeks under some conditions [41]. This continual renewal requires a population of multipotent ISCs (or progenitor cells) to self-renew and give rise to the rapidly proliferating population of transit-amplifying (TA) cells (Figure 37.1), precursors to terminally differentiated intestinal epithelial lineages [14].

As a result of the early work of Cheng and Leblond and others [42–44], ISCs have long been known to reside in the base of the crypts. The precise identity and location of these stem cells, however, has proven difficult to elucidate. Early studies by Potten and colleagues postulated that a population of slowly cycling, label-retaining cells (LRCs) existed in the " + 4" position relative to the base of the crypt (Figure 37.1) and formed the ISC population [44], while Leblond and colleagues [43] focused on a more rapidly cycling population of cells located in the crypt base which they referred to as the crypt-based columnar cells (CBCs). Despite much work in the field, little progress was made in definitively isolating these cells for confirmation of their stem cell characteristics. However, recent works by a number of laboratories have now discovered cellular and genetic markers expressed in ISCs located at both the +4 position and the crypt base [32–40]. Identification and verification of stem cell activity within these populations have improved our understanding of the location and function of these important cells. Here we review the three primary intestinal epithelial stem/progenitor populations and a number of intriguing minor stem cell populations in the context of proliferative potential for epithelial expansion.

37.2.1 Lgr5

Using lineage-tracing techniques in mouse model systems, Hans Clevers' group first identified a population of cells interspersed among Paneth cells in the crypt base that express the Wnt signaling target gene, leucine-rich-repeat-containing G-protein coupled receptor 5 (Lgr5) [32]. He went on to demonstrate that these Lgr5-expressing cells are capable of generating all epithelial lineages of the small intestine and colon [32]. Interestingly, these cells correspond to the original population of CBCs first described by Cheng and Leblond [43]. These Lgr5-expressing cells were found to be a rapidly proliferating population with cycling times on the order of 24 h and thought to undergo symmetric division [13]. This finding was somewhat unexpected as somatic stem cells in other tissues are believed to divide asymmetrically and be relatively quiescent, with long cell cycle times. Further, cells located in the " + 4" position were largely negative for Lgr5 expression, suggesting that these two stem cell compartments may differentially contribute to tissue homeostasis [13]. Importantly, the Clevers' group has also shown that single Lgr5-expressing mouse intestinal cells can be cultured *in vitro* to expand and generate complete crypt-villus structures [30]. These protocols have now been extended to human cells [29,31,45] as well as to the colon [27,29].

37.2.2 Bmi1

Mario Capecchi's group employed a similar lineage-tracing strategy to characterize the role of the Polycomb-repressor complex 1(PRC1) gene *Bmi1* in intestinal self-renewal. Bmi1 is a well known regulator of stem cell behavior in neuronal, hematopoietic, and leukemia cells [36,46−48]. Interestingly, using a tamoxifen-inducible reporter system, they demonstrated that Bmi1 was expressed primarily at the +4−5 position in ∼10% of crypts [36]. Consistent with the definition of a stem cell population, Bmi1-expressing cells were also able to give rise to all epithelial cell lineages and their presence persisted throughout the life of the mice. Furthermore, ablation of the Bmi1-expressing cells using induced diphtheria toxin-mediated death resulted in rapid epithelial denuding and subsequent death of the animals in 2−3 days, further confirming the essential nature of the Bmi1-expressing stem cell population.

Building on this work, de Sauvage and colleagues recently demonstrated that ablation of Lgr5-expressing intestinal epithelial cells resulted in the complete loss of Lgr5-expressing cells from the crypt base but did not result in significant histological changes within the crypts [49]. A compensatory increase in Bmi1-expressing ISCs was responsible for epithelial repopulation in the absence of Lgr5-expressing cells. Interestingly, although Lgr5-expressing cells were ablated throughout both the small and large intestine, Bmi1-expressing cells contributed to epithelial repopulation only in the duodenum and jejunum

with little to no expression in the ileum and colon, suggesting that the Bmi1-expressing population is regionally limited. Finally, they also demonstrated that Bmi1-expressing ISCs could give rise to Lgr5-expressing CBCs, supporting the notion of at least two distinct (although potentially overlapping) stem cell populations within the intestine. Under homeostatic conditions—at least in the proximal small intestine—slowly cycling Bmi1-expressing ISCs are upstream of the more rapidly cycling Lgr5-expressing cells. These experiments have important consequences for understanding which populations could respond more rapidly to injury and adequately repopulate the intestinal epithelium.

37.2.3 mTert

A third population of ISCs identified by David Breault's group is characterized by expression of mouse telomerase reverse transcriptase (mTert) [34]. Montgomery et al. found that mTert expression marks a population of slowly cycling LRCs, primarily in the +5 to +8 crypt cell position (consistent with historical reports of the " + 4" position). These mTert-expressing cells were identified as single cells in only 1/150 crypts on average, more infrequent than Bmi1-expressing cells, but unlike the Bmi1-ISC population, they are found in both the proximal and distal small intestine. While these cells are distinct from the Lgr5-expressing cells, they display some overlap with the Bmi1-ISC population and similarly contribute to all mature intestinal epithelial lineages.

Resistance to radiation injury has traditionally been considered a hallmark of stem cells. Paradoxically, cells in the +4 position have long been known to be exquisitely sensitive to low-dose (1 Gy) radiation damage. Lgr5-expressing cells, while resistant to low-dose radiation damage, are sensitive to higher dose (10 Gy) irradiation [32]. However, following both low- and high-dose radiation injury of the intestine, mTert$^+$ cells in the +4 position showed no evidence of apoptosis, suggesting functional heterogeneity of cells in this position. Furthermore, the number of mTert$^+$ crypts increased 12- to 15-fold following radiation damage, suggesting that mTert$^+$ ISCs contribute in the regenerative response and restoration of a radiation-sensitive ISC population following injury. Finally, as with the Bmi1-expressing ISC population, mTert ISCs can give rise to Lgr5-expressing cells, again suggesting that they exist upstream of the rapidly cycling Lgr5-ISC population under homeostatic conditions.

37.2.4 Additional ISC Markers

Courtney Houchen's group has characterized an additional discrete population of slowly cycling/quiescent

ISCs primarily located in the +4 position which are marked by expression of the doublecortin and CaM kinase-like-1 (*Dcamkl1*) gene [33]. While these cells often localize to the +4 area, a subset can be found within the Lgr5-expressing population. Unique to the Dcamkl1-expressing ISC population, however, is a spatial expression pattern weighted toward the distal small intestine. Furthermore, these cells are resistant to moderate- (6 Gy) but not lethal high-dose (12 Gy) radiation damage [33]. Whether Dcamkl1 cells can give rise to Lgr5-expressing cells is currently unknown.

CD24, a glycosylphosphatidyl-inositol-anchored membrane protein, which functions in cell–cell and cell–matrix adhesion and signal transduction has been identified by two different groups as a marker of ISCs [39,50]. This population marks a much broader domain and appears to overlap other ISC populations including those expressing the SRY-box containing gene 9 (*Sox9*) [51]. Additional markers of putative ISC populations include CD133 [37,40] and Musashi-1 (Msi1) [35]. The precise spatial, temporal, and situational relationship between the ISC populations expressing these various genes remains to be determined.

A recent and very intriguing report from Jonathan Epstein's group has identified a novel marker of ISCs located in the +4 position, the atypical homeobox gene *Hopx*[38]. Hopx is expressed in 85–90% of intestinal crypts under both homoestatic and injury conditions. However, unlike other +4 position ISC populations, it is expressed throughout the entire length of the small intestine. As expected for a stem cell marker, Hopx$^+$ cells are able to produce all differentiated intestinal epithelial cell types and form organoids under *in vitro* culture conditions. Interestingly, although under homeostatic conditions Lgr5$^+$ and Hopx$^+$ cell populations are distinct, they are able to interconvert— that is, Hopx-expressing cells can give rise to Lgr5-expressing cells and vice versa. There is clearly much work to be done to further elucidate the role of Hopx$^+$ ISCs in maintaining intestinal homeostasis and response to injury, but the identification of a bifunctional stem cell population in the intestine provides experimental proof of a mathematical model of ISC function [13] and helps to reconcile divergent behaviors of ISC populations.

It appears clear from the available data that multiple subpopulations of stem cells in the small intestine and colon exist and that more remain to be discovered. Under homeostatic conditions, the Lgr5-expressing CBC ISC population appears to be responsible for the majority of epithelial tissue regeneration. Following tissue damage however, a second population of normally slow cycling ISCs characterized by the expression of Bmi1, mTert, Dcamkl1, Hopx, and possibly other genes that have yet to be identified, is able to rapidly proliferate in response to injury and repopulate both the homeostatic, Lgr5-expressing ISC population as well as the entire intestinal epithelium. Critical issues that remain to be addressed include the spatial, temporal, and hierarchical relationship between the various +4/slow cycling ISC populations and the rapidly cycling Lgr5-expressing CBC population as well as markers of the +4/slow cycling ISC population in the colon. Ongoing experiments in our laboratory and others are expected to answer these questions and provide additional insight into basic ISC biology as well as improving methods for the *in vitro* generation of viable intestinal epithelium.

37.3 ADVANCES IN *IN VITRO* EXPANSION OF INTESTINAL EPITHELIUM

For nearly 100 years, the prospect of culturing intestinal epithelium to successfully mimic the growth and lineage differentiation that occurs *in vivo* has been an enigmatic challenge due to the inherent nature of this rapidly renewing cell population and the elusiveness of the ISC. Recently, groundbreaking work from the laboratories of both Calvin Kuo and Hans Clevers reported successful long-term culture of intestinal epithelium [28,30]. While both groups successfully generated *in vitro* systems to grow and study intestinal epithelium, they took different approaches. Kuo's group relied upon a mesenchymal niche to drive stem cell maintenance and epithelial differentiation, deriving intestinal "organoids" composed of both epithelial and mesenchymal compartments. Clevers' group used single epithelial cells conditioned with signaling factors devoid of mesenchymal components to derive "enteroids." For the purpose of this review, enteroids will refer to *in vitro* derived intestinal epithelium devoid of mesenchymal components, while organoids will refer to 3D intestinal epithelial cultures that have mesenchymal or stromal components (see Table 37.1). While both of these reports have advanced in the field of intestinal biology immensely, each system has advantages and disadvantages for addressing the ultimate goal of generating intestine for therapeutic repair in patients.

The work from Sato and colleagues suggests that programs that dictate intestinal epithelial proliferation and differentiation are cell autonomous [30]. They eloquently demonstrate that only four factors are required to sustain stem cell maintenance, stem cell division, and progenitor differentiation in culture [30]. Their study describes a unique long-term *in vitro* culture system that can be derived and maintained from mouse isolated small intestinal crypts as well as single intestinal epithelial cells that give rise to intestinal epithelial enteroids [30]. Further, their system suggests that mesenchymal components are

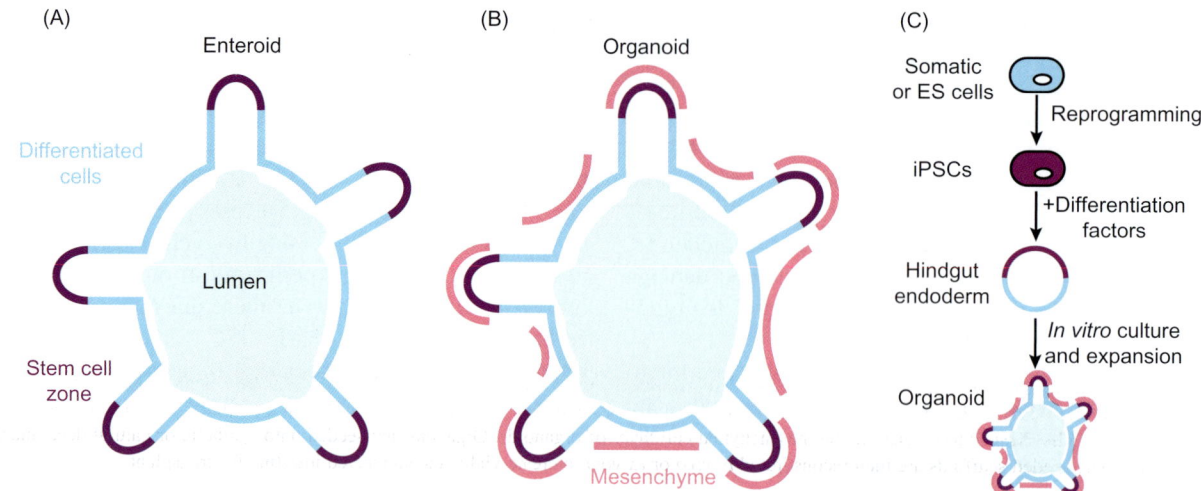

FIGURE 37.2 Schematic representation of *in vitro* cultured intestinal epithelia. (A) Enteroids generated by the method of Sato et al. include differentiated villus-like regions (blue) and proliferative crypt-like regions containing a stem cell zone (purple). Apoptotic cells are shed into the inner lumen. (B) Organoids created by [28] contain similar villus and crypt-like domains as the enteroids, but also have an outer mesenchymal layer (pink). (C) The method of [31] generates organoids from reprogrammed human embryonic stem (ES) cells or somatic cells through targeted differentiation into intestinal epithelium, complete with mesenchyme.

not an absolute requirement for the *in vitro* propagation of intestinal epithelium. However, it has not been determined if epithelia alone could effectively seed and survive on an intestinal scaffold. Most certainly, intestine bioengineered and grown *ex vivo* would require mesenchymal cell components, however, scaffolds seeded and grown *in vivo* may be able to utilize native stroma to reseed the mesenchyme to support epithelial growth.

The culture conditions described by Sato et al. were designed based on the vast amount of knowledge accumulated over the years regarding important factors necessary to sustain epithelial cells. Knowing the critical role for Wnt signaling in maintaining the proliferative crypt [52–54], the culturing conditions included supplementation with the Wnt agonist, R-spondin-1. In line with the need for effective proliferation, epidermal growth factor was added to the culture milieu [55]. For long-term culture growth and successful passaging, an expansion of crypts was achieved by the addition of Noggin, an inhibitor of bone morphogenetic protein [56]. Finally, because laminin is expressed near the crypt base [57], the authors chose to use an element that has had success in supporting the growth of mammary epithelium, Matrigel, enriched with laminin. Further, since the initial report, additional factors including Wnt3A produced by Paneth cells in the base of the crypt [58] and a Rho inhibitor to inhibit cell death due to anoikis [59] have fine-tuned the culture system.

In the cultures initiated from whole crypts, the isolated mouse intestinal crypts underwent fission events to expand into enteroids. Sustained culturing of these isolated crypts resulted in enteroids comprised of numerous crypt domains, encircling a luminal space and interspersed with differentiated "villus-like" epithelium (Figure 37.2A). Consistent with an *in vivo* environment, apoptotic cells were shed into the luminal space and Paneth cells were observed to be associated with progenitor cells where new crypt buds formed. Strikingly, the epithelium was organized into a single layer as marked by E-cadherin. Each week, the enteroids could be dissociated and a portion replated, allowing their maintenance in culture for over 8 months.

In contrast to the enteroid culture system, Ootani and colleagues approached their culturing conditions in a different manner. They did not specifically isolate the crypts, but rather, they embedded minced tissue from neonatal mouse small or large intestine directly into a 3D collagen gel [28]. Importantly, their preparation included stromal cells, known to be important for establishing a successful crypt niche, and the culture was exposed to an air–liquid interface for robust growth. The structures that developed were sphere-like organoids that exhibit proliferation and multilineage differentiation (Figure 37.2B). Similar to the culturing conditions of Sato et al., long-term growth of organoid cultures is also improved by the inclusion of a Wnt agonist. The *in vitro* organoids can grow for >350 days and are established not only with epithelial cells but also with supporting mesenchymal cells, such as fibroblasts, that aid in the development of a functional gut that can periodically contract, indicating the presence of working muscle cells and neurons.

Enteroid cultures produced by Clevers and colleagues lack stromal cells, which were thought to be instrumental in providing a functional ISC niche. Remarkably, their culture conditions promote long-term growth of intestinal epithelium beginning from just a single sorted ISC, although the

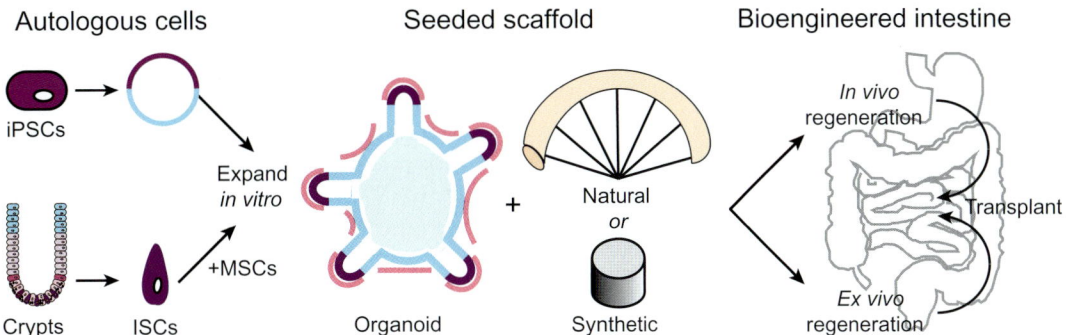

FIGURE 37.3 Potential strategies utilizing stem cells for bioengineering intestine. (Left to right) Autologous patient-derived iPSCs or crypt stem cells are cultured and expanded *in vitro* to generate sufficient intestinal epithelium in the form of organoids. Single ISC may require the addition of mesenchymal stem cells (MSCs) to regenerate the mesenchymal cell layer of organoids. Organoids are seeded onto synthetic or natural decellularized tissue scaffolds. The seeded scaffolds are then reconstituted *in vivo* or *ex vivo* to create viable bioengineered intestine for transplant.

efficiency is quite low (6%). It is interesting that, with the appropriate exogenously supplied signals, ISCs can be cultured and maintained without an explicitly supplied niche. The inclusion of stroma, however, is an advantage of the culture system developed by Kuo and colleagues. This *in vitro* niche environment allows for the study of the ISC in the context of the niche that exists *in vivo*. Further, it may also allow for the study of interactions of stem cells with other cell types, such as smooth muscle, endothelium, or neurons, all relevant *in vivo* interactions, and may prove to be a more viable expansion model for seeding scaffolds for intestinal bioengineering.

Finally, the observation that a structural niche is not required for the enteroid culture to thrive *in vitro* shatters the previous notion that developmental morphogen gradients dictate differential proliferation versus differentiation function among cell populations. It is typically believed that the stem cell must be maintained in a protective structural niche at the crypt base, supplied with subepithelial myofibroblasts and other mesenchymal elements, for structural and signaling support. Sato and colleagues clearly defined culture conditions that allow for a self-renewing intestinal epithelium in the absence of a mesenchymal niche by adding in the necessary mesenchymal components (possibly mesenchymal stem cells capable of regenerating multiple stromal elements or bone marrow–derived cells) to allow for adequate growth. Further, the enteroid culture system suggests that the ability to remain a stem cell versus a progenitor or a differentiated cell is a cell autonomous function and not dictated by levels of cell signaling factors. Clearly the ability to grow small organ-like cell clusters may be sufficient for seeding larger scaffolds. However, whether or not the intestinal ECM provides additional important cues for seeding of discrete cells types along the crypt-villus axis during reconstitution of a bioengineered intestine remains to be determined.

While a number of different intestinal culture systems bring the promise of expanding a small number of epithelial cells, the issue of transplant rejection and immunosuppression after transplantation remains if these cells are harvested from nonautologous donors. Recent development in generation of iPSCs and more excitingly, in coaxing these iPSCs toward an intestinal identity bring the promise of autologous bioengineered intestinal transplantation. Using human cells, James Wells' laboratory recently demonstrated a robust and efficient process to direct differentiation of iPSCs into intestinal tissue [31] (Figure 37.2C). These iPSC-derived organoids were composed of polarized columnar epithelium, surrounded by mesenchyme, which formed villus-like structures and crypt-like regions of proliferative cells and expressed ISC markers. Consistently, all intestinal lineages were represented in this iPSC system. The observation that intestinal mesenchyme differentiation was coordinated with epithelial differentiation indicates that crosstalk between these two regions is important for developing human intestinal organoids. This would agree with the notion that a mesenchymal niche is important for the ISC to reside and thrive in an *in vivo* setting.

The establishment of these culture systems has opened a vast area of research regarding mechanistic questions, exciting real-time imaging, and fine-tuned manipulation that has not been able to occur in the field of intestinal biology until now. The implications these systems have for effectively and successfully expanding intestinal tissue for regenerative therapeutic purposes is eminent. Although the human iPSC system has the advantage over the systems from Ootani and Sato because of the potential to use autologous patient-derived cells, it also has limitations. Components of the *in vivo* intestine, such as the enteric nervous, vascular, lymphatic, and immune systems are not represented. Thus, there is a need for further improvement to the system, or arguably to define the requirement of other markers and factors essential for establishing and maintaining a functional intestinal epithelium *in vitro* to provide expansive tissue growth. It does, however, provide an exceptional starting point to understand how to regenerate human tissue for therapeutic purposes.

37.4 CONCLUDING REMARKS

The recent advances in the ISC field have identified and characterized discrete stem cell populations that are able to be isolated and expanded *in vitro*. These technologies, coupled with the ability of iPSCs to form *de novo* (and potentially patient-derived) intestinal epithelium provide a foundation for which these cell types could be isolated from patients for the purpose of engineering autologous intestine for transplant (Figure 37.3). Either patient- or donor-derived iPSCs or ISCs would be isolated and cultured *in vitro* to create organoids. The organoids would be seeded onto either natural or synthetic tissue scaffolds and reconstituted *in vivo* or *ex vivo* for subsequent transplant. One important question remaining is what specific cell types are required to successfully seed a scaffold? Given the importance of mesenchymal and stromal cells in stem niche maintenance, these populations will undoubtedly be critical components. A major concern for using *in vitro* cultured cells is the potential for transformation, and thus a demonstration of the safety of such engineered tissues is essential. Further, future studies will likely center on optimization of the scaffolding materials and methods of *in vivo* or *ex vivo* manipulation of the seeded scaffolds. Demonstration of normal physiology and functionality of bioengineered intestine using newly developed assays will surely be a main focus in the near future and help move this technology toward the clinical realm. The effort of multiple groups employing a number of techniques to attain this goal brings excitement to the field and increases the likelihood of success.

REFERENCES

[1] Ueno T, Wada M, Hoshino K, Yonekawa Y, Fukuzawa M. Current status of intestinal transplantation in Japan. Transplant Proc 2011;43:2405—7.

[2] Orlando G, Baptista P, Birchall M, De Coppi P, Farney A, Guimaraes-Souza NK, et al. Regenerative medicine as applied to solid organ transplantation: current status and future challenges. Transpl Int 2011;24:223—32.

[3] Howell JC, Wells JM. Generating intestinal tissue from stem cells: potential for research and therapy. Regen Med 2011;6:743—55.

[4] Atala A, Bauer SB, Soker S, Yoo JJ, Retik AB. Tissue-engineered autologous bladders for patients needing cystoplasty. Lancet 2006;367:1241—6.

[5] Baptista PM, Siddiqui MM, Lozier G, Rodriguez SR, Atala A, Soker S. The use of whole organ decellularization for the generation of a vascularized liver organoid. Hepatology 2011;53:604—17.

[6] Baptista PM, Orlando G, Mirmalek-Sani SH, Siddiqui M, Atala A, Soker S. Whole organ decellularization—a tool for bioscaffold

fabrication and organ bioengineering. Conf Proc IEEE Eng Med Biol Soc 2009;2009:6526—9.

[7] Levin DE, Dreyfuss JM, Grikscheit TC. Tissue-engineered small intestine. Expert Rev Med Devices 2011;8:673—5.

[8] Thompson JS, Weseman R, Rochling FA, Mercer DF. Current management of the short bowel syndrome. Surg Clin North Am 2011;91:493—510.

[9] Garg M, Jones RM, Vaughan RB, Testro AG. Intestinal transplantation: current status and future directions. J Gastroenterol Hepatol 2011;26:1221—8.

[10] Bakker H, Bozzetti F, Staun M, Leon-Sanz M, Hebuterne X, Pertkiewicz M, et al. Home parenteral nutrition in adults: a European Multicentre Survey in 1997. ESPEN-Home artificial nutrition working group. Clin Nutr 1999;18:135—40.

[11] Wales PW, de Silva N, Kim J, Lecce L, To T, Moore A. Neonatal short bowel syndrome: population-based estimates of incidence and mortality rates. J Pediatr Surg 2004;39:690—5.

[12] Peyret B, Collardeau S, Touzet S, Loras-Duclaux I, Yantren H, Michalski MC, et al. Prevalence of liver complications in children receiving long-term parenteral nutrition. Eur J Clin Nutr 2011;65:743—9.

[13] Li L, Clevers H. Coexistence of quiescent and active adult stem cells in mammals. Science 2010;327:542—5.

[14] Potten CS, Booth C, Pritchard DM. The intestinal epithelial stem cell: the mucosal governor. Int J Exp Pathol 1997;78:219—43.

[15] Rizvi AZ, Wong MH. Epithelial stem cells and their niche: there's no place like home. Stem Cell 2005;23:150—65.

[16] Choi RS, Vacanti JP. Preliminary studies of tissue-engineered intestine using isolated epithelial organoid units on tubular synthetic biodegradable scaffolds. Transplant Proc 1997;29:848—51.

[17] Sala FG, Matthews JA, Speer AL, Torashima Y, Barthel ER, Grikscheit TC. A multicellular approach forms a significant amount of tissue-engineered small intestine in the mouse. Tissue Eng Part A 2011;17:1841—50.

[18] Grikscheit TC, Siddique A, Ochoa ER, Srinivasan A, Alsberg E, Hodin RA, et al. Tissue-engineered small intestine improves recovery after massive small bowel resection. Ann Surg 2004;240:748—54.

[19] Grikscheit T, Ochoa ER, Srinivasan A, Gaissert H, Vacanti JP. Tissue-engineered esophagus: experimental substitution by onlay patch or interposition. J Thorac Cardiovasc Surg 2003;126:537—44.

[20] Grikscheit TC, Ochoa ER, Ramsanahie A, Alsberg E, Mooney D, Whang EE, et al. Tissue-engineered large intestine resembles native colon with appropriate *in vitro* physiology and architecture. Ann Surg 2003;238:35—41.

[21] Speer AL, Sala FG, Matthews JA, Grikscheit TC. Murine tissue-engineered stomach demonstrates epithelial differentiation. J Surg Res 2011;171:6—14.

[22] Agopian VG, Chen DC, Avansino JR, Stelzner M. Intestinal stem cell organoid transplantation generates neomucosa in dogs. J Gastrointest Surg 2009;13:971—82.

[23] Sala FG, Kunisaki SM, Ochoa ER, Vacanti J, Grikscheit TC. Tissue-engineered small intestine and stomach form from autologous tissue in a preclinical large animal model. J Surg Res 2009;156:205—12.

[24] Gupta A, Dixit A, Sales KM, Winslet MC, Seifalian AM. Tissue engineering of small intestine—current status. Biomacromolecules 2006;7:2701—9.

[25] Chen DC, Avansino JR, Agopian VG, Hoagland VD, Woolman JD, Pan S, et al. Comparison of polyester scaffolds for bioengineered intestinal mucosa. Cell Tissues Org 2006;184:154—65.

[26] Wulkersdorfer B, Kao KK, Agopian VG, Dunn JC, Wu BM, Stelzner M. Growth factors adsorbed on polyglycolic acid mesh augment growth of bioengineered intestinal neomucosa. J Surg Res 2011;169:169—78.

[27] Jung P, Sato T, Merlos-Suarez A, Barriga FM, Iglesias M, Rossell D, et al. Isolation and *in vitro* expansion of human colonic stem cells. Nat Med 2011;17:1225—7.

[28] Ootani A, Li X, Sangiorgi E, Ho QT, Ueno H, Toda S, et al. Sustained *in vitro* intestinal epithelial culture within a Wnt-dependent stem cell niche. Nat Med 2009;15:701—6.

[29] Sato T, Stange DE, Ferrante M, Vries RG, Van Es JH, Van den Brink S, et al. Long-term expansion of epithelial organoids from human colon, adenoma, adenocarcinoma, and Barrett's epithelium. Gastroenterology 2011;141:1762—72.

[30] Sato T, Vries RG, Snippert HJ, van de Wetering M, Barker N, Stange DE, et al. Single Lgr5 stem cells build crypt-villus structures *in vitro* without a mesenchymal niche. Nature 2009;459:262—5.

[31] Spence JR, Mayhew CN, Rankin SA, Kuhar MF, Vallance JE, Tolle K, et al. Directed differentiation of human pluripotent stem cells into intestinal tissue *in vitro*. Nature 2011;470:105—9.

[32] Barker N, van Es JH, Kuipers J, Kujala P, van den Born M, Cozijnsen M, et al. Identification of stem cells in small intestine and colon by marker gene Lgr5. Nature 2007;449:1003—7.

[33] May R, Riehl TE, Hunt C, Sureban SM, Anant S, Houchen CW. Identification of a novel putative gastrointestinal stem cell and adenoma stem cell marker, doublecortin and CaM kinase-like-1, following radiation injury and in adenomatous polyposis coli/multiple intestinal neoplasia mice. Stem Cell 2008;26:630—7.

[34] Montgomery RK, Carlone DL, Richmond CA, Farilla L, Kranendonk ME, Henderson DE, et al. Mouse telomerase reverse transcriptase (mTert) expression marks slowly cycling intestinal stem cells. Proc Natl Acad Sci USA 2011;108:179—84.

[35] Potten CS, Booth C, Tudor GL, Booth D, Brady G, Hurley P, et al. Identification of a putative intestinal stem cell and early lineage marker; Musashi-1. Differentiation 2003;71:28—41.

[36] Sangiorgi E, Capecchi MR. Bmi1 is expressed *in vivo* in intestinal stem cells. Nat Genet 2008;40:915—20.

[37] Snippert HJ, van Es JH, van den Born M, Begthel H, Stange DE, Barker N, et al. Prominin-1/CD133 marks stem cells and early progenitors in mouse small intestine. Gastroenterology 2009;136:2187—94 e1.

[38] Takeda N, Jain R, Leboeuf MR, Wang Q, Lu MM, Epstein JA. Interconversion between intestinal stem cell populations in distinct niches. Science 2011;334:1420—4.

[39] von Furstenberg RJ, Gulati AS, Baxi A, Doherty JM, Stappenbeck TS, Gracz AD, et al. Sorting mouse jejunal epithelial cells with CD24 yields a population with characteristics of intestinal stem cells. Am J Physiol Gastrointest Liver Physiol 2011;300: G409—17.

[40] Zhu L, Gibson P, Currle DS, Tong Y, Richardson RJ, Bayazitov IT, et al. Prominin 1 marks intestinal stem cells that are susceptible to neoplastic transformation. Nature 2009;457:603—7.

[41] Cheng H. Origin, differentiation and renewal of the four main epithelial cell types in the mouse small intestine. IV. Paneth cells. Am J Anat 1974;141:521—35.

[42] Cheng H, Leblond CP. Origin, differentiation and renewal of the four main epithelial cell types in the mouse small intestine. I. Columnar cell. Am J Anat 1974;141:461—79.

[43] Cheng H, Leblond CP. Origin, differentiation and renewal of the four main epithelial cell types in the mouse small intestine. V. Unitarian theory of the origin of the four epithelial cell types. Am J Anat 1974;141:537—61.

[44] Potten CS, Kovacs L, Hamilton E. Continuous labelling studies on mouse skin and intestine. Cell Tissue Kinet 1974;7:271—83.

[45] McCracken KW, Howell JC, Wells JM, Spence JR. Generating human intestinal tissue from pluripotent stem cells *in vitro*. Nat Protoc 2011;6:1920—8.

[46] Lessard J, Sauvageau G. Bmi-1 determines the proliferative capacity of normal and leukaemic stem cells. Nature 2003;423:255—60.

[47] Leung C, Lingbeek M, Shakhova O, Liu J, Tanger E, Saremaslani P, et al. Bmi-1 is essential for cerebellar development and is overexpressed in human medulloblastomas. Nature 2004;428:337—41.

[48] Molofsky AV, Pardal R, Iwashita T, Park IK, Clarke MF, Morrison SJ. Bmi-1 dependence distinguishes neural stem cell self-renewal from progenitor proliferation. Nature 2003;425:962—7.

[49] Tian H, Biehs B, Warming S, Leong KG, Rangell L, Klein OD, et al. A reserve stem cell population in small intestine renders Lgr5-positive cells dispensable. Nature 2011;478:255—9.

[50] Gracz AD, Ramalingam S, Magness ST. Sox9 expression marks a subset of CD24-expressing small intestine epithelial stem cells that form organoids *in vitro*. Am J Physiol Gastrointest Liver Physiol 2010;298.

[51] Formeister EJ, Sionas AL, Lorance DK, Barkley CL, Lee GH, Magness ST. Distinct SOX9 levels differentially mark stem/progenitor populations and enteroendocrine cells of the small intestine epithelium. Am J Physiol Gastrointest Liver Physiol 2009;296:G1108—18.

[52] Korinek V, Barker N, Willert K, Molenaar M, Roose J, Wagenaar G, et al. Two members of the Tcf family implicated in Wnt/beta-catenin signaling during embryogenesis in the mouse. Mol Cell Biol 1998;18:1248—56.

[53] Kuhnert F, Davis CR, Wang HT, Chu P, Lee M, Yuan J, et al. Essential requirement for Wnt signaling in proliferation of adult small intestine and colon revealed by adenoviral expression of Dickkopf-1. Proc Natl Acad Sci USA 2004;101:266—71.

[54] Pinto D, Gregorieff A, Begthel H, Clevers H. Canonical Wnt signals are essential for homeostasis of the intestinal epithelium. Genes Dev 2003;17:1709—13.

[55] Dignass AU, Sturm A. Peptide growth factors in the intestine. Eur J Gastroenterol Hepatol 2001;13:763—70.

[56] Haramis AP, Begthel H, van den Born M, van Es J, Jonkheer S, Offerhaus GJ, et al. *De novo* crypt formation and juvenile polyposis on BMP inhibition in mouse intestine. Science 2004;303:1684—6.

[57] Sasaki T, Giltay R, Talts U, Timpl R, Talts JF. Expression and distribution of laminin alpha1 and alpha2 chains in embryonic and adult mouse tissues: an immunochemical approach. Exp Cell Res 2002;275:185—99.

[58] Sato T, van Es JH, Snippert HJ, Stange DE, Vries RG, van den Born M, et al. Paneth cells constitute the niche for Lgr5 stem cells in intestinal crypts. Nature 2011;469:415—8.

[59] Hofmann C, Obermeier F, Artinger M, Hausmann M, Falk W, Schoelmerich J, et al. Cell—cell contacts prevent anoikis in primary human colonic epithelial cells. Gastroenterology 2007;132:587—600.

Part VI

Endocrine Pancreas and Islets of Langerhans

Chapter 38

Current Status of Pancreas Transplantation

Gaetano Ciancio[a], Junichiro Sageshima[a], Linda Chen[a], Alberto Pugliese[b], and George W. Burke[a,b]

[a]The DeWitt Daughtry Family Department of Surgery, The Lillian Jean Kaplan Renal Transplant Center, Division of Transplantation and The Miami Transplant Institute, University of Miami, Miami, FL, [b]Diabetes Research Institute, University of Miami, Miller School of Medicine, Miami, FL

Chapter Outline

The goal of pancreas transplantation (PT) is to restore euglycemia providing long-term insulin independence, increase patient survival, stabilize or improve diabetic retinopathy and neuropathy, and in combination with kidney transplantation, eliminate the need for dialysis. Simultaneous kidney—pancreas (SPK) transplant results have improved consistently over the past decade [1,2] as a consequence of advances in immunosuppressive strategies [3,4], improvements in surgical technique, and better understanding of intra- and postoperative care [5]. At the University of Miami (UM), PT has been performed since 1990. Most (>95%) have been SPK transplants. SPK transplantation (SPKT) is considered as the best treatment option for patients with type 1 diabetes (T1D) and end-stage renal disease (ESRD) [1,2,6].

38.1 IMPORTANCE OF GLYCEMIC CONTROL

Nearly every recipient of an SPKT (c-peptide defined T1D/ESRD) at our center has become euglycemic, off insulin therapy with good renal function (creatinine/ eGFR). Physiologically normal renal function is critical to cardiovascular health, and renal dysfunction [7] and KT graft loss [8] are significant risk factors for CV death. Interestingly, while optimal glycemic control is intuitively a key component to the health and welfare of patients with T1D, the benefit from maintaining blood glucose/ HbA1c as close to normal as possible has been challenging to demonstrate. The Diabetes Control and Complication Trial (DCCT) was the landmark, multicenter clinical study [1983–1993] in which patients with T1D were randomized to receive intensive glycemic control therapy or conventional therapy [9]. Patients after an average of 6.5 years in the intensive therapy group were found to have a lower risk for progression of markers of early microvascular complications of diabetes, specifically retinopathy, nephropathy, and neuropathy [9]. This was the first study to clearly show benefit to better glycemic control in patients with T1D, although it took 3–4 years for the benefit to become manifest and there was a higher risk of significant hypoglycemic events.

Subsequent studies, in which those patients in the conventional therapy arm were offered the opportunity to

Regenerative Medicine Applications in Organ Transplantation.

receive intensive glycemic control therapy, have now provided up to 22 years of follow-up. As part of the EDIC (Epidemiology of Diabetes Interventions and Complications) Study, intense therapy resulted in decreased progression of carotid intima-media thickness 6 years after the end of this trial [10]. Since carotid intima-media thickness is a marker of atherosclerosis that correlates with clinical cardiovascular events, e.g., coronary artery disease/stroke, it was anticipated that these clinical developments would be reduced over time in the intensively treated group. Consistent with the idea that improvement in outcome in patients with T1D is dependent on lasting duration of improved glycemic control, another landmark follow-up paper demonstrated long-term beneficial effect on the risk of CV disease [11]. After a mean follow-up of 17 years (EDIC/DCCT) a reduced risk of any cardiovascular disease event, as well as non-fatal myocardial infarction, stroke, or death from CV disease was identified in the intensively treated group [11]. The improvement in glycemic control in the intensively treated group was associated with a higher number of severe hypoglycemic events. However, when evaluated (mean 18 years later) there did not appear to be a decline in cognitive function related to the hypoglycemic events [12]. In fact, the greater concern is the negative effect of chronic hyperglycemia on cognitive function [13]. Finally, there is sustained improvement in retinopathy and nephropathy at 4 years [14] and nephropathy (GFR) at 22 years [15] associated with better glycemic control achieved in the intensively treated group.

Taken together, this remarkable series of reports confirms the importance of long-term glycemic control positively impacting critical parameters in the health of patients with T1D. This effect may take years to decades to become evident.

38.2 HYPERCOAGULABLE STATE IN T1D/ESRD

The pancreas transplant portion of SPK has historically been more prone to thrombosis than other solid organ transplants, accounting for up to 15−20% of pancreas graft loss. When viewed in the context of Virchow's triad (hypercoagulability, endothelial cell damage, and venous stasis), thrombosis may in fact be predicted, since all three criteria are met: (i) hypercoagulability is clearly defined by the thromboelastogram (TEG, *vide infra*) [16]; (ii) endothelial cell damage, which results from an ischemia/reperfusion injury and mediated by cytokines, O_2 radicals, nitric oxide, etc.; furthermore, immunosuppression, particularly calcineurin inhibitors (tacrolimus, cyclosporine A), can induce endothelial cell damage by eliciting procoagulants [17]; (iii) venous stasis occurs

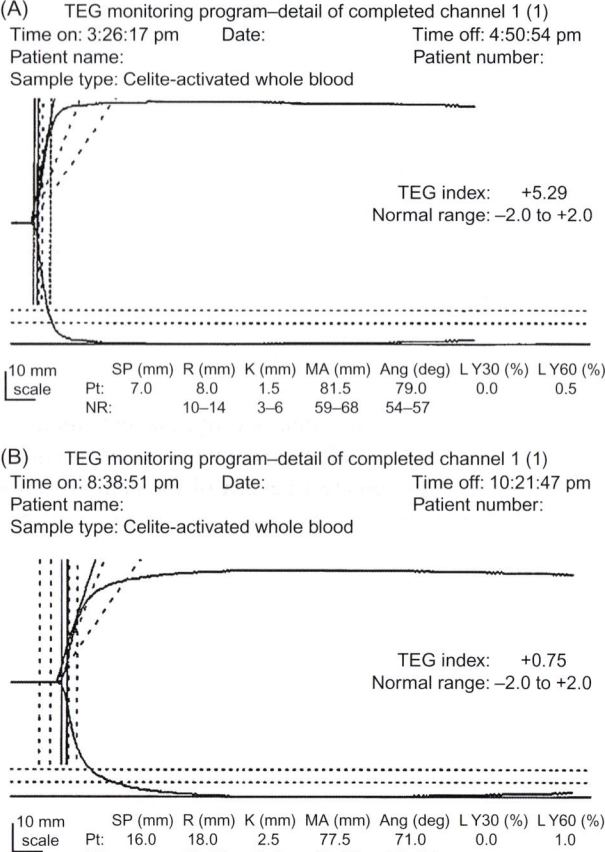

FIGURE 38.1 (A) TEG of patient intraoperatively demonstrating severe hypercoagulable pattern (see text for details). (B) TEG of same patient after heparinization (therapeutic) showing normal or near normalization of coagulation parameters (see text for details).

when the spleen is removed from the tail of the pancreas, as the major source of blood return through the splenic vein is lost. The pancreas portion of the splenic vein remains with its high capacitance, but there is only limited flow from the pancreas. The superior mesenteric vein similarly no longer receives venous return from the small bowel, but is limited to small pancreatic venous radicals. Thus the pancreas transplant fulfills Virchow's triad for propensity to venous thrombosis.

Indeed, at UM we perform a TEG at the time of SPK transplant surgery which consistently demonstrates a hypercoagulable pattern [16] (Figure 38.1A). Generally, rheologic assessment including the combination of (i) shortened prothrombin time, INR, and partial thromboplastin time and (ii) elevated platelet count, fibrinogen, and hematocrit along with hyperlipidemia, which are all features associated with hypercoagulability, are conceptually integrated in the TEG. Since each of these factors may vary over time, the performance of the intraoperative TEG allows a "real time" evaluation of the degree of hypercoagulability at the time of transplantation.

Although the uremic effect on platelets could offset the hypercoagulability associated with T1D, our experience suggests it does not. This has led us to use heparin intraoperatively, when the degree of hypercoagulability is matched with the degree of operative field hemostasis. As seen in Figure 38.1B, heparin can be titrated to decrease the degree of hypercoagulability offering protection from pancreas transplant thrombosis. Dextran sulfate is used for those patients with less pronounced hypercoagulability [18]. The PT loss rate of 1% from thrombosis shows that this has been an effective strategy, while reoperating from bleeding is also low (2%) [16,19].

When seen in the context of other risk factors for atherosclerosis, e.g., hypertension, obesity, diabetes mellitus, insulin resistance, dyslipidemia (components of the metabolic syndrome), the greatest benefit of the demonstration of the hypercoagulable state may lie in its long-term therapy. While SPK transplant prolongs patient survival [6,19,20], recognition and treatment of the hypercoagulable state along with new approaches to inflammation and atherosclerosis may allow further improvement.

38.3 PANCREAS TRANSPLANT, TECHNICAL ASPECTS

38.3.1 Exocrine Drainage

PT is performed primarily for its endocrine function, based on the vascularized islet cell, with the exocrine function, that is, digestive enzymes including amylase and lipase, etc., nearly irrelevant in most cases. Drainage of the pancreas/duodenum graft has evolved over time. It is performed either into the bladder or the small bowel. Bladder drainage is used in our center foremost for safety reasons, since it avoids intra-abdominal enteric spill. Bladder drainage also allows monitoring of urine amylase which provides insight into pancreatic graft function typically useful in diagnosing acute rejection (AR). After the administration of antirejection therapy, the need for repeat pancreatic biopsy can also be avoided if urine amylase returns to baseline levels. However, bladder-drained pancreas transplants are associated with multiple urological and metabolic complications, requiring enteric conversion in 14−50% of most reported series, although this is 8−12% in our experience [2,19]. When exocrine pancreas secretions are enterically drained, metabolic acidosis and dehydration are avoided. However, reliance on rising levels of serum amylase and lipase, without the ability to detect a fall in urine levels, may result in rejection episodes that may progress undiagnosed before treatment is started, and this delay increases the likelihood of allograft loss. The most serious complication of the enteric-drained pancreas transplant is a leak from the anastomotic site with resultant intra-abdominal sepsis.

38.3.2 Venous Drainage

Venous effluent from the portal vein of the pancreas transplant is usually through the systemic circulation (external iliac vein), which results in loss of first pass insulin through the liver, and typically hyper-c-peptidemia. Alternatively, the pancreas transplant portal vein can be anastomosed to a mesenteric venous radical leading to the liver, a more physiologic placement, avoiding hyper-c-peptidemia. The combination of exocrine enteric drainage and venous portal drainage was attempted to achieve a more physiologic result, but has not demonstrated clinical benefit and has not gained wide acceptance [21].

38.3.3 Arterial Approaches

Patients with T1D/ESRD generally are hypertensive and often have significant atherosclerosis involving coronary arteries, carotid arteries, as well as the intra-abdominal used for arterial inflow for the kidney and pancreas transplants. This atherosclerotic process can be particularly severe in patients with a significant cigarette smoking history. Our standard approach is to use the external iliac artery (EIA) on the right for pancreas transplant inflow (through the Y graft, fashioned from the donor common, internal and external iliac arteries to the splenic and SMA arteries of the pancreas transplant, on the back table); with the left EIA used for inflow to the kidney transplant. Subsequent options, if there is severe atherosclerosis of the EIA, include proximal dissection to include the internal iliac artery (although typically this is involved in the atherosclerotic process), and common iliac artery and/or aorta if necessary. Further approaches include endarteroctomy of the involved arterial segment; replacement of diseased EIA segment with donor EIA [22]; long donor Carrel arterial patch (patch angioplasty); and the use of single donor arterial conduit for both kidney and pancreas transplants. The use of an occluding Foley catheter/balloon has been described to occlude inflow in severely atherosclerotic arteries when placement of an external vascular clamp is not prudent [23]. The placement of both KT and PT on the same (ipsilateral right) side has been introduced to preserve the left side for future transplants [24]. We have recently described the use of a donor arterial jump graft (EIA) from the distal SMA of the PT (inflow from the recipient EIA) on the right side, with bladder-duodenal drainage to the KT (donor renal artery on the left side) after inflow from the severely atherosclerotic left EIA was insufficient [25].

38.4 IMMUNOSUPPRESSION IN PT

Recent protocols in PT include attempts to: (i) reduce calcineurin inhibitor short- and long-term nephrotoxicity, (ii)

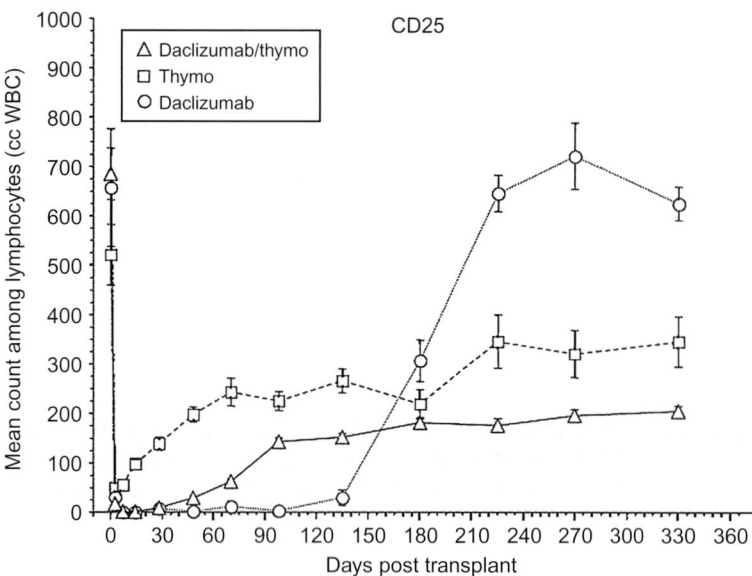

FIGURE 38.2 Mean CD25$^+$ lymphocyte count in the peripheral blood for patients following transplantation. The induction immunosuppression groups included daclizumab/thymoglobulin, thymoglobulin alone, and daclizumab alone.

reduce or avoid corticosteroids, (iii) use adjunctive maintenance antiproliferative agents or mTOR inhibitors, and (iv) utilize new induction antibodies. These include nondepleting anti-CD25 (IL-2R) monoclonals (daclizumab or basiliximab) or depleting monoclonals (alemtuzumab) and polyclonal antibodies (thymoglobulin). The goal is to reduce the incidence and severity of AR as well as to prevent long-term chronic allograft dysfunction [3].

38.4.1 Induction with Daclizumab

The results of a multicenter survey from 2001 using daclizumab as induction therapy revealed a low incidence of AR in combination with tacrolimus, mycophenolate mofetil (MMF), and steroids in SPK recipients [26]. MMF as primary immunosuppression was later investigated in a multicenter, open label, comparative trial [27]. SPK recipients were randomized to one of three groups. The rate of renal allograft AR was similar in the two daclizumab arms, however, significantly higher in the no antibody arm. Although the follow-up was short, this study emphasized the important role of induction antibody in reducing AR [28]. Subsequently, in a 3-year study comparing no induction with anti-IL-2R and T-cell-depleting antibody, actual 3-year kidney survival was significantly better in the induction groups, but cytomegalovirus (CMV) infection was higher in the T-cell depletion arm [29].

38.4.2 Induction with Daclizumab in Combination with Thymoglobulin

In our center, SPK recipients were randomized prospectively to MMF or rapamycin (Rapa) starting in September 2000. Patients received daclizumab and thymoglobulin as induction immunosuppression [30] as well as tapering steroids and tacrolimus. This results in a delay of over 45 days of peripheral blood CD25$^+$ cells compared to patients receiving thymoglobulin alone (Figure 38.2), and may affect a reduction in allogeneic (donor specific) effector T cells [30]. Patient, kidney, and pancreas survival rates were similar in the two arms. However, kidney-specific (biopsy-proven) AR occurred more often in the MMF versus the Rapa arm. Similarly, pancreas-specific AR (biopsy-proven or clinically suspected) occurred more often in the MMF versus Rapa arm. GI symptoms led to dose reductions or withholding MMF in the first year, significantly more often than in the Rapa group. Excellent kidney and pancreas graft survival was achieved using MMF or Rapa-based immunosuppression in this long-term study. Rates of AR for pancreas and biopsy-proven kidney transplantation were statistically significantly better with Rapa than MMF. It appears that Rapa is better tolerated than MMF from the GI standpoint in this patient population with T1D, in whom 60% have gastroparesis [19].

38.4.3 Induction with Alemtuzumab

A nonrandomized study of 75 SPK and pancreas recipients who received alemtuzumab (anti-CD52) (four doses for induction and up to 12 doses monthly maintenance within the first year) and MMF ($\geq$2 g/day) monotherapy was reported from Minnesota in 2005 [31]. Patient and pancreas survival rates were compared to a historical group of 266 consecutive pancreas recipients on thymoglobulin (induction) and tacrolimus (maintenance). The interim conclusion was that while it eliminated undesired calcineurin inhibitor and steroid-related side effects, the combination of

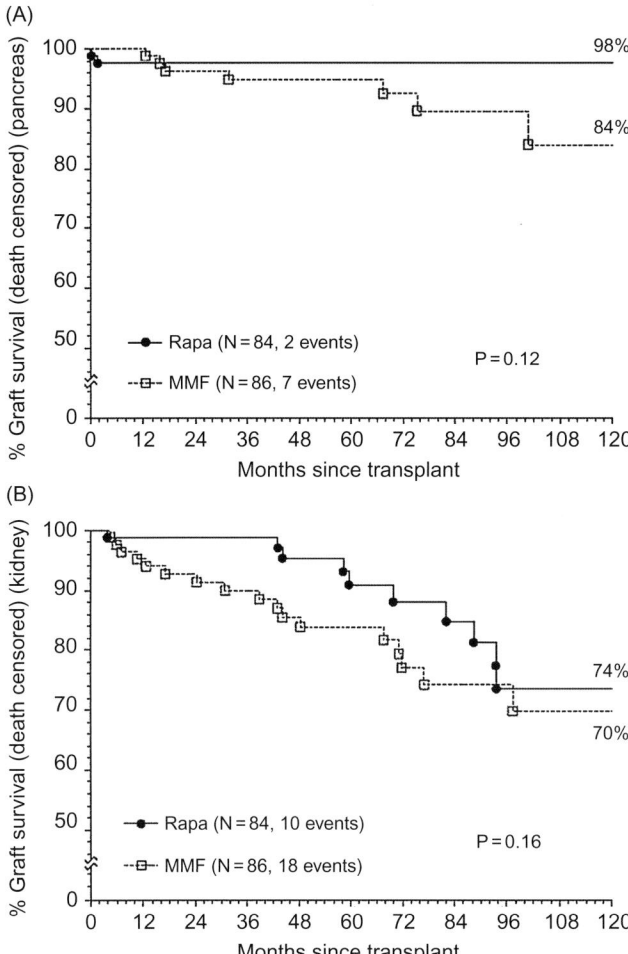

FIGURE 38.3 SPKT recipient treated with Rapa versus MMF: (A) % pancreas death-censored graft survival and (B) % kidney death-censored graft survival.

alemtuzumab and MMF was associated with a very high rejection rate (higher than expected for SPKT), albeit with acceptable (graft and native) kidney function. Kaufman et al. [32] compared the effects of alemtuzumab to those of thymoglobulin as induction agents given a steroid-free regimen in combination with tacrolimus/sirolimus-based maintenance therapy. The 1-year actual patient survival rates and AR were not statistically different [32]. Viral infectious complications were statistically significantly lower in the alemtuzumab group. Although steroid-free protocols for PT are effective, questions remain regarding optimal patient selection and long-term outcome [33].

38.5 PT: PATIENT, PANCREAS, AND KIDNEY TRANSPLANT SURVIVAL

PT restores euglycemia and provides long-term insulin independence in most patients. In our early (first 10

years—1990–1999) experience, 10-year survival rates for patients, pancreas, and kidney transplants were 84%, 76%, and 51%, respectively, among SPK recipients with T1D and ESRD [20]. In our more recent 10-year experience, with recipients, 25% of whom had experienced a cardiovascular event (myocardial infarction, CABG, angioplasty, CVA) before the transplant, the patient survival combining both Rapa and MMF groups, was about 70%. Death-censored pancreas transplant survival was 98% in the patients treated with Rapa and 84% in patients treated with MMF (Figure 38.3B) [19]. Death-censored kidney transplant survival was 74% in those patients treated with Rapa and 70% in patients treated with MMF [19] (Figure 38.3A). Fifteen-year actuarial patient survival was recently reported to be 56% for SPK, 42% for PAK, and 59% for PTA with pancreas survival rates of 36%, 18%, and 16%, respectively [6]. In our more recent study, where biopsy- proven AR was lower for both kidney and pancreas transplant recipients in the Rapa arm, and there was less rejection-related kidney transplant graft loss in the rapa arm, it is anticipated that better cardiovascular protection may result, but that it may require two decades to become apparent [2,11,19].

Other findings from our recent 10-year experience included relatively low rates of specific, transplant-related complications [19]. There were no lymphoproliferative disorders in either arm (Rapa or MMF). The incidence of serious (requiring hospital admission) viral infections included CMV (2 Rapa, 1 MMF arm), EBV (1 Rapa, 0 MMF), and BK (polyoma) virus (1 Rapa, 0 MMF) with no significant difference between arms. The proportion of high-risk CMV SPK recipients (i.e., donor CMV seropositive into recipient seronegative) was about 1/3 in each arm, and prophylaxis with gancyclovir was used as described.

Other potential complications, for example, wound infections (5/84 Rapa; 5/86 MMF) and incisional hernia repairs (12/84 Rapa; 8/86 MMF; $P = 0.31$) were similar in both arms. There were no patients in either arm who required hospitalization for mouth ulceration or lymphoceles [19].

From a metabolic standpoint, there was evidence for Rapa-related insulin resistance, with higher levels of HbA1C (5.52 Rapa, 5.40 MMF; $P = 0.00004$) despite no difference in c-peptide levels. Nonetheless the HbA1C was normal in both groups. The incidence of T2D (elevated HbA1C with normal to high c-peptide levels) was slightly greater in the Rapa arm (5 Rapa, 3 MMF) though this was not statistically significant. Similarly, there were significantly higher levels of both cholesterol and triglycerides in the Rapa than the MMF arm, however, levels were within normal limits in both groups. Furthermore, levels of HDL and LDL were not significantly different between groups after 3 months and the percent of SPKT recipients receiving lipid-lowering agents was similar after 1 year [19].

38.6 PT: IMPACT ON DIABETES-RELATED COMPLICATIONS

PT also ameliorates some diabetic complications. It can reverse pre-existing histological lesions of diabetic nephropathy in the native kidneys, but similar to findings from the DCCT/EDIC studies, reversal requires more than 5 years of normoglycemia [34–36]. There is also evidence for improvement of renal function after PT documented by reduction of proteinuria and stable creatinine clearance [37]. The majority of patients that undergo SPK transplant have already developed some degree of retinopathy and most have received laser therapy. For those patients without ESRD, the percentage of patients with improved or stabilized retinopathy was significantly higher among those receiving pancreas transplantation alone (PTA) than those treated with intensive insulin [38]. For SPK recipients with longer duration of T1D (mean 24.6 years) and ESRD, more than 90% of patients have stable diabetic retinopathy following transplantation [39]. Some improvement in gastroparesis, as well as postural hypotension, generally occurs after PT [40]. Sensory and motor neuropathy, as shown by nerve conduction studies, have also improved after SPK [41].

38.7 T1D RECURRENCE AFTER PT

At our center we are monitoring our patients for the possible recurrence of autoimmune diabetes, and have identified several patients who have developed T1D recurrence (T1DR) in the transplanted pancreas in the absence of rejection [42,43]. These patients present generally 5–10 years after SPKT with hyperglycemia often with measurable c-peptide and stable levels of urine amylase with normal/unchanged creatinine levels. Thus, they have evidence of endocrine pancreas dysfunction, while exocrine pancreas and KT function remain unchanged. This suggests a process involving the islet cell specifically, i.e., recurrent autoimmunity and not rejection. Biopsy of the PT has typically revealed insulitis (islet cell infiltrate) with T cells (CD4 and CD8) and B cells (CD20), and variable insulin staining. Retrospective analysis of levels of autoantibodies (GAD65, IA-2, and ZnT8) has revealed a rise in levels 3½ months to 2½ years prior to the development of hyperglycemia. Analysis of peripheral blood T cells has also shown the presence of autoreactive T cells [42,43]. The demonstration of autoantigen-reactive memory T-cell involvement [42,43] may provide insight into future therapies for T1DR that may possibly translate to therapy for spontaneous T1D as well [44]. Our attempts

at therapy for T1DR, which have included a combination of anti-T and anti-B-cell therapy, as well as plasmapheresis have resulted in transient increase in c-peptide, but not insulin independence. Furthermore, the reappearance of autoreactive memory T cells in the peripheral blood 1–3 years after treatment has heralded the continued fall and ultimately loss of c-peptide secretion [43].

Interestingly, in those patients with severe T1DR who no longer secrete c-peptide and lack insulin staining in the islets of their pancreas transplant biopsies, we have demonstrated the presence of insulin staining in the ductal cells [45]. This finding may have relevance to the field of regenerative medicine, since the ductal epithelial cells are presumably receiving signals to synthesize insulin, yet other stimuli that may encourage differentiation into beta cells are lacking. Elucidation of the signaling pathways involved in this differentiation may provide clues to both the field of stem cell research and regenerative medicine.

38.8 CHALLENGES AND FUTURE PERSPECTIVES IN PT

The present challenge is finding the balance between benefit (protection from rejection and resultant long-term graft function) and risk (side effects, infection, cancers) for patients with T1D and ESRD who receive an SPK transplant. Advances in immunosuppression protocols based on novel induction antibodies and effective maintenance agents have resulted in low rates of AR and excellent graft survival. Some centers, including our own, have also witnessed a low rate of viral sequelae and lymphoproliferative disorders [19]. The major hurdle is the requirement for chronic immunosuppression and the inability of current regimens to achieve transplantation tolerance and prevent recurrent autoimmunity. Improved immune monitoring for both rejection and autoimmunity may provide information that will help in the development of tolerogenic protocols that would ideally provide freedom from chronic immunosuppression.

ACKNOWLEDGMENT

The authors wish to acknowledge the pleasant and patient pursuit of perfection on the part of Ms. Maruja Chavez in the preparation of this pancreas transplant manuscript.

REFERENCES

[1] Leichtman AB, Cohen D, Keith D, O'Connor K, Goldstein M, McBride V, et al. Kidney and pancreas transplantation in the United States, 1997–2006: the HRSA breakthrough collaboratives and the 58 DSA challenge. Am J Transplant 2008;8:946–57.

[2] Burke GW, Ciancio G, Sollinger HW. Advances in pancreas transplantation. Transplantation 2004;77:62–7.

[3] Ciancio G, Burke GW, Miller J. Current treatment practices in immunosuppression. Expert Opin Pharmacother 2000;1:1307−30.

[4] Ciancio G, Mattiazzi A, Roth D, Kupin W, Miller J, Burke GW. The use of daclizumab as induction therapy in combination with tacrolimus and mycophenolate mofetil in recipients with previous transplants. Clin Transplant 2003;17:428−32.

[5] Burke GW, Ciancio G. Critical care issues in the renal and pancreatic allograft recipient. In: Civetta JM, Taylor RW, Kirby RR, editors. Critical care. 3rd ed. Philadelphia: J.B. Lippincott Company; 1997. p. 1311−5.

[6] White SA, Shaw JA, Sutherland DE. Pancreas transplantation. Lancet 2009;373:1808−17.

[7] Ninomiya T, Perkovic V, de Galan BE, Zoungas S, Pillai A, Jardine M, et al. ADVANCE Collaborative Group. Albuminuria and kidney function independently predict cardiovascular and renal outcomes in diabetes. J Am Soc Nephrol 2009;20:1813−21.

[8] Holdaas H, Fellström B, Cole E, Nyberg G, Olsson AG, Pedersen TR, et al. Assessment of LEscol in Renal Transplantation (ALERT) Study Investigators Long-term cardiac outcomes in renal transplant recipients receiving fluvastatin: the ALERT extension study. Am J Transplant 2005;5:2929−36.

[9] The Diabetes Control and Complications Trial Research Group. The effect of intensive therapy of diabetes on the development and progression of long-term complications in insulin-dependent diabetes mellitus. N Engl J Med 1993;329:977−86.

[10] Nathan DM, Lachin J, Cleary P, Orchard T, Brillon DJ, Backlund JY, et al. Diabetes control and complications trial; epidemiology of diabetes interventions and complications research group. Intensive diabetes therapy and carotid intima-media thickness in type 1 diabetes mellitus. N Engl J Med 2003;348:2294−303.

[11] Nathan DM, Cleary PA, Backlund JY, Genuth SM, Lachin JM, Orchard TJ, et al. Diabetes Control and Complications Trial/Epidemiology of Diabetes Interventions and Complications (DCCT/EDIC) study research group. Intensive diabetes treatment and cardiovascular disease in patients with type 1 diabetes. N Engl J Med 2005;353:2643−53.

[12] Study Research Group, Jacobson AM, Musen G, Ryan CM, Silvers N, Cleary P, et al. Diabetes Control and Complications Trial/Epidemiology of Diabetes Interventions and Complications Long-term effect of diabetes and its treatment on cognitive function. N Engl J Med 2007;356:1842−52.

[13] McCrimmon RJ, Ryan CM, Frier BM. Diabetes and cognitive dysfunction. Lancet 2012;379:2291−9.

[14] The Diabetes Control and Complications Trial/Epidemiology of Diabetes Interventions and Complications Research Group. Retinopathy and nephropathy in patients with type 1 diabetes four years after a trial of intensive therapy. N Engl J Med 2000;342:381−9.

[15] DCCT/EDIC Research Group, de Boer IH, Sun W, Cleary PA, Lachin JM, Molitch ME, et al. Intensive diabetes therapy and glomerular filtration rate in type 1 diabetes. N Engl J Med 2011;365:2366−76.

[16] Burke 3rd. GW, Ciancio G, Figueiro J, Buigas R, Olson L, Roth D, et al. Hypercoagulable state associated with kidney-pancreas transplantation. Thromboelastogram-directed anti-coagulation and implications for future therapy. Clin Transplant 2004;18:423−8.

[17] Burke GW, Ciancio G, Cirocco R, Markou M, Olson L, Contreras N, et al. Microangiopathy in kidney and simultaneous pancreas/

kidney recipients treated with tacrolimus: evidence of endothelin and cytokine involvement. Transplantation 1999;68:1336−42.

[18] Cola C, Ansell J. Heparin-induced thrombocytopenia and arterial thrombosis: alternative therapies. Am Heart J 1990;119:368−74.

[19] Ciancio G, Sageshima J, Chen L, Gaynor JJ, Hanson L, Tueros L, et al. Advantage of rapamycin over mycophenolate mofetil when used with tacrolimus for simultaneous kidney transplant recipients: randomized, single-center trial at 10 years. Am J Transplant 2012;12:3363−76.

[20] Burke GW, Ciancio G, Olson L, Roth D, Miller J. Ten-year survival after simultaneous pancreas/kidney transplantation with bladder drainage and tacrolimus-based immunosuppression. Transplant Proc 2001;33:1681−3.

[21] Stratta RJ, Shokouh-Amiri MH, Egidi MF, Grewal HP, Lo A, Kizilisik AT, et al. Long-term experience with simultaneous kidney-pancreas transplantation with portal-enteric drainage and tacrolimus/mycophenolate mofetil-based immunosuppression. Clin Transplant 2013;17:69−77.

[22] Moon JI, Ciancio G, Burke GW. Arterial reconstruction with donor iliac vessels during pancreas transplantation: an intraoperative approach to arterial injury or inadequate flow. Clin Transplant 2005;19:286−90.

[23] Fridell JA, Gage E, Goggins WC, Powelson JA. Complex arterial reconstruction for pancreas transplantation in recipients with advanced arteriosclerosis. Transplantation 2007;83:1385−8.

[24] Fridell JA, Shah A, Milgrom ML, Goggins WC, Leapman SB, Pescovitz MD. Ipsilateral placement of simultaneous pancreas and kidney allografts. Transplantation 2004;78:1074−6.

[25] Gorin MA, Garcia-Roig M, Burke 3rd GW, Ciancio G. Emergent renal revascularization of a simultaneous pancreas-kidney transplant recipient. Transplantation 2012;93:16−7.

[26] Bruce DS, Sollinger HW, Humar A, Sutherland DE, Light JA, Kaufman DB, et al. Multicenter survey of daclizumab induction in simultaneous kidney−pancreas transplant recipients. Transplantation 2001;72:1637−43.

[27] Stratta RJ, Alloway RR, Hodge E, Lo A. A multicenter, open-label, comparative trial of two daclizumab dosing strategies vs. no antibody induction in combination with tacrolimus, mycophenolate mofetil, and steroids for the prevention of acute rejection in simultaneous kidney-pancreas transplant recipients: interim analysis. Clin Transplant 2002;16:60−8.

[28] Stratta RJ, Alloway RR, Hodge E, Lo A. A multicenter, open-label, comparative trial of two daclizumab dosing strategies versus no antibody induction in simultaneous kidney−pancreas transplantation: 6-month interim analysis. Transplant Proc 2002;34:1903−5.

[29] Burke 3rd GW, Kaufman DB, Millis JM, Gaber AO, Johnson CP, Sutherland DE, et al. Prospective, randomized trial of the effect of antibody induction in simultaneous pancreas and kidney transplantation: three-year results. Transplantation 2004;77:1269−75.

[30] Sageshima J, Ciancio G, Gaynor JJ, Chen L, Guerra G, Kupin W, et al. Addition of anti-CD25 to thymoglobulin for induction therapy: delayed return of peripheral blood CD25-positive population. Clin Transplant 2011;25:132−5.

[31] Gruessner RW, Kandaswamy R, Humar A, Gruessner AC, Sutherland DE. Calcineurin inhibitor- and steroid-free immunosuppression in pancreas−kidney and solitary pancreas transplantation. Transplantation 2005;79:1184−9.

[32] Kaufman DB, Leventhal JR, Gallon LG, Parker MA. Alemtuzumab induction and prednisone-free maintenance immunotherapy in simultaneous pancreas—kidney transplantation comparison with rabbit antithymocyte globulin induction—long-term results. Am J Transplant 2006;6:331—9.

[33] Mineo D, Sageshima J, Burke GW, Ricordi C. Minimization and withdrawal of steroids in pancreas and islet transplantation. Transpl Int 2009;22:20—37.

[34] Fioretto P, Mauer SM, Bilous RW, Goetz FC, Sutherland DE, Steffes MW. Effects of pancreas transplantation on glomerular structure in insulin-dependent diabetic patients with their own kidneys. Lancet 1993;342:1193—6.

[35] Fioretto P, Steffes MW, Sutherland DE, Goetz FC, Mauer M. Reversal of lesions of diabetic nephropathy after pancreas transplantation. N Eng J Med 1998;339:69—75.

[36] Fioretto P, Sutherland DE, Najafian B, Mauer M. Remodeling of renal interstitial and tubular lesions in pancreas transplant recipients. Kidney Int 2006;69:907—12.

[37] Coppelli A, Giannarelli R, Vistoli F, Del Prato S, Rizzo G, Mosca F, et al. The beneficial effects of pancreas transplant alone on diabetic nephropathy. Diabetes Care 2005;28:1366—70.

[38] Giannarelli R, Coppelli A, Sartini MS, Del Chiaro M, Vistoli F, Rizzo G, et al. Pancreas transplant alone has beneficial effects on retinopathy in type 1 diabetic patients. Diabetologia 2006;49:2977—82.

[39] Pearce IA, Ilango B, Sells RA, Wong D. Stabilisation of diabetic retinopathy following simultaneous pancreas and kidney transplant. Br J Ophthalmol 2000;84:736—40.

[40] Hathaway DK, Hartwig MS, Milstead J, Elmer D, Evans S, Gaber AO. Improvement in quality of life reported by diabetic recipients of kidney-only and pancreas—kidney allografts. Transplant Proc 1994;26:512—4.

[41] Navarro X, Sutherland DE, Kennedy WR. Long-term effects of pancreatic transplantation on diabetic neuropathy. Ann Neurol 1997;42:727—36.

[42] Laughlin E, Burke G, Pugliese A, Falk B, Nepom G. Recurrence of autoreactive antigen-specific CD4$^+$ T cells in autoimmune diabetes after pancreas transplantation. Clin Immunol 2008;128:23—30.

[43] Vendrame F, Pileggi A, Laughlin E, Allende G, Martin-Pagola A, Molano RD, et al. Recurrence of type 1 diabetes after simultaneous pancreas-kidney transplantation, despite immunosuppression, is associated with autoantibodies and pathogenic autoreactive CD4 T cells. Diabetes 2010;59:947—57.

[44] Burke 3rd GW, Vendrame F, Pileggi A, Ciancio G, Reijonen H, Pugliese A. Recurrence of autoimmunity following pancreas transplantation. Curr Diab Rep 2011;11:413—9.

[45] Martin-Pagola A, Sisino G, Allende G, Dominguez-Bendala J, Gianani R, Reijonen H, et al. Insulin protein and proliferation in ductal cells in the transplanted pancreas of patients with type 1 diabetes and recurrence of autoimmunity. Diabetologia 2008;51:1803—13.

Chapter 39

Living-Donor Pancreas Transplantation: Progress, Pitfalls, and Promise

Rainer W.G. Gruessner, David E.R. Sutherland, and Angelika C. Gruessner

Department of Surgery, University of Arizona, Tucson, AZ

39.1 INTRODUCTION

The concept of solid-organ donation for transplantation is virtually unique among surgical procedures in that a healthy volunteer is exposed to the risks of surgery solely for the benefit of another individual. The first successful living-donor (LD) kidney transplant (between monozygous twins) was performed by Joseph Murray in Boston in 1954 [1], for which he was awarded the Nobel Prize in medicine in 1990.

The first extrarenal organ to be successfully transplanted using an LD was the pancreas. Since then, LDs have also successfully been used for liver [2,3], lung [4,5], and intestine [6,7] transplants.

The first LD pancreas transplant took place on June 20, 1979, at the University of Minnesota, Minneapolis [8]. Most pancreas transplants in the 1970s were segmental transplants using deceased donors, so at that time, it was already known that a segmental pancreas graft's physiologic reserve was capable of making the recipient insulin-independent. Unknown at that time was whether an LD's remaining pancreas had enough physiologic reserve to sustain long-term insulin independence. The concept of donating the distal pancreas was based on the observation that patients with benign or malignant pancreatic diseases can undergo a distal hemipancreatectomy without demonstrable changes in endocrine function. Nonetheless, the concern over potentially serious surgical and metabolic complications in LDs has hampered their widespread use. According to the International Pancreas Transplant Registry (IPTR), as of October 2012, a total of 156 LD pancreas transplants have been done worldwide (United States, 136; non-United States, 20), representing only 0.4% of all pancreas transplants [9].

39.2 RATIONALE FOR PANCREAS TRANSPLANTS

A successful pancreas transplant, the definitive treatment for insulin-dependent diabetes mellitus (IDDM), (i) restores normoglycemia without exposing patients to the risk of severe hypoglycemia and (ii) prevents, halts, or reverses the development or progression of secondary complications of IDDM [10–12]. Although improved glucose control reduces the long-term complications of IDDM, as shown in the Diabetes Control and Complications Trial, intensive insulin regimens can cause severe iatrogenic hypoglycemia [13,14].

Regenerative Medicine Applications in Organ Transplantation.

A pancreas transplant is primarily done to improve quality of life [15]. Recipients do not need to inject insulin and monitor glucose concentrations several times a day. In contrast to intensive exogenous insulin therapy, hypoglycemic episodes and/or unawareness basically do not occur after a successful transplant. The short- and long-term advantages of a pancreas transplant have to be balanced against the potential morbidity and mortality associated with the surgical procedure and the side effects from the long-term immunosuppression that is needed to prevent alloimmunity and autoimmune recurrence [16].

39.3 RATIONALE FOR LD PANCREAS TRANSPLANTS

The rationale for LD pancreas transplants has shifted over time. Initially, in the azathioprine (AZA) and early cyclosporin A (CSA) eras, the use of LDs (as compared with deceased donors) was associated with significantly better graft survival rates. In the world's largest series at the University of Minnesota, the 1-year graft survival rate for technically successful pancreas after kidney (PAK) transplants in the AZA and CSA eras (January 1, 1979, through March 31, 1994) was 78% with LDs versus 57% with deceased donors [17,18]. Better graft outcome with LDs in the 1970s and 1980s was mainly due to the significantly lower rate of graft loss from rejection.

With the introduction of tacrolimus (TAC) and mycophenolate mofetil (MMF) in the 1990s, and their combined use, graft survival improved markedly for deceased donor pancreas recipients, because of a significantly lower graft loss rate from rejection. The immunologic advantage of LD pancreas transplants in the TAC era is no longer as distinct as it was in the AZA and CSA eras [19]. For that reason, the incentive for using LDs has waned. Further, in contrast to kidney and liver transplants, a shortage of deceased donors for pancreas transplants does not exist. Thus, in the TAC era, LDs for solitary pancreas transplants are now used only if the recipient (i) is highly sensitized (panel-reactive antibody [PRA] > 80%) and has a low probability of receiving a deceased donor graft; (ii) must avoid high-dose immunosuppression; or (iii) has a nondiabetic identical twin or a six-antigen-matched sibling.

In contrast to solitary pancreas transplants (for which, again, no shortage of deceased organs exists), the demand for deceased donor kidneys is continuously increasing [20]. As a consequence, the average waiting times for a kidney transplant alone (KTA) or a simultaneous pancreas–kidney (SPK) transplant using a deceased donor have markedly increased in the United States: for deceased donor SPK transplants, from 255 days in 1995 to 386 days in 2011 [9].

Given this high, unmet demand for deceased SPK donors, pancreas transplant professionals now recommend use of an LD for the kidney transplant, followed later by a deceased donor pancreas transplant (thus shifting emphasis from the SPK to the PAK category) [21]. That way, however, the recipient must undergo two operations, including receiving anesthesia twice (although, in some high-risk candidates, two operations are preferred, because each procedure is smaller than a combined transplant). Long-term pancreas graft outcome with a deceased donor in the PAK category is (slightly) less favorable than in the SPK category, even though long-term kidney graft outcome is better with an LD than with a deceased donor. For those two reasons—the shortage of deceased donor kidneys and the less favorable long-term pancreas graft outcome in the PAK (vs. SPK) category—the use of LDs for SPK recipients has been advocated [22,23]. The SPK option allows the donor and recipient to undergo only one procedure each, and the recipient has the great (immunologic) benefit of receiving two LD organs. Despite the apparent advantages for the recipient, the SPK procedure has not been widely applied, given the surgical and metabolic risks to the LD, so it remains relatively rare, done under very specific circumstances.

39.4 THE DONOR

39.4.1 Selection

The principles for accepting a potential LD for a pancreas transplant are not different than for other solid-organ transplants. Potential LDs must understand the complex nature of the procedure and the surgical and metabolic risks to his or her health, must not be coerced, must provide voluntary consent, must be mentally competent, and must be of legal age. They must undergo a thorough medical, social, and, frequently, psychological evaluation. Initial screening usually rules out volunteers with major health problems, such as current or previous disorders of the pancreas, active infections or malignancies, major personality disorders, and drug or alcohol dependence. Single parents of minor children are not accepted. The social and psychological evaluation assesses each potential LD's voluntarism and altruism, as well as the dynamics of the donor–recipient relationship [24].

Strict exclusion criteria were developed in 1996 at the University of Minnesota, in an effort to minimize the risk of hemipancreas LDs developing diabetes after donation (Table 39.1). These criteria recognized the importance of both personal and family history of endocrine disease in establishing risk, along with the role of obesity and other conditions associated with insulin resistance in unmasking β-cell dysfunction after donation. As a result, the exclusion criteria (as listed in Table 39.1) are considered standard and

TABLE 39.1 Exclusion Criteria for Living Pancreas Donors

Historical and clinical criteria

1. Additional first-degree relative with type 1 diabetes (other than proposed recipient)
2. History of type 2 diabetes in any first-degree relative (parent, sibling, child)
3. Personal history of gestational diabetes
4. Age of the donor within 10 years of the age at which type 1 diabetes was diagnosed in the proposed recipient
5. Clinical evidence of diseases associated with insulin resistance (e.g., polycystic ovarian syndrome, hypertension)
6. History of active diseases of the exocrine pancreas (e.g., active or chronic pancreatitis)
7. Personal history of an autoimmune endocrine disorder involving the thyroid, adrenal or pituitary gland or gonads
8. Body mass index >27 kg/m^2
9. Age >50 years
10. Active or uncontrolled psychiatric disorders
11. Heavy smoking, alcoholism, or excessive alcohol use
12. Hypertension, cardiac disease
13. Active infections or malignant disorders

Metabolic criteria

1. Any glucose value above 150 mg/dL during standard OGTTs
2. Hb A$_{1c}$ >6%
3. Glucose disposal rate calculated from data collected during intravenous glucose tolerance tests <1.0%
4. Presence of elevated titer of islet cell autoantibodies or anti-GAD antibodies
5. Acute insulin response to intravenous glucose or intravenous arginine of <300% of basal value
6. GPAIS of <300% of basal value

rule out potential LDs whose risk of developing diabetes after donation is deemed to be too high. In addition, potential LDs with metabolic abnormalities uncovered during preoperative testing are also excluded, because of uncertainty over whether their hemipancreas would provide adequate function for the recipient [25].

39.4.2 Workup

The medical evaluation of potential LDs includes both pancreas-nonspecific tests and pancreas-specific tests. Pancreas-nonspecific tests are the same as for kidney LDs, including an electrocardiogram (ECG) and chest radiograph; ABO blood typing and tissue typing; leukocyte crossmatch and PRA tests; a biochemistry profile; liver function tests; a lipid profile; a complete blood count; a coagulation profile; hepatitis A, B, and C tests; cytomegalovirus (CMV), Epstein−Barr virus (EBV), human immunodeficiency virus (HIV), and rapid plasma reagin (RPR) tests; urine analysis and urine culture; in

women ≤55 years old, a serum pregnancy test; in women ≥40 years old, a mammogram and Pap smear; in all women, a pelvic and breast examination; and, in men >50 years, a prostate-specific antigen (PSA) test. In addition, all potential LDs must undergo a history and physical examination; SPK LDs must also undergo serial blood pressure measurements [26].

The metabolic testing for potential pancreas LDs includes a standard oral glucose tolerance test (OGTT), hemoglobin A$_{1c}$ (Hb A$_{1c}$) assessment, and measurement of antibodies associated with type 1 diabetes. In addition, a comprehensive insulin secretory test is done in the fasting state. To measure both arginine- and glucose-induced insulin secretion, the glucose potentiation of arginine-induced insulin secretion (GPAIS) test is done, which repeats the arginine stimulation test after glucose has been administered intravenously at a rate of 900 mg/min for 60 min. The secretory responses on the GPAIS test are determined by calculating the mean of the three highest consecutive insulin values obtained in the first 5 min after the acute injection of glucose or arginine, and then subtracting the basal values obtained before the acute injection from the mean value. A normal response should be >300% of the basal value. Potential LDs with a secretory response <300% of the basal value are informed that their risk of developing diabetes after a hemipancreatectomy is high [26].

A potential LD who has cleared all of the above tests still needs to undergo a radiographic study to determine the anatomic suitability of the pancreas. In contrast to kidney LDs, in whom arterial variations are common, the blood supply to the distal pancreas via the splenic artery varies little. But even with an anatomic variation (such as a splenic artery off the suprarenal aorta), a distal pancreatectomy usually is technically feasible. Until the mid-1990s, aortography was the gold standard for assessing the vascular anatomy of the LD's pancreas. Since the mid-1990s, magnetic resonance imaging (MRI) and angiography (MRA) have become increasingly popular because of their less invasive nature. In addition, MRI/MRA studies provide details of parenchymal structure and allow 3D reconstruction of not only arterial but also venous anatomy. An alternative to MRI/MRA is computed tomography (CT) and angiography (CTA), which also allow 3D vascular reconstruction.

If several medically suitable pancreas LDs are available, the final selection is based on the histocompatibility result: an identical twin or a human leukocyte antigen (HLA)-identical sibling is the ideal choice (providing all other criteria for pancreas donation are met) [26].

39.4.3 Surgery

Open or laparoscopic procurement of the distal pancreas from an LD has been described in detail in the literature

[27−29]. It is important to emphasize that, in contrast to a standard distal pancreatectomy for benign or malignant disorders, the blood supply via the splenic artery and veins needs to be preserved until shortly before graft procurement, in order to not jeopardize the oxygen supply to both endocrine and exocrine tissues. The pancreas is usually divided slightly to the left of the portal vein to leave ≥ 50% of pancreatic tissue with the LD. Whenever possible, a spleen-preserving distal pancreatectomy is done. The introduction of the laparoscopic (hand-assisted) distal pancreatectomy has shortened hospitalization and recovery time, making the use of LDs safer and more attractive. The median duration of hospitalization after an open distal pancreatectomy is 8 days (range 6−24 days); after a laparoscopic distal pancreatectomy, 4−6 days [29,30].

Postoperative care of LDs is similar to that of any patient undergoing a major abdominal procedure. Serum amylase levels are initially obtained daily to assess endocrine function of the remaining pancreas; levels usually return to a normal range within 3 days after donation. Because the spleen is preserved in most LDs, those with postoperative shoulder or left flank pain undergo splenic radionuclide or CT scanning to ensure viability of the spleen and to rule out abscess formation. In the early postoperative period, sequential technetium (Tc) 99m-sulfur colloid scans of the spleen have shown markedly decreased or absent uptake. But, in most LDs, over a period of 2 weeks, splenic blood flow and function return to normal or near-normal [31].

39.4.4 Outcome

In contrast to kidney or liver LDs, the mortality rate of pancreas LDs, according to the IPTR, has been 0% [32]. For pancreas LDs, morbidity includes both surgical and medical complications as well as adverse metabolic changes.

In general, surgical complications are rare; relaparotomies are required in <5% of pancreas LDs. The most common surgical complication is a splenectomy, either at the time of the distal pancreatectomy (most commonly due to bleeding from the spleen) or postoperatively (most commonly due to splenic ischemia). The incidence of splenectomy (intra- and postoperatively) ranges from 10% to 15%. Postoperative development of pancreatitis, pancreatic fistulas, and pancreatic or splenic abscesses occurs in <5% of LDs. The incidence of relaparotomies for such complications is <3% [33,34].

The effects of a distal pancreatectomy on insulin secretion, glucose metabolism, and pancreatic α- and β-cell function have been studied at various points after donation [35−43]. A detailed discussion with the potential LD regarding the potential metabolic complications is required.

The first metabolic study by Kendall et al. [35] in 28 LDs demonstrated deterioration of insulin secretion and

of glucose tolerance at 1 year after donation. Of the 28 LDs studied, 25% had abnormal glucose tolerance, with modest impairment of early insulin secretion during the OGTT at 1 year. Interestingly, their mean postoperative 24-h glucose-profile value did not exceed the range established in nondiabetic individuals [36].

Given that LDs generally maintain normoglycemia after a distal pancreatectomy, despite diminished insulin secretion, Seaquist and Robertson suggested that healthy humans may compensate for a distal pancreatectomy by increasing glucose disposal [37]. Their study demonstrated that first-phase insulin secretion in response to glucose, the insulin secretory response to arginine, and first-phase glucagon secretion in response to intravenous arginine were significantly decreased in LDs, as compared with controls matched for age, sex, and body mass index. As expected, LDs of a distal pancreas had significantly less insulin secretory reserve than the matched controls [37].

The observation of diminished β-cell mass and function in LDs of a distal pancreas who nonetheless maintained normal plasma glucose levels (fasting state) and normal Hb A_{1c} levels led to this hypothesis: to compensate for decreased β-cell mass, insulin sensitivity may increase in the liver and/or soft tissues (muscle or fat tissue). However, neither insulin- nor glucose-mediated glucose uptake increases after a distal pancreatectomy [36,38]. A follow-up study by Seaquist et al. supported this hypothesis instead: disproportionate hyperproinsulinemia may result from increased β-cell demand in LDs [39].

The relationship between diabetes and obesity 9−18 years after a distal pancreatectomy in LDs was studied by Robertson et al. in 8 donor−recipient pairs [40]. No differences between the two groups (LDs and recipients) were noted in any of these values: fasting plasma glucose, Hb A_{1c}, fasting insulin or C-peptide, acute insulin or C-peptide response to arginine and to glucose, or β-cell secretory reserve. Robertson et al. concluded that obesity should be a contraindication to donation of a distal pancreas and that LDs should assiduously avoid becoming obese.

In another study, Robertson et al. assessed glucagon, catecholamines, and symptom responses to hypoglycemia in eight LDs of a distal pancreas. Glucagon responses to arginine, as well as insulin responses to glucose and arginine, were diminished, yet no deficiencies in glucagon responses were detected during hypoglycemia. These results contrast with previous findings that LDs have diminished β-cell responses to glucose and arginine, as well as diminished α-cell responses to arginine [41].

A larger study in 46 LDs assessed Hb A_{1c} levels at various points after donation. In 10 LDs who underwent a distal pancreatectomy through August 31, 1996, elevated Hb A_{1c} levels were noted; of those 10, three required insulin >6 years after donation. The causes of elevated Hb A_{1c} levels in all three LDs are now considered contraindications to donation (gestational diabetes, increased body mass index, undisclosed history of alcohol abuse). In all LDs who underwent a distal pancreatectomy after September 1, 1996, Hb A_{1c} levels remained normal [42].

Clearly, a meticulous metabolic evaluation of potential LDs is important. Over the last two decades, modifications have frequently been necessary. Over time, acceptance criteria have become more stringent, in order to avert any negative impact on the LD's metabolic outcome. Ultimately, nowadays, only 15–20% of all potential LDs qualify for donation. Avoidance of obesity after donation appears to be key in diminishing the risk of long-term metabolic complications.

As with LDs of other organs, the vast majority of pancreas LDs stand by their decision to donate. In a retrospective study, of 46 pancreas LDs, 43 stated that they had made the correct decision. About 20% of LDs, however, have experienced social, marital, financial, and/or employment problems. A more detailed analysis of psychosocial outcome in pancreas LDs after donation is warranted [43].

39.5 THE RECIPIENT

39.5.1 Selection

An LD pancreas transplant is an option for virtually all pancreas transplant candidates. The rate-limiting aspect is the possible unavailability of a suitable LD. Because deceased donor pancreases are *not* in short supply, the use of LDs is primarily considered for these five subgroups [44]: (i) candidates with an identical twin or an HLA-identical sibling, even though recurrent disease is a concern; (ii) diabetic candidates with a rare tissue type and a well-matched LD; (iii) candidates with a high PRA level and negative crossmatch results with an LD; (iv) candidates who strongly prefer an elective transplant for medical or personal reasons; and (v) diabetic candidates for whom standard immunosuppression is not recommended.

LD pancreas transplants can be performed in three recipient categories: (i) an SPK transplant for diabetic and uremic candidates, given the long waiting time for a deceased donor SPK transplant (because of the kidney); (ii) a PAK transplant for diabetic and posturemic candidates with a previous kidney graft, ideally from the same LD; and (iii) a pancreas transplant alone (PTA) for nonuremic candidates with brittle diabetes.

39.5.2 Surgery

LD pancreas transplants vary from deceased donor pancreas transplants in that only the LD's splenic artery and vein are used for the vascular anastomoses with the recipient's external iliac vessels. But like deceased donor pancreas transplants, LD pancreas transplants are preferably placed intra-abdominally on the right side of the pelvis, because the external iliac vessels are in a more shallow position there than on the left side. Detailed descriptions of the surgical procedure have been published [18,45].

The technical aspects and considerations with regard to managing exocrine pancreatic secretions are similar for LD and deceased donor pancreas transplants. Bladder drainage and enteric drainage are most common. Bladder drainage (Figure 39.1A) allows measurement of urinary amylase levels and thus timely diagnosis of rejection, yet causes complications (such as metabolic acidosis and urinary tract infections). Enteric drainage (Figure 39.1B) is more physiologic, but early detection of rejection is not possible in the absence of exocrine urinary monitoring.

With LD pancreas transplants (unlike with deceased donor pancreas transplants), the use of portal vein drainage is not an alternative. Portal vein drainage would require the use of vascular extension graft(s); this modification and drainage into the low-flow portal circulation might increase the risk of graft thrombosis. Thus, systemic vein drainage is the technique of choice for LD segmental grafts. No convincing evidence exists today that systemic vein drainage places pancreas recipients at a higher risk for developing arteriosclerosis due to peripheral hyperinsulinemia [46,47]. Furthermore, comparable metabolic control is achieved with portal and systemic vein drainage [48,49].

The spectrum and management of posttransplant complications does not differ for LD and deceased donor pancreas transplants. The first 24–48 h posttransplant are the most crucial: the recipient undergoes the physiologic response to surgical trauma; the transplanted pancreas is in a varying degree of ischemic or reperfusion injury; and immunosuppression begins. Bicarbonate replacement is important for bladder-drained recipients. An insulin drip is started if blood sugar levels exceed 130 mg/dL, so as not to further stress the insulin-producing cells on top of the preservation and ischemic injury.

One of the most crucial elements in the immediate posttransplant care of LD segmental pancreas recipients is anticoagulation [50,51]. During the transplant procedure, such recipients receive 20–40 U/kg of heparin. The target partial thromboplastin time (PTT) within the first 24 h posttransplant is 50–70 s. The heparin drip is usually discontinued on posttransplant day 5. In addition, recipients are started on Coumadin on posttransplant day 3 (and continue on it for up to 6 months).

Immunosuppression is initiated in the operating room; the first dose of antibody therapy is usually given before graft reperfusion. Standard immunosuppressive protocols consist of quadruple immunosuppression for induction

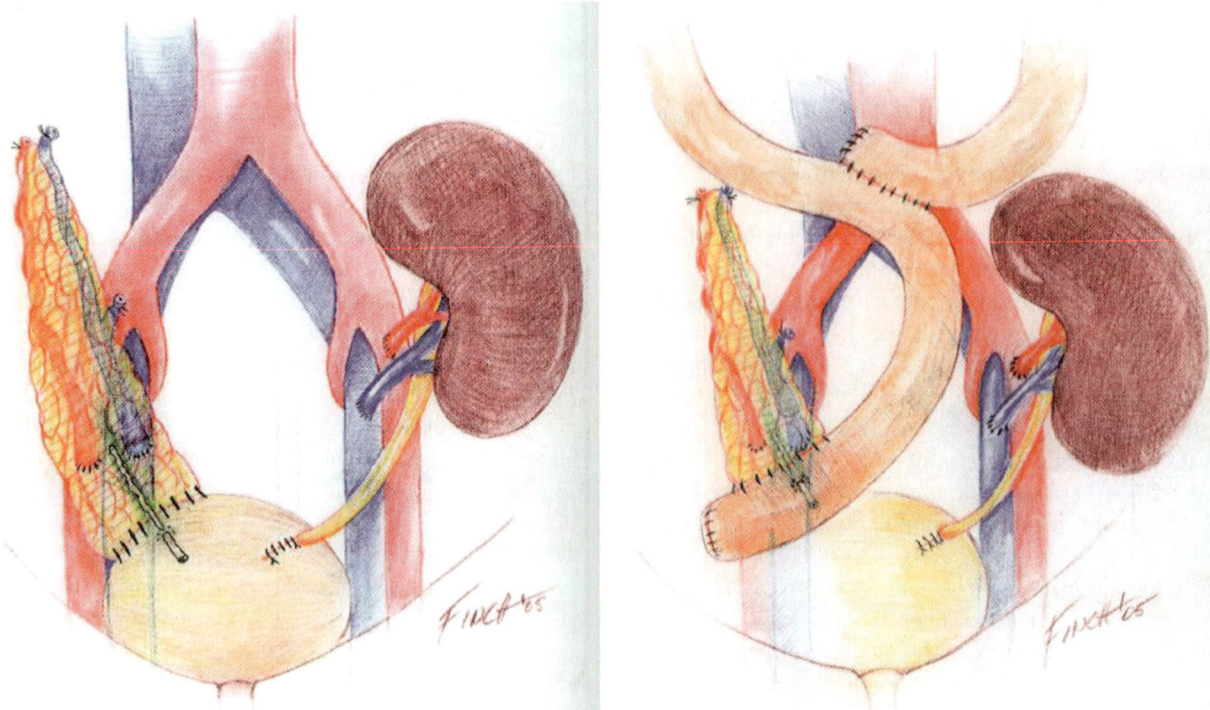

FIGURE 39.1 (A) LD segmental pancreas transplant with systemic vein and bladder exocrine drainage (color plate, Figure PA-9, in Gruessner RWG, Benedetti E. *Living Donor Organ Transplantation*). (B) LD segmental pancreas transplant with systemic vein and bowel exocrine drainage (color plate, Figure PA-10, in Gruessner RWG, Benedetti E. *Living Donor Organ Transplantation*).

and triple immunosuppression for maintenance therapy; because the rejection rate after LD pancreas transplants is lower than after deceased donor pancreas transplants, some LD recipients may be able to avoid steroids altogether and instead take only TAC and MMF.

39.5.3 Complications

Surgical complications after LD and deceased donor pancreas transplants are similar. The four most common complications in LD pancreas recipients are (i) bleeding (a rate of 6−8%) [52], (ii) thrombosis (5−13%), (iii) pancreatic leaks (5−20%); and (iv) intra-abdominal infections (4−10%). Other surgical complications that may also require a laparotomy include severe pancreatitis, pseudoaneurysms, arteriovenous (AV) fistulas in the graft, and wound dehiscence.

Note that the risk of thrombosis is higher in LD pancreas recipients than in deceased donor pancreas recipients because of the smaller caliber of LD blood vessels [53,54]. Thrombosis usually occurs in the first week posttransplant and is manifested by a sudden increase in the recipient's insulin requirement or by a sharp drop in urinary amylase levels. Venous thrombosis may be associated with a swollen and tender graft, hematuria, and lower extremity edema ipsilaterally. Diagnosis is usually

confirmed by Doppler ultrasound. Exploration usually requires a graft pancreatectomy.

With regard to pancreatic leaks, the absence of donor duodenum in LD pancreas recipients increases the incidence. A leak from an enteric anastomosis almost always leads to a laparotomy and may require a graft pancreatectomy.

Given the advent of advanced interventional radiologic procedures to drain intra-abdominal abscesses, the incidence of reoperations has markedly decreased [53].

The most common *nonsurgical* complication after LD pancreas transplants is rejection. The incidence is about 30% within the first year posttransplant. In bladder-drained recipients, diagnosis of rejection is usually based on an increase in serum amylase and lipase levels and on a decrease in urinary amylase levels. A sustained drop in urinary amylase levels >25% from baseline should prompt a pancreas biopsy to rule out rejection. In enterically drained recipients, only serum amylase and lipase levels can be relied on [52]. Other signs and symptoms include tenderness over the graft, unexplained fever, and hyperglycemia (usually a late finding). Diagnosis can be confirmed by a percutaneous pancreas biopsy [52,55]. If a percutaneous biopsy is not possible for technical reasons, empiric therapy with antibodies may be started.

Other nonsurgical complications include infections, such as CMV, hepatitis C virus (HCV), and extra-abdominal

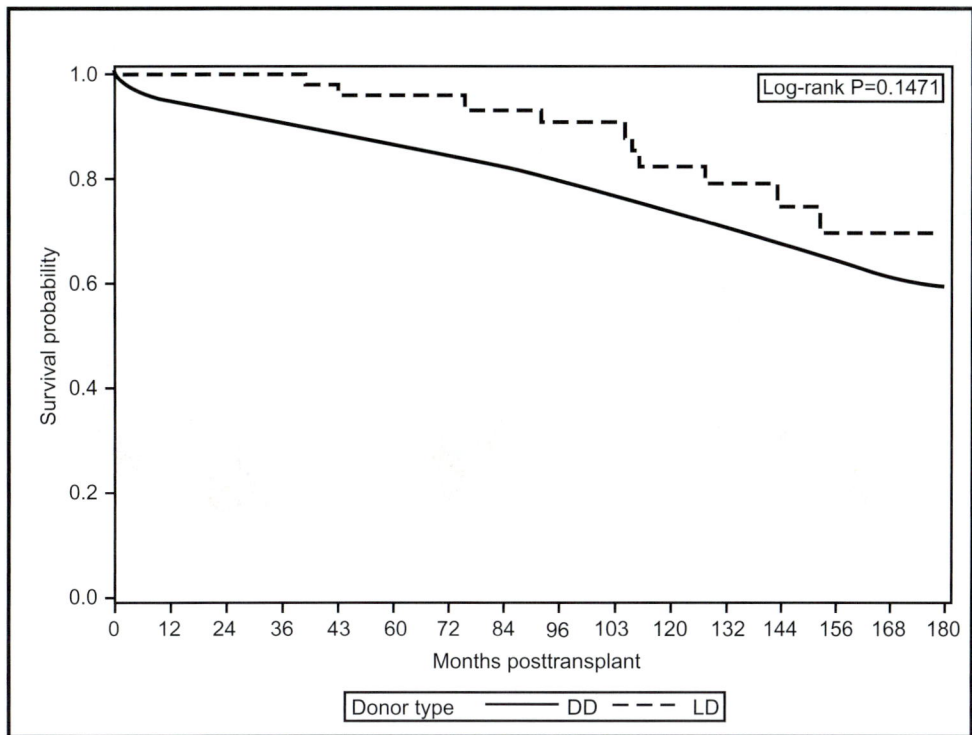

FIGURE 39.2 Patient survival rates for LD SPK recipients.

bacterial or fungal infections; posttransplant malignancies, such as posttransplant lymphoproliferative disorder; and, rarely, graft-versus-host disease. The diagnosis and management of these complications is similar in pancreas and other solid-organ transplant recipients.

39.5.4 Outcome

By December 31, 2011, a total of 156 segmental pancreas transplants using LDs had been reported to the IPTR from six countries. Most reports came from the United States (88%). Of the LD transplants in the United States, 124 (86%) were done at the University of Minnesota [9].

The first LD pancreas transplant was in the PAK category, done at the University of Minnesota on June 20, 1979. The first LD PTA was done at Huddinge Hospital, Stockholm, Sweden, on March 18, 1980; the first LD SPK transplant, at Jackson Memorial Hospital, Miami, FL, on November 12, 1980 [56].

The short- and long-term *patient* survival rates for LD pancreas recipients are high: for PAK recipients, 92% at 5 years, 92% at 10 years, and 79% at 20 years; for PTA recipients, 92% at 5 years and 86% at 10 years; and for SPK recipients, 94% at 5 years and 81% at 10 years [9]. Because LD pancreas transplants in the SPK category were not performed in higher numbers until the second half of the 1990s, follow-up time is shorter than for the PTA and PAK categories.

Pancreas *graft* survival rates, during the 33 years that LD pancreas transplants have been performed, have significantly improved, largely thanks to two developments: (i) the introduction of new immunosuppressants (TAC and MMF) in the 1990s and (ii) the aggressive use of anticoagulation. In the past decade, 1-year graft survival rates have ranged from 51% for solitary transplant recipients (PAK, PTA) to 89% for SPK recipients (the 1-year kidney graft survival rate in the SPK category is 99%) [9].

In all, two LD pancreas recipients have had functioning grafts for >20 years; eight for >15 years; 41 for >10 years; and 63 for >5 years [9].

Since the first successful LD SPK transplant in 1994, short- and long-term patient (Figure 39.2), pancreas graft (Figure 39.3), and kidney graft (Figure 39.4) survival rates have become excellent.

39.6 THE IDENTICAL-TWIN EXPERIENCE

The autoimmune cause of diabetes mellitus is based on several independent observations, including (i) the presence of a lymphocytic infiltrate in the islets ("isletitis"), (ii) the appearance of a series of autoantibodies, coupled with a progressive loss of insulin secretion, (iii) the specificity of pancreatic β-cell destruction, and (iv) the recurrence of type 1 diabetes mellitus in identical-twin pancreas recipients in the absence of immunosuppressive therapy [57,58].

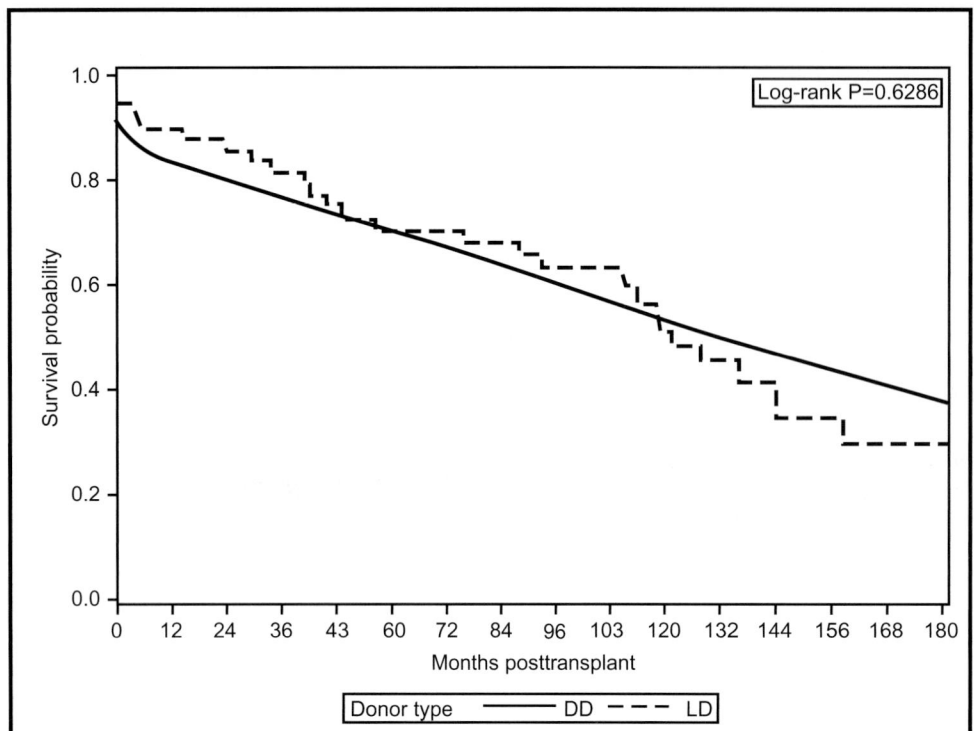

FIGURE 39.3 Pancreas graft survival rates for LD SPK recipients.

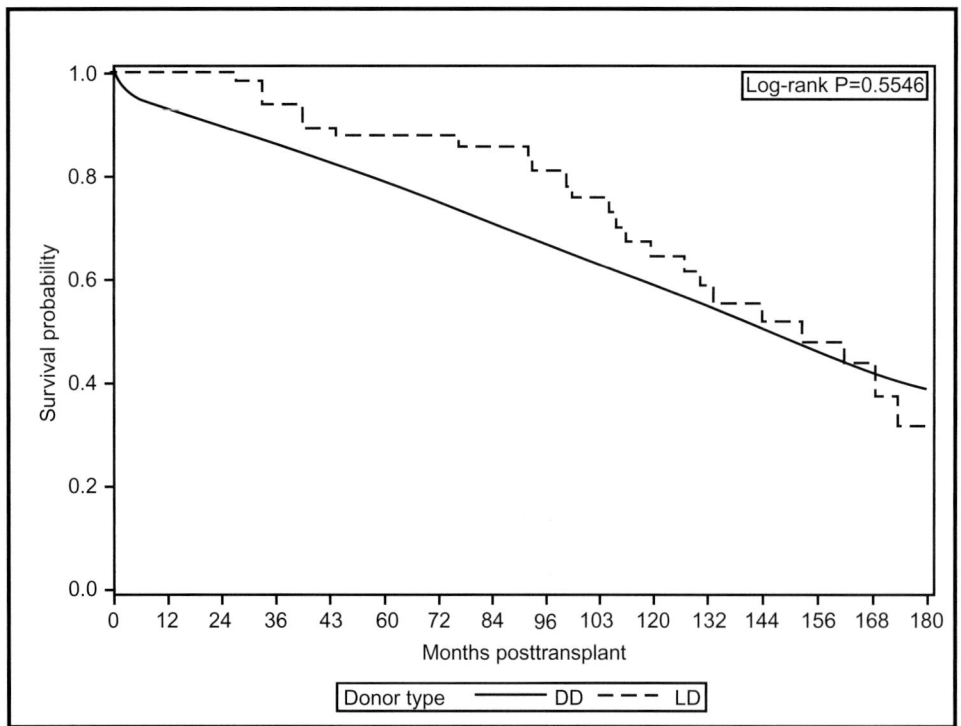

FIGURE 39.4 Kidney graft survival rates for LD SPK recipients.

The University of Minnesota experience is unique in the world: nine pancreas transplants between monozygous twins. Of those transplants, seven were technically successful. The first four were done in the pre-cyclosporine (pre-CSA) era (October 1, 1980, through July 31, 1983) [59].

The first three such recipients were not given any induction or maintenance immunosuppression. Each of them demonstrated normal glucose metabolism early posttransplant, but remained insulin-independent for only 5–12 weeks. In Recipient 1, once progressive hyperglycemia was diagnosed, no attempt was made at graft salvage. In Recipient 2, AZA was started 6 weeks posttransplant, but in the absence of clinical improvement, the drug was stopped 6 weeks later. In Recipient 3, antirejection treatment was started with Minnesota antilymphocyte globulin (ALG), and AZA was given, but efforts were discontinued in the absence of clinical improvement [59,60].

In those three recipients, pancreas graft biopsies at the time of the decline in graft function revealed mononuclear cell infiltrates in the islets consisting of T11 (pan T), OKT8 (suppressor/killer), OKT9 (transferrin receptor), OKT10 (activated), and HLA-DR-reactive mononuclear cells, as well as 63D3 and OKM1 reactive monocytes [61]. Further immunohistopathologic analysis showed that the isletitis was mostly constituted by CD8$^+$/T-lymphocyte receptor α, β (TCR$_{\alpha\beta}^+$) T lymphocytes surrounding and infiltrating the affected islets. CD4$^-$/CD8$^-$/TCR$_{\gamma\delta}^+$ T lymphocytes were observed within the islets [62].

In two of those three recipients, pancreas graft biopsies obtained after loss of graft function revealed resolution of the inflammatory process and selective destruction of all β cells, but no infiltrate was noted in islets without β cells. Because no immunohistologic evidence of a humorally mediated immune reaction was seen in any of the biopsies, selected β-cell destruction was deemed a consequence of β-cell-mediated immunity leading to recurrent diabetes mellitus [60,62].

Recipient 4 was prophylactically given AZA posttransplant. A biopsy at 6 weeks posttransplant, however, showed mild isletitis without β-cell destruction. At 36 months posttransplant, mild hyperglycemia developed. A biopsy showed resolution of isletitis, but destruction of β cells in 70% of the islets. CSA was added to the immunosuppressive regimen. The recipient temporarily required only a relatively low amount of exogenous insulin, but 5 years posttransplant, became fully insulin-dependent [52,54].

The last three identical-twin pancreas transplants at the University of Minnesota were done from May 1, 1987, through September 30, 1990. Of those three recipients, two were given induction therapy with Minnesota ALG; all three were maintained on CSA-based maintenance therapy. Of note, two have remained normoglycemic for over 15 years. The third recipient, on lower-dose immunosuppressive therapy, developed biopsy-proven isletitis at 1 year posttransplant and became fully insulindependent at 8 years posttransplant, despite intermittent anti-T-cell therapy.

This unique identical-twin transplant experience lends strong clinical evidence to diabetes mellitus being an autoimmune disease. It also demonstrates that disease recurrence in the absence of immunosuppressive therapy can occur as early as several weeks posttransplant and that low-dose immunosuppressive therapy cannot prevent disease recurrence. Furthermore, immunosuppression given late (that is, months or years after the clinical onset of diabetes mellitus) does not allow β-cell regeneration. The fact that immunosuppression prevents recurrence of disease in pancreas transplant recipients means that it should also prevent progression of disease in patients with *de novo* autoimmune diabetes mellitus if applied early enough. Such a study, preferably in children with a genetically high risk of developing diabetes mellitus, has not been done, because of concern over immunosuppressive side effects.

AUTHOR CONTRIBUTIONS

Rainer Gruessner, MD, University of Arizona, wrote the manuscript and performed data analysis. Angelika Gruessner, PhD, University of Arizona, researched data and coauthored the manuscript. David Sutherland, MD, PhD, University of Minnesota, also coauthored the manuscript.

The authors have no relevant conflicts of interest to disclose.

ACKNOWLEDGMENTS

The authors would like to thank Mary Knatterud, PhD, University of Arizona Department of Surgery, for her editorial assistance and Jack Roberts, University of Arizona Department of Surgery, for help with preparing the manuscript.

REFERENCES

[1] Murray JE, Merrill JP, Harrison JH. Renal homotransplantation in identical twins. Surg Forum 1955;6:432–6.

[2] Raia S, Nery JR, Mies S. Liver transplantation from live donors. Lancet 1989;2:497.

[3] Broelsch CE, Whitington PF, Emond JC, Heffron TG, Thistlethwaite JR, Stevens L, et al. Liver transplantation in children from living related donors: surgical techniques and results. Ann Surg 1991;214:428–39.

[4] Starnes VA, Lewiston NJ, Luikart H, Theodore J, Stinson EB, Shumway NE. Current trends in lung transplantation: lobar transplantation and expanded use of single lungs. J Thorac Cardiovasc Surg 1992;104:1060–5.

[5] Starnes V, Barr M, Cohen R. Lobar transplantation: indications, technique, and outcome. J Thorac Cardiovasc Surg 1994;108:403–11.

[6] Fortner JG, Sichuk G, Litwin SD, Beattie Jr. EJ. Immunological responses to an intestinal allograft with HL-A-identical donor-recipient. Transplantation 1972;14:531–5.

[7] Gruessner RWG, Sharp HL. Living related intestinal transplantation—first report of a standardized surgical technique. Transplantation 1997;64:1605–7.

[8] Sutherland DER, Goetz FC, Najarian JS. Living-related donor segmental pancreatectomy for transplantation. Transplant Proc 1980;12:19—25.

[9] Gruessner A. International Pancreas Transplant Registry (IPTR) analysis. Personal communication, October 20, 2012.

[10] White SA, Shaw JA, Sutherland DER. Pancreas transplantation. Lancet 2009;373:1808—17.

[11] Luzi L. Pancreas transplantation and diabetic complications. N Engl J Med 1998;339:115—7.

[12] Nathan DM, Fogel H, Norman D, Russell PS, Tolkoff-Rubin N, Delmonico FL, et al. Long-term metabolic and quality of life results with pancreatic/renal transplantation in insulin-dependent diabetes mellitus. Transplantation 1991;52:85—91.

[13] Diabetes Control and Complications Trial Research Group. The effect of intensive treatment of diabetes on the development and progression of long-term complications in insulin-dependent diabetes mellitus. N Engl J Med 1993;329:977—86.

[14] Diabetes Control and Complications Trial/Epidemiology of Diabetes Interventions and Complications Research Group. Retinopathy and nephropathy in patients with type 1 diabetes four years after a trial of intensive therapy. N Engl J Med 2000;342:381—9.

[15] Gruessner RWG. Living donor pancreas transplantation. In: Gruessner RWG, Sutherland DER, editors. Transplantation of the pancreas. New York: Springer-Verlag; 2004. p. 423—40 [chapter 14].

[16] Sutherland DER, Najarian JS, Gruessner RWG. History of living donor pancreas transplantation. In: Gruessner RWG, Benedetti E, editors. Living donor organ transplantation. New York: McGraw-Hill; 2008. p. 369—83 [chapter 18].

[17] Gruessner RWG, Najarian JS, Gruessner AC, Sutherland DER. Pancreas transplants from living related donors. In: Touraine JL, Traeger J, Bétuel H, et al., editors. Organ shortage—the solutions. Dordrecht, the Netherlands: Kluwer Academic; 1995. p. 77—83.

[18] Sutherland DE, Gruessner R, Dunn D, Moudry-Munns K, Gruessner A, Najarian JS. Pancreas transplants from living-related donors. Transplant Proc 1994;26:443—5.

[19] Gruessner RWG. Rationale for living donor pancreas transplants. In: Gruessner RWG, Benedetti E, editors. Living donor organ transplantation. New York: McGraw-Hill; 2008. p. 383—4 [chapter 18].

[20] Gruessner RWG, Sutherland DER, Gruessner AC. Mortality assessment for pancreas transplants. J Transplant 2004;4:2018—26.

[21] Gruessner AC, Sutherland DER, Dunn DL, Najarian JS, Humar A, Kandaswamy R, et al. Pancreas after kidney transplants in posturemic patients with type 1 diabetes mellitus. J Am Soc Nephrol 2001;12:2490—9.

[22] Gruessner RWG, Kendall DM, Drangstveit MB, Gruessner AC, Sutherland DE. Simultaneous pancreas—kidney transplantation from live donors. Ann Surg 1977;226:471—82.

[23] Benedetti E, Dunn T, Massad MG, Raofi V, Bartholomew A, Gruessner RWG, et al. Successful living related simultaneous pancreas—kidney transplant between identical twins. Transplantation 1999;67:915—8.

[24] Gruessner RWG, Sutherland DER. Simultaneous kidney and segmental pancreas transplants from living related donors—the first two successful cases. Transplantation 1996;61:1265—8.

[25] Sutherland DER, Radosevich D, Gruessner RWG, Gruessner AC, Kandaswamy R. Pushing the envelope: living donor pancreas transplantation. Curr Opin Organ Transplant 2012;17:106—15.

[26] Seaquist ER, Gruessner RWG. Pancreas transplantation: the donor. Selection and workup. In: Gruessner RWG, Benedetti E, editors. Living donor organ transplantation. New York: McGraw-Hill; 2008. p. 385—9 [chapter 19].

[27] Sutherland DER, Ascher NL. Distal pancreas donation from a living relative. In: Simmons RL, Finch ME, Ascher NL, Najarian JS, editors. Manual of vascular access, organ donation, and transplantation. New York: Springer-Verlag; 1984. p. 153—64.

[28] Gruessner RWG. Laparoscopic donor distal pancreatectomy. In: Gruessner RWG, Benedetti E, editors. Living donor organ transplantation. New York: McGraw-Hill; 2008. p. 393—8 [chapter 19].

[29] Gruessner RWG, Kandaswamy R, Denny R. Laparoscopic simultaneous nephrectomy and distal pancreatectomy from a live donor. J Am Coll Surg 2001;193:333—7.

[30] Tan M, Kandaswamy R, Sutherland DER, Gruessner RWG. Laparoscopic donor distal pancreatectomy for living donor pancreas and pancreas—kidney transplantation. Am J Trans 2005;5:1966—70.

[31] Oberholzer J, Avila JG, Benedetti E. Standard open distal pancreatectomy. In: Gruessner RWG, Benedetti E, editors. Living donor organ transplantation. New York: McGraw-Hill; 2008. p. 389—93 [chapter 19].

[32] Sutherland DER, Gruessner RWG, Dunn EDL, Matas AJ, Humar A, Kandaswamy R, et al. Lessons learned from more than 1000 pancreas transplants at a single institution. Ann Surg 2001;233:463—501.

[33] Gruessner RWG. Morbidity, mortality, and long-term outcome. In: Gruessner RWG, Benedetti E, editors. Living donor organ transplantation. New York: McGraw-Hill; 2008. p. 398—400 [chapter 19].

[34] Reynoso JF, Gruessner CE, Sutherland DER, Gruessner RWG. Short and long-term outcome for living pancreas donors. J Hepatobiliary Pancreat Sci 2010;17:92—6.

[35] Kendall DM, Sutherland DER, Najarian JS, Goetz FC, Robertson RP. Effects of hemipancreatectomy on insulin secretion and glucose tolerance in healthy humans. N Engl J Med 1990;322:898—903.

[36] Robertson RP. Endocrine function and metabolic outcomes in pancreas and islet transplantation. In: Gruessner RWG, Sutherland DER, editors. Transplantation of the pancreas. New York: Springer-Verlag; 2004. p. 441—54 [chapter 15].

[37] Seaquist ER, Robertson RP. Effects of hemipancreatectomy on pancreatic alpha and beta cell function in healthy human donors. J Clin Invest 1992;89:1761—6.

[38] Seaquist ER, Pyzdrowski K, Moran A, Teuscher AU, Robertson RP. Insulin-mediated and glucose-mediated glucose uptake following hemipancreatectomy in healthy human donors. Diabetologia 1994;37:1036—43.

[39] Seaquist ER, Kahn SE, Clark PM, Hales CN, Porte Jr. D, Robertson RP. Hyperproinsulinemia is associated with increased β cell demand after hemipancreatectomy in humans. J Clin Invest 1996;97:455—60.

[40] Robertson RP, Lanz KJ, Sutherland DER, Seaquist ER. Relationship between diabetes and obesity 9 to 18 years after

hemipancreatectomy and transplantation in donors and recipients. Transplantation 2002;73:736–41.

[41] Robertson RP, Sutherland DER, Seaquist ER, Lanz KL. Glucagon, catecholamine, and symptom responses to hypoglycemia in living donors of pancreas segments. Diabetes 2003;52:1689–94.

[42] Gruessner RWG, Sutherland DER, Drangstveit MB, Bland BJ, Gruessner AC. Pancreas transplants from living donors: short- and long-term outcome. Transplant Proc 2001;33:819–20.

[43] Sutherland D, Najarian J, Gruessner R. Living donor pancreas transplantation. In: Hakim NS, Canelo R, Papalois V, editors. Living related transplantation. London: Imperial College Press; 2010. p. 95–117.

[44] Kandaswamy R. Pancreas transplantation: the recipient. Selection and workup. In: Gruessner RWG, Benedetti E, editors. Living donor organ transplantation. New York: McGraw-Hill; 2008. p. 401–3 [chapter 20].

[45] Gruessner RWG. Pancreas transplantation: the recipient. Surgical procedures. In: Gruessner RWG, Benedetti E, editors. Living donor organ transplantation. New York: McGraw-Hill; 2008. p. 403–8 [chapter 20].

[46] Diem P, Abid M, Redmon JB, Sutherland DE, Robertson RP. Systemic venous drainage of pancreas allografts as independent cause of hyperinsulinemia in type 1 diabetic recipients. Diabetes 1990;39:534–40.

[47] Hughes TA, Gaber AO, Amiri HS, Wang X, Elmer DS, Winsett RP, et al. Kidney-pancreas transplantation. The effect of portal versus systemic venous drainage of the pancreas on the lipoprotein composition. Transplantation 1995;60:1406–12.

[48] Stratta RJ, Shokouh-Amiri MH, Egidi MF, Grewal HP, Kizilisik AT, Nezakatgoo N, et al. A prospective comparison of simultaneous kidney–pancreas transplantation with systemic-enteric versus portal-enteric drainage. Ann Surg. 2001;233:740–51.

[49] Cattral MS, Bigam DL, Hemming AW, Carpentier A, Greig PD, Wright E, et al. Portal venous and enteric exocrine drainage versus systemic venous and bladder exocrine drainage of pancreas grafts: clinical outcome of 40 consecutive transplant recipients. Ann Surg 2000;232:688–95.

[50] Sutherland DER, Goetz FC, Najarian JS. Surgery and possible complications in the living donor for the pancreas transplant, short and long term. In: Touraine JL, editor. Transplantation and clinical Immunology XVI. Amsterdam: Elsevier Science Publishers B. V. 1985. p. 7–15.

[51] Leone JP, Christensen K. Postoperative management: uncomplicated course. In: Gruessner RWG, Sutherland DER, editors. Transplantation of the pancreas. New York: Springer-Verlag; 2004. p. 179–90 [chapter 9].

[52] Gruessner RWG, Sutherland DER. Clinical diagnosis in pancreas allograft rejection. In: Solez K, Racusen LC, Billingham ME, editors. Solid organ transplant rejection: mechanisms, pathology, and diagnosis. New York: Marcel Dekker; 1996. p. 455–99.

[53] Kandaswamy R. Perioperative care, immunosuppressive therapy, and posttransplant complications. In: Gruessner RWG, Benedetti E, editors. Living donor organ transplantation. New York: McGraw-Hill; 2008. p. 408–12 [chapter 20].

[54] Kandaswamy R, Humar A, Gruessner AC, Harmon JV, Granger DK, Lynch S, et al. Vascular graft thrombosis after pancreas transplantation: comparison of the FK 506 and cyclosporine eras. Transplant Proc 1999;31:602–3.

[55] Kaplan AJ, Valente JF, First MR, Demmy AM, Munda R. Early operative intervention for urologic complications of kidney-pancreas transplantation. World J Surg 1998;22:890–4.

[56] Sutherland DER. Pancreas and islet transplantation. II. Clinical trials. Diabetologia 1981;20:435–50.

[57] Gruessner RWG. The identical twin transplant experience: recurrence of disease. In: Gruessner RWG, Benedetti E, editors. Living donor organ transplantation. New York: McGraw-Hill; 2008. p. 412–3 [chapter 20].

[58] Gruessner RWG. Immunology in pancreas transplantation—autoimmune reactivation. In: Gruessner RWG, Sutherland DER, editors. Transplantation of the pancreas. New York: Springer-Verlag; 2004. p. 393–7.

[59] Sutherland DER, Sibley RK, Xu XZ, Michael A, Srikanta AM, Taub F, et al. Twin-to-twin pancreas transplantation: reversal and reenactment of the pathogenesis of type 1 diabetes. Trans Assoc Am Phys 1984;97:80–7.

[60] Sutherland DER, Goetz FC, Sibley RK. Recurrence of disease in pancreas transplants. Diabetes 1989;38:85–7.

[61] Sibley RK, Sutherland DER, Goetz F, Michael AF. Recurrent diabetes mellitus in the pancreas iso- and allograft. A light and electron microscopic and immunohistochemical analysis of four cases. Lab Invest 1985;53:132–44.

[62] Santamaria P, Nakhleh RE, Sutherland DER, Barbosa JJ. Characterization of T lymphocytes infiltrating human pancreas allograft affected by isletitis and recurrent diabetes. Diabetes 1992;41:53–61.

Chapter 40

Current Status of Islet Transplantation

Paolo Cravedi[a], Piero Ruggenenti[a,b], Andrea Remuzzi[a], and Giuseppe Remuzzi[a,b]

[a]Mario Negri Institute for Pharmacological Research, Bergamo, Italy, [b]Unit of Nephrology, Azienda Ospedaliera Ospedali Riuniti di Bergamo, Bergamo, Italy

Chapter Outline

40.1 INTRODUCTION

Type 1 diabetes is the clinical consequence of immune-mediated destruction of insulin-producing pancreatic β cells. Hence from its first description, the main goal of treatment has been the identification of strategies to replace insulin deficiency, such as exogenous insulin and, later on, pancreas transplantation.

More recently, islet transplantation has been proposed as an alternative, relatively safer way to replace β-cell function. However, despite the extreme interest of this supposedly safer and more easily tolerated procedure compared to whole pancreas transplantation, results have been unsatisfactory for many years. In 2000, research in the field of pancreatic islet cell transplantation was boosted by a key paper reporting insulin independence in seven out of seven patients with type 1 diabetes mellitus over a median follow-up of 12 months [1]. The two major novelties of this protocol were the administration of increased doses of pancreatic islets taken from at least two pancreas donors, and an immunosuppressive protocol devoid of steroids. Until then, clinical outcomes had been disappointing. Of the 267 islet preparations transplanted since 1990, <10%

had resulted in insulin independence for more than 1 year [2]. With the new protocol, success rates have increased in parallel with significant improvements in the technical procedure and medical management of islet transplantation. However, true insulin independence rates for a prolonged period of time are still very low, and patients are required to take immunosuppressive medication as long as there is evidence of remaining graft function. By and large, despite the high number of islet infused and the avoidance of steroids, long-term results of the Edmonton protocol are far less successful than what is expected from short-term outcomes.

Moreover, islet transplantation remains a very complex procedure, whose planning and execution require dedicated facilities, highly trained personnel, and large economical investments. Another issue concerns the relative inadequacy of cadaveric donor organ availability, whose requirements are particularly stringent for islets transplantation because of the persistent variability of islet cell yield per organ. Thus, in spite of the initial enthusiasm, high costs and unsatisfactory outcomes make island transplantation a less than ideal treatment for type 1 diabetes.

40.2 THE BURDEN OF TYPE 1 DIABETES MELLITUS

Type 1 diabetes is the most common metabolic disease in childhood with incidence rates ranging from 8 to >50 per 100,000 population per year in Western countries [3]. For children aged 0–14 years, the prevalence of type 1 diabetes is estimated to be at least one million worldwide by the year 2025 [4]. Great variability, however, exists in the prevalence of type 1 diabetes across different countries. According to the Multinational Project for Childhood Diabetes (DIAMOND) project [5], a >350-fold difference in the incidence of type 1 diabetes among the 100 populations worldwide is present with age-adjusted incidences ranging from a low of 0.1/100,000 per year in China and Venezuela to a high of 36.5/100,000 in Finland and 36.8/100,000 per year in Sardinia. Recent forecast analyses predict that the number of patients with type 1 diabetes will triplicate by the year 2050 [6], a trend that is shared by most countries worldwide [7].

Children with type 1 diabetes usually present with a several day history of typical symptoms, such as frequent urination, excessive thirst, and weight loss, which appear when about 80% of the pancreatic β cells are already destroyed. If those symptoms are misinterpreted, progressive insulin deficiency leads to a potentially life-threatening condition in the form of diabetic ketoacidosis. Over time, micro- and macrovascular complications also occur, especially if glycemic control is suboptimal. The relative risks for cardiovascular disease and total mortality associated with type 1 diabetes have declined relative to earlier studies, but type 1 diabetes continues to be associated with higher cardiovascular and death rates than the nondiabetic population. For example at the attained age of 60–69 years, there are approximately three extra deaths per 100 per year in men (28.51/1000 person years at risk) and two per 100 per year for women (17.99/1000 person years at risk) with type 1 diabetes [8].

40.3 PATHOPHYSIOLOGY OF TYPE 1 DIABETES MELLITUS

Pancreatic tissue is composed of two cell types: acinar cells that excrete digestive enzymes into pancreatic ducts (exocrine function) and the cells contained in the islets of Langerhans that release various hormones into the blood (endocrine function). The islets of Langerhans are composed of α cells secreting glucagon, β cells secreting insulin, δ cells secreting somatostatin, and PP cells secreting pancreatic polypeptide. Pancreatic islets in type 1 diabetic patients show insulitis, which is characterized by the infiltration of predominantly CD8 T lymphocytes [9]. Although many individuals may have autoreactive T cells

specific for β-cell autoantigens, only a selected number of people develop type 1 diabetes, probably because among subjects with autoreactive T cells only those with permissive combinations of predisposing genetic and environmental factors will eventually manifest the diabetic phenotype. The most important genes associated with an increased risk of type 1 diabetes are those located within the major histocompatibility complex human leukocyte antigen (HLA) class II region. Of the non-HLA associated genes involved in type 1 diabetes pathogenesis, the insulin gene confers the highest risk [10]. Environmental factors triggering the onset of the disease are thought to be infectious agents, dietary factors, and environmental toxins, although no unique causal factor has consistently been identified [11,12].

The humoral response may also play a role in the destruction of β cells in type 1 diabetes, and this may be especially important during the first year after the appearance of autoantibodies [9]. Autoantibodies that have been implicated in the development of type 1 diabetes target insulin, glutamic acid decarboxylase (GAD, an enzyme produced primarily by islet cells), and the transmembrane protein tyrosine kinase IA-2. Presence of a single autoantibody usually does not predict progression to overt type 1 diabetes, but combined positivity confers a significantly increased risk [13]. Recently, a new autoantigen was detected in the form of the zinc transporter Slc30A8. The presence of antibodies against the transporter improves the accuracy with which future occurrence of type 1 diabetes can be predicted [14]. On the other hand, the pathogenic role of autoantibodies in type 1 diabetes is debated and may depend on a variety of host factors [15]. Indeed, it has been shown that type 1 diabetes can occur in the absence of a functional B-cell compartment and thus those autoantibodies are not a prerequisite for the inception of the disease [16]. In islet transplant patients, the presence of autoreactive T cells at the time of transplantation has been associated with graft failure, whereas the impact of autoantibodies on graft outcome has not been consistently found [17–19]. Importantly, a recent report described recurrence of autoimmunity on a pancreatic graft in a patient with type 1 diabetes even in the absence of a sizeable increase in GAD and/or IA-2 autoantibody titers [20]. This suggests that autoantibodies may participate in diabetes pathogenesis, but are not a cardinal feature of this condition.

Whether human β cells have the potential to regenerate has been a matter of lively debate between diabetologists. This is an issue with major clinical implications since hyperglycemia of type 1 diabetes is caused by progressive autoimmune disruption of insulin-producing β cells of pancreatic islets. Thus, provided the autoimmune process is inhibited, the possibility to induce surviving cells to proliferate and replenish the β-cell compartment might allow restoring normal glucose homeostasis and achieving freedom from exogenous insulin dependency for millions of type 1 diabetics worldwide. In experimental animals mature β cells can proliferate [21] but, so far, failure to demonstrate that reproduction is possible also in the diabetic "milieu," has challenged the possibility to apply this approach in clinical practice and cooled down diabetologists' enthusiasm on this line of research. This issue has been elegantly addressed by Nir et al. [22] who—in genetically modified mice made diabetic by inducing 80% β-cell ablation via endogenous diphtheria toxin production—showed that cell mass and glucose homeostasis can be fully restored in a few weeks after exhaustion of toxin production. Finding that this was accompanied by a dramatically increased proliferation rate even in the face of severe hyperglycemia challenged the common belief that glucotoxicity is a major impediment to β-cell survival.

40.4 STANDARD MANAGEMENT OF PATIENTS WITH TYPE 1 DIABETES

Glycemic control is the cornerstone of diabetes care, which requires a meticulous balance of insulin replacement with diet and exercise. In 1993, the Diabetes Control and Complications Trial (DCCT) showed that a system of intensive diabetes management aimed at near-normal glycemic control dramatically reduces the risk of microvascular complications and favorably affects the risk of macrovascular complications as compared to a less strict control approach [23–25]. However, the treatment regimens used by subjects randomized to the intensive treatment arm of the DCCT also significantly increased their risk of severe hypoglycemia and led to more weight gain [26].

Since the publication of the DCCT results, a variety of insulin analogs, better and more sophisticated insulin pumps, and faster and more accurate glucose meters have become widely used in the treatment of type 1 diabetes, making the prospects for patients with type 1 diabetes far better than they were in the past [27,28].

In most type 1 diabetes patients, however, the goal of near-normalization of glycated hemoglobin (HbA1c, a parameter of glucose control over the last 3 months) remains elusive. Several large, multicenter studies demonstrated a persistent gap between attained and target HbA1c levels. Successful implementation of intensive diabetes management in routine clinical practice continues to be a major challenge. The unremitting daily task of controlling blood glucose while avoiding hypoglycemia is arduous and often frustrating. Recent meta-analyses show that the use of insulin analogs and pump therapy, when compared with conventional insulins and injection-based regimens, respectively, have had only a modest impact on glycemic control and rates of adverse events.

Glycemic control is particularly challenging in adolescent patients. In the DCCT the mean HbA1c for adolescents as compared to adults was 1−2% higher in both the intensive and conventionally treated arms. Despite this, rates of hypoglycemia were higher in adolescents than in adults [29]. Studies published after DCCT have shown that mean levels of HbA1c have remained higher than current glycemic goals [30]. Management of type 1 diabetes requires many lifelong daily tasks that the child and/or family must perform to maintain a relatively healthy metabolism and glycemic control. Although in younger children these tasks are performed primarily by the care givers, in the teenage years the burden of diabetes management falls on the adolescents themselves. These patients more than others also require considerable psychosocial support, ongoing education, and guidance from a cohesive diabetes team working with each patient to set and achieve individualized treatment goals.

40.5 INDICATIONS TO ISLET TRANSPLANTATION

A minority of patients with type 1 diabetes have a difficult glycemic control and are prone to experiencing unaware hypoglycemic episodes that can threaten their lives. These patients are generally thought to have an indication for whole pancreas or pancreatic islet transplantation, a procedure that may lead to an improvement in the quality of life or they may even be saved from fatal hypoglycemia when provided with functionally active β cells [31]. In addition, whole pancreas or pancreatic islet transplantation may be considered in patients with severe clinical and emotional problems with exogenous insulin therapy [32].

The Edmonton group has proposed two scores to quantify the severity of labile diabetes. The HYPO score quantifies the extent of the problem of hypoglycemia by assigning scores to capillary glucose readings from a 4-week observation period in combination with a score for self-reported hypoglycemic episodes in the previous year. The lability index (LI) quantifies the extent of glucose excursions over time and is calculated using the formula proposed by this group [33].

The 2006 American Diabetes Association (ADA) guidelines acknowledged the advantages of islet transplantation over whole pancreas transplantation in terms of morbidity and mortality associated with the operative procedure. However, they clearly stated that islet transplantation is an experimental procedure, only to be performed in the setting of controlled research studies. As for type 1 diabetes patients who will also be receiving a kidney transplantation, simultaneous pancreas transplantation, rather than islet transplantation, is the treatment of choice, because it may improve kidney survival and will provide insulin independence in the majority of patients [32].

40.6 THE ORIGINS OF ISLET TRANSPLANTATION

The first evidence that islet transplantation could represent a cure for type 1 diabetes was published in 1972, when experiments in rodents showed that artificially induced diabetes mellitus could be reversed by transplanted pancreatic islets [34]. The advantage of islet transplantation over whole pancreas transplantation is avoidance of the major surgery needed and the theoretical possibility to manipulate these cells to make them more easily tolerized by the host immune system.

In the 90s, research activity into islet transplantation greatly increased. Clinical success rates, however, were generally low, with <10% of patients being insulin independent at 1 year after transplantation. More encouraging results were obtained in patients who already had a kidney transplant, with higher rates of insulin independency and graft function as defined by C-peptide secretion [35,36]. In 2000, a report by the Edmonton group was published describing seven type 1 diabetics with a history of severe hypoglycemia and poor metabolic control who underwent islet transplantation using a modified, steroid-free immunosuppressive protocol. In addition, each patient received at least two different islet transplantations, thus the total transplanted islet mass per patient was remarkably higher than in previous series. Over a median follow-up of 11.9 months (range 4.4–14.9), all patients were insulin free [1]. The so-called Edmonton protocol was subsequently adopted and modified by many centers and is now still the most widely adopted strategy worldwide.

40.7 CLINICAL OUTCOMES OF ISLET TRANSPLANTATION

40.7.1 Insulin Independence and Improved Glycemic Control

Many centers have published the results of their islet transplant programs [37–49]. In 2005, the Edmonton group reported 5-year results of the first 65 islet transplant patients treated according to their protocol. According to this report, 44 (68%) transplanted patients had become insulin independent, with a median duration of insulin independency of 15 months (IQR 6.2–25.5). Five of these patients received only a single islet infusion, 33 received two infusions, and six received three infusions. Insulin independency after 5 years was 10%. Nonetheless, after 5 years, some residual graft function

could be demonstrated in about 80% of patients on the basis of detectable serum C-peptide levels. Diabetic lability and the occurrence of severe hypoglycemia were effectively diminished [50].

Following the initial Edmonton results in 2000, a large international trial in nine centers in the United States and Europe was initiated by the Immune Tolerance Network to examine the feasibility and reproducibility of islet transplantation using the Edmonton protocol. The primary endpoint, defined as insulin independency with adequate glycemic control 1 year after the final transplantation, was met by 16 out of 36 subjects (44%). Only five of these patients were still insulin independent after 2 years (14%). Of note, the considerable differences in results obtained by the various participating sites emphasize the difficulties related to the procedure and suggests the need for concentrating this procedure in highly experienced centers. Again, graft function as defined by detectable C-peptide levels and associated improvements in diabetic control were preserved in a higher percentage of patients (70% after 2 years) [51].

The Groupe de Recherche Rhin Rhone Alpes Geneve pour la transplantation d'Ilots de Langerhans (GRAGIL) reported results obtained in 15 patients who received one ($n = 8$) or two islet infusions ($n = 7$). Median insulin independence duration was 4.7 (3.1–15.2) months after one infusion versus 19 (9.6–20.8) months after two infusions (not significant). Only 37.5% of single-graft patients had a β score ≥ 4 compared with 100% of double-graft patients, further supporting the idea that islet mass represents a major determinant for transplant success [52].

A report from the Japanese Trial of Islet Transplantation showed that only three out of 18 recipients of islet transplantation achieved insulin independency and only for a period of 2 weeks to 6 months. Graft function was preserved in 63% after 2 years. As in the other reports, HbA1c levels decreased and blood glucose levels stabilized with disappearance of hypoglycemia unawareness. In this report, no information was provided about the amount of islet equivalents (IEQ; number of islets in a preparation adjusted for size of the islet, 1 IEQ equals a single islet of 150 μm in diameter) per kg body weight infused. Of note, in Japan all pancreata are obtained from nonheart-beating donors, since pancreata from brain dead donors are usually allocated to whole pancreas or pancreas/kidney transplantation. In addition, the presence of brain death is frequently not examined because of cultural reasons, and invasive procedures are usually not allowed even in brain dead donors before cardiac arrest occurs. This may lead to decreased viability of pancreatic tissue when compared with pancreata from brain dead donors [53].

The largest registry of islet transplant data is the Collaborative Islet Transplant Registry (CITR), which retrieves its data mainly from US and Canadian medical institutions and two European centers. In their 2008 update considering 279 recipients of an islet transplantation reported between 1999 and 2007, the registry reported 24% insulin independence after 3 years. Graft function as defined by detectable C-peptide levels after 3 years was 23–26%. The prevalence of hypoglycemic events decreased dramatically and mean HbA1c levels substantially improved. Predictors of better islet graft function were: higher number of islet infusions, greater number of total IEQ infused, older recipient age, lower recipient HbA1c levels, whether the processing center was affiliated with the transplantation center, higher islet viability, larger islet size, and use of daclizumab, etanercept, or calcineurin inhibitors in the immunosuppressive regimens. In-hospital administration of steroids was associated with a negative outcome [54,55].

Recently, the CITR reported the outcome data of 627 islet transplants performed between 1999 and 2010. Insulin independence at 3 years after transplant improved from 27% in patients transplanted between 1999 and 2002 to 44% in patients who received a transplant between 2007 and 2010 (Figure 40.1). This success, however, was probably more the result of a careful selection of recipients (lower serum creatinine and donor-specific antibody titer) than of a real improvement in the transplant procedure. Consistently, the number of islet transplant performed each year did decline during the last era [56].

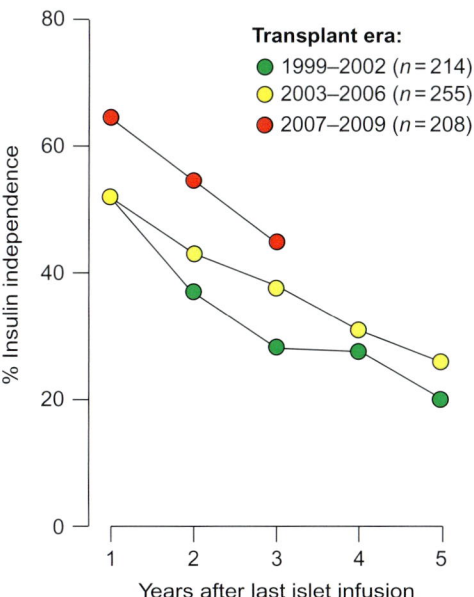

FIGURE 40.1 Rate of insulin independence at different time points after islet infusion according to the transplant era. *Data derived from CITR 2012 [52].*

40.7.2 Long-Term Diabetic Complications

Until now, it has not been sufficiently established whether pancreatic islet transplantation can prevent diabetic complications or halt their progression [57,58].

40.7.2.1 Cardiovascular Complications

In a retrospective study, cardiovascular function was compared between a group of 17 patients who received an islet-after-kidney transplantation and a group of 25 patients with previous kidney transplantation who were still on the waiting list for an islet transplantation or who had experienced early islet graft failure. Baseline characteristics for both groups were similar. Islet transplantation was associated with an improvement in ejection fraction and left ventricular diastolic function compared to baseline. Moreover, arterial intima-media thickness was stable in the islet transplant group, but worsened in the kidney-only group [59]. Recently, data have been published showing that, in a cohort of 15 consecutive islet transplant recipients who reached insulin independence, carotid intima-media thickness did actually decrease after the procedure, suggesting that optimal glycemic control may lead to a regression of atherosclerotic lesions [60].

40.7.2.2 Nephropathy, Retinopathy, and Neuropathy

Increased kidney graft survival rates and stabilization of microalbuminuria have been reported after islet transplantation [61]. Conversely, an uncontrolled observational study by the Edmonton group suggested an overall decline in estimated glomerular filtration rate (GFR) during 4 years of follow-up after islet transplantation alone, and an increase in albuminuria in a significant proportion of patients [62]. Subsequently, Maffi et al. showed that even a mildly decreased renal function pretransplantation should be considered as a contraindication for the currently used immunosuppressive regimen of sirolimus in combination with tacrolimus, since it was associated with progression to end-stage renal disease [47]. In patients with renal impairment, nephrotoxicity of immunosuppressive drugs like calcineurin and mTOR inhibitors might offset the benefits of improved metabolic control. Renal impairment progressively worsens even in those selected patients with type 1 diabetes who benefit from a 5-year normoglycemia period after a single pancreas transplantation, both as a result of immunosuppressive drug toxicity and, probably, of the marginal effect of glycemia control on already damaged diabetic kidney [63,64]. The same applies also to the rare subjects with prolonged normoglycemia after islet transplantation, which limits the indications for the procedure to those type 1 diabetes patients with normal renal function. In this cohort, aggressive treatment of risk factors for nephropathy, such as blood pressure, low-density lipoprotein cholesterol, together with careful tacrolimus level monitoring, have been associated with preserved renal function after islet transplantation, according to a retrospective series of 35 patients [65].

The Edmonton and the Miami series reported ocular problems posttransplantation in 8.5% and 15% of patients, respectively. Adverse events included retinal bleeds, tractional retinal detachment, and central retinal vein occlusion [42,50]. However, after 1−2 years, diabetic retinopathy seems to stabilize [66]. Moreover, at 1 year after transplantation, arterial and venous retinal blood flow velocities are significantly increased, possibly indicating improved retinal microcirculation [67]. The acute adverse effects on retinopathy may be due to the sudden improvement in glycemic control after islet transplantation. The DCCT also reported initial deterioration of diabetic retinopathy in patients with pre-existing disease who were treated in the intensive insulin treatment arm as compared to those in the conventional treatment arm; however, after 1 year differences between treatment arms disappeared, and after 36 months of follow-up, intensive treatment was consistently associated with significantly less progression of diabetic retinopathy [68]. Consistently, a study on 44 patients showed that, at 36 months after transplant, progression of retinopathy was slower in subjects who received islet transplantation compared to those on intensive insulin therapy [69]. More studies are needed to assess whether the overall effect of islet transplantation on diabetic retinopathy is beneficial in the long term.

Reports on the effect of islet transplantation on diabetic neuropathy suggest that the procedure has only marginal effect on this microvascular complication. Lee et al. performed nerve conduction studies in eight patients with at least 1 year of follow-up after transplantation. They concluded that peripheral neuropathy stabilized or maybe even improved, although no formal statistical analysis was provided and conclusions were based on clinical observations by a single neurologist [66]. Del Carro et al. compared nerve conduction studies in patients who had received an islet-after-kidney transplantation to patients having received kidney transplantation only. In their interpretation of the results, they suggested that worsening of diabetic neuropathy seemed to be halted by islet transplantation, but no statistically significant differences between the two groups could be demonstrated [70]. A prospective cohort study compared the progression of microvascular complications between 31 patients who received an islet transplant and 11 who remained on the waiting list. Islet transplantation was associated with lower HbA1c levels and slower progression of retinopathy, whereas neuropathy and GFR were not remarkably affected [71].

By and large, the potential benefits of the islet transplant procedure have been identified only in the minority of patients who reach prolonged insulin independence. Moreover, lack of comparisons with control groups on insulin therapy prevents any consideration on real superiority of islet transplantation. Therefore, large, multicenter, randomized trials are needed to assess the role of islet transplantation in slowing the progression of diabetic complications over conventional supportive therapy.

40.7.3 Adverse Events in Islet Transplantation

Adverse events related to islet transplantation are principally related to the procedure itself and to the adverse effects of the immunosuppressive regimen. During the procedure, a large mass of β cells is percutaneously and transhepatically injected into the portal vein. This may lead to portal vein thrombosis or thrombosis of segmental branches. On the other hand, incidence rates of up to 14% have been reported for intraperitoneal bleeding, which may require blood transfusion or even surgical intervention. This complication can be effectively prevented by sealing the catheter tract using thrombostatic coils and tissue fibrin glue [72]. Other relatively frequent procedure-related complications are abdominal pain from puncturation of the peritoneum or gallbladder, and a transient rise of hepatic enzymes [73]. Posttransplantation focal hepatic steatosis occurs in approximately 20% of patients, possibly due to a local paracrine effect of insulin, but its significance with regard to graft function is not clear yet [74,75]. According to the recent CITR report of islet transplants performed between 1999 and 2010, the incidence of life-threatening events has significantly declined over time. The incidence of any clinically reportable adverse events in year 1 declined from 50−53% in 1999−2006 to 38% in 2007−2010. Peritoneal hemorrhage or gallbladder perforation declined from 5.4% in 1999−2003 to 3.1% in 2007−2010 [56].

Type 1 diabetes patients receiving pancreatic islet transplantation may need an additional kidney and/or whole pancreas transplantation later in life. Therefore, posttransplantation alloimmunization in roughly 10−30% of patients using immunosuppression is a cause for concern [76,77]. Of note, up to 100% of patients develop HLA alloreactivity, with 71% having HLA panel-reactive antibodies (PRA) ≥ 50%, after withdrawal of immunosuppression because of islet graft failure or side effects [76,77]. Pre- or posttransplantation alloreactivity against HLA class I and II may also be associated with reduced pancreatic islet graft survival itself [78,79], although some authors suggested that increased PRA had no clinical significance under adequate immunosuppression [77].

As opposed to solid organ transplantation, pretransplantation testing of PRA is currently not performed in pancreatic islet transplantation. Thus, the impact of PRA positivity on clinical outcome after islet transplantation or on future whole organ transplantation has to be further investigated.

40.8 IMMUNOSUPPRESSIVE REGIMENS FOR ISLET TRANSPLANTATION

Islet transplantation currently requires potent immunosuppression to prevent immune-mediated attrition, and this must be sustained indefinitely if allo- or autoimmune-mediated injury is to be avoided. The goal of immunosuppression is to provide the minimal effective amount for effective and sustained immunologic protection, but without nonimmune-related side effects.

After the Edmonton experience published in 2000 [1], the steroid-free immune-suppressive protocol this group used was adopted by many centers, although it was not the only change being introduced. Changes with regard to recipient and donor selection, the technical procedure, and the infusion of a large number of pancreatic islets from multiple donors will all have contributed to the favorable short-term results. The Edmonton immunosuppressive regimen consists of induction therapy with a monoclonal antibody against the interleukin-2 receptor (daclizumab), and maintenance therapy with a calcineurin inhibitor (tacrolimus) and a mammalian mTOR inhibitor (sirolimus). Sirolimus has been shown to display significant synergy with calcineurin inhibitors, control autoimmunity, induce apoptosis of T cells and other inflammatory cells, and induce generation of regulatory T cells (Tregs). However, data have also emerged showing its potentially harmful effects on β-cell regeneration, similarly to what observed with calcineurin inhibitors [22,80]. In mice with induced diabetes, β-cell proliferation was fully blunted and hyperglycemia persisted when β cells were exposed to circulating concentrations of rapamicine and tacrolimus similar to those achieved and maintained in human recipients of pancreatic islets with the aim of preventing graft rejection. These findings questioned the rationale of the steroid-free sirolimus- and tacrolimus-based antirejection regimens proposed by the Edmonton group in the early 2000s, and subsequently adopted by the large majority of Transplant Centers worldwide, to avoid β-cell steroid toxicity [51].

Moreover, sirolimus and tacrolimus exert direct nephrotoxic effects and they often induce the development of hyperlipidemia and hypertension, which may further increase the risk of micro- and macrovascular complications [81]. Therefore, the combined use of sirolimus and tacrolimus to prevent acute rejection of transplanted pancreatic islets is certainly not ideal.

To increase islet transplantation success rates and diminish the often severe side effects associated with chronic use of immunosuppressive drugs [42], various centers are implementing new immunosuppressive regimens, both for the induction and for the maintenance phase [43,49,82−84].

T-cell-directed induction immunodepletion or modulation is emerging as a valuable adjunct in islet transplantation. Outcomes from the CITR showed that induction therapy with thymoglobulin or alternative non-Fc-binding humanized anti-CD3 mAb teplizumab (hOKT3-ala-ala) results into significantly higher 5-year insulin independence rates [40]. In an attempt to promote a protolerogenic state, Froud et al. tested induction therapy with alemtuzumab in three islet transplant recipients [83]. Alemtuzumab is a humanized monoclonal antibody against CD-52, which is present on the surface of mature lymphocytes. Its administration leads to severe lymphocyte depletion and may favorably influence the Treg versus effector T-cell ratio during T-cell repopulation [85]. Indeed, in these three patients, glucose metabolism seemed to be better than in historic controls, with no major infectious complications. However, other changes in the immunosuppressive regimen, such as the use of steroids on the day before islet infusion, the early switch from tacrolimus to mycophenolate mofetil (MMF) during the maintenance phase, and the use of etanercept, may all have contributed to improved outcomes in this study. In preliminary unpublished studies, the Edmonton group found >50% insulin independence at 5 years, when alemtuzumab was combined with tacrolimus and MMF, but when alemtuzumab was combined only with sirolimus, they found irreversible islet rejection in all cases. Thus, alemtuzumab induction was necessary, but not sufficient on its own in our hands, to prevent islet rejection over time [86].

Costimulation blockade with belatacept (LEA29Y) is currently the subject of a clinical islet transplant trial [86]. Posselt et al. studied the effect of combined induction with thymoglobulin and belatacept, using sirolimus or MMF maintenance in five islet-alone recipients, all of whom achieved single-donor islet engraftment success, with one receiving a late retransplant after return to insulin at 445 days posttransplant. These latter results are very promising and ongoing comparative experience with CIT-04 will help determine whether or not upfront T-cell depletion is needed for costimulation to work [87].

Tumor necrosis factor (TNF)α is a regulator of the immune response, and its activity is inhibited by etanercept, a recombinant TNFα receptor protein. From the University of Minnesota came an interesting report of high success rates in eight patients using a protocol in which etanercept was administered as induction therapy, combined with prednisone, daclizumab, and rabbit antithymocyte globulin. Of the eight patients, five were still insulin independent after 1 year. Of note, patients received an islet graft from a single donor [43]. More centers are now using etanercept as additional induction therapy, a strategy which is supported by the fact that the CITR found an association between etanercept use and graft survival [48,88]. Experimental data showed that anakinra, a competitive antagonist of IL-1, improves etenercept efficacy in preventing islet rejection [89] and is well tolerated in humans [90].

Some studies investigated the combination of etanercept induction with long-term use of subcutaneous exenatide, a glucagon-like peptide-1 (GLP-1) analog. GLP-1 is a hormone derived from the gut, which stimulates insulin secretion, suppresses glucagon secretion, and inhibits gastric emptying [91]. Combined treatment with etanercept and exenatide in addition to the Edmonton immunosuppressive protocol was shown to reduce the number of islets needed to achieve insulin independence [48]. In addition, combined etanercept and exenatide use improved glucose control and graft survival in patients who needed a second transplantation because of progressive graft dysfunction [88]. In two studies with islet transplantation patients, exenatide reduced insulin requirements, although in one study they tended to rise again at the end of the 3-month study period, possibly due to exhaustion of β cells [82,83,92]. However, these studies were very small and nonrandomized. Of note, exenatide needs to be administered subcutaneously twice-a-daily, causes severe nausea, and may lead to hypoglycemia. Therefore, randomized controlled trials are needed to define whether its use confers additional benefit over immunosuppressive therapy alone in islet transplantation recipients [93].

Despite immunosuppressive therapy aimed at preventing rejection (i.e., alloimmunity), outcomes of islet transplantation may also be adversely influenced by autoimmune injury. A recent study showed delayed graft function in patients with pretransplant cellular autoreactivity to β-cell autoantigens; in four out of 10 patients with recurrence of autoreactivity posttransplantation, insulin independence was never achieved. Moreover, in five out of eight patients in whom cellular autoreactivity occurred *de novo* after transplantation, time to insulin independence was prolonged [17]. In the international trial of the Edmonton protocol, patients with one or two autoantibodies in the serum before the final infusion had a significantly lower insulin independence rate than those without autoantibodies [51].

40.8.1 Autologous Islet Transplantation

The concept of autologous islet transplantation after pancreatectomy arose at the University of Minnesota in the late 1970s [94]. Pancreatectomy had already been performed as a treatment for patients with chronic, unrelentingly painful pancreatitis [95]. However, it was seen as an undesirable method, in part because removal of the gland inevitably causes insulin-dependent diabetes. This drawback led to the concept of not disposing of the resected pancreas, but using it as a source of islet tissue that could be used for autologous transplant. In contrast to alloimmune islet transplantation in patients with type 1 diabetes mellitus, no immunosuppression is needed for patients receiving autologous islet infusion, due to the autologous source of the islets and lack of pre-existing autoimmune reactivity.

Sutherland et al. first reported autologous islet transplantation after pancreatectomy in 1978 [94]. Since then, islet autotransplantation has become a widely used strategy to prevent diabetes after pancreatectomy. A recent report of 85 total pancreatectomy patients from the United Kingdom showed that the group of 50 patients receiving concomitant autologous islet transplantation had a significantly lower median insulin requirement than those without concomitant transplantation, although only five patients remained insulin independent [96]. Of 173 recipients of autologous islet transplantation postpancreatectomy at the University of Minnesota, 55 (32%) were insulin independent and 57 (33%) had partial islet function as defined by the need of only once-daily long-acting insulin at some posttransplant point. Importantly, the rate of decline of insulin independence was remarkably limited, with 46% insulin independence at 5 years follow-up and 28% at 10 years [97]. Despite the lower number of islets needed to provide insulin independence in autologous compared to allogeneic transplantation, islet cell mass is an important predictor of success in both transplant procedures [51,98]. Therefore, improvements in islet yields from fibrotic and inflamed pancreata are expected to further improve outcomes of autologous islet transplantation [99].

40.8.2 Cost-Efficacy of Islet Transplantation

According to the 2000 French National Cost Study [100], the costs of hospitalization for pancreas transplantation (DRG 279) were €25,674. The processing of the 5.6 pancreata used for a single islet transplantation costs €23,755; with the hospitalization, the total costs are about €34,178. Since about 80% of patients are still insulin independent at 3 years after a pancreas transplantation, compared to only 44% of those who receive an islet transplantation [56,101], pancreas transplantation is likely to be remarkably more cost-effective compared to islet transplantation, even taking into account the higher risk of surgical complications related to whole organ grafting.

A recent cost-effectiveness exploratory analysis comparing islet transplantation with standard insulin therapy in type 1 diabetes adult patients affected by hypoglycemia unawareness showed that islet transplantation becomes cost-saving at about 9−10 years after the procedure [102]. However, since the vast majority of islet transplant recipients lose islet independence within 5 years after transplantation, this result in fact means that islet transplantation is less cost-effective than standard insulin therapy. Therefore, despite the absence of *ad hoc* studies formally addressing the issue of the cost-effectiveness of islet transplantation compared to other treatments for type 1 diabetes, available data suggest that such procedure is hardly superior to pancreas transplantation or insulin therapy.

40.9 FUTURE DEVELOPMENTS

40.9.1 New Sources of Islets

Islet cell harvest from whole pancreas remains a limiting step, as the efficiency of the procedure and cell viability postharvest is relatively poor. There are several steps in the whole procedure of islet transplantation which may be targeted in order to improve islet recovery and posttransplantation protection. Pretransplantation procedures related to pancreas preservation, enzymatic digestion, purification, culture, and shipment may be further refined [103]. While many laboratories are developing methods to improve these processes, given the severe shortage of donor pancreases, other investigators are exploring alternative sources of β cells.

Other sources of pancreatic β cells are mesenchymal stem cells (MSCs) that display the capability to transdifferentiate into insulin-producing cells [104,105]. Of even more importance are the immunomodulatory and anti-inflammatory properties of MSCs, which might control the autoimmune response preventing immune injury of newly proliferating cells [106].

Alternative sources for β-cell replacement include human embryonic stem cells (hESCs). Recently, it has been shown that the small molecule (−)-indolactam V induces differentiation of hESCs into pancreatic progenitor cells *in vitro* [107]. The more plentiful pancreatic ductal cells isolated from human donor pancreases can be trans-differentiated into the more scarce β cells [108]. Similarly, mouse experiments have shown that bile duct epithelial cells [109], acinar cells [110], and hepatic cells [111] can also be trans-differentiated into β cells. The differentiation of human fibroblast-derived induced pluripotent stem cells (iPSCs) into β cells provides another alternative that is particularly enticing due to its potential avoidance of allogeneic rejection [112].

Another possible source of pancreatic islets is xenotransplantation, with which some experience has been gained in humans. Six groups have independently reported that pig islets transplanted into nonhuman primates can maintain normoglycemia for more than 6 months and more positive results are expected by genetic manipulation of transplanted islets [113].

In 1994, a Swedish group reported xenotransplantation with fetal porcine pancreatic islets in 10 diabetic patients. Although insulin requirements did not decrease, the procedure was well-tolerated and there was no evidence of transmission of porcine endogenous retroviruses after 4−7 years of follow-up [114,115]. More recently, xenotransplantation has been performed in China, Russia, and Mexico [116,117]. In 2005, the group from Mexico reported a 4-year follow-up of 12 diabetic patients not taking immunosuppressive therapy who had received one to three subcutaneous implantations of a device containing porcine pancreatic islets and Sertoli cells. Sertoli cells, being immune-privileged, were added because they may confer immunoprotection to transplanted endocrine tissue. Follow-up showed a decreased insulin requirement in 50% of patients, but the decrease in HbA1c was lower than in the 50% of patients not having a favorable response to the transplantation. Porcine C-peptide was not detectable in the urine, and the significance of this study remains to be determined. Importantly, severe ethical issues have been raised with regard to xenotransplantation as it is currently being performed. The program in China was suspended and the International Xenotransplantation Association has seriously objected to the Mexican and Russian studies, as they feel that the safety of the patient and of the general public (especially with regard to the spread of porcine endogenous retroviruses) are not sufficiently guaranteed [118−120]. More experimental studies are needed before clinical trials in humans can be initiated [121].

In 2005, Matsumoto et al. performed the first islet transplantation from a living-related donor in a patient who had brittle diabetes due to chronic pancreatitis. The procedure resulted in good glycemic control and no major complications in both the donor and the recipient [122,123]. However, results cannot be generalized to the type 1 diabetes population, as diabetic disease in the recipient did not result from an autoimmune process. Moreover, partial pancreatectomy in the donor implies major surgery with associated risks of morbidity and mortality. In the long term, donors may be at increased risk of developing diabetes mellitus themselves [124].

40.9.2 Improving the Transplant Procedure

Efficacy of islet transplant procedure is hampered by massive cell loss shortly after infusion because of an

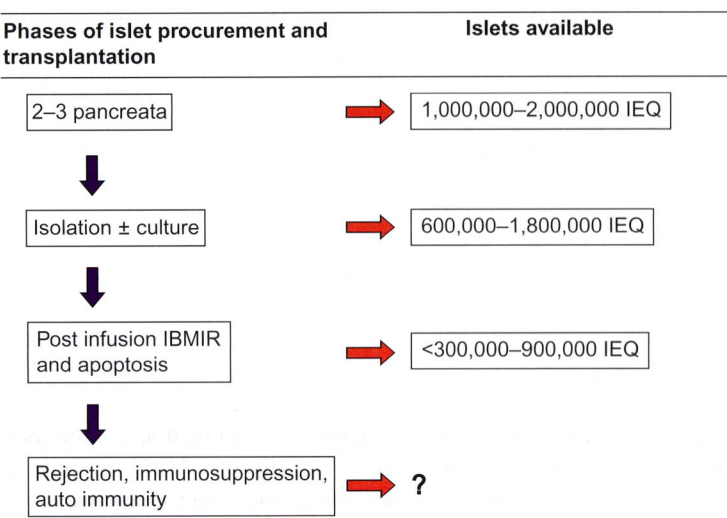

Phases of islet procurement and transplantation	Islets available
2–3 pancreata	1,000,000–2,000,000 IEQ
Isolation ± culture	600,000–1,800,000 IEQ
Post infusion IBMIR and apoptosis	<300,000–900,000 IEQ
Rejection, immunosuppression, auto immunity	?

FIGURE 40.2 Loss of pancreatic islet mass, from graft preparation to postinfusion degradation. *Data derived from CITR 2008 fifth annual report [55].* IEQ, islet equivalents; IBMIR, instant blood-mediated inflammatory reaction. The IBMIR reduces islet mass by 50–70% [125]. *Modified from [126].*

inflammatory reaction termed instant blood-mediated inflammatory reaction (IBMIR). This reaction involves activation of the complement and coagulation cascades, ultimately resulting in clot formation and infiltration of leukocytes into the islets, which leads to disruption of islet integrity and islet destruction (Figure 40.2). Different strategies have been proposed to prevent this phenomenon, since heparin can prevent clotting and decrease complement activation. Peritransplant heparinization of either the patient or, to prevent bleeding complications, the pancreatic islets themselves has been proposed as a strategy to improve outcomes [127–129]. This may prevent the immediate and significant postprocedural islet loss. Moreover, it is now possible to visualize islets in the peri-transplantation phase using 18F-fluorodeoxyglucose positron-emission tomography combined with computed tomography in order to assess islet survival and distribution, which may also be used to evaluate alternative sites of implantation [130].

For the transplantation procedure itself, it has been recognized that the liver is not the ideal site for transplantation because of the procedure-related complications, the relatively low oxygen supply in the liver, the exposure to toxins absorbed from the gastrointestinal tract, and the IBMIR, which causes substantial islet loss shortly after infusion. Many alternative sites have been explored, including the omentum, pancreas, gastrointestinal tract, and muscular tissue. However, these alternative approaches have so far remained experimental, with none of them being convincingly superior to the currently used method [131,132].

Islet encapsulation as a strategy to improve graft survival is one of the main areas in experimental research. The use of semipermeable encapsulation material, such as alginate gel or membrane devices, should protect the islets against the alloimmune response while at the same

time allowing them to sense glucose levels and secrete insulin [133,134]. This technique must allow adequate diffusion of oxygen and nutrients to maintain islet viability and function but, at the same time, must be selective enough to prevent the permeation of host immune proteins. Islet encapsulation in alginate gel and in polysulfone hollow fibers allows adequate transport of nutrients to maintain islet function and viability, making the use of these immunoisolation strategies for transplantation a potentially important field of investigation to also transplant xenogeneic islets [135,136].

Recently, four patients received an intraperitoneal infusion of encapsulated allogeneic islets. Despite no immunosuppression, all patients turned positive for serum C-peptide response, both in basal and after stimulation, and anti-MHC class I–II and GAD65 antibodies all tested negative at 3 years after transplant. Daily mean blood glucose, as well as HbA1c levels, significantly improved after transplant, with daily exogenous insulin consumption declining in all cases, but with full insulin independence reached, just transiently, in one single patient [137].

40.9.3 Induction of Immune Tolerance

Over the past several decades, the generation of a large array of immunosuppressive agents has increased the number of therapeutic tools available to prevent acute rejection. However, as detailed above, toxicity of immunosuppressive drugs may offset their benefits, hence the final goal of transplant medicine is to achieve T- and B-cell tolerance that is antigen specific without the need for long-term generalized immunosuppression. Many strategies have been proposed to promote transplant tolerance in rodent model of transplantation, including infusion of Tregs, MSCs, or immature dendritic cells (DCs). Most strategies, however, failed when transferred to nonhuman primate models. A treatment that selectively destroys activated cytopathic donor reactive T cells while sparing resting and immunoregulatory T cells has been tested in a model of islet transplantation. In particular, short-term sirolimus and IL-2.Ig plus mutant antagonist-type IL-15.Ig cytolytic fusion proteins posttransplantation resulted in prolonged, drug-free engraftment [138].

Recently, an isolated case with more than 11 years of insulin independency after islet transplantation was described [139]. The intriguing question is which factors have contributed to the outcome in this particular patient. The patient had previously received a kidney transplant and was on an immunosuppressive regimen comprising antithymocyte globulin as induction therapy followed by prednisone (which was rapidly tapered), cyclosporine, and azathioprine, which was later switched to MMF. Interestingly, the authors investigated the cellular immune response and found that the patient was hyporesponsive toward donor antigens, possibly as a result of the expanded Treg pool. Consistently, Huurman et al. examined cytokine profiles in patients who achieved insulin independence and found that they were skewed towards a Treg phenotype. In particular, expression of the Treg cytokine IL-10 was associated with low alloreactivity and superior islet function [140]. The role of Tregs in allograft tolerance has long been recognized in solid organ and bone marrow transplantation, and much research is devoted to translating this knowledge into therapeutic options, which may also benefit islet transplantation [141].

40.9.4 Novel Therapeutic Perspectives for Type 1 Diabetes Mellitus

New therapeutic approaches for patients with type 1 diabetes are underway. Indeed, refinement of insulin pumps in combination with continuous glucose monitoring systems may lead to better glycemic control [142]. A recent randomized, controlled study in 12 patients showed that day-and-night closed-loop basal insulin delivery can improve glucose control in adolescents over conventional pump therapy. However, unannounced moderate-intensity exercise and excessive prandial boluses pose challenges to hypoglycemia-free closed-loop basal insulin delivery [143].

Apart from symptomatic treatment, trials are being conducted using immunosuppressive drugs to control autoimmune response and preserve β cells. Indeed, regeneration of the existing β-cell pool is possible and happens in humans under physiological conditions, such as pregnancy and insulin resistance [104]. However, in diabetic patients, this process is thought to be impaired by autoimmune injury. In this regard, trials in patients with new-onset disease evaluating the effects of the combined use of MMF and daclizumab (www.clinicaltrials.gov;

NCT00100178) are underway. More intriguingly, attempts have been carried out also for the induction of tolerance. Along this line, compelling evidence has been accumulated to suggest that in addition to their immuno-suppressive properties, CD3-specific antibodies can also promote immune tolerance especially in the context of ongoing immune responses [144]. Starting from this background, studies have shown that this therapy may, at least partially, preserve β-cell mass in newly diagnosed type 1 diabetics [144]. An alternative approach is targeting B cells, in light of the importance of humoral response in the pathogenesis of type 1 diabetes and in consideration of their role of antigen presenting cells. Starting from the results with B-cell depleting monoclonal antibody obtained in a mouse model of diabetes [145], a randomized, double-blind study tested the effect of rituximab in 87 patients with newly diagnosed type 1 diabetes in preserving islet function [146]. One year after infusion, the mean area under the curve (AUC) for the serum C-peptide is significantly higher in the rituximab group than in the placebo group. A more drastic approach is autologous nonmyeloablative hematopoietic stem cell transplantation, which may reset autoreactive T cells and prevent the clinical onset of the disease in new-onset type 1diabetics [147]. According to this protocol, sustained normoglycemia was maintained for a mean of 2.5 years in approximately 50% of patients. However, acute drug toxicity, risk of infections, and sterility may outweigh the benefits of this protocol.

Alternative approaches for the induction of tolerance include strategies of molecular biology. In particular, evidence has been provided that "immature" DCs can promote tolerance. To this end, CD40, CD80, and CD86 cell surface molecules were specifically down-regulated by treating *ex vivo* DCs from mice with a mixture of specific antisense oligonucleotides. This promoted the emergence of Tregs that might possibly prevent the occurrence of diabetes [148]. Intriguingly, to circumvent the technical issues of *ex vivo* DC manipulation, a mouse study showed that the same immature phenotype can be induced by using a microsphere-based vaccine injected subcutaneously [149]. This approach effectively prevented new-onset diabetes or even reversed it, forming the basis for also testing this approach in humans.

40.10 CONCLUSION

Despite the enthusiasm that welcomed, more than 40 years ago, islet transplantation as a noninvasive strategy to normalize glucose metabolism in patients with type 1 diabetes, this procedure is still far from representing the ideal treatment for this condition. Rates of long-term insulin independence are disappointingly low and no clear evidence exists of any superiority of islet transplantation over insulin therapy in preventing diabetes-related complications. Thus, widespread applicability of pancreatic islet transplantation in clinical practice should be reassessed in view of uncertain benefits, well-documented risks, and considerable costs [150]. Long-term comparative cost/effectiveness analyses versus whole organ transplantation and insulin therapy in the setting of randomized trials should be implemented to understand whether islet pancreas transplantation may offer real advancements in the treatment of type 1 diabetics. In the meantime, only a highly selected group of patients with brittle diabetes may benefit from the procedure, which requires a high degree of expertise. Research studies are ongoing to identify novel sources of islets, improve transplant procedure, and prevent their rejection by the host immune system. At the same time, researchers are looking for strategies to prevent β-cell destruction in the early phases of diabetes.

The ADA recommends performing islet transplantation only in the context of controlled research studies [32]. This advice has been largely disregarded so far and the search for strategies to improve outcomes of the procedure has distracted attention from designing properly controlled trials comparing islet transplantation with medical therapy in type 1 diabetes. Until results of such studies are available, islet transplantation should be considered as just an experimental procedure [150,151].

REFERENCES

[1] Shapiro AM, Lakey JR, Ryan EA, Korbutt GS, Toth E, Warnock GL, et al. Islet transplantation in seven patients with type 1 diabetes mellitus using a glucocorticoid-free immunosuppressive regimen. N Engl J Med 2000;343:230–8.

[2] Bretzel RG, Brandhorst D, Brandhorst H, Eckhard M, Ernst W, Friemann S, et al. Improved survival of intraportal pancreatic islet cell allografts in patients with type-1 diabetes mellitus by refined peritransplant management. J Mol Med 1999;77:140–3.

[3] Daneman D. State of the world's children with diabetes. Pediatr Diabetes 2009;10:120–6.

[4] Green A. Descriptive epidemiology of type 1 diabetes in youth: incidence, mortality, prevalence, and secular trends. Endocr Res 2008;33:1–15.

[5] WHO Diamond Project Group. WHO multinational project for childhood diabetes. Diabetes Care 1990;13(10):1062–8.

[6] Imperatore G, Boyle JP, Thompson TJ, Case D, Dabelea D, Hamman RF, et al. Projections of type 1 and type 2 diabetes burden in the U.S. population aged <20 years through 2050: dynamic modeling of incidence, mortality, and population growth. Diabetes Care 2012;35:2515–20.

[7] DIAMOND Project Group. Incidence and trends of childhood type 1 diabetes worldwide 1990–1999. Diabet Med 2006;23:857–66.

[8] Livingstone SJ, Looker HC, Hothersall EJ, Wild SH, Lindsay RS, Chalmers J, et al. Risk of cardiovascular disease and total mortality in adults with type 1 diabetes: scottish registry linkage study. PLoS Med 2012;9:e1001321.

[9] Knip M, Siljander H. Autoimmune mechanisms in type 1 diabetes. Autoimmun Rev 2008;7:550–7.

[10] Concannon P, Rich SS, Nepom GT. Genetics of type 1A diabetes. N Engl J Med 2009;360:1646–54.

[11] Atkinson MA, Eisenbarth GS. Type 1 diabetes: new perspectives on disease pathogenesis and treatment. Lancet 2001;358:221–9.

[12] Gianani R, Eisenbarth GS. The stages of type 1A diabetes: 2005. Immunol Rev 2005;204:232–49.

[13] Brezar V, Carel JC, Boitard C, Mallone R. Beyond the hormone: insulin as an autoimmune target in type 1 diabetes. Endocr Rev 2011;32:623–69.

[14] Wenzlau JM, Juhl K, Yu L, Moua O, Sarkar SA, Gottlieb P, et al. The cation efflux transporter ZnT8 (Slc30A8) is a major autoantigen in human type 1 diabetes. Proc Natl Acad Sci USA 2007;104:17040–5.

[15] Csorba TR, Lyon AW, Hollenberg MD. Autoimmunity and the pathogenesis of type 1 diabetes. Crit Rev Clin Lab Sci 2010;47:51–71.

[16] Martin S, Wolf-Eichbaum D, Duinkerken G, Scherbaum WA, Kolb H, Noordzij JG, et al. Development of type 1 diabetes despite severe hereditary B-lymphocyte deficiency. N Engl J Med 2001;345:1036–40.

[17] Huurman VA, Hilbrands R, Pinkse GG, Gillard P, Duinkerken G, van de Linde P, et al. Cellular islet autoimmunity associates with clinical outcome of islet cell transplantation. PLoS One 2008;3: e2435.

[18] Jaeger C, Brendel MD, Hering BJ, Eckhard M, Bretzel RG. Progressive islet graft failure occurs significantly earlier in autoantibody-positive than in autoantibody-negative IDDM recipients of intrahepatic islet allografts. Diabetes 1997;46:1907–10.

[19] Bosi E, Braghi S, Maffi P, Scirpoli M, Bertuzzi F, Pozza G, et al. Autoantibody response to islet transplantation in type 1 diabetes. Diabetes 2001;50:2464–71.

[20] Assalino M, Genevay M, Morel P, Demuylder-Mischler S, Toso C, Berney T. Recurrence of type 1 diabetes after simultaneous pancreas-kidney transplantation in the absence of GAD and IA-2 autoantibodies. Am J Transplant 2012;12:492–5.

[21] Dor Y, Brown J, Martinez OI, Melton DA. Adult pancreatic beta-cells are formed by self-duplication rather than stem-cell differentiation. Nature 2004;429:41–6.

[22] Nir T, Melton DA, Dor Y. Recovery from diabetes in mice by beta cell regeneration. J Clin Invest 2007;117:2553–61.

[23] The DCCT Research Group. The effect of intensive treatment of diabetes on the development and progression of long-term complications in insulin-dependent diabetes mellitus. N Engl J Med 1993;329:977–86.

[24] Nathan DM, Cleary PA, Backlund JY, Genuth SM, Lachin JM, Orchard TJ, et al. Intensive diabetes treatment and cardiovascular disease in patients with type 1 diabetes. N Engl J Med 2005;353:2643–53.

[25] Lachin JM, Genuth S, Nathan DM, Zinman B, Rutledge BN. Effect of glycemic exposure on the risk of microvascular complications in the diabetes control and complications trial—revisited. Diabetes 2008;57:995–1001.

[26] Group H. The diabetes control and complications trial research group. Diabetes 1997;46:271–86.

[27] Nordwall M, Bojestig M, Arnqvist HJ, Ludvigsson J. Declining incidence of severe retinopathy and persisting decrease of nephropathy

in an unselected population of Type 1 diabetes-the linkoping diabetes complications study. Diabetologia 2004;47:1266−72.

[28] Nathan DM, Zinman B, Cleary PA, Backlund JY, Genuth S, Miller R, et al. Modern-day clinical course of type 1 diabetes mellitus after 30 years duration: the diabetes control and complications trial/epidemiology of diabetes interventions and complications and Pittsburgh epidemiology of diabetes complications experience (1983−2005). Arch Intern Med 2009;169:1307−16.

[29] Trial Research Group. Effect of intensive diabetes treatment on the development and progression of long-term complications in adolescents with insulin-dependent diabetes mellitus. J Pediatr 1994;125:177−88.

[30] Danne T, Mortensen HB, Hougaard P, Lynggaard H, Aanstoot HJ, Chiarelli F, et al. Persistent differences among centers over 3 years in glycemic control and hypoglycemia in a study of 3805 children and adolescents with type 1 diabetes from the hvidore study group. Diabetes Care 2001;24:1342−7.

[31] Ryan EA, Bigam D, Shapiro AM. Current indications for pancreas or islet transplant. Diabetes Obes Metab 2006;8:1−7.

[32] Robertson RP, Davis C, Larsen J, Stratta R, Sutherland DE. Pancreas and islet transplantation in type 1 diabetes. Diabetes Care 2006;29:935.

[33] Ryan EA, Shandro T, Green K, Paty BW, Senior PA, Bigam D, et al. Assessment of the severity of hypoglycemia and glycemic lability in type 1 diabetic subjects undergoing islet transplantation. Diabetes 2004;53:955−62.

[34] Ballinger WF, Lacy PE. Transplantation of intact pancreatic islets in rats. Surgery 1972;72:175−86.

[35] Secchi A, Socci C, Maffi P, Taglietti MV, Falqui L, Bertuzzi F, et al. Islet transplantation in IDDM patients. Diabetologia 1997;40:225−31.

[36] Benhamou PY, Oberholzer J, Toso C, Kessler L, Penfornis A, Bayle F, et al. Human islet transplantation network for the treatment of type I diabetes: first data from the Swiss-French GRAGIL consortium (1999−2000). Groupe de Recherche Rhin Rhjne Alpes Geneve pour la transplantation d'Ilots de Langerhans. Diabetologia 2001;44:859−64.

[37] Hirshberg B, Rother KI, Digon 3rd BJ, Lee J, Gaglia JL, Hines K, et al. Benefits and risks of solitary islet transplantation for type 1 diabetes using steroid-sparing immunosuppression: the National Institutes of Health experience. Diabetes Care 2003;26:3288−95.

[38] Frank A, Deng S, Huang X, Velidedeoglu E, Bae YS, Liu C, et al. Transplantation for type I diabetes: comparison of vascularized whole-organ pancreas with isolated pancreatic islets. Ann Surg 2004;240:631−43.

[39] Goss JA, Goodpastor SE, Brunicardi FC, Barth MH, Soltes GD, Garber AJ, et al. Development of a human pancreatic islet-transplant program through a collaborative relationship with a remote islet-isolation center. Transplantation 2004;77:462−6.

[40] Hering BJ, Kandaswamy R, Harmon JV, Ansite JD, Clemmings SM, Sakai T, et al. Transplantation of cultured islets from two-layer preserved pancreases in type 1 diabetes with anti-CD3 antibody. Am J Transplant 2004;4:390−401.

[41] Froud T, Ricordi C, Baidal DA, Hafiz MM, Ponte G, Cure P, et al. Islet transplantation in type 1 diabetes mellitus using cultured islets and steroid-free immunosuppression: Miami experience. Am J Transplant 2005;5:2037−46.

[42] Hafiz MM, Faradji RN, Froud T, Pileggi A, Baidal DA, Cure P, et al. Immunosuppression and procedure-related complications in 26 patients with type 1 diabetes mellitus receiving allogeneic islet cell transplantation. Transplantation 2005;80:1718−28.

[43] Hering BJ, Kandaswamy R, Ansite JD, Eckman PM, Nakano M, Sawada T, et al. Single-donor, marginal-dose islet transplantation in patients with type 1 diabetes. JAMA 2005;293:830−5.

[44] Warnock GL, Meloche RM, Thompson D, Shapiro RJ, Fung M, Ao Z, et al. Improved human pancreatic islet isolation for a prospective cohort study of islet transplantation vs. best medical therapy in type 1 diabetes mellitus. Arch Surg 2005;140:735−44.

[45] Keymeulen B, Gillard P, Mathieu C, Movahedi B, Maleux G, Delvaux G, et al. Correlation between beta cell mass and glycemic control in type 1 diabetic recipients of islet cell graft. Proc Natl Acad Sci USA 2006;103:17444−9.

[46] O'Connell PJ, Hawthorne WJ, Holmes-Walker DJ, Nankivell BJ, Gunton JE, Patel AT, et al. Clinical islet transplantation in type 1 diabetes mellitus: results of Australia's first trial. Med J Aust 2006;184:221−5.

[47] Maffi P, Bertuzzi F, De Taddeo F, Magistretti P, Nano R, Fiorina P, et al. Kidney function after islet transplant alone in type 1 diabetes: impact of immunosuppressive therapy on progression of diabetic nephropathy. Diabetes Care 2007;30:1150−5.

[48] Gangemi A, Salehi P, Hatipoglu B, Martellotto J, Barbaro B, Kuechle JB, et al. Islet transplantation for brittle type 1 diabetes: the UIC protocol. Am J Transplant 2008;8:1250−61.

[49] Gillard P, Ling Z, Mathieu C, Crenier L, Lannoo M, Maes B, et al. Comparison of sirolimus alone with sirolimus plus tacrolimus in type 1 diabetic recipients of cultured islet cell grafts. Transplantation 2008;85:256−63.

[50] Ryan EA, Paty BW, Senior PA, Bigam D, Alfadhli E, Kneteman NM, et al. Five-year follow-up after clinical islet transplantation. Diabetes 2005;54:2060−9.

[51] Shapiro AM, Ricordi C, Hering BJ, Auchincloss H, Lindblad R, Robertson RP, et al. International trial of the Edmonton protocol for islet transplantation. N Engl J Med 2006;355:1318−30.

[52] Borot S, Niclauss N, Wojtusciszyn A, Brault C, Demuylder-Mischler S, Muller Y, et al. Impact of the number of infusions on 2-year results of islet-after-kidney transplantation in the GRAGIL network. Transplantation 2011;92:1031−8.

[53] Kenmochi T, Asano T, Maruyama M, Saigo K, Akutsu N, Iwashita C, et al. Clinical islet transplantation in Japan. J Hepatobiliary Pancreat Surg 2009;16:124−30.

[54] Alejandro R, Barton FB, Hering BJ, Wease S. 2008 update from the collaborative islet transplant registry. Transplantation 2008;86:1783−8.

[55] Collaborative Islet Transplantation Registry. Fifth annual report. Available from: <http://citregistry.com>; 2008.

[56] Barton FB, Rickels MR, Alejandro R, Hering BJ, Wease S, Naziruddin B, et al. Improvement in outcomes of clinical islet transplantation: 1999−2010. Diabetes Care 2012;35:1436−45.

[57] Lee TC, Barshes NR, Agee EE, O'Mahoney CA, Brunicardi FC, Goss JA. The effect of whole organ pancreas transplantation and PIT on diabetic complications. Curr Diab Rep 2006;6:323−7.

[58] Fiorina P, Shapiro AM, Ricordi C, Secchi A. The clinical impact of islet transplantation. Am J Transplant 2008;8:1990−7.

[59] Fiorina P, Gremizzi C, Maffi P, Caldara R, Tavano D, Monti L, et al. Islet transplantation is associated with an improvement of

cardiovascular function in type 1 diabetic kidney transplant patients. Diabetes Care 2005;28:1358−65.

[60] Danielson KK, Hatipoglu B, Kinzer K, Kaplan B, Martellotto J, Qi M, et al. Reduction in carotid intima-media thickness after pancreatic islet transplantation in patients with type 1 diabetes Diabetes Care 2012 [published ahead of print November 19, 2012] . Available from: http://dx.doi.org/10.2337/dc12-0679

[61] Fiorina P, Folli F, Zerbini G, Maffi P, Gremizzi C, Di Carlo V, et al. Islet transplantation is associated with improvement of renal function among uremic patients with type I diabetes mellitus and kidney transplants. J Am Soc Nephrol 2003;14:2150−8.

[62] Senior PA, Zeman M, Paty BW, Ryan EA, Shapiro AM. Changes in renal function after clinical islet transplantation: four-year observational study. Am J Transplant 2007;7:91−8.

[63] Fioretto P, Mauer SM, Bilous RW, Goetz FC, Sutherland DE, Steffes MW. Effects of pancreas transplantation on glomerular structure in insulin-dependent diabetic patients with their own kidneys. Lancet 1993;342:1193−6.

[64] Fioretto P, Steffes MW, Sutherland DE, Goetz FC, Mauer M. Reversal of lesions of diabetic nephropathy after pancreas transplantation. N Engl J Med 1998;339:69−75.

[65] Leitao CB, Cure P, Messinger S, Pileggi A, Lenz O, Froud T, et al. Stable renal function after islet transplantation: importance of patient selection and aggressive clinical management. Transplantation 2009;87:681−8.

[66] Lee TC, Barshes NR, O'Mahony CA, Nguyen L, Brunicardi FC, Ricordi C, et al. The effect of pancreatic islet transplantation on progression of diabetic retinopathy and neuropathy. Transplant Proc 2005;37:2263−5.

[67] Venturini M, Fiorina P, Maffi P, Losio C, Vergani A, Secchi A, et al. Early increase of retinal arterial and venous blood flow velocities at color Doppler imaging in brittle type 1 diabetes after islet transplant alone. Transplantation 2006;81:1274−7.

[68] The Diabetes Control and Complications Trial Research Group. The effect of intensive treatment of diabetes on the development and progression of long-term complications in insulin-dependent diabetes mellitus. N Engl J Med 1993;329:977−86.

[69] Thompson DM, Begg IS, Harris C, Ao Z, Fung MA, Meloche RM, et al. Reduced progression of diabetic retinopathy after islet cell transplantation compared with intensive medical therapy. Transplantation 2008;85:1400−5.

[70] Del Carro U, Fiorina P, Amadio S, De Toni Franceschini L, Petrelli A, Menini S, et al. Evaluation of polyneuropathy markers in type 1 diabetic kidney transplant patients and effects of islet transplantation: neurophysiological and skin biopsy longitudinal analysis. Diabetes Care 2007;30:3063−9.

[71] Warnock GL, Thompson DM, Meloche RM, Shapiro RJ, Ao Z, Keown P, et al. A multi-year analysis of islet transplantation compared with intensive medical therapy on progression of complications in type 1 diabetes. Transplantation 2008;86:1762−6.

[72] Villiger P, Ryan EA, Owen R, O'Kelly K, Oberholzer J, Al Saif F, et al. Prevention of bleeding after islet transplantation: lessons learned from a multivariate analysis of 132 cases at a single institution. Am J Transplant 2005;5:2992−8.

[73] Ryan EA, Paty BW, Senior PA, Shapiro AM. Risks and side effects of islet transplantation. Curr Diab Rep 2004;4:304−9.

[74] Markmann JF, Rosen M, Siegelman ES, Soulen MC, Deng S, Barker CF, et al. Magnetic resonance-defined periportal steatosis

following intraportal islet transplantation: a functional footprint of islet graft survival?. Diabetes 2003;52:1591−4.

[75] Bhargava R, Senior PA, Ackerman TE, Ryan EA, Paty BW, Lakey JR, et al. Prevalence of hepatic steatosis after islet transplantation and its relation to graft function. Diabetes 2004;53:1311−7.

[76] Campbell PM, Senior PA, Salam A, Labranche K, Bigam DL, Kneteman NM, et al. High risk of sensitization after failed islet transplantation. Am J Transplant 2007;7:2311−7.

[77] Cardani R, Pileggi A, Ricordi C, Gomez C, Baidal DA, Ponte GG, et al. Allosensitization of islet allograft recipients. Transplantation 2007;84:1413−27.

[78] Lobo PI, Spencer C, Simmons WD, Hagspiel KD, Angle JF, Deng S, et al. Development of anti-human leukocyte antigen class 1 antibodies following allogeneic islet cell transplantation. Transplant Proc 2005;37:3438−40.

[79] Campbell PM, Salam A, Ryan EA, Senior P, Paty BW, Bigam D, et al. Pretransplant HLA antibodies are associated with reduced graft survival after clinical islet transplantation. Am J Transplant 2007;7:1242−8.

[80] Berney T, Secchi A. Rapamycin in islet transplantation: friend or foe? Transpl Int 2009;22:153−61.

[81] Halloran PF. Immunosuppressive drugs for kidney transplantation. N Engl J Med 2004;351:2715−29.

[82] Ghofaili KA, Fung M, Ao Z, Meloche M, Shapiro RJ, Warnock GL, et al. Effect of exenatide on beta cell function after islet transplantation in type 1 diabetes. Transplantation 2007;83:24−8.

[83] Froud T, Baidal DA, Faradji R, Cure P, Mineo D, Selvaggi G, et al. Islet transplantation with alemtuzumab induction and calcineurin-free maintenance immunosuppression results in improved short- and long-term outcomes. Transplantation 2008;86: 1695−701.

[84] Mineo D, Sageshima J, Burke GW, Ricordi C. Minimization and withdrawal of steroids in pancreas and islet transplantation. Transpl Int 2009;22:20−37.

[85] Weaver TA, Kirk AD. Alemtuzumab. Transplantation 2007;84:1545−7.

[86] Shapiro AM. State of the art of clinical islet transplantation and novel protocols of immunosuppression. Curr Diab Rep 2011;11:345−54.

[87] Posselt AM, Szot GL, Frassetto LA, Masharani U, Tavakol M, Amin R, et al. Islet transplantation in type 1 diabetic patients using calcineurin inhibitor-free immunosuppressive protocols based on T-cell adhesion or costimulation blockade. Transplantation 2010;90:1595−601.

[88] Faradji RN, Tharavanij T, Messinger S, Froud T, Pileggi A, Monroy K, et al. Long-term insulin independence and improvement in insulin secretion after supplemental islet infusion under exenatide and etanercept. Transplantation 2008;86:1658−65.

[89] McCall M, Pawlick R, Kin T, Shapiro AM. Anakinra potentiates the protective effects of etanercept in transplantation of marginal mass human islets in immunodeficient mice. Am J Transplant 2012;12:322−9.

[90] Takita M, Matsumoto S, Shimoda M, Chujo D, Itoh T, Sorelle JA, et al. Safety and tolerability of the T-cell depletion protocol coupled with anakinra and etanercept for clinical islet cell transplantation. Clin Transplant 2012;26:471−84.

[91] Gentilella R, Bianchi C, Rossi A, Rotella CM. Exenatide: a review from pharmacology to clinical practice. Diabetes Obes Metab 2009;11:544—56.

[92] Froud T, Faradji RN, Pileggi A, Messinger S, Baidal DA, Ponte GM, et al. The use of exenatide in islet transplant recipients with chronic allograft dysfunction: safety, efficacy, and metabolic effects. Transplantation 2008;86:36—45.

[93] Rickels MR, Naji A. Exenatide use in islet transplantation: words of caution. Transplantation 2009;87:153.

[94] Sutherland DE, Matas AJ, Najarian JS. Pancreatic islet cell transplantation. Surg Clin North Am 1978;58:365—82.

[95] Braasch JW, Vito L, Nugent FW. Total pancreatectomy of end-stage chronic pancreatitis. Ann Surg 1978;188:317—22.

[96] Garcea G, Weaver J, Phillips J, Pollard CA, Ilouz SC, Webb MA, et al. Total pancreatectomy with and without islet cell transplantation for chronic pancreatitis: a series of 85 consecutive patients. Pancreas 2009;38:1—7.

[97] Sutherland DE, Gruessner AC, Carlson AM, Blondet JJ, Balamurugan AN, Reigstad KF, et al. Islet autotransplant outcomes after total pancreatectomy: a contrast to islet allograft outcomes. Transplantation 2008;86:1799—802.

[98] Wahoff DC, Papalois BE, Najarian JS, Kendall DM, Farney AC, Leone JP, et al. Autologous islet transplantation to prevent diabetes after pancreatic resection. Ann Surg 1995;222:562—75 [discussion 575—9].

[99] Naziruddin B, Matsumoto S, Noguchi H, Takita M, Shimoda M, Fujita Y, et al. Improved pancreatic islet isolation outcome in autologous transplantation for chronic pancreatitis. Cell Transplant 2012;21:553—8.

[100] Guignard AP, Oberholzer J, Benhamou PY, Touzet S, Bucher P, Penfornis A, et al. Cost analysis of human islet transplantation for the treatment of type 1 diabetes in the Swiss-French Consortium GRAGIL. Diabetes Care 2004;27:895—900.

[101] Boggi U, Vistoli F, Egidi FM, Marchetti P, De Lio N, Perrone V, et al. Transplantation of the pancreas. Curr Diab Rep 2012;12:568—79.

[102] Beckwith J, Nyman JA, Flanagan B, Schrover R, Schuurman HJ. A health economic analysis of clinical islet transplantation. Clin Transplant 2012;26:23—33.

[103] Ichii H, Ricordi C. Current status of islet cell transplantation. J Hepatobiliary Pancreat Surg 2009;16:101—12.

[104] Porat S, Dor Y. New sources of pancreatic beta cells. Curr Diab Rep 2007;7:304—8.

[105] Claiborn KC, Stoffers DA. Toward a cell-based cure for diabetes: advances in production and transplant of beta cells. Mt Sinai J Med 2008;75:362—71.

[106] English K. Mechanisms of mesenchymal stromal cell immunomodulation. Immunol Cell Biol 2013;91:19—26.

[107] Chen S, Borowiak M, Fox JL, Maehr R, Osafune K, Davidow L, et al. A small molecule that directs differentiation of human ESCs into the pancreatic lineage. Nat Chem Biol 2009;5:258—65.

[108] Bonner-Weir S, Taneja M, Weir GC, Tatarkiewicz K, Song KH, Sharma A, et al. In vitro cultivation of human islets from expanded ductal tissue. Proc Natl Acad Sci USA 2000;97:7999—8004.

[109] Nagaya M, Katsuta H, Kaneto H, Bonner-Weir S, Weir GC. Adult mouse intrahepatic biliary epithelial cells induced in vitro to become insulin-producing cells. J Endocrinol 2009;201:37—47.

[110] Zhou Q, Brown J, Kanarek A, Rajagopal J, Melton DA. In vivo reprogramming of adult pancreatic exocrine cells to beta cells. Nature 2008;455:627—32.

[111] Aviv V, Meivar-Levy I, Rachmut IH, Rubinek T, Mor E, Ferber S. Exendin-4 promotes liver cell proliferation and enhances the PDX-1-induced liver to pancreas transdifferentiation process. J Biol Chem 2009;284:33509—335020.

[112] Tateishi K, He J, Taranova O, Liang G, D'Alessio AC, Zhang Y. Generation of insulin-secreting islet-like clusters from human skin fibroblasts. J Biol Chem 2008;283:31601—7.

[113] van der Windt DJ, Bottino R, Kumar G, Wijkstrom M, Hara H, Ezzelarab M, et al. Clinical islet xenotransplantation: how close are we? Diabetes 2012;61:3046—55.

[114] Groth CG, Korsgren O, Tibell A, Tollemar J, Moller E, Bolinder J, et al. Transplantation of porcine fetal pancreas to diabetic patients. Lancet 1994;344:1402—4.

[115] Heneine W, Tibell A, Switzer WM, Sandstrom P, Rosales GV, Mathews A, et al. No evidence of infection with porcine endogenous retrovirus in recipients of porcine islet-cell xenografts. Lancet 1998;352:695—9.

[116] Valdes-Gonzalez RA, Dorantes LM, Garibay GN, Bracho-Blanchet E, Mendez AJ, Davila-Perez R, et al. Xenotransplantation of porcine neonatal islets of Langerhans and Sertoli cells: a 4-year study. Eur J Endocrinol 2005;153: 419—27.

[117] Wang W. A pilot trial with pig-to-man islet transplantation at the 3rd Xiang-Ya Hospital of the Central South University in Changsha. Xenotransplantation 2007;14:358.

[118] Sykes M, Cozzi E, d'Apice A, Pierson R, O'Connell P, Cowan P, et al. Clinical trial of islet xenotransplantation in Mexico. Xenotransplantation 2006;13:371—2.

[119] Groth C. Towards developing guidelines on Xenotransplantation in China. Xenotransplantation 2007;14:358—9.

[120] Sykes M, Pierson 3rd RN, O'Connell P, D'Apice A, Cowan P, Cozzi E, et al. Reply to "Critics slam Russian trial to test pig pancreas for diabetes". Nat Med 2007;13:662—3.

[121] Rajotte RV. Moving towards clinical application. Xenotransplantation 2008;15:113—5.

[122] Matsumoto S, Okitsu T, Iwanaga Y, Noguchi H, Nagata H, Yonekawa Y, et al. Insulin independence after living-donor distal pancreatectomy and islet allotransplantation. Lancet 2005;365:1642—4.

[123] Matsumoto S, Okitsu T, Iwanaga Y, Noguchi H, Nagata H, Yonekawa Y, et al. Follow-up study of the first successful living donor islet transplantation. Transplantation 2006;82:1629—33.

[124] Hirshberg B. Can we justify living donor islet transplantation? Curr Diab Rep 2006;6:307—9.

[125] Korsgren O, Nilsson B, Berne C, Felldin M, Foss A, Kallen R, et al. Current status of clinical islet transplantation. Transplantation 2005;79:1289—93.

[126] Cravedi P, van der Meer IM, Cattaneo S, Ruggenenti P, Remuzzi G. Successes and disappointments with clinical islet transplantation. Adv Exp Med Biol 2010;654:749—69.

[127] Johansson H, Lukinius A, Moberg L, Lundgren T, Berne C, Foss A, et al. Tissue factor produced by the endocrine cells of the islets of Langerhans is associated with a negative outcome of clinical islet transplantation. Diabetes 2005;54:1755—62.

[128] Cabric S, Sanchez J, Lundgren T, Foss A, Felldin M, Kallen R, et al. Islet surface heparinization prevents the instant blood-

mediated inflammatory reaction in islet transplantation. Diabetes 2007;56:2008−15.

[129] Koh A, Senior P, Salam A, Kin T, Imes S, Dinyari P, et al. Insulin-heparin infusions peritransplant substantially improve single-donor clinical islet transplant success. Transplantation 2010;89:465−71.

[130] Eich T, Eriksson O, Lundgren T. Visualization of early engraftment in clinical islet transplantation by positron-emission tomography. N Engl J Med 2007;356:2754−5.

[131] Merani S, Toso C, Emamaullee J, Shapiro AM. Optimal implantation site for pancreatic islet transplantation. Br J Surg 2008;95:1449−61.

[132] van der Windt DJ, Echeverri GJ, Ijzermans JN, Coopers DK. The choice of anatomical site for islet transplantation. Cell Transplant 2008;17:1005−14.

[133] Figliuzzi M, Plati T, Cornolti R, Adobati F, Fagiani A, Rossi L, et al. Biocompatibility and function of microencapsulated pancreatic islets. Acta Biomater 2006;2:221−7.

[134] Beck J, Angus R, Madsen B, Britt D, Vernon B, Nguyen KT. Islet encapsulation: strategies to enhance islet cell functions. Tissue Eng 2007;13:589−99.

[135] Cornolti R, Figliuzzi M, Remuzzi A. Effect of micro- and macroencapsulation on oxygen consumption by pancreatic islets. Cell Transplant 2009;18:195−201.

[136] O'Sullivan ES, Vegas A, Anderson DG, Weir GC. Islets transplanted in immunoisolation devices: a review of the progress and the challenges that remain. Endocr Rev 2011;32:827−44.

[137] Basta G, Montanucci P, Luca G, Boselli C, Noya G, Barbaro B, et al. Long-term metabolic and immunological follow-up of non-immunosuppressed patients with type 1 diabetes treated with microencapsulated islet allografts: four cases. Diabetes Care 2011;34:2406−9.

[138] Koulmanda M, Qipo A, Fan Z, Smith N, Auchincloss H, Zheng XX, et al. Prolonged survival of allogeneic islets in cynomolgus monkeys after short-term triple therapy. Am J Transplant 2012;12:1296−302.

[139] Berney T, Ferrari-Lacraz S, Buhler L, Oberholzer J, Marangon N, Philippe J, et al. Long-term insulin-independence after allogeneic islet transplantation for type 1 diabetes: over the 10-year mark. Am J Transplant 2009;9:419−23.

[140] Huurman VA, Velthuis JH, Hilbrands R, Tree TI, Gillard P, van der Meer-Prins PM, et al. Allograft-specific cytokine profiles associate with clinical outcome after islet cell transplantation. Am J Transplant 2009;9:382−8.

[141] Schiopu A, Wood KJ. Regulatory T cells: hypes and limitations. Curr Opin Organ Transplant 2008;13:333−8.

[142] The Juvenile Diabetes Research Foundation Continuous Glucose Monitoring Study Group. Continuous glucose monitoring and intensive treatment of type 1 diabetes. N Engl J Med 2008;359:1464−76.

[143] Elleri D, Allen JM, Kumareswaran K, Leelarathna L, Nodale M, Caldwell K, et al. Closed-Loop basal insulin delivery over 36 hours in adolescents with type 1 diabetes: randomized clinical trial. Diabetes Care 2012.

[144] Chatenoud L, Bluestone JA. CD3-specific antibodies: a portal to the treatment of autoimmunity. Nat Rev Immunol 2007;7:622−32.

[145] Hu CY, Rodriguez-Pinto D, Du W, Ahuja A, Henegariu O, Wong FS, et al. Treatment with CD20-specific antibody prevents and reverses autoimmune diabetes in mice. J Clin Invest 2007;117:3857−67.

[146] Pescovitz MD, Greenbaum CJ, Krause-Steinrauf H, Becker DJ, Gitelman SE, Goland R, et al. Rituximab, B-lymphocyte depletion, and preservation of beta-cell function. N Engl J Med 2009;361:2143−52.

[147] Couri CE, Oliveira MC, Stracieri AB, Moraes DA, Pieroni F, Barros GM, et al. C-peptide levels and insulin independence following autologous nonmyeloablative hematopoietic stem cell transplantation in newly diagnosed type 1 diabetes mellitus. JAMA 2009;301:1573−9.

[148] Machen J, Harnaha J, Lakomy R, Styche A, Trucco M, Giannoukakis N. Antisense oligonucleotides down-regulating costimulation confer diabetes-preventive properties to nonobese diabetic mouse dendritic cells. J Immunol 2004;173:4331−41.

[149] Phillips B, Nylander K, Harnaha J, Machen J, Lakomy R, Styche A, et al. A microsphere-based vaccine prevents and reverses new-onset autoimmune diabetes. Diabetes 2008;57:1544−55.

[150] Ruggenenti P, Remuzzi A, Remuzzi G. Decision time for pancreatic islet-cell transplantation. Lancet 2008;371:883−4.

[151] Cravedi P, Remuzzi A, Remuzzi G. Comment on: Robertson (2010) Islet transplantation a decade later and strategies for filling a half-full glass. Diabetes 2010;59:1285−91 [Diabetes 59, 13].

Chapter 41

Pancreatic Islets Regeneration: The Bioengineering Approach

Timil Patel[a], Marcus Salvatori[b], Sij Hemal[a], Andrea Peloso[a,c], Ravi Katari[a], Joao Paulo Zambon[a], Shay Soker[a], and Giuseppe Orlando[a]

[a]Wake Forest School of Medicine, Winston-Salem, [b]Newcastle School of Medicine, University of Newcastle-upon-Tyne, UK, [c]School of Medicine, University of Pavia, Pavia, Italy

41.1 INTRODUCTION

The pancreas is a mixed exocrine and endocrine organ that functions in the digestion, metabolism, and energy utilization for the body [1]. One of the most important roles of the pancreas is the physiologic control of glucose homeostasis, which is achieved by the release of two key hormones: insulin and glucagon [1]. The pancreatic organ is located in the retroperitoneum and its mass is largely composed of exocrine cells clustered in lobules (acini) [1]. Within these acini are further clusters of endocrine cells called the islets of Langerhans of which two types predominate: beta cells and alpha cells [1]. Alpha cells of the pancreatic islets secrete glucagon and beta cells secrete insulin [2]. Both of these hormones coordinate fuel storage and utilization in the body [2].

Diabetes mellitus is a metabolic disorder of the endocrine pancreas that fails to produce sufficient insulin to maintain glucose homeostasis [2]. Worldwide, it is projected that 366 million individuals will be afflicted by diabetes in 2030, drastically up from 171 million in 2000 [3]. An aging population, obesity and physical inactivity are some of the major driving factors that are drastically increasing the prevalence of diabetes [3]. Patients with diabetes exhibit high risks for severe complications involving nearly every major organ, heart failure being the leading cause of death [4].

The treatment of diabetes mellitus, however, remains inadequate. When Holman and his team followed two groups of diabetic patients receiving either dietary restriction or intensive hypoglycemic medications to control blood glucose over 10 years, it was reported that the patients undergoing intensive therapy had a relative reduction in risk for all-cause mortality of only 27% [5]. In fact, intensive glycemic control in combination with dietary modification, increased physical activity and compliant use of hypoglycemics and exogenous insulin has not been shown to eliminate microvascular and macrovascular complications of diabetes [6]. In addition, although exogenous insulin therapy prevents acute metabolic decompensation in type 1 diabetes, <40% of patients actually achieve and maintain therapeutic targets [7]. When one looks cumulatively, therefore, multisystem organ damage in diabetes still remains a leading cause of morbidity and mortality. The current management guidelines of diabetics, however, stress continued use of pharmaceuticals and improving lifestyle choices. While medications and lifestyle changes certainly reduce the incidence of serious diabetic complications, they do not pave the way for complete remission or cure.

Beta-cell replacement through whole pancreas or islet cell transplantation is the sole treatment able to establish long-term, stable euglycemia in diabetic patients, particularly in type 1 diabetics. Whole pancreas transplantation was first performed by Kelly et al. [8] and has been shown to produce higher rates of insulin independence than islet transplantation. Navarro et al. studied the influence of pancreas transplantation and found that patients with abnormal autonomic function showed long-term improved survival with transplantation than those patients without transplantation [9]. Zehr et al. further surveyed quality of life posttransplantation and found high patient satisfaction fueled largely by improved secondary diabetic complications like neuropathy [10]. Nonetheless, the mentioned benefits of pancreas transplantation are overshadowed by the lifelong immunosuppression and its subsequent complications and the dire shortage of available organs for transplantation. As a result, clinical application is limited.

Pancreatic islet transplantation was introduced later than its whole pancreas counterpart. To test the benefits of pancreas-islet transplantation, the "Edmonton protocol" used an immunosuppressive regimen of sirolimus or tacrolimus to prevent graft rejection and track posttransplant insulin independence in selected patients [11]. The Edmonton protocol, however, has been heavily criticized for its underperformance, namely because average posttransplant insulin dependence did not exceed 15 months [11]. In fact, at 2 years of follow up only 24% remained free of exogenous insulin requirements [12]. Although the mechanisms that regulate pancreatic beta-cell mass are poorly understood, there are several proposed mechanisms by which to explain the disappointing performance of therapeutic islet transplantation. Nir et al. sought to study beta-cell regeneration in a diabetic state and administered doxycycline in transgenic mice [13]. They found giving doxycycline actually resulted in apoptosis of nearly 80% of beta cells. Similarly, treatment with sirolimus and tacrolimus also contributed to the inhibition of beta-cell regeneration [13]. These results thus point out that immunosuppression negatively affects B-cell proliferation. Furthermore, Jansson and Carlsson found that blood perfusion and tissue oxygen tension of grated islets were notably decreased thus highly suggesting that vascular integrity is disturbed with the transplantation of isolated islets [14]. Other physiological stressors proposed to decrease the efficacy of islet transplantation include low isolation efficiency [12], disruption of the basement membrane (BM) [15], and inflammatory-mediated host response [16]. In addition, each recipient of an islet transplantation procedure requires 2−4 cadaveric pancreata, thus worsening an already small organ pool [17]. Taking into account all of these weaknesses of therapeutic islet transplantation, the American Diabetes Association does not foresee a viable future in this practice [18].

New and pioneering advances in the field of regenerative medicine strive to overcome the limitations of existing diabetes treatment strategies. Collective efforts across islet encapsulation technology, investigating the importance of the extracellular membrane and refining methods for decellularization−recellularization promises to create a bioengineered pancreas capable of restoring function in patients with diabetes.

41.2 ISLET ENCAPSULATION

Regenerative medicine promises to contribute to the advancement of islet transplantation through the development of microencapsulation technology and the exploitation of bioengineered microenvironments [19]. The fundamental goal of encapsulation of cells is to get around the immune-mediated destruction of the graft without the use of toxic immunosuppression [20]. Encapsulation is a means of immune isolation which serves to "camouflage" the foreign antigens of the islet allo- or xeno-graft from host immune surveillance [19]. The scheme of encapsulation comprises placing islets within a semipermeable membrane composed of inert material [20]. The membrane has pores small enough to deter the entrance of immune cells and antibodies but allow the passage of smaller molecules, such as insulin and glucose [20].

Theoretically, successful encapsulation eliminates the need for aggressive, lifelong immunosuppression, with consequent improvements in β-cell viability and host morbidity [19]. There are several elements, however, that impact the survival and function of microencapsulated human islet cells. Although promising results have been obtained in early animal studies, the clinical value of islet encapsulation has been limited by the following obstacles, recently reviewed by Vaithilingam and Tuch [20]: (i) poor biocompatibility of capsule material, (ii) inadequate immune isolation due to the penetration of small immune mediators, and (iii) hypoxia secondary to failed revascularization [19,20].

Biocompatibility of the microcapsules is essential for the vitality and success of the microencapsulated islet cells. Bioincompatibility, on the other hand, leads to islet cell death through extensive pericapsular fibrosis [21]. The failure of biocompatibility is a failure of the quality and purity of the encapsulation material [20]. In order to increase biocompatibility and concurrently decrease fibrotic overgrowth, exclusion chromatography alongside barium bead formation to purify protein contaminants has resulted in less fibrotic growth [22].

The definition of immune protection rests in the capacity of the microcapsule to shield the encapsulated islets from the human immune system and the notion of immune isolation can be traced all the way back to the 1930s [20]. The size of the membrane pores on the microcapsule protects the inner islets from large antibodies and immune cells [20]. However, small molecules, such as chemokines, cytokines, and nitric oxide, can still cause damage to the encapsulated islet cells by attracting macrophages (chemokines) and direct cellular damage (nitrous oxide) [20]. Novel techniques, however, have been studied to immunologically protect the encapsulated islets including coencapsulation with agents such as erythrocytes and Sertoli cells, which release immunosuppressive factors [20,23−25].

Native human islets are vested in a profoundly intrinsic network of capillaries and are thus continuously receiving plenty of nutrients and oxygen [26]. In an attempt to isolate islet cells, capillary networks are compromised and the islets undergo hypoxic stress [20]. A similar strategy to overcome chemokines and nitrous oxide for immune protection has been used to the issue of hypoxia. Various coencapsulation agents like fibroblast growth factor (FGF)-1, however, have been shown to increase neovascularization [27].

41.3 THE IMPORTANCE OF THE PANCREATIC EXTRACELLULAR MATRIX

Biomaterial carriers can certainly fulfill the structural and anchorage requirements of the seeded cells; however, they cannot extend and accomplish the breadth of functions performed by the native extracellular matrix (ECM). Over the last decades, research has shown that the ECM plays a crucial role in the welfare of cells, tissues, and organs. The role of the ECM is extensive and dynamic. One of the long established functions of the ECM is its ability to provide structural support for organs and tissues. Recently, however, more attention has focused on the ECM's unique role in the determination, differentiation, proliferation, survival, polarity migration, and behavior of cells [28].

The ECM contains large and complex proteins that exhibit distinct domains whose sequences and arrangements are highly conserved [28]. These domains serve various functions. Some domains have been found to bind to adhesion receptors like integrins that mediate cell−matrix adhesions while other domains have been implicated in the binding of growth factors and transduction of signaling into cells. [28]. Several growth factors, notably FGFs and vascular endothelial growth factors (VEGFs), bind strongly to heparin and to heparan sulfate, an element of many ECM proteoglycans (PGs) [28]. It has been proposed that growth factors are released from the ECM by degradation of ECM proteins and of the glycosaminoglycan components of PGs thus furthering the notion that the heparin sulfate PGs essentially serve as a basin for various growth factors and that the presentation of these growth factor signals by the ECM proteins is an important function of the ECM itself [28]. However, it has also been proposed that heparan sulfate serves as a key cofactor for some growth factors [29]. For example, the binding of FGF to its receptor fibroblast growth factor receptor (FGFR) is contingent upon a heparan sulfate of the ECM binding simultaneously [29]. As a result, there is no doubt of the significance of the native ECM in the function and downstream effectiveness of growth factors, such as FGF and VEGF.

The native pancreatic (ECM) is a three-dimensional, structural framework of proteins in a state of "dynamic reciprocity" with the cells of the endocrine pancreas [30]. The ECM comes in two forms—interstitial matrix and BM [31]. The islet BM is composed of collagen IV, laminin, and fibronectin (FN) [31]. The ECM is a vital component of the islets' microenvironment because an interference of the cell-ECM relationship leads to the induction of apoptosis [31].

The ECM regulates essential aspects of islet biology. To begin with, it has been demonstrated that cultured pancreatic endocrine cells placed within a three-dimensional collagen matrix have the capacity to reorganize into islet-like organoids in vitro thus strongly implying that the ECM plays the role in this process [32]. The precise mechanism of the ordered distribution of the various cell types remains elusive, however, one theory is that non-B cells (which are concentrated peripherally) have a greater affinity for interaction with the ECM compared with the B cells (which are concentrated centrally) [32].

The ECM also has been shown to be involved in the development of the pancreas with strong evidence that the matrix plays a key role in the epithelial−mesenchymal interactions during the developmental process of the pancreatic tissues [33,34]. It has been proposed that the presence of BM in the developing embryonic pancreatic epithelium is essential for the formation of the pancreatic ducts [33]. In addition, when investigating the rat pancreatic morphology at various developmental stages in utero, Hisaoka et al. determined that morphological changes in the pancreas were closely related to parallel changes in the ECM organization especially at critical junctures of pancreatic development, such as elongation of the branching epithelium and acinar cell differentiation [34].

Furthermore, it has been noted that specific components of the ECM are capable of inducing the differentiation of pancreatic B cells [35]. In vitro, Jiang et al. revealed that a component of the ECM called laminin-1 stimulated selective differentiation of pancreatic precursor cells to become insulin-producing B cells [35]. In fact, when laminin-1 was experimentally removed in the medium of the ECM, the differentiation rate of pancreatic B cells decreased by nearly two-thirds [35]. The importance of laminin-1 was further affirmed when it was discovered that the BM of fetal pancreatic epithelium highly expressed laminin-1 thus implying it also plays a role in B-cell differentiation *in vivo*[35].

The ECM-expressed laminins also impact cellular processes by binding with integrins, a family of cell surface receptors largely responsible for cell—matrix adhesion [36] and the transduction of external signals to the cytoskeleton [37]. Integrins have been found to be involved in the initiation of several pancreatic intracellular signaling cascades such as focal adhesion kinase (FAK), paxillin, extracellular signal-regulated kinases (ERK) 1/2 [38], phosphatidylinositol (PI), 3-kinase and mitogen-activated protein kinase (MAP kinase ERK), protein kinase B (PKB/Akt) [39] and NF-κB [40]. Since laminins are a component of the ECM and are also involved in the activation of various pro-survival signaling pathways, the role of ECM is thereby inferred in the importance of pancreatic islet cell survival and functionality.

The ECM has also been shown to be involved in intracellular signaling. Rondas and his team found that by blocking the interaction between B1 integrins and the ECM with an anti-B1 integrin antibody, there was corresponding inhibition of glucose-induced phosphorylation of FAK (Tyr-397), paxillin (Tyr-118) and ERK1/2 (Thr-202/Tyr-204) [36]. Of grave importance, studies done by Weber et al. have also indicated that the ECM is vital in the survival of pancreatic beta cells [41]. When pancreatic beta cells were placed in either an ECM-containing gel or a non-ECM-containing gel, Weber et al. discovered that survival and functionality were markedly greater in beta cells in the ECM gels [41]. Hammer et al. similarly found that pancreatic beta cells cultured on ECM had decreased activity Caspase-8 (proapoptosis protein) when compared to beta cells cultured without the native ECM [40]. As the pace of discovery in the functional breadth of the ECM continues to expand, the native ECM undoubtedly is becoming the preferred platform for methods in pancreatic bioengineering.

41.4 DECELLULARIZATION—RECELLULARIZATION TECHNOLOGY

The regeneration of whole organs can be a big step in hopes to alleviate and someday even eliminate the hurdles that currently face organ transplantation: a dire shortage of available organs and lifelong immunosuppression post-transplantation [30]. The myriad of functions exerted by native ECM, all essential to cells welfare, has led scientists to deem this native matrix as—possibly—the ideal supporting scaffold for pancreatic bioengineering [30].

The justification for using a native matrix is to sequester important ECM proteins that provide protein "roadmaps" of previous resident cells [30]. Additionally, the benefit of matrix proteins is they are highly conserved, thus removing all cellular contents via decellularization has the potential of producing a promising minimally immunogenic scaffold with an intact structure for new tissue regeneration [30]. Pioneering and cutting-edge technologies in regenerative medicine have allowed researchers to recognize the advantages of preserving innate ECM for organ bioengineering and regeneration [42−45]. The innate ECM represents a biochemical, geometric, and spatially ideal platform for regeneration and implantation because it is both biocompatible [46] and has basic proteins and polysaccharides combined with matrix-bound growth factors and cytokines already preserved at physiological levels [47]. Also, the native ECM preserves an undamaged and patent vasculature capable of sustaining physiologic blood pressures when implanted *in vivo*[46]. As a result, regenerative medicine researchers around the world are now investigating the use of animal-derived ECM scaffolds as a platform for organ bioengineering and regeneration studies.

The technique of producing scaffolds of the ECM from whole animal or human-cadaveric organs involves removing all cells from an organ (decellularization) through exposing the organ parenchyma to detergents, proteases and chemicals by perfusion of the native vasculature [42−44,48]. Concurrently, the decellularization process preserves the native composition of and structure of the ECM (see Figure 2 in Chapter 19) [48]. The fundamental objective of organ decellularization is to eradicate all cellular material without adversely disturbing the composition, biological activity and mechanical integrity of the three-dimensional matrix, which is critical for successful cell regeneration [48].

The protocols for successful and robust decellularization comprise a blend of physical, chemical, and enzymatic approaches [49]. The initial step in decellularization is the lysis of the cell membrane using physical treatments, chemical, and ionic solutions [49].

Common methods for physical treatment involve snap freezing, which forms intracellular ice crystals thereby leading to cell membrane disruption and rupture [49,50]. Another physical method involves applying direct pressure to the tissues; however, tissues such as liver and lung that do not have a densely organized ECM can typically undergo this method [49].

Alkaline and acid treatments, such as acetic acid, peracetic acid (PAA), hydrochloric acid, sulfuric acid, and ammonium hydroxide (NH_4OH), can disrupt cell membranes and intracellular organelles and such treatments can be used in decellularization protocols to eradicate cytoplasmic components of the cell as well as remove nucleic acids, such as RNA and DNA [49−52]. Ionic solutions, such as sodium dodecyl sulfate (SDS) and sodium deoxycholate, work by solubilizing cytoplasmic and nuclear cellular membranes and thus removing nuclear and cytoplasmic proteins [49,53].

The lysis of the cell membrane is followed by separating the cellular components from the ECM using enzymatic treatments [49]. Enzymes, such as trypsin, endonucleases, and exonuclease, are commonly used [49]. Trypsin cleaves peptide bonds while endonucleases and exonucleases catalyze the hydrolysis of the interior and terminal bonds of ribonucleotide and deoxyribonucleotide chains, respectively [49,54,55].

Proper decellularization protocols are intended to reach a succession of key outcomes: disruption of the cell membrane, cell lysis and removal, and degradation of cytoplasmic and nuclear materials. A thorough decellularization is critical since residual cellular material may contain antigenic epitopes that can trigger an inflammatory response [56]. The completely decellularized ECM functions as an excellent skeleton onto which pluripotent cells are seeded with the intent of reconstituting the cellular compartment (recellularization). The successful recellularization of ECM scaffolds has been reported in several organ systems: liver [57], respiratory tract [58], nerve [59], tendon [60], valve [61], bladder [62], and mammary gland [27]. There rests tremendous potential and promise that building upon the aforementioned successes in various organ systems, scientists in the future may regenerate more complex, modular, and functional organs capable of performing the same as native human organs. As a result, success in bioengineered organs can address the two major obstacles facing organ transplantation: (i) shortage of available native organs and (ii) lifelong immunosupression posttransplantation.

The development of suitable bioreactors has revolutionized organ bioengineering. In principle, bioreactors provide a suitable environment where the seeded scaffolds can mature under conditions very similar to *in vivo*. After maturation in the bioreactors, these scaffolds can be prepared for implantation. Bioreactors offer several benefits for seeded scaffolds: (i) expedite the uniform distribution of seed populations onto scaffolds and (ii) provide a robust and vigorous nutrient supply, waste removal, and hydrodynamic shear stress [63,64].

Bioreactors provide two important functions for effective tissue engineering: flow-dependent shear stress and mechanical stimulation [65]. The flow-dependent shear stress enhances ensuing cellular development while persistent mechanical stimulation stimulates several signaling pathways that influence the newly engrafted cells to differentiate and assume function [65]. An ideal bioreactor device allows for adequate and precise perfusion of innate channels with the seed population, modifiable

levels of sheer stress, and real-time monitoring. Studies have shown that static recellularization as opposed to uniform cellular dispersion by perfusion bioreactors is associated with uneven deposition cells, namely increased cells around the organ exterior with a subsequent lack of cells within the parenchyma [66]. The lack of cells within the parenchyma compromises organ integrity largely due to poor deposition of cells around important vascular channels [66].

41.5 WHOLE-ORGAN PANCREAS BIOENGINEERING

There are only a handful of studies that report successful repopulation of decellularized pancreatic ECM. As a result, there is no doubt that pancreatic bioengineering lags significantly behind other organs in the field. De Carlo et al. reported that a homologous acellular pancreatic matrix was a viable scaffold for rat islet cultures long-term sustainability and function [67]. De Carlo et al. proved that islets they paired to the pancreatic matrix showed a predictive glucose-induced release all through a long-term *in vitro* incubation as opposed to islet cultures paired with nonmatrix material or with liver matrix [67]. Furthermore, the scientists aimed to implant the cultured islets into diabetic rats by placing the islet/matrix cultures into poly(vinyl alcohol) (PVA)/poly(ethylene glycol) (PEG) tubes [67]. When these tubes were then implanted into diabetic rats, De Carlo et al. noted a modest reduction of hyperglycemia [67]. Conrad et al. [68] also reported abbreviated findings describing the successful recellularization of murine pancreatic matrix with human islet cells. The islets displayed preserved glucose-stimulated insulin release and cell viability; however, complete findings have yet to be published. In another study, Omer et al. assessed porcine neonatal pancreatic cell clusters (NPCCs) microencapsulated in barium alginate that were transplanted into immunocompetent mice [69]. It was reported that blood glucose levels were normalized in 81% of the transplanted animals and remained normal throughout the 20-week experiment [69]. Omer et al. reported that barium alginate may serve as a possible agent for transplantation due to its success in evading immunity in animal models [69].

However, one serious drawback of note among these investigations is that the scaffolds described in these studies do not hold the physiological and structural capacity to support the critical mass of beta cells necessary to meet human insulin demands [67]. There is a challenge in applying decellularization−recellularization protocols to larger, more complex organs largely because perfusion decellularization works in part on diffusion [46]. Large organs with greater parenchymal mass demand higher perfusion pressures, much stronger and prolonged

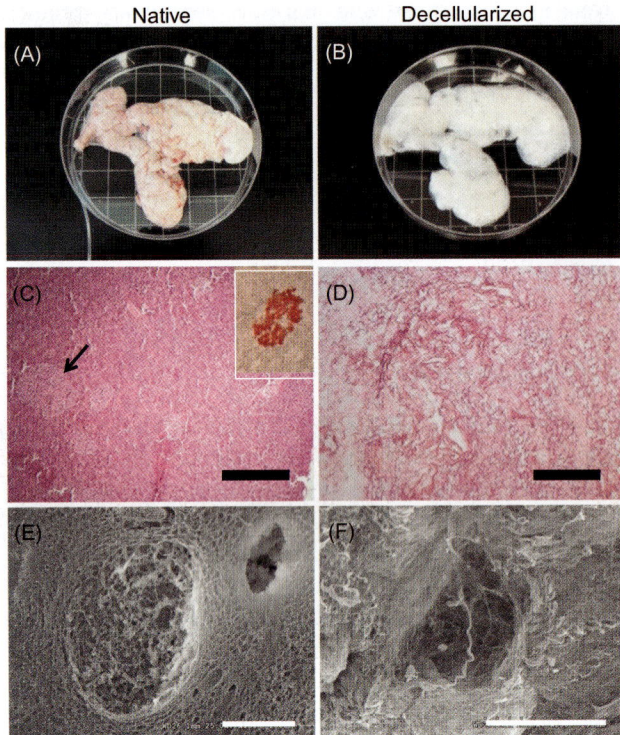

FIGURE 41.1 Decellularization of porcine pancreas. (A) Native porcine pancreas. (B) The pancreas after decellularization, characterized by whitish, translucent appearance. H&E staining before (C) and after decellularization (D); in panel D, staining shows pink eosinophilic staining typical of collagen while no basophilic staining indicative of cellular nuclear material was detected. Scanning electron micrographs of an islet in the native pancreatic matrix (E) versus putative site of an islet in the decellularized pancreatic matrix (F). All scale bars = 200 μm. *From Mirmalek-Sani SH et al. [70], with permission.*

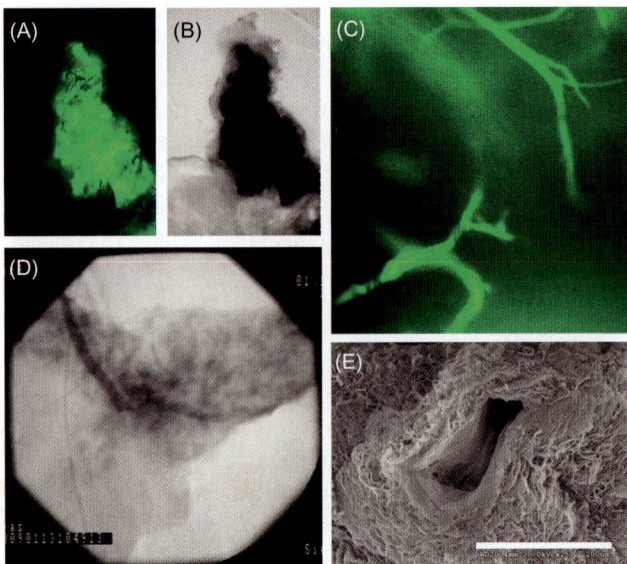

FIGURE 41.2 Preservation of intact vasculature in the decellularized porcine pancreas. (A and B) Perfusion of fluorescein isothiocyanate (FITC)-labeled dextran beads inside the decellularized pancreas, shown under fluorescent and bright light microscopy, respectively. (C) High magnification of panel A shows intact vessels inside the decellularized pancreas perfused with FITC-labeled dextran beads. (D) Fluoroangiograph of acellular porcine pancreas vasculature following perfusion of Conray® contrast agent. (F) Scanning electron micrograph of a preserved blood vessel with intact and smooth basal lamina layer. Scale bar = 200 μm. *From Mirmalek-Sani SH et al. [70], with permission.*

detergent exposures that have the potential to damage native architecture and ECM proteins [46].

Our group at Wake Forest School of Medicine recently described the generation of intact, whole-organ ECM scaffolds from a more clinically relevant model of porcine pancreas [70]. We were able to reach optimal decellularization results with continuous perfusion of Triton® X-100- and DNase-based solutions (Figure 41.1). Advanced imaging studies showed the standard decellularization protocol kept the vascular network intact, thus permitting subsequent recellularization (Figure 41.2). The decellularization protocol also did not harm or destroy the essential structural matrix proteins, such as collagen, elastic, FN, and laminin—all of which were confirmed by characterization studies of the acellular porcine pancreas (Figure 41.3). When seeded with human amniotic fluid stem cells (hAFSCs) and porcine pancreatic islets, our scaffolds did allow attachment and supported cell growth and welfare. Interestingly, porcine pancreatic ECM was able to promote glucose-mediated insulin release when

reintroduced with differentiated islets in culture (Figure 41.4). Overall, our study validates the idea that porcine pancreatic ECM is adept at supporting the growth of human amniotic progenitor cells and may represent a valuable platform for the bioengineering and regeneration of insulin-producing cells.

41.6 FUTURE CHALLENGES

The use of regenerative technologies still faces numerous challenges in the treatment of insulin-dependent diabetes. Further research needs to be done to understand exactly how growth factors enhance proliferation and insulin response. In addition, a deeper knowledge on a molecular level needs to occur to answer questions regarding the definite role of ECM proteins [71]. Furthermore, there are several unanswered questions in the field of whole-organ bioengineering. Although protocols have been written for whole-organ pancreatic decellularization in pigs and rats, analogous studies and protocols involving nonhuman primates have not been attempted. In hopes to advance the field of regenerative medicine, studies progressing to nonhuman primates are essential as there are marked differences in porcine pancreata compared to human or murine pancreata [72].

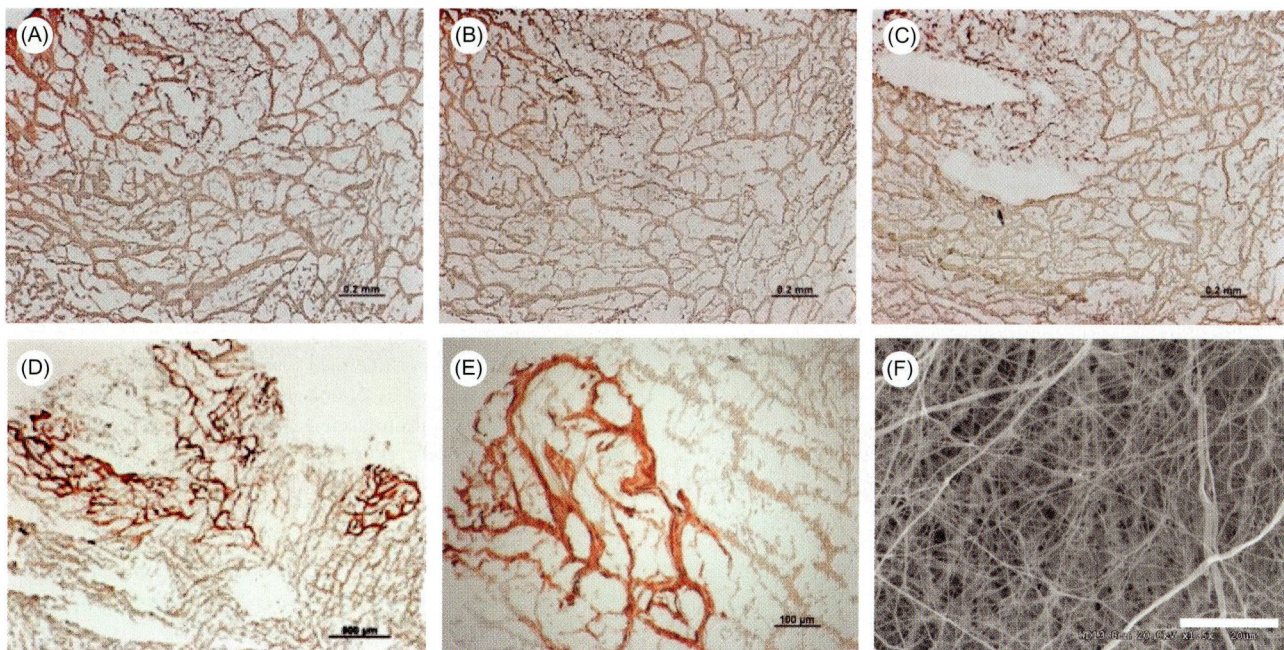

FIGURE 41.3 Characterization of ECM components in decellularized porcine pancreas. Immunohistochemical staining shows widespread expression of native ECM proteins, namely collagen type I (A), collagen type III (B), and collagen type IV (C). (D and E) Discreet staining for BM protein, laminin, presumably at the site of an islet. (F) Scanning electron micrograph shows dense and fibrous arrangement of matrix proteins. Scale bars = 200 μm (A−C), 500 μm (D), 100 μm (E), and 20 μm (F). *From Mirmalek-Sani SH et al. [70], with permission.*

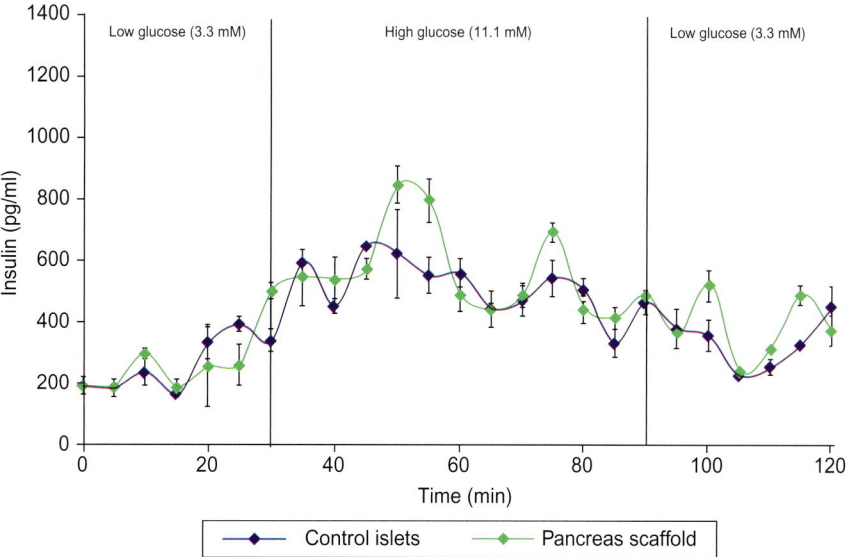

FIGURE 41.4 Porcine islet glucose challenge. Time course data from perfusion of control (unseeded) porcine islets (blue) or islets seeded onto acellular pancreatic matrix (green). Groups were cultured for 3 days then subjected to a glucose challenge, being preperfused for 30 min in low glucose medium prior to collection of samples for insulin secretion analysis. Low glucose levels (3.3 mM) represented physiological euglycemia of 60 mg/dL, and the transition to high glucose (11.1 mM) represented upper limit postprandial levels of 200 mg/dL. Both groups displayed increased insulin secretion during high glucose perfusion, with islets seeded onto scaffolds demonstrating higher peak insulin secretion values, observed specifically at 50, 55, and 75 min. Both groups showed a reduction of insulin secretion after a return to low (basal) glucose concentration. *From Mirmalek-Sani SH et al. [70], with permission.*

Moreover, even after the achievement of a bioengineer organ there remains the need for extensive collaborative effort to assess long-term viability and function.

REFERENCES

[1] Molina PE. Endocrine pancreas.. In: Molina PE, editor. Endocrine physiology. 3rd ed. New York: McGraw-Hill; 2010 [chapter 7].

[2] Funk JL. Disorders of the endocrine pancreas. In: McPhee SJ, Hammer GD, editors. Pathophysiology of disease. 6th ed. New York: McGraw-Hill; 2010 [chapter 18].

[3] Wild S, Roglic G, Green A, Sicree R, King H. Global prevalence of diabetes: estimates for the year 2000 and projections for 2030. Diabetes Care 2004;27:1047−53.

[4] Chan JC, Malik V, Jia W, Kadowaki T, Yajnik CS, Yoon KH, et al. Diabetes in Asia: epidemiology, risk factors, and pathophysiology. JAMA 2009;301:2129.

[5] Holman RR, Paul SK, Bethel MA, Matthews DR, Neil HA. 10-year follow-up of intensive glucose control in type 2 diabetes. N Engl J Med 2008;359:1577.

[6] The absence of a glycemic threshold for the development of long-term complications: the perspective of the diabetes control and complications trial. Diabetes 1996;45(10):1289−98.

[7] Orlando G, Stratta RJ, Light J. Pancreas transplantation for type 2 diabetes mellitus. Curr Opin Organ Transplant 2010;16(1):110−5.

[8] Kelly WD, Lillehei RC, Merkel FK, Idezuki Y, Goetz FC. Allotransplantation of the pancreas and duodenum along with the kidney in diabetic nephropathy. Surgery 1967;61(6):827−37.

[9] Navarro X, Kennedy WR, Loewenson RB, Sutherland DE. Influence of pancreas transplantation on cardiorespiratory reflexes, nerve conduction, and mortality in diabetes mellitus. Diabetes 1990;39(7):802−6.

[10] Zehr PS, Milde FK, Hart LK, Corry RJ. Pancreas transplantation: assessing secondary complications and life quality. Diabetologia 1991;34:S138−40.

[11] Ruggenenti P, Remuzzi A, Remuzzi G. Decision time for pancreatic islet-cell transplantation. Lancet 2008;371(9616):883−4.

[12] Shapiro AM, Ricordi C, Hering BJ, Auchincloss H, Lindblad R, Robertson RP, et al. International trial of the Edmonton protocol for islet transplantation. N Engl J Med 2006;355(13):1318−30.

[13] Nir T, Melton DA, Dor Y. Recovery from diabetes in mice by beta cell regeneration. J Clin Invest 2007;117(9):2553−61.

[14] Jansson L, Carlsson PO. Graft vascular function after transplantation of pancreatic islets. Diabetologia 2002;45(6):749−63.

[15] Nagata NA, Inoue K, Tabata Y. Co-culture of extracellular matrix suppresses the cell death of rat pancreatic islets. J Biomater Sci Polym Ed 2002;13(5):579−90.

[16] Shimizu H, Ohashi K, Utoh R, Ise K, Gotoh M, Yamato M, et al. Bioengineering of a functional sheet of islet cells for the treatment of diabetes mellitus. Biomaterials 2009;30(30):5943−9.

[17] Shapiro AM, Lakey JR, Ryan EA, Korbutt GS, Toth E, Warnock GL, et al. Islet transplantation in seven patients with type 1 diabetes mellitus using a glucocorticoid-free immunosuppressive regimen. N Engl J Med 2000;343(4):230−8.

[18] Robertson RP, Davis C, Larsen J, Stratta R, Sutherland DE. Pancreas and islet transplantation in type 1 diabetes. Diabetes Care 2006;29(4):935.

[19] Orlando G. Immunosuppression-free transplantation reconsidered from a regenerative medicine perspective. Expert Rev Clin Immunol 2012;8(2):179−87.

[20] Vaithilingam V, Tuch BE. Islet transplantation and encapsulation: an update on recent developments. Rev Diabet Stud 2011;8(1):51−67.

[21] de Vos P, Straaten JF, Nieuwenhuizen AG, de Groot M, Ploeg RJ, De Haan BJ, et al. Why do microencapsulated islet grafts fail in the absence of fibrotic overgrowth? Diabetes 1999;48(7):1381−8.

[22] Menard M, Dusseault J, Langlois G, Baille WE, Tam SK, Yahia L, et al. Role of protein contaminants in the immunogenicity of alginates. J Biomed Mater Res B Appl Biomater 2010;93(2):333−40.

[23] Wiegand F, Kroncke KD, Kolb-Bachofen V. Macrophage-generated nitric oxide as cytotoxic factor in destruction of alginate-encapsulated islets. Protection by arginine analogs and/or coencapsulated erythrocytes. Transplantation 1993;56(5):1206−12.

[24] Yang H, Wright JR. Co-encapsulation of sertoli enriched testicular cell fractions further prolongs fish-to-mouse islet xenograft survival. Transplantation 1999;67(6):815−20.

[25] Chae SY, Lee M, Kim SW, Bae YH. Protection of insulin secreting cells from nitric oxide induced cellular damage by crosslinked hemoglobin. Biomaterials 2004;25(5):843−50.

[26] Carlsson PO, Palm F, Andersson A, Liss P. Markedly decreased oxygen tension in transplanted rat pancreatic islets irrespective of the implantation site. Diabetes 2001;50(3):489−95.

[27] Wicha MS, Lowrie G, Kohn E, Bagavandoss P, Mahn T. Extracellular matrix promotes mammary epithelial growth and differentiation *in vitro*. Proc Natl Acad Sci USA 1982;79 (10):3213−7.

[28] Hynes RO. The extracellular matrix: not just pretty fibrils. Science 2009;326(5957):1216−9.

[29] Mohammadi M, Olsen SK, Goetz R. A protein canyon in the FGF−FGF receptor dimer selects from an à la carte menu of heparan sulfate motifs. Curr Opin Struct Biol 2005;15:506−16.

[30] Song JJ, Ott HC. Organ engineering based on decellularized matrix scaffolds. Trends Mol Med 2011;17(8):424−32.

[31] Wang RN, Rosenberg L. Maintenance of beta-cell function and survival following islet isolation requires re-establishment of the islet-matrix relationship. J Endocrinol 1999;163(2):181−90.

[32] Montesano R, Mouron P, Amherdt M, Orci L. Collagen matrix promotes reorganization of pancreatic endocrine cell monolayers into islet-like organoids. J Cell Biol 1983;97(3):935−9.

[33] Gittes GK, Galante PE, Hanahan D, Rutter WJ, Debase HT. Lineage-specific morphogenesis in the developing pancreas: role of mesenchymal factors. Development 1996;122(2):439−47.

[34] Hisaoka M, Haratake J, Hashimoto H. Pancreatic morphogenesis and extracellular matrix organization during rat development. Differentiation 1993;53(3):163−72.

[35] Jiang FX, Cram DS, DeAizpurua HJ, Harrison LC. Laminin-1 promotes differentiation of fetal mouse pancreatic beta-cells. Diabetes 1999;48(4):722−30.

[36] Hynes RO. Integrins: versatility, modulation, and signaling in cell adhesion. Cell 1992;69(1):11−25.

[37] Stendahl JC, Kaufman DB, Stupp SI. Extracellular matrix in pancreatic islets: relevance to scaffold design and transplantation. Cell Transplant 2009;18(1):1−12.

[38] Rondas D, Tomas A, Soto-Riberio M, Wehrle-Haller B, Halban PA. Novel mechanistic link between focal adhesion

remodeling and glucose-stimulated insulin secretion. J Biol Chem 2012;287(4):2423—36.

[39] Hammar E, Parnaud G, Bosco D, Perriraz N, Maedler K, Donath M, et al. Extracellular matrix protects pancreatic beta cells against apoptosis: role of short- and long-term signaling pathways. Diabetes 2004;53(8):2034—41.

[40] Hammar EB, Irminger JC, Rickenbach K, Parnaud G, Ribaux P, Bosco D, et al. Activation of NF-kappaB by extracellular matrix is involved in spreading and glucose-stimulated insulin secretion of pancreatic beta cells. J Biol Chem 2005;280(34):30630—7.

[41] Weber LM, Hayda KN, Anesth KS. Cell-matrix interactions improve beta-cell survival and insulin secretion in three-dimensional culture. Tissue Eng Part A 2008;14(12):1959—68.

[42] Orlando G, Baptist P, Birchall M, De Coppi P, Farney A, Guimaraes-Souza NK, et al. Regenerative medicine as applied to solid organ transplantation: current status and future challenges. Transpl Int 2011;24(3):223—32.

[43] Orlando G, Wood KJ, Stratta RJ, Yoo JJ, Atala A, Soker S. Regenerative medicine and organ transplantation: past, present, and future. Transplantation 2011;91(12):1310—7.

[44] Orlando G, Wood KJ, De Coppi P, Baptista PM, Binder KW, Bitar KN, et al. Regenerative medicine as applied to general surgery. Ann Surg 2012;255(5):867—80.

[45] Badylak SF, Weiss DJ, Caplan A, Macchiarini P. Engineered whole organs and complex tissues. Lancet 2012;379(9819):943—52.

[46] Orlando G, Farney AC, Iskandar SS, Mirmalek-Sani SH, Sullivan DC, Moran E, et al. Production and implantation of renal extracellular matrix scaffolds from porcine kidneys as a platform for renal bioengineering investigations. Ann Surg 2012;256(2):363—70.

[47] Wang Y, Cui CB, Yamauchi M, Miguez P, Roach M, Malavarca R, et al. Lineage restriction of human hepatic stem cells to mature fates is made efficient by tissue-specific biomatrix scaffolds. Hepatology 2011;53(1):293—305.

[48] Badylak SF, Taylor D, Uygun K. Whole-organ tissue engineering: decellularization and recellularization of three-dimensional matrix scaffolds. Annu Rev Biomed Eng 2011;13:27—53.

[49] Gilbert TW, Sellaro TL, Badylak SF. Decellularization of tissues and organs. Biomaterials 2006;27(19):3675—83.

[50] Jackson DW, Grood ES, Arnoczky SP, Butler DL, Simon TM. Cruciate reconstruction using freeze dried anterior cruciate ligament allograft and a ligament augmentation device (LAD). An experimental study in a goat model. Am J Sports Med 1987;15:528—38.

[51] Freytes DO, Badylak SF, Webster TJ, Geddes LA, Rundell AE. Biaxial strength of multilaminated extracellular matrix scaffolds. Biomaterials 2004;25:2353—61.

[52] De Filippo RE, Yoo JJ, Atala A. Urethral replacement using cell seeded tubularized collagen matrices. J Urol 2002;168:1789—92 [discussion 1792—3].

[53] Lin P, Chan WC, Badylak SF, Bhatia SN. Assessing porcine liver-derived biomatrix for hepatic tissue engineering. Tissue Eng 2004;10:1046—53.

[54] Bader A, Schilling T, Teebken OE, Brandes G, Herden T, Steinhoff G, et al. Tissue engineering of heart valves—human endothelial cell seeding of detergent acellularized porcine valves. Eur J Cardiothorac Surg 1998;14:279—84.

[55] Dahl SL, Koh J, Prabhakar V, Niklason LE. Decellularized native and engineered arterial scaffolds for transplantation. Cell Transplant 2003;12:659—66.

[56] Badylak SF, Gilbert TW. Immune response to biologic scaffold materials. Semin Immunol 2008;20(2):109—16.

[57] Baptista PM, Siddiqui MM, Lozier G, Rodriguez SR, Atala A, Soker S. The use of whole organ decellularization for the generation of a vascularized liver organoid. Hepatology 2011;53(2):604—17.

[58] Song JJ, Kim SS, Liu Z, Madsen JC, Mathisen DJ, Vacanti JP, et al. Enhanced in vivo function of bioartificial lungs in rats. Ann Thorac Surg 2011;92(3):998—1005. discussion 1005—6.

[59] Crapo PM, Medberry CJ, Reing JE, Tottey S, van der Merwe Y, Jones KE, et al. Biologic scaffolds composed of central nervous system extracellular matrix. Biomaterials 2012;33(13):3539—47.

[60] Martinello T, Bronzini I, Volpin A, Vindigni V, Maccatrozzo L, Caporale G, et al. Successful recellularization of human tendon scaffolds using adipose-derived mesenchymal stem cells and collagen gel. J Tissue Eng Regen Med 2012 Jun 19 [Epub ahead of print].

[61] Honge JL, Funder J, Hansen E, Dohmen PM, Konertz W, Hasenkam JM. Recellularization of aortic valves in pigs. Eur J Cardiothorac Surg 2011;39(6):829—34.

[62] Loai Y, Yeger H, Coz C, Antoon R, Islam SS, Moore K, et al. Bladder tissue engineering: tissue regeneration and neovascularization of HA-VEGF-incorporated bladder acellular constructs in mouse and porcine animal models. J Biomed Mater Res A 2010;94(4):1205—15.

[63] Baiguera S, Birchall MA, Macchiarini P. Tissue-engineered tracheal transplantation. Transplantation 2010;89(5):485—91.

[64] Rauh J, Milan F, Gunther KP, Stiehler M. Bioreactor systems for bone tissue engineering. Tissue Eng Part B Rev 2011;17(4):263—80.

[65] Chen HC, Hu YC. Bioreactors for tissue engineering. Biotechnol Lett 2006;28(18):1415—23.

[66] Martin I, Obradovic B, Freed LE, Vunjak-Novakovic G. Method for quantitative analysis of glycosaminoglycan distribution in cultured natural and engineered cartilage. Ann Biomed Eng 1999;27(5):656—62.

[67] De Carlo E, Baiguera S, Conconi MT, Vigolo S, Grandi C, Lora S, et al. Pancreatic acellular matrix supports islet survival and function in a synthetic tubular device: in vitro and in vivo studies. Int J Mol Med 2010;25(2):195—202.

[68] Conrad C, Schuetz C, Clippinger B, Vacanti J, Markmann J, Ott H. Bio-engineered endocrine pancreas based on decellularized pancreatic matrix and mesenchymal stem cell/islet cell coculture. J Am Coll Surg 2010;211(3):S62.

[69] Omer A, Duvivier-Kali VF, Trivedi N, Wilmot K, Bonner-Weir S, Weird GC. Survival and maturation of microencapsulated porcine neonatal pancreatic cell clusters transplanted into immunocompetent diabetic mice. Diabetes 2003;52:69—75.

[70] Mirmalek-Sani SH, Orlando G, McQuilling JP, Pareta R, Mack D, Salvatori M, et al. Porcine pancreas extracellular matrix as a platform for endocrine pancreas engineering. Biomaterials 2013;34(22):5488—95.

[71] Otonkoski T, Banerjee M, Korsgren O, Thornell LE, Virtanen I. Unique basement membrane structure of human pancreatic islets: implications for beta-cell growth and differentiation. Diabetes Obes Metab 2008:119—27.

[72] Van Deijnen JH, Hulstaert CE, Wolters GH, van Schilfgaarde R. Significance of the peri-insular extracellular matrix for islet isolation from the pancreas of rat, dog, pig, and man. Cell Tissue Res 1992;267(1):139—46.

Chapter 42

Pancreatic Islet Regeneration: The Developmental and Stem Cell Biology Approach

Giacomo Lanzoni, Camillo Ricordi, Luca Inverardi, and Juan Domínguez-Bendala

Diabetes Research Institute, University of Miami Miller School of Medicine, Miami, FL

Chapter Outline

42.1 INTRODUCTION

Type 1 diabetes (T1D) results from the loss of insulin-producing cells. In patients with a predisposing genetic background, undefined environmental factors trigger an autoimmune attack that ultimately results in pancreatic islet β-cell destruction [1]. After disease presentation, a life-saving but chronic exogenous insulin administration regimen must be followed. This hormonal substitutive therapy cannot replicate the physiologic glycemic control exerted by native β cells. The long-term use of insulin therapy is often associated with the development of debilitating and sometimes life-threatening complications, such as vascular degeneration, blindness, amputations, and kidney failure. The quest for alternative approaches that could provide independence from exogenous insulin therapies is therefore of paramount clinical interest. Therapies that normalized blood glucose control would impact the life of millions of patients with type 1 diabetes and a significant subset of patients with type 2 diabetes (T2D) who have a functional deficiency in insulin production. Islet transplantation has proven effective at restoring normoglycemia in T1D patients [2], and it has established proof of principle that cell therapies can revert, sometimes for years, the symptoms of the disease [3]. It can slow or prevent the disease progression in recipients [4] and has been shown to markedly ameliorate the patients' quality of life [5].

A major hurdle for this approach to become widely used is the scarcity of transplantable insulin-producing cells: issues range from the limited number of available pancreata to the variability in islet yield, quality and engraftment potential [6]. Thus, we need to identify better sources of transplantable insulin-producing cells [7]. The derivation of β-like cells from stem cells is arguably the foremost strategy currently under investigation. Other promising avenues of research are based on the study of cells that already reside in the pancreas, namely residual β cells for expansion, progenitors for endogenous regeneration and exocrine cells (acinar/ductal) for reprogramming.

A second hurdle is represented by the need of immunosuppression to prevent transplant rejection and recurrence of autoimmunity (in the case of T1D). The side effects of

Regenerative Medicine Applications in Organ Transplantation.

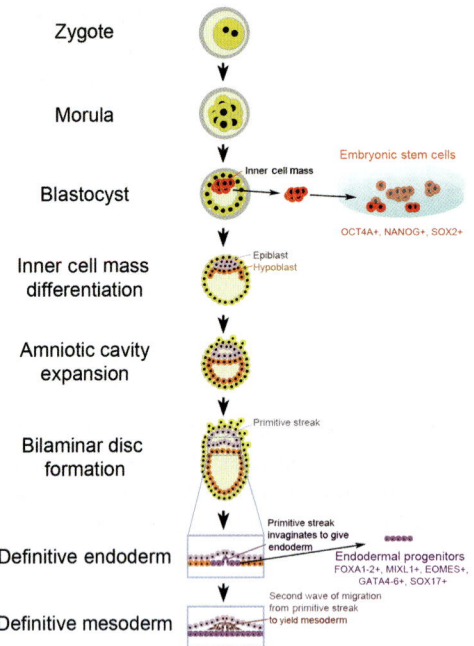

FIGURE 42.1 Generation of definitive endoderm: developmental events from the zygote to the trilaminar gastrula.

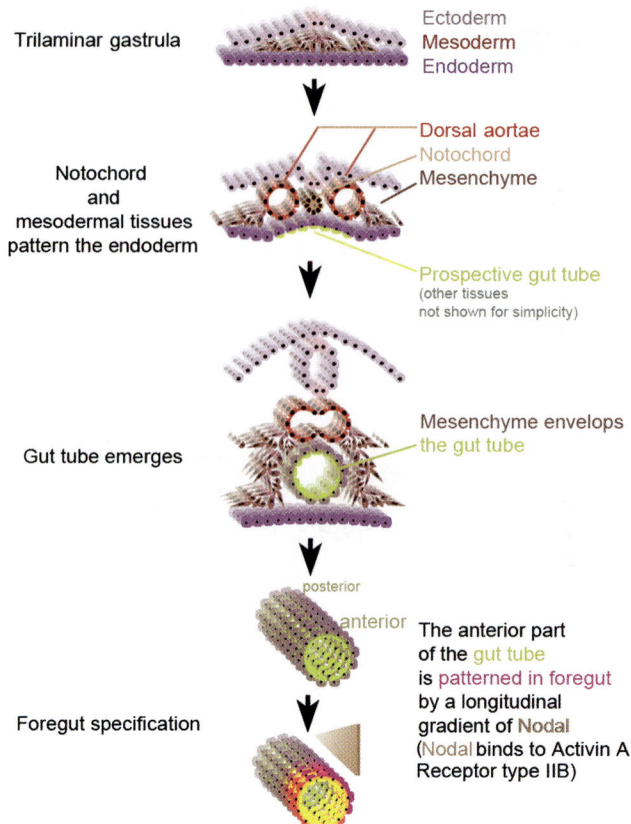

FIGURE 42.2 Patterning and specification of the foregut.

these treatments can have a serious impact on the recipient's health and quality of life [8]. Therefore, it is imperative that we refine treatments to maintain efficacy while reducing or eliminating unwanted complications, while having in mind the ultimate goal of transplantation immunology, namely the induction of tolerance, through safe and clinically applicable protocols [6].

42.2 EMBRYONIC DEVELOPMENT OF PANCREAS AND ISLETS

Most of our understanding of the basic "genetic blueprint" of pancreatic development, and particularly the lineage conducting to endocrine β cells, derives from gain- and loss-of-function studies conducted in mice. We will start our discussion by reviewing the cascade of events that takes place from the onset of the pancreatic program in the foregut epithelium to the specification of ductal, exocrine, and endocrine cell types. The objective of most stem cell differentiation protocols is to obtain functional endocrine β cells, the ones destroyed in type 1 diabetes, but other endocrine lineages are of great interest too, since the pancreatic islet exerts its functions as a coordinated system [9,10].

The most important molecular players in pancreatic development are highly conserved between mouse and human, but differences between both systems continue to emerge and thus have to be taken into account [11]. Two decades of research have shaped a basic "roadmap" of the

major molecular events driving mouse β-cell development from the early blastocyst [12–14]. Critical developmental milestones are: (a) generation of definitive endoderm/gut epithelium (Figures 42.1 and 42.2); (b) pancreatic differentiation (Figure 42.3); (c) endocrine specification (Figure 42.4); (d) β-cell differentiation (Figure 42.5); to reach (e) the functional β-cell phenotype (Figure 42.6). The process is orchestrated by a number of factors in a wide network of interactions. For the purpose of this discussion we will provide a simplified view focused on the steps where master regulators act, but the cascade of events is extremely intricate and far from being fully understood [15].

42.2.1 Generation of Definitive Endoderm/Gut Epithelium

After the formation of the zygote, the morula becomes a blastocyst through the segregation of an inner cell mass from an external trophoblastic cell layer. Embryonic stem cells (ESCs) can be isolated from the inner cell mass at this stage (Figure 42.1). The initial waves of migration and maturation from the inner cell mass give rise to the germ layers found in the subsequent embryonic stage, the

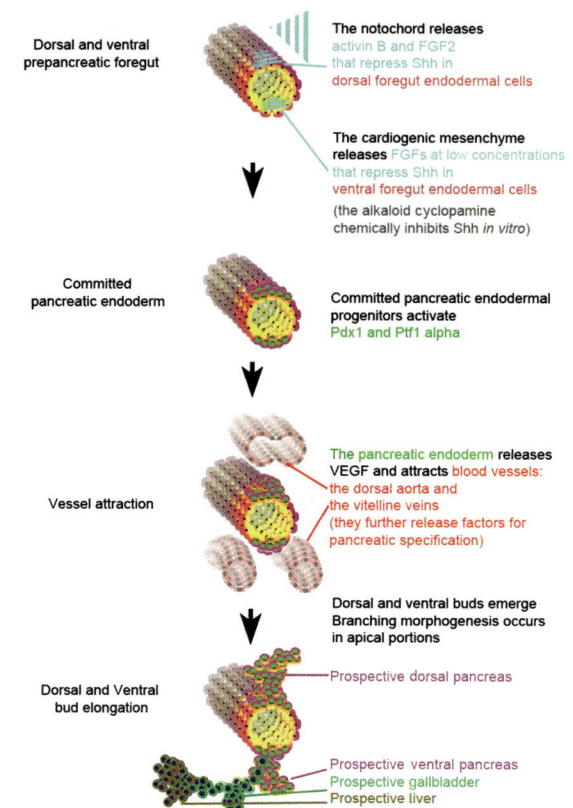

Dorsal and ventral prepancreatic foregut

The notochord releases activin B and FGF2 that repress Shh in dorsal foregut endodermal cells

The cardiogenic mesenchyme releases FGFs at low concentrations that repress Shh in ventral foregut endodermal cells (the alkaloid cyclopamine chemically inhibits Shh *in vitro*)

Committed pancreatic endoderm

Committed pancreatic endodermal progenitors activate Pdx1 and Ptf1 alpha

Vessel attraction

The pancreatic endoderm releases VEGF and attracts blood vessels: the dorsal aorta and the vitelline veins (they further release factors for pancreatic specification)

Dorsal and ventral buds emerge Branching morphogenesis occurs in apical portions

Dorsal and Ventral bud elongation

Prospective dorsal pancreas

Prospective ventral pancreas
Prospective gallbladder
Prospective liver

FIGURE 42.3   Pancreatic differentiation: evolution of the dorsal and ventral buds.

gastrula. Definitive endoderm emerges during gastrulation and is characterized by the expression of the transcription factors *Foxa1-2/HNF3beta, Mixl1, Eomes, Gata4-6, SOX17*, and other members of the *Sox* family [16,17]. The definitive endoderm-derived gut tube is subsequently patterned under the influence of signals from the surrounding developing tissues, namely the mesenchyme and the notochord. The *anterior* part of the gut tube will evolve into the foregut, from which pancreas, liver, and lungs will bud out. The initial patterning of the gut epithelium is controlled by Nodal, a member of the transforming growth factor beta (TGFβ) family [18].

42.2.2 Pancreatic Differentiation

Sonic Hedgehog (*Shh*) is highly expressed throughout the gut, where it mediates interactions between the gut endoderm and the surrounding mesoderm. However, as of e8.5 in the developing mouse, there is a region of the epithelium where *Shh* is specifically absent: the *Pdx1 + /Ptf1α+* prepancreatic endoderm [17,19]. Knockout experiments showed that the entire pancreas arises from *Pdx1+* progenitor cells [20]. The concerted action of the transcription factors *Pdx1* and *Ptf1α* is necessary for the initiation of the pancreatic program. The ability of these factors to prime

this commitment is evidenced by the fact that their simultaneous ectopic expression in the posterior endoderm induces a stable conversion into pancreas [21]. Thus, *Shh* repression and activation of *Pdx1* and *Ptf1a* are fundamental events of pancreatic specification. The notochord suppresses Shh expression in the dorsal portion of the prepancreatic region through pro-pancreatic factors like activin β and fibroblast growth factor 2 (FGF2). Chemical inhibition of *Shh* by the steroid alkaloid cyclopamine stimulates pancreatic differentiation, as Pdx1 expression is no longer restricted throughout the posterior foregut [22] (Figure 42.3). In the ventral portion of the prepancreatic region, signals deriving from the cardiogenic mesenchyme (chiefly of the FGF family), repress *Shh* at first and subsequently induce liver formation, whereas a low concentration of these signals leads to a default differentiation into ventral pancreas. A dorsal endodermal bud emerges from the foregut into the overlying mesenchyme at e9.5 in mice (around 4 weeks in human development) and a ventral bud evaginates 0.5 days later in mice (6 days later in humans). The first will give rise to the dorsal portion of the pancreas, whereas the second one will mature into liver, gallbladder, biliary tree, and ventral portion of the pancreas. After the patterning signals from the mesenchyme [23] and the notochord [24], blood vessels impart inductive instructions to the developing endoderm [25]. The early endoderm and the prepancreatic region, indeed, express vascular endothelial growth factor (VEGF) and could thus attract and induce maturation of nearby vessels [26]. The dorsal bud emerges in proximity of the dorsal aorta, whereas the ventral bud appears in proximity of the vitelline veins. Thus, blood vessels and factors derived from the endothelium represent major morphogenetic agents in pancreatic specification [27]. The buds elongate giving rise to primordia with a stalk region connected to an apical region where branching morphogenesis occurs (Figure 42.3). Minimal ultrastructural differences are seen among cells during this initial period of development, but an early wave of endocrine granule maturation occurs, predominantly positive for glucagon. The rotation of the gut brings the two developing portions of the organ into close contact and fusion occurs around e12 in the mouse (week 5 in human development). The ventral bud duct fuses with the distal portion of the dorsal bud duct, giving rise to the major pancreatic duct or duct of Wirsung. The portion of the dorsal bud duct proximal to the duodenum persists as the accessory pancreatic duct (or duct of Santorini).

42.2.3 Endocrine, Ductal, and Acinar Specification

The cellular architecture in the pancreas changes dramatically around e13 in the mouse. The development of lineage-tracing technologies, chiefly the Cre recombinase

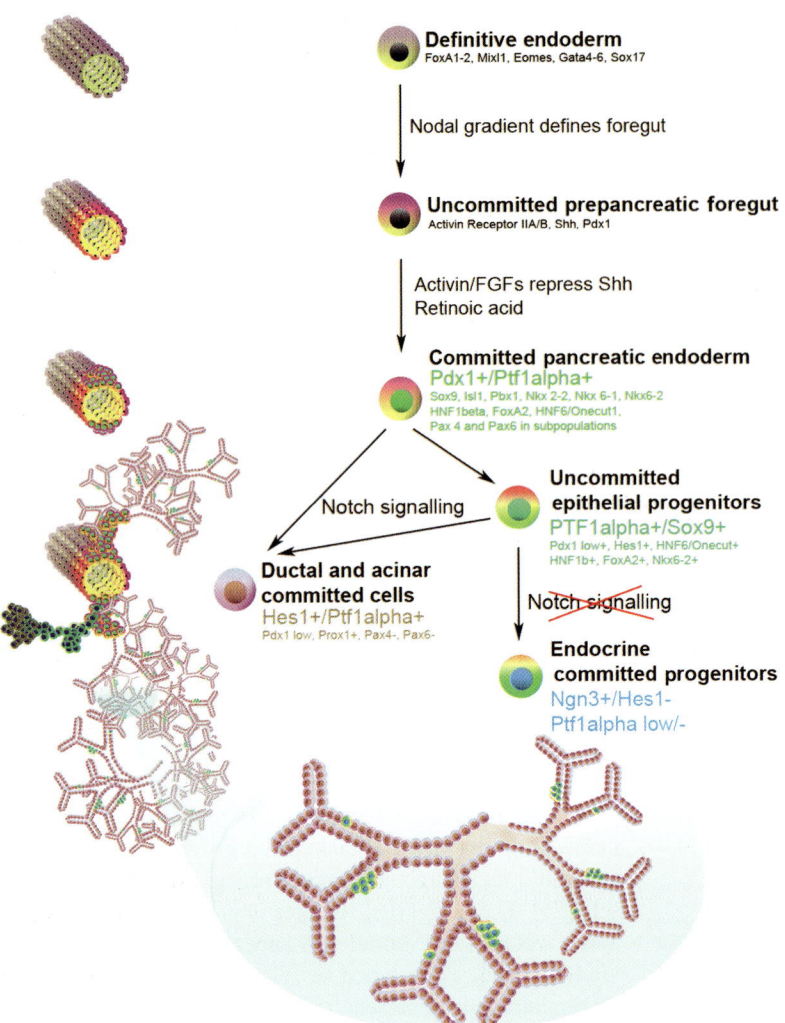

FIGURE 42.4 Endocrine, ductal and acinar specification from the definitive endoderm.

Definitive endoderm
FoxA1-2, Mixl1, Eomes, Gata4-6, Sox17

Nodal gradient defines foregut

Uncommitted prepancreatic foregut
Activin Receptor IIA/B, Shh, Pdx1

Activin/FGFs repress Shh
Retinoic acid

Committed pancreatic endoderm
Pdx1+/Ptf1alpha+
Sox9, Isl1, Pbx1, Nkx 2-2, Nkx 6-1, Nkx6-2
HNF1beta, FoxA2, HNF6/Onecut1,
Pax 4 and Pax6 in subpopulations

Notch signalling

Uncommitted epithelial progenitors
PTF1alpha+/Sox9+
Pdx1 low+, Hes1+, HNF6/Onecut+
HNF1b+, FoxA2+, Nkx6-2+

Notch signalling

Ductal and acinar committed cells
Hes1+/Ptf1alpha+
Pdx1 low, Prox1+, Pax4-, Pax6-

Endocrine committed progenitors
Ngn3+/Hes1-
Ptf1alpha low/-

(CRE)- and CRE^ER-loxP systems [28,29] has fostered major advancements in the determination of the origin of each cell type within the pancreas. Thus, Gu et al. [30,31] marked the progeny of cells expressing either *Ngn3* or *Pdx1* at different stages of development, and observed a key sequence of events in the development of the three major commitments of the pancreatic endoderm. Not only did they confirm that descendants of early *Pdx1+* progenitors give rise to every pancreatic epithelial tissue (endocrine, exocrine, and ductal) [20], but they also determined that *Ngn3+* cells represent the progenitors of islet endocrine cells, and reported that ductal progenitors are specified in a time window around day 10.5. Endocrine cells are specified from *Pdx1+* cells after activation of *Ngn3*, enabled by a switch-off of the *Notch* pathway. Acinar cells originate from *Pdx1+/Ngn3-* cells where both *Notch* signaling and *Ptf1α* expression persist. Finally, ductal cells arise from committed progenitors that

acquire Pdx1 expression between e9.5 and e11.5 and that do not express *Ngn3*.

The key instructive component driving epithelial progenitors to enter the endocrine lineage is thought to be the loss of *Notch* signaling through a *lateral specification* process [32−36]. The binding of ligands, such as *delta* or *serrate* to the *Notch* receptor stimulates the expression of high levels of *Hes1 (hairy enhancer of split 1)*, a potent repressor of the pro-endocrine transcription factor *Ngn3*. Active Notch signaling at this stage is associated with progenitor cell proliferation. Acinar and ductal lineage represent the progeny of *Hes1+* cells. However, *Hes1* repression upon Notch interruption brings about a transient and robust expression of *Ngn3*, paralleled by a downregulation of *Ptf1α*, which instructs the cell toward endocrine commitment [33−36]. The concerted action of the transcription factors Pdx1, Sox9, Hnf6, Hnf1b, and Foxa2 coordinates the activation of Ngn3 progenitors [37].

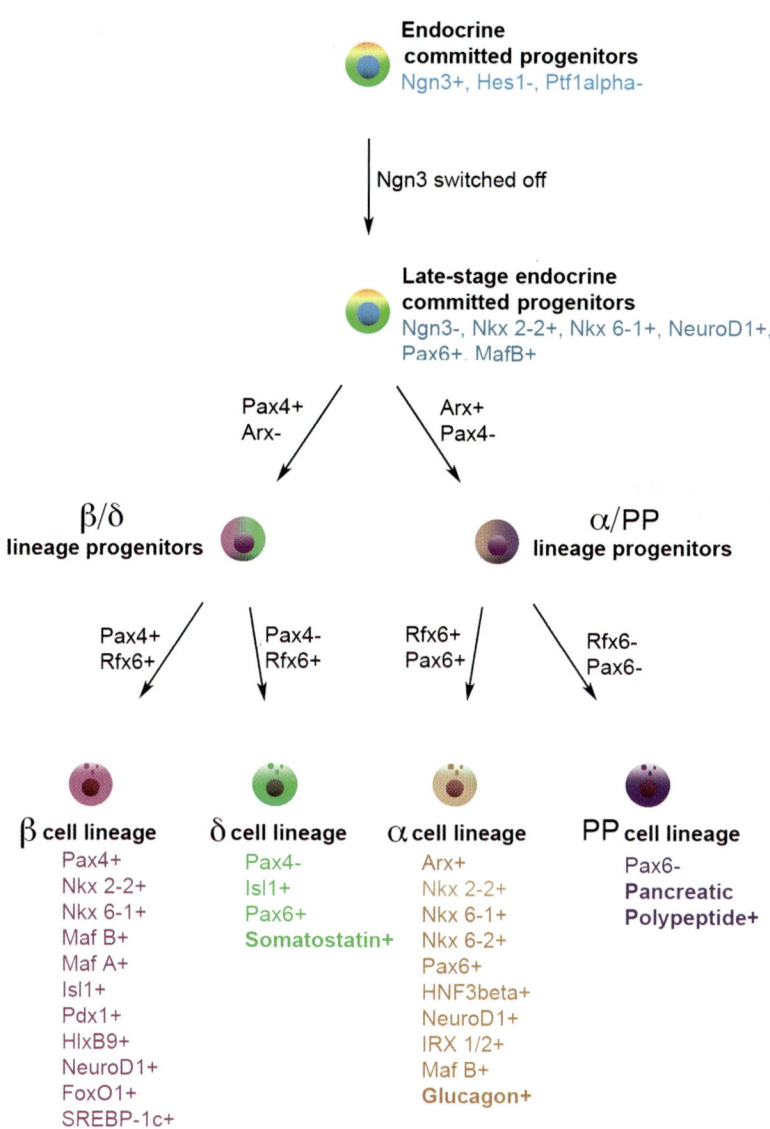

FIGURE 42.5 Endocrine-committed progenitor maturation: pancreatic endocrine lineages and β-cell differentiation.

The cross-talk between endocrine cells and blood vessels represents a key inductive interaction that, begun at the earliest stages of pancreatic differentiation [27], continues at this stage.

42.2.4 β-Cell Differentiation

The pathway controlling β-cell specification from *Ngn3+*/Hes1-/*Ptf1α-* endocrine-committed progenitors is still a field of active investigation [15]. Several transcription factors have been identified with a prominent role in the steps leading to β-cell differentiation. Animals lacking *Nkx2.2* [38] and *Nkx6.1* [39], two members of the NK family of homeodomain proteins, have defects in β-cell formation. *Nkx2.2* is expressed in the early pancreatic epithelium and then becomes restricted to *Ngn3+* progenitors. Later on, it can be found in endocrine cells with the exception of δ cells [40]. *Nkx2.2* null mutants develop with no detectable β cells, a major reduction in α cells, minor effects on PP cells and no effects on δ cells [38]. Thus, *Nkx2.2* seems to be a marker of multipotent progenitors early in pancreatic commitment and with a subsequent role in α, β, and PP lineages. Similarly, *Nkx6.1* is expressed in the committed pancreatic endoderm and its expression is subsequently restricted to β cells [41], supporting a role in multipotent and endocrine-committed progenitors [39]. *Nkx6.2* seems to have a partial redundancy with *Nkx6.1*, and *Nkx6.1/Nkx 6.2* double mutants cause a 92% reduction in the generation of β cells [42]. The central role of another transcription factor, the

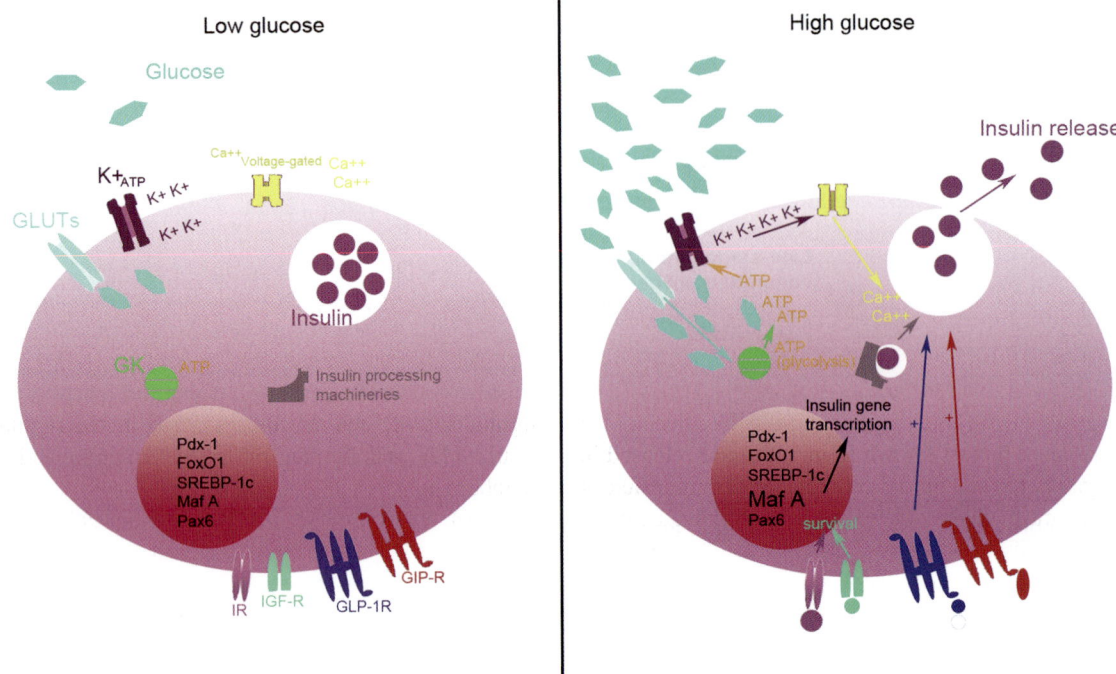

FIGURE 42.6 The functional β-cell phenotype: glucose-stimulated insulin release, transcriptional machineries, and fate-controlling network.

insulin-promoter binding *BETA2/NeuroD*, is evidenced by the fact that the knockout murine model develops a pancreas with a dramatic reduction in the number of β cells, impaired islet morphogenesis and additional abnormalities in the exocrine pancreas [43]. *Pax6* is another transcription factor with a major role in multipotent endocrine-committed progenitors and in α-cell commitment, since knockout mice exhibit an almost complete absence of α cells and a significant reduction and abnormal distribution of all other hormone-producing cells within the islet [44,45]. *Rfx6* is a transcription factor downstream of *Ngn3* involved in the commitment of progenitors for all endocrine cells except polypeptide-producing PP cells [46]. *Pax4* is a member of the *Pax* family that appears shortly after endocrine specification, colocalizing with *Ngn3* [47] and probably representing one of its targets [48,49]. *Pax4* acts downstream of *Ngn3*, *Nkx2.2*, and *Nkx6.1* as a major switch in the definition of β/δ cell progenitors, inhibiting *Arx* and the lineage selection toward α/PP cells [50–52]. *Pax4* is subsequently a hallmark of the commitment to the β lineage, since it is repressed in δ cells. The knockout of this gene results in a total absence of β cells but not α cells [53]; its expression reaches its maximum between e13.5 and e15.5 in the mouse, coinciding with an extraordinary wave of endocrine cell differentiation known as secondary transition [53,54]. Endocrine-committed progenitors and immature β cells express *MafB*, and as they mature and turn on *Pdx1* at high levels, they switch off *MafB* and start *MafA*

production [55]. *MafA* has been identified as specific to β-cell differentiation: it acts in the β-lineage reactivation of Pdx1 [55] and it is a critical regulator of the insulin gene in the mature β-cell. MafA is dispensable for the differentiation of insulin expressing cells during development, but its role is key in the functional regulation of glucose-mediated insulin secretion in mature β cell [56].

42.2.5 The Functional β-Cell Phenotype

Nature designed pancreatic islet β cells to sense glucose in the extracellular environment and to secrete insulin in response to increases in glucose concentration [57–59]. The cellular machinery synthesizes, processes, and stores insulin in granules. The β cell is dependent on glucose as a substrate for energy metabolism, and neither fatty acids nor amino acids can act as substrates to support substantial adenosine triphosphate (ATP) production. Glucose uptake from the extracellular environment is mediated by glucose transporters of different types (glucose transporter 2 (GLUT2) in the mouse, GLUT1, 2, and 3 in humans) [60]. Unlike most other tissues, the glucose metabolism in β cells is initiated by glucokinase (GK): the activity of this enzyme increases dramatically in response to minute increases in blood glucose concentrations. Glycolysis subsequently leads to an increase in the ATP/ADP ratio. High ATP levels close potassium channels and this prevents the efflux of K^+ ions from the cells. The accumulation of positive charge (K^+ ions) in the cytosol brings about the opening of voltage-gated calcium channels, resulting in an action potential. An influx of calcium ions (Ca^{++}) stimulates the exocytosis of the insulin granules and the hormone promptly reaches the bloodstream. Multiple waves of action potentials and depolarization occur in islets exposed to glucose [61]. There are two phases of insulin release. The first phase is fast and is characterized by the release of preexisting insulin granules. The second is prolonged and it involves the synthesis and release of insulin as long as blood glucose levels remain high. After lowering blood glucose to normal levels, the pool of insulin granules is replenished to be ready for the next fast phase. Oral glucose administration induces a more rapid and extensive secretion of insulin than intravenous administration [62] due to the fact that intestinal factors known as incretins are released in the bloodstream and signal to β cells the upcoming influx of glucose. The most characterized incretin is the glucagon-like peptide 1 (GLP-1), released by intestinal L cells, that can bind to the GLP-1 receptor (GLP-1R) on β cells. GLP-1 is a powerful incretin that enhances glucose-stimulated insulin secretion (GSIS) through GLP-1R activation in β cells. This probably results from the modulation of three types of ion channels: ATP-sensitive K^+ (K_{ATP}) channels, voltage-dependent Ca^{++} channels and voltage-dependent K^+ (K_V) channels, all potent glucose-dependent regulators of insulin secretion [63].

β cell-specific functions, and in particular GSIS, are regulated by the concerted action of the transcription factors *Pdx1*, *FoxO1*, *SREBP-1c*, and *MafA* [64]. A central role for Pax6 has been recently reported [65]. As described earlier, *Pdx1* plays a central role in pancreatic and β-cell development, and its degree of activation can be reduced under diabetic conditions [66]. Downregulation of *Pdx1* was shown to suppress the expression of GLP-1R, impairing the GLP-1 insulinotropic effect, and to perturb insulin processing [67]. *FoxO1* is thought to integrate cell proliferation with adaptive β-cell function. It exerts an effect on the survival and replication of mature β cells, since it is a target of pathways converging on the receptor IGF-1R, which relay survival signals from ligands, such as insulin and insulin-like growth factor-1 (IGF-1) [64,68]. *FoxO1* acts as a regulator of *Pdx1* by binding to its promoter, and it shows a differential localization in the cytoplasm or in the nuclei of, respectively, *Pdx1+* or *Pdx1−* insulin-producing cells [68]. Nuclear translocation is also stimulated by stressors, such as hyperglycemia. *FoxO1* is necessary to drive correct β-cell regeneration, as mice bearing a β cell-specific *FoxO1* mutation exhibited a reduced β-cell mass and hyperglycemia following physiologic stress. The reduction in β-cell mass was not due to death or apoptosis, but due to dedifferentiation and/or failed maturation to the β-cell phenotype [69]. *SREBP-1c* is a regulator of lipogenic enzymes in the liver: it is upregulated by and mediates the adaptation to diets rich in carbohydrates, sugars, and saturated fatty acids, and it is downregulated by dietary intake of polyunsaturated fatty acids. Diets of the first type can cause a metabolic disturbance in β cells termed glucolipotoxicity. In these cells, the activation of *SREBP-1c* has been associated to impaired insulin secretion [70]. *SREBP-1c* could repress the transcription of *Pdx1* and of the major substrate of insulin receptor signaling, IRS2, causing a decrease in ATP due to energy consumption through lipogenesis and uncoupling protein-2 (UCP-2) activation, and possibly impairing ionic channels pathways [71]. Thus, *SREBP-1c* is involved in multiple pathways regulating β-cell function. Recent findings highlighted the central role of *Pax6* in the regulation of mouse [72] and human [65] islet function and glucose metabolism.

MafA controls the transcription of insulin by binding the C1 element of the insulin promoter in humans (or the RIPE3b element in rat). Intriguingly, *MafA* expression is glucose-regulated [73]. In the mouse, this protein specifically localizes in the nuclei of insulin-positive cells, where it synergistically interacts with Pdx1 and Beta2/NeuroD1 to activate the transcription of the insulin gene. Upregulation of *MafA* alone was able to elevate insulin messenger RNA (mRNA) levels [74]. On the other hand, considering the range of pathways affected by its activity (Insulin1, Insulin2, Pdx1, NeuroD1, Nkx6.1, Glut2, GLP-1R, and pyruvate carboxylase), it is not surprising that a reduction in MafA activity is paralleled by a dysfunction of GSIS in β cells [56,75].

Pax6 regulates in the mouse the mRNA levels of a number of genes involved in β-cell function, such as insulin 1 and 2, Pdx1, MafA, GLUT2, PC1/3, GK, Nkx6.1, cMaf, PC2, GLP-1R, and gastric Inhibitory Polypeptide Receptor (GIPR) [72]. In humans, a common allelic variant of the gene is associated with reduced expression of the transcription factor itself and of its target PCSK1, a proprotein convertase. This translates in increased fasting insulin, insulin resistance, and a reduction of arginine-, glucose-, and potassium-stimulated insulin release [65].

42.2.6 Differences between Human and Mouse Pancreatic Development

Most transcription factors discussed here are highly conserved across species, but striking differences remain between human and mouse development. Early waves of endocrine differentiation occur rapidly in the mouse after pancreatic commitment, whereas in humans the first endocrine cells can be detected around 7 weeks postconception, 3 weeks after pancreatic specification [76]. In the mouse pancreas the early endocrine population is predominantly glucagon-positive, whereas in humans the first endocrine cells that appear are insulin-positive, and glucagon-expressing cells appear only after 8.5 weeks postconception. The organization into islets occurs just prior to birth in the mouse, whereas islets can be observed in the human pancreas around 12 weeks postconception [77,78].

42.3 RECAPITULATING DEVELOPMENT: STRATEGIES FOR β-CELL DIFFERENTIATION

The general aim of β-cell differentiation protocols is to obtain clinically relevant numbers of insulin-producing, glucose-responsive cells that could be used to restore β-cell mass in diabetic patients. The major strategies applied to induce *in vitro* commitment of stem cells in this direction make use of a combination of endocrine-promoting factors and culture conditions found to have a beneficial effect on islets or β-cell mass. After the tentativeness of initial approaches, which yielded questionable results, current efforts are focused on refining culture conditions designed to more accurately mimic pancreatic development.

42.4 STEM CELLS AND PROGENITORS FOR THE β-CELL PHENOTYPE

The possibility of instructing stem cells toward functional insulin-releasing cells is very promising. Over the last few years, several differentiation protocols have been developed for a plethora of different stem and progenitor cells. However, current protocols have limitations that range from a low reliability of the protocol to a limited yield in β-like cells. Some of the most successful protocols were designed to recapitulate *in vitro* several stages of pancreatic organogenesis. During development, the progression from multipotent progenitor cells to mature pancreatic islet cells is coordinated by the intricate interaction of inductive signals and regulatory elements discussed in the previous sections. Thus, mimicking that sequence of events in the bench is a challenging and cumbersome task. Two central questions still remain to be answered, namely, what is the most reliable stem/progenitor cell for the procurement of pancreatic endocrine cell phenotypes? And what is the most trustworthy and clinically translatable protocol to obtain β cells? We will now discuss the steps that have been taken to answer these two questions—which will surely keep investigators busy for years to come.

42.4.1 Pancreatic Progenitors

β cells were once thought to be incapable of proliferating in their terminally mature form. Recent findings have made us reconsider this "dogma". Now it is generally accepted that β cells can proliferate, and that this activity dynamically responds to systemic needs throughout life [79]. In adult rodents, insulin-producing cells retain a significant proliferative capacity [80]. β-cell mass grows in a number of circumstances including hyperglycemia [81−84], obesity [85], pregnancy [86], and experimental interventions, such as cellophane wrapping, streptozotocin treatment [87], partial duct ligation, and partial pancreatectomy [88]. In humans, the regenerative potential of β cells is less well documented [89], but physiological β-cell replication undoubtedly occurs in infancy and it gradually declines thereafter to adulthood [90]. Signs of β-cell regeneration and function were found in patients with type 1 diabetes not only after onset [89], but also even decades after diagnosis [91,92]. Differences among the rodent and human systems may account for discrepancies in several findings. The normal turnover for mouse β cells is 30 days, whereas for adult human β cells it can be measured in decades [93,94]. Obesity is a known inducer of β-cell mass expansion, with up to 30-fold increases in rodents and 0.5-fold in humans [93,94], but pregnancy and hyperglycemia exert relatively less marked effects in human β-cell mass. In contrast to the observations in rodents, regeneration does not seem to occur in humans after partial pancreatectomy [95]. Regardless of the model of islet regeneration, the origin of the newly formed β cells in the adult organ remains actively debated. The main pathway for β-cell mass expansion is replication from other insulin-positive cells [80,96], but the contribution from progenitors has been shown with lineage-marking approaches [97]. It is now well known that a limited number of insulin-positive cells can survive and function in the long term in type 1 diabetic patients [92,98], but unfortunately, in the majority of cases, β cells are almost completely lost, and their replicative capacity is thought to decline with age. This may ultimately hamper the success of strategies aimed at stimulating the replication of existent β cells in patients [99].

Thus, major efforts are focused in β-cell neogenesis, transdifferentiation and reprogramming from different cell subsets. Progenitor cells with the potential to undergo β-cell differentiation have been described in adult pancreatic ductal/acinar tissue. The sight of a single islet cell or small endocrine cell clusters sprouting from the ducts is not a rare finding in sections of adult pancreas. Such observations become more frequent in a number of experimental or pathological conditions [100−102]. The appearance of islet complexes in association with ducts is particularly impressive in cases of chronic pancreatitis and pancreatic fibrosis [103]. The conclusion has been that islets might emerge from ductal structures, migrating to the acinar surroundings. Studies by Bonner-Weir et al. [104] supported the hypothesis that pancreatic regeneration in the partially pancreatectomized rat can occur not only through replication of existing endocrine and exocrine cells, but also through proliferation and maturation from the ductal epithelium into new pancreatic lobules consisting of endocrine and exocrine cells. Adult murine or human ductal cells transduced with adenoviral vectors carrying the transcription factors Pdx1, Ngn3, Pax4, and NeuroD strongly upregulated the expression of the insulin gene [105].

The "ductal origin" hypothesis was put in question by the findings of Dor et al. [80] in 2004, who showed that adult β-cell regeneration chiefly occurs through self-replication rather than differentiation from noninsulin-producing pancreatic progenitors. However, recent reports suggest that cells with ductal markers do indeed give rise to new endocrine and exocrine cells both in physiological turnover, after ductal ligation, in partial pancreatectomy and under *in vitro* stimulation [106−108]. This is in line with our own findings that Pdx1 and insulin are activated in the ductal epithelium of transplanted human pancreata upon recurrence of autoimmunity [109]. However, these cells frequently show a hybrid ductal-β-cell phenotype of unknown significance. Even if the ductal hypothesis of β-cell regeneration continues to be hotly debated [110], a subset of multipotent cells residing in ducts may be instructed to commit to the β-cell fate, thus representing a major candidate for regeneration strategies. Other progenitors with pancreatic endocrine differentiation potential were described within islets [87,111−113]. Thus, Fernandes et al. [87] studied islet regeneration after streptozotocin treatment and identified a population of resident somatostatin+/Pdx1+ cells that can commit to the insulin-producing fate. Among islet resident populations, murine α cells unexpectedly showed potential to differentiate into β-like cells after ectopic expression of the β-cell commitment transcription factor Pax4 [114]. Similarly, lineage tracing the progeny of glucagon-producing cells in mice showed that α cells contributed to the regeneration of insulin-producing cells after extreme β-cell ablation, resulting in improved glycemic control [115]. After pancreatic duct ligation and alloxan-mediated ablation of β cells, α cells robustly proliferate, possibly reverting to a precursor phenotype, but can also directly convert in insulin-producing cells without cell division [116]. These findings are of great interest, since α cells represent the major component of pancreatic islets after depletion of β cells in patients with type 1 diabetes, and they are not affected by immune-mediated destruction. Moreover, they are certainly available in large numbers also in T2D patients.

Could such plasticity also exist in humans? The question remains open. Intriguingly, anecdotal reports describe the existence of insulin and glucagon double positive cells detected as rare elements in the fibrotic human pancreas [117], which may be the result of transdifferentiation attempts, putative progenitors activated in adult life, or merely an example of a disease-related, physiologically insignificant occurrence. Still, the *in vivo* streptozotocin-induced β-cell ablation does not seem to induce α-cell transdifferentiation or β-cell regeneration in nonhuman primates.

The differentiation of pancreatic acinar cells to β cells is actively investigated, and contrasting reports are fueling a debate on the general feasibility of the approach. Rodent acinar cells were reported to convert to endocrine cells after pancreatic duct ligation, with endocrine-acinar intermediates showing amylase and insulin granules [118]. More recently, it was found in rodents that antagonizing the Ptf1a-mediated control of the acinar phenotype is sufficient to convert acinar cells into endocrine and insulin-producing cells. As we have seen, Ptf1a acts as a major player in the commitment of epithelial progenitors to the acinar fate during development, and these findings support an additional role for this factor in the maintenance of their differentiated state after embryogenesis. On the other hand, no acinar contribution to β-cell replenishment was observed in a lineage-tracing mouse model designed to track the progeny of elastase I-expressing acinar cells in several injury models (pancreatectomy, pancreatic duct ligation, and caerulein-induced pancreatitis) [119]. An innovative transgenic approach led to the reprogramming of murine exocrine tissue *in vivo* through the use of viral vectors carrying Pdx1, Ngn3, and MafA [120]. As we have seen in previous sections, these transcription factors govern key developmental steps of pancreatic patterning, endocrine commitment, and β-cell maturation, respectively. The effect in the pancreatic exocrine tissue was striking, with the appearance of cells expressing several markers and features of bona fide β cells as well as amelioration of hyperglycemia. This method is thought to directly reprogram adult exocrine cells into β cells without reversion to a pluripotent state, but it is actually unknown if the transcription factors reprogram mature acinar/ductal cells, or rather a subset of cells with progenitor features. Given the large availability of acinar tissue, regeneration and reprogramming from this source remains a fertile field of research.

It was recently observed that niches in the biliary tree, namely peribiliary glands, harbor highly undifferentiated stem cells with the potential to commit to pancreatic and hepatic fates [121,122]. These stem cells niches seem to be related and anatomically connected to committed progenitor cell structures, namely pancreatic duct glands, that show signs of progressive maturation toward pancreatic fates, including Ngn3+ endocrine progenitors and insulin+ β-like cells ([123]). This led to the hypothesis of the existence of a ramified network of stem and progenitor cell niches, starting in the biliary tree and ending

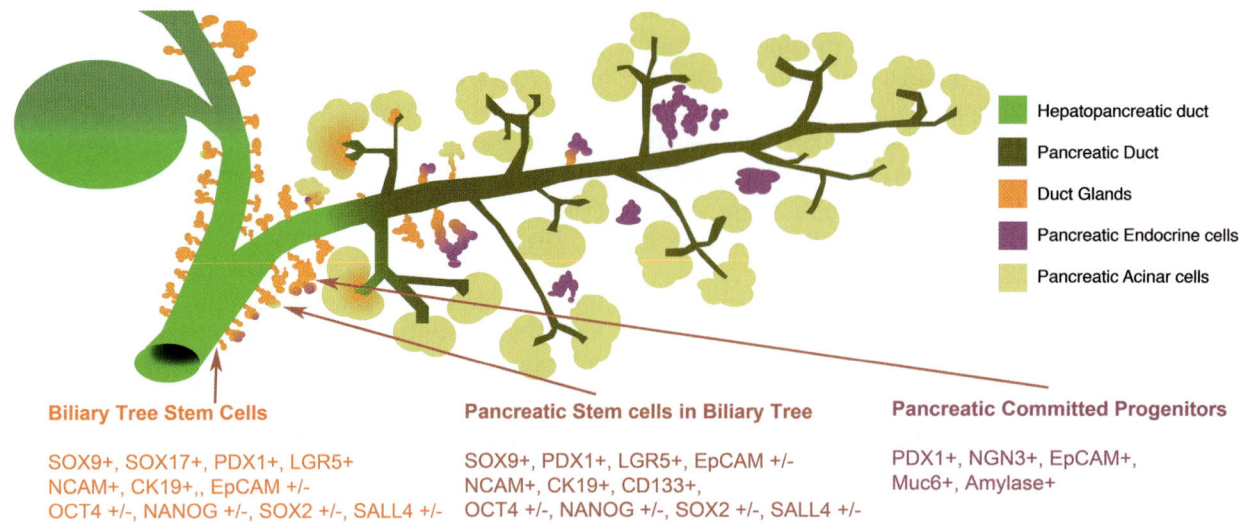

Hepatopancreatic duct

Pancreatic Duct

Duct Glands

Pancreatic Endocrine cells

Pancreatic Acinar cells

Biliary Tree Stem Cells

SOX9+, SOX17+, PDX1+, LGR5+
NCAM+, CK19+,, EpCAM +/-
OCT4 +/-, NANOG +/-, SOX2 +/-, SALL4 +/-

Pancreatic Stem cells in Biliary Tree

SOX9+, PDX1+, LGR5+, EpCAM +/-
NCAM+, CK19+, CD133+,
OCT4 +/-, NANOG +/-, SOX2 +/-, SALL4 +/-

Pancreatic Committed Progenitors

PDX1+, NGN3+, EpCAM+,
Muc6+, Amylase+

FIGURE 42.7   Biliary tree stem cells and pancreatic committed progenitor niches in the adult pancreas (a simplified scheme).

in the mature pancreas. Such a lineage structure would mirror at the anatomical level the sequence of events that takes place during embryogenesis. These niches could represent remnants of the embryonic development or actual progenitors at different stages of differentiation, contributing to a slow-rate organogenesis throughout life ([123]) (Figure 42.7). Whatever their nature, the β-cell differentiation potential of biliary tree stem cells is supported by a growing body of evidence and represents an exciting new avenue of research for β-cell regeneration.

Mesenchymal stem cells (MSCs) have also been extensively studied in the field: not only have they proven their worth as immune regulators, but also they exhibit a broad differentiation potential that was thought to encompass even the β-cell fate [124]. MSCs from several different sources (bone marrow, adipose tissue, cord blood, amniotic fluid, amnios, etc.) have been transdifferentiated to β-like cells with different degrees of success [125−129]. The distinct advantage of these cells resides in their safety profile, solidly established through numerous clinical trials for hematologic conditions and immune-based disorders. However, the induction of these precursors to an islet fate shows a low efficiency, and success to date has resulted in cells that are hybrids of islet cell and mesenchymal cell phenotypes.

Due to their extensive multipotency and indefinite expansion potential, ESCs represent the gold standard in stem cell research and are one of the most investigated sources for regenerative approaches in the diabetes field. Several protocols have been developed to induce ESC differentiation to endocrine progenitors and β cells. Major advancements were accomplished by recapitulating key developmental events [130]. As stated already, mimicking embryonic events *in vivo* is a cumbersome task and

advanced stage maturation to functional insulin-producing cells *in vitro* has not been conclusively established yet. Given the low yield of differentiation, earlier approaches made use of selectable markers under the control of the insulin promoter to obtain murine ESC-derived insulin-expressing cells. A simple differentiation protocol, based on nicotinamide and cell clustering, was sufficient to yield functional maturation in cells that ameliorated glycemia in diabetic mice [131]. The first report on the derivation of insulin-producing cells and other types of pancreatic endocrine cells from nongenetically manipulated murine ESCs appeared in 2001 [132]. The multistep differentiation strategy started with the generation of embryoid bodies, followed by the enrichment of nestin-positive cells, selected in a serum-free culture medium that was not permissive to most other cell types; subsequently, a transient induction step with bFGF was performed and maturation was finally allowed to happen in N2 medium containing B27 supplements. This protocol yielded clusters that somewhat resembled pancreatic islets. However, additional research on this method revealed that the cells generated in this manner were not of endodermal, but rather ectodermal, origin; and that their insulin positivity was merely a reflection of the uptake of insulin from the culture medium [133].

The first successful steps in human ESC maturation toward the β-cell fate were the result of the careful definition of the conditions required to generate definitive endoderm. A critical factor for this to happen was the use of medium containing Activin A and minimal or no serum [134,135]. The first protocol showing convincing differentiation of human ESC into insulin- and other endocrine hormone-producing cells was developed by a team from Novocell [136] formerly CyThera [135], now

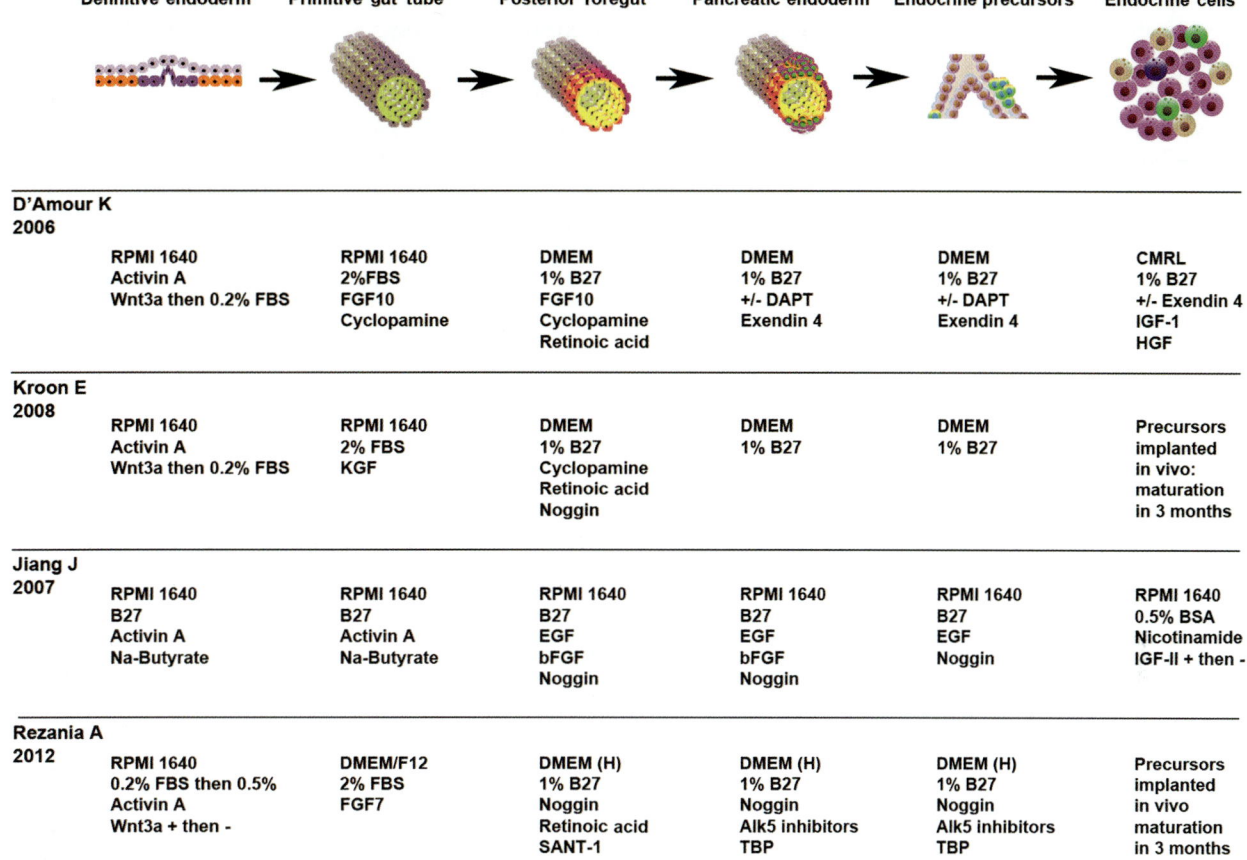

Definitive endoderm	Primitive gut tube	Posterior foregut	Pancreatic endoderm	Endocrine precursors	Endocrine cells
D'Amour K 2006					
RPMI 1640 Activin A Wnt3a then 0.2% FBS	RPMI 1640 2%FBS FGF10 Cyclopamine	DMEM 1% B27 FGF10 Cyclopamine Retinoic acid	DMEM 1% B27 +/- DAPT Exendin 4	DMEM 1% B27 +/- DAPT Exendin 4	CMRL 1% B27 +/- Exendin 4 IGF-1 HGF
Kroon E 2008					
RPMI 1640 Activin A Wnt3a then 0.2% FBS	RPMI 1640 2% FBS KGF	DMEM 1% B27 Cyclopamine Retinoic acid Noggin	DMEM 1% B27	DMEM 1% B27	Precursors implanted in vivo: maturation in 3 months
Jiang J 2007					
RPMI 1640 B27 Activin A Na-Butyrate	RPMI 1640 B27 Activin A Na-Butyrate	RPMI 1640 B27 EGF bFGF Noggin	RPMI 1640 B27 EGF bFGF Noggin	RPMI 1640 B27 EGF Noggin	RPMI 1640 0.5% BSA Nicotinamide IGF-II + then -
Rezania A 2012					
RPMI 1640 0.2% FBS then 0.5% Activin A Wnt3a + then -	DMEM/F12 2% FBS FGF7	DMEM (H) 1% B27 Noggin Retinoic acid SANT-1	DMEM (H) 1% B27 Noggin Alk5 inhibitors TBP	DMEM (H) 1% B27 Noggin Alk5 inhibitors TBP	Precursors implanted in vivo maturation in 3 months

FIGURE 42.8 Differentiation protocols for the differentiation of human embryonic stem cells (hESC) into β cells.

Viacyte [137]). The multistage protocol was developed after testing a number of factors involved in pancreatic differentiation, such as activins, Wnt3a, TGFβ1, TGFβ inhibitors, FGFs, FGF inhibitors, retinoids, γ-secretase inhibitors, GLPs and analogs, IGFs, VEGF, nicotinamide and hedgehog inhibitors. The main features of the sequential induction are shown in Figure 42.8, Amour protocol. It was determined that specification of ESC to definitive endoderm, obtained with Activin A and Wnt3a (stage 1), was crucial to efficiently obtain pancreatic endocrine cells. After Activin A removal, the cells acquired a phenotype consistent with that of the gut tube. The addition of FGF10, the hedgehog-signaling inhibitor KAAD-cyclopamine (stage 2) and subsequently retinoic acid (stage 3) provided further induction to pancreatic epithelial progenitors. Pancreatic endocrine commitment became evident after the addition of the GLP-1 agonist Exendin 4 (stage 4), and late-stage maturation was stimulated by IGF1 and HGF (stage 5). The cells obtained with this protocol resembled fetal β cells, as they released insulin in response to a number of stimuli but only minimally in response to glucose. It was immediately clear that the protocol, developed for the CyT203 cell line,

may not be easily replicated with other cell lines. Later developments of the protocol included the substitution of FGF10 and cyclopamine with KGF in the differentiation cocktail for stage 2 or the use of Noggin for stage 3, but the key change was the use of a final maturation step *in vivo* [138] (Figure 42.8, Kroon protocol). In a first set of experiments, human ESC-derived progenitors at stage 4 were transplanted in immunodeficient mice and matured into functional glucose-responsive insulin-producing cells. In a different set of experiments, the animals were rendered diabetic by streptozotocin after transplantation with ESC-derived cells. Streptozotocin enters murine β cells through the GLUT2 transporter, and since GLUT2 is expressed at low levels in human islets, this chemical was not expected to be significantly harmful to the graft. The ESC-derived progenitors matured into functional glucose-responsive insulin-producing cells that were able to ameliorate and nearly normalize glycemia in hosts. Other groups developed or modified the differentiation cocktails with intriguing results, expanding the applicability of the differentiation protocols to other ESC cell lines and to induced pluripotent stem (iPS) cell lines [139,140] (Figure 42.8, Jiang protocol). The manufacturing process for the most recent evolution of the Viacyte protocol, applied to the human ESC CyT49 line, has been scaled up and seems to be approaching clinical trials [137]. Another protocol, evolved from previous strategies [141], was developed for the widely available H1 ESC line (Figure 42.8, Rezania protocols) [142].

There is no consensus yet on a single robust and reliable protocol, and the strategies keep evolving. Several groups reported a lack of reproducibility of current protocols with different ESC lines, reinforcing the need for improvements in current strategies [130]. Safety remains the major concern for clinical translatability. The problem of transplanting a mixed population of immature human embryonic stem cell (hESC)-derived progenitors is evidenced by the formation of teratomas in the grafts in animal models. Promising strategies to remove progenitors with tumor-forming potential rely on selections based on surface markers [143].

The long-term stability of the phenotype and best strategies to avoid rejection are other open questions. It has been argued, however, that the use of encapsulation and immunoisolation devices will enable the use of allogeneic material in fully immunocompetent recipients, and that any tumorigenic escapees would be promptly destroyed by the host's immune response. The encapsulation in retrievable devices would enable the safe removal of the graft in case of tumor formation.

Recent techniques to generate iPS cells through the reprogramming of differentiated adult cells [144,145] have rapidly become a staple of many regenerative medicine laboratories. iPS cells share major characteristics with ESC, including an indefinite expansion potential in the undifferentiated state and the ability to differentiate in derivatives of all germ layers. There are no ethical concerns in their procurement, since they are not derived from human embryos. Moreover, they could be generated from the prospective recipients, a very appealing feature for the design of autologous therapies. One of the first reports in the diabetes field showed that iPS cells generated from skin fibroblasts could be induced to generate clusters of insulin-producing cells [146]. A major difference in comparison to classical ESCs lies in the fact that iPS cells bear an epigenetic memory of the tissue of origin [147]. This was initially considered a problem [148], but it was recently brilliantly exploited to derive β cell-specific iPS cells that are prone to commit to the β-cell phenotype [149].

42.5 miRNA AND siRNA IN DIFFERENTIATION

Small RNA molecules, such as microRNAs (miRNAs), can act as gene expression regulators. The study of their mechanism of action is unveiling novel regulatory networks that control cell development and physiology. The manipulation of the gene expression profile with miRNAs or synthetic analogs, such as small interfering RNAs (siRNAs) represents an emerging tool in the field of stem cell differentiation [150]. miRNAs are small noncoding RNAs that control gene expression at the posttranscriptional level [151]. After transcription, miRNAs can bind mRNA molecules with partial complementarity, and their primary function seems to be the downregulation of gene expression [152,153]. miRNAs can control very complex gene expression regulatory networks [154,155], regulate embryonic development and display tissue-/cell-specific patterns [156,157]. They also control pancreatic and endocrine differentiation, and their regulatory function is not dispensable [158]. mir-375 inhibition was shown to have a deleterious effect on pancreatic development [159]. mir-7 is an islet-specific miRNA [160] expressed in the human developing pancreas from week 9. Intriguingly, the peak of expression was observed between weeks 14 and 18, coincident with an exponential phase of hormone-producing cell differentiation [161]. The elucidation of miRNA regulatory networks in the mouse and human pancreas is redefining the way we understand pancreatic development [150,162], and novel tools to harness stem cell differentiation are rapidly emerging [163].

42.6 CHALLENGES AND PERSPECTIVES

Unlike other conditions potentially treatable with stem cells, diabetes can be reversed today with an effective form of cell therapy (islet transplantation). The development and progressive refinement of this method over the past two decades, aided by advances in immune regulation and immunoisolation, has helped us envision the most likely embodiments of the next-generation stem cell-based therapies for diabetes. Regulatory hurdles in the way of the clinical use of ESCs, once thought daunting, have now been largely overcome in the context of clinical trials for other conditions, such as macular degeneration or spinal cord injury. In short, the field appears to be poised for a quantum leap in our ability to take islet transplantation to a new level. We expect to witness significant progress within the next 5−10 years in our ability to reproduce *ex vivo* the microenvironment leading to functional maturation of stem cells *in vivo*. It is our belief that a fully mature cell product, able to correct hyperglycemia right after transplantation, would be preferable to the current state of the art, which entails the transplantation of immature progenitors whose maturation within the host may take several months. Safety is another concern that, fortunately, is being addressed by means of continuous refinement of the protocols (resulting in a lower likelihood of transplanting carryover undifferentiated cells) as well as the use of markers for the selective destruction of tumorigenic cells and/or positive selection of the nonreplicative ones. The development of physical barriers, such as immunoisolation devices and encapsulation is also contributing to lessen our initial fears that even a marginal incidence of teratogenic lesions may hamper the therapeutic prospects of these approaches. The first clinical trials for diabetes using (hES) cells are around the corner, and, if proven safe and effective, could revolutionize the treatment of the disease. Simultaneously, scientists have made great strides in other alternative means to replenish islet mass, exploiting either the natural ability of the pancreas to regenerate its own β cells or the close kinship between such cells and their surrounding tissues—the latter by means of novel *reprogramming* techniques that could potentially "recycle" into β cells the acinar tissue that is routinely discarded after each islet isolation. This new strategy, whose advent and rapid development has taken the field by surprise, may very well become a viable alternative to the use of stem cells down the line, although its review is out of the scope of this chapter.

REFERENCES

[1] Bluestone JA, Herold K, Eisenbarth G. Genetics, pathogenesis and clinical interventions in type 1 diabetes. Nature 2010;464:1293–300.

[2] Dominguez-Bendala J, Inverardi L, Ricordi C. Stem cell-derived islet cells for transplantation. Curr Opin Organ Transplant 2011;16(1):76–82.

[3] Ricordi C, Strom TB. Clinical islet transplantation: advances and immunological challenges. Nat Rev Immunol 2004;4:259–68.

[4] Thompson DM, Meloche M, Ao Z, Paty B, Keown P, Shapiro RJ, et al. Reduced progression of diabetic microvascular complications with islet cell transplantation compared with intensive medical therapy. Transplantation 2011;91:373–8.

[5] Poggioli R, Faradji RN, Ponte G, Betancourt A, Messinger S, Baidal DA, et al. Quality of life after islet transplantation. Am J Transplant 2006;6:371–8.

[6] Harlan DM, Kenyon NS, Korsgren O, Roep BO. Current advances and travails in islet transplantation. Diabetes 2009;58:2175–84.

[7] Leitao CB, Cure P, Tharavanij T, Baidal DA, Alejandro R. Current challenges in islet transplantation. Curr Diab Rep 2008;8:324–31.

[8] Van Belle T, von Herrath M. Immunosuppression in islet transplantation. J Clin Invest 2008;118:1625–8.

[9] Caicedo A. Paracrine and autocrine interactions in the human islet: more than meets the eye. Semin Cell Dev Biol 2013;24(1):11–21.

[10] Marchetti P, Lupi R, Bugliani M, Kirkpatrick CL, Sebastiani G, Grieco FA, et al. A local glucagon-like peptide 1 (GLP-1) system in human pancreatic islets. Diabetologia 2012;55(12):3262–72.

[11] Barker CJ, Leibiger IB, Berggren PO. The pancreatic islet as a signaling hub. Adv Biol Regul 2013;53(1):156–63.

[12] Edlund H. Pancreatic organogenesis—developmental mechanisms and implications for therapy. Nat Rev Genet 2002;3:524–32.

[13] Edlund H. Developmental biology of the pancreas. Diabetes 2001;50:5–9.

[14] Kumar M, Melton D. Pancreas specification: a budding question. Curr Opin Genet Dev 2003;13:401–7.

[15] Gittes GK. Developmental biology of the pancreas: a comprehensive review. Dev Biol 2009;326:4–35.

[16] Tam PP, Kanai-Azuma M, Kanai Y. Early endoderm development in vertebrates: lineage differentiation and morphogenetic function. Curr Opin Genet Dev 2003;13:393–400.

[17] de Santa Barbara P, van den Brink GR, Roberts DJ. Development and differentiation of the intestinal epithelium. Cell Mol Life Sci 2003;60:1322–32.

[18] Lowe LA, Yamada S, Kuehn MR. Genetic dissection of nodal function in patterning the mouse embryo. Development 2001;128:1831–43.

[19] Roberts DJ, Smith DM, Goff DJ, Tabin CJ. Epithelial-mesenchymal signaling during the regionalization of the chick gut. Development 1998;125:2791–801.

[20] Jonsson J, Carlsson L, Edlund T, Edlund H. Insulin-promoter-factor 1 is required for pancreas development in mice. Nature 1994;371:606–9.

[21] Afelik S, Chen Y, Pieler T. Combined ectopic expression of Pdx1 and Ptf1a/p48 results in the stable conversion of posterior endoderm into endocrine and exocrine pancreatic tissue. Genes Dev 2006;20:1441–6.

[22] Kim SK, Melton DA. Pancreas development is promoted by cyclopamine, a hedgehog signaling inhibitor. Proc Natl Acad Sci USA 1998;95:13036–41.

[23] Wells JM, Melton DA. Early mouse endoderm is patterned by soluble factors from adjacent germ layers. Development 2000;127:1563–72.

[24] Kim SK, Hebrok M, Melton DA. Notochord to endoderm signaling is required for pancreas development. Development 1997;124:4243–52.

[25] Lammert E, Cleaver O, Melton D. Induction of pancreatic differentiation by signals from blood vessels. Science 2001;294:564–7.

[26] Dumont DJ, Fong GH, Puri MC, Gradwohl G, Alitalo K, Breitman ML. Vascularization of the mouse embryo: a study of flk-1, tek, tie, and vascular endothelial growth factor expression during development. Dev Dyn 1995;203:80–92.

[27] Lammert E, Cleaver O, Melton D. Role of endothelial cells in early pancreas and liver development. Mech Dev 2003;120:59–64.

[28] Brocard J, Feil R, Chambon P, Metzger D. A chimeric Cre recombinase inducible by synthetic,but not by natural ligands of the glucocorticoid receptor. Nucl Acids Res 1998;26:4086–90.

[29] Feil R, Brocard J, Mascrez B, LeMeur M, Metzger D, Chambon P. Ligand-activated site-specific recombination in mice. Proc Natl Acad Sci USA 1996;93:10887–90.

[30] Gu G, Dubauskaite J, Melton DA. Direct evidence for the pancreatic lineage: NGN3+ cells are islet progenitors and are distinct from duct progenitors. Development 2002;129:2447–57.

[31] Gu G, Brown JR, Melton DA. Direct lineage tracing reveals the ontogeny of pancreatic cell fates during mouse embryogenesis. Mech Dev 2003;120:35–43.

[32] Chitnis AB. The role of Notch in lateral inhibition and cell fate specification. Mol Cell Neurosci 1995;6:311–21.

[33] Apelqvist A, Li H, Sommer L, Beatus P, Anderson DJ, Honjo T, et al. Notch signalling controls pancreatic cell differentiation. Nature 1999;400:877–81.

[34] Gradwohl G, Dierich A, LeMeur M, Guillemot F. neurogenin3 is required for the development of the four endocrine cell lineages of the pancreas. Proc Natl Acad Sci USA 2000;97:1607–11.

[35] Jensen J, Pedersen EE, Galante P, Hald J, Heller RS, Ishibashi M, et al. Control of endodermal endocrine development by Hes-1. Nat Genet 2000;24:36–44.

[36] Edlund H. Factors controlling pancreatic cell differentiation and function. Diabetologia 2001;44:1071−9.

[37] Oliver-Krasinski JM, Kasner MT, Yang J, Crutchlow MF, Rustgi AK, Kaestner KH, et al. The diabetes gene Pdx1 regulates the transcriptional network of pancreatic endocrine progenitor cells in mice. J Clin Invest 2009;119:1888−98.

[38] Sussel L, Kalamaras J, Hartigan-O'Connor DJ, Meneses JJ, Pedersen RA, Rubenstein JL, et al. Mice lacking the homeodomain transcription factor Nkx2.2 have diabetes due to arrested differentiation of pancreatic beta cells. Development 1998;125:2213−21.

[39] Sander M, Sussel L, Conners J, Scheel D, Kalamaras J, Dela Cruz F, et al. Homeobox gene Nkx6.1 lies downstream of Nkx2.2 in the major pathway of beta-cell formation in the pancreas. Development 2000;127:5533−40.

[40] Watada H, Scheel DW, Leung J, German MS. Distinct gene expression programs function in progenitor and mature islet cells. J Biol Chem 2003;278:17130−40.

[41] Oster A, Jensen J, Serup P, Galante P, Madsen OD, Larsson LI. Rat endocrine pancreatic development in relation to two homeobox gene products (Pdx-1 and Nkx 6.1). J Histochem Cytochem 1998;46:707−15.

[42] Henseleit KD, Nelson SB, Kuhlbrodt K, Hennings JC, Ericson J, Sander M. NKX6 transcription factor activity is required for alpha- and beta-cell development in the pancreas. Development 2005;132:3139−49.

[43] Naya FJ, Huang HP, Qiu Y, Mutoh H, DeMayo FJ, Leiter AB, et al. Diabetes, defective pancreatic morphogenesis, and abnormal enteroendocrine differentiation in BETA2/neuroD-deficient mice. Genes Dev 1997;11:2323−34.

[44] St-Onge L, Sosa-Pineda B, Chowdhury K, Mansouri A, Gruss P. Pax6 is required for differentiation of glucagon-producing alpha-cells in mouse pancreas. Nature 1997;387:406−9.

[45] Dohrmann C, Gruss P, Lemaire L. Pax genes and the differentiation of hormone-producing endocrine cells in the pancreas. Mech Dev 2000;92:47−54.

[46] Smith SB, Qu HQ, Taleb N, Kishimoto NY, Scheel DW, Lu Y, et al. Rfx6 directs islet formation and insulin production in mice and humans. Nature 2010;463:775−80.

[47] Wang J, Elghazi L, Parker SE, Kizilocak H, Asano M, Sussel L, et al. The concerted activities of Pax4 and Nkx2.2 are essential to initiate pancreatic beta-cell differentiation. Dev Biol 2004;266:178−89.

[48] Heremans Y, Van De Casteele M, in't Veld P, Gradwohl G, Serup P, Madsen O, et al. Recapitulation of embryonic neuroendocrine differentiation in adult human pancreatic duct cells expressing neurogenin 3. J Cell Biol 2002;159:303−12.

[49] Smith SB, Gasa R, Watada H, Wang J, Griffen SC, German MS. Neurogenin3 and hepatic nuclear factor 1 cooperate in activating pancreatic expression of Pax4. J Biol Chem 2003;278:38254−9.

[50] Collombat P, Hecksher-Sorensen J, Broccoli V, Krull J, Ponte I, Mundiger T, et al. The simultaneous loss of Arx and Pax4 genes promotes a somatostatin-producing cell fate specification at the expense of the alpha- and beta-cell lineages in the mouse endocrine pancreas. Development 2005;132:2969−80.

[51] Collombat P, Mansouri A, Hecksher-Sorensen J, Serup P, Krull J, Gradwohl G, et al. Opposing actions of Arx and Pax4 in endocrine pancreas development. Genes Dev 2003;17:2591−603.

[52] Heller RS, Jenny M, Collombat P, Mansouri A, Tomasetto C, Madsen OD, et al. Genetic determinants of pancreatic epsilon-cell development. Dev Biol 2005;286:217−24.

[53] Sosa-Pineda B, Chowdhury K, Torres M, Oliver G, Gruss P. The Pax4 gene is essential for differentiation of insulin-producing beta cells in the mammalian pancreas. Nature 1997;386:399−402.

[54] Sosa-Pineda B. The gene Pax4 is an essential regulator of pancreatic beta-cell development. Mol Cell 2004;18:289−94.

[55] Nishimura W, Kondo T, Salameh T, El Khattabi I, Dodge R, Bonner-Weir S, et al. A switch from MafB to MafA expression accompanies differentiation to pancreatic beta-cells. Dev Biol 2006;293:526−39.

[56] Zhang C, Moriguchi T, Kajihara M, Esaki R, Harada A, Shimohata H, et al. MafA is a key regulator of glucose-stimulated insulin secretion. Mol Cell Biol 2005;25:4969−76.

[57] Goke B. Islet cell function: alpha and beta cells—partners towards normoglycaemia. Int J Clin Pract Suppl 2008;159:2−7.

[58] Leibiger IB, Leibiger B, Berggren PO. Insulin signaling in the pancreatic beta-cell. Annu Rev Nutr 2008;28:233−51.

[59] Nolan CJ, Prentki M. The islet beta-cell: fuel responsive and vulnerable. Trends Endocrinol Metab 2008;19:285−91.

[60] van de Bunt M, Gloyn AL. A tale of two glucose transporters: how GLUT2 re-emerged as a contender for glucose transport into the human beta cell. Diabetologia 2012;55:2312−5.

[61] Dean PM, Matthews EK. Electrical activity in pancreatic islet cells. Nature 1968;219:389−90.

[62] McIntyre N, Holdsworth CD, Turner DS. New interpretation of oral glucose tolerance. Lancet 1964;2:20−1.

[63] MacDonald PE, Salapatek AM, Wheeler MB. Glucagon-like peptide-1 receptor activation antagonizes voltage-dependent repolarizing K(+) currents in beta-cells: a possible glucose-dependent insulinotropic mechanism. Diabetes 2002;51:443−7.

[64] Shao S, Fang Z, Yu X, Zhang M. Transcription factors involved in glucose-stimulated insulin secretion of pancreatic beta cells. Biochem Biophys Res Commun 2009;384:401−4.

[65] Ahlqvist E, Turrini F, Lang ST, Taneera J, Zhou Y, Almgren P, et al. A common variant upstream of the PAX6 gene influences islet function in man. Diabetologia 2012;55:94−104.

[66] Habener JF, Stoffers DA. A newly discovered role of transcription factors involved in pancreas development and the pathogenesis of diabetes mellitus. Proc Assoc Am Phys 1998;110:12−21.

[67] Wang H, Iezzi M, Theander S, Antinozzi PA, Gauthier BR, Halban PA, et al. Suppression of Pdx-1 perturbs proinsulin processing, insulin secretion and GLP-1 signalling in INS-1 cells. Diabetologia 2005;48:720−31.

[68] Kitamura T, Nakae J, Kitamura Y, Kido Y, Biggs 3rd WH, Wright CV, et al. The forkhead transcription factor Foxo1 links insulin signaling to Pdx1 regulation of pancreatic beta cell growth. J Clin Invest 2002;110:1839−47.

[69] Talchai C, Xuan S, Lin HV, Sussel L, Accili D. Pancreatic beta cell dedifferentiation as a mechanism of diabetic beta cell failure. Cell 2012;150:1223−34.

[70] Kato T, Shimano H, Yamamoto T, Ishikawa M, Kumadaki S, Matsuzaka T, et al. Palmitate impairs and eicosapentaenoate restores insulin secretion through regulation of SREBP-1c in pancreatic islets. Diabetes 2008;57:2382−92.

[71] Shimano H, Amemiya-Kudo M, Takahashi A, Kato T, Ishikawa M, Yamada N. Sterol regulatory element-binding protein-1c and pancreatic beta-cell dysfunction. Diabetes Obes Metab 2007;9 (Suppl. 2):133−9.

[72] Gosmain Y, Katz LS, Masson MH, Cheyssac C, Poisson C, Philippe J. Pax6 is crucial for beta-cell function, insulin biosynthesis, and glucose-induced insulin secretion. Mol Endocrinol 2012;26:696−709.

[73] Kataoka K, Han SI, Shioda S, Hirai M, Nishizawa M, Handa H. MafA is a glucose-regulated and pancreatic beta-cell-specific transcriptional activator for the insulin gene. J Biol Chem 2002;277:49903−10.

[74] Matsuoka TA, Artner I, Henderson E, Means A, Sander M, Stein R. The MafA transcription factor appears to be responsible for tissue-specific expression of insulin. Proc Natl Acad Sci USA 2004;101:2930−3.

[75] Wang H, Brun T, Kataoka K, Sharma AJ, Wollheim CB. MAFA controls genes implicated in insulin biosynthesis and secretion. Diabetologia 2007;50:348−58.

[76] Piper K, Brickwood S, Turnpenny LW, Cameron IT, Ball SG, Wilson DI, et al. Beta cell differentiation during early human pancreas development. J Endocrinol 2004;181:11−23.

[77] Falin LI. The development and cytodifferentiation of the islets of Langerhans in human embryos and foetuses. Acta Anat (Basel) 1967;68:147−68.

[78] Slack JM. Developmental biology of the pancreas. Development 1995;121:1569−80.

[79] Rhodes CJ. Type 2 diabetes-a matter of beta-cell life and death?. Science 2005;307:380−4.

[80] Dor Y, Brown J, Martinez OI, Melton DA. Adult pancreatic beta-cells are formed by self-duplication rather than stem-cell differentiation. Nature 2004;429:41−6.

[81] Bonner-Weir S, Deery D, Leahy JL, Weir GC. Compensatory growth of pancreatic beta-cells in adult rats after short-term glucose infusion. Diabetes 1989;38:49−53.

[82] Bruning JC, Winnay J, Bonner-Weir S, Taylor SI, Accili D, Kahn CR. Development of a novel polygenic model of NIDDM in mice heterozygous for IR and IRS-1 null alleles. Cell 1997;88:561−72.

[83] Nir T, Melton DA, Dor Y. Recovery from diabetes in mice by beta cell regeneration. J Clin Invest 2007;117:2553−61.

[84] Porat S, Weinberg-Corem N, Tornovsky-Babaey S, Schyr-Ben-Haroush R, Hija A, Stolovich-Rain M, et al. Control of pancreatic beta cell regeneration by glucose metabolism. Cell Metab 2011;13:440−9.

[85] Butler AE, Janson J, Bonner-Weir S, Ritzel R, Rizza RA, Butler PC. Beta-cell deficit and increased beta-cell apoptosis in humans with type 2 diabetes. Diabetes 2003;52:102−10.

[86] Buchanan TA, Kjos SL. Gestational diabetes: risk or myth? J Clin Endocrinol Metab 1999;84:1854−7.

[87] Fernandes A, King LC, Guz Y, Stein R, Wright CV, Teitelman G. Differentiation of new insulin-producing cells is induced by injury in adult pancreatic islets. Endocrinology 1997;138:1750−62.

[88] Dominguez-Bendala J, Inverardi L, Ricordi C. Regeneration of pancreatic beta-cell mass for the treatment of diabetes. Expert Opin Biol Ther 2012;12:731−41.

[89] Butler PC, Meier JJ, Butler AE, Bhushan A. The replication of beta cells in normal physiology, in disease and for therapy. Nat Clin Pract Endocrinol Metab 2007;3:758−68.

[90] Meier JJ, Butler AE, Saisho Y, Monchamp T, Galasso R, Bhushan A, et al. Beta-cell replication is the primary mechanism subserving the postnatal expansion of beta-cell mass in humans. Diabetes 2008;57:1584−94.

[91] Meier JJ, Lin JC, Butler AE, Galasso R, Martinez DS, Butler PC. Direct evidence of attempted beta cell regeneration in an 89-year-old patient with recent-onset type 1 diabetes. Diabetologia 2006;49:1838−44.

[92] Keenan HA, Sun JK, Levine J, Doria A, Aiello LP, Eisenbarth G, et al. Residual insulin production and pancreatic ss-cell turnover after 50 years of diabetes: Joslin Medalist study. Diabetes 2010;59:2846−53.

[93] Saisho Y, Butler AE, Manesso E, Elashoff D, Rizza RA, Butler PC. Beta-cell mass and turnover in humans: effects of obesity and aging. Diabetes Care 2013;36(1):111−7.

[94] Cnop M, Igoillo-Esteve M, Hughes SJ, Walker JN, Cnop I, Clark A. Longevity of human islet alpha- and beta-cells. Diabetes Obes Metab 2011;13:39−46.

[95] Menge BA, Tannapfel A, Belyaev O, Drescher R, Muller C, Uhl W, et al. Partial pancreatectomy in adult humans does not provoke beta-cell regeneration. Diabetes 2008;57:142−9.

[96] Teta M, Rankin MM, Long SY, Stein GM, Kushner JA. Growth and regeneration of adult beta cells does not involve specialized progenitors. Dev Cell 2007;12:817−26.

[97] Abouna S, Old RW, Pelengaris S, Epstein D, Ifandi V, Sweeney I, et al. Non-beta-cell progenitors of beta-cells in pregnant mice. Organogenesis 2010;6:125−33.

[98] Gianani R, Campbell-Thompson M, Sarkar SA, Wasserfall C, Pugliese A, Solis JM, et al. Dimorphic histopathology of long-standing childhood-onset diabetes. Diabetologia 2010;53:690−8.

[99] Chung CH, Levine F. Adult pancreatic alpha-cells: a new source of cells for beta-cell regeneration. Rev Diabet Stud 2010;7:124−31.

[100] Rosenberg L, Vinik AI. Trophic stimulation of the ductular-islet cell axis: a new approach to the treatment of diabetes. Adv Exp Med Biol 1992;321:95−104 [discussion 105−109].

[101] Gepts W. Pathologic anatomy of the pancreas in juvenile diabetes mellitus. Diabetes 1965;14:619−33.

[102] Weaver CV, Sorenson RL, Kaung HC. Immunocytochemical localization of insulin-immunoreactive cells in the pancreatic ducts of rats treated with trypsin inhibitor. Diabetologia 1985;28:781−5.

[103] Gianani R, Putnam A, Still T, Yu L, Miao D, Gill RG, et al. Initial results of screening of nondiabetic organ donors for expression of islet autoantibodies. J Clin Endocrinol Metab 2006;91:1855−61.

[104] Bonner-Weir S, Baxter LA, Schuppin GT, Smith FE. A second pathway for regeneration of adult exocrine and endocrine pancreas. A possible recapitulation of embryonic development. Diabetes 1993;42:1715−20.

[105] Noguchi H, Xu G, Matsumoto S, Kaneto H, Kobayashi N, Bonner-Weir S, et al. Induction of pancreatic stem/progenitor cells into insulin-producing cells by adenoviral-mediated gene transfer technology. Cell Transplant 2006;15:929−38.

[106] Bonner-Weir S, Inada A, Yatoh S, Li WC, Aye T, Toschi E, et al. Transdifferentiation of pancreatic ductal cells to endocrine beta-cells. Biochem Soc Trans 2008;36:353−6.

[107] Li WC, Rukstalis JM, Nishimura W, Tchipashvili V, Habener JF, Sharma A, et al. Activation of pancreatic-duct-derived progenitor cells during pancreas regeneration in adult rats. J Cell Sci 2010;123:2792−802.

[108] Yatoh S, Dodge R, Akashi T, Omer A, Sharma A, Weir GC, et al. Differentiation of affinity-purified human pancreatic duct cells to beta-cells. Diabetes 2007;56:1802−9.

[109] Martin-Pagola A, Sisino G, Allende G, Dominguez-Bendala J, Gianani R, Reijonen H, et al. Insulin protein and proliferation in ductal cells in the transplanted pancreas of patients with type 1 diabetes and recurrence of autoimmunity. Diabetologia 2008;51:1803−13.

[110] Kushner JA, Weir GC, Bonner-Weir S. Ductal origin hypothesis of pancreatic regeneration under attack. Cell Metab 2010;11:2−3.

[111] Guz Y, Nasir I, Teitelman G. Regeneration of pancreatic beta cells from intra-islet precursor cells in an experimental model of diabetes. Endocrinology 2001;142:4956−68.

[112] Kodama S, Toyonaga T, Kondo T, Matsumoto K, Tsuruzoe K, Kawashima J, et al. Enhanced expression of PDX-1 and Ngn3 by exendin-4 during beta cell regeneration in STZ-treated mice. Biochem Biophys Res Commun 2005;327:1170−8.

[113] Joglekar MV, Hardikar AA. Isolation, expansion, and characterization of human islet-derived progenitor cells. Methods Mol Biol 2012;879:351−66.

[114] Collombat P, Xu X, Ravassard P, Sosa-Pineda B, Dussaud S, Billestrup N, et al. The ectopic expression of Pax4 in the mouse pancreas converts progenitor cells into alpha and subsequently beta cells. Cell 2009;138:449−62.

[115] Thorel F, Nepote V, Avril I, Kohno K, Desgraz R, Chera S, et al. Conversion of adult pancreatic alpha-cells to beta-cells after extreme beta-cell loss. Nature 2010;464:1149−54.

[116] Chung CH, Hao E, Piran R, Keinan E, Levine F. Pancreatic beta-cell neogenesis by direct conversion from mature alpha-cells. Stem Cell 2010;28:1630−8.

[117] Gianani R. Beta cell regeneration in human pancreas. Semin Immunopathol 2011;33:23−7.

[118] Bertelli E, Bendayan M. Intermediate endocrine-acinar pancreatic cells in duct ligation conditions. Am J Physiol 1997;273:1641−9.

[119] Desai BM, Oliver-Krasinski J, De Leon DD, Farzad C, Hong N, Leach SD, et al. Preexisting pancreatic acinar cells contribute to acinar cell, but not islet beta cell, regeneration. J Clin Invest 2007;117:971−7.

[120] Zhou Q, Brown J, Kanarek A, Rajagopal J, Melton DA. *In vivo* reprogramming of adult pancreatic exocrine cells to beta-cells. Nature 2008;455:627−32.

[121] Cardinale V, Wang Y, Carpino G, Cui CB, Gatto M, Rossi M, et al. Multipotent stem/progenitor cells in human biliary tree give rise to hepatocytes, cholangiocytes, and pancreatic islets. Hepatology 2011;54:2159−72.

[122] Cardinale V, Wang Y, Carpino G, Mendel G, Alpini G, Gaudio E, et al. The biliary tree—a reservoir of multipotent stem cells. Nat Rev Gastroenterol Hepatol 2012;9:231−40.

[123] Wang YL, Carpino G, Cui CB, Dominguez-Bendala J, Wauthier E, Cardinale V, et al. Biliary tree stem cells, precursors to pancreatic committed progenitors: evidence for possible life-long pancreatic organogenesis. Stem Cells 2013 Jul 11. Available from: http://dx.doi.org/10.1002/stem.1460 [Epub ahead of print] in press. PMID: 23847135.

[124] Domínguez-Bendala JL, Inverardi L, Ricordi C. Concise review: mesenchymal stem cells for diabetes. Stem Cell Trans Med 2012;1:59−63.

[125] Moriscot C, de Fraipont F, Richard MJ, Marchand M, Savatier P, Bosco D, et al. Human bone marrow mesenchymal stem cells can express insulin and key transcription factors of the endocrine pancreas developmental pathway upon genetic and/or microenvironmental manipulation *in vitro*. Stem Cell 2005;23:594−603.

[126] Seo MJ, Suh SY, Bae YC, Jung JS. Differentiation of human adipose stromal cells into hepatic lineage *in vitro* and *in vivo*. Biochem Biophys Res Commun 2005;328:258−64.

[127] Chiou SH, Chen SJ, Chang YL, Chen YC, Li HY, Chen DT, et al. MafA promotes the reprogramming of placenta-derived multipotent stem cells into pancreatic islets-like and insulin + cells. J Cell Mol Med 2011;15:612−24.

[128] Prabakar KR, Dominguez-Bendala J, Molano RD, Pileggi A, Villate S, Ricordi C, et al. Generation of glucose-responsive, insulin-producing cells from human umbilical cord blood-derived mesenchymal stem cells. Cell Transplant 2012;21 (6):1321−39.

[129] Furth ME, Atala A. Stem cell sources to treat diabetes. J Cell Biochem 2009;106:507−11.

[130] Mfopou JK, Chen B, Sui L, Sermon K, Bouwens L. Recent advances and prospects in the differentiation of pancreatic cells from human embryonic stem cells. Diabetes 2010;59:2094−101.

[131] Soria B, Roche E, Berna G, Leon-Quinto T, Reig JA, Martin F. Insulin-secreting cells derived from embryonic stem cells normalize glycemia in streptozotocin-induced diabetic mice. Diabetes 2000;49:157−62.

[132] Lumelsky N, Blondel O, Laeng P, Velasco I, Ravin R, McKay R. Differentiation of embryonic stem cells to insulin-secreting structures similar to pancreatic islets. Science 2001;292:1389−94.

[133] Hansson M, Tonning A, Frandsen U, Petri A, Rajagopal J, Englund MC, et al. Artifactual insulin release from differentiated embryonic stem cells. Diabetes 2004;53:2603−9.

[134] Segev H, Fishman B, Ziskind A, Shulman M, Itskovitz-Eldor J. Differentiation of human embryonic stem cells into insulin-producing clusters. Stem Cell 2004;22:265−74.

[135] D'Amour KA, Agulnick AD, Eliazer S, Kelly OG, Kroon E, Baetge EE. Efficient differentiation of human embryonic stem cells to definitive endoderm. Nat Biotechnol 2005;23:1534−41.

[136] D'Amour KA, Bang AG, Eliazer S, Kelly OG, Agulnick AD, Smart NG, et al. Production of pancreatic hormone-expressing endocrine cells from human embryonic stem cells. Nat Biotechnol 2006;24:1392−401.

[137] Schulz TC, Young HY, Agulnick AD, Babin MJ, Baetge EE, Bang AG, et al. A scalable system for production of functional pancreatic progenitors from human embryonic stem cells. PLoS One 2012;7:e37004.

[138] Kroon E, Martinson LA, Kadoya K, Bang AG, Kelly OG, Eliazer S, et al. Pancreatic endoderm derived from human embryonic stem cells generates glucose-responsive insulin-secreting cells *in vivo*. Nat Biotechnol 2008;26:443−52.

[139] Jiang J, Au M, Lu K, Eshpeter A, Korbutt G, Fisk G, et al. Generation of insulin-producing islet-like clusters from human embryonic stem cells. Stem Cell 2007;25:1940−53.

[140] Zhang D, Jiang W, Liu M, Sui X, Yin X, Chen S, et al. Highly efficient differentiation of human ES cells and iPS cells into mature pancreatic insulin-producing cells. Cell Res 2009;19:429–38.

[141] Rezania A, Riedel MJ, Wideman RD, Karanu F, Ao Z, Warnock GL, et al. Production of functional glucagon-secreting alpha-cells from human embryonic stem cells. Diabetes 2011;60:239–47.

[142] Rezania A, Bruin JE, Riedel MJ, Mojibian M, Asadi A, Xu J, et al. Maturation of human embryonic stem cell-derived pancreatic progenitors into functional islets capable of treating pre-existing diabetes in mice. Diabetes 2012;61:2016–29.

[143] Kelly OG, Chan MY, Martinson LA, Kadoya K, Ostertag TM, Ross KG, et al. Cell-surface markers for the isolation of pancreatic cell types derived from human embryonic stem cells. Nat Biotechnol 2011;29:750–6.

[144] Takahashi K, Yamanaka S. Induction of pluripotent stem cells from mouse embryonic and adult fibroblast cultures by defined factors. Cell 2006;126:663–76.

[145] Yu J, Hu K, Smuga-Otto K, Tian S, Stewart R, Slukvin II, et al. Human induced pluripotent stem cells free of vector and transgene sequences. Science 2009;324:797–801.

[146] Tateishi K, He J, Taranova O, Liang G, D'Alessio AC, Zhang Y. Generation of insulin-secreting islet-like clusters from human skin fibroblasts. J Biol Chem 2008;283:31601–7.

[147] Kim K, Doi A, Wen B, Ng K, Zhao R, Cahan P, et al. Epigenetic memory in induced pluripotent stem cells. Nature 2010;467:285–90.

[148] Pera MF. Stem cells: the dark side of induced pluripotency. Nature 2011;471:46–7.

[149] Bar-Nur O, Russ HA, Efrat S, Benvenisty N. Epigenetic memory and preferential lineage-specific differentiation in induced pluripotent stem cells derived from human pancreatic islet beta cells. Cell Stem Cell 2011;9:17–23.

[150] Hinton A, Hunter S, Reyes G, Fogel GB, King CC. From pluripotency to islets: miRNAs as critical regulators of human cellular differentiation. Adv Genet 2012;79:1–34.

[151] Bartel DP. MicroRNAs: target recognition and regulatory functions. Cell 2009;136:215–33.

[152] Bagga S, Pasquinelli AE. Identification and analysis of microRNAs. Genet Eng (N Y) 2006;27:1–20.

[153] Massirer KB, Pasquinelli AE. The evolving role of microRNAs in animal gene expression. Bioessays 2006;28:449–52.

[154] Kim VN, Nam JW. Genomics of microRNA. Trends Genet 2006;22:165–73.

[155] Griffiths-Jones S. The microRNA Registry. Nucleic Acids Res 2004;32:109–11.

[156] Wienholds E, Kloosterman WP, Miska E, Alvarez-Saavedra E, Berezikov E, de Bruijn E, et al. MicroRNA expression in zebrafish embryonic development. Science 2005;309:310–1.

[157] Lagos-Quintana M, Rauhut R, Yalcin A, Meyer J, Lendeckel W, Tuschl T. Identification of tissue-specific microRNAs from mouse. Curr Biol 2002;12:735–9.

[158] Lynn FC, Skewes-Cox P, Kosaka Y, McManus MT, Harfe BD, German MS. MicroRNA expression is required for pancreatic islet cell genesis in the mouse. Diabetes 2007;56(12):2938–45.

[159] Kloosterman WP, Lagendijk AK, Ketting RF, Moulton JD, Plasterk RH. Targeted inhibition of miRNA maturation with morpholinos reveals a role for miR-375 in pancreatic islet development. PLoS Biol 2007;5:203.

[160] Bravo-Egana V, Rosero S, Molano RD, Pileggi A, Ricordi C, Dominguez-Bendala J, et al. Quantitative differential expression analysis reveals miR-7 as major islet microRNA. Biochem Biophys Res Commun 2008;366:922–6.

[161] Correa-Medina M, Bravo-Egana V, Rosero S, Ricordi C, Edlund H, Diez J, et al. MicroRNA miR-7 is preferentially expressed in endocrine cells of the developing and adult human pancreas. Gene Expr Patterns 2009;9(4):193–9.

[162] Rosero S, Bravo-Egana V, Jiang Z, Khuri S, Tsinoremas N, Klein D, et al. MicroRNA signature of the human developing pancreas. BMC Genomics 2010;11:509.

[163] Dominguez-Bendala J, Alvarez-Cubela S, Nieto M, Vargas N, Espino-Grosso P, Sacher VY, et al. Intracardial embryonic delivery of developmental modifiers in utero. Cold Spring Harb Protoc 2012;2012(9):962–8.

Chapter 43

Microencapsulation Technology

Rajesh A. Pareta, John P. McQuilling, Alan C. Farney, and Emmanuel C. Opara

Wake Forest Institute for Regenerative Medicine, Wake Forest School of Medicine, Winston-Salem, NC

Chapter Outline

43.1 INTRODUCTION

The hallmark of type 1 diabetes is destruction of the insulin-producing β-cells of the pancreas by an autoimmune disease [1] resulting in the obligatory need for exogenous insulin to regulate blood glucose in the afflicted individuals. The bioartificial pancreas (BAP) involves a bioengineering approach to treating diabetes with islet transplants, which will secrete pancreatic hormones in response to the host blood glucose like a real pancreas. Presently, islet transplantation is limited by a shortage of pancreas for human islets isolation and allogenic islet recipients also require long-term immunosuppression to prevent rejection and recovery of autoimmunity. Design of BAP should take this into account. In type 1 diabetes, the patients have preexisting antibodies and immune cells primed against β-cell surface markers and insulin [2], and hence a simple islet transplantation without immunosuppression is not viable. In the BAP construct illustrated in Figure 43.1, islets require a protective semipermeable coating to immunoisolate them and preserve their viability and functionality upon transplantation. This approach opens up possibilities for allo- and xenotransplants, and thus has the potential to overcome the shortage of islets while addressing the issue of transplant rejection. To summarize briefly, the BAP construct is not only a viable option to address the human islet shortage but also offers tremendous benefits to the transplant recipient such as relief from long-term use of immunosuppressant drugs.

In this review, we will focus on two technologies which may enhance the development of a viable BAP. The first is the use of alginate to microencapsulate islets. Alginate is a widely used naturally occurring biopolymer for islet microencapsulation [3]. The second technology, which is currently in its infancy, is the use of extracellular matrix (ECM) to influence the viability and function of islets in BAP constructs. ECM is a three-dimensional meshwork of proteins and polysaccharides that impart structure and mechanical stability to tissues [3], which can be obtained after decellularization of tissue and organs. Alginate has been extensively used because it has unique properties and advantages for islet microencapsulation such as its mechanical strength, hydrophilic nature, and ability to cross-link at physiological conditions, ECM is a relatively new and upcoming technology where the tissue's own natural scaffold can be used to promote the interactions between the islets and a construct matrix.

43.2 ISLET ISOLATION

In fabricating a BAP, a critical starting point is the procurement of pancreatic islets from an abundant source using reliable techniques to preserve β-cell function. Thus, isolation of whole islets without inflicting any

Regenerative Medicine Applications in Organ Transplantation.

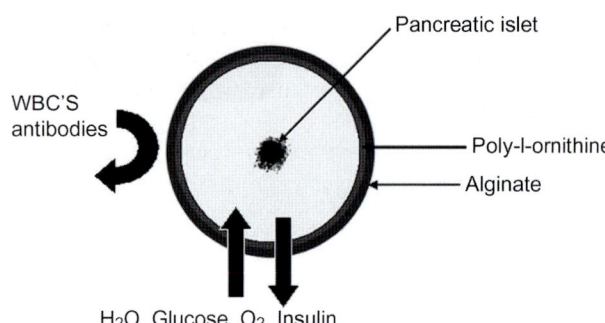

FIGURE 43.1 Concept of a BAP. A protective semipermeable coating to immunoisolate islets (iso-, allo-, or xenosource) and preserve their viability and functionality upon transplantation [1].

significant damage to the cells is a key component of developing a viable BAP. A critical balance of composition, process, and duration of collagenase digestion is required for isolating islets with integrity, viability, and high purity with a significant yield. This overall process has tremendous impact on the clinical outcome of islet transplants [4]. The pancreas is digested with combined collagenase and protease action, which disintegrates the intercellular matrix of collagen, releasing islets. These islets are isolated, purified, tested for viability, and sometimes cultured before being transplanted in the patient. Collagenase digestion disrupts islet−exocrine tissue adhesive contacts [5]. Thus, shorter duration or lower concentration of collagenase would lead to incomplete digestion of islets from exocrine tissue, leading to reduced yield on purification. On the other hand, extended duration of incubation or higher concentration of collagenase would adversely affect the islet cell−cell adhesion, leading to loss of islet integrity and viability. Intra-islet cell−cell adhesion is protease sensitive, while extra-islet cell−matrix adhesion is collagenase sensitive. In the pig pancreas, very little periinsular capsule is present, and the structural integration of the porcine islet in the exocrine pancreas is almost exclusively cell−cell adhesion. In canine, the islets are almost exclusively encapsulated with very little exocrine−endocrine cell−cell contact. In rodent and human, the situation is intermediate with a tendency toward predominance of cell−matrix adhesion. The presence of protease in the collagenase preparations has been reported to reduce the yield and quality of isolated islets from rats [6], however, it is more efficient for the isolation of pig islets [7].

43.3 ALGINATE-BASED MICROENCAPSULATION OF ISLETS

It is crucial that the biomaterial used to encapsulate islets must be biocompatible and permeable (for hormonal, nutrient, and oxygen exchange). Hydrogels are such semisolid materials, which not only are soft but remain stable under mechanical stress. Hydrogels are very attractive for making microcapsules. Hydrogels also provide higher permeability for low-molecular weight nutrients and metabolites. Furthermore, the soft and pliable features of the gel reduce the mechanical or frictional irritations to surrounding tissue [8]. Alginate is one such hydrogel which has been widely used owing to its many excellent properties conducive to islet transplants.

Alginate molecules are linear block copolymers of β-D-mannuronic (M) and α-L-guluronic acids (G). It forms a gel in the presence of divalent ions like Ca^{2+} and Ba^{2+}. Recent studies have shown that divalent ions cross-link not only G blocks but also blocks of alternating M and G (M−G blocks) [9]. Mainly calcium is used for gelling, as barium is known to be toxic and concerns have been raised about patients' safety if it is used as the cross-linking agent. Alginate is one of the few biopolymers that allow cell encapsulation at physiological conditions. The encapsulation can be done at room or body temperature, at physiological pH, and in isotonic solutions. Alginate-based capsules have been shown to be stable for years in both animals and humans [10−12]. Also, since alginate materials are negatively charged, the attachment of immune cells to the microcapsule is limited due to the negative charge on the cell surface thus making alginate very highly biocompatible [13].

In most tissues it has been shown that maximum diffusion distance for effective oxygen and nutrient diffusion from blood capillary to cells is about 200 μm. Absence of this convection inside a capsule induces a nutrient gradient from the capsule surface to center of cells. Present insights suggest microcapsules as a preferable system over macrocapsules due to their high surface to volume ratio for fast exchange of hormones and nutrients. Microencapsulation uses the interfacial precipitation predominantly, where a polyanionic polymer (alginate) gels with a divalent cation (Ca^{2+}, Ba^{2+}). Cells are suspended in an alginate solution and its droplets are generated by air jet spray method [14], electrostatic generators [15,16], submerged oscillating coaxial extrusion nozzles [17], conformal coatings [18], and spinning disk atomization [19].

Of these methods, the air jet spray method, which uses a two-channel air droplet microencapsulator, is most commonly used. Two-channel air droplet microencapsulators operate by allowing the alginate cell suspension to drip through an inner channel of the device while the outer channel uses an air jacket to shear off the alginate droplet. Using this method, the diameters of the inner and outer channels, the flow rate of the alginate, and air pressure of the outer channel can be adjusted to vary the microcapsule size [14]. In order to prevent hypoxic damage to cells, microencapsulation must be done relatively quickly even at lower temperature such as 4°C.

A reduction in capsule size would benefit the cells and also exponentially decrease the total transplant volume. Therefore, much work has been done with various new technologies to make beads as small as 185 μm (diameter) which is about four times smaller than conventional beads (800 μm). The smaller the diameter of the capsules the better the diffusion of nutrients and oxygen to the cells, and it has been shown that microcapsules with a diameter of 600 ± 100 μm had improved stability *in vivo* compared to larger capsules with diameters of 1000 ± 100 μm [20].

Uncoated non-permselective alginate microbeads have been reported to have a high permeability (>600 kD). Uptake studies with IgG (150 kD) and thyroglobulin (669 kD) suggested that they permeated uncoated alginate microbeads. Similarly, uncoated alginate microbeads implanted in peritoneum were positive for both IgG and C3 components after only 1 week [21]. Therefore, various immunoreactive molecules from macrophages and T-cells to smaller cytokine molecules such as IL-1β, TNF-α, and IFN-γ can easily penetrate into the microcapsules and can damage or destroy the encapsulated islets [22]. The role of permselective coating of alginate microcapsules cannot be overemphasized, and this issue can be illustrated with studies of encapsulated islet xenografts in the spontaneously diabetic nonobese diabetic (NOD) mouse. While studies performed with pig islet xenografts encapsulated with permselective alginate microcapsules showed prolonged reversal of hyperglycemia in immunocompetent diabetic NOD mice [23,24], another study using uncoated alginate microcapsules to encapsulate fish islets showed rapid destruction of the xenografts in NOD mice [25], thus highlighting the need to provide immunoisolation for islets within alginate microcapsules designated for transplant studies.

Encapsulated islets may incite a host inflammatory response. There are two general targets of the host-derived responses:

1. Inflammatory reaction against the capsule material: With the present technology these reactions can be successfully prevented by applying purification steps to the materials to be used [8].
2. Host response against the allogenic or xenogenic cell-derived bioactive factors or antigens that leak out of the capsules. It results in overgrowth by macrophages and lymphocytes on a small portion ($\sim 10\%$) of the capsules and in a humoral immune response against the encapsulated tissue. It has long been known that islets secrete cytokines upon stress [26]. Encapsulated islets have been shown to produce the cytokines MCP-1, MIP, nitric oxide (NO), and IL-6 under stress (stress induced by adding IL-1β and TNF-α), and these cytokines are well known to contribute to the recruitment

and activation of inflammatory cells [27]. Also, it has been demonstrated that activated macrophages on the 2−10% microcapsules with overgrowth do secrete the cytokines IL-1β and TNF-α when cultured with encapsulated islets but not with empty capsules [28]. This activation of inflammatory cells results in the production of cytokines, which are deleterious not only to the islet cells in the overgrown capsules but also the islets in the vast majority of transplanted, clean, and non-overgrown capsules.

43.3.1 Semipermeable Membrane Coating Techniques for Alginate

To provide immunoisolation for the microcapsules, it is essential to apply a permeability barrier between the encapsulated cells and the host immune system. Applying a polyamino acid layer, followed by an additional outer coating of alginate, creates an adequate barrier from the host system. The positively charged polyamino molecules will readily bind to the negatively charged alginate molecules forming a complex membrane [29,30], which significantly reduces the pore size of the microcapsule and prevents immune cells from entering into it [31−33]. In order to prevent interactions of nonbound polyamines to host tissue, a thin second layer of alginate is added. This polyamino acid barrier also acts as a shell, providing mechanical stability to the microcapsule, allowing for the liquefaction of the inner alginate core [34]. The thickness and pore size of this barrier can be varied through adjustments in incubation time and concentration of the polymer used [35].

The most researched permselective biomaterial is poly-L-lysine (PLL) which was the first material used to generate this barrier [36]; however, more recent research has shown that poly-L-ornithine (PLO) has markedly reduced immune response and provides more mechanical support to the microcapsules. Like PLL, PLO is a positively charged polyamine which, when applied to alginate microcapsules, forms a semipermeable membrane which significantly reduces the porosity of the microcapsules, allowing for immunoisolation without impairing oxygen and nutrient diffusion. PLO has been shown to evoke less of an immune response as well as to have improved mechanical properties in comparison to PLL [37−41].

When compared to alginate−PLL microcapsules, alginate−PLO microcapsules have been shown to better resist swelling and bursting under osmotic stress [37]. Bead swelling is an important factor to take into consideration because it can cause increases in pore size and permeability, as well as in shear stress, leading to decreased islet viability [42,43]. It has been hypothesized that the improved mechanical properties of alginate−PLO microcapsules over alginate—PLL microcapsules are due to the improved bonding of PLO to alginate owing to the shorter monomer structure of PLO [37,44]. Also, while PLL seems to bind to M—G sequences, PLO has been shown to prefer M—M sequences [45]. Long-term studies, in which empty alginate—PLO microcapsules were injected intraperitoneally in rodents, dogs, or pigs have always resulted in retrieval of intact and overgrowth-free microcapsules up to 1-year postimplant [46]. Usually after the coating with a cationic poly(amino acid), e.g., PLL or PLO is followed by a surface coating of low viscosity alginate, resulting in a microcapsule morphology that presents encapsulated cells in a sol layer of alginate, followed by PLL/PLO coating and gel layer of alginate on exterior, thus creating an alginate—PLL/PLO—alginate construct known as APA microcapsules.

43.3.2 Studies of Microencapsulated Islet Grafts in Large Animals and Humans

The technique of microencapsulation of islets prior to transplantation has shown promise in both large animal trials and pilot clinical trials. Multiple canine and primate studies have been conducted and have demonstrated the ability of encapsulated islets to maintain insulin independence [47—49]. A study conducted by Sun et al. demonstrated the ability of encapsulated islet xenografts to reverse diabetes for periods of time greater than 800 days [11]. A more recent study by Dufrane et al. demonstrated the ability of encapsulated islets to survive and produce insulin in the kidney capsule of *Cynomolgus macacus* for up to 6 months [51].

Several pilot clinical studies [10,12,45,50] have been conducted in humans. While these trials have failed to establish long-term insulin independence in any of the subjects, they have shown that the implantation of viable encapsulated islets can stabilize the blood glucose levels and reduce the required amount of exogenous insulin required. In a study by Soon-Shiong et al., a long-term type 1 diabetic patient was implanted with 15,000 encapsulated islet equivalents per kilogram body weight and evaluated for up to 9 months posttransplantation. In this study, average blood glucose levels were maintained at 135 mg/dL, and daily insulin requirements decreased from 0.69 ± 0.01 U/kg to 0 U/kg, and hyperglycemic episodes (>200 mg/dL) decreased from 11.7% to 6.14% at 9 months. Furthermore, the patient's quality of life was evaluated and shown to have greatly improved over the duration of the study [11]. The study by Calafiore et al. evaluated two individuals 60 days after receiving the encapsulated allografts. Although insulin independence was not attained, there was a significant reduction in the daily insulin requirements as well as a significant

reduction in the number of hypoglycemic events [45]. A third human study by Elliot et al. evaluated the effectiveness of porcine xenograft encapsulated islets up to 9.5 years after implantation. In this study, immediately after implantation, the daily insulin dosage was reduced by 30%, and C-peptide was present in urine samples up to 14 months posttransplantation. Retrieval of the capsules 9.5 years later revealed that the islets were still capable of producing insulin, however, the levels of insulin were significantly reduced and C-peptide could not be measured [12]. Tuch et al. studied the safety and viability of human islets microencapsulated in non-permselective alginate microbeads and found that allografts of these encapsulated islets were safe but had no efficacy in diabetic patients [50]. In addition to these small pilot trials, larger clinical trials are underway in New Zealand and Russia by Living Cell Technologies Limited (LCT). LCT is currently performing phase I and II clinical trials with DIABECELL® which are encapsulated neonatal porcine islets that are injected into the peritoneal cavity via laparoscopy at doses of 10,000—20,000 islet equivalents/kg. Currently, the short-term and long-term safety and effectiveness as well as proper dosage are being evaluated [51].

43.3.3 Microencapsulation Devices

One major limitation to the development of the microencapsulated islet technology is the scarcity of high-throughput devices. Current available microencapsulation devices are incapable of efficiently encapsulating large numbers of islets in a reasonable amount of time. This may result in hypoxic stress and loss of islet viability or islet function [3]. A newly proposed alternative procedure for islet microencapsulation utilizes multichannel air jacket microfluidic devices. These devices have the advantage of rapidly encapsulating large numbers of islets into microcapsules at speeds in excess of eight times of those of conventional methods without affecting the functionality of the islets. Additionally, this microfluidic approach can be used to produce microcapsules in the size range of 300—500 μ in diameter and are easily scaled up to increase production rates and can be cost effectively produced using rapid prototyping technology [52]. Figure 43.2 is a picture of PLO-coated APA microcapsules made with a prototype microfluidic device in our laboratory. An important consideration in the evaluation of the devices is the ability to produce microcapsules with uniformity in shape and size as the morphology of the microcapsules used to encapsulate islets plays a critical role in the performance of the BAP construct. Spherical microcapsules are necessary for long-term functionality; irregularities or imperfections in the microcapsules can cause an immune response and result in loss of islet function [53].

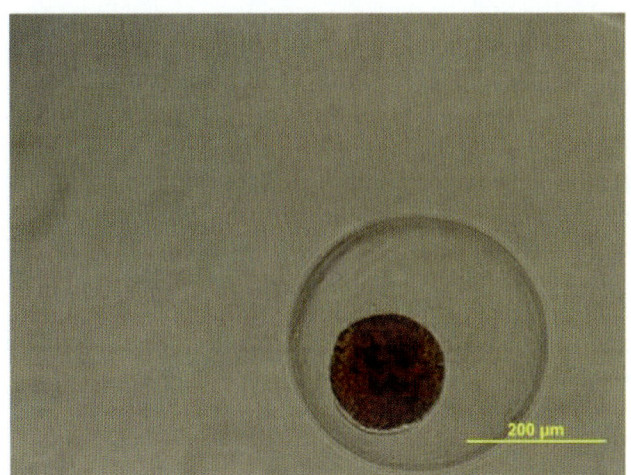

200 µm

FIGURE 43.2 Alginate—PLO—alginate (APA) microcapsules made with a prototype microfluidic device. Islet stained with dithizone.

43.3.4 Oxygen Requirements of Islets and the Effect of Transplantation Site on BAP Function

Although islets constitute approximately 1% of the pancreas, they receive about 6−10% of its blood flow [54], indicating a disproportionate level of perfusion in which islets receive and consume lots of oxygen. The usual high oxygen requirement of islets is interrupted during the process of islet isolation and processing when islets are used for transplantation, and studies have shown that hypoxia has significant deleterious effects on the survival and function of islets [55]. In the immediate posttransplant period, isolated islets are forced to depend upon diffusion of oxygen and nutrients through peripheral perfusion from the surrounding tissue within the site of transplantation [56], until revascularization by angiogenesis, a process that requires 7−10 days [57]. The peritoneal cavity is a commonly used site for implantation of microencapsulated islets. An advantage of the peritoneal site is the ease of transplantation, but the site has a number of disadvantages, including a low vascular density. Observations suggest that engraftment is slow and inefficient within the peritoneal cavity, probably because of the low vascular density and an extended period of hypoxia before engraftment occurs. Therefore, the death of most of the encapsulated islet grafts owing to severe hypoxia results in the need for large quantities of microencapsulated islets to achieve normoglycemia in studies performed in large animals and humans [3].

Considering the issue of adequate nutrient supply as discussed above, it is necessary to find a site where encapsulated islets are in close contact with the bloodstream. Unfortunately, it is difficult to find such a site since it should combine the capacity to bear a large graft volume in the immediate vicinity of blood vessels. Transplantation of encapsulated islets is most commonly done intraperitoneally, as it offers the advantages of laparoscopic implantation or through injection and allows ample room to implant numerous microcapsules [58]. In addition to the problem of avascular supply discussed earlier, another major disadvantage is that microcapsules that are implanted intraperitoneally are vulnerable to an immune response from intraperitoneal T-cells and macrophages [59−61] and have less access to the vasculature. This results in an increased likelihood of fibrotic growth over encapsulated islets, a loss of graft functionality, and a delay in insulin uptake into the blood circulation [62].

Consequently, alternative transplantation sites have been investigated, including transplanting into liver [63], kidney capsule, subcutaneously, and into an omentum pouch [3,51,64−66]. In the study conducted by Toso et al., microcapsules were injected into the portal veins of rats; however, the results of the study showed that immunosuppressants were necessary to prevent fibrotic overgrowth, and the risk of hepatic thrombosis makes this approach impractical. The studies by Dufrane et al. that investigated implant sites such as subcutaneous and the kidney capsule showed that encapsulated islets implanted in these two sites had less cellular overgrowth compared to encapsulated islets implanted intraperitoneally. The studies by Dufrane et al. demonstrated the functionality of encapsulated islets implanted within the kidney capsule of primates [51]; however, clinical application would be difficult given the limited space within this site [8]. The attraction for the omentum pouch is that like the kidney capsule, it offers a well-vascularized site for transplantation but has more space for microcapsules and is easier to access [67]. In addition, microencapsulated islets transplanted in the omentum pouch are easily retrievable for posttransplant evaluation [3].

Taking the oxygen requirements of islets into consideration, a recent study has described a promising approach that involves enclosure of microencapsulated islets in a macrochamber specifically engineered for islet transplantation. The subcutaneous implantable device allows for controlled and adequate oxygen supply and provides immunological protection of donor islets against the host immune system. This minimally invasive implantable BAP was shown to normalize blood sugar in streptozotocin-induced diabetic rodents for up to 3 months after subcutaneous transplantation [68]. In another study, investigators showed that encapsulation of solid calcium peroxide within hydrophobic polydimethylsiloxane resulted in sustained oxygen generation that lasted for more than 6 weeks and was enough to prevent hypoxia-induced cell dysfunction and death in insulin-producing cells [69].

43.4 POTENTIAL ROLE OF ECM-BASED TECHNOLOGY IN THE DEVELOPMENT OF THE BAP

In organ bioengineering, seeding of cells on supporting scaffolding material offers an exciting opportunity to enhance the clinical application of bioartificial organs [70−73]. In the last decade, more than 50 patients have received an organ manufactured from autologous cells which were seeded on supporting scaffolding material, with no use of immunosuppression at any time after the implantation. With this groundbreaking, history-making achievement tissue engineering has shown the potential to dramatically impact solid organ transplantation by successfully addressing the two major barriers to solid organ transplantation, namely the need for a new, ideally inexhaustible source of organs and for immunosuppression-free status posttransplantation.

Scaffolds may be synthetic or natural. Natural scaffolds consist of the innate ECM of animal or human organs and can now consistently be produced by perfusion of detergent solutions through the organ's vasculature, a process called decellularization [74]. Innate ECM represents a biochemically, geometrically, and spatially ideal platform for bioengineering investigations, because it is biocompatible [75], it has both basic components (proteins and polysaccharides) and matrix-bound growth factors and cytokines preserved [76]. It retains an intact and patent vasculature which—when implanted *in vivo*—sustains the physiologic blood pressure [75], and it is able to drive differentiation of progenitor cells into an organ-specific phenotype [77]. Importantly, when cells are seeded on ECM samples, they attach and expand well; when cells are seeded within whole, intact ECM scaffolds and allowed to mature into bioreactors, cells proliferate and show signs of active metabolism and effective function [78−81].

Pancreatic islet transplantation seems to be an ideal ground for the implementation of ECM-based bioengineering technology, because the role of matrix−integrin interactions on beta-cell survival and function is well known. This was demonstrated more than a decade ago by Wang and Rosenberg. In their classic study [82], the ability of canine islets in culture to attach to a collagen matrix was shown to decline progressively over 6 days. This decline was accompanied by a decrease in integrin expression and beta-cell function and an increase in apoptosis; yet it could be prevented or delayed by exposure of islets to matrix proteins. This observation provided evidence that the disruption of the cell−matrix relationship following pancreatic islets isolation can be prevented by restoration of a culture microenvironment that includes matrix proteins. Later, several investigations were focused on either increasing the survival of islets *in vitro* through

support in a solid matrix or on restoring the ECM environment and determining the effect of cell−matrix and cell−cell interaction on survival. In one study, Daoud et al. tried to identify the factors responsible for postisolation islet survival and promotion of function *in vitro* [83]. By investigating the effects of collagen I and IV, fibronectin, and laminin on human islet adhesion, survival, and functionality, the authors observed that collagen I/IV and fibronectin are essential for cell adhesion, while fibronectin is the only ECM protein capable of maintaining islet structural integrity and insulin content distribution.

Some groups are currently using pancreas ECM to support and enhance the islet viability *in vivo* in small animal models. Rat pancreata were minced and decellularized to obtain acellular ECM. When seeded on ECM patches, islets adhered well to the pancreatic matrix, maintained their long-term viability and function, and showed a constant glucose-induced insulin release during long-term *in vitro* incubation. In contrast, islets cultured on plastic or on non-pancreatic matrix showed a progressive reduction [84]. Moreover, when acellular matrix/islet cultures were inserted into poly(vinyl alcohol)/poly(ethylene glycol) tubes to obtain implantable devices, an *in vitro* constant insulin release could be detected. When the devices were implanted into diabetic rats, a reduced insulin requirement was noted suggesting insulin secretory activity of islets contained in the device. Later, immunofluorescence confirmed the presence of insulin- and glucagon-producing cells in the explanted devices.

From the foregoing illustration of the important role of ECM on islet function, we envisage a scenario where coencapsulation of micro/nanoparticles of ECM with islets in alginate microcapsules would result in significant enhancement of the function and longevity of encapsulated islet grafts. Indeed, it has recently been shown that the coencapsulation of islets with ECM proteins and mesenchymal stromal cells in a silk hydrogel resulted in enhanced function of the islets in *in vitro* studies [85]. Overall, ECM-based bioengineering technology holds a great promise for pancreatic islet transplantation research because of the essential role played by ECM−islet interactions for islet integrity. Nevertheless, knowledge of pancreas ECM biology and of the interactions with pancreatic islets remains inadequate and represents the biggest hurdle to overcome before it can be used in the development of a BAP for the treatment of diabetes. Acellular pancreas ECM can be produced effectively and consistently; however we still do not know how much damage is caused, and whether, in doing so, we destroy molecular domains that are essential for cells/islet to attach and grow. Also, in other scenarios, innate ECM has been used to drive differentiation of progenitor cells toward an organ-specific phenotype. Therefore, in the future, pancreas ECM may be used also to obtain insulin-producing cells from different strains of progenitor cells, a situation which would enhance the availability of islets for BAP constructs.

43.5 CONCLUSIONS

Alginate-based and ECM-based materials offer complementary technologies, which working together have the potential to advance the development of a viable BAP and provide metabolic function and a cure for type 1 diabetes. At present, we know more about the encapsulation biomaterials and semipermeable membrane materials. We can transplant more volume (smaller microcapsules size) as well as purified alginates/polymers which are biocompatible. With recent developments in technologies for the BAP as discussed in this chapter, there is tremendous hope that routine use of BAP constructs will become a clinical reality in the not so distant future.

ACKNOWLEDGMENT

The authors would like to acknowledge financial support from the National Institutes of Health (RO1 DK080897) and the Vila Rosenfeld Estate, Greenville, NC for the work in Dr. Opara's laboratory at the Wake Forest Institute for Regenerative Medicine.

LIST OF ABBREVIATIONS

BAP bioartificial pancreas
ECM extracellular matrix

REFERENCES

[1] Pugliese A, Eisenbarth GS. Type I diabetes mellitus: lessons for human autoimmunity. J Lab Clin Med 1992;120:363—6.

[2] Jaeger C, Brendel MD, Eckhard M, Bretzel RG. Islet autoantibodies as potential markers for disease recurrence in clinical islet transplantation. Exp Clin Endocrinol Diabetes 2000;108:328—33.

[3] Opara EC, Kendall WF. Immunoisolation techniques for islet cell transplantation. Expert Opin Biol Ther 2002;2:503—11.

[4] Lakey JRT, Kobayashi N, James Shapiro AM, Ricordi C, Okitsu T. Current human islet isolation protocol. Hiranomachi: Medical review Co., LTD; 2004.

[5] Wolters GHJ, Fritschy WM, Gerrits D, Vanschilfagaarde R. A versatile alginate droplet generator applicable for microencapsulation of pancreatic-islets. J Appl Biomater 1992;3:281—6.

[6] Vos-Scheperkeuter GH, van Suylichem PT, Vonk MW, Wolters GH, van Schilfgaarde R. Histochemical analysis of the role of class I and class II Clostridium histolyticum collagenase in the degradation of rat pancreatic extracellular matrix for islet isolation. Cell Transplant 1997;6:403—12.

[7] van Deijnen JH, Hulstaert CE, Wolters GH, van Schilfgaarde R. Significance of the peri-insular extracellular matrix for islet isolation from the pancreas of rat, dog, pig, and man. Cell Tissue Res 1992;267:139—46.

[8] de Vos P, Hoogmoed CG, Busscher HJ. Chemistry and biocompatibility of alginate-PLL capsules for immunoprotection of mammalian cells. J Biomed Mater Res 2002;60:252—9.

[9] Donati I, Holtan S, Morch YA, Borgogna M, Dentini M, Skjak-Braek G. New hypothesis on the role of alternating sequences in calcium-alginate gels. Biomacromolecules 2005;6:1031—40.

[10] Soon-Shiong P, Heintz RE, Merideth N, Yao QX, Yao Z, Zheng T, et al. Insulin independence in a type 1 diabetic patient after encapsulated islet transplantation. Lancet 1994;343:950—1.

[11] Sun Y, Ma X, Zhou D, Vacek I, Sun AM. Normalization of diabetes in spontaneously diabetic cynomologus monkeys by xenografts of microencapsulated porcine islets without immunosuppression. J Clin Invest 1996;98:1417—22.

[12] Elliott RB, Escobar L, Tan PL, Muzina M, Zwain S, Buchanan C. Live encapsulated porcine islets from a type 1 diabetic patient 9.5 yr after xenotransplantation. Xenotransplantation 2007;14:157—61.

[13] Smidsrod O, Skjak-Braek G. Alginate as immobilization matrix for cells. Trends Biotechnol 1990;8:71—8.

[14] Wolters GH, Fritschy WM, Gerrits D, van Schilfgaarde R. A versatile alginate droplet generator applicable for microencapsulation of pancreatic islets. J Appl Biomater 1991;3:281—6.

[15] Halle JP, Leblond FA, Pariseau JF, Jutras P, Brabant MJ, Lepage Y. Studies on small (< 300 microns) microcapsules: II—Parameters governing the production of alginate beads by high voltage electrostatic pulses. Cell Transplant 1994;3:365—72.

[16] Hsu BR, Chen HC, Fu SH, Huang YY, Huang HS. The use of field effects to generate calcium alginate microspheres and its application in cell transplantation. J Formos Med Assoc 1994;93:240—5.

[17] Dawson RM, Broughton RL, Stevenson WT, Sefton MV. Microencapsulation of CHO cells in a hydroxyethyl methacrylate-methyl methacrylate copolymer. Biomaterials 1987;8:360—6.

[18] Desmangles AI, Jordan O, Marquis-Weible F. Interfacial photopolymerization of beta-cell clusters: approaches to reduce coating thickness using ionic and lipophilic dyes. Biotechnol Bioeng 2001;72:634—41.

[19] Senuma Y, Lowe C, Zweifel Y, Hilborn JG, Marison I. Alginate hydrogel microspheres and microcapsules prepared by spinning disk atomization. Biotechnol Bioeng 2000;67:616—22.

[20] Omer A, Duvivier-Kali V, Fernandes J, Tchipashvili V, Colton CK, Weir GC. Long-term normoglycemia in rats receiving transplants with encapsulated islets. Transplantation 2005;79:52—8.

[21] Lanza RP, Kuhtreiber WM, Ecker DM, Marsh JP, Chick WL. Transplantation of porcine and bovine islets into mice without immunosuppression using uncoated alginate microspheres. Transplant Proc 1995;27:3321.

[22] van Schilfgaarde R, de Vos P. Factors influencing the properties and performance of microcapsules for immunoprotection of pancreatic islets. J Mol Med (Berl) 1999;77:199—205.

[23] Lum ZP, Tai IT, Krestow M, Norton J, Vacek I, Sun AM. Prolonged reversal of diabetic state in NOD mice by xenografts of microencapsulated rat islets. Diabetes 1991;40:1511—6.

[24] Cui H, Tucker-Burden C, Cauffiel SM, Barry AK, Iwakoshi NN, Weber CJ, et al. Long-term metabolic control of autoimmune diabetes in spontaneously diabetic nonobese diabetic mice by

nonvascularized microencapsulated adult porcine islets. Transplantation 2009;88:160−9.

[25] Xu BY, Yang H, Serreze DV, MacIntosh R, Yu W, Wright Jr JR. Rapid destruction of encapsulated islet xenografts by NOD mice is CD4-dependent and facilitated by B-cells: innate immunity and autoimmunity do not play significant roles. Transplantation 2005;80:402−9.

[26] Cardozo AK, Proost P, Gysemans C, Chen MC, Mathieu C, Eizirik DL. IL-1β and IFN-γ induce the expression of diverse chemokines and IL-15 in human and rat pancreatic islet cells, and in islets from pre-diabetic NOD mice. Diabetologia 2003;46:255−66.

[27] De Groot CJ, Woodroofe MN. The role of chemokines and chemokine receptors in CNS inflammation. Prog Brain Res 2001;132:533−44.

[28] de Vos P, de Haan BJ, de Haan A, van Zanten J, Faas MM. Factors influencing functional survival of microencapsulated islet grafts. Cell Transplant 2004;13:515−24.

[29] Bystrický S, Malovíková A, Sticzay T. Interaction of alginates and pectins with cationic polypeptides. Carbohydr Polym 1990;13:283−94.

[30] Thu B, Bruheim P, Espevik T, Smidsrod O, Soon-Shiong P, Skjak-Braek G. Alginate polycation microcapsules. II. Some functional properties. Biomaterials 1996;17:1069−79.

[31] King GA, Daugulis AJ, Faulkner P, Goosen MFA. Alginate-poly-lysine microcapsules of controlled membrane molecular weight cutoff for mammalian cell culture engineering. Biotechnol Prog 1987;3:231−40.

[32] Halle JP, Leblond FA, Pariseau JF, Jutras P, Brabant MJ, Lepage Y. Studies on small (< 300 microns) microcapsules: II–Parameters governing the production of alginate beads by high voltage electrostatic pulses. Cell Transplant 1994;3:365−72.

[33] Kulseng B, Thu B, Espevik T, Skjak-Braek G. Alginate polylysine microcapsules as immune barrier: permeability of cytokines and immunoglobulins over the capsule membrane. Cell Transplant 1997;6:387−94.

[34] Darrabie M, Freeman BK, Kendall Jr WF, Hobbs HA, Opara EC. Durability of sodium sulfate-treated polylysine-alginate microcapsules. J Biomed Mater Res 2001;54:396−9.

[35] Gugerli R, Cantana E, Heinzen C, von Stockar U, Marison IW. Quantitative study of the production and properties of alginate/poly-L-lysine microcapsules. J Microencapsul 2002;19:571−90.

[36] Lim F, Sun AM. Microencapsulated islets as bioartificial endocrine pancreas. Science 1980;210:908−10.

[37] Darrabie MD, Kendall WF, Opara EC. Characteristics of poly-L-ornithine-coated alginate microcapsules. Biomaterials 2005;26:6846−52.

[38] Brunetti P, Basta G, Faloerni A, Calcinaro F, Pietropaolo M, Calafiore R. Immunoprotection of pancreatic islet grafts within artificial microcapsules. Int J Artif Organs 1991;14:789−91.

[39] Calafiore R, Basta G, Boselli C, Bufalari A, Giustozzi GM, Luca G, et al. Effects of alginate/polyaminoacidic coherent microcapsule transplantation in adult pigs. Transplant Proc 1997;29:2126−7.

[40] Calafiore R, Basta G, Luca G, Calvitti M, Calabrese G, Racanicchi L, et al. Grafts of microencapsulated pancreatic islet cells for the therapy of diabetes mellitus in non-immunosuppressed animals. Biotechnol Appl Biochem 2004;39:159−64.

[41] Kizilel S, Garfinkel M, Opara E. The bioartificial pancreas: progress and challenges. Diabetes Technol Ther 2005;7:968−85.

[42] Thu B, Bruheim P, Espevik T, Smidsrod O, SoonShiong P, SkjakBraek G. Alginate polycation microcapsules .1. Interaction between alginate and polycation. Biomaterials 1996;17:1031−40.

[43] Ching CD, Harland RC, Collins BH, Kendall W, Hobbs H, Opara EC. A reliable method for isolation of viable porcine islet cells. Arch Surg 2001;136:276−9.

[44] Inaki Y, Tohnai N, Miyabayashi K, Miyata M. Isopoly-L-ornithine derivative as nucleic acid model. Nucleic Acids Symp Ser 1997;:25−6.

[45] Calafiore R, Basta G, Luca G, Lemmi A, Montanucci MP, Calabrese G, et al. Microencapsulated pancreatic islet allografts into nonimmunosuppressed patients with type 1 diabetes: first two cases. Diabetes Care 2006;29:137−8.

[46] de Vos P, Faas MM, Strand B, Calafiore R. Alginate-based microcapsules for immunoisolation of pancreatic islets. Biomaterials 2006;27:5603−17.

[47] Kendall Jr WF, Collins BH, Opara EC. Islet cell transplantation for the treatment of diabetes mellitus. Expert Opin Biol Ther 2001;1:109−19.

[48] Soon-Shiong P, Feldman E, Nelson R, Komtebedde J, Smidsrod O, Skjak-Braek G, et al. Successful reversal of spontaneous diabetes in dogs by intraperitoneal microencapsulated islets. Transplantation 1992;54:769−74.

[49] Wang T, Adcock J, Kuhtreiber W, Qiang D, Salleng KJ, Trenary I, et al. Successful allotransplantation of encapsulated islets in pancreatectomized canines for diabetic management without the use of immunosuppression. Transplantation 2008;85:331−7.

[50] Tuch BE, Keogh GW, Williams LJ, Wu W, Foster JL, Vaithilingam V, et al. Safety and viability of microencapsulated human islets transplanted into diabetic humans. Diabetes Care 2009;32:1887−9.

[51] Zukerman W. 'Pig Sushi' diabetes trial brings xenotransplant hope. New Sci 2010;:.

[52] Tendulkar S, McQuilling JP, Childers C, Pareta R, Opara EC, Ramasubramanian MK. A scalable microfluidic device for the mass production of microencapsulated islets. Transplant Proc 2011;43:3184−7.

[53] Hobbs HA, Kendall Jr WF, Darrabie M, Opara EC. Prevention of morphological changes in alginate microcapsules for islet xenotransplantation. J Investig Med 2001;49:572−5.

[54] Eddlestone GT, Oldham SB, Lipson LG, Premdas FH, Beigelman PM. Electrical activity, cAMP concentration, and insulin release in mouse islets of Langerhans. Am J Physiol 1985;248:C145−53.

[55] Dionne KE, Colton CK, Yarmush ML. Effect of hypoxia on insulin secretion by isolated rat and canine islets of Langerhans. Diabetes 1993;42:12−21.

[56] Davalli AM, Scaglia L, Zangen DH, Hollister J, BonnerWeir S, Weir GC. Vulnerability of islets in the immediate posttransplantation period - Dynamic changes in structure and function. Diabetes 1996;45:1161−7.

[57] Menger MD, Jaeger S, Walter P, Feifel G, Hammersen F, Messmer K. Angiogenesis and hemodynamics of microvasculature of transplanted islets of Langerhans. Diabetes 1989;38(Suppl. 1):199−201.

[58] Elliott RB, Escobar L, Garkavenko O, Croxson MC, Schroeder BA, McGregor M, et al. No evidence of infection with porcine

endogenous retrovirus in recipients of encapsulated porcine islet xenografts. Cell Transplant 2000;9:895–901.

[59] De Vos P, Van Straaten JF, Nieuwenhuizen AG, de Groot M, Ploeg RJ, De Haan BJ, et al. Why do microencapsulated islet grafts fail in the absence of fibrotic overgrowth? Diabetes 1999;48:1381–8.

[60] De Vos P, Smedema I, van Goor H, Moes H, van Zanten J, Netters S, et al. Association between macrophage activation and function of micro-encapsulated rat islets. Diabetologia 2003;46:666–73.

[61] Safley SA, Kapp LM, Tucker-Burden C, Hering B, Kapp JA, Weber CJ. Inhibition of cellular immune responses to encapsulated porcine islet xenografts by simultaneous blockade of two different costimulatory pathways. Transplantation 2005;79:409–18.

[62] De Vos P, De Haan B, Pater J, Van Schilfgaarde R. Association between capsule diameter, adequacy of encapsulation, and survival of microencapsulated rat islet allografts. Transplantation 1996;62:893–9.

[63] Toso C, Mathe Z, Morel P, Oberholzer J, Bosco D, Sainz-Vidal D, et al. Effect of microcapsule composition and short-term immunosuppression on intraportal biocompatibility. Cell Transplant 2005;14:159–67.

[64] McQuilling JP, Arenas-Herrera J, Childers C, Pareta RA, Khanna O, Jiang B, et al. New alginate microcapsule system for angiogenic protein delivery and immunoisolation of islets for transplantation in the rat omentum pouch. Transplant Proc 2011;43:3262–4.

[65] Kobayashi T, Aomatsu Y, Iwata H, Kin T, Kanehiro H, Hisanga M, et al. Survival of microencapsulated islets at 400 days post-transplantation in the omental pouch of NOD mice. Cell Transplant 2006;15:359–65.

[66] Moya ML, Garfinkel MR, Liu X, Lucas S, Opara EC, Greisler HP, et al. Fibroblast growth factor-1 (FGF-1) loaded microbeads enhance local capillary neovascularization. J Surg Res 2010;160:208–12.

[67] Kin T, Korbutt GS, Rajotte RV. Survival and metabolic function of syngeneic rat islet grafts transplanted in the omental pouch. Am J Transplant 2003;3:281–5.

[68] Ludwig B, Rotem A, Schmid J, Weir GC, Colton CK, Brendel MD, et al. Improvement of islet function in a bioartificial pancreas by enhanced oxygen supply and growth hormone releasing hormone agonist. Proc Natl Acad Sci U S A 2012;109:5022–7.

[69] Pedraza E, Coronel MM, Fraker CA, Ricordi C, Stabler CL. Preventing hypoxia-induced cell death in beta cells and islets via hydrolytically activated, oxygen-generating biomaterials. Proc Natl Acad Sci U S A 2012;109:4245–50.

[70] Orlando G, Bendala JD, Shupe T, Bergman C, Bitar KN, Booth C, et al. Cell and organ bioengineering technology as applied to gastrointestinal diseases. Gut 2013;62:774–86.

[71] Orlando G, Wood KJ, De Coppi P, Baptista PM, Binder KW, Bitar KN, et al. Regenerative medicine as applied to general surgery. Ann Surg 2012;255:867–80.

[72] Orlando G, Wood KJ, Stratta RJ, Yoo JJ, Atala A, Soker S. Regenerative medicine and organ transplantation: past, present, and future. Transplantation 2011;91:1310–7.

[73] Orlando G, Baptista P, Birchall M, De Coppi P, Farney A, Guimaraes-Souza NK, et al. Regenerative medicine as applied to solid organ transplantation: current status and future challenges. Transpl Int 2011;24:223–32.

[74] Badylak SF, Taylor D, Uygun K. Whole-organ tissue engineering: decellularization and recellularization of three-dimensional matrix scaffolds. Annu Rev Biomed Eng 2011;13:27–53.

[75] Orlando G, Farney AC, Iskandar SS, Mirmalek-Sani SH, Sullivan DC, Moran E, et al. Production and implantation of renal extracellular matrix scaffolds from porcine kidneys as a platform for renal bioengineering investigations. Ann Surg 2012;256:363–70.

[76] Wang Y, Cui CB, Yamauchi M, Miguez P, Roach M, Malavarca R, et al. Lineage restriction of human hepatic stem cells to mature fates is made efficient by tissue-specific biomatrix scaffolds. Hepatology 2011;53:293–305.

[77] Ross EA, Williams MJ, Hamazaki T, Terada N, Clapp WL, Adin C, et al. Embryonic stem cells proliferate and differentiate when seeded into kidney scaffolds. J Am Soc Nephrol 2009;20:2338–47.

[78] Ott HC, Clippinger B, Conrad C, Schuetz C, Pomerantseva I, Ikonomou L, et al. Regeneration and orthotopic transplantation of a bioartificial lung. Nat Med 2010;16:927–33.

[79] Ott HC, Matthiesen TS, Goh SK, Black LD, Kren SM, Netoff TI, et al. Perfusion-decellularized matrix: using nature's platform to engineer a bioartificial heart. Nat Med 2008;14:213–21.

[80] Uygun BE, Soto-Gutierrez A, Yagi H, Izamis ML, Guzzardi MA, Shulman C, et al. Organ reengineering through development of a transplantable recellularized liver graft using decellularized liver matrix. Nat Med 2010;16:814–20.

[81] Baptista PM, Siddiqui MM, Lozier G, Rodriguez SR, Atala A, Soker S. The use of whole organ decellularization for the generation of a vascularized liver organoid. Hepatology 2011;53:604–17.

[82] Wang RN, Rosenberg L. Maintenance of beta-cell function and survival following islet isolation requires re-establishment of the islet-matrix relationship. J Endocrinol 1999;163:181–90.

[83] Daoud J, Petropavlovskaia M, Rosenberg L, Tabrizian M. The effect of extracellular matrix components on the preservation of human islet function in vitro. Biomaterials 2010;31:1676–82.

[84] De Carlo E, Baiguera S, Conconi MT, Vigolo S, Grandi C, Lora S, et al. Pancreatic acellular matrix supports islet survival and function in a synthetic tubular device: in vitro and in vivo studies. Int J Mol Med 2010;25:195–202.

[85] Davis NE, Beenken-Rothkopf LN, Mirsoian A, Kojic N, Kaplan DL, Barron AE, et al. Enhanced function of pancreatic islets co-encapsulated with ECM proteins and mesenchymal stromal cells in a silk hydrogel. Biomaterials 2012;33:6691–7.

Chapter 44

Autologous Islets Transplantation

Alan C. Farney[a], Emmanuel C. Opara[b], and David E.R. Sutherland[c]

[a]Transplant Service, Department of General Surgery, Wake Forest Baptist Health Medical Center Blvd, Winston-Salem, NC, [b]Institute for Regenerative Medicine, Wake Forest School of Medicine, Medical Center Blvd., Winston-Salem, NC, [c]Division of Transplantation, Department of Surgery, University of Minnesota, Minneapolis, MN

Chapter Outline

44.1 INTRODUCTION

Regenerative medicine aims to repair or replace diseased or damaged tissue. Since metabolically active tissues require oxygen, transplanted healthy tissues must acquire a vascular supply, a process known as engraftment. Engraftment is the *sine qua non* of successful tissue or organ transplantation. For solid organ transplantation, revascularization is immediate (necessarily so) and occurs upon completion of vascular anastomosis, whereas cellular or tissue grafts must develop a vascular supply or rely upon diffusion. Pancreatic islet tissue has the interesting ability to engraft at ectopic sites as a free graft of tissue. The first successful tissue or organ transplants (whether vascularized or not) were done in the setting of either no immune barriers (autologous transplantation) or low immune barriers (isologous transplantation). Subsequently, the development of immunosuppression has allowed transplantation of vascularized organs in the setting of full alloimmunity. However, immunosuppression, which does spectacularly allow transplantation across certain immune barriers, is merely an adjunct to transplantation. The most common types of transplants, tissues such as blood or bone, require no immunosuppression for successful transplantation. The beneficial impact of autologous transplantation, such as islet autotransplantation (IAT), is threefold: autologous islet transplantation has significant clinical impact, it helps define the achievable limits of allogeneic islet grafts that must cross alloimmune barriers, and it is an important example of clinical regenerative medicine.

44.2 BACKGROUND

The fields of transplantation, diabetes mellitus, and chronic pancreatitis (CP) have an interesting, shared history. In 1889, Minkowski successfully performed total pancreatectomy (TP) in dogs to investigate the effect of pancreatic digestive enzymes on absorption of fats. The dogs quickly developed polyuria and polydipsia, leading Minkowski to test the urine for glucose (which of course confirmed the diagnosis of diabetes). These observations, published by Minkowski and von Mering, provided the first clear scientific link between the pancreas and diabetes [1]. But how was the pancreas mechanistically linked to diabetes? In 1892, Hedon performed an elegant experiment in dogs where he first performed a distal pancreatectomy, then ligated and divided the ductal attachments between the pancreatic head and duodenum, and transplanted the pancreatic head on its vascular pedicle to a subcutaneous site underneath the anterior abdominal wall [2]. Hedon noted that the dogs remained euglycemic, refuting one possible theory that conjectured that an exocrine secretion was responsible for proper maintenance of glucose levels. In a second operation, Hedon did a completion pancreatectomy, by resecting the autologously transplanted subcutaneous pancreatic head. The development of glycosuria following the second procedure confirmed the idea that the pancreas controlled glucose through production of an "internal secretion." The pancreas was known to be heterogeneous: from where did the "internal secretion" derive? Contemporaneous with the work by Minkowski and Hedon, a number of pathologists, including Opie and Dewitt, accrued evidence that chronic changes in the islets of Langerhans, not the exocrine pancreas, were associated with the development of diabetes [3,4]. In1920, Barron described a patient with pancreaticolithiasis and histology demonstrating severe CP but preservation of the islets of Langerhans where the "internal secretion" should reside. Reading this article piqued Banting's interest, launching a search for the "internal secretion," and culminating in the discovery of insulin and the shared Nobel prize between Banting and Macleod [5,6].

The first human IAT was done at the University Of Minnesota in 1977 [7,8] for a patient with severe CP. IAT was added to the near-TP ($>95\%$) procedure to avert surgical diabetes. At the time, allogeneic islet transplantation was an established laboratory model in rodents, and both allogeneic islet transplants and IATs had been performed successfully in larger animals, but allogeneic islet transplants for type 1 diabetes in humans were uniformly unsuccessful [7]. Experimentally, successful islet transplantation was easier to accomplish in autologous or isologous animal models. For the first patient who underwent IAT at Minnesota, the primary goal was treatment of debilitating pain by the pancreatectomy. Additionally, it was hoped that IAT might help avoid the typically brittle diabetes that often ensues following TP, and that successful IAT might also have implications for islet transplantation for treatment of type 1 diabetes. At the time, islet isolation was done by chopping of the pancreatic tissue and stationary collagenase digestion (only later, was intraductal collagenase injection introduced). Islets were not quantified by count or conversion to islet equivalents based on standardized 150 µm islets, but measurement of insulin by radioimmune assay indicated that the first IAT tissue preparation contained β-cells. The first IAT was successful and the patient remained insulin independent for 6 years until there was death with function from causes unrelated to the IAT [9].

44.3 PATIENT SELECTION AND SURGICAL CONSIDERATIONS

IAT is an adjunctive rather than primary treatment. The most common clinical scenario in which IAT is employed is recurring pain associated with CP. Multiple hospitalizations for pain, associated hyperamylasemia or hyperlipasemia, malnutrition, and narcotic use are clinical features of CP, but may not always be present. Many patients referred to the surgeon for management of CP have already undergone attempts to control pain, including nonsurgical (endoscopic sphincterotomy or stent placement) and surgical interventions. In some cases, partial pancreatectomy (for small duct disease) or decompression procedures (for large duct disease) have been attempted as a means to control pain, but proved unsuccessful. TP is the surgical procedure most likely to result in control of CP pain, but control or prevention of diabetes after TP and IAT is not assured. Nonresection surgical intervention is less likely to cause diabetes but is also less likely to control pain. Probably the best way to determine the proper surgical treatment for recurring pain associated with CP is to weigh the penalty of diabetes versus the benefits of controlling pain. If control of pain is paramount, justifying surgical diabetes, TP should be selected as the surgical treatment option (Figure 44.1). If TP is to be done, IAT should accompany it, as the risks of surgical diabetes outweigh the risks of IAT.

Other factors may bear on the decision to offer TP-IAT. Patients with untreated large duct CP may

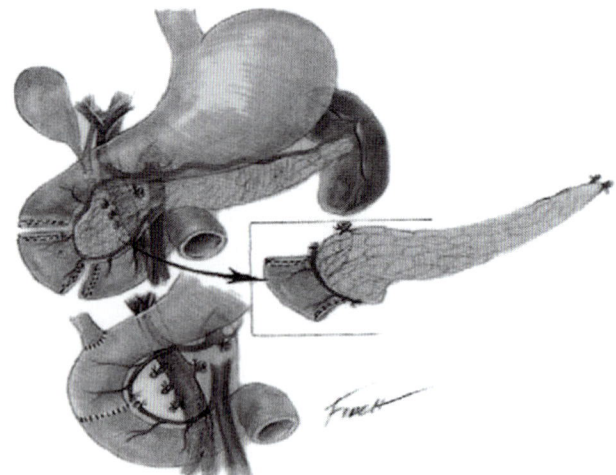

FIGURE 44.1 Surgical technique for TP as done for many patients in the University of Minnesota experience. The second portion of the duodenum is resected with the pancreas, sparing the pylorus and distal duodenum. A reconstruction is done by creating a duoduodenostomy and a choledochoduodenostomy. *Adapted from [9]; with permission. Copyright© 1991, Elsevier.*

respond to decompression procedures, either surgical or endoscopic. Pancreaticojejunostomy, particularly lateral pancreaticojejunostomy where the pancreatic duct is widely opened and drained into the small bowel, is an effective procedure for approximately two thirds of patients with large duct disease. However, for patients in whom pain recurs following surgical decompression, TP-IAT is a less effective salvage procedure (TP may control pain, but IAT is often not successful) because prior surgical opening of the pancreatic duct may preclude satisfactory pancreatic duct injection of collagenase during the islet isolation process. Thus, if endoscopic decompression is not successful, TP-IAT may reasonably be considered as the next procedure of choice for large duct disease.

Another factor that bears on whether to perform IAT is the presence of preexisting diabetes or glucose intolerance before TP. The natural history of CP includes diabetes. Candidates for TP-IAT should undergo metabolic testing to help predict the success of IAT and weigh the benefits versus the risks of islet infusion. Patients with impaired glucose tolerance are less likely to remain insulin independent following IAT, but IAT should still be considered since partial graft function may ameliorate the brittleness of surgical diabetes. Finally, another factor that may impact outcome of CP is the presence of malignancy. CP itself is probably a risk factor for pancreatic cancer, and on rare occasions, adenocarcinoma of the pancreas may result in pancreatitis. Abdominal imaging (computed tomography or magnetic resonance imaging) and modalities such as endoscopic ultrasound help to resolve concerns for malignancy.

Although CP is the most common disease for which TP-IAT has been applied, any benign disease of the pancreas for which TP is a treatment option warrants the consideration of IAT as an adjunct to TP. A number of authors have reported successful TP-IAT for benign neoplasms of the pancreas [10–13], and IAT has been used with pancreatectomy for traumatic disruption of the pancreatic duct [14]. Both TP and completion pancreatectomy and IAT following a Whipple procedure for ampullary carcinoma have also been reported [15,16].

Although CP most commonly occurs during adult years, children may also develop CP. In children, CP may result from injury or genetic disorders, but as in adults, the cause of CP is often unknown. The clinical indications for TP-IAT for CP are similar for adults and children; prevention of recurrent pain is the primary goal. IAT should accompany TP for children because surgical diabetes is often difficult to manage in children, and the procedure offers avoidance of diabetes in half or more of cases [17–20].

44.4 ISLET ISOLATION AND INTRAPORTAL INFUSION

Human islet isolation has evolved since the first clinical IAT. No two laboratories isolate islets in the exact same manner, but most standard operating procedures for islet tissue preparation share the same fundamental steps. Animal models of islet isolation, first in rodents then in dogs, have contributed significantly to current practice. In 1977, Kretschmer reported that dispersed (nonpurified) pancreatic tissue could prevent surgical diabetes in dogs [21], and it was essentially this technique that was applied in the first human IAT at the University of Minnesota. Subsequent improvements or changes in islet isolation technique include intraductal collagenase injection, methods for counting islets and standardizing counts of islets, density gradient purification, and culture of islets before transplantation [22–25].

Current state of the art for human islet isolation requires a processing facility that conforms to good manufacturing processes (GMPs), and the tissue should be handled with good tissue practices (GTPs), as set forth by the US Food and Drug Administration (FDA). In the United States, an investigational new drug application must be submitted, reviewed, and approved by the FDA for processing and transplantation of allogeneic islets before starting clinical work, but currently for IAT, meeting GMP and GTP guidelines and registration with the FDA is sufficient.

A variety of techniques may be utilized for pancreatectomy, but regardless of technique, warm ischemia should be limited by maintaining pancreatic blood supply (splenic arterial blood flow to the tail of the pancreas) until the pancreas is removed. Immediately after removal, the pancreas is cooled topically by immersion in cold (4C) Hanks' balanced saline or modified University of Wisconsin® solution and transported to the islet isolation laboratory. In the laboratory, the pancreatic duct is cannulated (some centers will perform duct cannulation and collagenase infusion in the operating room) [26] and the pancreas injected with a commercial clinical grade collagenase prepared according to GMP standards.

The pancreas is dispersed by collagenase digestion as the tissue is warmed to the effective temperature for collagenase enzymatic activity inside a Riccordi chamber and tissue circuit (Figure 44.2). A side port on the tissue circuit allows sampling of the tissue slurry to monitor the digestion process. Tissue samples are stained with dithizone and the islet and exocrine tissue assessed morphologically. Tissue morphology, including the number of free islets, is used to determine when to stop enzymatic dispersal of the pancreatic tissue; experience helps to recognize this point. At the termination of digestion, the tissue is rapidly cooled (stopping enzymatic activity),

FIGURE 44.2 Riccordi chamber and tissue circuit. Collagenase is infused into the pancreatic duct, and then the pancreatic tissue is dispersed by warming the tissue to temperatures allowing enzymatic activity. Once adequately dispersed (a judgment made by assessing dithizone stained samples), the tissue is rapidly cooled, washed, and pooled.

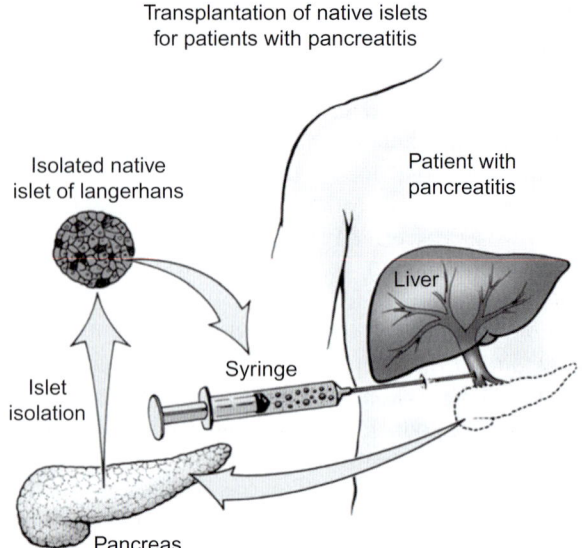

FIGURE 44.3 Intraportal IAT. The pancreas is dispersed by ductal collagenase, washed by centrifugation, pooled, and the dispersed tissue is infused into a branch of the portal vein. Embolized islet tissue engrafts within the liver. *Adapted from [29]; with permission. Copyright© 2007, Elsevier.*

collected, and then washed by repeated centrifugation. At this point, a decision must be made whether to further purify the dispersed pancreatic islet tissue. Some centers routinely purify, whereas others do so selectively depending on the final volume of tissue obtained at the end of collagenase dispersion [27]. Islet purification reduces tissue volume, thereby minimizing portal venous hypertension that may be encountered during intraportal venous infusion. Decreased islet yield and procedure length are clear drawbacks to purification, and it has been postulated that the process of purification may also remove stem cells or other cell types that improve engraftment or contribute to the endocrine mass [28]. If the tissue volume exceeds 15cc, the University of Minnesota islet program has purified all or part of the dispersed pancreatic tissue using a COBE 2991 cell processor [29]. After collection of either the purified or unpurified dispersed pancreatic tissue, the volume of tissue for transplantation is diluted to a fine tissue slurry and transported back to the operating room for infusion (Figure 44.3). Prior to intraportal infusion (the most common site), the patient is heparinized (70 units/kg), and the baseline portal venous pressure is measured. The dispersed or purified islet tissue is then embolized into a mesenteric venous tributary of the portal vein over 30–60 min while monitoring portal venous pressure. Some centers catheterize the umbilical vein and infuse the islet tissue while in the intensive care unit following the TP and others have utilized a percutaneous transhepatic route, radiologically, to access the portal system and infuse the islet tissue into the liver [30,31]. If the islet isolation process is prolonged, infusion of the islet tissue after surgery makes sense. Regardless of the technique chosen to access the portal system, portal venous

pressure should be monitored during the infusion. At the University of Minnesota, the infusion will be terminated if the portal venous pressure exceeds 30 cm water. If portal venous hypertension precludes infusion of the entire tissue volume, alternate intra-abdominal sites for transplantation may then be selected (an advantage of the intraoperative infusion technique).

44.5 SITES FOR ISLET ENGRAFTMENT

Islet tissue possesses an important characteristic of regenerative potential: it has the ability to engraft at a number of ectopic sites. Isologous islet transplant models in rodents and autologous islet transplant models in dogs have been used to explore the liver (intraportal), intraperitoneal site, omental pouch, renal subcapsule, and spleen (reflux via the splenic or gastric veins) as possible sites for transplantation of islets [32–41]. The intrahepatic site via the portal system seems to be the most efficient [38] and is the preferred site for islet transplantation in humans [9,42]. Although there was initial concern whether long-term intrahepatic function of islets was achievable [43], subsequently it has been shown that long-term function is possible. The mass of viable islet tissue transplanted appears to directly impact outcome [24,38], and factors affecting islet yield have an indirect bearing. In humans, long-term function (insulin independence) after TP and intrahepatic IAT is well established, demonstrating the stability of engraftment of islet tissue outside of the pancreas [27,44–46]. Although euglycemia is maintained long term, metabolic studies show abnormal α-cell (glucagon) responses to hypoglycemia (glucose counterregulation) from islets in the intrahepatic site of large animals and humans [40,47,48]. Other ectopic sites for islet engraftment also appear to have the same metabolic defect, and though the clinical significance appears modest, the observation supports the teleological supposition that islets belong to in the pancreas. Although there are no reports of organ dysfunction, local production of insulin may induce hepatic steatosis, indicating that ectopic engraftment of islet tissue may also have negative consequences for the local environment [49–51]. Stagner has reported successful intrapancreatic IAT in rodents and dogs [52], suggesting that the pancreatic site might be considered for allogeneic islet transplantation for diabetic patients, or as a possible IAT site in patients who undergo less than TP for benign conditions of the pancreas.

Complications of IAT include portal venous thrombosis and hepatic necrosis, both related to the volume of tissue infused into the portal system [53,54]. Bacterial contamination of the islet tissue (usually from prior endoscopic or surgical manipulation of the pancreatic duct rather than introduction through the isolation process) probably also increases the risk of portal thrombosis by increasing inflammation within the distal portal vasculature. Limiting the volume of transplanted tissue, monitoring portal venous pressure during the infusion, heparinization, and prophylactic antibiotics may all reduce the risk of portal thrombosis (see the section on complications of IAT).

44.6 IAT RESULTS

Successful IAT generally implies insulin independence after total or near-TP although minimal need for insulin or easily controlled blood glucose levels might also reasonably be considered a success given the difficulty of managing brittle surgical diabetes. Islet mass correlates with metabolic outcome after TP-IAT. Measurement of islet mass has evolved over time. Tissue insulin was measured as an estimate of islet mass for the first (and first successful) IAT in 1977 [7]. Subsequently, Wahoff reported that 74% of patients who received >300,000 islets remained insulin independent more than 2 years after total or near-TP. However, counting islets is not straightforward, as islets vary in size from 50 to 500 μm in diameter. Currently, islet yields are standardized to islet equivalent counts (IEQs) based upon a "standard" human islet diameter of 150 μm, and the mass of the transplant is referred to as the number of IEQ infused per body weight of the recipient in kilograms (IEQ/kg). IEQs predict the likelihood of insulin independence after TP-IAT, but not perfectly. Patients infused with "low" yields have displayed insulin independence and long-term

function, whereas others who have received very large numbers of islets (IEQs) have displayed little to no metabolic function after TP-IAT [27,55]. In recent analyses at the University of Minnesota, nearly half of the patients who received >2000 IEQ/kg were insulin independent after TP-IAT, and almost three quarters were insulin independent at 3 years who received >5000 IEQ/kg, but there was considerable variance [27,29,56]. Nearly all IAT recipients demonstrate circulating C-peptide. However, clinical control of blood glucose levels after IAT depends upon the engrafted mass of islets, which is always a fraction of the islet mass contained in the pancreas before the pancreatectomy. Teuscher studied patients who remained insulin independent long-term after IAT with glucose and arginine stimulation. Insulin secretory reserve correlated with the mass of infused islets but was significantly diminished even though the patients were insulin-free [57]. Factors that predict successful islet isolation, such as prior pancreatic surgery and degree of pancreatic fibrosis, also correlate with insulin independence after TP-IAT (Table 44.1) [9,42]. Although prior pancreatic resection does impact the chance of successful IAT after completion pancreatectomy, prior surgical pancreatic duct drainage procedures appear to be even more disadvantageous. Islet isolation after a prior drainage procedure results in lower islet yield (usually due to inability to infuse collagenase into the pancreatic duct) and is associated with a lower rate of insulin independence after IAT [29]. At the University of Minnesota, only 18% of patients who were initially treated with a Puestow procedure prior to TP-IAT achieved insulin independence, compared to a >70% insulin independence rate for patients who had no surgical intervention prior to TP-IAT [42]. In a more recent report with larger numbers of patients without prior surgery, insulin independence after TP-IAT is probably not as high as 70%, but islet yield, which correlates with an improved endocrine outcome, is significantly higher compared to patients who had direct surgery on the pancreas [56]. Of course, without IAT, 100% of patients who undergo a TP after a Puestow will be severely diabetic, also warranting IAT in these cases. Overall, approximately one quarter to one half of patients achieve insulin independence after TP-IAT [27,29,46,56,58]. Metabolic outcomes in children surpass adults. Of 53 pediatric IATs at the University of Minnesota, more than half were insulin independent at 3 years, and another quarter demonstrated partial function [56]. Patients without definitive insulin independence may still benefit from partial metabolic function. In the Minnesota series, up to one third of patients have partial function, based upon insulin usage and C-peptide measurements. Although patients with partial function may require a basal long-acting insulin dose, they have HbA1c levels consistently below 7.0%, produce C-peptide, and avoid the typical glucose level excursions

TABLE. 44.1 A Comparison of Clinical Factors That May Account for the Different Outcomes After Autologous and Allogeneic Islet Transplantation

Potential Negative Factors	IAT	Allogeneic Islet Transplant
Chronic pancreatitis	1. Severity decreases yield 2. Prior pancreatic surgery decreases yield	Not present
Single pancreas donor	Only one possible	Multiple pancreas donors possible
Alloimmunity	Not present	Rejection
Autoimmunity	Not present	Autoimmune injury
Immunosuppression	Not needed	β-Cell toxicity
Deceased donor	No (living donor)	1. Preservation injury decreases yield or islet viability 2. Adverse effects of brain death on cellular function and viability
Diabetes in recipient at time of transplant	Not present	1. Hyperglycemia β-cell toxic 2. Insulin resistance
Islet isolation purification step	Not always done	Unavoidable loss of islet tissue

associated with surgical diabetes [27]. Once achieved, persistence of function long-term is favorable. Graft failure may occur early, but patients who are insulin independent at 2 years often maintain full function more than 10 years and up to 20 years after TP-IAT [29,42,44,45]. Approximately half of patients who are insulin independent after TP-IAT remain insulin independent at 5 years [56]. The majority of patients with early partial function continue to display C-peptide and lower insulin usage in the long-term [27,46]. Several reports of maintenance of insulin independence through pregnancy after TP-IAT, suggest that metabolic function in some patients after IAT, is surprisingly clinically robust and durable [59–61].

Pain control is the primary goal of patients with CP who are treated with TP-IAT. Recurrent pain, often associated with eating, results in malnutrition and may necessitate hyperalimentation. In patients with the most severe form of the disease, the inability to eat and the long-term complications of intravenous hyperalimentation (infections, cirrhosis) are life threatening, clearly making TP justified. Many other patients have a significantly diminished quality of life, and for them TP is warranted if the estimated quality of life will be better even if they have surgical diabetes. Relief of pain associated with CP following TP is not 100%. For unclear reasons, even after TP, some patients continue to have pain. Some patients may continue to have pain following TP because the processing of pain has "centralized," akin to that described for phantom limb pain after amputation [62]. Other patients may have opioid induced hyperalgesia from prolonged exposure to narcotics. Even so, when queried about control of pain, patients report complete or partial

remission of pain after TP-IAT in the range of 80–90% in the long-term [9,42]. A clinically significant outcome after TP-IAT is the ability to discontinue narcotics. In a report from Cincinnati, approximately 60% of patients who had TP-IAT for severe chronic pancreatic were able to stop narcotics, and a quality of life assessment tool, the Short Form-36, indicated an improved quality of life [58,63].

44.7 COMPLICATIONS OF TP-IAT

Complications from IAT generally stem from intraportal tissue embolization and its effects on portal circulation. As noted in the section on intraportal infusion, portal venous pressure should be carefully monitored during IAT. Portal vein thrombosis, hepatic infarction, and disseminated intravascular coagulation (including a fatality associated with DIC) have been reported [53,54,64–67]. The majority of these complications occurred in the era before use of pancreatic duct collagenase injection and other modifications in the islet isolation procedure. More recently, White reported splenic infarction following an intrasplenic IAT and portal vein thrombosis after a combined intraportal and splenic IAT [53]. Risk factors for portal venous thrombosis include the volume of tissue for infusion and development of portal venous hypertension during infusion. Preexisting portal hypertension is considered a contraindication to IAT, but successful IAT has been carried out in the setting of cirrhosis without baseline portal hypertension [68]. Bacterial contamination of dispersed pancreatic tissue is common, probably because of prior pancreatic instrumentation or surgical drainage procedures. Care should be taken to wash tissue adequately (by centrifugation) prior to infusion, especially if the patient had a prior Puestow (or other enteric drainage) procedure. A gram stain of the final tissue preparation is done, and if positive, precludes transplantation. Culture results, which take time, are not obtained until after transplantation. However, positive tissue cultures do not necessarily portend poor clinical outcome as patients often do not display an inflammatory response nor develop infections with the same organisms cultured from the infused tissue [69,70]. Nonetheless, cultures of the final tissue preparation should be followed and the patient treated with antibiotics as determined by the clinical context. Prophylactic antibiotics given prior to IAT may reduce the risk of infection from adventitious organisms.

Development of intrahepatic metastasis after infusion of a malignancy contaminated IAT has remained a theoretical complication. Preoperative abdominal imaging and islet isolation, which greatly reduces pancreatic volume, probably contribute to the low risk of malignancy following IAT [29].

44.8 UNSUCCESSFUL IAT: SALVAGE BY ALLOGENEIC PANCREAS TRANSPLANTATION

Approximately one third of TP-IAT patients display overt surgical diabetes and 100% have exocrine failure and require oral enzyme replacement [29]. If IAT is not successful, then whole organ pancreas allotransplantation with exocrine drainage should be considered to treat both the endocrine and exocrine failure that results from TP. At the University of Minnesota, 26 pancreas allotransplants have been done for 18 patients who had previously been treated for CP by TP. Gruessner reported >70% 1 year graft survival (both endocrine and exocrine function) using tacrolimus-based maintenance immunosuppression [71].

44.9 WHAT DO THE RESULTS OF IAT MEAN FOR ALLOGENEIC ISLET TRANSPLANTATION?

Compared to allo–islet transplantation, IAT has higher early and long-term graft survival and achieves these results with a lower mass of transplanted islets. Results from the University of Minnesota suggest that if a patient has not had prior pancreatic surgery, an IAT maintains insulin independence in a higher fraction of patients, and if >5000 IEQ/kg are recovered, 75% may achieve insulin independence. There may be a negative effect of CP on islet isolation, but a 75% insulin independence rate from a single pancreas probably represents the achievable upper limit of state-of-the-art islet transplantation. Compared to allo–islet transplant, IAT has advantages

that increase the chance of success (Table 44.1). First, autologous transplantation faces no immunological barriers, either alloimmunity or autoimmunity. As a consequence, IAT does not require immunosuppression, and therefore can be done without exposure to diabetogenic agents such as calcineurin inhibitors and steroids [72]. Another significant and unique aspect of IAT is that the donor is living. Living donation shortens cold ischemia time, which is clearly associated with early graft function of other organ types, and avoids biological derangements associated with brain death that may affect early and late graft function [73]. The recipients of IAT also differ from recipients of allo–islet transplants in other ways. Before TP-IAT the patient is not overtly diabetic. Preexisting diabetes may impact allogeneic islet transplant outcome because of the toxic effect of hyperglycemia on β-cells, or metabolic stress in the setting of insulin resistance (Table 44.1). Strategies to improve allo–islet transplant results likely need to address some or all of the differences between IAT and allo–islet transplantation [27].

44.10 SUMMARY

Successful IAT following near-total or TP was the first substantial proof that human islet tissue could engraft at an ectopic site and maintain insulin independence [7]. This clinical observation holds promise for other regenerative medicine strategies aimed at treating diabetes. Follow-up observations have documented that if early insulin independence is achieved, then long-term function, 10 years or longer, is possible, indicating that intrahepatic β-cell residence is durable [9,29]. Patients who have not had prior pancreatic surgery and have high islet yields have the best chance of insulin independence after TP-IAT, as high as 75%; a rate that appears to be the upper limit of success when islets are isolated from a single pancreas from a live donor. If pancreatectomy is to be done for benign diseases of the pancreas, IAT should accompany it, as the risks of surgical diabetes outweigh the risks of IAT. Despite the negative effect of CP on islet isolation yield, IAT holds a number of advantages over allogeneic islet transplantation. Surmounting the differences between IAT and allogeneic islet transplant may improve the therapeutic potential of cellular based replacement treatments for type 1 diabetes.

REFERENCES

[1] Von Mering J, Minkowski O. Diabetes mellitus nach pankreasexstirpation. Arch Exp Pathol Pharmakol 1889;26:371–87.

[2] Hedon E. Greffe sous-cutanee du pancreas, son importance dans l'etude du diabete pancreatique. Archs Physiol 1892;4:617–28.

[3] Dewitt LM. Morphology and physiology of areas of Langerhans in some vertebrates. J Exp Med 1906;8:193−239.

[4] Opie EL. On the relation of chronic interstitial pancreatitis to the islands of Langerhans and to diabetes mellitus. J Exp Med 1901;5:397−428.

[5] Banting FG. The history of insulin. Edinb Med J 1929;36:1−18.

[6] Barron M. The relation of the islets of Langerhans to diabetes with special reference to cases of pancreatic lithiasis. Surgery Gynec Ostet 1920;31:437−48.

[7] Sutherland DE, Matas AJ, Najarian JS. Pancreatic islet cell transplantation. Surg Clin North Am 1978;58:365−82.

[8] Najarian JS, Sutherland DE, Matas AJ, Goetz FC. Human islet autotransplantation following pancreatectomy. Transplant Proc 1979;11:336−40.

[9] Farney AC, Najarian JS, Nakhleh RE, Lloveras G, Field MJ, Gores PF, et al. Autotransplantation of dispersed pancreatic islet tissue combined with total or near-total pancreatectomy for treatment of chronic pancreatitis. Surgery 1991;110:427−37; [discussion 437−439].

[10] Fournier B, Andereggen E, Buhler L, Cretin N, Mage R, Sinigaglia C, et al. Islands of Langerhans autotransplantation after pancreatic resection for benign pathology. Schweiz Med Wochenschr Suppl 1997;89:41S−5S.

[11] Oberholzer J, Triponez F, Mage R, Andereggen E, Buhler L, Cretin N, et al. Human islet transplantation: lessons from 13 autologous and 13 allogeneic transplantations. Transplantation 2000;69:1115−23.

[12] Lee BW, Jee JH, Heo JS, Choi SH, Jang KT, Noh JH, et al. The favorable outcome of human islet transplantation in Korea: experiences of 10 autologous transplantations. Transplantation 2005;79:1568−74.

[13] Ris F, Niclauss N, Morel P, Demuylder-Mischler S, Muller Y, Meier R, et al. Islet autotransplantation after extended pancreatectomy for focal benign disease of the pancreas. Transplantation 2011;91:895−901.

[14] Dardenne S, Sterkers A, Leroy C, Da Mata L, Zerbib P, Pruvot FR, et al. Laparoscopic spleen-preserving distal pancreatectomy followed by intramuscular autologous islet transplantation for traumatic pancreatic transection in a young adult. JOP 2012;13:285−8.

[15] Alsaif F, Molinari M, Al-Masloom A, Lakey JR, Kin T, Shapiro AM. Pancreatic islet autotransplantation with completion pancreatectomy in the management of uncontrolled pancreatic fistula after whipple resection for ampullary adenocarcinoma. Pancreas 2006;32:430−1.

[16] Iyegha UP, Asghar JA, Beilman GJ. Total pancreatectomy and islet auto-transplantation as treatment for ampullary adenocarcinoma in the setting of pancreatic ductal disruption secondary to acute necrotizing pancreatitis. A case report. JOP 2012;13:239−42.

[17] Bottino R, Bertera S, Grupillo M, Melvin PR, Humar A, Mazariegos G, et al. Isolation of human islets for autologous islet transplantation in children and adolescents with chronic pancreatitis. J Transplant 2012;2012: [642787]

[18] Schmulewitz N. Total pancreatectomy with autologous islet cell transplantation in children: making a difference. Clin Gastroenterol Hepatol 2011;9:725−6.

[19] Bellin MD, Freeman ML, Schwarzenberg SJ, Dunn TB, Beilman GJ, Vickers SM, et al. Quality of life improves for pediatric patients after total pancreatectomy and islet auto-transplant for chronic pancreatitis. Clin Gastroenterol Hepatol 2011;9:793−9.

[20] Bellin MD, Sutherland DE. Pediatric islet autotransplantation: indication, technique, and outcome. Curr Diab Rep 2010;10:326−31.

[21] Kretschmer GJ, Sutherland DE, Matas AJ, Steffes MW, Najarian JS. The dispersed pancreas: transplantation without islet purification in totally pancreatectomized dogs. Diabetologia 1977;13:495−502.

[22] Merrell RC, Marincola F, Maeda M, Cobb L, Basadonna G. The metabolic response of intrasplenic islet autografts. Surg Gynecol Obstet 1985;160:552−6.

[23] Munn SR, Kaufman DB, Meloche RM, Field MJ, Sutherland DE. Weight-corrected islet counts are predictive of outcome in the canine intrahepatic islet autograft model. Diabetes Res 1988;9:121−4.

[24] Warnock GL, Rajotte RV. Critical mass of purified islets that induce normoglycemia after implantation into dogs. Diabetes 1988;37:467−70.

[25] Scharp DW, Marchetti P, Swanson C, Newton M, McCullough CS, Olack B. The effect of transplantation site and islet mass on long-term survival and metabolic and hormonal function of canine purified islet autografts. Cell Transplant 1992;1:245−54.

[26] Naziruddin B, Matsumoto S, Noguchi H, Takita M, Shimoda M, Fujita Y, et al. Improved pancreatic islet isolation outcome in autologous transplantation for chronic pancreatitis. Cell Transplant 2012;21:553−8.

[27] Sutherland DE, Gruessner AC, Carlson AM, Blondet JJ, Balamurugan AN, Reigstad KF, et al. Islet autotransplant outcomes after total pancreatectomy: a contrast to islet allograft outcomes. Transplantation 2008;86:1799−802.

[28] Gmyr V, Kerr-Conte J, Vandewalle B, Proye C, Lefebvre J, Pattou F. Human pancreatic ductal cells: large-scale isolation and expansion. Cell Transplant 2001;10:109−21.

[29] Blondet JJ, Carlson AM, Kobayashi T, Jie T, Bellin M, Hering BJ, et al. The role of total pancreatectomy and islet autotransplantation for chronic pancreatitis. Surg Clin North Am 2007;87:1477−501.

[30] Morgan KA, Nishimura M, Uflacker R, Adams DB. Percutaneous transhepatic islet cell autotransplantation after pancreatectomy for chronic pancreatitis: a novel approach. HPB (Oxford) 2011;13:511−6.

[31] Pollard C, Gravante G, Webb M, Chung WY, Illouz S, Ong SL, et al. Use of the recanalised umbilical vein for islet autotransplantation following total pancreatectomy. Pancreatology 2011;11:233−9.

[32] Kemp CB, Knight MJ, Scharp DW, Ballinger WF, Lacy PE. Effect of transplantation site on the results of pancreatic islet isografts in diabetic rats. Diabetologia 1973;9:486−91.

[33] Lorenz D, Rosenbaum KD, Petermann J, Ziegler M, Beckert R, Dorn A. Transplantation of isologous islets of Langerhans in diabetic rats. Acta Diabetol Lat 1975;12:30−40.

[34] Griffith RC, Scharp DW, Hartman BK, Ballinger WF, Lacy PE. A morphologic study of intrahepatic portal-vein islet isografts. Diabetes 1977;26:201−14.

[35] Cobb LF, Merrell RC. Intrasplenic islet autografts: insulin response to intravenous glucose challenge. Curr Surg 1983;40:36−9.

[36] Toledo-Pereyra LH, Rowlett AL, Lodish M. Autotransplantation of pancreatic islet cell fragments into the renal capsule prepared without collagenase. Am Surg 1984;50:679—81.

[37] Hesse UJ, Sutherland DE, Gores PF, Sitges-Serra A, Najarian JS. Comparison of splenic and renal subcapsular islet autografting in dogs. Transplantation 1986;41:271—4.

[38] Kaufman DB, Morel P, Field MJ, Munn SR, Sutherland DE. Purified canine islet autografts. Functional outcome as influenced by islet number and implantation site. Transplantation 1990;50:385—91.

[39] Wahoff DC, Sutherland DE, Hower CD, Lloveras JK, Gores PF. Free intraperitoneal islet autografts in pancreatectomized dogs— impact of islet purity and posttransplantation exogenous insulin. Surgery 1994;116:742—8; [discussion 748—750].

[40] Gupta V, Wahoff DC, Rooney DP, Poitout V, Sutherland DE, Kendall DM, et al. The defective glucagon response from transplanted intrahepatic pancreatic islets during hypoglycemia is transplantation site-determined. Diabetes 1997;46:28—33.

[41] McQuilling JP, Arenas-Herrera J, Childers C, Pareta RA, Khanna O, Jiang B, et al. New alginate microcapsule system for angiogenic protein delivery and immunoisolation of islets for transplantation in the rat omentum pouch. Transplant Proc 2011;43:3262—4.

[42] Wahoff DC, Papalois BE, Najarian JS, Kendall DM, Farney AC, Leone JP, et al. Autologous islet transplantation to prevent diabetes after pancreatic resection. Ann Surg 1995;222:562—75 [discussion 575—579]

[43] Alejandro R, Cutfield RG, Shienvold FL, Polonsky KS, Noel J, Olson L, et al. Natural history of intrahepatic canine islet cell autografts. J Clin Invest 1986;78:1339—48.

[44] Farney AC, Hering BJ, Nelson L, Tanioka Y, Gilmore T, Leone J, et al. No late failures of intraportal human islet autografts beyond 2 years. Transplant Proc 1998;30:420.

[45] Robertson RP, Lanz KJ, Sutherland DE, Kendall DM. Prevention of diabetes for up to 13 years by autoislet transplantation after pancreatectomy for chronic pancreatitis. Diabetes 2001;50:47—50.

[46] Webb MA, Illouz SC, Pollard CA, Gregory R, Mayberry JF, Tordoff SG, et al. Islet auto transplantation following total pancreatectomy: a long-term assessment of graft function. Pancreas 2008;37:282—7.

[47] Portis AJ, Warnock GL, Finegood DT, Belcastro AN, Rajotte RV. Glucoregulatory response to moderate exercise in long-term islet cell autografted dogs. Can J Physiol Pharmacol 1990;68:1308—12.

[48] Pyzdrowski KL, Kendall DM, Halter JB, Nakhleh RE, Sutherland DE, Robertson RP. Preserved insulin secretion and insulin independence in recipients of islet autografts. N Engl J Med 1992;327:220—6.

[49] Desai CS, Khan KM, Megawa FB, Rilo H, Jie T, Gruessner A, et al. Influence of liver histopathology on transaminitis following total pancreatectomy and autologous islet transplantation. Dig Dis Sci 2013;58(5):1349—54.

[50] Bhargava R, Senior PA, Ackerman TE, Ryan EA, Paty BW, Lakey JR, et al. Prevalence of hepatic steatosis after islet transplantation and its relation to graft function. Diabetes 2004;53:1311—7.

[51] Markmann JF, Rosen M, Siegelman ES, Soulen MC, Deng S, Barker CF, et al. Magnetic resonance-defined periportal steatosis following intraportal islet transplantation: a functional footprint of islet graft survival?. Diabetes 2003;52:1591—4.

[52] Stagner JI, Rilo HL, White KK. The pancreas as an islet transplantation site. Confirmation in a syngeneic rodent and canine autotransplant model. JOP 2007;8:628—36.

[53] White SA, London NJ, Johnson PR, Davies JE, Pollard C, Contractor HH, et al. The risks of total pancreatectomy and splenic islet autotransplantation. Cell Transplant 2000;9:19—24.

[54] Memsic L, Busuttil RW, Traverso LW. Bleeding esophageal varices and portal vein thrombosis after pancreatic mixed-cell autotransplantation. Surgery 1984;95:238—42.

[55] Webb MA, Illouz SC, Pollard CA, Musto PP, Berry D, Dennison AR. Long-term maintenance of graft function after islet autotransplantation of less than 1000 IEQ/kg. Pancreas 2006;33:433—4.

[56] Sutherland DE, Radosevich DM, Bellin MD, Hering BJ, Beilman GJ, Dunn TB, et al. Total pancreatectomy and islet autotransplantation for chronic pancreatitis. J Am Coll Surg 2012;214:409—24 [discussion 424—426]

[57] Teuscher AU, Kendall DM, Smets YF, Leone JP, Sutherland DE, Robertson RP. Successful islet autotransplantation in humans: functional insulin secretory reserve as an estimate of surviving islet cell mass. Diabetes 1998;47:324—30.

[58] Ahmad SA, Lowy AM, Wray CJ, D'Alessio D, Choe KA, James LE, et al. Factors associated with insulin and narcotic independence after islet autotransplantation in patients with severe chronic pancreatitis. J Am Coll Surg 2005;201:680—7.

[59] Teuscher AU, Sutherland DE, Robertson RP. Successful pregnancy after pancreatic islet autotransplantation. Transplant Proc 1994;26:3520.

[60] Wahoff DC, Leone JP, Farney AC, Teuscher AU, Sutherland DE. Pregnancy after total pancreatectomy and autologous islet transplantation. Surgery 1995;117:353—4.

[61] Jung HS, Choi SH, Noh JH, Ohi SH, Ahn YR, Lee MK, et al. Healthy twin birth after autologous islet transplantation in a pancreatectomized patient due to a benign tumor. Transplant Proc 2007;39:1723—5.

[62] Woolf CJ. Central sensitization: implications for the diagnosis and treatment of pain. Pain 2011;152:2—15.

[63] Rodriguez Rilo HL, Ahmad SA, D'Alessio D, Iwanaga Y, Kim J, Choe KA, et al. Total pancreatectomy and autologous islet cell transplantation as a means to treat severe chronic pancreatitis. J Gastrointest Surg 2003;7:978—89.

[64] Cameron JL, Mehigan DG, Broe PJ, Zuidema GD. Distal pancreatectomy and islet autotransplantation for chronic pancreatitis. Ann Surg 1981;193:312—7.

[65] Mittal VK, Toledo-Pereyra LH, Sharma M, Ramaswamy K, Puri VK, Cortez JA, et al. Acute portal hypertension and disseminated intravascular coagulation following pancreatic islet autotransplantation after subtotal pancreatectomy. Transplantation 1981;31:302—4.

[66] Toledo-Pereyra LH, Rowlett AL, Cain W, Rosenberg JC, Gordon DA, MacKenzie GH. Hepatic infarction following intraportal islet cell autotransplantation after near-total pancreatectomy. Transplantation 1984;38:88—9.

[67] Froberg MK, Leone JP, Jessurun J, Sutherland DE. Fatal disseminated intravascular coagulation after autologous islet transplantation. Hum Pathol 1997;28:1295—8.

[68] Buhler L, Andereggen E, Fournier B, Cretin N, Deng S, Janjic D, et al. Islet autotransplantation in a cirrhotic liver. Swiss Surg 1997;3:35—8.

[69] Lloveras J, Farney AC, Sutherland DE, Wahoff D, Field J, Gores PF. Significance of contaminated islet preparations in clinical islet transplantation. Transplant Proc 1994;26:579—80.

[70] Wray CJ, Ahmad SA, Lowy AM, D'Alessio DA, Gelrud A, Choe KA, et al. Clinical significance of bacterial cultures from 28 autologous islet cell transplant solutions. Pancreatology 2005;5:562—9.

[71] Gruessner RW, Sutherland DE, Drangstveit MB, Kandaswamy R, Gruessner AC. Pancreas allotransplants in patients with a previous total pancreatectomy for chronic pancreatitis. J Am Coll Surg 2008;206:458—65.

[72] Shapiro AM, Hao E, Lakey JR, Finegood D, Rajotte RV, Kneteman NM. Diabetogenic synergism in canine islet autografts from cyclosporine and steroids in combination. Transplant Proc 1998;30:527.

[73] Kusaka M, Pratschke J, Wilhelm MJ, Ziai F, Zandi-Nejad K, Mackenzie HS, et al. Activation of inflammatory mediators in rat renal isografts by donor brain death. Transplantation 2000;69:405—10.

Part VII

Lung

Chapter 45

Current Status of Lung Transplantation

Dirk Van Raemdonck[a,b,c] and Toni Lerut[a,c]

[a]Division of Experimental Thoracic Surgery, Department of Clinical and Experimental Medicine, KU Leuven, Belgium, [b]Lung Transplant Program, University Hospitals Leuven, Leuven, Belgium, [c]Department of Thoracic Surgery, University Hospitals Leuven, Leuven, Belgium

Chapter Outline

45.1 INTRODUCTION

Lung transplantation (LTx) has come of age. It is no longer experimental therapy, but a successful treatment modality for carefully selected patients suffering from many different forms of end-stage lung disease. The aim of this replacement therapy is not only to prolong survival but also to improve the quality of life in patients at home or hospital who are often oxygen dependent and with limited exercise capacity and activities of daily life.

This ultimate therapy for respiratory insufficiency, just as transplantation for failure of any other solid organ, is limited by the shortage of suitable donors and acceptable grafts. In addition, LTx is hampered by early- and late-stage complications related to primary graft dysfunction (PGD), acute and chronic rejection (so-called bronchiolitis obliterans syndrome, BOS), opportunistic infections, malignancies, and other side effects of life-long immunosuppressive therapy. Further research to better understand the mechanisms behind these problems and to find new solutions to deal with limitations and obstacles of LTx are needed. Exciting developments in regenerative medicine and stem cell therapy in recent years may help to overcome these problems and finally conspire with transplantation to restore function of diseased organs.

This chapter aims to give an update on the current status of LTx and to give an insight into future practice to deal with current challenges. For a more detailed reading on specific issues, we refer the reader to a recent and more comprehensive textbook [1].

45.2 HISTORY OF LTX

The first LTx in a human was performed in 1963 by James Hardy at the University of Mississippi Medical Center in Jackson, MS, USA, in a patient with bronchial carcinoma, no longer a good indication for LTx nowadays [2]. It took the lung transplant community another two decades to better control the often fatal problems seen in the early days after LTx like acute rejection, invasive infections, and bronchial anastomotic dehiscence [3]. The first long-term successes were reported by the Stanford team [4] after heart−lung transplantation (HLTx) in a group of patients with idiopathic pulmonary arterial hypertension (IPAH) and by the Toronto team [5] after single-lung transplantation (SLTx) in a group of patients with pulmonary fibrosis. Originally, double-lung transplantation (DLTx) was believed to be required, not only for septic lung diseases but also for emphysematous lung disease. As a result, a technique of en bloc DLTx was developed in Toronto, first in the laboratory by John Dark and collaborators [6] and later in the clinic by Alec Patterson and coworkers [7]. In this original technique, both lungs remain connected by a left atrial cuff, main pulmonary artery, and trachea. Full cardiopulmonary bypass was needed to implant the double lung en bloc. After a few years, this technique was abandoned by the pioneering groups because of the complexity of a high incidence of lethal airway complications as a result of ischemic airway necrosis with dehiscence of the tracheal anastomosis [8]. To overcome this problem, a technique of bilateral bronchial anastomosis was developed by the group in Marseille [9]. Other groups in Bordeaux [10] and in Copenhagen [11] continued to use a tracheal anastomosis after revascularizing the donor airway by connecting a donor aortic patch including the orifices of bronchial arteries with the recipient's aorta.

In the late 1980s, the Hôpital Bichat group in Paris reported successful outcome in emphysema patients after SLTx [12]. The feared compression of the allograft by the hyperinflated native lung or the ventilation-perfusion imbalance on lung scans did not occur in these emphysema patients after SLTx. This was confirmed by other groups [13]. As a result of this observation, many transplant teams worldwide started their transplant program with SLTx in emphysema patients [14,15]. With more experience accrued over the years, it became evident from large institutional reports [16−18] and from analysis of data submitted to the Registry of the International Society for Heart and Lung Transplantation (ISHLT) [19,20] that emphysema patients did better after DLTx compared to SLTx. DLTx is now considered to be the procedure of choice in every patient with end-stage lung disease if no donor lung shortage would exist.

As a result of the airway complications observed after en bloc DLTx, a new technique for double-lung replacement was developed in the early 1990s known as bilateral single-lung (BLTx) or sequential single-lung transplantation [21]. This operation was originally performed using a bilateral anterior thoracotomy with transverse sternal division, the so-called "clamshell" incision [22]. This technique facilitated bilateral LTx as this operation can often be performed without cardiopulmonary bypass as is mandatory with en bloc DLTx. However, sternal complications such as chronic pain, sternal override, sternochondritis, and wound dehiscence are not uncommon. Therefore, the technique evolved over the years toward the use of an anterior thoracotomy on both sides without sternal division [23]. This surgical evolution has largely decreased early morbidity related to the incision.

Surgical techniques have further evolved over the years. Lobar LTx for small recipients, mainly cystic fibrosis (CF) patients, was introduced in the 1990s whereby the pulmonary graft can be recovered from cadaveric [24] or living donors [25,26]. Split LTx whereby one lung is divided in two lobes transplanted bilaterally in one small recipient spares the contralateral single lung that can be utilized for another recipient on the waiting list for SLTx [27]. Parenchymal resection following transplantation [28] can be utilized to downsize the volume of oversized pulmonary grafts to better fit into the chest cavity avoiding cardiac compression, especially in recipients with restrictive lung disease [29].

45.3 CURRENT ACTIVITIES WORLDWIDE

According to the 2012 report of the Registry of ISHLT, more than 43,000 lung and heart−lung transplant procedures have been performed worldwide until June 30, 2011 [19]. The annual number of procedures is still increasing with 3519 cases reported in the year 2010 worldwide. This growth has largely been the result of a constant rise in the number of bilateral lung transplants since the mid-1990s, while the annual number of single-lung transplants remains constant and the number of heart−lung transplants now stabilizes around 100 cases per year over the last 7 years. SLTx and BLTx currently make up more than 97% of all procedures performed worldwide [19].

The number of centers that report their activities to the ISHLT Registry reaches 178 worldwide. Nearly 47% of the centers have a volume of fewer than 10 transplants per year accounting for nearly 9% of the total transplant volume. Seven centers with an average activity of more than 50 transplants per year performed 20% of the procedures worldwide. Based on data from the United Network for Organ Sharing (UNOS) in the United States, center transplant volume has been identified as an important predictive factor with a 30-day mortality being lowest (4.1%) in centers with more than 20 lung transplants per year and a 2% increase in 30-day mortality for each decrease by 1 case/year [30]. Although high volume centers do not have significantly lower incidences of individual postoperative complications after LTx, they are best able to minimize the adverse effects of these complications [31].

45.4 INDICATIONS

The indications for LTx have broadened over time including a diverse spectrum of pulmonary diseases of the lung including airways, parenchyma, and vasculature. Chronic obstructive pulmonary disorder (COPD) remains the most common indication for which LTx is performed accounting for approximately 40% of all procedures performed to date [19]. The most common cause is smoking-induced emphysema (34.0%) followed by an inherited disorder in younger patients resulting from a deficiency of the protective enzyme alpha-1-antitrypsin causing emphysema (6.1%). More recently, the proportion of transplants performed for idiopathic pulmonary fibrosis (IPF) has steadily increased to 23.2%, particularly in the United States, where this disorder now represents the leading indication for LTx since the introduction of the lung allocation score (LAS) being higher in these patients. CF is the third major indication, accounting for 16.7% of all procedures. IPAH, once a leading indication for HLTx, currently constitutes only 3.1% of all procedures reflecting major advances in the medical management of these patients. Other less common indications include bronchiectasis (2.8%), sarcoidosis (2.5%), lymphangioleiomyomatosis (1.1%), retransplants for BOS (1.5%), connective tissue disorders (1.2%), and other rare indications (7.8%) [19]. Use of LTx as a definitive cure for patients with early-stage lung cancer (*in situ* or minimally invasive adenocarcinoma) [32], or with pulmonary metastases from low malignant primary tumors as done in liver transplantation [33], is not commonly accepted by most centers because of the high rate of cancer recurrence. Nevertheless, LTx can be performed to treat malignant diseases with results approaching those for nonneoplastic indications, provided patients are carefully selected and staged [34].

The distribution of indications may well vary between transplant centers according to the referral pattern of patients with end-stage pulmonary disease. Some centers transplant more emphysema patients while others have more recipients with IPF or CF on their waiting list. The recent introduction of the LAS in the United States and in some Eurotransplant countries has resulted in a shift with less emphysema patients being listed for LTx because of a smaller chance of being transplanted compared to recipients with other pulmonary diseases resulting in higher LASes [35].

45.5 RECIPIENT SELECTION

Because of the limited supply of good donor organs, it is important to select only those patients that qualify and have the best chance to benefit from the procedure. Contraindications to LTx are listed in Table 45.1. Patients with limited survival expectance (<12−24 months)

TABLE 45.1 Contraindications to LTx

- Absolute contraindications
  - Recent malignancy (other than nonmelanoma skin cancer or multiple locations of *in situ* lung adenocarcinoma)
  - Active extrapulmonary infection
  - Active Hepatitis B or C and HIV viral infections
  - Active or recent cigarette smoking
  - Severe psychiatric illness
  - Documented noncompliance with medical therapy
- Relative contraindications
  - Age >65 years
  - Extrapulmonary vital organ dysfunction (except combined transplant)
  - Obesity and poor nutritional status
  - Prior thoracic procedures
  - Medical comorbidities (CAD, DM, osteoporosis, GOR)
  - Ventilator-dependency or extracorporeal life support

CAD, coronary artery disease; DM, diabetes mellitus; GOR, gastrooesophageal reflux.
Adapted from [36].

resulting from progressive end-stage pulmonary disease despite optimal medical therapy, including supplementary oxygen or other surgical alternative strategies, should be referred for transplant screening. Disease-specific guidelines for referral defined by the pulmonary council of the ISHLT were updated and published in an expert consensus report in 2006 [36]. A detailed discussion of these criteria for each specific type of lung disease is beyond the scope of this chapter. It is of utmost importance to make referring lung physicians aware of these guidelines. Early referral for initial assessment at the transplant center should be encouraged, even before the need for listing is anticipated, to identify and to address potential barriers to transplantation, to initiate patient and family education, and to promote familiarity with the transplant team. Patients with progressive lung disorders like IPF, CF, and IPAH who continue to deteriorate despite optimal medical treatment should be screened early so that activation on the waiting list can be done with a realistic time perspective to find a suitable donor.

The upper age cutoff for LTx is usually set at 65 years as advanced age has been consistently identified as a risk factor for increased posttransplant mortality [19]. However, functional rather than chronological age should be considered when evaluating an individual candidate. Mean age at transplantation has risen over the years [19] with recent reports on successful LTx in patients over 70 years of age [37].

The presence of significant extrapulmonary vital organ dysfunction precludes isolated LTx, but combined organ procedures such as heart−lung, lung−liver, or lung−kidney can be considered in selected patients [38].

Prior thoracic procedures such as pleurodesis for pneumothorax or previous parenchymal resection is associated with an increased risk of perioperative bleeding, particularly when cardiopulmonary bypass is considered, but is not a contraindication to transplantation in experienced surgical hands [39,40].

The risk posed by other medical comorbidities such as coronary artery disease, osteoporosis, diabetes mellitus, and gastrooesophageal reflux must be assessed individually based on severity of disease and ease of control with standard therapies.

Ventilator-dependence before LTx has long been recognized as a risk factor for increased short-term posttransplant mortality [19]. These patients usually get priority in allocation of organs. Recent reports have demonstrated that survival in these patients is still inferior compared to nonventilated patients, but still better than patients who do not receive a transplant on time [41–43].

Even more controversial is LTx in patients who are being bridged on extracorporeal devices for severe hypoxemia, hypercarbia, or right heart failure. Extracorporeal life support with extracorporeal membrane oxygenation (ECMO) [44–48] or pumpless interventional lung assist [49,50] in these sick patients should be tailored to minimize morbidity and to provide the appropriate mode and level of cardiopulmonary support for each patient's physiologic requirements. Novel device refinements and further development of extracorporeal life support in an ambulatory and simplified manner will help maintain these patients in a better condition until transplantation [51–57].

45.6 TIMING OF LISTING AND ORGAN ALLOCATION

Listing for LTx is considered at a time when the lung disorder has advanced to a disabling and potentially life-threatening stage, such that survival with transplantation is deemed to be more likely than survival without. The presence of primary or secondary pulmonary hypertension, signs of cor pulmonale, hypercapnia, and hypoxemia are general indicators of poor prognosis, and these patients should get priority because of an increased risk of dying before a suitable donor is offered. In the United States, a system was implemented in 2007 that ranks patients on the waiting list according to a calculated LAS based on a dozen parameters that balances medical urgency versus posttransplant survival, so-called transplant benefit [58]. Recent studies have demonstrated that the implementation of LAS resulted in a profound impact on the dynamics of LTx in the United States with the number of patients on the active waiting list dropping to approximately one-half [35]. More importantly, median

waiting time has significantly decreased to less than 200 days and waiting list mortality has dropped significantly for IPF and CF patients since the introduction of LAS [59,60]. Other large organ-exchange organizations such as Eurotransplant have recently adopted a similar LAS in members states with low donor rates now prioritizing patients on their predicted survival benefit and no longer on total waiting time [61].

45.7 TRANSPLANT PROCEDURE

Four types of transplants have been described: single lung, double lung (or bilateral lung), heart–lung, and lobar lung. There has been an intense discussion and shift over the last three decades as to the most appropriate type for each individual form of lung disease.

45.7.1 Heart–Lung Versus Double Lung

In the 1980s, HLTx was widely performed in patients with Eisenmenger's syndrome, IPAH, and CF [62,63]. Following the growing success of heart transplantation and the increasing donor age with less suitable heart–lung blocs becoming available, HLTx nowadays accounts for less than 3% of all procedures [19] and is reserved by many transplant teams for patients with irreversible cardiopulmonary failure resulting from a surgically uncorrectable congenital heart disease leading to Eisenmenger's syndrome (Figure 45.1) [64]. Other teams prefer DLTx when the congenital cardiac defect (such as atrial septal defect, ventricular septal defect, or patent ductus arteriosus) can still be repaired simultaneously thereby optimizing organ allocation [65].

Patients with IPAH are currently treated either with HLTx or DLTx (Figure 45.2) depending on the preference of the team, the organ availability, and the estimation of ability of the dysfunctional right ventricle to recover once pulmonary artery pressures have normalized [66,67]. SLTx as opposed to DLTx for patients with IPAH or severe secondary pulmonary hypertension poses an increased risk of perioperative allograft edema since virtually the entire cardiac output must be borne by the freshly implanted lung.

45.7.2 Double Lung Versus Single Lung

Patients with CF or other forms of suppurative lung disease (e.g., bronchiectasis) should undergo DLTx (Figure 45.3) because of concerns related to leaving a chronically infected lung in place after SLTx [68,69]. Occasionally, in a CF patient with previous pneumonectomy, SLTx can be performed [70]. In such patients, lung implantation obviously can only be done with the recipient on full cardiopulmonary bypass.

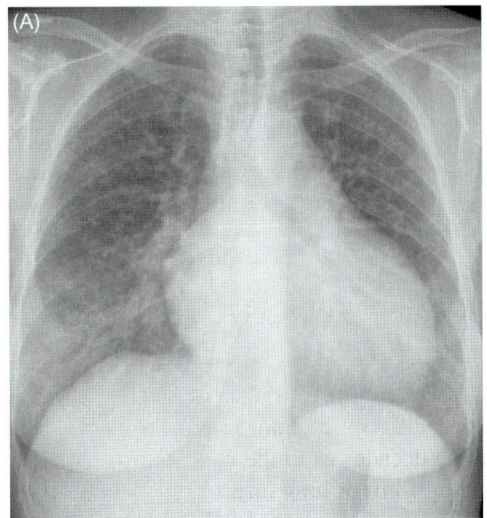
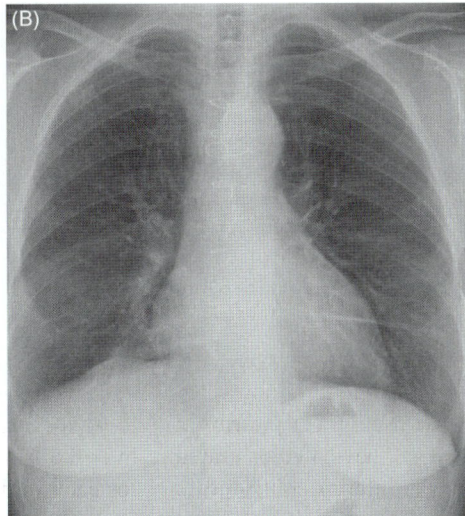

FIGURE 45.1 Chest X-ray 1 day before (A) and 1 year after (B) HLTx in a 45-year-old female patient (O positive, 158 cm, 64 kg) with Eisenmenger's syndrome from a 50-year-old female donor (O positive, 165 cm, 70 kg) after a waiting period of 76 days. Note the normal size of the heart and both lungs after transplantation.

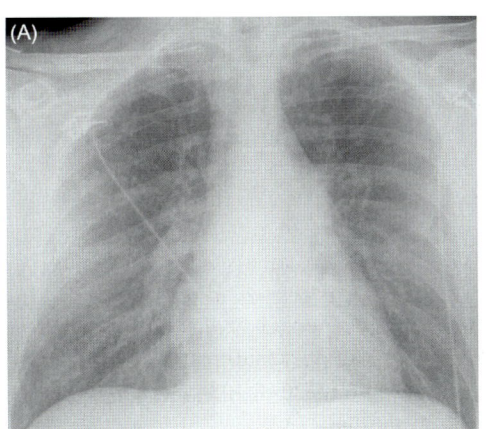
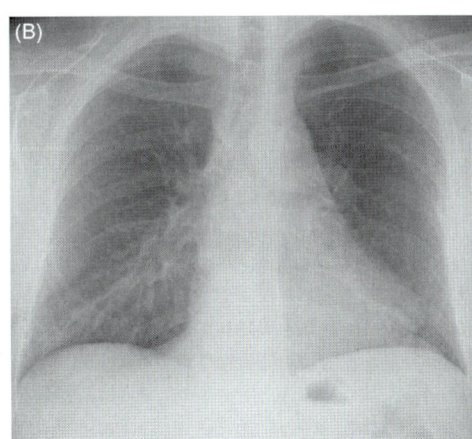

FIGURE 45.2 Chest X-ray 1 day before (A) and 1 year after (B) DLTx in a 43-year-old male patient (O positive, 183 cm, 73 kg) with IPAH from a 45-year-old male donor (O negative, 180 cm, 85 kg) after a waiting period of 58 days on the high-urgency waiting list. Note the reduced size of the previously dilated right atrium.

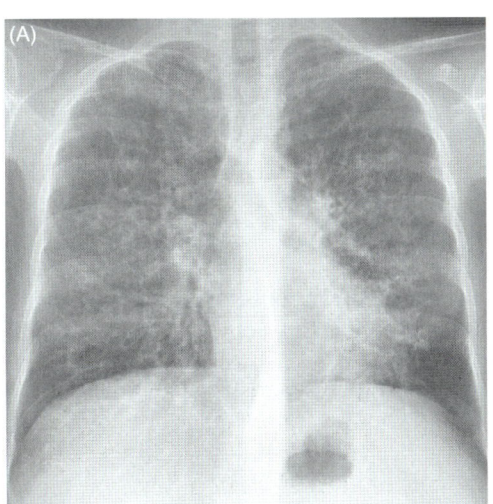
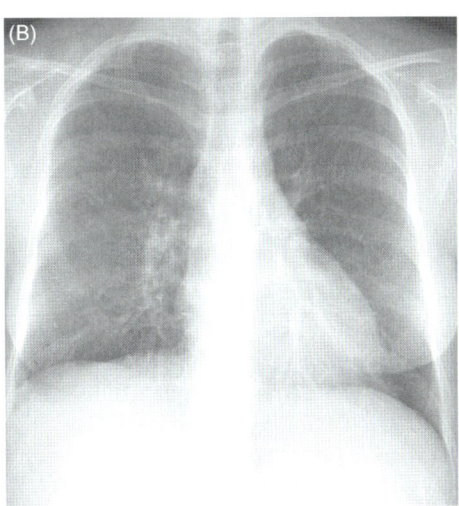

FIGURE 45.3 Chest X-ray 1 day before (A) and 1 year after (B) combined double-lung−liver transplantation in a 21-year-old female patient (A negative, 153 cm, 44 kg) with CF from a 27-year-old female donor (O positive, 167 cm, 57 kg) after a waiting period of 232 days. Note the smaller size of both pulmonary allografts relative to the hyperinflated native lungs.

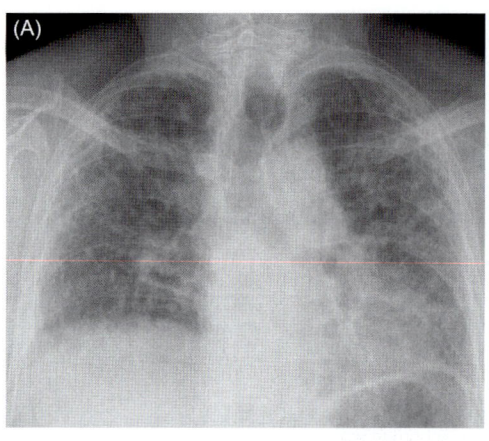
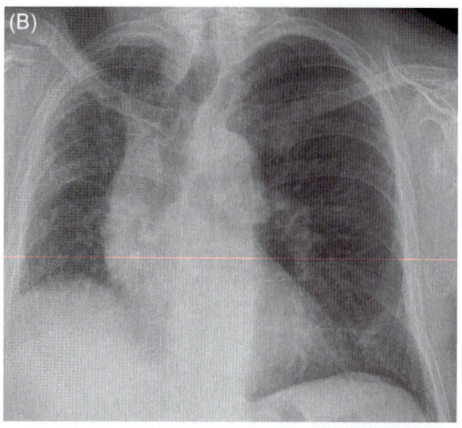

FIGURE 45.4 Chest X-ray 1 day before (A) and 1 year after (B) left SLTx in a 65-year-old male patient (O positive, 168 cm, 65 kg) with IPF from a 37-year-old male donor (O negative, 182 cm, 90 kg) after a waiting period of 112 days. Note the larger size of the left pulmonary allograft relative to the fibrotic native right lung.

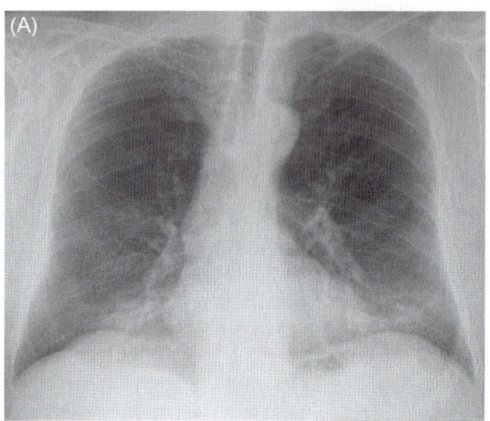
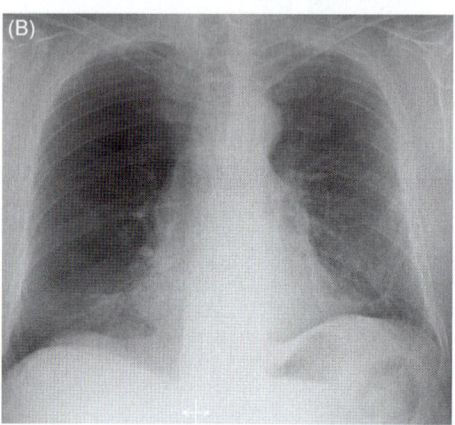

FIGURE 45.5 Chest X-ray 1 day before (A) and 1 year after (B) left SLTx in a 61-year-old male patient (A positive, 166 cm, 70 kg) with smoking-induced emphysema from a 19-year-old male donor (A positive, 185 cm, 70 kg) after a waiting period of 74 days. Note the smaller size of the left pulmonary allograft relative to the hyperinflated native right lung.

IPF patients are usually treated with SLTx (Figure 45.4). This type of LTx is believed to be ideal for patients with restrictive lung disease as the largest proportion of perfusion as well as ventilation will be diverted toward the allograft resulting from persistent high vascular resistance and low compliance in the remaining native lung [5]. The use of DLTx in this category of patients has been advocated in the last decade although the survival advantage of DLTx compared to SLTx has been conflicting between recent series [71–75].

For patients with emphysema, SLTx has been the procedure of choice for a long time (Figure 45.5) and is still performed at many transplant centers as this offers a more efficient use of the limited donor pool saving the contralateral lung for a twin recipient [12–14]. In recent years, however, DLTx has been advocated [18,20] in emphysema patients (Figure 45.6) as outcome after SLTx may be hampered by late complications affecting the native lung such as infection, malignancy, and hyperinflation [76,77]. Also, long-term survival after DLTx is superior especially when BOS develops [78]. Lungs of less ideal donors (older age, previous smokers, minimally infected or edematous organs) are best suited for patients with low-

perioperative risk such as those with COPD but ideally both lungs should then be transplanted. These lungs may otherwise be declined for higher risk patients such as those with CF, IPF, or IPAH. When transplanting a single lung for COPD or IPF, the side of the transplant (right versus left) does not seem to matter in terms of outcome and should not delay transplantation except when a specific side is preferred for anatomical reasons [79].

According to the ISHLT registry, in the year 2010, DLTx accounted for 74% of all lung transplants across all age groups and diagnoses [19]. The proportion of DLTx/SLTx has constantly risen since 1994 for all four major transplant indications. Nevertheless, until June 2011 in total more single-lung than double-lung transplant procedures have been reported for COPD and IPF (6048 versus 5539 and 4430 versus 3495, respectively). From all indications, emphysema accounted for 45.6% of all SLTx and 26.6% of all DLTx [19].

45.7.3 Lobar Lung

Finally, lobar-lung transplant from a taller donor (Figure 45.7) is reserved for small-sized recipients (e.g.,

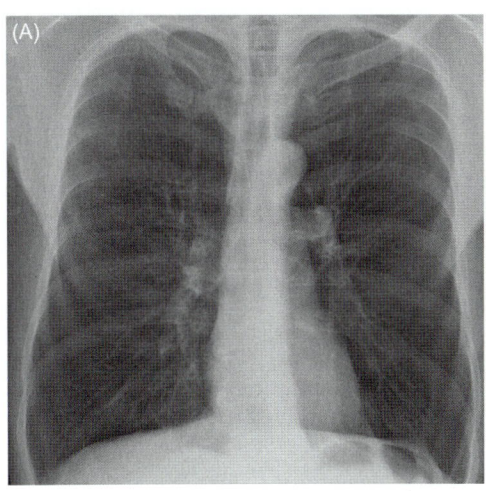
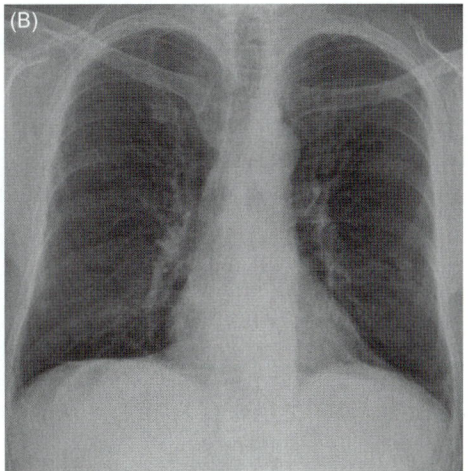

FIGURE 45.6 Chest X-ray 1 day before (A) and 1 year after (B) DLTx in a 56-year-old male patient (A positive, 160 cm, 46 kg) with smoking-induced emphysema from a 55-year-old male donor (A positive, 172 cm, 80 kg) after a waiting period of 454 days. Note the smaller size of both pulmonary allografts relative to the hyperinflated native lungs.

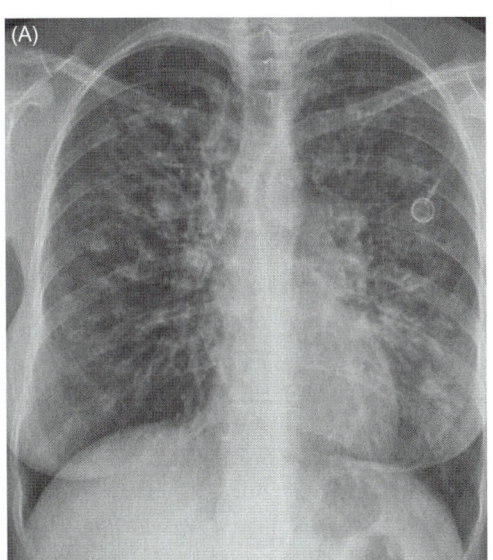
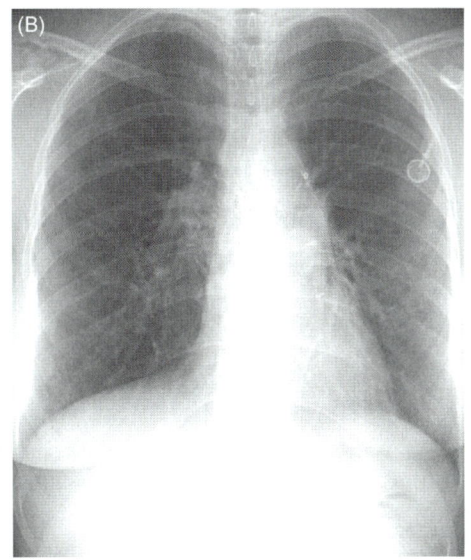

FIGURE 45.7 Chest X-ray 1 day before (A) and 1 year after (B) bilateral lobar LTx (right middle + lower lobes and left lower lobe) in a 22-year-old female patient (AB positive, 152 cm, 52 kg) with CF from a 44-year-old male donor (AB negative, 185 cm, 85 kg) after a waiting period of 263 days. Note the lobes fit nicely in the previously hyperinflated chest.

CF patients or small females) facing a long waiting time for a size-matched donor [28,80,81]. Successful outcome has also been reported with split lung transplants for bilateral lobar LTx saving the contralateral donor lung for a twin recipient [27,28]. Size matching between donor and recipient is usually based on donor and recipient height and/or predicted total lung capacity (pTLC). pTLC better than the real TLC of the recipient is a suitable indicator for decision making between whole lung and lobar LTx [82]. However, wide discrepancies in lung sizing did not affect overall posttransplant survival or pulmonary function in a recent series from the Cleveland Clinic [83]. If needed, lung volume can be reduced by shaving off part of the lung with staplers. The use of computed tomography volumetry in living lobar-lung donors is a valid method to estimate forced vital capacity of the graft and thus predict pulmonary function in the recipient [84].

45.8 DONOR TYPES, SELECTION, AND MANAGEMENT

45.8.1 Donor Types

Two types of solid organ donors can be distinguished: deceased donors and living donors (Table 45.2). The largest experience in LTx comes from postmortem donors. These donors die as a result of brain death (brain-dead donors or heart-beating donors) or following cardiac arrest (circulatory-arrested donors or nonheart-beating donors). The cause of brain death in the first category results from intracranial trauma, bleeding, ischemia, hypoxia, tumor, intoxication, or infection. Lungs from such donors get injured as a result of different hits that may affect the donor in the hours before and after brain death (lung contusion, aspiration, ventilation, infection). In addition, the increased intracranial pressure leading to

TABLE 45.2 Different Types of Solid Organ Donors

Deceased Donors	Living Donors
• Brain-dead donors (heart-beating)	• Related
• Circulatory-arrested donors (nonheart-beating)[a]	– Genetically
– I Death on arrival	– Emotionally
– II Failed resuscitation	• Unrelated
– III Awaiting cardiac arrest	– Samaritarian
– IV Cardiac arrest in brain-dead donor	– Paid
– V Euthanasia donor	

[a]According to proposed modification of the Maastricht Classification [85].

TABLE 45.3 Standard Lung Donor Selection Criteria

- No history of primary pulmonary disease
- No history of active pulmonary infection
- No history of previous malignancy (except skin cancer and some brain tumors)
- No history of previous chest surgery
- Age < 55 years
- Clear serial chest X-ray
- Normal gas exchange ($PaO_2 > 300$ mmHg on $FiO_2 = 1.0$, PEEP 5 cm H_2O)
- Short donor ventilation time (<72 h)
- No evidence of aspiration or sepsis
- Absence of purulent secretions at bronchoscopy
- Absence of organisms on sputum gram stain
- Absence of chest trauma
- Tobacco history <20 pack years
- ABO blood-group compatibility
- Appropriate size match with prospective recipient

Adapted from [103].

brain-stem compression and stop of cerebral perfusion, results in a sympathetic discharge leading to systemic hypertension followed by hypotension and loss of body control by the central nervous system (so-called Cushing response). All this will lead to a systemic inflammatory syndrome triggering neutrophilic lung infiltration. The increase of both hydrostatic pressure and vascular permeability may end up in neurogenic pulmonary edema. In the second category of postmortem donors who develop circulatory arrest resulting from myocardial hypoxia, arrhythmia, or hypovolemia, the delay between (unexpected) circulatory arrest and cold pulmoplegia, so-called warm ischemia, may also damage lung tissue prior to cold preservation. Research has shown that the tolerance of the lung to warm ischemia is around 60 min, leaving some time to organize organ recovery from these donors [86]. Four types of nonheart-beating donors have been described according to the 1995 Maastricht classification [87]. Recently, adding a fifth category with lung donors after euthanasia [85] was proposed for countries with an existing legal framework [88]. The Leuven Lung Transplant Group has reported good outcome in recipients transplanted with lungs recovered from this specific type of donor [89]. Recent studies have shown that lungs from both controlled [90–96] and uncontrolled [97] circulatory-arrested donors can be successfully transplanted, although the practice is not authorized by law in all countries. The incidence of severe PGD using lungs from uncontrolled donors, however, was reported to be somewhat higher compared to brain-dead donors [97]. This is an indication that pulmonary grafts with longer warm ischemic periods should be tested first by means of *ex vivo* lung perfusion (EVLP) prior to acceptance and subsequent transplantation [98–100].

The worldwide experience with LTx from living donors is much less, but the outcome may be superior as only carefully selected, healthy donors with noninjured lungs are accepted, and cold ischemic times are usually short. Two patients (relatives) serve as a donor, each giving one lobe to

the recipient usually suffering from CF or any other form of end-stage lung disease. Two well-established programs in the world (University of California at Los Angeles and University of Kyoto) have published good results after LTx from such living donors [25,26,101,102].

45.8.2 Donor Selection

The ideal or standard donor criteria are listed in Table 45.3 [104]. Adherence to these criteria will result in a recovery rate of only 10–15% of lungs from all effective multiorgan donors [105,106]. Similar to the guidelines for patient referral and listing [36], donor selection criteria applied in the early 1980s after the initial successes were not based on prospective, randomized, well-powered studies but rather on a combination of recommendations based on small and/or retrospective series and/or registry studies as well as expert opinion consensus. The evidence for adhering to these strict donor criteria was extensively reviewed some years ago by a panel of the pulmonary council of the ISHLT and was found to be very low [107]. Recommendations were made to relax the original acceptance criteria. In a national cohort study in the United Kingdom, Bonser and coauthors concluded that, although lungs from donors with a smoking history are associated with worse outcomes, the individual probability of survival is greater if they are accepted than if they are declined, and the patient chooses to wait for a potential transplant from a donor with a negative smoking history [108]. This scenario should be discussed at the time when accepting the patient on the lung transplant waiting list. Several cases of successful transplantation with ABO-incompatible lungs using specific treatment measures have been reported recently [109–111].

As a result of the large discrepancy between the number of patients listed for LTx and the number of donor lungs that become available, transplant teams have looked at ways to increase the donor pool [103,112] and have relaxed their donor requirements now using so-called extended-criteria donors. Reports from different institutions worldwide have demonstrated that immediate and long-term survival in lung recipients from such donors, if carefully selected, is similar to those from standard-criteria donors [113–115]. A large study based on Eurotransplant data demonstrated that the use of a lung donor score accurately reflected the likelihood of organ acceptance and also predicted patient mortality [116]. Its application at the time of donor reporting may facilitate donor risk assessment and patient selection. COPD patients with no signs of pulmonary hypertension are especially suited to receive lungs from these extended-criteria donors with older age (>55 years), smoking history (>20 pack years), inferior oxygenation (PaO_2/FiO_2 < 300 mmHg) because of the lower risk to develop severe PGD. For the same reason, patients with COPD are the best candidates to receive lungs from donors after circulatory arrest with an inherent period of warm ischemia prior to cold pulmonary flush and preservation.

45.8.3 Donor Management

Care of the organ donor starts before declaration of brain death. The general principles of intensive care medicine apply to the management of the potential organ donor. Active donor management is thought to preserve organ quality and improve retrieval rate [105,106]. Because as many organs as possible will need to be recovered from a given donor, the team in charge has to consider a treatment in the best interest of all organs. The general management focuses on maintaining body temperature, blood pressure, acid–base equilibrium, electrolytes balance, intravascular volume, and prevention of infection. Special attention on hemodynamic, hormonal, and respiratory disturbances are needed to preserve lung function before retrieval in brain-dead donors [105,106].

Fluid loading to compensate hypotension can contribute to further lung edema and impaired oxygenation if not well monitored. Central venous pressure alone to guide donor maintenance is not reliable and the use of pulmonary artery pressure catheter is advised to optimize filling pressures and monitor cardiac performance. Organ perfusion should be maintained with vasopressors and inotropes. Vasopressin has been shown to have a stabilizing effect on blood pressure and to reduce inotropic requirements [117]. In addition, vasopressin is effective against diabetes insipidus.

Brain death also results in endocrine and metabolic dysfunction that could affect graft survival. Replacement of hormones like cortisol, insulin, thyroid, and antidiuretic hormones before organ recovery may increase graft viability and early recipient outcome [118–120]. The use of inhaled β_2-adrenergic agents to stimulate alveolar liquid clearance [121] was recently studied in a randomized trial by the California Transplant Donor Network, but turned out to be negative in terms of improved donor oxygenation and increased donor lung utilization compared to placebo [122].

A ventilatory strategy with high tidal volume is potentially harmful and may exacerbate donor lung injury already triggered by the systemic inflammatory response. Recently, a randomized controlled trial demonstrated that a lung protective ventilator strategy (with tidal volumes of 6–8 mL/kg of predicted body weight, positive end-expiratory pressure, PEEP, of 8–10 cm H_2O, apnea tests performed by using continuous positive airway pressure, and closed circuit for airway suction) increased the number of eligible and recovered lungs compared with a conventional strategy [123]. Bronchoscopy should be routinely performed on all potential lung donors to remove mucous plugs, to assess anatomy, to detect presence of inflammation, aspiration, blood clots, and signs of infection, and to confirm correct endotracheal tube placement.

In a recent study by our group on a large number of lung donors, donor cause of brain death (traumatic versus vascular versus other causes) had no impact on survival or freedom from BOS after LTx. However, the time interval between the moment of brain death and lung retrieval was found to be important with a survival advantage for those recipients with such an interval longer than 10 h suggesting that longer donor management after brain death may improve lung quality [124].

45.9 LUNG PRESERVATION TECHNIQUES

Lung preservation and procurement are key steps in the transition from donor to recipient that may significantly influence the quality of the graft and thus the outcome in the transplanted patient [125]. Although no prospective trial has been carried out so far, clinical and experimental evidence suggest that cold flush and storage in an extracellular type of solution containing 5% dextran 40 kDA as colloid [126] is currently the method of choice for lung preservation resulting in a lower incidence of PGD compared to intracellular-type solutions [127,128]. Perfadex® (XVIVO Perfusion AB, Göteborg, Sweden) is the preservation solution clinically most used worldwide with safe preservation times up to 12 h [129].

Many transplant teams have adopted a combined flush technique with an antegrade flush through the pulmonary artery followed by a retrograde flush through each of the pulmonary veins after heart excision with the lung still ventilated inside the cadaver or on the backtable [130,131]. Strong evidence to support this practice, however, is lacking in the absence of controlled randomized studies.

EVLP in a circuit for several hours has become possible in recent years following extensive experimental work by Steen and colleagues in Lund [98,132] and Cypel and collaborators in Toronto [133]. EVLP is now on the horizon as a potential method to safely prolong the storage period outside the body [134] and to avoid the cold ischemic period [135]. A multicenter, international, randomized trial (Inspire) is ongoing comparing standard cold storage with Perfadex® versus normothermic perfusion in the Organ Care System Lung™ (Transmedics, Andover, MA, USA) [136]. EVLP is also useful to evaluate the quality of the lung prior to transplantation when the initial graft assessment in the donor was not possible or not satisfactory as previously described. Moreover, it is hoped that this technique will help to increase the number of acceptable grafts by reconditioning lungs of inferior or questionable quality prior to transplantation. Several groups worldwide have now reported successful transplantation of such lungs after EVLP with a recovery yield varying between 46% and 100% [100,137–141]. Finally, EVLP may become a way to get around the innate and adaptive immunity in the recipient by interacting with the allogenicity of the pulmonary graft while perfused on the device prior to transplantation. Much interest is directed toward the addition of mesenchymal stem cells during EVLP [142].

45.10 RESULTS

45.10.1 Overall Survival

According to the 2012 report of the ISHLT registry, the overall survival after SLTx and DLTx was 88% at 3 months, 79% at 1 year, 64% at 3 years, 53% at 5 years, and 30% at 10 years [19]. The overall median survival (half-life) is 5.5 years and for those who survived at least the first year, half-life survival is 7.7 years. Survival after combined HLTx is somewhat inferior with 71% at 3 months and 63% at 1 year. However, those recipients that survive the first year have generally good outcomes with a conditional half-life of 10.0 years.

Survival has consistently increased by era and is largely driven by better early survival. Three-month survival has improved between the earliest era starting in 1988 and the most recent era beginning in 2004 from 81% to 90%, and 1-year survival has improved from 70% to 81%. This better short-term survival in recent years is probably multifactorial related to better lung preservation methods with a decrease in and/or better control of PGD, improved surgical techniques, and earlier diagnosis and

more effective treatment of acute rejection and infections. The overall half-life survival has also improved in the same periods from 3.9 to 5.9 years and the conditional half-life among 1-year survivors has also improved (from 7.0 years in period 1988–1995 and not yet reached in period 2004–June 2010) [19]. These data demonstrate that longer-term survival has also improved over time, probably as a result of earlier diagnosis and better prevention and/or treatment of chronic rejection.

When stratified by pretransplant diagnosis, unadjusted 3-month mortality was reported to be lowest in COPD without A1ATD (9%) and highest in IPAH (23%), followed by sarcoidosis (16%), and IPF (15%). The conditional half-life survival among 1-year survivors was 10.4 years in recipients with CF, 10.0 years with IPAH, 8.4 years with sarcoidosis, 8.6 years with A1ATD, and 6.8 years with COPD and IPF [19]. These differences between early mortality and long-term survival between diagnostic groups are most likely related to early complications, including PGD occurring more frequently in patients with IPAH while more comorbidities enhanced by the side effects of long-term immunosuppressive therapy (arterial hypertension, renal insufficiency, diabetes, osteoporosis) may play a bigger role in older patients with COPD and IPF. In particular, data from the growing number of transplants for recipients aged older than 65 years consistently has shown worse survival: the half-life survival for patients aged older than 65 years was 3.6 years compared with 6.5 years for those aged 35–49 years.

In recipients of heart–lung transplants, overall and conditional survival is best in those with Eisenmenger's syndrome [19].

45.10.2 Survival After SLTx Versus DLTx

Patients with DLTx experience better long-term survival across all indications compared to SLTx (1-year conditional half-live survival 9.4 years versus 6.5 years, respectively) [19]. These differences should be interpreted with caution as they may be influenced by multiple clinical factors that inform the decision to perform a particular procedure type, including age of the recipient, comorbidities, preferences of the transplant center, and characteristics of the donor lungs. Moreover, the risk of secondary complications, related to the presence of the native lung that may impact survival, is avoided in double-lung recipients.

45.10.3 Survival Benefit

Determining whether and for whom transplantation actually extends survival is difficult in the absence of randomized trials [143]. Published studies examining the possible survival benefit were nicely summarized in a recent review [144].

All studies looking at IPF patients conferred a survival benefit after LTx [145–149]. Survival benefit in CF patients is a more debated issue. While multiple studies [145–148,150] have suggested a survival benefit, other authors [151–153] have challenged this stating that only a subset of patients may benefit. These studies were heavily criticized by other experts for methodological shortcomings [154,155]. This debate has resulted in an international conference for better understanding between leaders in the field of LTx and physicians taking care of CF patients [156]. Calculating survival benefit in COPD patients is more complex. Because protracted survival is possible even in the advanced stages of COPD and long-term survival after LTx is inferior, it has been difficult to determine whether LTx truly extends survival in this patient group. Results have been conflicting [145–148,157,158]. In an older study based on UNOS data, the risk of transplantation relative to the waiting time never dropped below 1 during 2 years of follow up [145]. This indicates that risk for mortality after LTx remains higher than the risk of dying on the waiting list. In contrast, another study based on data from Eurotransplant has shown that emphysema patients benefited in survival (relative risk of dying <1) by day 260 after LTx [146]. In a more recent study based on UNOS data between 1987 and 2004, the authors reported an overall survival benefit for emphysema patients that was larger after DLTx compared to SLTx (mean difference 307 days (95% CI 217–523 days)). With DLTx, 44.6% of patients would gain 1 year or more, 29.4% would gain or lose less than 1 year, and 26% would lose 1 year or more. In contrast, with SLTx, only 22% of patients would derive a survival benefit of a least 1 year [159]. Several factors such as FEV_1, body mass index, exercise capacity, functional status, and the need for continuous mechanical ventilation or supplemental oxygen were associated with survival benefit. A recent study by the Lausanne group analyzed survival impact of LTx for COPD according to the pretransplant BODE index based on Body mass index, degree of airflow Obstruction, Dyspnea, and Exercise capacity [160]. A global survival benefit (observed versus predicted survival) was seen only in patients with more severe disease (BODE ≥ 7). This supports the use of the BODE index [161] as a selection criterion for LTx candidates.

45.10.4 Cause of Death and Related Risk Factors

In the latest ISHLT report [19], the main causes of death in the first 30 days are primary graft failure and non-CMV (cytomegalovirus) infections. After the first year, BOS and non-CMV infections were the predominant identifiable reported causes of death. Death caused by malignancies increased over the years accounting for 15% of all reported deaths between 5 and 10 years after LTx and 16% beyond 10 years.

Categorical risk factors for 1-year mortality are the recipient's underlying disease and severity of illness at the time of transplant. Compared with SLTx for COPD as reference group in the multivariate analysis, all patients with other diagnosis and transplant type had an increased relative risk for 1-year mortality with retransplantation (RR 2.09 (1.71–2.56)) and IPAH (RR 1.73 (1.28–2.34)) having the highest risk. Continuous variables with a significant effect on 1-year mortality include older recipient age, lower transplant center volume, lower cardiac output, lower forced vital capacity, higher pretransplant bilirubin level, higher supplemental oxygen requirement at rest, and more negative donor-recipient height difference [19].

Categorical risk factors for 5-year mortality in 1-year survivors are early BOS and acute rejection occurring within the first year as well as dialysis and treatment for infections by discharge. Continuous variables associated with 5-year mortality among those surviving 1 year are limited to both lower and higher recipient age and lower transplant center volume, highlighting the importance of transplant center experience to factors that affect mortality beyond the transplant procedure itself [19].

45.10.5 Quality of Life

More than 85% of all survivors are reported to have no activity limitations up to 5 years after LTx. After 1 year, around 25% of patients are back at work part-time or full-time. Employment increases to 40% after 5 years [19]. Health-related quality of life significantly increases after LTx but will largely depend on the incidence of infections, rejections, and the onset of BOS [162].

Although survival benefit remains unclear for certain transplant indications, improvement in pulmonary function, hemodynamics, and quality of life may also justify LTx for these patients who would otherwise have to survive in poor condition. Therefore, better instruments to measure quality of life are needed [163] and the success of transplantation should be judged by the net gain in quality-adjusted life-years taking both length and quality of survival into consideration [164].

45.11 COMPLICATIONS AND MORBIDITIES

Both the effectiveness and the cost of LTx are greatly affected by complications and morbidities. These can be subdivided in early (surgical) problems often related to the transplant procedure itself and late (medical) problems related to the allogenicity of the graft and long-term immunosuppressive treatment obligatory to prevent rejection and prolong graft survival.

45.11.1 Surgical Complications

Among the early problems, postoperative bleeding is a known risk that is significantly higher in patients in whom cardiopulmonary support was used to safely replace both native lungs. This is often needed in patients with increased right ventricular afterload due to preexisting pulmonary hypertension, in patients with ventilatory problems and severe hypercapnia (e.g., CF), or in patients with preexisting hypoxemia (e.g., cystic or pulmonary fibrosis) or as a result of severe edema after reperfusion of the firstly implanted lung. The use of veno-arterial ECMO has been advocated as this technique is less invasive compared to conventional cardiopulmonary bypass and can be executed with low dose (or no) heparin when using a heparin-coated circuit [44,165]. In case of pleural adhesions, these are taken down before the patient is connected to the bypass circuit whenever possible. If postoperative bleeding occurs in the first hours after the transplant, it is better to reexplore the patient early to avoid massive blood transfusion, coagulopathy, and retained hemothorax.

PGD describes a form of acute lung injury by development of noncardiac pulmonary edema within 72 h after reperfusion of the allograft in the absence of other identifiable secondary causes like venous anastomotic obstruction, hyperacute rejection, or fluid overload. PGD is presumed to be a form of ischemia-reperfusion injury [166], but other factors may contribute like donor death–related injury, surgical trauma, lymphatic disruption, and recipient pulmonary hypertension. The impact of donor factors [167] appears predominant in the initial 24 h after reperfusion, whereas recipient risk factors [168] such as preexisting pulmonary hypertension become more important thereafter. An association between graft ischemic time and PGD has not been consistently demonstrated [169]. Different grades of PGD have been described by a working group of the ISHLT in relation to the oxygenation capacity and chest X-ray abnormalities (Table 45.4) [170]. This grading is important to describe the evolution over time (T0-T72) and to compare the incidence between different centers. PGD occurs in up to 50% of patients. In most cases, the injury is mild and transient, but in about 20% of cases it can result in severe hypoxemia (PGD 3) causing similar problems as described for acute respiratory distress syndrome (Figure 45.8). PGD is responsible for

TABLE 45.4 Grades of PGDa

PGD Grade	PaO$_2$/FiO$_2$ (mm Hg)	Radiographic Infiltrates Consistent with Pulmonary Edema
0	>300	Absent
1	>300	Present
2	200–300	Present
3	<200	Present

aAccording to ISHLT Classification [170].

nearly one-third of the early deaths in the first 30 days after LTx [19]. The risk of death in survivors of severe PGD remains higher even beyond the first year because of incomplete and protracted recovery despite sometimes normal lung function and exercise tolerance [19,171]. Some studies [17,171,172] have suggested an increased risk of BOS in survivors of PGD, but this was not confirmed by others and the mechanistic link between both events remains uncertain. In patients with preoperative severe pulmonary hypertension and a preconditioned, hypertrophic right ventricle, prolonged support with ECMO in the intensive care unit to prevent severe PGD has been recommended by the Vienna group [173]. Treatment of severe PGD is supportive initially with increased PEEPs, lung-protective ventilator settings, negative fluid balance, exogenous surfactant, inhaled nitric oxide, or other vasodilators to stabilize the patient until pulmonary edema resolves and oxygenation improves [174]. If the evolution, however, worsens in the first 24 h, early institution of extracorporeal support with veno-arterial ECMO is recommended with reported success in about 50% despite serious morbidity resulting from bleeding, cardiac tamponade, lower leg ischemia, renal failure, sepsis, and stroke [44,174]. The results of emergent retransplantation in this situation have been poor with nearly half of the patients not surviving the first 30 days [175].

Anastomotic complications form a serious risk of postoperative morbidity and mortality if not recognized early on. The healing of the bronchial anastomosis has been the Achilles' heel of LTx for nearly 20 years [3] as the bronchial circulation is interrupted, but the incidence has decreased over the last decade with better preservation and surgical techniques [176–180]. Airway ischemia can result in bronchial mucosal and cartilaginous necrosis, pseudomembrane formation, fungal superinfection [181], partial or total anastomotic dehiscence putting the patient at risk of bacterial mediastinitis, bronchopleural fistula with empyema, bronchovascular fistula [182,183], or can heal with sequellae such as bronchial stenosis or bronchomalacia.

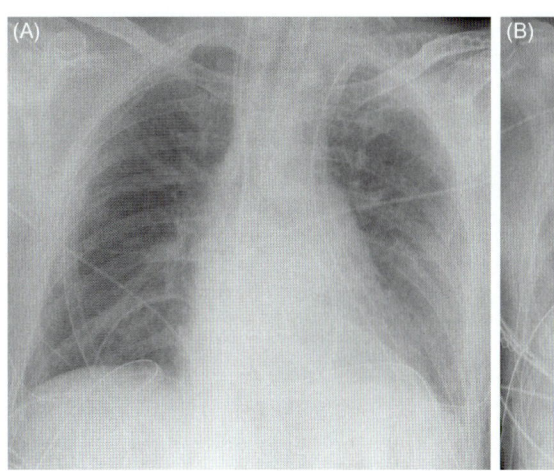 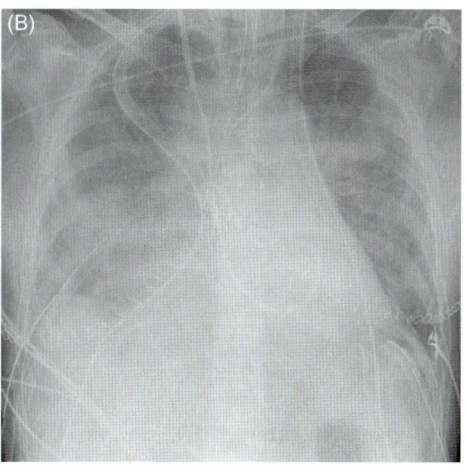

FIGURE 45.8 Chest X-ray on arrival in the intensive care unit in (A) a 45-year-old male patient with IPAH after receiving DLTx with intraoperative ECMO support. No radiographic signs of PGD at T0 (grade 0) [170] and (B) a 55-year-old female patient with IPF after receiving DLTx with intraoperative ECMO support. Note the bilateral infiltrates (R > L) and $PO_2/FiO_2 > 300$ mmHg compatible with PGD grade 1 at T0 [170].

The prevalence of bronchial anastomotic complications that need a bronchoscopic or surgical intervention is around 10%. The type of treatment depends on the presentation and should be tailored to the patient's condition [175,184−186]. Vascular anastomotic complications are rare after LTx but may occur at the level of the arterial as well as the atrial anastomosis [187,188]. This is usually the result of a technical failure with kinking, stenosis, or thrombosis as a result and can be diagnosed with the help of echocardiography [189], isotope scan, or selective arteriography. Early reintervention is needed to save the allograft. In case of chronic obstruction of the pulmonary artery, successful balloon dilation and stenting have been reported [190,191]. In case of lobar torsion [192−195] or parenchymal infarction, early reintervention with anatomical resection in the involved segment or lobe is warranted to prevent further complications such as bleeding, bronchopleural fistula, empyema, and pulmonary sepsis.

Direct injury to the phrenic [196,197] or vagal [198,199] nerves are well documented and may have a serious impact on posttransplant respiratory weaning and feeding prolonging hospital stay.

45.11.2 Medical Complications

Discussing all possible medical complications and morbidities after LTx is beyond the scope of this chapter and we refer the reader to more specific papers and text books on LTx [1].

Early on, bacterial (*Pseudomonas*, *Staphylococcus aureus*), viral (CMV), and fungal (*Aspergillus*, *Candida*) infections can be encountered in the immunosuppressed recipient and should be monitored carefully and treated adequately with antibiotic, antiviral, and antifungal agents. Acute cellular rejection occurs in about one-third of the recipients within the first year. Transbronchial lung biopsy is the gold standard for diagnosis if clinical

presentation on its own is not convincing. Treatment consists of a 3-day pulse of intravenous steroids followed by a tapering course. In recent years, a second form of acute humoral rejection based on donor-specific anti-HLA antibodies was recognized [200,201]. In addition to steroids, these patients may benefit from plasmapheresis, IV immunoglobulins, and anti-CD20 monoclonal antibodies.

Late complications include chronic allograft dysfunction (CLAD), malignancies [77,202], opportunistic infections, and other side effects of long-term immunosuppressants like hypertension, renal dysfunction, diabetes mellitus, dyslipidemia, and osteoporosis. CLAD remains the major obstacle to long-term graft and patient survival. This is characterized by a progressive and irreversible decline in FEV_1 at different grades [203] on pulmonary function testing and by a fibroproliferative process narrowing the lumen of small airways on histology. The disorder is clinically defined as BOS. According to the ISHLT registry, BOS in all 2 weeks survivors is reported in 48% and 76% of recipients by 5 and 10 years after transplantation, respectively. Different phenotypes have recently been identified [204]. Patients with the inflammatory form, so-called neutrophilic reversible allograft dysfunction, may benefit from a treatment with the macrolide azithromycin [205]. For patients with the fibroproliferative form, now called restrictive allograft syndrome (RAS) [206], retransplantation remains the only option to improve their pulmonary function.

45.12 EXPERIENCE WITH LTX AT THE UNIVERSITY HOSPITALS LEUVEN

The transplant program at the University Hospitals Leuven, Belgium, was initiated in 1991 under the leadership of Toni Lerut (Thoracic Surgery), Willem Daenen (Cardiac Surgery), and Maurits Demedts (Pneumology). Until the end of 2012, 711 LTx (505 DL; 159 SL; 47 HL)

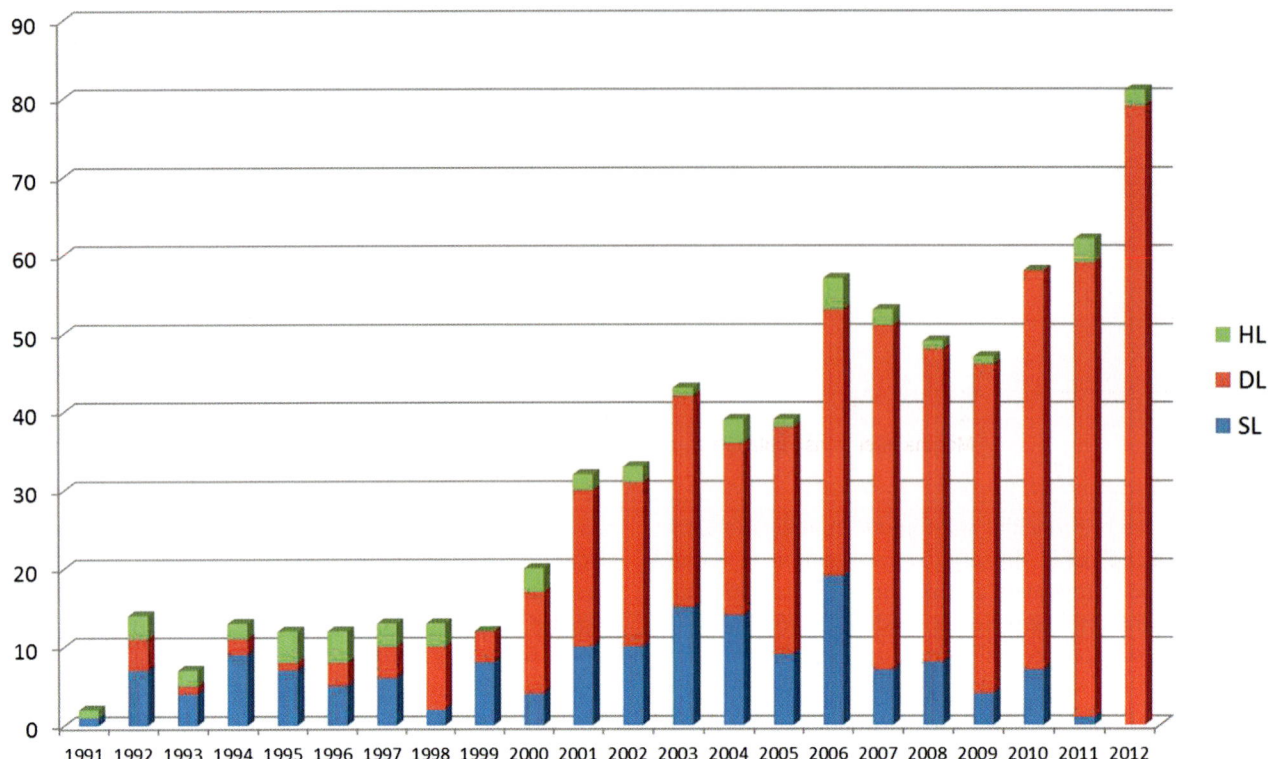

FIGURE 45.9   Number and types of lung-transplant procedures performed at the University Hospitals Leuven between 1991 and 2012—SL, single lung; DL, double lung; HL, heart—lung.

TABLE 45.5 Indications for LTx at the University Hospitals Leuven for 711 Procedures in 679 Recipients Between 1991 and 2012

Indication	Number	%
COPD	299	42
Alpha-1-antitrypsin deficiency	32	4.5
Pulmonary fibrosis	146	20.5
CF	92	13
IPAH	33	4.6
Eisenmenger	29	4.1
Bronchiectasis	14	2.0
Redo transplantation for graft failure	32[a]	4.5
Other	34	4.8
Total	711	100

[a]Including 1 for PGD and 31 for CLAD.

were performed in 679 patients, including 32 (4.5%) redo LTx. The growth in numbers and evolution in transplant type over the years is depicted in Figure 45.9. The indications for LTx are listed in Table 45.5. Overall hospital mortality was 5.8% (41/711). Survival in all SL + DL recipients transplanted between January 2000 and until end of 2010 ($n = 433$) is superior to the overall survival reported in the ISHLT registry in the period January 2000 till June 2009 (Figure 45.10).

45.13 CONCLUSIONS AND FUTURE DIRECTIONS

Since the introduction of successful LTx nearly three decades ago, this treatment has now become an effective and safe therapy for selected patients suffering from a variety of end-stage pulmonary diseases offering a prolonged and improved quality of life. Two major obstacles that remain limiting its clinical usefulness as standard therapy for more patients are donor organ shortage and CLAD. Alternative sources to expand the donor pool such as the use of donors after circulatory arrest and *ex vivo* perfusion for reconditioning of unacceptable donor lungs are emerging strategies. Macrolide treatment may be effective in patients with the neutrophilic phenotype of CLAD while retransplantation is the only effective therapy for patients with the RAS.

Further research is needed to find new ways to create bioartificial organs and to induce selective immune tolerance to accept the allograft in the absence of toxic immunosuppressants. These two steps would allow LTx to become an effective and safe therapy for many more patients suffering from end-stage pulmonary disease.

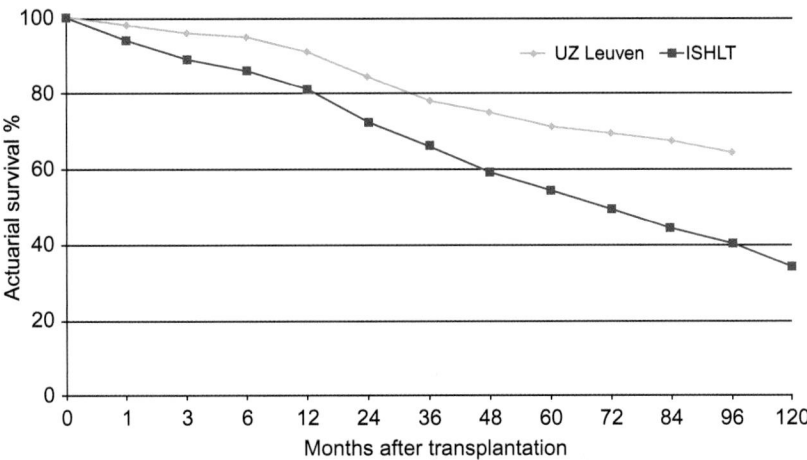

FIGURE 45.10 Overall survival in 433 single-lung and double-lung recipients transplanted at the University Hospitals Leuven (UZ Leuven) between January 2000 and until end 2010 in comparison to overall survival in patients in the Registry of the ISHLT transplanted between January 2000 and throughout June 2009 [19].

ACKNOWLEDGMENTS

Dirk Van Raemdonck is a senior clinical investigator supported by the Fund for Research—Flanders (G.3C04.99).

We acknowledge the large contribution to the success of our program by all members of the Leuven Lung Transplant Group (www.longtransplantatie.be).

Disclosures

Dirk Van Raemdonck is a consultant for Transmedics, Andover, MA, USA, currently acting as principal investigator of the Expand Registry. He received research support from XVIVO® Perfusion (Vitrolife), Göteborg, Sweden, in the past.

REFERENCES

[1] Vigneswaran WT, Garrity ER, editors. Lung Transplantation. New York, NY: Informa Health Care; 2010.

[2] Hardy JD, Webb WR, Dalton Jr MR, Walker Jr GR. Lung homotransplantation in man. JAMA 1963;186:1065−74.

[3] Wildevuur CR, Benfield JR. A review of 23 human lung transplantations by 20 surgeons. Ann Thorac Surg 1970;9:489−515.

[4] Reitz BA, Wallwork JL, Hunt SA, Pennock JL, Billingham ME, Oyer PE, et al. Heart−lung transplantation: successful therapy for patients with pulmonary vascular disease. N Eng J Med 1982;306:557−64.

[5] Toronto Lung Transplant Group. Unilateral transplantation for pulmonary fibrosis. N Engl J Med 1986;314:1140−5.

[6] Dark JH, Patterson GA, Al-Jilaihawi AN, Hsu H, Egan T, Cooper JD. Experimental en bloc double-lung transplantation. Ann Thorac Surg 1986;42:394−8.

[7] Patterson GA, Cooper JD, Goldman B, Weisel RD, Pearson FG, Waters PF, et al. Technique of successful clinical double lung transplantation. Ann Thorac Surg 1988;45:626−33.

[8] Patterson GA, Cooper JD, Dark JH, Jones MT. Experimental and clinical double lung transplantation. J Thorac Cardiovasc Surg 1988;95:70−4.

[9] Noirclerc MJ, Metras D, Vaillant A, Dumon JF, Zimmermann JM, Caamano A, et al. Bilateral bronchial anastomosis in double

lung and heart−lung transplantations. Eur J Cardiothorac Surg 1990;4:314−7.

[10] Baudet EM, Dromer C, Dubrez J, Jougon JB, Roques X, Velly JF, et al. Intermediate-term results after en bloc double-lung transplantation with bronchial arterial revascularization. Bordeaux lung and heart−lung transplant group. J Thorac Cardiovasc Surg 1996;112:1292−9.

[11] Pettersson G, Nørgaard MA, Arendrup H, Brandenhof P, Helvind M, Joyce F, et al. Direct bronchial artery revascularization and en bloc double lung transplantation—surgical techniques and early outcome. J Heart Lung Transplant 1997;16:320−33.

[12] Mal H, Andreassian B, Pamela F, Duchatelle JP, Rondeau F, Dubois F, et al. Unilateral lung transplantation in end-stage pulmonary emphysema. Am Rev Respir Dis 1989;140:797−802.

[13] Kaiser LR, Cooper JD, Trulock EP, Pasque MK, Triantafillou A, Haydock D. The evolution of single lung transplantation for emphysema. The Washington University Lung Transplant Group. J Thorac Cardiovasc Surg 1991;102:333−9.

[14] Klepetko W, Laufer G, Laczkovics A, Seitelberger R, Müller MR, Wollenek G, et al. Unilateral lung transplantation as an effective therapy in primary lung emphysema. Chirurg 1991;62:271−5.

[15] Van Raemdonck D, Verleden G, Coosemans W, Michiels E, De Leyn P, Buyse B, et al. Isolated lung transplantation; initial experience at the University Hospitals Leuven. Acta Chir Belg 1994;94:245−57.

[16] Cassivi SD, Meyers BF, Battafarano RJ, Guthrie TJ, Trulock EP, Lynch JP, et al. Thirteen-year experience in lung transplantation for emphysema. Ann Thorac Surg 2002;74:1663−9.

[17] Kreisel D, Krupnick AS, Puri V, Guthrie TJ, Trulock E, Meyers BF, et al. Short- and long-term outcome of 1000 adult lung transplant recipients at a single center. J Thorac Cardiovasc Surg 2011;141:215−22.

[18] Pochettino A, Kotloff RM, Rosengard BR, Arcasoy SM, Blumenthal NP, Kaiser LR, et al. Bilateral versus single lung transplantation for chronic obstructive pulmonary disease: intermediate results. Ann Thorac Surg 2000;70:1813−9.

[19] Christie JD, Edwards LB, Kucheryavaya AY, Benden C, Dipchand AI, Dobbels F, , et al.For the International Society for Heart and Lung Transplantation The registry of the International Society for Heart and Lung Transplantation: 29th official adult

lung and heart−lung transplant report—2012. J Heart Lung Transplant 2012;31:1073−86.

[20] Thabut G, Christie JD, Ravaud P, Castier Y, Brugière O, Fournier M, et al. Survival after bilateral versus single lung transplantation for patients with chronic obstructive pulmonary disease: a retrospective analysis of registry data. Lancet 2008;371:744−51.

[21] Bisson A, Bonette P. A new technique for double lung transplantation. "Bilateral single lung" transplantation. J Thorac Cardiovasc Surg 1992;103:40−6.

[22] Pasque MK, Cooper JD, Kaiser LR, Haydock DA, Triantafillou A, Trulock EP. Improved technique for bilateral lung transplantation: rationale and initial clinical experience. Ann Thorac Surg 1990;49:785−91.

[23] Meyers BF, Sundaresan RS, Guthrie T, Cooper JD, Patterson GA. Bilateral sequential lung transplantation without sternal division eliminates posttransplantation sternal complications. J Thorac Cardiovasc Surg 1999;117:358−64.

[24] Aigner C, Winkler G, Jaksch P, Ankersmit J, Marta G, Taghavi S, et al. Size-reduced lung transplantation: an advanced operative strategy to alleviate donor organ shortage. Transplant Proc 2004;36:2801−5.

[25] Starnes VA, Bowdish ME, Woo MS, Barbers RG, Schenkel FA, Horn MV, et al. A decade of living lobar lung transplantation: recipient outcomes. J Thorac Cardiovasc Surg 2004;127:114−22.

[26] Date H. Update on living-donor lobar lung transplantation. Curr Opin Organ Transplant 2011;16:453−7.

[27] Couetil JP, Tolan MJ, Loulmet DF, Guinvarch A, Chevalier PG, Achkar A, et al. Pulmonary bipartitioning and lobar transplantation: a new approach to donor organ shortage. J Thorac Cardiovasc Surg 1997;113:529−37.

[28] Aigner C, Mazhar S, Jaksch P, Seebacher G, Taghavi S, Marta G, et al. Lobar transplantation, split lung transplantation and peripheral segmental resection—reliable procedures for downsizing donor lungs. Eur J Cardiothorac Surg 2004;25:179−83.

[29] Shigemura N, Bermudez C, Hattler BG, Johnson B, Crespo M, Pilewski J, et al. Impact of graft volume reduction for oversized grafts after lung transplantation on outcome in recipients with end-stage restrictive pulmonary diseases. J Heart Lung Transplant 2009;28:130−4.

[30] Weiss ES, Allen JG, Meguid RA, Patel ND, Merlo CA, Orens JB, et al. The impact of center volume on survival in lung transplantation: an analysis of more than 10,000 cases. Ann Thorac Surg 2009;88:1062−70.

[31] Kilic A, George TJ, Beaty CA, Merlo CA, Conte JV, Shah AS. The effect of center volume on the incidence of postoperative complications and their impact on survival after lung transplantation. J Thorac Cardiovasc Surg 2012;144:1502−9.

[32] de Perrot M, Chernenko S, Waddell TK, Shargall Y, Pierre AF, Hutcheon M, et al. Role of lung transplantation in the treatment of bronchogenic carcinoma for patients with end-stage pulmonary disease. J Clin Oncol 2004;22:4351−6.

[33] Lerut JP, Orlando G, Sempoux C, Ciccarelli O, Van Beers BE, Danse E, et al. Hepatic haemangioendothelioma in adults: excellent outcome following liver transplantation. Transpl Int 2004;17:202−7.

[34] Machuca TN, Keshavjee S. Transplantation for lung cancer. Curr Opin Organ Transplant 2012;17:479−84.

[35] Takahashi SM, Garrity ER. The impact of the lung allocation score. Semin Respir Crit Care Med 2010;31:108−14.

[36] Orens JB, Estenne M, Arcasoy S, Conte JV, Corris P, Egan JJ, et al. Pulmonary Scientific Council of the International Society for Heart and Lung Transplantation International guidelines for the selection of lung transplant candidates: 2006 update—a consensus report from the pulmonary scientific council of the International Society for Heart and Lung Transplantation. J Heart Lung Transplant 2006;25:745−55.

[37] Kilic A, Merlo CA, Conte JV, Shah AS. Lung transplantation in patients 70 years old or older: have outcomes changed after implementation of the lung allocation score? J Thorac Cardiovasc Surg 2012;144:1502−9.

[38] Grannas G, Neipp M, Hoeper MM, Gottlieb J, Lück R, Becker T, et al. Indications for and outcomes after combined lung and liver transplantation: a single center experience on 13 consecutive cases. Transplantation 2008;85:524−31.

[39] Curtis HJ, Bourke SJ, Dark JH, Corris PA. Lung transplantation outcome in cystic fibrosis patients with previous pneumothorax. J Heart Lung Transplant 2005;24:865−9.

[40] Rolla M, Anile M, Venuta F, Diso D, Quattrucci S, De Giacomo T, et al. Lung transplantation for cystic fibrosis after thoracic surgical procedures. Transplant Proc 2011;43:1162−3.

[41] Baz MA, Palmer SM, Staples ED, Greer DG, Tapson VF, Davis DD. Lung transplantation after long-term mechanical ventilation: results and 1-year follow-up. Chest 2001;119:224−7.

[42] Mason DP, Thuita L, Nowicki ER, Murthy SC, Pettersson GB, Blackstone EH. Should lung transplantation be performed for patients on mechanical respiratory support? The US experience. J Thorac Cardiovasc Surg 2010;139:765−73.

[43] Gottlieb J, Warnecke G, Hadem J, Dierich M, Wiesner O, Fühner T, et al. Outcome of critically ill lung transplant candidates on invasive respiratory support. Intensive Care Med 2012;38:968−75.

[44] Aigner C, Wisser W, Taghavi S, Lang G, Jaksch P, Czyzewski D, et al. Institutional experience with extracorporeal membrane oxygenation in lung transplantation. Eur J Cardiothorac Surg 2007;31:468−73.

[45] Olsson KM, Simon A, Strueber M, Hadem J, Wiesner O, Gottlieb J, et al. Extracorporeal membrane oxygenation in nonintubated patients as a bridge to lung transplantation. Am J Transplant 2010;10:2173−8.

[46] Shigemura N, Bermudez C, Bhama J, Bonde P, Thacker J, Toyoda Y. Successful lung transplantation after extended use of extracorporeal membrane oxygenation as a bridge. Transplant Proc 2011;43:2063−5.

[47] Javidfar J, Brodie D, Iribarne A, Jurado J, Lavelle M, Brenner K, et al. Extracorporeal membrane oxygenation as a bridge to lung transplantation and recovery. J Thorac Cardiovasc Surg 2012;144:716−21.

[48] Bittner HB, Lehmann S, Rastan A, Garbade J, Binner C, Mohr F, et al. Outcome of extracorporeal membrane oxygenation as a bridge to lung transplantation and graft recovery. Ann Thorac Surg 2012;94:942−9.

[49] Fischer S, Simon AR, Welte T, Hoeper MM, Meyer A, Tessmann R, et al. Bridge to lung transplantation with the novel pumpless interventional lung assist device novalung. J Thorac Cardiovasc Surg 2006;131:719−23.

[50] Strueber M, Hoeper MM, Fischer S, Cypel M, Warnecke G, Gottlieb J, et al. Bridge to thoracic organ transplantation in patients with pulmonary arterial hypertension using a pumpless lung assist device. Am J Transplant 2009;9:853−7.

[51] Cypel M, Keshavjee S. Extracorporeal life support as a bridge to lung transplantation. Clin Chest Med 2011;32:245−51.

[52] Strueber M. Extracorporeal support as a bridge to lung transplantation. Curr Opin Crit Care 2010;16:69−73.

[53] Lang G, Taghavi S, Aigner C, Rényi-Vamos F, Jaksch P, Augustin V, et al. Primary lung transplantation after bridge with extracorporeal membrane oxygenation: a plea for a shift in our paradigms for indications. Transplantation 2012;93:729−36.

[54] Fuehner T, Kuehn C, Hadem J, Wiesner O, Gottlieb J, Tudorache I, et al. Extracorporeal membrane oxygenation in awake patients as bridge to lung transplantation. Am J Respir Crit Care Med 2012;185:763−8.

[55] Shafii AE, Mason DP, Brown CR, Vakil N, Johnston DR, McCurry KR, et al. Growing experience with extracorporeal membrane oxygenation as a bridge to lung transplantation. ASAIO J 2012;58:526−9.

[56] Nosotti M, Rosso L, Tosi D, Palleschi A, Mendogni P, Nataloni IF, et al. Extracorporeal membrane oxygenation with spontaneous breathing as a bridge to lung transplantation. Interact Cardiovasc Thorac Surg 2013;16:55−9.

[57] Schmidt F, Sasse M, Boehne M, Mueller C, Bertram H, Kuehn C, et al. Concept of "awake venovenous extracorporeal membrane oxygenation" in pediatric patients awaiting lung transplantation. Pediatr Transplant 2013;17:224−30.

[58] Egan TM, Murray S, Bustami RT, Shearon TH, McCullough KP, Edwards LB, et al. Development of the new lung allocation system in the United States. Am J Transplant 2006;6:1212−27.

[59] Gries C, Mulligan MS, Edelman JD, Raghu G, Curtis JR, Goss CH. Lung allocation score for lung transplantation: impact on disease severity and survival. Chest 2007;132:1954−61.

[60] Chen H, Shiboski SC, Golden JA, Gould MK, Hays SR, Hoopes CW, et al. The impact of the lung allocation score on lung transplantation for pulmonary arterial hypertension. Am J Respir Crit Care Med 2009;18:468−74.

[61] Smits JM, Nossent GD, de Vries E, Rahmel A, Meiser B, Strueber M, et al. Evaluation of the lung allocation score in highly urgent and urgent lung transplant candidates in Eurotransplant. J Heart Lung Transplant 2011;30:22−8.

[62] Stoica SC, McNeil KD, Perreas K, Sharples LD, Satchithananda DK, Tsui SS, et al. Heart−lung transplantation for Eisenmenger syndrome: early and long-term results. Ann Thorac Surg 2001;72:1887−91.

[63] Dennis C, Caine N, Sharples L, Smyth R, Higenbottam T, Stewart S, et al. Heart−lung transplantation for end-stage respiratory disease in patients with cystic fibrosis at Papworth Hospital. J Heart Lung Transplant 1993;12:893−902.

[64] Pigula FA, Gandhi SK, Ristich J, Stukus D, McCurry K, Webber SA, et al. Cardiopulmonary transplantation for congenital heart diseases in the adult. J Heart Lung Transplant 2001;20:297−303.

[65] Choong CK, Sweet SC, Guthrie TJ, Mendeloff EN, Haddad FJ, Schuler P, et al. Repair of congenital heart lesions combined with lung transplantation for the treatment of severe pulmonary hypertension: a 13-year experience. J Thorac Cardiovasc Surg 2005;129:661−9.

[66] Toyoda Y, Thacker J, Santos R, Nguyen D, Bhama J, Bermudez C, et al. Long-term outcome of lung and heart−lung transplantation for idiopathic pulmonary arterial hypertension. Ann Thorac Surg 2008;86:1116−22.

[67] Fadel E, Mercier O, Mussot S, Leroy-Ladurie F, Cerrina J, Chapelier A, et al. Long-term outcome of double-lung and heart−lung transplantation for pulmonary hypertension: a comparative retrospective study of 219 patients. Eur J Cardiothorac Surg 2010;38:277−84.

[68] Inci I, Stanimirov O, Benden C, Kestenholz P, Hofer M, Boehler A, et al. Lung transplantation for cystic fibrosis: a single center experience of 100 consecutive cases. Eur J Cardiothorac Surg 2012;41:435−40.

[69] Hayes Jr D, Meyer KC. Lung transplantation for advanced bronchiectasis. Semin Respir Crit Care Med 2010;31:123−38.

[70] Le Pimpec-Barthes F, Thomas PA, Bonnette P, Mussot S, DeFrancquen P, Hernigou A, et al. Single-lung transplantation in patients with previous contralateral pneumonectomy: technical aspect and results. Eur J Cardiothorac Surg 2009;36:927−32.

[71] Weiss ES, Allen JG, Merlo CA, Conte JV, Shah AS. Survival after single versus bilateral lung transplantation for high-risk patients with pulmonary fibrosis. Ann Thorac Surg 2009;88:1616−25.

[72] Meyers BF, Lynch JP, Trulock EP, Guthrie P, Cooper JD, Patterson GA. Single versus bilateral lung transplantation for idiopathic pulmonary fibrosis: a ten-year institutional experience. J Thorac Cardiovasc Surg 2000;120:99−107.

[73] Mason DP, Brizzio ME, Alster JM, McNeill AM, Murthy SC, Budev MM, et al. Lung transplantation for idiopathic pulmonary fibrosis. Ann Thorac Surg 2007;84:1121−8.

[74] Rinaldi M, Sansone F, Boffini M, El Qarra S, Solidoro P, Cavallo N, et al. Single versus double lung transplantation in pulmonary fibrosis: a debated topic. Transplant Proc 2008;40:2010−2.

[75] Thabut G, Christie JD, Ravaud P, Castier Y, Dauriat G, Jebrak G, et al. Survival after bilateral versus single-lung transplantation for idiopathic pulmonary fibrosis. Ann Intern Med 2009;151:767−74.

[76] King CS, Khandar S, Burton N, Shlobin OA, Ahmad S, Lefrak E, et al. Native lung complications in single-lung recipients and the role of pneumonectomy. J Heart Lung Transplant 2009;28:851−6.

[77] Minai OA, Shah S, Mazzone P, Budey MM, Sahoo D, Murthy S, et al. Bronchogenic carcinoma after lung transplantation: characteristics and outcomes. J Thorac Oncol 2008;3:1404−9.

[78] Hadjiliadis D, Chaparro C, Gutierrez C, Steele MP, Singer LG, Davis RD, et al. Impact of lung transplant operation on bronchiolitis obliterans syndrome in patients with chronic obstructive pulmonary disease. Am J Transplant 2006;6:183−9.

[79] Tsagkaropoulos S, Belmans A, Verleden GM, Coosemans W, Decaluwe H, De Leyn P, et al. Single lung transplantation: does side matter? Eur J Cardiothorac Surg 2011;40:83−92.

[80] Keating DT, Marasco SF, Negri J, Esmore D, Burton JH, Griffiths AP, et al. Long-term outcomes of cadaveric lobar lung transplantation: helping to maximize resources. J Heart Lung Transplant 2010;29:439−44.

[81] Inci I, Schuurmans MM, Kestenholz P, Schneiter D, Hillinger S, Opitz I, et al. Long-term outcomes of bilateral lobar lung transplantation. Eur J Cardiothorac Surg 2013;43:1220−5.

[82] Loizzi D, Aigner C, Jaksch P, Scheed A, Mora B, Sollitto F, et al. A scale for decision making between whole lung transplantation

or lobar transplantation. Eur J Cardiothorac Surg 2010;37:1122−5.

[83] Mason DP, Batizy LH, Wu J, Nowicki ER, Murthy SC, McNeill AM, et al. Matching donor to recipient in lung transplantation: How much does size matter? J Thorac Cardiovasc Surg 2009;137:1234−40.

[84] Chen F, Kubo T, Shoji T, Fujinaga T, Bando T, Date H. Comparison of pulmonary function test and computed tomography volumetry in living lung donors. J Heart Lung Transplant 2011;30:572−5.

[85] Detry O, Le Dinh H, Noterdaeme T, De Roover A, Honoré P, Squifflet JP, et al. Categories of donation after cardiocirculatory death. Transplant Proc 2012;44:1189−95.

[86] Van Raemdonck DEM, Rega FR, Neyrinck AP, Jannis N, Verleden GM, Lerut T. Non-heart-beating donors. Semin Thorac Cardiovasc Surg 2004;16:309−21.

[87] Kootstra G, Daemen JH, Oomen AP. Categories of non-heart-beating donors. Transplant Proc 1995;27:2893−4.

[88] Ysebaert D, Van Beeumen G, De Greef K, Squifflet JP, Detry O, De Roover A, et al. Organ procurement after euthanasia: Belgian experience. Transplant Proc 2009;41:585−6.

[89] Van Raemdonck D, Verleden GM, Dupont L, Ysebaert D, Monbaliu D, Neyrinck A, et al. Initial experience with transplantation of lungs recovered from donors after euthanasia. Appl Cardiopulm Pathophysiol 2011;15:38−48.

[90] Wigfield CH, Love RB. Donation after cardiac death lung transplantation outcomes. Curr Opin Organ Transplant 2011;16:462−8.

[91] Van De Wauwer C, Verschuuren EA, van der Bij W, Nossent GD, Erasmus ME. The use of non-heart-beating lung donors category III can increase the donor pool. Eur J Cardiothorac Surg 2011;39:175−80.

[92] De Vleeschauwer SI, Wauters S, Dupont LJ, Verleden SE, Willems-Widyastuti A, Vanaudenaerde BM, et al. Medium-term outcome after lung transplantation is comparable between brain-dead and cardiac-dead donors. J Heart Lung Transplant 2011;30:975−81.

[93] Mason DP, Brown CR, Murthy SC, Vakil N, Lyon C, Budev MM, et al. Growing single-center experience with lung transplantation using donation after cardiac death. Ann Thorac Surg 2012;94:406−12.

[94] Zych B, Popov AF, Amrani M, Bahrami T, Redmond KC, Krueger H, et al. Lungs from donation after circulatory death donors: an alternative source to brain-dead donors? Midterm results at a single institution. Eur J Cardiothorac Surg 2012;42:542−9.

[95] Puri V, Scavuzzo M, Guthrie T, Hachem R, Krupnick AS, Kreisel D, et al. Lung transplantation and donation after cardiac death: a single center experience. Ann Thorac Surg 2009;88:1609−15.

[96] De Oliveira NC, Osaki S, Maloney JD, Meyer KC, Kohmoto T, D'Alessandro AM, et al. Lung transplantation with donation after cardiac death donors: long-term follow-up in a single center. J Thorac Cardiovasc Surg 2010;139:1306−15.

[97] Gomez-de-Antonio D, Campo-Canaveral JL, Crowley S, Valdivia D, Cordoba M, Moradiellos J, et al. Clinical lung transplantation from uncontrolled non-heart-beating donors revisited. J Heart Lung Transplant 2012;31:349−53.

[98] Steen S, Sjöberg T, Pierre L, Liao Q, Eriksson L, Algotsson L. Transplantation of lungs from a non-heart-beating donor. Lancet 2001;357(9259):825−9.

[99] Moradiellos J, Naranjo M, Cordoba M, Salas C, Gōmez D, Campo-Caňaveral JL, et al. Clinical lung transplantation after ex vivo evaluation of uncontrolled non heart-beating donors lungs: initial experience. J Heart Lung Transplant 2011;30 (Suppl. 4):S38.

[100] Cypel M, Yeung JC, Machuca T, Chen M, Singer LG, Yasufuku K, et al. Experience with the first 50 ex vivo lung perfusions in clinical transplantation. J Thorac Cardiovasc Surg 2012;144:1200−6.

[101] Date H, Yamane M, Toyooka S, Okazaki M, Aoe M, Sano Y. Current status and potential of living-donor lobar lung transplantation. Front Biosci 2008;13:1433−9.

[102] Date H, Shiraishi T, Sugimoto S, Shoji T, Chen F, Hiratsuka M, et al. Outcome of living-donor lobar lung transplantation using a single donor. J Thorac Cardiovasc Surg 2012;144:710−5.

[103] Van Raemdonck DEM, Verleden GM, Coosemans W, Decaluwé H, Decker G, De Leyn P, et al. Increasing the donor pool. Eur Respir Mon 2009;45:104−27.

[104] Aigner C, Seebacher G, Klepetko W. Lung transplantation. Donor selection. Chest Surg Clin N Am 2003;13:429−42.

[105] Van Raemdonck D, Neyrinck A, Verleden GM, Dupont L, Coosemans W, Decaluwé H, et al. Donor selection and management. Proc Am Thorac Soc 2009;6:28−38.

[106] Snell G, Westall GP. Selection and management of the lung donor. Clin Chest Med 2011;32:223−32.

[107] Orens JB, Boehler A, de Perrot M, Estenne M, Glanville AR, Keshavjee S, , et al.Pulmonary Council, International Society for Heart and Lung Transplantation A review of lung transplant donor acceptability criteria. J Heart Lung Transplant 2003;22:1183−200.

[108] Bonser RS, Taylor R, Collett D, Thomas HL, Dark JH, Neuberger J, Cardiothoracic Advisory Group to NHS Blood and Transplant and the Association of Lung Transplant Physicians (UK). Effect of donor smoking on survival after lung transplantation: a cohort study of a prospective registry. Lancet 2012;380:747−55.

[109] Strüber M, Warnecke G, Hafer C, Goudeva L, Fegbeutel C, Fischer S, et al. Intentional ABO-incompatible lung transplantation. Am J Transplant 2008;8:2476−8.

[110] Shoji T, Bando T, Fujinaga T, Chen F, Yurugi K, Maekawa T, et al. ABO-incompatible living-donor lobar lung transplantation. J Heart Lung Transplant 2011;30:479−80.

[111] Grasemann H, de Perrot M, Bendiak GN, Cox P, van Arsdell GS, Keshavjee S, et al. ABO-incompatible lung transplantation in an infant. Am J Transplant 2012;12:779−81.

[112] Yeung JC, Cypel M, Waddell TK, Van Raemdonck D, Keshavjee S. Update on donor assessment, resuscitation, and acceptance criteria, including novel techniques—non-heart-beating donor lung retrieval and ex vivo donor lung perfusion. Thorac Surg Clin 2009;19:261−74.

[113] Aigner C, Winkler G, Jaksch P, Seebacher G, Lang G, Taghavi S, et al. Extended donor criteria for lung transplantation—a clinical reality. Eur J Cardiothorac Surg 2005;27:757−61.

[114] Meers C, Van Raemdonck D, Verleden GM, Coosemans W, Decaluwé H, De Leyn P, et al. The number of lung transplants can be safely doubled using extended criteria donors; a single-center review. Transplant Int 2010;23:628–35.

[115] Schiavon M, Falcoz PE, Santelmo N, Massard G. Does the use of extended criteria donors influence early and long-term results of lung transplantation? Interact Cardiovasc Thorac Surg 2012;14:183–7.

[116] Smits J, van der Bij W, Van Raemdonck D, de Vries E, Rahmel A, Laufer G, et al. Defining an extended criteria donor lung: an empirical approach based on the Eurotransplant experience. Transpl Int 2011;24:393–400.

[117] Rostron AJ, Avlonitis VS, Cork DMW, Grenade DS, Kirby JA, Dark JH. Hemodynamic resuscitation with arginine vasopressin reduces lung injury following brain death in the transplant donor. Transplantation 2008;85:597–606.

[118] Follette DM, Rudich SM, Babcock WD. Improved oxygenation and increased lung donor recovery with high-dose steroid administration after brain death. J Heart Lung Transplant 1998;17:423–9.

[119] Rosendale JD, Kauffman HM, McBride MA, Chabelewski FL, Zaroff JG, Garrity ER, et al. Aggressive pharmacologic donor management results in more transplanted organs. Transplantation 2003;75:482–7.

[120] Venkateswaran RV, Patchell VB, Wilson IC, Mascaro JG, Thompson RD, Quinn DW, et al. Early donor management increases the retrieval rate of lungs for transplantation. Ann Thorac Surg 2008;85:278–86.

[121] Ware LB, Fang X, Wang Y, Sakuma T, Hall TS, Matthay MA. Selected contribution: mechanisms that may stimulate the resolution of alveolar edema in the transplanted human lung. J Appl Physiol 2002;93:1869–74.

[122] Ware LB, Landeck M, Koyama T, Johnson E, Bernard GR, Lee JW, et al. A randomized trial of nebulized albuterol to enhance resolution of pulmonary edema in 506 brain dead organ donors. J Heart Lung Transplant 2012;31(Suppl.):S116 [abstract]

[123] Mascia L, Pasero D, Slutsky AS, Arguis MJ, Berardino M, Grasso S, et al. Effect of a lung protective strategy for organ donors on eligibility and availability of lungs for transplantation: a randomized controlled trial. JAMA 2010;304:2620–7.

[124] Wauters S, Verleden GM, Belmans A, Coosemans W, De Leyn P, Nafteux P, et al. Donor cause of brain death and related time intervals: does it affect outcome after lung transplantation? Eur J Cardiothorac Surg 2011;39:68–76.

[125] Van Raemdonck D. Thoracic organs: current preservation technology and future prospects; part 1: lung. Curr Opin Organ Transplant 2010;15:150–5.

[126] Keshavjee S, Yamazaki F, Yokomise H, Cardoso PF, Mullen JB, Slutsky AS, et al. The role of dextran 40 and potassium in extended hypothermic lung preservation for transplantation. J Thorac Cardiovasc Surg 1992;103:314–25.

[127] Thabut G, Vinatier I, Brugière O, Lesèche G, Loirat P, Bisson A, et al. Influence of preservation solution on early graft failure in clinical lung transplantation. Am J Respir Crit Care Med 2001;164:1204–8.

[128] Marasco SF, Bailey M, McGlade D, Snell G, Westall G, Oto T, et al. Effect of donor preservation solution and survival in lung transplantation. J Heart Lung Transplant 2011;30:414–9.

[129] Steen S, Kimblad PO, Sjöberg T, Lindberg L, Ingemansson R, Massa G. Safe lung preservation for twenty-four hours with Perfadex. Ann Thorac Surg 1994;57:450–7.

[130] Varela A, Cordoba M, Serrano-Fiz S, Burgos R, Montero CG, Téllez G, et al. Early lung allograft function after retrograde and anterograde preservation. J Thorac Cardiovasc Surg 1997;114:1119–20.

[131] Venuta F, Rendina EA, Bufi M, Della Rocca G, De Giacomo T, Costa MG, et al. Preimplantation retrograde pneumoplegia in clinical lung transplantation. J Thorac Cardiovasc Surg 1999;118:107–14.

[132] Steen S, Liao Q, Wierup PN, Bolys R, Pierre L, Sjöberg T. Transplantation of lungs from non-heart-beating donors after functional assessment *ex vivo*. Ann Thorac Surg 2003;76:244–52.

[133] Cypel M, Yeung JC, Hirayama S, Rubacha M, Fischer S, Anraku M, et al. Technique for prolonged normothermic *ex vivo* lung perfusion. J Heart Lung Transplant 2008;27:1319–25.

[134] Cypel M, Rubacha M, Yeung J, Hirayama S, Torbicki K, Madonik M, et al. Normothermic *ex vivo* perfusion prevents lung injury compared to extended cold preservation for transplantation. Am J Transplant 2009;9:2262–9.

[135] Warnecke G, Moradiellos J, Tudorache I, Kühn C, Avsar M, Wiegmann B, et al. Normothermic perfusion of donor lungs for preservation and assessment with the organ care system lung before bilateral transplantation: a pilot study of 12 patients. Lancet 2012;380:1851–8.

[136] Warnecke G, Weigmann B, Van Raemdonck D, Massard G, Santelmo N, Falcoz P-E, et al. The INSPIRE international lung trial with the organ care system technology (OCSTM). J Heart Lung Transplant 2013;32(Suppl.):S16 [abstract].

[137] Ingemansson R, Eyjolfsson A, Mared L, Pierre L, Algotsson L, Ekmehag B, et al. Clinical transplantation of initially rejected donor lungs after reconditioning *ex vivo*. Ann Thorac Surg 2009;87:255–60.

[138] Cypel M, Yeung JC, Liu M, Anraku M, Chen F, Karolak W, et al. Normothermic *ex vivo* lung perfusion in clinical lung transplantation. N Engl J Med 2011;364:1431–40.

[139] Zych B, Popov AF, Stavri G, Bashford A, Bahrami T, Amrani M, et al. Early outcomes of bilateral sequential single lung transplantation after *ex-vivo* lung evaluation and reconditioning. J Heart Lung Transplant 2012;31:274–81.

[140] Aigner C, Slama A, Hötzenecker K, Scheed A, Urbanek B, Schmid W, et al. Clinical *ex vivo* lung perfusion—pushing the limits. Am J Transplant 2012;12:1839–47.

[141] Wallinder A, Ricksten SE, Hansson C, Riise GC, Silverborn M, Liden H, et al. Transplantation of initially rejected donor lungs after *ex vivo* lung perfusion. J Thorac Cardiovasc Surg 2012;144:1222–8.

[142] Van Raemdonck D, Neyrinck A, Rega F, Devos T, Pirenne J. Machine perfusion in organ transplantation: a tool for *ex-vivo* graft conditioning with mesenchymal stem cells? Curr Opin Organ Transplant 2013;18(1):24–33.

[143] Thabut G, Fournier M. Assessing survival benefits from lung transplantation. Rev Mal Respir 2011;28:1–6.

[144] Kotloff RM, Thabut G. Lung transplantation. Am J Respir Crit Care Med 2011;184:159–71.

[145] Hosenpud JD, Bennett LE, Keck BM, Edwards EB, Novick RJ. Effect of diagnosis on survival benefit of lung transplantation for end-stage lung disease. Lancet 1998;351:24−7.

[146] De Meester J, Smits JM, Persijn GG, Haverich A. Listing for lung transplantation, stratified by type of end-stage lung disease, the Eurotransplant experience. J Heart Lung Transplant 2001;20:518−24.

[147] Charman SC, Sharples LD, McNeil KD, Wallwork J. Assessment of survival benefit after lung transplantation by patient diagnosis. J Heart Lung Transplant 2002;21:226−32.

[148] Titman A, Rogers CA, Bonser RS, Banner NR, Sharples LD. Disease-specific survival benefit of lung transplantation in adults: a national cohort study. Am J Transplant 2009;9:1640−9.

[149] Thabut G, Mal H, Castier Y, Groussard O, Brugiere O, Marrash-Chahla R, et al. Survival benefit of lung transplantation for patients with idiopathic pulmonary fibrosis. J Thorac Cardiovasc Surg 2003;126:469−75.

[150] Aurora P, Spencer H, Moreno-Galdo A. Lung transplantation and life extension in children with cystic fibrosis. Lancet 1999;354:1591−3.

[151] Liou TG, Adler FR, Cahill BC, FitzSimmons SC, Huang D, Hibbs JR, et al. Survival effect of lung transplantation among patients with cystic fibrosis. JAMA 2001;286:2683−9.

[152] Liou TG, Adler FR, Huang D. Use of lung transplantation survival models to refine patient selection in cystic fibrosis. Am J Respir Crit Care Med 2005;171:1053−9.

[153] Liou TG, Adler FR, Cox DR, Cahill BC. Lung transplantation and survival in children with cystic fibrosis. N Eng J Med 2007;357:2143−52.

[154] Sweet SC, Aurora P, Benden C, Wong JY, Goldfarb SB, Elidemir O, et al.International Pediatric Lung Transplant Collaborative Lung transplantation and survival in children with cystic fibrosis: solid statistics—flawed interpretation. Pediatr Transplant 2008;12:129−36.

[155] Aurora P, Spencer H, Moreno-Galdo A. Lung transplantation in children with cystic fibrosis: a view from Europe. Am J Respir Crit Care Med 2008;177:935−6.

[156] Adler FR, Aurora P, Barker DH, Barr ML, Blackwell LS, Bosma OH, et al. Lung transplantation for cystic fibrosis. Proc Am Thorac Soc 2009;6:619−33.

[157] Stavem K, Bjortuft O, Borgan O, Geiran O, Boe J. Lung transplantation in patients with chronic obstructive pulmonary disease in a national cohort is without obvious survival benefit. J Heart Lung Transplant 2006;25:75−84.

[158] Geertsma A, Ten Vergert EM, Bonsel GJ, de Boer WJ, van der Bij W. Does lung transplantation prolong life? A comparison of survival with and without transplantation. J Heart Lung Transplant 1998;17:511−6.

[159] Thabut G, Ravaud P, Christie JD, Castier Y, Fournier M, Mal H, et al. Determinants of the survival benefit of lung transplantation in patients with chronic obstructive pulmonary disease. Am J Respir Crit Care Med 2008;177:1156−63.

[160] Lahzami S, Bridevaux PO, Soccal PM, Wellinger J, Robert JH, Ris HB, et al. Survival impact of lung transplantation for COPD. Eur Respir J 2010;36:74−80.

[161] Celli BR, Cote CG, Marin JM, Casanova C, Montes de Oca M, Mendez RA, et al. The body mass index, airflow obstruction, dyspnea, and exercise capacity index in chronic obstructive pulmonary disease. N Engl J Med 2004;350:1005−12.

[162] Kugler C, Fischer S, Gottlieb J, Welte T, Simon A, Haverich A, et al. Health-related quality of life in two hundred-eighty lung transplant recipients. J Heart Lung Transplant 2005;24:2262−8.

[163] Choong CK, Meyers BF. Quality of life after transplantation. Thorac Surg Clin 2004;14:385−407.

[164] Snyder LD, Palmer SM. Quality, quantity, or both? Life after lung transplantation. Chest 2005;128:1086−7.

[165] Ius F, Kuehn C, Tudorache I, Sommer W, Avsar M, Boethig D, et al. Lung transplantation on cardiopulmonary support: venoarterial extracorporeal membrane oxygenation outperformed cardiopulmonary bypass. J Thorac Cardiovasc Surg 2012;144:1510−6.

[166] de Perrot M, Liu M, Waddell TK, Keshavjee S. Ischemia-reperfusion induced lung injury. Am J Respir Crit Care Med 2003;167:490−511.

[167] de Perrot M, Bonser RS, Dark J, Kelly RF, McGiffin D, Menza R, , et al.ISHLT Working Group on Primary Lung Graft Dysfunction Report of the ISHLT working group on primary lung graft dysfunction part III: donor-related risk factors and markers. J Heart Lung Transplant 2005;24:1460−7.

[168] Barr ML, Kawut SM, Whelan TP, Girgis R, Böttcher H, Sonett J, et al.ISHLT Working Group on Primary Lung Graft Dysfunction Report of the ISHLT working group on primary lung graft dysfunction part IV: recipient-related risk factors and markers. J Heart Lung Transplant 2005;24:1468−82.

[169] Thabut G, Mal H, Cerrina J, Dartevelle P, Dromer C, Velly JF, et al. Graft ischemic time and outcome of lung transplantation: a multicenter analysis. Am J Respir Crit Care Med 2005;171:786−91.

[170] Christie JD, Carby M, Bag R, Corris P, Hertz M, Weill D. Report of the ISHLT working group on primary lung graft dysfunction part II: definition. A consensus statement of the International Society for Heart and Lung Transplantation. J Heart Lung Transplant 2005;24:1454−9.

[171] Whitson BA, Prekker ME, Herrington CS, Whelan TP, Radosevich DM, Hertz MI, et al. Primary graft dysfunction and long-term pulmonary function after lung transplantation. J Heart Lung Transplant 2007;26:1004−11.

[172] Daud SA, Yusen RD, Meyers BF, Chakinala MM, Walter MJ, Aloush AA, et al. Impact of immediate primary lung allograft dysfunction on bronchiolitis obliterans syndrome. Am J Respir Crit Care Med 2007;175:507−13.

[173] Pereszlenyi A, Lang G, Steltzer H, Hetz H, Kocher A, Neuhauser P, et al. Bilateral lung transplantation with intra- and postoperatively prolonged ECMO support in patients with pulmonary hypertension. Eur J Cardiothorac Surg 2002;21:858−63.

[174] Shargall Y, Guenther G, Ahya VN, Ardehali A, Singhal A, Keshavjee S, ISHLT working group on primary graft dysfunction. Report of the ISHLT working group on primary lung graft dysfunction part IV: treatment. J Heart Lung Transplant 2005;24:1489−500.

[175] Aigner C, Jaksch P, Taghavi S, Lang G, Reza-Hoda MA, Wisser W, et al. Pulmonary retransplantation: is it worth the effort? A long-term analysis of 46 cases. J Heart Lung Transplant 2008;27:60−5.

[176] Meyers BF, de la Morena M, Sweet SC, Trulock EP, Guthrie TJ, Mendeloff EF, et al. Primary graft dysfunction and other selected complications of lung transplantation: a single-center experience of 983 patients. J Thorac Cardiovasc Surg 2005;129:1421—9.

[177] Van De Wauwer C, Van Raemdonck D, Verleden GM, Dupont L, De Leyn P, Coosemans W, et al. Risk factors for airway complications within the first year after lung transplantation. Eur J Cardiothorac Surg 2007;31:703—10.

[178] Weder W, Inci I, Korom S, Kestenholz PB, Hillinger S, Eich C, et al. Airway complications after lung transplantation: risk factors, prevention and outcome. Eur J Cardiothorac Surg 2009;35:293—8.

[179] Aigner C, Jaksch P, Seebacher G, Neuhauser P, Marta G, Wisser W, et al. Single running suture—the new standard technique for bronchial anastomoses in lung transplantation. Eur J Cardiothorac Surg 2003;23:488—93.

[180] Fitzsullivan E, Gries CJ, Phelan P, Farjah F, Gilbert E, Keech JC, et al. Reduction in airway complications after lung transplantation with novel anastomotic technique. Ann Thorac Surg 2011;92:309—15.

[181] Nunley DR, Gal AA, Vega JD, Perlino C, Smith P, Lawrence EC. Saprophytic fungal infections and complications involving the bronchial anastomosis following lung transplantation. Chest 2002;122:1185—91.

[182] Knight J, Elwing JM, Milstone A. Bronchovascular fistula formation: a rare airway complication after lung transplantation. J Heart Lung Transplant 2008;27:1179—85.

[183] Rea F, Marulli G, Loy M, Bortolotti L, Giacometti C, Schiavon M, et al. Salvage right pneumonectomy in a patient with bronchial-pulmonary artery fistula after bilateral sequential lung transplantation. J Heart Lung Transplant 2006;25:1383—6.

[184] Mughal MM, Gildea TR, Murthy S, Petterson G, DeCamp M, Mehta AS. Short-term deployment of self-expanding metallic stents facilitates healing of bronchial dehiscence. Am J Respir Crit Care Med 2005;172:768—71.

[185] Gottlieb J, Fuehner T, Dierich M, Wiesner O, Simon AR, Welte T. Are metallic stents really safe? A long-term analysis in lung transplant recipients. Eur Respir J 2009;34:1417—22.

[186] Schäfers HJ, Schäfer CM, Zink C, Haverich A, Borst HG. Surgical treatment of airway complications after lung transplantation. J Thorac Cardiovasc Surg 1994;107:1476—80.

[187] Griffith BP, Magee MJ, Gonzalez IF, Houel R, Armitage JM, Hardesty RL, et al. Anastomotic pitfalls in lung transplantation. J Thorac Cardiovasc Surg 1994;107:743—53.

[188] Clark SC, Levine AJ, Hasan A, Hilton CJ, Forty J, Dark JH. Vascular complications of lung transplantation. Ann Thorac Surg 1996;61:1079—82.

[189] Schulman LL, Anandarangam T, Leibowitz DW, Ditullio MR, McGregor CC, Galantowicz ME, et al. Four-year prospective study of pulmonary venous thrombosis after lung transplantation. J Am Soc Echocardiogr 2001;14:806—12.

[190] Banerjee SK, Santhanakrishnan K, Shapiro L, Dunning J, Tsui S, Parmar J. Successful stenting of anastomotic stenosis of the left pulmonary artery after single lung transplantation. Eur Respir Rev 2011;20:59—62.

[191] Waurick PE, Kleber FX, Ewert R, Pfitzmann R, Bruch L, Hummel L, et al. Pulmonary artery stenosis 5 years after single lung transplantation in primary pulmonary hypertension. J Heart Lung Transplant 1999;18:1243—5.

[192] Holloway B, Mukadam M, Thompson R, Bonser R. Cardiac herniation and lung torsion following heart and lung transplantation. Interact Cardiovasc Thorac Surg 2010;10:1044—6.

[193] Grazia TJ, Hodges TN, Cleveland Jr JC, Sheridan BC, Zamora MR. Lobar torsion complicating bilateral lung transplantation. J Heart Lung Transplant 2003;22:102—6.

[194] Shakoor H, Murthy S, Mason D, Johnston D, Shah SS, Carrillo MC, et al. Lobar torsion after lung transplantation—a case report and review of the literature. Artif Organs 2009;33:551—4.

[195] Nguyen JC, Manoley J, Kanne JP. Bilateral whole-lung torsion after bilateral lung transplantation. J Thorac Imaging 2011;26:17—9.

[196] Maziak DE, Maurer JR, Kesten S. Diaphragmatic paralysis: a complication of lung transplantation. Ann Thorac Surg 1996;61:170—3.

[197] Ferdinande P, Bruyninckx F, Van Raemdonck D, Daenen W, Verleden G, Leuven Lung Transplant Group. Phrenic nerve dysfunction after heart—lung and lung transplantation. J Heart Lung Transplant 2004;23:105—9.

[198] Au J, Hawkins T, Venables C, Morritt G, Scott CG, Gascoigne AD, et al. Upper gastrointestinal dysmotility in heart—lung transplant recipients. Ann Thorac Surg 1993;55:94—7.

[199] Paul S, Escareno CE, Clancy K, Jaklitsch MT, Bueno R, Lautz DB. Gastrointestinal complications after lung transplantation. J Heart Lung Transplant 2009;28:475 9.

[200] Morrell MR, Patterson GA, Trulock EP, Hachem RR. Acute antibody-mediated rejection after lung transplantation. J Heart Lung Transplant 2009;28:96—100.

[201] Glanville AR. Antibody-mediated rejection in lung transplantation: myth or reality? J Heart Lung Transplant 2010;29:395—400.

[202] Robbins HY, Arcasoy SM. Malignancies following lung transplantation. Clin Chest Med 2011;32:343—55.

[203] Estenne M, Maurer JR, Boehler A, Egan JJ, Frost A, Hertz M, et al. Bronchiolitis obliterans syndrome 2001: an update of the diagnostic criteria. J Heart Lung Transplant 2002;21:297—310.

[204] Verleden GM, Vos R, Verleden SE, De Wever W, De Vleeschauwer SI, Willems-Widyastuti A, et al. Survival determinants in lung transplant patients with chronic allograft dysfunction. Transplantation 2011;92:703—8.

[205] Vos R, Vanaudenaerde BM, Verleden SE, De Vleeschauwer SI, Willems-Widyastuti A, Van Raemdonck DE, et al. A randomized placebo-controlled trial of azithromycin to prevent bronchiolitis obliterans syndrome after lung transplantation. Eur Respir J 2011;37:164—72.

[206] Sato M, Waddell TK, Wagnetz U, Roberts HC, Hwang DM, Haroon A, et al. Restrictive allograft syndrome (RAS): a novel form of chronic lung allograft dysfunction. J Heart Lung Transplant 2011;30:735—42.

Additional Reading

Levvey BJ, Harkess M, Hopkins P, Chambers D, Merry C, Glanville AR, et al. Excellent clinical outcomes from a national donation-after-determination-of-cardiac-death lung transplant collaborative. Am J Transplant 2012;12:2406—13.

Minai OA, Shah S, Mazzone P, Budey MM, Sahoo D, Murthy S, et al. Bronchogenic carcinoma after lung transplantation: characteristics and outcomes. J Thorac Oncol 2008;3:1404—9.

Chapter 46

Living Related Lung Transplantation

Progress, Pitfalls, and Promise

Hiroshi Date

Department of Thoracic Surgery, Kyoto University Graduate School of Medicine, Kyoto, Japan

Chapter Outline

46.1 HISTORY AND CONCEPT

Living-donor lobar lung transplantation (LDLLT) was introduced by Starnes and his colleagues as an alternative form of treatment for patients who have a decline in physical condition with a limited life expectancy. The first patient was an 11-year-old girl with bronchopulmonary dysplasia [1]. In 1990, she underwent a right single lobe transplantation using the right upper lobe of her mother and survived. The second patient was a 3-year-old girl with Eisenmenger's syndrome. She underwent a right single lobe transplantation using the middle lobe of her father plus ventricular septal defect closure but died of primary graft dysfunction which developed immediately after reperfusion. After the unsuccessful result in the second case, they developed bilateral LDLLT in which two healthy donors donate their right or left lower lobes (Figure 46.1) [2,3]. Since then, bilateral LDLLT has been performed as a lifesaving procedure to deal with the shortage of cadaveric donors. Because only two lobes are transplanted, LDLLT seems to be best suited for children and small adults, and was applied most exclusively to cystic fibrosis in the beginning [3]. However, it is now well known that LDLLT can be applied to restrictive, obstructive, infectious, and hypertensive lung diseases for both pediatric and adult patients when the size matching is acceptable [4,5].

As of the end of 2012, LDLLT has been performed in approximately 400 patients worldwide. Although LDLLT began in the United States, it has decreased in the United States because of the recent change by the Organ Procurement and Transplantation Network to an urgency/benefit allocation system for cadaveric donor lungs. During the past several years, reports on LDLLT have been most exclusively from Japan where the average waiting time for a cadaveric lung is more than 2 years [6]. Other than Japanese experience, small practices in LDLLT have been reported from Brazil [7] and China [8]. The results of bilateral LDLLT have been equal to or better than conventional cadaveric lung transplantation (CLT).

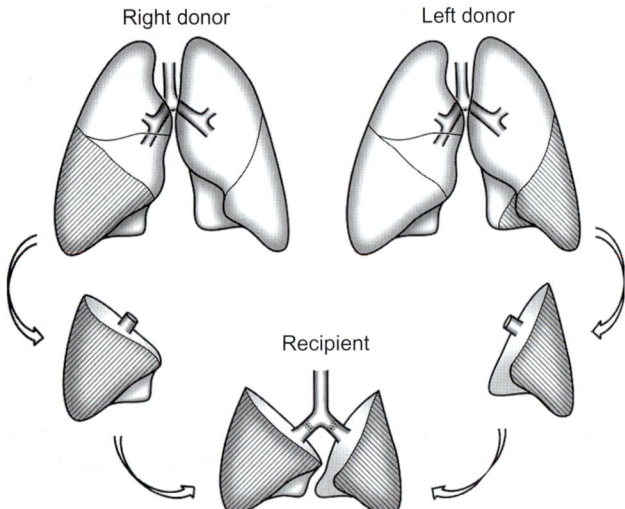

FIGURE 46.1 Bilateral LDLLT. Right and left lower lobes from two healthy donors are implanted in a recipient in place of whole right and left lungs, respectively.

46.2 PATIENT SELECTION

The candidate for LDLLT should be <65 years old with progressive lung disease. All recipients should fulfill the criteria for conventional CLT. Because of possible serious complications in the donor lobectomy, LDLLT should be indicated only for critically ill patients who are unlikely to survive the long wait for cadaveric lungs. On the other hand, when the recipient is too sick, it would not be justified to perform two lobectomies from two healthy donors. In our LDLLT experience, all patients were oxygen dependent and 62% of them were bed bound and 11% of them were on a ventilator at the time of transplantation. Controversy exists if LDLLT can be applied to patients already on a ventilator or requiring retransplantation. St. Louis group reported that LDLLT provided better survival than conventional CLT for retransplantation [9]. Perioperative mortality of retransplantation was only 7.7% in the patients who had LDLLT versus 42.3% in the CLT group. Okayama [10], Fukuoka [11], and Kyoto [12,13] universities reported successful LDLLT for ventilator-dependent patients. We have successfully performed LDLLT for all nine patients who had been on a ventilator for as long as 7 months. The University of Southern California (USC) group reported that patients on ventilator preoperatively had significantly worse outcomes, and those undergoing retransplantation had an increased risk of death among their 123 LDLLTs [14]. Successful LDLLT has been reported in two patients on extracorporeal membrane oxygenation (ECMO) by Okayama group [15]. In both of the patients, bridging time of ECMO to LDLLT was 2 days, and both could be weaned from cardiopulmonary bypass (CPB) support immediately after transplantation in the operating room (OR).

Because only two lobes are transplanted, cystic fibrosis represents the most common indication for LDLLT in the United States, as these patients are usually small in body size. The distribution of diagnoses is quite unique to Japan where cystic fibrosis is a very rare disease. We have accepted various lung diseases including hypertensive, restrictive, obstructive, and infectious lung diseases. In our experience, interstitial pneumonia (IP), bronchiolitis obliterans, and pulmonary hypertension (PH) were the three major indications. Most of the patients with IP were on systemic corticosteroid therapy. Most of the patients with bronchiolitis obliterans were after hematopoietic stem cell transplantation (HSCT) for various malignancies like leukemia. Patients with idiopathic pulmonary arterial hypertension (IPAH) were on high-dose epoprostenol therapy.

46.3 DONOR SELECTION

Preoperative workup consists of posterior–anterior and left lateral chest roentgenogram, high-resolution computed tomographic scan of the chest (at maximal inspiration and expiration), formal pulmonary function tests, measurement of room air blood gases, electrocardiogram, and Doppler echocardiogram. Three-dimensional multidetector computed tomography angiography is created for the confirmation of the pulmonary arterial and venous anatomy (Figure 46.2) [16]. The completeness of pulmonary fissures is carefully evaluated by high-resolution computed tomography. Although HLA matching is not required for donor selection, a prospective cross-match to rule out the presence of anti-HLA antibodies is performed.

Eligibility criteria for living lobar lung donation at Kyoto University are summarized in Table 46.1. Although immediate family members (relatives within the third degree or a spouse) have been the only donors in our institution, non-Japanese institutions have accepted extended family members and unrelated individuals [14]. Extracting more than one lobe from the donor should be prohibited.

Recently, many cases of ABO-incompatible organ transplantation, especially kidney and liver transplantation, have been performed to overcome the donor organ shortage. We recently reported successful ABO-incompatible LDLLT performed in a 10-year-old boy with bronchiolitis obliterans after bone marrow transplantation for recurrent acute myeloid leukemia [17]. His blood type had changed from AB to O by the bone marrow transplantation and received type B and AB donor lobar lungs.

Potential donors should be competent, willing to donate free of coercion, medically and psychosocially suitable, fully informed of the risks and benefits as a

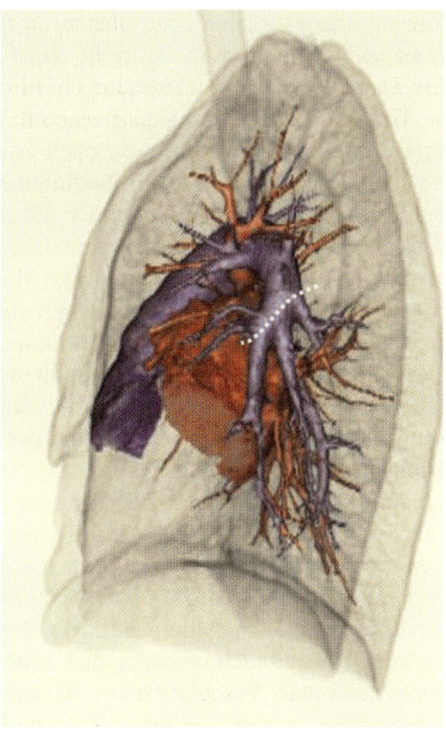

FIGURE 46.2 **3D-CT angiography in a typical left donor.** A white dotted line shows the planned cutting oblique line of the pulmonary artery, thus to preserve ligula branches.

FIGURE 46.2 **3D-CT angiography in a typical left donor.** A white dotted line shows the planned cutting oblique line of the pulmonary artery, thus to preserve ligula branches.

TABLE 46.1 **The Eligibility Criteria for Living Lung Donation (Kyoto University)**

Medical criteria
- Age 20−60 years
- ABO blood type compatible with recipient
- Blood-related relatives within the third degree or a spouse
- No significant past medical history
- No recent viral infection
- No significant abnormalities on echocardiogram and electrocardiogram
- No significant ipsilateral pulmonary pathology on computed tomography
- Arterial oxygen tension ≥ 80 mmHg (room air)
- FVC, FEV1 ≥ 85% of predicted
- No previous ipsilateral thoracic surgery
- No active tobacco smoking.

Social and ethical criteria
- No significant mental disorders proved by a psychiatrist
- No ethical issues or concerns about donor motivation.

donor, and fully informed of risks, benefits, and alternative treatment available to the recipient. In our institution, potential donors are interviewed at least three times to provide them with multiple opportunities to question, reconsider, or withdraw as a donor.

After a suitable donor pair is found, the larger donor with better vital capacity is selected for the donation of the right lower lobe and the other for the removal of the left lower lobe.

46.4 SIZE MATCHING

Appropriate size matching between the donor and recipient is important in LDLLT. It is often inevitable that small grafts are implanted in LDLLT in which only two lobes are implanted. Excessively small grafts may cause high pulmonary artery pressure, resulting in lung edema [18]. A pleural space problem may increase the risk of empyema. Overexpansion of the donor lobes may contribute obstructive physiology by early closure of small airways [19]. On the other hand, the adult lower lobe might be too big for small children. The use of oversized grafts could cause high airway resistance, atelectasis, and hemodynamic instability by the time of chest closure [20].

46.4.1 Functional Size Matching

For "functional size matching," we utilize graft forced vital capacity (FVC) [21]. We have previously proposed a formula to estimate the graft FVC based on the donor's measured FVC and the number of pulmonary segments implanted [5]. Given that the right lower lobe consists of 5 segments, the left lower lobe of 4, and the whole lung of 19, total FVC of the two grafts is estimated by the following equation:

Total FVC of the two grafts = Measured FVC of the right

donor $\times 5/19$ + measured FVC of the left donor $\times 4/19$

When the total FVC of the two grafts is more than 45% of the predicted FVC of the recipient (calculated from a knowledge of height, age, and sex), we accept the size disparity regardless of the recipient's diagnosis.

Total FVC of the two grafts/predicted FVC of the recipient > 0.45

The recipient's mean measured FVC at 6 months after LDLLT was well correlated with the estimated graft FVC [21]. In contrast, we found no significant correlation between the recipient's predicted FVC and the recipient's measured FVC. These results indicate that the amount of lung tissue implanted, not recipient factors like diagnosis, determines recipient FVC.

46.4.2 Anatomical Size Matching

For "anatomical size matching," three-dimensional computed tomography (3D-CT) volumetry is performed both for the donor and the recipient (Figure 46.3) [16,22,23]. CT images are obtained using a multidetector CT scanner

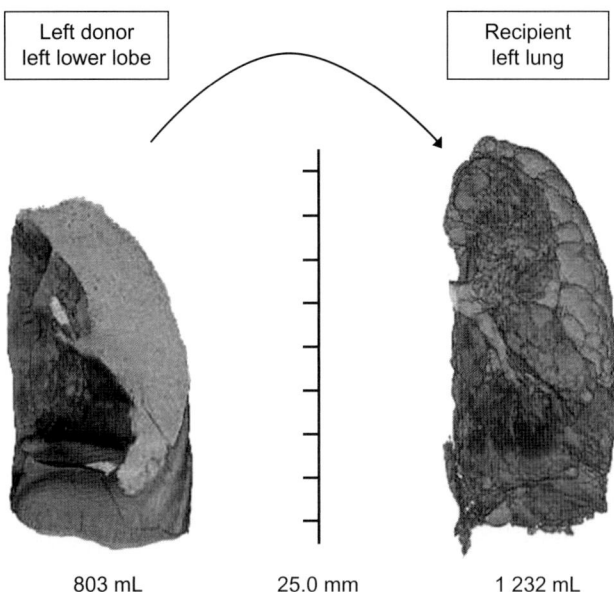

Left donor
left lower lobe

Recipient
left lung

803 mL   25.0 mm   1 232 mL

FIGURE 46.3 Anatomical size matching for the left donor graft and the recipient left hemithorax, using three-dimensional volumetry. The recipient was an adult male whose left hemithorax was 1232 mL. The left donor was his wife whose left lower lobe was 803 mL. The ratio of left donor graft to recipient left hemithorax was estimated to be 65.2%.

during a single respiratory pause at the end of maximum inspiratory effort. The upper and lower thresholds of anatomical size matching have not been determined yet. We have accepted a wide range of volume ratio between the donor's lower lobe graft and the corresponding recipient's chest cavity. When the ratio was within 40−160%, we found that recipient's adaptation ability to undersized or oversized grafts was remarkable.

46.5 SURGICAL TECHNIQUE

Three surgical teams and a back table team are required to perform bilateral LDLLT. They communicate with each other closely to minimize graft ischemic time. The recipient and the right-side donor are brought to OR at the same time. The left-side donor is brought to OR 30 min later.

46.5.1 Donor Lobectomy

The most common procedure involves a right lower lobectomy from a larger donor and a left lower lobectomy from a smaller donor. An epidural catheter for postoperative pain is placed the day before the surgery to avoid any complications related to heparinization. After induction of general anesthesia, donors are intubated with a left-sided double lumen endotracheal tube. Fiberoptic bronchoscopy was performed to determine if lower lobectomy was

feasible leaving adequate length for closure on the donor bronchus and length for anastomosis in the recipient.

The donors are placed in the lateral decubitus position and a posterolateral thoracotomy is performed through the fifth intercostal space. Fissures are developed using linear stapling devices. The pericardium surrounding the inferior pulmonary vein is opened circumferentially. Dissection in the fissure is carried out to isolate the pulmonary artery to the lower lobe, and to define the anatomy of the pulmonary arteries to the middle lobe in the right-side donor and to the lingular segment in the left-side donor. If the branches of middle lobe artery and lingular artery are small, they are ligated and divided. However, if such braches are large enough, arterioplasty using autopericardial patch should be performed [24].

Intravenous prostaglandin E1 is administered to decrease a systolic blood pressure by 10−20 mmHg. Five thousand units of heparin and 500 mg of methylprednisolone are administered intravenously. After placing vascular clamps in appropriate positions, the division of the pulmonary vein, the pulmonary artery and bronchus are carried out in this order. Vascular stamps are closed with 5−0 polypropylene running sutures. The bronchus is closed with 4−0 polypropylene interrupted sutures. The bronchial stamp is covered with pedicled pericardial fat tissue.

On the back table, the lobes are flushed with preservation solution both antegradely and retrogradely from a bag about 50 cm above the table. Lobes are gently ventilated with room air during the flush.

46.5.2 Recipient Implantation

Recipients are anesthetized and intubated with a single lumen endotracheal tube in children and with a left-sided double lumen endotracheal tube in adults. The "clamshell" incision is used and both chest cavities are entered through the fourth intercostal space. The sternum is notched at the level of transection by aiming the sternal saw at a 45° angle and cutting toward the midpoint to facilitate postoperative sternal adaptation.

Pleural and hilar dissections are performed as much as possible before heparinization to reduce blood loss. The ascending aorta and the right atrium are cannulated after heparinization and patients are placed on standard CPB. After bilateral pneumonectomy, hilar preparation is performed to facilitate subsequent implantation. The chest is irrigated with warm saline containing antibiotics.

The right lower lobe implantation is performed followed by the left lower lobe implantation. The bronchus, the pulmonary vein, and the pulmonary artery are anastomosed consecutively. The bronchial anastomosis is begun with a running 4−0 polydioxanone suture for membranous portion and completed with simple interrupted sutures or a running suture for cartilaginous portion. We use end-to-end anastomosis when the bronchial size is equivalent, and use telescoping technique when the discrepancy in bronchial size is obvious. The bronchial wrapping is not employed except for patients on high-dose steroid therapy. The venous anastomosis is conducted between the donor inferior pulmonary vein and the recipient superior pulmonary vein using a running 6−0 polypropylene suture. The pulmonary arterial anastomosis is completed in an end-to-end fashion using a running 6−0 polypropylene suture.

Just before completing the bilateral implantations, 500 mg to 1 g of methylprednisolone is given intravenously and nitric oxide inhalation is initiated at 20 ppm. After both lungs are reperfused and ventilated, CPB is gradually weaned and then removed.

The alternative strategy for cardiopulmonary support during recipient's operation for LDLLT is the use of ECMO via femoral artery and vein. ECMO allows lower heparin use, which seems to reduce perioperative bleeding [25]. It is especially useful when extensive pleural adhesion is found. Activated clotting time is maintained to be around 200 s. We have utilized ECMO instead of CPB in most of LDLLT procedures since 2012.

46.6 LDLLT USING OVERSIZED GRAFT

For small children, the adult lower lobe might be too big. The use of oversized grafts could cause high airway resistance, atelectasis, and hemodynamic instability by the time of chest closure [20]. To overcome these problems, we have developed several techniques including single lobe transplantation with or without contralateral pneumonectomy, delayed chest closure, and downsizing the graft.

Single LDLLT from a single living donor can be performed for selected small recipients. We retrospectively investigated 14 critically ill patients who had undergone single LDLLT at three lung transplant centers in Japan [26]. Three- and 5-year survival rates were 70% and 56%, respectively. Survival among these 14 patients was significantly worse than survival in a group of 78 patients undergoing bilateral LDLLT during the same period. Single LDLLT provides acceptable results for sick patients who would die soon otherwise. However, bilateral LDLLT appears to be a better option if two living donors are found.

We reported successful right lower lobe transplantation and simultaneous left pneumonectomy in an 8-year-old girl on a ventilator [13]. The graft donated by her mother was estimated to be 200% larger than the right chest cavity of the recipient.

It has been reported that delayed chest closure can be safely used after cadaveric bilateral lung transplantation. This technique can be applied to LDLLT [27]. The oversized graft volume is expected to decrease during the waiting period by the improvement of pulmonary edema and the dimensions of the recipient's right heart are expected to decrease because of the reduction in the afterload after LDLLT.

We reported another strategy for oversized graft by downsizing a graft on a back table. A 15-year-old boy with bronchiolitis obliterans successfully underwent bilateral LDLLT with segmentectomy of the superior segment of an oversized right lower lobe graft obtained from his father [16].

46.7 LDLLT USING UNDERSIZED GRAFT

When grafts are too small, a limited amount of vascular bed might cause high pulmonary artery pressure, resulting in lung edema [18]. Intrathracic dead space can remain and cause complications, such as postoperative bleeding, persistent air leakage, and empyema. Moreover, hyperinflation of the grafted lungs may result in insufficient respiratory dynamics or hemodynamic collapse after LDLLT [19].

We reported a successful LDLLT in which a very large size mismatch between donor lungs and recipient chest cavity was solved by sparing the bilateral native upper lobes [28]. A recipient, 44-year-old man with bronchiolitis obliterans, was 17 cm taller than his donors, his sister, and his wife. Regarding functional size matching, the estimated graft FVC was 45.7% of the recipient-predicted FVC. Regarding anatomical size matching, the volume ratio of the graft was only 22% in the right side and 36% in the left side. By sparing native upper lobes, adequate chest cavity for small grafts was provided. Candidates for this approach should have no infection in the spared lobes and minimum pleural adhesion with well-developed interlobar fissures. Considering these factors, a space occupying, noninfectious disease, like bronchiolitis obliterans, would be an ideal indication. Pulmonary fibrosis, pulmonary artery hypertension, emphysema, and lymphangioleiomyomatosis may also be possible indications.

46.8 POSTOPERATIVE MANAGEMENT

The patient is kept intubated for at least 3 days to maintain optimal expansion of the implanted lobes. We use pressure-limited ventilation and kept maximal ventilation pressure <25 cmH$_2$O. Fiber optic bronchoscopy is performed every 12 h during intubation to assess donor airway viability and to suction any retained secretions. Bedside postoperative pulmonary rehabilitation is initiated as soon as possible.

Postoperative immunosuppression consists of triple drug therapy with cyclosporine (CSA) or tacrolimus (FK), mycophenolate mofetil (MMF), and corticosteroids. Induction cytolytic therapy is not used. The combination of CSA + MMF + steroid is chosen for patients with infectious lung diseases, pediatric patients, and patients on steroid; the combination of FK + MMF + steroid for other patients. Except 125 mg of methylprednisolone during the first 3 days, all immunosuppressants are given via the nasal tube inserted to the proximal jejunum. Under careful monitoring of daily serum creatinine, CSA and FK trough levels are often reduced to below the target range.

We judge acute rejection on the basis of radiographic and clinical findings without transbronchial lung biopsy, because the risk of pneumothorax and bleeding after transbronchial lung biopsy may be greater after LDLLT. Because two lobes are donated by different donors, acute rejection is usually seen unilaterally. Early acute rejection episodes are characterized by dyspnea, low-grade fever, leukocytosis, hypoxemia, and diffuse interstitial infiltrate on chest radiographs and CT. A trial bolus dose of methylprednisolone 500 mg is administered and various clinical signs are carefully observed. If acute rejection is indeed the problem, two additional daily bolus doses of methylprednisolone are given. If acute rejection is encountered more than three times, CSA is switched to FK.

46.9 OUTCOME OF LIVING DONORS

Successful LDLLT largely depends on donor outcome. In our experience, all donors have returned to their previous life styles without any restrictions. However, long-term outcomes of live donors have not been well documented. It is because the donor follow-up continues generally through 1 year and then discontinues. More studies will be needed to understand the long-term results of living lung donors.

46.9.1 Perioperative Complications in Living Donors

Relatively high morbidity after lobectomy has been described in the previous reports, but there has been no reported perioperative mortality [29,30]. Morbidity rates varied from 20 to 60% depending on the definition of complications. Common complications are pleural effusion, bronchial stamp fistulas, hemorrhage, and arrhythmia. The Vancouver Forum Lung Group summarized the world experience on approximately 550 living lung donors in 2006 [31]. Approximately 5% of them have experienced complications requiring surgical or bronchoscopic intervention.

Relatively high morbidity after living-donor lobectomy as compared to standard lobectomy may be explained by three technical differences between the two surgical procedures. First, the circumferential pericardotomy surrounding the inferior pulmonary vein may increase the risk for arrhythmias and pericarditis. Second, an oblique transection of the right lower lobe bronchus may increase the risk for bronchial fistula and stenosis. Third, the administration of heparin may increase the risk of bleeding in the perioperative period.

46.9.2 Psychologic Outcome of Living Donors

The Massachusetts General Hospital (MGH) reported that living lung donors enjoyed generally satisfactory physical and emotional health [32]. Donors reported positive feelings about donation, but wished to be recognized and valued by the transplant team and the recipient. Okayama group reported that the average quality of life in the living lung donors was better than that of general population [33]. However, a fatal outcome in the recipient significantly impacted donor mental health. Interestingly, there was a significant correlation in mental health scores between the paired donors.

46.9.3 Pulmonary Function of Living Donors

The MGH group reported that mean donor FVC decreased by $16 \pm 3\%$ [32]. Post-donation FVC value was higher than preoperatively predicted value. We prospectively evaluated pulmonary function 3, 6, and 12 months after donor lobectomy [34]. FVC and FEV1 (forced expiratory volume in 1 s) recovered constantly up to more than 90% of the preoperative value 1 year after donor lobectomy.

46.10 OUTCOME OF LDLLT RECIPIENT

There are only three groups which have reported a summary of recipient outcome. The USC group recently published their 10-year experience on 123 LDLLT recipients including 39 children [14]. In their series, retransplantation and mechanical ventilation were identified as risk factors for mortality. One, 3-, and 5-year survivals were 70%, 54%, and 45%, respectively. St. Louis group reported similar results in 38 pediatric LDLLT recipients [35]. We (Okayama University group) recently published institutional results in 30 LDLLT recipients [6]. With mean follow-up period of 22 months, survival was 100%.

As of December 2012, the author has accumulated LDLLT experience in 79 patients (47 at Okayama University and 32 at Kyoto University). There were 57 females and 22 males with ages ranging from 6 to 64 years (average 32.7 years). Twenty of the patients were children and 59 were adults. Of note, there were only 13 adult men because of the size matching issue.

Recipient's diagnoses were listed in Table 46.2. We have accepted various diseases including restrictive, obstructive, vascular, and infectious lung diseases. IP, bronchiolitis obliterans, and PH were the three major indications which were summarized separately. All 79 patients were very sick and required oxygen inhalation preoperatively. Forty-six patients (62%) were bed bound and nine (11%) were on a ventilator.

Bilateral LDLLT was performed in 68 patients and single LDLLT was performed in 11 small patients. There were six early deaths for a hospital mortality of 7.6%. The causes of early death were graft failure due to excessive small grafts in two, infection in two, acute rejection in one, and heart failure in one. There were eight late

deaths during a follow-up period of 3–172 months. The causes of early death were bronchiolitis obliterans syndrome (BOS) in two, cachexia in two, post transplant lymphoproliferative disorder (PTLD) in one, encephalitis in one, and unknown causes in two. The 5- and 10-year survivals were 86.0% and 78.3%, respectively (Figure 46.4).

The question of whether two pulmonary lobes can provide a sufficient long-term pulmonary function and clinical outcome to recipients has been recently answered. The USC group reported that LDLLT provided comparable intermediate and long-term pulmonary function and exercise capacity to bilateral CLT in adult recipients surviving more than 3 months after transplantation [36]. We have observed similar results in our LDLLT recipients. The measured recipient FVC ultimately reached 123% of the estimated graft FVC of two donor lobes (calculated based on the donor FVC and number of segments implanted) at 36 months after LDLLT [37].

TABLE 46.2 Diagnoses for LDLLT

Diagnoses	Number
Interstitial pneumonia	25
Bronchiolitis obliterans	20
Pulmonary hypertension	19
Bronchiectasis	6
Lymphangioleiomyomatosis	4
Retransplantation	2
Cystic fibrosis	1
Emphysema	1
Eosinophilic granuloma	1
Total	79

46.11 LDLLT FOR IP

Among 25 patients with IP, 23 were idiopathic interstitial pneumonia (IIP) and two were associated with dermatomyositis [38]. Most patients diagnosed as having idiopathic interstitial fibrosis (IPF) show the features which fulfill the histopathological criteria for usual interstitial pneumonia (UIP), the most common histologic type of IIP. It was recently reported that a histologic diagnosis of fibrotic nonspecific interstitial pneumonia (NSIP) was associated with a significantly worse survival rate than

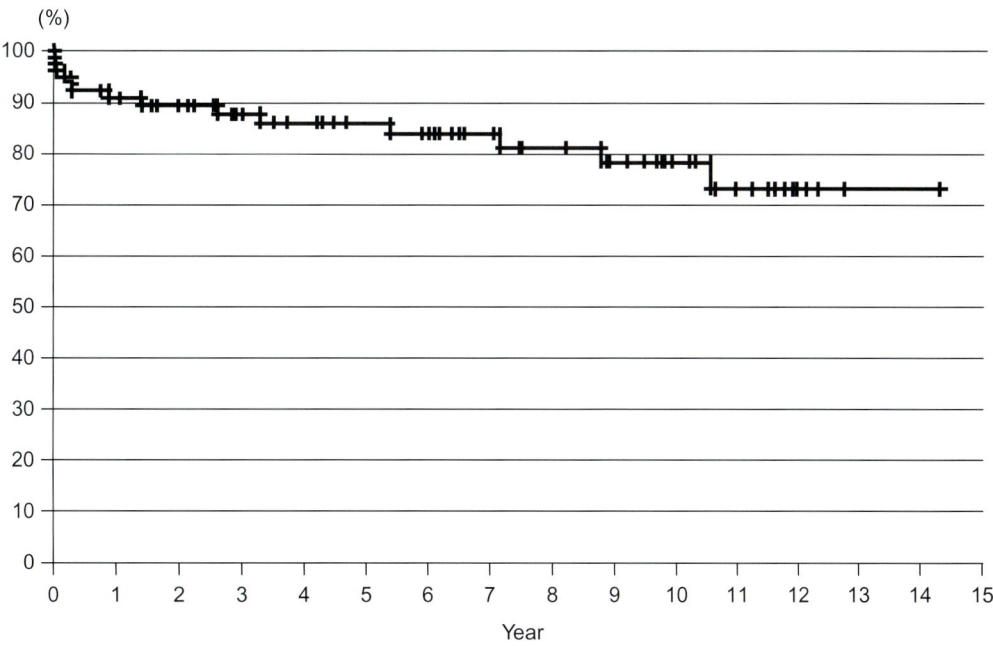

FIGURE 46.4 Survival after LDLLT (n = 79).   The 5- and 10-year survivals were 86.0% and 76.3%, respectively.

cellular NSIP, and a long-term survival rate similar to UIP [39]. Among patients awaiting lung transplant, patients with IPF have been demonstrated to have the highest mortality rate while awaiting cadaveric donors. Patients with IIPs have small chest cavities due to the nature of the restrictive lung disease, which we believe is rather beneficial for lobar implantation [40]. We have accepted patients on high-dose systemic corticosteroid therapy, as high as 50 mg/day of prednisone. Excellent bronchial healing was observed in all anastomoses. Various factors, such as short donor bronchial length, high blood flow in the small grafts implanted, and well-preserved lung parenchyma with short ischemic time, may contribute to the better oxygen supply to the donor bronchus resulting in excellent bronchial healing in LDLLT. The 5-year survival was 84.8% in 25 recipients with IP. These data support the option of LDLLT in patients with advanced IP, including UIP, fibrotic NSIP, and IP associated with collagen disease.

46.12 LDLLT FOR BRONCHIOLITIS OBLITERANS

In our 20 patients with bronchiolitis obliterans with or without pulmonary fibrosis, 17 were following HSCT for hematopoietic diseases like leukemia [41], 2 were after Steven–Johnson syndrome [12,42], and 1 was secondary to the ingestion of *Sauropus androgynus*.

Despite medical advances in the field of HSCT, chronic, progressive, and irreversible pulmonary complications, such as bronchiolitis obliterans and pulmonary fibrosis, remain significant and lead to cause of death. Of note, several patients have been reported who have undergone LDLLT after HSCT using the same living donor [43], which confers on the recipient advantages in terms of immunological predominance. The Japanese group summarized their 19 patients who had undergone LDLLT after HSCT [44]. Eight patients underwent LDLLT after HSCT in which one of the donors was the same living donor as in HSCT (SD group), whereas 11 received LDLLT from relatives who were not the HSCT donors (non-SD group). The 5-year survival was 100% and 58% in the SD and non-SD groups, respectively. For the three single LDLLTs in the SD group, immunosuppression was carefully tapered.

After HSCT, if the patient has received stem cell from a compatible but dissimilar ABO type, serum antibodies will not agree with red cell antigens. We reported a successful ABO-incompatible LDLLT in a 10-year-old boy with bronchiolitis obliterans after HSCT for recurrent acute myeloid leukemia [17]. His blood type had changed from AB to O since he underwent HSCT and he had no anti-A/B antibody, and received type B and AB donor

lobar lungs. There was no apparent acute cellular rejection or antibody-mediated rejection postoperatively.

Relapse of the recipient's hematologic malignancy has been reported in a young female 66 months after LDLLT. She was treated with intensive chemotherapy and has been in complete remission for 1 year since then [45].

The 5-year survival among the 17 patients undergoing LDLLT after HSCT was significantly worse than that of a group of 62 patients who underwent LDLLT without HSCT during the same period (67.2% vs. 91.2%, $p = 0.044$).

46.13 LDLLT FOR PH

Among 19 patients with PH, 13 were IPAH, three were pulmonary veno-occlusive disease (PVOD), two were Eisenmenger's syndrome, and one was pulmonary capillary hemagiomatosis (PCH).

Although the knowledge of the predictors of survival in patients with PH is helpful, ultimately the timetable must be set by the unique situation of each patient. In our LDLLT experience, all IPAH patients were on high-dose intravenous epoprostenol with inotropic support except for one patient. Most of them were in WHO class IV and were bed bound. There are obvious concerns regarding whether PH would develop in only two lobes implanted that would receive a patient's entire cardiac output. We first reported a successful LDLLT for an adult patient with IPAH [46]. Although limited amount of lung tissue is implanted, pulmonary arterial pressure becomes nearly normal soon after LDLLT, validating the functional capacity of the two adult lobes to handle the cardiac output of both adult and pediatric recipients with PH including IPAH, PVOD, PCH, and Eisenmenger's syndrome [47,48].

Because graft ischemic time is short in LDLLT, primary graft dysfunction is infrequently encountered as compared with conventional CLT in general. However, five of the 19 recipients (26%) with PH experienced severe lung edema associated with left ventricular dysfunction in the early postoperative period. Two of them required ECMO support. Clinical findings demonstrated by echocardiography helped to establish the diagnosis of left ventricular dysfunction. Contrary to the early right ventricular function recovery, the impaired left ventricular function persisted at 2 months despite findings that left ventricular geometry was restored earlier after the reversal of PH [49]. Chronic preoperative preload reduction may adversely affect left ventricular compliance and muscle stiffness. Four of five patients responded to therapy including steroid pulse, inotropic drugs, after-load reduction with vasodilators, and nitric oxide inhalation. We recommend that patients be kept on a ventilator for at least a week and weaned from the ventilator very slowly along with elective tracheostomy. A left atrial line should be placed through the left appendage at the time of transplantation to monitor left atrial pressure.

The 5-year survival was 94.7% in 19 recipients with PH. The survival was significantly better in patients receiving LDLLT than in patients medically treated [47].

46.14 COMPARISON WITH CLT

Advantages and disadvantages of LDLLT compared to CLT are summarized in Table 46.3. The current availability of cadaveric donor lungs has not been able to meet the increasing demand of potential recipients in most counties. The use of LDLLT has decreased in the United States, and the recent change by the Organ Procurement and Transplantation Network to an urgency/benefit allocation system for cadaveric donor lungs in patients of 12 years and older may further reduce the demand. In contrast, the average waiting time for a cadaveric lung is still more than 2 years in Japan.

In general, the ischemic time for LDLLT is much shorter than CLT. In our experience, the ischemic time of the right graft was 158 ± 6 min and that of the left graft was 113 ± 5 min. Although only two lobes are transplanted, LDLLT seems to be associated with less frequent primary graft failure. We believe that using a "small but perfect graft" is a great advantage in LDLLT.

Experienced centers have recently reported the incidence of bronchial complications in CLT to be about 5%. Contraindications to CLT include current high-dose systemic corticosteroid therapy, because it may increase airway complications, although low-dose pretransplantation corticosteroid therapy (≤ 20 mg/day prednisone) is acceptable. We have accepted high-dose systemic corticosteroid therapy, as high as 50 mg/day of prednisone, in LDLLT. Among 147 anastomoses, excellent bronchial healing was observed in 143 (97%) anastomoses [50]. Various factors,

such as short donor bronchial length, high blood flow in the small grafts implanted, and well-preserved lung parenchyma with short ischemic time, may contribute to better oxygen supply to the donor bronchus resulting in excellent bronchial healing in LDLLT.

BOS has been the major obstacle after CLT. The USC group suggested that LDLLT was associated with a lower incidence of BOS especially in pediatric patients. They also indicated that shorter ischemic time in LDLLT could explain the reduced incidence of BOS. In our first 40 LDLLT recipients who survived longer than 6 months, 10 recipients (25%) developed BOS. Interestingly, seven of the 10 recipients developed unilateral BOS and their FEV1 decline stopped within 9 months. Transplanting two lobes obtained from two different donors appears to be beneficial in long term, because contralateral unaffected lung may function as a reservoir in case of unilateral BOS [51].

Knowing that most of our patients were very sick at the time of transplantation, our LDLLT result, the 5-year survival of 86%, is very encouraging.

REFERENCES

[1] Starnes VA, Lewiston NJ, Luikart H, Theodore J, Stinson EB, Shumway NE. Current trends in lung transplantation: lobar transplantation and expanded use of single lungs. J Thorac Cardiovasc Surg 1992;104:1060—8.

[2] Starnes VA, Barr ML, Cohen RG. Lobar transplantation: indications, technique, and outcome. J Thorac Cardiovasc Surg 1994;108:403—11.

[3] Starnes VA, Barr ML, Cohen RG, Hagen JA, Wells WJ, Horn MV, et al. Living-donor lobar lung transplantation experience: intermediate results. J Thorac Cardiovasc Surg 1996;112:1284—91.

[4] Starnes VA, Barr ML, Schenkel FA, Horn MV, Cohen RG, Hagen JA, et al. Experience with living-donor lobar lung transplantation for indications other than cystic fibrosis. J Thorac Cardiovasc Surg 1997;114:917—21.

[5] Date H, Aoe M, Nagahiro I, Sano Y, Andou A, Matsubara H, et al. Living-donor lobar lung transplantation for various lung diseases. J Thorac Cardiovasc Surg 2003;126:476—81.

[6] Date H, Aoe M, Sano Y, Nagahiro I, Miyaji K, Goto K, et al. Improved survival after living-donor lobar lung transplantation. J Thorac Cardiovasc Surg 2004;128:933—40.

[7] Camargo SM, Camargo JJP, Schio SM, Sánchez LB, Felicetti JC, Moreira Jda S, et al. Complications related to lobectomy in living lobar lung transplant donors. J Bras Pneumol 2008;34:256—63.

[8] Chen QK, Jiang GN, Ding JA, Gao W, Chen C, Zhou X. First successful bilateral living-donor lobar lung transplantation in China. Chin Med J 2010;123:1477—8.

[9] Kozower BD, Sweeet SC, de la Morena M, Schuler P, Guthrie TJ, Patterson GA, et al. Living donor lobar grafts improve lung retransplantation survival. J Thorac Cardiovasc Surg 2006;131:1142—7.

[10] Toyooka S, Yamane M, Oto T, Sano Y, Okazaki M, Hanazaki M, et al. Favorable outcomes after living-donor lobar lung

TABLE 46.3 Comparison Between LDLLT and CLT

	LDLLT	CLT
Waiting time	Short	Long
Schedule	Controllable	Uncontrollable
Ischemic time	Short	Long
Graft size	Small	Full
Primary graft failure	Infrequent	10—20%
Infection transmitted from graft	Infrequent	Frequent
Number of teams	3	2
Bronchial complication	Rare	5%
Chronic rejection	Often unilateral	Major cause of death

LDLLT, living-donor lobar lung transplantation; CLT, cadaveric lung transplantation.

transplantation in ventilator-dependent patients. Surg Today 2008;38:1078−82.

[11] Shiraishi T, Hiratsuka M, Munakata M, Higuchi T, Makihata S, Yoshinaga Y, et al. Living-donor single-lobe lung transplantation for bronchiolitis obliterans in a 4-year-old boy. J Thorac Cardiovasc Surg 2007;134:1092−3.

[12] Shoji T, Bando T, Fujinaga T, Date H. Living-donor single-lobe lung transplant in a 6-year-old girl after 7-month mechanical ventilator support. J Thorac Cardiovasc Surg 2010;139:112−3.

[13] Sonobe M, Bando T, Kusuki S, Fujinaga T, Shoji T, Chen F, et al. Living-donor, single-lobe lung transplantation and simultaneous contralateral pneumonectomy in a child. J Heart Lung Transplant 2011;30:471−4.

[14] Starnes VA, Bowdish ME, Woo MS, Barbers RG, Schenkel FA, Horn MV, et al. A decade of living lobar lung transplantation. Recipient outcomes. J Thorac Cardiovasc Surg 2004;127:114−22.

[15] Miyoshi K, Oto T, Okazaki M, Yamane M, Toyooka S, Goto K, et al. Expracorporeal membrane oxygenation bridging to living-donor lobar lung transplantation. Ann Thorac Surg 2009;88:56−7.

[16] Chen F, Fujinaga T, Shoji T, Yamada T, Nakajima D, Sakamoto J, et al. Perioperative assessment of oversized lobar graft downsizing in living-donor lobar lung transplantation using three-dimensional computed tomographic volumetry. Transplant Int 2010;23:41−4.

[17] Shoji T, Bando T, Fujinaga T, Chen F, Yurugi K, Maekawa T, et al. ABO-incompatible living-donor lobar lung transplantation. J Heart Lung Transplant 2011;30:479−80.

[18] Fujita T, Date H, Ueda K, Nagahiro I, Aoe M, Andou A, et al. Experimental study on size matching in a canine living-donor lobar lung transplant model. J Thorac Cardiovasc Surg 2002;123:104−9.

[19] Haddy SM, Bremner RM, Moore-Jefferies EW, Thangathurai D, Schenkel FA, Barr ML, et al. Hyperinflation resulting in hemodynamic collapse following living donor lobar transplantation. Anesthesiology 2002;97:1315−7.

[20] Oto T, Date H, Ueda K, Hayama M, Nagahiro I, Aoe M, et al. Experimental study of oversized grafts in a canine living-donor lobar lung transplantation model. J Heart Lung Transplant 2001;20:1325−30.

[21] Date H, Aoe M, Nagahiro I, Sano Y, Matsubara H, Goto K, et al. How to predict forced vital capacity after living-donor lobar-lung transplantation. J Heart Lung Transplant 2004;23:547−51.

[22] Camargo JJP, Irion KL, Marchiori E, Hochhegger B, Porto NS, Moraes BG, et al. Computed tomography measurement of lung volume in preoperative assessment for living donor lung transplantation: volume calculation using 3D surface rendering in the determination of size compatibility. Pediatr Transplant 2009;13:429−39.

[23] Chen F, Kubo T, Shoji T, Fujinaga T, Bando T, Date H. Comparison of pulmonary function test and computed tomography volumetry in living lung donors. J Heart Lung Transplant 2011;30:572−5.

[24] Chen F, Miwa S, Bando T, Date H. Pulmonary arterioplasty for the remaining arterial stump of the donor and the arterial cuff of the donor graft in living-donor lobar lung transplantation. Eur J Cardiovasc Surg 2012;42:138−9.

[25] Ius F, Kuehn C, Tudorache I, Sommer W, Avsar M, Boethig D, et al. Lung transplantation on cardiopulmonary support: venoarterial extracorporeal membrane oxygenation outperformed cardiopulmonary bypass. J Thorac Cardiovasc Surg 2012;144:1510−6.

[26] Date H, Shiraishi T, Sugimoto S, Shoji T, Chen F, Hiratsuka M, et al. Outcome of living-donor lobar lung transplantation using a single donor. J Thorac Cardiovasc Surg 2012;144:710−5.

[27] Chen F, Matsukawa S, Ishii H, Ikeda T, Shoji T, Fujinaga T, et al. Delayed chest closure assessed by transesophageal echocardiogram in single-lobe lung transplantation. Ann Thorac Surg 2011;92:2254−7.

[28] Fujinaga T, Bando T, Nakajima D, Sakamoto J, Chen F, Shoji T, et al. Living-donor lobar lung transplantation with sparing of bilateral native upper lobes: a novel strategy. J Heart Lung Transplant 2011;30:351−3.

[29] Battafarano RJ, Anderson RC, Meyers BF, Guthrie TJ, Schuller D, Cooper JD, et al. Perioperative complications after living donor lobectomy. J Thorac Cardiovasc Surg 2000;120:909−15.

[30] Bowdish ME, Barr ML, Schenkel FA, Woo MS, Bremner RM, Horn MV, et al. A decade of living lobar lung transplantation. Perioperative complications after 253 donor lobectomies. Am J Transplant 2004;4:1283−8.

[31] Barr ML, Belghiti J, Villamil FG, Pomfret EA, Sutherland DS, Grussner RW, et al. A report of the Vancouver Forum on the care of the live organ donor. Lung, liver, pancreas, and intestine data and medical guidelines. Transplantation 2006;81:1373−85.

[32] Prager LM, Wain JC, Roberts DH, Ginns LC. Medical and psychologic outcome of living lobar lung transplant donors. J Heart Lung Transplant 2006;25:1206−12.

[33] Nishioka M, Yokoyama C, Iwasaki M, Inukai M, Sunami N, Oto T. Donor quality of life in living-donor lobar lung transplantation. J Heart Lung Transplant 2011;30:1348−51.

[34] Chen F, Fujinaga T, Shoji T, Sonobe M, Sato T, Sakai H, et al. Outcomes and pulmonary function in living lobar lung transplant donors. Transplant Int 2012;25:153−7.

[35] Sweet SC. Pediatric living donor lobar lung transplantation. Pediatr Transplant 2006;10:861−8.

[36] Bowdish ME, Pessotto R, Barbers RG, Schenkel FA, Starnes VA, Barr ML. Long-term pulmonary function after living-donor lobar lung transplantation in adults. Ann Thorac Surg 2005;79:418−25.

[37] Yamane M, Date H, Okazaki M, Toyooka S, Aoe M, Sano Y. Long-term improvement in pulmonary function after living-donor lobar lung transplantation. J Heart Lung Transplant 2007;26:687−92.

[38] Shoji T, Bando T, Fujinaga T, Okubo K, Yukawa N, Mimori T, et al. Living-donor lobar lung transplantation for interstitial pneumonia associated with dermatomyositis. Transplant Int 2010;23:10−1.

[39] Nicholson AG, Colby TV, du Bois RM, Hansell DM, Wells AU. The prognostic significance of the histologic pattern of interstitial pneumonia in patients presenting with the clinical entity of cryptogenic fibrosing alveolitis. Am J Respir Crit Care Med 2000;162:2213−7.

[40] Date H, Tanimoto Y, Yamadori I, Aoe M, Sano Y, Shimizu N. A new treatment strategy for advanced idiopathic interstitial pneumonia: living-donor lobar lung transplantation. Chest 2005;128:1364−70.

[41] Yamane M, Sano Y, Toyooka S, Okazaki M, Date H, Oto T. Living-donor lobar lung transplantation for pulmonary

complications after hematopoietic stem cell transplantation. Transplantation 2008;86:1767−70.

[42] Date H, Sano Y, Aoe M, Goto K, Tedoriya T, Sano S, et al. Living-donor lobar lung transplantation for bronchiolitis obliterans after Stevens−Johnson syndrome. J Thorac Cardiovasc Surg 2002;123:389−91.

[43] Oshima K, Kikuchi A, Mochizuki S, Yamane M, Date H, Hanada R. Living-donor single lobe lung transplantation for bronchiolitis obliterans from mother to child following previous allogeneic hematopoietic stem cell transplantation from the same donor. Int J Hematol 2009;90:540−2.

[44] Chen F, Yamane M, Inoue M, Shiraishi T, Oto T, Minami M, et al. Less maintenance immunosuppression in lung transplantation following hematopoietic stem cell transplantation from the same living donor. Am J Transplant 2011;11:1509−16.

[45] Ishiyama K, Okumra H, Yamazaki H, Kondo Y, Waseda Y, Kotani T, et al. Intensive chemotherapy for a relapsed ALL patient who received living-donor lobar lung transplantation. Bone Marrow Transplant 2011;47:135−6.

[46] Date H, Nagahiro I, Aoe M, Matsubara H, Kusano K, Goto K, et al. Living-donor lobar lung transplantation for primary pulmonary hypertension in an adult. J Thorac Cardiovasc Surg 2001;122:817−8.

[47] Date H, Kusano KF, Matsubara H, Ogawa A, Fujio H, Miyaji K, et al. Living-donor lobar lung transplantation for pulmonary arterial hypertension after failure of epoprostenol therapy. J Am Coll Cardiol 2007;50:523−7.

[48] Aokage K, Date H, Okazaki M, Sano Y, Oto T, Kusano K, et al. Living-donor lobar lung transplantation and closure of atrial septal defect for adult Eisenmenger's syndrome. J Heart Lung Transplant 2009;28:1107−9.

[49] Toyooka S, Kusano KF, Goto K, Yamane M, Oto T, Sano Y, et al. Right but left ventricular function recovers early after living-donor lobar lung transplantation in patients with pulmonary arterial hypertension. J Thorac Cardiovasc Surg 2009;138:222−6.

[50] Toyooka S, Yamane M, Oto T, Sano Y, Okazaki M, Date H. Bronchial healing after living-donor lobar lung transplantation. Surg Today 2009;39:938−43.

[51] Shinya T, Sato S, Kato K, Gobara H, Akaki S, Date H, et al. Assessment of mean transit time in the engrafted lung with 133Xe lung ventilation scintigraphy improves diagnosis of bronchiolitis obliterans syndrome in living-donor lobar lung transplant recipients. Ann Nucl Med 2008;22:31−9.

Chapter 47

Artificial Lung Support

Marcelo Cypel and Shaf Keshavjee

Division of Thoracic Surgery, Toronto Lung Transplant Program, Toronto General Hospital, University Health Network, University of Toronto, Toronto, ON, Canada

47.1 INTRODUCTION

Artificial lung devices are membranes made of synthetic material that are connected to blood vessels through tubes and cannulas of silicone. The blood passing through the device is oxygenated and cleared of carbon dioxide. The most well-known term for artificial lung is extracorporeal membrane oxygenation or ECMO. However, because this technology can be applied for patients who do not have oxygenation problems, such as patients with hypercapnic respiratory failure, or to patients with severe pulmonary hypertension (PAH) and right ventricular failure, the most current term is extracorporeal lung support or ECLS.

In the 1930s the concept of artificial organ support started with the work of Carrel and Lindbergh [1]. However, the first successful clinical use in humans was described by Hill et al. [2] using a heart–lung machine in 1972 on a young man with post-traumatic respiratory failure. The patient stayed 75 h on support and eventually recovered.

Since then, ECLS has been widely used in the neonate and pediatric population where survival rates approached 50–60%. In contrast, clinical trials in adults demonstrated dismal results (<20% survival) [3]. However, in the last decade, improvements in patient selection, a better understanding of ventilator-associated lung injury, and improvements in artificial lung device technologies have made it possible to successfully bridge patients to recovery or to lung transplantation (LTx) [4–8]. Recent studies have shown more promising results using ECLS for adults with acute respiratory distress syndrome (ARDS) with survival rates ranging from 50% to 80%. This includes the experience from Michigan in 100 patients [9], the UK CESAR trial [10], and H1N1/ARDS reports [11,12].

47.2 INDICATIONS AND CONTRAINDICATIONS FOR ECLS

The main indications for ECLS are:

1. Bridge to recovery in patients with severe acute lung injury—ARDS;
2. Bridge to recovery in patients with severe primary graft dysfunction after lung transplantation;
3. Bridge to lung transplantation in patients with end-stage lung diseases (as an extension of life until a suitable organ becomes available). Patients with end-stage lung disease who are not eligible for lung transplantation are generally not candidates for ECLS, although some patients may benefit from extracorporeal CO_2 removal as a weaning strategy.

Regenerative Medicine Applications in Organ Transplantation.

The three main physiological indications for the use of ECLS are:

1. Refractory hypoxemia—usually $PaO_2/FiO_2 < 80$ mmHg despite maximal conventional respiratory support;
2. Hypercapnia and respiratory acidosis—inability to maintain safe levels of pH despite high ventilatory plateau pressures;
3. PAH and right ventricular dysfunction.

Absolute contraindications for the use of ECLS are currently difficult to define as with advancements in technology, indications have expanded over the past few years. However, unfavorable prognostic factors include sepsis, multiorgan dysfunction, acute renal failure, high vasopressor requirements, a long preceding duration of mechanical ventilation, advanced age, and obesity [13].

47.3 NEW TECHNOLOGY

The main components of the ECLS system include a membrane oxygenator, a pump, and tubing circuits (Figure 47.1). In the last decade, several important advancements in technology have contributed to improved management and overall outcomes of these patients, as detailed below.

47.3.1 Development of Polymethylpentene Membranes

In the past, most adult ECLS circuits used silicone membrane oxygenators and the remainder used polypropylene microporous oxygenators. Both these oxygenators had drawbacks. The introduction of polymethylpentene (PMP) membranes provided several technical advantages. Compared to silicone membrane oxygenators, the PMP

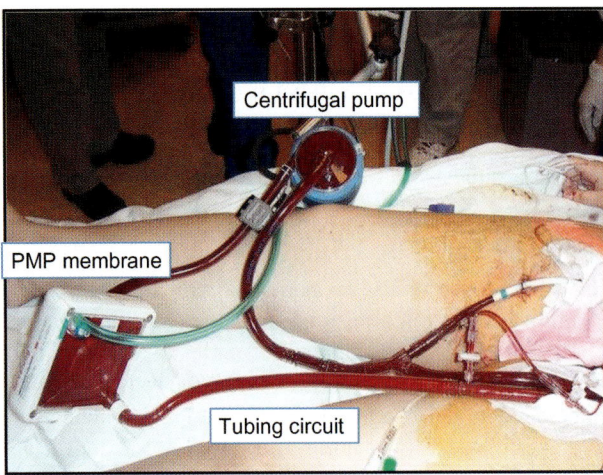

FIGURE 47.1 The main components of the ECLS system include a membrane oxygenator, a pump, cannulas, and tubing circuits.

oxygenator has reduced red blood cell and platelet transfusion requirements, significantly less plasma leakage, better gas exchange, lower resistance, and lower priming volume [14,15]. Compared to polypropylene microporous oxygenators, the PMP oxygenator has a reduced rate of oxygenator failure and can be functional for several weeks [4]. The PMP fibers are woven into a complex configuration of hollow fibers through which the oxygenated gas passes (Figure 47.2A). The hollow fibers themselves are then arranged into mats and stacked into a configuration that allows blood to pass between the fibers with low resistance. This provides maximum blood/gas mixing and gas transfer can take place without direct contact with blood. The low resistance also allows the use of these membranes without the need of an external pump (see below pumpless application of artificial lungs). Examples of new generation PMP membranes include the Quadrox D (Maquet), iLA (Novalung), and Hilite LT (Medos) (Figure 47.2B).

47.3.2 Introduction of Heparin-Coated Circuits

Heparin-coated circuits led to reduced rates of platelet, complement, and granulocyte activation and also significantly reduced heparin requirements [16,17]. Importantly, PMP oxygenators can also be readily heparin coated, whereas the silicone membrane oxygenators cannot. Thus, the modern ECLS circuit can be entirely heparin coated and requires less systemic heparinization or no heparin for few days if the patient is bleeding (recent history of trauma or surgery for example). In contrast, early ECLS circuits required full heparinization and consequently bleeding complications and daily blood product requirement were high.

47.3.3 Development of a New Generation of Centrifugal Pumps

Compared with traditional roller pumps, the centrifugal pumps have an improved performance and safety profile. They have virtually no risk of tubing rupture, require a smaller priming volume, do not require the use of a reservoir, and in general have a decreased incidence of hemolysis [18,19]. One of the most used pumps is the new generation Centrimag (Thoratec). This pump has no bearings but instead uses magnetic levitation (Figure 47.3). This approach diminishes both hemolysis and the likelihood of mechanical pump failure.

47.3.4 Development of New Cannulas

A major advance in recent year was the development of new cannulas that can be easily inserted using percutaneous Seldinger techniques. Moreover, dual lumen cannulas

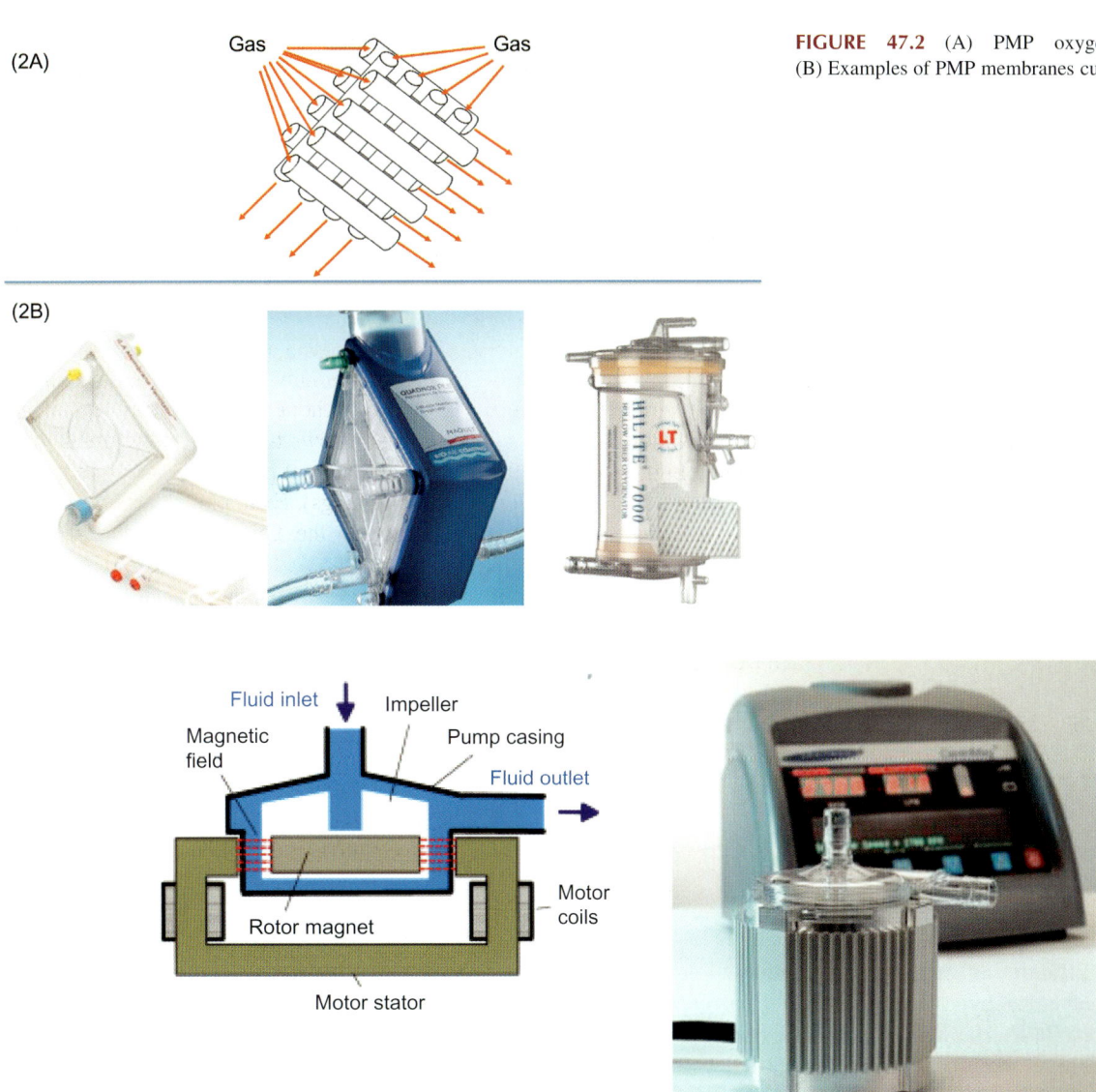

(2A)

Gas   Gas

FIGURE 47.2 (A) PMP oxygenator structure. (B) Examples of PMP membranes currently available.

(2B)

Fluid inlet   Impeller

Magnetic
field   Pump casing

               Fluid outlet

Rotor magnet   Motor
                           coils

Motor stator

FIGURE 47.2 (A) PMP oxygenator structure. (B) Examples of PMP membranes currently available.

FIGURE 47.3 Last generation magnetic levitation centrifugal pump (Thoratec).

for venous—venous ECLS have been developed and allow ECLS to be run through a single neck port decreasing risk of infection and allowing patient mobilization as the groin is kept free. The most common cannula used to date is the Avalon Elite (Maquet). The Novaport Twin (Novalung) and the Hemolung catether (Alung) are other dual lumen cannulas currently available (Figure 47.4).

47.4 MODES OF ECLS—CONFIGURATION OF DEVICE

In addition to the technical advances, device configuration can be individualized and tailored for specific patient ventilatory and hemodynamic requirements. The configuration and mode of ECLS will depend on the specific clinical scenario. A schematic of possible configurations per indication is shown in Figure 47.5.

47.4.1 Hypercapnic Respiratory Failure

Refractory hypercapnic respiratory failure and acidosis is a common scenario in patients with cystic fibrosis waiting for lung transplantation for example or patients with COPD. Noninvasive ventilation (NIV) has become an important option as a treatment modality in acute respiratory failure (ARF) in cystic fibrosis (CF), avoiding endotracheal intubation with its attendant complications [20]. If a suitable organ does not become available in time, respiratory failure progresses and mechanical ventilation

Avalon Elite (Maquet)   Novaport Twin (Novalung)

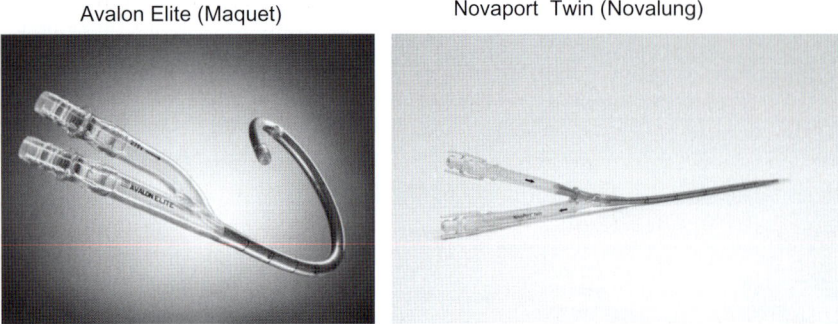

FIGURE 47.4 Dual lumen cannulas for venous–venous ECLS.

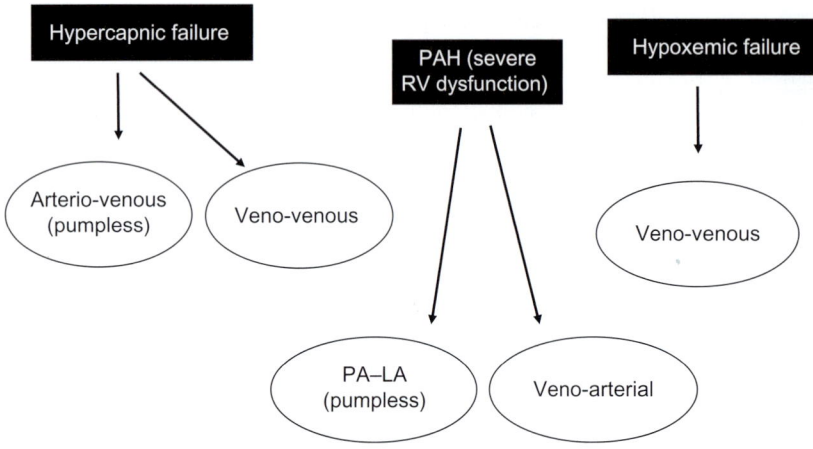

FIGURE 47.5 Algorithm for selection of ECLS configuration.

becomes necessary. At that stage, management becomes increasingly difficult as high-pressure ventilation is required and alveolar hypoventilation and hypercapnia often persist despite it. The large amounts of bronchopulmonary secretions in these patients make ventilation even more difficult. Traditionally, patient with hypercapnia and respiratory acidosis required ECLS with the use of a pump. However, with the advent of an interventional lung assist device (iLA, Novalung®, Germany), allowed patients to use the iLA in a *pumpless* arterio-venous (A-V) mode [4]. This low resistance (11 mmHg) hollow-fiber PMP membrane is attached to the systemic circulation (usually femoral artery) and receives only part of the cardiac output (15–20% of CO) for extracorporeal gas exchange. This allows prompt and effective CO_2 removal and correction of respiratory acidosis. CO_2 removal rates can be controlled by varying the sweep of gas flow up to 15 L/min. The usual recommended rate of CO_2 clearance is 20 mmHg/h. In order to use the pumpless device, the patient must have an adequate mean arterial blood pressure to be able to sustain good flows through the device (mean arterial pressure >80 mmHg). Since only a portion of the CO, in the range of one fifth of total, is oxygenated in the membrane, the PaO_2 is only augmented minimally with this mode of ECLS [4];

therefore, A-V iLA is not recommended in patients with severe hypoxemia. Cannulation is usually achieved percutaneously using a Seldinger technique (or an open modified Seldinger technique) in the femoral artery (13–15Fr) and femoral vein (17Fr). Insertion of traditional large bore long ECMO cannulas on the venous side provide more resistance to flow when A-V mode is being applied and therefore should be avoided. Since a centrifugal pump is not required, and the circuit is fully heparin coated, anticoagulation times (ACTs) are generally run in the range of 160–200 s, but if there is a concern about bleeding, an ACT of 150–180 s is acceptable.

In the initial publication from Fischer et al., 12 patients were bridged to transplantation using the iLA. The mean duration of iLA support in the 12 patients was 15 ± 8 days (4–32 days). Efficient CO_2 removal was rapidly achieved in all patients. Four patients died of multiorgan failure, two before LTx and two on days 16 and 30 after LTx. Thus, 10 of the 12 patients were successfully bridged to LTx, and eight of the 10 were alive 1 year post-transplantation [4].

One of the disadvantages of the pumpless iLA is the need for groin arterial cannulation which can lead to vascular complications and also immobilize the patient. As patient mobilization and rehabilitation seems to be an important component of the overall success of artificial lung devices in recent years, many investigators are now using low flow venous–venous ECLS using a single dual lumen cannula. The advantage here is simplicity of insertion and avoidance of groin and arterial cannulation. Most of the time, 1.5 L of veno-venous (VV) flows are sufficient to control CO_2 and maintain patients without mechanical ventilation. A concept of "respiratory dialysis" has also been proposed with even lower blood flows and smaller catheters using highly efficient membranes. The Hemolung (Alung) and Novalung Petit (Novalung) are examples of such technology.

47.4.2 Hypoxemic Respiratory Failure

Some patients will progress to hypoxemic respiratory failure and a different level of support is required. Whereas CO_2 removal can be achieved with low membrane flows (0.5–1 L/min) [21], substantial oxygenation requires more physiologic flows over the membrane (3–6 L/min). In order to achieve this, veno-venous (V-V) or veno-arterial (V-A) pump-driven ECLS support is required. V-V mode is the preferred choice if the patient is hypoxic, but hemodynamically stable. The advantages of the V-V mode in comparison to V-A mode are the decreased rate of complications, such as bleeding, arterial thrombosis, or neurologic complications. Generally, a 22Fr cannula is inserted into a femoral vein for drainage and a 17F single-stage cannula inserted into an internal jugular vein percutaneously for patient inflow. More recently, a dual-lumen single cannula system has been developed for V-V ECLS that has the advantage of simplicity, and importantly, allows for patient mobilization [22]. A 27Fr or 31Fr Avalon cannula can be inserted percutaneously. We usually use fluoroscopy as a guide for cannula insertion and positioning. Usual ACTs should range from 160 to 200 s.

47.4.3 Hypoxemic Respiratory Failure and Hemodynamic Compromise

For patients with respiratory failure and hemodynamic compromise, V-A ECLS is the recommended option since it provides both cardiac and pulmonary support. In fact, the initial experience with ECLS in LTx was using this mode [23]. Usually a femoral vein is cannulated for drainage and a femoral artery cannulated for blood return. Some authors also propose the use of the axillary artery with an interposition graft [24–26]. Although vascular access is a bit more difficult, the advantages of the axillary artery in this setting are the possibility of better patient mobilization and the low incidence of atherosclerosis in this vessel. Improved upper body oxygenated perfusion is also an important advantage. Another option to improve central oxygenation is to insert another cannula

into internal jugular vein and convert the circuit to a hybrid V-VA (V: femoral vein; VA: jugular vein and femoral artery) ECLS. This configuration of V-VA ECMO can also be used to provide partial cardiac support when cardiac function is depressed and does not improve with improved oxygenation on V-V support alone [27–29]. It is important to note that if the primary cause of hemodynamic shock is severe hypoxemia and respiratory acidosis, most of the time it can be readily reversed with V-V support.

The use of the traditional femoral V-A support often fails to correct hypoxemia in patients with somewhat preserved cardiac output. Thus this mode should be reserved for patients with primary hemodynamic failure, such as cardiomyopathy or severe right ventricle (RV) dysfunction due to PAH. A recent report demonstrates the application of V-A ECLS as bridge to LTx in awake and spontaneously breathing patients with PAH, avoiding the drawbacks and complications associated with intubation and prolonged mechanical ventilation. All five patients described in this series presented with cardiopulmonary failure due to PAH with or without concomitant lung disease. ECLS application was performed under local anesthesia without sedation and resulted in immediate stabilization of hemodynamics and gas exchange as well as recovery from secondary organ dysfunction. Two patients later required endotracheal intubation because of bleeding complications and both of them eventually died. The other three patients remained awake on ECMO support for 18–35 days until the time of transplantation after which full recovery was achieved [5].

47.4.4 PAH and Right Ventricular Failure

A novel mode of ECLS that we recently described is pulmonary artery to left atrium (PA to LA) ECLS configuration [7,30]. Although progress has been made for isolated lung failure, no truly effective solution existed for patients with primary PAH. Compared to patients with lung failure due to isolated lung parenchymal disorders, patients with end-stage PAH develop severe right heart failure. V-A or V-V ECLS does not effectively unload the RV. An atrial septostomy is sometimes performed as a last ditch effort, however, this leads to desaturated blood being systemically ejected as a result of the iatrogenic right to left shunt. In this scenario, we have demonstrated that the connection of a low-resistance gas exchange device (Novalung®) between the main trunk of the PA and the LA in a pumpless mode effectively creates an *oxygenating* shunt that pressure unloads the right ventricle much like an atrial septostomy. However, the important advantage in this case, is that the membrane oxygenates the blood and thus the central hypoxia seen with a simple septostomy is avoided. In our experience, patients improve dramatically as soon as flow across the Novalung® is instituted [7]. The elevated pressure in the pulmonary arteries serves as the driving force for the device and obviates the need for a pump. From a technical standpoint, patients are often so severely unstable such that they usually require femoral—femoral V-A ECLS support just prior to anesthetic induction. This is followed by median sternotomy and cannulation of the right superior pulmonary vein and PA. Extubation, physiotherapy and ambulation can be achieved while a patient is on pumpless PA-LA ECLS awaiting a compatible donor lung. Eight successful cases have been reported using this technique to bridge PAH patients for LTx [7,8,30]. In our experience, these patients become remarkably stable during PA-LA ECLS and three of these patients stayed on the device for more than 30 days without significant complications. Of note, in most cases, heart—lung transplantation is not required since the unloaded right ventricle recovers on the Novalung® and bilateral LTx provides ongoing remodeling and recovery of the right heart [7].

Another innovative ECLS approach to bridge patients with severe PAH to lung transplantation is V-V ECLS with added atrial septostomy. In two studies using adult sheep, right to left atrial shunting of oxygenated blood with VV-ECMO was capable of maintaining normal systemic hemodynamics and normal arterial blood gases during high right ventricular afterload dysfunction [31,32]. The theoretical advantage in comparison to PA-LA mode is the avoidance of sternotomy and central cannulation, however, the use of a pump is required.

47.5 CLINICAL MANAGEMENT

Patients receiving ECLS as a bridge to LTx require clinical management compared to patients with ARDS. Once ECLS is initiated, the ventilator should ideally be adjusted to "resting" lung settings. The ECLS flow should be maintained to sustain a venous blood saturation of 80—85% and an arterial saturation of 80—95%. Diuretics are given if required to maintain adequate urine output and remove excess fluid. If negative fluid balance cannot be achieved with diuretics, hemofiltration should be initiated early. Neurologic status is frequently checked and any deterioration should prompt further investigations. Cannulation sites and limb perfusion status are also frequently checked for bleeding and distal perfusion, respectively. Prophylactic antibiotics are given prior to insertion of cannulas.

47.6 CONCLUSIONS AND FUTURE OF ECLS

Despite significant advances and much improved outcomes with the use of ECLS, there are still significant challenges especially for the development of a long-term wearable artificial lung device. There has been ongoing research toward the development of a miniaturized, cell-coated artificial lung for long-term use—called the AmbuLung project.

REFERENCES

[1] Carrel A, Lindbergh CA. The culture of whole organs. Science 1935;81:621—3.

[2] Hill JD, O'Brien TG, Murray JJ, Dontigny L, Bramson ML, Osborn JJ, et al. Prolonged extracorporeal oxygenation for acute post-traumatic respiratory failure (shock-lung syndrome). Use of the Bramson membrane lung. N Engl J Med 1972;286:629—34.

[3] Zapol WM, Snider MT, Hill JD, Fallat RJ, Bartlett RH, Edmunds LH, et al. Extracorporeal membrane oxygenation in severe acute respiratory failure. A randomized prospective study. JAMA 1979;242:2193—6.

[4] Fischer S, Simon AR, Welte T, Hoeper MM, Meyer A, Tessmann R, et al. Bridge to lung transplantation with the novel pumpless interventional lung assist device NovaLung. J Thorac Cardiovasc Surg. 2006;131:719—23.

[5] Olsson KM, Simon A, Strueber M, Hadem J, Wiesner O, Gottlieb J, et al. Extracorporeal membrane oxygenation in nonintubated patients as bridge to lung transplantation. Am J Transplant 2010;10(9):2173—8.

[6] Aigner C, Wisser W, Taghavi S, Lang G, Jaksch P, Czyzewski D, et al. Institutional experience with extracorporeal membrane oxygenation in lung transplantation. Eur J Cardiothorac Surg 2007;31:468—73.

[7] Strueber M, Hoeper MM, Fischer S, Cypel M, Warnecke G, Gottlieb J, et al. Bridge to thoracic organ transplantation in patients with pulmonary arterial hypertension using a pumpless lung assist device. Am J Transplant 2009;9:853—7.

[8] Cypel M, Waddell TK, de Perrot M, Yeung JC, Chen F, Karolak W, et al. Safety and efficacy of the novalung interventional lung assist (iLA) device as a bridge to lung transplantation. J Heart Lung Transplant 2010;2:88.

[9] Kolla S, Awad SS, Rich PB, Schreiner RJ, Hirschl RB, Bartlett RH. Extracorporeal life support for 100 adult patients with severe respiratory failure. Ann Surg 1997;226:544—64.

[10] Peek GJ, Mugford M, Tiruvoipati R, Wilson A, Allen E, Thalanany MM, et al. Efficacy and economic assessment of conventional ventilatory support versus extracorporeal membrane oxygenation for severe adult respiratory failure (CESAR): a multicentre randomised controlled trial. Lancet 2009;374:1351—63.

[11] Freed DH, Henzler D, White CW, Fowler R, Zarychanski R, Hutchison J, et al. Extracorporeal lung support for patients who had severe respiratory failure secondary to influenza A (H1N1) 2009 infection in Canada. Can J Anaesth 2010;57:240—7.

[12] Davies A, Jones D, Bailey M, Beca J, Bellomo R, Blackwell N, et al. Extracorporeal membrane oxygenation for 2009 influenza A(H1N1) acute respiratory distress syndrome. JAMA 2009;302:1888—95.

[13] Fischer S, Bohn D, Rycus P, Pierre AF, de Perrot M, Waddell TK, et al. Extracorporeal membrane oxygenation for primary graft dysfunction after lung transplantation: analysis of the Extracorporeal Life Support Organization (ELSO) registry. J Heart Lung Transplant 2007;26:472—7.

[14] Peek GJ, Killer HM, Reeves R, Sosnowski AW, Firmin RK. Early experience with a polymethyl pentene oxygenator for adult extracorporeal life support. ASAIO J 2002;48:480—2.

[15] Khoshbin E, Roberts N, Harvey C, Machin D, Killer H, Peek GJ, et al. Poly-methyl pentene oxygenators have improved gas exchange capability and reduced transfusion requirements in adult extracorporeal membrane oxygenation. ASAIO J. 2005;51:281—7.

[16] Tamim M, Demircin M, Guvener M, Peker O, Yilmaz M. Heparin-coated circuits reduce complement activation and inflammatory response to cardiopulmonary bypass. Panminerva Med 1999;41:193—8.

[17] Moen O, Fosse E, Dregelid E, Brockmeier V, Andersson C, Hogasen K, et al. Centrifugal pump and heparin coating improves cardiopulmonary bypass biocompatibility. Ann Thorac Surg 1996; 62:1134—40.

[18] Lawson DS, Ing R, Cheifetz IM, Walczak R, Craig D, Schulman S, et al. Hemolytic characteristics of three commercially available centrifugal blood pumps. Pediatr Crit Care Med 2005;6:573—7.

[19] Valeri CR, MacGregor H, Ragno G, Healey N, Fonger J, Khuri SF. Effects of centrifugal and roller pumps on survival of autologous red cells in cardiopulmonary bypass surgery. Perfusion 2006;21:291—6.

[20] Noone PG. Non-invasive ventilation for the treatment of hypercapnic respiratory failure in cystic fibrosis. Thorax 2008;63:5—7.

[21] Zwischenberger BA, Clemson LA, Zwischenberger JB. Artificial lung: progress and prototypes. Expert Rev Med Devices 2006;3:485—97.

[22] Garcia JP, Iacono A, Kon ZN, Griffith BP. Ambulatory extracorporeal membrane oxygenation: a new approach for bridge-to-lung transplantation. J Thorac Cardiovasc Surg 2010;139:137—9.

[23] The Toronto Lung Transplant group. Sequential bilateral lung transplantation for paraquat poisoning. A case report. J Thorac Cardiovasc Surg 1985;89:734—42.

[24] Iglesias M, Jungebluth P, Sibila O, Aldabo I, Matute MP, Petit C, et al. Experimental safety and efficacy evaluation of an extracorporeal pumpless artificial lung in providing respiratory support through the axillary vessels. J Thorac Cardiovasc Surg 2007;133:339—45.

[25] Yokota K, Fujii T, Kimura K, Toriumi T, Sari A. Life-threatening hypoxemic respiratory failure after repair of acute type a aortic dissection: successful treatment with venoarterial extracorporeal life support using a prosthetic graft attached to the right axillary artery. Anesth Analg 2001;92:872—6.

[26] Mangi AA, Mason DP, Yun JJ, Murthy SC, Pettersson GB. Bridge to lung transplantation using short-term ambulatory extracorporeal membrane oxygenation. J Thorac Cardiovasc Surg 2010;140(3):713—5.

[27] Chou NK, Chen YS, Ko WJ, Huang SC, Chao A, Jan GJ, et al. Application of extracorporeal membrane oxygenation in adult burn patients. Artif Organs 2001;25:622—6.

[28] Madershahian N, Wittwer T, Strauch J, Franke UF, Wippermann J, Kaluza M, et al. Application of ECMO in multitrauma patients with ARDS as rescue therapy. J Card Surg 2007;22:180—4.

[29] Stohr F, Emmert MY, Lachat ML, Stocker R, Maggiorini M, Falk V, et al. Extracorporeal membrane oxygenation for acute respiratory distress syndrome: is the configuration mode an important predictor for the outcome? Interact Cardiovasc Thorac Surg 2011;12 (5):676—80.

[30] Camboni D, Philipp A, Arlt M, Pfeiffer M, Hilker M, Schmid C. First experience with a paracorporeal artificial lung in humans. ASAIO J 2009;55:304—6.

[31] Camboni D, Akay B, Pohlmann JR, Koch KL, Haft JW, Bartlett RH, et al. Veno-venous extracorporeal membrane oxygenation with interatrial shunting: a novel approach to lung transplantation for patients in right ventricular failure. J Thorac Cardiovasc Surg 2011;141:537—42.

[32] Camboni D, Akay B, Sassalos P, Toomasian JM, Haft JW, Bartlett RH, et al. Use of venovenous extracorporeal membrane oxygenation and an atrial septostomy for pulmonary and right ventricular failure. Ann Thorac Surg 2011;91:144—9.

Chapter 48

Lung Regeneration: The Bioengineering Approach

Joan Nichols and Joaquin Cortiella

Departments of Internal Medicine and Microbiology & Immunology, University of Texas Medical Branch at Galveston, Texas

Chapter Outline

48.1 INTRODUCTION

Lung transplantation is currently an appropriate treatment option for patients suffering from end-stage lung diseases when medical and surgical options have not been effective. Currently, lung transplantation is limited by the number of suitable donor organs available for use. The shortage of human lungs is exacerbated by the fact that the lung is easily compromised during the process of organ retrieval, and this results in rejection of the majority of procured organs [1,2]. Although *ex vivo* perfusion and reconditioning increase the number of lungs available for transplant, there are still too few organs available to meet the increasing need. One answer to the organ shortage problem is the development of patient-specific tissues or organs for transplantation. This has been a goal for many researchers working in biomedical engineering and regenerative medicine. Progress has been made in some areas of respiratory system organ engineering, such as for the trachea [3], but the field of lung tissue engineering has not progressed rapidly. This lack of progress is due to (i) the specialized needs of the lung in terms of strength and elasticity of the scaffold

necessary to support engineered tissues, (ii) the lack of identification of an appropriate cell source for tissue generation, and (iii) the problems associated with development of methods which promote adequate vascularization of engineered tissues which is needed in order to support gas exchange.

48.2 SCAFFOLDS FOR LUNG ENGINEERING

Although some attempts have been made to produce free-floating 3D matrix-independent, fluid-filled spheroids of airway epithelium [4], it is generally accepted that engineering lung tissue will require the development of a scaffold that meets the specialized needs of the lung. Of critical importance in selecting a scaffold for generation of lung tissue is the strength and elasticity of the material and the adsorption kinetics or capacity for cellular remodeling [5–7]. Since the primary physiological function to be replaced by engineered lung is gas exchange, an appropriate lung scaffold must allow gas exchange to occur unimpeded while providing adequate support to the cells and tissues. Lung scaffold design also needs to take into account the normal architecture and geometry of the lung as well as the need to support cell movement, transfer of nutrients into the tissues, and waste removal from tissues.

For development of lung tissue, scaffolding must remain long enough to support tissue development without impeding the elasticity or altering the elastic recoil of the engineered tissue or any adjoining normal tissue [6,7]. Scaffold components that are not as elastic as normal lung have the potential to contribute to development of restrictive conditions similar to the disease process which occurs in patients with idiopathic pulmonary fibrosis or sarcoidosis [6,7]. In the case of obstructive diseases, scaffold materials that limit the size of the airway could potentially add to existing obstructive problems.

Synthetic and natural polymers have been used as scaffolds to support efforts to engineer lung tissue. Synthetic polymers can be produced with a wide range of mechanical and chemical properties [8]. Degradable synthetic matrices that have been used to engineer lung tissue include polyglycolic acid (PGA) in the form of a felt sheet [9], PGA combined with pluronic F-127 [9], poly-lactic-*co*-glycolic acid (PLGA) [9] or poly-L-lactic-acid (PLLA) [9,10], poly-DL-lactic acid (PDLLA) [10], and polydimethylsiloxane [11].

Natural materials used in the past to engineer lung tissue include Matrigel™ [12], Gelfoam™ [13], Englebreth-Holms murine sarcoma basement membrane [14,15], and collagen [16–18]. Matrigel™ (Becton-Dickinson) is a solubilized basement membrane preparation extracted from Engelbreth-Holm-Swarm (EHS) mouse sarcoma, a tumor rich in extracellular matrix (ECM) proteins including laminin, collagen IV, and heparin sulfate proteoglycans [19]. EHS and a variety of component matrix proteins found in EHS are commercially available. Collagen (generally type I) has been used as a scaffold for engineering a variety of tissues including lung, and a number of collagen-based scaffolds are available for clinical use. Gelfoam™ (Pfizer) is a compressed sponge of porcine skin gelatin and was originally developed as a hemostatic device to arrest bleeding and promote clotting. There are limitations in the use of natural scaffold materials such as these due to their mechanical properties and variation in degradation rates. Scaffolds formed from these materials are also often very simple in terms of geometry, ECM composition, and lack the complexity as well as the elasticity or strength provided by normal lung ECM.

Acellular (AC) natural scaffolds are composed of the ECM secreted by the resident cells of the tissue or organ from which they are produced. As a result, organ-specific scaffolds already possess the correct anatomical, chemical, and morphological structure of the natural tissue, and this has been shown to facilitate the constructive remodeling of many different organs in both preclinical animal studies and human clinical applications [20,21]. The composition and structure of AC scaffolds or natural organ ECM depends on the origin of the tissues and the physiologic functions provided by the tissues [20–23]. Apart from meeting biological requirements, the biophysical cues originating from ECM microstructure and mechanical properties have been shown to be a major influence on the growth and differentiation of cells and regulation of cell behavior [8,23–26]. It has been suggested that organ-specific ECM structure and mechanics may actually guide tissue patterning and stem cell differentiation [27,28]. It is also important to remember that changes in ECM during remodeling, particularly of collagen and elastin fibers forming natural scaffolds, may in fact interfere with normal respiratory mechanics [29–31].

48.2.1 Development of AC Natural Lung Scaffolds for Lung Engineering

The structure of the lung is largely determined by the connective tissue network of the ECM and the organization of cells which generate the nonlinear mechanical properties which lead to the complex mechanical behavior of the lung which supports the process of breathing [31]. The ECM of normal lung parenchyma is composed primarily of collagen I, III, and elastin, and the principal function of these components is to form the mechanical scaffold that maintains the structure of the lung during the process of ventilation. Development of procedures for production of decellularized lung must allow for retention of key ECM components supporting lung functions while facilitating removal of cell debris and nucleic acids both of which could induce unwanted immune responses. Design of effective decellularization procedures are dictated by factors such as tissue density, tissue organization, or organ structure [22,23]. This is of particular importance for the lung since tissue density, and structure vary considerably among main stem bronchi, bronchioles, and distal lung. A variety of methods have been used to produce AC scaffolds from human tissues or organs [22,23], but these processes are not lung specific and do not take into account the delicate nature of lung structure. Techniques used for tissue and whole organ decellularization have been reviewed, including descriptions of solvents, detergents, physical agents, and enzymes [6,21–23].

Currently, there are only a few protocols for decellularization of lung [27,28,32–36], and these have been listed in Table 48.1. Protocols differ in the reagent used to facilitate cell removal, the physical conditions for decellularization, the duration of the process and, subsequently, in the ECM composition of the AC scaffold produced. Detergents used for the decellularization of lung include Triton X-100, sodium dodecyl sulfate (SDS),

TABLE 48.1 Comparison of Current Decellularization (DC) Methods and Outcomes after Recellularization

Protocol	Tissue Origin	Detergent Used	Condition For DC	Time for Process	ECM Remaining in AC Lung	Cell Source	Cell Responses or Cell Products
Lwebuga-Mukasa JS [1986] [37]	Human lung	Distilled water, then 0.1% Triton X, followed by 2% sodium deoxycholate	Immersion	48 + h	Collagen I, III, IV, V; laminin and fibronectin but no fibronectin and collagen V in basement membrane	Rat AE II cells	Loss of lamellar bodies and change in morphology of AE cells
Cortiella J [2010] [28]	Rat lung	1% SDS	Infusion followed by bioreactor immersion	3–5 weeks	Collagen I and elastin	mESC (C57BL6)	Prosurfactant protein C, surfactant protein A, CD31, cytokeratin-18 Clara cell secretory protein and TTF-1
Ott HC [2010] [38]	Rat lung	0.1% SDS	Infusion and immersion	120 min	Collagen, proteoglycans, and elastic fibers	Rat fetal lung cells	Surfactant protein A, SPC, and TTF-1 T1-α by FLCs
Petersen TH [2010] [39]	Rat lung	CHAPS	Infusion and immersion	180 min	Collagen I, elastin, and laminin	Neonatal rat epithelium and microvascular lung endothelial cells	Prosurfactant protein C and pro-SPB Clara cell secretory protein, aquaporin-5
Price AP [2010] [40]	Mouse lung	0.1% Triton X 100 followed by 2% sodium deoxycholate	Infusion and immersion	48 + h	Collagen I, elastin, glycosaminoglycans, and laminin	Murine fetal lung cells (E17)	Prosurfactant protein C Cytokeratin-18
Daly AB [2011] [41]	Mouse lung	0.1% Triton X 100 followed by 2% sodium deoxycholate	Infusion and immersion	3 days	Collagen I, collagen IV, laminin, fibronectin, but low levels of elastin and glycosaminoglycans	Mouse bone marrow MSCs	Transient expression of TTF-1 but no other lung lineage markers
Song JJ [2011] [42]	Rat lung	0.1% SDS	Perfusion	72 h	NA-Described in Ott HC [2010]	Rat fetal lung	TTF-1, pro-SPC, CC10, vimentin and ED-1 (CD68)
Bonvillan RW [2012] [43]	Rhesus macaque lung	0.1% Triton X 100 followed by 2% sodium deoxycholate	0.1% Triton X 100 followed by 2% sodium deoxycholate	3 days	GAGs are preserved. Collagen I, collagen IV, laminin, and some fibronectin	Rhesus macaque Bone marrow MSCs and adipose tissue	Cells adhered to scaffold

mESC, murine embryonic stem cells; pro-SPC, prosurfactant protein C; pro-SPB, prosurfactant protein B; CC10, Clara cell 10 kD protein; TTF-1, thyroid transcription factor-1; AE cells, alveolar epithelial cells; SDS, sodium dodecyl sulfate.

sodium deoxycholate, and 3-[(3-cholamidopropyl)dimethylammonio]-1-propanesulfonate (CHAPS). Triton X-100 ($C_{14}H_{22}O(C_2H_4O)_n$) is a nonionic surfactant which has a hydrophilic polyethylene oxide group and a hydrocarbon lipophilic or hydrophobic group. SDS is an organic compound with the formula $CH_3(CH_2)_{11}OSO_3Na$. It is an anionic surfactant used in both cleaning and hygiene products. Sodium deoxycholate (deoxycholic acid) is a water-soluble, bile-acid, ionic detergent generally used in methods for protein isolation or as a component of many cell lysis buffers (e.g., RIPA buffer). CHAPS is a zwitterionic detergent used in the laboratory to solubilize biological macromolecules such as proteins or as a nondenaturing solvent in some procedures for protein purification. Comparisons of matrix composition after decellularization of rat or mouse lungs have

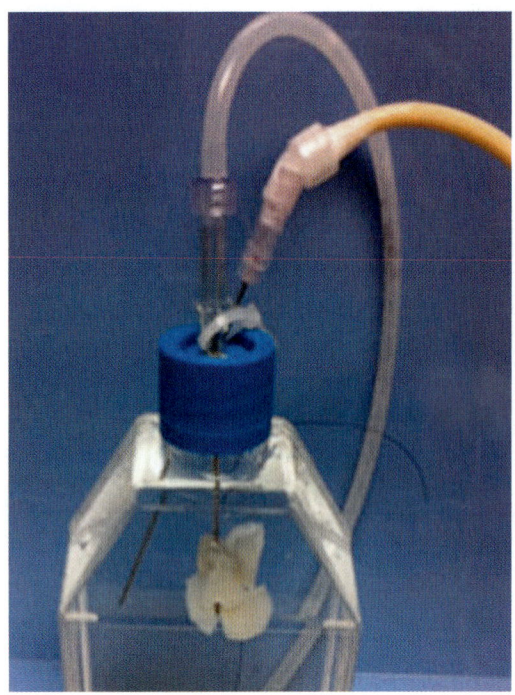

FIGURE 48.1 Perfusion system for decellularization of mouse or rat lung. Rat lungs were attached to a 16GX 3.5 in. spinal needle (Dyna Medical Corp, London, Ontario) and detergent was pumped through the trachea and continually removed from the decellularization chamber (a filtered 75 cm^2 tissue culture flask) using peristalic pumps.

shown in Figure 48.3A. In this decellularization system, catheters are attached to the trachea and pulmonary artery in order to allow fresh detergent to be flushed continually through the organ (Figure 48.3B) using peristaltic pumps (Figure 48.3C) to facilitate removal of cell debris and blood. Whole pig trachea-lungs or human trachea-lungs can be produced in this large perfusion system in hours to days depending on the detergent or detergent concentration used as well as the fluid flow rate (Figure 48.3D−F). Computer programs can also be written, which control pump speed and lung expansion (Figure 48.4B). Often our research group has used a bronchoscope during the decellularization process to perform a gross examination of the airways or vascular system as shown in Figure 48.4C and D.

48.2.2 Use of Nanotechnology to Design Scaffolds

In the future it is possible that lung scaffolds with high levels of complexity may be fabricated using nanotechnology [38,40]. Current nanofabricating techniques allow the creation of nanometer-sized scaffolds formed from combinations of ECM proteins with surface qualities that facilitate cell attachment and influence cell fate determination. These methodologies could be applied to produce scaffolds that are made from strong, highly elastic material with architecture and pore size similar to that found in the normal alveolus. Nanoparticle design of scaffolds has been attempted using a variety of different approaches. One scaffold with the potential to meet these requirements features a novel inverted colloidal crystal (ICC) geometry formed from modified hydrogel [41,43,45]. Primary colloidal crystals can form hexagonally packed lattices of spheres, with a wide range of diameters from nanometers to micrometers. ICCs are similarly organized structures where the spheres are replaced with cavities leaving open interstitial spaces within which cells can grow and form tissues. When ICC cavities exceed the diameter of cells, they can be used as 3D cell scaffolds [43,45]. The open geometry of the ICC lattice, high porosity (74% of free space), and large surface area make ICC an attractive structure to support the development of lung alveolus formation [41,43,45].

One promising option for scaffold development may be to use a combination of electrospray and spinning referred to as electrospinning. In the production of electrospun materials, a high electric field is applied to a droplet of fluid creating an electrically charged jet of polymer solution or melt, which dries or solidifies to leave a polymer fiber, typically in the micro- or nanoscale. This process has been extensively reviewed by others [42,46]. These techniques can regulate surface topography down to the nanometer range and include

indicated that variations in the ECM composition may occur depending on the detergent used [37,39,44]. The basement membrane is composed of collagen type IV, laminin, and proteoglycans. The interstitial matrix of the lung is composed of fibrilar collagens types I, II, and VII as well as elastin, proteoglycans, and hyaluron. Use of CHAPS or Triton X-100 combined with sodium deoxycholate in the decellularization process allowed for retention of varying amounts of basement membrane components such as collagen IV, laminin, and proteoglycans, as well as components of the interstitial matrix collagen 1 and elastin. The use of SDS resulted in removal of most basement membrane components but left interstitial matrix components collagen I and elastin.

Different physical systems used to facilitate decellularization include simple immersion in detergent, use of perfusion systems (Figure 48.1), or bioreactor chambers (Figure 48.2A−C). Use of a bioreactor to facilitate removal of cell debris is shown in Figure 48.2. Whole rat trachea-lungs were placed in a bioreactor chamber (Synthecon, Houston, TX) containing SDS (Figure 48.2A) and allowed to rotate with perfusion of fresh detergent for a period of days to weeks depending on the concentration of the detergent and the flow rates used (Figure 48.2B−D). Recently, more complex systems have been developed to produce large organ, AC pig or human trachea-lung such as the one

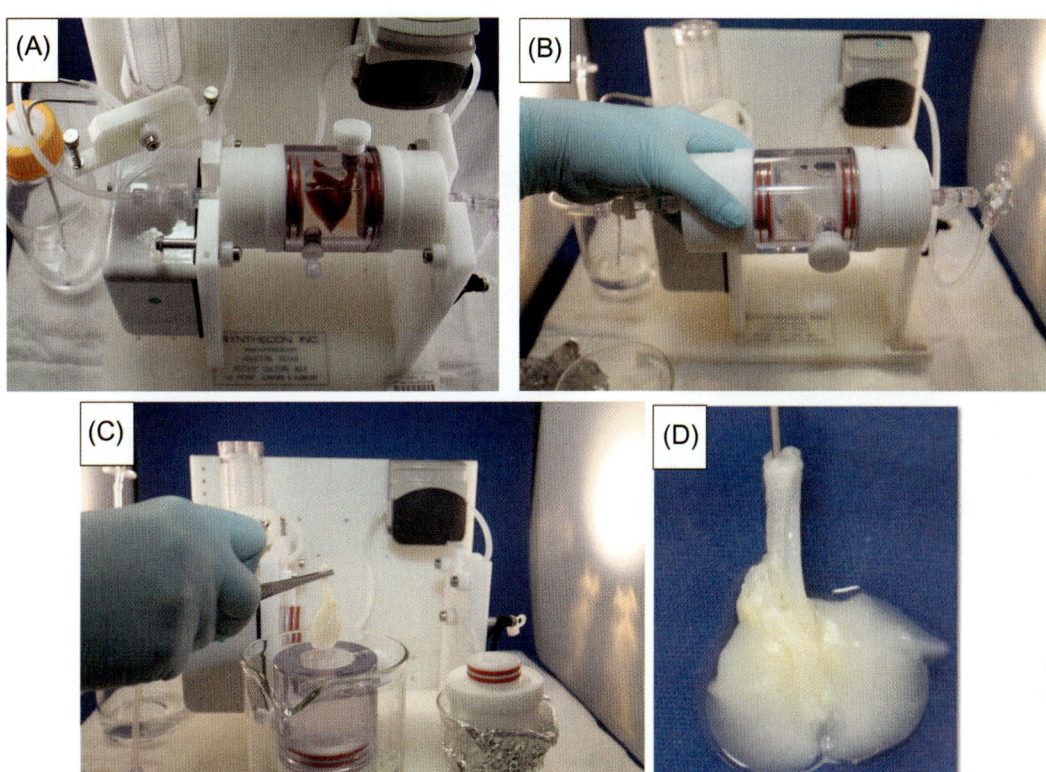

FIGURE 48.2   Bioreactor system for decellularization of mouse or rat trachea-lung. (A) Fresh lung in bioreactor chamber containing 2% SDS at the beginning of the decellularization process. (B) Decellularized lung in 1% SDS at the end of the decellularization process. (C) Acellular rat lung scaffold after washing in DNAase and PBS with antibiotics and antimycotics. (D) Acellular whole trachea−lung scaffold of rat. *Protocol from [29].*

nanoscale surface pattern fabrication, electrospinning, and self-assembly fabrication [42,46,47]. Electrospinning has been used to form 3D porous nonwoven mats or layers of biomaterials with geometries that support cell attachment, movement, and cell spreading. Electrospun meshes of collagen and elastin (1:1 ratio) have been produced from aqueous solution using this technology [48]. It is even possible to blend synthetic and natural polymers together to generate complex multicomponent scaffolds such as gelatin (degraded collagen) combined with either a conductive polymer, polyaniline, or with a mixture of PLA/PGA. These mixed scaffolds supported cell attachment and, depending on the pore size, induced cells to display different morphologies [48].

48.3 POTENTIAL CELL SOURCES FOR LUNG ENGINEERING

48.3.1 Endogenous Cell Sources

The lack of an adequate source of human cells for use in engineering lung remains a critical issue that continues to limit the development of functional engineered lung tissue

for transplantation. The generation of new lung tissue from lung progenitor cell populations is the most appealing option for a cell source since it would offer the possibility of autologous therapy or *in situ* therapy, which would minimize the risks of graft rejection and disease transmission. This is not the only option available, although, it is the least problematic one for the purpose of generation of an engineered lung for transplantation. The single most logical choice for a cell type to be used in the production of engineered lung tissue is the alveolar epithelial type II (AE II) cell of the distal lung. Regeneration of alveolar epithelia has been thought to be mediated by proliferation of existing AE II cells in the lung with their subsequent differentiation into alveolar epithelial type 1 (AE I) cell [49−55]. Adult AE II cells could be obtained from cadaveric human lungs not deemed suitable for transplantation. Currently, a variety of excellent procedures have been developed for the isolation of small numbers of AE II cells from human lung tissue [52−54]. Unfortunately, the current lack of isolation of sufficient numbers of endogenous lung progenitor cells or AE II cells as well as lack of proliferative capability of these cells *in vitro* make them an unsuitable choice for cell source for

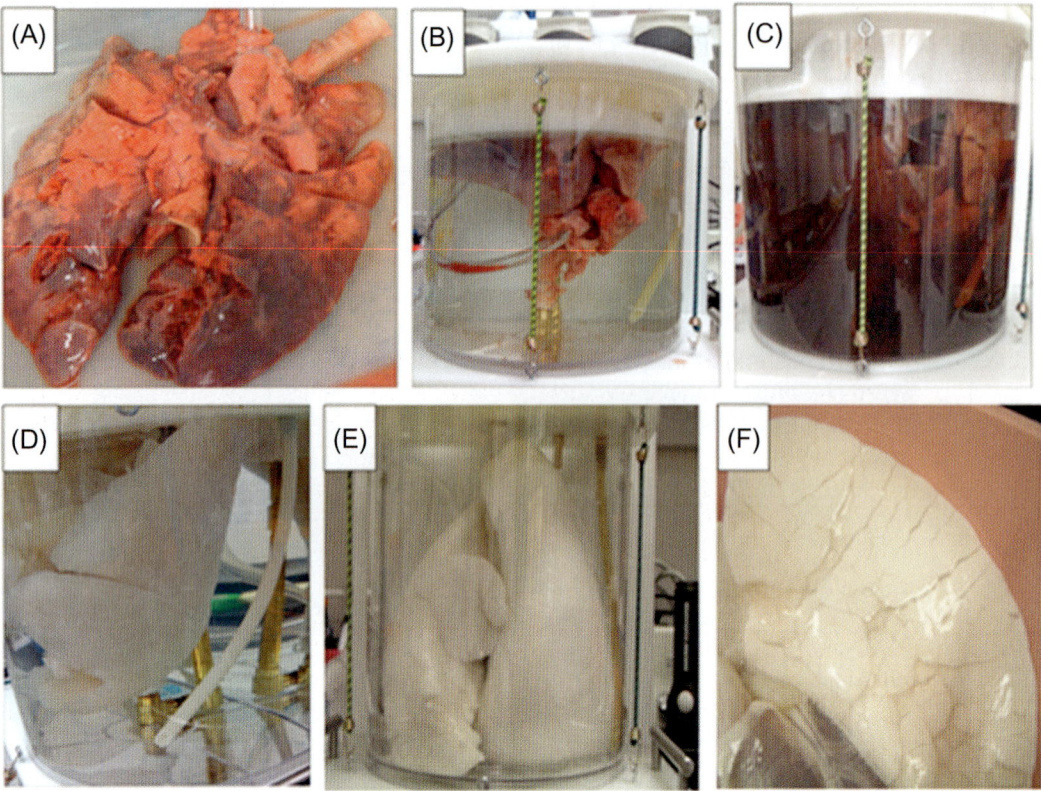

FIGURE 48.3 Large organ perfusion system for decellularization of pig lung. (A) Freshly isolated whole trachea-lung. (B) Picture of pig lung in the bioreactor chamber with cannula placed into pulmonary artery. (C) Image of chamber containing 2% SDS and cell debris on day 1 of the decellularization process. (D) Image of lung in chamber at end of decellularization process. (E) Lungs were expanded in the decellularization chamber to check for leaks in the pleura. (F) Acellular whole pig trachea-lung.

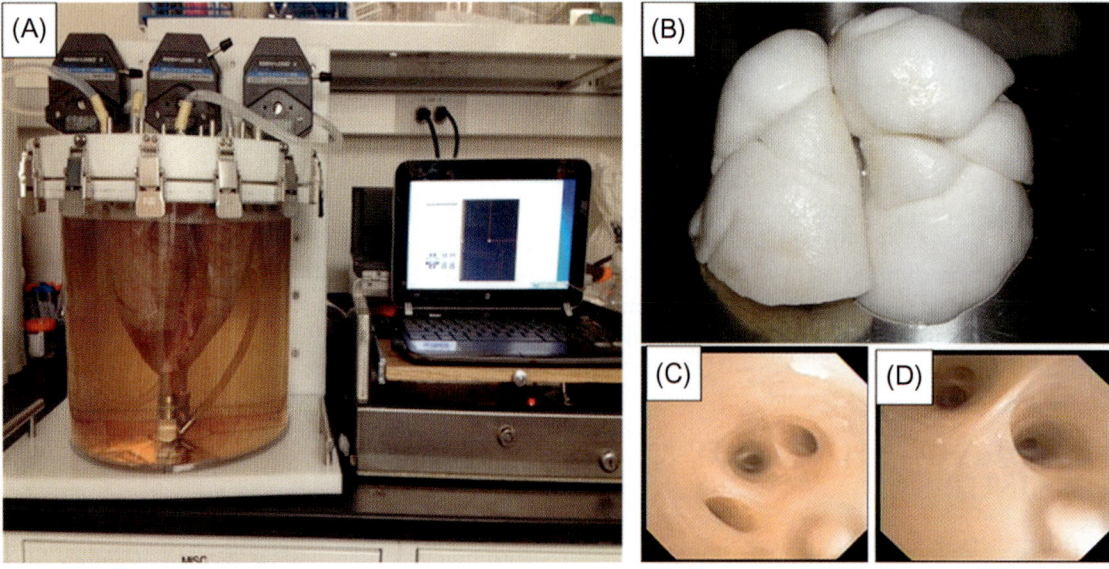

FIGURE 48.4 System for decellularization of human lung. (A) modified decellularization chamber showing the peristaltic pumps that were used to control flow into the trachea, pulmonary artery, and from the chamber into cell waste container. (B) Image of acellular lung scaffold produced from a pediatric lung. (C and D) Use of bronchoscope to examine the airways of the AC lung shown in panel B.

generation of lung tissue at this time. It is possible that culture procedures may be developed in the future which facilitate isolation and proliferation of these cells.

Defining endogenous stem or progenitor cell populations in animal or human lung has been difficult. This topic has been extensively reviewed [55−58]. Little is known regarding human endogenous lung stem and progenitor cells and their identity, organization, and regulation in the adult human lung. A number of excellent reviews on the topic exist [55−58], and although a number of putative stem or progenitor cell types have been identified in mice, many of these cell types have required that the lungs be damaged in order to identify cells involved in regeneration of epithelia [59−64]. Different epithelial cell types in distinct areas of the lung are sensitive to various agents and have been used in lung injury models to investigate stem cells in the adult lung [62−64]. The best approach for the identification of human lung stem or progenitor cells will require an approach based on isolation and clonal assay of cells similar to what has been done to identify mouse lung epithelial stem or progenitor cells [65].

There are few reports of lung-derived progenitor cells that have generated lung tissue *in vitro* or *in vivo*. One of the first reports regarding generation of lung tissue, by Nichols et al., described a mixed population of ovine c-kit positive somatic lung progenitor cells (SLPCs) that were used to produce surfactant protein C (SPC) as well as CC10 expressing cells [8]. Later work by other groups using human c-kit + lung stem cells showed that these cells were capable of self-renewal and multipotency *in vivo* with formation of thyroid transcription factor 1 (TTF-1) and other markers of lung lineage as well as development of lung tissue in injured mouse lungs [66]. Most recently, a population of putative human E-Cad/Lgr6þ stem cells isolated, and indefinitely expanded from human lungs, was shown to harbor both self-renewal capacity and the potency to differentiate *in vitro* and *in vivo*[67]. Single-cell injections of E-Cad/Lgr6þ cells in the kidney capsule of mice produce differentiated bronchioalveolar tissue, while retaining capacity for self-renewal, as demonstrated by serial transplantations under the kidney capsule or in the lung. Further work though is needed to examine the regenerative capacity of these progenitor cells. As stem cell advances make patient-specific stem cells easier to produce, maintain, and differentiate into lung lineages, we may begin to understand what will be required to produce tissues worthy of future clinical application.

48.3.2 Exogenous Cell Sources

Exogenous cells with the potential to be used to generate lung tissue include fetal lung-derived cells (FLC),

embryonic stem cells (ESC), and induced pluripotent stem cells (iPSC). Cells derived from rat or mouse fetal lung tissue have been used consistently by a number of research groups to produce mature lung cell lineages [13,27,28,32,34], but similar work has not been done using human fetal lung cells.

ESCs are pluripotent cells derived from the inner cell mass of the preimplantation embryos [68]. These cells can be maintained in culture in an undifferentiated state indefinitely, maintaining their proliferating capacity and have the unique potential to give rise to cells and tissues of all three embryonic germ layers [68]. ESCs have been shown to be able to differentiate into AE II cells as well as other lung lineage cells [32,69−75]. The first reports of derivation of a lung-specific cell from mouse ESCs showed production of a heterogeneous population of cells with low, differentiation efficiency [69−71]. Later studies where mouse ESCs were transferred in small airway growth medium (SAGM) showed a 20-fold increase in the expression of type II pneumocyte marker SPC [71]. The main disadvantages of directing ESC differentiation using a defined medium are the extensive assay time required and the low yields of target cells obtained [69−72]. One of the most promising studies using ESC circumvented the difficulties in generating a pure culture of ESC-derived AE II cells by generating stable transfected human embryonic stem cell (hESC) lines containing a single copy of an SPC-promoter-driven neomycin-resistant gene cassette [75]. In this study, production of AE II cells was accomplished without having to form embryotic bodies, and hESC lines were differentiated and cultured in the presence of neomycin to produce pure preparations of hESC-derived AE II cells with biological and phenotypic characteristics of AE II cells which demonstrated the ability to proliferate and differentiate into AE I cells [75].

iPSCs are derived by reprogramming adult cells into a primitive stem cell state. This process results in the creation of cells that are similar to ESCs in terms of their capability to differentiate into a variety of cell lineages, including endoderm that forms lung cell lineages [76]. Takahashi made the ground breaking discovery that four transcription factors (Klf4, Sox2, Oct4, and c-Myc), when introduced into mouse fibroblasts through retroviral transduction, led to the formation of cells with pluripotent properties [77]. Later, human pluripotent cells were generated by either the same set of transcriptional factors or another set of transcriptional factors that did not include the oncogenic transgene c-Myc [78]. There are few reports describing production of lung epithelial cells from iPSCs [78−80]. One report provides the most compelling argument for the use of a humanized version of a single lentiviral "stem cell cassette" vector to reprogram cells [79]. This single vector allowed for derivation of human iPSC containing a single excisable viral integration that on removal generates human iPSC free of integrated transgenes. This strategy was used to generate >100 lung disease-specific iPSC lines from individuals with a variety of diseases affecting the epithelial, endothelial, or interstitial compartments of the lung, including cystic fibrosis, α-1 antitrypsin deficiency-related emphysema, scleroderma, and sickle-cell disease. Use of this system was shown to robustly differentiate iPSC into definitive endoderm *in vitro* which is the developmental precursor tissue of lung epithelia [80]. The same techniques were used to generate lung and airway progenitor cells from mouse mesenchymal stromal cells (MSCs) and patient-specific cystic fibrosis iPSCs [80]. These disease-specific human lung progenitors formed respiratory epithelium when engrafted subcutaneously into immunodeficient mice [80].

Although the use of ESCs or fetal tissue as a source of progenitor cells has produced some promising results, tissues produced from these sources would have to be human leukocyte antigen (HLA) matched or engineered to express the graft recipient's HLA repertoire in order to be used for human organ transplantation. This would require the development of ESC or fetal tissue banks and, if a perfect HLA or tissue match was not available, immunosuppressive treatment of recipients of the engineered tissues. Although there are many problems that still must be addressed for ESC or iPSC technologies, such as the low efficiency of differentiation to lung lineages or the tendency for tumors to evolve after transplantation, these cell types may offer the best solutions for a renewable cell source for use in the future in the engineering of lung tissue.

48.3.3 Good Manufacturing Practice Considerations Related to Clinical Applications of Stem Cells

Production of engineered lung tissues from any cell source will require considerable cell expansion and manipulation *in vitro* prior to cell−scaffold construct formation. Because of potential contamination or damage to tissues, related compliance requirements for "good manufacturing practice" (GMP) and "good tissue practice" will have to be followed for cell sources used in the production of engineered lung [81,82]. Mechanisms to confirm that cell differentiation only occurs along desired lineages will also be necessary. Evaluation of karyotypic stability of the cells used and the possibility of formation of genetic alterations due exclusively to the manipulation of cells *in vitro* will also need to be addressed. The same potential for self-renewal and plasticity that makes embryonic and iPSCs or fetal germ tissue attractive

sources for production of engineered lung also raises concerns about their potential for tumorigenicity. Specific tumorigenic lung cancer stem cell populations [83] or transformed counterparts of lung-derived stem cells with the potential to give rise to carcinomas have been identified [84]. Regulatory frameworks may actually require the implantation of differentiated tissues rather than growth factor primed cell−matrix constructs in order to limit problems associated with using stem or progenitor cells for the engineering of tissues. The risk that adult growth factor primed endogenous lung progenitor cells or adult AE II cells have the capacity to give rise to cancer or form tumors is fairly low since the transforming events necessary to promote tumorigenic properties in stem cells requires more than the normal effects of growth factor activation and priming. It is also important to note that there was no indication of tumor formation from implantation of *in vitro* amplified, growth factor primed SLPCs into immunodeficient mice by Nichols et al. although development of mature lung lineage cells was shown [8]. Further work is needed to define the tumorigenic capabilities of endogenous lung progenitor cell populations or other potential cell sources.

48.4 ENGINEERING OF LUNG TISSUE

48.4.1 Early *In Vitro* Production of Lung Tissue

Early simple *in vitro* models of the lung, although very simplistic duplicated some rudimentary aspects of lung function. Information gained from simple one- or two-cell-type model systems paved the way for the design and execution of later complex multicell engineered tissue equivalents and has been extensively reviewed by the authors [5,6].

The first report describing transplantation of cell−scaffold constructs to generate lung tissue was done by Nichols et al. [8]. In this study ovine CD117 + SLPC-scaffold constructs were produced using a combination of PGA and/or PF-127. These constructs were implanted into both a large animal model (sheep) and a small animal model (nude mouse). In sheep, autologous cell−scaffold constructs were implanted directly into the right upper lobe of the lung into a pocket created by a wedge resection. In nude mice, similar cell constructs were placed on the animals' backs. Implanted constructs were well tolerated in these studies and lung tissue assembly was facilitated *in vivo* by the use of the PGA scaffold. In this same study, SLPC−PGA constructs formed on felt sheets of PGA were implanted into the thoracic cavity of three adult sheep with attachment of the construct to the right main stem bronchus site following a full pneumonectomy (Figure 48.5A and B). When harvested after 3 months,

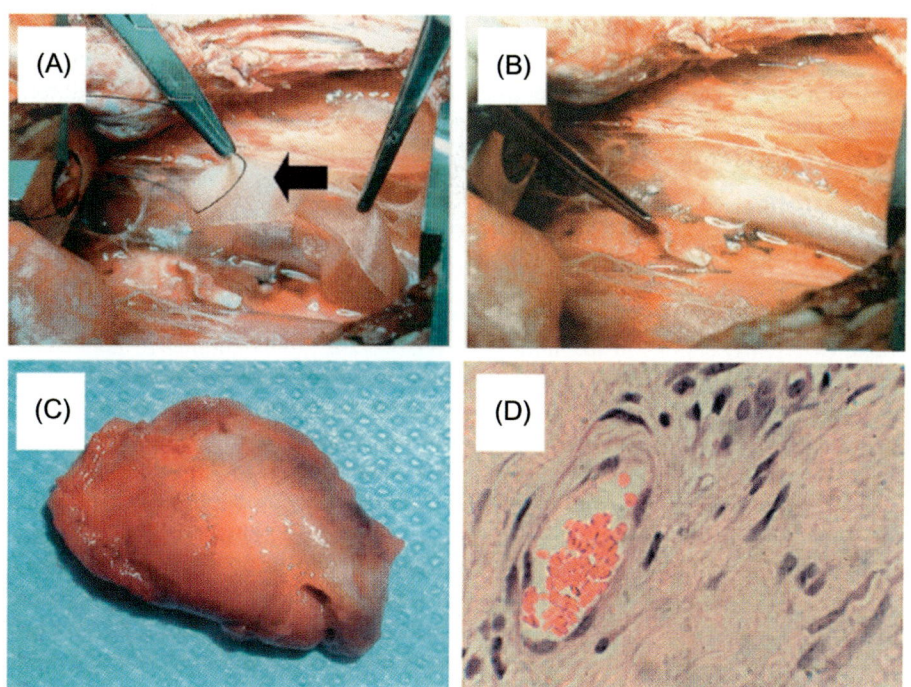

FIGURE 48.5 Implantation of somatic progenitor cell/PGA construct after a pneumonectomy. *In vivo* tissue-engineered ovine lung using SLPC/PGA construct. (A and B) Surgery showing implantation of construct at the right main stem bronchus site. Black arrow points to SLPC/PGA sheet being implanted at pneumonectomy site. (C) Engineered lung tissue after excision. (D) Section of engineered lung, produced 3 months after implantation, stained with hematoxylin and eosin (magnification 400X).

these implants were not shown to support development of lung epithelia but did form extremely well-vascularized fleshy tissue fragments (Figure 48.5C and D). Gelfoam™ constructs (sponge with cells) delivered into the lung by injection of the constructs (sponge with fetal rat lung-derived cells) into normal lung parenchyma has also been used to facilitate formation of lung tissue *in vivo*[13]. Most newly formed alveolar-like structures, though, were found close to the border between the sponge and the surrounding normal tissue with few found within the sponge itself. It is unclear why this occurred since the porosity of the Gelfoam™ scaffold should have provided cells with an adequate microenvironment to support cell movement and tissue development. Mondrinos et al. developed cell–scaffold constructs using fetal pulmonary cells (FPCs) grown on a Matrigel™ scaffold [9–12]. To examine the contribution of donor-derived endothelial cells to tissue construct vascularization, FPC–Matrigel™ constructs were injected into the anterior abdominal cavity of adult graft recipients (C57Bl6 mice) [12]. Implantation of Matrigel™ alone resulted in infiltration of host cells into the scaffold. In FPC–Matrigel™-FGF2 constructs the tissues were shown to develop ductal epithelial structures with development of pro-SPC expressing epithelial cells and patent vasculature. Vascularization of the construct was enhanced by addition of FGF2, and even in FGF2–Matrigel™ implantations alone there was significant increase in capillary density.

One of the first descriptions using a preparation of human AC lung scaffold evaluated AE II cell adherence and viability using strips of AC lung alveolar matrix seeded with rat AE II cells to examine the influence of ECM on cell attachment and morphology. The AC scaffolds described contained collagen I, II, IV, and V as well as laminin and fibronectin [33]. Rat AE II cells seeded on the strips of human AC lung scaffold took on some of the morphological characteristics of AE I cells such as loss of lamellar bodies and cytoplasmic flattening. From 1986 until 2010, there was no progress in the development of AC lung scaffolds for the engineering of lung tissue. This changed dramatically in 2010 when four research groups published their findings related to the production of engineered lung using whole trachea-lung mouse or rat AC scaffolds [27,28,32,34]. In these studies, whole AC trachea-lung scaffolds were seeded and cultured with mouse [34], rat fetal lung cells (FLCs) [28], neonatal rat [27], or murine embryonic stem cells (mESCs) [32]. All groups reported good cell attachment, survival of cells, and limited differentiation of cells into lung-specific cell phenotypes. All groups also reported that the recellularization of the scaffolds was not complete and that there were areas of the whole lung scaffolds that had not been adequately populated. Whole rat AC trachea-lung scaffold has also been shown to support attachment of murine bone marrow–derived MSCs or C10 mouse lung epithelial cells following intratracheal inoculation [35,36]. Although the MSCs were cultured in small airway growth medium (SAGM), the MSCs predominately expressed genes consistent with a mesenchymal or osteoblast phenotype, and no airway genes or vascular genes were expressed. Similar results were found when AC nonhuman primate lungs produced using a perfusion decellularization system were populated with rhesus bone marrow–derived MSCs and adipose tissue. [36].

48.4.2 *In Vivo* use of AC Lung Scaffolds

Use of whole AC trachea-lungs to engineer tissue makes the process of transplantation into animal models or potential clinical applications easier to work with since the lungs (i) fit properly into the thoracic cavity, (ii) support functions of the lung similar to normal lung, and (iii) possess the appropriate anatomical structures (trachea or bronchus) that allow it to be sutured in place in order to facilitate orthotopic transplantation. Engineered lung produced on AC scaffolds has been transplanted into animal models although graft recipient survival was limited in these early studies [27,28,37]. For the first of these studies, orthotopic implantation of an engineered left lung derived from neonatal rat lung cells cultured on whole AC rat lung matrix for up to 8 days was done following a left thoracotomy [27]. When implanted into rats for short time intervals, 45–120 min, the engineered lungs were shown to participate in gas exchange although there was some bleeding into the airways (Figure 48.3C). A second group engineered lung using rat neonatal lung cells cultured on AC rat scaffold for 5 days [28]. The engineered lung was transplanted orthotopically following a left-sided pneumonectomy. In these initial experiments by this research group, graft function *in vivo* was limited to a few hours due to development of pulmonary edema in the engineered lung. Later studies by the same group used engineered lungs derived from AC rat scaffolds seeded with human umbilical vein endothelial cells (HUVECS) and rat FLCs cultured for 7–10 days [37]. For these experiments, engineered lungs were transplanted into recipient athymic rats following a pneumonectomy. Athymic rats receiving a pneumonectomy with no transplant, or rats receiving transplantation of cadaveric lungs from Sprague Dawley rats were used as controls. Cadaveric lungs were shown to have slightly higher compliance levels than engineered lungs on postoperative day 7. Oxygenation levels for both engineered and cadaveric recipients was higher than for pneumonectomized animals up to day 7 but gradually declined in rats receiving an engineered lung between 7 and 14 days. Compliance and gas exchange of the engineered lung in this study gradually declined after 7 days due to progressive graft consolidation and inflammation. Engineered lung grafts but not cadaveric grafts also induced the formation of a thick fibrous scar surrounding the graft, causing restriction of graft expansion. This scarring was potentially due to a residual natural killer (NK) cell response in the athymic recipient [37]. Despite the problems encountered

following transplantation of these engineered tissues, this study provides hope that in the future we will be capable of engineering functional lung tissue capable of long term growth and survival *in vivo*.

48.4.3 Role of AC Matrix in Lung Lineage Differentiation

Lung ECM plays a major role in support of biomechanics in the lung [85], but we are just beginning to understand the influence of lung ECM on the differentiation of stem or progenitor cells and of subsequent tissue formation. Studies to examine this role have been limited although the field of tensegrity or mechanosensing has long suggested that (i) regional variation of ECM remodeling that occurs during embryogenesis leads to local differentials in ECM structure and mechanics, (ii) changes in matrix compliance (e.g., increased stiffness when the thinned basement membrane is stretched) alter mechanical force balance across membrane receptors that mediate cell-ECM adhesion, and (iii) altering the level of forces that are transmitted to the internal cytoskeleton will produce cell distortion and change intracellular biochemistry, thereby switching cells between growth, differentiation, and apoptosis [26,85–88]. As mentioned earlier, there is extensive regional variation in ECM composition and stiffness as you progress from the trachea to the bronchi and bronchioles and then to distal lung. Changes in ECM structure and composition could influence cell adhesion and provide critical cues that orchestrate tissue formation and cell function [88]. There have been a few reports supporting a role for lung-specific scaffold or at least cell–ECM interactions in influencing differentiation of stem or progenitor cells [10,32,86,87]. Early studies suggested that specific components of ECM materials could strongly influence the differentiation of ESC into pneumocytes [32]. One of the first reports regarding the influence of ECM on ESC differentiation into lung epithelial and endothelial lineages examined the efficiency of differentiation after allowing mESCs to attach to individual components of ECM. In these studies, ECM proteins collagen I, laminin, and fibronectin were shown to induce production of AE II cells from mESC cultured in 2D or 3D [10]. Production of surfactant proteins C and A and aquaporin-5 were enhanced by the presence of laminin. Similar results were found when culturing mESCs on whole AC rat lung scaffold [32,73]. Efficiency of differentiation was measured here by evaluation of CD31, cytokeratin-18, and pro-SPC expression and was shown to be increased by mESCs cultured on AC lung compared to commercially available matrices Gelfoam™, collagen I, or Matrigel™ [32]. Production of organized lung tissue as well as significant production of surfactant proteins A and

C were only seen for mESCs cultured on AC lung and not on any of the other matrices used. Other reports also support the concept that organ-specific stroma or ECM may even be required for proper site-specific differentiation and organization of lung tissues [87]. Comparison between liver- and lung-derived AC scaffolds indicated that liver derived scaffolds maintained the differentiation state of primary hepatocytes while lung-derived scaffolds allowed for both induction of lung lineage and maintenance of site-specific development of AE II cells [87].

48.4.4 Problems Associated with Use of AC Lung Scaffolds

The clinical application and success of decellularized trachea suggests that AC scaffolds may retain adequate strength to support both physiologic and anatomic functions of the trachea [3,89]. This may or may not be true for the lung. There are some limitations that must be considered before use of decellularized natural AC scaffold for development of lung tissues for clinical applications. Problems associated with the use of AC or natural scaffold material include the possibility that natural materials may harbor bacteria or viruses if adequate steps have not been taken to ensure the cleanliness of the materials produced. Retention of unwanted materials could lead to induction of immunological reactions leading to inflammation and graft rejection. The process of decellularization can also lead to degradation of the ECM, resulting in depletion of collagen, glycosaminoglycan, and elastin which compromises the mechanical integrity of the scaffold and any tissues generated on it. We must consider the effects of the decellularization process on the mechanical integrity of the lung ECM prior to and after engineering of lung tissue. As we have mentioned previously, AC ECM may be weakened, damaged, or degraded during the decellularization process. Quantitative evaluation of the composition of lung scaffolds suggests that there are great variations in AC ECM related to specific detergents used [41,43]. We do not know if the alteration in ECM composition will be detrimental to the production of tissues for the purpose of transplantation. What we know is that increased ECM production leading to deposition of collagen types I and III in fibrotic lungs results in lower lung compliance during assessment of pulmonary function tests (PFTs). Compliance levels are an important indicator of lung elasticity and, in disease states, correlate well with collagen composition of the lung ECM. Some groups reported lower compliance values for AC compared to normal lung in vitro[27,28,34,37] and after transplantation [37].

Our understanding of the link between lung tissue structure and function is incomplete due to the complexity of the problem and may not be easily understood or evaluated properly without examination of the ensemble behavior of all constituents of the lung which includes cells and ECM [90,91]. Culturing of selected cells on the AC lung scaffold will result in immediate changes to the ECM. Modifications to and remodeling of the ECM begins the moment cells attach. We may need to carefully examine the modifications made by cells to the AC ECM in order to determine what components of the ECM are truly critical to development of functional engineered tissues. For example, although SDS treatment removed basement membrane components during the process of decellularization, mESCs began to produce laminin and collagen IV relatively quickly following attachment and replaced the missing ECM components [32]. A recent comparative analysis of whole decellularized mouse lungs produced using Triton X-100/sodium deoxycholate, SDS or CHAPs also showed that although there were differences in gelatinase activation and protein composition of the AC scaffold, binding and initial growth following intratracheal inoculation with MSCs or C10 cells were similar [39].

Although cells and cell debris have been removed from the scaffold along with the HLAs, which are responsible for graft rejection, this does not mean that the scaffolds are no longer immunogenic. The host immune response to transplanted tissues or organs is an important determinant of graft function and survival. Natural scaffolds may induce immunogenic responses and invoke immunological reactions leading to inflammation and thereby causing graft rejection. A hallmark of tissue injury is increased turnover of ECM proteins [92]. Failure to remove ECM degradation products from the site of tissue injury or tissue remodeling can result in the induction of inflammation in the host. Hyaluronan is an important component of the lung interstitium and functions to help maintain the structural integrity of the lung. This protein also plays a major role in cell signaling following lung injury. Fragments of hyaluronan have been shown to trigger toll-like receptor (TLR), TLR-2- and TLR-4-dependent inflammation activation pathways resulting in initiation of inflammatory responses [92]. The immune response to AC scaffolds has not been studied extensively, and we need to develop a better understanding of the host response to the AC scaffold alone [93]. Investigation of the immunomodulatory effects of natural ECM scaffolds has recently indicated that tissue source, decellularization method, and chemical cross-linking modifications of scaffolds can affect the presence of damage associated molecular patterns (DAMPS) within the biologic scaffold. Presence of these DAMPS has correlated with differences in cell proliferation, cell death, secretion of proinflammatory chemokines, and upregulation of TLR-4 proinflammatory molecules and may influence remodeling of the ECM and associated tissue [94].

48.4.5 The Development of AC Lung Scaffolds for Clinical Use

Before discarded human lungs can be considered for use in clinical applications, there are some issues that must be addressed. Policies need to be developed related to organ procurement for development of AC scaffolds from discarded lungs. Production standards, processing and sterilization methods, evaluation of the product, and handling requirements prior to and after production of these scaffolds will have to be established. We will need to evaluate the influence of the manufacturing process on ECM composition and immunogenicity and determine what components of the AC scaffold ECM are critical for tissue engineering. Poorly manufactured lung scaffolds may also lead to an induction of immune responses in recipients, alter the process of tissue remodeling, and affect the functional outcome of the engineered lung. Natural materials may also harbor bacteria or viruses if adequate steps have not been taken to ensure the cleanliness of the materials produced. GMPs such as training and certification of personnel and design of clean room technologies in order to guarantee the safety and quality of each scaffold will also need to be developed. Sizing and long-term storage concerns need to be addressed as does the impact of potential complications related to the variability of each scaffold due to donor diversity. While xenografts are regulated as medical devices, most allograft tissue is classified as a Human Cell & Tissue/Product (HCT/P) by the FDA and not as a medical device. The FDA does not require a specific sterilization technique or Sterility Assurance Level (SAL) for allografts. Similar practices will have to be developed for the cell source to be used to engineer lung tissue for transplantation. 21 CFR Part 1271, became effective on April 4, 2001 for human tissues intended for transplantation that are regulated under section 361 of the PHS Act and 21 CFR Part 1270 [95]. This is a comprehensive plan for regulating human cells, tissues, or cellular and tissue based products that would include establishment of registration and product listing, donor suitability requirements, good tissue practice regulations, and other appropriate production requirements.

48.4.6 Vascularization of Engineered Lung

Generation of a functional vasculature is essential to the clinical success of any engineered tissue construct and remains a key challenge for the generation of clinically applicable engineered lung as well as for the field of regenerative medicine. Oxygen and nutrients cannot be delivered to cells residing in the interior of large-volume scaffolds via diffusion alone. The topic of vascular engineering has been extensively reviewed elsewhere [96,97] and will not be covered in detail in this chapter. In brief, a variety of scaffold materials and scaffold-free systems have been used to generate engineered vessels, but complete biomimetic constructs have not yet been produced. The same problems which limit the field of lung engineering also plague the field of blood vessel engineering such as lack of design and construction of appropriate scaffolds and lack of identification of adequate cell sources. A few novel approaches that could be used to produce vascular networks for lung engineering are worth mentioning at this time. Most recently developed vascularizing approaches for biomaterials attempt to mimic cellular communication that occurs in tissues and organs through the use of multiple cell types and biochemically modified scaffolds [97]. These attempts to match natural biological and mechanical cues through the use of ECM components have inspired the creation of scaffolds that incorporate growth factors, integrin-binding peptides, vasculature-forming cells, and physical features that promote diffusion and transport [97]. One novel approach to engineering blood vessels used sandwiched, striped patterns of endothelial cell and fibroblast sheets to control the formation of vasculature in the tissue, and in doing so identified that cell-to-cell interactions were critical in the fabrication of a specific 3D tissue structure with specific endothelial cell orientation [98].

Advances in fabrication technology have enabled the construction of microfluidic networks in scaffolds. Microscale technologies such as photolithography, micromachining, and micromolding have been utilized to fabricate 2D microfluidic networks on various biomaterials. These techniques allow for design of microfluidic networks by combining an oxygen transport simulation with biomimetic principles governing biological vascular trees [99]. One such system formed using a lost mold shapeforming process based on microstereolithography was used in the fabrication of a porous scaffold containing a microfluidic vascular network which was shown to provide oxygen and support cell proliferation under static and perfusion conditions [99]. This approach established a practical basis for designing an effective microfluidic network in a cell-seeded scaffold.

In vivo angiogenic sprouting occurs and generation of capillary beds occurs naturally and has been reproduced using 3D capillary beds in an *in vitro* microfluidic platform comprised of a biocompatible collagen I gel supported by a mechanical framework of alginate beads [100].The engineered vessels had patent lumens, formed robust ~1.5 mm capillary networks, and supported the perfusion of 1 μm fluorescent beads through them. Similar procedures may be used to produce a functional alveolar-endothelial junction which is the first of many steps that will be necessary in the production of a fully vascularized engineered lung construct. In the future, further research into mechanisms for regulating the spatial aspects of vascularization will be essential in order to produce vascularized lung capable of surviving after transplantation.

48.5 CONCLUSIONS

While the studies outlined in this chapter have generated great enthusiasm about the potential for regeneration of lung tissue for therapeutic use, the reality is that engineering of all of the critical component parts of lung tissue which support the physiologic function of the lung has been limited. In order to produce engineered lung tissue in the future for regenerative therapeutics, we will have to (i) select the cell source with the highest potential for use in transplantation (HLA matched) which may require tissue typing and HLA matching of banked cells; (ii) select the best biocompatible and degradable scaffold suitable for lung development, that takes into consideration the type of lung disease to be treated (restrictive or obstructive); (iii) determine the best combinations of growth factors and culture conditions which promote cellular differentiation and increase the efficiency of differentiation of ESCs, FLCs, or autologous stem cells in a manner suitable for *in vitro* production of 3D tissue engineered culture; (iv) select the best conditions overall which promote 3D production of lung tissue.

The progress toward development of individual components of engineered lung to date has been encouraging, particularly in the differentiation of ESCs and iPSCs into AE II cells. Although significant advances must be made before we will achieve engineered tissues worthy of clinical application we have progressed to the point of *in vivo* implantation of engineered tissues. We will need to develop a better understanding of factors promoting vascularization of engineered tissues before we can begin to engineer lung tissue capable of being used in clinical applications.

REFERENCES

[1] Zych B, Popov AF, Stavri G, Bashford A, Bahrami T, Amrani M, et al. Early outcomes of bilateral sequential single lung transplantation after *ex vivo* lung evaluation and reconditioning. J Heart Lung Transplant 2012;31(3):274–81.

[2] Medeiros IL, Pego-Fernandes PM, Mariani AW, Fernandes FG, do Vale Unterpertinger F, Canzian M, et al. Histologic and functional evaluation of lungs reconditioned by *ex vivo* lung perfusion.

J Heart Lung Transplant 2011; <http://dx.doi.org/10.1016/j.healun.2011.10.005>

[3] Macchiarini P, Jungebluth P, Go T, Asnaghi MA, Rees LE, Cogan TA, et al. Clinical transplantation of a tissue-engineered airway. Lancet 2008;372:2023–30.

[4] Pedersen PS, Frederiksen O, Holstein-Rathlou NH, Larson PL, Qvortrup K. Ion transport in epithelial spheroids derived from human airway cells. Am J Physiol Lung Cell Mol Physiol 1999;276:75–80.

[5] McAteer JA, Cavanagh TJ, Evan AP. Submersion culture of the intact fetal lung. In Vitro 1983;19:210–8.

[6] Nichols JE, Niles JA, Cortiella J. Design and development of tissue engineered lung: progress and challenges. Organogenesis 2009;5:57–61.

[7] Nichols JE, Cortiella J. Engineering of a complex organ: progress toward development of a tissue-engineered lung. Proc Am Thorac Soc 2008;5:723–30.

[8] Nichols JE, Niles JA, Cortiella J. Production and utilization of acellular lung scaffolds in tissue engineering. J Cell Biochem 2012;113:2185–92.

[9] Cortiella J, Nichols JE, Kojima K, Bonassar LJ, Dargon P, Roy AK, et al. Tissue-engineered lung: an *in vivo* and *in vitro* comparison of polyglycolic acid and pluronic F-127 hydrogel/somatic lung progenitor cell constructs to support tissue growth. Tissue Eng 2006;12:1213–25.

[10] Mondrinos MJ, Koutzaki S, Jiwanmall E, Li M, Dechadarevian JP, Lelkes PI, et al. Engineering three-dimensional pulmonary tissue constructs. Tissue Eng 2006;12:717–28.

[11] Lin YM, Zhang A, Rippon HJ, Bismarck A, Bishop AE. Tissue engineering of lung: the effect of extracellular matrix on the differentiation of embryonic stem cells to pneumocytes. Tissue Eng, Part A 2010;16:1515–26.

[12] Fritsche CS, Simsch O, Weinberg EJ, Orrick B, Stamm C, Kaazempur-Mofrad MR, et al. Pulmonary tissue engineering using dual-compartment polymer scaffolds with integrated vascular tree. Int J Artif Organs 2009;32:701–10.

[13] Mondrinos MJ, Koutzaki SH, Poblete HM, Crisanti MC, Lelkes PI, Finck CM. *In vivo* pulmonary tissue engineering: contribution of donor-derived endothelial cells to construct vascularization. Tissue Eng, Part A 2008;14:361–8.

[14] Andrade CF, Wong AP, Waddell TK, Keshavjee S, Liu M. Cell-based tissue engineering for lung regeneration. Am J Physiol Lung Cell Mol Physiol 2007;292:510–8.

[15] Blau H, Guzowski DE, Siddiqi ZA, Scarpelli EM, Bienkowski RS. Fetal type 2 pneumocytes form alveolar-like structures and maintain long-term differentiation on extracellular matrix. J Cell Physiol 1988;136:203–14.

[16] Shannon JM, Mason RJ, Jennings SD. Functional differentiation of alveolar type II epithelial cells *in vitro*: effects of cell shape, cell-matrix interactions and cell-cell interactions. Biochim Biophys Acta 1987;931:143–56.

[17] Sugihara H, Toda S, Miyabara S, Fujiyama C, Yonemitsu N. Reconstruction of alveolus-like structure from alveolar type II epithelial cells in three-dimensional collagen gel matrix culture. Am J Pathol 1993;142:783–92.

[18] Zhang WJ, Lin QX, Zhang Y, Liu CT, Qui LY, Wang HB, et al. The reconstruction of lung alveolus-like structure in collagen-matrigel/microcapsules scaffolds *in vitro*. J Cell Mol Med 2011;15:1878–86.

[19] Mondrinos MJ, Koutzaki S, Lelkes PI, Finck CM. A tissue-engineered model of fetal distal lung tissue. Am J Physiol Lung Cell Mol Physiol 2007;293:639–50.

[20] Kibby MC. Maintenance of the EHS sarcoma and matrigel preparation. J Tissue Cult Methods 1994;16:227–30.

[21] Badylak SF, Freytes DO, Gilbert TW. Extracellular matrix as a biological scaffold material: structure and function. Acta Biomater 2009;5:1–13.

[22] Crapo PM, Gilbert TW, Badylak SF. An overview of tissue and whole organ decellularization processes. Biomaterials 2011;32:3233–43.

[23] Gilbert TW, Sellaro TL, Badylak SF. Decellularization of tissues and organs. Biomaterials 2006;27:3675–83.

[24] Gilbert TW. Strategies for tissue and organ decellularization. J Cell Biochem 2012;113:2217–22.

[25] Pizzo AM, Kokini K, Vaughn LC, Waisner BZ, Voytik-Harbin SL. Extracellular matrix (ECM) microstructural composition regulates local cell-ECM biomechanics and fundamental fibroblast behavior: a multidimensional perspective. J Appl Physiol 2005;98:1909–21.

[26] Engler AJ, Sweeney HL, Disher DE. Matrix elasticity directs stem cell lineage specification. Cell 2006;126:667–89.

[27] Lwebuga-Mukasa JS, Ingbar DH, Madri JA. Repopulation of a human alveolar matrix by adult rat type II pneumocytes *in vitro*: a novel system for type II pneumocyte culture. Exp Cell Res 1986;162:423–35.

[28] Petersen TH, Calle EA, Zhao L, Lee EJ, Gui L, Raredon MB, et al. Tissue-engineered lungs for *in vivo* implantation. Science 2010;329:538–41.

[29] Cortiella J, Niles J, Cantu A, Brettler A, Pham A, Vargas G, et al. Influence of acellular natural lung matrix on murine embryonic stem cell differentiation and tissue formation. Tissue Eng, Part A 2010;16:2565–80.

[30] Gross TJ, Hunninghake GW. Idiopathic pulmonary fibrosis. N Engl J Med 2001;345:517–25.

[31] Phillips JE, Peng R, Burns L, Harris P, Garrido R, Tyagi G, et al. Bleomycin induced lung fibrosis increases work of breathing in the mouse. Pulm Pharmacol Ther 2012;25:281–5.

[32] Anciães AM, Olivo CR, Prado CM, Kagohara KH, Pinto TS, Moriya HT, et al. Respiratory mechanics do not always mirror pulmonary histological changes in emphysema. Clinics 2011;66:1797–803.

[33] Suki B, Ito S, Stamenović D, Lutchen KR, Ingenito EP. Biomechanics of the lung parenchyma: critical roles of collagen and mechanical forces. J Appl Physiol 2005;98:1892–9.

[34] Ott HC, Clippinger B, Conrad C, Schuetz C, Pomerantseva I, Ikonomou L, et al. Regeneration and orthotopic transplantation of a bioartificial lung. Nat Med 2010;16:927–33.

[35] Price AP, England KA, Matson AM, Blazar BR, Panoskaltsis-Mortari A. Development of a decellularized lung bioreactor system for bioengineering the lung: the matrix reloaded. Tissue Eng, Part A 2010;16:2581–91.

[36] Daly AB, Wallis JM, Borg ZD, Bonvillain RW, Deng B, Ballif BA, et al. Initial binding and recellularization of decellularized mouse lung scaffolds with bone marrow-derived mesenchymal stromal cells. Tissue Eng, Part A 2012;18:1–16.

[37] Peterson TH, Calle EA, Colehour MB, Niklason LE. Matrix composition and mechanics of decellularized lung scaffolds. Cells Tissues Organs 2012;195:222–31.

[38] Wallis JM, Borg ZD, Daly AB, Deng B, Ballif BA, Allen GB, et al. Comparative assessment of detergent-based protocols for mouse lung de-cellularization and re-cellularization. Tissue Eng, Part C Methods 2012;18:420–32.

[39] Song JJ, Kim SS, Liu Z, Madsen JC, Mathisen DJ, Vacanti JP, et al. Enhanced *in vivo* function of bioartificial lungs in rats. Ann Thorac Surg 2011;92:998–1005 [Discussion 1005–1006].

[40] Kelleher CM, Vacante JP. Engineering extracellular matrix through nanotechnology. J R Soc Interface 2010;7:717–29.

[41] Ayres CE, Jha BS, Sell SA, Bowlin GL, Simpson DG. Nanotechnology in the design of soft tissue scaffolds: innovations in structure and function. Wiley Interdiscip Rev Nanomed Nanobiotechnol 2010;2:20–34.

[42] Shanbhag S, Wang S, Kotov NA. Cell distribution profiles in three-dimensional scaffolds with inverted-colloidal-crystal geometry: modeling and experimental investigations. Small 2005;1:1208–14.

[43] Kotov NA, Liu Y, Wang S, Cumming C, Eghtedari M, Vagas G, et al. Inverted colloidal crystals as three-dimensional cell scaffolds. Langmuir 2004;20:7887–92.

[44] Bonvillain RW, Danchuk S, Sullivan DE, Betancourt AM, Semon JA, Eagle ME, et al. A nonhuman primate model of lung regeneration: detergent-mediated decellularization and initial *in vitro* recellularization with mesenchymal stem cells. Tissue Eng, Part A 2012;18:2437–52.

[45] Liu Y, Wang S, Krouse J, Kotov NA, Eghtedari M, Vargas G, et al. Rapid aqueous photo-polymerization route to polymer and polymer-composite hydrogel 3D inverted colloidal crystal scaffolds. J Biomed Mater Res A 2007;83:1–9.

[46] Li M, Mondrinos MJ, Gandhi MR, Ko FK, Weiss AS, Lelkes PI. Electrospun protein fibers as matrices for tissue engineering. Biomaterials 2005;26:5999–6008.

[47] Agarwal S, Wendorff JH, Greiner A. Progress in the field of electrospinning for tissue engineering applications. Adv Mater 2009;21:3343–51.

[48] Lelkes PI, Li M, Perets A, Mondrinos MJ, Gou Y, Chen X, et al. Designing intelligent polymeric scaffolds for tissue engineering: blending and co-electrospinning synthetic and natural fibers. In: Gdoutos EE, editor. Experimental analysis of nano and engineering materials and structures. Dordrecht, The Netherlands: Springer; 2007. p. 831–2.

[49] Buttafoco L, Kolkman NG, Engbers-Buijtenhuijs P, Poot AA, Dijkstra PJ, Vermes I, et al. Electrospinning of collagen and elastin for tissue engineering applications. Biomaterials 2006;27:724–34.

[50] Kapanci Y, Weibel ER, Kaplan HP, Robinson FR. Pathogenesis and reversibility of the pulmonary lesions of oxygen toxicity in monkeys. II. Ultrastructural and morphometric studies. Lab Invest 1969;20:101–17.

[51] Danto SI, Shannon JM, Borok Z, Zabski SM, Crandall ED. Reversible transdifferentiation of alveolar epithelial cells. Am J Respir Cell Mol Biol 1995;12:497–502.

[52] Campbell L, Hollins AJ, Al-Eid A, Newman GR, von Ruhland C, Gumbleton M. Caveolin-1 expression and caveolae biogenesis during cell transdifferentiation in lung alveolar epithelial primary cultures. Biochem Biophys Res Commun 1999;262:744–51.

[53] Driscoll B, Kikuchi A, Lau AN, Lee J, Reddy R, Jesudason E, et al. Isolation and characterization of distal lung progenitor cells. Methods Mol Biol 2012;879:109–22.

[54] Fujino N, Kubo H, Suzuki T, Ota C, Hegab AE, He M, et al. Isolation of alveolar epithelial type II progenitor cells from adult human lungs. Lab Invest 2011;91:363–78.

[55] Comhair SA, Xu W, Mavrakis L, Aldred MA, Asosingh K, Erzurum SC. Human primary lung endothelial cells in culture. Am J Respir Cell Mol Biol 2012;46:723–30.

[56] Bishop AE. Pulmonary epithelial stem cells. Cell Prolif 2004;37:89–96.

[57] McQualter JL, Bertoncello I. Concise review: deconstructing the lung to reveal its regenerative potential. Stem Cells 2012;30:811–6.

[58] Rock JR, Hogan BL. Epithelial progenitor cells in lung development, maintenance, repair, and disease. Annu Rev Cell Dev Biol 2011;27:493–512.

[59] Stripp BR, Reynolds SD. Maintenance and repair of the bronchiolar epithelium. Proc Am Thorac Soc 2008;5:328–33.

[60] Rawlins EL, Okubo T, Xue Y, Brass DM, Auten RL, Wang F, et al. The role of Scgb1a1+ Clara cells in the long-term maintenance and repair of lung airway, but not alveolar, epithelium. Cell Stem Cell 2009;4:525–34.

[61] Reynolds SD, Malkinson AM. Clara cell: progenitor for the bronchiolar epithelium. Int J Biochem Cell Biol 2010;42:1–4.

[62] Giangreco A, Reynolds SD, Stripp BR. Terminal bronchioles harbor a unique airway stem cell population that localizes to the bronchoalveolar duct junction. Am J Pathol 2002;161:173–82.

[63] Banerjee ER, Henderson Jr. WR. Characterization of lung stem cell niches in a mouse model of bleomycin-induced fibrosis. Stem Cell Res Ther 2012;3:21.

[64] Zheng D, Limmon GV, Yin L, Leung NH, Yu H, Chow VT, et al. Regeneration of alveolar type I and II cells from Scgb1a1-expressing cells following severe pulmonary damage induced by bleomycin and influenza. PLoS One 2012;7:e48451.

[65] Adamson IY. Pulmonary toxicity of bleomycin. Environ Health Perspect 1976;16:119–26.

[66] Bertoncello I, McQualter J. Isolation and clonal assay of adult lung epithelial stem/progenitor cells. Curr Protoc Stem Cell Biol 2011; [Chapter 2, Unit 2G.1].

[67] Kajstura J, Rota M, Hall SR, Hosoda T, D'Amario D, Sanada F, et al. Evidence for human lung stem cells. N Engl J Med 2011;364:1795–806.

[68] Oeztuerk-Winder F, Ventura JJ. Isolation, culture, and potentiality assessment of lung alveolar stem cells. Methods Mol Biol 2012;916:23–30.

[69] Thomson JA, Itskovitz-Eldor J, Shapiro SS, Waknitz MA, Swiergiel JJ, Marshall VS, et al. Embryonic stem cell lines derived from human blastocysts. Science 1998;282:1145–7 [Erratum in: Science 282, 1827].

[70] Ali NN, Edgar AJ, Samadikuchaksaraei A, Timson CM, Romanska HM, Polak JM, et al. Derivation of type II alveolar epithelial cells from murine embryonic stem cells. Tissue Eng 2002;8:541–50.

[71] Samadikuchaksaraei A, Cohen S, Isaac K, Rippon HJ, Polak JM, Bielby RC, et al. Derivation of distal airway epithelium from human embryonic stem cells. Tissue Eng 2006;12:867–75.

[72] Rippon HJ, Polak JM, Qin M, Bishop AE. Derivation of distal lung epithelial progenitors from murine embryonic stem cells using a novel three-step differentiation protocol. Stem Cells 2006;24:1389–98.

[73] Haviland DL, Burns AR, Zsigmond E, Wetsel RA. A pure population of lung alveolar epithelial type II cells derived from human embryonic stem cells. Proc Natl Acad Sci USA 2007;104:4449–54.

[74] Jensen T, Roszell B, Zang F, Girard E, Matson A, Thrall R, et al. A rapid lung de-cellularization protocol supports embryonic stem cell differentiation in vitro and following implantation. Tissue Eng, Part C Methods 2012;18:632–46.

[75] Denham M, Cole TJ, Mollard R. Embryonic stem cells form glandular structures and express surfactant protein C following culture with dissociated fetal respiratory tissue. Am J Physiol Lung Cell Mol Physiol 2006;290:1210–5.

[76] Wetsel RA, Wang D, Calame DG. Therapeutic potential of lung epithelial progenitor cells derived from embryonic and induced pluripotent stem cells. Annu Rev Med 2011;62:95–105.

[77] Takahashi K, Tanabe K, Ohnuki M, Narita M, Ichisaka T, Tomoda K, et al. Induction of pluripotent stem cells from adult human fibroblasts by defined factors. Cell 2007;131:861–72.

[78] Nakagawa M, Koyanagi M, Tanabe K, Takahashi K, Ichisaka T, Aoi T, et al. Generation of induced pluripotent stem cells without Myc from mouse and human fibroblasts. Nat Biotechnol 2008;26:101–6.

[79] Roszell B, Mondrinos MJ, Seaton A, Simons DM, Koutzaki SH, Fong GH, et al. Efficient derivation of alveolar type II cells from embryonic stem cells for in vivo application. Tissue Eng, Part A 2009;15:3351–65.

[80] Somers A, Jean JC, Sommer CA, Omari A, Ford CC, Mills JA, et al. Generation of transgene-free lung disease-specific human induced pluripotent stem cells using a single excisable lentiviral stem cell cassette. Stem Cells 2010;28:1728–40.

[81] Mou H, Zhao R, Sherwood R, Ahfeldt T, Lapey A, Wain J, et al. Generation of multipotent lung and airway progenitors from mouse ESCs and patient-specific cystic fibrosis iPSCs. Cell Stem Cell 2012;10:385–97.

[82] Halme DG, Kessler DA. FDA regulation of stem-cell-based therapies. N Engl J Med 2006;355:1730–5.

[83] Vats A, Tolley NS, Bishop AE, Polak JM. Embryonic stem cells and tissue engineering: delivering stem cells to the clinic. J R Soc Med 2005;98:346–50.

[84] Eramo A, Lotti F, Sette G, Pilozzi E, Biffoni M, Di Virgilio A, et al. Identification and expansion of the tumorigenic lung cancer stem cell population. Cell Death Differ 2008;15:504–14.

[85] Kim CF, Jackson EL, Woolfenden AE, Lawrence S, Babar I, Vogel S, et al. Identification of bronchioalveolar stem cells in normal lung and lung cancer. Cell 2005;121:823–35.

[86] Suki B, Ito S, Stamenovic D, Lutchen KR, Ingenito EP. Biomechanics of the lung parenchyma: critical roles of collagen and mechanical forces. J Appl Physiol 2005;98:1892–9.

[87] Chen SS, Fitzgerald W, Zimmerberg J, Kleinman HK, Margolis L. Cell-cell and cell-extracellular matrix interactions regulate embryonic stem cell differentiation. Stem Cells 2007;25:553–61.

[88] Ingber DE. Tensegrity II. How structural networks influence cellular information processing networks. J Cell Sci 2003;116:1397–408.

[89] Shamis Y, Hasson E, Soroker A, Bassat E, Shimoni Y, Ziv T, et al. Organ-specific scaffolds for in vitro expansion, differentiation, and organization of primary lung cells. Tissue Eng, Part C 2011;17:861–70.

[90] Bader A, Macchiarini P. Moving towards *in situ* tracheal regeneration: the bionic tissue engineered transplantation approach. J Cell Mol Med 2010;14:1877–89.

[91] Suki B, Bates JHT. Lung tissue mechanics as an emergent phenomenon. J Appl Physiol 2011;110:1111–8.

[92] Jiang D, Liang J, Fan J, Yu S, Chen S, Luo Y, et al. Regulation of lung injury and repair by toll-like receptors and hyaluronan. Nat Med 2005;11:1173–9.

[93] Keane TJ, Londono R, Turner NJ, Badylak SF. Consequences of ineffective decellularization of biologic scaffolds on the host response. Biomaterials 2012;33:1771–81.

[94] Daly KA, Liu S, Agrawal V, Brown BN, Huber A, Johnson SA, et al. The host response to endotoxin-contaminated dermal matrix. Tissue Eng, Part A 2012;18:1293–303.

[95] 361 of the PHS Act and 21 CFR Part 1270 [US Food and Drugs], 2001 at <www.fda.gov/CBER/tissue/tisreg.htm>; [accessed 01.01.13].

[96] Nemeno-Guanzon JG, Lee S, Berg JR, Jo YH, Yeo JE, Nam BM, et al. Trends in tissue engineering for blood vessels. J Biomed Biotechnol 2012; doi: 10.1155/2012/956345. Epub November 8, 2012.

[97] Bae H, Puranik AS, Gauvin R, Edalat F, Carrillo-Conde B, Peppas NA, et al. Building vascular networks. Sci Transl Med 2012;4:160ps23.

[98] Muraoka M, Shimizu T, Itoga K, Takahashi H, Okano T. Control of the formation of vascular networks in 3D tissue engineered constructs. Biomaterials 2013;34:696–703.

[99] Kang TY, Hong JM, Jung JW, Yoo JJ, Cho DW. Design and assessment of a microfluidic network system for oxygen transport in engineered tissue. Langmuir 2013;29(2):701–9.

[100] Chan JM, Zervantonakis IK, Rimchala T, Polacheck WJ, Whisler J, Kamm RD. Engineering of *in vitro* 3D capillary beds by self-directed angiogenic sprouting. PLoS One 2012;7: e50582.

Chapter 49

Lung Regeneration: The Developmental Biology Approach

Nicholas Hamilton, Colin R. Butler, Adam Giangreco, and Sam M. Janes
Lungs for Living Research Centre, Division of Medicine, University College London

Chapter Outline

49.1 INTRODUCTION

The respiratory system begins life during the fourth week of gestation as a respiratory diverticulum to the foregut, which divides into two bronchial buds. By 24 weeks, these buds have divided further to complete the basic structure of the lung. While the developing lung is structurally continuous, it is composed of distinct cellular regions from the onset of the canalicular period at 16 weeks. The proximal region of the trachea and bronchi represent the main conducting portion of the airway and are lined by a pseudostratified epithelium mainly consisting of ciliated, goblet, serous, and basal cells [1]. Distally the bronchiolar epithelium consists of columnar epithelium consisting of ciliated cells, secretory Clara cells, a few basal cells and no goblet cells. The conducting airways terminate in a transitional zone, known as the respiratory bronchioles, which are lined by cuboidal ciliated epithelium. These bronchioles then divide into the alveoli, which are lined by type I and type II pneumocytes. Maturation of the lung continues after birth with a reorganization of the pulmonary capillary bed between 2 and 5 years of age, at which point the lung is considered fully developed. Lung function begins to decline at around 18–25 years of age at a rate of around 20 mL in FEV1 per year in healthy individuals and 50–100 mL in FEV1 in those with chronic lung disease. This produces a theoretical

estimate of complete lung dysfunction at 130–140 years of age [2].

Throughout life the respiratory system carries out an amazing array of functions. It continually warms and humidifies inspired air, exchanges harmful gases for life-giving oxygen and neutralizes bacteria and other inhalants via the mucociliary escalator. In order to sustain this multitude of functions, bronchiolar epithelium turns over at a rate of around 1% per day with an estimated epithelial half-life of 6 months in the trachea and 17 months in the bronchioles [3,4]. It is therefore unsurprising that the mechanisms controlling lung homeostasis are complex and liable to failure with advancing age and airway-related disease.

Chronic lung disease represents a leading cause of global morbidity; the World Health Organization estimates that the prevalence of chronic obstructive pulmonary disease (COPD) will increase such that it will be the third leading cause of death worldwide by 2030 [5]. Chronic lung disease is characterized by respiratory cellular changes that include an impairment of mucociliary clearance, an increase in mucus producing cells and a reduced reparative ability. Unfortunately, current understanding of the complex cellular interactions that lead to these changes is largely inadequate and most modern medical treatments fail in terms of their ultimate goals: limiting disease progression and mortality. By improving our understanding of the pathways involved in lung homeostasis and repair it is hoped that novel therapies can be developed that target the root cause of these diseases and prevent their progression.

Traditionally, the homeostatic mechanisms of epithelial organs were believed to lie in a hierarchical arrangement based on slowly dividing, tissue-specific stem cells which produce highly proliferative, short-lived transit-amplifying (TA) daughter cells which are committed to differentiation. While a wealth of evidence exists supporting this arrangement in high turnover organs such as the intestine, it has recently been suggested that more slowly dividing epithelial organs, such as the lung and pancreas, may instead rely on a population of committed progenitor cells during homeostasis [6]. In this chapter we will present an overview of the signaling mechanisms that regulate murine lung development before discussing adult airway homeostasis and regeneration. Finally, we consider the implications of these findings for future improvements in tissue engineering of the human airways.

49.2 OVERVIEW OF LUNG DEVELOPMENT

Both the conducting airways and the lungs are derived from the anterior foregut endoderm during embryonic development. The endodermal gut tube is formed by a series of morphological changes after gastrulation. This primitive tube is then subdivided into the foregut, which goes on to form the thyroid, liver, lungs, pancreas and stomach, the midgut, which will become the small intestine, and the hindgut, which will become the colon [7].

The anterior foregut divides to give both the future esophagus and the future trachea. The earliest indicator of this process is localized differential expression of the transcription factors Nkx2.1, which is expressed in ventrally, and Sox2, which is highly expressed dorsally. The foregut lumen divides, giving rise to the future trachea from the Nkx2.1 + region and the future esophagus from the Sox2 + region [8]. The resultant tubular structures then undergo a rapid expansion process, although the nature of the signals that induce this phase of growth are currently unknown.

At approximately the same time as early tracheal development, the Nkx2.1+ region also gives rise to the primordial lung buds. Reciprocal interactions between the endoderm and the surrounding mesoderm induce the formation of two buds from the ventral surface of the foregut after 4 weeks of human embryonic development (mouse day 9.5). These buds undergo a process of branching morphogenesis during which they expand into the surrounding mesenchyme, forming the skeleton of the future airways.

These stages of branching can be divided into three distinct developmental processes: the pseudoglandular stage (mouse day 9.5—16.5, human weeks 5—17), the canalicular stage (mouse day 17.5—18.5, human weeks 17—25) and the saccular stage (mouse day 18.5—postnatal day 5, human weeks 25—40) [9]. The pseudoglandular stage is marked by the generation of the characteristic branching pattern of the airways and is also the stage during which epithelial cells begin to express differentiation markers. Elegant work by Metzger et al. [10] has recently demonstrated that, in mice, this early pattern formation is highly stereotyped and can be broken down into three local branching modes that are used in three different combinations to generate overall airway morphology. Once this branching pattern has been established, the canalicular stage sees the terminal buds narrow before, during the saccular stage, these buds generate saccules that are the precursors of mature alveoli. To summarize, increasingly branched airway structures are specified during branching morphogenesis, initially the main bronchi and bronchioles and ultimately, at the saccular stage, the distal epithelial saccules. Concurrently, branching morphogenesis sees progressive lineage restriction of the initially multipotent lung bud cells, which is achieved by virtue of differential exposure to paracrine signals in the proximodistal axis (Figure 49.1).

49.3 CELL SIGNALING IN LUNG DEVELOPMENT

Rather than being a cell intrinsic process, lung development is largely directed by reciprocal interactions between the endoderm and the surrounding mesenchyme. The distal tip of the mesoderm is a key regulator of this complex developmental program, producing many of the relevant signaling factors. One of the most important molecules in the developing murine airway is Fgf10, which is critical for the initiation of lung development through its effect on the Fgfr2-expressing endoderm [11,12]. Detailed characterization of Fgf10 mutant mouse phenotypes has shown that reduced levels of Fgf10 in the epithelium result in decreased proliferation during the pseudoglandular stage and decreased Wnt signaling pathway activation [13]. If Fgf10 is expressed ectopically in the airway epithelium, the reverse is true as a hyperproliferative epithelium with fewer differentiated cells emerges [14]. Fgf10 also has an important role in the branching program as Fgf10 or Fgfr2 -/- airways fail to initiate branching [11,15] and conditional deletion of Fgfr2 in the lung epithelium results in an abnormal branching pattern [16]. Fgf10 may also be important later during the specification of proximal versus distal cell types through interactions with the Notch signaling pathway [17]. Other groups have shown that mesothelial Fgf9 controls and activates Fgf10 from the mesenchyme via Fgfr2b, Shp2, and Ras and that sprouty2 is an inducible

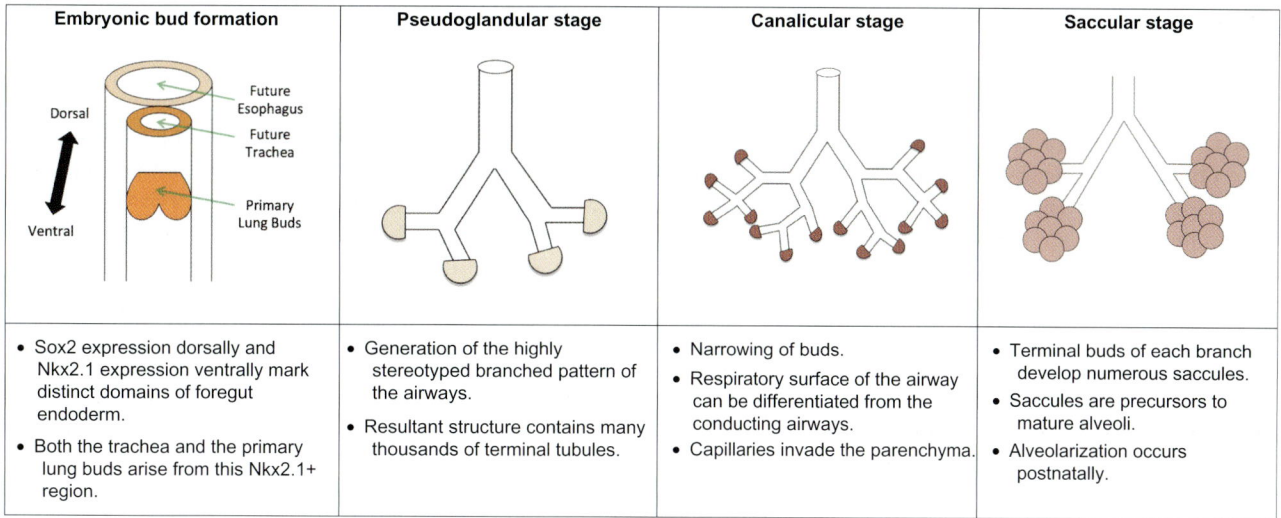

Embryonic bud formation	Pseudoglandular stage	Canalicular stage	Saccular stage
• Sox2 expression dorsally and Nkx2.1 expression ventrally mark distinct domains of foregut endoderm. • Both the trachea and the primary lung buds arise from this Nkx2.1+ region.	• Generation of the highly stereotyped branched pattern of the airways. • Resultant structure contains many thousands of terminal tubules.	• Narrowing of buds. • Respiratory surface of the airway can be differentiated from the conducting airways. • Capillaries invade the parenchyma.	• Terminal buds of each branch develop numerous saccules. • Saccules are precursors to mature alveoli. • Alveolarization occurs postnatally.

FIGURE 49.1 Overview of stages of lung development.

modulator of this pathway [18–20] Overall, these results support the existence of a system in which Fgf10-Fgfr2 signaling promotes self-renewal, inhibits differentiation and supports branching morphogenesis in the developing airway epithelium.

There is evidence that transforming growth factor beta (TGFβ) acts antagonistically to Fgf10 signaling during development. TGFβ binds to the receptors TβRI and TβRII, triggering a phosphorylation cascade that ultimately leads to the nuclear localization of the transcription factors Smad2 and Smad3 [21]. Xing et al. [22] have demonstrated that the inhibitory effect of TGFβ isoforms on branching depends upon TβRII and sees a concurrent decrease in cellular proliferation associated with increased phosphatase and tensin homolog (PTEN) expression. Consistent with these findings, the epithelium-specific deletion of PTEN during development leads to a hyperproliferative phenotype with expansion of stem/progenitor cell populations [23]. In addition to the TGFβ isoforms, another member of the TGFβ superfamily of ligands, Bmp4 is also expressed in the distal epithelial tip and in the mesenchyme surrounding the developing airways [24]. The receptor Bmpr1a is expressed throughout both the epithelium and the mesenchyme and epithelium-specific deletion of Bmpr1a reduces proliferation and decreases self-renewal of alveolar type II cells [25], suggesting that the physiological role of Bmp4 is to positively regulate epithelial proliferation and self-renewal. However, overall the contribution of BMP signaling to airway development remains ill-defined, particularly as *in vitro* data appears contradictory [24].

The Wnt/β-catenin signaling pathway is another important player in determining the levels of proliferation within the developing mouse airways. There is evidence that different members of the Wnt family have different effects on proliferation. Wnt7b and Wnt2 are both expressed in the mesenchyme that surrounds the lung buds and mouse knockout models show that these positively regulate proliferation via the canonical Wnt pathway [26,27]. The lungs in these animals are smaller in size but their patterning is unaffected. The knockout phenotype for Wnt5a, which is expressed in both distal lung epithelium and the surrounding mesenchyme, however, exhibits decreased proliferation, suggesting that it may act in opposition to Wnt7a and Wnt2 [28]. In addition to this role in proliferation, β-catenin, the transcriptional mediator of Wnt signaling, is necessary for specification of airway lineage fate. It appears that β-catenin acts early in development to specify the future respiratory lineages: inactivation of β-catenin in the ventral foregut endoderm halted the development of both the trachea and the lungs [29]. This effect was mediated through loss of respiratory cell fate rather than any possible decrease in progenitor proliferation [29]. Further, β-catenin is important later in the developmental process in determining proximal versus distal cell fate. If Wnt/β-catenin signaling is disrupted by deletion of β-catenin itself or expression of Dickkopf-1, a canonical Wnt pathway inhibitor, the developing proximal airway expands at the expense of the distal airways [30]. Results from experiments that have constitutively activated the Wnt-β-catenin pathway vary depending on the developmental stage at which activation occurs. Reynolds et al. [31] found that activation in late gestation embryos inhibited the differentiation of airway progenitor cells but had no effect on their proliferation. More recently, the same group have demonstrated that a similar intervention earlier in development leads to abnormal bronchial differentiation but spared alveolar development. It was shown that this Wnt pathway-driven effect occurs as a result of negative regulation of a Sox2-dependent bronchiolar differentiation program [32]. Overall it is clear that Wnt signaling plays multiple, stage-specific roles within the developing epithelium, although further work will be required to delineate these and determine the nature of interactions with other pathways.

Unlike the signaling pathways discussed thus far, Notch signaling is not thought to have an effect on proliferation in the developing murine airways [33]. However, several groups have now reported an effect of Notch on the differentiation status of the airway. When the C-terminal portion of Notch1, which is cleaved to activate downstream Notch responses under physiological conditions, is ectopically expressed in the proximal airway epithelium, mucosecretory cell differentiation is favored at the expense of ciliated cell fate [34]. Further, in the distal lung the same intervention led to a deficit in differentiation of alveolar cell types [34]. Conversely, when Notch signaling is conditionally inactivated, a ciliated phenotype characterized by an absence of mucosecretory cells and a more abundant neuroendocrine cell population is observed. Whether inactivation is achieved by interruption of Pofut1 or Rbpjk, alveolar development is unaffected in the knockout conditions [35]. These results indicate that low levels of Notch signaling are required for proper initiation of alveolar development and that Notch signaling regulates the cell fate choice between the ciliated and secretory lineages.

Correct branching morphogenesis is dependent on the organization of the vasculature plexus. Initially, a primitive capillary plexus surrounds the laryngo-tracheal groove. Once stimulated by VEGF, which is predominantly secreted by the primitive epithelium, hemangioblasts differentiate into the stereotypic capillary network that surrounds the branching airway, thus highlighting the importance of epithelial–endothelial crosstalk during lung morphogenesis [36].

Recent evidence suggests that retinoic acid (RA) may act as a regulatory molecule capable of integrating multiple pathways that affect early lung development. The Cardoso group [37] has shown that RA modulates the Fgf, TGFβ and Wnt pathways. During lung bud initiation, RA positively regulates the Wnt pathway and concurrently represses TGFβ signaling to drive proliferation. In the mesenchyme, Fgf10 expression is positively regulated by Wnt signals and inhibited by TGFβ; therefore, RA concentration indirectly determines the level and distribution of Fgf10 necessary for lung bud formation and branching morphogenesis.

49.4 ENDOGENOUS AIRWAY HOMEOSTASIS

The term "Committed Progenitor" (CP) refers to a cell with the ability to differentiate into multiple cell lineages within a given tissue [38]. These differ from "true" stem cells in that they have a finite capacity for self-renewal, lack pluripotency and exist in a differentiated state. CP cells were first reported in murine ear, tail and dorsal epidermis in experiments using *in vivo* genetic lineage tracing to show that epidermal basal cells were able to proliferate independently from regional stem cell niches and appeared to be under intrinsic as opposed to environmental control [6]. This gave rise to an alternative model for intrinsic homeostasis not characterized by obligate transit-amplifying cells in the resting state but instead a differentiated facultative transit-amplifying progenitor cell population with the capacity to proliferate and produce daughter cells capable of generating more differentiated lineages [1].

Substantial evidence now exists to suggest the presence and vital role of CP cells in the airway epithelium. In mice, *in vivo* injury models with noxious inhalants demonstrate the proliferation and dedifferentiation of epithelial basal cells within tracheobronchial tissue [39–41]. *In vitro* studies utilizing tracheal grafts with cell fractionation and retroviral lineage tracing distinguished tracheobronchial basal cells from other cell groups by their differentiation potential and clonogenicity [42,43]. More recently, *in vivo* lineage studies using GFP to label Keratin 5-positive mouse tracheal basal cells have shown tracheobronchial basal cells to exhibit multiple stable cell clones capable of growth and differentiation during homeostasis [44].

Within bronchiolar epithelium the use of oxidant gases to selectively damage ciliated cells demonstrated repair through a nonciliated cell with underlying Clara cells losing their resting state secretory granules and endoplasmic reticulum [40]. These cells are termed A Clara cells and have been shown to proliferative and differentiate into both mature Clara cells and ciliated epithelial cells in response to injury [39]. In addition, murine lineage-tracing studies using Scgb1a1 to tag bronchiolar Clara cells show their importance in proliferation and differentiation into mature epithelial cells in the noninjured state [45].

Within the alveoli, type II pneumocytes have been identified as potential progenitor cells and are known to contribute to renewal of alveolar type I cells and maintenance of the alveolar structure [46]. While this type II to type I pneumocyte relationship has been demonstrated in lineage-tracing studies, it is still unclear whether all type II pneumocytes behave in the same way or if individual subtypes exist that act as CP cells. Further, a population

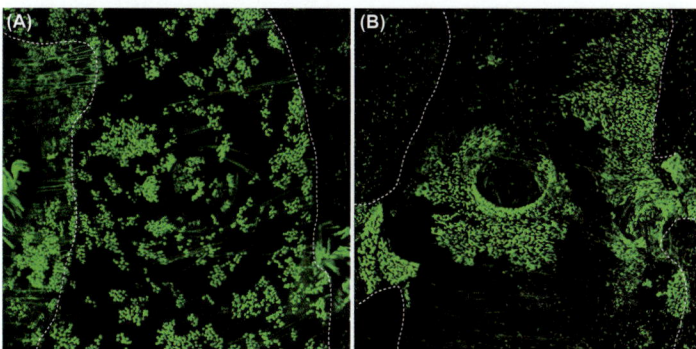

FIGURE 49.2 Confocal/IF images of airway epithelium.

of lineage negative, integrin α6β4-expressing cells has been shown to play a role in distal lung homeostasis [47]. Overall, however, the cellular basis of alveolar homeostasis is still poorly understood.

Evidence suggesting that CP cells, rather than traditional stem cells, are responsible for lung homeostasis is provided by Giangreco et al. [48]. An aggregation chimera murine model was used with the assumption that if CP cells were responsible for airway homeostasis then over time the airway would be colonized by frequent clonal cell patches with a varied size and inconsistent distribution between individuals inconsistent with the concept of lung stem cell niches. Conversely, the authors argued, if stem cells were responsible for lung homeostasis then over time the airways would contain large clonally derived cell patches with a consistent distribution relating to airway stem cells niches. In this model it was found that the airways display randomly distributed chimeric cell patches of variable size with no association with lung stem cell niches. As such, it was concluded that the cells responsible for airway homeostasis exhibit features of committed progenitor cells, rather than the traditional tissue-specific stem cell [48] (Figure 49.2).

49.5 AIRWAY REGENERATION FOLLOWING INJURY

While committed progenitor cells appear to play an important role in day-to-day epithelial homeostasis, their mitotic activity is believed to be too low to support the rapid proliferation of cells needed to repair the epithelium following injury [49]. Stem cells by nature are more suited to the task of producing a large number of cells that rapidly proliferate and differentiate in response to injury and have been observed to act alongside CP cells in other organs of foregut endoderm origin, such as the pancreas [50,51].

Early indications of lung stem cells came from the identification of a distinct population of cells expressing high levels of keratin within murine tracheal submucosal gland ducts [52]. Stem cells within the limbic system and hair follicles also express high levels of keratin suggesting the

submucosal glands as a lung stem cell niche. In further experiments, bromodeoxyuridine (BrdU) labeling was performed following intratracheal detergent and S02 inhalation. These methods are used to cause epithelial injury and therefore promote repair within the mouse trachea. Cells can be traced with BrdU due to its prolonged retention in cells that divide slowly, such as stem cells [41]. At days 3 and 6, BrdU positive epithelial cells were found along the length of the trachea but by days 20 and 95, BrdU cells were localized to the submucosal gland ducts in the upper trachea and at symmetrically arranged positions near the cartilage intercartilaginous junction in the lower trachea. Further studies have shown that submucosal gland duct associated basal cells are highly resistant to injury and demonstrate multipotent differentiation and high rates of clonal expansion [41,52]. In addition, these cells have a unique molecular signature with an upregulation of keratin genes that differentiates them from basal CP cells [52].

Within the bronchioles, stem cell niches are believed to exist adjacent to calcitonin gene-related peptides (CGRP)-expressing neuroepithelial bodies and terminal bronchioalveolar junctions. Variant Clara cell secretory protein (vCCSP)-expressing cells have been implicated as the stem/progenitor cells of these regions [3,48,53]. These variant Clara stem cells have a low turnover in normal physiological states but respond to injury by regenerating precursor secretory and ciliated epithelial cells. Furthermore, the destruction of vCCSP-expressing cells using a transgenic suicide gene model prevents effective epithelial regeneration following injury [48,54].

Within the alveoli, stem cell niches are difficult to identify but a number of cells appear to show stem cell-like behavior. Following acute oxygen injury, alveolar epithelial cells (AECs) upregulate telomerase activity (regarded as a stem cell marker) suggesting that a subpopulation of AECs are progenitor cells or that all AECs can potentially respond by reactivating progenitor-associated genetic programs [55]. These telomerase positive alveolar cells show stem cell characteristics: they are relatively resistant to injury-induced apoptosis and have a higher proliferative activity. The abundance of these telomerase positive cells within the alveoli support the concept that, if these cells do exist, they must be widely distributed.

The concept of stem cells that can respond to injury is further supported by the identification of bronchioalveolar stem cells residing at the bronchioalveolar duct junction. These cells are resistant to distal airway injury and proliferate during regeneration of the epithelium [56]. Bronchioalveolar stem cells also exhibit the stem cell characteristics of multipotency and self-renewal in clonal assays [57]. Other regenerative cells include SPC (+) pneumocytes which are able to replenish type I and type II cells following injury and a subpopulation of SPC (-)/CPC (-) cells expressing integrin α6β4 have been shown to regenerate alveolar tissue following bleomycin injury [47]. Work by Kajstura et al. [58] used the stem cell marker c-kit to label cells within the terminal lung that exhibited pluripotency in vitro. However, the localization of these cells within the airway epithelium and their exact role in lung regeneration remains uncertain.

While CP cells are not suited to provide the rapidly proliferating cells needed for extensive regeneration, it is known that CP cells become activated to proliferate and differentiate following injury. This leads to questions about the extent to which injury regeneration is CP cell or stem cell mediated. Experiments in which chimeric mice were exposed to varying degrees of naphthalene injury showed that injury severity appeared to determine airway CP cell versus airway stem cell activation with more severe injuries

being stem cell led at the expense of CP cells and minor injuries being more dependent upon CP cells [48].

49.6 CELL SIGNALING IN THE ADULT AIRWAY EPITHELIUM

Roles for both CP cells and stem cells have been described in airway homeostasis and repair and an understanding of how these processes are regulated is vital if future therapies aiming to manipulate are to be successful. The Wnt/β-catenin signaling pathway is essential in both embryonic lung development and in the homeostasis of a number of other adult organs. Wnt ligands bind to the frizzled (Frz) family of receptors and activate disheveled, a second protein of the Wnt receptor complex. This process leads components of the destruction complex, which in the absence of Wnt signals phosphorylates cytoplasmic β-catenin and targets it for degradation, to complex with the membrane-associated Wnt receptor proteins. Stabilized β-catenin then accumulates and can enter the nucleus where it influences gene expression [59] (Figure 49.3).

Wnt signaling affects the proliferation and differentiation of postnatal airway progenitor cells [60]. Loss of the zinc finger transcription factor Gata6, which regulates Frz2 (a noncanonical Wnt receptor that inhibits the canonical Wnt pathway), leads to expansion of the progenitor pool [61]. Moreover, Gata6 is required for proper epithelial

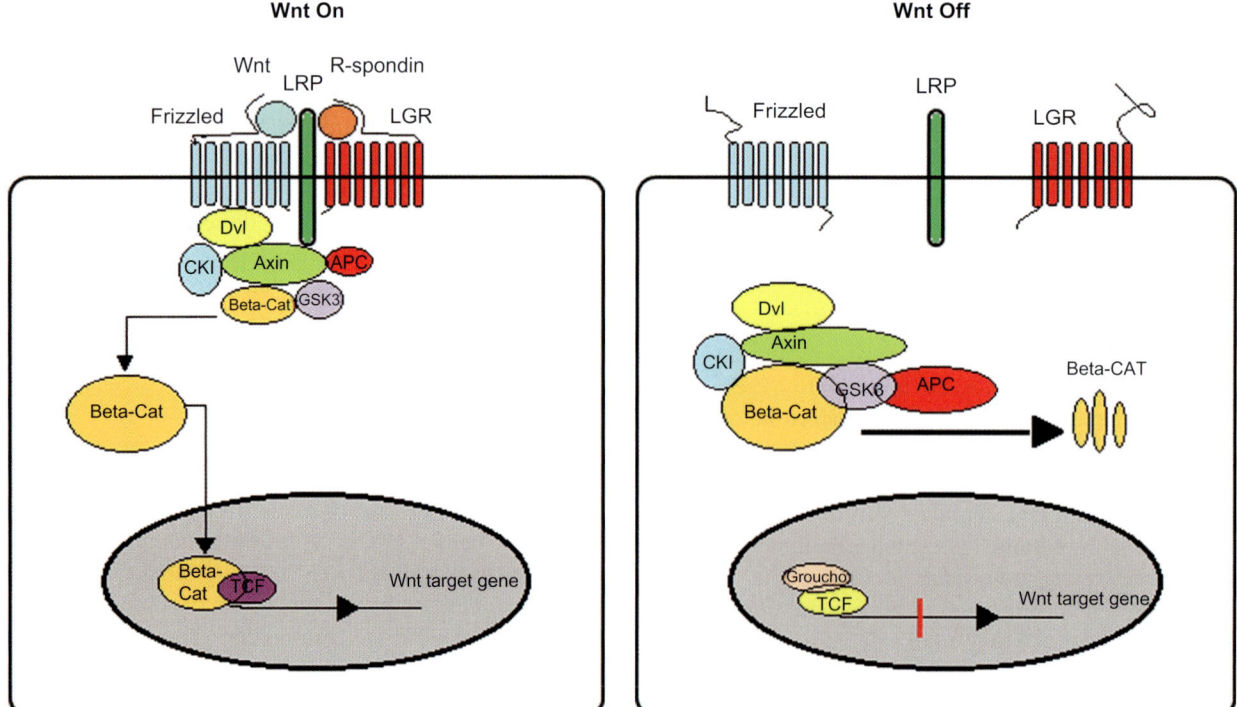

FIGURE 49.3   Wnt signaling pathway summary diagram.

regeneration following postnatal injury. Wnt signaling is activated in response to injury, thus expanding the population of progenitors capable of repopulating the damaged airway. When Gata6 is conditionally inactivated, this leads to a hyperproliferative response with increased numbers of progenitors but reduced airway differentiation [61]. The importance of Wnt signaling in the human airway epithelium is supported by the upregulation of several Wnt pathway genes during the differentiation of human bronchial epithelial cells *in vitro*[62].

TGFβ inhibits epithelial cell proliferation in a range of adult organs, including the airways [63], where it may be of particular significance as defective TGFβ signaling may lead to the abnormal epithelial phenotype seen in asthmatic patients [64].

Furthermore, the mitogen-activated protein kinase (MAPK) pathway has been demonstrated to affect the self-renewal properties of airway stem/progenitor cells [65]. Ventura et al. showed that deletion of p38-alpha from the mature airways leads to a lack of differentiated cells and increased progenitor cell proliferation. As such, it is proposed that p38-alpha counteracts proliferation signals, such as those provided by epidermal growth factor, while positively regulating CCAAT/enhancer-binding protein (C/EBP), which is required for airway differentiation [66]. More recently the same authors have described the regulation of a putative lung stem cell population, marked by E-Cadherin and Lgr6 expression, by a pathway involving p38-alpha, p53 and miR-17−92 [67]. When p38-alpha levels are reduced, miR-17−92 expression increases, reducing the levels of lung-specific transcription factors such as C/EBP and GATA6 and causing alterations in the expression of lung epithelial marker proteins and changes in integrin expression [67].

The Notch signaling pathway also appears to be of continued importance in the differentiation of basal cells in the adult airway epithelium. Cell−cell interactions appear to be of critical importance in determining whether basal cells self-renew or differentiate to form luminal progeny (ciliated cells, Clara cells, and mucosecretory cells) [33].

A more global understanding of the molecular and cellular processes that regulate lung homeostasis and regeneration must be a key objective of future research, particularly given that these processes are aberrant in airway diseases, including cancer, and as such offer an opportunity for early detection and intervention.

49.7 GENERATION OF AIRWAY CELLS FROM PLURIPOTENT CELLS USING A DEVELOPMENTAL APPROACH

Following extreme injury or in transplanted constructs, the physiological mechanisms for airway regeneration may be insufficient to restore the airway epithelium. A rapidly available source of cells suitable for transplantation is therefore desirable. Toward this goal several groups have investigated the possibility of generating airway cells from human embryonic (ES) or induced pluripotent stem (iPS) cells. Early attempts that suggested that this approach would be fruitful saw cells expressing a range of mature lung epithelial markers [68] and a pure population of type II cells characteristic of the distal airway epithelium derived from ES cells [69].

Given the ethical problems that surround the use of ES cells in clinical applications, the derivation of lung progenitors from iPS cells was another significant breakthrough [70]. Importantly, this work suggested that the application of developmentally important signaling molecules to pluripotent cells encouraged *in vivo*-like differentiation. While these studies demonstrated the potential of this approach for generating airway progenitor cells, it is necessary to achieve a mature airway epithelium for tissue engineering applications. Wong et al. [71] recently demonstrated proximal airway epithelial differentiation from both human ES and iPS cells using a similarly developmental approach. Initially cells were exposed to Activin-A, which mimics endogenous Nodal signaling during definitive endoderm specification, and Wnt3a, which is expressed at the initiation of gastrulation, in order to induce endodermal differentiation. After 4 days, cells began to express endodermal markers and addition of Shh and Fgf2 promoted anterior foregut identity as demonstrated by increased Sox2 and Nkx2.1 expression. Application of Fgf10, Fgf7, and BMP4, whose developmental importance is discussed above, directed the cells toward an airway phenotype and led cells to express markers of airway basal cells (CK5 and P63). Finally, proximal differentiation of these cells was encouraged by the addition of Fgf18, whose overexpression during development leads to airway proximalization [72]. This produced a population with increased expression of the aforementioned basal cells markers as well as markers of ciliated (FoxJ1) and mucosecretory (Muc5AC) cells. Culture of these cells for 5 weeks at an air−liquid interface confirmed their potential to form a polarized, well-differentiated epithelium.

Presently there are doubts over the clinical utility of ES/iPS cell-derived tissue as a result of ethical concerns, our incomplete understanding of the differences in gene expression and DNA methylation in iPS cells [73] and the similarities between induced pluripotency and the generation of tumor cells *in vitro* [74]. To this end, other potential sources of pluripotent cells have been investigated, which have included both fat- and bone marrow-derived mesenchymal stem cells. While these cells have been

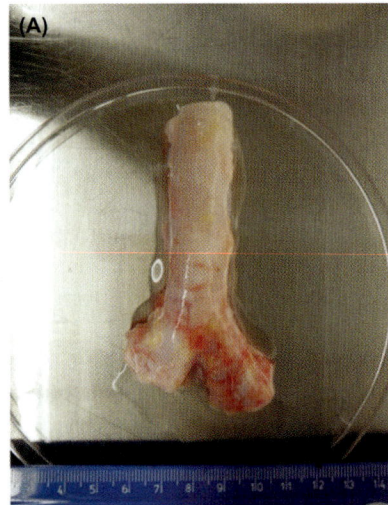

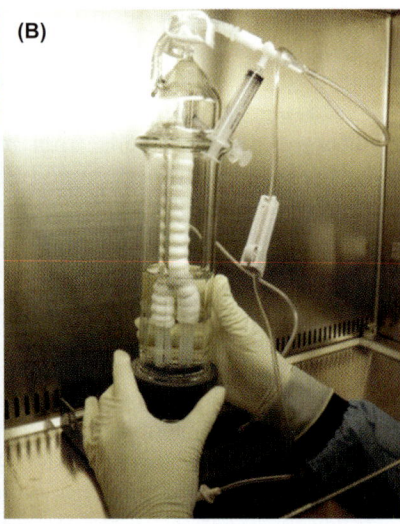

FIGURE 49.4 Bioreactor with trachea photo.

shown to improve lung repair following injury, the mechanism appears to be pseudopharmaceutical, with the release of anti-inflammatory cytokines and stimulation of endogenous repair, rather than engraftment [75].

Human amniotic fluid stem cells (hAFSC) are pluripotent and give rise to all three germ layers *in vitro*[76]. As an exogenous source of stems cells, these have clinical potential over ES cells as they do not form teratomas when injected *in vivo*. Furthermore, as they are derived from discarded amniocentesis specimens, they are considered ethically neutral. The potential for regenerative applications has been demonstrated by the ability of the cells to express makers of cell types of the niche in which they reside. Certainly in the developing embryonic lungs, hAFSCs will incorporate and express airway epithelial cell makers. In models of acute lung injury, hASCFs will integrate in a loco-regional manner according to the distribution of injury, and will adopt the cell lineages of CC10 positive Clara cells and Sp-C positive alveolar cells for naphthalene and oxygen-induced injury, respectively [77]. Whether they are truly involved in repair or whether their effect is more mediated by paracrine effects is still being debated, however, *in vitro* manipulation of pluripotent cells toward an airway fate represents a promising breakthrough and may provide a rapid alternative source of cells for airway tissue engineering in the future.

49.8 AIRWAY TISSUE ENGINEERING

The relative simplicity of the upper airway and failure of conventional treatments to effectively treat long-segment airway stenosis has driven the development of tissue-engineered upper airway transplants. While these therapies do not yet involve direct manipulation of the cells important in airway homeostasis and regeneration, the outcomes of early trials highlight a number of important

factors with regards to therapeutic airway regeneration. First, it has become clear that the extracellular matrix is of central importance in determining tissue-engineered airway graft survival [78]. Matrices that do not match the target tissue exhibit poor regeneration compared to decellularized scaffolds taken from the target tissue or synthetic scaffolds designed to replicate the target ECM. Moreover, the survival of regenerating airway mucosa *in vivo* appears to be influenced by factors produced outside the airway epithelium, as demonstrated in a series of pig trials, where survival was vastly improved with the addition of mesenchymal stem cell-derived chondrocytes. [79,80]. Other groups have reported enhanced airway mucosal repair with the engraftment of both endothelial and epithelial cell types. In these cases, engrafted cells were not in direct contact but localized adjacent to the site of injury suggesting paracrine effects to play a dominant role in regeneration [71]. Regardless of the mechanism, the engraftment of one or more cell types appears to be significant and delineating which combination of cells requires further investigation.

Tissue engineering of more distal airways has proven to be more challenging. The anatomical complexity of alveoli and the arrangement of the relevant cells and matrices are simply difficult to replicate. Despite this, various groups have risen to this challenge by demonstrating the feasibility of decellularized lung engrafted with cells in orthotopic transplant models [81,82]. These groups demonstrated that their tissue-engineered lungs were capable of maintaining gaseous exchange albeit with limited survival. Similar to upper airway grafts, the choice of engrafted cells is important in determining survival. Currently, the longest transplant survival has been observed in those seeded with fetal pneumocytes and human umbilical cord endothelial cells [83]. The improved survival seen with the addition of endothelial cells may reflect the crosstalk that exists between epithelium, mesenchyme, and endothelium, known to be crucial in normal airway development. Further investigation of these pathways and their application in a tissue-engineered setting would appear to be essential if lung regeneration is to be successful.

While tissue-engineered airway transplants have advanced significantly over the past decade and have already benefited the lives of chronic airway sufferers, limitations to the technology remain. On several occasions the transplants have suffered from poor graft mucosalization, partly due to either breakdown of the epithelial layer posttransplantation or an incomplete mucosal layer preimplantation [84,85]. These problems can result in infection, stasis of secretions and potentially life-threatening complications. The prospect of being able to manipulate airway CP or stem cells *ex vivo* to facilitate rapidly functional epithelization of the implantable graft is very attractive. In addition, the application of lung stem cells to monitor and regulate regeneration in response to injury would go further to producing a transplantable graft that is more phenotypically similar to human airway. In order to realize these goals, however, an improved understanding of the behavior of lung CP cells, stem cells, and the signals that regulate their behavior is essential (Figure 49.4).

REFERENCES

[1] Snyder JC, Teisanu RM, Stripp BR. Endogenous lung stem cells and contribution to disease. J Pathol 2009;217:254–64.

[2] Ito K, Barnes PJ. COPD as a disease of accelerated lung aging. Chest 2009;135:173–80.

[3] Hong KU, Reynolds SD, Giangreco A, Hurley CM, Stripp BR. Clara cell secretory protein-expressing cells of the airway neuroepithelial body microenvironment include a label-retaining subset and are critical for epithelial renewal after progenitor cell depletion. Am J Respir Cell Mol Biol 2001;24:671–81.

[4] Rawlins EL, Hogan BL. Ciliated epithelial cell lifespan in the mouse trachea and lung. Am J Physiol Lung Cell Mol Physiol 2008;295:231–4.

[5] Rock JR, Randell SH, Hogan BL. Airway basal stem cells: a perspective on their roles in epithelial homeostasis and remodeling. Dis Models Mech 2010;3:545–56.

[6] Butler C, Birchall M, Giangreco A. Interventional and intrinsic airway homeostasis and repair. Physiology 2012;27:140–7.

[7] Zorn AM, Wells JM. Vertebrate endoderm development and organ formation. Ann Rev Cell Dev Biol 2009;25:221–51.

[8] Que J, Okubo T, Goldenring JR, Nam KT, Kurotani R, Morrisey EE, et al. Multiple dose-dependent roles for Sox2 in the patterning and differentiation of anterior foregut endoderm. Development 2007;134:2521–31.

[9] Morrisey EE, Hogan BL. Preparing for the first breath: genetic and cellular mechanisms in lung development. Dev Cell 2010;18:8–23.

[10] Metzger RJ, Klein OD, Martin GR, Krasnow MA. The branching programme of mouse lung development. Nature 2008;453:745–50.

[11] Sekine K, Ohuchi H, Fujiwara M, Yamasaki M, Yoshizawa T, Sato T, et al. Fgf10 is essential for limb and lung formation. Nat Genetics 1999;21:138–41.

[12] Arman E, Haffner-Krausz R, Gorivodsky M, Lonai P. Fgfr2 is required for limb outgrowth and lung-branching morphogenesis. Proc Natl Acad Sci USA 1999;96:11895–9.

[13] Ramasamy SK, Mailleux AA, Gupte VV, Mata F, Sala FG, Veltmaat JM, et al. Fgf10 dosage is critical for the amplification of epithelial cell progenitors and for the formation of multiple mesenchymal lineages during lung development. Dev Biol 2007;307:237–47.

[14] Nyeng P, Norgaard GA, Kobberup S, Jensen J. FGF10 maintains distal lung bud epithelium and excessive signaling leads to progenitor state arrest, distalization, and goblet cell metaplasia. BMC Dev Biol 2008;8:2.

[15] De Moerlooze L, Spencer-Dene B, Revest JM, Hajihosseini M, Rosewell I, Dickson C. An important role for the IIIb isoform of fibroblast growth factor receptor 2 (FGFR2) in mesenchymal-epithelial signalling during mouse organogenesis. Development 2000;127:483–92.

[16] Abler LL, Mansour SL, Sun X. Conditional gene inactivation reveals roles for Fgf10 and Fgfr2 in establishing a normal pattern of epithelial branching in the mouse lung. Dev Dyn 2009;238:1999–2013.

[17] Tsao PN, Chen F, Izvolsky KI, Walker J, Kukuruzinska MA, Lu J, et al. Gamma-secretase activation of notch signaling regulates the balance of proximal and distal fates in progenitor cells of the developing lung. J Biol Chem 2008;283:29532–44.

[18] Del Moral PM, De Langhe SP, Sala FG, Veltmaat JM, Tefft D, Wang K, et al. Differential role of FGF9 on epithelium and mesenchyme in mouse embryonic lung. Dev Biol 2006;293:77–89.

[19] Tefft D, De Langhe SP, Del Moral PM, Sala F, Shi W, Bellusci S, et al. A novel function for the protein tyrosine phosphatase Shp2 during lung branching morphogenesis. Dev Biol 2005;282:422–31.

[20] Tefft D, Lee M, Smith S, Crowe DL, Bellusci S, Warburton D. mSprouty2 inhibits FGF10-activated MAP kinase by differentially binding to upstream target proteins. Am J Physiol Lung Cell Mol Physiol 2002;283:700–6.

[21] Santibanez JF, Quintanilla M, Bernabeu C. TGF-beta/TGF-beta receptor system and its role in physiological and pathological conditions. Clin Sci 2011;121:233–51.

[22] Xing Y, Li C, Hu L, Tiozzo C, Li M, Chai Y, et al. Mechanisms of TGFbeta inhibition of LUNG endodermal morphogenesis: the role of TbetaRII, Smads, Nkx2.1 and Pten. Dev Biol 2008;320:340–50.

[23] Tiozzo C, De Langhe S, Yu M, Londhe VA, Carraro G, Li M, et al. Deletion of pten expands lung epithelial progenitor pools and confers resistance to airway injury. Am J Respir Crit Care Med 2009;180:701–12.

[24] Weaver M, Dunn NR, Hogan BL. Bmp4 and Fgf10 play opposing roles during lung bud morphogenesis. Development 2000;127:2695–704.

[25] Eblaghie MC, Reedy M, Oliver T, Mishina Y, Hogan BL. Evidence that autocrine signaling through bmpr1a regulates the proliferation, survival and morphogenetic behavior of distal lung epithelial cells. Dev Biol 2006;291:67–82.

[26] Rajagopal J, Carroll TJ, Guseh JS, Bores SA, Blank LJ, Anderson WJ, et al. Wnt7b stimulates embryonic lung growth by

coordinately increasing the replication of epithelium and mesen-chyme. Development 2008;135:1625—34.

[27] Goss AM, Tian Y, Tsukiyama T, Cohen ED, Zhou D, Lu MM, et al. Wnt2/2b and beta-catenin signaling are necessary and suffi-cient to specify lung progenitors in the foregut. Dev Cell 2009;17:290—8.

[28] Li C, Xiao J, Hormi K, Borok Z, Minoo P. Wnt5a participates in distal lung morphogenesis. Dev Biol 2002;248:68—81.

[29] Harris-Johnson KS, Domyan ET, Vezina CM, Sun X. beta-Catenin promotes respiratory progenitor identity in mouse foregut. Proc Natl Acad Sci USA 2009;106:16287—92.

[30] Shu W, Guttentag S, Wang Z, Andl T, Ballard P, Lu MM, et al. Wnt/beta-catenin signaling acts upstream of N-myc, BMP4, and FGF signaling to regulate proximal-distal patterning in the lung. Dev Biol 2005;283:226—39.

[31] Reynolds SD, Zemke AC, Giangreco A, Brockway BL, Teisanu RM, Drake JA, et al. Conditional stabilization of beta-catenin expands the pool of lung stem cells. Stem Cell 2008;26:1337—46.

[32] Hashimoto S, Chen H, Que J, Brockway BL, Drake JA, Snyder JC, et al. beta-Catenin-SOX2 signaling regulates the fate of devel-oping airway epithelium. J Cell Sci 2012;125:932—42.

[33] Rock JR, Gao X, Xue Y, Randell SH, Kong YY, Hogan BL. Notch-dependent differentiation of adult airway basal stem cells. Cell Stem Cell 2011;8:639—48.

[34] Guseh JS, Bores SA, Stanger BZ, Zhou Q, Anderson WJ, Melton DA, et al. Notch signaling promotes airway mucous metaplasia and inhibits alveolar development. Development 2009;136:1751—9.

[35] Morimoto M, Liu Z, Cheng HT, Winters N, Bader D, Kopan R. Canonical notch signaling in the developing lung is required for determination of arterial smooth muscle cells and selection of Clara versus ciliated cell fate. J Cell Sci 2010;123:213—24.

[36] Del Moral PM, Sala FG, Tefft D, Shi W, Keshet E, Bellusci S, et al. VEGF-A signaling through flk-1 is a critical facilitator of early embryonic lung epithelial to endothelial crosstalk and branching morphogenesis. Dev Biol 2006;290:177—88.

[37] Chen F, Cao Y, Qian J, Shao F, Niederreither K, Cardoso WV. A retinoic acid-dependent network in the foregut controls formation of the mouse lung primordium. J Clin Invest 2010;120:2040—8.

[38] Weiss DJ, Bertoncello I, Borok Z, Kim C, Panoskaltsis-Mortari A, Reynolds S, et al. Stem cells and cell therapies in lung biology and lung diseases. Proc Am Thorac Soc 2011;8:223—72.

[39] Evans MJ, Cabral-Anderson LJ, Freeman G. Role of the Clara cell in renewal of the bronchiolar epithelium. Lab Invest J Tech Methods Pathol 1978;38:648—53.

[40] Evans MJ, Johnson LV, Stephens RJ, Freeman G. Renewal of the terminal bronchiolar epithelium in the rat following exposure to NO_2 or O_3. Lab Invest J Tech Methods Pathol 1976;35:246—57.

[41] Borthwick DW, Shahbazian M, Krantz QT, Dorin JR, Randell SH. Evidence for stem-cell niches in the tracheal epithelium. Am J Respir Cell Mol Biol 2001;24:662—70.

[42] Borthwick DW, West JD, Keighren MA, Flockhart JH, Innes BA, Dorin JR. Murine submucosal glands are clonally derived and show a cystic fibrosis gene-dependent distribution pattern. Am J Respir Cell Mol Biol 1999;20:1181—9.

[43] Engelhardt JF, Schlossberg H, Yankaskas JR, Dudus L. Progenitor cells of the adult human airway involved in submuco-sal gland development. Development 1995;121:2031—46.

[44] Rock JR, Onaitis MW, Rawlins EL, Lu Y, Clark CP, Xue Y, et al. Basal cells as stem cells of the mouse trachea and human air-way epithelium. Proc Natl Acad Sci USA 2009;106:12771—5.

[45] Rawlins EL, Okubo T, Xue Y, Brass DM, Auten RL, Hasegawa H, et al. The role of Scgb1a1+ Clara cells in the long-term main-tenance and repair of lung airway, but not alveolar, epithelium. Cell Stem Cell 2009;4:525—34.

[46] Adamson IY, Bowden DH. Derivation of type 1 epithelium from type 2 cells in the developing rat lung. Lab Invest J Tech Methods Pathol 1975;32:736—45.

[47] Chapman HA, Li X, Alexander JP, Brumwell A, Lorizio W, Tan K, et al. Integrin alpha6beta4 identifies an adult distal lung epithe-lial population with regenerative potential in mice. J Clin Invest 2011;121:2855—62.

[48] Giangreco A, Arwert EN, Rosewell IR, Snyder J, Watt FM, Stripp BR. Stem cells are dispensable for lung homeostasis but restore airways after injury. Proc Natl Acad Sci USA 2009;106:9286—91.

[49] Jones P, Simons BD. Epidermal homeostasis: do committed pro-genitors work while stem cells sleep? Nat Rev Mol Cell Biol 2008;9:82—8.

[50] Dor Y, Brown J, Martinez OI, Melton DA. Adult pancreatic beta-cells are formed by self-duplication rather than stem-cell differen-tiation. Nature 2004;429:41—6.

[51] Smukler SR, Arntfield ME, Razavi R, Bikopoulos G, Karpowicz P, Seaberg R, et al. The adult mouse and human pancreas contain rare multipotent stem cells that express insulin. Cell Stem Cell 2011;8:281—93.

[52] Hegab AE, Ha VL, Gilbert JL, Zhang KX, Malkoski SP, Chon AT, et al. Novel stem/progenitor cell population from murine tra-cheal submucosal gland ducts with multipotent regenerative potential. Stem Cell 2011;29:1283—93.

[53] Giangreco A, Reynolds SD, Stripp BR. Terminal bronchioles har-bor a unique airway stem cell population that localizes to the bronchoalveolar duct junction. Am J Pathol 2002;161:173—82.

[54] Reynolds SD, Hong KU, Giangreco A, Mango GW, Guron C, Morimoto Y, et al. Conditional Clara cell ablation reveals a self-renewing progenitor function of pulmonary neuroendocrine cells. Am J Physiol Lung Cel Mol Physiol 2000;278:1256—63.

[55] Reddy R, Buckley S, Doerken M, Barsky L, Weinberg K, Anderson KD, et al. Isolation of a putative progenitor subpopula-tion of alveolar epithelial type 2 cells. Am J Physiol Lung Cell Mol Physiol 2004;286:658—67.

[56] Kim CF, Jackson EL, Woolfenden AE, Lawrence S, Babar I, Vogel S, et al. Identification of bronchioalveolar stem cells in nor-mal lung and lung cancer. Cell 2005;121:823—35.

[57] Rock JR, Barkauskas CE, Cronce MJ, Xue Y, Harris JR, Liang J, et al. Multiple stromal populations contribute to pulmonary fibro-sis without evidence for epithelial to mesenchymal transition. Proc Natl Acad Sci USA 2011;108:1475—83.

[58] Kajstura J, Rota M, Hall SR, Hosoda T, D'Amario D, Sanada F, et al. Evidence for human lung stem cells. New Engl J Med 2011;364:1795—806.

[59] Clevers H. Wnt/beta-catenin signaling in development and dis-ease. Cell 2006;127:469—80.

[60] Giangreco A, Lu L, Vickers C, Teixeira VH, Groot KR, Butler CR, et al. beta-Catenin determines upper airway progenitor cell fate and preinvasive squamous lung cancer progression by modulating epithelial-mesenchymal transition. J Pathol 2012;226:575—87.

[61] Zhang Y, Goss AM, Cohen ED, Kadzik R, Lepore JJ, Muthukumaraswamy K, et al. A Gata6-Wnt pathway required for epithelial stem cell development and airway regeneration. Nat Genet 2008;40:862−70.

[62] Ross AJ, Dailey LA, Brighton LE, Devlin RB. Transcriptional profiling of mucociliary differentiation in human airway epithelial cells. Am J Respir Cell Mol Biol 2007;37:169−85.

[63] Lange AW, Keiser AR, Wells JM, Zorn AM, Whitsett JA. Sox17 promotes cell cycle progression and inhibits TGF-beta/Smad3 signaling to initiate progenitor cell behavior in the respiratory epithelium. PLoS One 2009;4:5711.

[64] Semlali A, Jacques E, Plante S, Biardel S, Milot J, Laviolette M, et al. TGF-beta suppresses EGF-induced MAPK signaling and proliferation in asthmatic epithelial cells. Am J Respir Cell Mol Biol 2008;38:202−8.

[65] Ventura JJ, Tenbaum S, Perdiguero E, Huth M, Guerra C, Barbacid M, et al. p38alpha MAP kinase is essential in lung stem and progenitor cell proliferation and differentiation. Nat Genet 2007;39:750−8.

[66] Basseres DS, Levantini E, Ji H, Monti S, Elf S, Dayaram T, et al. Respiratory failure due to differentiation arrest and expansion of alveolar cells following lung-specific loss of the transcription factor C/EBPalpha in mice. Mol Cell Biol 2006;26:1109−23.

[67] Oeztuerk-Winder F, Guinot A, Ochalek A, Ventura JJ. Regulation of human lung alveolar multipotent cells by a novel p38alpha MAPK/miR-17-92 axis. EMBO J 2012;31:3431−41.

[68] Van Haute L, De Block G, Liebaers I, Sermon K, De Rycke M. Generation of lung epithelial-like tissue from human embryonic stem cells. Respir Res 2009;10:105.

[69] Wang D, Haviland DL, Burns AR, Zsigmond E, Wetsel RA. A pure population of lung alveolar epithelial type II cells derived from human embryonic stem cells. Proc Natl Acad Sci USA 2007;104:4449−54.

[70] Mou H, Zhao R, Sherwood R, Ahfeldt T, Lapey A, Wain J, et al. Generation of multipotent lung and airway progenitors from mouse ESCs and patient-specific cystic fibrosis iPSCs. Cell Stem Cell 2012;10:385−97.

[71] Wong AP, Bear CE, Chin S, Pasceri P, Thompson TO, Huan LJ, et al. Directed differentiation of human pluripotent stem cells into mature airway epithelia expressing functional CFTRTR protein. Nat Biotechnol 2012;30:876−82.

[72] Whitsett JA, Clark JC, Picard L, Tichelaar JW, Wert SE, Itoh N, et al. Fibroblast growth factor 18 influences proximal programming during lung morphogenesis. J Biol Chem 2002;277:22743−9.

[73] Tobin SC, Kim K. Generating pluripotent stem cells: differential epigenetic changes during cellular reprogramming. FEBS Lett 2012;586:2874−81.

[74] Riggs JW, Barrilleaux BL, Varlakhanova N, Bush KM, Chan V, Knoepfler PS. Induced pluripotency and oncogenic transformation are related processes. Stem Cell Dev 2012;22(1):37−50.

[75] Ionescu L, Byrne RN, van Haaften T, Vadivel A, Alphonse RS, Rey-Parra GJ, et al. Stem cell conditioned medium improves acute lung injury in mice: in vivo evidence for stem cell paracrine action. Am J Physiol Lung Cell Mol Physiol 2012;303:967−77.

[76] De Coppi P, Bartsch Jr. G, Siddiqui MM, Xu T, Santos CC, Perin L, et al. Isolation of amniotic stem cell lines with potential for therapy. Nat Biotechnol 2007;25:100−6.

[77] Vosdoganes P, Wallace EM, Chan ST, Acharya R, Moss TJ, Lim R. Human amnion epithelial cells repair established lung injury. Cell Transplant 2012. Oct 4. [Epub ahead of print].

[78] Streuli C. Extracellular matrix remodelling and cellular differentiation. Curr Opin Cell Biol 1999;11:634−40.

[79] Go T, Jungebluth P, Baiguero S, Asnaghi A, Martorell J, Ostertag H, et al. Both epithelial cells and mesenchymal stem cell-derived chondrocytes contribute to the survival of tissue-engineered airway transplants in pigs. J Thorac Cardiovasc Surg 2010;139:437−43.

[80] Zani BG, Kojima K, Vacanti CA, Edelman ER. Tissue-engineered endothelial and epithelial implants differentially and synergistically regulate airway repair. Proc Natl Acad Sci USA 2008;105:7046−51.

[81] Ott HC, Clippinger B, Conrad C, Schuetz C, Pomerantseva I, Ikonomou L, et al. Regeneration and orthotopic transplantation of a bioartificial lung. Nat Med 2010;16:927−33.

[82] Petersen TH, Calle EA, Zhao L, Lee EJ, Gui L, Raredon MB, et al. Tissue-engineered lungs for in vivo implantation. Science 2010;329:538−41.

[83] Song JJ, Kim SS, Liu Z, Madsen JC, Mathisen DJ, Vacanti JP, et al. Enhanced in vivo function of bioartificial lungs in rats. Ann Thorac Surg 2011;92:998−1005.

[84] Macchiarini P, Jungebluth P, Go T, Asnaghi MA, Rees LE, Cogan TA, et al. Clinical transplantation of a tissue-engineered airway. Lancet 2008;372:2023−30.

[85] Birchall M, Macchiarini P. Airway transplantation: a debate worth having? Transplantation 2008;85:1075−80.

Chapter 50

Lung Regeneration: The Stem Cell Approach

Shigeo Masuda

Gene Expression Laboratory, The Salk Institute for Biological Studies, La Jolla, CA

Chapter Outline

50.1 LUNG STEM/PROGENITOR CELLS AND DIFFERENTIATED CELLS

In the trachea and bronchi, basal cells are located beneath the columnar airway epithelium. They are a less differentiated progenitor cell source for columnar epithelial cells and give rise to ciliated cells, secretory cells, and goblet cells.

In the terminal bronchioles, Clara cells, which are non-ciliated secretory cells, are found and are characterized by secretion of Clara cell secretory protein (CCSP) (also known as secretoglobin 1a1, Scgb1a1, or CC10). Clara cells are normally differentiated cells but are capable of serving as a self-regenerating local progenitor for ciliated cells and are essential in airway repair after injury [1]. Bronchioalveolar stem cells (BASCs) are progenitor cells found at the bronchioalveolar duct junction (BADJ). BASCs have been identified by expression of both CCSP and surfactant protein C (SPC) (also known as Sftpc). SPC is an alveolar type 2 (AT2) cell marker. In addition, BASCs express Sca1 and CD34 [2]. These cells have been proposed to be precursor cells for both Clara cells and AT2 cells; however, the definition and capability of this putative cell population remains debatable [1], because their lineage fate has not been followed.

In the alveoli, AT2 cells function to secrete surfactant, whereas alveolar type 1 (AT1) cells enable gas exchange between alveolar and capillary networks. Both AT1 and AT2 cells are differentiated cell types, although AT2 cells maintain some plasticity and can differentiate into AT1 cells.

Recently, Kajstura et al. [3] have identified lung stem cells (LSCs) from human lung tissue, using the stem cell antigen, c-kit. LSCs were shown to be self-renewing, clonogenic, and multipotent both *in vitro* and *in vivo*. Furthermore, from a single cell preparation, the cells could be expanded, administered to a mouse after lung injury, and result in the formation of human bronchioles, alveoli, and pulmonary vessels. It has been demonstrated that these human tissues were integrated into the damaged mouse lung, at the level of both airways and vessels. Therefore, human lung tissues of both endodermal and mesodermal origins could be generated in mice with a single precursor cell.

Kajstura et al. [3] tried to isolate human LSCs from samples of both adult and fetal lung tissues. Normal adult human lung tissues were obtained from unused donor organs, and fetal lungs were obtained after fetal death. Lung samples were digested and c-kit-positive cells were collected by means of immunosorting, followed by culture in F-12 medium plus 10% fetal-calf serum. For clonal analysis, c-kit-positive cells were subjected to sorting by fluorescence-activated cell sorting (FACS) to place single cells in each well of Terasaki plates or seeded at a limiting dilution. After 3 or 4 weeks, multicellular clones were obtained; cells in the clones continued to express c-kit. Cloning efficiency averaged 1%. Clonal human LSCs expressed NANOG, OCT3/4, SOX2, and KLF4, suggesting that adult human LSCs were multipotent cells with a high degree of plasticity. Regarding this finding, it would be of interest to see whether they have characteristics similar to those of induced pluripotent stem cells (iPSCs) [4]. Are human LSCs also expressing another factor (c-Myc) of Yamanaka factors for reprogramming? Moreover, expression of L-Myc (lung Myc) is indeed interesting [4], because L-Myc is expressed in mouse normal lung [5] and human lung cancer [6], it and is an alternative promising factor for reprogramming [7]. Another interest is whether human LSCs during proper *in vitro* culture for iPSCs would give rise to a colony similar to iPSCs [4]. If this were true, generated iPSCs may be free of exogenous factors, which would mean that they could be immaculate and have a great impact on regenerative medicine.

In order to confirm human LSCs, serial transplantation *in vivo* is required to establish whether these cells create functionally integrated structures in the relevant tissue microenvironment. Kajstura et al. [3] induced cryoinjury in the lung of immunosuppressed mice. Shortly after lung injury, human LSCs were administered in the region adjacent to the area of damage. After 10−14 days, human LSCs had formed human bronchioles, alveoli, and pulmonary vessels, partly restoring the structural integrity of the recipient parenchyma. Therefore, human LSCs showed self-renewal and multipotentiality *in vivo*. They also identified undifferentiated, cycling c-kit-positive human LSCs within the regenerated human lung parenchyma and in the adjacent, intact recipient mouse lung. After serial transplantation, they confirmed that the newly formed human lung structures could be derived from the serially transplanted human LSCs.

In adult human lung, LSCs were found at a frequency of 1 per 24,000 cells; counts in the bronchioles and alveoli were 1 per 6000 cells and 1 per 30,000 cells, respectively, suggesting that human LSCs showed a preferential localization in small bronchioles [3]. On the other hand, in fetal lung, the frequency of human LSCs varied from 1 per 11,000 cells to 1 per 600 cells, resulting in an average count of 1 per 4100 cells. There were no differences in the frequency of human LSCs among samples with gestational 12−36 weeks. If these results by Kajstura et al. [3] are truly reproducible, the discovery of human LSCs will advance the future potential for stem cell therapy in human lung disease.

50.2 LUNG DIFFERENTIATION FROM EMBRYONIC STEM CELLS AND IPSCS

During development, the lung arises from the anterior foregut endoderm region, which itself arises from the definitive endoderm (DE) that develops soon after gastrulation [8]. The lung field is first identified by expression of the transcription factor Nkx2.1. Recent studies have been trying to recapitulate in culture the key signaling events that direct early lineage commitment in the embryo. This mimicking has improved the efficiency in obtaining differentiated cells of endodermal origin. Similar to what is seen during gastrulation in the embryo,

high levels of nodal signaling (stimulated by activin A) induce formation of the DE in embryonic stem cells (ESCs) and iPSCs [9]. Treatment of activin-A-induced endoderm with inhibitors of bone morphogenetic protein (BMP) signaling and nodal signaling specifies a highly enriched population of cells with anterior foregut endoderm identity [10]. These cells can be directed to an early lung endoderm fate by timed treatment with signaling factors known to be required for lung development, including WNTs, fibroblast growth factors (FGFs), BMPs, endothelial growth factors (EGFs), and keratinocyte growth factors (KGFs). This stepwise differentiation protocol from undifferentiated ESCs to anteriorized endoderm greatly increases the efficiency of deriving anterior endoderm cell types such as the lung and thyroid. These results validate the approach of recapitulating key steps in early embryonic development in order to efficiently derive cells of endodermal origin from undifferentiated ESCs.

Two recent papers report advances in inducing formation of differentiated lung cells from ESCs and human iPSCs [11,12]. Both approaches use a stepwise application of inhibitors and activators of WNT, BMP, and FGF to induce lung endoderm formation. It has been shown that the timing and duration of application of these factors were found to be critical for proper specification of lung epithelial cells. In addition, both studies found that a stepwise process through the stages of lung development was required to obtain efficient production of differentiated lung epithelium.

Mou et al. [11] first tried to generate DE from mouse ESCs. It was reported recently that a high dose of activin, transient WNT activation, and staged BMP4 inhibition promote commitment of ESCs into DE with high efficiency [13] based on the dual expression of the endoderm transcription factors FOXA2 and SOX17. Next, because Nkx2.1 is the earliest marker of lung endoderm and distinguishes lung endoderm from the rest of the foregut endoderm, they focused on producing Nkx2.1 + lung endoderm. They investigated whether the newly generated DE cells have the competence to generate Nkx2.1 + cells. After 2-day culture in medium containing BMP4, FGF2, and GSK3iXV (a WNT agonist), they found that less than 1% of the cells were Nkx2.1 + . Instead, more than 60% of the cells were Cdx2 + , indicating that the majority of the cells were specified to a hindgut fate. Because the endoderm cells at this fate did not differentiate into Nkx2.1 + cells efficiently, they examined whether an "anteriorization" step would facilitate lung fate specification. It was reported that Noggin (BMP inhibitor) synergized with SB431542 (TGFβ inhibitor) to suppress a posterior endoderm fate (Cdx2 +) in favor of an anterior endoderm fate (Sox2 +) [10]. Mou et al. [11] found that TGFβ inhibition (A-83-01) alone for 2 days was sufficient to increase FoxA2 + Sox2 + anterior endoderm and enhanced the competence of endoderm cells to form Nkx2.1 + cells after exposure to the BMP4/FGF2/WNT agonist cocktail for another 2 days. Taken together, stage-specific TGFβ inhibition regionalizes naive endoderm to anterior foregut endoderm and facilitates the differentiation of Nkx2.1 + cells.

Because Nkx2.1 is not a specific marker of the lung, and its expression is also found in the thyroid and ventral forebrain, Mou et al. [11] determined the identity of the Nkx2.1 + cells generated in their culture system. The Nkx2.1 + cells were positive for endodermal marker FOXA2, negative for neuroectodermal marker TUJ1, and negative for thyroid marker PAX8. The Nkx2.1 + cells were also proliferative as demonstrated by costaining for KI67. Subpopulations of Nkx2.1 + cells were positive for SOX2 (an airway progenitor marker) and FOXP2/SOX9 (multipotent lung progenitor marker).

It has been shown that BMP4, FGF2, and WNT signaling are each required to specify lung endoderm from the anterior foregut based on mouse genetic studies [14–16]. Therefore, Mou et al. [11] examined whether they could recapitulate these findings in their in vitro mouse ESC differentiation system. They exposed the anteriorized endoderm cells to combinations of BMP4, FGF2, and WNT agonists and antagonists. They observed that BMP4, FGF2, and WNT signaling were each necessary for lung specification from foregut endoderm cells. The Nkx2.1 + immature lung progenitors were shown to be able to mature into Nkx2.1 + Sox2 + proximal progenitor cells and Nkx2.1 + p63 + airway basal stem cells in vitro, though production of these cells still needs to be optimized. When transplanted in vivo, ESC-derived Nkx2.1 + cell populations could differentiate into mature airway epithelium.

Furthermore, Mou et al. [11] determined whether a similar stepwise differentiation approach could be used to generate lung airway progenitors from cystic fibrosis disease-specific human iPSCs. This protocol was also successfully applied to several human iPSCs. Finally, human iPSC-derived Nkx2.1 + mixed cell populations were subcutaneously engrafted. Many spheres formed in engrafted tissues under the skin of immunodeficient recipient mice. In Nkx2.1 + spheres, some of the Nkx2.1 + cells coexpressed p63, indicating that these Nkx2.1 + cells had matured into airway basal stem cells. Differentiation into ciliated cells, Clara cells, and mucin-secreting goblet cells was not detected, suggesting that further optimization of differentiation conditions is required.

Longmire et al. [12] revealed the efficient derivation of purified lung and thyroid progenitors from mouse ESCs. Inhibition of TGFβ and BMP signaling, followed by activation of BMP and FGF signaling, was able to derive lung and thyroid progenitor cells efficiently from DE. Using Nkx2-1GFP knockin reporter ESCs,

Longmire et al. purified progenitor cells for expansion in culture and these progenitors had transcriptome that overlap with developing lung epithelium. Lung and thyroid progenitor cells expressed markers indicative of lung and thyroid lineages and could recellularize a 3D lung tissue scaffold. Longmire et al. derived a pure population of progenitors that could recapitulate the lung/thyroid development.

Although the functional capacity of many lineages differentiated from ESCs or iPSCs remains one of the most important issues in regenerative biology, there are few assays for determining functionality of, in particular, lung cells. Mou et al. [11] and Longmire et al. [12] do not claim to have generated fully differentiated lung epithelium from pluripotent stem cells, and further study is required to drive more mature and functional lung epithelial lineages. The structurally complex environment will be difficult to recapitulate in vitro and will likely require in vivo assays for full functional assessment.

Although Mou et al. [11] demonstrated stepwise generation of lung endoderm progenitors from human iPSCs, they failed to generate mature lung epithelial cells. No studies have been able to generate proximal conducting airway epithelial cells that have functional polarized CFTR (cystic fibrosis transmembrane conductance regulator). More recently, Wong et al. [17] have developed a method to recapitulate sequential processes that restrict progenitor cells from endoderm to proximal lung epithelia and have generated functional proximal conducting airway epithelia expressing CFTR from human pluripotent stem cells.

Wong et al. [17] first differentiated human ESCs toward DE using a method previously described by D'Amour [18]. Treatment with activin-A and WNT3A for 4 days was sufficient to induce cells into DE as determined by coexpression of CXCR4 and cKIT, and endoderm transcription factors FOXA2 and SOX17. After gastrulation, anterior–posterior signals pattern primitive gut tube into distinct regions. High concentration of FGF2 induces NKX2.1 expression typical of the early lung endoderm [15]. Sonic hedgehog (SHH) signaling promotes embryonic lung growth and suppresses pancreatic development [19]. Therefore, to promote anterior foregut identity and specify lung cell fate, DE cells were treated with FGF2 and SHH. After 5 days of exposure, the majority of cells expressed pan-endoderm transcription factor FOXA2 and transcription factor NKX2.1. Upregulation of genes associated with anterior foregut endoderm transcription factors—*SOX2* and *NKX2.1*—as well as pharyngeal endoderm *FOXG* and thyroid *TG* and *PAX9* was observed. Posterior hindgut gene *CDX2* was not detected. NKX2.1 is a good marker for lung endoderm but is also expressed in the thyroid and forebrain. Although other thyroid genes were detected, ectoderm

PAX6 was not detected, which excludes possibility of forebrain-derived *NKX2.1* expression. Collectively, FGF2 and SHH can efficiently induce anterior foregut derivatives from DE.

Previous studies in mouse embryonic lung organ cultures have identified specific growth factors essential for lung development from NKX2.1-expressing endoderm [20], including FGF and BMP4. FGF10 is key growth factor expressed by mesenchyme at the earliest stage of lung development and stimulates lung bud outgrowth and organogenesis [19]. FGF7, also expressed by mesenchyme, is mainly involved in stimulating fluid secretion in the lung but has a role in epithelial cell growth [21]. Wong et al. [17] observed that combination of FGF7 and FGF10 at 50 ng/ml augmented mRNA expression of transcription factor genes *NKX2.1* and *FOXA2* compared to DE levels. High concentrations of BMP4 stimulate distal cell fate, while low concentrations promote proximal cell fate [22]. Therefore, Wong et al. exposed human ESC-derived anterior foregut cells to FGF10, FGF7, and varying concentrations of BMP4 to determine optimal concentration of BMP4 that would induce proximal cell fate. They found that combination of FGF7, FGF10, and low concentrations of BMP4 could induce upregulation of some conducting airway cell lineages from anterior foregut endoderm.

FGF18 is known to enhance proximal but not distal airway formation and also known to play a role in increasing size of conducting airways following maturation of epithelium [23]. Wong et al. [17] added FGF18, which resulted in further upregulation of airway genes *KRT5*, *TRP63*, *FOXJ1*, *SOX17*, *MUC5AC*, and *CFTR*. Moreover, flow cytometric quantification showed that cells expressed pan-cytokeratin (pan-KRT, 33%) CFTR (30%), FOXJ1 (36%), and NKX2.1 (32%), suggesting that at least one-third of the cells in culture were of ciliated CFTR-expressing airway phenotype. More than 50% of the cells were P63 +, indicating that the vast majority of cells were potentially basal cell progenitors, previously shown to give rise to other proximal airway lineages [24]. Taken together, their findings support mechanism for airway development in which FGF7, FGF10, BMP4, and FGF18 are required to promote airway lineage development.

In order to mature cells toward functional airway epithelium *in vitro*, Wong et al. [17] employed commercially available media that support growth and differentiation of primary bronchial epithelial cells *ex vivo* along with air−liquid interface (ALI) growth condition to mimic postnatal airway epithelial niche *in vivo* and promote differentiation, maturation, and polarization of epithelium. For ALI culture, they exposed cells to air on apical side of transwell membrane while basolateral side was exposed to media. After 5 weeks of ALI, flow cytometry

showed that at least 50% of CFTR + cells coexpressed pan-KRT, FOXJ1, and LHS28 (basal bodies of cilia), suggesting enrichment of cells typical of ciliated epithelium. In addition, the majority of cells expressed other conducting airway epithelia markers: acetylated tubulin TUBA1A (cilia), MUC1 (Goblet cell), KRT14 (basal epithelia), and pan-endoderm marker FOXA2. A smaller percentage of cells expressed transcription factor NKX2.1, which in later stages of lung development regulates Clara cell and AT2 cell differentiation. Proximal airway epithelium was established as indicated by protein expression of mucin 16 (marker of tracheal epithelium) and cytokeratin 16. Gene expression analysis showed upregulation of proximal airway lineage genes (*SOX17*, *FOXJ1*, *MUC5AC*, *TRP63*, *KRT5*, *ARG2*, *SOX2*, *CFTR*, *KRT16*, *MUC16*, *NGFR*) comparable or higher than levels in total adult lung or tracheal tissue. Immunofluorescence staining and confocal analysis of 5-week culture of ALI demonstrated contiguous patches of epithelial cells typified by membrane expression of Zona Occludin-1 (ZO1), a protein associated with tight junctions, and costaining with pan-KRT and CFTR. Stained sections of cultures showed ciliated cells, confirmed with antibody staining for cilia (βIV tubulin) and apically localized CFTR. In addition, cultures stained positive for MUC5AC on surface of cells, indicative of mucin production. These findings suggest that ALI can induce maturation and polarization of ciliated large airway epithelium with proper localization of CFTR protein.

Next, Wong et al. [17] demonstrated as a proof of concept that treatment of cystic fibrosis (CF) patient iPSC-derived epithelial cells with small-molecule compound to correct for common CF processing mutation resulted in enhanced plasma membrane localization of mature CFTR protein. Overall, CF iPSC-derived airway cells may provide renewable source of patient-specific cells to identify new or validate existing CF therapeutic drugs.

This study by Wong et al. [17] is the first study demonstrating that human pluripotent stem cells can be directed to differentiate *in vitro* into CFTR-functional conducting airway epithelium. The great majority of cells express airway epithelia markers, with establishment of CFTR function observed in a third of cultures. Further refinement by isolating cells using positive and negative selection or cell-surface marker identification of lung progenitor populations would improve purity of lung epithelial cells.

Wang et al. [25] have demonstrated that cells derived from ES cells (i.e., AT2 cells) can have beneficial effects in lung injury, and it provides a platform for potential therapeutic use of ESCs in a variety of lung diseases. They have succeeded in induction of pure population of AT2 cells *in vitro*, and they have subsequently demonstrated that intratracheally transplanted AT2 cells derived from human ESCs differentiate and functionally repair the epithelium of injured alveolus in a mouse model of bleomycin-induced acute lung injury.

Recently, Cheng et al. [26] have established endodermal progenitor (EP) cell lines from human ESCs and iPSCs. Upon manipulation of their culture conditions *in vitro* or transplantation into mice, clonally derived EP cells differentiate into numerous endodermal lineages, including glucose-responsive pancreatic β-cells, hepatocytes, and intestinal epithelia both *in vitro* and *in vivo*. EP cells display a proliferative capacity similar to ESCs yet lack teratoma-forming ability.

There has been a stepwise differentiation protocol, previously shown to induce DE and its derivative hepatic lineages from mouse and human ESCs [27,28]. In this protocol, activin A was added to induce DE, and BMP4 and bFGF were subsequently used to specify a hepatic fate from DE. Cheng et al. [26] applied this protocol and could generate hepatic cells after 2 weeks of induction. Notably, after 3−4 weeks, these cultures contained two cell populations with distinct morphology; one population resembled immature hepatocytes, whereas the other population resembled undifferentiated ESC colonies. They used a pipette to manually isolate colonies of a given morphology and performed gene expression analysis. The ESC markers NANOG and OCT4 were not expressed in either cell type, indicating that these colonies were not contamination of ESCs in the culture. The early endoderm marker SOX17 was enriched in the putative progenitor colonies. Based on these findings, they hypothesized that the undifferentiated colonies might represent EP cells undergoing both self-renewal and differentiation in culture.

Cheng et al. [26] examined the bulk differentiation cultures by flow cytometry for expression of KIT (CD117) and CXCR4, which are expressed on early endoderm [18,28]. A subpopulation of CXCR4 + CD117 + cells (3−60%) was consistently present in the cultures. These double-positive cells also expressed the endoderm markers FOXA1 and SOX17. The CXCR4 + CD117 + FOXA1 + SOX17 + population likely marks putative EP cells from the undifferentiated colonies. To prove this, they performed FACS sorting to isolate CXCR4 + and CXCR4 − cells for further study. They observed that the CXCR4 + population expressed the immature endoderm marker SOX17 and was negative for the early hepatocyte marker AFP. Purified CXCR4 + cells could be expanded in culture, but to better expand the EP cell population while maintaining the FOXA1 + SOX17 + phenotype, they discovered culture conditions, which consist of media containing BMP4, bFGF, EGF, and vascular endothelial growth factor (VEGF) on matrigel and mouse embryonic fibroblast (MEF) feeders.

With this optimized culture condition, Cheng et al. [26] developed a simplified protocol for EP cell

production. Human ESCs or iPSCs were induced to differentiate with high activin A, which promotes endoderm formation [18]. EP cells were established from day 5 transient endoderm by sorting for CXCR4 + /CD117 high cells and by maintaining them in a 5% O_2 environment. Both ESC-derived and iPSC-derived EP cell lines proliferated extensively and displayed a homogenous EP cell phenotype (FOXA1 + SOX17 + FOXA2 +). They could establish EP cell lines from the human ESC lines H9 and CHB8 and from two human iPSC lines.

Cheng et al. [26] next investigated whether expanded EP cells retain the ability for multipotent endodermal differentiation. They found that EP cell lines could be differentiated under conditions known to promote hepatocyte, pancreatic, or intestinal specification *in vitro*, but whether EP cells could be differentiated into lung cells remains unanswered. These data suggest that EP cells are a robust stem cell population that can be generated from multiple stages of differentiation while maintaining multipotency except differentiation potential into lung. To determine if EP cell developmental potential was restricted to endodermal lineages, cultures were induced with conditions established to drive ESCs toward either neuroectoderm or mesoderm, but eventually it was suggested that EP cells were committed to the endoderm only. Furthermore, EP cells were proved to lack tumorigenicity and form endodermal tissues (gut epithelia and liver, but not lung) *in vivo*. Although differentiation into lung cells from EP cells has not been mentioned by Cheng et al. [26], further examination might enable either lung differentiation from existing EP cell lines or establishment of new EP cell lines capable of lung differentiation.

50.3 LUNG REGENERATION

Large airway regeneration has combined the culture of mesenchymal cells and epithelial cells using artificial or native scaffolds. It has been demonstrated that an engineered mainstem bronchus could be produced and successfully transplanted into a patient [29].

Distal lung regeneration approaches have followed two tracks: segmental approaches and whole lung approaches. Segmental approaches have focused on the growth of pulmonary cells on small pieces of scaffold material, which have successfully demonstrated that pulmonary cells can proliferate *in vitro* and can sometimes form alveolar structures *in vitro* and *in vivo*. Engineered tissue segments can be delivered in minimally invasive ways either via catheter or bronchoscope. But it has proven difficult to induce these segmental engineered tissues to integrate with vasculature or airway tree of host. Whole lung regeneration has focused on the growth of complete engineered lungs, which could then be transplanted into a patient to replace failing lung or lung lobe. Recent work has used decellularized native lung as starting scaffold material. These scaffolds are reseeded with cells, and then are cultured *ex vivo* to allow cells to repopulate the scaffolds. It was demonstrated that whole rodent lungs can be grown in the laboratory and can successfully exchange gas when transplanted into rats [30,31]. Furthermore, other studies have demonstrated the feasibility of using ESCs or iPSCs to repopulate decellularized scaffolds. Future research in human lung regeneration will focus on epithelial cells derived from progenitor populations, such as ESCs or iPSCs.

REFERENCES

[1] Rawlins EL, Okubo T, Xue Y, Brass DM, Auten RL, Hasegawa H, et al. The role of Scgb1a1 + Clara cells in the long-term maintenance and repair of lung airway, but not alveolar, epithelium. Cell Stem Cell 2009;4:525–34.

[2] Kim CF, Jackson EL, Woolfenden AE, Lawrence S, Babar I, Vogel S, et al. Identification of bronchioalveolar stem cells in normal lung and lung cancer. Cell 2005;121:823–35.

[3] Kajstura J, Rota M, Hall SR, Hosoda T, D'Amario D, Sanada F, et al. Evidence for human lung stem cells. N Engl J Med 2011;364:1795–806.

[4] Masuda S. Evidence for human lung stem cells. N Engl J Med 2011;365:464–5.

[5] Hatton KS, Mahon K, Chin L, Chiu FC, Lee HW, Peng D, et al. Expression and activity of L-Myc in normal mouse development. Mol Cell Biol 1996;16:1794–804.

[6] Nau MM, Brooks BJ, Battey J, Sausville E, Gazdar AF, Kirsch IR, et al. L-myc, a new myc-related gene amplified and expressed in human small cell lung cancer. Nature 1985;318:69–73.

[7] Nakagawa M, Takizawa N, Narita M, Ichisaka T, Yamanaka S. Promotion of direct reprogramming by transformation-deficient Myc. Proc Natl Acad Sci USA 2010;107:14152–7.

[8] Kadzik RS, Morrisey EE. Directing lung endoderm differentiation in pluripotent stem cells. Cell Stem Cell 2012;10:355–61.

[9] Murry CE, Keller G. Differentiation of embryonic stem cells to clinically relevant populations: lessons from embryonic development. Cell 2008;132:661–80.

[10] Green MD, Chen A, Nostro MC, d'Souza SL, Schaniel C, Lemischka IR, et al. Generation of anterior foregut endoderm from human embryonic and induced pluripotent stem cells. Nat Biotechnol 2011;29:267–72.

[11] Mou H, Zhao R, Sherwood R, Ahfeldt T, Lapey A, Wain J, et al. Generation of multipotent lung and airway progenitors from mouse ESCs and patient-specific cystic fibrosis iPSCs. Cell Stem Cell 2012;10:385–97.

[12] Longmire TA, Ikonomou L, Hawkins F, Christodoulou C, Cao Y, Jean JC, et al. Efficient derivation of purified lung and thyroid progenitors from embryonic stem cells. Cell Stem Cell 2012;10:398–411.

[13] Sherwood RI, Maehr R, Mazzoni EO, Melton DA. Wnt signaling specifies and patterns intestinal endoderm. Mech Dev 2011;128:387–400.

[14] Domyan ET, Ferretti E, Throckmorton K, Mishina Y, Nicolis SK, Sun X. Signaling through BMP receptors promotes respiratory identity in the foregut via repression of Sox2. Development 2011;138:971–81.

[15] Serls AE, Doherty S, Parvatiyar P, Wells JM, Deutsch GH. Different thresholds of fibroblast growth factors pattern the ventral foregut into liver and lung. Development 2005;132:35–47.

[16] Goss AM, Tian Y, Tsukiyama T, Cohen ED, Zhou D, Lu MM, et al. Wnt2/2b and beta-catenin signaling are necessary and sufficient to specify lung progenitors in the foregut. Dev Cell 2009;17:290–8.

[17] Wong AP, Bear CE, Chin S, Pasceri P, Thompson TO, Huan LJ, et al. Directed differentiation of human pluripotent stem cells into mature airway epithelia expressing functional CFTRTR protein. Nat Biotechnol 2012;30:876–82.

[18] D'Amour KA, Agulnick AD, Eliazer S, Kelly OG, Kroon E, Baetge EE. Efficient differentiation of human embryonic stem cells to definitive endoderm. Nat Biotechnol 2005;23:1534–41.

[19] Bellusci S, Furuta Y, Rush MG, Henderson R, Winnier G, Hogan BL. Involvement of Sonic hedgehog (Shh) in mouse embryonic lung growth and morphogenesis. Development 1997;124:53–63.

[20] Kimura J, Deutsch GH. Key mechanisms of early lung development. Pediatr Dev Pathol 2007;10:335–47.

[21] Shiratori M, Oshika E, Ung LP, Singh G, Shinozuka H, Warburton D, et al. Keratinocyte growth factor and embryonic rat lung morphogenesis. Am J Respir Cell Mol Biol 1996;15:328–38.

[22] Weaver M, Yingling JM, Dunn NR, Bellusci S, Hogan BL. BMP signaling regulates proximal–distal differentiation of endoderm in mouse lung development. Development 1999;126:4005–15.

[23] Whitsett JA, Clark JC, Picard L, Tichelaar JW, Wert SE, Itoh N, et al. Fibroblast growth factor 18 influences proximal programming during lung morphogenesis. J Biol Chem 2002;277:22743–9.

[24] Rock JR, Onaitis MW, Rawlins EL, Lu Y, Clark CP, Xue Y, et al. Basal cells as stem cells of the mouse trachea and human airway epithelium. Proc Natl Acad Sci USA 2009;106:12771–5.

[25] Wang D, Morales JE, Calame DG, Alcorn JL, Wetsel RA. Transplantation of human embryonic stem cell-derived alveolar epithelial type II cells abrogates acute lung injury in mice. Mol Ther 2010;18:625–34.

[26] Cheng X, Ying L, Lu L, Galvão AM, Mills JA, Lin HC, et al. Self-renewing endodermal progenitor lines generated from human pluripotent stem cells. Cell Stem Cell 2012;10:371–84.

[27] Gadue P, Huber TL, Paddison PJ, Keller GM. Wnt and TGF-beta signaling are required for the induction of an *in vitro* model of primitive streak formation using embryonic stem cells. Proc Natl Acad Sci USA 2006;103:16806–11.

[28] Gouon-Evans V, Boussemart L, Gadue P, Nierhoff D, Koehler CI, Kubo A, et al. BMP-4 is required for hepatic specification of mouse embryonic stem cell-derived definitive endoderm. Nat Biotechnol 2006;24:1402–11.

[29] Macchiarini P, Jungebluth P, Go T, Asnaghi MA, Rees LE, Cogan TA, et al. Clinical transplantation of a tissue-engineered airway. Lancet 2008;372:2023–30.

[30] Petersen TH, Calle EA, Zhao L, Lee EJ, Gui L, Raredon MB, et al. Tissue-engineered lungs for *in vivo* implantation. Science 2010;329:538–41.

[31] Ott HC, Clippinger B, Conrad C, Schuetz C, Pomerantseva I, Ikonomou L, et al. Regeneration and orthotopic transplantation of a bioartificial lung. Nat Med 2010;16:927–33.

Composite Tissues Allotransplantation

Chapter 51

Current Status of CTA

Palmina Petruzzo[a,b] and Jean Michel Dubernard[a]

[a]Department of Transplantation, Hopital Edouard Herriot, Lyon, France, [b]Department of Surgery, University of Cagliari, Italy

Chapter Outline

51.1 INTRODUCTION

Composite tissue allotransplantation (CTA) means simultaneous transplantation of skin, connective tissue, muscles, bones, nerves, vessels, and these multiple tissues are of ectodermal and mesodermal origin, determining a different antigenic load. Skin was considered the most antigenic tissue as proposed by Murray [1] in his relative scale of antigenicity of tissues and organs. However, some experimental studies concluded that no single tissue is dominant in primarily vascularized limb allografts; moreover they demonstrated that a whole limb allograft elicits a less intense immune response than each individual component of the CTAs [2].

The evolution of knowledge in solid organ transplantation and the development of surgical procedures in reconstructive surgery allowed the realization of CTA, also called vascularized composite allotransplantation, to reflect the similarity to solid organ transplantation or reconstructive transplantation.

CTAs share some characteristics with solid organ transplantation as both are vascularized allografts, recovered from human donors as an anatomical/structural unit (e.g., hand, face, lower extremity, larynx), which is then transplanted in human recipients who have to take immunosuppressants in order to avoid allograft rejection and loss. At the same time, CTAs are different from solid organ transplantation as they are mainly external, visible, constituted of tissues originating from several germ layers, and nerve regrowth is required to assure graft function.

The three main challenges in this new field of transplantation are:

1. immunological for the different antigenic load of the multiple tissues, including skin, which was always considered the most antigenic tissue [1];
2. technical for the complex and long surgery particularly in face transplantation, the necessity of a well-coordinated system and a multidisciplinary team being the success of a CTA based on a satisfactory function which depends on the quality of surgery, nerve regrowth, rehabilitation program, and cortical reorganization;
3. psychological for the importance of the aesthetic aspect of the CTAs and patients' capacity to accept and integrate in their own body image visible cadaveric tissues.

Despite the small clinical volume, there has been a considerable progress in this field and CTA is rapidly emerging from an experimental model to standard of care.

51.2 INDICATIONS AND OUTCOMES

CTAs are not life-saving transplants, but they allow the performance of reconstructions of parts of the body lost

Regenerative Medicine Applications in Organ Transplantation.

because of accidents, congenital malformations, and surgical excision when conventional reconstructive surgery is not possible. In contrast to solid organ transplantation, CTA recipients are usually healthy subjects except for the severe tissue defect and the final goal of this type of transplantation is to improve patients' quality of life. Indeed, patients suffering from severe tissue defects show an altered sense of body integrity; many of them do not have an integrated social life and live in isolation, without working and with a difficult economic status. It is common that these patients show depression and alcohol abuse.

The improvement of patients' quality of life is subjective, and it is based on graft function and on life changes not only due to graft function but also due to the immunosuppressive therapy, being that the side-effects in CTAs are the same as reported in solid organ transplantation.

The major controversies surrounding CTA are the uncertain long-term functional results and the unclear risk—benefit ratio.

Although patient survival is not considered an outcome in CTA, it is a critical point in nonlife-saving transplants. Until now the results reported by the International Registry on Hand and Composite Tissue Allotransplantation (IRHCTT) [3,4] are encouraging as global patient survival, considering together upper extremity transplantation and face transplantation, is 93.1% (96.5% at 1 year). When we consider graft survival, the results are superior to those achieved in solid organ transplantation, indeed graft survival is 100% in face transplantation and 90.48% in hand transplantation (92.8% at 1 year); although the initial success in several upper extremity transplantations were removed during the follow-up in "emerging countries" and graft loss was due to limited access to immunosuppressive drugs and to suboptimal follow-up.

51.3 IMMUNOSUPPRESSIVE TREATMENT

The large majority of CTA recipients have been maintained on immunosuppressive therapy similar to that used in solid organ transplantation, consisting of tacrolimus, steroids, and mycophenolate mofetil (MMF). The induction therapy included antithymocyte globulin, basiliximab, and more recently, Campath-1H.

Although steroid-sparing maintenance and a switch from tacrolimus to sirolimus have recently been utilized in CTA, in the early postoperative period, all recipients received tacrolimus because of the stimulatory effect of this drug on the synthesis of axotomy-induced growth-associated protein (GAP-43) that seems to promote nerve regeneration [5].

The side-effects reported in CTA are those of traditional immunosuppressive treatment and they are considered the major limit for a wider application of CTAs,

which are nonlife-saving transplants. Progress in this new field of transplantation is undoubtedly influenced by the necessity of long-term immunosuppression, for this reason different strategies aimed at minimizing maintenance immunosuppression or inducing donor-specific tolerance were performed.

At present it is difficult to demonstrate the superiority of one immunosuppressive regimen over another due to the lack of prospective randomized studies and to the limited number of grafted patients. Indeed, the superiority of the regimen based on low-dose steroids, tacrolimus, and MMF, which has been successfully used by the majority of teams, over that based on donor cell infusion associated to depleting induction therapy followed by calcineurin inhibitor monotherapy, has not been demonstrated; similarly, the superiority of an induction therapy, based on alemtuzumab instead of antithymocyte globulins, has not been shown [6].

Donor cell infusion associated with depleting induction therapy followed by calcineurin inhibitor monotherapy is considered an interesting approach in CTA, although up to now the induction of adequate tolerance has not been achieved [6,7]. At present no CTA recipient proved to be spontaneously tolerant; indeed, it was noted in the first-hand allotransplantation and in all recipients who discontinued the immunosuppressive therapy that consequent rejection of the graft inevitably occurred [3,4,8].

Mixed allogenic chimerism induces donor-specific tolerance in a wide spectrum of allografts [9,10], including some experimental composite tissue allografts [11], but the major obstacle to widespread application of bone marrow infusion to achieve mixed chimerism is graft-versus-host disease (GVHD), which would be unacceptable in a nonlife-saving transplantation [12].

The main complications reported in the IRHCTT [3,4] are metabolic ones (above all hyperglycemia, increased creatinine values with several cases of renal function deterioration, arterial hypertension), infections (above all CMV reactivation or infection, herpes virus infection, and bacterial infection), and malignancies (two lymphoproliferative diseases, one uterus carcinoma, and two basal cell carcinomas).

51.4 ACUTE AND CHRONIC REJECTIONS

The majority of CTA recipients experienced at least one episode of acute rejection (AR) in the first year after transplantation. According to the IRHCTT [3,4], 85% of the hand-grafted patients and 54.5% of the face-grafted patients presented at least one episode of AR in the first posttransplant year. On the other hand, the high rate of AR episodes reported in this field of transplantation might be due to the easy diagnosis of AR, as the corresponding lesions are easily seen and confirmed histologically on biopsy specimens taken from them. All AR episodes were completely reversible in compliant patients, provided they were promptly diagnosed and treated. Clinical experience seems to confirm the contention that skin is the main target of AR. Indeed, the first clinical signs of AR manifest on the skin: the suspicion is based on visual inspection and then confirmed by histological examination. In addition, it seems that even during severe rejection, the changes found in underlying tissues (muscles, nerves, bones, and tendons) are less severe than those present in the skin [8].

AR reactions manifest clinically as erythematous macules, diffuse redness (Figure 51.1), or asymptomatic papules over the allografted skin [8,13,14]. Microscopically, they show characteristic, although non-specific, changes mainly involving the dermis and the epidermis, that may extend to the hypodermis in the case of severe rejection (Figure 51.2). A specific score (Banff score 2007) to assess the severity of AR in CTA has been established. This system comprises the following five severity grades [15]:

Grade 0 (no rejection): no or rare inflammatory infiltrates.

Grade I (mild rejection): mild perivascular infiltration. No involvement of the overlying epidermis.

Grade II (moderate rejection): moderate-to-severe perivascular inflammation with or without mild epidermal and/or adnexal involvement (limited to spongiosis and exocytosis). No epidermal dyskeratosis or apoptosis.

Grade III (severe rejection): dense dermal inflammation associated with and epidermal involvement (basal keratinocyte vacuolization, keratinocyte apoptosis, and/or necrosis).

Grade IV (necrotizing AR): frank necrosis of epidermis or other skin structures.

The choice of AR treatment [3,4,6] is based on the Banff score grade, the frequency of the episodes, and the sensibility to steroid treatment. In the majority of cases, the episodes were reversed by increasing oral steroid dose (12.9%) or by using i.v. steroids (87%), or with administration of polyclonal (Thymoglobulin) or monoclonal antibodies (Campath-1H) in the other cases. In addition, local immunosuppressants (steroid and tacrolimus ointments) were often used although their efficacy remains unproven; these do not seem to be sufficient to reverse episodes of severe AR without additional systemic immunosuppressive treatment. For the first time in CTA, extracorporeal photochemotherapy (EPC) was employed in a few cases of face allotransplantation [16].

Despite a high incidence of AR, the occurrence of chronic rejection in CTA might be much rarer than in solid organ transplantation.

At present, insufficient data are available to define specific changes of chronic rejection in CTA. The Banff 2007 classification has not included features of chronic rejection yet [8]. Clinicopathologic features suggestive of chronic rejection could include myointimal proliferation

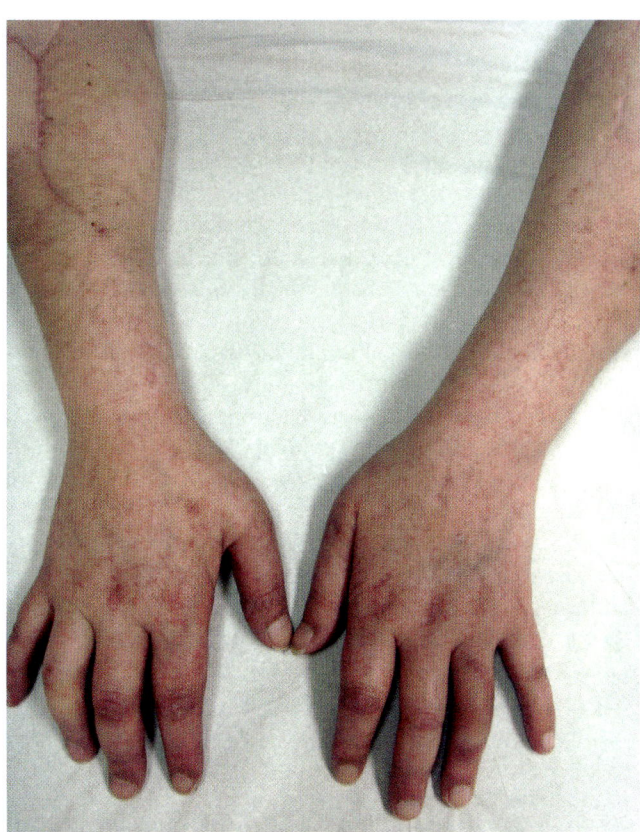

FIGURE 51.1 Maculopapular lesions of grafted upper extremities during an AR episode.

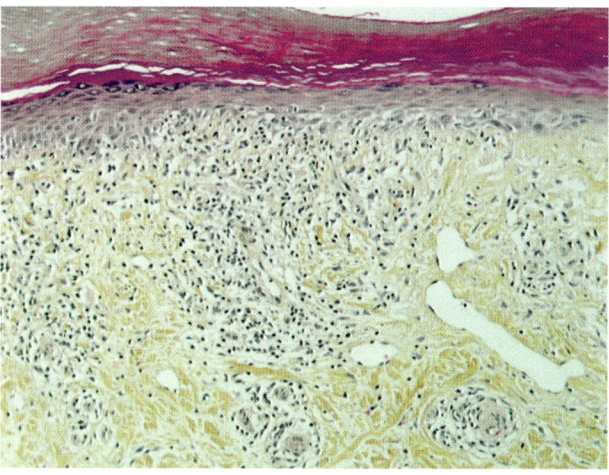

FIGURE 51.2 Histology of skin biopsy during an AR episode (grade III) shows a dense perivascular dermal infiltrate reaching the epidermis.

of arterioles, loss of adnexa, nail changes, skin and muscular atrophy, and fibrosis of deep tissues [15]. Although an experimental murine study [17] showed that graft vasculopathy is the last lesion to occur during the chronic rejection process, in a recent study, different degrees of vasculopathy were reported in six hand allotransplantations with loss of one of them [18]. That study claimed that large and small arteries are also affected in the early posttransplant period and that vessels could be the first unrecognized target of rejection in hand as in heart transplantation. All hand-graft tissues were investigated in one of our previous studies [19] searching possible signs of chronic rejection, especially vessels with Duplex ultrasounds, magnetic resonance imaging (MRI), and angiography without finding stenosis or significant alteration of the arterial silhouettes, without detecting any signs of chronic rejection in the examined patients.

All these data seem to suggest that composite tissue allografts are relatively resistant to chronic rejection and that the mechanisms involved in CTA might differ from those involved in solid organ transplantation.

51.5 UPPER EXTREMITY ALLOTRANSPLANTATION

After an attempt in Colombia in 1964, the first successfully human hand allotransplantation was performed in Lyon, France, in September 1998 by an international team of surgeons [20], who also performed the first bilateral hand allotransplantation in January 2000 [21]. The results achieved in these first cases showed the feasibility of the surgical technique, the efficacy of the immunosuppressive protocol, the limited adverse effects, and the importance of patient compliance. The first bilateral arm allotransplantation was successfully performed in Munich, Germany, in October 2008. Although hand allotransplantation is still considered experimental, lately it has been considered "ethical" when performed on properly selected patients, particularly on bilateral upper extremity amputees, as reported by the North American surgeons' survey, 2009.

Since then, upper extremity allotransplantation programs have been launched all over the world: 60 patients have been transplanted receiving single (32 patients) or bilateral (28 patients) upper extremity transplantation. A thumb transplantation was performed on one patient. The selection of donor and recipient is very important in hand transplantation as well as a detailed plan of surgical procedure, an efficient and safe immunosuppressive protocol, a rehabilitation program, and a severe monitoring of the recipient in the follow-up.

Donor and recipient selection is one of the crucial points of upper extremity transplantation and the basis for ethical considerations and debates. Although common criteria of selection do not exist and each team created and applied its own inclusion and exclusion criteria, the majority of the Western teams agree that hand transplantation candidates must be carefully selected firstly on the recipient's decision, who has to be able to evaluate the balance between his/her quality of life and all the possible complications correlated to the transplantation. Indeed, the first step of this process is psychological assessment of patient motivation and a prospective compliance with the immunosuppressive treatment and the rehabilitation program. The majority of patients selected for hand transplantation were young and suffered from severe disability owing to the loss of one or both hands. More than half of them used cosmetic or functional prostheses, but they were not satisfied with this alternative.

The surgery is based on bone consolidation, vessel anastomosis, nerve, tendon, and muscle sutures. The repair sequence of the different tissues varied considerably. However, bone fixation, arterial and venous anastomosis with limb reperfusion were performed in the majority of cases. Median and ulnar nerves were always repaired, while the radial nerve was reconstructed in some limbs. Tendon repair was achieved by suturing individual tendons, or in groups or by using a mixed individual/group technique. Despite the fact that the time between amputation and transplantation varied from 2 months to 30 years, it was possible to repair the major peripheral nerves and vessels in all cases, ensuring graft survival and nerve regeneration.

Then the patients must follow a rehabilitation program, which includes physiotherapy, electrostimulation, and occupational therapy. Physiotherapy usually starts as soon as swelling subsides, and it is performed at least once a day for 6–18 months, in some cases even after the patient returns to work. Physiotherapy includes a standard rehabilitation program for flexor and extensor tendons, sensory reeducation, and cortical reintegration. Occupational therapy focuses on sensory, visual, and motor stimulation of the grafted hand. Currently, there are no standardized protocols for hand-grafted patients, and the majority of teams apply the same protocols used after replantation procedures.

Hand amputation may lead to a modification in motor and sensitive cortical representation with activation of lateral motor cortex sites; functional MRI (fMRI) evidenced that following the graft, hand representation shifts from lateral to medial region in the motor cortex and reoccupies the normal hand region [22].

Functional recovery is the goal of hand allotransplantation and it is based on nerve regeneration and cortical reorganization. The evaluation of functional recovery remains one of the crucial points in hand transplantation because each team performs its own tests and scores, which are generally used for replantation procedures or to assess disabilities due to single or multiple disorders of the upper limb. On the contrary, in hand transplantation, teams have to evaluate both cosmetic and functional results as well as take into account "what really happened to the patient" following hand transplantation. This includes assessing his/her psychological outcome, social behavior, work status, subjective satisfaction, body image, and well-being [23].

Skin color and texture, hair, and nail growth were normal in the hand-grafted patients with viable hands (Figures 51.3 and 51.4). All of them achieved a certain degree of sensorimotor recovery [3,4]. Protective sensation recovery (i.e., the ability to detect pain, thermal stimuli, and gross tactile sensation) occurred in all grafted hands. Nerve regeneration allowed a certain degree of discriminative sensation, although this was not to the same degree at all parts of the graft.

Motor recovery began with extrinsic muscle function, allowing all patients to perform grasp and pinch activities. Function of intrinsic muscles was observed only at a later stage, starting between 9 and 12 months posttransplantation in the majority of patients in hand transplantation. Activation of intrinsic muscles was confirmed by electromyographic studies in several hands. Extrinsic and intrinsic muscle recovery enabled patients to perform most of the daily activities, including eating, driving, grasping objects, riding a bicycle or a motorbike, shaving, using the telephone, and writing.

At 2 years, the majority of hand-grafted patients have returned to work and improved manual skills allowed them not only to resume their previous jobs but also, in some cases, to find more suitable employment.

In conclusion, although the significant difference in the appearance of sensitivity and motion recovery related to the amputation level, the sensorimotor recovery enabled all patients to perform most daily activities allowing them to live a normal social life.

Despite the different immunosuppressive regimens, most transplanted hand-grafted patients (about 85%) experienced at least one episode of AR [3,4]. It is important to note that the diagnosis of AR in hand transplantation is easier than in solid organ transplantation as it is based on macroscopic observation of the skin and its biopsy and this feature could explain the high rate of AR episodes reported in this field of transplantation.

As reported by the IRHCTT [3,4] in most cases of hand transplantation, treatment of the first rejection episode included high-dose i.v. steroids followed by an increase in oral steroid dosage. In cases where no steroids were administered intravenously, oral steroid treatment was increased. Treatment of a second, third, or fourth rejection episode varied considerably from application of topical drugs only (steroid and/or tacrolimus creams) to i.v. steroids, to the use of antithymocyte globulins or Alemtuzumab or

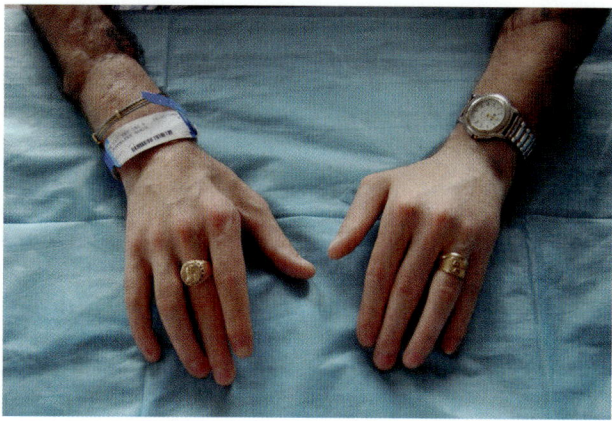

FIGURE 51.3 Macroscopic aspect of the first bilateral hand transplantation.

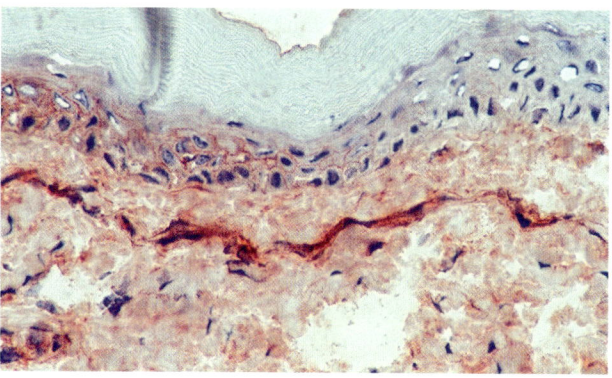

FIGURE 51.4 Histological aspect of normal skin in a hand allograft at the junction between recipient (left) and donor (right), highlighted by the donor's specific HLA-A23 antibody.

Basiliximab. It is interesting to note that the majority of AR episodes were easily reversed by a steroid treatment thanks perhaps to an early diagnosis.

Only one patient who underwent simultaneous face and bilateral hand allotransplantations died 45 days after transplantation. In Western countries five recipients and in China seven recipients lost their grafted hands. The principal cause of graft loss was non-compliance to the immunosuppressive regimen for different reasons. Only one patient has lost the grafted hand for acute ischemia due to vasculopathy [18].

51.6 FACE TRANSPLANTATION

The first partial face transplant (Figure 51.5) was performed by Amiens and Lyon teams in November 2005 [24] and since then 20 partial or total face allotransplantations have been performed all over the world.

At present patients with severe facial disfigurement not amenable to reconstruction are likely to benefit from face transplantation, which provides opportunities for functional improvement and better aesthetics currently impossible with other surgical techniques. The final appearance is likely to be a composite of donor and recipient faces, neither identical to original face nor to that of donor.

We can summarize the goals of this transplantation as physical, psychological, and social. The physical change is the easiest outcome to achieve while psychologically the situation is more complex as the recipient has to support the distress of the disfiguration before the transplantation, and then the new body image and the perception of how others perceive him/her. The capacity of graft integration into body image also conditions social reintegration (interaction with family, return to work, meeting new people). In conclusion, the final aim of face transplantation is to restore function and appearance to enable the individual to resume his/her normal life.

The inclusion criteria have to ensure that the recipient always understands the risk to benefit ratio. The recipient has to understand the limitations of the procedure, the source of the face graft (cadaveric donor), the compliance to the lifelong immunosuppressive regimen and to the rehabilitation program in the postoperative period, the possible side-effects, and AR episodes.

Surgical procedure is very complex and long (the entire procedures ranged from 12 to 24 h) as different anatomical structures and aesthetic units can be transplanted, such as scalp, forehead, eyelids, cheeks, chin, nose, lips, ear, neck, lacrimal glands, salivary glands, and tongue. In the majority of cases, vessels and nerves were firstly anastomosed, followed by bone fixation when it was necessary; muscles and all soft tissues were sutured in layers and finally skin suture was performed.

Until now, as reported by the IRHCTT at the XXIV International Congress of the Transplantation Society (Berlin, July 15 – 19, 2012), the immunosuppressive regimen in face transplantation has been based on antithymocyte globulins or humanized IL-2, tacrolimus, MMF, and corticosteroids as induction therapy and tacrolimus, MMF, and prednisone as maintenance therapy. In addition, in three cases prior to surgery, bone marrow was collected from the donor's iliac crests and then infused into the recipient on days 4 and 11 after face allotransplantation. During the follow-up, there was withdrawal of steroids in four cases and withdrawal of MMF in three cases; a switch from tacrolimus to sirolimus in two cases.

In face allotransplantation as well as in hand allotransplantation, there was a high incidence of AR episodes. In this type of CTA [4,25], there was an involvement of skin and oral mucosa in the rejection process and the episodes, at first suspected by visual inspection, were then confirmed by histological evaluation of skin and mucosa biopsies. Banff CTA score was always used to grade the severity of the rejection. All the AR episodes were successfully treated at least with i.v. bolus of steroid. For the first time in CTA, EPC was performed to avoid AR episodes in some cases of face transplantation [16,24].

The final goal of face allotransplantation is improving the patient's quality of life, which is based on functional and aesthetic aspect recovery. The face allografted patients were satisfied and the majority of them "very satisfied" with the new face, which was considered "new but own face," improving the confidence in their personal appearance. The functional recovery was based on recovery of discriminative sensibility, which was shown in 90% of recipients, and muscular tone with consequent motion recovery. Muscular tonus was normal in 75% of the grafted muscles, a hypertonus was shown in the other 20% and more rarely a certain degree of hypotonus. Motion recovery was impressive, although limited, particularly at level of orbicularis, myrtiformis, and buccinator muscles. One year after face allotransplantation, with a different degree of difficulty, the patients were able to perform the majority of daily activities, such as opening and closing eyelids, eating, drinking, swallowing, chewing, speaking, smiling, and blowing.

So far two face-grafted patients have died (with consequent graft loss). No other graft losses occurred.

51.7 LARYNX AND TRACHEAL TRANSPLANTATIONS

There is no good surgical, medical, or prosthetic solution to the problems faced by those with a larynx whose function is irreversibly damaged by trauma or tumor. Although in the last 10 years laryngeal transplantation has become a reality, it is still considered an experimental procedure and full-scale clinical trials have not started yet.

The first true successful laryngeal transplantation was performed in 1998 in Cleveland, OH [26,27], followed by

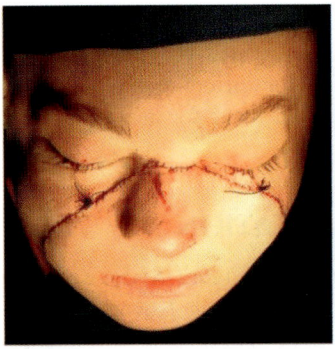
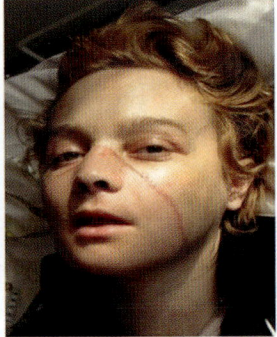

11/27/2005   Day 15

FIGURE 51.5 Macroscopic aspect of the first face transplantation after surgery and 15 days later.

another transplantation in 2011 in the University of California, Davis Medical Centre, which was performed on a woman who had already undergone pancreas kidney transplantation; in addition several cases of larynx and tracheal transplantations [28] have been performed in Colombia.

In larynx allotransplantation as well as for all the other CTAs, the purpose of the transplantation is to improve the quality of life of the patient. Patients have undergone a laryngectomy or with severe laryngeal damage or disease show loss of upper airway sphincter, loss of nasal airway, loss of phonation which has not been sufficiently replaced by devices yet, presence of a tracheostomy with poor healing, cosmetics, and recurrent infections. All these conditions affect the patient's quality of life.

The indications for laryngeal transplantation are severe traumatic or stenotic injuries causing a loss of laryngeal function or patients, who as a result of a large benign or low-grade malignant tumor, have undergone treatment by way of a total laryngectomy. At present it is not possible to propose laryngeal transplantation to the biggest pool of patients with locally advanced laryngeal cancer as an immunosuppression regimen has to be introduced.

Donor criteria included [28]: age of donor (18–50), gender and ABO blood type matching, similar weight between donor and recipient (± 20%), no smoking habit, not suffered intubation in excess of 48 h, and no visible structural damage evidenced by indirect fiber laryngoscopy. The insertion of catheters in jugular and subclavian veins has to be avoided. Donors' family consent has always been required.

Harvesting procedure usually includes a laryngo–pharyngeal complex constituted of six rings of trachea, thyroid, and parathyroids. Vascular anastomoses were performed between donor and recipient superior thyroid artery and between internal jugular vein and common facial vein. Both superior laryngeal nerves were located and sutured, while only the right recurrent laryngeal nerve was located and anastomosed.

The main barrier to successful laryngeal transplantation is restoration of normal laryngeal speech and breathing functions. The larynx has a very complex motor function: this complexity is reflected by the high density of motor fibers in the recurrent nerve as compared to the small size of the muscles it innervates. The superior laryngeal nerve supplies the cricothyroid muscle which assures sensation of the superior part preventing aspiration of saliva and food and restores patient's ability to shorten and lengthen vocal cords, with the resultant ability to vary the pitch of his/her voice; the recurrent laryngeal nerve provides most of the motor drive to the larynx and its restoration assures its function. Unfortunately, it also represents the principal barrier to laryngeal

transplantation as bilateral paralysis follows transplantation with impossibility to obtain volitional movement of vocal folds. On the basis of the experience of Cleveland's patient, although phonation was within the normal range at 36 months posttransplantation, it was not possible to close the tracheostomy as it lacked a significant volitional abduction of cords.

Immunosuppressive regimen was reported only by Cleveland team and it was based prior to surgery on cyclosporine A, azathioprine, steroids while in the postoperative period on muromonab-CD3, cyclosporine, MMF, and steroids. Eight years since transplantation, the immunosuppressive regimen includes steroids, MMF, and tacrolimus.

Sign of larynx rejection is acute laryngeal edema with decrease in quality of voice. Endoscopy and serial biopsies of graft tracheal mucosa show any signs of rejection. It seems possible to evidence laryngeal rejection monitoring the function of transplanted parathyroids.

After more than 10 years, the patient from Cleveland reports an improved quality of life being able to smell, to taste, to communicate through a "normal" voice, and to work.

Columbia team reported 90% of graft survival at 2 years.

In 2008, five patients received the first vascularized tracheal allotransplantations [29] in order to restore a permanently damaged airway (long stenosis and low-grade chondrosarcoma). The cartilaginous trachea was wrapped with host connective tissue (forearm) and vascularization based on the radial artery and veins; recipient mucosa was grafted into the transplant. Triple immunosuppression was stopped after a short period of time and the allotransplants were ready for the transplantation after a few months. At present the major obstacles are to assure a submucosal vascular supply and to determine if immunosuppression can be effectively stopped without loss of airway lumen.

51.8 ABDOMINAL WALL

Abdominal wall transplantation is a type of CTA that has been utilized to reconstitute the abdominal domain of adult and pediatric small bowel and multivisceral organ recipients.

Small intestine transplantation has become the treatment of choice for patients with chronic intestinal failure, but the application of this procedure is limited by the relatively high rate of complications that can occur. One of these is the difficulty or the impossibility of closing the abdominal wall. This problem is due to the prolonged absence of the midgut (prior to bowel resection surgery) with loss of volume of the abdominal domain or loss of surface coverage of the abdominal wall from repeated laparoscopies, stomas, fistulas, desmoids tumors, and so on. Moreover at the end of the intestinal transplantation, there is often organ edema, which prevents primary abdominal closure. Failure to close the abdomen leaves the recipient with an open wound or requires several surgical procedures. Both conditions can determine infections, fistulas, organ injury, and consequent increase in patient morbidity and mortality.

Abdominal wall transplantation differs from the other CTAs because it is performed as a secondary procedure for coverage purposes after a life-saving transplantation, such as intestinal transplant. The primary importance is to preserve intestinal or multivisceral graft.

Several cases of abdominal wall transplantations have been performed in the United States and Italy [30–32] after small bowel and multivisceral transplants. The abdominal walls were retrieved from the same multiorgan cadaveric donor, except for some cases when the abdominal wall was transplanted some days later. In these cases, it was retrieved from a separate donor.

The musculocutaneous flap consisting of the median oval cutaneous pan, the bilateral rectus abdominis muscles, a part of oblique muscles, the deep muscular sheet with the parietal peritoneum is harvested, flushed with University Wisconsin solution, and then stored in ice. The flap can be vascularized using a microsurgical technique between donor and recipient inferior epigastric vessels or an end-to-side anastomosis between donor and recipient iliac vessels. The fascia is sutured to the edges of the abdominal wall defect and the skin is closed in continuity with the recipient skin. No reinnervation has been performed with consequent hypotrophy of the muscular layer in the follow-up but without significant hernia formation.

More recently, the transplantation of nonvascularized rectus sheath fascia has been described with its preparation into the recipient on the back-table. In this series, the authors reported a 44% rate of abdominal infection [33].

In one case of liver and double kidney pediatric transplantation, the posterior rectus sheath fascia in continuity with the liver and falciform ligament was successfully used for fascia closure during the initial transplantation procedure [34].

In all adult grafted patients, the immunosuppression consisted of alemtuzumab induction and steroid-free tacrolimus-based maintenance therapy, while in pediatric transplantation the regimen included steroid, tacrolimus, and daclizumab.

Clinical inspection and biopsies were used to monitor rejection episodes of abdominal wall. Rejection episodes were successfully treated with steroid boluses. It is interesting to note that the intestine and the skin of the abdominal wall graft can reject independently from each other.

51.9 LOWER EXTREMITY TRANSPLANTATION

The first successful lower extremity transplantation was performed in 2006 in Toronto, Canada, when a functioning limb for one ischiopagus twin with a lethal cardiac anomaly was transplanted to the other [35]. The results are encouraging although the transplanted extremity is 6.5 cm shorter compared to the other one. The passive range of motion is almost normal. The muscular tone is normal, good power with hip flexion and knee extension and flexion while there is a reduction in the active dorsi- and plantar flexion of the foot. There is also an incomplete sensitivity recovery but the patient can perform the majority of physical activities with a specific lower extremity functional scale of 86.25%.

A bilateral lower extremity transplantation was performed in one adult recipient in Valencia in the summer of 2011 with encouraging results (data reported by Pedro Cavadas at the XXIV International Congress of the Transplantation Society, Berlin, July 2012), while a quadruple limb transplantation failed in February 2012 in Turkey.

51.10 PENIS TRANSPLANTATION

The first successful penis transplantation was performed in September 2005 in Guangzhou, China [36]. The recipient was a 44-year-old man who sustained the loss of most of his penis (1 cm long stump) in an accident. Urination became difficult and sexual intercourse impossible. His quality of life was severally affected.

The donor was a brain-dead 22-year-old man. Authorization of the donor's family was obtained.

Transplantation included anastomosis of urethra, corpus spongiosum and corpus cavernosum, and sutures of deep dorsal vein, dorsal artery, dorsal nerve, and superficial dorsal vein.

The immunosuppressive regimen included Zenopax, as induction treatment, steroid, cyclosporine A, and MMF as maintenance therapy.

Although successful, the patient and his wife suffered a psychological trauma as a result of the procedure and the surgery was reversed 15 days later. Histological examination of the amputated penis showed no sign of rejection.

51.11 UTERUS TRANSPLANTATION

The interest in uterine transplantation has been increasing since the mid-twentieth century as being the only possibility to overcome infertility problems linked to uterine absence or uncorrectable anomalies. New reproductive procedures are not useful in these situations and the only chance for women affected by congenital or acquired uterine disorders is a gestational surrogacy. It consists of using gametes of a genetic couple to produce embryos that are then transferred to the womb of a woman who agrees to act as a host for the pregnancy.

The first partially successful human uterine transplantation [37] was performed on April 6, 2000, in Saudi Arabia on a 26-year-old female who had lost her uterus 6 years earlier due to postpartum hemorrhage. The donor, a 46-year-old patient with multiloculated ovarian cysts, underwent a hysterectomy modified to preserve tissue and vascular integrity. The donor's uterus was harvested and flushed with cold modified Eurocollins solution.

The donor's uterus was connected in the orthotopic position to the recipient's vaginal vault and additional fixation was achieved by shortening the uterosacral ligament. The uterine arteries and veins were extended using reversed segments of the great saphenous vein, then connected to the external iliac arteries and veins, respectively.

Immunosuppression was maintained by oral cyclosporine A, azathioprine, and prednisolone, after preoperative administration of steroid and cyclosporine A.

An episode of AR was treated and controlled on the ninth day with antithymocytic globulin. The transplanted uterus responded well to combined estrogen and progesterone therapy, with endometrial proliferation up to 18 mm. The patient had two episodes of withdrawal bleeding upon cessation of the hormonal therapy.

Unfortunately, she developed acute vascular thrombosis 99 days after transplantation and hysterectomy was necessary. Macro- and microscopic histopathological examinations revealed acute thrombosis in the vessels of the uterine body, with resulting infarction. Both fallopian tubes remained viable, however, with no evidence of rejection. The acute vascular occlusion appeared to be caused by inadequate uterine structure support, which led to probable tension, torsion, or kinking of the connected vascular uterine grafts.

Another case of uterus transplantation (mother-to-daughter) was performed in Goteborg, Sweden, in September 2012.

51.12 TONGUE TRANSPLANTATION

The first tongue transplantation was performed in Austria in July 2003 [38,39]. The recipient was a 42-year-old man with tongue cancer. The surgical procedure started with the recipient total glossectomy followed by transplantation of the harvested tongue, including anastomosis of arteries and veins as well as suture of nerves and muscles. Although the patient was discharged from the hospital 1 month later with a tracheostomy for airway support and a gastrostomy for nutrition, after a few months he

presented some useful sensations, enabling him to swallow saliva and some fluids. The immunosuppressive regimen was efficient and 8 months after transplantation the recipient did not show any sign of rejection.

At 12 months after transplantation, a recurrence of the cancer occurred in the neck and 1 month later the patient died.

51.13 CONCLUSIONS

Despite initial skepticism and debate, CTA is a clinical reality. However, the requirement of long-term immunosuppression, the lack of clinical trials and consequently of clear indications and functional outcomes constitute the limits of this new field of transplantation which should be overcome in the near future.

REFERENCES

[1] Murray JE. Organ transplantation (skin, kidney, heart) and plastic surgeon. Plast Reconstr Surg 1971;47:425—31.

[2] Lee WP, Yaremchuk MJ, Pan YC, Randolph MA, Tan CM, Weiland AJ. Relative antigenicity of components of vascularized limb allograft. Plast Reconstr Surg 1991;87:401—11.

[3] Petruzzo P, Lanzetta M, Dubernard JM, Landin L, Cavadas P, Margreiter R, et al. The international registry on hand and composite tissue transplantation. Transplantation 2010;90:1590—4.

[4] Petruzzo P, Dubernard JM. The international registry on hand and composite tissue transplantation. ClinTranspl 2011;247—53 [Chapter 22].

[5] Gold BG, Yew JY, Zeleny-Pooley M. The immunosuppressant FK506 increases GAP-43 mRNA levels in axotomized sensory neurons. Neurosci Lett 1998;241:25—8.

[6] Morelon E, Kanitakis J, Petruzzo P. Immunological issues in clinical composite tissue allotransplantation: where do we stand today? Transplantation 2012;93:855—9.

[7] Brandacher G, Lee WA, Schneeberger S. Minimizing immunosuppression in hand transplantation. Expert Rev Clin Immunol 2012;8:673—84.

[8] Kanitakis J, Jullien D, Petruzzo P, Hakim N, Claudy A, Revillard JP, et al. Clinicopathologic features of graft rejection of the first human hand allograft. Transplantation 2003;76:688—93.

[9] Wood K, Sachs D. Chimerism and transplantation tolerance: cause and effect. Immunol Today 1996;17:584—7.

[10] Elwood ET, Larsen CP, Maurer DH, Routenberg KL, Neylan JF, Whelchel JD, et al. Microchimerism and rejection in clinical transplantation. Lancet 1997;34:1358—60.

[11] Siemionow M, Izycki D, Ozer K, Ozmen S, Klimczak A. Role of thymus in operational tolerance induction in limb allograft transplant model. Transplantation 2006;81:1568—76.

[12] Ramsamooj R, Llull R, Black KS, Hewitt CW. Composite tissue allografts in rats: IV. Graft-versus-host disease in recipients of vascularized bone marrow transplants. Plast Reconstr Surg 1999;104:1365—71.

[13] Landin L, Cavadas P, Nthumba P, Ibanez J, Vera-Sempere F. Preliminary results of bilateral arm transplantation. Transplantation 2009;88:749—51.

[14] Schneeberger S, Kreczy A, Brandacher G, Steurer W, Margreiter R. Steroid and ATG-resistant rejection after double forearm transplantation responds to Campath 1-H. Am J Transplant 2004;4:1372–4.

[15] Cendales LC, Kanitakis J, Schneeberger S, Burns C, Ruiz P, Landin L, et al. The Banff 2007 working classification of skin-containing composite tissue allograft pathology. Am J Transplant 2008;8:1396–400.

[16] Hivelin M, Siemionow M, Grimbert P, Lantieri L. Extracorporeal photopheresis: from solid organ to face transplantation. Transpl Immunol 2009;21:117–28.

[17] Unadkat JV, Schneeberger S, Horibe EH, Goldbach C, Solari MG, Washington KM, et al. Composite tissue vasculopathy and degeneration following multiple episodes of acute rejection in reconstructive transplantation. Am J Transplant 2010;10:251–61.

[18] Kaufman CL, Ouseph R, Blair B, Kutz JE, Tsai TM, Scheker LR, et al. Graft vasculopathy in clinical hand transplantation. Am J Transplant 2012;12:1004–16.

[19] Petruzzo P, Kanitakis J, Badet L, Pialat JB, Boutroy S, Charpulat R, et al. Long-term follow-up in composite tissue allotransplantation: in-depth study of five (hand and face) recipients. Am J Transplant 2011;11:808–16.

[20] Dubernard JM, Owen ER, Herzberg G, Lanzetta M, Martin X, Kapila H, et al. Human hand allograft: report on first 6 months. Lancet 1999;353:1315–20.

[21] Dubernard JM, Petruzzo P, Lanzetta MM, Parmentier H, Martin X, Dawahra M, et al. Functional results of the first human double-hand transplantation. Ann Surg 2003;238:128–36.

[22] Giraux P, Sirigu A, Schneider F, Dubernard JM. Cortical reorganization in motor cortex after graft of both hands. Nature Neurosci 2001;4:1.

[23] Lanzetta M, Petruzzo P. A comprehensive functional score system in hand transplantation. In: Lanzetta M, Dubernard JM, editors. Hand Transplantation. Italy: Springer-Verlag; 2007. p. 355–62.

[24] Dubernard JM, Lengele B, Morelon E, Testelin S, Badet L, Moure C, et al. Outcomes 18 months after the first human partial face transplantation. N Engl J Med 2007;357:2451–60.

[25] Siemionow M, Ozturk C. Face transplantation: outcomes, concerns, controversies, and future directions. J Craniofac Surg 2012;23:254–9.

[26] Strome M, Stein J, Esclamado R, Hicks D, Lorenz RR, Braun W, et al. Laryngeal transplantation and 40-month follow-up. N Engl J Med 2001;344:1676–9.

[27] Birchall MA, Lorenz RR, Berke GS, Genden EM, Haughey BH, Sieminow M, et al. Laryngeal transplantation in 2005: a review. Am J Transplant 2006;6:20–6.

[28] Duque E, Duque J, Nieves M, Mejía G, López B, Tintinago L. Larynx and trachea donors. Transplant Proc 2007;39:2076–8.

[29] Delaere PR, Vranckx JJ, Meulemans J, Vander Poorten V, Segers K, Van Raemdonck D, et al. Learning curve in tracheal allotransplantation. Am J Transplant 2012;12:2538–45.

[30] Levi DM, Tzakis AG, Kato T, Madariaga J, Mittal K, Nery J, et al. Transplantation of the abdominal wall. Lancet 2003;36:2173–6.

[31] Selvaggi G, Levi DM, Cipriani R, Sgarzani R, Pinna AD, Tzakis AG. Abdominal wall transplantation: surgical and immunologic aspects. Transplant Proc 2009;41:521–2.

[32] Cipriani R, Contedini F, Santoli M, Gelati C, Sgarzani R, Cucchetti A, et al. Abdominal transplantation with microsurgical technique. Am J Transplant 2007;7:1304–7.

[33] Gondolesi G, Selvaggi G, Tzakis A, Rodríguez-Laiz G, González-Campaña A, Fauda M, et al. Use of the abdominal rectus fascia as a nonvascularized allograft for abdominal wall closure after liver, intestinal, and multivisceral transplantation. Transplantation 2009;87:1884–8.

[34] Agarwal S, Dorafshar AH, Harland RC, Millis JM, Gottlieb LJ. Liver and vascularized posterior rectus sheath fascia composite tissue allotransplantation. Am J Transplant 2010;10:2712–6.

[35] Fattah A, Cypel T, Donner EJ, Wang F, Alman BA, Zuker RM. The first successful lower extremity transplantation: 6-year follow-up and implications for cortical plasticity. Am J Transplant 2011;11:2762–7.

[36] Hu W, Lu J, Zhang L, Wu W, Nie H, Zhu Y, et al. A preliminary report of penile transplantation: part 1. Eur Urol 2006;50:851–3.

[37] Fageeh W, Raffa H, Jabbad H, Marzouki A. Transplantation of the human uterus. Int J Gynecol Obstet 2002;76:245–51.

[38] Birchall M. Tongue transplantation. Lancet 2004;363:1663.

[39] Kermer C, Watzinger F, Oeckher M. Tongue transplantation: 10-month follow-up. Transplantation 2008;85:654–5.

Chapter 52

Bioengineering of the Lower Urinary Tract

Majid Mirzazadeh and Anthony Atala

Wake Forest School of Medicine, Winston-Salem, NC

52.1 BLADDER REGENERATION

52.1.1 Introduction

The properly functioning urinary bladder is composed of a compliant muscular wall and highly specialized urothelium. When operating effectively, it provides a low-pressure, high-capacity reservoir while simultaneously protecting the upper urinary tract from pressure, reflux of urine, and infection. Specialized characteristics of the bladder's smooth muscle fibers, nerves, and blood vessels permit repeated coordinated bladder contractions without compromising storage capability or upper tract protection. When this organ becomes damaged, any attempt at replacing it, in whole or part, must take into consideration all these properties, including reproducing or recreating them. The bladder can lose the ability to store and empty effectively as a result of numerous conditions, such as cancer, trauma, infection, inflammation, or iatrogenic injury. Additionally, congenital or neurologic conditions, such as spina bifida or spinal cord injury, can also cause progressive loss of bladder function resulting in debilitating urinary incontinence or renal impairment. When conservative measures fail to protect the upper tracts or lead to worsening quality of life, surgical reconstruction of the bladder is usually considered. While using native urologic tissue could have the best outcome, whenever there is a lack of native urologic tissue, reconstruction with native nonurologic tissues (skin, gastrointestinal segments, or mucosa from multiple body sites),

homologous tissues (cadaver fascia), heterologous tissues (bovine collagen), or artificial materials (silicone, polyurethane, Teflon) may be considered. None of them has optimum results in practice. The present gold standard is augmentation cystoplasty. It can be performed with the use of the small bowel, large bowel, or less often, stomach. Auto augmentation, described as a partial removal of the detrusor while sparing the urothelium, has also been advocated as a solution in subsets of patients. However, the potential complications from the use of a portion of the gastrointestinal tract to perform augmentation cystoplasty are well characterized. They include urinary tract infections, intestinal obstruction, mucus production, electrolyte abnormalities, perforation, change in fecal transit time, and carcinogenesis. All these issues beg for an alternate solution. Tissue engineering was proposed as an alternative for generating bladder tissue for reconstruction in the early 1990s, and tremendous progress has been made in this field since that time. Tissue engineering, as defined by Langer and Vacanti [1] in 1993, is "an interdisciplinary field that applies the principles of engineering and life sciences toward the development of biologic substitutes that restore, maintain or improve tissue function or a whole organ".

The goal of tissue engineering is to develop biologic substitutes that can restore and maintain normal function. Tissue engineering may involve matrices alone, wherein the body's natural ability to regenerate is used to orient or direct new tissue growth, or it may use matrices with cells. When cells are used for tissue engineering, donor tissue (heterologous, allogenic, or autologous) is dissociated into individual cells, which are implanted directly into the host or expanded in culture, attached to a support matrix, and reimplanted after expansion. Ideally, this approach allows lost tissue function to be restored or replaced completely with limited complications [2–7].

52.2 EMERGING RULES FOR INDUCING ORGAN REGENERATION

In the field of regenerative medicine no widely accepted paradigm is currently available that can guide formulation of new theories on the mechanism of regeneration in adults and open new directions for improved regeneration outcomes. The four rules have emerged from multiyear quantitative studies with skin and peripheral nerve regeneration using scaffold libraries based on a simple, well-defined collagen scaffold. These largely quantitative rules distinguish sharply between spontaneously regenerative and nonregenerative tissues, select the two reactants that are required for regeneration, recognize the essential modification of the wound healing process that must be realized prior to regeneration, and identify three structural features of scaffolds that are required for regenerative activity.

The four rules:

1. In an injured organ epithelial tissues and the basement membrane regenerate spontaneously while the stroma does not.
2. The required reactants for inducing regeneration are an appropriate scaffold and, optionally, epithelial cells.
3. Scaffolds are regeneratively active if they inhibit contraction and scar formation.
4. Structural features in scaffolds that are required for regenerative activity: pore structure, degradation rate, and surface chemistry [8].

52.2.1 Scaffolds and the Role of Biomaterials

A critical aspect of regenerating a large portion of bladder tissue is the bioscaffold selection. The scaffold must possess certain mechanical and physical properties to allow the organ to develop into a desired shape and maintain it. Tissue-engineered scaffolds should (i) facilitate the localization and delivery of tissue-specific cells to precise sites in the body, (ii) maintain a three-dimensional architecture that permits the formation of new tissues, and (iii) guide the development of new tissues with appropriate function. It has been demonstrated that tissue morphogenesis is heavily influenced by the interactions between cells and the extracellular matrix (ECM) during normal tissue development [9].

Currently, two categories of bladder bioscaffold are used in bladder tissue engineering: natural and synthetic matrices. Natural matrices were among the first to be used and they included skin, preserved bladder, omentum, peritoneum, lyophilized dura mater, chemically treated amniotic membrane, and calcium-treated pericardium. Graft rejection, urinary tract infection, and calculus formation are among the problems encountered with the use of natural matrices.

Decellularized small intestine submucosa has been extensively studied as a scaffold for bladder augmentation. The submucosa is harvested from the jejunum of pigs. However, the process of decellularizing the matrices affects their strength and quality and their ability to provide a reliable scaffold for regeneration and remodeling. A study [10] showed that large bladder augmentation with cell-seeded or unseeded small-intestinal submucosa (SIS) matrix did not improve bladder regeneration. A recent study focused on applying plastic compression to achieve higher concentrations of collagen, showing the scaffolds to have an increase in the desirable mechanical properties at the neo tissue level [11]. Furthermore, natural matrices are still liable to maintain a certain amount of cellularity, thus providing immunogenic components to the recipient. To overcome some of these challenges encountered in the use of natural matrices, synthetic polymers, such as polyglycolic acid (PGA) and polylactic coglycolic acid are the biomaterials that have been developed and used. Biomaterials in genitourinary tissue engineering function as an artificial ECM and elicit biologic and mechanical functions of native ECM found in body tissues. Biomaterials facilitate the localization and delivery of cells and/or bioactive factors (such as cell adhesion peptides and growth factors) to desired sites in the body; define a three-dimensional space for the formation of new tissues with appropriate structure; and guide the development of new tissues with appropriate function [12]. While direct injection of cell suspensions without biomaterial matrices has been used [13], it is difficult to control the localization of transplanted cells.

The ideal biomaterial should be biocompatible, promote cellular interaction and tissue development, and possess proper mechanical and physical properties. As mentioned, while naturally derived materials and acellular tissue matrices have the potential advantage of biologic recognition, synthetic polymers can be produced reproducibly on a large scale with controlled properties of strength, degradation rate, and microstructure.

52.2.2 Vascularization

A restriction of tissue engineering is that cells cannot be implanted in volumes exceeding $3 \, mL^3$ because of the limitations of nutrition and gas exchange [14]. To achieve the goals of engineering large complex tissues, and possibly internal organs, vascularization of the regenerating cells is essential. Three approaches have been used for vascularization of bioengineered tissue:

1. Incorporation of angiogenic factors in the bioengineered tissue
2. Seeding ECM with other cell types in the bioengineered tissue
3. Prevascularization of the matrix prior to cell seeding.

Angiogenic growth factors may be incorporated into the bioengineered tissue prior to implantation, in order to attract host capillaries and to enhance neovascularization of the implanted tissue. Many obstacles must be overcome before large entire tissue-engineered solid organs are produced. Recent developments in angiogenesis research may provide important knowledge and essential materials to accomplish this goal.

52.2.3 Sources of Cells

Harvesting cells from the organ of interest and culturing them has become an important tool in tissue engineering. Autologous bladder cells are harvested from the diseased organ, separated into cell lines of interest, and expanded *in vitro* to sufficient quantities. They can then be seeded onto matrices to create tissue for implantation back into the host. Previous work has demonstrated that in certain populations, such as in patients with a neurogenic bladder, cells can be safely and effectively harvested and cultured. However, the search for cells that possess more characteristics of healthy tissue and have more regenerative potential continues. Smooth muscle cells (SMCs) from neuropathic bladders have shown abnormal growth, less contractile ability, and inferior adherence compared with normal controls, making them less than ideal cells for culture in this particular population [15]. Other sources of cells are also important, because often, the diseased bladder does not contain enough normal cells to begin expansion. The patient's own cells may also not be desirable for creation of new bladder tissue, as is the case in patients with previously demonstrated bladder malignancy. Multiple possibilities for sources of cells have been investigated, including stem cells and differentiated cells from organs other than the bladder; however, to date, autologous bladder cells remain the gold standard for culture and seeding.

Stem cells are a potential source of tissue regeneration, and are of different types: totipotent stem cells that are formed when an egg and sperm merge can differentiate into embryonic and extraembryonic cell types and can construct a complete, viable organism. Stem cells that can differentiate into cells from all three germ layers are descendants of totipotent cells and are called pluripotent,

whereas the ones that are more restricted are multipotent stem cells. Pluripotent stem cells are of great interest in tissue engineering because of their unlimited self-renewal potential and their ability to differentiate into any cell type of endodermal, mesodermal, or ectodermal lineage [16]. Human embryonic stem cells are surrounded by ethical and moral concerns because of the need for destruction of human embryos to develop these cell lines. Therefore, search for alternative sources of stem cells continues. An alternate source of stem cells is the amniotic fluid and the placenta. Multiple partially differentiated cell types derived from the fetus have been identified in the amniotic fluid and are referred to as amniotic fluid and placental stem cells [17]. Indeed, these amniotic stem cells have been used in some animal trials with promising results.

It has been shown that cell therapy with human amniotic fluid stem cells and bone marrow-derived mesenchymal stem cells temporarily ameliorated bladder dysfunction in a Parkinson disease model. In contrast to integration, cells may act on the injured environment via cell signaling [18]. These cells lack the teratoma-forming characteristics of embryonic stem cells, although maintaining the ability to differentiate into cells of all three embryonic germ layers [17].

Urine has also been investigated as a potential source of cells and would be an attractive option, because it would provide cells for tissue regeneration in a noninvasive manner. These urine progenitor cells comprise only 0.2% of urine cells, but these cells that are found in sparse numbers may hold great promise for future tissue regeneration. The use of these cells also avoids ethical issues, because they are autologous somatic cells; however, their self-renewal potential has not yet been clearly defined [19]. Adult stem cells are another potential source of SMCs for bladder tissue engineering. They are particularly attractive, because they avoid the ethical problems associated with embryonic stem cells. However, they are found only in sparse numbers in the host, do not expand well in cultures, and have a more restricted differentiation potential. Adult stem cells have been isolated from bone marrow, skeletal muscle, adipose tissue, lung, testis, umbilical cord, and placenta. In a recent study, isolation and characterization of SMCs from porcine adipose and peripheral blood that are phenotypically and functionally indistinguishable from bladder-derived SMCs was described. The ability to create urologic structures *de novo* from scaffolds seeded by autologous adipose- or peripheral blood-derived SMCs may greatly facilitate the translation of urologic tissue engineering technologies into clinical practice, in the future.

Also, in another recent study, Bone marrow-derived cells implanted into radiation-injured urinary bladders could reconstruct functional bladder tissues in rats [20].

Therefore, the implantation of bone marrow-derived cells is a potentially useful treatment for radiotherapy-induced urinary dysfunctions [21]. Because the bone marrow is not a target for bladder cancer metastases, these cells are also safe for use in this patient population [22]. These cells could be especially helpful in engineering bladders for patients with bladder cancer who need bladder augmentation.

A new and exciting advancement in stem cell research involves genetic reprogramming. This technique involves the dedifferentiation of adult somatic cells to produce patient-specific pluripotent stem cells or induced pluripotent state cells. This technique was first described in 2006; however, cells were shown to be incompletely reprogrammed in that pilot study [23]. Subsequent studies were performed using human rather than mouse cells and revealed complete reprogramming of cells [24,25].

52.2.4 Clinical Application

As mentioned, because of problems with using GI segments in the bladder, investigators have attempted alternative reconstructive procedures for bladder replacement or repair such as the use of tissue expansion, matrices for tissue regeneration, and tissue engineering with cell transplantation.

52.2.4.1 Tissue Expansion

A system of progressive dilation for ureters and bladders has been proposed as a method of bladder augmentation but has not yet been attempted clinically [26]. Another method to progressively expand native bladder tissue has also been used for augmenting bladder volumes in animals. Within 30 days after progressive dilation, neo reservoir volume was expanded at least 10-fold [26].

52.2.4.2 Matrices

Nonseeded allogeneic acellular bladder matrices have served as scaffolds for the ingrowth of host bladder wall components. The matrices are prepared by mechanically and chemically removing all cellular components from bladder tissue [27,28]. The matrices serve as vehicles for partial bladder regeneration, and relevant antigenicity is not evident. For example, SIS (a biodegradable, acellular, xenogeneic collagen-based tissue-matrix graft) was first used in the early 1980s as an acellular matrix for tissue replacement in the vascular field. It has been shown to promote regeneration of a variety of host tissues, including blood vessels and ligaments [29]. Animal studies have shown that the nonseeded SIS matrix used for bladder augmentation can regenerate *in vivo*[30].

In multiple studies using various materials as nonseeded grafts for cystoplasty, the urothelial layer regenerated normally, but the muscle layer, although present, was not fully developed [27,28]. Often the grafts contracted to 60−70% of their original sizes [31], with little increase in bladder capacity or compliance [32]. Moreover, in a recent rat model study using bone marrow mesenchymal stem cells (BMSCs) seeded onto amniotic membranes fixed to Tachosil sponges as grafts for urinary bladder muscle layer augmentation, morphological and urodynamic evaluation of urinary bladder wall regeneration showed that muscle regeneration guarantee contraction but not proper function [33].

Studies involving acellular matrices that may provide the necessary environment to promote cell migration, growth, and differentiation are being conducted. With continued research, these matrices may have a clinical role in bladder replacement in the future.

52.2.4.3 Cell Transplantation

Cell-seeded allogeneic acellular bladder matrices have been used for bladder augmentation in dogs. Trigone-sparing cystectomy was performed in dogs randomly assigned to one of three groups. One group underwent closure of the trigone without a reconstructive procedure; another underwent reconstruction with a nonseeded bladder-shaped biodegradable scaffold; and the last underwent reconstruction using a bladder-shaped biodegradable scaffold that delivered seeded autologous urothelial cells and SMCs [6].

The cystectomy-only and nonseeded controls maintained average bladder capacities of 22% and 46% of preoperative values, respectively, compared with 95% in the cell-seeded tissue-engineered bladder replacements (Figure 52.1). The subtotal cystectomy reservoirs that were not reconstructed and the polymer-only reconstructed bladders showed a marked decrease in bladder compliance (10% and 42% total compliance). The compliance of the cell-seeded tissue-engineered bladders showed almost no difference from preoperative values that were measured when the native bladder was present (106%). Histologically, the nonseeded scaffold bladders presented a pattern of normal urothelial cells with a thickened fibrotic submucosa and a thin layer of muscle fibers. The retrieved tissue-engineered bladders showed a normal cellular organization, consisting of a trilayer of urothelium, submucosa, and muscle [6]. Preliminary clinical trials for the application of this technology have been performed and are under evaluation.

Although SMCs are typically used as a cell source for the reconstruction of hollow organs by conventional tissue engineering techniques, a recent study done by Huber and Badylak [34], detecting phenotypic changes in SMCs

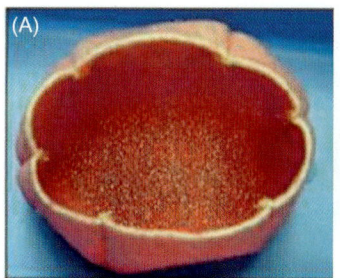 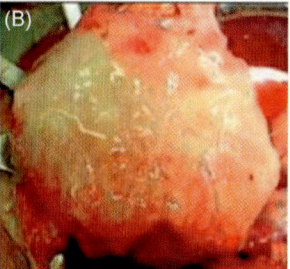 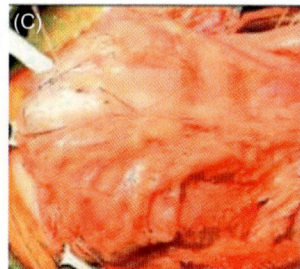

FIGURE 52.1 Construction of engineered bladder. (A) Scaffold seeded with cells and (B) engineered bladder anastamosed to native bladder with running 4−0 polyglycolic sutures. (C) Implant covered with fibrin glue and omentum.

which occur as a result of conventional cell isolation and expansion techniques, brought up questions about the necessity for a tissue-specific cell source for regenerative medicine applications.

The most complete clinical trial for tissue engineering in urology to date was published in 2006 [35]. It involved the engineering of human bladder tissue for young patients with end-stage bladder disease by isolating autologous bladder urothelial and muscle cells, expanding the cells, and attaching them to matrices. Seven patients with myelomeningocele who ranged in age from 4 to 19 years participated in this trial. They all had high-pressure, poorly compliant bladders, and were identified as candidates for augmentation cystoplasty. They each underwent a bladder biopsy for harvesting of urothelial and SMCs, which were then expanded and seeded on a biodegradable, bladder-shaped scaffold. Based on initial preclinical studies, a collaged matrix derived from decellularized bladder submucosa was used for seeding the cells of the first three patients without an omental wrap of the implanted tissue. Further animal studies demonstrated that wrapping the implanted tissue in omentum improved vascularization and those collagen-PGA composite matrices performed better in the long term. After realizing that omental wrapping was beneficial to the vascularization of the tissue *in vivo*, the protocol was revised, and one patient underwent implantation with a cell-seeded collagen matrix with an omental wrap. Protocol was revised once more, and the remaining three patients underwent implantation with cell-seeded composite collagen-PGA bladders wrapped in omentum. Approximately 7−8 weeks before planned implantation, the patients also underwent cystoscopic evaluation and a bladder biopsy through a suprapubic incision; 1−2 cm^2 of bladder tissue was harvested. Muscle cells were supplemented with 10% fetal bovine serum, whereas urothelial cells were grown with keratinocyte growth medium. The scaffolds used ranged in size from 70 to 150 cm^2 and were shaped like a bladder with polyglycolic sutures.

The exterior of the scaffolds was seeded with SMCs. The inside of the scaffolds was coated with urothelial cells 48 h later. The constructs were implanted in the patients 7 weeks after biopsy and after the patients had been on 7 days of oral antibiotics. The engineered bladder was sutured to native bladder tissue with 4-0 polyglycolic suture in a running fashion. One of the four patients who had collagen scaffolds implanted received an omental wrap of their construct, whereas all three patients receiving the composite scaffolds had their implants wrapped in omentum. In fact, the patients who had the collagen composite constructs implanted with an omental wrap had lower mean bladder leak point pressures, greater capacity, greater compliance, and longer dry periods. The bladder biopsies revealed adequate structural architecture and phenotype. On cystoscopic evaluation, the margins between composite matrix-based engineered segments and the native bladders were grossly indistinguishable (Figure 52.2) [35]. The use of composite scaffold made of PGA proved to support cell growth and survival better than collagen scaffolds and to be optimal for the engineering of bladder tissue in this trial. Also, omental wrapping proved to be beneficial. This clinical trial is overall the most complete clinical trial for tissue engineering to date and demonstrates the potential applications of tissue engineering to urology. This clinical experience is promising, since it showed that engineered tissues can be safely and effectively implanted in patients; however, further studies must be conducted if engineered bladders are to become the first-line option for bladder reconstruction.

In summary, with the great progress and success achieved in bladder regeneration, it should be considered as a field with great potential. On the other hand, there are following challenges that need to be considered.

The two areas that have been thoroughly studied in the tissue engineering field are biomaterial science and cell biology. Although we may be closer to identifying a suitable biomaterial, we have not yet taken into account the many factors that may predispose biomaterials to fail as a bladder substitute. The most common reason for failure is porosity, which is thought to be a valuable indicator for adequate cellularization. On the other hand, when we aim to replace an impermeable organ such as the bladder, permeability/porosity may hinder cellularization, with urine halting repair and causing fibrosis. Moreover, the

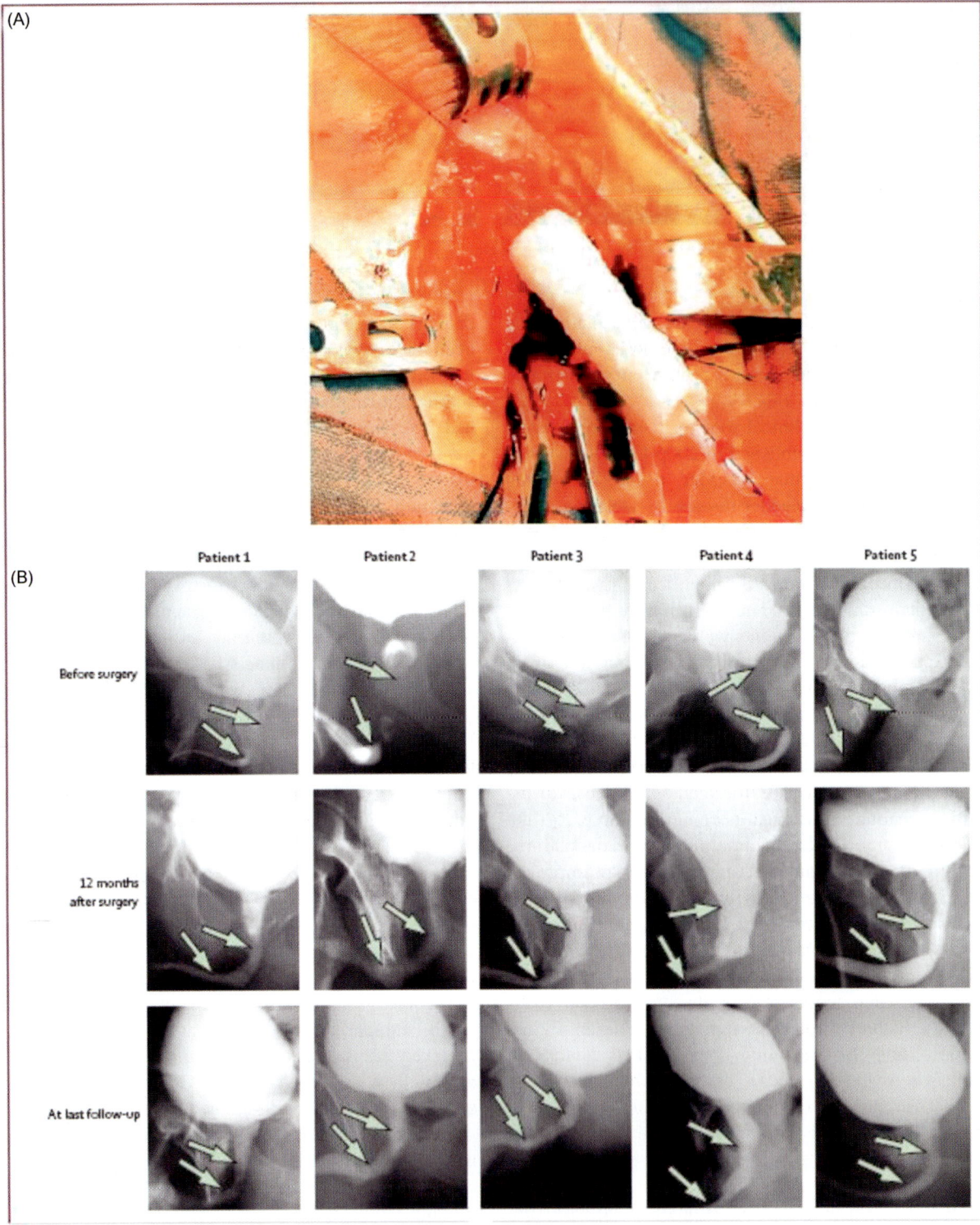

FIGURE 52.2   Neourethra implantation and clinical outcome.   (A) A cell-seeded graft sutured to the normal urethral margins from the first patient. (B) Voiding cystourethrograms of all five patients before surgery (arrows show the abnormal margins), 12 months after surgery (arrows show margins of tissue-engineered urethras), and at last follow-up (arrows show margins of tissue-engineered urethras). *Figures from Ref. [36].*

mechanical properties of the biomaterials may have been well investigated prior to implantation, but the minor and minute changes in the architecture of those biomaterials upon implantation will have major impact on their mechanical properties, thus limiting their use as a bladder-replacement technology. Finally, although biomaterials are noncellular and hence nonimmunogenic, recent literature suggests that allogenic natural scaffolds and possibly collagen-based synthetic scaffolds may induce an immune response that is innate in nature. This immune response may cause fibrosis and induce histological changes that are detrimental.

In terms of cell biology, with or without scaffolds, a plethora of literature is available on the adhesion properties, proliferation and differentiation characteristics of urinary bladder cells on different natural and synthetic scaffolds. On the other hand, it is well known that once cells are removed from an organ and are grown *in vitro*, they dedifferentiate and require complex environmental cues and factors to get them to differentiate into relevant cells types, whether it be urothelial cells as an impermeable layer or SMCs as a major component for the unique mechanical properties of the bladder. The importance of this differentiation process is emphasized by the fact that bladders that are not adequately cycled during fetal life tend to end up being abnormal when the children are born. Cell-seeded scaffolds need specific fine tuning not only with growth factors or cytokines, but also mechanical stimulation, which is a key factor to induce cells to differentiate normally. Bioreactors have been thoroughly investigated in more dynamic organs, such as cardiovascular tissues, and it will be very important to further investigate the role of mechanical stimulation and the impact of stretch on urinary bladder cells grown on suitable scaffolds and their ability to naturally differentiate. In addition, we need to be cognizant of the fact that patients who need bladder replacement have diseased bladders and their bladder cells may be abnormal, hence we need a suitable source of cells that may be easily harvested and differentiated to become normal urinary bladder cells. Cells that harbor stemness potential, such as embryonic stem cells or BMSCs, may be the ultimate cell source. These cells will need to be primed and driven to regenerate the different bladder cell constituents (smooth muscle, urothelium, vessels, and nerves). Finally, we need to provide the regenerated bladder with the nerves and blood supply that will allow it to function as a compliant contractile organ, rather than only a space-occupying cavity that passively fills.

Future work must be orchestrated between the cellular biologist, biomaterials scientist and developmental biologist to successfully tissue engineer urinary bladder *in vitro*. This construct must withstand implantation with minimal or no immune response, undergo remodeling, and continue to be as compliant and mechanically resilient as the normal bladder for a long period of time [37].

52.3 URETERAL REGENERATION

Ureteral nonseeded matrices have been used as a scaffold for the ingrowths of ureteral tissue in rats. On implantation, the acellular matrices promoted the regeneration of the ureteral wall components [38]. In a more recent study, nonseeded ureteral collagen acellular matrices were tabularized, but attempts to use them to replace 3-cm segments of canine ureters were unsuccessful [39].

Cell-seeded biodegradable polymer scaffolds have been used with more success to reconstruct ureteral tissues. In one study, urothelial and SMCs isolated from bladders and expanded *in vitro* were seeded onto PGA scaffolds with tubular configurations and implanted subcutaneously into athymic mice. After implantation, the urothelial cells proliferated to form a multilayered luminal lining of tubular structures, while the SMCs organized into multilayered structures surrounding the urothelial cells. Abundant angiogenesis was evident. Polymer scaffold degradation resulted in the eventual formation of natural urothelial tissues. This approach has also been used to replace ureters in dogs [40].

52.4 URETHRAL REGENERATION

Various strategies have been proposed to regenerate urethral tissue. Woven meshes of PGA, without cells, have been used to reconstruct urethras in dogs [41]. PGA has been used as a cell transplantation vehicle to engineer tubular urothelium *in vivo*. SIS without cells was used as an onlay patch graft for urethroplasty in rabbits [42], and a homologous free graft of acellular urethral matrix was also used in a rabbit model [43].

Bladder-derived acellular collagen matrix has proven to be a suitable graft for repairing urethral defects in rabbits. The created neourethras demonstrated a normal urothelial luminal lining and organized muscle bundles [44]. Results were confirmed clinically in a series of patients with a history of failed hypospadias reconstruction whose urethral defects were repaired with human bladder acellular collagen matrices (Figure 52.1) [45]. An advantage of this material over nongenital tissue grafts for urethroplasty is that it is "off the shelf," eliminating the need for additional surgical procedures for graft harvesting and decreasing operative time and potential morbidity from the harvest procedure.

The above techniques, using nonseeded acellular matrices, were successfully applied experimentally and clinically for onlay urethral repairs. However, when tubularized repairs were attempted experimentally, adequate urethral tissue regeneration was not achieved, and complications, such as graft contracture and stricture formation ensued [46].

Most recently, we were able to show that synthetic biomaterials can also be used in urethral reconstruction when they are tabularized and seeded with autologous cells [47]. This group used polygycolic acid: poly lactide-co-glycolic acid scaffolds seeded with autologous cells derived from bladder biopsies taken from each patient. The seeded scaffolds were then used to repair urethral defects in five boys. Upon follow-up, it was found that most of the boys had excellent urinary flow rates postoperatively, and voiding cystourethrograms indicated that these patients maintained wide urethral calibers. Urethral biopsies revealed that the grafts had developed a normal appearing architecture consisting of urothelial and muscular tissue.

52.5 CONCLUSION

Tissue engineering efforts are currently being undertaken for every type of tissue and organ in the urinary system. Most of the effort expended to engineer genitourinary tissues has occurred within the last decade. Tissue engineering techniques require a cell culture facility designed for human application. Personnel who have mastered the techniques of cell harvest, culture, and expansion as well as polymer design are essential for the successful application of this technology. Before these engineering techniques can be applied to humans, further studies need to be performed in many of the tissues described. Recent progress suggests that engineered urologic tissues and cell therapy may have clinical applicability.

REFERENCES

[1] Langer R, Vacanti JP. Tissue engineering. Science 1993;260:920–6.

[2] Atala A, Cima LG, Kim W, Paige KT, Vacanti JP, Retik AB, et al. Injectable alginate seeded with chondrocytes as a potential treatment for vesicoureteral reflux. J Urol 1993;150:745–7.

[3] Atala A. Future perspectives in reconstructive surgery using tissue engineering. Urol Clin North Am 1999;26:157–65.

[4] Amiel GE, Atala A. Current and future modalities for functional renal replacement. Urol Clin North Am 1999;26:235–46.

[5] Kershen RT, Atala A. New advances in injectable therapies for the treatment of incontinence and vesicoureteral reflux. Urol Clin North Am 1999;26:81–94.

[6] Oberpenning F, Meng J, Yoo JJ, Atala A. De novo reconstitution of a functional mammalian urinary bladder by tissue engineering. Nat Biotechnol 1999;17:149–55.

[7] Park HJ, Yoo JJ, Kershen RT, Moreland R, Atala A. Reconstitution of human corporal smooth muscle and endothelial cells in vivo. J Urol 1999;162:1106–9.

[8] Yannas IV. Emerging rules for inducing organ regeneration. Biomaterials 2013;34:321–30.

[9] Lee SJ, Atala A. Scaffold technologies for controlling cell behavior in tissue engineering. Biomed Mater 2013;8:010201.

[10] Zhang Y, Frimberger D, Cheng EY, Lin HK, Kropp BP. Challenges in a larger bladder replacement with cell-seeded and unseeded small intestinal submucosa grafts in a subtotal cystectomy model. BJU Int 2006;98:1100–5.

[11] Engelhardt EM, Stegberg E, Brown RA, Hubbell JA, Wurm FM, Adam M, et al. Compressed collagen gel: a novel scaffold for human bladder cells. J Tissue Eng Regen Med 2010;4:123–30.

[12] Kim BS, Mooney DJ. Engineering smooth muscle tissue with a predefined structure. J Biomed Mater Res 1998;41:322–32.

[13] Brittberg M, Lindahl A, Nilsson A, Ohlsson C, Isaksson O, Peterson L. Treatment of deep cartilage defects in the knee with autologous chondrocyte transplantation. N Engl J Med 1994;331:889–95.

[14] Folkman J, Hochberg M. Self-regulation of growth in three dimensions. J Exp Med 1973;138:745–53.

[15] Lin HK, Cowan R, Moore P, Zhang Y, Yang Q, Peterson Jr. JA, et al. Characterization of neuropathic bladder smooth muscle cells in culture. J Urol 2004;171:1348–52.

[16] Yu RN, Estrada CR. Stem cells: a review and implications for urology. Urology 2010;75:664–70.

[17] De Coppi P, Bartsch Jr. G, Siddiqui MM, Xu T, Santos CC, Perin L, et al. Isolation of amniotic stem cell lines with potential for therapy. Nat Biotechnol 2007;25:100–6.

[18] Soler R, Fullhase C, Hanson A, Campeau L, Santos C, Andersson KE. Stem cell therapy ameliorates bladder dysfunction in an animal model of Parkinson disease. J Urol 2012;187:1491–7.

[19] Zhang Y, McNeill E, Tian H, Soker S, Andersson KE, Yoo JJ, et al. Urine derived cells are a potential source for urological tissue reconstruction. J Urol 2008;180:2226–33.

[20] Basu J, Jayo MJ, Ilagan RM, Guthrie KI, Sangha N, Genheimer CW, et al. Regeneration of native-like neo-urinary tissue from nonbladder cell sources. Tissue Eng Part A 2012;18:1025–34.

[21] Imamura T, Ishizuka O, Lei Z, Hida S, Sudha GS, Kato H, et al. Bone marrow-derived cells implanted into radiation-injured urinary bladders reconstruct functional bladder tissues in rats. Tissue Eng Part A 2012;18:1698–709.

[22] Tian H, Bharadwaj S, Liu Y, Ma H, Ma PX, Atala A, et al. Myogenic differentiation of human bone marrow mesenchymal stem cells on a 3D nano fibrous scaffold for bladder tissue engineering. Biomaterials 2010;31:870–7.

[23] Takahashi K, Tanabe K, Ohnuki M, Narita M, Ichisaka T, Tomoda K, et al. Induction of pluripotent stem cells from adult human fibroblasts by defined factors. Cell 2007;131:861–72.

[24] Yu J, Vodyanik MA, Smuga-Otto K, Antosiewicz-Bourget J, Frane JL, Tian S, et al. Induced pluripotent stem cell lines derived from human somatic cells. Science 2007;318:1917–20.

[25] Takahashi K, Yamanaka S. Induction of pluripotent stem cells from mouse embryonic and adult fibroblast cultures by defined factors. Cell 2006;126:663–76.

[26] Satar N, Yoo JJ, Atala A. Progressive dilation for bladder tissue expansion. J Urol 1999;162:829–31.

[27] Yoo JJ, Meng J, Oberpenning F, Atala A. Bladder augmentation using allogenic bladder submucosa seeded with cells. Urology 1998;51:221–5.

[28] Probst M, Dahiya R, Carrier S, Tanagho EA. Reproduction of functional smooth muscle tissue and partial bladder replacement. Br J Urol 1997;79:505–15.

[29] Badylak SF, Lantz GC, Coffey A, Geddes LA. Small intestinal submucosa as a large diameter vascular graft in the dog. J Surg Res 1989;47:74—80.

[30] Kropp BP, Cheng EY, Lin HK, Zhang Y. Reliable and reproducible bladder regeneration using unseeded distal small intestinal submucosa. J Urol 2004;172:1710—3.

[31] Portis AJ, Elbahnasy AM, Shalhav AL, Brewer A, Humphrey P, McDougall EM, et al. Laparoscopic augmentation cystoplasty with different biodegradable grafts in an animal model. J Urol 2000;164:1405—11.

[32] Landman J, Olweny E, Sundaram CP, Andreoni C, Collyer WC, Rehman J, et al. Laparoscopic mid sagittal hemicystectomy and bladder reconstruction with small intestinal submucosa and reimplantation of ureter into small intestinal submucosa: 1-year followup. J Urol 2004;171:2450—5.

[33] Adamowicz J, Juszczak K, Bajek A, Tworkiewicz J, Nowacki M, Marszalek A, et al. Morphological and urodynamic evaluation of urinary bladder wall regeneration: muscles guarantee contraction but not proper function—a rat model research study. Transplant Proc 2012;44:1429—34.

[34] Huber A, Badylak SF. Phenotypic changes in cultured smooth muscle cells: limitation or opportunity for tissue engineering of hollow organs? J Tissue Eng Regen Med 2012;6:505—11.

[35] Atala A, Bauer SB, Soker S, Yoo JJ, Retik AB. Tissue-engineered autologous bladders for patients needing cystoplasty. Lancet 2006;367:1241—6.

[36] Atala A, Bauer SB, Yoo JJ, Retik AB. Tissue-engineering autologous bladder for patients needing cystoplasty. Lancet 2006;367:1241—6.

[37] Farhat WA. Bladder regeneration: great potential but challenges remain. Regen Med 2011;6:537—8.

[38] Dahms SE, Piechota HJ, Nunes L, Dahiya R, Lue TF, Tanagho EA. Free ureteral replacement in rats: regeneration of ureteral wall components in the acellular matrix graft. Urology 1997;50:818—25.

[39] Osman Y, Shokeir A, Gabr M, El-Tabey N, Mohsen T, El-Baz M. Canine ureteral replacement with long acellular matrix tube: is it clinically applicable? J Urol 2004;172:1151—4.

[40] Atala A, Vacanti JP, Peters CA, Mandell J, Retik AB, Freeman MR. Formation of urothelial structures *in vivo* from dissociated cells attached to biodegradable polymer scaffolds *in vitro*. J Urol 1992;148:658—62.

[41] Olsen L, Bowald S, Busch C, Carlsten J, Eriksson I. Urethral reconstruction with a new synthetic absorbable device. An experimental study. Scand J Urol Nephrol 1992;26:323—6.

[42] Kropp BP, Ludlow JK, Spicer D, Rippy MK, Badylak SF, Adams MC, et al. Rabbit urethral regeneration using small intestinal submucosa onlay grafts. Urology 1998;52:138—42.

[43] Sievert KD, Bakircioglu ME, Nunes L, Tu R, Dahiya R, Tanagho EA. Homologous acellular matrix graft for urethral reconstruction in the rabbit: histological and functional evaluation. J Urol 2000;163:1958—65.

[44] Chen F, Yoo JJ, Atala A. Acellular collagen matrix as a possible "off the shelf" biomaterial for urethral repair. Urology 1999;54:407—10.

[45] Atala A, Guzman L, Retik AB. A novel inert collagen matrix for hypospadias repair. J Urol 1999;162:1148—51.

[46] le Roux PJ. Endoscopic urethroplasty with unseeded small intestinal submucosa collagen matrix grafts: a pilot study. J Urol 2005;173:140—3.

[47] Raya-Rivera A, Esquiliano DR, Yoo JJ, Lopez-Bayghen E, Soker S, Atala A. Tissue-engineered autologous urethras for patients who need reconstruction: an observational study. Lancet 2011;377 (9772):1175—82.

Chapter 53

Upper Airways Regeneration and Bioengineering

Silvia Baiguera[a] and Paolo Macchiarini[a,b]

[a]BIOAIRLab, European Center for Thoracic Research (CERT), Florence, Italy, [b]Advanced Center for Translational Regenerative Medicine (ACTREM), Karolinska Institutet, Stockholm, Sweden

Chapter Outline

53.1 THE UPPER RESPIRATORY TRACT

The upper respiratory tract or upper airway primarily refers to the parts of the respiratory system lying outside of the thorax and it is composed of the nose, pharynx, larynx, and trachea (mainly the extrathoracic or cervical part). It is not just a simple conduit for air, food, and drink, since it warms, humidifies, and filters air passing into the lungs, playing a critical role in respiration. Moreover, it subserves a multiplicity of functions, such as phonation, olfaction, air conditioning, digestion, preservation of airway patency, and protection of the airways. As a consequence, the upper respiratory tract has a complex arrangement of muscle, soft tissue, cartilage, and bone in its walls. Among the different organs composing the upper airway, the larynx, or voice box, and the trachea play vital roles in the respiratory tract. The *larynx* is located immediately below the pharynx and extends vertically to the fourth, fifth, and sixth cervical vertebrae. It has a specialized constrictor–dilator mechanism in the airway. Its "oldest" function is to prevent food, liquid, and other foreign material from entering the lower airway. With evolution, the vocal cords developed to produce the sound used in voice production. Larynx primary function is, then, to orchestrate swallowing, breathing, coughing, and the voice. It is composed of three single cartilages (epiglottis, thyroid, and cricoid) and paired cartilages (arytenoid, corniculate, and cuneiform), all connected by ligaments and moved by various muscles (such as posterior cricoarytenoid, lateral cricoarytenoid, and thyroarytenoid). The epiglottis, feather-shaped fibroelastic cartilage, is a flap/lid at the base of the tongue, partly covering the opening of the larynx, and aids in airway protection during deglutition. Thyroid cartilage (the Adam's apple) is the largest one and it is composed of two superior horns that join in the midline and remain open posteriorly, forming an effective protective shield for the opening of the airway and supporting most of the soft tissue folds in the larynx. Cricoid cartilage, the only complete ring in the larynx, serves to support the posterior laryngeal structures. It has two surfaces that articulate with the thyroid cartilage; while by a dense cricotracheal ligament it attaches to the first tracheal ring. Arytenoid cartilages are pyramidal, articulate with the cricoid cartilage and act as the attachment for the vocal cords and the muscles of voice control. Cuneiform (on the top of arytenoid) and corniculate (fibroelastic cartilages, which lie laterally to the arytenoid cartilages) cartilages act as structural support to the ary-epiglottic fold. The vocal cords extend from the arytenoid cartilages posteriorly to the thyroid cartilage anteriorly and are composed of muscle, ligament, and submucosal soft tissue [1]. The larynx is lined with a respiratory epithelium (nonkeratinizing stratified squamous for the supraglottis and true vocal cords; and pseudostratified ciliated for the false vocal cords, ventricle, and subglottis) and a mucous membrane (seromucous glands are present in the lower epiglottis) which contains high levels of immunologically active cells, making the larynx an important immunological organ [2−4].

At its lower end, the larynx joins the trachea. The *trachea*, a cartilaginous/connective airtight tube, extends from the sixth cervical vertebra (cricoids cartilage) to the upper border of the fifth thoracic vertebra, where it divides into the bronchial bifurcation. Its main function is to conduct air from the nose to the lungs and it is characterized by specific structural and mechanical properties: the cartilaginous structure prevents collapse during respiration and provides flexibility to the tracheal pipe, maintaining patency of the lumen; the muscular tissues reduce lumen size during the cough reflex, facilitating airway clearance; while the mucosal membrane allows air conditioning (warming, moistening, and removal of particulate materials) and prevents the epithelium from dehydration by moving air. The trachea consists of 18−24 imperfect (horseshoe-like) rings of hyaline cartilage joined by fibroelastic tissue and closed posteriorly by a membranous structure, consisting of longitudinally oriented smooth muscles, named *pars membranacea*. The luminal part of the trachea is covered with a mucosal membrane, continuous above with that of larynx, that serves to protect against infection and aid in mucous clearance. It consists of areolar and lymphoid tissue and presents a well-marked basement membrane, supporting a stratified columnar and ciliated epithelium, containing basal, ciliated, secretory (goblet, serous, and Clara cells), neuroendocrine and less well-categorized "indeterminate" or "intermediate" cells [5,6]. The presence of multipotent basal stem cells, that play a role in airway epithelium homeostasis and regeneration after injury, has been recently demonstrated [6]. The submucous layer is composed of a loose meshwork of connective tissue, containing large blood vessels, nerves, and mucous glands, the main function of which is the secretion, through microvilli, of mucous that is moved upwards by the ciliated cells, allowing and facilitating airway clearance. Adult stem cells of mesenchymal derivation, which could play a pivotal role in cell repair, regeneration, and functional restoration, have been identified in the tracheal connective tissue, even if more research is necessary to clearly characterize them [7,8].

53.2 AIRWAY PATHOLOGIES: *THE CLINICAL NEED*

Diseases which compromise airways alter breathing, speech, and swallowing and the consequences are detrimental effects on psychology, socialization, and employment. A wide spectrum of pathologies may afflict the larynx and/or the trachea.

Among benign *laryngeal* diseases, laryngomalacia, vocal cord paralysis, stenosis, and laryngeal webs are the most common congenital disorders, while trauma secondary to endotracheal intubation (specially long-term intubation), tracheotomy or laryngotracheal reconstruction, caustic ingestion, local infections, gastroesophageal reflux, or accident are the main causes for acquired laryngotracheal stenosis and/or webs [9−11] (Table 53.1). Laryngeal carcinoma is the second most common malignancy of the head and neck and more than 90% is squamous cell carcinoma type [12]. Glottic cancer is most frequent, tends to be well differentiated, to grow slowly, and to metastasize late, whereas supraglottic cancer, being the lymphatic vascularity denser than in the glottis and subglottis, is characterized by a significantly higher incidence of cervical lymph node metastases. Most of the congenital laryngeal disorders are often managed by observation (in >90% of cases the lesions gradually improves by themselves), while for acquired laryngeal disorders, endoscopic procedures are most often used as treatment approach. However, moderate to severe pathologies generally require an open surgical procedure, such as anterior/posterior cricoid split (in association with stenting, cartilage grafting, or laryngofissure), laryngotracheal augmentation, cricotracheal resection, and laryngectomy. Anterior/posterior cricoids split procedures have considerably facilitated the management of congenital and acquired laryngeal pathologies, and remain the most popular techniques for reconstruction. However, the postoperative period can be the source of significant complications, such as airway obstruction, hematoma formation, pneumothorax, infection, recurrent stenosis, bleeding, infections, and many patients undergo reintubation, postoperative tracheostomy, or multiple surgical procedures. Autologous auricleor thyroid ala cartilage is the most common tissue used to expand stenotic areas, even if results useful only for anterior' grafting [13]. For more wide and complex augmentation, some successful results have been obtained using costal cartilage, however, limited tissue availability, donor site pain, scarring, infection, resorption, necrosis, and potentially pneumothorax are the main complications associated with this surgical strategy [14,15]. In case of severe circumferential stenosis, cricotracheal resection is the only possible surgical treatment. Although being a single-stage procedure with a high rate of success and minimal morbidity, postoperative complications, such as granulation tissue, separation of anastomosis (due to excessive tension), partial or complete restenosis (due to circumferential proliferating scar), prolapse of the arytenoids cartilage, brachiocephalic artery hemorrhage, trachea-esophageal fistula, vocal cord

TABLE 53.1 Main Laryngeal Pathologies

	Pathophysiology	Cause	Incidence	Mortality	Treatment
Congenital					
Laryngomalacia	Floppy cartilage which tends to prolapse over the larynx during inspiration, leading to airway obstruction	Delay in maturation of the supporting laryngeal structures	Frequency is unknown; the diagnosis is often presumed. 60% of all congenital laryngeal disorders	Rarely. Lesion may interfere with normal growth and development. In severe cases, feeding problems	Observation. In severe cases, surgical approach: tracheotomy, epiglottoplasty, supraglottoplasty, laser epiglottopexy
Vocal fold paralysis	Weakness (paralysis) of one (unilateral) or both (bilateral) vocal cords	Idiopathic or secondary to CNM immaturity or CNS lesions	15—20% of all congenital laryngeal disorders	Rarely (in extreme cases, excessive choking can lead to death)	Observation Endotracheal intubation (for bilateral vocal cord paralysis), tracheotomy, arytenoidectomy
Subglottic stenosis	Partial or circumferential narrowing of the airway	Failure of the laryngeal lumen to recanalize during embryogenesis	5% of congenital laryngeal disorders	Rarely 5.9% due to tracheotomy complications	Observation Endotracheal intubation, tracheotomy (for significant cases), laryngotracheoplasty (for severe cases)
Laryngeal webs	Presence of a membrane across the anterior portion of the glottis	Incomplete recanalization of the laryngotracheal tube	1 every 10,000 births 5% of congenital laryngeal anomalies	Rarely	Observation. Endoscopic division, staged dilations, laser management, tracheotomy, laryngotracheal reconstruction
Atresia	Complete absence of the laryngeal lumen	Presumably autosomal dominant inheritance for the partial form; extreme form of stenosis or webs	1:100,000 births. 50 reported cases of larynx atresia in the world literature	Very high Survivors are extremely rare, only when a surgical airway is immediately established	Emergent tracheotomy Use of the esophagus as a tissue replacement with the subsequent requirement for the reconstruction of the feeding tube
Acquired					
Stenosis	Partial or circumferential narrowing of the airway	Trauma secondary to endotracheal intubation, tracheotomy, laryngotracheal reconstruction Caustic ingestion, local infections, gastroesophageal reflux, foreign bodies, external trauma	The incidence of stenosis after intubation is reported to be 1—10%	2 — 5% due to tracheotomy complications	Conservative approach (for mild cases). Surgical intervention: endoscopic dilation, laser resection, stenting, tracheotomy, cricoid split, cartilage grafts, laryngofissure, laryngo reconstruction
Vocal cord paralysis	Weakness (paralysis) of one (unilateral) or both (bilateral) vocal cords	Secondary to: malignant disease (31%); surgical trauma (29%); idiopathic (24%); nonsurgical trauma (7%); inflammatory (4%); neurologic (1%)	0.42%	Rarely (in extreme cases, excessive choking can lead to death)	Unilateral: Speech therapy; intracordal injections; surgical medialization of the paralyzed cord; selective reinnervation; Bilateral: restrict activity; partial vocal cord resection or arytenoidectomy with laser; vocal cord lateralization; tracheotomy

(Continued)

TABLE 53.1 (Continued)

	Pathophysiology	Cause	Incidence	Mortality	Treatment
Laryngeal webs	Presence of a membrane across the anterior portion of the glottis	Secondary to: forceful or prolonged intubation, surgery, severe laryngeal infections, or accident	40–60%	Rarely	Surgery, laryngeal keel
Cancer	Mainly squamous cell carcinoma (>90%) 65% glottis; 35% supraglottic	Smoking, alcohol consumption	5 cases in 10,000 patients (12,500 new cases/year)	30% 5-year survival: 80–95% (for stages I and II); 10–50% (for stages III and IV)	Radiation therapy and/or combined with chemotherapy, laser excision, Surgical approach: partial or total laryngectomy

CNM: central neuromuscular immaturity; CNS: central nervous system.

paralysis, dyspnea, dysphonia, inability to decannulate, and recurrent laryngeal nerve damage, have been reported [16–18]. Concerning neoplastic patients, a small number of them undergo local resection, leaving permanent defects in the vocal cords and hoarseness, while the most advanced cases are treated by total laryngectomy [19]. The laryngectomy often requires tracheal anastomosis to the skin and the complete separation of the respiratory and digestive tracts, and results associated with severe mutilation, infections, pharyngo cutaneous fistula, tracheobronchitis, speech, and swallowing problems. The loss of the larynx leads to significant difficulties with social reintegration, because only palliative measurements are available for vocal rehabilitation and the tracheotomy must be maintained for life. Moreover the use of "organ-sparing" chemoradiation has high morbidity, 5% mortality, and can leave a functionless larynx.

Regarding *tracheal* disorders, most lead to central airway obstruction with subsequent respiratory insufficiency (Table 53.2). Others, such as fistulas between the trachea and innominate artery or esophagus, may also occur and be equally deleterious. Among benign lesions, atresia, congenital or acquired tracheomalacia, and stenosis are the most common anomalies of the trachea. Prolonged intubation, chest trauma, chronic tracheal infections, or inflammatory conditions are the most frequent causes of acquired tracheomalacia and stenosis [20]. Primary tracheal tumors, mainly adenoid cystic carcinoma and squamous cell carcinoma, are the least common neoplastic lesions of the airways [21,22], but their usually insidious onset often leads to a delay in diagnosis, making these potentially treatable lesions difficult to treat and often fatal [23–26]. The gold standard procedure for almost all the tracheal pathologies is tracheal resection followed by end-to-end anastomosis reconstruction, which yields a successful outcome (over 70%) [9,27]. However,

anastomotic complications (granulations, stenosis, separation, and poor wound healing), even if uncommon, lead to severe morbidity and mortality [28,29]. Moreover, extensive lesions (longer than half the trachea in adults or one-third in children) cannot be managed using this conventional procedure, the resection would not leave enough length of native airway for primary reconstruction and excessive anastomotic tension could cause severe and fatal postoperative complications [30], and palliative treatments (such as stenting, T tubes, radiation or a combination of these therapies) are the only solutions [28]. Different approaches have been proposed so far, ranging from autologous/allogenic tissue flaps and patches to synthetic solid or porous stents and prostheses, but only limited success has been achieved due to the lack of adequate vascularization and respiratory epithelium along the lumen, ischemia, and immune rejection that lead to inflammation, anastomotic dehiscence, stenosis, necrosis, and erosion of main vessels [28,31,32]. Moreover, as a result of the trachea not being located in a mesenchymal environment, but in direct contact with the breathing air, infection and contamination are more likely to occur. Malignant tracheal tumours can be completely resected in <60%, with a consequent low 5-year survival rate [33] while inoperable lesions are treated with palliation (such as stents or neoadjuvant radiotherapy) with a reported 5-year survival rate of about 5% [23–26].

53.3 AIRWAY RECONSTRUCTION: *A CHALLENGING SURGICAL ISSUE*

For more than 50 years, several approaches have been pursued to reconstruct the airways in cases when conventional surgical approaches were unsuitable. The resulting clinical outcome was inconsistent, incomplete, and controversial.

TABLE 53.2 Main Tracheal Pathologies

Tracheal Pathologies		Pathophysiology	Cause	Incidence	Mortality	Treatment
Benign						
Atresia		Partial or complete absence of the trachea	Congenital malformation; in 90% association with congenital anomalies	Uncommon (<100 have been reported)	Usually lethal	Immediate airway stabilization Reconstructive measures
Tracheomalacia	Congenital	Structural abnormality of the tracheal cartilage	Developmental defect. Can be associated with congenital anomalies (tracheoesophageal fistula or esophageal atresia)	1:2100 newborns	Extremely rare. Can cause cyanosis due to episodic severe airway obstruction	Allowing time to pass noninvasive ventilation, pharmacological treatment Surgical therapy: stenting, endotracheal, or tracheostomy tube Pharmacological and/or conservative treatments
	Acquired		Secondary to tracheal injury: intubation, endobronchial tube, chest trauma, chronic tracheal infections, inflammatory conditions	No definitive rates are available		Noninvasive, positive-pressure ventilation Surgical therapy: stenting, tracheostomy, aortopexy
Stenosis	Congenital	Fibrotic airway narrowing	Development defect: *pars membranacea* is deficient and the wall consists of complete or almost complete cartilaginous rings	<1:200,000 people	28% after surgical intervention: 53% in patients with intracardiac anomalies 73% in patients younger than 1 month	Tracheal dilation using rigid bronchoscope Laser surgery Tracheal resection and primary anastomosis (stenosis length <50% in adults and 30% in children)
	Acquired		Caused by different pathological processes: postintubation injury, trauma, or tracheostomy	10–22% (1–2% of the patients are symptomatic or have severe stenosis) Severe stenosis: 4.9 cases/million/year	2–5% after surgical intervention	
Tracheoesophageal fistula	Congenital	Communication between the trachea and esophagus	Developmental anomaly: 17–70% have associated anomalies (Down syndrome, duodenal atresia, cardiovascular defects)	1:2000/4000 live births	Usually fatal: in case of tracheal atresia 5–20% for infants with comorbidities 0% for healthy infants undergoing surgical repair	Surgical treatment (usually using tracheal prosthesis)

(Continued)

TABLE 53.2 (Continued)

Tracheal Pathologies	Pathophysiology	Cause	Incidence	Mortality	Treatment
	Acquired non malignant	Secondary to infection, ruptured diverticula, and trauma.	0.5% of patients undergoing tracheostomy	10.3%	For critically ill patients: conservative treatment Surgical repair
	Acquired malignant	Secondary to malignant disease	4.5% for primary esophageal tumor	33−61%	Palliative (esophageal exclusion, bypass, resection, endoprosthesis) or supportive (nasogastric drainage, tracheostomy, gastrostomy) treatment
			0.3% for primary lung tumor		Radiation therapy
Malign					
Tracheal tumor	Neoplastic lesions		0.1 person/ 100.00 (89−90% are malignant)	Median survival: 6 months (squamous cell carcinoma: median survival 44 months, 5-year survival 34% adenoid cystic: median survival 115 months, 5-year survival 78%; carcinoid: 5-year survival 95%)	Laser resection, stenting as palliative Radiochemo therapy Surgery therapy: resection

Muscle flaps, autologous cartilage, tracheal grafts, or different homografts have been employed for *laryngeal reconstruction* with clinically reliable outcomes. In 1965, partial laryngeal resection, was performed in 20 patients to remove advanced laryngeal cancer, by spreading external laryngeal muscles between the cricoids cartilage and the remaining half of the thyroid. The procedure allowed preservation of laryngeal speech at follow-up, even if in varying degree and with the presence of tracheotomy in some patients [34]. The ipsilateral thyrohyoid muscle has been used in 31 patients to reconstruct the glottis with some positive results (gratifying voice and glottis competence), even if reconstructed larynx shrinkage and nonmobile pseudocord were observed at follow-up [35]. The use of digastric tendon graft improved the functional outcome of patients treated by hemilaryngectomy [36]. Tracheal segments have also been used to reconstruct extended hemilaryngectomy defects after the removal of laryngeal cancers [37−40]. The initial "two-stage resection" strategy (neck dissection and

wrapping of the trachea with vascularized fascia performed at the first surgery, and resection of the primary tumor and tracheal autotransplantation at the second surgery), used in 38 patients with predictable, good functional results but with the major disadvantage of low reliability from an oncological viewpoint [37,38], has been improved by performing closure of the laryngeal defect simultaneously with tracheal revascularization [39,40]. This "one-stage" tracheal autotransplantation strategy provided a high reconstructive, functional, and oncological reliability but resulted, however, in a quite complicated procedure which required a later time definitive reconstructive procedure, and can be applied only in case of some type of laryngeal carcinomas. Laryngeal reconstruction with cryopreserved tracheal, thyroid and cricoids cartilaginous or aortic homografts have also been proposed as a reliable and versatile reconstructive option for performing conservation laryngeal surgery that allow for airway, swallowing, and voice preservation [41−43], however, the lack of a truly laryngeal architecture did not permit a real functional restoration or improved quality of life. Factors that contribute to unsatisfactory surgical outcomes include difficulty reconstructing (i) delicate tissue and structures (which as a whole are dynamic and influenced by movements necessary for swallowing and phonation) and (ii) the native contour of the laryngeal surface. To this end, the first attempt at replacing the larynx with a partial laryngeal transplantation was performed in 1969 in a patient with cancer, though it is associated with recurrence and poor recovery [44]. To date, 14 laryngeal transplantations [12,45] and 1 laryngeal and laryngotracheal transplantation [46] have been performed with clinical successful results. However, even if a near-total functional recovery of speech, swallowing, taste, and smell was obtained, the main critical issue was the necessity of a lifelong heavy immunosuppressive therapy, preventing the use of this strategy in a tumor context [47,48].

The various techniques adopted so far to identify the ideal *tracheal* graft resulted unsuccessfully, mainly due to the lack of adequate vascularization and respiratory epithelium along the lumen [28,32]. Prosthesis (solid or porous), allo (heterotopic or orthotopic, viable or nonviable), or auto (mainly vascular) grafts have been used for tracheal replacement, and in some cases clinically applied [28,32]. However, these approaches usually failed due to prosthesis migration, dislodgement, inflammation, rejection (viable allografts), insufficient epithelization with subsequent bacterial colonization, granulation, scar formation, stenosis, fistula development, liquefaction, necrosis, and erosion of main vessels [49]. Moreover, most of the grafts were tracheal patches rather than real tracheal replacements (the preservation of the recipients' posterior tracheal membranous wall was indeed necessary) and, in cases of pediatric patients, no graft grow with the recipient has been reported [50–56].

Successful tracheal allotransplantation has been recently performed in five patients (four affected by long-segment stenosis and one by low-grade chondrosarcoma) with successful results [57,58]. Before being transplanted into the tracheal defect, to achieve graft vascularization, the tracheal allotransplant was implanted for 9 months into the forearm of the patient. During this period, immunosuppressive therapy was necessary (stopped 7.5 months after implantation), the cartilaginous viability was maintained, while the membranous posterior wall of the allograft underwent avascular necrosis and a replacement with recipient buccal mucosa was necessary [57]. Even if clinically successful, this approach has limited therapeutic potential, since results are possible only for benign lesions, but not for malignant tumors, because of the risk of tumor progression during the period of pretransplant immunosuppression [58].

As reported above, long-segment airway reconstruction, for both advanced laryngeal and tracheal structural disorders, is still a challenging surgical issue.

In order to identify a functional substitute, displaying anatomical, physiological, and biomechanical properties equivalent to the native tissue, researchers and doctors have paid attention to the tissue engineering strategy, which has already provided functional human organ replacements in various clinical settings [59,60], as a solution for the airway functional reconstruction.

53.4 AIRWAY TISSUE ENGINEERING: *A POSSIBLE FUNCTIONAL SOLUTION*

According to the concept of tissue engineering, three components—cells, scaffold, and regulatory factors—are important for the regeneration of tissues. Several strategies, based on different cells (from autologous to allogenic, differentiated to stem/progenitor cells), different culturing conditions (static, dynamic, or *in vivo* approach, in which the body can be utilized as a bioreactor), various scaffold materials (natural and synthetic, biodegradable and permanent), and regulatory/growth/boosting factors (used as local and/or systemic pharmaceutical intervention to promote autologous progenitor cell mobilization and thus improve graft integrity), are currently investigated for airway functional regeneration [61–63]. However, the effectiveness of most approaches can be related to several limitations in the translation of results from *in vitro* studies, *in vivo* animal model experience to clinical cases. For this reason, in this chapter, we report only the tissue-engineered airway strategies which have been evaluated in preclinical studies and have been (or are to be) translated to clinical practice. In particular, for each organ, first strategies based on biological-derived scaffolds [which preserve the natural extracellular matrix (ECM) composition, do not produce toxic biodegradable products, do not induce inflammation, and their natural degradability, with release of growth factors and peptides, could stimulate constructive tissue remodelling] will be reported and then those performed evaluating synthetic-based scaffolds (which have the advantages that there are no issues related to a limited donor pool, graft structures can be tailored to the size and shape required for a particular patient, and material properties can be easily controlled).

Laryngeal tissue engineering aims mainly to restore the altered vibratory and respiratory functions of the larynx, however, the main problem is how to construct a bioengineered larynx with whole laryngeal framework and low immunogenicity. Indeed, the architecture of an entire larynx is complicated and its dynamic function requires the appropriate activation of the small muscles with respiration, in order to combine sphincter and breathing functions. Moreover, in comparison with trachea, the site of the larynx is more prone to infection, because of increased exposure to bacteria in the oral cavity from saliva and it is surgically difficult to affix the scaffold to the defect in the larynx, because of vocal fold movement, while maintaining the space needed for tissue regeneration [64]. So far, tissue engineering has only been used for partial laryngeal reconstruction.

Hyaluronic-based scaffolds (Hyalograft C), seeded with chondrocytes via a bioreactor and characterized by good *in vitro* biomechanical properties, did not succeed *in vivo*, because of a nonspecific foreign body response, leading to the degradation of engineered cartilage (with loss of the original three-dimensional shape and significant reduction of the cartilage mass) and graft failure [65]. Instead, cartilaginous sheets (scaffold-free grafts obtained by seeding autologous auricular cartilaginous cells on fibronectin-conditioned semipermeable polyester membrane via a bioreactor), implanted for laryngeal reconstruction, did not degrade but migration and buckling were reported due to graft biomechanical failure [66]. Decellularized matrices (mainly porcine decellularized urinary bladder matrix) have also been considered for laryngeal regeneration after partial hemilaryngectomies [67–69]: regeneration, superior to that observed using control standard procedure of thyroid cartilage, epithelium, connective tissue, glandular structures and of some skeletal muscles, was reported [67,68]. ECM scaffolds could then also be promising templates for constructive remodeling of laryngeal tissue [67,68]. Starting from this concept and based on the idea that the availability of substitutes displaying properties (anatomical, physiological, and biomechanical) equivalent to normal larynxes would provide the right, complex architecture and dynamics for normal voice production and sphincter action, Hou et al. [70] attempted to create a low-immune decellularized whole-larynx scaffold, comprising laryngeal muscles, matrix, and an integrated cartilage framework. After implantation (in the greater omentum) of the reseeded graft [with mesenchymal stem cells (MSCs)], muscle bundles and vessels were observed. However, a number of lymphocytes were seen around muscle bundles, indicating a serious inflammation reaction and suggesting an immunological reaction [70]. Human laryngeal acellular matrices have also been obtained in a brief and clinically useful time: bioengineered human grafts resulted structurally and mechanically similar to the native larynx (in all the different structures) and contained angiogenic factors which exert pro-angiogenic properties [71] (Figure 53.1). Even if this acellular laryngeal graft may provide the right, complex architecture and dynamics for normal voice production and sphincter action, no *in vivo* evaluation has to date been performed.

Among different synthetic materials, polypropylene mesh with a pore size of 260 μm, reinforced with a polypropylene supporting ring and coated with collagen (porcine dermal atelocollagen, types I and III) seems to be more

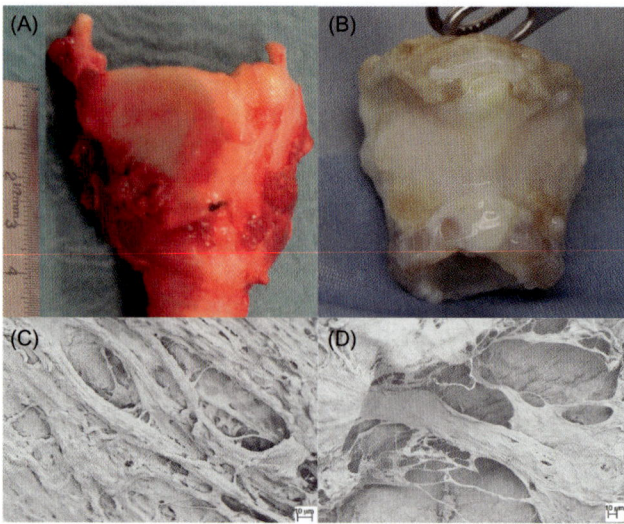

FIGURE 53.1 Larynx. (A and B) Representative images of human native (A) and decellularized (B) larynx. (C and D) Scanning electron microscopy micrographs of human native (C) and decellularized (D) larynx showing that the micro- and ultrastructural details of larynx architecture are retained after tissue decellularization.

suitable for laryngeal regeneration with clinical promising results. In preclinical studies, it has been preclotted with arterial blood and implanted for the treatment of subglottic stenosis: confluent epithelial regeneration, good incorporation of the scaffold mesh were observed, even if granulation tissue and mesh exposure were also reported [72]. In order to recreate the biological environment, to obtain a suitable regeneration, the same synthetic scaffold has been designed based on the replication of the luminal shape of a canine larynx. The graft, preclotted with a mixture of peripheral blood and bone marrow-derived stromal cells, was used to perform a hemilaryngectomy: soft tissue regeneration and the presence of mucosal cells were observed [73]. Considering the efficacy, the device, injected with autologous venous blood, has been implanted into four patients affected by airway stenosis ($n = 1$ subglottis stenosis) or cancer invasion ($n = 3$ thyroid cancers) [74]. It was reported that a good epithelialization on the luminal surface occurred in all the patients, as verified during the postoperative observation period (8–34 months), and in only one case air leak was revealed. These promising results demonstrate the ability to regenerate cricoid cartilage using scaffolds composed of polypropylene and collagen sponge in clinical applications [74]. Recently, the same artificial material, coated with different concentrations of collagen and wrapped with autologous fascia, resulted in a viable alternative for the regeneration of laryngeal defects, even if scar-like tissue in the middle of the regenerated vocal folds and a reduction of the treated vocal fold was observed, suggesting that additional approaches are required to regenerate a normal and functional vocal fold [64].

Most of the tracheal tissue engineering strategies, till now developed, seemed useless for whole long-segment (>6 cm) airway applications, because they were unable to resist collapse with consequent stenosis development [75]. Scaffolds developed joining the biocompatible properties of natural biomaterials with the mechanical properties of synthetic materials, characterized by good *in vivo* mechanical stability, are far from being translated into clinical applications because their functionality, biomechanics, viability, and adequate vascularization remain undetermined [76–78].

Increasing attention has been focused on naturally derived scaffold materials and different approaches with decellularized matrices have been tried to identify the more suitable scaffold for a functional, tissue-engineered trachea [79–83]. However, the main limitation was related to the decellularization protocol: most reagents used, indeed, damaged and/or disrupted the main ECM components (such as collagen, glycosaminoglycan, and elastin), compromising the ability of the scaffold to provide mechanical support during the *in vivo* remodeling process [82,84]. The first positive clinical success was reported in 2004: a porcine jejunal segment, decellularized using a detergent-enzymatic method (DEM) [85] and developed *in vitro* directly into a vascularized bioartificial matrix with all cellular tracheal functioning elements, was successfully used to repair a tracheal defect in a 58-year-old man [86]. The bioartificial patch, reseeded *in vitro* with autologous muscle cells and fibroblasts, functionally and morphologically integrated into the adjacent airway without signs of chronic inflammation, infection, or granulation tissue [86]. A decellularized human tracheal matrix (7-cm-long), obtained using the same DEM and characterized by structural and mechanical properties similar to native trachea, lack of immunogenicity and containing pro-angiogenic factors [87] (Figure 53.2), was used to replace, after *in vitro* recellularization (with autologous epithelial respiratory cells and MSC-derived chondrocytes via a bioreactor), the patient's left main bronchus (stenosed from tuberculosis) [88]. After 4 years, the patient is well, active with normal lung function and, more importantly, does not require immunosuppressive drugs [88]. To date the same approach, improved using an alternative cell technological approach (decellularized human tracheas were intraoperatively seeded with autologous bone marrow stromal cells and conditioned with growth and boosting factors), has been clinically applied for both benign ($n = 5$) and malignant ($n = 3$) airway diseases, resulting in a lasting and quicker recovery in comparison with classic surgery. Although initial promising results (the implanted scaffolds vascularized and lined with complete respiratory neomucosa), a partial collapse of the most proximal part of the graft was observed in about 30% of patients. This is probably related to a transmembrane migration of stem-

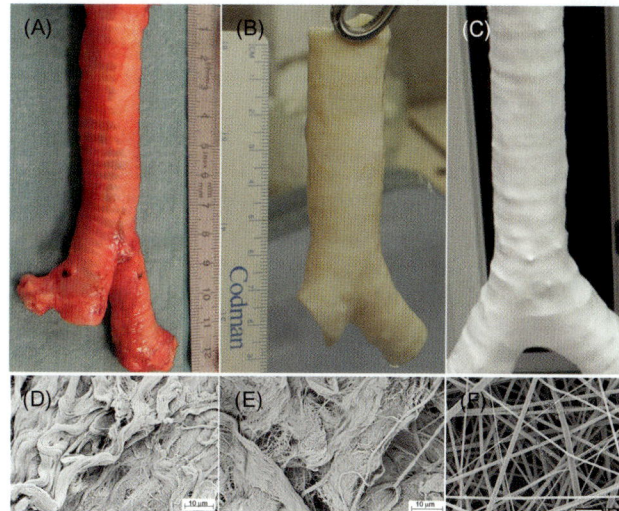

FIGURE 53.2 Trachea. (A–C) Representative images of human native (A), decellularized (B), and synthetic (C) trachea. (D–F) Scanning electron microscopy micrographs of human native (D) and decellularized (E) trachea showing that the micro- and ultrastructural details of tracheal architecture are retained after tissue decellularization. (F) Scanning electron microscopy micrographs of polyethylene terephthalate (PET) tracheal scaffold.

cell-derived chondrocytes from the outer (chondrocyte compartment) to the internal lumen (epithelial compartment) of the scaffold due to an oxygen concentration gradient developing throughout the engineered trachea. However, it is also possible that the decellularization process, affecting the scaffold properties and surface topography, could have an effect on the long-term graft properties [89]. In order to overcame the drawbacks related to the use of biologically-derived scaffolds, synthetic substitutes can represent a possible alternative, that is anyway characterized by pros and cons.

The same artificial material used for laryngeal regeneration, Marlex mesh tube covered by collagen sponge, has also been evaluated also as tracheal patch, resulting in favorable epithelialization, with proper reconstruction of original contours [73]. Using a similar prosthesis, soaked with peripheral blood, bone marrow aspirate, or bone marrow MSCs, Nakamura et al. [90] showed that the use of bone marrow and MSCs seemed to facilitate the healing and functional recovery (ciliar movement) of the reconstructed tracheal tissue. However, these approaches, even if promising, have been used only to repair tracheal defects and further long-term observation is required. The same synthetic scaffold, moistened with autologous peripheral blood, was first used in a clinical study in a 78-year-old woman affected by thyroid cancer [9]. The right half of three rings of the trachea was resected and the construct was properly trimmed to repair the defect. The artificial material was covered by epithelium after 2 months and was completely covered after 20 months with no complications. Clinical application of this tracheal prosthesis has begun for human patients with tracheal defects from 2002 [9,74], and although positive and promising results have been obtained, this approach resulted successful only in clinical cases where most of the trachea itself resulted normal; in cases of the presence of unfavorable conditions (such as infection and asthma in stenotic patients), this regenerative solution was not suitable [91]. Conditioning the scaffold with basic fibroblast growth factor (b-FGF), some positive results have been obtained also in patients affected by stenosis [91], however, the procedure involved two-staged operations (to enlarge the stenotic region and to implant tracheal scaffold) and b-FGF could not be applied to oncological patients because it may contribute to tumor recurrence. The majority of synthetic tracheal tissue-engineered approaches have been focused on developing scaffolds for patch or large circumferential defect repair, however, less common are the "Y"-shaped defects for bifurcation repair (to date only one clinical application has been reported). A Y-shaped scaffold made of Marlex mesh (polypropylene, 260 μm pore size), reinforced with polypropylene spiral and coated with porcine skin collagen types I and III, was implanted into beagle dogs [92]. A Y-shaped silicon stent was inserted into but not fixed to the prosthesis, and 8 weeks after surgery it was removed endoscopically. 14 out of 20 dogs died after experimentation due to obstruction of the main bronchus, omental necrosis, and air leakage. The same construct (60-mm-long, 18-mm-outer diameter), used for the replacement of the tracheobronchial bifurcation, was evaluated after a long-term follow-up. After 5 years, the prosthesis was infiltrated by the surrounding connective tissue, completely incorporated by the host trachea and bronchus, and neither stenosis nor dehiscence was observed. Moreover, the frequency of the epithelial cilia was maintained within the normal range, indicating functional recovery of the regenerating airway [93].

A significant application of tissue engineering concepts was clinically carried out by implanting a synthetic tracheobronchial substitute into a patient affected by a recurrence of a primary tracheal mucoepidermoid carcinoma involving the distal trachea and both main bronchi [94] (Figure 53.2). The tracheobronchial graft was realized by using a nanocomposite polymer (polyhedral oligomericsilsesquioxane covalently bonded to poly(carbonate-urea)urethane) and combining two different routes: a casted form for the cartilage "U" shaped rings and a coagulated form for the "connective" tracheal part. Before implantation, the bioartificial scaffold was seeded with autologous bone marrow mononuclear cells, via bioreactor. Postoperative bronchoscopy assessment verified, at 1 week, a normal and patent airway, while the biopsy

samples showed the presence of necrotic connective tissue associated with fungi contamination and neoformed vessels. Two months after transplantation, biopsy revealed large granulation areas with initial signs of epithelialization, more organized vessel formations, and no bacterial or fungi contamination. The patient is still alive and without evidence of malignancy 18 months after transplant [94].

53.5 CONCLUSIONS

Many steps forward have recently been made into laryngeal and tracheal tissue engineering, and the constructs that have been developed show promise for a clinically useful tissue-engineered airway replacement. However, many problems have arisen, such as how to maintain the long-term structural properties of bioengineered grafts, minimizing the rejection response, optimizing tracheal synthetic graft and culture conditions (like bioreactor design), adhesion, proliferation and function of the seeded cells, reepithelialization, and promoting revascularization. All these different aspects remain to be answered before a full clinical trial accreditation may be obtained. Moreover, the replacement of active movement, as would be required to replace the larynx, brings another order of the magnitude of complexity. The ability to produce functioning muscles, especially if these could be rapidly innervated and revascularized, would hugely extend the possible applications of regenerative medicine.

REFERENCES

[1] Isaacs RS, Sykes JM. Anatomy and physiology of the upper airway. Anesthesiol Clin North Am 2002;20:733–45.

[2] Barker E, Haverson K, Stokes CR, Birchall M, Bailey M. The larynx as an immunological organ: immunological architecture in the pig as a large animal model. Clin Exp Immunol 2006;143:6–14.

[3] Rees LE, Ayoub O, Haverson K, Birchall MA, Bailey M. Differential major histocompatibility complex class II locus expression on human laryngeal epithelium. Clin Exp Immunol 2003;134:497–502.

[4] Gorti GK, Birchall MA, Haverson K, Macchiarini P, Bailey M. A preclinical model for laryngeal transplantation: anatomy and mucosal immunology of the porcine larynx. Transplantation 1999;68(11): 1638–42.

[5] Mercer RR, Russell ML, Roggli VL, Crapo JD. Cell number and distribution in human and rat airways. Am J Respir Cell Mol Biol 1994;10:613–24.

[6] Rock JR, Randell SH, Hogan BL. Airway basal stem cells: a perspective on their roles in epithelial homeostasis and remodeling. Dis Model Mech 2010;3:545–56.

[7] Okubo T, Knoepfler PS, Eisenman RN, Hogan BL. Nmyc plays an essential role during lung development as a dosage-sensitive regulator of progenitor cell proliferation and differentiation. Development 2005;132:1363–74.

[8] Cardoso WV, Lü J. Regulation of early lung morphogenesis: questions, facts and controversies. Development 2006;133:1611−24.

[9] Omori K, Nakamura T, Kanemaru S, Asato R, Yamashita M, Tanaka S, et al. Regenerative medicine of the trachea: the first human case. Ann Otol Rhinol Laryngol 2005;114:429−33.

[10] Abdelkafy WM, El Atriby MN, Iskandar NM, Mattox DE, Mansour KA. Slide tracheoplasty applied to acquired subglottic and upper tracheal stenosis: an experimental study in a canine model. Arch Otolaryngol Head Neck Surg 2007;133:327−30.

[11] Younis RT, Lazar RH, Bustillo A. Revision single-stage laryngotracheal reconstruction in children. Ann Otol Rhinol Laryngol 2004;113:367−72.

[12] Duque E, Duque J, Nieves M, Mejía G, López B, Tintinago L. Management of larynx and trachea donors. Transplant Proc 2007;39:2076−8.

[13] de Jong AL, Park AH, Raveh E, Schwartz MR, Forte V. Comparison of thyroid, auricular, and costal cartilage donor sites for laryngotracheal reconstruction in an animal model. Arch Otolaryngol Head Neck Surg 2000;126:49−53.

[14] Ludemann JP, Hughes CA, Noah Z, Holinger LD. Complications of pediatric laryngotracheal reconstruction: prevention strategies. Ann Otol Rhinol Laryngol 1999;108:1019−26.

[15] Jacobs BR, Salman BA, Cotton RT, Lyons K, Brilli RJ. Postoperative management of children after single-stage laryngotracheal reconstruction. Crit Care Med 2001;29:164−8.

[16] Rutter MJ, Hartley BE, Cotton RT. Cricotracheal resection in children. Arch Otolaryngol Head Neck Surg 2001;127:289−92.

[17] Primov-Fever A, Talmi YP, Yellin A, Wolf M. Cricotracheal resection for airway reconstruction: the sheba medical center experience. IMAJ 2006;8:543−7.

[18] Hart CK, Richter GT, Cotton RT, Rutter MJ. Arytenoid prolapse: a source of obstruction following laryngotracheoplasty. Otolaryngol Head Neck Surg 2009;140:752−6.

[19] Effective head & neck cancer management: second consensus document. London: The Royal College of Surgeons of England, British Association of Otorhinolaryngologists, Head and Neck Surgeons; 2000.

[20] Nouraei SAR, Nouraei SM, Howard DJ, Sandhu GS. Estimating the population incidence of adult post-intubation laryngotracheal stenosis. Clin Otolaryngol 2007;32:407−9.

[21] Macchiarini P. Primary tracheal tumours. Lancet Onc 2006;7:83−91.

[22] Honings J, Gaissert HA, van der Heijden HF, Verhagen AF, Kaanders JH, Marres HA. Clinical aspects and treatment of primary tracheal malignancies. Acta Oto-Laryngol 2010;130:763−72.

[23] Gelder CM, Hetzel MR. Primary tracheal tumours: a national survey. Thorax 2006;48:688−92.

[24] Licht PB, Friis S, Pettersson G. Tracheal cancer in Denmark: a nationwide study. Eur J Cardiothorac Surg 2001;19:339−45.

[25] Bhattacharyya N. Contemporary staging and prognosis for primary tracheal malignancies: a population-based analysis. Otolaryngol Head Neck Surg 2004;131:639−42.

[26] Yang KY, Chen YM, Huang MH, Perng RP. Revisit of primary malignant neoplasms of the trachea: clinical characteristics and survival analysis. Jpn J Clin Oncol 1997;27:305−39.

[27] Grillo HC, Donahue DM, Mathisen DJ, Wain JC, Wright CD. Postintubation tracheal stenosis. Treatment and result. J Thorac Cardiovasc Surg 1995;109:486−92.

[28] Grillo HC. Tracheal replacement: a critical review. Ann Thorac Surg 2002;73:1995−2004.

[29] Wright CD, Grillo HC, Wain JC, Wong DR, Donahue DM, Gaissert HA, et al. Anastomotic complications after tracheal resection: prognostic factors and management. J Thorac Cardiovasc Surg 2004;128:731−9.

[30] Mulliken JB, Grillo HC. The limits of tracheal resection with primary anastomosis: further anatomical studies in man. J Thorac Cardiovasc Surg 1968;55:418−21.

[31] Kucera KA, Doss AE, Dunn SS, Clemson LA, Zwischenberger JB. Tracheal replacements: Part 1. ASAIO J 2007;53:497−505.

[32] Doss AE, Dunn SS, Kucera KA, Clemson LA, Zwischenberger JB. Tracheal replacements: Part 2. ASAIO J 2007;53(5):631−9.

[33] Chao MW, Smith JG, Laidlaw C, Joon DL, Ball D. Results of treating primary tumors of the trachea with radiotherapy. Int J Radiat Oncol Biol Phys 1998;41:779−85.

[34] Miodoński J, Sekula J, Olszewski E. Enlarged hemilaryngectomy (subtotal larygectomy) with immediate reconstruction for advanced cancer of the larynx. J Laryngol Otol 1965;79:1025−31.

[35] Quinn HJ. Free muscle transplant method of glottic reconstruction after hemilaryngectomy. Laryngoscope 1975;85:985−6.

[36] Park NH, Major Jr. JW, Sauers PL. Hemilaryngectomy and vocal cord reconstruction with digastric tendon graft. Surg Gynecol Obstet 1982;155:253−6.

[37] Delaere PR, Poorten VV, Goeleven A, Feron M, Hermans R. Tracheal autotransplantation: a reliable reconstructive technique for extended hemilaryngectomy defects. Laryngoscope 1998;108:929−34.

[38] Delaere PR, Poorten VV, Vanclooster C, Goeleven A, Hermans R. Results of larynx preservation surgery for advanced laryngeal cancer through tracheal autotransplantation. Arch Otolaryngol Head Neck Surg 2000;126:1207−15.

[39] Delaere PR, Hermans R. Tracheal autotransplantation as a new and reliable technique for the functional treatment of advanced laryngeal cancer. Laryngoscope 2003;113:1244−51.

[40] Delaere PR, Vranckx J, Dooms C, Meulemans J, Hermans R. Tracheal autotransplantation: guidelines for optimal functional outcome. Laryngoscope 2011;121:1708−14.

[41] Garozzo A, Rossi M. Glottic reconstruction by implant of homologous laryngeal cartilages. J Laryngol Otol 1993;107:427−9.

[42] Kunachak S, Kulapaditharom B, Vajaradul Y, Rochanawutanon M. Cryopreserved, irradiated tracheal homograft transplantation for laryngotracheal reconstruction in human beings. Otolaryngol Head Neck Surg 2000;122:911−6.

[43] Zeitels SM, Wain JC, Barbu AM, Bryson PC, Burns JA. Aortic homograft reconstruction of partial laryngectomy defects: a new technique. Ann Otol Rhinol Laryngol 2012;121:301−6.

[44] Kluyskens P, Ringoir S. Follow-up of a human larynx transplantation. Laryngoscope 1970;80:1244−50.

[45] Strome M, Stein J, Esclamado R, Hicks D, Lorenz RR, Braun W, et al. Laryngeal transplantation and 40-month follow-up. N Engl J Med 2001;344:1676−9.

[46] UC Newsroom. Patient gets successful larynx transplant. Oakland, CA: University of California; 2011 [Available from: <http://www.universityofcalifornia.edu/news/article/24844>; 2011.]

[47] Birchall MA. Tongue transplantation. Lancet 2004;363:1663.

[48] Birchall MA, Macchiarini P. Airway transplantation: a debate worth having? Transplantation 2008;85:1075−80.

[49] Baiguera S, Macchiarini P. Trachea. In: Steinhoff G, editor. Regenerative medicine—from protocol to patient. 1st ed. London, UK: Springer Publ.; 2011. p. 691−711.

[50] Bujia J, Pitzke P, Krombach F, Hammer C, Wilmes E, Herberhold C, et al. Immunological behavior of preserved human tracheal allografts: immunological monitoring of a human tracheal recipient. Clin Transplant 1991;5:376−80.

[51] Elliott MJ, Haw MP, Jacobs JP, Bailey CM, Evans JN, Herberhold C. Tracheal reconstruction in children using cadaveric homograft trachea. Eur J Cardiothorac Surg 1996;10:707−12.

[52] Jacobs JP, Elliott MJ, Haw MP, Bailey CM, Herberhold C. Pediatric tracheal homograft reconstruction: a novel approach to complex tracheal stenosis in children. J Thorac Cardiovasc Surg 1996;112:1549−60.

[53] Jacobs JP, Quintessenza JA, Andrews T, Burke RP, Spektor Z, Delius RE, et al. Tracheal allograft reconstruction: the total North American and worldwide pediatric experiences. Ann Thorac Surg 1999;68:1043−52.

[54] Dodge-Khatami A, Nijdam NC, Broekhuis E, Von Rosenstiel IA, Dahlem PG, Hazekamp MG. Carotid artery patch plasty as a last resort repair for long-segment congenital tracheal stenosis. J Thorac Cardiovasc Surg 2002;123:826−8.

[55] Hazekamp MG, Nijdam N. Use of autologous arterial patches for tracheal reconstruction in young infants. Ann Thorac Surg 2004;77:2259−64.

[56] Azorin JF, Bertin F, Martinod E, Laskar M. Tracheal replacement with an aortic autograft. Eur J Cardiothorac Surg 2006;29:261−3.

[57] Delaere PR, Vranckx J, Verleden G, De Leyn P, Van Raemdonck D, Leuven Tracheal Transplant Group. Tracheal allotransplantation after withdrawal of immunosuppressive therapy. N Engl J Med 2010;362:138−45.

[58] Delaere PR, Vranckx J, Meulemans J, Vander Poorten V, Segers K, Van Raemdonck D, et al. Learning curve in tracheal allotransplantation. Am J Transplant 2012;12:2538−45.

[59] Atala A, Bauer SB, Soker S, Yoo JJ, Retik AB. Tissue-engineered autologous bladders for patients needing cystoplasty. Lancet 2006;367:1241−6.

[60] Pham C, Greenwood J, Cleland H, Woodruff P, Maddern G. Bioengineered skin substitutes for the management of burns: a systematic review. Burns 2007;33:946−57.

[61] Kalathur M, Baiguera S, Macchiarini P. Translating tissue-engineered tracheal replacement from bench to bedside. Cell Mol Life Sci 2010;67:4185−96.

[62] Baiguera S, D'Innocenzo B, Macchiarini P. Current status of regenerative replacement of the airway. Expert Rev Respir Med 2011;5:487−94.

[63] Jungebluth P, Moll G, Baiguera S, Macchiarini P. Tissue-engineered airway: a regenerative solution. Clin Pharmacol Ther 2012;91:81−93.

[64] Kitani Y, Kanemaru S, Umeda H, Suehiro A, Kishimoto Y, Hirano S, et al. Laryngeal regeneration using tissue engineering techniques in a canine model. Ann Otol Rhinol Laryngol 2011;120:49−56.

[65] Weidenbecher M, Henderson JH, Tucker HM, Baskin JZ, Awadallah A, Dennis JE. Hyaluronan-based scaffolds to tissue-engineer cartilage implants for laryngotracheal reconstruction. Laryngoscope 2007;117:1745−9.

[66] Gilpin DA, Weidenbecher MS, Dennis JE. Scaffold-free tissue-engineered cartilage implants for laryngo tracheal reconstruction. Laryngoscope 2010;120:612−7.

[67] Huber JE, Spievack A, Simmons-Byrd A, Ringel RL, Badylak S. Extracellular matrix as a scaffold for laryngeal reconstruction. Ann Otol Rhinol Laryngol 2003;112:428−33.

[68] Ringel RL, Kahane JC, Hillsamer PJ, Lee AS, Badylak SF. The application of tissue engineering procedures to repair the larynx. J Speech Lang Hear Res 2006;49:194−208.

[69] Xu CC, Chan RW, Tirunagari N. A biodegradable, acellular xenogeneic scaffold for regeneration of the vocal fold lamina propria. Tissue Eng 2007;13:551−66.

[70] Hou N, Cui P, Luo J, Ma R, Zhu L. Tissue-engineered larynx using perfusion-decellularized technique and mesenchymal stem cells in a rabbit model. Acta Otolaryngol 2011;131:645−52.

[71] Baiguera S, Gonfiotti A, Jaus M, Comin CE, Paglierani M, Del Gaudio C, et al. Development of bioengineered human larynx. Biomaterials 2011;32:4433−42.

[72] Omori K, Nakamura T, Kanemaru S, Kojima H, Magrufov A, Hiratsuka Y, et al. Cricoid regeneration using in situ tissue engineering in canine larynx for the treatment of subglottic stenosis. Ann Otol Rhinol Laryngol 2004;113:623−7.

[73] Yamashita M, Omori K, Kanemaru S, Magrufov A, Tamura Y, Umeda H, et al. Experimental regeneration of canine larynx: a trial with tissue engineering techniques. Acta Otolaryngol Suppl 2007;557:66−72.

[74] Omori K, Tada Y, Suzuki T, Nomoto Y, Matsuzuka T, Kobayashi K, et al. Clinical application of in situ tissue engineering using a scaffolding technique for reconstruction of the larynx and trachea. Ann Otol Rhinol Laryngol 2008;117:673−8.

[75] Kojima K, Ignotz RA, Kushibiki T, Tinsley KW, Tabata Y, Vacanti CA. Tissue-engineered trachea from sheep marrow stromal cells with transforming growth factor beta2 released from biodegradable microspheres in a nude rat recipient. J Thorac Cardiovasc Surg 2004;128:147−53.

[76] Shi H, Xu Z, Qin X, Zhao X, Lu D. Experimental study of replacing circumferential tracheal defects with new prosthesis. Ann Thorac Surg 2005;79:672−6.

[77] Wu W, Cheng X, Zhao Y, Chen F, Feng X, Mao T. Tissue engineering of trachea-like cartilage grafts by using chondrocyte macroaggregate: experimental study in rabbits. Artif Organs 2007;31:826−34.

[78] Tatekawa Y, Kawazoe N, Chen G, Shirasaki Y, Komuro H, Kaneko M. Tracheal defect repair using a PLGA-collagen hybrid scaffold reinforced by a copolymer stent with bFGF-impregnated gelatin hydrogel. Pediatr Surg Int 2010;26:575−80.

[79] Park JW, Pavcnik D, Uchida BT, Timmermans H, Corless CL, Yamakado K, et al. Small intestinal submucosa covered expandable Z stents for treatment of tracheal injury: an experimental pilot study in swine. J Vasc Interv Radiol 2000;11: 1325−30.

[80] Gubbels SP, Richardson M, Trune D, Bascom DA, Wax MK. Tracheal reconstruction with porcine small intestine submucosa in a rabbit model. Otolaryngol Head Neck Surg 2006;134: 1028−35.

[81] Zhang L, Liu Z, Cui P, Zhao D, Chen W. SIS with tissue-cultured allogenic cartilages patch tracheoplasty in a rabbit model for tracheal defect. Acta Otolaryngol 2007;127:631−6.

[82] Gilbert TW, Gilbert S, Madden M, Reinolds SD, Badylak SF. Morphologic assessment of extracellular matrix scaffolds for patch tracheoplasty in a canine model. Ann Thorac Surg 2008;86: 967−74.

[83] Remlinger NT, Czajka CA, Juhas ME, Vorp DA, Stolz D, Badylak SF, et al. Hydrated xenogeneic tracheal matrix as scaffold for tracheal reconstruction. Biomaterials 2010;31:3520–6.

[84] Gilbert TW, Sellaro TL, Badylak SF. Decellularization of tissues and organs. Biomaterials 2006;27:3675–83.

[85] Meezan E, Hjelle JT, Brendel K. A simple, versatile, non disruptive method for the isolation of morphologically and chemically pure basement membranes from several tissues. Life Sci 1975; 17:1721–32.

[86] Macchiarini P, Walles T, Biancosino C, Mertsching H. First human transplantation of a bioengineered airway tissue. J Thorac Cardiovasc Surg 2004;128:638–41.

[87] Baiguera S, Jungebluth P, Burns A, Mavilia C, Haag J, De Coppi P, et al. Tissue engineered human tracheas for in vivo implantation. Biomaterials 2010;31:8931–8.

[88] Macchiarini P, Jungebluth P, Go T, Asnaghi MA, Rees LA, Cogan TA, et al. Clinical transplantation of a tissue-engineered airway. Lancet 2008;372:2023–30.

[89] Baiguera S, Del Gaudio C, Jaus MO, Polizzi L, Gonfiotti A, Comin CE, et al. Long-term changes to in vitro preserved bioengineered human trachea and their implications for decellularized tissues. Biomaterials 2012;33:3662–72.

[90] Nakamura T, Sato T, Araki M. In situ tissue engineering for tracheal reconstruction using a luminar remodeling type of artificial trachea. J Thorac Cardiovasc Surg 2009;138:811–9.

[91] Kanemaru S, Hirano S, Umeda H, Yamashita M, Suehiro A, Nakamura T, et al. A tissue-engineering approach for stenosis of the trachea and/or cricoid. Acta Otolaryngol Suppl 2010;563:79–83.

[92] Sekine T, Nakamura T, Ueda H, Matsumoto K, Yamamoto Y, Takimoto Y, et al. Replacement of the tracheobronchial bifurcation by a newly developed Y-shaped artificial trachea. ASAIO J 1999;45:131–4.

[93] Nakamura T, Teramachi M, Sekine T, Kawanami R, Fukuda S, Yoshitani M, et al. Artificial trachea and long term follow-up in carinal reconstruction in dogs. Int J Artif Organs 2000;23: 718–24.

[94] Jungebluth P, Alici E, Baiguera S, Le Blanc K, Blomberg P, Bozóky B, et al. Tracheobronchial transplantation using a stem cell-seeded bioartificial nanocomposite. Lancet 2011;378: 1997–2004.

Chapter 54

Skin Regeneration and Bioengineering

Dennis P. Orgill[a,b,c] and Ryan Gobble[b,c]

[a]Harvard Medical School, Boston, MA, [b]Division of Plastic Surgery, Brigham and Women's Hospital, Boston, MA, [c]Department of Surgery, Brigham and Women's Hospital, Boston, MA

Chapter Outline

54.1 INTRODUCTION

54.1.1 Skin Physiology

Skin is a remarkable organ that provides several critical functions for life. At a basic level, it can be considered as a bilayer structure made up of the epidermis and the dermis. The epidermis is separated from the dermis by the basement membrane, which is contiguous around the hair follicles and adnexal organs, which are epidermal derivatives. Depending on the sequence of maturation of keratinocytes, very different structures can result. Palmar skin is much different than skin just a few millimeters away. The epidermis is primarily composed of keratinocyte-derived cells, which originate next to the basement membrane where they proliferate and migrate towards the surface undergoing apoptosis and cell death. The cell bodies of keratinocytes form the stratum corneum, which provides a semipermeable water barrier and also a very effective barrier to microorganisms. The dermis is primarily a biopolymer comprised of Type I collagen and glycosaminoglycans (GAGs). The pigmentation of the skin largely resides in the epidermis through the excretion of melanin by melanocytes, which are neuroendocrine-derived cells. Important immune cells include Langerhans cells. The epidermis does not have an intrinsic blood supply but is nourished through the basement membrane by diffusion from the underlying superficial plexus.

The dermis is mechanically strong and has two layers. The deeper layer is the reticular dermis, which has larger collagen fibers. The superficial, or papillary, dermis has finer collagen fibers located just beneath the basement membrane. Topographically, the epidermal-derived adnexal organs invaginate into the dermis and often extend into the underlying fat. The thickness of skin varies tremendously throughout the body, being 2−3 mm thick on the back to 0.25 mm in thickness on the eyelids. Dermal thickness also varies with age, being thinner in the elderly and infants.

54.1.2 Need for Skin Regeneration

Autologous skin grafts are an effective method of skin restoration, but have some obvious drawbacks. Donor sites by necessity result in scarring. Additionally, for very large cases of skin loss, there may not be an adequate donor site. Finally, there can be color, contour, or texture mismatches with surrounding skin. Skin loss is common and can result from trauma, burns, congenital defects, metabolic diseases, vascular diseases, neuropathy, and infection.

Clearly, if there were a product or process available that would facilitate the regeneration of skin it would be of

Regenerative Medicine Applications in Organ Transplantation.

enormous benefit to patients with skin loss. A regenerative approach would minimize scarring and would restore the entire function of skin. Skin grafts by themselves can be thought of as partially regenerative therapies.

In the 1960s and 1970s, great advances in the treatment of burn victims occurred to the point that many with large burns [>50% total body surface area (TBSA)] were surviving. As larger burns became salvageable, there became a need for better solutions for skin replacement. The National Institutes of Health and the US Department of Defense recognized the need for skin replacement and funded several laboratories to study new technologies to create skin either from cells, biomaterials, or a combination of both.

Howard Green, MD, then at MIT in the Biology Department, was a pioneer in cell and skin culture and developed a method to culture keratinocytes to large surface areas from small skin biopsies. Eugene Bell, also in Biology at MIT, pioneered the concept of fibroblast-based collagen gels that could be covered with keratinocytes, recognizing the importance of dermal–epidermal interaction in the function of skin. Ioannis Yannas, PhD in the Department of Mechanical Engineering at MIT, felt that the dermis was the most critical problem presented by the large burn patient and developed biodegradable scaffolds for dermal regeneration. The ideas that came from these three laboratories provide the foundation for other innovations in skin regeneration present today.

54.1.3 Regeneration versus Scarring

Unlike amphibians which have developed a regenerative response to injury, mammals exhibit scarring and contraction. Mammals do, however, appear to have the genetic possibility of a regenerative response to skin injury. A large body of literature has found that skin injury during certain gestational periods results in minimal scarring. Similarly, organs like the liver are well known to regenerate after surgical excision. Superficial skin injury also has a regenerative response and superficial burns heal with little to no scarring. Dunkin studied incisional wounds on the forearms of volunteers and found that injuries superficial to 0.37 mm did not have a perceptible scar [1]. Therefore, it appears that the reticular dermis has a potential for forming scar tissue rather than regeneration.

In addition, humans have the ability to form excess scar tissue, such as hypertrophic scarring or keloid. Although animal models for this are not ideal, many investigators have used the Red Duroc pig and recently Gurtner's group has reported on the use of a skin tension device in rats to induce a hypertrophic response and have correlated this with the expression of the focal adhesion kinase pathway [2].

54.1.4 Engineering Principles of Design

In the process of designing products to facilitate the regeneration of skin, several engineering principles are critical.

1. Materials must contour to the complex surface of the body. The body has complex curvatures with movement that place high mechanical demands on skin. Strength, modulus of elasticity and drapeability are critical mechanical properties that must be considered.
2. Materials must control for water loss. Excessive evaporation will desiccate wounds, whereas a totally occlusive dressing will allow for fluid accumulation. Materials that approximate the water vapor transmission rate of intact stratum corneum are desirable.
3. Materials must be biocompatible: nontoxic, noninflammatory, nonimmunogenic and not contain microbiological or that could lead to infection from bacteria, virus, or prions.

54.2 IMPORTANT TECHNOLOGIES

54.2.1 Cell Culture

54.2.1.1 Cultured Epidermal Autografts

Based upon the pioneering work of Rheinwald and Green in 1975 that allowed keratinocytes to be cultured and clonally expanded on a feeder layer of lethally irradiated mouse fibroblasts [3], cultured epidermal autografts (CEAs) can be produced from a 2-cm^2 sample of skin and can produce enough epidermis to cover the entire body in approximately 3 weeks [4]. Epicel™ (Genzyme Corportation, Framingham, MA) is a CEA approved by the FDA as a Humanitarian Use Device for patients who have deep dermal or full thickness burns of greater than or equal to 30% of their body. Epicel employs murine fibroblasts as feeder cells and as such is considered a xenotransplantation product [5]. Histologic examination of full thickness biopsies of Epicel-treated wounds demonstrates that a complete stratified epithelium develops in 6 days, however, the development of a fully mature epidermis with rete ridges and hemidesmisome attachments may take 6–12 months [6,7]. Epicel is only able to produce an epidermal layer rather a true dual-layered dermal/epidermal template, which limits its clinical usefulness. The most commonly reported adverse events that were reported included death (9%), sepsis (3.7%) multisystem organ failure (3.3%), and skin graft failure (1.3%) [8].

CEAs have been used most extensively in the treatment of burns with mixed results. A nonrandomized trial of 64 patients that compared treatment with CEAs (22 patients) to conventional coverage with autologous graft (42 patients) demonstrated significantly reduced mortality in the CEA group (14% vs. 48%, $P < 0.01$). This improvement in mortality was observed despite a significantly higher percentage of burn in the CEA group (72% vs. 62%, $P < 0.001$) and presence of inhalational injury (91% vs. 62%, $P < 0.01$). There was an increased need for revisional surgery and costs were significantly higher in the CEA group [9]. Haith et al. reported on the use of CEA in nine patients with an average burn of 70% (40–93%) and felt that CEA was "a tremendous asset to the management of massive burn patients [10]." They noted that CEA take was highest in patients with early granulation tissue or freshly excised wounds. Carsin also reported good results using Epicel CEA in 30 patients with severe burns. Epicel was applied to $37 \pm 17\%$ of TBSA. In these patients, the average burn size was $78 \pm 10\%$ with $65 \pm 16\%$ third degree burns. 90% of these patients also had evidence of inhalational injury. They demonstrated permanent coverage of $26 \pm 15\%$ of TBSA with survival rate of 90% [11]. In a study of 84 patients with $\geq 50\%$ TBSA, 59 patients survived (70% survival). 88% of survivors were treated with CEA, and the authors noted that CEA had been integrated into the routine management of patients with large TBSA burns in the region [12].

In contrast, Williamson reported 28 patients who underwent CEA application over an average of 10.4% TBSA with a mean take of 26.9% of the grafted area (2.8% TBSA). When compared to the 5 years before CEAs were used, there was no significant difference in duration of hospital stay or number of autograft harvests. He concluded that CEAs are unpredictable in the treatment of burn wounds and should only be used as a biologic dressing and experimental adjunct to standard burn management using autograft [13]. Similar poor results were demonstrated in a study of 16 patients with a mean TBSA burn of 68% as definitive coverage of only 4.7% TBSA was achieved using CEA at a cost of $43,000 per patient [14].

Other reported indications for CEA include the treatment of chronic venous ulcers and rheumatoid ulcers in which they have been shown to increase ulcer healing rates [15–17]. The main disadvantages of CEA are related to the long cultivation time, variable take rate, high costs, and increased risk of infection [18,19]. Blistering may occur for up to 6–12 months due to the lack of a true dermis and dermoepidermal junction [20]. It is postulated that blistering occurs secondary to the abnormal anchoring fibrils that develop under culture conditions and due to the enzymatic treatment of the CEA sheets required to detach the CEA from the culture flasks prior to grafting [21].

To address some of the shortcomings observed with CEA, including fragility during cultivation and application, as well as the length of time required to cultivate, adaptations to the original technique have been made.

A composite biocompatible skin graft (CBSG) has been developed. CBSG consists of a pliable hyaluronate-derived membrane (Laserskin®), which has been previously seeded with allogenic dermal fibroblasts, upon which CEAs are grown [22]. This technique has several advantages over the traditional method of cultivation and application of CEA. CBSG require a relatively low seeding density ($2-4 \times 10^4$ cells/cm^2), can be grafted at a less-differentiated stage secondary to being grown on Laserskin, and is easily transferred from the culture dish directly onto the wound bed since it is grown on a transferable membrane. It also eliminates the need for enzymatic digestion and does not contract in cell culture [23]. However, despite these promising advantages, to date this technique has been limited to animal studies and has not been expanded to human studies.

CEA can also be placed directly onto meshed allograft with excellent take as popularized by Cuono et al. [24]. In this technique, CEAs are harvested from thin split thickness autografts from unburned areas and then cultured using the techniques of Rheiwald and Green [3]. Cryopreserved allograft skin is then meshed 1.5:1 and applied directly to the wound bed. Once the allograft has been determined to have completely taken, the epidermal layer is removed by dermabrasion and the CEA applied directly to the dermal layer of the allograft. Differentiation of the CEA into a stratified epidermis with stratum granulosum and stratum corneum was observed in patients treated with this method. Successful take rates of 75–90% using the Cuono method have been achieved [25,26].

A modification of the Cuono method has been used with excellent results when using widely meshed autograft [27]. In this study, 12 children who had deep burns to $60 \pm 16\%$ TBSA underwent early excision and coverage with as much area as possible with autograft. A small skin graft was taken at the time of excision and initial coverage to expand into CEA. When CEAs were available, approximately 3 weeks later, a widely meshed autograft (6X), which are usually associated with poor take rates, was then applied directly to the remaining wounds and then covered with CEA. The take rate using this method was $84 \pm 12\%$ and an additional $30 \pm 9\%$ TBSA was covered. This study was performed in France, where access to allograft was limited and thus expanded autografts were chosen in place of allograft.

54.2.1.2 Apligraf

Apligraf® (Organogenesis, Inc., Canton, MA) is a bilayer consisting of a dermal equivalent layer of neonatal foreskin fibroblasts in a bovine collagen matrix and an epidermal equivalent of neonatal allogenic keratinocytes. Apligraf is supplied as a circular disk with a diameter of 75 mm and a thickness of 0.75 mm and has a limited shelf life of 5–10 days. When Apligraf is applied to a wound, it does not take a skin graft, rather it promotes fibrovascular ingrowth and epitheliazation, which ultimately leads to wound healing [4]. Preparation of the wound bed including debriding necrotic tissue and treating local infection is of upmost importance if Apligraf is to be used. The most common adverse event with Apligraf treatment is wound infection. In a trial of Apligraf treatment of venous leg ulcers (VLUs), wound infection occurred in 29.2% of patients receiving Apligraf versus 14.0% in the control group [4].

Apligraf is approved by the FDA for the treatment of VLU and diabetic foot ulcers (DFUs). The approval for treatment of VLU was based upon a subgroup analysis of 120 patients with VLU >1 year in duration from a 15-center, randomized, controlled trial in 240 patients comparing the efficacy of Apligraf plus compression versus compression alone [28]. Apligraf was applied directly on the wound along with a three-layered compression wrap consisting of a nonadherent primary dressing followed by a gauze pressure dressing and a final layer of self adherent elastic bandage. The ulcers in the control group were treated with a standard multilayer compression dressing, which consisted of the nonadherent primary dressing, a gauze bolster, a zinc oxide-impregnated paste bandage, and a self-adherent elastic bandage. Apligraf plus compression was significantly more effective than compression alone in the percentage of patients healed by 6 months (47% vs. 19%; $P < 0.005$) and the median time to complete wound closure ($P < 0.005$). Multivariate regression analysis also showed that Apligraf treatment resulted in twice as many patients with complete wound closure by 6 months ($P < 0.005$).

The approval of Apligraf for the treatment of DFU was based upon a 24-center, randomized, controlled trial in 208 patients comparing the efficacy of Apligraf (112 patients) versus saline-moistened guaze (96 patients) [29]. Apligraf was applied directly to the wound after debridement and irrigation followed by application of a nonadherent primary dressing, dry guaze, petroleum guaze, and a final layer of Kling. Apligraf could be reapplied weekly for a maximum of five applications. Control groups were treated with saline-moistened guaze and the same secondary dressing as the Apligraf group. After 12 weeks of treatment, significantly more patients treated with Apligraf had achieved complete wound healing compared to controls (56% vs. 38%, $P < 0.005$). The odds ratio for complete healing with Apligraf treatment compared to controls was 2.14 (95% CI 1.23–3.74). Apligraf was also associated with a decreased amputation rate compared to controls (6% vs. 16%). Osteomyelitis was also demonstrated to be significantly reduced with Apligraf treatment compared to controls (3% vs. 10%). There was no difference in ulcer recurrence between treatment arms within the first 6 months posttreatment. The authors concluded that Apligraf should be considered an adjunct to the standard of care (saline moistened gauze) rather than a substitute for it.

Apligraf has been expanded beyond these indications and is currently being used for many different conditions in which there is a need for skin regeneration. It has been used to treat pressure ulcers with wound healing noted in 13 of 21 wounds in an average of 29 days. Seven of these wounds had complete healing. The diabetic wounds in this study healed in 86% of patients. Three diabetic wounds failed, of which two were expected pretreatment as the Apligraf was applied in a distal phalynx toe wound associated with ischemia. The expected outcome was amputation so Apligraf was placed as a salvage measure. The third failure was in a patient who had concomitant renal failure and had undergone prior contralateral below knee amputation [30].

Apligraf has been used in full-thickness burns in a multicenter, randomized clinical trial. 38 patients underwent meshed autograft followed by Apligraf placement. Control arms included meshed autograft covered with meshed allograft and meshed autograft not covered by a biological dressing. While there were no significant differences in the percentage of autograft take or in the number of days until 75% autograft take, investigators still rated the Apligraf arm superior in 22 patients (58%) with improvements specifically noted in pigmentation, vascularity, and pliability. By 24 months, patients treated with Apligraf had a significant improvement in pigmentation and vascularity compared to controls. Improvements in pliability were noted within the first week, and Vancouver Burn Scar Assessment Scale scores were significantly improved throughout the length of the study [31]. The authors theorized that the superior cosmetic performance observed with Apligraf versus controls might have been related to the Apligraf dermis covering the interstices of the autograft, potentially even allowing epidermal cells from the autograft to replace the neonatal keratinocytes. Furthermore, they questioned whether the neonatal fibroblasts, which make up the dermal component, may have been better able to generate a near normal dermis secondary to their immaturity.

In patients undergoing Mohs micrographic or excisional surgery, Apligraf has been shown to result in a more pliable and less-vascularized wound as assessed by the Vancouver Burn Scar Assessment Scale compared to healing by secondary intention [32]. There was no difference in healing at 6 months between the treatment arms (100% in both). Other rarer uses of Apligraf include the treatment of wounds related to epidermolysis bullosa [33–35], harlequin ichthyosis [36], bullous morphea [37], polyarteritis nodosa [38], and pyoderma gangrenosum [39].

54.2.2 Scaffolds

54.2.2.1 Collagen-GAG Scaffolds

There is a spectrum of biomaterials that can be considered for scaffold-based approaches. Totally synthetic biomaterials have the advantage of excellent quality control during manufacturing, high degrees of purity, and low processing costs. The disadvantage of their use is that our current manufacturing technologies do not allow for the rich biological interactions that semisynthetic or biologically derived polymers possess. In contrast, biologically derived materials tend to offer a high degree of biological activity, but less control over the manufacturing process and more heterogeneity in the final product. Sterilization and immunogenicity can sometimes be issues.

A compromise between these two approaches is the process of biologically derived materials. One example of this is collagen, which is the most common protein of the dermis providing substantial mechanical strength and resilience. It has a complex hierarchical structure based on a triple helical structure. The complex structure is largely formed in the extracellular space through self-assembly mechanisms. This self assembly results in a "banding" effect seen on electron microscopy which is pH dependent [40].

There have been many types of collagen described today, Type I being the most common found in the dermis. It has a rich biochemistry with multiple moieties that allow for binding and cross-linking of the structure. The mechanical properties of materials made of collagen vary substantially based on the pH of formation, the orientation of the polymers, the cross-linking of the polymer and the addition of copolymers like GAGs (Figure 54.1).

Collagen-based materials have been shown to favorably interact with biological materials and can be

chemically altered to change their reactivity. For example, the removal of telopeptides can reduce the antigenicity of collagen [42].

Manufacturing processes can be designed to provide oriented materials like tapes, using cross filtration devices to form blood vessels and using lyophilization methods to form a highly porous solid. The rich chemistry of collagen combined with the ability to alter mechanical properties and its biocompatibility allow it to be an excellent biomaterial for many applications. Biodegradation is a critical parameter in designing scaffold-based materials. An important concept is the fact that these materials will persist for some time while host cells lay down a replacement structure. Approximating the degradation rate with the rate of synthesis allows a new structure to replace the temporary scaffold.

Both the site of implantation and composition of the scaffold alter the biodegradation rate. Implantation into sites with high blood flow or enzymatic activity can hasten degradation. Physicochemical parameters, such as the thickness of the construct, the cross-link density, the porosity, the degree of collagen banding, and the orientation of collagen can all affect degradation. Yannas developed collagen-GAG scaffolds for skin regeneration and optimized the physicochemical parameters in a guinea pig model. Applied to an acutely formed wound surface, these scaffolds allowed infiltration by blood vessels, fibroblasts, and inflammatory cells within a few days of application. Critical pore sizes between 50 and 500 μm were shown to be optimal. Pore sizes <50 μm often resulted in lack of integration of the biomaterial. Pore sizes over 500 μm showed mechanical weakness resulting in faster wound contraction [43].

Degradation rates were primarily a function of the cross-link density, with low cross-linking materials degrading rapidly, and with high cross-linking a foreign body response occurring.

To control moisture transmission, a layer of silicone was added to the top. The silicone layer approximates the water vapor transmission rate of intact stratum conium and provides a barrier for bacterial invasion.

This material was first manufactured at MIT and used by Dr John Burke and colleagues at the Massachusetts General Hospital and Shiners' Burns Institute. They reported on 10 patients in 1981 in the Annals of Surgery [44]. Dr Burke was an advocate of early excision and grafting and the use of the meticulous technique. In these early studies, they placed the material over large areas after excision of burn eschar and then removed the silicone at sometime after 14 days and replaced this with a very thin autologous skin graft. When compared to standard expanded mesh grafts, both patients and physicians noted that this construct was softer more pliable and smoother.

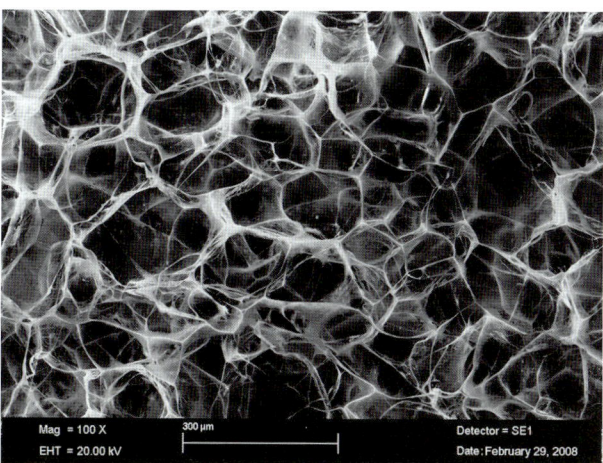

FIGURE 54.1 Porous collagen-GAG scaffold. *Originally appeared in [41].*

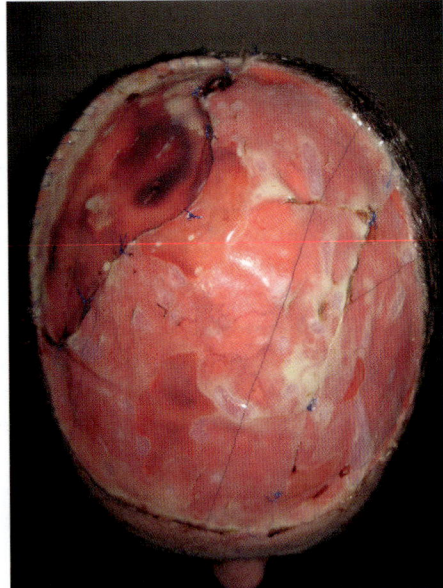

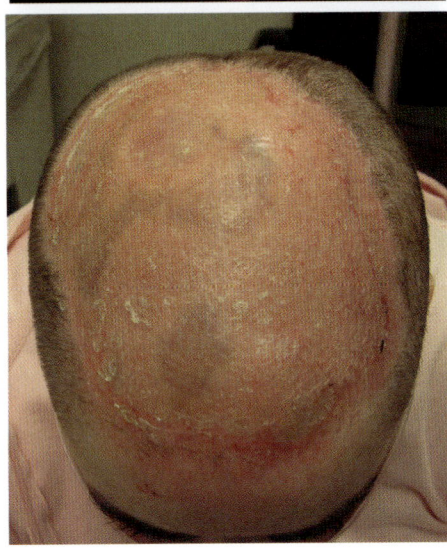

FIGURE 54.2 A 38-year-old woman with a squamous cell carcinoma of the scalp. She was treated with a collagen-GAG scaffold after excision and burning down of the calvaria to clear the tumor. This was followed by a skin graft 6 weeks later. This photo was taken 3 months postirradiation with 3500 cGy. *Originally appeared in [41].*

Later, this material became commercially manufactured and a multicenter trial was reported in Annals of Surgery [45]. They found similar findings that Burke et al. reported in 1981 [44]. They did find that it had a lower take rate than autologous skin grafts. This product is currently manufactured from bovine Achilles tendon as a source material (Integra®, Integra LifeSciences, Plainsboro, NJ).

Subsequent work has been performed in the areas of burn scar revision and the treatment of joint contractures [46,47]. The material is also used in cases of exposed bone or tendon (Figure 54.2).

54.2.2.2 Acellular Dermal Matrix

Acellular dermal matrixes have been developed for the treatment of a range of conditions in which an intact dermal schapolding is required. Acellular dermal matrixes are typically freeze-dried on an intact basement membrane obtained from donor skin. The donor skin is typically either of human or porcine origin. The dermal matrix is treated to remove all immunogenic materials including sweat glands, fibroblasts, smooth muscle, and vascular endothelium [48]. The process is completed with attention paid towards preserving the collagen and elastin fibers. Maintaining freedom from disease transmission is of the utmost importance. AlloDerm® (LifeCell, Branchburg, NJ), from donated human skin, is one of the most extensively studied acellular dermal matrixes and has been used either alone or with cultured autografts to successfully treat burn wounds and dermal defects [49,50].

When used in the treatment of burn reconstruction, AlloDerm is most typically meshed and placed directly on the wound bed. An ultrathin split thickness skin graft (0.004−0.008 in.) is then placed on the AlloDerm as a one-stage procedure [51]. A multicenter study demonstrated improved take (81.1% vs. 65.4%) of the AlloDerm when meshed rather than applied as a single solid sheet [52], and most commonly the AlloDerm is meshed at a 1:1 ratio, which allows plasmatic imbibition to support the overlying skin graft until neovascularization is able to occur. The graft is typically not expanded, however, as doing so could compromise the uniform dermal base on which the skin graft is applied. Immobilization of the skin graft is maintained for a minimum of 5 days and more commonly for 7−10 days, longer than typically done for a standard skin graft [49]. Outcomes were equivalent with allograft and ultrathin split thickness skin graft compared to thicker split thickness skin graft in the treatment of full thickness and deep partial thickness burns [52]. Application of ultrathin split thickness skin grafts allows for faster healing of the donor site.

AlloDerm has been used extensively in burn reconstruction with excellent outcomes. Calcutt et al. used AlloDerm along with thin split thickness skin graft (0.004−0.008 in.) in 21 patients with acute burns and six patients with traumatic skin lost. Successful take was observed in 26/27 patients and scar evaluation was noted to be favorable [51]. AlloDerm was used in the treatment of three patients with full thickness burns to the extremity. Ultrathin skin grafts were applied on meshed AlloDerm to the hand in two cases and the dorsum of the foot in one case. Functional and aesthetic outcomes were determined to be good to excellent [53]. Use of AlloDerm for burn reconstruction over joints has also been shown to have favorable outcomes. In a study of 31 patients with large burns (average 38.5% TBSA) >90% improvement of joint function was achieved using AlloDerm and skin grafting in patients with burns crossing the joints. Furthermore, the authors noted improved scar characteristics (elasticity, thickness, transepidermal water loss) [54]. Evaluation of scar assessment and skin quality after AlloDerm placement and ultrathin skin graft suggests that results are similar to those observed after standard skin grafts [52]. A study of two patients followed out 6 years after AlloDerm placement and split thickness skin graft demonstrated significantly improved viscoelastic properties compared to conventional split thickness skin grafts [55].

AlloDerm has also been used in the treatment of burn contractures. In a study of 38 patients with burn contractures at 72 sites, AlloDerm was used for contracture release with an improvement of in range of motion [56]. Motykie et al. demonstrated that the use of AlloDerm for the reconstruction of scar contractures in joints of the lower extremity [57] and the hand [58] resulted in less need for skin regrafting, fewer additional reconstructive surgeries and significantly improved range of motion compared to autografts alone. A two-stage approach for the treatment of burn contracture in the hand and wrist has been proposed. Nine patients with burn contracture underwent full thickness primary burn contracture release along with tenolysis and capsulotomy to ensure full mobility of each joint. AlloDerm was placed on the wound bed and maintained in position using a vacuum-assisted closure device. After 10–14 days, a split thickness skin graft at 0.012-inch thickness was applied onto the AlloDerm base. An 83% gain in passive range of motion at each joint and an 89% gain in length of each webspace were obtained at 1-year follow-up [59]. Initiation of physical therapy in the early postoperative period is essential in all burn contracture releases to maintain joint mobility and prevent contracture recurrence regardless of the type of reconstruction performed (e.g., full thickness skin graft, AlloDerm, and split thickness skin graft).

AlloDerm has also been used successfully in many different clinical applications, including breast reconstruction [60] and augmentation [61], cleft palate repair [62], abdominal wall reconstruction [63,64], and urethral reconstruction [65] amongst others and will continue to remain an early choice when complex reconstruction is required and native tissues are lacking.

54.2.3 Combination Products

54.2.3.1 Collagen-GAG Scaffolds Combined with Cells

Combining cells with scaffolds dates back to the 1980s, when keratinocytes were centrifuged into collagen-GAG scaffolds [66]. In this case, many cells did not go totally through the matrix but became entangled within the pore

structure. Interestingly, they began to divide, proliferate, and migrate towards the surface to produce a confluent epidermis in 10–14 days depending on the cell number. Cystic structures would sometimes form, migrate to the surface, fuse with the epidermis and then eventually extrude. When confluence occurred, the silicone elastomer would spontaneously separate from the underlying scaffold. Adding cells to the matrix also increased the rate that cells and blood vessels would enter the scaffold from the wound surface. In preclinical models, long-term contraction was reduced compared to both unseeded scaffolds and open wounds, but less than the contraction seen with full-thickness skin grafts. Histologic evaluation of the neodermis showed a structure that differed from scar tissue but lacked hair follicles or adnexal glands. Collagen morphology appeared wavier and with more of a random orientation compared to highly compacted and oriented collagen seen in open wound controls.

Wood has pioneered a cell spraying technology, that when used with widely meshed CG scaffolds results in faster epithelialization [67]. Melanocytes also appear to persist using the spray technology [68].

54.2.3.2 Collagen-GAG Scaffolds Combined with Suction

In order to increase the rate of cellular ingrowth into the scaffolds, these devices have been combined with microderformational therapy [69]. These devices provide a distributed suction to the scaffolds using highly porous polyurethane foam as an interface material. Early results in humans suggest that combining these two technologies increases the rate of ingrowth [70].

54.2.3.3 Collagen-GAG Scaffolds Combined with Cultured Keratinocytes

Vascularized collagen-GAG scaffolds can also be combined with keratinocytes in culture and have been used together both experimentally and in the clinic [71].

54.3 FUTURE

Ideally, replacement skin would perfectly replace what was lost. It would be identical in skin color, texture, and adnexal gland as well as have similar mechanical properties and mass transport properties. Current versions of CG scaffolds can provide a reasonable replacement of the dermis and often the epidermis as well. In many cases, the color match is also excellent (Figure 54.3). Hair follicles can be added to the scaffolds a few days after application and they not only provide hair growth but also are a source of keratinocytes that migrate and divide at the junction of the scaffold and silicone layer [72]. Better

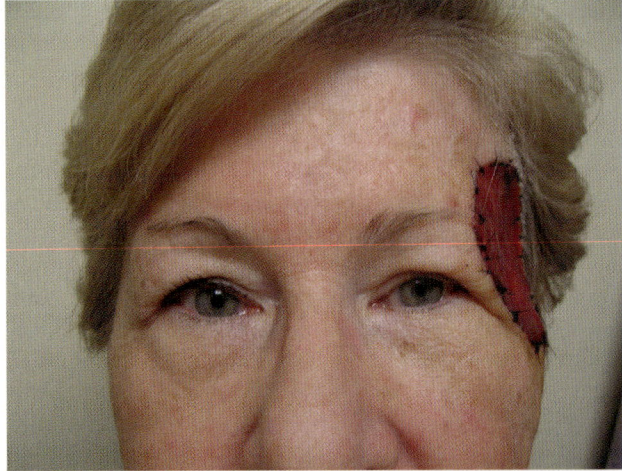

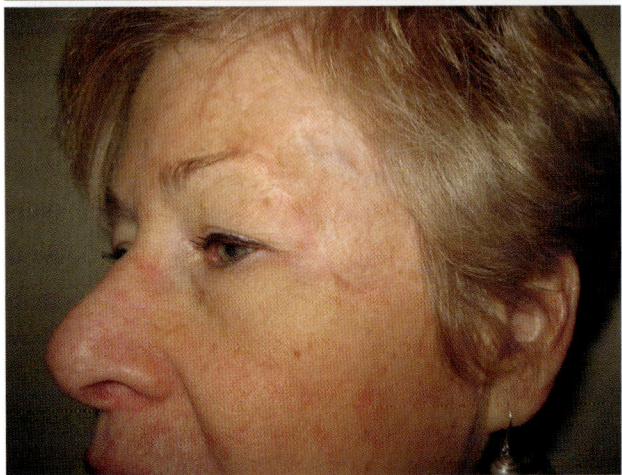

FIGURE 54.3 A 65-year-old woman with melanoma of the left temple was treated by resection and application of collagen-GAG scaffold. Once the silicone was removed, a thick skin graft was taken from the upper arm. At 1 year, there is little visible scar. *Originally appeared in [41].*

methods to regenerate hair follicles and adnexal organs as well as better methods to control skin color would be desirable. Finally, long-term function of skin including a natural aging process would be desired.

54.4 CONCLUSIONS

Advances in biodegradable scaffold design and in cell culture methodologies have added several new technologies that can be used to treat skin loss. These technologies, as currently developed treat thousands of patients each year with burns, trauma, malignancy, or infections. As these products and methodologies improve, we can look forward to a new era when areas of skin loss can be precisely treated to match the structural, microarchitectural, and visual characteristics of normal skin specific to the area of skin loss.

REFERENCES

[1] Dunkin CS, Pleat JM, Gillespie PH, Tyler MP, Roberts AH, McGrouther DA. Scarring occurs at a critical depth of skin injury: precise measurement in a graduated dermal scratch in human volunteers. Plast Reconstr Surg 2007;119:1722–32.

[2] Paterno J, Vial IN, Wong VW, Rustad KC, Sorkin M, Shi Y, et al. Akt-mediated mechanotransduction in murine fibroblasts during hypertrophic scar formation. Wound Repair Regen 2011;19:49–58.

[3] Rheinwald JG, Green H. Serial cultivation of strains of human epidermal keratinocytes: the formation of keratinizing colonies from single cells. Cell 1975;6:331–43.

[4] Ehrenreich M, Ruszczak Z. Update on tissue-engineered biological dressings. Tissue Eng 2006;12:2407–24.

[5] Wright KA, Nadire KB, Busto P, Tubo R, McPherson JM, Wentworth BM. Alternative delivery of keratinocytes using a polyurethane membrane and the implications for its use in the treatment of full-thickness burn injury. Burns 1998;24:7–17.

[6] Jones I, Currie L, Martin R. A guide to biological skin substitutes. Br J Plast Surg 2002;55:185–93.

[7] Compton CC. Current concepts in pediatric burn care: the biology of cultured epithelial autografts: an eight-year study in pediatric burn patients. Eur J Pediatr Surg 1992;2:216–22.

[8] Epicel Product Website. Genzyme, 2013. (Accessed 6/26/2013, 2013, at http://www.epicel.com/.)

[9] Munster AM. Cultured skin for massive burns. A prospective, controlled trial. Ann Surg 1996;224:372–5.

[10] Haith Jr. LR, Patton ML, Goldman WT. Cultured epidermal autograft and the treatment of the massive burn injury. J Burn Care Rehabil 1992;13:142–6.

[11] Carsin H, Ainaud P, Le Bever H, Rives J, Lakhel A, Stephanazzi J, et al. Cultured epithelial autografts in extensive burn coverage of severely traumatized patients: a five year single-center experience with 30 patients. Burns 2000;26:379–87.

[12] Wood FM, Kolybaba ML, Allen P. The use of cultured epithelial autograft in the treatment of major burn wounds: eleven years of clinical experience. Burns 2006;32:538–44.

[13] Williamson JS, Snelling CF, Clugston P, Macdonald IB, Germann E. Cultured epithelial autograft: five years of clinical experience with twenty-eight patients. J Trauma 1995;39:309–19.

[14] Rue 3rd LW, Cioffi WG, McManus WF, Pruitt Jr. BA. Wound closure and outcome in extensively burned patients treated with cultured autologous keratinocytes. J Trauma 1993;34:662–7.

[15] De Luca M, Albanese E, Cancedda R, Viacava A, Faggioni A, Zambruno G, et al. Treatment of leg ulcers with cryopreserved allogeneic cultured epithelium. A multicenter study. Arch Dermatol 1992;128:633–8.

[16] Leigh IM, Purkis PE, Navsaria HA, Phillips TJ. Treatment of chronic venous ulcers with sheets of cultured allogenic keratinocytes. Br J Dermatol 1987;117:591–7.

[17] Paquet P, Quatresooz P, Braham C, Pierard GE. Tapping into the influence of keratinocyte allografts and biocenosis on healing of chronic leg ulcers: split-ulcer controlled pilot study. Dermatol Surg 2005;31:431–5.

[18] Wood FM, Stoner M. Implication of basement membrane development on the underlying scar in partial-thickness burn injury. Burns 1996;22:459–62.

[19] Chester DL, Balderson DS, Papini RP. A review of keratinocyte delivery to the wound bed. J Burn Care Rehabil 2004;25:266–75.

[20] Compton CC, Gill JM, Bradford DA, Regauer S, Gallico GG, O'Connor NE. Skin regenerated from cultured epithelial autografts on full-thickness burn wounds from 6 days to 5 years after grafting. A light, electron microscopic and immunohistochemical study. Lab Invest 1989;60:600—12.

[21] Horch RE, Kopp J, Kneser U, Beier J, Bach AD. Tissue engineering of cultured skin substitutes. J Cell Mol Med 2005;9:592—608.

[22] Lam PK, Chan ES, To EW, Lau CH, Yen SC, King WW. Development and evaluation of a new composite Laserskin graft. J Trauma 1999;47:918—22.

[23] Lam PK, Chan ES, Liew CT, Lau C, Yen SC, King WW. Combination of a new composite biocampatible skin graft on the neodermis of artificial skin in an animal model. ANZ J Surg 2002;72:360—3.

[24] Cuono CB, Langdon R, Birchall N, Barttelbort S, McGuire J. Composite autologous-allogeneic skin replacement: development and clinical application. Plast Reconstr Surg 1987;80:626—37.

[25] Hickerson WL, Compton C, Fletchall S, Smith LR. Cultured epidermal autografts and allodermis combination for permanent burn wound coverage. Burns 1994;20(Suppl. 1):52—5.

[26] Krupp S, Benathan M, Meuli M, Deglise B, Holzer E, Wiesner L, et al. Current concepts in pediatric burn care: management of burn wounds with cultured epidermal autografts. Eur J Pediatr Surg 1992;2:210—5.

[27] Braye F, Oddou L, Bertin-Maghit M, Belgacem S, Damour O, Spitalier P, et al. Widely meshed autograft associated with cultured autologous epithelium for the treatment of major burns in children: report of 12 cases. Eur J Pediatr Surg 2000;10:35—40.

[28] Falanga V, Sabolinski M. A bilayered living skin construct (APLIGRAF) accelerates complete closure of hard-to-heal venous ulcers. Wound Repair Regen 1999;7:201—7.

[29] Veves A, Falanga V, Armstrong DG, Sabolinski ML. Graftskin, a human skin equivalent, is effective in the management of noninfected neuropathic diabetic foot ulcers: a prospective randomized multicenter clinical trial. Diabetes Care 2001;24:290—5.

[30] Brem H, Balledux J, Bloom T, Kerstein MD, Hollier L. Healing of diabetic foot ulcers and pressure ulcers with human skin equivalent: a new paradigm in wound healing. Arch Surg 2000;135:627—34.

[31] Waymack P, Duff RG, Sabolinski M. The effect of a tissue engineered bilayered living skin analog, over meshed split-thickness autografts on the healing of excised burn wounds. The Apligraf burn study group. Burns 2000;26:609—19.

[32] Gohari S, Gambla C, Healey M, Spaulding G, Gordon KB, Swan J, et al. Evaluation of tissue-engineered skin (human skin substitute) and secondary intention healing in the treatment of full thickness wounds after Mohs micrographic or excisional surgery. Dermatol Surg 2002;28:1107—14.

[33] Falabella AF, Schachner LA, Valencia IC, Eaglstein WH. The use of tissue-engineered skin (Apligraf) to treat a newborn with epidermolysis bullosa. Arch Dermatol 1999;135:1219—22.

[34] Falabella AF, Valencia IC, Eaglstein WH, Schachner LA. Tissue-engineered skin (Apligraf) in the healing of patients with epidermolysis bullosa wounds. Arch Dermatol 2000;136:1225—30.

[35] Fivenson DP, Scherschun L, Cohen LV. Apligraf in the treatment of severe mitten deformity associated with recessive dystrophic epidermolysis bullosa. Plast Reconstr Surg 2003;112:584—8.

[36] Jiang QJ, Izakovic J, Zenker M, Fartasch M, Meneguzzi G, Rascher W, et al. Treatment of two patients with Herlitz junctional epidermolysis bullosa with artificial skin bioequivalents. J Pediatr 2002;141:553—9.

[37] Martin LK, Kirsner RS. Ulcers caused by bullous morphea treated with tissue-engineered skin. Int J Dermatol 2003;42:402—4.

[38] Pennoyer JW, Susser WS, Chapman MS. Ulcers associated with polyarteritis nodosa treated with bioengineered human skin equivalent (Apligraf). J Am Acad Dermatol 2002;46:145.

[39] de Imus G, Golomb C, Wilkel C, Tsoukas M, Nowak M, Falanga V. Accelerated healing of pyoderma gangrenosum treated with bioengineered skin and concomitant immunosuppression. J Am Acad Dermatol 2001;44:61—6.

[40] Rainey JK, Wen CK, Goh MC. Hierarchical assembly and the onset of banding in fibrous long spacing collagen revealed by atomic force microscopy. Matrix Biol 2002;21:647—60.

[41] Yannas IV, Orgill DP, Burke JF. Template for skin regeneration. Plast Reconstr Surg 2011;127(Suppl. 1):60S—70S.

[42] Glowacki J, Mizuno S. Collagen scaffolds for tissue engineering. Biopolymers 2008;89:338—44.

[43] Yannas IV, Lee E, Orgill DP, Skrabut EM, Murphy GF. Synthesis and characterization of a model extracellular matrix that induces partial regeneration of adult mammalian skin. Proc Natl Acad Sci USA 1989;86:933—7.

[44] Burke JF, Yannas IV, Quinby Jr. WC, Bondoc CC, Jung WK. Successful use of a physiologically acceptable artificial skin in the treatment of extensive burn injury. Ann Surg 1981;194:413—28.

[45] Heimbach D, Luterman A, Burke J, Cram A, Herndon D, Hunt J, et al. Artificial dermis for major burns. A multi-center randomized clinical trial. Ann Surg 1988;208:313—20.

[46] Figus A, Leon-Villapalos J, Philp B, Dziewulski P. Severe multiple extensive postburn contractures: a simultaneous approach with total scar tissue excision and resurfacing with dermal regeneration template. J Burn Care Res 2007;28:913—7.

[47] Stiefel D, Schiestl C, Meuli M. Integra artificial skin for burn scar revision in adolescents and children. Burns 2010;36:114—20.

[48] Brusselaers N, Pirayesh A, Hoeksema H, Richters CD, Verbelen J, Beele H, et al. Skin replacement in burn wounds. J Trauma 2010;68:490—501.

[49] Wainwright DJ, Bury SB. Acellular dermal matrix in the management of the burn patient. Aesthet Surg J 2011;31:13—23.

[50] Kinsella Jr. CR, Grunwaldt LJ, Cooper GM, Mills MC, Losee JE. Scalp reconstruction: regeneration with acellular dermal matrix. J Craniofac Surg 2010;21:605—7.

[51] Callcut RA, Schurr MJ, Sloan M, Faucher LD. Clinical experience with Alloderm: a one-staged composite dermal/epidermal replacement utilizing processed cadaver dermis and thin autografts. Burns 2006;32:583—8.

[52] Wainwright D, Madden M, Luterman A, Hunt J, Monafo W, Heimbach D, et al. Clinical evaluation of an acellular allograft dermal matrix in full-thickness burns. J Burn Care Rehabil 1996;17:124—36.

[53] Lattari V, Jones LM, Varcelotti JR, Latenser BA, Sherman HF, Barrette RR. The use of a permanent dermal allograft in full-thickness burns of the hand and foot: a report of three cases. J Burn Care Rehabil 1997;18:147—155.

[54] Yim H, Cho YS, Seo CH, Lee BC, Ko JH, Kim D, et al. The use of AlloDerm on major burn patients: AlloDerm prevents post-burn joint contracture. Burns 2010;36:322—8.

[55] Sin P, Brychta P. Cutometrical measurement confirms the efficacy of the composite skin grafting using allogeneic acellular dermis in burns. Acta Chir Plast 2006;48:59—64.

[56] Chaudhari SRD, Zieger M, Sood CR. Use of AlloDerm to prevent recontracture following burn scar contracture release. J Burn Care Res 2007;28:85.

[57] Motykie GDWS, McCauley RL, Herndon DN. Reconstruction of the burned lower extremity utilizing AlloDerm underlay. J Burn Care Rehabil 2003;24:131.

[58] Motykie GD, Wolf SE, McCauley RL, Herndon, DN. Reconstruction of the burned hand utilizing AlloDerm underlay. J Burn Care Rehabil 2003;25:165.

[59] Askari M, Cohen MJ, Grossman PH, Kulber DA. The use of acellular dermal matrix in release of burn contracture scars in the hand. Plast Reconstr Surg 2011;127:1593−9.

[60] Brooke S, Mesa J, Uluer M, Michelotti B, Moyer K, Neves RI, et al. Complications in tissue expander breast reconstruction: a comparison of AlloDerm, DermaMatrix, and FlexHD acellular inferior pole dermal slings. Ann Plast Surg 2012;69:347−9.

[61] Namnoum JD, Moyer HR. The role of acellular dermal matrix in the treatment of capsular contracture. Clin Plast Surg 2012;39:127−36.

[62] Aldekhayel SA, Sinno H, Gilardino MS. Acellular dermal matrix in cleft palate repair: an evidence-based review. Plast Reconstr Surg 2012;130:177−82.

[63] Zhong T, Janis JE, Ahmad J, Hofer SO. Outcomes after abdominal wall reconstruction using acellular dermal matrix: a systematic review. J Plast Reconstr Aesthet Surg 2011;64:1562−71.

[64] Chivukula KK, Hollands C. Human acellular dermal matrix for neonates with complex abdominal wall defects: short- and long-term outcomes. Am Surg 2012;78:346−8.

[65] Carpenter CP, Daniali LN, Shah NP, Granick M, Jordan ML. Distal urethral reconstruction with AlloDerm: a case report and review of the literature. Can J Urol 2012;19:6207−10.

[66] Yannas IV, Burke JF, Orgill DP, Skrabut EM. Wound tissue can utilize a polymeric template to synthesize a functional extension of skin. Science 1982;215:174−6.

[67] Navarro FA, Stoner ML, Park CS, Huertas JC, Lee HB, Wood FM, et al. Sprayed keratinocyte suspensions accelerate epidermal coverage in a porcine microwound model. J Burn Care Rehabil 2000;21:513−8.

[68] Navarro FA, Stoner ML, Lee HB, Park CS, Wood FM, Orgill DP. Melanocyte repopulation in full-thickness wounds using a cell spray apparatus. J Burn Care Rehabil 2001;22:41−6.

[69] Orgill DP, Bayer L, Neuwalder J, Felter R. Microdeformational wound therapy—a new era in wound healing. Bus Brief: Glob Surg Future Dir 2005;1−3.

[70] Molnar JA, DeFranzo AJ, Hadaegh A, Morykwas MJ, Shen P, Argenta LC. Acceleration of Integra incorporation in complex tissue defects with subatmospheric pressure. Plast Reconstr Surg 2004;113:1339−46.

[71] Wood FM, Stoner ML, Fowler BV, Fear MW. The use of a non-cultured autologous cell suspension and Integra dermal regeneration template to repair full-thickness skin wounds in a porcine model: a one-step process. Burns 2007;33:693−700.

[72] Navsaria HA, Ojeh NO, Moiemen N, Griffiths MA, Frame JD. Reepithelialization of a full-thickness burn from stem cells of hair follicles micrografted into a tissue-engineered dermal template (Integra). Plast Reconstr Surg 2004;113:978−81.

Chapter 55

The Use of Skin Substitutes in the Treatment of Burns

Reema Chawla[a], Amelia Seifalian[a], Naiem S. Moiemen[b], Peter E. Butler[a,c], and Alexander M. Seifalian[a,c]

[a]UCL Centre for Nanotechnology and Regenerative Medicine, Division of Surgery and Interventional Science, London, United Kingdom, [b]Department of Burns and Plastic Surgery, Queen Elizabeth Hospital, University Hospitals Birmingham NHS Foundation Trust, Edgbaston, Birmingham, United Kingdom, [c]Department of Plastic and Reconstructive Surgery, Royal Free Hampstead NHS Trust Hospital, London, United Kingdom

55.1 INTRODUCTION

Burn injuries are a significant cause of mortality and morbidity and it is important to manage a patient that has been admitted with severe burns imminently. This is especially so if a particularly large percentage total body surface area (%TBSA) has been affected. Severe burns are notoriously complicated by massive fluid loss leading to a complicated disease state, characterized by high cardiac output, increased oxygen consumption, and fat and protein wasting. This hypermetabolic condition causes the immune system to become compromised and leads to attenuated wound healing [1]. Other complications arise from coexistent hypothermia and infection which both lead to impaired blood flow and tissue necrosis, which further aggravate morbidity [2,3]. Despite advances in the management of burns, mortality remains high. Skin substitutes are the mainstay of treatment involved in controlling fluid loss. The idea of covering the exposed burnt tissue to prevent and control fluid and heat loss is the mainstay of treatment; skin substitutes are the main method in which this is achieved [4]. The pursuit of an economically viable, nonimmunogenic, topical measure in order to control heat and fluid loss and prevent infection remains [5,6]. In addition, a gold standard method of coverage would also be easy to handle and apply, adhere well, reduce pain and be biodegradable, so as to integrate well with the patient's own epithelium [7].

55.2 AIM AND METHOD

The aim of this chapter is to review skin substitutes that are used primarily in the treatment of burn injuries. A Pubmed search was performed with the MeSH term "Artificial Skin" and "Skin Substitutes" concentrating on papers from 2005–2012.

55.3 TREATMENT OF BURNS

In addition to managing the complex fluid status of the patient with deep-partial and full-thickness burns, affected tissue is promptly excised in order to prepare the wound bed without delay to avoid consequences of poor healing [1,3]; this allows the application of a topical measure in order to control fluid and heat loss and allow reepithelialization to occur [8]. Subsequent treatment should facilitate a return to a state close to the patients' original

Regenerative Medicine Applications in Organ Transplantation.

appearance and function and can be split up into three principles depending on individual patient needs [9]:

1. Mimicking the normal wound healing environment using dressings.
2. Replicating the chemical environment of the wound using chemokines and growth factors.
3. Replacing the damaged skin, utilizing skin substitutes.

Following excision, continuity should be promptly restored; topical antimicrobial therapy followed by either temporary or permanent coverage of the wound achieves this [10]. Historically, autografts from healthy, unaffected skin have been the mainstay of treatment [1].

55.3.1 Autografts

The aim of using split-thickness skin grafts is to transfer residual keratinocyte stem cells that aid with regeneration and faster wound epithelialization; however, the meshing technique used to increase the surface area leads to the characteristic diamond-like scar [1,3,11]. Aesthetically, this is not ideal and may lead to the formation of keloids, hypertrophic scarring, and contractures due to the lack of dermal support and the fact that the body cannot replicate the distinct collagen pattern to lay down [11].

Results obtained with full-thickness skin grafts are much better aesthetically; however, their technical difficulties in combination with greater donor site morbidity can lead to severe complications [3]. In addition, full-thickness skin grafts by their very nature can only be restricted in their size as they can only be harvested from a few areas of the body (e.g., groin, thigh, lower abdomen) and therefore can only be performed for burned skin that has less than 2% TBSA [11].

The advantages of autografts are that they are immunologically compatible with the patient. Although these types of graft can often provide an average functional and cosmetic appearance, they are not ideal, especially as they increase the %TBSA affected and cause an additional insult to the already damaged skin leading to further scarring. Often use of autologous skin grafts can lead to the formation of contractures, requiring further surgery [1,3]. It has the added disadvantage that severely burned patients, who have a high percentage total burns surface area (%TBSA), will not have enough healthy skin left to be able to perform this procedure [11]. It is for this reason that alternative methods of substituting the skin have been explored over the previous three decades, many of which have been successful in increasing the odds of survival in burns patients [1,9].

55.3.2 Skin Substitutes

The ideal skin substitute should be a safe, easy to apply, topical material that promotes healing and leads to good aesthetic and functional recovery (i.e., it should have a barrier function that is similar to the stratum corneum in skin) [12−14]. It should have good mechanophysical properties in order to protect underlying tissues, control any burn wound infection, prevent heat and fluid loss, be nonimmunogenic, elastic, and adhere to the wound well [15]. An ideal substitute would provide analgesia, or minimize the pain and discomfort associated with burns treatment caused by changing dressings, etc. In addition, it should reduce length of rehabilitation and therefore hospital stay with minimal operative procedures required to produce an aesthetically and functionally viable result with decreased risk of contractures, stiffness, and scar hypertrophy [1]. It is imperative that any bioengineered matrices are not only chemically and environmentally stable but also economical [15].

There are several different types of skin substitute available. Skin substitutes can be either temporary or permanent; composed of natural or synthetic materials. Temporary skin substitutes aim to fulfill the first and sometimes second principle of burns treatment mentioned above; they can be biological or synthetic and are commonly draped onto the freshly excised wound bed in the interim, while waiting for a permanent skin substitute [2]. Permanent skin substitutes are surgically fixated and become incorporated into the patients skin, most often they are a combination of synthetic and biological materials [3]. The quality of the result is often better than that seen in a split-thickness skin graft [2].

The early types of skin substitute to be used are naturally occurring tissues that are derived directly from human or porcine sources and have not been further engineered or treated with the exception of disinfection and preservation. Their allogeneic and xenogeneic sources render them temporary skin substitutes and are either removed when healing is complete in a superficial partial-thickness burn or used as an intermediate before the permanent skin substitute is applied in a deep-partial or full-thickness burn [2].

55.3.3 Temporary Skin Substitutes

55.3.3.1 Cadaveric Skin

Historically, cadaveric skin has been very useful as a temporary skin substitute for burns. It contains cytokines and growth factors which provide an appropriate healing environment [3]. It is used when there is not enough donor site skin for autologous grafting or in order to cover split-thickness meshed grafts. Cadaveric skin can be used fresh or from a glycerol-preserved state; both show comparable metabolic activity of up to 95% [1,16]. Glycerol dehydrates the skin and prevents oxidative and hydrolytic reactions which are detrimental to the cells [8,17]. It also minimizes microbial growth and antigenicity of the skin and allows for easy handling [8]. However, there are disadvantages to using cadaveric skin, including low supply, substandard quality, infection and immune rejection, despite glycerolized cadaveric skin slowing onset of rejection by up to 5 weeks [16]. There are several methods to try and prevent rejection, none of which show any signs of promise. Immunosuppressants are very efficacious, but come with a myriad of side effects. Removal of the antigenic epidermal cells is technically challenging and has not been performed successfully to date [16]. To prevent infection, all cadaveric skin is pretreated with antibiotics; despite this, certain bacteria still persist. Additionally, there is also a worry that cadaveric skin may carry viruses or prion diseases [1,3].

Recent developments have led to advancements in decellularizing and sterilizing human allogeneic cadaveric skin [18]. This method eliminates any immune rejection that the patient may have faced in a conventional cadaveric skin substitute. The patient's own autologous keratinocytes and fibroblasts can then repopulate the scaffold in the laboratory before the skin can be transferred back to the patient [19]. Unfortunately, this method of preparing the cadaveric skin takes up to 4 weeks, meaning that the delay in preparation can lead to compromised wound healing and in the interim, another skin substitute is required. In addition, it does not eliminate the risks of viral transmission, despite having been sterilized [14,20,21].

55.3.3.2 Amniotic Membrane

Amniotic membrane is another type of biological wound dressing that has been successfully used worldwide [5]. The avascular nature of amnion and its thick basement membrane containing collagen, laminin, and proteinase inhibitors aid healing of burn wounds by adhering to principles one and two previously mentioned [22]. In some way, allogeneic amnion has been more successfully used internationally than human cadaveric skin at preserving the excised wound bed of a burn [5,22]. Its complex structure minimizes fluid and electrolyte loss from the wound bed and simultaneously reduces infection [8]. Amnion is extremely thin, which can be considered both an advantage and a disadvantage. It dresses the wound with a thin translucent layer, allowing the clinician to maintain the status of the wound bed underneath and additionally maintain an aerobic environment to encourage healing [8]. This is extremely advantageous especially when compared to the bulky dressings previously used. However, its fragility also means that it is extremely difficult to process, preserve, and handle. In addition, amnion does not need to be changed as frequently as dressings do, due to the fact that it acts as a barrier to

pathogens; therefore it minimizes pain associated with this [1].

One of the main aims in successful burns treatment is for rapid epithelialization to occur; amnion successfully achieves this in partial thickness burns [22]. However, in full-thickness burns, the amnion disintegrates before this is achieved. Although amnion can prevent bacterial penetration, it is often the case that bacteria can grow between the wound bed and the amniotic membrane [22]. Again there is also a risk that the amniotic membrane can be contaminated with bacteria or viruses especially when retrieved from vaginal deliveries [1].

Silver ions have been found to have strong antimicrobial properties due to their ions interacting with microbial respiratory enzymes causing instant death [2]. Incorporating 0.5% silver nitrate has been found to have a greater antimicrobial effect and prevents these sorts of complications [1]. However one of the main disadvantages of using silver nitrate is that it may inhibit epithelial and fibroblast proliferation which could potentially lead to impaired wound healing [2,23].

55.3.3.3 Xenografts

Xenografts are economical and widely available; fresh disinfected porcine dermal grafts are used as they are similar to human cadaveric skin in the result that they give the patient [9]. However, the antigenicity of these grafts is significantly higher than allogeneic human cadaveric grafts and rejection is commonly seen. Although in previous years, dermis from other mammals and reptiles has been utilized [1].

Similar to the developments in human cadaveric allogeneic skin, porcine dermis can also be decellularized to produce an acellular dermal matrix for use as a skin substitute, which will be discussed later on [1,24].

Porcine small intestinal submucosa is an acellular material and is similar in appearance and function to amniotic membrane. It provides a scaffold that can deliver a good wound healing response. Several companies have utilized this and have developed commercialized versions, such as Oasis™ (Healthpoint) and Fortaflex™ [9,25].

55.3.4 Permanent Skin Substitutes

55.3.4.1 Engineered Skin Substitutes

The remaining skin substitutes are modified and engineered specifically from either human, porcine, bovine, murine, or synthetic origins. In addition to the qualities of an ideal skin substitute described previously, ultimately these matrices should be able to be rapidly produced in large quantities and act as a complete autologous dermo-epidermal replacement [11].

Successful tissue engineered skin substitutes aim to emulate the microanatomy and physiology of human skin which can allow better integration with minimal scar formation [26]. The similarity to normal morphological properties of skin helps manifest epithelial—mesenchymal interactions which are key for good quality wound healing [1,11,27]. The scaffold should aid in promoting rapid vascularization and remodeling [11,20].

All tissue engineered skin substitutes consist of a basic porous matrix, which once applied to the wound bed allow cellular infiltration and extracellular matrix (ECM) deposition allowing reepithelialization to occur [3]. Engineered skin substitutes can contain additional cells or growth factors to exploit and encourage the body's capacity for wound healing [9]. Cells are obtained from a combination of autologous, allogeneic, or xenogeneic sources [28]. If the permanent substitutes contain autologous cells, they can take up to 4 weeks to grow after harvesting from patient biopsy, so in the interim, temporary skin substitutes have to be used [11].

Most engineered skin substitutes are intended for permanent use where the scaffold is biodegradable in order to permanently support a regenerating epidermis [11,20].

There are different types of skin substitute depending on the depth of the burn and the needs of the individual patient. There are epidermal substitutes, dermal substitutes, and bilayered substitutes, which are a combination of the two and are successfully used in full-thickness burns.

These specially engineered permanent skin substitutes can also treat chronic wounds, such as venous or diabetic ulcers, although it has to be taken into consideration that the wound in these situations has a considerably different pathology to the freshly excised skin in burns [29,30]. Chronic wounds often have an inflammatory pathology which means that there is an array of cytokines that the wound is exposed to [9,31].

Epidermal Substitutes

Epidermal substitutes aim to speed up the process by which reepithelialization occurs on a wound bed that includes a dermal layer; in this situation, it is important to provide a cellular-based substitute that gives the wound bed a smooth surface for scarless reepithelialization [11]. Keratinocytes are the predominant cell type seeded into the matrix of epidermal substitutes, mimicking the usual physiological environment [3,9]. A small biopsy is taken from healthy skin in which keratinocytes, melanocytes, fibroblasts, and endothelial cells can be harvested [11]. Keratinocytes are cultured with a layer of irradiated 3T3 murine fibroblasts which act as feeder layer cells in a

medium containing fetal calf serum (FCS) [32]. This method produces a cultured epithelial autograft (CEA) within 3—4 weeks [1,3,11]. However, these xenogeneic materials used in the culture may be a potential risk for the patient due to zoonoses and other potential hazards from using animal materials in human implants [32].

CEAs avoid the meshing that occurs in autografts, therefore reducing the abnormal scarring. Their use also decreases the discomfort involved in obtaining an autograft. Most importantly, an extremely large piece of tissue can be obtained from a small biopsy [1]. However, CEAs take up to 4 weeks to prepare and have a short shelf life. They are also difficult to handle due to their fragility [11]. Another major disadvantage is that the rate of engraftment is variable, due to the process of enzymatic detachment from the fibroblast feeder layer [1]. With the high production costs involved and no distinct method for quality control, their use may not be justified [11] (Figure 55.1).

Keratinocyte culture delivery systems have been developed to overcome these problems. This involves combining the keratinocytes with natural or synthetic carrier materials, such as petrolatum gauze, silicone membrane, polyurethane membrane, hyaluronic acid (HA) based membranes, collagen sponges, and fibrin glue [3]. Examples of commercially available CEAs are Epicel®, Epidex™, Myskin™, Laserskin™ (Vivoderm™), BioSeed-S™, CellSpray™, Cryoskin™ (CryoCeal™), Transcell, and ReCell® (Table 55.1).

Epicel® is an example of a conventional CEA. Confluent sheets of autologous keratinocytes are cultured in bovine serum with proliferation arrested murine 3T3

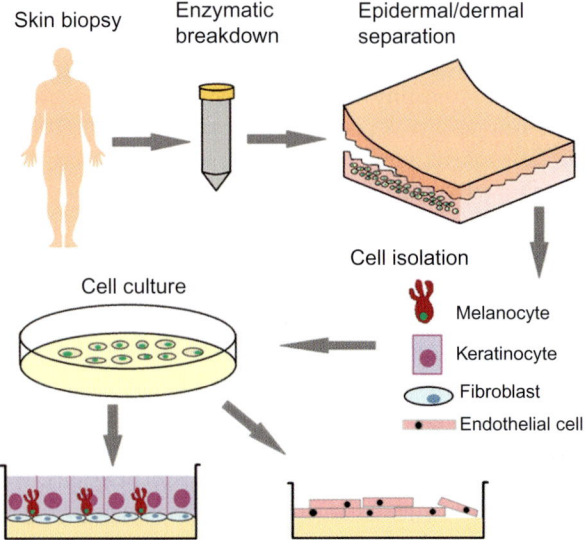

FIGURE 55.1 This image shows how keratinocytes are cultured from an initial skin biopsy into a confluent sheet of cells, used as a CEA to place on top of dermal substitutes. *Revised from [11].*

TABLE 55.1 All the Different Types of Skin Substitutes Commercially Available and Their Composition and Applications

Product	Company	Price (US $/cm²)	Source of Materials	Type of Cells Used	Temporary/ Permanent	Grown in	Application
Epidermal substitutes							
Epicel	Genzyme Corp	—	Human autologous and murine fibroblasts	Confluent keratinocytes from hair follicles grown in proliferation-arrested fibroblasts	Permanent	Petrolatum gauze backing	Full, partial-thickness burns and chronic ulcers
Epidex	Euroderm GmbH	—	Human autologous	Confluent keratinocytes from the outer root sheath of hair follicles	Permanent	Silicone membrane	Full, partial-thickness burns and chronic ulcers
Laserskin (VivodermTM)	(Fidia Advanced Polymers)	—	Human autologous	Subconfluent Keratinocytes	Permanent	Esterified laser-perforated hyaluronic acid matrix	Full, partial-thickness burns and chronic ulcers
BioSeed-S	BioTissue Technologies AG	—	Human autologous	Oral mucosal cells	Permanent	Fibrin matrix	Partial-thickness burns and chronic ulcers
Myskin	Celltran Ltd.	—	Human autologous	Subconfluent keratinocytes	Permanent	Specially treated silicone sheet	Partial-thickness burns and chronic ulcers
CellSpray	Avita Medical	—	Human autologous	Preconfluent keratinocytes	Permanent	Delivered into a suspension for spray	Partial-thickness burns and chronic ulcers
Cryoskin (CryoCeal)		—	Human allogeneic	Cryopreserved (viable cells) confluent keratinocytes	Permanent	Delivered on a gel-like proprietary chemical surface	Pending Medicines and Healthcare Products Regulatory Agency (MHRA) approval
ReCell	Clinical Cell Culture (C3), Ltd.	—	Human autologous	Epidermal cell suspension	. . .	Delivered into a suspension for spray	. . .
Dermal substitutes							
1. Acellular							
1.1 Bovine							
Matriderm	(Skin&Healthcare)	6	Bovine dermis and ligamentum nuchae	Acellular	Permanent	Collagen types I, III, V matrix covered with elastin fibers	Full- and partial-thickness wounds
Pelnac	(Kowa Co)	4	Bovine	Acellular	. . .	Collagen matrix and silicon layer	Full- and partial-thickness wounds
Terudermis	(Terumo Corp.)	5	Bovine	Acellular	. . .	Collagen matrix and silicon layer	. . .
Primatrix	(Tei Biosciences Inc)	—	Fetal bovine dermis	Acellular	. . .	Collagen matrix	Chronic ulcers
Renoskin	(Groupe Perouse Plastie)	8	Bovine	Acellular	. . .	Collagen matrix and silicon outer layer	Full- and partial-thickness wounds

(Continued)

TABLE 55.1 (Continued)

Product	Company	Price (US $/cm²)	Source of Materials	Type of Cells Used	Temporary/ Permanent	Grown in	Application
1.2 Porcine							
Permaco™	(Tissue Science Laboratories Plc.)	14	Porcine	Acellular	Permanent	Dermal collagen (and elastin matrix)	Dermal filler and full and partial thickness wounds
Oasis	(Health point)	4	Porcine	Acellular	Permanent	Small intestine submucosa, collagen matrix	Full- and partial-thickness burns, venous and diabetic ulcers
FortaFlex		–	Porcine	Acellular	Permanent	Small intestine submucosa	Full- and partial-thickness burns, venous and diabetic ulcers
EZ-Derm™	(Brennan Medical)	3	Porcine	Acellular	Permanent	Aldehyde-crosslinked dermal collagen matrix	Full- and partial-thickness wound
Biobrane™		–	Porcine	Acellular	Temporary	Collagen chemically bound to silicone/nylon membrane	Partial-thickness burns and wounds
1.3 Human collagen/elastin							
Alloderm™	(Life Cell Corp)	10	Human allogeneic	Acellular (freeze-dried)	Permanent	Allogeneic dermis	Full- and partial-thickness wound
Surederm	(Hans Biomed	3	Human allogeneic	Acellular	Permanent	Collagen and elastin matrix	

fibroblasts, until the sheets are between two and eight cell layers thick. Acquiring the keratinocytes in this particular skin substitute is noninvasive as they are obtained from the patients' hair follicles. Each cultured CEA is attached to petrolatum gauze backing with stainless steel surgical clips [1]. Epicel® is indicated for deep-partial and full-thickness burns of total body surface area (TBSA) greater than 30%.

A similar method of obtaining keratinocytes is used in Epidex™, where they are isolated from the outer root sheath of hair follicles in the anagen phase of growth. However, the confluent layers of keratinocytes have a silicone backing instead of the gauze that is used in Epicel® [33].

Although, the problems with fragility and handling remain as the CEAs are simply reinforced. Newer epidermal substitutes continue to utilize keratinocytes; however, they are grown into a subconfluent layer within sheets of stratified cells in an engineered matrix [11]. Laserskin™ (Fidia Advanced Biopolymers, Padua, Italy), also known as Vivoderm™, delivers the autologous keratinocytes into a benzyl esterified derivative of HA that contains laser drilled microperforations that allow the cells to migrate and proliferate in the wound bed [1].

Myskin™ is similar to Laserskin™ in that the keratinocytes are partially confluent; however, they are seeded into a silicone sheet. However, unlike Laserskin™, it cannot be used as a sole skin substitute for deep-partial thickness burns [34]. It is often used in conjunction with and on top of a meshed skin graft. It is for this reason that it is used mainly for the treatment of chronic ulcers seen commonly in diabetic patients.

BioSeed-S™ uses a two-component fibrin glue technology (Baxter International Inc, USA) to support the autologous keratinocytes. The function of the fibrin glue is to support the cells to the wound and allows better engraftment. The cells are delivered via a syringe which is much more convenient than handling the fragile CEAs seen above [35].

Another approach to an epidermal substitute is by using cultured autologous keratinocytes delivered into a suspension, so they can be applied directly to the prepared wound bed in the form of a spray. ReCell® and CellSpray™ are commercial examples of this method, which has been shown to promote faster epithelialization and epidermal maturation in wound models in partial thickness burns. However, the amount of attachment is dependent on adequate preparation of the wound bed [11].

Dermal Substitutes

Dermal substitutes generally consist of elements found in the ECM that make the dermis elastic and durable [2,3]. These include collagen, elastin, adhesion molecules, and HA. Unlike the predominantly cellular base of epidermal substitutes, dermal substitutes generally consist of a porous scaffold that allows infiltration of exudate from the freshly excised burns tissue.

Natural and artificial biomaterials have been widely used in dermal substitutes as scaffolds or additives; in order to mimic skin composition and create an appropriate microenvironment for cell proliferation, differentiation, and migration [36]. Examples of natural materials include collagen, chitosan, gelatin, HA, glycosaminoglycans (GAGs), fibrin, and fibronectin [17]. Although some of their dressings have been commercialized, their cellular skin grafts are still in the laboratory stage and further clinical data are needed for Food and Drug Administration (FDA) approval or commercialization [7]. Examples of artificial materials include among others PLG, poly(lactide-co-glycolide), PLA, polylactic acid, and silicone. Advantages of natural biomaterials include their ability for cell adhesion, interaction with their environment using integrins, focal adhesions, and stimulate downstream signaling pathways [37].

One of the most widespread naturally occurring proteins responsible for structural integrity is collagen. Collagens are secreted by different components of the skin and can be arranged in fibrils as well as in networks forming the major part of the ECM, stimulating wound healing [7,38,39]. The most commonly used dermal substitute clinically is Integra®, which is a bilayered, acellular matrix, composed of bovine type 1 collagen and chondroitin sulfate [4]. When applied to the patient, host fibroblasts migrate into the scaffold which is subsequently replaced by connective tissue [40,41]. The outer silastic layer is removed after 3 weeks and replaced by a split-thickness autograft. However, Integra® is notoriously difficult to handle; its fragility often causing failure of the graft by mechanical loss and separation of the silastic layer [4,42]. Apligraf™ also makes use of collagen mixed with dermal fibroblasts (obtained from neonatal foreskin) to obtain fibroblast−collagen interaction, which results in rearrangement and increased density of collagen fibrils with a dermal-lattice equivalent being formed [43−45]. The cost of Apligraf™ is higher than Integra® with the

added disadvantage of the use of allogeneic cells which may pose some immunogenicity [9,46].

Within the different forms of collagen used in dressings or matrices (hydrogel, sponge, and lattice), collagen sponge has proven to have stronger mechanical properties than hydrated collagen gel and several types of artificial skin have been developed in the form of sponge [7,47]. However, collagen alone proved not sufficient to support proper spatial organization of the endothelial cell network [48]. Therefore, crosslinking treatment or combining other natural or synthetic polymers such as GAGs, chitosan, polycarprolactone (PCL), and poly(lactic-co-glycolic acid) (PLGA) to enhance bridges between collagen molecules seems fit in order to prevent collagen matrix from contracting [7,49−51]. Even though collagen may seem a suitable and obvious natural polymer to be used in scaffolds, problems involving wound contraction and scarring discourage the use, as well as high costs of pure collagen [7,31,52].

Alloderm and Oasis are two other dermal substitutes that have been prepared from decellularized human cadaveric dermis and porcine intestinal submucosa, respectively [26,53]. Decellularization eliminates their antigenicity; however along with Integra® and Apligraf™, they continue to share the limitations that are involved when using xenografts or allografts with respect to costs and virus and prion disease transmission [7,41,54]. There have been a few reports of bovine spongiform encephalitis in the United States originating directly from bovine constituents [20].

HA (D-glucuronic acid $1->3$ N-acetyl-D-glucosamine) or hyaluronan in its sodium salt form has shown biocompatibility [7], the most important characteristic of a skin substitute [35,55]. With a high capacity for water sorption and water retention, influencing several cellular functions such as adhesion, migration, and proliferation, it has become a popular medical agent used in skin as well as in other medical applications [56,57]. HA is a major component of the skin ECM and is involved in tissue repair and wound healing. Even more so, scarless wound healing appears to occur with low levels of collagen and high levels of HA [58]. There are several studies done developing a scaffold incorporating HA, among which Wang et al. [18,36] have developed an adequate scaffold using a bilayered gelatin−chondroitin 6 sulfate−HA membrane seeded with fibroblasts and keratinocytes where HA proved to stimulate cellular migration, designing a well-formed skin equivalent in 21 days. Commercially this has been used in Hyalograft 3D™ (Fidia Advanced Polymers), where autologous fibroblasts have been seeded onto the esterified matrix. Cultured autologous fibroblasts sourced from a patient biopsy enhance the reepithelialization of a full-thickness burn, when compared to an acellular dermal substitute [1,59].

Again there is a 3—4 week delay in this procedure [60]. Though HA appears to be an excellent polymer, sufficiently low concentrations (0.01—0.1%) must be used since higher concentrations proved to inhibit its beneficial capacities [56].

Other materials in commercial use include polyglactic acid (Dermagraft) and fibrin (Bioseed) and those in development are chitosan, gelatin, and alginates [57,61—64]. However, the most common method for the fabrication of these scaffolds, despite the materials being different, is freeze drying [7,65]. Although, freeze drying can produce a porous scaffold, the mechanical strength and morphological structure of these scaffolds are not ideal [66]. Often a skin forms near the surface meaning that the cells cannot migrate and integrate into the whole graft resulting in graft failure. Coupled with the extremely high cost, and more importantly, the use of animal derivatives severely disadvantages tissue engineered constructs such as Integra® [20,67,68].

Another disadvantage with Integra® (and Matriderm) is that it requires at least two surgeries to apply it because both layers together are too thick to apply in one operation [12,69]. The greater the thickness of a skin substitute, the longer the dermis will take to vascularize and the more likely it is that the epidermis will undergo necrosis. The dermal component is applied initially, followed by the epidermal component once the dermis has had enough time to vascularize. It is imperative that the skin substitute is placed on a well-prepared, vascularized wound bed in order for best results for wound healing [11].

Despite the myriad of skin substitutes available, autografts remain the gold standard of treatment for burns patients; this very fact demonstrates that a fresh impetus is needed to progress the field. Investigating polymer-based scaffolds that overcome the shortcomings of current substitutes is one possible solution. The versatility offered by synthetic materials is highly appealing as they can be tailored to specific needs.

55.3.5 Cells

Keratinocytes and fibroblasts are most commonly used to repopulate scaffolds intended for epidermal and dermal replacement, respectively [70,71]. Often, other cells are cultured alongside the keratinocytes, such as stem cells and subconfluent noncontact inhibited cells that have a higher proliferative and wound healing capacity [72]. However, these studies are limited to animal models, which mean that the addition of these extra cells may not produce a measurable effect [73].

In order for these living cells to be cultured, each cell type is expanded in its appropriate culture medium, which then can be used in grafts. However, to determine the quality of such skin grafts, to date only a limited number of indicators or markers are available [68].

Cytokeratin 19 (K19) has proven to be an indicator of young, possibly laterally expanding skin, expressed in the stratum basale of skin substitutes, indicating potentially growing and functional epidermis [11]. However, this marker cannot be detected in skin of human individuals over 2 years of age, and may be a subtype of keratinocyte stem cell, exclusively expressed in very young skin.

Apart from the K19 marker, other keratinocyte stem cell markers include the highly expressed integrin α6 chain, which is in combination with low expression of the transferrin receptor, CD71. A reliable bioassay by which the usefulness of these markers can be determined is the formation of a stratified epidermis that remains fully functional for a minimum of 12 weeks posttransplantation [11].

Research indicates that if endothelial cells are co-cultured with fibroblasts that have been isolated from the patient, the rate of vascularization can be improved meaning that wound healing can occur much quicker. This is especially increased when angiogenic factors such as vascular endothelial growth factor (VEGF) and fibroblast growth factor (FGF) are used in the skin substitute [9]. Currently, the release of these angiogenic factors is not controlled. Future developments in this area would involve incorporating the angiogenic factors into the matrix so that they are released gradually and can effectively and continually promote vascularization [9,45,74]. Melanocytes may also be isolated from a skin biopsy in order to be cultured and added to a skin substitute to contribute to the full function of the epidermis, in order to mimic the innate skin construction, regulating skin pigmentation, and protecting the skin against UV radiation as well as enhancing aesthetics [75,76].

55.4 ECONOMIC ISSUES

The costs involved with engineered tissue production are high. In producing an off-the-shelf skin replacement, important considerations are cost of the research and development process, product safety, clinical trials, storage, shelf-life, and eventually the cost of producing, marketing and selling the actual product [37,71].

Published calculations of cost-effectiveness of treatments are problematic since evidence is limited and criteria for cost-effectiveness differ widely. Moreover, they are often fraught with subjectivity [77]. Further, some of the costs may be difficult or impossible to measure. Thus it is not surprising that various analyses of the cost-effectiveness of cell-based wound therapy or skin substitutes have reached conflicting conclusions [77].

Some authors are skeptical of high-cost treatments. Ho et al. [78] performed a meta-analysis on 23 trials that compared clinical outcomes for artificial skin grafts plus standard care against standard care alone, analyzing data on time to heal, adverse events and costs, from nearly 2000 patients. They concluded that the use of artificial skin grafts increases treatment costs. However, due to limited evidence, there is no evident conclusion about the long-term effect, suggesting that by year 1 there may be net cost savings [78].

Overall, the large amount of evidence supports the concept that despite their high initial costs, bioengineered skin products may be cost-effective or even cost saving for appropriately selected patients. Future economic analyses should explore a longer time horizon, balancing out initial high costs, and expecting lower average treatment costs [77]. Furthermore, not only the cost-effectiveness but the actual effectiveness of these novel bioengineered products deserves further research [77].

55.5 CHALLENGES AND FUTURE DIRECTIONS

Up until today, the ultimate goal of tissue engineering has not yet been reached. Even though a large amount of skin substitute products are commercially available, they still do not offer the complete regeneration of functional skin, including all the skin appendages (hair follicles, sweat glands, and sensory organs) and layers (epidermis and dermis). Neither have they been completely able to establish a fully functional vascular and nerve network and scar-free integration with the surrounding host tissue [19,79].

Though autologous cells have proven to add value being widely used in skin substitutes, they are still not readily available off-the-shelf for acute use, nor can they be easily commercialized, so composite of natural and synthetic polymers may provide a solution [19]. Recent developments in this area include biodegradable electrospun scaffolds in the replacement of human donor dermis or bovine collagen [20]. Considerable focus has already been placed on heterologous cells, e.g., stem cells, which could accelerate wound repair or even reconstitute the wound bed without being immunologically rejected, making them promising as an off-the-shelf therapy [37,79]. With that, incorporating growth factors in dermal matrix substitutes may promote further wound healing, though results are not unanimous and demand further investigation.

Furthermore, since our knowledge about the basic mechanisms of wound healing and the body's response to injury is expanding to the biomolecular level, an emerging prospect in the field of skin tissue engineering is gene therapy. Gene therapy in burnt skin involves particle-mediated gene transfer, introducing biological agents into injured tissue. These biological agents hold the capacity

for bacterial clearance and improving wound healing by enhancing the effect of "positive" growth hormones and suppressing the "negative" factors. Developing gene-activated matrices (biodegradable polymers incorporating a therapeutic gene) may be a promising field in tissue engineering [1].

Altogether these developments may provide a solution for overcoming the shortcomings of current products with appropriate biocompatibility and mechanical, physical, and chemical properties. The key in developing future generation enhanced skin substitutes is understanding the embryonic tissue development in adult tissue regeneration, in order to create a high standard, e.g., biomaterial scaffolds which are specifically enhanced to release growth factors and various signaling molecules, facilitating cell migration and adhesion. Substitutes could also be the carrier of carefully selected cells, which interact after biodegrading of the substitute, so as to regenerate all of the skin structures [19].

So additional research has to be done on the development of scaffolds, stem cell biology, and biomaterial engineering, through *in vitro* as well as *in vivo* randomized and controlled clinical trials, to provide us with the answers we need to succeed in creating these next generation skin substitutes.

REFERENCES

[1] Atiyeh BS, Hayek SN, Gunn SW. New technologies for burn wound closure and healing—review of the literature. Burns J Int Soc Burn Injuries 2005;31(8):944—56 [Review].

[2] DeSanti L. Pathophysiology and current management of burn injury. Adv Skin Wound Care 2005;18(6):323—32 [Comparative Study Review], quiz 32-4.

[3] Brusselaers N, Pirayesh A, Hoeksema H, Richters CD, Verbelen J, Beele H, et al. Skin replacement in burn wounds. J Trauma 2010 Feb;68(2):490—501 [Review].

[4] Burke JF, Yannas IV, Quinby Jr WC, Bondoc CC, Jung WK. Successful use of a physiologically acceptable artificial skin in the treatment of extensive burn injury. Ann Surg 1981 Oct;194 (4):413—28.

[5] Gajiwala K, Lobo Gajiwala A. Use of banked tissue in plastic surgery. Cell Tissue Bank 2003;4(2—4):141—6.

[6] Atiyeh BS, Gunn SW, Hayek SN. State of the art in burn treatment. World J Surg 2005;29(2):131—48.

[7] Zhong SP, Zhang YZ, Lim CT. Tissue scaffolds for skin wound healing and dermal reconstruction. Wiley Interdiscip Rev Nanomed Nanobiotechnol 2010;2(5):510—25 [Review].

[8] Cronin H, Goldstein G. Biologic skin substitutes and their applications in dermatology. Dermatolog Surg Off Publ Am Soc Dermatolog Surg [et al.] 2012;August(28).

[9] Auger FA, Lacroix D, Germain L. Skin substitutes and wound healing. Skin Pharmacol Physiol 2009;22(2):94—102.

[10] Papini R. Management of burn injuries of various depths. BMJ 2004;329(7458):158—60 [Review].

[11] Bottcher-Haberzeth S, Biedermann T, Reichmann E. Tissue engineering of skin. Burns J Int Soc Burn Injuries 2010;36(4):450−60 [Review].

[12] Atherton DD, Tang R, Jones I, Jawad M. Early excision and application of Matriderm with simultaneous autologous skin grafting in facial burns. Plast Reconstr Surg 2010;125(2):60e−61ee [Case Reports].

[13] Batheja P, Song Y, Wertz P, Michniak-Kohn B. Effects of growth conditions on the barrier properties of a human skin equivalent. Pharm Res 2009;26(7):1689−700 [Research Support, U.S. Gov't, Non-P.H.S.].

[14] Barai ND, Boyce ST, Hoath SB, Visscher MO, Kasting GB. Improved barrier function observed in cultured skin substitutes developed under anchored conditions. Skin Res Technol: Off J Int Soc Bioeng Skin 2008 Nov;14(4):418−24 [Evaluation Studies].

[15] Bellamy KE, Waters MG. Designing a prosthesis to simulate the elastic properties of skin. Bio-Med Mater Eng 2005;15 (1−2):21−7 [Clinical Trial Comparative Study].

[16] Choi HR, Kim SK, Kwon SB, Park KC. The fixation of living skin equivalents. Appl Immunohistochem Mol Morphol: AIMM/Off Pub Soc Appl Immunohistochem 2006;14(1):122−5 [Comparative Study Evaluation Studies Research Support, Non-U.S. Gov't].

[17] Adekogbe I, Ghanem A. Fabrication and characterization of DTBP-crosslinked chitosan scaffolds for skin tissue engineering. Biomaterials 2005;26(35):7241−50 [Evaluation Studies].

[18] Lee DY. Acellular dermal equivalent derived from fibroblast culture alone. Clin Exp Dermatol 2010;35(2):197−9 [Letter].

[19] Metcalfe AD, Ferguson MW. Tissue engineering of replacement skin: the crossroads of biomaterials, wound healing, embryonic development, stem cells and regeneration. J Royal Soc Interface/Royal Soc 2007;4(14):413−37 [Research Support, Non U.S. Gov't Review].

[20] Blackwood KA, McKean R, Canton I, Freeman CO, Franklin KL, Cole D, et al. Development of biodegradable electrospun scaffolds for dermal replacement. Biomaterials 2008;29(21):3091−104 [Research Support, Non-U.S. Gov't].

[21] Barai ND, Supp AP, Kasting GB, Visscher MO, Boyce ST. Improvement of epidermal barrier properties in cultured skin substitutes after grafting onto athymic mice. Skin Pharmacol Physiol 2007;20(1):21−8 [Research Support, Non-U.S. Gov't].

[22] Loeffelbein DJ, Baumann C, Stoeckelhuber M, Hasler R, Mucke T, Steinstrasser L, et al. Amniotic membrane as part of a skin substitute for full-thickness wounds: an experimental evaluation in a porcine model. J Biomed Mater Res Part B Appl Biomater 2012;100(5):1245−56.

[23] Boehnke K, Mirancea N, Pavesio A, Fusenig NE, Boukamp P, Stark HJ. Effects of fibroblasts and microenvironment on epidermal regeneration and tissue function in long-term skin equivalents. Eur J Cell Biol 2007;86(11−12):731−46 [Research Support, Non-U.S. Gov't].

[24] Jiong C, Jiake C, Chunmao H, Yingen P, Qiuhe W, Zhouxi F, et al. Clinical application and long-term follow-up study of porcine acellular dermal matrix combined with autoskin grafting. J Burn Care Res Off Publ Am Burn Assoc 2010;31(2):280−5.

[25] Widgerow AD. Bioengineered matrices—part 1: attaining structural success in biologic skin substitutes. Ann Plast Surg 2012;68 (6):568−73.

[26] Shakespeare PG. The role of skin substitutes in the treatment of burn injuries. Clinics Dermatol 2005;23(4):413−8 [Review].

[27] Black AF, Bouez C, Perrier E, Schlotmann K, Chapuis F, Damour O. Optimization and characterization of an engineered human skin equivalent. Tissue Eng 2005;11(5−6):723−33 [Research Support, Non-U.S. Gov't].

[28] Fioretti F, Lebreton-DeCoster C, Gueniche F, Yousfi M, Humbert P, Godeau G, et al. Human bone marrow-derived cells: an attractive source to populate dermal substitutes. Wound Repair Regen Off Publ Wound Healing Soc [and] Eur Tissue Repair Soc 2008;16(1):87−94 [Research Support, Non-U.S. Gov't].

[29] Canonico S, Campitiello F, Della Corte A, Fattopace A. The use of a dermal substitute and thin skin grafts in the cure of "complex" leg ulcers. Dermatol Surg: Off Publ Am Soc Dermatol Surg [et al.] 2009;35(2):195−200.

[30] Gibbs S, van den Hoogenband HM, Kirtschig G, Richters CD, Spiekstra SW, Breetveld M, et al. Autologous full-thickness skin substitute for healing chronic wounds. Br J Dermatol 2006;155 (2):267−74 [Evaluation Studies].

[31] DeCarbo WT. Special segment: soft tissue matrices—bilayered bioengineered skin substitute to augment wound healing. Foot Ankle Spec 2009;2(6):303−5.

[32] Coolen NA, Verkerk M, Reijnen L, Vlig M, van den Bogaerdt AJ, Breetveld M, et al. Culture of keratinocytes for transplantation without the need of feeder layer cells. Cell Transplantat 2007;16 (6):649−61.

[33] Ortega-Zilic N, Hunziker T, Lauchli S, Mayer DO, Huber C, Baumann Conzett K, et al. Epidex(R) swiss field trial 2004−2008. Dermatology 2010;221(4):365−72 [Multicenter Study].

[34] Seet WT, Maarof M, Khairul Anuar K, Chua KH, Ahmad Irfan AW, Ng MH, et al. Shelf-life evaluation of bilayered human skin equivalent, myderm. PLoS One 2012;7(8):e40978.

[35] Liu P, Deng Z, Han S, Liu T, Wen N, Lu W, et al. Tissue-engineered skin containing mesenchymal stem cells improves burn wounds. Artif Organs 2008;32(12):925−31 [Research Support, Non-U.S. Gov't].

[36] Wang TW, Sun JS, Wu HC, Tsuang YH, Wang WH, Lin FH. The effect of gelatin−chondroitin sulfate−hyaluronic acid skin substitute on wound healing in SCID mice. Biomaterials 2006;27 (33):5689−97.

[37] Metcalfe AD, Ferguson MW. Bioengineering skin using mechanisms of regeneration and repair. Biomaterials 2007;28 (34):5100−13 [Research Support, Non-U.S. Gov't Review].

[38] Frank L, Lebreton-Decoster C, Godeau G, Coulomb B, Jozefonvicz J. Dextran derivatives modulate collagen matrix organization in dermal equivalent. J Biomater Sci Polym Ed 2006;17 (5):499−517.

[39] Chen G, Sato T, Ohgushi H, Ushida T, Tateishi T, Tanaka J. Culturing of skin fibroblasts in a thin PLGA−collagen hybrid mesh. Biomaterials 2005;26(15):2559−66 [Comparative Study Evaluation Studies Research Support, Non-U.S. Gov't].

[40] Dainiak MB, Allan IU, Savina IN, Cornelio L, James ES, James SL, et al. Gelatin−fibrinogen cryogel dermal matrices for wound repair: preparation, optimisation and *in vitro* study. Biomaterials 2010;31(1):67−76.

[41] MacNeil S. Progress and opportunities for tissue-engineered skin. Nature 2007;445(7130):874−80 [Research Support, Non-U.S. Gov't Review].

[42] Corradino B, Di Lorenzo S, Leto Barone AA, Maresi E, Moschella F. Reconstruction of full thickness scalp defects after

tumour excision in elderly patients: our experience with integra dermal regeneration template. J Plast Reconstr Aesthet Surg JPRAS 2010;63(3):e245−7 [Case Reports].

[43] Hirt-Burri N, Scaletta C, Gerber S, Pioletti DP, Applegate LA. Wound-healing gene family expression differences between fetal and foreskin cells used for bioengineered skin substitutes. Artif Organs 2008;32(7):509−18 [Comparative Study Research Support, Non-U.S. Gov't].

[44] El Ghalbzouri A, Commandeur S, Rietveld MH, Mulder AA, Willemze R. Replacement of animal-derived collagen matrix by human fibroblast-derived dermal matrix for human skin equivalent products. Biomaterials 2009;30(1):71−8.

[45] Moravvej H, Hormozi AK, Hosseini SN, Sorouri R, Mozafari N, Ghazisaidi MR, et al. Comparison of the application of allogeneic fibroblast and autologous mesh grafting with the conventional method in the treatment of third-degree burns. J Burn Care Res Off Publ Am Burn Assoc 2012;7.

[46] Hu S, Kirsner RS, Falanga V, Phillips T, Eaglstein WH. Evaluation of Apligraf persistence and basement membrane restoration in donor site wounds: a pilot study. Wound Repair Regener Off Publ Wound Healing Soc [and] Eur Tissue Repair Soc 2006;14(4):427−33 [Comparative Study Multicenter Study Randomized Controlled Trial].

[47] Marx G, Hotovely-Salomon A, Levdansky L, Gaberman E, Snir G, Sievner Z, et al. Haptide-coated collagen sponge as a bioactive matrix for tissue regeneration. J Biomed Mater Res Part B Appl Biomater 2008;84(2):571−83 [Research Support, Non-U.S. Gov't].

[48] Ponec M, El Ghalbzouri A, Dijkman R, Kempenaar J, van der Pluijm G, Koolwijk P. Endothelial network formed with human dermal microvascular endothelial cells in autologous multicellular skin substitutes. Angiogenesis 2004;7(4):295−305.

[49] Chong EJ, Phan TT, Lim IJ, Zhang YZ, Bay BH, Ramakrishna S, et al. Evaluation of electrospun PCL/gelatin nanofibrous scaffold for wound healing and layered dermal reconstitution. Acta Biomater 2007;3(3):321−30 [Evaluation Studies].

[50] Harrison CA, Gossiel F, Layton CM, Bullock AJ, Johnson T, Blumsohn A, et al. Use of an *in vitro* model of tissue-engineered skin to investigate the mechanism of skin graft contraction. Tissue Eng 2006;12(11):3119−33 [Research Support, Non-U.S. Gov't].

[51] Ma L, Shi Y, Chen Y, Zhao H, Gao C, Han C. *In vitro* and *in vivo* biological performance of collagen−chitosan/silicone membrane bilayer dermal equivalent. J Mater Sci Mater Med 2007;18(11):2185−91 [Evaluation Studies Research Support, Non-U.S. Gov't].

[52] DeCarbo WT. Special segment: soft tissue matrices—Apligraf bilayered skin substitute to augment healing of chronic wounds in diabetic patients. Foot Ankle Specialist 2009;2(6):299−302.

[53] Aldekhayel SA, Sinno H, Gilardino MS. Acellular dermal matrix in cleft palate repair: an evidence-based review. Plast Reconstr Surg 2012;130(1):177−82.

[54] Gordley K, Cole P, Hicks J, Hollier L. A comparative, long term assessment of soft tissue substitutes: alloderm, enduragen, and dermamatrix. J Plast Reconstr Aesthet Surg JPRAS 2009;62 (6):849−50 [Comparative Study Evaluation Studies Letter].

[55] Potter MJ, Linge C, Cussons P, Dye JF, Sanders R. An investigation to optimize angiogenesis within potential dermal replacements. Plast Reconstr Surg 2006;117(6):1876−85 [Comparative Study Evaluation Studies].

[56] Liu H, Yin Y, Yao K. Construction of chitosan−gelatin−hyaluronic acid artificial skin in vitro. J Biomater Appl 2007;21(4):413−30.

[57] Mohamed A, Xing MM. Nanomaterials and nanotechnology for skin tissue engineering. Int J Burns Trauma 2012;2(1):29−41.

[58] Peramo A, Marcelo CL. Bioengineering the skin−implant interface: the use of regenerative therapies in implanted devices. Ann Biomed Eng 2010;38(6):2013−31 [Review].

[59] Gravante G, Delogu D, Giordan N, Morano G, Montone A, Esposito G. The use of Hyalomatrix PA in the treatment of deep partial-thickness burns. J Burn Care Res Off Publ Am Burn Assoc 2007;28(2):269−74.

[60] Myers SR, Partha VN, Soranzo C, Price RD, Navsaria HA. Hyalomatrix: a temporary epidermal barrier, hyaluronan delivery, and neodermis induction system for keratinocyte stem cell therapy. Tissue Eng 2007;13(11):2733−41 [Research Support, Non-U.S. Gov't].

[61] Anthony ET, Syed M, Myers S, Moir G, Navsaria H. The development of novel dermal matrices for cutaneous wound repair. Drug Disc Today Ther Strateg 2006;3(1):81−6.

[62] Boucard N, Viton C, Agay D, Mari E, Roger T, Chancerelle Y, et al. The use of physical hydrogels of chitosan for skin regeneration following third-degree burns. Biomaterials 2007;28(24):3478−88 [Research Support, Non-U.S. Gov't].

[63] Lee SB, Kim YH, Chong MS, Hong SH, Lee YM. Study of gelatin-containing artificial skin V: fabrication of gelatin scaffolds using a salt-leaching method. Biomaterials 2005;26(14):1961−8 [Comparative Study Evaluation Studies Research Support, Non-U. S. Gov't].

[64] Nagakura T, Hirata H, Tsujii M, Sugimoto T, Miyamoto K, Horiuchi T, et al. Effect of viscous injectable pure alginate sol on cultured fibroblasts. Plast Reconstr Surg 2005;116(3):831−8.

[65] Bottino MC, Jose MV, Thomas V, Dean DR, Janowski GM. Freeze-dried acellular dermal matrix graft: effects of rehydration on physical, chemical, and mechanical properties. Dent Mater Off Publ Acad Dent Mater 2009;25(9):1109−15.

[66] Chin CD, Khanna K, Sia SK. A microfabricated porous collagen-based scaffold as prototype for skin substitutes. Biomed Microdevices 2008;10(3):459−67 [Research Support, Non-U.S. Gov't].

[67] Reiffel AJ, Henderson PW, Krijgh DD, Belkin DA, Zheng Y, Bonassar LJ, et al. Mathematical modeling and frequency gradient analysis of cellular and vascular invasion into Integra and Strattice: toward optimal design of tissue regeneration scaffolds. Plast Reconst Surg 2012;129(1):89−99.

[68] Helary C, Zarka M, Giraud-Guille MM. Fibroblasts within concentrated collagen hydrogels favour chronic skin wound healing. J Tissue Eng Regen Med 2012;6(3):225−37.

[69] Cahn F, Kyriakides TR. Generation of an artificial skin construct containing a non-degradable fiber mesh: a potential transcutaneous interface. Biomed Mater 2008;3(3):034110 [Research Support, N.I.H., Extramural Research Support, U.S. Gov't, Non-P.H.S.].

[70] Morimoto N, Saso Y, Tomihata K, Taira T, Takahashi Y, Ohta M, et al. Viability and function of autologous and allogeneic fibroblasts seeded in dermal substitutes after implantation. J Surg Res 2005;125(1):56−67.

[71] Hata K. Current issues regarding skin substitutes using living cells as industrial materials. J Artif Organs Off J Japanese Soc Artif Organs 2007;10(3):129−32 [Review].

[72] Chan RK, Zamora DO, Wrice NL, Baer DG, Renz EM, Christy RJ, et al. Development of a vascularized skin construct using adipose-derived stem cells from debrided burned skin. Stem Cell Int 2012;2012:841203.

[73] Hernon CA, Harrison CA, Thornton DJ, MacNeil S. Enhancement of keratinocyte performance in the production of tissue-engineered skin using a low-calcium medium. Wound Repair Regener Off Publ Wound Healing Soc [and] Eur Tissue Repair Soc 2007;15(5):718−26.

[74] Han T, Wang H, Zhang YQ. Combining platelet-rich plasma and tissue-engineered skin in the treatment of large skin wound. J Craniofac Surg 2012;23(2):439−47.

[75] Lee JH, Kim JE, Kim BJ, Cho KH. *In vitro* phototoxicity test using artificial skin with melanocytes. Photodermatol,
Photoimmunol Photomed 2007;23(2−3):73−80 [Evaluation Studies In Vitro Research Support, Non-U.S. Gov't].

[76] Eves PC, Beck AJ, Shard AG, Mac Neil S. A chemically defined surface for the co-culture of melanocytes and keratinocytes. Biomaterials 2005;26(34):7068−81 [Comparative Study Evaluation Studies Research Support, Non-U.S. Gov't].

[77] Langer R. Perspectives and challenges in tissue engineering and regenerative medicine. Adv Mater 2009;21(32−33):3235−6.

[78] Ho G, Barbenel J, Grant MH. Effect of low-level laser treatment of tissue-engineered skin substitutes: contraction of collagen lattices. J Biomed Optics 2009;14(3):034002 [Research Support, Non-U.S. Gov't].

[79] Guerra L, Dellambra E, Panacchia L, Paionni E. Tissue engineering for damaged surface and lining epithelia: stem cells, current clinical applications, and available engineered tissues. Tissue Eng Part B Rev 2009;15(2):91−112 [Review].

Chapter 56

Bone Regeneration and Bioengineering

Rania M. Elbackly[a,b], Maddalena Mastrogiacomo[a], and Ranieri Cancedda[a]

[a]*Dipartimento di Medicina Sperimentale, Università di Genova & AOU San Martino-Istituto Nazionale per la Ricerca sul Cancro, Genova, Italy,*
[b]*Tissue Engineering Labs, Faculty of Dentistry, Alexandria University, El-Guish Road, Alexandria, Egypt*

Chapter Outline

56.1 BONE ENGINEERING IN THE LAST 20 YEARS

Owing to the limitations and drawbacks of available treatment options, almost 20 years ago, the concept of an engineered bone graft material for the treatment of large bone defects was introduced. This was based on the notion that an implant composed of osteogenically committed cells and an osteoconductive scaffold [1—6] could give rise to new bone tissue. Hydroxyapatite (HA) and other calcium phosphate—based ceramics have been of most interest among proposed biomaterials owing to their innate osteoconductivity and their ability to "integrate" with host bone [7—13].

In 1991, cell-seeded constructs implanted subcutaneously into immunocompromised mice were the first true demonstration of new bone deposition in porous bioceramic scaffolds. This work was then followed by several others who obtained similar results upon the implantation of bone marrow mesenchymal stem cells (BMSC) from different species and implanting them ectopically in syngeneic rats [14,15], in immunodeficient mice [16—19], or within small experimentally induced osseous defects [2,20].

Immunodeficient mouse models have also provided the advantage of qualitative and quantitative evaluation of the performance of different ceramic scaffolds engineered with BMSC [21—23], including kinetics of bone formation and scaffold resorption utilizing X-ray synchrotron radiation computed microtomography (microCT) and microdiffraction. Others have attempted to facilitate the vascularization of this newly formed bone by the implantation of osteoprogenitor cell/ceramic composites to create vascularized bone flaps [24,25].

Following these initial works with small animal models, tissue engineering approaches for bone regeneration have also been conducted in large animals. The development of these models has long provided the stepping stone to evaluate the feasibility of clinical translation of tissue engineering approaches. The defect model was in most instances a large segmental defect in long bone. This was then filled with a hollow cylinder of porous bioceramic seeded with *in vitro* expanded autologous osteogenic progenitors.

One of the earlier studies was conducted to compare the effect of empty porous ceramic constructs (65% HA and 35% beta-tricalcium phosphate, β-TCP) with those that were seeded with BMSC, on the healing of 21 mm segmental defects in the dog femur [26]. The seeded constructs maintained mechanical integrity, and radiographic union was established rapidly. On the other hand, defects receiving empty scaffolds were subject to multiple fractures.

Similar studies have also been conducted in sheep using highly resorbable coral-based scaffolds or porous hydroxyapatite ceramic scaffolds seeded with *in vitro* expanded BMSC [27,28] to repair critical size defects. Although bone formation in the latter study could be histologically observed in both cell-seeded and nonseeded implants, it could be seen in the internal macropore space only in the seeded implants. In the nonseeded implants, bone formation was limited to the periphery.

Other studies have searched for alternative resorbable ceramics that are clinically available, such as silicon-stabilized TCP/HA (Skelite™) in promoting the repair of a critical size (4.8 cm), experimentally induced defect in a weight-bearing tibia sheep model [8]. In this study, 10–20% of periosteum was deliberately left intact at the defect site as a source of osteogenic cells. The results showed that a progressive growth of bone from the periosteum remnants to the inner part of the scaffold was observed. At the same time, a progressive reduction in the quantity of the scaffold material was also observed, suggesting that these two processes were interrelated in a similar way to the normal bone remodeling process. The dynamic interaction between bone deposition and scaffold resorption was indeed demonstrated by the same group confirming that scaffold resorption was detected only when bone formation was promoted by BMSC previously seeded within the construct and never in the empty scaffolds. This coupling of scaffold resorption and bone deposition was assessed by means of X-ray computed microCT with synchrotron radiation [21–23,29].

Although all of the aforementioned studies utilized porous ceramic scaffolds, they were limited to the use of nonosteogenically induced BMSC. Osteogenically induced cells have also been used as in combination with coral scaffolds to restore 25 mm long defects, created in the middle third of goat femurs [30]. These implants showed bony union at 4 months, and the engineered bone was further remodeled into newly formed cortexed bone at 8 months. Moreover, the mechanical properties of the engineered bone were similar to those of the contralateral normal femur. Similar results were also shown when osteogenically induced BMSC seeded on porous β-HA were used to repair a 30-mm long mandibular segmental defect in a canine model [31].

In an approach to engineer an injectable bone graft material, the group at Nagoya University, Japan, evaluated the bone regenerative capacity of a composite made of BMSC in fibrin glue associated with a biodegradable (β-TCP) scaffold injected into the subcutaneous space on the dorsum of a rat [32]. The same group then used the injectable bone concept by mixing BMSC and platelet-rich plasma (PRP) in a dog mandible model [33,34]. They showed that while the control and PRP groups performed poorly, the BMSC/PRP group had well-formed mature bone and neovascularization, comparable with that achieved by the autologous cancellous bone group. Other researchers also tested the use of BMSC-PRP-fibrin gel as a grafting substitute for alveolar augmentation with simultaneous tooth implant placement in adult dogs [35].

The aforementioned studies and other similar works allowed the drawing of a number of general conclusions in spite of the differences in animal models, defects, chemical composition, geometry, and biodegradability of the scaffold used. These can be summed up as follows: (i) The addition of BMSC to the bioceramic scaffold significantly enhances its performance and thus healing of the defect. (ii) In larger sized implants, bone formation commenced from the periphery. This could reflect inadequate vascularization in the inner parts of the scaffold, thus affecting viability of seeded cells due to their exposure to ischemic circumstances [36,37]. Furthermore, vascularization may contribute to microenvironment homeostasis and to the recruitment of additional mesenchymal progenitors in the proximity of the lesion [38]. Indeed, transplantation models of live bone grafts from Rosa26A mice showed that only 70% of induced osteogenesis depended upon the donor progenitor cells [39]. Eventually, with the presence of advantageous conditions, bone formed also in the inner part of the scaffold starting from the peripheral cell/bone layer. (iii) Complete gap filling with new bone appears to be related to the use of resorbable ceramic scaffolds seeded with osteogenic cells since nonresorbable ceramics, such as 100% HA, were retained within the lesion gap at the end of the experimental phases. Although these were functionally effective, they could not fully reestablish bone integrity. It has been shown that a range of 12–18 months is the optimal scaffold resorption time to allow for new bone production. (iv) Osteogenic preinduction of BMSC prior to implantation also appears to favor bone formation in the constructs.

Following these initial experiments with large animal models, Quarto et al. reported the first human clinical cases in which the repair of massive bone defects (4–8 cm) by means of implantation of a porous ceramic scaffold seeded with *in vitro* expanded autologous BMSC was performed [40,41] (Figure 56.1). The study included four patients ranging from 16–41 years of age who presented with bone deficits in which alternative more "conventional" therapies had previously failed. Porous HA

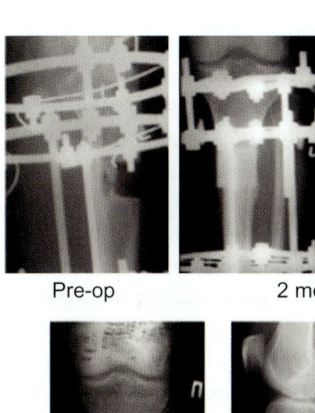
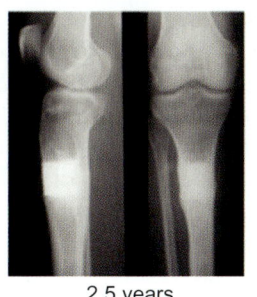

Pre-op   2 months

2 months   2.5 years

FIGURE 56.1 First human clinical case in which the repair of a massive bone defect (4–8 cm) was performed by implantation of a porous ceramic scaffold seeded with *in vitro* expanded autologous BMSC. Figure shows pre-operative and follow-up radiographs at 2 months and after 2.5 years as well as a photograph of the patient showing complete recovery.

ceramic scaffolds were custom made to match the bone defects in terms of size and shape, and external fixation was initially provided for mechanical stability. Patients were evaluated at different times postsurgery by conventional radiographs and CT scans. Abundant callus formation along the implant and a good integration at the interface with the host bone were observed by the second month after implantation. Complete fusion between the implant and the host bone occurred 5−7 months after surgery. All patients recovered limb function between 6 and 12 months. Late follow-up 6−7 years postoperation for two of the patients was also conducted confirming good implant integration with the absence of fractures.

In the same year, regeneration of bone after traumatic avulsion of the distal phalanx of the thumb was reported by Vacanti and co-workers [42]. They used autologous periosteal cells seeded onto a porous coral implant. Although debatable at the time, it was concluded that this paper represented an important effort in the advancement of tissue engineering to clinical translation [43].

Scientists at the National Institute of Advanced Industrial Science and Technology (AIST), Amagasaki City, Hyogo, Japan [44] then reported treatment of benign bone tumors in three patients using tissue-engineered implants. The reconstruction of a human mandible in a 56-year-old patient who had received a subtotal mandibulectomy [45] was also successfully carried out. This was done by a bone-muscle-flap *in vivo* prefabrication technique in which the outer scaffold of the mandible replacement was computer designed, according to tomography data of the defect region. A titanium mesh was chosen for the external scaffold and loaded with HA blocks coated with recombinant human bone morphogenetic protein-7 (rhBMP-7) and BMSC. The patient served as his own bioreactor as the scaffold was implanted into his latissimus dorsi muscle to allow for growth of heterotopic bone and ingrowth of vessels from the thoracodorsal artery. After 7 weeks, the mandible replacement was transplanted, along with the adjacent vessel pedicle, into the mandibular defect. Although long-term results could not be evaluated due to the unfortunate death of the patient 15 months after mandible replacement, initial data revealed that within 4 weeks after implantation the patient had regained masticatory function [46]. Due to the use of both rhBMP7 and whole bone marrow to maximize bone induction, no conclusions could be made on whether regeneration of bone tissue was attributable mostly to the bone marrow cells or BMP7 or both.

It has also been demonstrated that the use of a vascularized pedicled bone flap for reconstruction of a hemimandible obtained after intramuscular implantation of a HA/rhBMP7 composite could have favorable results without any addition of harvested bone, bone marrow, or stem cells [47].

The injectable bone approach using BMSC and PRP was used in the posterior maxilla or mandible of three patients with simultaneous implant placement in the defect area [33]. The results of this investigation indicated that injectable cells/PRP composites used for the plasty area, administered along with implant placement, provided stable and predictable results in terms of implant success. In the same way, this material was utilized for alveolar cleft osteoplasty [48] as well as for periodontal tissue regeneration [49]. Few other examples exist in the literature where maxillary sinus floor augmentation was carried out using a bone matrix derived from mandibular periosteum cells on a polymeric fleece [50,51] or mandibular periosteal cells seeded on a collagen matrix or onto natural bone mineral [52]. Both studies reported beneficial clinical outcomes; however, due to the lack or the inadequacy of the control group of patients, the authors were not able to clearly demonstrate whether the observed bone formation was the result of an osteoinduction or an osteoconduction process.

56.2 NEW REGENERATIVE MEDICINE APPROACH TO BONE REPAIR

Some 20 odd years ago, tissue engineering was introduced as a therapeutic strategy using the patient *ex vivo* expanded cells as the simple building blocks of any tissue to reconstruct an entire organ or "body part." Nevertheless, regardless of the significant successes obtained in the engineering of many tissues, this turned out to be a very difficult task.

The discovery of stem cells in virtually all tissues, the fact that the tissue itself is a reservoir of bioactive cues, and the acquired ability to trigger the mobilization of progenitor cells to home at injury sites have added a new perspective. Research is now focusing on harnessing naturally existent mechanisms and boosting them in order to stimulate the body to heal itself. Some studies have shown that progenitor cells are mobilized and accumulate in the circulation to home at the lesion site as a response to traumatic injuries. This suggests that the body itself is capable of triggering its own intrinsic healing capacity given the appropriate signals. As a result, a new therapeutic strategy that integrates biology and engineering has emerged making it possible to design biomimetic environments that will guide tissue development and regeneration *in vivo*.

The regeneration of bone is particularly challenging since it is a dynamic organ that is constantly subjected to all types of mechanical stresses and shows differences according to location both in the way it is formed and the way it heals and remodels. With the continuous evolution of our understanding of the basic mechanisms of healing, light has been shed on the amazing intrinsic healing capacity of the bone. In the context of bone regeneration strategies, it is also becoming clear that recruitment of host-derived progenitor cells is a major contributor to the final healing process. An example of a biomimetic concept for bone regeneration, changing our perception of the field today, is one where the trophic and immune-modulatory effects of stem cells, scaffolds resembling the natural extracellular matrix (ECM), and delivered multiple growth factors are combined in a delicate synchronized fashion.

For the regeneration of large bone defects, the scaffold remains a crucial part of the delivered strategy, yet its role is being revisited. New generations of scaffolds aim to mimic the morphology and function of the ECM. They will need the added requirement of being themselves "bioactive" and be capable alone of inducing host-cell recruitment and homing at the graft site.

In an effort to mimic the natural microenvironment during bone healing, the integration in the scaffold of a variety of growth factors and cytokines has also become mandatory. However, research conducted in the past 20 years has shown the limitations of using single or dual growth factor delivery approaches since they cannot reproduce the intricate synchrony between the several growth factors and cytokines and their inhibitors that takes place in nature. Hence, attention is now being redirected to using timed delivery of multiple growth factors, using molecules derived from natural ECM, and using cocktails of naturally derived growth factors such as platelet-rich plasma or the stem cell secretome ("biomimetic" approach).

When one considers regenerative medicine applications of these new strategies, rather than the often impossible mission of "building from scratch," clinical availability becomes more apparent and less futuristic, and these strategies will likely become much more rapidly integrated into everyday clinical practice.

56.3 THE PARADIGM SHIFT: BIOMIMETIC STIMULATION OF ENDOGENOUS BONE REPAIR AND REGENERATION

56.3.1 Bone Marrow—Derived Mesenchymal Stem Cells in Tissue Engineering: From Tissue Engraftment to Injury Drugstores

Mesenchymal stem cells have long been defined by their self-renewal and multipotential differentiation properties. The growing consensus is that mesenchymal stem cells and pericytes sharing a panel of common markers are actually one and the same [53]. The present theory is that pericytes are released from broken or inflamed blood vessels at the site of tissue damage to become mesenchymal stem cells (MSC). The latter are then activated by the injury releasing a myriad of bioactive molecules which first modulate the immune response and then secrete trophic factors, thereby creating a regenerative microenvironment [54] (Figure 56.2). Analysis of the *in vitro* secreting profile of MSC has since shown that they release a variety of cytokines that are antiapoptotic, immunosuppressive, proliferation enhancing, and angiogenesis modulating [55].

Different progenitor and stem cell subpopulations have different proteomic profiles and so secrete different factors. MSC secretome is further modulated by physiological, pharmacological, cytokine, or growth factor

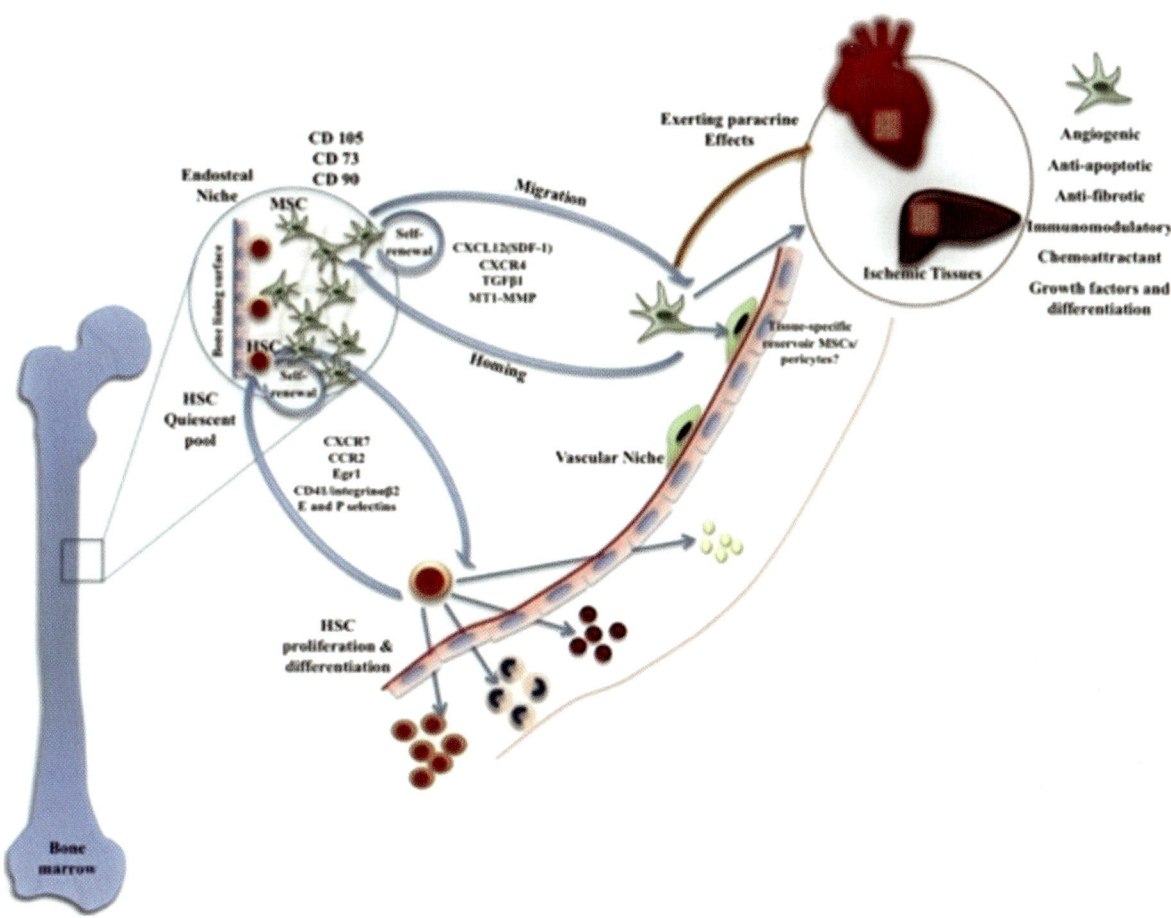

FIGURE 56.2 Schematic depicting some of the numerous stem cell niche interactions during homeostasis and injury. *(Reproduced from Elbackly and Cancedda, 123, 2010).*

preconditioning and/or genetic manipulation. However, the lesion microenvironment has a profound impact on the MSC secretome [56]. Adult stem cells may be able to monitor the microenvironment and respond to specific cues by their surroundings, thus producing different factors to respond to the dynamic microenvironment of the lesion [57]. Thus, it seems unlikely that MSC transplanted at an injury site contribute to the repair/regeneration process by simply replenishing differentiated cell populations. Indeed, bone marrow—derived MSC engraftment in the infarcted heart has been shown to be quite low as is their transdifferentiation posttransplantation into cardiomyocytes [56]. Apparently, infarct transplanted MSC secrete trophic factors that induce activation and proliferation of locally resident cardiac progenitor cells, thus mediating cardiovascular regeneration.

Transplanted bone marrow—derived MSC do not promote vascular growth by incorporating into vessel walls but rather by acting as cytokine factories to stimulate endothelial progenitors and endothelial cells [58].

It has been established that the cell culture conditioned medium of MSC is highly angiogenic. Cultured

MSC upregulate the secretion of prosurvival and angiogenic factors such as IGF-1, VEGF-A, angiopoietins, and HGF in addition to IL-6, IL-8, and CXCL1 [56]. These act both in an autocrine and paracrine fashion [59]. Interestingly, they also express a number of specific endothelial markers as Tie2/Tek, VEGFR2, PECAM1/CD31, and VE-Cadherin and have increased angiogenic potential. Indeed, MSC subjected to hypoxia show increased expression of a panel of genes encoding cytokines related to arteriogenesis such as VEGF, MCP-1, bFGF, IL-6, IL-8, PLGF, and osteopontin [37,58] while showing less effective osteogenic differentiation. This behavior is of utmost clinical relevance since physiological oxygen tension falls to 1% in the fracture hematoma. In fact, MSC react to the initial hostile environment for implanted MSC-seeded tissue engineering constructs, by secreting more angiogenic factors and modulating angiogenic processes to promote vascular invasion of the constructs. The increased secretion of osteopontin may point to their enhancing macrophage infiltration which also interacts with bone formation. The secretory profile of MSC largely changes during their osteogenenic differentiation

[60]. This further emphasizes the dynamic nature of the MSC secretome being influenced by external factors.

We recently reported that, when combinations of murine bone marrow MSC−scaffold constructs were grafted in immune-competent syngenic recipient mice (a well characterized ectopic model of bone formation), a trade between host and donor cells occurred: progenitor cells migrated from the recipient to the implant to form an engineered tissue, and the implanted cells left the scaffold and entered the circulatory flow. We demonstrated that seeded MSC trigger a cascade of events resulting in the mobilization of macrophages, the induction of their functional switch from a proinflammatory to a proresolving phenotype (unpublished results), and the subsequent formation of a bone regenerative niche through the recruitment into the scaffold of specific bone marrow-derived circulating endothelial and pluripotent mesenchymal progenitors with vasculogenic and osteogenic properties, respectively [61,62].

The intrinsic MSC capacity to activate endogenous regenerative mechanisms and to recruit host cells is critically dependent on MSC commitment level, highlighting the importance of carefully investigating the differences in the soluble morphogens secreted during the different stages of MSC differentiation. Future bone regeneration studies should aim at harnessing the full potential of implanted MSC to enhance their *in situ* survival not only to improve their engraftment and transdifferentiation but rather to enhance their autocrine and paracrine effects to trigger host endogenous regeneration cascades [54] avoiding the complication of cell delivery [56,57,63] (Figure 56.3). It would be appealing to customize the MSC secretome by preconditioning them to exert a more sustained effect at the site of injury. For example, preconditioning of MSC by TNF-α leads to activation of P38

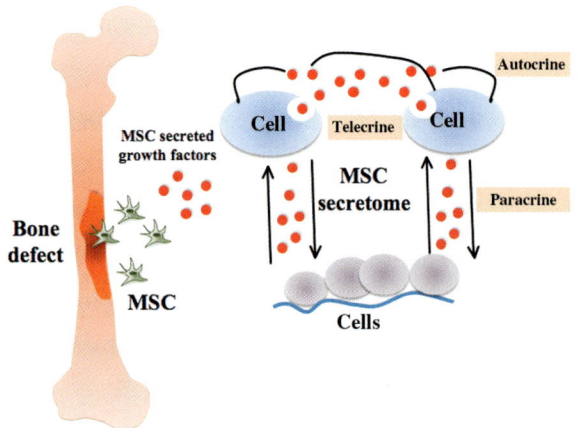

FIGURE 56.3   Schematic illustration of tissue regeneration via endogenous regenerative approaches. MSC secreted growth factors can produce their effects via autocrine, telecrine, or paracrine mechanisms.

MAPK enhancing their paracrine activity and causes them to secrete higher levels of VEGF, HGF, and IGF-1.

MSC have been primed to modify their secretome by internalizing 1−2 μm-sized biodegradable particles releasing bioactive agents [64]. Enhancing MSC homing at the injury site to display their autocrine and paracrine effects has been approached by engineering the surface of MSC using adhesion ligands to enhance cell rolling, which is a crucial step in cell homing [65].

56.3.2 Complementing the Wound Healing Cascade: Inflammation, Angiogenesis, and Tissue Repair

The earliest phase of fracture healing is characterized by local tissue hypoxia and an inflammatory response. It has been suggested that local inflammation initiates bone regeneration by stimulating the migration of MSC, fibroblasts, endothelial, and immune cells such as macrophages that, in turn, drive the formation of the soft callus during fracture repair [66]. Hypoxia and hypoxia inducible factor (HIF) reestablish the disturbed oxygen supply by promoting angiogenesis via increased secretion of VEGF, IL-6, and IL-8. In particular, IL-6 and IL-8 are strong proinflammatory factors that are upregulated during hypoxia and increase the migration of leukocytes. These and other factors were shown to be present in relatively high concentrations in fracture hematomas retrieved from patients. Indeed, data from animal studies have emphasized the role of the fracture hematoma in the healing process since its removal prolonged fracture healing [66].

Moreover, a common feature to both normal bone remodeling and fracture repair is the requirement for vascularization. Inflammatory cytokines such as IL-1β have been shown to potentiate endothelial cell responses increasing their proliferation and augmenting their ability to form tube-like structures *in vitro*[67]. Transplanted MSC also appear to react to the inflammatory environment by secreting MCP-1 and GCP-2, which lead to increased VEGF presence at the site of injury [68]. Some studies failed to show a significant contribution of the cells to the direct deposition of new bone, and confirmed the hypothesis that MSC act principally through secreted trophic factors which are mostly proangiogenic and work in synergy with the local inflammatory environment to promote vascular in growth and host progenitor cell recruitment [69].

56.3.3 Circulating Osteoprogenitor Cells: Nature's Response to Injury?

Fracture may induce mobilization of endothelial progenitor cells from the bone marrow to peripheral blood, and these cells themselves can promote both neovascularization and initiate the healing process in damaged bone tissue [70]. Peripheral blood of fracture patients presented an increase in the number of circulating CD133 + and CD34 + cells 48 h after fracture. Circulating osteoprogenitor cells have also been identified in fracture of mice. These cells may home to sites of injury by virtue of the inflammatory milieu containing high levels of chemoattractants such as SDF-1 and BMPs. Activation of the SDF-1/CXCR4 axis by hypoxia, angiogenic peptides, and inflammatory cytokines may play a significant role [71]. Owing to the intimate relationship between vasculature and bone, these circulating cells may indeed find a way to the injury site via blood vessels [71]. In animal models of BMP 2−induced ectopic bone formation, almost 50% of bone forming cells were found to be CD45⁻ cells that had homed from the bone marrow to the implantation site due to the attraction exerted by both tissue injury and the implanted BMP-2 [72,73]. High levels of VEGF, SDF-1, and HIF-1 were found highly expressed in the immediate vicinity of the implant pointing to a rather indirect role of BMP-2 in the process. It has been concluded that marrow cells in intact bone represent a major if not exclusive source of circulating osteoprogenitor cells.

A proportion of these cells expresses both CD34 and osteocalcin, suggesting the existence of a common circulating hematopoietic and osteoblast progenitor cell. Other studies have highlighted the possibility of two distinct populations of circulating osteogenic cells: one related to hematopoietic stem cells and the other to plastic adherent MSC, both of which originate from bone marrow [71]. Furthermore, analysis of heterotopic ossification lesions from fibrodysplasia ossificans progressiva (FOP) patients has shown the presence of endothelial marker expression in chondrocytes and osteoblasts possibly due to ALK2 activation, which may cause an endothelial to mesenchymal transition in these cells [71]. Although it appears that circulating osteogenic cells are major contributors to bone formation, levels of circulating cells fall to normal within a few days after fracture or BMP-2 implantation meaning that their response is transient provoked by the injury and the inflammatory nature of that site [70,72]. In principle, the presence of these circulating cells could hold great clinical promise. Strategies could be developed to enhance the migration of these cells thereby promoting natural endogenous repair mechanisms.

56.4 LEARNING NATURE'S LESSON: PROVIDING INSIGHT INTO BONE REGENERATION

Preclinical and clinical studies undoubtedly indicate that the use of MSC-based therapies for reconstruction and repair of bone tissue is indeed feasible [74]. The discovery of MSC sources, other than bone marrow, that are clinically accessible and ethically acceptable such as adipose tissue and umbilical cord blood and tissue offers a robust alternative to bone marrow−derived MSC since they have similar characteristics [75]. The now accepted concept of a perivascular niche and pericytes as predecessors to MSC hints to the fact that they can be isolated from virtually all tissues [76]. However, there are still numerous limitations such as culture conditions, cell commitment, tumorigenicity, lack of relevant *in vitro* studies, and *in vivo* in the immunodeficient mouse model [77]. Even more difficult is the identification of true MSC *in vivo* as well as distinguishing osteoprogenitors from more immature MSC due to the paucity of true stage-specific markers. It has recently been demonstrated that in a model of MSC hierarchy, osteoprogenitors could be distinguished from MSC with multilineage potential as being slowly growing cells as compared to more immature MSC [78].

56.4.1 Bone Repair: Going Back to the Roots

Although it has been recently shown that a rare population of circulating bone marrow MSC could home at the site of bone fracture, it remains to be seen whether these cells are not just a simple leakage from bone marrow at the fracture site or locally resident MSC moving from the areas surrounding the lesion [71,78]. It is important that, while studies using BMP-2 have documented that almost 50% of the bone-forming cells had arrived from the circulation, the origin of the rest of the cells remains largely unknown [72,73]. Additionally, the recruited cells have been shown to be present as bone lining cells on the newly formed bone and yet not integrated within new bone as osteocytes, so, it is still not clear whether they support repair by depositing new bone matrix or by producing osteoinductive factors [79]. While locally present cells including some in the bone marrow at the fractured bone are important to its repair, cells of bone marrow also give rise to inflammatory cells and osteoclast precursors migrating to the bone fracture although these cells do not appear to integrate into the callus itself as neither do the systemically injected MSC that home at the site of injury [79]. It is hence becoming clear that bone itself could hold the keys to its repair and that the local environment represents the true niche for the healing progenitors arriving from bone marrow, injured bone tissue, periosteum, and soft tissues close to the bone that are carried to the fracture site via blood vessels invading the callus [77] (Figure 56.4).

56.4.2 The Periosteum: An Underrated Player in Bone Engineering

Since Duhamel in 1739 noted that silver wires embedded under the periosteum became covered by osseous matrix, the periosteum has come to be looked upon as a crucial

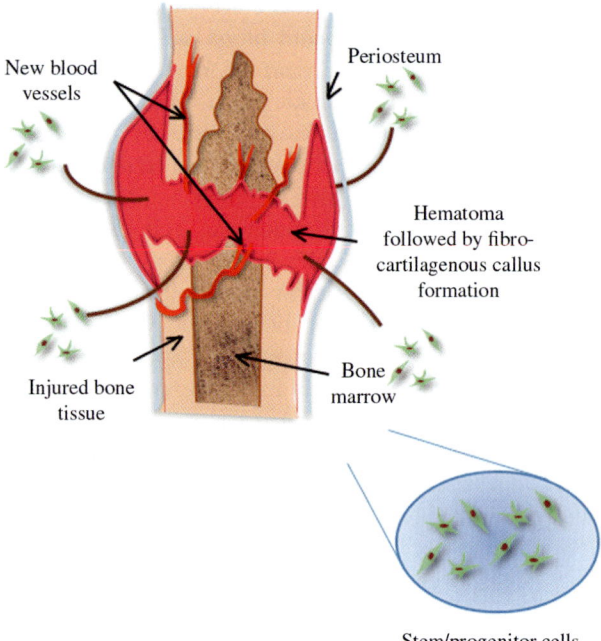

New blood vessels

Periosteum

Hematoma followed by fibro-cartilagenous callus formation

Injured bone tissue

Bone marrow

Stem/progenitor cells

FIGURE 56.4   Bone remodeling after fracture. Mesenchymal stem cells are recruited from various sources within the lesion and induced to form new bone via both endochondral and intramembranous pathways.

element of the angiogenesis and osteogenesis processes occurring during periosteal bone growth [80]. Angiogenesis and osteogenesis are two closely knit events, and vascularization is certainly a key factor for successful bone regeneration. Bone formation is most often limited due to insufficient oxygen supply, nutrient delivery, and inadequate waste removal. Bone is supplied by a hierarchical nourishing meshwork of arterioles and capillaries and forms only within vascularized sites [81]. Cell−scaffold constructs have been employed for bone engineering, but, despite the great potential of engineered bone substitutes, their clinical realization has been less than satisfactory due to poor graft integration which is a consequence of a lack of adequate blood supply [82,83]. In all cases, the most limiting factor was a reduced active blood vessel network for these constructs to survive and integrate with the existing host tissue [84]. To increase the vascular supply for enhancing bone healing and repair, scientists have attempted to engineer periosteum substitutes to more closely mimic the endogenous repair process and overcome current problems of traditional cell−scaffold-based approaches [70,85−88]. The periosteum is a microvascularized connective tissue covering the outer surface of cortical bone [79,88]. Its outer fibrous layer contributes to blood supply of bone while the inner "cambium" layer is highly cellular. It is within this layer that mesenchymal progenitor cells, differentiated osteogenic cells, and osteoblasts as well as fibroblasts are found. Being rich in vascular networks, the periosteum also

contains numerous endothelial pericytes which, in turn, may also serve as a reservoir of progenitor cells. Both endochondral bone repair and distraction osteogenesis primarily rely on the periosteum. Stripping of the periosteum and even the physical disruption between the bone and the overlying soft tissue impair bone healing possibly due to interference with the periosteal stem cell niche [89,90].

The current view supports that periosteum and bone marrow are the main local sources of skeletal stem/progenitor cells for bone repair with distinct but complementary roles [21,79,88,89,91]. The periosteum is essential for bone graft healing and remodeling. Removal of the periosteum from live bone grafts resulted in a 73% decrease in new bone and cartilage formation in the graft, a 10-fold reduction in neovascularization, and 75% decrease in the number of osteoclasts. On the other hand, removal of bone marrow had minimal effects on periosteal bone formation on the graft [88]. It is evident that these tissues hold specific stem/progenitor cell populations, the fate of which depends on the local tissue environment [90].

Indeed, periosteal cells possess a double role [82]. Murine periosteal cells have been shown to contain high number of MSC expressing both osteoprogenitor and chondroprogenitor markers. During repair, the periosteum gives rise to chondrocytes, osteoblasts, and osteocytes. Periosteal-mediated bone repair with its intricate repair mechanisms recapitulates chondrogenic and osteogenic events occurring during development [88]. Additionally, periosteal cells were shown to produce high amounts of VEGF, which enhanced the proliferation and survival of endothelial cells. *In vivo*, these cells were capable of forming mature bone with hematopoiesis-supportive stroma indicating the presence of a highly uncommitted skeletal stem cell in the periosteal cell population. *In vivo*, the implanted periosteal cells were also found to express αSMA and PDGFR-β and to be localized in perivascular locations in blood vessels further emphasizing their proangiogenic role by supporting stable blood vessel formation as well as by attraction of host endothelial cells. Newly formed vessels deliver osteoblast precursors for new bone formation besides nutrients and oxygen supply [92,93].

Designing a periosteal substitute that biomimics the fracture microenvironment and recapitulates a periosteal-mediated response by being both angiogenic and osteogenic may provide a changing concept for enhancing bone grafting and engineering.

56.5 BIOFUNCTIONALIZING THE SCAFFOLD FOR ENDOGENOUS BONE REPAIR

Skeletal repair requires the structural and mechanical support that can only be provided by an appropriate scaffold.

The use of MSC alone in skeletal repair without the structural and mechanical support provided by the scaffold would not be possible. Recent developments in development and healing biology are rapidly paving the way to the production of scaffolds that can truly mimic the natural ECM both in morphology and function and that are "primed" to stimulate, enhance, or control the tissue's innate capacity for repair [77].

The "optimal" scaffold should guide and promote tissue regeneration. For bone, it should have high porosity to allow cell infiltration, adequate vascularization for nutrient, and waste transfer while providing mechanical support and having degradation rates appropriate to the neotissue formation. The scaffold should support cells, and allow the release of signaling molecules to activate pathways that will, in turn, control the pattern and extent of bone formation [94].

In an attempt to combine such intricate processes, novel scaffolds have been designed such as bioactive glass-derived shell scaffolds with a bioresorbable gelatin coating to obtain a biomimetic porous scaffold. This scaffold could reproduce the complex morphology of bone by being composed of both natural apatite and collagen [94]. Additionally, the gelatin coating could possibly be used as a drug delivery medium, thereby enhancing its bioactivity.

Other recent efforts have focused on incorporating natural tissue components into the biomaterials to enhance osteoconduction, osteoinduction, and osteointegration [95]. Incorporating hydroxyapatite on the biomaterial surfaces and using simulated body fluids with similar ion composition to plasma are one such approach as it is supplementing the scaffold with proangiogenic factors to enhance angiogenesis. VEGF-incorporating biomimetic polylactic/polyglycolic acid sintered microspheres supported growth and attachment of endothelial cells and formation of bone-like mineral layers on their surface [95]. The scaffold modulated protein adsorption and controlled the release kinetics of VEGF allowing the delivery of the desirable amount of the growth factor in a sustained manner.

Biomimetic approaches strive to introduce new strategies such as the use of bioactive scaffolds for host cell recruitment. The goal of this new direction is to provide cell-free scaffold for large-scale clinical application. Indeed, the development of innovative biomimetic scaffolds that could stimulate the body's endogenous repair capacity could hold the key to bridging the gap between basic research and clinical application [96]. Using natural ECM molecules provides a supportive stroma and conduit for blood vessels and lymphatics. The decellularization of various organs and tissues to be used as naturally derived ECM scaffolds has been proposed [63] in addition to designing artificial ECM-mimicking materials such as self-assembling hydrogels.

Other solutions for clinical implementation will include the *in vivo* implantation of *in vitro* generated hypertrophic cartilage templates using human MSC. These implants can effectively remodel into mature bone tissue via endochondral ossification [97]. This strategy offers a route that recapitulates embryonic development, hence the "biomimetic" approach. Additionally, the implants offer greater vasculogenic potential due to VEGF production by hypertrophic chondrocytes, higher osteoinductivity due to production of BMPs, and higher chances of survival in the hypoxic microenvironment because of the nature of cartilage cells.

Intraoperative manufacturing employing freshly isolated stromal vascular fraction such as the ones in the human adipose tissue has been proposed [97]. These cells appear to contain 100-fold higher numbers of fibroblastic colony forming units per volume of tissue and have intrinsic vasculogenic potential.

56.6 SINGLE VERSUS MULTIPLE GROWTH FACTOR DELIVERY: PROVIDING THE RIGHT SIGNALS

In an effort to mimic the natural microenvironment during wound healing, the integration of a variety of growth factors and cytokines has become a valid concept in tissue engineering/regenerative medicine strategies. Bone itself is a reservoir of growth factors. During fracture repair, PDGF, FGF, IGF, and TGF-β are the most dominant growth factors. Various members of the BMP family show temporal differences in their appearance during fracture repair. Though present in all phases of fracture healing, BMP-2 expression appears to be accentuated both in the inflammatory phase (1−6 days) and the osteogenic phase (20−30 days) [98]. Similar dynamic patterns were also seen during tooth extraction socket healing. Therefore, designing growth factor delivery strategies will likely rely on multiple growth factors rather than a single one and should take into consideration their spatiotemporal patterns and the existence of both synergistic and inhibitory effects. Indeed, bolus delivery of single growth factors is ineffective since they tend to diffuse away from the wound and are deactivated. To overcome loss of bioactivity and low availability, higher doses are required resulting in increased risk of toxicity and again high cost.

To date, rhBMP-2 and rhBMP-7 (OP-1) are the only BMPs that have been FDA approved for clinical use for spinal fusion applications and tibial fractures, respectively. However, their use has been met with many obstacles. The need for very high doses to meet efficacy results in BMPs being very expensive remedies [98]. This is in addition to recent reports of immunogenic reaction to carriers, ectopic bone formation, and critical soft tissue swellings with high doses. Indeed, the FDA issued a warning regarding off-label use of rhBMP in cervical spine fusion due to severe life-threatening complications reported for a number of clinical cases (www.fda.gov). This is further complicated by the fact that BMP-mediated bone formation strongly depends on the local presence of various BMP activity regulating stimulators and inhibitors as the mere presence of BMPs does not guarantee efficient bone healing.

Delivering multiple growth factors such as TGF-β and BMP-7 for their synergistic effects or sequential release of BMP-2 followed by BMP-7 could enhance osteogenic differentiation. Combining VEGF and FGF to induce vascular growth during bone healing may mediate the crosstalk between osteoblasts and endothelial cells. VEGF secreted by osteoblasts attracts endothelial cells. In turn, endothelial cells have been shown to release BMPs [83]. Significant progress has been made in developing dual or multiple growth factor delivery systems via microsphere encapsulation or by chemical cross-linking with the scaffold to manipulate release profiles [99]. Another alternative to exogenous growth factor delivery would be the use of inexpensive compounds that could locally stimulate induction of endogenous BMPs, or by their blocking antagonists [98]. Possibly an effective strategy could be one that utilizes a cocktail of biologically active molecules such as the PRP that could more precisely recapitulate the cascade of signaling events [99].

56.6.1 PRP as an Endogenous Regenerative Tool for Bone Regeneration: Naturally Orchestrating Multiple Growth Factor Delivery

PRP has long been defined as autologous plasma that has a concentration of platelets much higher than in normal blood [100]. The effects of PRP rely on the unique biological activity of platelets and their involvement in the wound healing cascade. The main role of platelets is to create a hemostatic plug and to promote fibrin generation and blood clotting to prevent blood loss. However, they are also a vital part of the innate immune response. They combat infection and modulate inflammation [101], they promote cell chemotaxis and proliferation, and they promote wound healing, angiogenesis, and bone formation.

Within platelets, the most abundant source of secreted proteins is the α-granule [101]. Secreted proteins can be grouped in different families based on their biological activity. Factors such as PDGF, IGF-1, VEGF, CTGF, and several other chemokines and cytokines favor wound healing and promote angiogenesis [101,102] in cooperation with proangiogenic mediators such as SDF-1, MMP-

1, MMP-2, MMP-9, and angiopoietin that are also present. FGF-2 is a mitogenic factor as well as TGF-β1 that is also known to recruit inflammatory cells to the wound. IGF-1 stimulates matrix formation. Adhesion proteins which can act as cell adhesion molecules thereby mediating cell migration are also released by platelets [102]. The proinflammatory molecule CD40 ligand present on platelet membranes is believed to play an important role in stimulating angiogenesis by promoting endothelial cell proliferation. BMP-2, BMP-4, and BMP-6 are synthesized by megakaryocytes and released by platelets in the acidic hypoxic environment of bone fracture. Indeed, it has been suggested that PRP stimulates osteoblastic differentiation of myoblasts and osteoblasts in the presence of BMP-2, BMP-4, BMP-6, and BMP-7 possibly by playing a potentiating role in BMP-dependent osteoblastic differentiation [102].

Yet, though platelets secrete activators, they also secrete a large number of inhibitors of the osteogenic process. Thrombospondin-1 is a strong antiangiogenic factor known to inhibit cell proliferation *in vitro* at high concentrations. Both IL-1 and MIP-1α, also secreted by platelets, play a role in osteoclastogenesis. Indeed, a major puzzling fact is that platelets store and release a wide range of both activators and inhibitors for the same pathways. It appears that local conditions probably dictate the regulation of these factors and which side will eventually win indicating a response to a finely tuned process[101].

Platelets are activated by vessel rupture as a result of injury. In PRP, they are activated by addition of thrombin which serves to induce platelet degranulation and cleaves fibrinogen into fibrin. The presence of calcium further catalyzes the reaction resulting in completion of platelet degranulation. This activation in turn leads to the release of the contained bioactive factors (Figure 56.5). Activated platelets secrete 70% of their stored growth factors within 10 min of clotting and close to 100% within the first 24 h, and then they synthesize additional amounts of growth factors for about 8 days until they are depleted and die [100]. It has also been shown that lower concentrations of thrombin and calcium chloride can lead to sustained growth factor release for up to 6 days after activation and create a stronger fibrin matrix.

The appeal of PRP as therapeutic agent, stems from the healing trigger activity of growth factors and biologically active molecules that are released by platelets in their natural mix and in the right proportions to each other. The general consensus for clinical use of PRP is that it contain 5 times more than the concentration of platelets than in normal blood. Lower concentrations appear to have minimal effects while too high concentrations may have negative effects on the tissue repair process. However, the use of PRP has been associated with significant challenges. While many studies have reported beneficial effects of PRP on the wound healing process

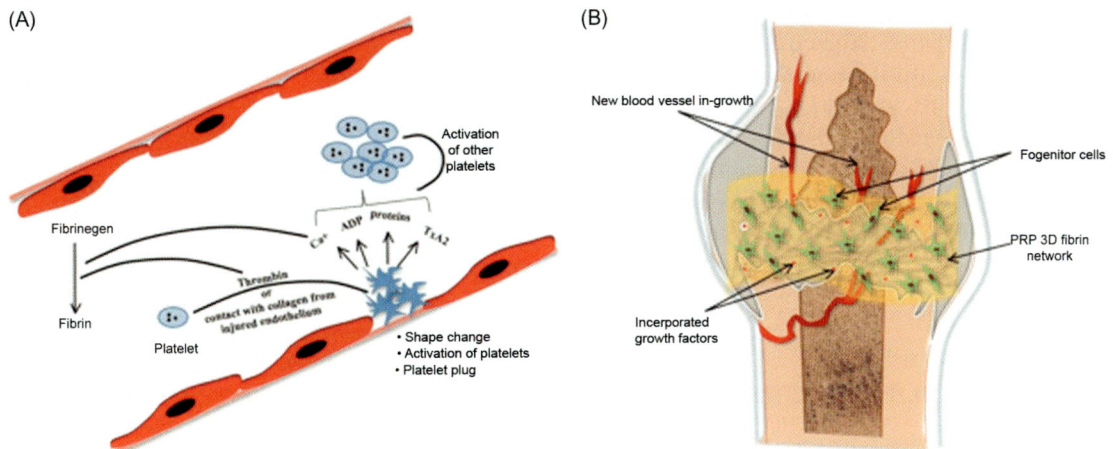

FIGURE 56.5 Platelet-rich plasma as an endogenous regenerative technology for bone regeneration. Cartoon in (A) depicts platelet activation and in (B) how PRP can contribute to bone healing.

especially in orthopedic, craniofacial, and dental fields, others have shown no positive role of PRP. Positive effects have included improved wound healing of both soft and hard tissues as well as reduction of postoperative infection, pain, and blood loss. The lack of controlled clinical trials has led to a difficulty in interpreting results. This is in addition to the wide variability of existing preparation protocols and the lack of standardization measures [103,104].

Different bone substitute materials also appear to influence the overall effects of PRP treatment. PRP could enhance *de novo* bone formation when combined with periosteum and Bio-Oss (deproteinized bovine bone) as the scaffold [104]. No such effects were seen when bioglass or phycogenic hydroxyapatite were used.

Although controversy still remains regarding the efficacy of PRP, many *in vitro* and *in vivo* animal and human studies support its use [105]. Single injections of PRP in monkey jaws enhanced late rather than early bone formation [106]. PRP was also beneficial in the treatment of periodontal defects as recently shown by a systematic review of literature [103]. An elegant approach has recently been proposed using PRP gel as a delivery scaffold for bone marrow-derived cell aggregates and ceramic microparticles by delivering them to the site of the bone defect using a minimally invasive technique. This is particularly appealing for cranio-maxillofacial defects [107]. A similar study also introduced a microinvasive approach using PRP and adipose-derived stem cells in alginate microspheres to improve both angiogenesis and osteogenesis [108]. Using this microencapsulation technique, the combination of both PRP and adipose-derived stem cells could have created a favorable environment to enhance bone regeneration. The authors reported enhanced cell viability when PRP was present and that PRP promoted

cell migration to the surface of some microspheres as well as increasing proliferation. They also showed that PRP could stimulate osteogenic differentiation of the cells. *In vivo* experiments further revealed increased vascularization and bone formation when PRP was present.

It has been hypothesized that although PRP may exert an early effect, its role is quickly taken over by the cells that it recruits to the wound microenvironment. These cells along with the seeded cells continue to secrete a myriad of growth factors and cytokines that take part in both angiogenesis and osteogenesis [108]. The role of PRP is probably acting as a reservoir of cytokines and growth factors that can promote the migration, proliferation, and differentiation of progenitor cells, especially osteoprogenitors. They can thus modulate the injury microenvironment, thereby accentuating the healing process indirectly especially when healing is delayed [106]. Recent studies on the effects of TGF-β1, which is present in high concentrations in PRP, have shown that it does not regulate osteoblastic differentiation but instead directs BMSC to resorption sites. PRP may act in a similar fashion by promoting BMSC migration rather than by having a direct effect on osteogenesis.

PRP seems to have a significant effect also on stimulating vascularization which is well known to be a prerequisite to bone formation. Angiogenic factor-enriched PRP increased both angiogenic and osteogenic effects when combined with a β-TCP scaffold for repair of rat cranial defects. It increased periosteal blood perfusion in the defects as demonstrated by laser Doppler flowsmetry [109]. It may well be possible that the overall improvement in bone regeneration is actually an indirect effect attributed to an enhanced angiogenic effect of PRP. It is hence important to no longer regard PRP as an agent that will induce increased bone formation but rather as a cocktail of signaling molecules that can modify the innate host response leading to a triggering of endogenous regenerative capacity [110]. By controlling and standardizing the protocol of PRP preparation and by combining PRP with biomaterials and/or cells, distinct composites for multiple growth factor delivery may contribute to the bone regeneration by creating a more favorable angiogenic and osteogenic environment [110] (Figure 56.5).

REFERENCES

[1] Matsumura G, Hibino N, Ikada Y, Kurosawa H, Shin'oka T. Successful application of tissue engineered vascular autografts: clinical experience. Biomaterials 2003;24:2303—8.

[2] Krebsbach PH, Mankani MH, Satomura K, Kuznetsov SA, Robey PG. Repair of craniotomy defects using bone marrow stromal cells. Transplantation 1998;66:1272—8.

[3] Mnaymneh W, Malinin TI, Makley JT, Dick HM. Massive osteoarticular allografts in the reconstruction of extremities following resection of tumors not requiring chemotherapy and radiation. Clin Orthop Relat Res 1985:76—87.

[4] Boyde A, Corsi A, Quarto R, Cancedda R, Bianco P. Osteoconduction in large macroporous hydroxyapatite ceramic implants: evidence for a complementary integration and disintegration mechanism. Bone 1999;24:579—89.

[5] Ohgushi H, Miyake J, Tateishi T. Mesenchymal stem cells and bioceramics: strategies to regenerate the skeleton. Novartis Found Symp 2003;249:118—27 [Discussion 27—32, 70—4, 239—41].

[6] Stevenson S. The immune response to osteochondral allografts in dogs. J Bone Joint Surg Am 1987;69:573—82.

[7] Kruyt MC, Dhert WJ, Oner C, van Blitterswijk CA, Verbout AJ, de Bruijn JD. Optimization of bone-tissue engineering in goats. J Biomed Mater Res B Appl Biomater 2004;69:113—20.

[8] Mastrogiacomo M, Corsi A, Francioso E, Di Comite M, Monetti F, Scaglione S, et al. Reconstruction of extensive long bone defects in sheep using resorbable bioceramics based on silicon stabilized tricalcium phosphate. Tissue Eng 2006;12:1261—73.

[9] Ge Z, Baguenard S, Lim LY, Wee A, Khor E. Hydroxyapatite-chitin materials as potential tissue engineered bone substitutes. Biomaterials 2004;25:1049—58.

[10] Elsinger EC, Leal L. Coralline hydroxyapatite bone graft substitutes. J Foot Ankle Surg 1996;35:396—9.

[11] Marcacci M, Kon E, Zaffagnini S, Giardino R, Rocca M, Corsi A, et al. Reconstruction of extensive long-bone defects in sheep using porous hydroxyapatite sponges. Calcif Tissue Int 1999;64:83—90.

[12] Sartoris DJ, Holmes RE, Resnick D. Coralline hydroxyapatite bone graft substitutes: radiographic evaluation. J Foot Surg 1992;31:301—13.

[13] Heise U, Osborn JF, Duwe F. Hydroxyapatite ceramic as a bone substitute. Int Orthop 1990;14:329—38.

[14] Goshima J, Goldberg VM, Caplan AI. Osteogenic potential of culture-expanded rat marrow cells as assayed in vivo with porous calcium phosphate ceramic. Biomaterials 1991;12:253—8.

[15] Goshima J, Goldberg VM, Caplan AI. The osteogenic potential of culture-expanded rat marrow mesenchymal cells assayed in vivo

in calcium phosphate ceramic blocks. Clin Orthop Relat Res 1991:298—311.

[16] Krebsbach PH, Kuznetsov SA, Satomura K, Emmons RV, Rowe DW, Robey PG. Bone formation in vivo: comparison of osteogenesis by transplanted mouse and human marrow stromal fibroblasts. Transplantation 1997;63:1059—69.

[17] Kadiyala S, Young RG, Thiede MA, Bruder SP. Culture expanded canine mesenchymal stem cells possess osteochondrogenic potential in vivo and in vitro. Cell Transplant 1997;6:125—34.

[18] Goshima J, Goldberg VM, Caplan AI. The origin of bone formed in composite grafts of porous calcium phosphate ceramic loaded with marrow cells. Clin Orthop Relat Res 1991:274—83.

[19] Martin I, Muraglia A, Campanile G, Cancedda R, Quarto R. Fibroblast growth factor-2 supports ex vivo expansion and maintenance of osteogenic precursors from human bone marrow. Endocrinology 1997;138:4456—62.

[20] Ohgushi H, Goldberg VM, Caplan AI. Repair of bone defects with marrow cells and porous ceramic. Experiments in rats. Acta Orthop Scand 1989;60:334—9.

[21] Mastrogiacomo M, Papadimitropoulos A, Cedola A, Peyrin F, Giannoni P, Pearce SG, et al. Engineering of bone using bone marrow stromal cells and a silicon-stabilized tricalcium phosphate bioceramic: evidence for a coupling between bone formation and scaffold resorption. Biomaterials 2007;28:1376—84.

[22] Komlev VS, Peyrin F, Mastrogiacomo M, Cedola A, Papadimitropoulos A, Rustichelli F, et al. Kinetics of in Vivo bone deposition by bone marrow stromal cells into porous calcium phosphate scaffolds: an X-Ray computed microtomography study. Tissue Eng 2006;12:3449—58.

[23] Mastrogiacomo M, Komlev VS, Hausard M, Peyrin F, Turquier F, Casari S, et al. Synchrotron radiation microtomography of bone engineered from bone marrow stromal cells. Tissue Eng 2004;10:1767—74.

[24] Mankani MH, Krebsbach PH, Satomura K, Kuznetsov SA, Hoyt R, Robey PG. Pedicled bone flap formation using transplanted bone marrow stromal cells. Arch Surg 2001;136:263—70.

[25] Casabona F, Martin I, Muraglia A, Berrino P, Santi P, Cancedda R, et al. Prefabricated engineered bone flaps: an experimental model of tissue reconstruction in plastic surgery. Plast Reconstr Surg 1998;101:577—81.

[26] Bruder SP, Kraus KH, Goldberg VM, Kadiyala S. The effect of implants loaded with autologous mesenchymal stem cells on the healing of canine segmental bone defects. J Bone Joint Surg Am 1998;80:985—96.

[27] Petite H, Viateau V, Bensaid W, Meunier A, de Pollak C, Bourguignon M, et al. Tissue-engineered bone regeneration. Nat Biotechnol 2000;18:959—63.

[28] Kon E, Muraglia A, Corsi A, Bianco P, Marcacci M, Martin I, et al. Autologous bone marrow stromal cells loaded onto porous hydroxyapatite ceramic accelerate bone repair in critical-size defects of sheep long bones. J Biomed Mater Res 2000;49:328—37.

[29] Mastrogiacomo M, Muraglia A, Komlev V, Peyrin F, Rustichelli F, Crovace A, et al. Tissue engineering of bone: search for a better scaffold. Orthod Craniofac Res 2005;8:277—84.

[30] Zhu L, Liu W, Cui L, Cao Y. Tissue-engineered bone repair of goat-femur defects with osteogenically induced bone marrow stromal cells. Tissue Eng 2006;12:423—33.

[31] Yuan J, Cui L, Zhang WJ, Liu W, Cao Y. Repair of canine mandibular bone defects with bone marrow stromal cells and porous beta-tricalcium phosphate. Biomaterials 2007;28:1005–13.

[32] Yamada Y, Boo JS, Ozawa R, Nagasaka T, Okazaki Y, Hata K, et al. Bone regeneration following injection of mesenchymal stem cells and fibrin glue with a biodegradable scaffold. J Craniomaxillofac Surg 2003;31:27–33.

[33] Yamada Y, Ueda M, Naiki T, Nagasaka T. Tissue-engineered injectable bone regeneration for osseointegrated dental implants. Clin Oral Implants Res 2004;15:589–97.

[34] Yamada Y, Ueda M, Hibi H, Nagasaka T. Translational research for injectable tissue-engineered bone regeneration using mesenchymal stem cells and platelet-rich plasma: from basic research to clinical case study. Cell Transplant 2004;13:343–55.

[35] Ito K, Yamada Y, Naiki T, Ueda M. Simultaneous implant placement and bone regeneration around dental implants using tissue-engineered bone with fibrin glue, mesenchymal stem cells and platelet-rich plasma. Clin Oral Implants Res 2006;17:579–86.

[36] Viateau V, Guillemin G, Calando Y, Logeart D, Oudina K, Sedel L, et al. Induction of a barrier membrane to facilitate reconstruction of massive segmental diaphyseal bone defects: an ovine model. Vet Surg 2006;35:445–52.

[37] Potier E, Ferreira E, Andriamanalijaona R, Pujol JP, Oudina K, Logeart-Avramoglou D, et al. Hypoxia affects mesenchymal stromal cell osteogenic differentiation and angiogenic factor expression. Bone 2007;40:1078–87.

[38] Lee SY, Miwa M, Sakai Y, Kuroda R, Matsumoto T, Iwakura T, et al. In vitro multipotentiality and characterization of human unfractured traumatic hemarthrosis-derived progenitor cells: a potential cell source for tissue repair. J Cell Physiol 2007;210:561–6.

[39] Shao X, Goh JC, Hutmacher DW, Lee EH, Zigang G. Repair of large articular osteochondral defects using hybrid scaffolds and bone marrow-derived mesenchymal stem cells in a rabbit model. Tissue Eng 2006;12:1539–51.

[40] Marcacci M, Kon K, Moukhachev V, Lavroukov A, Kutepov S, Quarto R, et al. Stem cells associated with macroporous bioceramics for long bone repair: 6 to 7 year outcome of a pilot clinical study. Tissue Eng 2007;13:947–55.

[41] Quarto R, Mastrogiacomo M, Cancedda R, Kutepov SM, Mukhachev V, Lavroukov A, et al. Repair of large bone defects with the use of autologous bone marrow stromal cells. N Engl J Med 2001;344:385–6.

[42] Vacanti CA, Bonassar LJ, Vacanti MP, Shufflebarger J. Replacement of an avulsed phalanx with tissue-engineered bone. N Engl J Med 2001;344:1511–4.

[43] Hentz VR, Chang J. Tissue engineering for reconstruction of the thumb. N Engl J Med 2001;344:1547–8.

[44] Morishita T, Honoki K, Ohgushi H, Kotobuki N, Matsushima A, Takakura Y. Tissue engineering approach to the treatment of bone tumors: three cases of cultured bone grafts derived from patients' mesenchymal stem cells. Artif Organs 2006;30:115–8.

[45] Gronthos S. Reconstruction of human mandible by tissue engineering. Lancet 2004;364:735–6.

[46] Warnke PH, Springer IN, Acil Y, Julga G, Wiltfang J, Ludwig K, et al. The mechanical integrity of in vivo engineered heterotopic bone. Biomaterials 2006;27:1081–7.

[47] Heliotis M, Lavery KM, Ripamonti U, Tsiridis E, di Silvio L. Transformation of a prefabricated hydroxyapatite/osteogenic protein-1 implant into a vascularised pedicled bone flap in the human chest. Int J Oral Maxillofac Surg 2006;35:265–9.

[48] Hibi H, Yamada Y, Ueda M, Endo Y. Alveolar cleft osteoplasty using tissue-engineered osteogenic material. Int J Oral Maxillofac Surg 2006;35:551–5.

[49] Yamada Y, Ueda M, Hibi H, Baba S. A novel approach to periodontal tissue regeneration with mesenchymal stem cells and platelet-rich plasma using tissue engineering technology: a clinical case report. Int J Periodontics Restorative Dent 2006;26:363–9.

[50] Schimming R, Schmelzeisen R. Tissue-engineered bone for maxillary sinus augmentation. J Oral Maxillofac Surg 2004;62:724–9.

[51] Schmelzeisen R, Schimming R, Sittinger M. Making bone: implant insertion into tissue-engineered bone for maxillary sinus floor augmentation-a preliminary report. J Craniomaxillofac Surg 2003;31:34–9.

[52] Springer IN, Nocini PF, Schlegel KA, De Santis D, Park J, Warnke PH, et al. Two techniques for the preparation of cell-scaffold constructs suitable for sinus augmentation: steps into clinical application. Tissue Eng 2006;12:2649–56.

[53] Caplan AI, Correa D. The MSC: an injury drugstore. Cell Stem Cell 2011;9:11–5.

[54] El Backly RM, Cancedda R. Bone marrow stem cells in clinical application: harnessing paracrine roles and niche mechanisms. Adv Biochem Eng Biotechnol 2010;123:265–92.

[55] Schinkothe T, Bloch W, Schmidt A. In vitro secreting profile of human mesenchymal stem cells. Stem Cells Dev 2008;17:199–206.

[56] Ranganath SH, Levy O, Inamdar MS, Karp JM. Harnessing the mesenchymal stem cell secretome for the treatment of cardiovascular disease. Cell Stem Cell 2012;10:244–58.

[57] Roche S, D'Ippolito G, Gomez LA, Bouckenooghe T, Lehmann S, Montero-Menei CN, et al. Comparative analysis of protein expression of three stem cell populations: models of cytokine delivery system in vivo. Int J Pharm 2012;440:72–82.

[58] Heil M, Ziegelhoeffer T, Mees B, Schaper W. A different outlook on the role of bone marrow stem cells in vascular growth: bone marrow delivers software not hardware. Circ Res 2004;94:573–4.

[59] Oskowitz A, McFerrin H, Gutschow M, Carter ML, Pochampally R. Serum-deprived human multipotent mesenchymal stromal cells (MSCs) are highly angiogenic. Stem Cell Res 2011;6:215–25.

[60] Kim JM, Kim J, Kim YH, Kim KT, Ryu SH, Lee TG, et al. Comparative secretome analysis of human bone marrow-derived mesenchymal stem cells during osteogenesis. J Cell Physiol 2012;228:216–24.

[61] Tasso R, Fais F, Reverberi D, Tortelli F, Cancedda R. The recruitment of two consecutive and different waves of host stem/progenitor cells during the development of tissue-engineered bone in a murine model. Biomaterials 2010;31:2121–9.

[62] Tasso R, Augello A, Boccardo S, Salvi S, Carida M, Postiglione F, et al. Recruitment of a host's osteoprogenitor cells using exogenous mesenchymal stem cells seeded on porous ceramic. Tissue Eng Part A 2009;15:2203–12.

[63] Chen FM, Sun HH, Lu H, Yu Q. Stem cell-delivery therapeutics for periodontal tissue regeneration. Biomaterials 2012;33:6320–44.

[64] Sarkar D, Spencer JA, Phillips JA, Zhao W, Schafer S, Spelke DP, et al. Engineered cell homing. Blood 2011;118:e184−91.

[65] Sarkar D, Ankrum JA, Teo GS, Carman CV, Karp JM. Cellular and extracellular programming of cell fate through engineered intracrine-, paracrine-, and endocrine-like mechanisms. Biomaterials 2011;32:3053−61.

[66] Kolar P, Gaber T, Perka C, Duda GN, Buttgereit F. Human early fracture hematoma is characterized by inflammation and hypoxia. Clin Orthop Relat Res 2011;469:3118−26.

[67] Rosell A, Arai K, Lok J, He T, Guo S, Navarro M, et al. Interleukin-1beta augments angiogenic responses of murine endothelial progenitor cells in vitro. J Cereb Blood Flow Metab 2009;29:933−43.

[68] Cho HH, Kim YJ, Kim JT, Song JS, Shin KK, Bae YC, et al. The role of chemokines in proangiogenic action induced by human adipose tissue-derived mesenchymal stem cells in the murine model of hindlimb ischemia. Cell Physiol Biochem 2009;24:511−8.

[69] Khosla S, Westendorf JJ, Modder UI. Concise review: insights from normal bone remodeling and stem cell-based therapies for bone repair. Stem Cells 2010;28:2124−8.

[70] Ma XL, Sun XL, Wan CY, Ma JX, Tian P. Significance of circulating endothelial progenitor cells in patients with fracture healing process. J Orthop Res 2012;30:1860−6.

[71] Pignolo RJ, Kassem M. Circulating osteogenic cells: implications for injury, repair, and regeneration. J Bone Miner Res 2011;26:1685−93.

[72] Otsuru S, Tamai K, Yamazaki T, Yoshikawa H, Kaneda Y. Circulating bone marrow-derived osteoblast progenitor cells are recruited to the bone-forming site by the CXCR4/stromal cell-derived factor-1 pathway. Stem Cells 2008;26:223−34.

[73] Otsuru S, Tamai K, Yamazaki T, Yoshikawa H, Kaneda Y. Bone marrow-derived osteoblast progenitor cells in circulating blood contribute to ectopic bone formation in mice. Biochem Biophys Res Commun 2007;354:453−8.

[74] Mastrogiacomo M, Scaglione S, Martinetti R, Dolcini L, Beltrame F, Cancedda R, et al. Role of scaffold internal structure on in vivo bone formation in macroporous calcium phosphate bioceramics. Biomaterials 2006;27:3230−7.

[75] Wen Y, Jiang B, Cui J, Li G, Yu M, Wang F, et al. Superior osteogenic capacity of different mesenchymal stem cells for bone tissue engineering. Oral Surg Oral Med Oral Pathol Oral Radiol 2012;xx:e1−9.

[76] Arthur A, Zannettino A, Gronthos S. The therapeutic applications of multipotential mesenchymal/stromal stem cells in skeletal tissue repair. J Cell Physiol 2009;218:237−45.

[77] Neman J, Hambrecht A, Cadry C, Jandial R. Stem cell-mediated osteogenesis: therapeutic potential for bone tissue engineering. Biologics 2012;6:47−57.

[78] Jones E, Yang X. Mesenchymal stem cells and bone regeneration: current status. Injury 2011;42:562−8.

[79] Colnot C. Cell sources for bone tissue engineering: insights from basic science. Tissue Eng Part B Rev 2011;17:449−57.

[80] Dwek JR. The periosteum: what is it, where is it, and what mimics it in its absence? Skeletal Radiol 2010;39:319−23.

[81] Schmid J, Wallkamm B, Hammerle CH, Gogolewski S, Lang NP. The significance of angiogenesis in guided bone regeneration. A case report of a rabbit experiment. Clin Oral Implants Res 1997;8:244−8.

[82] van Gastel N, Torrekens S, Roberts SJ, Moermans K, Schrooten J, Carmeliet P, et al. Engineering vascularized bone: osteogenic and pro-angiogenic potential of murine periosteal cells. Stem Cells 2012;30:2460−71.

[83] Nguyen LH, Annabi N, Nikkhah M, Bae H, Binan L, Park S, et al. Vascularized bone tissue engineering: approaches for potential improvement. Tissue Eng Part B Rev 2012;18:363−82.

[84] Giannoni P, Mastrogiacomo M, Alini M, Pearce SG, Corsi A, Santolini F, et al. Regeneration of large bone defects in sheep using bone marrow stromal cells. J Tissue Eng Regen Med 2008;2:253−62.

[85] Guo H, Li X, Yuan X, Ma X. Reconstruction of radial bone defects using the reinforced tissue-engineered periosteum: an experimental study on rabbit weight-bearing segment. J Trauma Acute Care Surg 2012;72:E94−100.

[86] Gassling V, Hedderich J, Acil Y, Purcz N, Wiltfang J, Douglas T. Comparison of platelet rich fibrin and collagen as osteoblast-seeded scaffolds for bone tissue engineering applications. Clin Oral Implants Res 2011;24:320−8.

[87] Zhao L, Zhao J, Wang S, Wang J, Liu J. Comparative study between tissue-engineered periosteum and structural allograft in rabbit critical-sized radial defect model. J Biomed Mater Res B Appl Biomater 2011;97:1−9.

[88] Zhang X, Awad HA, O'Keefe RJ, Guldberg RE, Schwarz EM. A perspective: engineering periosteum for structural bone graft healing. Clin Orthop Relat Res 2008;466:1777−87.

[89] Ozaki A, Tsunoda M, Kinoshita S, Saura R. Role of fracture hematoma and periosteum during fracture healing in rats: interaction of fracture hematoma and the periosteum in the initial step of the healing process. J Orthop Sci 2000;5:64−70.

[90] Colnot C. Skeletal cell fate decisions within periosteum and bone marrow during bone regeneration. J Bone Miner Res 2009;24:274−82.

[91] Zhang X, Xie C, Lin AS, Ito H, Awad H, Lieberman JR, et al. Periosteal progenitor cell fate in segmental cortical bone graft transplantations: implications for functional tissue engineering. J Bone Miner Res 2005;20:2124−37.

[92] Maes C, Kobayashi T, Selig MK, Torrekens S, Roth SI, Mackem S, et al. Osteoblast precursors, but not mature osteoblasts, move into developing and fractured bones along with invading blood vessels. Dev Cell 2010;19:329−44.

[93] Kanczler JM, Oreffo RO. Osteogenesis and angiogenesis: the potential for engineering bone. Eur Cell Mater 2008;15:100−14.

[94] Bellucci D, Sola A, Gentile P, Ciardelli G, Cannillo V. Biomimetic coating on bioactive glass-derived scaffolds mimicking bone tissue. J Biomed Mater Res A 2012;100:3259−66.

[95] Jabbarzadeh E, Deng M, Lv Q, Jiang T, Khan YM, Nair LS, et al. VEGF-incorporated biomimetic poly(lactide-co-glycolide) sintered microsphere scaffolds for bone tissue engineering. J Biomed Mater Res B Appl Biomater 2012;100:2187−96.

[96] Hollister SJ. Scaffold engineering: a bridge to where? Biofabrication 2009;1:1−14.

[97] Jakob M, Saxer F, Scotti C, Schreiner S, Studer P, Scherberich A, et al. Perspective on the evolution of cell-based bone tissue engineering strategies. Eur Surg Res 2012;49:1−7.

[98] Lissenberg-Thunnissen SN, de Gorter DJ, Sier CF, Schipper IB. Use and efficacy of bone morphogenetic proteins in fracture healing. Int Orthop 2011;35:1271—80.

[99] Chen FM, Zhang M, Wu ZF. Toward delivery of multiple growth factors in tissue engineering. Biomaterials 2010;31:6279—308.

[100] Marx RE. Platelet-rich plasma (PRP): what is PRP and what is not PRP? Implant Dent 2001;10:225—8.

[101] Nurden AT. Platelets, inflammation and tissue regeneration. Thromb Haemost 2011;105(Suppl. 1):S13—33.

[102] Alsousou J, Thompson M, Hulley P, Noble A, Willett K. The biology of platelet-rich plasma and its application in trauma and orthopaedic surgery: a review of the literature. J Bone Joint Surg Br 2009;91:987—96.

[103] Plachokova AS, Nikolidakis D, Mulder J, Jansen JA, Creugers NH. Effect of platelet-rich plasma on bone regeneration in dentistry: a systematic review. Clin Oral Implants Res 2008;19:539—45.

[104] Metzler P, von Wilmowsky C, Zimmermann R, Wiltfang J, Schlegel KA. The effect of current used bone substitution materials and platelet-rich plasma on periosteal cells by ectopic site implantation: an in-vivo pilot study. J Craniomaxillofac Surg 2012;40:409—15.

[105] Sheth U, Simunovic N, Klein G, Fu F, Einhorn TA, Schemitsch E, et al. Efficacy of autologous platelet-rich plasma use for orthopaedic indications: a meta-analysis. J Bone Joint Surg Am 2012;94:298—307.

[106] Iqbal J, Pepkowitz SH, Klapper E. Platelet-rich plasma for the replenishment of bone. Curr Osteoporos Rep 2011;9:258—63.

[107] Chatterjea A, Yuan H, Fennema E, Chatterjea S, Garritsen H, Renard A, et al. Engineering new bone via a minimally invasive route using human bone marrow derived stromal cell aggregates, micro ceramic particles and human platelet rich plasma gel. Tissue Eng Part A 2012;19:340—9.

[108] Man Y, Wang P, Guo Y, Xiang L, Yang Y, Qu Y, et al. Angiogenic and osteogenic potential of platelet-rich plasma and adipose-derived stem cell laden alginate microspheres. Biomaterials 2012;33:8802—11.

[109] Kim ES, Kim JJ, Park EJ. Angiogenic factor-enriched platelet-rich plasma enhances in vivo bone formation around alloplastic graft material. J Adv Prosthodont 2010;2:7—13.

[110] El Backly RM, Zaky SH, Muraglia A, Tonachini L, Brun F, Canciani B, et al. A platelet-rich plasma-based membrane as a periosteal substitute with enhanced osteogenic and angiogenic properties: a new concept for bone repair. Tissue Eng Part A 2012;19:152—65.

Chapter 57

Nerve Regeneration and Bioengineering

Tina Sedaghati[a], Gavin Jell[a], and Alexander M. Seifalian[a,b]

[a]UCL Centre for Nanotechnology and Regenerative Medicine, Division of Surgical and Interventional Sciences, University College London, London, UK, [b]Royal Free London NHS Foundation Trust Hospital, London, UK

Chapter Outline

57.1 INTRODUCTION

Peripheral nerve (PN) injury can be classified as a common injury with an estimated incidence of 300,000 reported cases per year in Europe [1]. Trauma is the major cause of the reported PN injury cases [2]. Congenital, mechanical, chemical, and thermal pathologies can be other sources of damage to the PN system. As a result of injury, a gap would be generated between two ends of the nerve. The size of gap determines recovery rate. Incomplete repair of the resulted gap can cause numbness, impairment of sensory or motor function, and increase in the possibility of developing chronic pain [3]. The majority of patients suffering from PN trauma are mainly at the peak of their employment; accordingly, any loss of function can have economic consequences. PN injuries can be classified into neuropraxia, axonotmesis, and neurotmesis depending on the type and severity of damage (Table 57.1) [4].

Accordingly, nerve cells respond differently to the damages ranging from repair by regeneration to cell death. Regeneration of the nerve across a gap occurs only if a suitable nerve growth supporting environment is provided. Regeneration is initiated by the interruption of the myelin sheaths causing Schwann cells (SCs) to detach from the axons known as Wallerian degeneration. The myelin sheath debris is then phagocytosed by resident and recruited macrophages. Later, the detached SCs proliferate and form columns of cells lining the endoneurial tubes known as Bands of Büngner. These newly formed columns serve as guidance for the regenerating axons (Figure 57.1) [5].

57.2 CURRENT TREATMENT OPTIONS AND THEIR LIMITATIONS

Numerous microsurgical techniques have been developed to repair nerve defects. They can be divided into three main groups, as shown in Figure 57.2: (1) coaptation, (2) nerve tissue grafting, and (3) commercial nerve conduits made of biomaterials.

57.2.1 Coaptation

End-to-end coaptation is the most common repair technique of nerve defects with a short gap (<8 mm). This technique can be performed in 82% of PN injury cases where the proximal nerve stump is directly sutured to the distal one [2]. However, this method of repair is not practical if the gap between the two ends of a nerve is >8 mm. In this situation, tension is generated at the

Regenerative Medicine Applications in Organ Transplantation.

TABLE 57.1 PN Injury Classification Based on the Extent and Type of the Damage

Injury Type	Description
Neuropraxia	Mildest nerve injury
	Nerve structure remains intact
	Complete recovery occurs within days to weeks
Axonotmesis	Severe nerve injury
	Loss of axon continuity with no or partially myelin intruption
	Maintenance of connective tissue structure surrounding the nerve
	Possible recovery without surgery
	Surgery requirement depending on severity of scar formation
	Regeneration occurs over weeks to years
Neurotmesis	Most severe nerve injury
	Complete disruption of the nerve, including axons, myelin sheath, and all surrounding nerve connective tissue structure
	Surgery intervention is required
	Partial recovery may be achieved

suturing site resulting in the reduction of blood flow to the repaired site [3,6], consequently, axonal regeneration is inhibited. Furthermore, neural regeneration could be impeded due to inflammation and fibrotic reactions induced by synthetic suture used in this repair technique [6]. The forced orientation given by surgeon during the direct suturing of nerve fibers may also cause morphomeric mismatch of the regenerating axons [2]. An alternative tension-free nerve repair method would therefore be the best option for the reconstruction of a completely transacted nerve.

57.2.2 Nerve Autograft

Up to date, nerve autografting remains the most frequent, successful, and "gold standard" approach to repair PN defects regardless of their gap size [3]. So far, autologous nerve graft has been the most successful biological material used as a nerve conduit. An autologous nerve graft provides a suitable scaffold which contains all the essential components for nerve regeneration, such as SCs and growth factors [7], without any risk of immune rejection. The cutaneous sural sensory nerve is the most commonly used donor nerve, which is composed of multiple small fascicles. Faster vascularization is the main advantage of using small and thin nerve grafts compared

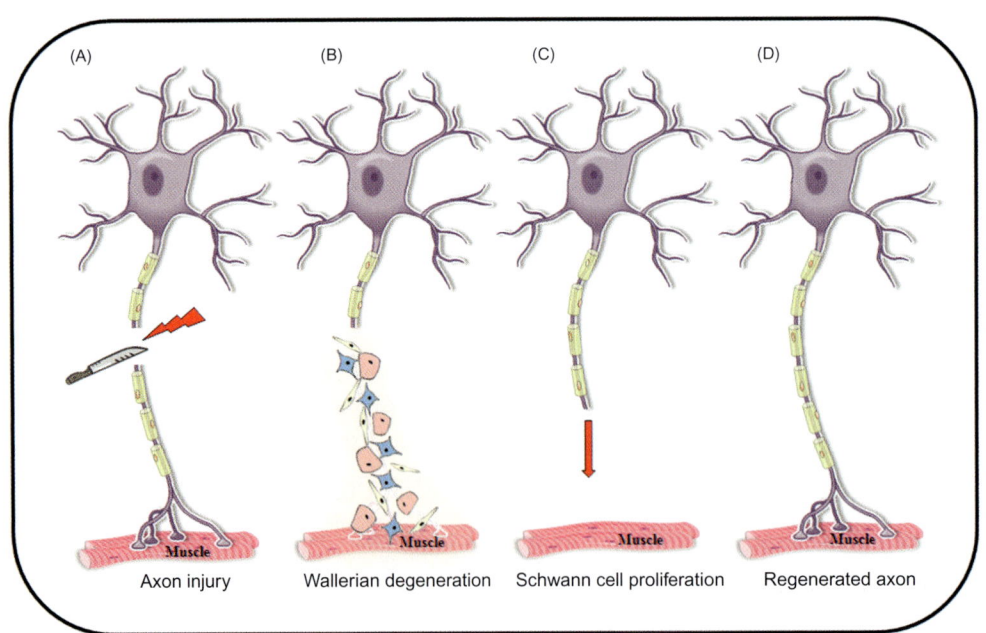

FIGURE 57.1 Axonal regeneration in the PNS: (A) neural injury, (B) detection of injury by resident fibroblasts and macrophages in response to myelin debris and further recruitment of repair cells, including SCs and macrophages, via cytokine gradients, (C) proliferating SCs form a column of cells, known as bands of Bunger, which act as a guide for directional regeneration of the axons, and (D) successful neuronal regeneration and target reinnervation.

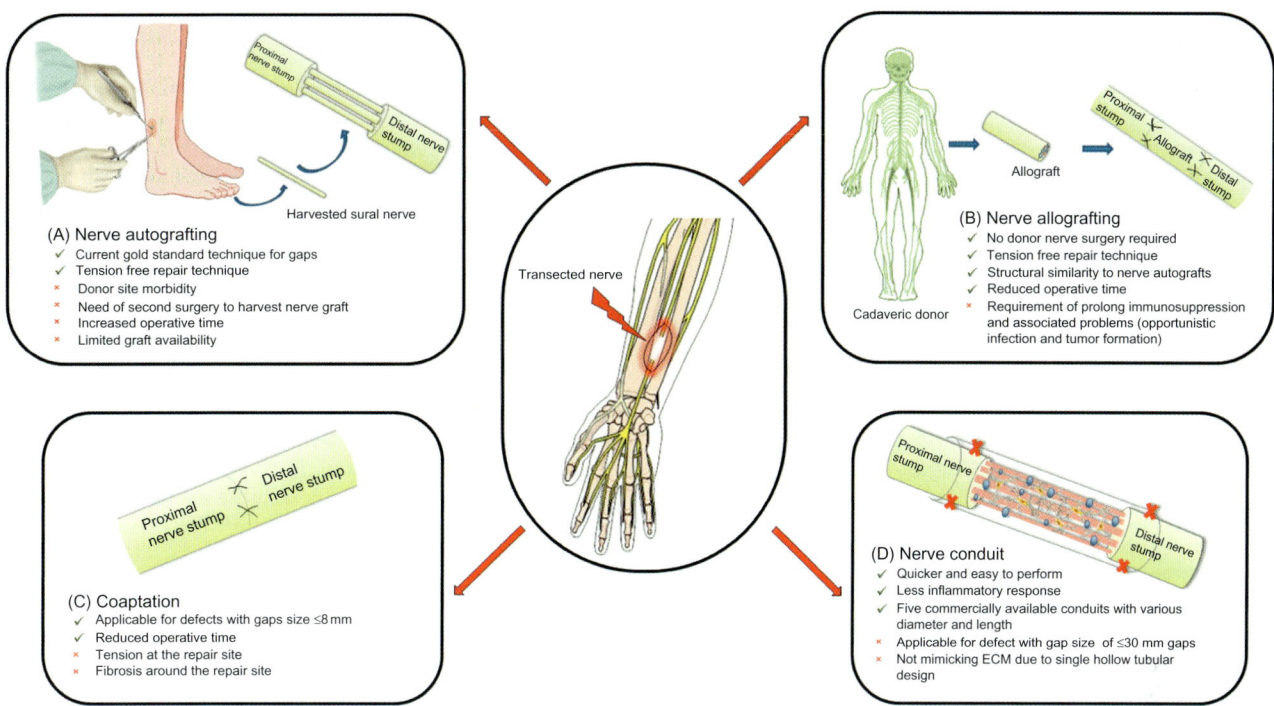

FIGURE 57.2 Advantages (green tick) and disadvantages (red cross) of current treatment options to repair PN defects: (**A**) nerve autografting, (B) nerve allografting, (C) coaptation, and (D) nerve conduit.

to larger and thicker ones [6]. Furthermore, cutaneous nerves of arm and forearm, dorsal sensory part of radial nerve, and distal segment of anterior interosseous nerve are other occasionally usable donor nerves [6]. Anesthesia, extensive disfiguring endoneural scars, painful neuroma formation, possibility of a hematoma or infection as well as denervation and numbness of the harvesting site are possible drawbacks associated with this repair technique [3,8]. It is important to mention that a full functional recovery is not achievable even by using this "gold standard" technique. Size and sensory nature of the donor nerve limit its ability to achieve full functional recovery. Insufficient availability of suitable donor nerve in elder patients as well as those suffering from diabetes complications are other limiting factors of this technique [9].

57.2.3 Nerve Allograft

There are several experimental reports in which allograft nerve has been used as an alternative to nerve autograft to bridge two ends of a nerve together both in nonhuman [10] and human [11–13] primates. Allograft tissue would serve as a temporary scaffold in which it enhances neural regeneration by providing the essential structural characteristics of the nerve tissue. It would be replaced by the host's own tissue over time. Moreover, being readily available as an unlimited source of graft material made this substrate an attractive choice for neural repair. Systemic and prolonged

immunosuppression is required for this technique to prevent rejection of the graft, in turn, increasing the possibility of opportunistic infections and tumor formation in treated patients [13].

AxoGen Company, manufacturer of Avance® nerve graft, claims their graft is nonimmunogenic due to their decellularizing and cleansing techniques. Thus, it avoids the need of systemic immune suppression associated with allografting [13]. This graft has been tested in accordance with ISO 10993 standards and it may be a suitable alternative option to autografting since a comparable rate of sensation return is achieved using this graft in nerve defects ranging from 5 to 50 mm without infection or rejection [14].

57.2.4 Commercially Available FDA and CE-Approved Nerve Conduits

To overcome the limitations associated with nerve autograft and allograft techniques, nerve conduits made of new biomaterial have emerged as an alternative option. This tensionless nerve repair technique is relatively rapid and easier to perform. It can also prevent the donor site morbidity associated with autografting as well as the prolonged immunosuppression associated with allografting. Furthermore, degradation and physiochemical properties of synthetic polymers can be easily tailored compared to the natural materials. To date, NeuraGen, NeuroFlex/ NeuroMatrix, Neurotube, Neurolac nerve conduits and

Box 1　PN Injury Classification Based on the Extent and Type of the Damage

NeuraGen Nerve Conduit

NeuraGen nerve conduit, commercialized by Integra NeuroSciences, is a semipermeable biocompatible tube made of purified collagen I isolated from Bovine Achilles tendon. It has defined permeability and great mechanical strength owing to the preservation of collagen's fibrillar structure during the manufacturing process [15]. It would be completely degraded via normal metabolic pathways by 36 months [15,16]. NeuraGen can be made in different inner diameters ranging from 1.5 to 7 mm with maximum length of 30 mm. It has been reported in several studies that the level of functional recovery achieved with this conduit was similar to direct coaptation technique [15,17]. FDA has approved this conduit in June 2001 despite a small risk of transmitting animal diseases, such as the new variant Creutzfeldt Jacob disease, to human [18].

Neurotube Nerve Conduit

The neurotube is a flexible, porous, and biodegradable nerve conduit made of woven mesh of polyglycolic acid (PGA) polymer. This conduit has been approved by FDA in 1999 and commercialized by Synovis Life Technologies Inc. PGA polymer is being commonly used in absorbable Vicryl sutures and sheaths for various applications. Neurotube conduit porosity allows the infiltration of nutrients and oxygen, which plays key roles during nerve regeneration. Luminal collapse could be prevented in this conduit due to designing suitable wall thickness with enough strength. Flexibility of conduit accommodates the movement of joints and associated tendon gliding. This conduit can be made with inner diameters of 2−8 mm and length of 20 or 40 mm. A majority of experimental studies using PGA conduit observed conduit's fragments around 1 month postimplantation [19]. The company claims complete absorption of conduit by hydrolysis in approximately 6 months after implantation.

Neurolac Nerve Conduit

Neurolac nerve conduit is another synthetic biodegradable nerve conduit composed of poly-DL-lactic-ε-caprolactone

(p(DLLA-ε-CL)). It has been approved for clinical use in PN and cranial nerve repairs by FDA in 2003 and commercialized by Polyganics Inc. This nerve conduit has been used in several studies to bridge short nerve gaps (several mm), which showed promising results. The Polyganics company states complete absorption of the conduit through hydrolysis within 16 months of implantation. However, Meek and Jansen found some remnants of Neurolac conduit and foreign body reactions present after 2 years follow-up after implantation [20]. Damage to the nerve's surrounding tissue could be reduced by using this nerve conduit as its degradation products are less acidic compared to NeuraGen tube's ones [21]. The transparency is another advantage of this conduit in which it allows the surgeon to observe and confirm the correct position of the nerve ends as well as eliminating any blood clots within the tube which may hamper PN regeneration.

Neuroflex and NeuroMatrix Cuffs

NeuroFlex and NeuroMatrix are resorbable, nonfriable, semipermeable tubular matrices with pore size ranging from 0.001 to 0.005 μm. Neuroflex is flexible with added kink resistance property, which allows the tube to bend without kinking or collapsing compared to nonflexible NeuroMatrix. Both of these matrices are engineered from highly purified type I collagen derived from bovine Achilles tendon. The lengths of currently available products are 25 mm with different internal diameter sizes ranging from 2 to 6 mm. However, the company claims that t has the ability of fabricating tubes with larger or smaller inner diameters. Based on the company statement, these matrices are nonpyrogenic and they would be reabsorbed over a period of about 4−8 months. These tubes obtained FDA approval in September 2001 and have been commercialized by Collagen Matrix Inc. To date, no publication is reported about the clinical implantation of Neuroflex or NeuroMatrix.

Avance® nerve graft have been produced and approved either by FDA or CE, to be used in clinic as an alternative to autologous nerve transplants (Box 1).

57.3 TISSUE ENGINEERING STRATEGIES

Since the late 19th century, a series of new techniques based on tissue engineering have been developed to repair PN defects known as entubulation. In this repair technique, a tubular construct would be used to bridge the transacted nerve ends together. This repair method provides an appropriate physiological microenvironment at the damaged site, which has a desirable influence on SC growth, maturation of nerve fibers throughout the regeneration procedure, proper fiber orientation, and minimizes the invasion of exogenous cells and fibrous tissue into the wound site [22]. This repair technique would avoid any axonal escape at the damaged site. Accordingly, the possibility of neuroma formation would be reduced. Additionally, this guidance of axons guarantees that the optimal number of fibers reaches their target quickly, as larger numbers of nerve fibers result in better end-organ reinnervation [9]. Currently, 18% of reconstruction of PN damages is performed by means of entubulation or grafting. With the evolution of tissue engineering, a large range of biological and synthetic materials have been investigated *in vitro* and *in vivo* for the development of potential nerve conduits (Table 57.2).

TABLE 57.2 Summary of Majority Natural and Synthetic Materials Used in Research as Nerve Conduit to Enhance PN Regeneration

Natural	Synthetic	
Bone	Biodegradable	Nonbiodegradable
Artery [23]	PGA [24]	Silicon [25]
Vein [26]	PCL [27]	PTFE [28]
Muscle [29]	PLLA [30]	
Purified collagen [31]	PHB [32]	
Gelatin [33]	PVA [34]	
Alginate [35]	Above copolymers [36,37]	
Chitosan [38]	POSS-PCL [39]	

Polyglycolic acid (PGA), poly-ε-caprolactone (PCL), poly-ʟ-lactic acid (PLLA), poly-β-hydroxybutyrate (PHB), polyvinyl alcohol (PVA), polyhedral oligomeric silsesquioxane-modified poly-caprolactone-urea urethane (POSS-PCL), and poly-4-fluoroethylene (PTFE).

57.3.1 Nerve Conduit Ideal Properties

The ideal nerve conduit should be biocompatible by means of being noncytotoxic, nonimmunogenic, and noncarcinogenic, biodegradable at a controllable rate according to the axonal growth rate, sufficiently porous to mimic the function of natural extracellular matrix (ECM) by allowing the diffusion of nutrients, whereas inhibiting the invasion of cells and factors that cause pathological processes, such as scar and edema, favorable for healing and regeneration processes, and flexible to facilitate constant protection of the regenerating axons as well as permitting surgical manipulation (Figure 57.3) [40].

Furthermore, the shape of the implanted nerve conduit is another important factor in which granulomatous inflammatory response was found to be more severe when the nerve conduit sample had sharper borders. There is a probability that sheer forces among the cell membrane and the nerve conduit would be possibly higher at the edges during movement of the conduit. These sites with higher sheer forces result in cell damage which may then enhance the inflammatory response as cited by Lam et al. [41]. Moreover, the volume of the nerve conduit, determined by its diameter, has an effect on the availability and gradient of diffusible proregenerative neurotrophic factors (NTFs) released by proximal end of

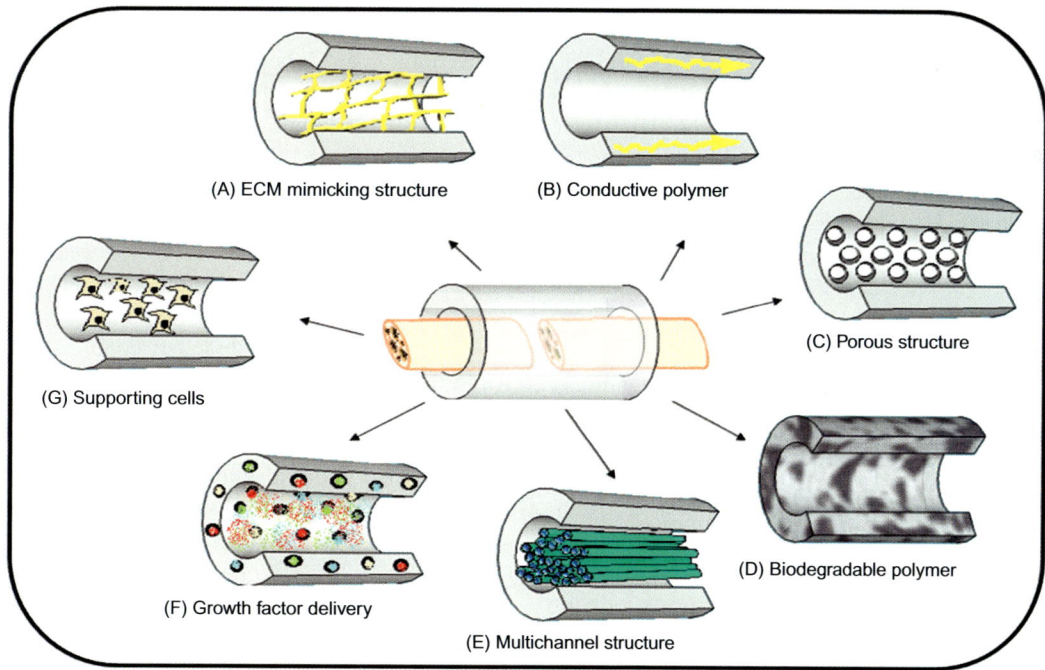

FIGURE 57.3 Design strategies for fabrication of a nerve conduit. (A) Intra-luminal ECM mimicking to enhance Schwann cell migration and guidance of the regenerating axons. (B) Conductive materials may allow retain of function by transferring the electrical signals during regeneration. (C) Porosity to optimize nutrient and oxygen exchange. (D) and (G) Suitable biodegradable material which support neuronal cells during regeneration. (E) Multi-channel design to control axonal dispersion by mimicking native nerve fascicular structure. (F) Incorporation of growth factors into the nerve conduit may accelerate neuronal regeneration.

injured nerve in which as volume is increased, the NTF concentration would be decreased as they become diluted within the space of conduit [42]. However, diameter of the nerve conduit should be slightly larger than the injured nerve diameter to avoid compression of the nerve [8].

57.3.2 Nerve Conduit Biomaterials

Usage of the decalcified bone to bridge a sciatic nerve gap in a canine model was one of the first attempts in the use of biological substances in the field of neural tissue engineering [43]. Tubes made of natural substances, such as bone, arteries, vein, skeletal muscle and ECM components, have been investigated in various studies. The use of a majority of these natural tissues as conduits was frequently complicated by scar formation and collapse due to the surrounding tissue pressure with final results of impairment of nerve regeneration and repair [3]. Furthermore, the clinical use of ECM components, such as laminin, collagen, and fibronectin, is currently limited as they are mostly derived from animal or cancerous sources.

Synthetic polymers used as conduits can be divided into two main groups: (1) nonbiodegradable and (2) biodegradable polymers. Originally, inert nondegradable materials, like silicon, had shown promising results due to the accumulation of a high concentration of NTFs by being impermeable. However, the clinical use of these nonbiodegradable conduits is limited as they cause chronic nerve compression and foreign body reaction following the regeneration of the nerve. Therefore, biodegradable materials, especially aliphatic polyesters, have been strongly considered as nerve scaffolding material as they eliminate the need of removing the conduit by second surgery associated with nonbiodegradable materials [44]. The degradation rate of conduits made of biodegradable polymers might vary depending on the material, the nerve type and size, as well as the defect size. For example, Den Dunnen taught that a nerve conduit must remain intact for a minimum of 3 months to repair a 10 mm gap of the rat's sciatic nerve [45]. Materials which would not remain intact with the nerve long enough and would degrade quickly as a function of time, would not allow full maturation of the regenerated nerve. As a result, quick degradation might result in more scar formation which has a negative effect on regeneration and maturation of the damaged nerve [45]. So, a longer degradation time guarantees that the regenerating nerve would be able to endure the stress of mobilization.

Furthermore, nanomaterials have shown great potential to enhance nerve regeneration. These nanomaterials provide a new dimension of interaction with biological systems that takes place on a subcellular level with a high degree of specificity. It has been well established that cells would respond differently to micro- or nano-clues presented by the scaffold. These clues could have significant effects on cellular behavior, such as growth, differentiation, and ECM production [46—48]. Unique properties of the nanocomposites make them a suitable substance for the next generation of nerve conduits. Incorporation of carbon nanotubes into various polymers is one example [49,50]. Fabrication of a conductive substrate could significantly improve nerve regeneration shown by several studies [51—53]. Furthermore, it has been reported that nanodiamond monolayers provide an excellent growth substrate on various materials for functional neuronal networks and bypass the necessity of protein coating, which could be a limiting factor as they are normally derived from animal sources [54].

In our laboratory, we have been working on the enhancement of nerve regeneration using innovative biodegradable nanocomposite polymers. The best example of such materials is a biodegradable nanocomposite polymer based on the incorporation of polyhedral oligomeric silsesquioxane (POSS) nanoparticles into poly(caprolactone) urethane/urea (PCL) [55]. This biodegradable nanocomposite polymer demonstrated steady degradation with maintenance of its mechanical properties after exposure to hydrolytic enzymes and plasma protein fractions. This nanocomposite showed dramatic degradation by oxidation [56]. In addition, its surface nanotopography enhanced the cellular adhesion, growth, and migration in several cell types [39,57,58]. Extensive work is currently being carried out in our laboratory to optimize the development of a single-channel nerve conduit using this nanocomposite polymer. Additionally, we are working to fabricate a multichannel nerve conduit to facilitate nerve regeneration over longer nerve gaps.

As you can see, the evolution of nerve regeneration lies in the improvement of the nanotechnology field. The majority of the experimental conduits have already achieved similar regenerative potential to autografting. In a little while, the nanofabricated conduits will find their way into the clinical realm as suitable alternatives to traditional autografting. In this way, nanotechnology shows great potential to alter the way in which reconstructive surgeons treat patients with traumatic nerve injury by significantly improving the surgical outcome as well as avoiding the need for multiple-staged operations.

57.3.3 Nerve Conduit Fabrication Techniques

It is crucial to highlight the influence of the fabrication technique on the nerve conduit properties. Different fabrication techniques have been investigated both *in vitro* and *in vivo* to develop an ideal nerve conduit. Solvent casting, particulate leaching, freeze-drying, solvent-induced phase separation, and electrospinning are some of the most popular methods used for fabricating nerve conduits.

Solvent casting is one of the oldest methods of producing high quality nonporous polymeric films with various thicknesses [30,59]. The method is a relatively easy and controllable technique, which involves evaporating the solvent of polymer solution from a casting surface leaving a polymeric film. Having control over the film's thickness and its uniformity is one of the most important advantages of the fabrication technique. The film's physiochemical properties depend on the solvent used, casting mould substrate, as well as casting temperature. Solvent casting can be combined with other common fabrication techniques, such as particulate leaching, to fabricate a porous three-dimensional scaffold for tissue engineering application. Porosity is one of the important factors during regeneration as it allows mass transport into the conduit as well as extrusion of debris following degeneration [60,61]. The freeze-drying method is another technique to produce porous constructs in which mechanical properties, porosity, pore size, interconnectivity, and distribution can be controlled. In this technique, the solvent, frequently water, would be sublimated in a specific condition with defined and controlled temperature and pressure [38,62]. Phase inversion is the most effective method to fabricate porous scaffolds. In this technique, polymer solution would coagulate upon inversion into a nonmiscible solvent. This technique can be merged with the particulate leaching method to control porosity and pore-related properties [63,64]. Electrospinning is one of the promising fabrication techniques in which nanofibers can be produced by the induction of electric charges into the polymer solution. These nanoscale fibers can mimic the natural ECM architecture. So they have shown promising effects on cell adhesion, mass transport, and orientation of newly formed tissue [51,65].

Although, the commercially available nerve conduits have revolutionized the treatment of PN injury, their clinical use is limited to small-diameter nerves with gaps <30 mm. This is due to their basic hollow tube design which fails to support axonal regeneration in defects with longer gaps and larger diameters. Modification of the nerve conduit's luminal environment has risen as a potential strategy to overcome this limitation. Addition of supportive cells or neurotropic and neurotrophic factors into conduit are examples of such modification.

57.3.4 Supportive Cells

Addition of supportive cells into the nerve conduit has shown promising results, however, the complete mechanism of their action is still unknown. A list of the common cells used in the PN regeneration studies is summarized in Table 57.3.

SCs are principal supporting cells essential for promoting axonal fibers regeneration following injury in both the PNS [35] and the CNS [73]. Following Wallerian degeneration, columns of cells would be formed by proliferating SCs, known as bands of Büngner, which would be served as a scaffold for regenerating axons by expressing specific adhesion molecules (CAMs) on their surface membrane [74]. Additionally, SCs would create a more favorable physiological environment for neurite outgrowth and axonal regeneration by synthesis and release of diffusible neurotrophic growth factors [75]. The use of autologous SC transplantation in clinic has been limited due to insufficient donor sources [74] as well as the need for time-consuming secondary surgery to isolate a sufficient amount of pure autologous SCs [35,76]. On the other side, immunological rejection is the main limitation of implanting allogenic SCs [74].

In recent years, there has been a significant increase of interest in using stem cells to repair PN defects. Promising results have been documented in which in some circumstances specific cell differentiation factor was not required as stem cells were capable to synthesize and release factors and support ECM components that promote their differentiation and nerve repair [77]. To this point, skin precursor cells [71], bone marrow [77−79], hair follicle stem cells [80], and adipose-derived stem cells (ADSCs) have been studied both *in vitro* and *in vivo* to enhance PN regeneration. Among the stem cell sources, there has been an exponential attraction to ADSCs in clinical cases [81]. Isolating a sufficient amount of ADSCs easily from adipose tissue in order to be injected immediately postisolation is one of the reasons which made ADSCs an attractive choice for cell therapy in nerve regeneration. Additionally, it eliminates the requirement of immunosuppression associated with allograft cell source by being autologous. Recently, full sciatic nerve regeneration has been achieved on rats treated with the ADSCs containing conduit in which the regenerated nerve fibers were thicker in size compared to the regenerated fibers achieved by the plain conduit [82].

57.3.5 Growth Factor Incorporation

As an alternative to cell therapy, NTF delivery has shown promising results on the enhancement of PN regeneration both *in vitro* and *in vivo*. Nerve injury results in substantial local cell-mediated synthesis of proteins, known as neurotrophic growth factors, such as glial cell-derived neurotrophic factor (GDNF) [83], brain-derived neurotrophic factor (BDNF) [84], transforming growth factor-beta (TGF-β) [85], and fibroblast growth factor (FBF) [7]. Following the injury, there is a biphasic release profile of

TABLE 57.3 Advantages and Disadvantage of Common Cells Incorporated into the Nerve Conduit to Enhance PN Regeneration

Cell Type	Advantages	Disadvantages

SCs [66]	— Secretion of NTFs
	— Participation in the myelination of regenerated axons
	— Nonimmunogenic
		— Difficult to harvest
		  — Invasive nerve biopsies to harvest cells
		  — Frequent contamination of fibroblasts
		  — Relatively small amount of glial cells obtained from primary cultures
		  — Slow expansion of cells *in vitro* and limited survival in culture

Olfactory ensheathing cells (OECs) [67]	— Secretion of NTFs — Participation in the myelination of regenerated axons — Homing in both peripheral and central nervous systems — Higher migratory potential to penetrate glial scars in comparison to SCs — Nonimmunogenic	— Same as Schwann cells
Neural stem cells (NSCs) [68]	— Nonimmunogenic	— High possibility of tumor formation following NSC transplantation
Bone marrow derived mesenchymal stem cells (BMSCs) [69]	— Easy to harvest and expansion in culture — Nonimmunogenic	— Relatively small amount of BMSCs obtained from primary cultures — High donor-dependent variations in quality — Painful invasive intervention — Short half-life
ADSCs [70]	— Participation in the myelination of regenerated axons — Requirement of minimally invasive approaches to harvest cells — Ability to promote motor neurite outgrowth *in vitro* — Nonimmunogenic	— Further evaluation is required to investigate the fate of transplanted cells — Clarify the mechanism of their influence on nerve regeneration
Skin-derived precursor cells [71] and hair follicle stem cells [72]	— Easily accessible source — A large population of neural crest stem cells — Nonimmunogenic	— Donor site morbidity

neurotrophic growth factors. The first rapid boost release takes place within the first 6 h of injury at both the proximal and distal nerve stumps. The second release phase commences 2−3 days postinjury. At this stage, there is a close association between the growth factor release and maximal SCs proliferation in the proximal nerve stump as well as blood borne macrophages penetration to the injury site [22]. Following the nerve injury, significant decline in availability of these NTFs may be one of the reasons for poor recovery [86]. This reduction happens due to the disruption of retrograde transport of growth factors from the distal end of the damaged nerve. Controlled release of NTFs combined with nerve conduit has shown promising outcomes *in vivo*. The delivery of growth factors can be obtained by methods, such as polymeric microspheres [87] or mini-pumps [88].

However, the use of these NTFs is not yet applicable for clinical cases due to their unknown optimal release concentration dosage, short half-life, and optimal release time which are crucial factors during nerve regeneration.

57.4 CONCLUSION AND FUTURE PROSPECTS

Unsatisfactory outcomes and limitations of two main PN repair techniques, coaptation or nerve autografting, have encouraged scientists to investigate alternative options to enhance PN regeneration. Recent advances in nanotechnology [89,90] and tissue engineering [47,91,92] have offered the most effective approach to repair neural defects as a broad range of applications in regenerative medicine has been covered by these two subjects. Development of a synthetic scaffold with structure mimicking to the natural ECM is the key to successful nerve regeneration. This three-dimensional scaffold can regulate tissue progression by managing several cellular behaviors, including proliferation and differentiation, through providing topographical, electrochemical cues. An ideal scaffold for nerve regeneration must be biocompatible, immunologically inert, porous, biodegradable with nontoxic byproducts, conductive, and enhance neurite outgrowth. Various nerve conduits have been fabricated from a wide range of biomaterials, such as natural, synthetic, and biohybrid substances. Conduits made of synthetic materials are widely chosen as ideal scaffold material for PN regeneration both *in vitro* and *in vivo* [93–96] as their physiochemical properties can be easily tailored through easy and economical fabrication processes. However, further tissue engineering improvements are needed as completely satisfactory outcomes have not yet been achieved in clinic.

In the future, neural tissue engineering will significantly profit from advances in biomaterials and nanotechnology sciences. These innovations promise the fabrication of implants with more rational designs, which significantly enhance larger nerve defects regeneration as well as elimination of donor site morbidity associated with nerve autografting. Integration of more self-sustained biological components like SCs or biomolecules such as Argenin–Glycin–Aspartic acid (RGD) peptide [97,98] can be examples of such improvements in which they affect cellular attachment and neurite outgrowth.

Further momentum in the neural tissue engineering field may potentially be the investigation of pioneering nanocomposite materials. For example, the University College London research team has recently developed a biodegradable nanocomposite polymer based on incorporating POSS nanoparticles into PCL polymer. Steady degradation of scaffold when it was in contact with hydrolytic enzymes and plasma protein fractions, while preserving its mechanical properties, is the best example of enhancement of PCL properties following the nanoparticle incorporation [55]. To conclude, the biocompatibility of this nanocomposite with various cell types has made this material a suitable scaffolding substrate for tissue engineering a range of tissues as well as biomedical devices.

REFERENCES

[1] Ciardelli G, Chiono V. Materials for peripheral nerve regeneration. Macromol Biosci 2006;6:13–26.

[2] Lohmeyer JA, Siemers F, Machens HG, Mailander P. The clinical use of artificial nerve conduits for digital nerve repair: a prospective cohort study and literature review. J Reconstr Microsurg 2009;25:55–61.

[3] Hood B, Levene HB, Levi AD. Transplantation of autologous Schwann cells for the repair of segmental peripheral nerve defects. Neurosurg Focus 2009;26.

[4] Pfister BJ, Gordon T, Loverde JR, Kochar AS, Mackinnon SE, Cullen DK. Biomedical engineering strategies for peripheral nerve repair: surgical applications, state of the art, and future challenges. Crit Rev Biomed Eng 2011;39:81–124.

[5] Terenghi G. Peripheral nerve regeneration and neurotrophic factors. J Anat 1999;194:1–14.

[6] Bhandari PS, Sadhotra LP, Bhargava P, Bath AS, Mukherjee MK, Bavdekar RD. What is new in peripheral nerve repair? Indian J Neurotrauma (IJNT) 2007;4:21–3.

[7] Lee AC, Yu VM, Lowe III JB, Brenner MJ, Hunter DA, Mackinnon SE, et al. Controlled release of nerve growth factor enhances sciatic nerve regeneration. Exp Neurol 2003;184:295–303.

[8] Mackinnon SE, Dellon AL. Clinical nerve reconstruction with a bioabsorbable polyglycolic acid tube. Plast Reconstr Surg 1990;85:419–24.

[9] Belkas JS, Munro CA, Shoichet MS, Midha R. Peripheral nerve regeneration through a synthetic hydrogel nerve tube. Restor Neurol Neurosci 2005;23:19–29.

[10] Bain JR, Mackinnon SE, Hudson AR, Wade J, Evans P, Makino A, et al. The peripheral nerve allograft in the primate immunosuppressed with Cyclosporin A. I. Histologic and electrophysiologic assessment. Plast Reconstr Surg 1992;90:1036–46.

[11] Moore AM, MacEwan M, Santosa KB, Chenard KE, Ray WZ, Hunter DA, et al. Acellular nerve allografts in peripheral nerve regeneration: a comparative study. Muscle Nerve 2011;44:221–34.

[12] Yang RG, Zhong HB, Zhu JL, Zuo TT, Wu KJ, Hou SX. Clinical safety about repairing the peripheral nerve defects with chemically extracted acellular nerve allograft. Zhonghua Wai Ke Za Zhi 2012;50:74–6.

[13] Mackinnon SE, Doolabh VB, Novak CB, Trulock EP. Clinical outcome following nerve allograft transplantation. Plast Reconstr Surg 2001;107:1419–29.

[14] Brooks DN, Weber RV, Chao JD, Rinker BD, Zoldos J, Robichaux MR, et al. Processed nerve allografts for peripheral nerve reconstruction: a multicenter study of utilization and outcomes in sensory, mixed, and motor nerve reconstructions. Microsurgery 2012;32:1–14.

[15] Archibald SJ, Krarup C, Shefner J, Li ST, Madison RD. A collagen-based nerve guide conduit for peripheral nerve repair: an electrophysiological study of nerve regeneration in rodents and nonhuman primates. J Comp Neurol 1991;306:685–96.

[16] Dellon AL, Maloney Jr. CT. Salvage of sensation in a hallux-to-thumb transfer by nerve tube reconstruction. J Hand Surg Am 2006;31:1495–8.

[17] Akassoglou K, Akpinar P, Murray S, Strickland S. Fibrin is a regulator of Schwann cell migration after sciatic nerve injury in mice. Neurosci Lett 2003;338:185–8.

[18] Mosahebi A, Simon M, Wiberg M, Terenghi G. A novel use of alginate hydrogel as Schwann cell matrix. Tissue Eng 2001;7:525–34.

[19] Rosen JM, Pham HN, Abraham G, Harold L, Hentz VR. Artificial nerve graft compared to autograft in a rat model. J Rehabil Res Dev 1989;26:1–14.

[20] Meek MF, Jansen K. Two years after in vivo implantation of poly (DL-lactide-epsilon-caprolactone) nerve guides: has the material finally resorbed?. J Biomed Mater Res A 2009;89:734−8.

[21] Luis AL, Rodrigues JM, Lobato JV, Lopes MA, Amado S, Veloso AP, et al. Evaluation of two biodegradable nerve guides for the reconstruction of the rat sciatic nerve. Biomed Mater Eng 2007;17:39−52.

[22] Chang CJ. The effect of pulse-released nerve growth factor from genipin-crosslinked gelatin in Schwann cell-seeded polycaprolactone conduits on large-gap peripheral nerve regeneration. Tissue Eng Part A 2009;15:547−57.

[23] Dahlin LB. Techniques of peripheral nerve repair. Scand J Surg 2008;97:310−6.

[24] Rosson GD, Williams EH, Dellon AL. Motor nerve regeneration across a conduit. Microsurgery 2009;29:107−14.

[25] Lundborg G, Rosen B, Dahlin L, Holmberg J, Rosen I. Tubular repair of the median or ulnar nerve in the human forearm: a 5-year follow-up. J Hand Surg Br 2004;29:100−7.

[26] Marcoccio I, Vigasio A. Muscle-in-vein nerve guide for secondary reconstruction in digital nerve lesions. J Hand Surg Am 2010;35:1418−26.

[27] Jiang X, Mi R, Hoke A, Chew SY. Nanofibrous nerve conduit-enhanced peripheral nerve regeneration. J Tissue Eng Regen Med 2012;:10.1002/term.1531.

[28] Mersa B, Agir H, Aydin A, Sen C. Comparison of expanded poly-tetrafluoroethylene (ePTFE) with autogenous vein as a nerve conduit in rat sciatic nerve defects. Kulak Burun Bogaz Ihtis Derg 2004;13:103−11.

[29] Gattuso JM, Glasby MA, Gschmeissner SE, Norris RW. A comparison of immediate and delayed repair of peripheral nerves using freeze-thawed autologous skeletal muscle grafts in the rat. Br J Plast Surg 1989;42:306−13.

[30] Li J, Shi R. Fabrication of patterned multi-walled poly-L-lactic acid conduits for nerve regeneration. J Neurosci Methods 2007;165:257−64.

[31] Taras JS, Jacoby SM, Lincoski CJ. Reconstruction of digital nerves with collagen conduits. J Hand Surg Am 2011;36:1441−6.

[32] Kalbermatten DF, Pettersson J, Kingham PJ, Pierer G, Wiberg M, Terenghi G. New fibrin conduit for peripheral nerve repair. J Reconstr Microsurg 2009;25:27−33.

[33] Martin-Lopez E, Nieto-Diaz M, Nieto-Sampedro M. Differential adhesiveness and neurite-promoting activity for neural cells of chitosan, gelatin, and poly-L-lysine films. J Biomater Appl 2010;26:791−809.

[34] Rutkowski GE, Heath CA. Development of a bioartificial nerve graft. II. Nerve regeneration in vitro. Biotechnol Prog 2002;18:373−9.

[35] Mosahebi A, Fuller P, Wiberg M, Terenghi G. Effect of allogeneic Schwann cell transplantation on peripheral nerve regeneration. Exp Neurol 2002;173:213−23.

[36] Sun M, Kingham PJ, Reid AJ, Armstrong SJ, Terenghi G, Downes S. In vitro and in vivo testing of novel ultrathin PCL and PCL/PLA blend films as peripheral nerve conduit. J Biomed Mater Res A 2010;93:1470−81.

[37] Li BC, Jiao SS, Xu C, You H, Chen JM. PLGA conduit seeded with olfactory ensheathing cells for bridging sciatic nerve defect of rats. J Biomed Mater Res A 2010;94:769−80.

[38] Simoes MJ, Amado S, Gartner A, Armada-da-Silva PA, Raimondo S, Vieira M, et al. Use of chitosan scaffolds for repairing rat sciatic nerve defects. Ital J Anat Embryol 2010;115:190−210.

[39] Sedaghati T, Yang SY, Mosahebi A, Alavijeh MS, Seifalian AM. Nerve regeneration with aid of nanotechnology and cellular engineering. Biotechnol Appl Biochem 2011;58:288−300.

[40] de Ruiter GC, Malessy MJ, Yaszemski MJ, Windebank AJ, Spinner RJ. Designing ideal conduits for peripheral nerve repair. Neurosurg Focus 2009;26:E5.

[41] Lam KH, Schakenraad JM, Nieuwenhuis P. The influence of hardness, wettability and fragmentation of polymers on the inflammatory response. Biocompatibility Degradable Biomaterials 1992; http://dissertations.ub.rug.nl/faculties/medicine/1992/k.h.lam/

[42] Shin RH, Friedrich PF, Crum BA, Bishop AT, Shin AY. Treatment of a segmental nerve defect in the rat with use of bioabsorbable synthetic nerve conduits: a comparison of commercially available conduits. J Bone Joint Surg Am 2008;91:2194−204.

[43] IJpma FF, Van De Graaf RC, Meek MF. The early history of tubulation in nerve repair. J Hand Surg Eur Vol 2008;33:581−6.

[44] Merle M, Dellon AL, Campbell JN, Chang PS. Complications from silicon-polymer intubation of nerves. Microsurgery 1989;10:130−3.

[45] Den Dunnen WF, Van der Lei B, Schakenraad JM, Blaauw EH, Stokroos I, Pennings AJ, et al. Long-term evaluation of nerve regeneration in a biodegradable nerve guide. Microsurgery 1993;14:508−15.

[46] Saracino GA, Cigognini D, Silva D, Caprini A, Gelain F. Nanomaterials design and tests for neural tissue engineering. Chem Soc Rev 2012;42:225−62.

[47] Zhang L, Webster T. Review: nanotechnology and nanomaterials: promises for improved tissue regeneration. Nano Today 2009;4:66−80.

[48] Lin YL, Jen JC, Hsu SH, Chiu IM. Sciatic nerve repair by microgrooved nerve conduits made of chitosan-gold nanocomposites. Surg Neurol 2008;70:1−18.

[49] Antoniadou EV, Ahmad RK, Jackman RB, Seifalian AM. Next generation brain implant coatings and nerve regeneration via novel conductive nanocomposite development. Conf Proc IEEE Eng Med Biol Soc 2011;2011:3253−7.

[50] Cellot G, Cilia E, Cipollone S, Rancic V, Sucapane A, Giordani S, et al. Carbon nanotubes might improve neuronal performance by favouring electrical shortcuts. Nat Nanotechnol 2009;4:126−133.

[51] Wang Y, Zhao Z, Zhao B, Qi HX, Peng J, Zhang L, et al. Biocompatibility evaluation of electrospun aligned poly (propylene carbonate) nanofibrous scaffolds with peripheral nerve tissues and cells in vitro. Chin Med J (Engl) 2011;124:2361−6.

[52] Huang YC, Hsu SH, Kuo WC, Chang-Chien CL, Cheng H, Huang YY. Effects of laminin-coated carbon nanotube/chitosan fibers on guided neurite growth. J Biomed Mater Res A 2011;99:86−93.

[53] Lee JY, Bashur CA, Goldstein AS, Schmidt CE. Polypyrrole-coated electrospun PLGA nanofibers for neural tissue applications. Biomaterials 2009;30:4325−35.

[54] Thalhammer A, Edgington RJ, Cingolani LA, Schoepfer R, Jackman RB. The use of nanodiamond monolayer coatings to

promote the formation of functional neuronal networks. Biomaterials 2010;31:2097–104.

[55] Raghunath J, Zhang H, Edirisinghe MJ, Darbyshire A, Butler PE, Seifalian AM. A new biodegradable nanocomposite based on polyhedral oligomeric silsesquioxane nanocages: cytocompatibility and investigation into electrohydrodynamic jet fabrication techniques for tissue-engineered scaffolds. Biotechnol Appl Biochem 2009;52:1–8.

[56] Raghunath J, Georgiou G, Armitage D, Nazhat SN, Sales KM, Butler PE, et al. Degradation studies on biodegradable nanocomposite based on polycaprolactone/polycarbonate (80:20%) polyhedral oligomeric silsesquioxane. J Biomed Mater Res A 2009;91:834–44.

[57] Gupta A, Vara DS, Punshon G, Sales KM, Winslet MC, Seifalian AM. In vitro small intestinal epithelial cell growth on a nanocomposite polycaprolactone scaffold. Biotechnol Appl Biochem 2009;54:221–9.

[58] Adwan H, Fuller B, Seldon C, Davidson B, Seifalian A. Modifying three-dimensional scaffolds from novel nanocomposite materials using dissolvable porogen particles for use in liver tissue engineering. J Biomater Appl 2012;24:1–12.

[59] Hsu SH, Su CH, Chiu IM. A novel approach to align adult neural stem cells on micropatterned conduits for peripheral nerve regeneration: a feasibility study. Artif Organs 2009;33:26–35.

[60] Bozkurt A, Lassner F, O'Dey D, Deumens R, Bocker A, Schwendt T, et al. The role of microstructured and interconnected pore channels in a collagen-based nerve guide on axonal regeneration in peripheral nerves. Biomaterials 2012;33:1363–75.

[61] Guo B, Sun Y, Finne-Wistrand A, Mustafa K, Albertsson AC. Electroactive porous tubular scaffolds with degradability and noncytotoxicity for neural tissue regeneration. Acta Biomater 2012;8:144–53.

[62] Chang CJ, Hsu SH. The effect of high outflow permeability in asymmetric poly(DL-lactic acid-co-glycolic acid) conduits for peripheral nerve regeneration. Biomaterials 2006;27:1035–42.

[63] Wen X, Tresco PA. Fabrication and characterization of permeable degradable poly(DL-lactide-co-glycolide) (PLGA) hollow fiber phase inversion membranes for use as nerve tract guidance channels. Biomaterials 2006;27:3800–9.

[64] Zhang N, Zhang C, Wen X. Fabrication of semipermeable hollow fiber membranes with highly aligned texture for nerve guidance. J Biomed Mater Res A 2005;75:941–9.

[65] Kijenska E, Prabhakaran MP, Swieszkowski W, Kurzydlowski KJ, Ramakrishna S. Electrospun bio-composite P(LLA-CL)/collagen I/collagen III scaffolds for nerve tissue engineering. J Biomed Mater Res B Appl Biomater 2012;100:1093–102.

[66] Hu J, Zhou J, Li X, Wang F, Lu H. Schwann cells promote neurite outgrowth of dorsal root ganglion neurons through secretion of nerve growth factor. Indian J Exp Biol 2011;49:177–82.

[67] Radtke C, Wewetzer K, Reimers K, Vogt PM. Transplantation of olfactory ensheathing cells as adjunct cell therapy for peripheral nerve injury. Cell Transplant 2011;20:145–52.

[68] Radtke C, Redeker J, Jokuszies A, Vogt PM. In vivo transformation of neural stem cells following transplantation in the injured nervous system. J Reconstr Microsurg 2010;26:211–2.

[69] Cai S, Shea GK, Tsui AY, Chan YS, Shum DK. Derivation of clinically applicable Schwann cells from bone marrow stromal cells for neural repair and regeneration. CNS Neurol Disord Drug Targets 2011;10:500–8.

[70] Lopatina T, Kalinina N, Karagyaur M, Stambolsky D, Rubina K, Revischin A, et al. Adipose-derived stem cells stimulate regeneration of peripheral nerves: BDNF secreted by these cells promotes nerve healing and axon growth de novo. PLoS One 2011;6: e17899.

[71] Walsh S, Biernaskie J, Kemp SW, Midha R. Supplementation of acellular nerve grafts with skin derived precursor cells promotes peripheral nerve regeneration. Neuroscience 2009;164:1097–107.

[72] Lin H, Liu F, Zhang C, Zhang Z, Kong Z, Zhang X, et al. Characterization of nerve conduits seeded with neurons and Schwann cells derived from hair follicle neural crest stem cells. Tissue Eng Part A 2011;17:1691–8.

[73] Guenard V, Kleitman N, Morrissey TK, Bunge RP, Aebischer P. Syngeneic Schwann cells derived from adult nerves seeded in semipermeable guidance channels enhance peripheral nerve regeneration. J Neurosci 1992;12:3310–20.

[74] Sinis N, Schaller HE, Schulte-Eversum C, Schlosshauer B, Doser M, Dietz K, et al. Nerve regeneration across a 2-cm gap in the rat median nerve using a resorbable nerve conduit filled with Schwann cells. J Neurosurg 2005;103:1067–76.

[75] Mosahebi A, Wiberg M, Terenghi G. Addition of fibronectin to alginate matrix improves peripheral nerve regeneration in tissue-engineered conduits. Tissue Eng 2003;9:209–18.

[76] Hadlock T, Sundback C, Hunter D, Cheney M, Vacanti JP. A polymer foam conduit seeded with Schwann cells promotes guided peripheral nerve regeneration. Tissue Eng 2000; 6:119–27.

[77] Chen CJ, Ou YC, Liao SL, Chen WY, Chen SY, Wu CW, et al. Transplantation of bone marrow stromal cells for peripheral nerve repair. Exp Neurol 2007;204:443–53.

[78] Keilhoff G, Goihl A, Stang F, Wolf G, Fansa H. Peripheral nerve tissue engineering: autologous Schwann cells vs. transdifferentiated mesenchymal stem cells. Tissue Eng 2006;12:1451–65.

[79] Hu J, Zhu QT, Liu XL, Xu YB, Zhu JK. Repair of extended peripheral nerve lesions in rhesus monkeys using acellular allogenic nerve grafts implanted with autologous mesenchymal stem cells. Exp Neurol 2007;204:658–66.

[80] Amoh Y, Li L, Campillo R, Kawahara K, Katsuoka K, Penman S, et al. Implanted hair follicle stem cells form Schwann cells that support repair of severed peripheral nerves. Proc Natl Acad Sci USA 2005;102:17734–8.

[81] Gir P, Oni G, Brown SA, Mojallal A, Rohrich RJ. Human adipose stem cells: current clinical applications. Plast Reconstr Surg 2012;129:1277–90.

[82] Santiago LY, Clavijo-Alvarez J, Brayfield C, Rubin JP, Marra KG. Delivery of adipose-derived precursor cells for peripheral nerve repair. Cell Transplant 2009;18:145–58.

[83] De Boer R, Borntraeger A, Knight AM, Hebert-Blouin MN, Spinner RJ, Malessy MJ, et al. Short- and long-term peripheral nerve regeneration using a poly-lactic-co-glycolic-acid scaffold containing nerve growth factor and glial cell line-derived neurotrophic factor releasing microspheres. J Biomed Mater Res A 2012;100:2139–46.

[84] Takemura Y, Imai S, Kojima H, Katagi M, Yamakawa I, Kasahara T, et al. Brain-derived neurotrophic factor from bone

marrow-derived cells promotes post-injury repair of peripheral nerve. PLoS One 2012;7:e44592.

[85] Zhang Y, Jin Y, Nie X, Wang Y, Liu P, Shen N. Tissue engineering peripheral nerve with TGF-beta repair sciatic nerve defect. Sheng Wu Yi Xue Gong Cheng Xue Za Zhi 2007;24:394—8.

[86] Boyd JG, Gordon T. A dose-dependent facilitation and inhibition of peripheral nerve regeneration by brain-derived neurotrophic factor. Eur J Neurosci 2002;15:613—26.

[87] Wood MD, Kim H, Bilbily A, Kemp SW, Lafontaine C, Gordon T, et al. GDNF released from microspheres enhances nerve regeneration after delayed repair. Muscle Nerve 2012;46:122—4.

[88] McGuinness SL, Shepherd RK. Exogenous BDNF rescues rat spiral ganglion neurons in vivo. Otol Neurotol 2005;26:1064—72.

[89] Jain KK. Role of nanotechnology in developing new therapies for diseases of the nervous system. Nanomedicine 2006;1:9—12.

[90] Ibrahim AM, Gerstle TL, Rabie AN, Song YA, Melik R, Han J, et al. Nanotechnology in Plastic Surgery. Plast Reconstr Surg 2012;130:879—87.

[91] Stabenfeldt SE, Garcia AJ, LaPlaca MC. Thermoreversible laminin-functionalized hydrogel for neural tissue engineering. J Biomed Mater Res A 2006;77:718—25.

[92] Muir D. The potentiation of peripheral nerve sheaths in regeneration and repair. Exp Neurol 2010;223:102—11.

[93] Carbonetto ST, Gruver MM, Turner DC. Nerve fiber growth on defined hydrogel substrates. Science 1982;216:897—9.

[94] Bellamkonda R, Ranieri JP, Bouche N, Aebischer P. Hydrogel-based three-dimensional matrix for neural cells. J Biomed Mater Res 1995;29:663—71.

[95] Geller HM, Fawcett JW. Building a bridge: engineering spinal cord repair. Exp Neurol 2002;174:125—36.

[96] Schmidt CE, Leach JB. Neural tissue engineering: strategies for repair and regeneration. Annu Rev Biomed Eng 2003;5:293—347.

[97] de Mel A, Punshon G, Ramesh B, Sarkar, S, Darbyshire A, Hamilton G, et al. In situ endothelialization potential of a biofunctionalised nanocomposite biomaterial-based small diameter bypass graft. Biomed Mater Eng 2009;19:317—31.

[98] de Mel A, Jell G, Stevens MM, Seifalian AM. Biofunctionalization of biomaterials for accelerated in situ endothelialization: a review. Biomacromolecules 2008;9:2969—79.

Chapter 58

Vessel Regeneration and Bioengineering[*]

Shuhei Tara[a], Ethan W. Dean[b], Kevin A. Rocco[c], Brooks V. Udelsman[b], Hirotsugu Kurobe[a], Toshiharu Shinoka[d], and Christopher K. Breuer[a]

[a]Tissue Engineering Program and Surgical Research, Nationwide Children's Hospital Columbus, OH, [b]Yale University School of Medicine, New Haven, CT, [c]Department of Biomedical Engineering, Yale University, New Haven, CT, [d]Department of Cardiothoracic Surgery, The Heart Center, Nationwide Children's Hospital, Columbus, OH

Chapter Outline

58.1 INTRODUCTION

In the field of cardiovascular medicine, transplantation and surgical reconstruction are frequently required in the care of patients with cardiac defects or vascular disease. Transplantation is becoming an increasingly impractical solution due to the severe shortage of donor tissue, as evidenced by the steady increase in the number of patients who die while on transplant waiting lists [1]. Surgical reconstruction with synthetic materials is an alternative strategy, but this approach has limited efficacy due to complications, such as progressive stenosis, thrombosis, deposition of calcium, rejection, risk of infection, and the need for persistent anticoagulation therapy [2]. To address the critical need for a viable long-term solution, cardiac

surgeons, cardiologists, and scientists have turned to the nascent field of tissue engineering.

Stated most simply, the goal of tissue engineering is to develop alternative materials that integrate with the patient's native tissue to restore physiologic function [3]. Often, this involves the use of synthetic or natural materials termed scaffolds to provide a three-dimensional surface for cellular proliferation and new tissue formation followed by degradation of the scaffold at a later point. Therefore, the traditional concept of tissue engineering consists of the following three components: (i) a tissue-inducing scaffold material, (ii) isolation and use of cells or cell substitutes, and (iii) the integration of the cells and the scaffold via a seeding technique [1]. In 1986, Weinberg and Bell generated what was widely regarded as the first tissue-engineered blood vessel substitute, consisting of cultures of bovine endothelial cells (ECs), smooth muscle cells (SMCs), and fibroblasts embedded in

---

[*] Shuhei Tara and Ethan W. Dean contributed equally to the preparation of this manuscript and should be listed as cofirst authors.

a collagen gel [4]. However, this tissue-engineered construct lacked adequate strength and required reinforcement with a Dacron® mesh. In the years since, tissue-engineered vascular grafts (TEVGs) have been greatly improved and refined, ultimately reaching clinical application in a 2001 trial that employed a bone marrow-derived mononuclear cell (BM-MNC) seeded biodegradable polymeric scaffold for use in pediatric patients undergoing extracardiac total cavopulmonary connection (EC-TCPC) procedures [5].

Advances in stem cell science have been integral to the development of tissue engineering. In addition to the ability of stem cells to self-renew and differentiate into various mature cell types, their crucial role in tissue remodeling and repair has led to the use of these cells in a variety of TEVG applications [6]. Adult stem cells—including mesenchymal stem cells (MSCs) and hematopoietic stem cells (HSCs), which have been extensively characterized—exist in a variety of adult tissues and are widely available. Alternatively, embryonic stem (ES) cells, best known for their potential to differentiate into cell types of all three germ layers, can be isolated from the inner cell mass of blastocysts [7]. More recently, induced pluripotent stem (iPS) cells were generated from adult human cells, thus providing a replacement to ES cells [8]. Characterization of these cell lines and the ongoing fusion of tissue engineering with advances in stem cell science hold great promise for vascular regeneration therapy.

In this chapter, we present the history, current progress, and future directions of TEVG research with a focus on three factors: scaffold materials, cell types, and seeding methods. We also introduce our clinical trial in which we employ a biodegradable scaffold seeded with BM-MNCs to treat patients with congenital heart disease.

58.2 MATERIALS

58.2.1 Current Methods for Vascular Repair

Coronary artery bypass surgery and peripheral vascular bypass procedures (e.g., femoropopliteal bypass) currently rely on grafts composed of autologous tissue isolated from either the internal mammary artery or saphenous vein with 10-year patency rates ranging from 90% to 50%, respectively [9,10]. However, such autografts have additional disadvantages including the inconvenience of harvesting and preparing the tissue graft as well as insufficient availability in patients with widespread vascular disease or in those whose vessels have been harvested for a previous procedure. Consequently, the development of synthetic grafts has been a major achievement in vascular repair.

Expanded-polytetrafluoroethylene (ePTFE, Gortex®), polyethylene terephthalate (PET, Dacron®), and polyurethane (PU) are the most commonly used synthetic graft materials for vascular bypass surgery due to the ease in adjusting the mechanical properties of the material as well as a history of relative success in applications that require grafts >6 mm in diameter in a high-flow, low-resistance circulation [11].

ePTFE (Gortex®) is an expansion-processed form of PTFE (Teflon®), achieving a highly microporous structure composed of PTFE nodes interconnected by fine PTFE fibrils, thereby permitting cellular infiltration without inducing a foreign body rejection [12]. These fluorocarbon polymers are unique and useful materials with properties that include hydrophobicity, low surface free energy, low coefficient of friction, and high chemical resistance; such characteristics impart biocompatibility and antithrombogenicity when used in vascular applications [13]. However, ePTFE is highly crystalline (>90%), with a reported elastic modulus of 0.5 GPa, and tensile strength of 14.0 MPa, resulting in a construct that is far more stiff than native artery [14]. While more than 95% of aorto-iliac ePTFE grafts are functional at 5 years [15], only 36% remain patent in femoropopliteal bypass procedures [16] compared to 75% patency when autologous vein grafts are used [17].

PET (Dacron®) grafts are composed of a nondegradable polyester that is first spun into filaments and then either woven or knitted to generate a vascular graft that meets the essential requirements of strength, durability, and functionality [18]. Reported tensile strength of a PET graft is around 175 MPa, indicating that the material is significantly stiffer than native artery [14]. Just as is true of clothing, the manufacturing process has a significant impact on the overall material properties as woven grafts typically have small pores and are stiffer whereas knitted grafts, typically velour, result in larger pores and are more compliant. As a result of the larger pore size, knitted PET grafts have improved neotissue infiltration but must first undergo preclotting with albumin, gelatin, or blood prior to implantation to prevent transmural bleeding. Approximately 50% of collagen-impregnated, knitted PET conduits are patent at 5 years when used in femoropopliteal bypass procedures [16].

Polyurethanes are copolymers that contain a urethane linkage typically formed by reacting two or more monomers with a hydroxyl group and isocyanate group. The use of different isocyanate-containing and hydroxyl-containing components allows for the synthesis of a variety of Polyurethanes with a wide range of material properties and a microstructure consisting of a hard domain and a soft domain. The hard domain, which is interspersed within the soft domain, is hydrogen bonded and has a high modulus thus conferring strength to the material. The soft domain is characterized by a low modulus, giving the composite material stretch and durability [19]. Varying the composition of this microstructure results in tensile strength values ranging from 20 MPa to 90 MPa with tensile modulus values of 5–1150 MPa [14]. Poly(ester urethane) and poly(ether urethane) both possess excellent material properties for vascular graft applications, but *in vivo* studies revealed that soft domain polyesters were susceptible to hydrolytic degradation while polyether segments were degraded by reactive oxygen species [20]. Currently, carbonate-based polyurethanes (CPUs) are gaining popularity due to better *in vivo* stability and compliance values five times greater than PET or ePTFE, nearly matching the compliance of native artery [21].

However, none of these synthetic materials have proven suitable for generating grafts <6 mm in diameter, as these smaller conduits are more vulnerable to failure secondary to thrombus formation or intimal hyperplasia. Furthermore, these nondegradable synthetic grafts lack the capacity for growth, resulting in unacceptable or unproven long-term outcomes. This is of particular concern in pediatric patients, who often undergo vascular graft surgery at a young age. TEVGs attempt to address these concerns by restoring native vascular structure and function through the combination of a biodegradable scaffold material seeded with one or more cell types prior to implantation.

58.2.2 Functional Role of TEVG Scaffolds

In the field of tissue engineering, the role of the scaffold is to provide a temporary three-dimensional structure for cellular attachment and proliferation *in vitro* or *in vivo*. These materials typically possess biomimetic properties and are highly porous, thereby facilitating cellular infiltration, stimulation of neotissue formation, and integration with native tissue [22]. For TEVGs in particular, scaffold materials should demonstrate mechanical properties akin to the native vessel, namely strength, durability, and compliance. Lastly, these materials must ensure biocompatibility so as not to induce an immunological response [23]. Both synthetic and naturally occurring polymers can be utilized to satisfy these requirements, enabling a tissue engineering approach for the development of a functional vascular graft that induces patient vessel regeneration *in situ*[2].

58.2.3 Natural Materials for TEVG Scaffolds

As previously mentioned, Weinberg and Bell constructed the first TEVG by employing a collagen gel as a natural material scaffold for neovessel growth. However, the graft lacked sufficient strength and was unsuitable for implantation [4]. Kanda et al. attempted a similar approach by constructing hybrid grafts composed of type I collagen gel and cultured SMCs and ECs [24]. These constructs were evaluated *in vivo* as venous [25] and arterial implants [26] but required Dacron® reinforcement due to the inferior mechanical properties of the collagen gel. Various methods have been investigated to improve the mechanical properties of collagen gels (e.g., crosslinking agents like glutaraldehyde), but none have been proven to yield a structurally stable TEVG [27]. Nonetheless, these pioneering studies demonstrated that natural materials can support three-dimensional tissue culture and induce positive vascular remodeling *in vivo*.

As an alternative to collagen for natural scaffolds, fibrin holds particular promise due to its ability to induce collagen and elastin synthesis and offer improved mechanical properties [28]. Fibrin can also be extracted noninvasively from a patient's own blood sample to create an autologous scaffold material [29,30]. Swartz et al. highlighted the feasibility of this approach by implanting SMC-impregnated fibrin gel constructs into the jugular veins of lambs, with later analyses demonstrating considerable remodeling and elastic fiber development by 15 weeks postimplantation [28]. Furthermore, by combining fibrin gels with biodegradable polymeric scaffolds (knitted polylactide) [31], endothelialized autologous vessels were successfully implanted in the carotid arteries of sheep [32]. After 6 months, all of the ovine grafts were patent with positive endothelialization and remodeling, and no evidence of thrombosis, calcification, or aneurysm in the midsection of the graft. Remodeling was ongoing at 6 months due to incomplete degradation of the synthetic scaffold, and collagen (hydroxyproline) content was 40% that of native tissue [32].

Finally, decellularized donor tissue, often in the form of a xenograft, can serve as a naturally available scaffold. Decellularized donor tissue contains intact, structurally organized, and mechanically competent collagen and elastin, but lacks cellular components and DNA [33]. During an optimized decellularization process, all cellular immunogenic components are removed using any combination of physical agitation, chemical surfactants, and enzymatic digestion, all of which work to disrupt cells and remove small proteins, lipids, and nucleotide remnants while leaving the extracellular matrix (ECM) intact [34]. Using these methods, Kaushal et al. decellularized porcine iliac vessels, seeded them with ECs, and implanted the constructs into ovine carotid arteries [35]. The TEVG constructs remained patent out to 130 days and remodeled into neovessel, whereas the unseeded control group occluded within 15 days. These results indicate that decellularized vascular scaffolds are susceptible to early failure unless first undergoing endothelialization or additional modification. Furthermore, elements of the ECM are exposed to physical and chemical stresses during the process of decellularization which can adversely affect the biomechanical properties of the ECM. This deterioration may ultimately lead to degenerative structural graft failure [36]. Additional drawbacks of decellularized donor tissue include the inability to modify matrix content and architecture, variability among donor sources, and risk of viral transmission from animal tissue.

58.2.4 Synthetic Biodegradable Scaffolds for TEVG

In addition to the previously discussed roles of the scaffold in tissue engineering, the ideal scaffold in vascular applications meets the following criteria: (i) provides a suitable surface for cell adhesion and proliferation, (ii) is

highly porous in bulk to allow for neotissue infiltration and exchange of nutrients and metabolic waste (Figure 58.1), and (iii) is biodegradable and bioresorbable to ultimately restore physiologic function without the necessity for synthetic material [37]. The selection of an appropriate biodegradable scaffold material is dependent on a variety of factors, such as biocompatibility, mechanical properties, and biodegradability (Table 58.1), and it is a crucial first step in designing constructs for vascular engineering.

Degradable polymers of the aliphatic polyester family, such as poly(glycolic acid) (PGA), poly(lactic acid) (PLA), and poly(caprolactone) (PCL), are commonly used materials for constructing TEVGs due to their history of successful clinical application, FDA approval for human implantation, and a broad range of material properties [38]. PGA is a highly crystalline, hydrophilic

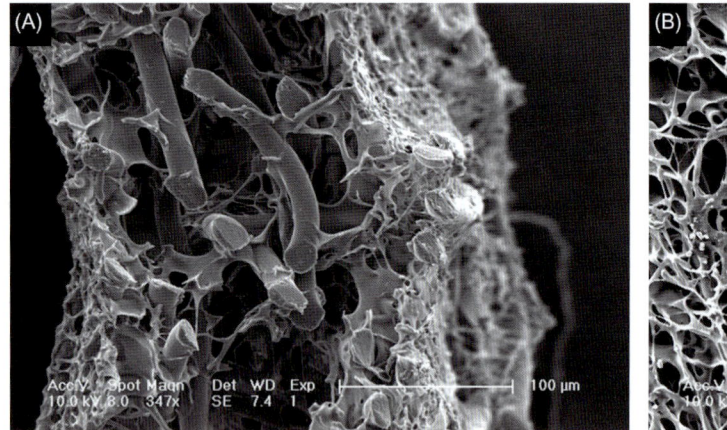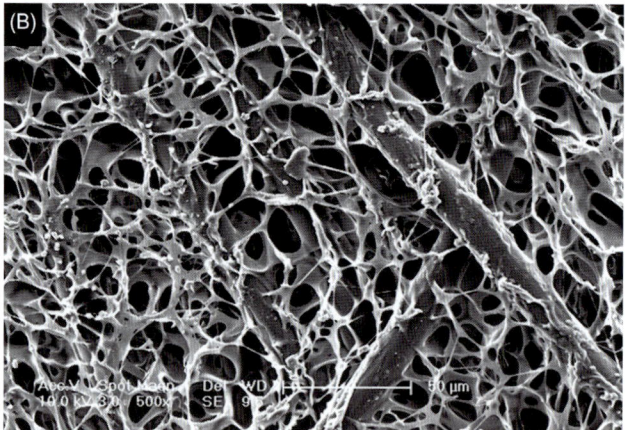

FIGURE 58.1   Scanning electron microscope images of a TEVG scaffold. (A) Cross-sectional view of the scaffold wall demonstrates tubular PGA mesh fibers coated with PCLA (a copolymer of PCL and PLA). Image at 347× magnification. (B) Increased magnification reveals the high porosity of the scaffold structure. This facilitates cellular infiltration into the scaffold and provides physical space for the formation of neotissue. Image at 500× magnification.

TABLE 58.1 Biodegradable Polymers for TEVGs

Polymer	Tm	Tg	Initial Tensile Strength (Mpa)	Elastic Modulus (GPa)	Elongation at Break (%)	Degradation Period
PGA	230	36	890	8.4	30	2−3 weeks
PLA	170	56	900	8.5	25	6−12 months
PCLA (75:25)	140	22	500	4.8	70	8−10 weeks
PCLA (50:50)	105	−17	12	0.9	600	4−6 weeks
PCL	60	−60	50	0.3	70	12 weeks
PHB	177	4	43	−	5	>6 weeks

Tm, melting temperature; Tg, glass-forming temperature; Mpa, mega pascal; GPa, giga pascal; PGA, polyglycolic acid; PLA, polylactic acid; PCLA, copolymer of L-lactide and ε-caprolactone; PCL, poly ε-caprolactone; PHB, poly-3-hydroxybutyrate.

polymer that degrades rapidly *in vivo*. PLA is a methylated version of PGA which is less hydrophilic and hence persists longer following implantation [14]. These polymers degrade via hydrolytic cleavage at the ester bond, producing fragments of diminishing molecular weight. Macroscopically, the degradation of these materials is evidenced first by a loss of mechanical properties, followed by a decrease in mass/volume. In the final phase of *in vivo* degradation, low molecular weight polymer fragments are phagocytosed by macrophages and cleaved into naturally occurring metabolites [39]. It is for this reason that PLA is often interchangeable with PLLA, the latter referring to the polymer of L-lactic acid, the biologically present enantiomer. For complete mass loss *in vivo*, the polymer degradation process may take anywhere from weeks for hydrophilic PGA to years for high molecular weight, hydrophobic PCL [40]. Therefore, PCL is most suitable for applications in which the mechanical properties of the scaffold must be maintained for long periods of time. PCL also exhibits unusual properties that are not found in PGA or PLA. Because of its low glass transition temperature of $-60°C$, PCL is always in a viscous state at room temperature [41].

The degradation rate for each of these polymers is further determined by initial molecular weight, exposed surface area, crystallinity, and ratio of constituent monomers [41]. Combining these materials with biopolymers like collagen or with additional synthetic polymers to create copolymers, such as poly(l-lactic acid-co-glycolic acid) (PLGA) and poly(caprolactone-co-l-lactic acid) (PCLA), allows for the tuning of mechanical properties and degradation rates through control of the composition ratios and molecular weights. It is for these reasons that copolymeric PLGA scaffolds have been deployed in a wide range of tissue engineering applications [42].

Polyhydroxyalkanoates (PHAs) like poly-3-hydroxybutyrate (PHB) are another noteworthy form of degradable polyesters that are assembled by a variety of microorganisms while also remaining biocompatible for human implantation [41]. Bacteria, such as *Azotobacter beijerinckii* and *Rhizobium* spp., synthesize these thermoplastic polymers to serve as carbon and energy reserves, with excess stores accumulating as insoluble spherical granules in the cytoplasm. These granules are readily available for isolation and can then be processed into biomaterials for tissue engineering applications [43].

58.3 CELL TYPES

The application of various cell types in tissue engineering has undergone a remarkable expansion and evolution, and many different cell sources are now used for TEVGs. Prior to the isolation and identification of human stem cells, matured somatic cells isolated from adult tissues were primarily employed. These cell types may be utilized in combination with natural or synthetic scaffold materials.

58.3.1 Matured Somatic Cells

ECs and SMCs have received the most attention in vessel engineering as these cells are the primary cellular constituents of a mature blood vessel. These cell types are most commonly derived from a tissue biopsy, followed by *ex vivo* culture to allow for proliferation prior to seeding. However, since matured cells have limited capacity for proliferation and because vessel engineering often requires *ex vivo* expansion, recent studies have shifted the focus of cell seeding from matured somatic cells to stem cells, which possess an unlimited capacity for proliferation. Matured cell seeding will be discussed first, followed by a review of current stem cell seeding strategies.

58.3.1.1 Endothelial Cells

The presence of a confluent monolayer of ECs on the luminal surface of a vascular graft greatly enhances its thromboresistance and prevents the development of neointimal hyperplasia through inhibition of bioactive substances responsible for SMC migration, proliferation, and production of ECM [44]. In 1978, Herring et al. introduced a technique in which ECs were harvested from venous tissue by scraping the luminal surface. Isolated cells were then seeded onto a nonbiodegradable prosthetic material and incubated prior to implantation in a femoropopliteal artery bypass [45]. The initial study justified a second, more comprehensive clinical trial, but this method did not demonstrate any significant advantage over autologous vein grafting [46]. However, subsequent randomized clinical trials revealed improved patency rates compared to autologous grafting and demonstrated that the implantation of EC-seeded ePTFE grafts resulted in significantly better outcomes compared to a nonseeded control group [47]. These results highlight the desirability of an EC layer in vascular grafts, but direct EC seeding of synthetic ePTFE conduits is challenging and may offer only a slight advantage to the patient.

58.3.1.2 Smooth Muscle Cells

It is widely accepted that SMCs are an integral component of a stable and efficient blood vessel. Throughout development, SMCs are the predominant source of the complex ECM that ultimately defines the mechanical behavior of a vessel [48]. Yue et al. explored the use of this cell type in TEVGs by seeding cultured SMCs onto a biodegradable scaffold and implanting the grafts into rat aortas. The implanted constructs demonstrated rapid neomedia formation in comparison to unseeded controls [49]. Also of interest, Niklason et al. developed a technique for creating a tissue-engineered graft by seeding SMCs onto a PGA scaffold followed by *in vitro* culture in a pulsatile radial stress environment for 8 weeks [50]. These grafts demonstrated physiologic and mechanical functions comparable to native human vessels and are being considered for clinical application [51].

58.3.2 Stem Cells

As a result of major advances in stem cell technology during the last decade, stem cells have drawn considerable attention for potential use in tissue engineering applications. The theoretical advantage of using stem cells as a cell source in vascular tissue engineering is their capacity for self-renewal and ability to differentiate into the various cellular constituents of a mature blood vessel. However, there is limited data available to suggest that stem cells serve as direct precursors to the formation of vascular neotissue *in vivo*[52].

In this text, we divide stem cells into three groups according to their origins and characteristics: adult stem cells (including MSCs and HSCs), ES cells, and iPS cells.

58.3.2.1 Adult Stem Cells

The primary advantage of cell seeding with adult stem cells is that they do not necessitate immunosuppression of the recipient after implantation, in contrast to many other stem cell subtypes. Additionally, gene manipulation of the seeded cells is not required and teratoma formation does not occur. The two subtypes of adult stem cells will be reviewed below.

58.3.2.1.1 Mesenchymal Stem Cells

MSCs originate from mesenchyme, the embryonic connective tissue that is derived from the mesoderm. Their ability to differentiate into multiple cell lineages coupled with their presence within multiple tissues of the adult human body have made MSCs the subject of intense research within the field of tissue engineering [23]. In humans, a MSC phenotype is suggested by the presence of cell surface markers CD29, CD44, CD71, CD90, CD106, CD120a, CD124, CD166, SH2, SH3, SH4, integrins α_1, α_5, and β_1, coupled with a deficiency of markers, such as CD14, CD34, CD45, HLA-DR, CD 11a, CD31, and integrins α_4 and β_2[53]. These cells are characterized by their adherence to extracellular material as well as their spindle-shaped, fibroblast-like appearance in culture dishes [53].

The use of MSCs in clinical applications requires an understanding of their unique biological characteristics that contribute to the desired therapeutic effect. The following four properties are considered to be the most important: (i) the ability to migrate to sites of inflammation when administrated systemically, (ii) the ability to differentiate into various cell types, (iii) the ability to secrete multiple bioactive molecules capable of inhibiting inflammation and stimulating cell recovery from injury, and (iv) the lack of immunogenicity coupled with the ability to perform immunomodulatory functions [54].

MSCs were initially described by Friedenstein et al. in 1966 as a multipotent population of nonhematopoietic cells in the bone marrow [55]; subsequent studies have shown that additional adult tissues also contain MSCs [56,57], and that these stem cells have enormous potential within the field of tissue engineering. These MSCs were once thought to be committed cell lines that could give rise to only a single type of cell, but are now known to have wider differentiation potential [58]. Mirza et al. demonstrated this key point through *in vivo* induction of bone marrow-derived MSCs (BM-MSCs) to the vascular lineage with green fluorescence protein-labeled rat BM-MSCs cultured on PU vascular prostheses [59]. Additionally, Hashi et al. conducted a 60-day *in vivo* study that demonstrated the antithrombogenic property of MSCs when seeded onto nanofibrous vascular grafts [60].

Recently, an additional human MSC cell type was described. These so-called multilineage-differentiating stress-enduring (Muse) cells can be isolated from cultured fibroblasts, bone marrow aspirates, or bone marrow stromal cells. The cells are doubly positive for SSEA-3 and CD105, thereby allowing for efficient isolation via cell separation techniques like flow cytometry [61]. Although Muse cells appear to demonstrate less proliferative capability compared to other stem cell subtypes, such as ES cells and iPS cells, they only rarely develop into teratomas, which is currently the most serious complication occurring with the clinical usage of stem cells [62]. This evidence indicates that, among stem cells, Muse cells may be a safer and more effective option for clinical translation, with clear potential for application in novel vascular bioengineering solutions.

58.3.2.1.2 Hematopoietic Stem Cells

HSCs represent another type of adult stem cell used for seeding in vascular engineering. Classically, HSCs are multipotent stem cells that give rise to all the blood cell types from the myeloid (monocytes/macrophages, neutrophils, basophils, eosinophils, erythrocytes, megakaryocytes/platelets, and dendritic cells), and lymphoid (T cells, B cells, NK cells) lineages. The definition of HSCs has undergone considerable revision over the last two decades. The cell surface marker CD34 was previously thought to be a marker specific for the hematopoietic lineage, but in 1997, Asahara et al. confirmed CD34 positivity on endothelial progenitor cells (EPCs) isolated from adult human peripheral blood [63]. It was subsequently revealed that these cells participated in postnatal neovascularization after their mobilization from the bone marrow [64]. Like MSCs, HSCs are now recognized as multipotent progenitors which can differentiate into multiple lineages. In addition to CD34, the cell surface markers AC133 and vascular endothelial growth factor receptor 2 (VEGFR-2) are regarded as specific for HSCs [65], although it has been suggested that these markers may also be expressed by the EPC subpopulation [66]. It is thought that early stage EPCs express markers in common with HSCs, whereas differentiated EPCs lose HSC markers and instead express mature EC markers, such as VE-cadherin, platelet/endothelial cell adhesion molecule 1, VEGFR-2, E-selectin, and von Willebrand factor [63,67].

Like HSCs, EPCs also possess the ability to endothelialize TEVGs. Kaushal et al. demonstrated that *ex vivo* expanded EPCs isolated from the peripheral blood of sheep effectively achieved luminal coverage after seeding onto decellularized arterial grafts [35]. Interestingly, the differentiation of naïve EPCs into mature ECs has been shown to be enhanced by cocultures with CD34 negative cells [63], indicating that a heterogeneous growth condition might be favorable for stem cell differentiation. Ultimately, the primary advantages of these hematopoietic EPCs include medical regulatory feasibility (straightforward isolation via noninvasive sampling of peripheral blood), as well as their ability to home to sites of neovascularization where they serve as potent mediators of vasculogenesis [68].

58.3.2.1.3 Sources of Adult Stem Cells

Several putative sources of adult stem cells have been identified or suggested within the adult human body. Bone marrow, peripheral blood, adipose tissue, umbilical cord blood, placental tissue, amniotic membrane, dermis, and skeletal muscle are examples of known stem cell sources [69]. In this section, we present three of the most common sources of adult stem cells: bone marrow, peripheral blood, and adipose tissue, each of which is easily harvested and readily available for clinical use.

Bone marrow cells have undergone the most successful translation in human studies that involve a stem cell approach. BM-MSCs were initially characterized *in vitro* through isolation of bone marrow stromal cells with multipotent differentiation capability. These BM-MSCs are uniformly positive for SH2, SH3, CD29, CD44, CD71, CD90, CD105, CD120a, CD124, and additional surface proteins characteristic of bone marrow cells [70]. One of the advantages in using BM as a source for cells in TEVG creation is that BM contains several lineages and differentiated stages of cells, as well as an abundance of cytokines that may enhance neovascular development. Multiple

animal studies have demonstrated that vascular grafts seeded with BM cells may be a reasonable therapeutic option. In 1996, Noishiki et al. demonstrated the presence of endothelialization in synthetic vessels created by seeding BM-derived cells onto a biodegradable scaffold [71]. Additionally, Matsumura et al. reported the results of a study in which scaffolds seeded with BM cells were implanted as canine inferior vena cava interposition grafts with subsequent development of neovascular tissue [72].

Mobilization of progenitor cells from bone marrow into the circulation is induced by a variety of physiological or pathological signals. EPCs can be easily isolated from peripheral blood using techniques based on magnetic beads coated with antibodies, such as CD34, AC133, or markers common to both early and more differentiated progenitors like VEGFR-2 [66]. In addition, the number of EPCs within peripheral blood can be enhanced via mobilization with granulocyte colony-stimulating factor [23]. Unfortunately, patients must undergo apheresis to isolate these cells from the bloodstream, and the safety of this technique compared to BM aspiration has not been established.

Adipose tissue, which can be harvested subcutaneously with minimally invasive techniques, contains MSCs in the stromal cell fraction that can give rise to several lineages, including ECs. Zuk et al. noted that these cells express CD13, CD29, CD44, CD71, CD90, CD105/SH2, SH3, and STRO-1. In contrast, no expression of the hematopoietic lineage markers CD14, CD16, CD31, CD34, CD45, CD56, CD61, CD 62E, CD 104, and CD106 was observed [73]. These adipose-derived MSCs have recently been termed adipose-derived stem cells/progenitor cells (ASCs) [74] and have been shown to secrete multiple angiogenic growth factors, including vascular endothelial growth factor (VEGF), hepatocyte growth factor, and chemokine stromal cell-derived factor 1 [75]. The combination of these biological properties (i.e., ease of isolation coupled with abundant growth factor release) suggests that adipose tissue will remain a promising candidate as a cell source for tissue engineering applications.

58.3.2.2 Embryonic Stem Cells

ES cells are pluripotent cells derived from the inner cell mass of the early embryo. ES cells can proliferate indefinitely while retaining the ability to differentiate into virtually any cell in the human body. They possess greater proliferative capacity compared with adult stem cells, suggesting that they might be a useful cell source for *ex vivo* expansion and seeding. Shen et al. demonstrated this concept by seeding ECs derived from mouse ES cells onto a PGA scaffold, and confirming the development of an EC monolayer [76]. Although mouse ES cell research has been ongoing for many years, data on human ES cells is limited. Research on human ES cells is still fairly novel, ES cells having first been discovered in 1998 [77]. Federal research funding restrictions on human embryonic cell lines in the United States have also arisen due to political and ethical concerns [78]. Additionally, there are further difficulties when ES cells are used as a cell source for TEVGs, including the requirement for homogeneous separation, risks of immunological reaction, untoward genetic changes, and overall technical difficulty [23].

58.3.2.3 iPS Cells

In 2006, Takahashi et al. successfully demonstrated the induction of adult fibroblasts into pluripotent stem cells by introducing four factors: Oct3/4, Sox2, c-Myc, and Klf4 [79]. The resulting cells were appropriately termed iPS cells. The iPS cells were similar to ES cells in morphology, proliferation, surface antigens, gene expression, epigenetic status of pluripotent cell-specific genes, and telomerase activity, but differed with respect to epigenetic modification, lifespan, and differentiation potential [8]. Importantly, the iPS cell approach offers distinct advantages over ES cell utilization, primarily in that iPS cells are autologous and therefore the transplantation does not require immunosuppressive therapy. Additionally, iPS cell research obviates the political and ethical dilemma associated with embryo destruction and ES cell harvest. Studies have yet to demonstrate that iPS cells differentiate into mature vascular cells on the scaffold *in vivo*. However, iPS cells differentiated *ex vivo* may prove to be an effective cell source for constructing TEVGs, especially when employed in the creation of cell sheets, and it has been proposed that seeded iPS cells may actually function in a paracrine manner to induce neovascular formation [44]. These developments are promising, but a number of obstacles must still be overcome prior to the implementation of iPS cells in TEVG applications, the most serious being the potential for undifferentiated iPS cells to form teratomas following implantation.

58.4 CELL SEEDING TECHNIQUE

A variety of cell seeding techniques have been developed in order to effectively and reproducibly apply cells onto the TEVG scaffold material. However, no single method has proved universally efficacious with regard to optimal seeding efficiency or improvement of long-term graft function. The expansion of the field of vascular tissue engineering has prompted the search for the most cost-effective, reliable, and efficient seeding technique [80]. The following text serves as a review of some of the most commonly used cell seeding methods in TEVG fabrication.

58.4.1 Passive Seeding

Passive seeding (also known as static seeding) is the simplest and most widely used method of cell delivery, but it is also the least efficient and least reproducible approach [81]. This method involves pipetting a cell suspension directly into the lumen of the scaffold or onto the outer surface of the scaffold. Statically seeded cells are incubated with the scaffold for several hours to several days prior to implantation. This seeding technique yields seeding efficiencies of approximately 10–25% [82]. Increasing the incubation time allows for the cells to obtain a more mature morphology and thus more efficient cell attachment, but it may also lead to higher incidences of contamination or unfavorable cellular changes.

58.4.2 Dynamic Seeding Techniques (Rotational Seeding and Vacuum Seeding)

To increase cell seeding efficiency, uniformity, and penetration into the scaffold, two methods of dynamic seeding have gained popularity: rotational seeding and vacuum seeding. Rotational seeding systems rely on centrifugal forces to enhance cell-scaffold interaction [80]. Briefly, scaffolds are affixed to a mandrel and placed in a spinner flask filled with the cell seeding suspension. The rotation of the cell suspension within the spinner flask drives cells into the scaffold resulting in seeding efficiencies ranging from 38% to 90% [83]. Low-speed rotational systems often require prolonged seeding time, whereas high-speed rotational systems increase seeding efficiency and graft wall penetration at the cost of potentially deleterious effects on cell morphology.

Alternatively, a vacuum system utilizes a pressure differential to drive cells into a porous scaffold material. This method relies on either internal or external vacuum pressure to force a cell suspension through the micropores of a TEVG scaffold (Figure 58.2), and is an extremely rapid means of cell seeding with seeding efficiencies ranging from 60% to 90% [80]. Moreover, the simplicity of some vacuum apparatuses allows for their use as an operator-independent disposable seeding device, reducing the risks of contamination and seeding variability while increasing clinical utility [84].

58.4.3 Sheet-Based Techniques

Sheet-based techniques often do not employ synthetic or exogenous materials at all; rather, the vessels are created through the use of autologous fibroblasts and ECs obtained via biopsy. The autologous cells are cultured on a substrate and can be lifted off in contiguous layers with preservation of the ECM. These robust cellular sheets can be used as

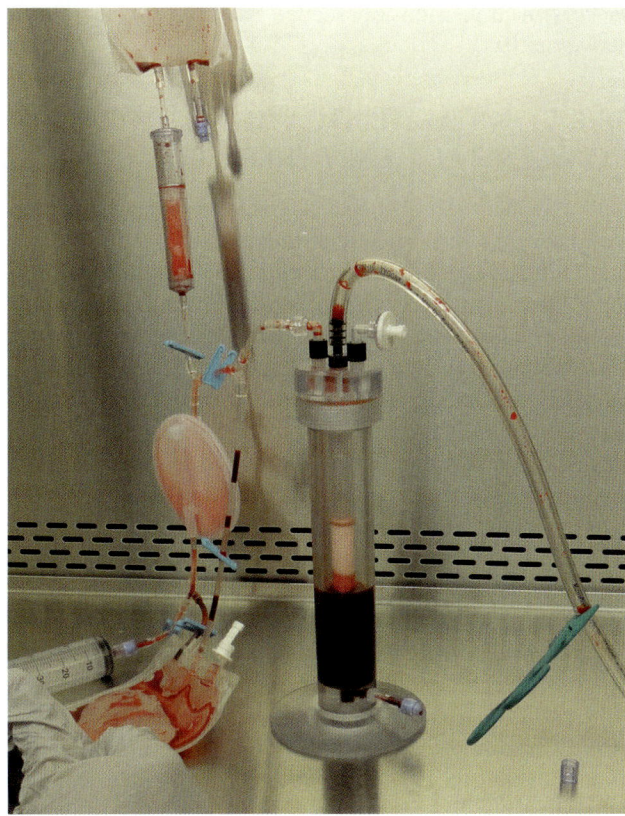

FIGURE 58.2   A vacuum seeding system utilizes a pressure differential to drive cells into a porous scaffold material, resulting in increased seeding efficiency compared to static methods. Newer vacuum seeding designs are completely enclosed and operator independent, reducing the risks of contamination and allowing for more reproducible results.

patch grafts, or they can be rolled onto the lumen of a tubular scaffold to form a TEVG. These approaches are termed "sheet-based tissue engineering" [85]. One study demonstrated the fabrication of 10 TEVGs using sheet-based tissue engineering methodology, followed by implantation as arteriovenous shunts for hemodialysis access. Primary patency was maintained in seven (78%) of the remaining nine patients 1 month after implantation and 5 (60%) of the remaining eight patients 6 months after implantation [86]. Researchers have also reported the development of a novel sheet-based technique to create three-dimensional cardiac tissue-engineered grafts [87]. Researchers employed temperature-responsive culture surfaces to obtain stable cell sheets of single-cell thickness. These sheets could be harvested without destruction of the intercellular connections, enabling overlaying onto a second and third cell sheet and subsequent implantation in a rat model. In our own study using iPS cells for TEVG fabrication, this new cell sheet technique improved seeding efficiency dramatically in comparison to traditional static seeding (86.5% versus 4.9%) [44].

58.4.4 Additional Methods of Seeding

Electrostatic cell seeding refers to cell seeding via manipulation of the electrostatic properties of vascular scaffolds [80]. This is accomplished by inducing a temporary positive surface charge on the luminal surface of an ePTFE scaffold in order to enhance negatively charged EC adhesion. Research has demonstrated that the utilization of this method for as little as 16 min results in seeding efficiencies of up to 90% [88].

Magnetic cell seeding techniques involve the use of magnetic forces to increase cell seeding efficiency. The basic principle relies on the use of a magnet to attract nanoparticles attached to the surface or inside of seeded cells [80]. The advantages of these techniques include short duration, seeding efficiency >90% [89,90], and reproducible results, all of which are essential for constructing grafts for clinical applications.

58.5 MECHANISM OF NEOTISSUE FORMATION IN TEVGS

It was previously believed that the stem cell fraction within the seeded BM-MNC population differentiated into the mature vascular cells of developing TEVGs. In past studies, we identified small populations of hematopoietic and vascular progenitor cells within the BM-MNC population used for seeding. However, we observed that the number of seeded cells in the graft decreased rapidly in the first few days after implantation, ultimately resulting in the absence of all BM-MNCs within 1 week post-implantation (Figure 58.3). As a result of these unique findings, we were the first to suggest that TEVGs transform into functional neovessel *in situ* via an inflammatory process of vascular remodeling [91].

Although the precise mechanism of neotissue formation in TEVGs remains to be fully elucidated, the following mechanism is presumed to be involved in the inflammation-mediated process of TEVG remodeling (Figure 58.4). After surgical implantation of the graft, seeded BM-MNCs release chemokines like monocyte chemoattractant protein 1 (MCP-1), which attract circulating monocytes into the scaffold. Monocytes differentiate into macrophages which release a variety of chemokines, cytokines, and growth factors, such as platelet-derived growth factor and VEGF. Macrophage-derived signaling molecules induce migration of adjacent host SMCs and ECs into the scaffold. Macrophages, fibroblasts, and SMCs begin to deposit ECM while the scaffold begins to degrade. The ECM undergoes continuous remodeling as scaffold degradation continues. Given sufficient time, the biomechanical properties of the neovessel are no longer determined by scaffold characteristics and are instead related to collagen and elastin

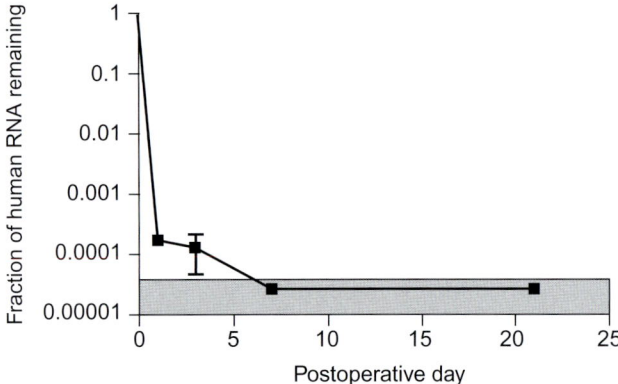

FIGURE 58.3 Dynamics of seeded BM-MNCs after implantation *in vivo*; RT-PCR of explanted cell-seeded scaffolds confirms no detectable human RNA expression after 1 week *in vivo*. *Adopted with permission from Roh et al. [91]*

content [52]. Finally, the seeded TEVG has completely transformed into a neovessel, with cellular, biomechanical, and physiologic characteristics similar to that of a native vessel. This model, in which seeded cells augment the host inflammatory response rather than serve as direct precursors in the formation of neotissue, contrasts with the classic tissue engineering theory of seeded cells as "building blocks" of neotissue. Rather, our findings suggest a regenerative medicine paradigm, in which seeded cells merely promote the host's natural ability to heal [91].

Additional studies have highlighted the critical importance of host-derived macrophages and monocytes in the process of TEVG remodeling, relating both to the formation of vascular neotissue and to the development of TEVG stenosis. It has been demonstrated that monocytes infiltrating a BM-MNC-seeded graft undergo transformation into M1 macrophages, whereas macrophages in unseeded TEVGs demonstrated phenotypic switching from the M1 to the M2 subtype [92,93]. M2 macrophages are considered to be a "healing macrophage" and have the ability to secrete cytokines and chemotactic factors while simultaneously degrading scaffold or ECM material in order to facilitate ingrowth of new tissue. Additional *in vivo* studies conducted by our lab involved implantation of TEVGs into a macrophage-depleted mouse model. While the absence of macrophages resulted in increased rates of patency and increased luminal diameters, we also observed a decrease in neotissue formation, with histological analysis revealing the absence of both the endothelial and smooth muscle layers, as well as reduced collagen deposition. These results highlight the importance of macrophage infiltration and activation for the successful transition of seeded scaffold to functional neovessel [94].

58.6 CLINICAL APPLICATION OF TEVGS IN CONGENITAL HEART DISEASE

Approximately 0.6% of live births are affected by moderate to severe forms of congenital heart disease [95], many of which require surgical intervention with various prosthetic materials to restore normal cardiac function. However, synthetic materials, such as Gortex® and Dacron®, lack growth potential and require reoperation to up-size the conduit as the child grows. The ability to engineer a vascular graft that remodels into a neovessel *in situ* is thus an attractive solution due to the restoration of autologous living tissue with growth potential, thereby obviating the need for surgical replacement [72].

58.6.1 Pilot Clinical Study

Following the demonstration of the safety and efficacy of our TEVG technique through *in vitro* studies and preclinical large animal experiments, we initiated the first human clinical trial investigating TEVG implantation in children with congenital heart defects in 2001. This study was performed under approval of the ethics committee at Tokyo Women's Medical University and was conducted as a prospective, nonrandomized clinical trial [72,96,97]. The following sections provide a review of the methods and results from this innovative clinical trial.

58.6.2 Patient Selection

The following inclusion criteria were used for patient selection: (i) elective surgery, (ii) age younger than 30 years, (iii) full understanding of the risks and benefits of the procedure by the patient or parent(s), and (iv) minimal extracardiac disease burden. Informed consent was obtained from the patient or parent(s) before proceeding. Between September 2001 and December 2004, 25 patients underwent an EC-TCPC with TEVG used as conduit (Table 58.2 and Figure 58.5) [97]. Mean patient age at the time of TEVG implantation was 5.5 years, and mean patient body weight was 19.5 kg. Anticoagulation and antiplatelet therapy with warfarin and aspirin were started 2 days postoperatively and continued for 3–6 months with a target INR of 1.5–2.0. After 6 months, patients were maintained on aspirin alone for an additional year. Patients were followed postoperatively in a multidisciplinary clinic. Graft patency and cardiac function were monitored with transthoracic echocardiography, multislice computed tomography (CT), angiography, or

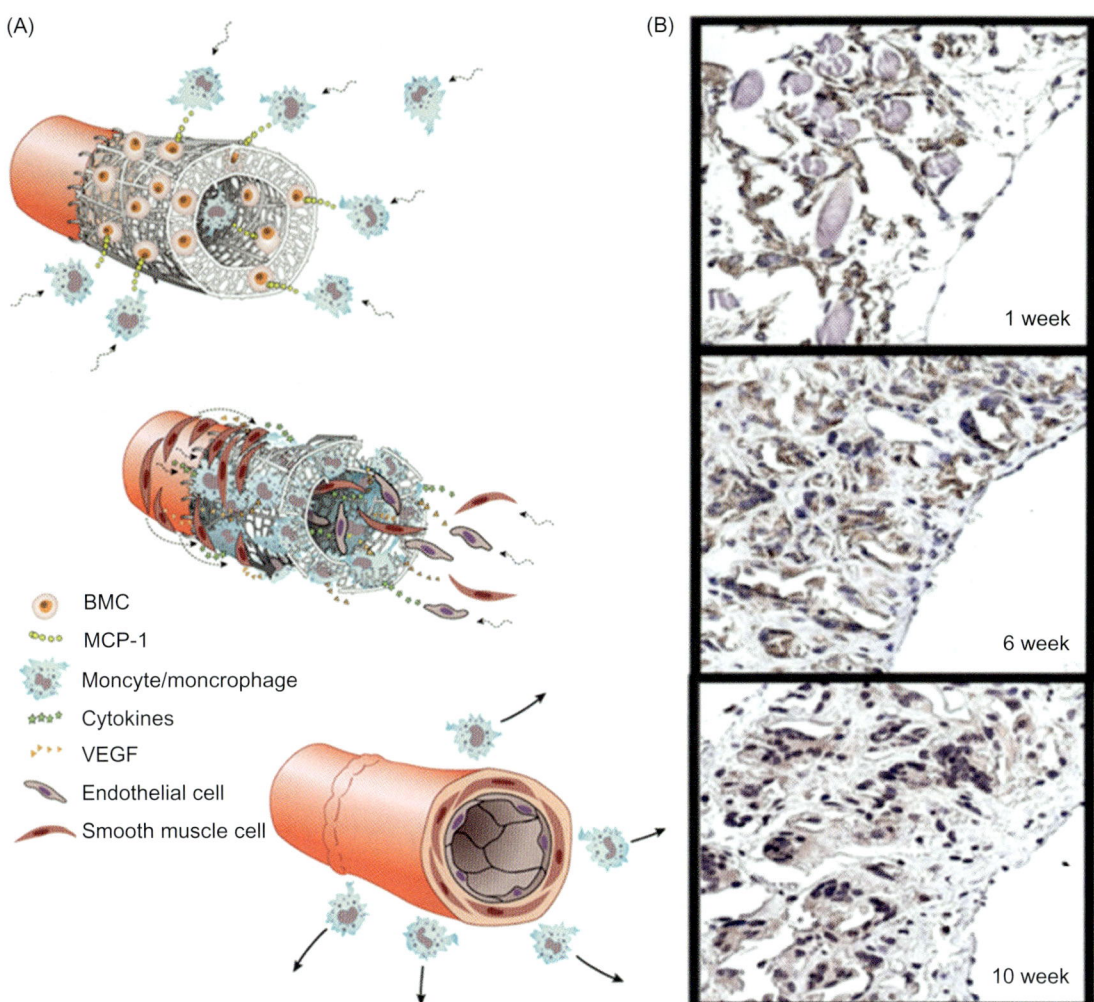

FIGURE 58.4 Proposed mechanism of neovessel formation after implantation of a cell-seeded biodegradable scaffold. (A) Seeded BM-MNCs release an early pulse of cytokines, including MCP-1, resulting in increased monocyte infiltration into the scaffold. Recruited monocytes release a variety of cytokines and growth factors (e.g., VEGF) that enhance angiogenesis and recruit SMCs and ECs to the scaffold. These components of the neovessel arise primarily from the proliferation and inward migration of mature vascular cells from adjacent native vessel segments. ECs and SMCs assemble into a mature vessel, consisting of an endothelial monolayer surrounded by a smooth muscle media. The scaffold degrades and monocytes migrate away, leaving behind a neovessel composed of autologous tissue. (B) Immunohistochemical staining for VEGF in scaffolds seeded with bone marrow cells at 1, 6, and 10 weeks after implantation demonstrates consistent VEGF expression during the transition from scaffold to neovessel (brown indicates positive VEGF expression). Images at 400× magnification. *Adapted with permission from Roh et al. [91].*

magnetic resonance image (MRI) angiography. Additionally, all patients were contacted by telephone to confirm their most recent clinical status.

58.6.3 Scaffold Preparation

Porous polymer scaffolds were constructed based on a copolymer of L-lactide acid and ε-caprolactone (50:50) (PCLA) reinforced with woven PGA fabric [96]. On the basis of patient vessel size, an appropriate diameter scaffold was selected (range: 12−24 mm). The scaffolds were

0.6−0.7 mm in thickness, and originally 13 cm in length prior to modification by the surgeon to accommodate unique patient anatomy. The scaffolds were designed to completely degrade within 6 months *in vivo*.

58.6.4 Cell Collection and Seeding

With the patient under general anesthesia and prior to the median sternotomy, bone marrow cells (5 mL/kg body weight) were aspirated from the anterior superior iliac spine via a puncture needle attached to a heparin-containing

TABLE 58.2 Patient Status After TEVG Implantation

Patient No.	Age at Operation (Year)	Graft Type	Graft Diameter (mm)	Patient Status	Graft Patency	Graft-Related Complications
1	2	PLA	16	Alive	Patent	None
2	1	PLA	20	Alive	Patent	None
3	7	PLA	18	Alive	Patent	Stenosis
4	21	PLA	24	Alive	Patent	None
5	4	PLA	20	Alive	Patent	None
6	12	PLA	24	Alive	Patent	None
7	17	PLA	24	Alive	Patent	None
8	19	PLA	22	Dead	Patent	None
9	3	PLA	12	Alive	Patent	Stenosis
10	2	PLA	16	Dead	Patent	None
11	2	PGA	20	Dead	Patent	None
12	13	PGA	16	Alive	Patent	Stenosis
13	2	PGA	16	Alive	Patent	Thrombosis
14	2	PGA	18	Alive	Patent	None
15	2	PGA	12	Alive	Patent	None
16	2	PGA	16	Alive	Patent	None
17	24	PGA	18	Alive	Patent	None
18	1	PGA	16	Alive	Patent	Stenosis
19	11	PGA	18	Alive	Patent	None
20	2	PGA	16	Alive	Patent	None
21	3	PGA	16	Alive	Patent	None
22	4	PGA	18	Alive	Patent	None
23	4	PGA	18	Alive	Patent	None
24	13	PGA	16	Alive	Patent	None
25	2	PGA	18	Dead	Patent	None

(100 units/mL) syringe [96]. Bone marrow aspirates were passed through a nylon cell strainer (Becton Dickinson, Franklin Lakes, NJ) to remove fat and bone fractions. The mononuclear cell fraction was collected via the Histopaque-1077 (Sigma Chemical Co., St. Louis, MO) density gradient centrifugation method. These cells were then seeded onto the scaffold by manual pipetting. The seeded scaffold was incubated intraoperatively in diluted autologous plasma for 2 h prior to implantation.

58.6.5 Early Results (Within 1 Month of Implantation)

Early results were defined as clinical and radiographic events within 30 days of patient discharge after

TEVG implantation. At this point, all patients were alive and symptom free. Angiography, ultrasonography, or CT revealed that all TEVGs were patent without stenosis, thrombosis, or aneurysmal dilatation [97].

58.6.6 Mid-Term Results (1-Year Postimplantation)

Patients continued to be monitored clinically and radiographically (angiography, ultrasonography, MRI angiography, or CT). At 1-year postimplantation, all grafts were patent with no thrombosis, stenosis, obstruction, or calcification on cineangiography or CT. Notably, the diameter of the implanted conduits increased in a nonaneurysmal

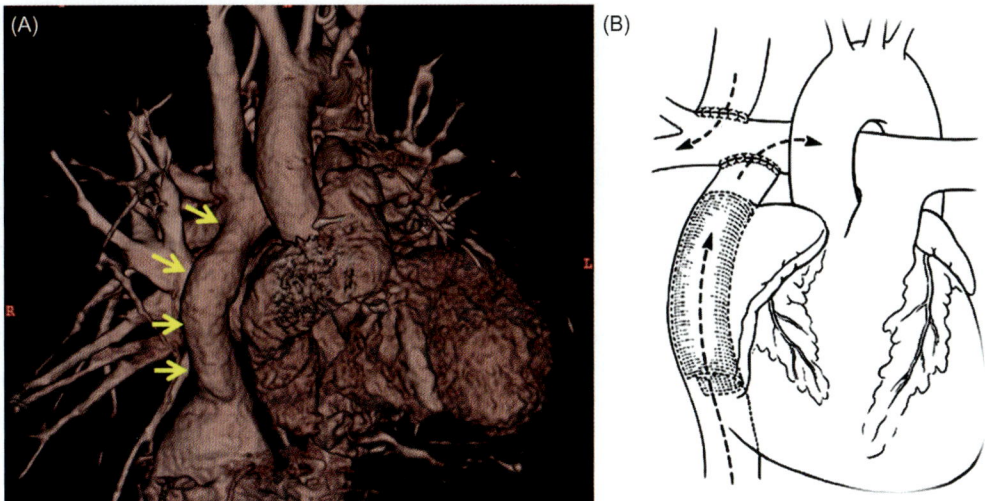

FIGURE 58.5 EC-TCPC utilizing a TEVG. (A) Three-dimensional CT image 1 year after TEVG implantation in a patient with congenital heart disease. Arrows indicate the location of the graft. (B) Schematic representation of the EC-TCPC operation. There are three major components in this procedure: (i) end-to-side anastomosis of the superior vena cava to the undivided right pulmonary artery; (ii) construction of a composite intraatrial tunnel with the use of the posterior wall of the right atrium; and (iii) use of a prosthetic vascular conduit to route the inferior vena cava to the enlarged orifice of the transected superior vena cava at the site of anastomosis with the main pulmonary artery.

fashion over time ($110 \pm 7\%$ of the implanted size). Partial mural thrombosis was identified in one patient (Patient 13 in Table 58.2) who was successfully treated with warfarin. One patient with hypoplastic left heart syndrome died of congestive heart failure secondary to severe tricuspid regurgitation 6 months after TEVG implantation (Patient 10 in Table 58.2), but this was unrelated to TEVG function [96].

58.6.7 Long-Term Results (Four or More Years Postimplantation)

Overall, there was no graft-related mortality during the follow-up period (range 4.3–7.3 years; mean 5.8 years). All patients received catheterization-based angiography, CT, or MRI, which demonstrated no evidence of aneurysm formation, graft rupture, or ectopic calcification. An additional three patients with known cardiovascular abnormalities died after TEVG implantation at long-term follow-up (total: four deaths postimplantation). Importantly, surveillance imaging in the months prior to all four patient deaths demonstrated a patent TEVG. Six of the patients (24%) had asymptomatic graft narrowing noted on routine surveillance imaging. Four of those six patients underwent successful balloon angioplasty, including one patient who required repeat balloon angioplasty and stent placement in the stenosed segment of the TEVG [2,97].

58.6.8 Summary

This first human clinical trial confirmed that our TEVGs functioned well without aneurysmal change or graft rupture. Three important observations from that study are as follows. First, asymptomatic graft stenosis developed in six patients with smaller diameter (<8 mm) conduits (four of whom required and were successfully treated with angioplasty and/or stenting). Second, there were no reported thromboembolic, hemorrhagic, or infectious complications that necessitated graft explantation or replacement. Third, there was significant growth of the TEVGs as demonstrated by serial imaging. These data support the overall feasibility and safety of a vascular tissue engineering approach. Despite the overall success of this initial clinical study, further translation of this technology remains limited by our incomplete understanding of the processes involved in transforming an implanted scaffold into a functional neovessel. The results of the trial indicated that stenosis is the primary means of graft-related failure. Therefore, our ongoing research has focused on gaining a better understanding of the mechanisms underlying TEVG stenosis in order to identify the potential targets and strategies that could prevent this complication. Ultimately, this would allow for the rational design of a second generation TEVG with improved safety and efficacy, while also expanding the range of applications in which this technology can be employed.

58.7 PERSPECTIVES FOR THE FUTURE

The technique of combining a stem cell source with a biodegradable scaffold material to achieve functional repair holds great promise for the patient with cardiac defects or vascular disease. However, many challenges remain to be met in both the basic science and clinical arenas before vascular tissue engineering becomes the new standard of care. For example, isolation of the appropriate mononuclear cell fraction for seeding is commonly performed using conventional density centrifugation methods. However, this method has several limitations including risk of contamination, the delay between bone marrow aspiration and graft readiness, and the need for a specialized class 10,000 clean room. To address this challenge, our lab has developed a new technique featuring an entirely closed filtration system [98]. In this system, harvested bone marrow is passed across a specialized filter to capture the mononuclear cells. The captured cells are then transferred to a chamber where vacuum suction draws the cells into the scaffold. The development of this unique system is only one response to the many challenges of clinical translation, but it is our hope that more physicians will be able to utilize TEVG technology safely and efficiently.

The development of various animal models has allowed the field of vascular tissue engineering to test many different combinations of scaffold materials, cell populations, seeding techniques, and bioreactor culturing conditions. However, due to the vast array of the aforementioned variables, a great number of methodologies remain untested. Moreover, the time requirements and cost of conducting animal experiments limit the number that can be undertaken. Put simply, the current trial-and-error approach to TEVG development inadequately addresses the complexities of neovessel formation. An alternative framework to TEVG construction lies in continuum mechanics and computational modeling. To this end, a multiaxial biomechanical testing device designed for mouse arteries has been implemented to increase the feasibility of creating models for TEVG development [99]. Indeed, this technology is already being incorporated in TEVGs cultured in bioreactors [100]. A rigorous and predictive modeling system will not only shed light onto the process of neovessel formation, but will also offer guidance toward the ideal mechanical properties of scaffold materials and conditions for neovascularization.

Despite its relative youth, the field of vascular tissue engineering has made great progress in the last two decades. The technology has reached the clinic in the form of a pilot clinical trial and additional clinical translation is a certainty in the coming years. However, to ensure the safety and efficacy of future treatments, all aspects of TEVG construction from scaffold materials and cell populations, to seeding methods and clinical release criteria need to be examined. To achieve successful translation of this complex multidisciplinary technology to the clinic, vascular tissue engineering requires the active participation of biologists, engineers, computer scientists, and clinicians in a concerted and directed effort.

REFERENCES

[1] Langer R, Vacanti JP. Tissue engineering. Science 1993;260:920—6.

[2] Kurobe H, Maxfield MW, Breuer CK, Shinoka T. Tissue engineering and regenerative medicine concise review: tissue-engineered vascular grafts for cardiac surgery: past, present, and future. Stem Cells Trans Med 2012;1:566—71.

[3] Vacanti JP, Langer R. Tissue engineering: the design and fabrication of living replacement devices for surgical reconstruction and transplantation. Lancet 1999;354:32—4.

[4] Weinberg CB, Bell E. A blood vessel model constructed from collagen and cultured vascular cells. Science 1986;231:397—400.

[5] Shinoka T, Imai Y, Ikada Y. Transplantation of a tissue-engineered pulmonary artery. N Engl J Med 2001;344:532—3.

[6] Verstappen J, Katsaros C, Torensma R, Von den Hoff JW. A functional model for adult stem cells in epithelial tissues. Wound Repair Regen 2009;17:296—305.

[7] Lerou PH, Daley GQ. Therapeutic potential of embryonic stem cells. Blood Rev 2005;19:321—31.

[8] Takahashi K, Okita K, Nakagawa M, Yamanaka S. Induction of pluripotent stem cells from fibroblast cultures. Nat Protoc 2007;2:3081—9.

[9] Bourassa MG. Long-term vein graft patency. Curr Opin Cardiol 1994;9:685—91.

[10] The VA Coronary Artery Bypass Surgery Cooperative Study Group. Eighteen-year follow-up in the veterans affairs cooperative study of coronary artery bypass surgery for stable angina. Circulation 1992;86:121—130.

[11] Ravi S, Chaikof EL. Biomaterials for vascular tissue engineering. Regen Med 2010;5:107—20.

[12] Florian A, Cohn LH, Dammin GJ, Collins Jr. JJ. Small vessel replacement with gore-tex (expanded polytetrafluoroethylene). Arch Surg 1976;111:267—70.

[13] Hunt MOJ, Belu AM, Linton RW, DeSimone JM. End-functionalized polymers. 1. Synthesis and characterization of perfluoroalkyl-terminated polymers via chlorosilane derivatives. Macromolecules 1993;26:7391.

[14] Kannan RY, Salacinski HJ, Butler PE, Hamilton G, Seifalian AM. Current status of prosthetic bypass grafts: a review. J Biomed Mater Res B Appl Biomater 2005;74:570—81.

[15] Prager M, Polterauer P, Bohmig HJ, Wagner O, Fugl A, Kretschmer G, et al. Collagen versus gelatin-coated Dacron versus stretch polytetrafluoroethylene in abdominal aortic bifurcation graft surgery: results of a seven-year prospective, randomized multicenter trial. Surgery 2001;130:408—14.

[16] van Det RJ, Vriens BH, van der Palen J, Geelkerken RH. Dacron or ePTFE for femoro-popliteal above-knee bypass grafting: short- and long-term results of a multicentre randomised trial. Eur J Vasc Endovasc Surg 2009;37:457—63.

[17] Taylor Jr. LM, Edwards JM, Porter JM. Present status of reversed vein bypass grafting: five-year results of a modern series. J Vasc Surg 1990;11:193—205, discussion 205—6.

[18] Debakey ME, Jordan Jr. GL, Abbott JP, Halpert B, O'Neal RM. The fate of dacron vascular grafts. Arch Surg 1964;89:757–82.

[19] Tiwari A, Salacinski H, Seifalian AM, Hamilton G. New prostheses for use in bypass grafts with special emphasis on polyurethanes. Cardiovasc Surg 2002;10:191–7.

[20] Christenson EM, Dadsetan M, Wiggins M, Anderson JM, Hiltner A. Poly(carbonate urethane) and poly(ether urethane) biodegradation: in vivo studies. J Biomed Mater Res A 2004;69:407–16.

[21] Tai NR, Salacinski HJ, Edwards A, Hamilton G, Seifalian AM. Compliance properties of conduits used in vascular reconstruction. Br J Surg 2000;87:1516–24.

[22] Rustad KC, Sorkin M, Levi B, Longaker MT, Gurtner GC. Strategies for organ level tissue engineering. Organogenesis 2010;6:151–7.

[23] Riha GM, Lin PH, Lumsden AB, Yao Q, Chen C. Review: application of stem cells for vascular tissue engineering. Tissue Eng 2005;11:1535–52.

[24] Kanda K, Matsuda T, Oka T. In vitro reconstruction of hybrid vascular tissue. Hierarchic and oriented cell layers. ASAIO J 1993;39:561–5.

[25] Hirai J, Matsuda T. Venous reconstruction using hybrid vascular tissue composed of vascular cells and collagen: tissue regeneration process. Cell Transplant 1996;5:93–105.

[26] Matsuda T, Miwa H. A hybrid vascular model biomimicking the hierarchic structure of arterial wall: neointimal stability and neoarterial regeneration process under arterial circulation. J Thorac Cardiovasc Surg 1995;110:988–97.

[27] Charulatha V, Rajaram A. Influence of different crosslinking treatments on the physical properties of collagen membranes. Biomaterials 2003;24:759–67.

[28] Swartz DD, Russell JA, Andreadis ST. Engineering of fibrin-based functional and implantable small-diameter blood vessels. Am J Physiol Heart Circ Physiol 2005;288:1451–60.

[29] Cummings CL, Gawlitta D, Nerem RM, Stegemann JP. Properties of engineered vascular constructs made from collagen, fibrin, and collagen-fibrin mixtures. Biomaterials 2004;25:3699–706.

[30] Haisch A, Loch A, David J, Pruss A, Hansen R, Sittinger M. Preparation of a pure autologous biodegradable fibrin matrix for tissue engineering. Med Biol Eng Comput 2000;38:686–9.

[31] Tschoeke B, Flanagan TC, Koch S, Harwoko MS, Deichmann T, Ella V, et al. Tissue-engineered small-caliber vascular graft based on a novel biodegradable composite fibrin-polylactide scaffold. Tissue Eng Part A 2009;15:1909–18.

[32] Koch S, Flanagan TC, Sachweh JS, Tanios F, Schnoering H, Deichmann T, et al. Fibrin-polylactide-based tissue-engineered vascular graft in the arterial circulation. Biomaterials 2010;31:4731–9.

[33] Dahl SL, Blum JL, Niklason LE. Bioengineered vascular grafts: can we make them off-the-shelf? Trends Cardiovasc Med 2011;21:83–9.

[34] Konuma T, Devaney EJ, Bove EL, Gelehrter S, Hirsch JC, Tavakkol Z, et al. Performance of CryoValve SG decellularized pulmonary allografts compared with standard cryopreserved allografts. Ann Thorac Surg 2009;88:849–54, discussion 554–5.

[35] Kaushal S, Amiel GE, Guleserian KJ, Shapira OM, Perry T, Sutherland FW, et al. Functional small-diameter neovessels created using endothelial progenitor cells expanded ex vivo. Nat Med 2001;7:1035–40.

[36] Simon P, Kasimir MT, Seebacher G, Weigel G, Ullrich R, Salzer-Muhar U, et al. Early failure of the tissue engineered porcine heart valve SYNERGRAFT in pediatric patients. Eur J Cardiothorac Surg 2003;23:1002–6, discussion 1006.

[37] Li L, Li Y, Li J, Yao L, Mak AFT, Ko F, et al. Antibacterial properties of nanosilver PLLA fibrous membranes. J Nanomat 2009; Available from: http://dx.doi.org/10.1155/2009/168041 [Article ID 168041, 5 pages].

[38] Athanasiou KA, Niederauer GG, Agrawal CM. Sterilization, toxicity, biocompatibility and clinical applications of polylactic acid/polyglycolic acid copolymers. Biomaterials 1996;17:93–102.

[39] Woodward SC, Brewer PS, Moatamed F, Schindler A, Pitt CG. The intracellular degradation of poly(epsilon-caprolactone). J Biomed Mater Res 1985;19:437–44.

[40] Gunatillake PA, Adhikari R. Biodegradable synthetic polymers for tissue engineering. Eur Cell Mater 2003;5:1–16 discussion 16.

[41] Pitt CG, Marks TA, Schindler A. Biodegradable drug delivery systems based on aliphatic polyesters: application to contraceptives and narcotic antagonists. NIDA Res Monogr 1981;28:232–53.

[42] Willerth SM, Sakiyama-Elbert SE. Combining stem cells and biomaterial scaffolds for constructing tissues and cell delivery. Cambridge (MA): StemBook; 2008.

[43] Dawes EA. Polyhydroxybutyrate: an intriguing biopolymer. Biosci Rep 1988;8:537–47.

[44] Hibino N, Duncan DR, Nalbandian A, Yi T, Qyang Y, Shinoka T, et al. Evaluation of the use of an induced puripotent stem cell sheet for the construction of tissue-engineered vascular grafts. J Thorac Cardiovasc Surg 2012;143:696–703.

[45] Herring M, Gardner A, Glover J. A single-staged technique for seeding vascular grafts with autogenous endothelium. Surgery 1978;84:498–504.

[46] Herring M, Smith J, Dalsing M, Glover J, Compton R, Etchberger K, et al. Endothelial seeding of polytetrafluoroethylene femoral popliteal bypasses: the failure of low-density seeding to improve patency. J Vasc Surg 1994;20:650–5.

[47] Meinhart JG, Deutsch M, Fischlein T, Howanietz N, Froschl A, Zilla P. Clinical autologous in vitro endothelialization of 153 infrainguinal ePTFE grafts. Ann Thorac Surg 2001;71:327–31.

[48] Wagenseil JE, Mecham RP. Vascular extracellular matrix and arterial mechanics. Physiol Rev 2009;89:957–89.

[49] Yue X, van der Lei B, Schakenraad JM, van Oene GH, Kuit JH, Feijen J, et al. Smooth muscle cell seeding in biodegradable grafts in rats: a new method to enhance the process of arterial wall regeneration. Surgery 1988;103:206–12.

[50] Niklason LE, Gao J, Abbott WM, Hirschi KK, Houser S, Marini R, et al. Functional arteries grown in vitro. Science 1999;284:489–93.

[51] Peck M, Gebhart D, Dusserre N, McAllister TN, L'Heureux N. The evolution of vascular tissue engineering and current state of the art. Cells Tissues Organs 2012;195:144–58.

[52] Naito Y, Shinoka T, Duncan D, Hibino N, Solomon D, Cleary M, et al. Vascular tissue engineering: towards the next generation vascular grafts. Adv Drug Deliv Rev 2011;63:312–23.

[53] Roberts I. Mesenchymal stem cells. Vox Sang 2004;87:38–41.

[54] Wang S, Qu X, Zhao RC. Clinical applications of mesenchymal stem cells. J Hematol Oncol 2012;5:19.

[55] Friedenstein AJ, Piatetzky II S, Petrakova KV. Osteogenesis in transplants of bone marrow cells. J Embryol Exp Morphol 1966;16:381–90.

[56] Tuli R, Tuli S, Nandi S, Wang ML, Alexander PG, Haleem-Smith H, et al. Characterization of multipotential mesenchymal progenitor cells derived from human trabecular bone. Stem Cells 2003;21:681−93.

[57] Asakura A, Komaki M, Rudnicki M. Muscle satellite cells are multipotential stem cells that exhibit myogenic, osteogenic, and adipogenic differentiation. Differentiation 2001;68:245−53.

[58] Young HE, Steele TA, Bray RA, Hudson J, Floyd JA, Hawkins K, et al. Human reserve pluripotent mesenchymal stem cells are present in the connective tissues of skeletal muscle and dermis derived from fetal, adult, and geriatric donors. Anat Rec 2001; 264:51−62.

[59] Mirza A, Hyvelin JM, Rochefort GY, Lermusiaux P, Antier D, Awede B, et al. Undifferentiated mesenchymal stem cells seeded on a vascular prosthesis contribute to the restoration of a physiologic vascular wall. J Vasc Surg 2008;47:1313−21.

[60] Hashi CK, Zhu Y, Yang GY, Young WL, Hsiao BS, Wang K, et al. Antithrombogenic property of bone marrow mesenchymal stem cells in nanofibrous vascular grafts. Proc Natl Acad Sci USA 2007;104:11915−20.

[61] Kuroda Y, Kitada M, Wakao S, Nishikawa K, Tanimura Y, Makinoshima H, et al. Unique multipotent cells in adult human mesenchymal cell populations. Proc Natl Acad Sci USA 2010;107:8639−43.

[62] Wakao S, Kitada M, Kuroda Y, Shigemoto T, Matsuse D, Akashi H, et al. Multilineage-differentiating stress-enduring (Muse) cells are a primary source of induced pluripotent stem cells in human fibroblasts. Proc Natl Acad Sci USA 2011;108:9875−80.

[63] Asahara T, Murohara T, Sullivan A, Silver M, van der Zee R, Li T, et al. Isolation of putative progenitor endothelial cells for angiogenesis. Science 1997;275:964−7.

[64] Asahara T, Masuda H, Takahashi T, Kalka C, Pastore C, Silver M, et al. Bone marrow origin of endothelial progenitor cells responsible for postnatal vasculogenesis in physiological and pathological neovascularization. Circ Res 1999;85:221−8.

[65] Ziegler BL, Valtieri M, Porada GA, De Maria R, Muller R, Masella B, et al. KDR receptor: a key marker defining hematopoietic stem cells. Science 1999;285:1553−8.

[66] Peichev M, Naiyer AJ, Pereira D, Zhu Z, Lane WJ, Williams M, et al. Expression of VEGFR-2 and AC133 by circulating human CD34(+) cells identifies a population of functional endothelial precursors. Blood 2000;95:952−8.

[67] Rehman J, Li J, Orschell CM, March KL. Peripheral blood "endothelial progenitor cells" are derived from monocyte/macrophages and secrete angiogenic growth factors. Circulation 2003; 107:1164−9.

[68] Asahara T, Kawamoto A, Masuda H. Concise review: Circulating endothelial progenitor cells for vascular medicine. Stem Cells 2011;29:1650−5.

[69] Steigman SA, Fauza DO. Autologous approaches to tissue engineering. Cambridge (MA): StemBook; 2008.

[70] Pittenger MF, Mackay AM, Beck SC, Jaiswal RK, Douglas R, Mosca JD, et al. Multilineage potential of adult human mesenchymal stem cells. Science 1999;284:143−7.

[71] Noishiki Y, Tomizawa Y, Yamane Y, Matsumoto A. Autocrine angiogenic vascular prosthesis with bone marrow transplantation. Nat Med 1996;2:90−3.

[72] Matsumura G, Hibino N, Ikada Y, Kurosawa H, Shin'oka T. Successful application of tissue engineered vascular autografts: clinical experience. Biomaterials 2003;24:2303−8.

[73] Zuk PA, Zhu M, Mizuno H, Huang J, Futrell JW, Katz AJ, et al. Multilineage cells from human adipose tissue: implications for cell-based therapies. Tissue Eng 2001;7:211−28.

[74] Kondo K, Shintani S, Shibata R, Murakami H, Murakami R, Imaizumi M, et al. Implantation of adipose-derived regenerative cells enhances ischemia-induced angiogenesis. Arterioscler Thromb Vasc Biol 2009;29:61−6.

[75] Rehman J, Traktuev D, Li J, Merfeld-Clauss S, Temm-Grove CJ, Bovenkerk JE, et al. Secretion of angiogenic and antiapoptotic factors by human adipose stromal cells. Circulation 2004;109:1292−8.

[76] Shen G, Tsung HC, Wu CF, Liu XY, Wang XY, Liu W, et al. Tissue engineering of blood vessels with endothelial cells differentiated from mouse embryonic stem cells. Cell Res 2003;13:335−41.

[77] Gearhart J. New potential for human embryonic stem cells. Science 1998;282:1061−2.

[78] Daley GQ. Missed opportunities in embryonic stem-cell research. N Engl J Med 2004;351:627−8.

[79] Takahashi K, Yamanaka S. Induction of pluripotent stem cells from mouse embryonic and adult fibroblast cultures by defined factors. Cell 2006;126:663−76.

[80] Villalona GA, Udelsman B, Duncan DR, McGillicuddy E, Sawh-Martinez RF, Hibino N, et al. Cell-seeding techniques in vascular tissue engineering. Tissue Eng Part B Rev 2010;16:341−50.

[81] Pawlowski KJ, Rittgers SE, Schmidt SP, Bowlin GL. Endothelial cell seeding of polymeric vascular grafts. Front Biosci 2004;9:1412−21.

[82] Roh JD, Nelson GN, Udelsman BV, Brennan MP, Lockhart B, Fong PM, et al. Centrifugal seeding increases seeding efficiency and cellular distribution of bone marrow stromal cells in porous biodegradable scaffolds. Tissue Eng 2007;13:2743−9.

[83] Hsu SH, Tsai IJ, Lin DJ, Chen DC. The effect of dynamic culture conditions on endothelial cell seeding and retention on small diameter polyurethane vascular grafts. Med Eng Phys 2005;27:267−72.

[84] Udelsman B, Hibino N, Villalona GA, McGillicuddy E, Nieponice A, Sakamoto Y, et al. Development of an operator-independent method for seeding tissue-engineered vascular grafts. Tissue Eng Part C Methods 2011;17:731−6.

[85] L'Heureux N, Dusserre N, Marini A, Garrido S, de la Fuente L, McAllister T. Technology insight: the evolution of tissue-engineered vascular grafts—from research to clinical practice. Nat Clin Pract Cardiovasc Med 2007;4:389−95.

[86] McAllister TN, Maruszewski M, Garrido SA, Wystrychowski W, Dusserre N, Marini A, et al. Effectiveness of haemodialysis access with an autologous tissue-engineered vascular graft: a multicentre cohort study. Lancet 2009;373:1440−6.

[87] Shimizu T, Yamato M, Isoi Y, Akutsu T, Setomaru T, Abe K, et al. Fabrication of pulsatile cardiac tissue grafts using a novel 3-dimensional cell sheet manipulation technique and temperature-responsive cell culture surfaces. Circ Res 2002;90:40.

[88] Bowlin GL, Meyer A, Fields C, Cassano A, Makhoul RG, Allen C, et al. The persistence of electrostatically seeded endothelial cells lining a small diameter expanded polytetrafluoroethylene vascular graft. J Biomater Appl 2001;16:157−73.

[89] Perea H, Aigner J, Hopfner U, Wintermantel E. Direct magnetic tubular cell seeding: a novel approach for vascular tissue engineering. Cells Tissues Organs 2006;183:156–65.

[90] Shimizu K, Ito A, Arinobe M, Murase Y, Iwata Y, Narita Y, et al. Effective cell-seeding technique using magnetite nanoparticles and magnetic force onto decellularized blood vessels for vascular tissue engineering. J Biosci Bioeng 2007;103:472–8.

[91] Roh JD, Sawh-Martinez R, Brennan MP, Jay SM, Devine L, Rao DA, et al. Tissue-engineered vascular grafts transform into mature blood vessels via an inflammation-mediated process of vascular remodeling. Proc Natl Acad Sci USA 2010;107:4669–74.

[92] Badylak SF, Valentin JE, Ravindra AK, McCabe GP, Stewart-Akers AM. Macrophage phenotype as a determinant of biologic scaffold remodeling. Tissue Eng Part A 2008;14:1835–42.

[93] Brown BN, Valentin JE, Stewart-Akers AM, McCabe GP, Badylak SF. Macrophage phenotype and remodeling outcomes in response to biologic scaffolds with and without a cellular component. Biomaterials 2009;30:1482–91.

[94] Hibino N, Yi T, Duncan DR, Rathore A, Dean E, Naito Y, et al. A critical role for macrophages in neovessel formation and the development of stenosis in tissue-engineered vascular grafts. FASEB J 2011;25:4253–63.

[95] Hoffman JI, Kaplan S. The incidence of congenital heart disease. J Am Coll Cardiol 2002;39:1890–900.

[96] Shin'oka T, Matsumura G, Hibino N, Naito Y, Watanabe M, Konuma T, et al. Midterm clinical result of tissue-engineered vascular autografts seeded with autologous bone marrow cells. J Thorac Cardiovasc Surg 2005;129:1330–8.

[97] Hibino N, McGillicuddy E, Matsumura G, Ichihara Y, Naito Y, Breuer C, et al. Late-term results of tissue-engineered vascular grafts in humans. J Thorac Cardiovasc Surg 2010; 139:431–6.

[98] Hibino N, Nalbandian A, Devine L, Martinez RS, McGillicuddy E, Yi T, et al. Comparison of human bone marrow mononuclear cell isolation methods for creating tissue-engineered vascular grafts: novel filter system versus traditional density centrifugation method. Tissue Eng Part C Methods 2011;17:993–8.

[99] Gleason RL, Gray SP, Wilson E, Humphrey JD. A multiaxial computer-controlled organ culture and biomechanical device for mouse carotid arteries. J Biomech Eng 2004;126:787–95.

[100] Niklason LE, Yeh AT, Calle EA, Bai Y, Valentin A, Humphrey JD. Enabling tools for engineering collagenous tissues integrating bioreactors, intravital imaging, and biomechanical modeling. Proc Natl Acad Sci USA 2010;107:3335–9.

Chapter 59

Corneal Bioengineering[*]

Francesca Corradini[a], Michela Zattoni[a], Paolo Rama[b], Michele De Luca[a], and Graziella Pellegrini[a]

[a]Center for Regenerative Medicine "Stefano Ferrari", University of Modena and Reggio Emilia, Modena, Italy, [b]San Raffaele Scientific Institute, Ophthalmology Unit, Milan, Italy

59.1 STRATIFIED EPITHELIA

The epithelial tissue covers the whole surface of the body. It is made up of cells closely packed and organized in one or more layers and rests on a basement membrane, which is subdivided into a basal lamina (thought to be produced by the epithelium) and a reticular lamina (produced by connective tissue cells). The basement membrane anchors the epithelium to the underlying connective tissue providing structural support for the epithelium.

The various types of human epithelia can be divided into simple and stratified epithelia, depending on the number of layers.

Simple epithelia are formed by a single layer of cells all attached to the basement membrane, while stratified epithelia consist of multiple layers of keratinocytes. Depending on the morphology of the apical layer, stratified epithelia are divided into squamous (flat and scale-like), cuboidal (cube shaped), and columnar (column shaped). Keratinocytes forming the basal layer rest on the basement membrane and are the only cells in a stratified epithelium endowed with proliferative capacity.

Squamous epithelia are constantly renewed. For instance, human epidermis and corneal epithelium are renewed approximately every 3–4 weeks and 9–12 months, respectively. Moreover, squamous epithelia represent the first protective barriers against the external environment. Thus, they experience daily assaults like wounds, which need timely repair.

Self-renewing epithelial stem cells are responsible for such regeneration and repair processes. Under normal homeostasis, stem cells divide infrequently and generate committed progenitors often referred to as transient amplifying (TA) cells, which in turn generate terminally differentiated cells after a limited number of cell division. Autologous cultures of keratinocyte stem cells are used worldwide for the regeneration of many types of squamous epithelia.

59.2 IDENTIFICATION OF HUMAN SQUAMOUS EPITHELIAL STEM CELLS: THE HOLOCLONE-FORMING CELL

The proliferative compartment of human squamous epithelia contains three types of keratinocytes with different proliferative capacity, referred to as holoclones,

---

[*] Francesca Corradini and Michela Zattoni contributed equally to this work.

meroclones, and paraclones. Under appropriate culture conditions, these clones can be isolated and cultured at a single cell level [1].

Holoclone-forming cell are the smallest colony-forming keratinocytes and their number declines during aging [1,2] (Figure 59.1). Holoclones have all the characteristics of stem cells, including self-renewal ability [3,4], telomerase activity [5], and an impressive proliferative potential [6,7]. The holoclone-forming keratinocyte is the stem cell of virtually all human squamous epithelia, including the epidermis, the hair follicle, and the corneal epithelium [6–11].

Paraclone-forming cells have a very limited proliferative ability and have the properties of TA progenitors. Meroclones have an intermediate proliferative potential and are a reservoir of paraclones [1,6]. The transition from a holoclone to a paraclone, known as clonal conversion, is an unidirectional process that occurs during serial cultivation.

While paraclones can be visually identified as small colonies with an irregular perimeter mainly composed of large and flattened keratinocyte, it is not possible to distinguish holoclones from meroclones based on their size or morphology. In short-term clonogenic assays, meroclone-forming cell can even generate larger colonies than holoclones but this is not an indicator of the long-term regenerative potential of the clone (unpublished).

Therefore, the most reliable tool to evaluate the stem cell content of a specific stratified epithelium is clonal analysis. The classification of clonal type is based on the frequency of terminal (aborted) colonies produced when the clone is transferred to an indicator dish. When 0–5% of colonies are terminal the founding clone is scored as a holoclone. When all colonies formed are terminal (or when no colonies formed), the clone is classified as a paraclone. When >5% but <100% of the colonies are terminal, the clone is classified as a meroclone [1].

59.3 STRUCTURE AND FUNCTION OF CORNEAL EPITHELIUM

The anterior part of the human eye is composed of the conjunctiva and the cornea. The conjunctival epithelium lies on a vascularized stroma and allows the movement of the eyelid over the cornea, the limbal (see below) vascular supply, and the maintenance of the normal lid-globe apposition. Human conjunctival stem cells are located in the basal layer of the entire conjunctival epithelium covering the bulb and the fornix [6]. Conjunctival stem cells are bipotent, in that they generate goblet cells (unicellular mucin-secreting glands) according to a rather specific cell doubling clock [6].

Our window to the world is represented by the cornea, which is transparent and allows refraction of light. The cornea is composed of five layers: the outermost nonkeratinized stratified epithelium, the Bowman's membrane, the fibroblast (keratocyte)-populated stroma, Descemet's membrane, and the inner endothelium (Figure 59.2). The corneal epithelium represents approximately 10% of the total corneal thickness.

The limbus is the narrow zone between the cornea and the bulbar conjunctiva. The limbal epithelium consists mainly of keratinocytes, contains Langerhans cells and melanocytes, and harbors the stem cells of the corneal epithelium.

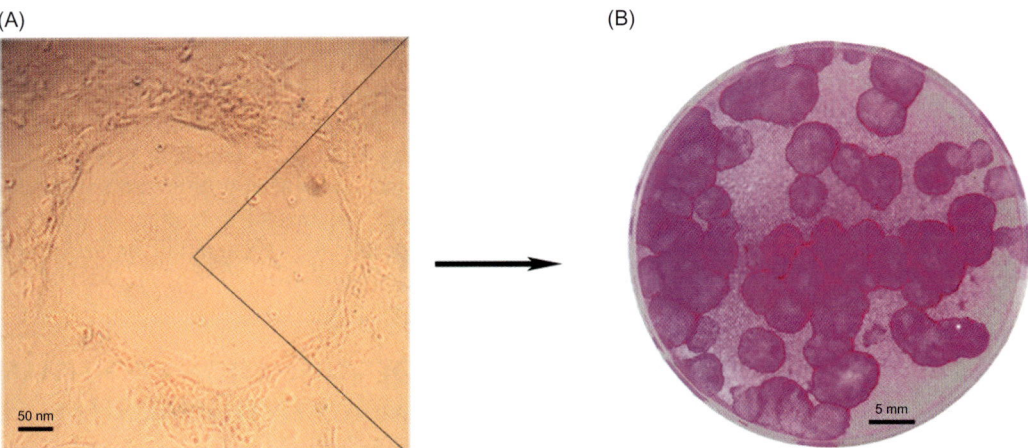

(A)   50 nm

(B)   5 mm

FIGURE 59.1 Daughter colonies produced by a holoclone. The clone was disaggregated and one-quarter of the cells were cultured into the indicator dishes. The cells were allowed to grow for 12 days, fixed and stained with rhodamine. The original clone was classified as holoclone based on the number of terminal colonies (<5%).

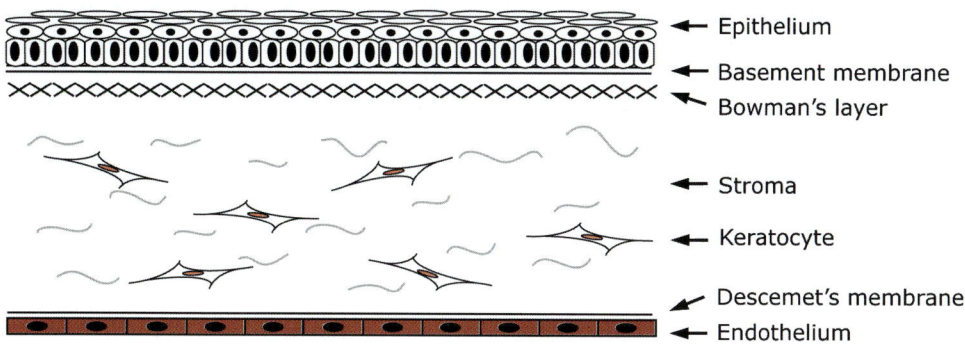

← Epithelium

← Basement membrane

← Bowman's layer

← Stroma

← Keratocyte

← Descemet's membrane

← Endothelium

FIGURE 59.2 Schematic representation of the human cornea in cross section.

Relatively undifferentiated and slow-cycling epithelial cells have been found in the limbal basal layer but not in the central cornea [12−14]. It is well documented that these cells migrate from the limbus towards a wounded cornea [9,13,15,16]. In addition, mathematical analysis of the maintenance of the corneal epithelial cell mass [17] and the mosaic analysis of stem cell function and corneal wound healing [18] strengthened the hypothesis that the corneal epithelium is regenerated by limbal stem cells. The regenerative capacity of the corneal epithelium is limited; in fact, in the absence of limbal epithelium, repeated corneal wounds lead to progressive vascularization and recurrent erosions of the cornea [19].

Clonal analysis of the human limbal−corneal epithelium have shown that limbal cells are highly clonogenic and endowed with a high proliferative potential. In contrast, central and paracentral corneal cells generated only aborted colonies with a negligible proliferative capacity. The peripheral cornea (namely, the area of the corneal adjacent to the limbus) contains some clonogenic cells, which, however, have a limited capacity for multiplication. Indeed, only limbal cells are able to generate holoclones with a frequency of approximately 5−10% [4,6].

Thus, the stem cells of the human corneal epithelium are located exclusively in the basal layer of the limbus. The transparency of the cornea, which depends on stromal avascularity and epithelial integrity, is essential to visual acuity. Integrity, constant renewal, and repair of the corneal epithelium rely on self-renewing limbal stem cells.

59.4 THE LIMBAL EPITHELIAL STEM CELLS

59.4.1 Location

Limbal stem cells are thought to reside in the palisades of Vogt, a well-defined stem cell niche providing a protecting microenvironment that controls self-renewal, proliferation, and differentiation potential of limbal stem cells

[20]. It has been proposed that limbal stem cells are located at the bottom of the epithelial papillae forming the Palisades of Vogt [21,22], specifically in crypts identified as solid cords of cells that extend from the peripheral end of the palisades of Vogts into the underlying stroma. The palisades of Vogt contain melanocytes [16], Langerhans cells [23], and T lymphocytes [24]. The melanin pigmentation protects limbal basal stem cells against ultraviolet light damage and the consequent generation of reactive oxygen species [25].

59.4.2 Characteristics and Markers

As with many other somatic stem cells, limbal stem cells are slow cycling, at least under normal homeostasis [14]. However, they retain the ability to rapidly proliferate in response to injury. Limbal stem cells also share a number of characteristics with other adult somatic stem cells, such as morphology and cell size. The basal cells of the limbus ($10.1 \pm 0.8\,\mu m$ in diameter) are significantly smaller than basal cells of the central and peripheral cornea ($17.1 \pm 0.8\,\mu m$ in diameter) [26]. They have a high nucleus/cytoplasm ratio and heterochromatin rich nuclei [27]. In addition, the absence of markers like involucrin [27], which is associated to a differentiation state of the cells, strengthens the notion that the limbal basal layer is the site of stem cell precursors. Finally, the basal layer of the limbus consists of cells rich in keratins but lacking keratins strictly associated to corneal terminal differentiation as cytokeratin 3 (CK3) and cytokeratin 12 (CK12), which are expressed in the limbal suprabasal cells and in both basal and suprabasal cells of the corneal epithelium [12,28].

Different approaches have been used to identify potential molecular markers distinguishing limbal stem cells from TA progenitors. Initially, integrins like α9, which is localized into small clusters of cells in the limbal epithelium, have been suggested as a possible marker [29]. The integrin β1, originally proposed as an epidermal stem cell marker [27], was also associated to limbal stem cells, however, β1 is also expressed in the basal layer of the cornea, suggesting that it also identifies corneal progenitors. Nerve growth factor (NGF) has been thought to promote limbal cell survival through the high affinity NGF-receptor TrkA [30]. This receptor is expressed in both corneal and limbal basal cells whereas the low affinity NGF-receptor p75 was found in suprabasal limbal and entire corneal epithelium. Intermediate filaments, like vimentin, are expressed by the majority of basal limbal cells (in some instances corneal) [21,31]. However, none of these proteins seems to specifically identify limbal stem cells.

The ATP-binding cassette subfamily G, member 2 (ABCG2), which is expressed in a wide variety of stem cells, has been found in (some) limbal basal cells but not in corneal epithelial cells, supporting the idea that it is a possible marker for limbal stem cells [32,33].

Notch-1 is a ligand-activated transmembrane receptor, which has been reported in other systems to maintain progenitor cells in an undifferentiated state. It has been shown that Notch-1 is expressed in the basal layer of the limbus, particularly in the cluster of cells contained in the palisades of Vogt, but it is not expressed in the central cornea [34]. Furthermore, ABCG2-positive limbal basal cells coexpress Notch-1, strengthening the notion that Notch-1 might mark the limbal stem cell compartment [34]. However, other groups provided evidence suggesting a role of Notch-1 in regulating differentiation [35], proliferation, and corneal repair [36] processes of the central corneal epithelium.

59.5 CULTURE METHOD OF LIMBAL STEM CELL FOR GRAFTABLE EPITHELIUM

Allogeneic corneal transplantation (keratoplasty) is used worldwide to restore the cornea in patients suffering from severe chemical burns of the eye. However, the keratoplasty is aimed only at the restoration of the stromal scar. Indeed, the clinical success of the keratoplasty relies on patient's spared limbal stem cells, which generate the host-derived corneal epithelium needed to resurface the donor corneal stroma. But ocular burns may destroy the limbus, causing limbal stem cell deficiency (LSCD). In case of LSCD, a functional corneal epithelium can no longer be formed, the keratoplasty is unsuccessful and the cornea reacquires an epithelium by invasion of bulbar conjunctival cells originating beyond the destroyed limbus. This process leads to corneal vascularization, inflammation, opacification, and loss of vision [20].

In patients with chemical or thermal burn-dependent LSCD, the only way to prevent the corneal conjunctivalization is to restore the limbus. In 1989, Kenyon and Tseng performed the first trial of large limbal fragments transplantation in patients with unilateral limbal/corneal destruction. They transplanted conjunctiva and limbus from the uninjured eye onto the diseased eye [37].

The notion that limbal stem cells, detected as holoclone-forming cells, can be cultivated [6,38], has fostered the clinical use of autologous limbal cultures for the regeneration of a destroyed corneal epithelium [39—41]. If the injury does not generate stromal scarring, limbal cultures are sufficient to restore a functional corneal epithelium and normal visual acuity, but if there is stroma destruction, the eye might recover only 10—30% of vision. To improve the recovery of visual acuity after grafting, the patient needs a subsequent keratoplasty to restore the stromal scar. The engrafted limbal stem cells are able to resurface the donor stroma and the eye can recover full visual acuity [40,41].

This said, the clinical success of an autologous limbal culture first depends on the quality of cultures used to prepare the graft. This does not mean that the cultures should contain a well-organized stratified epithelium resembling the corneal epithelium, but rather that they must contain a sufficient number of the keratinocyte stem cells essential for long-term corneal renewal. Only when this criterion is met, does success then depend exclusively on the clinical procedure. In the absence of an adequate number of limbal stem cells, failures of long-term corneal regeneration are inevitable and will entail not only suffering of the patients, but also general confusion as to what results are to be expected.

Several factors are required for the preservation of cultured stem cell number, amongst which appropriate feeder layer of lethally irradiated fibroblasts, substrate, and fetal calf serum is probably the most important.

Only few (dozens) stem cells are present in the 1—2 mm² limbal biopsy used to establish the culture. During primary culture this number increases as a consequence of the amplification of stem cell population, establishing the condition for the clinical success.

The first key factor essential for the preservation of limbal stem cells in culture is an appropriate feeder layer of lethally irradiated fibroblasts [3]. In particular, primary limbal stem cells need to be cultivated on a feeder layer of clinical grade [GMP (good manufacturing practice)-certified] 3T3-J2 cells, which have been used for decades worldwide to grow both epidermal and limbal stem cells. The use of a single clone of 3T3 assures an identical feeder layer among all cultures and eliminates the intrinsic variability of autologous fibroblasts, hence allowing rigorous in-process quality control. Moreover, since the 3T3 Master and Working Cell Banks are produced under rigorous GMP conditions they are safer than other allogeneic human cells.

Another key factor for the maintenance of limbal stem cells is the selection of an appropriate batch of fetal calf serum; this would assure the reproducibility of the culture eliminating the variability of autologous serum.

The first clinical application of cultured human limbal cells could not be reproduced on a large scale because of the fragility of the epithelium [39]. Limbal keratinocytes were grown directly on plastic and the resulting epithelial sheets were detached enzymatically. The lack of anchorage causes a substantial (approximately 50%) shrinking of the epithelium and a rapid loss of clonogenic cells, limiting long distance transportation of the graft and timing between graft preparation and transplantation [3].

The introduction of fibrin glue as supporting material for cell cultures greatly improved the clinical application of both epidermal and limbal stem cells [42,43].

The fibrin matrix is an ideal mechanical support since it is natural, nontoxic, and manageable, in that it adheres to the prepared wound bed without the need of sutures. Fibrin is naturally degraded within 24 h after transplantation, allowing the cultured epithelium to adhere and engraft on the wound bed.

Both epidermal and limbal keratinocytes grown on fibrin have the same clonogenic ability, proliferative potential, morphology, and differentiation properties as those cultured directly on plastic [40,42,43].

More importantly, the number of stem cells contained in both epidermal and limbal cultures is preserved by the fibrin substrate. This property is of paramount importance for the long-term clinical performance of the cultures.

The use of the fibrin substrate also allows use of subconfluent cultures since the detachment of cohesive epithelial sheets is no longer needed. This allows more flexibility in the coordination between cultivation and surgical procedures. In this respect, fibrin-cultured cells preserve their clonogenic ability for at least 3 days (as opposed to the 24 h of the detached epithelium) and can be safely transported at room temperature [3].

For clinical application, the subconfluent limbal keratinocyte primary culture is trypsinized and a portion is plated on a circular fibrin substrate of 3 cm diameter in the presence of feeder layer and the remainder is cryopreserved [40–42] (Figure 59.3). At confluence, fibrin-cultured epithelial sheet is grafted over the surgically prepared corneal and limbal wound region of the injured eye. The corneal wound bed is prepared by the removal of the abnormal epithelium and fibrovascular tissue as described by Rama et al. Conjunctiva is sutured over the peripheral fibrin sheet in order to help it to adhere on the surface [40,41].

59.6 LEARNING STEM CELL BIOLOGY FROM CLINICS

The clinical use of limbal cultures was first described by Pellegrini et al in 1997. Since then, many reports on the clinical use of this technology have been published [44–47]. Small numbers of cases, heterogeneous etiology, limited follow-up (often <1 year), heterogeneous sources of cells (autologous or allogeneic limbal cells or autologous oral cells), heterogeneous substrates and culture procedures, and absence of stringent quality criteria for limbal cultures hampered comparable and interpretable results.

Rama et al. have recently reported a 10-year follow-up study of transplantation of autologous cultures of limbal stem cells in 112 patients with severe chemical burn-dependent corneal destruction.

In this study, LSCD grading was based on clinical and cytological evaluation and eye examination, which included visual acuity, tonometry, slit-lamp examination, fundus, ecography, Shirmer test, and photographs. These parameters were evaluated before and after treatment providing a reliable objective assessment of clinical success. Moreover, clinical outcomes were evaluated in light of the percentage of holoclone-forming cells and by stringent statistical analysis [41]. The corneal surface was permanently restored in 76.6% of the eyes, which were covered by a transparent, renewing and functional corneal epithelium. The Kaplan–Meier survival chart showed that failures mainly occurred within the first year after grafting whereas successful cultures remained stable for up to 10 years of follow-up. 40% of patients presented stromal destruction and in order to improve visual acuity, they underwent successful keratoplasty between 4 and 6.5 years after grafting.

In this study, clinical success was not correlated to the total number of clonogenic cells nor to the morphology and colony size. The only correlation has been seen with the number of holoclones assessed by the expression of the transcription factor p63.

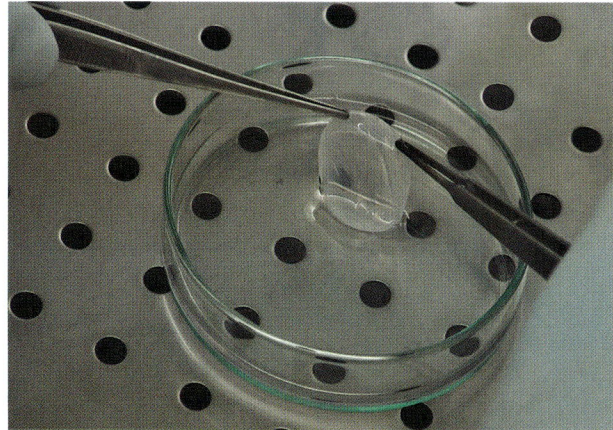

FIGURE 59.3 Circular fibrin matrix of approximately 3 cm diameter used for secondary culture of limbal cells.

p63 is a transcription factor belonging to the p53 gene family [48]. p63 is normally expressed in keratinocytes with proliferative potential, including skin, cervix, prostate, and cornea [11,48,49]. It plays a key role in the regulation of epithelial development, proliferation, and maintenance [50–53]; p63 knockout mice lack stratified epithelia and contain cluster of terminally differentiated keratinocytes on the exposed epidermis but not proliferative basal layer [51,52].

The p63 gene generates 10 p63 isoforms [54]: the transactivating isoforms, and the truncated ΔN isoforms. Alternatively splicing of each transcript gives rise to five different C termini, labeled as α, β, γ, δ, ϵ [54].

The $\Delta N\alpha$ isoform is strongly expressed in human limbal holoclones and weakly in meroclones but is undetectable in paraclones [9,11]. Depending on the condition, limbal and corneal keratinocyte may contain three ΔN isoforms. In human, resting cornea $\Delta Np63\alpha$ is present in the limbus but is absent from central cornea [9,11,27]. Both $\Delta Np63\beta$ and $\Delta Np63\gamma$ become relatively abundant (as compared to α isoform) during clonal conversion *in vitro* and correlate with cell migration and corneal regeneration during wounding [9].

$\Delta Np63\alpha$ has been linked to Bmi-1 and C/EBPδ. The coexpression of these three proteins identifies mitotically quiescent limbal stem cell and holoclones but not meroclones or paraclones [4].

Under normal homeostasis, quiescent limbal stem cells coexpress C/EBPδ and Bmi-1, which regulate mitotic quiescence and self-renewal, and $\Delta Np63\alpha$, which is responsible for the stem cell proliferative potential [9,11,49,55]. During an acute wound, $\Delta Np63\alpha$ positive cells are released from C/EBPδ-dependent mitotic constants, actively proliferate and migrate to the central cornea. This process leads to the irreversible loss of self-renewal properties and the progressive loss of $\Delta Np63\alpha$ because of the entrance in the TA compartment. Activated stem cells and TA cells increase the expression of $\Delta Np63\beta$ and $\Delta Np63\gamma$, which is associated to terminal differentiation during the regeneration of the corneal epithelium [4,9].

This said, the clinical success of limbal cultures strictly depends on the number of stem cells, detected as p63 bright holoclone-forming cells, contained in the graft. Indeed, primary limbal cultures containing more than 3% of p63 bright holoclones were successful in almost 80% of patients, whereas cultures containing 3% or less failed in almost 90% of the patients. Therefore, a minimum of approximately 3000 stem cells is required for the clinical success of the procedure (a primary limbal culture contains a minimum of 3×10^5 cells, about 30% of which are clonogenic; therefore, 3% of these clonogenic cells, namely 3000, should be holoclones) [41]. Of note, holoclones represent a tiny percentage ($\sim$1%) of the total

limbal cells grafted on the patient. Therefore, corneal regeneration cannot be ascribed to a nonspecific stimulatory effect of epithelial cultures (fibrin or surgical manipulation) on spared residual limbal cells [41]. Similarly, holoclones represent a tiny percentage ($\sim$3–5%) of clonogenic cells contained in the graft. Therefore, the colony-forming efficiency assay is not sufficient to evaluate the number of stem cells and to predict the performance of the graft [56], in that at least 95% of the clonogenic cells are meroclones and paraclones. Indeed, the number of total clonogenic cells contained in successful or unsuccessful cultures was almost identical.

Clonal conversion, hence disappearance of holoclone-forming cells, might occur very rapidly under nonoptimal culture conditions. A properly prepared feeder layer of lethally irradiated 3T3 cells together with adequate selected fetal calf serum [3,57,58] are necessary for the preservation of holoclones. Keratinocytes cultured using this method have been used worldwide since the 1980s [3,59] for life-saving treatment of massive full-thickness burns [42,43,60]. No adverse effects have been reported during the past 30 years and this culture method has been approved for use in the United States, Japan, Italy, and South Korea. In addition, appropriate substrates for the cultivation of stem cells are required for the retention of holoclones and both plastic and fibrin have been demonstrated to preserve holoclone-forming cells [40,42].

In summary, clinical results clearly demonstrated that (i) to achieve clinical success, the cultures must contain a sufficient number of stem cells essential for long-term epithelial renewal, (ii) such stem cells can be identified as p63 bright holoclones, (iii) the number of clonogenic cells cannot be used as a surrogate for a keratinocyte stem cell assay, (iv) the corneal regeneration achieved in these patients can be ascribed neither to a stimulation of resident spared stem cells nor to an aspecific stimulatory effect of the cultures, (v) appropriate culture conditions are of paramount importance to preserve stem cells and to assure a long-term regeneration of a functional self-renewing epithelium.

59.7 ALTERNATIVE METHODS OF AUTOLOGOUS GRAFT CULTIVATION ON DIFFERENT SUBSTRATES

The first patient suffering from LSCD treated with limbal stem cell therapy dates 15 years ago [39].

Since then alternative limbal culture methods for the clinical use of this technology have been proposed (reviewed in [3,44–46,61–65]), but the retention of stem cells has not been investigated.

Cells have been cultured on a variety of substrates and culture media in the presence or absence of feeder cells or FCS. All these studies (reviewed in [3,44−46,61−65]) have been limited by comparatively small numbers of cases, heterogeneous etiology, limited follow-up (often <1 year) and heterogeneous sources of transplant and cell culture techniques.

59.7.1 Amniotic Membrane

The human amniotic membrane (AM) consists of a single layer of columnar epithelial cells attached to a basement membrane [66] and its thickness can range from 0.02 mm to 0.5 mm [67]. It has been shown that the amniotic promotes epithelialization, inhibits fibrosis, and has anti-inflammatory ability and low immunogenicity. Thanks to these properties, the human AM has been widely used as a biological medication in ophthalmic surgery and has been proposed as a suitable substrate for limbal cell cultivation [67−70].

The first clinical use of limbal cells grown onto AM was reported in 1999 [71]. Since then, different cultivation protocols for the transplantation of AM-cultured limbal epithelial cells have been proposed [63,71−75]. The most used protocol is the "explant culture system" in which a biopsy is placed directly on the AM with or without trimming and limbal stem cells migrate out of the biopsy [72,74,76−78].

Moreover, different methods for preparing the AM prior to limbal culture have been described. AM has been used fresh or cryopreserved, with or without removal of the layer of columnar epithelial cells attached to its basement membrane, and AM limbal cultures have been performed with or without feeder layer. However, the best method for AM preparation and use is still unclear. Furthermore, the preservation of limbal stem cells onto the AM has never been formally demonstrated.

Gruterich et al. have shown that limbal epithelial cells expanded on intact AM retained their characteristics while if they have been cultured on denuded AM acquire a more differentiated corneal phenotype [79,80]. Other authors reported that denuded AM is colonized much quicker than intact AM and better supported the growth of limbal cells [81,82]. Recently, a 10-year clinical study of xeno-free autologous limbal cells cultured on de-epithelized AM has been reported. Despite the long follow-up and the large number of patients involved, the survival curve of success showed absence of stability of positive clinical outcome on long term [83]; moreover, a great limitation of the work is the absence of rigorous criteria for the diagnosis of LSCD and the clinical outcome of the cultures [84].

In all these studies, stringent quality criteria for limbal cultures were not adopted and the retention of stem cells has not been evaluated, making it quite difficult to evaluate the clinical performance of the graft.

59.7.2 Poly(N-isopropylacrylammide) Polymer

Poly(N-isopropylacrylamide) polymer is a newly developed temperature-responsive support, which reversibly alters its hydration properties with temperature. It is laid down on the culture surface in thin film allowing cell growth in normal culture condition at 37°C. The reduction of the temperature below 30°C leads to swelling and complete detachment of adherent cells without enzymes or EDTA. Nishida et al. first described the clinical application of this polymer for the transplantation of autologous oral mucosal epithelial cells into patients with bilateral LSCDs. The epithelial sheet was transplanted without sutures directly to the exposed transparent stromal bed of the diseased eye [85,86].

However, the effects of this support on the cell behavior have not been investigated. Neither holoclones preservation nor proliferative potential have been tested on cells grown on the polymer. Nevertheless clinical results were encouraging, as those patients have no real therapeutic alternatives.

59.8 ALTERNATIVE STEM CELL SOURCES

Autologous cultures of limbal cells can be applied to total unilateral LSCD or severe bilateral corneal destruction, provided that a tiny limbal area is spared. This technology cannot be used in case of total bilateral LSCD.

A possible alternative therapy for patients with total bilateral LSCD is represented by limbal-allograft transplantation [87]. In such cases, long-term immunosuppression is required with the consequence of possible systemic complications, including infection and liver and kidney dysfunction. Moreover, recurrent graft failure is quite common in patients with Stevens−Johnson syndrome or ocular pemphigoid, owing to different severe preoperative conditions (such as inflammation of the ocular surface, severe dry eye, abnormalities in epithelial differentiation of the ocular surface) [88,89].

59.8.1 Oral Mucosa

To avoid allograft rejection, alternative strategies have been studied for the replacement of destroyed corneal epithelium. Oral mucosa represents a valid and easily accessible source for the cultivation of squamous epithelial. Oral mucosal epithelial sheets and corneal epithelial sheets originating from limbal stem cells have been shown to possess the same optical transparency [85].

During the last decade, oral mucosa epithelial cells have been widely used to treat patients with various severe ocular surface disorders [85,90−92]. As with autologous limbal cells, many substrates have been used and many culture methods have been adopted [85,90,93–98].

In some of these studies, a substantial improvement (or even complete restoration) of the corneal epithelium has been reported, with a follow-up of up to 35 months. However, the means by which transplanted oral mucosa cells can improve or restore the ocular surface require clarification. Are these cells able to transdifferentiate into corneal epithelium? This seems unlikely and is not supported by any evidence. The effect of oral keratinocytes could be explained by what has been called functional, not anatomical, stem cell deficiency [99]. In other words, if the LSCD is only partial, the applied oral keratinocytes may stimulate remaining resident limbal cells to regenerate the corneal epithelium. A rigorous study of the supposed transdifferentiation capacity of human buccal keratinocytes onto authentic limbal–corneal cells is needed. A rigorous evaluation of the clinical performance of oral mucosa cells in total and complete LSCD is needed as well.

59.9 GMP AND IMPLICATIONS FOR REGULATION

In Europe, cultured epithelial cells for transplantation were recently classified as advanced therapy medicinal products (ATMPs) and they fall under the regulation of human medicinal products.

The competent authority in Europe is the European Medicinal Agency (EMA) and these products should be compliant with Tissue and Cells Directive 2004/23/EC, article 1 of Directive 2001/83/EC, article 2 of Regulations (EC) No. 726/2004.

Advanced Therapies include genes: gene therapy, cells: cell therapy, and tissues: tissue engineering.

The aim of this regulation is to guarantee a high level of health protection, harmonize and facilitate market access, foster competitiveness, and provide overall legal certainty.

Key elements are:

- No marketing without prior approval
- Demonstration of quality, safety, and efficacy against tailored technical requirements
- Scientific assessment by EMA (new scientific Committee)
- Risk management and long-term traceability.

Additional advantages are competitiveness aspects such as

- Direct access to the community market
- Harmonized data protection of 10 years
- Accelerated assessment
- Scientific advice at reduced fee
- Special incentives for small and medium enterprises (SMEs).

This regulation lays down *specific* rules concerning the authorization, supervision, and pharmacovigilance of ATMPs, such as culture of human cells *in vitro*, providing important insights into cell biology, disease processes, and potential therapies. *In vitro* cell culture brings a number of concerns: both cells and cell environment allow the growth of microorganisms and, depending on type and number of manipulations, cells become prone to genetic changes. Moreover, additional risks can result from cell culture interchanges, cross-contaminations, or mislabeling during laboratory manipulations.

Three specific characteristics of cell cultures are fundamental to the assurance of good-quality cell culture work: purity, identity, and stability.

- Purity means that cells are free from microbiological contamination.
- Identity means that cells are what they are claimed to be.
- Stability means that genotype, phenotype, and function remain stable, during growth and passage *in vitro*, for a defined time.

These requirements should be obtained on cell type-based approach, however, it is acknowledged that most of them come from conventional pharmacology and toxicology, and may not always be appropriate for cell-based medicinal product.

Human cell-based medicinal products are heterogeneous with regard to the origin and type of the cells and to the complexity of the product. Cells may be self-renewing stem cells, more committed progenitor cells, or terminally differentiated cells exerting a specific defined physiological function. Cells may be of autologous or allogeneic origin. In addition, the cells may also be genetically modified (Figure 59.4). The cells may be used alone or associated with biomolecules, chemical substances, and combined with structural materials that alone might be classified as medical devices, and they are defined as "combination products." All these differences introduce uncertainties and variability in product characterization. Moreover, purity depends on body site and environment where biopsy originating the tissue was obtained: different body sites have different levels of sterility, blood is sterile whereas epithelia are not, the operating theatre where biopsy is removed is not sterile as well.

Identity relies on cell characterization but frequently, mixed cell types are present in the original tissue as well as in the cultured cells and some impurities or contaminants coming from other cell types or surrounding tissues can remain in culture.

Stability can highlight some change in function or phenotype in culture; therefore, a range of acceptable changes should be defined. Unfortunately, the definitive answer to all these questions and the definition of a correct range

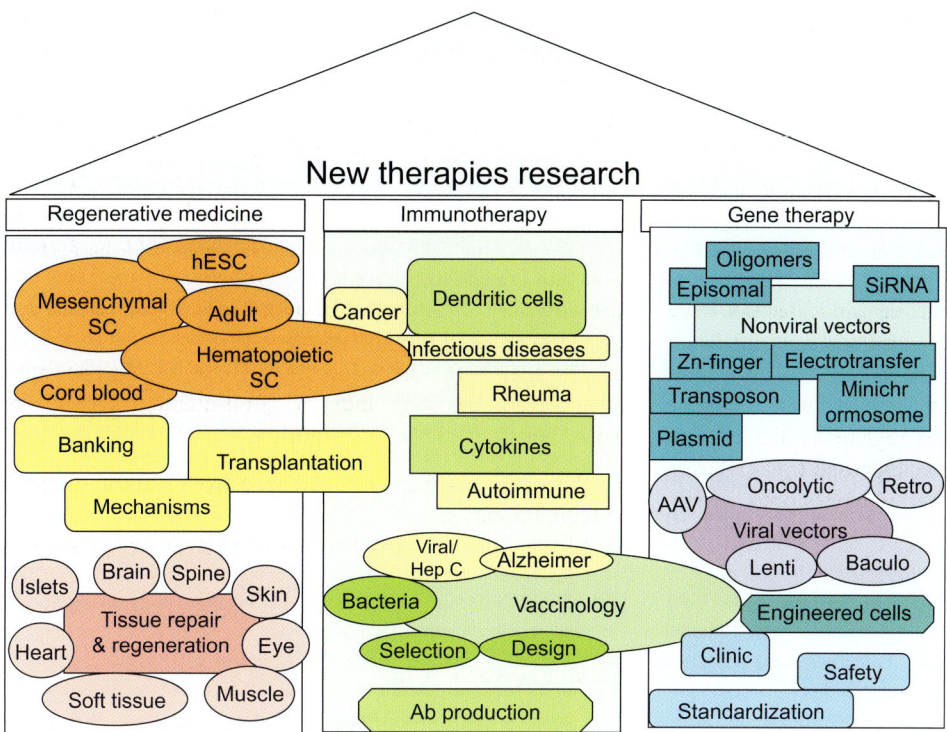

FIGURE 59.4 The diagram represents the new advanced therapies that must follow the regulation of human medicinal products edited by competent authorities.

come out only after long-term clinical experience, so a risk analysis approach should be used to justify all steps of development of a regenerative medicine protocol and all previous experience should be taken into consideration; this analysis will be the basis for the risk management plan, considering that no risk means no treatment but certainty of pathology does not increase safety.

59.10 CONCLUDING REMARKS

LSCDs are painful and highly disabling diseases caused by the loss or dysfunction of limbal stem cells. In such cases, the corneal epithelium integrity and function cannot be sustained and consequently, epithelial defects occur leading to chronic pain and severe visual impairment with significant patient invalidity. The management of these diseases is demanding and complex. Patients suffering from LSCD have contributed considerably to shed light on the biology of human limbal stem cells. Transplantation of cultured limbal stem cells represents a promising cell therapy in regenerative medicine. The difficulties encountered in limbal stem cell therapy have been great and include the patient and donor eye selection criteria, the culture method used, the transplantation technique, and the subjective and objective outcome measures. Much progress has been made and assimilated by scientists and physicians. Cell therapy with limbal stem cells is ready for worldwide use because the essential criteria is now well understood for graftable cultures and for their application in surgery. We

underline the relevance of a discipline for defining the quality and suitability of cultured limbal epithelial grafts. These careful considerations are relevant to the future use of any cultured cell type for therapeutic purposes.

ACKNOWLEDGMENTS

We thank Professor Howard Green for supplying 3T3-J2 cells. We thank all the collaborators of previously laboratories who supported us in the epithelia characterization; in particular, Dr Sergio Bondanza and Dr Patrizia Paterna (IDI, Rome), Dr Osvaldo Golisano, Dr Enzo Di Iorio, and Dr Vanessa Barbaro (FBOV, Venice); the ophthalmologists Dott. Carlo Traverso, Dott. Paolo Rama, Dott. Alessandro Lambiase, and Dott. Stefano Bonini who helped us in the definition of clinical protocols.

We thank: MIUR, Italian Ministry of Health and European Community for the research funding; Seventh Framework Program: Optimization of Stem Cell Therapy for Degenerative Epithelial and Muscle Diseases (OptiStem, HEALTH-F5-2009-223098); Regione Emilia–Romagna: area 1b, Regenerative Medicine and POR-FESR 2007-13-Tecnopolo.

REFERENCES

[1] Barrandon Y, Green H. Three clonal types of keratinocyte with different capacities for multiplication. Proc Natl Acad Sci USA 1987;84:2302–6.

[2] Barrandon Y, Green H. Cell size as a determinant of the clone-forming ability of human keratinocytes. Proc Natl Acad Sci USA 1985;82:5390–4.

[3] De Luca M, Pellegrini G, Green H. Regeneration of squamous epithelia from stem cells of cultured grafts. Regener Med 2006;1:45−57.

[4] Barbaro V, Testa A, Di Iorio E, Mavilio F, Pellegrini G, De Luca M. C/EBPdelta regulates cell cycle and self-renewal of human limbal stem cells. J Cell Biol 2007;177:1037−49.

[5] Dellambra E, Golisano O, Bondanza S, Siviero E, Lacal P, Molinari M, et al. Downregulation of 14-3-3sigma prevents clonal evolution and leads to immortalization of primary human keratinocytes. J Cell Biol 2000;149:1117−30.

[6] Pellegrini G, Golisano O, Paterna P, Lambiase A, Bonini S, Rama P, et al. Location and clonal analysis of stem cells and their differentiated progeny in the human ocular surface. J Cell Biol 1999;145:769−82.

[7] Rochat A, Kobayashi K, Barrandon Y. Location of stem cells of human hair follicles by clonal analysis. Cell 1994;76:1063−73.

[8] Mathor MB, Ferrari G, Dellambra E, Cilli M, Mavilio F, Cancedda R, et al. Clonal analysis of stably transduced human epidermal stem cells in culture. Proc Natl Acad Sci USA 1996;93:10371−6.

[9] Di Iorio E, Barbaro V, Ruzza A, Ponzin D, Pellegrini G, De Luca M. Isoforms of DeltaNp63 and the migration of ocular limbal cells in human corneal regeneration. Proc Natl Acad Sci USA 2005;102:9523−8.

[10] Oshima H, Rochat A, Kedzia C, Kobayashi K, Barrandon Y. Morphogenesis and renewal of hair follicles from adult multipotent stem cells. Cell 2001;104:233−45.

[11] Pellegrini G, Dellambra E, Golisano O, Martinelli E, Fantozzi I, Bondanza S, et al. p63 identifies keratinocyte stem cells. Proc Natl Acad Sci USA 2001;98:3156−61.

[12] Schermer A, Galvin S, Sun TT. Differentiation related expression of a major 64K corneal keratin in vivo and in culture suggests limbal location of corneal epithelial stem cells. J Cell Biol 1986;103:49−62.

[13] Lehrer MS, Sun TT, Lavker RM. Strategies of epithelial repair: modulation of stem cell and transit amplifying cell proliferation. J Cell Sci 1998;111:2867−75.

[14] Cotsarelis G, Cheng SZ, Dong G, Sun TT, Lavker RM. Existence of slow-cycling limbal epithelial basal cells that can be preferentially stimulated to proliferate: implications on epithelial stem cells. Cell 1989;57:201−9.

[15] Nagasaki T, Zhao J. Centripetal movement of corneal epithelial cells in the normal adult mouse. Invest Ophthalmol Visual Sci 2003;44:558−66.

[16] Davanger M, Evensen A. Role of the pericorneal papillary structure in renewal of corneal epithelium. Nature 1971;229:560−1.

[17] Sharma A, Coles WH. Kinetics of corneal epithelial maintenance and graft loss. A population balance model. Invest Ophthalmol Visual Sci 1989;30:1962−71.

[18] Mort RL, Ramaesh T, Kleinjan DA, Morley SD, West JD. Mosaic analysis of stem cell function and wound healing in the mouse corneal epithelium. BMC Dev Biol 2009;9:4.

[19] Huang AJ, Tseng SC. Corneal epithelial wound healing in the absence of limbal epithelium. Invest Ophthalmol Visual Sci 1991;32:96−105.

[20] Dua HS, Azuara-Blanco A. Limbal stem cells of the corneal epithelium. Surv Ophthalmol 2000;44:415−25.

[21] Schlotzer-Schrehardt U, Kruse FE. Identification and characterization of limbal stem cells. Exp Eye Res 2005;81:247−64.

[22] Dua HS, Shanmuganathan VA, Powell-Richards AO, Tighe PJ, Joseph A. Limbal epithelial crypts: a novel anatomical structure and a putative limbal stem cell niche. Br J Ophthalmol 2005;89:529−32.

[23] Baum JL. Melanocyte and Langerhans cell population of the cornea and limbus in the albino animal. Am J Ophthalmol 1970;69:669−76.

[24] Vantrappen L, Geboes K, Missotten L, Maudgal PC, Desmet V. Lymphocytes and Langerhans cells in the normal human cornea. Invest Ophthalmol Visual Sci 1985;26:220−5.

[25] Shimmura S, Tsubota K. Ultraviolet B-induced mitochondrial dysfunction is associated with decreased cell detachment of corneal epithelial cells in vitro. Invest Ophthalmol Visual Sci 1997;38:620−6.

[26] Romano AC, Espana EM, Yoo SH, Budak MT, Wolosin JM, Tseng SC. Different cell sizes in human limbal and central corneal basal epithelia measured by confocal microscopy and flow cytometry. Invest Ophthalmol Visual Sci 2003;44:5125−9.

[27] Chen Z, de Paiva CS, Luo L, Kretzer FL, Pflugfelder SC, Li DQ. Characterization of putative stem cell phenotype in human limbal epithelia. Stem Cells 2004;22:355−66.

[28] Sun TT, Tseng SC, Lavker RM. Location of corneal epithelial stem cells. Nature 2010;463:10−1.

[29] Pajoohesh-Ganji A, Ghosh SP, Stepp MA. Regional distribution of alpha9beta1 integrin within the limbus of the mouse ocular surface. Dev Dyn 2004;230:518−28.

[30] Lambiase A, Bonini S, Micera A, Rama P, Aloe L. Expression of nerve growth factor receptors on the ocular surface in healthy subjects and during manifestation of inflammatory diseases. Invest Ophthalmol Visual Sci 1998;39:1272−5.

[31] Chee KY, Kicic A, Wiffen SJ. Limbal stem cells: the search for a marker. Clin Experiment Ophthalmol 2006;34:64−73.

[32] Watanabe K, Nishida K, Yamato M, Umemoto T, Sumide T, Yamamoto K, et al. Human limbal epithelium contains side population cells expressing the ATP-binding cassette transporter ABCG2. FEBS Lett 2004;565:6−10.

[33] Budak MT, Alpdogan OS, Zhou M, Lavker RM, Akinci MA, Wolosin JM. Ocular surface epithelia contain ABCG2-dependent side population cells exhibiting features associated with stem cells. J Cell Sci 2005;118:1715−24.

[34] Thomas PB, Liu YH, Zhuang FF, Selvam S, Song SW, Smith RE, et al. Identification of Notch-1 expression in the limbal basal epithelium. Mol Vision 2007;13:337−44.

[35] Ma A, Boulton M, Zhao B, Connon C, Cai J, Albon J. A role for notch signaling in human corneal epithelial cell differentiation and proliferation. Invest Ophthalmol Visual Sci 2007;48:3576−85.

[36] Vauclair S, Majo F, Durham AD, Ghyselinck NB, Barrandon Y, Radtke F. Corneal epithelial cell fate is maintained during repair by Notch1 signaling via the regulation of vitamin A metabolism. Dev Cell 2007;13:242−53.

[37] Kenyon KR, Tseng SC. Limbal autograft transplantation for ocular surface disorders. Ophthalmology 1989;96:709−22.

[38] Lindberg K, Brown ME, Chaves HV, Kenyon KR, Rheinwald JG. In vitro propagation of human ocular surface epithelial cells for transplantation. Invest Ophthalmol Visual Sci 1993;34:2672−9.

[39] Pellegrini G, Traverso CE, Franzi AT, Zingirian M, Cancedda R, De Luca M. Long-term restoration of damaged corneal surfaces with autologous cultivated corneal epithelium. Lancet 1997;349:990−3.

[40] Rama P, Bonini S, Lambiase A, Golisano O, Paterna P, De Luca M, et al. Autologous fibrin-cultured limbal stem cells permanently restore the corneal surface of patients with total limbal stem cell deficiency. Transplantation 2001;72:1478–85.

[41] Rama P, Matuska S, Paganoni G, Spinelli A, De Luca M, Pellegrini G. Limbal stem-cell therapy and long-term corneal regeneration. N Engl J Med 2010;363:147–55.

[42] Pellegrini G, Ranno R, Stracuzzi G, Bondanza S, Guerra L, Zambruno G, et al. The control of epidermal stem cells (holo-clones) in the treatment of massive full-thickness burns with autologous keratinocytes cultured on fibrin. Transplantation 1999;68:868–79.

[43] Ronfard V, Rives JM, Neveux Y, Carsin H, Barrandon Y. Long-term regeneration of human epidermis on third degree burns transplanted with autologous cultured epithelium grown on a fibrin matrix. Transplantation 2000;70:1588–98.

[44] Notara M, Alatza A, Gilfillan J, Harris AR, Levis HJ, Schrader S, et al. In sickness and in health: corneal epithelial stem cell biology, pathology and therapy. Exp Eye Res 2010;90:188–95.

[45] Baylis O, Figueiredo F, Henein C, Lako M, Ahmad S. 13 years of cultured limbal epithelial cell therapy: a review of the outcomes. J Cell Biochem 2011;112:993–1002.

[46] Shortt AJ, Secker GA, Notara MD, Limb GA, Khaw PT, Tuft SJ, et al. Transplantation of ex vivo cultured limbal epithelial stem cells: a review of techniques and clinical results. Surv Ophthalmol 2007;52:483–502.

[47] O'Callaghan AR, Daniels JT. Concise review: limbal epithelial stem cell therapy: controversies and challenges. Stem Cells 2011;29:1923–32.

[48] Yang A, Kaghad M, Wang Y, Gillett E, Fleming MD, Dotsch V, et al. p63, a p53 homolog at 3q27–29, encodes multiple products with transactivating, death-inducing, and dominant-negative activities. Mol Cell 1998;1998(2):305–16.

[49] Parsa R, Yang A, McKeon F, Green H. Association of p63 with proliferative potential in normal and neoplastic human keratinocytes. J Invest Dermatol 1999;113:1099–105.

[50] Koster MI, Kim S, Mills AA, DeMayo FJ, Roop DR. p63 is the molecular switch for initiation of an epithelial stratification program. Genes Dev 2004;18:126–131.

[51] Yang A, Schweitzer R, Sun D, Kaghad M, Walker N, Bronson RT, et al. p63 is essential for regenerative proliferation in limb, craniofacial and epithelial development. Nature 1999;398:714–8.

[52] Mills AA, Zheng B, Wang XJ, Vogel H, Roop DR, Bradley A. p63 is a p53 homologue required for limb and epidermal morphogenesis. Nature 1999;398:708–13.

[53] Senoo M, Pinto F, Crum CP, McKeon F. p63 is essential for the proliferative potential of stem cells in stratified epithelia. Cell 2007;129:523–36.

[54] Mangiulli M, Valletti A, Caratozzolo MF, Tullo A, Sbisa E, Pesole G, et al. Identification and functional characterization of two new transcriptional variants of the human p63 gene. Nucleic Acids Res 2009;37:6092–104.

[55] McKeon F. p63 and the epithelial stem cell: more than status quo? Genes Dev 2004;18:465–9.

[56] Pellegrini G, Rama P, Mavilio F, De Luca M. Epithelial stem cells in corneal regeneration and epidermal gene therapy. J Pathol 2009;217:217–28.

[57] Rheinwald JG, Green H. Serial cultivation of strains of human epidermal keratinocytes: the formation of keratinizing colonies from single cells. Cell 1975;6:331–43.

[58] Green H, Kehinde O, Thomas J. Growth of cultured human epidermal cells into multiple epithelia suitable for grafting. Proc Natl Acad Sci USA 1979;76:5665–8.

[59] Green H. The birth of therapy with cultured cells. BioEssays 2008;30:897–903.

[60] Gallico 3rd GG, O'Connor NE, Compton CC, Kehinde O, Green H. Permanent coverage of large burn wounds with autologous cultured human epithelium. N Engl J Med 1984;311:448–51.

[61] Pellegrini G, De Luca M, Arsenijevic Y. Towards therapeutic application of ocular stem cells. Semin Cell Dev Biol 2007;18:805–18.

[62] Daniels JT, Notara M, Shortt AJ, Secker G, Harris A, Tuft SJ. Limbal epithelial stem cell therapy. Expert Opin Biol Ther 2007;7:1–3.

[63] Schwab IR, Reyes M, Isseroff RR. Successful transplantation of bioengineered tissue replacements in patients with ocular surface disease. Am J Ophthalmol 2000;130:543–4.

[64] Takacs L, Toth E, Berta A, Vereb G. Stem cells of the adult cornea: from cytometric markers to therapeutic applications. Cytometry A 2009;75:54–66.

[65] McIntosh Ambrose W, Schein O, Elisseeff J. A tale of two tissues: stem cells in cartilage and corneal tissue engineering. Curr Stem Cell Res Ther 2010;5:37–48.

[66] van Herendael BJ, Oberti C, Brosens I. Microanatomy of the human amniotic membranes. A light microscopic, transmission, and scanning electron microscopic study. Am J Obstet Gynecol 1978;131:872–80.

[67] Dua HS, Gomes JA, King AJ, Maharajan VS. The amniotic membrane in ophthalmology. Surv Ophthalmol 2004;49:51–77.

[68] Gomes JA, Romano A, Santos MS, Dua HS. Amniotic membrane use in ophthalmology. Curr Opin Ophthalmol 2005;16:233–40.

[69] Hao Y, Ma DH, Hwang DG, Kim WS, Zhang F. Identification of antiangiogenic and antiinflammatory proteins in human amniotic membrane. Cornea 2000;19:348–52.

[70] Fernandes M, Sridhar MS, Sangwan VS, Rao GN. Amniotic membrane transplantation for ocular surface reconstruction. Cornea 2005;24:643–53.

[71] Schwab IR. Cultured corneal epithelia for ocular surface disease. Trans Am Ophthalmol Soc 1999;97:891–986.

[72] Tsai RJ, Li LM, Chen JK. Reconstruction of damaged corneas by transplantation of autologous limbal epithelial cells. N Engl J Med 2000;343:86–93.

[73] Nakamura T, Koizumi N, Tsuzuki M, Inoki K, Sano Y, Sotozono C, et al. Successful regrafting of cultivated corneal epithelium using amniotic membrane as a carrier in severe ocular surface disease. Cornea 2003;22:70–1.

[74] Koizumi N, Inatomi T, Suzuki T, Sotozono C, Kinoshita S. Cultivated corneal epithelial stem cell transplantation in ocular surface disorders. Ophthalmology 2001;108:1569–74.

[75] Sangwan VS, Vemuganti GK, Iftekhar G, Bansal AK, Rao GN. Use of autologous cultured limbal and conjunctival epithelium in a patient with severe bilateral ocular surface disease induced by acid injury: a case report of unique application. Cornea 2003;22:478–81.

[76] Sangwan VS, Murthy SI, Vemuganti GK, Bansal AK, Gangopadhyay N, Rao GN. Cultivated corneal epithelial transplantation for severe ocular surface disease in vernal keratoconjunctivitis. Cornea 2005;24:426−30.

[77] Sangwan VS, Matalia HP, Vemuganti GK, Fatima A, Ifthekar G, Singh S, et al. Clinical outcome of autologous cultivated limbal epithelium transplantation. Indian J Ophthalmol 2006;54:29−34.

[78] Fatima A, Sangwan VS, Iftekhar G, Reddy P, Matalia H, Balasubramanian D, et al. Technique of cultivating limbal derived corneal epithelium on human amniotic membrane for clinical transplantation. J Postgrad Med 2006;52:257−61.

[79] Grueterich M, Espana EM, Tseng SC. Ex vivo expansion of limbal epithelial stem cells: amniotic membrane serving as a stem cell niche. Surv Ophthalmol 2003;48:631−46.

[80] Grueterich M, Tseng SC. Human limbal progenitor cells expanded on intact amniotic membrane ex vivo. Arch Ophthalmol 2002;120:783−90.

[81] Koizumi N, Fullwood NJ, Bairaktaris G, Inatomi T, Kinoshita S, Quantock AJ. Cultivation of corneal epithelial cells on intact and denuded human amniotic membrane. Invest Ophthalmol Visual Sci 2000;41:2506−2513.

[82] Koizumi N, Rigby H, Fullwood NJ, Kawasaki S, Tanioka H, Koizumi K, et al. Comparison of intact and denuded amniotic membrane as a substrate for cell-suspension culture of human limbal epithelial cells. Graefes Arch Clin Exp Ophthalmol 2007;245:123−34.

[83] Mason C, Manzotti E. Regenerative medicine cell therapies: numbers of units manufactured and patients treated between 1988 and 2010. Regener Med 2010;5:307−13.

[84] Sangwan VS, Basu S, Vemuganti GK, Sejpal K, Subramaniam SV, Bandyopadhyay S, et al. Clinical outcomes of xeno-free autologous cultivated limbal epithelial transplantation: a 10-year study. Br J Ophthalmol 2011;95:1525−9.

[85] Nishida K, Yamato M, Hayashida Y, Watanabe K, Yamamoto K, Adachi E, et al. Corneal reconstruction with tissue-engineered cell sheets composed of autologous oral mucosal epithelium. N Engl J Med 2004;351:1187−96.

[86] Nishida K, Yamato M, Hayashida Y, Watanabe K, Maeda N, Watanabe H, et al. Functional bioengineered corneal epithelial sheet grafts from corneal stem cells expanded ex vivo on a temperature-responsive cell culture surface. Transplantation 2004;77:379−85.

[87] Tsubota K, Satake Y, Kaido M, Shinozaki N, Shimmura S, Bissen-Miyajima H, et al. Treatment of severe ocular-surface disorders with corneal epithelial stem-cell transplantation. N Engl J Med 1999;340:1697−703.

[88] Shimazaki J, Shimmura S, Fujishima H, Tsubota K. Association of preoperative tear function with surgical outcome in severe Stevens−Johnson syndrome. Ophthalmology 2000;107:1518−23.

[89] Samson CM, Nduaguba C, Baltatzis S, Foster CS. Limbal stem cell transplantation in chronic inflammatory eye disease. Ophthalmology 2002;109:862−8.

[90] Ma DH, Kuo MT, Tsai YJ, Chen HC, Chen XL, Wang SF, et al. Transplantation of cultivated oral mucosal epithelial cells for severe corneal burn. Eye (Lond) 2009;23:1442−50.

[91] Satake Y, Dogru M, Yamane GY, Kinoshita S, Tsubota K, Shimazaki J. Barrier function and cytologic features of the ocular surface epithelium after autologous cultivated oral mucosal epithelial transplantation. Arch Ophthalmol 2008;126:23−8.

[92] Liu J, Sheha H, Fu Y, Giegengack M, Tseng SC. Oral mucosal graft with amniotic membrane transplantation for total limbal stem cell deficiency. Am J Ophthalmol 2011;152:739−47.

[93] Inatomi T, Nakamura T, Kojyo M, Koizumi N, Sotozono C, Kinoshita S. Ocular surface reconstruction with combination of cultivated autologous oral mucosal epithelial transplantation and penetrating keratoplasty. Am J Ophthalmol 2006;142:757−64.

[94] Nakamura T, Inatomi T, Sotozono C, Amemiya T, Kanamura N, Kinoshita S. Transplantation of cultivated autologous oral mucosal epithelial cells in patients with severe ocular surface disorders. Br J Ophthalmol 2004;88:1280−4.

[95] Ang LP, Nakamura T, Inatomi T, Sotozono C, Koizumi N, Yokoi N, et al. Autologous serum-derived cultivated oral epithelial transplants for severe ocular surface disease. Arch Ophthalmol 2006;124:1543−51.

[96] Inatomi T, Nakamura T, Koizumi N, Sotozono C, Yokoi N, Kinoshita S. Midterm results on ocular surface reconstruction using cultivated autologous oral mucosal epithelial transplantation. Am J Ophthalmol 2006;141:267−75.

[97] Chen HC, Chen HL, Lai JY, Chen CC, Tsai YJ, Kuo MT, et al. Persistence of transplanted oral mucosal epithelial cells in human cornea. Invest Ophthalmol Visual Sci 2009;50:4660−8.

[98] Nakamura T, Inatomi T, Cooper LJ, Rigby H, Fullwood NJ, Kinoshita S. Phenotypic investigation of human eyes with transplanted autologous cultivated oral mucosal epithelial sheets for severe ocular surface diseases. Ophthalmology 2007;114:1080−8.

[99] Pellegrini G. Changing the cell source in cell therapy? N Engl J Med 2004;351:1170−2.

Chapter 60

Esophagus Bioengineering

Giorgia Totonelli[a], Panagiotis Maghsoudlou[a], and Paolo De Coppi[a,b,c]

[a]Surgery Unit, Institute of Child Health and Great Ormond Street Hospital, University College London, London, [b]Wake Forest Institute for Regenerative Medicine, Wake Forest University School of Medicine, Winston-Salem, NC, [c]Centre for Stem Cells and Regenerative Medicine, University College London, London

60.1 INTRODUCTION

60.1.1 Anatomy and Function

The human esophagus is a tubular muscular conduit that connects the oral cavity to the stomach and allows the passage of food bolus and liquids, owing to its peristaltic activity that is caused by the alternate contraction and relaxation of smooth muscle cells (SMCs). This peristaltic activity is initiated by either extrinsic or intrinsic pathways. With a total length of 20–25 cm in an adult, the esophagus crosses three anatomical districts (neck, thorax, and abdomen) and along its entire length is composed of four main different layers: mucosa, submucosa, muscularis externa, and adventitia. The mucosal layer is composed of nonkeratinized squamous epithelium with cells at different states of differentiation, which produce mucus to protect the esophagus from the stress caused by the passage of the food bolus. The submucosa is represented by a loose network of collagen fibers, mainly collagen type I and III, constituting the connective tissue, with interposed esophageal glands and blood vessels. The muscularis externa consists of inner circular and outer longitudinal layers. The upper part of this muscle is composed of skeletal muscle that changes to smooth muscle distally while both types of cells are present in the middle. The adventitia represents the outer layer of the esophagus and is composed of loose connective tissue full of blood and lymphatic vessels, adipose tissue, and squamous cell epithelium. Hence, the reconstruction of the complex morphology of the esophageal conduit represents a challenge.

60.1.2 The Clinical Problem

Several conditions, both congenital and acquired, may require esophageal tissue replacement. Every year there are 5000–10,000 patients diagnosed with esophageal diseases for which esophageal replacement is needed [1]. In the pediatric population the primary indication for esophageal replacement is long-gap esophageal atresia (EA) with insufficient length for primary anastomosis. EA has an incidence of 1:3000 to 1:5000 in live births and in most cases a primary anastomosis of the upper and lower pouches is possible, with relatively minor complications. Patients with long-gap EA, which fail a primary repair, receive a denervated gastric pull-up or interposition graft using either jejunum or colon, with many associated early and late postoperative complications, such as stricture formation and the potentially carcinogenic effect of acid reflux [2,3]. In children, gastric transposition and intestinal interposition can also be used in esophageal strictures not responsive to dilatation following failed EA repair or caustic ingestion, or for rare neoplastic conditions such as inflammatory pseudotumor, leiomyosarcoma, and teratoma [4]. By contrast, the commonest indication for esophageal replacement in adults is cancer, a condition with an escalating incidence [5], and difficult treatment, mainly due to the poor regenerative capacity of the esophageal tissue. The two major types of esophageal cancer are the squamous carcinoma and the adenocarcinoma, whose increased risk is related to smoking, alcohol, obesity, esophageal reflux and Barrett's esophagus, a premalignant condition. In the case of diffuse Barrett's esophagus, colon interposition is sometimes indicated. Herein, treatment of esophageal disorders entails the removal of the affected portion, both patch or circumferential, and its substitution with functional tissue. Unfortunately, all of these methods of esophageal replacement severely impair the quality of life of recipient adults and children [6,7] and present problems related to donor site morbidity. Even recent developments in endoluminal resection, which removes the diseased inner layers of the esophagus through an endoscope, while reducing morbidity, still results in a high rate of stenosis and consequent dysphagia [8]. Despite its 60-year history, conventional organ transplantation is not a solution for the failure of every organ, due to technical and ethical issues, and is specifically unable to address the unmet needs of esophageal replacement. Thus, regenerative medicine techniques, which extend the boundaries of reconstruction and do not, in most applications, require immunosuppression, present attractive alternatives [9,10].

60.2 REGENERATIVE MEDICINE

Regenerative medicine has been used to describe the use of natural human substances, such as genes, proteins, cells, and biomaterials to regenerate diseased or damaged human tissue [11,12] in order to restore normal function [13]. Tissue engineering aims to either enhance repair in sites of esophageal damage or create constructs to replace the esophagus in cases of deficient tissue. The components of a tissue-engineered esophagus comprise of the scaffold and the appropriate cells seeded onto it (Figure 60.1). Enhancing repair of a damaged tissue may

necessitate only one of these two components. For example, the scaffold may be used for mechanical support to allow cellular penetration from the surrounding tissue as has been done in the bladder [14], urethra [15], skin [16] and trachea [17−20]. Alternatively the matrix has been used to promote regeneration and avoid stricture of the esophagus. Indeed, in a limited number of patients which underwent aggressive endoscopic resection of early-stage esophageal neoplasia, early implantation of a biological scaffold derived from extracellular matrix (ECM), was successfully used avoiding the need for esophagectomy [21]. On the other hand, cellular delivery without a scaffold may also be used to enhance regeneration following an acute injury such as an infarction in the heart and ischemic or toxic injury in the kidney. Tissue engineering with the endpoint of organogenesis has been successful through a combination of appropriate cells with a scaffold. Recently, we have published on a 2-year follow-up on the first pediatric patient, receiving a tissue-engineered trachea [22]. More complex, however, is the engineering of modular organs, such as the esophagus, due to their structural and functional characteristics.

Material science is concerned with the production of acellular scaffolds that can be seeded with cells, allowing and promoting their growth. General attributes that a scaffold must fulfill include:

1. biocompatibility that does not lead to an immunogenic response from the host;
2. biodegradation in a suitable time period that permits sufficient cellular growth while not producing harmful degradation products;
3. mechanical properties that are in line with tissue growth (i.e., a significant amount of three-dimensional support in the early stages without obstructing cellular ECM production in the later stages).

More specific scaffold attributes relate to the aim of recreating the nonmechanical attributes of the ECM. These properties include mediation of cell adhesion, as well as positive influence on cell survival and

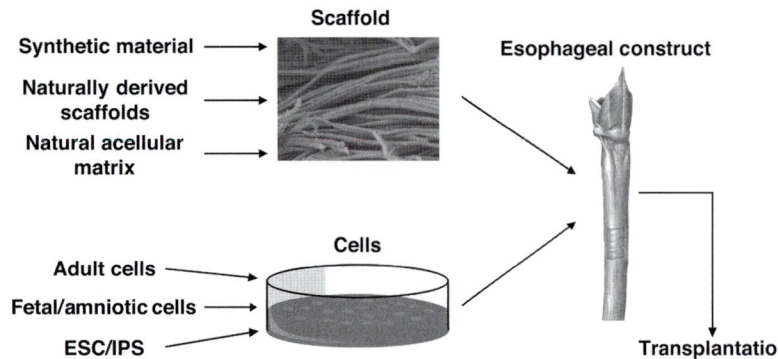

FIGURE 60.1 Esophageal tissue engineering. A tissue-engineered esophageal construct may be created by the combination of a scaffold and cells, grown in a bioreactor and transplanted in patients. A three-dimensional scaffold may be created from synthetic material, collagen, or a decellularized matrix. Cells for the use of tissue engineering are derived from a number of sources such as the adult, fetus, and the embryo. Additionally, nonseeded scaffolds may be transplanted with the aim of being repopulated by host cells; ESC: embryonic stem cells, IPS: induced pluripotent stem cells.

proliferation by means of growth factors and cytokines. Furthermore, it has been suggested that the ECM is involved in mechanochemical transduction, as shown by the differentiation of MSC to neurons, muscle cells and osteoblasts when seeded on substrate mimicking the elasticity of each of those host tissues, respectively [23].

The successful combination of scaffold, cells, and growth factors is crucial for the generation of a functional tissue. Attempts in tissue engineering of the esophagus have followed one of two methodologies. The majority of the studies have transplanted the scaffold without any cellular support expecting that the epithelium and smooth muscle will migrate to cover it. Cellular migration, however, has been more common in the luminal side by the epithelium when compared to the smooth muscle layer extraluminally. Acellular scaffolds that have been used fall into all three categories discussed below, namely: synthetic, naturally derived, and natural acellular matrix. Aiming to reduce complications arising from acellular approaches, scaffolds seeded with epithelial and SMCs prior to transplantation have also been used with better functional results.

60.2.1 Acellular Scaffolds

The majority of identified studies transplanted acellular scaffolds with the aim that host epithelial and SMCs will migrate to repopulate the new conduit. Initially, nonabsorbable materials were used as scaffold for the replacement of esophageal defect. However, several complications occurred due to the absence of biodegradation and to the long-term biocompatibility (i.e., leakage, infections, and strictures). Hence, natural acellular matrices and biodegradable scaffolds gradually replaced the nonbiodegradable materials. Acellular scaffolds studied to date conform to one of three categories: synthetic, naturally derived, and natural acellular matrices.

60.2.2 Synthetic Scaffolds

Both synthetic and natural polymers have been used to replace esophageal defect, with the advantage of being reproducible and the possibility to adapt their properties to the current necessities. Acellular synthetic scaffolds such as polyethylene plastic [24–26] and silicon [27,28] have been used for esophageal replacement, but the nature of the materials did not allow cellular migration and led to poor results in animal models. When polyvinylidene fluoride (PVDF) and polyglastin-910 (Vicryl®) were compared for the regeneration of patch defects in rabbits, PVDF was shown to lead to improved results with an absence of strictures and neoepithelialization [29]. In a different study, the combination of Vicryl® and collagen brought about positive results both for patch and

tubular defects in dogs, with a low mortality of 8.3% [30]. The successful use of synthetic polymers in other organs such as the trachea [31], suggests that this approach may appear attractive and further development of appropriate materials is needed.

60.2.3 Naturally Derived Scaffolds

In a series of experiments performed by a research group in Japan, porcine dermal collagen scaffolds were used to produce porous tubular structures (Table 60.1) [32–37]. The general methodology involved the use of these scaffolds to replace 5–10 cm tubular defects in the cervical or intrathoracic portion of the esophagus in dogs. A silicon tube was used as a stent to support the scaffold until repopulation occurred. Aiming to avoid complications such as stenosis [32,33], the research group compared whether this was related to the time for which the scaffold was supported by the stent. In an experiment where three groups of dogs had a 5-cm cervical surgically created defect, the stent was removed at either 2, 3, or 4 weeks. With increasing stent duration, it was observed that greater epithelial and muscle cell densities were achieved in the collagen scaffold, and this correlated with decreased stenosis and mortality [35]. However, when the collagen scaffold replaced 10-cm portions of the esophagus there was poor cellular migration in the muscular layer, suggesting that there are limitations to the size of defect that may be replaced by this methodology [34]. Moreover, when the same methods were used to replace intrathoracic portions of the esophagus in dogs, muscular regeneration was completely absent, something the authors attributed to the lack of a vascular supply in the thorax [36]. In an attempt to address this, the scaffold was wrapped in omentum [37], as recently described for trachea tissue engineering [22,47]. However, muscular regeneration remained absent, while an increase in mid-portion stenosis and mortality was observed [37].

60.2.4 Natural Acellular Matrices

Natural acellular matrices are derived from human and animal organs or tissues that have been treated to remove cells and immunogenic material [48]. Importantly, however, they retain the macro- and microarchitecture of the tissue of origin, with its three-dimensional morphology, which is believed to promote cell orientation, growth, and proliferation. Also, the molecular components of the native ECM are maintained in the decellularized tissue, being rich in collagen, elastin, fibronectin, laminin, and growth factors [20,49–51]. They have the added hypothetical advantages over synthetic scaffolds of not producing potentially toxic degradation products or inducing inflammation characteristics that may be important in the

TABLE 60.1 Overview of *In Vivo* Transplantation of Acellular Matrices

Animal Model (*n*)	Scaffold		Results		References
	Type	Size	Scaffold Regeneration	Clinical Course	
Canine (19)	Collagen with silicon stent (not removed)	5 cm circumferential gap, cervical esophagus	Partial epithelial regeneration	26% mortality	[32]
Canine (26)	Collagen with silicon stent (removed between 2 and 8 weeks)	5 cm circumferential gap, cervical esophagus	Epithelial regeneration, no stenosis	0% mortality when stent dislodged after 4 weeks (*n* = 4)	[33]
Canine (7)	Collagen with silicon stent (removed at 6 weeks)	10 cm circumferential gap, cervical esophagus	Epithelial and partial muscular regeneration, no stenosis	29% mortality	[34]
Canine (43)	Collagen with silicon stent (removed either at 2, 3, or 4 weeks)	5 cm circumferential gap, cervical esophagus	Epithelial and muscular regeneration, no stenosis	0% mortality when stent was removed at 4 weeks (*n* = 16)	[35]
Canine (9)	Collagen with silicon stent (removed at 4 weeks)	5 cm circumferential gap, thoracic esophagus	Epithelial but no muscular regeneration, mid-portion stenosis	11% mortality	[36]
Canine (14)	Collagen with silicon stent (removed at 4–8 weeks) ± OMPx	5 cm circumferential gap, thoracic esophagus	Epithelial regeneration, mid-portion stenosis	11% mortality in control group, 80% in OMPx group	[37]
Canine (15)	ECM scaffold from either small intestine (*n* = 12) or urinary bladder submucosa (*n* = 3)	5 cm semicircumferential or 5 cm circumferential, cervical esophagus	Mucosal and muscular regeneration. Stenosis in case of complete circumferential defects	0% mortality	[38]
Pigs (10)	Elastin-based acellular biomaterial patch (from porcine aorta)	2-cm circular defect, abdominal esophagus	Mucosal and muscular regeneration	0% mortality. No complications reported in treatment groups	[39]
Canine (12)	Urinary bladder matrix scaffold	Complete transection with replacement of endomucosa with matrix	Mucosal and muscular regeneration	0% mortality. No complications reported in treatment groups	[40]
Canine (12)	AlloDerm®	2 cm × 1 cm anterolateral patch defect, cervical esophagus	Epithelialization with blood capillaries formation	No stricture, leakage and dysphagia	[41]
Rats (67)	SIS patch graft	Semicircumferential defect, cervical or abdominal esophagus	Mucosal and muscular regeneration at 150 days	94% survival at 150 days	[42]
Rats (85)	SIS patch graft, or tube interposition	Semicircumferential defect or segmental esophageal excision	Tube interposition unsuccessful. Mucosal and muscular regeneration at 150 days in patch group	100% survival for patch group (and no complications reported), 0% survival for tube interposition group at 28 days	[43]
Rats (27)	Gastric acellular matrix scaffold	Patch defects, abdominal esophagus	Mucosal regeneration seen at 2 weeks. No muscular regeneration seen up to 18 months	11% complication rate	[44]
Pigs (14)	SIS (tubular)	4-cm defect, cervical esophagus	Prosthesis not found either macroscopically or histologically	Only 1 pig survived the full 4 week study. The other pigs have to be sacrificed prematurely due to severe stenosis	[45]

(Continued)

TABLE 60.1 (Continued)

Animal Model (*n*)	Scaffold		Results		References
	Type	Size	Scaffold Regeneration	Clinical Course	
Human (5)	Porcine small intestinal mucosa	8- to 13-cm *en bloc* resection of mucosa and submucosa for superficial carcinoma	Restoration of normal mucosa as early as 4 months	Strictures; perforation in one patient	[21]
Human (1)	Porcine small intestinal mucosa	5 cm × 3 cm defect cervical esophagus	Intact esophagus with normal calibre	No complications encountered	[46]

Fresh   Decellularized

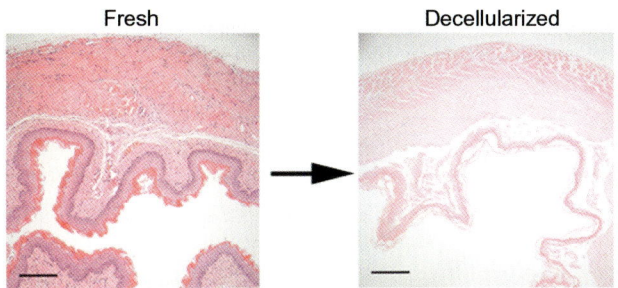

FIGURE 60.2 Production of esophageal natural acellular matrices. Decellularization involves treatment of fresh esophageal tissue with a combination of solutions that will remove the cells but maintain the structural characteristics of the native ECM. The optimal methodology of esophageal decellularization is currently under investigation. Our experience with the DET is illustrated here with hematoxylin and eosin staining of a representative decellularized esophagus demonstrating preservation of the native architecture (scale bar 100 μm).

prevention of stenosis [24,52,53]. Natural acellular matrices are usually obtained through different chemical or enzymatic treatment of the native tissue to remove the cellular components. These techniques include the use of sodium dodecyl sulfate (SDS), sodium deoxycholate, Triton X-100, DNA-se and trypsin digestion to generate a natural acellular matrix (Figure 60.2). One of the advantages of the decellularization process is to remove the human leukocyte antigen (HLA) in the acellular scaffold. This helps to reduce the inflammation and fibrosis reaction after implantation [50,54]. Decellularized scaffolds that have been used for esophageal organs originated from the esophagus as well as from other tissues such as the small intestinal submucosa (SIS) [21,32−35, 38,39,46,55] and the gastric acellular matrix [44].

Significant heterogeneity exists among studies, both with respect to the type of scaffold, extent of surgery and species used, which partly explains the range of results reported. However, the development of an ideal scaffold for the esophagus replacement still needs to be achieved, as scaffolds used so far showed their own advantages and

drawbacks. Thus, regeneration of the muscularis propria layer is seen to take place in some studies [38−40], but not others [44]. Isch et al. [41] performed a cervical esophagoplasty in a canine model using AlloDerm®, a human decellularized skin usually employed for abdominal wound reconstruction. All dogs survived without major complications like dysphagia, leakage, stricture, and infection, examined using barium esophagogram. The histological analysis of the cervical patch showed a good reepithelialization and blood capillary formation in the graft after 1 month postsurgery. The AlloDerm® patch demonstrated its potential to support tissue regeneration. Differently, studies that have attempted tube interposition with SIS reported the development of esophageal stenosis and increased mortality [38,42,45]. Alternatively, SIS may produce better results when used as patch to repair esophageal defects [38,42,43,45,56]. Badylak et al. [44] laid sheets of SIS onto the raw internal surface of esophagus following endoscopic submucosal resection in five patients with superficial cancers. With a follow-up of 4−24 months, the scaffold promoted physiological remodeling as evident by endoscopy and histological characterization following biopsy. Strictures still formed, but only at areas outside those lined by SIS, suggesting that possible technical improvements in scaffold delivery could ameliorate this. In fact, when SIS was used to completely cover a 3 cm × 5 cm mucosal defect in the cervical esophagus, there was no stenosis and endoscopy at 4 weeks demonstrated good integration of the scaffold [46]. Gastric acellular matrices (GAMs) have already been used for the replacement of urinary bladder and small intestine. Urita et al. [44] implanted a GAM patch in the abdominal esophagus of a rat, obtaining a good mucosal regeneration without stenosis or leakage. However, no muscular regeneration was observed up to 18 months.

Hypothetically, decellularized esophageal tissue should retain the signals, both chemical and structural, that will direct the appropriate migration and differentiation of host cells, in a way unlikely to occur with

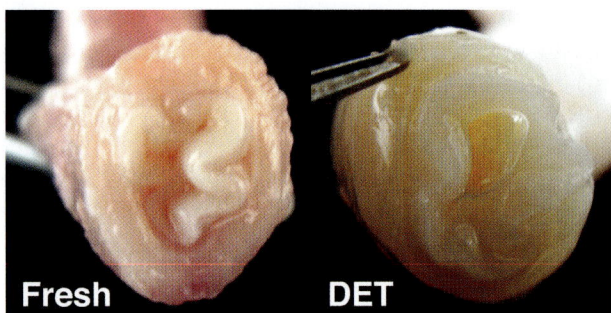

FIGURE 60.3 Macroscopic appearance of the esophagus during the decellularization process. Macroscopically, the decellularization treatment generates a translucent acellular matrix that maintains the whole structure of the native esophagus. The esophageal wall of the treated sample appears more translucent than the fresh tissue, without losing its structural characteristics.

scaffolds originating outside the esophagus, such as SIS. Ozeki et al. [54] compared two methods of decellularization of adult rat esophagus based on deoxycholate and Triton X-100, respectively, and assessed the resulting scaffolds using routine histology and biocompatibility. Those treated with deoxycholate showed superior mechanical properties, maintenance of the ECM and a lower DNA content than those treated with Triton X-100. Bhrany et al. [57] found a combination of 0.5% SDS and Triton X-100 to be effective in decellularization, albeit with a loss of tensile strength as measured by burst pressure studies. Our experience with the detergent-enzymatic treatment (DET) in the decellularization of the intestine [50] allowed us to use the same methodology in the esophagus (Figure 60.3), leading to an improved preservation of a multilayered macro- and microarchitecture [58]. This is of paramount importance for the orientation of the cells in the cell-seeded constructs.

60.2.5 Cell-Seeded Scaffolds

To reduce complications arising from acellular approaches, some authors have seeded the scaffolds prior to transplantation. As mentioned, the two main cell types that are important for esophageal tissue engineering are those that will reconstitute the epithelium and the muscle layer on the luminal and extraluminal sides, respectively. Also important in the formation of a functional esophagus are the vascular and neuronal cell components but we could locate no studies that have explored these in engineered esophagus.

A number of in vitro experiments have examined the seeding and culture of esophageal epithelial cells (EECs) and different scaffolds to assess the optimal combination. When a matrix composed of decellularized human skin

was compared to synthetic scaffolds in vitro for the capacity to support cultured epithelial cells, the decellularized scaffold proved superior. The decellularized skin showed a proliferating basal cell layer, six layers of stratification and a thick keratin layer. The porous nature of synthetic scaffolds led to areas of proliferating basal cells, three layers of stratification and keratinization. It was suggested that mimicking the characteristics of the natural scaffold in a synthetic environment could prove to be the way forward [59]. Another study compared the growth of human esophageal squamous cells on human decellularized esophagus, porcine decellularized esophagus, human decellularized dermis, and collagen [60]. Interestingly the porcine matrix and collagen gave better results leading to the formation of a mature stratified epithelium. When rat EECs were seeded onto three-dimensional (3-D) collagen scaffolds they were shown to be viable for up to 8 weeks in vitro but did not fully integrate within the scaffold, remaining on the surface as individual cells or small clusters [61]. Seeding of sheep EEC on the same 3-D collagen scaffold resulted in the absence of epithelium sheet formation, which was attributed to cellular penetration into the scaffold and loss of cell-to-cell contact [62]. However, when the same cells were seeded on the 2-D collagen scaffolds a single layer of epithelium was evident following 3 weeks of in vitro culture that remained viable up to 6 weeks. The same group has also performed in vivo studies of vascularization of the EEC-scaffold construct by stent tubularization and omental transplantation in lambs for 8–12 weeks [63]. Omental wrapping resulted in vascular penetration within and around the constructs. Following stent removal, the engineered conduit revealed a structure similar to the esophagus with areas of esophageal epithelium on the luminal side and vascular ingrowths on the outer perimeter. Positive selection of the epithelial population could increase proliferative capacity as demonstrated by Kofler et al. [64], who selected ovine EEC for expression of pancytokeratin-26 (PCK-26) using fluorescence-activated cell sorting (FACS) and seeded them on collagen scaffolds for a week. PCK-26 negative cells had a proliferative capability of 13%, which compared poorly to the 80% of the PCK-26 positive subpopulation. Accordingly, as evidenced by electron microscopy the PCK-26 negative subpopulation had minimal cell attachment on the collagen scaffolds, whereas the PCK-26 positive cells had a uniform distribution. However, in vivo experiments using EEC-scaffold constructs, similarly to results in acellular approaches, have shown more promise for regeneration of partial rather than circumferential defects in rats and dogs [65–67]. An innovative approach recently described seeded cells on a temperature-responsive dish that became hydrophilic at 20°C and allowed harvesting of a single-cell sheet [64]. Following harvest, the cell sheets were transplanted in dogs that had undergone endoscopic submucosal resection of 5 cm in length and 180° in circumference. The cell sheets adhered to and survived on the underlying muscle layers in the ulcer sites, providing an intact, stratified epithelium. Complete wound healing was observed at 4 weeks with no signs of stricture and an intact epithelium. In comparison, the animals that received ESD without the cell sheets, had fibrin mesh and inflammation in the corresponding sites. Wei et al. [66] obtained mucosal epithelial cells from oral biopsy (OMEC) that were seeded onto SIS scaffolds and implanted as esophageal substitutes for half-circumferential 5-cm defects. Four weeks after implantation, the SIS-OMEC group showed complete reepithelialization, whereas the SIS group showed only partial epithelialization. Eight weeks after surgery, the squamous epithelium was found to cover the entire graft surface in both groups, suggesting migration from the host. Interestingly, muscular regeneration was observed in the SIS-OMEC animals, but not in the SIS group. Prior to that, Grikscheit et al. [67] isolated esophageal organoid units (OUs) created following digestion of rat esophagi and consisting of mesenchymal cores with an epithelial lining. These were seeded on synthetic scaffold tubes, transplanted in the omentum and then used as a graft for the esophagus. Following 4 weeks of omental growth the construct consisted of mucosa, submucosa, and muscularis propria. While green fluorescent labeling confirmed the OU-origin of the neoesophagus, rats that received the construct to cover a 2.5 cm patch defect of 2 cm circumferential defect gained weight on a normal diet. However, the muscular wall, lack organization, and the constructs were associated to strictures and dilations.

To overcome the limitations of using EEC in isolation, esophageal constructs prepared using EEC-seeded collagen scaffolds were placed on the latissimus dorsi muscle of athymic mice with the intention to harvest and tubularize the muscle once the epithelial side has matured [68,69]. Up to a month after implantation, the luminal structure and epithelium was maintained and 8 days after grafting rat fibroblasts infiltrated from the muscle and neovascularization appeared in the collagen layer. Miki et al. [70] found an increase in the number of epithelial layers from two when EEC seeded alone, to 18 when coseeded with fibroblasts. This increased stratification with fibroblast coculture was correlated with the number of fibroblasts added. A more recent study by Hayashi et al. [71] cultured both epithelial and fibroblast cells on a bed of SMCs embedded in a collagen gel *in vitro*, prior to transplanting them on the latissimus dorsi of athymic rats. Nakase et al. [72] also aimed to combine different cell lines and scaffolds into one tubular structure in dogs. They used oral keratinocytes and fibroblasts cultured on human amniotic membrane and SMC seeded on PGA.

These two scaffolds were then rolled together and implanted into the omentum for 3 weeks, following which they were transplanted to cover a 3-cm intrathoracic esophageal defect. Both muscular and epithelial layers were present at 420 days of follow-up, and were of similar thickness to adjacent native esophagus, although no peristaltic activity was observed. In a control group, in which the same procedure was followed without addition of keratinocytes and fibroblasts, strictures developed and there was almost complete obstruction within 3 weeks.

Additionally, in more recent studies, the combination of synthetic and naturally derived scaffolds has been investigated. The aim of such an approach would be to bring together the precisely controlled biomechanical properties of synthetic scaffolds with the growth factors and ECM cues present in naturally derived materials such as collagen and fibronectin. Zhu et al. [73] used poly(L-lactide-*co*-caprolactone) (PLLC) scaffolds that were aminolysed to introduce free amino groups, which could be used as bridges to couple collagen (PLLC−Col) and fibronectin (PLLC-Fn) by the means of glutaraldehyde. Following a 12-day culture with porcine esophageal cells it was shown that the PLLC-Fn surface was more favorable to epithelium regeneration than PLLC−Col. In a subsequent study, the same group investigated a nanofibrous PLLC-Fn scaffold and demonstrated by means of SEM, immunostaining and Western Blotting to compare collagen type IV synthesis that fibronectin coating improved cell growth [74]. In a similar study, collagen IV was used instead of fibronectin. The epithelial cell culture demonstrated that the grafted collagen IV accelerated greatly the epithelium regeneration. SEM showed good distribution on the scaffold and extensive interconnection between each other, while immunostaining with anticytokeratin AE1/AE3 confirmed the squamous esophageal epithelial nature of the cells [75].

60.3 FUTURE PERSPECTIVES

Based on the above literature, it is clear that although tissue engineering has been proposed as a solution for the current treatments of esophageal defects, currently, there is no clear strategy for recreating all the portions of the esophagus in man [76,77]. The significant heterogeneity among studies and the developed complications like leakage, strictures, infections, inflammations, immune rejection, poor reepithelialization, and poor muscle regeneration in the grafts, suggests that several issues still need to be addressed. These are mainly related to the optimal scaffold, the cell sources for the epithelial and muscular components, peristalsis, and vascularization (Table 60.1). The stenotic changes that are the main complication encountered with esophageal constructs are likely related to poor regeneration of natural architecture.

Therefore, the main challenge is to develop an ideal scaffold that mimics the morphology and function of the native esophageal tissue, to treat esophageal disorders.

The recent trend in clinical tissue engineering has been to use decellularized scaffolds. It has been suggested that, in the short term, they would be an advantageous choice due to their enhancement of cellular proliferation, migration, and differentiation. However, the lack of positive results when trying to replace a tubular defect, confirms that the use of biomaterials alone as a means of esophageal repair is unsuccessful. We envisage a point where "smart polymers" may replace scaffolds of biological origin and facilitate an "off-the-shelf" approach to esophageal tissue engineering. POSS-PCU, a synthetic material used in clinical trials of vascular grafting, has been utilized as an alternative to biologic scaffolds in the generation of tracheal scaffolds [78]. These have the added advantages of being tailor-made and retain biomechanical properties indefinitely, while there is no need for an organ donor, with all the attendant convenience, infection and ethical issues of the latter. However, early experience shows that these scaffolds do not epithelialize or vascularize easily [31]. The study of cell-scaffold interactions is likely to substantially inform the development of better biomaterials for organ and tissue regeneration. Ritchie et al. [79] found that esophageal muscle cells seeded onto collagen membranes required mechanical stimulation to retain normal contractile properties in a bioreactor, showing the importance of a multidisciplinary engineering approach to this problem, but we could find no other references to the application of bioreactors to esophageal tissue engineering. *Ex vivo* models, such as bioreactors and microfluidic organotypic chambers, are urgently required in order to explore the effects of varying stem cell/cell-scaffold-signaling combinations in the generation of functional esophageal tissue preimplantation. The use of a bioreactor could provide the cells with a microenvironment similar to the biological one that could promote the migration, adhesion, proliferation and expression of surface markers for the cell-scaffold interaction. In order to reproduce the native conditions, the muscular layer of the esophagus should experience a passive and active stretch mimicking the passage of the bolus. Moreover, all the cellular layers should receive the adequate nutrients for their metabolic activities. Herein, the bioreactor device together with a 3-D scaffold, could promote a uniform, multilayered, and oriented organization of the cells to regenerate a functional tissue.

The general consensus indicates a significant advantage in repopulating scaffolds with cells prior to implantation. Studies that have seeded EEC have had positive results in repopulating the epithelial layer, both as an onlay patch [65,66] and as a total interposition graft [67]. Nevertheless, as with cell-free approaches, in cases in which only the lumen was seeded, there was a poor regeneration of the muscular layer, indicating a need for coseeding with SMC. This is not a surprise, since esophageal strictures can be managed clinically easily with an intestinal patch (free graft) as partial substitution while they have a very high chance of recurrence when such material is used to repair the whole circumference. Herein, both epithelial and SMCs, together with blood vessels and nerves are crucial components for the regeneration of the esophageal tissue. However, the difficulties encountered to coculture multiple cell lines which require different conditions to grow and proliferate, result in the need for a device that promotes the multilayered structure of the engineered tissue. Also, studies are required to identify the optimal cell types and sources to repopulate esophageal scaffolds. Ideally, cell sources should be autologous, easy to harvest, highly proliferative, and should have the ability to differentiate into many specialized cell types.

Equally important to muscular regeneration is the challenge of replicating peristaltic contractility and a vascular supply in an artificial esophagus. Watanabe et al. [80] developed nickel−titanium, shaped-memory, alloy coils, which were placed in an annular manner on a Gore-Tex vascular graft for esophageal replacement. Interestingly, low-voltage electrical current passing through the coils generated peristaltic movements in the artificial esophagus implanted in a goat model, suggesting that reprovision of appropriate muscular stimuli, either by enhanced neural regeneration or by electrical means, may be a profitable route for investigation if functionally normal swallowing is to be achieved. What is more, we propose that the physiological contribution of neural crest cells is a prerequisite for functional peristalsis.

Regarding the vascular component, the esophagus holds an additional challenge due to the tenuous intrinsic vascular anatomy of the esophagus in man and the association of stenosis with poor vascularization. Wrapping the engineered esophagus in the omentum prior to thoracic transplantation is one potential solution, as proposed by Nakase et al. [72]. This is also the experience of the authors, who have mobilized the omentum in the chest to vascularize the transplanted trachea [22]. The use of intraluminal stents is another solution to avoid stenosis. Where collagen scaffolds were used in the above studies, the stenosis and mortality was inversely correlated to the length of stay of the intraluminal stent [33−35]. The use of stents allows time for epithelial and muscular migration onto the cell-free scaffolds. This approach has also been tested clinically in the trachea and allows epithelial ingrowth and persists for about 6 weeks before complete degradation [22,81,82].

60.4 CONCLUSION

There is evidence that tissue engineering may represent a promising new route for treating severe congenital or acquired esophageal disorders. We present possible lines for investigation that could indicate what such products will look like, but propose that, in the short- to medium-term, a combination of decellularized scaffolds with muscle and epithelial cells of autologous (including autologous stem cell) origin are likely to be the most expeditious route. Major questions of vascularity, cell–cell and cell–scaffold interaction, and motility remain outstanding, however, before the bioengineered neoesophagus becomes an established, effective treatment for complex congenital and acquired malformations in adults and children.

REFERENCES

[1] Badylak SF, Meurling S, Chen M, Spievack A, Simmons-Byrd A, Lafayette W. Resorbable bioscaffold for esophageal repair in a dog model. J Pediatr Surg 2000;35:1097–103.

[2] Ludman L, Spitz L. Quality of life after gastric transposition for esophageal atresia. J Pediatr Surg 2003;38:53–7.

[3] Ure BM, Slany E, Eypasch EP, Gharib M, Holschneider AM, Troidl H. Long-term functional results and quality of life after colon interposition for long-gap esophageal atresia. Eur J Pediatr Surg 1995;5:206–10.

[4] Spitz L, Kiely E, Pierro A. Gastric transposition in children—a 21-year experience. J Pediatr Surg 2004;39:276–81.

[5] Allum WH, Blazeby JM, Griffin SM, Cunningham, D, Jankowski JA, Wong, R. Association of Upper Gastrointestinal Surgeons of Great Britain and Ireland, the British Society of Gastroenterology and the British Association of Surgical Oncology. Guidelines for the management of esophageal and gastric cancer. Gut 60:1449–72.

[6] Deurloo JA, Ekkelkamp S, Hartman EE, Sprangers MA, Aronson DC. Quality of life in adult survivors of correction of esophageal atresia. Arch Surg 2005;140:976–80.

[7] Somppi E, Tammela O, Ruuska T, Rahnasto J, Laitinen J, Turjanmaa V, et al. Outcome of patients operated on for esophageal atresia: 30 years' experience. J Pediatr Surg 1998;33:1341–6.

[8] van Vilsteren FG, Pouw RE, Herrero LA, Peters FP, Bisschops R, Houben M, et al. Learning to perform endoscopic resection of esophageal neoplasia is associated with significant complications even within a structured training program. Endoscopy 2012;44:4–12.

[9] Zani A, Pierro A, Elvassore N, De Coppi P. Tissue engineering: an option for esophageal replacement? Semin Pediatr Surg 2009;18:57–62.

[10] Orlando G, Wood KJ, De Coppi P, Baptista PM, Binder KW, Bitar KN, et al. Regenerative medicine as applied to general surgery. Ann Surg 2012.

[11] Kemp P. History of regenerative medicine: looking backwards to move forwards. Regen Med 2006;1:653–69.

[12] Atala A. Engineering organs. Curr Opin Biotechnol 2009;20:575–92.

[13] Mason C, Dunnill P. A brief definition of regenerative medicine. Regen Med 2008;3:1–5.

[14] Atala A, Bauer SB, Soker S, Yoo JJ, Retik AB. Tissue-engineered autologous bladders for patients needing cystoplasty. Lancet 2006;367:1241–6.

[15] Atala A, Guzman L, Retik AB. A novel inert collagen matrix for hypospadias repair. J Urol 1999;162:1148–51.

[16] Chalmers RL, Smock E, Geh JL. Experience of integra(®) in cancer reconstructive surgery. J Plast Reconstr Aesthet Surg 2010;63:2081–90.

[17] Macchiarini P, Jungebluth P, Go T, Asnaghi MA, Rees LE, Cogan TA, et al. Clinical transplantation of a tissue-engineered airway. Lancet 372:2023 e30.

[18] Laurance J. British boy receives trachea transplant built with his own stem cells. BMJ 2010;340:c1633.

[19] Fishman JM, De Coppi P, Elliott MJ, Atala A, Birchall MA, Macchiarini P. Airway tissue engineering. Expert Opin Biol Ther 2011;11:1623–35.

[20] Baiguera S, Jungebluth P, Burns A, Mavilia C, Haag J, De Coppi P, et al. Tissue engineered human tracheas for in vivo implantation. Biomaterials 2010;31:8931–8.

[21] Badylak SF, Hoppo T, Nieponice A, Gilbert TW, Davison JM, Jobe BA. Esophageal preservation in five male patients after endoscopic inner-layer circumferential resection in the setting of superficial cancer: a regenerative medicine approach with a biologic scaffold. Tissue Eng Part A 2011;17:1643–50.

[22] Elliott MJ, De Coppi P, Speggiorin S, Roebuck D, Butler CR, Samuel E, et al. Stem-cell-based, tissue engineered tracheal replacement in a child: a 2-year follow-up study. Lancet 2012 Sep 15;380(9846):994–1000.

[23] Engler AJ, Sen S, Sweeney HL, Discher DE. Matrix elasticity directs stem cell lineage specification. Cell 126:677–689.

[24] Freud E, Efrati I, Kidron D, Finally R, Mares AJ. Comparative experimental study of esophageal wall regeneration after prosthetic replacement. J Biomed Mater Res 1999;45:84–91.

[25] Berman EF. The experimental replacement of portions of the esophagus by a plastic tube. Ann Surg 1952;135:337–43.

[26] Lister J, Altman RP, Allison WA. Prosthetic substitution of thoracic esophagus in puppies: use of marlex mesh with collagen or anterior rectus sheath. Ann Surg 1965;162:812–24.

[27] Fryfogle JD, Cyrowski GA, Rothwell D, Rheault G, Clark T. Replacement of the middle third of the esophagus with a silicone rubber prosthesis. An experiment and clinical study. Dis Chest 1963;43:464–75.

[28] Watanabe K, Mark JB. Segmental replacement of the thoracic esophagus with a Silastic prosthesis. Am J Surg 1971;121:238–40.

[29] Lynen Jansen P, Klinge U, Anurov M, Titkova S, Mertens PR, Jansen M. Surgical mesh as a scaffold for tissue regeneration in the esophagus. Eur Surg Res 2004;36:104–11.

[30] Shinhar D, Finaly R, Niska A, Mares AJ. The use of collagen-coated vicryl mesh for reconstruction of the canine cervical esophagus. Pediatr Surg Int 1998;13:84–7.

[31] Jungebluth P, Alici E, Baiguera S, et al. Tracheobronchial transplantation with a stem-cell-seeded bioartificial nanocomposite: a proof-of-concept study. Lancet 2011;378:1997–2004.

[32] Natsume T, Ike O, Okada T, Takimoto N, Shimizu Y, Ikada Y. Porous collagen sponge for esophageal replacement. J Biomed Mater Res 1993;27:867–75.

[33] Takimoto Y, Okumura N, Nakamura T, Natsume T, Shimizu Y. Long-term follow-up of the experimental replacement of the esophagus with a collagen-silicone composite tube. ASAIO J 1993;39:736—9.

[34] Takimoto Y, Nakamura T, Teramachi M, Kiyotani T, Shimizu Y. Replacement of long segments of the esophagus with a collagen-silicone composite tube. ASAIO J 1995;41:605—8.

[35] Takimoto Y, Nakamura T, Yamamoto Y, Kiyotani T, Teramachi M, Shimizu Y. The experimental replacement of a cervical esophageal segment with an artificial prosthesis with the use of collagen matrix and a silicone stent. J Thorac Cardiovasc Surg 1998;116:98—106.

[36] Yamamoto Y, Nakamura T, Shimizu Y, Matsumoto K, Takimoto Y, Kiyotani T, et al. Intrathoracic esophageal replacement in the dog with the use of an artificial esophagus composed of a collagen sponge with a double-layered silicone tube. J Thorac Cardiovasc Surg 1999;118:276—86.

[37] Yamamoto Y, Nakamura T, Shimizu Y, Matsumoto K, Takimoto Y, Liu Y, et al. Intrathoracic esophageal replacement with a collagen sponge—silicone double layer tube: evaluation of omental-pedicle wrapping and prolonged placement of an inner stent. ASAIO J 2000;46:734—9.

[38] Badylak S, Meurling S, Chen M, Spievack A, Simmons-Byrd A. Resorbable bioscaffold for esophageal repair in a dog model. J Pediatr Surg 2000;35:1097—103.

[39] Kajitani M, Wadia Y, Hinds MT, Teach J, Swartz KR, Gregory KW. Successful repair of esophageal injury using an elastin based biomaterial patch. ASAIO J 2001;47:342—5.

[40] Nieponice A, Gilbert TW, Badylak SF. Reinforcement of esophageal anastomoses with an extracellular matrix scaffold in a canine model. Ann Thorac Surg 2006;82:2050—8.

[41] Isch JA, Engum SA, Ruble CA, Davis MM, Grosfeld JL. Patch esophagoplasty using alloderm as a tissue scaffold. J Pediatr Surg 2001;36:266—8.

[42] Lopes MF, Cabrita A, Ilharco J, Pessa P, Paiva-Carvalho J, Pires A, et al. Esophageal replacement in rat using porcine intestinal submucosa as a patch or a tube-shaped graft. Dis Esophagus 2006;19:254—9.

[43] Lopes MF, Cabrita A, Ilharco J, Pessa P, Patrício J. Grafts of porcine intestinal submucosa for repair of cervical and abdominal esophageal defects in the rat. J Invest Surg 2006;19:105—11.

[44] Urita Y, Komuro H, Chen G, Shinya M, Kaneko S, Kaneko M, et al. Regeneration of the esophagus using gastric acellular matrix: an experimental study in a rat model. Pediatr Surg Int 2007;23:21—6.

[45] Doede T, Bondartschuk M, Joerck C, Schulze E, Goernig M. Unsuccessful alloplastic esophageal replacement with porcine small intestinal submucosa. Artif Organs 2009;33:328—33.

[46] Clough A, Ball J, Smith GS, Leibman S. Porcine small intestine submucosa matrix (Surgisis) for esophageal perforation. Ann Thorac Surg 2011;91:15—6.

[47] Teramachi M, Okumura N, Nakamura T, Yamamoto Y, Kiyotani T, Takimoto Y, et al. Intrathoracic tracheal reconstruction with a collagen-conjugated prosthesis: evaluation of the efficacy of omental wrapping. J Thorac Cardiovasc Surg 1997;113:701—11.

[48] Badylak SF, Weiss DJ, Caplan A, Macchiarini P. Engineered whole organs and complex tissues. Lancet 2012;379:943—52.

[49] Fishman JM, Ansari T, Sibbons P, De Coppi P, Birchall MA. Decellularized rabbit cricoarytenoid dorsalis muscle for laryngeal regeneration. Ann Otol Rhinol Laryngol 2012;121:129—38.

[50] Totonelli G, Maghsoudlou P, Garriboli M, Riegler J, Orlando G, Burns AJ, et al. A rat decellularized small bowel scaffold that preserves villus-crypt architecture for intestinal regeneration. Biomaterials 2012;33:3401—10.

[51] Tan JY, Chua CK, Leong KF, Chian KS, Leong WS, Tan LP. Esophageal tissue engineering: an in-depth review on scaffold design. Biotechnol Bioeng 2012;109:1—15.

[52] Mertsching H, Schanz J, Steger V, Schandar M, Schenk M, Hansmann J, et al. Generation and transplantation of an autologous vascularized bioartificial human tissue. Transplantation 2009;27:203—10.

[53] Gilbert TW, Sellaro TL, Badylak SF. Decellularization of tissues and organs. Biomaterials 2006;27:3675—83.

[54] Ozeki M, Narita Y, Kagami H, Ohmiya N, Itoh A, Hirooka Y, et al. Evaluation of decellularized esophagus as a scaffold for cultured esophageal epithelial cells. J Biomed Mater Res A 2006;79:771—8.

[55] Freud E, Greif M, Rozner M, Finaly R, Efrati I, Kidron D, et al. Bridging of esophageal defects with lyophilized dura mater: an experimental study. J Pediatr Surg 1993;28:986—9.

[56] Badylak SF, Vorp DA, Spievack AR, Simmons-Byrd A, Hanke J, Freytes DO, et al. Esophageal reconstruction with ECM and muscle tissue in a dog model. J Surg Res 2005;128:87—97.

[57] Bhrany AD, Beckstead BL, Lang TC, Farwell DG, Giachelli CM, Ratner BD. Development of an esophagus acellular matrix tissue scaffold. Tissue Eng 2006;12:319—30.

[58] Totonelli G, Maghsoudlou P, Georgiades F, Garriboli M, Koshy K, Turmaine M, et al. Detergent enzymatic treatment for the development of a natural acellular matrix for oesophageal regeneration. Pediatr Surg Int 2013;29:87—95.

[59] Beckstead BL, Pan S, Bhrany AD, Bratt-Leal AM, Ratner BD, Giachelli CM. Esophageal epithelial cell interaction with synthetic and natural scaffolds for tissue engineering. Biomaterials 26:6217—28.

[60] Green N, Huang Q, Khan L, Battaglia G, Corfe B, MacNeil S, et al. The development and characterization of an organotypic tissue-engineered human esophageal mucosal model. Tissue Eng Part A 2010;16:1053—64.

[61] Saxena AK, Ainoedhofer H, Hollwarth ME. Esophagus tissue engineering: *in vitro* generation of esophageal epithelial cell sheets and viability on scaffold. J Pediatr 2009;44:896—901.

[62] Saxena AK, Ainoedhofer H, Hollwarth ME. Culture of ovine esophageal epithelial cells and *in vitro* esophagus tissue engineering. Tissue Eng Part C Methods 2010;16:109—14.

[63] Saxena AK, Baumgart H, Komann C, Ainoedhofer H, Soltysiak P, Kofler K, et al. Esophagus tissue engineering: in situ generation of rudimentary tubular vascularized esophageal conduit using the ovine model. J Pediatr Surg 2010;45:859—64.

[64] Kofler K, Ainoedhofer H, Hollwarth ME, Saxena AK. Fluorescence-activated cell sorting of PCK-26 antigen-positive cells enables selection of ovine esophageal epithelial cells with improved viability on scaffolds for esophagus tissue engineering. Pediatr Surg Int 2010;26:97—104.

[65] Ohki T, Yamato M, Murakami D, Takagi R, Yang J, Namiki H, et al. Treatment of esophageal ulcerations using endoscopic

transplantation of tissue-engineered autologous oral mucosal epithelial cell sheets in a canine model. Gut 2006;55:1704—10.

[66] Wei RQ, Tan B, Tan MY, Luo JC, Deng L, Chen XH, et al. Grafts of porcine small intestinal submucosa with cultured autologous oral mucosal epithelial cells for esophageal repair in a canine model. Exp Biol Med 2009;234:453—61.

[67] Grikscheit T, Ochoa ER, Srinivasan A, Gaissert H, Vacanti JP. Tissue-engineered esophagus: experimental substitution by onlay patch or interposition. J Thorac Cardiovasc Surg 2003;126:537—44.

[68] Sato M, Ando N, Ozawa S, Miki H, Kitajima M. An artificial esophagus consisting of cultured human esophageal epithelial cells, polyglycolic acid mesh, and collagen. ASAIO J 1994;40:389—92.

[69] Sato M, Ando N, Ozawa S, Nagashima A, Kitajima M. A hybrid artificial esophagus using cultured human esophageal epithelial cells. ASAIO J 1993;39:554—7.

[70] Miki H, Ando N, Ozawa S, Sato M, Hayashi K, Kitajima M. An artificial esophagus constructed of cultured human esophageal epithelial cells, fibroblasts, polyglycolic acid mesh, and collagen. ASAIO J 1999;45:502—8.

[71] Hayashi K, Ando N, Ozawa S, Kitagawa Y, Miki H, Sato M, et al. A neo-esophagus reconstructed by cultured human esophageal epithelial cells, smooth muscle cells, fibroblasts, and collagen. ASAIO J 2004;50:261—6.

[72] Nakase Y, Nakamura T, Kin S, Nakashima S, Yoshikawa T, Kuriu Y, et al. Intrathoracic esophageal replacement by in situ tissue-engineered esophagus. J Thorac Cardiovasc Surg 2008;136:850—9.

[73] Zhu Y, Chian KS, Chan-Park MB, Mhaisalkar PS, Ratner BD. Protein bonding on biodegradable poly(L-lactide-co-caprolactone) membrane for esophageal tissue engineering. Biomaterials 2006;27:68—78.

[74] Zhu Y, Leong MF, Ong WF, Chan-Park MB, Chian KS. Esophageal epithelium regeneration on fibronectin grafted poly(L-lactide-co-caprolactone) (PLLC) nanofiber scaffold. Biomaterials 2007;28:861—8.

[75] Zhu Y, Ong WF. Epithelium regeneration on collagen (IV) grafted polycaprolactone for esophageal tissue engineering. Mater Sci Eng C 2009;29:1046—50.

[76] Orlando G, García-Arrarás JE, Soker T, Booth C, Wang Z, Ross CL, et al. Regeneration and bioengineering of the gastrointestinal tract: current status and future perspectives. Dig Liv Dis 2012;44 (9):714—20.

[77] Orlando G, Domínguez-Bendala J, Shupe T, Bergman C, Bitar KN, Booth C, et al. Cell and organ bioengineering technology as applied to gastrointestinal diseases. Gut 2013;62 (5):774—86.

[78] Ahmed M, Ghanbari H, Cousins BG, Hamilton G, Seifalian AM. Small calibre polyhedral oligomeric silsesquioxane nanocomposite cardiovascular grafts: influence of porosity on the structure, haemocompatibility and mechanical properties. Acta Biomater 2011;7:3857—67.

[79] Ritchie AC, Wijaya S, Ong WF, Zhong SP, Chian KS. Dependence of alignment direction on magnitude of strain in esophageal smooth muscle cells. Biotechnol Bioeng 2009;102:1703—11.

[80] Watanabe M, Sekine K, Hori Y, Shiraishi Y, Maeda T, Honma D, et al. Artificial esophagus with peristaltic movement. ASAIO J 2005;51:158—61.

[81] Vondrys D, Elliott MJ, McLaren CA, Noctor C, Roebuck DJ. First experience with biodegradable airway stents in children. Ann Thorac Surg 2011;92:1870—4.

[82] Lischke R, Pozniak J, Vondrys D, Elliott MJ. Novel biodegradable stents in the treatment of bronchial stenosis after lung transplantation. Eur J Cardiothorac Surg 2011;40:619—24.

Part IX

Immunosuppression-free Transplantation in the Regenerative Medicine Era

Chapter 61

Stem Cell-Based Approach to Immunomodulation

Kathryn J. Wood[a], Karen English[b], and Ou Li[a]

[a]Transplantation Research Immunology Group, Nuffield Department of Surgical Sciences, John Radcliffe Hospital, University of Oxford, Oxford, UK,

[b]Cellular Immunology Group, Institute of Immunology, National University of Ireland Maynooth, Co. Kildare, Ireland

Chapter Outline

61.1 IMMUNOMODULATION

The human immune system has evolved to respond to challenges that pose a threat to the host in a precise and controlled way. A constant balance exists to ensure an effective, but not an excessive immune response is made to any unwanted stimuli. The mechanisms used by the immune system to prevent damage to the host and to maintain homeostasis are multifactorial and complex. Many of these mechanisms involve immune regulatory cell populations [1,2]. It may be possible to take advantage of these mechanisms to modulate the immune response to allogeneic tissues using stem cell-based therapies.

Despite the significant achievements accomplished over the past 60 years in solid organ transplantation (SOT), rejection remains a significant barrier to long-term graft survival [3]. While the advent of immunosuppressive drugs has facilitated improved outcomes, the toxicity and associated complications of lifelong nonspecific immunosuppression are substantial limiting factors [4]. Thus, there is a significant requirement for immunosuppressive therapies with increased specificity and lower levels of toxicity to improve both the quality and quantity of life of transplant recipients. Stem cell-based therapies may offer an opportunity in this context.

61.2 HEMATOPOIETIC STEM CELLS

Hematopoietic stem cells (HSCs) are self-renewing and multipotent, with the capacity to generate all of the cell populations of the hematopoietic system. HSCs also have the ability to modulate immune responsiveness and induce tolerance to donor alloantigens in the setting of SOT. The strategies for tolerance induction using HSCs that are being explored most actively at present involve one or more mechanisms of tolerance including the continuous deletion of donor reactive leukocytes by establishing the presence of high levels of donor-derived HSCs in the recipient (mixed chimerism), transient depletion, and/or deletion of donor reactive leukocytes followed by the establishment of immunoregulation following an infusion of HSCs to control or suppress responses to donor alloantigens [5].

61.2.1 Immunomodulation Through Mixed Chimerism

The coexistence of stable mixtures of donor and recipient cells resulting in a state of allospecific tolerance is an idea that was initially restricted to the field of bone marrow and HSC transplantation. A small number of bone marrow transplant recipients who subsequently required a renal transplant were transplanted with a kidney from their bone marrow donor. These transplant recipients exhibited tolerance to the donor alloantigens and required no long-term immunosuppression [6−8]. Obviously, bone marrow or HSC transplantation is not an appropriate approach that should be considered for most recipients on transplant waiting lists, as achieving fully allogeneic chimerism requires a conditioning regimen that is relatively toxic and has the drawback of reducing the immunocompetance of the recipient's immune system in some situations. However, these clinical case reports provided a foundation for the development of nonmyeloablative conditioning protocols wherein donor bone marrow cells/HSCs are introduced into recipients under conditions that allow the development of chimerism leading to long-term allograft survival in the absence of a to long-term requirement for nonspecific immunosuppression.

Many different approaches have been used to achieve chimerism *in vivo*. Total lymphoid irradiation (TLI) alone or in combination with bone marrow infusion has been shown to be effective at inducing tolerance in rodents, primates, and humans [9−12]. Despite the requirement for irradiation that has arguably inhibited the development and clinical application of these protocols to their fullest extent, strategies using TLI have continued to be refined leading to more recent reports that this approach can be used safely and effectively to induce tolerance to a kidney transplant. Notably, Scandling et al. have developed a conditioning regimen of TLI and antithymocyte globulin after transplantation that facilitates the engraftment of donor HSCs. The persistent mixed chimerism achieved modulates the immune response to donor alloantigens and enables immunosuppression to be discontinued without rejection episodes or clinical manifestations of graft-versus-host disease [12].

The limitations of myeloablative therapy prompted the development and refinement of alternate approaches in animal models. Nonmyeloablative conditioning regimens have been developed that promote deletion of donor reactive cells in the thymus following the infusion of a high dose of donor bone marrow. The demonstration that transient macrochimerism via nonmyeloablative conditioning could achieve tolerance to renal allograft transplanted along with donor bone marrow in nonhuman primate models provided the impetus for the translation of this approach to clinical transplantation [13−15].

Mixed chimerism was first used successfully to treat a patient with multiple myeloma who had end-stage renal failure and received a kidney from an HLA-identical living donor [16]. The pretransplant conditioning strategy included thymic irradiation, whole-body irradiation, splenectomy, and donor marrow infusion followed by the administration of a short course of cyclosporin posttransplantation. As in the primate studies, macrochimerism was only detectable in the early period after transplantation, and after the first 2 months the percentage of donor cells declined. Nevertheless, withdrawal of immunosuppressive therapy (cyclosporin) at 10 weeks posttransplantation did not trigger rejection and immunosuppression-free survival, that is, operational tolerance to donor alloantigens, persisted long term. At the time of writing, nine patients with end-stage renal failure who had an HLA-identical donor have been treated with this protocol [17], with only one recipient developing evidence of renal allograft rejection after stopping immunosuppression that was treated successfully by the temporary reintroduction of immunosuppression. Although chimerism was lost in all of these patients, data suggesting that there was enrichment for $CD25^+CD4$ T cells with regulatory activity after transplantation was obtained [18], suggesting that the maintenance of the macrochimeric state up to the time of transplantation may be sufficient to enable other mechanisms to be induced enabling the graft to function in the absence of long-term immunosuppressive drug therapy.

As a result of these encouraging data and further refinement of the protocols in experimental models, the mixed chimerism approach for tolerance induction was extended to HLA-mismatched kidney transplants [19]. The first two patients enrolled in this trial were treated successfully and immunosuppression was withdrawn. However, the third patient developed irreversible humoral rejection 10 days after transplantation. Careful review of this case showed that this patient had high levels of pre-existing alloantibody activity, but no detectable donor-specific antibody (DSA) before transplantation. The clinical protocol was therefore modified further to include Rituximab (anti-CD20). Stable renal allograft function has been maintained in seven of the eight subjects in whom immunosuppression was discontinued, although, very importantly, long-term monitoring of all of these patients continues to determine if the tolerant state is stable in the long term. The analysis of samples from patients enrolled in this trial showed that the chimeric state was achieved, but that it persisted only transiently. Evidence for unresponsiveness to donor alloantigens was obtained through *in vitro* assays and data supporting the idea that regulatory mechanisms were operating were obtained by analyzing FOXP3 expression by PCR [19].

There is therefore good evidence that HSCs can modulate immune responses in the setting of SOT by harnessing multiple mechanisms that include deletion and regulation of donor reactivity. However, the treatment regimens required to achieve mixed chimerism, albeit transiently, remain relatively aggressive and to date these approaches have not been adopted by the majority of clinical transplant centers. Studies are ongoing to refine mixed chimerism protocols and further clinical trials are planned.

61.3 MESENCHYMAL STROMAL CELLS

Mesenchymal stromal cells (MSCs) are a subpopulation of multipotent cells originally identified in the bone marrow [20]. MSCs are characterized by their fibroblast-like appearance, colony forming unit capacity and their rapid adherence to tissue culture plastic. While MSCs are relatively easy to isolate, culture, and expand, the lack of a unique marker or panel of markers to identify MSCs from other cell populations remains an issue that has led to a lack of uniformity in the cell populations being used by different teams both experimentally and in clinical studies. In 2006, the International Society for Cellular Therapy proposed a set of phenotypic and functional criteria to define MSCs [21], a process that has been helpful. However, the discovery of new markers which specifically identify MSCs remains a priority and is eagerly awaited.

MSCs are multifunctional. They have the capacity to differentiate into adipocytes, chondrocytes, and osteoblasts *in vitro* and *in vivo* [22]. Based on their differentiation potential, initial studies using MSCs focused on their regenerative capacity [23,24]. These effects are mediated predominantly as a result of the production of trophic factors [25,26]. Interestingly, some of these trophic factors can also facilitate modulation of immune responses by MSCs. Supporting the concept that MSCs can have both repair and immune modulatory properties.

The ability of MSCs to modulate or suppress immune responses was first described in a transplant model. Infusion of MSCs was found to facilitate the survival of allogeneic skin grafts [27], in part through secretion of soluble factors [28]. Initially, studies focused primarily on the suppression of the adaptive immune response [3]. MSCs were found to inhibit T-cell function, have the capacity to shift the T-helper lymphocyte balance, induce T-cell apoptosis and the generation of functional regulatory T cells (Treg) [29–31]. With respect to B cells, only limited data are available at present and in some cases they are contradictory. However, there is a suggestion that MSCs can also suppress B-cell proliferation and function [32]. As well as modulating adaptive immunity, MSCs have the capacity to modulate multiple components of the innate immune system, including complement, toll-like receptor (TLR) signaling, macrophages, dendritic cells (DCs), neutrophils, mast cells, and natural killer cells [33–38]. The therapeutic efficacy of the

anti-inflammatory effects of MSC has been established in a number of experimental models, including graft-versus-host disease, sepsis, inflammatory bowel disease, and allergic airway disease [29,37–41].

In the case of SOT, MSCs can exert their immune modulatory effects in two ways, by controlling ischemia reperfusion injury [42] and by suppressing allograft rejection [30,43,44]. In some situations, MSCs have been reported capable of inducing a state of specific immunological unresponsiveness or tolerance to defined sets of antigens [30,45]. Whether this is a direct property of MSCs or an indirect effect of their ability to deviate the immune response and generate Treg requires further investigation. Nevertheless this potential to induce tolerance to alloantigens clearly merits further study.

The *in vitro* immunosuppressive capacity combined with the proven therapeutic efficacy of MSCs in preclinical models has paved the way for clinical studies using MSCs as adjunctive cell therapy.

61.3.1 Modulation of Ischemia Reperfusion Injury by MSCs

Ischemia reperfusion injury causes sterile inflammation and results in the production of a number of damage-associated molecular patterns (DAMPs), including necrotic cells, cellular debris, heat shock proteins, and high mobility group box-1 (HMGB-1) [46]. DAMPs activate pattern recognition receptors (PRRs), such as TLRs, C-type lectin receptors, nucleotide-binding oligomerization domain (NOD) and NOD-like receptors, the receptor for advanced glycation end products (RAGE) and retinoic acid inducible gene I (RIG I) receptors. Signaling through these receptors results in activation of the inflammasome [47] and the complement system, upregulating gene transcription and production of micro-RNAs [46] involved in the inflammatory response. Together, these factors lead to the production of pro-inflammatory cytokines, the activation of platelets and endothelial cells, tissue hypoxia and the recruitment of innate and eventually adaptive immune cells [46].

MSCs express a number of pathogen recognition receptors, including TLR1–9 [48–51], NOD receptors [52,53], and RAGE [54]. These receptors are functionally active, and binding to their respective ligands leads to alterations in MSC function. For example, stimulation of NOD-like receptors on MSCs leads to the production of IL-8 and vascular endothelial growth factor [52,53], HMGB-1 signaling through RAGE-induced MSC migration and inhibited MSC production of IDO (indoleamine 2,3-dioxygenase) [55]. TLR3 and TLR4 activation of MSCs results in differential effects with TLR4 priming inducing a pro-inflammatory phenotype and secretion of IL-6, IL-8, and TGF-β whereas TLR3 priming induces anti-inflammatory MSCs producing IDO, prostaglandin E-2 (PGE-2), IL-4, and IL-1RA [56]. The data relating to immunomodulation by MSCs following PRR engagement are contradictory at present. While in some studies TLR3 and TLR4 signaling has been reported to enhance the functional activity of MSCs *in vitro* through mechanisms, including IDO induction via IFN-β and protein kinase R signaling [48], in others it has been found to attenuate the immunosuppressive effects of MSCs [57]. It is clear that MSCs respond to environmental cues that may be present during ischemia reperfusion injury. The exact combination of these cues may dictate the outcome of their response. Further research is required to understand how these DAMPs and the other environmental cues present in injured tissue impact MSC function.

MSCs are also responsive to complement and migrate in response to complement components, including C1q, C3a, and C5a [58,59]. The presence of high levels of C3 activation has been found to correlate with the enhanced immunosuppressive capacity of MSCs [60]. Importantly, MSCs express CD59, a complement regulatory protein, and also release complement factor H which protects them from complement-mediated lysis [60,61]. In addition, stimulation through the receptors for the small, soluble molecules C3a and C5a released during complement activation protects MSCs from oxidative damage [59]. This together with the finding that MSCs produce a number of antioxidants, including heme-oxygenase 1 and superoxide dismutase [62,63] links to the observations that MSCs have been shown to suppress oxidative stress and inflammation in ischemia reperfusion injury models *in vivo* [64−67]. The exact mechanisms of action in this situation are unclear. However, protection of tissue injury by MSCs was associated with increased expression of IL-10, heme-oxygenase 1, and hepatocyte growth factor, decreased expression of the pro-inflammatory cytokines IL-1β, TNF-α, and IFN-γ, reduced reactive oxygen species, reduced apoptosis, and decreased numbers of activated T cells and infiltrating immune cells [64,66−68].

Administration of MSCs *in vivo* can reduce intra-graft inflammatory gene expression and recruitment of antigen presenting cells into the allograft, as demonstrated in a prolonged cold ischemic kidney transplant model [68], resulting in long-term protection from chronic allograft nephropathy [69]. Importantly, ischemia reperfusion plays a key role in the recruitment of MSCs to transplanted organs [45], thus MSCs present in the recipient as well as those administered as a therapy will have a tendency to migrate to injured tissue, in this case the transplant. This experimental study provides insight into the findings from the first reported pilot study of MSCs in kidney transplant patients. Although, the pilot study reported the feasibility of the use of MSCs as a cell therapy in kidney transplantation, serum creatinine levels were increased 7−14 days

post-MSC infusion and a graft biopsy revealed the presence of a focal granulocyte inflammatory infiltrate [70]. Nevertheless, patients maintained stable graft function after 180 and 360 days, and on a positive note MSC administration was associated with decreased CD8[+] memory T cells and a progressive increase in Treg [70]. Overall, the data available to date suggest that MSCs can exert protective effects in ischemic reperfusion injuries through anti-inflammatory and paracrine factors and this likely plays an important part in MSC enhancement of allograft survival. However, much more information is needed on the effect of the ischemic environment on the function of MSCs and how this might impact the outcome of MSC therapy in SOT [71].

61.3.2 Modulation of Macrophage Function by MSCs

Macrophages and neutrophils are generally the first innate immune cells to infiltrate the graft postischemia reperfusion injury. While neutrophils are present only in the graft during inflammatory episodes, macrophages are present throughout the life of the graft albeit in reduced numbers after resolution of tissue injury in the absence of rejection. Neutrophils and macrophages play a role in rejection through the production of pro-inflammatory cytokines and by driving the activation of antigen-specific T cells [72]. MSCs have the capacity to reeducate monocytes/macrophages by inducing alternatively activated macrophages downregulating the production of TNF-α, IL-1α, IL-6, and IL-12p70, increasing the production of IL-10 and enhancing phagocytic activity [33,34,36,37] through production of IDO and PGE-2 [73,74]. There is no direct evidence at present that MSCs function in this way after transplantation, but clearly the possibility that they could promote the development of alternatively activated macrophages within the allograft is intriguing. Again, the microenvironmental cues present at the site of MSC activation seem to determine the particular mechanism of actions deployed by MSCs in modulating the immune response and resolving inflammation.

61.3.3 Modulation of DCs by MSCs

Both donor and recipient DCs play a critical role in triggering graft rejection through the direct, indirect, or semi-direct pathways of allorecognition [3]. MSCs can interfere with the key features of DC function; migration, maturation, and antigen presentation [35] and they mediate these effects through downregulation of cell surface molecules associated with immunostimulation, including MHC class II, CD40, CD80, and CD86 [75−78] and modulation of the lymph node homing chemokine receptor CCR7 *in vitro*[35] and *in vivo*[79]. The soluble factor IL-6 produced by MSCs has been shown to be involved in downregulation of maturation markers [35,75,77], whereas contact-dependent Notch signaling but not IL-6 was shown to be required for DC modulation in another study [76]. MSCs have been shown to block DC maturation and function in a kidney allograft model [30], however, the exact mechanisms utilized by MSCs to achieve this effect remain to be elucidated.

Analogous to the effects of MSCs on macrophage polarization, mentioned above, MSCs can also promote the generation of tolerogenic DCs [78,80–83]. MSC educated DCs have the capacity to suppress alloreactive responses and prolong islet allograft survival [84] and to induce a state of tolerance in the context of SOT (cardiac allograft) in the presence of low dose immunosuppression [80].

The mechanisms used by MSCs to generate tolerogenic DCs are thought to be influenced by the context in which MSCs interact with DCs. The key mediator in MSC modulation of DC maturation is IL-6 [35,75,77], however, the mechanisms involved in MSC promotion of tolerogenic DCs are less clear. A central role for PGE-2 [83] and cell contact-dependent activation of the Notch signaling pathway [76,78] but not IL-6 has been reported [76,83], as well as a contact-dependent mechanism involving activation of AKT and impaired NFκB signaling [79]. Finally, mouse embryonic fibroblast-derived MSCs have been shown to generate a novel population of IL-10-dependent tolerogenic DCs through an IL-10-activated SOCS3-dependent mechanism [82]. MSC induction of tolerogenic DCs is a key mechanism through which MSCs modulation of the immune response is achieved. However, there are still significant gaps in our understanding of exactly how MSCs promote the generation of tolerogenic DCs.

61.3.4 Modulation of T-Cell Responses in Rejection by MSCs

T-cell proliferation and activation are prerequisites for allograft rejection [2,85]. A large body of data demonstrate that MSCs can modulate T-cell proliferation, activation, and function both *in vitro* and *in vivo* [28,44,86–89]. Moreover, the capacity for MSCs to inhibit Th17 cell differentiation [90,91] or to shift the T-helper cell balance in favor of a more anti-inflammatory phenotype has been demonstrated *in vitro* [92–96]. The mechanisms utilized by MSCs in mediating these effects vary between *in vitro* and *in vivo* models. However, the secretion of soluble factors by MSCs is a common feature (English, 2012). IDO and PGE-2 have been implicated in MSC inhibition of Th17 differentiation [90,91]. In the case of PGE-2, the steps involved in the process require contact-dependent COX-2 induction of PGE-2 and direct

inhibition through EP4 [90]. MSCs can also mediate this effect through suppressing the Th17 transcription factor RORγt and upregulating Foxp3 to induce a Treg phenotype producing IL-10 [96]. MSC-derived TGF-β has been shown to play a partial role in shifting the balance of Th1/Th2/Th17 and Treg in an autoimmune disease model [31]. A role for matrix metalloproteinase (MMP)2 and MMP9 secreted by MSCs facilitating cleavage of CD25 expressed on CD4$^+$ T cells thereby inhibiting alloantigen driven proliferation and so preventing islet allograft rejection has also been described [44]. Other evidence suggests that MSC-derived MMPs also cleave CCL2 which subsequently inhibits Th17 activation via a STAT3-dependent pathway [97].

MSCs also have the capacity to expand or induce Treg in the setting of an alloimmune response [43,45,98,99] and in some cases can generate a state of Treg-dependent tolerance [30,45]. Both of these studies elegantly demonstrate the importance of Treg in MSC-induced tolerance using Treg depletion strategies with IDO potentially playing a significant role [30]. *In vitro*, MSC induction of Treg is thought to involve cell contact, PGE-2, and TGF-β [94]. *In vivo*, MSC-derived TGF-β was required for the generation of antigen-specific Treg and overall, TGF-β seems to be the major soluble factor involved in MSC promotion of Treg *in vivo* [29,31,100,101].

61.3.5 Modulation of B-Cell Responses by MSCs

The reported effects of MSCs on B-cell activation, proliferation, and function have been variable and in some cases contradictory. On the positive side, MSCs have been shown to inhibit B-cell proliferation [86,102–104] and Ig production [32,103,105] *in vitro*. The mechanisms of action involve contact-dependent factors [104], including programmed death-1/programmed death ligands-1/2 [102] as well as soluble factors like MMP-cleaved CCL2 [105]. The differences in the reported data may well be due to the purity of B cells (excluding T-cell help) and by further understanding the role that TLR ligands (LPS, CPG, Poly I.C.) have on MSC activation and function. *In vivo*, the effect of MSCs has been examined in the B-cell driven pathology of systemic lupus erythematosus both in mouse models and in patient samples. MSCs were found to enhance survival and reduce serum creatinine, blood urea nitrogen, proteinuria, C3 deposition and circulating dsDNA antibodies, as well as antigen-specific IgM and IgG secretion [33,86,106] in one study, but again other data are contradictory [104,107], with for example Schena et al. reporting no effect of MSCs on survival, proteinuria, or dsDNA antibodies [104]. In transplantation, MSCs have been reported to reduce intra-graft IgG deposits as well as circulating DSAs [69,80] providing protection from graft injury [69] and inducing allograft tolerance in the presence of immunosuppression [80]. Notably, failure of MSCs to prolong allograft survival has been associated with MSC promotion of intra-graft B-cell infiltration [108], and while administration of donor MSCs posttransplant leads to sensitization and premature graft dysfunction, this was not thought to be associated with antibody-mediated humoral rejection [45]. Rather, MSCs administered posttransplant (but not pretransplant) localized in the transplanted kidney in response to ischemia reperfusion injury and subsequently produced IL-6 and TNF-α promoting a pro-inflammatory environment facilitating neutrophil infiltration and C3 deposition [45]. These studies highlight the significant gap in our understanding of the effect of the microenvironment on MSC activation and function and particularly the effect this has on how MSCs see B cells and vice versa.

61.4 CLINICAL APPLICATION OF MSCS IN TRANSPLANTATION

The application of MSC therapy in conjunction with a reduced immunosuppressive regimen is theoretically very appealing as MSCs not only promote the resolution of inflammation and enhance graft repair but may also facilitate the induction of tolerance. Based on the safety and efficacy data generated in preclinical models and clinical trials utilizing MSC therapy for acute graft-versus-host disease [30,43,109], MSC therapy is currently being evaluated in SOT [110]. Perico et al. provided the first report on MSCs in kidney transplant patients in a pilot study examining safety and feasibility [70]. Culture expanded autologous bone marrow-derived MSCs (1.7×10^6–2.0×10^6/kg body weight) were administered intravenously on day 7 postkidney (living donor) transplant in addition to T-cell depletion induction therapy in two patients. This study reported an increase in serum creatinine in both patients 7−14 days after MSC infusion. In addition, a focal inflammatory (granulocyte) infiltrate was observed in a graft biopsy taken from one of the patients, but acute graft rejection was ruled out. Importantly, both patients maintained stable graft function which was associated with decreased memory $CD8^+$ T cells and increased Treg [70]. Importantly, similar effects were observed in a mouse model of kidney allograft transplantation, and elegantly demonstrated the correlation of premature graft injury with posttransplant (but not pretransplant) MSC infusion and localization of the MSCs primarily in the injured graft. Indeed these studies may help to resolve the disparate findings with regard to MSC efficacy in prolonging graft survival [108,111−115]. This highlights the importance of timing

of MSC administration and the requirement for a better understanding of the influence of the transplanted graft microenvironment (e.g., ischemia reperfusion) on MSC function. Patients in a large randomized controlled trial (106 patients over three arms) investigating the safety and efficacy of autologous bone marrow-derived MSCs (in kidney transplantation) in combination with standard dose calcineurin inhibitors (CNI) or low dose CNI were compared to control groups receiving anti-IL-2 receptor antibody therapy in combination with standard dose CNI [116]. This trial demonstrated safety and efficacy with decreased incidence of acute rejection and glucocorticoid-resistant rejection and increased estimated glomerular filtration rate and faster recovery of renal function in the first month posttransplant as well as better estimated renal function at year 1. In addition, MSC-treated groups revealed significantly reduced risk of opportunistic infections than the control group [116]. Importantly, this large randomized controlled trial did not observe increased creatinine levels in MSC-treated patients and this may be associated with the dose and/or timing ($1 - 2 \times 10^6$ on day 0 and day 14 versus $1.7 - 2 \times 10^6$ on day 7 posttransplant) of MSC administration or with differences in immunosuppressive regimen [70,116]. These are important factors which may significantly impact MSC efficacy in SOT and require careful consideration.

61.5 CONCLUDING REMARKS

The characteristics of MSCs make them an ideal candidate for modulating immune responsiveness *in vivo* at the level of both the innate and adaptive responses. Early data from clinical trials suggest that MSCs are safe and efficacious, but highlight the gap in our understanding of exactly how MSCs mediate their protective effects *in vivo*. More work is needed to elucidate the microenvironmental cues that drive MSC function *in vivo*; data that will hold the key to the successful therapeutic application of MSCs in the future.

ACKNOWLEDGMENTS

Karen English is supported by a HRB Translational Medicine Postdoctoral Fellowship and a Marie Curie Career Integration Grant. Work from the authors' own laboratories was supported by grants from The Wellcome Trust, European Union FP7 programme (OPTISTEM), and the Medical Research Council, UK.

LIST OF ABBREVIATIONS

CNI	calcineurin inhibitor
DAMPs	damage-associated molecular patterns
DC	dendritic cell
dsDNA	double-stranded deoxyribonucleic acid
HMGB-1	high mobility group protein box-1
HSCs	hematopoietic stem cells

IDO	indoleamine 2,3-dioxygenase
IFN-γ	interferon gamma
Ig	immunoglobulin
IL	interleukin
MMP	matrix metalloproteinase
MSCs	mesenchymal stromal cells
NK	natural killer
NOD	nucleotide oligomerization domain
PGE-2	prostaglandin E-2
Poly I.C.	polyinosinic:polycytidylic acid
PRR	pattern recognition receptor
RAGE	receptor for advanced glycation end products
SOT	solid organ transplantation
TGF-β	transforming growth factor beta
TLI	total lymphoid irradiation
TLR	toll-like receptor
TNF-α	tumor necrosis factor alpha
TNFR1	tumor necrosis factor receptor 1
Treg	T regulatory cell

REFERENCES

[1] Wood KJ, Sakaguchi S. Regulatory T cells in transplantation tolerance. Nat Immunol Rev 2003;3:199−210.

[2] Wood KJ, Bushell A, Hester J. Regulatory immune cells in transplantation. Nat Rev Immunol 2012;12:417−30.

[3] Wood KJ, Goto R. Mechanisms of rejection: current perspectives. Transplantation 2012;93:1−10. Available from: http://dx.doi.org/10.1097/TP.0b013e31823cab44.

[4] Halloran PF. Immunosuppressive drugs for kidney transplantation. N Engl J Med 2004;351:2715−29.

[5] Issa F, Wood KJ. Translating tolerogenic therapies to the clinic—where do we stand? Front Immunol 2012;3.

[6] Jacobsen N, Taaning E, Ladefoged J, Kristensen J, Pedersen F. Tolerance to an HLA-B,DR disparate kidney allograft after bone marrow transplantation from the same donor. Lancet 1994;343:800.

[7] Sayegh M, Fine N, Smith J, Renneke H, Milford E, Tilney N. Immunologic tolerance to renal allografts after bone marrow transplants from the same donors. Ann Inter Med 1991;114:954.

[8] Sorof J, Koerper M, Portale A, Potter D, DeSantes K, Morton C. Renal transplantation without chronic immunosuppression after T cell depleted HLA-mismatched bone marrow transplantation. Transplantation 1995;59:1633−5.

[9] Myburgh JA, Smit JA, Stark JH, Browde S. Total lymphoid irradiation in kidney and liver transplantation in the baboon: prolonged graft survival and alteration in cell subsets with low cumulative dose regimens. J Immunol 1984;132:1019−25.

[10] Slavin S, Strober S, Fuks Z, Kaplan HS. Induction of specific tissue transplantation tolerance using fractionated total lymphoid irradiation in adult mice: long-term survival of allogeneic bone marrow and skin grafts. J Exp Med 1977;146:34−48.

[11] Strober S, Dhillon M, Schubert M, Holm B, Engleman E, Benike C, et al. Acquired immune tolerance to cadaveric renal allografts: a study of three patients treated with total lymphoid irradiation. N Engl J Med 1989;321:28−33.

[12] Scandling JD, Busque S, Dejbakhsh-Jones S, Benike C, Millan MT, Shizuru JA, et al. Tolerance and chimerism after

renal and hematopoietic-cell transplantation. N Engl J Med 2008;358:362−8.

[13] Kawai T, Cosimi A, Colvin R, Powelson J, Eason J, Kozlowski T, et al. Mixed allogeneic chimerism and renal allograft tolerance in cynomolgus monkeys. Transplantation 1995;59:256−62.

[14] Monaco AP, Medawar P. Chimerism in organ transplantation: conflicting experiments and clinical observations. Transplantation 2003;75:13S−6S.

[15] Sachs DH, Sykes M, Kawai T, Cosimi AB. Immuno-intervention for the induction of transplantation tolerance through mixed chimerism. Semin Immunol 2011;23:165−73.

[16] Spitzer T, Delmonico F, Tolkoff-Rubin N, McAfee S, Sackstein R, Saidman S, et al. Combined histocompatibility leukocyte antigen-matched donor bone marrow and renal transplantation for multiple myeloma with end stage renal disease: the induction of allograft tolerance through mixed lymphohematopoietic chimersim. Transplantation 1999;68:480−4.

[17] Buhler L, Spitzer T, Sykes M, Sachs D, Delmonico F, Tolkoff-Rubin N, et al. Induction of kidney allograft tolerance after transient lymphohematopoietic chimerism in patients with multiple myeloma and end-stage renal disease. Transplantation 2002;74:1405−9.

[18] Fudaba Y, Spitzer TR, Shaffer J, Kawai T, Fehr T, Delmonico F, et al. Myeloma responses and tolerance following combined kidney and nonmyeloablative marrow transplantation: *in vivo* and *in vitro* analyses. Am J Transplant 2006;6:2121−33.

[19] Kawai T, Cosimi AB, Spitzer TR, Tolkoff-Rubin N, Suthanthiran M, Saidman SL, et al. HLA-mismatched renal transplantation without maintenance immunosuppression. N Engl J Med 2008;358:353−61.

[20] Friedenstein AJ, Gorskaja JF, Kulagina NN. Fibroblast precursors in normal and irradiated mouse hematopoietic organs. Exp Hematol 1976;4:267−74.

[21] Dominici M, Le Blanc K, Mueller I, Slaper-Cortenbach I, Marini FC, Krause DS, et al. Minimal criteria for defining multipotent mesenchymal stromal cells. The international society for cellular therapy position statement. Cytotherapy 2006;8:315−7.

[22] Pittenger MF, Mackay AM, Beck SC, Jaiswal RK, Douglas R, Mosca JD, et al. Multilineage potential of adult human mesenchymal stem cells. Science 1999;284:143−7.

[23] Mahmood A, Lu D, Lu M, Chopp M. Treatment of traumatic brain injury in adult rats with intravenous administration of human bone marrow stromal cells. Neurosurgery 2003;53:697−702 [discussion 702−3].

[24] Murphy JM, Fink DJ, Hunziker EB, Barry FP. Stem cell therapy in a caprine model of osteoarthritis. Arthritis Rheum 2003;48:3464−74.

[25] Caplan AI, Dennis JE. Mesenchymal stem cells as trophic mediators. J Cell Biochem 2006;98:1076−84.

[26] Prockop DJ. Repair of tissues by adult stem/progenitor cells (MSCs): controversies, myths, and changing paradigms. Mol Ther 2009;17:939−46.

[27] Bartholomew A, Sturgeon C, Siatskas M, Ferrer K, McIntosh K, Patil S, et al. Mesenchymal stem cells suppress lymphocyte proliferation *in vitro* and prolong skin graft survival *in vivo*. Exp Hematol 2002;30:42−8.

[28] Di Nicola M, Carlo-Stella C, Magni M, Milanesi M, Longoni PD, Matteucci P, et al. Human bone marrow stromal cells suppress

T-lymphocyte proliferation induced by cellular or nonspecific mitogenic stimuli. Blood 2002;99:3838–43.

[29] Akiyama K, Chen C, Wang D, Xu X, Qu C, Yamaza T, et al. Mesenchymal-stem-cell-induced immunoregulation involves FAS-ligand-/FAS-mediated T cell apoptosis. Cell Stem Cell 2012;10:544–55.

[30] Ge W, Jiang J, Arp J, Liu W, Garcia B, Wang H. Regulatory T-cell generation and kidney allograft tolerance induced by mesenchymal stem cells associated with indoleamine 2,3-dioxygenase expression. Transplantation 2010;90:1312–20.

[31] Kong QF, Sun B, Bai SS, Zhai DX, Wang GY, Liu YM, et al. Administration of bone marrow stromal cells ameliorates experimental autoimmune myasthenia gravis by altering the balance of Th1/Th2/Th17/Treg cell subsets through the secretion of TGF-beta. J Neuroimmunol 2009;207:83–91.

[32] Comoli P, Ginevri F, Maccario R, Avanzini MA, Marconi M, Groff A, et al. Human mesenchymal stem cells inhibit antibody production induced in vitro by allostimulation. Nephrol Dialysis Transplant 2008;23:1196–202.

[33] Choi H, Lee RH, Bazhanov N, Oh JY, Prockop DJ. Anti-inflammatory protein TSG-6 secreted by activated MSCs attenuates zymosan-induced mouse peritonitis by decreasing TLR2/NF-kappaB signaling in resident macrophages. Blood 2011;118:330–8.

[34] Cutler AJ, Limbani V, Girdlestone J, Navarrete CV. Umbilical cord-derived mesenchymal stromal cells modulate monocyte function to suppress T cell proliferation. J Immunol 2010;185:6617–23.

[35] English K, Barry FP, Mahon BP. Murine mesenchymal stem cells suppress dendritic cell migration, maturation and antigen presentation. Immunol Lett 2008;115:50–8.

[36] Kim J, Hematti P. Mesenchymal stem cell-educated macrophages: a novel type of alternatively activated macrophages. Exp Hematol 2009;37:1445–53.

[37] Nemeth K, Leelahavanichkul A, Yuen PST, Mayer B, Parmelee A, Doi K, et al. Bone marrow stromal cells attenuate sepsis via prostaglandin E-2-dependent reprogramming of host macrophages to increase their interleukin-10 production. Nat Med 2009;15:42–9.

[38] Spaggiari GM, Capobianco A, Becchetti S, Mingari MC, Moretta L. Mesenchymal stem cell-natural killer cell interactions: evidence that activated NK cells are capable of killing MSCs, whereas MSCs can inhibit IL-2-induced NK-cell proliferation. Blood 2006;107:1484–90.

[39] Kavanagh H, Mahon BP. Allogeneic mesenchymal stem cells prevent allergic airway inflammation by inducing murine regulatory T cells. Allergy 2011;66:523–31.

[40] Polchert D, Sobinsky J, Douglas G, Kidd M, Moadsiri A, Reina E, et al. IFN-gamma activation of mesenchymal stem cells for treatment and prevention of graft versus host disease. Eur J Immunol 2008;38:1745–55.

[41] Ren GW, Zhang LY, Zhao X, Xu GW, Zhang YY, Roberts AI, et al. Mesenchymal stem cell-mediated immunosuppression occurs via concerted action of chemokines and nitric oxide. Cell Stem Cell 2008;2:141–50.

[42] Liu H, Liu S, Li Y, Wang X, Xue W, Ge G, et al. The role of SDF-1-CXCR4/CXCR7 axis in the therapeutic effects of hypoxia-preconditioned mesenchymal stem cells for renal ischemia/reperfusion injury. PLoS One 2012;7:e34608.

[43] Casiraghi F, Azzollini N, Cassis P, Imberti B, Morigi M, Cugini D, et al. Pretransplant infusion of mesenchymal stem cells prolongs the survival of a semiallogeneic heart transplant through the generation of regulatory T cells. J Immunol 2008;181:3933–46.

[44] Ding YC, Xu DM, Feng G, Bushell A, Muschel RJ, Wood KJ. Mesenchymal stem cells prevent the rejection of fully allogenic Islet grafts by the immunosuppressive activity of matrix metalloproteinase-2 and-9. Diabetes 2009;58:1797–806.

[45] Casiraghi F, Azzollini N, Todeschini M, Cavinato RA, Cassis P, Solini S, et al. Localization of mesenchymal stromal cells dictates their immune or proinflammatory effects in kidney transplantation. Am J Transplant 2012;12:2373–83.

[46] Eltzschig HK, Eckle T. Ischemia and reperfusion--from mechanism to translation. Nat Med 2011;17:1391–401.

[47] Ogura Y, Sutterwala FS, Flavell RA. The inflammasome: first line of the immune response to cell stress. Cell 2006;126:659–62.

[48] Opitz CA, Litzenburger UM, Lutz C, Lanz TV, Tritschler I, Koppel A, et al. Toll-like receptor engagement enhances the immunosuppressive properties of human bone marrow-derived mesenchymal stem cells by inducing indoleamine-2,3-dioxygenase-1 via interferon-beta and protein kinase R. Stem Cell 2009;27:909–19.

[49] Pevsner-Fischer M, Morad V, Cohen-Sfady M, Rousso-Noori L, Zanin-Zhorov A, Cohen S, et al. Toll-like receptors and their ligands control mesenchymal stem cell functions. Blood 2007;109:1422–32.

[50] Romieu-Mourez R, Francois M, Boivin MN, Bouchentouf M, Spaner DE, Galipeau J. Cytokine modulation of TLR expression and activation in mesenchymal stromal cells leads to a proinflammatory phenotype. J Immunol 2009;182:7963–73.

[51] Tomchuck SL, Zwezdaryk KJ, Coffelt SB, Waterman RS, Danka ES, Scandurro AB. Toll-like receptors on human mesenchymal stem cells drive their migration and immunomodulating responses. Stem Cell 2008;26:99–107.

[52] Kim HS, Shin TH, Yang SR, Seo MS, Kim DJ, Kang SK, et al. Implication of NOD1 and NOD2 for the differentiation of multipotent mesenchymal stem cells derived from human umbilical cord blood. PLoS One 2010;5:e15369.

[53] Sioud M, Mobergslien A, Boudabous A, Floisand Y. Evidence for the involvement of galectin-3 in mesenchymal stem cell suppression of allogeneic T-cell proliferation. Scand J Immunol 2010;71:267–74.

[54] Kume S, Kato S, Yamagishi S, Inagaki Y, Ueda S, Arima N, et al. Advanced glycation end-products attenuate human mesenchymal stem cells and prevent cognate differentiation into adipose tissue, cartilage, and bone. J Bone Miner Res 2005;20:1647–58.

[55] Lotfi R, Eisenbacher J, Solgi G, Fuchs K, Yildiz T, Nienhaus C, et al. Human mesenchymal stem cells respond to native but not oxidized damage associated molecular pattern molecules from necrotic (tumor) material. Eur J Immunol 2011;41:2021–8.

[56] Waterman RS, Tomchuck SL, Henkle SL, Betancourt AM. A new mesenchymal stem cell (MSC) paradigm: polarization into a pro-inflammatory MSC1 or an Immunosuppressive MSC2 phenotype. PLoS One 2010;5:e10088.

[57] Liotta F, Angeli R, Cosmi L, Fili L, Manuelli C, Frosali F, et al. Toll-like receptors 3 and 4 are expressed by human bone marrow-derived mesenchymal stem cells and can inhibit their T-cell modulatory activity by impairing Notch signaling. Stem Cell 2008;26:279–89.

[58] Qiu Y, Marquez-Curtis LA, Janowska-Wieczorek A. Mesenchymal stromal cells derived from umbilical cord blood migrate in response to complement C1q. Cytotherapy 2012;14:285–95.

[59] Schraufstatter IU, Discipio RG, Zhao M, Khaldoyanidi SK. C3a and C5a are chemotactic factors for human mesenchymal stem cells, which cause prolonged ERK1/2 phosphorylation. J Immunol 2009;182:3827–36.

[60] Moll G, Jitschin R, von Bahr L, Rasmusson-Duprez I, Sundberg B, Lonnies L, et al. Mesenchymal stromal cells engage complement and complement receptor bearing innate effector cells to modulate immune responses. PLoS One 2011;6:e21703.

[61] Tu Z, Li Q, Bu H, Lin F. Mesenchymal stem cells inhibit complement activation by secreting factor H. Stem Cell Dev 2010;19:1803–9.

[62] Kemp K, Hares K, Mallam E, Heesom KJ, Scolding N, Wilkins A. Mesenchymal stem cell-secreted superoxide dismutase promotes cerebellar neuronal survival. J Neurochem 2010;114:1569–80.

[63] Mougiakakos D, Jitschin R, Johansson CC, Okita R, Kiessling R, Le Blanc K. The impact of inflammatory licensing on heme oxygenase-1-mediated induction of regulatory T cells by human mesenchymal stem cells. Blood 2011;117:4826–35.

[64] Chen S, Chen L, Wu X, Lin J, Fang J, Chen X, et al. Ischemia postconditioning and mesenchymal stem cells engraftment synergistically attenuate ischemia reperfusion-induced lung injury in rats. J Surg Res 2012;178:81–91.

[65] Chen YT, Sun CK, Lin YC, Chang LT, Chen YL, Tsai TH, et al. Adipose-derived mesenchymal stem cell protects kidneys against ischemia-reperfusion injury through suppressing oxidative stress and inflammatory reaction. J Transl Med 2011;9:51.

[66] Du T, Cheng J, Zhong L, Zhao XF, Zhu J, Zhu YJ, et al. The alleviation of acute and chronic kidney injury by human Wharton's jelly-derived mesenchymal stromal cells triggered by ischemia-reperfusion injury via an endocrine mechanism. Cytotherapy 2012;14:1215–27.

[67] Sun CK, Yen CH, Lin YC, Tsai TH, Chang LT, Kao YH, et al. Autologous transplantation of adipose-derived mesenchymal stem cells markedly reduced acute ischemia-reperfusion lung injury in a rodent model. J Transl Med 2011;9:118.

[68] Hara Y, Stolk M, Ringe J, Dehne T, Ladhoff J, Kotsch K, et al. *In vivo* effect of bone marrow-derived mesenchymal stem cells in a rat kidney transplantation model with prolonged cold ischemia. Transpl Int 2011;24:1112–23.

[69] Franquesa M, Herrero E, Torras J, Ripoll E, Flaquer M, Goma M, et al. Mesenchymal stem cell therapy prevents interstitial fibrosis and tubular atrophy in a rat kidney allograft model. Stem Cell Dev 2012;21:3125.

[70] Perico N, Casiraghi F, Introna M, Gotti E, Todeschini M, Cavinato RA, et al. Autologous mesenchymal stromal cells and kidney transplantation: a pilot study of safety and clinical feasibility. Clin J Am Soc Nephrol 2011;6:412–22.

[71] Sotiropoulou PA, Perez SA, Gritzapis AD, Baxevanis CN, Papamichail M. Interactions between human mesenchymal stem cells and natural killer cells. Stem Cell 2006;24:74–85.

[72] Wyburn KR, Jose MD, Wu H, Atkins RC, Chadban SJ. The role of macrophages in allograft rejection. Transplantation 2005;80:1641–7.

[73] Francois M, Romieu-Mourez R, Li M, Galipeau J. Human MSC suppression correlates with cytokine induction of indoleamine 2,3-dioxygenase and bystander M2 macrophage differentiation. Mol Ther 2012;20:187–95.

[74] Maggini J, Mirkin G, Bognanni I, Holmberg J, Piazzon IM, Nepomnaschy I, et al. Mouse bone marrow-derived mesenchymal stromal cells turn activated macrophages into a regulatory-like profile. PLoS One 2010;5:e9252.

[75] Djouad F, Charbonnier LM, Bouffi C, Louis-Plence P, Bony C, Apparailly F, et al. Mesenchymal stem cells inhibit the differentiation of dendritic cells through an interleukin-6-dependent mechanism. Stem Cell 2007;25:2025–32.

[76] Li YP, Paczesny S, Lauret E, Poirault S, Bordigoni P, Mekhloufi F, et al. Human mesenchymal stem cells license adult CD34(+) hemopoietic progenitor cells to differentiate into regulatory dendritic cells through activation of the notch pathway. J Immunol 2008;180:1598–608.

[77] Nauta AJ, Kruisselbrink AB, Lurvink E, Willemze R, Fibbe WE. Mesenchymal stem cells inhibit generation and function of both CD34$^+$-derived and monocyte-derived dendritic cells. J Immunol 2006;177:2080–7.

[78] Zhang B, Liu R, Shi D, Liu XX, Chen Y, Dou XW, et al. Mesenchymal stem cells induce mature dendritic cells into a novel Jagged-2-dependent regulatory dendritic cell population. Blood 2009;113:46–57.

[79] Chiesa S, Morbelli S, Morando S, Massollo M, Marini C, Bertoni A, et al. Mesenchymal stem cells impair *in vivo* T-cell priming by dendritic cells. Proc Natl Acad Sci USA 2011;108:17384–9.

[80] Ge W, Jiang J, Baroja ML, Arp J, Zassoko R, Liu W, et al. Infusion of mesenchymal stem cells and rapamycin synergize to attenuate alloimmune responses and promote cardiac allograft tolerance. Am J Transplant 2009;9:1760–72.

[81] Li H, Guo ZK, Jiang XX, Zhu H, Li XS, Mao N. Mesenchymal stem cells alter migratory property of t and dendritic cells to delay the development of murine lethal acute graft-versus-host disease. Stem Cell 2008;26:2531–41.

[82] Liu X, Qu X, Chen Y, Liao L, Cheng K, Shao C, et al. Mesenchymal stem/stromal cells induce the generation of novel IL-10-dependent regulatory dendritic cells by SOCS3 activation. J Immunol 2012;189:1182–92.

[83] Spaggiari GM, Abdelrazik H, Becchetti F, Moretta L. MSCs inhibit monocyte-derived DC maturation and function by selectively interfering with the generation of immature DCs: central role of MSC-derived prostaglandin E-2. Blood 2009;113:6576–83.

[84] Huang Y, Chen P, Zhang CB, Ko GJ, Ruiz M, Fiorina P, et al. Kidney-derived mesenchymal stromal cells modulate dendritic cell function to suppress alloimmune responses and delay allograft rejection. Transplantation 2010;90:1307–11.

[85] Issa F, Chandrasekharan D, Wood KJ. Regulatory T cells as modulators of chronic allograft dysfunction. Curr Opin Immunol 2011;23:648–54.

[86] Asari S, Itakura S, Ferreri K, Liu CP, Kuroda Y, Kandeel F, et al. Mesenchymal stem cells suppress B-cell terminal differentiation. Exp Hematol 2009;37:604–15.

[87] English K, Barry FP, Field-Corbett CP, Mahon BP. IFN-gamma and TNF-alpha differentially regulate immunomodulation by murine mesenchymal stem cells. Immunol Lett 2007;110:91–100.

[88] English K, Mahon BP. Allogeneic mesenchymal stem cells: agents of immune modulation. J Cell Biochem 2011;112:1963–8.

[89] Glennie S, Soeiro I, Dyson PJ, Lam EW, Dazzi F. Bone marrow mesenchymal stem cells induce division arrest anergy of activated T cells. Blood 2005;105:2821–7.

[90] Duffy MM, Pindjakova J, Hanley SA, McCarthy C, Weidhofer GA, Sweeney EM, et al. Mesenchymal stem cell inhibition of T-helper 17 cell-differentiation is triggered by cell–cell contact and mediated by prostaglandin E2 via the EP4 receptor. Eur J Immunol 2011;41:2840–51.

[91] Tatara R, Ozaki K, Kikuchi Y, Hatanaka K, Oh I, Meguro A, et al. Mesenchymal stromal cells inhibit Th17 but not regulatory T-cell differentiation. Cytotherapy 2011;13:686–94.

[92] Bai L, Lennon DP, Eaton V, Maier K, Caplan AI, Miller SD, et al. Human bone marrow-derived mesenchymal stem cells induce Th2-polarized immune response and promote endogenous repair in animal models of multiple sclerosis. Glia 2009;57:1192–203.

[93] Batten P, Sarathchandra P, Antoniw JW, Tay SS, Lowdell MW, Taylor PM, et al. Human mesenchymal stem cells induce T cell anergy and downregulate T cell allo-responses via the TH2 pathway: relevance to tissue engineering human heart valves. Tissue Eng 2006;12:2263–73.

[94] English K, Ryan JM, Tobin L, Murphy MJ, Barry FP, Mahon BP. Cell contact, prostaglandin E-2 and transforming growth factor beta 1 play non-redundant roles in human mesenchymal stem cell induction of CD4(+)CD25(High)forkhead box P3(+) regulatory T cells. Clin Exp Immunol 2009;156:149–60.

[95] Fiorina P, Jurewicz M, Augello A, Vergani A, Dada S, La Rosa S, et al. Immunomodulatory function of bone marrow-derived mesenchymal stem cells in experimental autoimmune type 1 diabetes. J Immunol 2009;183:993–1004.

[96] Ghannam S, Pene J, Torcy-Moquet G, Jorgensen C, Yssel H. Mesenchymal stem cells inhibit human Th17 cell differentiation and function and induce a T regulatory cell phenotype. J Immunol 2010;185:302–12.

[97] Rafei M, Campeau PA, Aguilar-Mahecha A, Buchanan M, Williams P, Birman E, et al. Mesenchymal stromal cells ameliorate experimental autoimmune encephalomyelitis by inhibiting CD4 Th17 T Cells in a CC chemokine ligand 2-dependent manner. J Immunol 2009;182:5994–6002.

[98] Jia Z, Jiao C, Zhao S, Li X, Ren X, Zhang L, et al. Immunomodulatory effects of mesenchymal stem cells in a rat corneal allograft rejection model. Exp Eye Res 2012;102:44–9.

[99] Wang Y, Zhang A, Ye Z, Xie H, Zheng S. Bone marrow-derived mesenchymal stem cells inhibit acute rejection of rat liver allografts in association with regulatory T-cell expansion. Transplant Proc 2009;41:4352–6.

[100] Nemeth K, Keane-Myers A, Brown JM, Metcalfe DD, Gorham JD, Bundoc VG, et al. Bone marrow stromal cells use TGF-beta to suppress allergic responses in a mouse model of ragweed-induced asthma. Proc Natl Acad Sci USA 2010;107:5652–7.

[101] Zhao W, Wang Y, Wang D, Sun B, Wang G, Wang J, et al. TGF-beta expression by allogeneic bone marrow stromal cells ameliorates diabetes in NOD mice through modulating the distribution of CD4$^+$ T-cell subsets. Cell Immunol 2008;253:23–30.

[102] Augello A, Tasso R, Negrini SM, Amateis A, Indiveri F, Cancedda R, et al. Bone marrow mesenchymal progenitor cells inhibit lymphocyte proliferation by activation of the programmed death 1 pathway. Eur J Immunol 2005;35:1482–90.

[103] Corcione A, Benvenuto F, Ferretti E, Giunti D, Cappiello V, Cazzanti F, et al. Human mesenchymal stem cells modulate B-cell functions. Blood 2006;107:367–72.

[104] Schena F, Gambini C, Gregorio A, Mosconi M, Reverberi D, Gattorno M, et al. Interferon-gamma-dependent inhibition of B cell activation by bone marrow-derived mesenchymal stem cells in a murine model of systemic lupus erythematosus. Arthritis Rheum 2010;62:2776–86.

[105] Rafei M, Hsieh J, Fortier S, Li MY, Yuan S, Birman E, et al. Mesenchymal stromal cell-derived CCL2 suppresses plasma cell immunoglobulin production via STAT3 inactivation and PAX5 induction. Blood 2008;112:4991–8.

[106] Zhou K, Zhang H, Jin O, Feng X, Yao G, Hou Y, et al. Transplantation of human bone marrow mesenchymal stem cell ameliorates the autoimmune pathogenesis in MRL/lpr mice. Cell Mol Immunol 2008;5:417–24.

[107] Youd M, Blickarz C, Woodworth L, Touzjian T, Edling A, Tedstone J, et al. Allogeneic mesenchymal stem cells do not protect NZB × NZW F1 mice from developing lupus disease. Clin Exp Immunol 2010;161:176–86.

[108] Seifert M, Stolk M, Polenz D, Volk HD. Detrimental effects of rat mesenchymal stromal cell pre-treatment in a model of acute kidney rejection. Front Immunol 2012;3:202.

[109] LeBlanc K, Frassoni F, Ball L, Locatelli F, Roelofs H, Lewis I, et al. Mesenchymal stem cells for treatment of steroid-resistant, severe, acute graft-versus-host disease: a phase II study. Lancet 2008;371:1579–86.

[110] Hoogduijn MJ, Popp FC, Grohnert A, Crop MJ, van Rhijn M, Rowshani AT, et al. Advancement of mesenchymal stem cell therapy in solid organ transplantation (MISOT). Transplantation 2010;90:124–6.

[111] Eggenhofer E, Renner P, Soeder Y, Popp FC, Hoogduijn MJ, Geissler EK, et al. Features of synergism between mesenchymal stem cells and immunosuppressive drugs in a murine heart transplantation model. Transpl Immunol 2011;25:141–7.

[112] Eggenhofer E, Steinmann JF, Renner P, Slowik P, Piso P, Geissler EK, et al. Mesenchymal stem cells together with mycophenolate mofetil inhibit antigen presenting cell and T cell infiltration into allogeneic heart grafts. Transpl Immunol 2011;24:157–63.

[113] Inoue S, Popp FC, Koehl GE, Piso P, Schlitt HJ, Geissler EK, et al. Immunomodulatory effects of mesenchymal stem cells in a rat organ transplant model. Transplantation 2006;81:1589–95.

[114] Popp FC, Eggenhofer E, Renner P, Slowik P, Lang SA, Kaspar H, et al. Mesenchymal stem cells can induce long-term acceptance of solid organ allografts in synergy with low-dose mycophenolate. Transpl Immunol 2008;20:55–60.

[115] Renner P, Eggenhofer E, Rosenauer A, Popp FC, Steinmann JF, Slowik P, et al. Mesenchymal stem cells require a sufficient, ongoing immune response to exert their immunosuppressive function. Transplant Proc 2009;41:2607–11.

[116] Tan J, Wu W, Xu X, Liao L, Zheng F, Messinger S, et al. Induction therapy with autologous mesenchymal stem cells in living-related kidney transplants: a randomized controlled trial. JAMA 2012;307:1169–77.

Immunosuppression-Free Renal Transplantation

Fred Fändrich

Department of Applied Cellular Medicine, University Schleswig-Holstein, Kiel, Germany

Chapter Outline

62.1 GRAFT TOLERANCE: IMMUNOLOGICAL BACKGROUND AND DEFINITION

Graft tolerance has been—and still is—considered the ultimate goal for patients undergoing transplantation for end-stage organ diseases. There is now accumulating evidence that tolerance can be observed in a clinical setup as well. For instance, organ transplantation between identical twins does not need immunosuppressive treatment since the immunologic response of the recipient will recognize donor-derived major histocompatibility complex (MHC)-restricted antigens as "self." This was first described by Dr. Barrett Brown who achieved permanent survival of skin grafts when these were transplanted between identical twins [1]. As such, genetic disparity between organ donor and recipient plays a major role in initiating a cell-mediated immune response. But even allogeneic renal allografts successfully acquired immunological tolerance if recipients had been engrafted with bone marrow transplants from the same donor some time before [2]. At this point, it is important to make the following distinction between immunological and transplantation tolerance as we miss an agreed on definition of the term "tolerance." More or less a stringent definition of immunological tolerance can be formulated for experimental transplant models, which comprise four central components: (i) prolonged allograft survival in absence of immunosuppressive medication, (ii) normal organ function in absence of measurable antidonor immune responses, (iii) normal histologic appearance of the graft, and (iv) acceptance of a second-set transplanted graft (e.g., skin) but rejection of a third-party graft [3]. For obvious ethical reasons, these criteria cannot be applied to humans following transplantation. From an immunologist's point of view, tolerance is considered as a state of immune unresponsiveness specific to a particular antigen or set of antigens after previous exposure to that antigen (or set of antigens) whereas transplant tolerance describes a state of donor-specific unresponsiveness without the need for ongoing treatment with immunosuppressive compounds. This distinction is necessary as the scope of *in vitro* assays which attempt to assess antidonor immune responses are neither predictive to determine the future development of tolerance nor do they correlate with maintenance of tolerance. The clinical variables to assess graft integrity are insensitive and often fail to indicate graft damage. Significant injury is frequently evident long before biochemical or physiological markers of function are observed. Cellular infiltrates from graft biopsies do not—under routine circumstances—distinguish between harmful cytotoxic effector cells and protective immune cells, which are deactivating and regulative in

Regenerative Medicine Applications in Organ Transplantation.

character. Moreover, the underlying microenvironment whether inflammatory and promoting immune responses or noninflammatory is not tested and evaluated for clinical decision making. A further confounding variable in this scenario is nonimmunological parameters of graft deterioration, for instance hypertension, metabolic syndrome, infections, and disease recurrence. Subsequently, the categorization of an individual as "tolerant" is limited to the observation of stable graft function in the absence of immunosuppression. A practical approach to define graft acceptance, now termed "clinical operational tolerance" (COT), relates to organ recipients being completely off all immunosuppressives [IS] for at least 1 year and without histological signs of rejection injury [4]. Moreover, a normal immune response to immune stimuli, such as infections and tumors (third-party antigens) can be expected from an operationally tolerant patient.

In order to develop potential tolerance promoting strategies it is important to understand what causes rejection of vascularized organs in an allogeneic setup. Under normal circumstances, transplantation of a vascularized graft or a tissue will raise an unspecific activation of innate immune effector cells, e.g., macrophages, neutrophils, mast cells, and eosinophils, which solicit secondary cell-mediated immune responses, including cytotoxic and T-helper lymphocytes. In this scenario, tissue injury due to hypoxia and cold and warm ischemic events sets free soluble inflammatory cytokines, which stimulate and recruit effector cells of innate immunity (macrophages, NK cells, etc.) to clear up damaged tissue components. Secondary to this unspecific immune response, lymphocytes are chemotactically recruited to the site of organ/tissue damage where they have direct access to upregulated donor antigens, directly presented on the graft, to elicit cytotoxic immune responses defined as the direct way of antigen recognition. Alternatively, opsonized donor antigens are processed and subsequently presented as peptides within the MHC of recipient antigen-presenting cells (APCs), so-called indirect antigen presentation. Whereas direct antigen recognition primarily activates cytotoxic T cells, indirect antigen presentation triggers T-helper cells which give rise to both T- and B-cell-mediated cellular and humoral rejection responses, respectively. Conversely, unspecific immune-stimulating triggers, such as organ retrieval, cold- and warm ischemic time, or organ reperfusion, do not cause immune rejection on their own but can only enhance and accelerate specific immune responses in the context of graft antigens.

Without doubt, during the last two decades a remarkable improvement for short-term results following organ transplantation has been achieved whereas long-term outcome is still less satisfactory. In 2009, more than 20,000 organ transplants were performed in the United States alone to treat patients suffering from kidney, liver, heart, lung, and intestine organ dysfunction or failure. At current state, more than 170,000 organ recipients live with donated grafts, over 80,000 patients are listed for organ transplantation and waiting for those [3]. Although transplantation medicine has made great progress in terms of technical skills, we are well aware that lifelong suppression of global T- and B-cell immunity has extensive side effects and requires a high level of patient compliance. It must be stated as a matter of fact that long-term immunosuppressive treatment shortens life expectancy, limits survival of the graft, and impairs quality of life of patients. As such, there is a strong need to achieve a state of graft acceptance which will allow no or individually tailored immunosuppressive medication.

62.2 LESSONS LEARNED FROM IMMUNOLOGY

T lymphocytes are central to cause organ rejections. T cells mature in the thymus and central tolerance describes a chief mechanism of deletion of autoreactive T cells in the thymus which—following positive and negative selection of specific T-cell subsets—creates a milieu of T lymphocyte self-tolerance to an individual's peripheral antigen repertoire. However, many potentially reactive T cells escape intrathymic deletion as many antigens encountered in the peripheral body are absent intrathymically. Those potentially reactive T cells which escape intrathymic deletion are regulated in peripheral lymphatic tissues to prevent autoimmunity. Other nondeletional defined immunologic mechanisms which prevent T-cell activation distinguish anergy, immune deviation (ignorance), exhaustion, and suppression of T lymphocytes [5].

Nevertheless, our understanding of how immune regulation is in fact maintained and to which extent cellular, soluble, receptor/ligand engagement, tissue-specific, and other physiologic features (physical barriers, vasculature, tight junctions, immunologic barrier) play a role in sustaining immune homeostasis is still not completely understood. In 1994, Matzinger was the first to postulate a more general conceptual framework of "danger signals" which suggested that the immune system of vertebrates responds to substances that cause damage, rather than to those that are simply "foreign" (nonself antigens) [6]. She claimed that a broad scope of intrinsic and extrinsic substrates or molecular structures may activate APCs to express costimulatory molecules which in turn initiate immune responses. As such, Matzinger took a more general view on immune regulatory networks, which do not narrow our perspective on mere genetically discrepant mismatches but included unspecific immunologic "danger events," such as tissue inflammation, as well. However, this "danger-signal" concept still is not able to explain the full complexity of immunologic response patterns, as I briefly want to outline.

Let me illustrate the complexity of immune reactions with a simple observation, which can be made in a model of rat liver transplantation. Grafting of LEW inbred rat strain liver organs expressing the haplotype RT1.[l] to DA inbred rats will result in long-term graft acceptance in most circumstances, whereas the reversed direction, namely DA [RT1.[av1]] donor liver to LEW recipient triggers an acute rejection response and subsequent liver failure within 7 days following transplantation. This experiment clearly contradicts Matzinger's danger/nondanger paradigm [7], as there is an identical constellation of danger signals and an identical mismatch of alloantigens in both experimental setups. Nevertheless, the clinical outcome represents both arms of possible immune responses, tolerance versus rejection. To contribute even a higher degree of complexity to this immunological paradigm, let me emphasize that this paradoxical observation is not a mere liver-related organ specific phenomenon. Liver allografts in pigs and rodents are uniquely capable of acquiring a state of tolerance to themselves as well as to other grafts and tissue components [7]. Microchimerism, the coexistence of donor- and recipient-derived immune cells in the same host has been postulated as a necessary prerequisite for tolerance development in this scenario [8]. Following liver transplantation, donor leukocytes emigrate from the graft and disseminate extrahepatic tissues. This phenomenon has also been reported for successful outcome in long-term accepted renal transplants [9].

Beyond these considerations, immunologic memory poses another significant barrier to achieving tolerance. As outlined by Billingham and Medawar, neonatal tolerance can be overcome by transfer of immunologic memory [10]. This observation can be confirmed as memory T cells demonstrate a lower level of activation threshold, possess less stringent hurdles for activation of T-cell proliferation via costimulatory signals, do not need APCs for antigen presentation, differ in migration pattern, and are less susceptible to T-cell depletion following treatment with compounds, such as alemtuzumab or thymoglobulin [3,5,11–13]. In animal models, adoptive transfer of memory T cells is able to break tolerance induced with anti-DC40L antibodies [13]. Moreover, the specificity of those memory T cells does not need to be donor-antigen-related as memory T cells resulting from exposure to bacterial or viral pathogens can prevent tolerance development via cross-reactive immune responses [11,12]. In humans, the pool of memory T cells is much more expanded than in models of inbred rats, as exposure to environmental and infectious pathogens is substantially stronger. The putative size of T-cell repertoire capable to respond to alloantigens has been calculated to encompass 10^{12} T cells of which 10^{10} bear specificity for alloantigen recognition. Based on these numbers, it can be assumed that 5–10% of T lymphocytes are endowed with receptors to elicit an alloorgan response. Experimental and clinical approaches to achieve a state of operational tolerance aimed to deplete the large number of alloreactive T cells. The underlying rational assumes that reemerging T-cell clones will accommodate donor-specific antigens and unresponsiveness to the graft will ensue. However, widespread proliferation of residual T lymphocytes—especially with memory function—will follow this form of medically induced lymphopenia, a process termed "homeostatic proliferation." This reconstitution of lymphocyte clones with functional properties of T memory cells has been reported to block tolerance induction in both experimental and human transplantation settings [13,14]. A further confounding variable which impacts tolerance development is age. There is solid evidence that age-related decline in immunity weakens cell-mediated host responses to infections, malignancy, and in acute allograft rejection. Whereas costimulatory blockade treatment extended survival of allogeneic skin grafts in young mice, it failed to prolong graft survival of aged recipients [15]. This observation was associated with higher numbers of CD8 memory T cells and stronger IFN-γ release in response to allostimulation measured for aged CD8$^+$ T cells when compared with young naïve T lymphocytes. Enhanced CD8$^+$ T-cell alloreactivity impaired the ability of costimulatory blockade-based tolerance induction in the chosen experimental mouse model. Moreover, transplantation tolerance needs thymic participation. Functional and morphologic involution of the thymus gland in aged animals poses a significant clinical barrier to organ transplantation. Old mice demonstrate a general refractoriness to mount adequate immune responses to novel antigens and become resistant to tolerance to cardiac allografts. Interestingly, this blockade could be overcome by surgical castration or temporary disruption of gonadal function of male mice, a treatment which refurnished thymic regeneration and priming of thymocytes [16].

As such, questions arise as to what extent we are able to predict immune tolerance for a given donor/ recipient pair and which are the strongest parameters in play which determine immunologic response patterns? How do those patterns change with an ongoing phase of immune adaptation or accommodation? Which external and internal events, for instance infection episodes or metabolic inflammatory disturbances are prone to shift a given immunologic equilibrium to one or the other side? At current state, we routinely determine the strength of graft-specific (directed against the transplant) immune reactions from graft biopsies, differential blood laboratory parameters, clinical course, and functional graft status.

62.3 ETHICAL CONSIDERATIONS AND CONCERNS REGARDING TRANSPLANTATION TOLERANCE-INDUCING REGIMENS

These considerations bring up ethical concerns involved in applying tolerance strategies to human transplantation. Opinion leaders in the field of transplantation medicine have voiced their concerns in an editorial report, which addresses tolerance-inducing regimen in renal transplantation [17]. First of all, they fear that publications on successful transplantation without the need of maintenance immunosuppression might encourage patients and physicians to consider drug adherence with lower priority. Second, they ask whether successfully weaned off patients were fully immunocompetent at all. In other words, do "tolerant" patients mount adequate immune responses or do they exhibit a global immune defect and related unresponsiveness to allostimulation. Even this aspect is still a matter of debate, there is some clinical evidence that tolerant patients being off all immunosuppression display comparable IFNγ-producing T cells to healthy volunteers and immunosuppressed graft recipients. However, whereas

immunosuppressed patients have impaired antibody responses to flu antigens following flu vaccinations, tolerant patients show similar responses to healthy individuals [17]. Third, the question arises whether "tolerant" patients demonstrate a better long-term graft outcome relative to contemporary immunosuppressive regimens. So far, there is little clinical evidence to promote this assumption. First publications on this topic report a higher incidence of drug (immunosuppression)-related morbidity in terms of new-onset posttransplant diabetes mellitus, infectious complications, and cardiovascular disease [18] of patients with maintenance immunosuppressives.

Nevertheless, keeping these concerns in mind, considerable progress has been made in the field of COT in recent years.

62.4 CLINICAL TRANSPLANTATION TOLERANCE INDUCTION

COT in renal transplantation must still be considered as a rare event [19,20] (Table 62.1). Immunosuppression-free renal transplantation has been achieved clinically in three different ways. Clinical observations of spontaneous graft

TABLE 62.1 Overview on Observed COT in IS Free Renal Transplantation[a]

Nonadherence to IS	Number of Patients	Type of Study	Documented COT Period	Tolerance Mechanism/Strategy	Percentage of COT Patients
Spontaneous withdrawal Survey	25	Case reports	Few months to 27 years	Unknown/none	24% (6/25)
	124	Descriptive, investigative	Few months to 20 years	Unknown/none	81% (101/124)
Intentional tolerance induction (clinical trials)	58	Prospective, comparative, nonrandomized	Not achieved	Molecule based	0% (58/58)
	6	Case reports	15 months to >7 years	Central tolerance, previous donor-BMT, full donor chimerism	100% (6/6)
	117	Prospective, noncomparartive	28 months to 14 years	Central tolerance, donor-HSC myelosuppressive, transient/stable microchimerism	8% (9/117)
	8	Prospective noncomparative	1 year	Central and peripheral tolerance, stable microchimerism by use of HSC and FSC[b]	63% (5/8)
	1	Prospective, noncomparative	Not achieved	Unknown/ESC transfer	0% (0/1)
	17	Prospective, noncomparative	Few weeks to 9 months, COT not achieved	Peripheral donor M2-polarized macrophages	0% (0/17)

[a]For more detailed information, please see Orlando et al. [20].
[b]Published by Leventhal [21].

acceptance in the absence of all immunosuppression (IS) were first reported by Owens et al. [22] in a series of 203 renal transplant recipients. Among those, six patients had completely dropped their graft protecting medication and demonstrated stable function for periods between 17 and 52 months. However, two patients eventually suffered from acute kidney graft rejection. More so-called sporadic cases, in which patients discontinued IS for reasons of non-adherence and without consulting the advice of a transplant physician are published to date [23]. Conversely, there are reported cases of kidney allograft recipients who, due to severe drug-related or otherwise life-threatening complications, were gradually weaned off immunosuppressive medication under close supervision of their physicians [19]. A third cohort of patients was enrolled into clinical protocols in which COT was the planned objective before kidney transplantation (KTx) was performed, which will be discussed in detail later on.

As mentioned above, there are published cases of COT following KTx which comprise the ideal situation of identical twins [24] or in who the causes are unknown [25]. Moreover, there are a number of publications of patients who accepted their renal allograft free of IS but documentation did not exceed 1-year follow-up [26,27] or other anecdotal reports [22,23], not further addressed here. Without doubt, most cases of COT following KTx relate to patients who spontaneously stopped IS due to nonadherence to their immunosuppressive drugs. One more recently published review examined the medical history of 10 kidney recipients who displayed COT for a period between 8 and 10 years follow-up [23]. Given the subtle and detailed online monitoring of these case reports, a couple of clinically relevant observations can be made. First, COT occurs in the presence of antidonor class II antibodies. Kidney graft failure may happen as late as after 9 and 13 years postwithdrawal, respectively, even in the absence of histologic signs of chronic rejection. Moreover, weaning of IS over a long period (until 4 years posttransplantation) and young donor age may favor the development of operational tolerance. Donor age (graft <30 years of age) is possibly linked with more robust regenerative reserves to facilitate COT. One major clinical observation from these sporadically tolerant patients is the finding that graft dysfunction can develop at any time after transplantation irrespective of evident lesions specific for acute or chronic rejection.

Sporadic development of COT can imply a variety of different clinical and immunologically related and unrelated mechanisms which are difficult to discern. As such, the implementation of clinical protocols which aim at achieving operational tolerance in a prospective way is strongly favorable. While numerous protocols have addressed the induction of tolerance in various animal settings it is frustrating that clinical transplantation is still practiced today without application of these approaches.

Even tolerance strategies have been described in a broad scope of small and large animal experiments, translation of these approaches into clinical application has not proven to warrant induction of operational tolerance, not even for a minor fraction of recipients as only 2.5% of intentionally weaned off patients did not experience kidney rejection or graft failure [20]. As such, the gap between the medical science of transplantation immunology and the practice of clinical medicine becomes more than apparent. Nevertheless, the following section will discern different clinically applied strategies to achieve COT based on the underlying immunological regulating mechanisms. These comprise deletional and regulatory approaches to assimilate immune responses from donor and recipient, respectively.

62.5 CLINICAL TOLERANCE STRATEGIES BASED ON T-CELL DEPLETION

As outlined before, exchange of donor- and recipient-derived passenger leukocytes in the host and graft, respectively, has strong implications to tolerize graft-versus-host and host-versus-graft immune reactions [8]. The mutual engagement and engraftment between those effector cells has been described as microchimerism, defined as <1% donor cells which circulate in the recipient peripheral blood compartment or are detectable from bone marrow aspirates. The mixed chimerism approach looks back on a long history of rodent models which successfully demonstrated coexistence of hematopoietic cells of donor and recipient origin (>1% and <100%) as the basis for graft acceptance. In contrast to full-lineage donor chimerism where all hematopoietic cells are donor derived (100% chimerism), mixed chimerism benefits from less stringent conditioning regimens for its development. Original regimens to induce chimerism included myeloablative doses of irradiation in combination with T-cell depletion and thymic radiation [28,29]. Nowadays, less toxic nonmyeloablative protocols include costimulation blockers, irradiation of the thymic area, and nonmyeloablative conditioning to promote engraftment of donor bone marrow or blood-borne hematopoietic stem cells [30–35]. Further strategies aimed to avoid irradiation and cytoreductive treatment at all [36] and instead combined conventional IS [37–39] with NK-cell depletion [40], nondepleting antibodies [41], and facilitating cells (FCs) [42]. However, none of these approaches were sufficiently potent to develop mixed chimerism if devoid of any myelosuppressive drug. Most recently, a small pilot clinical trial of eight kidney-transplanted patients who received MHC-mismatched living unrelated renal allografts was reported to have achieved operational tolerance in five of the eight recipients [21]. The conditioning regimen applied consisted of a combination of mobilized cells enriched for hematopoietic stem cells, along with so-called "graft-facilitating cells"—a plasmacytoid precursor like dendritic cell (DC)—and simultaneous nonmyeloablative conditioning. The latter consisted of two doses of cyclophosphamide, 200 cGy total body irradiation (TBI), and three doses of preoperative fludarabine. Immunosuppression after transplant comprised the administration of mycophenolate mofetil and tacrolimus. Immunologic monitoring of these patients found a reduction in CD3$^+$ T cells early on but return of CD4$^+$ T lymphocytes throughout the first posttransplant year, whereas CD8$^+$ T-cell counts remained low and inversion of the CD4/CD8 ratio was an evident finding in all patients. Hematopoietic reconstitution was further characterized by recovery of CD19$^+$ B cells, CD56$^+$ NK cells, and monocytes. Stable chimerism was associated with a proportional increase in CD4$^+$/CD25$^+$/CD127$^-$ regulatory T cells (T$_{regs}$), not observed in recipients with transient mixed chimerism. Stable tolerant patients showed donor-specific alloantigen unresponsiveness and adequate third-party immune responses, even if donor chimerism was lost. Five of the eight patients have sustained full donor chimerism without graft-versus-host disease (GvHD) and absence of *in vitro* T-cell proliferation against pretransplant isolated recipient stimulator cells, whereas peripheral blood leukocytes (PBLs) obtained from the living donor vigorously proliferated following stimulation with recipient stimulators. These findings let us assume that donor-derived immune cells underwent a tolerizing mechanism to accept host alloantigens after transplantation. Accordingly, intentional induction of mixed macrochimerism to induce operational tolerance was applied in other clinical pilot studies which also reported to induce donor-specific unresponsiveness by use of hematopoietic stem cell transplantation (HSCT) [43–45]. However, based on a different conditioning regimen, kidney allograft recipients did not develop ongoing macrochimerism (>1% donor cells) beyond day 21 following HSCT. Two of 10 recipients lost their grafts due to rejections, one defined as humoral in nature. Nine patients developed a so-called "engraftment syndrome," namely a capillary leak syndrome with simultaneous antidonor antibody production and elevated creatinine levels [46]. Biopsy proven C4d complement deposition and prevailing antidonor HLA class II antibodies has been associated with an increased risk of chronic allograft nephropathy [43] and impaired long-term survival [47,48]. A further clinical trial included 12 HLA-identical kidney/CD34-selected peripheral blood stem cell transplanted patients, which underwent nonmyeloablative conditioning [48]. Eight of 12 patients could successfully be tapered from IS, only one of these patients with measurable donor macrochimerism. One patient suffered from disease recurrence, a further three patients from rejection crisis which demanded ongoing immunosuppression [48]. No durable donor chimerism was

achieved when the same conditioning regimen was applied to HLA-mismatched recipients [49]. Subsequently, the question arises whether FCs are crucial and obligatory to avoid antibody formation, C4d graft deposition, and GvHD on one side and durable chimerism on the other. As a direct clinical comparison between FC-depleted and FC-repleted or unmodified grafts in humans is missing, this question is open to debate and must be addressed in appropriate prospective randomized trials in the future.

In summary, there is rising evidence that optimized nonmyeloablative recipient conditioning protocols in conjunction with donor-derived stem cell grafting can induce tolerance in HLA-mismatched hosts without concomitant risk to trigger GvHD or humoral immune responses which impair graft outcome in the long run. However, so far all applied conditioning strategies have included nonmyeloablative cytotoxic compounds and some form of irradiation therapy. Most recently, experimental work from Wekerle's group has investigated a myeloablative-free combination of host conditioning. This consisted of polyclonal recipient-type T$_{regs}$, which enabled fully allogeneic bone marrow of donor origin to engraft when administered in combination with costimulatory blockade and rapamycin [50,51].

In this context, additional transfusion of T$_{regs}$ hypothetically might enhance engraftment of transferred hematopoietic stem cells as previously reported [52]. These observations indicate that regulatory mechanisms are important players which contribute to the development of operational tolerance. It can be assumed that there exists a fine-tuned balance between necessary deletional events and regulatory pathways which both solicit mutual acceptance of donor- and recipient-related immune response during the process of tolerance induction.

62.6 CLINICAL TOLERANCE STRATEGIES BASED ON REGULATORY MECHANISMS

Hematopoietic cells of several different lineages display immune regulatory properties, which have been tested positive to promote tolerance to auto- and alloantigens. Ever since the successful outcome of Medawar's experiments which in part can be explained by parabiosis, the linking together of the circulatory system of two genetically different individuals, as described by Owen [53], it became apparently clear that whole blood cellular fractions mediate tolerant induction in one or the other way. Clinical evidence for a tolerogenic property of intentionally administered donor-specific blood transfusions (DSTs) was first shown by Newton and Anderson [54] in the context of living-related renal transplantation. Burlingham et al. [25] reported a rare case of a patient who had successfully withdrawn himself from all immunosuppression at 1 year following

DST and KTx from his mother. At this point, however, it is not yet clear which cell type among a broad arsenal of different tolerogenic cell entities and underlying tolerance-inducing strategies is best with regard to safety, efficacy, and related costs.

T_{regs} are known to play an unequivocal role in modulating the immune response. Selective depletion of T_{regs} evokes severe autoimmune disturbances in neonates [28] and can lead to lethal autoimmunity in adults [55]. In 1990, Hall et al. reported successful tolerance induction in an MHC-mismatched PVG to DA strain combination if T cells from tolerant animals (100 days posttransplant) were harvested and tested in adoptive transfer models [56]. The phenomenon that T cells from a tolerant animal are able to specifically disable naïve T lymphocytes to respond to alloantigens appropriately was first described by Herman Waldmann and termed "infectious tolerance" [57] (Figure 62.1). The putative effector cell which conveys these tolerizing properties to naïve lymphocytes is based on the expression of the nuclear transcription factor Forkhead box protein 3 (Foxp3), a key regulator of T_{reg} development in both mice and humans [58,59]. Among $CD4^+/Foxp3^+$ T_{regs}, several subpopulations can be distinguished: type 1 (TR1) cells which produce mainly interleukin (IL)-10 [60], T-helper 3 cells which release transforming growth factor (TGF)β [61], and T_{regs} which produce IL-35 [62]. Moreover, $CD3^+/CD8^-/CD4^-$ (double negative) T cells [63], $CD8^+/CD28^-$ [64], NKT cells [65], and $\gamma\delta$ T lymphocytes can exert suppressive immune properties [66]. However, most important and best studied in the context of peripheral immune regulation appear to be $CD4^+/CD25^+/CD127$low T_{regs} which exist in two forms, a naturally occurring thymic-primed regulatory form (nT_{reg}) and the adaptive

T_{reg} which results from *de novo* stimulation of naïve T cells [67]. If we assume that T_{regs} initiate and sustain the process of tolerance development and infectious tolerance, respectively, we could certainly benefit from the knowledge of exactly which regulatory cell type is responsible and where and when they act [68]. Currently, evidence for $Foxp3^+$ T cells as essential regulator cells for tolerance induction is mainly based on descriptive evidence. Circumstantial evidence has been categorized in three ways: (i) adoptive transfer of $Foxp3^+$ T_{regs} prolonged allograft survival [69–71], (ii) tolerance induction in rodents is associated with induction of peripheral $Foxp3^+$ T_{regs} [72–74], and (iii) tolerated allografts harbor small numbers of $Foxp3^+$ T_{regs} [75–77]. The demonstration that $Foxp3^+$ cells constantly suppress persistent cytotoxic T cells which otherwise inevitably would cause graft damage is still lacking. In part, this is due to the lack of a natural cell surface marker which would allow *in vivo* manipulation of $Foxp3^+$ T_{regs}. As recently reported, a transgenic mouse strain was created such that all $Foxp3^-$ expressing cells coexpress a human CD2 molecule, which is linked to the CD52 epitope on the cell surface. This CD2_CD52-fusion protein binds CD2 antibodies which in turn can be used to ablate $Foxp3^+$ cells in these transgenic animals [68,78]. Results obtained from this transgenic model undoubtedly proofed that $Foxp3^+$ T_{reg} cells prevent cytotoxic graft infiltrated T lymphocytes from tissue rejection. Moreover, peripherally induced T_{regs} sustain tolerance by an "infectious" mechanism as they were able to convey their tolerizing properties to naïve T cells [68]. Based on these proof-of-concept experiments empowering $Foxp3^+$ regulatory T lymphocytes appear to be an attractive route for COT induction.

In this scenario, one might ask what role do T_{regs} play in an intact human immune system and what immunosuppressive drug will be more or less permissive for T_{reg} development and function? Moreover, Foxp3 is of limited utility to isolate T_{regs} as it is expressed intracellularly requiring permeabilization of the cell for its detection. Beyond these limitations, in contrast to mice, Foxp3 may be expressed only transiently in humans and $Foxp3^+$ T_{regs} can convert to turn into IL-17 producing T-helper cells, known to accentuate immune responses [79]. In this context, it must be emphasized that both naturally occurring T_{regs} (nT_{regs}) and TGFβ-induced T_{regs} (iT_{regs}) can be converted into T-helper 17 (Th17) cells in the presence of inflammatory cytokines [80–82]. Th17 cells are associated with enhancement of the inflammatory environment which usually aggravates autoimmune diseases, such as rheumatoid arthritis, systemic lupus erythematodes, and rejection crisis, seen in allografted patients. This might explain the paradox why T_{regs} may be associated with kidney transplant rejection under certain circumstances. Expression and detection of Foxp3 in urine sediment and biopsies could be correlated—not with quiescence—but

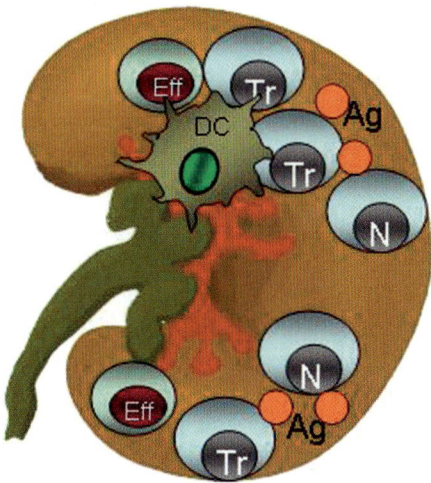

FIGURE 62.1 The phenomenon that T cells from a tolerant animal are able to specifically disable naïve lymphocytes to respond to alloantigens appropriately was first described by Herman Waldmann and termed "infectious tolerance"

rather graft rejection [83,84]. We can surmise that T_{regs} infiltrate rejecting allografts in an attempt to block inflammatory immune processes. Consequently, elevated levels of Foxp3 gene expression or T_{reg} numbers do not necessarily indicate a tolerance promoting microenvironment [85,86] (see Figure 62.2a and b).

As far as their clinical use is considered, a number of studies have shown a positive association between circulating T_{regs} and good graft function in liver [87,88], kidney [89,90], and lung [91,92] transplantation. At this point, more than 100 clinical trials assess the role of T_{regs} for various diseases, including hematological cancer, GvHD, type 1 diabetes mellitus and others (for a comprehensive overview, see www.clinical.trials.gov). T_{regs} can be separated into either adoptive transfer of *ex vivo* expanded T cells and *de novo* T_{reg} induction or conversion of naïve T cells into T_{regs}*in vivo*. Conversely, in the presence of TGFβ all-trans-retinoic acid efficiently converts adult human peripheral blood naïve $CD4^+$ T cells into T_{regs} with suppressive ability [87,89].

If T_{regs}, which by now can be expanded in therapeutically relevant numbers, are to be applied clinically, questions concerning their homing and trafficking in immunosuppressed hosts must be addressed with care. As recently outlined, T_{regs}—if *ex vivo* expanded and transferred back to a patient—should maintain their tolerogenic properties life long. As mentioned above, $Foxp3^+$ T_{regs} can convert to proinflammatory cytokine-producing cells and lose their suppressive capacity [93,94], which has aroused considerable controversy. In fact, an IFNγ-producing T_{reg} type has been identified in the peripheral blood of diabetic [95] and multiple sclerosis patients [96]. However, these Th1-like T_{regs} underwent reconversion to classical T_{regs} if IL12 was removed from affected individuals. As such, the cytokine

profile and epigenetic modifications of the Foxp3 locus mainly contribute to Foxp3 transcription. Demethylation of key regions of the Foxp3 locus results in long-term stability of its expression and promotes their suppressive feature. Concomitant treatment with azacytidine after HSCT has yielded higher numbers of T_{regs} without loss of antitumor immunity and warrants further investigation in the setting of organ transplantation [97]. In addition, histone deacetylase (HDAC) inhibitors increase T_{regs} which do not release proinflammatory cytokines [98]. The same report used systemic application of HDAC *in vivo* after transplantation which prolonged allograft survival. Infectious tolerance conferred by T_{regs} is crucial to sustain graft tolerance as indicated before. Infectious tolerance depends on ongoing TGFβ production from T_{regs}, a prerequisite to prevent allograft rejection [68,99]. A recent trial of T_{reg} therapy in HSCT demonstrated that transferred cells were no longer detected in circulation after 2 weeks. It is unclear whether injected T_{regs} died or migrated to other tissues [100]. In cancer patients, infused $CD8^+$ T cells persisted for at least 6 months [101]. Adjuvant treatment with low doses of IL-2 promotes T_{reg} expansion in an inflammatory environment and ameliorated GvHD [102].

62.7 IMMUNOSUPPRESSIVE DRUGS IN THE CONTEXT OF CLINICAL ORGAN TOLERANCE INDUCTION

This brings up the question in which way immunosuppression may affect the longevity and function of transferred cells. It is generally accepted that calcineurin inhibitors, likewise cyclosporine and tacrolimus are detrimental to T_{regs} and tailoring immunosuppression to use other compounds, such as sirolimus and antithymocyte globulin, is thought to preserve and enhance their *in vivo* suppressor

(A)
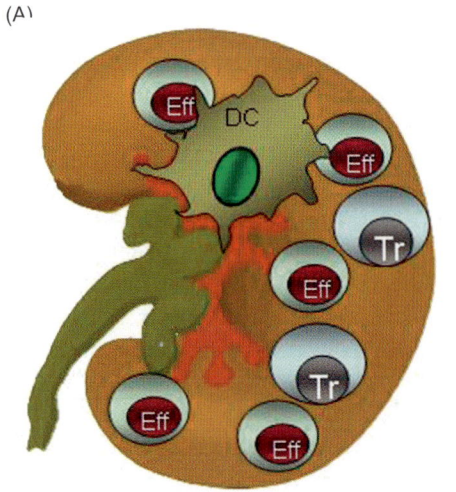

(B)
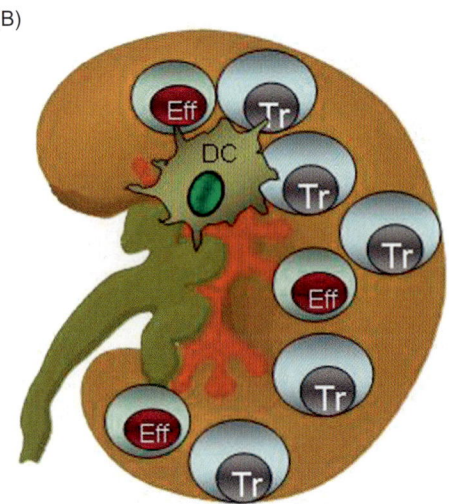

FIGURE 62.2   Tregs infiltrate rejecting allografts in an attempt to block inflammatory immune processes. Consequently, elevated levels of Foxp3 gene expression or Treg numbers do not necessarily indicate a tolerance promoting microenvironment

function [103]. Whereas sirolimus may impair T_{reg} expansion [104] antithymocyte globulin selectively spares this adverse effect and selectively enhances T_{reg} function [105].

A further concern addresses the number of T_{regs} needed in the clinical setting of organ tolerance. Assuming that no lymphodepletion treatment is used, a dose of $49-79 \times 10^9$ T_{regs} will be required [106] to reach a calculated ratio of 1:1 to 1:2, likewise $33-50\%$ T_{regs} in relation to conventional T cells (T_{conv}). Even now it seems possible to manufacture such extensive numbers by repeated stimulation [104], T_{regs} tend to loose $Foxp3^+$ expression under this procedure [107]. Subsequently, a combination of lymphodepletion and short-term *ex vivo* T_{reg} expansion appears to be the most effective approach to positively balance T_{reg}/T_{conv} ratios. In combination with thymoglobulin induction, a single dose of $3-5 \times 10^9$ T_{regs} can effectively surmount the required 33% level needed, as this number is easily produced from one single unit of blood within 14 days of *ex vivo* expansion time [108]. Other depleting agents useful for induction therapy comprise alemtuzumab and basiliximab. It is beyond the scope of this chapter to go into details describing the large panel of clinical study outcome after their use (for an extensive review, see Wagner et al. [109]).

62.8 CHEMOKINE RECEPTOR-MEDIATED TRAFFICKING OF T_{REGS}

We must keep in mind that different subtypes of T_{regs} may exert distinct therapeutic effects depending on which immune cells mediate the rejection process. Moreover, chemokine receptors and cognate ligands determine appropriate trafficking of T_{regs} which is critical for their function [90]. Expression of trafficking and homing receptors on T_{regs} varies with respect to their individual fate and state of activation [110]. Normally, T_{regs} are endowed with CD62L and CCR7 to allow entry to lymph nodes where they encounter APCs, which in turn modulate the expression of homing receptors as a prerequisite to migrate to sites in which they exert their regulatory and suppressive functions [111]. Interference with proper homing receptor expression can break tolerance as demonstrated in a model of cardiac and islet allograft mouse model, respectively [112,113]. Since $Foxp3^+$ T_{regs} variously express CCR2, CCR4, CCR5, CCR8, $\alpha 4\beta 7$ integrin, CCR9, CXCR1, CXCR3, CXCR4, CXCR6, and other integrins and addressins [90], they display a functional diversity which might promote and impair tolerance development. As recently exemplified, overexpression of CCL22 on transplanted islets enforced recruitment of CCR4 positive T_{regs} to the site of transplantation and ameliorated rejection damage considerably [114].

Much attention is currently attributed to mesenchymal stem cells (MSCs) as another potential source of cells with tolerogenic properties [115–117]. MSCs are thought to downregulate overshooting immune responses in various clinical settings. Subsequently, they are currently applied to circumvent GvHD after bone marrow transplantation and to limit rejection episodes after cell or organ transplantation [118]. A detailed description of their suppressive properties and how to possibly use them in a clinical setting is given elsewhere in this compendium.

Another cell type which bears inherent tolerogenic properties and regulatory functions are monocyte-derived DCs. DCs initiate either robust immunity against pathogens as well as maintaining immunological tolerance to self antigens. Undoubtedly, subpopulations of DCs are able to prime and expand T_{regs} from naïve DCs [119,120]. Vice versa, accumulating *in vivo* and *in vitro* evidence suggests that T_{regs} by binding to DCs interfere with activation of $CD25^-$ T cells and confer tolerogenic properties to DCs in place [121,122], also exemplified in Figure 62.1. In order to safeguard the generation and clinical use of tolerogenic DCs one key target has now been described for the Wnt-β-catenin signaling pathway [123]. Intracellular β-catenin signaling in intestinal DCs is required for the expression of anti-inflammatory mediators, such as retinoic acid metabolizing enzymes, IL-10, and TGFβ. An anti-inflammatory cytokine profile released by β-catenin activated DCs promotes T_{reg} induction while suppressing inflammatory effector T cells. Alternative efforts to prevent *in vivo* activation of DCs attempt to modulate DCs using a variety of different agents which inhibit DC maturation *in vitro*. Among those, rapamycin [124], dexamethasone [119], LF 15-0195 (a 15-deoxyspergualin analogue) [120], and ITF 2357 (an HDAC inhibitor) [122] [79/88] are tested.

Further insight into peripheral regulatory immune networks comes from experimental and clinical trials, which investigated the tolerogenic properties of macrophages. Macrophages belong to the innate immunity and exert a dichotomic behavior as they can accentuate and attenuate immune responses, extensively reviewed elsewhere [125,126]. Initial experiments with rat-derived embryonic stem cell-like cells (RESCs) identified RESC-derived monocytes as the crucial cell population for long-term cardiac allograft acceptance in a rat strain model where host animals were preconditioned with intraportally injected RESCs, 7 days before transplantation ensued [123]. Based on these observations, a regulatory-type macrophage termed M_{reg} was characterized and specified with the effort to use it as a medical product to promote tolerance in renal transplantation [127]. Subtle functional analysis of M_{regs} demonstrates their IDO- and contact-dependent deletion of activated T cells, their ability to suppress T-cell proliferation, their anti-inflammatory potential, and their capacity to expand regulatory lymphocytes in coculture experiments [128,129]. A recently published series of manuscripts describes this type of immunoregulatory macrophages for an autoimmune animal model of DSS-induced inflammatory bowel disease and in the setting of living-related and unrelated KTxs [127,130–133]. In kidney allograft recipients, these M_{regs} induced a semistable state of alloantigen-specific graft acceptance which allowed for stable minimization of immunosuppressive drugs. Five patients were enrolled in a pilot clinical trial of living-related and unrelated KTx, to investigate the potential of M-CSF and IFNγ-pretreated monocytes to induce COT. One patient was weaned off immunosuppression for a period of 36 weeks after which he experienced acute kidney rejection which was successfully reversed with conventional immunosuppressive drugs. Two patients also suffered from acute rejection crisis when IS was discontinued, one after 2 and another after 34 weeks. Both rejections were reversible. Two patients kept on minimalized tacrolimus monotherapy (trough levels between 4 and 6 ng/mL) after an initial 3-month triple therapy (prednisone plus MMF) demonstrated stable kidney function for more than 5 years. Of note, all patients exhibited donor-specific unresponsiveness but normal responses to third-party antigens [130,134]. Based on optimized methodological and technical manufacturing technologies, two living-donor KTxs, which included preconditioning recipients with donor-derived M_{regs}, 6 days prior to transplantation, were performed. Patient MM, a 23-year-old female, had one single HLA-B and DR mismatch, whereas patient CA, a 47-year-old man, received a fully mismatched kidney from a 40-year-old unrelated donor. Both patients received donor-derived M_{regs} under cover of azathioprine 7 days prior to transplantation. Hence, with graft transfer to the recipient conventional immunosuppression with steroids and tacrolimus was started. Steroids were tapered to cessation after 2 weeks and tacrolimus monotherapy maintained. At 3 years, both patients kept on low-level tacrolimus <5 mg had stable function and no biopsy proven signs of rejection, tacrolimus trough levels ranging between 4.5 (MM) and 2.5 (CA) ng/mL, respectively. Expression of the strongest 10 discriminatory gene markers of tolerance as identified by the IOT-RISET (European) and ITN (USA) tolerance networks, respectively [87,88], were serially measured and gradually converged upon the genetic signature associated with COT [127]. Of note, trafficking of indiodine-labeled M_{regs} in patient MM revealed sequential migration to lungs, liver, spleen, and bone marrow, where cells were detectable by gamma-scinthigraphy after 32 h. At this time point, cells were still vital and not the target of host-mediated rejection, since no indiodine radioactivity along the urinary tract was detectable.

62.9 OBSTACLES AND PERSPECTIVES IN COT

One major concern raised by the implementation of tolerogenic strategies after SOT is the risk for graft rejection and subsequent graft loss, once IS has been withdrawn. Although we still lack methods capable of identifying recipients who may be amenable candidates for complete IS withdrawal, a number of potential assays to address operational tolerant patients are now at hand in the field of KTx, reviewed extensively elsewhere in this compendium. Unless validated and reliable clinical biomarkers prove correct, one major challenge for transplant investigators is to decide when and in which steps weaning from IS could be achieved to relieve organ recipients from the heavy burden of IS-related adverse effects thereby improving life quality and patient's adherence to treatment [135]. As a matter of fact, graft dysfunction usually does not manifest clinically before substantial immunological damage has occurred in the graft. Until recently, pharmacological monitoring and graft histology were still the only diagnostic tools routinely at hand in most transplant centers and not useful to decide on IS weaning and/or minimization. A recent publication found that specific mRNA expressed in urinary cells anticipates kidney allograft rejection 60–90 days before it is diagnosed clinically [136]. If this diagnostic mean holds upto its promise, it could open a new door to prospective clinical weaning trials as therapeutic interventions before severe onset of graft damage appears to be possible.

Nevertheless, COT must still be considered as a courageous endeavor. The main clinical arguments against such an attempt can be summarized as follows: (i) in contrast to liver transplanted patients, COT is an exceptional finding in patients who sporadically stop IS intake, (ii) the risk of graft loss is considerably higher if rejection ensues IS withdrawal. (iii) Conversely, all molecule-based strategies have failed even though they worked successfully in various animal models. At the current stage, there are still a lot of open questions concerning future tolerance-inducing strategies. These comprise patient selection, the therapeutic combination of IS, costimulatory blockade agents, the source of suppressor cell type, immunomonitoring, pretransplant (memory cell status?), underlying disease (patients with autoimmune and vasculitis diseases are much more difficult to regulate), age, gender, donor-recipient HLA compatibility, and many more. A thoughtful mean to approximate this goal might take advantage of Roy Calne's termed "probe" tolerance strategy which largely, but not completely withdraws immunosuppressive drugs [9]. "Almost tolerance" with minimized but sufficient immunosuppression, as recently recommended by Knechtle et al. [137] would create a therapeutic safety corridor, ameliorating IS-related adverse effects while considerably improving life quality of patients.

In summary, there is a clear recognizable common endeavor of numerous academic transplant centers to launch clinical tolerance protocols in the future. Of course, these attempts must safeguard the highest ethical standards and demonstrate that COT development neither puts at risk grafts nor patients. Based on the high number of graft losses due to chronic rejection and on IS-related mortality, this endeavor seems justified to me. Hopefully, the transplant community will encourage transplant surgeons, physicians, and immunologists to go this way. It will still be, without doubt, a long and winding road.

REFERENCES

[1] Brown JB. Homografting of skin: with report of success in identical twins. Surgery 1937;102:508.

[2] Sayegh MH, Fine NA, Smith JL, Rennke HG, Milford EL, Tilney NL. Immunological-tolerance to renal-allografts after bone-marrow transplants from the same donors. Ann Intern Med 1991;114:954−5.

[3] Newell KA. Clinical transplantation tolerance. Semin Immunopathol 2011;33:91−104.

[4] Orlando G. Immunosuppression-free transplantation reconsidered from a regenerative medicine perspective. Expert Rev Clin Immunol 2012;8:179−87.

[5] Bishop GA, Sharland AF, Ierino FL, Sandrin MS, Hall BM, Alexander SI, et al. Operational tolerance in organ transplantation versus tissue engineering: into the future. Transplantation 2011;92:E39.

[6] Matzinger P. Tolerance, danger, and the extended family. Annu Rev Immunol 1994;12:991−1045.

[7] Sriwatanawongsa V, Davies HFS, Calne RY. The essential roles of parenchymal tissues and passenger leukocytes in the tolerance induced by liver grafting in rats. Nat Med 1995;1:428−32.

[8] Starzl TE. Chimerism and tolerance in transplantation. Proc Natl Acad Sci USA 2004;101:14607−14.

[9] Calne R, Friend P, Moffatt S, Bradley A, Hale G, Firth J, et al. Prope tolerance, perioperative campath 1H, and low-dose cyclosporin monotherapy in renal allograft recipients. Lancet 1998;351:1701−2.

[10] Medawar PB. Immunological tolerance—phenomenon of tolerance provides a testing ground for theories of immune response. Science 1961;133:303.

[11] Adams AB, Pearson TC, Larsen CP. Heterologous immunity: an overlooked barrier to tolerance. Immunol Rev 2003;196:147−60.

[12] Welsh RM, Markees TG, Woda BA, Daniels KA, Brehm MA, Mordes JP, et al. Virus-induced abrogation of transplantation tolerance induced by donor-specific transfusion and anti-CD154 antibody. J Virol 2000;74:2210−8.

[13] Wu ZH, Bensinger SJ, Zhang JD, Chen CQ, Yuan XL, Huang XL, et al. Homeostatic proliferation is a barrier to transplantation tolerance. Nat Med 2004;10:87−92.

[14] Pearl JP, Parris J, Hale DA, Hoffmann SC, Bernstein WB, Mccoy KL, et al. Immunocompetent T cells with a memory-like phenotype are the dominant cell type following antibody-mediated T-cell depletion. Am J Transplant 2005;5:465−74.

[15] Du W, Shen H, Galan A, Goldstein DR. An age-specific CD8(+) T cell pathway that impairs the effectiveness of strategies to prolong allograft survival. J Immunol 2011;187:3631−40.

[16] Zhao GP, Moore DJ, Kim JI, Lee KM, O'Connor MR, Duff PE, et al. Inhibition of transplantation tolerance by immune senescence is reversed by endocrine modulation. Sci Transl Med 2011;3:.

[17] Halloran PF, Bromberg J, Kaplan B, Vincenti F. Tolerance versus immunosuppression: a perspective. Am J Transplant 2008;8: 1365−6.

[18] Orlando G, Manzia T, Baiocchi L, Sanchez-Fueyo A, Angelico M, Tisone G. The tor vergata weaning off immunosuppression protocol in stable HCV liver transplant patients: the updated follow up at 78 months. Transpl Immunol 2008;20:43−7.

[19] Orlando G. Finding the right time for weaning off immunosuppression in solid organ transplant recipients. Expert Rev Clin Immunol 2010;6:879−92.

[20] Orlando G, Hematti P, Stratta RJ, Burke GW, Di Cocco P, Pisani F, et al. Clinical operational tolerance after renal transplantation current status and future challenges. Ann Surg 2010;252:915−28.

[21] Leventhal J, Abecassis M, Miller J, Gallon L, Ravindra K, Tollerud DJ, et al. Chimerism and tolerance without GvHD or engraftment syndrome in HLA-mismatched combined kidney and hematopoietic stem cell transplantation. Sci Transl Med 2012;4 (124):124ra28.

[22] Owens ML, Maxwell JG, Goodnight J, Wolcott MW. Discontinuance of immunosuppression in renal-transplant patients. Arch Surg 1975;110:1450−1.

[23] Zoller KM, Cho SI, Cohen JJ, Harrington JT. Cessation of immunosuppressive therapy after successful transplantation—a national survey. Kidney Int 1980;18:110−4.

[24] Weil R, Starzl TE, Porter KA, Kershaw M, Schroter GPJ, Koep LJ. Renal isotransplantation without immunosuppression. Ann Surg 1980;192:108−10.

[25] Burlingham WJ, Grailer AP, Fechner JH, Kusaka S, Trucco M, Kocova M, et al. Microchimerism linked to cytotoxic T-lymphocyte functional unresponsiveness (Clonal Anergy) in a tolerant renal-transplant recipient. Transplantation 1995;59:1147−55.

[26] Burke GW, Ciancio G, Cirocco R, Markou M, Coker D, Roth D, et al. Association of interleukin-10 with rejection-sparing effect in septic kidney transplant recipients. Transplantation 1996;61: 1114−6.

[27] Trivedi HL, Mishra VV, Vanikar AV, Modi PR, Shah VR, Shah PR, et al. Embryonic stem cell derived and adult hematopoietic stem cell transplantation for tolerance induction in a renal allograft recipient: a case report. Transplant Proc 2006;38:3103−8.

[28] Cobbold SP, Martin G, Qin S, Waldmann H. Monoclonal antibodies to promote marrow engraftment and tissue graft tolerance. Nature 1986;323:164−6.

[29] Ildstad ST, Sachs DH. Reconstitution with syngeneic plus allogeneic or xenogeneic bone-marrow leads to specific acceptance of allografts or xenografts. Nature 1984;307:168−70.

[30] Adams AB, Durham MM, Kean L, Shirasugi N, Ha JW, Williams MA, et al. Costimulation blockade, busulfan, and bone marrow promote titratable macrochimerism, induce transplantation tolerance, and correct genetic hemoglobinopathies with minimal myelosuppression. J Immunol 2001;167:1103−11.

[31] Kean LS, Durham MM, Adams AB, Hsu LL, Perry JR, Dillehay D, et al. A cure for murine sickle cell disease through stable mixed chimerism and tolerance induction after nonmyeloablative conditioning and major histocompatibility complex-mismatched bone marrow transplantation. Blood 2002;99:1840−9.

[32] Seung E, Iwakoshi N, Woda BA, Markees TG, Mordes JP, Rossini AA, et al. Allogeneic hematopoietic chimerism in mice treated with sub-lethal myeloablation and anti-CD154 antibody: absence of graft vs. host disease (GvHD), induction of skin

allograft tolerance, and prevention of recurrent autoimmunity in islet-allografted NOD mice. FASEB J 2000;14:A1071.

[33] Sharabi Y, Sachs DH. Mixed chimerism and permanent specific transplantation tolerance induced by a nonlethal preparative regimen. J Exp Med 1989;169:493–502.

[34] Wekerle T, Sayegh MH, Hill J, Zhao Y, Chandraker A, Swenson KG, et al. Extrathymic T cell deletion and allogeneic stem cell engraftment induced with costimulatory blockade is followed by central T cell tolerance. J Exp Med 1998;187:2037–44.

[35] Wekerle T, Sayegh MH, Ito H, Hill J, Chandraker A, Pearson DA, et al. Anti-CD154 or CTLA4lg obviates the need for thymic irradiation in a non-myeloablative conditioning regimen for the induction of mixed hematopoietic chimerism and tolerance. Transplantation 1999;68:1348–55.

[36] Pilat N, Wekerle T. Transplantation tolerance through mixed chimerism. Nat Rev Nephrol 2010;6:594–605.

[37] Blaha P, Bigenzahn S, Koporc Z, Schmid M, Langer F, Selzer E, et al. The influence of immunosuppressive drugs on tolerance induction through bone marrow transplantation with costimulation blockade. Blood 2003;101:2886–93.

[38] Blaha P, Bigenzahn S, Koporc Z, Sykes M, Muehlbacher F, Wekerle T. Short-term immunosuppression facilitates induction of mixed chimerism and tolerance after bone marrow transplantation without cytoreductive conditioning. Transplantation 2005;80:237–43.

[39] Taylor PA, Lees CJ, Wilson JM, Ehrhardt MJ, Campbell MT, Noelle RJ, et al. Combined effects of calcineurin inhibitors or sirolimus with anti-CD40L mAb on alloengraftment under nonmyeloblative conditions. Blood 2002;100:3400–7.

[40] Westerhuis G, Maas WGE, Willemze R, Toes REM, Fibbe WE. Long-term mixed chimerism after immunologic conditioning and MHC-mismatched stem-cell transplantation is dependent on NK-cell tolerance. Blood 2005;106:2215–20.

[41] Graca L, Daley S, Fairchild PJ, Cobbold SP, Waldmann H. Co-receptor and co-stimulation blockade for mixed chimerism and tolerance without myelosuppressive conditioning. BMC Immunol 2006;7.

[42] Gandy KL, Domen J, Aguila H, Weissman IL. CD8(+)TCR(+) and CD8(+)TCR(−) cells in whole bone marrow facilitate the engraftment of hematopoietic stem cells across allogeneic barriers. Immunity 1999;11:579–90.

[43] Kawai T, Cosimi AB, Sachs DH. Preclinical and clinical studies on the induction of renal allograft tolerance through transient mixed chimerism. Curr Opin Organ Transplant 2011;16:366–71.

[44] Kawai T, Cosimi AB, Spitzer TR, Tolkoff-Rubin N, Suthanthiran M, Saidman SL, et al. Brief report: HLA-mismatched renal transplantation without maintenance immunosuppression. N Engl J Med 2008;358:353–61.

[45] Sachs DH, Sykes M, Kawai T, Cosimi AB. Immuno-intervention for the induction of transplantation tolerance through mixed chimerism. Semin Immunol 2011;23:165–73.

[46] Farris AB, Taheri D, Kawai T, Fazlollahi L, Wong W, Tolkoff-Rubin N, et al. Acute renal endothelial injury during marrow recovery in a cohort of combined kidney and bone marrow allografts. Am J Transplant 2011;11:1464–77.

[47] Mao Q, Terasaki PI, Cai J, Briley K, Catrou P, Haisch C, et al. Extremely high association between appearance of HLA antibodies and failure of kidney grafts in a five-year longitudinal study. Am J Transplant 2007;7:864–71.

[48] Scandling JD, Busque S, Shizuru JA, Engleman EG, Strober S. Induced immune tolerance for kidney transplantation. N Engl J Med 2011;365:1359–60.

[49] Millan TLT, Shizuru JA, Hoffmann P, Dejbakhsh-Jones S, Scandling JD, Grumet FC, et al. Mixed chimerism and immuno-suppressive drug withdrawal after HLA-mismatched kidney and hematopoietic progenitor transplantation. Transplantation 2002;73:1386–91.

[50] Pilat N, Baranyi U, Klaus C, Jaeckel E, Mpofu N, Wrba F, et al. Treg-therapy allows mixed chimerism and transplantation tolerance without cytoreductive conditioning. Am J Transplant 2010;10:751–62.

[51] Pilat N, Wekerle T. Mechanistic and therapeutic role of regulatory T cells in tolerance through mixed chimerism. Curr Opin Organ Transplant 2010;15:725–30.

[52] Fujisaki J, Wu J, Carlson AL, Silberstein L, Putheti P, Larocca R, et al. *In vivo* imaging of T-reg cells providing immune privilege to the haematopoietic stem-cell niche. Nature 2011;474:216–9.

[53] Owen RD. Immunogenetic consequences of vascular anastomoses between bovine twins. Science 1945;102:400–1.

[54] Newton WT, Anderson CB. Planned preimmunization of renal-allograft recipients. Surgery 1973;74:430–6.

[55] Kim JM, Rasmussen JP, Rudensky AY. Regulatory T cells prevent catastrophic autoimmunity throughout the lifespan of mice. Nat Immunol 2007;8:191–7.

[56] Hall BM, Jelbart ME, Dorsch SE. Suppressor T cells in rats with prolonged cardiac allograft survival after treatment with cyclosporine. Transplantation 1984;37:595–600.

[57] Qin SX, Cobbold SP, Pope H, Elliott J, Kioussis D, Davies J, et al. Infectious transplantation tolerance. Science 1993;259:974–7.

[58] Fontenot JD, Gavin MA, Rudensky AY. Foxp3 programs the development and function of CD4(+)CD25(+) regulatory T cells. Nat Immunol 2003;4:330–6.

[59] Hori S, Nomura T, Sakaguchi S. Control of regulatory T-cell development by the transcription factor Foxp3. Science 2003;299:1057–61.

[60] Groux H, OGarra A, Bigler M, Rouleau M, Antonenko S, deVries JE, et al. A CD4(+) T-cell subset inhibits antigen-specific T-cell responses and prevents colitis. Nature 1997;389:737–42.

[61] Faria AM, Weiner HL. Oral tolerance. Immunol Rev 2005;206:232–59.

[62] Collison LW, Workman CJ, Kuo TT, Boyd K, Wang Y, Vignali KM, et al. The inhibitory cytokine IL-35 contributes to regulatory T-cell function. Nature 2007;450:566–9.

[63] Zhang ZX, Yang L, Young KJ, DuTemple B, Zhang L. Identification of a previously unknown antigen-specific regulatory T cell and its mechanism of suppression. Nat Med 2000;6:782–9.

[64] Vlad G, Cortesini R, Suciu-Foca N. CD8(+) T suppressor cells and the ILT3 master switch. Hum Immunol 2008;69:681–6.

[65] Seino K, Fukao K, Muramoto K, Yanagisawa K, Takada Y, Kakuta S, et al. Requirement for natural killer T (NKT) cells in the induction of allograft tolerance. Proc Natl Acad Sci USA 2001;98:2577–81.

[66] Hayday A, Tigelaar R. Immunoregulation in the tissues by gamma delta T cells. Nat Rev Immunol 2003;3:233–42.

[67] McMurchy AN, Bushell A, Levings MK, Wood KJ. Moving to tolerance: clinical application of T regulatory cells. Semin Immunol 2011;23:304–13.

[68] Kendal AR, Chen Y, Regateiro FS, Ma JB, Adams E, Cobbold SP, et al. Sustained suppression by Foxp3(+) regulatory T cells is vital for infectious transplantation tolerance. J Exp Med 2011;208:2043–53.

[69] Feng G, Wood KJ, Bushell A. Interferon-gamma conditioning *ex vivo* generates CD25(+)CD62L(+)Foxp3(+) regulatory T cells that prevent allograft rejection: potential avenues for cellular therapy. Transplantation 2008;86:578−89.

[70] Graca L, Thompson S, Lin CY, Adams E, Cobbold SP, Waldmann H. Both CD4(+)CD25(+) and CD4(+)CD25(−) regulatory cells mediate dominant transplantation tolerance. J Immunol 2002;168:5558−65.

[71] Xia GL, He J, Leventhal JR. Prevention of allograft rejection in wild-type mice by amplifying suppressive activity of allospecific Foxp3(+)CD4(+)CD25(+) regulatory T (Treg) cells. Am J Transplant 2009;9:210.

[72] Battaglia M, Stabilini A, Migliavacca B, Horejs-Hoeck J, Kaupper T, Roncarolo MG. Rapamycin promotes expansion of functional CD4(+)CD25(+)FOXP3(+) regulatory T cells of both healthy subjects and type 1 diabetic patients. J Immunol 2006;177:8338−47.

[73] Cobbold SP, Castejon R, Adams E, Zelenika D, Graca L, Humm S, et al. Induction of foxP3(+) regulatory T cells in the periphery of T cell receptor transgenic mice tolerized to transplants. J Immunol 2004;172:6003−10.

[74] Turnquist HR, Raimondi G, Zahorchak AF, Fischer RT, Wang ZL, Thomson AW. Rapamycin-conditioned dendritic cells are poor stimulators of allogeneic CD4(+) T cells, but enrich for antigen-specific Foxp3(+) T regulatory cells and promote organ transplant tolerance. J Immunol 2007;178:7018−31.

[75] Fan ZG, Spencer JA, Lu Y, Pitsillides CM, Singh G, Kim P, et al. *In vivo* tracking of "color-coded" effector, natural and induced regulatory T cells in the allograft response. Nat Med 2010;16:718−22.

[76] Lee I, Wang LQ, Wells AD, Dorf ME, Ozkaynak E, Hancock WW. Recruitment of Foxp3(+) T regulatory cells mediating allograft tolerance depends on the CCR4 chemokine receptor. J Exp Med 2005;201:1037−44.

[77] Semiletova NV, Shen XD, Baibakov B, Andakyan A. Intensity of transplant chronic rejection correlates with level of graft-infiltrating regulatory cells. J Heart Lung Transplant 2010;29:335−41.

[78] Komatsu N, Mariotti-Ferrandiz ME, Wang Y, Malissen B, Waldmann H, Hori S. Heterogeneity of natural Foxp3(+) T cells: a committed regulatory T-cell lineage and an uncommitted minor population retaining plasticity. Proc Natl Acad Sci USA 2009;106:1903−8.

[79] Allan SE, Crome SQ, Crellin NK, Passerini L, Steiner TS, Bacchetta R, et al. Activation-induced FOXP3 in human T effector cells does not suppress proliferation or cytokine production. Int Immunol 2007;19:345−54.

[80] Bestard O, Cruzado JM, Rama I, Torras J, Freixanet M, Seron D, et al. Presence of FoxP3(+) regulatory T cells predicts outcome of subclinical rejection of renal allografts. J Am Soc Nephrol 2008;19:2020−6.

[81] Meloni F, Vitulo P, Bunco AM, Paschetto E, Morosini M, Cascina A, et al. Regulatory CD4⁺CD25⁺ T cells in the peripheral blood of lung transplant recipients: correlation with transplant outcome. Transplantation 2004;77:762−6.

[82] Radhakrishnan S, Cabrera R, Schenk EL, Nava-Parada P, Bell MP, Van Keulen VP, et al. Reprogrammed FoxP3(+) T regulatory cells become IL-17(+) antigen-specific autoimmune effectors *in vitro* and *in vivo* (retraction of vol. 181, p. 3137, 2008). J Immunol 2010;184:6556.

[83] Li BG, Hartono C, Ding RC, Sharma VK, Ramaswamy R, Qian B, et al. Noninvasive diagnosis of renal-allograft rejection by measurement of messenger RNA for perforin and granzyme B in urine. N Engl J Med 2001;344:947−54.

[84] Muthukumar T, Dadhania D, Ding RC, Snopkowski C, Naqvi R, Lee JB, et al. Messenger RNA for FOXP3 in the urine of renal-allograft recipients. N Engl J Med 2005;353:2342−51.

[85] Baan CC, Dijke IE, Weimar W. Regulatory T cells in alloreactivity after clinical heart transplantation. Curr Opin Organ Transplant 2009;14:577−82.

[86] Neujahr DC, Cardona AC, Ulukpo O, Rigby M, Pelaez A, Ramirez A, et al. Dynamics of human regulatory T cells in lung lavages of lung transplant recipients. Transplantation 2009;88:521−7.

[87] Afzali B, Mitchell P, Lechler RI, John S, Lombardi G. Translational mini-review series on Th17 Cells: induction of interleukin-17 production by regulatory T cells. Clin Exp Immunol 2010;159:120−30.

[88] Schiopu A, Wood KJ. Regulatory T cells: hypes and limitations. Curr Opin Organ Transplant 2008;13:333−8.

[89] Koenen HJPM, Smeets RL, Vink PM, van Rijssen E, Boots AMH, Joosten I. Human CD25(high)Foxp3(pos) regulatory T cells differentiate into IL-17-producing cells. Blood 2008;112:2340−52.

[90] Sagoo P, Lombardi G, Lechler RI. Regulatory T cells as therapeutic cells. Curr Opin Organ Transplant 2008;13:645−53.

[91] Demirkiran A, Kok A, Kwekkeboom J, Kusters JG, Metselaar HJ, Tilanus HW, et al. Low circulating regulatory T-cell levels after acute rejection in liver transplantation. Liver Transpl 2006;12:277−84.

[92] Wang J, Huizinga TWJ, Toes REM. *De novo* generation and enhanced suppression of human CD4(+)CD25(+) regulatory T cells by retinoic acid. J Immunol 2009;183:4119−26.

[93] d'Hennezel E, Piccirillo CA. Functional plasticity in human FOXP3(+) regulatory T cells. Implications for cell-based immunotherapy. Hum Vaccin Immunother 2012;8(7):1001−5.

[94] Hamann A. Regulatory T cells stay on course. Immunity 2012;36:161−3.

[95] McClymont SA, Putnam AL, Lee MR, Esensten JH, Liu WH, Hulme MA, et al. Plasticity of human regulatory T cells in healthy subjects and patients with type 1 diabetes. J Immunol 2011;186:3918−26.

[96] Dominguez-Villar M, Baecher-Allan CM, Hafler DA. Identification of T-helper type 1-like, Foxp3(+) regulatory T cells in human autoimmune disease. Nat Med 2011;17:673−5.

[97] Goodyear OC, Dennis M, Jilani NY, Loke J, Siddique S, Ryan G, et al. Azacitidine augments expansion of regulatory T cells after allogeneic stem cell transplantation in patients with acute myeloid leukemia (AML). Blood 2012;119:3361−9.

[98] Beier UH, Wang LQ, Bhatti TR, Liu YJ, Han RX, Ge GH, et al. Sirtuin-1 targeting promotes Foxp3(+) T-regulatory cell function and prolongs allograft survival. Mol Cell Biol 2011;31:1022−9.

[99] Andersson J, Tran DQ, Pesu M, Davidson TS, Ramsey H, O'Shea JJ, et al. CD4(+)FoxP3(+) regulatory T cells confer infectious tolerance in a TGF-beta-dependent manner. J Exp Med 2008;205:1975−81.

[100] Brunstein CG, Miller JS, Cao Q, McKenna DH, Hippen KL, Curtsinger J, et al. Infusion of *ex vivo* expanded T regulatory cells in adults transplanted with umbilical cord blood: safety profile and detection kinetics. Blood 2011;117:1061−70.

[101] Kalos M, Porter DL, June CH. Chimeric antigen receptor-modified T cells in CLL Reply. N Engl J Med 2011;365:1938.

[102] Koreth J, Matsuoka K, Kim HT, McDonough SM, Bindra B, Alyea EP, et al. Interleukin-2 and regulatory T cells in graft-versus-host disease. N Engl J Med 2011;365:2055—66.

[103] Ma AL, Qi SJ, Song LJ, Hu YX, Dun H, Massicotte E, et al. Adoptive transfer of CD4(+)CD25(+) regulatory cells combined with low-dose sirolimus and anti-thymocyte globulin delays acute rejection of renal allografts in Cynomolgus monkeys. Int Immunopharmacol 2011;11:618—29.

[104] Hippen KL, Merkel SC, Schirm DK, Sieben CM, Sumstad D, Kadidlo DM, et al. Massive *ex vivo* expansion of human natural regulatory T cells (Tregs) with minimal loss of *in vivo* junctional activity. Sci Transl Med 2011;18:83ra41.

[105] Scandling JD, Busque S, Dejbakhsh-Jones S, Benike C, Sarwal M, Millan MT, et al. Tolerance and withdrawal of immunosuppressive drugs in patients given kidney and hematopoietic cell transplants. Am J Transplant 2012;12:1133—45.

[106] Tang QZ, Lee K. Regulatory T-cell therapy for transplantation: how many cells do we need?. Curr Opin Organ Transplant 2012;17:349—54.

[107] Hoffmann P, Boeld TJ, Eder R, Huehn J, Floess S, Wieczorek G, et al. Loss of FOXP3 expression in natural human CD4(+) CD25(+) regulatory T cells upon repetitive *in vitro* stimulation. Eur J Immunol 2009;39:1088—97.

[108] Putnam AL, Brusko TM, Lee MR, Liu WH, Szot GL, Ghosh T, et al. Expansion of human regulatory T cells from patients with type 1 diabetes. Diabetes 2009;58:652—62.

[109] Wagner SJ, Brennan DC. Induction therapy in renal transplant recipients how convincing is the current evidence? Drugs 2012;72:671—83.

[110] Waldmann H, Adams E, Fairchild P, Cobbold S. Regulation and privilege in transplantation tolerance. J Clin Immunol 2008;28:716—25.

[111] Long E, Wood KJ. Regulatory T cells in transplantation: transferring mouse studies to the clinic. Transplantation 2009;88: 1050—6.

[112] Hara M, Kingsley CI, Niimi M, Read S, Turvey SE, Bushell AR, et al. IL-30 is required for regulatory T cells to mediate tolerance to alloantigens *in vivo*. J Immunol 2001;166:3789—96.

[113] Tang QZ, Bluestone JA, Kang SM. CD4(+)Foxp3(+) regulatory T cell therapy in transplantation. J Mol Cell Biol 2012;4:11—21.

[114] Ganusov VV, De Boer RJ. Do most lymphocytes in humans really reside in the gut? Trends Immunol 2007;28:514—8.

[115] Crop M, Baan C, Weimar W, Hoogduijn M. Potential of mesenchymal stem cells as immune therapy in solid-organ transplantation. Transpl Int 2009;22:365—76.

[116] Le Blanc K, Pittenger MF. Mesenchymal stem cells: progress toward promise. Cytotherapy 2005;7:36—45.

[117] Le Blanc K, Ringden O. Immunomodulation by mesenchymal stem cells and clinical experience. J Intern Med 2007;262:509—25.

[118] Ding YC, Xu DM, Feng G, Bushell A, Muschel RJ, Wood KJ. Mesenchymal stem cells prevent the rejection of fully allogenic islet grafts by the immunosuppressive activity of matrix metalloproteinase-2 and-9. Diabetes 2009;58:1797—806.

[119] Stax AM, Gelderman KA, Schlagwein N, Essers MC, Kamerling SWA, Woltman AM, et al. Induction of donor-specific T-cell hyporesponsiveness using dexamethasone-treated dendritic cells in two fully mismatched rat kidney transplantation models. Transplantation 2008;86:1275—82.

[120] Zhang XS, Li M, Lian DM, Zheng XF, Zhang ZX, Ichim TE, et al. Generation of therapeutic dendritic cells and regulatory T cells for preventing allogeneic cardiac graft rejection. Clin Immunol 2008;127:313—21.

[121] Brem-Exner BG, Sattler C, Hutchinson JA, Koehl GE, Kronenberg K, Farkas S, et al. Macrophages driven to a novel state of activation have anti-inflammatory properties in mice. J Immunol 2008;180:335—49.

[122] Reddy P, Sun YP, Toubai T, Duran-Struuck R, Clouthier SG, Weisiger E, et al. Histone deacetylase inhibition modulates indoleamine 2,3-dioxygenase-dependent DC functions and regulates experimental graft-versus-host disease in mice. J Clin Invest 2008;118:2562—73.

[123] Fandrich F, Lin XB, Chai GX, Schulze M, Ganten D, Bader M, et al. Preimplantation-stage stem cells induce long-term alloneic graft acceptance without supplementary host conditioning. Nat Med 2002;8:171—8.

[124] Reichardt W, Durr C, von Elverfeldt D, Juttner E, Gerlach UV, Yamada M, et al. Impact of mammalian target of rapamycin inhibition on lymphoid homing and tolerogenic function of nanoparticle-labeled dendritic cells following allogeneic hematopoietic cell transplantation. J Immunol 2008;181:4770—9.

[125] Geissmann F, Gordon S, Hume DA, Mowat AM, Randolph GJ. Unravelling mononuclear phagocyte heterogeneity. Nat Rev Immunol 2010;10:453—60.

[126] Gordon S, Taylor PR. Monocyte and macrophage heterogeneity. Nat Rev Immunol 2005;5:953—64.

[127] Hutchinson JA, Riquelme P, Sawitzki B, Tomiuk S, Miqueu P, Zuhayra M, et al. Cutting edge: immunological consequences and trafficking of human regulatory macrophages administered to renal transplant recipients. J Immunol 2011;187:2072—8.

[128] Broichhausen C, Riquelme P, Geissler EK, Hutchinson JA. Regulatory macrophages as therapeutic targets and therapeutic agents in solid organ transplantation. Curr Opin Organ Transplant 2012;17:332—42.

[129] Hutchinson JA, Riquelme P, Geissler EK. Human regulatory macrophages as a cell-based medicinal product. Curr Opin Organ Transplant 2012;17:48—54.

[130] Hutchinson JA, Brem-Exner BG, Riquelme P, Roelen D, Schulze M, Ivens K, et al. A cell-based approach to the minimization of immunosuppression in renal transplantation. Transpl Int 2008;21:742—54.

[131] Hutchinson JA, Riquelme P, Brem-Exner BG, Schulze M, Matthai M, Renders L, et al. Transplant acceptance-inducing cells as an immune-conditioning therapy in renal transplantation. Transpl Int 2008;21:728—41.

[132] Hutchinson JA, Riquelme P, Geissler EK, Fandrich F. Macrophages driven to a novel state of activation have tolerogenic properties in kidney transplant recipients. Am J Transplant 2009;9:657.

[133] Hutchinson JA, Roelen D, Riquelme P, Brem-Exner BG, Witzke O, Philipp T, et al. Preoperative treatment of a presensitized kidney transplant recipient with donor-derived transplant acceptance-inducing cells. Transpl Int 2008;21:808—13.

[134] Hutchinson JA, Govert F, Riquelme P, Brasen JH, Brem-Exner BG, Matthai M, et al. Administration of donor-derived transplant acceptance-inducing cells to the recipients of renal transplants from deceased donors is technically feasible. Clin Transplant 2009;23:140—5.

[135] Kirk AD. Clinical Tolerance 2008. Transplantation 2009;87: 953—5.

[136] Suthanthiran M, Ding R, Sharma V, Abecassis M, Dadhania D, Samstein B, et al. Urinary cell messenger RNA expression signatures anticipate acute cellular rejection: a report from CTOT-04. Am J Transplant 2011;11:29.

[137] Knechtle SJ, Pascual J, Bloom DD, Torrealba JR, Jankowska-Gan E, Burlingham WJ, et al. Early and limited use of tacrolimus to avoid rejection in an alemtuzumab and sirolimus regimen for kidney transplantation: clinical results and immune monitoring. Am J Transplant 2009;9:1087—98.

Chapter 63

Immunosuppression-Free Liver Transplantation

Tommaso M. Manzia[a], Leonardo Baiocchi[b], and Giuseppe Tisone[c]

[a]*Tor Vergata University of Rome, U.O.C. Chirurgia dei Trapianti, Policlinico Tor Vergata, Viale Oxford, Rome, Italy,* [b]*Tor Vergata University of Rome, U.O.C. Medicina Interna, Policlinico Tor Vergata, Viale Oxford, Rome, Italy,* [c]*Head of Transplantation Surgery, Tor Vergata University of Rome, U.O.C. Chirurgia dei Trapianti, Policlinico Tor Vergata, Viale Oxford, Rome, Italy*

Chapter Outline

63.1 INTRODUCTION

Transplantation is the treatment of choice for end-stage organ failure. The improving results have shifted clinicians' interest from prevention of acute rejection and short-term patient and graft survival toward long-term survival and quality of life. Posttransplant morbidity and mortality are mainly immunosuppression (IS) related, e.g., infections, diabetes, hyperlipidaemia, hypertension, renal failure, and malignancy [1−3]. The European Liver Transplant Registry (ELTR, www.eltr.org) and the United Network for Organ Sharing (UNOS, www.optm.transplant.hrsa.gov) report a 5-year patient survival rate after liver transplantation (LT) of 71% and 72%, respectively, and there can be no doubt that advances in IS management and postoperative care have contributed to this improvement. Currently, one of the chief interests of the scientific transplant community is to attempt the complete discontinuation of IS with organ acceptance, namely clinical operational tolerance (COT). Successful weaning from IS has already been reported in kidney and LT [4]. In this chapter, we will focus on IS-free LT analyzing recipients worldwide and reporting their outcome and future prospects. A fundamental step toward understanding IS-free LT is the concept of immunological tolerance. This chapter will also try to underline the difference between immunological tolerance and COT and explain why immunological tolerance is not achievable in humans at present. We will also assess the clinical predictive factors for COT and how the discovery of biomarkers and their clinical applications could, in the near future, improve the weaning success rate.

63.1 DEFINITION OF TRANSPLANTATION TOLERANCE

63.1.1 Immunological Tolerance

In 1953, Medawar was the first to describe transplantation tolerance in mice that underwent skin transplantation and received an injection of mononuclear cells from the same donor [5]. From the immunological point of view, tolerance was defined as *a functioning graft without histological sign of rejection in the absence of IS and in an immunocompetent host that can accept a second graft from the same donor but reject a graft from a different third-party donor*[6]. In other words, it is an immune nonreactivity toward a specific set of antigens that is indefinitely maintained in the absence of ongoing IS. In this setting we can recognize a central and a peripheral tolerance. *Central tolerance* is a natural process which takes place in the thymus in which autoreactive T-cells go through apoptotic deletion to avoid autoreactive T-cell development. On the other hand, *peripheral tolerance* consists of deletion during hematic circulation of those autoreactive/alloreactive T-cells that escaped thymic regulation.

63.1.2 Lessons from Animal Models

Liver allografts in some animal models are often spontaneously accepted in completely unrelated individuals without IS. Several studies on LT rat models have shown that the liver might be accepted also in the presence of major histocompatibility complex (MHC) mismatch. This is valid, for instance, between Piebald Viral Glaxo (PVG) donor and Dark Agouti recipients [7,8].

The reason for spontaneous liver allograft acceptance is still under debate even if there are a number of proposed mechanisms for which there is considerable experimental support: (i) possibly the liver produces large amount of class I MHC antigen that could modulate the recipients' immune response [9]; (ii) the liver is rich in donor hematopoietic stem cells that could be important in the establishment of chimerism in the recipients [10]; (iii) a similar role may be played by donor liver leukocytes that migrate to the recipients during LT as suggested by experiments on donor leukocytes depleted by irradiation [11]; (iv) the antigen-presenting cells of the rat liver show the indoleamine 2,3-dioxygenase (IDO) gene expression; this is an enzyme present in the placenta during pregnancy that is involved in preventing rejection of the fetus by the maternal T-cells [12]. In liver-transplanted rats, IDO seems to inhibit T-cell activation and proliferation protecting the transplanted cells, tissues, and organs from immune attack [13]. (v) Rat livers show low levels of interleukin (IL)-4, a cytokine-promoting humoral immunity. The inoculation of IL-4 in these animals significantly decreases allograft survival and causes death due to allograft rejection [14]. Interestingly, treatment with IL-4 of the donor PVG strain seems to protect Lewis rat recipients from the rejection that normally occurs. In fact, donor IL-4 treatment converts rejection to acceptance in the majority of Lewis recipients, inducing IDO and INF-gamma (normally produced by NK-cells) and increasing leukocytes transfer to the donor liver [15]. (vi) In the liver transplant rat model, the common gamma chain of the cytokine receptors that are involved in the T-cell proliferation and their survival are expressed at a lower rate; as a consequence the potential signaling by the cytokine microenvironment is blocked, thus inducing the apoptotic death of the infiltrating alloreactive T-cells and permitting the survival of the liver graft [16]. (vii) Finally, the liver is located in a peculiar microenvironment as it is constantly exposed to different antigens deriving from food and/or bacteria. This continuous exposure might be responsible for an increased acceptance of the liver allograft in comparison with other grafts [17,18].

63.1.3 COT in Liver

Since the concept of immunological tolerance is not clinically applicable, the term COT was introduced to better define the condition whereby a *transplant retains function without features of acute or chronic rejection in the absence of any IS*[19].

The first case of COT in solid organ transplantation was described in 1956 by Dr. J. Murray who performed the first renal transplant on the Herrick twins; as the result of genetic identity between the brothers, the recipient did not receive any IS [20].

While in experimental models, tolerance is induced using therapies aimed at the deletion of alloreactive cytopathic T-cells and generation of T-reg creating "tolerized animal models" [21], in human COT we cannot assume this, because our understanding of the mechanism responsible for transplant tolerance is incomplete and we lack a detailed picture of the nature of donor-specific immune responses. Several studies on IS weaning in LT have been published so far employing COT as the clinical end point. These trials focused more on the feasibility and safety of IS withdrawal than on the mechanism responsible for the maintenance of the tolerant status. Even if protocol liver biopsies have not been universally conducted and we cannot exclude minimal graft damage in weaned patients, COT is the condition where the transplant recipient is not receiving any IS while maintaining good graft function without signs of rejection.

63.1.3.1 Worldwide COT Experience in the Liver

Spontaneous COT in LT from Cadaveric Donors

We define as *spontaneous COT*, the IS withdrawal achieved in transplant recipients who do not receive any molecules or tolerogenic cells.

The very first series of spontaneous COT after LT was described by the Starzl group in 1995. They found that 11 LT recipients were IS free as a consequence of noncompliance or posttransplant lymphoproliferative disorders. They therefore designed a prospective trial in which IS was withdrawn in 95 patients with IS-related chronic toxicity. The weaning process was attempted and complete IS withdrawal was achieved within 3 years. Twenty eight (29%) recipients were weaned off IS, and after 10 years no episodes of rejection were reported [22−27]. A similar experience was reported by King's College, where the weaning protocol was successful in five out of 18 patients enrolled (28%), but only two remained completely IS free in the long term. In fact, one patient resumed IS for late acute rejection, one for kidney transplantation and one for retransplantation due to chronic rejection. The King's College team also identified as independent predictors of a sustained IS-free state, the nonautoimmune disease liver disorders, fewer donor—recipients HLA mismatches, and no previous acute rejection [28,29].

Pons et al. enrolled nine patients in an IS weaning study, investigating whether chimerism of the liver endothelium could influence the success of weaning. Three (33%) patients were successfully weaned, however, the hypothesis was not confirmed. In fact, chimerism of endothelium did not predict the COT [30].

Other authors have reported single cases of COT that were undertaken due to malignancy or as an accidental finding [31—34].

In 2006, the Tor Vergata University group published the results of an observational prospective clinical trial in which IS withdrawal was attempted in 34 long-term LT recipients with HCV-recurrent disease [35]. Complete and permanent IS withdrawal was achieved in eight patients (23.4%) (TOL), whereas 14 (41.2%) developed rejection within 8 months (non-TOL). Twelve (35.2%) rejected during tapering. After 4 years, the weaned patients showed stabilization/improvement of histological fibrosis, lower necro-inflammation, and improved liver function. Low blood cyclosporine (CsA) trough levels during the first posttransplant week and initial steroid-free IS were found to be independent predictors of weaning achievement. After 78 months of follow-up, seven of the eight originally tolerant patients remain alive and in good condition, while one died due to severe HCV recurrence (10 years post-LT and 6 years after complete removal of IS). Four out of 26 non-TOL individuals died due to: HCV recurrence (*n*: 2), lung carcinoma (*n*: 1), and acute myocardial infarction (*n*: 1) after a mean time of 10 years from LT. Of the non-TOL individuals, 50% developed new onset diabetes and about 60% suffered from either cardiovascular or infectious diseases [36]. After a longer follow-up (more than 10 years) out of eight TOL, six remain COT while one underwent a kidney transplant and required IS resumption [37] (Table 63.1).

The European Consortium of transplant tolerance, conceived and led by Prof. Sanchez-Fueyo, recently presented preliminary results of a prospective trial regarding 102 LT recipients who were enrolled in an IS weaning study. Forty patients (39%), with a median follow-up from LT of 8 years, were sustainably weaned off IS (30 of them were between 12 and 34 months of follow-up and 10 were completing the 12 months) [38].

Since early 2005, 29 patients of our Center (17M, median age 57 (SD 9.6) years) with a mean follow-up from LT of 110 months were included in this multicentre trial. Immediately after LT, all recipients were under calcineurin inhibitors (CNIs). At the time of weaning, the IS regimen was as follows, nine (31%) were under mycophenolate mofetil (MMF) and 20 (69%) under CNI. All patients underwent liver biopsies: before entry, after 12 months of IS withdrawal, and when rejection was clinically suspected. Patients were assessed every 4 weeks and IS was gradually discontinued with the aim of complete withdrawal by month 12 after the beginning of the study. Following IS discontinuation, patients were followed up monthly for 1 year. Liver functional tests (LFTs) were assessed at every clinical follow-up. Complete and permanent IS withdrawal was achieved in 14 (48%) LT recipients, whereas 14 had to resume IS during tapering. One, after an initial response, showed deterioration of LFTs and resumed IS without further complications. Interestingly, of the 15 non-TOL recipients only three showed histological signs of acute rejection at liver biopsy, and only one required corticosteroid bolus administration. Finally, the study identified: (i) long-term follow-up from LT, (ii) the lower dose of CsA at LT, (iii) use of MMF, as independent predictive factors of sustainable weaning off IS [39].

In summary, at the time of writing, the Transplant Unit of Tor Vergata University follows 23 individuals weaned off IS from a mean time of 60 months. At the last follow-up, the mean values of LFTs in this cohort showed only minimal fluctuations that could be justified by the large presence of HCV-positive patients in this series. None, at the most recent biopsy, showed histological signs of either acute or chronic rejection. Excluding those with HCV recurrence, the mean inflammation and staging Ishak score was 2 and 1, respectively.

Analyzing the quality of life of TOL recipients versus non-TOL, we realized that TOL patients are less prone to developing feelings of inadequacy and inferiority. In fact, the non-TOL cohort display a medium—high level of interpersonal sensitivity and develop stress which may be attributed to both the awareness of the failure of the weaning attempt and the lifelong need for IS [40].

For all the reasons above, the current policy of our center is to maintain our LT long-term cohort on very low doses of IS. In other words, we always try to achieve a *tailored IS therapy* conforming to each single patient in order to avoid both rejection and morbidity and when possible to achieve IS avoidance.

Spontaneous COT in LT from Parental Donors

COT in living donor LT has been reported mainly in pediatric studies.

Kyoto University reported 88 of 581 pediatric living donor liver recipients (15%) who were operationally tolerant 2 years after transplant. Among the tolerant Kyoto children, 38% underwent IS withdrawal because of EBV infection or other complications. They reported normal

TABLE 63.1 Spontaneous COT

Year of Publication/Journal	Author	Study Type	No. of Patients	Adult/Pediatric	Mean Age at Time of Weaning (Years)	Selection Criteria	Baseline IS	Mean Time Between LT and Weaning (Months)	Tolerant	Mean Follow-Up from IS Withdrawal (Months)	Acute Rejection	Chronic Rejection
1995 Transpl Proc 2007 Transpl Immunol	Mazariegos [23–27]	Prospective, single arm	95	Both (30% children)	11.8	>5 years post-LT, >2 years without rejection episode, evidence of compliance	CsA or Tac + Aza	70	28 (29%)	180	25 (26%)	None
1998 Hepatology 2005 Transpl Proc	Devlin [28] Girlanda [29]	Prospective, single arm	18	Adult	47.7	>5 years post-LT, chronic IS-related toxicity	CsA, Aza, prednisolone	60	2 (11%) of the initial 5 (28%)	120	5 (27.7%)	1 (5.6%)
2003 Transplantation	Pons [30]	Prospective, single arm	9	Adult	53.1	>2 years post-LT, normal graft function	CsA	>60	3 (33%)	20	2 (22%)	None
2000 J Clin Invest	VanBuskirk [34]	Observational	1	NA	NA	–	NA	36	1	36	None	None
2006 J Hepatol 2008 Transpl Immunol	Tisone [35,36]	Prospective, controlled	34 (only HCV)	Adult	62.2	>1 year post-LT, established HCV recurrence disease, normal graft function, evidence of compliance	CsA	63.5	8 (24%)	122[a]	26 (76.4%)	None
2007 Liver transpl	Gras [31]	Prospective, placebo-controlled, randomized	1	Pediatric	2	First single transplant, age <15 years, no autoimmune disease	Tac	2	1	55	None	None
2007 Transpl Immunol	Hsu [32]	Observational	1	NA	NA	–	NA	60	1	60	None	None
2008 N Engl J Med	Alexander [33]	Observational	1	Pediatric	9	–	CsA, prednisone	14	1	46	None	None
2010 Am J Transpl (Suppl)	RISET Consortium [38]	Prospective	102 (24 HCV)	Adult	56	<3 years from tx, no liver autoimmune disease, no rejection episode in the previous 12 months	Tac, CsA, MMF	35.8	40 (39%)	25.7	56 (58%), 89% biopsy proved	None

NA, not available; LDLT, living donor liver transplant.

[a]Follow-up to July 2011: seven recipients remain alive with stable graft function, one required IS resumption after 7 years of IS-off due to kidney transplant, while one died for HCV recurrence 10 years post-LT and 6 years after complete IS withdrawal.

LFTs in COT individuals; however, an increase in the number of bile ducts, and a higher degree of fibrosis was evidenced. As liver function tests themselves are not considered sufficient to detect the development of subclinical rejection, they therefore suggested careful histological monitoring of these individuals [41−43].

Enthusiastic results among pediatric recipients of parental living donor liver transplant were recently published by Feng et al.; of the 20 recipients, 12 (60%) were successfully withdrawn from IS and after 3 years maintained normal graft function. Of the recipients who experienced rejection, all returned to normal allograft function after IS resumption [44]. In contrast with the Kyoto group, no histological inflammation or fibrosis progression was observed over time. In addition they suggested the long-term follow-up as the main predictor for IS-free state.

A Korean transplant group has also reported encouraging results. In their experience, IS withdrawal was attempted in five pediatric liver transplant recipients who received parental donation 4 years before (no treatment compliance (n: 3); posttransplant lymphoproliferative disorder (n: 1); no evidence of chronic rejection after 4 years was reported) [45].

These stimulating preliminary results show that achievement of COT is strictly related to the careful selection of the candidate and, as mentioned below, long-term follow-up and living related LT seem to be the most important clinical predictors of sustained weaning off IS (Table 63.2).

Intentional Induction of COT in LT

Since IS withdrawal has a low success rate (less than 30%), most groups have tried to induce COT by medication, either molecular or cell based.

The Pittsburgh group worked on the principle that the strong depletion of effector T-cells prior to the engraftment, followed by a low Tacrolimus (Tac) dose after transplantation, could lead to graft acceptance [46]. The authors administered *ab initio* an IS protocol supposed to be tolerogenic, in 82 adult kidney, pancreas, liver, or intestinal transplant recipients. The goal of the study was to eliminate those effector T-cells (namely Th1 and Th2) responsible for graft rejection [47−54]. In addition, minimization of IS was attempted in order to find a balance between irreversible immune damage to the graft and donor-specific clonal exhaustion−deletion. In fact, experimental and clinical data have shown that multidrug IS usage at the time of transplantation does not favor the development of tolerance compared to the use of low-dose monodrug therapy [55]. The 1-year patient and graft survival rates were 95% and 82%, respectively. IS morbidity was eliminated and 48/72 survivors were receiving a reduced dose of Tac. Even though none had IS

withdrawal, it must be highlighted that the very low dose of IS was obtained early after transplantation in organs considered heavily immunogenic such as kidney, pancreas, and intestine, where IS minimization was obtained in 25/39 (64%), 5/12 (42%), and 6/11 (54%), respectively. The highest percentage of success was, as expected, obtained in LT recipients where 12/17 (70%) underwent low-dose Tac monotherapy.

Due to the good results shown by the Starzl study, Eason et al. attempted the complete weaning of IS in 18 LT recipients who had undergone transplantation at least 6 months earlier. These patients received rabbit antithymocyte globulin for induction in combination with antimetabolites (namely MMF and Tac). In this case, weaning was achieved in just one recipient (5%) whereas most of the patients had biopsy proven acute rejection either during tapering (n: 11, 61%) or after withdrawal (n: 3, 16%). (The fate of the remaining three patients is difficult to ascertain) [56].

These results, although discouraging, introduce an important point of debate. While some concerns may arise due to the fact that no patients could be weaned off IS, the striking reduction in the immunosuppressant daily dose was an outstanding achievement. Moreover, it should be underscored that it was obtained after transplantation of highly immunogenic organs.

In 2007, Assy et al. proposed ursodeoxycholic acid as a tolerogenic molecule in LT recipients [57]; the rationale was that ursodeoxycholic acid reduces the incidence and severity of acute rejection in liver allografts reducing the expression of MHC class molecules in biliary epithelia and central vein endothelia of the liver [58]. However, the results were discouraging, in fact 15/26 (58%) patients developed an episode of acute rejection within 12−21 weeks from the beginning of tapering.

The Ghent-Brussels group described three patients who underwent LT for malignancy outside of the general criteria (patients suffering from multifocal hepatocarcinoma (×2) and cholangiocarcinoma (×1)). They all received a right lobe living donor graft and CD34 + stem cells harvested from the same donor. The major aim was to attempt withdrawal of IS. The IS protocol consisted of steroids, rapamycin, and antithymocite globulin. Two patients were weaned off IS 3 weeks after LT and remained stable until their death due to tumor recurrence (after 1½ years). The other patient could not be weaned off IS due to acute rejection and is still alive under Tac monotherapy [59]. Apart from the concern of tumor recurrence that precludes including patients affected by advanced malignancies in LT programs, this protocol is promising due to the good success rate of weaning off IS by using stem cells.

Some case reports of bone marrow transplantation for hematologic disease and successful weaning off IS after

TABLE 63.2 Spontaneous COT in LT from Parental Donation

Year of Publication/Journal	Author	Study Type	No. of Patients	Adult/Pediatric	Mean Age at Time of Weaning (Years)	Selection Criteria	Baseline IS	Mean Time Between LT and Weaning (Months)	Tolerant	Mean Follow-Up from IS Withdrawal (Months)	Acute Rejection	Chronic Rejection
2001 Transplantation 2007 Transpl Immunol	Takatsuki [42] Koshiba [43]	Mixed, single arm	581 (overall LT population)	Pediatric	1.1	>2 years post-LT, no previous acute rejection, normal graft function	Tac	>24	88 (15%)	24	16 (25.4%)	None
2009 Yonsei Med J	Lee JH [45]	Observational	5	Pediatric	–	Noncompliance to medication, PTLD	Tac	45	5 (100%)	32	–	None
2012 JAMA	Feng [44]	Prospective	20	Pediatric	8.5	Tx from parental graft, age >4 years, stable graft function on CNI monotherapy	Tac or CsA	88.8	12 (63%)	Range: 18–39	7(37%), 10.5% biopsy proved	None

NA, not available; LDLT, living donor liver transplant.
*Follow-up to July 2011: seven recipients remain alive with stable graft function, one required IS resumption after 7 years of IS-off due to kidney transplant, while one died for HCV recurrence 10 years post-LT and 6 years after complete IS withdrawal.

LT are present in literature; Margreiter et al. described a case of a young patient who developed liver-based chronic graft versus host disease after allogenic bone marrow transplantation for sideroblastic anemia. The patient was treated by LT from the same HLA identical donor, and IS was gradually discontinued and finally withdrawn after 71 months [60]. A similar experience was reported by Clavien [61] and Andreoni [62] who described two cases of COT after LT in two patients who received allogenic bone marrow transplantation a few years earlier and underwent LT for cholangiocarcinoma and HCV-related cirrhosis, respectively.

Based on these reports, Tzakis hypothesized that the perioperative infusion of donor bone marrow cells might favor COT after LT [63]. A group of 104 LT recipients were allocated to receive (n: 45) or not to receive (n: 59) donor bone marrow cells at the time of LT. IS was reduced by one-third upon enrollment, by another one-third the second year, and was completely withdrawn the third year. Twenty three recipients (22%) were off IS after 8 years of follow-up [64], but the working hypothesis was not confirmed as the donor bone marrow cells infusion did not increase the likelihood of COT. Although the data from bone marrow and LT patients would support this study, the results were discouraging as the incidence of acute rejection and COT were similar in both groups. The immunomodulatory properties of the bone marrow—derived stem cells, as shown in in vitro studies [65], are dependent on the infused dose [66]. We could therefore argue that the number of the cells infused in these subjects was maybe insufficient to achieve a sustainable IS withdrawal (Table 63.3).

63.1.4 The Impact of IS-Free Status on HCV Recurrence

Since HCV recurrence represents the main indication to LT worldwide [67], this part of the chapter is dedicated to the impact of IS withdrawal on the natural history of HCV-recurrent disease after transplant.

HCV infection of the allograft occurs at the reperfusion during transplantation in all patients with chronic hepatitis C who have detectable serum HCV-RNA levels prior to LT. Up to 30% of these recipients develop cirrhosis within 5 years. It has been well documented that the course of HCV infection is accelerated in LT patients when compared to an immune competent host. In fact, cirrhosis occurs after about 30 years in HCV immunocompetent patients while only 10 years are usually sufficient in LT recipients [68]. Although several factors contribute to the faster progression of HCV-related liver disease after LT (donor age >65 years, macrosteatosis of the donor liver, prolonged cold ischemic time, the age of

the recipients >50, insulin-dependent diabetes, and the high viral load at the time of LT) [69], the IS regimen remains one of the main posttransplant variables responsible to HCV disease recurrence [70] that has the possibility to be modified or adjusted.

Rejection treatment with corticosteroid boluses and OKT3 has been associated with high severity of HCV recurrence because of increased HCV-RNA viral load and accelerated fibrosis progression [71]. Although CsA has shown good antiviral response on HCV replication, when compared to Tac, in in vitro studies [72,73], clinical researches failed to support this finding as a significant difference on clinical outcomes with either CsA or Tac was not evidenced [74], and both drugs equally mitigate the immune response against the HCV.

Data on azathioprine and MMF in HCV-transplanted patients are characterized by conflicting results. A systematic review comparing the impact of AZA and MMF on HCV recurrence showed that AZA administration might be associated with a lower HCV recurrence rate than MMF [75]. On the other hand, MMF use has been shown to improve fibrosis progression in HCV recipients either alone [76] or in association with low-dose CNIs administration [77]. There are no data, at present, to support the use of mTOR inhibitors' in HCV recipients [78].

It must be highlighted that specific IS agents may elicit negative effects on the disease with an indirect mechanism. In particular, it has been reported that insulin resistance and diabetes, common side effects of long-term IS administration, are associated with higher fibrosis progression after LT [79,80]. Therefore, the use of IS regimen with reduced or absent metabolic side effects is a possible strategy that may slow posttransplant disease progression [81].

These data suggest that even though the ideal IS regimen in the HCV transplant setting is as yet undefined, it is likely that the less potent is the drug, the slower is the natural course of HCV.

Literature reports 21 cases of COT in HCV LT recipients [82]. Unfortunately, in most studies protocol liver biopsies were not performed so histological data are not available. Only the Tor Vergata group reported the pathological findings over a period of 10 years. Of the 34 HCV patients originally enrolled to the weaning study, six were off IS over 10 years. These recipients have been followed and compared to those who could not achieve IS withdrawal. Ten consecutive yearly biopsies were available for each patient. When baseline biopsies were compared with 10-year biopsies, the TOL-HCV patients showed an improvement in grading and no differences in staging meanwhile in non-TOL-HCV group, staging increased. In terms of the fibrosis progression rate at 10 years, the TOL-HCV patients showed a slower progression of tissue damage than the non-TOL patients (-0.06 ± 0.12 vs

TABLE 63.3 Intentional Induction of Operational Tolerance

Year of Publication/ Journal	Author	Study Type	No. of Patients	Adult/ Pediatric	Mean Age at Time of Weaning (Years)	Selection Criteria	Donor Bone Marrow Transplant Before LT	Induction of Operational Tolerance	Mean time Between LT and Weaning (Months)	Tolerant	Follow-Up from IS Withdrawal (Months)	Acute Rejection	Chronic Rejection
2002 Bone Marrow Transplant	Urban [60]	Observational	1	Pediatric	7.5	–	Yes	Aza, prednisone, CsA	71	1	30	None	None
2003 Lancet	Starzl [46]	Prospective, single arm	82	Adult	NA	First transplant	35 patients (43%)	Anti-thymocyte globulin + Tac	*Ab initio*	0 (0%)	15	NA	NA
2003 Transplantation	Kadry [61]	Prospective	1	Adult	41	–	Yes	–	*Ab initio*	1	12	None	None
2004 N Engl J Med	Andreoni [62]	Observational	1	Adult	42	–	Yes	–	6	1	24	None	None
2004 Transpl Immunol	Donckier [59]	Prospective, single arm	3	Adult	54	Liver cancer ineligible for LT or other curative options, living donor graft	No	Anti-thymocyte globulin, steroids, rapamycin + stem cells	*Ab initio*	3 (100%)	498 days (alive), 561 and 356 days (dead for tumor recurrence)	3 (100%) during tapering (resolved)	None
2005 Transplantation	Eason [21]	Prospective, single arm	18	Adult	NA	First transplant, >6 months post-LT, not previous acute rejection	No	Anti-thymocyte globulin + MMF + Tac	>6	1 (5.6%)	12	11 (61%) during tapering, 3 (3.16%) after withdrawal	None
2005 Am J Transplant 2010 Transplantation	Tryphonopoulos [63,64]	Prospective, controlled	104	Adult	NA	>36 of follow-up, no previous acute rejection, no autoimmune diseases	45 patients (43%)	–	>12	23 (22%)	7.27 ± 0.28 years	20 (67%)	2 (1.9%)
2007 Transplantation	Assy [57]	Prospective, controlled	26	Adult	53	Rejection-free for >24 months	No	UDCA	56	2 (8%)	12	15 (58%)	None

NA, not available; DBMC, donor bone marrow cells; UDCA, ursodeoxycholic acid.

0.1 ± 0.2, respectively, $p = 0.04$). At the last biopsy taken, 63% of non-TOL-HCV patients showed features of advanced fibrosis (defined as a fibrosis >4 according to the Ishak score) [83] and 40% frank cirrhosis. On the contrary, none of the patients in the TOL-HCV group presented either advanced fibrosis or cirrhosis. No evidence of early or late chronic rejection was ever observed during the follow-ups [37].

Even if, is not possible to draw a definitive conclusion on the basis of these data due to the small number of patients included, the major strength of this study derives from the availability of yearly protocol liver biopsies in all patients. Since the nearly 300 biopsies performed in 34 patients over a 10-year period were studied by the same pathologist, this allowed us to closely monitor the histological evolution of recurrent HCV-related disease while minimizing the potential biases attributable to sampling errors or interpretation by different specialists.

63.1.5 Prediction of COT

As mentioned above, about 25% of carefully selected LT recipients may be successfully weaned off IS for sustained periods of time. More recent preliminary results from multicentre trials conducted in Europe [38] and the United States [44] in adult and pediatric patients have shown that COT may occur more frequently than originally estimated. Clinical predictor factors, such as stable graft function, length of follow-up after LT, IS type and blood levels, close matching of HLA type between donor and recipient, absence of autoimmune liver diseases, no history of rejection, were used as inclusion criteria for patient selection [27,35,63,64].

Long-term results of IS weaning so far reported have shown that the longer the follow-up after LT the higher the likelihood of weaning off IS. In this setting, the European Consortium of transplant tolerance showed that 60% of successful IS weanings were followed up for over 10 years from LT and almost the total cohort of the TOL patients were followed up for over 6 years from LT [38].

IS regime might also determine the successful achievement of COT [34]. In the early posttransplant period, IS seems to play a key role in tolerance achievement after LT. The low dose of CNI in the first postoperative weeks is considered as an independent factor in sustained IS withdrawal due to interference, with a currently unknown mechanism, in the pathway of T-cell activation, T-receptor downregulation and in the cytokine microenvironment [84]. Recently, our group reported that CsA monotherapy and corticosteroid avoidance during the first post-LT week, together with the switch from CsA to MFM in the medium-term after LT, might predict sustained tolerance in up to 50% of patients [39]. Moreover, 90% of the TOL group who were treated with a low dose of CsA

at the time of transplantation were switched to MMF during the follow-up, due to either renal failure or hyperlipidaemia. This group also had over 10 years of follow-up at the time of the beginning of tapering.

Having determined some important clinical predictors of COT, in recent years, clinicians and scientists have focused their attention on determining the immunological signature of the tolerant recipients with the aim of identifying those patients who could successfully be weaned off IS. The development of noninvasive biomarker testing in transplantation currently assumes an important role in LT due to the increasing likelihood of identifying COT recipients who might be able to undergo a reduction of IS or be withdrawn from it. Biomarkers are commonly defined as *objective and measurable indicators of physiological processes, pathological conditions, or pharmacological responses*[85]. In the transplant setting, their principal role is to evaluate the immune status of recipients, either phenotypic or genotypic, and to identify those associated with COT.

The two strategies that have been successfully adopted mainly on peripheral blood samples are the multiparameter flow cytometry (capable of analyzing in detail the number and the surface of phenotypes of peripheral blood APCs and lymphocytes) and microarrays or real-time PCR gene expression profiling (employed to determine which genes are differentially expressed in tolerant versus nontolerant).

In recent years, several studies have focused on the immunophenotypic and genotypic pathway of peripheral blood mononuclear cells (PBMCs), comparing the phenotypic expression of liver-tolerant recipients with those who require IS. In 2003, Mazariegos et al. quantified the human dendritic cell (representing a part of APCs) subsets in six TOL LT recipients and in 23 non-TOL LT recipients. They showed that the ratio between *monocytoid dendritic cells inducing Th1* responses and *plasmacytoid dendritic cell inducing Th2* responses was significantly lower in the TOL group [86]. One year later, Li et al. phenotyped PBMCs in 12 COT LT recipients and in 19 pediatric patients who received living donor LT and were under IS. In the TOL group an increasing number of CD4$^+$CD25^{high+} subsets of the B-cells, a high ratio of $\gamma\delta1/\gamma\delta2$ T-cells, and a decreasing number of NK-cells were observed [87].

Based on these reports, we also focused our study [88] on the gene expression profiling and immunophenotyping of the PBMCs. In this setting, 16 COT LT recipients, 16 recipients requiring ongoing IS therapy, and 10 healthy individuals were enrolled in an immunological study. We confirmed, as in the previous report, that CD4$^+$CD25$^+$ subsets, $\gamma\delta1$ T-cells, and $\gamma\delta1/\gamma\delta2$ ratio increased in TOL, but no differences were found in B-cell, NK, and dendritic cells among the three groups. This study was also the first to employ oligonucleotide microarray analysis on

PBMCs samples to identify gene encoding for γδ T-cells, NK receptors, and for proteins involved in cell proliferation.

Recently, Pons et al. [89] showed an increasing subset of CD4$^+$CD25$^+$ T-cells during the tapering process of the five patients who successfully discontinued IS. These results opened an important debate because they showed, firstly, that the phenotypic pathway changes during the IS tapering. This finding strongly suggests that a prospective trial on IS weaning together with biomarker studies are needed to better understand the changing of the immunological signature over time.

The largest immunological study on COT so far reported was conducted by the European Consortium of transplant tolerance in 2008 [90]. We analyzed transcriptional patterns in the peripheral blood of 80 LT recipients and 16 nontransplanted healthy individuals by employing oligonucleotide microarrays and quantitative real-time PCR. Functional analysis of the whole expression data set identified the NK-cell signaling pathway as the most significant prediction factor for COT. Additional flow cytometry studies showed that specific PBMC subsets (CD4$^+$CD25$^+$foxp3 T-cells, Vδ1$^+$-T-cells, plasmacytoid dendritic cells) were also increased in tolerant recipients, confirming previous reports. Pons et al. [89] and Li [87] also showed that the high expression of *FOXP3* gene, the master regulator of CD4$^+$CD25$^+$ T-reg cells, is strongly related to tolerance status.

These findings suggest that transcriptional profiling of peripheral blood might be used as a biomarker of operational tolerance to successfully identify LT recipients who can discontinue IS therapy.

A recent study conducted by our European group also analyzed the liver tissue samples collected from liver recipients enrolled in a prospective multicenter clinical trial on IS withdrawal. In TOL and non-TOL recipients, substantial differences were found in intra-graft gene expression of messages responsible for the regulation of iron homeostasis. Furthermore, there was also an increase in hepcidin and ferritin serum levels in TOL recipients. These preliminary results open new perspectives for a novel set of biomarkers that may guide clinicians in IS withdrawal after LT [91].

From all the data collected, it is clear that the main limit of these studies is the small sample of patients due to the paucity of COT population. In this setting, the European Union (RISET Consortium) and the Immune Tolerance Network are conducting large clinical trials to better identify COT features (Figure 63.1).

63.1.6 COT from the Practical Point of View

While IS weaning in the clinical field remains within a research framework, careful attempts should be made by expert hepatologists or transplant surgeons. The following paragraphs provide guidelines, on the basis of the literature currently available, on how successful IS withdrawal may be implemented. In addition, a paragraph on how to manage liver impairment during the weaning process is included.

63.1.6.1 Possible Candidates for IS

Patients exhibiting severe side effects from immunosuppressive drugs (diabetes, renal failure, hyperlipidaemia, hypertension), and/or risk of neoplasm development (history of previous non-hepatocarcinoma neoplasms, tobacco or alcohol consumption), and/or HCV disease recurrence represent the majority of the liver-transplanted population [92]; and are those recipients who could most benefit from IS withdrawal. Unfortunately, not all LT recipients can successfully undergo IS withdrawal due to the high rate of IS withdrawal failure [35]. In order to attempt the weaning process, it is important to carefully select those recipients with the main predictors of operational tolerance; for instance, patients under low-dose IS monotherapy are more likely to be operationally tolerant than those requiring double therapy because they are already "prope tolerant"[39].

As first shown by the King's College group [28,29], the history of liver autoimmune disease (autoimmune hepatitis, primary liver cirrhosis, primary sclerosing cholangitis) seems to be a contraindication to attempting weaning due to the impossibility of predicting the operational tolerance status and also because the disease itself requires IS to avoid recurrence. Since half of the patients cannot be withdrawn from therapy due to liver impairment and/or rejection, starting IS weaning in those recipients who experienced graft rejection in the preceding 12 months is not recommended. Furthermore, the liver graft function should be normal in the preceding 6 months, or there should be only minor alterations in the liver function tests that have remained unchanged over the previous 6 months (AST/ALT <2-fold normal levels; ALP <1.5-fold normal levels; GGT <2-fold normal levels; bilirubin <2 mg/dL) [28,29,35,37,38,44].

To attempt the weaning process, IS must be tapered slowly (see below) and safely; therefore, clinical follow-up visits every 2−3 weeks for these patients are essential [35].

63.1.6.2 When to Attempt IS

Since long-term follow-up seems to be the most important clinical predictor for COT, most authors prefer to attempt weaning 12−36 months after LT. In fact, the longer the time lapse, the higher the likelihood of operational tolerance. As mentioned above, the IS-free state seems to benefit HCV LT recipients but HCV damage takes place

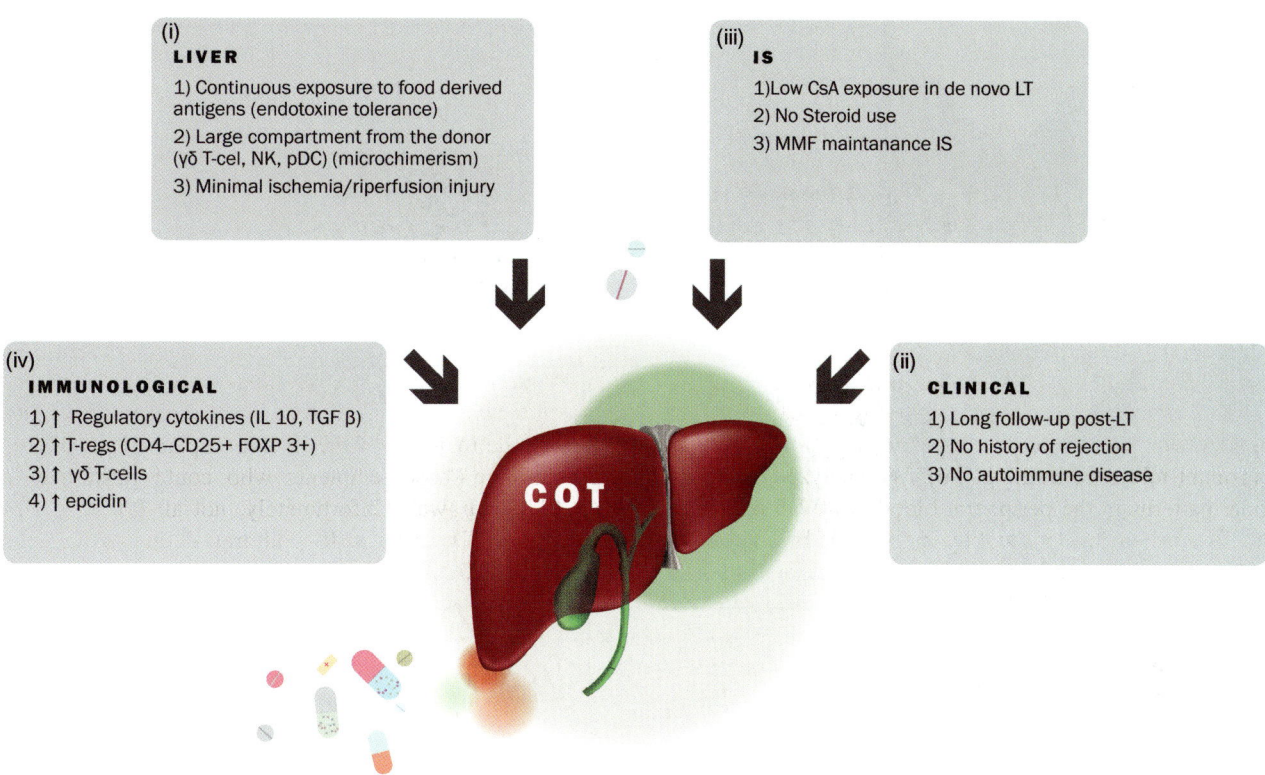

NK: natural killer cells; pDC: plasmocitoid dendritic cells; CsA: Cyclosporine; MMF: Mycophenolate Mofetil; COT:clinical operational tolerance

FIGURE 63.1 Components associated with liver COT. COT in liver transplantation could be influenced by several factors: (i) intrinsic liver factors such as the continuous antigen exposure due to portal flow or the hematopoietic cell transfer from the donor; (ii) clinical factors such as the length of follow-up or the absence of a history of rejection; (iii) immunosuppressive drug use, e.g., low-dose CNI in *de novo* recipients, steroid avoidance, and antimetabolite use in long term; (iv) immunological pathway: The presence of IL-10 and TGF-β microenvironment turns donor-activated CD4 + T-cells into tissue protective T-regs. NK, natural killer cells; pDC, plasmocitoid dentritic cells.

shortly after LT. In this cohort, immune surveillance should be resumed early to counteract the disease. In this setting, we need to intervene at the right time to contrast disease recurrence without exposing recipients to an unjustified rejection due to IS withdrawal. We think that the development and spread of new biological markers as early identifiers of the *hidden operational tolerant recipient* could lead to the anticipation of the weaning process [4,24,25].

63.1.6.3 Method of Weaning

It is highly advisable to perform a liver biopsy before attempting weaning, to exclude subclinical rejection. After that, IS can be gradually discontinued with the aim of achieving a 50% decrease in drug dosages by month 3 and complete withdrawal by month 6. After drug discontinuation, patients should be followed up every 2–3 weeks for 3 months and monthly thereafter until month 12 after initiation of the study. After 1 year, patients can

return at the standard follow-up schedule. Liver function tests should be obtained at every clinical follow-up visit [26,28,29,35] (Figure 63.2).

63.1.6.4 How to Manage Liver Impairment

If IS tapering/withdrawal results in an increase in liver function test values under 2-fold normal levels for AST/ALT/GGT, 1.5-fold normal levels for ALP, or 2 mg/dL for bilirubin, it is advisable to stop decreasing drug dosages or to resume IS; patients should repeat liver function tests within 1 week. Worsening or persistence of liver function test alterations will constitute an indication for liver biopsy. Increases in liver function tests beyond 2-fold normal levels for AST/ALT/GGT, 1.5-fold normal levels for ALP, or 2 mg/dL for bilirubin will result in liver biopsy. Usually, liver function recovers after the resumption of IS. In those rare cases in which IS resumption does not result in patient recovery, corticosteroids are required [38,44] (Figure 63.3).

FIGURE 63.2   Patients selection and weaning method. It is highly recommended to carefully select those recipients who could safely undergo weaning off IS. Recipients who show IS side effects could be of greater benefit of IS withdrawal. It is advisable to attempt weaning gradually with the aim of achieving a 50% decrease in drug dosages by month 3 and complete withdrawal by month 6.

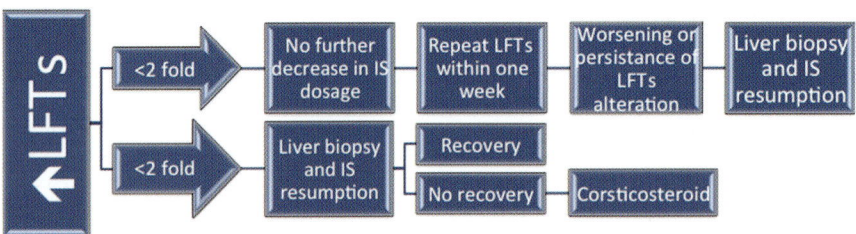

FIGURE 63.3   How to manage liver impairment during or after IS withdrawal. If IS tapering/withdrawal results in an increase in LFTs under 2-fold normal levels, it is advisable to stop decreasing drug dosages or to resume IS; patients should repeat LFTs within 1 week. Worsening or persistence of LFT's alterations will constitute an indication for liver biopsy. Increases in LFTs beyond 2-fold normal levels will result in liver biopsy and IS resumption. Corticosteroids are required if baseline-IS does not restore the liver function.

63.1.7 Current Prospect and Future Perspectives

The data reported above show that COT in LT is currently achievable in about 25–30% of carefully selected cases. The likelihood of TOL is higher in comparison to that identified earlier because several clinical predictor factors such as long-term follow-up were identified and used for patient selection. The pathways of immune response triggered by the engraftment were also studied, and the researchers identified several immunological factors that could predict COT. In this scenario, we are almost ready to identify those patients who could

successfully discontinue IS, but, on the other hand, we have nothing to offer those who cannot discontinue it. Several papers have demonstrated that the tolerogenic protocols adopted in the last few years have failed to induce COT. Therefore, new clinical trials are in progress to better clarify the "immunological jungle" present in the transplantation setting with the aim of identifying the immunological printing of COT. Therefore, it is possible to hypothesize that those phenotypes which are not associated with COT can also be identified and maybe blocked or inhibited by new tolerogenic drugs.

It may be time to consider combining both immunology and regenerative medicine approaches to achieve

COT. In fact, we believe that the harnessing of the immunomodulatory capabilities of stem cells—with their ability for tissue repair and regeneration—could provide a stimulating opportunity for further research in the field of LT, with the potential to address the two major issues of modern transplantation: COT and an inexhaustible source of organs [92,93]. In other words, the bioengineering platform may be able to supply the transplant field with new strategies to solve the shortage of organs and to attempt COT avoiding IS *ab initio*[94].

REFERENCES

[1] Fridell JA, Jain A, Reyes J, Beiderman R, Green M, Sindhi R, et al. Causes of mortality beyond 1 year after primary pediatric liver transplant under tacrolimus. Transplantation 2002;74:1721−4.

[2] McDiarmid SV. Management of the pediatric liver transplant patient. Liver Transpl 2001;7:77−86.

[3] Wallot MA, Mathot M, Janssen M, Holter T, Paul K, Paul Buts J, et al. Long-term survival and late graft loss in pediatric liver transplant recipients—a 15-year single-center experience. Liver Transpl 2002;8:615−22.

[4] Orlando G. Finding the right time for weaning off immunosuppression in solid organ transplant recipients. Expert Rev Clin Immunol 2010;6:879−92.

[5] Billingham R, Brent L, Medawar P. Acquired immunological tolerance to foreign cells. Nature 1953;172:603−6.

[6] Ashton-Chess J, Giral M, Brouard S, Soulillou JP. Spontaneous operational tolerance after immunosuppressive drug withdrawal in clinical renal allotransplantation. Transplantation 2007;84:1215−9.

[7] Houssin D, Gigou M, Franco D, Bismuth H, Charpentier B, Lang P, et al. Specific transplantation tolerance induced by spontaneously tolerated liver allograft in inbred strains of rats. Transplantation 1980;29:418.

[8] Zimmermann FA, Davies HS, Knoll PP, Gokel JM, Schmidt T. Orthotopic liver allografts in the rat: the influence of strain combination on the fate of the graft. Transplantation 1984;37:406.

[9] Sumimoto R, Kamada N. Specific suppression of allograft rejection by soluble class I antigen and complexes with monoclonal antibody. Transplantation 1990;50:678.

[10] Qian S, Demetris A, Murase N, Rao A, Fung J, Starzl T. Murine liver allograft transplantation: tolerance and donor cell chimerism. Hepatology 1994;19:916.

[11] Sriwatanawongsa V, Davies HS, Calne RY. The essential roles of parenchymal tissues and passenger leukocytes in the tolerance induced by liver grafting in rats. Nat Med 1995;1(5):428−32.

[12] Munn DH, Zhou M, Attwood JT, Bondarev I, Conway SJ, Marshall B, et al. Prevention of allogeneic fetal rejection by tryptophan catabolism. Science 1998;281:1191−3.

[13] Lin YC, Chen CL, Nakano T, Goto S, Kao YH, Hsu LW, et al. Immunological role of indoleamine 2,3-dioxygenase in rat liver allograft rejection and tolerance. J Gastroenterol Hepatol 2008;23:243−50.

[14] Wang C, Li J, Cordoba SP, Tran G, Hodgkinson SJ, Hall BM, et al. Post-transplant IL-4 treatment converts rat liver allograft tolerance to rejection. Transplantation 2005;79:1116−20.

[15] Wang C, Tay SS, Tran GT, Hodgkinson SJ, Allen RD, Hall BM, et al. Donor IL-4-treatment induces alternatively activated liver macrophages and IDO-expressing NK cells and promotes rat liver allograft acceptance. Transpl Immunol 2010;22:172−8.

[16] Ganbold A, Andersen S, Tay SS, Cunningham E, Ilie V, Krishnan S, et al. Expression of common gamma chain signalling cytokines and their receptors distinguishes rejection from tolerance in a rat organ transplant model. Transpl Immunol 2012;27:89−94.

[17] Crispe IN. The liver as a lymphoid organ. Annu Rev Immunol 2009;27:147−63.

[18] Kern M, Popov A, Kurts C, Schultze JL, Knolle PA. Taking off the brakes: T cell immunity in the liver. Trends Immunol 2010;31:311−7.

[19] Lerut J, Sanchez-Fueyo A. An appraisal of tolerance in liver transplantation. Am J Transplant 2006;6:1774−80.

[20] Merrill JP, Murray JE, Harrison JH. Successful homotransplantation of the human kidney between identical twins. JAMA 1956;160:277−82.

[21] Li XC, Strom TB, Turka LA, Wells AD. T cell death and transplantation tolerance. Immunity 2001;14:407−16.

[22] Reyes J, Zeevi A, Ramos H, Tzakis A, Todo S, Demetris AJ, et al. Frequent achievement of a drug-free state after orthotopic liver transplantation. Transplant Proc 1993;25:3315−9.

[23] Mazariegos GV, Ramos H, Shapiro R, Zeevi A, Fung JJ, Starzl TE. Weaning of immunosuppression in long-term recipients of living related renal transplants: a preliminary study. Transpl Proc 1995;27:207−9.

[24] Ramos HC, Reyes J, Abu-Elmagd K, Zeevi A, Reinsmoen N, Tzakis A, et al. Weaning of immunosuppression in long-term liver transplant recipients. Transplantation 1995;59:212−7.

[25] Mazariegos GV, Reyes J, Marino IR, Demetris AJ, Flynn B, Irish W, et al. Weaning of immunosuppression in liver transplant recipients. Transplantation 1997;63:243−9.

[26] Tzakis AG, Reyes J, Zeevi A, Ramos H, Nour B, Reinsmoen N, et al. Early tolerance in pediatric liver allograft recipients. J Pediatr Surg 1994;29:754−6.

[27] Mazariegos GV, Sindhi R, Thomson AW, Marcos A. Clinical tolerance following liver transplantation: long term results and future prospects. Transpl Immunol 2007;17:114−9.

[28] Devlin J, Doherty D, Thomson L, Wong T, Donaldson P, Portmann B, et al. Defining the outcome of immunosuppression withdrawal after liver transplantation. Hepatology 27:926−933.

[29] Girlanda R, Rela M, Williams R, O'Grady JG, Heaton ND. Long-term outcome of immunosuppression withdrawal after liver transplantation. Transplant Proc 2005;37:1708−9.

[30] Pons JA, Yelamos J, Ramirez P, Oliver-Bonet M, Sanchez A, Rodriguez-Gago M, et al. Endothelial cell chimerism does not influence allograft tolerance in liver transplant patients after withdrawal of immunosuppression. Transplantation 2003;75:1045−7.

[31] Gras J, Wieers G, Vaerman JL, Truong DQ, Sokal E, Otte JB, et al. Early immunological monitoring after paediatric liver transplantation: cytokine immune deviation and graft acceptance in 40 recipients. Liver Transpl 2007;13:426−33.

[32] Hsu LW, Goto S, Nakano T, Lai CY, Lin YC, Kao YH, et al. Immunosuppressive activity of serum taken from a liver transplant recipient after withdrawal of immunosuppressants. Transpl Immunol 2007;17:137−46.

[33] Alexander SI, Smith N, Hu M, Verran D, Shun A, Dorney S, et al. Chimerism and tolerance in a recipient of a deceased-donor liver transplant. N Engl J Med 2008;358:369–74.

[34] VanBuskirk AM, Burlingham WJ, Jankowska-Gan E, Chin T, Kusaka S, Geissler F, et al. Human allograft acceptance is associated with immune regulation. J Clin Invest 2000;106:145–55.

[35] Tisone G, Orlando G, Cardillo A, Palmieri G, Manzia TM, Baiocchi L, et al. Complete weaning off immunosuppression in HCV liver transplant recipients is feasible and favourably impacts on the progression of disease recurrence. J Hepatol 2006;44:702–9.

[36] Orlando G, Manzia T, Baiocchi L, Sanchez-Fueyo A, Angelico M, Tisone G. The Tor Vergat weaning off immunosuppression in HCV liver transplant patients: the updated follow up at 78 months. Transpl Immunol 2008;20:43–7.

[37] Manzia TM, Angelico R, Baiocchi L, Toti L, Ciano P, Palmieri G, et al. The Tor Vergata weaning of immunosuppression protocols in stable HCV liver transplant patients: the 10 year followup. Transpl Int 2013;26(3):259–66.

[38] Benitez CE, Lozano JJ, Martinez Llordella M, Puig-Pey I, Lopez M, Tisone G, et al. Use of transcriptional biomarkers to identify liver transplant recipients who can successfully discontinue immunosuppressive therapy. Am J Transpl 2010;10:A-517 [Abstract]

[39] Manzia TM, Toti L, Angelico R, Sorge R, Manuelli M, Iaria G, et al. Weaning off immunosuppression after liver transplantation: predictive factors and short-term results. Liver Transpl 2010;16:0–16 [Abstract]

[40] Manzia TM, Dazzi E, Angelico R, Cillis A, Ciano A, Manuelli M, et al. Quality of life of successfully immunosuppression weaned off transplant recipients. Liver Transpl 2011;17: [Abstract]

[41] Takatsuki M, Uemoto S, Inomata Y, Sakamoto S, Hayashi M, Ueda M, et al. Analysis of alloreactivity and intragraft cytokine profiles in living donor liver transplant recipients with graft acceptance. Transpl Immunol 2001;8:279–86.

[42] Takatsuki M, Uemoto S, Inomata Y, Egawa H, Kiuchi T, Fujita S, et al. Weaning of immunosuppression in living donor liver transplant recipients. Transplantation 2001;72:449–54.

[43] Koshiba T, Li Y, Takemura M, Wu Y, Sakaguchi S, Minato N, et al. Clinical, immunological, and pathological aspects of operational tolerance after pediatric living-donor liver transplantation. Transpl Immunol 2007;17:94–7.

[44] Feng S, Ekong UD, Lobritto SJ, Demetris AJ, Roberts JP, Rosenthal P, et al. Complete immunosuppression withdrawal and subsequent allograft function among pediatric recipients of parental living donor liver transplants. JAMA 2012;307:283–93.

[45] Lee JH, Lee SK, Lee HJ, Seo JM, Joh JW, Kim SJ, et al. Withdrawal of immunosuppression in pediatric liver transplant recipients in Korea. Yonsei Med J 2009;50:784–8.

[46] Starzl TE, Murase N, Abu-Elmagd K, Gray EA, Shapiro R, Eghtesad B, et al. Tolerogenic immunosuppression for organ transplantation. Lancet 2003;361:1502–10.

[47] Stefanova I, Dorfman JR, Tsukamoto M, Germain RN. On the role of self-recognition in T cell responses to foreign antigen. Immunol Rev 2003;191:97–106.

[48] Afzali B, Lechler RI, Hernandez-Fuentes MP. Allorecognition and the alloresponse: clinical implications. Tissue Antigens 2007;69:545–56.

[49] Afzali B, Lombardi G, Lechler RI. Pathways of major histocompatibility complex allorecognition. Curr Opin Organ Transplant 2008;13:438–44.

[50] Strom TB, Tilney NL, Carpenter CB, Busch GJ. Identity and cytotoxic capacity of cells infiltrating renal allografts. N Engl J Med 1975;292:1257–63.

[51] Sarwal M, Chua MS, Kambham N, Hsieh SC, Setterwhite T, Masek M, et al. Molecular heterogeneity in acute renal allograft rejection identified by DNA microarray profiling. N Engl J Med 2003;349:125–38.

[52] Kroemer A, Edtinger K, Li XC. The innate natural killer cells in transplant rejection and tolerance induction. Curr Opin Organ Transplant 2008;13:339–43.

[53] Li F, Atz ME, Reed EF. Human leukocyte antigen antibodies in chronic transplant vasculopathy-mechanisms and pathways. Curr Opin Immunol 2009;21:557–62.

[54] Seetharam A, Tiriveedhi V, Mohanakumar T. Alloimmunity and autoimmunity in chronic rejection. Curr Opin Organ Transplant 2010;15:531–6.

[55] Yoshizawa A, Ito A, Li Y, Koshiba T, Sakaguchi S, Wood KJ, et al. The roles of CD25 + CD4 + regulatory T cells in operational tolerance after living donor liver transplantation. Transplant Proc 2005;37:37–9.

[56] Eason JD, Cohen AJ, Nair S, Alcantera T, Loss GE. Tolerance: is it worth the risk? Transplantation 2005;79:1157–9.

[57] Assy N, Adams PC, Myers P, Simon V, Minuk GY, Wall W, et al. Randomized controlled trial of total immunosuppression withdrawal in liver transplant recipients: role of ursodeoxycholic acid. Transplantation 2007;83:1571–6.

[58] Persson H, Friman S, Schersten T, Svanvik J, Karlberg I. Ursodeoxycholic acid for prevention of acute rejection in liver transplant recipients. Lancet 1990;336:52–3.

[59] Donckier V, Troisi R, Toungouz M, Colle I, Van Vlierberghe H, Jacquy C, et al. Donor stem cell infusion after non-myeloablative conditioning for tolerance induction to HLA mismatched adult living-donor liver graft. Transpl Immunol 2004;13:139–46.

[60] Urban CH, Deutschmann A, Kerbl R, Lackner H, Schwinger W, Konigsrainer A, et al. Organ tolerance following cadaveric liver transplantation for chronic graft-versus-host disease after allogenic bone marrow transplantation. Bone Marrow Transplant 2002;30:535–7.

[61] Kadry Z, Mullhaupt B, Renner EL, Bauerfeind P, Schanz U, Pestalozzi BC, et al. Living donor liver transplantation and tolerance: a potential strategy in cholangiocarcinoma. Transplantation 2003;76:1003–6.

[62] Andreoni KA, Lin JI, Groben PA. Liver transplantation 27 years after bone marrow transplantation from the same living donor. N Engl J Med 2004;350:2624–5.

[63] Tryphonopoulos P, Tzakis AG, Weppler D, Garcia-Morales R, Kato T, Madariaga JR, et al. The role of donor bone marrow infusions in withdrawal of immunosuppression in adult liver allotransplantation. Am J Transplant 2005;5:608–13.

[64] Tryphonopoulos P, Ruiz P, Weppler D, Nishida S, Levi DM, Moon J, et al. Long-term follow-up of 23 operational tolerant liver transplant recipients. Transplantation 2010;90:1556–61.

[65] Hematti P. Role of mesenchymal stromal cells in solid organ transplantation. Transplant Rev 2008;22:262–73.

[66] Le Blanc K, Ringden O. Immunomodulation by mesenchymal stem cells and clinical experience. J Intern Med 2007;262:509–25.

[67] US Department of Health and Human Services. OPTN/SRTR annual report. <www.srtr.org/annual_reports/2010>; 2010 [accessed November 2012].

[68] Berenguer M, López-Labrador FX, Wright TL. Hepatitis C and liver transplantation. J Hepatol 2001;35:666–78.

[69] Berenguer M, Ferrell L, Watson J, Prieto M, Kim M, Rayón M, et al. HCV-related fibrosis progression following liver transplantation: increase in recent years. J Hepatol 2000;32:673–84.

[70] Lake JR. The role of immunosuppression in recurrence of hepatitis C. Liver Transpl 2003;9:63–6.

[71] Berenguer M, Aguilera V, Prieto M, San Juan F, Rayón JM, Benlloch S, et al. Significant improvement in the outcome of HCV-infected transplant recipients by avoiding rapid steroid tapering and potent induction immunosuppression. J Hepatol 2006;44:717–22.

[72] Nakagawa M, Sakamoto N, Tanabe Y, Koyama T, Itsui Y, Takeda Y, et al. Suppression of hepatitis C virus replication by cyclosporin a is mediated by blockade of cyclophilins. Gastroenterology 2005;129:1031–41.

[73] Watashi K, Hijikata M, Hosaka M, Yamaji M, Shimotohno K. Cyclosporin A suppresses replication of hepatitis C virus genome in cultured hepatocytes. Hepatology 2003;38:1282–8.

[74] Berenguer M, Royuela A, Zamora J. Immunosuppression with calcineurin inhibitors with respect to the outcome of HCV recurrence after liver transplantation: results of a meta-analysis. Liver Transpl 2007;13:21–9.

[75] Germani G, Pleguezuelo M, Villamil F, Vaghjiani S, Tsochatzis E, Andreana L, et al. Azathioprine in liver transplantation: a reevaluation of its use and a comparison with mycophenolate mofetil. Am J Transplant 2009;9:1725–31.

[76] Manzia TM, Angelico R, Toti L, Bellini MI, Sforza D, Palmieri G, et al. Long-term, maintenance MMF monotherapy improves the fibrosis progression in liver transplant recipients with recurrent hepatitis C. Transpl Int 2011;24:461–8.

[77] Bahra M, Neumann UI, Jacob D, Puhl G, Klupp J, Langrehr JM, et al. MMF and calcineurin taper in recurrent hepatitis C after liver transplantation: impact on histological course. Am J Transplant 2005;5:406–11.

[78] Samonakis DN, Germani G, Burroughs AK. Immunosuppression and HCV recurrence after liver transplantation. J Hepatol 2012;56:973–83.

[79] Veldt BJ, Poterucha JJ, Watt KD, Wiesner RH, Hay JE, Rosen CB, et al. Insulin resistance, serum adipokines and risk of fibrosis progression in patients transplanted for hepatitis C. Am J Transplant 2009;9:1406–13.

[80] Foxton MR, Quaglia A, Muiesan P, Heneghan MA, Portmann B, Norris S, et al. The impact of diabetes mellitus on fibrosis progression in patients transplanted for hepatitis C. Am J Transplant 2006;6:1922–9.

[81] Trotter JF. Hot-topic debate on hepatitis C virus: the type of immunosuppression matters. Liver Transpl 2011;17:20–3.

[82] Manzia TM, Angelico R, Toti L, Lai Q, Ciano P, Angelico M, et al. Hepatitis C virus recurrence and immunosuppression-free state after liver transplantation. Expert Rev Clin Immunol 2012;8:635–44.

[83] Hui AY, Liew CT, Go MY, Chim AM, Chan HL, Leung NW, et al. Quantitative assessment of fibrosis in liver biopsies from patients with chronic hepatitis B. Liver Int 2004;24:611–8.

[84] Sánchez-Fueyo A, Strom TB. Immunologic basis of graft rejection and tolerance following transplantation of liver or other solid organs. Gastroenterology 2011;140:51–64.

[85] Biomarker Definition Working Group. Biomarkers and surrogate endpoints: preferred definitions and conceptual framework. Clin Pharmacol Ther 2001;69:89–95.

[86] Mazariegos GV, Zahorchak AF, Reyes J, Ostrowski L, Flynn B, Zeevi A, et al. Dendritic cell subset ratio in peripheral blood correlates with successful withdrawal of immunosuppression in liver transplant patients. Am J Transplant 2003;3:689–96.

[87] Li Y, Koshiba T, Yoshizawa A, Yonekawa Y, Masuda K, Ito A, et al. Analyses of peripheral blood mononuclear cells in operational tolerance after pediatric living donor liver transplantation. Am J Transplant 2004;4(12):2118–25.

[88] Martinez-Llordella M, Puig-Pey I, Orlando G, Ramoni M, Tisone G, Rimola A, et al. Multiparameter immune profiling of operational tolerance in liver transplantation. Am J Transplant 2007;7:309–19.

[89] Pons JA, Revilla-Nuin B, Baroja-Mazo A, Ramírez P, Martínez-Alarcón L, Sánchez-Bueno F, et al. FoxP3 in peripheral blood is associated with operational tolerance in liver transplant patients during immunosuppression withdrawal. Transplantation 2008;86:1370–8.

[90] Martinez-Llordella M, Lozano JJ, Puig-Pey I, Orlando G, Tisone G, Lerut J, et al. Using transcriptional profiling to develop a diagnostic test of operational tolerance in liver transplant recipients. J Clin Invest 2008;118:2845–57.

[91] Bohne F, Martínez-Llordella M, Lozano JJ, Miquel R, Benítez C, Londoño MC, et al. Intra-graft expression of genes involved in iron homeostasis predicts the development of operational tolerance in human liver transplantation. J Clin Invest 2012;122:368–82.

[92] Geissler EK, Schlitt HJ. Immunosuppression for liver transplantation. Gut 2009;58:452–63.

[93] Orlando G, Baptista P, Birchall M, De Coppi P, Farney A, Guimaraes-Souza NK, et al. Regenerative medicine as applied to solid organ transplantation: current status and future challenges. Transpl Int 2011;24:223–32.

[94] Orlando G, Cocco P, Corona L, Manzia TM, Clemente C, Famulari A. Operational tolerance in kidney transplantation in the regenerative medicine era. In: Trzcinska M, editor. Kidney transplantation new perspectives. InTech; 2011. p. 193–212.

[95] Orlando G, Wood KJ, Stratta RJ, Yoo JJ, Atala A, Soker S. Regenerative medicine and organ transplantation: past, present and future. Transplantation 2011;91(12):1310–7.

Chapter 64

Biomarkers of Operational Tolerance in Liver Transplantation

Takaaki Koshiba[a], Hidenori Ohe[b,c], and Alex G.Bishop[d]

[a]Department of Disaster and Comprehensive Medicine, Fukushima Medical University, Fukushima, Japan, [b]Department of Surgery, Graduate School of Medicine, Kyoto University, Kyoto, Japan, [c]Terasaki Foundation Laboratory, Los Angeles, CA, [d]Transplantation Laboratory, Royal Prince Alfred Hospital, Sydney, Australia

Chapter Outline

64.1 INTRODUCTION

Billingham and Medawar in 1953 first described a state of acquired immunologic tolerance in rodents injected *in utero* or during the neonatal period with bone marrow−derived donor cells [1]. These animals later accepted an allograft taken from the same donor strain from which the original cells had been harvested, while maintaining the ability to reject third-party grafts. Up to now, tolerance has been defined in rodents as (i) long-term survival of primary grafts in the absence of immunosuppression (IS); (ii) acceptance of donor-specific secondary grafts; (iii) rejection of third-party secondary grafts [2]. In addition, many studies in rodents have demonstrated that tolerance is characterized by normal histology of accepted grafts [3−8] that cannot usually be distinguished from syngeneic transplants (Tx). There is thus no histological marker of tolerance.

Graft acceptance, in combination with *in vivo* immunological hyporesponsiveness, is more frequently achieved after liver transplantation (LTx) compared to other solid organ grafts [9−12]. This occurs both in animals (pigs, rodents) and humans, although in humans, demonstration of tolerance by challenge with a subsequent donor graft is not feasible for ethical reasons. In addition, confirmation of normal liver histology by routine biopsy after transplant has usually not been performed, due to the increased risk for recipients [13]. Therefore, clinical tolerance, defined functionally *in vivo* as persisting normal liver function tests for greater than 1 year after complete cessation of IS is referred to as "operational" tolerance (OT), to distinguish it from the complete immunological tolerance identified in animal models [2−8].

In several centers, substantial efforts have been made to identify reliable biomarkers to monitor patients who may discontinue IS without risk of rejection. In addition, identification of biomarkers of OT might provide a significant clue towards development of a novel therapy for its induction. Several potential biomarkers of OT have been identified mainly by retrospective studies using peripheral blood or liver graft biopsy tissues (Table 64.1).

Regenerative Medicine Applications in Organ Transplantation.

TABLE 64.1 Potential Biomarkers of OT in Human Liver Transplant

Parameter	Tissue Assayed	Assay Type	Description	Reference
NK Cells				
Peripheral NK cells	Blood	Flow cytometry	The frequency of peripheral NK cells down regulated	[14]
NK-cell transcript	Blood	Microarray/RT-PCR	NK cell-related gene expression	[15,16]
γδ T Cells				
Vδ1/Vδ2 T-cell ratio	Blood Graft	Flow cytometry RT-PCR	An increase of Vδ1/Vδ2 T-cell ratio An increase of Vδ1/Vδ2 T-cell ratio	[14,17] [18]
TCR sequence of Vδ1 T-cells	Graft	RT-PCR/autosequence	Skewing of TCR sequence of Vδ1 T-cells	[18]
Regulatory T-Cells				
$CD4^+CD25^{high+}$T-cells	Blood	Flow cytometry	The frequency of peripheral $CD4^+CD25^{high+}$T cells upregulated	[14,17,19]
$CD4^+CD25^{high+}CD45RA^+$ T cells	Blood	Flow cytometry	The frequency of peripheral $CD4^+CD25^{high+}$ $CD45RA^+$ T cells and $CD4^+CD25^{very\ high+}$ CD45RA-T cells up regulated	[20]
$CD4^+CD25^{very\ high+}CD45RA^-$ T cells	Blood	Cell sorting/MLR	Donor antigen-specific property of $CD4^+CD25^{high+}CD45RA^+$ T-cells and $CD4^+CD25^{very\ high+}CD45RA^-$ T-cells	[20]
$CD4^+CD25^{high+}CD127lo$ T-cells	Blood	Flow cytometry	The frequency of peripheral $CD4^+CD25^{high+}CD127lo$ upregulated	[21,22]
$Foxp3^+$ cells	Graft	Immunohistochemistry	The number of $Foxp3^+$ cells upregulated	[23]
Dendritic Cells				
pDC/mDC ratio	Blood	Flow cytometry	An increase of pDC/mDC ratio An elevated PD-L1/CD86 expression ratio on pDC	[19]
HLA-G				
Serum HLA-G	Blood	ELISA	Increased serum HLA-G levels	[24]
HLA-G expression on mDC	Blood	Flow cytometry	Increased HLA-G expression on mDC	[21]
B-cells				
$CD19^+$ cells	Blood	Flow cytometry	The frequency of $CD19^+$ cells upregulated	[14]
Cytokine				
Cytokine gene polymorphisms	Blood	PCR-SSP	Low TNF-α profiles and high/intermediate IL-10 profiles	[25]
Cytokine gene expressions	Graft	RT-PCR	Intra-graft Th2 immune deviation	[26]
Humoral Responses				
DSA	Blood	ELISA	The absence of HLA DSA in the serum derived from operationally tolerant recipient	[27]
HLA antibody	Blood	LABScreening single antigen beads	The absence of HLA Abs early post-Tx and subsequent operational tolerance	[28]
C4d deposition	Graft	Immunohistochemistry	The absence of C4d deposition	[29][a]
Others				
Iron homeostasis	Graft	Microarray/RT-PCR	Intragraft gene expression associated with iron homeostasis	[30][a]

pDC, plasmacytoid dendritic cell; mDC, monocytoid dendritic cell.
[a]*Prospective study.*

64.2 POTENTIAL BIOMARKERS OF OPERATIONAL TOLERANCE IN LTx

64.2.1 Regulatory T Cells

Since the demonstration by Sakaguchi et al. that regulatory T cells (Tregs) are characterized by co-expression of CD4 and CD25 in rodents, there has been considerable interest in the role of regulatory cells in nondeletional tolerance [31,32]. The description by Hori et al. that Foxp3 is the master control gene of Tregs, and can thus more accurately identify these cells, has increased interest in their role in OT of LTx patients [33].

Li et al. reported that the frequency of circulating $CD4^+CD25^{high}$ cells was increased in OT recipients compared to recipients on maintenance IS after haploidentical (parent to child) pediatric LTx [14] (Figure 64.1). Subsequently, Martinez-Llordella et al. found that the frequency of $CD4^+Foxp3^+$ cells was increased in OT recipients after cadaveric adult LTx [17]. Furthermore, $Foxp3^+$ cells were present in biopsies of IS-free recipients after pediatric LTx [23]. Immunofluorescent staining of biopsy sections showed that these cells were localized in the portal area, and their number was significantly increased in accepted grafts compared to chronically rejecting grafts, grafts of fully immunosuppressed recipients or normal livers. Double staining for both CD4 and Foxp3 revealed that most $Foxp3^+$ cells were $CD4^+$. This raises the possibility that Tregs function as suppressors of rejection within grafts *in situ*. Overall, the several studies of OT patients showing increased frequency of Tregs in the peripheral blood and graft suggest that these cells are efficient biomarkers of OT and may play a role in nondeletional tolerance of liver transplants.

64.2.2 Memory and Naïve Tregs

Murine Foxp3-expressing Tregs first described by Sakaguchi's original work have a memory phenotype (memory Tregs) [31]. Recently, his group and others identified a distinct subset of Foxp3-expressing CD4CD25 cells with naïve phenotype in humans (naïve Tregs) [34−40]. Human memory Tregs are likely to be a counterpart of murine Tregs [41−45] while naïve Tregs appear to be specific for humans. Of interest, the latter exert only weak suppressive properties but can proliferate after antigenic stimulation in the presence of IL-2, unlike memory Tregs [46,47]. Expression of Foxp3 is weak in naïve Tregs but is increased once these cells are activated. The degree by which the frequency of naïve Tregs predominates over memory Tregs in the peripheral blood of children gradually decreases with age and conversion between the two occurs in adults [34−37,39,48−50]. One possibility is that naïve Tregs are a functionally immature form of memory Tregs and differentiate into mature memory Tregs after subsequent antigenic stimulation. However, recent human studies found decreased frequency and impaired function of naïve Tregs in the blood of patients with autoimmune diseases [37,39,51−54].

Nafady-Hego et al. examined the frequency and function of both memory Tregs (referred to as conventional Tregs in their work) and naïve Tregs [20]. They found that the frequencies of both memory and naïve Tregs were increased in OT recipients after pediatric LTx compared to intolerant recipients who failed to taper IS (Figure 64.2). OT recipients exhibited donor antigen-specific *in vitro* hyporesponsiveness. Hyporesponsiveness was abrogated by depletion of either naïve or memory Tregs. Thus, the authors concluded that both subsets appeared to contribute to establishment of OT. Of interest, Foxp3 expression of naïve Tregs was significantly higher in OT recipients than in healthy individuals or intolerant recipients. In addition, the number of naïve Tregs gradually increased with time after complete cessation of IS. Such a gradual increase was not observed in the number of memory Tregs.

These results for naïve and memory T cells in OT pediatric LTx patients imply that naïve Tregs, similar to memory Tregs, can play a crucial role in OT once they are activated, although naïve Tregs are normally in a resting state. Also, the results raise the possibility that novel tolerance induction strategies might be better to target expansion of naïve Tregs rather than memory Tregs which are refractory to expansion.

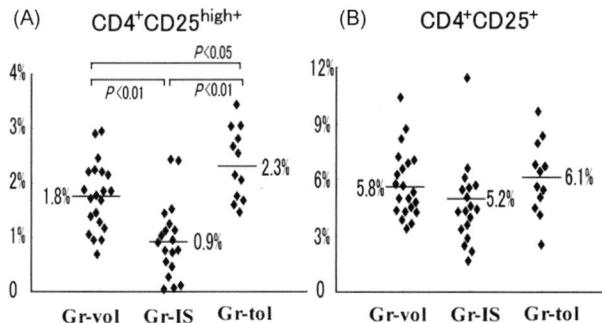

FIGURE 64.1 Analysis of $CD4^+CD25^+$ Tregs in operationally tolerant liver transplant patients, patients on IS, and age-matched normal volunteers. Twelve pediatric LTx patients, who exhibited normal or nearly normal liver function tests for more than 1 year after cessation of IS, were examined for circulating $CD4^+CD25^{high+}$ cells by FACS analysis (group-tolerance: Gr-tol). Nineteen age-matched LTx patients on IS (group-IS: Gr-IS) and 24 age-matched healthy volunteers (group-volunteers: Gr-vol) were used as controls. (A) The frequency of $CD4^+CD25^{high+}$ lymphocytes was significantly higher in Gr-tol compared to those in Gr-IS and Gr-vol. (B) The frequency of $CD4^+$ lymphocytes showing low/intermediate expression of CD25 ($CD4^+CD25^+$) did not differ among the three groups.

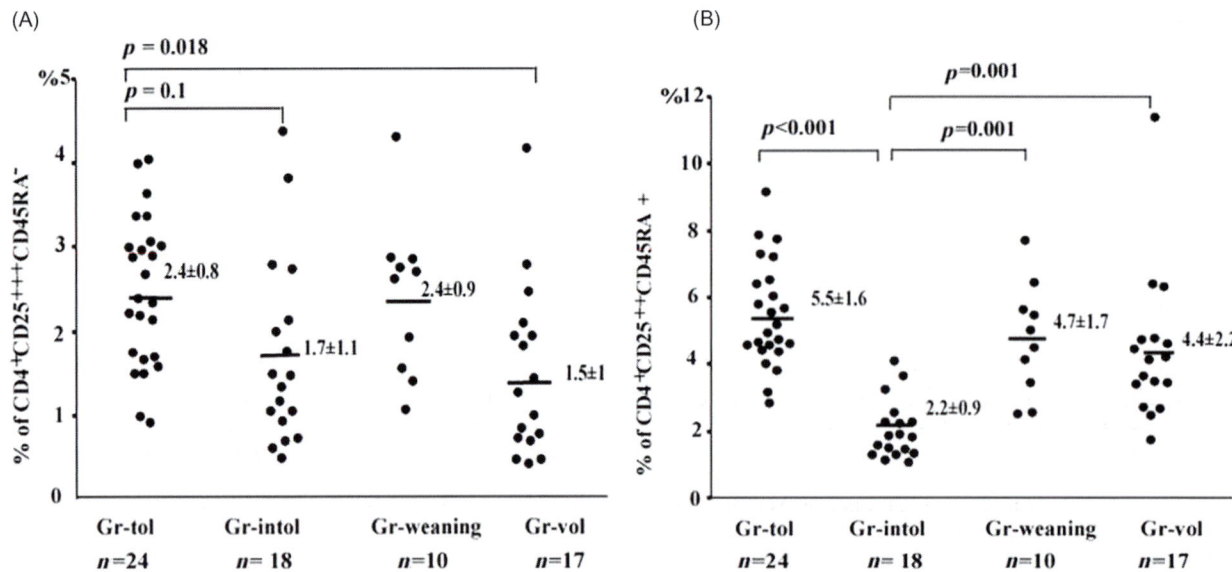

FIGURE 64.2 The frequency of memory and naïve Tregs. (A) The frequency of CD4$^+$ memory Tregs (CD4$^+$CD25^{+++}CD45RA$^-$) was significantly higher in tolerant liver transplant recipients (Gr-tol) compared to healthy volunteers (Gr-vol)(Gr-tol vs Gr-vol $p = 0.018$). A trend toward an increased frequency of conventional Tregs was observed in Gr-tol compared to intolerant recipients (Gr-intol)(Gr-tol vs Gr-intol $p = 0.1$). (B) The frequency of naïve CD4 + Tregs (CD4$^+$CD25^{++}CD45RA$^+$) was significantly reduced in Gr-intol compared with the other groups (Gr-intol vs Gr-tol $p < 0.001$, Gr-intol vs Gr-weaning $p = 0.001$, Gr-intol vs Gr-vol $p = 0.001$).

One problem was the lack of specific cell surface biomarkers to define and separate Tregs from effector cells. Recently, there has been technical progress in isolation of Tregs. Two groups have reported that CD127 (IL-7 receptor α) is downregulated in a subset of CD4$^+$ cells and a substantial proportion ($>90\%$) of this cell fraction expresses Foxp3. Lack of CD127 expression efficiently quantified Tregs in subjects with type 1 diabetes [55,56] and CD127$^{low+/-}$ CD4$^+$CD25$^+$ cells have been confirmed to accurately identify Tregs [57], which offers a practical advantage over identification of intracellular Foxp3 expression.

In two separate studies, identification of Tregs using CD4$^+$CD25^{++}CD127$^{low+/-}$ found significantly more in OT pediatric LTx recipients compared to intolerant recipients [21,22]. CD127 is potentially a more efficient marker of Tregs than CD25 as CD4$^+$CD25$^+$ T cells in peripheral blood display a range of intensities of CD25$^+$ expression due to the presence of effector cells (activated/memory-type cells) which express CD25 at intermediate levels. In contrast, in cord blood, most CD4$^+$CD25$^+$ cells exhibit high CD25 intensity since the fetus is not normally exposed to pathogens so that effector cells are not yet activated. Nafady-Hego et al. have confirmed by flow cytometry that CD127 efficiently separates Tregs from activated effector cells in the setting of pediatric LTx. Functional analysis revealed that the use of CD127 in combination with CD4$^+$ and CD25^{++} was an efficient marker of OT after pediatric LTx [58].

64.2.3 γδ T Cells

γδ T cells are highly conserved and are the likely ancestors of αβ T cells and B cells [59,60]. They localize to epithelial barriers and traffic to inflamed tissue [61,62]. Although little is known about their function, there is emerging evidence that they participate in innate immunity and may regulate adaptive immune responses [63]. In humans, the subsets of γδ T cells have been defined based upon δ chain usage. Of the two prominent subsets in peripheral blood, denoted Vδ1 and Vδ2 [64–66], Vδ2 cells predominate over Vδ1 cells. Vδ2 T cells respond to inflammation/infection by producing inflammatory cytokines [67–69].

In marked contrast, Vδ1 cells predominate in the intestinal mucosa and the spleen [64,66,67,70] and may be potentially immunoregulatory [71]. Peng et al. identified a dominant Vδ1 cell population in tumor-infiltrating lymphocytes and found it to possess potent immunosuppressive capacity [72]. Vδ1 cells are significantly increased in the peripheral blood as well as in deciduas during normal but not abortive pregnancy [73]. In addition, Vδ1 cells in the deciduas produced IL-10 and transforming growth factor-beta (TGF-β) which are both immunoregulatory cytokines [74]. Thus, it is tempting to consider that Vδ1 cells may contribute to feto-maternal tolerance.

Li et al. have found that Vδ1 cells, which otherwise reside mainly within the intestine, emerge into peripheral blood and predominate over Vδ2 cells in OT recipients

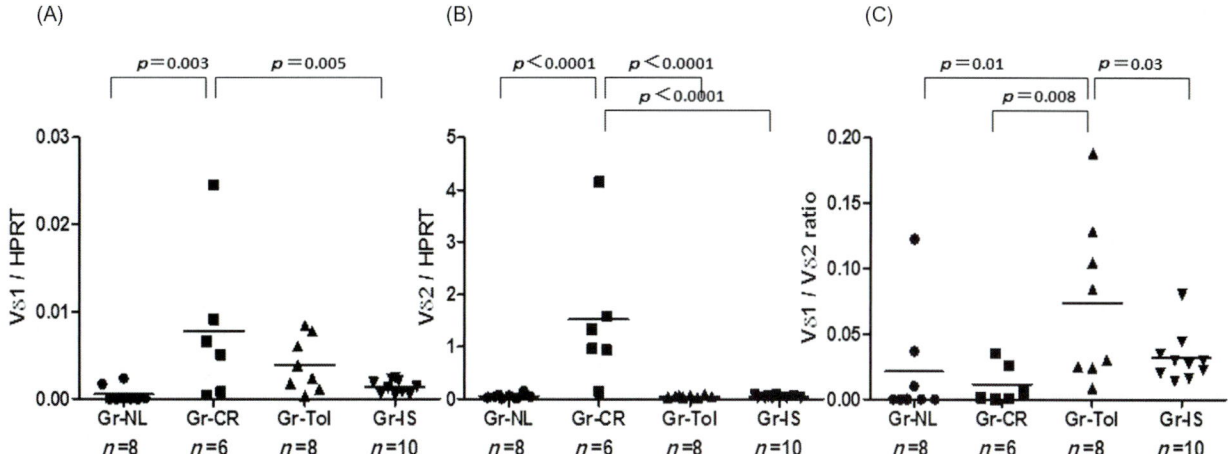

FIGURE 64.3 Vδ1/Vδ2 ratio is increased within grafts of living-donor (semi-allogeneic) LTx recipients. Expression of Vδ1 cells and Vδ2 cells in liver tissue was quantified by real-time PCR. (A) Extent of infiltration with Vδ1 cells analyzed by ratio of Vδ1 mRNA to HPRT mRNA expression. (B) Extent of infiltration with Vδ2 cells analyzed by ratio of Vδ2 mRNA to HPRT mRNA expression. (C) Ratio of Vδ1/Vδ2 mRNA. The Vδ1 and Vδ2 levels were assessed in the tolerance (or almost tolerance) group (Gr-Tol), maintenance immunosuppression group (Gr-IS), chronic rejection group (Gr-Cr), and normal liver group (Gr-NL). The Vδ1/Vδ2 ratio was significantly increased in Gr-Tol, compared to Gr-IS, Gr-CR, and Gr-NL (Gr-Tol vs Gr-IS $p = 0.03$, Gr-Tol vs Gr-NL $p = 0.01$, Gr-Tol vs Gr-CR $p = 0.008$). It did not differ among Gr-IS, Gr-CR, and Gr-NL (Gr-IS vs Gr-CR or Gr-NL NS).

after haploidentical pediatric LTx [14]. This was confirmed by Martínez-Llordella et al. in the peripheral blood of OT recipients after cadaveric adult LTx [17]. They extended this in a study of the transcriptional pattern of peripheral blood showing that increased mRNA expression of several Vδ1-associated genes correlated well with OT [15].

Recently, Zhao et al. have investigated the Vδ1/Vδ2 ratio at the transcriptional level in biopsies from liver allografts after living-donor semi-allogeneic (parent to child) pediatric LTx. They also examined the complementarity determining region 3 (CDR3) sequence of the δ chain of Vδ1 cells within the graft to determine the diversity of T-cell receptors [18]. Data obtained from OT or almost OT (Gr-Tol) were compared with those of fully immunosuppressed recipients' grafts (Gr-IS), chronically rejecting grafts (Gr-CR), and the normal liver (Gr-NL). The Vδ1/Vδ2 ratio was the highest in Gr-Tol compared with Gr-IS, Gr-CR, and Gr-NL. The Vδ1/Vδ2 ratio did not differ among Gr-IS, Gr-CR, and Gr-NL (Figure 64.3).

There was an identical CDR3 sequence (100% homologous) for Vδ1 cells exhibited by all recipients in Gr-Tol. In Gr-Tol, this sequence was dominant in a substantial proportion of recipients as it was the most frequent sequence detected in all PCR product clones derived from each biopsy. It was observed in most samples in Gr-NL, although it was dominant in only a single sample. In contrast, this specific sequence was not observed in any recipient in Gr-IS or Gr-CR, showing that it very effectively discriminates between accepted and rejected livers.

These findings show that accumulation of a unique Vδ1 T-cell clone within graft infiltrates closely correlates

with good graft function in OT (or almost OT) semi-allogeneic LTx recipients. If these results can be confirmed in other OT patient populations, it raises the possibility that Vδ1 T cells might function as regulatory cells, although their relative importance compared to conventional $CD4^+CD25^+Foxp3^+$ Tregs in transplant tolerance is yet to be determined. It also suggests that the oligoclonal Vδ1 response in the OT patients was antigen-driven. If this is indeed the case, the identification of a specific ligand for these Vδ1 cells may help us to fully understand how they accumulate within accepted grafts and provide a novel therapy for induction of tolerance.

The unique Vδ1 T-cell clone identified in these OT patients requires confirmation whether it might be the first qualitative (rather than quantitative) biomarker of OT. Of note, the Barcelona group demonstrated that CDR3 sequences of Vδ1 cells in the peripheral blood derived from OT adult recipients after cadaveric LTx were polyclonal with no evidence of oligoclonal expansion [75]. Whether this difference is due to the different patient populations or to the source of cells analyzed (peripheral blood vs graft biopsy) has not been established.

This raises the issue of whether graft biopsies are likely to provide a more accurate analysis of the tolerant state than peripheral blood. There has been general reluctance to perform routine biopsy before commencing weaning of IS due to the possibility of adverse effects of the biopsy procedure. In contrast, there is considerable current incentive to perform LTx biopsies as the Banff Working Group on Liver Allograft Pathology strongly recommends routine preweaning biopsy [76]. If all groups adopted this policy, it would provide greater opportunity to evaluate the effectiveness of graft biopsy versus peripheral blood for prospective identification of markers of OT. This issue is discussed further in the section on biomarkers of fibrosis.

64.2.4 Cytokines: The Th1/Th2 Paradigm

After the description of the Th1/Th2 paradigm by Mosmann and Coffman, there was much interest in the role of Th1 and Th2 cytokines in inducing rejection and protecting from rejection, respectively [77]. In experimental transplantation, increase of Th1 cytokines and/or decrease of Th2 cytokine upregulation was shown to play a central role in Tx rejection [78–87].

Cytokine gene loci are highly polymorphic, and examination of polymorphism can distinguish between low and high producers. Mazariegos et al. examined several Th1 and Th2 cytokine polymorphisms by using the peripheral blood derived from OT LTx recipients versus intolerant recipients [25]. Polymorphisms associated with high producers of IL-10 and low producers of TNF-α were found in OT. Takatsuki et al. examined intragraft Th1/Th2 cytokines at the transcriptional level and responsiveness to donor antigens by peripheral blood in mixed lymphocyte reaction of OT living-related LTx recipients [26]. OT was characterized by *in vivo* hyporesponsiveness and the lack of intragraft Th1 cytokine expression. On the other hand, IL-10 expression was equivalent to that in the normal liver. Thus, it is likely that Th1/Th2 paradigm theory appears to be applicable for human OT.

64.2.5 Humoral Responses: Donor-Specific Antibody, C4d Deposition and B Cells

There is currently considerable interest in the role of humoral responses in rejection and tolerance after organ transplantation. The pathogenic role of antihuman leukocyte antigen (HLA) antibodies (Abs) after renal transplantation (RTx) is well established [88–90]. In contrast, their role in LTx has long been controversial [91–95]. Girnita et al. examined the prevalence of circulating HLA Ab and donor-specific HLA Ab after LTx using ELISA [27]. Most were pediatric patients or young adults and received livers from cadaver donors. Anti-HLA Abs were detected in a subset of OT recipients, recipients who could not taper IS (intolerant), and recipients during the weaning process. In addition, donor-specific antibodies (DSAs) were detected in some intolerant recipients and recipients during the weaning process. Of note, however, no donor-specific HLA Ab was found in tolerant recipients. They concluded that a tolerant state was characterized by the lack of donor-specific HLA Ab. In support, Ohe et al. have shown

that donor-recipient HLA-A matching positively impacts OT in the setting of haploidentical pediatric LTx [22]. More recently, Waki et al. reported that the amount of circulating HLA Abs even early post-Tx negatively correlated with subsequent development of OT after pediatric living-donor LTx [28]. These findings in OT of LTx are despite the fact that a positive impact of donor-recipient HLA matching on outcome after LTx in general has not been established [96–100].

As discussed above in the section on γδ T cells, there is now justification for routine preweaning biopsies for identification of recipients who can successfully stop IS. Very recently, Feng et al. have conducted a prospective, multicenter, open label pilot trial of IS withdrawal in pediatric living-related LTx patients [29]. In most cases, the primary disease was biliary atresia, and age at Tx was less than 1 year. All recipients received haploidentical grafts. Twenty recipients were enrolled of whom 12 became OT. These had less portal inflammation and less C4d deposition in the preweaning biopsy than recipients who could not be removed from IS. The authors concluded that active and inadequately regulated antidonor cellular and humoral responses interfered with tolerance. Consequently, C4d status in preweaning biopsy appears to be a promising quantitative biomarker of tolerance.

In 2004, when Li et al. systemically analyzed phenotypes of peripheral blood mononuclear cells (PBMCs) derived from OT recipients after pediatric LTx, they found an increased frequency of CD19[+] B cells in tolerant recipients [14]. In this work, they speculated that increased frequency of B cells was merely a consequence of Th2 immune deviation but was not directly responsible for OT. However, recently, several groups found that a tolerant state was accompanied by increased circulating B cells after RTx [101,102]. In addition, they reported that those B cells produced IL-10 and TGF-β and thereby, B cells could function as immune regulators. In the setting of LTx, it remains elusive whether B cells play a role in development of OT.

64.2.6 Dendritic Cell Subset Ratio and HLA-G

Recently, there has been accumulating evidence that dendritic cells (DC), traditionally regarded as being responsible for rejection [103,104], can regulate immune systems [105]. In humans, classical myeloid (monocytoid) DC (mDC) and plasmacytoid DC (pDC) circulate in the peripheral blood [19]. They are distinguishable by differential cell surface antigens. In experimental settings, pDC appear to be tolerogenic since freshly isolated pDC of donor origin can markedly prolong organ transplant and hematopoietic cell engraftment [106,107]. Of interest, pDC that present donor Ag within secondary lymphoid tissue appear to promote organ transplant tolerance by inducing antigen-specific Tregs [108]. Tokita et al. found an elevated pDC/mDC ratio in the peripheral blood of pediatric liver Tx recipients who could successfully stop IS, or were in the process of weaning IS, versus recipients that required continued maintenance IS [19]. The authors also examined costimulatory molecule expression on pDC and mDC. The ratio of the inhibitory molecule, programmed death ligand-1 (PD-L1) to the costimulatory molecule CD86 (B7-2) on pDC was significantly higher in OT recipients compared to that in recipients on maintenance IS [19]. This was not the case on mDC. Of note, the frequency of CD4$^+$CD25^{high+} Tregs was increased in OT recipients and correlated with the PD-L1/CD86 ratio on pDC. These findings suggest that there is a functional interplay between pDC and Tregs [109].

HLA-G is a nonclassical HLA class I molecule with immunoregulatory properties that plays an important role in feto-maternal tolerance [110]. Compared to classical HLA class I, HLA-G is expressed on a much more restricted range of cells and tissues, including trophoblast, thymus, cornea, erythroid lineage, and mesenchymal stem cells [111,112]. It has been reported that HLA-G plays an important role in regulating both the maturation and function of DC. Interaction between HLA-G and its receptor leads to inhibition of DC maturation [113]. Moreover, HLA-G-expressing antigen-presenting cells induce CD4$^+$ T-cell anergy and differentiation of Tregs [114].

In pediatric LTx recipients, Castellaneta et al. found increased HLA-G expression on circulating mDC in OT compared to recipients on maintenance IS [21]. This was not the case on pDC. Of interest, there was a positive correlation between HLA-G expression on mDC and the expression of Foxp3 within Tregs. A separate study reported that serum soluble HLA-G was significantly higher in OT and stable IS pediatric liver Tx recipients compared to those who experienced at least two rejection episodes after 1 year post-Tx [24] although the source of soluble HLA-G was not identified.

Thus, it remains to be examined whether there is linkage between increased soluble HLA-G levels and increased HLA-G expression on mDC in OT LTx recipients. Collectively, three studies showed that in LTx OT, (i) pDC to mDC ratio was upregulated, (ii) PD-L1 expression to CD86 expression ratio on pDC was upregulated accompanied by positive correlation with Tregs frequency, (iii) HLA-G expression on mDC was upregulated accompanied by positive correlation with Tregs frequency, and (iv) increased soluble HLA-G [19,21,24]. Functional interaction between Tregs and specific molecules on DC subsets remains to be elucidated.

64.2.7 Transcriptional Profiling of Liver Transplant Tolerance

Martinez-Llordella et al. used transcriptional profiling of PBMCs derived from LTx recipients to identify differential expression of genes between OT and intolerant recipients [15]. Gene expression indicated that NK cells and $\gamma\delta$ T cells appear to preferentially associate with OT [15]. In accordance, a subsequent transcriptional study demonstrated high expression of NK cell-related genes in peripheral blood from OT pediatric and adult LTx recipients [16]. These gene expression studies contrast to the findings of decreased circulating NK cells in OT by FACS analysis [14].

Bohne et al. conducted a prospective multicenter IS withdrawal trial of LTx patients in Europe [30]. They extensively examined intragraft gene expression in biopsies and found that iron homeostasis-related genes were associated with OT [30]. These transcriptional profiling approaches have the potential to identify novel biochemical and immunological mechanisms of OT, and it will be interesting to determine whether the candidate biomarkers identified to date can be independently validated.

64.2.8 Organ Specificity of Tolerance Biomarkers: Liver versus Kidney

In both clinical and experimental transplantation, successful IS withdrawal is more frequently achievable in LTx compared to other organ Tx including RTx [9–12]. Therefore, it is not surprising that there might be organ-specific as well as organ-independent biomarkers of OT. A study comparing LTx and RTx by transcriptional profiling of PBMCs [115] showed that in LTx OT, genes preferentially expressed by CD56$^+$ lymphocytes (NK cells) were increased, while in RTx OT, genes associated with CD19$^+$ cells (B cells) were increased. There was no significant overlap. Monitoring of OT in LTx and RTx has recently been compared [116], and it appears that $\gamma\delta$ T cells and NK cells appear to be associated with LTx while B cells are associated with RTx. Whether this reflects organ-specific differences in the mechanism of tolerance induction is yet to be determined.

64.2.9 Limitations of Existing Data

Several researchers have made substantial efforts to develop reliable biomarkers of OT after LTx (Table 64.1). To our knowledge, however, most studies were retrospective except the two studies listed here [29,30]. There is a possibility that the biomarkers of OT identified in a retrospective study might be merely a consequence of withdrawal of IS. If so, they are likely to be the effects rather than the causes of tolerance. In reviewing these studies, the size of each was relatively small and the time post-Tx when the sample was obtained was very different among studies and even among subjects within each study. Furthermore, the IS-weaning protocol differed between studies. It is also possible that tolerance mechanisms may differ between haploidentical LTx and fully mismatched LTx or between children and adults. These factors are likely to affect the validity of existing biomarkers shown in Table 64.1. For instance, tolerance appeared to be characterized by the lack of DSA in a retrospective study conducted by Girnita et al. [27]. In contrast, pre-weaning determination of DSA did not efficiently predict tolerance in a prospective study conducted by Feng et al. One explanation of this discrepancy was that the data generated by the retrospective study was not validated by the prospective study [29]. Another explanation was that sample size of the prospective study was not large enough to obtain a definitive conclusion. Finally, most subjects enrolled in the retrospective study received fully mismatched grafts, whereas all subjects in the prospective study received haploidentical grafts. Altogether, it is tempting to conclude that large-scale prospective studies are needed to confirm the validity of candidate biomarkers listed in Table 64.1.

64.3 NOVEL BIOMARKERS

64.3.1 Non-HLA Antibodies

It is well known that rejection can occur in HLA-identical sibling transplants. Thus, more than three decades ago, it was assumed that antibodies to non-HLA systems might be responsible for rejection [117]. Several data suggest that such non-HLA antibodies may occur as alloantibodies or autoantibodies [118]. In reviewing the literature, non-HLA antibodies, which are potentially related with rejection, include major histocompatibility class I chain-related gene A (MICA) and gene B (MICB) [119–121], glutathione S-transferase T1 [122,123], angiotensin II type 1-receptor (AT_1R) [118,124,125], endothelial cell antigens [126,127], vimentin [128,129], and non-HLA IgM antibodies [130]. Details of those non-HLA antibodies and rejection after organ Tx were extensively reviewed by Dragun [118].

64.3.2 Major Histocompatibility Class I Chain-Related Gene A and Gene B

MICA and MICB loci are located on chromosome 6 and are closely linked to the HLA-B locus. They are highly polymorphic [131]. Due to their polymorphism, sensitization is likely to occur similarly to anti-HLA antibodies. However, in the absence of HLA antibodies, antibodies to MICA or MICB are not always accompanied by C4d

positivity [132]. Therefore, one possibility is that their presence may merely reflect enhancement of NK cell cytotoxicity [133] as MICA and MICB are ligands for NKG2D on NK cells. Their binding results in upregulation of NK cell function [134]. Tumor cells that overexpress MICA are more sensitive to NK cells, but this is abolished if the interaction with NKG2D is blocked [134].

Similar mechanisms may take place in organ allografts when transplanted allografts increase expression of MICA and MICB during rejection. A positive correlation between rejection episodes, and the presence of MICA and MICB antibodies was reported in the setting of renal, pancreas, and heart Tx [119–121,135–143]. However, it remains elusive whether MICA and MICB antibodies initiate antibody-mediated rejection.

64.3.3 Angiotensin II Type I Receptor

AT_1R mediates most physiologic and pathophysiologic actions of its endogenous ligand, angiotensin II, including regulation of arterial pressure and water–salt balance [144]. Overactivity of the angiotensin II-AT_1R axis leads to hypertension, cardiac, renal, and vascular remodeling [145]. Dragun et al. found circulating agonistic antibodies to the AT_1R (AT_1R Ab) in renal allograft recipients who experienced severe vascular rejection and simultaneously malignant hypertension in the absence of anti-HLA antibodies [124]. Of interest, this malignant hypertension resembled pre-eclampsia in pregnancy, suggesting a common immunological feature between pre-eclampsia and this form of graft rejection [146]. In addition, removal of AT_1R Ab by plasmapheresis and pharmacological blockade of AT_1R improved renal function in Tx recipients with AT_1R Ab. Transfer of human AT_1R Ab into transplanted animals caused a transmural arteritis and hypertension similar to humans. [124].

It remains to be clarified whether non-HLA antibodies are the cause or the effect of rejection. Therefore, further studies are needed to clarify whether non-HLA antibody-related pathologies represent "typical" rejection or whether they are autoimmune phenomena induced by Tx. Consequently, it remains to be established whether non-HLA antibodies are reliable biomarkers of rejection and therefore of the absence of OT.

64.4 BIOMARKERS OF FIBROSIS

As previously described, OT is defined by clinical criteria rather than by graft histology [147,148]. Thus, the question arises whether establishment of OT, defined by normal liver function tests and the absence of clinical rejection, offers protection against the antigen-dependent histologic changes of grafts characterized by cellular infiltrates, fibrosis, and damage to vessels and bile ducts

[149]. Studies from Birmingham, Groningen, Chicago, and Kyoto independently reported the presence of fibrosis within grafts in clinically stable liver Tx recipients [149–152]. Of note, the Kyoto group found that a subset of IS-free recipients, although exhibiting normal liver function tests, showed more fibrosis compared to fully immunosuppressed recipients [149]. Fibrosis present in OT recipients may represent a variant of rejection. "Clinically overt" rejection after cessation of IS or during the process of weaning IS can usually be resolved by reintroduction of IS without any penalty [153]. In contrast, "Clinically silent" progression of fibrosis may lead to graft dysfunction over the long term and can only be detected by biopsy with its attendant risks. There is thus considerable need for development of noninvasive methods to identify the extent of liver fibrosis.

64.5 CONCLUSIONS

There is a need to establish reliable biomarkers of liver transplant tolerance that can prospectively identify patients who are most likely to be successfully weaned from IS. This will enable conduct of weaning in a more informed manner than current clinical practice. One approach is to validate existing candidate biomarkers identified by retrospective studies in large-scale multicenter prospective studies of patients about to commence weaning of IS. In addition, noninvasive biomarkers of graft fibrosis are required, since this can occur in a "clinically silent" manner that can lead to long-term liver damage.

ACKNOWLEDGMENTS

The authors thank Dr. Ying Li, Dr. Hanaa Nafady-Hego, Dr. Xiangdong Zhao, Dr. Naoki Satoda, Dr. Mami Yoshitomi, and Dr. Kayo Waki for their contribution to this chapter and Prof. Koichi Tanaka, Prof. Jacques Pirenne, Prof. Kathryn Wood, and Prof. Shimon Sakaguchi for all they taught us.

REFERENCES

[1] Billingham RE, Brent L, Medawar PB. Actively acquired tolerance of foreign cells. Nature 1953;172:603–6.

[2] Cobbold SP, Adams E, Marshall SE, Davies JD, Waldmann H. Mechanisms of peripheral tolerance and suppression induced by monoclonal antibodies to CD4 and CD8. Immunol Rev 1996;149:5–33.

[3] Cuturi MC, Josien R, Douillard P, Pannetier C, Cantarovich D, Smit H, et al. Prolongation of allogeneic heart graft survival in rats by administration of a peptide (a.a. 75–84) from the alpha 1 helix of the first domain of HLA-B7 01. Transplantation 1995;59:661–9.

[4] Knechtle SJ. Knowledge about transplantation tolerance gained in primates. Curr Opin Immunol 2000;12:552–6.

[5] Levisetti MG, Padrid PA, Szot GL, Mittal N, Meehan SM, Wardrip CL, et al. Immunosuppressive effects of human CTLA4Ig in a non-human primate model of allogeneic pancreatic islet transplantation. J Immunol 1997;159:5187–91.

[6] Remuzzi G, Rossini M, Imberti O, Perico N. Kidney graft survival in rats without immunosuppressants after intrathymic glomerular transplantation. Lancet 1991;337:750–2.

[7] Sablinski T, Hancock WW, Tilney NL, Kupiec-Weglinski JW. CD4 monoclonal antibodies in organ transplantation—a review of progress. Transplantation 1991;52:579–89.

[8] Subbotin V, Sun H, Aitouche A, Valdivia LA, Fung JJ, Starzl TE, et al. Abrogation of chronic rejection in a murine model of aortic allotransplantation by prior induction of donor-specific tolerance. Transplantation 1997;64:690–5.

[9] Calne RY, Sells RA, Pena JR, Davis DR, Millard PR, Herbertson BM, et al. Induction of immunological tolerance by porcine liver allografts. Nature 1969;223:472–6.

[10] Zimmermann FA, Butcher GW, Davies HS, Brons G, Kamada N, Türel O. Techniques for orthotopic liver transplantation in the rat and some studies of the immunologic responses to fully allogeneic liver grafts. Transplant Proc 1979;11:571–7.

[11] Qian S, Demetris AJ, Murase N, Rao AS, Fung JJ, Starzl TE. Murine liver allograft transplantation: tolerance and donor cell chimerism. Hepatology 1994;19:916–24.

[12] Lerut J, Sanchez-Fueyo A. An appraisal of tolerance in liver transplantation. Am J Transplant 2006;6:1774–80.

[13] Sánchez-Fueyo A. Identification of tolerant recipients following liver transplantation. Int Immunopharmacol 2010;10:1501–4.

[14] Li Y, Koshiba T, Yoshizawa A, Yonekawa Y, Masuda K, Ito A, et al. Analyses of peripheral blood mononuclear cells in operational tolerance after pediatric living donor liver transplantation. Am J Transplant 2004;4:2118–25.

[15] Martínez-Llordella M, Lozano JJ, Puig-Pey I, Orlando G, Tisone G, Lerut J, et al. Using transcriptional profiling to develop a diagnostic test of operational tolerance in liver transplant recipients. J Clin Invest 2008;118:2845–57.

[16] Li L, Wozniak LJ, Rodder S, Heish S, Talisetti A, Wang Q, et al. A common peripheral blood gene set for diagnosis of operational tolerance in pediatric and adult liver transplantation. Am J Transplant 2012;12:1218–28.

[17] Martínez-Llordella M, Puig-Pey I, Orlando G, Ramoni M, Tisone G, Rimola A, et al. Multiparameter immune profiling of operational tolerance in liver transplantation. Am J Transplant 2007;7:309–19.

[18] Zhao X, Li Y, Ohe H, Nafady-Hego H, Uemoto S, Bishop GA, et al. Intragraft Vδ1 γδ T cells with a unique T-cell receptor are closely associated with pediatric semiallogeneic liver transplant tolerance. Transplantation 2013;95:192–202.

[19] Tokita D, Mazariegos GV, Zahorchak AF, Chien N, Abe M, Raimondi G, et al. High PD-L1/CD86 ratio on plasmacytoid dendritic cells correlates with elevated T-regulatory cells in liver transplant tolerance. Transplantation 2008;85:369–77.

[20] Nafady-Hego H, Li Y, Ohe H, Zhao X, Satoda N, Sakaguchi S, et al. The generation of donor-specific CD4$^+$CD25$^+$CD45RA$^+$ naive regulatory T cells in operationally tolerant patients after pediatric living-donor liver transplantation. Transplantation 2010;90:1547–55.

[21] Castellaneta A, Mazariegos GV, Nayyar N, Zeevi A, Thomson AW. HLA-G level on monocytoid dendritic cells correlates with

regulatory T-cell Foxp3 expression in liver transplant tolerance. Transplantation 2011;91:1132−40.

[22] Ohe H, Waki K, Yoshitomi M, Morimoto T, Nafady-Hego H, Satoda N, et al. Factors affecting operational tolerance after pediatric living-donor liver transplantation: impact of early post-transplant events and HLA match. Transpl Int 2012;25:97−106.

[23] Li Y, Zhao X, Cheng D, Haga H, Tsuruyama T, Wood K, et al. The presence of Foxp3 expressing T cells within grafts of tolerant human liver transplant recipients. Transplantation 2008;86:1837−43.

[24] Zarkhin V, Talisetti A, Li L, Wozniak LJ, McDiarmid SV, Cox K, et al. Expression of soluble HLA-G identifies favorable outcomes in liver transplant recipients. Transplantation 2010;90:1000−5.

[25] Mazariegos GV, Reyes J, Webber SA, Thomson AW, Ostrowski L, Abmed M, et al. Cytokine gene polymorphisms in children successfully withdrawn from immunosuppression after liver transplantation. Transplantation 2002;73:1342−5.

[26] Takatsuki M, Uemoto S, Inomata Y, Sakamoto S, Hayashi M, Ueda M, et al. Analysis of alloreactivity and intragraft cytokine profiles in living donor liver transplant recipients with graft acceptance. Transpl Immunol 2001;8:279−86.

[27] Girnita A, Mazariegos GV, Castellaneta A, Reyes J, Bentlejewski C, Thomson AW, et al. Liver transplant recipients weaned off immunosuppression lack circulating donor-specific antibodies. Hum Immunol 2010;71:274−6.

[28] Waki K, Sugawara Y, Mizuta K, Taniguchi M, Ozawa M, Hirata M, et al. Predicting operational tolerance in pediatric living-donor liver transplantation by absence of HLA antibodies. Transplantation 2013;95:177−83.

[29] Feng S, Ekong UD, Lobritto SJ, Demetris AJ, Roberts JP, Rosenthal P, et al. Complete immunosuppression withdrawal and subsequent allograft function among pediatric recipients of parental living donor liver transplants. JAMA 2012;307:283−93.

[30] Bohne F, Martínez-Llordella M, Lozano JJ, Miquel R, Benítez C, Londoño MC, et al. Intra-graft expression of genes involved in iron homeostasis predicts the development of operational tolerance in human liver transplantation. J Clin Invest 2012;122:368−82.

[31] Sakaguchi S, Sakaguchi N, Asano M, Itoh M, Toda M. Immunologic self-tolerance maintained by activated T cells expressing IL-2 receptor alpha-chains (CD25). Breakdown of a single mechanism of self-tolerance causes various autoimmune diseases. J Immunol 1995;155:1151−64.

[32] Wood KJ, Sakaguchi S. Regulatory T cells in transplantation tolerance. Nat Rev Immunol 2003;3:199−210.

[33] Hori S, Nomura T, Sakaguchi S. Control of regulatory T cell development by the transcription factor Foxp3. Science 2003;299:1057−61.

[34] Valmori D, Merlo A, Souleimanian NE, Hesdorffer CS, Ayyoub M. A peripheral circulating compartment of natural naive CD4 Tregs. J Clin Invest 2005;115:1953−62.

[35] Seddiki N, Santner-Nanan B, Tangye SG, Alexander SI, Solomon M, Lee S, et al. Persistence of naive CD45RA$^+$ regulatory T cells in adult life. Blood 2006;107:2830−8.

[36] Hoffmann P, Eder R, Boeld TJ, Doser K, Piseshka B, Andreesen R, et al. Only the CD45RA$^+$ subpopulation of CD4$^+$CD25^{high} T cells gives rise to homogeneous regulatory T-cell lines upon *in vitro* expansion. Blood 2006;108:4260−7.

[37] Haas J, Fritzsching B, Trübswetter P, Korporal M, Milkova L, Fritz B, et al. Prevalence of newly generated naive regulatory T

cells (Treg) is critical for Treg suppressive function and determines Treg dysfunction in multiple sclerosis. J Immunol 2007;179:1322−30.

[38] Hoffmann P, Boeld TJ, Eder R, Huehn J, Floess S, Wieczorek G, et al. Loss of Foxp3 expression in natural human CD4$^+$CD25$^+$ regulatory T cells upon repetitive *in vitro* stimulation. Eur J Immunol 2009;39:1088−97.

[39] Miyara M, Yoshioka Y, Kitoh A, Shima T, Wing K, Niwa A, et al. Functional delineation and differentiation dynamics of human CD4$^+$ T cells expressing the Foxp3 transcription factor. Immunity 2009;30:899−911.

[40] Putnam AL, Brusko TM, Lee MR, Liu W, Szot GL, Ghosh T, et al. Expansion of human regulatory T cells from patients with type 1 diabetes. Diabetes 2009;58:652−62.

[41] Dieckmann D, Plottner H, Berchtold S, Berger T, Schuler G. *Ex vivo* isolation and characterization of CD4(+)CD25(+) T cells with regulatory properties from human blood. J Exp Med 2001;193:1303−10.

[42] Taams LS, Smith J, Rustin MH, Salmon M, Poulter LW, Akbar AN. Human anergic/suppressive CD4(+)CD25(+) T cells: a highly differentiated and apoptosis-prone population. Eur J Immunol 2001;31:1122−31.

[43] Baecher-Allan C, Brown JA, Freeman GJ, Hafler DA. CD4$^+$CD25^{high} regulatory cells in human peripheral blood. J Immunol 2001;167:1245−53.

[44] Jonuleit H, Schmitt E, Stassen M, Tuettenberg A, Knop J, Enk AH. Identification and functional characterization of human CD4 (+)CD25(+) T cells with regulatory properties isolated from peripheral blood. J Exp Med 2001;193:1285−94.

[45] Yagi H, Nomura T, Nakamura K, Yamazaki S, Kitawaki T, Hori S, et al. Crucial role of Foxp3 in the development and function of human CD25$^+$CD4$^+$ regulatory T cells. Int Immunol 2004;16:1643−56.

[46] Sereti I, Imamichi H, Natarajan V, Imamichi T, Ramchandani MS, Badralmaa Y, et al. *In vivo* expansion of CD4CD45RO$^-$CD25 T cells expressing Foxp3 in IL-2-treated HIV-infected patients. J Clin Invest 2005;115:1839−47.

[47] Fujimaki W, Takahashi N, Ohnuma K, Nagatsu M, Kurosawa H, Yoshida S, et al. Comparative study of regulatory T cell function of human CD25CD4 T cells from thymocytes, cord blood, and adult peripheral blood. Clin Dev Immunol 2008;30:58−9.

[48] Brusko T, Atkinson M. Treg in type 1 diabetes. Cell Biochem Biophys 2007;48:165−75.

[49] Booth NJ, McQuaid AJ, Sobande T, Kissane S, Agius E, Jackson SE, et al. Different proliferative potential and migratory characteristics of human CD4$^+$ regulatory T cells that express either CD45RA or CD45RO. J Immunol 2010;184:4317−26.

[50] Mikulkova Z, Praksova P, Stourac P, Bednarik J, Michalek J. Imbalance in T-cell and cytokine profiles in patients with relapsing-remitting multiple sclerosis. J Neurol Sci 2011;300:135−41.

[51] Dau PC, Callahan JP. Immune modulation during treatment of systemic sclerosis with plasmapheresis and immunosuppressive drugs. Clin Immunol Immunopathol 1994;70:159−65.

[52] Abdulahad WH, van der Geld YM, Stegeman CA, Kallenberg CG. Persistent expansion of CD4$^+$ effector memory T cells in Wegener's granulomatosis. Kidney Int 2006;70:938−47.

[53] Darmochwal-Kolarz D, Saito S, Rolinski J, Tabarkiewicz J, Kolarz B, Leszczynska-Gorzelak B, et al. Activated T lymphocytes in pre-eclampsia. Am J Reprod Immunol 2007;58:39−45.

[54] Venken K, Hellings N, Broekmans T, Hensen K, Rummens JL, Stinissen P. Natural naive CD4$^+$CD25$^+$CD127^{low} regulatory T cell (Treg) development and function are disturbed in multiple sclerosis patients: recovery of memory Treg homeostasis during disease progression. J Immunol 2008;180:6411−20.

[55] Liu W, Putnam AL, Xu-Yu Z, Szot GL, Lee MR, Zhu S, et al. CD127 expression inversely correlates with Foxp3 and suppressive function of human CD4$^+$T reg cells. J Exp Med 2006;203:1701−11.

[56] Seddiki N, Santner-Nanan B, Martinson J, Zaunders J, Sasson S, Landay A, et al. Expression of interleukin (IL)-2 and IL-7 receptors discriminates between human regulatory and activated T cells. J Exp Med 2006;203:1693−700.

[57] Ardon H, Verbinnen B, Maes W, Beez T, Van Gool S, De Vleeschouwer S. Technical advancement in regulatory T cell isolation and characterization using CD127 expression in patients with malignant glioma treated with autologous dendritic cell vaccination. J Immunol Methods 2010;352:169−73.

[58] Nafady-Hego H, Li Y, Ohe H, Elgendy H, Sakaguchi S, Uemoto S, et al. Significance of CD127 versus Foxp3 to define regulatory T cells in tolerant recipients after pediatric living-donor liver transplantation. Transpl Int 2011;24:225.

[59] Ferrick DA, King DP, Jackson KA, Braun RK, Tam S, Hyde DM, et al. Intraepithelial gamma delta T lymphocytes: sentinel cells at mucosal barriers. Springer Semin Immunopathol 2000;22:283−96.

[60] Richards MH, Nelson JL. The evolution of vertebrate antigen receptors: a phylogenetic approach. Mol Biol Evol 2000;17:146−55.

[61] Hayday AC. [gamma][delta] Cells: a right time and a right place for a conserved third way of protection. Annu Rev Immunol 2000;18:975−1026.

[62] Wilson E, Aydintug MK, Jutila MA. A circulating bovine gamma delta T cell subset, which is found in large numbers in the spleen, accumulates inefficiently in an artificial site of inflammation: correlation with lack of expression of E-selectin ligands and L-selectin. J Immunol 1999;162:4914−9.

[63] Holtmeier W, Kabelitz D. gammadelta T cells link innate and adaptive immune responses. Chem Immunol Allergy 2005;86:151−83.

[64] De Rosa SC, Andrus JP, Perfetto SP, Mantovani JJ, Herzenberg LA, Herzenberg LA, et al. Ontogeny of gamma delta T cells in humans. J Immunol 2004;172:1637−45.

[65] De Rosa SC, Mitra DK, Watanabe N, Herzenberg LA, Herzenberg LA, Roederer M. Vdelta1 and Vdelta2 gammadelta T cells express distinct surface markers and might be developmentally distinct lineages. J Leukoc Biol 2001;70:518−26.

[66] Holtmeier W, Rowell DL, Nyberg A, Kagnoff MF. Distinct delta T cell receptor repertoires in monozygotic twins concordant for coeliac disease. Clin Exp Immunol 1997;107:148−57.

[67] Kamath AB, Wang L, Das H, Li L, Reinhold VN, Bukowski JF. Antigens in tea-beverage prime human Vgamma 2Vdelta 2 T cells in vitro and in vivo for memory and nonmemory antibacterial cytokine responses. Proc Natl Acad Sci USA 2003;100:6009−14.

[68] Wang L, Das H, Kamath A, Li L, Bukowski JF. Human V gamma 2V delta 2 T cells augment migration-inhibitory factor secretion and counteract the inhibitory effect of glucocorticoids on IL-1 beta and TNF-alpha production. J Immunol 2002;168:4889−96.

[69] Shen Y, Zhou D, Qiu L, Lai X, Simon M, Shen L, et al. Adaptive immune response of Vgamma2Vdelta2 + T cells during mycobacterial infections. Science 2002;295:2255−8.

[70] Das H, Sugita M, Brenner MB. Mechanisms of Vdelta1 gamma delta T cell activation by microbial components. J Immunol 2004;172:6578−86.

[71] Kress E, Hedges JF, Jutila MA. Distinct gene expression in human Vdelta1 and Vdelta2 gamma delta T cells following non-TCR agonist stimulation. Mol Immunol 2006;43:2002−11.

[72] Peng G, Wang HY, Peng W, Kiniwa Y, Seo KH, Wang RF. Tumor-infiltrating gamma delta T cells suppress T and dendritic cell function via mechanisms controlled by a unique toll-like receptor signaling pathway. Immunity 2007;27:334−48.

[73] Barakonyi A, Polgar B, Szekeres-Bartho J. The role of gamma/delta T-cell receptor-positive cells in pregnancy: part II. Am J Reprod Immunol 1999;42:83−7.

[74] Nagaeva O, Jonsson L, Mincheva-Nilsson L. Dominant IL-10 and TGF-beta mRNA expression in gammadeltaT cells of human early pregnancy decidua suggests immunoregulatory potential. Am J Reprod Immunol 2002;48:9−17.

[75] Puig-Pey I, Bohne F, Benítez C, López M, Martínez-Llordella M, Oppenheimer F, et al. Characterization of $\gamma\delta$ T cell subsets in organ transplantation. Transpl Int 2010;23:1045−55.

[76] Banff Working Group on Liver Allograft Pathology. Importance of liver biopsy findings in immunosuppression management: biopsy monitoring and working criteria for patients with operational tolerance. Liver Transpl 2012;18:1154−70.

[77] Mosmann TR, Cherwinski H, Bond MW, Giedlin MA, Coffman RL. Two types of murine helper T cell clone. I. Definition according to profiles of lymphokine activities and secreted proteins. J Immunol 1986;136:2348−57.

[78] Babcock GF, Alexander JW. The effects of blood transfusion on cytokine production by TH1 and TH2 lymphocytes in the mouse. Transplantation 1996;61:465−8.

[79] Chen N, Gao Q, Field EH. Prevention of Th1 response is critical for tolerance. Transplantation 1996;61:1076−83.

[80] Ganschow R, Broering DC, Nolkemper D, Albani J, Kemper MJ, Rogiers X, et al. Th2 cytokine profile in infants predisposes to improved graft acceptance after liver transplantation. Transplantation 2001;72:929−34.

[81] Hancock WW, Lord RH, Colby AJ, Diamantstein T, Rickles FR, Dijkstra C, et al. Identification of IL 2R$^+$T cells and macrophages within rejecting rat cardiac allografts, and comparison of the effects of treatment with anti-IL 2R monoclonal antibody or cyclosporin. J Immunol 1987;138:164−70.

[82] Hara M, Kingsley CI, Niimi M, Read S, Turvey SE, Bushell AR, et al. IL-10 is required for regulatory T cells to mediate tolerance to alloantigens in vivo. J Immunol 2001;166:3789−96.

[83] Onodera K, Hancock WW, Graser E, Lehmann M, Sayegh MH, Strom TB, et al. Type 2 helper T cell-type cytokines and the development of "infectious" tolerance in rat cardiac allograft recipients. J Immunol 1997;158:1572−81.

[84] Sayegh MH, Akalin E, Hancock WW, Russell ME, Carpenter CB, Linsley PS, et al. CD28-B7 blockade after alloantigenic challenge in vivo inhibits Th1 cytokines but spares Th2. J Exp Med 1995;181:1869−74.

[85] Suthanthiran M, Strom TB. Renal transplantation. N Engl J Med 1994;331:365−76.

[86] Takeuchi T, Lowry RP, Konieczny B. Heart allografts in murine systems. The differential activation of Th2-like effector cells in peripheral tolerance. Transplantation 1992;53:1281−94.

[87] Zhai Y, Ghobrial RM, Busuttil RW, Kupiec-Weglinski JW. Th1 and Th2 cytokines in organ transplantation: paradigm lost? Crit Rev Immunol 1999;19:155−72.

[88] Cardarelli F, Pascual M, Tolkoff-Rubin N, Delmonico FL, Wong W, Schoenfeld DA, et al. Prevalence and significance of anti-HLA and donor-specific antibodies long-term after renal transplantation. Transpl Int 2005;18:532−40.

[89] Campos EF, Tedesco-Silva H, Machado PG, Franco M, Medina-Pestana JO, Gerbase-DeLima M. Post-transplant anti-HLA class II antibodies as risk factor for late kidney allograft failure. Am J Transplant 2006;6:2316−20.

[90] Terasaki P, Mizutani K. Antibody mediated rejection: update 2006. Clin J Am Soc Nephrol 2006;1:400−3.

[91] Gordon RD, Fung JJ, Markus B, Fox I, Iwatsuki S, Esquivel CO, et al. The antibody crossmatch in liver transplantation. Surgery 1986;100:705−15.

[92] Castillo-Rama M, Castro MJ, Bernardo I, Meneu-Diaz JC, Elola-Olaso AM, Calleja-Antolin SM, et al. Preformed antibodies detected by cytotoxic assay or multibead array decrease liver allograft survival: role of human leukocyte antigen compatibility. Liver Transpl 2008;14:554−62.

[93] Goh A, Scalamogna M, De Feo T, Poli F, Terasaki PI. Human leukocyte antigen crossmatch testing is important for liver retransplantation. Liver Transpl 2010;16:308−13.

[94] Fontana M, Moradpour D, Aubert V, Pantaleo G, Pascual M. Prevalence of anti-HLA antibodies after liver transplantation. Transpl Int 2010;23:858−9.

[95] Colvin RB. Dimensions of antibody-mediated rejection. Am J Transplant 2010;10:1509−10.

[96] Neumann UP, Guckelberger O, Langrehr JM, Lang M, Schmitz V, Theruvath T, et al. Impact of human leukocyte antigen matching in liver transplantation. Transplantation 2003;75:132−7.

[97] Navarro V, Herrine S, Katopes C, Colombe B, Spain CV. The effect of HLA class I (A and B) and class II (DR) compatibility on liver transplantation outcomes: an analysis of the OPTN database. Liver Transpl 2006;12:652−8.

[98] Sieders E, Hepkema BG, Peeters PM, TenVergert EM, de Jong KP, Porte RJ. The effect of HLA mismatches, shared cross-reactive antigen groups, and shared HLA-DR antigens on the outcome after pediatric liver transplantation. Liver Transpl 2005;11:1541−9.

[99] Devlin J, Doherty D, Thomson L, Wong T, Donaldson P, Portmann B, et al. Defining the outcome of immunosuppression withdrawal after liver transplantation. Hepatology 1998;27:926−33.

[100] Tryphonopoulos P, Ruiz P, Weppler D, Nishida S, Levi DM, Moon J, et al. Long-term follow-up of 23 operational tolerant liver transplant recipients. Transplantation 2010;90:1556−61.

[101] Sagoo P, Perucha E, Sawitzki B, Tomiuk S, Stephens DA, Miqueu P, et al. Development of a cross-platform biomarker signature to detect renal transplant tolerance in humans. J Clin Invest 2010;120:1848−61.

[102] Newell KA, Asare A, Kirk AD, Gisler TD, Bourcier K, Suthanthiran M, et al. Identification of a B cell signature associated with renal transplant tolerance in humans. J Clin Invest 2010;120:1836−47.

[103] Lechler RI, Batchelor JR. Restoration of immunogenicity to passenger cell-depleted kidney allografts by the addition of donor strain dendritic cells. J Exp Med 1982;155:31−41.

[104] Larsen CP, Austyn JM, Morris PJ. The role of graft-derived dendritic leukocytes in the rejection of vascularized organ allografts. Recent findings on the migration and function of dendritic leukocytes after transplantation. Ann Surg 1990;212:308−15.

[105] Morelli AE, Thomson AW. Tolerogenic dendritic cells and the quest for transplant tolerance. Nat Rev Immunol 2007;7:610−21.

[106] Abe M, Wang Z, de Creus A, Thomson AW. Plasmacytoid dendritic cell precursors induce allogeneic T-cell hyporesponsiveness and prolong heart graft survival. Am J Transplant 2005;5:1808−19.

[107] Fugier-Vivier IJ, Rezzoug F, Huang Y, Graul-Layman AJ, Schanie CL, Xu H, et al. Plasmacytoid precursor dendritic cells facilitate allogeneic hematopoietic stem cell engraftment. J Exp Med 2005;201:373−83.

[108] Ochando JC, Homma C, Yang Y, Hidalgo A, Garin A, Tacke F, et al. Alloantigen-presenting plasmacytoid dendritic cells mediate tolerance to vascularized grafts. Nat Immunol 2006;7:652−62.

[109] Tang Q, Bluestone JA. Plasmacytoid DCs and T(reg) cells: casual acquaintance or monogamous relationship? Nat Immunol 2006;7:551−3.

[110] Kovats S, Main EK, Librach C, Stubblebine M, Fisher SJ, DeMars R. A class I antigen, HLA-G, expressed in human trophoblasts. Science 1990;248:220−3.

[111] Le Rond S, Azéma C, Krawice-Radanne I, Durrbach A, Guettier C, Carosella ED, et al. Evidence to support the role of HLA-G5 in allograft acceptance through induction of immunosuppressive/regulatory T cells. J Immunol 2006;176:3266−76.

[112] Selmani Z, Naji A, Gaiffe E, Obert L, Tiberghien P, Rouas-Freiss N, et al. HLA-G is a crucial immunosuppressive molecule secreted by adult human mesenchymal stem cells. Transplantation 2009;87 (9 Suppl):S62−66.

[113] Liang S, Baibakov B, Horuzsko A. HLA-G inhibits the functions of murine dendritic cells via the PIR-B immune inhibitory receptor. Eur J Immunol 2002;32:2418−26.

[114] LeMaoult J, Krawice-Radanne I, Dausset J, Carosella ED. HLA-G1-expressing antigen-presenting cells induce immunosuppressive CD4$^+$ T cells. Proc Natl Acad Sci USA 2004;101:7064−9.

[115] Lozano JJ, Pallier A, Martinez-Llordella M, Danger R, López M, Giral M, et al. Comparison of transcriptional and blood cell-phenotypic markers between operationally tolerant liver and kidney recipients. Am J Transplant 2011;11:1916−26.

[116] Bishop GA, Ierino FL, Sharland AF, Hall BM, Alexander SI, Sandrin MS, et al. Approaching the promise of operational tolerance in clinical transplantation. Transplantation 2011;91:1065−74.

[117] Ahern AT, Artruc SB, DellaPelle P, Cosimi AB, Russell PS, Colvin RB, et al. Hyperacute rejection of HLA-AB-identical renal allografts associated with B lymphocyte and endothelial reactive antibodies. Transplantation 1982;33:103−6.

[118] Dragun D. Humoral responses directed against non-human leukocyte antigens in solid-organ transplantation. Transplantation 2008;86:1019−25.

[119] Sumitran-Karuppan S, Tyden G, Reinholt F, Berg U, Moller E. Hyperacute rejections of two consecutive renal allografts

and early loss of the third transplant caused by non-HLA antibodies specific for endothelial cells. Transpl Immunol 1997;5:321–7.

[120] Sumitran-Holgersson S, Wilczek HE, Holgersson J, Söderström K. Identification of the nonclassical HLA molecules, mica, as targets for humoral immunity associated with irreversible rejection of kidney allografts. Transplantation 2002;74:268–77.

[121] Zou Y, Heinemann FM, Grosse-Wilde H, Sireci G, Wang Z, Lavingia B, et al. Detection of anti-MICA antibodies in patients awaiting kidney transplantation, during the post-transplant course, and in eluates from rejected kidney allografts by Luminex flow cytometry. Hum Immunol 2006;67:230–7.

[122] Rodriguez-Mahou M, Salcedo M, Fernandez-Cruz E, Tiscar JL, Bañares R, Clemente G, et al. Antibodies against glutathione S-transferase T1 (GSTT1) in patients with GSTT1 null genotype as prognostic marker: long-term follow-up after liver transplantation. Transplantation 2007;83:1126–9.

[123] Aguilera I, Alvarez-Marquez A, Gentil MA, Fernandez-Alonso J, Fijo J, Saez C, et al. Anti-glutathione S-transferase T1 antibody-mediated rejection in C4d-positive renal allograft recipients. Nephrol Dial Transplant 2008;23:2393–8.

[124] Dragun D, Müller DN, Bräsen JH, Fritsche L, Nieminen-Kelhä M, Dechend R, et al. Angiotensin II type 1-receptor activating antibodies in renal-allograft rejection. N Engl J Med 2005;352:558–69.

[125] Amico P, Hönger G, Bielmann D, Lutz D, Garzoni D, Steiger J, et al. Incidence and prediction of early antibody-mediated rejection due to non-human leukocyte antigen-antibodies. Transplantation 2008;85:1557–63.

[126] Rose ML. Role of endothelial cells in allograft rejection. Vasc Med 1997;2:105–14.

[127] Praprotnik S, Blank M, Levy Y, Tavor S, Boffa MC, Weksler B, et al. Anti-endothelial cell antibodies from patients with thrombotic thrombocytopenic purpura specifically activate small vessel endothelial cells. Int Immunol 2001;13:203–10.

[128] Jurcevic S, Ainsworth ME, Pomerance A, Smith JD, Robinson DR, Dunn MJ, et al. Antivimentin antibodies are an independent predictor of transplant-associated coronary artery disease after cardiac transplantation. Transplantation 2001;71:886–92.

[129] Carter V, Shenton BK, Jaques B, Turner D, Talbot D, Gupta A, et al. Vimentin antibodies: a non-HLA antibody as a potential risk factor in renal transplantation. Transplant Proc 2005;37:654–7.

[130] Lawson C, Holder AL, Stanford RE, Smith J, Rose ML. Anti-intercellular adhesion molecule-1 antibodies in sera of heart transplant recipients: a role in endothelial cell activation. Transplantation 2005;80:264–71.

[131] Bahram S, Bresnahan M, Geraghty DE, Spies T. A second lineage of mammalian major histocompatibility complex class I genes. Proc Natl Acad Sci USA 1994;91:6259–63.

[132] Scornik JC, Guerra G, Schold JD, Srinivas TR, Dragun D, Meier-Kriesche HU. Value of posttransplant antibody tests in the evaluation of patients with renal graft dysfunction. Am J Transplant 2007;7(7):1808–14.

[133] Bauer S, Groh V, Wu J, Steinle A, Phillips JH, Lanier LL, et al. Activation of NK cells and T cells by NKG2D, a receptor for stress-inducible MICA. Science 1999;285:727–9.

[134] Jinushi M, Hodi FS, Dranoff G. Therapy-induced antibodies to MHC class I chain-related protein A antagonize immune suppression and stimulate antitumor cytotoxicity. Proc Natl Acad Sci USA 2006;103:9190–5.

[135] Hankey KG, Drachenberg CB, Papadimitriou JC, Klassen DK, Philosophe B, Bartlett ST, et al. MIC expression in renal and pancreatic allografts. Transplantation 2002;73:304–6.

[136] Quiroga I, Salio M, Koo DD, Cerundolo L, Shepherd D, Cerundolo V, et al. Expression of MHC class I-related Chain B (MICB) molecules on renal transplant biopsies. Transplantation 2006;81:1196–203.

[137] Suárez-Alvarez B, López-Vázquez A, Díaz-Molina B, Bernardo-Rodríguez MJ, Alvarez-López R, Pascual D, et al. The predictive value of soluble major histocompatibility complex class I chain-related molecule A (MICA) levels on heart allograft rejection. Transplantation 2006;82:354–61.

[138] Suárez-Alvarez B, López-Vázquez A, Gonzalez MZ, Fdez-Morera JL, Díaz-Molina B, Blanco-Gelaz MA, et al. The relationship of anti-MICA antibodies and MICA expression with heart allograft rejection. Am J Transplant 2007;7:1842–8.

[139] Mizutani K, Terasaki P, Bignon JD, Hourmant M, Cesbron-Gautier A, Shih RN, et al. Association of kidney transplant failure and antibodies against MICA. Hum Immunol 2006;9:683–91.

[140] Panigrahi A, Siddiqui JA, Rai A, Margoob A, Khaira A, Bhowmik D, et al. Allosensitization to HLA and MICA is an important measure of renal graft outcome. Clin Transpl 2007;211–7.

[141] Terasaki PI, Ozawa M, Castro R. Four-year follow-up of a prospective trial of HLA and MICA antibodies on kidney graft survival. Am J Transplant 2007;7:408–15.

[142] Mizutani K, Terasaki P, Rosen A, Esquenazi V, Miller J, Shih RN, et al. Serial ten-year follow-up of HLA and MICA antibody production prior to kidney graft failure. Am J Transplant. 2005;5:2265–72.

[143] Zou Y, Stastny P, Süsal C, Döhler B, Opelz G. Antibodies against MICA antigens and kidney-transplant rejection. N Engl J Med 2007;357:1293–300.

[144] Hunyady L, Catt KJ. Pleiotropic AT1 receptor signaling pathways mediating physiological and pathogenic actions of angiotensin II. Mol Endocrinol 2006;20:953–70.

[145] Dzau VJ, Bernstein K, Celermajer D, Cohen J, Dahlöf B, Deanfield J, et al. The relevance of tissue angiotensin-converting enzyme: manifestations in mechanistic and endpoint data. Am J Cardiol 2001;88(9A):1L–20L.

[146] Mellor AL, Munn DH. Immunology at the maternal-fetal interface: lessons for T cell tolerance and suppression. Annu Rev Immunol 2000;18:367–91.

[147] Matthews JB, Ramos E, Bluestone JA. Clinical trials of transplant tolerance: slow but steady progress. Am J Transplant 2003;3:794–803.

[148] Tisone G, Orlando G, Angelico M. Operational tolerance in clinical liver transplantation: emerging developments. Transpl Immunol 2007;17:108—13.

[149] Yoshitomi M, Koshiba T, Haga H, Li Y, Zhao X, Cheng D, et al. Requirement of protocol biopsy before and after complete cessation of immunosuppression after liver transplantation. Transplantation 2009;87:606—14.

[150] Evans HM, Kelly DA, McKiernan PJ, Hübscher S. Progressive histological damage in liver allografts following pediatric liver transplantation. Hepatology 2006;43:1109—17.

[151] Scheenstra R, Peeters PM, Verkade HJ, Gouw AS. Graft fibrosis after pediatric liver transplantation: ten years of follow-up. Hepatology 2009;49:880—6.

[152] Ekong UD, Melin-Aldana H, Seshadri R, Lokar J, Harris D, Whitington PF, et al. Graft histology characteristics in long-term survivors of pediatric liver transplantation. Liver Transpl 2008;14:1582—7.

[153] Koshiba T, Li Y, Takemura M, Wu Y, Sakaguchi S, Minato N, et al. Clinical, immunological, and pathological aspects of operational tolerance after pediatric living-donor liver transplantation. Transpl Immunol 2007;17:94—7.

Chapter 65

Biomarkers of Tolerance in Renal Transplantation

Faouzi Braza[a,b,c], Nicolas Degauque[a,b,c], Jean-Paul Soulillou[a,b,c], and Sophie Brouard[a,b,c]

[a]INSERM, UMR 1064, Nantes, France, [b]CHU de Nantes, ITUN, Nantes, France, [c]Université de Nantes, Faculté de Médecine, Nantes, France

65.1 INTRODUCTION

Inducing transplant tolerance (i.e., stable graft function after stopping immunosuppression in recipients who remain immunocompetent) remains the major goal in clinical transplantation. Numerous strategies to induce tolerance in experimental models had been reported but very few were transferable to human settings. The clinical trial by D. Sachs group accounts for the few reported successes and relies on mixed chimerism induction to HLA-mismatched transplant in end-stage renal disease patients [1]. Nevertheless, the conditioning regimen precludes its generalization in clinical transplantation and adverse events have been reported. The development of monoclonal antibodies has emerged as an interesting alternative to control graft rejection [2]. Monoclonal antibodies gave the possibility to target particular processes of the immune response. Many antibodies had been tested to deplete T cells, including alloreactive T cells, and to neutralize cytokines signalization in experimental transplantation or clinical trials and have demonstrated their capacity to improve long-term graft survival in combination with classic immunosuppression [3]. More recently, strategies have focused on the possibility to deplete memory T cells [4] and block costimulatory signals [5]. Nevertheless, monoclonal antibodies need to pass through robust clinical trial in order to avoid any deleterious and dangerous side

effects [6] and ensure a safe clinical application. Finally, a lot of recent studies have shed light on the potential of immune regulatory cells to control graft rejection [7–9]. However, this approach needs validation and a better understanding of regulatory processes. Tolerogenic dendritic cells, natural and induced regulatory T cells, suppressive macrophages and regulatory B cells are heterogeneous and plastic populations that could shift into effectors cells in an inflammatory microenvironment. Consequently, there is a high need to develop isolation and expansion technology to obtain stable regulatory cells in order to ensure the safety of transplanted recipients [8].

Rather than a proactive modification of the immune system, an intense area of research is devoted to the identification of feature characteristics of a graft outcome (such as chronic rejection or operational tolerance). The quest for biomarkers aimed to adjust the treatment of the patients according to their graft outcome. As a correlate, the identification of biomarkers requires large cohort of patients and a stringent statistical filter to identify relevant diagnostic and prognostic biomarkers. A biomarker is defined as "a characteristic that is objectively measured and evaluated as an indicator of normal biologic processes, pathogenic processes, or pharmacologic responses to a therapeutic intervention" [10]. It can be a gene, protein or peptide, RNA product (mRNA and miRNA), or a metabolite involved in a biological process. To be relevant, a biomarker must be indicative of a clinical state and easily identified and monitored. In kidney clinical transplantation, a good biomarker must be correlated with the clinical outcome and be assessed in a noninvasive manner (e.g., blood or urine samples). Currently, proteinuria and serum creatinine are used as surrogate markers to monitor graft function. Proteinuria is an early and sensitive marker of kidney damage [11]. Proteinuria that is inappropriately high is an indication of possible graft dysfunction [12–14] and is a good marker of the current clinical state of patients. Serum creatinine is closely monitored in the first days and weeks after transplantation in order to detect acute change of kidney function [15]. However, estimation of the glomerular filtration rate from plasma creatinine using the Cockroft or MDRD formulas is not a reliable marker of renal function loss in kidney transplant patients nor a good predictive marker of graft loss [16–18]. More recently, it has been shown that serum cystatin C was a filtration marker in kidney transplantation and the newly developed CKD-EPI (Chronic Kidney Disease Epidemiology Collaboration) equations performed better than measured GFR using urinary clearance of inulin [19]. Nevertheless, there is a strong need for the biodetection of noninvasive biomarkers, more sensitive and specific of the clinical state of the recipient. It is obvious that a biomarker cannot be suitable to all the situations in transplantation and specific studies are needed to answer specific needs: prognostic versus diagnostic graft outcome, identification of kidney dysfunction versus specific graft rejection (acute or chronic rejection), and identification of state of operational tolerance. Consequently, in addition to classical allograft histology and creatinine/proteinuria follow-up, a lot of strategies have been developed as noninvasive ones for the diagnosis and monitoring of transplanted recipients. This would allow us to move into personalized medicine approach in order to adapt immunosuppressive therapy in function of the recipient profile and target patients with potential molecular phenotype of chronic rejection or tolerance.

65.2 BIOMARKERS OF CHRONIC REJECTION

65.2.1 Chronic Allograft Rejection: One State Encompassing Many Phenotypes

According to the Banff classification, diagnosis of antibody-mediated chronic allograft rejection relied on renal dysfunction, histological features, and serum donor-specific antibodies (DSAs). The biological mechanisms leading to CAMR are poorly defined despite the identification of anti-donor antibodies as a negative prognostic parameter for graft survival. Despite the large spectrum of factors contributing to allograft fibrosis (hypertension,

drug toxicity, viral infections, and allo- and autoimmune inflammatory injuries) [20], many efforts have been made to standardize renal allograft biopsy interpretation in order to guide therapy. Consequently, all these factors are associated with the same state of "chronic allograft rejection." This term is not a true diagnosis but a descriptive histologic term not associated with a specific cause. Microcirculation inflammation was the first histological feature used to diagnose ABMR followed by the discovery of C4d staining [21,22]. With more than 10 years of studies, the usefulness of C4d is now in question as C4d has a low sensitivity for ABMR, missing at least 50% of chronic ABMR cases [23–25]. Recent studies have tried to identify specific factors of kidney allograft loss in order to understand the cause of injury to reverse and ameliorate graft outcome. A recent study has followed the future of 1317 kidney allograft transplanted over 10 years. In this large prospective study, 330 grafts were lost during this period, including 153 (11.6%) due to graft failure [26]. More than 90% of recipients with rejected graft had an identifiable diagnosis, including 37% due to glomerular diseases, 31% for fibrosis/atrophy, and 12% of acute rejection. Most cases of IF/TA could be attributed to a specific cause encompassing BK polyomavirus nephropathy, recurrent pyelonephritis, and immune-mediated rejection [26]. Indeed, most recipients exhibit features of humoral and/or cellular-mediated rejection strongly suggesting that inflammation is a major feature of chronic rejection. Cluster analysis of 240 kidney allograft biopsies demonstrated a strong link between inflammation and chronic rejection [27]. Indeed, recipients exhibiting fibrosis and atrophy had more inflammation and more graft failure when compared with patients with less IF/TA. Analysis of protocol biopsies has shown that recipients with fibrosis but no inflammation have better long-term survival than patients with both fibrosis and inflammation [28,29]. This inflammation is associated with cellular infiltration in 80% of renal allograft and correlates with reduced long-term graft survival [28,29]. Molecular study by DNA chips highlights the activation of endothelial cells as a mechanism of rejection [30]. Finally, microcirculation inflammation score predicts graft failure independently of time, C4d, and transplant glomerulopathy [31].

65.2.2 Monitoring Graft Survival

As described above, late graft failure is characterized by an ongoing alloimmune response. This inflammation can be cellular or antibody mediated with expression of proinflammatory cytokines and chemokines. Consequently, many targets could be used to monitor the graft survival. Various biological compartments are available, such as plasma, serum, peripheral blood mononuclear cells, urine, and biopsy. One key point is the use of a noninvasive method for the safety of the patients. The recent development of high-throughput technologies, including genomics, proteomics, and metabolomics, have permit to identify mechanisms of late graft injuries at a large scale.

The detection and monitoring of alloantibodies is an important clinical parameter to follow the anti-donor response. Systematic monitoring of anti-HLA antibodies has shown to be useful in the prediction of late graft loss [32,33]. The detection of anti-HLA antibodies is associated with an increase rate of graft dysfunction but the time of the occurrence of late graft loss varies from month to years. The detection of anti-HLA antibodies is now routinely performed in transplant centers using HLA-coated beads [34]. The use of Luminex technology enables assessment of a broad range of HLA class I and class II molecules (91 and 97, respectively). Pre-existing and *de novo* anti-HLA antibodies have both been shown to be associated with worse kidney survival. For instance, 8-year graft survival was significantly worse among patients with preformed HLA-DSA compared with sensitized patients without HLA-DSA and nonsensitized patients [35,36]. Similar observation has been reported for *de novo* HLA-DSA, indeed a large cohort of kidney transplant recipients included 5-year posttransplantation and monitored for 5.5-year postinclusion, patients with HLA-DSA exhibit a significantly lower graft survival as compared to patients without anti-HLA antibodies (49% versus 83% graft survival, respectively) [37]. Contradictory studies that do not report difference in graft survival between patients with and without anti-HLA antibodies have also been published [38]. To address this discrepancy, standardization of the definition of anti-HLA antibodies (especially the threshold of detection) is needed.

In addition to anti-HLA antibodies, immunoglobulin against non-HLA antigens can prime allogeneic injuries leading to chronic humoral rejection [39]. The specificities against non-HLA antigens are very diverse, including endothelial cells and MICA molecules, and can lead to endothelial cell injuries and complement activation [40,41]. These alloantibodies must be taken in consideration to monitor graft survival.

The combination of microarray and histological data from biopsies of transplanted recipients permits us to classify and identify different subgroups of acute graft rejection [42]. This study identified some molecular patterns correlated to distinct phenotypes of acute rejection, which could not previously be clearly defined on the basis of clinical or pathological criteria [42]. Transcriptomic pattern of acute rejection was associated with strong infiltration of CD20[+] B cells. No inflammatory infiltration and no inflammatory transcriptomic pattern were reported on patients with chronic allograft nephropathy [42]. The team of Halloran had extensively analyzed kidney biopsies using microarray and proposed pathogenesis-based transcript sets (PBTs) to

assign modification in the expression of genes to biological processes: IFNG effects for rejection; T cell and macrophage transcripts for T-cell-mediated rejection; endothelial and NK transcripts for antibody-mediated rejection [23,43]. Additional studies put on light specific patterns in both peripheral blood and biopsies of acute rejection and various biological functions were assigned to these gene signatures (immune signal transduction, cytoskeletal reorganization, and apoptosis). Immune and inflammatory mechanisms were identified in several independent studies, such as apoptosis and immune infiltration, using either renal biopsies or PBL [44,45]. Granzyme B and FOXP3 transcripts quantification in urine of patients under acute rejection have been shown to have diagnostic and prognostic values [46,47]. In the case of chronic graft injury, attention has been paid to genome-wide profiling in the search for biomarkers. Indeed, the development of IT/TA was associated with significant changes in the graft transcriptome. Among differentially expressed genes, higher levels of molecules involved in complement activation, leukocyte homing, T- and B-cell infiltration and activation, and epithelio-mesenchymal transition (EMT) were associated with chronic changes in graft [48]. Although such studies have identified gene pathways involved in chronic allograft nephropathy, the use of biopsies precludes the generalization of their use as biomarkers due to its invasiveness [49].

Transcriptomic profiling in peripheral blood has shown that immunoproteasome beta subunit 10 [50] and TRIB1 [51] mRNA were specifically increased in the graft and blood samples during chronic active antibody-mediated rejection as compared to patients with stable graft function. In addition, MyD88 and TLR4 mRNA are overexpressed in the blood of patients with chronic antibody-mediated rejection when compared to patients with stable graft function [52]. Interestingly overexpression of TLR4 was further confirmed in kidney graft biopsies from patients with chronic antibody-mediated rejection [52].

65.3 BIOMARKERS OF TRANSPLANTATION TOLERANCE

65.3.1 Tolerance in Kidney Transplantation

In immunology, tolerance is currently defined as an absence of deleterious response against self-antigens. Different mechanisms of central and peripheral tolerance are involved in this process. Alteration in self-tolerance can lead to autoimmune diseases with destruction of self-tissues. The problem is different in transplantation, as the immune system must tolerate foreign antigen to ensure graft survival. In experimental transplantation, the first report on neonatal tolerance proved the possibility to induce long-term graft survival without the use of immunosuppressors [53]. This first study has laid the ground of the definition of "true tolerance" in transplantation. True tolerance is thus defined by a well-functioning graft without histological lesions of rejection and the acceptance of a second graft from the same donor but rejecting a third-party graft. Obviously, this definition cannot be applied to clinic. The term of operational tolerance is thus frequently used to describe patients who, after stopping their immunosuppressive therapy, keep a good and stable function of their graft [54–60]. Clinical "operational tolerance" was defined as stable kidney transplant function with <150 mol/L creatinemia and <1 g/day proteinuria in the absence of immunosuppressive drugs for at least 1 year [54]. The withdrawal of immunosuppressive drugs was not intended on purpose but was undertaken for either personal noncompliance or adverse events [60]. Thus, the observation of several operational tolerant recipients (more than 100 described in the literature) is a living proof that such a state could exist but the achievement of such states of tolerance remains an unsolved question. Importantly, kidney operational tolerant are immune-competent hosts able to respond to other immune challenges, including viral infections [54] and appeared to be less exposed to viral or bacterial infections than immunosuppressed patients [60]. Up to date, three cohorts of operational tolerant patients had been extensively studied [58–60]. The American cohort has a majority of HLA-matched graft (20/25) with only one patient developing a specific anti-donor HLA response. The French cohort of 27 tolerant recipients reported distinct characteristics. Only five patients received an HLA-matched graft. Tolerant patients are older and stopped progressively their treatment mainly by noncompliance. Operational tolerance state is not a stable state as eight of the 27 patients have rejected their graft [60]. Surprisingly the deterioration of graft function was not due to the development of a *de novo* humoral response as only one of them presented specific antibodies against the donor in the serum and the majority of the patients who lost their state of operational tolerance displayed pretransplant immunization [60]. This study emphasizes the complexity and the high heterogeneity of the tolerance phenotype, but despite this, tolerant recipients offer the opportunity (i) to understand mechanisms responsible for tolerance in human kidney transplantation and (ii) to identify molecules that can induce, predict, or diagnose tolerance.

65.3.2 Looking for the Hidden Phenotype of Tolerance: B Cells are the Answer?

First attempts to characterize tolerance were based on immunophenotyping studies of immune cells, starting with studies on effectors and regulatory T cells. Although experimental models have demonstrated the potency of regulatory T cells to prevent graft rejection and inducing transplantation tolerance [61–63], their role in clinical settings remains to be defined. Tolerant recipients exhibit a similar number of Tregs as compared to healthy volunteers but a higher number as compared to patients under chronic rejection suggesting that clinically operational tolerance may be due to a maintained phenomenon of natural or acquired tolerance that is lacking in patients with chronic rejection [64]. A more important infiltrate of Foxp3$^+$ cells was noticed in biopsy of kidney graft from tolerant patients when compared to chronic rejection and stable patients [65]. Maintenance of regulatory T cells is associated with a decreased frequency of cytotoxic CD8$^+$ T cells in the blood of tolerant patients suggesting a global quiescent immune system in these recipients [66]. In addition, high-throughput study by DNA chips confirms a downregulation of inflammatory genes in the blood of tolerant patients ($n = 17$) compared to chronic rejection ($n = 11$) and healthy volunteers ($n = 8$) [55]. Indeed, classical markers for early and late T-cell activation (CD69, TACTILE, LAG3, or SLAM), expression of cytotoxicity-associated genes (granzyme, perforin, FAS, and granulysin) and prototypic T_H1/T_H2 genes (TNF-alpha, IL-4, and IL-10) were consistently underexpressed in tolerant patients. Approximately 90% of known pro-inflammatory cytokines were reduced in tolerant recipients, supporting evidence for immune quiescence and ignorance of donor antigen [55]. Although TGF-beta expression was not increased in tolerant recipients, 27% of the peripheral blood genes that differentiate tolerance from chronic rejection was regulated by this protein [55]. In accordance with this, a recent study highlights a dominant regulatory profile in tolerant recipients with higher expression of FOXP3, TGF-beta, and TGF-beta receptor [67]. This regulatory profile was stable over time and specific of the tolerance phenotype. Nevertheless, only five tolerant patients were included in this study and validation in a larger cohort must be performed to confirm these results [67]. Importantly, the study of Brouard et al. identified for the first time a tolerance footprint in clinical kidney transplantation [55]. This transcriptomic signature composed of 49 genes was specific of the tolerance state and could predict and detect potential tolerant recipients in stable patients under immunosuppression [55]. Most interestingly, most of the genes differentially expressed in the blood of tolerant recipients were related to B-cell functions suggesting a B-cell signature of kidney transplantation tolerance [55,57]. This was in accordance with previous results showing that tolerant recipients exhibit a higher number of B cells in their blood [64]. This B-cell phenotypic and transcriptomic signature was confirmed by two international and independent studies [58,59]. Thus, a naive/transitional B-cell phenotype was reported as a specific feature of operational tolerance associated with a specific

B-cell transcriptomic pattern. In the top 30 differentially expressed genes, enrichment in B-cell-related gene involved in B-cell signaling and Ig rearrangement were reported in these studies. Moreover, extensive statistical analysis highlighted three genes that predictably and specifically distinguish tolerant patients from stable patients and healthy volunteers [58]. New analysis demonstrated that overexpression of IGKV1D-13 was stable over time and accurately discriminated operational tolerant patients (TOL) from stable patients under immunosuppressive therapy (STA) (Newell, personal communication, ATC 2012). By adding eight B-cell-related genes, authors were able to distinguish tolerant from stable recipients with high positive and negative predictive values confirming the potency of B-cell genes as biomarkers for operational tolerance (Newell, personal communication, ATC 2012). An upregulation of CD20 mRNA, a specific marker for B cells, was noticed in the urine sediment cells of tolerant patients when compared to stable patients and healthy volunteers reflecting a B-cell accumulation in the tolerated graft [58]. Interestingly, comparison between liver and kidney tolerant recipients showed an absence of enriched B-cell markers in liver transplantation tolerance arguing for an organ-specific signature of tolerance [68].This B-cell signature raises important questions about the mechanisms responsible for allograft tolerance. Indeed recent evidence in rodents indicates that regulatory B cells can modulate and control inflammation in autoimmune disease [69], asthma [70], and transplantation [71]. In humans, many reports focused their study on the characterization of IL-10 secreting regulatory B cells expressing some transitional and memory B-cell markers [72−74]. Interestingly an enrichment of IL-10 secreting transitional B cells [58] and an accumulation of B cells expressing high level of FcgammaRIIb and BANK-1 [57], two inhibitory molecules, were found in blood from operationally tolerant patients suggesting a particular inhibitory profile of B cells. But for now, no report has yet described a functional role of B cells in the regulation of allogeneic response in human kidney transplantation. Very recently, Haynes et al. propose an interesting hypothesis in the establishment of kidney transplant tolerance [75]. This work suggests that immunosuppressive therapy would favor the emergence of transitional and naive B cells with regulatory capacities leading to transplant tolerance. This confirms recent observations in which patients undergoing progressive diminution of immunosuppressors are characterized by a "colonization" of transitional and naive B cells in peripheral blood [76].

However, these current observations hilight one fundamental question: Is the B-cell signature indicative of tolerance or simply a manifestation of an absence of immunosuppression? To answer this question, results must be replicated in a larger cohort of TOL patients and tested in weaning protocol. Thus the validity of these biomarkers and their predictive capacity must be tested in drug minimization trials conducted in subgroups of kidney recipients with low "immunological risk." One such study is currently in progress in which calcium inhibitor (CNI) weaning is being performed (http://clinicaltrials.gov/ct2/show/NCT01292525). This study is being conducted in the context of a European consortium (FP7 BioDrim) and with centers from the CENTAURE network (http://www.fondation-centaure.org/) [77].

65.3.3 Biomarkers in Renal Transplantation...What's Next?

Many hopes have been placed in the evolution and interpretation of high-throughput technology to identify specific biomarkers of chronic rejection and tolerance in kidney transplantation. The study of proteome, transcriptome, and immunome has allowed scientists and physicians to decipher mechanisms and physiological variations related to graft survival or rejection. However, these studies notably in kidney transplantation tolerance have been performed on small cohorts with highly selected patients. To be confirmed, the performance and clinical utility of biological signatures need to be validated in large patient cohorts that are representative of the population in which they will be used thereafter. In cardiac transplantation, one study described and developed the Allomap test based on the expression of 11 genes in the peripheral blood to identify heart transplant patients with a low probability of moderate-to-severe rejection [78]. The test was discovered and validated on independent sets of patients amounting to hundreds of samples. A clinical trial demonstrated the potency to use this signature as an alternative to routine biopsies for patients at low risk [79]. Consequently, many efforts have to be made in order to move toward large-scale clinical validation in the field of kidney transplantation. Such approaches are costly but cannot be evaded. To reach this goal, there is a strong necessity to properly design and standardize experimental procedures from cell isolation and RNA extraction to data analysis in order to validate these biomarkers in larger cohorts of transplanted patients [77,80]. Thankfully, the use of bioinformatics tools permits us today to re-analyze complete sets of data from different experiments and different platforms and confirm biological signatures notably in transplantation tolerance [68]. Powerful mathematical and bioinformatic tools permit us to establish and study biological functions, links between genes and proteins and characterize biomarkers and human disease [80−84]. This gives us the possibility to switch from a reductionist to a more global integrated approach consisting of the incorporation of many data from transcriptome, proteome, immunome, and clinicome, providing signatures of pathology and links to clinical research [85]. This field of "system medicine" is emerging and could give new insights in the quest for biomarkers in transplantation [80]. As a correlate of discovery studies, there is a reciprocal relationship between the criteria of inclusion of the patients and the output of the studies that leads to the modification of the clinical classification of the patients. The combination of biological markers and clinical features should improve the stratification of the patients and thus lead to true personalized medicine. It is now time to validate the biomarkers to switch from potential biomarkers into useful biomarkers leading to an improvement of follow-up of recipients.

REFERENCES

[1] Kawai T, Cosimi AB, Spitzer TR, Tolkoff-Rubin N, Suthanthiran M, Saidman SL, et al. HLA-mismatched renal transplantation without maintenance immunosuppression. N Engl J Med 2008; 358:353–61.

[2] Köhler G, Milstein C. Continuous cultures of fused cells secreting antibody of predefined specificity. Nature 1975;256:495–7.

[3] Soulillou JP, Cantarovich D, Le Mauff B, Giral M, Robillard N, Hourmant M, et al. Randomized controlled trial of a monoclonal antibody against the interleukin-2 receptor (33B3.1) as compared with rabbit antithymocyte globulin for prophylaxis against rejection of renal allografts. N Engl J Med 1990;322:1175–82.

[4] Vincenti F, Larsen C, Durrbach A, Wekerle T, Nashan B, Blancho G, et al. Costimulation blockade with belatacept in renal transplantation. N Engl J Med 2005;353:770–81.

[5] Poirier N, Azimzadeh AM, Zhang T, Dilek N, Mary C, Nguyen B, et al. Inducing CTLA-4-dependent immune regulation by selective CD28 blockade promotes regulatory T cells in organ transplantation. Sci Transl Med 2010;2(17):17ra10.

[6] Suntharalingam G, Perry MR, Ward S, Brett SJ, Castello-Cortes A, Brunner MD, et al. Cytokine storm in a phase 1 trial of the anti-CD28 monoclonal antibody TGN1412. N Engl J Med 2006;355: 1018–28.

[7] Wood KJ, Bushell A, Jones ND. Immunologic unresponsiveness to alloantigen *in vivo*: a role for regulatory T cells. Immunol Rev 2011;241:119–32.

[8] Wood KJ, Bushell A, Hester J. Regulatory immune cells in transplantation. Nat Publ Group 2012;12:417–30.

[9] Long E, Wood KJ. Regulatory T cells in transplantation: transferring mouse studies to the clinic. Transplantation 2009;88:1050–6.

[10] Biomarkers Definitions Working Group. Biomarkers and surrogate endpoints: preferred definitions and conceptual framework. Clin Pharmacol Ther 2001;69:89–95.

[11] Morath C, Zeier M. When should post-transplantation proteinuria be attributed to the renal allograft rather than to the native kidney?. Nat Clin Pract Nephrol 2007;3:18–9.

[12] Nankivell BJ, Borrows RJ, Fung CLS, O'Connell PJ, Allen RDM, Chapman JR. The natural history of chronic allograft nephropathy. N Engl J Med 2003;349:2326–33.

[13] Roodnat JI, Mulder PG, Rischen-Vos J, van Riemsdijk IC, van Gelder T, Zietse R, et al. Proteinuria after renal transplantation affects not only graft survival but also patient survival. Transplantation 2001;72:438–44.

[14] Roodnat JI, Mulder PG, Rischen-Vos J, van Riemsdijk IC, van Gelder T, Zietse R, et al. Proteinuria and death risk in the renal transplant population. Transplant Proc 2001;33:1170–1.

[15] Perrone RD, Madias NE, Levey AS. Serum creatinine as an index of renal function: new insights into old concepts. Clin Chem 1992;38:1933–53.

[16] Mariat C, Maillard N, Phayphet M, Thibaudin L, Laporte S, Alamartine E, et al. Estimated glomerular filtration rate as an end point in kidney transplant trial: where do we stand? Nephrol Dial Transplant 2008;23:33–8.

[17] Mariat C, Alamartine E, Afiani A, Thibaudin L, Laurent B, Berthoux P, et al. Predicting glomerular filtration rate in kidney transplantation: are the K/DOQI guidelines applicable? Am J Transplant 2005;5:2698–703.

[18] Gera M, Slezak JM, Rule AD, Larson TS, Stegall MD, Cosio FG. Assessment of changes in kidney allograft function using creatinine-based estimates of glomerular filtration rate. Am J Transplant 2007;7:880–7.

[19] Masson I, Maillard N, Tack I, Thibaudin L, Dubourg L, Delanaye P, et al. GFR estimation using standardized cystatin C in kidney transplant recipients. Am J Kidney Dis 2013;61(2):279–84.

[20] Racusen LC, Colvin RB, Solez K, Mihatsch MJ, Halloran PF, Campbell PM, et al. Antibody-mediated rejection criteria—an addition to the Banff 97 classification of renal allograft rejection. Am J Transplant 2003;3:708–14.

[21] Feucht HE, Felber E, Gokel MJ, Hillebrand G, Nattermann U, Brockmeyer C, et al. Vascular deposition of complement-split products in kidney allografts with cell-mediated rejection. Clin Exp Immunol 1991;86:464–70.

[22] Mauiyyedi S, Pelle PD, Saidman S, Collins AB, Pascual M, Tolkoff-Rubin NE, et al. Chronic humoral rejection: identification of antibody-mediated chronic renal allograft rejection by C4d deposits in peritubular capillaries. J Am Soc Nephrol 2001;12: 574–82.

[23] Halloran PF, de Freitas DG, Einecke G, Famulski KS, Hidalgo LG, MengeL M, et al. An integrated view of molecular changes, histopathology and outcomes in kidney transplants. Am J Transplant 2010;10:2223–30.

[24] Einecke G, Sis B, Reeve J, MengeL M, Campbell PM, Hidalgo LG, et al. Antibody-mediated microcirculation injury is the major cause of late kidney transplant failure. Am J Transplant 2009;9:2520–31.

[25] Sis B, Jhangri GS, Bunnag S, Allanach K, Kaplan B, Halloran PF. Endothelial gene expression in kidney transplants with alloantibody indicates antibody-mediated damage despite lack of C4d staining. Am J Transplant 2009;9:2312–23.

[26] El-Zoghby ZM, Stegall MD, Lager DJ, Kremers WK, Amer H, Gloor JM, et al. Identifying specific causes of kidney allograft loss. Am J Transplant 2009;9:527–35.

[27] Matas AJ, Leduc R, Rush D, Cecka JM, Connett J, Fieberg A, et al. Histopathologic clusters differentiate subgroups within the nonspecific diagnoses of CAN or CR: preliminary data from the DeKAF study. Am J Transplant 2010;10:315–23.

[28] Cosio FG, Grande JP, Wadei H, Larson TS, Griffin MD, Stegall MD. Predicting subsequent decline in kidney allograft function

from early surveillance biopsies. Am J Transplant 2005;5: 2464—72.

[29] Moreso F, Ibernon M, Gomà M, Carrera M, Fulladosa X, Hueso M, et al. Subclinical rejection associated with chronic allograft nephropathy in protocol biopsies as a risk factor for late graft loss. Am J Transplant 2006;6:747—52.

[30] Akalin E, Dinavahi R, Dikman S, de Boccardo G, Friedlander R, Schroppel B, et al. Transplant glomerulopathy may occur in the absence of donor-specific antibody and C4d staining. Clin J Am Soc Nephrol 2007;2:1261—7.

[31] Sis B, Jhangri GS, Riopel J, Chang J, de Freitas DG, Hidalgo L, et al. A new diagnostic algorithm for antibody-mediated microcirculation inflammation in kidney transplants. Am J Transplant 2012;12:1168—79.

[32] Lee PC, Terasaki PI, Takemoto SK, Lee PH, Hung CJ, Chen YL, et al. All chronic rejection failures of kidney transplants were preceded by the development of HLA antibodies. Transplantation 2002;74:1192—4.

[33] Terasaki PI, Ozawa M. Predicting kidney graft failure by HLA antibodies: a prospective trial. Am J Transplant 2004;4:438—43.

[34] Pei R, Lee JH, Shih NJ, Chen M, Terasaki PI. Single human leukocyte antigen flow cytometry beads for accurate identification of human leukocyte antigen antibody specificities. Transplantation 2003;75:43—9.

[35] Lefaucheur C, Suberbielle-Boissel C, Hill GS, Nochy D, Andrade J, Antoine C, et al. Clinical relevance of preformed HLA donor-specific antibodies in kidney transplantation. Am J Transplant 2008;8:324—31.

[36] Lefaucheur C, Nochy D, Hill GS, Suberbielle-Boissel C, Antoine C, Charron D, et al. Determinants of poor graft outcome in patients with antibody-mediated acute rejection. Am J Transplant 2007;7:832—41.

[37] Lachmann N, Terasaki PI, Budde K, Liefeldt L, Kahl A, Reinke P, et al. Anti-human leukocyte antigen and donor-specific antibodies detected by luminex posttransplant serve as biomarkers for chronic rejection of renal allografts. Transplantation 2009;87: 1505—13.

[38] Aubert V, Venetz J-P, Pantaleo G, Pascual M. Low levels of human leukocyte antigen donor-specific antibodies detected by solid phase assay before transplantation are frequently clinically irrelevant. Hum Immunol 2009;70:580—3.

[39] Thaunat O. Humoral immunity in chronic allograft rejection: puzzle pieces come together. Transpl Immunol 2012;26:101—6.

[40] Dragun D, Philippe A, Catar R. Role of non-HLA antibodies in organ transplantation. Curr Opin Organ Transplant 2012;17: 440—5.

[41] Zhang Q, Reed EF. Non-MHC antigenic targets of the humoral immune response in transplantation. Curr Opin Immunol 2010;22:682—8.

[42] Sarwal M, Chua MS, Kambham N, Hsieh SC, Satterwhite T, Masek M, et al. Molecular heterogeneity in acute renal allograft rejection identified by DNA microarray profiling. N Engl J Med 2003;349:125—38.

[43] Hidalgo LG, Sis B, Sellares J, Campbell PM, Mengel M, Einecke G, et al. NK cell transcripts and NK cells in kidney biopsies from patients with donor-specific antibodies: evidence for NK cell involvement in antibody-mediated rejection. Am J Transplant 2010;10:1812—22.

[44] Mueller TF, Einecke G, Reeve J, Sis B, Mengel M, Jhangri GS, et al. Microarray analysis of rejection in human kidney transplants using pathogenesis-based transcript sets. Am J Transplant 2007; 7:2712—22.

[45] Reeve J, Einecke G, Mengel M, Sis B, Kayser N, Kaplan B, et al. Diagnosing rejection in renal transplants: a comparison of molecular- and histopathology-based approaches. Am J Transplant 2009;9:1802—10.

[46] Muthukumar T, Dadhania D, Ding R, Snopkowski C, Naqvi R, Lee JB, et al. Messenger RNA for FOXP3 in the urine of renal-allograft recipients. N Engl J Med 2005;353:2342—51.

[47] Veale JL, Liang LW, Zhang Q, Gjertson DW, Du Z, Bloomquist EW, et al. Noninvasive diagnosis of cellular and antibody-mediated rejection by perforin and granzyme B in renal allografts. Hum Immunol 2006;67:777—86.

[48] Scherer A, Gwinner W, Mengel M, Kirsch T, Raulf F, Szustakowski JD, et al. Transcriptome changes in renal allograft protocol biopsies at 3 months precede the onset of interstitial fibrosis/tubular atrophy (IF/TA) at 6 months. Nephrol Dial Transplant 2009;24:2567—75.

[49] Ashton-Chess J, Giral M, Soulillou JP, Brouard S. Can immune monitoring help to minimize immunosuppression in kidney transplantation? Transpl Int 2009;22:110—9.

[50] Ashton-Chess J, Mai HL, Jovanovic V, Renaudin K, Foucher Y, Giral M, et al. Immunoproteasome beta subunit 10 is increased in chronic antibody-mediated rejection. Kidney Int 2010;77:880—90.

[51] Ashton-Chess J, Giral M, Mengel M, Renaudin K, Foucher Y, Gwinner W, et al. Tribbles-1 as a novel biomarker of chronic antibody-mediated rejection. J Am Soc Nephrol 2008;19: 1116—27.

[52] Braudeau C, Ashton-Chess J, Giral M, Dugast E, Louis S, Pallier A, et al. Contrasted blood and intragraft toll-like receptor 4 mRNA profiles in operational tolerance versus chronic rejection in kidney transplant recipients. Transplantation 2008;86:130—6.

[53] Billingham RE, Brent L, Medawar PB. Actively acquired tolerance of foreign cells. Nature 1953;172:603—6.

[54] Roussey-Kesler G, Giral M, Moreau A, Subra JF, Legendre C, Noël C, et al. Clinical operational tolerance after kidney transplantation. Am J Transplant 2006;6:736—46.

[55] Brouard S, Mansfield E, Braud C, Li L, Giral M, Hsieh SC, et al. Identification of a peripheral blood transcriptional biomarker panel associated with operational renal allograft tolerance. Proc Natl Acad Sci USA 2007;104:15448—53.

[56] Orlando G, Hematti P, Stratta RJ, Burke GW, Di Cocco P, Cocco PD, et al. Clinical operational tolerance after renal transplantation: current status and future challenges. Ann Surg 2010;252:915—28.

[57] Pallier A, Hillion S, Danger R, Giral M, Racape M, Degauque N, et al. Patients with drug-free long-term graft function display increased numbers of peripheral B cells with a memory and inhibitory phenotype. Kidney Int Nat Publ Group 2010;78:503—13.

[58] Newell KA, Asare A, Kirk AD, Gisler TD, Bourcier K, Suthanthiran M, et al. Identification of a B cell signature associated with renal transplant tolerance in humans. J Clin Invest 2010;120:1836—47.

[59] Sagoo P, Perucha E, Sawitzki B, Tomiuk S, Stephens DA, Miqueu P, et al. Development of a cross-platform biomarker signature to detect renal transplant tolerance in humans. J Clin Invest 2010;120:1848—61.

[60] Brouard S, Pallier A, Renaudin K, Foucher Y, Danger R, Devys A, et al. The natural history of clinical operational tolerance after kidney transplantation through twenty-seven cases. Am J Transplant 2012.

[61] Lin CY, Graca L, Cobbold SP, Waldmann H. Dominant transplantation tolerance impairs CD8^{+} T cell function but not expansion. Nat Immunol 2002;3:1208–13.

[62] Waldmann H, Cobbold S. Exploiting tolerance processes in transplantation. Science 2004;305:209–12.

[63] Graca L, Cobbold SP, Waldmann H. Identification of regulatory T cells in tolerated allografts. J Exp Med 2002;195:1641–6.

[64] Louis S, Braudeau C, Giral M, Dupont A, Moizant F, Robillard N, et al. Contrasting CD25hiCD4^{+}T cells/FOXP3 patterns in chronic rejection and operational drug-free tolerance. Transplantation 2006;81:398–407.

[65] Becker LE, de Oliveira Biazotto F, Conrad H, Schaier M, Kihm LP, Gross-Weissmann ML, et al. Cellular infiltrates and NFκB subunit c-Rel signaling in kidney allografts of patients with clinical operational tolerance. Transplantation 2012;94:729–37.

[66] Baeten D, Louis S, Braud C, Braudeau C, Ballet C, Moizant F, et al. Phenotypically and functionally distinct CD8^{+} lymphocyte populations in long-term drug-free tolerance and chronic rejection in human kidney graft recipients. J Am Soc Nephrol 2006; 17:294–304.

[67] Moraes-Vieira PMM, Takenaka MCS, Silva HM, Monteiro SM, Agena F, Lemos F, et al. GATA3 and a dominant regulatory gene expression profile discriminate operational tolerance in human transplantation. Clin Immunol 2012;142:117–26.

[68] Lozano JJ, Pallier A, Martinez-Llordella M, Danger R, López M, Giral M, et al. Comparison of transcriptional and blood cell-phenotypic markers between operationally tolerant liver and kidney recipients. Am J Transplant 2011;11:1916–26.

[69] Mauri C. Regulation of immunity and autoimmunity by B cells. Curr Opin Immunol 2010;22:761–7.

[70] Natarajan P, Singh A, McNamara JT, Secor ER, Guernsey LA, Thrall RS, et al. Regulatory B cells from hilar lymph nodes of tolerant mice in a murine model of allergic airway disease are CD5^{+} express TGF-beta, and co-localize with CD4^{+}Foxp3^{+} T cells. Nat Publ Group 2012;1–11.

[71] Ding Q, Yeung M, Camirand G, Zeng Q, Akiba H, Yagita H, et al. Regulatory B cells are identified by expression of TIM-1 and can be induced through TIM-1 ligation to promote tolerance in mice. J Clin Invest 2011;121:3645–56.

[72] Blair PA, NoreNa LY, Flores-Borja F, Rawlings DJ, Isenberg DA, Ehrenstein MR, et al. CD19^{+}CD24hiCD38hi B cells exhibit regulatory capacity in healthy individuals but are functionally impaired in systemic lupus erythematosus patients. Immunity 2010;32:129–40.

[73] Lemoine S, Morva A, Youinou P, Jamin C. Human T cells induce their own regulation through activation of B cells. J Autoimmun 2011;36:228–38.

[74] Iwata Y, Matsushita T, Horikawa M, DiLillo DJ, Yanaba K, Venturi GM, et al. Characterization of a rare IL-10-competent B-cell subset in humans that parallels mouse regulatory B10 cells. Blood 2011;117:530–41.

[75] Haynes LD, Jankowska-Gan E, Sheka A, Keller MR, Hernandez-Fuentes MP, Lechler RI, et al. Donor-specific indirect pathway analysis reveals a B-cell-independent signature which reflects outcomes in kidney transplant recipients. Am J Transplant 2012; 12:640–8.

[76] Porcheray F, Wong W, Saidman SL, De Vito J, Girouard TC, Chittenden M, et al. B-cell immunity in the context of T-cell tolerance after combined kidney and bone marrow transplantation in humans. Am J Transplant 2009;9:2126–35.

[77] Londoño MC, Danger R, Giral M, Soulillou JP, Sánchez-Fueyo A, Brouard S. A need for biomarkers of operational tolerance in liver and kidney transplantation. Am J Transplant 2012;12:1370–7.

[78] Deng MC, Eisen HJ, Mehra MR, Billingham M, Marboe CC, Berry G, et al. Noninvasive discrimination of rejection in cardiac allograft recipients using gene expression profiling. Am J Transplant 2006;6:150–60.

[79] Pham MX, Teuteberg JJ, Kfoury AG, Starling RC, Deng MC, Cappola TP, et al. Gene expression profiling for rejection surveillance after cardiac transplantation. N Engl J Med 2010;362: 1890–900.

[80] Auffray C, Chen Z, Hood L. Systems medicine: the future of medical genomics and healthcare. Genome Med 2009;1:2.

[81] Giallourakis C, Henson C, Reich M, Xie X, Mootha VK. Disease gene discovery through integrative genomics. Annu Rev Genomics Hum Genet 2005;6:381–406.

[82] Goh KI, Cusick ME, Valle D, Childs B, Vidal M, Barabási AL. The human disease network. Proc Natl Acad Sci USA 2007;104: 8685–90.

[83] Lamb J, Crawford ED, Peck D, Modell JW, Blat IC, Wrobel MJ, et al. The connectivity map: using gene-expression signatures to connect small molecules, genes, and disease. Science 2006;313: 1929–35.

[84] Hori SS, Gambhir SS. Mathematical model identifies blood biomarker-based early cancer detection strategies and limitations. Sci Transl Med 2011;3:109–16.

[85] Pepperkok R, Wiemann S. Integrating systems biology with clinical research. 2008;9(7):314.

Chapter 66

Immunocloaking

Lauren Brasile

BREONICS, Inc., Albany, NY

Chapter Outline

Since the time of Medawar's ground breaking work demonstrating the genetic control of the immune response in 1949, clinical transplantation has focused on immunosuppressive regimens that impact the effector arm of the recipient's immune system. If an organ-specific immunosuppressive treatment can be developed that prevents the initial allorecognition that occurs upon reperfusion of a transplanted organ, it would revolutionize transplantation. More importantly, it would help to minimize, or ideally eliminate, the toxic side effects of today's immunosuppressive regimens. A logical target for an organ-specific treatment would be the monolayer of endothelial cells lining the blood vessels within a vascularized allograft since the allograft vasculature is the primary target of immune-mediated rejection. Protecting the endothelium from immune effector mechanisms could theoretically ameliorate allograft rejection. Such an approach in essence constitutes the tissue engineering of an allograft's vasculature and can be viewed as "immunocloaking."

66.1 THE IMMUNE RESPONSE TO A VASCULARIZED ALLOGRAFT

Upon reperfusion when recipient immune cells encounter the foreign antigens of an allograft, allorecognition occurs and from that point onward there develops the uphill battle to prevent immunologic rejection of the tissue. The initial point of contact between the recipient immune cells in circulation and the donor tissue is the vascular endothelium. During the approximately 50 years of clinical transplantation, the corresponding intricate mechanisms involved in both cellular and humoral immune responses have been largely elucidated. Yet, despite the essentially eliminated occurrence of hyperacute rejection and the development of an impressive arsenal of immunosuppressive drugs, acute rejection episodes still occur and chronic rejection remains a major obstacle to transplantation. Chronic rejection is thought to be the consequence of continual damage mediated by cellular and particularly humoral immune responses initiated by indirect recognition of alloantigens [1].

Our understanding of the nature of alloantigens has led to discerning the underlying mechanisms involved in T-cell recognition, activation, and differentiation. This knowledge has led to our persistent focus on drug regimens to treat the effector arm of the immune response. From the earliest days of transplantation using glucocorticoids and azathioprine to the now used arsenal of calcineurin inhibitors, monoclonal antibodies, rapamycin, mycophenolate mofetil, and new therapeutics in clinical trials, toxicity has represented a continued problem. The continuous multidrug regimens routinely used to prevent rejection results in significant side effects, predominantly infections and malignancies. The exclusive focus on drugs that directly inhibit effector cell functions has not provided the requisite specificity to prevent these devastating toxicities. The major limitation of today's immunosuppressive drug regimens is the direct result of our inability to limit suppression to the immune cells involved in the alloresponses without adversely affecting other immune functions as well. Therefore, the development of an organ-specific therapy that protects the allograft vasculature from immune-mediated rejection by providing a physical barrier to inhibit the initial allorecognition could represent a major advancement.

66.1.1 Direct and Indirect Allorecognition

Clinical transplantation is an unique immunologic situation because both donor and recipient antigen presenting cells are present at reperfusion. Direct allorecognition is our inherent capacity to recognize foreign tissues. Much focus has been placed on the passenger leukocytes within a vascularized allograft with antigen presentation occurring in the secondary lymphoid tissues. However, the passenger leukocytes that are trapped in the allograft due to the normal transmigration that is requisite to normal immune surveillance at the time of organ recovery are known to leave the allograft soon after reperfusion and survive for only a short period [4,5]. In experiments where the direct pathway of allorecognition is absent the grafts are still rejected. The rejection is mediated by CD4$^+$ T cells that are primed by an indirect pathway [3,4]. Supporting this experimental observation is the case report where a human renal allograft underwent immunologic rejection despite the absence of allogeneic passenger leukocytes [6]. The report involved a male recipient who received a kidney from his HLA identical sister. The recipient had previously donated his bone marrow to his renal donor whose bone marrow 11 years later was identified as being 100% of the XY karyotype. The absence of allogeneic passenger leukocytes did not prevent the occurrence of acute cellular rejection. Although the living-related kidney donor was tolerant of the transplanted bone marrow, the recipient was not tolerant to his donor's kidney after weaning from the monotherapy consisting of Sirolimus. While non-HLA antigens were in all probability the target of the alloimmunity, this case supports the concept that the elimination of donor passenger leukocytes does not prevent immunologically mediated rejection.

It has been postulated that the indirect pathway of allorecognition represents a more relevant role clinically in the initiation of late acute rejection and chronic rejection [7–9]. This perspective is supported by the turnover

of graft tissues over time with the corresponding continual uptake of donor alloantigen by the recipient antigen presenting cells. The result of such a scenario would be recurrent activation of effector T cells both Th1 and Th2 [10–12]. The inevitable development of donor-specific antibody, particularly with isotope switching to Fc binding IgG, would be dependent upon the indirect pathway of alloreactivity. T cells activated by indirect allorecognition are predominantly stimulated following engraftment and then become the major mediator initiating alloresponses [13]. This may be due to the observation that allogeneic passenger leukocytes are susceptible to recipient NK cell attack within hours of trafficking to the secondary lymphoid tissues, thereby limiting their ability to serve an antigen presenting cells via the direct pathway of allorecognition [14].

An important consideration is that direct and indirect pathways of allorecognition are not necessarily mutually exclusive pathways of the rejection process since both pathways can be simultaneously involved in rejection. In fact, T cells can respond to alloantigen either directly or indirectly by both donor and recipient antigen presenting cells. What is most important is the long known observation that vascularized allografts are rejected by cellular-mediated and humoral mechanisms that are directed against the alloantigens expressed on the vascular endothelial cells within the allograft [15].

66.1.2 Vascular Endothelium as Antigen Presenting Cells

The luminal surfaces within the vasculature consist of a single monolayer of vascular endothelial cells (VECs). *In vivo*, VECs express both MHC class I and II antigens. Inasmuch, VECs are capable of serving as antigen presenting cells. Using a mixed lymphocyte–VEC proliferative assay, VECs were first identified to be able to stimulate T-cell proliferation 30 years ago [16]. VECs derived from human blood vessels of different anatomic sites, including arteries, veins, capillaries, and venules, express a variety of co-stimulating molecules resulting in increased receptor expression on activated and memory T cells. These inducibly upregulated receptors include LFA-3, CD58, elucidation of the OX40L, CD137, ICOS-L, and GITR-L [17–21]. Overall, arterial blood vessels tend to be less reactive to the inflammatory cytokines than are VECs located in the venous system [22]. Some of these variations in response can be attributed to differences in shear forces, PaO$_2$, vascular flow rates as well as local tissue signaling. Most importantly, VECs have been shown to initiate activation of CD4$^+$ and CD8$^+$ T cells leading to secretion of IL-2 and effector cytokines resulting in T-cell proliferation. While it has been reported that interactions between naïve T cells and VECs can lead to tolerance, VECs can also prime naïve and antigen-specific alloreactive T cells [22]. This continued VEC stimulation of effector pathways throughout the period of engraftment likely contributes to the development of chronic graft rejection [23]. The persistent secretion of chemokines by the graft VECs, such as Rantes and fractalkine, impacts leukocyte expression of integrins [24,25]. The upregulation of integrins, such as LFA-1, MAC-1, and VLA-4, facilitates the firm adherence of the activated T cells to the surface of the graft VECs [26,27].

Consequently, the VEC is a major site of inflammation, allorecognition, T-cell alloreactivity, leukocyte recruitment, diapedesis, and the target of both cellular- and humoral-mediated immune responses.

66.1.3 The Primary Target of Allograft Rejection

In the case of vascularized allografts, the heterogeneous cell types that constitute organ-specific functions make the assessment of the primary targets of immunologically mediated rejection more complex. Since the damage may be mediated by cellular-mediated mechanisms and/or humoral damaged with Cd4 deposition or reoccurrence of the original disease causal to the native organ failure, a clear diagnosis can be obscured. This is particularly the case if the original disease in the native kidney was mediated by autoantibodies. Therefore, the observed patterns of renal injury can cover a wide spectrum. The targets of rejecting allografts were described 32 years ago by focusing on whether there was principal damage to parenchymal cells or the vascular cells [28]. The morphological evidence supported the finding that whether the nature of the rejection process was early, intermediate, or late, the vasculature was the primary target [29,30]. These observed morphological patterns of injury included widespread VEC destruction. Interestingly, parenchymal damage that was the result of cytotoxic injury was observed in cases where there was an absence of immunosuppression.

Overall the blood vessels are the principle target of the immune response [31]. This observed vascular injury can be mediated by donor-specific antibody to the MHC antigens as well as non-MHC antigens with arteritis characterized by mononuclear cell infiltration [30]. More recent work has demonstrated that complement-dependent cytotoxicity, antibody-dependent cell-mediated cytotoxicity, and inflammation are causal processes in the resulting vasculitis. The vasculitis is primarily associated with humoral rejection with immune cell infiltration. The donor-specific antibodies that directly damage the VECs lining the vasculature results in necrosis due to the antibody deposition along with C4d complement breakdown product and fibrin,

neutrophil infiltration and fibrosis. Similarly the endothelitis is caused by the infiltration of $CD8^+$ T cells. This interpretation of the immunogenicity of the vascular endothelium as the primary target of rejection of vascularized allografts is supported by the results of engrafting small diameter vascular grafts where the VECs serve as antigen presenting cells leading to acute rejection. This result has led to the current popular approach of decellularizing vascular allografts to reduce the inherent antigenicity. The removal of the VECs with detergents resulted in reduced rejection of the vascular allograft [32].

The VECs are not only the primary target of the rejection process but also active participants. In addition to initiating rejection by antigen presentation to circulating T cells, VECs play an active role in the development of inflammation and thrombosis. The microvasculature is the major site of the rejection process where the damage can be mediated by both acute and chronic mechanisms and is targeted to the capillary network in renal allografts [33–35]. It has been postulated that microvasculature VECs are distinct from the macrovasculature in several important areas. In addition to antigen presentation, microvessel VECs demonstrate distinct co-stimulatory and adhesion molecules [36]. These can become upregulated following VEC activation mediated by IL-1 and TNF-α following allorecognition. The upregulation of adhesion molecules on the surface of VECs leads to increased leukocyte adhesion resulting in enhanced diapedesis into the allograft [37,38]. This extravasation of the activated leukocytes occurs through the VEC monolayer both within individual cells and between the VEC junctions [39]. Engagement of the T-cell receptor by microvascular VECs results in the formation of transendothelial protrusions on the lymphocytes that facilitates the transendothelial migration [40]. In another study using high-resolution confocal imaging of the distribution of adhesion receptors known to play a role in diapedesis, a transcellular route was identified as well as the established paracellular route of transmigration. Both routes of extravasation involved endothelial "cuplike" transmigratory structures that were enriched with ICAM-1 and VCAM-1 projections following stimulation by cytokines [41].

In response to injury, the VECs produce cytokines that in turn further increase MNC infiltration. The further upregulation of IFN-γ and TNF-α is associated with increased MHC class II antigen expression and results in further allograft injury mediated by both $CD4^+$ and $CD8^+$ effector cells [41,42]. The kinetics involved with immune cell rolling and firm adhesion to the VECs along with the diapedesis can be measured in seconds and minutes [43,44]. Therefore, the development of an organ-specific therapy to prevent allograft rejection must not only protect the VEC but also prevent direct T-cell interactions during the immediate posttransplant period.

66.1.4 Immunogenicity and Function of the Vascular Endothelium

66.1.4.1 VEC Immunogenicity

Normal microvascular VECs express MHC class I and II antigens on their cell surfaces at concentrations significantly higher than most other cells with the exclusion of the dendritic cell [45]. As long known, MHC antigens are highly immunogenic and their biologic function is to present foreign antigen for recognition by T cells. The VECs also express co-stimulatory and adhesion molecules that attract T cells, although allotypes have not been described. In addition to the well-known MHC antigens, VECs are known to express tissue-specific antigens. Identification of tissue-specific antigens had been expected because cell differentiation results in the development of specialized proteins that provide the biologic functions. Some of these tissue-specific antigens have demonstrated allotypes and are distinct from systemic antigens, such as the ABH and MHC antigens. The relevance of the tissue-specific antigens in clinical transplantation is supported by the observations of differential allograft survival between organs from the same donor transplanted into the single recipient. Further support for the relevance of tissue-specific alloantigens is provided by the well-established observation that retransplantation results are inferior to that of primary transplants. Similarly, HLA identical living-related renal allografts do develop irreversible rejection, albeit in low frequency, but would all undergo rejection in the absence of systemic immunosuppression except in the case of identical twins [46,47].

Polymorphic alloantigens with restrictive expression limited to the VECs, and in some cases expressed on the peripheral blood monocytes, were first demonstrated more than 25 years ago [48–50]. More recently, there has been a renewed interest in alloantigens with expression restricted to the VECs of an allograft [51,52]. The salient feature of the tissue-specific alloantigens on the allograft VECs is that current technology cannot detect presensitization of anti-VEC antibody to the donor VECs using standard crossmatching techniques.

66.1.4.2 Endothelial Cell Biology

Not only are VECs in a vascularized allograft the major transplant barrier but they also function as an endocrine-like organ capable of synthesizing clotting factors and inactivating hormones. A basic and critical role of the VECs is to control the exchange of metabolites, nutrients, and secreted molecules between blood and the interstitial fluid. The endothelium has been categorized into three broad groups: continuous, fenestrated, and sinusoidal [53,54]. Continuous endothelium occurs in large vessels, arterioles, and most venules. Fenestrated endothelium is observed in organs with function that necessitates a high rate of fluid exchange. The sinusoidal endothelium is found restricted to predominately liver, spleen, and bone marrow, and most notably lacks a basement membrane. The microvasculature displays dynamic features that render the regulation of physiologic functions, such as permeability, transport, molecule uptake, biosynthesis, and metabolism [54].

VECs are also able to phagocytize and contribute to the normal function of platelets by preventing activation and secreting tissue factor pathway inhibitors [55,56]. The net negative charge of the VECs in part provided by the surface expression of heparin proteoglycans and sulfates modulates a cell surface that accounts for its anticoagulant properties and contributes to its barrier functions. The endothelium also provides regional modulation of vascular dynamics via the endothelial isoform of nitric oxide synthase production of NO and other enzymatic activity like angiotensin-converting enzyme function. Therefore, the VECs play an essential role in maintaining homeostasis.

While the allograft VECs are capable of directly activating T cells, they can also process alloantigen contributing to the allorecognition process. VECs can stimulate monocyte differentiation into competent antigen presenting cells and are involved in the differentiation of dendritic cells [57,58]. Therefore, the development of an organ-specific therapy to prevent allograft rejection must not only prevent allorecognition but also protect the mandatory VEC functions. None of the current systemic immunosuppressive agents used clinically protect the VECs. Development of such an organ-specific therapy would hold the potential to not only protect the allograft from immune-mediated rejection but also help to eliminate the cardiovascular complications that are comorbidities in clinical transplantation. However, an organ-specific technology could not interfere with the critical VEC-related events that constitute vascular biology.

66.1.4.3 T-cell–Endothelial Cell Interactions

The luminal diameter of the microvasculature can be narrower than the diameter of the average T cell, particularly in the postcapillary venules [59]. Microvessel VECs do not interact with circulating T cells under normal conditions [60]. However, T cells once activated can modulate vascular endothelial cell functions mediated by both direct contact and soluble signals. The modulated effects can include remodeling of blood vessels, alterations in blood flow, recruitment of inflammatory cells, altered vascular wall permeability, and antigen presentation. Therefore, the interactions between T cells and the vascular endothelium inherently represent a linking of the immune and vascular systems and are bidirectional. If an organ-specific treatment that targets the allograft vasculature is to be achieved, it will be mandatory to prevent the allorecognition that is inherent in the bidirectional interaction between the immune and vascular systems.

66.2 POTENTIAL OF MODULATING THE IMMUNOGENICITY OF VASCULARIZED ALLOGRAFTS: CONCEPT OF IMMUNOCLOAKING

A huge body of research has focused on the encapsulation of cells to render them protected from an immune response. The encapsulation approach using various chemicals, such as liposomes, (lactide-co-glycolide) microspheres, alginate hydrogels, hyperbranched polyglycerol, and biotin/streptavidin, has been viewed as essential in protecting dispersed cells such as the insulin producing cells in the pancreas following isolation from the intact organ. Encapsulating suspensions of cells, while similar in attempting to provide protection from immune attack, is quite distinct from immunocloaking a complex three-dimensional vascularized organ containing approximately 3 trillion heterogeneous cells. Inasmuch, the following sections will focus on the potential of immunocloaking a whole vascularized allograft. As an introduction to the structural and functional consequences of immunocloaking to achieve antigenic modulation, the following will contain a brief review of pertinent work involving isolated cells.

66.2.1 Immunocloaking of Isolated Cell Suspensions

Hydrophilic macromolecules have been utilized to covalently modify the surface of cells with the goal of rendering the cells immunologically protected. Such covalent bonding of polymers to the cell surface has resulted in minimized immune responses, reduced degradation of enzymes and cell survival *in vivo*[61−63]. Polyethylene glycol (PEG) has been used to modify the surface of isolated and enriched pancreatic islet cells, red blood cells (RBCs), and a number of additional cell types. PEGylation is achieved by covalent bonding of the PEG to amines of proteins or carbohydrates on the surface of the treated cells or alternatively by inserting a conjugate consisting of PEG and lipid directly into the cell membrane [64]. PEG is a hydrated polymer with flexible chains which allows the polymer to function as steric barrier on the cell membrane [65].

Islet cell suspensions were covalently bonded with PEG polymers to eliminate their inherent immunogenicity [66,67]. The immunocloaked islet cells were shown to retain functionality for 12 months when administered in

conjunction with low dose cyclosporine [68]. However, in large animal models the period of insulin independence has been more limited and few clinical trials have been performed using immunocloaked cells [69,70]. The limited period of functionality is not specific for islet cells since similar results have been demonstrated using other cell types [71,72]. Additionally, PEGylation did not prevent recipient immune cell recruitment to the site of the grafted cells [73].

Major and minor RBC surface antigens have also been immunocloaked using polymers [74,75]. PEG-modified RBC surfaces provided masking of the ABH, RhD, and the C, c, E, S, and s RBC antigens. The modified RBCs were found to maintain their normal structure as well as function and did not activate complement. The immunocloaked RBCs have been used for targeted drug delivery of tissue plasminogen activator and urokinase plasminogen activator [76,77]. Therefore, immunocloaking of isolated cell surfaces can entail using macromolecules that can be hyperbranched and multifunctional. Theoretically such side chains could provide a mechanism for delivery for any number of therapeutic agents. Most importantly, immunocloaking did not adversely affect RBC lysis, hemoglobin oxygen carrier functions, survival in circulation, or homeostasis. However, a limitation to PEGylation of a particular cell surface, particularly RBCs with a relatively short life span, is that the treated cell would soon be replaced with new cells that would be without the protective treatment. The clinical efficacy of PEGylation may also be of limited use because of its dependence upon a steric exclusion to provide an immunocloaking barrier. In studies using additional biomaterials to immunocloak, implantation of the treated cells resulted in the accumulation of macrophage and neutrophils demonstrating an inflammatory response despite the use of the biomaterials [78,79]. Indirect antigen presentation has been proposed to be involved in this response since encapsulated allogeneic cells are rejected more quickly in comparison to control autologous encapsulated cells [80].

66.2.2 Embedding of Isolated Cells Within Immunocloaking Constructs

Relevant to immunocloaking vascular cells, matrix-embedded VECs have been shown to result in reduced responses to allogeneic activation. VECs that were embedded in Gelfoam blocks demonstrated reduced expression of MHC class II, co-stimulatory, and adhesion molecules, both *in vitro* and *in vivo* [81]. However, the embedding did not eliminate MHC class II responsiveness or co-stimulatory expression and had no effect on MHC class I or PD-L1 expression. It was proposed that matrix seeding of a potential cell-based therapy would provide benefit beyond simply providing a barrier to direct VEC/ T-cell interactions but rather provides a recapitulation of physiologic control [82].

This work supports the concept and principles that an immunocloaking approach focused on targeting the VECs within an allograft could provide protection from early phase allorecognition. However, the relevance of the studies is limited by the matrix used and by the lack of a vascular connection in the *in vivo* studies performed. Furthermore, the matrix used was Gelfoam. Gelfoam is a water-insoluble, nonelastic product made of purified pork skin gelatin. It is used as an absorbable hemostatic device. Therefore, its use in immunocloaking technologies represents a limitation.

To date the focus of immunocloaking technologies has been to treat predominately dissociated cells with the goal of limiting the allorecognition that occurs with the transplantation of allogeneic cells. This focus has resulted in the development of barriers that are selected based upon their physical dimensions to prevent the infiltration of cytotoxic cells and anti-donor antibodies. This focused approach has not lead to widespread clinical utility; partly because the barriers do not provide adequate nutrient delivery to the isolated cells. Additionally, inflammatory and immune responses still occur resulting in compromised cell function and viability.

The ability to successfully immunocloak a vascularized allograft will require a technology that not only interferes with allorecognition upon reperfusion but also supports the normal transfer of nutrients, oxygen, etc. to the subendothelium tissue that the vasculature provides. Similarly, any such technology cannot interfere with the normal physiologic functions of the vasculature.

66.2.3 Immunocloaking of an Organ

66.2.3.1 The Role of Ex Vivo Warm Perfusion

Requisite to tissue engineering the vasculature of an allograft with the goal of immunocloaking is an *ex vivo* near-normothermic perfusion in order to efficiently support continued synthetic functions. Such a warm perfusion technology represents a delivery platform for immunomodification that would ideally be an acellular technology based upon the principles of cell culture. Unlike a traditional bioreactor designed to expand populations of cells isolated by dissociative procedures, the warm perfusion must deliver all of the required molecules to maintain metabolism in the approximately three trillion heterogeneous cells that constitute an organ. Since no cell in an organ is more than a few microns away from a blood vessel, the perfusion should use the directional flow of the vascular bed. Physiologic parameters, such as temperature, pH, O_2 tension and perfusion pressures, must be

kept within a narrow range to support optimized *ex vivo* oxidative metabolism. Therefore, the perfusion technology must be based upon tissue culture principles that incorporate dynamic oxidative/reduction functions involving individual perfusate components that are interactive in order to support adequate basal metabolism [83–86]. Similarly, the vascular function as a refined charge barrier must be preserved.

Continued metabolism during the period of *ex vivo* immunocloaking is required in order to obtain optimal application of material in terms of adequate integrin binding, polymerization, and adequate coverage of the vasculature. The continued metabolism and corresponding synthesis are important since application of too much immunocloaking material results in the potential obstruction of microvessels. Similarly, inadequate application of immunocloaking material would leave parts of the vasculature untreated and at risk of immune activation. An additional consideration is that the warm perfusion must protect the normal barrier functions of the vascular wall and support adequate vasodilation to appropriately treat the microvessel bed.

These mandatory functions of a warm perfusion technology cannot be provided by hypothermic perfusion technology. At the temperatures used clinically to perfuse renal allografts, oxidative metabolism is inhibited by as much as 96% [87]. In the absence of the required level of continued metabolism, the translocation of the surface molecules, particularly the integrins, from the abluminal side of the endothelium would not occur. Without adequate integrin binding, the ability to retain an immunocloaking material on the luminal surface of the vasculature would be severely compromised.

66.2.3.2 Hypothesis

The subendothelial basement membrane is nonthrombogenic and nonimmunogenic. This extracellular basement membrane beneath the VECs contains a pore size of approximately 50 nm [88,89]. Such a small pore size restricts passive diffusion across the membrane to only small proteins, oxygen, and nutrients. Because circulating cells are limited in their ability to effectively transmigrate pores of <2.0 μm, there is a spatial inhibition that surpasses the basement membrane porosity by a factor of approximately 40-fold [90]. Cell trafficking associated with normal surveillance is dependent upon circulating leukocytes being in direct contact with the overlying vascular endothelium [91]. VECs function as the gate keepers for transmigration from the circulation, where traversing across the basement membrane only occurs in the presence of its overlaying vascular endothelium. Therefore, immunocloaking the vasculature of an allograft with a natural membrane that is not immunogenic or thrombogenic could provide organ-specific protection from the allorecognition that occurs upon reperfusion. Moreover, the prevention of allorecognition provided by the immunocloaking would not be at the expense of compromised VEC viability and function.

66.2.3.3 Potential Sources of Biomaterials

A wide array of artificial materials have been used in a variety of tissue engineering applications. These artificial materials have demonstrated limited efficacy in their ability to adhere to VEC [92]. In tissue engineering studies, VECs do not reestablish normal monolayers and are susceptible to thrombosis leading to the resulting loss of patency [93–94]. Further restricting the clinical potential of the artificial materials, is the limitation of the bioburden involved in using them *in vivo*. Overall the synthetic materials are deficient in bioactivity and do not possess the natural structural characteristics of extracellular matrices (ECMs)-derived components. The limitation of using synthetic materials is the all too frequent development of fibrous encapsulation that occurs following reimplantation. However, using ECM derivatives, such as collagen, gelatin, fibronectin, glycosaminoglycans, and even the commercially available Matrigel™, that is the product of the Engelbreth–Holm–Swarm murine sarcoma cell line, results in poor mechanical properties and immunogenicity [95]. Natural polymers, such as alginate, fibrin, chitosan, and hyaluronan, provide better mechanical properties but do not provide ECM architecture and functions [96].

A logical starting point for the selection of biomaterials to accomplish immunocloaking would entail using natural ECMs. ECMs are very complex three-dimensional scaffoldings that cells secrete and rest upon in tissues. Cells interact intimately with their specialized ECM that regulates and maintains fundamental cellular functions. This interdependence provides the resulting molecular composition, structure, and mechanical properties of tissues. The cell–cell and cell–ECM interactions, in large part, contribute to the differentiated functions of various tissues. The types and concentrations of the components of ECM vary widely throughout the body. ECM predominantly consists of an insoluble network of molecules with self-assembly consisting of a laminin template for the scaffolding, glycoproteins, proteoglycans, carbohydrate moieties, vimentin, fibronectin, elastin, and a network of collagen fibrils. The resident cells' ability to attach and embed within its ECM affects its phenotype and functions. The cell–ECM interactions are affected by the number and intensity of the adhesions between the cell surface integrins and specific ECM peptides [97]. A large number of integrins that facilitate cell binding to its ECM have been described. However, the β1 integrin subfamily represents a major class of integrins that interact with a number of individual ECM components, including laminins and type IV collagen [98,99]. These integrins provide cell anchorage to the ECM that facilitates activated self-assembly [100]. The peptide sequence Arg–Gly–Asp, commonly referred to as "RGD" facilitates cellular binding via their respective adhesion molecules [101]. *In vitro* studies using ECM prevented cell penetration into the engineered tissues while also providing a compatible environment for confluent cell populations [102].

66.2.3.4 Application of ECM to the Target Tissue

In order to apply ECM materials to the vasculature, it is necessary to solubilize the material prior to application. By applying the solubilized material prepolymerization, it is feasible to control its administration in a targeted fashion. Once the solubilized ECM is delivered into a physiologic environment, self-assembly for repolymerization occurs. How well the repolymerized ECM functions to provide its normal biologic role depends upon how the material is originally isolated and handled. For example, if detergents are used to decellularize the ECM scaffolding, individual components like the glycosaminoglycans can be lost [103]. Therefore, the processing necessary to extract ECM from intact tissue requires distinct processing steps that can impact the ending structure and function when reimplanted *in vivo*. The distinct processing steps can involve decellularization, sonication, freeze/thaw, dehydration, disinfection, lyophilization, and terminal sterilization [104,105]. The removal of cell remnants and contaminating DNA removes antigenic epitopes associated with cell membranes and their immunogenicity. Isolation processes that preserve the hydration state through decellularization and sterilization result in functional ECM [106].

66.3 INITIAL IMMUNOCLOAKING STUDIES

Recently, a bioengineered barrier membrane that is predominantly comprised of type IV collagen, vitrogen, fibronectin, laminin, entactin, glycosaminoglycans, and proteoglycans has been developed [107]. The components are polymerized to result in a tri-dimensional transparent membrane. The transparent membrane is referred to as NB-LVF4, a nanobarrier membrane. The interaction between the VEC lining the graft vasculature and the recognition domains within the NB-LVF4 membrane is receptor specific via the laminin and type IV collagen portions of the membrane. Lysine-derived cross-links and disulfide bonds stabilize the components in the barrier membrane. In its present biosynthetic form the membrane appears as a fine mesh lacking the banded fibrillar structure of other collagen types. Each molecule has a globular region at one end and a disulfide region at the other end. Additionally, each molecule is interconnected with others through disulfide bonds. The application process involves solubilizing the synthesized barrier membrane followed by application to the kidneys during a near-normothermic acellular perfusion at 32°C for 3 h. The solubilized membrane can be slowly introduced into the arterial line. With the continued oxidative metabolism during *ex vivo* warm perfusion the self-assembly of the NB-LVF4 membrane is localized to the luminal surface by the concordant translocation of the endothelial cell surface integrins.

Since the barrier membrane is permeable to small molecular weight compounds, free transport of nutrients and oxygen is unaffected and the graft tissue remains viable. Likewise, the barrier membrane supports cellular functions similar to the natural role of ECMs in substrata tissues *in vivo*.

66.3.1 Efficacy of Immunocloaking ECM in a Renal Transplant Model

A number of important initial research questions have been addressed in order to test the proposed hypothesis. These research questions included:

- Would immunocloaking of the vascular endothelium prevent allorecognition as assessed in immunologic assays *in vitro* studies?
- Is it feasible to apply the barrier membrane to the vasculature of the kidneys, to result in adequate coverage of the luminal surfaces?
- If successfully applied, would covering the renal vasculature adversely affect renal function?
- If application of the barrier membrane to the renal vasculature did not adversely affect function, would such immunocloaking provide protection against early allograft rejection?

66.3.1.1 NB-LVF4 Application Prevents Immune Cell Activation

Mixed Lymphocyte/VEC Reactions

Mixed lymphocyte/VEC reactions (MLERs) were performed where porcine donor VECs strongly stimulated recipient peripheral blood mononuclear cells (MNCs) resulting in a mean stimulation index of 49.8 compared with the negative control consisting of responding cells alone. Immunocloaking the same donor VEC with NB-LVF4 inhibited these *in vitro* alloresponses by 99.9% (Table 66.1). Further evidence supporting the nonimmunogenicity of the NB-LVF4 barrier membrane is provided by the resulting mean stimulation index of only 0.7 when

TABLE 66.1 Mixed Lymphocyte−Vascular Endothelial Cell Reaction[a]

Combination	CPM[b]	SI[c]	% Inhibition
Mononuclear cells	385 (±25)	–	–
Mononuclear cells + allogeneic VECs	19,200 (±345)	49.8	–
Mononuclear cells + NB-LVF4-treated allogeneic VECs	409 (±18)	1.1	99.9
Mononuclear cells + NB-LVF4	280 (±12)	0.7	–

[a]Results represented as the mean of the three experiments.
[b]CPM, counts per minute of the mean.
[c]SI (stimulation index) = mean CPM test/mean CPM negative control.

responding MNCs were stimulated with the NB-LVF4 membrane alone. These results demonstrate that immunocloaking with the NB-LVF4 prevents the T-cell proliferation that normally occurs when immune cells are stimulated with VECs.

NB-LVF4 Application Prevents Antigen Presentation and T-Cell Activation

Additional immunological assays have been performed using human responding MNCs. The responding MNCs were stimulated with autologous lymphocytes (negative controls) and untreated confluent monolayers of donor umbilical vein VECs (positive controls). The test groups consisted of the same confluent donor VECs that were immunocloaked with the NB-LVF4 barrier membrane.

Early T-cell activation was measured using $CD4^+$, $CD69^+$ detection by flow cytometry following 24 h ($n = 10$). Measurements were also made for the cytokines IL-2, IL-6, and MIG using the Luminex xMap platform. In addition, following stimulation, supernatants were collected from each triplicate well and tested for additional cytokine and chemokine responses to evaluate antigen presentation. These tested cytokines and chemokines that are associated with antigen presentation included IL-1β, TNF-α, MIP-1α, and γ-IFN.

Immunocloaking of the VECs resulted in statistically significant inhibition of the cytokines and chemokines: IL-1β, IL-6, γ-IFN, IL-2, TNF-α, CD-69, MIG, and MIP-1α ($P < 0.05$) (Figure 66.1). The finding of the inhibition of the cytokines produced by antigen presenting cells, MIP-1α, IL-1β, γ-IFN, and TNF-α suggests that immunocloaking VECs with NB-LVF4 prevents antigen presentation. Since IL-1, γ-IFN, and TNF-α are pro-inflammatory cytokines, the ability of the NB-LVF4 barrier membrane to inhibit their release provides further evidence of the protective effect of immunocloaking. Similarly, the

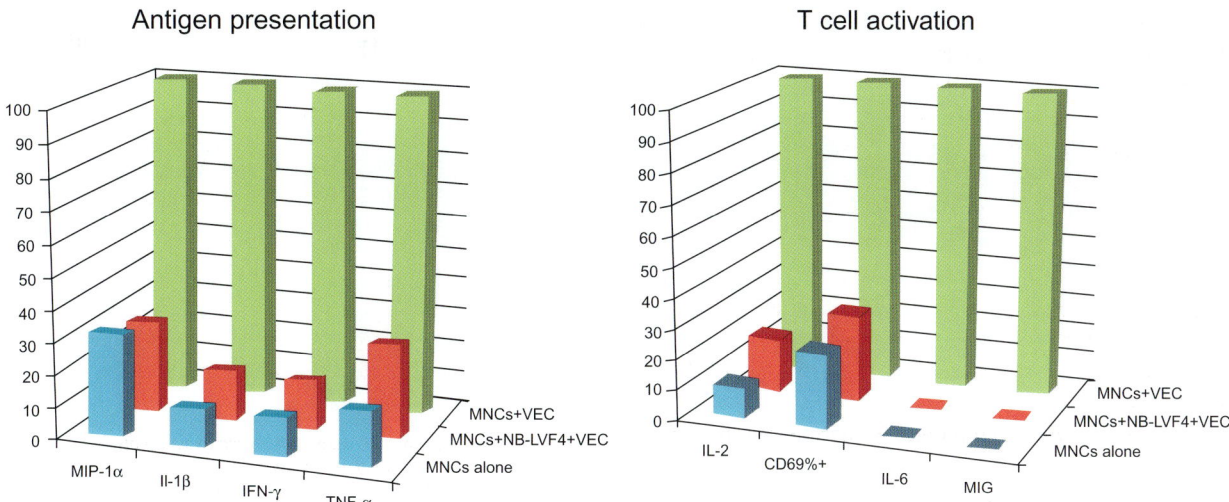

FIGURE 66.1 Inhibition of antigen presentation and the prevention of T-cell activation. MNCs: Responding peripheral blood mononuclear cells; VECs: confluent monolayers of vascular endothelial cells; NB-LVF4 + VECs: NB-LVF4 treatment of confluent monolayers of vascular endothelial cells. Y axis: Values represented as a percent of the positive controls tested in triplicate of $n = 5$.

inhibition of markers of T-cell activation, IL-6, IL-2, CD-69, and MIG suggests the blockade of T-cell-mediated responses when the VECs are immunocloaked with NB-LVF4 membrane. It is reasonable to assume that the immunocloaking NB-LVF4 provides that primary antigen recognition would also prevent endothelial cell activation. Preventing endothelial cell activation would likewise prevent the externalization of the preformed Weibel−Palade bodies that contain P-selectin adhesion molecules. Without a pro-inflammatory signal, the multistep leukocyte extravasation process would be adverted while the NB-LVF4 masks the allograft. The results of these immunologic screenings support the hypothesis that immunocloaking can be successfully used in an organ-specific manner to prevent the allorecognition that normally occurs upon reperfusion.

Feasibility of Targeting Application of the NB-LVF4 Barrier Membrane to the Vasculature of an Allograft

Deposition of the barrier membrane within the vasculature of canine kidneys was evaluated during a period of *ex vivo* near-normothermic perfusion. Following administration of the solubilized membrane, the kidneys continued on warm perfusion for a 3-h subsequent period of perfusion to allow for optimal polymerization. The treated kidneys were then sectioned and evaluated histologically using an antibody to laminin to determine the location of the membrane within the renal vasculature. Various protein concentrations of the solubilized NB-LVF4 membrane were evaluated per gram of kidney to determine optimal dosing. At low doses, inadequate coverage of the luminal surfaces was observed with

deposition predominately in the large vessels. Higher doses were found to result in thick deposition that in some cases obstructed the glomeruli and microvessels. At moderate concentrations, NB-LVF4 treatment of kidneys provided ubiquitous coverage in approximately 90% of the vascular luminal surfaces, including both small and large blood vessels. No occlusion of vessels was observed by immunohistochemical evaluation at an optimized range of dosing. This observation was supported by the perfusion data where the perfusion pressures and vascular flow rates remained stable throughout the 3-h period following NB-LVF4 membrane application. Similarly, while there were some isolated areas of kidney vasculature not covered, these areas were focal and randomly dispersed. All other areas appeared to be uniformly immunocloaked. In contrast, the untreated control kidneys that were warm perfused without application of the NB-LVF4 membrane uniformly stained negative for extracellular basement within the renal vasculature [107].

Electron Microscopy of Immunocloaked Vascular Endothelium

An electron micrograph of the vascular endothelium of a normal renal arcuate artery is shown in Figure 66.2A. The vessel wall is composed of a single monolayer of endothelial cells. No more than two endothelial cells are needed to cover the circumference of microvessels, while five or more are needed to cover larger blood vessels. VECs have irregular borders forming a delicate mosaic that appears to bulge into the lumens of the blood vessels in areas where the nuclei are located. The VECs have a thin glycoprotein layer over their luminal surface. In

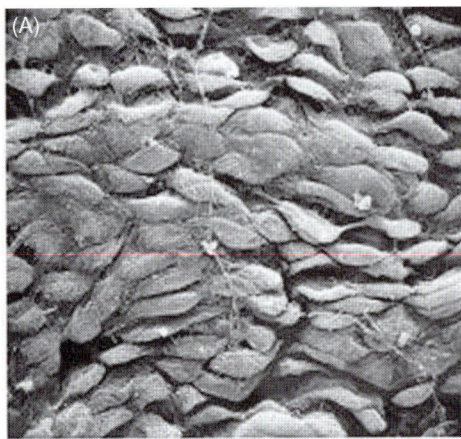
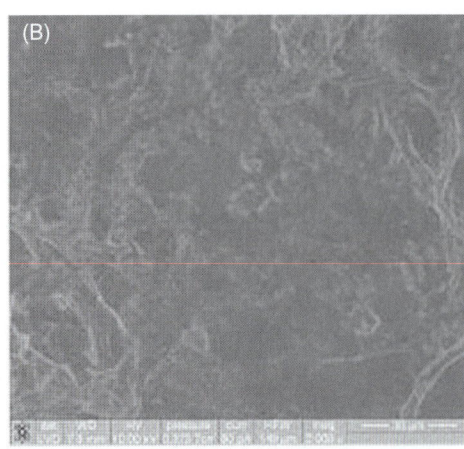

FIGURE 66.2 Electron microscopy of immunocloaked endothelium: (A) VECs lining the luminal surface of the renal arcuate artery and (B) NB-LVF4 membrane applied to the VECs lining the renal artery.

Figure 66.2A it can be seen that adjacent VECs tend to meet in a simple end-to-end pattern but can frequently be seen overlapping with one cell overlaying another along an oblique course.

Applying the NB-LVF4 barrier membrane to the luminal surface of blood vessels results in an uniform surface coating beneath which the VECs remain in their normal configuration (Figure 66.2B). While the normal barrier functions of the endothelium remain unaffected, the entire luminal surface appears to become immunocloaked to immune cells in the circulation. The NB-LVF4-coated surface remains nonthrombogenic and nonimmunogenic.

Immunocloaking the Renal Vasculature Does not Adversely Affect Renal Function

Using a canine autotransplant model, the effect of applying the immunocloaking membrane on the resulting renal function was determined. The left kidneys were nephrectomized, flushed of blood, and then placed on *ex vivo* near-normothermic perfusion. The solubilized NB-LVF4 membrane was administered and once applied, the perfusion was continued for an additional 3 h. Following the NB-LVF4 barrier membrane application and polymerization, the kidneys were reimplanted with contralateral nephrectomy of the untreated kidneys.

Sequential testing posttransplant in the autotransplanted kidneys included serum chemistries, urinalysis, hematology, and calculations of GFR (Table 66.2). All four treated autotransplanted kidneys provided normal serum chemistries accompanied by normal urine with no evidence of proteinuria. The dogs were followed for several months and the normal renal function was maintained throughout the posttreatment period. The results demonstrate that application of an immunocloaking material within the renal vasculature during a period of *ex vivo* near-normothermic perfusion does not adversely affect renal function.

Immunocloaking Provides Protection Against Early Allograft Rejection

A canine allotransplant model was used to determine the effect of treatment with NB-LVF4 on early allograft immune responses. The solubilized NB-LVF4 membrane was again slowly administered during the *ex vivo* near-normothermic perfusion. The membrane was allowed to polymerize during the period of continued *ex vivo* perfusion and the treated kidneys were allotransplanted with nephrectomy of the native kidneys. No systemic immunosuppression was given. Untreated controls were similarly warm perfused and allotransplanted with nephrectomy of the native kidneys but without NB-LVF4 treatment or systemic immunosuppression. In the untreated controls, the mean onset of rejection occurred on day 6 ($\pm$1). In contrast, in the dogs transplanted with a NB-LVF4 membrane treated kidney the mean onset of rejection was delayed approximately fivefold occurring on day 30 ($\pm$1.0) [107].

While the retention times have not been definitively delineated, it appears that the NB-LVF4 membrane is present within the luminal surfaces for approximately 21 days. The deterioration of the membrane integrity in subsequent days would occur most likely first in the large vessels and lastly in the microvessel bed. This hypothesis is supported by the differential pressures and blood flow in the microvasculature compared to the large vessels and by the known role of the microvessels in diapedesis. If this estimate proves to be accurate, it would explain the delayed allorecognition observed in the allotransplanted kidneys that were treated with the NB-LVF4 membrane and the subsequent rejection of the allograft by day 30.

66.3.1.2 Concluding Remarks

This body of evidence taken together demonstrates that an organ-specific treatment using a subendothelial basement membrane applied in a manner that covers the

TABLE 66.2 NB-LVF4 Treated Canine Kidney Autotransplants

Blood Chemistries

	Pre-Tx	Post-Tx D1	Post-Tx D7	Post-Tx D14	Post-Tx D180	Normal Range
BUN (mg/dL)	16.3 ± 3.5	33.3 ± 6.9	17.8 ± 2.6	16.1 ± 3.2	16 ± 1.4	6–27
Total protein (g/dL)	6.1 ± 0.3	4.5 ± 0.3	6.2 ± 0.3	6.3 ± 0.4	6.3 ± 0.4	5.0–7.5
Sodium (mEq/L)	148.3 ± 2.2	139 ± 2.9	148.8 ± 3.3	148.8 ± 1.0	149.8 ± 2.2	145–158
Chloride (mEq/L)	113.8 ± 1.5	101.5 ± 2.4	113 ± 1.2	113.5 ± 1.0	113.5 ± 1.3	105–122
Potassium (mmol/L)	5.1 ± 0.4	5.2 ± 0.4	5.2 ± 0.4	5.2 ± 0.3	5.1 ± 0.3	3.5–5.8
Creatinine (mg/dL)	1.8 ± 0.1	2.4 ± 0.3	1.9 ± 0.2	1.6 ± 0.1	1.7 ± 0.2	<20
Calcium (Eq/L)	10.3 ± 0.1	10.4 ± 0.1	10.4 ± 0.1	10.4 ± 0.1	10.4 ± 0.1	9.0–11.8

Hematoloqy

	Pre-Tx	Post-Tx D7	Normal Range
RBC (cells/µL)	5.9 ± 0.8	5.2 ± 1.0	4.0–6.0
HGB (g/dL)	13.5 ± 1.4	12.1 ± 2.0	11.0–18.0
HCT (%)	40.3 ± 5.1	35.9 ± 7.7	35.0–60.0
MCVfl	67.6 ± 0.7	67.6 ± 0.4	60.0–80.0
MCHC (%)	33.6 ± 2.0	34.5 ± 1.3	33.0–37.0
RDW (%)	13.2 ± 0.8	12.6 ± 0.3	11.6–13.7
PLT (K/µL)	300.8 ± 149	273.3 ± 153	150–450
WBC (K/µL)	10.7 ± 3.8	10.4 ± 3.7	4.5–10.5

Urinalysis

	Pre-Tx	Post-Tx D1	Post-Tx D7	Post-Tx D180	Normal Range
Glucose (mg/dL)	Negative	Negative	Negative	Negative	Negative
Bilirubin	Negative	Negative	Negative	Negative	0–1
Specific gravity	1.04 ± 0.01	1.02 ± 0.01	1.039 ± 0.01	1.04 ± 0.0	>1.025
Blood	Negative	1 ± 1	Negative	Negative	0–trace
Total protein (mg/dL)	Negative	0.17 ± 0.2	Negative	Negative	<10 mg/dL
Nitrite	Negative	Negative	Negative	Negative	Negative
GFR (mL/min/kg)	3.1 ± 0.2	–	3.1 ± 0.2	1.5 ± 0.1	1.78–3.87
Urine creat (mg/dL)	70 ± 25	–	7.3 ± 17	55 ± 23	>20

vascular luminal surfaces does not adversely affect renal function. A single treatment of the barrier membrane provides protection against the initial allorecognition that occurs upon reperfusion; despite any role that passenger leukocytes might play. The results of immunologic studies provide the initial evidence supporting the protective effect of the barrier membrane as functioning by preventing antigen presentation, T-cell activation and proliferation. While the immunocloaking technology is in its infancy, organ-specific treatments that could minimize, or ideally eliminate, the need for the toxic systemic immunosuppression regimens that clinical transplantation is dependent upon today could revolutionize the field. Clearly much more research and development is essential to moving this technology forward to become a clinical reality. Next steps should include modifying the chemistry of the barrier membrane to enhance the intravascular retention times. Also important will be the development of minimally invasive technology to readminister the solubilized barrier membrane during the posttransplant period. One could envision the potential utility of receiving an injection approximately every 3 weeks instead of the daily multidrug immunosuppressive regimen occurring today. A most tantalizing concept worthy of intense investigation would be to use the window of opportunity where allorecognition does not occur to induce donor-specific tolerance. Alternatively, developing strategies to optimize co-stimulatory blockade in combination with low dose systemic immunosuppression during the period the barrier membrane is beginning to degrade within the allograft vasculature could positively impact long-term outcomes. Certainly eliminating the need for calcineurin inhibitors during the early posttransplant period could help to make transplanting ischemically damaged allografts more feasible.

REFERENCES

[1] Joosten SA, Yvo WJ, Sijpkens YW, Van Kooten C, Paul LC. Chronic renal allograft rejection: pathophysiologic considerations. Kidney Int 2005;68:1—13.

[2] Holt PG, Haining S, Nelson DJ, Sedgewick JD. Origin and steady-state turnover of class II MHC-bearing dendritic cells in the epithelium of the conducting airways. J Immunol 1994; 153:256—61.

[3] Kamath AT, Pooley J, O'Keeffe MA, Vremec D, Zhan Y, Lew AM, et al. The development, maturation, and turnover rate of mouse spleen dendritic cell populations. J Immunol 2000; 165:6762—70.

[4] Steele DJ, Laufer TM, Smiley ST, Ando Y, Grusby MJ, Glimcher LH, et al. Two levels of help for B cell alloantibody production. J Exp Med 1996;183:699—703.

[5] Dalloul AH, Chmouzis E, Ngo K, Fung-Leung WP. Adoptively transferred CD4$^+$ lymphocytes from CD8$^{-/-}$ mice are sufficient to mediate rejection of MHC class II or class I disparate skin grafts. J Immunol 1996;156:4114—9.

[6] Preston EH, Light JA, Kampen RL, Kirk AD. Human renal allograft rejection despite the absence of allogeneic passenger leukocytes. Am J Transplant 2004;4:283—5.

[7] Divate SA. Acute renal allograft rejection: progress in understanding cellular and molecular mechanisms. J Postgrad Med 2000; 46:293.

[8] Lechler R, Batchelor J. Immunogenicity of retransplanted rat kidney allografts. Effect of inducing chimerism in the first recipient and quantitative studies on immunosuppression of the second recipient. J Exp Med 1982;156:1835—41.

[9] Watschinger B. Indirect recognition of allo MHC peptides—potential role in human transplantation. Nephr Dial Transplant 1999;14:8—11.

[10] Shoskes DA, Wood KJ. Indirect presentation of MHC antigens in transplantation. Immunol Today 1994;15:32—8.

[11] Baker RJ, Hernandez-Fuentes MP, Brookes PA, Chaudhry AN, Cook HT, Lechler RI. Loss of direct and maintenance of indirect alloresponses in renal allograft recipients: implications for the pathogenesis of chronic allograft nephropathy. J Immunol 2001; 167:7199—206.

[12] Vella JP, Spadafora-Ferreira M, Murphy B, Alexander SI, Harmon W, Carpenter CB, et al. Indirect allorecognition of major histocompatibility complex allopeptides in human renal transplant recipients with chronic graft dysfunction. Transplantation 1997; 64:795—800.

[13] Opelz G. Impact of HLA compatibility on survival of kidney transplants from unrelated live donors. Transplantation 1997;10:1473—5.

[14] Brennan TV, Jaigirdar A, Hoang V, Hayden T, Liu FC, Zaid H, et al. Preferential priming of alloreactive T cells with indirect reactivity. Am J Transplant 2009;9:709—18.

[15] Van Breda Vriesman PJC, Feldman JD. IgM alloantibody elicited by first set renal allografts: *in vivo* and *in vitro* studies. J Immunol 1972;108:1188—9.

[16] Hirschberg H, Evensen SA, Henriksen T, Thorsby E. Stimulation of human lymphocytes by allogeneic endothelial cells *in vitro*. Tissue Antigens 1974;4:257—61.

[17] Hirschberg H, Evensen SA, Henriksen T, Thorsby E. Stimulation of human lymphocytes by cultured allogeneic skin and endothelial cells *in vitro*. Transplantation 1975;19:191—4.

[18] Omari KI, Dorovini-Zis K. Expression and function of the costimulatory molecules B7-1 (CD80) and B7-2 (CD86) in an *in vitro* model of the human blood—brain barrier. J Neuro Immunol 2001;113:129—41.

[19] Epperson DE, Pober JS. Antigen-presenting function of human endothelial cells. Direct activation of resting CD8 T Cells. J Immunol 1994;153:5402—12.

[20] Khayyamian S, Hutloff A, Büchner K, Gräfe M, Henn V, Kroczek RA, et al. ICOS-ligand, expressed on human endothelial cells, costimulates Th1 and Th2 cytokine secretion by memory CD4$^+$ T cells. Proc Natl Acad Sci USA 2002;99:6198—203.

[21] Nocentini G, Riccardi C. GITR: a multifaceted regulator of immunity belongining to the tumor necrosis factor receptor superfamily. Eur J Immunol 2005;35:1016—22.

[22] Valujskikh A, Heeger PS. Emerging roles of endothelial cells in transplant rejection. Curr Opin Immunol 2003;15:493—8.

[23] Berg LP, James MJ, Alvarez-Iglesias M, Gelnnie S, Lechler RI, Marelli-Berg FM. Functional consequences of noncognate interactions between CD4$^+$ memory T lymphocytes and the endothelium. J Immunol 2002;168:3227—34.

[24] Von Hundelshausen P, Weber KS, Huo Y, Proudfoot AE, Nelson PJ, Ley K, et al. RANTES deposition by platelets triggers monocyte arrest on inflamed and artherosclerotic endothelium. Immunity 2001;14:377—86.

[25] Nishimura M, Umehjara H, Nakayama T, Yoneda O, Hieshima K, Kakizaki M, et al. Dual functions of fractalkine/CX3C ligand 1 in trafficking of perforin$^+$/granzyme B$^+$ cytotoxic effector lymphocytes that are defined by CX3CR1 expression. J Immunol 2002; 168:6173—80.

[26] Butcher EC, Picker LJ. Lymphocyte homing and homeostasis. Science 1996;272:60—6.

[27] Nelson PJ, Krensky AM. Chemokines, chemokine receptors, and allograft rejection. Immunity 2001;14:377—86.

[28] McCluskey RT. Comments on targets in rejecting allografts. Transplant Proc 1980;12:22−5.

[29] Busch GJ, Reynolds ES, Galvanek EG, Braun WE, Dammin GJ. Human renal allografts. The role of vascular injury in early graft failure. Medicine 1971;50:29−83.

[30] Busch GJ, Garovoy MR, Tilney NL. Variant forms of arteritis in human renal allografts. Transplantation 1979;11:100−3.

[31] Porter KA, Joseph NH, Rendall JM, Stolinski C, Hoehn RJ, Calne RY. The role of lymphocytes in the rejection of canine renal homotransplants. Lab Invest 1964;13:1080−98.

[32] Rose ML. Endothelial cells as antigen-presenting cells: role in human transplant rejection. Cell Mol Life Sci 1998;54:965−78.

[33] Trpkov K, Campbell P, Pazderka F, Cockfield S, Solez K, Halloran PF. Pathologic features of acute renal allograft rejection associated with donor-specific antibody. Analysis using the Banff grading schema. Transplantation 1996;61:1586−92.

[34] Halloran PF, Wadgymar A, Ritchie S, Falk J, Soiez K, Srinvasa NS. The significance of the anti-class I antibody response. I. Clinical and pathologic features of antibody class I-mediated rejection. Transplantation 1990;49:85−91.

[35] Halloran PF, Schlaut J, Soiez K, Srinivasa NS. The significance of the anti-class I response. II. Clinical and pathological features of renal transplants with anti-class I-like antibody. Transplantation 1992;53:550−5.

[36] Racusen LC, Colvin RB, Solex K, Mihatsch MJ, Halloran PF, Campbell PM, et al. Antibody-mediated rejection criteria-an addition to the Banff 97 classification of renal allograft rejection. AMJ Transplant 2003;3:708−14.

[37] Taflin C, Charron D, Glotz D, Mooney N. Immunological function of the endothelial cell within the setting of organ transplantation. Immunol Lett 2001;1−6.

[38] Bevilaqua MP. Endothelial−leukocyte adhesion molecules. Annu Rev Immunol 1993;11:767−804.

[39] Shivizu Y, Shaw S, Graver N. Activation-independent binding of human memory T cells to adhesion molecule ELAM-1. Nature 1996;349:786−99.

[40] Muller WA, Berman ME, Newman PJ, Delisserj HM, Albelda SM. A heterophilic adhesion mechanism for platelet/endothelial cell adhesion molecule 1. J Exp Med 1992;175:1401−4.

[41] Carman CV, Springer TA. A transmigratory cup in leukocyte dispedesis both through individual vascular endothelial cells and between them. J Cell Biol 2004;167:377−88.

[42] Faull RJ, Russ GR. Tubular expression of intercellular adhesion molecules-1 during renal allograft rejection. Transplant Int 1991;4:141−3.

[43] Carman CV, Jun CD, Salas A, Springer TA. Endothelial cells proactively form microvilli-like membrane projections upon ICAM-1 engagement of leukocyte LFA-11. J Immunol 2003;171:6135−44.

[44] Luu NT, Rainger GE, Nash GB. Kinetics of the different steps during neutrophil migration through cultured endothelial monolayers treated with tumour necrosis factor-alpha. J Vasc Res 1999;36:477−85.

[45] Hancock WW, Kraft N, Atkins RC. The immunohistochemical demonstration of major histocompatibility antigens in the human kidney using monoclonal antibodies. Pathology 1982;31:75−8.

[46] Brasile L, Clarke J, Galouzis T, Cerilli J. The clinical significance of the vascular endothelial cell antigen system: evidence for genetic linkage between the endothelial cell antigen system and the major histocompatibility complex. Transplant Proc 1985;17: 741−3.

[47] Cerilli J, Brasile L, Galouzis T, Lempert N, Clarke J. The vascular endothelial cell antigen system. Transplantation 1985;39:286−9.

[48] Brasile L, Zerbe T, Rabin B, Clarke J, Abrams A, Cerilli J. The identification of antibody to vascular endothelial cells (VEC) in patients undergoing cardiac transplantation. Transplantation 1985;40:672−5.

[49] Moraes R, Stastny P. A new antigen system expressed in human endothelial cells. Clin Invest 1977;60:449−54.

[50] Paul LC, Claas FH, van Es LA, Kalff MW, de Graeff J. Accelerated rejection of a renal allograft associated with pretransplantation antibodies directed against donor antigens on endothelium and monocytes. N Engl J Med 1979;11:427.

[51] Lucchiarai N, Panajotopoulos N, Xu C, Rodrigues H, Ianhez LE, Kalil J, et al. Antibodies eluted from acutely rejected renal allografts bind to and activate human endothelial cells. Human Immunol 2000;61:518−27.

[52] Zhang Q, Reed EF. Non-MHC antigenic targets of the humoral immune response in transplantation. Curr Opin Immunol 2010;22:682−8.

[53] Taylor AE, Granger DN. Exchange of macromolecules across the microcirculation. In: Renkin EM, Michel CC, editors. Handbook of physiology—the cardiovascular system, vol. IV. Bethesda, MD: American Physiological Society; 1984. p. 467−520. Section 2.

[54] Simionescu N, Simionescu M. Ultrastructure of the microvessel wall. In: Renkin EM, Michel CC, editors. Handbook of physiology—the cardiovascular system, vol. IV. Bethesda, MD: American Physiological Society; 1984. p. 41−101. Section 2.

[55] Pober JS, Gimbrone Jr MA, Collins T, Cotran RS, Ault KA, Fiers W, et al. Interactions of T lymphocytes with human vascular endothelial cells: role of endothelial cell surface antigens. Immunobiology 1984;168:483−94.

[56] Manes TD, Pober JS, Lkuger MS. Endothelial cell−T lymphocyte interations: IP-10 stimulates rapid transendothelial migration of human effort but not central memory CD4$^+$ T cells. Requirements for shear stress and adhesion molecules. Transplantation 2006;82: S9−14.

[57] Denton MD, Geehan CS, Alexander SI, Sayegh MH, Briscoe DM. Endothelial cells modify the costimulatory capacity of transmigration leukocytes and promote CD28-mediated CD4(+) T-cell alloactivation. J Exp Med. 1999;190:555−66.

[58] Randolph GJ, Beaulieu S, Lebecque S, Steinman RM, Muller WA. Differentiation of monocytes into dendritic cells in a model of transendothelial trafficking. Science 1998;282:480−3.

[59] Pober JS, Cotran RS. Overview: the role of endothelial cells in inflammation. Transplantation 1990;50:537−44.

[60] Pober JS. Immunobiology of human vascular endothelium. Immunol Res 1999;19:225−32.

[61] Scott MD, Murad KL, Koumpouras F, Talbot M, Eaton JW. Chemical camouflage of antigenic determinants: "Stealth" erythrocytes. Proc Natl Acad Sci USA 1997;94:7566−71.

[62] Veronese FM, Pasut G. PEGylation, successful approach to drug delivery. Drug Discov Today 2005;10:1451−8.

[63] Williams DF. On the mechanisms of biocompatibility. Biomaterials 2008;29:2941−53.

[64] Kellam B, De Bank PA, Shakeshaff KM. Chemical modification of mammalian cell surfaces. Chem Soc Rev 2003;32:327−37.

[65] Chen AM, Scott MD. Current and future applications of immunological attenuation via pegylation of cells and tissue. BioDrugs 2001;15:833−47.

[66] Hume PS, Bowman CN, Anseth KS. Functional PEG hydrogels through reactive dip-coating for the formation of immunoactive barriers. Biomaterials 2011;32:6204−12.

[67] Kizilel S, Scavone A, Liu XA, Nothias JM, Ostrega D, Witkowski P, et al. Encapsulation of pancreatic islets within nano-thin functional polyethylene glycol coatings for enhanced insulin secretion. Tissue Eng Part A 2010;16:2217−28.

[68] Lee DY, Nam JH, Byun Y. Functional and histological evaluation of transplanted pancreatic islets immunoprotected by PEGylation and cycloporin for 1 year. Biomaterials 2007;28:1957−66.

[69] Soon-Shiong P, Heintz RE, Merideth N, Yao QX, Yao Z, Zheng T, et al. Insulin independence in a type 1 diabetic patient after encapsulated islet transplantation. Lancet 1994;343:950−1.

[70] Calafiore R, Basta G, Luca G, Lemmi A, Montanucci MP, Calabrese G, et al. Microencapsulated pancreatic islet allografts into nonimmunosuppressed patients with type 1 diabetes: first two cases. Diabetes Care 2006;29:137−8.

[71] Tibell A, Rafael E, Wennberg L, Nordenstrom J, Berstrom M, Geller RL, et al. Survival macroencapsulated allogeneic parathyroid tissue one year after transplantation in nonimmunosuppressed humans. Cell Transplant 2001;10:591−9.

[72] Bloch J, Bachoud-Levi AC, Deglon N, Lefaucheur JP, Winkel L, Boisse S, et al. Neuroprotective gene therapy for Huntington's disease, using polymer-encapsulated cells engineered to secrete human ciliary neurotrophic factor: results of a phase I study. Hum Gene Ther 2004;15:968−75.

[73] Lee DY, Park SJ, Lee S, Nam JH, Byun Y. Highly poly(ethylene) glycolylated islets improve long-term islet allograft survival without immunosuppressive medication. Tissue Eng 2007;13: 2133−41.

[74] Murad KT, Mahany KI, Brugnara C, Kuypers FA, Eaton JW, Scott MD. Structural and functional consequences of antigenic modulation of red blood cells with methoxypoly(ethylene glycol). Blood 1999;93:2121−7.

[75] Bradley AJ, Murad KI, Regan KL, Scott MD. Biophysical consequences of linker chemistry and polymer size on stealth erythrocytes: size does matter. Biochem Biophys Acta 2002;1561: 147−58.

[76] Murciano JC, Medinilla S, Eslin D, Atochina E, Cines DB, Muzykantov VR. Prophylactic fibrinolysis through selective dissolution of nascent clots by tPA-carrying erythrocytes. Nat Biotechnol 2003;21:891−6.

[77] Murciano JC, Higazi AAR, Cines DB, Muzykantov VR. Soluble urokinase receptor conjugated to carrier red blood cells binds latent pro-urokinase and alters its functional profile. J Control Release 2009;139:190−6.

[78] Rihova B. Immunocompatibility and biocompatibility of cell delivery systems. Adv Drug Deliv Rev 2000;42:65−80.

[79] Ziats NP, Miller KM, Anderson JM. *In vitro* and *in vivo* interactions of cells with biomaterials. Biomaterials 1988;9:5−13.

[80] Omer A, Duvivier-Kali V, Fernandes J, Tchipashvili V, Colton CK, Weir GC. Long-term normoglycemia in rats receiving transplants with encapsulated islets. Transplantation 2005;79: 52−8.

[81] Methe H, Nugent HM, Groothuis A, Seifert P, Sayegh MH, Edelman ER. Matrix embedding alters immune response against endothelial cells *in vitro* and *in vivo*. Circulation 2005;112: 189−95.

[82] Methe H, Groothuis A, Sayegh MH, Edelman ER. Matrix adherence of endothelial cells attenuates immune reactivity: induction of hyporesponsiveness in allo- and xenogeneic models. FASEB J 2007;21(7):1515−26.

[83] Brasile L, Stubenitsky B, Haisch C, Kon M, Kootstra G. Repair of damaged organs *in vitro*. Am J Transplant 2005;5:300−6.

[84] Brasile L, Stubenitsky B, Haisch C, Kon M, Kootstra G. Induction of heme oxygenase-1 in kidneys during *ex vivo* warm perfusion. Transplantation 2003;76:1145−9.

[85] Brasile L, Stubenitsky B, Booster M, Haisch C, Kon M, Kootstra G. NOS: the underlying mechanism preserving vascular integrity and during *ex vivo* warm perfusion. Am J Transplant 2003;3: 674−9.

[86] Brasile L, Stubenitsky B, Booster M, Lindell S, Arenada D, Buck C, et al. Overcoming severe renal ischemia: the role of *ex vivo* warm perfusion. Transplantation 2002;73:897−901.

[87] Bickford RG, Winton FR. The influence of temperature on the isolated dog kidney. J Physiol 1937;89:198−219.

[88] Kalluri R. Basement membranes: structure, assembly and role in tumour angiogenesis. Nat Rev Cancer 2003;3:422−33.

[89] Abrams GA, Goodman SL, Nealey PF, Franco M, Murphy CJ. Nanoscale topography of the basement membrane underlying the corneal epithelium of the rhesus macaque. Cell Tissue Res 2000;299:39−46.

[90] Kuntz RM, Saltzman WM. Neutrophil motility in extracellular matrix gels: mesh size and adhesion affect speed of migration. Biophys J 1997;72:1472−80.

[91] Huber AR, Weiss SJ. Disruption of the subendothelial basement membrane during neutrophil diapedesis in an *in vitro* construct of a blood vessel wall. J Clin Invest 1989;83:1122−36.

[92] Meinhart JG, Deutsch M, Fischlein T, Howanietz N, Froschl A, Zilla P. Clinical autologous *in vitro* endothelialization of 153 infrainguinal ePTFE grafts. Ann Thorac Surg 2001;71:S327−31.

[93] Bos GW, Poot AA, Beugeling T, van Aken WG, Feijen J. Small-diameter vascular graft prostheses: current status. Arch Physiol Biochem 1998;106:100−15.

[94] Conte MS. The ideal small arterial substitute: a search for the Holy Grail? FASEB J 1998;12:43−5.

[95] Boccaccini AR, Maquet V. Bioresorbable and bioactive polymer/bioglass composites with tailored pore structure for tissue engineering applications. Compos Sci Technol 2003;63:2417−29.

[96] Wang H, Zhou J, Liu Z, Wang C. Injectable cardiac tissue engineering for the treatment of myocardial infarction. J Cell Mol Med 2010;14:1044−55.

[97] Panitch A, Yamaoka T, Fournier MJ, Mason TL, Tirrell DA. Design and biosynthesis of elastin-like artificial extracellular matrix proteins containing periodically spaced fibronectin CS5 domains. Macromolecules 1999;32:1701−3.

[98] Salmivirta K, Talts J, Olsson M, Sasaki T, Timple R, Ekblom P. Binding of mouse nidogen-2 to basement membrane components and cells and its expression in embryonic and adult tissues suggest complementary functions of the two nidogens. Exp Cell Res 2002;279:188.

[99] Hyashi K, Madri JA, Yurchenco PD. Endothelial cells interact with the core protein of basement membrane perlecan through beta 1 and beta 3 integrins: an adhesion modulated by glycosaminolglycan. J Cell Biol 1992;119:945−59.

[100] Colognato H, Winkelmann DA, Yurchenco PD. Laminin polymerization induces a receptor-cytoskeleton network. J Cell Biol 1999;145:619−31.

[101] Ruoslahti E, Pierschbacher MD. New perspectives in cell-adhesion—RGD and integrins. Science 1987;238:491−7.

[102] Brown B, Lindberg K, Reing J, Stolz DB, Badylak SF. The basement membrane component of biological scaffolds derived from extracellular matrix. Tissue Eng 2006;12:519−26.

[103] Gilbert TW, Sellaro TL, Badylak SF. Decellularization of tissues and organs. Biomaterials 2006;27:3675−83.

[104] Woods T, Gratzer PF. Effectiveness of three extraction techniques in the development of a decellularized bone-anterior cruciate ligament-bone graft. Biomaterials 2005;26:7339−49.

[105] Freytes DO, Stoner RM, Badylak SF. Uniaxial and biaxial properties of terminally sterilized porcine urinary bladder matrix scaffolds. J Biomed Mater Res B Appl Biomater 2008;84:205−17.

[106] Freytes DO, Tullius RS, Valentin JE, Stewart-Akers AM, Badylak SF. Hydrated versus lyophilized forms of porcine extracellular matrix derived from the urinary bladder. J. Biomed Mater Res A 2008;87:862−72.

[107] Brasile L, Glowacki P, Castracane J, Stubenitsky BM. Pretransplant kidney-specific treatment to eliminate the need for systemic immunosuppression. Transplantation 2010;90:1294−8.

Chapter 67

The Need for Immune Modulation Despite Regenerative Medicine

Damelys Calderon[a,b], Michel Pucéat[b,c], Sylvaine You[a,b], Philippe Menasché[b,c], and Lucienne Chatenoud[a,b]

[a]Institut National de la Santé et de la Recherche Médicale, Paris, France, [b]Université Paris Descartes, Sorbonne Paris Cité, Faculté de Médecine, Paris, France, [c]Institut National de la Santé et de la Recherche Médicale, UMR 633, Hôpital Européen Georges Pompidou, Paris, France

67.1 INTRODUCTION

As remarkably illustrated by the extremely comprehensive list of chapters in this book, the clinical fields where clinical regenerative medicine can find its place in transplantation are numerous and diverse. The basic requirements that apply to most of them are the presence, to replace the damaged tissue, of a scaffold (of biological or artificial origin) that will support the survival, growth, and homeostasis of cellular components, which vary depending on the situation [1,2]. These cellular components, in terms of histocompatibility with the host, may be allogeneic (as are for instance the derivatives of embryonic stem cells (ESC)) [3], autologous (as in the case of hematopoietic or mesenchymal cells) [4], or autologous yet "transformed" cells as "induced pluripotent stem cells" or iPS cells that originate from adult differentiated cells which through genetic engineering are reprogrammed toward embryonic pluripotency to then be derived into various organ-specific cell progenitors [5–7].

Challenging the initial prevailing dogma, compelling evidence is being accumulated to show that both ESC and iPS are susceptible to rejection by the host immune system [8–12]. In the case of ESCs among the underlying molecular mechanisms triggering this immune response are the expression of conventional major histocompatibility antigens (MHC), especially class I MHC alloantigens, as well as of minor alloantigens [8]. In the case of iPS cells the situation is also complicated. On the one hand one cannot exclude the development upon implantation of an autoimmune reaction and, on the other hand, some of the products of the genes that induce the pluripotency, in particular *OCT4*, appear to be potential targets for natural immunity [13].

Taking a step back to better assess the overall results we believe that presently the more sensible conclusion is to admit that depending on the cell progenitor population used and the existing mismatch with the recipient an immune response of variable intensity will be inevitably elicited which must be taken into consideration to achieve any efficient clinical translation of the strategy.

We know very well, from experience in transplantation and autoimmunity, that immense progress has been made in the field of immunosuppressive therapy over the past 30 years. A large panel of biological and chemical agents has been introduced in the clinic targeting either selective lymphocyte subsets involved in the rejection process or intracellular signaling pathways that are essential for the functional capacity of lymphocytes.

A major problem remains, however, in that the vast majority of these agents depress immunity globally and are therefore devoid of any specificity for the antigen(s) triggering the pathogenic reaction, i.e., autoantigens or alloantigens. This is why these immunosuppressive strategies present the important pitfalls of their incomplete effectiveness (i.e., in transplantation it is well accepted that current conventional treatments are clearly effective for the prevention of acute allograft rejection but are much less effective to curb chronic rejection) and, second, that their effectiveness involves chronic administration which unfortunately leads in too many cases to a state of overimmunosuppression, characterized by an increased frequency of infections and tumors, that are often a consequence of uncontrolled viral infections.

The only effective way to overcome this problem would be to induce a state of "operational immune tolerance" that is to say, to reprogram the immune system aiming at controlling pathogenic immune responses versus alloantigens, autoantigens, and virtually any neoantigen expressed by the "manipulated" tissue without affecting the ability of the recipient to react effectively against various exogenous antigens, including those expressed by infectious agents.

Our goal here is to present a set of arguments suggesting that the ability to induce immune tolerance is no longer a myth or a possibility exclusively confined to the field of experimental medicine but is becoming a reality in the clinic through new immune intervention strategies. These strategies take advantage of concepts coming from basic immunology and involve, in particular, our improved knowledge of the immune mechanisms underlying the physiological "self-tolerance," which operate in all normal individuals to control unwanted autoimmune responses. For the sake of clarity of presentation, we will therefore first briefly provide some important definitions and historical notes to then describe what are the immune mechanisms involved in the maintenance of self-tolerance and then, finally, discuss how some of these mechanisms can be exploited to induce tolerance to alloantigens and autoantigens, thus representing an interesting avenue in regenerative medicine.

67.2 DEFINITIONS AND HISTORICAL NOTES

The definition of immune tolerance is not univocal and may vary in its formulation depending on the experts one interrogates.

For the fundamental immunologist, the concept of immune tolerance gathers all the mechanisms allowing that, under physiological conditions, the immune system does not develop aggressive or pathogenic reactions against the host tissues that harbor it, although it is now well established that the B- and T-lymphocyte repertoire is autoreactive due to the processes that underlie their generation in the bone marrow and the thymus, respectively. It is this state of "peaceful" coexistence of an autoreactive immune system and its host which is termed immune tolerance or physiological self-tolerance.

For the specialist in transplantation or autoimmunity, tolerance is defined as the absence of a destructive reaction against the target tissue while specific immune responses to foreign antigens or tumors are preserved. This definition, entirely valid in the experimental context, must be applied with caution to the clinical situation. Indeed, because of the difficulty, not to say the present impossibility to directly test for the state of specific tolerance to alloantigens or autoantigens (even though major efforts are made to identify markers of tolerance [14]), the habit is more to refer to "operational tolerance" that is a situation where long-term graft survival is observed in the absence of chronic immunosuppression. Sir R. Calne coined the term "prope" or almost tolerance to include under this umbrella the situations where immunosuppression is minimal yet present [15–17]. It seems obvious that if a particular therapeutic strategy allowed us to induce long-term survival of a transplant or remission of an established autoimmune disease in the total absence of immunosuppressive treatment, the patient would be immunologically "tolerant" although for practical reasons (lack of validated tests to directly measure the immune mechanisms underlying the tolerance) it may turn out impossible to "diagnose" the state of tolerance.

At this point it is crucial to recall that immune tolerance is not genetically determined. It is not an innate status, it is acquired during embryonic development for self-tolerance and in the postnatal period for foreign antigens such as transplantation alloantigens, following their introduction under specific conditions (be it early in the postnatal period or in utero or after appropriate manipulations in the adult individual). The seminal experiences of "neonatal tolerance" conducted in the 1950s by Billingham, Brent, and Medawar, who earned the Nobel Prize, provided the first clear-cut demonstration of the acquired nature of immune tolerance [18]. The model consisted of the administration to newborn mice of allogeneic bone marrow and cells in the absence of any other immunosuppressive treatment. Such a manipulation established a state of tolerance, assessed by the fact that when the mice became adults, skin allografts histocompatible with the cells injected at birth survived indefinitely while "third-party" grafts (histoincompatible with both the recipient and the cells injected at birth) were normally rejected, thereby confirming that the mice treated at birth were not immunocompromised [18].

Obviously, on the basis of these results, major efforts were devoted to examine if was it possible, and if yes under which conditions, to reproduce such phenomenon in an adult host? Today, with the hindsight of several decades, we can say that in the experimental field, the answer to this question is certainly positive. In animals, not only in rodents but also in large mammals (nonhuman primates), different therapeutic strategies have resulted in operational allograft tolerance. Interestingly, it was found that the most effective modalities to achieve transplantation tolerance consisted in the possibility of using some immunological mechanisms underlying physiological immune tolerance that can "reprogram" the immune system so that alloantigens are recognized but do not induce an "aggressive" immune response [19–37].

67.3 THE IMMUNE MECHANISMS THAT SUSTAIN SELF-TOLERANCE: CENTRAL AND PERIPHERAL TOLERANCE

The original prevailing dogma stated that self-tolerance exclusively relied on the complete elimination of all autoreactive lymphocytes clones during their differentiation in the thymus, for T-cells, or the bone marrow, for B-cells. This process was termed "negative selection" or central tolerance (since the thymus and bone marrow are central lymphoid organs versus the spleen, lymph nodes, tonsils, and intestinal lymphoid tissue that are the peripheral lymphoid organs). This negative selection involves apoptotic cell death, that is to say, a programmed cell death secondary to intracellular signal transduction leading to the activation of specialized enzymes called caspases, which will fragment the DNA. Challenging this concept, robust data have been accumulated over the past 40 years to show that the negative selection filter is, in fact, far from complete and only eliminates lymphocytes bearing receptors recognizing self-antigens with high affinity explaining, as we mentioned above, that in every normal individual autoreactive T- and B-lymphocytes emerge at the periphery. The fact that autoimmune diseases do not normally develop despite the presence of these autoreactive T- and B-cells has for long been a central paradox of immunology now explained by various mechanisms, grouped under the term of peripheral tolerance. Among these peripheral tolerance mechanisms are lymphocyte ignorance or indifference, lymphocyte anergy, peripheral deletion, and immune regulation also defined active tolerance mediated by regulatory T-cells.

We shall concentrate in more detail on immune regulation because of the many data showing the central role of this mechanism of peripheral tolerance in the situations where immune tolerance is induced toward autoantigens or alloantigens both in the experimental setting and in the clinic. Immune regulation involves specialized subsets of T-lymphocytes initially called suppressor T-cells and more recently renamed as regulatory T-cells (Tregs) [38–42]. Regulatory T-cells have been shown to control most immune responses including autoimmune responses as well as immune responses to alloantigens, tumor, and infectious agents. Two broad categories of Tregs, natural Treg and induced or adaptive Treg were identified [43,44].

Natural Treg cells of thymic origin constitute a separate lineage of $CD4^+$ thymocytes, characterized by the expression of CD25 (the α chain of human interleukin (IL)-2) and of the transcription factor FoxP3, a member of the family winged-helix/forkhead transcription factors [45,46]. It is the work of the S. Sakaguchi group who showed back in 1995 that if, in normal mice, the exportation from the thymus to the periphery of $CD4^+CD25^+$ Tregs was prevented, following thymectomy performed on day 3 after birth, a polyautoimmune syndrome was generated including gastritis, thyroiditis, orchitis, or oophoritis (depending on the sex of the mouse) [38,39]. The identification of the *FoxP3* gene occurred independently from the study of Tregs, through the identification of gene mutations responsible in humans for a rare X-linked disease, described in 1982, and called immune dysregulation, polyendocrinopathy, enteropathy, X linked, or IPEX. This syndrome is a collection of autoimmune manifestations and severe inflammatory bowel disease (enteropathy). It was only years after its description that it became clear that the IPEX syndrome resulted from single mutations associated with loss of function of the *FoxP3* gene, located on the Y-chromosome, that are in their majority within the area forkhead which binds to DNA. In 2003, it was clearly established that the expression of FoxP3 was a marker for the natural $CD4^+CD25^+$ Treg lineage, associated with their suppressive function [45,46]. Indeed, if one transduced $CD4^+CD25^-$ conventional nonregulatory lymphocytes with a retrovirus encoding FoxP3 they acquire the regulatory capacity and express different surface molecules typical of Tregs namely, CD25, CTLA4, and GITR.

Induced or adaptive Tregs are derived from mature $CD4^+$ T-cells at the periphery and acquire their regulatory function when activated by various antigens in appropriate conditions, particularly in an environment that includes key cytokines such as IL10 or TGF-β (for "transforming growth factor"-β) [43,44,47]. It is now well recognized that there are subsets of adaptive Tregs that do not express FoxP3 such as IL10-producing Tr1 cells (for T regulatory 1), initially described in the bone marrow transplant setting [48], and TGF-β-producing Th3 cells described in oral tolerance models [49].

67.4 STRATEGIES TO PROMOTE TOLERANCE APPLICABLE TO REGENERATIVE MEDICINE

One strategy widely studied to induce allograft tolerance has been the use of polyclonal antibodies that upon short-term administration favored, in mice, the induction of immune tolerance to histoincompatible skin allografts [19,20]. Data also confirm this capacity in autoimmunity [50]. The use of monoclonal antibodies from the early 1980s has confirmed that adequate treatment of the recipient by antibodies or fusion proteins targeting functionally important lymphocyte surface receptors (CD3, CD4, CD8, and CD4) or co-stimulatory molecules (CD28, B7 family, LFA-1, ICAM-1) could reproduce satisfactorily the observations initially made with polyclonal anti-T-cell antisera [36,51−59]. In rodents it has been well demonstrated that immune tolerance can be induced by this type of strategy in adult thymectomized hosts proving that mechanisms of peripheral tolerance sustain the effect. More recent data emphasize the important role of Treg cells, which are found preferentially in tolerated grafts or in the draining lymph nodes [60]. These Tregs are essential mediators of immune tolerance as well demonstrated by adoptive transfer experiments in which following their purification Tregs from tolerant animals transfer the protection into syngeneic naive recipients [36,61].

We can therefore say that, contrary to conventional immunosuppressive acting either by eliminating immune cells or by inhibiting their functional abilities that biological products (monoclonal antibodies, fusion proteins) have a much broader spectrum of pharmacological and biological activities. Thus, according to their fine specificity, these agents will not only eliminate their targets or inhibit their function but also act as agonists on lymphocyte subpopulations transducing activation signals or effectively neutralizing the action of key cytokines or chemokines. It is interesting to add that massive lymphocyte depletion is not a fundamental prerequisite for the induction of tolerance. Polyclonal anti-lymphocyte sera are potent depleting agents which is not the case for many monoclonal anti-T-cell antibodies expressing tolerogenic properties.

67.4.1 The Special Case of CD3 Monoclonal Antibodies

It is important to focus at this point on a class of biological agents that are CD3 monoclonal antibodies for which presently available data are quite encouraging in terms of their potentiality to induce operational immune tolerance in the clinic.

The history of CD3 antibody is paradoxical well illustrating how empirical the development of important drugs may be. OKT3, a mouse IgG2a, was the first monoclonal introduced in clinical transplantation in the early 1980s, even before the complex structure of the CD3 molecule and its functional significance were elucidated [62−64]. Indeed, the target of OKT3 turned out to be one of the chains (ε) of the CD3 molecular complex that is nothing less than the signal transducing element of the T-cell receptor for antigen recognition [65,66]. Over the years a series of controlled studies have clearly demonstrated that OKT3 was an extremely powerful immunosuppressant, highly effective for the treatment of acute renal allograft rejection, an indication for which it was quickly marketed in the United States and Europe. Monitoring of patients treated with OKT3 provided an amount of data on the mode of action and side effects of murine monoclonal antibodies. Over the last 10 years, the use of OKT3 was gradually abandoned because of side effects resulting from the mitogenic capacity of the antibody [67−70].

Another interesting historical note is that CD3 antibodies directed to rodent T-cells were developed well after OKT3 driving then the development of a number of experimental models to get insight into the therapeutic potentialities of CD3 antibodies [55,71,72]. In particular, it rapidly became apparent that in addition to their quite potent immunosuppressive capacity, CD3 antibodies could also induce immune tolerance to organ allografts and also in autoimmunity [54,55,73]. As an example, we may cite the work demonstrating that a nonmitogenic CD3 antibody when administered short-term-induced permanent survival of fully mismatched vascularized heart grafts [54,55]. In these tolerant animals, third-party skin allografts were rejected as compared to donor-matched skin grafts that survived indefinitely. In autoimmune diabetes, data showed that a single injection of a CD3 antibody in NOD mice (that develop spontaneous autoimmune type 1 diabetes) induces immune tolerance to β-cell antigens. Thus, the treatment prevented disease development, an effect associated with the inhibition of the development of the infiltration of the islets of Langerhans by mononuclear cells (e.g., insulitis) [73].

A step further was reached when we reported that a short five-day low-dose treatment with CD3 antibodies when applied to overtly diabetic (e.g., hyperglycemic) NOD mice induced long-standing remission of the disease by restoring self-tolerance [74−77]. Well demonstrating that immune tolerance to β-cell antigens had been restored NOD, in mice showing remission following CD3 antibody treatment, syngeneic islet grafts survived indefinitely as compared to untreated diabetic NOD where such islet grafts were rapidly destroyed by recurrence of autoimmunity. Importantly, skin allografts were normally rejected in tolerant NOD mice confirming the absence of chronic immunosuppression in these animals [75].

Identical data were reported in experimental allergic encephalomyelitis (EAE), an autoimmune disease that models human multiple sclerosis, which is induced following the administration of proteolipid protein or myelin oligodendrocyte glycoprotein with adjuvant [78]. Treatment was effective at stopping ongoing disease while no preventive effect was observed [78,79]. Work from Ochi et al. showed that oral administration of CD3 antibody could both prevent and treat EAE [80].

Data have also been reported in a model of inflammatory bowel disease [81] where severe colitis develops in IL2-deficient ($IL2^{-/-}$) mice upon administration of TNP-KLH (2,4,6-trinitrophenol-conjugated keyhole limpet hemocyanin). A single administration of CD3 antibody inhibited disease, an effect that was linked to blockade of T helper (Th)1 IFNγ-producing cells and appearance of TGFβ secretion by lamina propria T-cells.

Based on these data that concordantly showed that CD3 antibodies promote immune tolerance exclusively when they are administered in the context of a primed immune system (e.g., established/ongoing autoimmunity), we extended the observation to transplantation [82]. Using an islet transplant model (islets from BALB/c mice grafted under the kidney capsule of C57BL/6 recipients rendered diabetic following administration of streptozotocin), we showed that the short (5 days) CD3 antibody treatment applied at the time of transplantation (e.g., while no activation of alloreactive cells is present) significantly delayed rejection (42 days in treated mice versus 18 days in controls) yet all grafts were eventually rejected. Thus, only immunosuppression had been induced. In contrast, if the treatment was delayed once activation of alloreactive cells had occurred, day 7 posttransplant in this particular model, about 85% of the recipients accepted their grafts indefinitely. We proved that antigen-specific tolerance had been induced by performing, in mice with long-term surviving grafts, second allografts from the original donor (BALB/c) that were accepted indefinitely or from third-party donors (C3H) that were normally rejected. Here again CD4 + FoxP3 + T-cells from tolerant animals transferred the protection to naive recipients [82]. In collaboration with the group of KJ Wood, we recently confirmed that the same type of effect is obtained with vascularized grafts namely, heart allografts implanted in a completely mismatched situation [83].

Our understanding of the mechanisms underlying this interesting therapeutic effect has tremendously progressed over these last years. CD3 antibody induces partial T-cell depletion that affects about 50% of host's CD3/TCR$^+$ cells when Fc receptor-binding mitogenic CD3 antibodies, such as OKT3 in humans or the 145 2C11 antibody in mice, are used and 20−30% with nonmitogenic antibodies (e.g., F(ab)'2 fragments of 145 2C11 in mice) [76,77,84]. This CD3 antibody-induced depletion is not consequent to complement fixation or antibody-mediated cell-mediated cytotoxicity but rather to redirected T-cell lysis whereby cytotoxic T-cells "bridging" to other T-cells destroy them [85] by apoptosis [86]. Recent data showed that antigen-activated T-cells are particularly sensitive to the CD3 antibody-mediated effect while Tregs are more resistant [82,87,88].

Remnant T-cells undergo antigenic modulation of CD3/TCR, upon antibody binding followed by internalization or shedding [89,90]. In NOD mice presenting with recent onset diabetes, within the first days of CD3 antibody treatment, near-complete clearing of insulitis is observed. Treated mice show a transient Th2 polarization yet IL4-deficient ($IL4^{-/-}$) NOD mice are sensitive to the CD3 antibody treatment [74]. Pathogenic cells disappear during treatment but recover in tolerant hosts albeit in a variable fashion depending on their fine specificity [91]. By 2−6 weeks following the end of treatment, mononuclear cell infiltrates are observed in tolerant NOD mice at the periphery of the islets (e.g., peripheral insulitis). In parallel, clear-cut signs of an "active" transferable tolerance develop involving TGFβ-dependent Tregs [74,91]. The central role of TGFβ in the therapeutic effect is shown by an experiment where neutralizing antibodies to TGFβ completely abrogate CD3 antibody-induced tolerance [74,91].

In the islet transplant model, reduced anti-donor reactivity is found in the spleen of tolerant recipients, as assessed by IFNγ ELISPOT, and alloantigen-specific CD8$^+$ T-cells in tolerant mice cannot mediate a secondary response [82]. In addition, Tregs are increased in the long-term surviving grafts, a finding also observed in the cardiac allograft model.

We recently developed a new preclinical model which is better adapted to preclinical studies to test the therapeutic activity of antihuman CD3 antibodies. In fact, antihuman CD3 antibodies are species specific and only cross-react with chimpanzee T-cells. We established NOD mice expressing the human CD3ε chain as a transgene. The T-cells of these mice are sensitive *in vitro* and *in vivo* to antihuman CD3 antibodies [91]. Thus, this model is an interesting tool to get further insights into the mode of action of antihuman CD3 antibodies and to protocols based on combination strategies.

67.4.2 Clinical Applications of CD3 Monoclonal Antibodies

Clinical studies using CD3 antibodies in recent onset of insulin-dependent type 1 diabetes were started in 2000, based on the experimental results presented above.

Two humanized Fc-mutated CD3 antibodies were used that are, ChAglyCD3 (Otelixizumab), derived from the rat YTH 12.5 antibody [92] and OKT3γ1 Ala-Ala (Teplizumab) [93]. Phase I safety trials were carried out in renal allograft recipients presenting acute rejection episodes [94,95].

A phase I open trial used Teplizumab for 14 days included a total of 48 patients (24 antibody-treated and 24 untreated) [96,97]. Results suggested that in antibody-treated patients, progression of disease had been significantly reduced as assessed by C-peptide production following mixed meal test stimulation and exogenous insulin requirements. In parallel, a European phase II randomized double-blind controlled trial including 80 patients was conducted [98,99]. Otelixizumab significantly preserved the endogenous insulin-secreting capacity as assessed at 6, 12 and 18, and even 48 months by measuring C-peptide after parenteral glucose stimulation [98,99]. A significant decrease in insulin needs was also observed in treated patients as compared to placebo. As a whole, minor acute side effects linked to a still persisting, though limited, cytokine release were observed after the first infusions. These were dose dependent and did not, however, necessitate any pretreatment nor hampered the normal enrollment of patients.

Phase III trials were initiated, and very promising data were obtained from the one named Protégé, which used Teplizumab. This was a randomized placebo-controlled study including 554 patients, aged 8−35, with new onset type 1 diabetes treated with insulin since not more than 3 months. Patients received a 14 day course twice at day 0 and 6 months with three different dose regimens (cumulated dose 5, 6, or 17 mg × 2). Results reported a significant therapeutic effect when analyzing the C-peptide production and insulin needs [100].

67.4.3 The Immunogenicity of Progenitor Cells

As we mentioned in the introduction, contrasting with initial expectations, it rapidly became apparent that ESC, with their derived precursors, and iPS are "visible" to the immune system and therefore susceptible to rejection [8−12]. Major attention has been focused, for obvious reasons, on the expression of MHC molecules. Although mouse and human ESC express low levels of MHC class I and mostly undetectable class II antigens, expression of MHC class I increases when ESC differentiates into embryoid bodies and increases even further in more differentiated tissues. Progenitor cells derived from ESC also generally express low MHC class I levels that may undergo a significant increase of this expression following in vivo implantation within a pro-inflammatory environment where cytokines such as IFNγ may be present. Evidence for the differential epigenetic control of MHC

and antigen processing molecules in ESCs and differentiated ESCs has also been described. MHC class II antigens appear not to be expressed by ESC even when they are exposed to IFNγ [8]. Concerning the derived progenitors, MHC class II expression may depend on the lineage: it appears to be absent for instance at the surface of cardiac progenitors, but especially following in vivo implantation it may appear on differentiating endothelial cells which may "contaminate" the inoculums. This is relevant not only in terms of being a source of expression of "donor" MHC class II but also because endothelial cells, especially within an inflammatory environment, become efficient APCs. Therefore, even if in principle there are no donor APCs in ESC-derived preparations, unless they are derived on purpose [101], the differentiation following implantation of unwanted APC-like cell types is a possibility that must be considered and looked for.

Aside to MHC molecules, expression of minor transplantation antigens may lead to inevitable rejection of the ESC-derived progenitors. All polymorphic proteins that differ between the recipient and ESC are potential minor antigens, the importance of which in promoting rejection must be considered very seriously. From the molecular point of view, minor antigens are peptides presented by the recipient antigen presenting cell (e.g., indirect allorecognition). The best studied and characterized minor transplantation antigens are encoded by Y-chromosome genes.

All this well explains the data showing that in xenogeneic situations (human to rat) as well as in meaningful and stringent preclinical allogeneic models (Rhesus monkey progenitors into Rhesus recipients), it has been impossible to obtain the survival of ESC-derived cardiac cell progenitors in the absence of immunosuppression [3].

However, very relevant to our present discussion is the data from the group of P. Fairchild showing that in the case of ESCs the modalities of this rejection appear easier to control as compared to those typical of organ transplants making them, at least in rodents, more sensitive to tolerance induction protocols [102,103]. These authors derived a panel of ESC lines that differed in histocompatibility from the recipient strain (CBA/Ca) at defined loci. They could demonstrate that only a disparity in minor MHC is sufficient to provoke acute rejection of tissues differentiated from ESC. However, transplantation tolerance could be effectively established using minimal treatment with a combination of nondepleting monoclonal antibodies directed to CD4 and CD8 [102]. Importantly, these authors also emphasized the major role of TGFβ and FoxP3 + Tregs in this tolerance [102,103]. The very important conclusion from these results is that immunogenicity of ESC is real but that it may be controlled quite easily by means other than immunosuppressive agents namely, by biologically promoting immune tolerance that depends on Tregs. Biological agents aiming at blockade of costimulation have shown a similar effectiveness in significantly prolonging the survival of ESC-derived progenitors [11,103].

67.5 CONCLUSIONS

Despite enormous progress, especially in the characterization of pluripotent cell progenitors and their derivatives for use in different clinical situations, there is still much to do to achieve large-scale clinical use of these fascinating tools. The immunogenicity of these cells is a major problem, and the definition of therapeutic strategies to overcome this pitfall while obviating serious side effects remains a major challenge.

We hope to have provided clear evidence showing the interest and the major benefit that can be expected from the induction of immune tolerance as compared to conventional immunosuppressive therapy to circumvent this immunogenicity. We also hope we conveyed specific information on the large amount of data we have on CD3 antibodies showing that they presently are the first biologicals for which proof of concept of the capacity to achieve operational tolerance in the clinic is available. CD3 antibodies are currently in clinical development and would represent ideal candidates to be used in different indications related to regenerative medicine.

ACKNOWLEDGMENTS

The authors are indebted to the funding agencies that support the work in the field they address. This work was supported by grants from the Foundation Leducq (Grant number: 11 CVD 02), the European Union FP7 (TRIAD), the Foundation CENTAURE, the Juvenile Diabetes Research Foundation (JDRF), the Foundation Day Solvay, INSERM.

REFERENCES

[1] Orlando G. Regenerative medicine technology applied to gastroenterology: current status and future perspectives. World J Gastroenterol 2012;18:6874–5.

[2] Orlando G, Baptista P, Birchall M, De Coppi P, Farney A, Guimaraes-Souza NK, et al. Regenerative medicine as applied to solid organ transplantation: current status and future challenges. Transpl Int 2011;24:223–32.

[3] Blin G, Nury D, Stefanovic S, Neri T, Guillevic O, Brinon B, et al. A purified population of multipotent cardiovascular progenitors derived from primate pluripotent stem cells engrafts in postmyocardial infarcted nonhuman primates. J Clin Invest 2010;120:1125–39.

[4] Roemeling-van Rhijn M, Weimar W, Hoogduijn MJ. Mesenchymal stem cells: application for solid-organ transplantation. Curr Opin Organ Transplant 2012;17:55–62.

[5] Takahashi K, Tanabe K, Ohnuki M, Narita M, Ichisaka T, Tomoda K, et al. Induction of pluripotent stem cells from adult human fibroblasts by defined factors. Cell 2007;131:861–72.

[6] Yamanaka S. A fresh look at iPS cells. Cell 2009;137:13–7.

[7] Yamanaka S. Induced pluripotent stem cells: past, present, and future. Cell Stem Cell 2012;10:678–84.

[8] de Rham C, Villard J. How to cross immunogenetic hurdles to human embryonic stem cell transplantation. Semin Immunopathol 2011;33:525–34.

[9] Fairchild PJ. The challenge of immunogenicity in the quest for induced pluripotency. Nat Rev Immunol 2010;10:868–75.

[10] Lui KO, Fairchild PJ, Waldmann H. Prospects for ensuring acceptance of ES cell-derived tissues. (September 30, 2010), StemBook, ed. The Stem Cell Research Community, StemBook, http://dx.doi.org/10.3824/stembook.1.54.1.

[11] Pearl JI, Kean LS, Davis MM, Wu JC. Pluripotent stem cells: immune to the immune system? Sci Transl Med 2012;4:164ps25.

[12] Pearl JI, Lee AS, Leveson-Gower DB, Sun N, Ghosh Z, Lan F, et al. Short-term immunosuppression promotes engraftment of embryonic and induced pluripotent stem cells. Cell Stem Cell 2011;8:309–17.

[13] Dhodapkar KM, Feldman D, Matthews P, Radfar S, Pickering R, Turkula S, et al. Natural immunity to pluripotency antigen OCT4 in humans. Proc Natl Acad Sci USA 2010;107:8718–23.

[14] Newell KA, Asare A, Kirk AD, Gisler TD, Bourcier K, Suthanthiran M, et al. Identification of a B cell signature associated with renal transplant tolerance in humans. J Clin Invest 2010;120:1836–47.

[15] Calne R. "Prope" tolerance: induction, lymphocyte depletion with minimal maintenance. Transplantation 2005;80:6–7.

[16] Calne R, Watson CJ. Some observations on prope tolerance. Curr Opin Organ Transplant 2011;16:353–8.

[17] Calne RY. Prope tolerance—the future of organ transplantation from the laboratory to the clinic. Int Immunopharmacol 2005;5: 163–7.

[18] Billingham RE, Brent L, Medawar PB. Actively acquired tolerance to foreign cells. Nature 1953;172:603–6.

[19] Monaco AP, Wood ML, Russell PS. Studies on heterologous antilymphocyte serum in mice. III. Immunological tolerance and chimerism produced across the H2-locus with adult thymectomy and antilymphocyte serum. Ann NY Acad Sci 1966;129: 190–209.

[20] Wood ML, Monaco AP, Gozzo JJ, Liegeois A. Use of homozygous allogeneic bone marrow for induction of tolerance with antilymphocyte serum: dose and timing. Transplant Proc 1971;3: 676–9.

[21] Thomas J, Alqaisi M, Cunningham P, Carver M, Rebellato L, Gross U, et al. The development of a posttransplant TLI treatment strategy that promotes organ allograft acceptance without chronic immunosuppression. Transplantation 1992;53:247–58.

[22] Thomas J, Carver M, Cunningham P, Park K, Gonder J, Thomas F. Promotion of incompatible allograft acceptance in rhesus monkeys given posttransplant antithymocyte globulin and donor bone marrow. I. *In vivo* parameters and immunohistologic evidence suggesting microchimerism. Transplantation 1987;43:332–8.

[23] Thomas JM, Carver FM, Cunningham PR, Olson LC, Thomas FT. Kidney allograft tolerance in primates without chronic immunosuppression—the role of veto cells. Transplantation 1991;51: 198–207.

[24] Thomas JM, Carver FM, Foil MB, Hall WR, Adams C, Fahrenbruch GB. Renal allograft tolerance induced with ATG and donor bone marrow in outbred rhesus monkeys. Transplantation 1983;36:104−6.

[25] Thomas JM, Neville DM, Contreras JL, Eckhoff DE, Meng G, Lobashevsky AL, et al. Preclinical studies of allograft tolerance in rhesus monkeys: a novel anti-CD3-immunotoxin given peritransplant with donor bone marrow induces operational tolerance to kidney allografts. Transplantation 1997;64:124−35.

[26] Sykes M, Sachs DH. Bone marrow transplantation as a means of inducing tolerance. Semin Immunol 1990;2:401−17.

[27] Sykes M, Sheard M, Sachs DH. Effects of T cell depletion in radiation bone marrow chimeras. I. Evidence for a donor cell population which increases allogeneic chimerism but which lacks the potential to produce GVHD. J Immunol 1988;141:2282−8.

[28] Wekerle T, Kurtz J, Bigenzahn S, Takeuchi Y, Sykes M. Mechanisms of transplant tolerance induction using costimulatory blockade. Curr Opin Immunol 2002;14:592−600.

[29] Wekerle T, Kurtz J, Ito H, Ronquillo JV, Dong V, Zhao G, et al. Allogeneic bone marrow transplantation with co-stimulatory blockade induces macrochimerism and tolerance without cytoreductive host treatment. Nat Med 2000;6:464−9.

[30] Bushell A, Karim M, Kingsley CI, Wood KJ. Pretransplant blood transfusion without additional immunotherapy generates CD25 + CD4 + regulatory T cells: a potential explanation for the blood-transfusion effect. Transplantation 2003;76:449−55.

[31] Bushell A, Niimi M, Morris PJ, Wood KJ. Evidence for immune regulation in the induction of transplantation tolerance: a conditional but limited role for IL-4. J Immunol 1999;162:1359−66.

[32] Darby CR, Morris PJ, Wood KJ. Evidence that long-term cardiac allograft survival induced by anti-CD4 monoclonal antibody does not require depletion of CD4 + T cells. Transplantation 1992;54:483−90.

[33] Cobbold S, Waldmann H. Infectious tolerance. Curr Opin Immunol 1998;10:518−24.

[34] Cobbold SP, Adams E, Graca L, Daley S, Yates S, Paterson A, et al. Immune privilege induced by regulatory T cells in transplantation tolerance. Immunol Rev 2006;213:239−55.

[35] Cobbold SP, Qin S, Leong LY, Martin G, Waldmann H. Reprogramming the immune system for peripheral tolerance with CD4 and CD8 monoclonal antibodies. Immunol Rev 1992;129:165−201.

[36] Qin S, Cobbold SP, Pope H, Elliott J, Kioussis D, Davies J, et al. "Infectious" transplantation tolerance. Science 1993;259:974−7.

[37] Knechtle SJ, Vargo D, Fechner J, Zhai Y, Wang J, Hanaway MJ, et al. FN18-CRM9 immunotoxin promotes tolerance in primate renal allografts. Transplantation 1997;63:1−6.

[38] Sakaguchi S. Naturally arising CD4 + regulatory T cells for immunologic self-tolerance and negative control of immune responses. Annu Rev Immunol 2004;22:531−62.

[39] Sakaguchi S, Fukuma K, Kuribayashi K, Masuda T. Organ-specific autoimmune diseases induced in mice by elimination of T cell subset. I. Evidence for the active participation of T cells in natural self-tolerance; deficit of a T cell subset as a possible cause of autoimmune disease. J Exp Med 1985;161:72−87.

[40] Takahashi T, Kuniyasu Y, Toda M, Sakaguchi N, Itoh M, Iwata M, et al. Immunologic self-tolerance maintained by CD25 + CD4 + naturally anergic and suppressive T cells: induction of autoimmune disease by breaking their anergic/suppressive state. Int Immunol 1998;10:1969−80.

[41] Suri-payer E, Amar AZ, Thornton AM, Shevach EM. CD4 + CD25 + T cells inhibit both the induction and effector function of autoreactive T cells and represent a unique lineage of immunoregulatory cells. J Immunol 1998;160:1212−8.

[42] Thornton AM, Shevach EM. Suppressor effector function of CD4 + CD25 + immunoregulatory T cells is antigen nonspecific. J Immunol 2000;164:183−90.

[43] Bluestone JA, Abbas AK. Natural versus adaptive regulatory T cells. Nat Rev Immunol 2003;3:253−7.

[44] Chen W, Jin W, Hardegen N, Lei KJ, Li L, Marinos N, et al. Conversion of peripheral CD4 + CD25 − naive T cells to CD4 + CD25 + regulatory T cells by TGF-beta induction of transcription factor Foxp3. J Exp Med 2003;198:1875−86.

[45] Fontenot JD, Gavin MA, Rudensky AY. Foxp3 programs the development and function of CD4 + CD25 + regulatory T cells. Nat Immunol 2003;4:330−6.

[46] Hori S, Nomura T, Sakaguchi S. Control of regulatory T cell development by the transcription factor Foxp3. Science 2003;299:1057−61.

[47] Walker MR, Kasprowicz DJ, Gersuk VH, Benard A, Van Landeghen M, Buckner JH, et al. Induction of FoxP3 and acquisition of T regulatory activity by stimulated human CD4 + CD25 − T cells. J Clin Invest 2003;112:1437−43.

[48] Battaglia M, Gregori S, Bacchetta R, Roncarolo MG. Tr1 cells: from discovery to their clinical application. Semin Immunol 2006;18:120−7.

[49] Weiner HL, da Cunha AP, Quintana F, Wu H. Oral tolerance. Immunol Rev 2011;241:241−59.

[50] Like AA, Rossini AA, Guberski DL, Appel MC, Williams RM. Spontaneous diabetes mellitus: reversal and prevention in the BB/W rat with antiserum to rat lymphocytes. Science 1979;206:1421−3.

[51] Pearson TC, Madsen JC, Larsen CP, Morris PJ, Wood KJ. Induction of transplantation tolerance in adults using donor antigen and anti-CD4 monoclonal antibody. Transplantation 1992;54:475−83.

[52] Larsen CP, Elwood ET, Alexander DZ, Ritchie SC, Hendrix R, Tuckerburden C, et al. Long-term acceptance of skin and cardiac allografts after blocking CD40 and CD28 pathways. Nature 1996;381:434−8.

[53] Shizuru JA, Gregory AK, Chao CT, Fathman CG. Islet allograft survival after a single course of treatment of recipient with antibody to L3T4. Science 1987;237:278−80.

[54] Plain KM, Chen J, Merten S, He XY, Hall BM. Induction of specific tolerance to allografts in rats by therapy with non-mitogenic, non-depleting anti-CD3 monoclonal antibody: association with TH2 cytokines not anergy. Transplantation 1999;67:605−13.

[55] Nicolls MR, Aversa GG, Pearce NW, P6spinelli A, Berger MF, Gurley KE, et al. Induction of long-term specific tolerance to allografts in rats by therapy with an anti-CD3-like monoclonal antibody. Transplantation 1993;55:459−68.

[56] Gao W, Demirci G, Strom TB, Li XC. Stimulating PD-1-negative signals concurrent with blocking CD154 co-stimulation induces long-term islet allograft survival. Transplantation 2003;76:994−9.

[57] Nanji SA, Hancock WW, Luo B, Schur CD, Pawlick RL, Zhu LF, et al. Costimulation blockade of both inducible costimulator and CD40 ligand induces dominant tolerance to islet allografts and prevents spontaneous autoimmune diabetes in the NOD mouse. Diabetes 2006;55:27–33.

[58] Truong W, Plester JC, Hancock WW, Merani S, Murphy TL, Murphy KM, et al. Combined coinhibitory and costimulatory modulation with anti-BTLA and CTLA4Ig facilitates tolerance in murine islet allografts. Am J Transplant 2007;7:2663–74.

[59] Zhang QW, Rabant M, Schenk A, Valujskikh A. ICOS-dependent and -independent functions of memory CD4 T cells in allograft rejection. Am J Transplant 2008;8:497–506.

[60] Graca L, Cobbold SP, Waldmann H. Identification of regulatory T cells in tolerated allografts. J Exp Med 2002;195:1641–6.

[61] Kendal AR, Chen Y, Regateiro FS, Ma J, Adams E, Cobbold SP, et al. Sustained suppression by Foxp3 + regulatory T cells is vital for infectious transplantation tolerance. J Exp Med 2011;208: 2043–53.

[62] Cosimi AB, Colvin RB, Burton RC, Rubin RH, Goldstein G, Kung PC, et al. Use of monoclonal antibodies to T-cell subsets for immunologic monitoring and treatment in recipients of renal allografts. N Engl J Med 1981;305:308–14.

[63] Ortho Multicenter Transplant Study Group. A randomized clinical trial of OKT3 monoclonal antibody for acute rejection of cadaveric renal transplants. N Engl J Med 1985;313:337–42.

[64] Vigeral P, Chkoff N, Chatenoud L, Campos H, Lacombe M, Droz D, et al. Prophylactic use of OKT3 monoclonal antibody in cadaver kidney recipients. Utilization of OKT3 as the sole immunosuppressive agent. Transplantation 1986;41:730–3.

[65] Clevers H, Alarcon B, Wileman T, Terhorst C. The T cell receptor/CD3 complex: a dynamic protein ensemble. Annu Rev Immunol 1988;6:629–62.

[66] Davis MM, Chien YH. T cell antigen receptors. In: Paul W, editor. Fundamental immunology. New York, NY: Raven Press; 1999. p. 341–66.

[67] Van Wauwe JP, De Mey JR, Goossens JG. OKT3: a monoclonal anti-human T lymphocyte antibody with potent mitogenic properties. J Immunol 1980;124:2708–13.

[68] Chatenoud L, Ferran C, Legendre C, Thouard I, Merite S, Reuter A, et al. In vivo cell activation following OKT3 administration. Systemic cytokine release and modulation by corticosteroids. Transplantation 1990;49:697–702.

[69] Abramowicz D, Schandene L, Goldman M, Crusiaux A, Vereerstraeten P, De Pauw L, et al. Release of tumor necrosis factor, interleukin-2, and gamma-interferon in serum after injection of OKT3 monoclonal antibody in kidney transplant recipients. Transplantation 1989;47:606–8.

[70] Herbelin A, Chatenoud L, Roux-lombard P, De Groote D, Legendre C, Dayer JM, et al. In vivo soluble tumor necrosis factor receptor release in OKT3-treated patients. Differential regulation of TNF-sR55 and TNF-sR75. Transplantation 1995;59:1470–5.

[71] Leo O, Foo M, Sachs DH, Samelson LE, Bluestone JA. Identification of a monoclonal antibody specific for a murine T3 polypeptide. Proc Natl Acad Sci USA 1987;84:1374–8.

[72] Tomonari K. A rat antibody against a structure functionally related to the mouse T-cell receptor/T3 complex. Immunogenetics 1988;28:455–8.

[73] Hayward AR, Shreiber M. Neonatal injection of CD3 antibody into nonobese diabetic mice reduces the incidence of insulitis and diabetes. J Immunol 1989;143:1555–9.

[74] Belghith M, Bluestone JA, Barriot S, Megret J, Bach JF, Chatenoud L. TGF-beta-dependent mechanisms mediate restoration of self-tolerance induced by antibodies to CD3 in overt autoimmune diabetes. Nat Med 2003;9:1202–8.

[75] Chatenoud L, Thervet E, Primo J, Bach JF. Anti-CD3 antibody induces long-term remission of overt autoimmunity in nonobese diabetic mice. Proc Natl Acad Sci USA 1994;91:123–7.

[76] Chatenoud L. CD3-specific antibody-induced active tolerance: from bench to bedside. Nat Rev Immunol 2003;3:123–32.

[77] Chatenoud L, Primo J, Bach JF. CD3 antibody-induced dominant self tolerance in overtly diabetic NOD mice. J Immunol 1997;158:2947–54.

[78] Kohm AP, Williams JS, Bickford AL, McMahon JS, Chatenoud L, Bach JF, et al. Treatment with nonmitogenic anti-CD3 monoclonal antibody induces CD4 + T cell unresponsiveness and functional reversal of established experimental autoimmune encephalomyelitis. J Immunol 2005;174:4525–34.

[79] Perruche S, Zhang P, Liu Y, Saas P, Bluestone JA, Chen W. CD3-specific antibody-induced immune tolerance involves transforming growth factor-beta from phagocytes digesting apoptotic T cells. Nat Med 2008;14:528–35.

[80] Ochi H, Abraham M, Ishikawa H, Frenkel D, Yang K, Basso AS, et al. Oral CD3-specific antibody suppresses autoimmune encephalomyelitis by inducing CD4(+)CD25(−)LAP(+) T cells. Nat Med 2006;12:627–35.

[81] Ludviksson BR, Ehrhardt RO, Strober W. TGF-beta production regulates the development of the 2,4,6-trinitrophenol-conjugated keyhole limpet hemocyanin-induced colonic inflammation in IL-2-deficient mice. J Immunol 1997;159:3622–8.

[82] You S, Zuber J, Kuhn C, Baas M, Valette F, Sauvaget V, et al. Induction of allograft tolerance by monoclonal CD3 antibodies: a matter of timing. Am J Transplant 2012;12:2909–19.

[83] Riochy G, Sylvaine Y, Lucienne C, Kathryn W. Am J Transplant 2013; Am J Transplant 2013;13:1655–64.

[84] Hirsch R, Bluestone JA, De Nenno L, Gress RE. Anti-CD3 F(ab′) 2 fragments are immunosuppressive in vivo without evoking either the strong humoral response or morbidity associated with whole mAb. Transplantation 1990;49:1117–23.

[85] Wong JT, Colvin RB. Selective reduction and proliferation of the CD4 + and CD8 + T cell subsets with bispecific monoclonal antibodies: evidence for inter-T cell-mediated cytolysis. Clin Immunol Immunopathol 1991;58:236–50.

[86] Wesselborg S, Janssen O, Kabelitz D. Induction of activation-driven death (apoptosis) in activated but not resting peripheral blood T cells. J Immunol 1993;150:4338–45.

[87] Penaranda C, Tang Q, Bluestone JA. Anti-CD3 therapy promotes tolerance by selectively depleting pathogenic cells while preserving regulatory T cells. J Immunol 2011;187:2015–22.

[88] Chatenoud L. Immune therapy for type 1 diabetes mellitus—what is unique about anti-CD3 antibodies? Nat Rev Endocrinol 2010;6:149–57.

[89] Chatenoud L, Baudrihaye MF, Kreis H, Goldstein G, Schindler J, Bach JF. Human in vivo antigenic modulation induced by the anti-T cell OKT3 monoclonal antibody. Eur J Immunol 1982;12:979–82.

[90] Chatenoud L, Bach JF. Antigenic modulation: a major mechanism of antibody action. Immunol Today 1984;5:20—5.

[91] Kuhn C, You S, Valette F, Hale G, van Endert P, Bach JF, et al. Human CD3 transgenic mice: preclinical testing of antibodies promoting immune tolerance. Sci Transl Med 2011;3: 68ra10.

[92] Bolt S, Routledge E, Lloyd I, Chatenoud L, Pope H, Gorman SD, et al. The generation of a humanized, non-mitogenic CD3 monoclonal antibody which retains *in vitro* immunosuppressive properties. Eur J Immunol 1993;23:403—11.

[93] Alegre ML, Peterson LJ, Xu D, Sattar HA, Jeyarajah DR, Kowalkowski K, et al. A non-activating "humanized" anti-CD3 monoclonal antibody retains immunosuppressive properties *in vivo*. Transplantation 1994;57:1537—43.

[94] Friend PJ, Hale G, Chatenoud L, Rebello P, Bradley J, Thiru S, et al. Phase I study of an engineered aglycosylated humanized CD3 antibody in renal transplant rejection. Transplantation 1999;68:1632—7.

[95] Woodle ES, Xu D, Zivin RA, Auger J, Charette J, O'laughlin R, et al. Phase I trial of a humanized, Fc receptor nonbinding OKT3 antibody, huOKT3gamma1(Ala-Ala) in the treatment of acute renal allograft rejection. Transplantation 1999;68:608—16.

[96] Herold KC, Hagopian W, Auger JA, Poumian Ruiz E, Taylor L, Donaldson D, et al. Anti-CD3 monoclonal antibody in new-onset type 1 diabetes mellitus. N Engl J Med 2002;346:1692—8.

[97] Herold KC, Gitelman SE, Masharani U, Hagopian W, Bisikirska B, Donaldson D, et al. A single course of anti-CD3 monoclonal antibody hOKT3gamma1(Ala-Ala) results in improvement in C-peptide responses and clinical parameters for at least 2 years after onset of type 1 diabetes. Diabetes 2005;54:1763—9.

[98] Keymeulen B, Vandemeulebroucke E, Ziegler AG, Mathieu C, Kaufman L, Hale G, et al. Insulin needs after CD3-antibody therapy in new-onset type 1 diabetes. N Engl J Med 2005;352: 2598—608.

[99] Keymeulen B, Walter M, Mathieu C, Kaufman L, Gorus F, Hilbrands R, et al. Four-year metabolic outcome of a randomised controlled CD3-antibody trial in recent-onset type 1 diabetic patients depends on their age and baseline residual beta cell mass. Diabetologia 2010;53:614—23.

[100] Sherry N, Hagopian W, Ludvigsson J, Jain SM, Wahlen J, Ferry Jr. RJ, et al. Teplizumab for treatment of type 1 diabetes (Protege study): 1-year results from a randomised, placebo-controlled trial. Lancet 2011;378:487—97.

[101] Silk KM, Tseng SY, Nishimoto KP, Lebkowski J, Reddy A, Fairchild PJ. Differentiation of dendritic cells from human embryonic stem cells. Methods Mol Biol 2011;767:449—61.

[102] Robertson NJ, Brook FA, Gardner RL, Cobbold SP, Waldmann H, Fairchild PJ. Embryonic stem cell-derived tissues are immunogenic but their inherent immune privilege promotes the induction of tolerance. Proc Natl Acad Sci USA 2007;104:20920—5.

[103] Lui KO, Boyd AS, Cobbold SP, Waldmann H, Fairchild PJ. A role for regulatory T cells in acceptance of ESC-derived tissues transplanted across an major histocompatibility complex barrier. Stem Cells 2010;28:1905—14.

Chapter 68

Stem Cells Approach to I/R Injury

Takumi Teratani[a], Eiji Kobayashi[a], and Lauren Brasile[b]

[a]Division of Development of Advanced Therapy, Center for Development of Advanced Medical Technology, Jichi Medical University, Japan, [b]Breonics Inc, Albany, New York, USA

Chapter Outline

68.1 CANDIDATE STEM CELLS FOR REGENERATIVE MEDICINE

In recent years, much interest has accompanied the discovery that adult stem cells are capable of contributing to regeneration processes and repair of damaged tissues. The field of stem cell biology is advancing at an incredible pace with new discoveries being reported in the scientific literature on a weekly basis. Although adult hematopoietic stem cells from the bone marrow have long been recognized as being capable of developing into blood and immune cells, recent reports have revealed that adult stem cells from one tissue type appear to be capable of developing into cells having characteristics of other tissues.

Stem cells are endowed with the capacity to self-renew and to differentiate into various cell types, depending on the stimuli that they receive. For ease of discussion, we have classified stem cells as embryonic stem cells and adult stem cells. Embryonic stem cells are believed to hold great promise for clinical applications for the replacement of diseased or degenerating cell populations, tissues, and organs. However, embryonic stem cells are not the only stem cell candidates for generation of differentiated cell types [1]. Other pluripotent cell types include stem cells derived from the primordial germ cells of the gonadal ridge and possibly cancer stem cells, which recently have been identified in leukemia and several solid tumors [2,3].

In a breakthrough study, the Yamanaka group screened a combination of 24 candidate genes and surprisingly found that viral transudation of four previously known transcription factors can convert mouse embryonic fibroblasts and human fibroblasts into embryonic stem-like cells, also called "induced pluripotent stem (iPS) cells" [4,5]. In addition to circumventing ethical problems associated with human embryonic stem cells, this discovery could also address immunologic rejection associated with cell therapies because iPS cells lines could be created with human leukocyte antigen (HLA) haplotypes matching those of individual patients [6]. However, similar to embryonic stem cells, iPS cells form teratomas, leading to a problem that remains to be solved.

In the past decade, researchers have also defined committed stem cells or progenitor cells from various tissues in both adult animals and humans. These so-called adult stem cells can give rise to cells of a particular tissue and include, for example, bone marrow cells, one of the earliest clinically used and most widely studied stem cell populations. In 1957, based on more than 10 years of experimental research, the Thomas group reported the findings of six patients diagnosed with aplastic anemia or hematologic neoplasia, treated with irradiation and intravenous infusion of bone marrow—derived stem cell suspensions obtained from healthy donors [7,8]. The concept that bone marrow may contain some stem cells has been postulated by several investigators, and as previously mentioned, the best evidence that bone marrow—derived stem cells are in fact heterogeneous was provided by experiments showing that bone marrow—derived cells could support regeneration of various tissues [9].

There are two main groups of stem cells in the bone marrow, hematopoietic stem cells and mesenchymal stem cells (MSCs). Adult bone marrow—derived hematopoietic stem cells have long been recognized as giving rise to all blood cell lineages, including erythrocytes, platelets, and white blood cells. Under normal physiologic conditions, hematopoietic homeostasis is maintained by a delicate balance between processes such as self-renewal, proliferation, and differentiation and apoptosis or cell-cycle arrest in hematopoietic stem cells. Furthermore, the hematopoietic stem cell population is typically defined by surface expression of CD34 and represents heterogeneous cells, multipotent progenitor cells, and uncommitted differentiating cells. Hematopoietic stem cells have the potential to proliferate indefinitely and can differentiate into mature hematopoietic linage-specific cells [10,11]. Bone marrow—derived stem cells are currently considered a promising source for cell therapies targeting several diseases. Donor bone marrow—derived stem cells transplanted into animal models of various diseases have been shown to differentiate into different cell types, including bone, cartilage, cardiac muscle, vascular endothelial, neuronal, kidney, and liver cells [12—18]. In addition, differentiation of hematopoietic stem cells into parenchymal cells appears to be stimulated by injury, suggesting that these adult stem cells may also contribute to tissue repair.

On the other hand, MSCs have unique phenotypic markers. The minimal criteria established by the International Society of Cellular Therapy to define MSCs are as follows: plastic-adherent in culture; expression of CD105, CD73, CD29, and CD90; lack of expression of hematopoietic markers such as CD45, CD34, CD14, CD11b, CD19, CD79a, and HLA-DR; and ability to differentiate into osteoblasts, adipocytes, and chondrocytes [19]. When cells from bone marrow aspirate are cultured in plastic flasks, hematopoietic cells and hematopoietic stem cells do not adhere to the plastic and are removed with the change of media. The ratio of MSCs to bone marrow mononuclear cells is estimated to be only 10 MSCs per million bone marrow cells. Despite relatively low numbers, a 2 mL aspirate of bone marrow can be expanded 500-fold *ex vivo* to 12—35 billion MSCs within 3 weeks [20]. As the question of what MSCs do physiologically cannot currently be answered, the pragmatic question of what we can make them do is far more approachable. There has been a great deal of work, some of which is quite promising, suggesting that MSCs can be therapeutically useful in specific settings.

68.2 FUNCTIONS OF MSCs

MSCs have two basic properties: one is the secretion of various kinds of cytokines that promote organ regeneration and the other is immunomodulation [21—27]. Clinical applications of MSCs in solid organ transplantation take advantage of the immunologic properties of these cells [28—30]. MSCs possess an arsenal of immunosuppressive mechanisms, which can be deployed to modulate inflammation. Two very interesting paradigms have recently been proposed, suggesting that (i) MSCs have sentinel functions that allow them to sense their microenvironment and act accordingly and (ii) MSCs become polarized toward either a pro-inflammatory phenotype or an immunosuppressive phenotype depending on the toll-like receptor signals received [31,32]. Together, these concepts help in resolving some of the conflicting data stating that in some cases, MSCs enhance immune cell survival and function, and in others, they inhibit inflammation and encourage repair. The potential of MSCs to regenerate damaged tissue has also been attributed to the presence of chemokine receptors on the surface, which enables these cells to migrate toward gradients of growth factors secreted by damaged tissues [33] or tumors [34]. This migration has recently been exploited to mark breast cancer for radionuclide treatment using sodium iodide symporter-transduced MSCs as the targeting vehicle for tissue destruction [35]. Despite various mechanisms enabling MSC homing, the majority of the implanted cells are trapped in draining organs (liver, lung, and spleen) and only a very small fraction migrate to sites of tissue damage, with rates of engraftment depending on the method of administration and disease models [36—38]. Indeed, the inherent variability in the biodistribution of MSCs postimplantation has been described as a potential hindrance in the translation of MSC-based therapies to clinical practice. Thus, several cell-targeting efforts have been examined to improve the delivery and retention of MSCs at the desired tissue/site of injury postinjection. Based on our experimental data, we review these with a focus on novel strategies that include engineered cell-homing mechanisms and incorporation of nanomaterials.

68.3 ISCHEMIA/REPERFUSION INJURY FRAMEWORK

Ischemia/reperfusion (I/R) injury is a phenomenon in which cellular damage in a hypoxic organ is accentuated following restoration of oxygen delivery [39—41]. Increased production of reactive oxygen species, necrosis, vascular injury, and increase in mucosal permeability are some of the prominent features of I/R injury [42—52].

The study of I/R injury and its components is essential considering that this type of injury plays a major role in the outcome of several clinical conditions, such as trauma, hemorrhagic shock, organ transplantation, and revascularization processes, as well as autoimmune diseases manifesting as various forms of vasculitis and vasculopathies [53]. An in-depth investigation of the molecular and cellular events leading to I/R injury is needed to explore more efficient treatments to protect organs from injury.

Oxygen deprivation leads to mitochondrial dysfunction, formation of reactive oxygen species, and eventually cell death. Complement activation, neutrophil recruitment, T cell activation, and platelet—leukocyte interactions are some of the crucial players in I/R injury [53]. Increasing data derived from ongoing studies show that I/R injury is a very complex process that literally recruits all components of the immune response. Clever therapeutic strategies are needed to address various aspects of the I/R associated pathology.

68.4 PREVENTION OF SOLID ORGAN I/R INJURY WITH MSCs

68.4.1 Kidney

Many experimental studies have reported the therapeutic injection of MSCs to the ischemic kidney [54—56]. However, injection of MSCs has a risk of embolism at the site of ischemia site. In a recent study by Gao et al. [57], thermosensitive chitosan chloride hydrogel was explored as an injectable adipose tissue-derived MSC (AT-MSC) delivery scaffold for I/R-induced acute kidney injury. Thermosensitive chitosan chloride hydrogels with/without AT-MSCs were injected into the I/R site of a rat acute kidney injury model. Dihydroethidium staining was used to detect the number of reactive oxygen species *in vivo*. The AT-MSCs were transfected with firefly luciferase and monomeric red fluorescent protein reporter genes for assessment of their retention and survival using bioluminescence imaging. Differentiation behaviors of AT-MSCs were investigated using immunofluorescent and immunohistochemical staining. Proliferation and apoptosis *in vivo* of host renal cells were characterized by proliferating cell nuclear antigen and terminal deoxynucleotidyl transferase dUTP nick end labeling (TUNEL) staining. Results suggested that chitosan chloride hydrogels could improve the retention and survival of grafted AT-MSCs. Moreover, chitosan chloride hydrogels could enhance the proliferative activity and reduce apoptosis of host renal cells. At 4 weeks, significant improvement in renal function, microvessel density, and tubular cell proliferation were observed in chitosan chloride hydrogels with AT-MSCs. Therefore, application of thermosensitive

chitosan chloride hydrogel as a scaffold for AT-MSC delivery into renal areas could resolve the main obstacle of cell transplantation for acute kidney injury. As such, chitosan chloride hydrogel may be a potential cell carrier for treatment of acute kidney injury.

68.4.2 Intestine

Graft viability prior to implantation is a key factor influencing outcomes after organ transplantation. Along with brain death of the donor, surgical manipulation, and I/R injury, preservation damage is one of the many factors that affect the quality of transplanted organs. Especially, small intestine grafts associated barrier functions [58]. Control of bacterial translocation through the small intestine mucosa is the most important aspect of patient management after small intestine transplantation [59]. Small intestine transplantation using segmental grafts from living related donors has recently been proposed to be advantageous [60—63]. Graft length affected on immunological and nonimmunological reaction [64]. The ischemic storage time can also be greatly reduced with these grafts [62,63]. On the other hand, small intestinal free grafts have often been used for reconstruction after pharyngolaryngectomy or cervical eshophagectomy [65,66]. In such procedures, the small intestinal free flap does not need cold preservation solution and is usually preserved using moistened gauze during warm ischemia time of less than 6 h.

The ability to successfully preserve graft viability during the short period of ischemic storage in intestinal graft is still uncertain. Preventing hypothermia-induced cellular swelling is fundamental to successful organ preservation [67,68]. However, no solutions have repeatedly proven to be superior for small intestine preservation [69—71]. Multiple studies have documented that a variety of solutions, including University of Wisconsin (UW) solution, cannot prevent clinically unacceptable degrees of morphologic injury beyond 6—10 h of cold storage [72—75]. Unfortunately, extension of small intestine graft preservation time is needed to protect the mucosal layer because progressive subepithelial edema appears within minutes of the onset of ischemia and eventually may lead to mucosal breakdown [76].

In an MSC study by Jiang et al. [77], the therapeutic potential of bone marrow—MSC therapy in I/R-injured rats was recently evaluated with regard to intestinal mechanical barrier function. Their results suggested that MSCs are effective in reducing both the intestinal permeability and the pathologic damage associated with I/R. It has been much focused on whether MSCs acceleration leads to recovery of the structure and barrier function of the small intestine after I/R injury.

68.4.3 Liver

Liver transplantation is one of the most efficient treatments available for various end-stage hepatic diseases. However, one of the major limitations of liver transplantation is the scarcity of donor organs. To overcome this limitation, split liver transplantation or living donor transplantation shows the most promising outcomes. However, when these treatments are performed in adult recipients with adult living donors, size mismatch between graft and recipient becomes a critical problem. The first priority is donor safety; therefore, smaller-sized grafts, such as the left lobe of the liver, appear to be the optimal choice [78].

Liver transplantation may inevitably lead to hepatic I/R injury. Primary graft nonfunction or dysfunction, which occurs as a result of combined I/R injury and secondary tissue regeneration impairment, remains a serious complication in clinical practice and is especially true in cases of small-sized liver transplantation [79,80]. Many studies have been conducted in an attempt to elucidate the mechanisms of I/R injury using appropriate *in vivo* models. Effective treatment strategies aimed at reducing hepatic I/R injury and accelerating liver regeneration could offer major benefits for liver transplantations with size mismatch between the graft and recipient.

Recent reports have demonstrated the capacity of MSCs to be specifically involved in the repair of damaged liver. These results indicate that MSCs are an attractive cell source for regenerative medicine. As for the liver, most animal studies have been performed in drug-induced rodent models [81−83]. However, the role of MSCs in hepatic I/R injury remains to be established. We reported that transplanted MSCs are able to ameliorate hepatic I/R injury and significantly improve liver regeneration [84]. Male rats were separated into two groups: an MSC group given MSCs after reperfusion as treatment and a Control group given phosphate-buffered saline after reperfusion as placebo. The results of liver function tests, pathologic changes in the liver, and the remnant liver regeneration rate were assessed. The fate of transplanted MSCs in the luciferase expressing rats was examined by *in vivo* luminescent imaging. The MSC group showed peak luciferase activity of transplanted MSCs in the remnant liver 24 h after reperfusion, after which luciferase activity gradually declined. The elevation of serum alanine transaminase levels was significantly reduced by MSC injection. Histopathological findings showed that vacuolar change was lower in the MSC group compared to the Control group. Additionally, remnant liver regeneration rate was accelerated in the MSC group.

68.4.4 Limb

Muscle tissues are often destroyed by high-energy injuries such as those encountered in motor vehicle accidents, traumatic limb amputation, compartment syndrome, and crush injuries. Prolonged muscle ischemia causes irreversible necrosis and apoptosis, resulting in complications such as muscle atrophy and functional loss. Because ischemic muscle tissues degenerate irreversibly within 6−8 h [85−89], ischemic time is critical during limb replantation. Attempts at reperfusion more than 8 h after ischemic injury often result in failed replantation. Moreover, anaerobic metabolic products or potassium from the necrotic tissue can cause acute renal failure and fatal shock [90−94]. In addition, the larger the muscle volume involved, the more severe the systemic effects become [95].

Zhang et al. [96] reported that (i) bone marrow-derived MSC transplantation was capable of engraftment in ischemic muscle and engrafted MSCs participate in neovascular formation; (ii) a low dose of simvastatin promoted angiogenesis and improved blood perfusion after hindlimb ischemia in mice; and (iii) combined administration of low-dose simvastatin and bone marrow-derived MSCs was more efficient than simvastatin administration or MSC transplantation alone.

68.4.5 Heart

Myocardial ischemia and reperfusion-induced heart injury accompanied by reparative or replacement heart surgery such as coronary artery bypass grafts can result in myocardial stunning (reversible injury), myocardial infarction (necrosis), and myocardial apoptosis. Related symptoms are associated clinically with acute hypotension and low cardiac output or chronic heart failure, and ventricular remodeling [97]. Although the exact mechanisms underlying I/R heart injury remain obscure, accumulating evidence indicates that its etiology resides in intracellular Ca^{2+} overload during I/R and oxidative stress induced by reactive oxygen species released at the onset of reperfusion [97−101].

Stem cell therapy also has the potential to be a key component in the field of regenerative cardiovascular medicine [102]. MSC therapy in particular may be a leading candidate for cell-based treatment of the ischemic heart. These cells are an unique subset of stem cells that can be isolated from the bone marrow, adipose tissue, and even umbilical cord blood [103,104]. The MSCs are multipotent, and in response to I/R injury, they secrete a variety of cytokines that are cardioprotective or angiogenic [105,106]. In addition, MSCs engrafted into injured myocardium can potentially differentiate into cardiac myocytes [105]. However, engraftment and functional results are extremely variable in the literature and modest at best in a number of clinical trials [105,107−112].

68.5 CONCLUSIONS

In researching the regeneration and cell differentiation processes, many important findings involving stem cells have been discovered so far. New information is expected to be continually forthcoming as research and experimentation in this field continues to progress. The primary mechanisms involved in the ability of MSCs to maintain undifferentiated and their application in the treatment of various diseases are very important research topics. The development of new materials and technology is still in its initial stages, and many problems remain to be solved prior to practical use; however, the biological features of stem cells uncovered so far justify their continued exploration and the promise of future successful clinical applications.

ACKNOWLEDGMENTS

We wish to thank Drs. Yasuhiro Fujimoto, Naoya Kasahara, Junshi Doi, Yuku Iijima, Junji Iwasaki, Hiroyuki Kanazawa, and Taizen Urahashi from the Division of Development of Advanced Treatment, Center for Development of Advanced Medical Technology (CDAMTec), Jichi Medical University, Japan.

COMPETING INTERESTS

Author Eiji Kobayashi has been a visiting professor at CDAMTec and a chief scientific advisor to Otsuka Pharmaceutical Factory, Inc. (Naruto Japan) from 2009.

REFERENCES

[1] Anderson DJ, Gage FH, Weissman IL. Can stem cells cross linage boundaries? Nat Med 2001;7:393−5.

[2] Lin H. The tao of stem cells in the germline. Annu Rev Genet 1997;31:455−91.

[3] La PC. Cancer stem cells: lessons from melanoma. Stem Cell Rev 2009;5:61−5.

[4] Takahashi K, Yamanaka S. Induction of pluripotent stem cells from mouse embryonic and adult fibroblast cultures by defined factors. Cell 2006;126:663−76.

[5] Takahashi K, Tanabe K, Ohnuki M, Narita M, Ichisaka T, Tomoda K, et al. Induction of pluripotent stem cells from adult human fibroblasts by defined factors. Cell 2007;131:861−72.

[6] Yamanaka S. Strategies and new developments in the generation of patient-specific pluripotent stem cells. Cell Stem Cell 2007;1:39−49.

[7] Jacobson L, Marks E, Robson M. Effect of spleen protection on mortality following X-irradiation. J Lab Clin Med 1949;34:1538−43.

[8] Thomas ED, Lochte Jr. HL, Lu WC, Ferrebee JW. Intravenous infusion of bone marrow in patients receiving radiation and chemotherapy. N Engl J Med 1957;257:491−6.

[9] Mariusz ZR, Ewa KZ, Boguslaw M, Magdalena K. Bone marrow-derived stem cells—our key to longevity? J Appl Genet 2007;48:307−19.

[10] Banerjee P, Crawford L, Samuelson E, Feuer G. Hematopoietic stem cells and retroviral infection. Retrovirology 2010;4:7−8.

[11] Huang S, Terstappen LW. Lymphoid and myeloid differentiation of single human CD34+, HLA-DR+, CD38− hematopoietic stem cells. Blood 1994;83:1515−26.

[12] Pereira RF, Halford KW, O'Hara MD, Leeper DB, Sokolov BP, Pollard MD, et al. Cultured adherent cells from marrow can serve as long-lasting precursor cells for bone, cartilage, and lung in irradiated mice. Proc Natl Acad Sci USA 1995;92:4857−61.

[13] Orlic D, Kajstura J, Chimenti S, Jakoniuk I, Anderson SM, Li B, et al. Bone marrow cells regenerate infarcted myocardium. Nature 2001;410:701−5.

[14] Asahara T, Masuda H, Takahashi T, Kalka C, Pastore C, Silver M, et al. Bone marrow origin of endothelial progenitor cells responsible for postnatal vasculogenesis in physiological and pathological neovascularization. Circ Res 1999;85:221−8.

[15] Brazelton TR, Rossi FM, Keshet GI, Blau HM. From marrow to brain: expression of neuronal phenotypes in adult mice. Science 2000;290:1775−9.

[16] Krause DS, Theise ND, Collector MI, Henegariu O, Hwang S, Gardner R, et al. Multi-organ, multi-lineage engraftment by a single bone marrow-derived stem cell. Cell 2001;105:369−77.

[17] Lin F, Cordes K, Li L, Hood L, Couser WG, Shankland SJ, et al. Hematopoietic stem cells contribute to the regeneration of renal tubules after renal ischemia-reperfusion injury in mice. J Am Soc Nephrol 2003;14:1188−99.

[18] Terai S, Ishikawa T, Omori K, Aoyama K, Marumoto Y, Urata Y, et al. Improved liver function in patients with liver cirrhosis after autologous bone marrow cell infusion therapy. Stem Cell 2006;24:2292−8.

[19] Dominici M, Le, Blanc K, et al. Minimal criteria for defining multipotent mesenchymal stromal cells. The international society for cellular therapy position statement. Cytotherapy 2006;8:315−7.

[20] Pittenger MF, Mackay AM, Beck SC, Jaiswal RK, Douglas R, Mosca JD, et al. Multilineage potential of adult human mesenchymal stem cells. Science 1999;284:143−7.

[21] Banas A, Teratani T, Yamamoto Y, Tokuhara M, Takeshita F, Osaki M, et al. Rapid hepatic fate specification of adipose-derived stem cells and their therapeutic potential for liver failure. J Gastroenterol Hepatol 2009;24:70−7.

[22] Blaber SP, Webster RA, Hill CJ, Breen EJ, Kuah D, Vesey G, et al. Analysis of in vitro secretion profiles from adipose-derived cell populations. J Transl Med 2012;10:172.

[23] Engela AU, Baan CC, Dor FJ, Weimar W, Hoogduijn MJ. On the interactions between mesenchymal stem cells and regulatory T cells for immunomodulation in transplantation. Front Immunol 2012;3:126.

[24] Marigo I, Dazzi F. The immunomodulatory properties of mesenchymal stem cells. Semin Immunopathol 2011;33:593−602.

[25] Yagi H, Soto-Gutierrez A, Parekkadan B, Kitagawa Y, Tompkins RG, Kobayashi N, et al. Mesenchymal stem cells: mechanisms of immunomodulation and homing. Cell Transplant 2010;19: 667−79.

[26] Ishikawa T, Banas A, Hagiwara K, Iwaguro H, Ochiya T. Stem cells for hepatic regeneration: the role of adipose tissue derived mesenchymal stem cells. Curr Stem Cell Res Ther 2010;5:182−9.

[27] Otto WR, Wright NA. Mesenchymal stem cells: from experiment to clinic. Fibrogenesis Tissue Repair 2011;4:20.

[28] Dahlke MH, Hoogduijn M, Eggenhofer E, Popp FC, Renner P, Slowik P, et al. Toward MSC in solid organ transplantation: 2008 position paper of the MISOT study group. Transplantation 2009;88:614−9.

[29] Hoogduijn MJ, Popp FC, Grohnert A, Crop MJ, van Rhijn M, Rowshani AT, et al. Advancement of mesenchymal stem cell therapy in solid organ transplantation (MISOT). Transplantation 2010;90:124−6.

[30] Popp FC, Renner P, Eggenhofer E, Slowik P, Geissler EK, Piso P, et al. Mesenchymal stem cells as immunomodulators after liver transplantation. Liver Transpl 2009;15:1192−8.

[31] Waterman RS, Tomchuck SL, Henkle SL, Betancourt AM. A new mesenchymal stem cell (MSC) paradigm: polarization into a pro-inflammatory MSC1 or an immunosuppressive MSC2 phenotype. PLoS One 2010;5:e10088.

[32] Auletta JJ, Deans RJ, Bartholomew AM. Emerging roles for multipotent, bone marrow-derived stromal cells in host defense. Blood 2012;119:1801−9.

[33] Mirotsou M, Jayawardena TM, Schmeckpeper J, Gnecchi M, Dzau VJ. Paracrine mechanisms of stem cell reparative and regenerative actions in the heart. J Mol Cell Cardiol 2011;50: 280−9.

[34] Song C, Li G. CXCR4 and matrix metalloproteinase-2 are involved in mesenchymal stromal cell homing and engraftment to tumors. Cytotherapy 2011;13:549−61.

[35] Dwyer RM, Ryan J, Havelin RJ, Morris JC, Miller BW, Liu Z, et al. Mesenchymal stem cell-mediated delivery of the sodium iodide symporter supports radionuclide imaging and treatment of breast cancer. Stem Cells 2011;29:1149−57.

[36] Freyman T, Polin G, Osman H, Crary J, Lu M, Cheng L, et al. A quantitative, randomized study evaluating three methods of mesenchymal stem cell delivery following myocardial infarction. Eur Heart J 2006;27:1114−22.

[37] Wilson T, Stark C, Holmbom J, Rosling A, Kuusilehto A, Tirri T. Fate of bone marrow-derived stromal cells after intraperitoneal infusion or implantation into femoral bone defects in the host animal. J Tissue Eng 2010;:345806.

[38] Curley GF, Hayes M, Ansari B, Shaw G, Ryan A, Barry F, et al. Mesenchymal stem cells enhance recovery and repair following ventilator-induced lung injury in the rat. Thorax 2012;67: 496−501.

[39] Jaeschke H. Mechanisms of reperfusion injury after warm ischemia of the liver. J Hepatobiliary Pancreat Surg 1998;5:402−8.

[40] Teoh NC, Farrell GC. Hepatic ischemia reperfusion injury: pathogenic mechanisms and basis for hepatoprotection. J Gastroenterol Hepatol 2003;18:891−902.

[41] Jaeschke H. Molecular mechanisms of hepatic ischemia−reperfusion injury and preconditioning. Am J Physiol Gastrointest Liver Physiol 2003;284:G15−26.

[42] Chang JX, Chen S, Ma LP, Jiang LY, Chen JW, Chang RM, et al. Functional and morphological changes of the gut barrier during the restitution process after hemorrhagic shock. World J Gastroenterol 2005;11:5485−91.

[43] Lipton P. Ischemic cell death in brain neurons. Physiol Rev 1999;79:1431−568.

[44] Gao C, Xu L, Chai W, Sun X, Zhang H, Zhang G. Amelioration of intestinal ischemia−reperfusion injury with intraluminal hyperoxygenated solution: studies on structural and functional changes of enterocyte mitochondria. J Surg Res 2005;129:298−305.

[45] Madesh M, Bhaskar L, Balasubramanian KA. Enterocyte viability and mitochondrial function after graded intestinal ischemia and reperfusion in rats. Mol Cell Biochem 1997;167:81−7.

[46] Solligård E, Juel IS, Spigset O, Romundstad P, Grønbech JE, Aadahl P. Gut luminal lactate measured by microdialysis mirrors permeability of the intestinal mucosa after ischemia. Shock 2008;29:245−51.

[47] Szabó A, Vollmar B, Boros M, Menger MD. *In vivo* fluorescence microscopic imaging for dynamic quantitative assessment of intestinal mucosa permeability in mice. J Surg Res 2008;145: 179−85.

[48] Bodwell W. Ischemia, reperfusion, and reperfusion injury: role of oxygen free radicals and oxygen free radical scavengers. J Cardiovasc Nurs 1989;4:25−32.

[49] Ikeda H, Suzuki Y, Suzuki M, Koike M, Tamura J, Tong J, et al. Apoptosis is a major mode of cell death caused by ischaemia and ischaemia/reperfusion injury to the rat intestinal epithelium. Gut 1998;42:530−7.

[50] Kaminski KA, Bonda TA, Korecki J, Musial WJ. Oxidative stress and neutrophil activation—the two keystones of ischemia/reperfusion injury. Int J Cardiol 2002;86:41−59.

[51] McCord JM. Oxygen-derived free radicals in postischemic tissue injury. N Engl J Med 1985;312:159−63.

[52] Rose S, Floyd RA, Eneff K, Bühren V, Massion W. Intestinal ischemia: reperfusion-mediated increase in hydroxyl free radical formation as reported by salicylate hydroxylation. Shock 1994;1:452−6.

[53] Ioannou A, Dalle, Lucca J, Tsokos GC. Immunopathogenesis of ischemia/reperfusion-associated tissue damage. Clin Immunol 2011;141:3−14.

[54] Cao H, Qian H, Xu W, Zhu W, Zhang X, Chen Y, et al. Mesenchymal stem cells derived from human umbilical cord ameliorate ischemia/reperfusion-induced acute renal failure in rats. Biotechnol Lett 2010;32:725−32.

[55] Hara Y, Stolk M, Ringe J, Dehne T, Ladhoff J, Kotsch K, et al. *In vivo* effect of bone marrow-derived mesenchymal stem cells in a rat kidney transplantation model with prolonged cold ischemia. Transpl Int 2011;24:1112−23.

[56] Liu H, McTaggart SJ, Johnson DW, Gobe GC. Original article anti-oxidant pathways are stimulated by mesenchymal stromal cells in renal repair after ischemic injury. Cytotherapy 2012;14:162−72.

[57] Gao J, Liu R, Wu J, Liu Z, Li J, Zhou J, et al. The use of chitosan based hydrogel for enhancing the therapeutic benefits of adipose-derived MSCs for acute kidney injury. Biomaterials 2012;33: 3673−81.

[58] Roskott AM, Nieuwenhuijs VB, Dijkstra G, Koudstaal LG, Leuvenink HG, Ploeg RJ. Small bowel preservation for intestinal transplantation: a review. Transpl Int 2011;24:107−31.

[59] Ueno T, Fukuzawa M. Current status of intestinal transplantation. Surg Today 2010;40:1112–22.

[60] Cicalese L, Sileri P, Asolati M, Rastellini C, Abcarian H, Benedetti E. Low infectious complications in segmental living related small bowel transplantation in adults. Clin Transplant 2000;14:567–71.

[61] Ueno T, Wada M, Hoshino K, Yonekawa Y, Fukuzawa M. Current status of intestinal transplantation in Japan. Transplant Proc 2011;43:2405–7.

[62] Tzvetanov IG, Oberholzer J, Benedetti E. Current status of living donor small bowel transplantation. Curr Opin Org Transplant 2010;15:346–8.

[63] Ji G, Chu D, Wang W, Dong G. The safety of donor in living donor small bowel transplantation—an analysis of four cases. Clin Transplant 2009;23:761–4.

[64] Fujishiro J, Tahara K, Inoue S, Kaneko T, Kaneko M, Hashizume K, et al. Immunologic benefits of longer graft in rat allogenic small bowel transplantation. Transplantation 2005;79: 190–5.

[65] Moradi P, Glass GE, Atherton DD, Eccles S, Coffey M, Majithia A, et al. Reconstruction of pharyngolaryngectomy defects using the jejunal free flap: a 10-year experience from a single reconstructive center. Plast Reconstr Surg 2010; 126:1960–6.

[66] Zhao D, Gao X, Guan L, Su W, Gao J, Liu C, et al. Free jejunal graft for reconstruction of defects in the hypopharynx and cervical esophagus following the cancer resections. J Gastrointest Surg 2009;13:1368–72.

[67] Southard JH, van Gulik TM, Ametani MS, Vreugdenhil PK, Lindell SL, Pienaar BL, et al. Important components of the UW solution. Transplantation 1990;49:251–7.

[68] Schlachter K, Kokotilo MS, Carter J, Thiesen A, Ochs A, Khadaroo RG, et al. Redefining the properties of an osmotic agent in an intestinal-specific preservation solution. World J Gastroenterol 2010;16:5701–9.

[69] Porte RJ, Ploeg RJ, Hansen B, van, Bockel JH, Thorogood J, Persijn GG, et al. Long-term graft survival after liver transplantation in the UW era: late effects of cold ischemia and primary dysfunction. European multicentre study group. Transpl Int 1998;11: S164–7.

[70] Jamieson NV, Sundberg R, Lindell S, Claesson K, Moen J, Vreugdenhil PK, et al. Preservation of the canine liver for 24–48 hours using simple cold storage with UW solution. Transplantation 1988;46:517–22.

[71] Wahlberg JA, Love R, Landegaard L, Southard JH, Belzer FO. 72-hour preservation of the canine pancreas. Transplantation 1987;43:5–8.

[72] Taguchi T, Zorychta E, Guttman FM. Evaluation of UW solution for preservation of small intestinal transplants in the rat. Transplantation 1992;53:1202–5.

[73] Burgmann H, Reckendorfer H, Sperlich M, Spieckermann PG. Small bowel tissue high-energy phosphate alterations during hypothermic storage using different protecting solutions. Eur Surg Res 1992;24:84–8.

[74] Olson D, Stewart B, Carle M, Chen M, Madsen K, Zhu J, et al. The importance of impermeant support in small bowel preservation: a morphologic, metabolic and functional study. Am J Transplant 2001;1:236–342.

[75] Olson DW, Fujimoto Y, Madsen KL, Stewart BG, Carle M, Zeng J, et al. Potentiating the benefit of vascular-supplied glutamine during small bowel storage: importance of buffering agent. Transplantation 2002;73:178–85.

[76] Oltean M, Joshi M, Herlenius G, Olausson M. Improved intestinal preservation using an intraluminal macromolecular solution: evidence from a rat model. Transplantation 2010;89: 285–90.

[77] Jiang H, Qu L, Li Y, Gu L, Shi Y, Zhang J, et al. Bone marrow mesenchymal stem cells reduce intestinal ischemia/reperfusion injuries in rats. J Surg Res 2011;168:127–34.

[78] Ben-Haim M, Emre S, Fishbein TM, Sheiner PA, Bodian CA, Kim-Schluger L, et al. Critical graft size in adult-to-adult living donor liver transplantation: impact of the recipient's disease. Liver Transpl 2001;7:948–53.

[79] Azoulay D, Astarcioglu I, Bismuth H, Castaing D, Majno P, Adam R, Johann M. Split-liver transplantation. The Paul Brousse policy. Ann Surg 1996;224:737–46.

[80] Goss JA, Yersiz H, Shackleton CR, Seu P, Smith CV, Markowitz JS, et al. *In situ* splitting of the cadaveric liver for transplantation. Transplantation 1997;64:871–7.

[81] Banas A, Teratani T, Yamamoto Y, Tokuhara M, Takeshita F, Osaki M, et al. IFATS collection: *in vivo* therapeutic potential of human adipose tissue mesenchymal stem cells after transplantation into mice with liver injury. Stem Cells 2008;26: 2705–12.

[82] Tsai PC, Fu TW, Chen YM, Ko TL, Chen TH, Shih YH, et al. The therapeutic potential of human umbilical mesenchymal stem cells from Wharton's jelly in the treatment of rat liver fibrosis. Liver Transplant 2009;15:484–95.

[83] Oyagi S, Hirose M, Kojima M, Okuyama M, Kawase M, Nakamura T, et al. Therapeutic effect of transplanting HGF-treated bone marrow mesenchymal cells into CCl4-injured rats. J Hepatol 2006;44:742–8.

[84] Kanazawa H, Fujimoto Y, Teratani T, Iwasaki J, Kasahara N, Negishi K, et al. Bone marrow-derived mesenchymal stem cells ameliorate hepatic ischemia reperfusion injury in a rat model. PLoS One 2011;6:e19195.

[85] Morrison WA, O'Brien BW, MacLeod AM. Evaluation of digital replantation—a review of 100 cases. Orthop Clin N Am 1977;8:295–308.

[86] Gold AH, Lee GW. Upper extremity replantation: current concepts and patient selection. J Trauma 1981;21:551–7.

[87] Eckert P, Schnackerz K. Ischemic tolerance of human skeletal muscle. Ann Plast Surg 1991;26:77–84.

[88] Beyersdorf F, Unger A, Wildhirt A, Kretzer U, Deutschländer N, Krüger S, et al. Studies of reperfusion injury in skeletal muscle: preserved cellular viability after extended periods of warm ischemia. J Cardiovasc Surg 1991;32:664–76.

[89] Zdeblick TA, Shaffer JW, Field GA. An ischemia-induced model of revascularization failure of replanted limbs. J Hand Surg 1985;10A:125–31.

[90] Eiken O, Nabaseth DC, Mayer RF, Deterling Jr. RA. Limb replantation. The pathophysiological effects. Arch Surg 1964;88: 54–65.

[91] McCutcheon C, Hennessy B. Systemic reperfusion injury during arm replantation requiring intraoperative amputation. Anaesth Intensive Care 2002;30:71–3.

[92] Al-Qattan MM. Ischemia—reperfusion injury. Implications for the hand surgeon. J Hand Surg 1998;23:570—3.

[93] Kerrigan CL, Stotland MA. Ischemia reperfusion injury: a review. Microsurgery 1993;14:165—75.

[94] Zimmerman BJ, Granger DN. Mechanism of reperfusion injury. Am J Med Sci 1994;307:284—92.

[95] Akimau P, Yoshiya K, Hosotsubo H, Takakuwa T, Tanaka H, Sugimoto H. New experimental model of crush injury of the hindlimbs in rats. J Trauma 2005;58:51—8.

[96] Zhang Y, Zhang R, Li Y, He G, Zhang D, Zhang F. Simvastatin augments the efficacy of therapeutic angiogenesis induced by bone marrow-derived mesenchymal stem cells in a murine model of hindlimb ischemia. Mol Biol Rep 2012;39:285—93.

[97] Mentzer Jr. RM, Lasley RD, Jessel A, Karmazyn M. Intracellular sodium hydrogen exchange inhibition and clinical myocardial protection. Ann Thorac Surg 2003;75:S700—8.

[98] Piper HM, Garcia-Dorado D, Ovize M. A fresh look at reperfusion injury. Cardiovasc Res 1998;38:291—300.

[99] Sabri A, Byron KL, Samarel AM, Bell J, Lucchesi PA. Hydrogen peroxide activates mitogen-activated protein kinases and $Na^+—H^+$ exchange in neonatal rat cardiac myocytes. Circ Res 1998;82:1053—62.

[100] Myers ML, Farhangkhoee P, Karmazyn M. Hydrogen peroxide induced impairment of post-ischemic ventricular function is prevented by the sodium—hydrogen exchange inhibitor HOE 642 (cariporide). Cardiovasc Res 1998;40:290—6.

[101] Snabaitis AK, Hearse DJ, Avkiran M. Regulation of sarcolemmal Na^+/H^+ exchange by hydrogen peroxide in adult rat ventricular myocytes. Cardiovasc Res 2002;53:470—80.

[102] Jolicoeur EM, Granger CB, Fakunding JL, Mockrin SC, Grant SM, Ellis SG, et al. Bringing cardiovascular cell-based therapy to clinical application: perspectives based on a National Heart, Lung, and Blood Institute cell therapy working group meeting. Am Heart J 2007;153:732—42.

[103] Kumar S, Chanda D, Ponnazhagan S. Therapeutic potential of genetically modified mesenchymal stem cells. Gene Ther 2008; 15:711—5.

[104] Kestendjieva S, Kyurkchiev D, Tsvetkova G, Mehandjiev T, Dimitrov A, Nikolov A, et al. Characterization of mesenchymal stem cells isolated from the human umbilical cord. Cell Biol Int 2008;32:724—32.

[105] Amado LC, Saliaris AP, Schuleri St KH, John M, Xie JS, Cattaneo S, et al. Cardiac repair with intramyocardial injection of allogeneic mesenchymal stem cells after myocardial infarction. Proc Natl Acad Sci USA 2005;102:11474—9.

[106] Kinnaird T, Stabile E, Burnett MS, Lee CW, Barr S, Fuchs S, et al. Marrow-derived stromal cells express genes encoding a broad spectrum of arteriogenic cytokines and promote *in vitro* and *in vivo* arteriogenesis through paracrine mechanisms. Circ Res 2004;94:678—85.

[107] Berry MF, Engler AJ, Woo YJ, Pirolli TJ, Bish LT, Jayasankar V, et al. Mesenchymal stem cell injection after myocardial infarction improves myocardial compliance. Am J Physiol 2006;290:2196—203.

[108] Iso Y, Spees JL, Serrano C, Bakondi B, Pochampally R, Song YH, et al. Multipotent human stromal cells improve cardiac function after myocardial infarction in mice without long-term engraftment. Biochem Biophys Res Commun 2007;354:700—6.

[109] Carr CA, Stuckey DJ, Tatton L, Tyler DJ, Hale SJ, Sweeney D, et al. Bone marrow-derived stromal cells home to and remain in the infarcted rat heart but fail to improve function: an *in vivo* cine-MRI study. Am J Physiol 2008;295:533—42.

[110] Grauss RW, Winter EM, van Tuyn J, Pijnappels DA, Steijn RV, Hogers B, et al. Mesenchymal stem cells from ischemic heart disease patients improve left ventricular function after acute myocardial infarction. Am J Physiol 2007;293:H2438—47.

[111] Grauss RW, van Tuyn J, Steendijk P, Winter EM, Pijnappels DA, Hogers B, et al. Forced myocardin expression enhances the therapeutic effect of human mesenchymal stem cells after transplantation in ischemic mouse hearts. Stem Cells 2008;26:1083—93.

[112] Abdel-Latif A, Bolli R, Tleyjeh IM, Montori VM, Perin EC, Hornung CA, et al. Adult bone marrow-derived cells for cardiac repair: a systematic review and meta-analysis. Arch Intern Med 2007;167:989—97.

Chapter 69

Xenotransplantation: Past, Present, and Future

Pierre Gianello

Laboratory of Experimental Surgery and Transplantation, Institut de Recherche Expérimentale et Clinique, Université Catholique de Louvain, Brussels, Belgium

Chapter Outline

69. XENOTRANSPLANTATION

69.1 PAST, PRESENT, AND FUTURE

Animal-derived tissues, such as skin, tendons, and heart valves, are used daily in human medicine. In fact, porcine- and bovine-derived heart valves represent over 60% of the valve-replacement procedures in humans annually, and porcine insulin has been used for decades to treat diabetic patients. Although rejection of primarily vascularized xenografts remains a major hurdle to overcome, cell xenotransplantation could become the next major breakthrough in human medicine.

69.1.1 Past: Where Are We Coming from?

The beginning of xenotransplantation history involved cells and tissues (blood, bone, skin), and not primarily vascularized organs [1,2]. In the seventeenth and eighteenth centuries, blood xenotransfusion [3] as well as bone and skin xenografts in humans was in fact reported [2]. In the early nineteenth century, G. Voronoff proposed transplanting ape testicle slices to rejuvenate human males and applied this technique in more than 2000 human patients [4]. In the early 1900s, Princeteau treated a young boy with acute renal failure by transplanting two slices of a rabbit's kidney into his kidney, but the child died 16 days later [2]. In 1906, Jaboulay attempted a double-kidney xenotransplantation using the technique of vascular anastomosis, but the xenograft never functioned [5]. A few years later, E. Unger tried to graft an animal kidney into a 32-year-old man, but vascular thrombosis occurred [6]. In 1923 in New York, H. Neuhof grafted a lamb kidney into a human, but the patient died 9 days later [7]. Following these disappointing results, xenograft attempts in humans were stopped for a long period of time; in fact, no xenotransplants were reported for 40 years.

The real start of the xenotransplantation story was in 1963, when K. Reemtsma, working in New York, transplanted chimpanzee kidneys into thirteen humans [8]. These xenografts were quite successful since one patient survived 9 months with normal renal function and died from an unrelated cause. In Europe, the first primate xenograft was performed in 1966 by Cortesini, who transplanted a kidney into a young patient who eventually rejected this xenograft after 40 days [9]. During the same period in Colorado, T. Starzl performed six baboon kidney xenografts into patients with end-stage renal failure [10]. In each case, patient death occurred due to clinical complications associated with rejection. The first attempt to accomplish a heart xenograft was made by J. Hardy and C. M. Chavez at the University of Mississippi in 1964; the patient survived 2 h [11]. The second attempt was made by Christian Barnard on a 25-year-old woman; the baboon heart maintained a modest circulation for about 6 h prior to the onset of rejection [12].

At that time the concept of "brain death" was described in humans and the era of cadaveric allotransplantation was quickly underway. The rapid success of allotransplantation and the later discovery of several new immunosuppressive drugs demonstrated that allotransplantation was the best treatment for several life-threatening diseases. Organ shortage again became a major problem though, and xenotransplantation was therefore seen as a possible solution to the lack of donors. However, virologists were very reluctant to use superior primates as donors for xenotransplant tissue in humans due to possible parallelism with HIV retrovirus infection, which may have been transmitted from apes to humans. This was true especially after the case of Baby Fae in 1984, who was transplanted with a baboon's heart by L. Bailey in Loma Linda [13] and died 20 days later from rejection. At that time, the US Food and Drug Administration (FDA) published a guideline on infectious disease issues in xenotransplantation, but several scientists wrote an open letter declaring these guidelines insufficient. This debate generated a European moratorium on the use of nonhuman primate organs in xenotransplantation.

Since primates were no longer being considered as possible organ donors for humans, intense experimental research began to focus on pigs as potential donors. The pig is currently regarded as the most suitable source of organs for future xenotransplantation for several reasons. First, the pig has many anatomical and physiological similarities with man; studies have demonstrated that porcine organs are approximately the same size as human organs and have similar efficiency and renal/heart physiology. Pigs have a short gestation period and produce large litters of 10−15 offspring, or more, that rapidly grow to the size necessary for xenografting into human adults. Additionally, pigs can easily be bred under qualified pathogen-free conditions such that the majority of zoonotic agents can be eliminated. Since pigs are widely used for food, social and ethical issues are not anticipated to be problems. Finally, progress in genetic technology allows production of transgenic pigs expressing several membrane factors or knockout of specific pig genes. Therefore, as a source of organs for humans, the pig has considerable advantages over any other species, but the major problem of rapid hyperacute rejection (HAR) after grafting of vascularized pig organs remained and became the problem to solve.

Since both humans and superior apes have preformed antibodies against pig endothelial antigens, the most useful experimental model therefore became the pig-to-primate model. In 1984, G.P.J. Alexandre and P. Gianello reported for the first time that a pig renal transplant could survive and correct renal function for 23 days in a baboon with the use of an immune treatment comparable with that used in ABO-incompatible allotransplant recipients [14]. Afterward, several laboratories around the world focused on the physiopathology of hyperacute and vascular rejections occurring in pig-to-primate transplants.

U. Galilli soon identified the epitopes that are recognized on pig endothelial cells as a $Gal\alpha1-3Gal\alpha1-4GlcNAc-R$ (Gal) antigen synthesized by the enzyme α-galactosyltransferase (GT) [15]. This epitope is evident only in mammals but not in Old World primates and humans who have invalidated the enzyme. As a consequence, Old World primates and humans have developed anti-Gal antibodies, probably due to cross-reaction with Gal present on microbial intestinal flora. Without any treatment, these xenonatural antibodies (XNA) lead to HAR due to a very rapid binding to the Gal epitopes, which activates the classical complement pathway and leads to a complete destruction of the xenograft within minutes or hours. The binding of XNA to endothelial pig galactosyl residues is directly cytotoxic to endothelial cells by triggering a release of pro-inflammatory and chemoattractive factors that cause a massive graft inflammation. Due to the subsequent endothelial activation responsible for upregulation of P-selectin and the endothelial retraction exposing the subendothelial matrix, von Willebrand factor may initiate platelet aggregation. This platelet aggregation is rapidly followed by fibrin formation, leading to diffused thrombosis of the capillaries and consequent massive ischemic and hemorrhagic graft necrosis. This rejection can be transiently prevented by strategies such as antibody depletion (plasma exchange, immunoadsorption) or by complement blockage. In fact, by using successive plasmapheresis and clinical immune suppression in baboons, G.P.G. Alexandre and P. Gianello demonstrated in several primates that pig renal xenograft could function normally for 10−23 days.

A better understanding of the pathophysiology of HAR pushed D. White and E. Cozzi to produce transgenic pigs expressing a human decay-accelerating factor (hDAF) complement regulator (i.e., CD55). After the successful production of hDAF pigs, subsequent in vivo studies demonstrated that pig organs expressing the complement regulatory molecule were resistant to HAR [16]. The best results in the 1990s demonstrated kidney survival up to 78 and 90 days by implanting a transgenic

TABLE 69.1 Renal Xenograft from Transgenic Pigs

Author	Recipient	Transgene	Protocol		Graft Survival
Cozzi [17]	Cynomolgus	hCD55	CsA, CyP, MMF, steroids		2−51 days
Zaidi [19]	Cynomolgus	hCD55	CsA, CyP, steroids		6−35 days
Cozzi [20]	Cynomolgus	hCD55	CyA, CyP, steroids		5−78 days
Ghanekar [21]	Baboon	hCD55	CyP, CyA, steroids	Gal-conjugate, ATG, Rapa	Average 19 days
Vangerow [22]	Cynomolgus	hCD55	CyP, C1-inh, steroids		3−68 days
Lam [23]	Cynomolgus	hCD55	CyA, CyP, MMF, steroids	Gal-conjugate	3−27 days
Richards [24]	Cynomolgus	hCD55	CyP, CsA, steroids	RAD, MPS, TP10	4−60 days
Barth [25]	Baboon	hCD55	CyP, CVF, ATG, MMF	Thymokidneys, Gal-conjugate, anti-CD3	24−32 days
Ashton-Chess [26]	Baboon	hCD55	CyP, CsA, MMF, steroids		5−12 days
Cozzi [27]	Cynomolgus	hCD55	CyA, CyP, MMF, steroids	Gal-conjugate	2−37 days
Shimizu [28]	Baboon	hCD55	CVF, MMF, ATG, CyP	Thymokidneys, Gal-conjugate, anti-CD40, antimonkey CD3	9−30 days
Chen [29]	Baboon	hCD55	Tac, MMF, ATG, CVF, steroids	Gal-conjugate, LF15-0195, Rituxan	7−75 days
Tu [30]	*Macaca cyclopis*	hCD55	None		24−104 h
Cowan [31]	Baboon	hCD55/hCD59, HT			30 h−5 days
Baldan [18]	Cynomolgus	Hdaf	CyP, CyA steroids	MPS, MTX	1−90 days

CD55 pig kidney into a cynomolgus monkey and using clinically applicable immunosuppression (composed mainly of cyclosporin, cyclophosphamid, and steroids) [17,18]. Table 69.1 summarizes the main results obtained in the late 1990s by using transgenic pig kidneys from animals expressing either CD55, CD46, or CD59. During the same period, several laboratories showed that an orthotopic transgenic hCD46 pig heart could survive up to 57 days (Table 69.2), and others showed survival of a heterotopic transgenic hCD55 pig heart up to 139 days with the use of a clinically applicable immunosupppression (Table 69.3).

However, even if HAR is controlled, another type of rejection eventually develops. Delayed xenogeneic rejection (DXR), or acute humoral xenogeneic rejection (AHXR), involves both preformed and induced XNA and complements activation as well as cellular responses including cytokine production. In the 1990s and 2000s, different strategies were used to evade HAR and AHXR.

Several groups used organs transgenic for hCD55 in pig-to-nonhuman primate transplants under strong immuno-suppression, but kidney survival did not extend beyond 51 days, with evidence of IgM binding, complement deposition, and macrophage infiltrate [17,26] suggesting that acute cellular xenogeneic rejection also occurred. A second strategy, based on induction of T-cell tolerance, was to simultaneously transplant porcine thymus and kidney (called thymokidney) in a nonhuman primate recipient. This strategy resulted in graft survival up to 30 days with no evidence of cellular response, but humoral damage was exhibited [25].

During the same period, some physicians were also assessing the potential of pig xenografts in humans. In 1990 in Poland, J. Czapliscki grafted a pig's heart into a man who survived for less than 24 h; the reported cause of death was not organ rejection but the small size of the heart [50]. At Cedar Sinai Hospital in 1993, L. Makowka grafted a pig's liver into a woman with fulminant hepatic

TABLE 69.2 Orthotopic Heart Xenograft from Transgenic Pigs

Author	Recipient	Pig Organ	Transgene	Protocol		Graft Survival
Schmoeckel [32]	Baboon	Heart (ortho)	hCD55	CyP, CsA, steroids		<1−9 days
Vial [33]	Baboon	Heart (ortho)	hCD55	Cyp, CyA MMF, steroids		2−39days
Brandl [34]	Baboon	Heart (ortho)	hCD55, hCD46	Tac, Siro, ATG, steroids	Gal-conjugate, Rituxan, anti-HLA-DR antibody	1−9 days
Brandl [35]	Baboon	Heart (ortho)	hCD55	CyP, Tac, ATG, steroids	Gal-conjugate, Rapa,	1−25 days
Waterworth [36]	Baboon	Heart (ortho, Hetero)	hCD55	CsA, CyP, steroids		2−21 days

TABLE 69.3 Heterotopic Heart Xenograft from Transgenic Pigs

Author	Recipient	Transgene	Protocol		Graft Survival
Cozzi [37]	Cynomolgus	hCD55	CsA, CyP, steroids		6−62 days
Waterworth [38]	Cynomolgus	hCD55	CsA, CyP, steroids		6−62 days
McCurry [39]	Baboon	hCD55/hCD59	SP, CyP, steroids	AZA, IA	4−30 h
Lin [40],	Baboon	hCD55/hCD59	CsA, steroids	IA, MTX	<29 days
Adams [41]	Baboon	hCD46/IA			16 days
Chen [42]	Baboon	hCD55/hCD59			<3 h
McGregor [43]	Baboon	hCD46	tac, ATG, siro, steroids	Gal-conjugate, anti-CD20, enaxaparin	56−113
McGregor [44]	Baboon	hCD46	Tac, rapa, ATG, steroids	Rituximab	15−137 days
Byrne [45]	Baboon	hCD46	CyP, Tac, Rapa, ATG, steroids	TPC, Rituximab	0−139 days
Wu [46]	Baboon	hCD46	CsA, CyP, MM	Gal-conjugate NEX1285, anti-CD155	2 h−1 day
Lam [23]	Cynomolgus	hCD55	CsA, CyP, MMF, steroids	Gal-conjugate	Mean 27 days
Houser [47]	Baboon	hCD55	ATG, MMF, CVF, steroids	Gal-conjugate, thymic irradiation, anti-CD154	4−139 days
Kuwaki [48]	Baboon	hCD55	ATG, MMF, CVF, steroids	Gal-conjugate, thymic irradiation, anti-CD154	4−139 days
Wu [46]	Baboon	hCD55	CsA, CyP, MMF	Gal-conjugate, NEX1285, anti-CD154	0−36 days
Bhatti [49]	Baboon	hCD55	CsA, CyP, MMP, steroids		10−99 days

failure. The patient died after a few hours [51]. At the Karolinska Institute in 1994, C. Groth transplanted porcine fetal islet-like cell clusters into 10 insulin-dependent diabetic kidney-transplant patients under a standard immunosuppression regimen. The results were encouraging as demonstrated by the urinary excretion of small amounts of porcine C-peptide in some patients and the identification of some intact insulin-staining cells in a biopsy specimen [52,53].

Although the human data for primarily vascularized xenografts were not very impressive, the experimental data convinced scientists that using organs from

TABLE 69.4 Renal Xenograft from GTKO Pigs

Author	Recipient	Pig Organ	Transgene	Protocol		Graft Survival
Chen [56]	Baboon	Kidney	GTKO	tac, ATG, CVF, MMF, steroids		8−16 days
Yamada [57]	Baboon	Kidney	GTKO	MMF, ATG, steroids	Anti-CD2, anti-CD154	4−83 days
Ezzelarab [58]	Baboon	kidney	GTKO	MMF, ATG, steroids CVF	Anti-CD154	3−5 days
Le Bas-Bernardet [59]	Baboon	Kidney	GTKO/hCD55, hCD59, hCD39/hHT	CyP, Tac, steroids	MMF, hC1 inhibitor	12−15 days
Nishimura [60]	Baboon	Kidney (thymo)	GTKO	ATG, MMF, CVF, steroids	LoCD2b/anti-CD3 rIT, anti-CD154	LoCD2b: 15−23 days; anti-CD3 rIT: 14 days
Cowan [31]	Baboon	Kidney	GTKO, hCD55, hCD59	None		3−5 days
Griesemer [61]	Baboon	Kidney (thymo)	GTKO	ATG, MMF, FK 506	LoCD2b, anti-CD20, anti-CD154	28−83 days

TABLE 69.5 Heterotopic Heart Xenograft from GTKO Pigs

Author	Recipient	Transgene	Protocol		Graft Survival
Kuwaki [62]	Baboon	GTKO	ATG, CVF, MMF, steroids	LoCD2b, anti-CD154	16−179 days
Shimizu [63]	Baboon	GTKO	ATG, CVF, MMF	Thymic irradiation, anti-CD154	16−179 days
Hisashi [64]	Baboon	GTKO	ATG, CVF, MMF, steroids	Thymic irradiation, anti-CD154	16−179 days
Ezzelarab [58]	Baboon	GTKO	ATG, CVF, MMF, steroids	Anti-CD154	6 days−8 weeks
Bauer [65]	Baboon	GTKO, hCD46, hTM	Steroids, MMMF, ATG	Anti-CD20	2−19 days
Mohiuddin [66]	Baboon	GTKO, hCD46Tg	ATG, MMF, CVF, steroids	Anti-CD20, anti-CD154	36−236 days
McGregor [67]	Baboon	GTKO/hCD56	ATG, tac, siro	Anti-CD20	15−52 days

transgenic pigs to transplant humans was possible and may be used in the very near future. Progress was based on the main antigenic epitope (i.e., galactosyl) being identified and success in terms of large animal cloning. Hypothetically, knockout of the *GT* gene responsible for the presence of the Gal epitope on pig cells could be a major advantage to control HAR and early vascular rejection. In 2002, PPL reported that pigs with one allele of the α-1,3-galactosyltransferase (*GT*) gene rendered nonfunctional had been produced [54]. By subsequent modifications to the other allele [55], pigs with homozygous knockout of the *GT* gene were produced both in PPL Therapeutics and at Massachusetts General Hospital (MGH) in Boston.

69.1.2 Present

This success fueled great hope in the xenotransplant community for major progress in the field. However, few laboratories had access to the GalT-KO pigs and therefore it took several years before scientific data could be obtained from GalT-KO pig organs.

As shown in Table 69.4, kidney xenograft survival up to 83 days was obtained with GalT-KO pig organs in the context of a tolerance-induction protocol. By using GalT-KO pig hearts that also expressed an hCD46 complement regulator, researchers demonstrated that an heterotopic heart xenograft could survive up to 236 days in primates (Table 69.5). Immunosuppression required multiple strong treatments, and anti-CD154 monoclonal antibody (mAb) (against CD40L) was constantly needed to prevent rejection. Thus, by comparing the data obtained with GalT-KO organs and the results from wild-type transgenic pigs expressing similar complement regulators, GalT-KO kidney xenografts did not appear to provide a major advantage. No comparison was possible between orthotopic GalT-KO hearts and wild-type pigs since no data have been reported for orthotopic GalT-KO hearts, but use of heterotopic GalT-KO pig hearts seemed to provide a substantial advantage in terms of graft survival. The maximal survival obtained with transgenic hCD55 pig hearts reached 139 days, but use of heterotopic GalT-KO pig hearts yielded a maximal survival of 236 days. A persistent problem, however, was that the immune regimen required anti-CD154 mAb to attain the results, and this mAb is not clinically useful because it produces life-threatening thrombosis. We thus do not know the real impact of the GalT-KO pig hearts in xenotransplantation since no reports are available with orthotopic or heterotopic GalT-KO hearts without the use of anti-CD154 mAb or if it was the case, then the survival was never better or prolonged in comparison with transgenic organs from wild-type pigs.

Pig liver xenotransplantation has the advantage of being less susceptible to HAR and AHXR than other organs. However, using conventional immunosuppression, a liver from a hDAF pig enabled survival for only up to 8 days [68]. More recently, Ekser reported a study using GalT-KO pigs transgenic for CD46. Graft survival ranged between 4 and 7 days [69]. Pig liver-specific proteins, such as albumin, fibrinogen, haptoglobulin, and plasminogen, were produced following pig liver transplantation. However, while several coagulation factors produced by the porcine liver remained in a normal range and functioned satisfactorily, porcine albumin production was not sufficient. These current data suggest that pig livers may sustain the life of a nonhuman primate long enough to form a bridge to liver allotransplantation.

The intrinsic structure of porcine lungs makes them particularly susceptible to injury. Results from pig lung xenotransplantation report that graft survival ranged from hours to days, even following the xenotransplantation of genetically modified porcine organs [70]. In a clinically relevant *ex vivo* perfusion model comparing wild-type, hDAF, or GalT-KO lungs, GalT-KO lungs exhibited reduced complement activation, delayed acute lung injury, and prolonged survival [71]. However, GalT-KO lungs still failed within hours, with platelet sequestration, thrombin activation, and intravascular thrombosis. These data suggest that major efforts are still needed to understand and control the mechanisms behind lung xenograft rejection.

What we did learn with transgenic pig organs and GalT-KO organs is that, although HAR is controlled, AHXR is still a major problem since antibodies against proteins other than Gal are also powerful and this innate reaction has an impact on other system such as coagulation. Cowan et al. [72] investigated the problem of dysregulated coagulation and vascular injury in xenotransplantation and concluded that careful consideration of the tightly integrated and regulated processes of blood clotting and inflammation would be needed to solve the associated problems in xenografting. In this regard, it appears that tissue factor (TF) may play a key role in initiating the thrombotic microangiopathy and consumptive coagulopathy after xenotransplantation [73]. In order to control the biological effects of coagulation incompatibilities and overcome inflammation and thrombotic events, generation of genetically modified pigs expressing modulators of the clotting cascade has been suggested as an adjunct strategy to prolong xenograft survival. In this context, approaches aiming to overexpress human molecules, such as TF inhibitors [74], thrombomodulin [75], and CD39 [76], or to engineer pigs with low/no levels of procoagulant molecules on the vascular porcine endothelium were investigated. At this stage, however, it appears that a combined approach with appropriately genetically modified pig donors together with systemic treatment of the recipient will be necessary to overcome the coagulation dysregulation in xenotransplantation.

69.1.3 What Is the Future for Primarily Vascularized Organ Xenotransplantation?

In recent year, priority has been given to produce new GalT-KO pigs that also express several molecules previously identified as being crucial to avoiding HAR, coagulopathy, and AHXR. Unfortunately, the first multiple transgenic GaLT-KO pigs used in a kidney xenotransplantation model did not demonstrate major advantages, with survival of only 2–3 weeks being recently reported for GalT-KO/CD55/CD59/H-transferase pig kidneys grafted in both cynomolgus monkeys or baboons [59,77]. It should be noted that anti-CD154 mAb was not used in these experiments.

GalT-KO pigs are considered to be the basic platform for introducing novel genetic modifications into the pig genome, and the first step has been the production of GalT-KO animals expressing human complement regulatory proteins (i.e., CD55, CD46, CD59) [59,67]. Since human natural killer (NK) cells, macrophages, dendritic cells, and T-cells have an important cytotoxic role against porcine cells, other modifications have taken place to tackle cell-mediated rejection. Transgenic pigs expressing HLA-E, a molecule inhibiting NK cell adhesion and cytotoxicity, have recently been produced [78]. Likewise, transgenic pig fetuses expressing human CD47, which can inhibit the activity of human macrophages, have been generated [79]. To reduce pig cell immunogenicity, a mutant form of the *CIITA* gene has been introduced on a Gal-KO/CD46/CD55 background, resulting in down-regulation of swine leukocyte class II antigen expression and inhibition of CD4$^+$ T-proliferation in Mixed Lymphocyte Reaction (MLR) studies [80]. Pigs transgenic for LEA29Y, an inhibitory agent of the costimulatory B7/CD28 pathway under the control of porcine insulin promoter, have recently been developed. Similarly, porcine and human CTLA4-Ig transgenic pigs, with the gene under the control of an enolase promoter, have been obtained [81,82] with the aim of blocking T-cell activity [82,83]. Moreover, since constitutive expression of pCTLA4-Ig generates highly immunosuppressed pigs, pigs with inducible pCTLA4-Ig expression have recently been obtained [84]. Other approaches currently under investigation include pigs expressing the anti-inflammatory and antiapoptotic molecules HO-1 and A20 [85,86] and pigs transgenic for TRAIL (tumor necrosis factor-α—related apoptosis-inducing ligand), a molecule able to induce plasma cell apoptosis and reduce neutrophil lifespan [87]. More recently, a method using zinc finger nuclease that considerably increases the efficiency of gene invalidation has been described as applicable in pigs [88]. Ekser et al. [89] suggested that 25 additional genetic modifications, including expression of multiple complement regulatory proteins and/or thromboregulatory genes combined with novel immunosuppressive regimens, are needed to improve pig organ survival.

To identify antibody responses to non-Gal endothelial antigens involved in DXR of pig-to-primate cardiac xenorejection, Byrne et al. [90] analyzed primate sera by assessing the fixation of non-Gal Ab on GalT-KO porcine aortic endothelial cells. Results of proteomic analysis revealed 14 potential target antigens without defining any immunodominant targets. In a complementary experiment, they screened a retrovirus expression library to identify endothelial cell (EC) membrane antigens detected in sera from sensitized cardiac xenograft recipients [91]. Sequence analysis identified porcine CD9, CD46, CD59, and EC protein C, proteins included in EC functions, which suggests antinon-Gal responses. They also identified porcine annexin A2 and a GT that could potentially generate a new carbohydrate antigen.

As seen in Table 69.5, existing transgenic pigs on a GalT-KO background are numerous, and the wait is on to see if prolonged survival with clinically applicable immune regimen is obtained in the near future. At the same time, new genetically engineered pigs will be produced (Table 69.6).

TABLE 69.6 Actual Nonexhaustive List of Available GalT-KO and Transgenic Pigs[a]

Transgenic Wild-Type Pigs	Transgenic GTKO Pigs
CD55	GTKO/CD55/CD59/HT/TBM/CD39
CD59	GTKO/CD46/**CD39/TFPI/pCTLA4-Ig**
HT	GTKO/CD55/CD59/HT/CD39
CD55/59	GTKO/CD55/CD59/HT/TBM
CD55/59/HT	GTKO/CD46/CD55/CIITA-DN
GnT-III	GTKO/CD46/**TFPI/pCTLA4-Ig**
CD46	GTKO/CD46/CD55/**CD39**
CD39	GTKO/CT46/**TMB**
TRAIL	GTKO/CD55/CD39
CTLA4-Ig (islets/neurons)	GTKO/CD46/CD55/**TBM**
TBM (islets)	GTKO/CD46/CD55
shTNFR1Fc	GTKO/CD46/**CD39**
A20	GTKO/CD47
PERV siRNA	
HO-1	
HLA-E/β2m	
LEA29Y	

Bold means they are tissue specific.
[a]*Microinjection and somatic cell nuclear transfer (SNCT) cloning.*

69.1.4 Tolerance Induction in Pig-to-Nonhuman Primate Xenotransplantation

T-cell—mediated xenogeneic responses likely initiate a myriad of other cellular players including B-cells and NK-cells as well as the innate immune response. Consequently, the level of immunosuppression needed to prolong solid-organ xenograft survival is prohibitive and associated with significant morbidity and mortality. Given these factors, it is likely that the induction of tolerance across xenogeneic barriers will be necessary for successful xenotransplantation.

Indefinite tolerance may be obtained either by concomitant vascularized thymic transplantation or microchimerism or by using transgenic pig organs. Thymic transplantation is thought to lead to tolerance by a central or deletional mechanism. K. Yamada developed two methods of transplanting vascularized thymic tissue, which allow "thymic grafts" to function immediately after allogeneic/xenogeneic transplantation in MGH miniature swine, either by direct vascular anastomosis of the thymic blood supply [92] or as a thymokidney [93]. The authors demonstrated functional thymopoiesis in transplanted thymic grafts and donor-specific tolerance induction across a full MHC mismatch barrier in an allogeneic miniature swine model [94,95]. When this strategy was applied to the transplantation of GalT-KO pig kidneys to baboons, attempts resulted in markedly prolonged renal xenograft survival with normal renal function for up to 83 days; mean recipient survival exceeded 50 days with an anti-CD154—based, steroid-free tolerance-inducing regimen [57,61]. Thymic grafts supported thymopoiesis, and donor-specific unresponsiveness with normal antithird-party allogeneic responses in Cytotoxic T Lymphocyte (CTL) assays was observed in the maintenance period, suggesting that the baboon was on the path to tolerance [57,61]. More recently, the levels of T-cell depletion (TCD) required for prolonged GalT-KO thymokidney survival in baboons has been evaluated [60], and these data suggest that an optimal level of TCD is required for xenogeneic thymokidney graft survival and avoidance of lethal infection.

Another approach for induction of tolerance is through the creation of a chimeric state in the recipient. Hematopoietic stem cells obtained from miniature swine, GalT-KO swine, or hDAF pigs have been transplanted into baboons [96]. Thus far, even with the GalT-KO bone marrow, the majority of the cells (>90%) are cleared from circulation within hours, and peripheral chimerism is generally undetectable after 1 week [97]. These results suggest that especially in bone marrow models, an innate immune response, possibly mediated by macrophages, is responsible for the cell loss. Macrophage depletion has been shown to enhance the engraftment of porcine hematopoietic cells in baboons. Controlling macrophages may be possible through the exploitation of the species-specific interaction of CD47 with signal regulatory protein-alpha [98]. Recent experiments have evaluated the efficacy of human-CD47—transduced porcine GalT-KO bone marrow cells, and it is anticipated that these cells will be "protected" from host macrophages because they express host CD47 and thus will not be cleared by the reticuloendothelial system.

The third approach to induce tolerance to xenoantigens is to utilize human CD39 transgenic GalT-KO pigs as donors. Robson et al. have defined and characterized the role of signaling by extracellular nucleotides and nucleoside derivatives in transplant rejection and have examined purinergic mechanisms of inflammation in transplantation [94]. One of the key players in these processes is CD39, an ectonucleoside triphosphate diphosphohydrolase that hydrolyzes extracellular ATP and ADP to AMP and is uniquely expressed at high levels by vascular endothelium and also on T-regulatory cells [99,100]. Given species differences with low intrinsic CD39 basal functional expression in porcine tissues [95], it might be predicted that transgenic pig cells (bone marrow, passenger leukocytes, or vascular cells) overexpressing human CD39 would exert immunosuppressive effects locally. CD39 expression by vasculature and regulatory T-immune cells allows the integration of host responses by controlling both ATP-activatory and adenosine-mediated suppressive purinergic responses within the xenograft. CD39 and purinergic signaling pathways provide a potentially crucial bridge between vascular inflammation, thrombosis, and tolerance induction in xenotransplantation. Such integrated mechanisms are also possible in the setting of thymic xenotransplantation in which "xenothymic-educated or deleted" T-cells may ameliorate vascular injury of the graft.

69.1.5 Cellular Xenotransplantation—the Future of Xenotransplantation?

During the last 5 years, major developments and advancements in xenotransplantation have occurred in the field of cellular transplantation, both in terms of xenograft duration and function.

69.1.5.1 Pig Islet Xenotransplantation

Until recently, the maximum reported time of pig islet survival (insulin-positive cells, no function) was 53 days after xenotransplantation under the kidney capsule in non-diabetic cynomolgus monkeys utilizing immunosuppression with ATG, anti-IL-2R mAb, cyclosporin, and steroids [101]. In March 2006, however, two major studies were reported in which neonate or adult pig islet xenotransplantation in primates resulted in more than 180 days of survival (up to 344 days) and functionality [102,103]. The use of a heavy immunosuppressive regimen was, however, necessary to achieve these results. In particular, the use of anti-CD154 mAb, which induces thromboembolic events that preclude its clinical use [104], was a prerequisite in both studies. Such an immunosuppressive regimen seems unlikely to be acceptable in humans, but the results are still encouraging because an alternative immune regimen could be developed. In fact, the same teams recently showed interesting data obtained by infusing neonatal pig islets in primates and replacing anti-CD154 mAb with anti-CD40 mAb [105]. All these data were obtained by using wild-type pigs. In 2010, one report showed survival of 360 days for one animal expressing CD46 at the islet level. However, since anti-CD154 mAb was part of the immune regimen it was difficult to know whether the long survival was obtained because of the CD46 transgenesis or the anti-CD40L mAb [106]. In another recent experiment, Thompson et al. [107] showed that function and survival for neonatal pig islets from GalT-KO pigs were significantly better than for wild-type pig islets. These results clearly suggest that a lot of new data will be generated during the coming years, especially since specific expression of multiple transgenes is already a reality for complement, coagulation, and T-cell response at the level of pig islets. In fact, in 2011, Ayares et al. [108] reported at IXA in Miami the recent production of pigs that are GalT-KO and specifically express insulin promoter molecules such as CD46, TFPI, CD39, and CTLA4-Ig at the level of islets. Xenotransplantation of the islets from such pigs could have enormous potential for generating new information with a clinically applicable immunosuppressive regimen. With immunosuppression, it seems that these multiple GalT-KO transgenic pig islets could soon be used for pilot clinical trials.

These important results nevertheless suggested that alternatives, such as encapsulation of pig islets, need to be evaluated in the pig-to-primate model. In fact, encapsulation of pig (or human) islets in an alginate membrane permeable for glucose, insulin, and nutrients, but not for humoral and cellular components of the immune system, could allow successful transplantation in the absence of immunosuppressive medication. The microencapsulation process consists of placing one to three islets in a semipermeable immunoprotective capsule and simply injecting large numbers of these microcapsules, which are durable and difficult to disrupt mechanically [109]. Only a few studies have been reported regarding the biocompatibility of this type of microencapsulated pig islets in diabetic and nondiabetic primates, respectively [110,111]. Recently, it has been demonstrated that encapsulated adult pig islets can survive up to 6 months after transplantation in nondiabetic primates without immunosuppression when capsules are made with a very pure alginate (low level of endotoxin) and cultivated for 18 or 24 h in a serum-free medium containing a concentration of 1.8 mM of $CaCl_2$ prior to transplantation. Additionally, demonstration that the ratio of well-formed capsules must be over 90% to obtain a long-term *in vivo* biocompatibility in the pig-to-primate model was done [112]. Graft failure and strong fibrosis occurred when any of these conditions was not met. The survival of the pig islets for up to 6 months in the most stringent xenogeneic pig-to-primate model without any immunosuppression, despite an ongoing antipig-IgG response, confirmed that encapsulation protects the islets long term. Although the site under the kidney capsule is interesting for studying encapsulated islet grafts, more clinically useful sites need to be investigated. The absence of revascularization of the encapsulated islets interferes with both the functional performance and the longevity of the grafts. Reported

sites allowing for successful nonencapsulated islet transplantation, such as the liver and spleen, do not meet requirements since they do not tolerate the large volumes (>16 mL) of capsules (with a diameter over $600 \, \mu m$) required for primates; transplantation with such capsules constitutes a risk of severe thrombosis. Therefore, most transplantation of encapsulated pig islets into primates have been done intraperitoneally [110,111]. While this technique seems easy, the peritoneal site does not appear optimal. It is a preferential site for inflammation and immunological reactions [113], and peritoneal mesothelial cells facilitate the action of powerful innate immune mechanisms [113]. The subcutaneous space was thus investigated since it can be considered clinically useful. Subcutaneous spaces offered good alternatives to the peritoneal sites since biocompatibility, viability, and function of encapsulated pig islets were improved without any immunosuppression [112]. These results were directly translated to primates for preclinical studies.

There are few reports of transplantation of encapsulated pig islets in primates. Sun et al. reported normalization of blood glucose and insulin independence after transplantation of encapsulated adult pig islets in spontaneous diabetic cynomolgus monkeys for periods ranging from 120 to 804 days [110]. Although these results were very encouraging for the transplantation of encapsulated adult pig islets, several criticisms about the diabetic status of the recipients, the exact formulation of capsules, and the immune response against pig islets were raised. Only one additional, recent study describing xenotransplantation of microencapsulated neonatal pig islets in primates partially confirmed this result; however, there was only a mean reduction of 43% of the daily exogenous insulin in comparison with control animals and no changes in weekly blood glucose levels in either animal [111]. The lack of solid, consistent data on glucose metabolism (glycosylated hemoglobin evolution, glycosuria, intravenous glucose tolerance testing, etc.) renders this study difficult to interpret. However recently demonstration that macroencapsulated pig islets in a monolayer cellular device transplanted in the subcutaneous tissues of streptozotocin-diabetic primates can cure diabetes up to 6 months without any immunosuppression was obtained. Total diabetes correction was revealed by a glycosylated hemoglobin below 7% [114]. In addition, Veriter et al. also showed in the same diabetic primate model that a mixture of either bone marrow—derived Mesenchymal Stem Cells (MSCs) or adipose-derived MSCs may significantly improve the correction of the diabetes up to 7—8 months in numerous primates [131].

In the field of pig islet xenotransplantation, it therefore seems possible that major progress will be made in the next 5 years by using GalT-KO pig islets expressing multiple molecules involved in the complement cascade, coagulation, and T-cell response, and these grafts may function with a clinically applicable immune regimen (excluding anti-CD154 mAb). In the field of micro/ macroencapsulation, results now obtained without any immunosuppression may allow progression to pilot clinical trials as long as DPF pigs are used under GMP conditions. LCT in New Zealand has, in fact, begun a clinical trial by implanting microencapsulated pig islets in the peritoneum of eight patients. The actual results show an avoidance of uncontrolled hypoglycemia but no correction of the daily hyperglycemia.

69.1.5.2 Neuron Xenotransplantation

In neurodegenerative disorders, the cell replacement strategy, in which grafted cells develop and integrate, replacing the lost neurons within the host circuitry, is possible alone or in association with a neurotrophic or neuroprotective strategy, in which the grafted cells (genetically modified) can induce functional improvements by releasing *in situ* therapeutic molecules such as trophic factors or antiapoptotic/antioxidant agents. One of the prototypical neurodegenerative disorders amenable to such a cell therapy approach is Parkinson's disease. In this context, a primate model was developed and transgenic pigs specifically expressing CTLA4-Ig$^{+/+}$ at the level of fetal pig neurons were evaluated in the presence of peripheral immunosuppression. In this study, macaques (*Macaca fascicularis*) were rendered parkinsonian by systemic administration of the 1-methyl-4-phenyl-1,2,3,6-tetrahydropyridine (MPTP) toxin, a well-characterized model of Parkinson's disease in nonhuman primates that closely resemble the symptomatology in humans through depletion of the dopaminergic system. The reduction of dopaminergic innervations in the striatum severely affects locomotion and causes bradykinesia and/or akinesia, postural abnormalities, and tremor. Thus, the rationale was to transplant parkinsonian monkeys and assess their spontaneous locomotor activity before and after transplantation longitudinally. Xenografted animals were monitored for up to 320 days posttransplantation. Behavioral studies showed very significant recovery of spontaneous locomotion in most grafted animals that was still apparent at 11 months posttransplantation in one case. Such recovery is most probably due to the partial restoration of dopaminergic activity, which was detected by PET scans in at least five primates. Histological analysis of the brain from clinically improved animals revealed the existence of large porcine grafts composed of dopaminergic, serotoninergic, and GABAergic differentiated neurons and various glial components. These preliminary studies demonstrate that transplantation of CTLA4-Ig fetal porcine grafts in the striatum of immunosuppressed parkinsonian primates may enable long-term xenograft survival and differentiation, associated with considerable improvement of locomotor activity [92].

69.1.5.3 Cornea Xenotransplantation

Cornea transplantation has become a classical treatment for many types of blindness resulting from diseases affecting the anterior part of the eye. Allotransplantation of the cornea is well accepted, and with a local treatment most cases lead to excellent functional results. The main limitation of cornea grafting is access to donor corneas. Interestingly, animal experiments conducted in discordant species and some anecdotal accounts of xenotransplants in humans performed almost half a century ago have suggested that xenogenic corneal grafts are neither hyperacutely nor acutely rejected. Furthermore, local treatment (steroids) provides minor immunosuppression that allows lamellar transplants to remain functional for months. The immunoprivileged behavior of corneal transplantation, despite xenoantigens being expressed in the cornea, may be related to it being a nonvascularized graft and to specific local immune regulatory properties. Interestingly, and in contrast to vascularized organs, the rejection is mostly "cellular" when it occurs, despite a strong antibody response against xenoantigens (anti-Gal and non-Gal determinants). These unusual outcomes and characteristics have spurred a new wave of research in this field, and despite only a handful of laboratories working in this area in the world, old observations have been confirmed and the advantages of genetically modified animals and of new immunosuppressive treatment are currently being tested. The fact that pig-to-primate corneal grafts can work for months with only a minor local immunosuppressive regimen has to be appreciated to understand that corneal graft may become a major and an early clinical application of xenotransplantation.

69.1.6 Potential Zoonotic Risks

The risk of zoonosis is increased in xenotransplantation because normal host defenses are bypassed when human and animal tissues are placed in close contact *in vivo*. Animal pathogens that would normally not be of concern to human health may become infectious in a xenotransplantation setting. In addition, the use of immunosuppressive therapy to minimize graft rejection further exacerbates the risk of infection from otherwise noninfectious or latent animal pathogens. Cross-species infection is often unpredictable and, in recent years, the human population has been afflicted by animal-derived diseases such as avian flu, SARS, West Nile and Dengue fevers, bovine spongiform encephalopathy, AIDS, typhus, and plague [115]. Viruses are the predominant etiological agents for most of the outbreaks. For example, HIV is believed to have infected humans from a nonhuman primate species through rare zoonotic events, adapted to the new host, and silently spread in the human population [116]. The consequence of cross-species viral infection giving rise to a new disease and possible epidemic can be serious for both individual patients and the public. Xenotransplantation, unlike allotransplantation, which uses grafts sourced from a "free-range" human population, can provide clean organs/tissues free from known pathogens. Donor animals can be bred in special animal facilities under a strict health surveillance system, commonly referred to as specific pathogen-free (SPF) conditions. Indeed, the FDA and the World Health Organization have produced public health service guidelines for xenotransplantation covering animal facilities, donor animal selection, and screening for known infectious agents (http://www.fda.gov/BiologicsBloodVaccines/GuidanceComplianceRegulatory Information/Guidances/Xenotransplantation/default.htm). Pigs are considered the most promising species to be used as an animal donor. Among several advantages, the risk of transmission of infectious diseases from pigs to humans is lower than, for example, nonhuman primates. Although farming pigs in SPF conditions can eliminate most swine pathogens, viruses, bacteria, and parasites, certain types of viruses are more problematic: (i) porcine endogenous retroviruses (PERVs), which can infect human cells *in vitro* and are difficult to eliminate from porcine cells [117], and (ii) unknown or poorly characterized viruses.

69.1.7 PERV Transmission

Approximately 8% of the human genome and other vertebrates consists of retroviral sequences, or proviruses, that have been incorporated over millions of years. The majority of these proviral elements are defective, but in pigs, PERV, a member of the gamma-retrovirus (C-type retrovirus) family, is known to be transcriptionally active and results in the production of virions capable of infecting cells from a number of species *in vitro*[117,118]. PERVs were initially shown to be spontaneously produced from certain porcine cell lines and did not appear to infect human cells [119]. PERVs were first identified as being human-tropic in 1997 by Patience et al. [117] who demonstrated that an immortal porcine cell line could release a type-C retrovirus that infected human epithelial cells *in vitro*[117]. PERV was subsequently divided into three subtypes within the porcine genome, A, B, and C, according to host-cell tropism and the envelope sequence [118,120]. PERV A and B were found to infect selected human and pig cell lines, in addition to a number of other mammalian cell types, while PERV C was restricted to replication in porcine cell lines [10,16]. However, PERV C can recombine with PERV A and produce high-titer, human-tropic A/C recombinants [121]. Thus, if

xenotransplantation resulted in the expression of PERV *in vivo*, the risk of transmission of PERV to the recipient would raise the potential for selection and spread of new variants of PERV adapted to growth in the human host, first to immediate contacts and then to the general public. With regard to the development of organ-source pigs for humans, PERVs present a particularly difficult problem since little is known about their potential to infect human cells *in vivo* and their capacity to cause disease. Many studies have shown that cells derived from a number of animal species can be successfully infected *in vitro* [117,120,122,123]; however, productive infection of humans and nonhuman primates with PERV *in vivo* has yet to be shown [124–127]. This raises the question as to whether PERV is of real significance or if we have not yet found a suitable animal model to assess the relevance of PERV transmission *in vivo*. Support for the lack of a suitable animal model is evidenced by limited productive infection in certain species [128]. It was initially demonstrated that PERV could infect severe combined immunodeficiency (SCID) mice *in vivo*; however, later analysis revealed that this was not true infection but a cooperation between PERV and mouse retroviral elements [129]. In addition, there are some incidences that have clearly demonstrated the transient transmission of PERV *in vivo* without productive replication [130]. One potential problem is the inconsistency of methods used for PERV detection in recipients, and this should be improved.

In addition to PERV, zoonotic viruses have been identified that require specific diagnostic detection methods to confirm their absence. Even in the presence of careful monitoring of designated/SPF herds, there is the potential for viral contaminants to be present in pig tissues used in xenotransplantation. Therefore, porcine cytomegalovirus and porcine lymphotropic herpes virus (PLHV)-1, -2 and -3, porcine hepatitis E virus, and emerging viruses such as lymphocytic choriomeningitis virus or torque teno viruses of the genus *Anellovirus* need to be considered. A comprehensive list specifying which pathogens must be excluded by definition would clarify the level of appropriate testing required and build more consistency in source animals produced for these purposes.

69.1.8 Overall Conclusions

Despite the remaining challenges, the potential benefit offered by xenotransplantation, namely a solution for the current shortage of human organs for transplantation, makes it an area of research that should be strongly pursued. Important experimental data have been obtained with pig islet and neuronal cell xenotransplantation. This is especially true for pig islet cells that can be used in humans without immunosuppression if encapsulated. As far as solid-organ xenotransplantation is concerned, to date, genetic modifications in donors have not yet provided the necessary survival to envisage clinical use, but the existence of transgenic pigs on a background of GalT-KO should provide significant improvements in the near future.

ACKNOWLEDGMENTS

Members of FP6 EU project XENOME participated in the writing of this chapter through a monograph: M. Vadori, M.L. Lavitrano, E. Cozzi, J.P. Soulillou, G. Blancho, L. Scobie, Y. Takeuchi, Y. Yamada, S. Le Bas-Bernardet, R. Aron Badin. "Scientific, ethical, social and legal aspects of xenotransplantation" C.R. Casabona, M. Jorqui Azofra, and E. Cozzi (Eds.).

REFERENCES

[1] De Vito DA, Dauber JH, Hoffman LA. Rejection after organ transplantation: a historical review. Am J Crit Care 2000;9:419—29.

[2] Deschamps JY, Roux FA, Sai P, Gouin E. History of xenotransplantation. Xenotransplantation 2005;12:91—109.

[3] Farr AD. The first human blood transfusion. Med Hist 1980;24:143—62.

[4] Voronoff S, Imianitoff F. Rejuvenation by grafting. New York, NY: Adelphi; 1925.

[5] Jaboulay M. Greffe de reins au pli de coude par soudures artérielles et veineuse. Lyon Med 1906;10:575.

[6] Unger E. Nierentransplantationem. Klin Wochenschr 1910;47:573.

[7] Anzani A. Trapianto d'organo: problemi etici, aspetti sociali. Milano: Lauri; 1996.

[8] Reemtsma K, Mccracken BH, Schlegel JU, Pearl MA, Pearce CW, Dewitt CW, et al. Renal heterotransplantation in man. Ann Surg 1964;160:384—410.

[9] Stefanini P, Cortesini R, Casciani C, Arullani A, Cucchiara G, Ancarani E, et al. Kidney transplantation. Int Surg 1968;49:181—6.

[10] Starzl TE, Tzakis A, Fung JJ, Todo S, Demetris AJ, Manez R, et al. Prospects of clinical xenotransplantation. Transplant Proc 1994;26:1082—8.

[11] Hardy JD, Chavez CM. The first heart transplant in man. Developmental animal investigations with analysis of the 1964 case in the light of current clinical experience. Am J Cardiol 1968;22:772—81.

[12] Barnard CN, Wolpowitz A, Losman JG. Heterotopic cardiac transplantation with a xenograft for assistance of the left heart in cardiogenic shock after cardiopulmonary bypass. S Afr Med J 1977;52:1035—8.

[13] Bailey LL, Nehlsen-Cannarella SL, Concepcion W, Jolley WB. Baboon-to-human cardiac xenotransplantation in a neonate. JAMA 1985;254:3321—9.

[14] Gianello P, Latinne D, Alexandre GPJ. Pig to baboon xenograft. Xeno 1995;3(3):26—30.

[15] Galili U, Gregory CR, Morris RE. Contribution of anti-Gal to primate and human IgG binding to porcine endothelial cells. Transplantation 1995;60:210—3.

[16] Schuurman HJ, Pino-Chavez G, Phillips MJ, Thomas L, White DJ, Cozzi E. Incidence of hyperacute rejection in pig-to-primate transplantation using organs from hDAF-transgenic donors. Transplantation 2002;73:1146—51.

[17] Cozzi E, Vial C, Ostlie D, Farah B, Chavez G, Smith KG, et al. Maintenance triple immunosuppression with cyclosporin A, mycophenolate sodium and steroids allows prolonged survival of primate recipients of hDAF porcine renal xenografts. Xenotransplantation 2003;10:300—10.

[18] Baldan N, Rigotti P, Calabrese F, Cadrobbi R, Dedja A, Iacopetti I, et al. Ureteral stenosis in HDAF pig-to-primate renal xenotransplantation: a phenomenon related to immunological events? Am J Transplant 2004;4:475—81.

[19] Zaidi A, Bhatti F, Schmoeckel M, Cozzi E, Chavez G, Wallwork J, et al. Kidneys from HDAF transgenic pigs are physiologically compatible with primates. Transplant Proc 1998;30:2465—6.

[20] Cozzi E, Bhatti F, Schmoeckel M, Chavez G, Smith KG, Zaidi A, et al. Long-term survival of nonhuman primates receiving life-supporting transgenic porcine kidney xenografts. Transplantation 2000;70:15—21.

[21] Ghanekar A, Luo Y, Yang H, Garcia B, Luke P, Chakrabarti S, et al. The alpha-Gal analog GAS914 ameliorates delayed rejection of hDAF transgenic pig-to-baboon renal xenografts. Transplant Proc 2001;33:3853—4.

[22] Vangerow B, Hecker JM, Lorenz R, Loss M, Przemeck M, Appiah R, et al. C1-Inhibitor for treatment of acute vascular xenograft rejection in cynomolgus recipients of h-DAF transgenic porcine kidneys. Xenotransplantation 2001;8:266—72.

[23] Lam TT, Hausen B, Squiers E, Cozzi E, Morris RE. Cyclophosphamide-induced postoperative anemia in cynomolgus monkey recipients of hDAF-transgenic pig organ xenografts. Transplant Proc 2002;34:1451—2.

[24] Richards AC, Davies HF, McLaughlin ML, Copeman LS, Holmes BJ, Dos Santos CG, et al. Serum anti-pig antibodies as potential indicators of acute humoral xenograft rejection in pig-to-cynomolgus monkey kidney transplantation. Transplantation 2002;73:881—9.

[25] Barth RN, Yamamoto S, LaMattina JC, Kumagai N, Kitamura H, Vagefi PA, et al. Xenogeneic thymokidney and thymic tissue transplantation in a pig-to-baboon model: I. Evidence for pig-specific T-cell unresponsiveness. Transplantation 2003;75:1615—24.

[26] Ashton-Chess J, Roussel JC, Bernard P, Barreau N, Karam G, Dantal J, et al. The effect of immunoglobulin immunoadsorptions on delayed xenograft rejection of human CD55 transgenic pig kidneys in baboons. Xenotransplantation 2003;10:552—61.

[27] Cozzi E, Simioni P, Boldrin M, Seveso M, Calabrese F, Baldan N, et al. Alterations in the coagulation profile in renal pig-to-monkey xenotransplantation. Am J Transplant 2004;4:335—45.

[28] Shimizu A, Yamada K, Yamamoto S, Lavelle JM, Barth RN, Robson SC, et al. Thrombotic microangiopathic glomerulopathy in human decay accelerating factor-transgenic swine-to-baboon kidney xenografts. J Am Soc Nephrol 2005;16:2732—45.

[29] Chen G, Sun H, Yang H, Kubelik D, Garcia B, Luo Y, et al. The role of anti-non-Gal antibodies in the development of acute humoral xenograft rejection of hDAF transgenic porcine kidneys in baboons receiving anti-Gal antibody neutralization therapy. Transplantation 2006;81:273—83.

[30] Tu CF, Tai HC, Wu CP, Ho LL, Lin YJ, Hwang CS, et al. The *in vitro* protection of human decay accelerating factor and hDAF/heme oxygenase-1 transgenes in porcine aortic endothelial cells against sera of Formosan macaques. Transplant Proc 2010;42:2138−41.

[31] Cowan PJ, Aminian A, Barlow H, Brown AA, Chen CG, Fisicaro N, et al. Renal xenografts from triple-transgenic pigs are not hyperacutely rejected but cause coagulopathy in non-immunosuppressed baboons. Transplantation 2000;69:2504−15.

[32] Schmoeckel M, Bhatti FN, Zaidi A, Cozzi E, Waterworth PD, Tolan MJ, et al. Orthotopic heart transplantation in a transgenic pig-to-primate model. Transplantation 1998;65:1570−7.

[33] Vial CM, Ostlie DJ, Bhatti FN, Cozzi E, Goddard M, Chavez GP, et al. Life supporting function for over one month of a transgenic porcine heart in a baboon. J Heart Lung Transplant 2000;19:224−9.

[34] Brandl U, Michel S, Erhardt M, Brenner P, Burdorf L, Jockle H, et al. Transgenic animals in experimental xenotransplantation models: orthotopic heart transplantation in the pig-to-baboon model. Transplant Proc 2007;39:577−8.

[35] Brandl U, Michel S, Erhardt M, Brenner P, Bittmann I, Rossle M, et al. Administration of GAS914 in an orthotopic pig-to-baboon heart transplantation model. Xenotransplantation 2005;12:134−41.

[36] Waterworth PD, Dunning J, Tolan M, Cozzi E, Langford G, Chavez G, et al. Life-supporting pig-to-baboon heart xenotransplantation. J Heart Lung Transplant 1998;17:1201−7.

[37] Cozzi E, Langford GA, Wright L, Tucker A, Yannoutsos N, Richards A, et al. Comparative analysis of human DAF expression in the tissues of transgenic pigs and man. Transplant Proc 1995;27:319−20.

[38] Waterworth PD, Cozzi E, Tolan MJ, Langford G, Braidley P, Chavez G, et al. Pig-to-primate cardiac xenotransplantation and cyclophosphamide therapy. Transplant Proc 1997;29:899−900.

[39] McCurry KR, Kooyman DL, Alvarado CG, Cotterell AH, Martin MJ, Logan JS, et al. Human complement regulatory proteins protect swine-to-primate cardiac xenografts from humoral injury. Nat Med 1995;1:423−7.

[40] Lin SS, Hanaway MJ, Gonzalez-Stawinski GV, Lau CL, Parker W, Davis RD, et al. The role of anti-Galalpha1-3Gal antibodies in acute vascular rejection and accommodation of xenografts. Transplantation 2000;70:1667−74.

[41] Adams DH, Kadner A, Chen RH, Farivar RS. Human membrane cofactor protein (MCP, CD 46) protects transgenic pig hearts from hyperacute rejection in primates. Xenotransplantation 2001;8:36−40.

[42] Chen RH, Naficy S, Logan JS, Diamond LE, Adams DH. Hearts from transgenic pigs constructed with CD59/DAF genomic clones demonstrate improved survival in primates. Xenotransplantation 1999;6:194−200.

[43] McGregor CG, Teotia SS, Byrne GW, Michaels MG, Risdahl JM, Schirmer JM, et al. Cardiac xenotransplantation: progress toward the clinic. Transplantation 2004;78:1569−75.

[44] McGregor CG, Davies WR, Oi K, Teotia SS, Schirmer JM, Risdahl JM, et al. Cardiac xenotransplantation: recent preclinical progress with 3-month median survival. J Thorac Cardiovasc Surg 2005;130:844−51.

[45] Byrne GW, Davies WR, Oi K, Rao VP, Teotia SS, Ricci D, et al. Increased immunosuppression, not anticoagulation, extends cardiac xenograft survival. Transplantation 2006;82:1787−91.

[46] Wu G, Pfeiffer S, Schroder C, Zhang T, Nguyen BN, Kelishadi S, et al. Coagulation cascade activation triggers early failure of pig hearts expressing human complement regulatory genes. Xenotransplantation 2007;14:34−47.

[47] Houser SL, Kuwaki K, Knosalla C, Dor FJ, Gollackner B, Cheng J, et al. Thrombotic microangiopathy and graft arteriopathy in pig hearts following transplantation into baboons. Xenotransplantation 2004;11:416−25.

[48] Kuwaki K, Knosalla C, Dor FJ, Gollackner B, Tseng YL, Houser S, et al. Suppression of natural and elicited antibodies in pig-to-baboon heart transplantation using a human anti-human CD154 mAb-based regimen. Am J Transplant 2004;4:363−72.

[49] Bhatti FN, Schmoeckel M, Zaidi A, Cozzi E, Chavez G, Goddard M, et al. Three-month survival of HDAFF transgenic pig hearts transplanted into primates. Transplant Proc 1999;31:958.

[50] Czaplicki J, Blonska B, Religa Z. The lack of hyperacute xenogeneic heart transplant rejection in a human. J Heart Lung Transplant 1992;11:393−7.

[51] Makowa L, Cramer DV, Hoffman A, Breda M, Sher L, Eiras-Hreha G, et al. The use of a pig liver xenograft for temporary support of a patient with fulminant hepatic failure. Transplantation 1995;59:1654−9.

[52] Groth CG, Korsgren O, Tibell A, Tollemar J, Moller E, Bolinder J, et al. Transplantation of porcine fetal pancreas to diabetic patients. Lancet 1994;344:1402−4.

[53] Groth CG, Tibell A, Wennberg L, Korsgren O. Xenoislet transplantation: experimental and clinical aspects. J Mol Med 1999;77:153−4.

[54] Lai L, Kolber-Simonds D, Park KW, Cheong HT, Greenstein JL, Im GS, et al. Production of alpha-1,3-galactosyltransferase knockout pigs by nuclear transfer cloning. Science 2002;295:1089−92.

[55] Kolber-Simonds D, Lai L, Watt SR, Denaro M, Arn S, Augenstein ML, et al. Production of alpha-1,3-galactosyltransferase null pigs by means of nuclear transfer with fibroblasts bearing loss of heterozygosity mutations. Proc Natl Acad Sci USA 2004;101:7335−40.

[56] Chen G, Qian H, Starzl T, Sun H, Garcia B, Wang X, et al. Acute rejection is associated with antibodies to non-Gal antigens in baboons using Gal-knockout pig kidneys. Nat Med 2005;11:1295−8.

[57] Yamada K, Yazawa K, Shimizu A, Iwanaga T, Hisashi Y, Nuhn M, et al. Marked prolongation of porcine renal xenograft survival in baboons through the use of alpha1,3-galactosyltransferase gene-knockout donors and the cotransplantation of vascularized thymic tissue. Nat Med 2005;11:32−4.

[58] Ezzelarab M, Garcia B, Azimzadeh A, Sun H, Lin CC, Hara H, et al. The innate immune response and activation of coagulation in alpha1,3-galactosyltransferase gene-knockout xenograft recipients. Transplantation 2009;87:805−12.

[59] Le Bas-Bernardet S, Tillou X, Poirier N, Dilek N, Chatelais M, Devalliere J, et al. Xenotransplantation of galactosyl-transferase knockout, CD55, CD59, CD39, and fucosyl-transferase transgenic pig kidneys into baboons. Transplant Proc 2011;43:3426−30.

[60] Nishimura H, Scalea J, Wang Z, Shimizu A, Moran S, Gillon B, et al. First experience with the use of a recombinant CD3

immunotoxin as induction therapy in pig-to-primate xenotransplantation: the effect of T-cell depletion on outcome. Transplantation 2011;92:641−7.

[61] Griesemer AD, Hirakata A, Shimizu A, Moran S, Tena A, Iwaki H, et al. Results of gal-knockout porcine thymokidney xenografts. Am J Transplant 2009;9:2669−78.

[62] Kuwaki K, Tseng YL, Dor FJ, Shimizu A, Houser SL, Sanderson TM, et al. Heart transplantation in baboons using alpha1,3-galactosyltransferase gene-knockout pigs as donors: initial experience. Nat Med 2005;11:29−31.

[63] Shimizu A, Hisashi Y, Kuwaki K, Tseng YL, Dor FJ, Houser SL, et al. Thrombotic microangiopathy associated with humoral rejection of cardiac xenografts from alpha1,3-galactosyltransferase gene-knockout pigs in baboons. Am J Pathol 2008;172:1471−81.

[64] Hisashi Y, Yamada K, Kuwaki K, Tseng YL, Dor FJ, Houser SL, et al. Rejection of cardiac xenografts transplanted from alpha1,3-galactosyltransferase gene-knockout (GalT-KO) pigs to baboons. Am J Transplant 2008;8:2516−26.

[65] Bauer A, Postrach J, Thormann M, Blanck S, Faber C, Wintersperger B, et al. First experience with heterotopic thoracic pig-to-baboon cardiac xenotransplantation. Xenotransplantation 2010;17:243−9.

[66] Mohiuddin MM, Corcoran PC, Singh AK, Azimzadeh A, Hoyt Jr RF, Thomas ML, et al. B-cell depletion extends the survival of GTKO.hCD46Tg pig heart xenografts in baboons for up to 8 months. Am J Transplant 2012;12:763−71.

[67] McGregor CG, Ricci D, Miyagi N, Stalboerger PG, Du Z, Oehler EA, et al. Human CD55 expression blocks hyperacute rejection and restricts complement activation in Gal knockout cardiac xenografts. Transplantation 2012;93:686−92.

[68] Ramirez P, Chavez R, Majado M, Munitiz V, Munoz A, Hernandez Q, et al. Life-supporting human complement regulator decay accelerating factor transgenic pig liver xenograft maintains the metabolic function and coagulation in the nonhuman primate for up to 8 days. Transplantation 2000;70:989−98.

[69] Ekser B, Echeverri GJ, Hassett AC, Yazer MH, Long C, Meyer M, et al. Hepatic function after genetically engineered pig liver transplantation in baboons. Transplantation 2010;90:483−93.

[70] Nguyen BN, Azimzadeh AM, Zhang T, Wu G, Schuurman HJ, Sachs DH, et al. Life-supporting function of genetically modified swine lungs in baboons. J Thorac Cardiovasc Surg 2007;133:1354−63.

[71] Nguyen BN, Azimzadeh AM, Schroeder C, Buddensick T, Zhang T, Laaris A, et al. Absence of Gal epitope prolongs survival of swine lungs in an *ex vivo* model of hyperacute rejection. Xenotransplantation 2011;18:94−107.

[72] Cowan PJ, Robson SC, d'Apice AJ. Controlling coagulation dysregulation in xenotransplantation. Curr Opin Organ Transplant 2011;16:214−21.

[73] Lin CC, Cooper DK, Dorling A. Coagulation dysregulation as a barrier to xenotransplantation in the primate. Transpl Immunol 2009;21:75−80.

[74] Lee HJ, Lee BC, Kim YH, Paik NW, Rho HM. Characterization of transgenic pigs that express human decay accelerating factor and cell membrane-tethered human tissue factor pathway inhibitor. Reprod Domest Anim 2011;46:325−32.

[75] Miwa Y, Yamamoto K, Onishi A, Iwamoto M, Yazaki S, Haneda M, et al. Potential value of human thrombomodulin and DAF

expression for coagulation control in pig-to-human xenotransplantation. Xenotransplantation 2010;17:26−37.

[76] Dwyer KM, Robson SC, Nandurkar HH, Campbell DJ, Gock H, Murray-Segal LJ, et al. Thromboregulatory manifestations in human CD39 transgenic mice and the implications for thrombotic disease and transplantation. J Clin Invest 2004;113:1440−6.

[77] Cozzi E, Simioni P, Nottle M, Vadori M, De Benedictis GM, Baldan N, et al. Late Breaking IXA Oral Presentations : preliminary study in a life supporting pig to primate xenotransplantation model using GAL KO pigs transgenic for human CD39, CD55, CD59 and fucosyltransferase. Xenotransplantation 2009;16:544.

[78] Weiss EH, Lilienfeld BG, Muller S, Muller E, Herbach N, Kessler B, et al. HLA-E/human beta2-microglobulin transgenic pigs: protection against xenogeneic human anti-pig natural killer cell cytotoxicity. Transplantation 2009;87:35−43.

[79] Tena A, Turcotte N, Leto Barone A, Arn S, Terlouw S, Dobrinski J, et al. Parallel session 3: genetic engineering: miniature swine expressing human CD47 to enhance bone marrow engraftment in non-human primates. Xenotransplantation 2011;18:271.

[80] Ayares D, Phelps C, Vaught TD, Ball S, Mendicino M, Ramsoondar JJ, et al. Parallel session 3: genetic engineering: multitransgenic pigs for vascularized organ xenografts. Xenotransplantation 2011;18:269.

[81] Phelps CJ, Ball SF, Vaught TD, Vance AM, Mendicino M, Monahan JA, et al. Production and characterization of transgenic pigs expressing porcine CTLA4-Ig. Xenotransplantation 2009;16:477−85.

[82] Martin C, Plat M, Nerriere-Daguin V, Coulon F, Uzbekova S, Venturi E, et al. Transgenic expression of CTLA4-Ig by fetal pig neurons for xenotransplantation. Transgenic Res 2005;14:373−84.

[83] Aron Badin R, Padoan A, Vadori M, Boldrin M, Cavicchioli L, De Benedictis GM, et al. Late Breaking IXA Oral Presentations: transgenic porcine embryonic xenografts as a treatment for Parkinson's disease in non-human primates. Xenotransplantation 2009;16:543.

[84] Klymiuk N, Bocker W, Schonitzer V, Bahr A, Radic T, Frohlich T, et al. First inducible transgene expression in porcine large animal models. FASEB J 2012;26:1086−99.

[85] Oropeza M, Petersen B, Carnwath JW, Lucas-Hahn A, Lemme E, Hassel P, et al. Transgenic expression of the human A20 gene in cloned pigs provides protection against apoptotic and inflammatory stimuli. Xenotransplantation 2009;16:522−34.

[86] Petersen B, Ramackers W, Lucas-Hahn A, Lemme E, Hassel P, Queisser AL, et al. Transgenic expression of human heme oxygenase-1 in pigs confers resistance against xenograft rejection during *ex vivo* perfusion of porcine kidneys. Xenotransplantation 2011;18:355−68.

[87] Klose R, Kemter E, Bedke T, Bittmann I, Kelsser B, Endres R, et al. Expression of biologically active human TRAIL in transgenic pigs. Transplantation 2005;80:222−30.

[88] Hauschild J, Petersen B, Santiago Y, Queisser AL, Carnwath JW, Lucas-Hahn A, et al. Efficient generation of a biallelic knockout in pigs using zinc-finger nucleases. Proc Natl Acad Sci USA 2011;108:12013−7.

[89] Ekser B, Kumar G, Veroux M, Cooper DK. Therapeutic issues in the treatment of vascularized xenotransplants using gal-knockout donors in nonhuman primates. Curr Opin Organ Transplant 2011;16:222−30.

[90] Byrne GW, Stalboerger PG, Davila E, Heppelmann CJ, Gazi MH, McGregor HC, et al. Proteomic identification of non-Gal antibody targets after pig-to-primate cardiac xenotransplantation. Xenotransplantation 2008;15:268—76.

[91] Byrne GW, Stalboerger PG, Du Z, Davis TR, McGregor CG. Identification of new carbohydrate and membrane protein antigens in cardiac xenotransplantation. Transplantation 2011;91:287—92.

[92] Kamano C, Vagefi PA, Kumagai N, Yamamoto S, Barth RN, LaMattina JC, et al. Vascularized thymic lobe transplantation in miniature swine: thymopoiesis and tolerance induction across fully MHC-mismatched barriers. Proc Natl Acad Sci USA 2004;101:3827—32.

[93] Yamada K, Shimizu A, Utsugi R, Ierino FL, Gargollo P, Haller GW, et al. Thymic transplantation in miniature swine. II. Induction of tolerance by transplantation of composite thymokidneys to thymectomized recipients. J Immunol 2000;164:3079—86.

[94] Robson SC, Wu Y, Sun X, Knosalla C, Dwyer K, Enjyoji K. Ectonucleotidases of CD39 family modulate vascular inflammation and thrombosis in transplantation. Semin Thromb Hemost 2005;31:217—33.

[95] Khalpey Z, Yuen AH, Lavitrano M, McGregor CG, Kalsi KK, Yacoub MH, et al. Mammalian mismatches in nucleotide metabolism: implications for xenotransplantation. Mol Cell Biochem 2007;304:109—17.

[96] Sablinski T, Gianello PR, Bailin M, Bergen KS, Emery DW, Fishman JA, et al. Pig to monkey bone marrow and kidney xenotransplantation. Surgery 1997;121:381—91.

[97] Tseng YL, Dor FJ, Kuwaki K, Ryan D, Wood J, Denaro M, et al. Bone marrow transplantation from alpha1,3-galactosyltransferase gene-knockout pigs in baboons. Xenotransplantation 2004;11:361—70.

[98] Ide K, Wang H, Tahara H, Liu J, Wang X, Asahara T, et al. Role for CD47-SIRPalpha signaling in xenograft rejection by macrophages. Proc Natl Acad Sci USA 2007;104:5062—6.

[99] Robson SC, Kaczmarek E, Siegel JB, Candinas D, Koziak K, Millan M, et al. Loss of ATP diphosphohydrolase activity with endothelial cell activation. J Exp Med 1997;185:153—63.

[100] Shalev I, Schmelzle M, Robson SC, Levy G. Making sense of regulatory T cell suppressive function. Semin Immunol 2011;23:282—92.

[101] Rijkelijkhuizen JK, Haanstra KG, Wubben J, Tons A, Roos A, van Gijlswijk-Janssen DJ, et al. T-cell-specific immunosuppression results in more than 53 days survival of porcine islets of Langerhans in the monkey. Transplantation 2003;76:1359—68.

[102] Hering BJ, Wijkstrom M, Graham ML, Hardstedt M, Aasheim TC, Jie T, et al. Prolonged diabetes reversal after intraportal xenotransplantation of wild-type porcine islets in immunosuppressed nonhuman primates. Nat Med 2006;12:301—3.

[103] Cardona K, Korbutt GS, Milas Z, Lyon J, Cano J, Jiang W, et al. Long-term survival of neonatal porcine islets in nonhuman primates by targeting costimulation pathways. Nat Med 2006;12:304—6.

[104] Schuler W, Bigaud M, Brinkmann V, Di PF, Geisse S, Gram H, et al. Efficacy and safety of ABI793, a novel human anti-human CD154 monoclonal antibody, in cynomolgus monkey renal allotransplantation. Transplantation 2004;77:717—26.

[105] Thompson P, Cardona K, Russell M, Badell IR, Shaffer V, Korbutt G, et al. CD40-specific costimulation blockade enhances neonatal porcine islet survival in nonhuman primates. Am J Transplant 2011;11:947—57.

[106] van der Windt DJ, Bottino R, Casu A, Campanile N, Smetanka C, He J, Murase N, Hara H, Ball S. Long-term controlled normoglycemia in diabetic non-human primates after transplantation with hCD46 transgenic porcine islets. Am J Transplant 2009;9:2716—26.

[107] Thompson P, Badell IR, Lowe M, Cano J, Song M, Leopardi F, et al. Islet xenotransplantation using gal-deficient neonatal donors improves engraftment and function. Am J Transplant 2011;11:2593—602.

[108] Ayares D, Vaught TD, Ball S, Ramsoondar JJ, Monahan JA, Mendicino M, et al. Parallel session 3: genetic engineering: islet-specific expression of TFPI, CD39, and CTLA4Ig in transgenic pigs designed for xenoislet transplantation. Xenotransplantation 2011;18:269.

[109] Omer A, Duvivier-Kali VF, Trivedi N, Wilmot K, Bonner-Weir S, Weir GC. Survival and maturation of microencapsulated porcine neonatal pancreatic cell clusters transplanted into immunocompetent diabetic mice. Diabetes 2003;52:69—75.

[110] Sun Y, Ma X, Zhou D, Vacek I, Sun AM. Normalization of diabetes in spontaneously diabetic cynomologus monkeys by xenografts of microencapsulated porcine islets without immunosuppression. J Clin Invest 1996;98:1417—22.

[111] Elliott RB, Escobar L, Tan PL, Garkavenko O, Calafiore R, Basta P, et al. Intraperitoneal alginate-encapsulated neonatal porcine islets in a placebo-controlled study with 16 diabetic cynomolgus primates. Transplant Proc 2005;37:3505—8.

[112] Dufrane D, Goebbels RM, Saliez A, Guiot Y, Gianello P. Six-month survival of microencapsulated pig islets and alginate biocompatibility in primates: proof of concept. Transplantation 2006;81:1345—53.

[113] Hall JC, Heel KA, Papadimitriou JM, Platell C. The pathobiology of peritonitis. Gastroenterology 1998;114:185—96.

[114] Dufrane D, Goebbels RM, Gianello P. Alginate macroencapsulation of pig islets allows correction of streptozotocin-induced diabetes in primates up to 6 months without immunosuppression. Transplantation 2010;90:1054—62.

[115] Weiss RA. Cross-species infections. Curr Top Microbiol Immunol 2003;278:47—71.

[116] Gao F, Bailes E, Robertson DL, Chen Y, Rodenburg CM, Michael SF, et al. Origin of HIV-1 in the chimpanzee *Pan troglodytes troglodytes*. Nature 1999;397:436—41.

[117] Patience C, Takeuchi Y, Weiss RA. Infection of human cells by an endogenous retrovirus of pigs. Nat Med 1997;3:282—6.

[118] Takeuchi Y, Patience C, Magre S, Weiss RA, Banerjee PT, Le TP, Stoye JP. Host range and interference studies of three classes of pig endogenous retrovirus. J Virol 1998;72:9986—91.

[119] Moennig V, Frank H, Hunsmann G, Ohms P, Schwarz H, Schafer W. C-type particles produced by a permanent cell line from a leukemic pig. II. Physical, chemical, and serological characterization of the particles. Virology 1974;57:179—88.

[120] Wilson CA, Wong S, VanBrocklin M, Federspiel MJ. Extended analysis of the *in vitro* tropism of porcine endogenous retrovirus. J Virol 2000;74:49—56.

[121] Scobie L, Taylor S, Wood JC, Suling KM, Quinn G, Meikle S, et al. Absence of replication-competent human-tropic porcine endogenous retroviruses in the germ line DNA of inbred miniature swine. J Virol 2004;78:2502—9.

[122] Mattiuzzo G, Matouskova M, Takeuchi Y. Differential resistance to cell entry by porcine endogenous retrovirus subgroup A in rodent species. Retrovirology 2007;4:93.

[123] Martin U, Kiessig V, Blusch JH, Haverich A, von der HK, Herden T, et al. Expression of pig endogenous retrovirus by primary porcine endothelial cells and infection of human cells. Lancet 1998;352:692—4.

[124] Moscoso I, Hermida-Prieto M, Manez R, Lopez-Pelaez E, Centeno A, Diaz TM, et al. Lack of cross-species transmission of porcine endogenous retrovirus in pig-to-baboon xenotransplantation with sustained depletion of anti-alphaGal antibodies. Transplantation 2005;79:777—82.

[125] Garkavenko O, Dieckhoff B, Wynyard S, Denner J, Elliott RB, Tan PL, et al. Absence of transmission of potentially xenotic viruses in a prospective pig to primate islet xenotransplantation study. J Med Virol 2008;80:2046—52.

[126] Paradis K, Langford G, Long Z, Heneine W, Sandstrom P, Switzer WM, et al. Search for cross-species transmission of porcine endogenous retrovirus in patients treated with living pig tissue. The XEN 111 Study Group. Science 1999;285:1236—41.

[127] Levy MF, Argaw T, Wilson CA, Brooks J, Sandstrom P, Merks H, et al. No evidence of PERV infection in healthcare workers exposed to transgenic porcine liver extracorporeal support. Xenotransplantation 2007;14:309—15.

[128] Denner J, Specke V, Schwendemann J, Tacke SJ. Porcine endogenous retroviruses (PERVs): adaptation to human cells and attempts to infect small animals and non-human primates. Ann Transplant 2001;6:25—33.

[129] Martina Y, Kurian S, Cherqui S, Evanoff G, Wilson C, Salomon DR. Pseudotyping of porcine endogenous retrovirus by xenotropic murine leukemia virus in a pig islet xenotransplantation model. Am J Transplant 2005;5:1837—47.

[130] Martina Y, Marcucci KT, Cherqui S, Szabo A, Drysdale T, Srinivisan U, et al. Mice transgenic for a human porcine endogenous retrovirus receptor are susceptible to productive viral infection. J Virol 2006;80:3135—46.

[131] Veriter et al., Cell Transplantation 2013 Feb 4. http://dx.doi.org/10.3727/096368913X663550 [Epub ahead of print].

Chapter 70

Regenerative Medicine as an Industry

Alessandro Sannino[a] and Paolo A. Netti[b]

[a]Department of Engineering for Innovation, University of SalentoVia per Monteroni—Complesso Ecotekne, Lecce, Italy
[b]Centre for Advanced Biomaterials for Health Care, IIT@CRIB, Istituto Italiano di Tecnologia, Largo Barsanti e Matteucci, Napoli, Italy

Chapter Outline

70.1 INTRODUCTION

Under the Tissue Engineering (TE) realm a rather wide number of approaches is usually assembled—all devoted to tissue and organ regeneration—stemming from an even wider technological platforms, from medical devices to stem cells. A detailed history of the genesis and primordial TE approach has been provided by Lysaght et al. [1]. Ancient Egyptians might have been the first to apply TE principles to wound care around 1500 BC [2,3] even if, businesswise, it is not clear whether their products were actually sold in ancient Egypt and whether they had any competitors.

In our time, the modern concept of TE was enunciated and technologically proposed around the mid-1980s from a group of researchers working in the Boston area between MIT and Harvard. In the middle of the 80s, Langer and Vacanti [4] defined and formalized TE approach as an "interdisciplinary field that applies the principles of engineering and life sciences toward the development of biological substitutes that restore, maintain, or improve tissue function or a whole organ". Yannas and Burke [5] in the early 80s developed the first artificial skin for dermal regeneration based on a collagen-GAG matrix known today under the commercial name of Integra®. In the same period there were also other products for skin regeneration, including Transcyte®, Dermagraft®, Apligraf®, and Epicel®. TE began to emerge as an industry and by 1994 there were $246 million invested in the private sector and about 40 companies [6]. Transcyte®, albeit produced using cells, was an acellular product. Dermagraft®, on the other hand, was a dermal equivalent realized from scaffold seeded fibroblasts derived from foreskin [7]. Apligraf®, produced by Organogenesis, was made with collagen and included both a dermal equivalent and an epidermis [7]. Epicel® was presented by Genzyme which also presented Carticel for cartilage replacement. Epicel® was based on the work of Green [8], while Carticel was based on the work of Brittberg [9].

By the end of the millennium, total private sector investment experienced a remarkable increase to $600 million with total companies around 70 [10].

However, at the beginning of the twenty-first century, the analysis of the TE industry turned into a less exciting picture. A reality check showed a scenario where it was difficult to get any kind of real return on investment. Advanced Tissue Sciences, one of the most active companies with Transcyte® and Dermagraft® products, declared bankruptcy and liquidated with a loss of more than $300 million. Organogenesis also experienced severe financial problems, and it also went bankrupt. Several other companies turned into service businesses instead of product businesses.

More recent data reported by Lysaght et al. [1] indicated that the TE industry bounced back to a more positive track pointing again to great expectations. However, the strategy of the companies operating in TE sectors has significantly changed compared to the original scheme of the mid-90s in that most of them present products that do not contain cells. These include recombinant morphogenic products such as INFUSE or regenerative biomaterials based on decellularized matrices [1]. Also old products were represented and valorized. Organogenesis was able to recover from bankruptcy and Apligraf® has been further developed. Transcyte® and Dermagraft® have been acquired and represented by a new company, Advanced BioHealing.

There have been several factors that have contributed to the alternating financial and economical outcome of the TE industry sector and their understanding can help this industry to meet its promising potential and address the clinical shortage that is still unmet.

70.2 TISSUE ENGINEERING AND REGENERATIVE MEDICINE

An important aspect that should be mentioned to fully appreciate the TE industry performance and development is the contemporary emergence of stem cells technology and the progressive morphing of TE into regenerative medicine. With the aim of defining the borders of different tissue regeneration strategies and helping in creating a more systematic classification, different approaches were classified under two main titles "Tissue Engineering" and "Regenerative Medicine." TE approaches imply the implantation of fully or partially *in vitro* developed homologous tissue obtained by seeding cells into a biomaterial matrix and culturing it *in vitro* with the use of bioreactors. Regenerative medicine (RM), on the other hand, includes approaches consisting of the implant of the biomaterial matrix with or without seeded cells into the body, in order to facilitate regeneration of the tissue *in vivo*.

The TE approach offers the advantage that it relies on the potential evaluation of tissue prior to implantation but, on the other hand, to produce a fully functional tissue *in vitro*, the complex microenvironment occurring *in vivo*—including molecular and biophysical milieu—should be accurately reproduced. However, the complex interplay between microenvironmental condition and tissue-genesis and maturation is at the moment not completely understood and the contemporary reproduction of the basic molecular milieu (e.g., presentation of growth factors and matricellular cues) and biophysical stimuli (e.g., mechanical stress and deformation) is at present technologically not feasible.

Generally, the term TE is now accepted as inclusive of regenerative medicine and stem cell therapeutics. A Google search of these two terms generates 7,690,000 hits for TE and 4,550,000 for RM. Even if these terms have become rather commonplace in today's mass media culture, stem cells do not seem to display—yet—the same market potential of the first two, and timelines from lab to market seem to further postpone this potential.

The border between TE and other types of medical technologies is often unclear and quite fragmented, including approaches and markets that are not always fully related to tissue and organ regeneration, such as bioaesthetic products. Moreover, there are other fields sometimes perceived as proximate to TE but that, in fact, are not. For instance, not-for-profit cord blood banks, veterinary firms, clinical services, organ or tissue allografts, conventional bone marrow transplantation for blood-borne cancers, transfusion medicine, and media-based or financial services are all not to be considered included in the TE approach. Furthermore, cell-based immunotherapies, such as those for cancer treatment, which are recently experiencing some clinical success and are sometimes reported as TE-related therapies, should be excluded from the TE umbrella definition, and related market analysis. Stem cell banking is also often included in the TE enlarged field, but it should not be part of a business-oriented overview. This is because, even if stem cells represent a solid potential for future approaches toward tissue regeneration, it is also true that a bank providing their storage is a support for future activities, but not its core.

70.3 THE MARKET OF TE, MECHANISMS AND PERSPECTIVES

For a structured analysis of the market in terms of results overview and potentiality of TE-related products, it is perhaps important to underline the path that, in general, an advanced technology should follow to go from the benchside to the bedside, and correlate it to the mechanism of investment to support such a path. This analysis would provide a possible interpretation for the present status of the TE-related product market that has not fully fulfilled the originally promised potential and represents a possible suggestion as to where to concentrate future efforts.

Any high-tech product that is born as an idea in a research center laboratory must overcome a series of steps or barriers before it can hit the market. It is widely recognized that one of the most delicate steps of this path is the transition between a successful research project carried out on a lab bench to a prototype that enters the development pipeline of a company. This step is often exacerbated by the cultural, knowledge, and financial gap existing between the academic and industrial communities that need to be bridged. The cultural gap is radicated deeply in the motivation of the two different organizations and possibly in the difference of governance model. Whereas the industry tends to be working under extreme time pressure and budget constraints, the academic community, also by virtue of its support from the public sector, can work on long-term challenges and sometimes with less time pressure. Time to market, however, may represent a huge barrier for a product, or a whole class of products, such as TE-related ones, dividing them up into successful or not existing at all.

The financial and knowledge gap are depicted graphically in Figure 70.1. Technology developed in the research community is typically taken no further than proof of concept. In contrast, commercial production can rarely start before a pilot scale manufacturing plant is set up. The gap between these two is generally addressed as the "Valley of Death" of technology. In general, start-up companies—either universities or large company start-ups—and large companies, rather than small and medium enterprises (SMEs), seem to be better equipped to cross this gap.

The chain of actions generally undertaken to support the gap transition, with all the variations depending on the funding and managing player, including venture capital (VC), large companies, reported in Figure 70.2.

It can be viewed as a five-step path, going from pre-early stage lab research all the way through to the market. Intermediate stages may be variable in time and amount of funding rounds, but in general for medical device products it is generally between 5 and 7 years and for pharma over 10, and usually at least double the expense. Valorization and seed capital stages are generally the ones during which the IP is generated and expanded,

together with the know-how, preclinical studies and human resources hiring and formation. For biotech-related activities, the VC stage is generally devoted to human studies, process scale-up including quality assessment (QA) and quality control (QC), setting procedures and all the other premarket-related activities.

Risk profile is generally high in the early stages, and has a sharp drop after a successful phase I human trial. Safety profile, in fact, may add a lot of value to the potential product even if its efficacy still needs to be optimized. It is thus important to keep time and expenses within a range consistent with the high-risk profile during the first steps of the path.

Counting all companies operating in the field of TE, both with medical devices alone or integrated with cells, the total sector of activities shows in the period of 2007–2011 a remarkable 1.5-fold increase. The overall business volume moved from $2.4 to $3.6 billion with commercial stage spending almost doubling from $1.6 to $2.8 billion in the same period [11]. However, the total development stage spending decreases in the same period, from $0.86 to $0.78 billion. This decrease, which apparently might be seen as relatively slight is, on the contrary, very relevant if compared to the aforementioned 1.5-fold increase in total sector activities and twofold increase of commercial stage spending, resulting in over a 50% decrease from 2008 to 2011. This may be one of the

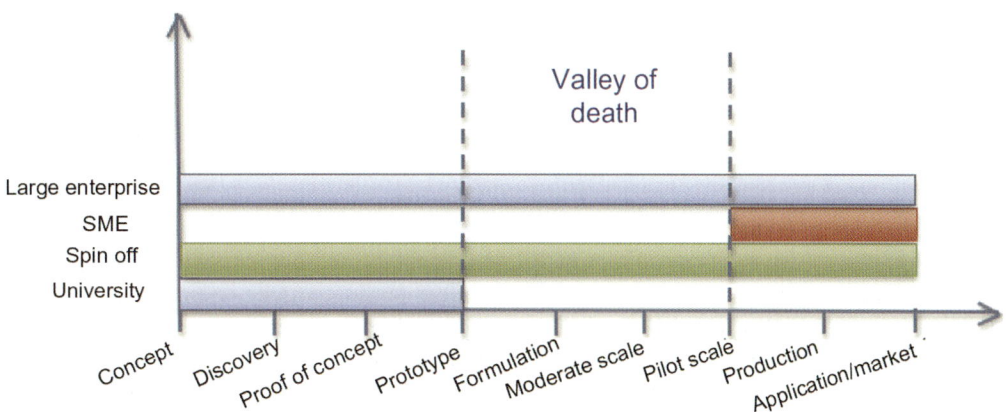

FIGURE 70.1 Typical phases involved in the development of a TE product with indication of synergy between public research organizations and SMEs, large companies and spin-offs.

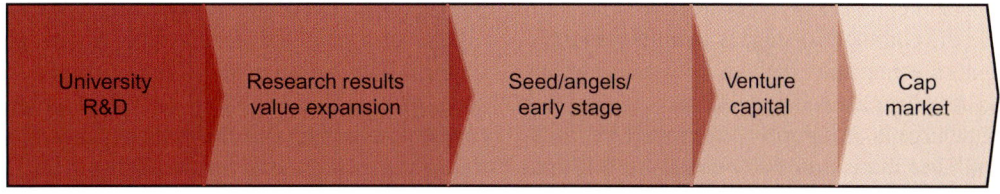

FIGURE 70.2 Typical five-step process of funding of a TE product from laboratory stage to market.

consequences of a reassessment of the market on a more developmental expense sensitive trend of the industry, moving also toward cheaper products in terms of production and development-related costs, which is more in line with other biomedical products. The need for cheaper products, as stated in a recent report in *Nature*[12] seems to be reflected in this specific financial trend.

In addition to this, there has been a consistent growth of the companies operating in the field at the commercial stage. Indeed, from 2007 to 2011 there has been an 18% increase of the total number of companies operating in the field (from 171 to 202) [11] while in the same period a remarkable 32% increase of the number of companies at the commercial stage has been registered (from 47 to 62). These trends indicate the need for a faster time to market of the developmental activity necessary to make the investments in the field more attractive. It is noteworthy that the marked decrease in spending for the developmental stage allowed the amount of sales in 2011 to match the total expenditures, making this market increasingly appealing.

A detailed breakdown by TE industry overall spending and stage of product development is reported in a recent report [11]. The important indication that emerges from the analysis is that out of the 47 companies operating in the field with medical devices without the use of cells, 31 are already at commercial stage. Remarkably, almost 70% of the overall companies have been able to run the whole translational path from lab to market. On the contrary, out of the 44 using cells and biomaterials, only 21 are at the commercial stage. Therefore, it appears that there is a potential difference in the commercial success rate between companies approaching the TE strategies including cells compared to those that do not use cells. This conclusion is also corroborated by the small fraction of stem cell companies, eight out of 104, that hit the commercial stage. Further support comes from a comparison between the breakdown of spending by industry segment (i.e., product platform: cells alone, cells + biomaterials, biomaterials alone) and stage of development (preclinical trial, clinical trial, and marketing) [11]. Indeed, the large majority of expenses for preclinical (81%) and clinical trials (91%) are encountered in companies dealing with cells, either alone or in combination with biomaterials. On the contrary, the majority of commercial expenses (76%) are encountered in companies dealing with biomaterials alone.

Preclinical and clinical trial stages dominate the financial and time-to-market target of companies dealing with stem cells, while the commercial stage is more dominant in the development of companies dealing with biomaterials alone. Overall, these data are encouraging and indicate that the TE industry at large is on a positive track with 167 business units overall and more than 6000 employee full-time equivalents. A larger investment in the biomaterial side focused approaches, intrinsically less risky than approaches including cell therapy, would probably push the market to a more attractive platform for large companies and investors. Furthermore, commercial success will also depend on companies understanding the FDA approval processes better [13,14]. This data confirms the longer time to market, and larger expenses, required by the products using cells. Even if FDA compliance with such trials actually on track allows their entering the market in the near future, the issue of a large investment and time to market for cell-including products still seems to make them less attractive than the fast track of the cheapest cell-free devices. In conclusion, the above analysis suggests that at the present stage and with the current scientific and technological knowledge, the use of autologous *ex vivo* cell manipulation may extend the amount of time and investment required precisely in the section where the risk profile is high, making the technology for the market less attractive, and sometimes unaffordable.

70.4 A CULTURAL DICHOTOMY BETWEEN BIOMATERIALS AND CELLS: WHERE TO PLACE THE BET?

A relevant issue regulating the business development of TE products is the major gap existing in the cultural approach between two well-known development paths: medical devices and pharma products. In fact, time to market and investment profiles in these two fields are completely different, and usually a player operating in one of them knows almost nothing about the other. "A biomaterial seeded with cells is something new, which is usually envisioned as a *Drug Device Combination*," explains Mr. Graziano Seghezzi, senior partner in Sofinnova, a major pan-European VC operating in the field of Life Science. He continues: "Looking for a potential market (i.e., buy out with a large market player) for such a product is like trying to sit between two chairs: medical device companies have a time to market of approximately 5 or 6 years, coupled with a target of investment that does not match with the time and investment profiles required by a drug-like product. Pharma companies, which have a time to market of approximately 10 or 12 years and larger investment scales, on the other hand have a strict rule on the manufacturing line control, and usually do not understand the *Device* component of the product." In such a way, a cell-seeded scaffold is born orphan of a potential market, and would require a longer commitment by an early stage investor. This does not seem to be a viable option for VC or general investment funds considering that the present Biotech market potential is not as high as it has been in the last 5 or 10 years. Alternatively, a biomaterial without cells, which would follow the same path as a medical device, both in terms of time to market and investment, seems to be most promising, especially if the efficacy/cost ratio of the product is relevant.

The above scenario is exacerbated by the contingency of the progressive reduction of potential attractivity of the biotech market by funding and VCs. Indeed, last year Mark Kessel called for the establishment of a biotech superfund funded by pharma, outlining the capital crisis faced by the biopharma industry and its negative impact on innovative therapies. He wrote that a superfund, modeled after the successful SEMATECH superfund, is a viable tool to address the need to fund innovation. In a panel held at the Biotech Showcase 2013, the capital crisis, its effect on innovation and the possibility of creating a superfund was discussed. The moderator, Brady Huggett (Nature Publishing Group), presented the current scenario of financing for private Biotech: "for the year 2012—Brady said— 14 companies went public, in 2011 there were 13, in 2010 there were 19, and over the past 5 years there were only 62 companies that we consider to be innovative that went public. So, if you are a VC and want to get some money back from your investment, now chances are quite slim." According to data published annually by "Nature," looking at the public funding to Biotech companies in the last 10 years, there was an average investment of $4.5 billion on an annual basis, while in the last 3 years there was a significant drop to $3.8 billion. In 2007, there were 362 financing rounds into innovative Biotech companies, which turned to 254 in 2011, and money falls right along with it: $5.3 billion of private investment in 2007, turned into $3.5 in 2011. Nature Publishing Group also looked at the number of VCs actively investing in innovative Biotech companies, which fell from 330 in 2007 to 194 in 2011. Thus, the general evidence of Biotech investments shows fewer VCs working in the space, less money invested and a relevant fall in number of IPOs.

Daphne Zohar, funding member and manager of one of the most innovative venture funds in the field, Puretech Ventures, said: "The landscape has changed because the exit environment for life science companies has been rather bleak over the last 10 years. The IPO market has not really provided an exit; it was rather more of a financing event due to the impossibility from the investors to sell their shares after the company was publicly traded. On the other hand, merger and acquisition (M&A) transactions have not been as lucrative with strategic acquirers, now structuring primarily milestone or 'biobucks' or 'earnout' transactions, that often do not return that much to the investors. As a result, VC funds have not provided good returns to their investors and many of them are struggling to raise new funds."

In such a scenario, Dr. Zohar explained what to expect in the near future: "The landscape may get better because pharma companies have been cutting internal research and focusing on externalization. Since there are not as many innovative projects moving forward, the scarcity of projects will force those companies to make earlier bets if they want to bring innovative projects into their pipelines." Once again, the opportunity represented by a shorter time to market, lower budgeted approach as for biomaterials without cells, if compared to stem cells or cell-powered structures, seems to be the most promising in terms of risk/opportunity ratio and fund-raising capability.

Apart from the financial and funding matters outlined above, technical issues related to regeneration patterns could also contribute to depict and anticipate the possible evolutive scenario of the TE industry. "About the difficulties in getting TE products to become commercially successful" Prof. I.V. Yannas, a pioneer in the field and leading scientist at MIT in Boston, stated "two issues have to be considered":

1. TE products are typically based on *in vitro* processes, which aim to prepare tissues or organs *in vitro*, to be later implanted in the organism. Development of these products has tacitly assumed that *in vitro* preparations will promptly "*take*" as soon as implanted. In fact, they rarely, if ever, perfom in this direction and are consequently considered failures that do not achieve market success. The vast majority of TE products have been prepared in this manner. A very small minority of products have been prepared as true regeneration devices, which are prepared *in vitro* in the form of a scaffold with biological activity but then adapt to the wound-healing environment of the organism and promote regeneration by inhibiting the normal wound-healing process. To confirm this point, it is sufficient to simply check the literature and see what percentage of methodologies are based on *in vitro* preparations and how many are prepared as regeneration devices which function *in vivo*.

2. Unlike *in vitro* preparations characteristic of TE devices, which require a laborious treatment of *ex vivo* autologous cells and, due to the presence of cells, present severe limitation and constrains in the packaging and shipping conditions, regeneration templates are free of cells. That is because the basis of regenerative activity is the regeneration of stroma, not the regeneration of epithelial cells—which regenerate spontaneously in the absence of scaffolds. Stroma regeneration requires simply a scaffold, not stem cells or cells of any kind. An exception to this statement is the need to seed a scaffold with autologous epithelial cells, in order to either speed up the process of regeneration or to regenerate epithelial tissues over very large distances.

In a recent paper Yannas [15] supports the central role of biomaterial scaffold in guiding the tissue regeneration process and the questionable use of *ex vivo* manipulation of autologous cells. In particular, biomaterials scaffolds are strictly necessary for stroma repair since epithelial tissues and the basement membrane regenerate spontaneously while the stroma does not. For example, in skin, the three tissues are arranged in an approximately simple, planar symmetry: the epidermis (epithelia comprising keratinocytes) is attached to the basement membrane which is in contact with the stroma (dermis) underneath, all in the same plane. In nerve fibers, the basement membrane separates the Schwann-cell axon units from the stroma (endoneurium). In skin, a mild injury which destroys the epidermis but nothing else heals with full regeneration of the epithelial cells [16−18]. Full-thickness excision of skin leads to a dramatically different response: the edges of the wound contract toward the center and eventually join, aided by formation of scar. However, the dermis does not regenerate in the adult mammal [19,20] nerve trunk and, following release of the crushing force, the crush site rapidly fills with the displaced tissues and structural recovery follows [21]. By 4−6 weeks regeneration of the myelin sheath has been reported to be complete [22,23]. Normal function was recovered following mild crushing [24]. In contrast, following a complete nerve transection (cut through) and separation of the two stumps by several millimeters following injury, each stump heals separately by contraction and scar formation (neuroma) [25]. The epithelial part of nerve fibers (Schwann cells with myelin sheath) regenerates spontaneously and independently of the intensity of injury. In contrast, the endoneurial stroma is not spontaneously regenerated either in the proximal or the distal stump. These observations are not limited to skin and peripheral nerves but have been independently extended to blood vessels and other internal organs. Injury of the epithelia in the urinary bladder and the gall bladder was followed by reepithelialization; however, injury to the underlying stroma was not regenerated [26].

The studies performed so far on skin and peripheral nerves show that the simplest conditions reported for synthesizing a desired tissue or organ at the injured site do not require the presence of cells [27]. This analysis showed that epithelial tissues and the basement membrane were both regenerated, or synthesized *de novo* in an *in vitro* process with no requirement for the presence of cells of another type or for extracellular matrix [27]. Regeneration of the stroma in skin and nerves required the exogenous addition only of a scaffold. This is because soluble regulators and other signaling molecules are naturally present in the wound exudate (endogenous reactants) and apparently suffice to provide the necessary signaling that eventually leads to regeneration. Several studies of skin wound-healing in which several kinds of growth factors and cytokines were exogenously added [27,28] have failed to induce regeneration of the stroma (dermis) either

in vitro or *in vivo*. In contrast, regeneration both of the dermis [29] and peripheral nerves [30] was accomplished using a collagen scaffold free of soluble factors.

The reported evidence supports the idea that investing in a proper biomaterial design may be more promising than adding the complexity represented by the addition of cells to biomaterials. Recently, a relevant effort has been devoted to the study and production of decellularized matrices. Decellularized extracellular matrix, mostly constituted by collagen, presents ameliorate antigenicity concerns, since the constituent molecules are conserved across species and are therefore tolerated immunologically [31]. Financial and time to market issues, related to the synthesis and authorization associated with the use of extracellular matrices of other kinds, show a preference for its use instead of decellularized scaffolds. Additionally, implants based upon acellular matrices are occasionally seeded with epithelial cells. However, the spontaneous response of epithelial cells to injury is best appreciated with matrices that have not been seeded prior to implantation. For example, in a clinical study of urethral repair using unseeded conduits based on an acellular bladder submucosa matrix, the biopsy specimens showed the typical urethral stratified epithelium, undoubtedly the result of epithelial tissue regeneration [32].

As regards biomaterial design and development, while keeping in mind that complexity is a drawback in terms of the financial and time to market issues already discussed, greater effort should be devoted to identifying the structural features of scaffolds that are required for regenerative activity. One is pore structure, and in particular pore dimension and degree of interconnectivity, other than appropriate micropatterns which may instruct specific regeneration patterns [33]. Degradation rate is another parameter, especially after it has been demonstrated that the half life of the implanted biomaterial seems to be independent of the anatomical site of the defect [29,30,34,35]. Also in terms of surface chemistry, there is direct evidence that collagens, together with a variety of GAGs, such as chondroitin sulfate, heparin sulfate, and hyaluronic acid [31] have regenerative activity in a variety of tissues and organs.

Clear commercial success examples of the TE industry have been either morphogenic molecules, such as INFUSE by Medtronic, or biomaterials matrices without cells. On these premises, as a possible future scenario of the TE industry, one may envisage an increased attention to engineering novel biomaterials that—enriched with a bioactive compound—might improve their regenerative potential without the use of autologous cells. Growth factor enriched scaffolds may transition to an *in situ* cell programming scheme that aims to recruit stem cells into the scaffold from the surrounding tissue and expand and differentiate them to promote functional tissue regeneration.

The ability to recruit and program the resident stem cell *in situ* to aid tissue regeneration is very promising, since it is technologically feasible and would finally lead to a regeneration process that does not rely on the *ex vivo* manipulation of cells. It has been proven that cell fate can be programmed using biological cues [36,37] albeit this approach has never been systematically tested by using scaffolds integrated with growth factor delivery devices.

The study of the optimal combination of these components on the quality and extent of the regeneration process would lead to the definition of a wide number of biomaterials, which may be promising both in terms of performance and cost.

70.5 CONCLUSION

In conclusion, efforts to promote the growth of the TE market should be devoted to the definition of the guidelines, which on one hand minimize the financial, and time to market efforts. This would make the investment in the field profitable, even in the current not-so-brilliant scenario. On the other hand, this approach seems to be perfectly in line with the scientific evidence pointed out by Yannas, that the simplest set of reactants that can induce regeneration (a scaffold and, optionally, autologous epithelial cells), added exogenously at the injured site, should be defined. Results reported in the literature seem to allow investigators to exclude the use of exogenous reactants that complicate experiments without being useful. However, in addition to inhibition of wound contraction (and the secondary process of scar formation) organ regeneration also requires the synthesis of normal organ tissue, comprising stroma and tissues formed by differentiated cells. Since the evidence shows that an active scaffold, with optional epithelial cells, suffices for both steps in the regenerative process, it appears that the scaffold further provides the required information for stroma synthesis [27].

ACKNOWLEDGMENT

The authors wish to thank Prof. Robert Langer and Prof. Ioannidis V. Iannas from MIT Boston for fruitful discussion and suggestions and Dr. Daphne Zohar of Puretech Ventures and Dr. Graziano Seghezzi of Sofinnova for their sapient insights from VC point of view.

REFERENCES

[1] Lysaght MJ, Jaklenec A, Deweerd E. Great expectations: private sector activity in tissue engineering, regenerative medicine, and stem cell therapeutics. Tissue Eng Part A 2008;14:305–15.

[2] Nahmias Y, et al. Tissue engineering application in general surgery. Fundamentals of tissue engineering and regenerative medicine. 1st ed. Berlin: Springer; 2009.

[3] Meyer U, et al. Fundamentals of tissue engineering and regenerative medicine. Springer; 2009.

[4] Langer R, Vacanti JP. Tissue Eng Sci 1993;260:920–6.

[5] Yannas I, Burke JF. Design of an artificial skin. I. Basic design principles. J Biomed Mater Res 1980;14:65–81.

[6] Lysaght MJ. Product development in tissue engineering. Tissue Eng 1995;1:221–8.

[7] Parenteau N, Naughton G. Skin: the first tissue-engineered products. Sci Am Am Ed 1999;280:83–5.

[8] Green H. Cultured-cells for the treatment of disease. Sci Am 1991;265:64–70.

[9] Brittberg M, et al. Treatment of deep cartilage defects in the knee with autologous chondrocyte transplantation. New Engl J Med 1994;331:889–95.

[10] Lysaght MJ, Reyes J. The growth of tissue engineering. Tissue Eng 2001;7:485–93.

[11] Jaklenec A, et al. Progress in the tissue engineering and stem cell industry "Are we there yet?" Tissue Eng Part B Rev 2012;18:155–66.

[12] Place ES, Evans ND, Stevens MM. Complexity in biomaterials for tissue engineering. Nat Mater 2009;8:457–70.

[13] Lee MH, et al. Considerations for tissue-engineered and regenerative medicine product development prior to clinical trials in the United States. Tissue Eng Part B Rev 2009;16:41–54.

[14] Johnson PC, et al. Hurdles in tissue engineering/regenerative medicine product commercialization: a survey of North American academia and industry. Tissue Eng Part A 2010;17:5–15.

[15] Yannas IV. Emerging rules for inducing organ regeneration. Biomaterials 2013;34:321–30.

[16] Briggaman RA, Dalldorf FG, Wheeler Jr. CE. Formation and origin of basal lamina and anchoring fibrils in adult human skin. J Cell Biol 1971;51:384–95.

[17] Schaffer CJ, Nanney LB. Cell biology of wound healing. Int Rev Cytol 1996;169:151–81.

[18] Nanney LB, King Jr. LE. Epidermal growth factor and transforming growth factor-a. In: Clark R, editor. The molecular and cellular biology of wound repair. Springer; 1996. p. 171.

[19] Billingham RE, Medawar PB. Contracture and intussusceptive growth in the healing of extensive wounds in mammalian skin. J Anat 1955;89:114–23.

[20] Berry DP, et al. Human wound contraction: collagen organization, fibroblasts, and myofibroblasts. Plast Rec Surg 1998;102:124–31.

[21] Haftek J, Thomas PK. Electron-microscope observations on the effects of localized crush injuries on the connective tissues of peripheral nerve. J Anat 1968;103:233–43.

[22] Goodrum JF, et al. Fatty acids from degenerating myelin lipids are conserved and reutilized for myelin synthesis during regeneration in peripheral nerve. J Neurochem 1995;65:1752–9.

[23] Goodrum JF, et al. Peripheral nerve regeneration and cholesterol reutilization are normal in the low-density lipoprotein receptor knockout mouse. J Neurosci Res 2000;59:581–6.

[24] Madison R, Archibald S, Krarup C. Peripheral nerve injury. In: Clark RA, editor. The molecular and cellular biology of wound repair. Springer; 1992. p. 450–87.

[25] Chamberlain LJ, et al. Connective tissue response to tubular implants for peripheral nerve regeneration: the role of myofibroblasts. J Comp Neurol 1997;417:415–30.

[26] Goss RJ. Adaptive growth. London: Logos Press; 1964.

[27] Yannas IV. Tissue and organ regeneration in adults. Springer; 2001.

[28] Fu XB, et al. Engineered growth factors and cutaneous wound healing: success and possible questions in the past 10 years. Wound Repair Regen 2005;13:122–30.

[29] Yannas IV, et al. Synthesis and characterization of a model extracellular-matrix that induces partial regeneration of adult mammalian skin. Proc Natl Acad Sci USA 1989;86:933–7.

[30] Soller EC, et al. Common features of optimal collagen scaffolds that disrupt wound contraction and enhance regeneration both in peripheral nerves and in skin. Biomaterials 2012;33:4783–91.

[31] Badylak SF, Freytes DO, Gilbert TW. Extracellular matrix as a biological scaffold material: structure and function. Acta Biomater 2009;5:1–13.

[32] El-Kassaby AW, et al. Urethral stricture repair with an off-the-shelf collagen matrix. J Urol 2003;169:170–3.

[33] Harley BA, et al. Fabricating tubular scaffolds with a radial pore size gradient by a spinning technique. Biomaterials 2006;27:866–74.

[34] Gilbert TW, et al. Degradation and remodeling of small intestinal submucosa in canine Achilles tendon repair. J Bone Joint Surg Am 2007;89A:621–30.

[35] Record RD, et al. *In vivo* degradation of C-14-labeled small intestinal submucosa (SIS) when used for urinary bladder repair. Biomaterials 2001;22:2653–9.

[36] Discher DE, Mooney DJ, Zandstra PW. Growth factors, matrices, and forces combine and control stem cells. Science 2009;324:1673–7.

[37] Sands RW, Mooney DJ. Polymers to direct cell fate by controlling the microenvironment. Curr Opin Biotechnol 2007;18:448.

Chapter 71

Ethics in Regenerative Medicine and Transplantation

Nancy M.P. King

Department of Social Sciences and Health Policy and Wake Forest Institute for Regenerative Medicine, Wake Forest School of Medicine, Center for Bioethics, Health, and Society and Graduate Program in Bioethics, Wake Forest University

Chapter Outline

71.1 INTRODUCTION

As all of the chapters in this volume have shown, the relationship between regenerative medicine and organ transplantation is characterized by a high degree of complexity and variety, an extraordinary range of disciplines and technologies, and a very broad spectrum of translational development, from the gathering of basic information to promising laboratory findings to clinical success. Considered by itself, regenerative medicine research raises many ethical questions. Some are common to early stage research involving novel biotechnologies, and several are relevant to regenerative medicine in particular [1−3].

Considered by itself, organ transplantation presents an even more ethically complex picture. From both a historical and a conceptual perspective, surgery is something of a latecomer to organized research [4−8]; as a result, organ transplantation combines features of standard surgical treatment, treatment innovations in surgery that have been improvised for individual patients, and organized research involving not only surgical techniques but also organ procurement, preservation, and distribution, as well

as new drugs and devices. Organ transplantation also raises issues of distributive justice, by virtue of the need to develop equitable distribution schemes for naturally scarce resources. Because transplantation is both a familiar technology and one that is continually advancing, the ethical issues it presents are likewise both familiar and constantly changing, along with changes in the science and in the social context of transplantation policy and practice.

Taken together, then, regenerative medicine and organ transplantation offer an unprecedented mix of ethical challenges. Some problems may diminish or disappear when these two technologies are combined, and others may increase. For example, the problem of immunosuppression, which complicates the harm−benefit calculus in traditional transplantation, is anticipated to disappear when regenerative medicine makes the regeneration of solid organs feasible, but the challenges of zoonosis may be heightened with the use of natural scaffolding derived from animal organs [9]. Some problems may reappear in a different guise. For instance, if the organ shortage decreases greatly as a result of developments in autologous regeneration and transplantation, then fair

Regenerative Medicine Applications in Organ Transplantation.

distribution of a naturally scarce resource may no longer be at issue, and the complexities of organ procurement may thus diminish or disappear entirely—or they may vanish in resource-rich countries that can afford to make regenerative technologies available, but persist in resource-poor countries, perhaps in unexpected ways.

Moreover, if regeneration remains costly and time-consuming, some justice questions may persist, and new ones may arise. The high cost of regeneration may remain a barrier for those who are not wealthy or well insured, and getting on the regeneration list early enough may become a significant concern for those who lack physician-advocates to help ensure that a new organ will be ready in time.

Finally, each of the above speculations is based on the assumption that organ regeneration will become part of the standard of care, or even replace traditional transplantation as the standard. However, as this volume has shown, despite its enormous promise, organ regeneration still has a long way to go along the translational trajectory. Indeed, it has moved from preclinical to human research in only a few instances to date [10,11]. Thus, for some time to come, organ regeneration will continue to be early stage organ regeneration research, sharing many of the characteristic features of first-in-human trials, but also expanding and updating our views of the ethical considerations relevant to translational research.

Organ regeneration research differs in many but not all respects from the old pharmacological model of discrete phases in research design and ethics. The focus of this chapter is on those similarities and differences, and on how they affect our understanding of biotechnological innovations that are continually developing. The combination of regenerative medicine and organ transplantation exemplifies modern biotechnology, and ethics and policy are challenged to keep pace. Importantly, many of the ethical issues raised by this combined technology are not truly novel. Instead, they represent long-standing questions that have never been perfectly addressed, but that reappear, as science advances, to renew old debates.

Many health-care providers and researchers have a basic knowledge of the major ethical principles in medical treatment and research. Those principles are autonomy and protection of those whose autonomy is diminished (combined, in research ethics, into the principle of respect for persons); beneficence (doing good) and nonmaleficence (avoiding harm); and justice [12,13]. Just as the scientific components of organ regeneration connect to basic knowledge of human growth and development, the ethical issues raised by regenerative medicine and organ transplantation connect to these basic building blocks of bioethics and human relationships in medicine and research.

These major principles provide a conceptual foundation for the practical work of research ethics, which often goes hand in hand with attention to the regulatory requirements that significantly shape the progress of both preclinical and clinical research. This second level of research ethics addresses clearly related but more fine-grained, applied precepts, including informed consent, the balancing of harms and benefits, the mandate to ensure that research is well designed and therefore likely to yield useful data, and the equitable selection of research subjects. This realm of middle-level principles and practices is familiar territory and thus a helpful starting point.

71.2 BLAZING THE TRAIL

71.2.1 The Translational Pathway

Everyone who undertakes preclinical research in an area of translational science must consider the research trajectory. Investigators must repeatedly assess the direction and promise of the line of research by asking themselves questions: what is the long-term goal of this line of research? What will it take to get there, and what data will be needed to demonstrate whether we have arrived? Although accurate prediction about where a new line of scientific inquiry may lead is difficult or impossible, each experimental step in the research trajectory must be designed so that the data gathered are likely to reduce the many uncertainties associated with the intervention being studied. Without this, the line of research cannot move forward; even with it, progress toward (and in) human trials is often decidedly nonlinear, in large part because generalizable knowledge can take many forms, including contributions to basic developmental biology, the discovery of a need to change directions, and the development of entirely new ideas [14].

Research design and ethics are linked in many ways. Because the design of translational research in regenerative medicine is complex, situation-specific, and governed by a plethora of regulatory requirements, the easiest way to highlight the ethics connection is to continue to ask provocative, difficult-to-answer questions, like this one: When is the right time to move from laboratory and animal studies into humans?

71.2.2 Are We Ready to Blaze This Trail?

This profoundly important question of design and ethics requires a careful balancing between the desire for knowledge and the hope of rapid progress against disease and disability: "Basic scientists want to know; clinicians want it to work" [15, p. 433]. Both the potential to decrease human suffering and pressures for commercialization and product development may affect the speed of bench to bedside translation. How, then, should researchers in regenerative medicine and organ transplantation go about deciding whether an intervention in the preclinical stages is ready for human subjects research? It is essential to remember that moving from preclinical to clinical trials is scientifically and ethically justifiable only when three conditions are met: (1) No more can reasonably be learned without delivering the experimental intervention to human research subjects. (2) The risks of harm and amount of uncertainty have been reduced as far as feasible. (3) The remaining uncertainties and risks of harm are not excessive under the circumstances [15].

The operative terms here—reasonable, feasible, not excessive—are imprecise. Because precision with regard to evaluative terms is impossible, the goal of using these terms is to promote habits of discussion and decision-making among researchers, oversight bodies, regulatory authorities, and research ethics scholars. Attending to the process goal of thoughtful and productive examination and discussion of how these terms and concepts apply to the unique circumstances of a particular line of research can help guide decision-making and promote reasoned consensus.

71.2.3 Ethics in Clinical Trial Design

The reality of clinical research has always been far more fluid than the traditional phase designations first developed for pharmaceutical research. This fluidity is especially applicable to regenerative medicine/organ transplantation research. When experimental interventions moved from pharmaceuticals to biotechnologies, and when surgery began to be widely and systematically viewed as subject to the requirements of research, traditional phase designations were recognized as often fitting poorly with the realities of research design. The best questions to ask, then, do not include "What phase is this study?" Instead, it may make far more sense to ask "What value does this trial have?" and "Will this trial produce valid data to help guide us toward the next research step?"

Every clinical trial must meet these basic requirements of *value* and *validity* [16]. First, to be valuable, every trial must ask a scientifically and socially meaningful question. There are no agreed-upon standards for determining value; that determination is made in the course of ongoing discussion among researchers, funding authorities, regulatory bodies, and the public [17], and researchers have an important role in this justificatory conversation. A familiar example of controversy about validity is the labeling of some drug trials as "me-too" studies; the label challenges the value of the contributions made by the development of additional pharmaceutical variants to treat common, mild conditions. Me-too studies can raise

ethical questions because the value of the results may be considered insufficient to justify the potential harm and inconvenience to research subjects.

Second, every trial, to be valid, must be designed so that the data gathered have the capacity to answer the question it poses, whether the answer is positive or negative. Trial data must be able to point to the appropriate next step in the research trajectory, able to guide researchers in designing the study that should come next, whatever it may be—one step farther along the research trajectory, a step back or rerouting along another path, or, in the case of later-stage research, a result sufficient to change practice if the intervention is deemed successful [14]. Validity determinations can be quite complex, taking account of statistical design, inclusion and exclusion criteria, and factors affecting the conduct of the trial, such as meeting or failing to meet enrollment goals, and the accuracy and completeness of data collection [16].

Enrolling subjects in a trial that cannot produce valid data raises ethical concerns because it is unfair to expose subjects to any risk of harm, discomfort, or inconvenience if generalizable knowledge does not emerge from the trial. Without generalizable knowledge, neither science nor society can benefit. This is why institutional review boards (IRBs) and funders look carefully at statistical considerations in trials, and why data and safety monitoring boards have the power to recommend that trials be stopped because of "futility"—i.e., inability to prove or disprove the trial's hypotheses. This is also one of the reasons why the failure to publish the results of negative studies causes ethical concern.

For regenerative medicine and organ transplantation research, as for other novel biotechnologies, validity poses a particular challenge because the population of potential subjects may be quite small, so that study sample sizes are also small. Careful attention to the design and conduct of small studies, and to the reporting of data from them, is needed to ensure validity and render the contributions of subjects worth their time and risk [18].

71.2.4 Subject Selection

Once a line of research is considered ready to take into human trials, a key decision must be made: Who should be the first research subjects? At every stage of clinical research, the choice of subjects must balance two criteria that may be at odds: first, subjects must be those who can provide the best data to enable determination of the appropriate next research step [14,15]. Second, subjects must also be at the least risk of harm from research participation. Harm here includes both harm arising directly from receiving the research intervention and the harm that may arise if any standard interventions must be detrimentally forgone or postponed as a result of research participation [15,19].

Many first-in-human safety trials of pharmaceuticals and other experimental interventions have conventionally enrolled healthy volunteers, but it has become common, in many areas of research, to enroll as first subjects patients who have the disease or condition of interest. Healthy volunteers may not provide the best data under the circumstances, as they are often poor models for the disease states under study. Moreover, many interventions, including chemotherapy, surgery, and novel biotechnologies, pose significant risks of harm to subjects. For these reasons, it is customary and logical to turn first to patients as subjects in nearly all research involving regenerative medicine and organ transplantation.

Even after patients are chosen as subjects, researchers must consider which among them should be enrolled first. Patients who are younger and/or whose disease has not progressed very far often represent a significantly different population from older patients. The same is true of the so-called treatment-naive patients in comparison with those who have tried many treatments and can derive no further benefit from standard therapies. For regenerative medicine and organ transplantation, as for many novel biotechnologies, patient-subjects who can provide good data may also be those who can experience the greatest amount of direct benefit if the intervention is successful. This reality raises important concerns, not only about research design but also about the "therapeutic misconception," discussed more fully below.

Because the subjects in regenerative medicine and organ transplantation research are patients, considerations of research design and ethics suggest that great care must be taken not only to select patient-subjects from whom the most can be learned, but also to assemble a thorough and compelling data-portrait of the course of the disease and the efficacy and adverse effects of available standard treatments, so that the data gathered from the experimental intervention can be usefully compared with that data-portrait. If beneficial effects are seen in patient-subjects, but the normal course of the disease is unpredictable, it may be difficult to determine whether the intervention is responsible for good outcomes. This could be the case, for instance, in an intervention after a recent diagnosis, if it is not yet known how much permanent impairment a particular patient-subject would experience without the experimental intervention [20]. Similarly, if adverse effects are seen in patients experiencing severe disease, it may not be possible to distinguish the adverse effects of the experimental intervention from the ordinary course of severe disease. This could be a concern when enrolling treatment-naive patients with a serious condition in a trial where there is concern that the intervention itself could produce a serious adverse effect.

Such possibilities do not argue against doing research on such conditions or enrolling such subjects. Rather, they serve as reminders that research and treatment, though increasingly related, are meaningfully different, and that the justification for every trial must take all such factors into consideration [15]. The complexity of these justifications will have a significant impact on regenerative medicine and organ transplantation research, at least in some instances, making thorough preclinical research and careful preparation for the transition to human subjects of signal importance.

71.3 TRAIL MARKERS

When the translational research trajectory has been mapped out as far as possible, it is time to begin traveling along it. Although the process of decision-making by the patient who is a potential subject captures much ethical and regulatory attention, decisions by investigators, sponsors, and oversight bodies have ethical dimensions too. Signposts all along the trail mark the places where key issues must be addressed and make it easy for others to follow.

71.3.1 Decision-making

It is the task of the IRB and other oversight bodies to determine whether and when a first-in-humans trial may justifiably move forward. All must agree that the potential benefits to society from the likely results of the trial, and the potential benefits to individual patient-subjects in the trial—if any are possible, of course—outweigh the risks of harm to patient-subjects in the trial [21]. The assessment and balancing of benefits and harms is especially challenging in the early stage clinical research, when what is uncertain and what is unknown loom particularly large. Genuine conversation between researchers and IRBs about this balancing helps to clarify what risks of harm and potential benefits are at issue, and to determine whether the offer of research participation to potential subjects is fair and justifiable under the circumstances. This conversation logically precedes determination of what information should be provided to potential subjects in the consent form and process. Nonetheless, it can be informed by asking "How should this research be described to potential subjects?"

71.3.2 The Harm—Benefit Calculus

How should the risks of harm and the potential benefits presented by a clinical trial be assessed and balanced? Once again, the guidance provided on this critical issue is more suggestive than precise. The earliest significant clinical research guidance document, the Nuremberg Code, puts it this way: "The degree of risk to be taken should never exceed that determined by the humanitarian importance of the problem to be solved by the experiment" [22]. An international guidance document that is more oriented toward research with human subjects who are also patients, the Declaration of Helsinki, makes a similar statement: "Medical research involving human subjects may only be conducted if the importance of the objective outweighs the inherent risks and burdens to the research subjects" [23].

And finally, the most modern guidance document, the Belmont Report, takes a small but significant next step, declaring: "It is commonly said that benefits and risks must be 'balanced' and shown to be 'in a favorable ratio.' The metaphorical character of these terms draws attention to the difficulty of making precise judgments. However, the idea of systematic, nonarbitrary analysis of risks and benefits should be emulated insofar as possible" [13]. Thus, the best available guidance notes that analysis of risks of harm and potential benefits is imprecise and non-formulaic, but that at the same time, oversight bodies must endeavor to be consistent and systematic. Even though reasonable people may reach different conclusions about the harm−benefit balance of a given clinical trial, researchers and oversight bodies must engage in analysis that requires a reasonable justification for any conclusion, so that others can examine their reasoning, reason together about other trials, and perpetuate the difficult practice of justificatory conversation under conditions of uncertainty.

71.3.3 Risks of Harm

In every early stage clinical trial, all anticipated risks of harm from the experimental intervention must be clearly identified, and what is known or expected about their nature, magnitude (size/intensity and duration), and likelihood must be described. If alternatives exist (such as standard treatments, however imperfect), it is equally important to acknowledge when harm could result from forgoing or delaying standard treatment. Finally, consideration of the possibility of unknown effects is essential [24]. Identifying the risks of harm in research is familiar territory to investigators, even in early phase trials. Potential benefit, in contrast, has not been as thoroughly addressed.

71.3.4 Potential for Direct Benefit: Ambiguous Expectations

In a study of how potential benefit is understood, described, and discussed in early stage gene transfer research, interviewers asked investigators whether they expected the patient-subjects to receive direct benefit

from the experimental intervention. One principal investigator's answer provoked this revealing exchange [25]:

> PI: "Oh, it's a long shot. It's a long shot."
> Interviewer: "If you were just to say yes or no what would you say?"
> PI: "Ah that's tough, that's actually, I'm really conflicted about that. I guess if you really push me, I'd have to say no, but I would like to say yes, but I don't think that would be honest at this point. It's a little bit too early... to work out."
> Interviewer: "I can also punch here 'don't know'."
> PI: "No I'll put no. It's the moral response."

Determining whether any potential benefits may reasonably be expected from an experimental intervention is a pervasive challenge in all research, but it has special significance in early stage research involving novel biotechnologies like regenerative medicine/organ transplantation, for two opposing reasons. On the one hand, direct benefit is often unlikely or impossible in early stage research [26,27]. On the other hand, direct benefit to patient-subjects is likely to be anticipated even in early stage regenerative medicine/organ regeneration research, because the expected harms to subjects from surgical procedures are substantial enough to make it ethically troubling to offer such procedures without the potential for direct benefit from the experimental intervention. Although societal benefit from the results of the research is the overriding goal of all research with human subjects, societal benefit alone can seem inadequate when the anticipated harms to individual subjects are significant and irreducible, even though not serious or permanent.

For research involving regenerative medicine and organ transplantation, these opposing factors demonstrate the need to take several steps addressing potential direct benefit: (1) conduct extensive preclinical research to develop and substantiate the potential for direct benefit to human subjects based on the results of laboratory and animal studies; (2) carefully explore the alternatives to study participation for each potential patient-subject, in order to minimize the risks of harm from forgoing standard treatment; and (3) ensure a thoughtful application of technical skill and knowledge during each procedure, including the standard surgical perspective of learning from each procedure how to revise technique for the next, so that it becomes possible to maximize whatever direct benefit may materialize.

71.3.5 Informed Consent

Anticipated risks of harm and potential benefits should be thoroughly and clearly described, and the basis for any expectations should be explained, for all clinical trials, as well as for all offers of treatment [26−28]. Because the work of regenerative medicine and organ transplantation is so complex and multidisciplinary, combining conventional surgery, surgical innovation, tissue engineering, drug–device combinations, and more, it is useful to compare informed consent in treatment and research for insights into best practices.

It may surprise researchers that the requirements for informed consent to treatment are, in general, less detailed and less robust than in research. There are several reasons for this difference. First, because the primary goal of research is to contribute to generalizable knowledge, research subjects must be thoroughly informed in order to understand that research participation is not focused on their personal benefit. In contrast, in clinical care, physicians and patients share the same goal—benefiting the patient (even though the parties to the relationship may have different definitions of benefit). Second, there is often (but not always) more uncertainty about the effects and outcomes of receiving an experimental intervention than exists about standard treatment. And finally, the informed consent requirements in research are applied prospectively, through regulations, whereas informed consent in the clinical context is enforced after the fact through litigation. Moreover, informed consent requirements for clinical care are delineated in state statutes and court decisions, generally with less precision and detail than are found in the Common Rule [29] or the FDA regulations [30].

It is noteworthy, however, that despite the vagueness and variability of informed consent requirements in clinical care, informed consent in surgery represents a more highly developed body of guidance and practice. Historically, consent arose as an issue in surgery even before *informed* consent was a recognized concept, because anesthetized patients are unable to object contemporaneously if something is done during surgery to which they did not agree beforehand [28, p. 120–3]. And although the standard hospital "op permit" does not usually contain a wealth of information, there is a pervasive expectation that the surgeon will discuss with the patient the nature and consequences of the procedure, the risks of harm, and the available alternatives and their consequences, including the consequences of not doing the procedure, before the op permit is signed. This is a more formal and thorough process than many physicians undertake when they prescribe a medication or propose a treatment less invasive than surgery. Thus, the requirements for informed consent in surgery and research, though far from identical, both reflect genuine effort to inform patients and subjects about what to expect from an intervention.

Several components of the discussion of potential benefit are particularly important in early stage research, including regenerative medicine research. First, it is essential to distinguish direct, inclusion, and societal benefits [26,27,31]. Direct benefits are those arising from the experimental intervention and are probably most important to patient-subjects. Inclusion benefits arise from simply participating in the research, whether or not one receives the experimental intervention or is benefited by it. Inclusion benefits are provided to all subjects and are sometimes considered inducements to participate. Examples include a free physical examination, medical testing and monitoring beyond what is required by the research, and other nonmonetary benefits. Sometimes, the additional testing and monitoring required by the research is described as beneficial: "Patients do better on study because we monitor them so closely." This claim is not well supported by the literature [32]. Close monitoring and additional testing may just as easily pose risks of harm, as when research-related testing gives a "false positive" result that produces anxiety and requires additional testing at the patient-subject's expense. Finally, societal benefits stem not from research participation but from the outcomes of the line of research.

Second, discussion of the potential for direct benefit must be more specific and detailed than this common boilerplate statement in research consent forms: "You may or may not benefit." Direct benefit can and should be described in terms that resemble the more familiar description of risks of harm: the nature of any anticipated benefit, its magnitude (i.e., its size and duration), and its likelihood. In research with high levels of uncertainty, these dimensions of direct benefit will necessarily be difficult to quantify; nonetheless, addressing them, even when precision is impossible, at least signals to potential subjects that there is more to the potential for benefit than "Either I will benefit or I won't" [27].

71.3.6 Long-Term Follow-Up

Understanding how the role of patient-subject differs from the patient's role includes understanding why long-term follow-up (LTFU) is sought—both to gather optimal study data and to properly protect the welfare and interests of patient-subjects. Thus, LTFU is an important component of disclosure in the consent form and process. Investigators need to design good LTFU and incorporate it into their protocols, and funding agencies need to support it. Investigators also need to address the practicalities of LTFU, in order to make adherence to recommended follow-up easier for patient-subjects.

In regenerative medicine/transplantation research, LTFU may be necessary over many years simply in order to determine whether an experimental intervention is a success or a failure. It may not be known right away how long successful regeneration and restoration of a minimum level of organ function will take; thus, LTFU may help determine both whether the experimental intervention is "working" and how long monitoring should last before success or failure is declared [1].

For these reasons, it will be wise to ensure that patient-subjects are given ample information about what to expect from an experimental organ regeneration intervention and the anticipated challenges of LTFU. Periodic revisiting of discussion about research participation and LTFU should be a component of an ongoing researcher—subject relationship [31], so that patient-subjects recognize that their role extends beyond receipt of the experimental intervention. Taking LTFU seriously helps to emphasize patient-subjects' role in knowledge production.

71.3.7 Therapeutic Misconception?

In research enrolling patients as subjects, the therapeutic misconception is the tendency to view research as treatment, to blur the distinction between research and treatment, and/or to have unreasonably high expectations of direct benefit from receiving the experimental intervention or unreasonably to discount the risks of harm. First identified by Paul Appelbaum and colleagues some 30 years ago [33], the therapeutic misconception is most often attributed to patient-subjects, but it is vital to recognize that it is also common in investigators and oversight bodies [34]. The therapeutic misconception is of concern because it may adversely affect understanding about the nature of the research and the likelihood that the experimental intervention will be beneficial for subjects. It thus might, but does not necessarily, compromise decision-making by patient-subjects. More importantly, it might also influence how investigators describe the research to potential subjects in the informed consent process, as well as how oversight bodies like IRBs, and even study sponsors, view the research [35—37].

In early stage research, available information about potential benefit is limited at best, but the goal of the line of research is to demonstrate clinical benefit. When potential subjects are patients with the disease or condition that the experimental intervention is ultimately intended to treat, an unintended consequence may unfortunately result: the mistaken belief that, if healthy volunteers are not enrolled, it is because they cannot benefit, but because patients will be enrolled, they stand at least some chance to benefit, almost by definition—especially when no good standard treatment exists [15,27,38]. In such cases, all involved are hoping that a new, untried intervention will offer some benefit that standard treatment cannot provide [34]. Although this hope is understandable, the resulting therapeutic misconception may have significant distorting effects on decision-making.

The surgical component of regenerative medicine/transplantation research may increase the risk of therapeutic misconception. Surgery, like chemotherapy, is generally not practiced on healthy volunteers, and a surgical intervention is difficult to limit to safety considerations in an early stage trial [4]. Moreover, randomized trials are unlikely to be feasible even at later stages of research involving regenerative medicine and organ transplantation, with the possible exception of some comparative effectiveness trials. Thus, hopes of benefit are likely to exist for all involved in early stage organ regeneration research. To reduce the therapeutic misconception, those hopes must be acknowledged and directly addressed.

It is not yet clear how best to identify the therapeutic misconception and assess its effects on decision-making in clinical research [39]. However, the likelihood of the therapeutic misconception in patient-subjects enrolled in early stage research can be considerably reduced if it is addressed and reduced in investigators and IRB members, so that the consent form and process provide clear, accurate, and realistic information about the potential for direct benefit [27,36]. Second, the possibility of the therapeutic misconception should never automatically disqualify patients as potential subjects, especially when vague or misleading information about potential benefit has contributed to their views. Hope for benefit is not always therapeutic misconception; it is acceptable, even desirable to hope for benefit if you do not expect it. Patient-subjects who are confident that they will experience benefit may be expressing a degree of optimism that is unproblematic in context [40,41].

71.4 MAKING (AND BREAKING) THE BANK

Novel biotechnologies are extremely expensive, and because biotechnology research has become a key focus of medical progress, the cost of medical care has become an overwhelming issue globally. A notable aspect of regenerative medicine is its recognition of the need to standardize and streamline production. To make organ regeneration feasible, accessible, and affordable outside the research context, it may be necessary to move from individually tailored interventions toward the development of production methods that can reduce the investment of time, labor, and cost. Attention to this goal distinguishes regenerative medicine from new technologies like gene transfer, where standardization, the development of platform technologies, and attempts at large-scale, cost-reducing production have been quite challenging [2]. This production perspective is an important step in lowering costs and increasing access to new technologies, but it could also have interesting ethical implications.

Just how economies of scale might be applied to organ regeneration remains to be determined, but two possibilities are already being explored: standardization of scaffolding, and banking of cells and tissues. Scaffolding materials can be produced in standard shapes and sizes rather than being individually tailored to fit; i.e., small, medium, and large organ scaffolds could be prepared, and perhaps even stockpiled until needed. Stem cell banking is already well under way in many places. Some biobanks are public, even nationwide, whereas others are private and for-profit. If a sufficient number of specimens can be publicly banked, reasonably good matches may be available for many whose own cells cannot reliably be used to reseed an organ scaffold [42,43]. Without a sufficient number of banked specimens, however, justice questions could arise [43].

One unknown that will continue to complicate and increase the expense of organ regeneration is the time factor. Regeneration takes time, regardless of the type of scaffold or the cell source. A minimal amount of time must be devoted to reseeding an organ scaffold, and a certain amount of time must be spent by each new organ in a bioreactor to ensure that it is ready for transplant. A great deal of research must still be done to determine the optimal amount of time to regenerate and condition a new organ—and to reduce the time, when possible.

Although the worldwide need for transplantable organs is a significant innovation driver, the costs of organ regeneration are still extremely high, and economies of scale may not significantly lower those costs for some time. Yet the costs associated with standard treatments—conventional organ transplantation and lifelong immunosuppression, or lifelong reliance on organ substitutes like hemodialysis or insulin therapy—are also extremely high. For many patients, the costs and complications of these "halfway technologies" [44] may exceed the projected costs of organ regeneration. Thus, in resource-rich countries where halfway technologies are available, organ regeneration may come to be seen as a relative bargain for the health-care system. In resource-poor countries, however, if standard treatments for organ failure are not reasonably available, the availability of organ regeneration technologies can scarcely be viewed as an advance unless costs are reduced significantly. Even the least expensive and most effective of organ replacement technologies must be critically compared with the costs of meeting basic health needs and preventing disease.

71.5 SUMMARY: BEYOND THE TRAIL'S END

The task of the clinical researcher, in essence, is to make a fair offer of research participation to potential subjects under conditions of uncertainty, where the goals are two-fold: to contribute to generalizable knowledge and to keep subjects as safe as possible under the circumstances. In regenerative medicine/transplantation research, the combination of promise, complexity, and uncertainty makes research design and conduct especially challenging. Only early, continual, and ongoing critical reflection by physicians, patients, researchers, and research subjects [45] can bring science and ethics together in organ regeneration research and treatment.

Perhaps it should not come as a surprise, then, that organ regeneration researchers can contribute to a policy-level conversation about research directions and priorities. Recognizing that research helps to shape policy decisions about health-care expenditures, researchers can critically examine and discuss how research targets are and should be identified, and how research activities are and should be prioritized. What should drive organ regeneration research:

- need (severity, prevalence, lack of effective standard treatment or symptom control)?
- science (easy to study, materials available, high generalizability)?
- society (advocacy group interest, available funding)?

Researchers represent only one voice in this policy discussion, but it is a voice from which others should learn. Similarly, researchers themselves stand to learn a great deal from policy makers and the public.

It may be more surprising, however, to suggest that organ regeneration researchers also have a role to play in a deeper and more philosophical conversation about the nature of normal human growth, development, and aging. Consider the probable effect of organ regeneration on how we understand organ failure. Some technologies involve artificial organ substitutes or medical therapies that supply what the organ cannot; these are usually procedures that are repeated and necessitate close monitoring and correction of adverse effects. Conventional organ transplantation requires ongoing immunosuppression and monitoring, and may need to be repeated because allotransplants often have relatively short productive life spans. Organ regeneration technologies would ideally require only a single surgical intervention at one time point. Such interventions may take more time to become effective, but are hoped to be significantly less invasive, with fewer risks of harm and adverse effects than conventional treatment.

Both of these factors could lead to earlier intervention with regenerative medicine technologies than with conventional treatments. Over time, the definition of "normal" organ function—the level of function at which treatment would be recommended, which will always be higher than the level that would cause serious illness or irreversible damage—would probably then change. The line between what is normal and abnormal would probably begin to shift, because the balance of harms and benefits would change as the potential harms of treatment are lessened. If treatment is unproblematic, then the condition or conditions in need of treatment may expand, and the threshold of need for treatment may be reached sooner.

This is one aspect of a very familiar phenomenon in health care. It is well known that the indications for new treatments tend to expand as the treatments are perfected and better understood. An even more common type of example is found in the broadening of indications for novel interventions from serious conditions to a wide range of minor indications. This path has been followed with human growth hormone injections for idiopathic short stature, botulinum toxin (Botox) injections for minor medical and cosmetic conditions, and testosterone injections for older men, to name just a few examples [46,47].

Changes in how we think about concepts like "normal function" and "need for treatment" are thus by no means unprecedented. The contested terrain of concepts of health and disease has never been truly fixed, nor truly fluid. Nonetheless, the potential for permanent alteration in function that is so attractive in novel biotechnologies may accelerate, and even broaden, such conceptual shifts. In addition to moving the line between normal and diseased, regenerative medicine technologies might change the relationship between prevention and treatment. Moving the "normal" line means recategorizing people who were formerly viewed as having normal organ function as now having abnormal function, in large part because treatment is improved. But we might alternatively consider earlier intervention as a way of maintaining normal function—and thus categorize regenerative interventions not as *treatment*, but as *prevention*.

The relationship between prevention and treatment is another distinction that seems sharp and clear but upon examination turns out to be quite nuanced [48]. If successful regenerative medicine interventions change the definition of disease so that it is diagnosed earlier, then treatment and secondary prevention could converge.

And finally, in the future, there might arise a cumulative effect from changes like these in the conceptualization of health and disease and in the practice of medicine. Over time, our understanding of healthy aging could change significantly. This could include changes in what we consider normal functioning as we age and ultimately a lengthening of the projected life span. Both changes are characteristic of what has been called human enhancement and even "transhumanism" [49—52]. Regenerative medicine is likely to play a signally important role in medical progress—and the expected concomitants of medical progress are corresponding changes in how we think and how we live. It is therefore part of the responsibility of both science and ethics to anticipate possible changes like this, and to help direct their examination and discussion.

REFERENCES

[1] Atala A, Lanza R, Thomson JA, Nerem R, editors. Principles of regenerative medicine. 2nd ed. Amsterdam: Elsevier Science; 2010.

[2] King NMP, Coughlin C, Furth M. Ethical issues in regenerative medicine. Wake Forest Intell Prop L J 2010;9:215—37.

[3] King NMP, Coughlin CN, Atala A. Pluripotent stem cells: in search of the "perfect" source. Minn J Law Sci Technol 2011;12:715—30.

[4] King NMP. The line between clinical innovation and human experimentation. Seton Hall Law Rev 2003;32:573—82.

[5] Mastroianni AC. Liability, regulation and policy in surgical innovation: the cutting edge of research and therapy. Health Matrix 2006;16:351—442.

[6] Angelos P. The ethical challenges of surgical innovation for patient care. Lancet 2010;376:1046—7.

[7] Fins JJ. Surgical innovation and ethical dilemmas: precautions and proximity. Cleveland Clinic J Med 2008;75:7—12.

[8] Morreim H, Mack MJ, Sade RM. Surgical innovation: too risky to remain unregulated? Ann Thorac Surg 2006;82:1957—65.

[9] Orlando G. Transplantation as a subfield of regenerative medicine. Expert Rev Clin Immunol 2011;7:137—41.

[10] Atala A, Bauer SB, Soker S, Yoo JJ, Retik AB. Tissue-engineered autologous bladders for patients needing cystoplasty. Lancet 2006;367:1241—6.

[11] Orlando G. Immunosuppression-free transplantation considered from a regenerative medicine perspective. Expert Rev Clin Immunol 2012;8:179—87.

[12] Beauchamp TL, Childress JF. Principles of biomedical ethics. 7th ed. New York, NY: Oxford University Press; 2013.

[13] National Commission for the Protection of Human Subjects of Biomedical and Behavioral research. The Belmont Report: Ethical principles and guidelines for the protection of human subjects of research. 1979 [Cited October 23, 2012] Available from: <http://www.hhs.gov/ohrp/humansubjects/guidance/belmont.html>.

[14] Kimmelman J. Gene transfer and the ethics of first-in-human trials: lost in translation. Cambridge, UK: Cambridge University Press; 2009.

[15] King NMP, Cohen-Haguenauer O. En route to ethical recommendations for gene transfer clinical trials. Mol Ther 2008;16:432—8.

[16] Emanuel EJ, Wendler D, Grady C, Warren G. What makes clinical research ethical? JAMA 2000;283:2701—11.

[17] Dresser R. When science offers salvation: patient advocacy and research ethics. New York, NY: Oxford University Press; 2001.

[18] Halpern SD, Karlawish JHT, Berlin JA. The continuing unethical conduct of underpowered clinical trials. JAMA 2002;288:358—62.

[19] Kimmelman J. Recent developments in gene transfer: risk and ethics. BMJ 2005;330:79—82.

[20] Bretzner F, Gilbert F, Baylis F, Brownstone RM. Target populations for first-in-human embryonic stem cell research in spinal cord injury. Cell Stem Cell 2011;8:468—75.

[21] King NMP, Churchill LR. Assessing and comparing potential benefits and risks of harm. In: Emanuel E, Grady C, Crouch RA, Lie RK, Miller FG, Wendler D, editors. Oxford textbook of clinical research ethics. New York, NY: Oxford University Press; 2008. p. 514—26.

[22] US Department of Defense. The Nuremberg code: Directives for human experimentation, U.S. v. Karl Brandt, Trials of war

criminals before Nuremberg military tribunals under control law 10. 1947 [Cited October 23, 2012]. Available from: <http://ori.dhhs.gov/education/products/RCRintro/c03/b1c3.html>.

[23] World Medical Association. Declaration of Helsinki. Helsinki: Eighteenth World Medical Assembly. 1964, and amended 1975−2008. [Cited October 23, 2012]. Available from: <http://www.wma.net/en/30publications/10policies/b3/>.

[24] Kimmelman J, London AJ. Predicting harms and benefits in translational trials: ethics, evidence, and uncertainty. PLoS Med 2011;8(3):e1001010. Available from: <http://dx.doi.org/10.1371/journal.pmed.1001010>.

[25] Henderson GE, King NM, P. Informed consent in gene transfer research. Unpublished data; 2002.

[26] King NMP. Defining and describing benefit appropriately in clinical trials. J Law Med Ethics 2000;28:332−43.

[27] King NMP, Henderson GE, Churchill LR, Davis AM, Hull SC, Nelson DK, et al. Consent forms and the therapeutic misconception: the example of gene transfer research. IRB 2005;27(1):1−8.

[28] Faden RR, Beauchamp TL, King NMP. A history and theory of informed consent. New York, NY: Oxford University Press; 1986.

[29] Department of Health and Human Services. Protection of human subjects: Title 45 CFR Part 46. Available from: http://www.gpo.gov/fdsys/browse/collectionCfr.action?collectionCode=CFR>; 2011.

[30] Food and Drug Administration. Protection of human subjects, Institutional review boards: Title 21 CFR Parts 50, 56. Available from: http://www.gpo.gov/fdsys/browse/collectionCfr.action?collectionCode = CFR>; 2012.

[31] National Institutes of Health. NIH guidance on informed consent for gene transfer research. [Cited October 23, 2012] Available from: <http://oba.od.nih.gov/oba/rac/ic/index.html>; 2012.

[32] Peppercorn JM, Weeks JC, Cook EF, Joffe S. Comparison of outcomes of patients treated within and outside clinical trials: conceptual framework and structured review. Lancet 2004;363:263−70.

[33] Appelbaum PS, Roth LH, Lidz C. The therapeutic misconception: informed consent in psychiatric research. Int J Law Psychiat 1982;5:319−29.

[34] Dresser R. The ubiquity and utility of the therapeutic misconception. Soc Philos Policy 2002;19:271−94.

[35] Churchill LR, Nelson DK, Henderson GE, King NMP, Davis AM, Leahey E, et al. Assessing benefits in clinical research: why diversity in benefit assessment can be risky. IRB 2003;25:1−8.

[36] Henderson GE, Easter MM, Zimmer C, King NMP. Therapeutic misconception in early phase gene transfer trials. Soc Sci Med 2006;62:239−53.

[37] Miller M. Phase I cancer trials: a collusion of misunderstanding. Hastings Cent Rep 2000;30:34−43.

[38] Dresser R. First-in-human trial participants: not a vulnerable population, but vulnerable nonetheless. J Law Med Ethics 2009;37:38−50.

[39] Henderson GE, Churchill LR, Davis AM, Grady C, Joffe S, Kass N, et al. Clinical trials and medical care: defining the therapeutic misconception. PLoS Medicine 2007;4:1735−8.

[40] Horng S, Grady C. Misunderstanding in clinical research: distinguishing therapeutic misconception, therapeutic misestimation, and therapeutic optimism. IRB 2003;25:11−6.

[41] Sulmasy DP, Astrow AB, He MK, Sells DM, Meropol NJ, Micco E, et al. The culture of faith and hope: patients' justifications for their high estimates of expected therapeutic benefit when enrolling in early phase oncology trials. Cancer 2010;116:3702−11.

[42] De Coppi PG, Bartsch Jr. G, Siddiqui MM, Xu T, Santos CC, Perin L, et al. Isolation of amniotic stem cell lines with potential for therapy. Nat Biotechnol 2007;25:100−6.

[43] Faden RR, Dawson L, Bateman-House AS, Agnew DM, Bok H, Brock DW, et al. Public stem cell banks: considerations of justice in stem cell research and therapy. Hastings Center Rep 2003;33:13−27.

[44] Thomas L. The technology of medicine. Lives of a cell: notes of a biology watcher. New York, NY: Viking; 197431−6

[45] Capron AM. Informed consent in catastrophic disease research and treatment. Univ PA Law Rev 1974;123:340−438.

[46] Groopman J. Hormones for men: is male menopause a question of medicine or of marketing? New Yorker July 29, 2002;34−38.

[47] Leschek EW, Rose SR, Yanovski JA, Troendle JF, Quigley CA, Chipman JJ, et al. Effect of growth hormone treatment on adult height in peripubertal children with idiopathic short stature: a randomized, double-blind, placebo-controlled trial. J Clin Endocrinol Metab 2004;89:3140−8.

[48] Juengst ET. Can enhancement be distinguished from prevention in genetic medicine? J Med Philos 1997;22:125−42.

[49] Fukuyama F. Our posthuman future: consequences of the biotechnology revolution. New York, NY: Farrar, Straus and Giroux.; 2002.

[50] Kurzweil R. The singularity is near: when humans transcend biology. New York, NY: Penguin; 2005.

[51] McKibben B. Enough: staying human in an engineered age. New York, NY: Henry Holt; 2003.

[52] Mehlman MJ. The price of perfection. Baltimore: Johns Hopkins University Press; 2009.

Chapter 72

Epilogue: Organ Bioengineering and Regeneration as The New Holy Grail of Organ Transplantation

Giuseppe Orlando[a,b]

[a]*Department of General Surgery, Section of Transplantation, Wake Forest School of Medicine, Winston Salem, NC,* [b]*Wake Forest Institute for Regenerative Medicine, Wake Forest School of Medicine, Winston Salem, NC*

Chapter Outline

Organ transplantation has fallen victim to the remarkable success noted from the extraordinary results achieved to date. As a result, the demand for organs is rising, unable to be met by the plateaued and limited supply of organs. Ultimately, this constraint has placed a proliferating number of candidate patients on a waiting list and increased mortality due to the long waiting times. Achievements in organ bioengineering and regeneration (OBR) have offered the quickest potential solution by allowing the application of OBR technologies to manufacture organs for transplantation purposes. The ideal solution will be the equivalent of an autograft (produced via the exploitation and manipulation of autologous cells) such that the recipient will not require antirejection medication. The immunosuppression-free state (IFS) has long been the ideal end product of transplant, and yet to this day, the goal of achieving stable, immediate, durable and reproducible IFS remains intangible. OBR has immense potential and capacity to revolutionize the field of organ transplantation by fulfilling the increasing organ demands and achieving IFS in transplanted patients. Hence, advances in the field of OBR could become the new Holy Grail for transplant sciences.

72.1 INTRODUCTION

In the past 20 years, over 160 patients have received organs created from autologous cells, which were seeded and expanded on a supportive scaffold and transplanted into a patient, without any need for immunosuppression [1–6] (Table 72.1). The success of such pioneering discoveries has shown that organ bioengineering and regeneration (OBR) has enormous potential to revolutionize organ transplantation and address two of the most important issues—the demand for an inexhaustible source of organs and the accomplishment of an immunosuppression-free state (IFS) posttransplantation [4,6,9]. Based on these preliminary yet pivotal results, we believe that OBR should become a major focus of transplant science in the coming years, and it should be referred to as the new Holy Grail for the field instead of plain clinical tolerance. The purpose of this chapter is to elucidate upon this rationale from the perspective of transplant surgeon—scientists specializing in OBR and attempt to bridge the gap between organ transplantation and regenerative medicine under the collective definition of OBR.

TABLE 72.1 Cases of Bioengineered Organs Implanted in Patients

	Study Organ	Indication to implantation	Bioengineering method	Type of scaffolds used	No. of patients	Patient's age	Length of follow-up post-implantation	Reported Outcome
Romagnoli[46]	Urethra	Posterior hypospadias	Cell sheet technology. Autologous urethral mucosal cell sheets generated by serial cultivation	No scaffolds. When ready for implantation, cell sheets were either mounted on a petrolatum gauze before implantation.	2	4- and 13-year old	18 and 6 month	Excellent
Pellegrini[47]	Cornea	Complete loss of the corneal-limbal epithelium	Cell sheet technology. Autologous corneal epithelial cell sheets generated by serial cultivation of limbal cells	No scaffolds. When ready for implantation, cell sheets were either mounted on a petrolatum gauze or on a soft contact lens before implantation.	2	32- and 69-year old	2 years	Excellent under the form of striking improvement in patients' comfort and visual acuity.
Rama[48]	Cornea	Severe or total, unilateral or partial burn-related lateral limbal stem-cell deficiency	Cell sheet technology. Autologous corneal epithelial cell sheets generated by serial cultivation of limbal cells		106	46.5 ± 14.4 (range, 14 to 80) years	Up to 10 years	76.6% success rate
Quarto[49]	Bone	Defect of long bones	Culture-expanded osteoprogenitor cells grown on porous bioceramic scaffolds	Synthetic	3	41-, 16-, 22-year old	6.5, 6 and 13 months, respectively	Excellent
Warnke[50]	Bone	Extended mandibular discontinuity following ablative tumor surgery	Titanium mesh cage filled with bone mineral blocks and infiltrated with recombinant human bone morphogenetic protein and bone marrow cells. To allow maturation, the construct was implanted into the latissimus dorsi muscle for 7 weeks	Synthetic	1	56-year old	4 weeks	Excellent, yet short term
Shinoka[35]	Right intermediate pulmonary artery	Pulmonary atresia	Autologous adult cells seeded on a tube that served as a scaffold, composed of a polycaprolactone—polylactic acid copolymer	Synthetic	1	4-year old	9 years	Excellent

Study	Site	Indication	Technique	Scaffold type	Number of patients	Age	Follow-up	Outcome
Atala[51]	Bladders	High-pressure or poorly compliant bladders complicating myelomeningocele	Autologous adult cells seeded on a biodegradable bladder-shaped scaffold made of collagen, or a composite of collagen and polyglycolic acid ... reinforced with woven polyglycolic acid	Synthetic and bladder derived	7	11 (range 4–19) years	46 (range 22–61) months	Excellent
Hibino[52]	Extra-cardiac cavopulmonary conduits	Single ventricle physiology	Autologous bone marrow mononuclear cells were seeded onto a biodegradable scaffold composed of either polyglycolic acid or poly-L lactic acid	Synthetic	25	5.5 (1–24 range) years	5.8 years	Five patients developed either thrombosis or stenosis of the graft. There was no evidence of aneurysm formation, graft rupture, ectopic calcification or death related to the procedure
McAllister[53]	Arteriovenous fistula for dialysis access	End-stage renal disease in patients requiring vascular access for hemodialysis. All patients had previous hemodialysis access failure, had no suitable vein for a new arteriovenous fistula, and would therefore be good candidates for the implantation of synthetic vascular grafts	Sheet-based technology with no scaffolds. Autologous adult cells were cultured in conditions that promote the deposition of ECM. The end product of such process is the production of cell sheets that can be detached intact from the culture dishes and wrapped around a stainless steel mandrel to allow fusion among sheets and eventually maturation.	No supporting scaffolds were ever used	9	60 (29–78) months	3 year	Cumulatively primary patency rate was 60% at 6 months from surgery. This figure is comparable to data about patency rate of native arteriovenous fistula from Dialysis Outcomes Quality Initiative
Macchiarini[36]	Windpipe	End-stage bronchomalacia	Decellularization-recellularization scaffold technology. Autologous adult and bone marrow stromal cells were seeded on a decellularized trachea retrieved from a deceased donor	Natural ECM scaffolds from deceased donor	1	30-year old	18 months	Eight months after the implantation, a proximal ventral collapse of the proximal end of the graft required the placement of a temporary endoluminal stent
Raya-Rivera[54]	Posterior urethra	Complete posterior urethral disruption caused by pelvic trauma	Autologous adult cells were seeded onto tabularized polyglycolic acid: poly (lactide-co-glycolide acid) scaffolds	Synthetic	5	11.6 (10–14) years	71 (36–76) months	Excellent
Jungebluth[37]	The upper airway defect remaining following the removal of the tumor, extended from 5 cm above the	Recurrent primary trachea cancer	Airway was replaced with a tailored bioartificial nanocomposite previously seeded with autologous bone-marrow mononuclear	Synthetic	1	36-year old	5 month	Patient asymptomatic and tumor-free

(Continued)

TABLE 72.1 (Continued)

Study	Organ	Indication to implantation	Bioengineering method	Type of scaffolds used	No. of patients	Patient's age	Length of follow-up post-implantation	Reported Outcome
	right tracheobronchial angle into the first 1.3 cm of the right main bronchus		cells via a bioreactor for 36 h. Postoperative granulocyte colony-stimulating factor filgrastim (10 μg/kg) and epoetin beta (40,000 UI) were given over 14 days					
Olausson[55]	Meso-Rex bypass, namely an interposition graft between the superior mesenteric vein and the left portal vein	Portal vein thrombosis in the context of idiopathic thrombocytopenic purpura	A 9 cm segment of allogenic donor iliac vein was decellularised and subsequently recellularised with endothelial and smooth muscle cells differentiated from stem cells obtained from the bone marrow of the recipient.	Natural ECM scaffolds obtained by decellularization of iliac vessel from a deceased donor	1	10-year old	12 months	The patient had normal laboratory values for 9 months. However, at 1 year the shunt was narrowed and damaged by mechanical obstruction of tissue in the mesocolon. Therefore, patient was re-operated to remove obstruction and allow vessel lengthening with a second stem-cell populated vein graft. After this second operation, portal blood flow normalized and patient improved physical and mental function and growth.
Elliott[37]	Trachea	Congenital tracheal stenosis and pulmonary sling	Innate human ECM scaffolds were seeded with bone marrow mesenchymal stem cells, with patches of autologous epithelium. Topical human recombinant erythropoietin was applied to encourage angiogenesis, and transforming growth factor β to support chondrogenesis. Intravenous human recombinant erythropoietin was continued postoperatively.	Natural ECM scaffold obtained from decellularization of a segment of upper airway from a deceased donor	1	12-year old	2 years	Excellent

72.2 ORGAN TRANSPLANTATION AS A *HALFWAY TECHNOLOGY*

The concept of halfway technology, introduced by Lewis Thomas [10], in its application to medicine describes treatments that improve upon symptoms without removing the cause of a specific clinical condition. Furthermore, the goal of halfway technology is to manage disease without proposing a cure in medical situations, which are deemed life-threatening, noncurable, or end-stage disease settings, or when a disease drastically deteriorates quality of life. Organ transplantation, although one of the greatest achievements in the history of modern medicine, remains in most cases a "halfway technology" for two reasons: firstly, it does not eliminate the cause of the standard disease that may relapse posttransplantation, and secondly, it requires lifelong antirejection therapy, which may potentially lead to severe acute or chronic toxicity, which further leads to additional clinical syndromes. Long-term management must be put into place to minimize side effects arising from drug toxicity, and also to optimize the quality of the patient's life and prevent untimely graft failure or premature death.

An example that illustrates organ transplantation as a halfway technology is in a scenario in which liver transplantation is performed in a patient with hepatitis C virus (HCV)-related end-stage liver disease. The transplant is lifesaving, however, the virus is not removed, and it may immediately reinfect the graft at reperfusion [11]. Currently, no effective antiviral therapy can reliably prevent reinfection of the graft, which unfortunately is a universal outcome. The onset of disease recurrence just happens to be a matter of time. What is even more vexing is histological evidence of liver graft damage detected as early as 9 days posttransplant and advancement to cirrhosis and allograft failure, which requires retransplantation and occurs in 25% and 10% of cases, respectively, within 5−10 years following the original transplant surgery [11]. Subsequently, HCV relapse is one of the most frustrating problems of modern liver transplantation, and it represents one of the most recurrent sources of allograft failure secondary to relapse of disease, regardless of the type of organ transplanted in the patient. Additionally, HCV relapse is favored by factors such as the degree of impairment of the host immune system (mainly due to the net state of immunosuppression), defined by the intensity of immunosuppressive therapy. Perplexingly, evidence in literature has shown that complete weaning of immunosuppression in HCV liver transplant defers disease relapse and thus, improves the global outcome [12−14]. Currently, approaches to establish safe, stable, and durable IFS are being examined [15] (see http://www.immunetolerance.org/studies/gradual-withdrawal-immunosuppression-patients-receiving-liver-transplant-awish).

Transplantation may be considered to be a complete technology in instances where it resolves not only the clinical symptoms but also gets rid of the source of disease without risk of relapse posttransplant. Such a scenario is feasible when a candidate patient receives an organ from a syngeneic donor, while the origin of the disease is removed. The candidate recipient's immune system accepts the new organ because of identical antigen match of donor. Furthermore, no immunosuppression is needed at any point of time posttransplant. An ideal example was noted in the case of kidney transplantation occurring between identical twins. The transplant community celebrated the longest living renal transplant survivor [16], Johanna Nightingale, who was recipient to a new kidney from her identical twin in December 1960 following chronic kidney disease occurring secondary to glomerulonephritis complication of *Streptococcus pyogenes* throat infection which originally occurred in 1955. Dialysis was not available to the patient and, consequently, she received supportive care for a few years until her attending physicians learnt that plastic surgeon, Dr. Joseph Murray, had been performing kidney transplants in Boston. Dr. Murray, realizing that Johanna was a perfect transplant candidate for having a healthy identical twin, performed a successful renal transplantation, and more than 50 years later, Johanna is doing well with a perfectly functioning graft, which never required immunosuppressive therapy due to genetic identity. Therefore, for three reasons organ transplantation among genetically identical individuals should be considered a full treatment rather than a halfway technology [17]: firstly, the graft reestablishes the condition of wellness; secondly, the source of the disease (in the case of Johanna, the streptococcal infection) is either removed, absent, or does not relapse; and finally, no immunosuppression is required posttransplantation.

72.1.1 The Escalating Organ Demand

As stated earlier organ transplantation has become a victim of its own success. For instance, the gold standard for renal replacement therapy in patients with end-stage renal disease is kidney transplantation because not only does it radically optimize patient survival and quality of life, but it is also cost-effective in comparison to maintenance dialysis. As per the United States Renal Data System report (http://www.usrds.org/), life expectancy from the time renal dialysis is started is approximately 8 years and 4.5 years for patients in the age groups of 40−44 and 60−64 years, respectively. These figures are exceeded by increased survival following kidney transplantation, which are 85%, 70%, and 44% after 5-, 10-, and 20 years, respectively (information based on the Organ Procurement and Transplantation Network data as of July

27, 2012; http://optn.transplant.hrsa.gov/) (Table 72.2). In recognition of the beneficial outcomes of transplantation compared to patients affected by the same disease who did not receive liver transplant, such as delay in disease recurrence, waiting lists for transplantation are increasing rapidly [12–14]. Additionally, immunosuppression weaning approaches are being studied to promote a safe and stable IFS [15] (see http://www.immunetolerance.org/studies/gradual-withdrawal-immunosuppression-patients-receiving-liver-transplant-awish). To meet the urgent and rapidly rising demands, efforts to expand the donor organ pool are being implanted, which unfortunately still have not been able to keep pace with the demand (Figure 72.1). Some of these efforts include extension of living and deceased donor acceptance criteria, transplants occurring across immunological barriers, paired donations, Good Samaritan or altruistic donations, etc. Despite such efforts the cumulative probability to receive an organ in the critical time period is decreasing rapidly (Table 72.3), while the waiting list mortality is increasing (http://www.medscape.org/viewarticle/488926). For instance, in the United States, only 16,813 kidney transplants were performed in the year 2011 (*source*: http://optn.transplant.hrsa.gov/) out of 96,574 patient candidates. This figure corresponds to the probability of receiving a renal graft at 1 year (from the time of registration on waiting list) of 9.65%. Although this figure increases at 3 and 5 years to 21.65% and 36%, respectively (http://unos.org/), the rate is not adequate to meet the demand for transplantable kidneys. As the number of uremic patients reaches epidemic proportions as a result of chronic noncommunicable diseases such as diabetes and hypertension in the developed world, as well as infectious diseases and malnutrition in developing countries, it becomes very important to point out the rising disparity between organ supply and demand [19]. Currently, research shows that only 6% of dialysis patients will ever receive a kidney transplant. Xenotransplantation, as a potential solution, remains highly unlikely in terms of bridging the gap between organ supply and demand; however, OBR technological developments and regenerative medicine have shown enormous potential to be successfully translated to clinical solutions [1–6,9]. The reported series of such successful OBR transplants are very rare and information about long-term follow up is deficient. Despite these drawbacks, it is important to recognize that it is possible to bioengineer body parts and perhaps even develop complex modular organs such as the kidneys or livers, theoretically at least. The triumph of OBR technologies will certainly depend on available resources and research investment in a field that is destined to revolutionize medical transplantation.

72.3 WHY PURSUING AN IFS?

Chronic immunosuppression, which is needed to prevent graft rejection by the recipient, places the patient at risk for acquiring infectious and metabolic complications, malignancies, as well as drug-specific toxicities [15,21–23]. In fact, infection is one of the leading causes of hospitalization in patients receiving kidney transplants within the first 2 years of the transplantation, and in pediatric kidney transplantation, infection is the leading cause of hospitalization [24]. Furthermore, immunosuppression contributes to cardiovascular events, which represent the major cause of death [25]. Transplant recipients have a greater incidence of cancer compared to an age-match healthy population, proven by studies stating that after 20 years of immunosuppression, about 40% of kidney transplant patients will develop cancer and moreover, the risk is correlated to the amount of immunosuppressive drugs received [26,27]. About 50% of the kidney grafts are lost within 10 years posttransplant secondary to chronic allograft nephropathy, a process dependent on alloantigen-dependent and independent aspects; the latter includes toxic effects of nephrotoxicity, hypertension, and hyperlipemia [28]. Thus, we believe that the overarching goal of transplantation should be to accomplish IFS as it avoids complications and costs stemming from lifelong immunosuppression and optimizing the patient's quality of life.

Not only is the pursuit of IFS free transplantation driven by the medical complications arising from immunosuppression, but also economic factors such as the high cost of immunosuppressive drugs and the impact on the patient's quality of life are playing a critical role in fueling these initiatives. Organ transplantation is one of the most expensive therapies, which has a major financial impact on health care systems globally. Unfortunately, countries, where the national health services are overloaded by the needs of an aging population, choose to adopt strict cost-optimization policies to limit expenses and access [29]. The challenge lies in developing strategies to lower expense of transplantation, improving quality of life, and reducing posttransplant complications such as acute rejection and infection rates. The safety and efficacy of immunosuppressive regimens are assessed by the outcomes discussed above and act as a guide for other treatments [30]. Patient expectations, mainly the anticipation of complete healing and similar quality of life as nontransplant individuals is also considered as a benchmark to assess outcomes. The frequency and dosage of prescribed medications have the most bearing towards the patient's lower perceived quality of life, i.e., the more the drug dosage and frequency, the lower the perception of the patient's quality of life. Additionally, patient nonadherence is one of the major causes of organ failure and an important factor in determining the outcome. Ultimately,

TABLE 72.2 Actuarial Patient Survival Rates After KT. Data Are Stratified by Year of Transplantation (based on open access OPTN data as of July 27, 2012).

A) Patient survival rate after 1, 5 and 10-year for transplants performed between 01/01/1991 and 12/31/1999 by Organ

	Year Survival					
	1		**5**		**10**	
Organ	**Sample Size**	**Survival Rate and 95% CI**	**Sample Size**	**Survival Rate and 95% CI**	**Sample Size**	**Survival Rate and 95% CI**
Heart	20469	83.83 [83.31,84.34]	20469	69.74 [69.10,70.38]	20469	51.57 [50.85,52.29]
Heart-Lung	498	65.79 [61.49,70.10]	498	42.14 [37.71,46.57]	498	26.73 [22.62,30.84]
Intestine	393	64.44 [59.39,69.49]	393	44.61 [39.33,49.89]	393	35.08 [29.90,40.27]
Kidney	99844	95.29 [95.15,95.42]	99844	84.90 [84.66,85.13]	99844	69.90 [69.56,70.24]
Kidney-Pancreas	6900	93.41 [92.81,94.02]	6900	83.68 [82.76,84.59]	6900	69.09 [67.86,70.33]
Liver	34594	83.28 [82.87,83.69]	34594	71.42 [70.91,71.92]	34594	58.36 [57.78,58.94]
Lung	6712	74.36 [73.30,75.42]	6712	44.91 [43.70,46.13]	6712	24.28 [23.19,25.37]
Pancreas	1434	91.92 [90.39,93.44]	1434	79.63 [77.19,82.07]	1434	62.91,59.45,66.37]

B) Patient survival rate after 1 and 5-year for transplants performed between 01/01/2000 and 12/31/2005 by Organ

	Year Survival			
	1		**5**	
Organ	**Sample Size**	**Survival Rate and 95% CI**	**Sample Size**	**Survival Rate and 95% CI**
Heart	12753	86.78 [86.18,87.38]	12753	73.74 [72.96,74.52]
Heart-Lung	212	68.42 [61.80,75.04]	212	47.92 [41.00,54.84]
Intestine	748	76.60 [73.42,79.79]	748	56.75 [52.99,60.51]
Kidney	90309	95.69 [95.56,95.83]	90309	85.41 [85.17,85.66]
Kidney-Pancreas	5367	94.73 [94.11,95.35]	5367	86.11 [85.13,87.09]
Liver	33816	86.14 [85.76,86.52]	33816	72.92 [72.42,73.42]
Lung	6723	81.35 [80.40,82.29]	6723	53.09 [51.87,54.31]
Pancreas	3116	94.64 [93.79,95.48]	3116	83.00 [81.49,84.51]

C) Patient survival rate after 20-year for transplants performed between 10/01/1987 and 12/31/1990 by Organ

	Year Survival	
	20	
Organ	**Sample Size**	**Survival Rate and 95% CI**
Heart	5843	18.45 [17.29,19.61]
Heart-Lung	202	14.16 [8.35,19.98]
Kidney	28987	43.69 [42.74,44.64]
Kidney-Pancreas	993	38.41 [34.37,42.45]
Liver	6926	33.95 [32.51,35.40]
Pancreas	238	24.71 [15.03,34.39]

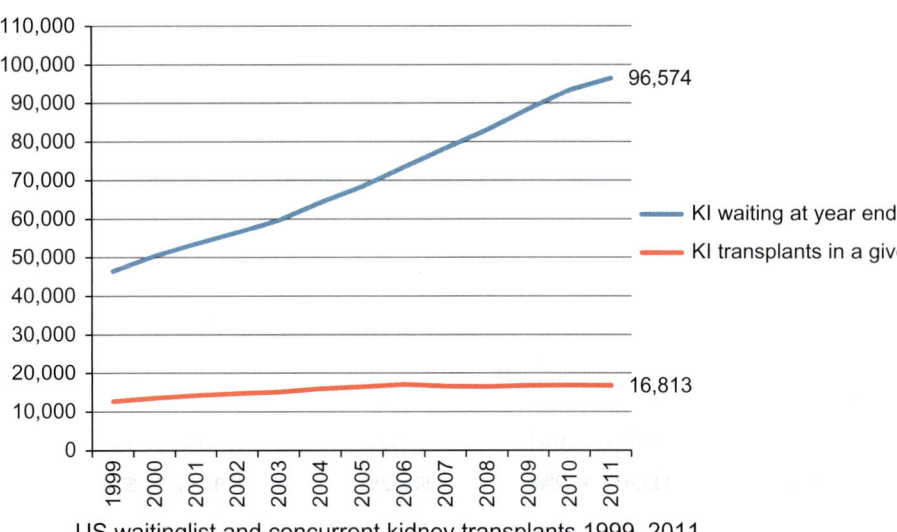

US waitinglist and concurrent kidney transplants 1999–2011

FIGURE 72.1 Representation of the comparison between the number of renal transplants performed yearly since 1999 and the number of patients on the waiting list for a kidney from a deceased donor. The gap between the two variables is dramatically increasing as a consequence of the constantly escalating number of patients registered on the waiting list and an unprecedented attenuation of the growth of kidney transplantation. (*Based on Organ Procurement and Transplantation Network data as of August 24, 2012*).

TABLE 72.3 Cumulative Probability to Receive Organs at 1, 3 and 5 Years From the Time of Registration on the Waiting List, in Two Different Time Frames: 01/01/1995-31/12/2006 (A) and 01/01/2007-12/31/2012 (B). Percentages are trending towards lower figures as years goes by due to the increasing gap between demand and offer (based on open acces Organ Procurement and Transplantation Network data as of July 27, 2012).

A)

		Year Post-Listing					
		1		3		5	
Organ	Number of Registrations	Percent (Probability*100) Transplanted In One Year	95% Confidence Interval	Percent (Probability*100) Transplanted In One Year	95% Confidence Interval	Percent (Probability*100) Transplanted In One Year	95% Confidence Interval
Heart	41101	53.98%	[53.5%, 54.5%]	61.92%	[61.5%, 62.4%]	63.19%	[62.7%, 63.6%]
Heart-Lung	1267	22.65%	[20.4%, 25.0%]	32.68%	[30.1%, 35.3%]	34.57%	[32.0%, 37.2%]
Intestine	2252	48.49%	[46.4%, 50.5%]	54.00%	[51.9%, 56.0%]	54.66%	[52.6%, 56.7%]
Kidney	287194	14.21%	[14.1%, 14.3%]	28.88%	[28.7%, 29.0%]	36.00%	[35.8%, 36.2%]
Kidney-Pancreas	19832	35.49%	[34.8%, 36.2%]	52.80%	[52.1%, 53.5%]	54.97%	[54.3%, 55.7%]
Liver	117489	39.93%	[39.7%, 40.2%]	48.67%	[48.4%, 49.0%]	50.49%	[50.2%, 50.8%]
Lung	22800	32.63%	[32.0%, 33.2%]	50.28%	[49.6%, 50.9%]	53.07%	[52.4%, 53.7%]
Pancreas	9489	38.28%	[37.3%, 39.2%]	49.00%	[48.0%, 50.0%]	50.34%	[49.3%, 51.3%]
Pancreas Islets	920	19.78%	[17.3%, 22.4%]	26.41%	[23.6%, 29.3%]	27.28%	[24.5%, 30.2%]
Total	502344	25.80%	[25.7%, 25.9%]	38.62%	[38.5%, 38.8%]	43.47%	[43.3%, 43.6%]

B)

Organ	Number of Registrations	Year Post-Listing			
		1		3	
		Percent (Probability*100) Transplanted In One Year	95% Confidence Interval	Percent (Probability*100) Transplanted In One Year	95% Confidence Interval
Heart	13534	59.23%	[58.4%, 60.0%]	67.34%	[66.5%, 68.1%]
Heart-Lung	224	46.43%	[39.9%, 52.7%]	0	[*, *]
Intestine	1049	59.49%	[56.5%, 62.4%]	65.51%	[62.5%, 68.3%]
Kidney	138803	9.65%	[9.49%, 9.80%]	21.65%	[21.4%, 21.9%]
Kidney-Pancreas	6335	34.98%	[33.8%, 36.2%]	50.32%	[49.1%, 51.6%]
Liver	45517	45.17%	[44.7%, 45.6%]	51.84%	[51.4%, 52.3%]
Lung	8710	63.88%	[62.9%, 64.9%]	72.13%	[71.2%, 73.1%]
Pancreas	3288	34.91%	[33.3%, 36.5%]	44.02%	[42.3%, 45.7%]
Pancreas Islets	311	31.83%	[26.8%, 37.0%]	44.29%	[38.6%, 49.9%]
Total	217771	23.75%	[23.6%, 23.9%]	34.26%	[34.1%, 34.5%]

*denotes value cannot be calculated due to less than 10 at risk at time point

the goal should be to achieve an optimal regimen that uses the least amount of drugs at the lowest effective doses to decrease toxicity and costs, but still avert disease relapse, graft rejection, and maintain graft function while increasing patient satisfaction.

At this point of time, IFS can be achieved only in genetically identical donors or by achieving immunological tolerance. True or complete tolerance is only seen in a small proportion of liver transplant recipients and continues to remain an ideal that the field of transplantation seeks to accomplish [15,21,22]. When liver transplantation is considered, IFS can only be attempted safely in a small subset of recipients who have to meet very strict criteria, about one in four patients who undergo weaning immunosuppression, in addition to having stable graft function at least for a year following transplantation. In the case of liver transplantation, tolerance is never achieved right away, nor is it stable [9]. The situation further deteriorates, when other organs are considered. For instance, clinical tolerance has never been stated for intestinal, islet, or whole pancreas transplantation, whereas two rare cases of IFS have been reported in lung and heart transplantation [22]. In renal transplantation, the accomplishment of a stable IFS is unique and exceptional; however, efforts to mimic such a state have involved complex treatment with nonreproducible results. Current tolerogenic strategies are not available for daily clinical practice as they are not practical, safe, or effective. Furthermore, our understanding of the immune mechanisms underlying tolerance is lacking [9,15,21,22].

Currently, OBR has bioengineered relatively simple hollow organs (using autologous cells derived from the patent), which have been implanted in over 50 patients suffering from a variety of medical conditions without the need for antirejection treatment. Moreover, these figures surpass the number of organ recipients, who were successfully weaned off immunosuppression following the immediate postoperative period [9,15,21,22]. The production of more complex organs such as kidney, liver, heart, lung, pancreas, and intestine will be a definite challenge, to say the least and has not been documented thus far. However, it is important to recognize the tremendous potential of regenerative medicine to offer a new perspective in solving this baffling puzzle.

72.4 PRINCIPLES OF OBR

The chief concept driving the manufacturing of hollow organ implants is the idea, known as cell-scaffold technology, that organs can be regenerated by seeding cells on supporting scaffold materials. The cells, either adult cells, progenitor cells *per se*, or progenitor cells that were induced to differentiate into specific cell types, are seeded on a synthetic or natural scaffold; the latter created by detergent-based decellularization of animal or human organ, a process that ensures clearance of cellular components of most organs. Pioneered at Harvard by Vacanti's group in the 1980s, cell-scaffold technology is based on the idea that cells require a supporting structure to attach, grow, and exhibit their function. The first experiments performed by Vacanti's group consisted of fetal and adult rat cells, mouse hepatocytes, mice pancreatic islet cells, and mesenteric cells from the small intestine, which were seeded onto synthetic scaffolds [4,32]. The synthetic scaffolds were made of polymers organized into fiber networks mimicking the intertwined branching patterns present in all organs, allowing for cell viability through diffusion, vascularization, and cell proliferation [33]. These seeded scaffolds underwent 4 days of culture, and were then implanted in different regions of diverse animal model species. Six such cases of successful engraftment have been demonstrated by viable cells, mitotic activity, and vascularization in cell mass. This discovery represents the first efforts by scientists to produce organs *ex vivo* and implant living cells on supporting artificial scaffold. These innovative and pioneering results laid the foundation for the development of tissue-engineered vascular grafts (TEVG), which were introduced to clinical application to replace the right intermediate pulmonary artery in a child afflicted from single right ventricle and pulmonary atresia [4,34]. In this case, as well as in several other cases reported to date, autologous cells from patients were collected and cultured *in vitro*. In a few cases, stem cells were isolated and differentiated into specific cell types before being seeded on a supporting scaffold [35−37]. These seeded scaffolds were allowed to further develop in bioreactors or were implanted immediately after seeding without undergoing any maturation [37]. Cell-scaffold technology has made it possible to produce and implant simple hollow organs like TEVG, segments of upper airways, urethras, and neourinary conducts. The drawbacks associated with this technology are the small sample size, short follow up, and discussion lacking on the current and future complications. With more complex organs, the circumstances change as current studies are in early stages, and unfortunately, clinical translation may not be feasible on the basis of current knowledge and data. Ultimately, we believe that the time has now come to elevate the initial excitement and tackle the issues that are hampering success, as highlighted in the outstanding chapters of the present book.

72.4.1 Major Issues to Address

a. What Is the Mechanism of Organ Development, Regeneration, and Healing?

The basic mechanism underlying regeneration begins with cell seeding on a scaffold, followed by attachment, growth, and expansion of the cells. The maturation phase, which is normally performed in bioreactors, is not effective enough to mature the construct and needs to be adjusted dynamically in order to imitate the cellular environment following implantation. In theory, the bioreactor environment imitates the *in utero* physiological conditions during organ ontogenesis, which still remain unknown. Certainly, we must acknowledge that our lack of comprehensive understanding of the mechanisms of organ development and mechanisms underlying regeneration, repair, and healing are obstacles toward achieving our goal of bioengineering organs.

b. Bioreactors

"Are currently available bioreactors adequate?" A contemporary case from Great Ormond Street, London, reported a novel bioengineered segment of upper airway [37], which was implanted immediately postproduction, without undergoing any maturation and probed upon the quality of currently available bioreactors. In all the previous cases, in which a bioengineered segment of upper airway was implanted, the new construct had undergone maturation in a bioreactor prior to implantation.

But why are bioreactors important? An important step toward achieving our goals is the development of appropriate bioreactors, which can offer a dynamic, manageable and reproducible culture environment created to hold constructs capable of taking part in physiological process and from which the products can be harvested or extracted [38]. The key function of a bioreactor is to replicate the *in vivo* environment (temperature, oxygen concentration, pH, nutrient concentration, biochemical, and biomechanical stimuli) for a time period, considered to be short, for stimulating the maturation phase of the construct following scaffold cell seeding. Additionally, bioreactors also allow to monitor and control the whole biological, biochemical and biochemical process, and permit uniform scaffold cell distribution, ideal nutrient application, and waste metabolite removal, along with hydrodynamic shear stress stimulus [38,39]. In the case of bioreactors, shear stress is defined by the forces native to bioreactor-like rotation and other mechanical stimuli essential to promote tissue development, which overall contributes to sustain the metabolic activity of the cell seeded construct and promote differentiation. With regard to OBR, seeding techniques, types, and number of cells, mimicry of the *in vivo* biophysical environment, biophysical forces, composition, and architecture of biomaterials play a chief role in driving success.

c. Vascularizing Bioengineered Constructs

To date, all bioengineered constructs were implanted without being relinked to the systemic vascular system, jeopardizing the newly transplanted organ to a risk of ischemia and graft failure. Basic physiology demonstrates that all organs require a vascular supply, although it is known that cells can live within a zone of roughly $1-3$ mm away from a source of nutrients, waste metabolite removal, and oxygen [40]. Hence, it is very important to take into account the key role of vascularization in providing oxygen and nutrient supply unless the size of the body part is less than 3 mm. In a case performed by Macchiarini's group, proximal ventral collapse of the most proximal 1 cm of the tracheal graft was reported after 8 months postimplantation of the group's first bioengineered trachea. The group believed that this complication may have been as a result of the pulsatile compression from the aortic arch superiorly and from the movement of stem-cell derived chondrocytes into the endoluminal surface of the graft. Another feasible hypothesis pertaining to vascularization would be that this complication was a result of ischemia to the graft, as the graft had not been reconnected to systemic circulation. For even more complex modular organs, which we would like to manufacture in the future using OBR, it is certainly important to understand the principal role of reconnecting the vascular pedicles of the OBR organs to systemic circulation. Natural extracellular matrix (ECM) scaffolds acquired from decellularized animal and human organs embody the ideal foundation as they have a well-preserved framework of the inherent vascular tree. For instance, the injection of regular contrast media within the renal artery of renal ECM scaffolds acquired from discarded kidneys in human transplantation [41], or from pigs demonstrates an intact vasculature encompassing the innate hierarchical branching structures. The contrast flows from the larger blood vessels to the smaller capillary bed and eventually drains out of the renal vein without extravasation into the parenchyma of the scaffold.

d. The Prominent Role of the ECM

It has become apparent from research done in the last decade that ECM plays a very important role in supporting cells, tissues, and organs. In particular, the function of the ECM is to not only provide mechanical support and architecture but also exert its effects on cell adhesion, signaling, molecular composition, and binding of growth factors. Furthermore, the mechanical characteristics of ECM, mainly the stiffness and deformability, contribute primarily to the determination, differentiation, proliferation, survival, polarity, migration, and behavior of cells [42]. The importance of understanding the pivotal role of the ECM and its interaction with cells has been demonstrated by the proliferation of numerous organizations focusing on

ECM biology such as the American Society for Matrix Biology (www.asmb.net), the British Society for Matrix Biology (www.bsmb.ac.uk), the Federation of European Connective Tissue Societies, the International Society of Matrix Biology (www.ismb. org/), and the Matrix Biology Society of Australia and New Zealand (www.mbsanz.org/). We believe that the understanding of interactions between ECM and cells along with knowledge of interactions' mechanisms will promote the development of OBR technology and regenerative medicine.

e. Time to Establish an International Registry of Implanted Bioengineered Organs

Registries are vital to the success of OBR as they permit the strict monitoring and inspection of the outcomes of bioengineered implanted organs, the follow ups, potential complications, and thus represent a critical method to qualitatively and quantitatively evaluate the risk-to-benefit ratio [43]. Additionally, registries provide an ideal interface with the public and are the most suitable mode to convey accurate information to the scientific community and to the media [44]. Currently and to the best of our knowledge, over 50 patients have received a bioengineered organ. A registry can unite detailed information of every single case and offer institutions performing such procedures the opportunity to share expertise and deliver data needed for critical analysis. Thus, we believe that it is time to establish an International Registry of Implanted Bioengineered Organs, which can globally bridge the gap between the scientific communities and promote collaboration toward achieving success in the field of OBR.

72.5 OBR AS THE NEW HOLY GRAIL FOR TRANSPLANTATION

Regenerative medicine, OBR, and organ transplantation arose from the same family of brilliant ideas and extraordinary thought processes. Alexis Carrel is known as the father of both OBR and vascular and transplant surgery (Figure 72.2). His farsighted and imaginative studies of cell culture and investigations of *ex vivo* organ preservation and growth provided an important understanding and a critical foundation for tissue and OBR, which were not translated until decades later. Carrel, via his collaborative projects with engineer and legendary aviator, Charles Lindbergh, was able to develop the first hybrid device meshed from two disciplines, which complemented each other in an exceptional manner. For instance, the Carrel and Lindbergh perfusion pump permitted living organs to exist outside the body during surgery, which ultimately led to the development of not only perfusion systems for

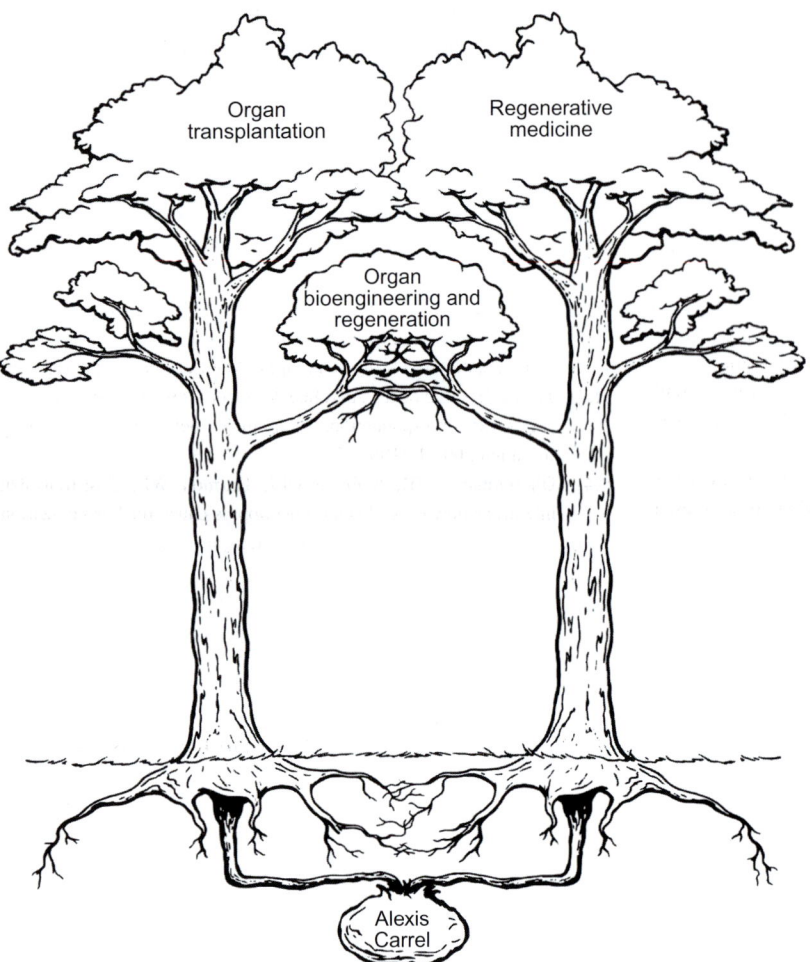

FIGURE 72.2 Organ transplantation and OBR share the same parenthood and generate from the same idea. Alexis Carrel's visionary work paved the ground of both fields. Organ transplantation progressed to clinical translation earlier than OBR and should *de facto* be considered, with blood transfusion, the most formidable example of cell therapy.

cardiac and transplant surgery but also of bioreactors that are currently used in regenerative medicine and OBR investigations. Perhaps, what is astonishing is that these developments took place decades before organ transplantation was clinically achievable, and almost a century before the implantation of bioengineered vessels, bladders, segments of upper airways, and urethras had become possible.

For Christians, the Holy Grail is believed to own phenomenal powers and has thus been the object of quest and pursuit for centuries, despite the fact that no one has ever found it. Thus, one questions the existence of the Holy Grail and ponders whether it was simply a product of human fantasy. It is important, however, to recognize the metaphor of the Holy Grail describing an exceptionally rare object or a near-unattainable ideal. With regard to transplantation, the establishment of IFS has long been the primary ambition of a plethora of investigations, which have required massive investments since the early pioneering years, and yet IFS remains an elusive goal.

As the present book has attempted to provide evidence, modern developments in novel OBR technologies are revealing the groundbreaking and revolutionary potential of OBR and regenerative medicine to transform transplantation and reinforce the extraordinary legacy of Alexis Carrel's innovative ideas. Therefore, OBR is a strong candidate to supplant IFS as the new Holy Grail of clinical transplantation mainly because of its capacity to act as a source of inexhaustible organs and as a primary "IFS" in clinical transplantation.

This book has summarized the latest developments in OBR technology, demonstrating the potential of OBR and regenerative medicine to revolutionize transplantation science through the actualization of Alexis Carrel's genius. Through these innovations, OBR promises to supplant IFS as the "Holy Grail" of clinical transplantation, as it offers an inexhaustible source of immunotolerant organs. OBR thus represents the missing link between regenerative medicine and organ transplantation, serving as a catalyst for the evolution and advancement of both fields. I believe that transplantation science will drive the progress of regenerative medicine in upcoming years, as necessity breeds innovation, and transplant surgeons are desperate for a new, inexhaustible source of organs. In return, regenerative medicine is destined to revolutionize transplantation for the reasons outlined by the outstanding research summarized in this book.

REFERENCES

[1] Orlando G, Soker S, Stratta RJ. Organ bioengineering and regeneration as the new Holy Grail for organ transplantation. Ann Surg 2013;258(2):221−32.

[2] Orlando G, Domínguez-Bendala J, Shupe T, Bergman C, Bitar KN, Booth C, et al. Cell and organ bioengineering technology as applied to gastrointestinal diseases. Gut 2012;62:774−86.

[3] Orlando G, Wood KJ, De Coppi P, Baptista PM, Binder KW, Bitar KN, et al. Regenerative medicine as applied to general surgery. Ann Surg 2012;255:867−80.

[4] Orlando G, Wood KJ, Stratta RJ, Yoo J, Atala A, Soker S. Regenerative medicine and organ transplantation: past, present and future. Transplantation 2011;91:1310−7.

[5] Orlando G, Baptista P, Birchall M, De Coppi P, Farney A, Opara E, et al. Regenerative medicine as applied to solid organ transplantation: current status and future development. Transpl Int 2011;24:223−32.

[6] Orlando G. Transplantation as a subfield of regenerative medicine. An interview by Lauren Constable. Expert Rev Clin Immunol 2011;7:137−41.

[7] Lechler RI, Sykes M, Thomson AW, Turka LA. Organ transplantation—how much of the promise has been realized? Nat Med 2005;11:605−13.

[8] Orlando G. Immunosuppression-free transplantation reconsidered from a regenerative medicine perspective. Exp Rev Clin Immun 2012;8:179−87.

[9] Orlando G, Wood KJ, Soker S, Stratta RJ. How regenerative medicine may contribute to the achievement of an immunosuppression-free state. Transplantation 2011;92:36−8.

[10] Thomas L. Notes of a biology-watcher: the technology of medicine. N Eng J Med 1971;285:1366−8.

[11] Rubín A, Aguilera V, Berenguer M. Liver transplantation and hepatitis C. Clin Res Hepatol Gastroenterol 2011;35:805−12.

[12] Tisone G, Orlando G, Cardillo A, Palmieri G, Manzia TM, Baiocchi L, et al. Complete weaning off immunosuppression reduces the progression rate of fibrosis in HCV liver transplant patients with recurrence of disease. Mid-term results. J Hepatol 2006;44:702−9.

[13] Orlando G, Manzia T, Baiocchi L, Sanchez-Fueyo A, Angelico M, Tisone G. The Tor Vergata weaning off immunosuppression protocol in stable HCV liver transplant patients. The updated follow up at 78 months. Transpl Immunol 2008;20:43−7.

[14] Manzia TM, Angelico R, Toti L, Angelico M, Orlando G, Tisone G. The Tor Vergata weaning of immunosuppression protocols in stable HCV liver transplant patients: the 10-year follow-up. Transpl Int 2013;26:259−66.

[15] Orlando G, Soker S, Wood K. Clinical operational tolerance after liver transplantation. J Hepatol 2009;50:1247−57.

[16] Tullius SG, Rudolf JA, Malek SK. Moving boundaries—the Nightingale twins and transplantation science. N Engl J Med 2012;366:1564−5.

[17] Cohen AB. Introduction—Technology in American Health Care. Ann Arbor, MI: The University of Michigan Press; 2007.

[18] Delmonico FL, McBride MA. Analysis of the wait list and deaths among candidates waiting for a kidney transplant. Transplantation 2008;86:1678−83.

[19] Perico N, Remuzzi G. Chronic kidney disease: a research and public health priority. Nephrol Dial Transplant 2012;27:19−26.

[20] Cooper DK. A brief history of cross-species organ transplantation. Proc (Bayl Univ Med Cent) 2012;25:49−57.

[21] Orlando G. Finding the right time for weaning off immunosuppression in solid organ transplant recipients. Expert Rev Clin Immunol 2010;6:879−92.

[22] Orlando G, Hematti P, Stratta RJ, Burke GW, Di Cocco P, Pisani F, et al. Clinical operational tolerance after renal transplantation: current status and future challenges. Ann Surg 2010;252:913−26.

[23] Tisone G, Orlando G, Angelico M. Operational tolerance in clinical liver transplantation. Emerging developments. Transpl Immunol 2007;17:108−13.

[24] Dharnidharka VR, Stablein DM, Harmon WE. Post-transplant infections now exceed acute rejection as cause for hospitalization: a report of the NAPRTCS. Am J Transplant 2004;4:384−9.

[25] U.S. Renal Data system, USRDS 2011 annual data report: atlas of chronic kidney disease and end-stage renal disease in the United States. Bethesda, MD: National Institutes of Health, National Institute of Diabetes and Digestive and Kidney Diseases; 2011.

[26] Dantal J, Soulillou JP. Immunosuppressive drugs and the risk of cancer after organ transplantation. N Eng J Med 2005;352:1371−3.

[27] Dantal J, Hourmant M, Cantarovich D, Giral M, Blancho G, Dreno B, et al. Effect of long-term immunosuppression in kidney-graft recipients on cancer incidence: randomised comparison of two cyclosporin regimens. Lancet 1998;351:623−8.

[28] Coulson MT, Jablonski P, Howden BO, Thomson NM, Stein AN. Beyond operational tolerance: effect of ischemic injury on development of chronic damage in renal grafts. Transplantation 2005;80:353−60.

[29] Filipponi F, Pisati R, Cavicchini G, Ulivieri MI, Ferrara R, Mosca F. Cost and outcome analysis and cost determinants of liver transplantation in a European National Health Service hospital. Transplantation 2003;75:1731−6.

[30] Karam VH, Gasquet I, Delvart V, Hiesse C, Dorent R, Danet C, et al. Quality of life in adult survivors beyond 10 years after liver, kidney, and heart transplantation. Transplantation 2003;76:1699−704.

[31] Galbraith CA, Hathaway D. Long-term effects of transplantation on quality of life. Transplantation 2004;77:84−7.

[32] Vacanti JP, Morse MA, Saltzman WM, Domb AJ, Perez-Atayde A, Langer R. Selective cell transplantation using bioabsorbable artificial polymers as matrices. J Pediatr Surg 1988;23:3−9.

[33] Vacanti JP. Beyond transplantation. Third Annual Samuel Jason Mixture lecture. Arch Surg 1988;123:545−9.

[34] Shinoka T, Imai Y, Ikada Y. Transplantation of a tissue-engineered pulmonary artery. N Engl J Med 2001;344:532−3.

[35] Macchiarini P, Jungebluth P, Go T, Asnaghi MA, Rees LE, Cogan TA, et al. Clinical transplantation of a tissue-engineered airway. Lancet 2008;372:2023−30.

[36] Jungebluth P, Alici E, Baiguera S, Le Blanc K, Blomberg P, Bozóky B, et al. Tracheobronchial transplantation with a stem-cell-seeded bioartificial nanocomposite: a proof-of-concept study. Lancet 2011;378:1997−2004.

[37] Elliott MJ, De Coppi P, Speggiorin S, Roebuck D, Butler CR, Samuel E, et al. Stem-cell-based, tissue engineered tracheal replacement in a child: a 2-year follow-up study. Lancet 2012;380: 994–1000.

[38] Baiguera S, Birchall MA, Macchiarini P. Tissue-engineered tracheal transplantation. Transplantation 2010;89:485–91.

[39] Rauh J, Milan F, Günther KP, Stiehler M. Bioreactor systems for bone tissue engineering. Tissue Eng Part B Rev 2011;17:263–80.

[40] Folkman J, Hochberg M. Self-regulation of growth in three dimensions. J Exp Med 1937;138:745–53.

[41] Orlando G, Farney A, Sullivan DC, Abouschwareb T, Iskandar S, Wood KJ, et al. Production and implantation of renal extracellular matrix scaffolds from porcine kidneys as a platform for renal bioengineering investigations. Ann Surg 2012;256:363–70.

[42] Hynes RO. The extracellular matrix: not just pretty fibrils. Science 2009;326:1216–9.

[43] Petruzzo P, Lanzetta M, Dubernard JM, Landin L, Cavadas P, Margreiter R, et al. The International Registry of hand and composite tissue transplantation. Transplantation 2010;12:1590–4.

[44] Petruzzo P, Lanzetta M, Dubernard JM, Margreiter R, Schuind F, Breidenbach W, et al. The International Registry of hand and composite tissue transplantation. Transplantation 2008;86:487–92.

[45] Romagnoli G, De Luca M, Faranda F, Bandelloni R, Franzi AT, Cataliotti F, et al. Treatment of posterior hypospadias by the autologous graft of cultured urethral epithelium. N Engl J Med 1990;323:527–30.

[46] Pellegrini G, Traverso CE, Franzi AT, Zingirian M, Cancedda R, De Luca M. Long-term restoration of damaged corneal surfaces with autologous cultivated corneal epithelium. Lancet 1997;349:990–3.

[47] Rama P, Matuska S, Paganoni G, Spinelli A, De Luca M, Pellegrini G. Limbal stem-cell therapy and long-term corneal regeneration. N Engl J Med 2010;363:147–55.

[48] Quarto R, Mastrogiacomo M, Cancedda R, Kutepov SM, Mukhachev V, Lavroukov A, et al. Repair of large bone defects with the use of autologous bone marrow stromal cells. N Engl J Med 2001;344:385–6.

[49] Warnke PH, Springer IN, Wiltfang J, Acil Y, Eufinger H, Wehmöller M, et al. Growth and transplantation of a custom vascularised bone graft in a man. Lancet 2004;364:766–70.

[50] Atala A, Bauer SB, Soker S, Yoo JJ, Retik AB. Tissue-engineered autologous bladders for patients needing cystoplasty. Lancet 2006;367:1241–6.

[51] Hibino N, McGillicuddy E, Matsumura G, Ichihara Y, Naito Y, Breuer C, et al. Late-term results of tissue-engineered vascular grafts in humans. J Thorac Cardiovasc Surg 2010;139: 431–6.

[52] McAllister TN, Maruszewski M, Garrido SA, Wystrychowski W, Dusserre N, Marini A, et al. Effectiveness of haemodialysis access with an autologous tissue-engineered vascular graft: a multicentre cohort study. Lancet 2009;373:1440–6.

[53] Raya-Rivera A, Esquiliano DR, Yoo JJ, Lopez-Bayghen E, Soker S, Atala A. Tissue-engineered autologous urethras for patients who need reconstruction: an observational study. Lancet 2011;377:1175–82.

[54] Olausson M, Patil PB, Kuna VK, Chougule P, Hernandez N, Methe K, et al. Transplantation of an allogeneic vein bioengineered with autologous stem cells: a proof-of-concept study. Lancet 2012;380:230–7.

Index